30 *Years*
Previous Solved Papers

GATE 2025

Electronics Engineering

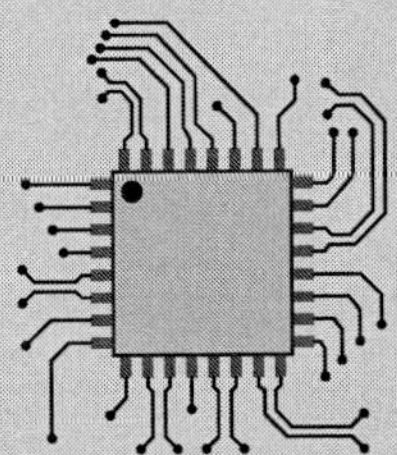

- Fully solved with explanations
- Analysis of previous papers
- Topicwise presentation
- Thoroughly revised & updated

MADE EASY Publications Pvt. Ltd.

Corporate Office: 44-A/4, Kalu Sarai (Near Hauz Khas Metro Station), New Delhi-110016 | **Ph.:** 9021300500

Email : infomep@madeeasy.in | **Web :** www.madeeasypublications.org

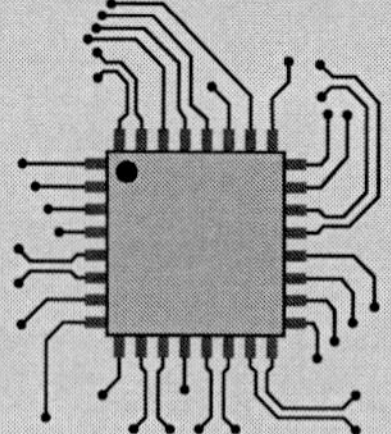

GATE - 2025

Electronics Engineering

Topicwise Previous GATE Solved Papers (1995-2024)

Editions

1st Edition	:	2007
2nd Edition	:	2008
3rd Edition	:	2009
4th Edition	:	2010
5th Edition	:	2011
6th Edition	:	2012
7th Edition	:	2013
8th Edition	:	2014
9th Edition	:	2015
10th Edition	:	2016
11th Edition	:	2017
12th Edition	:	2018
13th Edition	:	2019
14th Edition	:	2020
15th Edition	:	2021
16th Edition	:	2022
17th Edition	:	2023
18th Edition	**:**	**2024**

Preface

Over the period of time the GATE examination has become more challenging due to increasing number of candidates. Though every candidate has ability to succeed but competitive environment, in-depth knowledge, quality guidance and good source of study is required to achieve high level goals.

B. Singh (Ex. IES)

The new edition of **GATE 2025 Solved Papers : Electronics Engineering** has been fully revised, updated and edited. The whole book has been divided into topicwise sections.

At the beginning of each subject, analysis of previous papers are given to improve the understanding of subject.

I have true desire to serve student community by way of providing good source of study and quality guidance. I hope this book will be proved an important tool to succeed in GATE examination. Any suggestions from the readers for the improvement of this book are most welcome.

B. Singh (Ex. IES)
Chairman and Managing Director
MADE EASY Group

GATE-2025

Electronics Engineering

CONTENTS

Network Theory

UNIT

I

CONTENTS

Network Theory

Syllabus

Circuit analysis: Node and mesh analysis, superposition, Thevenin's theorem, Norton's theorem, reciprocity. Sinusoidal steady state analysis: phasors, complex power, maximum power transfer. Time and frequency domain analysis of linear circuits: RL, RC and RLC circuits, solution of network equations using Laplace transform. Linear 2-port network parameters, wye-delta transformation.

Analysis of Previous GATE Papers

Exam Year	1 Mark Ques.	2 Marks Ques.	3 Marks Ques.	Total Marks
1995	8	–	–	**8**
1996	3	1	–	**5**
1997	4	2	1	**11**
1998	6	–	–	**6**
1999	2	3	–	**8**
2000	3	2	–	**7**
2001	3	4	–	**11**
2002	2	2	–	**6**
2003	4	8	–	**20**
2004	5	5	–	**15**
2005	5	6	–	**17**
2006	6	–	–	**6**
2007	2	4	–	**10**
2008	2	7	–	**16**
2009	3	4	–	**11**
2010	2	4	–	**10**
2011	3	3	–	**9**
2012	4	4	–	**12**
2013	3	6	–	**15**

Exam Year	1 Mark Ques.	2 Mark Ques.	Total Marks
2014 Set-1	2	4	**10**
2014 Set-2	2	2	**6**
2014 Set-3	2	4	**10**
2014 Set-4	2	4	**10**
2015 Set-1	4	3	**10**
2015 Set-2	3	3	**9**
2015 Set-3	3	2	**7**
2016 Set-1	1	2	**5**
2016 Set-2	4	3	**10**
2016 Set-3	1	3	**7**
2017 Set-1	1	2	**5**
2017 Set-2	2	2	**6**
2018	1	3	**7**
2019	1	2	**5**
2020	3	2	**7**
2021	2	4	**10**
2022	2	3	**8**
2023	4	1	**6**
2024	2	2	**6**

1 Basics of Network Analysis

1.1 Two 2 H inductance coils are connected in series and are also magnetically coupled to each other the coefficient of coupling being 0.1. The total inductance of the combination can be

(a) 0.4 H (b) 3.2 H
(c) 4.0 H (d) 4.4 H

[1995 : 1 M]

1.2 The current i_4 in the circuit of figure is equal to

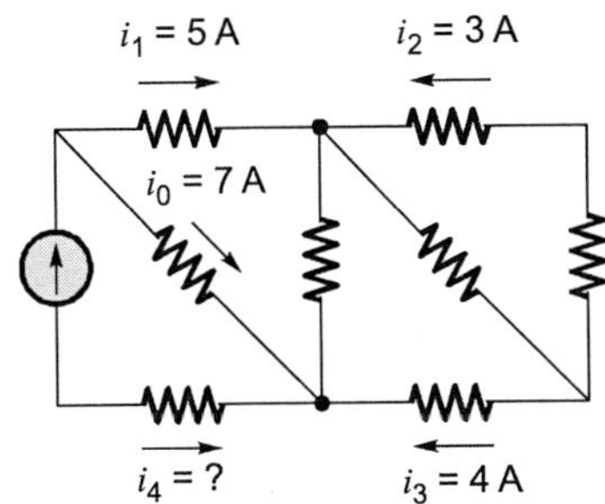

(a) 12 A (b) – 12 A
(c) 4 A (d) None of these

[1997 : 1 M]

1.3 The voltage V in figure is equal to

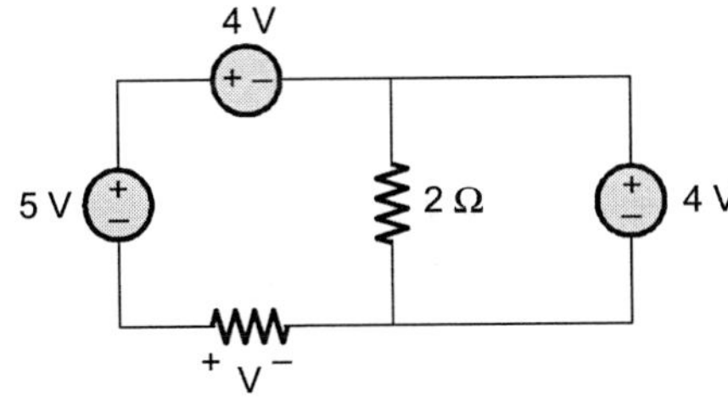

(a) 3 V (b) –3 V
(c) 5 V (d) None of these

[1997 : 1 M]

1.4 The voltage V in figure is always equal to

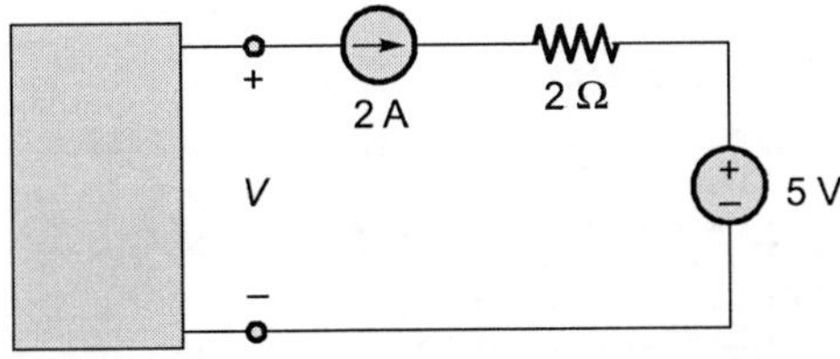

(a) 9 V (b) 5 V
(c) 1 V (d) None of these

[1997 : 1 M]

1.5 The voltage V in figure is

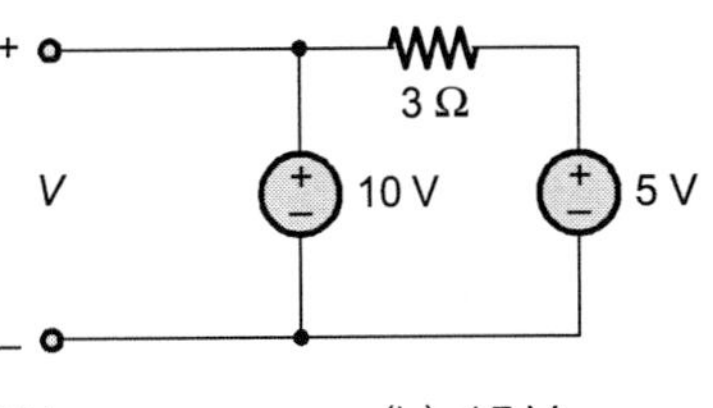

(a) 10 V (b) 15 V
(c) 5 V (d) None of the these

[1997 : 1 M]

1.6 In the circuit of Fig. the equivalent impedance seen across terminals A, B is

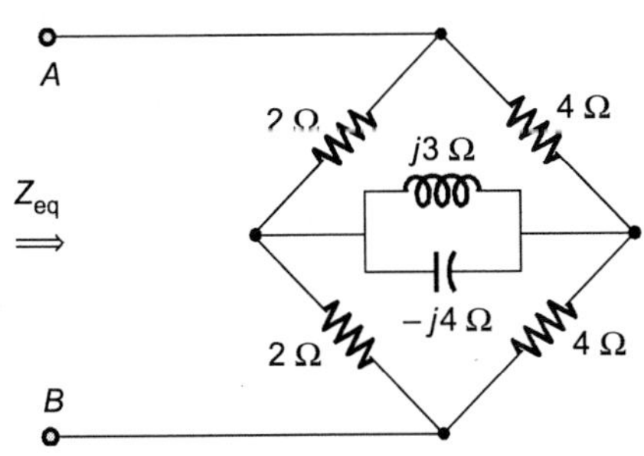

(a) (16/3) Ω (b) (8/3) Ω
(c) (8/3 + 12 j) Ω (d) None of the above

[1997 : 2 M]

1.7 In the circuit shown in the figure the current i_D through the ideal diode (zero cut in voltage and zero forward resistance) equals

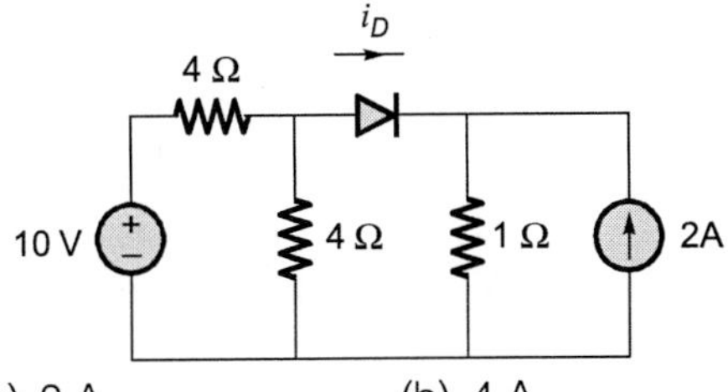

(a) 0 A (b) 4 A
(c) 1 A (d) None of these

[1997 : 3 Marks]

1.8 The nodal method of circuit analysis is based on

(a) KVL and Ohm's law
(b) KCL and Ohm's law
(c) KCL and KVL
(d) KCL, KVL and Ohm's law

[1998 : 1 M]

1.9 The voltage across the terminals a and b in Fig. is

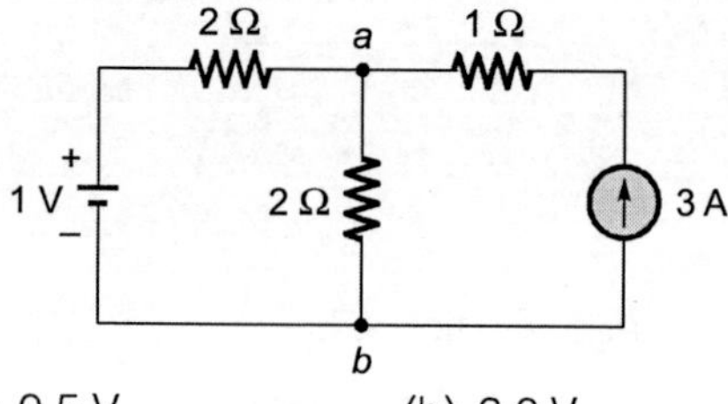

(a) 0.5 V (b) 3.0 V
(c) 3.5 V (d) 4.0 V

[1998 : 1 M]

1.10 A Delta-connected network with its Wye-equivalent is shown in the figure. The resistances R_1, R_2 and R_3 (in ohms) are respectively

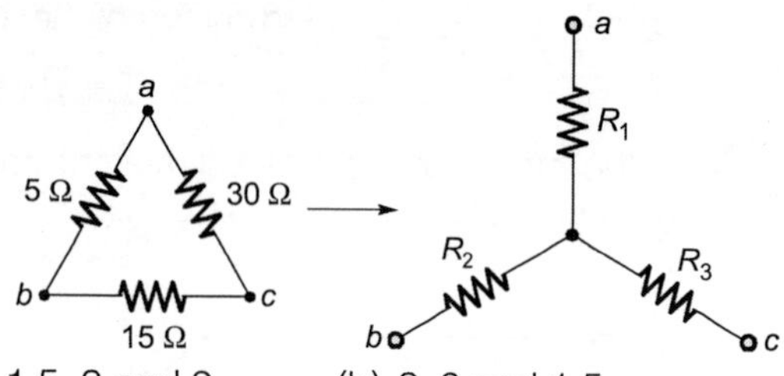

(a) 1.5, 3 and 9 (b) 3, 9 and 1.5
(c) 9, 3 and 1.5 (d) 3, 1.5 and 9

[1999 : 2 M]

1.11 In the circuit of the figure, the voltage $v(t)$ is

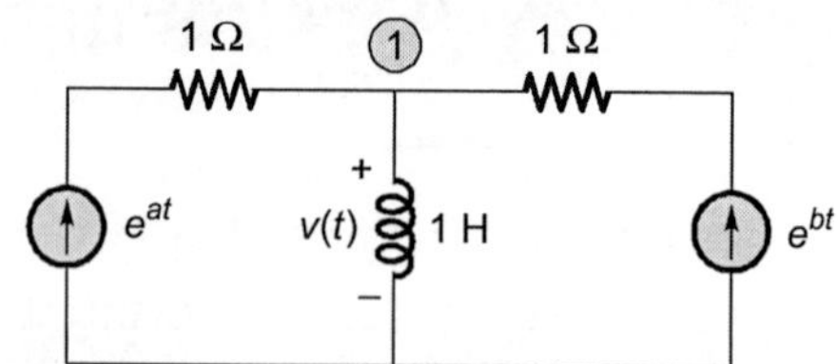

(a) $e^{at} - e^{bt}$ (b) $e^{at} + e^{bt}$
(c) $ae^{at} - be^{bt}$ (d) $ae^{at} + be^{bt}$

[2000 : 1 M]

1.12 In the circuit of the figure, the value of the voltage source E is

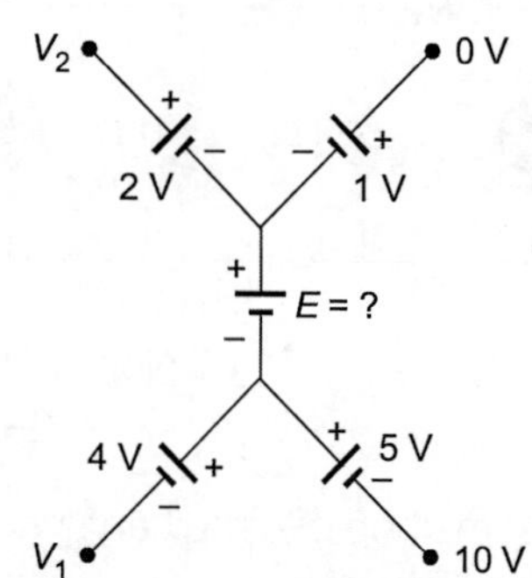

(a) −16 V (b) 4 V
(c) −6 V (d) 16 V

[2000 : 1 M]

1.13 For the circuit in the figure, the voltage V_o is

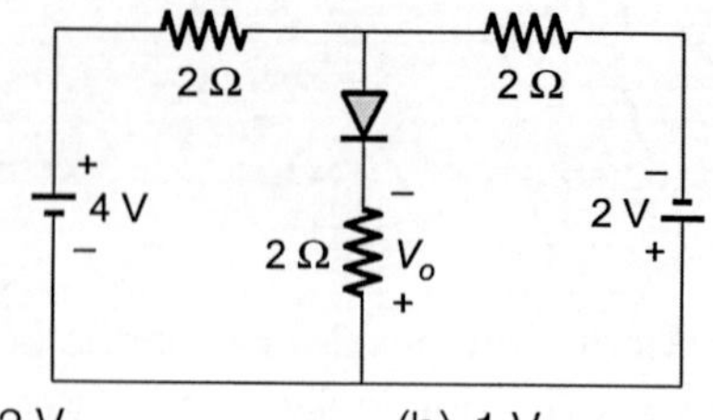

(a) 2 V (b) 1 V
(c) −1 V (d) None of these

[2000 : 2 M]

1.14 The voltage e_o in the figure is

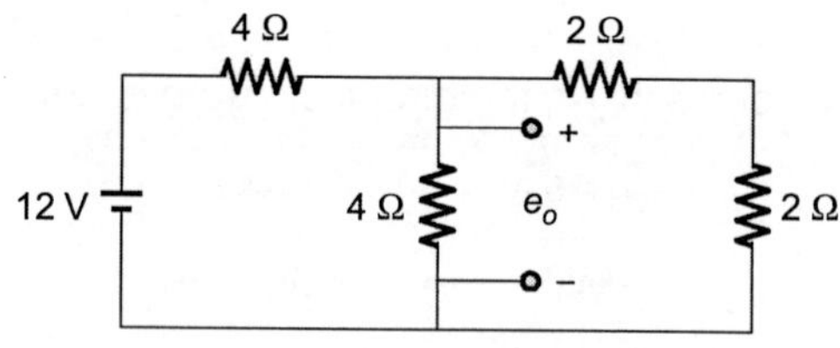

(a) 2 V (b) $\frac{4}{3}$ V
(c) 4 V (d) 8 V

[2001 : 1 M]

1.15 If each branch of a Delta circuit has impedance $\sqrt{3}\,Z$, then each branch of the equivalent Wye circuit has impedance.

(a) $\frac{Z}{\sqrt{3}}$ (b) $3Z$
(c) $3\sqrt{3}\,Z$ (d) $\frac{Z}{3}$

[2001 : 1 M]

1.16 The voltage e_o in the figure is

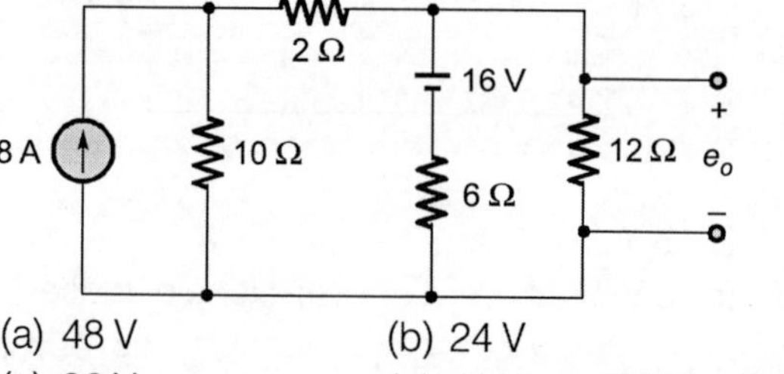

(a) 48 V (b) 24 V
(c) 36 V (d) 28 V

[2001 : 2 M]

1.17 The dependent current source shown in the figure

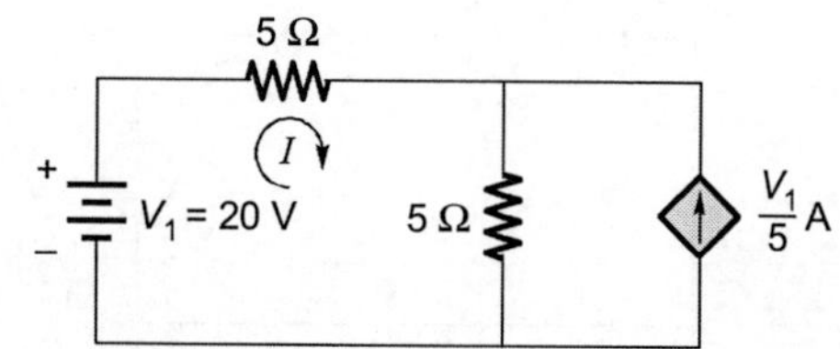

(a) delivers 80 W (b) absorbs 80 W
(c) delivers 40 W (d) absorbs 40 W

[2002 : 1 M]

1.18 If the 3-phase balanced source in the figure delivers 1500 W at a leading power factor 0.844, then the value of Z_L (in ohm) is approximately

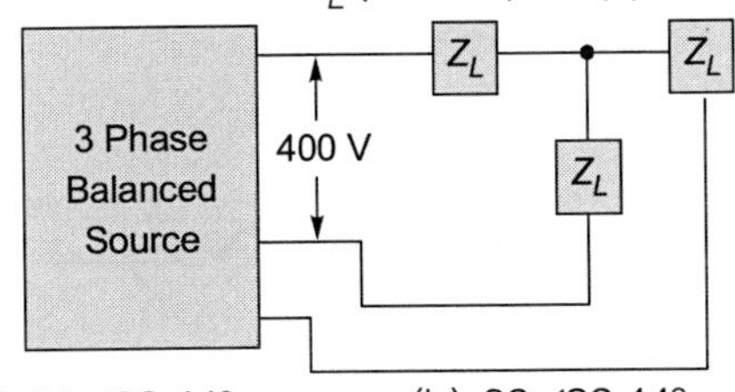

(a) $90\angle 32.44°$ (b) $80\angle 32.44°$
(c) $80\angle -32.44°$ (d) $90\angle -32.44°$

[2002 : 2 M]

1.19 The minimum number of equations required to analyze the circuit shown in the figure is

(a) 3
(b) 4
(c) 6
(d) 7

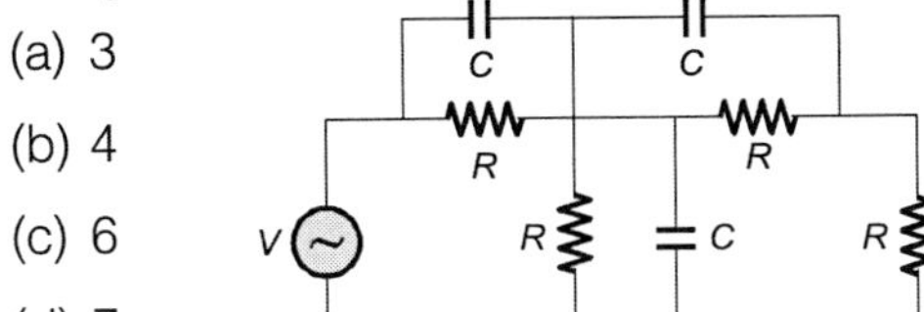

[2003 : 1 M]

1.20 Twelve 1 Ω resistances are used as edges to form a cube. The resistance between two diagonally opposite corners of the cube is

(a) $\frac{5}{6}$ Ω (b) 1 Ω
(c) $\frac{6}{5}$ Ω (d) $\frac{3}{2}$ Ω [2003 : 2 M]

1.21 The current flowing through the resistance R in the circuit in the figure has the form $P\cos 4t$, where P is

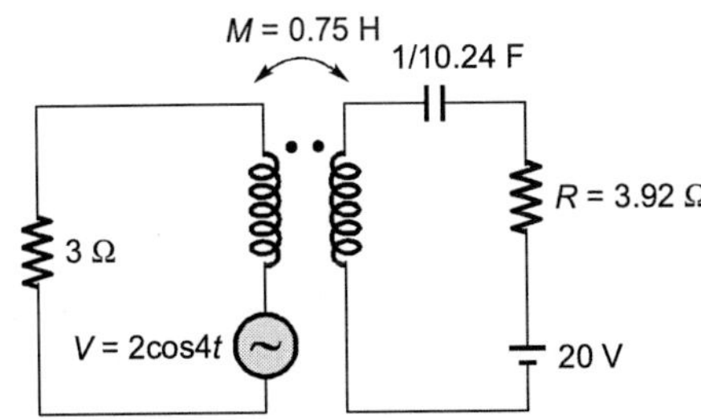

(a) $(0.18 + j0.72)$ (b) $(0.46 + j1.90)$
(c) $-(0.18 + j1.90)$ (d) $-(0.192 + j0.144)$

[2003 : 2 M]

1.22 An ideal sawtooth voltage waveform of frequency 500 Hz and amplitude 3 V is generated by charging a capacitor of 2 μF in every cycle. The charging requires

(a) constant voltage source of 3 V for 1 ms.
(b) constant voltage source of 3 V for 2 ms.
(c) constant current source of 3 mA for 1 ms.
(d) constant current source of 3 mA for 2 ms.

[2003 : 2 M]

1.23 The equivalent inductance measured between the terminals 1 and 2 for the circuit shown in the figure is

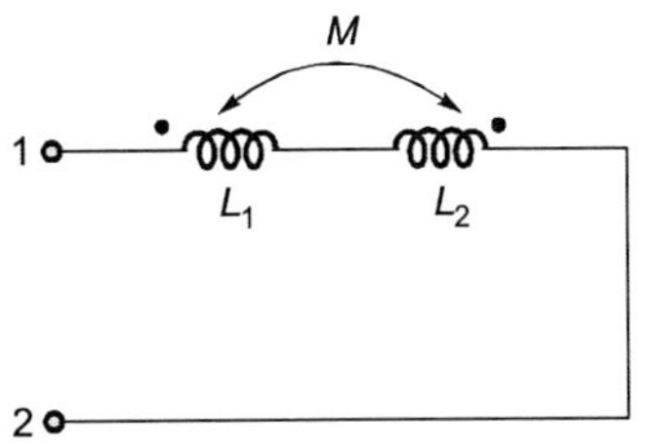

(a) $L_1 + L_2 + M$ (b) $L_1 + L_2 - M$
(c) $L_1 + L_2 + 2M$ (d) $L_1 + L_2 - 2M$

[2004 : 1 M]

1.24 For the circuit shown in the figure, the initial conditions are zero. Its transfer function $H(s) = \frac{V_c(s)}{V_i(s)}$ is

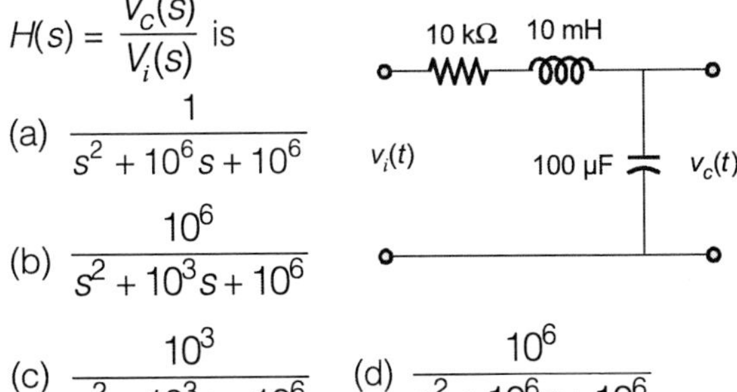

(a) $\frac{1}{s^2 + 10^6 s + 10^6}$
(b) $\frac{10^6}{s^2 + 10^3 s + 10^6}$
(c) $\frac{10^3}{s^2 + 10^3 s + 10^6}$ (d) $\frac{10^6}{s^2 + 10^6 s + 10^6}$

[2004 : 2 M]

1.25 Impedance Z as shown in the given figure is

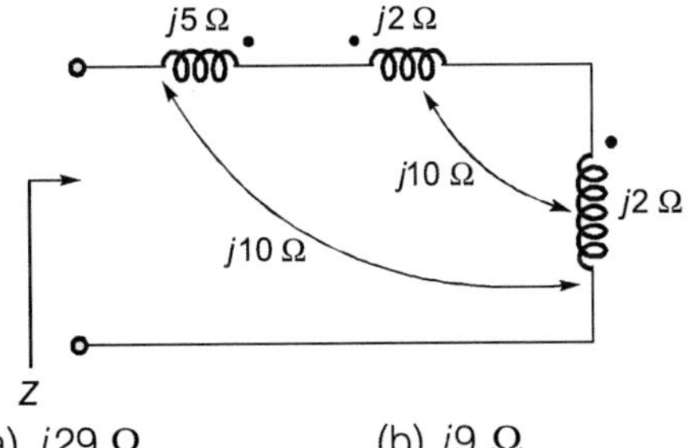

(a) $j29$ Ω (b) $j9$ Ω
(c) $j19$ Ω (d) $j39$ Ω [2005 : 2 M]

1.26 If $R_1 = R_2 = R_4 = R$ and $R_3 = 1.1\,R$ in the bridge circuit shown in the figure, then the reading in the ideal voltmeter connected between a and b is

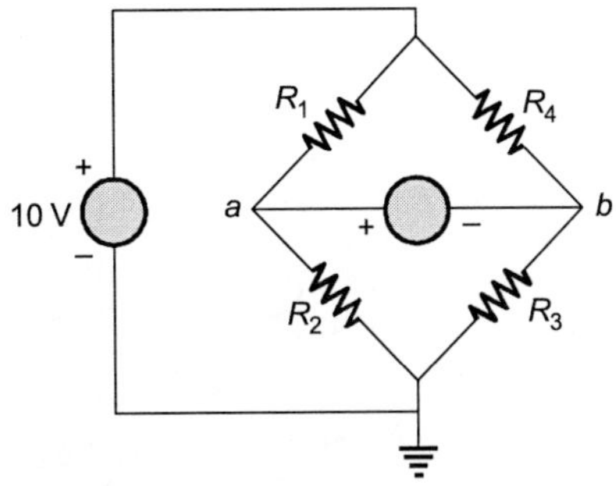

(a) 0.238 V (b) 0.138 V
(c) −0.238 V (d) 1 V [2005 : 2 M]

1.27 In the interconnection of ideal sources shown in the figure, it is known that the 60 V source is absorbing power.

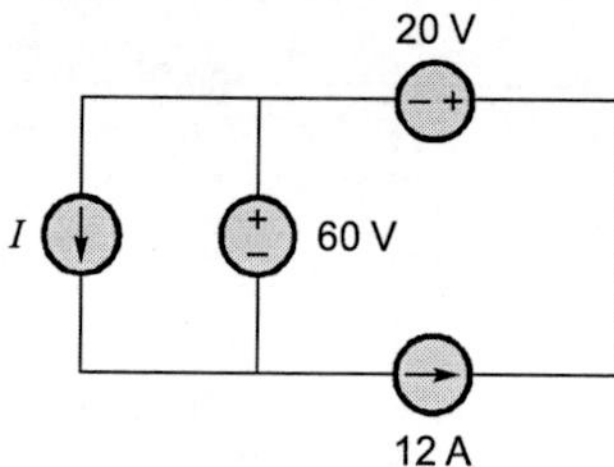

Which of the following can be the value of the current source I?

(a) 10 A (b) 13 A
(c) 15 A (d) 18 A [2009 : 1 M]

1.28 A fully charged mobile phone with a 12 V battery is good for a 10 minute talk-time. Assume that, during the talk-time the battery delivers a constant current of 2 A and its voltage drops linearly from 12 V to 10 V as shown in the figure. How much energy does the battery deliver during this talk-time?

(a) 220 J
(b) 12 kJ
(c) 13.2 kJ
(d) 14.4 J

[2009 : 1 M]

1.29 In the circuit shown, the power supplied by the voltage source is

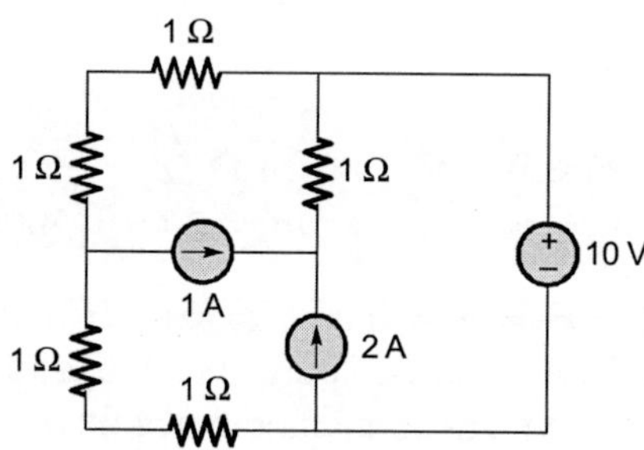

(a) 0 W (b) 5 W
(c) 10 W (d) 100 W [2010 : 2 M]

1.30 In the circuit shown below, the current I is equal to

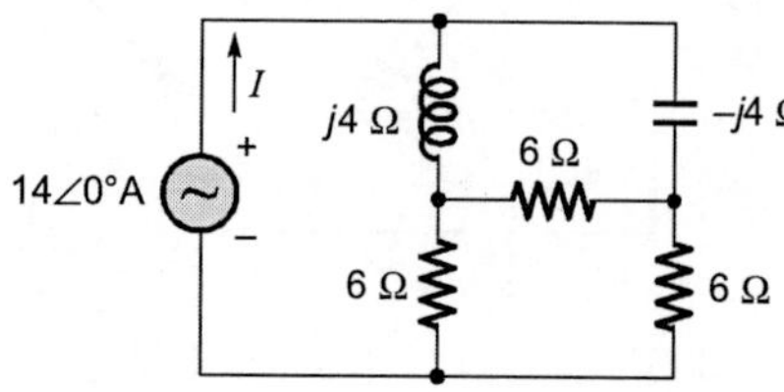

(a) $1.4\angle 0°$ A (b) $2.0\angle 0°$ A
(c) $2.8\angle 0°$ A (d) $3.2\angle 0°$ A [2011 : 2 M]

1.31 In the circuit shown below, the current through the inductor is

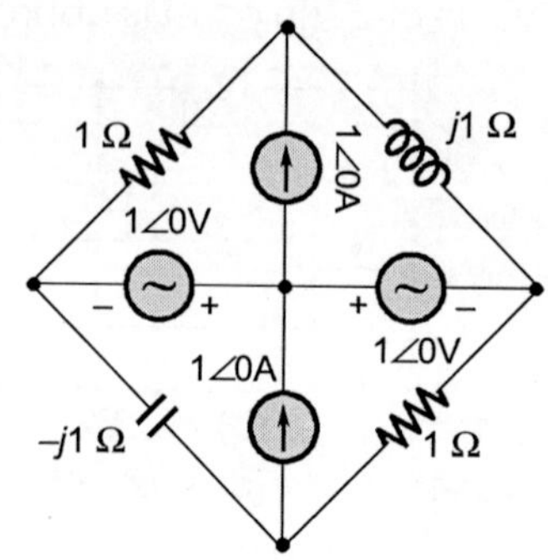

(a) $\frac{2}{1+j}$ A (b) $\frac{-1}{1+j}$ A
(c) $\frac{1}{1+j}$ A (d) 0 A

[2012 : 1 M]

1.32 The average power delivered to an impedance $(4 - j3)\Omega$ by a current $5\cos(100\pi t + 100)$ A is

(a) 44.2 W (b) 50 W
(c) 62.5 W (d) 125 W

[2012 : 1 M]

1.33 If $V_A - V_B = 6$ V, then $V_C - V_D$ is

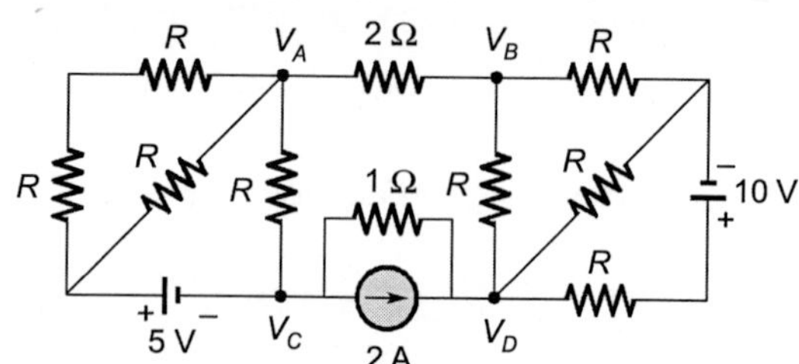

(a) – 5 V (b) 2 V
(c) 3 V (d) 6 V [2012 : 2 M]

1.34 Consider a delta connection of resistors and its equivalent star connection as shown below. If all elements of the delta connection are scaled by a factor k, $k > 0$, the elements of the corresponding star equivalent will be scaled by a factor of

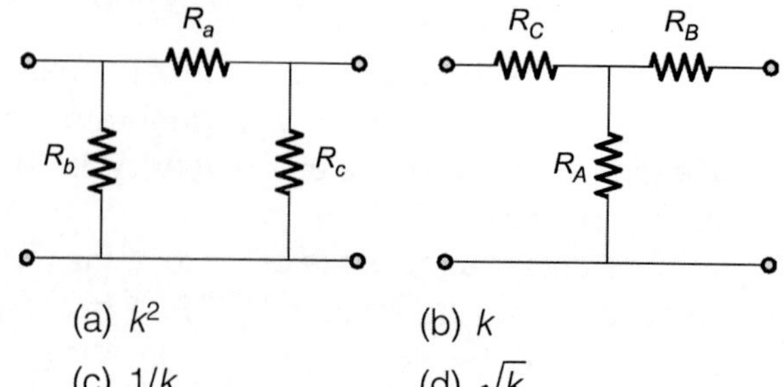

(a) k^2 (b) k
(c) $1/k$ (d) $\sqrt{k}$

[2013 : 1 M]

1.35 The following arrangement consists of an ideal transformer and an attenuator which attenuates by a factor of 0.8. An ac voltage $V_{WX_1} = 100$ V is applied across WX to get an open circuit voltage V_{YZ_1} across YZ. Next, an ac voltage $V_{YZ_2} = 100$ V is applied across YZ to get an open circuit voltage V_{WX_2} across WX. Then, V_{YZ_1}/V_{WX_1}, V_{WX_2}/V_{YZ_2} are respectively,

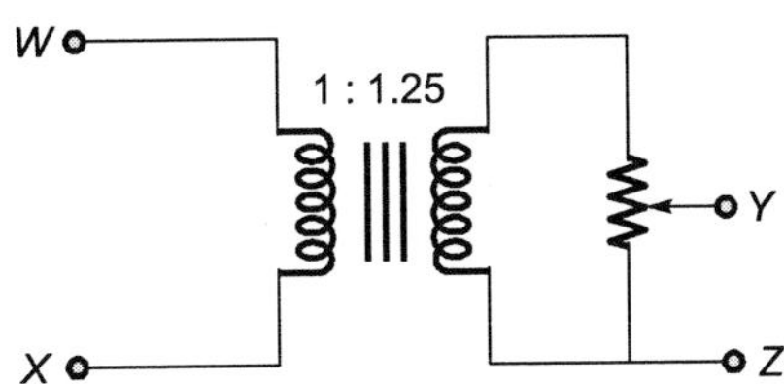

(a) 125/100 and 80/100
(b) 100/100 and 80/100
(c) 100/100 and 100/100
(d) 80/100 and 80/100

[2013 : 2 M]

1.36 Three capacitors C_1, C_2 and C_3 whose values are 10 μF, 5 μF, and 2 μF respectively, have breakdown voltages of 10 V, 5 V and 2 V respectively. For the interconnection shown below, the maximum safe voltage in volts that can be applied across the combination, and the corresponding total charge in μC stored in the effective capacitance across the terminals are, respectively

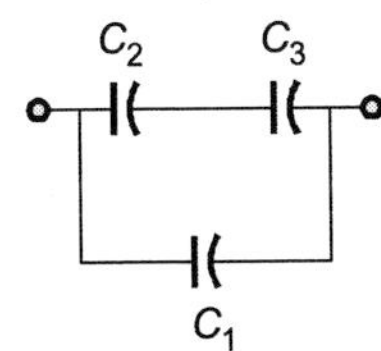

(a) 2.8 and 36 (b) 7 and 119
(c) 2.8 and 32 (d) 7 and 80

[2013 : 2 M]

Common Data For Questions 1.37 and 1.38:

Consider the following figure:

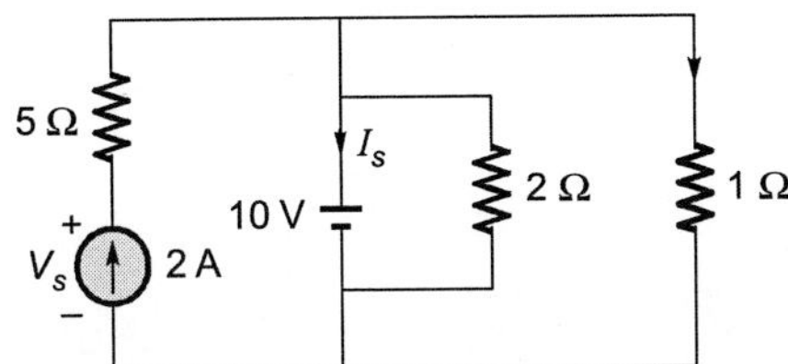

1.37 The current I_s in Amps in the voltage source, and voltage V_s in volts across the current source respectively, are

(a) 13, −20 (b) 8, −10
(c) −8, 20 (d) −13, 20 [2013 : 2 M]

1.38 The current in the 1 Ω resistor in Amps is

(a) 2 (b) 3.33
(c) 10 (d) 12

[2013 : 2 M]

1.39 Consider the configuration shown in the figure which is a portion of a larger electrical network

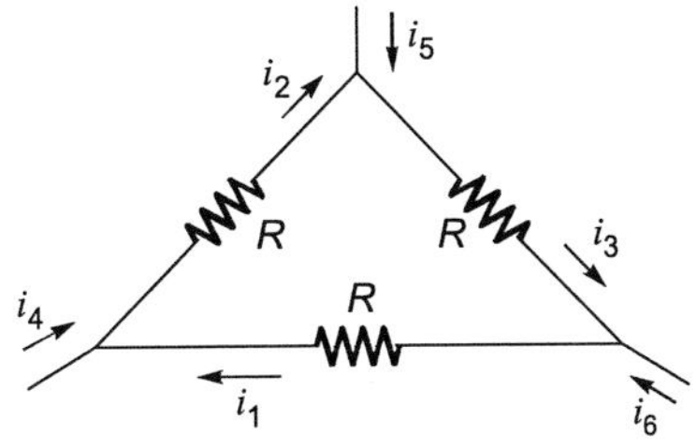

For $R = 1\ \Omega$ and currents $i_1 = 2$ A, $i_4 = -1$ A, $i_5 = -4$ A, which one of the following is TRUE?

(a) $i_6 = 5$ A
(b) $i_3 = -4$ A
(c) Data is sufficient to conclude that the supposed currents are impossible
(d) Data is insufficient to identify the currents i_2, i_3 and i_6

[2014 : 1 M, Set-1]

1.40 A Y-network has resistances of 10 Ω each in two of its arms, while the third arm has a resistance of 11 Ω. In the equivalent Δ-network, the lowest value (in Ω) among the three resistances is ______.

[2014 : 2 M, Set-1]

1.41 For the Y-network shown in the figure, the value of R_1 (in Ω) in the equivalent D-network is ____.

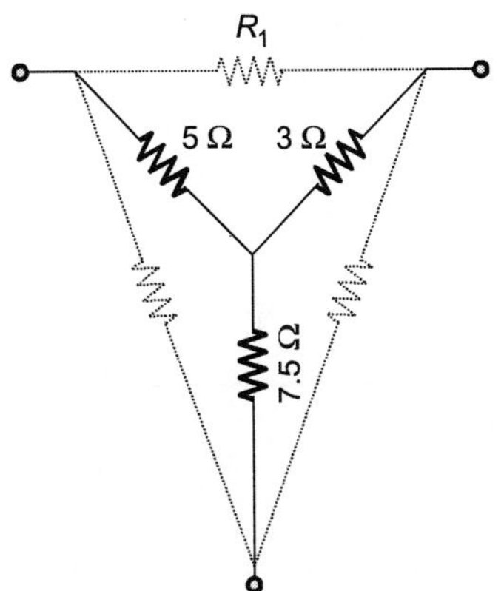

[2014 : 2 M, Set-3]

1.42 In the circuit shown in the figure, the value of node voltage V_2 is

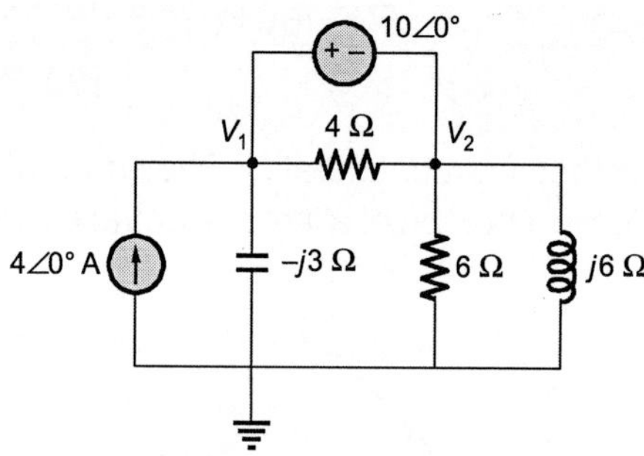

(a) $22 + j2$ V
(b) $2 + j22$ V
(c) $22 - j2$ V
(d) $2 - j22$ V

[2014 : 2 M, Set-3]

1.43 The circuit shown in the figure represents a

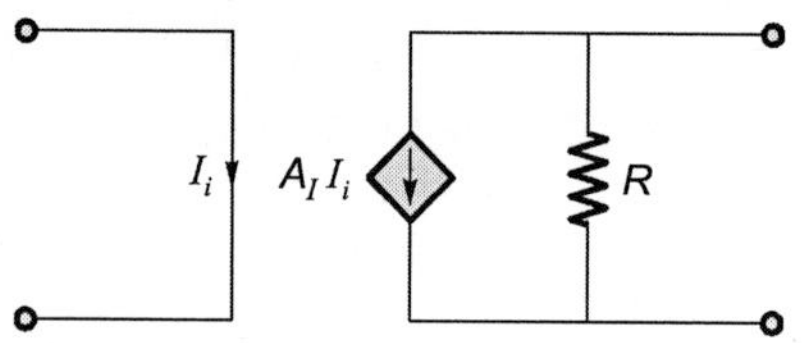

(a) voltage controlled voltage source
(b) voltage controlled current source
(c) current controlled current source
(d) current controlled voltage source

[2014 : 1 M, Set-4]

1.44 The magnitude of current (in mA) through the resistor R_2 in the figure shown is_______.

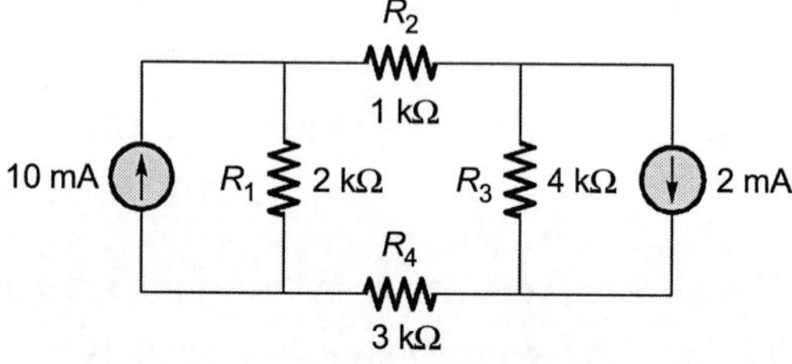

[2014 : 1 M, Set-4]

1.45 The equivalent resistance in the infinite ladder network shown in the figure, is R_e.

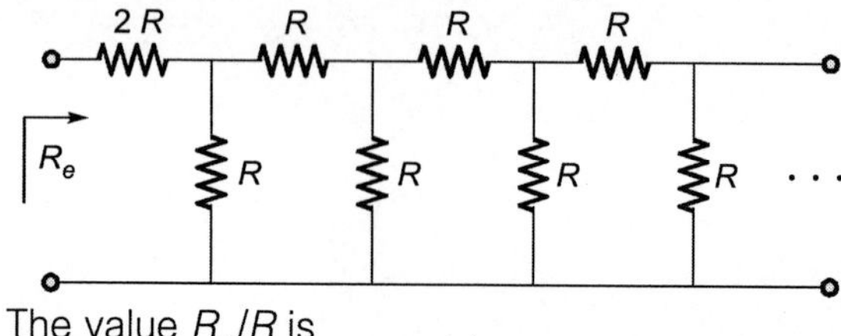

The value R_e/R is _______.

[2014 : 2 M, Set-4]

1.46 In the network shown in the figure, all resistors are identical with R = 300 Ω. The resistance R_{ab} (in Ω) of the network is _____________.

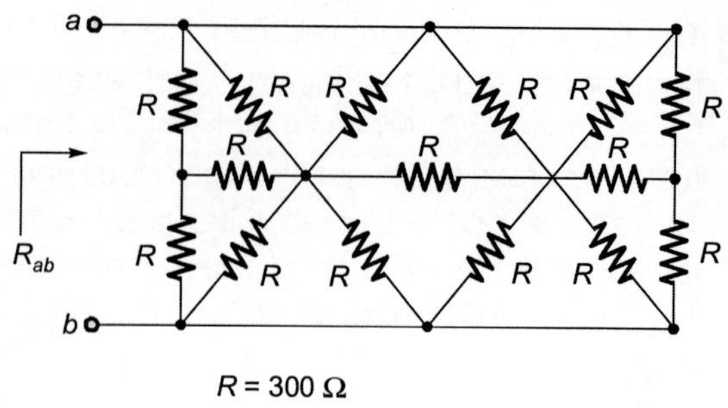

[2015 : 1 M, Set-1]

1.47 In the given circuit, the values of V_1 and V_2 respectively are

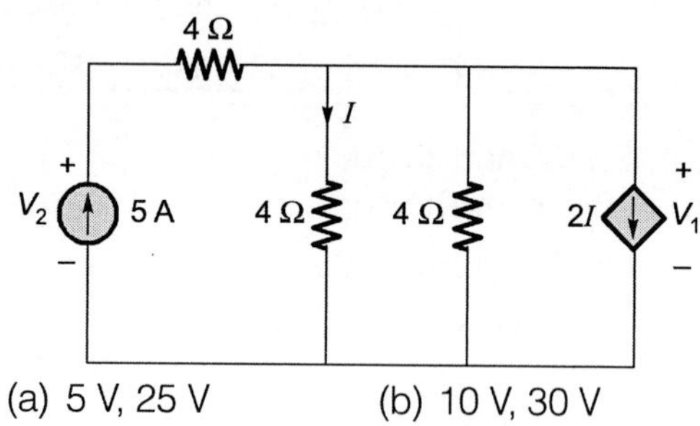

(a) 5 V, 25 V
(b) 10 V, 30 V
(c) 15 V, 35 V
(d) 0 V, 20 V

[2015 : 1 M, Set-1]

1.48 In the circuit shown, the voltage V_x (in Volts) is ____.

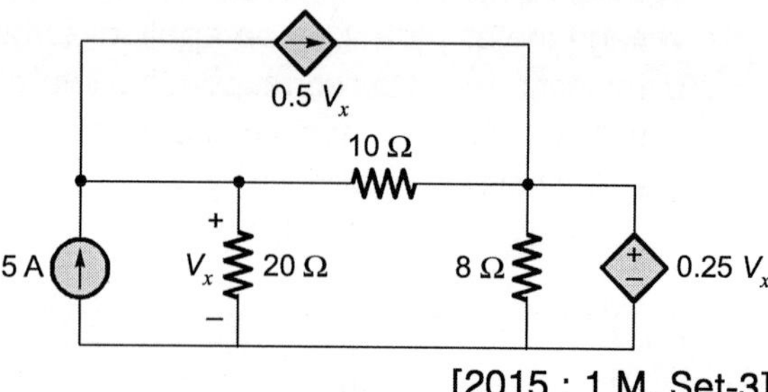

[2015 : 1 M, Set-3]

1.49 An AC voltage source $V = 10\sin(t)$ volts is applied to the following network. Assume that R_1 = 3 kΩ, R_2 = 6 kΩ and R_3 = 9 kΩ, and that the diode is ideal.

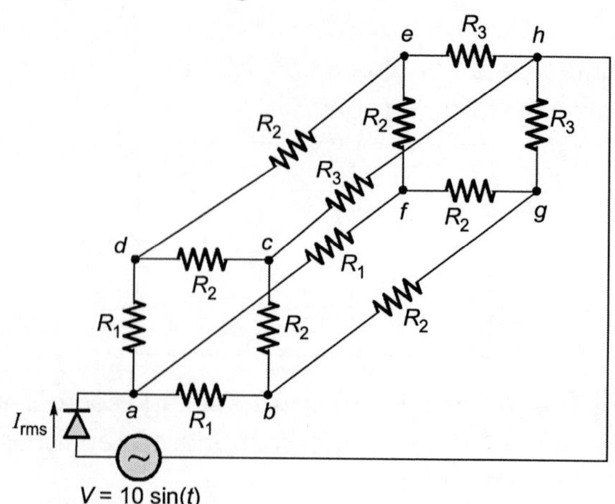

RMS current I_{rms} (in mA) through the diode is ____.

[2016 : 2 M, Set-1]

1.50 In the given circuit, each resistor has a value equal to 1 Ω.

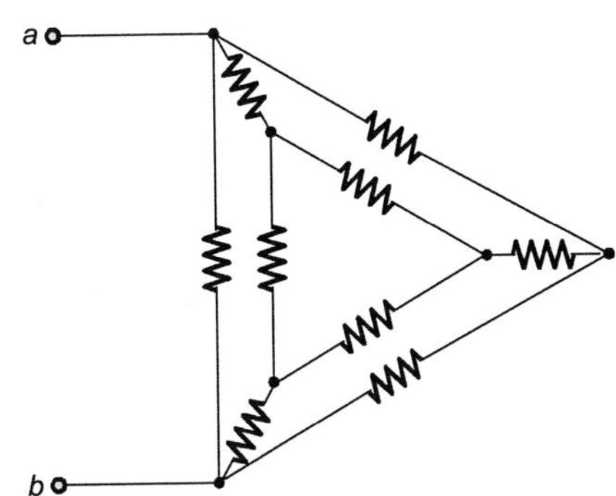

What is the equivalent resistance across the terminals *a* and *b*?

(a) 1/6 Ω
(b) 1/3 Ω
(c) 9/20 Ω
(d) 8/15 Ω

[2016 : 2 M, Set-2]

1.51 In the circuit shown in the figure, the magnitude of the current (in amperes) through R_2 is ____

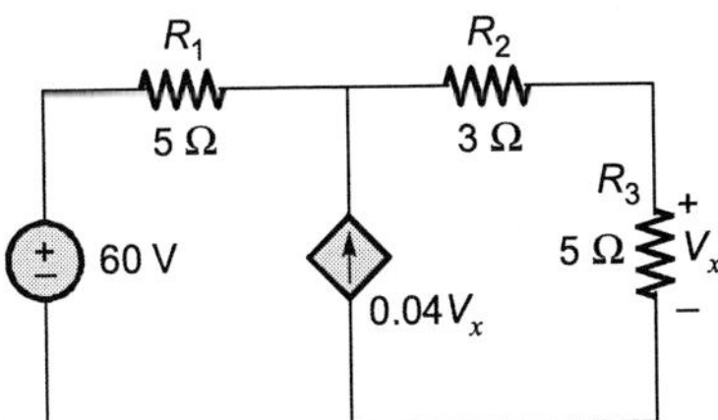

[2016 : 2 M, Set-2]

1.52 In the figure shown, the current *i* (in ampere) is ____.

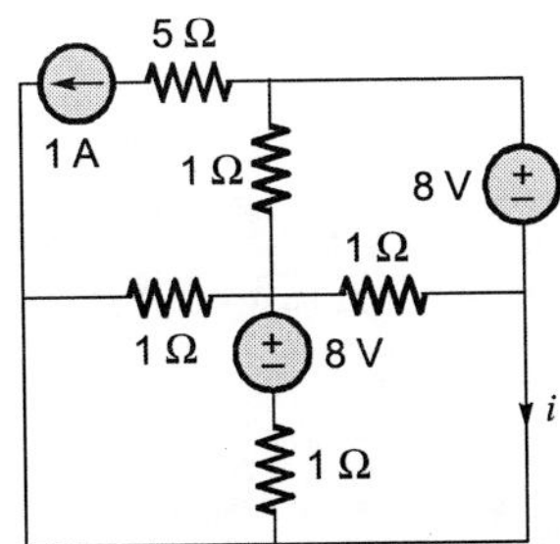

[2016 : 2 M, Set-3]

1.53 A connection is made consisting of resistance *A* in series with a parallel combination of resistances *B* and *C*. Three resistors of value 10 Ω, 5 Ω, 2 Ω are provided. Consider all possible permutations of the given resistors into the positions A, B, C and identify the configurations with maximum possible overall resistance, and also the ones with minimum possible overall resistance. The ratio of maximum to minimum values of the resistances (up to second decimal place) is _____ .

[2017 : 1 M, Set-2]

1.54 Consider the network shown below with $R_1 = 1\ \Omega$, $R_2 = 2\ \Omega$ and $R_3 = 3\ \Omega$. The network is connected to a constant voltage source of 11 V.

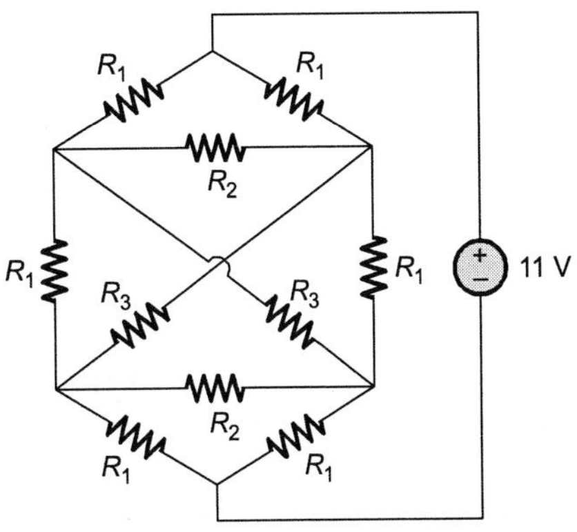

The magnitude of the current (in amperes, accurate to two decimal places) through the source is ____.

[2018 : 2 M]

1.55 Consider the circuit shown in the figure.

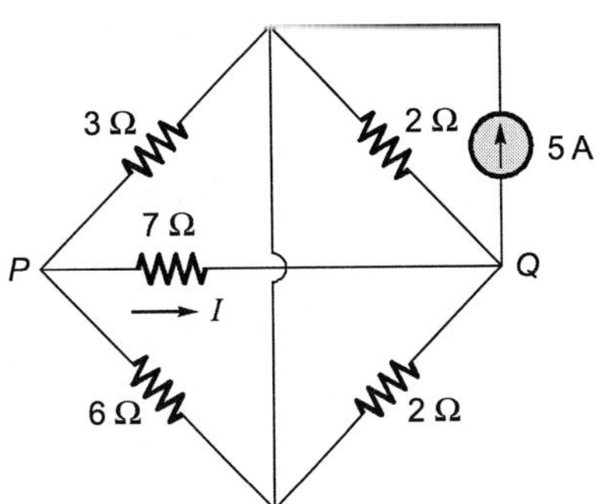

The current I flowing through the 7 Ω resistor between *P* and *Q* (rounded off to one decimal place) is _____ A.

[2021 : 1 M]

1.56 Consider the circuit shown in the figure.

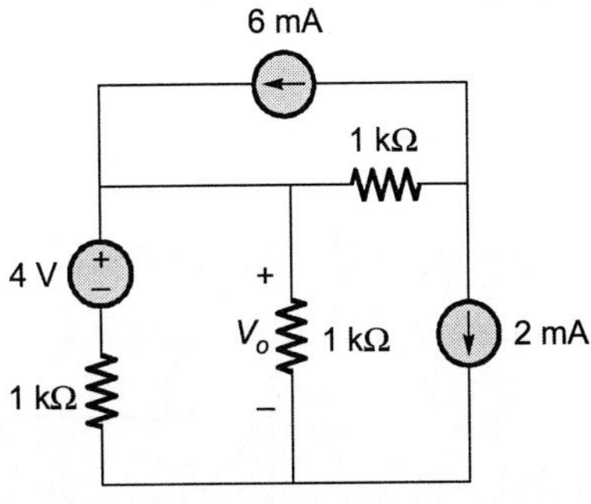

The value of V_o (rounded off to one decimal place) is _________ V.

[2021 : 1 M]

1.57 The current I in the circuit shown is ________.

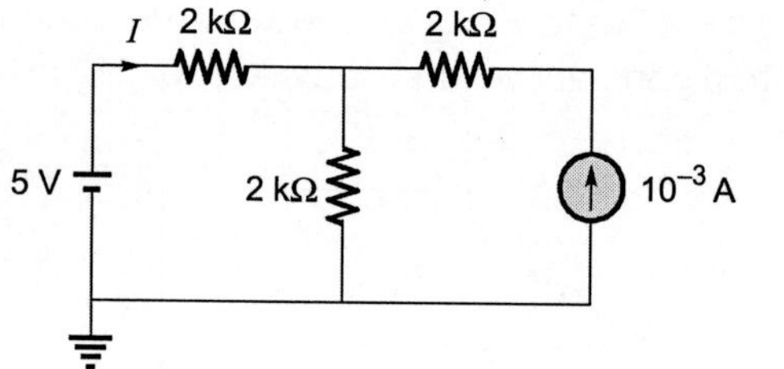

(a) 1.25×10^{-3} A (b) 0.75×10^{-3} A
(c) -0.5×10^{-3} A (d) 1.16×10^{-3} A

[2022 : 1 M]

1.58 Consider the circuit shown in the figure. The current I flowing through the 10 Ω resistor is ________.

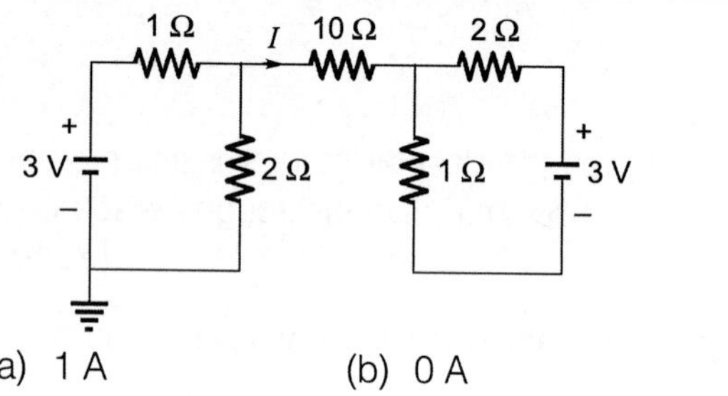

(a) 1 A (b) 0 A
(c) 0.1 A (d) −0.1 A

[2022 : 1 M]

1.59 In the circuit shown below, the current i flowing through 200 Ω resistor is ______ mA. (rounded off to two decimal places).

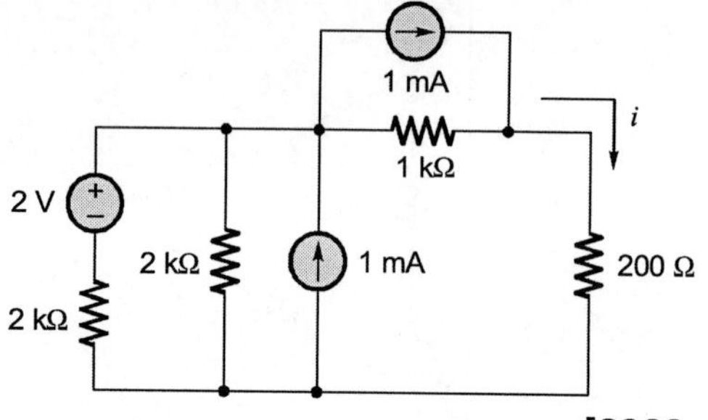

[2023 : 1 M]

1.60 In the given circuit, the current I_x (in mA) is ____.

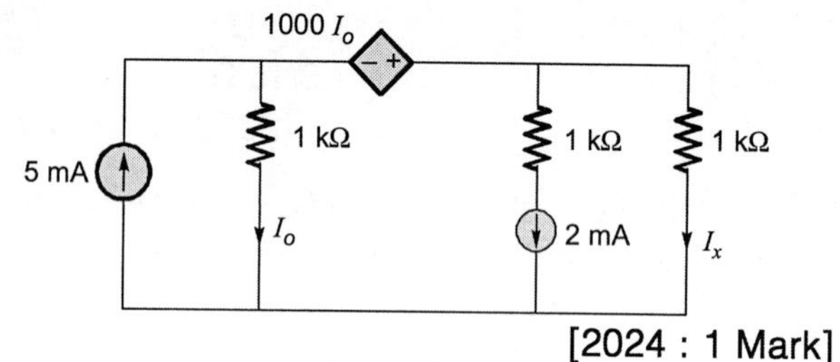

[2024 : 1 Mark]

Answers Basics of Network Analysis

1.1	(d)	1.2	(b)	1.3	(a)	1.4	(d)	1.5	(a)	1.6	(b)	1.7	(c)
1.8	(b)	1.9	(c)	1.10	(d)	1.11	(d)	1.12	(a)	1.13	(d)	1.14	(c)
1.15	(a)	1.16	(d)	1.17	(a)	1.18	(d)	1.19	(a)	1.20	(a)	1.21	(*)
1.22	(d)	1.23	(d)	1.24	(d)	1.25	(b)	1.26	(c)	1.27	(a)	1.28	(c)
1.29	(a)	1.30	(b)	1.31	(c)	1.32	(b)	1.33	(a)	1.34	(b)	1.35	(b)
1.36	(c)	1.37	(d)	1.38	(c)	1.39	(a)	1.40	(29.09)	1.41	(10)	1.42	(d)
1.43	(c)	1.44	(2.8)	1.45	(2.62)	1.46	(100)	1.47	(a)	1.48	(8)	1.49	(1)
1.50	(d)	1.51	(5)	1.52	(−1)	1.53	(2.143)	1.54	(8)	1.55	(0.5)	1.56	(1)
1.57	(b)	1.58	(b)	1.59	(1.36)	1.60	(2)						

Explanations Basics of Network Analysis

1.1 (d)

$$L = L_1 + L_2 \pm 2M = L_1 + L_2 \pm 2k\sqrt{L_1 L_2}$$

$$L = 2 + 2 \pm 2(0.1)\sqrt{2 \times 2} = 4 \pm 0.4$$

$$L = 3.6 \text{ H and } 4.4 \text{ H}$$

1.2 (b)

Apply KCL at node A

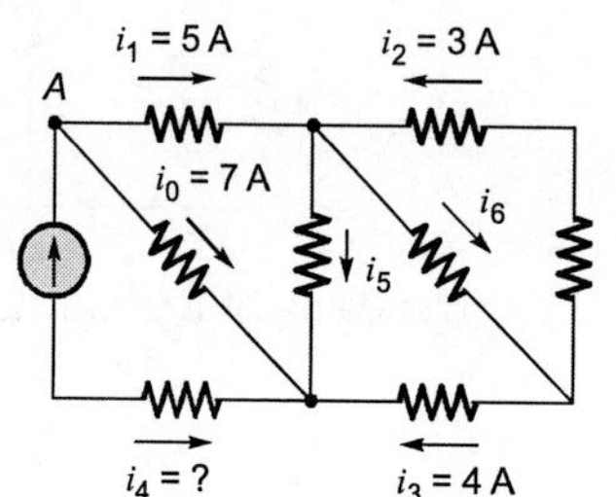

$$i_0 + i_1 + i_4 = 0$$

$$7 + 5 + i_4 = 0$$

$$i_4 = -12 \text{ A}$$

1.3 (a)

Apply KVL, $V + 5 - 4 - 4 = 0$

$$V = 3 \text{ V}$$

1.4 (d)

$$V = V_{2\text{A}} + 2 \times 2 + 5 = V_{2\text{A}} + 9$$

Since the voltage of 2 A current source is not known. So, it is not possible to find the value of voltage V.

1.5 (a)

Voltage in parallel is always equal.

1.6 (b)

The bridge is balanced

$$Z_{eq} = (2 \parallel 4) + (2 \parallel 4)$$

$$Z_{eq} = \frac{2\times4}{2+4} + \frac{2\times4}{2+4} = \frac{4}{3} + \frac{4}{3} = \frac{8}{3}\ \Omega$$

1.7 (c)

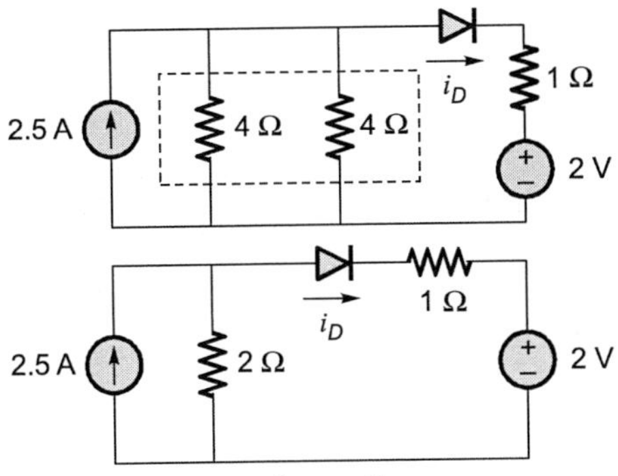

Using source transformation,

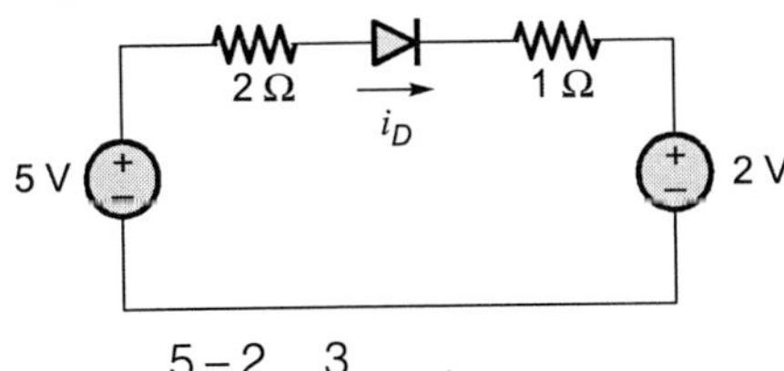

$$i_D = \frac{5-2}{2+1} = \frac{3}{3} = 1\ \text{Amp.}$$

1.8 (b)

The nodal or mesh method is based on KCL and Ohm's law.

1.9 (c)

Apply superposition theorem

For 1 volt source

$$V_{ab_1} = 1\times\frac{2}{2+2} = 0.5\ \text{V}$$

For 3 A source,

$$V_{ab_3} = 3\times\left(\frac{2}{2+2}\right)\times 2 = 3\ \text{V}$$

$$V_{ab} = V_{ab_1} + V_{ab_3} = 0.5 + 3 = 3.5\ \text{V}$$

1.10 (d)

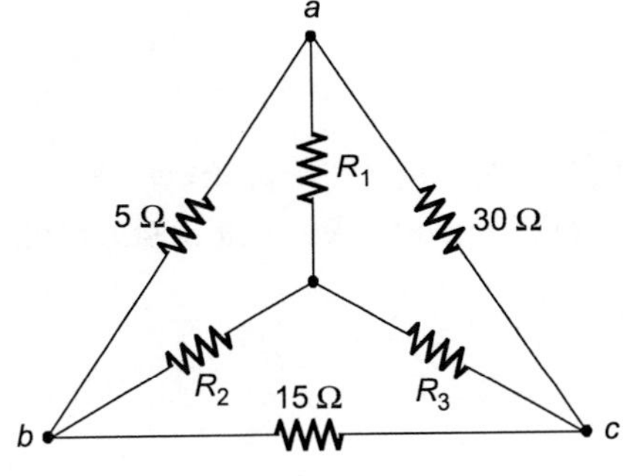

$$R_1 = \frac{5\times30}{5+30+15} = 3 \quad ; \quad R_2 = \frac{15\times5}{50} = 1.5$$

$$R_3 = \frac{15\times30}{50} = 9$$

1.11 (d)

Applying KCL at the node (1),

$$e^{at} + e^{bt} = i_L(t)$$

$$\Rightarrow v(t) = L\frac{d}{dt}\left[e^{at} + e^{bt}\right] = ae^{at} + be^{bt}$$

1.12 (a)

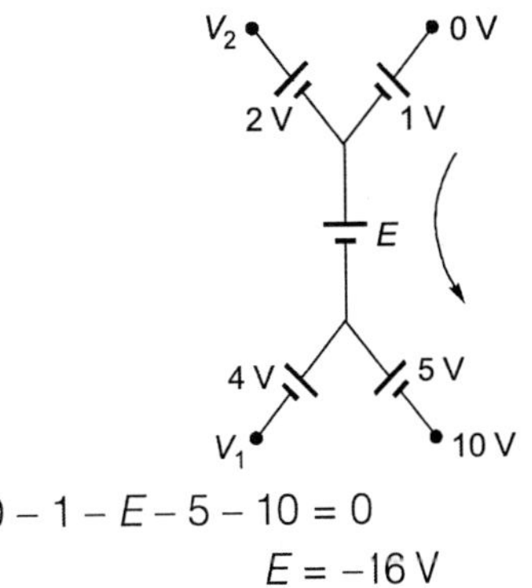

$$0 - 1 - E - 5 - 10 = 0$$

$$E = -16\ \text{V}$$

1.13 (d)

Since diode is forward bias it is taken as short circuit.

Applying KCL,

$$\frac{V-4}{2} + \frac{V}{2} + \frac{V+2}{2} = 0$$

$$3\ V = 2 \qquad \Rightarrow V = 2/3$$

$$\Rightarrow \qquad V_o = -V = -\frac{2}{3}$$

1.14 (c)

Applying KCL, $\dfrac{e_o - 12}{4} + \dfrac{e_o}{4} + \dfrac{e_o}{4} = 0$

$$\Rightarrow \qquad 3e_o = 12 \qquad \therefore\ e_o = 4\ \text{V}$$

1.15 (a)

$$Z_\Delta = 3Z_Y \quad \Rightarrow \sqrt{3}Z_\Delta = 3Z_Y \quad ; \quad Z_Y = \frac{Z_\Delta}{\sqrt{3}}$$

1.16 (d)

Applying source conversion

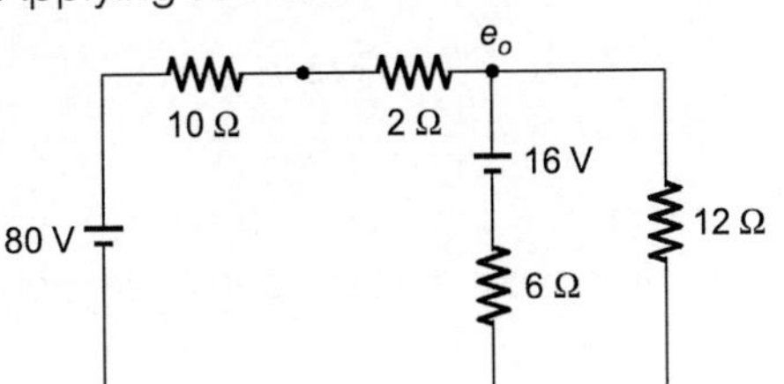

$$\frac{e_o - 80}{12} + \frac{e_o}{12} + \frac{e_o - 16}{6} = 0$$

$$4e_o = 112$$

$$e_o = \frac{112}{4} = 28 \text{ V}$$

1.17 (a)

Applying KVL, $20 - 5I - 5\left(I + \frac{V_1}{5}\right) = 0$

$20 - 10I - 20 = 0$

$\Rightarrow \quad I = 0$

$\therefore$ Only dependent source acts.

$$\frac{V_1}{5} = 4 \text{ A}$$

Power delivered $= I^2R = 16 \times 5 = 80$ W

1.18 (d)

$$3V_pI_p\cos\theta = 1500$$

$$3\left(\frac{V_L}{\sqrt{3}}\right)\left(\frac{V_L}{\sqrt{3}Z_L}\right)\cos\theta = 1500$$

$$Z_L = \frac{V_L^2 \cdot \cos\theta}{1500} = \frac{400^2 \times 0.844}{1500} = 90\ \Omega$$

$$\theta = \cos^{-1}(0.844) = 32.44$$

As power factor is leading, load is capacitive so angle will be negative.

$$\theta = -32.44°$$

1.19 (a)

As voltage at 1 node is known

$\therefore$ using nodal analysis only 3 equations required.

1.20 (a)

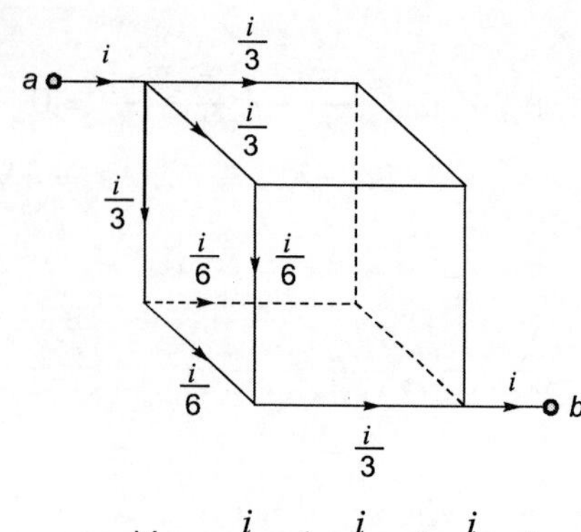

$$V_{ab} = \frac{i}{3}\times 1 + \frac{i}{6}\times 1 + \frac{i}{3}\times 1$$

$$\Rightarrow \quad R_{eq} = \frac{V_{ab}}{i} = \frac{5}{6}\ \Omega$$

1.21 (*)

Question is incomplete as L_1 and L_2 are not given.

1.22 (d)

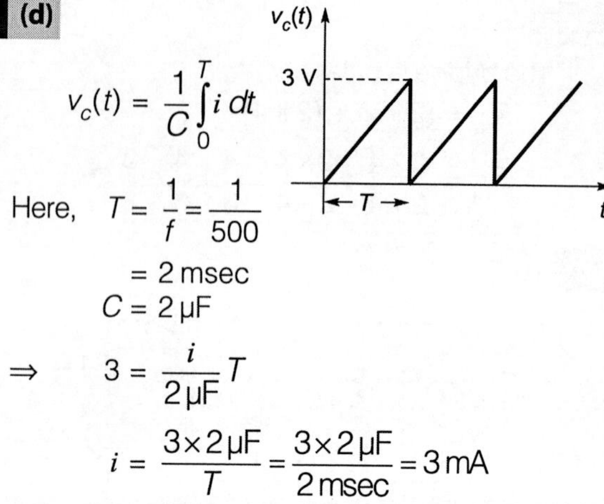

$$v_c(t) = \frac{1}{C}\int_0^T i\,dt$$

Here, $T = \frac{1}{f} = \frac{1}{500}$

$= 2$ msec

$C = 2\ \mu$F

$$\Rightarrow \quad 3 = \frac{i}{2\ \mu\text{F}}T$$

$$i = \frac{3\times 2\ \mu\text{F}}{T} = \frac{3\times 2\ \mu\text{F}}{2\text{ msec}} = 3\text{ mA}$$

Hence, the charging requires constant current source of 3 mA for 2 msec.

1.23 (d)

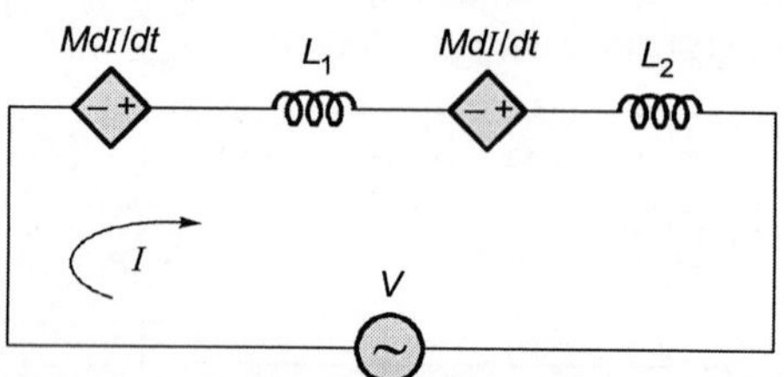

If current enters the dotted terminals of coil 1 then a voltage is developed across coil 2 whose higher potential is at dotted terminals.

$$V = \frac{-MdI}{dt} + \frac{L_1dI}{dt} - \frac{MdI}{dt} + L_2\frac{dI}{dt}$$

$$= (L_1 + L_2 - 2M)\frac{dI}{dt} \qquad V = L_{eq}\frac{dI}{dt}$$

1.24 (d)

$$H(s) = \frac{\frac{1}{sC}}{R + sL + \frac{1}{sC}} = \frac{1}{s^2LC + sCR + 1}$$

$$= \frac{1}{s^2(10\times 10^{-3}\times 100\times 10^{-6}) + s(10\times 10^3\times 100\times 10^{-6}) + 1}$$

$$H(s) = \frac{1}{10^{-6}s^2 + s + 1} = \frac{10^6}{s^2 + 10^6 s + 10^6}$$

1.25 (b)

$X = X_1 + X_2 + X_3 + 2X_m - 2X_m$

$= (j5 + j2 + j2 + j20 - j20)\ \Omega$

$= j9\ \Omega$ (one additive & other subtractive)

1.26 (c)

$$V_a = 5 \qquad (R_1 = R_2)$$

$$V_b = \frac{R_3}{R_3 + R_4} \times 10 = \frac{1.1}{2.1} \times 10$$

$$V = V_a - V_b = -0.238 \text{ V}$$

1.27 (a)

Since, the power is absorbed by 60 V source

$$I' = 12 - I$$

$$\Rightarrow \quad I' > 0$$

$$12 - I > 0$$

$$I < 12 \text{ A}$$

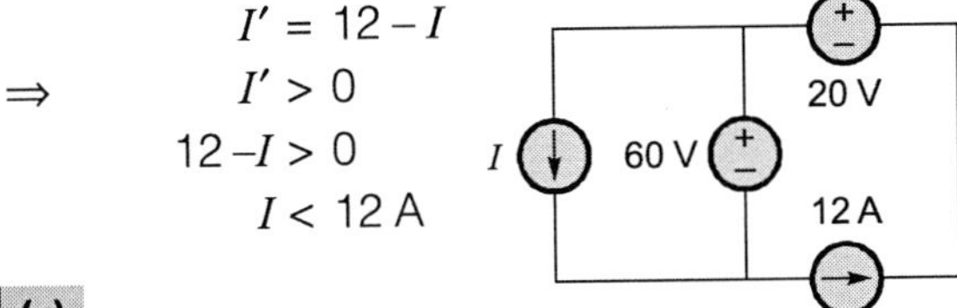

1.28 (c)

$$P = VI$$

$$\text{Energy} = P \cdot t = V \cdot It = (V.t)I$$

$$I = 2\text{A} \quad \text{(given)}$$

$$V \cdot t = \text{Area under } V - t \text{ curve}$$

$$V \cdot t = \left(\frac{1}{2} \times 2 \times 600\right) + (10 \times 600)$$

$$= 600 + 6000$$

$$V \cdot t = 6600$$

$$E = (6600) \times 2 = 13200 = 13.2 \text{ kJ}$$

1.29 (a)

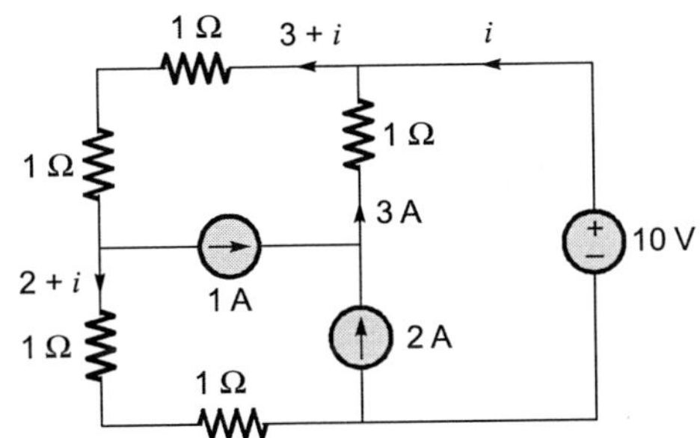

Applying KVL in outer loop,

$$(3 + i)2 + (2 + i)2 = 10$$

$$\Rightarrow 6 + 2i + 4 + 2i = 10$$

$$\Rightarrow \quad 4i = 0$$

$$\Rightarrow \quad i = 0$$

Power supplied by the voltage source,

$$P = Vi = 10 \times 0 = 0 \text{ W}$$

1.30 (b)

Converting delta into star, the circuit can be redrawn as below:

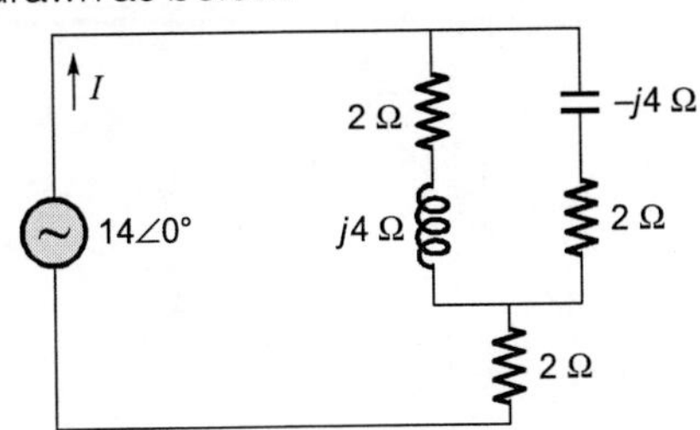

Equivalent impedance of the circuit,

$$Z = (2 + j4) || (2 - j4) + 2$$

$$\Rightarrow Z = \frac{(2 + j4)(2 - j4)}{2 + j4 + 2 - j4} + 2 = \frac{4 + 16}{4} + 2 = 7\,\Omega$$

Therefore,

$$\text{Current,} \quad I = \frac{14\angle 0^\circ}{7} = 2\angle 0^\circ \text{ A}$$

1.31 (c)

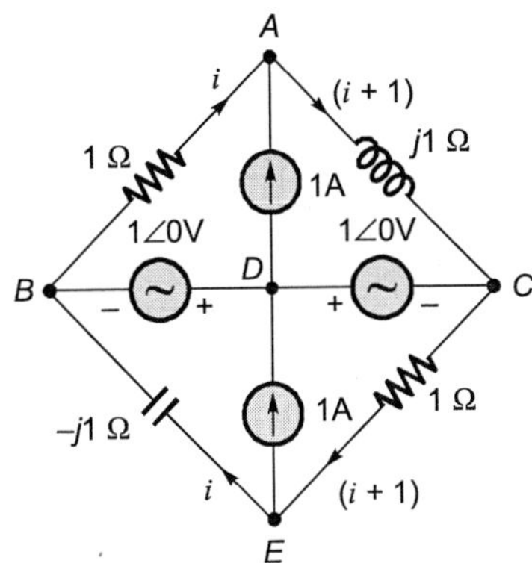

According to KCL at node D there will be no current in voltage sources.

According to KCL at node A current through inductor will be

$$i_1 = i + 1 \qquad \text{...(1)}$$

Applying KVL in loop $ACDBA$ we have

$$1 \times i + (i+1)\, j1 + 1\angle 0 - 1\angle 0 = 0$$

$$i + (i + 1)j = 0$$

$$(1 + j)i = -j$$

$$i = \frac{-j}{1 + j} \qquad \text{...(2)}$$

Therefore from (1) and (2) we have

$$i_1 = i + 1 = \frac{-j}{j + 1} + 1 = \frac{1}{1 + j}$$

1.32 (b)

Average power is same as RMS power.

$$P = I_{rms}^2 R = \left(\frac{5}{\sqrt{2}}\right)^2 \times 4 = \frac{25}{2} \times 4 = 50 \text{ W}$$

Note: Power is consumed only by resistance i.e. by real part of impedance.

1.33 (a)

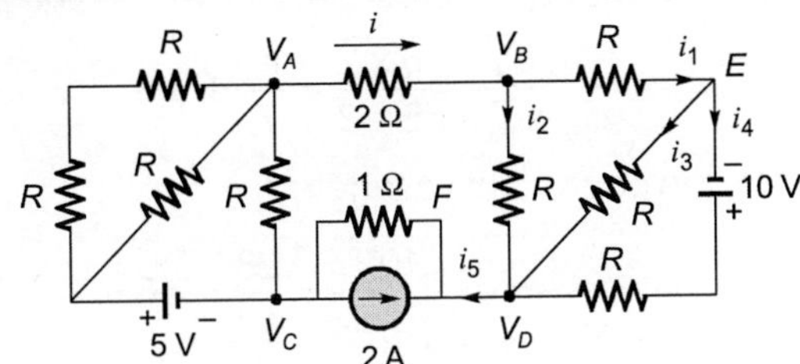

$$i = \frac{V_A - V_B}{2} = \frac{6}{2} = 3\text{ A} \quad ...(i)$$

KCL at node *B*, we have

$$i = i_2 + i_1$$

$$i_2 + i_1 = 3\text{ A} \quad ...(ii)$$

KCL at node *E*, we have

$$i_1 = i_3 + i_4 \quad ...(iii)$$

KCL at node *D* we have

$$i_5 = i_2 + i_3 + i_4 = i_2 + i_1 = 3\text{ A}$$

KCL at node *F*, we have

$$i_6 + 2 + i_5 = 0$$

$$i_6 = -2 - i_5 = -5\text{ A}$$

So, $V_C - V_D = 1 \times i_6 = -5\text{ V}$

1.34 (b)

$$R_A = \frac{R_b R_c}{R_a + R_b + R_c}$$

$$R'_a = kR_a; \quad R'_b = kR_b; \quad R'_c = kR_c$$

$$R'_A = \frac{kR_b \cdot kR_c}{kR_a + kR_b + kR_c} = \frac{k^2 R_b R_c}{k(R_A + R_b + R_c)}$$

$$= k \times \frac{R_b R_c}{R_a + R_b + R_c}$$

$$R'_A = kR_A$$

1.35 (b)

$$V_{YZ_1} = 100 \times 1.25 \times 0.8 = 100\text{ V}$$

In second case, when 100 V is applied at *YZ* terminals, this whole 100 V will appear across the secondary winding.

Hence, $V_{WX_2} = \dfrac{100}{1.25} = 80\text{ V}$

$$\Rightarrow \quad \frac{Y_{YZ_1}}{Y_{WX_1}} = \frac{100}{100}, \frac{V_{WX_2}}{V_{YZ_2}} = \frac{80}{100}$$

1.36 (c)

$$Q = CV$$

$$Q_1 = C_1 V_1 = 10 \times 10^{-6} \times 10 = 100\ \mu\text{C}$$

$$Q_2 = C_2 V_2 = 5 \times 10^{-6} \times 5 = 25\ \mu\text{C}$$

$$Q_3 = C_3 V_3 = 2 \times 10^{-6} \times 2 = 4\ \mu\text{C}$$

Capacitors C_2 and C_3 are in series.

In series charge is same.

So, the maximum charge on C_2 and C_3 will be minimum of $(Q_2, Q_3) = \min(25\ \mu\text{C}, 4\ \mu\text{C}) = 4\ \mu\text{C} = Q_{23}$.

In series the equivalent capacitance of C_2 and C_3 is

$$C_{23} = \frac{C_2 C_3}{C_2 + C_3} = \frac{5 \times 2}{5 + 2} = \frac{10}{7}\ \mu\text{F}$$

So, the equivalent voltage

$$V_{23} = \frac{Q_{23}}{C_{23}} = \frac{4 \times 10^{-6}}{\frac{10}{7} \times 10^{-6}} = \frac{28}{10} = 2.8\text{ V}$$

In parallel, the voltage is same.

$$V_1 = V_{23} = 2.8\text{ V}$$

Charge in capacitor C_1,

$$Q_1 = C_1 V_1 = 10 \times 10^{-6} \times 2.8 = 28\ \mu\text{C}$$

In parallel, the total charge

$$Q = Q_1 + Q_{23} = 4 + 28 = 32\ \mu\text{C}$$

1.37 (d)

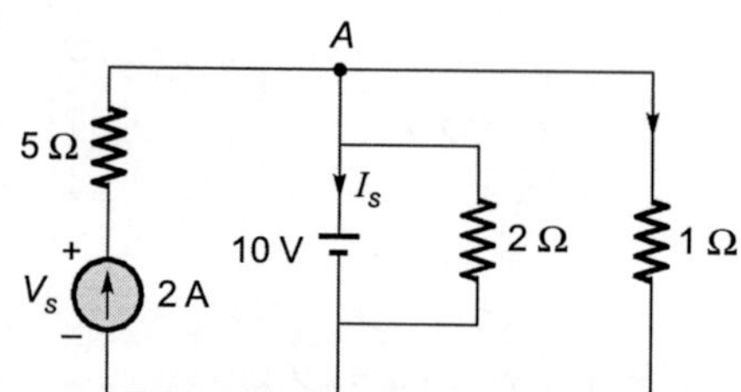

Voltage across 1 Ω resistance,

$$V_1 = 10\text{ V}$$

Current through 1 Ω resistance,

$$I_1 = \frac{10}{1} = 10\text{ Amp.}$$

Voltage across 2 Ω resistance,

$$V_2 = 10\text{ V}$$

Current through 2 Ω resistance,

$$I_2 = \frac{10}{2} = 5\text{ Amp.}$$

Apply KCL at node *A*,

$$-2 + I_s + I_2 + I_1 = 0$$

$$I_s = 2 - I_1 - I_2 = 2 - 10 - 5$$

$$I_s = -13\text{ Amp}$$

Voltage at node *A*,

$$V_A = 10\text{ V}$$

$$V_s - 10 = 10\text{ V}$$

$$V_s = 10 + 10 = 20\text{ V}$$

1.38 (c)

The current in the 1 Ω resistor

$$I_1 = \frac{10}{1} = 10\text{ Amp.}$$

1.39 (a)

Given data: $i_1 = 2\text{ A}, i_4 = -1\text{ A}, i_5 = -4\text{ A}$

$R = 1\ \Omega$

To calculate:

$i_6 = ?$

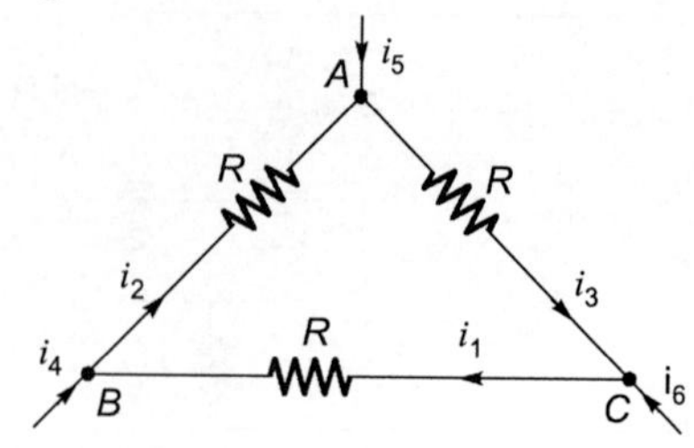

Using KVL at all the three nodes we get,

At node A,

$$i_5 - i_3 + i_2 = 0 \quad ...(i)$$

At node B,

$$i_4 + i_1 - i_2 = 0 \quad ...(ii)$$

At node C,

$$i_6 + i_3 - i_1 = 0 \quad ...(iii)$$

By putting the value of i_3 and i_2 from equation (i) and (ii) in equation (iii) we get,

$$i_6 + (i_2 + i_5) - i_1 = 0$$

$$i_6 + (i_1 + i_4 + i_5) - i_1 = 0$$

$$\therefore \quad i_6 + (2 - 1 - 4) - 2 = 0$$

$$i_6 = 5 \text{ A}$$

1.40 (29.09)

According to the question

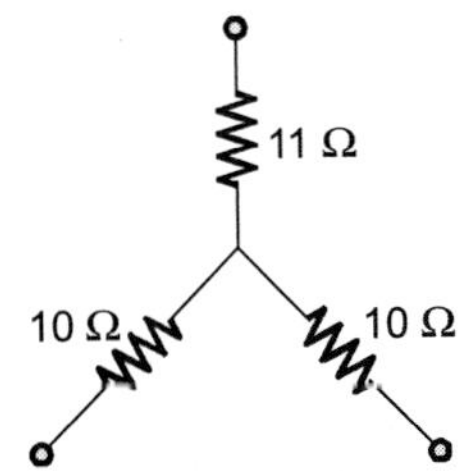

The equivalent Δ-network of the above Y-network is

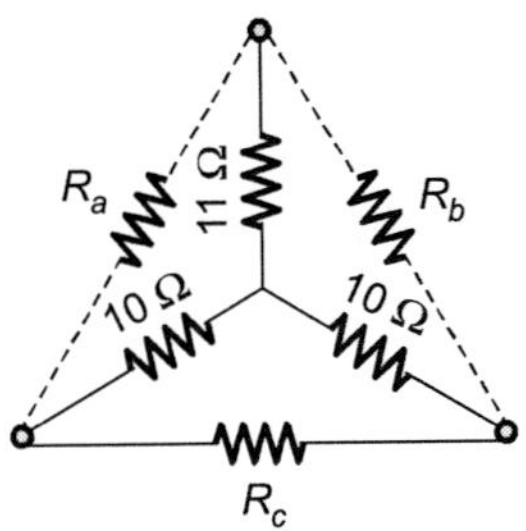

$$\text{Here, } R_a = 10 + 11 + \frac{10 \times 11}{10} = 32\ \Omega$$

$$R_b = 10 + 11 + \frac{10 \times 11}{10} = 32\ \Omega$$

$$R_c = 10 + 10 + \frac{10 \times 10}{11} = 29.09\ \Omega$$

Hence, the lowest value among the three resistances is 29.09 Ω.

1.41 (10)

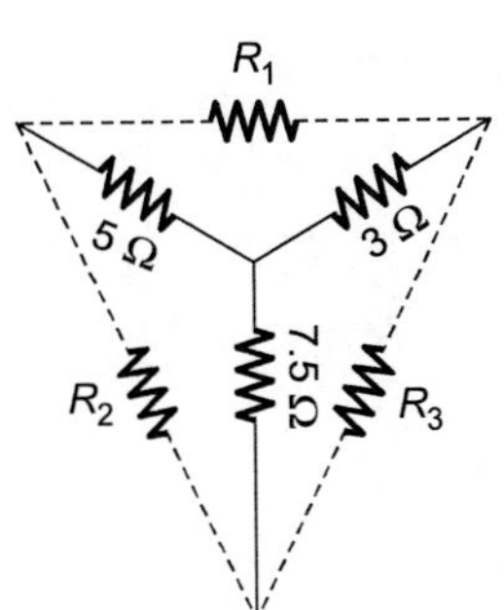

Using star-delta conversions,

The value of R_1 is given by

$$= 5 + 3 + \frac{5 \times 3}{7.5} = 10$$

1.42 (d)

Using the concept of super node, we get

$$V_1 - V_2 = 10\angle 0^\circ \quad ...(i)$$

$$= \frac{V_1}{-3j} + \frac{V_2}{6j} + \frac{V_2}{6} = 4\angle 0^\circ \quad ...(ii)$$

$$= \frac{-2V_1 + V_2 + jV_2}{6j} = 4\angle 0^\circ \quad ...(iii)$$

From equation (i) and (iii)

$$V_2 = \frac{20 + j24}{(-1 + j)} = \frac{31.241\angle 50.194}{\sqrt{2}\angle 135^\circ}$$

$$= 22.091\ \angle -84.806$$

$$\text{or} \quad V_2 = 2 - 22j$$

1.44 (2.8)

Using source transformation, we get,

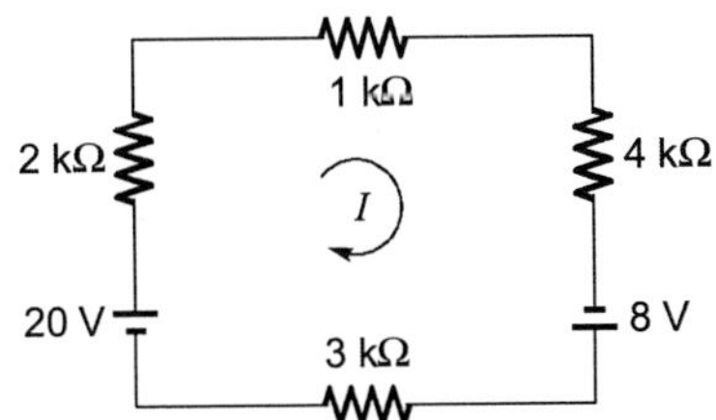

Applying KVL in above circuit, we get,

$$20 - 2I - I - 4I + 8 - 3I = 0$$

$$\text{or} \quad 28 = 10I \qquad \text{or} \quad I = 2.8 \text{ mA}$$

1.45 (2.62)

For an infinite ladder network, if all the resistance are comprises of same value R, then

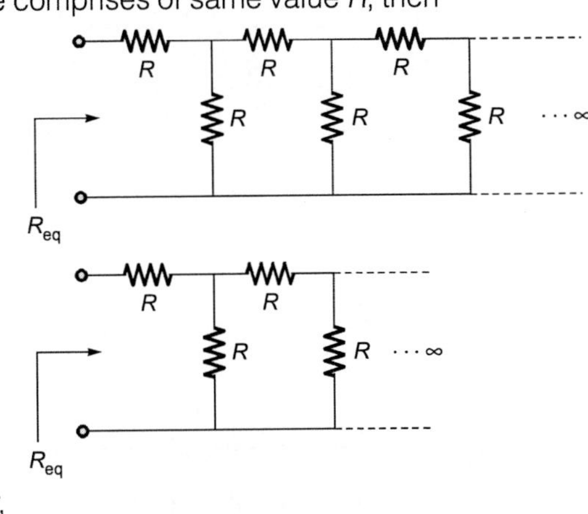

Or,

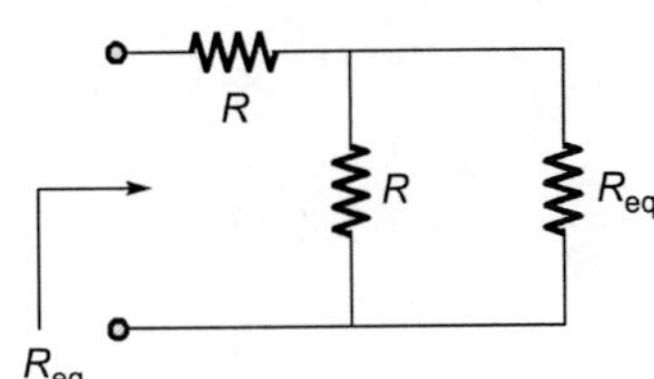

$$\therefore \quad R_{eq} = R + \frac{R \cdot R_{eq}}{R + R_{eq}} \qquad \text{...(i)}$$

After solving equation (i) we get,

$$R_{eq} = \left(\frac{1+\sqrt{5}}{2}\right)R \qquad \text{...(ii)}$$

From the given question, the circuit can be redraw as

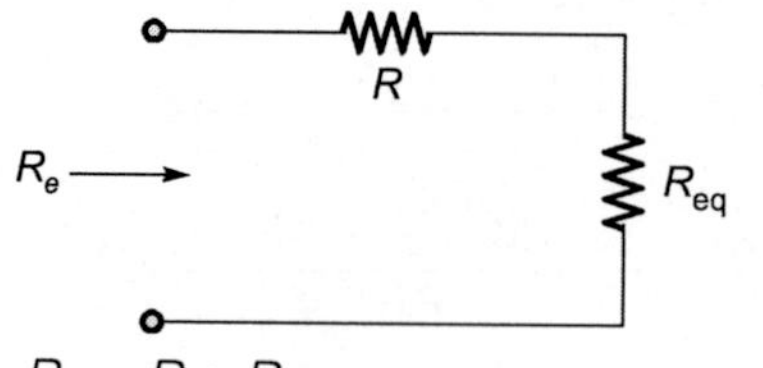

$$\therefore \quad R_e = R + R_{eq} \qquad \text{...(iii)}$$

From equation (ii) and (iii) we get,

$$R_e = R + \left(\frac{1+\sqrt{5}}{2}\right)R = 2.618\,R \qquad \text{...(iv)}$$

$$\text{or} \quad \frac{R_e}{R} = 2.618 = 2.62$$

1.46 (100)

Modifying the given circuit

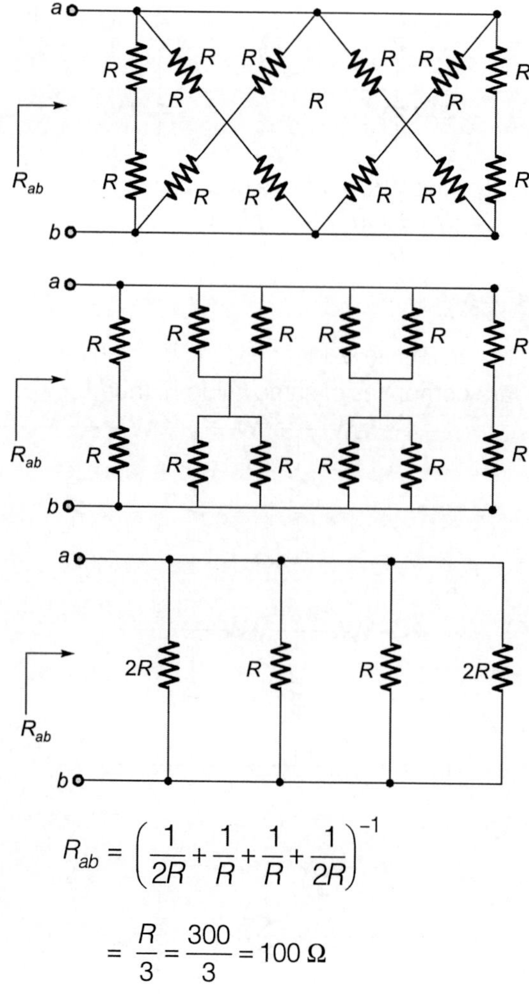

$$R_{ab} = \left(\frac{1}{2R} + \frac{1}{R} + \frac{1}{R} + \frac{1}{2R}\right)^{-1}$$

$$= \frac{R}{3} = \frac{300}{3} = 100\ \Omega$$

1.47 (a)

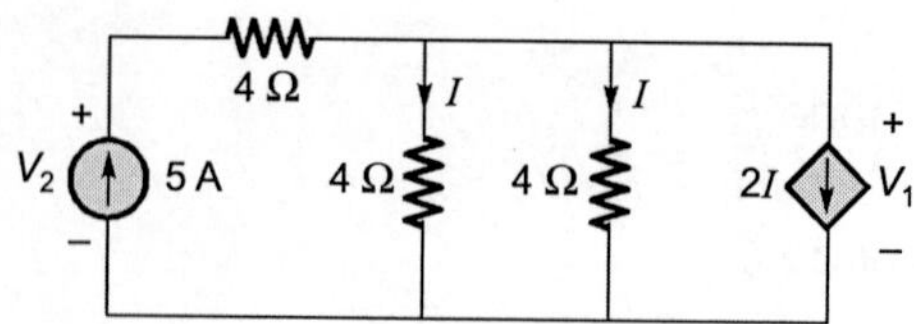

Current flowing through both the parallel 4 Ω will be I.

So, $V_2 = 4(I + I + 2I) + 4I$ by KVL

$I + I + 2I = 5$ by KCL

$$I = \frac{5}{4}A$$

$$V_2 = 4 \times 5 + \frac{4 \times 5}{4} = 25\text{ V}$$

$$V_1 = 4I = \frac{4 \times 5}{4} = 5\text{ V}$$

1.48 (8)

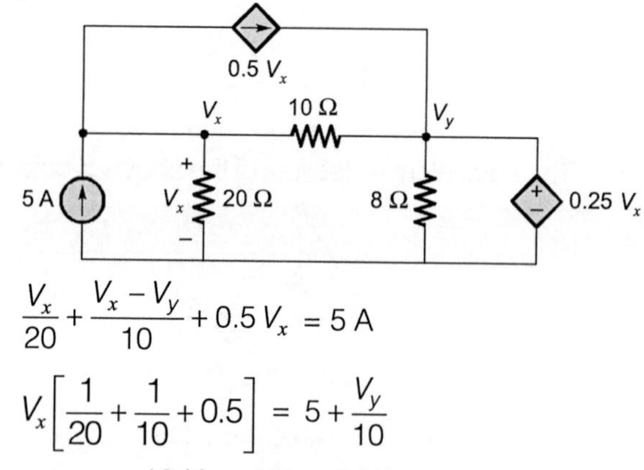

$$\frac{V_x}{20} + \frac{V_x - V_y}{10} + 0.5\,V_x = 5\text{ A}$$

$$V_x\left[\frac{1}{20} + \frac{1}{10} + 0.5\right] = 5 + \frac{V_y}{10}$$

$$13\,V_x = 100 + 2\,V_y \qquad \text{...(i)}$$

and also, $V_y = 0.25\,V_x$ ...(ii)

Solving equations (i) and (ii), we have

$$52\,V_y = 100 + 2\,V_y$$

$$\Rightarrow \quad 50\,V_y = 100 \Rightarrow V_y = 2\text{ V}$$

$$V_x = 4\,V_y = 8\text{ V}$$

1.49 (1)

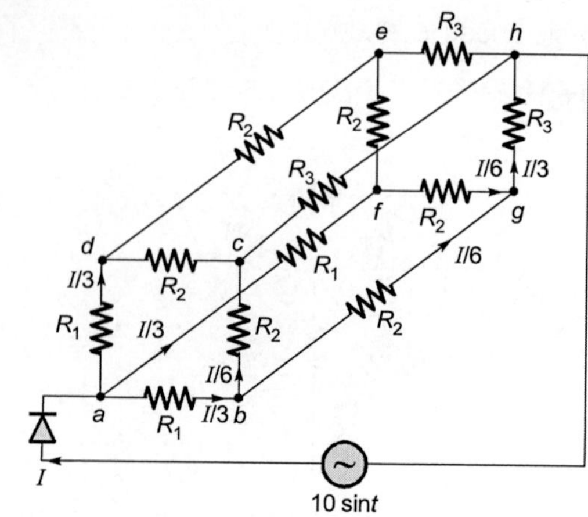

The equivalent resistance across terminal *ah* (outer loop) is

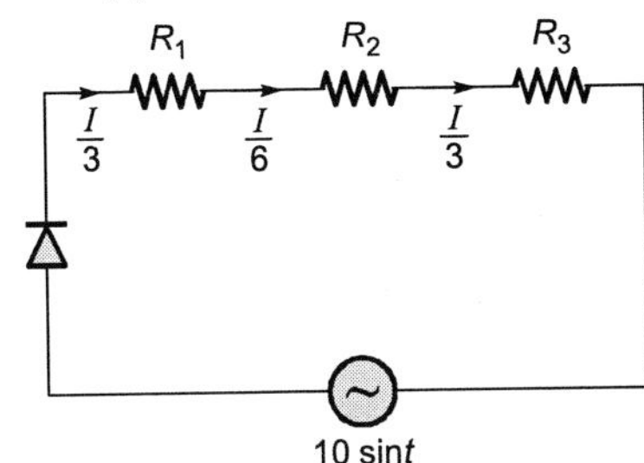

$$V = \frac{I}{3}\times 3\text{ k}\Omega + \frac{I}{6}\times 6\text{ k}\Omega + \frac{I}{3}\times 9\text{ k}\Omega = 5I$$

or $$\frac{V}{I} = 5\text{ k}\Omega$$

For half wave rectifier

$$I_{\text{rms}} = \frac{I_m}{(2)} = \frac{10\sin t}{5\text{ k}\Omega} = 2\sin t\text{ mA} = \frac{I_m}{2} = 1\text{ mA}$$

1.50 (d)

$R_{eq} \Rightarrow$

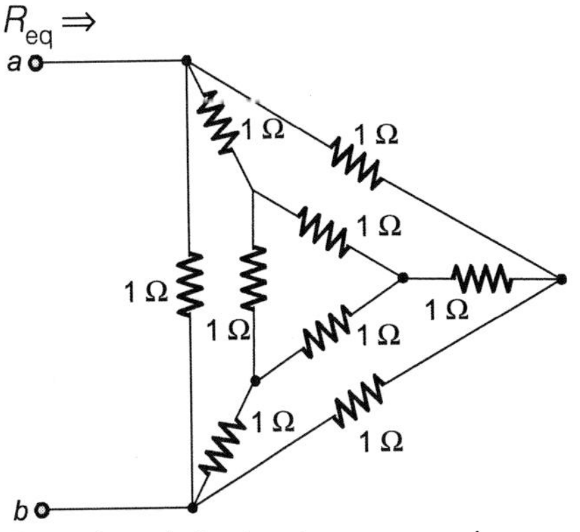

By using delta to star conversion

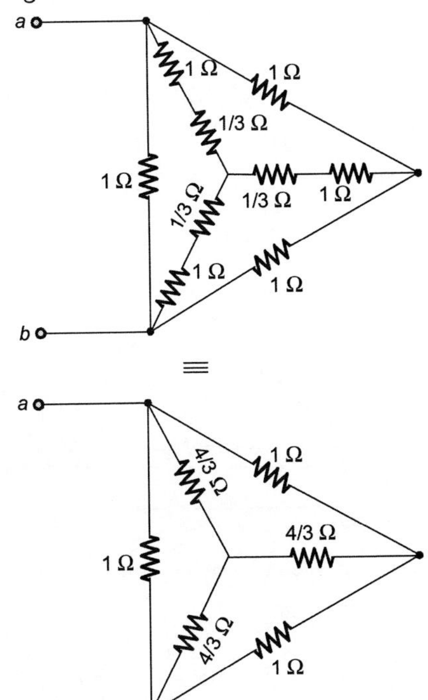

Again by star to delta conversion

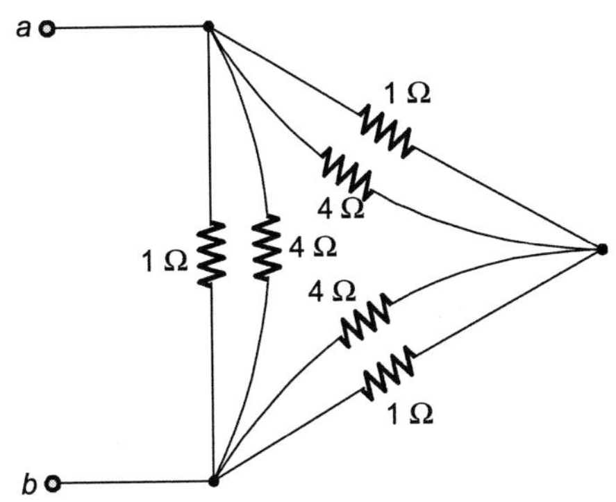

$$R_{ab} = \{4||1) + (4||1)\}||\{1||4)\}$$

$$= \left(\frac{4}{5}+\frac{4}{5}\right)\Bigg\|\frac{4}{5} = \frac{8}{15}\ \Omega$$

1.51 (5)

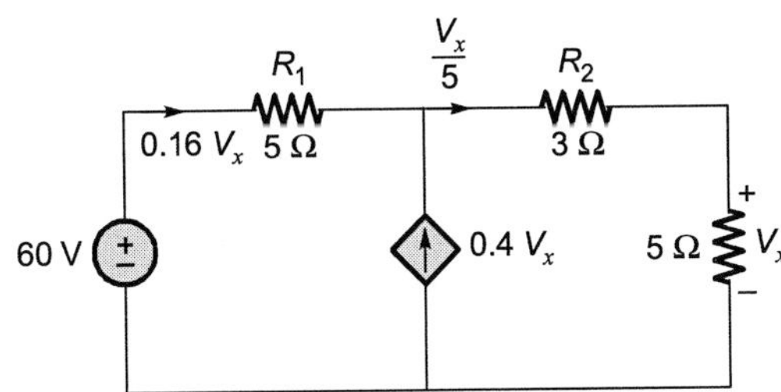

Using KVL in the outer loop

$$60 - 5(0.16\,V_x) - \frac{V_x}{5}\times 3 - V_x = 0$$

or $$V_x = 25\text{ V}$$

∴ The current flowing through

$$R_2 = \frac{V_x}{5} = \frac{25}{5} = 5\text{ A}$$

1.52 (–1)

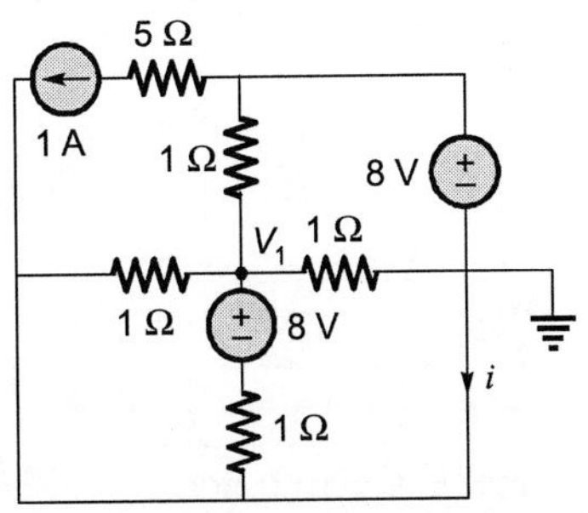

Using KCL at V_1

$$\frac{V_1}{1} + \frac{V_1 - 8}{1} + \frac{V_1 - 8}{1} + \frac{V_1}{1} = 0$$

or $$V_1 = 4\text{ V}$$

considering KVL, we get

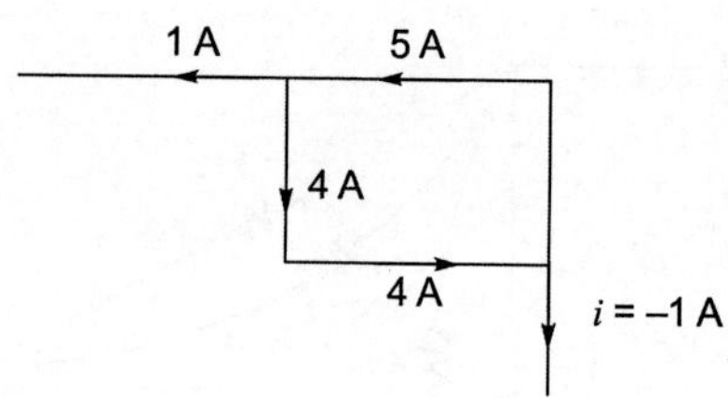

1.53 (2.143)

The connection of resistors is as shown below:

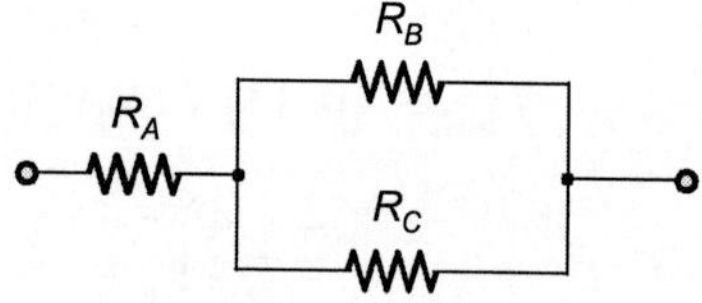

Given resistor values are : 10 Ω, 5 Ω, 2 Ω

The maximum resistance possible is,

$$R_{T(\max)} = 10\ \Omega + (5\ \Omega \parallel 2\ \Omega)$$

$$= \left(10+\frac{10}{7}\right)\Omega = \frac{80}{7}\ \Omega$$

The minimum resistance possible is,

$$R_{T(\min)} = 2\ \Omega + (10\ \Omega \parallel 5\ \Omega)$$

$$= \left(2+\frac{10}{3}\right)\Omega = \frac{16}{3}\ \Omega$$

$$\frac{R_{T(\max)}}{R_{T(\min)}} = \frac{80/7}{16/3} = \frac{15}{7} = 2.143$$

1.54 (8)

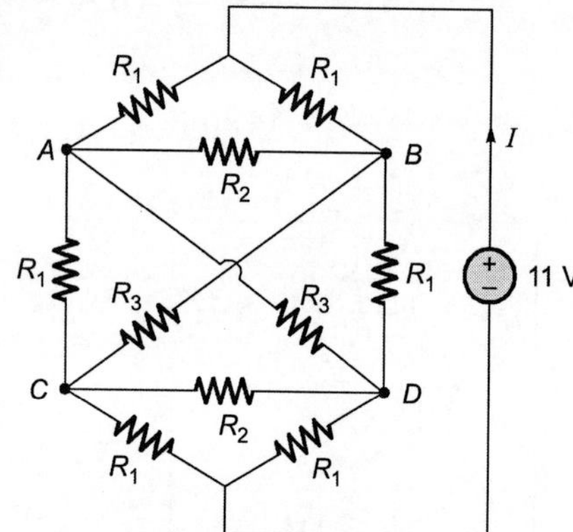

As the network is symmetric,

$$V_A = V_B \text{ and } V_C = V_D$$

So, current through R_2 resistors is zero and as $V_A = V_B$ and $V_C = V_D$, electrically the circuit can be reduced as,

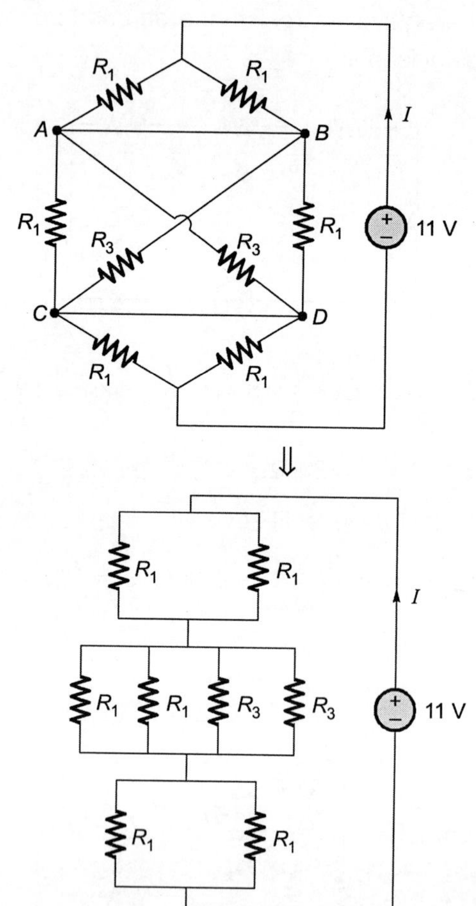

Total resistance,

$$R_T = 2(R_1 \parallel R_1) + (R_1 \parallel R_1 \parallel R_3 \parallel R_3) = R_1 + \left(\frac{R_1}{2} \middle\| \frac{R_3}{2}\right)$$

Given that, $R_1 = 1\ \Omega$ and $R_3 = 3\ \Omega$

So, $$R_T = 1+\left(\frac{1}{2} \middle\| \frac{3}{2}\right)\Omega = 1+\frac{3/2}{4} = \frac{11}{8}\ \Omega$$

$$I = \frac{11\text{V}}{R_T} = \frac{11}{(11/8)} = 8\text{ A}$$

1.55 (0.5)

Redraw the circuit,

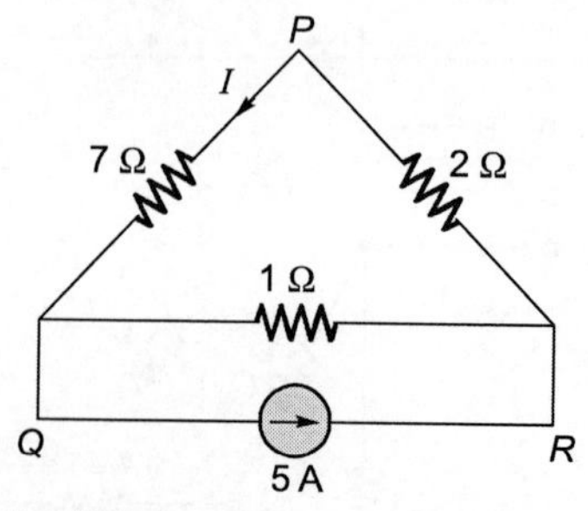

$$3\Omega \parallel 6\Omega = 2\Omega$$

$$2\Omega \parallel 2\Omega = 1\Omega$$

$$I = 5\times\frac{1}{10} = 0.5\text{ A}$$

1.56 (1)

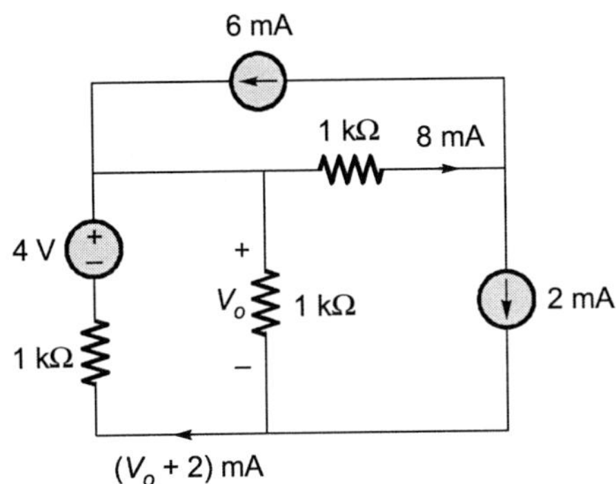

Write KVL equation first loop,

$$V_o - 4 + 1(V_o + 2) = 0$$

$$V_o = 1\text{ V}$$

1.57 (b)

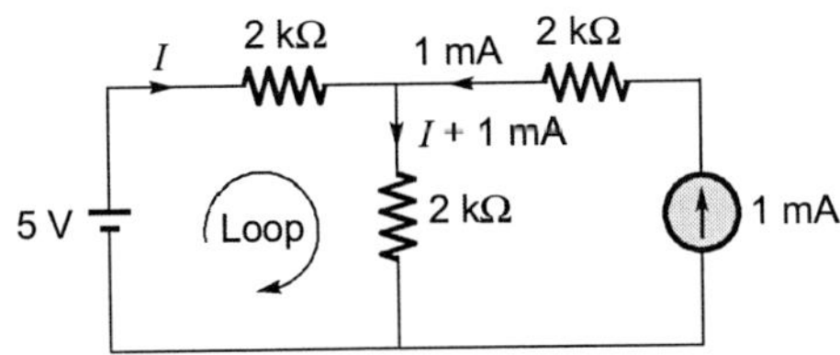

Applying KVL in the loop, we get,

$$5 = 2kI + 2k(I + 10^{-3})$$

$$5 = I(4k) + 2 \times 10^3 \times 10^{-3}$$

$$I(4k) = 3$$

$$I = \frac{3}{4k} = 0.75\text{ mA}$$

$$I = 0.75 \times 10^{-3}\text{ A}$$

1.58 (b)

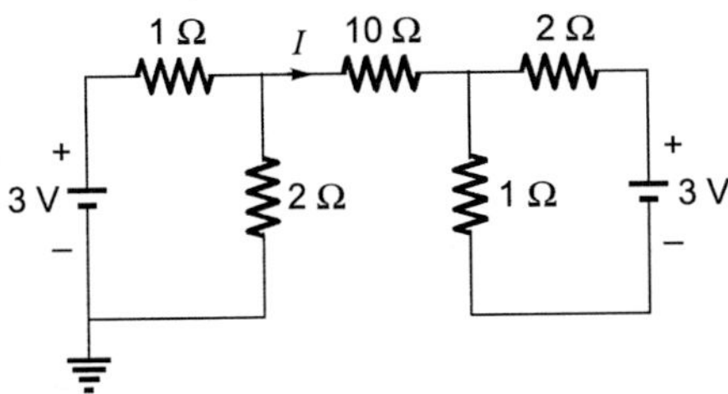

As there is no closed (or return) path for the current I to flow.

Therefore, $I = 0$

1.59 (1.36)

By applying source transformation,

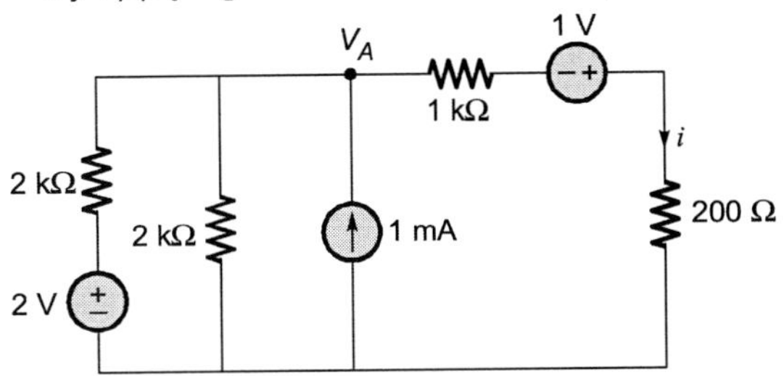

By applying nodal at node V_A,

$$\frac{V_A - 2}{2\text{ k}\Omega} + \frac{V_A}{2\text{ k}\Omega} + \frac{V_A + 1}{1.2\text{ k}\Omega} = 1\text{ mA}$$

$$V_A\left[\frac{1}{2} + \frac{1}{2} + \frac{1}{1.2}\right] = 1 + \frac{2}{2} - \left(\frac{1}{1.2}\right)$$

$$\therefore \quad V_A = 0.636\text{ V}$$

The current through 200 Ω resistor,

$$i = \frac{V_A + 1}{1.2\text{ k}\Omega} = \frac{0.636 + 1}{1.2} = 1.36\text{ mA}$$

1.60 (2)

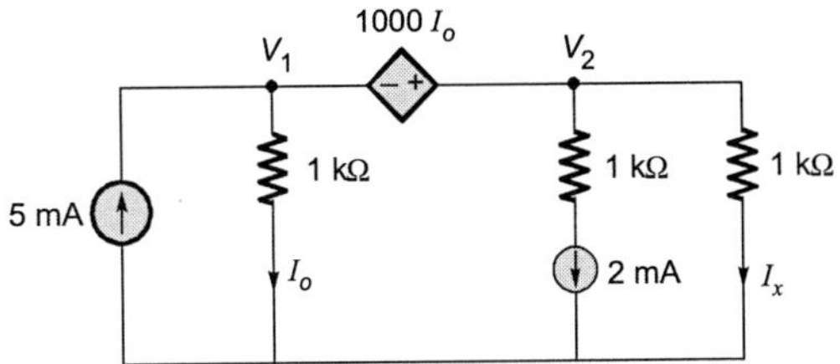

V_1 and V_2 form a super node. Applying KCL at V_1 and V_2 gives

$$\frac{V_1}{10^3} + \frac{V_2}{1 \times 10^3} + 2 \times 10^{-3} = 5 \times 10^{-3} \quad \text{...(i)}$$

Here, $$I_0 = \frac{V_1}{10^3}$$

and $$V_2 - V_1 = 10^3 I_0 = 10^3 \times \frac{V_1}{10^3} = V_1$$

$$V_2 = 2V_1$$

$$V_1 = \frac{V_2}{2} \quad \text{...(ii)}$$

Put in equation (i)

$$\frac{1}{2}V_2 + V_2 = 3$$

$$V_2 = 2\text{ V}$$

$$I_x = \frac{V_2}{10^3} = 2\text{ mA} = 2\text{ mA}$$

■■■

2 Sinusoidal Steady State

2.1 A DC voltage source is connected across a series *R-L-C* circuit. Under steady-state conditions, the applied DC voltage drops entirely across the

(a) *R* only (b) *L* only
(c) *C* only (d) *R* and *L* combination

[1995 : 1 M]

2.2 Consider a DC voltage source connected to a series *R-C* circuit. When the steady-state reaches, the ratio of the energy stored in the capacitor to the total energy supplied by the voltage source, is equal to

(a) 0.362 (b) 0.500
(c) 0.632 (d) 1.000 [1995 : 1 M]

2.3 The current, $i(t)$, through a 10-Ω resistor in series with an inductance, is given by

$$i(t) = 3 + 4\sin(100t + 45°) + 4\sin(300t + 60°) \text{ Amperes.}$$

The RMS value of the current and the power dissipated in the circuit are:

(a) $\sqrt{41}$ A, 410 W, respectively
(b) $\sqrt{35}$ A, 350 W, respectively
(c) 5 A, 250 W, respectively
(d) 11 A, 1210 W, respectively [1995 : 1 M]

2.4 A series *R-L-C* circuit has a *Q* of 100 and an impedance of $(100 + j0)\ \Omega$ at its resonant angular frequency of 10^7 radians/sec. The values of *R* and *L* are:

R = ____ ohms. *L* = ____ mH. [1995 : 1 M]

2.5 The rms value of a rectangular wave of period *T*, having a value of $+V$ for a duration, T_1 ($<T$) and $-V$ for the duration, $T - T_1 = T_2$, equals

(a) V (b) $\frac{T_1 - T_2}{T}V$
(c) $\frac{V}{\sqrt{2}}$ (d) $\frac{T_1}{T_2}V$ [1995 : 1 M]

2.6 In Fig., A_1, A_2 and A_3 are ideal ammeters. If A_2 and A_3 read 3 A and 4 A respectively, then A_1 should read

(a) 1 A
(b) 5 A
(c) 7 A
(d) None of the above

[1996 : 1 M]

Sinusoidal voltage source

2.7 In the circuit of figure, assume that the diodes are ideal and the meter is an average indicating ammeter. The ammeter will read

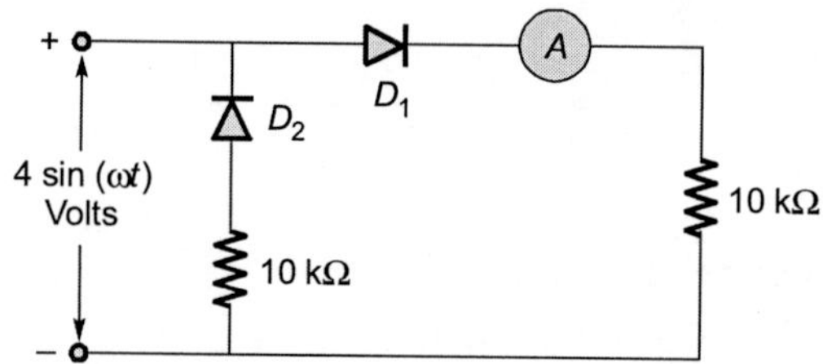

(a) $0.4\sqrt{2}$ mA (b) 0.4 mA
(c) $\frac{0.8}{\pi}$ mA (d) $\frac{0.4}{\pi}$ mA

[1996 : 1 M]

2.8 The parallel *RLC* circuit shown in figure is in resonance. In this circuit

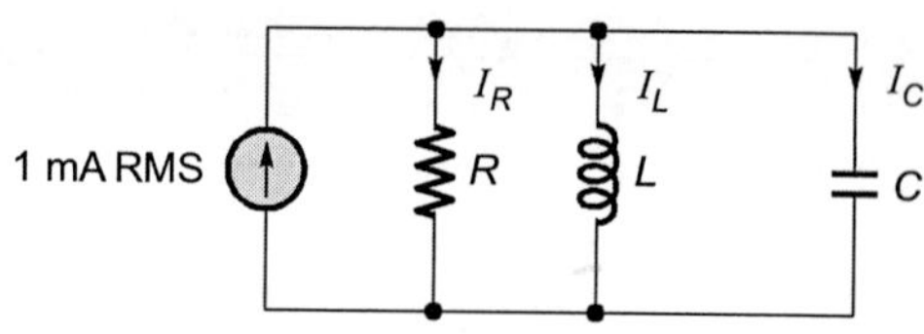

(a) $|I_R| < 1\text{mA}$ (b) $|I_R + I_L| > 1\text{mA}$
(c) $|I_R + I_C| < 1\text{mA}$ (d) $|I_L + I_C| > 1\text{mA}$

[1998 : 1 M]

2.9 When the angular frequency ω in the figure is varied from 0 to ∞, the locus of the current phasor I_2 is given by

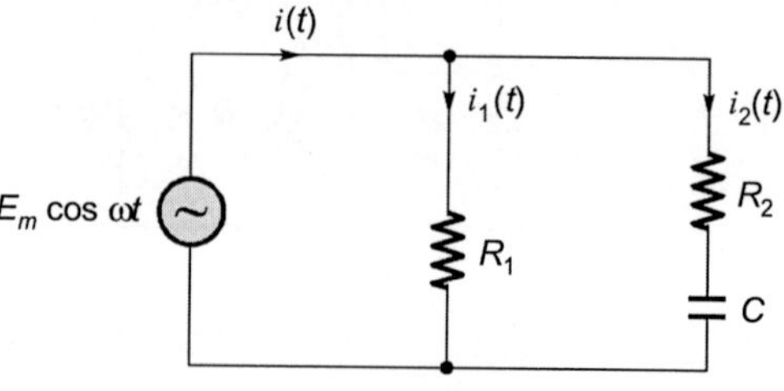

(a)

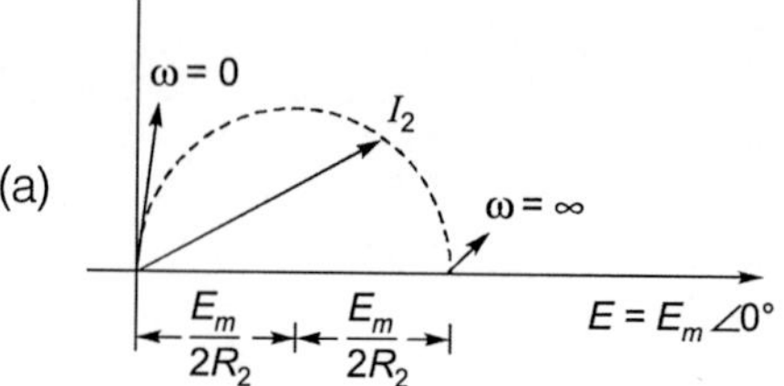

(b)

$\frac{E_m}{2R_2}$ $\frac{E_m}{2R_2}$ $\omega = 0$ $E = E_m \angle 0°$ I_2 $\omega = \infty$

(c)

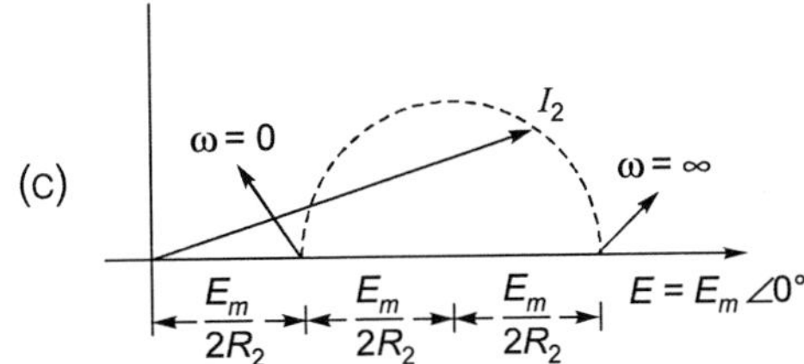

(d)

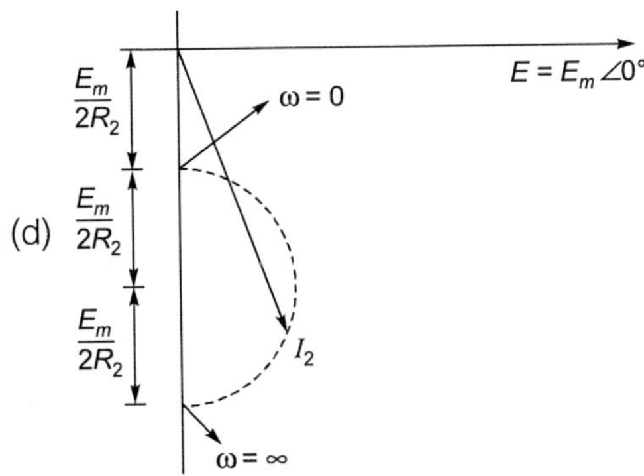

[2001 : 2 M]

2.10 A series RLC circuit has a resonance frequency of 1 kHz and a quality factor $Q = 100$. If each of R, L and C is doubled from its original value, the new Q of the circuit is

(a) 25 (b) 50

(c) 100 (d) 200 [2003 : 1 M]

2.11 An input voltage $v(t) = 10\sqrt{2}\cos(t + 10°) + 10\sqrt{5}\cos(2t + 10°)$ V is applied to a series combination of resistance $R = 1\ \Omega$ and an inductance $L = 1$ H. The resulting steady-state current $i(t)$ in ampere is

(a) $10\cos(t + 55°) + 10\cos(2t + 10° + \tan^{-1}2)$

(b) $10\cos(t + 55°) + 10\sqrt{\frac{3}{2}}\cos(2t + 55)$

(c) $10\cos(t - 35°) + 10\cos(2t + 10° - \tan^{-1}2)$

(d) $10\cos(t - 35) + 10\sqrt{\frac{3}{2}}\cos(2t - 35°)$

[2003 : 2 M]

2.12 The circuit shown in the figure, with $R = \frac{1}{3}\Omega$, $L = \frac{1}{4}$ H, $C = 3$ F has input voltage $V(t) = \sin 2t$. The resulting current $i(t)$ is

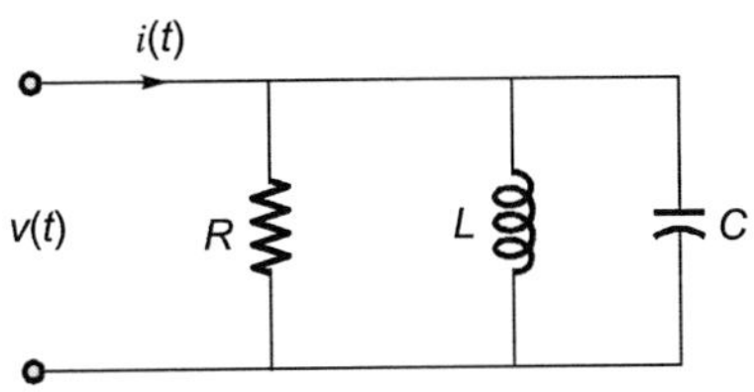

(a) $5\sin(2t + 53.1°)$

(b) $5\sin(2t - 53.1°)$

(c) $25\sin(2t + 53.1°)$

(d) $25\sin(2t - 53.1°)$ [2004 : 1 M]

2.13 For the circuit shown in the figure, the time constant $RC = 1$ ms. The input voltage is $v_i(t) = \sqrt{2}\sin 10^3 t$. The output voltage $v_0(t)$ is equal to

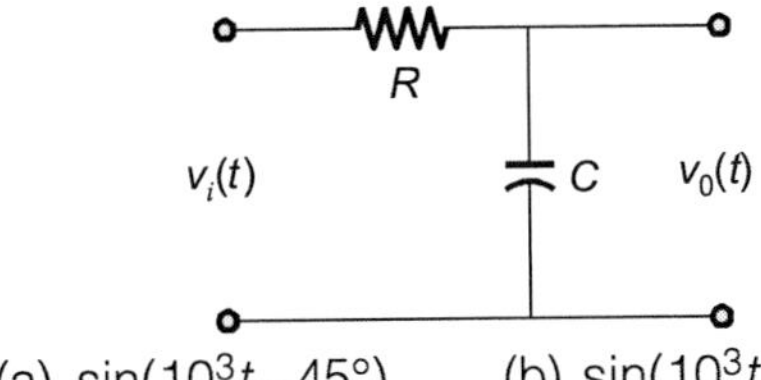

(a) $\sin(10^3t - 45°)$ (b) $\sin(10^3t + 45°)$

(c) $\sin(10^3t - 53°)$ (d) $\sin(10^3t + 53°)$

[2004 : 1 M]

2.14 The transfer function $H(s) = \frac{V_0(s)}{V_i(s)}$ of an R-L-C circuit is given by $H(s) = \frac{10^6}{s^2 + 20s + 10^6}$. The Quality factor ($Q$-factor) of this circuit is

(a) 25 (b) 50

(c) 100 (d) 5000 [2004 : 2 M]

2.15 Consider the following statements S_1 and S_2

S_1: At the resonant frequency the impedance of a series R-L-C circuit is zero.

S_2: In a parallel G-L-C circuit, increasing the conductance G results in increase in its Q factor.

Which one of the following is correct?

(a) S_1 is FALSE and S_2 is TRUE

(b) Both S_1 and S_2 are TRUE

(c) S_1 is TRUE and S_2 is FALSE

(d) Both S_1 and S_2 are FALSE [2004 : 2 M]

2.16 The condition on R, L and C such that the step response $y(t)$ in the figure has no oscillations, is

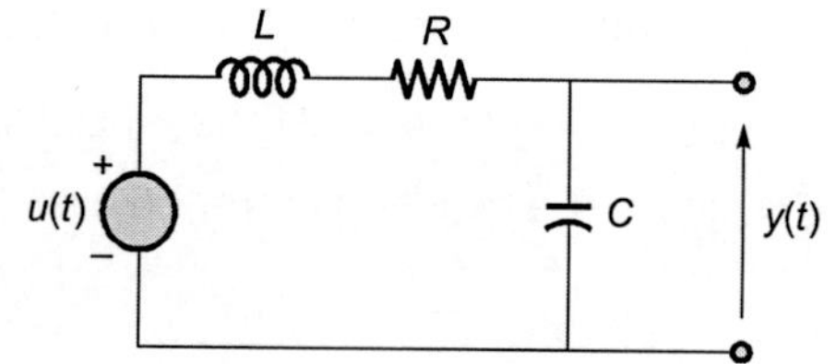

(a) $R \geq \frac{1}{2}\sqrt{\frac{L}{C}}$ (b) $R \geq \sqrt{\frac{L}{C}}$

(c) $R \geq 2\sqrt{\frac{L}{C}}$ (d) $R = \sqrt{\frac{1}{LC}}$

[2005 : 1 M]

2.17 In a series RLC circuit, R = 2 kΩ, L = 1 H, and $C = \frac{1}{400}\mu F$.

The resonant frequency is

(a) 2×10^4 Hz (b) $\frac{1}{\pi} \times 10^4$ Hz

(c) 10^4 Hz (d) $2\pi \times 10^4$ Hz

[2005 : 1 M]

2.18 For the circuit shown in the figure, the instantaneous current $i_1(t)$ is

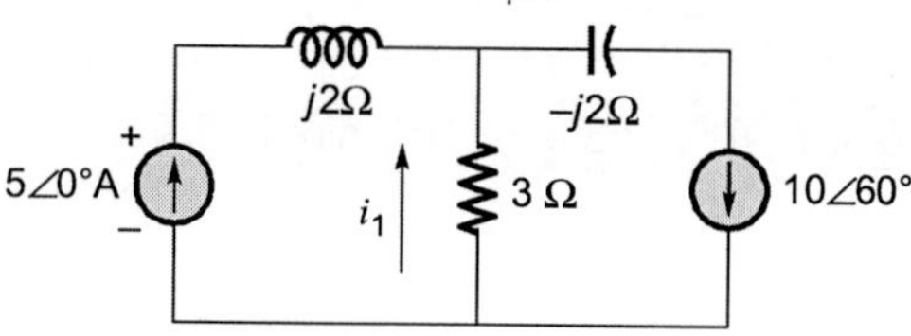

(a) $\frac{10\sqrt{3}}{2}\angle 90°$ A (b) $\frac{10\sqrt{3}}{2}\angle -90°$ A

(c) $5\angle 60°$ A (d) $5\angle -60°$ A

[2005 : 2 M]

2.19 In the AC network shown in the figure, the phasor voltage V_{AB} (in Volts) is

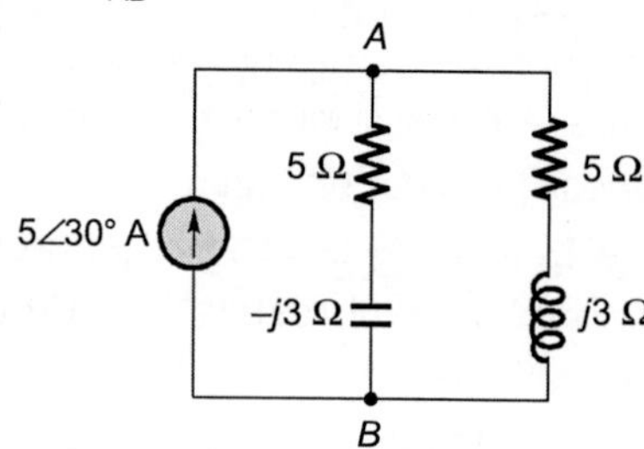

(a) 0 (b) $5\angle 30°$

(c) $12.5\angle 30°$ (d) $17\angle 30°$

[2007 : 2 M]

2.20 An AC source of RMS voltage 20 V with internal impedance Z_s = (1 + 2j) Ω feeds a load of impedance Z_L = (7 + 4j) Ω in the figure below. The reactive power consumed by the load is

(a) 8 VAR

(b) 16 VAR

(c) 28 VAR

(d) 32 VAR

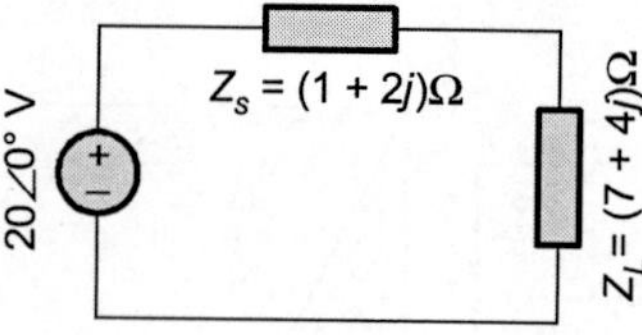

[2009 : 2 M]

2.21 For a parallel RLC circuit, which one of the following statements is NOT correct?

(a) The bandwidth of the circuit decreases if R is increased

(b) The bandwidth of the circuit remains same if L is increased

(c) At resonance, input impedance is a real quantity

(d) At resonance, the magnitude of input impedance attains its minimum value

[2010 : 1 M]

2.22 The current I in the circuit shown is

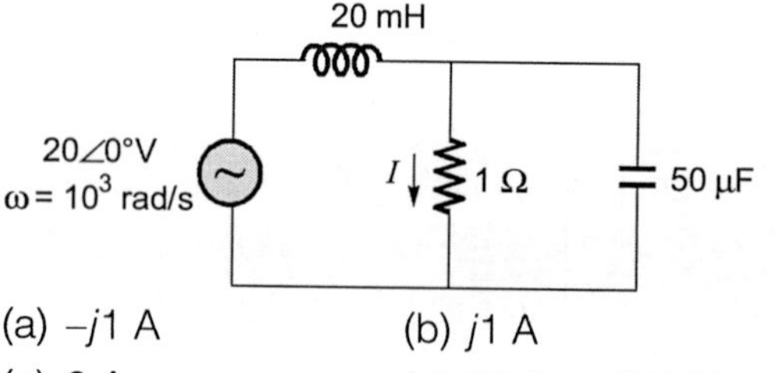

(a) $-j1$ A (b) $j1$ A

(c) 0 A (d) 20 A [2010 : 2 M]

2.23 The circuit shown below is driven by a sinusoidal input $v_i = V_p \cos(t/RC)$. The steady output v_0 is

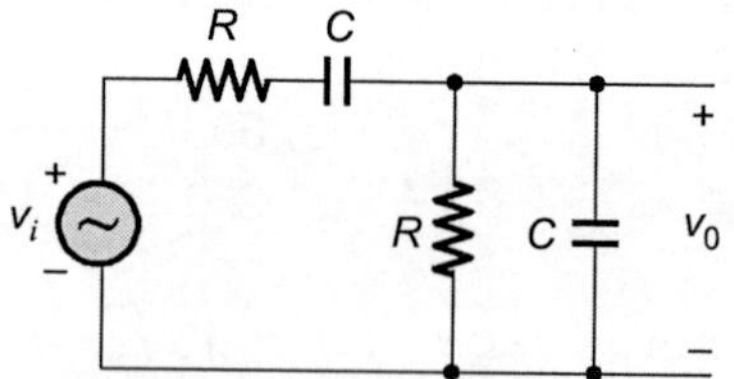

(a) $(V_p/3)\cos(t/RC)$ (b) $(V_p/3)\sin(t/RC)$

(c) $(V_p/2)\cos(t/RC)$ (d) $(V_p/2)\sin(t/RC)$

[2011 : 1 M]

2.24 Two magnetically uncoupled inductive coils have Q factors q_1 and q_2 at the chosen operating frequency. Their respective resistance are R_1 and R_2. When connected in series, their effective Q factor at the same operating frequency is

(a) $q_1 + q_2$

(b) $(1/q_1) + (1/q_2)$

(c) $(q_1R_1 + q_2R_2)/(R_1 + R_2)$

(d) $(q_1R_2 + q_2R_1)/(R_1 + R_2)$ [2013 : 2 M]

2.25 A 230 V rms source supplies power to two loads connected in parallel. The first load draws 10 kW at 0.8 leading power factor and the second one draws 10 kVA at 0.8 lagging power factor. The complex power delivered by the source is

(a) (18 + j 1.5) kVA (b) (18 – j 1.5) kVA
(c) (20 + j 1.5) kVA (d) (20 – j 1.5) kVA

[2014 : 2 M, Set-1]

2.26 A periodic variable x is shown in the figure as a function of time. The root-mean-square (rms) value of x is ________.

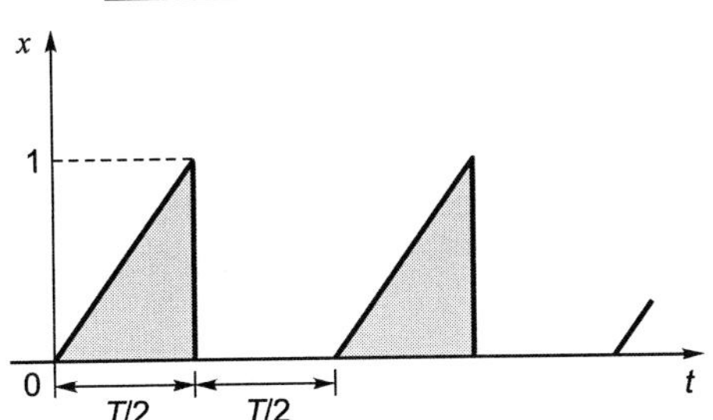

[2014 : 2 M, Set-1]

2.27 A series RC circuit is connected to a DC voltage source at time t = 0. The relation between the source voltage V_S, the resistance R, the capacitance C, and the current $i(t)$ is given below:

$$V_s = R\,i(t) + \frac{1}{C}\int_0^t i(t)dt$$

Which one of the following represents the current $i(t)$?

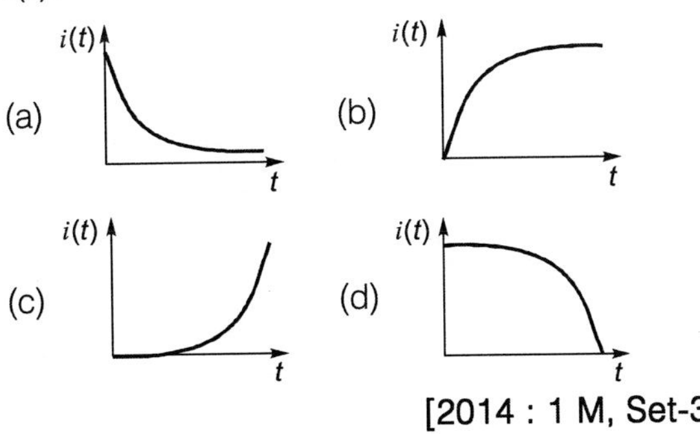

[2014 : 1 M, Set-3]

2.28 The steady state output of the circuit shown in the figure is given by $y(t) = A(\omega)\sin(\omega t + \phi(\omega))$. If the amplitude $|A(\omega)| = 0.25$, then the frequency ω is

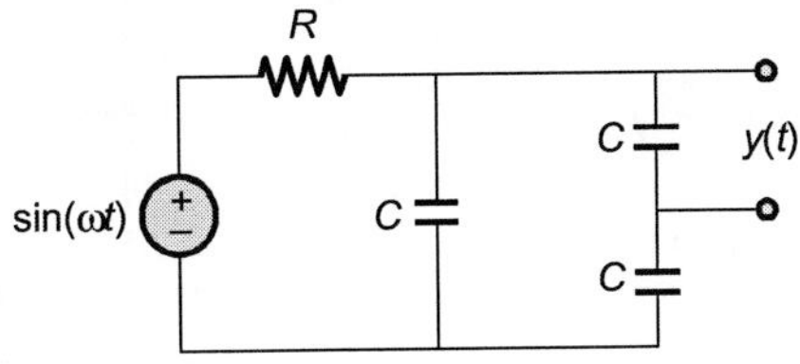

(a) $\frac{1}{\sqrt{3}RC}$ (b) $\frac{2}{\sqrt{3}RC}$

(c) $\frac{1}{RC}$ (d) $\frac{2}{RC}$

[2014 : 2 M, Set-4]

2.29 In the circuit shown, at resonance, the amplitude of the sinusoidal voltage (in Volts) across the capacitor is ________.

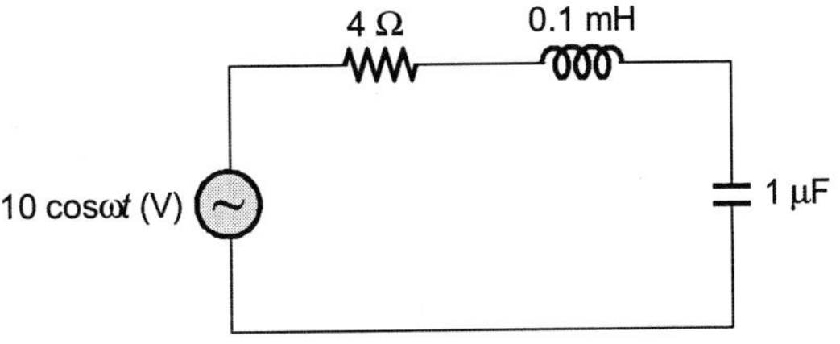

[2015 : 1 M, Set-1]

2.30 The damping ratio of a series RLC circuit can be expressed as

(a) $\frac{R^2C}{2L}$ (b) $\frac{2L}{R^2C}$

(c) $\frac{R}{2}\sqrt{\frac{C}{L}}$ (d) $\frac{2}{R}\sqrt{\frac{L}{C}}$

[2015 : 2 M, Set-1]

2.31 In the circuit shown, the average value of the voltage V_{ab} (in Volts) in steady state condition is _____.

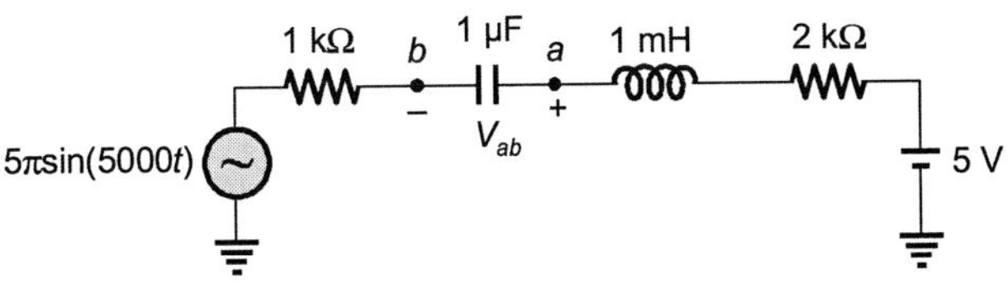

[2015 : 1 M, Set-2]

2.32 The voltage (V_c) across the capacitor (in Volts) in the network shown in ____ .

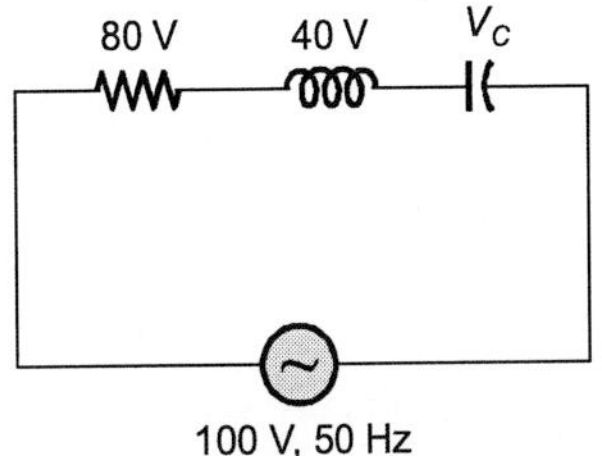

[2015 : 1 M, Set-2]

2.33 An LC tank circuit consists of an ideal capacitor C connected in parallel with a coil of inductance L having an internal resistance R. The resonant frequency of the tank circuit is

(a) $\dfrac{1}{2\pi\sqrt{LC}}$ (b) $\dfrac{1}{2\pi\sqrt{LC}}\sqrt{1-R^2\dfrac{C}{L}}$

(c) $\dfrac{1}{2\pi\sqrt{LC}}\sqrt{1-\dfrac{L}{R^2C}}$ (d) $\dfrac{1}{2\pi\sqrt{LC}}\left(1-R^2\dfrac{C}{L}\right)$

[2015 : 2 M, Set-2]

2.34 At very high frequencies, the peak output voltage V_o (in Volts) is ______.

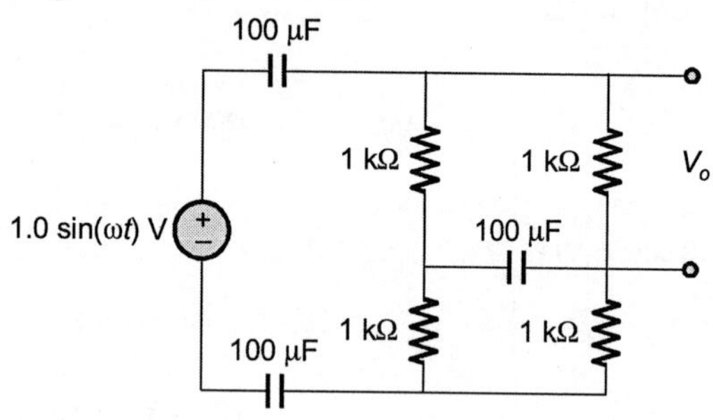

[2015 : 1 M, Set-3]

2.35 In the circuit shown, the current I flowing through the 50 Ω resistor will be zero if the value of capacitor C (in μF) is_____.

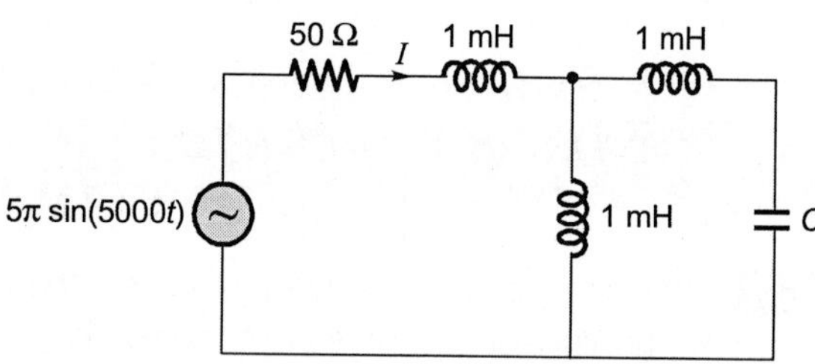

[2015 : 2 M, Set-3]

2.36 The figure shows an RLC circuit with a sinusoidal current source.

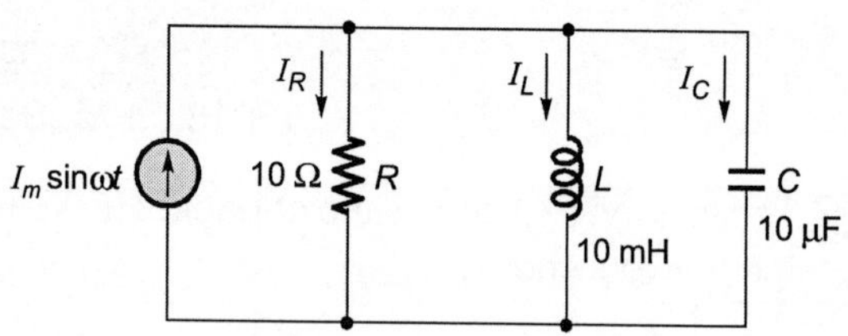

At resonance, the ratio $\dfrac{|I_L|}{|I_R|}$, i.e., the ratio of the magnitudes of the inductor current phasor and the resistor current phasor, is_____

[2016 : 1 M, Set-2]

2.37 In the RLC circuit shown in the figure, the input voltage is given by

$$V_i(t) = 2\cos(200\,t) + 4\sin(500\,t)$$

The output voltage $V_0(t)$ is

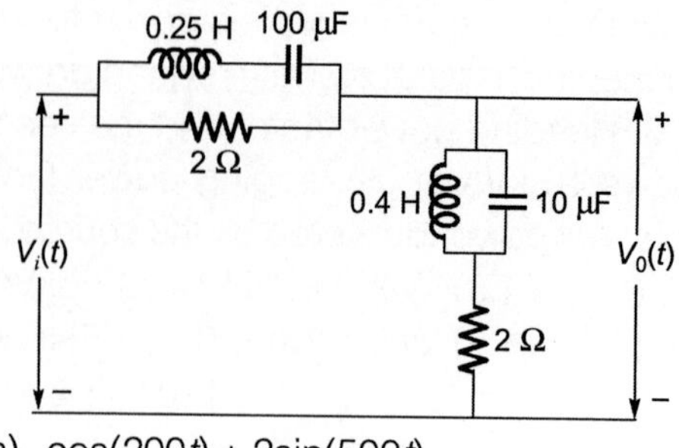

(a) $\cos(200t) + 2\sin(500t)$
(b) $2\cos(200t) + 4\sin(500t)$
(c) $\sin(200t) + 2\cos(500t)$
(d) $2\sin(200t) + 4\cos(500t)$

[2016 : 1 M, Set-3]

2.38 In the circuit shown, the positive angular frequency ω (in radians per second) at which the magnitude of the phase difference between the voltages V_1 and V_2 equals $\dfrac{\pi}{4}$ radians, is ________.

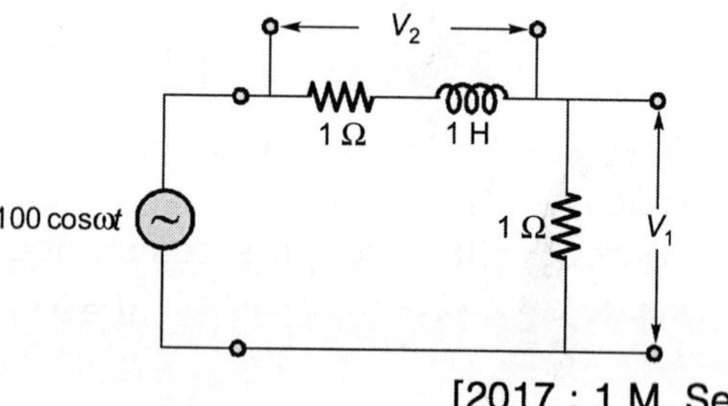

[2017 : 1 M, Set-1]

2.39 The figure shows an RLC circuit excited by the sinusoidal voltage 100cos(3t) Volts, where t is in seconds. The ratio $\dfrac{\text{amplitude of } V_2}{\text{amplitude of } V_1}$ is ________.

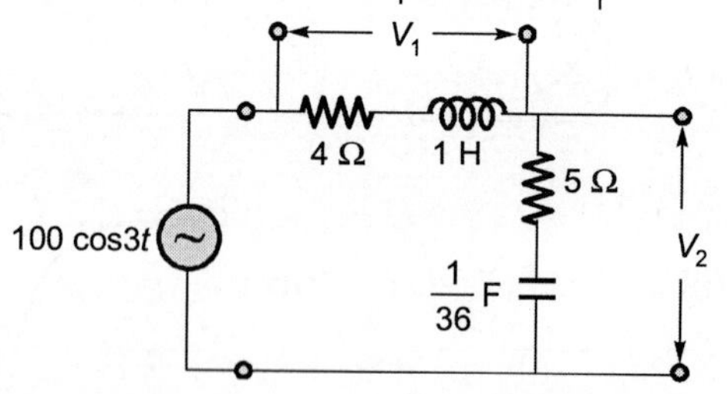

[2017 : 2 M, Set-1]

2.40 In the circuit shown, V is a sinusoidal voltage source. The current I is in phase with voltage V. The ratio $\dfrac{\text{amplitude of voltage across the capacitor}}{\text{amplitude of voltage across the resistor}}$ is ___.

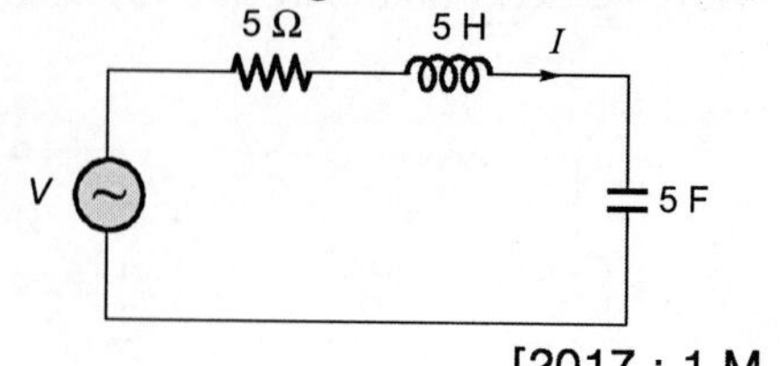

[2017 : 1 M, Set-2]

2.41 For the circuit given in the figure, the voltage V_C (in volts) across the capacitor is

(a) $1.25\sqrt{2}\sin(5t - 0.25\pi)$
(b) $1.25\sqrt{2}\sin(5t - 0.125\pi)$
(c) $2.5\sqrt{2}\sin(5t - 0.25\pi)$
(d) $2.5\sqrt{2}\sin(5t - 0.125\pi)$

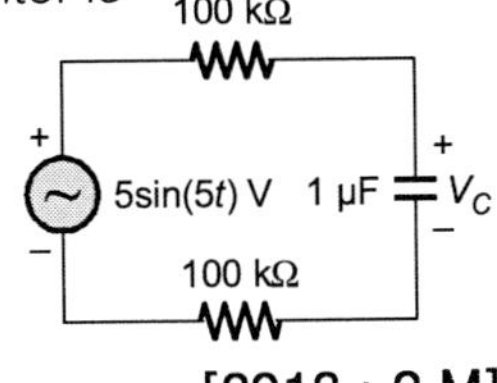

[2018 : 2 M]

2.42 In the circuit shown, if $v(t) = 2\sin(1000t)$ volts. $R = 1$ kΩ and $C = 1$ μF. then the steady-state current $i(t)$, in milliamperes (mA), is

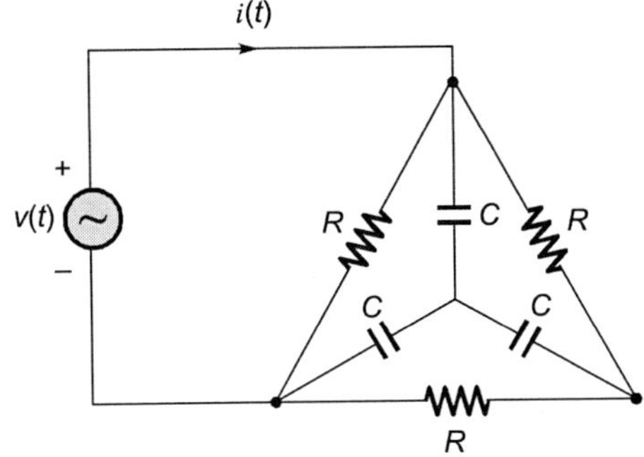

(a) $3\sin(1000t) + \cos(1000t)$
(b) $\sin(1000t) + \cos(1000t)$
(c) $\sin(1000t) + 3\cos(1000t)$
(d) $2\sin(1000t) + 2\cos(1000t)$

[2019 : 2 M]

2.43 The current in the RL-circuit shown below is $i(t) = 10\cos(5t - \pi/4)A$. The value of the inductor (rounded off to two decimal places) is ______ H.

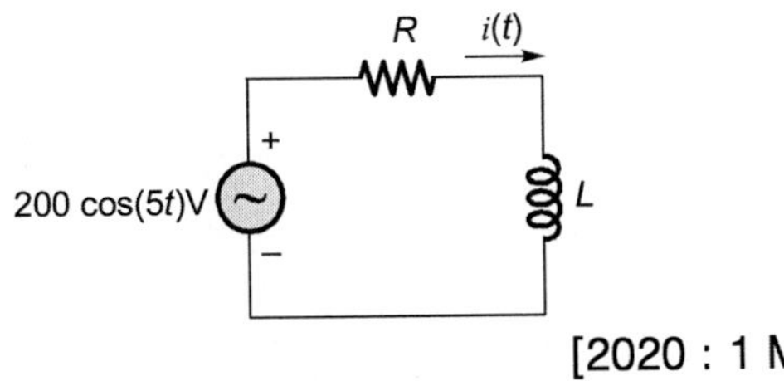

[2020 : 1 M]

2.44 The current I in the given network is

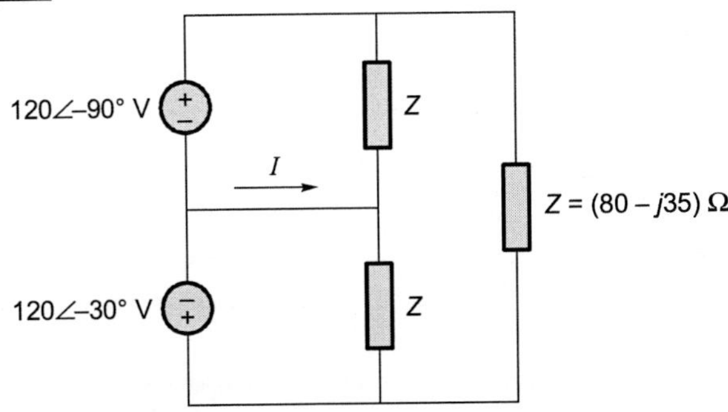

(a) $2.38\angle -23.63°$ A (b) 0 A
(c) $2.38\angle -96.37°$ A (d) $2.38\angle 143.63°$ A

[2020 : 2 M]

2.45 For the circuit shown, the locus of the impedance $Z(j\omega)$ is plotted as ω increases from zero to infinity. The values of R_1 and R_2 are:

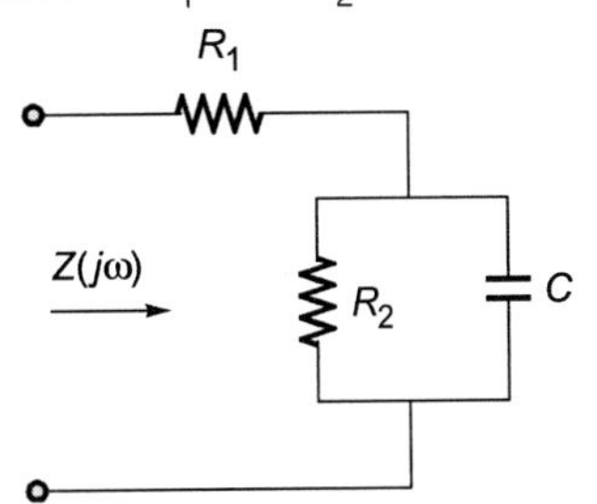

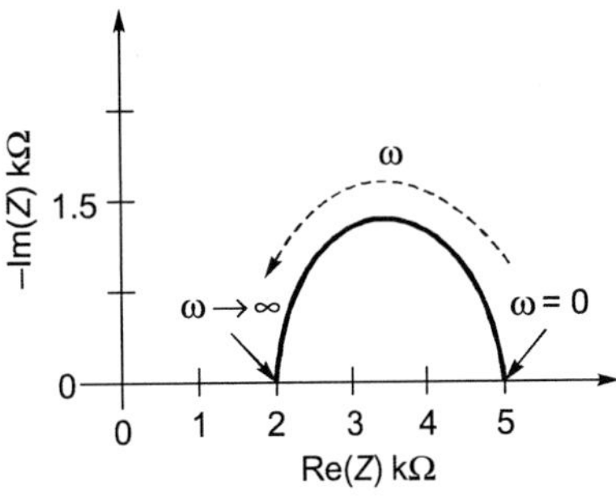

(a) $R_1 = 2$ kΩ, $R_2 = 3$ kΩ
(b) $R_1 = 5$ kΩ, $R_2 = 2$ kΩ
(c) $R_1 = 5$ kΩ, $R_2 = 2.5$ kΩ
(d) $R_1 = 2$ kΩ, $R_2 = 5$ kΩ

[2022 : 2 M]

2.46 Consider the circuit shown in the figure with input $V(t)$ in volts. The sinusoidal steady state current $I(t)$ flowing through the circuit is shown graphically (where t is in seconds). The circuit element Z can be ________.

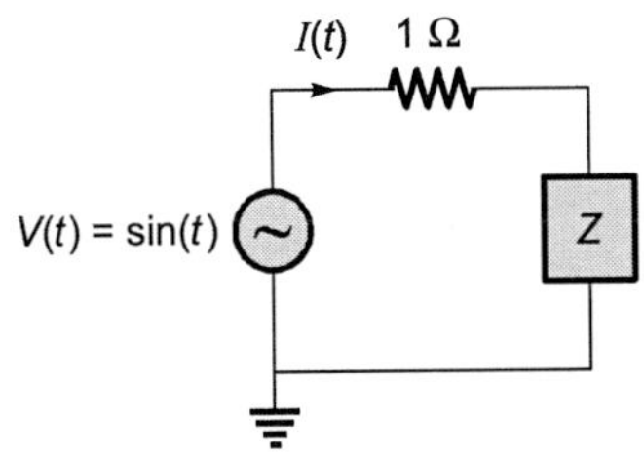

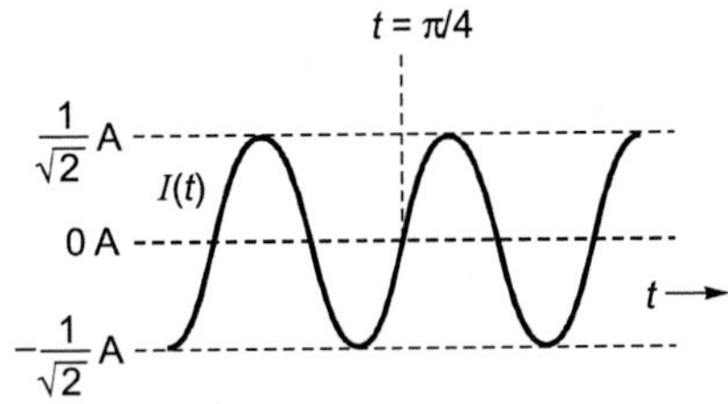

(a) a capacitor of 1 F
(b) an inductor of 1 H
(c) a capacitor of $\sqrt{3}$ F
(d) an inductor of $\sqrt{3}$ H

[2022 : 2 M]

2.47 A series RLC circuit has a quality factor Q of 1000 at a center frequency of 10^6 rad/s. The possible values of R, L and C are

(a) $R = 1\ \Omega$, $L = 1\ \mu$H and $C = 1\ \mu$F

(b) $R = 0.1\ \Omega$, $L = 1\ \mu$H and $C = 1\ \mu$F

(c) $R = 0.01\ \Omega$, $L = 1\ \mu$H and $C = 1\ \mu$F

(d) $R = 0.001\ \Omega$, $L = 1\ \mu$H and $C = 1\ \mu$F

[2023 : 1 M]

■■■

Answers Sinusoidal Steady State

2.1	(c)	2.2	(b)	2.3	(c)	2.4	(100, 1)	2.5	(a)	2.6	(b)	2.7	(d)
2.8	(b)	2.9	(a)	2.10	(b)	2.11	(c)	2.12	(a)	2.13	(a)	2.14	(b)
2.15	(d)	2.16	(c)	2.17	(b)	2.18	(a)	2.19	(d)	2.20	(b)	2.21	(d)
2.22	(a)	2.23	(a)	2.24	(c)	2.25	(b)	2.26	(0.408)	2.27	(a)	2.28	(b)
2.29	(25)	2.30	(c)	2.31	(5)	2.32	(100)	2.33	(b)	2.34	(0.5)	2.35	(20)
2.36	(0.316)	2.37	(b)	2.38	(1)	2.39	(2.6)	2.40	(0.2)	2.41	(c)	2.42	(a)
2.43	(2.828)	2.44	(d)	2.45	(a)	2.46	(b)	2.47	(d)				

Explanations Sinusoidal Steady State

2.1 (c)

For DC supply
Inductor behave as short circuit
capacitor behave as open circuit
So, under steady state conditions, the applied dc voltage drops entirely across the capacitor (c) only.

2.2 (b)

$$W_s = CV_s^2 \quad ; \quad W_c = \frac{1}{2}CV_s^2$$

$$\frac{W_c}{W_s} = 0.5$$

2.3 (c)

$$I_{rms} = \sqrt{3^2 + \left(\frac{4}{\sqrt{2}}\right)^2 + \left(\frac{4}{\sqrt{2}}\right)^2} = 5 \text{ Amp.}$$

Power is dissipated only in the 10 Ω resistor.

$$P = I_{rms}^2 R = (5)^2 \times 10 = 250 \text{ W}$$

2.4 (100, 1)

$$Z = R + j(X_L - X_C) = 100 + j0$$

Compare the real part

$$R = 100\ \Omega$$

$$Q = \frac{\omega L}{R}$$

$$L = \frac{QR}{\omega} = \frac{100 \times 100}{10^7} = 1 \text{ mH}$$

2.5 (a)

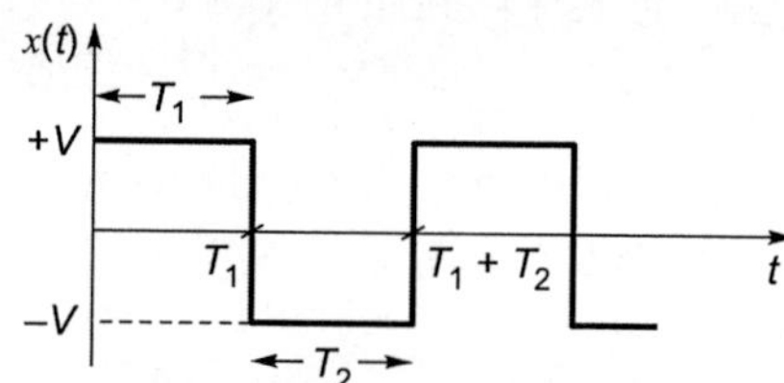

RMS value of $x(t)$

$$= \sqrt{\frac{1}{T}\int_0^T x^2(t)\,dt} = \sqrt{\frac{1}{T_1+T_2}\left[\int_0^{T_1}(V)^2 dt + \int_{T_1}^{T_1+T_2}(-V)^2 dt\right]}$$

$$= \sqrt{\frac{1}{T_1+T_2}[V^2T_1 + V^2T_2]} = \sqrt{V^2} = V$$

2.6 (b)

$$A_1^2 = A_2^2 + A_3^2 = 3^2 + 4^2$$

$$A_1 = 5 \text{ Amp.}$$

2.7 (d)

Diode D_1 will conduct for the positive half cycle of the input

The ammeter will read the average value

$$I_{avg} = \frac{V_m}{\pi} \times \frac{1}{R} = \frac{4}{\pi} \times \frac{1}{10 \times 10^3} = \frac{0.4}{\pi} \text{ mA}$$

2.8 (b)

At resonance

$$I = I_R = 1 \text{ mA}$$

$$|I_R + I_L| = \sqrt{I_R^2 + I_L^2} = \sqrt{1^2 + I_L^2} > 1\text{mA}$$

$$|I_R + I_L| > 1 \text{ mA}$$

2.9 (a)

$$i_2(t) = \frac{E_m \cos\omega t}{R_2 + \frac{1}{j\omega C}} = E_m\angle 0 \frac{j\omega C}{1 + j\omega CR_2}$$

$$\angle i_2(t) = \frac{E_m\angle 0\ \omega C\angle 90°}{\sqrt{1+\omega^2C^2R^2}\angle\tan^{-1}\omega CR_2}$$

$$i_2(t) = \frac{E_m\omega C}{\sqrt{1+\omega^2C^2R^2}}\angle 90° - \tan^{-1}\omega CR$$

at $\omega = 0$, $i_2(t) = 0$, $\omega = \infty$, $i_2(t) = \frac{E_m}{R_2}$

Option (a) satisfies both conditions.

2.10 (b)

$$Q = \frac{f_o}{BW} \quad ; \quad f_0 = \frac{1}{2\pi\sqrt{LC}}$$

$$BW = \frac{R}{L}$$

$$\left(\text{Characteristic equation} = s^2 + \frac{Rs}{L} + \frac{1}{LC}\right)$$

or $$Q = \frac{1}{R}\sqrt{\frac{L}{C}}$$

When R, L, C are doubled,

$$Q' = \frac{1}{2}Q = 50$$

2.11 (c)

$$i(t) = \frac{v(t)}{R + j\omega L} = \frac{10\sqrt{2}\cos(t+10°)}{1+1j} + \frac{10\sqrt{5}\cos(2t+10°)}{1+2j}$$

$$i(t) = \frac{10\sqrt{2}\cos(t+10°)}{\sqrt{2}\angle 45°} + \frac{10\sqrt{5}\cos(2t+10°)}{\sqrt{5}\angle\tan^{-1}2}$$

$$\therefore i(t) = 10\cos(t-35°) + 10\cos(2t + 10 - \tan^{-1}2)$$

2.12 (a)

$$i(t) = V(t).Y$$

$$i(t) = V(t)\left[\frac{1}{R_1} + \frac{1}{j\omega L} + j\omega c\right]$$

$$= \sin 2t\left[3 + \frac{4}{2j} + j\times 2\times 3\right]$$

$$= \sin 2t[3 - 2j + 6j] = \sin 2t[3 + 4j]$$

$$= 5\sin 2t\angle\tan^{-1}\frac{4}{3} = 5\sin(2t + 53.1°)$$

2.13 (a)

$$v_0(t) = \frac{\frac{1}{j\omega C}}{R + \frac{1}{j\omega C}}\ v_i(t) = \frac{1}{1 + j\omega CR}\sqrt{2}\sin 10^3 t$$

$$= \frac{1}{1 + j\times 10^3\times 10^{-3}}\sqrt{2}\sin 10^3 t$$

$$v_0(t) = \sin(10^3 t - 45°)$$

2.14 (b)

Characteristic equation = $s^2 + 20s + 10^6$

$$Q = \frac{\omega_o}{BW},\ \omega_o = \sqrt{10^6}\ ; \quad Q = \frac{10^3}{20} = \frac{1000}{20} = 50$$

2.15 (d)

S_1 : Impedance of series RLC circuit at resonant frequency is minimum

$$Z = R + j\left(\omega L - \frac{1}{\omega C}\right) \quad ; \quad \omega L - \frac{1}{\omega C} = 0$$

$Z = R$ (Purely resistive)

$$S_2 : Q = R\sqrt{\frac{C}{L}}$$

$$G = \frac{1}{R} \Rightarrow Q = \frac{1}{G}\sqrt{\frac{C}{L}}$$

$G\uparrow$ then $Q\downarrow$ if C & L are same

2.16 (c)

$$\text{Transfer function} = \frac{\frac{1}{sC}}{R + sL + \frac{1}{sC}} = \frac{1}{s^2LC + sCR + 1}$$

$$\frac{Y(s)}{U(s)} = \frac{\frac{1}{LC}}{s^2 + \frac{R}{L}s + \frac{1}{LC}}$$

$$2\xi\omega_n = \frac{R}{L}$$

$$\omega_n = \frac{1}{\sqrt{LC}}$$

$$\xi = \frac{R}{2L}\sqrt{LC} = \frac{R}{2}\sqrt{\frac{C}{L}}$$

For no oscillations, $\xi \geq 1$

$$\frac{R}{2}\sqrt{\frac{C}{L}} \geq 1 \quad ; \quad R \geq 2\sqrt{\frac{L}{C}}$$

2.17 (b)

$$f_0 = \frac{1}{2\pi\sqrt{LC}} = \frac{1}{2\pi\sqrt{1\times\frac{1}{400}\times 10^{-6}}}$$

$$= \frac{10^3 \times 20}{2\pi} = \frac{10^4}{\pi}\text{Hz}$$

2.18 (a)

When $5\angle 0°$ is acting along,

$$i_1(t) = -5\angle 0°$$

(as $10\angle 60°$ is kept open)

When $10\angle 60°$ is acting alone.

$$i_1(t) = 10\angle 60°$$

(as $5\angle 0°$ is kept open)

$$i_1(t) = 10\angle 60° - 5\angle 0° = 5 + 8.66j - 5$$

$$i_1(t) = 8.66j$$

$$i(t) = 5\sqrt{3}\angle 90° = \frac{10}{2}\sqrt{3}\angle 90°$$

2.19 (d)

$$V_{AB} = \text{Current} \times \text{impedance}$$

$$= 5\angle 30° \times (5-3j) \parallel (5+3j)$$

$$= 5\angle 30° \times \frac{(5-3j)\times(5+3j)}{5-3j+5+3j}$$

$$= 5\angle 30° \times \frac{25+9}{10}$$

$$\Rightarrow \quad = 5\angle 30° \times 3.4 = 17\angle 30°$$

2.20 (b)

1 + 2j

$20\angle°$V I 7 + 4j

$$I = \frac{20}{8+6j} = \frac{10}{4+3j} = 2\angle -36.87°$$

Reactive power

$$Q = I^2 X_L = 4 \times 4 = 16 \text{ VAR}$$

2.21 (d)

Characteristic equation for a parallel *RLC* circuit is

$$s^2 + \frac{1}{RC}s + \frac{1}{LC} = 0$$

where, Bandwidth $= \frac{1}{RC}$

(i) It is clear that the bandwidth of a parallel *RLC* circuit is independent of *L* and decreases if *R* is increased.

(ii) At resonance, imaginary part of input impedance is zero. Hence, at resonance input impedance is a real quantity.

(iii) In parallel *RLC* circuit, the admittance is minimum at resonance. Hence magnitude of input impedance attains its maximum value at resonance.

2.22 (a)

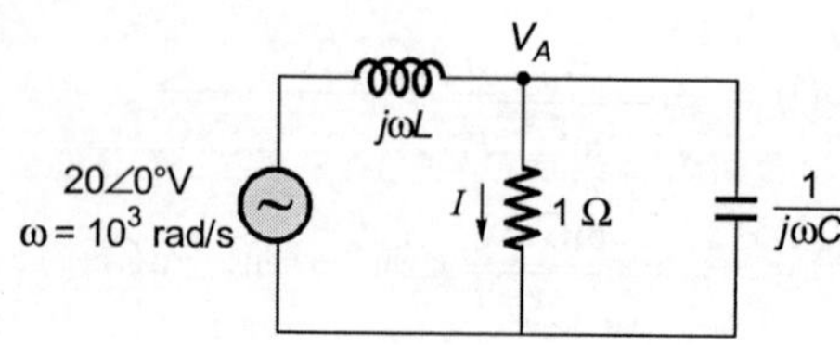

($L = 20$ mH, $C = 50$ μF)

Nodal analysis at node *A*,

$$\frac{V_A - 20}{j\omega L} + \frac{V_A}{1} + \frac{V_A}{1/j\omega C} = 0$$

$$V_A\left[\frac{1}{j10^3 \times 20 \times 10^{-3}} + 1 + j10^3 \times 50 \times 10^{-6}\right] = \frac{20}{j10^3 \times 20 \times 10^{-3}}$$

$$V_A\left[\frac{-j}{20} + 1 + \frac{j}{20}\right] = -j1$$

$$V_A = -j1 \text{ V}$$

$$I = \frac{V_A}{1} = -j1 \text{ A}$$

2.23 (a)

Redrawing the circuit s-domain

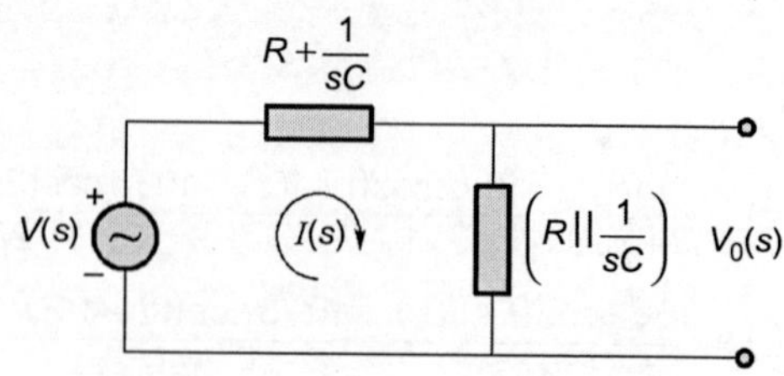

$$V_i(s) = \left(R + \frac{1}{sC}\right)I(s) + \frac{R\cdot\frac{1}{sC}}{R+\frac{1}{sC}}I(s)$$

$$V_i(s) = \frac{1+sCR}{sC}I(s) + \frac{R}{1+sCR}I(s) \quad \text{...(i)}$$

$\because \quad v_i = V_p\cos(t/RC)$

so here, $\omega = \frac{1}{RC}$

Now, $V_i(s) = \frac{(1+j\omega CR)}{j\omega C} I(s) + \frac{R}{(1+j\omega CR)} I(s)$

Put, $\omega = \frac{1}{RC}$

So, $V_i(s) = \left[\frac{(1+j)R}{j} + \frac{R}{1+j}\right] I(s)$

$$\frac{V_i(s)}{I(s)} = \frac{3R}{(1+j)} \quad ...(ii)$$

$$I(s) = \frac{V_i(s)}{3R} \times (1+j)$$

Now, $V_0(s) = \left(R \,||\, \frac{1}{sC}\right) I(s)$

$$\Rightarrow \quad V_0(s) = \frac{R \cdot \frac{1}{sC}}{R + \frac{1}{sC}} I(s)$$

$$\Rightarrow \quad V_0(s) = \frac{R}{1+sCR} \cdot \frac{V_i(s)}{3R}(1+j)$$

$$\Rightarrow \quad V_0(s) = \frac{R}{1+j} \cdot \frac{V_i(s)}{3R}(1+j)$$

$$\Rightarrow \quad V_0(s) = \frac{V_i(s)}{3}$$

In time domain,

$$v_0(t) = \frac{1}{3} v_i(t)$$

$$\Rightarrow \quad v_0(t) = \frac{V_p}{3} \cos\left(\frac{t}{RC}\right)$$

2.24 (c)

$$q_1 = \frac{\omega L_1}{R_1} \text{ and } q_2 = \frac{\omega L_2}{R_2}$$

$$\omega L_1 = q_1 R_1 \text{ and } \omega L_2 = q_2 R_2$$

Coils are connected in series.

So, $q \cdot R = \omega L_1 + \omega L_2 = q_1 R_1 + q_2 R_2$

$$q = \frac{q_1 R_1 + q_2 R_2}{R}$$

$$R = R_1 + R_2$$

$$q = \frac{q_1 R_1 + q_2 R_2}{R_1 + R_2}$$

2.25 (b)

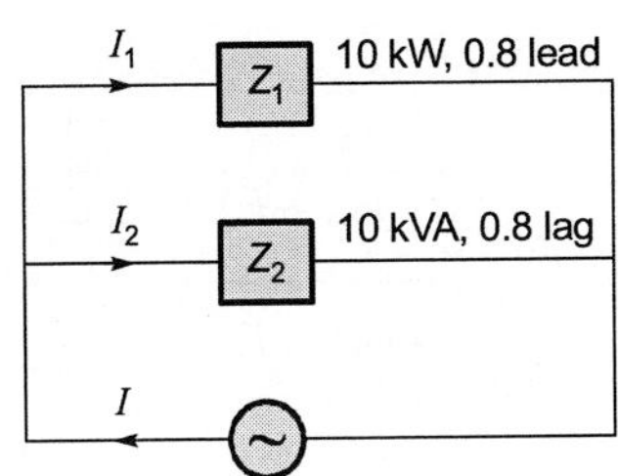

Real power (kW) $P = VI\cos\phi$ ...(i)

$\therefore \quad P_1 = V \times I_1 \cos\phi$

Thus, $I_1 = \frac{10 \times 10^3}{230 \times 0.8} = 54.34\angle 36.86$ ...(ii)

Total power (kVA)

$$P = V \cdot I_2$$

$\therefore \quad I_2 = 43.47\angle -36.86$ ...(iii)

Therefore,

$$\vec{I} = \vec{I_1} + \vec{I_2} = 78.25 + j6.25 \quad ...(iv)$$

So, the complex power delivered by source

$$S = VI^* = 230(78.25 - j6.25)$$

or $S = (18 - j1.5)$ kVA

2.26 (0.408)

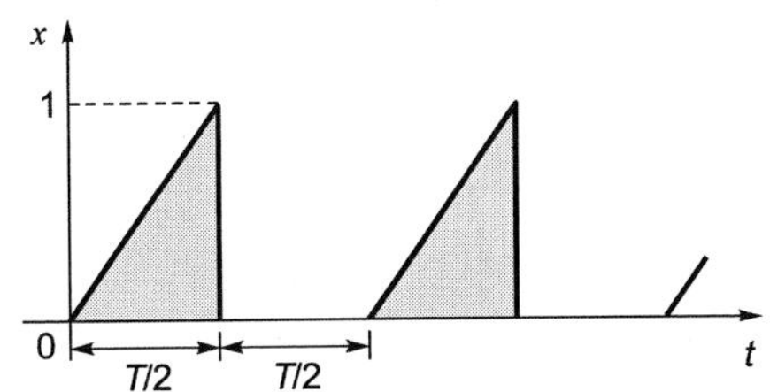

$$\text{RMS value} = \sqrt{\frac{1}{T}\int_0^T x^2(t)\,dt}$$

where, T = time period

for the given signal,

$$\text{RMS value} = \sqrt{\frac{1}{T}\int_0^{T/2}\left(\frac{2}{T}t\right)^2 dt + \int_{T/2}^{T}(0)^2 dt}$$

$$= \sqrt{\frac{1}{T}\int_0^{T/2}\frac{4}{T^2}t^2\,dt} = \sqrt{\frac{1}{T} \times \frac{4}{T^2} \times \left(\frac{t^3}{3}\right)_0^{T/2}}$$

$$= \sqrt{\frac{4}{T^3}\left(\frac{T^3}{24}\right)} = \sqrt{\frac{1}{6}} = 0.408$$

2.27 (a)

Given that:

$$V_s = Ri(t) + \frac{1}{C}\int_0^t i(t)\,dt \quad ...(i)$$

Using Laplace transform,

$$V(s) = RI(s) + \frac{1}{Cs} I(s) \quad ...(ii)$$

or $I(s) = \frac{V(s)}{\left(R + \frac{1}{Cs}\right)}$ ...(iii)

For, $V(s) = \frac{1}{s}$ ...(iv)

From equation (iii) and (iv),

$$I(s) = \frac{C}{(RCs+1)} = \frac{1}{R\left(s+\frac{1}{RC}\right)} \quad \text{...(v)}$$

Using inverse Laplace transform in equation (v), we get,

$$i(t) = \frac{1}{R}e^{-t/RC}$$

Thus, option (a) is correct.

2.28 (b)

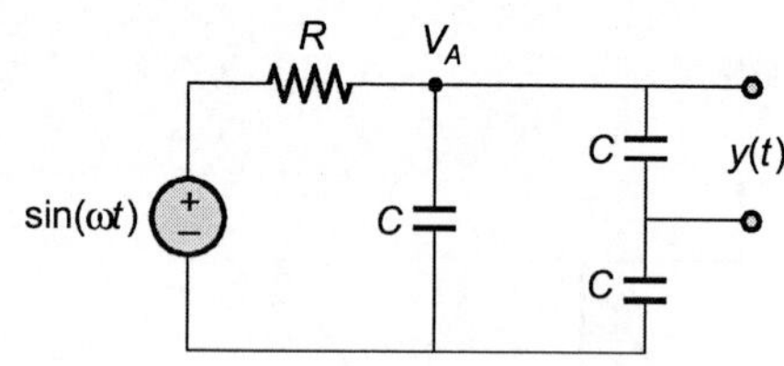

Applying KCL at node A, we get

$$= \frac{V_A - \sin\omega t}{R} + \frac{V_A}{\frac{1}{j\omega C}} + \frac{\frac{V_A}{2}}{\frac{2}{j\omega C}} = 0 \quad \text{...(i)}$$

$$= V_A\left[\frac{1}{R} + j\omega C + \frac{j\omega C}{2}\right] = \frac{\sin\omega t = 1\angle 0°}{R}$$

$$= V_A = \frac{2}{2+3RC\cdot j\omega} \quad \text{...(ii)}$$

Also $\quad Y = \frac{V_A}{2} = \frac{1}{2+3j\omega RC} \quad$ ...(iii)

$$\because \quad |A(\omega)| = \frac{1}{4}$$

$$\therefore \quad \frac{1}{4} = \frac{1}{\sqrt{4+9\omega^2(RC)^2}}$$

or $\quad \omega = \frac{2}{\sqrt{3}\,RC}$

2.29 (25)

At resonance, $I_{max} = \frac{10}{4} = 2.5$ A

$$\omega = \frac{1}{\sqrt{0.1\times10^{-3}\times10^{-6}}} = 10^5 \text{ rad/sec}$$

$$X_C = \frac{1}{\omega C} = \frac{1}{10^5\times1\times10^{-6}} = 10\ \Omega$$

Amplitude of the sinusoidal voltage across capacitor

$$V_{C_{max}} = IX_C = 10\times2.5 = 25 \text{ V}$$

2.30 (c)

$$\xi = \frac{1}{2Q}\ ;\ Q = \frac{1}{R}\sqrt{\frac{L}{C}}$$

$$\therefore \text{ Damping ratio } = \xi = \frac{R}{2}\sqrt{\frac{C}{L}}$$

2.31 (5)

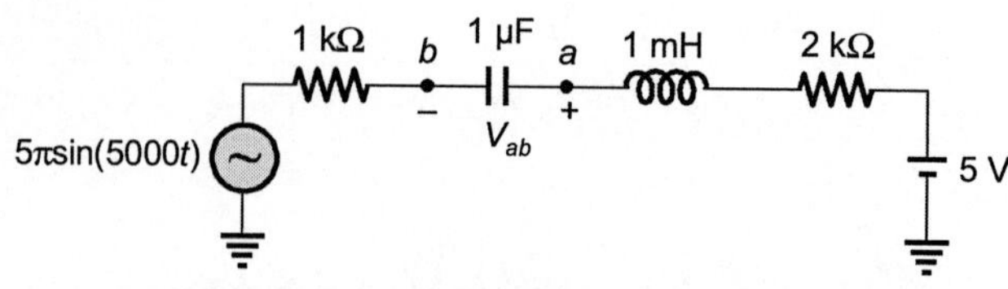

Applying superposition:

V_{ab} = 5 V [open circuited in steady state]

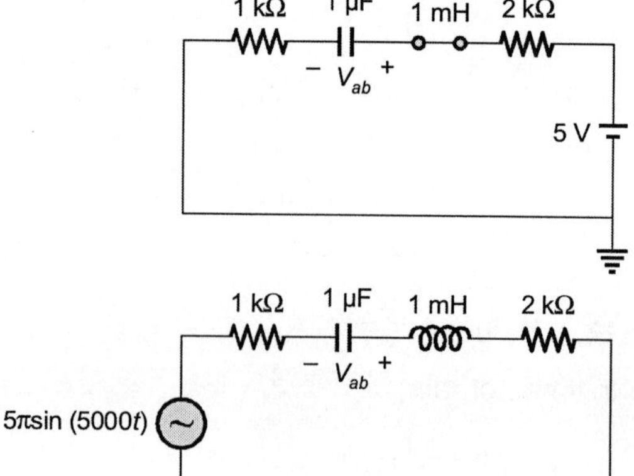

V_{ab} will be sinusoid with average value zero

⇒ Average V_{ab} = 5 V.

2.32 (100)

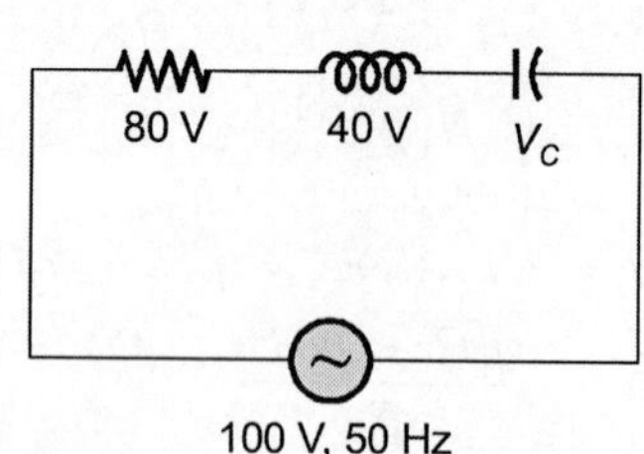

$$(80)^2 + (40 - V_C)^2 = 100^2$$

$$(40 - V_C)^2 = 100^2 - 80^2 = 3600$$

$$|40 - V_C| = 60$$

$$V_C = 100 \text{ V}$$

2.33 (b)

$$Z_{eq} = \frac{\frac{1}{j\omega C}\times(j\omega L + R)}{\frac{1}{j\omega C} + j\omega L + R}$$

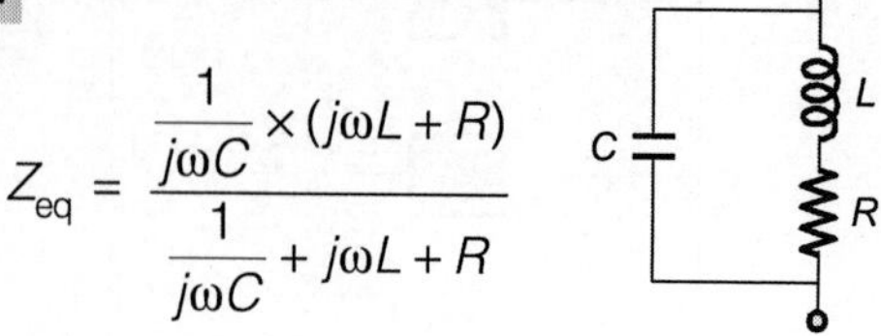

$$Z_{eq} = \frac{\left(\frac{R}{j\omega C}+\frac{L}{C}\right)}{R+j\left(\omega L-\frac{1}{\omega C}\right)} \times \frac{R-j\left(\omega L-\frac{1}{\omega C}\right)}{R-j\left(\omega L-\frac{1}{\omega C}\right)}$$

Equating imaginary part to zero.

$$\text{Img} = -\frac{R^2}{\omega C}-\frac{L}{C}\left(\omega L-\frac{1}{\omega C}\right) = 0$$

$$\frac{R^2}{\omega C}+\frac{L}{C}\left(\omega L-\frac{1}{\omega C}\right) = 0$$

$$\frac{CR^2+\omega^2L^2C-L}{\omega C^2} = 0$$

$$\omega^2 = \frac{L-R^2C}{L^2C} \quad \Rightarrow \omega = \frac{1}{\sqrt{LC}}\sqrt{1-\frac{R^2C}{L}}$$

$$f = \frac{1}{2\pi\sqrt{LC}}\sqrt{1-\frac{R^2C}{L}}$$

2.34 (0.5)

Circuit contains balanced Wheat-stone bridge. Also at high frequencies capacitor can be considered as short circuits.

Redrawing the circuit:

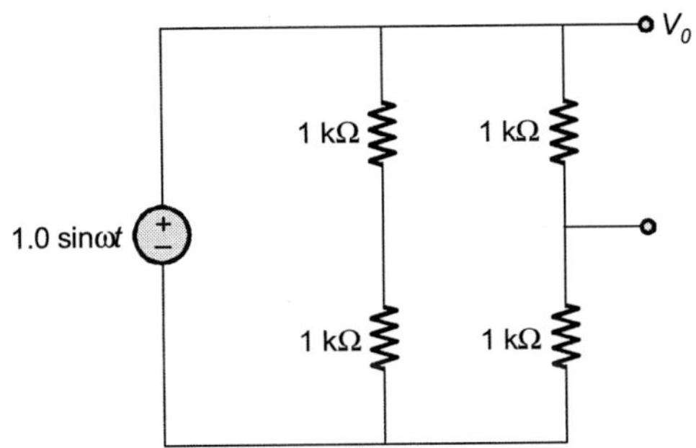

$$V_0 = 1\sin\omega t \times 1\text{ k}\Omega = 0.5\sin\omega t$$

Peak output = 0.5 V

2.35 (20)

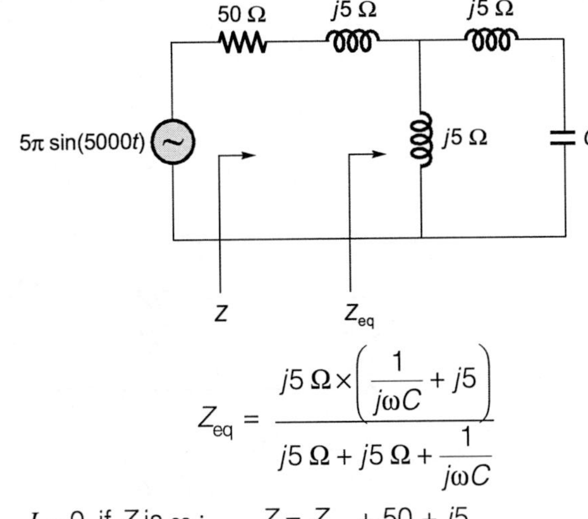

$$Z_{eq} = \frac{j5\,\Omega \times \left(\frac{1}{j\omega C}+j5\right)}{j5\,\Omega + j5\,\Omega + \frac{1}{j\omega C}}$$

$I = 0$ if Z is ∞; $\quad Z = Z_{eq} + 50 + j5$

For z to be ∞, $Z_{eq} = \infty$

$$\Rightarrow j5+j5+\frac{1}{j\omega C} = 0$$

$$10 = \frac{1}{5000 \times C} \quad \Rightarrow C = \frac{1}{5\times10^3\times10} = 20\ \mu\text{F}$$

2.36 (0.316)

At resonance (for parallel RLC circuit)

$$I_R = I$$

$$I_L = QI\angle{-90^\circ}$$

$$I_C = QI\angle 90^\circ$$

For parallel RLC circuit

$$\frac{|I_L|}{|I_R|} = \frac{IQ}{I} = Q = R\sqrt{\frac{C}{L}}$$

$$= 10\sqrt{\frac{10\times10^{-6}}{10\times10^{-3}}} = 0.316$$

2.37 (b)

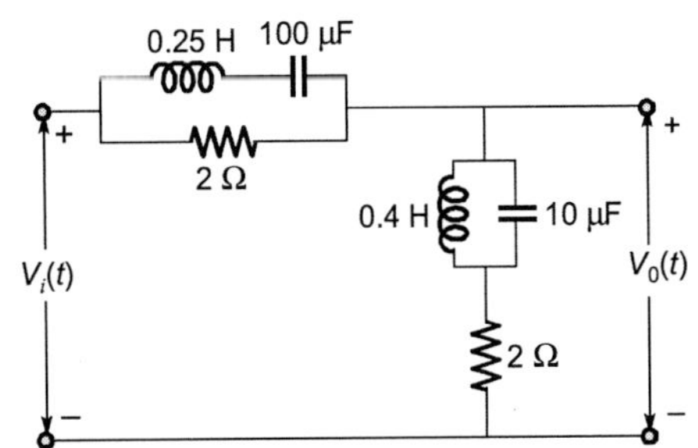

Where, $\quad V_i(t) = 2\cos(200t) + 4\sin(500t)$

As different frequencies are operating, using superposition theorem, we get

for ω = 200 rad/sec

$X_L = \omega L = (200)(0.25) = 50\ \Omega$;

$$X_C = \frac{1}{\omega C} = \frac{1}{200\times100\times10^{-6}} = 50\ \Omega$$

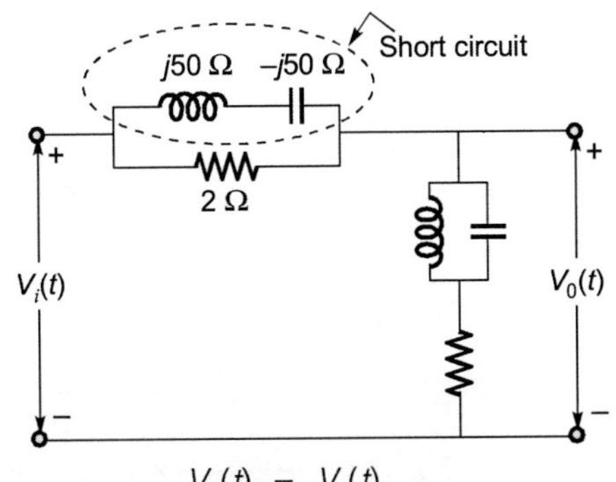

$$V_0(t) = V_i(t)$$

for ω = 500 rad/sec

$$X_L = 0.4\times500 = 200\ \Omega$$

$$X_C = \frac{1}{10\times10^{-6}\times500} = 200\ \Omega$$

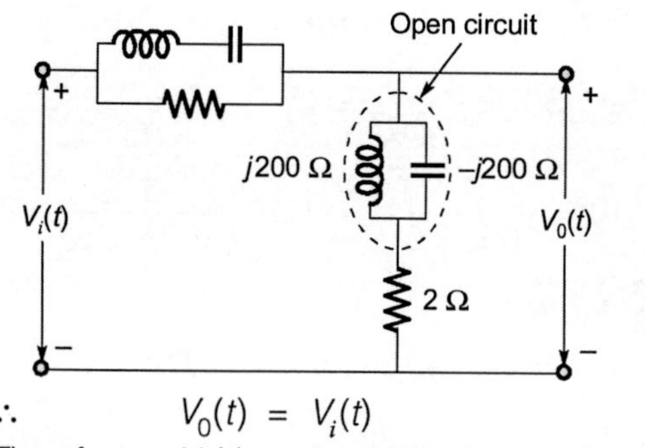

$\therefore$ $\quad V_0(t) = V_i(t)$

Therefore $\quad V_0(t) = 2\cos(200t) + 4\sin(500t)$

2.38 (1)

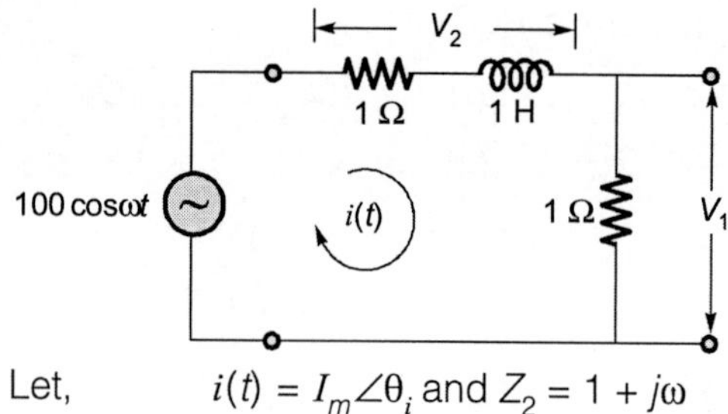

Let, $\quad i(t) = I_m\angle\theta_i$ and $Z_2 = 1 + j\omega$

$$= \sqrt{1+\omega^2}\,\angle\theta_2$$

$$\theta_2 = \tan^{-1}\left(\frac{\omega}{1}\right)$$

$$V_1 = i(t)\,(1\ \Omega) = I_m\angle\theta_i$$

$$V_2 = i(t)Z_2 = I_m\sqrt{1+\omega^2}\,\angle\theta_2 + \theta_i$$

From the given data,

$$(\theta_i + \theta_2) - (\theta_i) = \frac{\pi}{4}$$

$$\theta_2 = \frac{\pi}{4}$$

$$\tan^{-1}\left(\frac{\omega}{1}\right) = \frac{\pi}{4}$$

$$\omega = 1 \text{ rad/sec}$$

2.39 (2.6)

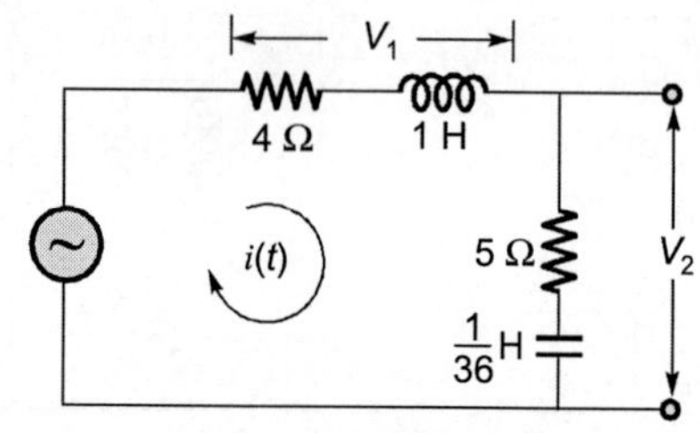

$$\omega = 3 \text{ rad/sec} ;\ Z_1 = (4 + j3)\ \Omega$$

and $\quad Z_2 = (5 - j12)\ \Omega$

$$|V_2| = |i|\,|Z_2| = |i|\sqrt{5^2+12^2} = 13|i|$$

$$|V_1| = |i|\,|Z_1| = |i|\sqrt{4^2+3^2} = 5\,|i|$$

$$\frac{|V_2|}{|V_1|} = \frac{13|i|}{5|i|} = \frac{13}{5} = 2.6$$

2.40 (0.2)

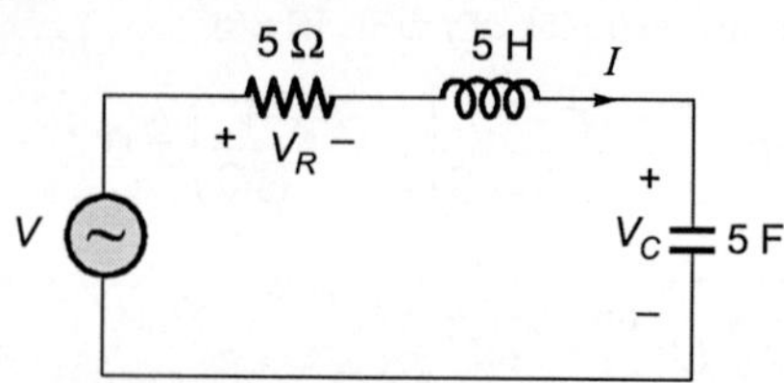

Given that, V and I have same phase. So, the circuit is in resonance.

At resonance,

$$V_C = QV_R$$

So, $\quad \dfrac{\text{Amplitude of } V_C}{\text{Amplitude of } V_R} = Q = \dfrac{1}{R}\sqrt{\dfrac{L}{C}} = \dfrac{1}{5}\sqrt{\dfrac{5}{5}} = 0.2$

2.41 (c)

$$\frac{1}{\omega C} = \frac{1}{5\times10^{-6}} = 200\text{ k}\Omega$$

$$V_C = \frac{5\angle0^\circ}{200 - j200}\times(-j200)\text{ V}$$

$$= \frac{5\angle0^\circ\times1\angle-90^\circ}{\sqrt{2}\angle-45^\circ}\text{ V}$$

$$= \frac{5}{\sqrt{2}}\angle-45^\circ\text{ V} = 2.5\sqrt{2}\sin\left(5t - \frac{\pi}{4}\right)\text{V}$$

$$= 2.5\sqrt{2}\sin(5t - 0.25\pi)\text{ V}$$

2.42 (a)

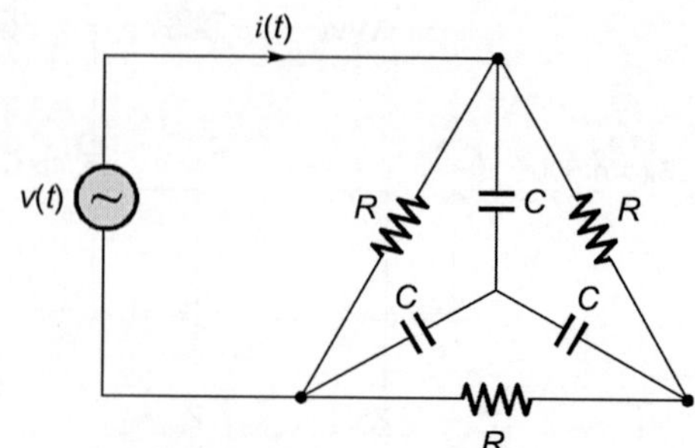

Here, $\quad X_C = \dfrac{1}{\omega C} = \dfrac{1}{10^3\times10^{-6}} = \dfrac{1}{10^{-3}}$

$$X_C = 10^3\ \Omega$$

$$R = 10^3\ \Omega \quad \text{(Given)}$$

$$v(t) = 2\sin1000t\text{ V} = 2\angle0^\circ\text{ V}$$

Redrawing the given network, we get,

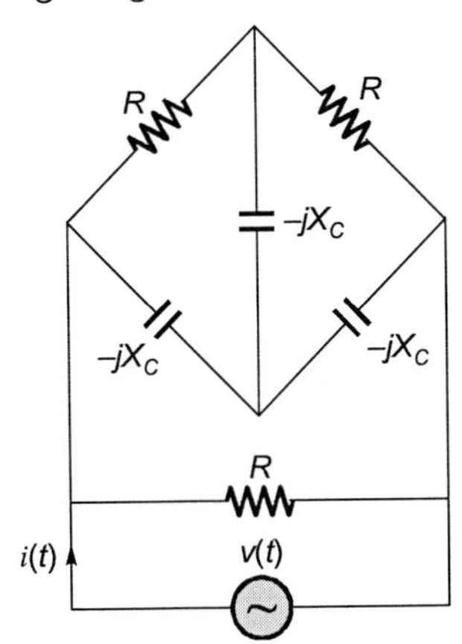

As the bridge is balanced, it can be redrawn as

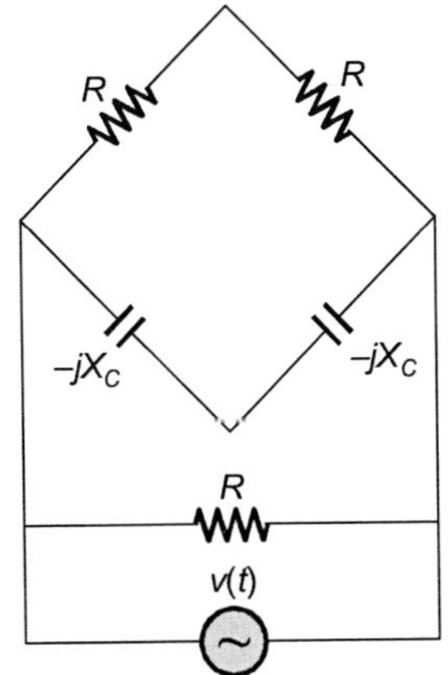

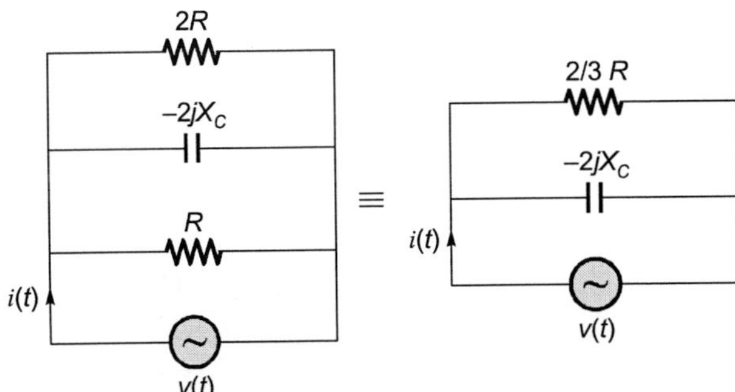

$$\therefore \quad Y_{eq} = Y_1 + Y_2$$

$$= \frac{3}{2}R + \frac{1}{-2jX_C}$$

$$= \frac{3}{2}\times 10^{-3} + j\frac{1}{2}\times 10^{-3}$$

$$\therefore \quad i(t) = v(t)\times Y_{eq} = 2\angle 0^\circ \left[\frac{3}{2} + j\frac{1}{2}\right] \text{mA}$$

$$= (3 + j1) \text{ mA}$$

$$= 3\sin(1000t) + \cos(1000t) \text{ mA}$$

2.43 (2.828)

$$Z = \frac{V}{I} = \frac{200\angle 0^\circ}{10\angle -45^\circ} = 20\angle 45^\circ = 10\sqrt{2} + j10\sqrt{2}$$

$$X_L = 10\sqrt{2}$$

$$\omega L = 10\sqrt{2}$$

$$L = \frac{10\sqrt{2}}{5} = 2.828\ H$$

2.44 (d)

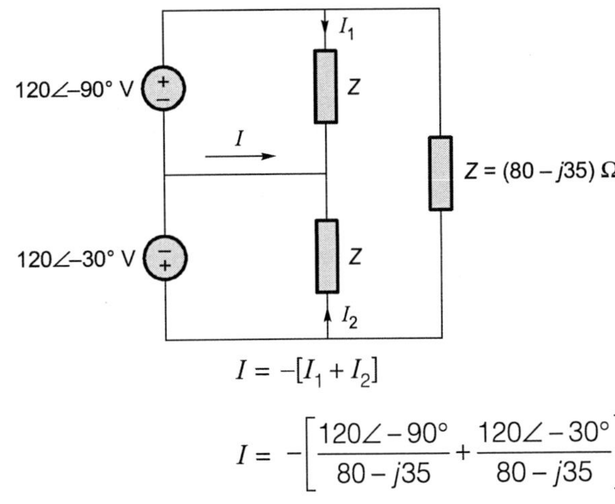

$$I = -[I_1 + I_2]$$

$$I = -\left[\frac{120\angle -90^\circ}{80 - j35} + \frac{120\angle -30^\circ}{80 - j35}\right]$$

$$I = 2.38\angle 143.7^\circ$$

2.45 (a)

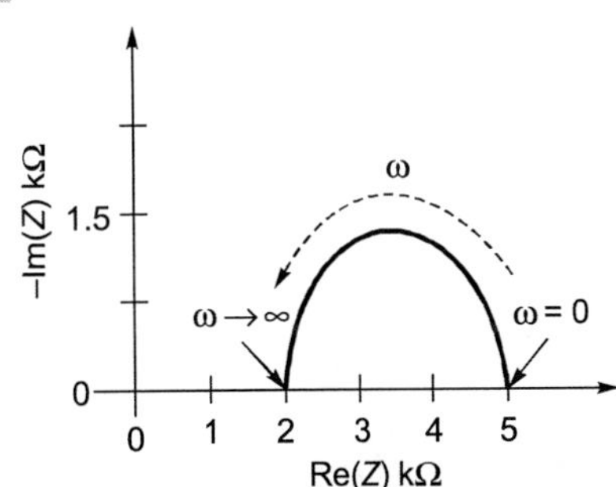

At ω = 0 rad/sec,

$$X_C = \frac{1}{\omega C} = \frac{1}{0} = \infty$$

The equivalent circuit at ω = 0 rad/sec is as drawn below :

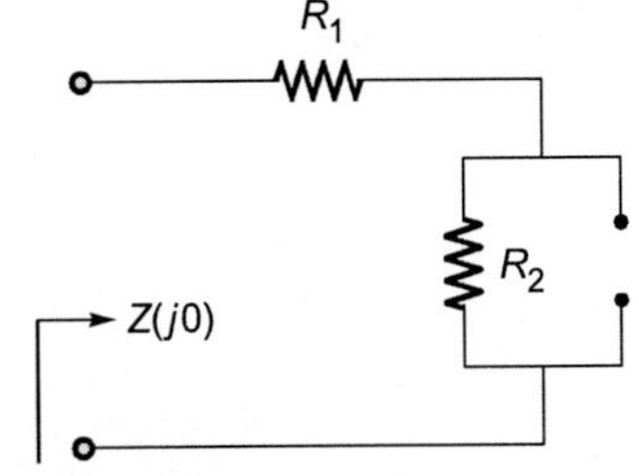

At ω = 0 rad/sec

$$Z(j0) = R_1 + R_2$$

From the plot of the locus of impedance, we get,

$$Z(j0) = 5 \text{ k}\Omega$$

$\therefore \quad R_1 + R_2 = 5\ k\Omega$

At $\omega = \infty$ rad/sec,

$$X_c = \frac{1}{j\omega C} = \frac{1}{\infty} = 0\ \Omega$$

The equivalent circuit at $\omega = \infty$ rad/sec is as drawn below:

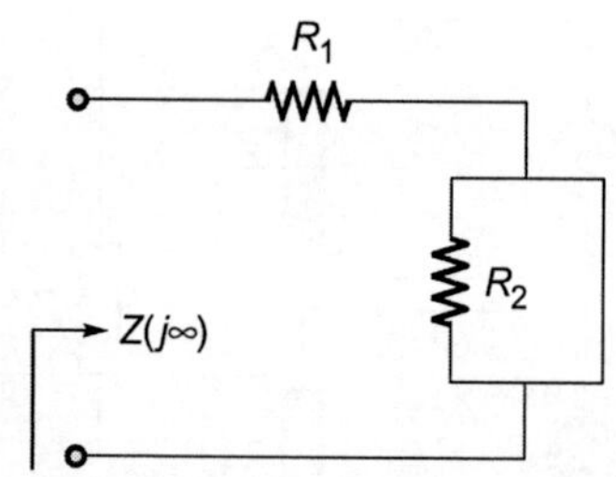

At $\omega = \infty$ rad/sec,

$$Z(j\infty) = R_1$$

From the plot of the locus of impedance, we get,

$$Z(j\infty) = 2\ k\Omega$$

$$\therefore \quad R_1 = 2\ k\Omega$$

Using equation (i),

$$R_1 + R_2 = 5\ k\Omega$$

$$\therefore \quad R_2 = 5 - 2 = 3\ k\Omega$$

Hence, $R_1 = 2\ k\Omega$ and $R_2 = 3\ k\Omega$.

2.46 (b)

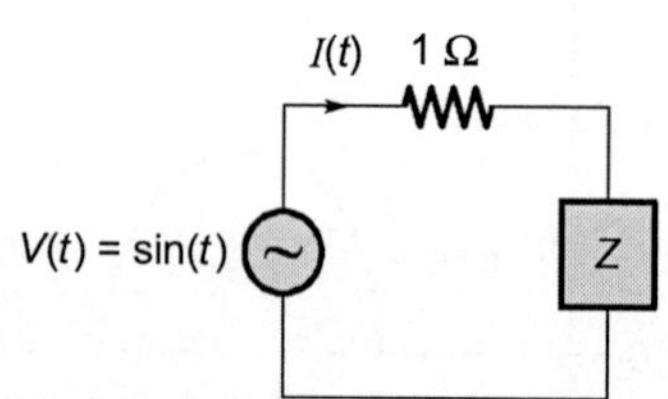

From the given waveform of steady state current $I(t)$, we can write

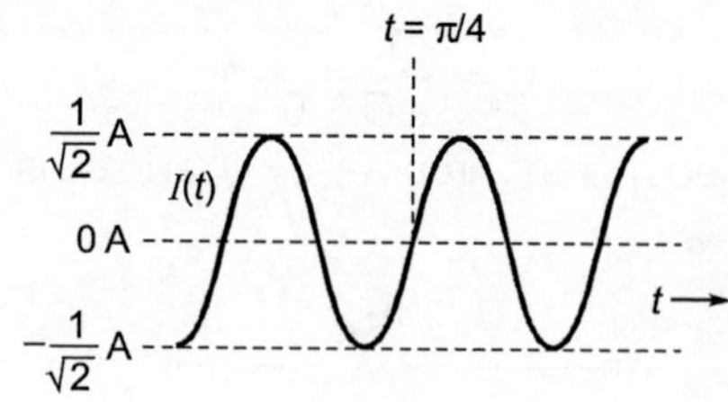

$$I(t) = \frac{1}{\sqrt{2}}\sin\left(t - \frac{\pi}{4}\right) A$$

Hence, the current $I(t)$ is lagging. Therefore, element Z is an inductor ($Z = j\omega L$).

$$\therefore \quad I(t) = \frac{V(t)}{1+Z} = \frac{\sin(t)}{Z+1} A \quad \text{...(i)}$$

From the waveform, the maximum value of current

$$|I(t)|_{max} = \frac{1}{\sqrt{2}} A$$

Using equation (i)

$$|I(t)|_{max} = \frac{1}{|Z+1|} = \frac{1}{\sqrt{2}}$$

$$\Rightarrow \quad |j\omega L + 1| = \sqrt{2}$$

$$\Rightarrow \quad \sqrt{1+\omega^2 L^2} = \sqrt{2}$$

$$\Rightarrow \quad 1 + \omega^2 L^2 = 2$$

Given : $\omega = 1$ rad/sec

$$\therefore \quad L^2 = 1$$

$$\Rightarrow \quad L = 1\ H$$

2.47 (d)

Given: Quality factor, $Q = 1000$ and center frequency $\omega_o = 10^6$ rad/sec

We know, for series RLC circuit,

$$\text{Quality factor, } Q = \frac{\omega_o L}{R}$$

Also, center frequency,

$$\omega_o = \sqrt{\frac{1}{LC}}$$

$$Q = \frac{1}{\sqrt{LC}} \times \frac{1}{R} = \frac{1}{R}\sqrt{\frac{L}{C}}$$

So, $L = 1\ \mu H$, $C = 1\ \mu F$ and $R = 0.001\ \Omega$

■■■

3 Network Theorems

3.1 The value of the resistance, R, connected across the terminals, A and B, (ref. Fig.), which will absorb the maximum power, is

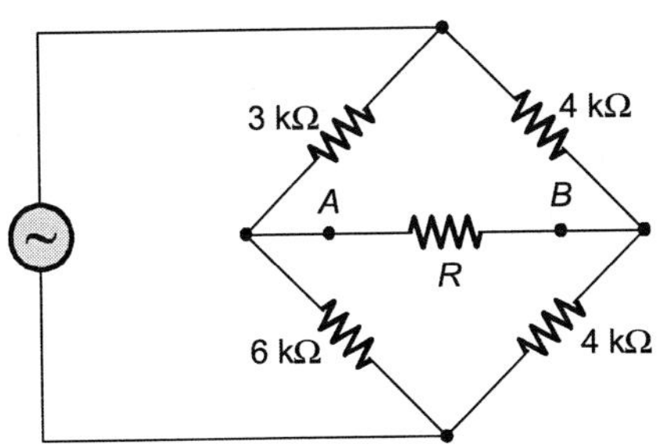

(a) 4.00 kΩ
(b) 4.11 kΩ
(c) 8.00 kΩ
(d) 9.00 kΩ [1995 : 1 M]

3.2 Superposition theorem is NOT applicable to networks containing

(a) non linear elements
(b) dependent voltage sources
(c) dependent current sources
(d) transformers [1998 : 1 M]

3.3 The Thevenin equivalent voltage V_{TH} appearing between the terminals A and B of the network shown in the figure is given by

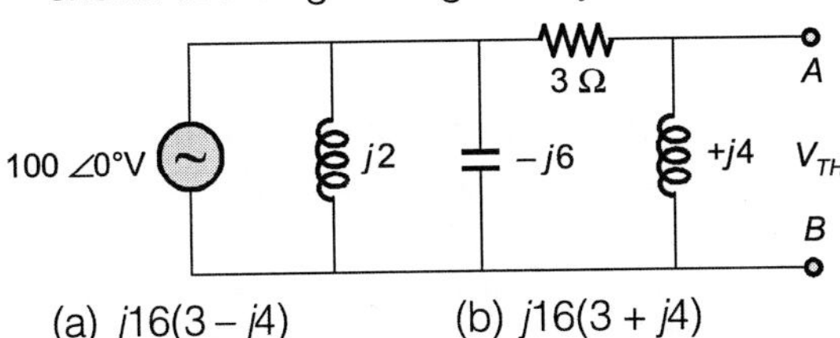

(a) $j16(3 - j4)$
(b) $j16(3 + j4)$
(c) $16(3 + j4)$
(d) $16(3 - j4)$

[1999 : 2 M]

3.4 The value of R (in ohms) required for maximum power transfer in the network shown in the figure is

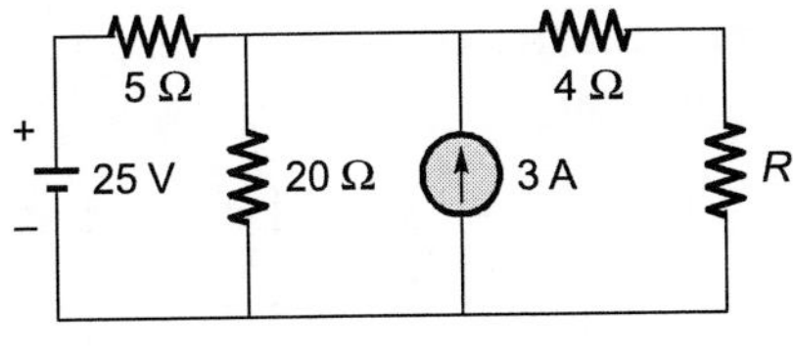

(a) 2
(b) 4
(c) 8
(d) 16

[1999 : 2 M]

3.5 Use the data of the figure (a). The current i in the circuit of the figure (b)

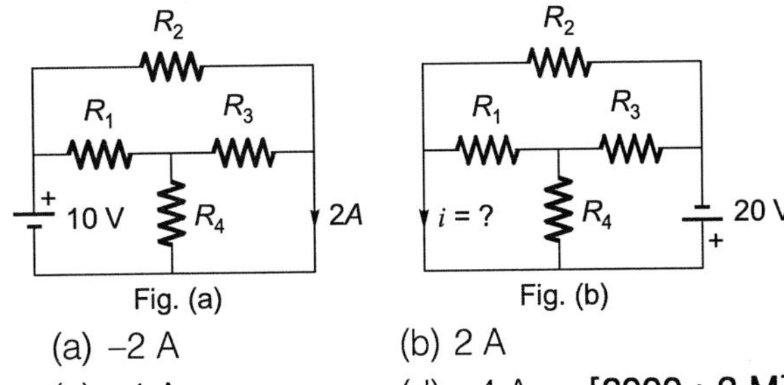

(a) –2 A
(b) 2 A
(c) –4 A
(d) +4 A [2000 : 2 M]

3.6 In the figure, the value of the load resistor R which maximizes the power delivered to it is

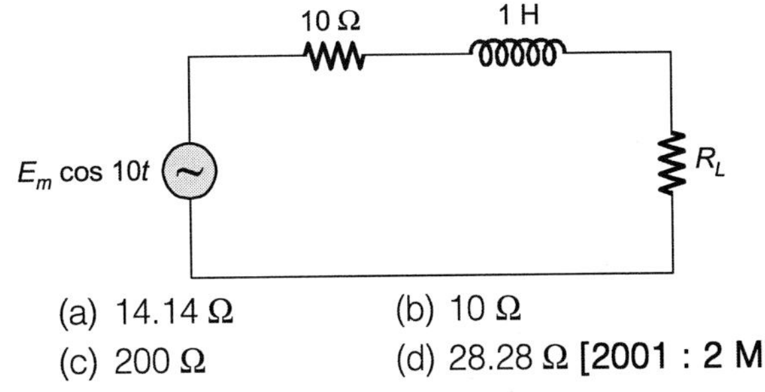

(a) 14.14 Ω
(b) 10 Ω
(c) 200 Ω
(d) 28.28 Ω [2001 : 2 M]

3.7 In the network of the figure, the maximum power is delivered to R_L if its value is

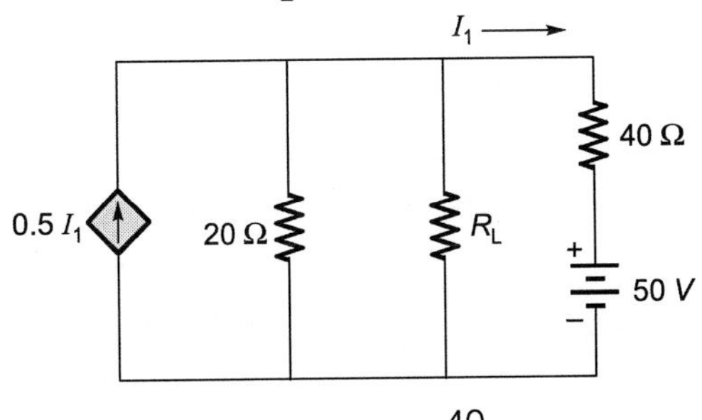

(a) 16 Ω
(b) $\frac{40}{3}$ Ω
(c) 60 Ω
(d) 20 Ω [2002 : 2 M]

3.8 A source of angular frequency 1 rad/sec has a source impedance consisting of 1 Ω resistance in series with 1 H inductance. The load that will obtain the maximum power transfer is

(a) 1 Ω resistance
(b) 1 Ω resistance in parallel with 1 H inductance
(c) 1 Ω resistance in series with 1 F capacitor
(d) 1 Ω resistance in parallel with 1 F capacitor

[2003 : 1 M]

3.9 The maximum power that can be transferred to the load resistor R_L from the voltage source in the figure is

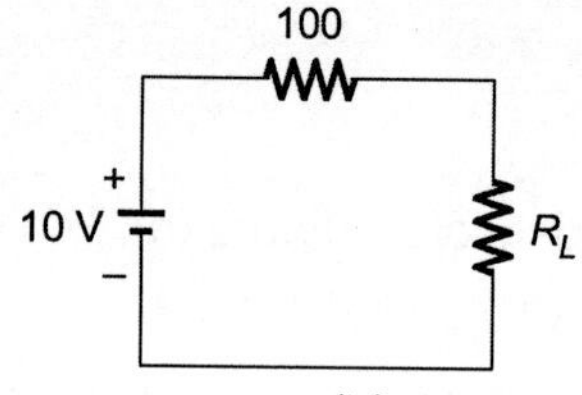

(a) 1 W (b) 10 W
(c) 0.25 W (d) 0.5 W **[2005 : 1 M]**

3.10 For the circuit shown in the figure, Thevenin's voltage and Thevenin's equivalent resistance at terminals *a-b* is

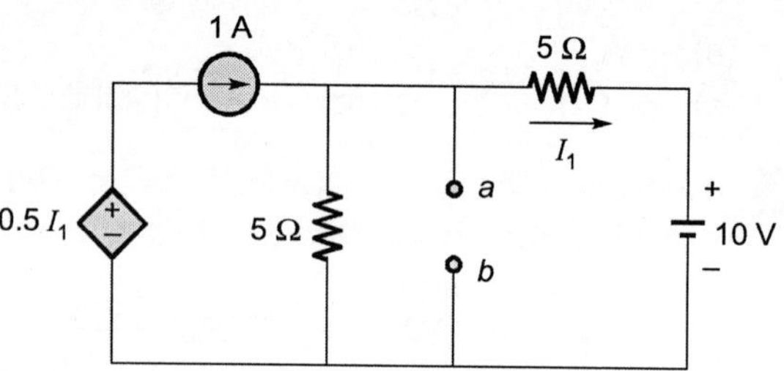

(a) 5 V and 2 Ω (b) 7.5 V and 2.5 Ω
(c) 4 V and 2 Ω (d) 3 V and 2.5 Ω

[2005 : 2 M]

3.11 An independent voltage source in series with an impedance $Z_s = R_s + jX_s$ delivers a maximum average power to a load impedance Z_L when

(a) $Z_L = R_s + jX_s$ (b) $Z_L = R_s$
(c) $Z_L = jX_s$ (d) $Z_L = R_s - jX_s$

[2007 : 1 M]

3.12 For the circuit shown in the figure, the Thevenin voltage and resistance looking into *X-Y* are

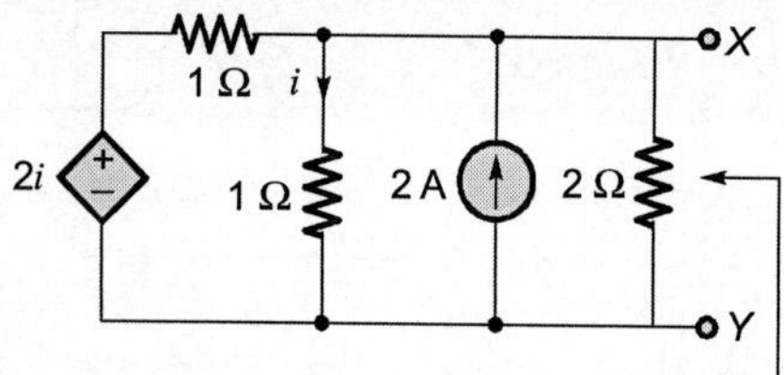

(a) 4/3 V, 2 Ω (b) 4 V, 2/3 Ω
(c) 4/3 V, 2/3 Ω (d) 4 V, 2 Ω **[2007 : 2 M]**

3.13 The Thevenin equivalent impedance Z_{Th} between the nodes *P* and *Q* in the following circuit is

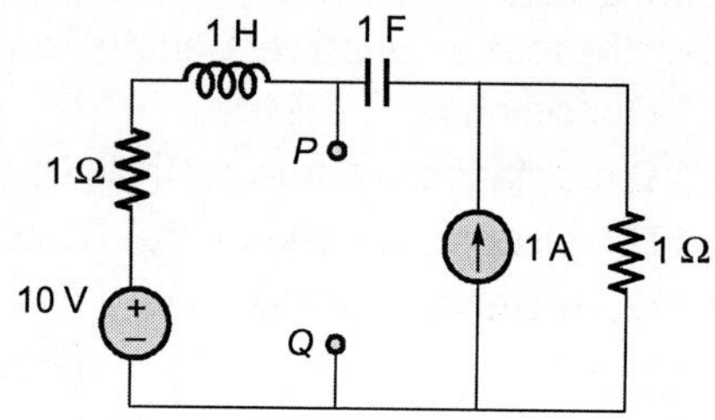

(a) 1 (b) $1 + s + \frac{1}{s}$
(c) $2 + s + \frac{1}{s}$ (d) $\frac{s^2 + s + 1}{s^2 + 2s + 1}$

[2008 : 2 M]

3.14 In the circuit shown, what value of R_L maximizes the power delivered to R_L?

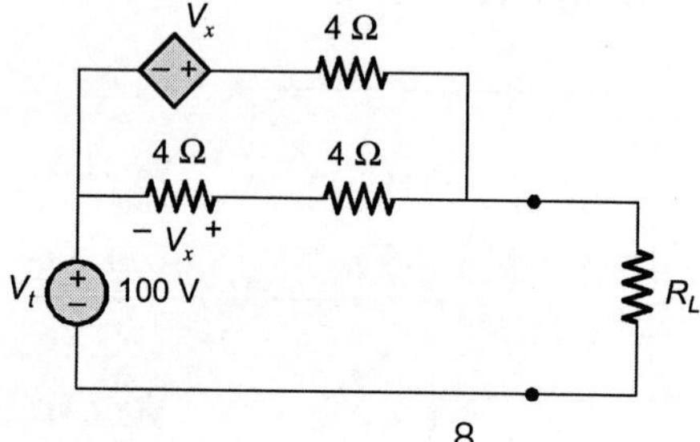

(a) 2.4 Ω (b) $\frac{8}{3}$ Ω
(c) 4 Ω (d) 6 Ω **[2009 : 2 M]**

3.15 In the circuit shown below, the Norton equivalent current in amperes with respect to the terminals *P* and *Q* is

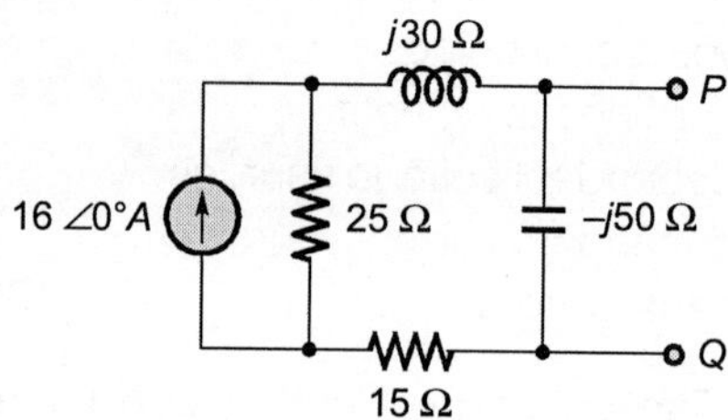

(a) $6.4 - j4.8$ (b) $6.56 - j7.87$
(c) $10 + j0$ (d) $16 + j0$ **[2011 : 1 M]**

3.16 In the circuit shown below, the value of R_L such that the power transferred to R_L is maximum is

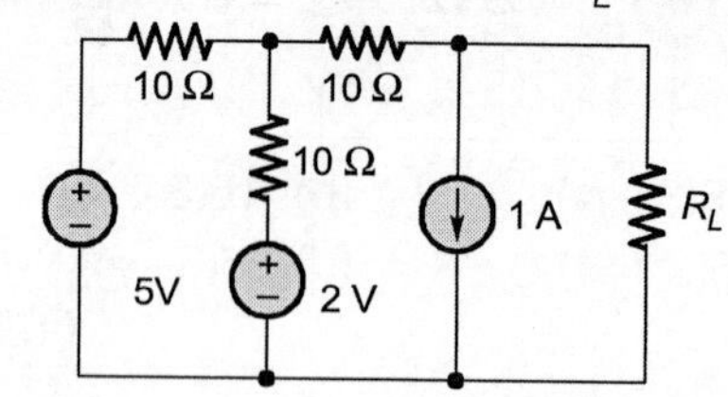

(a) 5 Ω (b) 10 Ω
(c) 15 Ω (d) 20 Ω **[2011 : 1 M]**

3.17 The impedance looking into nodes 1 and 2 in the given circuit is

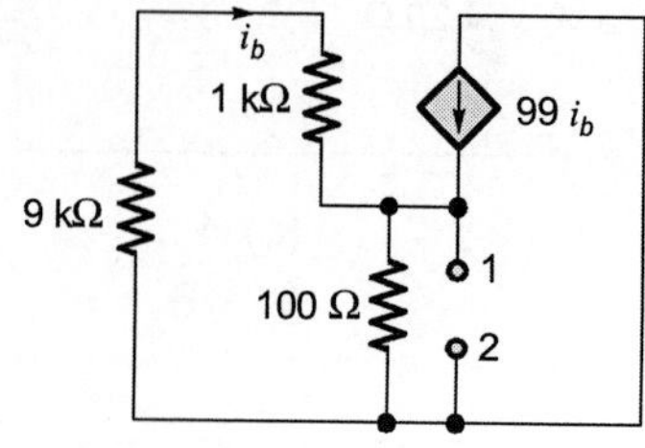

(a) 50 Ω (b) 100 Ω
(c) 5 kΩ (d) 10.1 kΩ [2012 : 1 M]

3.18 Assuming both the voltage sources are in phase, the value of R for which maximum power is transferred from circuit A to circuit B is

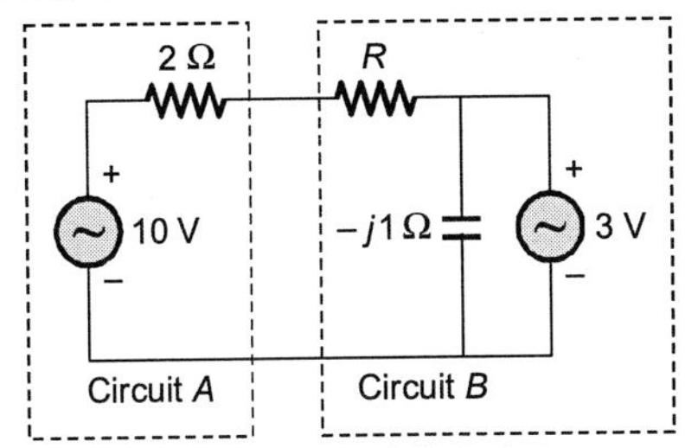

(a) 0.8 Ω (b) 1.4 Ω
(c) 2 Ω (d) 2.8 Ω [2012 : 2 M]

3.19 A source $v_s(t) = V\cos 100\pi t$ has an internal impedance of $(4 + j3)\Omega$. If a purely resistive load connected to this source has to extract the maximum power out of the source, its value in Ω should be
(a) 3 (b) 4
(c) 5 (d) 7 [2013 : 1 M]

3.20 In the circuit shown below, if the source voltage $V_s = 100\angle 53.13°$V then the Thevenin's equivalent voltage in Volts as seen by the load resistance R_L is

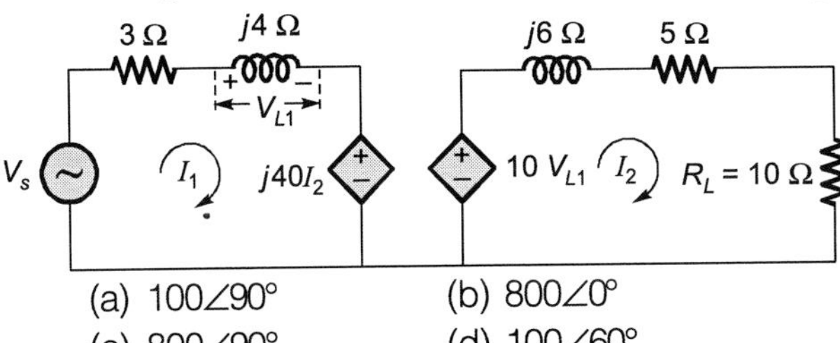

(a) $100\angle 90°$ (b) $800\angle 0°$
(c) $800\angle 90°$ (d) $100\angle 60°$
[2013 : 2 M]

3.21 Norton's theorem states that a complex network connected to a load can be replaced with an equivalent impedance
(a) in series with a current source
(b) in parallel with a voltage source
(c) in series with a voltage source
(d) in parallel with a current source
[2014 : 1 M, Set-2]

3.22 In the figure shown, the value of the current I (in Amperes) is ______.

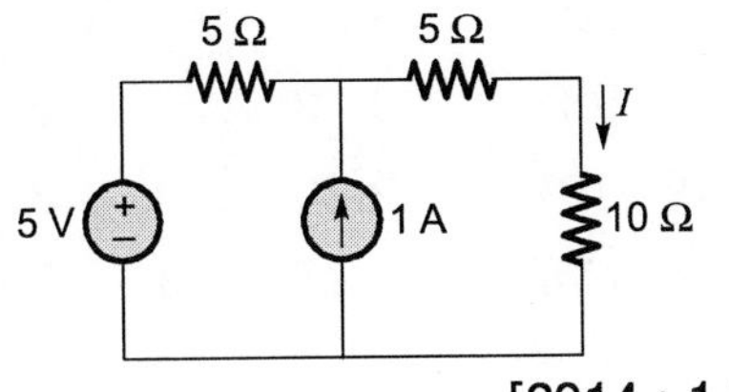

[2014 : 1 M, Set-3]

3.23 In the circuit shown in the figure, the angular frequency ω (in rad/s), at which the Norton equivalent impedance as seen from terminals b-b' is purely resistive, is ______.

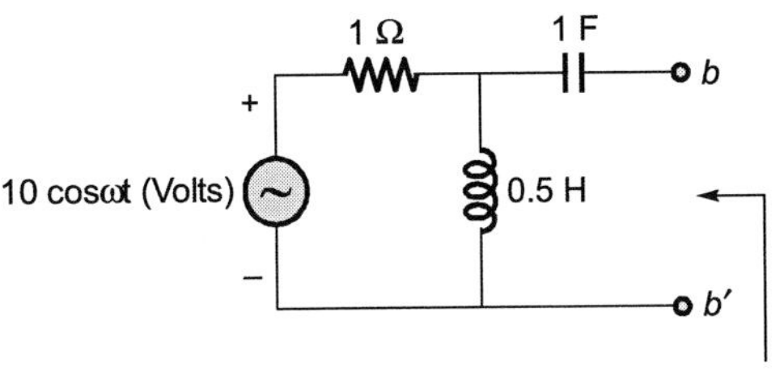

[2014 : 2 M, Set-3]

3.24 In the given circuit, the maximum power (in Watts) that can be transferred to the load R_L is

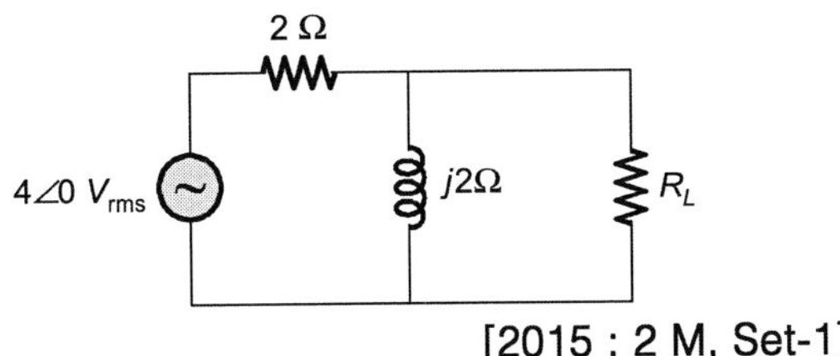

[2015 : 2 M, Set-1]

3.25 In the circuit shown, the Norton equivalent resistance (in Ω) across terminals a-b is______.

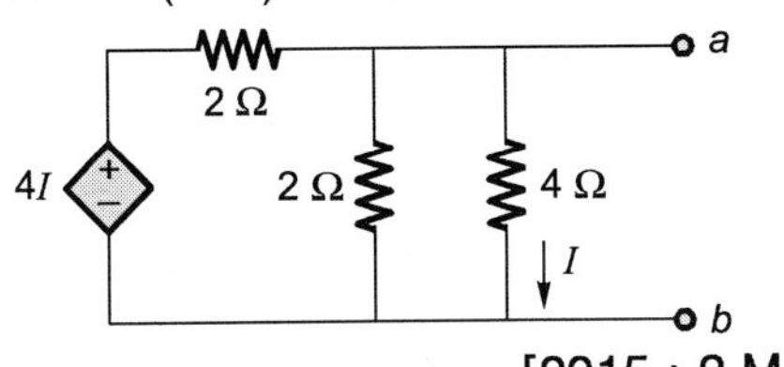

[2015 : 2 M, Set-2]

3.26 For the current shown in the figure, the Thevenin equivalent voltage (in Volts) across terminals a-b is ______.

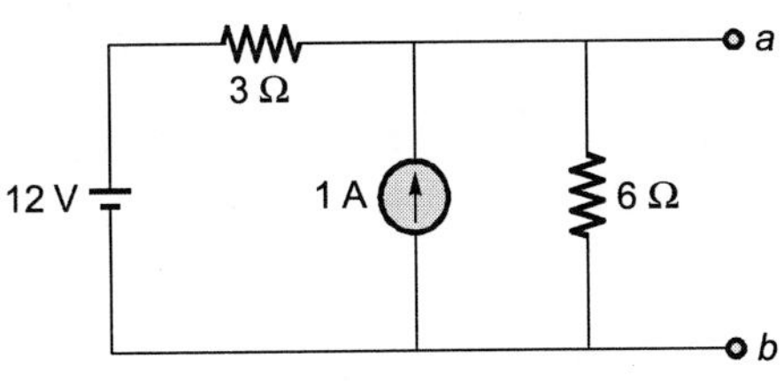

[2015 : 1 M, Set-3]

3.27 In the circuit shown in the figure, the maximum power (in watt) delivered to the resistor R is ______.

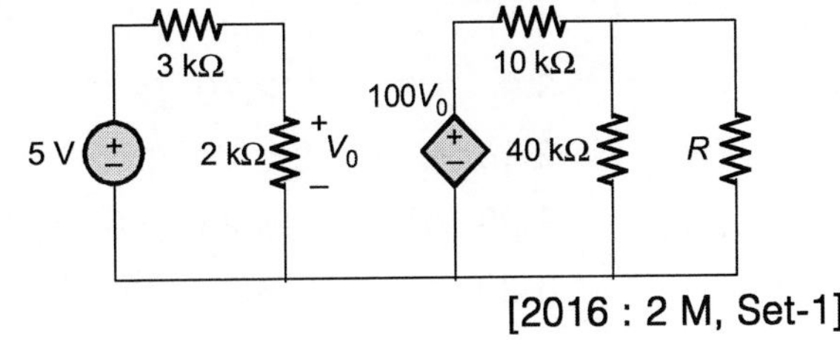

[2016 : 2 M, Set-1]

3.28 In the circuit shown below, V_s is constant voltage source and I_L is a constant current load.

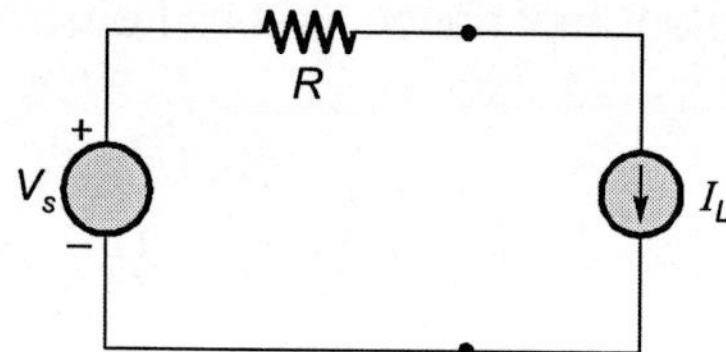

The value of I_L that maximizes the power absorbed by the constant current load is

(a) $\frac{V_s}{4R}$
(b) $\frac{V_s}{2R}$
(c) $\frac{V_s}{R}$
(d) ∞ [2016 : 1 M, Set-2]

3.29 Consider the circuit shown in the figure.

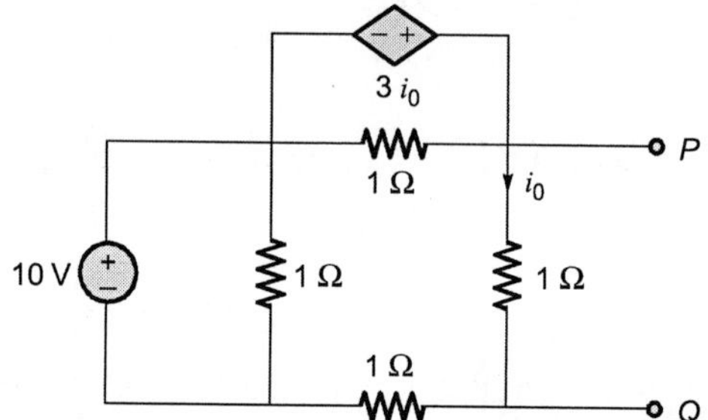

The Thevenin equivalent resistance (in Ω) across P-Q is _____.

[2017 : 2 M, Set-2]

3.30 Consider the two-port resistive network shown in the figure. When an excitation of 5 V is applied across Port 1, and Port 2 is shorted, the current through the short circuit at Port 2 is measured to be 1 A (see (a) in the figure).
Now, if an excitation of 5 V is applied across Port 2, and Port 1 is shorted (see (b) in the figure), what is the current through the short circuit at Port 1?

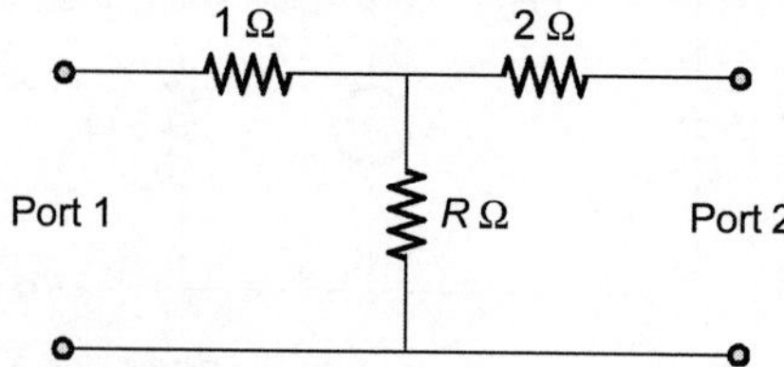

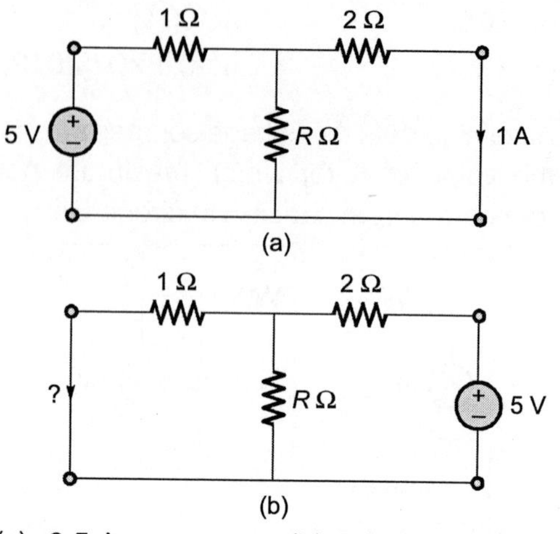

(a) 0.5 A
(b) 2.5 A
(c) 1 A
(d) 2 A [2019 : 1 M]

3.31 In the circuit shown below, the Thevenin voltage V_{Th} is

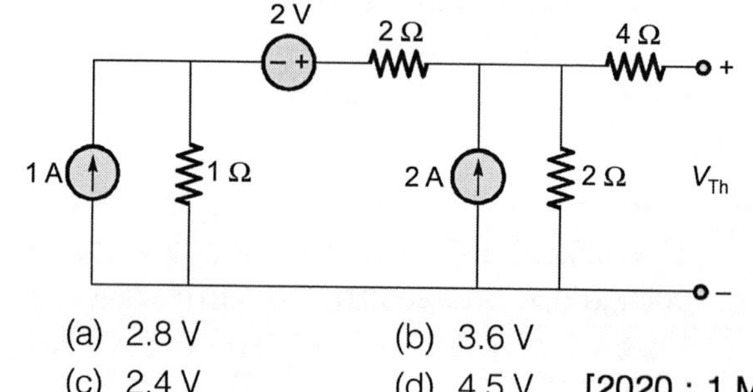

(a) 2.8 V
(b) 3.6 V
(c) 2.4 V
(d) 4.5 V [2020 : 1 M]

3.32 In the network shown below, maximum power is to be transferred to the load R_L.

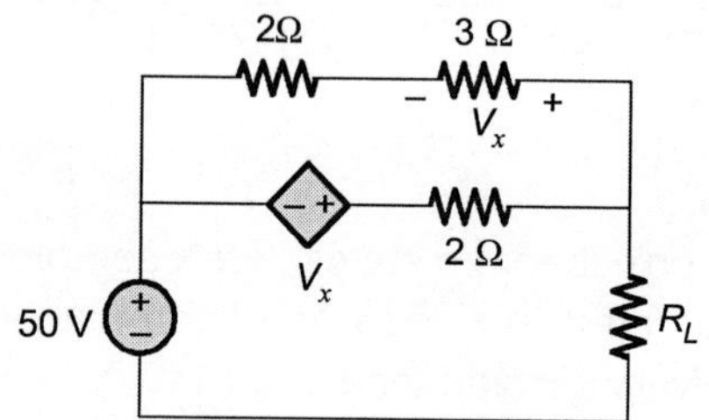

The value of R_L (in Ω) is _______.

[2024 : 2 Marks]

Answers Network Theorems

3.1 (a)	3.2 (a)	3.3 (a)	3.4 (c)	3.5 (c)	3.6 (a)	3.7 (a)
3.8 (c)	3.9 (c)	3.10 (b)	3.11 (d)	3.12 (d)	3.13 (a)	3.14 (c)
3.15 (a)	3.16 (c)	3.17 (a)	3.18 (a)	3.19 (c)	3.20 (c)	3.21 (d)
3.22 (0.5)	3.23 (2)	3.24 (1.649)	3.25 (1.33)	3.26 (10)	3.27 (0.8)	3.28 (b)
3.29 (–1)	3.30 (c)	3.31 (b)	3.32 (2.5)			

Explanations Network Theorems

3.1 (a)

Maximum power will be absorbed by R when $R = R_{Th}$.

$$R_{AB} = R_{Th} = (3 \parallel 6) + (4 \parallel 4)$$

$$R_{Th} = R = 2 + 2 = 4\ \text{k}\Omega$$

3.2 (a)

Superposition theorem is applicable for linear network.

3.3 (a)

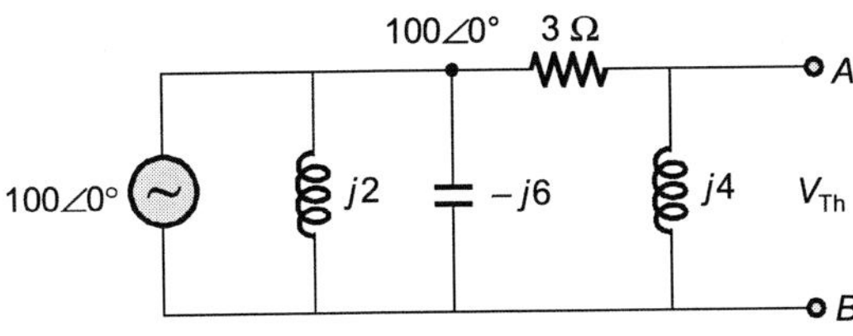

$$V = 100\angle 0^\circ$$

$$\therefore\ V_{TH} = 100\angle 0^\circ \cdot \frac{4j}{3+4j} = \frac{100 \times 4j(3-4j)}{25}$$

$$V_{TH} = 16j(3-4j)$$

3.4 (c)

For MPT, R should be equal to R_{eq} of the circuit seen from the terminal after removing R. Deactivating voltage and current sources.

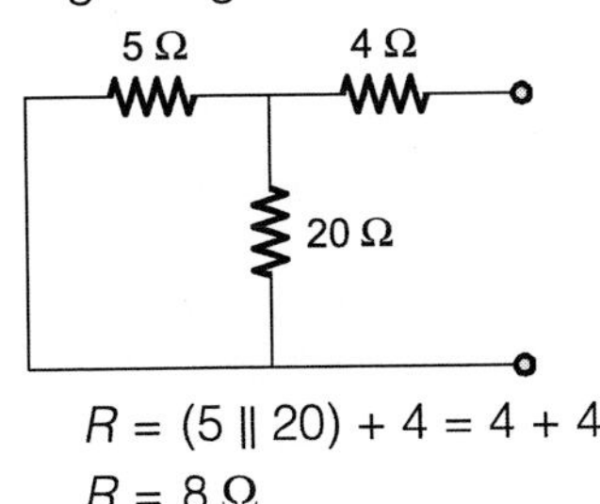

$$R = (5 \parallel 20) + 4 = 4 + 4$$

$$R = 8\ \Omega$$

3.5 (c)

This is a reciprocal and linear network. According to reciprocity theorem which states "Two loops A and B of a network N and if an ideal voltage source E in loop A produces a current I in loop B, then interchanging positions an identical source in loop B produces the same current in loop A". Since network is linear, principle of homogeneity holds and so when volt source is doubled, current also doubles with opposite direction.

3.6 (a)

$$X_S = \omega L = 10\ \Omega$$

$$Z_S = 10 + j10$$

$$\text{R for max power transfer} = |Z_s| = 10\sqrt{2}$$

$$= |10 + 10j| = 10\sqrt{2}\angle 45^\circ$$

$$= 14.14\ \Omega$$

3.7 (a)

For maximum power delivered to R_L, open circuit R_L,

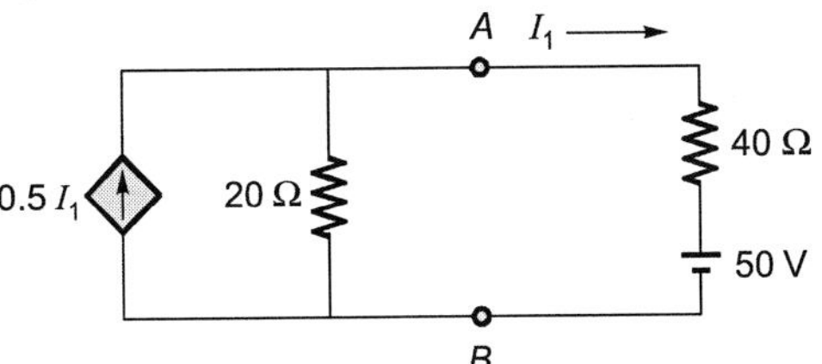

R_{Th} across AB,

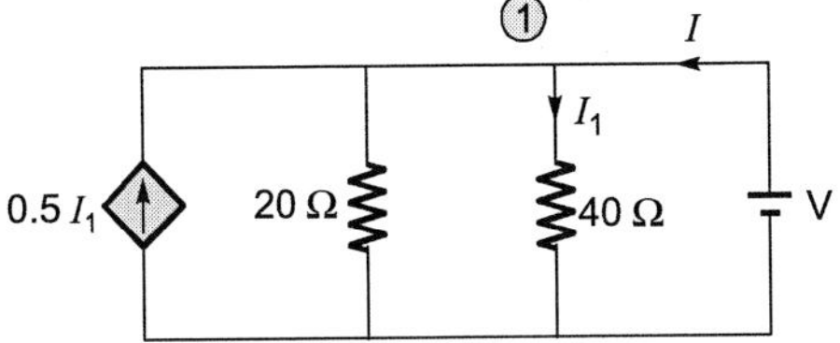

KCL at Node 1,

$$0.5I_1 + I = \frac{V}{20} + \frac{V}{40}$$

$$\Rightarrow 0.5.\frac{V}{40} + I = \frac{V}{20} + \frac{V}{40}$$

$$I = V\left(\frac{1}{20} + \frac{1}{40} - \frac{1}{80}\right)$$

$$\Rightarrow\ \frac{V}{I} = R_{TH} = 16$$

$$\therefore\ R_L = R_{TH} = 16$$

3.8 (c)

$$Z_L = R_S - jX_s$$

$$\therefore\ Z_L = 1 - 1j$$

3.9 (c)

For maximum power transfer,

$$R_L = R_s = 100\ \Omega$$

$$\therefore\ P_{max} = \frac{V^2}{4R_L} = \frac{5\times 5}{100} = 0.25\text{ W}$$

(V across R_L = 5V)

3.10 (b)

For V_{Th}:

$$\frac{V_{ab}}{5}+\frac{V_{ab}-10}{5}=1$$

(when current source is in series with voltage source effect of current source is taken)

$$2V_{ab} = 15 \Rightarrow V_{ab} = 7.5\text{ V}$$

$$\frac{0.5I_1}{I_1}=\frac{5}{R}\Rightarrow R=10\,\Omega$$

(Impedance connected to $0.5I_1$ current source)

For R_{Th}, Make independent sources dead,

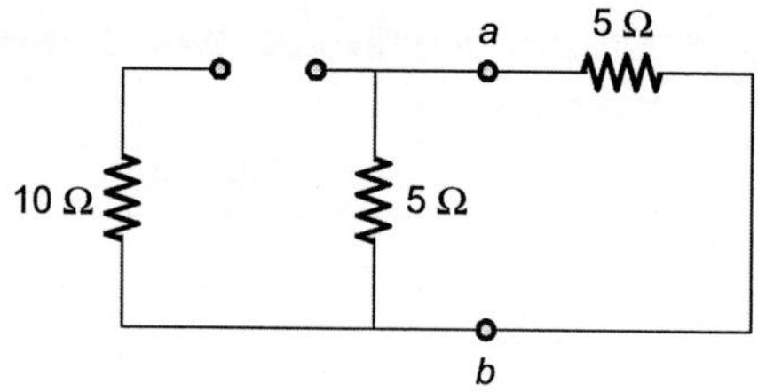

$$R_{Th} = 5\,||\,5 = 2.5\,\Omega$$

3.11 (d)

$$Z_L = R_s - jX_s$$

For maximum power transfer

$$Z_L = Z_S^* = R_s - jX_s$$

3.12 (d)

For V_{Th}, applying KCL at node X

$$2 = \frac{V_{Th}}{2}+\frac{V_{Th}}{1}+\frac{V_{Th}-2i}{1}$$

where, $$i = \frac{V_{Th}}{1}$$

So $$2 = V_{Th}\left(\frac{1}{2}+\frac{1}{1}+\frac{1}{1}-\frac{2}{1}\right)$$

$$V_{Th} = 4\text{ volt}$$

From the figure $I_{sc} = 2$ Amp

So $$R_{Th} = \frac{V_{Th}}{I_{sc}} = \frac{4}{2} = 2\,\Omega$$

3.13 (a)

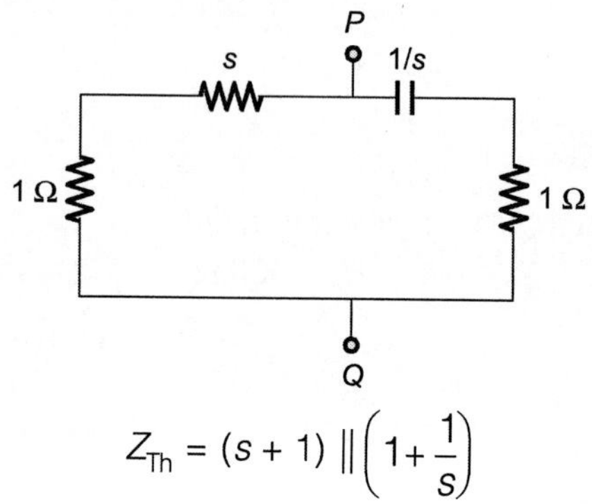

$$Z_{Th} = (s+1)\,||\left(1+\frac{1}{s}\right)$$

$$= \frac{(s+1)\left(1+\frac{1}{s}\right)}{(s+1)+1+\frac{1}{s}} = \frac{(s+1)^2/s}{(s+1)(s+1)/s} = 1$$

3.14 (c)

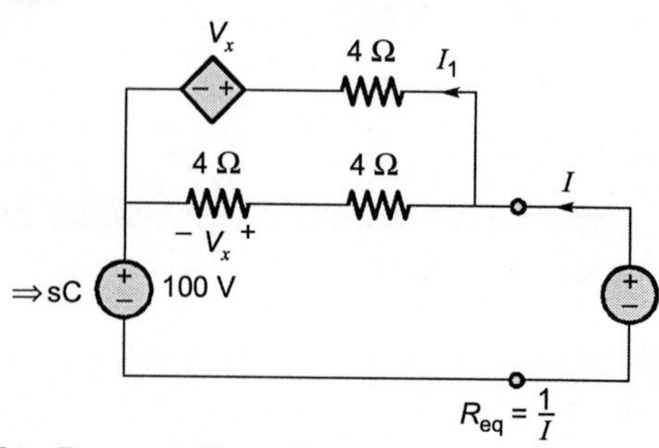

For P_{max}; $R_L = R_{eq}$

KVL $\quad 1 = 4I_1 + V_x \quad$ ... (1)

$V_x = 4(I - I_1) \quad$ ... (2)

Simplifying $$R_{eq} = \frac{1}{I} = 4\,\Omega$$

3.15 (a)

For Norton equivalent current short circuiting the terminal PQ.

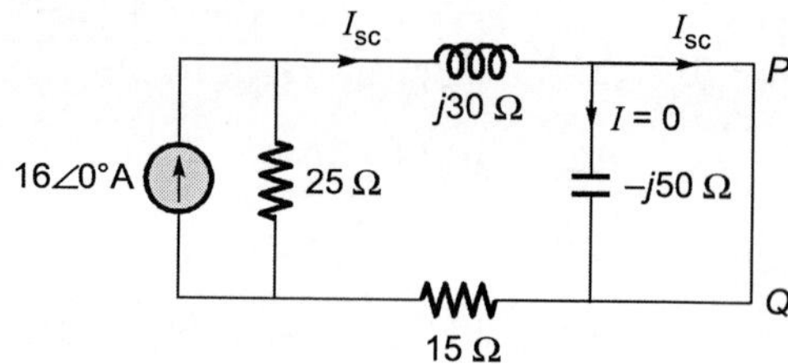

Short circuit current,

$$I_{SC} = \frac{25}{15+j30+25}\times 16\angle 0^\circ$$

$$= \frac{25}{40+j30}\times 16\angle 0^\circ = \frac{(25\times 16)\angle 0^\circ}{50\angle 36.86}$$

$$= 8\angle -36.86^\circ$$

Hence Norton current is

$$I_N = I_{SC} = 8\angle{-36.86^\circ}$$

$$I_N = (6.4 - j4.8)\text{ A}$$

3.16 (c)

For maximum power transfer

$$R_L = R_{Th}$$

To calculate R_{Th} deactivate all the energy sources.

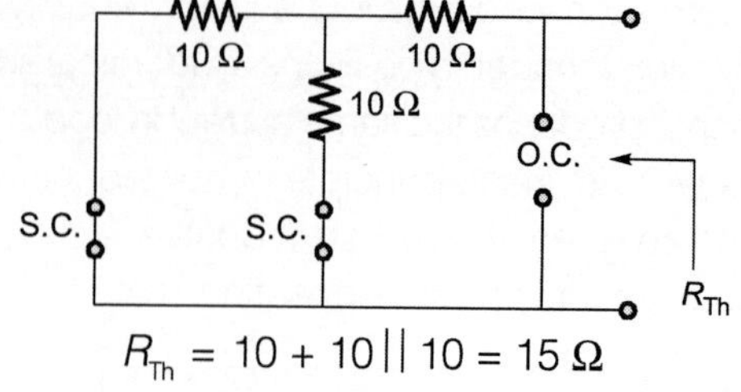

$$R_{Th} = 10 + 10\,||\,10 = 15\,\Omega$$

3.17 (a)

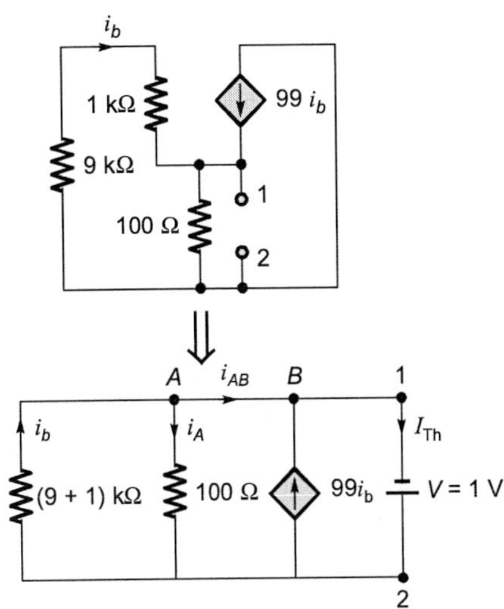

To find Thevenin impedance across node 1 and 2. Connect a 1 V source and find the current through voltage source.

Then, $$Z_{Th} = \frac{1}{I}$$

By applying KCL at node B and A

$$i_{AB} + 99i_b = I$$
$$i_b = i_A + i_{AB}$$
$$\Rightarrow \quad i_b - i_A + 99i_b = I$$
$$\Rightarrow \quad 100i_b - i_A = I \quad \text{...(i)}$$

By applying KVL in outer loop

$$10 \times 10^3\, i_b = 1$$
$$i_b = 10^{-4}\text{A}$$

and $$10 \times 10^3\, i_b = -100\, i_A$$
$$\Rightarrow \quad i_A = -100\, i_b$$

∴ From equation (i)

$$100i_b + 100i_b = I$$
$$\Rightarrow \quad I = 200i_b = 200 \times 10^{-4} = 0.02$$
$$\therefore \quad Z_{Th} = \frac{1}{I} = \frac{1}{0.02} = 50\ \Omega$$

3.18 (a)

Redrawing the diagram

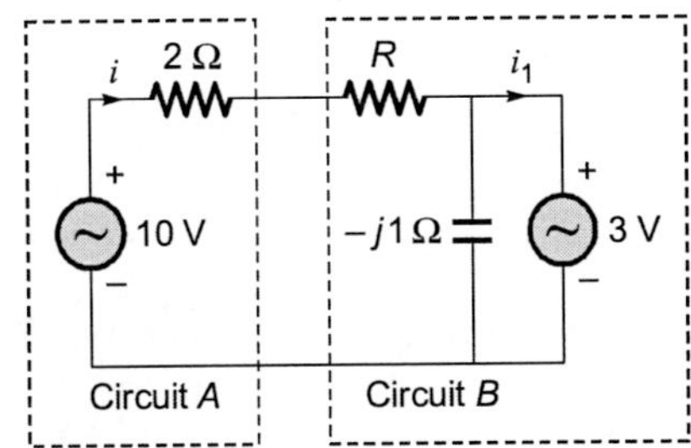

Current through R will be

$$i = \frac{10-3}{2+R} = \left(\frac{7}{2+R}\right) \text{A}$$

Current through 3 V source is

$$i_1 = i - \frac{3}{-j1} = i - 3j$$

So power delivered to circuit B by circuit A is

$$P = i^2R + i_1 \times 3$$
$$P = \left(\frac{7}{2+R}\right)^2 \cdot R + \left(\frac{7}{2+R} - 3j\right)3$$

for P to be maximum $\frac{\partial P}{\partial R}$ will be zero.

$$\frac{\partial P}{\partial R} = 0$$
$$\left(\frac{7}{2+R}\right)^2 - \frac{98R}{(2+R)^3} - \frac{21}{(2+R)^2} = 0$$
$$49\,(2 + R) - 98\,R - 21\,(2 + R) = 0$$
$$98 - 42 = 49\,R + 21\,R$$
$$R = \frac{56}{70} = 0.8\ \Omega$$

3.19 (c)

For pure resistive load to extract the maximum power

$$R_L = |Z_s| = \sqrt{R_s^2 + X_s^2} = \sqrt{4^2 + 3^2} = 5\ \Omega$$

3.20 (c)

To find V_{Th}, open circuit the load voltage R_L then

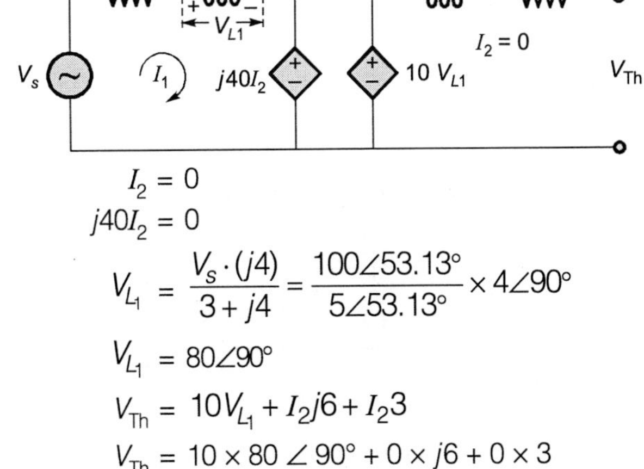

$$I_2 = 0$$
$$j40I_2 = 0$$
$$V_{L_1} = \frac{V_s \cdot (j4)}{3 + j4} = \frac{100\angle 53.13^\circ}{5\angle 53.13^\circ} \times 4\angle 90^\circ$$
$$V_{L_1} = 80\angle 90^\circ$$
$$V_{Th} = 10V_{L_1} + I_2 j6 + I_2 3$$
$$V_{Th} = 10 \times 80\angle 90^\circ + 0 \times j6 + 0 \times 3$$
$$V_{Th} = 800\angle 90^\circ\ \text{V}$$

3.21 (d)

Norton's theorem states that a complex network connected to a load can be replaced with an equivalent impedance in parallel with a current source

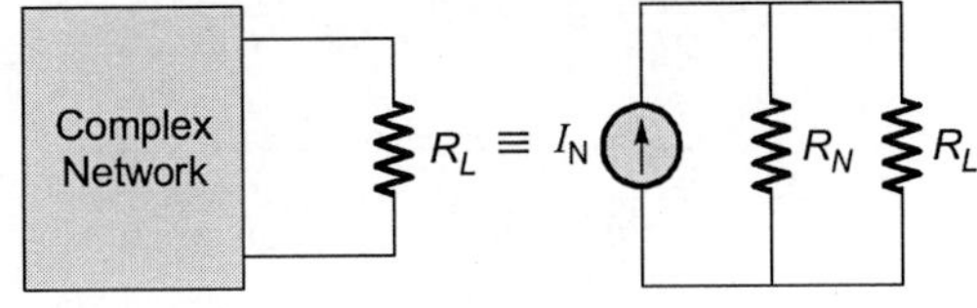

3.22 (0.5)

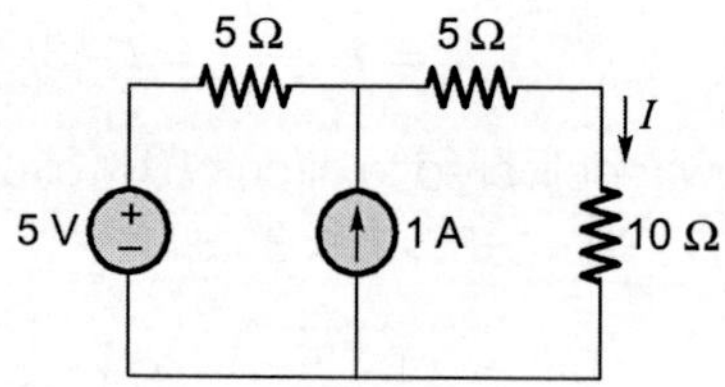

Using super position theorem,
when 5 V source acting alone, we get

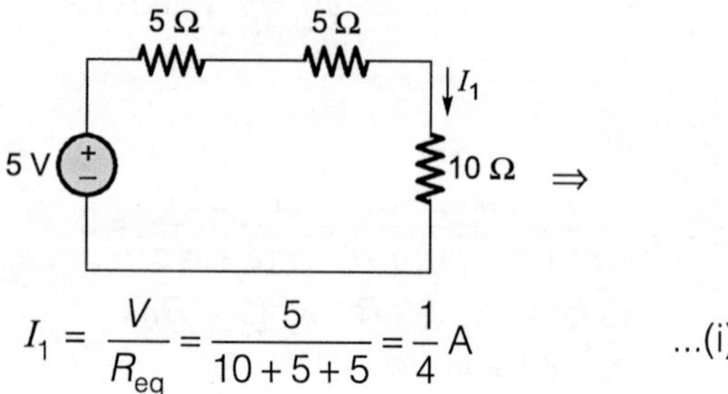

$$I_1 = \frac{V}{R_{eq}} = \frac{5}{10+5+5} = \frac{1}{4}\text{ A} \qquad ...(i)$$

When 1 A source acting alone, we get

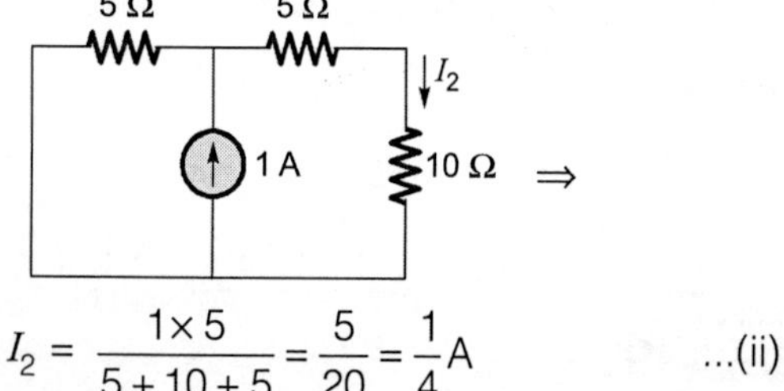

$$I_2 = \frac{1\times 5}{5+10+5} = \frac{5}{20} = \frac{1}{4}\text{A} \qquad ...(ii)$$

Therefore,

$$I = I_1 + I_2 = \frac{1}{2}\text{ A} = 0.5\text{ A}$$

3.23 (2)

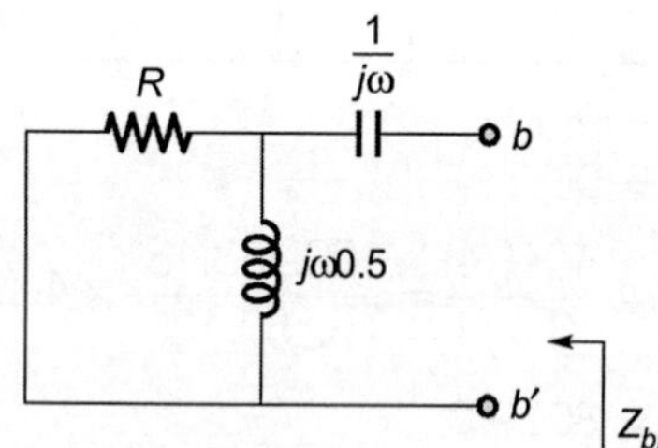

Finding Z_N:

$$Z'_{bb} = \frac{1\times j0.5\omega}{1+j0.5\omega} + \frac{1}{j\omega} = \frac{j\omega}{2+j\omega} + \frac{1}{j\omega}$$

$$\text{or } Z'_{bb} = \frac{2-\omega^2+j\omega}{2j\omega-\omega^2} \qquad ...(i)$$

Rationalizing equation (i), we get,

$$Z'_{bb} = \frac{(2-\omega^2)+j\omega}{2j\omega-\omega^2} \times \frac{-\omega^2-j2\omega}{-\omega^2-j2\omega}$$

$$= -\frac{2\omega^2+\omega^4+2\omega^2}{\omega^4+2\omega^2} + j\frac{(\omega^3-4\omega)}{\omega^4+2\omega^2}$$

In order to have a purely resistive impedance Z'_{bb}, the imaginary part of equation (ii) will be equaled to zero.

$$\therefore \quad \frac{-4\omega+\omega^3}{\omega^4+2\omega^2} = 0$$

or $\omega^3 = 4\omega$

or $\omega = \sqrt{4} = 2$ rad/sec

3.24 (1.649)

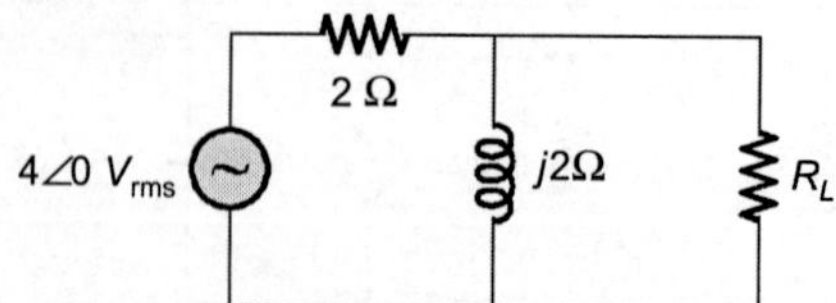

For maximum power transfer

$$R_L = |Z_{Th}| = |2 \parallel j2| = \left|\frac{2\times j2}{2+j2}\right| = 1.414\ \Omega$$

$$V_{Th} = \frac{8\angle 90^\circ}{2+j2} = 2.828\angle 45^\circ$$

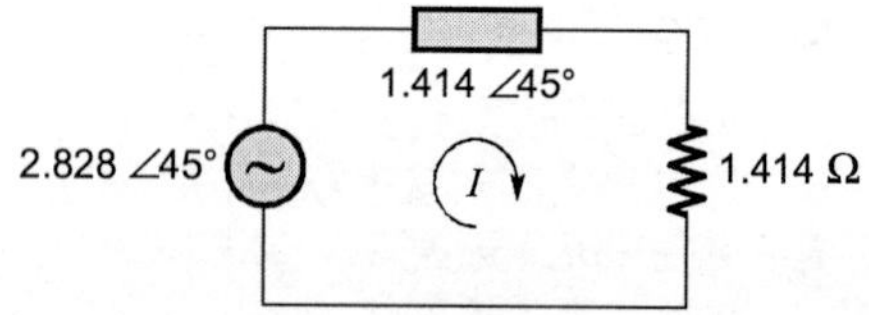

$$I = \frac{2.828\angle 45^\circ}{1.414\angle 45^\circ + 1.414} = 1.08\angle 22.5^\circ$$

$$\text{Power} = I^2R = (1.08)^2 \times \sqrt{2} = 1.649\text{ W}$$

3.25 (1.33)

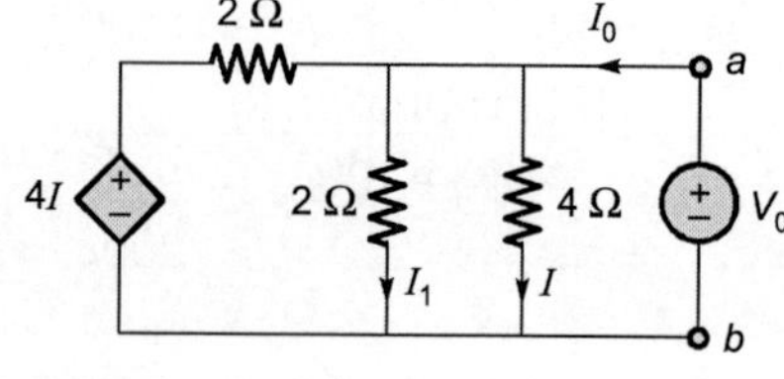

$$I = \frac{V_o}{4}$$

$$I_1 = \frac{V_o}{2}$$

Applying KCL

$$\frac{V_o - 4I}{2} + \frac{V_o}{2} + \frac{V_o}{4} = I_0$$

From here, $V_o \cdot \frac{3}{4} = I_o$

$$R_N = \frac{V_o}{I_o} = \frac{4}{3} = 1.33\ \Omega$$

3.26 (10)

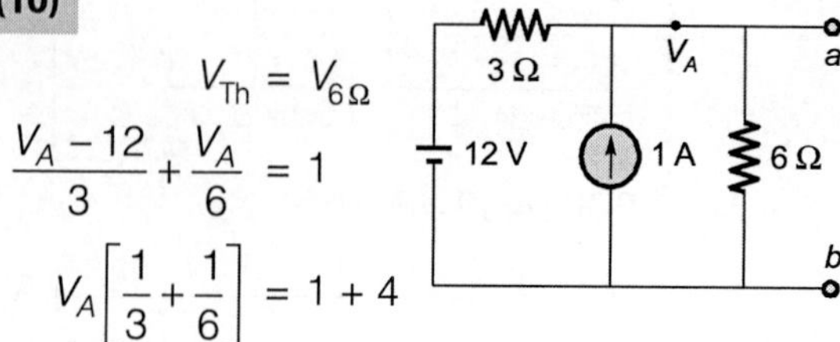

$$V_{Th} = V_{6\Omega}$$

$$\frac{V_A - 12}{3} + \frac{V_A}{6} = 1$$

$$V_A\left[\frac{1}{3}+\frac{1}{6}\right] = 1 + 4$$

$$V_A \frac{3}{6} = 5$$

$$V_{Th} = V_{6\Omega} = V_A = 10 \text{ V}$$

3.27 (0.8)

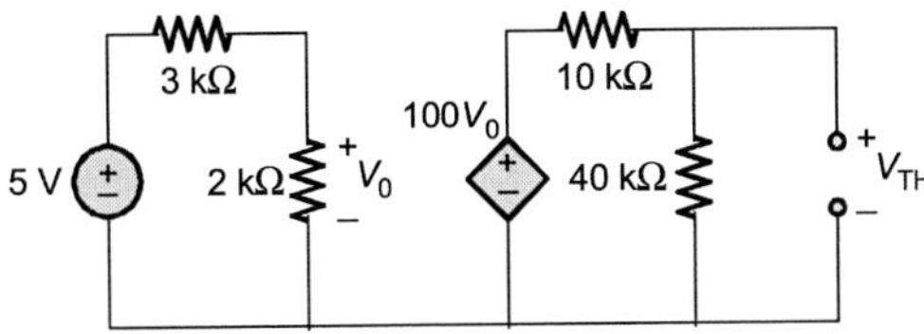

For maximum power transfer,

$$R = R_{Th} \ ; \quad V_0 = 5 \times \frac{2 \text{ k}\Omega}{5 \text{ k}\Omega} = 2 \text{ V}$$

From output loop, $V_{TH} = 100 \times 2 \times \frac{40 \text{ k}\Omega}{50 \text{ k}\Omega}$

$$V_{TH} = 160 \text{ V}$$

and $R_{TH} = 10 \text{ k}\Omega \,||\, 40 \text{ k}\Omega = \frac{10 \times 40}{50} = 8 \text{ k}\Omega$

$$\therefore \text{ Maximum power } = \frac{V_{Th}^2}{4R_{Th}} = \frac{160 \times 160}{4 \times 8000} = 0.8 \text{ W}$$

3.28 (b)

In maximum power transformation, half of the voltage drops across source resistance, remaining half across the load.

$\therefore$ voltage across source (R)

$$I_L R = \frac{V_s}{2} \ ; \quad I_L = \frac{V_s}{2R}$$

3.29 (-1)

- The equivalent circuit to calculate the Thevenin equivalent resistance (R_{Th}) is as follows:

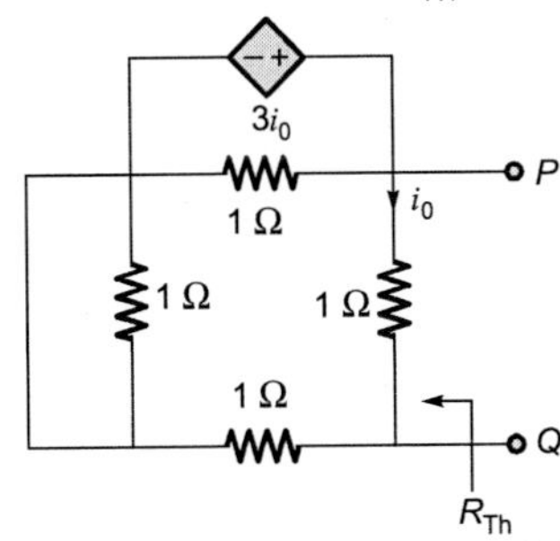

- It can be further reduced as follows:

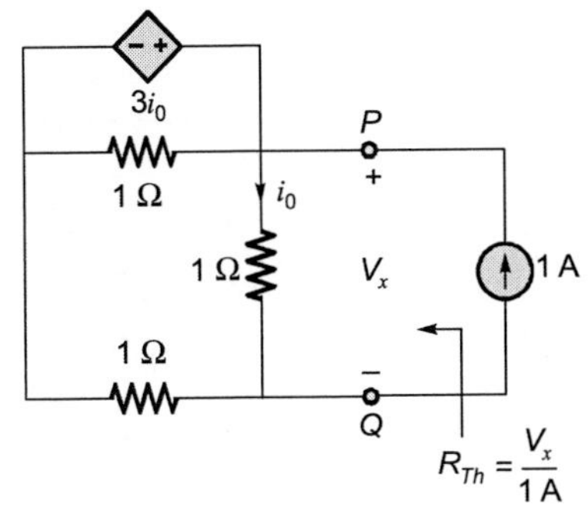

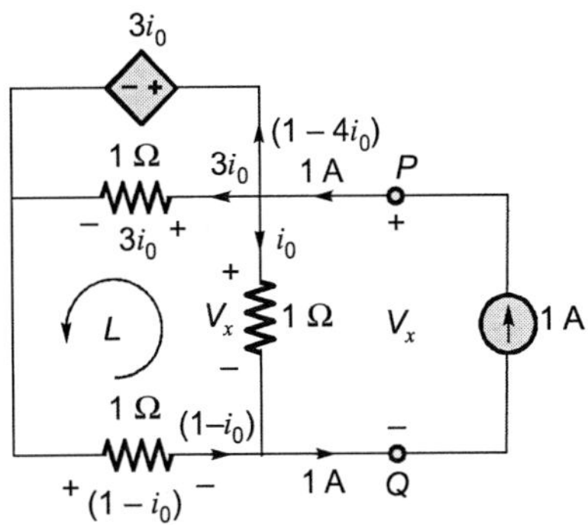

- By applying KVL in the Loop L,

$$V_x = 3i_0 + (1 - i_0) = 2i_0 + 1$$

Also, $\quad V_x = i_0(1\ \Omega)$

- So, $2i_0 + 1 = i_0$

$$i_0 = -1 \text{ A}$$

And $\quad V_x = -1 \text{V}$

So, $\quad R_{Th} = \frac{V_x}{1\text{A}} = -1\ \Omega$

3.30 (c)

According to reciprocity theorem,

In a linear bilateral single source network the ratio of response to excitation remains the same even after their positions get interchanged.

$$\therefore \quad \frac{I}{5} = \frac{1}{5} \Rightarrow I = 1 \text{ A}$$

3.31 (b)

By applying source transformation

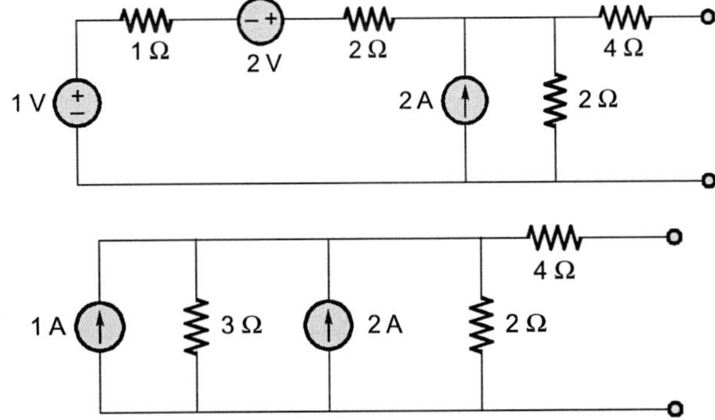

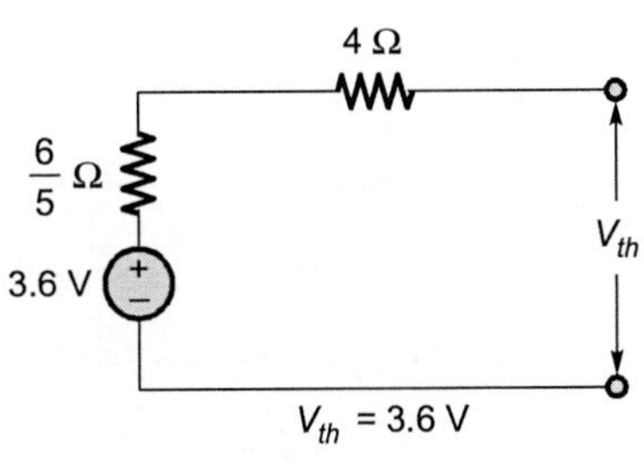

$$V_{th} = 3.6 \text{ V}$$

3.32 (2.5)

For maximum power transfer, the load resistance R_L should be equal to the Thevenin equivalent resistance across the load terminals. To obtain the Thevenin equivalent resistance, we deactivate the independent voltage source and connect a voltage V across the load terminal.

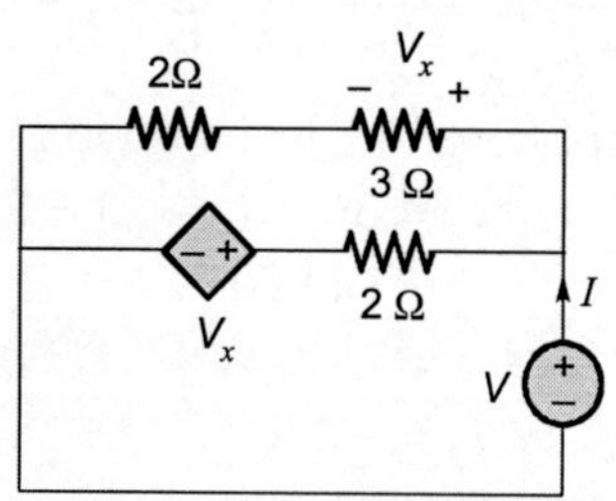

Using KCL, we can write,

$$I = \frac{V - V_x}{2} + \frac{V}{5} = \frac{V}{2} - \frac{V_x}{2} + \frac{V}{5}$$

But, $$V_x = \frac{V}{5} \times 3$$

$$\therefore \quad I = \frac{V}{2} - \frac{1}{2} \times \frac{3V}{5} + \frac{V}{5} = \frac{V}{2} - \frac{3V}{10} + \frac{V}{5}$$

$$I = \frac{5V - 3V + 2V}{10} = \frac{4V}{10}$$

$$\therefore \quad R_{Th} = \frac{V}{I} = \frac{10}{4} = 2.5\ \Omega$$

Hence, for maximum power transfer,

$$R_L = R_{Th} = 2.5\ \Omega$$

■■■

4 Transient Analysis

4.1 The RMS value of a rectangular wave of period T, having a value of $+V$ for a duration, T_1 ($<T$) and $-V$ for the duration, $T - T_1 = T_2$ equals

(a) V (b) $\frac{T_1 - T_2}{T}$ V

(c) $\frac{V}{\sqrt{2}}$ (d) $\frac{T_1}{T_2}$ V

[1995 : 1 M]

4.2 The voltage V_{C1}, V_{C2} and V_{C3} across the capacitors in the circuit in Fig., under steady state, are respectively.

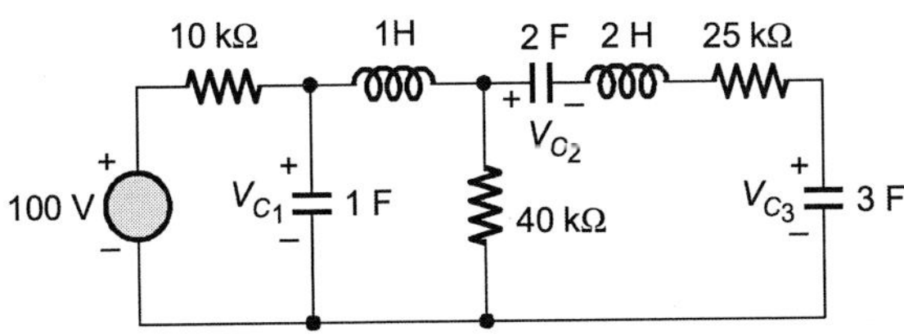

(a) 80 V, 32 V, 48 V (b) 80 V, 48 V, 32 V
(c) 20 V, 8 V, 12 V (d) 20 V, 12 V, 8 V

[1996 : 2 M]

4.3 In the circuit of Fig. the energy absorbed by the 4 Ω resistor in the time interval (0, ∞) is

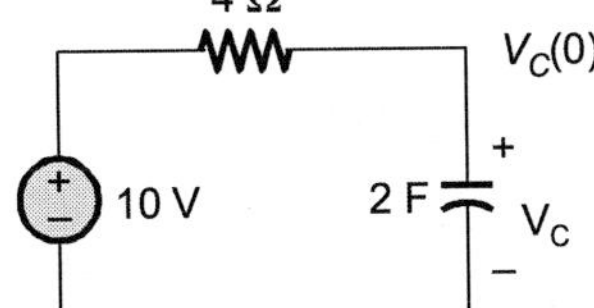

(a) 36 Joules

(b) 16 Joules

(c) 256 Joules

(d) None of the above

[1997 : 2 M]

4.4 In the figure, the switch was closed for a long time before opening at $t = 0$. The voltage V_x at $t = 0^+$ is

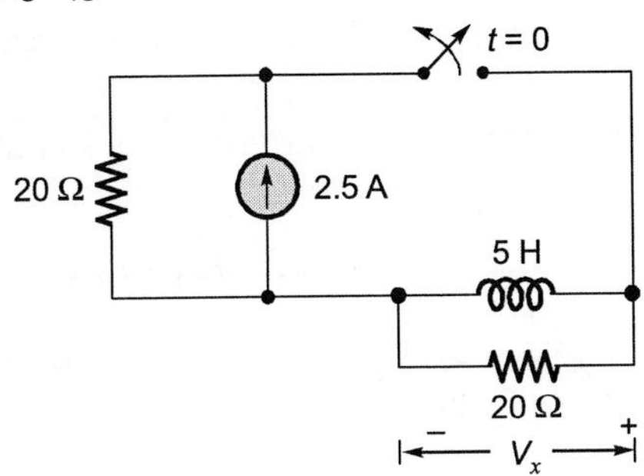

(a) 25 V (b) 50 V
(c) –50 V (d) 0 V

[2002 : 1 M]

The circuit for 4.6 and 4.7 is given. Assume that the switch S is in position 1 for a long time and thrown to position 2 at $t = 0$.

4.5 At $t = 0^+$, the current i_1 is

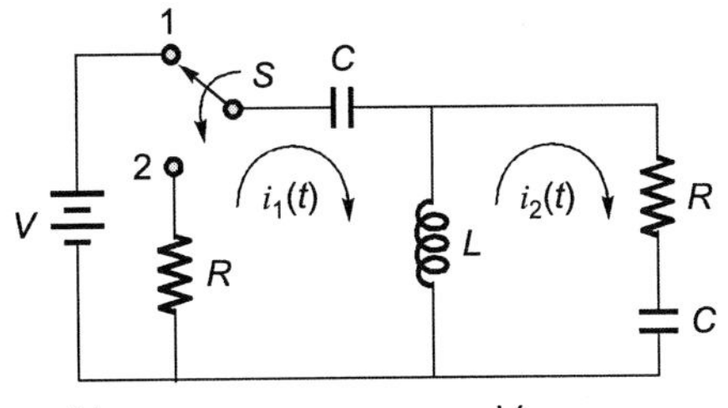

(a) $\frac{-V}{2R}$ (b) $\frac{-V}{R}$

(c) $\frac{-V}{4R}$ (d) zero

[2003 : 2 M]

4.6 $I_1(s)$ and $I_2(s)$ are the Laplace transforms of $i_1(t)$ and $i_2(t)$ respectively. The equations for the loop currents $I_1(s)$ and $I_2(s)$ for the circuit shown in the figure, after the switch is brought from position 1 to position 2 at $t = 0$, are

(a) $\begin{bmatrix} R + Ls + \frac{1}{Cs} & -Ls \\ -Ls & R + \frac{1}{Cs} \end{bmatrix} \begin{bmatrix} I_1(s) \\ I_2(s) \end{bmatrix} = \begin{bmatrix} \frac{V}{s} \\ 0 \end{bmatrix}$

(b) $\begin{bmatrix} R + Ls + \frac{1}{Cs} & -Ls \\ -Ls & R + \frac{1}{Cs} \end{bmatrix} \begin{bmatrix} I_1(s) \\ I_2(s) \end{bmatrix} = \begin{bmatrix} -\frac{V}{s} \\ 0 \end{bmatrix}$

(c) $\begin{bmatrix} R + Ls + \frac{1}{Cs} & -Ls \\ -Ls & R + Ls + \frac{1}{Cs} \end{bmatrix} \begin{bmatrix} I_1(s) \\ I_2(s) \end{bmatrix} = \begin{bmatrix} -\frac{V}{s} \\ 0 \end{bmatrix}$

(d) $\begin{bmatrix} R + Ls + \frac{1}{Cs} & -Ls \\ -Ls & R + Ls + \frac{1}{Cs} \end{bmatrix} \begin{bmatrix} I_1(s) \\ I_2(s) \end{bmatrix} = \begin{bmatrix} \frac{V}{s} \\ 0 \end{bmatrix}$

[2003 : 2 M]

4.7 For the R-L circuit shown in the figure, the input voltage $v_i(t) = u(t)$. The current $i(t)$ is

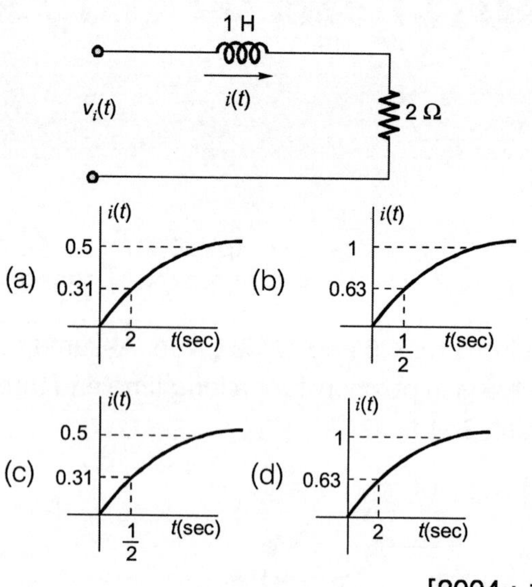

[2004 : 1 M]

4.8 The circuit shown in the figure has initial current $i_L(0^-) = 1$ A through the inductor and an initial voltage $v_C(0^-) = -1$ V across the capacitor. For input $v(t) = u(t)$, the Laplace transform of the current $i(t)$ for $t \geq 0$ is

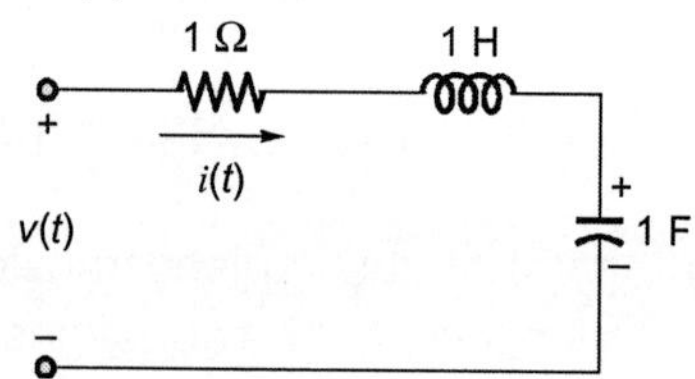

(a) $\dfrac{s}{s^2+s+1}$ (b) $\dfrac{s+2}{s^2+s+1}$

(c) $\dfrac{s-2}{s^2+s-1}$ (d) $\dfrac{s-2}{s^2+s+1}$

[2004 : 2 M]

4.9 A square pulse of 3 volts amplitude is applied to C-R circuit shown in the figure. The capacitor is initially uncharged. The output voltage V_2 at time $t = 2$ sec is

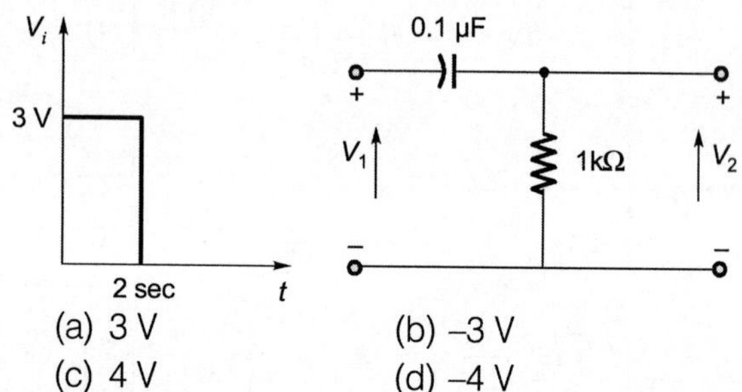

(a) 3 V (b) –3 V

(c) 4 V (d) –4 V

[2005 : 2 M]

4.10 A 2 mH inductor with some initial current can be represented as shown below, where s is the Laplace Transform variable. The value of initial current is

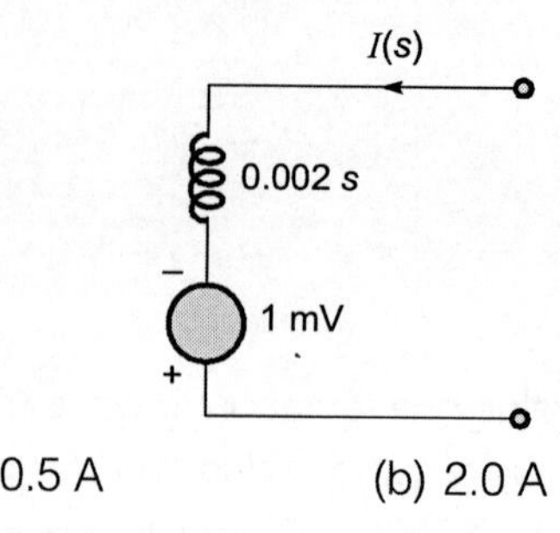

(a) 0.5 A (b) 2.0 A

(c) 1.0 A (d) 0.0 A

[2006 : 1 M]

4.11 In the figure shown below, assume that all the capacitors are initially uncharged. If $v_i(t) = 10u(t)$ Volts, $v_0(t)$ is given by

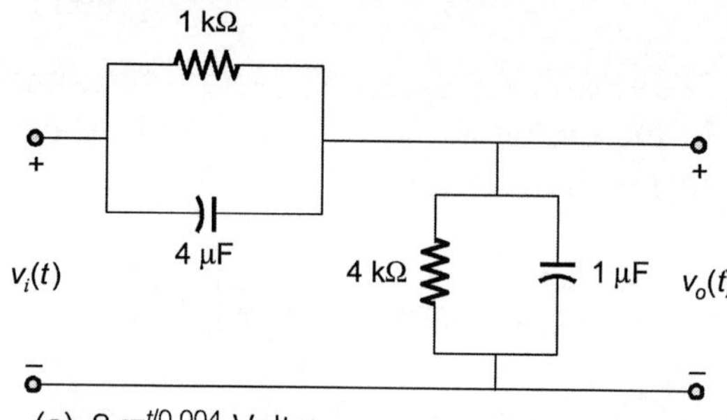

(a) $8e^{-t/0.004}$ Volts

(b) $8(1 - e^{-t/0.004})$ Volts

(c) $8u(t)$ Volts

(d) 8 Volts

[2006 : 1 M]

4.12 In the circuit shown, V_C is 0 volts at $t = 0$ sec. For $t > 0$, the capacitor current $i_c(t)$, where t is in seconds, is given by

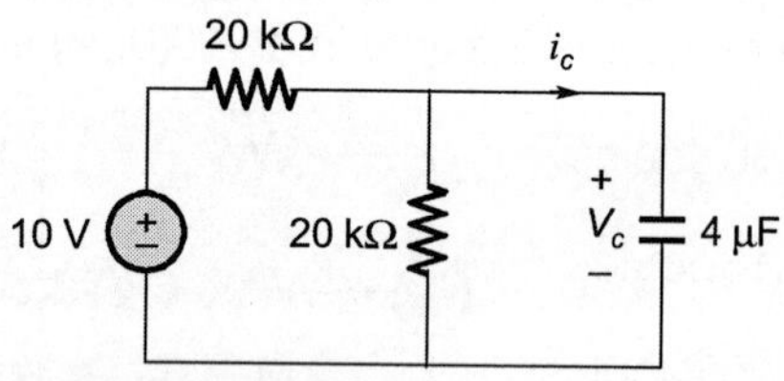

(a) 0.50 exp(–25t) mA

(b) 0.25 exp(–25 t) mA

(c) 0.50 exp(–12.5t) mA

(d) 0.25 exp(–6.25 t) mA

[2007 : 2 M]

4.13 In the following circuit, the switch S is closed at $t = 0$.

The rate of change of current $\dfrac{di}{dt}(0^+)$ is given by

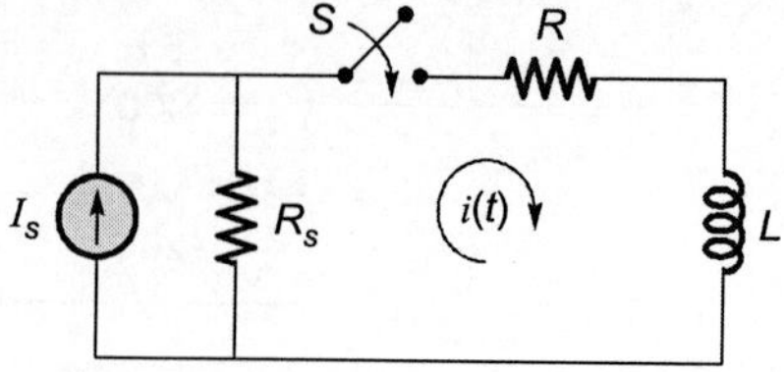

(a) 0 (b) $\frac{R_s I_s}{L}$

(c) $\frac{(R+R_s)I_s}{L}$ (d) ∞

[2008 : 1 M]

4.14 The circuit shown in the figure is used to charge the capacitor C alternately from two current sources as indicated. The switches S_1 and S_2 are mechanically coupled and connected as follows:
For $2nT \le t < (2n+1)T$, $(n = 0,1,2...)$ S_1 to P_1 and S_2 to P_2.
For $(2n+1)T \le t < (2n+2)T$, $(n = 0,1,2. ...)$ S_1 to Q_1 and S_2 to Q_2

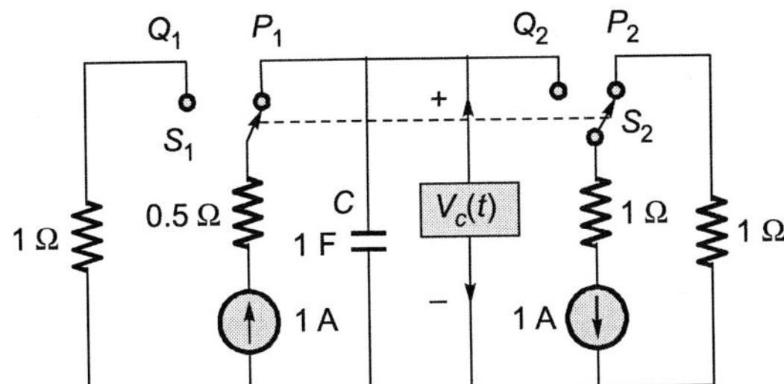

Assume that the capacitor has zero initial charge. Given that $u(t)$ is a unit step function, the voltage $V_c(t)$ across the capacitor is given by

(a) $\sum_{n=0}^{\infty}(-1)^n tu(t-nT)$

(b) $u(t) + 2\sum_{n=1}^{\infty}(-1)^n u(t-nT)$

(c) $tu(t) + 2\sum_{n=1}^{\infty}(-1)^n (t-nT)u(t-nT)$

(d) $\sum_{n=0}^{\infty}\left[0.5 - e^{-(t-2nT)} + 0.5e^{-(t-2nT-T)}\right]$

[2008 : 2 M]

Common Data for Questions 4.15 and 4.16:

The following series RLC circuit with zero initial conditions is excited by a unit impulse function $\delta(t)$.

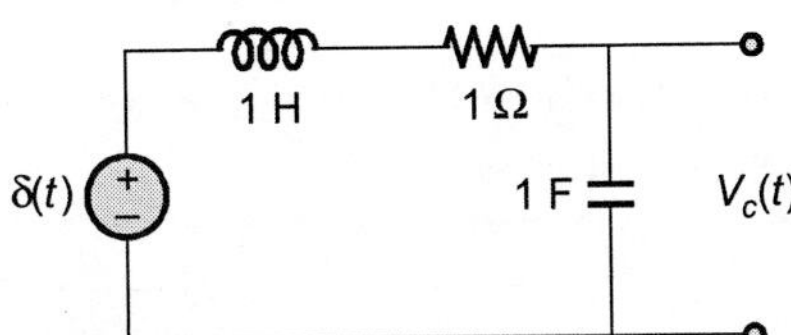

4.15 For $t > 0$, the output voltage $V_c(t)$ is

(a) $\frac{2}{\sqrt{3}}\left(e^{-\frac{1}{2}t} - e^{-\frac{\sqrt{3}}{2}t}\right)$

(b) $\frac{2}{\sqrt{3}}te^{-\frac{1}{2}t}$

(c) $\frac{2}{\sqrt{3}}e^{-\frac{1}{2}t}\cos\left(\frac{\sqrt{3}}{2}t\right)$

(d) $\frac{2}{\sqrt{3}}e^{-\frac{1}{2}t}\sin\left(\frac{\sqrt{3}}{2}t\right)$

[2008 : 2 M]

4.16 For $t > 0$, the voltage across the resistor is

(a) $\frac{1}{\sqrt{3}}\left(e^{-\frac{\sqrt{3}}{2}t} - e^{-\frac{1}{2}t}\right)$

(b) $e^{-\frac{1}{2}t}\left[\cos\left(\frac{\sqrt{3}t}{2}\right) - \frac{1}{\sqrt{3}}\sin\left(\frac{\sqrt{3}t}{2}\right)\right]$

(c) $\frac{2}{\sqrt{3}}e^{-\frac{1}{2}t}\sin\left(\frac{\sqrt{3}t}{2}\right)$

(d) $\frac{2}{\sqrt{3}}e^{-\frac{1}{2}t}\cos\left(\frac{\sqrt{3}t}{2}\right)$

[2008 : 2 M]

4.17 The switch in the circuit shown was on position a for a long time, and is moved to position b at time $t = 0$. The current $i(t)$ for $t > 0$ is given by

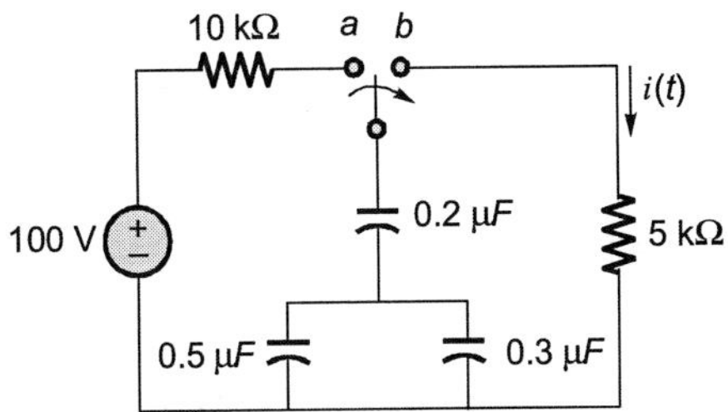

(a) $0.2e^{-125t}u(t)$ mA (b) $20e^{-1250t}u(t)$ mA
(c) $0.2e^{-1250t}u(t)$ mA (d) $20e^{-1000t}u(t)$ mA

[2009 : 2 M]

4.18 The time domain behaviour of an RL circuit is represented by

$$L\frac{di}{dt} + Ri = V_0\left(1 + Be^{-Rt/L}\sin t\right)u(t)$$

For an initial current of $i(0) = \frac{V_0}{R}$, the steady state value of the current is given by

(a) $i(t) \to \frac{V_0}{R}$

(b) $i(t) \to \frac{2V_0}{R}$

(c) $i(t) \to \frac{V_0}{R}(1+B)$

(d) $i(t) \to \frac{2V_0}{R}(1+B)$

[2009 : 2 M]

4.19 In the circuit shown, the switch S is open for a long time and is closed at $t = 0$. The current $i(t)$ for $t \geq 0^+$ is

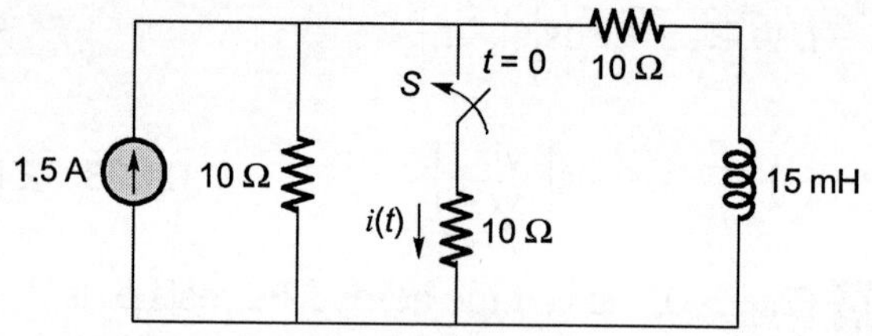

(a) $i(t) = 0.5 - 0.125\, e^{-1000t}$ A
(b) $i(t) = 1.5 - 0.125\, e^{-1000t}$ A
(c) $i(t) = 0.5 - 0.5\, e^{-1000t}$ A
(d) $i(t) = 0.375\, e^{-1000t}$ A **[2010 : 2 M]**

4.20 In the circuit shown below, the initial charge on the capacitor is 2.5 mC, with the voltage polarity as indicated. The switch is closed at time $t = 0$. The current $i(t)$ at a time t after the switch is closed is

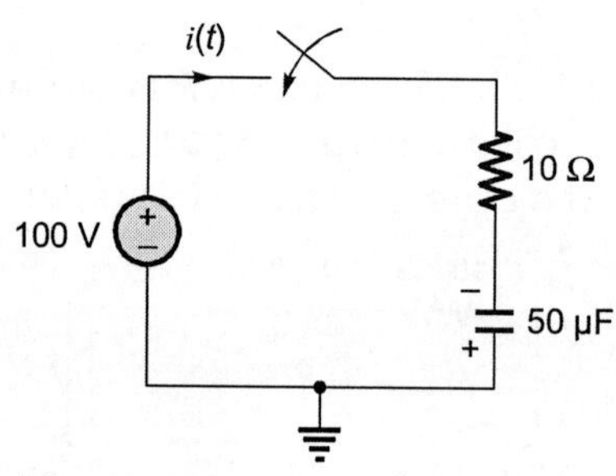

(a) $i(t) = 15 \exp(-2 \times 10^3 t)$ A
(b) $i(t) = 5 \exp(-2 \times 10^3 t)$ A
(c) $i(t) = 10 \exp(-2 \times 10^3 t)$ A
(d) $i(t) = -5 \exp(-2 \times 10^3 t)$ A **[2011 : 2 M]**

4.21 In the following figure, C_1 and C_2 are ideal capacitors. C_1 had been charged to 12 V before the ideal switch S is closed at $t = 0$. The current $i(t)$ for all t is

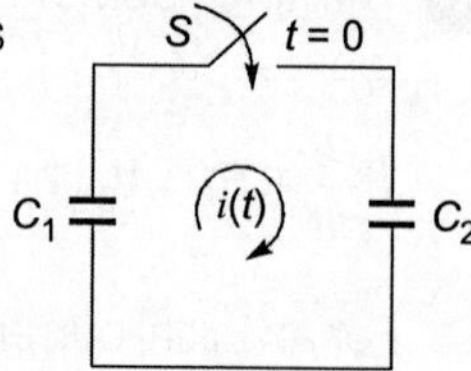

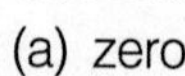
(a) zero
(b) a step function
(c) an exponentially decaying function
(d) an impulse function
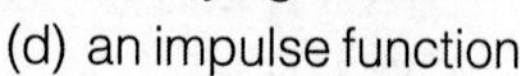

[2012 : 1 M]

4.22 For maximum power transfer between two cascaded sections of an electrical network, the relationship between the output impedance Z_1 of the first section to the input impedance Z_2 of the second section is

(a) $Z_2 = Z_1$ (b) $Z_2 = -Z_1$
(c) $Z_2 = Z_1^*$ (d) $Z_2 = -Z_1^*$

[2014 : 1 M, Set-1]

4.23 In the circuit shown in the figure, the value of capacitor C (in mF) needed to have critically damped response $i(t)$ is ______.

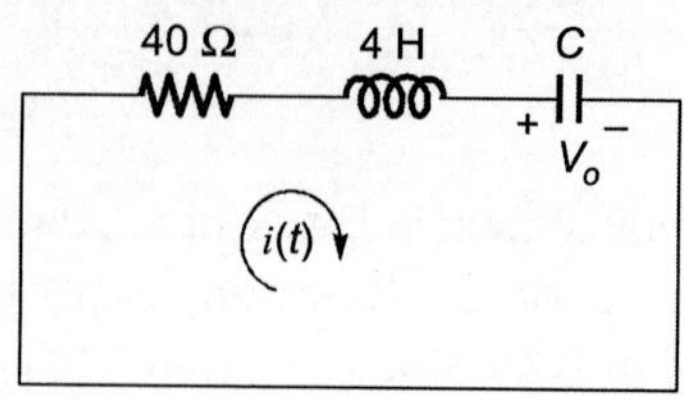

[2014 : 2 M, Set-1]

4.24 In the figure shown, the ideal switch has been open for a long time. If it is closed at $t = 0$, then the magnitude of the current (in mA) through the 4 kΩ resistor at $t = 0^+$ is ______.

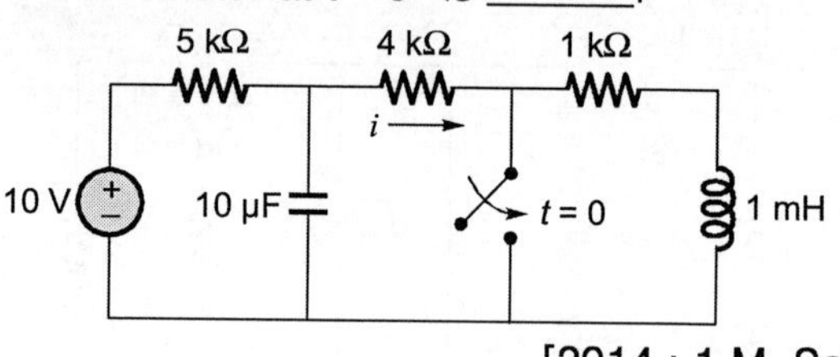

[2014 : 1 M, Set-2]

4.25 In the figure shown, the capacitor is initially uncharged. Which one of the following expressions describes the current $I(t)$ (in mA) for $t > 0$?

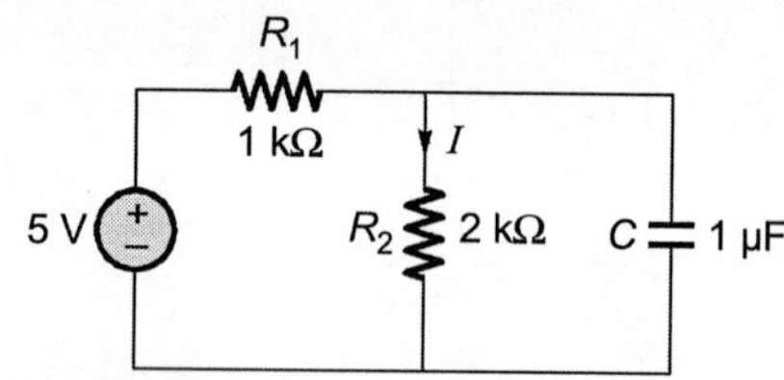

(a) $I(t) = \frac{5}{3}(1 - e^{-t/\tau}),\ \tau = \frac{2}{3}$ msec
(b) $I(t) = \frac{5}{2}(1 - e^{-t/\tau}),\ \tau = \frac{2}{3}$ msec
(c) $I(t) = \frac{5}{3}(1 - e^{-t/\tau}),\ \tau = 3$ msec
(d) $I(t) = \frac{5}{2}(1 - e^{-t/\tau}),\ \tau = 3$ msec

[2014 : 2 M, Set-2]

4.26 In the circuit shown in the figure, the value of $v_o(t)$ (in volts) for $t \to \infty$ is ___.

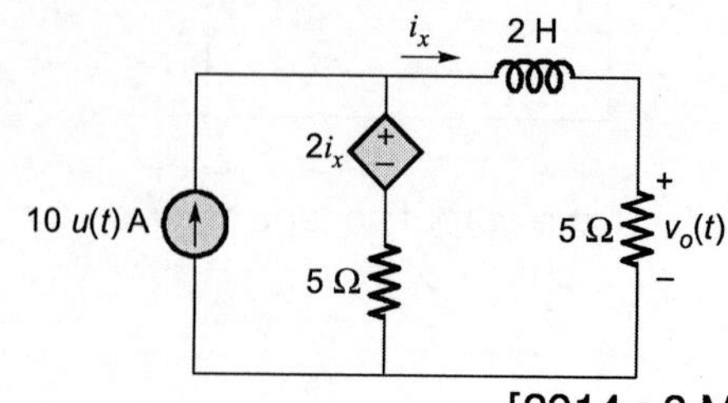

[2014 : 2 M, Set-4]

4.27 In the circuit shown, the switch SW is thrown from position A to position B at time $t = 0$. The energy (in μJ) taken from the 3 V source to charge the 0.1 μF capacitor form 0 V to 3 V is

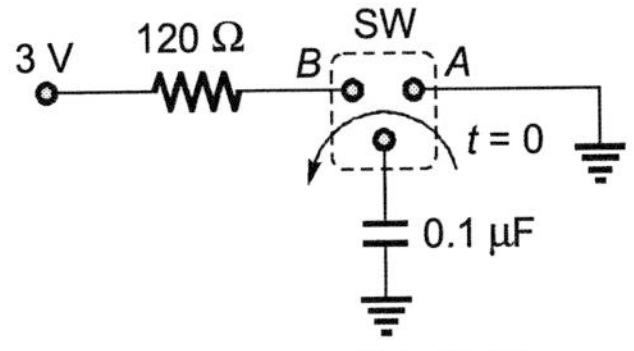

(a) 0.3 (b) 0.45
(c) 0.9 (d) 3 [2015 : 1 M, Set-1]

4.28 In the circuit shown, switch SW is closed at $t = 0$. Assuming zero initial conditions, the value of $v_c(t)$ (in Volts) at $t = 1$ sec is ________.

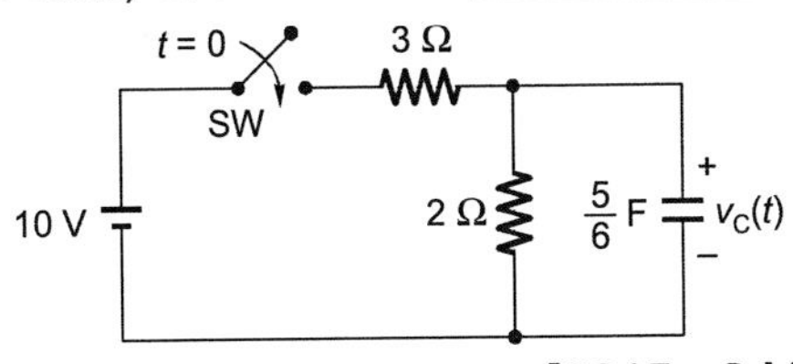

[2015 : 2 M, Set-1]

4.29 In the circuit shown, the initial voltages across the capacitors C_1 and C_2 and 1 V and 3 V, respectively. The switch is closed at time $t = 0$. The total energy dissipated (in Joules) in the resistor R until steady state is reached, is _____.

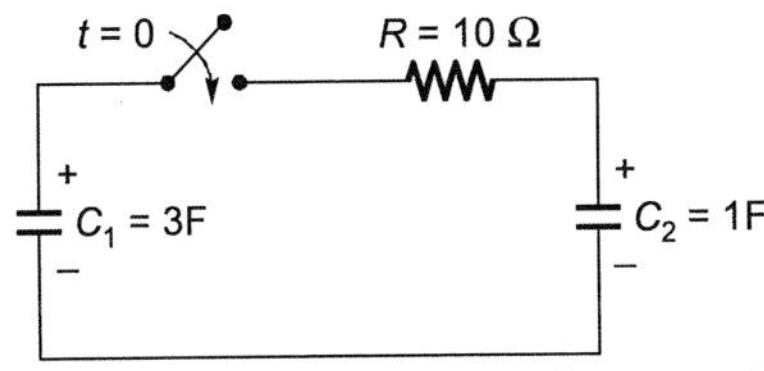

[2015 : 2 M, Set-2]

4.30 The switch has been in position 1 for a long time and abruptly changes to position 2 at $t = 0$.

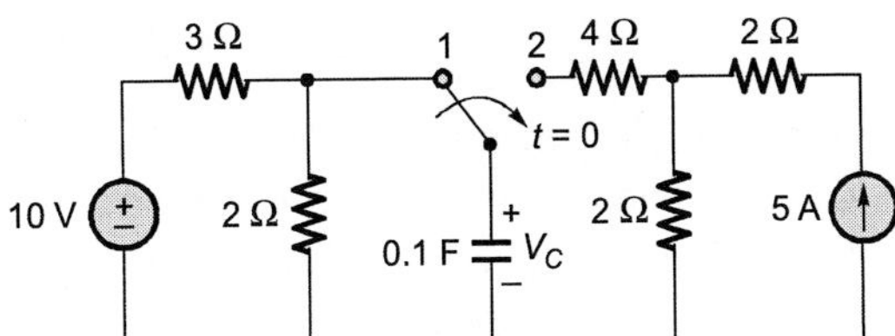

If time t is in seconds, the capacitor voltage V_C (in volts) for $t > 0$ is given by
(a) $4(1 - \exp(-t/0.5))$ (b) $10 - 6\exp(-t/0.5)$
(c) $4(1 - \exp(-t/0.6))$ (d) $10 - 6\exp(-t/0.6)$
[2016 : 1 M, Set-2]

4.31 The switch S in the circuit shown has been closed for a long time. It is opened at time t=0 and remains open after that. Assume that the diode has zero reverse current and zero forward voltage drop.

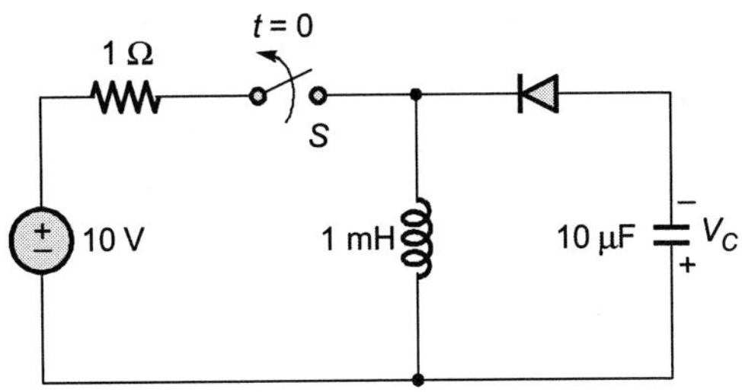

The steady state magnitude of the capacitor voltage V_C (in volts) is _____.

[2016 : 2 M, Set-2]

4.32 Assume that the circuit in the figure has reached the steady state before time $t = 0$ when the 3 Ω resistor suddenly burns out, resulting in an open circuit. The current $i(t)$ (in ampere) at $t = 0^+$ is ________.

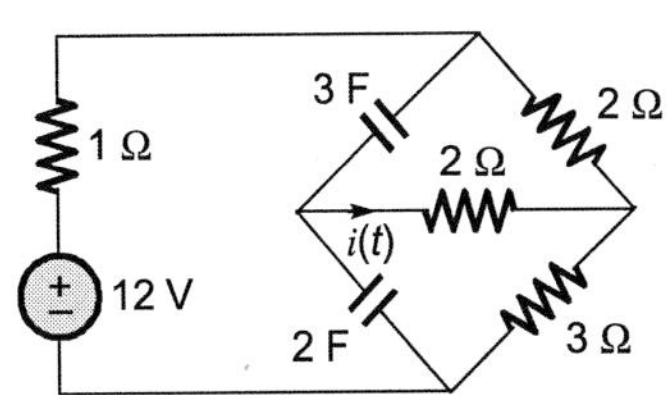

[2016 : 2 M, Set-3]

4.33 In the circuit shown, the voltage $V_{IN}(t)$ is described by:

$$V_{IN}(t) = \begin{cases} 0, & \text{for } t < 0 \\ 15 \text{ Volts}, & \text{for } t \geq 0 \end{cases}$$

where t is in seconds. The time (in seconds) at which the current I in the circuit will reach the value 2 Amperes is ________.

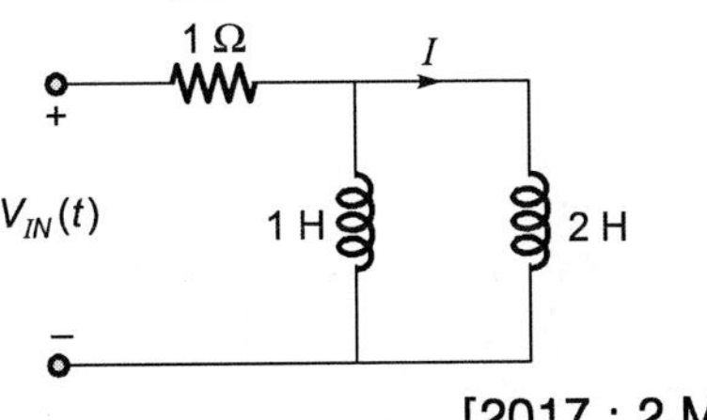

[2017 : 2 M, Set-1]

4.34 The switch in the circuit, shown in the figure, was open for a long time and is closed at $t = 0$.

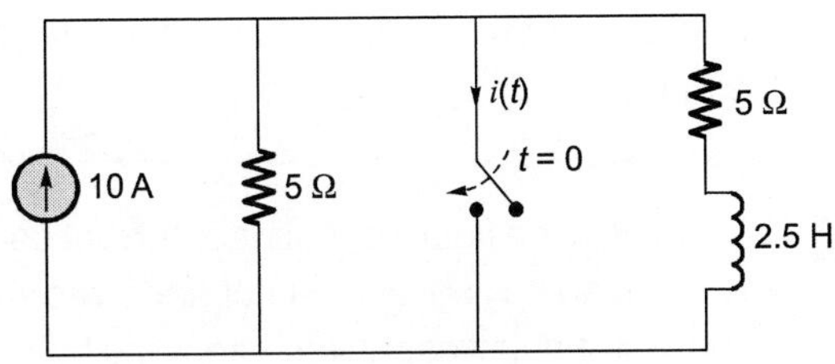

The current $i(t)$ (in ampere) at $t = 0.5$ seconds is _____.

[2017 : 2 M, Set-2]

4.35 For the circuit given in the figure, the magnitude of the loop current (in amperes, correct to three decimal places) 0.5 second after closing the switch is _______.

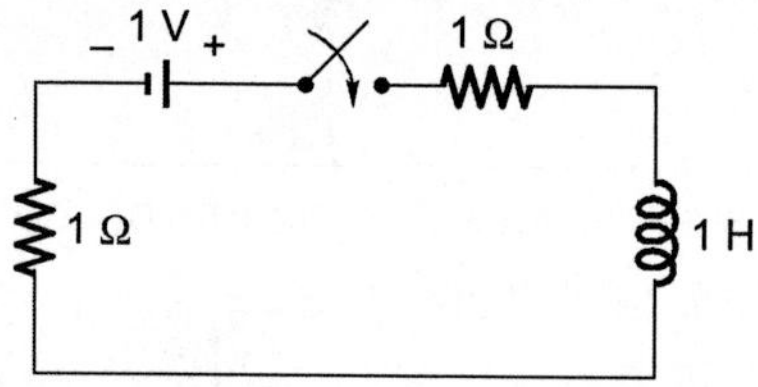

[2018 : 2 M]

4.36 The RC circuit shown below has a variable resistance $R(t)$ given by the following expression:

$$R(t) = R_o\left(1 - \frac{t}{T}\right) \text{ for } 0 \le t < T$$

where R_o = 1 Ω, and C = 1 F. We are also given that $T = 3R_oC$ and the source voltage is V_s = 1 V. If the current at time $t = 0$ is 1 A. then the current $I(t)$, in amperes, at time t = $T/2$ is _______ (rounded off to 2 decimal places).

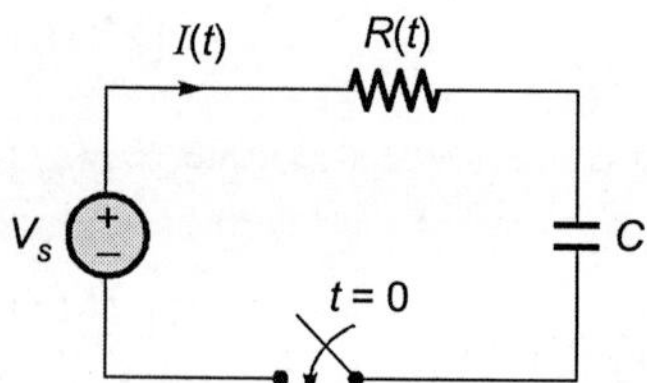

[2019 : 2 M]

4.37 In the circuit shown in the figure, the switch is closed at time $t = 0$, while the capacitor is initially charged to –5 V (i.e., $v_c(0)$ = –5 V).

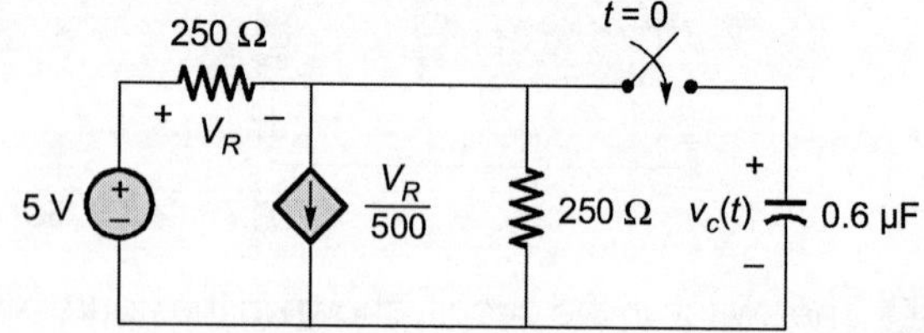

The time after which the voltage across the capacitor becomes zero (rounded off to three decimal places) is _____ ms.

[2021 : 2 M]

4.38 The circuit in the figure contains a current source driving a load having an inductor and a resistor in series, with a shunt capacitor across the load. The ammeter is assumed to have zero resistance. The switch is closed at time $t = 0$.

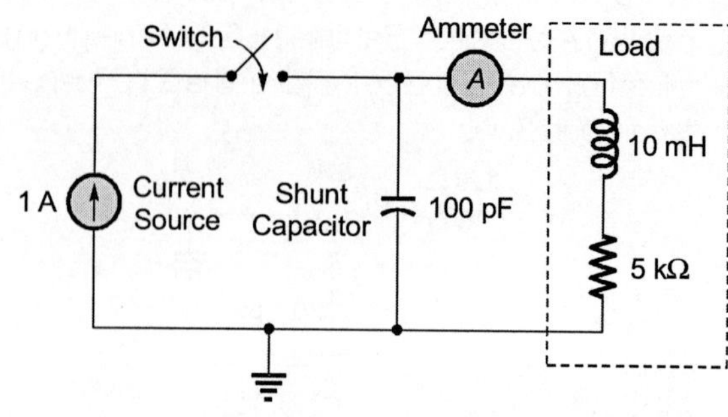

Initially, when the switch is open, the capacitor is discharged and the ammeter reads zero ampere. After the switch is closed, the ammeter reading keeps fluctuating for some time till it settles to a final steady value. The maximum ammeter reading that one will observe after the switch is closed (rounded off to two decimal places) is _____ A.

[2021 : 2 M]

4.39 The switch in the circuit in the figure is in position P for a long time and then moved to position Q at time $t = 0$.

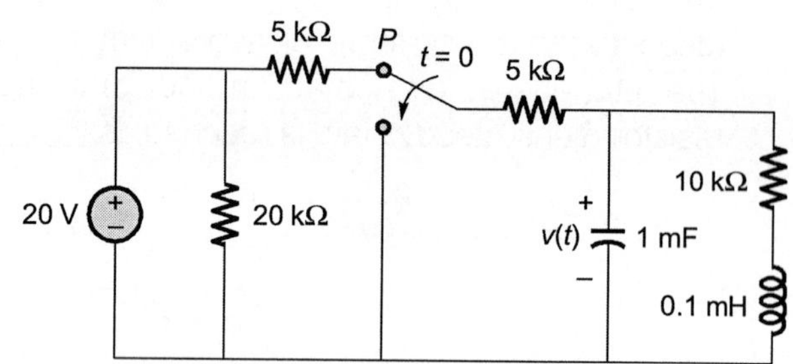

The value of $\frac{dv(t)}{dt}$ at $t = 0^+$ is

(a) 3 V/s (b) –5 V/s
(c) 0 V/s (d) –3 V/s

[2021 : 2 M]

4.40 In the circuit shown below, switch S was closed for a long time. If the switch is opened at $t = 0$, the maximum magnitude of the voltage V_R, in volts, is ________ (rounded off to the nearest integer).

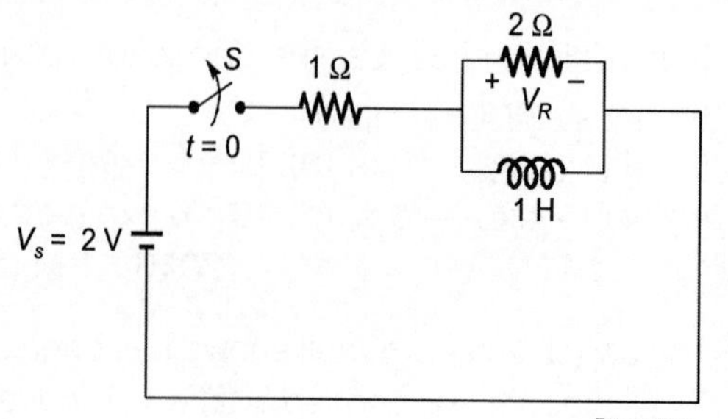

[2023 : 1 M]

4.41 The switch S_1 was closed and S_2 was open for a long time. At $t = 0$, switch S_1 is opened and S_2 is closed, simultaneously. The value of $i_c(0^+)$, in amperes, is

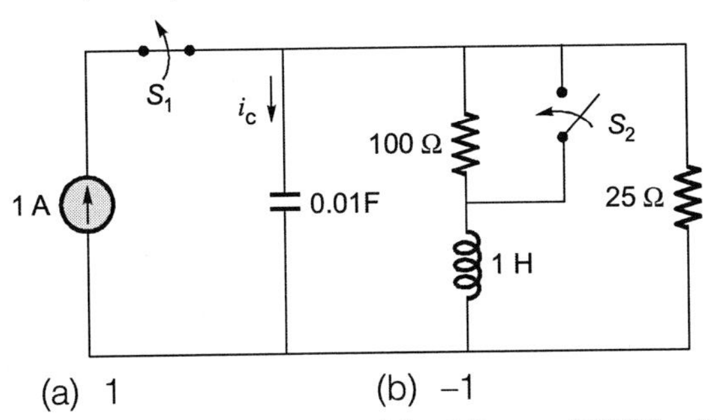

(a) 1
(b) –1
(c) 0.2
(d) 0.8 [2023 : 2 M]

4.42 In the circuit given below, the switch S was kept open for a sufficiently long time and is closed at time $t = 0$. The time constant (in seconds) of the circuit for $t > 0$ is ______.

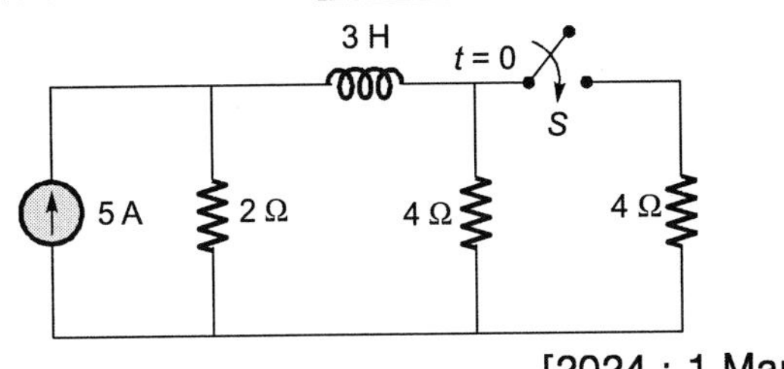

[2024 : 1 Mark]

Answers Transient Analysis

4.1	(a)	4.2	(b)	4.3	(b)	4.4	(c)	4.5	(a)	4.6	(c)	4.7	(c)
4.8	(b)	4.9	(b)	4.10	(a)	4.11	(c)	4.12	(a)	4.13	(b)	4.14	(c)
4.15	(d)	4.16	(b)	4.17	(b)	4.18	(a)	4.19	(a)	4.20	(a)	4.21	(d)
4.22	(c)	4.23	(10)	4.24	(1.25)	4.25	(a)	4.26	(31.25)	4.27	(c)	4.28	(2.528)
4.29	(1.5)	4.30	(d)	4.31	(100)	4.32	(–1)	4.33	(0.3405)	4.34	(8.16)	4.35	(0.316)
4.36	(0.25)	4.37	(0.1386)	4.38	(1.44)	4.39	(d)	4.40	(4)	4.41	(b)	4.42	(0.75)

Explanations Transient Analysis

4.1 (a)

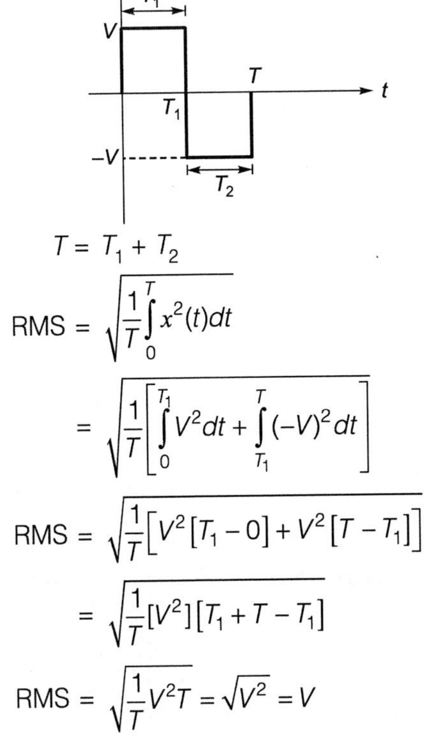

$$T = T_1 + T_2$$

$$\text{RMS} = \sqrt{\frac{1}{T}\int_0^T x^2(t)dt}$$

$$= \sqrt{\frac{1}{T}\left[\int_0^{T_1} V^2 dt + \int_{T_1}^{T} (-V)^2 dt\right]}$$

$$\text{RMS} = \sqrt{\frac{1}{T}\left[V^2[T_1 - 0] + V^2[T - T_1]\right]}$$

$$= \sqrt{\frac{1}{T}[V^2][T_1 + T - T_1]}$$

$$\text{RMS} = \sqrt{\frac{1}{T}V^2 T} = \sqrt{V^2} = V$$

4.2 (b)

At steady state

Inductor behave as short circuit

Capacitor behave as open circuit

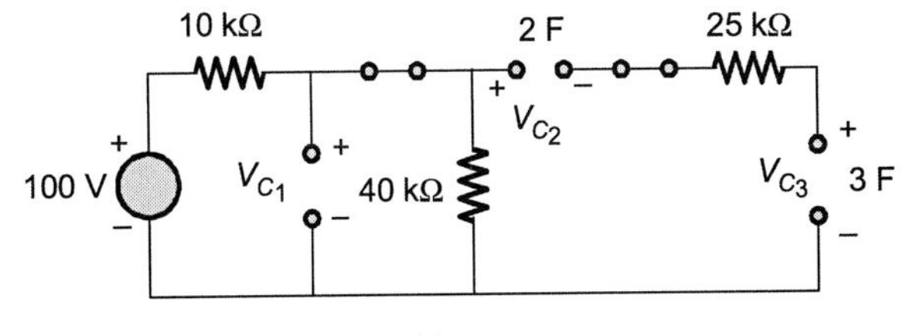

$$V_{C_1} = 100 \times \frac{40}{10+40} = 80 \text{ V}$$

$$V_{C_2} = 80 \times \frac{3}{2+3} = 48 \text{ V}$$

$$V_{C_3} = 80 \times \frac{2}{2+3} = 32 \text{ V}$$

4.3 (b)

$$V_{(0^-)} = V_{(0^+)} = 6 \text{ V}$$

So, at $t = 0^+ \rightarrow$

$$V_R = 10 - 6 = 4 \text{ V}$$

$$I_{R(0^+)} = \frac{4}{4} = 1 \text{ Amp.}$$

$$I_{R(\infty)} = 0 \text{ Amp.}$$

$$\tau = RC = 8 \text{ sec.}$$

$$i(t) = i_\infty + (i_{0^+} - i_\infty)e^{-t/\tau}$$

$$= 0 + (1-0)e^{-t/8}$$

$$i(t) = e^{-t/8}$$

Energy absorbed by 4 Ω resistor in (0, ∞)

$$E = \int_0^\infty i^2 R\,dt = \int_0^\infty 4e^{-t/4}dt = 4\int_0^\infty e^{-t/4}dt$$

$$E = 4\left.\frac{e^{-t/4}}{-1/4}\right|_0^\infty = -16[e^{-t/4}]_0^\infty$$

$$E = -16[0-1] = 16 \text{ J}$$

4.4 (c)

When switch was closed circuit was in steady state,

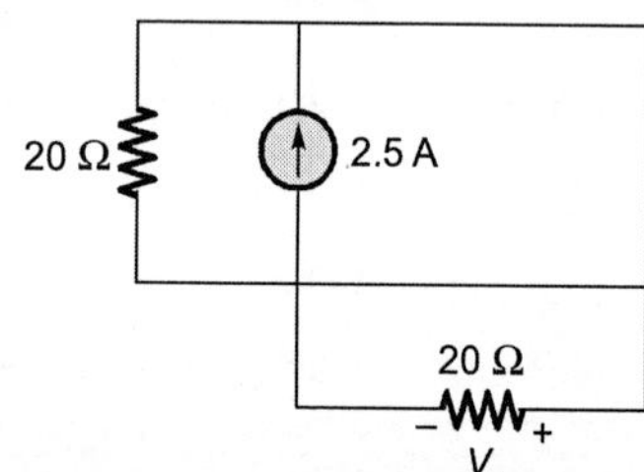

$$i_L(0^-) = 2.5 \text{ A}$$

At, $t = 0^+$

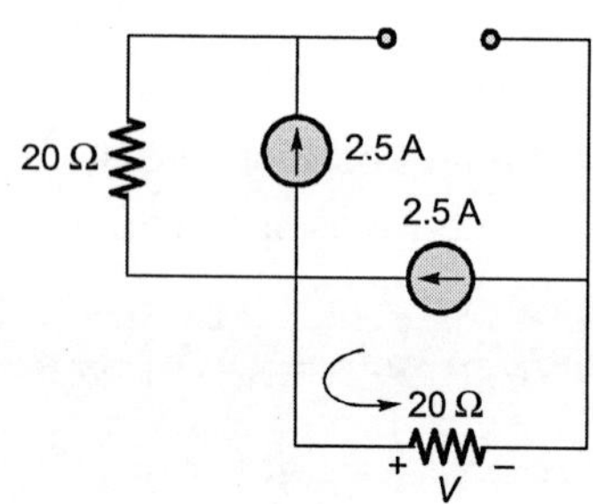

$$\Rightarrow \quad V = IR$$

$$= 2.5 \times 20 = 50 \text{ V}$$

$$\therefore \quad V_x = -50 \text{ V}$$

(Polarity of V_x is given reverse of V)

4.5 (a)

At $t = 0^-$, in steady state,

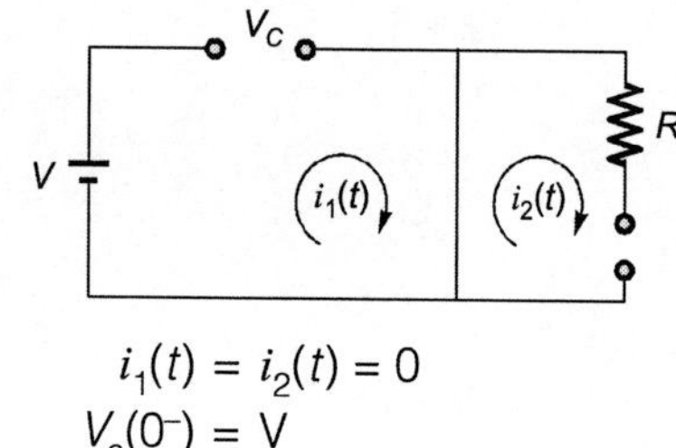

$$i_1(t) = i_2(t) = 0$$

$$V_c(0^-) = V$$

At $t = 0^+$

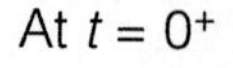

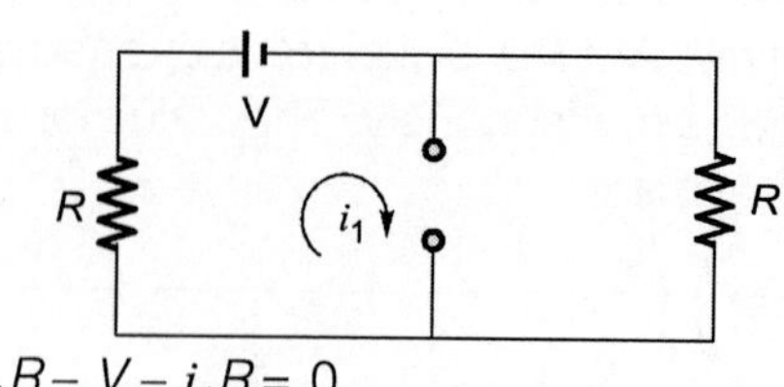

$$-i_1 R - V - i_1 R = 0$$

$$\therefore \quad i_1 = -\frac{V}{2R}$$

4.6 (c)

When switch is in position 2,

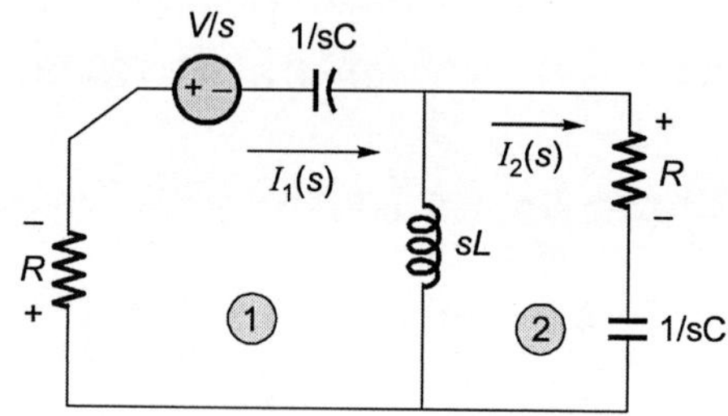

KVL in loop (1),

$$I_1(s)\cdot R + \frac{V}{s} + I_1(s)\cdot\frac{1}{sC} + [I_1(s) - I_2(s)]sL = 0$$

$$\Rightarrow I_1(s)\left[R + \frac{1}{sC} + sL\right] - I_2(s)\cdot sL = \frac{-V}{s}$$

KVL in loop (2),

$$[I_2(s) - I_1(s)]sL + I_2(s)R + I_2(s)\cdot\frac{1}{sC} = 0$$

$$\Rightarrow -I_1(s)\cdot sL + I_2(s)\left[R + sL + \frac{1}{sC}\right] = 0$$

$$\begin{bmatrix} R + sL + \frac{1}{sC} & -sL \\ -sL & R + sL + \frac{1}{sC} \end{bmatrix}\begin{bmatrix} I_1(s) \\ I_2(s) \end{bmatrix} = \begin{bmatrix} -\frac{V}{s} \\ 0 \end{bmatrix}$$

4.7 (c)

$$I(s) = \frac{V(s)}{s+2} = \frac{1}{s(s+2)} = \frac{1}{s(s+2)} = \frac{1}{2}\left[\frac{1}{s} - \frac{1}{s+2}\right]$$

$$i(t) = \frac{1}{2}(1 - e^{-2t})$$

At $t = 0,\ i(t) = 0$

$t = \infty,\ i(t) = 0.5$

$t = \frac{1}{2},\ i(t) = 0.31$

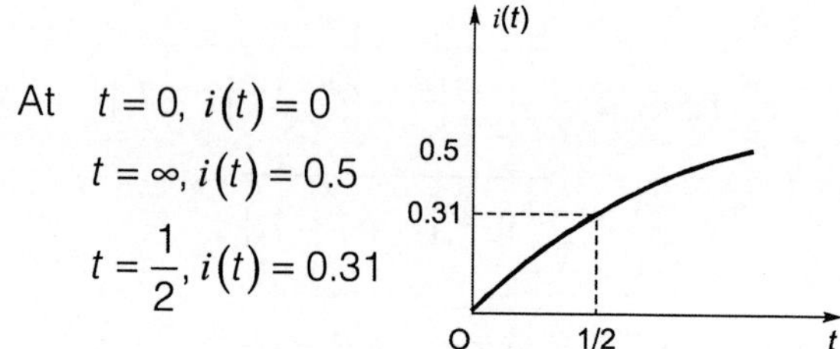

Graph (c) satisfies all conditions.

4.8 (b)

KVL, $v(t) = Ri(t) + \frac{Ldi(t)}{dt} + \frac{1}{C}\int_0^{\infty} i(t)dt$

Taking Laplace transform on both sides,

$$V(s) = RI(s) + LsI(s) - LI(0^+) + \frac{I(s)}{sC} + \frac{V_C(0^+)}{s}$$

$$\Rightarrow \quad \frac{1}{s} = I(s) + sI(s) - 1 + \frac{I(s)}{s} - \frac{1}{s}$$

$$\frac{2}{s} + 1 = \frac{I(s)}{s}[s^2 + s + 1]$$

$$I(s) = \frac{s+2}{s^2+s+1}$$

4.9 (b)

$$RC = 0.1 \times 10^{-6} \times 10^3 = 10^{-4} = 100\ \mu s$$

As RC is very small, so steady state will be reached in 2 sec.

$$V_c = 3\text{ V}$$
$$V_2 = -V_C = -3\text{ V}$$

4.10 (a)

$$V = \frac{LdI}{dt}$$
$$V(s) = sLI(s) - LI(0^+)$$
$$\Rightarrow \quad -LI(0^+) = -1\text{ mV (given in question)}$$
$$I(0^+) = \frac{1\text{mV}}{2\text{mH}} = 0.5\text{ A}$$

4.11 (c)

$$Z_1 = \frac{R_1 \times \frac{1}{C_1 s}}{R_1 + \frac{1}{C_1 s}} = \frac{R_1}{R_1 C_1 s + 1} = \frac{1\text{k}}{4 \times 10^{-3}s + 1}$$

$$Z_2 = \frac{R_2 \times \frac{1}{C_2 s}}{R_2 + \frac{1}{C_2 s}}$$

$$= \frac{R_2}{R_2 C_2 s + 1} = \frac{4\text{k}}{4 \times 10^{-3}s + 1}$$

$$V_o(t) = \left(\frac{Z_2}{Z_1 + Z_2}\right) V_i(t)$$

$$Z_2 = 4Z_1$$

$$\frac{Z_2}{Z_1 + Z_2} = \frac{4Z_1}{Z_1 + 4Z_1} = \frac{4Z_1}{5Z_1} = 0.8$$

$$v_o(t) = 0.8\, v_i(t) = 0.8 \times 10\, u(t)$$
$$v_o(t) = 8u(t)$$

4.12 (a)

At $t = 0^+$ Capacitor is short circuit and at $t = \infty$ Capacitor is open circuit

So, $\quad I_c(0^+) = \frac{10\text{ V}}{20\text{ k}\Omega} = 0.5\text{ mA}$

$$I_c(\infty) = 0$$

Time constant of the circuit $\tau = R_{eq}C$

$$\tau = 4\ \mu\text{F} \times 20\text{ k}\Omega \parallel 20\text{ k}\Omega$$
$$\Rightarrow \quad \tau = 40\text{ msec}$$

Using direct formula

$$I_c(t) = I_c(\infty) - [I_c(\infty) - I_c(0)]\, e^{-t/\tau}$$
$$I_c(t) = 0 - (0 - 0.5)\, e^{-t/40\text{ msec}}$$
$$I_c(t) = 0.5e^{-25t}\text{ mA}$$

4.13 (b)

At $t = 0$, the inductor behaves as an open circuit

So, $\quad V_L = I_S R_S$

$$V_L = L\frac{di}{dt}(0^+)$$

$$\Rightarrow \quad \frac{di}{dt}(0^+) = \frac{V_L}{L} - \frac{I_s R_s}{L}$$

4.14 (c)

The waveform of voltage $V_c(t)$ is shown below.

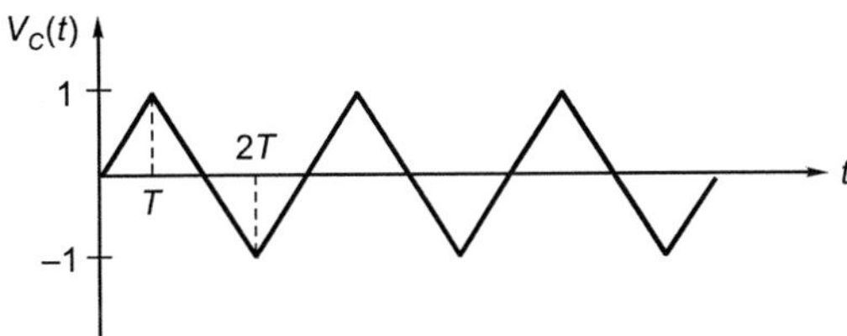

In mathematical form,

$0 < t < T$,

$$C = 1\text{ F}, I = 1\text{ A}$$
$$V_c = \int_0^t dt = t$$

at $t = T$; $\quad V_c = T$

$T < t < 2T$

$$V_c = T - \int_T^t dt = 2T - t$$

at $t = 2T$; $\quad V_C = 0$

$2T < t < 3T$

$$V_C = \int_{2T}^t dt = t - 2T$$

at $t = 3T$; $\quad V_C = T$

$3T < t < 4T$

$$V_C = T - \int_{3T}^t dt = 4T - t$$

$$\therefore \quad V_c(t) = tu(t) - 2(t-T)u(t-2T) + 2(t-2T)u(t-2T)\ldots$$

$$= tu(t) + 2\sum_{n=1}^{\infty}(-1)^n(t-nT)u(t-nT)$$

4.15 (d)

$$V_C(s) = \frac{1}{\left(s+1+\frac{1}{s}\right)}\cdot\left(\frac{1}{s}\right)$$

$$= \frac{1}{s^2+s+1}$$

$$= \frac{1}{\left(s+\frac{1}{2}\right)^2+\left(\frac{\sqrt{3}}{2}\right)^2}$$

$$V_C(t) = \frac{2}{\sqrt{3}}e^{-\frac{t}{2}}\sin\left(\frac{\sqrt{3}}{2}t\right)$$

4.16 (b)

$$V_R(s) = \frac{1}{\left(s+1+\frac{1}{s}\right)}\cdot 1 = \frac{s}{s^2+s+1}$$

$$= \frac{\left(s+\frac{1}{2}\right)}{\left(s+\frac{1}{2}\right)^2+\left(\frac{\sqrt{3}}{2}\right)^2} - \frac{1/2\cdot\frac{\sqrt{3}}{2}\cdot\frac{2}{\sqrt{3}}}{\left(s+\frac{1}{2}\right)^2+\left(\frac{\sqrt{3}}{2}\right)^2}$$

$$V_R(t) = e^{-t/2}\cos\left(\frac{\sqrt{3}}{2}t\right) - \frac{1}{\sqrt{3}}e^{-t/2}\sin\left(\frac{\sqrt{3}}{2}t\right)$$

$$V_R(t) = e^{-t/2}\left[\cos\left(\frac{\sqrt{3}}{2}t\right) - \frac{1}{\sqrt{3}}\sin\left(\frac{\sqrt{3}}{2}t\right)\right]$$

4.17 (b)

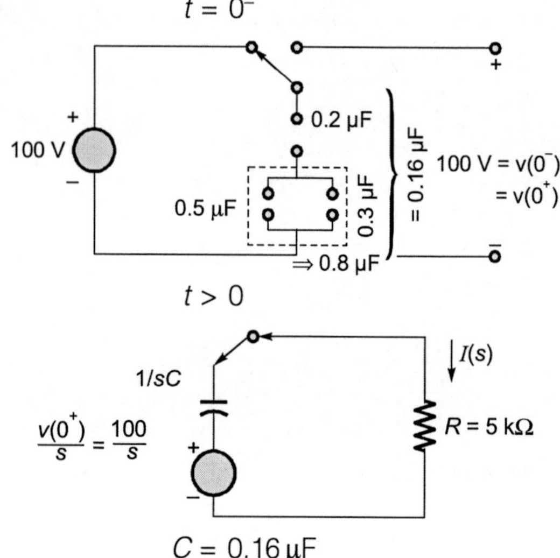

$$i(t) = \frac{v(0^+)}{R}e^{-t/RC}\cdot u(t)$$

$$v(0^+) = 100\text{ V}$$

$$\frac{1}{RC} = \frac{1}{5\times10^3\times0.16\times10^{-6}}$$

$$R = 5\text{ k}\Omega$$

$$i(t) = 20\,e^{-1250t}u(t)\text{ mA}$$

4.19 (a)

Method 1:

At $t = 0^-$, the circuit is shown below:

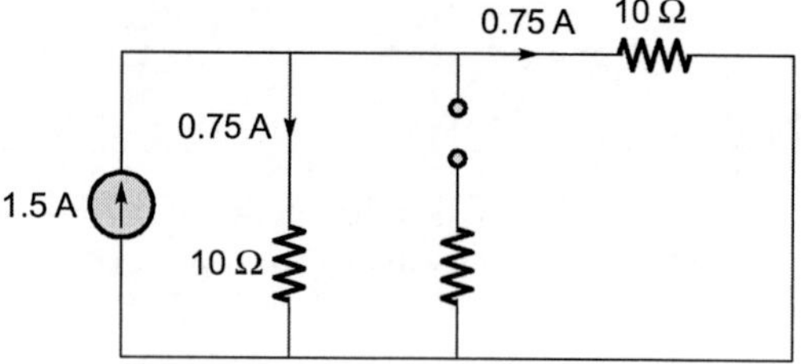

Current will divide equally in both 10 Ω resistors as shown in diagram.

At $t \geq 0^+$, the circuit in s-domain is shown below:

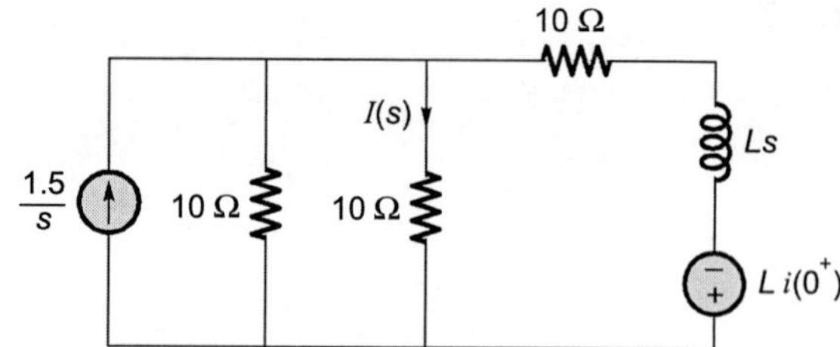

[$L = 15$ mH, $i(0^+) = i(0^-) = 0.75$ A]

The above circuit can be rearranged as given below:

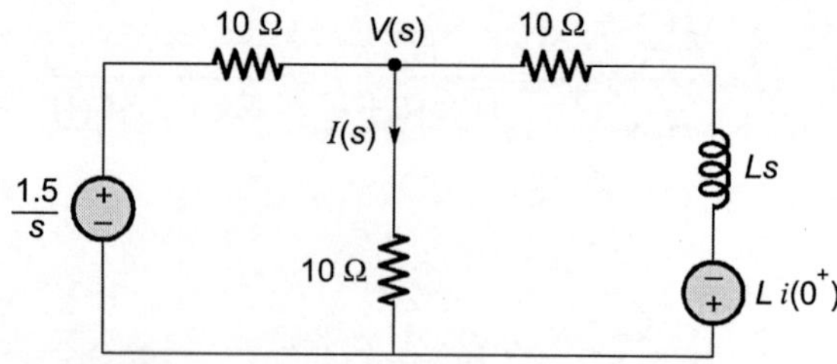

Applying nodal analysis,

$$\frac{V(s)-\frac{1.5}{s}}{10} + \frac{V(s)}{10} + \frac{V(s)+15\times10^{-3}\times0.75}{10+15\times10^{-3}s} = 0$$

$$\frac{V(s)}{10} + \frac{V(s)}{10} + \frac{V(s)}{10+15\times10^{-3}s}$$

$$= \frac{15}{10s} - \frac{15\times10^{-3}\times0.75}{10+15\times10^{-3}s}$$

$$V(s)\left[\frac{10+15\times10^{-3}s+5}{5(10+15\times10^{-3}s)}\right]$$

$$= 15\left[\frac{10 + 15\times10^{-3}\,s - 10\,s\times10^{-3}\times0.75}{10\,s(10 + 15\times10^{-3}\,s)}\right]$$

$$(1 + 10^{-3}\,s)\,V(s) = \frac{(10 + 7.5\times10^{-3}\,s)}{2\,s}\left[\frac{s+1000}{1000}\right]$$

$$V(s) = \frac{10000 + 7.5\,s}{2000\,s} = \frac{10000 + 7.5\,s}{2\,s(s+1000)}$$

$$= \frac{10000}{2\times1000\,s} + \frac{\{10000 + 7.5(-1000)\}}{2(-1000)\,(s+1000)}$$

$$= \frac{5}{s} - \frac{1.25}{s+1000}$$

Taking inverse Laplace transform,

$$v(t) = 5 - 1.25\,e^{-1000t}\text{ V}$$

$$i(t) = \frac{v(t)}{10} = 0.5 - 0.125\,e^{-1000t}\text{ A}$$

Method 2:

$$i(0^+) = \frac{0.75}{2} = 0.375\text{ A}$$

$$i(\infty) = 0.5\text{ A}$$

$$i(t) = i(\infty) - \{i(\infty) - i(0^+)\}\,e^{-Rt/L}$$

where, R = equivalent resistance seen across L with current source opened

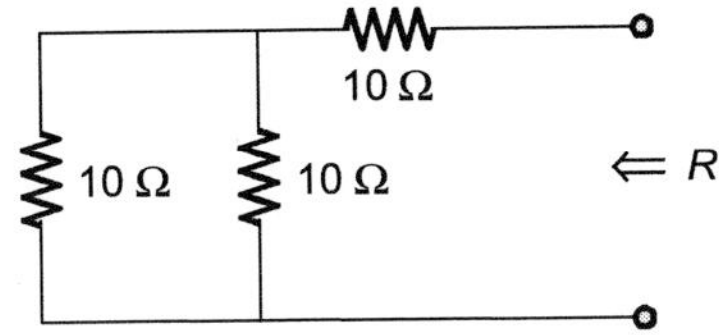

$$R = 10 + (10\,||\,10) = 15\,\Omega$$

$$i(t) = 0.5 - \{0.5 - 0.375\}e^{-15t/15\times10^{-3}}$$

$$= 0.5 - 0.125\,e^{-1000t}\text{ A}$$

4.20 (a)

$$V(0^-) = V(0^+) = \frac{Q}{C} = \frac{2.5\times10^{-3}}{50\times10^{-6}} = -50\text{ V}$$

$\because$ (direction given is opposite)

$$V(\infty) = 100\text{ V}$$

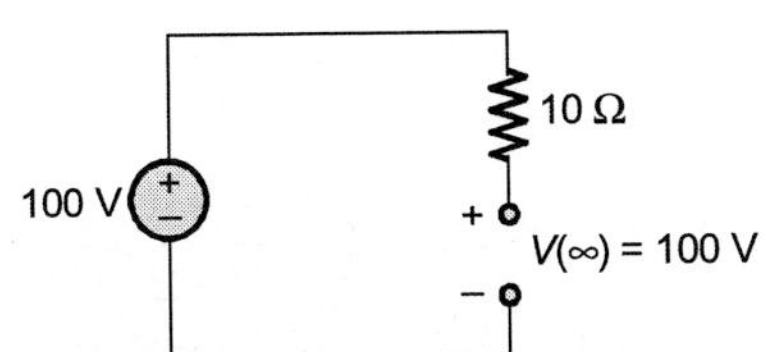

$$V(t) = V(\infty) + [V(0^+) - V(\infty)]e^{-t/\tau}$$

$$= 100 + (-50 - 100)e^{-t/\tau}$$

$$V(t) = 100 - 150e^{-t/\tau}$$

$$i(t) = C\frac{dV}{dt} = C\frac{150}{\tau}e^{-t/\tau}$$

and $\tau = 10\times50\times10^{-6} = 5\times10^{-4}$

$$i(t) = 50\times10^{-6}\times\frac{150}{10\times50\times10^{-6}}e^{-2\times10^3 t}$$

$$= 15e^{-2\times10^3 t}\text{ Amp}$$

4.21 (d)

Since there is no resistance so time constant will be zero. That means as soon as the switch will be closed voltages at C_1 and C_2 will become equal and capacitor allows sudden change of voltage only if impulse of current will pass through it.

4.23 (10)

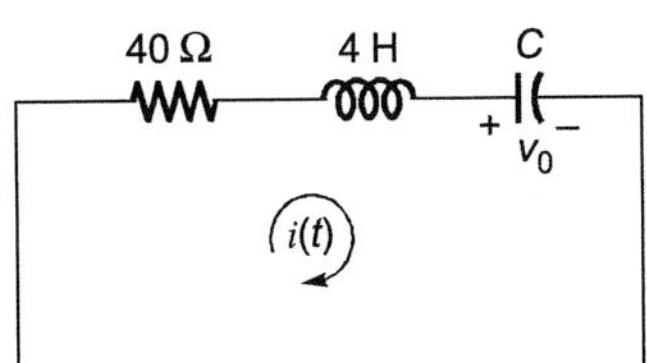

For critically damped system

$$\xi = 1 = \frac{1}{2Q} \quad \text{...(i)}$$

where ξ = Damping factor

Q = Quality factor

For series RLC circuit

$$Q = \frac{1}{R}\sqrt{\frac{L}{C}} \quad \text{...(ii)}$$

From equation (i) and (ii),

$$\frac{1}{\frac{2}{R}\sqrt{\frac{L}{C}}} = 1$$

or $$C = \left(\frac{2}{R}\right)^2 L = \left(\frac{2}{40}\right)^2\times4 = 10\text{ mF}$$

4.24 (1.25)

At steady state $t = 0^-$

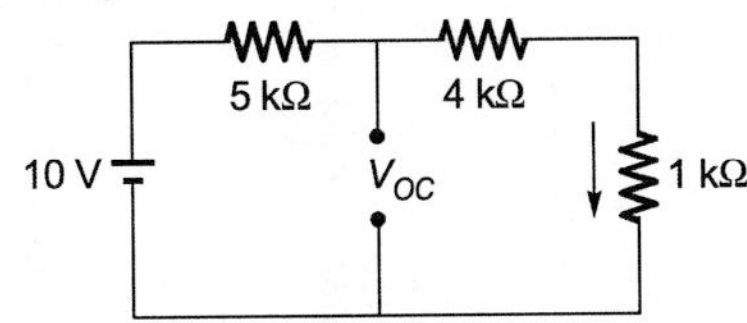

$$\therefore V_c(0^-) = V_c(0^+) = 5\text{ V}$$

$$I_L(0^-) = I_L(0^+) = 1\text{ mA}$$

At, $t = 0$, switch get closed

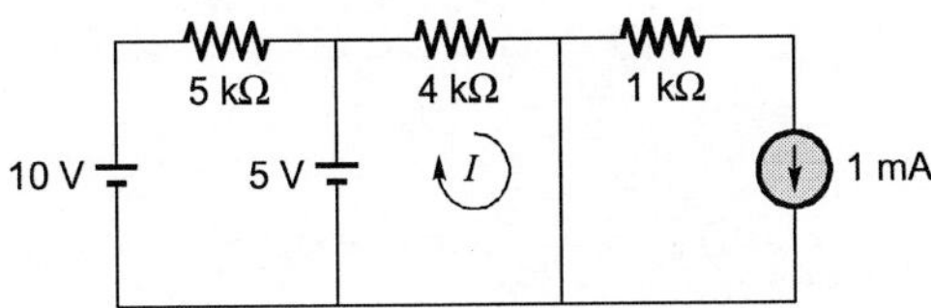

Thus, the current through 4 Ω resistance is

$$I = \frac{5}{4\times10^3} = 1.25\text{ mA}$$

4.25 (a)

Converting the given circuit into frequency domain and applying KCL at $V(s)$

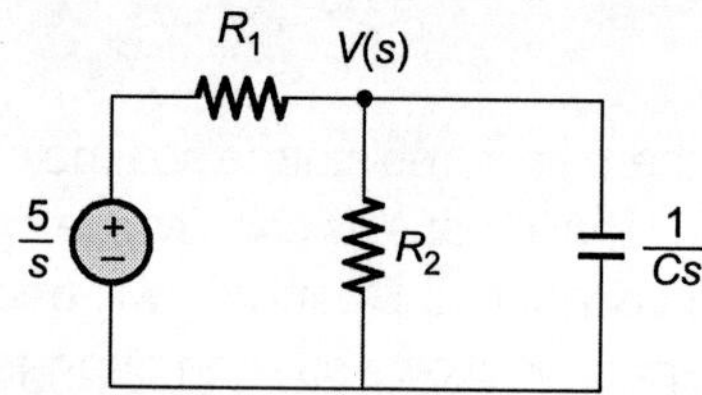

We get,

$$\frac{V(s)-\dfrac{5}{s}}{R_1}+\frac{V(s)}{R_2}+\frac{V(s)}{\dfrac{1}{Cs}} = 0 \qquad \text{...(i)}$$

$\because$ R_1 = 1 kΩ, R_2 = 2 kΩ and C = 1 μF

Using the components value we get,

$$V(s)\left[\frac{1}{1}+\frac{1}{2}+s\right] = \frac{5}{s}$$

$$\text{or} \quad V(s) = \frac{5}{s\left(s+\dfrac{3}{2}\right)} \qquad \text{...(ii)}$$

Using partial fraction on equation (ii)

$$V(s) = \frac{10}{3s}-\frac{10}{3\left(s+\dfrac{3}{2}\right)} \qquad \text{...(iii)}$$

Using inverse Laplace transform

$$v(t) = \frac{10}{3}\left[1-e^{-\frac{3}{2}t}\right] \qquad \text{...(iv)}$$

$$\therefore \text{ Current } I = \frac{V(t)}{R_2} = \frac{5}{3}\left[1-e^{-\frac{3}{2}t}\right]\text{mA}$$

4.26 (31.25)

At steady state the inductor act as a short circuit.

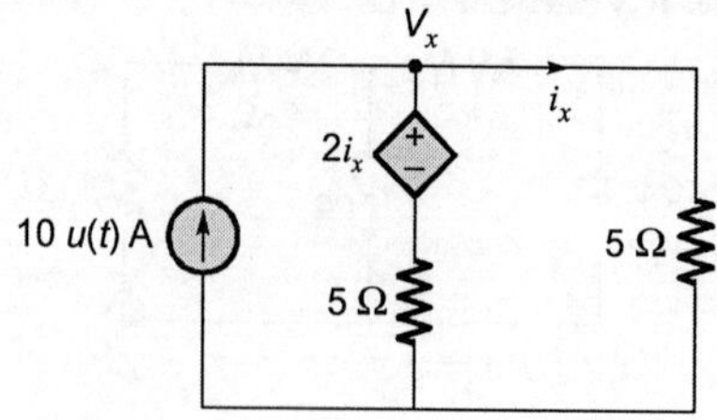

$\therefore$ $V_x = 5i_x$

By KCL

$$-10+\frac{V_x-2i_x}{5}+i_x = 0$$

$$-10+\frac{5i_x-2i_x}{5}+i_x = 0$$

or

$$\frac{8i_x}{5} = 10$$

$$\Rightarrow \quad i_x = \frac{50}{8}\text{ A}$$

$$\therefore v_o(t) = 5i_x(t) = \frac{5\times 50}{8} = 31.25 \text{ V}$$

4.27 (c)

120 Ω
3 V
B
0.1 μF

$v_c(0^-) = 0$ V

$v_c(0^+) = 0$ V

$v_c(\infty) = 3$ V

Time constant,

$$\tau = RC = 120 \times 0.1 \times 10^{-6}$$

$$v_c(t) = 3+(0-3)e^{-t/\tau}$$

$$= 3(1-e^{-t/\tau})$$

$$I_c(t) = \frac{Cdv_c(t)}{dt} = \frac{0.1\times 10^{-6}\times 3\times e^{-t/\tau}}{\tau}$$

$$= \frac{0.1\times 10^{-6}\times 3e^{-t/\tau}}{120\times 0.1\times 10^{-6}} = \frac{1}{40}e^{-t/\tau}$$

$$\text{Energy} = \int_0^\infty VIdt$$

$$= \int_0^\infty 3\cdot\frac{1}{40}\times e^{-t/\tau} = -\frac{3}{40}\times\tau e^{-t/\tau}\Big|_0^\infty$$

$$= \frac{3}{40}\times 12\times 10^{-6} = 0.9\ \mu\text{J}$$

4.28 (2.528)

$v_c(0^-) = 0$ V

$v_c(0^+) = 0$ V

At $t = \infty$

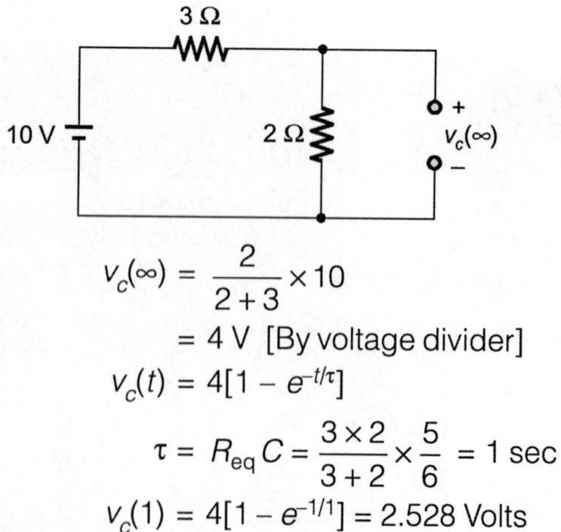

$$v_c(\infty) = \frac{2}{2+3}\times 10$$

$$= 4 \text{ V [By voltage divider]}$$

$$v_c(t) = 4[1-e^{-t/\tau}]$$

$$\tau = R_{eq}C = \frac{3\times 2}{3+2}\times\frac{5}{6} = 1 \text{ sec}$$

$$v_c(1) = 4[1-e^{-1/1}] = 2.528 \text{ Volts}$$

4.29 (1.5)

$$\text{Initial energy} = \frac{1}{2}(C_1V_1^2+C_2V_2^2)$$

$$= \frac{1}{2}(3\times 1^2+1\times 3^2) = 6 \text{ J}$$

Final energy stored in capacitor

$$= \frac{1}{2}(C_1 + C_2)V^2$$

$$C_1 V_1 + C_2 V_2 = (C_1 + C_2)V$$
$$1 \times 3 + 3 \times 1 = (1 + 3)V$$
$$V = 1.5 \text{ V}$$

$$\text{Final energy} = \frac{1}{2}(1+3)\times(1.5)^2 = 4.5 \text{ J}$$

Energy dissipated = 6 – 4.5 = 1.5 J

4.30 (d)

at $t = 0^-$, Switch is at position-1

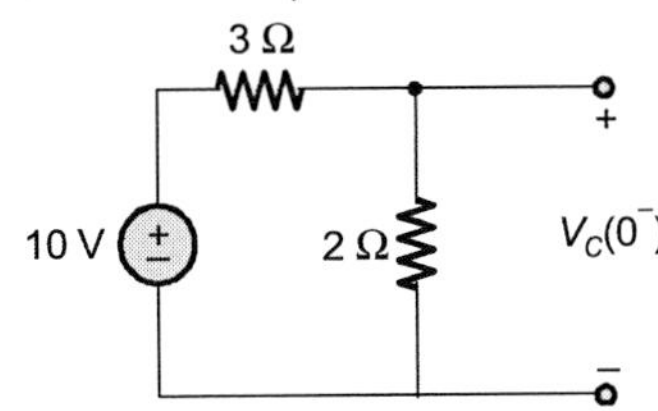

where, $V_c(0^-) = \frac{10\times 2}{2+3} = 4 \text{ V}$ ...(1)

$\therefore$ $V_c(0^-) = V_c(0^+) = 4 \text{ V}$

at $t = \infty$

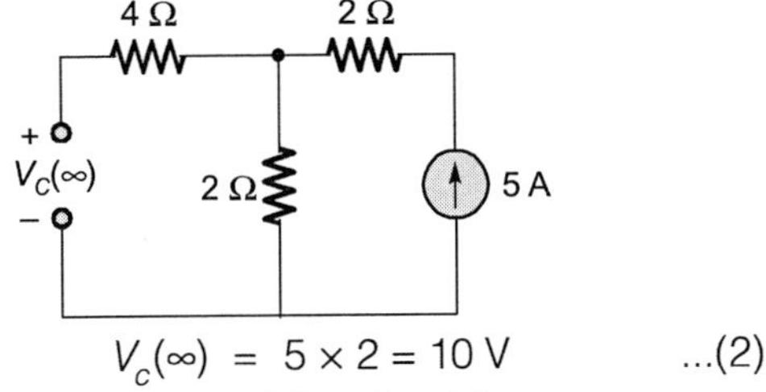

$$V_c(\infty) = 5 \times 2 = 10 \text{ V} \quad ...(2)$$

The time constant of the circuit is

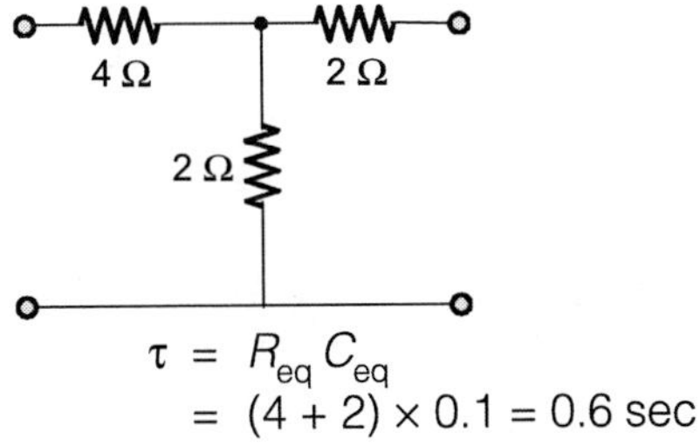

$$\tau = R_{eq} C_{eq}$$
$$= (4 + 2) \times 0.1 = 0.6 \text{ sec}$$

$$\therefore \quad V_c(t) = V_c(\infty) + [V_c(0^+) - V_c(\infty)]e^{-t/\tau}$$
$$= 10 + (4 - 10)e^{-t/0.6}$$
$$V_c(t) = (10 - 6e^{-t/0.6}) \text{ V}$$

4.31 (100)

At $t = 0^-$

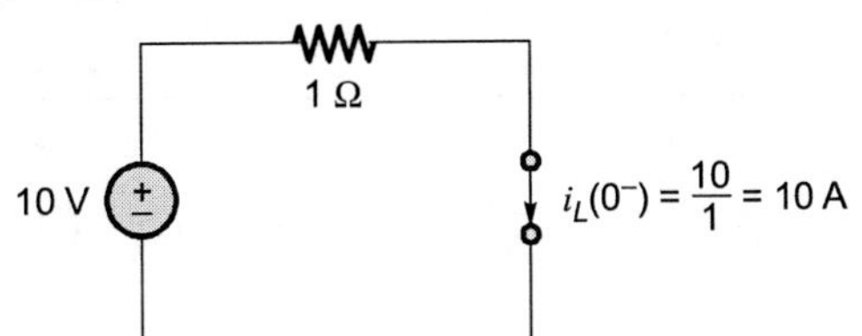

$$i_L(0^-) = \frac{10}{1} = 10 \text{ A}$$

For $t > 0$ (using Laplace transform)

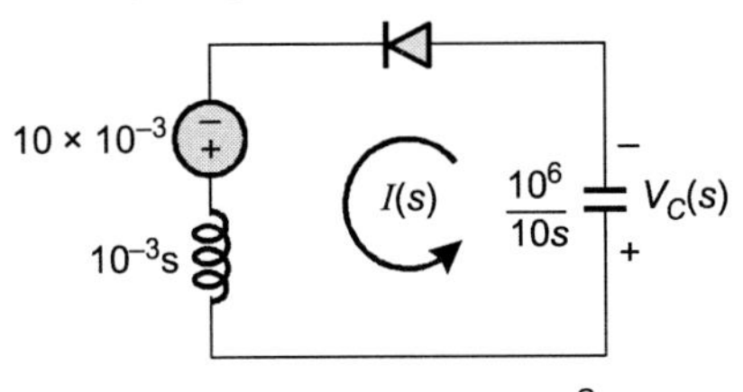

$$I(s) = \frac{10\times 10^{-3}}{10^{-3}s + \frac{10^6}{10s}}$$

$$V_c(s) = I(s)\times\frac{10^6}{10s} = \frac{10^6}{s^2+10^8}$$

Taking inverse Laplace, we get

$$V_c(t) = 100\sin 10^4 t \text{ V}$$

$\therefore$ Steady state magnitude voltage across capacitor is 100 V.

4.32 (–1)

at $t = 0^-$

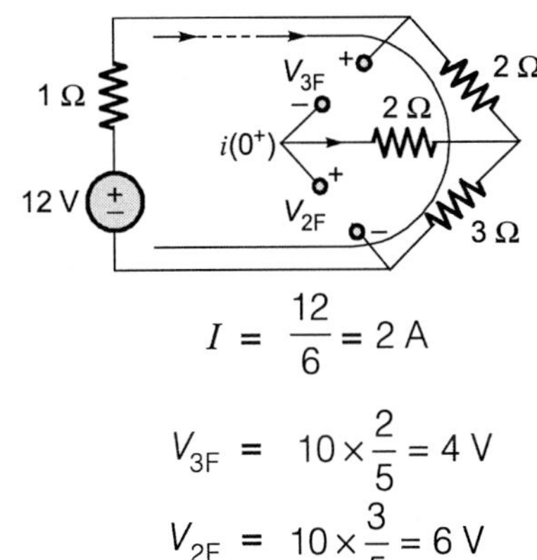

$$I = \frac{12}{6} = 2 \text{ A}$$

$$V_{3F} = 10\times\frac{2}{5} = 4 \text{ V}$$

$$V_{2F} = 10\times\frac{3}{5} = 6 \text{ V}$$

at $t = 0^+$

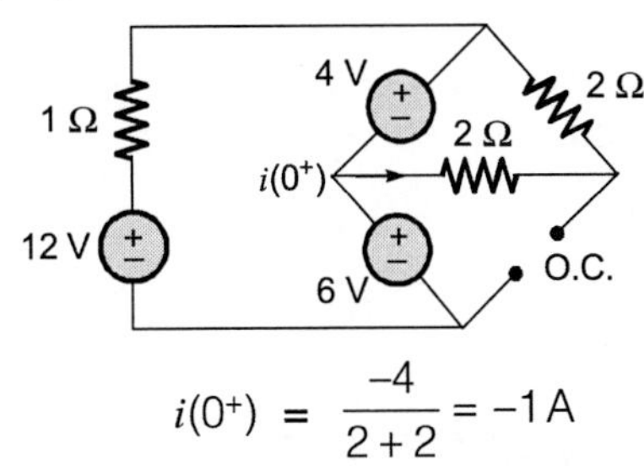

$$i(0^+) = \frac{-4}{2+2} = -1 \text{ A}$$

Note : As the current direction is not mentioned in the question, thus by reversing the current direction 1 A can also be the answer.

4.33 (0.3405)

$$i_s(t) = \frac{V}{R}\left[1-e^{\frac{-Rt}{L}}\right] = \frac{15}{1}\left[1-e^{\frac{-3t}{2}}\right] \text{ A}$$

Current through 2 H,

$$i(t) = i_s(t)\frac{1}{1+2} = 5\left[1-e^{\frac{-3t}{2}}\right] \text{A}$$

At $i(t) = 2$ A, $\quad 2 = 5\left[1-e^{\frac{-3t}{2}}\right]$

By solving, $\quad t = 0.3405$ sec

4.34 (8.16)

- The equivalent circuit at $t = 0^-$ is as follows:

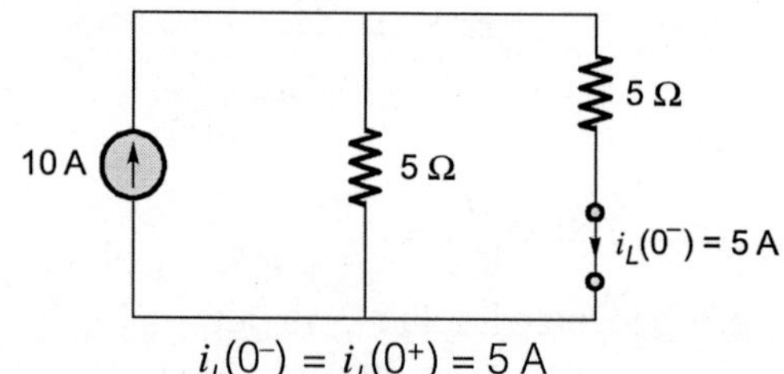

$$i_L(0^-) = i_L(0^+) = 5 \text{ A}$$

- The Laplace transform model of the circuit for $t > 0$ is as follows:

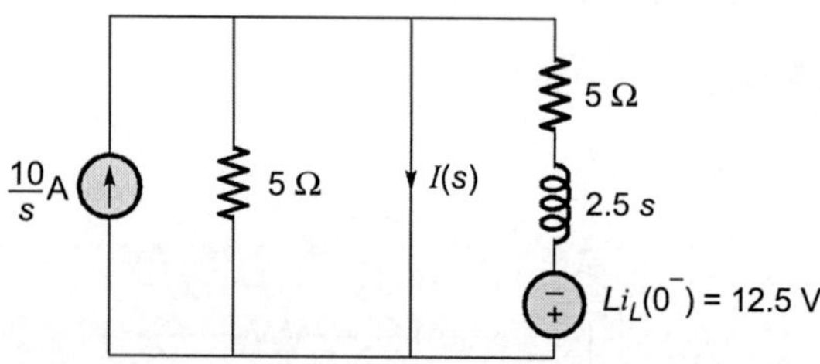

$$I(s) = \frac{10}{s} - \frac{12.5}{5+2.5s} = \frac{10}{s} - \frac{5}{s+2}$$

- By taking inverse Laplace transform,

$$i(t) = (10 - 5e^{-2t})\, u(t) \text{ A}$$

- At $t = 0.5$ seconds,

$$i(t) = \left(10 - \frac{5}{e}\right)\text{A} = 8.16 \text{ A}$$

4.35 (0.316)

$$\text{Loop current, } i(t) = \frac{1}{1+1}(1-e^{-t/\tau}) \text{ A} \quad ; \quad t>0$$

$$\tau = \frac{L}{R_{eq}} = \frac{1}{1+1} = \frac{1}{2} \text{ sec}$$

$$i(t) = \frac{1}{2}(1-e^{-2t}) \text{ A} \ ; \ t>0$$

At $t = 0.5$ sec,

$$i(t) = \frac{1}{2}(1-e^{-1}) \text{ A} = 0.316 \text{ A}$$

4.36 (0.25)

$$T = 3R_oC = 3\text{sec}$$

$$R(t) = \left(1-\frac{t}{3}\right) \ ; \ 0 \le t \le 3 \text{ sec}$$

$$R(t)\, i(t) + \frac{1}{C}\int i(t)dt = 1$$

$$\left(1-\frac{t}{3}\right) i(t) + \int i(t)dt = 1$$

Differentiating both sides, we get,

$$\left(1-\frac{t}{3}\right)\frac{di}{dt} - \frac{i}{3} + i = 0$$

$$(3-t)\frac{di}{dt} + 2i = 0$$

$$\frac{di}{i} = -\frac{2}{(3-t)}dt$$

Integrating on both sides, we get,

$$\ln(i) = 2\ln(3-t) + \ln(c)$$

$$i(t) = c(3-t)^2 \ ; \ t \ge 0$$

Given that, $i(0) = 1$ A.

So, $\quad c(3-0)^2 = 1$ A

$$c = \frac{1}{9} \text{ A}$$

$$i(t) = \frac{1}{9}(3-t)^2 \text{ A}$$

At $t = \frac{T}{2} = 1.5$ sec,

$$i(1.5) = \frac{1}{9}(1.5)^2 = 0.25 \text{ A}$$

4.37 (0.1386)

$$V_c(0^-) = -5\text{V}$$

$$V_c(0^+) = -5\text{V}$$

$t \to \infty$, capacitor acts as a O.C.

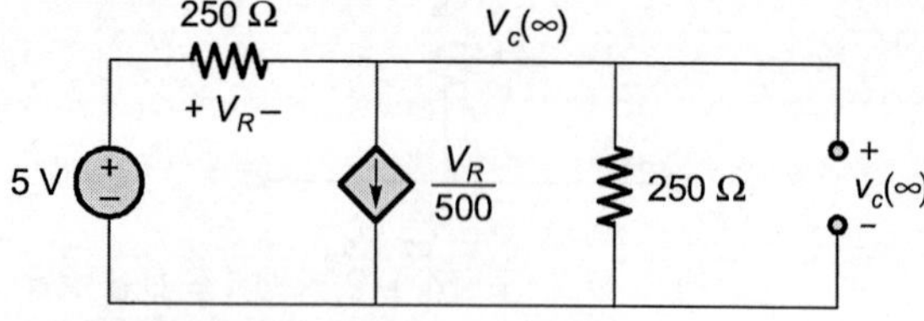

Write KCL at node

$$\frac{V_c(\infty)-5}{250} + \frac{V_R}{500} + \frac{V_c(\infty)}{250} = 0$$

$$V_c(\infty) = \frac{5}{3} \text{ Volts}$$

Time constant $(\tau) = R_{eq}C$.

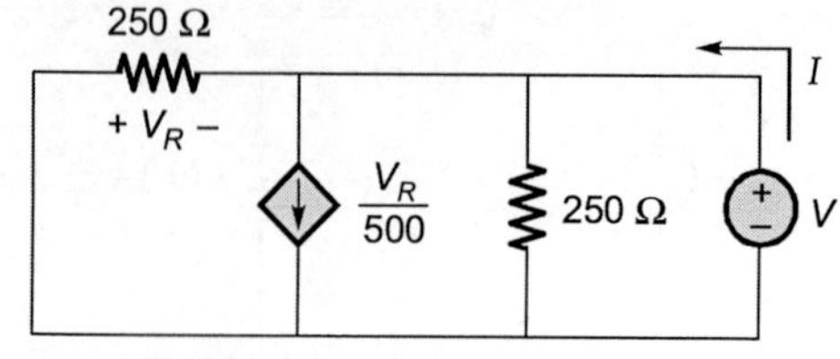

$$I = \frac{V}{250} + \frac{V_R}{500} + \frac{V}{250}$$

$$V_R = -V$$

$$I = \frac{V}{250} - \frac{V}{500} + \frac{V}{250}$$

$$\frac{V}{I} = \frac{500}{3}\Omega; \ R_{eq} = \frac{500}{3}\Omega$$

$$\tau = \frac{500}{3} \times 0.6\mu = 0.1 \times 10^{-3}$$

$$v_c(t) = v_c(\infty) + (v_c(0) - v_c(\infty))e^{-t/\tau}$$

$$v_c(t) = \frac{5}{3} + \left(-5 - \frac{5}{3}\right)e^{-t/0.1\times10^{-3}}$$

$$0 = \frac{5}{3} - \frac{20}{3}e^{-10000t}$$

$$t = 0.1386 \text{ m sec}$$

4.38 (1.44)

Apply Laplace transform,

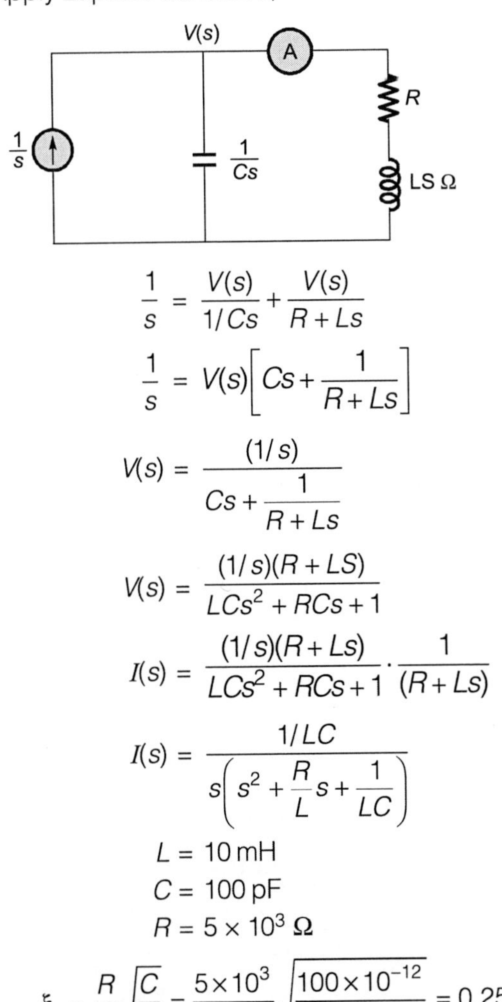

$$\frac{1}{s} = \frac{V(s)}{1/Cs} + \frac{V(s)}{R+Ls}$$

$$\frac{1}{s} = V(s)\left[Cs + \frac{1}{R+Ls}\right]$$

$$V(s) = \frac{(1/s)}{Cs + \frac{1}{R+Ls}}$$

$$V(s) = \frac{(1/s)(R+LS)}{LCs^2 + RCs + 1}$$

$$I(s) = \frac{(1/s)(R+Ls)}{LCs^2 + RCs + 1} \cdot \frac{1}{(R+Ls)}$$

$$I(s) = \frac{1/LC}{s\left(s^2 + \frac{R}{L}s + \frac{1}{LC}\right)}$$

$$L = 10 \text{ mH}$$

$$C = 100 \text{ pF}$$

$$R = 5 \times 10^3 \ \Omega$$

$$\xi = \frac{R}{2}\sqrt{\frac{C}{L}} = \frac{5\times10^3}{2}\sqrt{\frac{100\times10^{-12}}{10\times10^{-3}}} = 0.25$$

Max. over shoot = $e^{-\pi\xi/\sqrt{1-\xi^2}} = 0.44$

Maximum value= Steady state + Max. overshoot

$$= 1 + 0.44 = 1.44 \text{ A}$$

4.39 (d)

Inductor and capacitors are connected to the inductance source for a long time, so these elements have reached steady state.

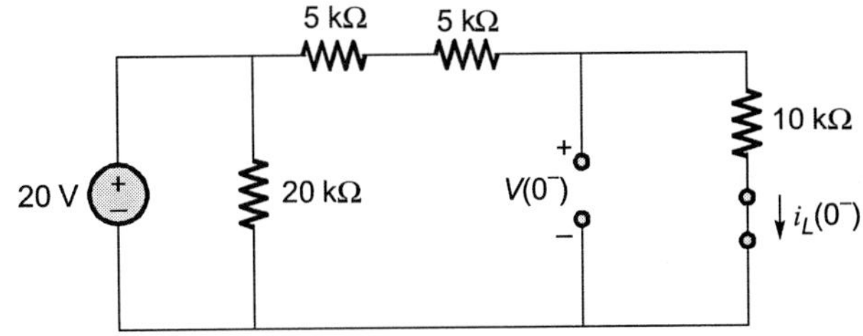

$$i_L(0^-) = \frac{20}{5\,k\Omega + 5\,k\Omega + 10\,k\Omega} = 1 \text{ mA}$$

$$V(0^-) = 10 \text{ V}$$

$$t = 0^+$$

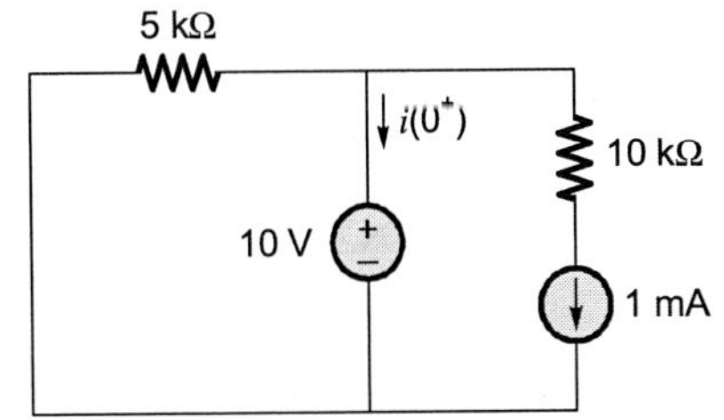

$$i(0^+) + \frac{10}{5k} + 1 \text{ mA} = 0$$

$$i(0^+) = -3 \text{ mA}$$

$$C\frac{dV(0^+)}{dt} = -3 \text{ mA}$$

$$1\times10^{-3}\frac{dV(0^+)}{dt} = -3 \text{ mA}$$

$$\frac{dV(0^+)}{dt} = -3 \text{ V/s}$$

4.40 (4)

At $t = 0^-$, the switch was closed, hence the circuit is in steady state.

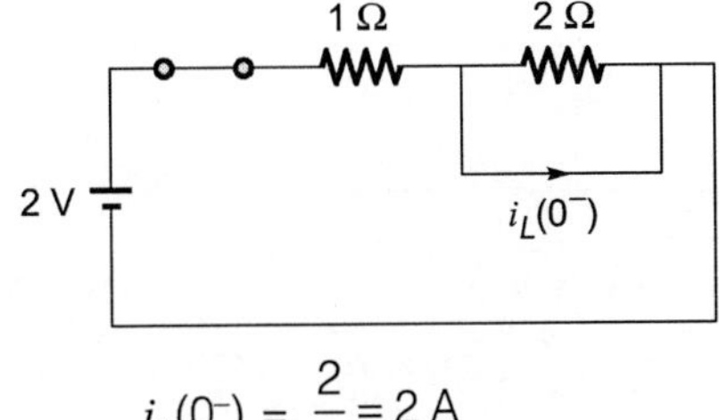

$$i_L(0^-) = \frac{2}{1} = 2 \text{ A}$$

At $t = 0^+$, the switch is open, hence the circuit becomes,

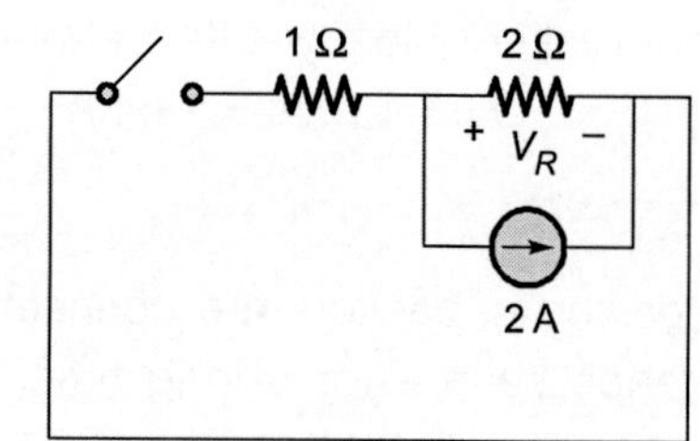

$$V_R = -2 \times 2 = -4$$

Magnitude of voltage V_R,

$$|V_R| = 4$$

4.41 (b)

At $t = 0^-$; switch $S_1 \rightarrow$ closed, switch $S_2 \rightarrow$ opened, hence the circuit is in steady state.

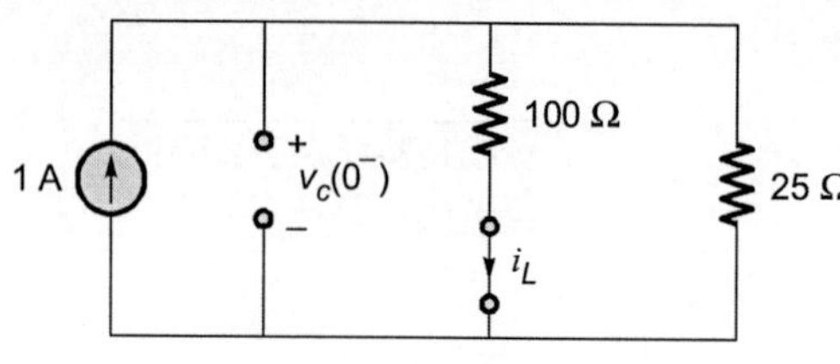

$$i_L(0^-) = \frac{1 \times 25}{100 + 25} = 0.2 \text{ A}$$

$$v_c(0^-) = \frac{1}{5} \times 100 = 20 \text{ V}$$

At $t = 0^+$; switch $S_1 \rightarrow$ opened, switch $S_2 \rightarrow$ closed, the circuit can be redrawn as

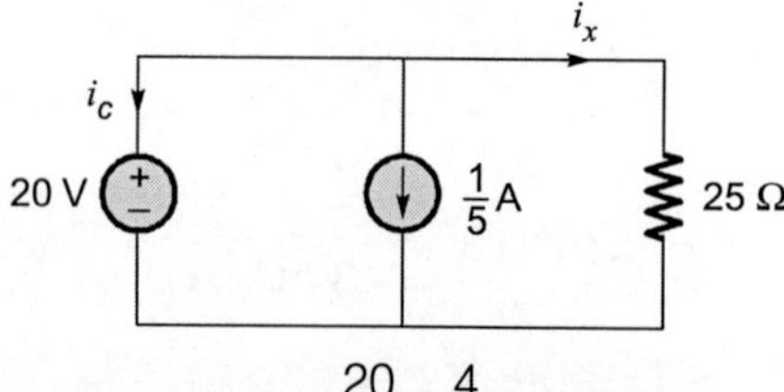

$$i_x = \frac{20}{25} = \frac{4}{5} \text{ A} = 0.8 \text{ A}$$

By applying KCL:

$$-i_c = i_x + 0.2 = 0.8 + 0.2$$

$$\Rightarrow \quad i_c = -1 \text{ A}$$

4.42 (0.75)

After the switch is closed, the circuit for $t > 0$ is given as below:

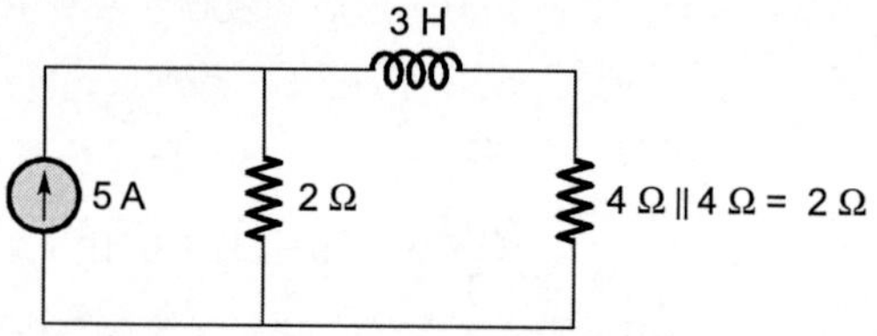

To calculate the time constant, the current source is considered as open-circuit.

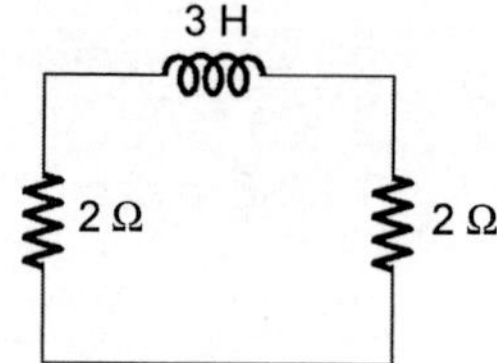

It is an R-L circuit with time-constant given as:

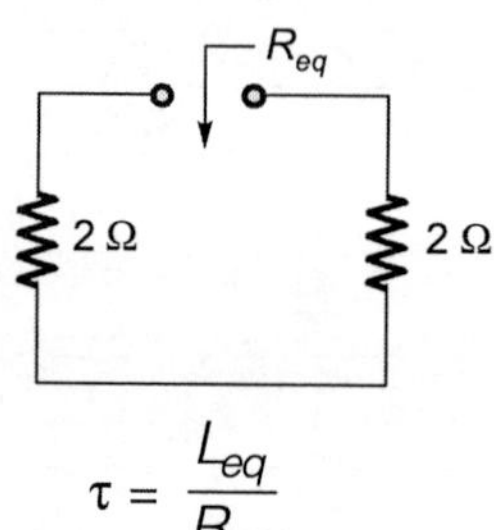

$$\tau = \frac{L_{eq}}{R_{eq}}$$

We have, $L_{eq} = 3$ H and $R_{eq} = 2 + 2 = 4\ \Omega$

$$\therefore \quad \tau = \frac{3}{4} = 0.75 \text{ sec}$$

■■■

5 Two Port Networks

5.1 The short-circuit admittance matrix of a two-port network is

$$\begin{bmatrix} 0 & -1/2 \\ 1/2 & 0 \end{bmatrix}$$

The two-port network is
(a) non-reciprocal and passive
(b) non-reciprocal and active
(c) reciprocal and passive
(d) reciprocal and active

[1998 : 1 M]

5.2 A 2-port network is shown in the figure. The parameter h_{21} for this network can be given by

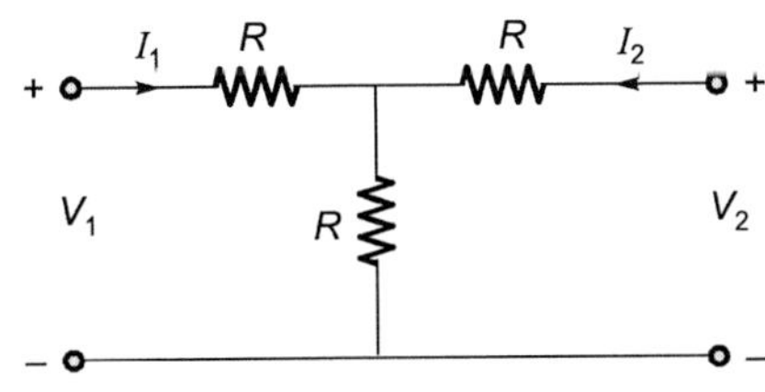

(a) −1/2 (b) +1/2
(c) −3/2 (d) +3/2

[1999 : 1 M]

5.3 The admittance parameter y_{12} in the 2-port network in Figure is

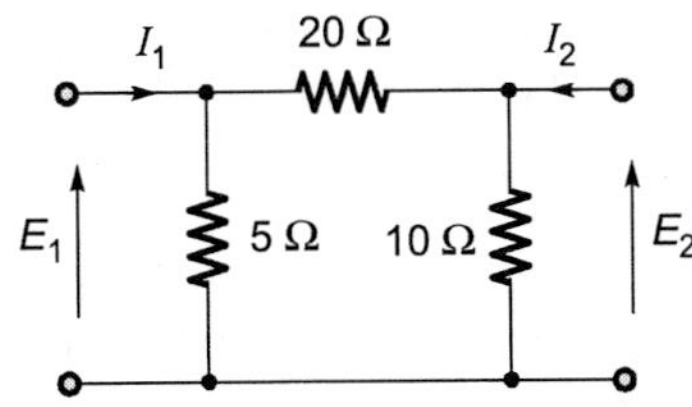

(a) −0.2 mho (b) 0.1 mho
(c) −0.05 mho (d) 0.05 mho[2001 : 1 M]

5.4 The Z parameters Z_{11} and Z_{21} for the 2-port network in the figure are

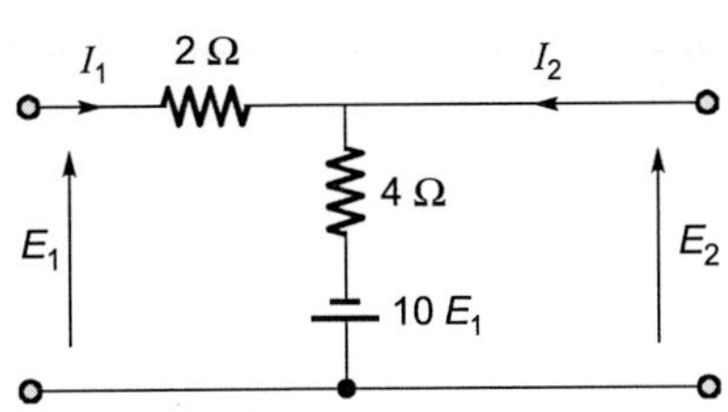

(a) $Z_{11} = \frac{-6}{11}\Omega$; $Z_{21} = \frac{16}{11}\Omega$

(b) $Z_{11} = \frac{6}{11}\Omega$; $Z_{21} = \frac{4}{11}\Omega$

(c) $Z_{11} = \frac{6}{11}\Omega$; $Z_{21} = \frac{-16}{11}\Omega$

(d) $Z_{11} = \frac{4}{11}\Omega$; $Z_{21} = \frac{4}{11}\Omega$

[2001 : 2 M]

5.5 The impedance parameters Z_{11} and Z_{12} of the two-port network in the figure are

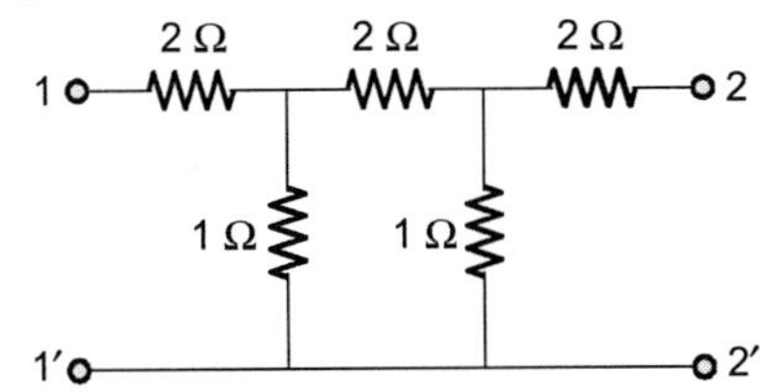

(a) Z_{11} = 2.75 Ω and Z_{12} = 0.25 Ω
(b) Z_{11} = 3 Ω and Z_{12} = 0.5 Ω
(c) Z_{11} = 3 Ω and Z_{12} = 0.25 Ω
(d) Z_{11} = 2.25 Ω and Z_{12} = 0.5 Ω

[2003 : 2 M]

5.6 For the lattice circuit shown in the figure, $Z_a = j2\ \Omega$ and Z_b = 2 Ω. The values of the open circuit impedance parameters $Z = \begin{bmatrix} Z_{11} & Z_{12} \\ Z_{21} & Z_{22} \end{bmatrix}$ are

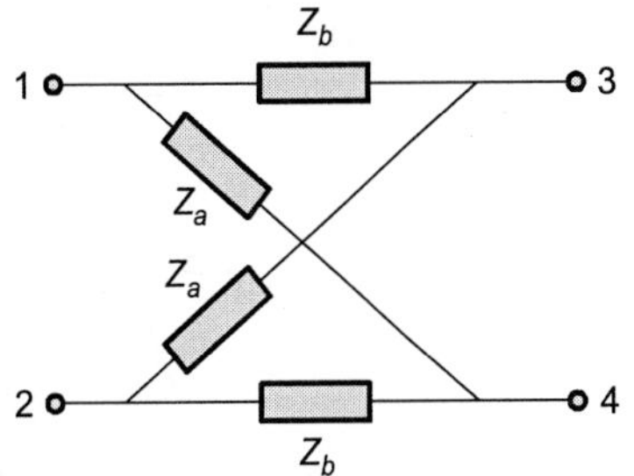

(a) $\begin{bmatrix} 1-j & 1+j \\ 1+j & 1+j \end{bmatrix}$ (b) $\begin{bmatrix} 1-j & 1+j \\ -1+j & 1-j \end{bmatrix}$

(c) $\begin{bmatrix} 1+j & 1+j \\ 1-j & 1-j \end{bmatrix}$ (d) $\begin{bmatrix} 1+j & -1+j \\ -1+j & 1+j \end{bmatrix}$

[2004 : 2 M]

5.7 The ABCD parameters of an ideal $n:1$ transformer shown in the figure are $\begin{bmatrix} n & 0 \\ 0 & X \end{bmatrix}$. The value of X will be

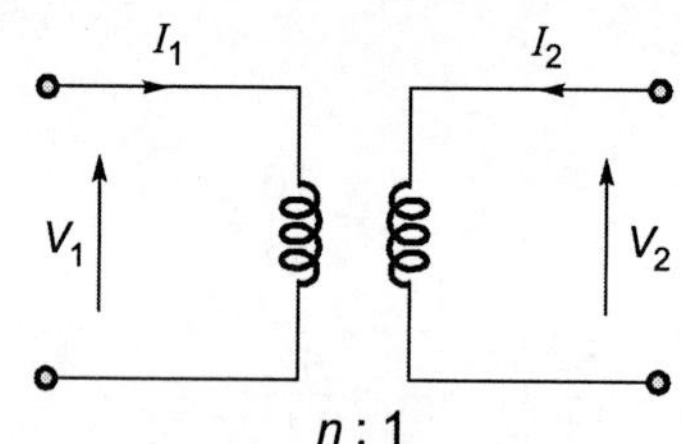

(a) n (b) $\frac{1}{n}$

(c) n^2 (d) $\frac{1}{n^2}$ [2005 : 1 M]

5.8 The h-parameters of the circuit shown in the figure are

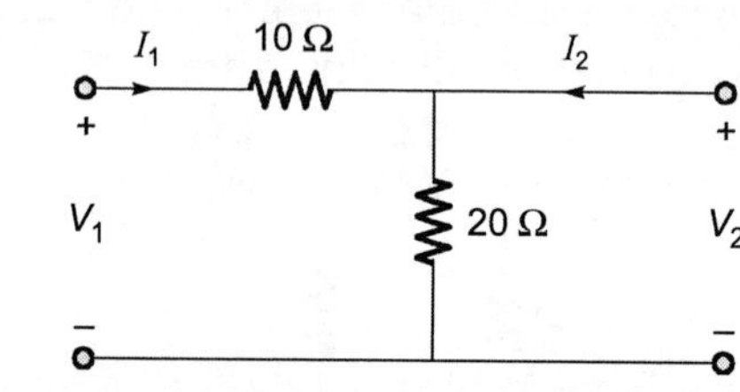

(a) $\begin{bmatrix} 0.1 & 0.1 \\ -0.1 & 0.3 \end{bmatrix}$ (b) $\begin{bmatrix} 10 & -1 \\ 1 & 0.05 \end{bmatrix}$

(c) $\begin{bmatrix} 30 & 20 \\ 20 & 20 \end{bmatrix}$ (d) $\begin{bmatrix} 10 & 1 \\ -1 & 0.05 \end{bmatrix}$

[2005 : 2 M]

5.9 A two-port network is represented by $ABCD$ parameters given by

$$\begin{bmatrix} V_1 \\ I_1 \end{bmatrix} = \begin{bmatrix} A & B \\ C & D \end{bmatrix} \begin{bmatrix} V_2 \\ -I_2 \end{bmatrix}$$

If port-2 is terminated by R_L, the input impedance seen at port-1 is given by

(a) $\frac{A + BR_L}{C + DR_L}$ (b) $\frac{AR_L + C}{BR_L + D}$

(c) $\frac{DR_L + A}{BR_L + C}$ (d) $\frac{B + AR_L}{D + CR_L}$

[2006 : 1 M]

5.10 In the two port network shown in the figure below, Z_{12} and Z_{21} are, respectively

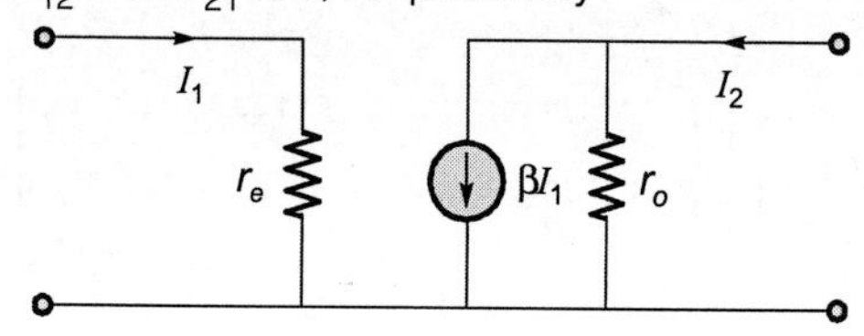

(a) r_e and βr_o (b) 0 and $-\beta r_o$

(c) 0 and βr_o (d) r_e and $-\beta r_o$

[2006 : 1 M]

Linked Answer Questions 5.11 to 5.12:

A two-port network shown below is excited by external dc sources. The voltages and currents are measured with voltmeters V_1, V_2 and ammeters A_1, A_2 (all assumed to be ideal) as indicated. Under following switch conditions, the readings obtained are:

(i) S_1 – Open, S_2 – closed $A_1 = 0$ A, $V_1 = 4.5$ V, $V_2 = 1.5$ V, $A_2 = 1$ A

(ii) S_1 – Closed, S_2-Open $A_1 = 4$ A, $V_1 = 6$ V, $V_2 = 6$V , $A_2 = 0$ A

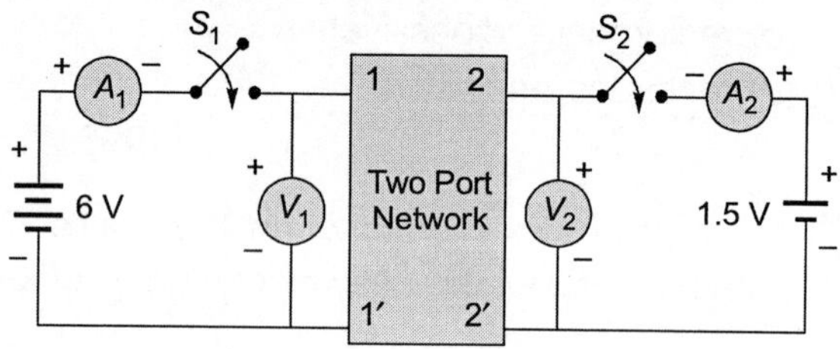

5.11 The Z-parameter matrix for this network is

(a) $\begin{bmatrix} 1.5 & 1.5 \\ 4.5 & 1.5 \end{bmatrix}$ (b) $\begin{bmatrix} 1.5 & 4.5 \\ 1.5 & 4.5 \end{bmatrix}$

(c) $\begin{bmatrix} 1.5 & 4.5 \\ 1.5 & 1.5 \end{bmatrix}$ (d) $\begin{bmatrix} 4.5 & 1.5 \\ 1.5 & 4.5 \end{bmatrix}$

[2008 : 2 M]

5.12 The h-parameter matrix for this network is

(a) $\begin{bmatrix} -3 & 3 \\ -1 & 0.67 \end{bmatrix}$ (b) $\begin{bmatrix} -3 & -1 \\ 3 & 0.67 \end{bmatrix}$

(c) $\begin{bmatrix} 3 & 3 \\ 1 & 0.67 \end{bmatrix}$ (d) $\begin{bmatrix} 3 & 1 \\ -3 & -0.67 \end{bmatrix}$

[2008 : 2 M]

5.13 For the two-port network shown below, the short-circuit admittance parameter matrix is

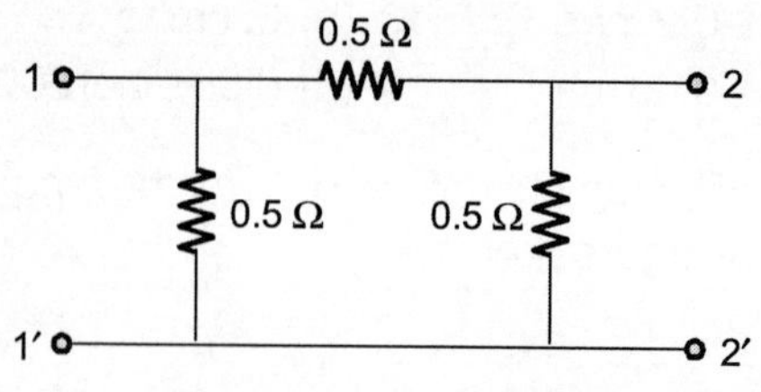

(a) $\begin{bmatrix} 4 & -2 \\ -2 & 4 \end{bmatrix} S$ (b) $\begin{bmatrix} 1 & -0.5 \\ -0.5 & 1 \end{bmatrix} S$

(c) $\begin{bmatrix} 1 & 0.5 \\ 0.5 & 1 \end{bmatrix} S$ (d) $\begin{bmatrix} 4 & 2 \\ 2 & 4 \end{bmatrix} S$

[2010 : 1 M]

5.14 In the circuit shown below, the network N is described by the following Y matrix:

$Y = \begin{bmatrix} 0.1S & -0.01S \\ 0.01S & 0.1S \end{bmatrix}$. The voltage gain $\frac{V_2}{V_1}$ is

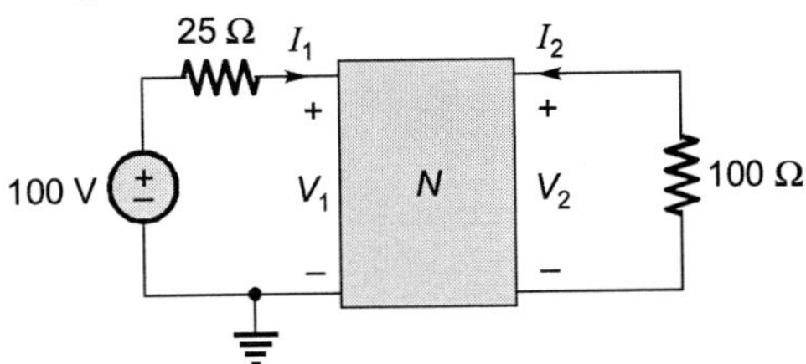

(a) 1/90 (b) –1/90
(c) –1/99 (d) –1/11

[2011 : 2 M]

Common Data for 5.15 and 5.16:

With 10 V dc connected at port A in the linear nonreciprocal two-port network shown below, the following were observed.

(i) 1 Ω connected at port B draws a current of 3 A.
(ii) 2.5 Ω connected at port B draws a current of 2 A.

5.15 For the same network, with 6 V dc connected at port A, 1 Ω connected at port B draws 7/3 A. If 8 V dc is connected to port A, the open circuit voltage at port B is

(a) 6 V (b) 7 V
(c) 8 V (d) 9 V

[2012 : 2 M]

5.16 With 10 V dc connected at port A, the current drawn by 7 Ω connected at port B is

(a) 3/7 A (b) 5/7 A
(c) 1 A (d) 9/7 A

[2012 : 2 M]

5.17 In the h-parameter model of the 2-port network given in the figure shown, the value of h_{22} (in S) is ______.

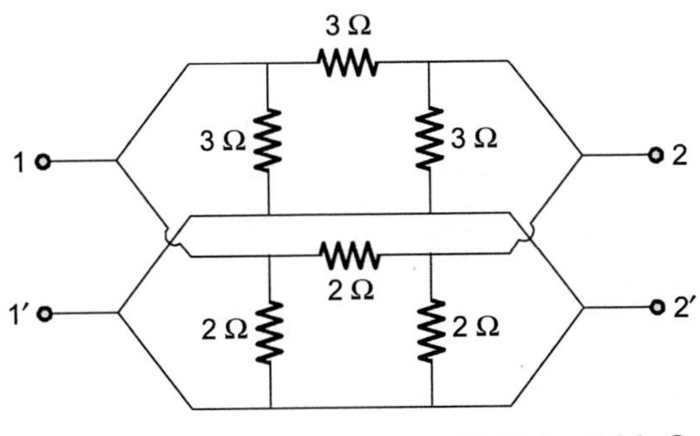

[2014 : 2 M, Set-2]

5.18 Consider the building block called 'Network N' shown in the figure.

Let C = 100 μF and R = 10 kΩ,

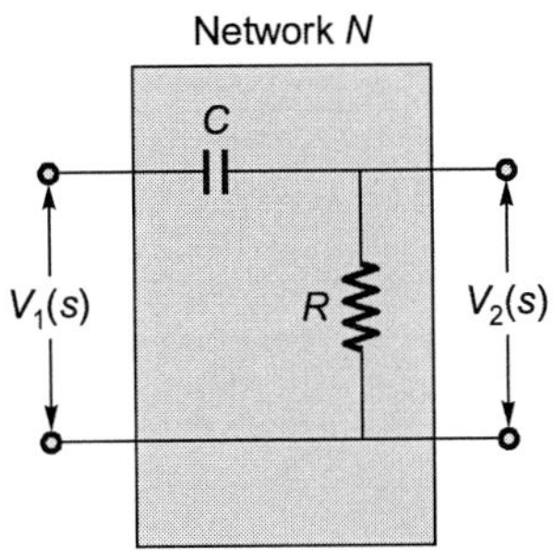

Two such blocks are connected in cascade, as shown in the figure.

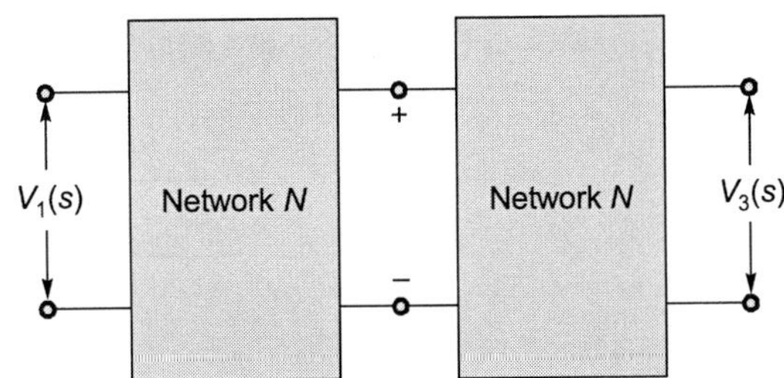

The transfer function $\frac{V_3(s)}{V_1(s)}$ of the cascaded network is

(a) $\frac{s}{1+s}$ (b) $\frac{s^2}{1+3s+s^2}$

(c) $\left(\frac{s}{1+s}\right)^2$ (d) $\frac{s}{2+s}$

[2014 : 2 M, Set-3]

5.19 For the two-port network shown in the figure, the impedance (Z) matrix (in Ω) is

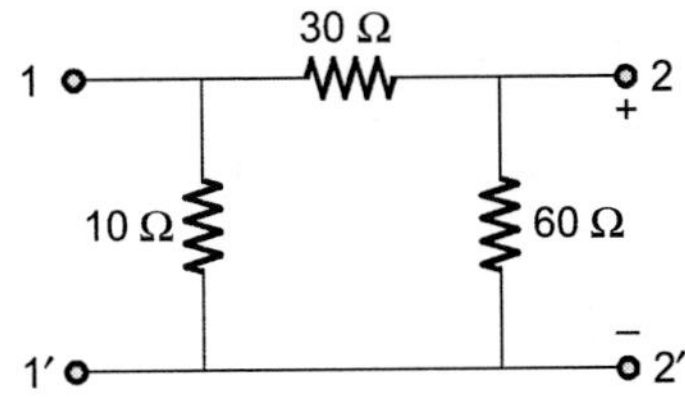

(a) $\begin{bmatrix} 6 & 24 \\ 42 & 9 \end{bmatrix}$ (b) $\begin{bmatrix} 9 & 8 \\ 8 & 24 \end{bmatrix}$

(c) $\begin{bmatrix} 9 & 6 \\ 6 & 24 \end{bmatrix}$ (d) $\begin{bmatrix} 42 & 6 \\ 6 & 60 \end{bmatrix}$

[2014 : 2 M, Set-4]

5.20 The 2-port admittance matrix of the circuit shown is given by

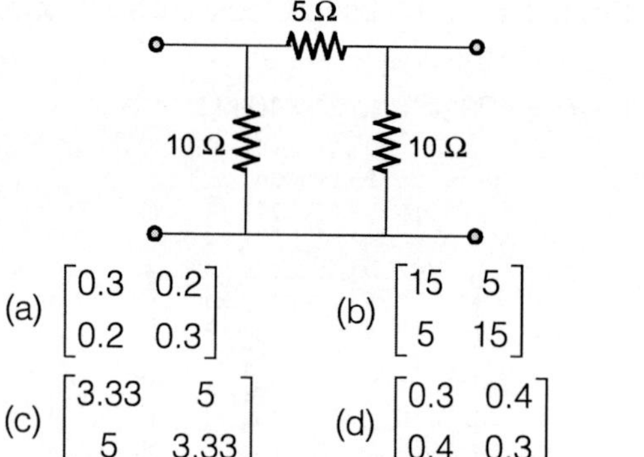

(a) $\begin{bmatrix} 0.3 & 0.2 \\ 0.2 & 0.3 \end{bmatrix}$ (b) $\begin{bmatrix} 15 & 5 \\ 5 & 15 \end{bmatrix}$

(c) $\begin{bmatrix} 3.33 & 5 \\ 5 & 3.33 \end{bmatrix}$ (d) $\begin{bmatrix} 0.3 & 0.4 \\ 0.4 & 0.3 \end{bmatrix}$

[2015 : 1 M, Set-2]

5.21 The *ABCD* parameters of the following 2-port network are

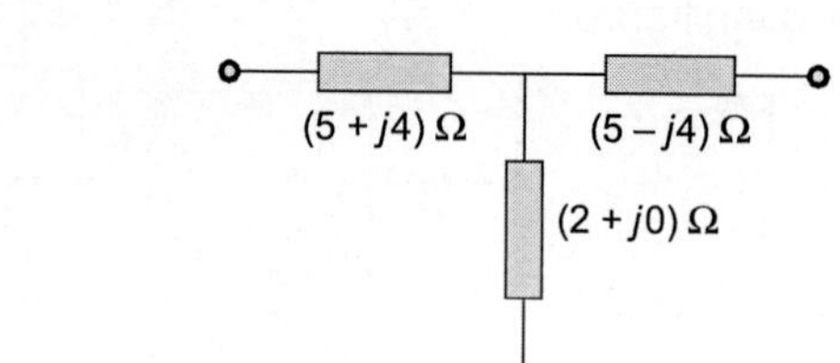

(a) $\begin{bmatrix} 3.5+j2 & 20.5 \\ 20.5 & 3.5-j2 \end{bmatrix}$

(b) $\begin{bmatrix} 3.5+j2 & 0.5 \\ 0.5 & 3.5-j2 \end{bmatrix}$

(c) $\begin{bmatrix} 10 & 2+j0 \\ 2+j0 & 10 \end{bmatrix}$

(d) $\begin{bmatrix} 7+j4 & 0.5 \\ 30.5 & 7-j4 \end{bmatrix}$

[2015 : 2 M, Set-3]

5.22 Consider a two-port network with the transmission matrix : $T = \begin{bmatrix} A & B \\ C & D \end{bmatrix}$.

If the network is reciprocal, then

(a) $T^{-1} = T$

(b) $T^2 = T$

(c) Determinant $(T) = 0$

(d) Determinant $(T) = 1$

[2016 : 1 M, Set-1]

5.23 The *Z*-parameter matrix for the two-port network shown is

$$\begin{bmatrix} 2j\omega & j\omega \\ j\omega & 3+2j\omega \end{bmatrix},$$

Where the entries are in Ω.

Suppose $Z_b(j\omega) = R_b + j\omega$

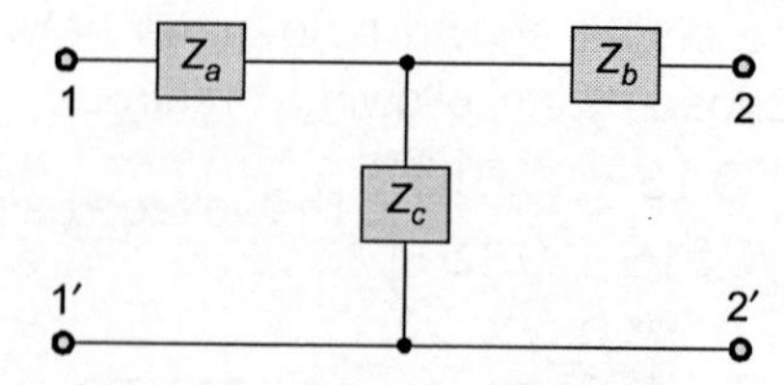

Then the value of R_b (in Ω) equals _____

[2016 : 1 M, Set-2]

5.24 The *Z*-parameter matrix $\begin{bmatrix} Z_{11} & Z_{12} \\ Z_{21} & Z_{22} \end{bmatrix}$ for the two-port network shown is

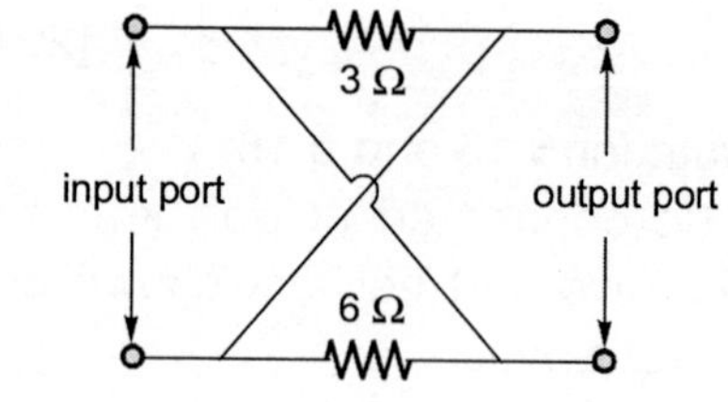

(a) $\begin{bmatrix} 2 & -2 \\ -2 & 2 \end{bmatrix}$ (b) $\begin{bmatrix} 2 & 2 \\ 2 & 2 \end{bmatrix}$

(c) $\begin{bmatrix} 9 & -3 \\ 6 & 9 \end{bmatrix}$ (d) $\begin{bmatrix} 9 & 3 \\ 6 & 9 \end{bmatrix}$

[2016 : 2 M, Set-3]

5.25 The *ABCD* matrix for a two-port network is defined by

$$\begin{bmatrix} V_1 \\ I_1 \end{bmatrix} = \begin{bmatrix} A & B \\ C & D \end{bmatrix} \begin{bmatrix} V_2 \\ -I_2 \end{bmatrix}$$

I_1 2 Ω 2 Ω I_2

V_1 5 Ω V_2

The parameter *B* for the given two-port network (in ohms, correct to two decimal places) is _____.

[2018 : 1 M]

5.26 In the given circuit, the two-port network has the impedance matrix $[Z] = \begin{bmatrix} 40 & 60 \\ 60 & 120 \end{bmatrix}$. The value of Z_L for which maximum power is transferred to the load is ________ Ω.

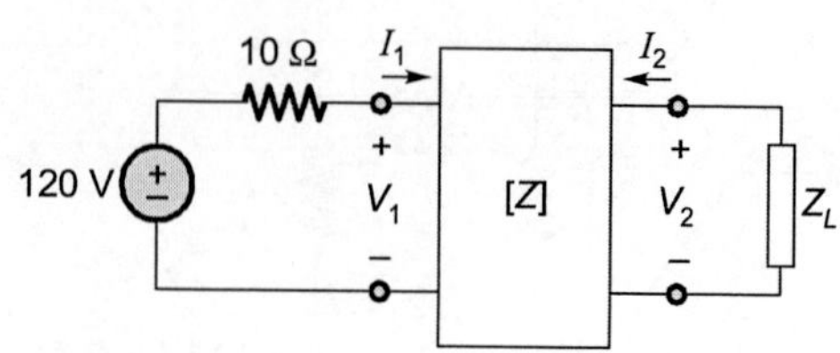

[2020 : 1 M]

5.27 For a 2-port network consisting of an ideal lossless transformer, the parameter S_{21} (rounded off to two decimal places) for a reference impedance of 10 Ω, is ______.

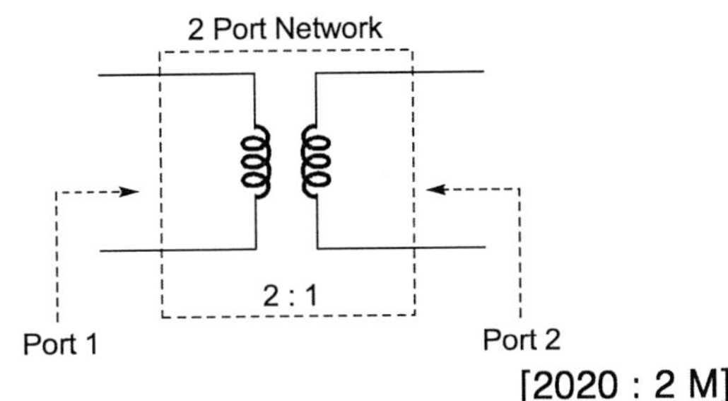

[2020 : 2 M]

5.28 Consider the two-port network shown in the figure.

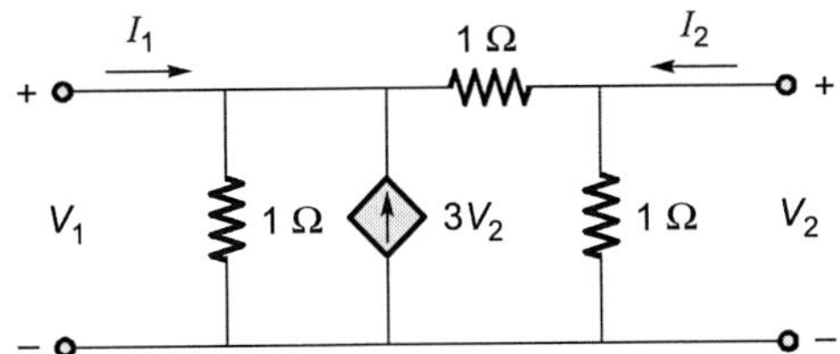

The admittances parameters, in Siemens, are

(a) $y_{11} = 2$, $y_{12} = -4$, $y_{21} = -4$, $y_{22} = 2$
(b) $y_{11} = 2$, $y_{12} = -4$, $y_{21} = -4$, $y_{22} = 3$
(c) $y_{11} = 2$, $y_{12} = -4$, $y_{21} = -1$, $y_{22} = 2$
(d) $y_{11} = 1$, $y_{12} = -2$, $y_{21} = -1$, $y_{22} = 3$

[2021 : 2 M]

5.29 A linear 2-port network is shown in Fig. (a). An ideal DC voltage source of 10 V is connected across Port 1. A variable resistance R is connected across Port 2. As R is varied, the measured voltage and current at Port 2 is shown in Fig. (b) as a V_2 versus $-I_2$ plot. Note that for $V_2 = 5$ V, $I_2 = 0$ mA, and for $V_2 = 4$ V, $I_2 = -4$ mA.

When the variable resistance R at Port 2 is replaced by the load shown in Fig. (c), the current I_2 is ______ mA (rounded off to one decimal place).

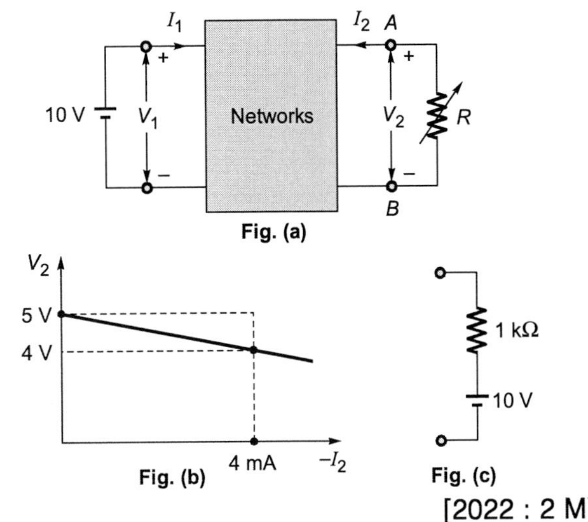

Fig. (a)

Fig. (b)

Fig. (c)

[2022 : 2 M]

5.30 For the two port network shown below, the [Y]-parameters is given as

$$[Y] = \frac{1}{100}\begin{bmatrix} 2 & -1 \\ -1 & 4/3 \end{bmatrix} S$$

The value of load impedance Z_L, in Ω, for maximum power transfer will be ______.
(rounded off to the nearest integer).

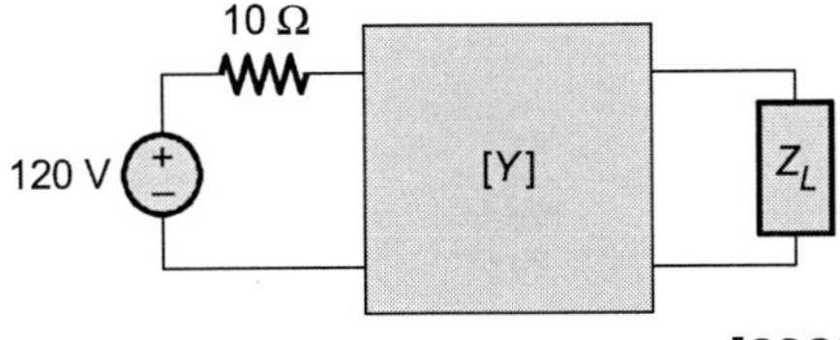

[2023 : 1 M]

5.31 For the two port network shown below, the value of the Y_{21} parameter (in Siemens) is ______.

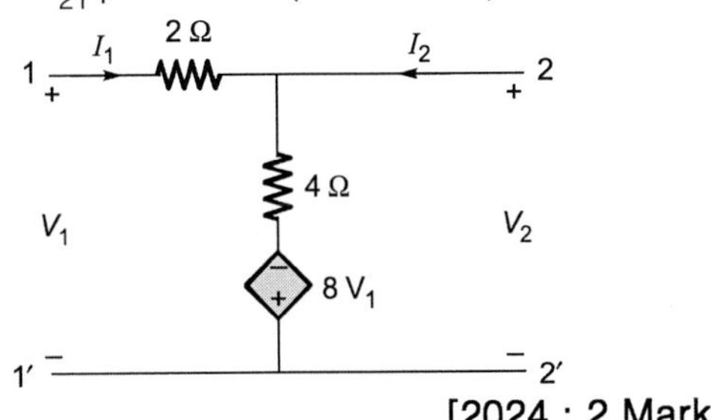

[2024 : 2 Marks]

■■■

Answers Two Port Networks

5.1 (b)	5.2 (a)	5.3 (c)	5.4 (c)	5.5 (a)	5.6 (d)	5.7 (b)
5.8 (d)	5.9 (d)	5.10 (b)	5.11 (c)	5.12 (a)	5.13 (a)	5.14 (d)
5.15 (c)	5.16 (c)	5.17 (1.25)	5.18 (b)	5.19 (c)	5.20 (a)	5.21 (b)
5.22 (d)	5.23 (3)	5.24 (a)	5.25 (4.80)	5.26 (48)	5.27 (0.8)	5.28 (c)
5.29 (4)	5.30 (80)	5.31 (1.5)				

Explanations Two Port Networks

5.1 (b)

$Y_{12} \neq Y_{21}$

So, the given two-port network is non-reciprocal and active.

5.2 (a)

$$I_2 = h_{21}I_1 + h_{22}V_2 \quad ; \quad h_{21} = \left.\frac{I_2}{I_1}\right|_{V_2=0}$$

when $\quad V_2 = 0 = I_2R(I_1 + I_2)_R$

$$I_2 = -\frac{I_1}{2}$$

$$\Rightarrow \quad \frac{I_2}{I_1} = -\frac{1}{2}$$

5.3 (c)

$$\begin{bmatrix} y_1 + y_3 & -y_3 \\ -y_3 & y_2 + y_3 \end{bmatrix} = \begin{bmatrix} y_{11} & y_{12} \\ y_{21} & y_{22} \end{bmatrix}$$

$$y_{12} = -y_3$$

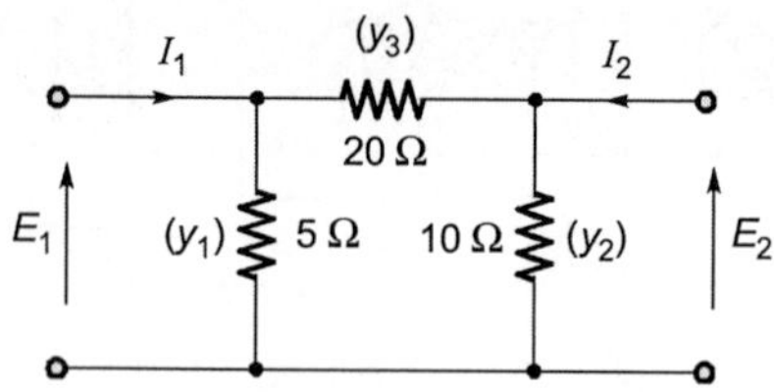

$$y_{12} = -\frac{1}{20} = -0.05 \text{ mho}$$

5.4 (c)

$$V_1 = Z_{11}I_1 + Z_{12}I_2$$
$$V_2 = Z_{21}I_1 + Z_{22}I_2$$
$$Z_{11} = \left.\frac{V_1}{I_1}\right|_{I_2=0}$$

⇒ Applying KVL in LHS loop

$E_1 - 2I_1 - 4I_1 + 10E_1 = 0$

$$11E_1 = 6I_1$$

$$\Rightarrow \quad \frac{E_1}{I_1} = \frac{6}{11}\Omega$$

$$Z_{21} = \left.\frac{V_2}{I_1}\right|_{I_2=0}$$

KVL in RHS loop,

$E_2 - 4I_1 + 10E_1 = 0$

$$\Rightarrow \quad E_2 - 4I_1 + 10 \times \frac{6}{11}I_1 = 0 \qquad \left(E_1 = \frac{b}{11}I_1\right)$$

$$\Rightarrow \quad 11E_2 - 44I_1 + 60I_1 = 0$$

$$\frac{E_2}{I_1} = -\frac{16}{11}\Omega$$

5.5 (a)

Using $\Delta - Y$ conversion,

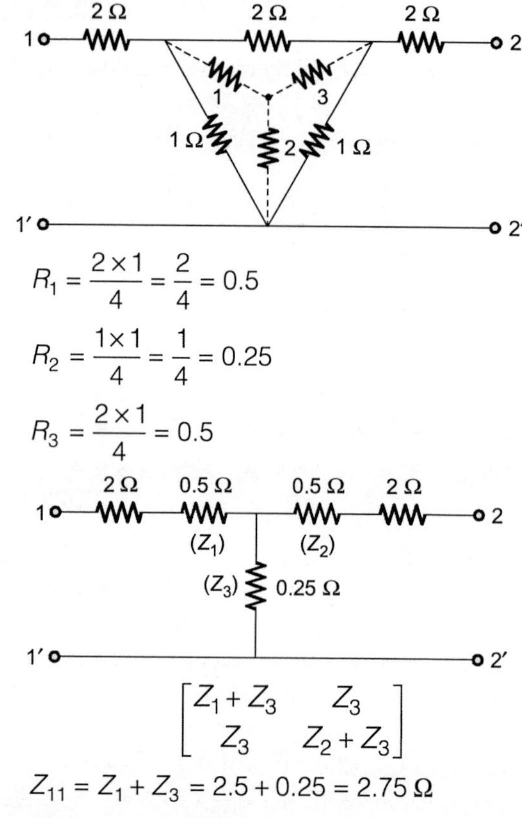

$$R_1 = \frac{2 \times 1}{4} = \frac{2}{4} = 0.5$$

$$R_2 = \frac{1 \times 1}{4} = \frac{1}{4} = 0.25$$

$$R_3 = \frac{2 \times 1}{4} = 0.5$$

$$\begin{bmatrix} Z_1 + Z_3 & Z_3 \\ Z_3 & Z_2 + Z_3 \end{bmatrix}$$

$Z_{11} = Z_1 + Z_3 = 2.5 + 0.25 = 2.75\ \Omega$

$Z_{12} = Z_3 = 0.25\ \Omega$

5.6 (d)

For Lattice network, Z-parameter is given as

$$\begin{bmatrix} \dfrac{Z_a + Z_b}{2} & \dfrac{Z_a - Z_b}{2} \\ \dfrac{Z_a - Z_b}{2} & \dfrac{Z_a + Z_b}{2} \end{bmatrix}$$

$Z_a = 2j$, $Z_b = 2\ \Omega$

$$\begin{bmatrix} 1 + j & -1 + j \\ -1 + j & 1 + j \end{bmatrix}$$

5.7 (b)

$$\begin{bmatrix} V_1 \\ I_1 \end{bmatrix} = \begin{bmatrix} A & B \\ C & D \end{bmatrix} \begin{bmatrix} V_2 \\ -I_2 \end{bmatrix}$$

$$-\frac{I_2}{I_1} = \frac{V_1}{V_2} = \frac{n}{1}$$

$$V_1 = AV_2 - BI_2\,;\quad I_1 = CV_2 - DI_2$$

$$A = \left.\frac{V_1}{V_2}\right|_{I_2=0} = n$$

$$D = -\left.\frac{I_1}{I_2}\right|_{V_2=0} = \frac{V_2}{V_1} = \frac{1}{n}$$

5.8 (d)

$$V_1 = h_{11}I_1 + h_{12}V_2$$

$$I_2 = h_{21}I_1 + h_{22}V_2$$

$$h_{11} = \left.\frac{V_1}{I_1}\right|_{V_2=0},\quad h_{12} = \left.\frac{V_1}{V_2}\right|_{I_1=0}$$

$$h_{21} = \left.\frac{I_2}{I_1}\right|_{V_2=0},\quad h_{22} = \left.\frac{I_2}{V_2}\right|_{I_1=0}$$

When $V_2 = 0$

$$I_1 = -I_2$$

$$\frac{I_2}{I_1} = -1 = h_{21}$$

$$V_1 = 10I_1$$

$$\Rightarrow \quad \frac{V_1}{I_1} = 10$$

when $\quad I_1 = 0$

$$V_1 = V_2$$

$$\Rightarrow \quad \frac{V_1}{V_2} = h_{12} = 1$$

(As no drop in 10 Ω resistance)

$$V_2 = 20I_2$$

$$\Rightarrow \quad \frac{I_2}{V_2} = h_{22} = \frac{1}{20} = 0.05$$

$$\begin{bmatrix} 10 & 1 \\ -1 & 0.05 \end{bmatrix}$$

5.9 (d)

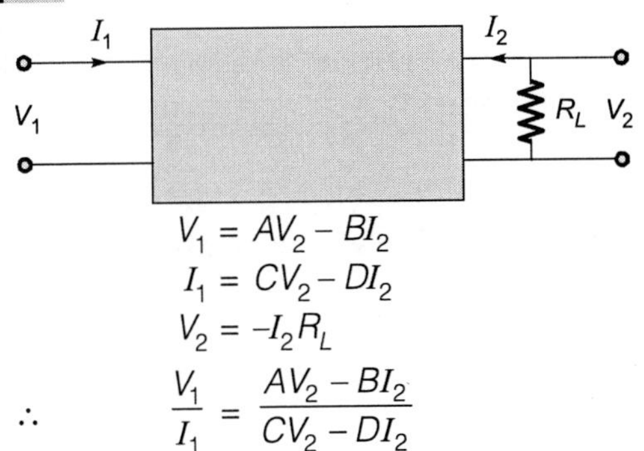

$$V_1 = AV_2 - BI_2$$

$$I_1 = CV_2 - DI_2$$

$$V_2 = -I_2R_L$$

$$\therefore \quad \frac{V_1}{I_1} = \frac{AV_2 - BI_2}{CV_2 - DI_2}$$

$$= \frac{-A\cdot I_2R_L - BI_2}{-C\cdot I_2R_L - DI_2}$$

$$\text{Input impedance} = \frac{AR_L + B}{CR_L + D}$$

5.10 (b)

$$V_1 = Z_{11}I_1 + Z_{12}I_2$$

$$V_2 = Z_{21}I_1 + Z_{22}I_2$$

$$Z_{12} = \left.\frac{V_1}{I_2}\right|_{I_1=0}\;;\quad Z_{12} = \left.\frac{V_2}{I_1}\right|_{I_2=0}$$

When, $\quad I_1 = 0$

$\Rightarrow \quad V_1 = 0$

$$\Rightarrow \quad \frac{V_1}{I_2} = 0 = Z_{12}$$

When $\quad I_2 = 0,\ V_2 = -\beta I_1 r_o$

$$\Rightarrow \quad \frac{V_2}{I_1} = -\beta r_o = Z_{21}$$

5.11 (c)

When, $\quad I = 0$

then, $\quad V_1 = 4.5\text{ V}$

$V_2 = 1.5\text{ V}$

$I_2 = 1\text{ A}$

$$V_1 = Z_{11}I_1 + Z_{12}I_2$$

$$V_2 = Z_{21}I_1 + Z_{22}I_2$$

$$Z_{12} = \left.\frac{V_1}{I_2}\right|_{I_1=0} = \frac{4.5}{1} = 4.5$$

$$Z_{22} = \left.\frac{V_2}{I_2}\right|_{I_1=0} = \frac{1.5}{1} = 1.5$$

When, $\quad I_2 = 0$

then, $\quad I_1 = 4\text{ A}$

$V_1 = 6\text{ V}$

$V_2 = 6\text{ V}$

$$Z_{11} = \left.\frac{V_1}{I_1}\right|_{I_2=0} = \frac{6}{4} = 1.5$$

$$Z_{21} = \left.\frac{V_2}{I_1}\right|_{I_2=0} = \frac{6}{4} = 1.5$$

So, Z-parameter matrix is $\begin{bmatrix} 1.5 & 4.5 \\ 1.5 & 1.5 \end{bmatrix}$

5.12 (a)

$$V_1 = h_{11}I_1 + h_{12}V_2$$

$$I_2 = h_{21}I_1 + h_{22}V_2$$

$$h_{12} = \left.\frac{V_1}{V_2}\right|_{I_1=0} = \frac{4.5}{1.5} = 3$$

$$h_{22} = \left.\frac{I_2}{V_2}\right|_{I_1=0} = \frac{1}{1.5} = 0.67$$

$$h_{11} = \left.\frac{V_1}{I_1}\right|_{V_2=0}$$

When , $V_2 = 0, Z_{21}I_1 + Z_{22}I_2 = 0$

$\Rightarrow$ $$I_2 = -\frac{Z_{21}I_1}{Z_{22}}$$

$\Rightarrow$ $$V_1 = Z_{11}I_1 + Z_{12}\left(-\frac{Z_{21}I_1}{Z_{22}}\right)$$

$\Rightarrow$ $$\frac{V_1}{I_1} = \left(Z_{11} - \frac{Z_{12}Z_{21}}{Z_{22}}\right) = h_{11}$$

$\Rightarrow$ $$h_{11} = 1.5 - \frac{4.5 \times 1.5}{1.5} = -3$$

$\Rightarrow$ $$h_{21} = \left.\frac{I_2}{I_1}\right|_{V_2=0} = \frac{-Z_{21}}{Z_{22}}$$

$$= \frac{-1.5}{1.5} = -1$$

So, h-parameter matrix is $\begin{bmatrix} -3 & 3 \\ -1 & 0.67 \end{bmatrix}$

5.13 (a)

Short-circuit admittance parameters for a two-port π-network are,

$$Y_{11} = Y_a + Y_b$$
$$Y_{12} = Y_{21} = -Y_b$$
$$Y_{22} = Y_b + Y_c$$

For the given network,

$$Y_a = Y_b = Y_c = \frac{1}{0.5} = 2\ \mho$$

So, $Y_{11} = 2 + 2 = 4\ \mho$

$$Y_{12} = Y_{21} = -2\ \mho$$
$$Y_{22} = 2 + 2 = 4\ \mho$$

5.14 (d)

$$I_2 = Y_{21}V_1 + Y_{22}V_2$$

$\therefore$ $I_2 = 0.01\ V_1 + 0.1\ V_2$ ...(i)

From given figure,

$$V_2 = -I_2R_L = -100\ I_2$$

$\therefore$ $$I_2 = -\frac{V_2}{100}$$

Putting value of I_2 in equation (i),

$$-\frac{V_2}{100} = 0.01\ V_1 + 0.1\ V_2$$

$\Rightarrow$ $-0.01\ V_2 - 0.1\ V_2 = 0.01\ V_1$

$\Rightarrow$ $-0.11\ V_2 = 0.01\ V_1$

$\therefore$ $$\frac{V_2}{V_1} = -\frac{0.01}{0.11}$$

or, $$\frac{V_2}{V_1} = -\frac{1}{11}$$

5.15 (c)

Case (i):

$$V_{DC} = 10\text{ V}, R_L = 1\ \Omega, I = 3\text{ A}$$
$$V_{Th} = I(R_{Th} + R_L) = 3(R_{Th} + 1)$$

$\Rightarrow$ $V_{Th} = 3\ R_{Th} + 3$ ...(i)

Case (ii):

$$R_L = 2.5\ \Omega, I = 2\text{ A}, V_{DC} = 10\text{ V}$$
$$V_{Th} = 2(R_{Th} + 2.5)$$

$V_{Th} = 2R_{Th} + 5$ ...(ii)

From equations (i) and (ii)

$V_{Th} = 9\text{ V}$ and $R_{TH} = 2\ \Omega$ ...(iii)

Now, V_{Th} depends on independent voltage source and varies with applied voltage. R_{Th} does not depend on voltage source and is same for any applied voltage source, since voltage source is short circuited while calculating R_{Th}.

For $V_{DC} = 6$ V, $R_L = 1\ \Omega$, $I = 7/3$

$$V_{Th} = I(R_{Th} + R_L)$$

$V_{Th} = \frac{7}{3}(2+1) = 7\text{ V}$ ...(iv)

$\because$ The network is linear and non reciprocal, it may contain dependent voltage source.

$\therefore$ $V_{Th} = aV + b$ ...(v)

$\Rightarrow$ $9 = a10 + b$ from equation (iii)

and $7 = a6 + b$ from equation (iv)

Solving, we get $a = 1/2$ and $b = 4$

$\therefore$ $$V_{Th} = \frac{V}{2} + 4 = \frac{8}{2} + 4 = 8\text{ V}$$

For 8 V source.

5.16 (c)

From the above solution

When, $V_{DC} = 10\text{ V}$

$V_{Th} = 9\text{ V}$ and $R_{Th} = 2\ \Omega$

When, $R_L = 7\ \Omega, I = ?$

$V_{Th} = I(R_{Th} + R_L)$

$$I = \frac{V_{Th}}{R_{Th} + R_L} = \frac{9}{2+7} = 1\text{ A}$$

5.17 (1.25)

When two, two port networks are connected in parallel then individual Y-parameters are added. Therefore from the given network,

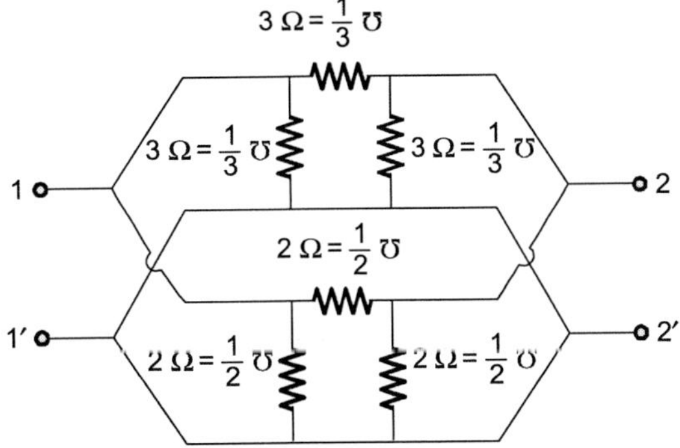

For network (1) Y-parameter is,

$$Y_1 = \begin{bmatrix} \frac{1}{3}+\frac{1}{3} & -\frac{1}{3} \\ -\frac{1}{3} & \frac{1}{3}+\frac{1}{3} \end{bmatrix}$$

Similarly for network (2),

$$Y_2 = \begin{bmatrix} 1 & -\frac{1}{2} \\ -\frac{1}{2} & 1 \end{bmatrix}$$

Thus, $$Y = Y_1 + Y_2 = \begin{bmatrix} \frac{5}{3} & -\frac{5}{6} \\ -\frac{5}{6} & \frac{5}{3} \end{bmatrix}$$

$\therefore$ $$I_1 = Y_{11} V_1 + Y_{12} V_2$$

$$= \frac{5}{3} Y_1 - \frac{5}{6} V_2 \quad \text{...(i)}$$

$$I_2 = Y_{21} V_1 + Y_{22} V_2$$

$$= -\frac{5}{6} V_1 + \frac{5}{3} V_2 \quad \text{...(ii)}$$

Also $$h_{22} = \left.\frac{I_2}{V_2}\right|_{I_1=0}$$

$\therefore$ From equation (i), we get,

$$V_1 = \frac{1}{2} V_2 \quad \text{...(iii)}$$

and from equation (ii) and (iii), we get,

$$h_{22} = \frac{I_2}{V_2} = \frac{15}{12} = 1.25$$

5.18 (b)

The cascaded network is,

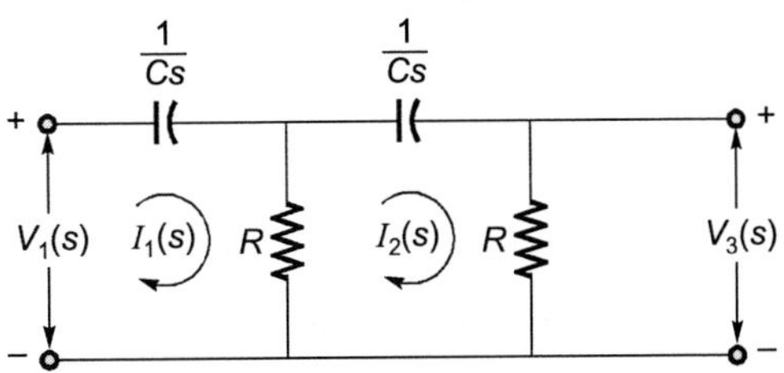

Applying mesh analysis to determine the current $I_2(s)$.

We get,

$$\left(R + \frac{1}{Cs}\right) I_1(s) = RI_2(s) = V_1(s) \quad \text{...(i)}$$

$$\left(R + \frac{1}{Cs}\right) I_2(s) - RI_2(s) = 0 \quad \text{...(ii)}$$

Also,

$$V_3(s) = I_2(s) \times R \quad \text{...(iii)}$$

From equation (i) and (ii), we get

$$= \left(R + \frac{1}{Cs}\right) \times \left(1 + \frac{1}{RCs}\right) I_2(s) - RI_2(s) = V_1(s)$$

$$= \left(\frac{RCs+1}{Cs}\right)\left(\frac{RCs+1}{RCs}\right) I_2(s) - RI_2(s) = V_1(s)$$

or $$I_2(s) = \frac{RC^2 s^2 V_1(s)}{(1+RCs)^2 - R^2 C^2 s^2} \quad \text{...(iv)}$$

Using equation (iii) and (iv), we get

$$\frac{V_3(s)}{V_1(s)} = \frac{s^2 R^2 C^2}{1 + 3RCs + s^2 R^2 C^2} \quad \text{...(v)}$$

$\because$ $R = 10\text{ k}\Omega$, $C = 100\ \mu\text{F}$ and $RC = 1$

$$\frac{V_3(s)}{V_1(s)} = \frac{s^2}{1 + 3s + s^2}$$

5.19 (c)

Converting Π-network to Y-network, we get,

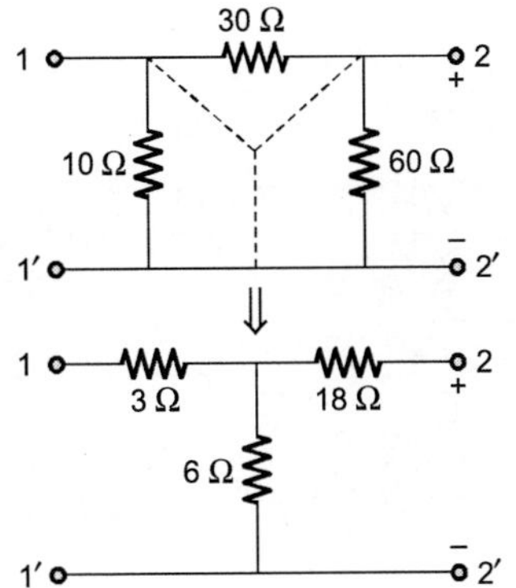

$\therefore$ Z-parameter

$$Z_{11} = 3\,\Omega + 6\,\Omega = 9\,\Omega$$
$$Z_{12} = Z_{21} = 6\,\Omega$$
$$Z_{22} = 18\,\Omega + 6\,\Omega = 24\,\Omega$$
$$[Z] = \begin{bmatrix} 9 & 6 \\ 6 & 24 \end{bmatrix}$$

5.20 (a)

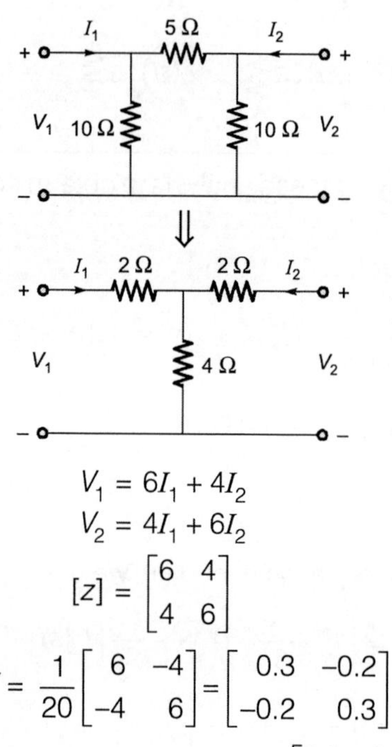

$$V_1 = 6I_1 + 4I_2$$
$$V_2 = 4I_1 + 6I_2$$
$$[z] = \begin{bmatrix} 6 & 4 \\ 4 & 6 \end{bmatrix}$$
$$Y = \frac{1}{20}\begin{bmatrix} 6 & -4 \\ -4 & 6 \end{bmatrix} = \begin{bmatrix} 0.3 & -0.2 \\ -0.2 & 0.3 \end{bmatrix}$$

Ignoring negative sign: $[Y] = \begin{bmatrix} 0.3 & 0.2 \\ 0.2 & 0.3 \end{bmatrix}$

5.21 (b)

$$\begin{bmatrix} V_1 \\ I_1 \end{bmatrix} = \begin{bmatrix} A & B \\ C & D \end{bmatrix}\begin{bmatrix} V_2 \\ I_2 \end{bmatrix}$$

When $I_2 = 0$,

$$V_1 = (5 + j4 + 2)\,I_1$$
$$V_2 = 2I_1$$
$$A = \left.\frac{V_1}{V_2}\right|_{I_2=0} = \frac{7+j4}{2} = 3.5 + j2$$
$$C = \left.\frac{I_1}{V_2}\right|_{I_2=0} = \frac{1}{2} = 0.5$$

From here only (b) option matches.

5.22 (d)

For reciprocal network $AD - BC = 1$

$|T| = 1$

5.23 (3)

For T-network,

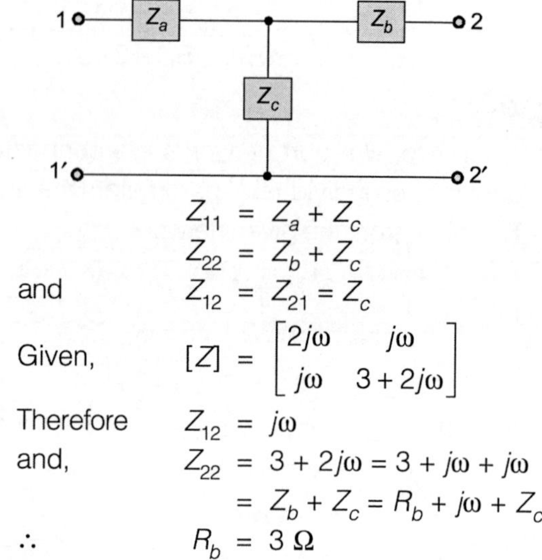

$$Z_{11} = Z_a + Z_c$$
$$Z_{22} = Z_b + Z_c$$

and $\quad Z_{12} = Z_{21} = Z_c$

Given, $\quad [Z] = \begin{bmatrix} 2j\omega & j\omega \\ j\omega & 3+2j\omega \end{bmatrix}$

Therefore $\quad Z_{12} = j\omega$

and, $\quad Z_{22} = 3 + 2j\omega = 3 + j\omega + j\omega$

$$= Z_b + Z_c = R_b + j\omega + Z_c$$

$\therefore \quad R_b = 3\,\Omega$

5.24 (a)

Redrawing the circuit,

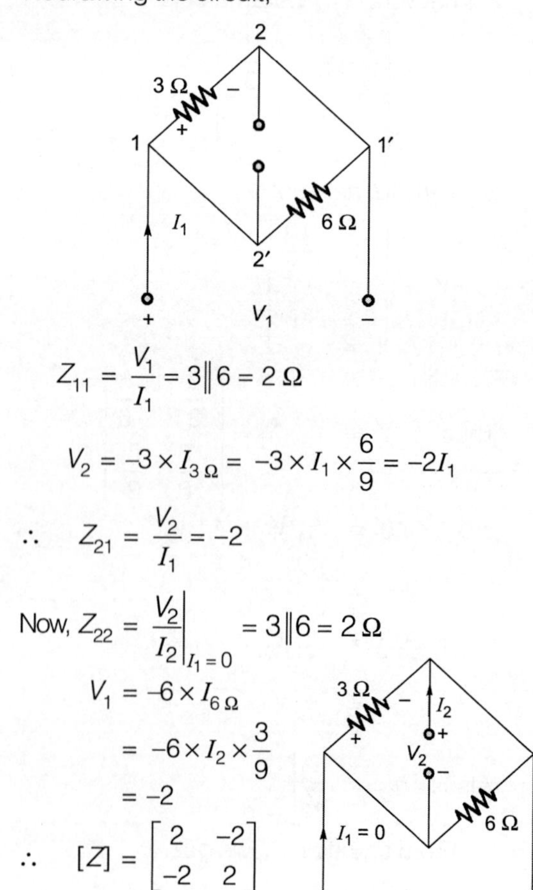

$$Z_{11} = \frac{V_1}{I_1} = 3\|6 = 2\,\Omega$$
$$V_2 = -3 \times I_{3\,\Omega} = -3 \times I_1 \times \frac{6}{9} = -2I_1$$
$$\therefore \quad Z_{21} = \frac{V_2}{I_1} = -2$$

Now, $Z_{22} = \left.\frac{V_2}{I_2}\right|_{I_1=0} = 3\|6 = 2\,\Omega$

$$V_1 = -6 \times I_{6\,\Omega}$$
$$= -6 \times I_2 \times \frac{3}{9}$$
$$= -2$$
$$\therefore \quad [Z] = \begin{bmatrix} 2 & -2 \\ -2 & 2 \end{bmatrix}$$

5.25 (4.80)

$$B = -\left.\frac{V_1}{I_2}\right|_{V_2=0}$$

When $V_2 = 0$ (i.e., when port-2 is short circuited),

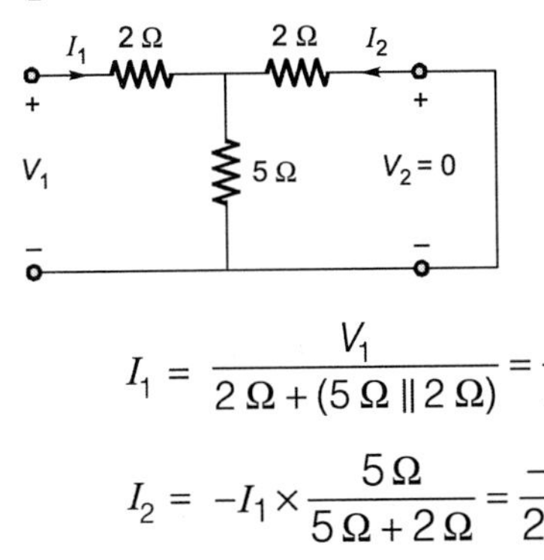

$$I_1 = \frac{V_1}{2\,\Omega + (5\,\Omega \parallel 2\,\Omega)} = \frac{7V_1}{24\,\Omega}$$

$$I_2 = -I_1 \times \frac{5\,\Omega}{5\,\Omega + 2\,\Omega} = \frac{-5V_1}{24\,\Omega}$$

So, $$B = -\frac{V_1}{I_2} = \frac{24}{5}\,\Omega = 4.80\,\Omega$$

5.26 (48)

From maximum power transfer theorem

$$Z_L = Z_{th}$$

$$Z_{th} = Z_{22} - \frac{Z_{12} \times Z_{21}}{R_S + Z_{11}}$$

For given data,

$$Z_{th} = 120 - \frac{60 \times 60}{10 + 40} = 48\,\Omega$$

$$Z_L = 48\Omega$$

5.27 (0.8)

For ideal transformer of $n:1$, the scattering matrix is

$$\begin{bmatrix} S_{11} & S_{12} \\ S_{21} & S_{22} \end{bmatrix} = \begin{bmatrix} \dfrac{n^2-1}{n^2+1} & \dfrac{2n}{n^2+1} \\ \dfrac{2n}{n^2+1} & \left(\dfrac{1-n^2}{1+n^2}\right) \end{bmatrix}$$

$$S_{21} = \frac{2n}{n^2+1} = \frac{2(2)}{2^2+1} = \frac{4}{5} = 0.8$$

5.28 (c)

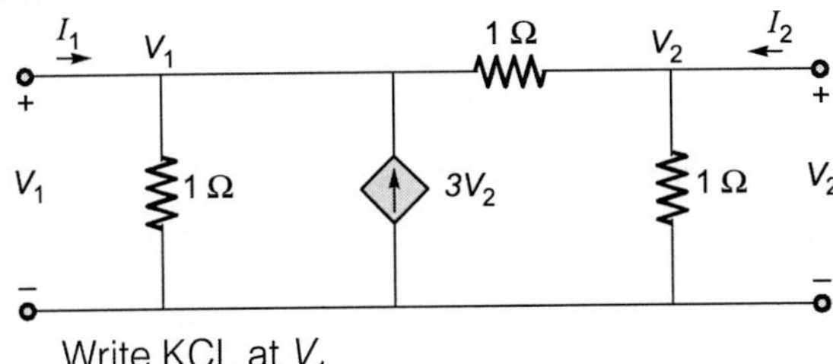

Write KCL at V_1

$$I_1 = \frac{V_1}{1} + \frac{V_1 - V_2}{1} - 3V_2$$

$$I_1 = 2V_1 - 4V_2 \quad \text{... (i)}$$

Write KCL at V_2

$$I_2 = \frac{V_2}{1} + \frac{V_2 - V_1}{1}$$

$$I_2 = -V_1 + 2V_2 \quad \text{... (ii)}$$

$$[y] = \begin{bmatrix} 2 & -4 \\ -1 & 2 \end{bmatrix} \mho$$

5.29 (4)

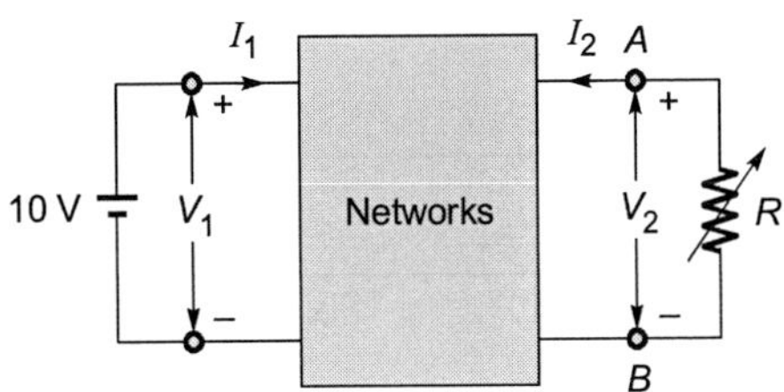

Given,

For $V_2 = 5$ V, $I_2 = 0$ mA

For $V_2 = 4$ V, $I_2 = -4$ mA

Calculation of Thevenin voltage across terminals A-B,

$$V_{th} = \left.V_2\right|_{I_2=0} = 5\text{ V (given)}$$

Calculation of Thevenin resistance across terminals A-B,

Short-circuit current,

$$I_{SC} = -\left.I_2\right|_{V_2=0}$$

From the given plot of V_2 versus $-I_2$

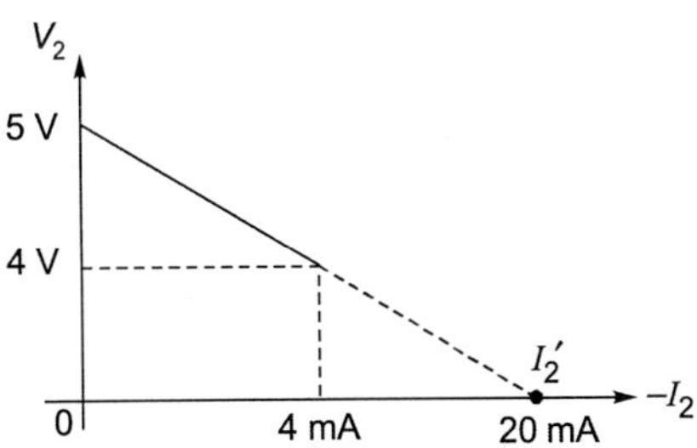

Slope of the line,

$$m = \frac{5-4}{0 - 4\times10^{-3}} = -250$$

For $V_2 = 0$, we can calculate I_2 as:

$$m = -250 = \frac{4-0}{4\times10^{-3} - I_2'}$$

$\Rightarrow$ $-1 + 250I_2' = 4$

$\Rightarrow$ $250I_2' = 5$

$\Rightarrow$ $$I_2' = \frac{1}{50} = 20\text{ mA}$$

Hence, $$I_{SC} = -\left.I_2\right|_{V_2=0} = I_2' = 20\text{ mA}$$

The Thevenin resistance,

$$R_{th} = \frac{V_{OC}}{I_{SC}} = \frac{V_{th}}{I_{SC}} = \frac{5\text{ V}}{20\text{ mA}}$$

$$\therefore \quad R_{th} = 250\ \Omega$$

When the variable resistance R is replaced with the given load, the Thevenin equivalent circuit can be drawn as below:

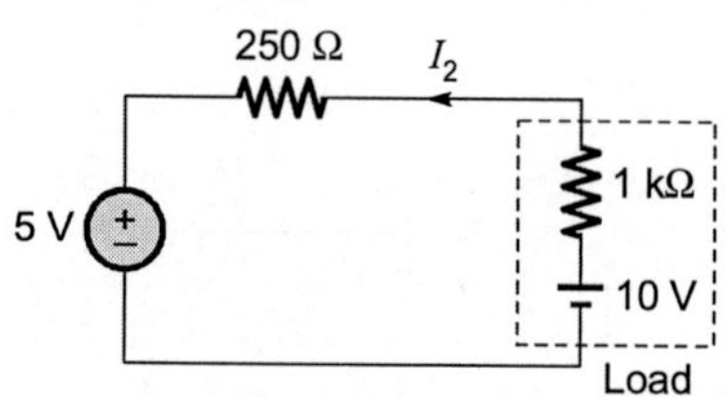

$$I_2 = \frac{10-5}{1000+250} = \frac{5}{1250} = \frac{5}{1.25} \times 10^{-3}\text{A} = 4\text{ mA}$$

5.30 (80)

Given, Y-parameters,

$$[Y] = \begin{bmatrix} \frac{2}{100} & -\frac{1}{100} \\ -\frac{1}{100} & \frac{4}{300} \end{bmatrix}$$

For the given Y-parameter the two-port network is

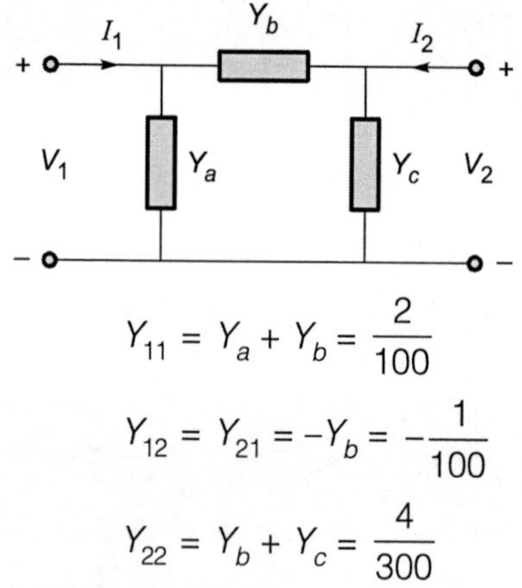

$$Y_{11} = Y_a + Y_b = \frac{2}{100}$$

$$Y_{12} = Y_{21} = -Y_b = -\frac{1}{100}$$

$$Y_{22} = Y_b + Y_c = \frac{4}{300}$$

On solving, $Y_b = \frac{1}{100} S$

$$Y_a = \frac{1}{100} S$$

$$Y_c = \frac{1}{300} S$$

The network becomes,

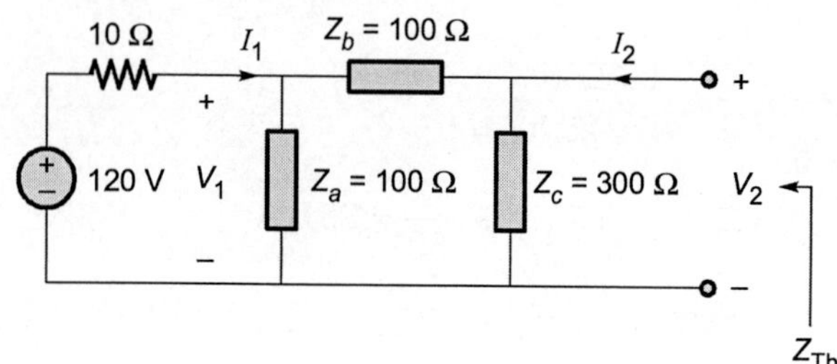

Converting delta to star,

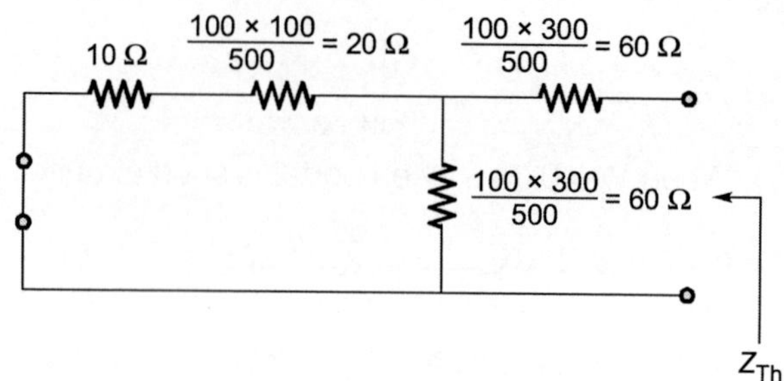

$$Z_{Th} = 60 + [(20+10) \parallel 60]$$

$$= 60 + \frac{30 \times 60}{30+60} = 80\ \Omega$$

For maximum power transfer,

$$Z_L = Z_{Th} = 80\ \Omega$$

5.31 (1.5)

Given two-port network,

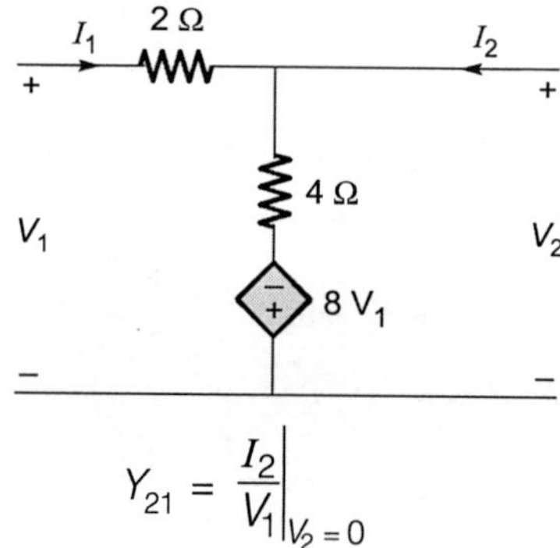

$$Y_{21} = \left.\frac{I_2}{V_1}\right|_{V_2=0}$$

Considering $V_2 = 0$, the circuit can be drawn as below:

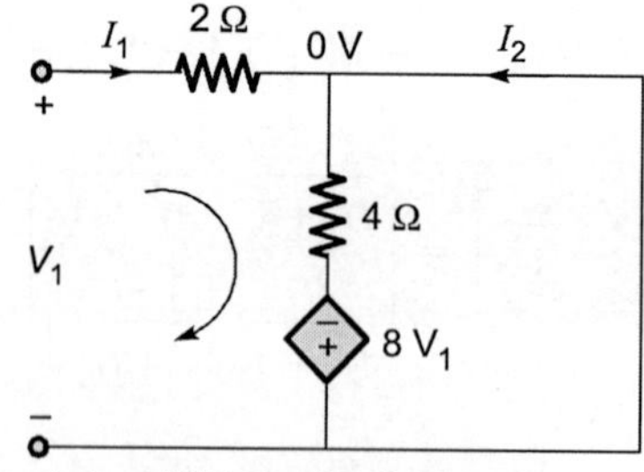

By applying KVL in the loop,

$$V_1 - 2I_1 - 4(I_1 + I_2) + 8V_1 = 0$$

$$9V_1 - 6I_1 - 4I_2 = 0$$

But $\quad I_1 = \frac{V_1 - 0}{2} \Rightarrow \frac{V_1}{2} = I_1$

$$\therefore \quad 9V_1 - 6\left(\frac{V_1}{2}\right) - 4I_2 = 0$$

$$6V_1 = 4I_2$$

$$\therefore \quad Y_{21} = \frac{I_2}{V_1} = \frac{6}{4} = \frac{3}{2} = 1.5\text{ S}$$

■■■

6 Graph Theory and State Equations

6.1 The number of independent loops for a network with n nodes and b branches is
(a) $n - 1$
(b) $b - n$
(c) $b - n + 1$
(d) independent of the number of nodes

[1996 : 1 M]

6.2 A network has 7 nodes and 5 independent loops. The number of branches in the network is
(a) 13 (b) 12
(c) 11 (d) 10

[1998 : 1 M]

6.3 Identify which of the following is NOT a tree of the graph shown in the figure

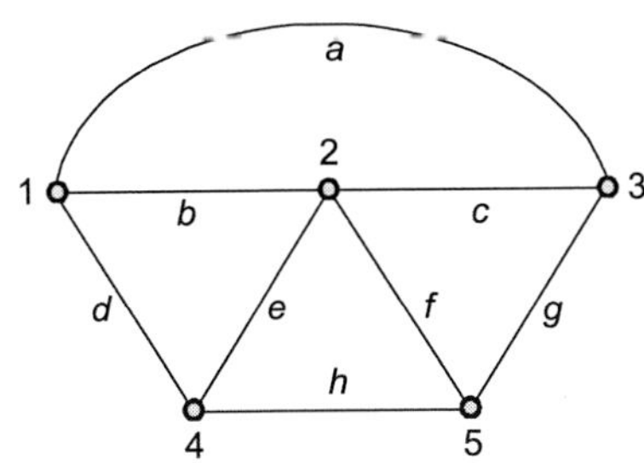

(a) *begh* (b) *defg*
(c) *adhg* (d) *aegh*

[1999 : 1 M]

6.4 The differential equation for the current $i(t)$ in the circuit of the figure is

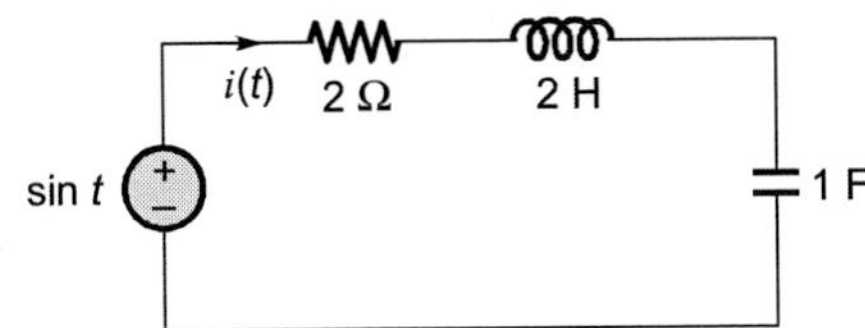

(a) $2\frac{d^2i}{dt^2} + 2\frac{di}{dt} + i(t) = \sin t$

(b) $\frac{d^2i}{dt^2} + 2\frac{di}{dt} + 2i(t) = \cos t$

(c) $2\frac{d^2i}{dt^2} + 2\frac{di}{dt} + i(t) = \cos t$

(d) $\frac{d^2i}{dt^2} + 2\frac{di}{dt} + 2i(t) = \sin t$

[2003 : 1 M]

6.5 Consider the network graph shown in the figure. Which one of the following is NOT a 'tree' of this graph?

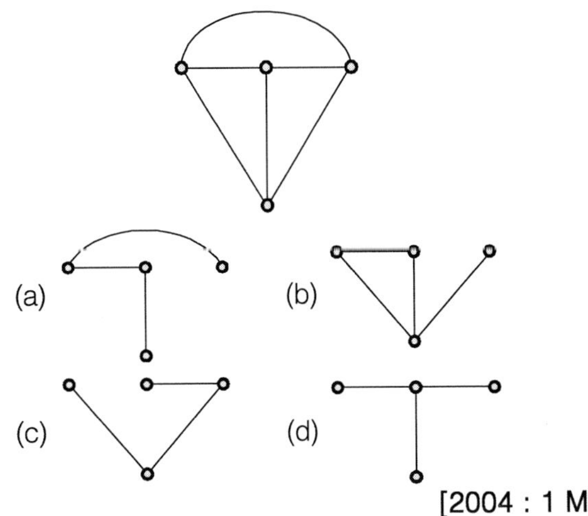

[2004 : 1 M]

6.6 In the following graph, the number of trees (P) and the number of cut-sets(Q) are

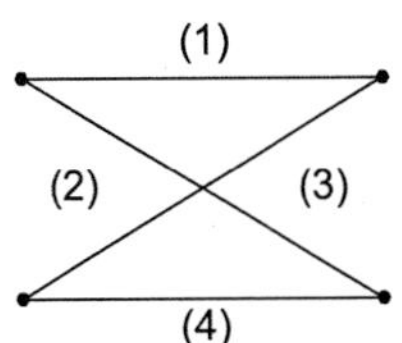

(a) $P = 2, Q = 2$ (b) $P = 2, Q = 6$
(c) $P = 4, Q = 6$ (d) $P = 4, Q = 10$

[2008 : 1 M]

■■■

Answers Graph Theory and State Equations

6.1 (c) 6.2 (c) 6.3 (c) 6.4 (c) 6.5 (b) 6.6 (c)

Explanations Graph Theory and State Equations

6.1 (c)

The number of independent loops for a network with n nodes and b branches is

$b-(n-1)=b-n+1$

6.2 (c)

$$l = b-(n-1)$$
$$b = l+(n-1)=5+(7-1)=5+6$$
$$b = 11$$

6.3 (c)

Tree does not contain any loop.

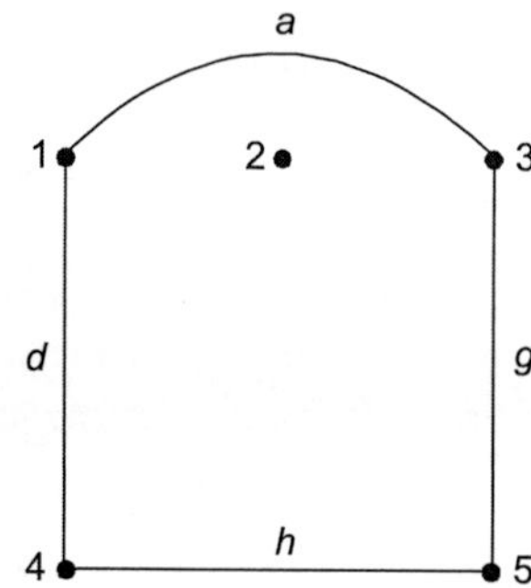

6.4 (c)

Applying KVL,

$$\sin t = i(t)\times 2+L\frac{di(t)}{dt}+\frac{1}{c}\int i(t)dt$$

$$\sin t = 2i(t)+2\frac{di(t)}{dt}+\int i(t)dt$$

Differentiating with respect to t,

$$\cos(t)=\frac{2di(t)}{dt}+\frac{2d^2i(t)}{dt^2}+i(t)$$

6.5 (b)

It is forming a closed loop. So it can't be a tree.

6.6 (c)

Different trees (P) are shown below

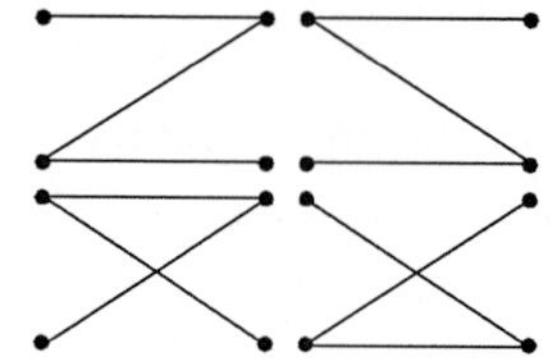

Different cut-sets (Q) are shown below:

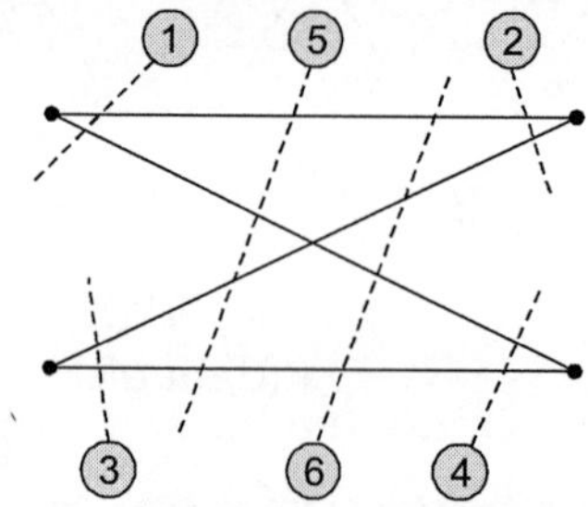

7 Network Functions

7.1 The circuit of the figure represents a

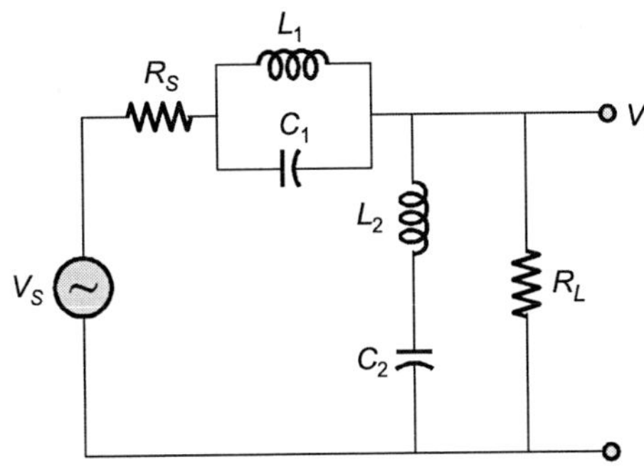

(a) low pass filter (b) high pass filter
(c) band pass filter (d) band reject filter

[2000 : 1 M]

7.2 The driving-point impedance $Z(s)$ of a network has the pole-zero locations as shown in the figure. If $Z(0) = 3$, then $Z(s)$ is

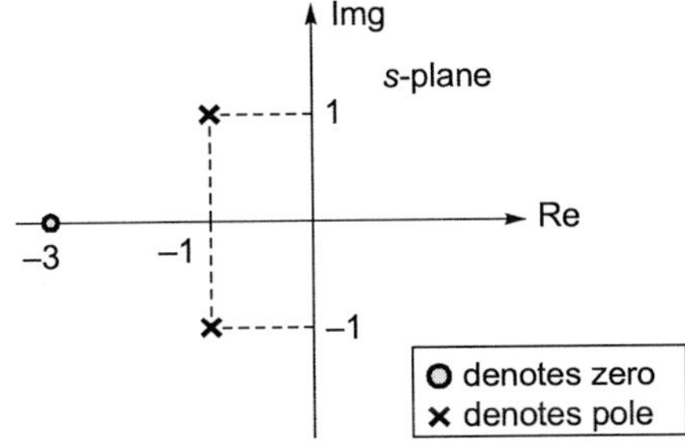

(a) $\frac{3(s+3)}{s^2+2s+3}$ (b) $\frac{2(s+3)}{s^2+2s+2}$

(c) $\frac{3(s-3)}{s^2-2s-2}$ (d) $\frac{2(s-3)}{s^2-2s-3}$

[2003 : 2 M]

7.3 The first and the last critical frequency of an *RC*-driving point impedance function must respectively be

(a) *a* zero and *a* pole (b) *a* zero and *a* zero
(c) *a* pole and *a* pole (d) *a* pole and *a* zero

[2005 : 1 M]

7.4 The first and the last critical frequencies (singularities) of a driving point impedance function of a passive network having two kinds of elements, are a pole and a zero respectively. The above property will be satisfied by

(a) *RL* network only
(b) *RC* network only
(c) *LC* network only
(d) *RC* as well as *RL* networks

[2006 : 1 M]

7.5 A negative resistance R_{neg} is connected to a passive network *N* having driving point impedance as shown below. For $Z_2(s)$ to be positive real

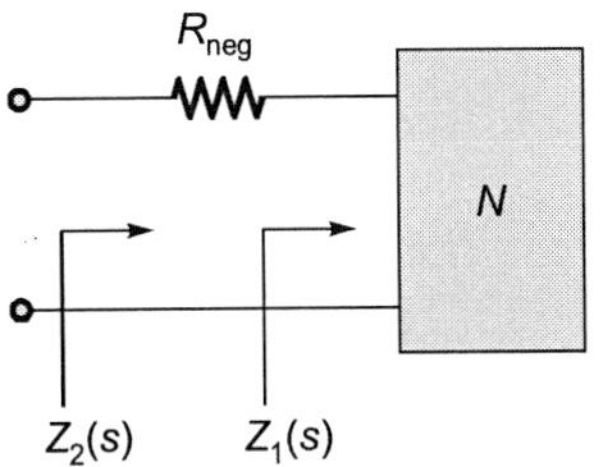

(a) $|R_{neg}| \le \text{Re}\, Z_1(j\omega),\ \forall\omega$

(b) $|R_{neg}| \le |Z_1(j\omega)|,\ \forall\omega$

(c) $|R_{neg}| \le |\text{Im}\, Z_1(j\omega)|,\ \forall\omega$

(d) $|R_{neg}| \le |\le\angle Z_1(j\omega)|,\ \forall\omega$

[2006 : 1 M]

7.6 The *RC* circuit shown in the figure is

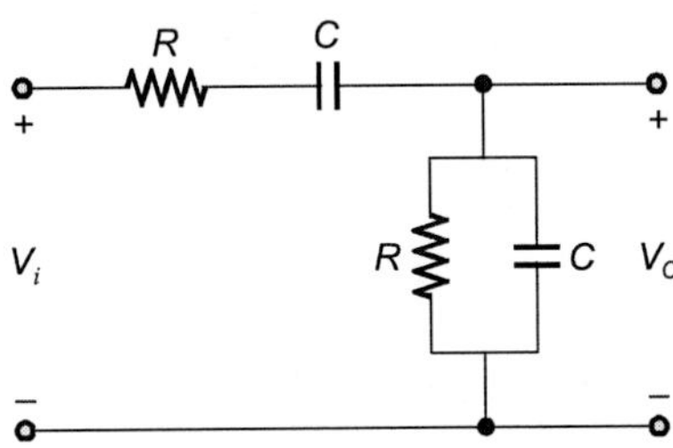

(a) a low-pass filter (b) a high-pass filter
(c) a band-pass filter (d) a band-reject filter

[2007 : 1 M]

7.7 Two series resonant filters are as shown in the figure. Let the 3-dB bandwidth of Filter 1 be B_1 and that of Filter 2 be B_2. The value of B_1/B_2 is

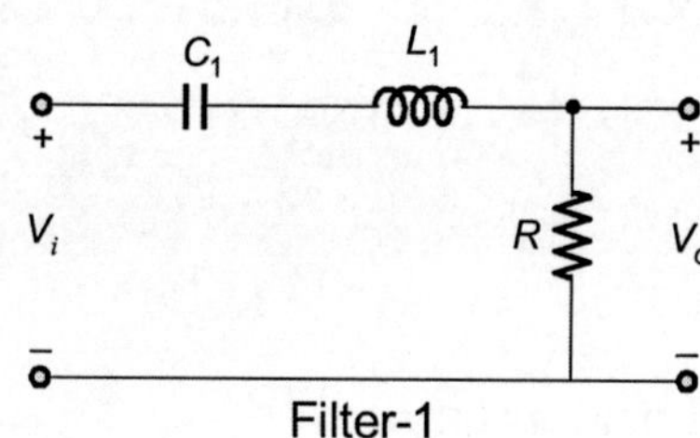

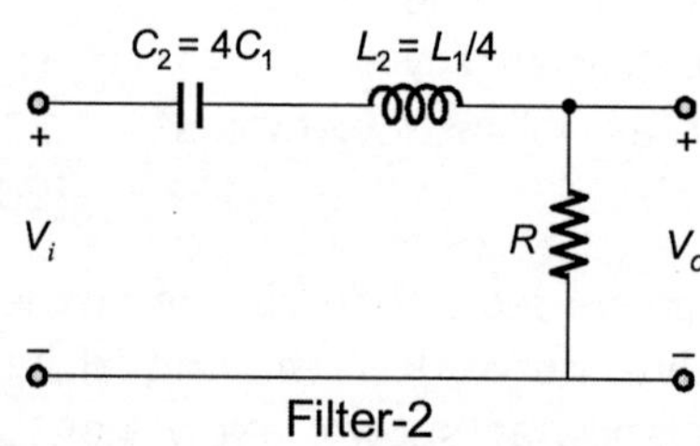

(a) 4
(b) 1
(c) $\frac{1}{2}$
(d) $\frac{1}{4}$

[2007 : 2 M]

7.8 The driving point impedance of the following network

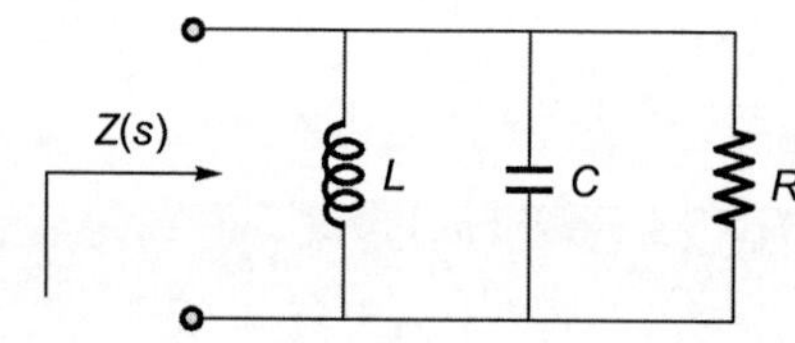

is given by $Z(s) = \frac{0.2s}{s^2 + 0.1s + 2}$. The component values are

(a) $L = 5$ H, $R = 0.5\ \Omega$, $C = 0.1$ F
(b) $L = 0.1$ H, $R = 0.5\ \Omega$, $C = 5$ F
(c) $L = 5$ H, $R = 2\ \Omega$, $C = 0.1$ F
(d) $L = 0.1$ H, $R = 2\ \Omega$, $C = 5$ F

[2008 : 2 M]

7.9 If the transfer function of the following network is

$$\frac{V_0(s)}{V_i(s)} = \frac{1}{2 + sCR}$$

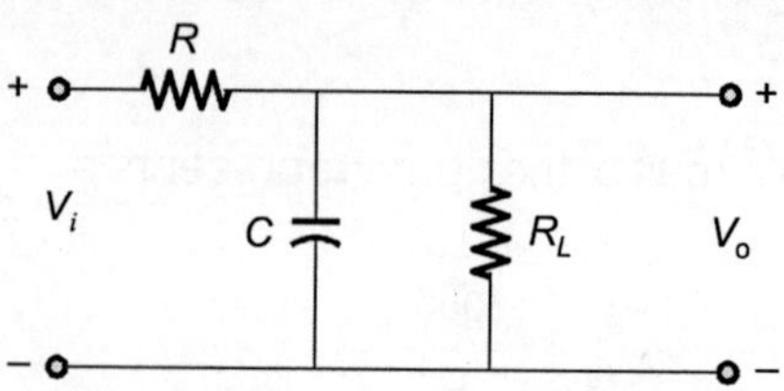

the value of the load resistance R_L is

(a) $R/4$
(b) $R/2$
(c) R
(d) $2R$

[2009 : 1 M]

7.10 The transfer function $\frac{V_2(s)}{V_1(s)}$ of the circuit shown below is

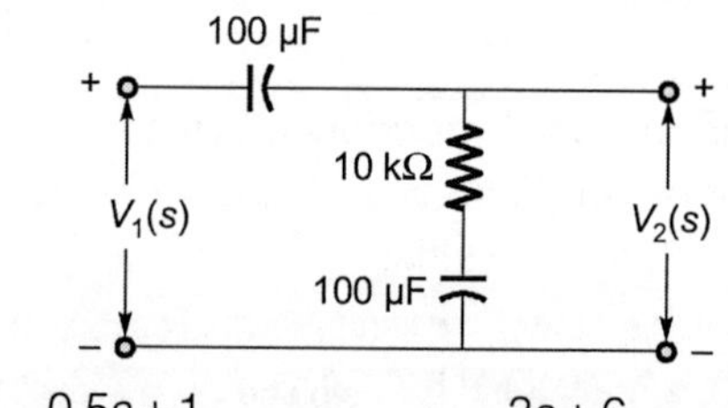

(a) $\frac{0.5s + 1}{s + 1}$
(b) $\frac{3s + 6}{s + 2}$
(c) $\frac{s + 2}{s + 1}$
(d) $\frac{s + 1}{s + 2}$

[2013 : 1 M]

■■■

Answers Network Functions

7.1	(d)	7.2	(b)	7.3	(d)	7.4	(b)	7.5	(a)	7.6	(c)	7.7	(d)
7.8	(d)	7.9	(c)	7.10	(d)								

Explanations Network Functions

7.1 (d)

Analyzing the circuit for $\omega = 0$ and $\omega = \infty$

At $\omega = 0$,

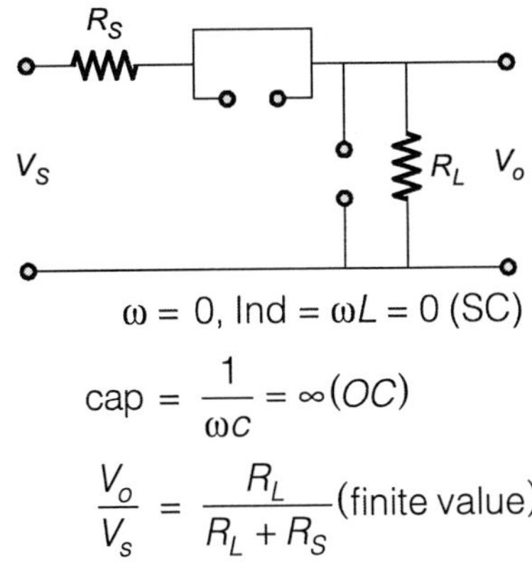

$\omega = 0$, Ind $= \omega L = 0$ (SC)

$$\text{cap} = \frac{1}{\omega C} = \infty (OC)$$

$$\frac{V_o}{V_s} = \frac{R_L}{R_L + R_S} \text{(finite value)}$$

At $\omega = \infty$

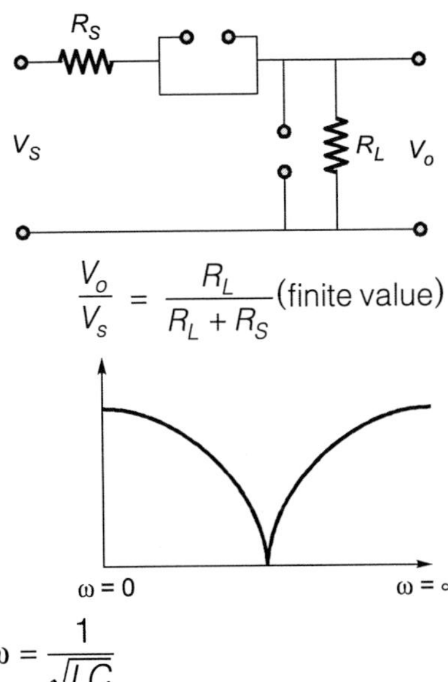

$$\frac{V_o}{V_s} = \frac{R_L}{R_L + R_S} \text{(finite value)}$$

At $\omega = \dfrac{1}{\sqrt{LC}}$

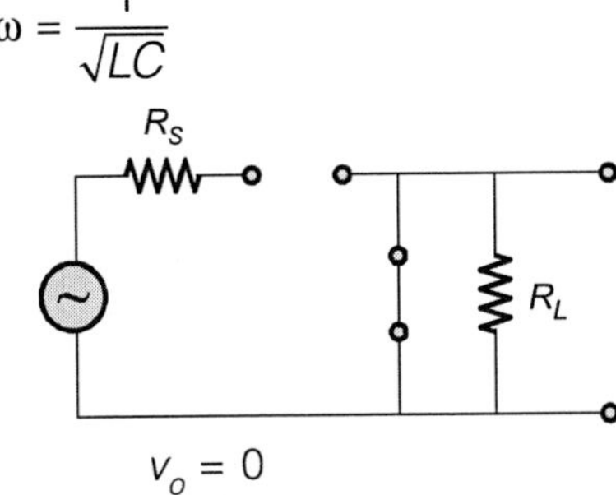

$V_o = 0$

7.2 (b)

$$Z(s) = \frac{K(s-z)}{(s-p_1)(s-p_2)} = \frac{K(s+3)}{(s+1+j)(s+1-j)}$$

$$Z(s) = \frac{K(s+3)}{(s+1)^2 - j2} = \frac{K(s+3)}{(s+1)^2 + 1}$$

$$Z(0)|_{\omega=0} = 0$$

$$\Rightarrow \quad \frac{3K}{2} = 3$$

$$\Rightarrow \quad K = 2$$

$$\therefore \quad Z(s) = \frac{2(s+3)}{s^2 + 2s + 2}$$

7.3 (d)

For stability poles and zero interlace on real axis. Since its *RC*, first pole should come and zero at last.

7.4 (b)

RC impedance function has

(1) 1st critical frequency due to pole.

(2) Last critical frequency due to zero.

7.5 (a)

For $Z_2(s)$ to be positive real,

$$\text{Re}\{Z(s)\} \geq |R_{neg}|$$

$$\Rightarrow \quad |R_{neg}| \leq \text{Re}\{Z_1(j\omega)\}, \text{ for all } \omega$$

7.6 (c)

At $\omega \to \infty$, Capacitor $\to$ short circuited

Circuit looks like,

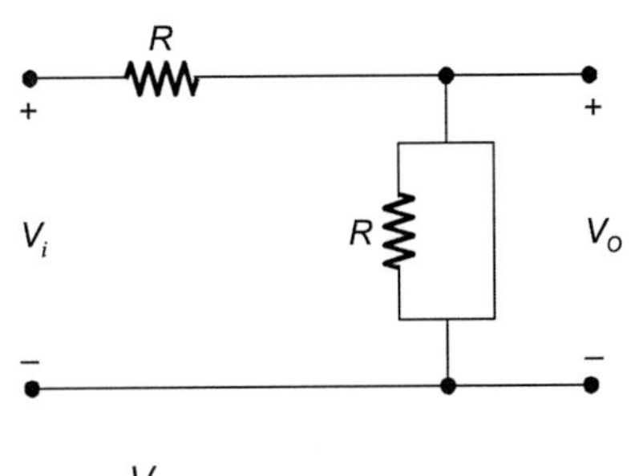

$$\frac{V_o}{V_i} = 0$$

at $\omega \to 0$, Capacitor $\to$ open circuited

Circuit looks like

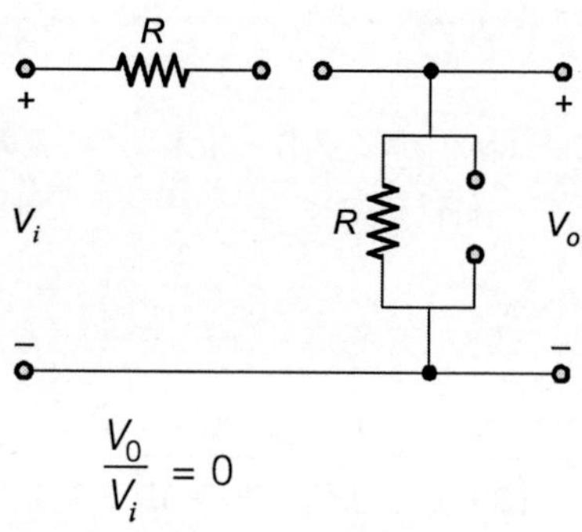

$$\frac{V_0}{V_i} = 0$$

So frequency response of the circuit will be,

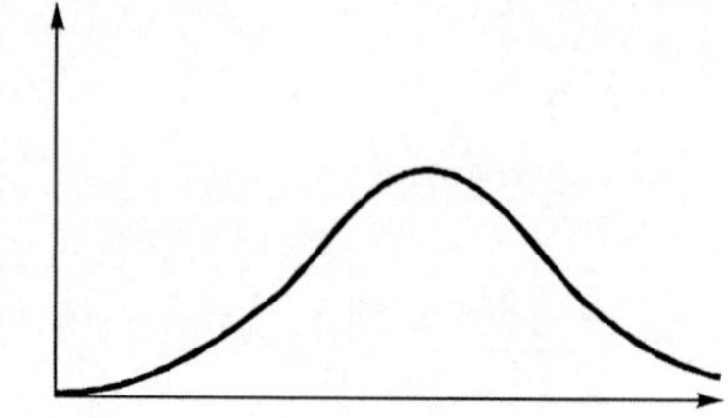

So the circuit is Band pass filter

7.7 (d)

Bandwidth of series RLC circuit is R/L

Bandwidth of filter 1; $B_1 = \frac{R}{L_1}$

Bandwidth of filter 2; $B_2 = \frac{R}{L_2}$

$$= \frac{R}{L_1/4} = \frac{4R}{L_1}$$

So, $\frac{B_1}{B_2} = \frac{1}{4}$

7.8 (d)

$$Z(s) = \frac{0.2\,s}{s^2 + 0.1s + 2}$$

$$Y(s) = \frac{s^2 + 0.1s + 2}{0.2\,s}$$

$$= \frac{s}{0.2} + \frac{1}{2} + \frac{2}{0.2\,s}$$

$$= 5s + 0.5 + \frac{10}{s}$$

Comparing with

$$Y(s) = Cs + \frac{1}{R} + \frac{1}{Ls}$$

$$C = 5\text{ F},\ R = \frac{1}{0.5} = 2\ \Omega$$

$$L = \frac{1}{10} = 0.1\text{ H}$$

7.9 (c)

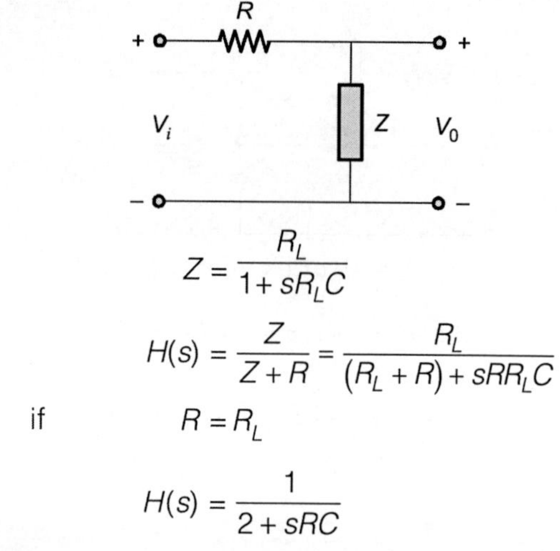

$$Z = \frac{R_L}{1 + sR_LC}$$

$$H(s) = \frac{Z}{Z+R} = \frac{R_L}{(R_L + R) + sRR_LC}$$

if $R = R_L$

$$H(s) = \frac{1}{2 + sRC}$$

7.10 (d)

$$\frac{V_2(s)}{V_1(s)} = \frac{10\times 10^3 + \frac{1}{100\times 10^{-6}s}}{10\times 10^3 + \frac{1}{100\times 10^{-6}s} + \frac{1}{100\times 10^{-6}s}}$$

$$\frac{V_2(s)}{V_1(s)} = \frac{s\times 10^4 + 10^4}{s\times 10^4 + 10^4 + 10^4} = \frac{10^4(1+s)}{10^4(s+2)}$$

$$\frac{V_2(s)}{V_1(s)} = \frac{s+1}{s+2}$$

■■■

Electromagnetics

UNIT

II

CONTENTS

Electromagnetics

Syllabus

Maxwell's equations: differential and integral forms and their interpretation, boundary conditions, wave equation, Poynting vector.

Plane waves and properties: reflection and refraction, polarization, phase and group velocity, propagation through various media, skin depth.

Transmission lines: equations, characteristic impedance, impedance matching, impedance transformation, S-parameters, Smith chart. Rectangular and circular waveguides, light propagation in optical fibers, dipole and monopole antennas, linear antenna arrays.

Analysis of Previous GATE Papers

Exam Year	1 Mark Ques.	2 Marks Ques.	Total Marks
1995	5	–	**5**
1996	3	4	**11**
1997	2	1	**4**
1998	12	–	**12**
1999	3	4	**11**
2000	4	4	**12**
2001	4	4	**12**
2002	4	4	**12**
2003	2	7	**16**
2004	2	6	**14**
2005	2	6	**14**
2006	2	8	**18**
2007	2	7	**16**
2008	2	5	**12**
2009	2	3	**8**
2010	3	2	**7**
2011	3	3	**9**
2012	4	4	**12**
2013	1	2	**5**

Exam Year	1 Mark Ques.	2 Mark Ques.	Total Marks
2014 Set-1	2	3	**8**
2014 Set-2	2	3	**8**
2014 Set-3	2	2	**6**
2014 Set-4	2	2	**6**
2015 Set-1	2	3	**8**
2015 Set-2	2	3	**8**
2015 Set-3	2	4	**10**
2016 Set-1	2	4	**10**
2016 Set-2	3	4	**11**
2016 Set-3	2	4	**10**
2017 Set-1	2	3	**8**
2017 Set-2	2	3	**8**
2018	2	3	**8**
2019	3	3	**9**
2020	2	2	**6**
2021	3	4	**11**
2022	2	2	**6**
2023	1	5	**11**
2024	2	3	**8**

1 Basics of Electromagnetics

1.1 The electric field strength at distant point, P, due to a point charge, $+q$, located at the origin, is 100 μV/m. If the point charge is now enclosed by a perfectly conducting metal sheet sphere whose center is at the origin, then the electric field strength at the point, P, outside the sphere, becomes

(a) zero (b) 100 μV/m
(c) – 100 μV/m (d) 50 μV/m

[1995 : 1 M]

1.2 In the infinite plane, $y = 6$ m, there exists a uniform surface charge density of $(1/6000\pi)$ μC/m^2. The associated electric field strength is

(a) $30\,\hat{i}$ V/m (b) $3\,\hat{j}$ V/m
(c) $30\,\hat{k}$ V/m (d) $60\,\hat{j}$ V/m

[1995 : 1 M]

1.3 A metal sphere with 1 m radius and a surface charge density of 10 Coulombs/m^2 is enclosed in a cube of 10 m side. The total outward electric displacement flux normal to the surface of the cube is

(a) $40\,\pi$ Coulombs
(b) $10\,\pi$ Coulombs
(c) $5\,\pi$ Coulombs
(d) None of the above

[1996 : 1 M]

1.4 An electric field on a plane is described by its potential $V = 20\left(r^{-1} + r^{-2}\right)$ where r is the distance from the source. The field is due to

(a) a monopole
(b) a dipole
(c) both a monopole and a dipole
(d) a quadrupole

[1999 : 1 M]

1.5 The electric field on the surface of a perfect conductor is 2 V/m. The conductor is immersed in water with $\epsilon = 80\epsilon_0$. The surface charge density on the conductor is

(a) 0 C/m^2
(b) 2 C/m^2
(c) 1.8×10^{-11} C/m^2
(d) 1.41×10^{-9} C/m^2
($\epsilon = 10^{-9})/(36\,\pi)F/m$)

[2002 : 2 M]

1.6 The unit of $\nabla \times \vec{H}$ is

(a) Ampere (b) Ampere/meter
(c) Ampere/meter^2 (d) Ampere-meter

[2003 : 1 M]

1.7 If the electric field intensity is given by $\vec{E} = (x\hat{u}_x + y\hat{u}_y + z\hat{u}_z)$ Volt/m the potential difference between $X(2, 0, 0)$ and $Y(1, 2, 3)$ is

(a) +1 volt (b) – 1 volt
(c) + 5 volt (d) + 6 volt

[2003 : 2 M]

1.8 Medium 1 has the electrical permittivity $\varepsilon_1 = 1.5\varepsilon_0$ farad/m and occupies the region to the left of $x = 0$ plane. Medium 2 has the electrical permittivity $\varepsilon_2 = 2.5\varepsilon_0$ farad/m and occupies the region to the right of $x = 0$ plane. If E_1 in medium 1 is $E_1 = (2u_x - 3u_y + 1u_z)$ volt/m, then E_2 in medium 2 is

(a) $(2.0u_x - 7.5u_y + 2.5u_z)$ volt/m
(b) $(2.0u_x - 2.0u_y + 0.6u_z)$ volt/m
(c) $(1.2u_x - 3.0u_y + 1.0u_z)$ volt/m
(d) $(1.2u_x - 2.0u_y + 0.6u_z)$ volt/m

[2003 : 2 M]

1.9 A medium is divided into regions I and II about $x = 0$ plane, as shown in the figure below. An electromagnetic wave with electric field $\vec{E}_1 = 4\hat{a}_x + 3\hat{a}_y + 5\hat{a}_z$ is incident normally on the interface from region-I. The electric field $\vec{E}_2$ in region-II at the interface is

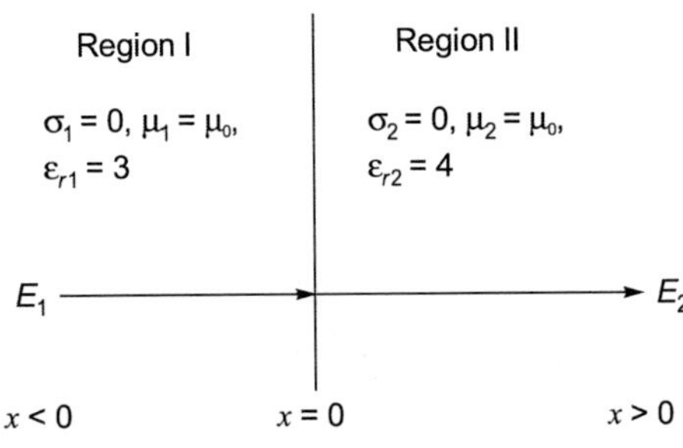

(a) $\vec{E}_2 = \vec{E}_1$
(b) $4\hat{a}_x + 0.75\hat{a}_y - 1.25\hat{a}_z$
(c) $3\hat{a}_x + 3\hat{a}_y + 5\hat{a}_z$
(d) $-3\hat{a}_x + 3\hat{a}_y + 5\hat{a}_z$

[2006 : 2 M]

1.10 If C is a closed curve enclosing a surface S, then the magnetic field intensity $\vec{H}$, the current density $\vec{J}$ and the electric flux density $\vec{D}$ are related by

(a) $\iint_S \vec{H}\cdot\vec{ds} = \oint_C \left(\vec{J}+\frac{\partial \vec{D}}{\partial t}\right)\cdot\vec{dl}$

(b) $\int_C \vec{H}\cdot\vec{dl} = \oiint_S \left(\vec{J}+\frac{\partial \vec{D}}{\partial t}\right)\cdot\vec{ds}$

(c) $\oiint_S \vec{H}\cdot\vec{ds} = \int_C \left(\vec{J}+\frac{\partial \vec{D}}{\partial t}\right)\cdot\vec{dl}$

(d) $\oint_C \vec{H}\cdot\vec{dl} = \iint_S \left(\vec{J}+\frac{\partial \vec{D}}{\partial t}\right)\cdot\vec{ds}$ [2007 : 1 M]

1.11 For static electric and magnetic fields in an inhomogeneous source-free medium, which of the following represents the correct form of two of Maxwell's equations?

(a) $\nabla\cdot\vec{E}=0$, $\nabla\times\vec{B}=0$

(b) $\nabla\cdot\vec{E}=0$, $\nabla\cdot\vec{B}=0$

(c) $\nabla\times\vec{E}=0$, $\nabla\times\vec{B}=0$

(d) $\nabla\times\vec{E}=0$, $\nabla\cdot\vec{B}=0$

[2008 : 1 M]

1.12 Two infinitely long wires carrying current are as shown in the figure below. One wire is in the y-z plane and parallel to the y-axis. The other wire is in the x-y plane and parallel to the x-axis. Which components of the resulting magnetic field are non-zero at the origin?

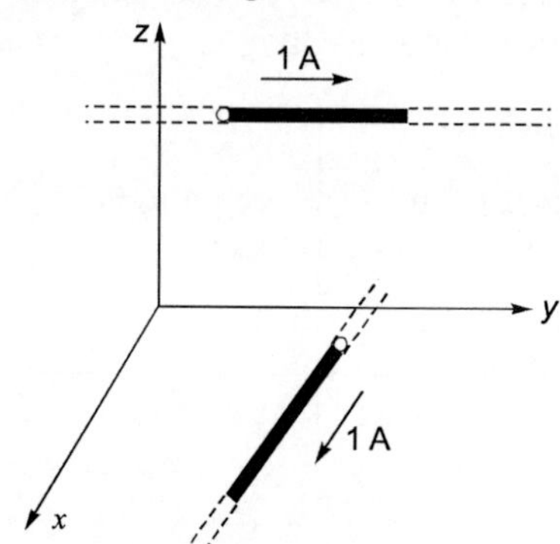

(a) x, y, z components
(b) x, y components
(c) y, z components
(d) x, z components [2009 : 1 M]

1.13 If a vector field $\vec{V}$ is related to another vector field $\vec{A}$ through $\vec{V}=\nabla\times\vec{A}$, which of the following is true? Note: C and S_C refer to any closed contour and any surface whose boundary is C.

(a) $\oint_C \vec{V}\cdot\vec{dl} = \iint_{S_C} \vec{A}\cdot\vec{dS}$

(b) $\oint_C \vec{A}\cdot\vec{dl} = \iint_{S_C} \vec{V}\cdot\vec{dS}$

(c) $\oint_C \nabla\times\vec{V}\cdot\vec{dl} = \iint_{S_C} \nabla\times\vec{A}\cdot\vec{dS}$

(d) $\oint_C \nabla\times\vec{A}\cdot\vec{dl} = \iint_{S_C} \vec{V}\cdot\vec{dS}$ [2009 : 2 M]

1.14 A magnetic field in air is measured to be

$$\vec{B} = B_0\left(\frac{x}{x^2+y^2}\hat{y} - \frac{y}{x^2+y^2}\hat{x}\right)$$

What current distribution leads to this field? [Hint: the algebra is trivial in cylindrical coordinates].

(a) $\vec{J} = -\frac{B_0\hat{z}}{\mu_0}\left(\frac{1}{x^2+y^2}\right), r\neq 0$

(b) $\vec{J} = -\frac{B_0\hat{z}}{\mu_0}\left(\frac{2}{x^2+y^2}\right), r\neq 0$

(c) $\vec{J} = 0, r\neq 0$

(d) $\vec{J} = \frac{B_0\hat{z}}{\mu_0}\left(\frac{1}{x^2+y^2}\right), r\neq 0$ [2009 : 2 M]

1.15 A current sheet $\vec{J} = 10\,\hat{u}_y$ A/m lies on the dielectric interface $x = 0$ between two dielectric media with $\varepsilon_{r_1}=1$, $\mu_{r_1}=1$ in Region-1 ($x<0$) and $\varepsilon_{r_2}=2$, $\mu_{r_2}=2$ in Region-2 ($x>0$). If the magnetic field in Region-1 at $x = 0^-$ is $\vec{H}_1 = 3\hat{u}_x + 30\hat{u}_y$ A/m, the magnetic field in Region-2 at $x = 0^+$ is

$x > 0$ (Region-2): $\varepsilon_{r_2} = 2$, $\mu_{r_2} = 2$

$\vec{J}$ $x = 0$

$x < 0$ (Region-1): $\varepsilon_{r_1} = 1$, $\mu_{r_1} = 1$

(a) $\vec{H}_2 = 1.5\,\hat{u}_x + 30\hat{u}_y - 10\hat{u}_z$ A/m
(b) $\vec{H}_2 = 3\,\hat{u}_x + 30\hat{u}_y - 10\hat{u}_z$ A/m
(c) $\vec{H}_2 = 1.5\,\hat{u}_x + 40\,\hat{u}_y$ A/m
(d) $\vec{H}_2 = 3\,\hat{u}_x + 30\,\hat{u}_y + 10\,\hat{u}_z$ A/m [2011 : 2 M]

Statement for Linked Answer Questions 1.16 and 1.17:

An infinitely long uniform solid wire of radius a carries a uniform dc current of density $\vec{j}$.

1.16 The magnetic field at a distance r from the center of the wire is proportional to
(a) r for $r < a$ and $1/r^2$ for $r > a$
(b) 0 for $r < a$ and $1/r$ for $r > a$
(c) r for $r < a$ and $1/r$ for $r > a$
(d) 0 for $r < a$ and $1/r^2$ for $r > a$

[2012 : 2 M]

1.17 A hole of radius $b(b < a)$ is now drilled along the length of the wire at a distance d from the center of the wire as shown below.

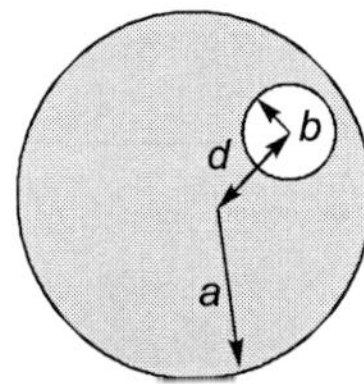

The magnetic field inside the hole is
(a) uniform and depends only on d
(b) uniform and depends only on b
(c) uniform and depends on both b and d
(d) non uniform

[2012 : 2 M]

1.18 The force on a point charge $+q$ kept at a distance d from the surface of an infinite grounded metal plate in a medium of permittivity $\in$ is
(a) 0
(b) $\dfrac{q^2}{16\pi \in d^2}$ away from the plate
(c) $\dfrac{q^2}{16\pi \in d^2}$ towards the plate
(d) $\dfrac{q^2}{4\pi \in d^2}$ towards the plate

[2014 : 1 M, Set-1]

1.19 A region shown below contains a perfect conducting half-space and air. The surface current $\vec{K}_s$ on the surface of the perfect conductor is $\vec{K}_s = \hat{x}2$ amperes per meter. The tangential $\vec{H}$ field in the air just above the perfect conductor is

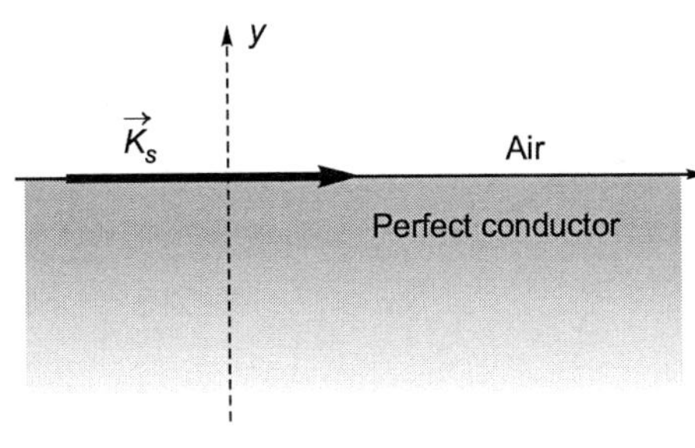

(a) $(\hat{x} + \hat{z})$ amperes per meter
(b) $\hat{x}2$ amperes per meter
(c) $-\hat{z}2$ amperese per meter
(d) $\hat{z}2$ amperes per meter

[2014 : 2 M, Set-3]

1.20 The electric field (assumed to be one-dimensional) between two points A and B is shown. Let ψ_A and ψ_B be the electrostatic potentials at A and B, respectively. The value of $\psi_A - \psi_B$ in Volts is ________.

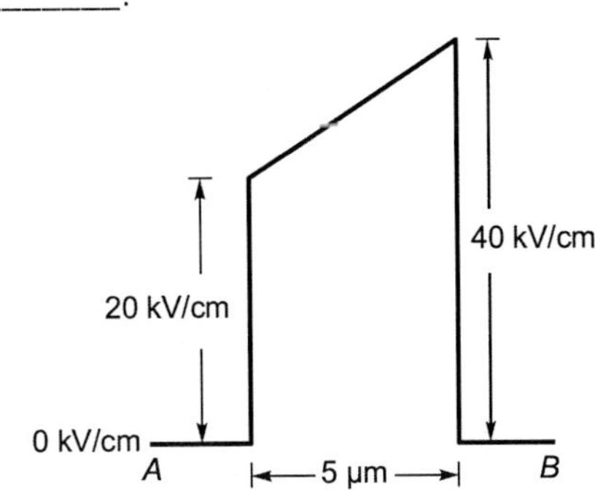

[2014 : 2 M, Set-4]

1.21 If $\vec{E} = -(2y^3 - 3yz^2)\hat{x} - (6xy^2 - 3xz^2)\hat{y} + (6xyz)\hat{z}$ is the electric field in a source free region, a valid expression for the electrostatic potential is
(a) $xy^3 - yz^2$
(b) $2xy^3 - xyz^2$
(c) $y^3 + xyz^2$
(d) $2xy^3 - 3xyz^2$

[2014 : 2 M, Set-4]

1.22 Consider a straight, infinitely long, current carrying conductor lying on the z-axis. Which one of the following plots (in linear scale) qualitatively represents the dependence of H_ϕ on r, where H_ϕ is the magnitude of the azimuthal component of magnetic field outside the conductor and r is the radial distance from the conductor?

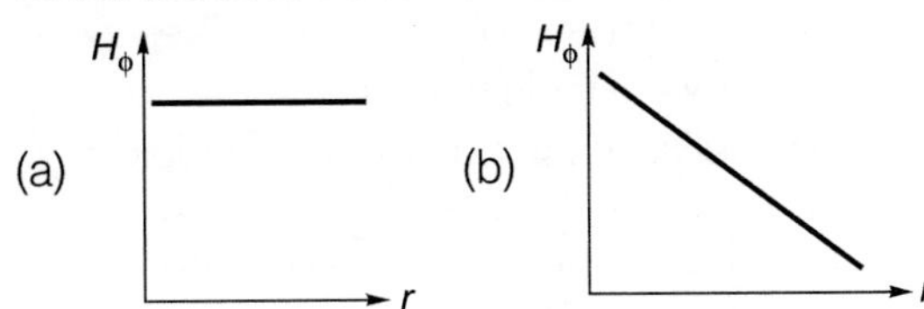

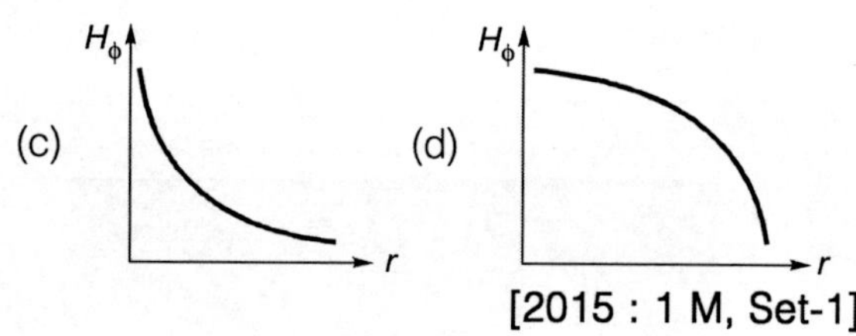

[2015 : 1 M, Set-1]

1.23 In a source free region in vacuum, if the electrostatic potential $\varphi = 2x^2 + y^2 + cz^2$, the value of constant c must be _____.

[2015 : 1 M, Set-2]

1.24 A vector field $\vec{D} = 2\rho^2\,\hat{a}_\rho + z\hat{a}_z$ exists inside a cylindrical region enclosed by the surfaces $\rho = 1$, $z = 0$ and $z = 5$. Let S be the surface bounding this cylindrical region. The surface integral of this field on S $\oiint_s \vec{D}\cdot\vec{ds}$ is _____.

[2015 : 2 M, Set-3]

1.25 Concentric spherical shells of radii 2 m, 4 m, and 8 m carry uniform surface charge densities of 20 nC/m^2, –4 nC/m^2 and ρ_s, respectively. The value of ρ_s(nC/m^2) required to ensure that the electric flux density $\vec{D} = \vec{0}$ at radius 10 m is_____.

[2016 : 1 M, Set-1]

1.26 The current density in a medium is given by

$$\vec{J} = \frac{400\sin\theta}{2\pi(r^2+4)}\hat{a}_r \text{ Am}^{-2}$$

The total current and the average current density flowing through the portion of a spherical surface $r = 0.8$ m, $\frac{\pi}{12} \le \theta \le \frac{\pi}{4}, 0 \le \phi \le 2\pi$ are given, respectively, by

(a) 15.09 A, 12.86 Am^{-2}
(b) 18.73 A, 13.65 Am^{-2}
(c) 12.86 A, 9.23 Am^{-2}
(d) 10.28 A, 7.56 Am^{-2}

[2016 : 2 M, Set-1]

1.27 A uniform and constant magnetic field $B = \hat{z}B$ exists in the $\hat{z}$ direction in vacuum. A particle of mass m with a small charge q is introduced into this region with an initial velocity $v = \hat{x}v_x + \hat{z}v_z$. Given that B, m, q, v_x and v_z are all non-zero, which one of the following describes the eventual trajectory of the particle?

(a) Helical motion in the $\hat{z}$ direction.
(b) Circular motion in the xy plane.
(c) Linear motion in the $\hat{z}$ direction.
(d) Linear motion in the $\hat{x}$ direction.

[2016 : 1 M, Set-2]

1.28 A positive charge q is placed at $x = 0$ between two infinite metal plates placed at $x = -d$ and at $x = +d$ respectively. The metal plates lie in the yz plane.

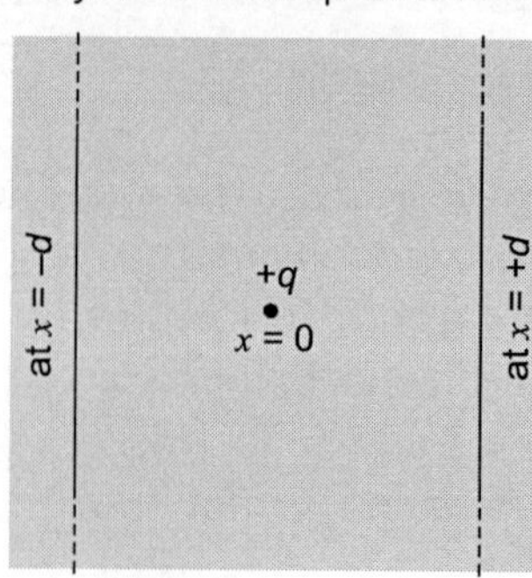

The charge is at rest at $t = 0$, when a voltage $+V$ is applied to the plate at $-d$ and voltage $-V$ is applied to the plate at $x = +d$. Assume that the quantity of the charge q is small enough that it does not perturb the field set up by the metal plates. The time that the charge q takes to reach the right plate is proportional to

(a) d/V
(b) $\sqrt{d}/V$
(c) $d/\sqrt{V}$
(d) $\sqrt{d/V}$

[2016 : 2 M, Set-2]

1.29 The parallel-plate capacitor shown in the figure has movable plates. The capacitor is charged so that the energy stored in it is E when the plate separation is d. The capacitor is then isolated electrically and the plates are moved such that the plate separation becomes $2d$.

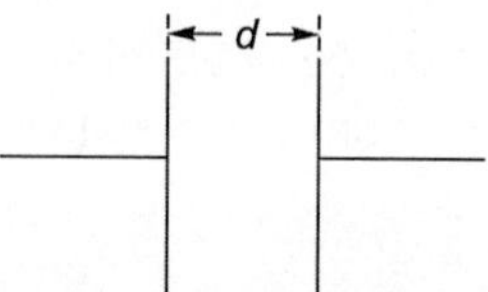

At this new plate separation, what is the energy stored in the capacitor, neglecting fringing effects?

(a) $2E$
(b) $\sqrt{2}E$
(c) E
(d) $E/2$

[2016 : 2 M, Set-2]

1.30 Consider the charge profile shown in the figure. The resultant potential distribution is best described by

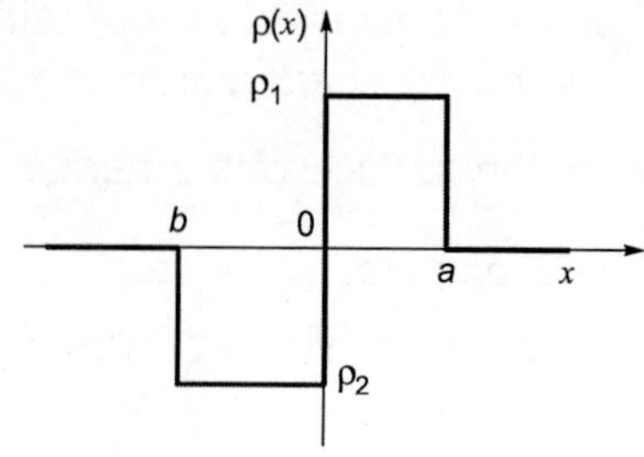

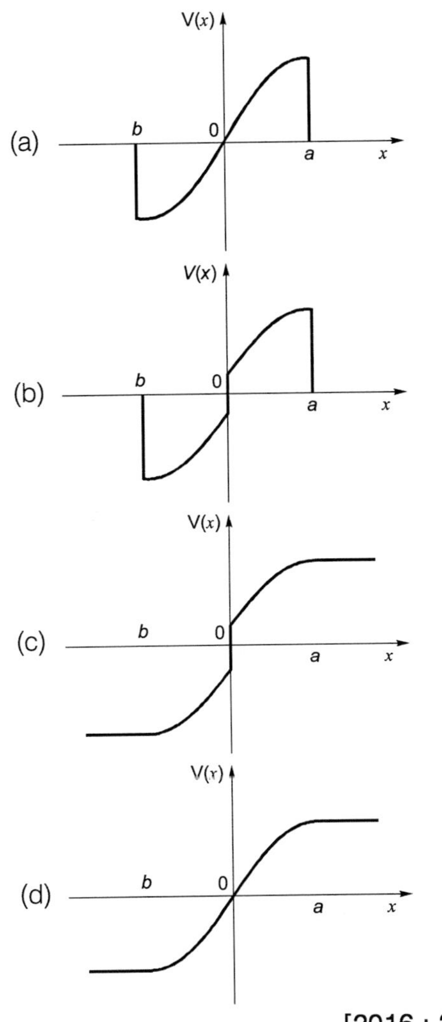

[2016 : 2 M, Set-3]

1.31 Two conducting spheres S1 and S2 of radii a and b ($b > a$) respectively, are placed far apart and connected by a long, thin conducting wire, as shown in the figure.

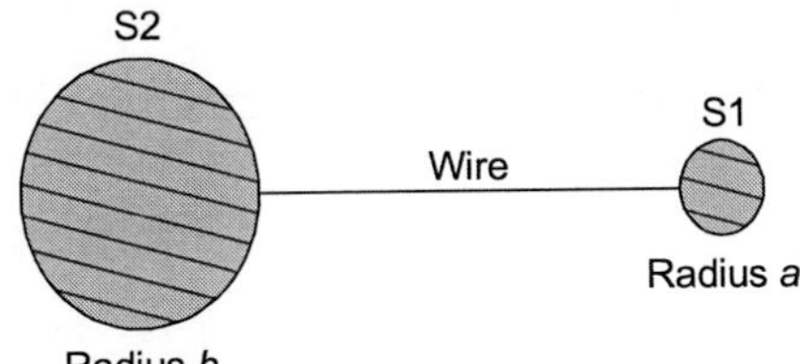

For some charge placed on this structure, the potential and surface electric field on S1 are V_a and E_a, and that on S2 are V_b and E_b respectively. Then, which of the following is CORRECT?

(a) $V_a = V_b$ and $E_a < E_b$
(b) $V_a > V_b$ and $E_a > E_b$
(c) $V_a = V_b$ and $E_a > E_b$
(d) $V_a > V_b$ and $E_a = E_b$

[2017 : 1 M, Set-2]

1.32 An electron (q_1) is moving in free space with velocity 10^5 m/s towards a stationary electron (q_2) far away. The closest distance that this moving electron gets to the stationary electron before the repulsive force diverts its path is _____ $\times 10^{-8}$ m. [Given, mass of electron $m = 9.11 \times 10^{-31}$ kg, charge of electron $e = -1.6 \times 10^{-19}$C, and permittivity $\varepsilon_0 = (1/36\pi) \times 10^{-9}$ F/m]

[2017 : 2 M, Set-2]

1.33 In the table shown, List-I and List-II, respectively, contain terms appearing on the left-hand side and the right-hand side of Maxwell's equations (in their standard form). Match the left-hand side with the corresponding right-hand side.

List-I	List-II
1. $\nabla \cdot \vec{D}$	P. 0
2. $\nabla \times \vec{E}$	Q. ρ
3. $\nabla \cdot \vec{B}$	R. $-\dfrac{\partial \vec{B}}{\partial t}$
4. $\nabla \times \vec{H}$	S. $\bar{J} + \dfrac{\partial \vec{D}}{\partial t}$

(a) 1-R, 2-Q, 3-S, 4-P
(b) 1-Q, 2-S, 3-P, 4-R
(c) 1-P, 2-R, 3-Q, 4-S
(d) 1-Q, 2-R, 3-P, 4-S

[2019 : 1 M]

1.34 What is the electric flux $\left(\int \vec{E} \cdot d\hat{a}\right)$ through a quarter-cylinder of height H (as shown in the figure) due to an infinitely long line charge along the axis of the cylinder with a charge density of Q?

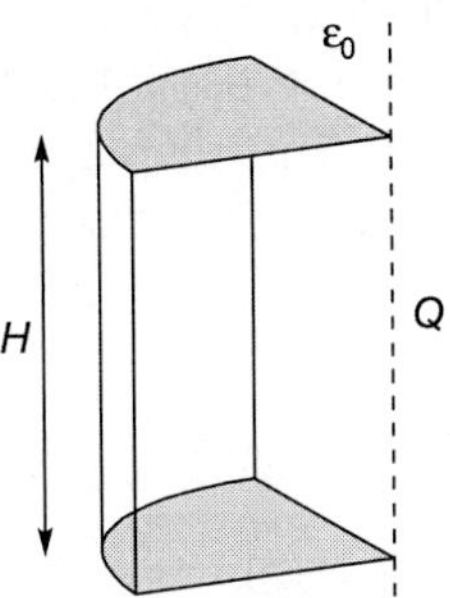

(a) $\dfrac{4H}{Q\varepsilon_0}$ (b) $\dfrac{H\varepsilon_0}{4Q}$

(c) $\dfrac{HQ}{\varepsilon_0}$ (d) $\dfrac{HQ}{4\varepsilon_0}$

[2019 : 1 M]

1.35 Two identical copper wires W_1 and W_2 placed in parallel as shown in the figure, carry currents I and $2I$, respectively, in opposite directions. If the two wires are separated by a distance of $4r$, then the magnitude of the magnetic field $\vec{B}$ between the wires at a distance r from W_1 is

W_1 ↕ r

W_2

(a) $\frac{\mu_0^2 I^2}{2\pi r^2}$ (b) $\frac{6\mu_0 I}{5\pi r}$

(c) $\frac{\mu_0 I}{6\pi r}$ (d) $\frac{5\mu_0 I}{6\pi r}$

[2019 : 2 M]

1.36 Consider the vector field

$\vec{F} = \hat{a}_x(4y - c_1 z) + \hat{a}_y(4x + 2z) + \hat{a}_z(2y + z)$

in a rectangular coordinate system (x, y, z) with unit vectors $\hat{a}_x, \hat{a}_y,$ and $\hat{a}_z$. If the field $\vec{F}$ is irrotational (conservative), then the constant c_1 (in integer) is ________.

[2021 : 1 M]

1.37 For a vector field $\vec{D} = \rho\cos^2\varphi\hat{a}_\rho + z^2\sin^2\varphi\hat{a}_\varphi$ in a cylindrical coordinate system (ρ, φ, z) with unit vectors $\hat{a}_\rho, \hat{a}_\varphi$ and $\hat{a}_z$, the net flux of $\vec{D}$ leaving the closed surface of the cylinder $(\rho = 3, 0 \le z \le 2)$ (rounded off to two decimal places) is ______.

[2021 : 2 M]

1.38 In a circuit, there is a series connection of an ideal resistor and an ideal capacitor. The conduction current (in Amperes) through the resistor is $2\sin(t + \pi/2)$. The displacement current (in Amperes) through the capacitor is ________.

(a) $2\sin(t)$ (b) $2\sin(t + \pi)$

(c) $2\sin(t + \pi/2)$ (d) 0

[2022 : 1 M]

1.39 In an electrostatic field, the electric displacement density vector, $\vec{D}$, is given by

$$\vec{D}(x, y, z) = (x^3\vec{i} + y^3\vec{j} + xy^2\vec{k})\ \text{C/m}^2,$$

where $\vec{i}, \vec{j}, \vec{k}$ are the unit vectors along x-axis, y-axis, and z-axis, respectively. Consider a cubical region R centered at the origin with each side of length 1 m, and vertices at (±0.5 m, ±0.5 m, ±0.5 m). The electric charge enclosed within R is ________ C (rounded off to two decimal places).

[2022 : 2 M]

1.40 Let $\hat{i}$ and $\hat{j}$ be the unit vectors along x and y axes, respectively and let A be a positive constant. Which one of the following statements is true for the vector fields $\vec{F}_1 = A(\hat{i}y + \hat{j}x)$ and $\vec{F}_2 = A(\hat{i}y - \hat{j}x)$?

(a) Neither $\vec{F}_1$ nor $\vec{F}_2$ is an electrostatic field.

(b) Only $\vec{F}_2$ is an electrostatic field

(c) Only $\vec{F}_1$ is an electrostatic field.

(d) Both $\vec{F}_1$ and $\vec{F}_2$ are electrostatic fields.

[2024 : 1 Mark]

1.41 Consider the Earth to be a perfect sphere of radius R. Then the surface area of the region, enclosed by the 60°N latitude circle, that contains the north pole in its interior is ________.

(a) $(2-\sqrt{3})\pi R^2$ (b) $\frac{(\sqrt{2}-1)\pi R^2}{2}$

(c) $\frac{(2+\sqrt{3})\pi R^2}{8\sqrt{2}}$ (d) $\frac{2\pi R^2}{3}$

[2024 : 2 Marks]

■■■

Answers Basics of Electromagnetics

1.1 (b)	1.2 (b)	1.3 (a)	1.4 (c)	1.5 (d)	1.6 (c)	1.7 (c)
1.8 (c)	1.9 (c)	1.10 (d)	1.11 (d)	1.12 (d)	1.13 (b)	1.14 (c)
1.15 (a)	1.16 (c)	1.17 (a)	1.18 (c)	1.19 (d)	1.20 (15)	1.21 (d)
1.22 (c)	1.23 (–3)	1.24 (78.54)	1.25 (–0.25)	1.26 (*)	1.27 (a)	1.28 (c)
1.29 (a)	1.30 (d)	1.31 (c)	1.32 (5.06)	1.33 (d)	1.34 (d)	1.35 (d)
1.36 (0)	1.37 (56.55)	1.38 (c)	1.39 (0.5)	1.40 (c)	1.41 (a)	

Explanations Basics of Electromagnetics

1.1 (b)

The radially outward field is normal to conductor and no induction effects E is same.

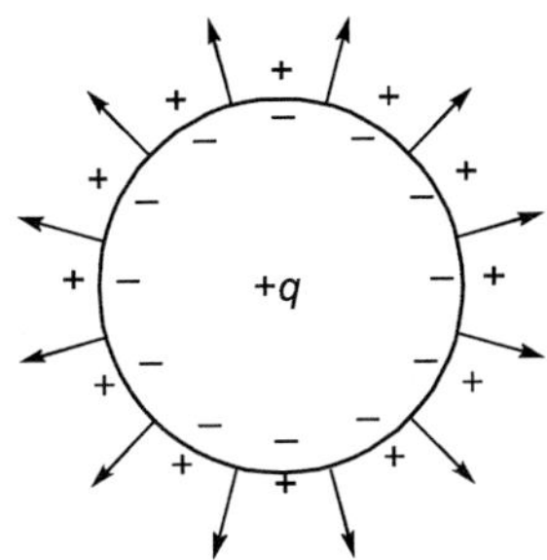

1.2 (b)

$$\vec{E} = \frac{\rho_s}{2\epsilon}\hat{a}_n$$

Where, $\hat{a}_n = \hat{a}_y$

$$\rho_s = \frac{1}{6000\pi}\times 10^{-6}\ \text{C/m}^2$$

$$\vec{E} = \frac{10^{-6}\times\pi}{2\times 8.854\times 10^{-12}\times 6000\pi}\hat{a}_y$$

$$\vec{E} = 3\hat{a}_y = 3j\ \text{V/m}$$

1.3 (a)

Net flux leaving = Charged enclosed

Cube encloses the sphere completely.

$$\oiint \vec{D}\cdot\vec{ds} = Q_{\text{enclosed}}$$

$$Q_{\text{enclosed}} = \rho_s A = \rho_s(4\pi r^2) = 10(4\pi\times 1^2) = 40\pi$$

1.4 (c)

$$V = \frac{Q}{4\pi\epsilon_o r}\text{(monopole)}$$

$$V = \frac{Qd\cos\theta}{4\pi\epsilon_o r^2}\text{(dipole)}$$

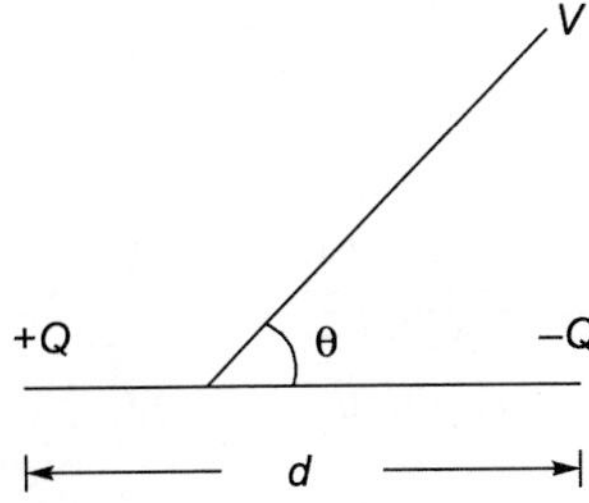

1.5 (d)

The surface change density is equally to normal flux density

$$D_{\text{normal}} = \rho_s$$

$$D = \rho_s = \epsilon E = 80\cdot\epsilon\cdot 2$$

1.6 (c)

$$\nabla\times\vec{H} = \frac{\partial\bar{D}}{\partial t}+\vec{J}$$

$$\frac{1}{m}\times\frac{\text{Ampere}}{\text{m}} = \frac{\text{A}}{\text{m}^2}$$

1.7 (c)

$$V = -\int\vec{E}\cdot\vec{dl} = -\left[\int_1^2 x\,dx+\int_2^0 y\,dy+\int_3^0 z\,dz\right]$$

$$= -\left[\frac{x^2}{2}\Big|_1^2+\frac{y^2}{2}\Big|_2^0+\frac{z^2}{2}\Big|_3^0\right]$$

$$= -\frac{1}{2}[2^2-1^2+0^2-2^2+0^2-3^2]$$

$$= -\frac{1}{2}\times -10 = 5\text{V}$$

1.8 (c)

$$\vec{E}_1 = 2u_x-3u_y+1u_z$$

$$\vec{E}_{1t} = -3u_y+u_z = E_{2t}\qquad (x = 0\text{ plane})$$

$$\vec{E}_{1n} = 2u_x$$

$$\vec{D}_{1n} = \vec{D}_{2n} = \epsilon\vec{E}$$

$$\Rightarrow\quad \epsilon_1\vec{E}_{1n} = \epsilon_2\vec{E}_{2n}$$

$$= 1.5\epsilon_o.2u_x = 2.5\epsilon_o.\vec{E}_{2n}$$

$$\vec{E}_{2n} = \frac{3}{2.5}u_x = 1.2\,u_x$$

$$\vec{E}_2 = \vec{E}_{2t}+\vec{E}_{2n}$$

$$\vec{E}_2 = -3u_y+u_z+1.2\,u_x$$

1.9 (c)

$$\vec{E}_{t_1} = 3\hat{a}_y+5\hat{a}_z = \vec{E}_{t_2}$$

$$\vec{D}_{n_1} = \vec{D}_{n_2}$$

$$\epsilon_2\vec{E}_{n_2} = \epsilon_1\vec{E}_{n_1}$$

$$\vec{E}_{n_2} = \frac{3\times 4\hat{a}_x}{4} = 3\hat{a}_x$$

$$\therefore\quad \vec{E}_2 = \vec{E}_{t_2}+\vec{E}_{n_2} = 3\hat{a}_y+5\hat{a}_z+3\hat{a}_x$$

1.11 (d)

Intensity - curl - line integral
Density - divergence - surface integral
This is order of understanding Maxwell's equations.
For any material

$$\nabla \times E = 0 \qquad \nabla \cdot B = 0$$
$$\nabla \times H = J \qquad \nabla \cdot D = \rho_v$$

For source free like charge or current free conditions

$$\nabla \times E = 0 \qquad \nabla \cdot B = 0$$
$$\nabla \times H = 0 \qquad \nabla \cdot D = 0$$

If $D = \in E$ and $B = \mu H$ and $\in$ or μ are constant with distance or direction then

$$\nabla \cdot (\in E) = 0 \qquad \nabla \times \frac{B}{\mu} = 0$$
$$\nabla \cdot E = 0 \qquad \nabla \times B = 0$$

These are called as homogeneous/isotropic medium.
For non-homogeneous material. However they are not valid.

1.12 (d)

$H_{direction}$ due to any current wire
= I flow direction × Radial vector from current to point of consideration
For wire in the y-z plane,

$$H_{direction} = \hat{a}_y \times (-\hat{a}_z) = -\hat{a}_x$$

For wire in the x-y plane,

$$H_{direction} = \hat{a}_x \times (-\hat{a}_y) = -\hat{a}_z$$

Therefore, resultant magnetic field has x and z components at the origin.

1.14 (c)

Applying $\nabla \cdot H = J$

$$\nabla \times B = \begin{vmatrix} \hat{a}_x & \hat{a}_y & \hat{a}_z \\ \frac{\partial}{\partial x} & \frac{\partial}{\partial y} & \frac{\partial}{\partial z} \\ \frac{-y}{x^2+y^2} & \frac{x}{x^2+y^2} & 0 \end{vmatrix} = 0$$

$$J = 0$$

1.15 (a)

For magnetic field boundary relations are

$$\vec{B}_{n_1} = \vec{B}_{n_2}$$

and $\vec{H}_{t_1} - \vec{H}_{t_2} = -\vec{J}_s \times \vec{a}_n$

$$\Rightarrow \quad B_{x1} = B_{x2} \quad (x \text{ is normal component})$$
$$\mu_1 H_{x1} = \mu_2 H_{x2}$$
$$\Rightarrow \quad 1 \times 3 = 2 \times H_{x2}$$
$$H_{x2} = 1.5$$
$$\Rightarrow \quad \vec{H}_{t_1} - \vec{H}_{t_2} = -10\hat{u}_y \times \hat{u}_x = +10\hat{u}_z$$
$$\vec{H}_{t_2} = H_{t_1} - 10\hat{u}_z$$
$$\vec{H}_{t_2} = 30\hat{u}_y - 10\hat{u}_z$$
$$\Rightarrow \quad \vec{H}_2 = 1.5\,\hat{u}_x + 30\hat{u}_y - 10\hat{u}_z \text{ A/m}$$

1.16 (c)

We know that magnetic flux density at a distance r from the wire is,

$$|B| = \frac{\mu_0 I}{2\pi r}$$

For $r < a \rightarrow \quad I = J \cdot \pi r^2$

$$\text{so} \quad |B| = \frac{\mu_0 Jr}{2}$$

$|B| \propto r$ for $r < a$

for $r > a$, $\quad I = J \times \pi a^2$

$$\text{So,} \quad |B| = \frac{\mu_0 J \pi a^2}{2\pi r}$$

$$|B| \propto \frac{1}{r} \text{ for } r > a.$$

1.17 (a)

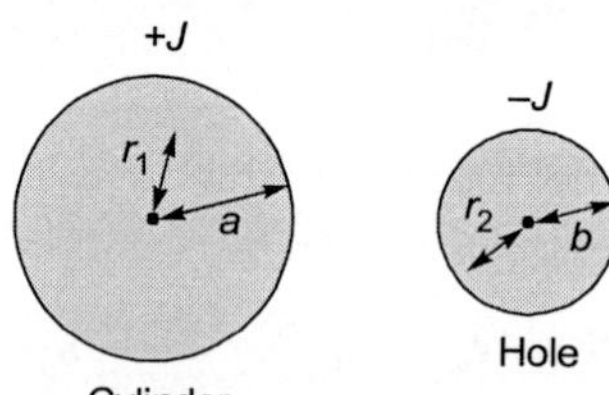

The H field due to entire cylinder without hole.

$$H(r_1) = \frac{Jr_1}{2}$$

The H field due to hole only.

$$H(r_2) = \frac{Jr_2}{2}$$

Current inside the hole is zero means

$$H(r) = \frac{Jr_1}{2} - \frac{Jr_2}{2} \simeq \frac{Jd}{2}$$

1.19 (d)

Since, we know from the property of magnetic boundary conditions

$$\vec{H}_{t2} - \vec{H}_{t1} = \hat{K}_s \times a_N$$

In the given question the tangential field $\vec{H}$ in the air just above the perfect conductor is asked with $H = 0$ inside the conductor.

$$\vec{H}_{t1} = 0$$

$$\vec{H}_{t2} = \hat{x}\,2 \times \hat{y}$$

$$\boxed{\vec{H}_{t2} = \hat{z}\,2 \text{ A/m}}$$

1.20 (15)

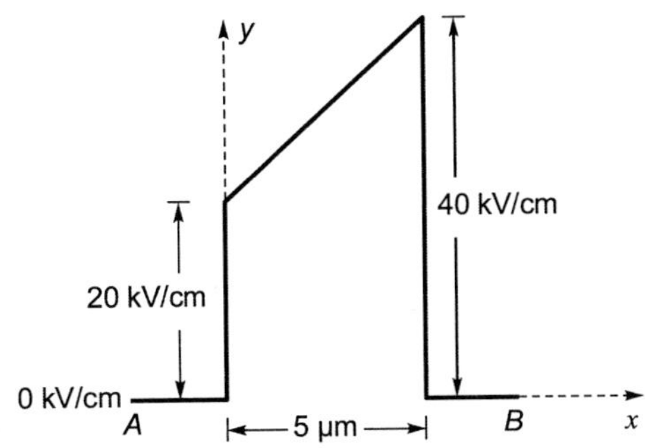

E is linear from A to B and $E = mx + c$

$$V = -\int_B^A (mx + c)\,dx$$

At point A: $x = 0$, $E = c = 2 \times 10^6$ V/m

At point B: $x = 5 \times 10^{-6}$

$$E = 4 \times 10^6 = m \times 5 \times 10^{-6} + 2 \times 10^6$$

$$m = 4 \times 10^{11}$$

$$\Psi_A - \Psi_B = -\int_B^A E \cdot dl$$

$$= -\int_{5\,\mu m}^{0} (mx + c)dx = -\left[cx + \frac{mx^2}{2}\right]_{5\,\mu m}^{0}$$

$$\Psi_A - \Psi_B = \left[(20 \times 10^5 \times 5 \times 10^{-6}) + \frac{4 \times 10^{11}}{2}[(5 \times 10^{-6})^2 - 0^2]\right]$$

$$= [10 + (2 \times 10^{11} \times 25 \times 10^{-12})]$$

$$= 15 \text{ V}$$

1.21 (d)

$$E = -\frac{\partial V}{\partial x}\hat{a}_x - \frac{\partial V}{\partial y}\hat{a}_y - \frac{\partial V}{\partial z}\hat{a}_z$$

$$\frac{\partial V}{\partial x} = 2y^3 - 3yz^2 \qquad V = 2xy^3 - 3xyz^2 + K_1$$

$$\frac{\partial V}{\partial y} = 6xy^2 - 3xz^2 \qquad V = 2xy^3 - 3xyz^2 + K_2$$

$$\frac{\partial V}{\partial z} = -6xyz \qquad V = -3xyz^2 + K_3$$

Comparing all three for uniqueness solution of potential

$$K_1 = K_2 = 0, \qquad K_3 = 2xy^3$$

$$\vec{E} = -(2y^3 - 3yz^2)\hat{x} - (6xy^2 - 3xz^2)\hat{y} + (6xyz)\hat{z}$$

as we know that $E = -\nabla V$

$$E = -\left(\frac{\partial}{\partial x}\hat{x} + \frac{\partial}{\partial y}\hat{y} + \frac{\partial}{\partial z}\hat{z}\right)V \qquad \text{...(i)}$$

Clearly option (d) satisfies the given relation.

1.22 (c)

$$H_\phi = \frac{I}{2\pi r}\hat{a}_\phi$$

$$\Rightarrow \quad \boxed{H_\phi \propto \frac{1}{r}} \quad \text{option (c) is satisfied.}$$

1.23 (-3)

$$\phi = 2x^2 + y^2 + cz^2$$

In source-free region,

$$\nabla \cdot \vec{D} = 0$$

$$\Rightarrow \quad \in(\nabla \cdot \vec{E}) = 0 \quad \Rightarrow \nabla \cdot \vec{E} = 0$$

where, $\vec{E} = -\nabla V = -\nabla\phi$

Now, $$\nabla V = \frac{\partial\phi}{\partial x}\hat{a}_x + \frac{\partial\phi}{\partial y}\hat{a}_y + \frac{\partial\phi}{\partial z}\hat{a}_z$$

$$\Rightarrow \quad -\vec{E} = 4x\,\hat{a}_x + 2y\,\hat{a}_y + 2cz\,\hat{a}_z$$

Again $\nabla \cdot \vec{E} = 0$

$$\Rightarrow \quad \frac{\partial E_x}{\partial x} + \frac{\partial E_y}{\partial y} + \frac{\partial E_z}{\partial z} = 4 + 2 + 2c = 0$$

$$c = -3$$

1.24 (78.54)

$$\vec{D} = 2\rho^2\hat{a}_\rho + z\hat{a}_z,$$

surface: $\rho = 1$, $z = 0$ and $z = 5$

$$\oiint_s \vec{D} \cdot \vec{ds} = \int_{\text{Top surface}} \vec{D} \cdot \vec{ds} + \int_{\text{Bottom surface}} \vec{D} \cdot \vec{ds} + \int_{\text{Curved surface}} \vec{D} \cdot \vec{ds}$$

$$= 5\pi + 0 + 20\pi = 25\pi = 78.54$$

1.25 (-0.25)

$$\oint \vec{D} \cdot \vec{ds} = Q \quad \text{(charge enclosed)}$$

$$Q_1 + Q_2 + Q_3 = 0$$

For $D = 0$

$$\rho_{s1} \cdot 4\pi 2^2 + \rho_{s2} \cdot 4\pi \cdot 4^2 + \rho_{s3} \cdot 4\pi \cdot 8^2 = Q = 0$$

$$20 \cdot 4 - 4.4^2 + \rho_{s3} \cdot 8^2 = 0$$

$$80 - 64 + \rho_{s3} \cdot 8^2 = 0$$

$$\rho_{s3} = \frac{-16}{64} = -0.25 \text{ nC/m}^2$$

1.26 (*)

$$I = \int \vec{J}\cdot\vec{ds}$$

$$= \int_{\theta=\frac{\pi}{12}}^{\pi/4}\int_{\phi=0}^{2\pi} \frac{400\sin\theta}{2\pi(r^2+4)}.r^2\sin\theta d\theta\, d\phi$$

$$= \frac{400}{2\pi(r^2+4)}\cdot r^2\cdot\phi\Big|_0^{2\pi}\int_{\pi/12}^{\pi/4}\sin^2\theta\, d\theta$$

$$= \frac{400r^2}{(r^2+4)}\int_{\pi/12}^{\pi/4}\left(\frac{1-\cos 2\theta}{2}\right)d\theta$$

$$= \frac{400\cdot r^2}{(r^2+4)}\left(\frac{\left(\frac{\pi}{4}-\frac{\pi}{12}\right)}{2}-\left(\frac{\sin 2\theta}{4}\right)_{\pi/12}^{\pi/4}\right)$$

$$= \frac{400\cdot r^2}{(r^2+4)}\left(\frac{\pi}{12}-\left(\frac{1-1/2}{4}\right)\right)\Bigg|_{r=0.8}$$

$$= \frac{400\times0.8\times0.8}{4.64}\times0.13 = 7.17 \text{ Amp}$$

Total area $= \int ds = \iint r^2\sin\theta d\theta\, d\phi$

$$= r^2\int_{\theta=\frac{\pi}{12}}^{\pi/4}\sin\theta d\theta\cdot2\pi$$

$$= r^2\cdot2\pi\cdot0.259\big|_{r=0.8}$$

$$= 0.8^2\times0.5\times2\pi\times\frac{1}{4} = 1.041 \text{ m}^2$$

Average current $= \frac{7.17}{1.041} = 6.88 \text{ A/m}^2$

Note : GATE key mentioned (MTA - marks to all)

1.27 (a)

Ba_z magnetic field, $v_xa_x + v_za_z$ velocity

$$F = Q(v\times B) \quad \text{by Lorentz's law}$$

$$= Q(v_xa_x + v_za_z)\times Ba_z$$

$$F_y = Q\,v_x\,.\,B(-a_y)$$

This results in a circular path in the XY plane with v_za_z component causing a linear path.

Both result in a helical path along Z axis.

1.28 (c)

Velocity being free velocity,

$$\frac{1}{2}mv^2 = qV$$

$$v = \frac{d}{t} = \sqrt{\frac{2qV}{m}}$$

$$\Rightarrow \quad t \propto \frac{d}{\sqrt{V}}$$

1.29 (a)

Let, $E = E_1$, Energy $E_1 = \frac{Q_1^2}{2C_1}$

Electrically isolated $\Rightarrow Q_2 = Q_1$

$d_2 = 2d_1 \quad \Rightarrow \quad C_2 = \frac{C_1}{2}$

$$E_2 = \frac{Q_2^2}{2C_2} = \frac{Q_1^2}{2\frac{C_1}{2}} = 2\left(\frac{Q_1^2}{2C_1}\right)$$

$$= 2E_1 = 2E$$

1.30 (d)

Applying Poisson's equations

$$\nabla^2V = \frac{\partial^2V}{\partial x^2} = -\frac{\rho_v}{\epsilon} = K$$

Constant charge density

$$\frac{\partial V}{\partial x} = -Kx + K'$$

$$V = \frac{-Kx^2}{2} + K'x + K''$$

Towards positive x or negative side.

It is a second order parabolic increase.

Due to symmetry of + and – charges $K'' = 0$ is expected with $V = 0$ at centre and graph passing through origin.

Beyond $x > a$ or $x < b$, $E = 0$ due to capacitive nature of + and – charges

$$V = -\int 0\cdot\vec{dl} = \text{Constant}$$

This constant is same at $x = a$ or $x = b$.

1.31 (c)

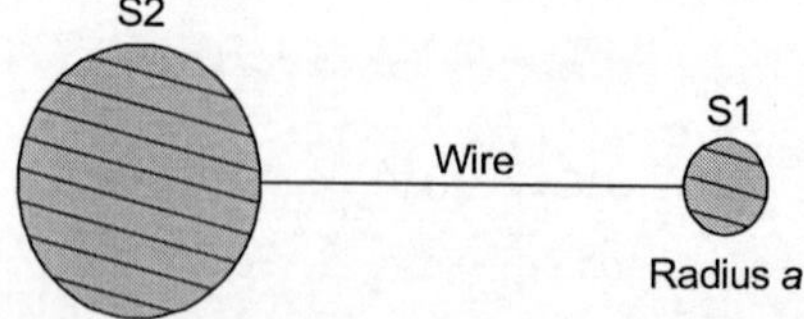

- When charge is placed on this structure, equilibrium is established such that both spheres are at same potential i.e.

$$V_a = V_b$$

- $V_a = V_b$

So, $$\frac{Q_a}{4\pi\epsilon_o a} = \frac{Q_b}{4\pi\epsilon_o b}$$

$$\frac{Q_b}{Q_a} = \frac{b}{a}$$

- Now, surface electric fields,

$$\frac{E_a}{E_b} = \left[\frac{Q_a/4\pi\varepsilon_o a^2}{Q_b/4\pi\varepsilon_o b^2}\right] = \frac{Q_a \times b^2}{Q_b \times a^2} = \frac{b}{a} > 1$$

So, $E_a > E_b$.

1.32 (5.06)

q_1 $v = 10^5$ m/s q_2

moving ← r → stationary

r is the distance at which kinetic energy of q_1 becomes zero [(because kinetic energy (KE) is converted into potential energy (PE)].

When q_1 reaches 'r', it starts diverting.

Kinetic energy, $KE = \frac{1}{2}mv^2$ and work done in moving q_1 charge to distance 'r' is

$$q_1 V_2 = q_1 \frac{q_2}{4\pi\varepsilon_0 r},$$

$$(q_1 = q_2 = -1.6 \times 10^{-19}\text{ C})$$

Now, $$\frac{1}{2}mv^2 = \frac{q_1 q_2}{4\pi\varepsilon_0 r}$$

$$\Rightarrow r = \frac{(2\times -1.6\times 10^{-19})\times(-1.6\times 10^{-19})}{4\pi\times\frac{10^{-9}}{36\pi}\times 9.11\times 10^{-31}\times(10^5)^2}$$

$$\simeq 5.06 \times 10^{-8}\text{ m}$$

1.33 (d)

$$\nabla\cdot\vec{D} = \rho_v$$

$$\nabla\times\vec{E} = -\frac{\partial\vec{B}}{\partial t}$$

$$\nabla\cdot\vec{B} = 0$$

$$\nabla\times\vec{H} = \vec{J} + \frac{\partial\vec{D}}{\partial t}$$

1.34 (d)

Electric field intensity ($\vec{E}$) at 'ρ' distance due to infinite long line having line charge density Q is

$$\vec{E} = \frac{Q}{2\pi\varepsilon_0\rho}\hat{a}_\rho$$

$$\int\vec{E}\cdot\overrightarrow{da} = \iint\frac{Q}{2\pi\varepsilon_0\rho}\hat{a}_\rho\cdot\rho\, d\phi\, dz\hat{a}_\rho$$

$$= \frac{Q}{2\pi\varepsilon_0}\int_{<\pi/2>} d\phi\int_{z=0}^{H} dz = \frac{Q}{2\pi\varepsilon_0}\left(\frac{\pi}{2}\right)H = \frac{HQ}{4\varepsilon_0}$$

1.35 (d)

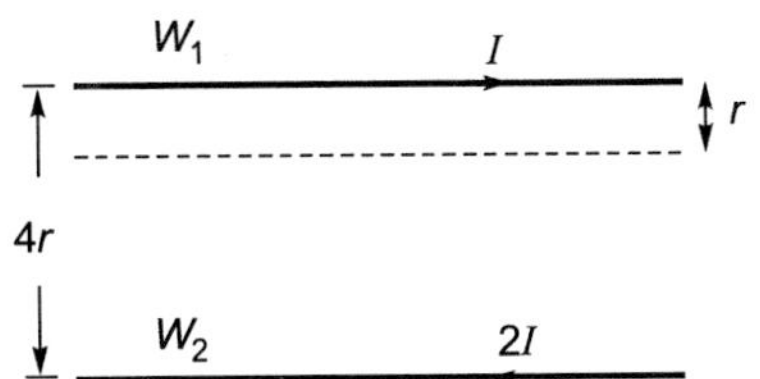

Magnetic flux density ($\vec{B}$) at r distance due to infinite line carrying current I is $\left|\vec{B}\right| = \frac{\mu_0 I}{2\pi r}$.

- $\vec{B}$ at r distance due to W_1 wire

$$= \left|\vec{B}_1\right| = \frac{\mu_0 I}{2\pi r} \quad \text{...(i)}$$

- $\vec{B}$ at $3r$ distance due to W_2 wire

$$= \left|\vec{B}_2\right| = \frac{\mu_0(2I)}{2\pi(3r)} \quad \text{...(ii)}$$

From right hand thumb rule, $\vec{B}$ due to both lines add in between conductors.

So, $$\left|\vec{B}\right| = \left|\vec{B}_1\right| + \left|\vec{B}_2\right|$$

$\therefore$ $$\left|\vec{B}\right| = \frac{\mu_0 I}{2\pi r} + \frac{2\mu_0 I}{6\pi r} = \frac{5\mu_0 I}{6\pi r}$$

1.36 (0)

$$\nabla\times\vec{F} = 0$$

$$\begin{vmatrix} i & j & k \\ \frac{\partial}{\partial x} & \frac{\partial}{\partial y} & \frac{\partial}{\partial z} \\ 4y - c_1 z & 4x + 2z & 2y + z \end{vmatrix} = 0$$

$$= i(2-2) - j(0 + c_1) + k(4-4) = 0$$

$$c_1 = 0$$

1.37 (56.55)

Method 1:

$$\vec{D} = \rho\cos^2\phi\hat{a}_\rho + z^2\sin^2\phi\hat{a}_\phi$$

Electric flux crossing the closed surface is

$$\psi = \oiint\vec{D}\cdot\overrightarrow{dS} = \iiint(\vec{\nabla}\cdot\vec{D})dv$$

$$\vec{\nabla}\cdot\vec{D} = \frac{1}{\rho}\frac{\partial}{\partial\rho}(\rho D_\rho) + \frac{1}{\rho}\frac{\partial D_\phi}{\partial\phi} + \frac{\partial D_z}{\partial z}$$

$$= \frac{1}{\rho}\frac{\partial}{\partial\rho}\left(\rho\rho\cos^2\phi\right) + \frac{1}{\rho}\frac{\partial}{\partial\phi}(z^2\sin^2\phi) + 0$$

$$= \frac{1}{\rho}(2\rho)\cos^2\phi + \frac{1}{\rho}z^2 2\sin\phi\cos\phi$$

$$= 2\cos^2\phi + \frac{z^2}{\rho}\sin 2\phi$$

$$\iiint(\vec{\nabla}\cdot\vec{D})dv = \iiint 2\cos^2\phi(\rho d\rho d\phi dz) + \iiint\left(\frac{z^2}{\rho}\sin 2\phi\right)\rho d\rho d\phi dz$$

$$= 2\int_{\rho=0}^{3}\rho d\rho\int_{\phi=0}^{2\pi}\left(\frac{1+\cos 2\phi}{2}\right)d\phi\int_{z=0}^{2}dz + \int_{\rho=0}^{2}d\rho\int_{\phi=0}^{2\pi}\sin 2\phi d\phi\int_{z=0}^{2}z^2dz$$

$$= 2\left(\frac{\rho^2}{2}\right)_{\rho=0}^{3}\frac{1}{2}(2\pi)(z)_{z=0}^{2} + 0$$

$$= 2\left(\frac{3^2}{2}\right)\pi(2) = 18\pi \text{ (Coulomb)}$$

= 56.55 (Coulomb)

Method 2: Electric flux crossing the closed surface is

$$\psi = \oint\!\!\oint \vec{D}\cdot\vec{dS}$$

Electric flux crossing $\rho = 3$ cylindrical surface is

$$\psi|_{\rho=3} = \oint\!\!\oint(\rho\cos^2\phi\hat{a}_\rho)\cdot(\rho d\phi dz)\hat{a}_\rho$$

$$= 3^2\int_{\phi=0}^{2\pi}\cos^2\phi d\phi\int_{z=0}^{2}dz$$

$$= 9\frac{1}{2}(2\pi)(2) = 18\pi\text{(Coulomb)}$$

= 56.55 (Coulomb)

1.38 (c)

For a series connection of an ideal resistor and ideal capacitor shown below,

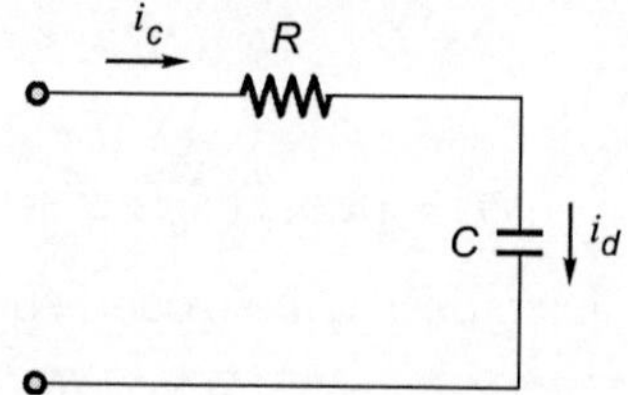

the current through both the elements remains same. Hence,

$$i_d = i_c$$

$$\therefore \quad i_d = 2\sin\left(t+\frac{\pi}{2}\right)\text{A}$$

1.39 (0.5)

The electric displacement density vector,

$$\vec{D}(x,y,z) = (x^3\vec{i} + y^3\vec{j} + xy^2\vec{k})\text{ C/m}^2$$

For a cubical region R, the charge enclosed,

$$Q_{enc} = \int_V \rho_V \cdot dV$$

where ρ_v = volume charge density

Using Gauss's law, $\vec{\nabla}\cdot\vec{D} = \rho_v$, we get,

$$Q_{enc} = \int_V(\vec{\nabla}\cdot\vec{D})\,dV$$

$$\vec{\nabla}\cdot\vec{D} = \frac{\partial}{\partial x}(x^3) + \frac{\partial}{\partial y}(y^3) + \frac{\partial}{\partial z}(xy^2)$$

$$\vec{\nabla}\cdot\vec{D} = 3x^2 + 3y^2 = 3(x^2+y^2)$$

Hence, $$Q_{enc} = \int_V 3(x^2+y^2)\,dx\,dy\,dz$$

$$Q_{enc} = 3\int_{x=-0.5}^{0.5}\int_{y=-0.5}^{0.5}\int_{z=-0.5}^{0.5}(x^2+y^2)\,dz\,dy\,dz$$

$$= 3\left[\left\{\frac{x^3}{3}\right\}_{x=-0.5}^{0.5}\{y\}_{y=-0.5}^{0.5}\{z\}_{z=-0.5}^{0.5} + \{x\}_{x=-0.5}^{0.5}\left\{\frac{y^3}{3}\right\}_{y=-0.5}^{0.5}\{z\}_{z=-0.5}^{0.5}\right]$$

$$Q_{enc} = \left[((0.5)^3-(-0.5)^3)(0.5+0.5)(0.5+0.5)\times 2\right]$$

$$Q_{enc} = 0.50\text{ C}$$

1.40 (c)

For an electrostatic field, the field lines do not circulate back on themselves; therefore, the curl of the electrostatic field is zero i.e. $\overline{\nabla}\times\overline{F} = 0$

$$\overline{F}_1 = A\left[y\hat{i} + x\hat{j}\right]$$

$$\Rightarrow \quad \overline{\nabla}\times\overline{F}_1 = A\begin{vmatrix}\hat{i} & \hat{j} & \hat{k}\\ \partial/\partial x & \partial/\partial y & \partial/\partial z\\ y & x & 0\end{vmatrix}$$

$$= A\left[0\hat{i} - 0\hat{j} + (1-1)\hat{k}\right]$$

$$= 0$$

$$\overline{F}_2 = A\left[y\hat{i} - x\hat{j}\right]$$

$$\Rightarrow \quad \overline{\nabla}\times\overline{F}_2 = A\begin{vmatrix}\hat{i} & \hat{j} & \hat{k}\\ \partial/\partial x & \partial/\partial y & \partial/\partial z\\ y & -x & 0\end{vmatrix}$$

$$= A\left[0\hat{i} - 0\hat{j} + (-1-1)\hat{k}\right]$$

$$= -2A\hat{k}$$

Hence, $\overline{F}_1$ is electrostatic field and $\overline{F}_2$ is not electrostatic field.

1.41 (a)

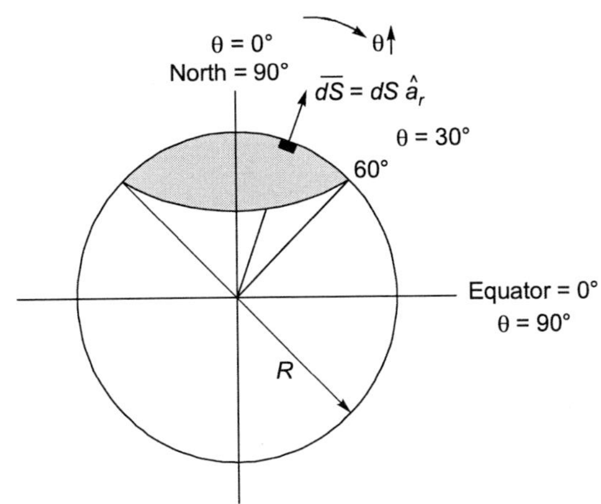

From the figure

Earth's North pole = 90°N Latitude

Earth's equator = 0° Latitude

In spherical co-ordinate system

Earth's North pole, θ = 0°

Earth's equator, θ = 90°

60°N latitude circle corresponds to θ = 30°

The area of the shaded portion,

$$\overline{dS} = dS\,\hat{a}_r$$

Hence, in spherical coordinate system

$$\overline{dS} = r^2 \sin\theta\, d\theta\, d\phi\, \hat{a}_r$$

$$\Rightarrow \quad S = \int dS = \int r^2 \sin\theta\, d\theta\, d\phi$$

$$= r^2 \int_{\theta=0^\circ}^{30^\circ} \sin\theta\, d\theta \int_{\phi=0}^{2\pi} d\phi \quad \text{at } r = R$$

$$= R^2 \left[-\cos\theta\Big|_0^{30^\circ}\right] \cdot \phi\Big|_0^{2\pi}$$

$$= R^2 \{-[\cos 30^\circ - \cos 0^\circ]\} \cdot 2\pi$$

$$= R^2 \left\{-\left[\frac{\sqrt{3}}{2} - 1\right]\right\} \cdot 2\pi = R^2 \cdot \frac{2-\sqrt{3}}{2} \cdot 2\pi$$

$$= (2-\sqrt{3})\pi R^2$$

■■■

2 Uniform Plane Waves

2.1 The intrinsic impedance of a lossy dielectric medium is given by

(a) $\frac{j\omega\mu}{\sigma}$ (b) $\frac{j\omega\epsilon}{\mu}$

(c) $\sqrt{\frac{j\omega\mu}{(\sigma+j\omega\epsilon)}}$ (d) $\sqrt{\frac{\mu}{\epsilon}}$

[1995 : 1 M]

2.2 Copper behaves as a

(a) conductor always

(b) conductor or dielectric depending on the applied electric field strength

(c) conductor or dielectric depending on the frequency

(d) conductor or dielectric depending on the electric current density

[1995 : 1 M]

2.3 A uniform plane wave in air is normally incident on an infinitely thick slab. If the refractive index of the glass slab is 1.5, then the percentage of the incident power that is reflected from the air-glass interface is

(a) 0% (b) 4%

(c) 20% (d) 100% [1996 : 2 M]

2.4 Some unknown material has a conductivity of 10^6 mho/m and a permeability of $4\pi \times 10^{-7}$ H/m. The skin depth for the material at 1 GHz is

(a) 15.9 μm (b) 20.9 μm

(c) 25.9 μm (d) 30.9 μm

[1996 : 2 M]

2.5 The intrinsic impedance of copper at high frequency is

(a) purely resistive

(b) purely inductive

(c) complex with a capacitive component

(d) complex with an inductive component

[1998 : 1 M]

2.6 The wavelength of a wave with propagation constant $(0.1\,\pi + j\,0.2\,\pi)\ \text{m}^{-1}$ is

(a) $\frac{2}{\sqrt{0.05}}$ m (b) 10 m

(c) 20 m (d) 30 m

[1998 : 1 M]

2.7 The depth of penetration of a wave in a lossy dielectric increases with increasing

(a) conductivity (b) permeability

(c) wavelength (d) permittivity

[1998 : 1 M]

2.8 The polarization of a wave with electric field vector $\vec{E} = E_o e^{j(\omega t-\beta z)}(\vec{a}_x + \vec{a}_y)$ is

(a) linear (b) elliptical

(c) left hand circular (d) right hand circular

[1998 : 1 M]

2.9 The time averaged Poynting vector, in W/m^2, for a wave with $\vec{E} = 24e^{j(\omega t+\beta z)}\vec{a}_y$ V/m in free space is

(a) $-\frac{2.4}{\pi}\vec{a}_z$ (b) $\frac{2.4}{\pi}\vec{a}_z$

(c) $\frac{4.8}{\pi}\vec{a}_z$ (d) $-\frac{4.8}{\pi}\vec{a}_z$

[1998 : 1 M]

2.10 A loop is rotating about the *y*-axis in a magnetic field $\vec{B} = B_0\cos(\omega t+\phi)\hat{a}_x T$. The voltage in the loop is

(a) zero

(b) due to rotation only

(c) due to transformer action only

(d) due to both rotation and transformer action

[1998 : 1 M]

2.11 The Maxwell's equation, $\nabla \times \vec{H} = \vec{J} + \frac{\partial \vec{D}}{\partial t}$ is based on

(a) Ampere's law (b) Gauss's law

(c) Faraday's law (d) Coulomb's law

[1998 : 1 M]

2.12 Identify which one of the following will NOT satisfy the wave equation.

(a) $50\, e^{j(\omega t-3z)}$ (b) $\sin[\omega(10z+5t)]$

(c) $\cos(y^2+5t)$ (d) $\sin(x)\cos(t)$

[1999 : 1 M]

2.13 A plane wave propagating through a medium [ϵ_r = 8, μ_r = 2, and σ = 0] has its electric field given by $\vec{E} = 0.5\sin(10^8 t - \beta z)$V/m. The wave impedance, in ohms is
(a) 377 (b) 198.5∠180°
(c) 182.9∠14° (d) 188.5 [1999 : 2 M]

2.14 A TEM wave is incident normally upon a perfect conductor. The *E* and *H* fields at the boundary will be, respectively,
(a) minimum and minimum
(b) maximum and maximum
(c) minimum and maximum
(d) maximum and minimum [2000 : 1 M]

2.15 A uniform plane wave in air impinges at 45° angle on a lossless dielectric material with dielectric constant ϵ_r. The transmitted wave propagates in a 30° direction with respect to the normal. The value of ϵ_r is
(a) 1.5 (b) $\sqrt{1.5}$
(c) 2 (d) $\sqrt{2}$ [2000 : 2 M]

2.16 Two coaxial cables 1 and 2 are filled with different dielectric constants ϵ_{r_1} and ϵ_{r_2} respectively. The ratio of the wavelengths in the two cables, (λ_1/λ_2) is
(a) $\sqrt{\epsilon_{r_1}/\epsilon_{r_2}}$ (b) $\sqrt{\epsilon_{r_2}/\epsilon_{r_1}}$
(c) $\epsilon_{r_1}/\epsilon_{r_2}$ (d) $\epsilon_{r_2}/\epsilon_{r_1}$
[2000 : 2 M]

2.17 If a plane electromagnetic wave satisfies the equation $\frac{\partial^2 E_x}{\partial z^2} = \frac{1}{c^2}\frac{\partial^2 E_x}{\partial t^2}$, the wave propagates in the
(a) x-direction
(b) z-direction
(c) y-direction
(d) xy plane at an angle of 45° between the x and z directions
[2001 : 1 M]

2.18 A material has conductivity of 10^{-2} mho/m and a relative permittivity of 4. The frequency at which the conduction current in the medium is equal to the displacement current is
(a) 45 MHz (b) 90 MHz
(c) 450 MHz (d) 900 MHz
[2001 : 2 M]

2.19 A plane wave is characterized by $\vec{E} = \left(0.5\hat{x} + \hat{y}e^{j\pi/2}\right)e^{j\omega t - jkz}$. This wave is
(a) linearly polarized
(b) circularly polarized
(c) elliptically polarized
(d) unpolarized
[2002 : 2 M]

2.20 Distilled water at 25°C is characterized by
$$\sigma = 1.7 \times 10^{-4} \text{ mho/m and } \epsilon = 78\epsilon_0$$
at a frequency of 3 GHz. Its loss tangent tan δ is
(a) 1.3×10^{-5} (b) 1.3×10^{-3}
(c) $1.7 \times 10^{-4}/78$ (d) $1.7 \times 10^{-4}/(78\epsilon_0)$
$\left(\epsilon = 10^{-9}/(36\pi)\text{F/m}\right)$ [2002 : 2 M]

2.21 The depth of penetration of electromagnetic wave in a medium having conductivity σ at a frequency of 1 MHz is 25 cm. The depth of penetration at a frequency of 4 MHz will be
(a) 6.25 cm (b) 12.50 cm
(c) 50.00 cm (d) 100.00 cm
[2003 : 1 M]

2.22 A uniform plane wave travelling in air is incident on the plane boundary between air and another dielectric medium with $\varepsilon_r = 4$. The reflection coefficient for electric field at normal incidence, is
(a) Zero (b) 0.5∠180°
(c) 0.333∠0° (d) 0.333∠180°
[2003 : 2 M]

2.23 If the electric field intensity associated with a uniform plane electromagnetic wave travelling in a perfect dielectric medium is given by $E(z, t) = 10\cos(2\pi \times 10^7 t - 0.1\pi z)$ volt/m, the velocity of the travelling wave is
(a) 3.00×10^8 m/sec (b) 2.00×10^8 m/sec
(c) 6.28×10^7 m/sec (d) 2.00×10^7 m/sec
[2003 : 2 M]

2.24 A parallel plate air-filled capacitor has plate area of 10^{-4} m^2 and plate separation of 10^{-3} m. It is connected to a 0.5 V, 3.6 GHz source. The magnitude of the displacement current is
$(\varepsilon_0 = 1/36\pi \times 10^{-9}$ F/m)
(a) 10 mA (b) 100 mA
(c) 10 A (d) 1.59 mA
[2004 : 2 M]

2.25 If $\vec{E}=(\hat{a}_x + j\hat{a}_y)e^{jkz-j\omega t}$ and $\vec{H}=\left(\frac{k}{\omega\mu}\right)(\hat{a}_y + j\hat{a}_x)e^{jkz-j\omega t}$, the time averaged Pointing vector is

(a) null vector
(b) $\left(\frac{k}{\omega\mu}\right)\hat{a}_z$
(c) $\left(\frac{2k}{\omega\mu}\right)\hat{a}_z$
(d) $\left(\frac{k}{2\omega\mu}\right)\hat{a}_z$

[2004 : 2 M]

2.26 The magnetic field intensity vector of a plane wave is given by

$\bar{H}(x,t)=10\sin(50000t+0.004x+30)\hat{a}_y$, where $\hat{a}_y$ denotes the unit vector in y direction. The wave is propagating with a phase velocity

(a) 5×10^4 m/s
(b) 3×10^8 m/s
(c) 1.25×10^7 m/s
(d) 3×10^6 m/s

[2005 : 1 M]

2.27 Refractive index of glass is 1.5. Find the wavelength of a beam of light with frequency of 10^{14} Hz in glass. Assume velocity of light is 3×10^8 m/s in vacuum

(a) 3 μm
(b) 3 mm
(c) 2 μm
(d) 1 mm

[2005 : 1 M]

2.28 The electric field of an electromagnetic wave propagating in the positive z-direction is given by $E=\hat{a}_x\sin(\omega t-\beta z)+\hat{a}_y\sin(\omega t-\beta z+\pi/2)$.

The wave is

(a) linearly polarized in the z-direction
(b) elliptically polarized
(c) left-hand circularly polarized
(d) right-hand circularly polarized

[2006 : 1 M]

2.29 A medium of relative permittivity $\epsilon_{r_2}=2$ forms an interface with free-space. A point source of electromagnetic energy is located in the medium at a depth of 1 meter from the interface. Due to the total internal reflection, the transmitted beam has a circular cross-section over the interface. The area of the beam cross-section at the interface is given by

(a) 2π m²
(b) π^2 m²
(c) $\pi/2$ m²
(d) π m²

[2006 : 2 M]

2.30 When a plane wave traveling in free-space is incident normally on a medium having $\epsilon_r=4.0$, the fraction of power transmitted into the medium is given byx

(a) 8/9
(b) 1/2
(c) 1/3
(d) 5/6

[2006 : 2 M]

2.31 The $\vec{H}$ field (in A/m) of a plane wave propagating in free space is given by

$\vec{H}=\hat{x}\frac{5\sqrt{3}}{\eta_0}\cos(\omega t-\beta z)+\hat{y}\frac{5}{\eta_0}\sin\left(\omega t-\beta z+\frac{\pi}{2}\right)$.

The time average power flow density in watts is

(a) $\frac{\eta_0}{100}$
(b) $\frac{100}{\eta_0}$
(c) $50\eta_0^2$
(d) $\frac{50}{\eta_0}$

[2007 : 2 M]

2.32 A Right Circularly Polarized (RCP) plane wave is incident at an angle of 60° to the normal, on an air-dielectric interface. If the reflected wave is linearly polarized, the relative dielectric constant ϵ_{r_2} is

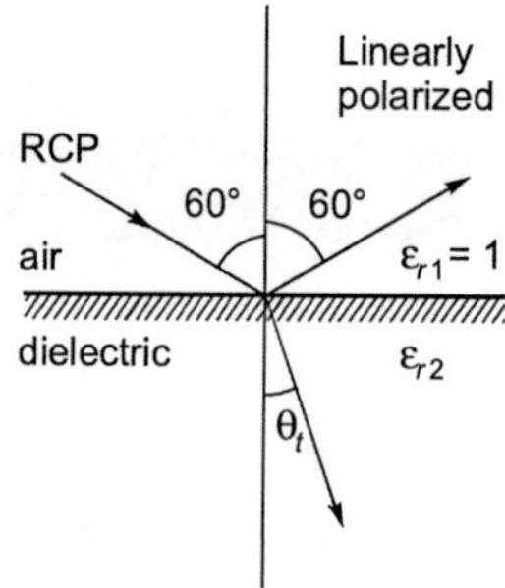

(a) $\sqrt{2}$
(b) $\sqrt{3}$
(c) 2
(d) 3

[2007 : 2 M]

2.33 A plane wave of wavelength λ is travelling in a direction making an angle 30° with positive x-axis and 90° with positive y-axis. The $\vec{E}$ field of the plane wave can be represented as (E_0 is a constant)

(a) $\vec{E}=\hat{y}E_0e^{j\left(\omega t-\frac{\sqrt{3}\pi}{\lambda}x-\frac{\pi}{\lambda}z\right)}$

(b) $\vec{E}=\hat{y}E_0e^{j\left(\omega t-\frac{\pi}{\lambda}x-\frac{\sqrt{3}\pi}{\lambda}z\right)}$

(c) $\vec{E}=\hat{y}E_0e^{j\left(\omega t+\frac{\sqrt{3}\pi}{\lambda}x+\frac{\pi}{\lambda}z\right)}$

(d) $\vec{E}=\hat{y}E_0e^{j\left(\omega t-\frac{\pi}{\lambda}x+\frac{\sqrt{3}\pi}{\lambda}z\right)}$

[2007 : 1 M]

2.34 A uniform plane wave in the free space is normally incident on an infinitely thick dielectric slab (dielectric constant $\varepsilon_r = 9$). The magnitude of the reflection coefficient is
(a) 0 (b) 0.3
(c) 0.5 (d) 0.8 [2008 : 2 M]

2.35 The electric field component of a time harmonic plane *EM* wave traveling in a nonmagnetic lossless dielectric medium has an amplitude of 1 V/m. If the relative permittivity of the medium is 4, the magnitude of the time average power density vector (in W/m²) is
(a) $\frac{1}{30\pi}$ (b) $\frac{1}{60\pi}$
(c) $\frac{1}{120\pi}$ (d) $\frac{1}{240\pi}$
[2010 : 1 M]

2.36 A plane wave having the electric field component $\vec{E}_i = 24\cos\left(3\times10^8 t - \beta y\right)\hat{a}_z$ V/m and traveling in free space is incident normally on a lossless medium with $\mu = \mu_0$ and $\epsilon = 9\epsilon_0$ which occupies the region $y \geq 0$. The reflected magnetic field component is given by
(a) $\frac{1}{10\pi}\cos\left(3\times10^8 t + \beta y\right)\hat{a}_x$ A/m
(b) $\frac{1}{20\pi}\cos\left(3\times10^8 t + \beta y\right)\hat{a}_x$ A/m
(c) $-\frac{1}{20\pi}\cos\left(3\times10^8 t + \beta y\right)\hat{a}_x$ A/m
(d) $-\frac{1}{10\pi}\cos\left(3\times10^8 t + \beta y\right)\hat{a}_x$ A/m
[2010 : 2 M]

2.37 The electric and magnetic fields for a TEM wave of frequency 14 GHz in a homogeneous medium of relative permittivity ε_r and relative permeability $\mu_r = 1$ are given by

$$\vec{E} = E_p e^{j(\omega t - 280\pi y)}\hat{u}_z \text{ V/m}$$

$$\vec{H} = 3e^{j(\omega t - 280\pi y)}\hat{u}_x \text{ A/m}$$

Assuming the speed of light in free space to be 3×10^8 m/s, intrinsic impedance of free space to be 120π, the relative permittivity ε_r of the medium and the electric field amplitude E_p are
(a) $\varepsilon_r = 3$, $E_p = 120\pi$
(b) $\varepsilon_r = 3$, $E_p = 360\pi$
(c) $\varepsilon_r = 9$, $E_p = 360\pi$
(d) $\varepsilon_r = 9$, $E_p = 120\pi$ [2011 : 2 M]

2.38 The electric field of a uniform plane electromagnetic wave in free space, along the positive x direction, is given by $\vec{E} = 10(\hat{a}_y + j\hat{a}_z)e^{-j25x}$. The frequency and polarization of the wave, respectively, are
(a) 1.2 GHz and left circular
(b) 4 Hz and left circular
(c) 1.2 GHz and right circular
(d) 4 GHz and right circular [2012 : 1 M]

2.39 A plane wave propagating in air with $\vec{E} = (8\hat{a}_x + 6\hat{a}_y + 5\hat{a}_z)\, e^{j(\omega t + 3x - 4y)}$ V/m is incident on a perfectly conducting slab positioned at $x \leq 0$. the $\vec{E}$ field of the reflected waves is
(a) $(-8\hat{a}_x - 6\hat{a}_y - 5\hat{a}_z)\, e^{j(\omega t + 3x + 4y)}$ V/m
(b) $(-8\hat{a}_x + 6\hat{a}_y - 5\hat{a}_z)\, e^{j(\omega t + 3x + 4y)}$ V/m
(c) $(8\hat{a}_x - 6\hat{a}_y - 5\hat{a}_z)\, e^{j(\omega t - 3x - 4y)}$ V/m
(d) $(8\hat{a}_x + 6\hat{a}_y - 5\hat{a}_z)\, e^{j(\omega t - 3x - 4y)}$ V/m
[2012 : 1 M]

Statement for Linked Answer Questions 2.40 and 2.41:
A monochromatic plane wave of wavelength $\lambda = 600$ μm is propagating in the direction as shown in the figure below. $\vec{E}_i$, $\vec{E}_r$, and $\vec{E}_t$ denote incident, reflected, and transmitted electric field vectors associated with the wave.

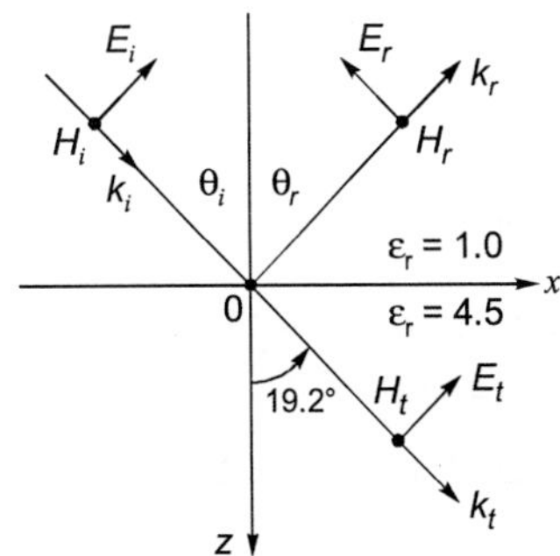

2.40 The angle of incidence θ_i and the expression for $\vec{E}_i$ are
(a) 60° and $\frac{E_o}{\sqrt{2}}(\hat{a}_x - \hat{a}_z)e^{-j\frac{\pi\times10^4(x+z)}{3\sqrt{2}}}$ V/m
(b) 45° and $\frac{E_o}{\sqrt{2}}(\hat{a}_x + \hat{a}_z)e^{-j\frac{\pi\times10^4 z}{3}}$ V/m

(c) $45°$ and $\frac{E_o}{\sqrt{2}}(\hat{a}_x - \hat{a}_z)e^{-j\frac{\pi\times10^4(x+z)}{3\sqrt{2}}}$ V/m

(d) $60°$ and $\frac{E_o}{\sqrt{2}}(\hat{a}_x - \hat{a}_z)e^{-j\frac{\pi\times10^4 z}{3}}$ V/m

[2013 : 2 M]

2.41 The expression for $\vec{E}_r$ is

(a) $0.23\frac{E_o}{\sqrt{2}}(\hat{a}_x + \hat{a}_z)e^{-j\frac{\pi\times10^4(x-z)}{3\sqrt{2}}}$ V/m

(b) $-\frac{E_o}{\sqrt{2}}(\hat{a}_x + \hat{a}_z)e^{j\frac{\pi\times10^4 z}{3}}$ V/m

(c) $0.44\frac{E_o}{\sqrt{2}}(\hat{a}_x + \hat{a}_z)e^{-j\frac{\pi\times10^4(x-z)}{3}}$ V/m

(d) $\frac{E_o}{\sqrt{2}}(\hat{a}_x + \hat{a}_z)e^{-j\frac{\pi\times10^4(x+z)}{3}}$ V/m

[2013 : 2 M]

2.42 Which one of the following field patterns represents a *TEM* wave travelling in the positive x direction?

(a) $E = +8\hat{y}, H = -4\hat{z}$

(b) $E = -2\hat{y}, H = -3\hat{z}$

(c) $E = +2\hat{z}, H = +2\hat{y}$

(d) $E = -3\hat{y}, H = +4\hat{z}$

[2014 : 1 M, Set-2]

2.43 If the electric field of a plane wave is

$$\vec{E}(z,t) = \hat{x}3\cos(\omega t - kz + 30°) - \hat{y}4\sin(\omega t - kz + 45°)\ (\text{mV/m}),$$

the polarization state of the plane wave is

(a) left elliptical (b) left circular

(c) right elliptical (d) right circular

[2014 : 2 M, Set-2]

2.44 Assume that a plane wave in air with an electric field $\vec{E} = 10\cos(\omega t - 3x - \sqrt{3}z)\hat{a}_y$ V/m is incident on a non-magnetic dielectric slab of relative permittivity 3 which covers the region $z > 0$. The angle of transmission in the dielectric slab is ________ degrees.

[2014 : 2 M, Set-3]

2.45 The electric field component of a plane wave travelling in a lossless dielectric medium is given by $\vec{E}(z,t) = \hat{a}_y 2\cos\left(10^8 t - \frac{z}{\sqrt{2}}\right)$ V/m. The wavelength (in m) for the wave is__________.

[2015 : 1 M, Set-1]

2.46 Consider a uniform plane wave with amplitude (E_0) of 10 V/m and 1.1 GHz frequency travelling an air, and incident normally on a dielectric medium with complex relative permittivity (ε_r) and permeability (μ_r) as shown in the figure.

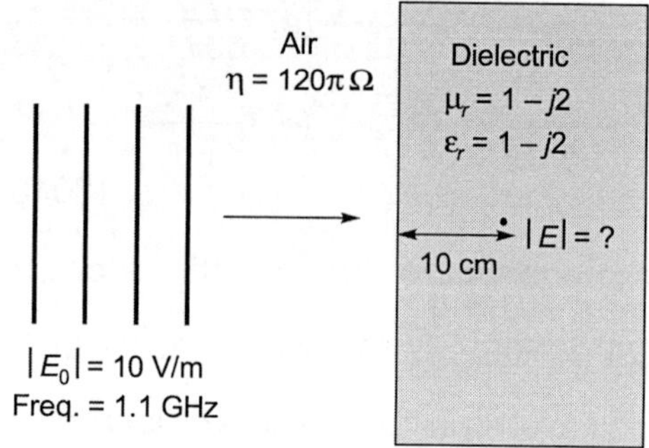

The magnitude of the transmitted electric field component (in V/m) after it has travelled a distance of 10 cm inside the dielectric region is____________.

[2015 : 2 M, Set-1]

2.47 The electric field intensity of a plane wave traveling in free space is given by the following expression

$$\vec{E}(x,t) = \hat{a}_y 24\pi\cos(\omega t - k_0 x)\ (\text{V/m})$$

In this field, consider a square area 10 cm × 10 cm on a plane $x + y = 1$. The total time-averaged power (in mW) passing through the square area is _____.

[2015 : 2 M, Set-1]

2.48 The electric field of a uniform plane electromagnetic wave is

$$\vec{E} = (\hat{a}_x + j2\hat{a}_y)\exp[j(2\pi\times10^7 t + 0.2z)].$$

The polarization of the wave is

(a) right handed circular

(b) right handed elliptical

(c) left handed circular

(d) left handed elliptical [2015 : 1 M, Set-2]

2.49 The electric field of a plane wave propagating in a lossless non-magnetic medium is given by the following expression

$$\vec{E}(z,t) = \hat{a}_x\ 5\cos(2\pi \times 10^9 t + \beta z) + \hat{a}_y\ 3\cos\left(2\pi \times 10^9 t + \beta z - \frac{\pi}{2}\right)$$

(a) Right Hand Circular
(b) Left Hand Elliptical
(c) Right Hand Elliptical
(d) Linear **[2015 : 2 M, Set-2]**

2.50 Let the electric field vector of a plane electromagnetic wave propagating in a homogenous medium be expressed as $E = \hat{x}E_x e^{-j(\omega t - \beta z)}$, where the propagation constant β is a function of the angular frequency ω. Assume that $\beta(\omega)$ and E_x are known and are real. From the information available, which one of the following CANNOT be determined?

(a) The type of polarization of the wave.
(b) The group velocity of the wave.
(c) The phase velocity of the wave.
(d) The power flux through the $z = 0$ plane.

[2016 : 1 M, Set-2]

2.51 If a right-handed circularly polarized wave is incident normally on a plane perfect conductor, then the reflected wave will be

(a) right-handed circularly polarized
(b) left-handed circularly polarized
(c) elliptically polarized with a tilt angle of 45°
(d) horizontally polarized

[2016 : 1 M, Set-3]

2.52 Faraday's law of electromagnetic induction is mathematically described by which one of the following equations?

(a) $\nabla \cdot \vec{B} = 0$ (b) $\nabla \cdot \vec{D} = \rho_v$

(c) $\nabla \times \vec{E} = -\dfrac{\partial \vec{B}}{\partial t}$ (d) $\nabla \times \vec{H} = \sigma \vec{E} + \dfrac{\partial \vec{D}}{\partial t}$

[2016 : 1 M, Set-3]

2.53 The expression for an electric field in free space is $\vec{E} = E_0(\hat{x} + \hat{y} + j2\hat{z})e^{-j(\omega t - kx + ky)}$, where x, y, z represent the spatial coordinates, t represents time, and ω, k are constants. This electric field

(a) does not represent a plane wave.
(b) represents a circularly polarized plane wave propagating normal to the z-axis.
(c) represents an elliptically polarized plane wave propagating along the x - y plane.
(d) represents a linearly polarized plane wave.

[2017 : 2 M, Set-1]

2.54 The permittivity of water at optical frequencies is $1.75\,\varepsilon_0$. It is found that an isotropic light source at a distance d under water forms an illuminated circular area of radius 5 m as shown in the figure. The critical angle is θ_C.

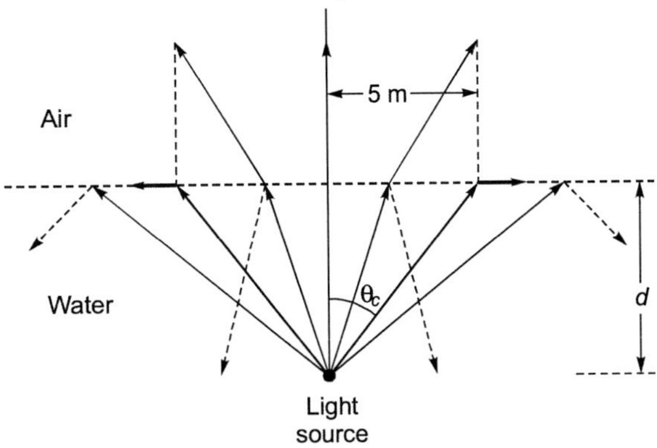

The value of d (in meter) is _____.

[2017 : 2 M, Set-2]

2.55 A uniform plane wave traveling in free space and having the electric field

$$\vec{E} = (\sqrt{2}\hat{a}_x - \hat{a}_z)\cos\left[6\sqrt{3}\pi \times 10^8 t - 2\pi(x + \sqrt{2}z)\right] \text{ V/m}$$

is incident on a dielectric medium (relative permittivity > 1, relative permeability = 1) as shown in the figure and there is no reflected wave.

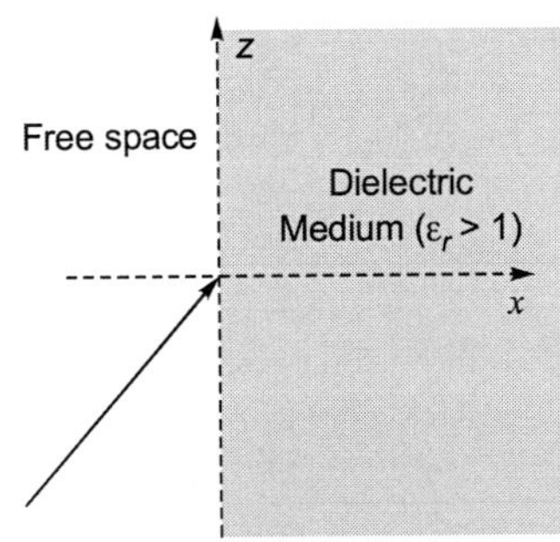

The relative permittivity (correct to two decimal places) of the dielectric medium is ____.

[2018 : 2 M]

2.56 The distance (in meters) a wave has to propagate in a medium having a skin depth of 0.1 m so that the amplitude of the wave attenuates by 20 dB, is

(a) 0.12 (b) 0.23
(c) 0.46 (d) 2.3

[2018 : 2 M]

2.57 The magnetic field of a uniform plane wave in vacuum is given by

$$\vec{H}(x, y, z, t) = (\hat{a}_x + 2\hat{a}_y + b\hat{a}_z)\cos(\omega t + 3x - y - z).$$

The value of b is ____.

[2020 : 2 M]

2.58 Consider a rectangular coordinate system (x, y, z) with unit vectors $\hat{a}_x, \hat{a}_y$ and $\hat{a}_z$. A plane wave traveling in the region $z \geq 0$ with electric field vector $E = 10\cos(2\times10^8 t + \beta z)\hat{a}_y$ is incident normally on the plane at $z = 0$, where β is the phase constant. The region $z \geq 0$ is in free space and the region $z < 0$ is filled with a lossless medium (permittivity $\epsilon = \epsilon_0$, permeability $\mu = 4\mu_0$, where $\epsilon_0 = 8.85 \times 10^{-12}$ F/m and $\mu_0 = 4\pi \times 10^{-7}$ H/m). The value of the reflection coefficient is

(a) $\frac{1}{3}$ (b) $\frac{3}{5}$

(c) $\frac{2}{5}$ (d) $\frac{2}{3}$

[2021 : 1 M]

2.59 Consider the following wave equation,

$$\frac{\partial^2 f(x,t)}{\partial t^2} = 10000\frac{\partial^2 f(x,t)}{\partial x^2}$$

Which of the given options is/are solution(s) to the given wave equation?

(a) $f(x,t) = e^{-(x-100t)^2} + e^{-(x+100t)^2}$

(b) $f(x,t) = e^{-(x-100t)} + 0.5e^{-(x+1000t)}$

(c) $f(x,t) = e^{-(x-100t)} + \sin(x+100t)$

(d) $f(x,t) = e^{j100\pi(-100x+t)} + e^{j100\pi(100x+t)}$

[2022 : 1 M]

2.60 The electric field of a plane electromagnetic wave is

$E = a_x C_{1x}\cos(\omega t - \beta z) + a_y C_{1y}\cos(\omega t - \beta z + \theta)$ V/m

Which of the following combination(s) will give rise to a left handed elliptically polarized (LHEP) wave?

(a) $C_{1x} = 1, C_{1y} = 1, \theta = \pi/4$

(b) $C_{1x} = 2, C_{1y} = 1, \theta = \pi/2$

(c) $C_{1x} = 1, C_{1y} = 2, \theta = 3\pi/2$

(d) $C_{1x} = 2, C_{1y} = 1, \theta = 3\pi/4$

[2023 : 2 M]

2.61 A uniform plane wave with electric field $\vec{E}(x) = A_y\hat{a}_y e^{-j\frac{2\pi x}{3}}$ V/m is travelling in the air (relative permittivity, $\epsilon_r = 1$ and relative permeability, $\mu_r = 1$) in the $+x$ direction (A_y is a positive constant, $\hat{a}_y$ is the unit vector along the y axis). It is incident normally on an ideal electric conductor (conductivity, $\sigma = \infty$) at $x = 0$. The position of the first null of the total magnetic field in the air (measured from $x = 0$, in metres) is _______.

(a) -6 (b) $-\frac{3}{2}$

(c) -3 (d) $-\frac{3}{4}$

[2024 : 2 Marks]

■■■

Answers Uniform Plane Waves

2.1 (d)	2.2 (a)	2.3 (b)	2.4 (a)	2.5 (d)	2.6 (b)	2.7 (d)
2.8 (a)	2.9 (a)	2.10 (d)	2.11 (a)	2.12 (c)	2.13 (d)	2.14 (c)
2.15 (c)	2.16 (b)	2.17 (b)	2.18 (a)	2.19 (c)	2.20 (a)	2.21 (b)
2.22 (d)	2.23 (b)	2.24 (a)	2.25 (a)	2.26 (c)	2.27 (c)	2.28 (c)
2.29 (d)	2.30 (a)	2.31 (d)	2.32 (d)	2.33 (a)	2.34 (c)	2.35 (c)
2.36 (a)	2.37 (d)	2.38 (a)	2.39 (c)	2.40 (c)	2.41 (a)	2.42 (b)
2.43 (a)	2.44 (30)	2.45 (8.88)	2.46 (0.1)	2.47 (53.29)	2.48 (b)	2.49 (b)
2.50 (d)	2.51 (b)	2.52 (c)	2.53 (c)	2.54 (4.33)	2.55 (2)	2.56 (b)
2.57 (1)	2.58 (a)	2.59 (a, c)	2.60 (a, b, d)	2.61 (d)		

Explanations Uniform Plane Waves

2.2 (a)

A material behave as conductor if,

$$\frac{\sigma}{\omega \in} >> 1$$

$$\sigma >> \omega\epsilon$$

For copper, conductivity $\sigma = 5.8 \times 10^7$ mho/m

$$\epsilon = 8.856 \times 10^{-12} \text{ F/m}$$

So, even at very large frequencies,

$$\sigma >> \omega\epsilon$$

So, copper always behave as conductor for practical frequencies.

2.3 (b)

Reflection coefficient,

Given: $\sqrt{\frac{\epsilon_2}{\epsilon_1}} = 1.5$

$$\Gamma = \frac{1.5-1}{1.5+1} = \frac{0.5}{2.5} = \frac{1}{5}$$

$$|\Gamma| = \frac{1}{5}$$

$$\frac{P_r}{P_i} = \left(\frac{1}{5}\right)^2 = \frac{1}{25}$$

$$P_r = \frac{P_i}{25} \times 100\% = 4\% \text{ of } P_i$$

2.4 (a)

$$\frac{\sigma}{\omega \in} = 18 \times 10^6 >> 1$$

Hence material is a good conductor.

Skin depth,

$$\delta = \frac{1}{\sqrt{\pi f \mu \sigma}} = \frac{1}{\sqrt{\pi \times 10^9 \times 4\pi \times 10^{-7} \times 10^6}} = \frac{10^{-4}}{2\pi}$$

$$\delta = 15.9 \ \mu\text{m}$$

2.5 (d)

We know that,

$$\text{Intrinsic impedance} = \eta = \sqrt{\frac{j\omega\mu}{\sigma + j\omega\in}}$$

For copper,

$$\sigma >> \omega\epsilon$$

$$\eta = \sqrt{\frac{j\omega\mu}{\sigma}} = \sqrt{\frac{\mu\omega}{\sigma}}\, e^{j45^\circ}$$

$$\theta = +45^\circ$$

So, the intrinsic impedance of copper at high frequency is complex with an inductive component.

2.6 (b)

Propagation constant

$$\gamma = \alpha + j\beta = 0.1\pi + j0.2\pi$$

Comparing the real and imaginary part

$$\beta = 0.2\pi = \frac{2\pi}{\lambda} \quad ; \quad \lambda = \frac{2\pi}{0.2\pi} = 10 \text{ m}$$

2.7 (d)

In a lossy dielectric, attenuation constant

$$\alpha = \frac{\sigma}{2}\sqrt{\frac{\mu}{\epsilon}}$$

Skin depth, $\delta = \frac{1}{\alpha} = \frac{2}{\sigma}\sqrt{\frac{\epsilon}{\mu}}$

So, the depth of penetration of a wave in a lossy dielectric increases with increasing permittivity.

2.8 (a)

$$E = E_o e^{j(\omega t - \beta z)}(\vec{a}_x + \vec{a}_y)$$

$$E = E_o e^{j(\omega t - \beta z)}\vec{a}_x + E_o e^{j(\omega t - \beta z)}\vec{a}_y$$

$$\vec{E} = E_x \vec{a}_x + E_y \vec{a}_y$$

So, $\vec{E}_x = E_o e^{j(\omega t - \beta z)}$

$$\vec{E}_y = E_o e^{j(\omega t - \beta z)}$$

$$\vec{E}_x = \vec{E}_y \text{ at any time } t$$

So, the polarization of the wave is linear.

2.9 (a)

$$\vec{E} = 24 e^{j(\omega t + \beta z)}\hat{a}_y \text{ V/m}$$

Wave is travelling in $-z$ direction

$$|\vec{E}| = 24$$

Direction of power $= -z = -\vec{a}_z$

In free space

$$\eta = \eta_o = 120\pi\ \Omega$$

$$P_{avg} = \frac{1}{2\eta_o}|E|^2(-\vec{a}_z)$$

$$P_{avg} = \frac{(24)^2}{2(120\pi)}(-\vec{a}_z)$$

$$P_{avg} = -\frac{2.4}{\pi}\vec{a}_z \text{ W/m}^2$$

2.10 (d)

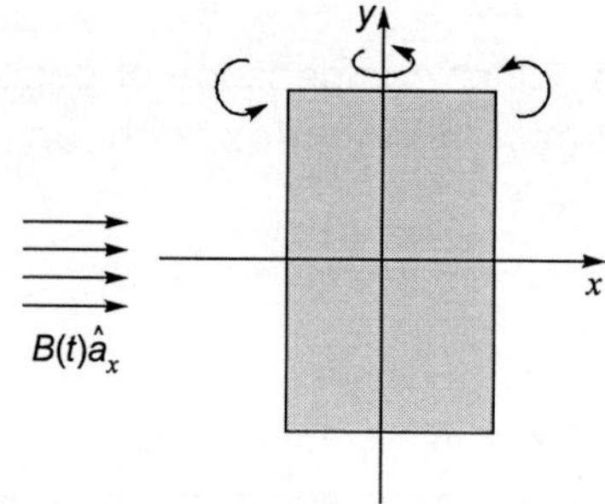

$$V = -\frac{d\psi_m}{dt} = -\frac{d}{dt}(BA\cos\theta)$$

A = Area of the loop

with θ = angle between B and area of loop. This changes due to rotation

$B(t) = \cos\omega t$ Harmonic

$$V = \underbrace{-AB\frac{d\cos\theta}{dt}}_{\text{Rotation Motional EMF}} - \underbrace{A\cos\theta\frac{dB}{dt}}_{\text{Harmonic Transformer Action}}$$

2.11 (a)

According to Ampere's law,

$$\nabla \times \vec{H} = \vec{J} + \frac{\partial \vec{D}}{\partial t}$$

2.12 (c)

$\cos(y^2 + 5t)$ does not satisfy the wave equation due to non-linearity of phase with distance.

2.13 (d)

$$\eta = \sqrt{\frac{j\omega\mu}{\sigma + j\omega\in}}, \quad \sigma = 0$$

$$= \sqrt{\frac{\mu}{\in}}\angle 0° = \sqrt{\frac{\mu_o\mu_r}{\in_o\in_r}} = \sqrt{\frac{\mu_o}{\in_o}}\sqrt{\frac{\mu_r}{\in_r}}$$

$$= 377\sqrt{\frac{2}{8}} = \frac{377}{2} = 188.5$$

2.14 (c)

Reflection coefficient for electric fields

$$\Gamma_E = \frac{\sqrt{\frac{j\omega\mu}{\sigma}} - 120\pi}{\sqrt{\frac{j\omega\mu}{\sigma}} + 120\pi} = -1 \text{ for conductors}$$

As $\eta = \sqrt{\frac{j\omega\mu}{\sigma}} \simeq 0$ with σ being large

when E reflection completely out of phase, H reflects in phase

$$E_r - E_i = 0, \quad E_{tang} \text{ should be zero}$$

$$H_r + H_i = 2H_t, \; H_{tang} \text{ should be maximum}$$

2.15 (c)

$$\frac{\sin\theta_t}{\sin\theta_i} = \sqrt{\frac{1}{\in_r}}$$

$$\frac{\sin 30°}{\sin 45°} = \frac{1}{\sqrt{\in_r}}$$

$$\Rightarrow \quad \frac{1/2}{1/\sqrt{2}} = \frac{1}{\sqrt{\in_r}}$$

$$\Rightarrow \quad \in_r = 2$$

2.16 (b)

$$v_1 = \frac{1}{\sqrt{\mu_1\in_1}}$$

$$v_2 = \frac{1}{\sqrt{\mu_2\in_2}}$$

$$\frac{v_1}{v_2} = \sqrt{\frac{\in_2}{\in_1}}$$

$v = f\lambda$; f = frequency of operation

$v \propto \lambda$

$$\therefore \quad \frac{\lambda_1}{\lambda_2} = \sqrt{\frac{\in_2}{\in_1}} = \sqrt{\frac{\in_0\in_{r_2}}{\in_0\in_{r_1}}} = \sqrt{\frac{\in_{r_2}}{\in_{r_1}}}$$

2.17 (b)

The E field is a function of z and directed in X. i.e. harmonic function of z or travelling in z direction.

2.18 (a)

$$|\sigma E| = |j\omega\in E| \quad ; \quad \sigma = 2\pi f\in_0\in_r$$

$$\Rightarrow \quad f = \frac{\sigma}{2\pi\times\in_0\in_r} = \frac{9\times10^9\times2\times10^{-2}}{4}$$

$$f = 45\times10^6 = 45 \text{ MHz}$$

2.19 (c)

E_x and E_y out of phase by 90°, unequal amplitudes. This is elliptical polarization.

2.20 (a)

$$\text{Loss tangent} = \frac{\sigma}{\omega\in}$$

$$= \frac{1.7\times10^{-4}}{2\pi\times3\times10^9\times78\in_0} = \frac{1.7\times10^{-4}\times9\times10^9}{3\times10^9\times39}$$

$$= 0.13\times10^{-4} = 1.3\times10^{-5}$$

2.21 (b)

For conductors, $\delta = \sqrt{\frac{2}{\omega\mu\sigma}}$

$$\delta \propto \frac{1}{\sqrt{f}} \quad ; \quad \frac{\delta_2}{\delta_1} = \sqrt{\frac{f_1}{f_2}}$$

$$\delta_2 = \sqrt{\frac{1\,\text{MHz}}{4\,\text{MHz}}} \times 25\text{ cm} = 12.5\text{ cm}$$

2.22 (d)

$$\Gamma = \frac{\eta_2 - \eta_1}{\eta_2 + \eta_1}$$

$$= \frac{\sqrt{\frac{\mu_o}{\epsilon_o \epsilon_r}} - \sqrt{\frac{\mu_o}{\epsilon_o}}}{\sqrt{\frac{\mu_o}{\epsilon_o \epsilon_r}} + \sqrt{\frac{\mu_o}{\epsilon_o}}} = \frac{\frac{1}{\sqrt{\epsilon_r}} - 1}{\frac{1}{\sqrt{\epsilon_r}} + 1}, \epsilon_r = 4$$

$$= \frac{-1/2}{3/2} = \frac{-1}{3} \Rightarrow 0.333\angle 180°$$

2.23 (b)

$$\omega = 2\pi \times 10^7$$
$$\beta = 0.1\pi$$

Comparing with $A\cos(\omega t - \beta z)$

$$v = \frac{\omega}{\beta} = \frac{2\pi \times 10^7}{0.1\pi} = 2 \times 10^8 \text{ m/s}$$

2.24 (a)

Displacement current,

$$I_d = AJ_d$$
$$= A\frac{\partial D}{\partial t} = A\in\frac{\partial E}{\partial t}$$
$$|I_d| = |A\in\omega E| = A\omega\in\frac{V}{d}$$

After putting values, we get,

$$I_d = 10\text{ mA}$$

2.25 (a)

$$\vec{P}_{avg} = \frac{1}{2}\text{Re}\left[\vec{E} \times \vec{H}^*\right]$$
$$= \frac{1}{2}\text{Re}\left[(\hat{a}_x + j\hat{a}_y)e^{jkz - j\omega t} \times \frac{k}{\omega\mu}(\hat{a}_y - j\hat{a}_x)e^{-jkz + j\omega t}\right]$$
$$\vec{P}_{avg} = \frac{k}{2\mu\omega}[0] = 0$$
$$\begin{bmatrix} 1 & j \\ -j & 1 \end{bmatrix} = 1 + j^2 = 1 - 1 = 0$$

2.26 (c)

$$\omega = 50{,}000 \ ; \ \beta = 0.004$$
$$v_p = \frac{\omega}{\beta} = \frac{5 \times 10^4}{4 \times 10^{-3}} = 1.25 \times 10^7 \text{ m/s}$$

2.27 (c)

Refractive index $= \sqrt{\frac{\epsilon_2}{\epsilon_1}}$

For glass with respective air $= \sqrt{\epsilon_R}$

$$V_p = \frac{3 \times 10^8}{\sqrt{\epsilon_R} \times f} = \frac{3 \times 10^8}{1.5 \times 10^{14}} = 2\,\mu\text{m}$$

2.28 (c)

Equal amplitudes and out of phase by 90°, wave is circularly polarized. Tracing the time advancing E on the $z = 0$ plane at $t = 0$, $t = T/4$, $t = T/2$ gives left/right notations.

2.29 (d)

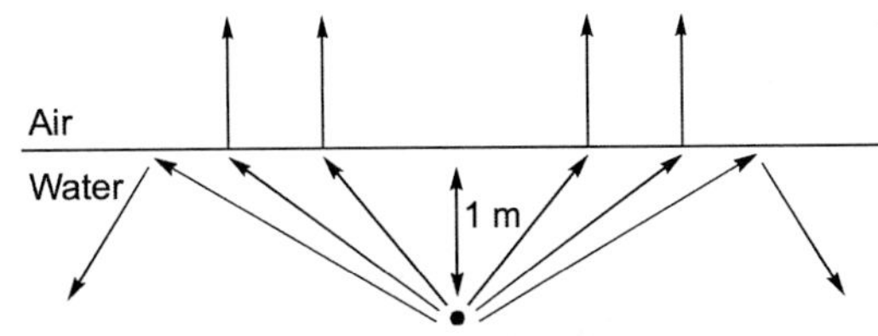

Total internal reflection beyond critical angle

Circular cross-section beam for rays incident below critical angle.

Calculating critical angle $\sin\theta_c = \sqrt{\frac{\epsilon_o}{2\epsilon_o}}$

$$\theta_c = 45°$$

1 m- depth ; 1 m-radius

$$\text{Area} = \pi R^2 = \pi$$

2.30 (a)

$$P_t = (1 - \Gamma^2)P_i$$

$$\Gamma = \frac{\eta_2 - \eta_1}{\eta_2 + \eta_1} = \frac{\sqrt{\frac{\mu_o}{\epsilon_o \epsilon_r}} - \sqrt{\frac{\mu_o}{\epsilon_o}}}{\sqrt{\frac{\mu_o}{\epsilon_o \epsilon_r}} + \sqrt{\frac{\mu_o}{\epsilon_o}}} = \frac{\frac{1}{2} - 1}{\frac{1}{2} + 1} = \frac{-1}{3}$$

$$\tau = 1 - \Gamma^2 = 1 - \frac{1}{9} = \frac{8}{9}$$

$$\therefore \quad P_t = \tau \cdot P_i$$

$$\frac{P_t}{P_i} = \frac{8}{9}$$

2.31 (d)

For free space,

$$P = \frac{E^2}{2\eta_0},\ E = \eta_0 H$$

$$P = \frac{\eta_0^2 H^2}{2\eta_0} = \frac{\eta_0 H^2}{2}$$

$$H = \frac{1}{\eta_0}\sqrt{(5\sqrt{3})^2 + (5)^2} = \frac{10}{\eta_0}$$

$$P = \frac{\eta_0}{2} \times \frac{100}{\eta_0^2} = \frac{50}{\eta_0} \text{ watts}$$

2.32 (d)

The reflected wave is linearly polarized when circularly polarized wave is incident under Brewster's angle of incidence.

$$\tan\theta_B = \sqrt{\frac{\epsilon_{r_2}}{\epsilon_{r_1}}} = \tan 60°$$

$$\epsilon_{r_2} = 3$$

2.33 (a)

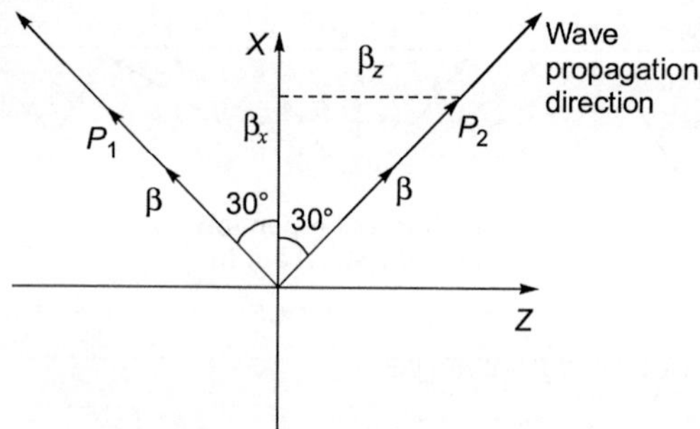

Wave is travelling in the *ZX*-plane (90° to *Y*-axis) and 30° to *X*-axis.

$$\beta_x = \beta\cos 30° = \frac{2\pi}{\lambda}\cdot\frac{\sqrt{3}}{2} = \frac{\sqrt{3}\pi}{\lambda}$$

$$\beta_z = \beta\sin 30° = \frac{\pi}{\lambda}$$

The two solutions possible are shown as P_1 and P_2 paths in the diagrams. Option (a) is P_2.

2.34 (c)

Dielectric constant of free space, $\epsilon_{r_1} = 1$

Dielectric constant of dielectric slab, $\epsilon_{r_2} = 9$

Magnitude of reflection coefficient,

$$|\Gamma| = \left|\frac{\sqrt{\epsilon_{r_1}} - \sqrt{\epsilon_{r_2}}}{\sqrt{\epsilon_{r_1}} + \sqrt{\epsilon_{r_2}}}\right| = \left|\frac{1-3}{1+3}\right| = \frac{2}{4} = 0.5$$

2.35 (c)

Magnitude of the time-average power density vector,

$$P_{av} = \frac{E^2}{2\eta}$$

where, $\eta = \sqrt{\mu/\epsilon}$

$$\eta = \sqrt{\frac{\mu_0}{4\epsilon_0}} = \frac{\eta_0}{2} = \frac{120\pi}{2}$$

$$P_{av} = \frac{1^2}{2\left(\frac{120\pi}{2}\right)} = \frac{1}{120\pi} \text{ W/m}^2$$

2.36 (a)

$$\vec{E_i} = 24\cos(3\times 10^8 t - \beta y)\,\hat{a}_z \text{ V/m}$$

The wave is travelling in +y direction.

$$H_i = \frac{1}{\eta}(\hat{a}_y \times \vec{E_i})$$

$$= \frac{24\cos}{120\pi}(3\times 10^8 t - \beta y)\,\hat{a}_x$$

$$= \frac{1}{5\pi}\cos(3\times 10^8 t - \beta y)\,\hat{a}_x$$

$$\frac{H_r}{H_i} = \frac{\eta_1 - \eta_2}{\eta_1 + \eta_2} = \frac{\sqrt{\frac{\mu_0}{\epsilon_1}} - \sqrt{\frac{\mu_0}{\epsilon_2}}}{\sqrt{\frac{\mu_0}{\epsilon_1}} + \sqrt{\frac{\mu_0}{\epsilon_2}}}$$

$$= \frac{\sqrt{\epsilon_2} - \sqrt{\epsilon_1}}{\sqrt{\epsilon_2} + \sqrt{\epsilon_1}} = \frac{\sqrt{9} - \sqrt{1}}{\sqrt{9} + \sqrt{1}} = \frac{3-1}{3+1} = \frac{2}{4} = \frac{1}{2}$$

Therefore, reflected magnetic field component,

$$\vec{H_r} = \frac{1}{10\pi}\cos(3\times 10^8 t + \beta y)\,\hat{a}_x$$

Due to change in propagation direction to $-y$.

2.37 (d)

Given: $\vec{E} = E_p\, e^{j(\omega t - 280\pi y)}\,\hat{u}_z$ V/m

$\vec{H} = 3\, e^{j(\omega t - 280\pi y)}\,\hat{u}_x$ A/m

$v = 3\times 10^8$ m/s

From given expression we conclude that,

$$\beta = 280\pi = \frac{2\pi}{\lambda}$$

$$\therefore \quad \lambda = \frac{1}{140} \text{ meter}$$

$$v = f\lambda = 14\times 10^9 \times \frac{1}{140} \text{ m/sec}$$

$$v = 1 \times 10^8 \text{ m/sec}$$

but, $$v = \frac{c}{\sqrt{\epsilon_r \mu_r}}$$

$$1 \times 10^8 = \frac{3 \times 10^8}{\sqrt{1 \times \epsilon_r}}$$

$\therefore$ $$\epsilon_r = 9$$

$$\frac{E_P}{H_P} = \eta = \sqrt{\frac{\mu}{\epsilon}} = \sqrt{\frac{\mu_0 \times 1}{\epsilon_0 \times 9}} = \frac{1}{3}\sqrt{\frac{\mu_0}{\epsilon_0}}$$

$\Rightarrow$ $$\frac{E_P}{3} = \frac{1}{3} \times 120\pi$$

$\therefore$ $$E_P = 120\pi$$

2.38 (a)

Out of phase by 90° and equal amplitude the wave is circularly polarized.

Check the trace of E with the time which gives left circularly polarized.

$$a_y = \sin \omega t$$
$$a_z = \cos \omega t$$

2.39 (c)

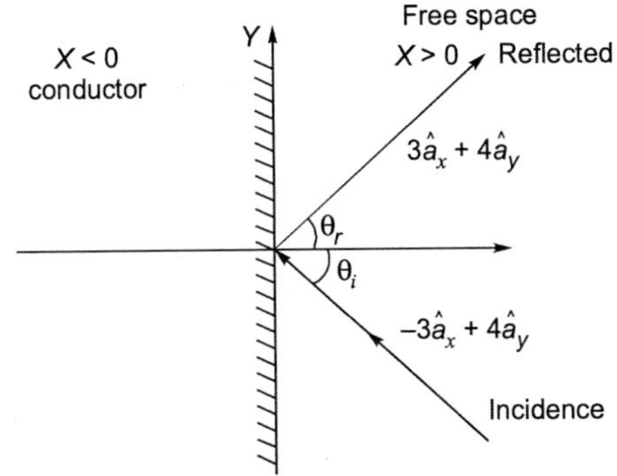

Incident wave propagation $\rightarrow -3\hat{a}_x + 4\hat{a}_y$

Reflected wave propagation $\rightarrow 3\hat{a}_x + 4\hat{a}_y$

E_i incident direction $\rightarrow 8\hat{a}_x + 6\hat{a}_y + 5\hat{a}_z$

$$\left.\begin{array}{l} E_i \text{ tangential} \rightarrow 6\hat{a}_y + 5\hat{a}_z \\ E_r \text{ tangential} \rightarrow -6\hat{a}_y - 5\hat{a}_z \end{array}\right\} E_{tang} = 0$$

$$\left.\begin{array}{l} E_i \text{ normal} = 8\hat{a}_x \\ E_r \text{ normal} = 8\hat{a}_x \end{array}\right\} E_{normal} = E_{max}$$

E_r reflected direction $= 8\hat{a}_x - 6\hat{a}_y - 5\hat{a}_z$

2.40 (c)

E_i propagates in ZX-plane.

E_i directed in ZX-plane.

It is a p-plane of incidence polarized wave.

Using Snell's law,

$$\frac{\sin\theta_i}{\sin 19.2^\circ} = \sqrt{\frac{4.5\epsilon_o}{\epsilon_o}} \qquad \theta_i = 45^\circ$$

$$\beta_x = \beta_z = \beta\cos 45^\circ = \frac{\beta}{\sqrt{2}}$$

(for propagation direction)

For field direction $= \dfrac{\hat{a}_x - \hat{a}_z}{\sqrt{2}}(45^\circ)$

2.41 (a)

For reflected wave.

E_r propagation direction changes without change in angle 45°.

E_r propagation direction $= \dfrac{\hat{a}_x - \hat{a}_z}{\sqrt{2}}$

Γ reflection coefficient

$$\Gamma = -0.23$$

E_r field direction suitably changes satisfying

$$E \cdot P = 0$$

2.42 (b)

Since, the TEM wave or plane wave is travelling in positive x-direction, then conditions should be

$$P = \vec{E} \times \vec{H}$$
$$\hat{a}_x = \hat{a}_y \times \hat{a}_z$$

or, $$\hat{a}_x = (\hat{a}_y) \times (-\hat{a}_z)$$

So, the option (b) satisfies the condition.

2.43 (a)

$$\vec{E}(z,t) = 3\hat{x}\cos(\omega t - kz + 30^\circ) - 4\hat{y}\sin(\omega t - kz + 45^\circ)$$

The plane wave is travelling in $+z$ direction

$$\hat{E}_x = 3\cos(\omega t - kz + 30^\circ)$$
$$\hat{E}_y = -4\sin(\omega t - kz + 45^\circ)$$

At $z = 0$, tracing E locus for various times

At $t = 0$, $E_x = 3\cos 30^\circ = 3 \cdot \frac{1}{2} = \frac{3}{2} = 1.5$

$$E_y = -4\sin 45^\circ = -2\sqrt{2} = -2.83$$

At $t = \dfrac{T}{6}$,

$$E_x = 3\cos\left(\frac{2\pi}{T} \times \frac{T}{6} + 30^\circ\right) = 0$$

$$E_y = -4\sin(60^\circ + 45^\circ) = -3.86$$

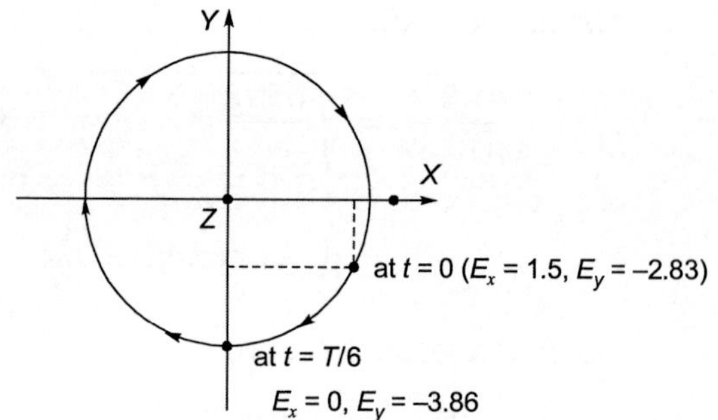

This is left elliptical polarized plane-wave.

2.44 (30)

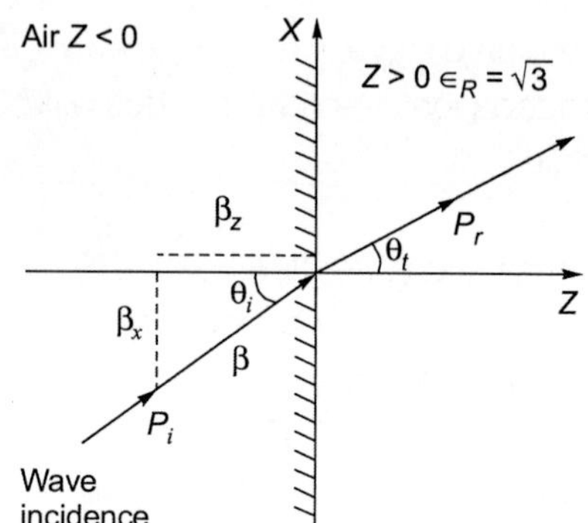

$$\beta_x = 3,\ \beta_z = \sqrt{3} \text{ from } E_i \text{ of the wave}$$
$$\theta_i = 60^\circ$$
$$\frac{\sin 60^\circ}{\sin\theta_t} = \sqrt{3}$$
$$\theta_t = 30^\circ$$

2.45 (8.88)

$$\vec{E}(z,t) = \hat{a}_y 2\cos\left(10^8 t - \frac{z}{\sqrt{2}}\right) \text{ V/m}$$

General form of

$$\vec{E}(z,t) = \hat{a}_y E_o \cos(\omega t - \beta z) \text{ V/m}$$

Comparing both, we get $\beta = \dfrac{1}{\sqrt{2}}$

Since, $\beta = \dfrac{2\pi}{\lambda}$

$$\Rightarrow \lambda = \frac{2\pi}{\beta} = \frac{2\pi}{\frac{1}{\sqrt{2}}} = 2\sqrt{2}\pi m$$

$$\Rightarrow \lambda = 8.88 \text{ m}$$

2.46 (0.1)

Material has μ_R and ϵ_R complex.

$$\eta \text{ of material} = \sqrt{\frac{j\omega\mu}{\sigma + j\omega\epsilon}} = \sqrt{\frac{\mu_o\mu_R}{\epsilon_o\epsilon_R}} = 120\pi$$

where, $\sigma = 0$ (dielectric)

No reflections at the boundary.

$$\gamma = \text{Propagation constant}$$
$$= \sqrt{j\omega\mu_o\mu_R(0 + j\omega\epsilon_o\epsilon_R)}$$
$$= j\omega\sqrt{\mu_o\epsilon_o}\cdot\sqrt{(1-j2)^2}$$
$$= j\omega\sqrt{\mu_o\epsilon_o} + 2\omega\sqrt{\mu_o\epsilon_o}$$
$$\alpha = 2\omega\sqrt{\mu_o\epsilon_o} = 46.07 \text{ Np/m}$$

Magnitude of transmitted electric field component

$$|E| = E_o e^{-\alpha z} \text{ with } Z = 0.1 \text{ m}$$
$$|E| = 0.1 \text{ V/m}$$

2.47 (53.29)

$$\vec{E}(x,t) = \hat{a}_y 24\pi\cos(\omega t - K_o x) \text{ V/m}$$

Power density

$$\vec{P} = \frac{1}{2}\frac{|E|^2}{\eta} = \frac{1}{2}\frac{(24\pi)^2}{120\pi}\hat{a}_x$$

Time-averaged power passing through square area

$$P_{avg} = \vec{P}\cdot\vec{ds}$$
$$= \left[\frac{1}{2}\frac{(24\pi)^2}{120\pi}\hat{a}_x\right]\cdot\left[\frac{\hat{a}_x + \hat{a}_y}{\sqrt{2}}\times 10^{-2}\right]$$
$$P_{avg} = \frac{24\pi}{\sqrt{2}} = 53.29 \text{ mW}$$

2.48 (b)

$\Rightarrow$ Wave has orthogonal components with unequal amplitudes and checking the time trace.

$\therefore$ Wave is right hand elliptically polarized.

2.49 (b)

$$\vec{E}(z,t) = \hat{a}_x 5\cos(2\pi\times 10^9 t + \beta z) + \hat{a}_y 3\cos\left(2\pi\times 10^9 t + \beta z - \frac{\pi}{2}\right)$$

$\Rightarrow$ Wave is travelling in $-\hat{a}_z$ direction.

$\Rightarrow$ Wave has orthogonal components with unequal amplitudes and checking the time trace.

$\therefore$ Wave is left hand elliptically polarized.

2.50 (d)

$\beta(\omega)$ is known so V_g can be calculated.

$v_p = \omega/\beta$ can be calculated.

Polarization can be identified.

μ_r and ϵ_r cannot be found, due to which power flux cannot be calculated as power flux

$$P = \frac{1}{2}\frac{|E|^2}{\eta}, \text{ where } \eta = 120\pi\times\sqrt{\frac{\mu_r}{\epsilon_r}}$$

2.51 (b)

Left circularly polarized, due to direction change after the reflection from conductor.

2.52 (c)

Rate of change of magnetic field results in induced voltage,

$$\nabla \times \vec{E} = -\frac{\partial \vec{B}}{\partial t}$$

2.53 (c)

$\vec{E}$ field direction $\Rightarrow (\hat{a}_x + \hat{a}_y + j2\hat{a}_z)$

Propagation direction $\Rightarrow k\hat{a}_x - k\hat{a}_y$

$\vec{E}$ is perpendicular to propagation

$$\vec{E} \cdot \vec{P} = 0$$

Component in $\hat{a}_z$ has magnitude of 2.

Component in X-Y plane has magnitude of $\sqrt{2}$. These two components are out of phase by 90° and have unequal amplitudes. So, it is elliptically polarized wave.

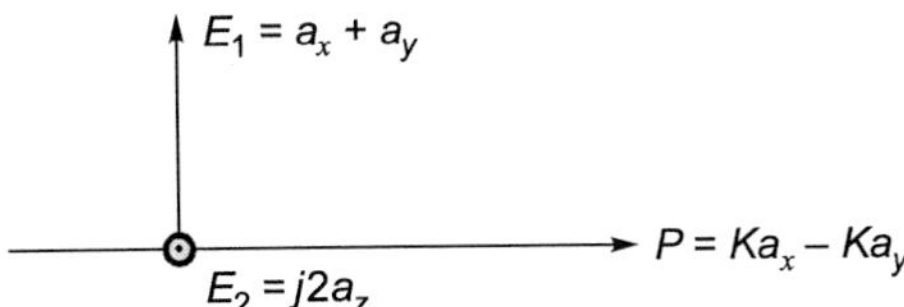

2.54 (4.33)

Critical angle,

$$\theta_c = \sin^{-1}\sqrt{\frac{\epsilon_2}{\epsilon_1}} = \sin^{-1}\sqrt{\frac{\epsilon_0}{1.75\,\epsilon_0}} = 49.1°$$

$$\tan 49.1 = 1.1547 = \frac{5}{d}$$

$$\Rightarrow \quad d = \frac{5}{1.1547} = 4.33 \text{ m}$$

5 m, d, θ_c

2.55 (2)

The wave has E_i direction and propagation direction both in same plane (ZX).

The wave is plane of incidence (P) polarized.

$$\beta_x = 1, \quad \beta_z = \sqrt{2}$$

$$\tan\theta_i = \frac{\beta_z}{\beta_x} = \sqrt{2}$$

No reflection in P polarized wave under Brewster's angle of incidence.

$$\theta_i = \theta_B$$

$$\tan\theta_B = \sqrt{2} = \sqrt{\epsilon_r}$$

$$\epsilon_r = 2$$

2.56 (b)

Attenuation constant,

$$\alpha = \frac{1}{\text{skin depth}} = 10 \text{ Np/m}$$

$$20\log_{10}\left(\frac{E_o}{E_x}\right) = 20 \text{ dB}$$

$$\frac{E_o}{E_x} = 10 \Rightarrow E_x = \frac{E_o}{10}$$

$$E_x = E_o e^{-\alpha x} = E_o e^{-10x} = \frac{E_o}{10}$$

$$e^{-10x} = \frac{1}{10}$$

$$x = \frac{1}{10} ln(10) = 0.23 \text{ m}$$

2.57 (1)

For uniform plane wave

$$\hat{a}_H \cdot \hat{a}_\rho = 0$$

$\hat{a}_H$ is unit vector in magnetic field direction $\hat{a}_\rho$ is unit vector in power flow direction

$$\hat{a}_H = \frac{1\hat{a}_x + 2\hat{a}_y + b\hat{a}_z}{\sqrt{1^2 + 2^2 + b^2}}$$

$$\hat{a}_\rho = \frac{-3\hat{a}_x + \hat{a}_y + \hat{a}_z}{\sqrt{3^2 + 1^2 + 1^2}}$$

$$\hat{a}_H \cdot \hat{a}_\rho = 0$$

$$(\hat{a}_x + 2\hat{a}_y + b\hat{a}_z) \cdot (-3\hat{a}_x + \hat{a}_y + \hat{a}_z) = 0$$

$$-3 + 2 + b = 0$$

$$b = 1$$

2.58 (a)

Given $\vec{E} = 10\cos(2\pi \times 10^8 t + \beta z)\hat{a}_y$ for $z \geq 0$ having free space. For $z < 0$ medium has $\epsilon_{r2} = 1$; $\mu_{r2} = 4$.

z = 0

z < 0 (medium-2)

$\mu_{r_2} = 4$; $\epsilon_{r_2} = 1$

$\eta_2 = 120\pi\sqrt{\frac{\mu_{r2}}{\epsilon_{r2}}}$

$= 120\pi\sqrt{\frac{4}{1}} = 120\pi(2)$

z ≥ 0 (air) (medium-1)

$\vec{E} = 10\cos(2\pi \times 10^8 t + \beta z)\hat{a}_y$

EM wave

$\eta_1 = 120\pi$

$$\Gamma = \frac{\eta_2 - \eta_1}{\eta_2 + \eta_1} = \frac{120\pi(2) - 120\pi}{120\pi(2) + 120\pi} = \frac{1}{3}$$

2.59 (a, c)

The given wave equation is

$$\frac{\partial^2 f(x,t)}{\partial t^2} = 10000\frac{\partial^2 f(x,t)}{\partial x^2}$$

The solution of the wave equation is given as

$$f(x, t) = f_1(x - vt) + f_2(x + vt)$$

where, $v = \sqrt{10000} = 100$

For option (a),

$$f_1(x - vt) = e^{-(x-100t)^2}$$

$$f_2(x + vt) = e^{-(x+100t)^2}$$

For option (c),

$$f_1(x - vt) = e^{-(x-100t)}$$

$$f_2(x + vt) = \sin(x + 100t)$$

For the above functions,

$$\frac{\partial^2 f(x,t)}{\partial t^2} = 10000\frac{\partial^2 f(x,t)}{\partial x^2}$$

Hence, options (a) and (c) are correct.

2.60 (a, b, d)

Given,

$$\vec{E} = \hat{a}_x C_{1x}\cos(\omega t - \beta z) + \hat{a}_y C_{1y}\cos(\omega t - \beta z + \theta)$$

At $z = 0$

$$\vec{E} = C_{1x}\cos\omega t\,\hat{a}_x + C_{1y}\cos(\omega t + \theta)\hat{a}_y$$

Going by options,

Option (a): $\vec{E} = \cos\omega t\,\hat{a}_x + \cos(\omega t + \pi/4)\hat{a}_y$

at $t = 0$, $\omega t = 0$, $\vec{E} = \hat{a}_x + \frac{1}{\sqrt{2}}\hat{a}_y$

at $t = T/4$, $\omega t = \pi/2$, $\vec{E} = 0 - \frac{1}{\sqrt{2}}\hat{a}_y$

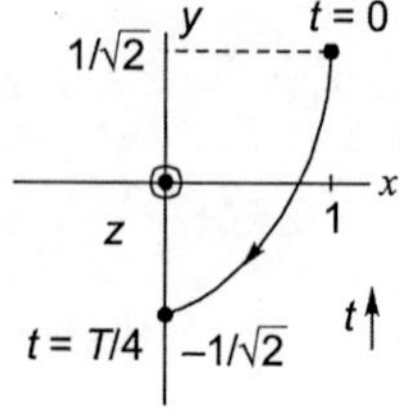

⇒ Hence, it is LHEP.

Option (b): $\vec{E} = 2\cos\omega t\,\hat{a}_x + \cos(\omega t + \pi/2)\hat{a}_y$

at $t = 0$, $\omega t = 0$, $\vec{E} = 2\hat{a}_x$

at $t = T/4$, $\omega t = \pi/2$, $\vec{E} = -1\hat{a}_y$

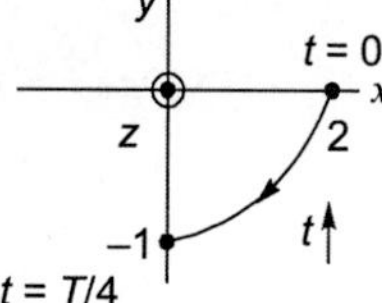

⇒ Hence, it is LHEP.

Option (c): $\vec{E} = \cos\omega t\,\hat{a}_x + 2\cos(\omega t + 3\pi/2)\hat{a}_y$

at $t = 0$, $\omega t = 0$, $\vec{E} = \hat{a}_x$

at $t = T/4$, $\omega t = \pi/2$, $\vec{E} = 2\hat{a}_y$

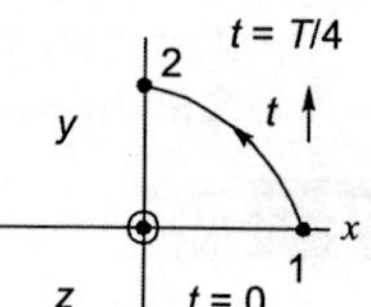

⇒ Hence, it is RHEP.

Option (d): $\vec{E} = 2\cos\omega t\,\hat{a}_x + \cos(\omega t + 3\pi/4)\hat{a}_y$

at $t = 0$, $\omega t = 0$, $\vec{E} = 2\hat{a}_x - \frac{1}{\sqrt{2}}\hat{a}_y$

at $t = T/4$, $\omega t = \pi/2$, $\vec{E} = 0 - \frac{1}{\sqrt{2}}\hat{a}_y = \frac{-1}{\sqrt{2}}\hat{a}_y$

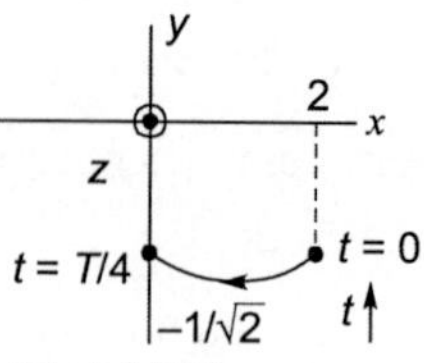

⇒ Hence, it is LHEP.

∴ Option (a), (b) and (d) are correct.

2.61 (d)

At perfect conductor,

$$\Gamma = -1$$

$$\Rightarrow \quad \Gamma = 1\angle\pi$$

air | Perfect conductor | $x = 0$

As we know that H_{min} occurs at E_{max}.

Hence, $E_{max} = E_0[1 + |\Gamma|]$

$$H_{min} = \frac{E_0}{\eta}\left[1 - |\Gamma|\right] \text{ at } 2\beta x_{max} = 2n\pi + \theta_\Gamma$$

So, $2\beta x_{max} = 2n\pi + \theta_\Gamma$

$$\Rightarrow \quad \frac{4\pi}{\lambda}x_{max} = 2n\pi + \pi$$

$$\Rightarrow \quad x_{max} = (2n+1)\frac{\lambda}{4};\ n = 0, 1, 2, 3...$$

So, the 1st value of x at which H-field will be minimum (i.e. zero as $|\Gamma| = 1$) is given by

$$x_{max} = \frac{\lambda}{4} \quad [\text{at } n = 0]$$

Now, from the equation given,

$$\beta = \frac{2\pi}{3}\text{ rad/m}$$

$$\Rightarrow \quad \frac{2\pi}{\lambda} = \frac{2\pi}{3} \quad \Rightarrow \lambda = 3\text{ m}$$

Hence, $x_{max} = \frac{3}{4}$ m

Since for the air $x < 0$, hence, the first null of the magnetic field occurs at

$$x = -\frac{3}{4}\text{ m}$$

■■■

3 Transmission Lines

3.1 A lossless transmission line having 50 Ω characteristic impedance and length λ/4 is short circuited at one end and connected to an ideal voltage source of 1 V at the other end. The current drawn from the voltage sources is

(a) 0 (b) 0.02 A
(c) ∞ (d) none of the these

[1996 : 1 M]

3.2 The capacitance per unit length and the characteristic impedance of a lossless transmission line are C and Z_0 respectively. The velocity of a travelling wave on the transmission line is

(a) Z_0C (b) $1/(Z_0C)$
(c) Z_0/C (d) C/Z_0 [1996 : 1 M]

3.3 A transmission line of 50 Ω characteristic impedance is terminated with a 100 Ω resistance. The minimum impedance measured on the line is equal to

(a) 0 Ω (b) 25 Ω
(c) 50 Ω (d) 100 Ω [1997 : 1 M]

3.4 All transmission line section in Figure, have a characteristic impedance $R_0 + j0$. The input impedance Z_{in} equals

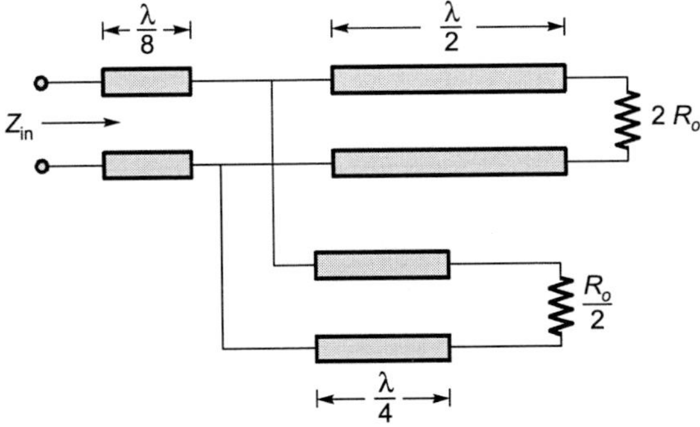

(a) $\frac{2}{3}R_0$ (b) R_0
(c) $\frac{3}{2}R_0$ (d) $2R_0$ [1998 : 1 M]

3.5 In a twin-wire transmission line in air, the adjacent voltage maxima are at 12.5 cm and 27.5 cm. The operating frequency is

(a) 300 MHz (b) 1 GHz
(c) 2 GHz (d) 6.28 GHz

[1999 : 2 M]

3.6 In air, a lossless transmission line of length 50 cm with $L = 10 \mu H/m$, $C = 40 pF/m$ is operated at 25 MHz. Its electrical path length is

(a) 0.5 meters (b) λ meters
(c) π/2 radians (d) 180 degrees

[1999 : 2 M]

3.7 The magnitudes of the open-circuit and short-circuit input impedances of a transmission line are 100 Ω and 25 Ω respectively. The characteristic impedance of the line is,

(a) 25 Ω (b) 50 Ω
(c) 75 Ω (d) 100 Ω

[2000 : 1 M]

3.8 A transmission line is distortionless if

(a) $RL = \frac{1}{GC}$ (b) $RL = GC$
(c) $LG = RC$ (d) $RG = LC$

[2001 : 1 M]

3.9 A uniform plane electromagnetic wave incident normally on a plane surface of a dielectric material is reflected with a VSWR of 3. What is the percentage of incident power that is reflected?

(a) 10% (b) 25%
(c) 50% (d) 75%

[2001 : 2 M]

3.10 The VSWR can have any value between

(a) 0 and 1 (b) −1 and +1
(c) 0 and ∞ (d) 1 and ∞

[2002 : 1 M]

3.11 In an impedance Smith chart, a clockwise movement along a constant resistance circle gives rise to

(a) a decrease in the value of reactance
(b) an increase in the value of reactance
(c) no change in the reactance value
(d) no change in the impedance value

[2002 : 1 M]

3.12 A short-circuited stub is shunt connected to a transmission line as shown in the figure. If $Z_0 = 50\ \Omega$, the admittance Y seen at the junction of the stub and the transmission line is

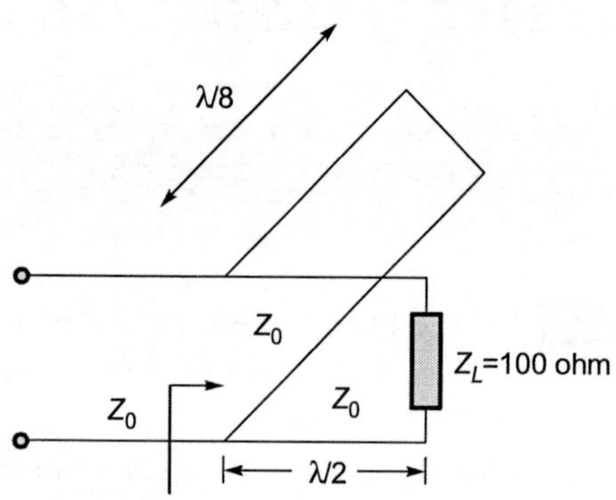

(a) $(0.01 - j0.02)$ mho
(b) $(0.02 - j0.01)$ mho
(c) $(0.04 - j0.02)$ mho
(d) $(0.02 + j0)$ mho [2003 : 2 M]

3.13 Consider a 300 Ω, quarter-wave long (at 1 GHz) transmission line as shown in the figure. It is connected to a 10 V, 50 Ω source at one end and is left open circuited at the other end. The magnitude of the voltage at the open circuit end of the line is

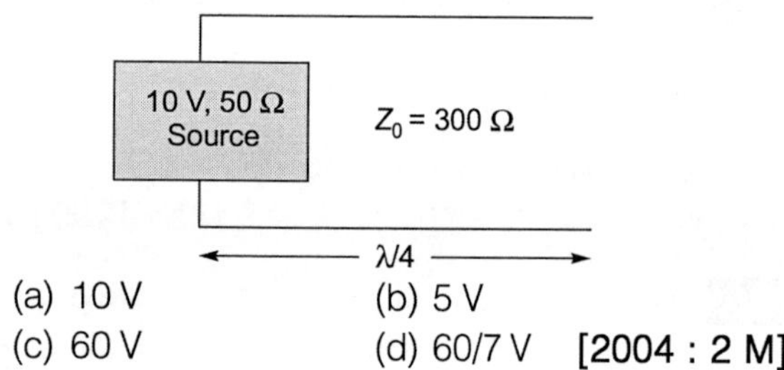

(a) 10 V (b) 5 V
(c) 60 V (d) 60/7 V [2004 : 2 M]

3.14 Consider an impedance $Z = R + jX$ marked with point P in an impedance Smith chart as shown in the figure. The movement from point P along a constant resistance circle in the clockwise direction by an angle 45° is equivalent to

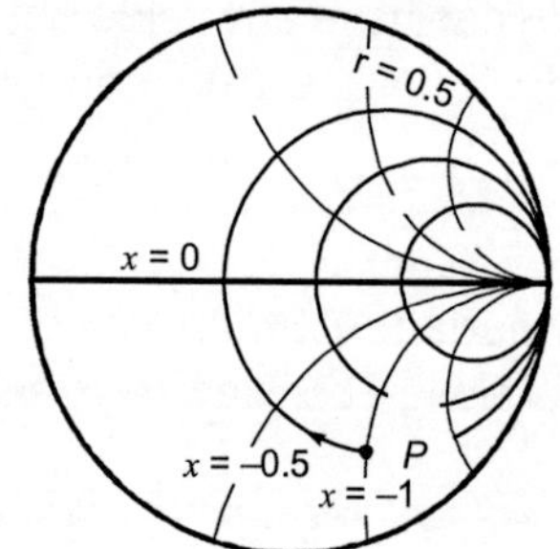

(a) adding an inductance in series with Z
(b) adding a capacitance in series with Z
(c) adding an inductance in shunt across Z
(d) adding a capacitance in shunt across Z

[2004 : 2 M]

3.15 A plane electromagnetic wave propagating in free space is incident normally on a large slab of loss-less, non-magnetic, dielectric material with $\epsilon > \epsilon_0$. Maxima and minima are observed when the electric field is measured in front of the slab. The maximum electric field is found to be 5 times the minimum field. The intrinsic impedance of the medium should be

(a) 120 πΩ (b) 60 πΩ
(c) 600 πΩ (d) 24 πΩ

[2004 : 2 M]

3.16 A lossless transmission line is terminated in a load which reflects a part of the incident power. The measured VSWR is 2. The percentage of the power that is reflected back is

(a) 57.73 (b) 33.33
(c) 0.11 (d) 11.11

[2004 : 2 M]

3.17 Characteristic impedance of a transmission line is 50 Ω. Input impedance of the open-circuited line is $Z_{OC} = 100 + j150$ Ω. When the transmission line is short-circuited, then value of the input impedance will be

(a) 50 Ω (b) $100 + j150$ Ω
(c) $7.69 + j11.54$ Ω (d) $7.69 - j11.54$ Ω

[2005 : 2 M]

Common data for Questions 3.18 and 3.19

Voltage standing wave pattern in a lossless transmission line with characteristic impedance 50 Ω and a resistive load is shown in the figure.

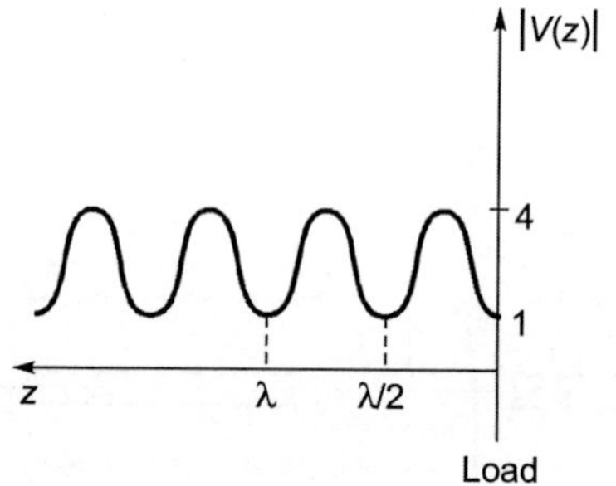

3.18 The value of the load resistance is

(a) 50 Ω (b) 200 Ω
(c) 12.5 Ω (d) 0 Ω [2005 : 2 M]

3.19 The reflection coefficient is given by

(a) –0.6 (b) –1
(c) 0.6 (d) 0

[2005 : 2 M]

3.20 Many circles are drawn in a Smith chart used for transmission line calculations. The circles shown in the figure represent

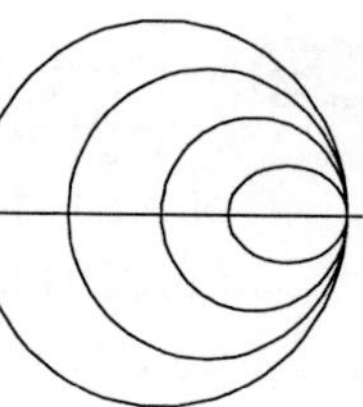

(a) unit circles
(b) constant resistance circles
(c) constant reactance circles
(d) constant reflection coefficient circles

[2005 : 2 M]

Common data for Questions 3.21 & 3.22:

A 30-Volts battery with zero source resistance is connected to a coaxial line of characteristic impedance of 50 Ohms at t = 0 second and terminated in an unknown resistive load. The line length is such that it takes 400 μs for an electromagnetic wave to travel from source end to load end and vice-versa. At t = 400 μs, the voltage at the load end is found to be 40 Volts.

3.21 The load resistance is

(a) 25 Ohms (b) 50 Ohms
(c) 75 Ohms (d) 100 Ohms

[2006 : 2 M]

3.22 The steady-state current through the load resistance is

(a) 1.2 Amps (b) 0.3 Amps
(c) 0.6 Amps (d) 0.4 Amps

[2006 : 2 M]

3.23 The parallel branches of a 2-wire transmission line are terminated in 100 Ω and 200 Ω resistors as shown in the figure. The characteristic impedance of the line is Z_0 = 50 Ω and each section has a length of λ/4. The voltage reflection coefficient Γ at the input is

(a) 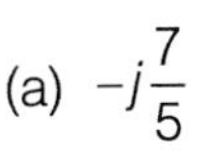

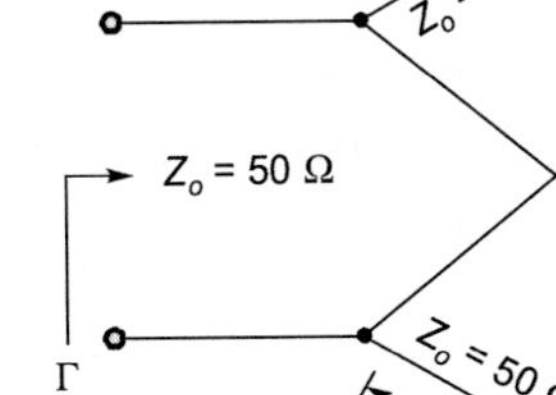

(b) $\frac{-5}{7}$

(c) 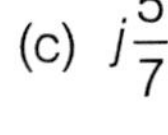

(d) $\frac{5}{7}$

[2007 : 2 M]

3.24 One end of a loss-less transmission line having the characteristic impedance of 75 Ω and length of 1 cm is short-circuited. At 3 GHz, the input impedance at the other end of the transmission line is

(a) 0 (b) Resistive
(c) Capacitive (d) Inductive

[2008 : 2 M]

3.25 A transmission line terminates in two branches, each of length λ/4, as shown. The branches are terminated by 50 Ω loads. The lines are lossless and have the characteristic impedances shown. Determine the impedance Z_i as seen by the source.

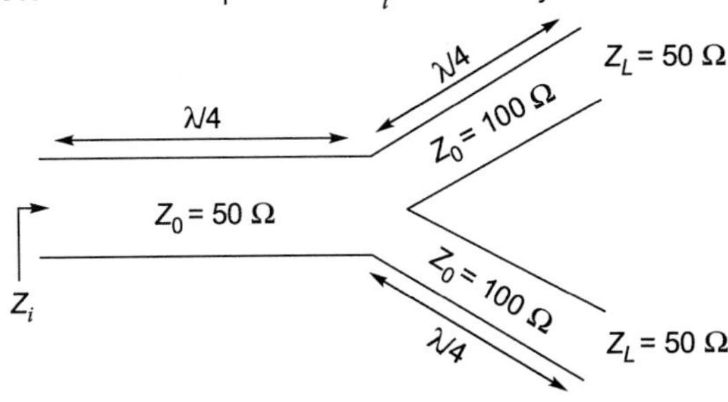

(a) 200 Ω (b) 100 Ω
(c) 50 Ω (d) 25 Ω [2009 : 2 M]

3.26 A transmission line has a characteristic impedance of 50 Ω and a resistance of 0.1 Ω/m. If the line is distortionless, the attenuation constant (in Np/m) is

(a) 500 (b) 5
(c) 0.014 (d) 0.002 [2010 : 1 M]

3.27 In the circuit shown, all the transmission line sections are lossless. The Voltage Standing Wave Ratio (VSWR) on the 60 Ω line is

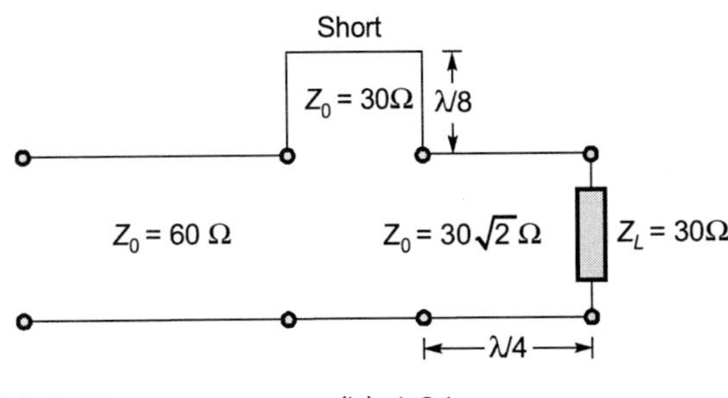

(a) 1.00 (b) 1.64
(c) 2.50 (d) 3.00 [2010 : 2 M]

3.28 A transmission line of characteristic impedance 50 Ω is terminated by a 50 Ω load. When excited by a sinusoidal voltage source at 10 GHz, the phase difference between two points spaced 2 mm apart on the line is found to be π/4 radians. The phase velocity of the wave along the line is

(a) 0.8×10^8 m/s (b) 1.2×10^8 m/s
(c) 1.6×10^8 m/s (d) 3×10^8 m/s

[2011 : 1 M]

3.29 A transmission line of characteristic impedance 50 Ω is terminated in a load impedance Z_L. The VSWR of the line is measured as 5 and the first of the voltage maxima in the line is observed at a distance of λ/4 from the load. The value of Z_L is

(a) $10\ \Omega$
(b) $250\ \Omega$
(c) $(19.23 + j46.15)\ \Omega$
(d) $(19.23 - j46.15)\ \Omega$ [2011 : 2 M]

3.30 A coaxial-cable with an inner diameter of 1 mm and outer diameter of 2.4 mm is filled with a dielectric of relative permittivity 10.89. Given $\mu_0 = 4\pi \times 10^{-7}$ H/m, $\varepsilon_0 = \frac{10^{-9}}{36\pi}$ F/m, the characteristic impedance of the cable as

(a) $330\ \Omega$
(b) $100\ \Omega$
(c) $143.3\ \Omega$
(d) $43.4\ \Omega$ [2012 : 1 M]

3.31 A transmission line with a characteristic impedance of $100\ \Omega$ is used to match a $50\ \Omega$ section to a $200\ \Omega$ section. If the matching is to be done both at 429 MHz and 1 GHz, the length of the transmission line can be approximately

(a) 82.5 cm
(b) 1.05 m
(c) 1.58 m
(d) 1.75 m [2012 : 2 M]

3.32 The return loss of a device is found to be 20 dB. The voltage standing wave ratio (VSWR) and magnitude of reflection coefficient are respectively

(a) 1.22 and 0.1
(b) 0.81 and 0.1
(c) −1.22 and 0.1
(d) 2.44 and 0.2

[2013 : 1 M]

3.33 For a parallel plate transmission line, let v be the speed of propagation and Z be the characteristic impedance. Neglecting fringe effects, a reduction of the spacing between the plates by a factor of two results in

(a) halving of v and no change in Z
(b) no changes in v and halving of Z
(c) no change in both v and z
(d) halving of both v and z

[2014 : 2 M, Set-1]

3.34 The input impedance of a $\frac{\lambda}{8}$ section of a lossless transmission line of characteristic impedance $50\ \Omega$ is found to be real when the other end is terminated by a load $Z_L = (R + jX)\ \Omega$. If X is $30\ \Omega$, the value of R (in Ω) is ______.

[2014 : 2 M, Set-1]

3.35 To maximize power transfer, a lossless transmission line is to be matched to a resistive load impedance via a $\lambda/4$ transformer as shown.

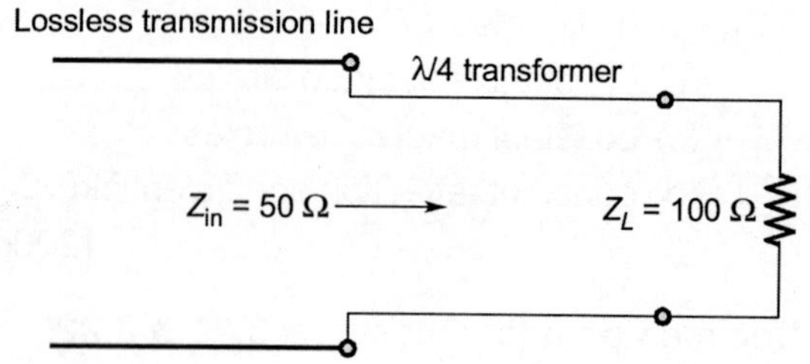

The characteristic impedance (in Ω) of the $\lambda/4$ transformer is ______.

[2014 : 1 M, Set-2]

3.36 In the transmission line shown, the impedance Z_{in} (in ohms) between node A and the ground is

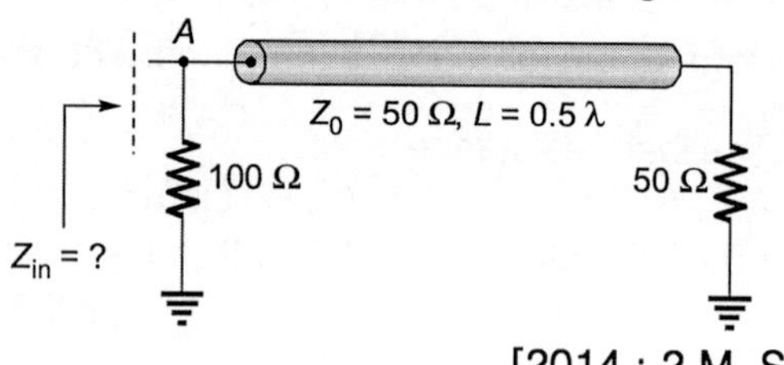

[2014 : 2 M, Set-2]

3.37 In the following figure, the transmitter T_x sends a wideband modulated *RF* signal via a coaxial cable to the receiver R_x. The output impedance Z_T of T_x, the characteristic impedance Z_0 of the cable and the input impedance Z_R of R_x are all real.

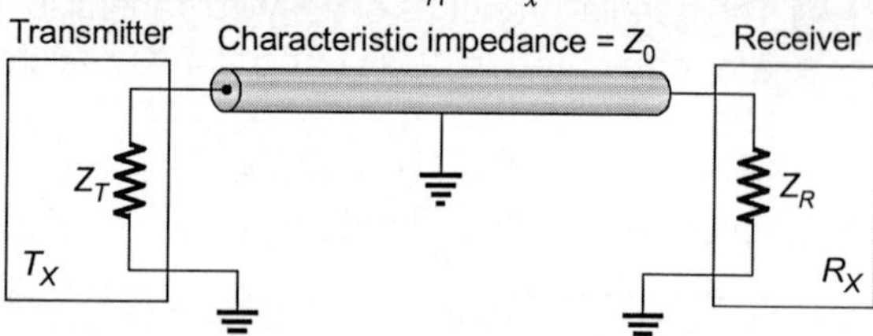

Which one of the following statements is **TRUE** about the distortion of the received signal due to impedance mismatch?

(a) The signal gets distorted if $Z_R \neq Z_0$, irrespective of the value of Z_T.
(b) The signal gets distorted if $Z_T \neq Z_0$, irrespective of the value of Z_R.
(c) Signal distortion implies impedance mismatch at both ends: $Z_T \neq Z_0$ and $Z_R \neq Z_0$.
(d) Impedance mismatches do NOT result in signal distortion but reduce power transfer efficiency.

[2014 : 1 M, Set-3]

3.38 A coaxial cable is made of two brass conductors. The spacing between the conductors is filled with Teflon ($\varepsilon_r = 2.1$, $\tan\delta = 0$). Which one of the following circuits can represent the lumped element model of a small piece of this cable having length Δz?

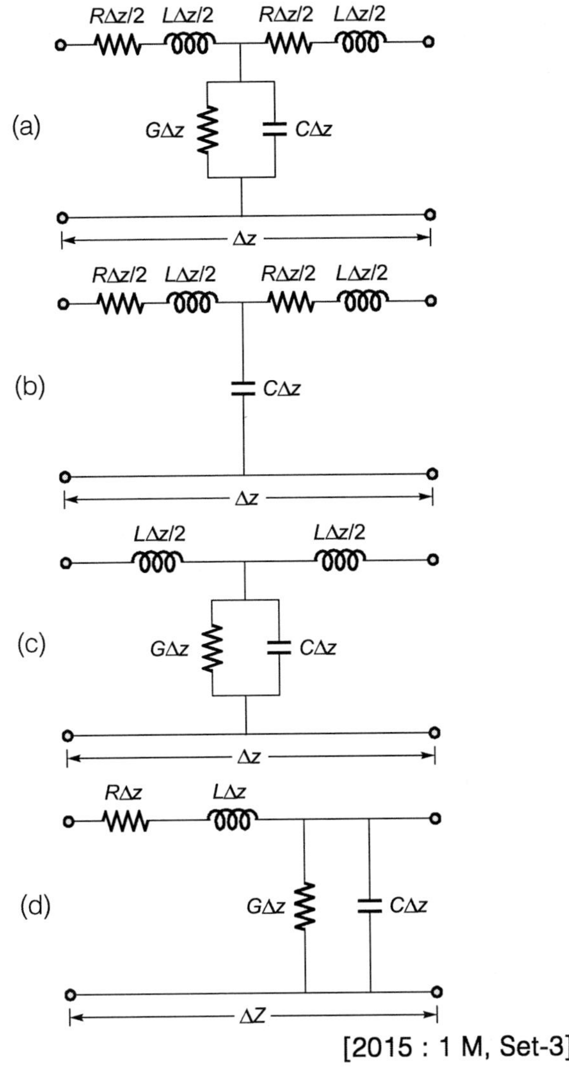

[2015 : 1 M, Set-3]

3.39 Consider the 3 m long lossless air-filled transmission line shown in the figure. It has a characteristic impedance of $120\,\pi\Omega$, is terminated by a short circuit, and is excited with a frequency of 37.5 MHz. What is the nature of the input impedance (Z_{in})?

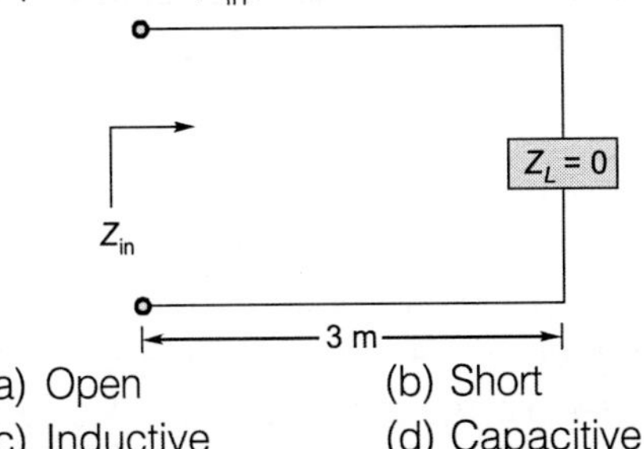

(a) Open
(b) Short
(c) Inductive
(d) Capacitive

[2015 : 2 M, Set-3]

3.40 A coaxial capacitor of inner radius 1 mm and outer radius 5 mm has a capacitance a per unit length of 172 pF/m. If the ratio of outer radius to inner is doubled, the capacitance per unit length (in pF/m) is_____. [2015 : 2 M, Set-3]

3.41 A 200 m long transmission line having parameters shown in the figure is terminated into a load R_L. The line is connected to a 400 V source having source resistance R_S through a switch which is closed at $t = 0$. The transient response of the circuit at the input of the line ($z = 0$) is also drawn in the figure. The value of R_L (in Ω) is _____.

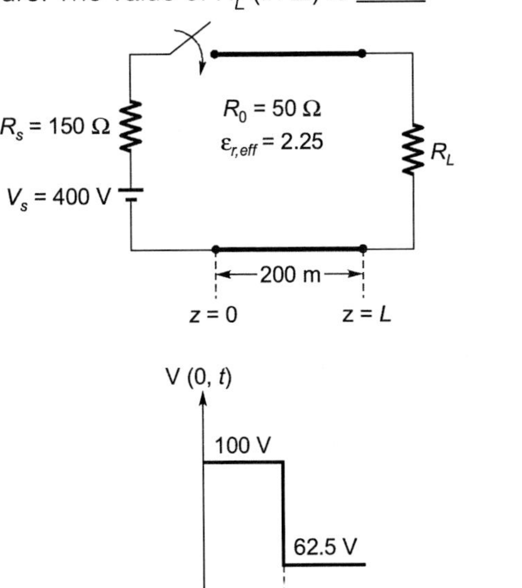

[2015 : 2 M, Set-3]

3.42 The propagation constant of a lossy transmission line is $(2 + j5)\ \text{m}^{-1}$ and its characteristic impedance is $(50 + j0)\Omega$ at $\omega = 10^6$ rad s^{-1}. The value of the line constants L, C, R, G are respectively,

(a) $L = 200\ \mu$H/m, $C = 0.1\ \mu$F/m, $R = 50\ \Omega$/m, $G = 0.02$ S/m
(b) $L = 250\ \mu$H/m, $C = 0.1\ \mu$F/m, $R = 100\ \Omega$/m, $G = 0.04$ S/m
(c) $L = 200\ \mu$H/m, $C = 0.2\ \mu$F/m, $R = 100\ \Omega$/m, $G = 0.02$ S/m
(d) $L = 250\ \mu$H/m, $C = 0.2\ \mu$F/m, $R = 50\ \Omega$/m, $G = 0.04$ S/m

[2016 : 1 M, Set-1]

3.43 A lossless microstrip transmission line consists of a trace of width w. It is drawn over a practically infinite ground plane and is separated by a dielectric slab of thickness t and relative permittivity $\varepsilon_r > 1$. The inductance per unit length and the characteristic impedance of this line are L and Z_0, respectively.

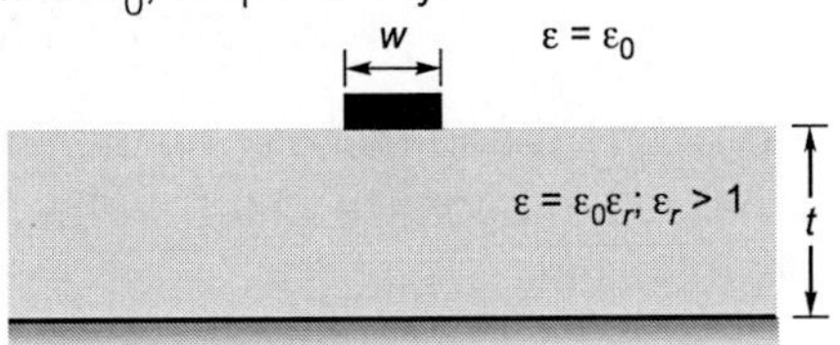

Which one of the following inequalities is always satisfied?

(a) $Z_0 > \sqrt{\dfrac{Lt}{\varepsilon_0\varepsilon_r w}}$ (b) $Z_0 < \sqrt{\dfrac{Lt}{\varepsilon_0\varepsilon_r w}}$

(c) $Z_0 > \sqrt{\dfrac{Lw}{\varepsilon_0\varepsilon_r t}}$ (d) $Z_0 < \sqrt{\dfrac{Lw}{\varepsilon_0\varepsilon_r t}}$

[2016 : 2 M, Set-2]

3.44 A microwave circuit consisting of lossless transmission lines T_1 and T_2 is shown in the figure. The plot shows the magnitude of the input reflection coefficient Γ as a function of frequency f. The phase velocity of the signal in the transmission lines is 2×10^8 m/s.

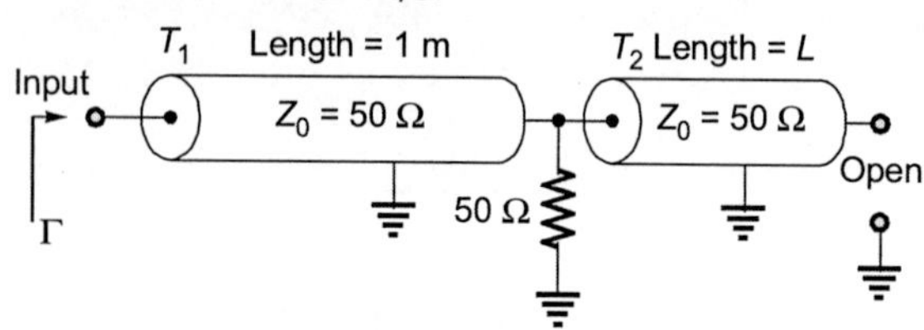

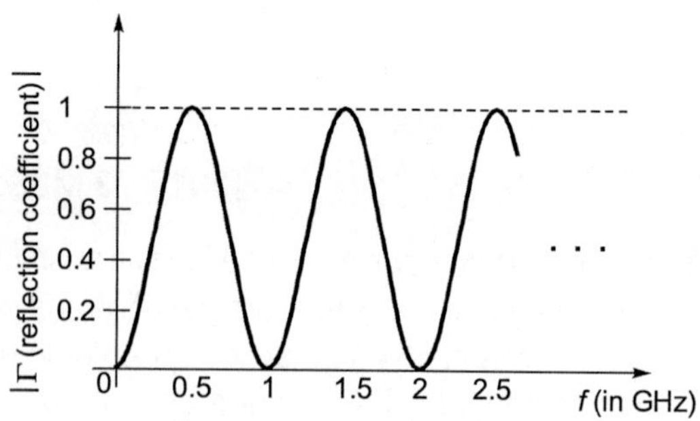

The length L (in meters) of T_2 is ______

[2016 : 2 M, Set-2]

3.45 The voltage of an electromagnetic wave propagating in a coaxial cable with uniform characteristic impedance is $V(l) = e^{-\gamma l + j\omega t}$ Volts, where l is the distance along the length of the cable in metres, $\gamma = (0.1 + j40)$ m^{-1} is the complex propagation constant, and $\omega = 2\pi \times 10^9$ rad/s is the angular frequency. The absolute value of the attenuation in the cable in dB/metre is

[2017 : 1 M, Set-1]

3.46 A two-wire transmission line terminates in a television set. The VSWR measured on the line is 5.8. The percentage of power that is reflected from the television set is _____.

[2017 : 1 M, Set-2]

3.47 The points P, Q and R shows on the Smith chart (normalized impedance chart) in the following figure represent:

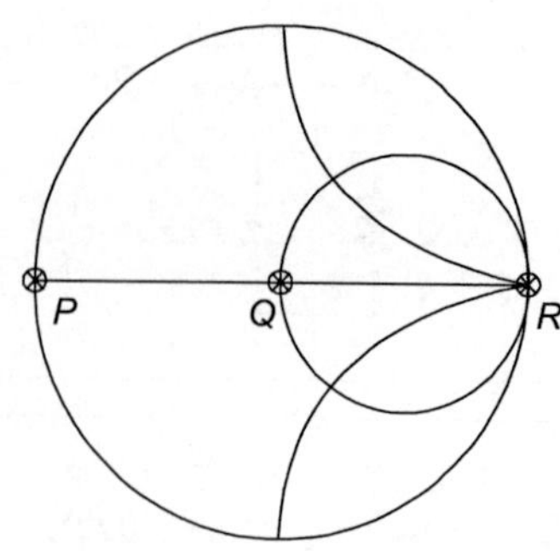

(a) P: Open Circuit, Q: Short Circuit, R: Matched Load

(b) P: Open Circuit, Q: Matched Load, R: Short Circuit

(c) P: Short Circuit, Q: Matched Load, R: Open Circuit

(d) P: Short Circuit, Q: Open Circuit, R: Matched Load

[2018 : 1 M]

3.48 A lossy transmission line has resistance per unit length R = 0.05 Ω/m. The line is distortionless and has characteristic impedance of 50 Ω. The attenuation constant (in Np/m correct to three decimal places) of the line is ________.

[2018 : 1 M]

3.49 A transmission line of length 3λ/4 and having a characteristic impedance of 50 Ω is terminated with a load of 400 Ω. The impedance (rounded off to two decimal places) seen at the input end of the transmission line is __________ Ω.

[2020 : 1 M]

3.50 The impedances $Z = jX$, for all X in the range $(-\infty, \infty)$, map to the Smith chart as

(a) a circle of radius 0.5 with centre at (0.5, 0).

(b) a point at the centre of the chart.

(c) a line passing through the centre of the chart.

(d) a circle of radius 1 with centre at (0, 0),

[2020 : 1 M]

3.51 The impedance matching network shown in the figure is to match a lossless line having characteristic impedance Z_0 = 50 Ω with a load impedance Z_L. A quarter-wave line having a characteristic impedance Z_1 = 75 Ω is connected to Z_L. Two stubs having characteristic impedance of 75 Ω each are connected to this quarter-wave line. One is a short-circuited (S.C.) sub of length 0.25λ connected across PQ and the other one is an open-circuited (O.C.) stub of length 0.5λ connected across RS.

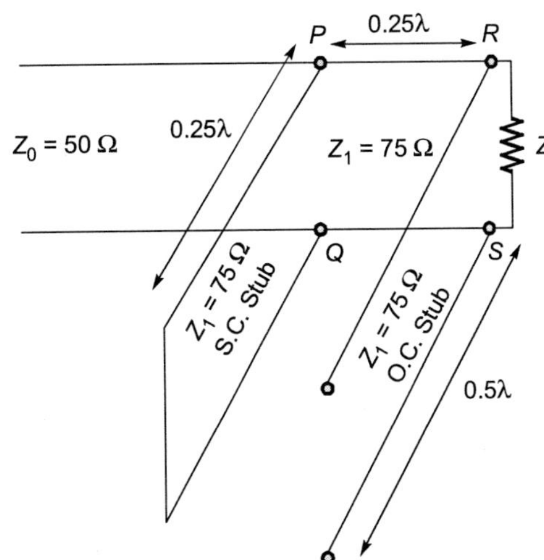

The impedance matching is achieved when the real part of Z_L is

(a) 75.0 Ω
(b) 112.5 Ω
(c) 33.3 Ω
(d) 50.0 Ω

[2021 : 2 M]

3.52 The S-parameters of a two port network is given as

$$[S] = \begin{bmatrix} S_{11} & S_{12} \\ S_{21} & S_{22} \end{bmatrix}$$

with reference to Z_0. Two lossless transmission line sections of electrical lengths $\theta_1 = \beta l_1$ and $\theta_2 = \beta l2$ are added to the input and output ports for measurement purposes, respectively. The S-parameters $[S']$ of the resultant two port network is

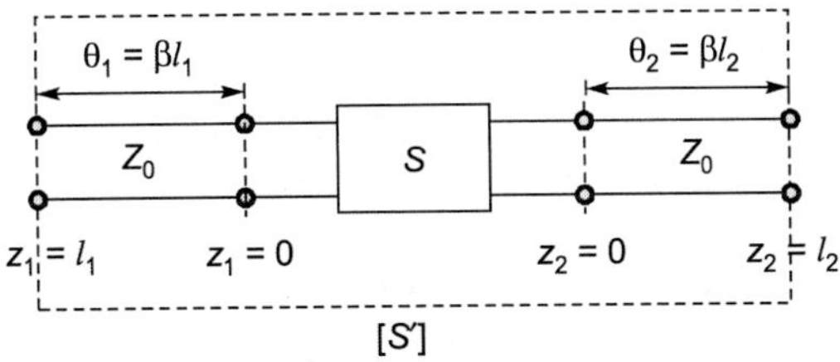

(a) $\begin{bmatrix} S_{11}e^{-j2\theta_1} & S_{12}e^{-j(\theta_1+\theta_2)} \\ S_{21}e^{-j(\theta_1+\theta_2)} & S_{22}e^{-j2\theta_2} \end{bmatrix}$

(b) $\begin{bmatrix} S_{11}e^{j2\theta_1} & S_{12}e^{-j(\theta_1+\theta_2)} \\ S_{21}e^{-j(\theta_1+\theta_2)} & S_{22}e^{j2\theta_2} \end{bmatrix}$

(c) $\begin{bmatrix} S_{11}e^{j2\theta_1} & S_{12}e^{ej(\theta_1+\theta_2)} \\ S_{21}e^{j(\theta_1+\theta_2)} & S_{22}e^{j2\theta_2} \end{bmatrix}$

(d) $\begin{bmatrix} S_{11}e^{-j2\theta_1} & S_{12}e^{ej(\theta_1+\theta_2)} \\ S_{21}e^{j(\theta_1+\theta_2)} & S_{22}e^{-j2\theta_2} \end{bmatrix}$

[2023 : 2 M]

3.53 The standing wave ratio on a 50 Ω lossless transmission line terminated in an unknown load impedance is found to be 2.0. The distance between successive voltage minima is 30 cm and the first minimum is located at 10 cm from the load. Z_L can be replaced by an equivalent length l_m and terminating resistance R_m of the same line. The value of R_m and l_m, respectively, are

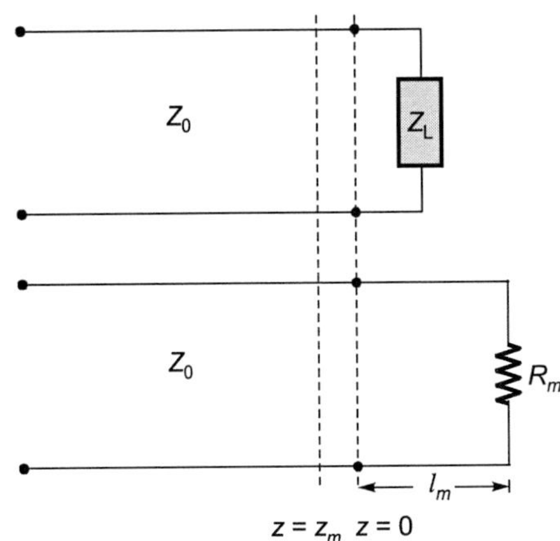

(a) R_m = 100 Ω, l_m = 20 cm
(b) R_m = 25 Ω, l_m = 20 cm
(c) R_m = 100 Ω, l_m = 5 cm
(d) R_m = 25 Ω, l_m = 5 cm

[2023 : 2 M]

3.54 The following circuit(s) representing a lumped element equivalent of an infinitesimal section of a transmission line is/are

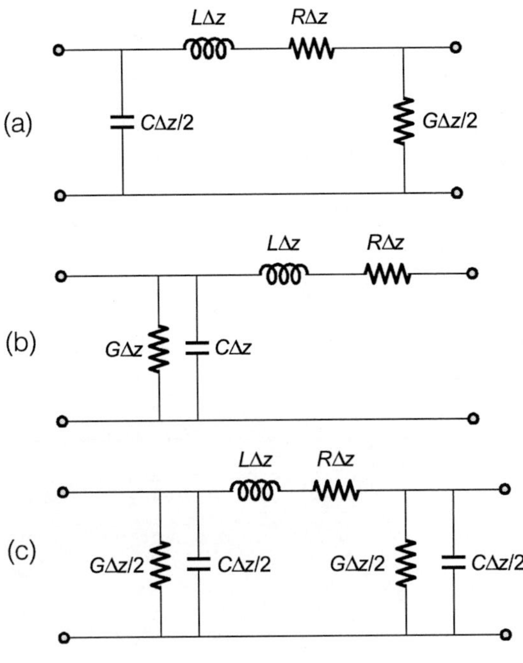

(d) 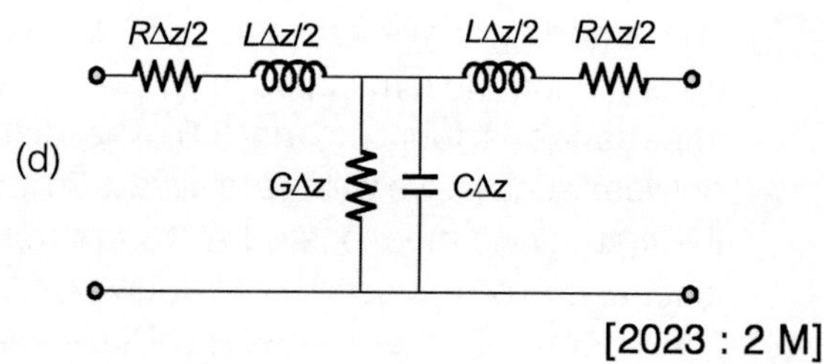

[2023 : 2 M]

3.55 A transparent dielectric coating is applied to glass ($\epsilon_r = 4$, $\mu_r = 1$) to eliminate the reflection of red light ($\lambda_0 = 0.75$ μm). The minimum thickness of the dielectric coating, in μm, that can be used is ________ (rounded off to two decimal places).

[2023 : 2 M]

3.56 Consider a lossless transmission line terminated with a short circuit as shown in the figure below. As one moves towards the generator from the load, the normalized impedances z_{inA}, z_{inB}, z_{inC} and z_{inD} (indicated in the figure) are ________.

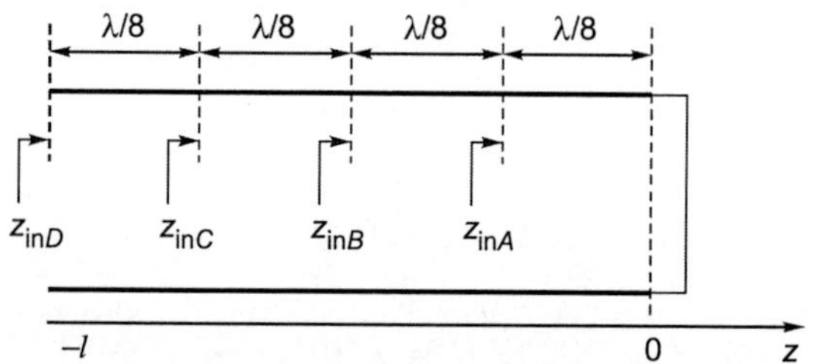

(a) $z_{inA} = +0.4j\,\Omega$, $z_{inB} = \infty$, $z_{inC} = -0.4j\,\Omega$, $z_{inD} = 0$

(b) $z_{inA} = -1j\,\Omega$, $z_{inB} = 0$, $z_{inC} = +1j\,\Omega$, $z_{inD} = \infty$

(c) $z_{inA} = \infty$, $z_{inB} = +0.4j\,\Omega$, $z_{inC} = 0$, $z_{inD} = +0.4j\,\Omega$

(d) $z_{inA} = +1j\,\Omega$, $z_{inB} = \infty$, $z_{inC} = -1j\,\Omega$, $z_{inD} = 0$

[2024 : 1 Mark]

3.57 A lossless transmission line with characteristic impedance $Z_0 = 50\ \Omega$ is terminated with an unknown load. The magnitude of the reflection coefficient is $|\Gamma| = 0.6$. As one moves towards the generator from the load, the maximum value of the input impedance magnitude looking towards the load (in Ω) is ________.

[2024 : 2 Marks]

■■■

Answers Transmission Lines

3.1 (a)	3.2 (b)	3.3 (b)	3.4 (b)	3.5 (b)	3.6 (c)	3.7 (b)
3.8 (c)	3.9 (b)	3.10 (d)	3.11 (b)	3.12 (a)	3.13 (c)	3.14 (a)
3.15 (d)	3.16 (d)	3.17 (d)	3.18 (c)	3.19 (a)	3.20 (b)	3.21 (d)
3.22 (b)	3.23 (d)	3.24 (d)	3.25 (d)	3.26 (d)	3.27 (b)	3.28 (c)
3.29 (a)	3.30 (*)	3.31 (c)	3.32 (a)	3.33 (b)	3.34 (40)	3.35 (70.72)
3.36 (33.3)	3.37 (c)	3.38 (b)	3.39 (d)	3.40 (120.22)	3.41 (30)	3.42 (b)
3.43 (b)	3.44 (0.1)	3.45 (0.868)	3.46 (49.83)	3.47 (c)	3.48 (0.001)	3.49 (6.25)
3.50 (d)	3.51 (b)	3.52 (a)	3.53 (b, c)	3.54 (b, c, d)	3.55 (0.133)	3.56 (d)
3.57 (200)						

Explanations Transmission Lines

3.1 (a)

For quarter wave transformer (λ/4),

$$Z_{in} = \frac{Z_0^2}{Z_L}$$

$$Z_L = 0\ \Omega$$

$$Z_{in} = \frac{Z_0^2}{0} = \infty = \text{open circuit}$$

$$I_s = \frac{V_s}{Z_{in}} = \frac{V_s}{\infty} = 0$$

3.2 (b)

$$Z_0 = \sqrt{\frac{L}{C}} \quad ; \quad Z_0^2 = \frac{L}{C}$$

$$L = Z_0^2 C$$

$$v = \frac{1}{\sqrt{LC}} = \frac{1}{\sqrt{(Z_0^2C)(C)}} = \frac{1}{Z_0 C}$$

3.3 (b)

$$R_L = 100\ \Omega \quad ; \quad R_0 = 50\ \Omega$$

$$SWR = \frac{100}{50} = 2$$

$$Z_{min} = \frac{Z_o}{\text{SWR}} = \frac{50}{2} = 25\ \Omega$$

3.4 (b)

For λ/4 line

$$Z_{in_1} = \frac{Z_{01}^2}{Z_{L_1}} = \frac{R_0^2}{R_0/2} = 2R_0$$

For λ/2 line

$$Z_{in_2} = Z_{L_2} = 2R_0$$

For λ/8 line

$$Z_L = (Z_{in_1}) \| (Z_{in_2}) = (2R_0) \| (2R_0) = R_0$$

For λ/8 line,

$$Z_{in} = Z_0\left[\frac{Z_L + jZ_0}{Z_0 + jZ_L}\right] = R_0\left[\frac{R_0 + jR_0}{R_0 + jR_0}\right] = R_0$$

3.5 (b)

Adjacent volt maximum are at λ/2 distance.

$$\therefore 27.5 - 12.5 = \frac{\lambda}{2}$$

$$\lambda = 30 \text{ cm} = 0.3 \text{ m}$$

$$f = \frac{c}{\lambda} = \frac{3\times10^8}{0.3} = 1 \text{ GHz}$$

3.6 (c)

Electrical path length = βl deg or rad.

$$v = \frac{1}{\sqrt{LC}} = \frac{1}{\sqrt{10\times10^{-6}\times40\times10^{-12}}}$$

$$= 0.5\times10^8$$

$$\lambda = \frac{v}{f} = \frac{0.5\times10^8}{25\times10^6}$$

$$\beta l = \frac{2\pi}{\lambda}\times l$$

$$= \frac{2\pi\times0.5\times25\times10^6}{0.5\times10^8} = \frac{\pi}{2}\text{rad.}$$

3.7 (b)

$$Z_0 = \sqrt{Z_{OC}\cdot Z_{SC}} = \sqrt{100\times25}$$

$$Z_0 = 10\times5 = 50\ \Omega$$

3.9 (b)

$$s = \frac{1+|\Gamma|}{1-|\Gamma|}$$

$$\Rightarrow \quad 3 = \frac{1+|\Gamma|}{1-|\Gamma|}$$

$$|\Gamma| = 0.5$$

$$\frac{P_r}{P_i} = |\Gamma|^2 = 0.25$$

∴ 25% of incident power is reflected.

3.11 (b)

Moving along clockwise direction is an inductive increase in the load impedance .

3.12 (a)

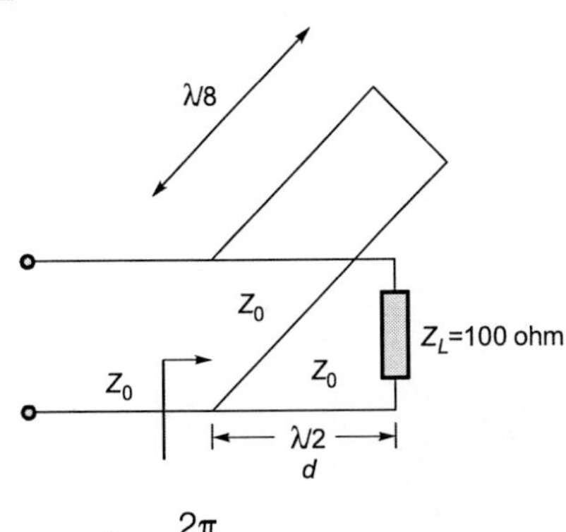

$$\beta = \frac{2\pi}{\lambda}$$

For Y_d $\quad \beta d = \frac{2\pi}{\lambda}\times\frac{\lambda}{2} = \pi$

$$Z_d = \frac{Z_0[Z_L + jZ_0\tan\beta d]}{Z_0 + jZ_L\tan\beta d} = \frac{50[100 + j50\tan\pi]}{(50 + j100\tan\pi)} = 100$$

$$Y_d = \frac{1}{Z_d} = \frac{1}{100} = 0.01$$

For Y_s $\beta d = \frac{2\pi}{\lambda} \times \frac{\lambda}{8} = \frac{\pi}{4}$, $Z_L \to 0$

$$Z_s = \frac{Z_0\left[Z_L + jZ_o\tan\frac{\pi}{4}\right]}{\left[Z_0 + jZ_L\tan\frac{\pi}{4}\right]} = jZ_o$$

$$Y_s = \frac{1}{Z_s} = \frac{1}{jZ_0} = -0.02j$$

$\therefore$ $Y = Y_d + Y_s = 0.01 - 0.02j$

3.13 (c)

The $\lambda/4$ line has $Z'_{in} = 0$ as $Z_L = \infty$

Current I_{in} into the line $= \frac{V}{50} = 0.2$ A

Using current equation on the line

$$I(x) = I_{in} = I_L\cos\beta x + \frac{jV_L}{Z_o}\sin\beta x$$

with $x = \lambda/4$, $I_L = 0$ as $Z_L = \infty$

Open circuit end voltage

$$V_L = 60 \text{ V}$$

3.14 (a)

Movement on constant resistance circle by an $\angle 45°$ in clockwise direction.

R is same and reactance increases i.e. addition of an inductance in series with Z.

3.15 (d)

$$\text{SWR} = \frac{E_{max}}{E_{min}}$$

$$\Rightarrow \quad s = \frac{5E_{min}}{E_{min}} = 5$$

$$5 = \frac{1+|\Gamma|}{1-|\Gamma|} \qquad |\Gamma| = \frac{2}{3}$$

$$\eta_2 = \frac{120\pi}{\sqrt{\epsilon_r}} \text{ and } \eta_2 = 120\pi$$

As $\eta_2 < \eta_1$

$$\Rightarrow -\frac{2}{3} = \frac{\eta_2 - 120\pi}{\eta + 120\pi}$$

$$\Rightarrow \eta_2 = 24\pi$$

3.16 (d)

$$\text{SWR} = \frac{1+\Gamma}{1-\Gamma}$$

$$\Rightarrow \quad 2 = \frac{1+\Gamma}{1-\Gamma}$$

Voltage reflection coefficient

$$\Rightarrow \quad \Gamma = \frac{1}{3}$$

Power reflection coefficient

$$\therefore \quad \frac{P_{ref}}{P_{inc}} = \Gamma^2 = \frac{1}{9}$$

$\Rightarrow$ 11.11% of P_i is reflected.

3.17 (d)

$$Z_0^2 = Z_{oc} \cdot Z_{sc}$$

$$Z_{sc} = \frac{50 \times 50}{100 + j150} = \frac{50}{2+3j} = \frac{50(2-3j)}{13}$$

$$Z_{SC} = 7.69 - 11.54j$$

3.18 (c)

$$S = \frac{V_{max}}{V_{min}} = \frac{4}{1} = 4$$

$$Z_{max} = Z_0 \cdot S$$

$$Z_{min} = Z_0/S$$

The diagram shows that minima is at load

$$\therefore \quad Z_L = Z_{min} = \frac{50}{4} = 12.5\ \Omega.$$

3.19 (a)

$$\Gamma = \frac{Z_L - Z_0}{Z_L + Z_0} = \frac{12.5 - 50}{12.5 + 50} = -0.6$$

3.21 (d)

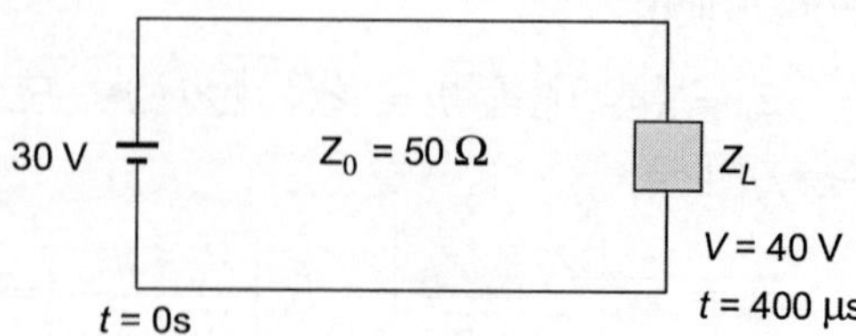

10 V reflects to add to 30 V in phase and gives 40 V.

$$\Gamma = \frac{V_r}{V_i} = \frac{10}{30} = \frac{1}{3} \quad ; \quad \Gamma = \frac{Z_L - Z_0}{Z_L + Z_0}$$

$$\Rightarrow \quad \frac{1}{3} = \frac{Z_L - 50}{Z_L + 50}$$

$$\Rightarrow \quad Z_L = 100\ \Omega$$

3.22 (b)

In steady state, $V_L = 30$ V

$$I_L = \frac{30}{100} = 0.3 \text{ A}$$

3.23 (d)

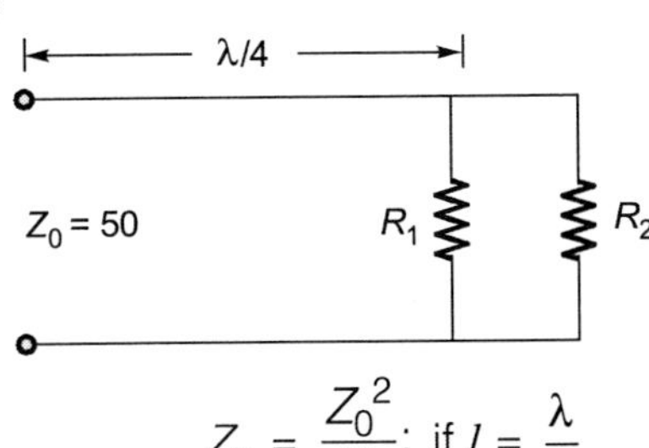

$$Z_{in} = \frac{Z_0^2}{Z_L}; \text{ if } l = \frac{\lambda}{4}$$

$$R_1 \text{ due to } 100\ \Omega = \frac{50^2}{100} = 25$$

$$R_2 \text{ due to } 200\ \Omega = \frac{50^2}{200} = \frac{25}{2}$$

$$\Rightarrow \quad R_1 \parallel R_2 = 25 \parallel \frac{25}{2} = \frac{25}{3}$$

$$Z_S = \frac{(50)^2}{25/3} = 300\ \Omega$$

$$\Rightarrow \quad \Gamma = \frac{Z_S - Z_0}{Z_S + Z_0} = \frac{300 - 50}{300 + 50} = \frac{5}{7}$$

3.24 (d)

It is a short circuit line with $l = 1$ cm.

As $$\lambda = \frac{3 \times 10^8}{3 \times 10^9} = 0.3 = 30 \text{ cm}$$

$$l < \frac{\lambda}{4}$$

The line has inductive nature.

As $Z_{SC} = jZ_o \tan \beta l$

3.25 (d)

$$Z_1 = \frac{Z_0^2}{Z_{L_1}} = \frac{(100)^2}{50} = 200\ \Omega$$

$$Z_2 = \frac{Z_0^2}{Z_2} = \frac{(100)^2}{50} = 200\Omega$$

$$Z_L = Z_1 \parallel Z_2 = 100\ \Omega$$

$$Z_i = \frac{Z_0^2}{Z_L} = \frac{(50)^2}{100} = 25\ \Omega$$

3.26 (d)

For distortionless transmission line,

$$LG = RC \qquad \Rightarrow \frac{L}{C} = \frac{R}{G}$$

Characteristic impedance,

$$Z_0 = \sqrt{\frac{L}{C}} = \sqrt{\frac{R}{G}}$$

Attenuation constant,

$$\alpha = \sqrt{RG} = \sqrt{R} \cdot \frac{\sqrt{R}}{Z_0} = \frac{R}{Z_0} = \frac{0.1}{50}$$

$$= 0.002 \text{ Np/m}$$

3.27 (b)

$$Z_{in} = Z_0\left(\frac{Z_L + jZ_0 \tan \beta l}{Z_0 + jZ_L \tan \beta l}\right)$$

Short
$Z_0 = 30\ \Omega$ λ/8
Z_1
$Z_0 = 60\ \Omega$ Z_2 $Z_0 = 30\sqrt{2}\ \Omega$ $Z_L = 30\ \Omega$
λ/4

Input impedance for λ/8 line,

$$Z_1 = 30\left[\frac{0 + j30 \tan\left(\frac{2\pi}{\lambda}\right)\left(\frac{\lambda}{8}\right)}{30 + 0}\right]$$

$$\Rightarrow \quad Z_1 = j30$$

Input impedance for λ/4 line,

$$Z_2 = 30\sqrt{2}\left[\frac{30 + j30\sqrt{2} \tan\left(\frac{2\pi}{\lambda}\right)\left(\frac{\lambda}{4}\right)}{30\sqrt{2} + j30 \tan\left(\frac{2\pi}{\lambda}\right)\left(\frac{\lambda}{4}\right)}\right]$$

$$= 30\sqrt{2}\left[\frac{\frac{30}{\tan \pi/2} + j30\sqrt{2}}{\frac{30\sqrt{2}}{\tan \pi/2} + j30}\right] = 60\ \Omega$$

or, $$Z_2 = \frac{(30\sqrt{2})^2}{30} = 60\ \Omega$$

Load impedance,

$$Z_L = Z_1 + Z_2 = j30 + 60$$

Magnitude of reflection coefficient,

$$|\Gamma| = \left|\frac{Z_L - Z_0}{Z_L + Z_0}\right| = \left|\frac{60 + j30 - 60}{60 + j30 + 60}\right|$$

$$= \left|\frac{j30}{120 + j30}\right| = \left|\frac{j1}{4 + j}\right|$$

$$= \frac{1}{\sqrt{16 + 1}} = \frac{1}{\sqrt{17}}$$

VSWR on 60 Ω line,

$$\text{VSWR} = \frac{1+|\Gamma|}{1-|\Gamma|} = \frac{1+\frac{1}{\sqrt{17}}}{1-\frac{1}{\sqrt{17}}} = 1.64$$

3.28 (c)

$$\text{Phase difference} = \frac{2\pi}{\lambda}\text{ (path difference)}$$

$$\Rightarrow \quad \frac{\pi}{4} = \frac{2\pi}{\lambda}(2\times 10^{-3})$$

$$\therefore \quad \lambda = 8\times 2\times 10^{-3}$$

$$= 16\times 10^{-3}\text{ m}$$

$$f = 10\text{ GHz} = 10\times 10^{9}\text{ Hz}$$

Hence the phase velocity of wave along the line is,

$$v = f\lambda = 10\times 10^{9}\times 16\times 10^{-3}\text{ m/sec}$$

$$\therefore \quad v = 1.6\times 10^{8}\text{ m/s}$$

3.29 (a)

$$\text{VSWR} = 5 = \frac{1+|\Gamma|}{1-|\Gamma|}$$

$$\therefore \quad |\Gamma| = \frac{2}{3} \Rightarrow \Gamma = \pm\frac{2}{3}$$

If maximum is at $\lambda/4$ from load then minimum will be at load itself.

For the phase of Γ,

$$2\beta x_{max} = 2n\pi + \theta$$

where x_{max} is distance of V_{max} from load.

$$2\cdot\frac{2\pi}{\lambda}\cdot\frac{\lambda}{4} = \theta$$

$$\Gamma = -\frac{2}{3}$$

$$\text{Now,} \quad -\frac{2}{3} = \frac{Z_L - Z_0}{Z_L + Z_0} = \frac{Z_L - 50}{Z_L + 50}$$

$$\Rightarrow \quad Z_L = 10\ \Omega$$

3.30 (*)

Characteristics impedance of the co-axial cable is given by

$$Z_o = \frac{138}{\sqrt{\epsilon_r}}\log\frac{D}{d}$$

$$= \frac{138}{\sqrt{10.89}}\log\left(\frac{2.4}{1}\right) = 15.89\ \Omega$$

3.31 (c)

The 100 Ω device used to match 50 Ω and 200 Ω devices is a quarter wave transformer.

Length of matching device

$$= (2n+1)\frac{\lambda}{4} = \text{odd multiple of } \frac{\lambda}{4}$$

Where n-depends on frequency of $\lambda/4$.

λ_1 of frequency 429 MHz = 0.7 m

λ_2 of frequency 1 GHz = 0.3 m

Common multiple = 2.1 m

Dividing by 4 = 0.525 m (not available in options)

Another odd multiple of 0.525 × 3 = 1.58 m

3.32 (a)

$$\text{Return loss} = -20\log|\Gamma|$$

$$-20\log|\Gamma| = 20$$

$$\Rightarrow \quad |\Gamma| = 10^{-1} = 0.1$$

$$\text{VSWR} = \frac{1+|\Gamma|}{1-|\Gamma|} = \frac{1+0.1}{1-0.1} = 1.22$$

3.33 (b)

$$Z_o = \sqrt{\frac{L}{C}} \text{ for parallel plate line}$$

$$\frac{C}{l} = \frac{\epsilon W}{d} \text{ and } \frac{L}{l} = \frac{\mu d}{W}$$

where, W = Width of the line

d = Separation between the lines

$Z_o \propto d$ (Z_o changes by half)

$$\text{Velocity} = \frac{1}{\sqrt{LC}} = \frac{1}{\sqrt{\mu_o\epsilon_o}} = 3\times 10^{8}\text{ m/s}$$

(No change in V)

3.34 (40)

$$Z_{in} = Z_o\left(\frac{Z_L + jZ_o\tan\beta l}{Z_o + jZ_L\tan\beta l}\right)$$

$$\text{at} \quad l = \frac{\lambda}{8}$$

$$\Rightarrow \quad \beta l = \frac{2\pi}{\lambda}\times\frac{\lambda}{8} = \frac{\pi}{4}$$

$$\therefore \tan\frac{\pi}{4} = 1$$

$$Z_{in} = Z_o\left[\frac{Z_L + j50}{Z_o + jZ_L}\right]$$

Given, $Z_L = R + jX = R + j30$ and $Z_o = 50\ \Omega$

$$30R = 80(50 + R)$$

$$Z_{in} = 50\left[\frac{R + j30 + j50}{50 + j(R + j30)}\right]$$

$$= 50\left[\frac{R + j80}{(50 - 30) + jR}\right] = 50\left[\frac{R + j80}{20 + jR}\right]$$

$$= 50\left[\frac{(R + j80)(20 - jR)}{20^2 + R^2}\right]$$

Since, only real part of Z_{in} exists, so imaginary part of $Z_{in} = 0$.

$$\therefore \quad Z_{in} = 50\left[\frac{20R + 1600j - jR^2 + 50R}{20^2 + R^2}\right]$$

$$Z_{imaginary} = 0 = \frac{1600 - R^2}{20^2 + R^2}$$

$$\therefore \quad R^2 = 1600$$

$$R = 40\ \Omega$$

3.35 (70.72)

Input impedance for quarter wave transformer is given as,

$$Z_{in} = \frac{Z_0^2}{Z_L}$$

Given: $Z_{in} = 50\ \Omega$, $Z_L = 100\ \Omega$

$$\therefore \quad Z_0^2 = 50 \times 100$$

$$Z_0 = \sqrt{50 \times 100}$$

$$Z_0 = 70.72\ \Omega$$

3.36 (33.3)

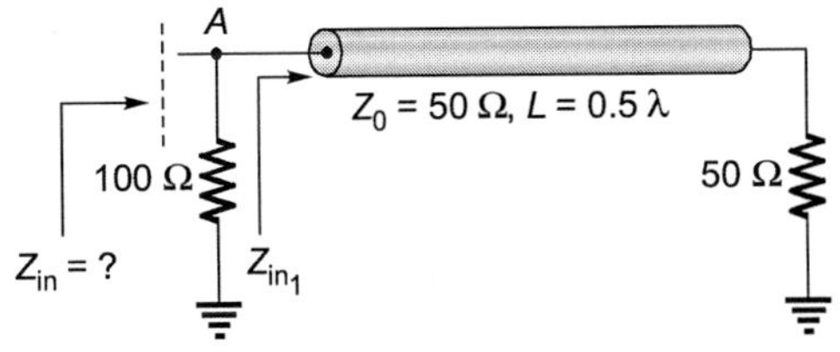

A; $Z_0 = 50\ \Omega$, $L = 0.5\ \lambda$; 100 Ω; 50 Ω; $Z_{in} = ?$; Z_{in_1}

$$Z_{in_1} = Z_0\left(\frac{Z_L + jZ_0 \tan\beta l}{Z_0 + jZ_L \tan\beta l}\right)$$

$$\beta l = \frac{2\pi}{\lambda} \times \frac{\lambda}{2} = \pi$$

$$Z_{in_1} = Z_0\left(\frac{Z_L + j0}{Z_0 + j0}\right)$$

$$Z_{in_1} = Z_L = 50\ \Omega$$

$$\therefore \quad Z_{in} = (Z_{in_1} \| 100\ \Omega) = \frac{100 \times 50}{150} = 33.3\ \Omega$$

3.37 (c)

The signal wave gets distorted if there is any impedance mismatch between the characteristic impedance Z_0 and the transmitting impedance Z_T and Z_0 and receiving impedance Z_R.

i.e. $Z_T \neq Z_0$ and $Z_R \neq Z_0$

then only distortion occurs.

As standing waveforms the signal has multi harmonics and hence distortion occurs.

3.38 (b)

The Teflon dielectric between the lines has

$\tan\delta$ = Loss tangent = 0

For the medium between the lines

$$G = 0$$

3.39 (d)

$$l = 3\text{ m}; \quad f = 37.5\text{ MHz}$$

$$\Rightarrow \quad \lambda = \frac{3 \times 10^8}{37.5 \times 10^6} = 8\text{ m}$$

$$Z_{in}(SC) = jZ_0 \tan\beta l = j120\pi \tan\frac{2\pi}{8} \cdot 3$$

$$= -j120\pi\Omega$$

3.40 (120.22)

$$C = \frac{2\pi \in}{\ln\frac{b}{a}}$$

$$172 \times 10^{-12} = \frac{2\pi \in}{\ln 5}$$

$$\Rightarrow \quad \in = \frac{172 \times 10^{-12} \times \ln 5}{2\pi}$$

If ratio of $\frac{b}{a}$ is doubled then

$$C = \frac{2\pi \in}{\ln 10} = \frac{2\pi \times \frac{172 \times 10^{-12} \times \ln 5}{2\pi}}{\ln 10}$$

$$C = 120.22\text{ pF}$$

3.41 (30)

At $t = 0$,

$$V_{in} = 400 \times \frac{R_o}{R_o + R_s} = 400 \times \frac{50}{200} = 100\text{ V}$$

At $t = T_0$ Γ_L 100 volts reflects

At $t = 2T_0$

$$V_{in} = 100 + (1 + \Gamma_{in})\,\Gamma_L 100 = 62.5$$

$$\Gamma_{in} = \frac{R_s - R_o}{R_s + R_o} = \frac{1}{2}$$

$$\text{Therefore,} \quad \Gamma_L = -\frac{1}{4} = \frac{R_L - 50}{R_L + 50}$$

$$R_L = 30\ \Omega$$

3.42 (b)

$$\gamma = \sqrt{(R+j\omega L)(G+j\omega C)}$$

$$Z_0 = \sqrt{\frac{R+j\omega L}{G+j\omega C}}$$

$$\gamma \cdot Z_0 = R + j\omega L = (2 + j5)(50 + j0) = 100 + j250$$

$$R = 100\ \Omega/\text{m}$$

$$L = \frac{250}{\omega} = \frac{250}{10^6} = 250\ \mu\text{H/m}$$

$$\frac{\gamma}{Z_0} = \frac{2+j5}{50} = G + j\omega C = 0.04 + j0.1$$

$$G = 0.04\ \text{S/m}$$

$$C = \frac{0.1}{\omega} = \frac{0.1}{10^6} = 0.1\ \mu\text{F/m}$$

3.44 (0.1)

At frequency of 1 GHz the line is matched and $|\Gamma| = 0$ as seen in the graph.

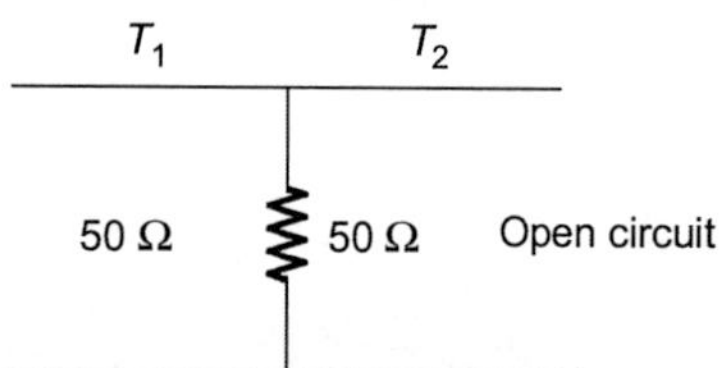

The line T_1 is having 50 Ω load parallel to T_2 input impedance.

This load is 50 Ω when T_2 input impedance is infinite.

$$Z_{in\,2} = \infty = Z_{L2}\ \text{(open circuit)}$$

T_2 has to be a $\frac{\lambda}{2}$ line.

$$f = 1\ \text{GHz}$$

$$\lambda = \frac{2\times 10^{10}}{1\times 10^9}\ \text{cm}$$

$$\frac{\lambda}{2} = L = 10\ \text{cm} = 0.1\ \text{m}$$

3.45 (0.868)

$$V(l) = V_o e^{-\alpha l} e^{-j\beta l} e^{j\omega t}$$

$$\text{Attenuation} = \frac{|\text{Input}|}{|\text{Output}|} = \frac{|V_o(0)|}{|V_o(l)|}$$

$$\text{Attenuation per meter} = \frac{|V_o|}{|V_o(1\text{m})|} = e^{\alpha}$$

Attenuation in dB/m = $(20\log e^{\alpha})$dB/m

$= 20(0.1)\log e = 0.868$ dB/m

3.46 (49.83)

Given that, VSWR (or) $s = 5.8$

$$s = \frac{1+|\Gamma|}{1-|\Gamma|} = 5.8$$

Γ = Reflection coefficient of television set

$$\Gamma = \frac{s-1}{s+1} = \frac{4.8}{6.8} = 0.70588$$

$$\frac{P_{\text{reflected}}}{P_{\text{incident}}} = \Gamma^2 = 0.4983 \text{ or } 49.83\%$$

3.47 (c)

For Short circuit,

$r = x = 0$ ⇒ Point "P"

For Open circuit,

$r = x = \infty$ ⇒ Point "R"

For Matched load,

$r = 1$ and $x = 0$ ⇒ Point "Q"

P: Short Circuit, Q: Matched Load

R: Open circuit

3.48 (0.001)

For a distortionless transmission line,

$$\frac{L}{R} = \frac{C}{G}$$

Propagation constant,

$$\gamma = \alpha + j\beta = \sqrt{(R+j\omega L)(G+j\omega C)} = \sqrt{RG}\left(1 + j\omega\frac{L}{R}\right)$$

Attenuation constant,

$$\alpha = \sqrt{RG}$$

Characteristic impedance,

$$Z_o = \sqrt{\frac{(R+j\omega L)}{(G+j\omega C)}} = \sqrt{\frac{R}{G}}$$

$$\sqrt{G} = \frac{\sqrt{R}}{Z_o}$$

So,

$$\alpha = \sqrt{R}\cdot\frac{\sqrt{R}}{Z_o} = \frac{R}{Z_o} = \frac{0.05}{50} = \frac{0.01}{10} = 0.001\ \text{Np/m}$$

3.49 (6.25)

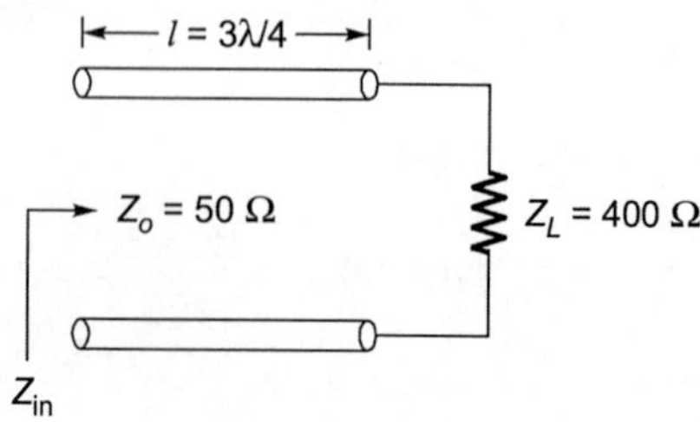

$$Z_{in} \text{ for } (l = \lambda/4) = \frac{Z_0^2}{Z_L} = \frac{50^2}{400} = \frac{25}{4} = 6.25\ \Omega$$

3.50 (d)

For given impedance Normalized impedance is

$$\frac{Z}{Z_0} = \frac{jX}{Z_0}$$

$$Z = jX$$

$$\Rightarrow \quad Z = 0 + jX$$

Normalized resistance = 0 ⇒ r = 0

$X = -\infty$ to ∞

$r = 0$ and X from $-\infty$ to ∞ is a unit circle (radius 1) and centre (0, 0) on a complex reflection coefficient plane:

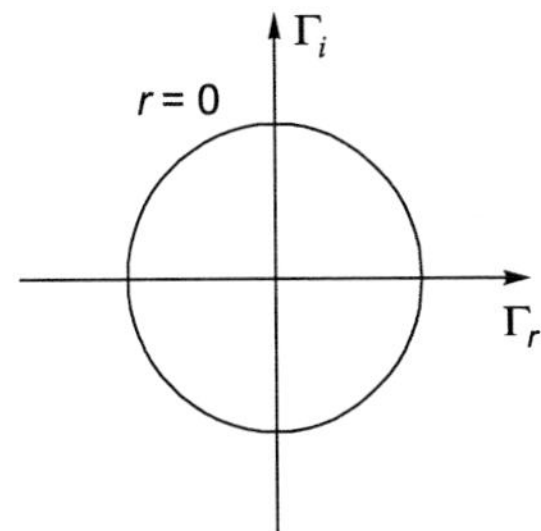

3.51 (b)

$$Z_{in_{\lambda/4}} = \frac{Z_0^2}{Z_L} = \frac{(75)^2}{0} = \infty \quad \left[\text{for } \frac{\lambda}{4} \text{ } Tx \text{ line}\right]$$

Given Z_L of $\frac{\lambda}{4}$ line is 0(SC) $\quad Z_{in_{\lambda/2}} = Z_L = \infty \quad \left[\text{for } \frac{\lambda}{2} \text{ } Tx \text{ line}\right]$

Given Z_L of $\frac{\lambda}{2}$ line is ∞(O.C)

The input impedance of $\frac{\lambda}{4}$ transmission line, as well as $\frac{\lambda}{2}$ transmission line is ∞, and they are in parallel with main transmission line, so they are not effective for main. Transmission line.

Final configuration of given line is

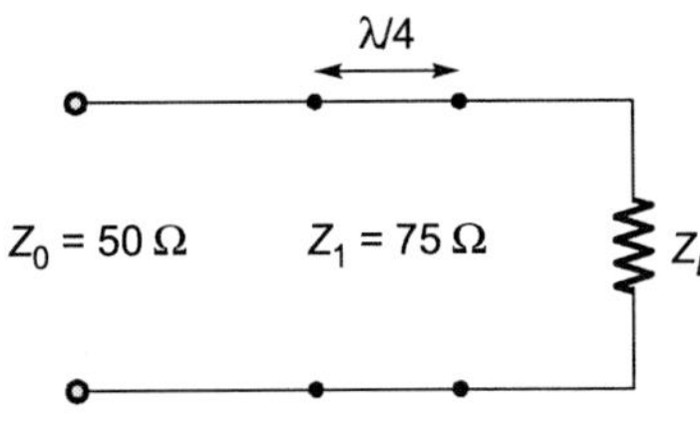

For impedance matching,

$$Z_L = \frac{Z_1^2}{Z_0} = \frac{(75)^2}{50} = 112.5$$

3.52 (a)

Let us evaluate S_{11} and S_{21} first at $V_2^+ = 0$

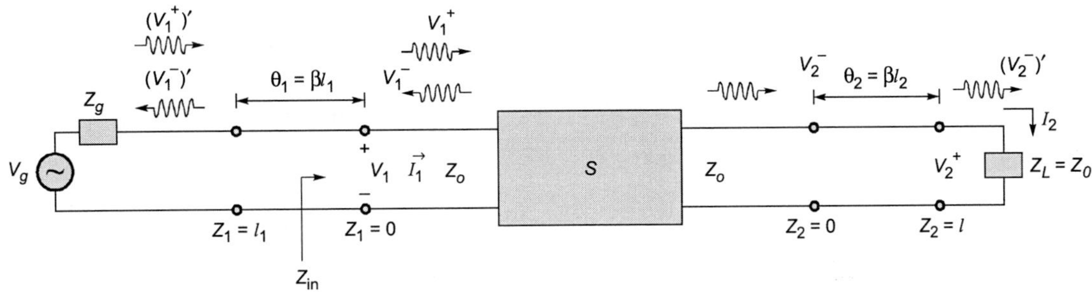

(a) S_{11}:

$$V_1 = V_1^+ + V_1^- = (V_1^+)' e^{-j\beta l_1} + (V_1^-) e^{+j\beta l_1}$$

$$\therefore \quad V_1^+ = (V_1^+)' e^{-j\beta l_1} \quad ; \quad V_1^- = (V_1^-) e^{+j\beta l_1}$$

$$S_{11} = \frac{V_1^-}{V_1^+} = \frac{(V_1^-) e^{+j\beta l_1}}{(V_1^+)' e^{-j\beta l_1}} = S'_{11} e^{+j2\beta l_1} \quad \Rightarrow S'_{11} = S_{11} e^{-j2\beta l_1} = S_{11} e^{-j2\theta_1}$$

$$\therefore \quad S'_{11} = S_{11} e^{-j2\theta_1}$$

(b) S_{21} :

$$V_2 = V_2^+ e^{+j\beta l_2} + V_2^- e^{-j\beta l_2} = (V_2^+)' + (V_2^-)'$$

Here, $V_2^+ = 0$

Hence, $V_2 = (V_2^-)' = V_2^- e^{-j\beta l_2} \Rightarrow V_2^- = (V_2^-)' e^{+j\beta l_2}$

From previous discussion in S_{11},

$$V_1^+ = (V_1^+)\, e^{-j\beta l_1}$$

$$\therefore\ S_{21} = \frac{V_2^-}{V_1^+} = \frac{(V_2^-)' e^{+j\beta l_2}}{(V_1^+)' e^{-j\beta l_1}} = S_{21}' e^{+j(\beta l_2 + \beta l_1)}$$

$$\Rightarrow\ S_{21}' = S_{21} e^{-j(\beta l_2 + \beta l_1)} = S_{21} e^{-j(\theta_1 + \theta_2)}$$

(looking at options (a) is correct)

3.53 (b, c)

Given $S = 2$, $Z_{min} = 10$ cm, $Z_0 = 50\ \Omega$

As we know that,

$$|\Gamma| = \frac{S-1}{S+1} = \frac{2-1}{2+1} = \frac{1}{3}$$

Now, distance between successive voltage minima = 30 cm

$$\Rightarrow\ \frac{\lambda}{2} = 30 \text{ cm}$$

$$\Rightarrow\ \lambda = 60 \text{ cm}$$

Also, for minima,

$$2\beta Z_{min} = (2n+1)\pi + \theta_\Gamma$$

At $n = 0$, 1st minima, $Z_{min} = 10$ cm

$$\frac{4\pi}{\lambda} Z_{min} = \pi + \theta_\Gamma$$

$$\Rightarrow\ \frac{4\pi}{60} * 10 = \pi + \theta_\Gamma$$

$$\Rightarrow\ \frac{2\pi}{3} - \pi = \theta_\Gamma$$

$$\Rightarrow\ \theta_\Gamma = \frac{-\pi}{3} \qquad \therefore\ \Gamma = \frac{1}{3}\angle -60^\circ$$

Now, $$\Gamma = \frac{Z_L - Z_0}{Z_L + Z_0}$$

$$\Rightarrow\ Z_L = Z_0\left[\frac{1+\Gamma}{1-\Gamma}\right]$$

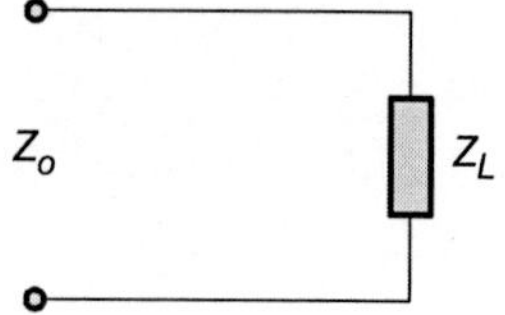

$$\Rightarrow\ Z_L = 50\left[\frac{1+0.33e^{-j\frac{\pi}{3}}}{1-0.33e^{-j\frac{\pi}{3}}}\right]$$

$$\Rightarrow\ Z_L = 67.97\angle -32.67^\circ$$

Now, $$Z_{in} = Z_0\left[\frac{Z_L + jZ_0\tan\beta l}{Z_0 + jZ_L\tan\beta l}\right]$$

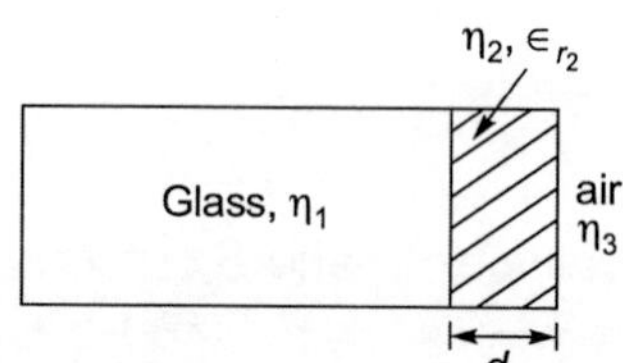

$$\Rightarrow\ Z_{in} = 50\left[\frac{R_m + j50\tan\beta l_m}{50 + jR_m\tan\beta l_m}\right]$$

Here, $Z_{in} = Z_L = 67.97\angle -32.67^\circ$

Going through options,

$R_m = 100\ \Omega$ and $l_m = 5$ cm and $R_m = 25\ \Omega$ and $l_m = 20$ cm } satisfy this identity, hence option (b) and (c) are correct.

3.55 (0.133)

For no reflection, impedance must be matched. Hence, η_2 acts like a quarter wave impedance transformer.

η_2, ϵ_{r_2}

Glass, η_1	(hatched layer, thickness d)	air η_3

So,

(i) $\eta_2 = \sqrt{\eta_1 \cdot \eta_3} \Rightarrow \epsilon_{r_2} = \sqrt{\epsilon_{r_1} \cdot \epsilon_{r_3}} \Rightarrow \epsilon_{r_2} = 2$

(ii) For impedance matching,

$$d = (2n+1)\frac{\lambda}{4};\ n = 0, 1, 2 \ldots$$

$$\lambda = \frac{\lambda_0}{\sqrt{\epsilon_r}} = \frac{\lambda_0}{\sqrt{\epsilon_{r_2}}}$$

Here, $$\lambda = \frac{0.75\times10^{-6}}{\sqrt{2}} = 0.53\times10^{-6}$$

Hence, for minimum distance, $n = 0$

So, $$d = \frac{\lambda}{4} = \frac{0.53\times10^{-6}}{4} = 0.133\ \mu\text{m}$$

3.56 (d)

The input impedance of a short-circuit transmission line of length 'l' is given by

$Z_{s/c} = jZ_o \tan \beta l.$

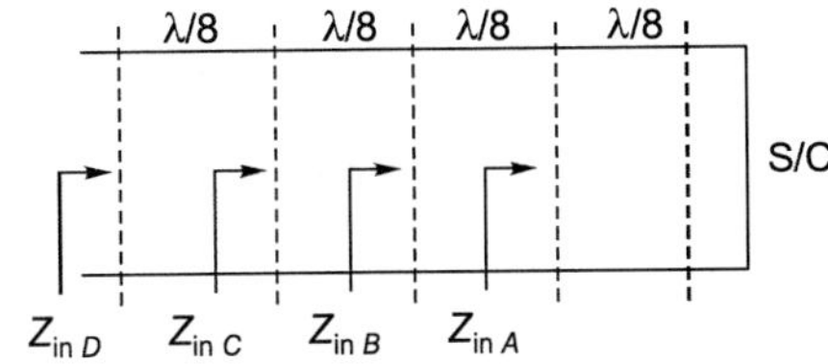

⇒ Normalized impedance,

$$\bar{Z}_{s/c} = \frac{Z_{S/C}}{Z_0} = j\tan\beta l$$

$$Z_{inA} = j\tan\left(\frac{2\pi}{\lambda}\cdot\frac{\lambda}{8}\right) = j\tan(\pi/4) = j1\ \Omega$$

$$Z_{inB} = j\tan\left(\frac{2\pi}{\lambda}\cdot\frac{\lambda}{4}\right) = \infty$$

$$Z_{inC} = j\tan\left(\frac{2\pi}{\lambda}\cdot\frac{3\lambda}{8}\right) = -j\,1\,\Omega$$

$$Z_{inD} = j\tan\left(\frac{2\pi}{\lambda}\cdot\frac{\lambda}{2}\right) = 0$$

3.57 (200)

As we know that

$$Z_{max} = SZ_0$$

Now,

$$S = \frac{1+|\Gamma|}{1-|\Gamma|} = \frac{1+0.6}{1-0.6} = \frac{1.6}{0.4} = 4$$

$$\therefore \quad Z_{max} = 4 \times 50 = 200\ \Omega.$$

■■■

4 Waveguides

4.1 A rectangular air filled waveguide has a cross section of 4 cm × 10 cm. The minimum frequency which can propagate in the waveguide is

(a) 1.5 GHz (b) 2.0 GHz
(c) 2.5 GHz (d) 3.0 GHz

[1997 : 1 M]

4.2 Assuming perfect conductors of a transmission line, pure *TEM* propagation is NOT possible in

(a) coaxial cable
(b) air-filled cylindrical waveguide
(c) parallel twin-wire line in air
(d) semi-infinite parallel plate waveguide

[1999 : 1 M]

4.3 A rectangular waveguide has dimensions 1 cm × 0.5 cm. Its cut-off frequency is

(a) 5 GHz (b) 10 GHz
(c) 15 GHz (d) 12 GHz

[2000 : 2 M]

4.4 The phase velocity of waves propagating in a hollow metal waveguide is

(a) greater than the velocity of light in free space.
(b) less than the velocity of light in free space
(c) equal to the velocity of light in free space
(d) equal to the group velocity

[2001 : 1 M]

4.5 The dominant mode in a rectangular waveguide is TE_{10}, because this mode has

(a) no attenuation
(b) no cut-off
(c) no magnetic field component
(d) the highest cut-off wavelength

[2001 : 1 M]

4.6 The phase velocity for the TE_{10} mode in an air-filled rectangular waveguide is

(a) less than c
(b) equal to c
(c) greater than c
(d) none of the above

Note: (c) is the velocity of plane waves in free space)

[2002 : 1 M]

4.7 A rectangular metal wave guide filled with a dielectric material of relative permittivity $\epsilon_r = 4$ has the inside dimensions 3.0 cm × 1.2 cm. The cut-off frequency for the dominant mode is

(a) 2.5 GHz (b) 5.0 GHz
(c) 10.0 GHz (d) 12.5 GHz

[2003 : 2 M]

4.8 The phase velocity of an electromagnetic wave propagating in a hollow metallic rectangular waveguide in the TE_{10} mode is

(a) equal to its group velocity
(b) less than the velocity of light in free space
(c) equal to the velocity of light in free space
(d) greater than the velocity of light in free space

[2004 : 1 M]

4.9 Which one of the following does represent the electric field lines for the TE_{02} mode in the cross-section of a hollow rectangular metallic waveguide?

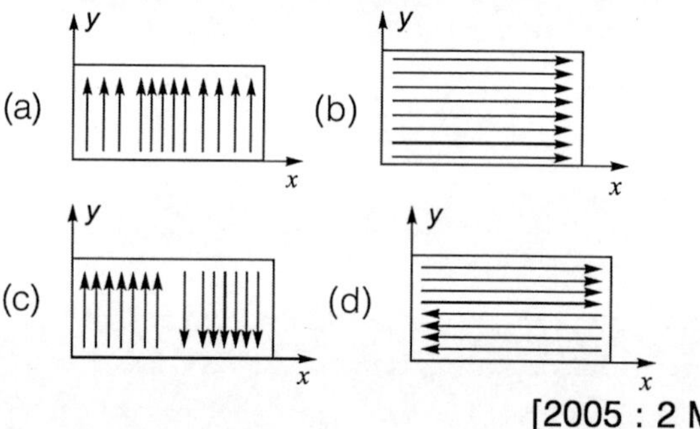

[2005 : 2 M]

4.10 A rectangular waveguide having TE_{10} mode as dominant mode is having a cutoff frequency of 18-GHz for the TE_{30} mode. The inner broad-wall dimension of the rectangular waveguide is

(a) 5/3 cms (b) 5 cms
(c) 5/2 cms (d) 10 cms

[2006 : 2 M]

4.11 An air-filled rectangular waveguide has inner dimensions of 3 cm × 2 cm. The wave impedance of the TE_{20} mode of propagation in the waveguide at a frequency of 30 GHz is (free space impedance $\eta_0 = 377\ \Omega$)

(a) 308 Ω (b) 355 Ω
(c) 400 Ω (d) 461 Ω [2007 : 2 M]

4.12 The $\vec{E}$ field in a rectangular waveguide of inner dimensions a × b is given by

$$\vec{E} = \frac{\omega\mu}{h^2}\left(\frac{\pi}{a}\right)H_0 \sin\left(\frac{2\pi x}{a}\right)\sin(\omega t - \beta z)\,\hat{y}$$

where H_0 is a constant, and a and b are the dimensions along the x-axis and the y-axis respectively. The mode of propagation in the waveguide is

(a) TE_{20} (b) TM_{11}
(c) TM_{20} (d) TE_{10} [2007 : 2 M]

4.13 A rectangular waveguide of internal dimensions (a = 4 cm and b = 3 cm) is to be operated in TE_{11} mode. The minimum operating frequency is

(a) 6.25 GHz (b) 6.0 GHz
(c) 5.0 GHz (d) 3.75 GHz

[2008 : 2 M]

4.14 Which of the following statements is true regarding the fundamental mode of the metallic waveguides shown?

P: Coaxial

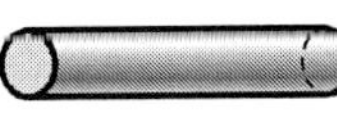
Q: Cylindrical

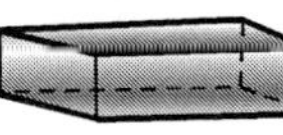
R: Rectangular

(a) Only P has no cutoff-frequency
(b) Only Q has no cutoff-frequency
(c) Only R has no cutoff-frequency
(d) All three have cutoff-frequencies

[2009 : 1 M]

4.15 The modes in a rectangular waveguide are denoted by TE_{mn}/TM_{mn} where m and n are the eigen numbers along the larger and smaller dimensions of the waveguide respectively. Which one of the following statements is **TRUE**?

(a) The TM_{10} mode of the waveguide does not exist
(b) The TE_{10} mode of the waveguide does not exist
(c) The TM_{10} and the TE_{10} modes both exist and have the same cut-off frequencies
(d) The TM_{10} and the TM_{01} modes both exist and have the same cut-off frequencies

[2011 : 1 M]

4.16 The magnetic field along the propagation direction inside a rectangular waveguide with the cross-section shown in figure is

$H_z = 3\cos(2.094 \times 10^2 x)\cos(2.618 \times 10^2 y)\cos(6.283 \times 10^{10} t - \beta z)$

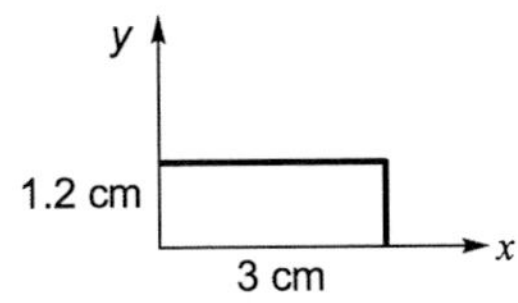

The phase velocity v_p of the wave inside the waveguide satisfies

(a) $v_p > c$ (b) $v_p = c$
(c) $0 < v_p < c$ (d) $v_p = 0$ [2012 : 2 M]

4.17 For a rectangular waveguide of internal dimensions $a \times b$ ($a > b$), the cut-off frequency for the TE_{11} mode is the arithmetic mean of the cut-off frequencies for TE_{10} mode and TE_{20} mode. If $a = \sqrt{5}$ cm, the value of b (in cm) is ______.

[2014 : 2 M, Set-2]

4.18 Consider an air filled rectangular waveguide with a cross-section of 5 cm × 3 cm. For this waveguide, the cut-off frequency (in MHz) of TE_{21} mode is ________. [2014 : 1 M, Set-3]

4.19 The longitudinal component of the magnetic field inside an air-filled rectangular waveguide made of a perfect electric conductor is given by the following expression

$$H_z(x, y, z, t) = 0.1\cos(25\pi x)\cos(30.3\,\pi y)\cos(12\pi \times 10^9\, t - \beta z)\ \text{(A/m)}$$

The cross-sectional dimensions of the waveguide are given as a = 0.08 m and b = 0.033 m. The mode of propagation inside the waveguide is

(a) TM_{12} (b) TM_{21}
(c) TE_{21} (d) TE_{12}

[2015 : 2 M, Set-1]

4.20 An air-filled rectangular waveguide of internal dimensions a cm × b cm ($a > b$) has a cutoff frequency of 6 GHz for the dominant TE_{10} mode. For the same waveguide, if the cutoff frequency of the TM_{11} mode is 15 GHz, the cutoff frequency of the TE_{01} mode in GHz is_____.

[2015 : 2 M, Set-2]

4.21 Consider an air-filled rectangular waveguide with dimensions a = 2.286 cm and b = 1.016 cm. At 10 GHz operating frequency, the value of the propagation constant (per meter) of the corresponding propagating mode is ________.

[2016 : 2 M, Set-3]

4.22 Consider an air-filled rectangular waveguide with dimensions a = 2.286 cm and b = 1.016 cm. The increasing order of the cut-off frequencies for different modes is

(a) $TE_{01} < TE_{10} < TE_{11} < TE_{20}$
(b) $TE_{20} < TE_{11} < TE_{10} < TE_{01}$
(c) $TE_{10} < TE_{20} < TE_{01} < TE_{11}$
(d) $TE_{10} < TE_{11} < TE_{20} < TE_{01}$

[2016 : 2 M, Set-3]

4.23 Standard air-filled rectangular waveguides of dimensions $a = 2.29$ cm and $b = 1.02$ cm are designed for radar applications. It is desired that these waveguides operate only in the dominant TE_{10} mode with the operating frequency at least 25% above the cutoff frequency of the TE_{10} mode but not higher than 95% of the next higher cutoff frequency. The range of the allowable operating frequency f is _____.

(a) 8.19 GHz $\le f \le$ 13.1 GHz
(b) 8.19 GHz $\le f \le$ 12.45 GHz
(c) 6.55 GHz $\le f \le$ 13.1 GHz
(d) 1.64 GHz $\le f \le$ 10.24 GHz

[2017 : 2 M, Set-2]

4.24 The cut-off frequency of TE_{01} mode of an air filled rectangular waveguide having inner dimensions a cm × b cm ($a > b$) is twice that of the dominant TE_{10} mode. When the waveguide is operated at a frequency which is 25% higher than the cut-off frequency of the dominant mode, the guide wavelength is found to be 4 cm. The value of b (in cm, correct to two decimal places) is ______.

[2018 : 2 M]

4.25 A rectangular waveguide of width w and height h has cut-off frequencies for TE_{10} and TE_{11} modes in the ratio 1 : 2. The aspect ratio w/h, rounded off to two decimal places, is _____.

[2019 : 2 M]

4.26 The dispersion equation of a waveguide, which relates the wave-number k to the frequency ω, is

$$k(\omega) = \left(\frac{1}{c}\right)\sqrt{\omega^2 - \omega_0^2}$$

where the speed of light $c = 3 \times 10^8$ m/s, and ω_0 is a constant. If the group velocity is 2×10^8 m/s, then the phase velocity is

(a) 2×10^8 m/s
(b) 3×10^8 m/s
(c) 1.5×10^8 m/s
(d) 4.5×10^8 m/s

[2019 : 2 M]

4.27 A standard air-filled rectangular waveguide with dimension $a = 8$ cm, $b = 4$ cm, operates at 3.4 GHz. For the dominant mode of wave propagation, the phase velocity of the signal is v_p. The value (rounded off to two decimal places) of v_p/c, where c denotes the velocity of light, is _______.

[2021 : 2 M]

4.28 A waveguide consists of two infinite parallel plates (perfect conductors) at a separation of 10^{-4} cm, with air as the dielectric. Assume the speed of light in air to be 3×10^8 m/s. The frequency/frequencies of TM waves which can propagate in this waveguide is/are _______.

(a) 6×10^{15} Hz
(b) 0.5×10^{12} Hz
(c) 8×10^{14} Hz
(d) 1×10^{13} Hz

[2022 : 2 M]

4.29 Consider a narrow band signal, propagating in a lossless dielectric medium ($\epsilon_r = 4$, $\mu_r = 1$), with phase velocity v_p and group velocity v_g. Which of the following statement is true?

(c is the velocity of light in vacuum.)

(a) $v_p > c$, $v_g > c$
(b) $v_p < c$, $v_g > c$
(c) $v_p > c$, $v_g < c$
(d) $v_p < c$, $v_g < c$

[2023 : 1 M]

■■■

Answers Waveguides

4.1	(a)	4.2	(b)	4.3	(c)	4.4	(a)	4.5	(d)	4.6	(c)	4.7	(a)
4.8	(d)	4.9	(d)	4.10	(c)	4.11	(c)	4.12	(a)	4.13	(a)	4.14	(a)
4.15	(a)	4.16	(a)	4.17	(2)	4.18	(7810)	4.19	(c)	4.20	(13.75)	4.21	(158)
4.22	(c)	4.23	(b)	4.24	(0.75)	4.25	(1.732)	4.26	(d)	4.27	(1.198)	4.28	(a, c)
4.29	(d)												

Explanations Waveguides

4.1 (a)

$a = 10$ cm ; $f_c = \dfrac{v_o}{2a}$; $v_o = 3 \times 10^{10}$ cm/sec

$$f_c = \frac{3\times10^{10}}{2\times10} = 1.5 \times 10^9 = 1.5 \text{ GHz}$$

4.2 (b)

A pure *TEM* mode is not possible in single conductor waveguide such as rectangular or cylindrical waveguide.

4.3 (c)

$\lambda_c = 2a$
(For TE_{10} dominant mode)
$= 2 \times 1$ cm $= 2 \times 10^{-2}$ m

$$f_c = \frac{c}{\lambda_C} = \frac{3\times10^8}{2\times10^{-2}} = 15 \text{ GHz}$$

4.4 (a)

$$V_p = \frac{c}{\cos\theta} = \frac{c}{\sqrt{1-\left(\frac{f_c}{f}\right)^2}}$$

with $f > f_c$, $V_p > c$ always.

4.5 (d)

Dominant mode in waveguide has lowest cut-off frequency and hence the highest cut-off wavelength.

4.6 (c)

$$V_p = \frac{c}{\cos\theta} = \frac{c}{\sqrt{1-\left(\frac{f_c}{f}\right)^2}}$$

with $f > f_c$, $V_p > c$ always.

4.7 (a)

$$v = \frac{c}{\sqrt{\epsilon_r}} = \frac{3\times10^8}{2} = 1.5\times10^8$$

$$f_c = \frac{v}{\lambda_C} = \frac{1.5\times10^8}{6\times10^{-2}} = 2.5 \text{ GHz}$$

or, $$f_c = \frac{v}{2a} = \frac{1.5\times10^8}{6\times10^{-2}} = 2.5 \text{ GHz}$$

4.9 (d)

Two maximas in the field along *y*-direction identifies TE_{02}.

4.10 (c)

TE_{30}, $\lambda_c = \dfrac{2a}{m} = \dfrac{2a}{3}$

$$\lambda_c = \frac{c}{f} = \frac{3\times10^8}{18\times10^9} = \frac{1}{60}$$

$$\frac{1}{60} = \frac{2a}{3}$$

$$\Rightarrow \quad a = \frac{1}{40} = 2.5 \text{ cm} = \frac{5}{2}\text{cm}$$

4.11 (c)

For TE_{20}, $\lambda_C = a = 3 \times 10^{-2}$m

$$f_c = \frac{c}{\lambda_C} = \frac{3\times10^8}{3\times10^{-2}} = 10^{10} \text{ Hz}$$

$$\eta_{TE} = \frac{\eta}{\sqrt{1-\left(\frac{f_c}{f}\right)^2}} = \frac{377}{\sqrt{1-\left(\frac{10^{10}}{3\times10^{10}}\right)^2}}$$

$\eta_{TE} = 400\ \Omega$

4.12 (a)

The wave is $E(x, z, t)$ with no propagation in *y*-direction.
This is possible only in TE waves when $n = 0$

$$\left(\cos\frac{n\pi}{b}y = 1\right)$$

4.13 (a)

Minimum operating frequency,

$$f_0 = \frac{c}{2\pi}\sqrt{\left(\frac{m\pi}{a}\right)^2 + \left(\frac{n\pi}{b}\right)^2}$$

For TE_{11} mode,

$$f_0 = \frac{3\times10^8}{2}\sqrt{\left(\frac{1}{0.04}\right)^2 + \left(\frac{1}{0.03}\right)^2}$$

$$= \frac{3\times10^8}{2}\times\frac{0.05}{0.04\times0.03} = 6.25 \text{ GHz}$$

4.14 (a)

P is coaxial line and support *TEM* wave

∴ P has no cutoff frequency

Q and *R* are waveguides and cutoff frequency of each depends upon their dimensions.

4.15 (a)

In case of rectangular wave guide TE_{mn} exists for all values of *m* and *n* except *m* = 0 and *n* = 0. For TM_{mn} to exist both values of *m* and *n* must be non zero.

4.16 (a)

From the given data

$$\omega_c = \left(\sqrt{\left(\frac{m\pi}{a}\right)^2 + \left(\frac{n\pi}{b}\right)^2}\right)c$$

$$f_c = \left(\sqrt{(2.094)^2 + (2.618)^2}\right)\times\frac{3\times10^{10}}{2\times3.14} = 16 \text{ GHz}$$

$$f = \frac{6.283\times10^{10}}{2\times3.14} = 10 \text{ GHz}$$

Hence, $f < f_c$, $\gamma = \alpha + j0$

$$\beta = 0, \quad V_p = \frac{\omega}{\beta} = \infty$$

4.17 (2)

Cut-off frequency of a rectangular waveguide is

$$f_c = \frac{c}{2\pi}\sqrt{\left(\frac{m\pi}{a}\right)^2 + \left(\frac{n\pi}{b}\right)^2}$$

$$f_c \text{ for } TE_{10} = \frac{c}{2\pi}\sqrt{\left(\frac{1\pi}{a}\right)^2} = \frac{c}{2a}$$

$$f_c \text{ for } TE_{20} = \frac{c}{2\pi}\sqrt{\left(\frac{2\pi}{a}\right)^2} = \frac{c}{a}$$

$$f_c \text{ for } TE_{11} = \frac{c}{2\pi}\sqrt{\left(\frac{\pi}{a}\right)^2 + \left(\frac{\pi}{b}\right)^2} = \frac{c}{2}\sqrt{\left(\frac{1}{a}\right)^2 + \left(\frac{1}{b}\right)^2}$$

Given, $$f_{c_{TE_{11}}} = \frac{f_{c_{TE_{10}}} + f_{c_{TE_{20}}}}{2}$$

$$\frac{c}{2}\sqrt{\left(\frac{1}{a}\right)^2 + \left(\frac{1}{b}\right)^2} = \frac{\frac{c}{2a} + \frac{c}{a}}{2}$$

$$\sqrt{\left(\frac{1}{a}\right)^2 + \left(\frac{1}{b}\right)^2} = \frac{1}{2a} + \frac{1}{a} = \frac{3}{2a}$$

$$\left(\frac{1}{a}\right)^2 + \left(\frac{1}{b}\right)^2 = \frac{9}{4a^2}$$

$$\Rightarrow \quad \left(\frac{1}{b}\right)^2 = \frac{5}{4a^2}$$

$$b = \frac{2a}{\sqrt{5}} = \frac{2\times\sqrt{5}}{\sqrt{5}}$$

(given $a = \sqrt{5}$ cm)

$$= 2 \text{ cm}$$

4.18 (7810)

Given $a = 5 \text{ cm} = 5\times10^{-2}$ m

$b = 3 \text{ cm} = 3\times10^{-2}$ m

$$f_c = \frac{c}{2\pi}\sqrt{\left(\frac{m\pi}{a}\right)^2 + \left(\frac{n\pi}{b}\right)^2}$$

$$f_{c_{TE_{21}}} = \frac{c}{2}\sqrt{\left(\frac{2}{5\text{ cm}}\right)^2 + \left(\frac{1}{3\text{ cm}}\right)^2}$$

$$= \frac{3\times10^8}{2}\sqrt{\left(\frac{2\times100}{5}\right)^2 + \left(\frac{100}{3}\right)^2}$$

$$f_c = 7810 \text{ MHz}$$

4.19 (c)

$$H_z(x, y, z, t) = 0.1\cos(25\pi x)\cos(30.3\pi y)\cos(12\pi\times10^9 t - \beta z)\text{A/m}$$

$$a = 0.08 \text{ m}, \; b = 0.033 \text{ m}$$

Axial component is $H \Rightarrow$ the propagating mode is TE_{mn}, *m*, *n* can be found by

$$\frac{m\pi}{a}x = 25\pi x$$

$$\Rightarrow \quad \frac{m}{0.08} = 25$$

$$\Rightarrow \quad m = 2$$

$$\frac{n\pi}{b}y = 30.3\pi y$$

$$\Rightarrow \quad \frac{n}{0.033} = 30.3$$

$$\Rightarrow \quad n = 1$$

$\therefore$ The mode of propagation is TE_{21}.

4.20 (13.75)

Given, f_c for TE_{10} mode is 6 GHz

$$\Rightarrow \quad \frac{c}{2a} = \frac{3\times10^{10}}{2\times a} = 6\times10^9$$

$$\Rightarrow \quad a = 2.5 \text{ cm}$$

Given f_c for TM_{11} mode is 15 GHz

$$\Rightarrow \frac{c}{2}\sqrt{\frac{1}{a^2}+\frac{1}{b^2}} = 15\times10^9$$

$$\frac{3\times10^{10}}{2}\sqrt{\frac{1}{6.25}+\frac{1}{b^2}} = 15\times10^9$$

$$\Rightarrow \quad b = \frac{2.5}{\sqrt{5.25}}$$

Now f_c for TE_{01} mode is,

$$f_c = \frac{c}{2b} = \frac{3\times10^{10}}{2.5}\sqrt{5.25}$$

$$= 13.75 \text{ GHz}$$

4.21 (158)

Operation frequency = 10 GHz

Cut off frequency for TE_{10}, $f_c = \frac{c}{2a}$

$$f_c = \frac{3\times10^8}{2\times2.286\times10^{-2}} = 6.57 \text{ GHz}$$

Next mode TE_{20} has f_c = 13 GHz

10 GHz operates only in dominant mode.

$$\gamma = \sqrt{\left(\frac{m\pi}{a}\right)^2 - \omega^2\mu\in} = \sqrt{\left(\frac{\pi}{a}\right)^2 - \frac{(2\pi f)^2}{(3\times10^8)^2}}$$

$$= \pi\sqrt{\left(\frac{1}{2.286\times10^{-2}}\right)^2 - \left(\frac{2\times10\times10^9}{3\times10^8}\right)^2}$$

$$= j158 \,(\text{m}^{-1})$$

4.22 (c)

$$a = 2.286 \text{ cm}; \quad b = 1.016 \text{ cm}$$

$$f_{c_{TE10}} = \frac{c}{2a}$$

$$f_{c_{TE01}} = \frac{c}{2b}$$

$$f_{c_{TE20}} = \frac{c}{a}$$

$$f_{c_{TE11}} = \frac{c}{2}\sqrt{\left(\frac{1}{a}\right)^2 + \left(\frac{1}{b}\right)^2}$$

As $a > 2b \Rightarrow f_{c_{TE10}} < f_{c_{TE20}} < f_{c_{TE01}} < f_{c_{TE11}}$

4.23 (b)

a = 2.29 cm, b = 1.02 cm

Waveguide is operating in (TE_{10}) dominant mode

- $$f_c(TE_{10}) = \frac{c}{2a} = \frac{3\times10^{10}}{2\times2.29} = 6.55 \text{ GHz}$$

  $\Rightarrow$ 25% above cut-off frequency

  $= 1.25\times6.55$ GHz $= 8.1875$ GHz

- Next higher order mode is TE_{20} ($\because a > 2b$)

  $$f_c(TE_{20}) = \frac{c}{a} = \frac{3\times10^{10}}{2.29} = 13.1 \text{GHz}$$

  $\Rightarrow$ 95% of $f_c(TE_{20}) = 0.95\times13.1$

  $= 12.445$ GHz

- Range of allowable operating frequency f is

  8.19 GHz $\le f \le$ 12.45 GHz

4.24 (0.75)

$$f_{c(01)} = 2f_{c(10)} = \frac{2c}{2a} = \frac{c}{a}$$

$$\frac{c}{2b} = \frac{c}{a} \Rightarrow a = 2b \quad \Rightarrow b = \frac{a}{2}$$

Operating frequency,

$$f = 1.25 f_{c(10)}$$

$$f_{c(10)} < 1.25 f_{c(10)} < [f_{c(01)} = 2f_{c(10)}]$$

So, for the given frequency, the waveguide will work in TE_{10} mode.

$$\text{So, } \lambda_g = \frac{\lambda_0}{\sqrt{1-\left(\frac{f_{c(10)}}{f}\right)^2}} = \frac{c/f}{\sqrt{1-\left(\frac{1}{1.25}\right)^2}} = \frac{c/f}{0.6}$$

$$\frac{c}{(1.25)f_{c(10)}\,(0.6)} = \lambda_g = 4 \text{ cm}$$

$$\frac{c}{f_{c(10)}} = 3\times10^{-2} = 2a$$

$$a = 1.5 \text{ cm}; \quad b = \frac{a}{2} = 0.75 \text{ cm}$$

4.25 (1.732)

$$f_{cmn} = \frac{c}{2}\sqrt{\left(\frac{m}{a}\right)^2 + \left(\frac{n}{b}\right)^2}$$

For TE_{10} mode,

$$f_{c10} = \frac{c}{2w} \quad \text{...(i)}$$

and For TE_{11} mode,

$$f_{c11} = \frac{c}{2}\sqrt{\left(\frac{1}{w}\right)^2 + \left(\frac{1}{h}\right)^2} = \frac{c}{2w}\sqrt{1+\left(\frac{w}{h}\right)^2} \quad ...(ii)$$

given, $$\frac{f_{c10}}{f_{c11}} = \frac{1}{2} \quad ...(iii)$$

put (i), (ii) in (iii)

$$\Rightarrow \frac{\frac{c}{2w}}{\frac{c}{2w}\sqrt{1+\left(\frac{w}{h}\right)^2}} = \frac{1}{2} \Rightarrow \sqrt{1+\left(\frac{w}{h}\right)^2} = 2$$

On solving above equation, we get,

$$\frac{w}{h} = \sqrt{3} = 1.732$$

4.26 (d)

By definition $v_p = \frac{\omega}{\beta} = \frac{\omega}{k}$

where, $k(\omega) = \left(\frac{1}{c}\right)\sqrt{\omega^2 - \omega_0^2}$ (given)

$$\therefore \quad v_p = \frac{c}{\sqrt{1-\left(\frac{\omega_0}{\omega}\right)^2}}$$

by definition, $v_g = \frac{d\omega}{d\beta} = \frac{d\omega}{dk}$

$$\frac{dk}{d\omega} = \frac{1}{c}\frac{1}{2\sqrt{\omega^2-\omega_0^2}} \times 2\omega$$

or $$v_g = c\sqrt{1-\left(\frac{\omega_0}{\omega}\right)^2}$$

$$\because \quad v_p \cdot v_g = c^2$$

$$\therefore \quad v_p = \frac{c^2}{v_g} = \frac{(3\times10^8)^2}{2\times10^8} = 4.5 \times 10^8 \text{ m/sec}$$

4.27 (1.198)

$$f_{c10} = \frac{c}{2a} = \frac{3\times10^8}{2(8\times10^{-2})} = 1.875 \text{ GHz}$$

Guide phase velocity,

$$V_p = \frac{c}{\sqrt{1-\left(\frac{f_{c10}}{f}\right)^2}}$$

$$\frac{V_p}{c} = \frac{1}{\sqrt{1-\left(\frac{f_{c10}}{f}\right)^2}} = \frac{1}{\sqrt{1-\left(\frac{1.875}{3.4}\right)^2}} = 1.198$$

4.28 (a, c)

Given, speed of light in air,

$$c = 3 \times 10^8 \text{ m/s}$$

a 10^{-4} cm

The TM mode that propagates at the lowest cut-off frequency in a parallel plate waveguide is TM_1 mode. Hence, cut-off frequency,

$$f_c\big|_{TM_1} = \frac{mc}{2a}; \quad m = 1$$

$$\therefore \quad f_c = \frac{3\times10^{10}\text{ cm/s}}{2\times10^{-4}\text{ cm}} = 1.5 \times 10^{14} \text{ Hz}$$

All the frequencies above f_c can propagate in this waveguide. Hence, option (a) and (c) are the correct answer.

4.29 (d)

- Phase velocity,

$$V_p = \frac{\omega}{\beta} = \frac{\omega}{\omega\sqrt{\mu\epsilon}} = \frac{1}{\sqrt{\mu_0\epsilon_0}}, \frac{1}{\sqrt{\mu_r\epsilon_r}} = \frac{c}{\sqrt{\mu_r\epsilon_r}}$$

$$\therefore \quad V_p < c$$

- Group velocity,

$$V_g = \frac{d\omega}{d\beta} = \frac{V_p}{1-\frac{\omega}{V_p}\frac{dV_p}{d\omega}}$$

Here, $V_p \neq f(\omega)$

$\therefore \quad V_g = V_p < c$

Hence, $V_p < c$

$V_g < c$

5 Antennas

5.1 An antenna, when radiating, has a highly directional radiation pattern. When the antenna is receiving, its radiation pattern

(a) is more directive
(b) is less directive
(c) is the same
(d) exhibits no directivity at all [1995 : 1 M]

5.2 A transverse electromagnetic wave with circular polarization is received by a dipole antenna. Due to polarization mismatch, the power transfer efficiency from the wave to the antenna is reduced to about

(a) 50% (b) 35.3%
(c) 25% (d) 0% [1996 : 2 M]

5.3 A 1 km long microwave link uses two antennas each having 30 dB gain. If the power transmitted by one antenna is 1 W at 3 GHz, the power received by the other antenna is approximately

(a) 98.6 μW (b) 76.8 μW
(c) 63.4 μW (d) 55.2 μW [1996 : 2 M]

5.4 A parabolic dish antenna has conical beam 2° wide. The directivity of the antenna is approximately

(a) 20 dB (b) 30 dB
(c) 40 dB (d) 50 dB [1997 : 2 M]

5.5 The vector $\vec{H}$ in the far field of an antenna satisfies

(a) $\nabla \cdot \vec{H} = 0$ and $\nabla \times \vec{H} = 0$
(b) $\nabla \cdot \vec{H} \neq 0$ and $\nabla \times \vec{H} \neq 0$
(c) $\nabla \cdot \vec{H} = 0$ and $\nabla \times \vec{H} \neq 0$
(d) $\nabla \cdot \vec{H} \neq 0$ and $\nabla \times \vec{H} = 0$ [1998 : 1 M]

5.6 The radiation resistance of a circular loop of one turn is 0.01 Ω. The radiation resistance of five turns of such a loop will be

(a) 0.002 Ω (b) 0.01 Ω
(c) 0.05 Ω (d) 0.25 Ω [1998 : 1 M]

5.7 An antenna in free space receives 2 μW of power when the incident electric field is 20 mV/m rms. The effective aperture of the antenna is

(a) 0.005 m^2 (b) 0.05 m^2
(c) 1.885 m^2 (d) 3.77 m^2 [1998 : 1 M]

5.8 The far field of an antenna varies with distance r as

(a) $1/r$ (b) $1/r^2$
(c) $1/r^3$ (d) $1/\sqrt{r}$ [1998 : 1 M]

5.9 A transmitting antenna radiates 251 W isotropically. A receiving antenna, located 100 m away from the transmitting antenna, has an effective aperture of 500 cm^2. The total power received by the antenna is

(a) 10 μW (b) 1 μW
(c) 20 μW (d) 100 μW [1999 : 2 M]

5.10 If the diameter of a λ/2 dipole antenna is increased from $\frac{\lambda}{100}$ to $\frac{\lambda}{50}$, then its

(a) bandwidth increases
(b) bandwidth decreases
(c) gain increases
(d) gain decreases [2000 : 1 M]

5.11 For an 8 feet (2.4 m) parabolic dish antenna operating at 4 GHz, the minimum distance required for far field measurement is closest to

(a) 7.5 cm (b) 15 cm
(c) 15 m (d) 150 m [2000 : 2 M]

5.12 A medium wave radio transmitter operating at a wavelength of 492 m has a tower antenna of height 124 m. What is the radiation resistance of the antenna?

(a) 25 Ω (b) 36.5 Ω
(c) 50 Ω (d) 73 Ω [2001 : 2 M]

5.13 In a uniform linear array, four isotropic radiating elements are spaced λ/4 apart. The progressive phase shift between the elements required for forming the main beam at 60° off the end-fire is:

(a) −π radians (b) −π/2 radians
(c) −π/4 radians (d) −π/8 radians
[2001 : 2 M]

5.14 The line-of-sight communication requires the transmit and receive antennas to face each other. If the transmit antenna is vertically polarized, for best reception the receiver antenna should be

(a) horizontally polarized
(b) vertically polarized
(c) at 45° with respect to horizontal polarization
(d) at 45° with respect to vertical polarization
[2002 : 1 M]

5.15 A person with a receiver is 5 km away from the transmitter. What is the distance that this person must move further to detect a 3-dB decrease in signal strength?
(a) 942 m (b) 2070 m
(c) 4978 m (d) 5320 m [2002 : 2 M]

5.16 Two identical antennas are placed in the $\theta = \pi/2$ plane as shown in the figure. The elements have equal amplitude excitation with 180° polarity difference, operating at wavelength λ. The correct value of the magnitude of the far-zone resultant electric field strength normalized with that of a single element, both computed for $\phi = 0$, is

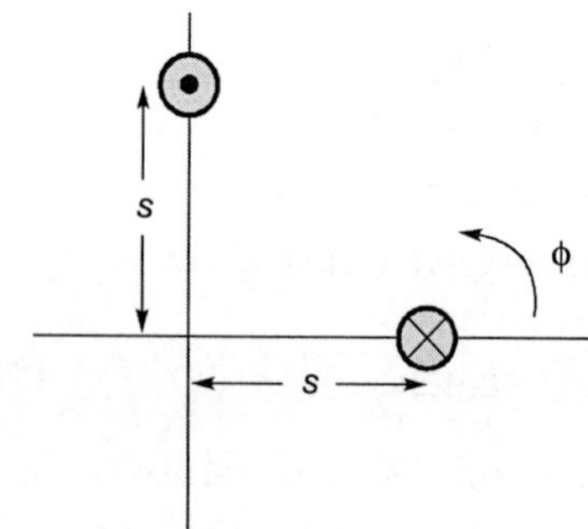

(a) $2\cos\left(\frac{2\pi s}{\lambda}\right)$ (b) $2\sin\left(\frac{2\pi s}{\lambda}\right)$
(c) $2\cos\left(\frac{\pi s}{\lambda}\right)$ (d) $2\sin\left(\frac{\pi s}{\lambda}\right)$
[2003 : 2 M]

5.17 Consider a lossless antenna with a directive gain of +6 dB. If 1 mW of power is fed to it the total power radiated by the antenna will be
(a) 4 mW (b) 1 mW
(c) 7 mW (d) 1/4 mW [2004 : 1 M]

5.18 Two identical and parallel dipole antennas are kept apart by a distance of $\lambda/4$ in the H-plane. They are fed with equal currents but the right most antenna has a phase shift of +90°. The radiation pattern is given as

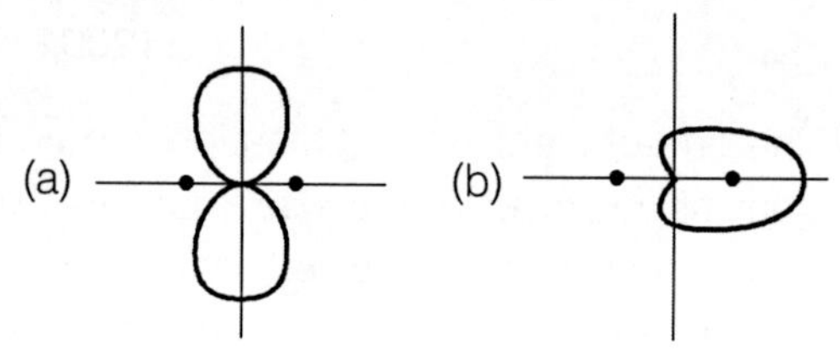

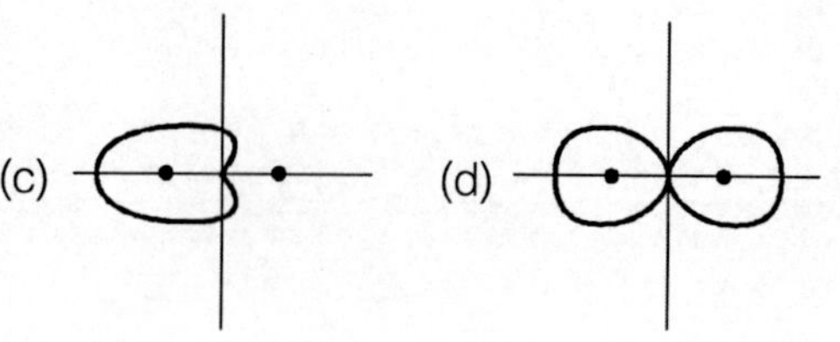

[2005 : 2 M]

5.19 A transmission line is feeding 1 Watt of power to a horn antenna having a gain of 10 dB. The antenna is matched to the transmission line. The total power radiated
(a) 10 Watts (b) 1 Watt
(c) 0.1 Watt (d) 0.01 Watt
[2006 : 1 M]

5.20 A mast antenna consisting of a 50 meter long vertical conductor operates over a perfectly conducting ground plane. It is base-fed at a frequency of 600 kHz. The radiation resistance of the antenna in Ohms is
(a) $\frac{2\pi^2}{5}$ (b) $\frac{\pi^2}{5}$
(c) $\frac{4\pi^2}{5}$ (d) $20\pi^2$ [2006 : 2 M]

5.21 A $\frac{\lambda}{2}$ dipole is kept horizontally at a height of $\frac{\lambda_0}{2}$ above a perfectly conducting infinite ground plane. The radiation pattern in the plane of the dipole ($\vec{E}$ plane) looks approximately as

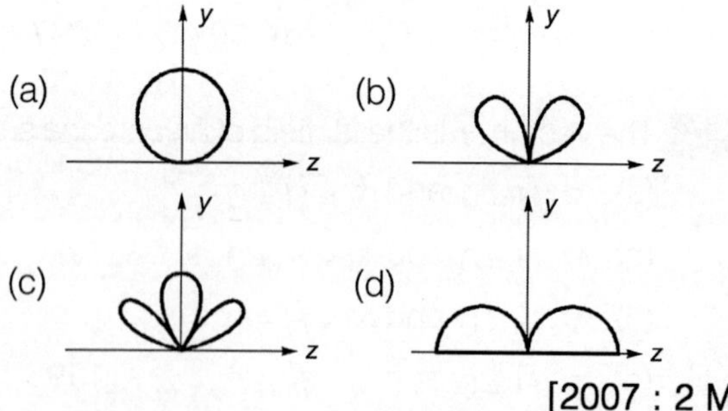

[2007 : 2 M]

5.22 For a Hertz dipole antenna, the Half Power Beam Width (HPBW) in the E-plane is
(a) 360° (b) 180°
(c) 90° (d) 45° [2008 : 1 M]

5.23 At 20 GHz, the gain of a parabolic dish antenna of 1 meter diameter and 70% efficiency is
(a) 15 dB (b) 25 dB
(c) 35 dB (d) 45 dB
[2008 : 2 M]

5.24 Consider the following statements regarding the complex Pointing vector $\vec{P}$ for the power radiated by a point source in an infinite homogenous and lossless medium. Re $(\vec{P})$ denotes the real part of $\vec{P}$, S denotes a spherical surface whose centre is at the point source, and $\hat{n}$ denotes the unit surface normal on S. Which of the following statements is TRUE?

(a) Re $(\vec{P})$ remains constant at any radial distance from the source

(b) Re $(\vec{P})$ increases with increasing radial distance from the source

(c) $\oiint_S \text{Re}(\vec{P}).\hat{n}\,dS$ remains constant at any radial distance from the source

(d) $\oiint_S \text{Re}(\vec{P}).\hat{n}\,dS$ decreases with increasing radial distance from the source

[2011 : 1 M]

5.25 The radiation pattern of an antenna in spherical co-ordinates is given by

$$F(\theta) = \cos^4\theta,\ 0 \le \theta \le \pi/2$$

The directivity of the antenna is

(a) 10 dB
(b) 12.6 dB
(c) 11.5 dB
(d) 18 dB

[2012 : 1 M]

5.26 In spherical coordinates, let $\hat{a}_\theta, \hat{a}_\phi$ denote unit vectors along the θ, ϕ directions.

$$\vec{E} = \frac{100}{r}\sin\theta\cos(\omega t - \beta r)\hat{a}_\theta \text{ V/m} \quad \text{and}$$

$$\vec{H} = \frac{0.265}{r}\sin\theta\cos(\omega t - \beta r)\hat{a}_\phi \text{ A/m}$$

represent the electric and magnetic field components of the EM wave at large distances r from a dipole antenna, in free space. The average power (W) crossing the hemispherical shell located at $r = 1$ km, $0 \le \theta \le \pi/2$ is _____.

[2014 : 2 M, Set-1]

5.27 For an antenna radiating in free space, the electric field at a distance of 1 km is found to be 12 mV/m. Given that intrinsic impedance of the free space is $120\pi\ \Omega$, the magnitude of average power density due to this antenna at a distance of 2 km from the antenna (in nW/m^2) is _____.

[2014 : 1 M, Set-4]

5.28 Match **Column A** with **Column B**.

Column-A
1. Point electromagnetic source
2. Dish antenna
3. Yagi-Uda antenna

Column-A
P. Highly directional
Q. End fire
R. Isotropic

	P	Q	R
(a)	1	2	3
(b)	2	3	1
(c)	2	1	3
(d)	3	2	1

[2014 : 1 M, Set-4]

5.29 Two half-wave dipole antennas placed as shown in the figure are excited with sinusoidally varying currents of frequency 3 MHz and phase shift of $\pi/2$ between them (the element at the origin leads in phase). If the maximum radiated E-field at the point P in the x-y plane occurs at an azimuthal angle of 60°, the distance d (in meters) between the antennas is_____.

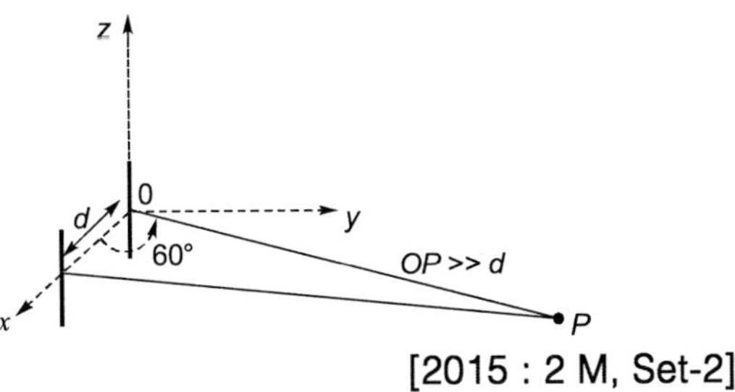

[2015 : 2 M, Set-2]

5.30 The directivity of an antenna array can be increased by adding more antenna elements, as a larger number of elements

(a) improves the radiation efficiency
(b) increases the effective area of the antenna
(c) results in a better impedance matching
(d) allows more power to be transmitted by the antenna

[2015 : 1 M, Set-3]

5.31 Two lossless X-band horn antennas are separated by a distance of 200λ. The amplitude reflection coefficients at the terminals of the transmitting and receiving antennas are 0.15 and 0.18, respectively. The maximum directivities of the transmitting and receiving antennas (over the isotropic antenna) are 18 dB and 22 dB, respectively. Assuming that the input power in the lossless transmission line connected to the antenna is 2 W, and that the antennas are perfectly aligned and polarization matched, the power (in mW) delivered to the load at the receiver is _____.

[2016 : 2 M, Set-1]

5.32 The electric field of a uniform plane wave travelling along the negative z direction is given by the following equation:

$\vec{E}_w^i = (\hat{a}_x + j\hat{a}_y)E_0 e^{jkz}$

This wave is incident upon a receiving antenna placed at the origin and whose radiated electric field towards the incident wave is given by the following equation:

$$\vec{E}_a = (\hat{a}_x + 2\hat{a}_y)E_I \frac{1}{r} e^{-jkr}$$

The polarization of the incident wave, the polarization of the antenna and losses due to the polarization mismatch are, respectively,

(a) Linear, Circular (clockwise), –5 dB
(b) Circular (clockwise), Linear, –5 dB
(c) Circular (clockwise), Linear, –3 dB
(d) Circular (anticlockwise), Linear, –3 dB

[2016 : 2 M, Set-1]

5.33 The far-zone power density radiated by a helical antenna is approximated as:

$$\vec{W}_{rad} = \vec{W}_{\text{average}} \approx \hat{a}_r C_0 \frac{1}{r^2} \cos^4 \theta$$

The radiated power density is symmetrical with respect to ϕ and exists only in the upper hemisphere $0 \le \theta \le \frac{\pi}{2}; 0 \le \phi \le 2\pi$; C_0 is a constant. The power radiated by the antenna (in watts) and the maximum directivity of antenna, respectively, are

(a) 1.5 C_0, 10 dB (b) 1.256 C_0, 10 dB
(c) 1.256 C_0, 12 dB (d) 1.5 C_0, 12 dB

[2016 : 2 M, Set-1]

5.34 Consider a wireless communication link between a transmitter and a receiver located in free space, with finite and strictly positive capacity. If the effective areas of the transmitter and the receiver antennas, and the distance between them are all doubled, and everything else remains unchanged, the maximum capacity of the wireless link

(a) increases by a factor of 2
(b) decreases by a factor of 2
(c) remains unchanged
(d) decreases by a factor of $\sqrt{2}$

[2017 : 1 M, Set-1]

5.35 A half wavelength dipole is kept in the x-y plane and oriented along 45° from the x-axis. Determine the direction of null in the radiation pattern for $0 \le \phi \le \pi$. Here the angle $\theta(0 \le \theta \le \pi)$ is measured from the z-axis, and the angle $\phi(0 \le \phi \le 2\pi)$ is measured from the x-axis in the x - y plane.

(a) $\theta = 90°, \phi = 45°$ (b) $\theta = 45°, \phi = 90°$
(c) $\theta = 90°, \phi = 135°$ (d) $\theta = 45°, \phi = 135°$

[2017 : 2 M, Set-1]

5.36 Radiation resistance of a small dipole current element of length l at a frequency of 3 GHz is 3 ohms. If the length is changed by 1%, then the percentage change in the radiation resistance, rounded off to two decimal places, is _____%.

[2019 : 1 M]

5.37 For an infinitesimally small dipole in free space, the electric field E_θ in the far field is proportional to $(e^{-jkr}/r)\sin\theta$, where $k = 2\pi/\lambda$. A vertical infinitesimally small electric dipole ($\delta l << \lambda$) is placed at a distance $h(h > 0)$ above an infinite ideal conducting plane, as shown in the figure. The minimum value of h, for which one of the maxima in the far field radiation pattern occurs at $\theta = 60°$. is

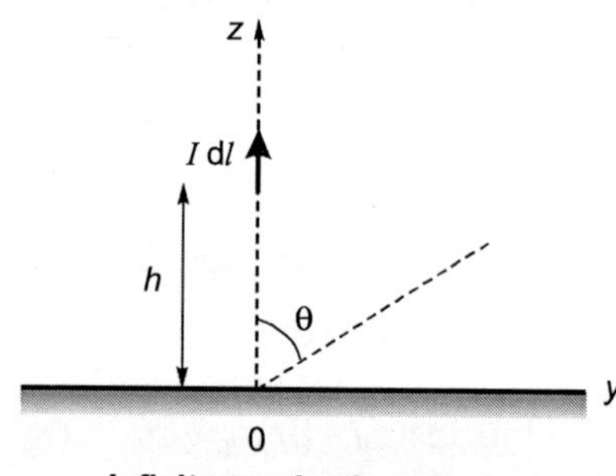

Infinite conduction plane

(a) 0.75λ (b) λ
(c) 0.25λ (d) 0.5λ [2020 : 2 M]

5.38 An antenna with a directive gain of 6 dB is radiating a total power of 16 kW. The amplitude of the electric field in free space at a distance of 8 km from the antenna in the direction of 6 dB gain (rounded off to three decimal places) is ___ V/m.

[2021 : 2 M]

■■■

Answers Antennas

5.1 (b)	5.2 (d)	5.3 (c)	5.4 (c)	5.5 (c)	5.6 (d)	5.7 (c)
5.8 (a)	5.9 (d)	5.10 (b)	5.11 (d)	5.12 (b)	5.13 (c)	5.14 (b)
5.15 (b)	5.16 (d)	5.17 (b)	5.18 (c)	5.19 (b)	5.20 (a)	5.21 (b)
5.22 (c)	5.23 (d)	5.24 (c)	5.25 (a)	5.26 (55.5)	5.27 (47.7)	5.28 (b)
5.29 (50)	5.30 (b)	5.31 (3)	5.32 (c, d)	5.33 (b)	5.34 (c)	5.35 (a)
5.36 (2.01)	5.37 (b)	5.38 (0.244)				

Explanations Antennas

5.1 (b)

An antenna, when radiating, has a highly directional radiation pattern. When the antenna is receiving, its radiation pattern is the same. Antenna is a reciprocal device, whose characteristics are the same when it is transmitting or receiving.

5.2 (d)

$$\text{Polarization loss} = |\cos\theta|^2$$

where θ is angle between received and transmission E fields of antennas.

$$\text{Polarization loss} = \left|\frac{E_T \cdot E_R}{|E_T||E_R|}\right|^2$$

$E_T = (\hat{a}_x + j\hat{a}_y)$ circular polarization

$E_R = (\hat{a}_x \text{or } \hat{a}_y)$ linear polarization

$$\text{Polarization loss} = \left|\frac{(\hat{a}_x + j\hat{a}_y)}{\sqrt{2}} \cdot \hat{a}_x\right|^2$$

$$= \frac{1}{2} = 50\% = -3\text{ dB}$$

5.3 (c)

$R = 1000$ m

$G_t = G_r = 30$ dB $= 1000$

$f = 3$ GHz

$P_t = 1$ W

$$\lambda = \frac{C}{f} = \frac{3\times10^{10}}{3\times10^{9}} = 10\text{ cm}$$

Power density P_d at 1 km

$$= \frac{P_t G_t}{4\pi R^2} = \frac{1\times1000}{4\pi(10^3)^2} = \frac{1}{4000\pi}$$

$$P_r = (A_e)_r P_d = \frac{G_r\lambda^2}{4\pi}\,\frac{1}{4000\pi}$$

$$= \frac{1000(0.1)^2}{4\pi}\times\frac{1}{4000\pi}$$

$P_r = 63.3\ \mu$W

5.4 (c)

$$\text{Directivity} = \frac{4\pi}{\Omega_A}\text{ (Steradian)}$$

$$\theta_{3dB} = 2^\circ$$

$$\text{Directivity} = \frac{41,253^\circ}{\theta_{3dB}\,\phi_{3dB}} = \frac{41,253^\circ}{(2)(2)} \cong 10,000$$

$$D = 40\text{ dB}$$

5.5 (c)

For field free space conditions have $J_c = 0$

$$\nabla\times\vec{H} = J_C + J_D$$

$$\nabla\times\vec{H} = J_c$$

$$\nabla\cdot\vec{H} = 0$$

So, $\nabla\times\vec{H} \neq 0$ and $\nabla\cdot\vec{H} = 0$

5.6 (d)

$$R_r = 80\pi\left(\frac{dl}{\lambda}\right)^2 \text{ for short loops}$$

$$= 20\pi\left(\frac{dl}{\lambda}\right)^2 \text{ for longer loops}$$

$$R_r \propto l^2$$

When turns are increased l increases as Nl

$$R_r \propto N^2$$

$$R_r = 0.25\ \Omega$$

5.7 (c)

$$\text{Power density, } P_d = \frac{E^2}{\eta} = \frac{(20\times10^{-3})^2}{120\pi}$$

$$= 1.061\times10^{-6}\text{ W/m}^2$$

$$A_e = \frac{P_r}{P_d} = \frac{2\times10^{-6}}{1.06\times10^{-6}} = 1.885\text{ m}^2$$

5.9 (d)

Intensity of *EM* wave at 100 m away

$$= \frac{251}{4\pi \times 100^2} \text{W/m}^2$$

Effective aperture = 500 cm^2 = 500 × 10^{-4} m^2

$$\text{Power} = \frac{251}{4\pi \times 100 \times 100} \times 500 \times 10^{-4} \text{watt}$$

$$= 100\ \mu\text{W}$$

5.10 (b)

Diameter of the conductor of the dipole effects only losses and loss resistance

$$R_l \propto \frac{1}{A}$$

Losses reduce when area increases and gain increases. Bandwidth is unaffected as the gain increase is due to loss aspects but not beam angle aspects.

Unlike an amplifier we cannot relate gain × bandwidth relationship in antennas.

5.11 (d)

To measure the antennas performance in any aspect the near field or $\frac{1}{r^2}$ and $\frac{1}{r^3}$ have to be neglected. The radiation field $\frac{1}{r}$ term should be strong enough ignoring $\frac{1}{r^2}$ and $\frac{1}{r^3}$.

This distance is called as Fraunhofer zone where measurements are carried.

$$r > \frac{2D^2}{\lambda}$$

Where *D* is the biggest dimension of the antenna.

$$\lambda = \frac{3 \times 10^8}{4 \times 10^9} = 7.5$$

$$r = \frac{2 \times 2.4 \times 2.4}{7.5} = 1.536 \text{ m}$$

5.12 (b)

$$\frac{\lambda}{4} = \frac{492}{4} = 123 \simeq L$$

∴ It is a quarter wave monopole antenna.

So, $R_a = 36.5\ \Omega$

5.13 (c)

$\theta_{max} = 60°$ from end fire means 60° from array axis.

Condition for maximum, $\psi \to 0$

$$E_T = E_o \frac{\sin(N\psi/2)}{\sin(\psi/2)}$$

$$E_T = NE_o = E_{max}$$

$$\psi = \alpha + \beta d \cos 60° = 0$$

$$\alpha + \frac{2\pi}{\lambda} \cdot \frac{\lambda}{4} \cdot \frac{1}{2} = 0$$

$$\alpha = -\frac{\pi}{4}$$

5.14 (b)

Both transmitting and receiving antennas should be oriented correctly for maximum induction in the receiver.

Transmitting antenna is vertical produced *E* field is vertical inducing *E* field is vertical so receiving antenna has to be vertical.

5.15 (b)

$$\text{Signal strength} = \frac{P}{4\pi R^2}$$

$$P_1 = \frac{P}{4\pi(5000)^2} \text{ and } P_2 = \frac{P}{4\pi(x)^2}$$

Given, $\frac{P_2}{P_1} = \frac{1}{2}$ (3 dB decrease)

$$\Rightarrow \left(\frac{5000}{x}\right)^2 = \frac{1}{2}$$

$$\Rightarrow x = 5000\sqrt{2}$$

∴Required distance

$$= 5000\sqrt{2} - 5000 = 2071 \text{ m}$$

5.16 (d)

$\theta = 90°$ means *XY* plane means $Z = 0$ plane

Far zone calculations of field at $\phi = 0$ means point on *X*-axis.

Angle made by the array axis to the point of consideration, if taken as θ.

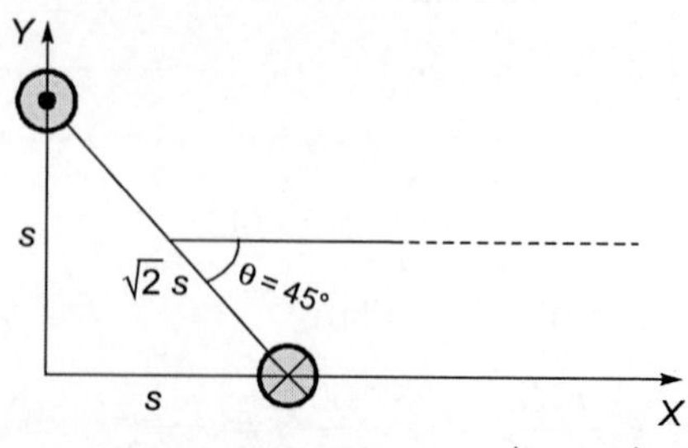

$$\text{Normalized array factor} = 2\left|\cos\frac{\psi}{2}\right|$$

$$\psi = \beta d \cos\theta + \alpha$$

$$d = \sqrt{2}s,\ \theta = 45°,\ \alpha = 180°$$

$$\Rightarrow \quad 2\left|\cos\frac{\psi}{2}\right| = 2\cos\left[\frac{\beta d\cos\theta + \alpha}{2}\right]$$

$$= 2\cos\left[\frac{2\pi}{\lambda.2}\sqrt{2}\,s\cos 45° + \frac{180°}{2}\right]$$

$$= 2\cos\left[\frac{\pi s}{\lambda} + 90°\right] = 2\sin\left(\frac{\pi s}{\lambda}\right)$$

5.17 (b)

When an antenna is lossless then the antenna efficiency is 100%. In such case whole of the power fed to antenna is radiated in the form of radiation. Therefore, if power fed to antenna is 1 mW then total radiated power is also 1 mW.

5.18 (c)

H-plane means ϕ pattern.

Both dipoles are horizontally isotropic in the *H*-plane.

$$E = 2E_o\cos(\psi/2)$$

$$\psi = \alpha + \beta d\cos\theta$$

$$= \frac{\pi}{2} + \frac{2\pi}{\lambda}\cdot\frac{\lambda}{4}\cos\theta$$

$$\psi = \frac{\pi}{2} + \frac{\pi}{2}\cos\theta$$

$\psi = 0$ $\quad \cos\theta = -1$ $\quad \theta_{max} = 180°$

$\psi = \pi$ $\quad \cos\theta = 1$ $\quad \theta_{NP} = 0°$

$\psi = \frac{\pi}{2}$ $\quad \cos\theta = 0$ $\quad \theta_{HPP} = 90°/270°$

Option (c) when θ is considered as shown in above calculations

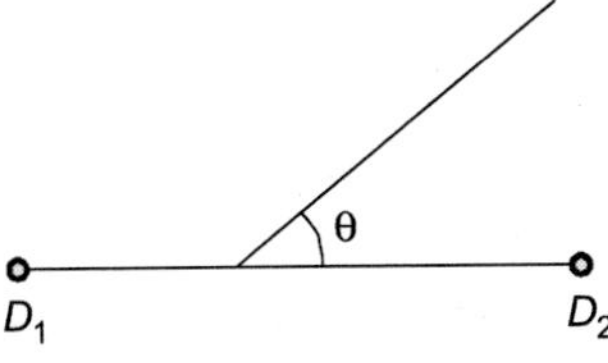

5.19 (b)

Gain of antenna is directive gain i.e. different from that of amplifier. It radiates same power. So 1 W.

5.20 (a)

A mast antenna is a tower antenna or monopole vertically erected on the ground for low frequency ground wave communication.

$l = 50$ m, $\quad \lambda = 500$ m

$$l = \frac{\lambda}{10} \text{ electrically short monopole}$$

$$R_r = 20\pi^2\left(\frac{dl}{\lambda}\right)^2 \times 2$$

This 2 in the multiplication is due to image in conducting earth.

$$R_r = \frac{2\pi^2}{5} \simeq 4\,\Omega$$

5.21 (b)

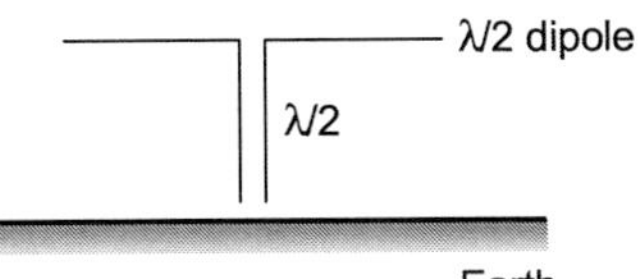

A horizontally placed dipole has an image in the earth and interference with the image formed gives the final pattern of radiation

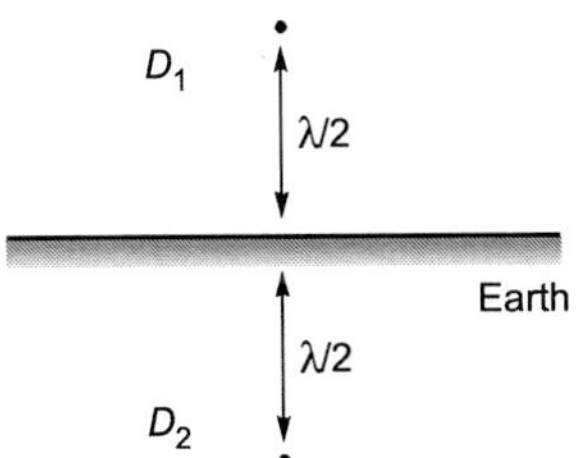

Dipole pattern

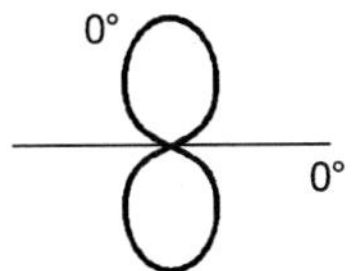

Array factor between D_1 and D_2

$$\psi = \alpha + \beta d\cos\theta$$

$d = \lambda$, $\lambda = 180°$ (image is out of phase)

$$\psi = \pi + 2\pi\cos\theta$$

Array pattern

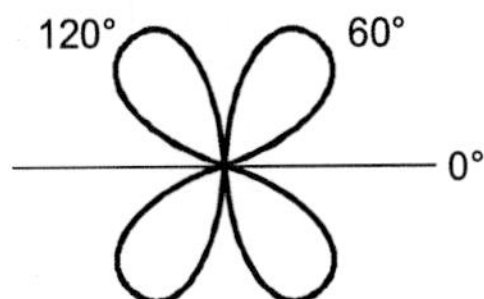

Resultant pattern can be obtained by multiplication of patterns.

5.22 (c)

E-plane mean θ-plane or vertical plane

$E \propto \sin\theta$ in Hertzian dipole

E_{max} when $\theta = 90°$

E_{min} when $\theta = 0°$

E has half power of maximum when $\theta = 45°$ and $135°$

HPBW in θ side $= 135° - 45° = 90°$

5.23 (d)

$$\text{Gain} = \frac{4\pi}{\lambda^2}A_e = \frac{4\pi}{\lambda^2}\pi R^2$$

Along with illumination or efficiency

$$\text{Gain} = \eta\pi^2\frac{(2R)^2}{\lambda^2} = 7\left(\frac{D}{\lambda}\right)^2 = 0.7\,\pi^2\left(\frac{1\times 20\times 10^9}{3\times 10^8}\right)^2$$

$$= 30705.4359$$

$$G(\text{dB}) = 10\log_{10}G$$

$$= 10\log_{10}(30705.4359) = 45\text{ dB}$$

5.24 (c)

Power density of any point source decreases with distance i.e. the density decreases and area of cross-over increases with the product being constant.

5.25 (a)

$F(\theta) = \cos^4\theta,\ 0 \le \theta \le \pi/2$

we know that directivity D is

$$D = \frac{4\pi u_{max}}{W_{rad}} \quad \text{...(1)}$$

$F(\theta)$ is nothing but radiation intensity $u(\theta, \phi)$ and W_{rad} is radiated power.

$$W_{rad} = \int_0^{2\pi}\int_0^{\pi/2}\cos^4\theta\sin\theta\, d\theta\, d\phi \quad \text{...(2)}$$

above equation is written from the formula

$$W_{rad} = \iint_{\theta\ \phi} u(\theta,\phi)\, d\Omega$$

where $d\Omega$ is solid angle and $d\Omega = \sin\theta\, d\theta\, d\phi$.

So from (2)

$$W_{rad} = \int_0^{2\pi}\int_0^{\pi/2}\cos^4\theta\sin\theta\, d\theta\, d\phi$$

$$W_{rad} = 2\pi\int_0^{\pi/2}\cos^4\theta\sin\theta\, d\theta$$

$$W_{rad} = -2\pi\left[\frac{\cos^5\theta}{5}\right]_0^{\pi/2}$$

$$W_{rad} = \frac{2\pi}{5} \quad \text{...(3)}$$

So, $$D = \frac{4\pi u_{max}}{\left(\frac{2\pi}{5}\right)} = 10\, u_{max} = 10$$

In dB directivity = $10\log_{10}D = 10$ dB.

5.26 (55.5)

$$\vec{E} = \text{Re}\left[\frac{100}{r}\sin\theta e^{-j\beta r}e^{j\omega t}\right]\hat{a}_\theta\text{ V/m}$$

$$\vec{H} = \text{Re}\left[\frac{0.265}{r}\sin\theta e^{-j\beta r}e^{j\omega t}\right]\hat{a}_\phi\text{ A/m}$$

Poynting vector,

$$\vec{P} = \frac{1}{2}\text{Re}[\vec{E}\times\vec{H}^*] = \frac{26.5}{2r^2}\sin^2\theta\hat{a}_r\text{ W/m}^2$$

The average power crossing the given hemispherical shell is,

$$\text{Power} = \int_{\theta=0}^{\pi/2}\int_{\phi=0}^{2\pi}\vec{P}\cdot r^2\sin\theta\, d\theta d\phi\hat{a}_r$$

$$= \frac{26.5}{2}\times 2\pi\int_{\theta=0}^{\pi/2}\sin^3\theta\, d\theta$$

$$= \frac{26.5}{2}\times 2\pi\frac{2}{3} = 55.5\text{ W}$$

5.27 (47.7)

Given electric field, E at a distance of 1 km = 12 mV/m

Also we know that $E \propto \frac{1}{r}$, where r is the distance where electric field is to be measured.

So, electric field, E at a distance of 2 km

$$= \frac{12\text{ mV/m}}{2} = 6\text{ mV/m}$$

Also, power density due to the antenna is given as,

$$P_{avg} = \frac{1}{2}\frac{E^2}{\eta} = \frac{1}{2}\times\frac{6\times 6\times 10^{-6}}{120\pi}$$

$$= 47.7\text{ nW/m}^2$$

5.28 (b)

- Point electromagnetic source radiates in all direction
- Dish antenna radiates any electromagnetic energy in any particular direction with narrow beam-width and high directivity.
- Yagi-uda antenna is high gain antenna used for T.V. reception. Its radiation is along the axis of the antenna.

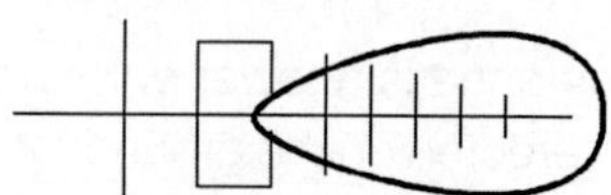

5.29 (50)

Both dipole antennas have isotropic pattern in ϕ view or H plane view.

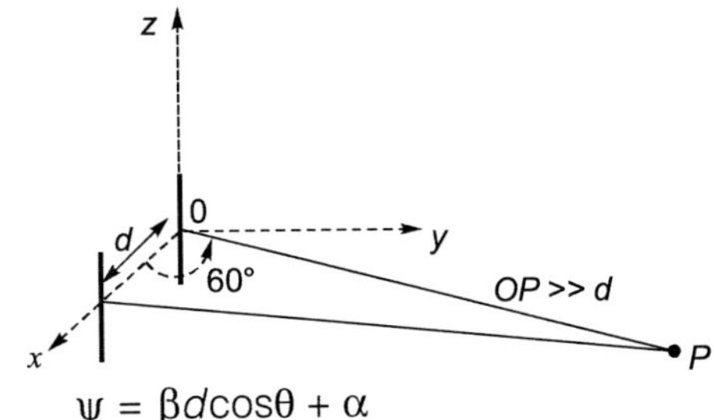

$\psi = \beta d\cos\theta + \alpha$

where, $\alpha = -90°, \beta = \dfrac{2\pi}{\lambda} = \dfrac{2\pi}{\dfrac{3\times10^8}{3\times10^6}} = \dfrac{2\pi}{100}$

Since maximum radiation ($\psi = 0$) is at $\theta = 60°$

$$\Rightarrow \quad 0 = \frac{2\pi}{100}d\cos60° - 90°$$

$$\Rightarrow \quad 90 = \frac{360}{100}d\frac{1}{2}$$

$$\Rightarrow \quad d = 50\text{ m}$$

5.30 (b)

Directivity (D) is increased by adding more antenna elements in an antenna array. Effective area (A_e) and directivity (D) are related by

$$A_e = \frac{\lambda^2}{4\pi}D$$

$\Rightarrow D\uparrow A_e\uparrow.$

5.31 (3)

$d = 200\,\lambda$

TX	RX
$\lvert\Gamma_t\rvert = 0.15$	$\lvert\Gamma_r\rvert = 0.18$
18 dB	22 dB
$P_t = 2$ W	$P_r = ?$

$G_t = 10^{1.8}$, $G_r = 10^{2.2}$

$$P_r = \frac{\left(1-|\Gamma_t|^2\right)\left(1-|\Gamma_r|^2\right)G_tG_r}{\left(\dfrac{4\pi d}{\lambda}\right)^2}\cdot P_t$$

$$= \frac{\left(1-|0.15|^2\right)\left(1-|0.18|^2\right)10^{1.8}\cdot10^{2.2}}{\left(\dfrac{4\pi200\lambda}{\lambda}\right)^2}\times2$$

$= 2.99476\times10^{-3}$ W $= 2.995$ mW ≈ 3 mW

5.32 (c, d)

$\vec{E}_T^i = (\hat{a}_x + j\hat{a}_y)E_0e^{jkz}$

$\Rightarrow$ Wave contains two orthogonal components and Y component leads X component leads by 90° and also wave is travelling in negative z-direction.

$\vec{E}_R = (\hat{a}_x + 2\hat{a}_y)E_I\dfrac{1}{r}e^{-jkr}$

$\Rightarrow$ Wave contains two orthogonal components with unequal amplitudes and both are in-phase.

$\Rightarrow$ Linear polarization.

Polarization Loss Factor, PLF $= \left|\hat{E}_T\cdot\hat{E}_R\right|^2$

Where, $E_T = \dfrac{\hat{a}_x + j\hat{a}_y}{\sqrt{2}}$

$E_R = \dfrac{\hat{a}_x + 2\hat{a}_y}{\sqrt{5}}$

$$\text{PLF} = \left|\frac{1+j2}{\sqrt{10}}\right|^2 = \frac{5}{10} = \frac{1}{2}$$

$$\Rightarrow \quad \text{PLF (dB)} = 10\log\frac{1}{2} = -3\text{ dB}$$

5.33 (b)

Power radiated $= \int P_{rad}\cdot ds$

$$= \int_{\theta=0}^{\pi/2}\int_{\phi=0}^{2\pi} C_0\frac{1}{r^2}\cos^4\theta\cdot r^2\sin\theta d\theta d\phi$$

$$= 2\pi\cdot\frac{C_0}{r^2}\cdot r^2\int_{\theta=0}^{\pi/2}\cos^4\theta\cdot d(-\cos\theta)$$

$$= 2\pi\cdot C_0\left(-\frac{\cos^5\theta}{5}\right)\Bigg|_0^{\pi/2}$$

$$= \frac{2\pi}{5}C_0 = 1.256\,C_0$$

$$\text{Directivity} = \frac{4\pi\cdot U(\theta,\phi)}{\int U(\theta,\phi)\cdot d\Omega}$$

$$= \frac{4\pi.C_0\cdot\cos^4\theta}{\int_{\theta=0}^{\pi/2}\int_{\phi=0}^{2\pi}C_0\cdot\cos^4\theta\cdot\sin\theta d\theta d\phi}$$

$$= \frac{2\cos^5\theta}{1/5} = 10\cos^5\theta$$

max value = 10

$\Rightarrow$ max value (in dB) $= 10\log_{10}10 = 10$ dB

5.34 (c)

As per Friis free space propagation equation,

$$W_r = \frac{W_tA_{er}A_{et}}{(\lambda d)^2}$$

when A_{er} and A_{et} are doubled and d also doubled, W_r is same. Hence capacity is also same.

5.35 (a)

XY plane means $Z = 0$ with $Z = r\cos\theta$ (spherical coordinates) $\theta = 90°$

With null points along the axis of the dipole it is $\phi = 45°$ and $225°$.

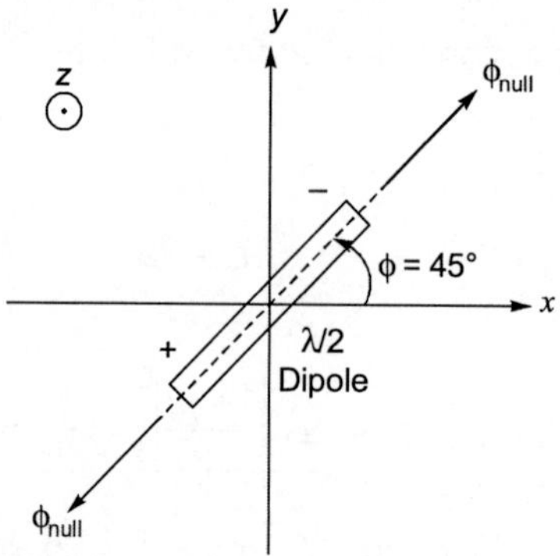

5.36 (2.01)

Radiation resistance of a small dipole current element of length 'l' is

$$R_{rad} = 80\pi^2\left(\frac{l}{\lambda}\right)^2 \Rightarrow R \propto l^2$$

$$\frac{R_2}{R_1} = \left(\frac{l_2}{l_1}\right)^2$$

If length is changed by 1% then percentage change in the radiation resistance.

$$\frac{R_2}{R_1} = \left(\frac{1.01l}{l}\right)^2 = 1.0201$$

Percentage change in radiation resistance

$$= \frac{R_2 - R_1}{R_1} \times 100$$

$$= 0.0201 \times 100 = 2.01\%$$

5.37 (b)

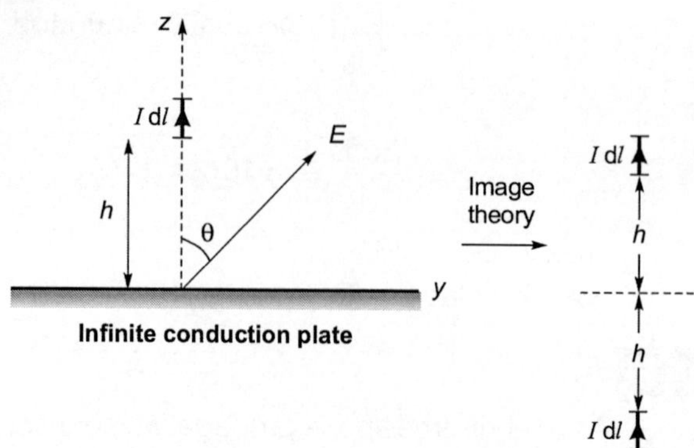

$$|\text{Total E}| = |(E_{\text{single element}})||(\text{A.F.})|$$

$$|(\text{A.F.})| = \frac{\sin\left(N\frac{\psi}{2}\right)}{\sin\left(\frac{\psi}{2}\right)} = \frac{\sin\left(2\frac{\psi}{2}\right)}{\sin\left(\frac{\psi}{2}\right)}$$

$$= \frac{2\sin\left(\frac{\psi}{2}\right)\cos\left(\frac{\psi}{2}\right)}{\sin\left(\frac{\psi}{2}\right)} = 2\cos\left(\frac{\psi}{2}\right)$$

$$|\text{A.F}_N| = \frac{\text{A.F}}{\text{A.F}_{max}} = \frac{2\cos\left(\frac{\psi}{2}\right)}{2} = \left|\cos\left(\frac{\psi}{2}\right)\right|$$

where, $\psi = \beta d\cos\theta = \frac{2\pi}{\lambda}(2h)\cos\theta$

$$\left|\text{A.F}_N|_{\theta=60°}\right| = \left|\cos\left(\frac{2\pi}{\lambda}h\cos 60°\right)\right|$$

$$= \left|\cos\left(\frac{\pi h}{\lambda}\right)\right|$$

$\cos\theta$ is maximum, whenever $\theta = n\pi$; n = 0, 1, 2 ...

$$\frac{\pi h}{\lambda} = n\pi \Rightarrow h = n\lambda$$

$\Rightarrow$ For $n = 1$, $h_{min} = \lambda$

5.38 (0.244)

$$G_d(\text{dB}) = 10\log_{10}(G_d)$$

$$\Rightarrow 6 = 10\log_{10} G_d$$

$$\Rightarrow G_d = 10^{0.6}$$

$$G_d = 3.981$$

$$E_m = \frac{\sqrt{60 G_d P_{rad}}}{r}$$

$$E_m = \frac{\sqrt{60(16\times10^3)(3.981)}}{8\times10^3}$$

$$= 0.244\ (\text{V/m})$$

■■■

6 Miscellaneous Topics

6.1 The frequency range for satellite communication is
(a) 1 kHz to 100 kHz (b) 100 kHz to 10 kHz
(c) 10 MHz to 30 MHz (d) 1 GHz to 30 GHz

[2000 : 1 M]

6.2 In a microwave test bench, why is the microwave signal amplitude modulated at 1 kHz?
(a) To increase the sensitivity of measurement
(b) To transmit the signal to a far-off place
(c) To study amplitude modulation
(d) Because crystal detector fails at microwave frequencies

[2006 : 2 M]

6.3 A load of 50 Ω is connected in shunt in a 2-wire transmission line of Z_0 = 50 Ω as shown in the figure. The 2-port scattering parameter matrix (*S*-matrix) of the shunt element is

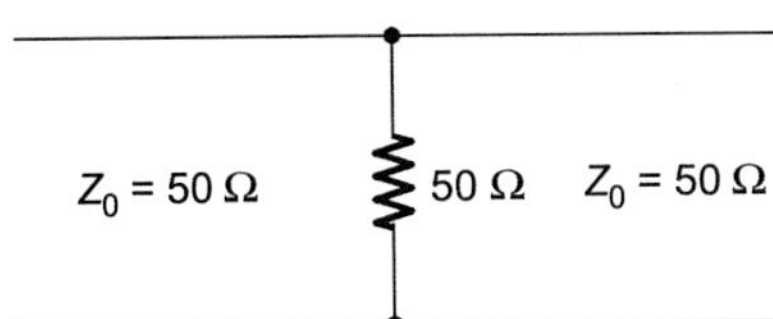

(a) $\begin{bmatrix} -\frac{1}{2} & \frac{1}{2} \\ \frac{1}{2} & -\frac{1}{2} \end{bmatrix}$ (b) $\begin{bmatrix} 0 & 1 \\ 1 & 0 \end{bmatrix}$

(c) $\begin{bmatrix} -\frac{1}{3} & \frac{2}{3} \\ \frac{2}{3} & -\frac{1}{3} \end{bmatrix}$ (d) $\begin{bmatrix} \frac{1}{4} & -\frac{3}{4} \\ -\frac{3}{4} & \frac{1}{4} \end{bmatrix}$

[2007 : 2 M]

6.4 In the design of a single mode step index optical fiber close to upper cut-off, the single-mode operation is NOT preserved if
(a) radius as well as operating wavelength are halved
(b) radius as well as operating wavelength are doubled
(c) radius is halved and operating wavelength is doubled
(d) radius is doubled and operating wavelength is halved

[2008 : 2 M]

6.5 If the scattering matrix [*S*] of a two port network is

$$[S] = \begin{bmatrix} 0.2\angle 0^\circ & 0.9\angle 90^\circ \\ 0.9\angle 90^\circ & 0.1\angle 90^\circ \end{bmatrix}$$

then the network is
(a) lossless and reciprocal
(b) lossless but not reciprocal
(c) not lossless but reciprocal
(d) neither lossless nor reciprocal

[2010 : 1 M]

6.6 A two-port network has scattering parameters given by $[S] = \begin{bmatrix} s_{11} & s_{12} \\ s_{21} & s_{22} \end{bmatrix}$. If the port-2 of the two-port is short circuited, the s_{11} parameter for the resultant one-port network is

(a) $\dfrac{s_{11} - s_{11}s_{22} + s_{12}s_{21}}{1 + s_{22}}$

(b) $\dfrac{s_{11} + s_{11}s_{22} - s_{12}s_{21}}{1 + s_{22}}$

(c) $\dfrac{s_{11} + s_{11}s_{22} + s_{12}s_{21}}{1 - s_{22}}$

(d) $\dfrac{s_{11} - s_{11}s_{22} + s_{12}s_{21}}{1 - s_{22}}$

[2014 : 1 M, Set-1]

6.7 A radar operating at 5 GHz uses a common antenna for transmission and reception. The antenna has a gain of 150 and is aligned for maximum directional radiation and reception to a target 1 km away having radar cross-section of 3 m^2. If it transmits 100 kW, then the received power (in μW) is ________.

[2016 : 2 M, Set-3]

6.8 Light from free space is incident at an angle θ_i to the normal of the facet of a step-index large core optical fibre. The core and cladding refractive indices are n_1=1.5 and n_2=1.4, respectively.

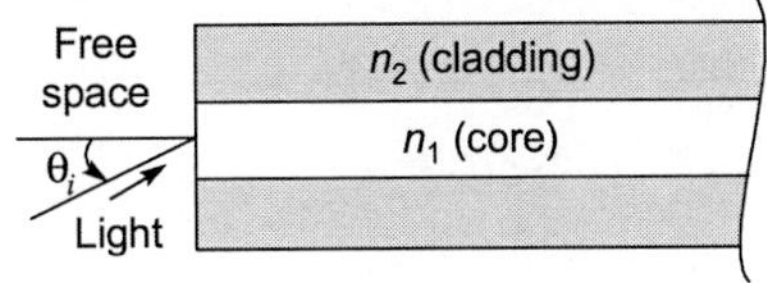

The maximum value of θ_i (in degrees) for which the incident light will be guided in the core of the fibre is _______.

[2016 : 1 M, Set-2]

6.9 An optical fiber is kept along the $\hat{z}$ direction. The refractive indices for the electric fields along $\hat{x}$ and $\hat{y}$ directions in the fiber are n_x = 1.5000 and n_y = 1.5001, respectively ($n_x \neq n_y$ due to the imperfection in the fiber cross-section). The free space wavelength of a light wave propagating in the fiber is 1.5 μm. If the lightwave is circularly polarized at the input of the fiber, the minimum propagation distance after which it becomes linearly polarized, in centimetres, is ______.

[2017 : 2 M, Set-1]

6.10 The refractive indices of the core and cladding of an optical fibre are 1.50 and 1.48, respectively. The critical propagation angle, which is defined as the maximum angle that the light beam makes with the axis of the optical fibre to achieve the total internal reflection, (rounded off to two decimal places) is _______ degree.

[2021 : 1 M]

■■■

Answers Miscellaneous Topics

6.1	(d)	6.2	(d)	6.3	(b)	6.4	(d)	6.5	(c)	6.6	(b)	6.7	(0.012)
6.8	(32.58)	6.9	(0.375)	6.10	(14.13)								

Explanations Miscellaneous Topics

6.1 (d)

The frequency range for satellite communication is 1 GHz to 30 GHz.

6.3 (b)

The line is terminated with 50 Ω at the center and so, matched on both the sides.

6.5 (c)

For reciprocal networks, $S_{12} = S_{21}$

For symmetrical networks, $S_{11} = S_{22}$

For antisymmetrical networks, $S_{11} = -S_{22}$

For lossless reciprocal networks,

$$|S_{11}| = |S_{22}|$$

and $|S_{11}|^2 + |S_{12}|^2 = 1$

6.6 (b)

$$b_1 = s_{11} \cdot a_1 + s_{12} \cdot a_2 \quad \text{...(i)}$$

$$b_2 = s_{21} \cdot a_1 + s_{22}\, a_2 \quad \text{...(ii)}$$

When output is short, $a_2 = -b_2$

$$-a_2 = s_{21} \cdot a_1 + s_{22}\, a_2$$

$$\Rightarrow \quad a_2 = \frac{-s_{21} \cdot a_1}{1 + s_{22}}$$

Substituting in first equation,

$$b_1 = s_{11} \cdot a_1 + s_{12}\left(\frac{-s_{21} \cdot a_1}{1 + s_{22}}\right)$$

$$= \frac{s_{11} \cdot a_1 + s_{11} \cdot s_{22} a_1 - s_{12} \cdot s_{21} a_1}{1 + s_{22}}$$

$$\frac{b_1}{a_1} = s_{11} = \frac{s_{11} + s_{11} \cdot s_{22} - s_{12} \cdot s_{21}}{1 + s_{22}}$$

6.7 (0.012)

f = 5 GHz, G = 150, R = 1 km,

P_t = 100 kW, σ = 3 m^2

$$P_r = \frac{(G\lambda)^2 \cdot \sigma}{(4\pi)^3 R^4} \times P_t$$

$$= \frac{\left(150 \times \frac{3 \times 10^8}{5 \times 10^9}\right)^2 \cdot 3}{(4\pi)^3 \cdot 10^{12}} \times 10^5$$

$$= \frac{81 \times 3 \times 10^5}{(4\pi)^3 \times 10^{12}}$$

$$= 1.224 \times 10^{-8} = 0.012\ \mu\text{W}$$

6.8 (32.58)

$$\sin\alpha_{max} = \sqrt{n_1^2 - n_2^2} = \sqrt{1.5^2 - 1.4^2}$$

$$\Rightarrow \quad \alpha_{max} = \sin^{-1}(0.5385) = 32.58°$$

6.9 (0.375)

Initially the wave is circularly polarized. So, the initial phase difference between field components in $\hat{a}_x$ direction and $\hat{a}_y$ direction is $\pi/2$.

To become linearly polarized, the wave must travel a minimum distance, such that, the phase difference at that point between the field components in $\hat{a}_x$ direction and $\hat{a}_y$ direction is π (i.e., the travel of this minimum distance should provide an additional phase difference of $\pi/2$ between $\hat{a}_x$ and $\hat{a}_y$ field components).

So,

$$z_{min} k_x \sim z_{min} k_y = \frac{\pi}{2}$$

$$z_{min}\left[\frac{\omega}{v_{px}} \sim \frac{\omega}{v_{py}}\right] = \frac{\pi}{2}$$

$$2\pi z_{min}\left[\frac{f}{c}\sqrt{\varepsilon_{rx}} \sim \frac{f}{c}\sqrt{\varepsilon_{ry}}\right] = \frac{\pi}{2}$$

$$\frac{4 z_{min}}{\lambda_0}\left[n_x \sim n_y\right] = 1$$

$$z_{min} = \frac{\lambda_0}{4(n_x \sim n_y)} = \frac{1.5}{4(1.5 \sim 1.5001)}\,\mu\text{m}$$

$$= \frac{1.5}{4(0.0001)}\,\mu\text{m} = \frac{1.5}{4}\,\text{cm}$$

$$z_{min} = 0.375\ \text{cm}$$

6.10 (14.13)

Given that

Refractive index of core $\eta_1 = 1.50$

Refractive index of clad $\eta_2 = 1.48$

Critical propagation angle (θ_a)

$$\theta_a = \sin^{-1}\left[\sqrt{\eta_1^2 - \eta_2^2}\right]$$

$$= \sin^{-1}\sqrt{1.5^2 - 1.48^2} = 14.13$$

■■■

Control Systems

CONTENTS

Control Systems

Syllabus

Basic control system components; Feedback principle; Transfer function; Block diagram representation; Signal flow graph; Transient and steady-state analysis of LTI systems; Frequency response; Routh-Hurwitz and Nyquist stability criteria; Bode and root-locus plots; Lag, lead and lag-lead compensation; State variable model and solution of state equation of LTI systems.

Analysis of Previous GATE Papers

Exam Year	1 Mark Ques.	2 Marks Ques.	Total Marks
1995	7	–	**7**
1996	–	–	–
1997	–	2	**4**
1998	9	1	**11**
1999	4	3	**10**
2000	1	1	**3**
2001	4	3	**10**
2002	4	4	**12**
2003	2	6	**14**
2004	2	9	**20**
2005	3	6	**15**
2006	2	8	**18**
2007	1	9	**19**
2008	1	8	**17**
2009	2	4	**10**
2010	3	3	**9**
2011	3	3	**9**
2012	1	5	**11**
2013	2	5	**12**

Exam Year	1 Mark Ques.	2 Mark Ques.	Total Marks
2014 Set-1	2	3	**8**
2014 Set-2	2	3	**8**
2014 Set-3	1	3	**7**
2014 Set-4	2	3	**8**
2015 Set-1	3	3	**9**
2015 Set-2	3	3	**9**
2015 Set-3	3	4	**11**
2016 Set-1	2	3	**8**
2016 Set-2	2	3	**8**
2016 Set-3	2	3	**8**
2017 Set-1	3	3	**9**
2017 Set-2	3	3	**9**
2018	1	3	**7**
2019	1	4	**9**
2020	2	3	**8**
2021	2	2	**6**
2022	2	3	**8**
2023	1	3	**7**
2024	2	3	**8**

1 Basics of Control Systems, Block Diagram and SFG's

1.1 Signal flow graph is used to find
(a) Stability of the system
(b) Controllability of the system
(c) Transfer function of the system
(d) Poles of the system.

[1995 : 1 M]

1.2 The transfer function of a linear system is the
(a) ratio of the output, $v_0(t)$ and input $v_i(t)$.
(b) ratio of the derivatives of the output and the input.
(c) ratio of the Laplace transform of the output and that of the input with all initial conditions zeros.
(d) none of these.

[1995 : 1 M]

1.3 In the signal flow graph of figure, y/x equals

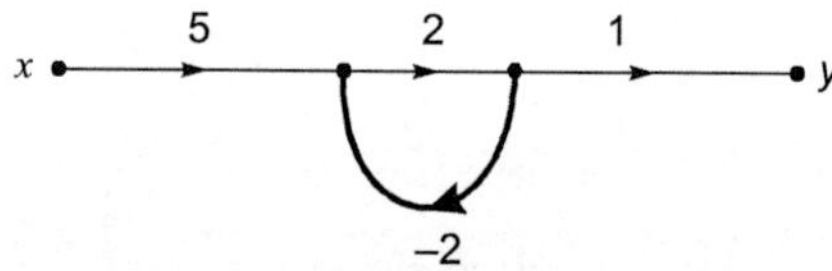

(a) 3 (b) 5/2
(c) 2 (d) None of the above

[1997 : 2 M]

1.4 The equivalent of the block diagram in the figure is given as

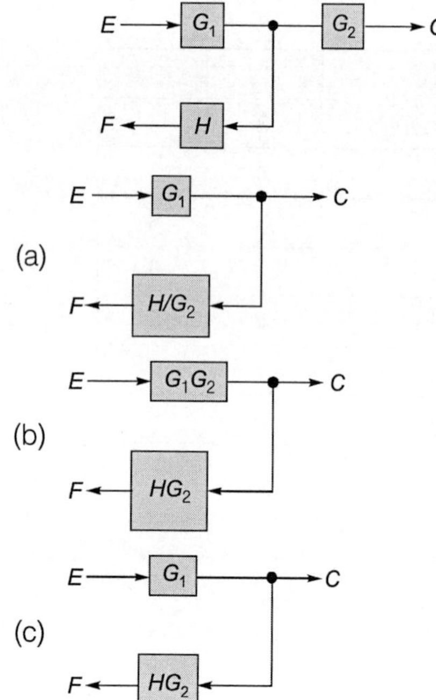

(d)

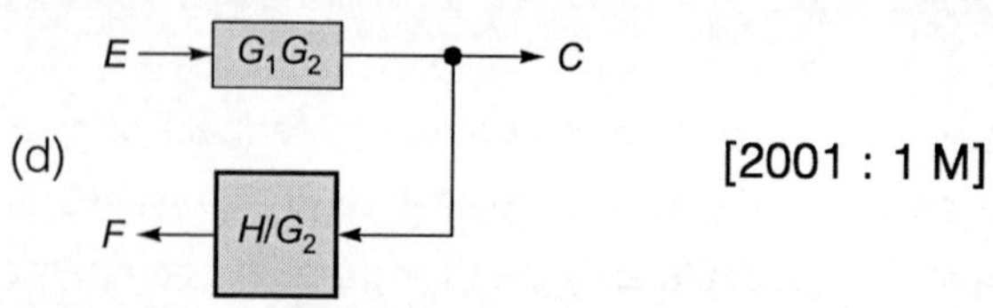

[2001 : 1 M]

1.5 An electrical system and its signal-flow graph representations are shown in the figure (a) and (b) respectively. The values of G_2 and H, respectively, are

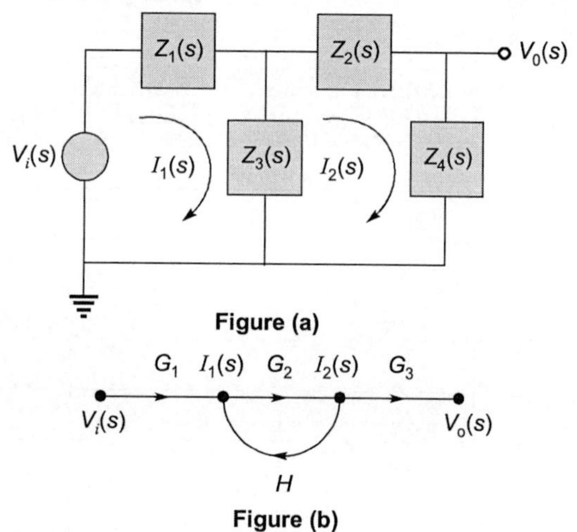

Figure (a)

Figure (b)

(a) $\dfrac{Z_3(s)}{Z_2(s)+Z_3(s)+Z_4(s)}, \dfrac{-Z_3(s)}{Z_1(s)+Z_3(s)}$

(b) $\dfrac{-Z_3(s)}{Z_2(s)-Z_3(s)+Z_4(s)}, \dfrac{-Z_3(s)}{Z_1(s)+Z_3(s)}$

(c) $\dfrac{Z_3(s)}{Z_2(s)+Z_3(s)+Z_4(s)}, \dfrac{Z_3(s)}{Z_1(s)+Z_3(s)}$

(d) $\dfrac{-Z_3(s)}{Z_2(s)-Z_3(s)+Z_4(s)}, \dfrac{Z_3(s)}{Z_1(s)+Z_3(s)}$

[2001 : 2 M]

1.6 The signal flow graph of a system is shown in the figure. The transfer function $\dfrac{C(s)}{R(s)}$ of the system is

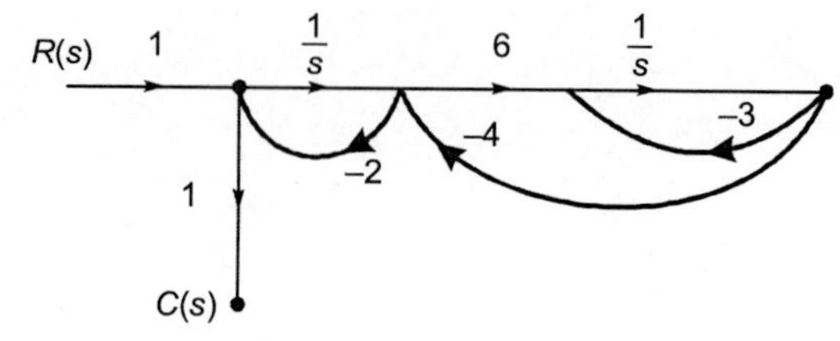

(a) $\dfrac{6}{s^2+29s+6}$ (b) $\dfrac{6s}{s^2+29s+6}$

(c) $\dfrac{s(s+2)}{s^2+29s+6}$ (d) $\dfrac{s(s+27)}{s^2+29s+6}$

[2003 : 2 M]

1.7 Consider the signal flow graph shown in the figure. The gain x_5/x_1 is

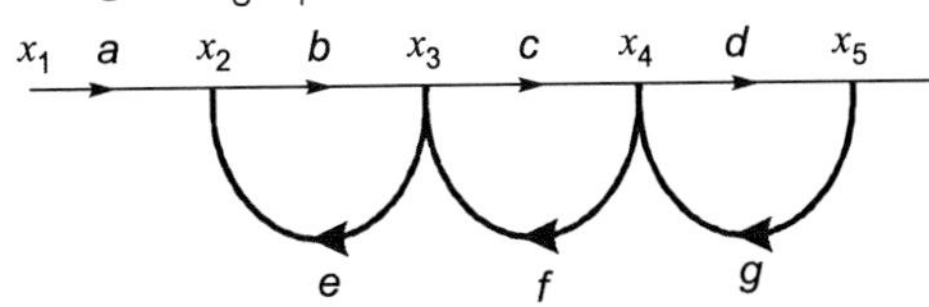

(a) $\dfrac{1-(be+cf+dg)}{abc}$

(b) $\dfrac{bedg}{1-(be+cf+dg)}$

(c) $\dfrac{abcd}{1-(be+cf+dg)+bedg}$

(d) $\dfrac{1-(be+cf+dg)+bedg}{abcd}$

[2004 : 2 M]

1.8 The transfer function $Y(s)/R(s)$ of the system shown is

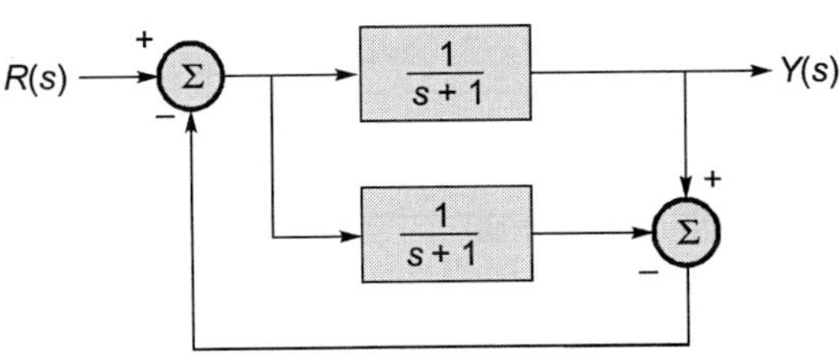

(a) 0 (b) $\dfrac{1}{s+1}$

(c) $\dfrac{2}{s+1}$ (d) $\dfrac{2}{s+3}$ [2010 : 1 M]

1.9 The input-output transfer function of a plant $H(s)=\dfrac{100}{s(s+10)^2}$. The plant is placed in a unity negative feedback configuration as shown in the figure below.

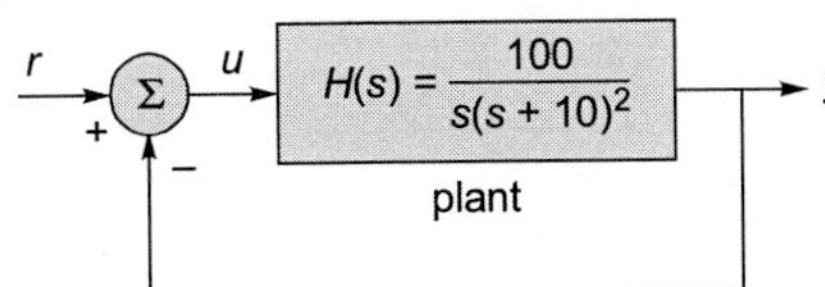

The signal flow graph that **DOES NOT** model the plant transfer function $H(s)$ is

(a)

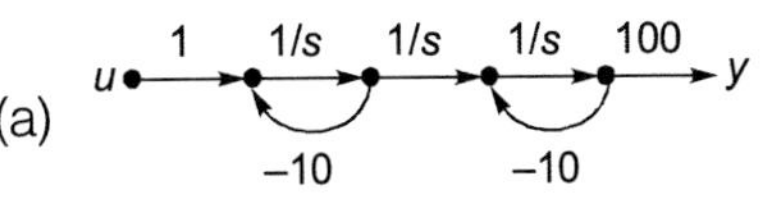

(b)

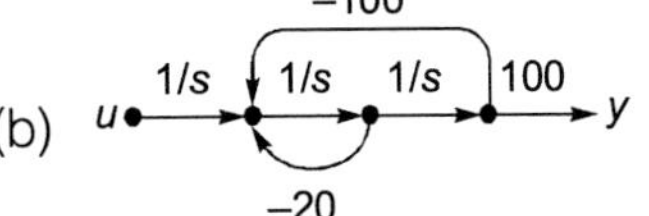

(c)

(d)

[2011 : 2 M]

1.10 The signal flow graph for a system is given below. The transfer function $\dfrac{Y(s)}{U(s)}$ for this system is

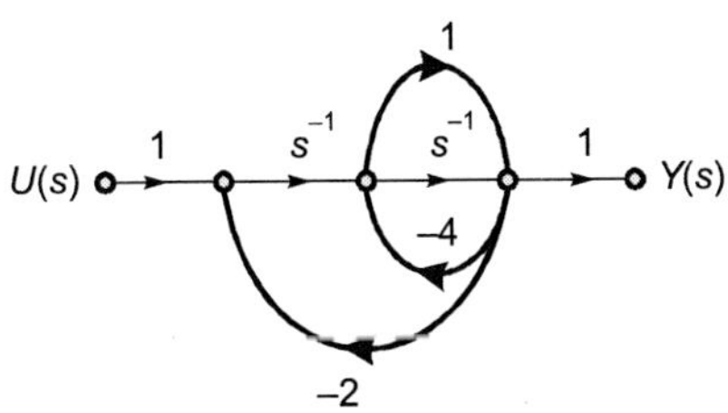

(a) $\dfrac{s+1}{5s^2+6s+2}$ (b) $\dfrac{s+1}{s^2+6s+2}$

(c) $\dfrac{s+1}{s^2+4s+2}$ (d) $\dfrac{1}{5s^2+6s+2}$

[2013 : 2 M]

1.11 For the following system,

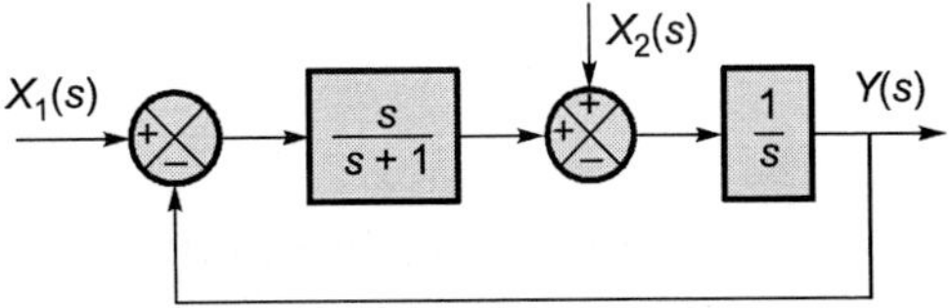

when $X_1(s)=0$, the transfer function $\dfrac{Y(s)}{X_2(s)}$ is

(a) $\dfrac{s+1}{s^2}$ (b) $\dfrac{1}{s+1}$

(c) $\dfrac{s+2}{s(s+1)}$ (d) $\dfrac{s+1}{s(s+2)}$

[2014 : 1 M, Set-2]

1.12 Consider the following block diagram in the figure.

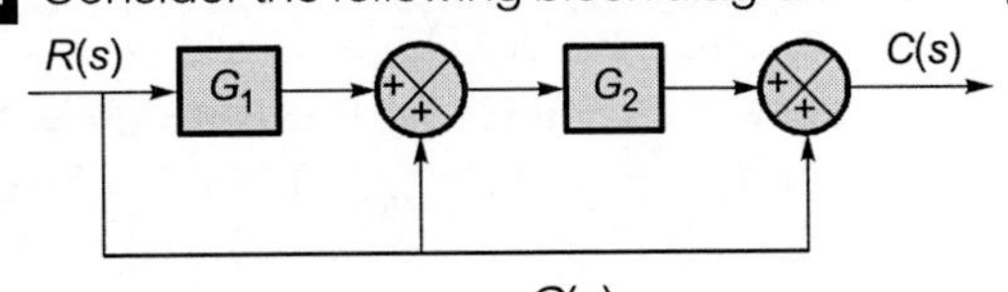

The transfer function $\dfrac{C(s)}{R(s)}$ is

(a) $\dfrac{G_1G_2}{1+G_1G_2}$ (b) $G_1G_2+G_1+1$

(c) $G_1G_2+G_2+1$ (d) $\dfrac{G_1}{1+G_1G_2}$

[2014 : 1 M, Set-3]

1.13 Negative feedback in a closed-loop control system **DOES NOT**

(a) reduce the overall gain
(b) reduce bandwidth
(c) improve disturbance rejection
(d) reduce sensitivity to parameter variation

[2015 : 1 M, Set-1]

1.14 For the signal flow graph shown in the figure, the value of $\dfrac{C(s)}{R(s)}$ is

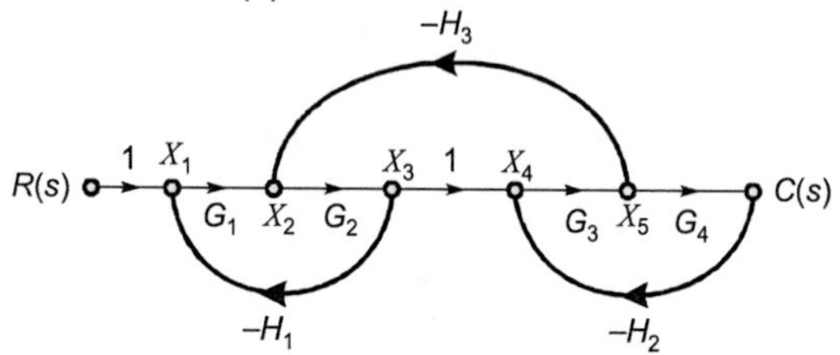

(a) $\dfrac{G_1G_2G_3G_4}{1-G_1G_2H_1-G_3G_4H_2-G_2G_3H_3+G_1G_2G_3G_4H_1H_2}$

(b) $\dfrac{G_1G_2G_3G_4}{1+G_1G_2H_1+G_3G_4H_2+G_2G_3H_3+G_1G_2G_3G_4H_1H_2}$

(c) $\dfrac{1}{1+G_1G_2H_1+G_3G_4H_2+G_2G_3H_3+G_1G_2G_3G_4H_1H_2}$

(d) $\dfrac{1}{1-G_1G_2H_1-G_3G_4H_2-G_2G_3H_3+G_1G_2G_3G_4H_1H_2}$

[2015 : 1 M, Set-2]

1.15 By performing cascading and/or summing/differencing operations using transfer function blocks $G_1(s)$ and $G_2(s)$, one CANNOT realize a transfer function of the form

(a) $G_1(s)\,G_2(s)$ (b) $\dfrac{G_1(s)}{G_2(s)}$

(c) $G_1(s)\left(\dfrac{1}{G_1(s)}+G_2(s)\right)$ (d) $G_1(s)\left(\dfrac{1}{G_1(s)}-G_2(s)\right)$

[2015 : 1 M, Set-2]

1.16 The position control of a DC servo-motor is given in the figure. The values of the parameters are K_T = 1 N-m A, R_a = 1 Ω, L_a = 0.1 H. J = 5 kg-m², B = 1 N-m (rad/sec) and K_b = 1 V/(rad/sec). The steady-state position response (in radians) due to unit impulse disturbance torque T_d is ________.

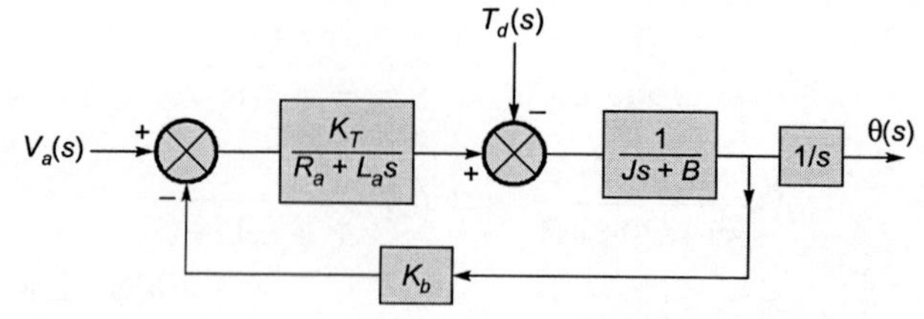

[2015 : 2 M, Set-3]

1.17 The block diagram of a feedback control system is shown in the figure. The overall closed-loop gain G of the system is

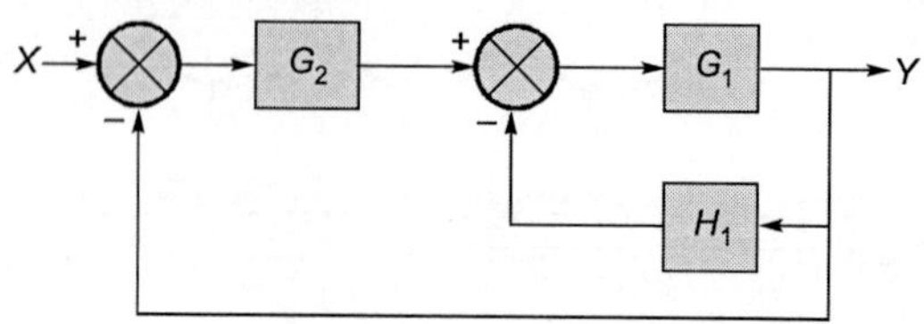

(a) $G=\dfrac{G_1G_2}{1+G_1H_1}$ (b) $G=\dfrac{G_1G_2}{1+G_1G_2+G_1H_1}$

(c) $G=\dfrac{G_1G_2}{1+G_1G_2H_1}$ (d) $G=\dfrac{G_1G_2}{1+G_1G_2+G_1G_2H_1}$

[2016 : 1 M, Set-3]

1.18 For the system shown in the figure, $Y(s)/X(s)$ = ____.

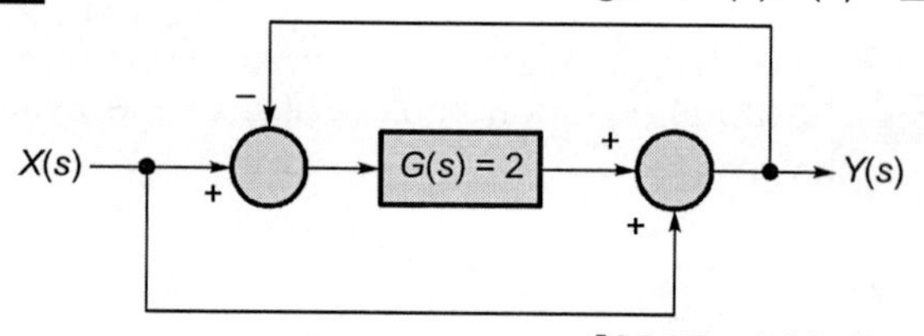

[2017 : 1 M, Set-2]

1.19 The block diagram of a system is illustrated in the figure shown, where $X(s)$ is the input and $Y(s)$ is the output. The transfer function $H(s)=\dfrac{Y(s)}{X(s)}$ is

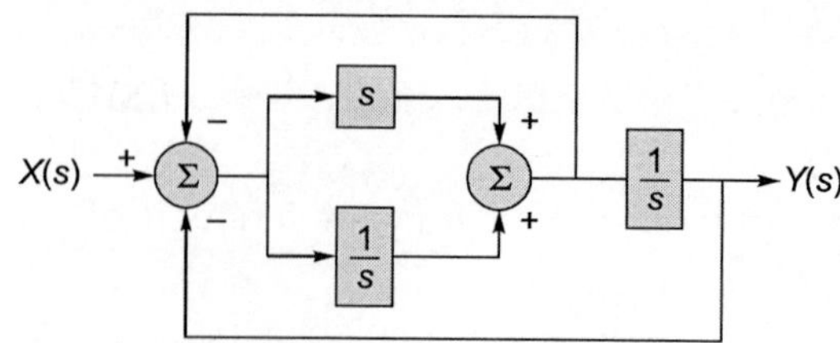

(a) $H(s)=\dfrac{s^2+1}{2s^2+1}$

(b) $H(s)=\dfrac{s^2+1}{s^3+2s^2+s+1}$

(c) $H(s)=\dfrac{s+1}{s^2+s+1}$

(d) $H(s)=\dfrac{s^2+1}{s^3+s^2+s+1}$

[2019 : 2 M]

1.20 The block diagram of a feedback control system is shown in the figure.

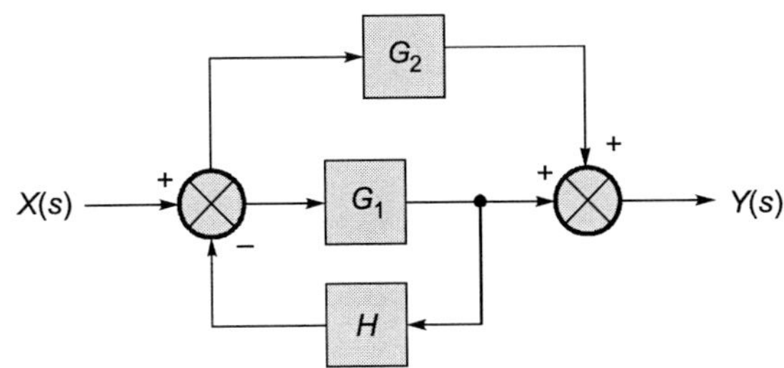

The transfer function $\frac{Y(s)}{X(s)}$ of the system is

(a) $\frac{G_1+G_2}{1+G_1H}$
(b) $\frac{G_1+G_2+G_1G_2H}{1+G_1H+G_2H}$
(c) $\frac{G_1+G_2+G_1G_2H}{1+G_1H}$
(d) $\frac{G_1+G_2}{1+G_1H+G_2H}$

[2021 : 1 M]

1.21 In the following block diagram, $R(s)$ and $D(s)$ are two inputs. The output $Y(s)$ is expressed as $Y(s) = G_1(s)R(s) + G_2(s)D(s)$.

$G_1(s)$ and $G_2(s)$ are given by

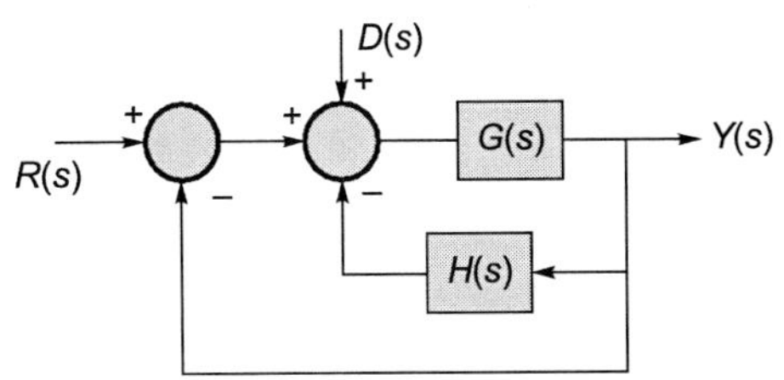

(a) $G_1(s)=\frac{G(s)}{1+G(s)+G(s)H(s)}$ and

$G_2(s)=\frac{G(s)}{1+G(s)+G(s)H(s)}$

(b) $G_1(s)=\frac{G(s)}{1+G(s)+H(s)}$ and

$G_2(s)=\frac{G(s)}{1+G(s)+H(s)}$

(c) $G_1(s)=\frac{G(s)}{1+G(s)+H(s)}$ and

$G_2(s)=\frac{G(s)}{1+G(s)+G(s)H(s)}$

(d) $G_1(s)=\frac{G(s)}{1+G(s)+G(s)H(s)}$ and

$G_2(s)=\frac{G(s)}{1+G(s)+H(s)}$

[2023 : 2 M]

■■■

Answers Basics of Control Systems, Block Diagram and SFG's

1.1	(c)	1.2	(c)	1.3	(c)	1.4	(d)	1.5	(c)	1.6	(d)	1.7	(c)
1.8	(b)	1.9	(d)	1.10	(a)	1.11	(d)	1.12	(c)	1.13	(b)	1.14	(b)
1.15	(b)	1.16	(–0.5)	1.17	(b)	1.18	(1)	1.19	(b)	1.20	(a)	1.21	(a)

Explanations Basics of Control Systems, Block Diagram and SFG's

1.1 (c)

The signal flow graph can not be used to comment on stability, controllability, observability or pole zero locations. It only gives transfer function.

1.2 (c)

The transfer function of a linear time invariant system is defined as the ratio of the Laplace transform of the output and that of the input with all initial conditions zero.

1.3 (c)

$$\frac{C}{R} = \frac{\Sigma P_K \Delta_K}{\Delta}$$

$$\frac{C}{R} = \frac{5\times2\times1}{1-(-4)} = \frac{10}{1+4} = \frac{10}{5} = 2$$

1.4 (d)

Take off point is moved after G_2 so H/G_2.

1.5 (c)

From KVL in both loops

In first loop,

$$V_i(s) = I_1(s)Z_1(s) + [I_1(s) - I_2(s)]Z_3(s)$$

$$V_i(s) = I_1(s)[Z_1(s) + Z_3(s)] - I_2(s)Z_3(s)$$

$$\frac{V_i(s)}{Z_1(s)+Z_3(s)} = I_1(s) - \frac{I_2(s)\cdot Z_3(s)}{Z_1(s)+Z_3(s)} \quad ...(i)$$

In second loop,

$$[I_2(s) - I_1(s)]Z_3(s) + I_2(s)Z_2(s) + I_2(s)Z_4(s) = 0$$

$$I_2(s)(Z_2(s) + Z_3(s) + Z_4(s)) = I_1(s)\cdot Z_3(s)$$

$$G_2 = \frac{I_2(s)}{I_1(s)} = \frac{Z_3(s)}{Z_2(s)+Z_3(s)+Z_4(s)}$$

From SFG, $I_1(s) = V_iG_1(s) + I_2(s)H(s)$

$$I_1(s) = V_i(s)\frac{1}{Z_1(s)+Z_3(s)} + I_2(s)\frac{Z_3(s)}{Z_1(s)+Z_3(s)}$$

... from (i)

$$\therefore \quad H(s) = \frac{Z_3(s)}{Z_1(s)+Z_3(s)}$$

(comparing above two equations)

1.6 (d)

$$P_1 = 1$$

$$\Delta_1 = 1 + \frac{3}{s} + \frac{24}{s} = \frac{s+27}{s}$$

$$G(s) = \frac{P_1\Delta_1}{1-(\text{loop gain}) + \text{pair of non-touching loops}}$$

$$L_1 = \frac{-3}{s}, L_2 = \frac{-24}{s}, L_3 = \frac{-2}{s}$$

L_1 and L_3 are non-touching.

$$\therefore \quad G(s) = \frac{\frac{(s+27)}{s}}{1-\left(\frac{-3}{s} - \frac{24}{s} - \frac{2}{s}\right) + \frac{-2}{s}\times\frac{-3}{s}}$$

$$= \frac{\frac{(s+27)}{s}}{1+\frac{29}{s}+\frac{6}{s^2}}$$

$$\frac{C(s)}{R(s)} = \frac{s(s+27)}{s^2+29s+6}$$

1.7 (c)

$P_1 = abcd, \quad \Delta_1 = 1$

$L_1 = be, L_2 = cf, L_3 = dg$

Non-touching loops are L_1 & $L_3 = bedg$

$$\therefore \quad \frac{X_5}{X_1} = \frac{abcd}{1-(be+cf+dg)+bedg}$$

1.8 (b)

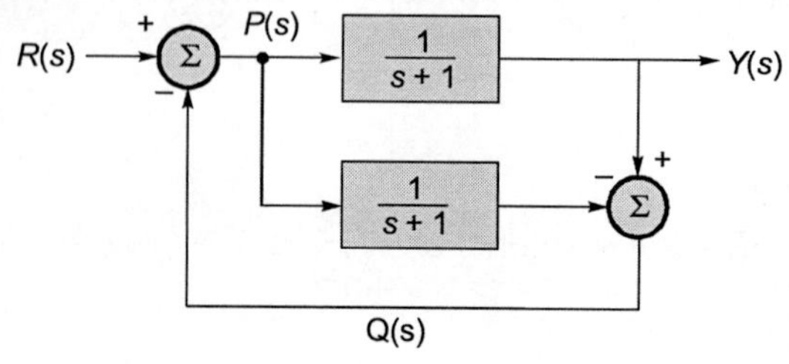

$$Q(s) = P(s)\left[\frac{1}{s+1} - \frac{1}{s+1}\right]$$

$$= 0$$

So, $\quad P(s) = R(s) - 0 = R(s)$

$$Y(s) = \frac{P(s)}{s+1} = \frac{R(s)}{s+1}$$

Therefore, $\dfrac{Y(s)}{R(s)} = \dfrac{1}{s+1}$

1.9 (d)

For option (d),

$$\frac{Y(s)}{U(s)} = \frac{100/s^3}{1+\frac{100}{s^2}}$$

Which is not transfer function of $H(s)$.

1.10 (a)

Using Mason's gain formula,

$$\Delta = 1 - [-2s^{-1} - 2s^{-2} - 4 - 4s^{-1}]$$

$$\Delta = 1 + \frac{2}{s} + \frac{2}{s^2} + 4 + \frac{4}{s} = \frac{5s^2+6s+2}{s^2}$$

$$P_1 = s^{-2} = \frac{1}{s^2}$$

$$P_2 = s^{-1} = \frac{1}{s}$$

$$\Delta_1 = 1$$

$$\Delta_2 = 1$$

$$\frac{Y(s)}{U(s)} = \frac{\sum P_k\Delta_k}{\Delta} = \frac{\frac{1}{s^2}\times 1 + \frac{1}{s}\times 1}{\frac{5s^2+6s+2}{s^2}}$$

$$\frac{Y(s)}{U(s)} = \frac{s+1}{5s^2+6s+2}$$

1.11 (d)

Redrawing the block diagram with $X_1(s) = 0$

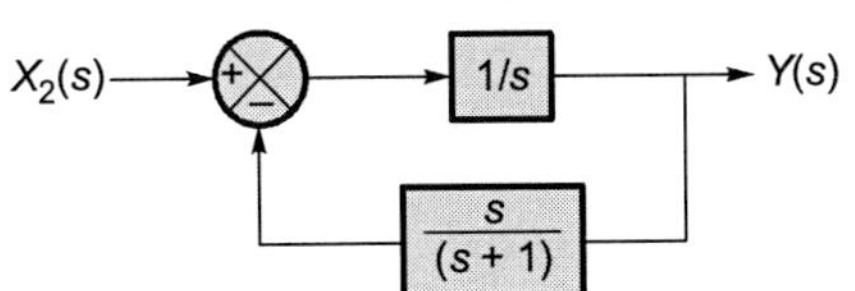

The transfer function

$$T(s) = \frac{Y(s)}{X_2(s)} = \frac{G(s)}{1+G(s)H(s)} \quad ...(i)$$

Here, $G(s) = \frac{1}{s}$ and $H(s) = \frac{s}{s+1}$

$$\frac{Y(s)}{X_2(s)} = \frac{1/s}{1+\frac{1}{s}\times\frac{s}{s+1}} = \frac{(s+1)}{s(s+2)}$$

1.12 (c)

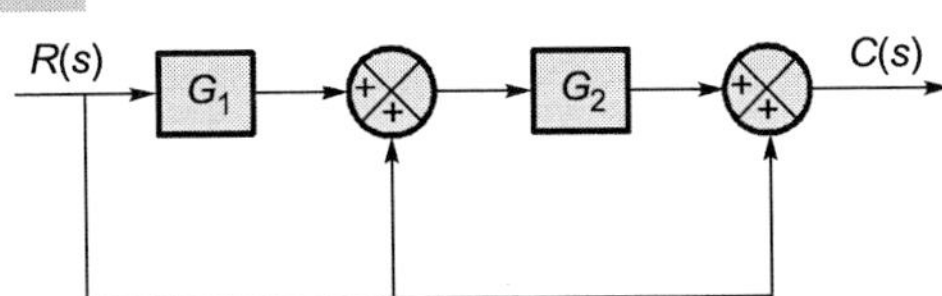

Converting the block diagram into signal flow graph on

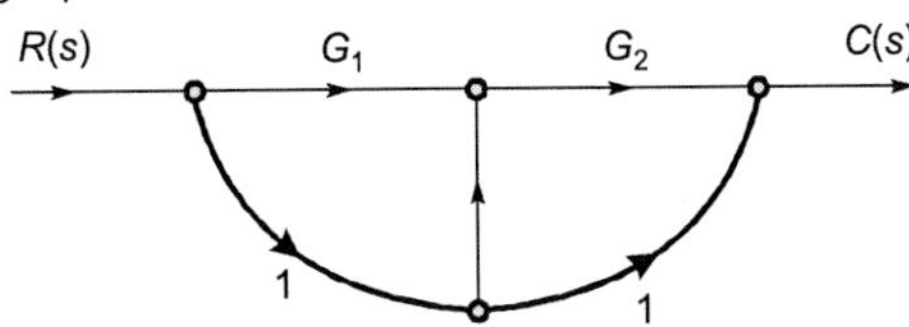

Forward paths,

$$P_1 = G_1\, G_2 \;;\; P_2 = G_2 \cdot 1 \;;\; P_3 = 1 \cdot 1 = 1$$

So, the transfer function is

$$\frac{C(s)}{R(s)} = G_1\, G_2 + G_2 + 1$$

1.14 (b)

Transfer function

$$\frac{C(s)}{R(s)} = \frac{\Sigma P_k \Delta_k}{\Delta}$$

$$= \frac{G_1G_2G_3G_4}{1-(-G_1G_2H_1 - G_2G_4H_2 - G_2G_3H_3) + G_1G_2H_1 \cdot G_3G_4H_2}$$

$$\frac{C(s)}{R(s)} = \frac{G_1G_2G_3G_4}{1+G_1G_2H_1 + G_2G_4H_2 + G_2G_3H_3 + G_1G_2H_1 \cdot G_3G_4H_2}$$

1.15 (b)

Given blocks has transfer function $G_1(s)$ and $G_2(s)$:

In cascade connection $G_1(s) \cdot G_2(s)$

In parallel connection $G_1(s) + G_2(s)$

From the given options:

(For cascading/ summing/differencing)

(a) $G_1(s)\, G_2(s) \Rightarrow$ Realization is possible.

(b) $\frac{G_1(s)}{G_2(s)} \Rightarrow$ Realization is not possible because the gain of blocks given as $G_1(s)$ and $G_2(s)$.

(c) $G_1(s)\left(\frac{1}{G_1(s)} + G_2(s)\right) = 1 + G_1(s)\, G_2(s)$

Realization is possible.

(d) $G_1(s)\left(\frac{1}{G_1(s)} - G_2(s)\right) = 1 - G_1(s)\, G_2(s)$

Realization is possible.

1.16 (-0.5)

The transfer function due to the disturbance torque $T_d(s)$ is

$$\frac{\theta(s)}{T_d(s)} = \frac{-\frac{1}{(Js+B)}\times\frac{1}{s}}{1+\left(\frac{1}{Js+B}\right)\left(\frac{K_b K_T}{R_a + L_a s}\right)}$$

$$= \frac{-(R_a + L_a s)\cdot\frac{1}{s}}{(R_a + L_a s)(Js+B) + K_b K_T}$$

The steady value of response for unit impulse input

$$\theta(0) = \lim_{s\to 0} \frac{-s(R_a + L_a s)\cdot\frac{1}{s}}{(R_a + L_a s)(Js+B) + K_b K_T}\cdot T_d$$

$$= -\frac{R_a}{R_a B + K_b K_T}\cdot 1$$

Given:

$K_T = 1$ N-m/A $R_a = 1\ \Omega$,

$B = 1$ N-m/rad/sec

and $K_b = 1$ V/rad/sec

Substituting the given values into above equation, we get

$$\theta(0) = -\frac{1}{2} = -0.5 \text{ rad}$$

1.17 (b)

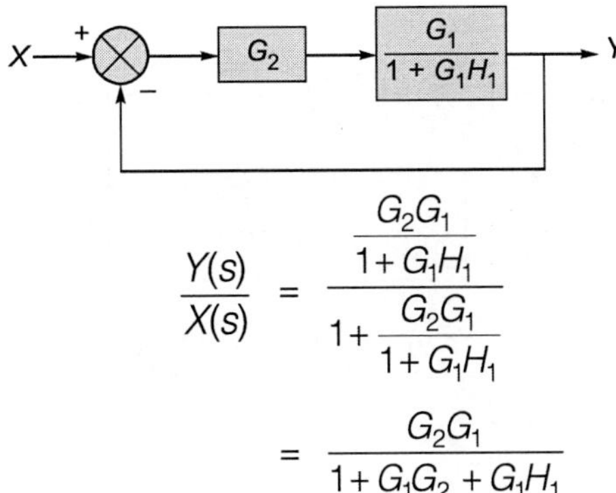

$$\frac{Y(s)}{X(s)} = \frac{\frac{G_2G_1}{1+G_1H_1}}{1+\frac{G_2G_1}{1+G_1H_1}}$$

$$= \frac{G_2G_1}{1+G_1G_2+G_1H_1}$$

1.18 (1)

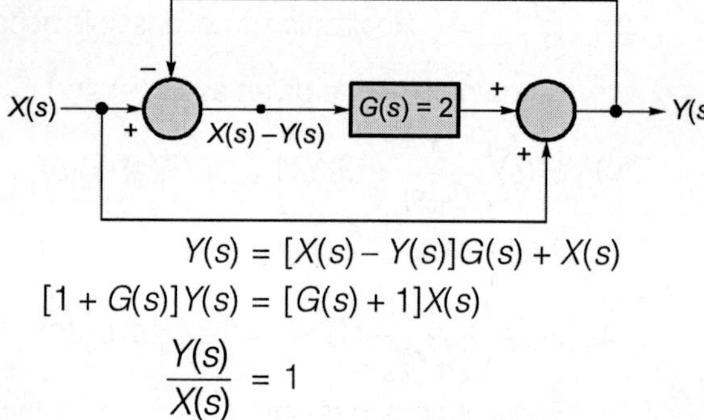

$$Y(s) = [X(s) - Y(s)]G(s) + X(s)$$

$$[1 + G(s)]Y(s) = [G(s) + 1]X(s)$$

$$\frac{Y(s)}{X(s)} = 1$$

1.19 (b)

Using block diagram reduction, we get,

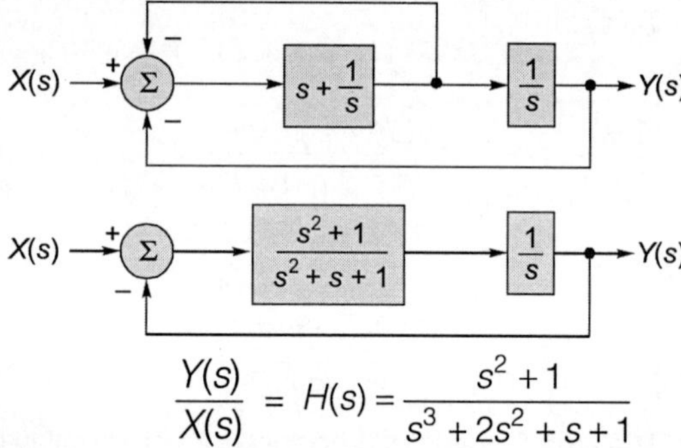

$$\frac{Y(s)}{X(s)} = H(s) = \frac{s^2+1}{s^3+2s^2+s+1}$$

1.20 (a)

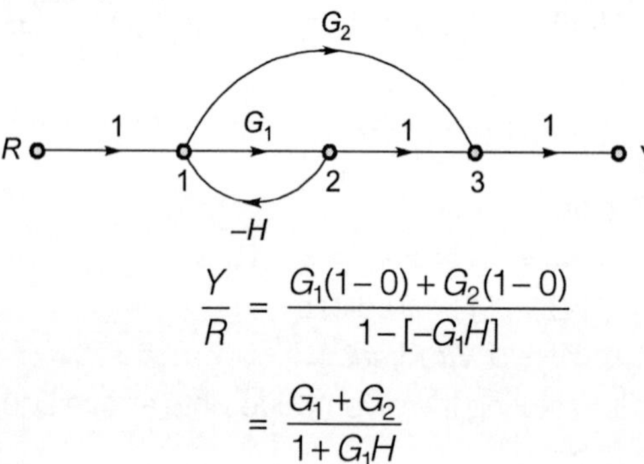

$$\frac{Y}{R} = \frac{G_1(1-0)+G_2(1-0)}{1-[-G_1H]}$$

$$= \frac{G_1+G_2}{1+G_1H}$$

1.21 (a)

Given, block diagram,

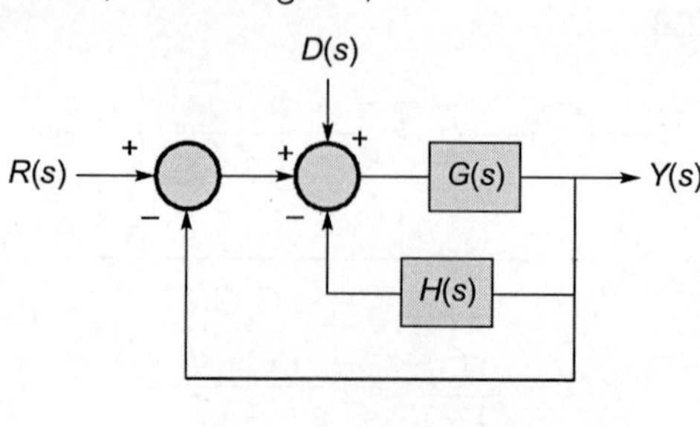

$$Y(s) = \underbrace{G_1(s)R(s)}_{Y_1(s)} + \underbrace{G_2(s)D(s)}_{Y_2(s)}$$

Considering first $R(s)$ only, then $Y(s)$ is $Y_1(s)$

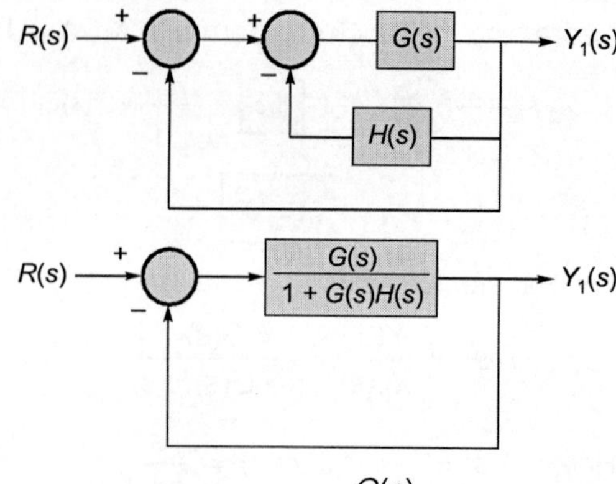

$$\frac{Y_1(s)}{R(s)} = \frac{\dfrac{G(s)}{1+G(s)H(s)}}{1+\dfrac{G(s)}{1+G(s)H(s)}}$$

$$\frac{Y_1(s)}{R(s)} = \frac{G(s)}{1+G(s)H(s)+G(s)}$$

$$Y_1(s) = \left[\frac{G(s)}{1+G(s)+G(s)H(s)}\right]R(s)$$

$$\therefore \quad G_1(s) = \frac{G(s)}{1+G(s)+G(s)H(s)}$$

Now considering $D(s)$ only, then $Y(s)$ is $Y_2(s)$

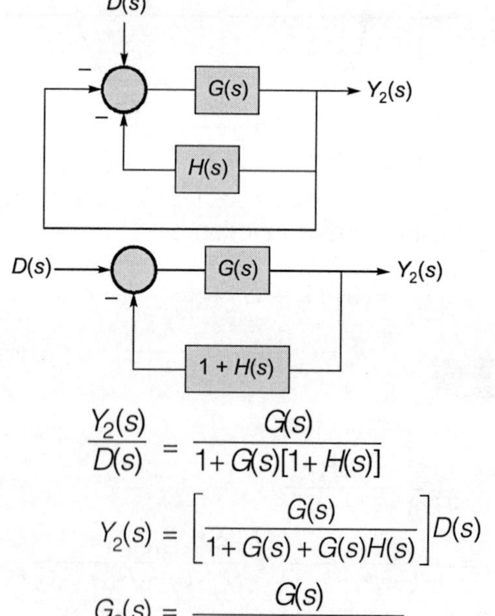

$$\frac{Y_2(s)}{D(s)} = \frac{G(s)}{1+G(s)[1+H(s)]}$$

$$Y_2(s) = \left[\frac{G(s)}{1+G(s)+G(s)H(s)}\right]D(s)$$

$$\therefore \quad G_2(s) = \frac{G(s)}{1+G(s)+G(s)H(s)}$$

Hence, $G_1(s)$ and $G_2(s)$ both are equal.

■■■

2 Compensators and Controllers

2.1 The transfer function of a tachometer is of the form

(a) Ks (b) $\frac{K}{s}$

(c) $\frac{K}{(s+1)}$ (d) $\frac{K}{s(s+1)}$

[1998 : 1 M]

2.2 The transfer function of a phase lead controller is $\frac{1+3Ts}{1+Ts}$. The maximum value of phase provided by this controller is

(a) 90° (b) 60°
(c) 45° (d) 30° [1998 : 1 M]

2.3 A PD controller is used to compensate a system. Compared to the uncompensated system, the compensated system has

(a) a higher type number
(b) reduced damping
(c) higher noise amplification
(d) larger transient overshoot [2003 : 1 M]

2.4 A double integrator plant, $G(s) = \frac{K}{s^2}$, $H(s) = 1$ is to be compensated to achieve the damping ratio $\zeta = 0.5$, and an undamped natural frequency, $\omega_n = 5$ rad/s. Which one of the following compensator $G_e(s)$ will be suitable?

(a) $\frac{s+3}{s+9.9}$ (b) $\frac{s+9.9}{s+3}$

(c) $\frac{s-6}{s+8.33}$ (d) $\frac{s+6}{s}$

[2005 : 2 M]

2.5 The transfer function of a phase-lead compensator is given by

$G_e(s) = \frac{1+3Ts}{1+Ts}$ where $T > 0$

The maximum phase-shift provided by such a compensator is

(a) $\pi/2$ (b) $\pi/3$
(c) $\pi/4$ (d) $\pi/6$

[2006 : 2 M]

2.6 A control system with a PD controller is shown in the figure. If the velocity error constant $K_V = 1000$ and the damping ratio $\zeta = 0.5$, then the values of K_P and K_D are

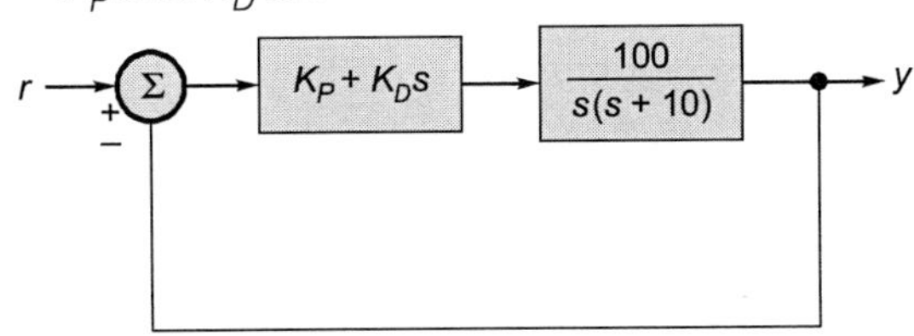

(a) $K_P = 100, K_D = 0.09$
(b) $K_P = 100, K_D = 0.9$
(c) $K_P = 10, K_D = 0.09$
(d) $K_P = 10, K_D = 0.9$

[2007 : 2 M]

2.7 The open-loop transfer function of a plant is given as $G(s) = \frac{1}{s^2-1}$. If the plant is operated in a unity feedback configuration, then the lead compensator that can stabilize this control system is

(a) $\frac{10(s-1)}{(s+2)}$ (b) $\frac{10(s+4)}{(s+2)}$

(c) $\frac{10(s+2)}{(s+10)}$ (d) $\frac{2(s+2)}{(s+10)}$

[2007 : 2 M]

2.8 Group I gives two possible choices for the impedance Z in the diagram. The circuit elements in Z satisfy the condition $R_2C_2 > R_1C_1$. The transfer function $\frac{V_0}{V_i}$ represents a kind of controller. Match the impedances in Group I with the types of controllers in Group II.

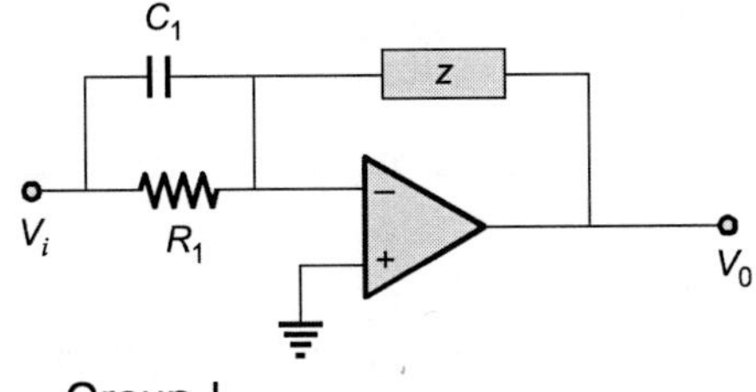

Group-I

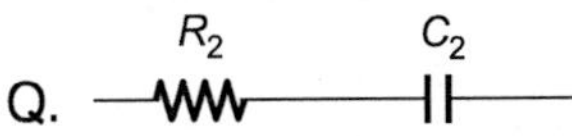

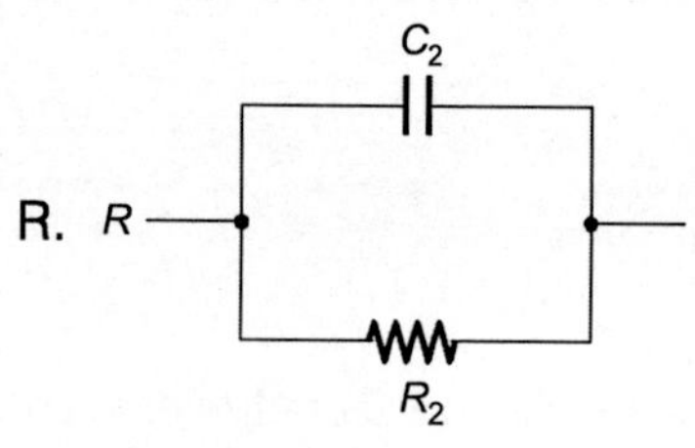

Group-II

1. PID controller
2. Lead compensator
3. Lag compensator

(a) Q–1, R–2 (b) Q–1, R–3

(c) Q–2, R–3 (d) Q–3, R–2

[2008 : 2 M]

2.9 The magnitude plot of a rational transfer function $G(s)$ with real coefficients is shown below. Which of the following compensators has such a magnitude plot?

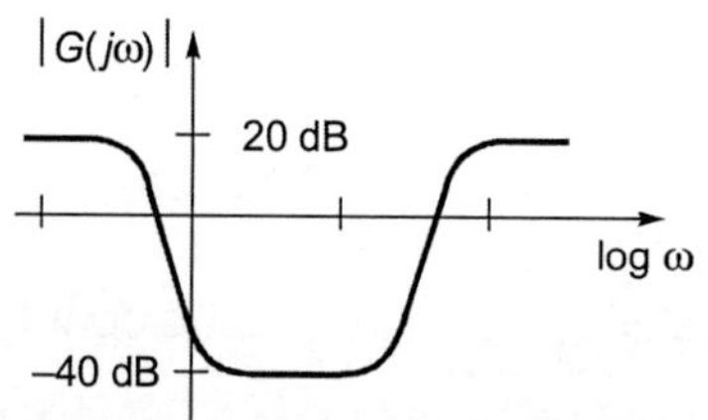

(a) Lead compensator (b) Lag compensator

(c) PID compensator (d) Lead-lag compensator

[2009 : 1 M]

2.10 A unity negative feedback closed loop system has a plant with the transfer function $G(s) = \frac{1}{s^2+2s+2}$ and a controller $G_c(s)$ in the feed forward path. For a unit step input, the transfer function of the controller that gives minimum steady state error is

(a) $G_c(s) = \frac{s+1}{s+2}$

(b) $G_c(s) = \frac{s+2}{s+1}$

(c) $G_c(s) = \frac{(s+1)(s+4)}{(s+2)(s+3)}$

(d) $G_c(s) = 1+\frac{2}{s}+3s$

[2010 : 2 M]

Linked Answer Questions

Statement for Linked Answer Questions 2.11 and 2.12:

The transfer function of a compensator is given as

$$G_c(s) = \frac{s+a}{s+b}$$

2.11 $G_c(s)$ is a lead compensator if

(a) $a = 1, b = 2$ (b) $a = 3, b = 2$

(c) $a = -3, b = -1$ (d) $a = 3, b = 1$

[2012 : 2 M]

2.12 The phase of the above lead compensator is maximum at

(a) $\sqrt{2}$ rad/s (b) $\sqrt{3}$ rad/s

(c) $\sqrt{6}$ rad/s (d) $1/\sqrt{3}$ rad/s

[2012 : 2 M]

2.13 For the following feedback system $G(s) = \frac{1}{(s+1)(s+2)}$. The 2%-settling time of the step response is required to be less than 2 seconds.

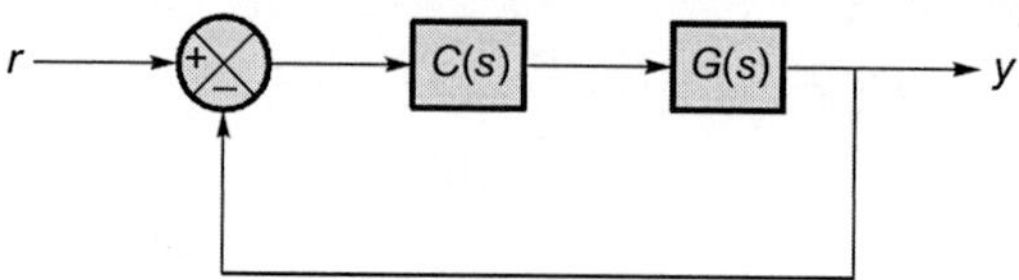

Which one of the following compensators $C(s)$ achieves this?

(a) $3\left(\frac{1}{s+5}\right)$ (b) $5\left(\frac{0.03}{s}+1\right)$

(c) $2(s+4)$ (d) $4\left(\frac{s+8}{s+3}\right)$

[2014 : 2 M, Set-1]

2.14 A lead compensator network includes a parallel combination of R and C in the feed-forward path. If the transfer function of the compensator is $G_c(s) = \frac{s+2}{s+4}$, the value of RC is________.

[2015 : 2 M, Set-1]

2.15 The transfer function of a first order controller is given as

$$G_C(s) = \frac{K(s+a)}{s+b}$$

where, K, a and b are positive real numbers. The condition for this controller to act as a phase lead compensator is

(a) $a < b$ (b) $a > b$

(c) $K < ab$ (d) $K > ab$

[2015 : 1 M, Set-3]

2.16 Which of the following can be the pole-zero configuration of a phase-lag controller (lag compensator)?

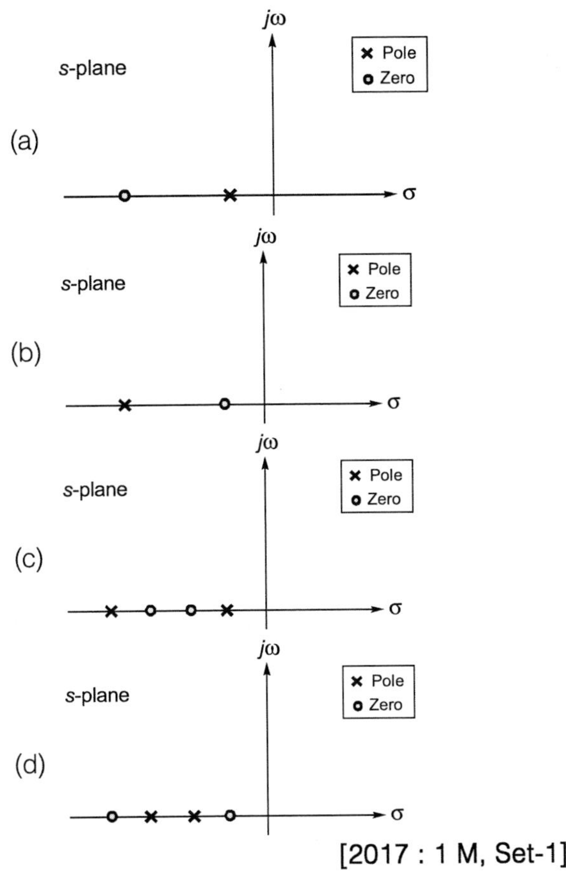

[2017 : 1 M, Set-1]

2.17 Which of the following statements is incorrect?

(a) Lead compensator is used to reduce the settling time.

(b) Lag compensator is used to reduce the steady state error.

(c) Lead compensator may increase the order of a system.

(d) Lag compensator always stabilizes an unstable system.

[2017 : 1 M, Set-2]

2.18 A unity feedback system that uses proportional-integral (PI) control is shown in the figure.

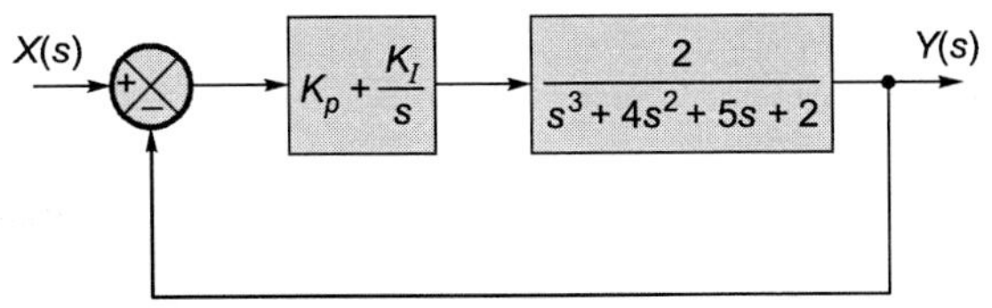

The stability of the overall system is controlled by tuning the PI control parameters K_p and K_I. The maximum value of K_I that can be chosen so as to keep the overall system stable or, in the worst case, marginally stable (rounded off to three decimal places) is ________.

[2021 : 2 M]

2.19 A satellite attitude control system, as shown below, has a plant with transfer function $G(s) = \frac{1}{s^2}$ cascaded with a compensator $C(s) = \frac{K(s+\alpha)}{s+4}$, where K and α are positive real constants.

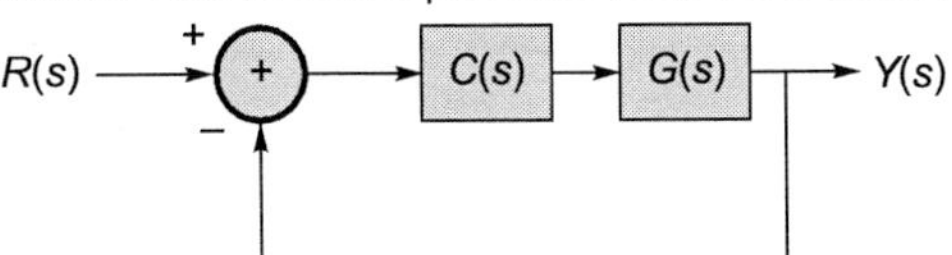

In order for the closed-loop system to have poles at $-1 \pm j\sqrt{3}$, the value of α must be ________.

(a) 0 (b) 3

(c) 1 (d) 2

[2024 : 2 Marks]

■■■

Answers Compensators and Controllers

2.1 (a)	2.2 (d)	2.3 (c)	2.4 (a)	2.5 (d)	2.6 (b)	2.7 (c)
2.8 (b)	2.9 (d)	2.10 (d)	2.11 (a)	2.12 (a)	2.13 (c)	2.14 (0.5)
2.15 (a)	2.16 (a)	2.17 (d)	2.18 (3.125)	2.19 (c)		

Explanations Compensators and Controllers

2.1 (a)

Tachometer is basically a differentiator. So, the transfer function of a tachometer is of the form Ks.

2.2 (d)

For lead compensator

$$\text{T.F.} = \frac{\alpha(1+T's)}{1+\alpha T's} = \frac{1+3Ts}{1+Ts}$$

So, $T' = 3T$

$$\alpha T' = T$$

$$\alpha 3T = T$$

$$\alpha = \frac{1}{3}$$

$$\phi_{max} = \sin^{-1}\left(\frac{1-\alpha}{1+\alpha}\right) = \sin^{-1}\left(\frac{1-1/3}{1+1/3}\right)$$

$$\phi_{max} = \sin^{-1}\left(\frac{2/3}{4/3}\right) = \sin^{-1}\left(\frac{1}{2}\right) = 30°$$

2.3 (c)

- A PD controller does not effect the type of the system.
- It reduces the overshoot and increases the damping.
- It increases the bandwidth therefore SNR decreases.
- System become more prone to noise

2.4 (a)

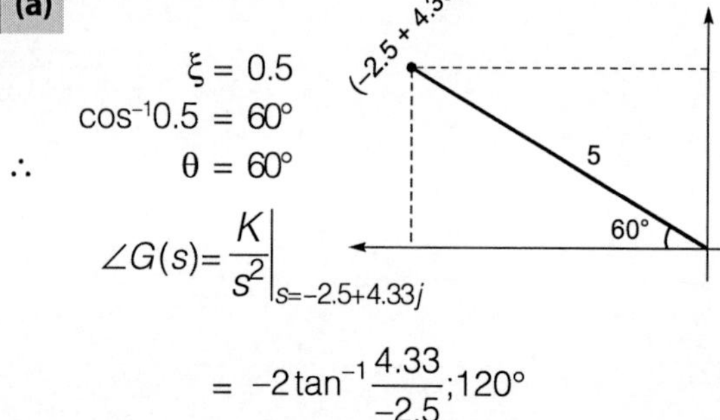

$$\xi = 0.5$$

$$\cos^{-1}0.5 = 60°$$

$\therefore$ $\theta = 60°$

$$\angle G(s) = \frac{K}{s^2}\bigg|_{s=-2.5+4.33j}$$

$$= -2\tan^{-1}\frac{4.33}{-2.5};120°$$

$\therefore$ For compensated system $\angle = 180 - 120$; 60°
(b) & (d) are lag network & for compensating lag network K/s^2, a lead network is required

$\therefore$ Putting s = −2.5 + 4.33j in (a) gives

$$\frac{K(s+3)}{s^2(s+9.9)} = \frac{0.5+j4.33}{7.4+j4.33} = 53°;60°$$

2.5 (d)

Maximum phase shift

$$\phi_m = \angle G_e(s)$$

$$\phi = \tan^{-1} 3\omega T - \tan^{-1}\omega T$$

For maximum phase shift

$$\frac{d\phi}{d\omega} = 0$$

$$\Rightarrow \frac{3T}{1+(3T\omega)^2} = \frac{T}{1+(T\omega)^2}$$

$$3[1+(T\omega)^2] = 1+(3T\omega)^2$$

$$3+3T^2\omega^2 = 1+9T^2\omega^2$$

$$2 = 6(\omega T)^2$$

$$(\omega T)^2 = \frac{1}{3}$$

$$\omega T = \frac{1}{\sqrt{3}}$$

$$\phi_{max} = \tan^{-1}3\times\frac{1}{\sqrt{3}} - \tan^{-1}\frac{1}{\sqrt{3}}$$

$$= \frac{\pi}{3} - \frac{\pi}{6} = \frac{\pi}{6}$$

2.6 (b)

$$K_v = \lim_{s\to 0} sG(s)H(s)$$

$$1000 = \lim_{s\to 0} s \times \frac{(K_P+K_D s)100}{s(s+10)}$$

$$\Rightarrow K_p = 100$$

Now characteristics eq. $(1 + G(s)H(s)) = 0$

$$1 + \frac{(K_P+K_D s)100}{s(s+10)} = 0$$

Putting $K_p = 100$

$$s^2 + 10s + 10^4 + 100K_D s = 0$$

$$s^2 + (10 + 100K_D)s + 10^4 = 0$$

Comparing with standard second order eq.

i.e. $s^2 + 2\xi\omega_n s + \omega_n^2 = 0$

So, $\omega_n = 100;\ 2\xi\omega_n = 10 + 100\,K_D$

Given, $\xi = 0.5$

$2 \times 0.5 \times 100 = 10 + 100\,K_D$

$K_D = 0.9$

2.7 (c)

Lead compensator is required for stability. Controllers given in option (a) and (b) are not lead compensator. From option (c) and (d), with option (d) system becomes unstable hence option (c) is correct.

2.8 (b)

$$\frac{V_i\,(R_1C_1s + 1)1}{R_1} = -\frac{V_0}{Z}$$

$$\Rightarrow \frac{V_0}{V_i} = -\frac{Z\,(R_1C_1s + 1)}{R_1}$$

In case of Q, $Z = \dfrac{R_2C_2s + 1}{C_2s + 1}$

In case of R, $Z = \dfrac{R_2}{R_2C_2s + 1}$

Considering Q,

$$\frac{V_0}{V_i} = -\frac{(R_1C_1s + 1)}{R_1}\cdot\frac{(R_2C_2s + 1)}{C_2s}$$

Considering R,

$$\frac{V_0}{V_i} = -\frac{(R_1C_1s + 1)}{R_1}\cdot\frac{R_2}{(R_2C_2s + 1)}$$

$\because$ Given that $R_2C_2 > R_1C_1$,

$\therefore$ Considering R, controller is lag compensator.

and Considering Q, Controller is PID controller.

2.10 (d)

Steady state error,

$$e_{ss} = \lim_{s\to 0}\frac{s\,R(s)}{1 + G(s)\,G_c(s)}$$

$$r(t) = u(t)$$

$$R(s) = \frac{1}{s}$$

$$e_{ss} = \lim_{s\to 0}\frac{s\,.\,\frac{1}{s}}{1 + G(s)\,G_c(s)}$$

$$e_{ss} = \lim_{s\to 0}\frac{1}{1 + G(s)\,G_c(s)}$$

Taking, $G_c(s) = \dfrac{s+1}{s+2},\ e_{ss} = \dfrac{2}{3}$

Taking, $G_c(s) = \dfrac{s+2}{s+1},\ e_{ss} = \dfrac{1}{3}$

Taking, $G_c(s) = \dfrac{(s+1)(s+4)}{(s+2)(s+3)},\ e_{ss} = \dfrac{3}{5}$

Taking, $G_c(s) = 1 + \dfrac{2}{s} + 3s,\ e_{ss} = 0$

2.11 (a)

$$G_c(s) = \left(\frac{s+a}{s+b}\right)$$

Phase $P = \tan^{-1}(\omega/a) - \tan^{-1}\omega/b$

for lead comparators phase must be +ve.

for this $\dfrac{\omega}{a} > \dfrac{\omega}{b} \Rightarrow a < b$.

So $a = 1$, $b = 2$.

2.12 (a)

For phase to be maximum,

$$\frac{\partial P}{\partial \omega} = 0$$

$$\frac{1/a}{1 + \frac{\omega^2}{a^2}} - \frac{1/b}{1 + \frac{\omega^2}{b^2}} = 0$$

$$\frac{1}{1 + \omega^2} - \frac{1/b}{1 + \frac{\omega^2}{4}} = 0$$

$$\frac{1}{1 + \omega^2} - \frac{2}{\omega^2 + 4} = 0$$

$$\omega^2 + 4 - 2 - 2\omega^2 = 0$$

$$\omega^2 = 2$$

$$\omega = \sqrt{2}\ \text{rad/sec}$$

2.13 (c)

Given open loop transfer function is

$$G(s) = \frac{1}{(s+1)(s+2)}$$

$$\therefore T(s) = \frac{G(s)}{1 + G(s)} = \frac{1}{(s+1)(s+2) + 1}$$

$$T(s) = \frac{1}{s^2 + 3s + 3} \qquad \text{...(i)}$$

Comparing equation (i) with standard transfer function

$$\xi\omega_n = \frac{3}{2} = 1.5$$

$\therefore$ 2% settling time

$$\tau_s = \frac{4}{\xi\omega_n} = \frac{4}{1.5} = 2.67 > 2 \text{ sec}$$

Thus, in order to make settling time (τ_s) less than 2 sec, PD controller should be used. Hence, option (c) is the correct choice.

Where, $C(s) = 2(s+4)$

$\therefore$ New transfer function = $T'(G)$

$$= \frac{C(s)G(s)}{1+C(s)G(s)} = \frac{2(s+4)}{s^2+3s+2+2s+8}$$

or $$T'(s) = \frac{2(s+4)}{s^2+5s+10}$$

$$\therefore \quad \tau_s = \frac{4}{\xi\omega_n} = \frac{4}{5/2} = \frac{4}{2.5} = 1.6$$

2.14 (0.5)

$$G_C(s) = \frac{s+2}{s+4} \quad \text{...(i)}$$

For lead compensator,

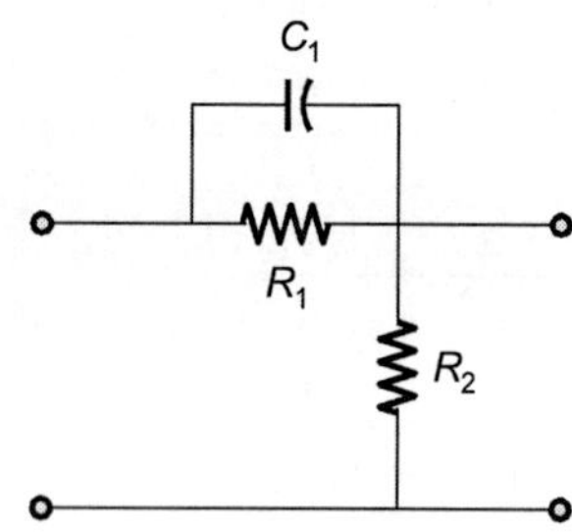

$$\text{Transfer function} = \frac{1+s\tau}{1+\alpha s\tau} \quad \text{...(ii)}$$

Where,

τ = Lead time constant = $R_1 C$

and $\alpha = \dfrac{R_2}{R_1+R_2}$

Comparing equation (i) and (ii), we get

$$\tau = \frac{1}{2} \quad \text{and} \quad \alpha\tau = \frac{1}{4}$$

or, $$\alpha = \frac{1}{2}$$

$\therefore$ RC time constant = 0.5

2.15 (a)

For phase lead compensator

$$G_c(s) = \frac{(1+s\tau)}{(1+\alpha s\tau)} \quad ; \quad \alpha < 1$$

Here, $\tau = \dfrac{1}{a}$

and $\alpha\tau = \dfrac{1}{b}$

or, $\alpha = \dfrac{a}{b} < 1$

or, $a < b$

2.16 (a)

Phase lag controller transfer function is

$$G_C(s) = \frac{s+Z}{s+P};\ |Z| > |P|$$

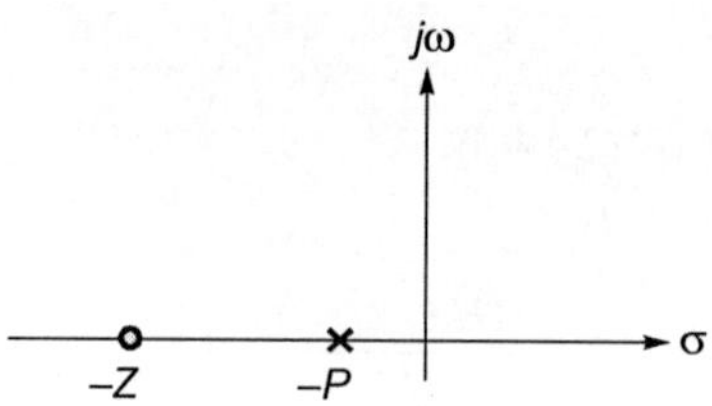

2.17 (d)

In case of high type systems Lag compensator fails to give stability.

2.18 (3.125)

$$GH = \left(\frac{sK_p+K_I}{s}\right)\left(\frac{2}{s^3+4s^2+5s+2}\right)$$

$$q(s) = s^4 + 4s^3 + 5s^2 + s(2+2k_p) + 2k_I$$

Necessary: $K_p > -1;\ K_I > 0$

s^4	1	5	$2K_I$
s^3	4	$2+2K_p$	
s^2	$\frac{18-2K_p}{4}$	$2K_I$	
s^1	$(9-K_p)(1+K_p)-8K_I$		
s^0	$2K_I$		

Sufficient:

$$\frac{18-2K_p}{4} > 0$$

$$\Rightarrow \quad K_p < 9$$

$$\therefore \quad -1 < K_p < 9$$

$$(18-2K_p)(2+2K_p) - 32K_I > 0$$

$$32K_I < 36 + 32K_p - 4K_p^2$$

$$\therefore \quad 0 < K_I < \frac{36+32K_p-4K_p^2}{32}$$

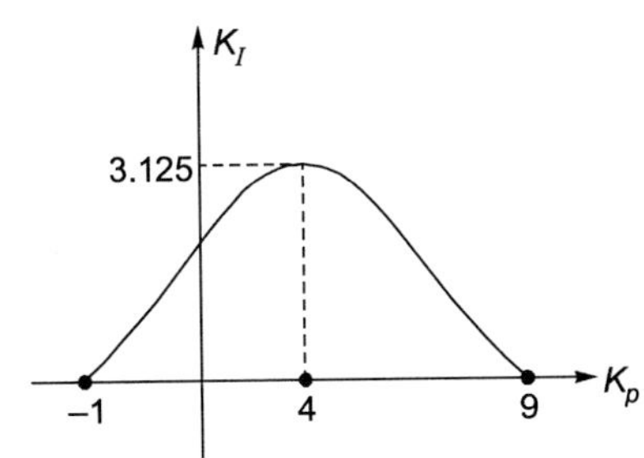

If $K_p = -1 \Rightarrow k_I = 0$
If $K_p = 9 \Rightarrow k_I = 0$

$$\therefore \quad \frac{dK_I}{dK_p} = 0$$

$$\Rightarrow \quad 32 - 8K_p = 0 = 0 \Rightarrow K_p = 4$$

$\therefore$ For $K_p = 4$, K_I is maximum, which is

$$K_I = \frac{36 + 32 \times 4 - 64}{32} = 3.125$$

For $K_p = 4$, $K_I < 3.125$ for stability

$$\therefore \quad K_{I\max} = 3.125$$

2.19 (c)

The open-loop transfer function of the system is given by

$$G(s) \cdot C(s) = \frac{K(s+\alpha)}{s^2(s+4)}$$

The characteristic equation of the system is

$$1 + G(s)\,C(s) = 0$$

$$\Rightarrow \quad 1 + \frac{K(s+\alpha)}{s^2(s+4)} = 0$$

$$\Rightarrow \quad s^3 + 4s^2 + Ks + K\alpha = 0 \quad \text{...(i)}$$

If the two poles of the closed-loop system are located at $s = -1 \pm j\sqrt{3}$ and assuming the third pole at $s = -a$, the characteristic equation of the system can be written as

$$(s+a)\left[(s+1)^2 + (\sqrt{3})^2\right] = 0$$

$$\Rightarrow \quad (s + a)\,[s^2 + 2s + 4] = 0$$

$$\Rightarrow s^3 + (2 + a)s^2 + (4 + 2a)s + 4a = 0 \quad \text{...(ii)}$$

Comparing equation (i) and (ii), we get,

$$2 + a = 4 \quad \Rightarrow a = 2$$

$$4 + 2a = K \quad \Rightarrow K = 8$$

$$K\alpha = 4a \quad \Rightarrow \alpha = \frac{4a}{K} = \frac{4 \times 2}{8} = 1$$

■■■

3 Time Response Analysis

3.1 For a second order system, damping ratio (ξ), is $0 < \xi < 1$, then the roots of the characteristic polynomial are
(a) real but not equal
(b) real and equal
(c) complex conjugates
(d) imaginary

[1995 : 1 M]

3.2 If $L[f(t)] = \dfrac{2(s+1)}{s^2+2s+5}$ then $f(0^+)$ and $f(\infty)$ are given by
(a) 0, 2 respectively (b) 2, 0 respectively
(c) 0, 1 respectively (d) 2/5, 0 respectively.
[**Note** : 'L' stands for 'Laplace Transform of']

[1995 : 1 M]

3.3 The step error coefficient of a system $G(s) = \dfrac{1}{(s+6)(s+1)}$ with unity feedback is
(a) 1/6 (b) ∞
(c) 0 (d) 1 [1995 : 1 M]

3.4 The final value theorem is used to find the
(a) steady state value of the system output
(b) initial value of the system output
(c) transient behavior of the system output
(d) none of these.

[1995 : 1 M]

3.5 If $F(s) = \dfrac{\omega}{s^2+\omega^2}$, then the value of $\lim_{t\to\infty} f(t)$, {where $F(s)$ is the $L[f(t)]$}
(a) cannot be determined
(b) is zero
(c) is unity
(d) is infinite [1998 : 1 M]

3.6 Consider a feedback control system with loop transfer function

$$G(s)H(s) = \frac{K(1+0.5s)}{s(1+s)(1+2s)}$$

The type of the closed loop system is
(a) zero (b) one
(c) two (d) three [1998 : 1 M]

3.7 Consider a unity feedback control system with open-loop transfer function $G(s) = \dfrac{K}{s(s+1)}$ the steady state error of the system due to a unit step input is
(a) zero (b) K
(c) $1/K$ (d) infinite [1998 : 1 M]

3.8 The unit impulse response of a linear time invariant system is the unit step function $u(t)$. For $t > 0$, the response of the system to an excitation $e^{-at}u(t)$, $a > 0$ will be
(a) ae^{-at} (b) $(1/a)(1-e^{-at})$
(c) $a(1-e^{-at})$ (d) $1-e^{-at}$

[1998 : 2 M]

3.9 For a second-order system with the closed-loop transfer function

$T(s) = \dfrac{9}{s^2+4s+9}$ the settling time for 2-percent band, in seconds, is
(a) 1.5 (b) 2.0
(c) 3.0 (d) 4.0 [1999 : 1 M]

3.10 If the closed-loop transfer function $T(s)$ of a unity negative feedback system is given by

$$T(s) = \frac{a_{n-1}s + a_n}{s^n + a_1 s^{n-1} + \ldots + a_{n-1}s + a_n}$$

then the steady state error for a unit ramp input is
(a) $\dfrac{a_n}{a_{n-1}}$ (b) $\dfrac{a_n}{a_{n-2}}$
(c) $\dfrac{a_{n-2}}{a_{n-1}}$ (d) zero [1999 : 2 M]

3.11 If the characteristic equation of a closed-loop system is $s^2 + 2s + 2 = 0$, then the system is
(a) overdamped (b) critically damped
(c) underdamped (d) undamped

[2001 : 1 M]

3.12 The open-loop DC gain of a unity negative feedback system with closed-loop transfer function $\dfrac{s+4}{s^2+7s+13}$ is

(a) $\frac{4}{13}$ (b) $\frac{4}{9}$

(c) 4 (d) 13

[2001 : 2 M]

3.13 Consider a system with the transfer function $G(s) = \frac{s+6}{Ks^2 + s + 6}$. Its damping ratio will be 0.5 when the value of K is

(a) 2/6 (b) 3

(c) 1/6 (d) 6

[2002 : 1 M]

3.14 The transfer function of a system is $G(s) = \frac{100}{(s+1)(s+100)}$. For a unit-step input to the system the approximate settling time for 2% criterion is

(a) 100 sec (b) 4 sec

(c) 1 sec (d) 0.01 sec

[2002 : 2 M]

3.15 A second-order system has the transfer function

$$\frac{C(s)}{R(s)} = \frac{4}{s^2 + 4s + 4}$$. With $r(t)$ as the unit-step function, the response $c(t)$ of the system is represented by

(a)
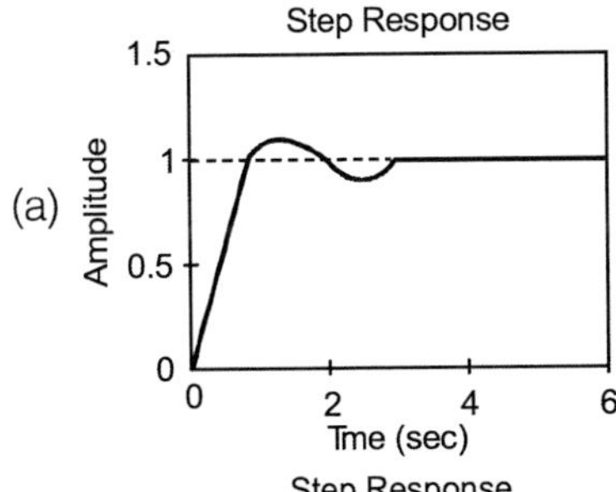

(b)
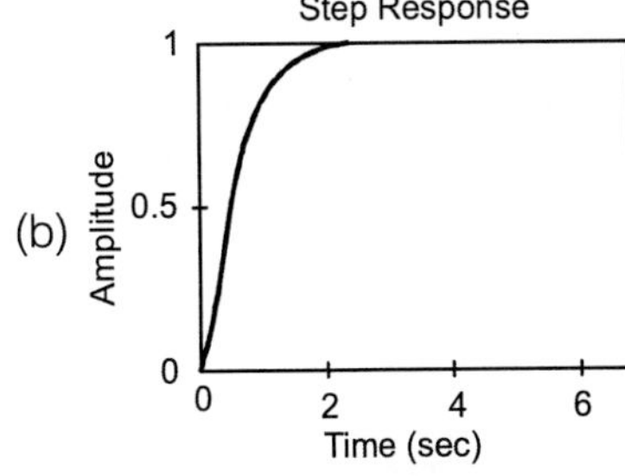

(c)
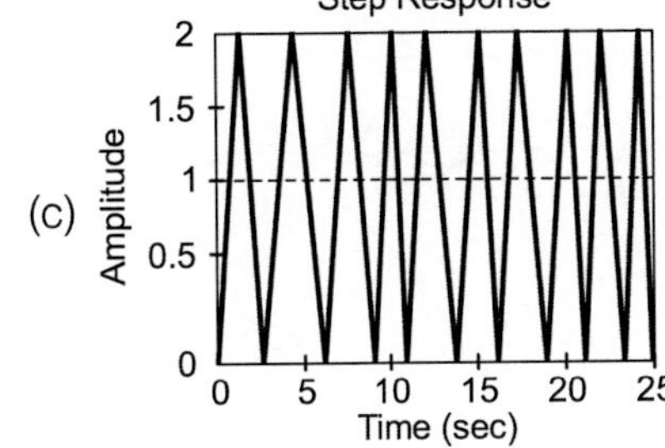

(d)
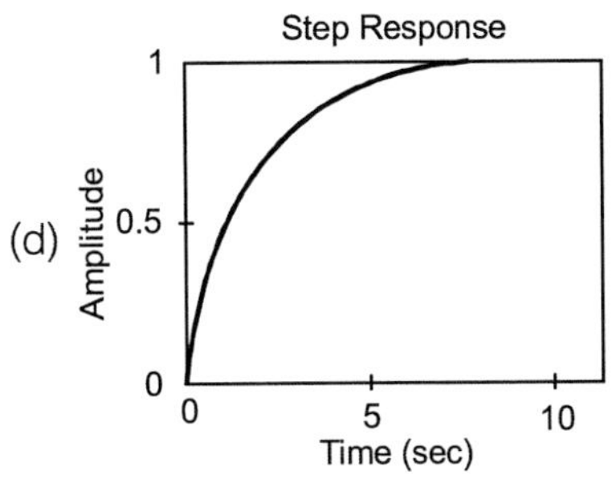

[2003 : 2 M]

3.16 A causal system having the transfer function $H(s) = \frac{1}{s+2}$ is excited with $10u(t)$. The time at which the output reaches 99% of its steady state value is

(a) 2.7 sec (b) 2.5 sec

(c) 2.3 sec (d) 2.1 sec

[2004 : 2 M]

3.17 A system described by the following differential equation $\frac{d^2y}{dt^2} + 3\frac{dy}{dt} + 2y = x(t)$ is initially at rest. For input $x(t) = 2u(t)$, the output $y(t)$ is

(a) $(1 - 2e^{-t} + e^{-2t})u(t)$

(b) $(1 + 2e^{-t} - 2e^{-2t})u(t)$

(c) $(0.5 + e^{-t} + 1.5e^{-2t})u(t)$

(d) $(0.5 + 2e^{-t} + 2e^{-2t})u(t)$

[2004 : 2 M]

3.18 Despite the presence of negative feedback, control systems still have problems of instability because the

(a) components used have nonlinearities

(b) dynamic equations of the systems are not known exactly

(c) mathematical analysis involves approximations

(d) system has large negative phase angle at high frequencies

[2005 : 1 M]

3.19 In the derivation of expression for peak percent overshoot, $M_p = \exp\left(\frac{-\pi\xi}{\sqrt{1-\xi^2}}\right) \times 100$ which one of the following conditions is NOT required?

(a) System is linear and time invariant

(b) The system transfer function has a pair of complex conjugate poles and no zeroes

(c) There is no transportation delay in the system

(d) The system has zero initial conditions

[2005 : 2 M]

3.20 A ramp input applied to an unity feedback system results in 5% steady state error. The type number and zero frequency gain of the system are respectively

(a) 1 and 20 (b) 0 and 20
(c) 0 and 1/20 (d) 1 and 1/20

[2005 : 2 M]

3.21 The unit-step response of a system starting from rest is given by

$c(t) = 1 - e^{-2t}$ for $t \geq 0$

The transfer function of the system is

(a) $\dfrac{1}{1+2s}$ (b) $\dfrac{2}{2+s}$

(c) $\dfrac{1}{2+s}$ (d) $\dfrac{2s}{1+2s}$ [2006 : 2 M]

3.22 The unit impulse response of a system is $h(t) = e^{-t}$, $t \geq 0$

For this system, the steady-state value of the output for unit step input is equal to

(a) –1 (b) 0
(c) 1 (d) ∞ [2006 : 2 M]

3.23 The transfer function of a plant is

$T(s) = \dfrac{5}{(s+5)(s^2+s+1)}$. The second-order approximation of $T(s)$ using dominant pole concept is

(a) $\dfrac{1}{(s+5)(s+1)}$ (b) $\dfrac{5}{(s+5)(s+1)}$

(c) $\dfrac{5}{s^2+s+1}$ (d) $\dfrac{1}{s^2+s+1}$

[2007 : 2 M]

3.24 Step responses of a set of three second-order underdamped systems all have the same percentage overshoot. Which of the following diagrams represents the poles of three systems?

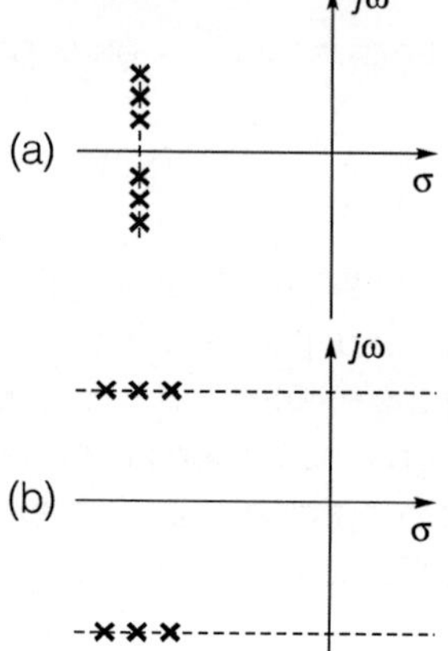

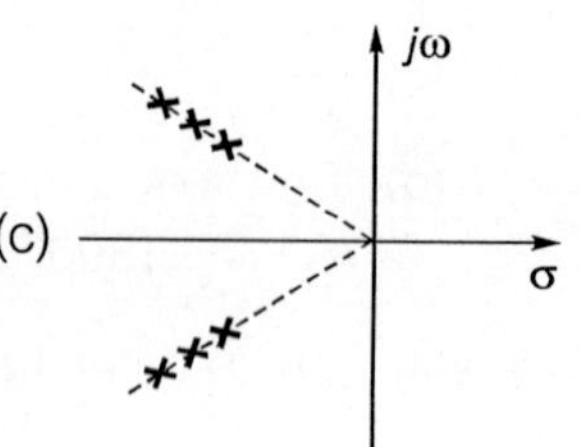

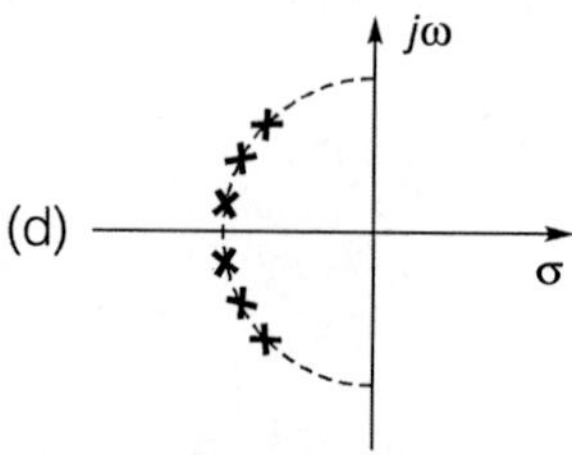

[2008 : 1 M]

3.25 A linear, time-invariant, causal continuous time system has a rational transfer function with simple poles at $s = -2$ and $s = -4$, and one simple zero at $s = -1$. A unit step $u(t)$ is applied at the input of the system. At steady state, the output has constant value of 1. The impulse response of this system is

(a) $[\exp(-2t) + \exp(-4t)]\, u(t)$
(b) $[-4\exp(-2t) + 12\exp(-4t) - \exp(-t)]\, u(t)$
(c) $[-4\exp(-2t) + 12\exp(-4t)]\, u(t)$
(d) $[-0.5\exp(-2t) + 1.5\exp(-4t)]\, u(t)$

[2008 : 2 M]

3.26 **Group I** lists a set of four transfer functions. **Group II** gives a list of possible step responses $y(t)$. Match the step responses with the corresponding transfer functions.

Group I

$P = \dfrac{25}{s^2+25}$; $Q = \dfrac{36}{s^2+20s+36}$

$R = \dfrac{36}{s^2+12s+36}$; $S = \dfrac{36}{s^2+7s+49}$

Group II

1.
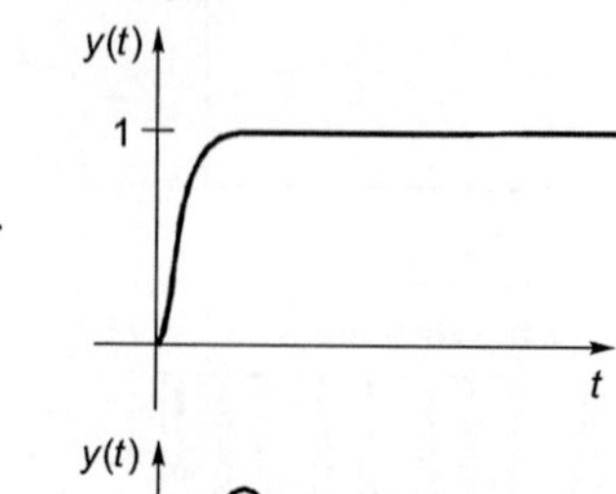

2.
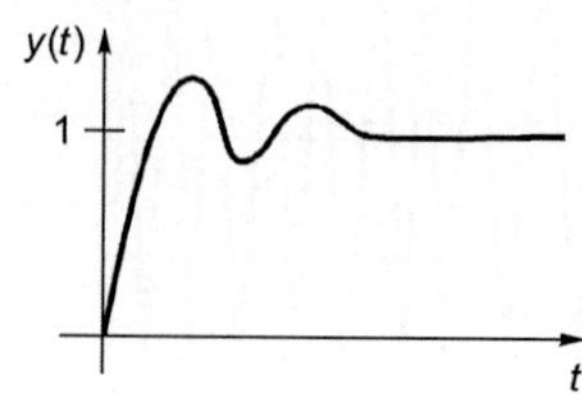

3.

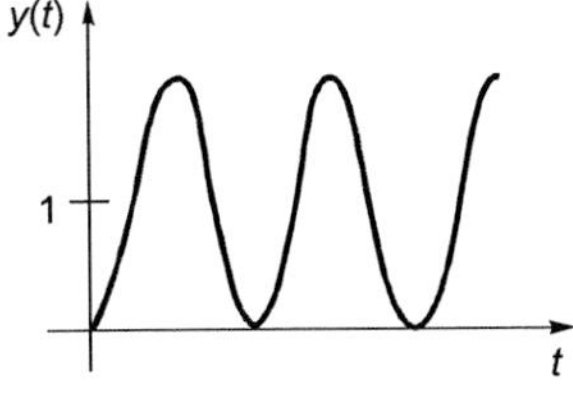

4.

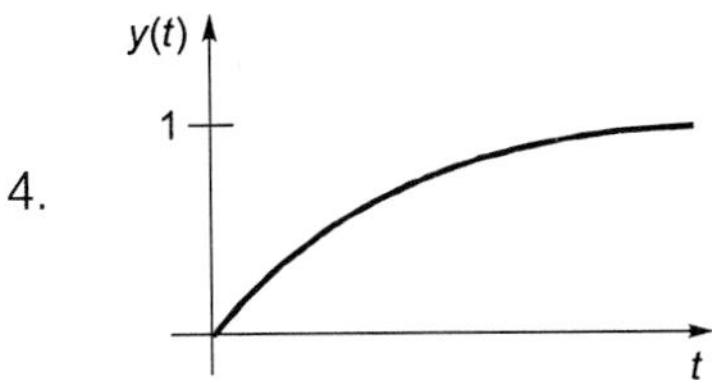

(a) P–3, Q–1, R–4, S–2
(b) P–3, Q–2, R–4, S–1
(c) P–2, Q–1, R–4, S–3
(d) P–3, Q–4, R–1, S–2 [2008 : 2 M]

3.27 The unit step response of an under-damped second order system has steady state value of –2. Which one of the following transfer functions has these properties?

(a) $\dfrac{-2.24}{s^2+2.59s+1.12}$ (b) $\dfrac{-3.82}{s^2+1.91s+1.91}$

(c) $\dfrac{-2.24}{s^2-2.59s+1.12}$ (d) $\dfrac{-3.82}{s^2-1.91s+1.91}$

[2009 : 2 M]

3.28 The differential equation $100\dfrac{d^2y}{dt^2}-20\dfrac{dy}{dt}+y=x(t)$ describes a system with an input $x(t)$ and an output $y(t)$. The system, which is initially relaxed, is excited by a unit step input. The output $y(t)$ can be represented by the waveform

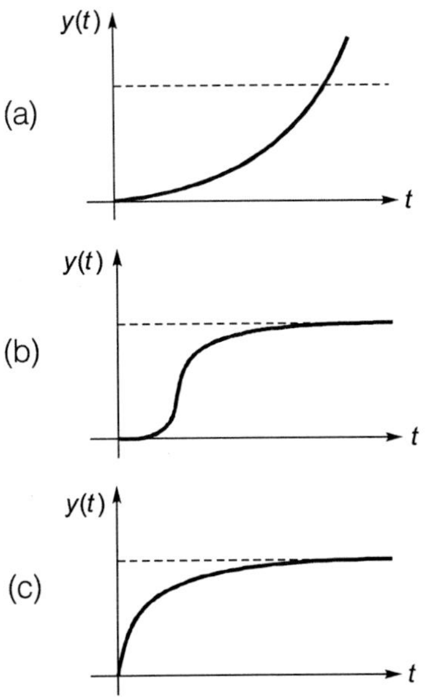

(d)

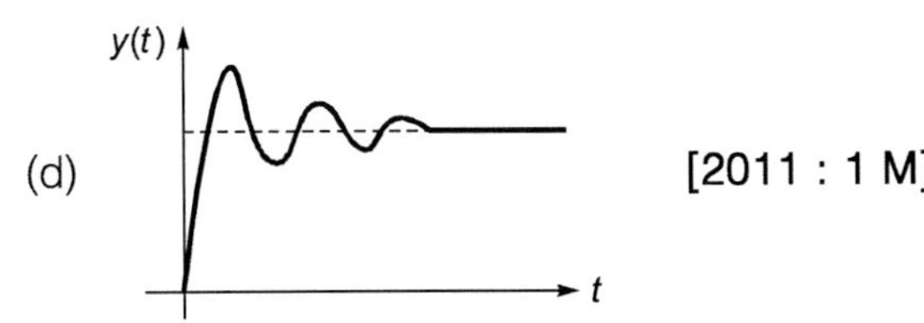

[2011 : 1 M]

3.29 A system described by a linear, constant coefficient, ordinary, first order differential equation has an exact solution given by $y(t)$ for $t > 0$, when the forcing function is $x(t)$ and the initial condition is $y(0)$. If one wishes to modify the system so that the solution becomes $-2y(t)$ for $t > 0$, we need to

(a) change the initial condition to $-y(0)$ and the forcing function to $2x(t)$
(b) change the initial condition to $2y(0)$ and the forcing function to $-x(t)$
(c) change the initial condition to $j\sqrt{2}\,y(0)$ and the forcing function to $j\sqrt{2}\,x(t)$
(d) change the initial condition to $-2y(0)$ and the forcing function to $-2x(t)$ [2013 : 2 M]

3.30 The open-loop transfer function of a dc motor is given as $\dfrac{\omega(s)}{V_a(s)}=\dfrac{10}{1+10s}$. When connected in feedback as shown below, the approximate value of K_a that will reduce the time constant of the closed loop system by one hundred times as compared to that of the open-loop system is

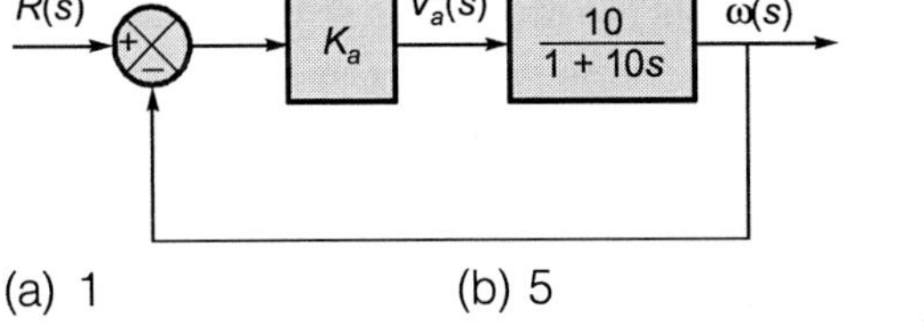

(a) 1 (b) 5
(c) 10 (d) 100 [2013 : 2 M]

3.31 The natural frequency of an undamped second-order system is 40 rad/s. If the system is damped with a damping ratio 0.3, the damped natural frequency in rad/s is ______.

[2014 : 1 M, Set-2]

3.32 The steady state error of the system shown in the figure for a unit step input is ______.

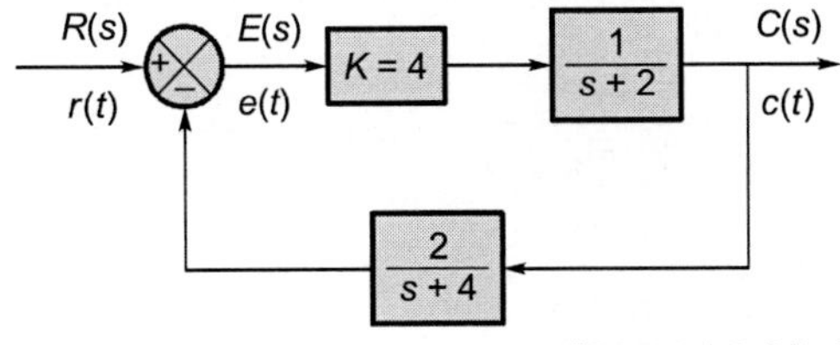

[2014 : 2 M, Set-3]

3.33 For the second order closed-loop system shown in the figure, the natural frequency (in rad/s) is

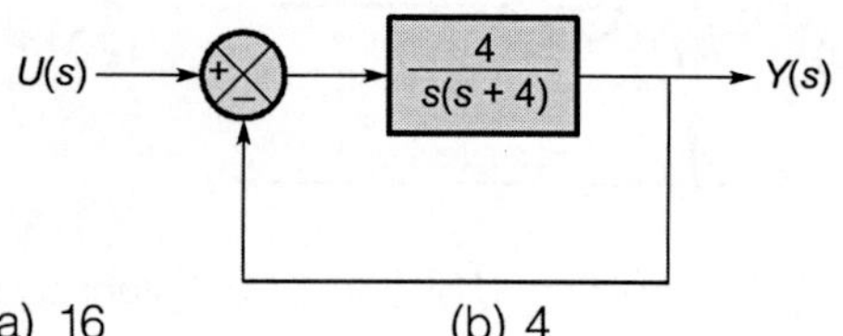

(a) 16 (b) 4
(c) 2 (d) 1

[2014 : 1 M, Set-4]

3.34 The characteristic equation of a unity negative feedback system is $1 + KG(s) = 0$. The open loop transfer function $G(s)$ has one pole at 0 and two poles at -1. The root locus of the system for varying K is shown in the figure.

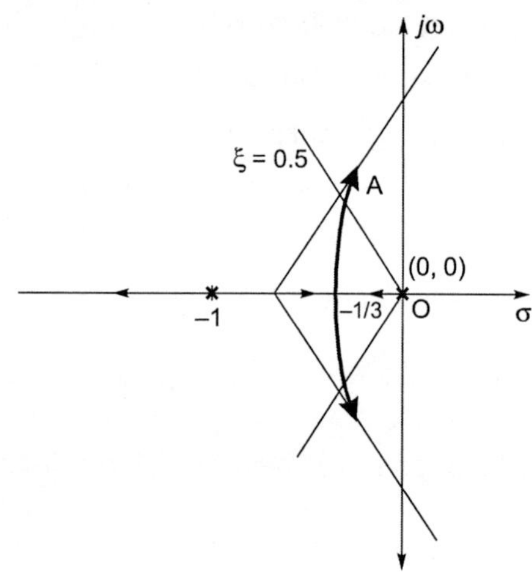

The constant damping ratio line, for $\xi = 0.5$, intersects the root locus at point A. The distance from the origin to point A is given as 0.5. The value of K at point A is ______.

[2014 : 2 M, Set-4]

3.35 A unity negative feedback system has an open-loop transfer function $G(s) = \frac{K}{s(s+10)}$. The gain K for the system to have a damping ratio of 0.25 is _____.

[2015 : 1 M, Set-2]

3.36 The output of a standard second-order system for a unit step input is given as $y(t) = 1 - \frac{2}{\sqrt{3}} e^{-t} \cos\left(\sqrt{3}t - \frac{\pi}{6}\right)$. The transfer function of the system is

(a) $\frac{2}{(s+2)(s+\sqrt{3})}$ (b) $\frac{1}{s^2+2s+1}$

(c) $\frac{3}{s^2+2s+3}$ (d) $\frac{4}{s^2+2s+4}$

[2015 : 2 M, Set-2]

3.37 The open-loop transfer function of unity-feedback control system is given by

$$G(s) = \frac{K}{s(s+2)}$$

For the peak overshoot of the closed-loop system to a unit step input to be 10%, the value of K is ___.

[2016 : 2 M, Set-1]

3.38 For the unity feedback control system shown in the figure, the open-loop transfer function $G(s)$ is given as

$$G(s) = \frac{2}{s(s+1)}$$

The steady state error e_{ss} due to a unit step input is

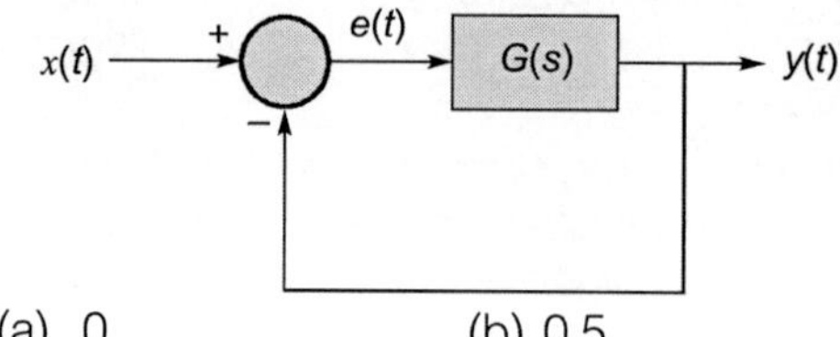

(a) 0 (b) 0.5
(c) 1.0 (d) ∞

[2016 : 1 M, Set-3]

3.39 The response of the system $G(s) = \frac{s-2}{(s+1)(s+3)}$ to the unit step input $u(t)$ is $y(t)$. The value of $\frac{dy}{dt}$ at $t = 0^+$ is ____

[2016 : 1 M, Set-2]

3.40 In the feedback system shown below

$$G(s) = \frac{1}{(s^2+2s)}.$$

The step response of the closed-loop system should have minimum settling time and have no overshoot.

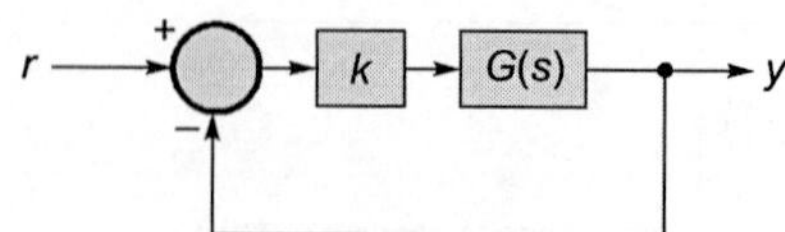

The required value of gain k to achieve this is _______.

[2016 : 2 M, Set-2]

3.41 The open loop transfer function

$$G(s) = \frac{(s+1)}{s^p(s+2)(s+3)}$$

where p is an integer, is connected in unity feedback configuration as shown in the figure.

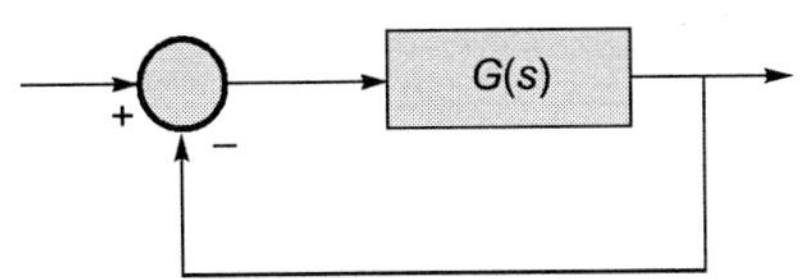

Given that the steady state error is zero for unit step input and is 6 for unit ramp input, the value of the parameter p is ________.

[2017 : 1 M, Set-1]

3.42 Consider a causal second-order system with the transfer function

$$G(s) = \frac{1}{1 + 2s + s^2}$$

with a unit-step $R(s) = \frac{1}{s}$ as an input. Let $C(s)$ be the corresponding output. The time taken by the system output $c(t)$ to reach 94% of its steady-state value $\lim_{t \to \infty} c(t)$, rounded off to two decimal places, is

(a) 5.25
(b) 4.50
(c) 2.81
(d) 3.89

[2019 : 2 M]

3.43 Consider the following closed loop control system

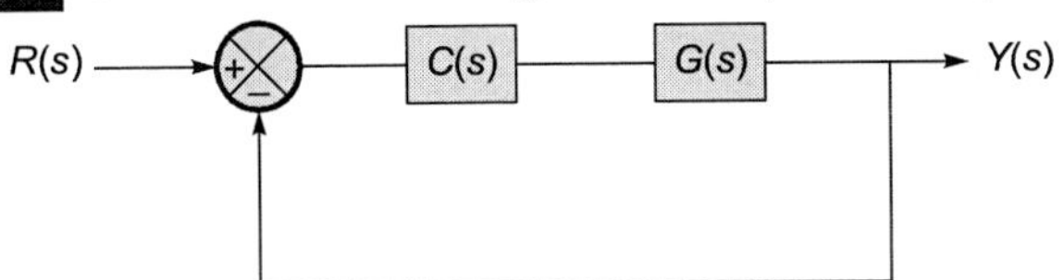

where $G(s) = \frac{1}{s(s+1)}$ and $C(s) = K\frac{s+1}{s+3}$. If the steady state error for a unit ramp input is 0.1, then the value of K is ______.

[2020 : 2 M]

3.44 Two linear time-invariant systems with transfer functions

$$G_1(s) = \frac{10}{s^2 + s + 1} \text{ and } G_2(s) = \frac{10}{s^2 + s\sqrt{10} + 10}$$

have unit step responses $y_1(t)$ and $y_2(t)$, respectively. Which of the following statements is/are true?

(a) $y_1(t)$ and $y_2(t)$ have the same percentage peak overshoot.
(b) $y_1(t)$ and $y_2(t)$ have the same steady-state value.
(c) $y_1(t)$ and $y_2(t)$ have the same damped frequency of oscillation.
(d) $y_1(t)$ and $y_2(t)$ have the same 2% settling time.

[2022 : 2 M]

3.45 The block diagram of a closed-loop control system is shown in the figure. $R(s)$, $Y(s)$, and $D(s)$ are the Laplace transforms of the time-domain signals $r(t)$, $y(t)$, and $d(t)$, respectively. Let the error signal be defined as $e(t) = r(t) - y(t)$. Assuming the reference input $r(t) = 0$ for all t, the steady-state error $e(\infty)$, due to a unit step disturbance $d(t)$, is ________ (rounded off to two decimal places).

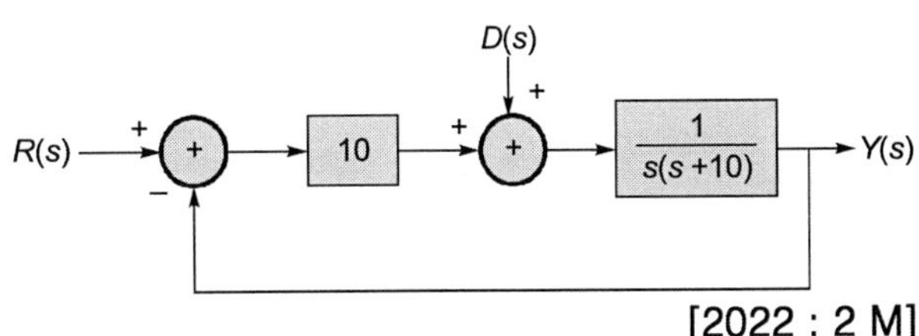

[2022 : 2 M]

3.46 A closed loop system is shown in the figure where $k > 0$ and $\alpha > 0$. The steady state error due to a ramp input ($R(s) = \alpha/s^2$) is given by

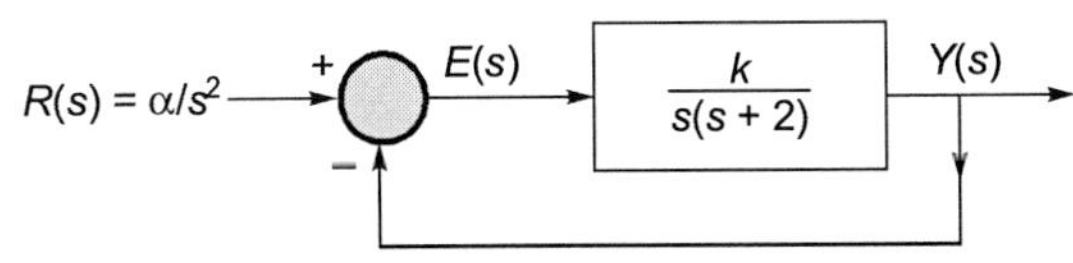

(a) $\frac{2\alpha}{k}$
(b) $\frac{\alpha}{k}$
(c) $\frac{\alpha}{2k}$
(d) $\frac{\alpha}{4k}$

[2023 : 2 M]

3.47 In the feedback control system shown in the figure below $G(s) = \frac{6}{s(s+1)(s+2)}$.

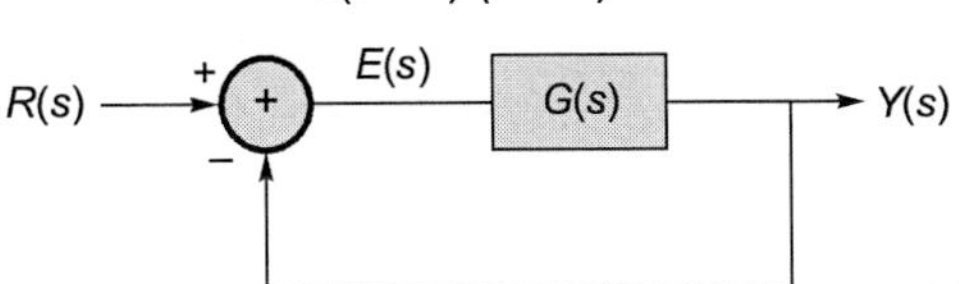

$R(s)$, $Y(s)$, and $E(s)$ are the Laplace transforms of $r(t)$, $y(t)$, and $e(t)$, respectively. If the input $r(t)$ is a unit step function, then ______.

(a) $\lim_{t \to \infty} e(t) = 0$
(b) $\lim_{t \to \infty} e(t)$ does not exist, $e(t)$ is oscillatory
(c) $\lim_{t \to \infty} e(t) = \frac{1}{3}$
(d) $\lim_{t \to \infty} e(t) = \frac{1}{4}$

[2024 : 1 Mark]

■■■

Answers Time Response Analysis

3.1 (c)	3.2 (b)	3.3 (a)	3.4 (a)	3.5 (a)	3.6 (b)	3.7 (a)
3.8 (b)	3.9 (b)	3.10 (d)	3.11 (c)	3.12 (b)	3.13 (c)	3.14 (b)
3.15 (b)	3.16 (c)	3.17 (a)	3.18 (d)	3.19 (c)	3.20 (a)	3.21 (b)
3.22 (c)	3.23 (d)	3.24 (c)	3.25 (c)	3.26 (d)	3.27 (b)	3.28 (a)
3.29 (d)	3.30 (c)	3.31 (38.16)	3.32 (0)	3.33 (c)	3.34 (0.37)	3.35 (400)
3.36 (d)	3.37 (2.8)	3.38 (a)	3.39 (1)	3.40 (1)	3.41 (1)	3.42 (b)
3.43 (30)	3.44 (a)	3.45 (–0.10)	3.46 (a)	3.47 (b)		

Explanations Time Response Analysis

3.1 (c)

$\xi < 1$

For underdamped system, roots or poles are:

$$s = -\xi\omega_n \pm j\omega_n\sqrt{1-\xi^2}$$

3.2 (b)

$$f(0^+) = \lim_{s\to\infty} sF(s) = \frac{2s^2+2s}{s^2+2s+5} = 2$$

$$f(\infty) = \lim_{s\to 0} sF(s) = \frac{2s^2+2s}{s^2+2s+5} = 0$$

3.3 (a)

$$K_p = \lim_{s\to 0} G(s)H(s)$$

$$= \lim_{s\to 0}\frac{1}{(s+6)(s+1)} = \frac{1}{6}$$

3.4 (a)

The final value theorem is used to find the steady state value of the system.

3.5 (a)

$s = \pm j\omega$

So, poles are pure imaginary, hence final value can not be determined.

3.6 (b)

Number of poles at origin of open loop transfer function gives the type of the system.

3.7 (a)

$$K_p = \lim_{s\to 0} G(s)H(s) = \lim_{s\to 0}\frac{K}{s(s+1)} = \infty$$

$$e_{ss} = \frac{A}{1+K_p} = \frac{A}{1+\infty} = 0$$

3.8 (b)

$$H(s) = \frac{1}{s}\ ;\quad R(s) = \frac{1}{s+a}$$

$$C(s) = R(s)\,H(s) = \frac{1}{s(s+a)} = \frac{1}{a}\left[\frac{1}{s} - \frac{1}{s+a}\right]$$

$$C(t) = \frac{1}{a}\left[1-e^{-at}\right]$$

3.9 (b)

$$t_s = \frac{4}{\xi\omega_n}$$

$$2\xi\omega_n = 4$$

$$\Rightarrow \quad \xi\omega_n = 2$$

$$\therefore \quad t_s = \frac{4}{2} = 2$$

3.10 (d)

$$T(s) = \frac{\dfrac{a_{n-1}s+a_n}{s^n + a_1 s^{n-1} + \ldots a_{n-2}s^2}}{1+\dfrac{a_{n-1}s+a_n}{s^n + a_1 s^{n-1} + \ldots a_{n-2}s^2}}$$

$$G(s) = \frac{a_{n-1}s+a_n}{s^n + a_1 s^{n-1} + \ldots a_{n-2}s^2}$$

For type 2, Ramp input

$$K_v = \lim_{s\to 0} sG(s)$$

$$K_v = \infty$$

$$e_{ss} = \frac{1}{K_v} = 0$$

3.11 (c)

$$s^2 + 2\xi\omega_n s + \omega_n^2 = 0$$

$$2\xi\omega_n = 2,\ \xi = \frac{1}{\omega_n}$$

$$\omega_n = \sqrt{2}$$

$$\therefore \quad \xi = \frac{1}{\sqrt{2}}$$

$\because \xi < 1$, hence response will be underdamped.

3.12 (b)

$$\text{CLTF} = \frac{G(s)}{1+G(s)H(s)} = \frac{s+4}{s^2+7s+13}$$

$$\frac{1+G(s)H(s)}{G(s)} = \frac{s^2+7s+13}{s+4}$$

$H(s) = 1$ for unity feedback.

$$\frac{1}{G(s)} = \frac{s^2+7s+13}{s+4} - 1 = \frac{s^2+6s+9}{s+4}$$

$$\therefore \quad G(s) = \frac{s+4}{s^2+6s+9}$$

For D.C. gain, $\omega = 0$

$$\therefore \quad G(0) = \text{open loop gain} = \frac{4}{9}$$

3.13 (c)

$$\frac{s+6}{K\left(s^2+\frac{s}{K}+\frac{6}{K}\right)}$$

Comparing with $s^2 + 2\xi\omega_n s + \omega_n^2 = 0$

$$\omega_n = \sqrt{\frac{6}{K}}$$

$$2\xi\omega_n = \frac{1}{K}$$

$$2 \times 0.5 \times \sqrt{\frac{6}{K}} = \frac{1}{K}$$

$$\Rightarrow \quad \frac{6}{K} = \frac{1}{K^2}$$

$$K = \frac{1}{6}$$

3.14 (b)

$$G(s) = \frac{100}{(s+1)(s+100)}$$

Taking dominant pole consideration,

$s = -100$ pole is not taken.

$$\therefore \quad G(s) = \frac{100}{s+1}$$

Now it is 1st order system

$$\therefore \quad t_s = 4T = 4 \times 1 = 4 \text{ sec}$$

3.15 (b)

$$\frac{C(s)}{R(s)} = \frac{4}{s^2+4s+4}$$

$$2\xi\omega_n = 4, \ \omega_n = 2$$

$$\therefore \quad \xi = 1 \qquad \text{(Critical damping)}$$

$$t_s = \frac{4}{\xi\omega_n} = \frac{4}{1\times 2} = 2$$

3.16 (c)

$$H(s) = \frac{1}{s+2}$$

$$r(t) = 10u(t)$$

$$R(s) = \frac{10}{s}$$

$$C(s) = H(s)\cdot R(s) = \frac{1}{s+2}\cdot\frac{10}{s}$$

$$\Rightarrow \quad \frac{10}{s(s+2)} = \frac{A}{s} + \frac{B}{s+2}$$

$$10 = A(s+2) + Bs$$

$$s = 0, \ 10 = 2A$$

$$\Rightarrow \quad A = 5$$

$$s = -2, \ 10 = -2B$$

$$\Rightarrow \quad B = -5$$

$$\therefore \quad C(s) = \frac{5}{s} - \frac{5}{s+2}$$

$$c(t) = 5[1 - e^{-2t}]\,u(t)$$

Steady state value when $t = 0$ is 5. 99% of steady state value reaches at

$$5[1 - e^{-2t}] = 0.99 \times 5$$

$$\Rightarrow \quad 1 - e^{-2t} = 0.99$$

$$e^{-2t} = 0.01$$

$$\Rightarrow \quad -2t = \ln 0.01$$

$$\Rightarrow \quad t = 2.3 \text{ sec}$$

3.17 (a)

$$\frac{d^2y}{dt^2} + \frac{3dy}{dt} + 2y = X(t)$$

$$s^2Y(s) + 3sY(s) + 2Y(s) = X(s)$$

$$x(t) = 2u(t)$$

$$X(s) = \frac{2}{s}$$

$$\therefore \quad (s^2 + 3s + 2)Y(s) = \frac{2}{s}$$

$$Y(s) = \frac{2}{s(s+2)(s+1)}$$

$$\frac{2}{s(s+2)(s+1)} = \frac{A}{s} + \frac{B}{s+2} + \frac{C}{s+1}$$

$$2 = A(s+2)(s+1) + Bs(s+1) + C(s+2)s$$

$s = 0, 2 = 2A \Rightarrow A = 1$

$s = -1, 2 = -C \Rightarrow C = -2$

$s = -2, 2 = 2B \Rightarrow B = 1$

$$Y(s) = \frac{1}{s} + \frac{1}{s+2} - \frac{2}{s+1}$$

$$y(t) = [1 + e^{-2t} - 2e^{-t}]u(t)$$

3.20 (a)

$$e_{ss} = \lim_{s\to 0} sE(s),\ R(s) = \frac{1}{s^2}$$

$$= \lim_{s\to 0} s\frac{R(s)}{1+G(s)}$$

$$= \lim_{s\to 0}\frac{1}{s+sG(s)} = \text{finite (given)}$$

$$K_v = \lim_{s\to 0} sG(s)$$

$$e_{ss} = \lim_{s\to 0}\frac{1}{sG(s)} = \text{finite} = 5\% = \frac{1}{20}$$

$\therefore\ K = 20$

K_v is finite for type 1 system having ramp input.

3.21 (b)

$$C(s) = \frac{1}{s} - \frac{1}{s+2} = \frac{2}{s(s+2)}$$

$$R(s) = \frac{1}{s} \qquad H(s) = \frac{C(s)}{R(s)} = \frac{2}{s+2}$$

3.22 (c)

$$h(t) = e^{-t}$$

$$H(s) = \frac{1}{s+1}$$

$$R(s) = \frac{1}{s}$$

$$\text{Output} = H(s)\cdot R(s) = \frac{1}{(s+1)}\cdot\frac{1}{s}$$

$$\frac{1}{s(s+1)} = \frac{A}{s} + \frac{B}{s+1}$$

$1 = A(s+1) + Bs$

$s = 0,\ 1 = A$

$s = -1,\ 1 = -B$

$$\therefore\quad \text{Output} = \frac{1}{s} - \frac{1}{s+1}$$

$$= (1 - e^{-t})u(t)$$

When $t = \infty$ at steady state.

Output = 1

3.23 (d)

In dominant pole concept, the factor that has to be eliminated should be in time constant form.

$$\frac{5}{(s+5)(s^2+s+1)} = \frac{5}{5\left(1+\frac{s}{5}\right)(s^2+s+1)}$$

$$= \frac{1}{s^2+s+1}$$

3.24 (c)

Peak overshoot depends on ξ as

$$M_p = e^{-\xi\pi/\sqrt{1-\xi^2}}$$

where $\quad \xi = \cos^{-1}\theta$

where θ is the angle made by pole from negative real axis.

To make M_p same, θ should be the same.

3.25 (c)

Transfer functions can be written as,

$$H(s) = \frac{K(s+1)}{(s+2)(s+4)}$$

Input, $\quad R(s) = \frac{1}{s}$

Output, $\quad C(s) = R(s)\,H(s)$

Given: $\lim_{s\to 0} sC(s) = 1$

or, $\lim_{s\to 0}\frac{s\cdot K(s+1)}{s(s+2)(s+4)} = 1$

or $\quad \frac{K}{8} = 1$

$\Rightarrow \quad K = 8$

$$H(s) = \frac{8(s+1)}{(s+2)(s+4)}$$

$$= -\frac{4}{s+2} + \frac{12}{s+4}$$

$$h(t) = (-4e^{-2t} + 12e^{-4t})\,u(t)$$

Which is also the required impulse response of the system.

3.26 (d)

Comparing the given transfer functions with

$$\frac{\omega_n^2}{s^2 + 2\xi\omega_n s + \omega_n^2}$$

$$P = \frac{25}{s^2+25} \Rightarrow \xi = 0$$

So, P is undamped.

$$Q = \frac{36}{s^2+20s+36} \Rightarrow \xi = \frac{20}{2\times 6} = 1.67$$

So, Q is overdamped.

$$R = \frac{36}{s^2+12s+36} \Rightarrow \xi = \frac{12}{2\times 6} = 1$$

So, R is critically damped.

$$S = \frac{49}{s^2+7s+49} \Rightarrow \xi = \frac{7}{2\times 7} = 0.5$$

So, S is underdamped.

3.27 (b)

Steady state Value = –2

denominator:

$$2\xi\omega_n = 1.91,\ \omega_n^2 = 1.91$$

$$\Rightarrow \omega_n \cong 1.4$$

$$\xi = \frac{1.91}{2\omega_n} = \frac{1.91}{2.8} < 1$$

...underdamped

3.28 (a)

$$100\frac{d^2y}{dt^2} - 20\frac{dy}{dt} + y = x(t)$$

Taking Laplace transform of both sides

$$100\, s^2\, Y(s) - 20\, sY(s) + Y(s) = \frac{X(s)}{s}$$

$$\Rightarrow \frac{Y(s)}{X(s)} = \frac{1/s}{(100s^2 - 20s + 1)} = \frac{1}{s(10s-1)^2}$$

Poles are at $s = \frac{1}{10}, \frac{1}{10}, 0$

As poles are on the right-hand side of s-plane so given system is unstable system.

Only option (a) represents unstable system.

3.29 (d)

$$\frac{dy(t)}{dt} + ky(t) = x(t)$$

Taking Laplace transform of both sides, we have

$$sY(s) - y(0) + kY(s) = X(s)$$

$$Y(s)[s+k] = X(s) + y(0)$$

$$\Rightarrow Y(s) = \frac{X(s)}{s+k} + \frac{y(0)}{s+k}$$

Taking inverse Laplace transform, we have

$$y(t) = e^{-kt}x(t) + y(0)e^{-kt}$$

So if we want $-2y(t)$ as a solution both $x(t)$ and $y(0)$ has to be multiplied by –2; hence change $x(t)$ by $-2x(t)$ and $y(0)$ by $-2y(0)$.

3.30 (c)

$$\frac{\omega(s)}{V_a(s)} = \frac{10}{1+10s}$$

$$1 + Ts = 1 + 10s \quad \Rightarrow Ts = 10s$$

$$\therefore \quad T = 10$$

$$T' = \frac{T}{100} = \frac{10}{100} = \frac{1}{10}$$

$$\frac{\omega(s)}{R(s)} = \frac{K_a \times \frac{10}{1+10s}}{1+K_a\frac{10}{1+10s}}$$

$$\frac{\omega(s)}{R(s)} = \frac{10K_a}{1+10s+10K_a} = \frac{10K_a}{(1+10K_a)+10s}$$

$$\frac{\omega(s)}{R(s)} = \frac{10K_a}{(1+10K_a)\left[1+\frac{10s}{1+10K_a}\right]}$$

$$T' = \frac{10}{1+10K_a} = \frac{1}{10}$$

$$1 + 10\,K_a = 100$$

$$10\,K_a = 99$$

$$K_a = 9.9 \cong 10$$

3.31 (38.16)

Given, $\omega_n = 40$ rad/sec. ; $\xi = 0.3$

$$\therefore \quad \omega_d = \omega_n\sqrt{1-\xi^2} = 38.16 \text{ rad/sec}$$

3.32 (0)

After converting the system into equivalent unity feedback system.

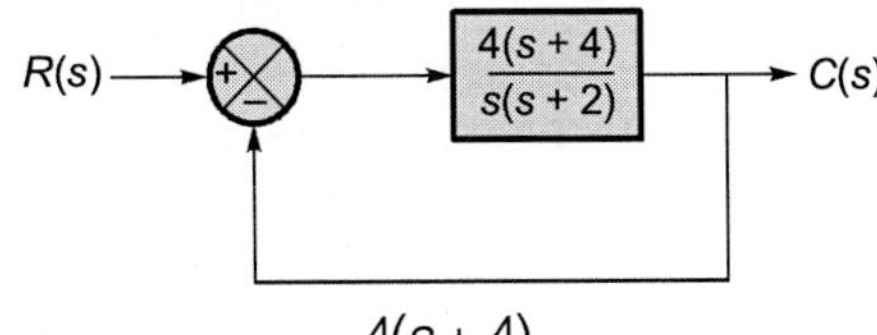

$$\therefore \quad \text{OLTF} = \frac{4(s+4)}{s(s+2)}$$

Type = 1

e_{ss} for step input = 0

3.33 (c)

The closed loop transfer function of the given system is

$$\frac{C(s)}{R(s)} = \frac{4}{s(s+4)+4} = \frac{4}{s^2+4s+4}$$

Comparing with standard second order transfer function $\dfrac{\omega_n^2}{s^2 + 2\xi\omega_n s + \omega_n^2}$

we get, $\omega_n^2 = 4$ or $\omega_n = \sqrt{4} = 2$ rad/sec.

3.34 (0.37)

According to the question

$$G(s)\,H(s) = \frac{K}{s(s+1)^2} \quad \text{...(i)}$$

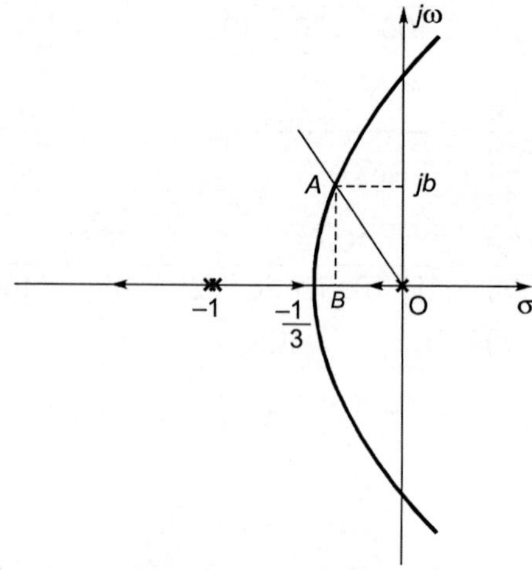

The damping ratio is given as $\xi = 0.5$

$\therefore \quad \theta = \cos^{-1}\xi = 60°$

Let the co-ordinates of point '*A*' is $(-a + jb)$

where, $\cos 60° = \dfrac{OB}{OA}$ ...(ii)

Also, $OA = 0.5$ (Given)

$\therefore \quad \boxed{OB = 0.25}$

Similarly,

$$\sin 60° = \frac{BA}{OA} \quad \text{...(iii)}$$

$\therefore \quad \boxed{BA = jb = 0.433j}$

Hence, co-ordinates of $\boxed{A = -0.25 + 0.43j}$

$\therefore$ The value of system gain K can be obtained as

$$|G(s)H(s)| = 1 \quad \text{...(iv)}$$

From (i) and (iv) we get $s = -0.25 + j4.33$

$$\boxed{K = 0.37}$$

3.35 (400)

$$T(s) = \frac{G(s)}{1 + G(s)\,H(s)} = \frac{K}{s^2 + 10s + K}$$

Comparing with standard second order transfer function

$$T(s) = \frac{\omega_n^2}{s^2 + 2\xi\omega_n s + \omega_n^2}$$

We have, $\omega_n^2 = K$

and $2\xi\omega_n = 10$

$\because \quad \xi = 0.25$ (given)

$\therefore \quad \omega_n = \dfrac{10}{2 \times 0.25} = 20$

and $K = \omega_n^2 = (20)^2 = 400$

3.36 (d)

The output of a second order system for unit step input

$$y(t) = 1 - \frac{e^{-\xi\omega_n t}}{\sqrt{1-\xi^2}}\sin(\omega_d t - \theta) \quad \text{...(i)}$$

From given data

$$y(t) = 1 - \frac{2}{\sqrt{3}}e^{-t}\cos\left(\sqrt{3}\,t - \frac{\pi}{6}\right)$$

$\therefore \quad \alpha = \xi\omega_n = 1$ and $\omega_d = \sqrt{3}$

$$\omega_n = \sqrt{\alpha^2 + \omega_d^2} = 2$$

$$\xi = \frac{1}{2}$$

Therefore, the transfer function

$$T(s) = \frac{\omega_n^2}{s^2 + 2\xi\omega_n s + \omega_n^2} = \frac{4}{s^2 + 2s + 4}$$

3.37 (2.8)

$$G(s) = \frac{K}{s(s+2)}\ ;\ H(s) = 1$$

Characteristic equation $= 1 + G(s)\,H(s) = 0$

$$1 + \frac{K}{s(s+2)} = 0$$

$$s^2 + 2s + K = 0 \Rightarrow \omega_n = \sqrt{K}$$

$$2\xi\omega_n = 2\ ;\ \xi = \frac{1}{\sqrt{K}}$$

$$M_p = e^{-\pi\xi/\sqrt{1-\xi^2}} = 0.1$$

$$-\frac{\pi\xi}{\sqrt{1-\xi^2}} = ln(0.1)$$

$$\Rightarrow \quad \frac{\pi\xi}{\sqrt{1-\xi^2}} = 2.3$$

$$\pi^2\xi^2 = (2.3)^2(1-\xi^2)$$

$$15.16\,\xi^2 = (2.3)^2$$

$$\Rightarrow \quad \xi = 0.59$$

$$\Rightarrow \text{ Also,} \quad K = \frac{1}{\xi^2} = 2.8$$

3.38 (a)

$$E(s) = \frac{R(s)}{1+G(s)H(s)}$$

$$R(s) = \frac{1}{s};\; G(s) = \frac{2}{s(s+1)};$$

$$H(s) = 1$$

$$E(s) = \frac{\frac{1}{s}}{1+\frac{2}{s(s+1)}} = \frac{s+1}{s^2+s+2}$$

$$e_{ss} = \lim_{s\to 0} sE(s) = \lim_{s\to 0}\frac{s(s+1)}{s^2+s+2} = 0$$

3.39 (1)

$$L\left(\frac{dy}{dt}\right) = (sY(s) - y(0))$$

$$Y(s) = G(s)\times\frac{1}{s} = \frac{s-2}{s(s+1)(s+3)}$$

$$y(0) = \lim_{s\to\infty} sY(s) \quad \text{(Applying initial value theorem)}$$

$$= \lim_{s\to\infty}\frac{s-2}{(s+1)(s+3)} = \frac{\left(1-\frac{2}{s}\right)}{s\left(1+\frac{1}{s}\right)\left(1+\frac{3}{s}\right)}$$

$$y(0) = 0$$

$$L\left(\frac{dy}{dt}\right) = sY(s) = \frac{s\times(s-2)}{s(s+1)(s+3)} = \frac{s-2}{(s+1)(s+3)}$$

$$\left.\frac{dy}{dt}\right|_{t=0} = \lim_{s\to\infty} sL\left(\frac{dy}{dt}\right)$$

$$= \lim_{s\to\infty}\frac{s\times(s-2)}{(s+1)(s+3)} = \frac{\left(1-\frac{2}{s}\right)}{\left(1+\frac{1}{s}\right)\left(1+\frac{3}{s}\right)} = 1$$

3.40 (1)

$$G(s) = \frac{1}{s^2+2s}$$

$$\frac{Y(s)}{R(s)} = \frac{\frac{K}{s^2+2s}}{1+\frac{K}{s^2+2s}} = \frac{K}{s^2+2s+K}$$

Minimum settling time and no overshoot implies critical damping i.e.

$$\xi = 1;\quad \omega_n = \sqrt{K}$$

$$\therefore \quad 2\times\xi\cdot\omega_n = 2$$

$$\omega_n = 1$$

$$\Rightarrow \quad \sqrt{K} = 1 \text{ or } K = 1$$

3.41 (1)

Steady state error of type 1 system, for step input is zero, for ramp input is $\frac{1}{K_v}$ and for parabolic input is infinity.

So, the given system must be of type-1 and $p = 1$.

3.42 (b)

$$G(s) = \frac{1}{s^2+2s+1} = \frac{1}{(s+1)^2}$$

$$R(s) = \frac{1}{s}$$

$$C(s) = G(s)R(s) = \frac{1}{s(s+1)^2}$$

Using partial fraction expansion, we get,

$$C(s) = \frac{A}{s} + \frac{B}{(s+1)} + \frac{C}{(s+1)^2}$$

$$A(s^2+2s+1) + B(s^2+s) + Cs = 1$$

$$A = 1$$

$$A + B = 0 \Rightarrow B = -1$$

$$2A + B + C = 0 \Rightarrow C = -1$$

$$\therefore \quad C(s) = \frac{1}{s} - \frac{1}{(s+1)} - \frac{1}{(s+1)^2}$$

and $\quad c(t) = (1 - e^{-t} - te^{-t})\,u(t)$

$$\lim_{t\to\infty} c(t) = 1$$

In order to reach 94% of its steady-state value,

$$(1 - e^{-t} - te^{-t}) = 0.94$$

By trial and error, we get,

$$t \approx 4.50 \text{ sec}$$

3.43 (30)

Open loop transfer function for the system

$$= C(s)\times G(s) = \frac{K(s+1)}{(s+3)}\times\frac{1}{s(s+1)}$$

Since the system is type-1 so for a given unit ramp input, steady state error

$$e_{ss} = \frac{1}{K_V}$$

where, $\quad K_V = \lim_{s\to 0} s\times\frac{K}{s(s+3)} = \frac{K}{3}$

So, $e_{ss} = \frac{1}{K/3} = \frac{3}{K}$

Given that, $e_{ss} = 0.1$

So, $0.1 = \frac{3}{K} \Rightarrow K = 30$

3.44 (a)

System-1, $G_1(s) = \frac{10}{s^2+s+1}$

The characteristics equation of the system is

$s^2 + s + 1 = 0$

Comparing with standard second order characteristic equation,

$s^2 + 2\xi\omega_n s + \omega_n^2 = 0$

We get, $\omega_{1n} = 1, 2\xi_1\omega_{1n} = 1$

$\Rightarrow \quad \xi_1 = \frac{1}{2} = 0.5$

The damped frequency of oscillations for system-1,

$$\omega_{d1} = \omega_{n1}\sqrt{1-\xi_1^2} = 1\sqrt{1-(0.5)^2} = 0.866 \text{ rad/sec}$$

Settling time, $t_{s1} = \frac{4}{\xi_1\omega_{n1}} = \frac{4}{0.5\times 1} = 8 \text{ sec}$

For a unit-step input, the steady value

$$y_1(\infty) = \lim_{s\to 0} sY_1(s) = \lim_{s\to 0} sG_1(s)R(s)$$

For unit-step input, $R(s) = \frac{1}{s}$. Hence,

$$y_1(\infty) = \lim_{s\to 0} \cdot s \times \frac{10}{s^2+s+1} \times \frac{1}{s} = 10$$

System-2, $G_2(s) = \frac{10}{s^2+\sqrt{10}s+10}$

The characteristic equation of the system is

$s^2 + \sqrt{10}s + 10 = 0$

Comparing with standard second order characteristic equation, we get

$\omega_{2n} = \sqrt{10}$,

$2\xi_2\omega_{2n} = \sqrt{10}$

$\Rightarrow \quad \xi_2 = \frac{1}{2} = 0.5$

The damped frequency of oscillations for system-2,

$$\omega_{d2} = \omega_{n2}\sqrt{1-\xi_2^2} = \sqrt{10}\cdot\sqrt{1-(0.5)^2} = 2.739 \text{ rad/sec}$$

Settling time,

$$t_{s2} = \frac{4}{\xi_2\omega_{n2}} = \frac{4}{0.5\sqrt{10}} = \frac{8}{\sqrt{10}}$$

$t_{s2} = 2.53$ sec

For a unit-step input, the steady value

$$y_2(\infty) = \lim_{s\to 0} sY_2(s)$$

$$y_2(\infty) = \lim_{s\to 0} sG_2(s)R(s)$$

$$y_2(\infty) = \lim_{s\to 0} \cdot s \times \frac{10}{s^2+\sqrt{10}s+10} \times \frac{1}{s}$$

$$y_2(\infty) = 1$$

From above, we can see that $y_1(t)$ and $y_2(t)$ have different steady-state values, damped frequency os oscillations and settling time.

Since 'ξ' i.e., damping ratio for both the system is same with $\xi_1 = \xi_2 = 0.5$, hence the percentage peak overshoot $\%M_P$,

$$\%M_P = e^{-\pi/\xi/\sqrt{1-\xi^2}} \times 100\%$$

is same for $y_1(t)$ and $y_2(t)$. Hence, option (a) is the correct answer.

3.45 (–0.10)

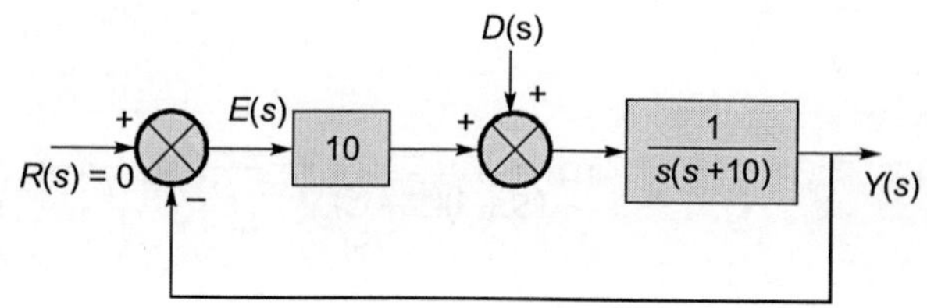

Assume $G_1(s) = 10, \quad G_2(s) = \frac{1}{s(s+10)}$

Given, $e(t) = r(t) - y(t)$

For $r(t) = 0, e(t) = -y(t)$

$E(s) = -Y(s)$ ...(i)

With $R(s) = 0$, the block diagram can be redrawn as below:

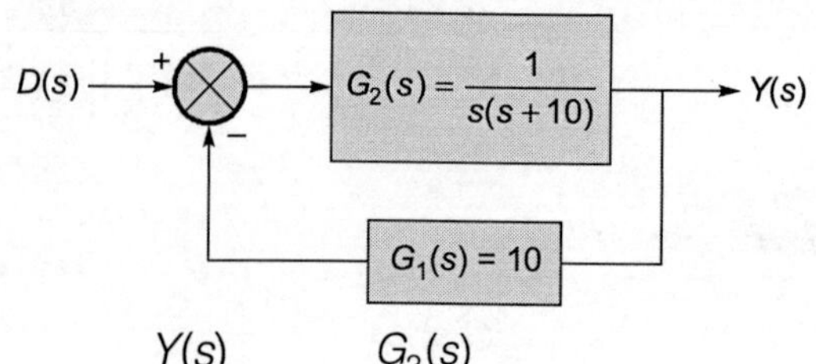

$$\frac{Y(s)}{D(s)} = \frac{G_2(s)}{1+G_1(s)G_2(s)}$$

Using equation (i)

$$\frac{E(s)}{D(s)} = \frac{-G_2(s)}{1+G_1(s)G_2(s)}$$

$$\frac{E(s)}{D(s)} = \frac{\frac{-1}{s(s+10)}}{1+\left(10\times\frac{1}{s(s+10)}\right)}$$

$$\frac{E(s)}{D(s)} = \frac{-1}{s^2+10s+10}$$

For a unit-step disturbance $d(t)$,

$$D(s) = \frac{1}{s}. \text{ Hence,}$$

$$E(s) = \frac{-1}{s(s^2+10s+10)}$$

Using final value theorem to obtain the steady-state error $e(\infty)$,

$$e(\infty) = \lim_{s\to 0} sE(s)$$

$$e(\infty) = \lim_{s\to 0} s \times \frac{-1}{s(s^2+10s+10)}$$

$$e(\infty) = \frac{-1}{10} = -0.1$$

3.46 (a)

Given, input is $r(t) = \alpha t u(t)$

$$R(s) = \frac{\alpha}{s^2}$$

From the figure,

$$G(s)H(s) = \frac{K}{s(s+2)}$$

Now steady state error for Ramp input is

$$e_{ss} = \frac{\alpha}{K_v}, \text{ where } \alpha \text{ is the magnitude}$$

of Ramp input

Velocity error constant,

$$K_v = \lim_{s\to 0}[sG(s)H(s)]$$

$$K_v = \lim_{s\to 0}\left[\frac{s\times K}{s(s+2)}\right] = \frac{K}{2}$$

$$\therefore \quad e_{ss} = \frac{\alpha\times 2}{K}$$

$$e_{ss} = \frac{2\alpha}{K}$$

3.47 (b)

Given system is

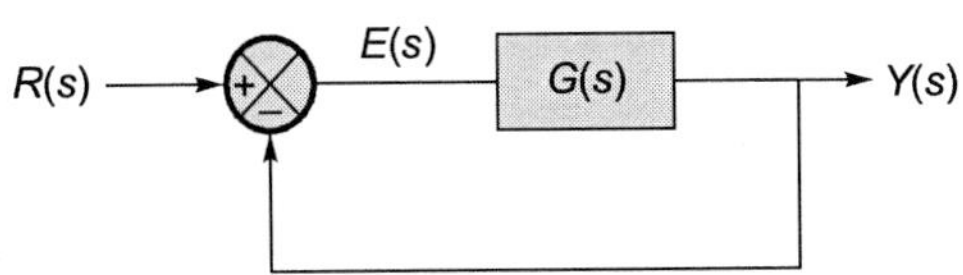

$$G(s) = \frac{6}{s(s+1)(s+2)}$$

The characteristic equation,

$$1 + G(s) = 0$$

$$1 + \frac{6}{s(s+1)(s+2)} = 0$$

$$s(s+1)(s+2) + 6 = 0$$

$$s^3 + 3s^2 + 2s + 6 = 0$$

Clearly, here the internal coefficient product is equal to external coefficient product i.e.,

$$3\times 2 = 6\times 1$$

Hence, the given system is marginally stable system.

$\therefore$ $\lim_{t\to\infty} e(t)$ does not exist because system is oscillatory with a pair of complex poles on the imaginary axis.

■■■

4 Stability Analysis

4.1 The number of roots of $s^3 + 5s^2 + 7s + 3 = 0$ in the left half of the s-plane is

(a) zero (b) one
(c) two (d) three

[1998 : 1 M]

4.2 The open loop transfer function of an unity feedback open loop system is $\dfrac{2s^2+6s+5}{(s+1)^2(s+2)}$. The characteristic equation of the closed loop system is

(a) $2s^2 + 6s + 5 = 0$
(b) $(s + 1)^2 (s + 2) = 0$
(c) $2s^2 + 6s + 5 + (s + 1)^2 (s + 2) = 0$
(d) $2s^2 + 6s + 5 - (s + 1)^2 (s + 2) = 0$

[1998 : 1 M]

4.3 An amplifier with resistive negative feedback has two left half plane poles in its open-loop transfer function. The amplifier

(a) will always be unstable at high frequency
(b) will be stable for all frequency
(c) may be unstable, depending on the feedback factor
(d) will oscillate at low frequency

[2000 : 1 M]

4.4 A system described by the transfer function

$$H(s) = \frac{1}{s^3 + \alpha s^2 + Ks + 3}$$ is stable.

The constraints on α and K are,

(a) $\alpha > 0, \alpha K < 3$ (b) $\alpha > 0, \alpha K > 3$
(c) $\alpha < 0, \alpha K > 3$ (d) $\alpha < 0, \alpha K < 3$

[2000 : 2 M]

4.5 The feedback control system in the figure is stable.

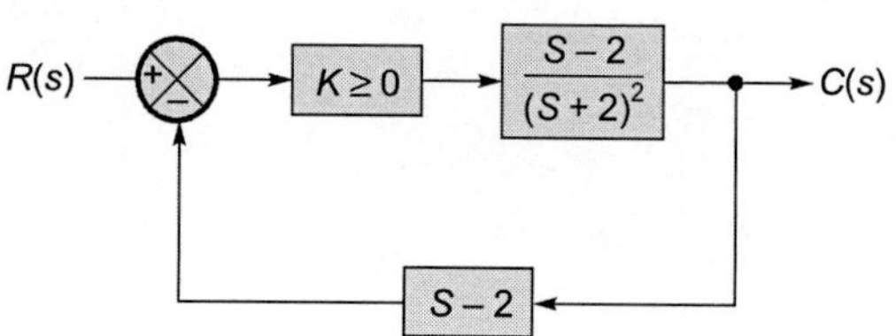

(a) for all $K \geq 0$ (b) only is $K \geq 0$
(c) only if $0 \leq K < 1$ (d) only if $0 \leq K \leq 2$

[2001 : 2 M]

4.6 The system shown in the figure remains stable when

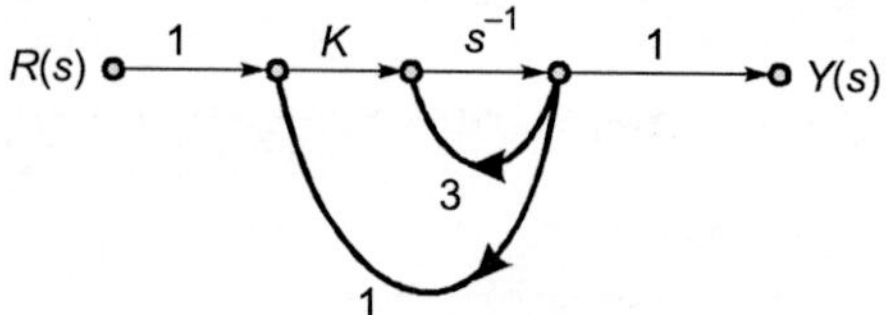

(a) $K < -1$ (b) $-1 < K < 1$
(c) $1 < K < 3$ (d) $K < -3$ [2002 : 2 M]

4.7 The characteristic polynomial of a system is $q(s) = 2s^5 + s^4 + 4s^3 + 2s^2 + 2s + 1$. The system is

(a) stable (b) marginally stable
(c) unstable (d) oscillatory

[2002 : 2 M]

4.8 The gain margin for the system with open-loop transfer function $G(s)H(s) = \dfrac{2(1+s)}{s^2}$, is

(a) ∞ (b) 0
(c) 1 (d) $-\infty$ [2004 : 1 M]

4.9 The open-loop transfer function of a unity feedback system is

$$G(s) = \frac{K}{s(s^2+s+2)(s+3)}$$

The range of K for which the system is stable is

(a) $\frac{21}{4} > K > 0$ (b) $13 > K > 0$
(c) $\frac{21}{4} < K < \infty$ (d) $-6 < K < \infty$

[2004 : 2 M]

4.10 For the polynomial

$P(s) = s^5 + s^4 + 2s^3 + 2s^2 + 3s + 15$, the number of roots which lie in the right half of the s-plane is

(a) 4 (b) 2
(c) 3 (d) 1 [2004 : 2 M]

4.11 The positive values of "K" and "a" so that the system shown in the figure below oscillates at a frequency of 2 rad/sec respectively are

R(s) → [$\frac{K(s+1)}{(s^3 + as^2 + 2s + 1)}$] → C(s)

(a) 1, 0.75 (b) 2, 0.75
(c) 1, 1 (d) 2, 2 [2006 : 2 M]

Common Data for Questions 4.12 & 4.13:
Consider a unity-gain feedback control system whose open-loop transfer function is $G(s) = \frac{as+1}{s^2}$

4.12 The value of "a" so that the system has a phase-margin equal to $\pi/4$ is approximately equal to
(a) 2.40 (b) 1.40
(c) 0.84 (d) 0.74 [2006 : 2 M]

4.13 With the value of "a" set for phase-margin of $\pi/4$, the value of unit-impulse response of the open-loop system at $t = 1$ second is equal to
(a) 3.40 (b) 2.40
(c) 1.84 (d) 1.74 [2006 : 2 M]

4.14 If the closed-loop transfer function of a control system is given as

$$T(s) = \frac{s-5}{(s+2)(s+3)}, \text{ then it is}$$

(a) an unstable system
(b) an uncontrollable system
(c) a minimum phase system
(d) a non-minimum phase system [2007 : 1 M]

4.15 A certain system has transfer function $G(s) = \frac{s+8}{s^2 + \alpha s - 4}$, where α is a parameter. Consider the standard negative unity feedback configuration as shown below.

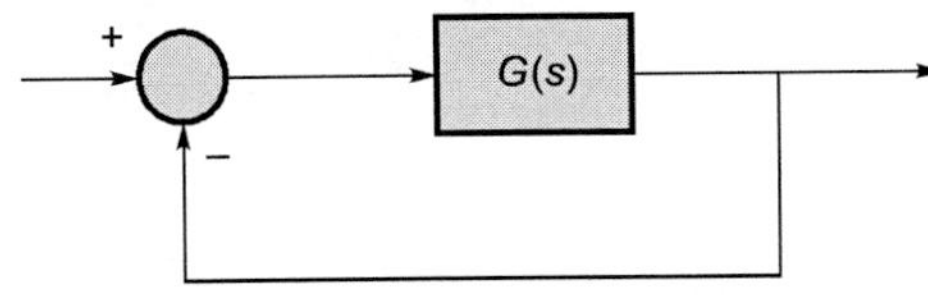

Which of the following statements is true?
(a) The closed loop system is never stable for any value of α
(b) For some positive values of α, the closed loop system is stable, but not for all positive values
(c) For all positive values of α, the closed loop system is stable
(d) The closed loop system is stable for all values of α, both positive and negative
[2008 : 2 M]

4.16 The number of open right half plane poles of $G(s)$

$$= \frac{10}{s^5 + 2s^4 + 3s^3 + 6s^2 + 5s + 3} \text{ is}$$

(a) 0 (b) 1
(c) 2 (d) 3 [2008 : 2 M]

4.17 The feedback system shown below oscillates at 2 rad/s when

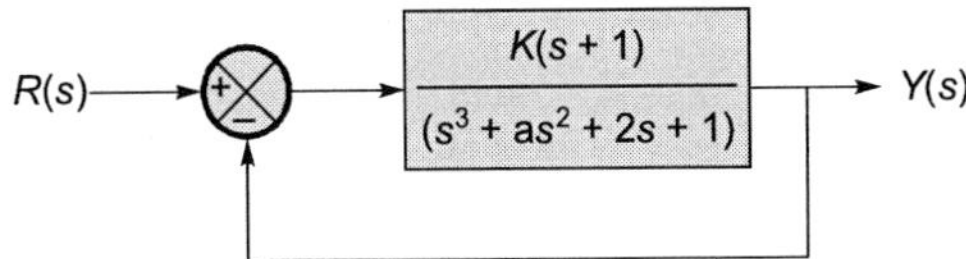

(a) $K = 2$ and $a = 0.75$
(b) $K = 3$ and $a = 0.75$
(c) $K = 4$ and $a = 0.5$
(d) $K = 2$ and $a = 0.5$ [2012 : 2 M]

4.18 The forward path transfer function of a unity negative feedback system is given by

$$G(s) = \frac{K}{(s+2)(s-1)}$$

The value of K which will place both the poles of the closed-loop system at the same location, is ____.
[2014 : 1 M, Set-1]

4.19 Consider a transfer function

$$G_p(s) = \frac{ps^2 + 3ps - 2}{s^2 + (3+p)s + (2-p)}$$

with p a positive real parameter. The maximum value of p until which G_p remains stable is ____.
[2014 : 2 M, Set-4]

4.20 A plant transfer function is given as

$$G(s) = \left(K_P + \frac{K_I}{s}\right)\frac{1}{s(s+2)}$$

When the plant operates in a unity feedback configuration, the condition for the stability of the closed loop system is

(a) $K_P > \frac{K_I}{2} > 0$ (b) $2K_I > K_P > 0$
(c) $2K_I < K_P$ (d) $2K_I > K_P$
[2015 : 2 M, Set-1]

4.21 The characteristic equation of an LTI system is given by

$$F(s) = s^5 + 2s^4 + 3s^3 + 6s^2 - 4s - 8 = 0$$

The number of roots that lie strictly in the left half s-plane is ____. [2015 : 2 M, Set-3]

4.22 Match the inferences X, Y and Z about a system, to the corresponding properties of the elements of first column in Routh's Table of the system characteristic equation.

List - I
X. The system is stable..
Y. The system is unstable..
Z. The test breaks down..

List - II
P. .. when all elements are positive
Q. ..when any one element is zero
R. ..when there is a change in sign of coefficients

(a) X – P; Y – Q; Z – R
(b) X – Q; Y – P; Z – R
(c) X – R; Y – Q; Z – P
(d) X – P; Y – R; Z – Q [2016 : 1 M, Set-1]

4.23 The transfer function of a linear time invariant system is given by
$H(s) = 2s^4 - 5s^3 + 5s - 2$
The number of zeroes in the right half of the s-plane is ___. [2016 : 2 M, Set-1]

4.24 The first two rows in the Routh table for the characteristic equation of a certain closed-loop control system are given as

s^3	1	$(2K + 3)$
s^2	$2K$	4

The range of K for which the system is stable is
(a) $-2.0 < K < 0.5$
(b) $0 < K < 0.5$
(c) $0 < K < \infty$
(d) $0.5 < K < \infty$
[2016 : 2 M, Set-3]

4.25 Which one of the following options correctly describes the locations of the roots of the equation $s^4 + s^2 + 1 = 0$ on the complex plane?
(a) Four left half plane (LHP) roots.
(b) One right half plane (RHP) root, one LHP root and two roots on the imaginary axis.
(c) Two RHP roots and two LHP roots.
(d) All four roots are on the imaginary axis
[2017 : 2 M, Set-1]

4.26 A unity feedback control system is characterized by the open-loop transfer function

$$G(s) = \frac{2(s+1)}{s^3 + Ks^2 + 2s + 1}$$

The value of K for which the system oscillates at 2 rad/s is _____.

[2017 : 2 M, Set-2]

4.27 Consider a unity feedback system, as in the figure shown, with an integral compensator $\frac{K}{s}$ and open-loop transfer function

$$G(s) = \frac{1}{s^2 + 3s + 2}$$

where $K > 0$. The positive value of K for which there are exactly two poles of the unity feedback system on the $j\omega$ axis is equal to _____ (rounded off to two decimal places).

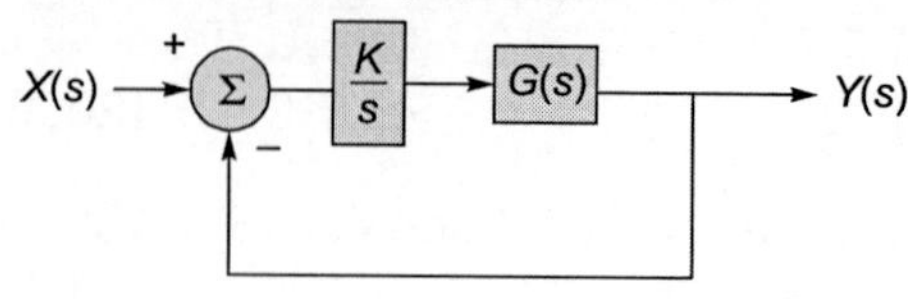

[2019 : 2 M]

4.28 The loop transfer function of a negative feedback system is

$$G(s)H(s) = \frac{K(s+11)}{s(s+2)(s+8)}$$

The value of K, for which the system is marginally stable, is ________. [2020 : 1 M]

4.29 The characteristic equation of a system is
$$s^3 + 3s^2 + (K + 2)s + 3K = 0$$
In the root locus plot for the given system, as K varies from 0 to ∞, the break-away or break-in point(s) lie within
(a) $(-2, -1)$
(b) $(-1, 0)$
(c) $(-3, -2)$
(d) $(-\infty, -3)$ [2020 : 2 M]

4.30 Consider an even polynomial $p(s)$ given by
$$p(s) = s^4 + 5s^2 + 4 + K$$
where K is an unknown real parameter. The complete range of K for which $p(s)$ has all its roots on the imaginary axis is ________.

(a) $-4 \le K \le \frac{9}{4}$
(b) $-3 \le K \le \frac{9}{2}$
(c) $-6 \le K \le \frac{5}{4}$
(d) $-5 \le K \le 0$

[2022 : 2 M]

4.33 Consider a unity negative feedback control system with forward path gain $G(s) = \frac{K}{(s+1)(s+2)(s+3)}$ as shown.

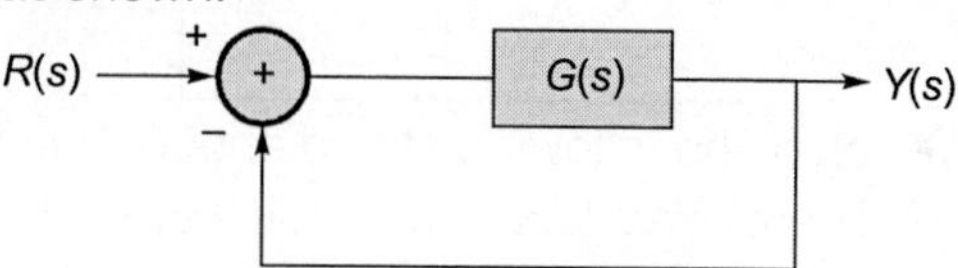

The impulse response of the closed-loop system decays faster than e^{-t} if ________.
(a) $-4 \le K \le -1$
(b) $1 \le K \le 5$
(c) $7 \le K \le 21$
(d) $-24 \le K \le -6$
[2024 : 2 Marks]

■■■

Answers Stability Analysis

4.1 (d)	4.2 (c)	4.3 (b)	4.4 (b)	4.5 (c)	4.6 (d)	4.7 (c)
4.8 (a)	4.9 (a)	4.10 (b)	4.11 (b)	4.12 (c)	4.13 (c)	4.14 (d)
4.15 (c)	4.16 (c)	4.17 (a)	4.18 (2.25)	4.19 (1.99)	4.20 (a)	4.21 (2)
4.22 (d)	4.23 (3)	4.24 (d)	4.25 (c)	4.26 (0.75)	4.27 (6)	4.28 (160)
4.29 (b)	4.30 (a)	4.33 (b)				

Explanations Stability Analysis

4.1 (d)

Using R-H criterion:

s^3	1	7
s^2	5	3
s^1	6.4	0
s^0	3	

There is no sign change in the first column of R-H array. So, there is no roots lie in R.H.S. of *s*-plane. So, all the 3 roots will lie in the left half of *s*-plane.

4.2 (c)

$2s^2 + 6s + 5 + (s + 1)^2 (s + 2) = 0$

Characteristic equation

$1 + GH = 0$

$$1+\frac{2s^2+6s+5}{(s+1)^2(s+2)}=0$$

$2s^2 + 6s + 5 + (s + 1)^2 (s + 2) = 0$

4.3 (b)

Second order system is stable.

4.4 (b)

The characteristic equation is,

$s^3 + \alpha s^2 + Ks + 3 = 0$

Routh table is

s^3	1	K
s^2	α	3
s^1	$\dfrac{\alpha K-3}{\alpha}$	0
s^0	3	0

For system to be stable

$\alpha > 0, \quad \dfrac{\alpha K - 3}{\alpha} > 0$

$\alpha K > 3$

4.5 (c)

$$\text{T.F.} = \frac{G_1G_2}{1+G_1G_2H} = \frac{\dfrac{K(s-2)}{(s+2)^2}}{1+\dfrac{K(s-2)(s-2)}{(s+2)^2}} = \frac{K(s-2)}{(s+2)^2+K(s-2)^2}$$

∴ Characteristic equation

$$= s^2 + 4 + 4s + Ks^2 - 4Ks + 4K$$

$(1+K)s^2 + K(4-4K)s + 4K + 4 = 0$

Routh table is

s^2	$(1+K)$	$(4K+4)$
s^1	$K(4-4K)$	0
s^0	$4(K+1)$	

For system to be stable,

$1 - K > 0$

$1 > K$ and $K \geq 0$ (given in question)

From third row,

$K > -1$

∴ $0 \leq K < 1$

4.6 (d)

$$\frac{Y(s)}{R(s)} = \frac{K/s}{1-\left(\dfrac{3}{s}+\dfrac{K}{s}\right)} = \frac{K}{s-(3+K)}$$

For system to be stable,

$3 + K < 0$

$\Rightarrow \quad K < -3$

4.7 (c)

Routh table is

s^5	2(1)	4(2)	2(1)
s^4	1	2	1
s^3	0	0	0
s^2			
s^1			
s^0			

$$\therefore \frac{d}{ds}(s^4 + 2s^2 + 1) = 0$$

$$4s^3 + 4s = 0$$

$$s = \pm j,\ s = \pm j$$

$$\frac{d}{ds}(s^2 + 1) = 0$$

$$2s = 0$$

$$\Rightarrow \quad s = 0$$

Double roots on imaginary axis so system is unstable.

4.8 (a)

$$\angle G(s)H(s) = -180° + \tan^{-1}\omega$$

For $\omega_\phi = -180° + \tan^{-1}\omega = -180°$

$$\omega = 0$$

$$|G(s)H(s)| = \frac{2\sqrt{1+\omega^2}}{\omega^2} = \infty$$

$$\text{G.M.} = \frac{1}{\infty} = 0$$

In dB G.M. $= \infty$

4.9 (a)

$$G(s) = \frac{K}{s(s^2+s+2)(s+3)} \quad ; \quad H(s) = 1$$

$$1 + G(s)H(s) = 1 + \frac{K}{s(s^3+3s^2+s^2+3s+2s+6)}$$

$$= \frac{s^4+4s^3+5s^2+6s+K}{s(s^3+4s^2+5s+6)}$$

$$1 + G(s)H(s) = 0$$

$$\Rightarrow s^4 + 4s^3 + 5s^2 + 6s + K = 0$$

s^4	1	5	K
s^3	4	6	
s^2	$\frac{7}{2}$	K	
s^1	$\dfrac{\frac{7}{2}\times 6 - 4K}{\frac{7}{2}}$	0	
s^0	K		

For system to be stable, $K > 0$

$$(21-4K).\frac{2}{7} > 0 \quad ; \quad \frac{21}{4} > K \Rightarrow K < \frac{21}{4}$$

$$\therefore \quad \frac{21}{4} > K > 0$$

4.10 (b)

$$P(s) = s^5 + s^4 + 2s^3 + 2s^2 + 3s + 15$$

s^5	1	2	3
s^4	1	2	15
s^3	$0(\epsilon)$	-12	0
s^2	$\frac{2\epsilon+12}{\epsilon}$	15	0
s^1	$\dfrac{\frac{-12(2\epsilon+12)}{\epsilon} - 15\epsilon}{\left(\frac{2\epsilon+12}{\epsilon}\right)}$		
s^0	15		

Let $\frac{2\epsilon+12}{\epsilon} = t$

Let ϵ be a small positive no.

$$\frac{-12t - 15\epsilon}{t}$$

$$s \Rightarrow -12 - \frac{15\epsilon}{t}$$

$\therefore$ Two sign change from s^2 to s and s to s^0

$\therefore$ 2 roots on RHS of s plane.

4.11 (b)

$$1 + G(s)H(s) = 1 + \frac{K(s+1)}{s^3+as^2+2s+1} = 0$$

$$\Rightarrow \frac{s^3+as^2+2s+1+Ks+K}{s^3+as^2+2s+1} = 0$$

$$s^3 + as^2 + (2+K)s + K + 1 = 0$$

s^3	1	$2+K$
s^2	a	$K+1$
s^1	$\frac{a(2+K)-(K+1)}{a}$	
s^0	$K+1$	

For oscillation, $\frac{a(2+K)-(K+1)}{a} = 0$

$$a = \frac{K+1}{K+2} \;;\quad as^2 + K + 1 = 0$$

$$s = j\omega,\ s^2 = -\omega^2 = -4 \quad \Rightarrow -4a + K + 1 = 0$$

$$a = \frac{K+1}{4} \Rightarrow \frac{K+1}{4} = \frac{K+1}{K+2} \Rightarrow K = 2$$

$$a = 0.75$$

4.12 (c)

$$\text{PM} = \frac{\pi}{4}$$

$$\Rightarrow 180 + \tan^{-1} a\omega - 180° = \frac{\pi}{4}$$

$$\tan\frac{\pi}{4} = a\omega \quad \Rightarrow a\omega = 1$$

Now for gain crossover frequency,

$$|G(s)| = 1$$

$$\Rightarrow \quad \frac{\sqrt{1+a^2\omega^2}}{\omega^2} = 1$$

$$\sqrt{1+1} = \omega^2 \qquad (\text{as } a\omega = 1)$$

$$\omega^2 = \sqrt{2} \quad \Rightarrow \quad \omega = (2)^{1/4}$$

$$a = \frac{1}{2^{1/4}} = 0.84$$

4.13 (c)

$$G(s) = \frac{0.84s+1}{s^2}$$

$$H(s) = 1,\ R(s) = 1$$

$$\therefore \quad C(s) = G(s)\cdot R(s) = \frac{0.84s+1}{s^2}$$

$$c(t) = L^{-1}\left[\frac{1+0.84s}{s^2}\right] = L^{-1}\left[\frac{1}{s^2}+\frac{0.84}{s}\right]$$

$$c(t) = [t+0.84]\,U(t)$$

At $t = 1,\ c(t) = 1 + 0.84 = 1.84$

4.14 (d)

As there is a right half zero, the system is a non-minimum phase system.

4.15 (c)

Closed loop gain is

$$\frac{G(s)}{1+G(s)} = \frac{s+8}{s^2+\alpha s-4+s+8}$$

Characteristic equation

$$q(s) = s^2 + (\alpha+1)s + 4$$

Closed loop system is stable only for $\alpha > -1$. Therefore, for all positive of α, the closed loop system is stable.

4.16 (c)

Characteristic equation

$$q(s) = s^5 + 2s^4 + 3s^3 + 6s^2 + 5s + 3$$

Putting $s = \frac{1}{z}$

$$q'(z) = 3z^5 + 5z^4 + 6z^3 + 3z^2 + 2z + 1$$

Routh array is

z^5	3	6	2
z^4	5	3	1
z^3	21/5	7/5	
z^2	$\dfrac{\frac{63}{5}-\frac{35}{5}}{21} = \dfrac{4}{3}$	1	
z^1	$\dfrac{\frac{28}{15}-\frac{21}{5}}{4/3} = \dfrac{-35/15}{4/3} = \dfrac{-7}{4}$		
z^0	1		

Since sign changes twice in Routh-array, therefore, there are two poles on right half plane.

4.17 (a)

Characteristic equation is

$$1 + G(s)H(s) = 0$$

$$1+\frac{K(s+1)}{s^3+as^2+2s+1} = 0$$

$$s^3 + as^2 + (2+K)s + (1+K) = 0$$

Routh array for this is

s^3	1	$(2+K)$
s^2	a	$(1+K)$
s^1	$\dfrac{a(2+K)-(1+K)}{a}$	
s^0	$(1+K)$	

For oscillation $\dfrac{a(2+K)-(1+K)}{a} = 0$

$$\Rightarrow \quad a = \left(\frac{1+K}{2+K}\right)$$

Now

$$as^2 + (1+K) = 0$$

$$-a\omega^2 + (1+K) = 0$$

given $\omega = 2$ rad/sec

$$-4a + (1+K) = 0$$

$$-4\frac{(1+K)}{(2+K)} + (1+K) = 0$$

$$-4(1+K) + (2+K)(1+K) = 0$$

$$(1+K)[(2+K)-4] = 0$$

$$K = -1,\ 2$$

but $K = -1$ is not possible as system will not oscillate for this as $a = 0$

so $K = 2$

$$a = \frac{1+K}{2+K} = \frac{3}{4} = 0.75$$

4.18 (2.25)

Given that, $G(s) = \dfrac{K}{(s+2)(s-1)}$

Using root locus method, the break point can be obtain as

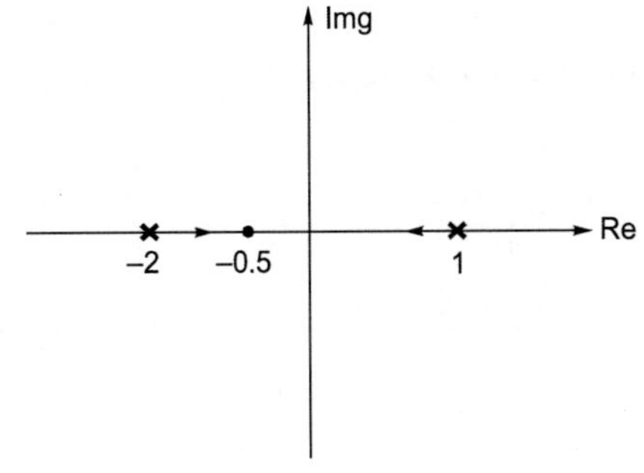

$$\Rightarrow \quad 1 + G(s) = 0$$

$$1+\frac{K}{(s+2)(s-1)} = 0$$

or $$K = -(s+2)(s-1)$$

$$\frac{dK}{ds} = -2s - 1 = 0$$

or $$s = -0.5$$

To have, both the poles at the same directions

$$|G(s)|_{s=-0.5} = 1$$

$$K = 2.25$$

4.19 (1.99)

Given transfer function

$$G_p(s) = \frac{ps^2+3ps-2}{s^2+(3+p)s+(2-p)} \quad ...(i)$$

The characteristic equation is

$$= s^2 + (3+p)s + (2-p) \quad ...(ii)$$

Using Routh Hurwitz criterion,

s^2	1	$(2-P)$
s^1	$(3+p)$	0
s^0	$(2-p)$	0

For system to be remain stable

$2 - p \geq 0$

$p_{max} = 1.99$

4.20 (a)

$$G(s) = \left(K_p + \frac{K_I}{s}\right)\left(\frac{1}{s(s+2)}\right)$$

The closed loop transfer function for unity feedback

$$\frac{G(s)}{1+G(s)} = \frac{(K_p s + K_I)}{s^2(s+2)+(K_p s + K_I)}$$

$$= \frac{K_p s + K_I}{s^3 + 2s^2 + K_p s + K_I}$$

Using Routh's tabular form:

s^3	1	K_p
s^2	2	K_I
s^1	$\frac{2K_p - K_I}{2}$	0
s^0	K_p	

For system to be stable;

$$K_P > 0$$

and $$\frac{2K_p - K_I}{2} > 0$$

or $$2K_p - K_I > 0$$

or $$K_p > \frac{K_I}{2} > 0$$

4.21 (2)

Using Routh's tabular form:

Auxiliary equation:

$$A = 2s^4 + 6s^2 - 8 = 0$$

$$\frac{dA}{ds} = 8s^3 + 12s = 0$$

s^5	1	3	−4
s^4	2	6	−8
s^3	0(8)	0(12)	0
s^2	3	−8	0
s^1	$\frac{100}{3}$	0	0
s^0	−8		

As one time row of zero occur in Routh's table therefore a pair of imaginary pole exists, also number of sign change in Routh's table is one. Hence, two poles lie strictly in the left half of s-plane.

4.22 (d)

When all elements are positive, the system is stable. When any element is zero, the test breaks down. When there is change in sign of coefficients, the system is unstable.

4.23 (3)

$$2s^4 - 5s^3 + 5s - 2 = 0$$

By Routh Array,

s^4	2	0	−2
s^3	−5	5	
s^2	2	−2	
s^1	0(2)		
s^0	−2		

Number of sign changes = number of roots (zeros) in right half of s-plane = 3

4.24 (d)

s^3	1	$(2K+3)$
s^2	$2K$	4
s^1	$\frac{4K^2+6K-4}{2K}$	0
s^0	4	

For stability, $K > 0$ ...(i)

$$4K^2 + 6K - 4 > 0$$

$$2K^2 + 3K - 2 > 0$$

$$(2K-1)(K+2) > 0$$

$$\Rightarrow \quad K > \frac{1}{2} \text{ or, } K < -2 \quad ...(ii)$$

Hence, $0.5 < K < \infty$

4.25 (c)

$$q(s) = s^4 + s^2 + 1 = 0$$

s^4	1	1	1
s^3	4	2	0
s^2	0.5	1	0
s^1	−6	0	0
s^0	1	0	0

$$\frac{dA(s)}{ds} = 4s^3 + 2s^1$$

There are two sign changes in the first column of the R-H table and the order of auxiliary equation is 4. So, four poles are symmetric about origin.

∴ 2 RHP roots and 2 LHP roots.

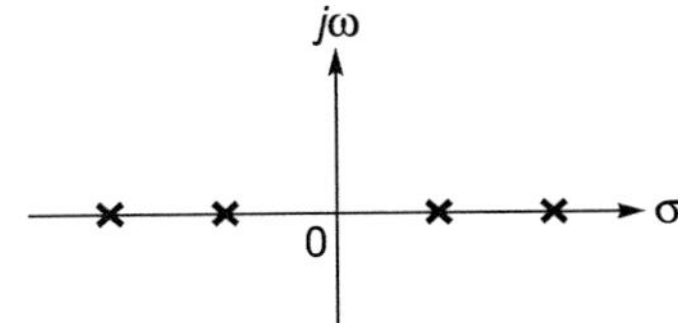

4.26 (0.75)

The given open loop transfer function is,

$$G(s) = \frac{2(s+1)}{s^3 + Ks^2 + 2s + 1}$$

The closed loop transfer function is,

$$T(s) = \frac{G(s)}{1+G(s)} = \frac{2(s+1)}{s^3 + Ks^2 + 4s + 3}$$

When system oscillates, i.e., when system is marginally stable,

$$(1)(3) = K(4)$$

So, $K = 0.75$

4.27 (6)

$$\frac{Y(s)}{X(s)} = \frac{K}{s^3 + 3s^2 + 2s + K}$$

Two poles of this system lie on the $j\omega$ axis when the system is marginally stable.

$$k_{mar} = 3 \times 2 = 6$$

4.28 (160)

Characteristic equation $q(s)$ for the given open loop system will be

$$q(s) = s^3 + 10s^2 + 16s + Ks + 11K = 0$$

Using *R-H* criteria,

s^3	1	$16 + K$
s^2	10	$11K$
s^1	$\frac{10(16+K) - 11K}{10}$	0
s^0	$11K$	0

For system to be marginally stabled

$$\frac{10(16+K) - 11K}{10} = 0$$

$$160 + 10K - 11K = 0$$

$$K = 160$$

4.29 (b)

$$Q(s) = 1 + G(s)\,H(s) = 0$$

$$s^3 + 3s^2 + 2s + ks + 3k = 0$$

$$-k = \frac{s^3 + 3s^2 + 2s}{s+3}$$

$$-\frac{dk}{ds} = \frac{(s+3)(3s^2 + 6s + 2) - (s^3 + 3s^2 + 2s)}{(s+3)^2} = 0$$

$$3s^3 + 6s^2 + 2s + 9s^2 + 18s + 6 - s^3 - 3s^2 - 2s = 0$$

$$2s^3 + 12s^2 + 18s + 6 = 0$$

$$s = -0.46, -3.87, -1.65$$

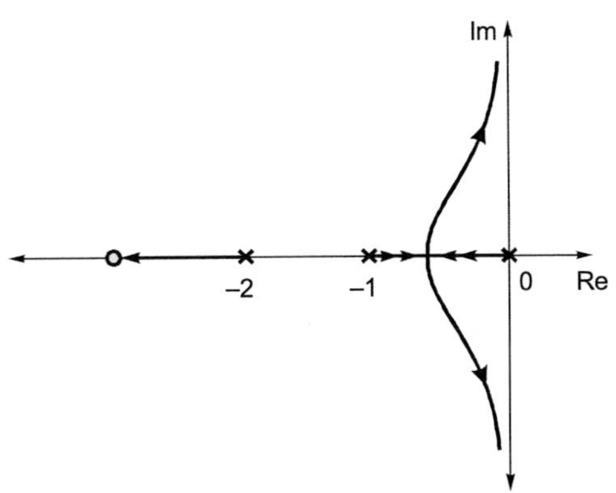

∴ Break-away point lies between (0, −1), i.e. (−1, 0).

4.30 (a)

$$p(s) = s^4 + 5s^2 + (4 + K)$$

Using Routh-Hurwitz criterion,

s^4	1	5	$(4+K)$
s^3	0	0	0
s^2			
s^1			
s^0			

As all rows of s^3 are zero. The auxiliary polynomial

$$A(s) = s^4 + 5s^2 + (4 + K)$$

$$\frac{dA(s)}{ds} = 4s^3 + 10s$$

The Routh array can be written as below:

s^4	1	5	$(4+K)$
s^3	4	10	0
s^2	$\frac{20-10}{4}=\frac{5}{2}$	$(4+K)$	
s^1	$\dfrac{\frac{50}{2}-4(4+K)}{\frac{5}{4}}$		
s^0	$4+K$		

Since, the odd row s_3 is zero. All the three roots lie on imaginary axis. The first-column of the *R-H* table must be positive as the system is marginally stable with all roots lying on imaginary axis. Therefore,

$$\frac{50}{2}-4(4+K) > 0$$

$$\Rightarrow \quad 4 + K < \frac{50}{8}$$

$$\Rightarrow \quad K < \frac{50}{8} - 4$$

$$\Rightarrow \quad K < \frac{9}{4} \quad \text{...(i)}$$

Also, $4 + K > 0 \Rightarrow K > -4$ ...(ii)

From equations (i) and (ii),

Range of $K: -4 < K < \frac{9}{4}$

4.33 (b)

Given, $G(s) = \dfrac{K}{(s+1)(s+2)(s+3)}$

The characteristic equation of the system is

$$1 + G(s) = 0$$

$$\Rightarrow \quad 1+\frac{K}{(s+1)(s+2)(s+3)} = 0$$

$$\Rightarrow \quad (s + 1)(s + 2)(s + 3) + K = 0$$

The impulse response of the closed-loop system decays faster than e^{-t} if the poles of the system has $\text{Re}\{s\} < -1$ i.e., the poles lie on the left of the $s = -1$ axis.

To analyze the poles lying to the left of $s = -1$ axis using Routh Hurwitz criterion, we substitute $s = z - 1$. The modified characteristic equation, so obtained is given by

$$(z - 1 + 1)(z - 1 + 2)(z - 1 + 3) + K = 0$$

$$\Rightarrow \quad z(z + 1)(z + 2) + K = 0$$

$$\Rightarrow \quad z^3 + 3z^2 + 2z + K = 0$$

Using Routh-Hurwitz criterion,

z^3	1	2
z^2	3	K
z^1	$\frac{6-K}{3}$	
z^0	K	

For the poles to lie on left of $s = -1$ axis,

$$6 - K > 0 \Rightarrow K < 6 \text{ and } K > 0$$

$$\therefore \quad 0 < K < 6$$

Option (b) satisfies the above range of K.

■■■

5 Root Locus

5.1 Consider the points $s_1 = -3 + j4$ and $s_2 = -3 - j2$ in the s-plane. Then, for a system with the open-loop transfer function

$$G(s)H(s) = \frac{K}{(s+1)^4}$$

(a) s_1 is on the root locus, but not s_2
(b) s_2 is on the root locus, but not s_1
(c) both s_1 and s_2 are on the root locus
(d) neither s_1 nor s_2 is on the root locus

[1999 : 2 M]

5.2 The root-locus diagram for a closed-loop feedback system is shown in the figure. The system is overdamped.

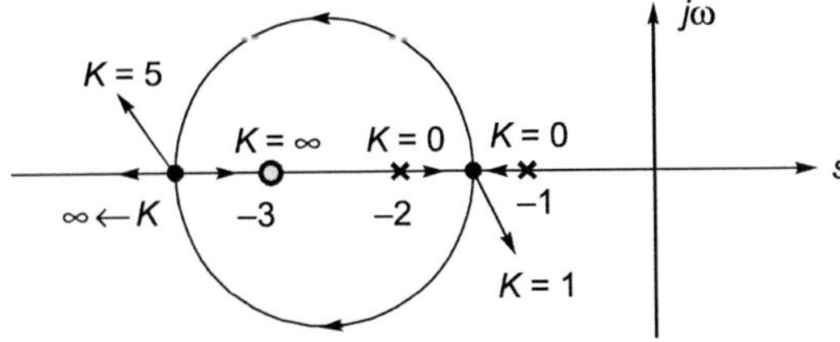

(a) only if $0 \le K \le 1$ (b) only if $1 < K < 5$
(c) only if $K > 5$ (d) if $0 \le K < 1$ or $K > 5$

[2001 : 1 M]

5.3 Which of the following points is NOT on the root locus of a system with the open-loop transfer function $G(s)H(s) = \dfrac{K}{s(s+1)(s+3)}$?

(a) $s = -j\sqrt{3}$ (b) $s = -1.5$
(c) $s = -3$ (d) $s = -\infty$ [2002 : 1 M]

5.4 The root locus of the system

$$G(s)H(s) = \frac{K}{s(s+2)(s+3)},$$

has the break-away point located at
(a) (–0.5, 0) (b) (–2.548, 0)
(c) (–4, 0) (d) (–0.784, 0)

[2003 : 2 M]

5.5 Given $G(s)H(s) = \dfrac{K}{s(s+1)(s+3)}$, the point of intersection of the asymptotes of the root loci with the real axis is

(a) –4 (b) 1.33
(c) –1.33 (d) 4 [2004 : 1 M]

5.6 An unity feedback system is given as,

$$G(s) = \frac{K(1-s)}{s(s+3)}$$

Indicate the correct root locus diagram

(a)
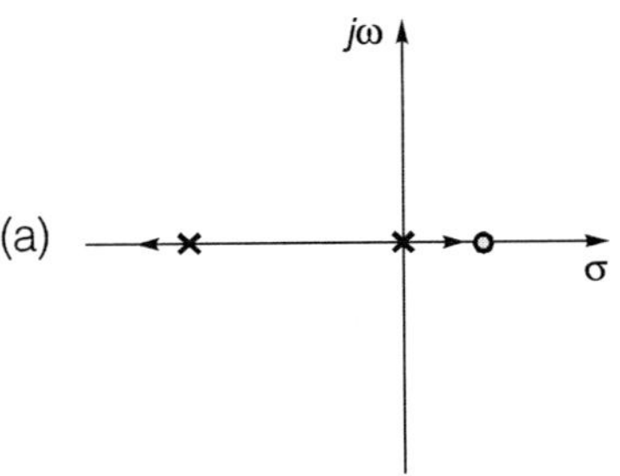

(b)
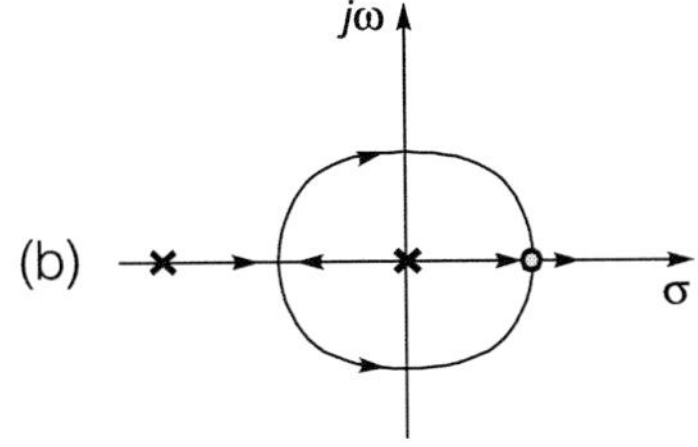

(c)
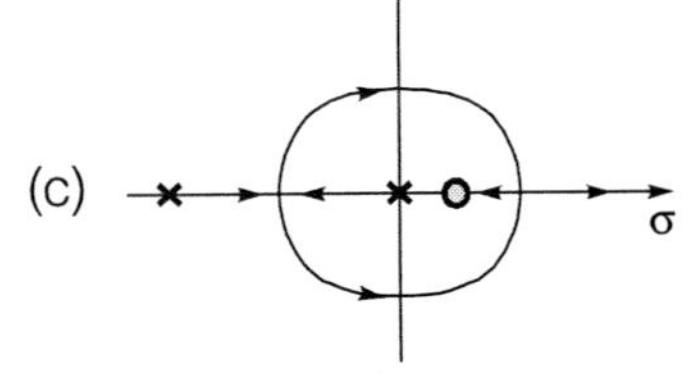

(d)
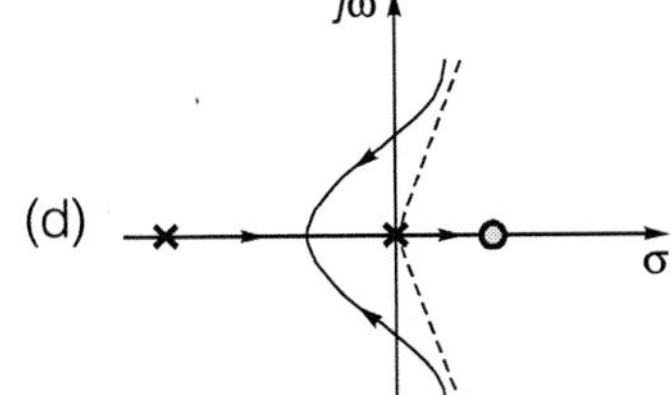

[2005 : 2 M]

5.7 A unity feedback control system has an open-loop transfer function $G(s) = \dfrac{K}{s(s^2+7s+12)}$. The gain K for which $s = -1 + j1$ will lie on the root locus of this system is
(a) 4 (b) 5.5
(c) 6.5 (d) 10 [2007 : 2 M]

5.8 The feedback configuration and the pole-zero locations of

$G(s) = \dfrac{s^2 - 2s + 2}{s^2 + 2s + 2}$ are shown below. The root locus for Negative values of K, i.e. for $-\infty < K < 0$, has breakaway/break-in points and angle of departure at pole P (with respect to the positive real axis) equal to

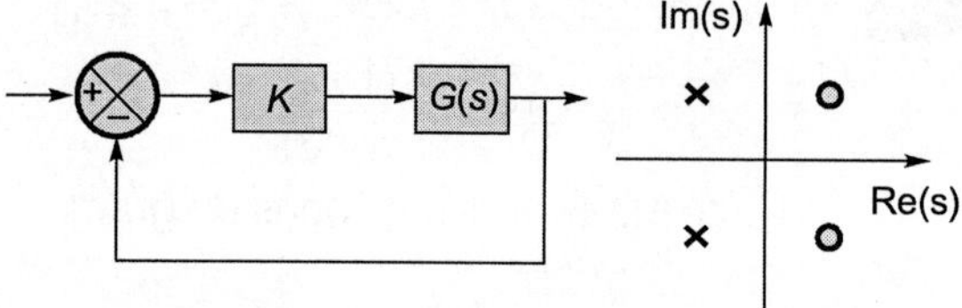

(a) $\pm\sqrt{2}$ and 0° (b) $\pm\sqrt{2}$ and 45°
(c) $\pm\sqrt{3}$ and 0° (d) $\pm\sqrt{3}$ and 45°

[2009 : 2 M]

5.9 The root locus plot for a system is given below. The open loop transfer function corresponding to this plot is given by

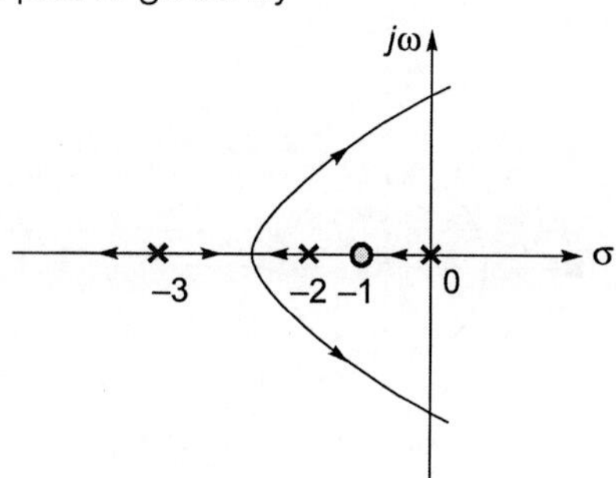

(a) $G(s)H(s) = K\dfrac{s(s+1)}{(s+2)(s+3)}$

(b) $G(s)H(s) = K\dfrac{(s+1)}{s(s+2)(s+3)^2}$

(c) $G(s)H(s) = K\dfrac{1}{s(s-1)(s+2)(s+3)}$

(d) $G(s)H(s) = K\dfrac{(s+1)}{s(s+2)(s+3)}$ [2011 : 1 M]

5.10 In the root locus plot shown in the figure, the pole/zero marks and the arrows have been removed. Which one of the following transfer functions has this root locus?

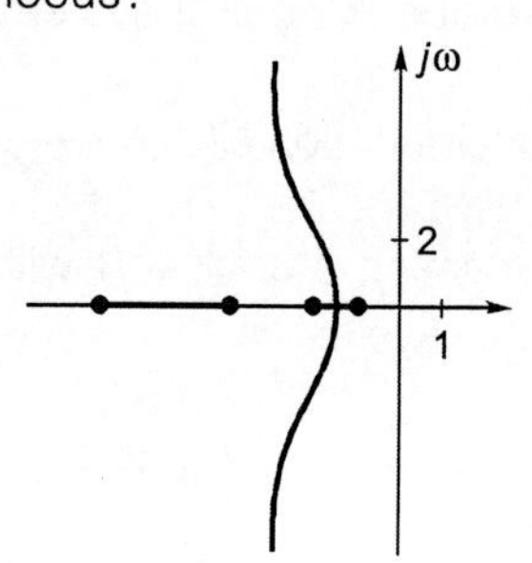

(a) $\dfrac{s+1}{(s+2)(s+4)(s+7)}$

(b) $\dfrac{s+4}{(s+1)(s+2)(s+7)}$

(c) $\dfrac{s+7}{(s+1)(s+2)(s+4)}$

(d) $\dfrac{(s+1)(s+2)}{(s+7)(s+4)}$ [2014 : 2 M, Set-3]

5.11 A unity negative feedback system has the open-loop transfer function $G(s) = \dfrac{K}{s(s+1)(s+3)}$. The value of the gain K (>0) at which the root locus crosses the imaginary axis is ______.

[2015 : 1 M, Set-1]

5.12 The open-loop transfer function of a unity feedback configuration is given as $G(s) = \dfrac{K(s+4)}{(s+8)(s^2-9)}$.

The value of a gain K(>0) for which $-1 + j2$ lies on the root locus is ________.

[2015 : 2 M, Set-1]

5.13 For the system shown in figure, $s = -2.75$ lies on the root locus if K is_____.

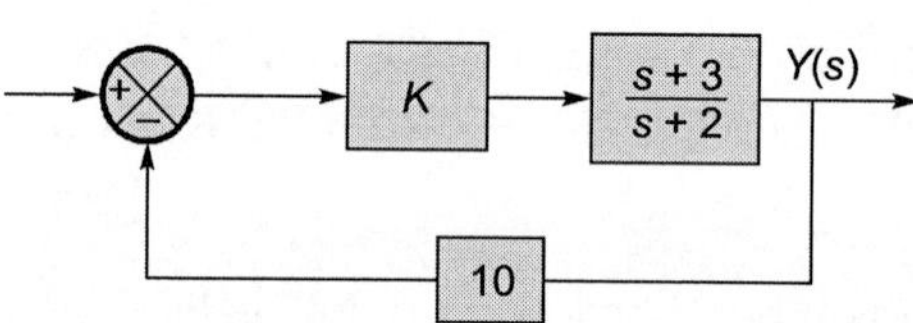

[2015 : 2 M, Set-3]

5.14 The open-loop transfer function of a unity-feedback control system is

$$G(s) = \frac{K}{s^2 + 5s + 5}$$

The value of K at the breakaway point of the feedback control system's root-locus plot is _____.

[2016 : 2 M, Set-1]

5.15 The forward-path transfer function and the feedback-path transfer function of a single loop negative feedback control system are given as

$$G(s) = \frac{K(s+2)}{s^2 + 2s + 2} \text{ and } H(s) = 1,$$

respectively. If the variable parameter K is real positive, then the location of the breakaway point on the root locus diagram of the system is ______.

[2016 : 2 M, Set-3]

5.16 A linear time invariant (LTI) system with the transfer function

$$G(s) = \frac{K(s^2 + 2s + 2)}{(s^2 - 3s + 2)}$$

is connected in unity feedback configuration as shown in the figure.

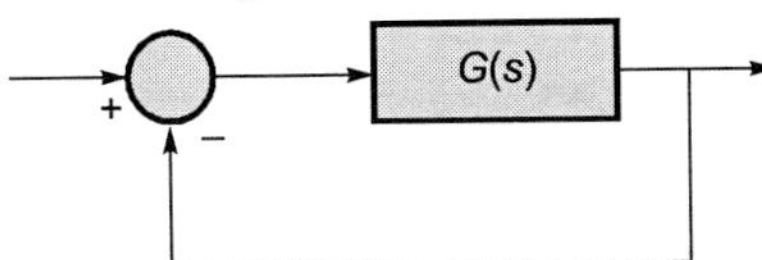

For the closed loop system shown, the root locus for $0 < K < \infty$ intersects the imaginary axis for $K = 1.5$. The closed loop system is stable for

(a) $K > 1.5$
(b) $1 < K < 1.5$
(c) $0 < K < 1.5$
(d) no positive value of K

[2017 : 2 M, Set-1]

5.17 The root-locus plot of a closed-loop system with unity negative feedback and transfer function $KG(s)$ in the forward path is shown in the figure. Note that K is varied from 0 to ∞.

Select the transfer function $G(s)$ that results in the root-locus plot of the closed-loop system as shown in the figure.

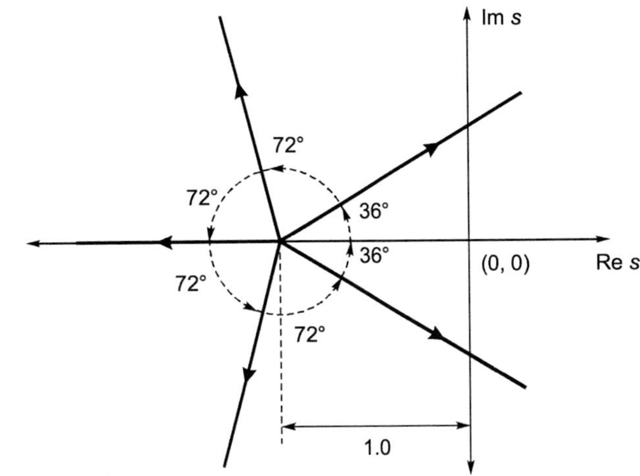

(a) $G(s) = \frac{1}{(s+1)^5}$
(b) $G(s) = \frac{1}{s^5 + 1}$
(c) $G(s) = \frac{s-1}{(s+1)^6}$
(d) $G(s) = \frac{s+1}{s^6 + 1}$

[2022 : 1 M]

5.18 The open loop transfer function of a unity negative feedback system is $G(s) = \frac{k}{s(1+sT_1)(1+sT_2)}$, where k, T_1 and T_2 are positive constants. The phase cross-over frequency, in rad/s, is

(a) $\frac{1}{\sqrt{T_1 T_2}}$
(b) $\frac{1}{T_1 T_2}$
(c) $\frac{1}{T_1\sqrt{T_2}}$
(d) $\frac{1}{T_2\sqrt{T_1}}$

[2023 : 1 M]

■■■

Answers Root Locus

5.1 (b)	5.2 (d)	5.3 (b)	5.4 (d)	5.5 (c)	5.6 (c)	5.7 (d)
5.8 (*)	5.9 (b)	5.10 (b)	5.11 (12)	5.12 (25.54)	5.13 (0.3)	5.14 (1.25)
5.15 (–3.41)	5.16 (a)	5.17 (a)	5.18 (a)			

Explanations Root Locus

5.1 (b)

$s_1 = -3 + 4j$

$$G(s)H(s) = \frac{K}{(-3+4j+1)^4} = \frac{K}{(-2+4j)^4}$$

$$\angle G(s)H(s) = -4[\pi - \tan^{-1} 2]$$

$$= -466° \neq \text{odd multiple of } \pi$$

So s_1 does not lie on root locus

$s_2 = -3 - j2$

$$G(s)H(s) = \frac{K}{(-3-j2+1)^4} = \frac{K}{(-2-2j)^4}$$

$$\angle G(s)H(s) = -4\left[\tan^{-1} 1\right] = -4 \times \frac{\pi}{4} = -\pi$$

Odd multiple of π

$\therefore$ s_2 lies on root locus

5.2 (d)

For overdamping roots of characteristic equation should lie on negative axis and be unequal.

5.3 (b)

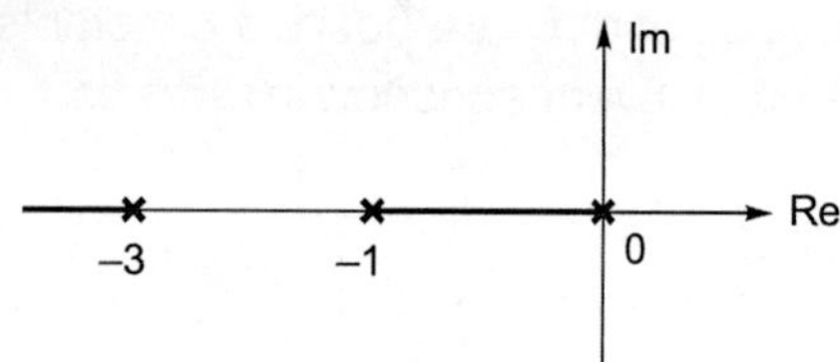

RL lies where no. of poles and zeroes to the right of the pole is odd.

$\therefore$ $s = -1.5$ doesn't lie on RL.

5.4 (d)

$$1 + G(s)H(s) = 0$$

$$\Rightarrow 1 + \frac{K}{s(s+2)(s+3)} = 0$$

$$\Rightarrow 1 + \frac{K}{s(s^2+5s+6)} = 0$$

$$\Rightarrow K = -1 \cdot (s^3 + 5s^2 + 6s)$$

$$\Rightarrow \frac{dK}{ds} = -(3s^2 + 10s + 6)$$

For break away point

$$\frac{dK}{ds} = 0$$

$$\Rightarrow 3s^2 + 10s + 6 = 0$$

$$s = \frac{-10 \pm \sqrt{100-72}}{6}$$

$$= \frac{-10 \pm 5.3}{6} = -0.784, -2.55$$

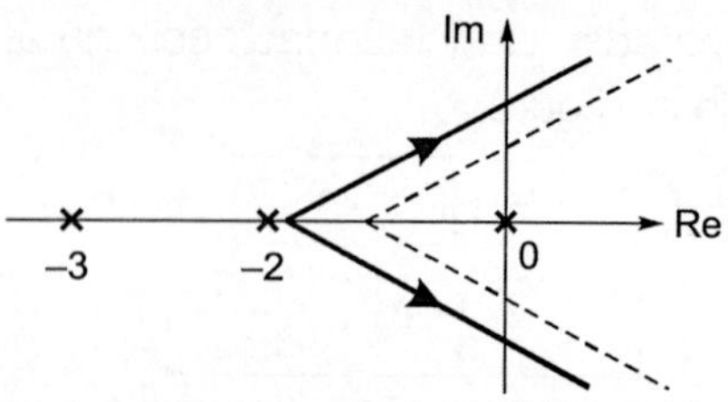

$$\text{Centroid} = \frac{-2-3}{3} = \frac{-5}{3} = -1.66$$

$$\text{angle of asymtotes} = (2K+1)\frac{\pi}{P-Z}$$

$$= (2K+1)\frac{\pi}{3}$$

$$\frac{\pi}{3}, \pi, \frac{5\pi}{3}$$

$\therefore$ Break away point is $(-0.784, 0)$ as -2.55 does not lie on RL.

5.5 (c)

$$\text{Centroid} = \frac{\Sigma P - \Sigma Z}{P - Z} = \frac{-1-3}{3} = -1.33$$

5.6 (c)

$$1 + G(s)H(s) = 0$$

$$K = \frac{s^2 + 3s}{1-s}$$

For break away and break-in point

$$\frac{dK}{ds} = (1-s)(2s+3) + s^2 + 3s = 0$$

$$= -s^2 + 2s + 3 = 0$$

$$\Rightarrow s^2 - 2s - 3 = 0$$

$$\Rightarrow (s-3)(s+1) = 0$$

$$s = 3, -1$$

-1 is the break-away point and 3 is the break-in point.

5.7 (d)

$$\text{T.F.} = \frac{G(s)}{1+G(s)}$$

As $H(s) = 1$

For the point $s = -1 + j1$ to lie on root locus

$$1 + G(s) = 0$$

$$\Rightarrow 1 + \frac{K}{s(s^2 + 7s + 12)} = 0$$

$$s(s^2 + 7s + 12) + K = 0$$

Putting $s = -1 + j$

$$(-1 + j)(1 - 2j - 1 - 7 + 7j + 12) + K = 0$$

$$\Rightarrow K = +10$$

5.8 (*)

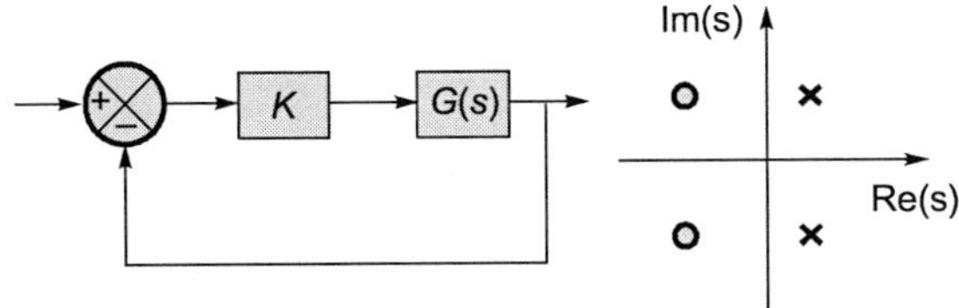

$$1 + G(s)H(s) = 0$$

$$1 + \frac{K(s^2 - 2s + 2)}{s^2 + 2s + 2} = 0$$

$$K = -\frac{s^2 + 2s + 2}{s^2 - 2s + 2}$$

put $\frac{\partial K}{\partial s} = 0$ we have

$$(s^2 - 2s + 2)(s + 1) - (s^2 + 2s + 2)$$
$$(s - 1) = 0$$
$$2s^2 - 4s^2 + 4 = 0$$
$$2s^2 = +4$$
$$s = \pm\sqrt{2}$$

Angle of departure is

$$\phi_D = \phi$$

where $\phi = \Sigma\phi_Z - \Sigma\phi_P$

$$\phi = 225°$$
$$\phi_D = \pm 225°$$

No option is matching.

5.9 (b)

From plot we can observe that one pole terminates at one zero at position –1 and three poles terminates to ∞. It means there are four poles and 1 zero. Pole at –3 goes on both sides. It means there are two poles at –3.

5.10 (b)

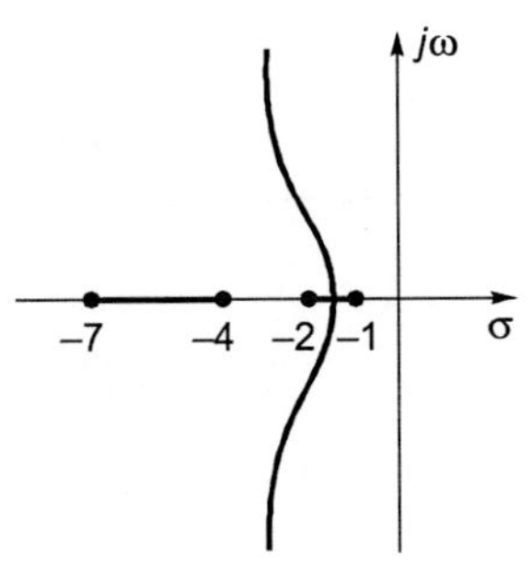

Since the root locus always emerges from the break-away points i.e. when two poles are at a part of same root locus lie. So σ = –1 and –2 will surely be poles.

Also we know that locus emerges from poles and terminates at zero. So, $s = -4$ is a zero and $s = -7$ is a pole.

$$\therefore T(s) = \frac{(s+4)}{(s+1)(s+2)(s+7)}$$

5.11 (12)

$$G(s) = \frac{K}{s(s+1)(s+3)}$$

The characteristic equation

$$= 1 + G(s)H(s) = 0$$
$$= s(s + 1)(s + 3) + K = 0$$
$$= s^3 + 4s^2 + 3s + K = 0$$

Using Routh's tabular form

s^3	1	3
s^2	4	K
s^1	$\frac{12-K}{4}$	0
s^0	K	

In order to cross the imaginary axis, system should be marginally stable

$$\therefore \quad \frac{12-K}{4} = 0$$

or $K = 12$

5.12 (25.54)

$$G(s) = \frac{K(s+4)}{(s+8)(s^2-9)} = \frac{K(s+4)}{(s+8)(s+3)(s-3)}$$

For the point $(-1 + 2j)$ to lie on the root locus the angle condition must be satisfy

i.e. $\angle G(s)H(s) = \pm(2Q + 1)\,180°$

Taking LHS:

$$\angle G(s)H(s)\big|_{s=-1+2j}$$

$$= \frac{\angle K + \angle(s+4)}{\angle(s+8) + \angle(s+3) + \angle(s-3)}$$

$$= \frac{\angle K + \angle(-1+2j+4)}{\angle(-1+2j+8) + \angle(-1+2j+3) + \angle(-1+2j-3)}$$

$$= \frac{0 + \tan^{-1}\left(\frac{2}{3}\right)}{\tan^{-1}\left(\frac{2}{7}\right) + \tan^{-1}(1) + 180° - \tan^{-1}\left(\frac{1}{2}\right)}$$

$$= \tan^{-1}\frac{2}{3} - \tan^{-1}\frac{2}{7} - \tan^{-1}1 - 180° + \tan^{-1}\frac{1}{2}$$

$= 33.69° - 15.945° - 45° - 180° + 26.565°$

$\simeq 180°$

As angle condition is satisfied the value of system gain K can be obtained by using magnitude condition

i.e. $|G(s)H(s)|_{s=-1+2j} = 1$

$$= \frac{K\sqrt{(-1+4)^2+2^2}}{\sqrt{(-1+8)^2+2^2}\cdot\sqrt{(-1+3)^2+2^2}\cdot\sqrt{(-1-3)^2+2^2}} = 1$$

$$= \frac{K\sqrt{9+4}}{\sqrt{49+4}\sqrt{4+4}\sqrt{16+4}} = 1$$

$$= \frac{K\sqrt{13}}{\sqrt{53}\sqrt{8}\sqrt{20}} = 1$$

$$K = \frac{\sqrt{53\times 8\times 20}}{\sqrt{13}} = 25.54$$

5.13 (0.3)

The open loop transfer function of the system:

$$G(s)\text{H}(s) = \frac{K(s+3)\times 10}{(s+2)}$$

In order to lie a point $s = -2.75$, the angle condition must be satisfy and therefore the magnitude condition is

$$|G(s)\,H(s)|\big|_{s=-2.75} = 1$$

$$= \left.\frac{10K\sqrt{s^2+3^2}}{\sqrt{s^2+2^2}}\right|_{s=-2.75} = 1$$

$$= \frac{10K\sqrt{(3-2.75)^2}}{\sqrt{(2-2.75)^2}} = 1$$

$$= \frac{0.25\times 10K}{0.75} \quad \text{or} \quad K = 0.3$$

5.14 (1.25)

Characteristic equation is $1 + G(s)\,H(s) = 0$

$$1+\frac{K}{s^2+5s+5} = 0$$

$$K = -s^2 - 5s - 5$$

For break away point $\frac{dK}{ds} = 0$

$$\frac{dK}{ds} = -2s - 5 = 0 \Rightarrow s = -2.5$$

Acc. to magnitude condition,

$$|G(s)\,H(s)|_{s=-2.5} = 1$$

$$|G(s)\,H(s)|_{s=-2.5} = \frac{|K|}{|(-2.5)^2+5\times -2.5+5|} = 1$$

$$\Rightarrow \quad |K| = |(6.25+5-12.5)| = 1.25$$

$$K = \pm 1.25$$

5.15 (-3.41)

$$G(s) = \frac{K(s+2)}{s^2+2s+2};\ H(s) = 1$$

$$1 + G(s)H(s) = 0$$

$$1+\frac{K}{s^2+2s+2}(s+2) = 0$$

$$K = -\frac{(s^2+2s+2)}{(s+2)}$$

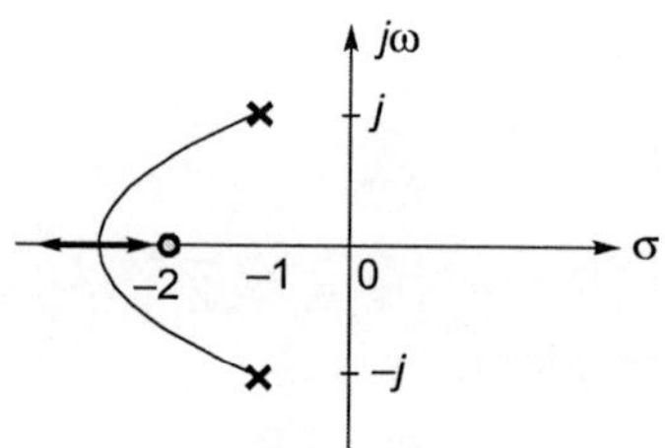

$$\frac{dK}{ds} = -\left(\frac{(2s+2)(s+2)-(s^2+2s+2)}{(s+2)^2}\right)$$

For break away points, $\frac{dK}{ds} = 0$

$(2s + 2)(s + 2) - (s^2 + 2s + 2) = 0$

$2s^2 + 6s + 4 - s^2 - 2s - 2 = 0$

$s^2 + 4s + 2 = 0$

$$s = \frac{-4\pm\sqrt{16-8}}{2}$$

$$= \frac{-4\pm 2\sqrt{2}}{2} = -2\pm\sqrt{2}$$

$s = -0.58$; $s = -3.41$

But, $s = -3.41$ lies on root locus

Hence $s = -3.41$

5.16 (a)

Given that,

$$G(s) = \frac{K(s^2+2s+2)}{s^2-3s+2}$$

The root locus plot of the given system is as follows:

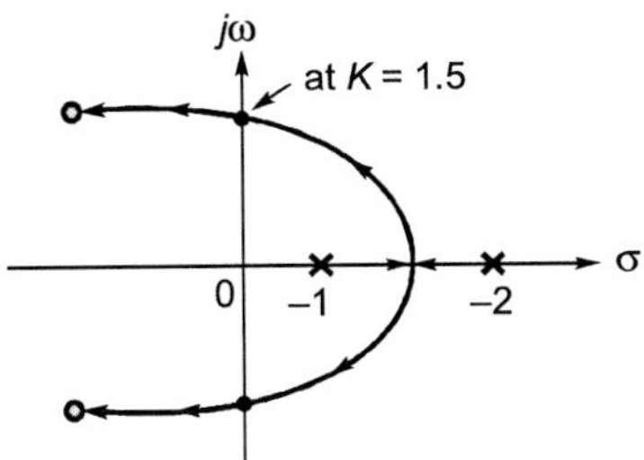

∴ System is stable for $K > 1.5$.

5.17 (a)

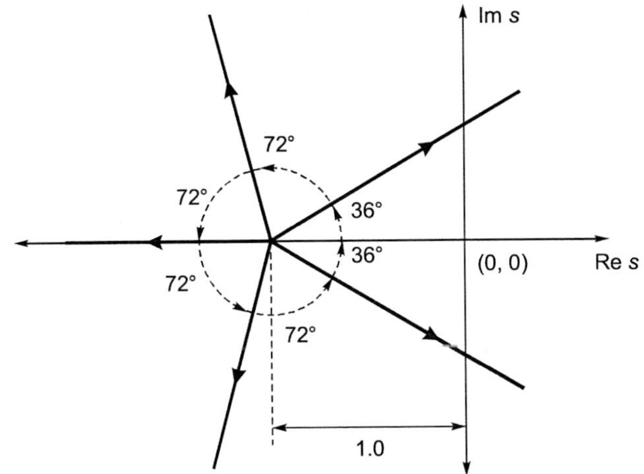

The given root locus plot has 5 branches originating from the same point and all terminating at infinity. So, there are five real multiple poles located at $s = -1$ and there are no zeros in the forward path transfer function $G(s)$. Hence,

$$G(s) = \frac{1}{(s+1)^5}$$

5.18 (a)

We know phase crossover frequency is that frequency at which phase of the open loop transfer function is −180°.

$$\therefore \quad G(s) = \frac{K}{s(1+sT_1)(1+sT_2)}$$

$$G(j\omega) = \frac{K}{(j\omega)(1+j\omega T_1)(1+j\omega T_2)}$$

Phase of $G(j\omega) = \phi = -90 - \tan^{-1}(\omega T_1) - \tan^{-1}(\omega T_2)$

∴ At $\omega = \omega_{pc}$, $\phi = -180°$

$$\therefore \quad -180° = -90° - \tan^{-1}(\omega_{pc}T_1) - \tan^{-1}(\omega_{pc}T_2)$$

$$90° = \tan^{-1}(\omega_{pc}T_1) + \tan^{-1}(\omega_{pc}T_2)$$

$$\tan^{-1}\left(\frac{\omega_{pc}T_1 + \omega_{pc}T_2}{1-\omega_{pc}^2 T_1 T_2}\right) = 90°$$

$$1 - \omega_{pc}^2 T_1 T_2 = 0$$

$$\omega_{pc} = \frac{1}{\sqrt{T_1 T_2}}$$

■■■

6 Frequency Response Analysis

6.1 The 3-dB bandwidth of a typical second-order system with the transfer function.

$$\frac{C(s)}{R(s)} = \frac{\omega_n^2}{s^2 + 2\xi\omega_n s + \omega_n^2}, \text{ is given by}$$

(a) $\omega_n\sqrt{1-2\xi^2}$

(b) $\omega_n\sqrt{(1-\xi^2)+\sqrt{\xi^4-\xi^2+1}}$

(c) $\omega_n\sqrt{(1-2\xi^2)+\sqrt{4\xi^4-4\xi^2+2}}$

(d) $\omega_n\sqrt{(1-2\xi^2)-\sqrt{4\xi^4-4\xi^2+2}}$ [1995 : 1 M]

6.2 Non-minimum phase transfer function is defined as the transfer function

(a) which has zeros in the right-half *s*-plane
(b) which has zeros only in the left-half *s*-plane
(c) which has poles in the right-half *s*-plane
(d) which has poles in the left-half *s*-plane

[1995 : 1 M]

6.3 In the Bode-plot of a unity feedback control system, the value of phase of $G(j\omega)$ at the gain cross over frequency is –125°. The phase margin of the system is

(a) –125° (b) –55°
(c) 55° (d) 125° [1998 : 1 M]

6.4 The Nyquist plot of a loop transfer function $G(j\omega)H(j\omega)$ of a system encloses the (–1, 0) point. The gain margin of the system is

(a) less than zero (b) zero
(c) greater than zero (d) infinity

[1998 : 1 M]

6.5 The gain margin (in dB) of a system having the loop transfer function

$$G(s)H(s) = \frac{\sqrt{2}}{s(s+1)} \text{ is}$$

(a) 0 (b) 3
(c) 6 (d) ∞ [1999 : 1 M]

6.6 The phase margin (in degrees) of a system having the loop transfer function

$$G(s)H(s) = \frac{2\sqrt{3}}{s(s+1)} \text{ is}$$

(a) 45° (b) –30°
(c) 60° (d) 30° [1999 : 1 M]

6.7 The Nyquist plot for the open-loop transfer function $G(s)$ of a unity negative feedback system is shown in the figure, if $G(s)$ has no pole in the right-half of *s*-plane, the number of roots of the system characteristic equation in the right-half of *s*-plane is

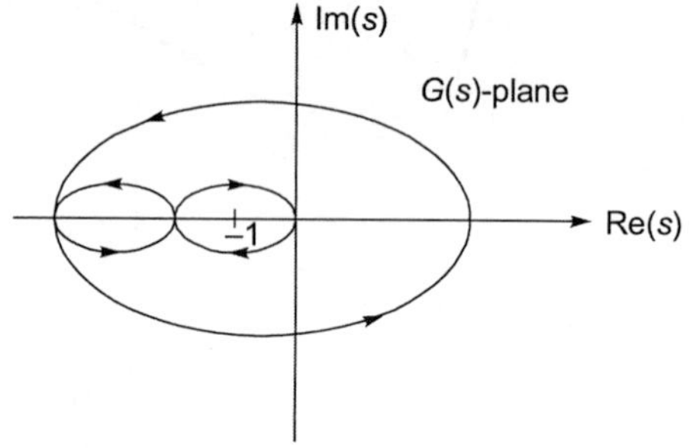

(a) 0 (b) 1
(c) 2 (d) 3 [2001 : 1 M]

6.8 The system with the open loop transfer function

$$G(s)H(s) = \frac{1}{s(s^2+s+1)} \text{ has a gain margin of}$$

(a) –6 dB (b) 0 dB
(c) 3.5 dB (d) 6 dB

[2002 : 2 M]

6.9 The phase margin of a system with the open-loop transfer function

$$G(s)H(s) = \frac{(1-s)}{(1+s)(2+s)}$$

(a) 0° (b) 63.4°
(c) 90° (d) ∞ [2002 : 1 M]

6.10 The gain margin and the phase margin of a feedback system with

$$G(s)H(s) = \frac{s}{(s+100)^3} \text{ are}$$

(a) 0 dB, 0° (b) ∞, ∞
(c) ∞, 0° (d) 88.5 dB, ∞

[2003 : 2 M]

6.11 In the figure, the Nyquist plot of the open-loop transfer function $G(s)H(s)$ of a system is shown. If $G(s)H(s)$ has one right-hand pole, the closed-loop system is

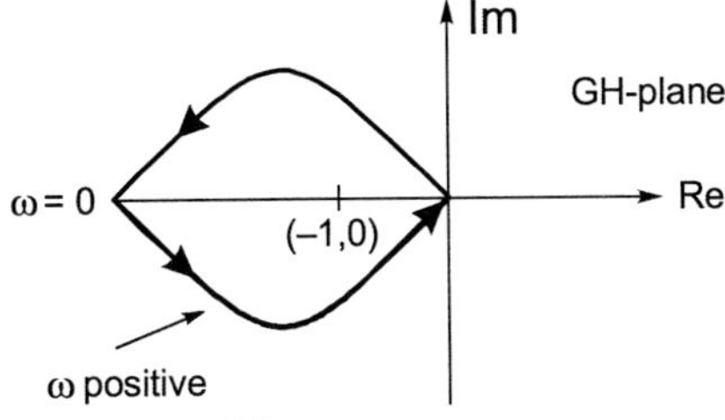

(a) always stable
(b) unstable with one closed-loop right hand pole
(c) unstable with two closed-loop right hand poles
(d) unstable with three closed-loop right hand poles

[2003 : 1 M]

6.12 The approximate Bode magnitude plot of a minimum-phase system is shown in the figure. The transfer function of the system is

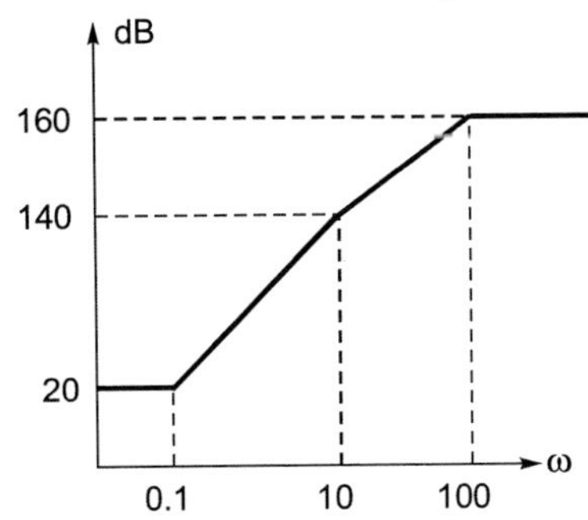

(a) $10^8 \dfrac{(s+0.1)^3}{(s+10)^2(s+100)}$

(b) $10^7 \dfrac{(s+0.1)^3}{(s+10)(s+100)}$

(c) $10^8 \dfrac{(s+0.1)^2}{(s+10)^2(s+100)}$

(d) $10^9 \dfrac{(s+0.1)^3}{(s+10)(s+100)^2}$

[2003 : 2 M]

6.13 Consider the Bode magnitude plot shown in the figure. The transfer function $H(s)$ is

dB
20 log |H(jω)|
0
–20
–20dB/decade
1 10 100 ω

(a) $\dfrac{(s+10)}{(s+1)(s+100)}$ (b) $\dfrac{10(s+1)}{(s+10)(s+100)}$

(c) $\dfrac{10^2(s+1)}{(s+10)(s+100)}$ (d) $\dfrac{10^3(s+100)}{(s+1)(s+10)}$

[2004 : 2 M]

6.14 A system has poles at 0.01 Hz, 1 Hz and 80 Hz; zeros at 5 Hz, 100 Hz and 200 Hz. The approximate phase of the system response at 20 Hz is

(a) –90° (b) 0°
(c) 90° (d) –180°

[2004 : 2 M]

6.15 Which one of the following polar diagrams corresponds to a lag network?

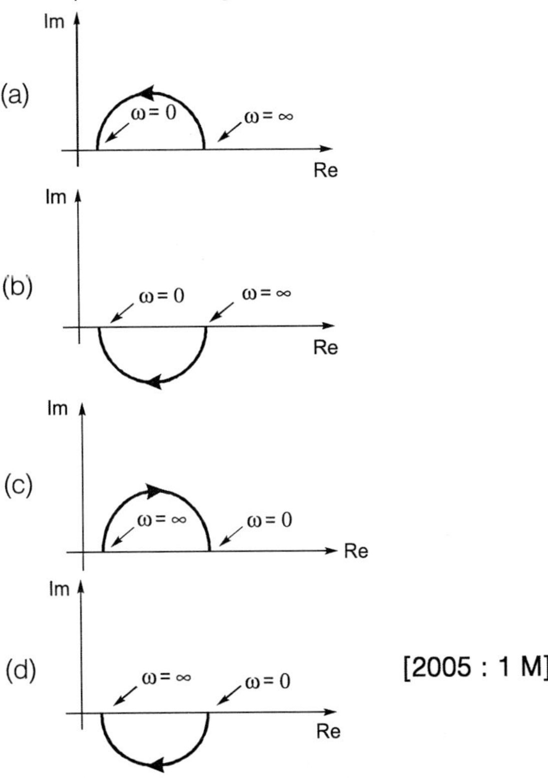

[2005 : 1 M]

6.16 The polar diagram of a conditionally stable system for open loop gain K = 1 is shown in the figure. The open loop transfer function of the system is known to be stable. The closed loop system is stable for

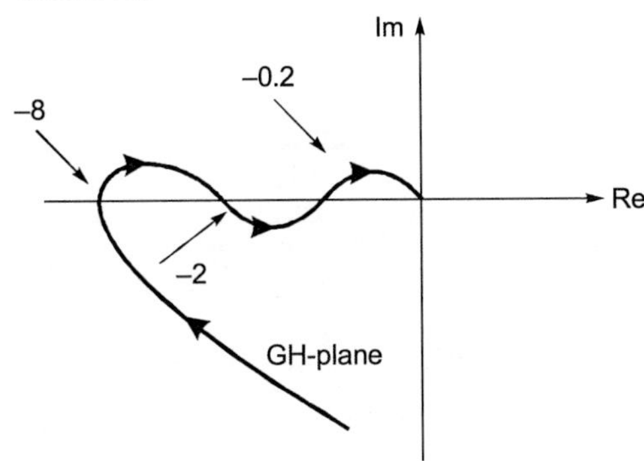

(a) $K < 5$ and $\frac{1}{2} < K < \frac{1}{8}$

(b) $K < \frac{1}{8}$ and $\frac{1}{2} < K < 5$

(c) $K < \frac{1}{8}$ and $5 < K$

(d) $K < \frac{1}{8}$ and $K < 5$ [2005 : 2 M]

Common Data for Questions 6.17 and 6.18

The open loop transfer function of a unity feedback system is given by $G(s) = \frac{3e^{-2s}}{s(s+2)}$.

6.17 The gain and phase crossover frequencies in rad/sec are, respectively

(a) 0.632 and 1.26 (b) 0.632 and 0.485
(c) 0.485 and 0.632 (d) 1.26 and 0.632

[2005 : 2 M]

6.18 Based on the above results, the gain and phase margins of the system will be

(a) –7.09 and 87.5°
(b) 7.09 and 87.5°
(c) 7.09 dB and –87.5°
(d) –7.09 dB and –87.5° [2005 : 2 M]

6.19 The open-loop transfer function of a unity-gain feedback control system is given by

$$G(s) = \frac{K}{(s+1)(s+2)}$$

The gain margin of the system in dB is given by

(a) 0 (b) 1
(c) 20 (d) ∞ [2006 : 1 M]

6.20 The Nyquist plot of $G(j\omega)\,H(j\omega)$ for a closed loop control system, passed through (–1, j0) point in the GH plane. The gain margin of the system in dB is equal to

(a) infinite (b) greater than zero
(c) less than zero (d) zero [2006 : 2 M]

6.21 In the system shown below, $x(t) = (\sin t)u(t)$. In steady-state, the response $y(t)$ will be

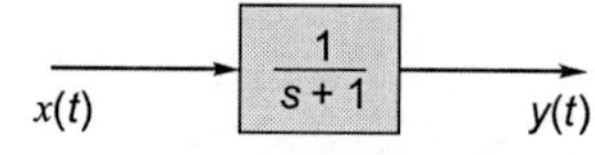

(a) $\frac{1}{\sqrt{2}}\sin\left(t - \frac{\pi}{4}\right)$ (b) $\frac{1}{\sqrt{2}}\sin\left(t + \frac{\pi}{4}\right)$

(c) $\frac{1}{\sqrt{2}}e^{-t}\sin t$ (d) $\sin t - \cos t$

[2006 : 1 M]

6.22 The frequency response of a linear, time-invariant system is given by $H(f) = \frac{5}{1 + j10\pi f}$. The step response of the system is

(a) $5(1 - e^{-5t})\,u(t)$ (b) $5(1 - e^{-t/5})u(t)$

(c) $\frac{1}{5}(1 - e^{-5t})u(t)$ (d) $\frac{1}{(s+5)(s+1)}$

[2007 : 2 M]

6.23 The asymptotic Bode plot of a transfer function is as shown in the figure. The transfer function $G(s)$ corresponding to this Bode plot is

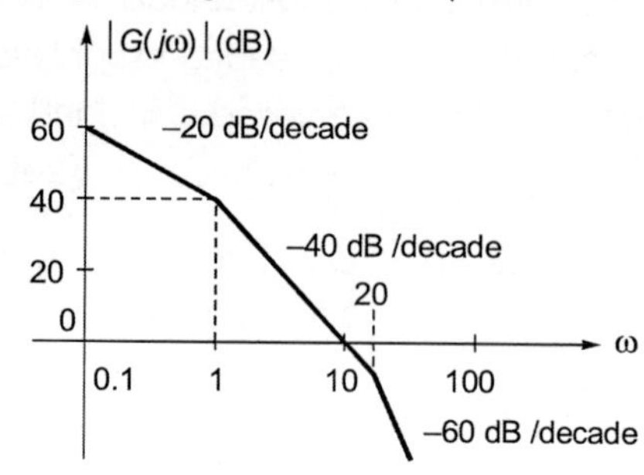

(a) $\frac{1}{(s+1)(s+20)}$ (b) $\frac{1}{s(s+1)(s+20)}$

(c) $\frac{100}{s(s+1)(s+20)}$ (d) $\frac{100}{s(s+1)(1+0.05s)}$

[2007 : 2 M]

6.24 The magnitude of frequency response of an underdamped second order system is 5 at 0 rad/sec and peaks to $\frac{10}{\sqrt{3}}$ at $5\sqrt{2}$ rad/sec. The transfer function of the system is

(a) $\frac{500}{s^2 + 10s + 100}$ (b) $\frac{375}{s^2 + 5s + 75}$

(c) $\frac{720}{s^2 + 12s + 144}$ (d) $\frac{1125}{s^2 + 25s + 225}$

[2008 : 2 M]

Common Data for Question 6.25 and 6.26:

The Nyquist plot of a stable transfer function $G(s)$ is shown in the figure. We are interested in the stability of the closed loop system in the feedback configuration shown.

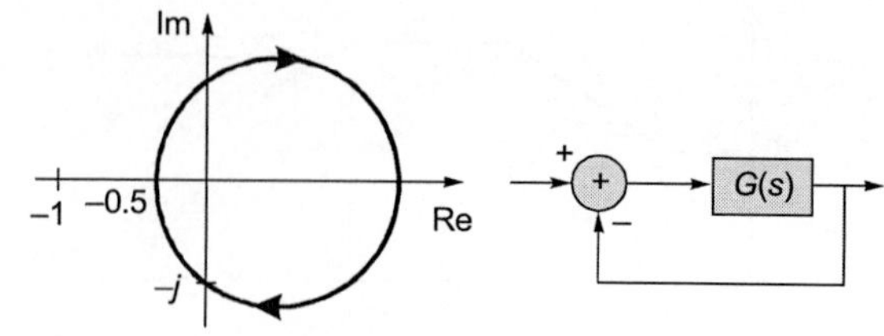

6.25 Which of the following statement is true?
(a) $G(s)$ is an all-pass filter
(b) $G(s)$ has a zero in the right-half plane
(c) $G(s)$ is the impedance of a passive network
(d) $G(s)$ is marginally stable [2009 : 2 M]

6.26 The gain and phase margins of $G(s)$ for closed loop stability are
(a) 6 dB and 180° (b) 3 dB and 180°
(c) 6 dB and 90° (d) 3 dB and 90°
[2009 : 2 M]

6.27 For the asymptotic Bode magnitude plot shown below, the system transfer function can be

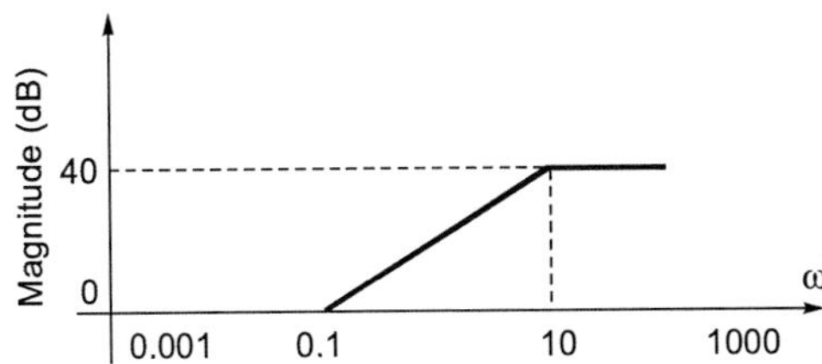

(a) $\dfrac{10s+1}{0.1s+1}$ (b) $\dfrac{100s+1}{0.1s+1}$
(c) $\dfrac{100s}{10s+1}$ (d) $\dfrac{0.1s+1}{10s+1}$
[2010 : 1 M]

6.28 A system with the transfer function $\dfrac{Y(s)}{X(s)} = \dfrac{s}{s+p}$ has an output $y(t) = \cos\left(2t - \dfrac{\pi}{3}\right)$ for the input signal $x(t) = p\cos\left(2t - \dfrac{\pi}{2}\right)$. Then, the system parameter 'p' is
(a) $\sqrt{3}$ (b) $\dfrac{2}{\sqrt{3}}$
(c) 1 (d) $\dfrac{\sqrt{3}}{2}$
[2010 : 1 M]

6.29 For the transfer function $G(j\omega) = 5 + j\omega$, the corresponding Nyquist plot for positive frequency has the form

(a)
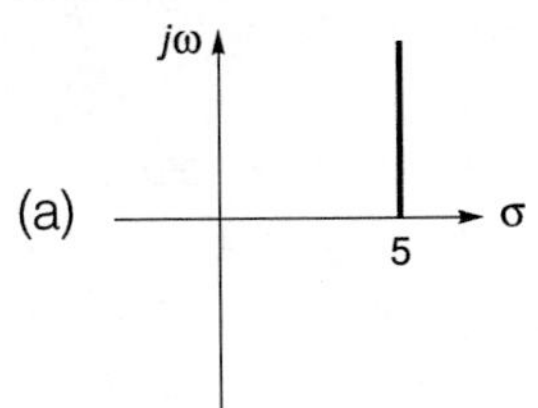

(b)
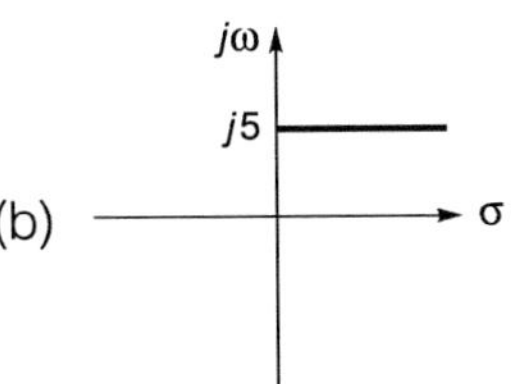

(c)
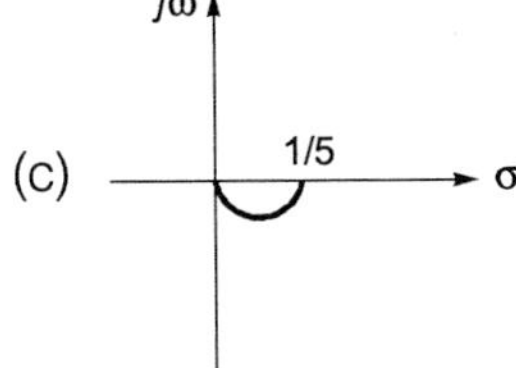

(d)
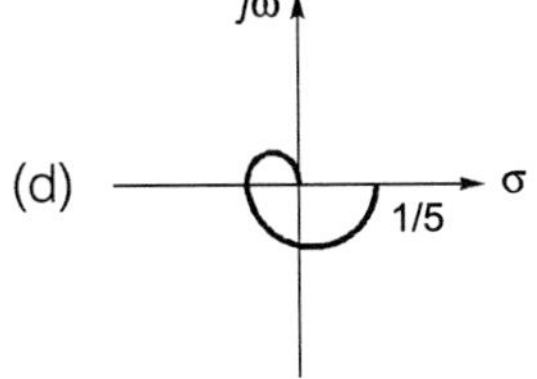

[2011 : 1 M]

6.30 The gain margin of the system under closed loop unity negative feedback is

$$G(s)\,H(s) = \frac{100}{s(s+10)^2}$$

(a) 0 dB (b) 20 dB
(c) 26 dB (d) 46 dB
[2011 : 2 M]

6.31 A system with transfer function

$$G(s) = \frac{(s^2+9)(s+2)}{(s+1)(s+3)(s+4)}$$

is excited by sin (ωt). The steady-state output of the system is zero at
(a) $\omega = 1$ rad/s (b) $\omega = 2$ rad/s
(c) $\omega = 3$ rad/s (d) $\omega = 4$ rad/s
[2012 : 1 M]

6.32 The Bode plot of a transfer function $G(s)$ is shown in the figure below:

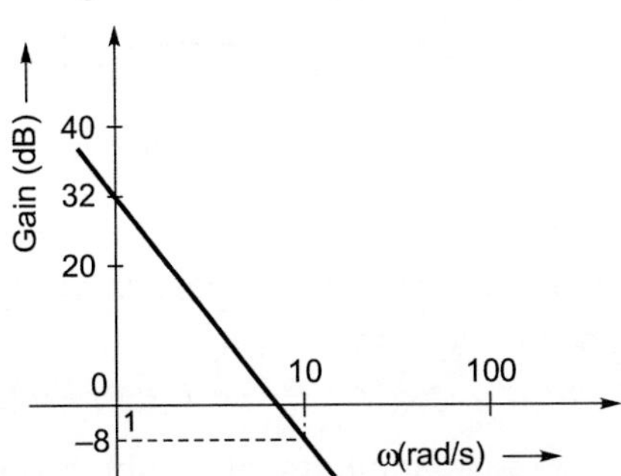

The gain $(20\log|G(s)|)$ is 32 dB and –8 dB at 1 rad/s and 10 rad/s respectively. The phase is negative for all ω. Then $G(s)$ is

(a) $\frac{39.8}{s}$ (b) $\frac{39.8}{s^2}$

(c) $\frac{32}{s}$ (d) $\frac{32}{s^2}$ [2013 : 1 M]

6.33 Consider the feedback system shown in the figure. The Nyquist plot of $G(s)$ is also shown. Which one of the following conclusions is correct?

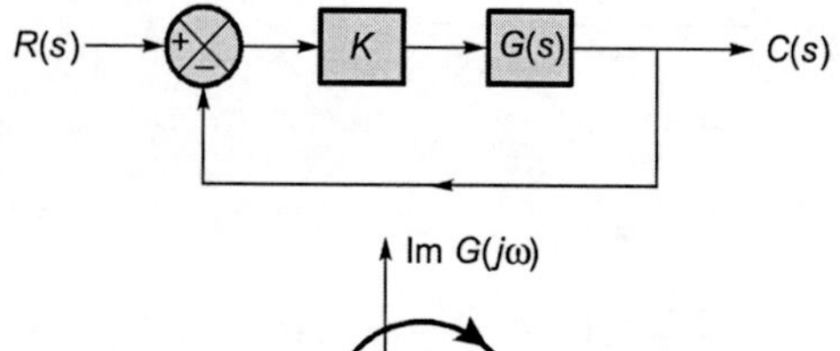

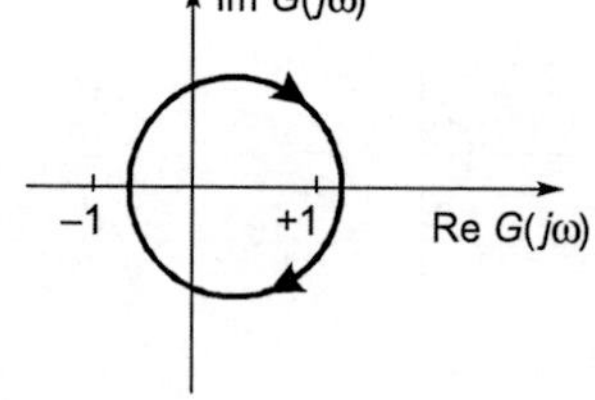

(a) $G(s)$ is an all-pass filter
(b) $G(s)$ is a strictly proper transfer function
(c) $G(s)$ is a stable and minimum-phase transfer function
(d) The closed-loop system is unstable for sufficiently large and positive K

[2014 : 1 M, Set-1]

6.34 The phase margin in degrees of

$$G(s) = \frac{10}{(s+0.1)(s+1)(s+10)}$$ calculated using the asymptotic Bode plot is ______.

[2014 : 2 M, Set-1]

6.35 The Bode asymptotic magnitude plot of a minimum phase system is shown in the figure.

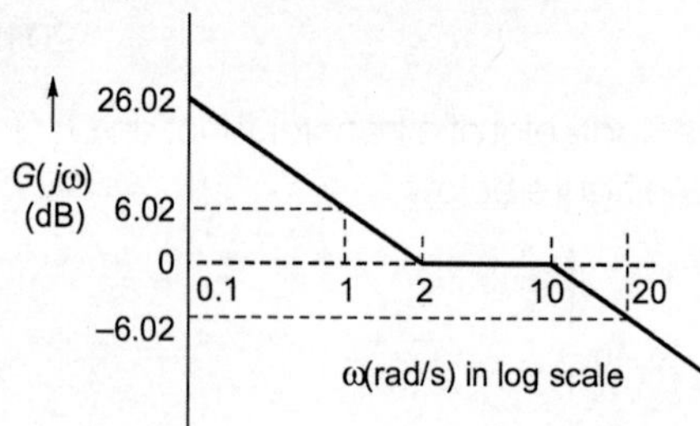

If the system is connected in a unity negative feedback configuration, the steady state error of the closed loop system, to a unit ramp input, is ______.

[2014 : 2 M, Set-2]

6.36 In a Bode magnitude plot, which one of the following slopes would be exhibited at high frequencies by a 4th order all-pole system?

(a) −80 dB/decade (b) −40 dB/decade
(c) +40 dB/decade (d) +80 dB/decade

[2014 : 1 M, Set-4]

6.37 The polar plot of the transfer function $G(s) = \frac{10(s+1)}{s+10}$ for $\leq \omega \leq \infty$ will be in the

(a) first quadrant (b) second quadrant
(c) third quadrant (d) fourth quadrant

[2015 : 1 M, Set-1]

6.38 The transfer function of a mass-spring-damper system is given by

$$G(s) = \frac{1}{Ms^2 + Bs + K}$$

The frequency response data for the system are given in the following table.

ω in rad/s	|G(jω)| in dB	arg (G(jω)) in deg
0.01	−18.5	−0.2
0.1	−18.5	−1.3
0.2	−18.4	−2.6
1	−16	−16.9
2	−11.4	−89.4
3	−21.5	−151
5	−32.8	−167
10	−45.3	−174.5

The unit step response of the system approaches a steady state value of ______.

[2015 : 2 M, Set-2]

6.39 Consider the Bode plot shown in figure. Assume that all the poles and zeros are real valued.

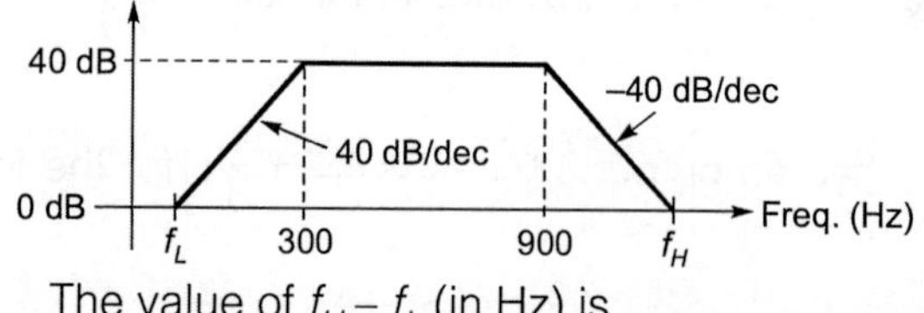

The value of $f_H - f_L$ (in Hz) is ______.

[2015 : 1 M, Set-3]

6.40 The phase margin (in degrees) of the system $G(s) = \frac{10}{s(s+10)}$ is ______. [2015 : 1 M, Set-3]

6.41 A closed-loop control system is stable if the Nyquist plot of the corresponding open-loop transfer function

(a) encircles the s-plane point $(-1 + j0)$ in the counterclockwise direction as many times as the number of right-half s-plane poles.
(b) encircles the s-plane point $(0 - j1)$ in the clockwise direction as many times as the number of right-half s-plane poles.
(c) encircles the s-plane point $(-1 + j0)$ in the counterclockwise direction as many times as the number of left-half s-plane poles.

(d) encircles the s-plane point $(-1 + j0)$ in the counterclockwise direction as many times as the number of right-half s-plane zeros.

[2016 : 1 M, Set-1]

6.42 The number and direction of encirclements around the point $-1 + j0$ in the complex plane by the Nyquist plot of $G(s) = \frac{1-s}{4+2s}$ is

(a) zero.
(b) one, anti-clockwise.
(c) one, clockwise.
(d) two, clockwise.

[2016 : 1 M, Set-2]

6.43 In the feedback system shown below

$$G(s) = \frac{1}{(s+1)(s+2)(s+3)}$$

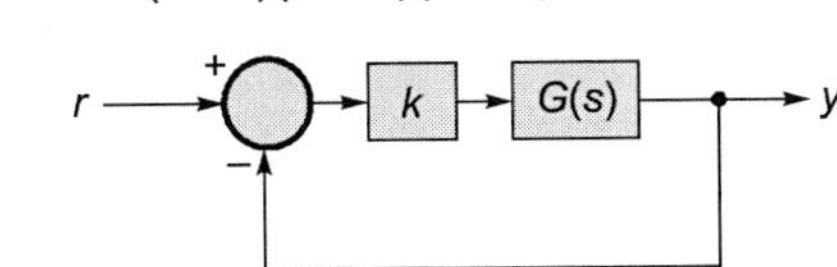

The positive value of k for which the gain margin of the loop is exactly 0 dB and the phase margin of the loop is exactly zero degree is ________

[2016 : 2 M, Set-2]

6.44 The asymptotic Bode phase plot of

$$G(s) = \frac{k}{(s+0.1)(s+10)(s+p_1)},$$ with k and p_1 both positive, is shown below.

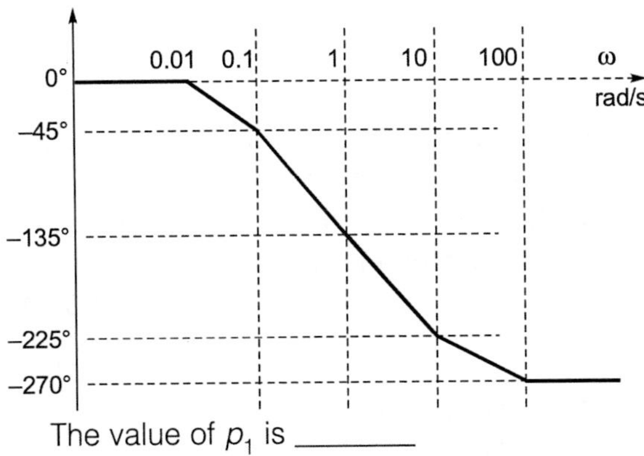

The value of p_1 is ________

[2016 : 2 M, Set-2]

6.45 Consider a stable system with transfer function

$$G(s) = \frac{s^p + b_1 s^{p-1} + \ldots + b_p}{s^q + a_1 s^{q-1} + \ldots + a_q}$$

where $b_1, \ldots, b_p$ and $a_1, \ldots, a_q$ are real valued constants. The slope of the Bode log magnitude curve of $G(s)$ converges to −60 dB/decade as $\omega \to \infty$. A possible pair of values for p and q is

(a) $p = 0$ and $q = 3$
(b) $p = 1$ and $q = 7$
(c) $p = 2$ and $q = 3$
(d) $p = 3$ and $q = 5$

[2017 : 1 M, Set-1]

6.46 The Nyquist plot of the transfer function

$$G(s) = \frac{K}{(s^2 + 2s + 2)(s + 2)}$$

does not encircle the point $(-1 + j0)$ for $K = 10$ but does encircle the point $(-1 + j0)$ for $K = 100$. Then the closed loop system (having unity gain feedback) is

(a) stable for $K = 10$ and stable for $K = 100$
(b) stable for $K = 10$ and unstable for $K = 100$
(c) unstable for $K = 10$ and stable for $K = 100$
(d) unstable for $K = 10$ and unstable for $K = 100$

[2017 : 2 M, Set-1]

6.47 A unity feedback control system is characterized by the open-loop transfer function

$$G(s) = \frac{10K(s+2)}{s^3 + 3s^2 + 10}$$

The Nyquist path and the corresponding Nyquist plot of $G(s)$ are shown in the figures below.

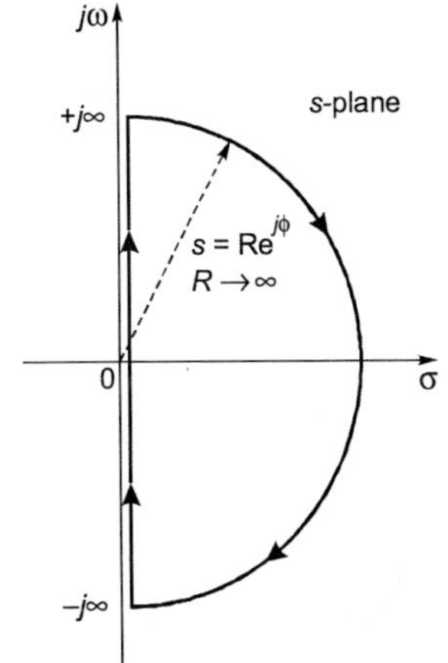

Nyquist path for $G(s)$

j Im G
G(s)-plane
+j 5.43 K
−jω
ω = 0
Re G
−K
2K
+jω
−j 5.43 K

Nyquist plot for $G(s)$

If $0 < K < 1$, then the number of poles of the closed-loop transfer function that lie in the right-half of the s-plane is

(a) 0 (b) 1
(c) 2 (d) 3 **[2017 : 2 M, Set-2]**

6.48 The Nyquist stability criterion and the Routh criterion both are powerful analysis tools for determining the stability of feedback controllers. Identify which of the following statements is FALSE:

(a) Both the criteria provide information relative to the stable gain range of the system.
(b) The general shape of the Nyquist plot is readily obtained from the Bode magnitude plot for all minimum-phase systems.
(c) The Routh criterion is not applicable in the condition of transport lag, which can be readily handled by the Nyquist criterion.
(d) The closed-loop frequency response for a unity feedback system cannot be obtained from the Nyquist plot. **[2018 : 1 M]**

6.49 For a unity feedback control system with the forward path transfer function $G(s) = \dfrac{K}{s(s+2)}$.
The peak resonant magnitude M_r of the closed-loop frequency response is 2. The corresponding value of the gain K (correct to two decimal places) is ___. **[2018 : 2 M]**

6.50 The figure below shows the Bode magnitude and phase plots of a stable transfer function $G(s) = \dfrac{n_o}{s^3 + d_2 s^2 + d_1 s + d_0}$.

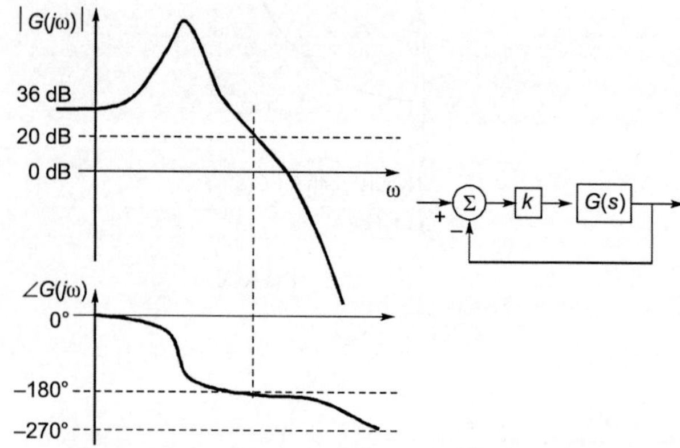

Consider the negative unity feedback configuration with gain k in the feedforward path. The closed loop is stable for $k < k_0$. The maximum value of k_0 is ______. **[2018 : 2 M]**

6.51 For an LTI system, the Bode plot for its gain is as illustrated in the figure shown. The number of system poles N_p and the number of system zeros N_z in the frequency range 1 Hz ≤ f ≤ 10^7 Hz is

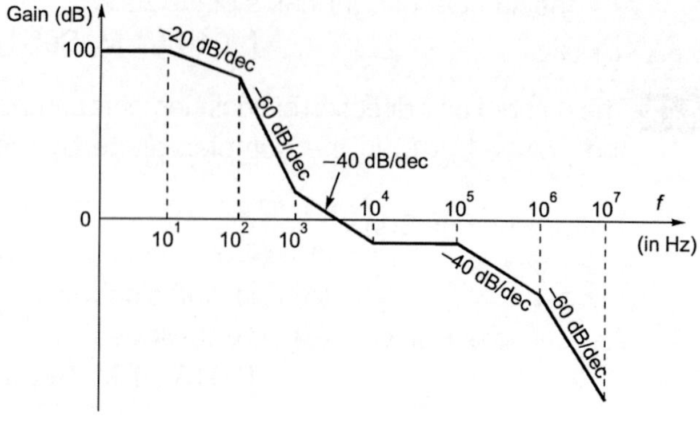

(a) $N_p = 6, N_z = 3$ (b) $N_p = 7, N_z = 4$
(c) $N_p = 5, N_z = 2$ (d) $N_p = 4, N_z = 2$
[2019 : 1 M]

6.52 The pole-zero map of a rational function $G(s)$ is shown below. When the closed counter Γ is mapped into the $G(s)$-plane, then the mapping encircles.

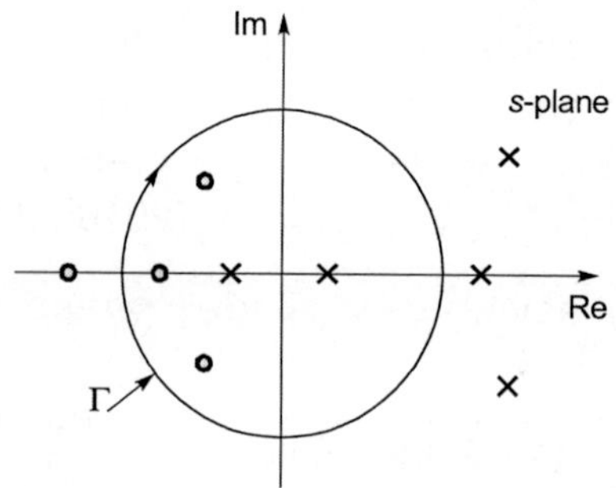

(a) the point $-1 + j0$ of the $G(s)$-plane once in the counter-clockwise direction.
(b) the origin of the $G(s)$-plane once in the clockwise direction.
(c) the origin of the $G(s)$-plane once in the counter-clockwise direction.
(d) the point $-1 + j0$ of the $G(s)$-plane once in the clockwise direction.

[2020 : 1 M]

6.53 The complete Nyquist plot of the open-loop transfer function $G(s) H(s)$ of a feedback control system is shown in the figure.

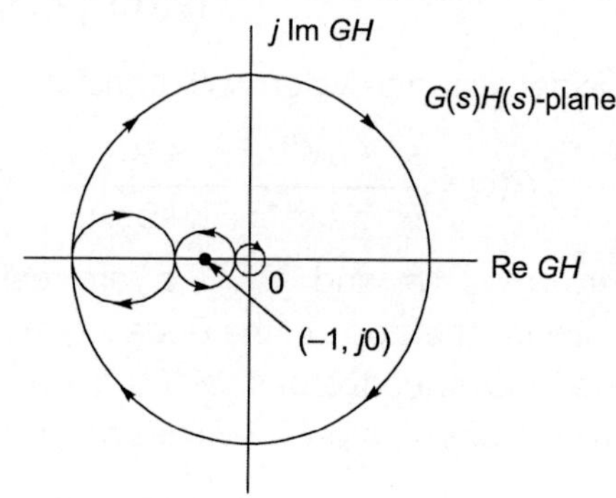

If $G(s)H(s)$ has one zero in the right-half of the s-plane, the number of poles that the closed-loop system will have in the right-half of the s-plane is

(a) 4 (b) 0
(c) 3 (d) 1 [2021 : 1 M]

6.54 Consider a closed-loop control system with unity negative feedback and $KG(s)$ in the forward path, where the gain $K = 2$. The complete Nyquist plot of the transfer function $G(s)$ is shown in the figure. Note that the Nyquist contour has been chosen to have the clockwise sense. Assume $G(s)$ has no poles on the closed right-half of the complex plane. The number of poles of the closed-loop transfer function in the closed right-half of the complex plane is ________.

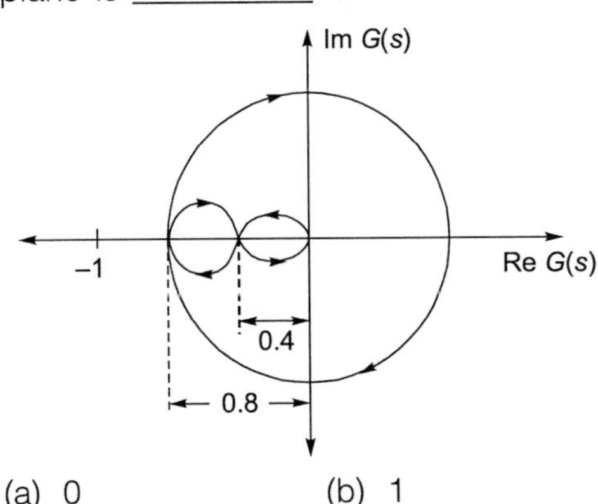

(a) 0 (b) 1
(c) 2 (d) 3

[2022 : 1 M]

6.55 The asymptotic magnitude Bode plot of a minimum phase system is shown in the figure. The transfer function of the system is

$(s) = \dfrac{k(s+z)^a}{s^b(s+p)^c}$, where k, z, p, a, b and c are positive constants. The value of $(a + b + c)$ is ________.

(rounded off to the nearest integer).

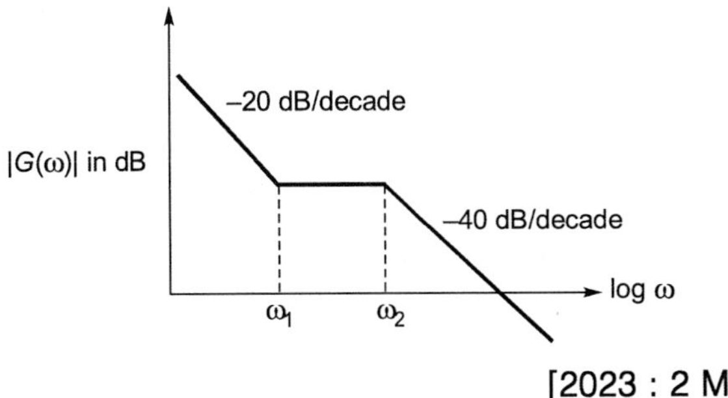

[2023 : 2 M]

6.56 In the context of Bode magnitude plots, 40 dB/decade is the same as

(a) 12 dB/octave (b) 10 dB/octave
(c) 6 dB/octave (d) 20 dB/octave

[2024 : 1 Mark]

■■■

Answers Frequency Response Analysis

6.1 (c)	6.2 (a & c)	6.3 (c)	6.4 (a)	6.5 (d)	6.6 (d)	6.7 (a)
6.8 (b)	6.9 (d)	6.10 (b)	6.11 (a)	6.12 (a)	6.13 (c)	6.14 (a)
6.15 (d)	6.16 (b)	6.17 (d)	6.18 (d)	6.19 (d)	6.20 (d)	6.21 (a)
6.22 (b)	6.23 (d)	6.24 (a)	6.25 (b)	6.26 (c)	6.27 (a)	6.28 (b)
6.29 (a)	6.30 (c)	6.31 (c)	6.32 (b)	6.33 (d)	6.34 (45)	6.35 (0.5)
6.36 (a)	6.37 (a)	6.38 (0.12)	6.39 (8970)	6.40 (84.36)	6.41 (a)	6.42 (a)
6.43 (60)	6.44 (1)	6.45 (a)	6.46 (b)	6.47 (c)	6.48 (d)	6.49 (14.92)
6.50 (0.10)	6.51 (a)	6.52 (b)	6.53 (*)	6.54 (c)	6.55. (4)	6.56 (a)

Explanations Frequency Response Analysis

6.1 (c)

Given that

Transfer function, $H(s) = \frac{C(s)}{R(s)} = \frac{\omega_n^2}{s^2 + 2\xi\omega_n s + \omega_n^2}$

Put $s = j\omega$

$$H(j\omega) = \frac{\omega_n^2}{(j\omega)^2 + 2\xi\omega_n(j\omega) + \omega_n^2}$$

$$= \frac{\omega_n^2}{(\omega_n^2 - \omega^2) + j2\xi\omega_n\,\omega}$$

$$H(j\omega) = \frac{1}{\left(1 - \frac{\omega^2}{\omega_n^2}\right) + j\frac{2\xi\omega}{\omega_n}}$$

Let $x = \frac{\omega}{\omega_n}$

$$H(j\omega) = \frac{1}{(1-x^2) + j2\xi x}$$

$$|H(j\omega)| = \frac{1}{\sqrt{(1-x^2)^2 + (2\xi x)^2}}$$

at 3dB frequency $H(j\omega_c) = \frac{1}{\sqrt{2}}$.

So, $\frac{1}{\sqrt{2}} = \frac{1}{\sqrt{(1-x^2)^2 + (2\xi x)^2}}$

$$\sqrt{(1-x^2)^2 + (2\xi x)^2} = \sqrt{2}$$

$$(1 - x^2)^2 + (2\xi x)^2 = 2$$

$$1 + x^4 - 2x^2 + 4\xi^2 x^2 = 2$$

$$x^4 - 2x^2 + 4\xi^2 x^2 = 1$$

$$x^4 + x^2(4\xi^2 - 2) - 1 = 0$$

$$x^2 = -\frac{(4\xi^2 - 2) \pm \sqrt{(4\xi^2 - 2)^2 - 4(1)(-1)}}{2}$$

$$x^2 = (1 - 2\xi^2) \pm \sqrt{4\xi^4 - 4\xi^2 + 2}$$

$$x = \sqrt{(1 - 2\xi^2) \pm \sqrt{4\xi^4 - 4\xi^2 + 2}} = \frac{\omega_c}{\omega_n}$$

$$\frac{\omega_c}{\omega_n} = \sqrt{(1 - 2\xi^2) \pm \sqrt{4\xi^4 - 4\xi^2 + 2}}$$

$$\omega_c = \omega_n\sqrt{(1 - 2\xi^2) \pm \sqrt{4\xi^4 - 4\xi^2 + 2}}$$

Since ω_c cannot be negative.

$$\omega_c = \omega_n\sqrt{(1 - 2\xi^2) + \sqrt{4\xi^4 - 4\xi^2 + 2}}$$

6.2 (a) & (c)

The non-minimum phase transfer function is defined as the transfer function which has zeros (or) poles in right side of s-plane.

6.3 (c)

$$\phi = -125°$$

$$\text{P.M.} = 180° + \phi = 180° - 125° = 55°$$

6.4 (a)

A system is unstable when Nyquist plot of $G(j\omega)\,H(j\omega)$ encloses the point $(-1 + j0)$ and gain margin of unstable system is always negative or less than zero. Therefore, the gain margin of given system is less than zero.

6.5 (d)

This is a second order function and so its gain margin is infinity.

6.6 (d)

Gain crossover frequency where gain is 1 is ω_g.

$$\left|\frac{2\sqrt{3}}{j\omega(1+j\omega)}\right| = 1 \quad \Rightarrow \quad \frac{2\sqrt{3}}{\omega\sqrt{1+\omega^2}} = 1$$

$$\Rightarrow \quad 2\sqrt{3} = \omega\sqrt{1+\omega^2} \quad \Rightarrow \omega = \sqrt{3}$$

$$\angle G(j\omega)H(j\omega) = -90° - \tan^{-1}\omega$$

$$= -90° - \tan^{-1}\sqrt{3} = -150°$$

$\therefore \quad PM = 180° - 150° = 30°$

6.7 (a)

$$N = 0$$

(1 encirclement in CW direction and other in ACW)

$$P = 0 \text{ (no pole in right half)}$$

So, $\quad N = P - Z$

$$Z = P - N = 0$$

$\therefore$ No roots on RH of s-plane.

6.8 (b)

$$G(s)H(s) = \frac{1}{s(s^2+s+1)}$$

ω when $\angle G(s)H(s) = -180°$

$$-180 = -90 - \tan^{-1}\frac{\omega}{1-\omega^2}$$

$$-90 = -\tan^{-1}\frac{\omega}{1-\omega^2}$$

$$1 - \omega^2 = 0$$

$$\omega_{pc} = 1 \text{ rad/sec}$$

Value of gain at $\omega_{pc} = 1$

$$|G(s)H(s)| = \frac{1}{\sqrt{|(1-\omega^2)|^2 + |j\omega|^2}}$$

$\therefore \quad$ G.M. $= -20 \log 1 = 0$

6.9 (d)

$$PM = 180° + \angle GH(j\omega)|_{\omega_{gc}}$$

For ω_{gc}, $\quad |G(s)H(s)|_{\omega = \omega_{gc}} = 1$

$$\left|\frac{1-s}{(1+s)(2+s)}\right| = \frac{\sqrt{1+\omega^2}}{\sqrt{1+\omega^2}\sqrt{4+\omega^2}} = 1$$

$$\sqrt{4+\omega^2} = 1$$

$$\Rightarrow \quad 4 + \omega^2 = 1$$

$$\omega^2 = -3 \text{ (imaginary)}$$

So no gain crossover frequency

$\therefore \quad PM = \infty$

6.10 (b)

$$G(s)H(s) = \frac{s}{(s+100)^3}$$

Calculating ω_{pc},

$$-180° = +90 - 3\tan^{-1}\frac{\omega_{pc}}{100}$$

$$-270° = -3\tan^{-1}\frac{\omega_{pc}}{100}$$

or, $\quad \omega_{pc} = \infty$

$\therefore \quad$ G.M. $= \infty$

Hence cannot be determined.

Calculating ω_{gc}

From, $\quad |G(s)H(s)| = 1$

$$\omega_{gc} = \text{negative value}$$

G.M. and P.M. of the system cannot be determined.

6.11 (a)

The encirclement of critical point (–1, 0) in A.C.W. direction is once.

$\therefore \quad N = 1, P = 1 \quad$ (given)

$$Z = P - N = O$$

No zero in RH of *s*-plane

So system is stable.

6.12 (a)

ω = 0.1 to 10, + 120 dB change

$\therefore$ 3 zeroes at 0.1

ω = 10 to 100, –40 dB

[i.e. +60 to +20] change

So two poles at ω = 10

ω = 100, –20 dB change.

$\therefore$ one pole at ω = 100

$$\therefore \quad T(s) = \frac{K(s+0.1)^3}{(s+10)^2(s+100)}$$

$$20\log(K/\omega)|_{\omega=10} = 140$$

$$\left.\frac{K}{\omega}\right|_{\omega=10} = 10^7$$

$\therefore \quad K = 10^8$

6.13 (c)

$$20 \log K = -20 \text{ dB}$$

$$\Rightarrow \quad K = 10^{-1} = 0.1$$

Zero at ω = 1 & poles at ω = 10, 100

$$H(s) = \frac{K(s+1)}{\left(\frac{s}{10}+1\right)\left(\frac{s}{100}+1\right)}$$

$$= \frac{0.1(s+1)\times 10 \times 100}{(s+10)(s+100)} = \frac{100(s+1)}{(s+10)(s+100)}$$

6.14 (a)

Pole at 0.01 and 1 Hz gives –180° phase.
Zero at 5 Hz gives +90° phase
∴ at 20 Hz –90° phase shift is provided.

6.15 (d)

Let $\frac{1}{s+1}$ be a lag network .

At $\omega = 0, \text{Mag} = \infty, \angle = -\tan^{-1} 0 = 0$

$\omega = \infty, \text{Mag} = 0, \angle = -\tan^{-1} \infty = -90°$

If in the direction of ω increasing phase shift is decreasing system is lag network.

6.16 (b)

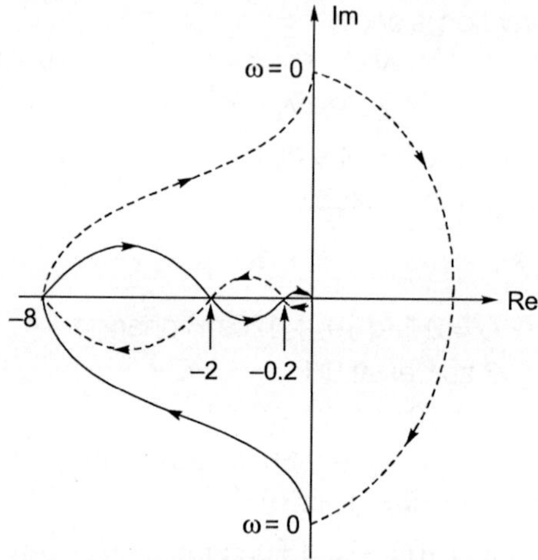

System is stable in region –0.2 to –2 & on the left side of –8 as number of encirclement there is zero.

$$0.2\,K < 1 \Rightarrow K < 5$$
$$2\,K > 1 \Rightarrow K > 0.5$$
$$\therefore \quad 0.5 < K < 5$$
$$1 > 8\,K$$

$K < \frac{1}{8}$ (negative sign only shows that it is on negative axis)

6.17 (d)

Gain cross over frequency where gain is 1.

$$|G(s)| = 1$$

$$\Rightarrow \quad \frac{3}{\omega(\omega^2+4)^{1/2}} = 1$$

$$\Rightarrow \quad \frac{9}{\omega^2(\omega^2+4)} = 1$$

$$\Rightarrow \quad \omega_{gc} = 1.26$$

Phase cross over frequency
where $\quad \angle GH = -180°$

$$\Rightarrow \quad \omega_{pc} = 0.632\left(\frac{\omega_\phi}{2} = \cot 2\omega_\phi\right)$$

6.18 (d)

$$GM\big|_{\omega=\omega_\phi} = \frac{3}{0.632(0.632^2+4)^{1/2}}$$

$$a = 2.26$$

$$\text{G.M.} = 20\log\frac{1}{a}$$

$$\Rightarrow \quad 20\log\frac{1}{2.26} = -7.09$$

Since, G.M. is negative system is unstable.
∴ P.M. is also negative.

$$\text{P.M.} = -87.5°$$

6.19 (d)

For 2nd order system G.M. = ∞

6.20 (d)

$$\text{G.M.} = 20\log\frac{1}{a}\text{dB}$$

$a = 1 \quad \therefore$ G.M. = 0

6.21 (a)

$$y(t) = x(t) * h(t)$$
$$Y(s) = X(s) \cdot H(s)$$
$$|H(\omega)| = \left|\frac{1}{s+1}\right| = \frac{1}{\sqrt{2}}\angle -45° = \frac{1}{\sqrt{2}}\angle -\frac{\pi}{4}$$
$$x(t) = \sin(t)\,u(t)$$
$$\therefore \quad y(t) = \frac{1}{\sqrt{2}}\sin\left(t-\frac{\pi}{4}\right)$$

6.22 (b)

$$H(f) = \frac{5}{1+j10\pi f}$$

$$H(s) = \frac{5}{1+5s} = \frac{5}{5\left(s+\frac{1}{5}\right)} = \frac{1}{s+\frac{1}{5}}$$

$$\text{Step response} = \frac{1}{s}\frac{1}{\left(s+\frac{1}{5}\right)}$$

$$\frac{1}{s} \times \frac{1}{\left(s+\frac{1}{5}\right)} = \frac{A}{s} + \frac{B}{s+\frac{1}{5}}$$

$$\Rightarrow A\left(s+\frac{1}{5}\right) + Bs = 1$$

when $s = 0$, $A = 5$

when $s = \frac{-1}{5}$, $B = -5$

$$Y(s) = \frac{5}{s} - \frac{5}{s + \frac{1}{5}}$$

$$\Rightarrow \quad y(t) = 5[1 - e^{-t/5}]\, u(t).$$

6.23 (d)

$$G(s) = \frac{K}{s(s+1)(s+20)} = \frac{K/20}{s(1+s)\left(1+\frac{s}{20}\right)}$$

Bode plot is in $(1 + sT)$ form

$$-20\log\omega + 20\log K = |60\ dB|_{\omega = 0.1}$$

$$-20\log 0.1 + 20\log K = 60 \text{ dB}$$

$$20\log K = 40\text{dB}$$

$$\Rightarrow \quad K = 100$$

$$\therefore \quad G(s) = \frac{100}{s(s+1)(1+.05s)}$$

6.24 (a)

$$M_r = \frac{5}{2\xi\sqrt{1-\xi^2}} = \frac{10}{\sqrt{3}}$$

or $\xi = \frac{1}{2}$

Also, $\omega_r = \omega_n\sqrt{1-2\xi^2}$

$$5\sqrt{2} = \omega_n\sqrt{1-2x\left(\frac{1}{4}\right)}$$

or $\omega_n = 10$ rad/sec

Only option (a) satisfies the conditions.

6.25 (b)

Nyquist plot will encircle the origin as many number of times as the difference between the number of right side poles and zeros of open loop transfer function.

i.e. $N = P - Z$

From the plot, $N = -1$

and $P = 0$ (Given)

$\therefore$ $Z = 1$

Thus, one open loop zero in right half of s-plane.

6.26 (c)

$$\text{Gain Margin} = 20\log\frac{1}{X} = 20\log\frac{1}{0.5} = 20\log 2 = 6 \text{ dB}$$

and phase margin = 90°

6.27 (a)

System transfer function,

$$G(s)\,H(s) = \frac{K\left(1+\frac{s}{0.1}\right)}{\left(1+\frac{s}{10}\right)}$$

Here, $20\log K = 0$

$\Rightarrow$ $K = 1$

Therefore, $G(s)\,H(s) = \frac{10s+1}{0.1s+1}$

6.28 (b)

Phase difference between input and output,

$$\phi = -\frac{\pi}{3} - \left(-\frac{\pi}{2}\right) = \frac{\pi}{6} = 30°$$

and $\omega = 2$ rad/s

From the transfer function,

$$\phi = 90° - \tan^{-1}\frac{\omega}{p}$$

$$90° - \tan^{-1}\frac{2}{p} = 30°$$

$$\tan^{-1}\frac{2}{p} = 60°$$

$$\frac{2}{p} = \tan 60° = \sqrt{3}$$

$$p = \frac{2}{\sqrt{3}}$$

6.29 (a)

$$G(j\omega) = 5 + j\omega$$

$$|G(j\omega)| = \sqrt{25+\omega^2}$$

At $\omega = 0$

$$|G(0)| = \sqrt{25+0} = 5$$

At $\omega = \infty$

$$|G(\infty)| = \infty$$

6.30 (c)

$$G(s)\,H(s) = \frac{100}{s(s+10)^2}$$

$$\phi = -90° - 2\tan^{-1}(\omega/10)$$

For phase cross-over frequency, $\phi = -180°$

$\therefore$ $-180 = -90° - 2\tan^{-1}(\omega/10)$

$\Rightarrow$ $\omega = 10$ rad/sec

Put, $s = j\omega$

$$G(j\omega)H(j\omega) = \frac{100}{j\omega(j\omega+100)^2}$$

$$|G(j\omega)H(j\omega)|_{\omega=10} = \frac{100}{\omega(\omega^2+100)} = \frac{100}{10(200)} = \frac{1}{20}$$

$$\text{G.M.} = \frac{1}{|G(j\omega)H(j\omega)|} = 20$$

G.M. in dB = 20 log 20 = 26 dB.

6.31 (c)

Transfer function is

$$G(s) = \frac{(s^2+9)(s+2)}{(s+1)(s+3)(s+4)}$$

Input $X(s) = \text{L.T}[\sin \omega t] = \dfrac{\omega}{s^2+\omega^2}$

So output, $Y(s) = \dfrac{(s^2+9)(s+2)}{(s+1)(s+3)(s+4)} \cdot \dfrac{\omega}{(s^2+\omega^2)}$

Now,

$$\lim_{t\to\infty} y(t) = \lim_{s\to 0} sY(s) = \lim_{s\to 0} s\frac{(s^2+9)(s+2)}{(s+1)(s+3)(s+4)} \cdot \frac{\omega}{(s^2+\omega^2)}$$

this will be zero if $s^2 + \omega^2$ will cancel $(s^2 + 9)$ term as only then final value theorem will be applicable. So at $\omega^2 = 9 \Rightarrow \omega = 3$ rad/sec, steady state output will be zero.

6.32 (b)

10 rad/s to 1 rad/s is 1 decade.

32 – (–8) = 40 dB

So, the slope is 40 dB/decade it means there are two poles at origin, it means either option (b) or option (d) is correct put $\omega = 1$ rad/sec in both the options.

$$20\log\left[\frac{39.8}{(1)^2}\right] = 32 \text{ dB}$$

$$20\log\left[\frac{32}{(1)^2}\right] = 30.1 \text{ dB}$$

So, option (b) is correct option $= \dfrac{39.8}{s^2}$.

6.33 (d)

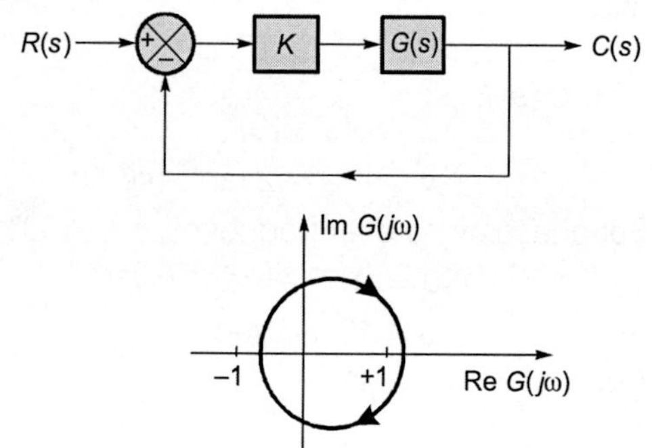

From the Nyquist plot, it can be seen that it does not encircle the critical point (–1, 0)

$N = P - Z, \quad N = 0$

- The closed loop-system is unstable for sufficiently large and positive K.

6.34 (45)

$$\frac{10}{0.1\times 10\left(\frac{s}{0.1}+1\right)\left(\frac{s}{1}+1\right)\left(\frac{s}{10}+1\right)}$$

$$= \frac{10}{\left(\frac{s}{0.1}+1\right)\left(\frac{s}{1}+1\right)\left(\frac{s}{10}+1\right)}$$

Amount of shift = 20log K = 20 log10 = 20 dB

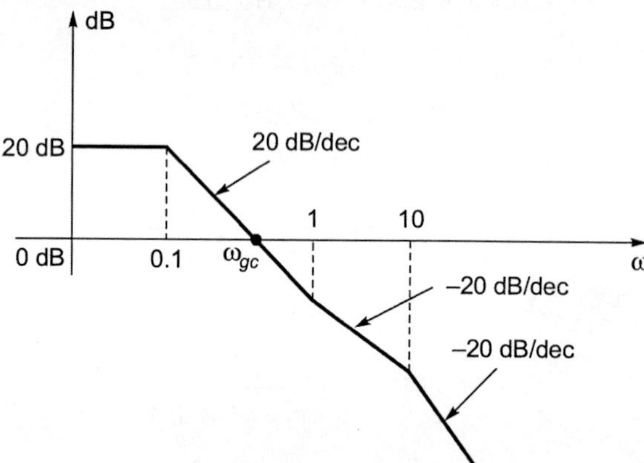

$$-20 \text{ dB} = \frac{20-0}{\log 0.1 - \log \omega_{gc}}$$

or, ω_{gc} = 1 rad/sec ...(i)

$$\text{PM} = 180° + \angle GH(j\omega)|_{\text{at } \omega = \omega_{gc}} \quad \text{...(ii)}$$

From the given transfer function,

$$\angle G(j\omega) = -\tan^{-1}\left(\frac{\omega_{gc}}{0.1}\right) - \tan^{-1}\left(\frac{\omega_{gc}}{1}\right) - \tan^{-1}\left(\frac{\omega_{gc}}{10}\right)$$

$$\text{or} \angle G(j\omega) = -84.289° - 45° - 5.711°$$

$$= -135° \quad \text{...(iii)}$$

∴ From equation (ii) and (iii) we get,

Phase margin PM = 180° – 135° = 45°

6.35 (0.5)

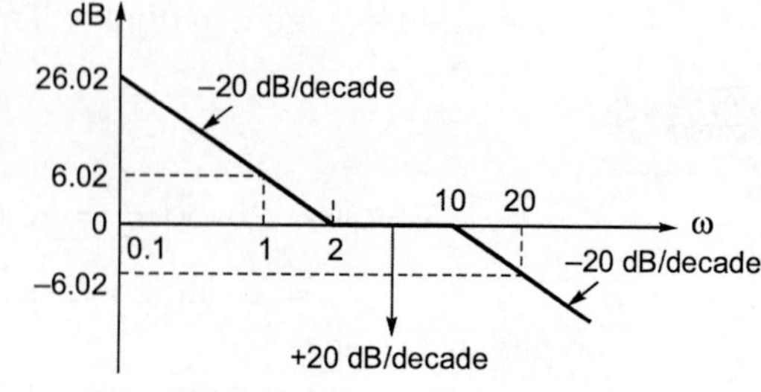

The open loop transfer function of the system is

$$G(s)\,H(s) = \frac{K\left(\frac{s}{2}+1\right)}{s\left(\frac{s}{10}+1\right)} \quad \text{...(i)}$$

where,

$$26.02\big|_{\omega=0.1} = 20\log K - 20\log \omega$$

or, $K = 1.99 \simeq 2$ ...(ii)

From equation (i) and (ii), we get

$$G(s) = \frac{20(s+2)}{2s(s+10)} \quad \text{...(iii)}$$

Steady state error for ramp input is

$$e_{ss} = \frac{A}{K_V} \text{ where } K_V = \lim_{s\to 0} sG(s)$$

From equation (iii),

$$K_V = \lim_{s\to 0} s \times \frac{10(s+2)}{s(s+10)} = 2$$

$$\therefore \quad e_{ss} = \frac{1}{2} = 0.5$$

6.36 (a)

A 4th order all-pole system means that the system must be having no zero or s-terms in numerator and s^4 terms in denominator or 4-poles. i.e.

$$H(s) \propto \frac{1}{s^4}$$

One pole exhibits –20 dB/decade slope, so 4 pole exhibits a slope of –80 dB/decade.

6.37 (a)

$$G(s) = \frac{10(s+1)}{(s+10)}$$

For

$\omega = 0 \Rightarrow M_1 = 1$ and $\phi_1 = 0°$

$\omega = 0 \Rightarrow M_2 = 10$

and $\phi_2 = -\tan^{-1}\infty + \tan^{-1}\infty = 0°$

∴ Polar plot

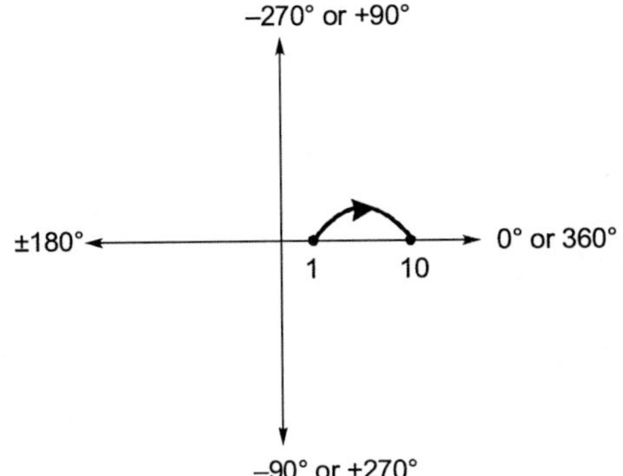

As the zero is nearer to the imaginary axis hence the direction of polar plot is clockwise.

6.38 (0.12)

$$\frac{C(s)}{R(s)} = G(s) = \frac{1}{Ms^2 + Bs + K}$$

Steady state value of response for unit step input

$$C(s) = \lim_{s\to 0} s \times \frac{1}{Ms^2+Bs+K} \times \frac{1}{s} = \frac{1}{K}$$

Also

$$G(\approx 0) = \frac{1}{K} = -18.5 \text{ dB}$$

(as per the given table)

$$\therefore \quad \frac{1}{K} = 10^{\left(-\frac{18.5}{20}\right)} = 0.12$$

6.39 (8970)

As we know,

$$\text{Slope} = \frac{y_2 - y_1}{x_2 - x_1}$$

$$\therefore \quad 40 \text{ dB/dec} = \frac{40-0}{\log 300 - \log f_L}$$

or, $\log\left(\frac{300}{f_L}\right) = \frac{40}{40} = 1$

or, $f_L = 30$ Hz

Similarly,

$$-40 \text{ dB/dec} = \frac{40-0}{\log 900 - \log f_H}$$

$$\log\left(\frac{900}{f_H}\right) = -1$$

or, $f_H = 9000$ Hz

Therefore, $f_H - f_L = 9000 - 30 = 8970$ Hz

6.40 (84.36)

$$G(s) = \frac{10}{s(s+10)}$$

$$\text{PM} = 180° + \angle G(\omega)\,H(\omega)\big|_{\omega=\omega_{gc}}$$

The gain cross over frequency can be calculated by

$$|G(\omega)H(\omega)| = 1 \;; \quad \frac{10}{\omega\sqrt{\omega^2+100}} = 1$$

$$\omega_{gc}^4 + 100\omega_{gc}^2 - 100 = 0$$

or $\omega_{gc}^2 = 0.99, \quad -100.99$

or $\omega_{gc} = 0.99$ rad/sec.

$$\therefore \text{PM} = 180° + \left(-90° - \tan^{-1}\frac{\omega_{gc}}{10}\right)$$

$$= 180° - 90° - \tan^{-1}\frac{0.99}{10} \simeq 84.364°$$

6.41 (a)

$N = P - Z$

N = Number of encirclements of $(-1 + j0)$. It is positive if Nyquist plot encircles the point $-1 + j0$ in counterclockwise direction.

Z = Number of closed loop poles lying in the right half of s-plane

P = Number of open loop poles lying in right half of s-plane

For stability $Z = 0 \Rightarrow N = P$

6.42 (a)

$$G(s) = \frac{1-s}{4+2s}$$

and $\angle G(s) = -\tan^{-1}\omega - \tan^{-1}\frac{\omega}{2}$

at $s = 0$, $|G(s)| = \frac{1}{4} = 0.25$

and $\angle G(s)|_{\omega=0} = 0°$

at $s = \infty$, $|G(s)| = -\frac{1}{2} = -0.5$

and $\angle G(s)|_{\omega=\infty} = -180°$

Nyquist plot is

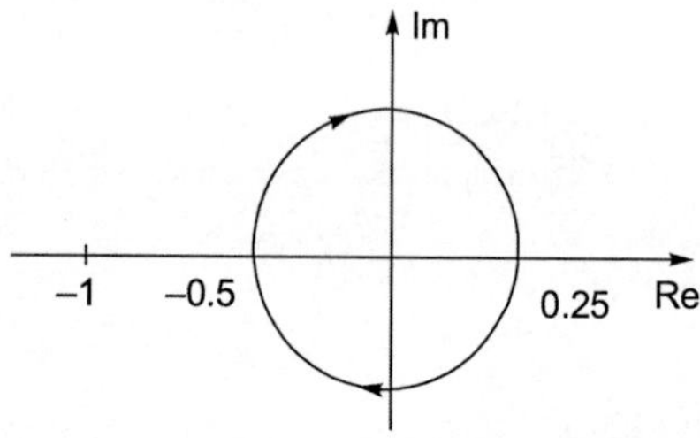

Hence, number of encirclements of $(-1 + j0) = 0$

6.43 (60)

$$1 + G(s)H(s) = 1 + \frac{K}{(s+1)(s+2)(s+3)} = 0$$

$(s + 1)(s^2 + 5s + 6) + K = 0$

$s^3 + 5s^2 + 6s + s^2 + 5s + 6 + K = 0$

$\Rightarrow s^3 + 6s^2 + 11s + 6 + K = 0$

Gain margin = 0 dB and phase margin = 0°

It implies marginal stable system

By Routh Array

s^3	1	11
s^2	6	$(6+K)$
s	$\frac{66-6-K}{6}$	0
s^0	$6+K$	

For marginal stable system,

$$60 - K = 0$$

$$\Rightarrow K = 60$$

6.44 (1)

$$G(s) = \frac{K}{(s+0.1)(s+10)(s+p_1)}$$

$$\angle G(s) = -\tan^{-1}\frac{\omega}{0.1} - \tan^{-1}\frac{\omega}{10} - \tan^{-1}\frac{\omega}{p_1}$$

$$\angle G(s)|_{\omega=1} = -\tan^{-1}\frac{1}{0.1} - \tan^{-1}\frac{1}{10} - \tan^{-1}\frac{1}{p_1} = -135°$$

$$-\tan^{-1}10 - \tan^{-1}0.1 - \tan^{-1}\frac{1}{p_1} = -135°$$

$$-84.28° - 5.71° - \tan^{-1}\frac{1}{p_1} = -135°$$

$$-\tan^{-1}\frac{1}{p_1} = -135° + 90°$$

$$\tan^{-1}\frac{1}{p_1} = 45°$$

$$\frac{1}{p_1} = 1 \quad \Rightarrow p_1 = 1$$

6.45 (a)

Final slope = –60 dB/decade, which indicates that, $q - p = 3$

Among the given options, option (a) satisfies this condition.

6.46 (b)

Given that, $G(s) = \dfrac{K}{(s^2+2s+2)(s+2)}$

P = open loop poles in RHS of s-plane

Z = closed loop poles in RHS of s-plane

N = number of encirclements about the point $(-1 + j0)$

For $K = 10$; $N = P - Z = 0 \Rightarrow Z = 0$: stable

For $K = 100$; $N = P - Z = 1 \Rightarrow Z \neq 0$: unstable

6.47 (c)

Given that, the open loop transfer function of a unity feedback system is,

$$G(s) = \frac{10k(s+2)}{s^3+3s^2+10}$$

From the given Nyquist plot, for $0 < k < 1$, the encirclements about the point $(-1 + j0)$ is,

$$N = 0$$
$$N = P - Z$$

P = number of open loop poles in the right half of s-plane

Z = number of closed loop poles in the right half of s-plane

To determine the value of P:

Applying R-H criteria to $G(s)$,

$$G(s) = \frac{10k(s+2)}{s^3+3s^2+10}$$

s^3	1	0
s^2	3	10
s^1	$-10/3$	0
s^0	10	0

Two sign changes are there in the first column of the RH table. So, two open loop poles are there in right half of s-plane.

So, $P = 2$

$$N = P - Z$$
$$0 = 2 - Z$$
$$Z = 2$$

$Z = 2$ indicates, there are two closed loop poles in right half of s-plane.

6.49 (14.92)

Maximum resonant peak,

$$M_r = \frac{1}{2\xi\sqrt{1-\xi^2}} = 2$$

$$2\xi\sqrt{1-\xi^2} = \frac{1}{2}$$

$$\xi^2(1-\xi^2) = \frac{1}{16}$$

$$\xi^4 - \xi^2 + \frac{1}{16} = 0$$

$$\xi^2 = \frac{1}{2} \pm \sqrt{\frac{1-\frac{1}{4}}{4}} = \frac{1}{2} \pm \frac{\sqrt{3}}{4}$$

As $M_r = 2 > 1$, $\xi < \frac{1}{\sqrt{2}}$ and $\xi^2 < \frac{1}{2}$

So, $$\xi^2 = \frac{1}{2} - \frac{\sqrt{3}}{4}$$

Given, $$G(s) = \frac{K}{s(s+2)} = \frac{\omega_n^2}{s(s+2\xi\omega_n)}$$

So, $$\omega_n = \sqrt{K}$$

$$2\xi\sqrt{K} = 2 \;;\; \sqrt{K} = \frac{1}{\xi}$$

$$K = \frac{1}{\xi^2} = \frac{1}{\left(\frac{1}{2} - \frac{\sqrt{3}}{4}\right)} = \frac{4}{2-\sqrt{3}} = 14.928$$

6.50 (0.10)

For $G(s)$,

$$M_{dB}(\omega_{pc}) = 20 \text{ dB}$$

When cascaded with k,

$$GM_{dB} = -20 \text{ dB} - 20\log_{10}(k) > 0 \text{ dB}$$

$$20 + 20\log_{10}(k) < 0$$
$$20\log_{10}(k) < -20$$
$$k < 10^{-1} = 0.10$$

So, $k_0 = 0.10$

6.51 (a)

Corner frequency (in Hz)	No. of poles (or) zeros
10	1pole
10^2	2 poles
10^3	1 zero
10^4	2 zeros
10^5	2 poles
10^6	1pole

Number of poles $(N_p) = 6$

Number of zeros $(N_z) = 3$

6.52 (b)

s-plane contour is encircling 2-poles and 3-zeros in clockwise direction hence the corresponding $G(s)$ plane contour encircles origin 2-times in anti-clockwise direction and 3-times in clockwise direction.

∴ Effectively once in clockwise direction.

6.53 (*)

The given Nyquist plot is not matched according to the data given in the question.

6.54 (c)

For $K=1$, the Nyquist plot for forward path transfer function is given as below:

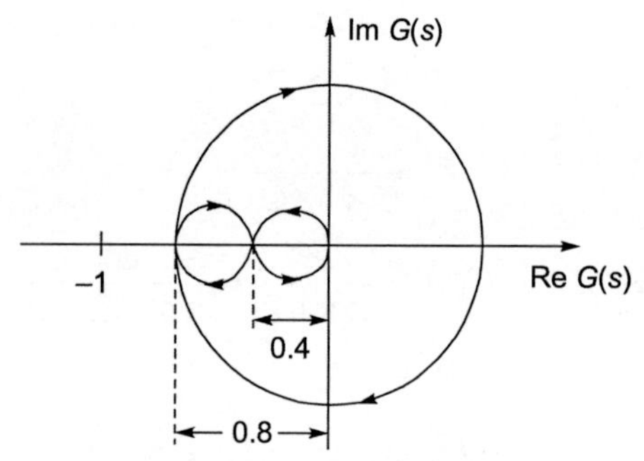

For K = 2, the Nyquist plot for $KG(s)$ will be as below:

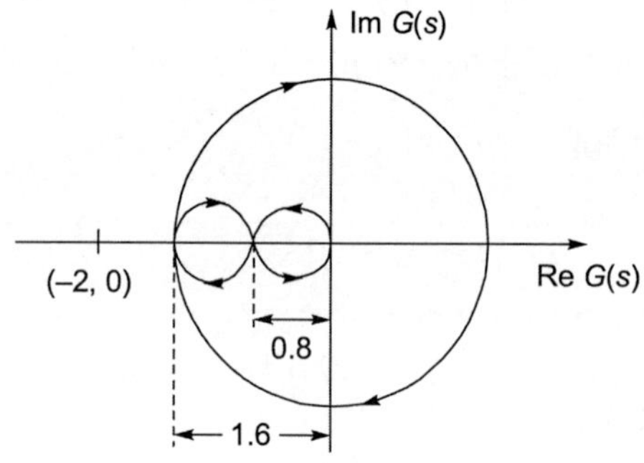

According to the Nyquist criterion,

$$N = P - Z$$

where, N = Number of encirclements about (−1, 0) in anticlockwise direction.

P = Number of open loop poles in RHS of complex plane.

Z = Number of closed-loop poles in the RHS of complex plane

From the Nyquist plot for $K = 2$,

$$N = -2$$

$$P = 0 \quad \text{(Given)}$$

Hence, $-2 = 0 - Z$

$\therefore$ $Z = 2$

Hence, two poles of the closed loop transfer function lies in the closed right half of the complex plane.

6.55 (4)

From the given Bode magnitude plot, it is clear that there is one pole at origin,

$\therefore$ $b = 1$

and at frequency ω_1, system has a zero

$\therefore$ $a = 1$

and at frequency ω_2, system have two poles

$\therefore$ $c = 2$

$\therefore$ $a + b + c = 1 + 1 + 2$

$a + b + c = 4$

6.56 (a)

Decade means ratio of frequencies = 10 i.e., 10 fold increase and octave means ratio of frequencies = 2 i.e., 2-fold increase.

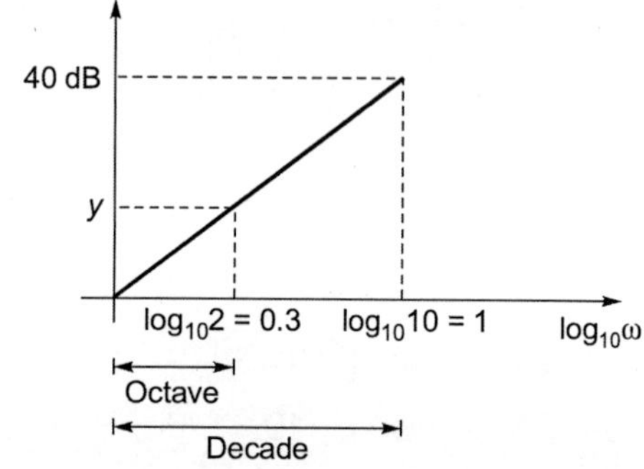

From the figure,

$$40 = \frac{y}{0.3}$$

$\Rightarrow$ $y = 12$ dB

Hence, 40 dB/decade is same as 12 dB/octave.

■■■

7 State Space Analysis

7.1 A certain linear time invariant system has the state and the output equations given below:

$$\begin{bmatrix} \dot{x}_1 \\ \dot{x}_2 \end{bmatrix} = \begin{bmatrix} 1 & -1 \\ 0 & 1 \end{bmatrix}\begin{bmatrix} x_1 \\ x_2 \end{bmatrix} + \begin{bmatrix} 0 \\ 1 \end{bmatrix}u$$

$$y = [1 \quad 1]\begin{bmatrix} x_1 \\ x_2 \end{bmatrix} \text{ is}$$

$x_1(0) = 1$, $x_2(0) = -1$, $u(0) = 0$ then $\left.\dfrac{dy}{dt}\right|_{t=0}$ is

(a) 1 (b) −1
(c) 0 (d) None of the above

[1997 : 2 M]

7.2 The system mode described by the state equations

$$\dot{x} = \begin{bmatrix} 0 & 1 \\ 2 & -3 \end{bmatrix}x + \begin{bmatrix} 0 \\ 1 \end{bmatrix}u, \; Y = [1 \quad 1]x \quad \text{is}$$

(a) controllable and observable
(b) controllable, but not observable
(c) observable, but not controllable
(d) neither controllable nor observable

[1999 : 1 M]

7.3 For the system described by the state equation

$$\dot{x} = \begin{bmatrix} 0 & 1 & 0 \\ 0 & 0 & 1 \\ 0.5 & 1 & 2 \end{bmatrix}x + \begin{bmatrix} 0 \\ 0 \\ 1 \end{bmatrix}u$$

If the control signal u is given by $u = [-0.5 \; -3 \; -5]x + v$, then the eigen values of the closed-loop system will be

(a) 0, −1, −2 (b) 0, −1, −3
(c) −1, −1, −2 (d) 0, −1, −1 [1999 : 2 M]

7.4 The transfer function $Y(s)/U(s)$ of a system described by the state equations $\dot{x}(t) = -2x(t) + 2u(t)$ and $y(t) = 0.5x(t)$ is

(a) $0.5/(s-2)$ (b) $1/(s-2)$
(c) $0.5/(s+2)$ (d) $1/(s+2)$

[2002 : 1 M]

7.5 The zero-input response of a system given by the state-space equation

$$\begin{bmatrix} \dot{x}_1 \\ \dot{x}_2 \end{bmatrix} = \begin{bmatrix} 1 & 0 \\ 1 & 1 \end{bmatrix}\begin{bmatrix} x_1 \\ x_2 \end{bmatrix} \text{ and } \begin{bmatrix} x_1(0) \\ x_2(0) \end{bmatrix} = \begin{bmatrix} 1 \\ 0 \end{bmatrix} \text{ is}$$

(a) $\begin{bmatrix} te^t \\ t \end{bmatrix}$ (b) $\begin{bmatrix} e^t \\ t \end{bmatrix}$

(c) $\begin{bmatrix} e^t \\ te^t \end{bmatrix}$ (d) $\begin{bmatrix} t \\ te^t \end{bmatrix}$ [2003 : 2 M]

7.6 The state variable equations of a system are:

1. $\dot{x}_1 = -3x_1 - x_2 + u$
2. $\dot{x}_2 = 2x_1$; $y = x_1 + u$

The system is

(a) controllable but not observable
(b) observable but not controllable
(c) neither controllable nor observable
(d) controllable and observable [2004 : 2 M]

7.7 Given $A = \begin{bmatrix} 1 & 0 \\ 0 & 1 \end{bmatrix}$, the state transition matrix e^{At} is given by

(a) $\begin{bmatrix} 0 & e^{-t} \\ e^{-t} & 0 \end{bmatrix}$ (b) $\begin{bmatrix} e^t & 0 \\ 0 & e^t \end{bmatrix}$

(c) $\begin{bmatrix} e^{-t} & 0 \\ 0 & e^{-t} \end{bmatrix}$ (d) $\begin{bmatrix} 0 & e^t \\ e^t & 0 \end{bmatrix}$

[2004 : 2 M]

7.8 A linear system is equivalently represented by two sets of state equations.

$X = AX + BU$ and $W = CW + DU$

The eigen values of the representations are also computed as $[\lambda]$ and $[\mu]$. Which one of the following statements is true?

(a) $[\lambda] = [\mu]$ and $X = W$
(b) $[\lambda] = [\mu]$ and $X \neq W$
(c) $[\lambda] \neq [\mu]$ and $X = W$
(d) $[\lambda] \neq [\mu]$ and $X \neq W$ [2005 : 1 M]

7.9 A linear system is described by the following state equation

$$\dot{X}(t) = AX(t) + BU(t), \; A = \begin{bmatrix} 0 & 1 \\ -1 & 0 \end{bmatrix}$$

The state-transition matrix of the system is

(a) $\begin{bmatrix} \cos t & \sin t \\ -\sin t & \cos t \end{bmatrix}$ (b) $\begin{bmatrix} -\cos t & \sin t \\ -\sin t & -\cos t \end{bmatrix}$

(c) $\begin{bmatrix} -\cos t & -\sin t \\ -\sin t & \cos t \end{bmatrix}$ (d) $\begin{bmatrix} \cos t & -\sin t \\ \cos t & \sin t \end{bmatrix}$

[2006 : 2 M]

7.10 The state space representation of a separately excited DC servo motor dynamics is given as

$$\begin{bmatrix} \frac{d\omega}{dt} \\ \frac{di_a}{dt} \end{bmatrix} = \begin{bmatrix} -1 & 1 \\ -1 & -10 \end{bmatrix}\begin{bmatrix} \omega \\ i_a \end{bmatrix} + \begin{bmatrix} 0 \\ 10 \end{bmatrix} u$$

where ω is the speed of the motor, i_a is the armature current and u is the armature voltage. The transfer function $\frac{\omega(s)}{U(s)}$ of the motor is

(a) $\frac{10}{s^2+11s+11}$ (b) $\frac{1}{s^2+11s+11}$

(c) $\frac{10s+10}{s^2+11s+11}$ (d) $\frac{1}{s^2+s+1}$

[2007 : 2 M]

Statement for Linked Answer Questions 7.11 & 7.12:

Consider a linear system whose state space representation is $\dot{x}(t) = Ax(t)$. If the initial state vector of the system is $x(0) = \begin{bmatrix} 1 \\ -2 \end{bmatrix}$, then the system response is $x(t) = \begin{bmatrix} e^{-2t} \\ -2e^{-2t} \end{bmatrix}$. If the initial state vector of the system changes to $x(0) = \begin{bmatrix} 1 \\ -1 \end{bmatrix}$, then the system response becomes $x(t) = \begin{bmatrix} e^{-t} \\ -e^{-t} \end{bmatrix}$.

7.11 The eigen-value and eigen-vector pairs (λ_i, v_i) for the system are

(a) $\left(-1, \begin{bmatrix} 1 \\ 1 \end{bmatrix}\right)$ and $\left(-2, \begin{bmatrix} 1 \\ 2 \end{bmatrix}\right)$

(b) $\left(-2, \begin{bmatrix} 1 \\ -1 \end{bmatrix}\right)$ and $\left(-1, \begin{bmatrix} 1 \\ -2 \end{bmatrix}\right)$

(c) $\left(-1, \begin{bmatrix} 1 \\ -1 \end{bmatrix}\right)$ and $\left(-2, \begin{bmatrix} 1 \\ -2 \end{bmatrix}\right)$

(d) $\left(-2, \begin{bmatrix} 1 \\ -1 \end{bmatrix}\right)$ and $\left(1, \begin{bmatrix} 1 \\ -2 \end{bmatrix}\right)$

[2007 : 2 M]

7.12 The system matrix A is

(a) $\begin{bmatrix} 0 & 1 \\ -1 & 1 \end{bmatrix}$ (b) $\begin{bmatrix} 1 & 1 \\ -1 & -2 \end{bmatrix}$

(c) $\begin{bmatrix} 1 & 1 \\ -1 & -2 \end{bmatrix}$ (d) $\begin{bmatrix} 0 & 1 \\ -2 & -3 \end{bmatrix}$

[2007 : 2 M]

7.13 A signal flow graph of a system is given below.

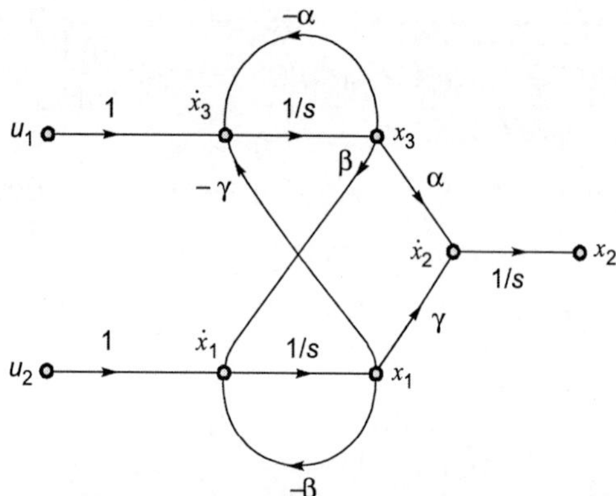

The set of equations that correspond to this signal flow graph is

(a) $\frac{d}{dt}\begin{pmatrix} x_1 \\ x_2 \\ x_3 \end{pmatrix} = \begin{bmatrix} \beta & -\gamma & 0 \\ \gamma & \alpha & 0 \\ -\alpha & -\beta & 0 \end{bmatrix}\begin{pmatrix} x_1 \\ x_2 \\ x_3 \end{pmatrix} + \begin{bmatrix} 1 & 0 \\ 0 & 0 \\ 0 & 1 \end{bmatrix}\begin{pmatrix} u_1 \\ u_2 \end{pmatrix}$

(b) $\frac{d}{dt}\begin{pmatrix} x_1 \\ x_2 \\ x_3 \end{pmatrix} = \begin{bmatrix} 0 & \alpha & \gamma \\ 0 & -\alpha & -\gamma \\ 0 & \beta & -\beta \end{bmatrix}\begin{pmatrix} x_1 \\ x_2 \\ x_3 \end{pmatrix} + \begin{bmatrix} 0 & 0 \\ 0 & 1 \\ 1 & 0 \end{bmatrix}\begin{pmatrix} u_1 \\ u_2 \end{pmatrix}$

(c) $\frac{d}{dt}\begin{pmatrix} x_1 \\ x_2 \\ x_3 \end{pmatrix} = \begin{bmatrix} -\alpha & \beta & 0 \\ -\beta & -\gamma & 0 \\ \alpha & \gamma & 0 \end{bmatrix}\begin{pmatrix} x_1 \\ x_2 \\ x_3 \end{pmatrix} + \begin{bmatrix} 1 & 0 \\ 0 & 1 \\ 0 & 0 \end{bmatrix}\begin{pmatrix} u_1 \\ u_2 \end{pmatrix}$

(d) $\frac{d}{dt}\begin{pmatrix} x_1 \\ x_2 \\ x_3 \end{pmatrix} = \begin{bmatrix} -\gamma & 0 & \beta \\ \gamma & 0 & \alpha \\ -\beta & 0 & -\alpha \end{bmatrix}\begin{pmatrix} x_1 \\ x_2 \\ x_3 \end{pmatrix} + \begin{bmatrix} 0 & 1 \\ 0 & 0 \\ 1 & 0 \end{bmatrix}\begin{pmatrix} u_1 \\ u_2 \end{pmatrix}$

[2008 : 2 M]

7.14 Consider the matrix $P = \begin{bmatrix} 0 & 1 \\ -2 & -3 \end{bmatrix}$. The value of e^P is

(a) $\begin{bmatrix} 2e^{-2} - 3e^{-1} & e^{-1} - e^{-2} \\ 2e^{-2} - 2e^{-1} & 5e^{-2} - e^{-1} \end{bmatrix}$

(b) $\begin{bmatrix} e^{-1}+e^{-2} & 2e^{-2}-e^{-1} \\ 2e^{-1}-4e^{2} & 3e^{-1}+2e^{-2} \end{bmatrix}$

(c) $\begin{bmatrix} 5e^{-2}-e^{-1} & 3e^{-1}-e^{-2} \\ 2e^{-2}-6e^{-1} & 4e^{-2}+e^{-1} \end{bmatrix}$

(d) $\begin{bmatrix} 2e^{-1}-e^{-2} & e^{-1}-e^{-2} \\ -2e^{-1}+2e^{-2} & -e^{-1}+2e^{-2} \end{bmatrix}$ [2008 : 2 M]

7.15 Consider the system $\frac{dx}{dt} = Ax + Bu$ with $A = \begin{bmatrix} 1 & 0 \\ 0 & 1 \end{bmatrix}$ and $B = \begin{bmatrix} p \\ q \end{bmatrix}$ where p and q are arbitrary real numbers. Which of the following statements about the controllability of the system is true?

(a) The system is completely state controllable for any nonzero values of p and q

(b) Only $p = 0$ and $q = 0$ result in controllability

(c) The system is uncontrollable for all values of p and q

(d) We cannot conclude about controllability from the given data [2009 : 1 M]

Common Data for Questions 7.16 and 7.17:

The signal flow graph of a system is shown below:

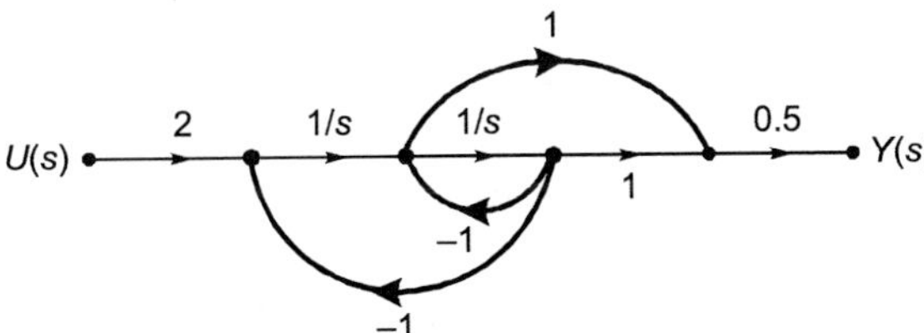

7.16 The state variable representation of the system can be

(a) $\dot{x} = \begin{bmatrix} 1 & 1 \\ -1 & 0 \end{bmatrix} x + \begin{bmatrix} 0 \\ 2 \end{bmatrix} u$

$y = \begin{bmatrix} 0 & 0.5 \end{bmatrix} x$

(b) $\dot{x} = \begin{bmatrix} -1 & 1 \\ -1 & 0 \end{bmatrix} x + \begin{bmatrix} 0 \\ 2 \end{bmatrix} u$

$y = \begin{bmatrix} 0 & 0.5 \end{bmatrix} x$

(c) $\dot{x} = \begin{bmatrix} 1 & 1 \\ -1 & 0 \end{bmatrix} x + \begin{bmatrix} 0 \\ 2 \end{bmatrix} u$

$y = \begin{bmatrix} 0.5 & 0.5 \end{bmatrix} x$

(d) $\dot{x} = \begin{bmatrix} -1 & 1 \\ -1 & 0 \end{bmatrix} x + \begin{bmatrix} 0 \\ 2 \end{bmatrix} u$ [2010 : 2 M]

$y = \begin{bmatrix} 0.5 & 0.5 \end{bmatrix} x$

7.17 The transfer function of the system is

(a) $\frac{s+1}{s^2+1}$ (b) $\frac{s-1}{s^2+1}$

(c) $\frac{s+1}{s^2+s+1}$ (d) $\frac{s-1}{s^2+s+1}$

[2010 : 2 M]

7.18 The block diagram of a system with one input u and two outputs y_1 and y_2 is given below.

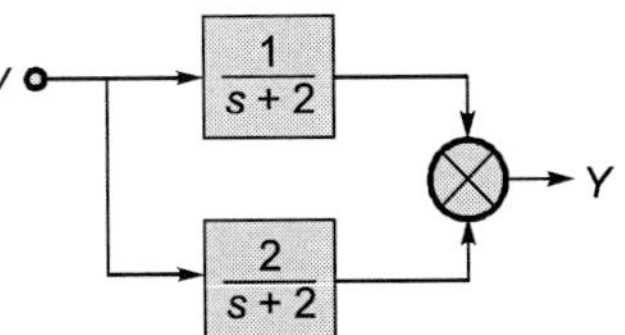

A state space model of the above system in terms of the state vector x and the output vector $y = [y_1 \;\; y_2]^T$ is

(a) $\dot{x} = [2]\,x + [1]\,u$; $y = [1 \;\; 2]\,x$

(b) $\dot{x} = [-2]\,x + [1]\,u$; $y = \begin{bmatrix} 1 \\ 2 \end{bmatrix} x$

(c) $\dot{x} = \begin{bmatrix} -2 & 0 \\ 0 & -2 \end{bmatrix} x + \begin{bmatrix} 1 \\ 1 \end{bmatrix} u$; $y = [1 \;\; 2]\,x$

(d) $\dot{x} = \begin{bmatrix} 2 & 0 \\ 0 & 2 \end{bmatrix} x + \begin{bmatrix} 1 \\ 2 \end{bmatrix} u$; $y = \begin{bmatrix} 1 \\ 2 \end{bmatrix} x$

[2011 : 2 M]

7.19 The state variable description of an LTI system is given by

$$\begin{pmatrix} \dot{x}_1 \\ \dot{x}_2 \\ \dot{x}_3 \end{pmatrix} = \begin{pmatrix} 0 & a_1 & 0 \\ 0 & 0 & a_2 \\ a_3 & 0 & 0 \end{pmatrix} \begin{pmatrix} x_1 \\ x_2 \\ x_3 \end{pmatrix} + \begin{pmatrix} 0 \\ 0 \\ 1 \end{pmatrix} u$$

$$y = (1 \;\; 0 \;\; 0) \begin{pmatrix} x_1 \\ x_2 \\ x_3 \end{pmatrix}$$

where y is the output and u is the input. The system is controllable for

(a) $a_1 \neq 0, a_2 = 0, a_3 \neq 0$

(b) $a_1 = 0, a_2 \neq 0, a_3 \neq 0$

(c) $a_1 = 0, a_2 \neq 0, a_3 = 0$

(d) $a_1 \neq 0, a_2 \neq 0, a_3 = 0$ [2012 : 2 M]

Statement for Linked Answer Questions 7.20 and 7.21:

The state diagram of a system is shown below is described by the state-variable equations:

$$\dot{X} = AX + Bu; \quad y = CX + Du$$

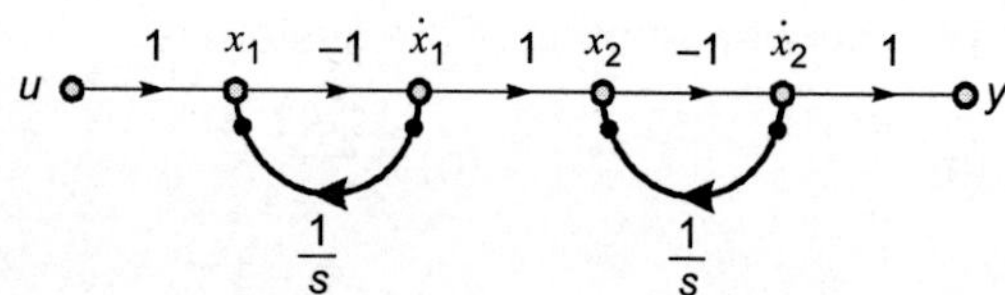

7.20 The state-variable equations of the system in the figure above are

(a) $\dot{X} = \begin{bmatrix} -1 & 0 \\ 1 & -1 \end{bmatrix} X + \begin{bmatrix} -1 \\ 1 \end{bmatrix} u$

$y = [1 \ -1]X + u$

(b) $\dot{X} = \begin{bmatrix} -1 & 0 \\ -1 & -1 \end{bmatrix} X + \begin{bmatrix} -1 \\ 1 \end{bmatrix} u$

$y = [-1 \ -1]X + u$

(c) $\dot{X} = \begin{bmatrix} -1 & 0 \\ -1 & -1 \end{bmatrix} X + \begin{bmatrix} -1 \\ 1 \end{bmatrix} u$

$y = [-1 \ -1]X - u$

(d) $\dot{X} = \begin{bmatrix} -1 & -1 \\ 0 & -1 \end{bmatrix} X + \begin{bmatrix} -1 \\ 1 \end{bmatrix} u$

$y = [1 \ -1]X - u$

[2013 : 2 M]

7.21 The state transition matrix e^{At} of the system shown in figure above is

(a) $\begin{bmatrix} e^{-t} & 0 \\ te^{-t} & e^{-t} \end{bmatrix}$ (b) $\begin{bmatrix} e^{-t} & 0 \\ -te^{-t} & e^{-t} \end{bmatrix}$

(c) $\begin{bmatrix} e^{-t} & 0 \\ e^{-t} & e^{-t} \end{bmatrix}$ (d) $\begin{bmatrix} e^{-t} & -te^{-t} \\ 0 & e^{-t} \end{bmatrix}$

[2013 : 2 M]

7.22 Consider the state space model of a system, as given below:

$$\begin{bmatrix} \dot{x}_1 \\ \dot{x}_2 \\ \dot{x}_3 \end{bmatrix} = \begin{bmatrix} -1 & 1 & 0 \\ 0 & -1 & 0 \\ 0 & 0 & -2 \end{bmatrix} \begin{bmatrix} x_1 \\ x_2 \\ x_3 \end{bmatrix} + \begin{bmatrix} 0 \\ 4 \\ 0 \end{bmatrix} u; \ y = [1 \ 1 \ 1] \begin{bmatrix} x_1 \\ x_2 \\ x_3 \end{bmatrix}.$$

The system is

(a) controllable and observable
(b) uncontrollable and observable
(c) uncontrollable and unobservable
(d) controllable and unobservable

[2014 : 2 M, Set-1]

7.23 An unforced linear time invariant (LTI) system is represented by

$$\begin{bmatrix} \dot{x}_1 \\ \dot{x}_2 \end{bmatrix} = \begin{bmatrix} -1 & 0 \\ 0 & -2 \end{bmatrix} \begin{bmatrix} x_1 \\ x_2 \end{bmatrix}$$

If the initial conditions are $x_1(0) = 1$ and $x_2(0) = -1$, the solution of the state equation is

(a) $x_1(t) = -1, x_2(t) = 2$
(b) $x_1(t) = -e^{-t}, x_2(t) = 2e^{-t}$
(c) $x_1(t) = e^{-t}, x_2(t) = -e^{-2t}$
(d) $x_1(t) = -e^{-t}, x_2(t) = -2e^{-t}$

[2014 : 2 M, Set-2]

7.24 Consider the state space system expressed by the signal flow diagram shown in the figure.

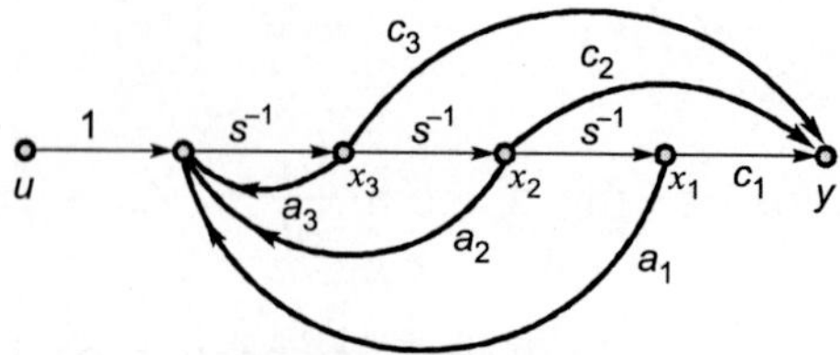

The corresponding system is

(a) always controllable
(b) always observable
(c) always stable
(d) always unstable

[2014 : 2 M, Set-2]

7.25 The state equation of a second-order linear system is given by,

$$\dot{x}(t) = Ax(t), \ x(0) = x_0$$

For $x_0 = \begin{bmatrix} 1 \\ -1 \end{bmatrix}$, $x(t) = \begin{bmatrix} e^{-t} \\ -e^{-t} \end{bmatrix}$ and for $x_0 = \begin{bmatrix} 0 \\ 1 \end{bmatrix}$,

$x(t) = \begin{bmatrix} e^{-t} - e^{-2t} \\ -e^{-t} + 2e^{-2t} \end{bmatrix}$.

When $x_0 = \begin{bmatrix} 3 \\ 5 \end{bmatrix}$, $x(t)$ is

(a) $\begin{bmatrix} -8e^{-t} + 11e^{-2t} \\ 8e^{-t} - 22e^{-2t} \end{bmatrix}$ (b) $\begin{bmatrix} 11e^{-t} - 8e^{-2t} \\ -11e^{-t} + 16e^{-2t} \end{bmatrix}$

(c) $\begin{bmatrix} 3e^{-t} - 5e^{-2t} \\ -3e^{-t} + 10e^{-2t} \end{bmatrix}$ (d) $\begin{bmatrix} -5e^{-t} + 6e^{-2t} \\ -5e^{-t} + 6e^{-2t} \end{bmatrix}$

[2014 : 2 M, Set-3]

7.26 The state transition matrix $\phi(t)$ of a system

$\begin{bmatrix} \dot{x}_1 \\ \dot{x}_2 \end{bmatrix} = \begin{bmatrix} 0 & 1 \\ 0 & 0 \end{bmatrix} \begin{bmatrix} x_1 \\ x_2 \end{bmatrix}$ is

(a) $\begin{bmatrix} t & 1 \\ 1 & 0 \end{bmatrix}$ (b) $\begin{bmatrix} 1 & 0 \\ t & 1 \end{bmatrix}$

(c) $\begin{bmatrix} 0 & 1 \\ 1 & t \end{bmatrix}$ (d) $\begin{bmatrix} 1 & t \\ 0 & 1 \end{bmatrix}$

[2014 : 2 M, Set-4]

7.27 The state variable representation of a system is given as

$$x = \begin{bmatrix} 0 & 1 \\ 0 & -1 \end{bmatrix} x, \quad x(0) = \begin{bmatrix} 1 \\ 0 \end{bmatrix}$$

$$y = [0 \ 1]x$$

The response $y(t)$ is

(a) $\sin(t)$ (b) $1 - e^t$
(c) $1 - \cos(t)$ (d) 0

[2015 : 2 M, Set-2]

7.28 A network is described by the state model as

$$\dot{x}_1 = 2x_1 - x_2 + 3u$$

$$\dot{x}_2 = -4x_2 - u$$

$$y = 3x_1 - 2x_2$$

The transfer function $H(s)\left(= \dfrac{Y(s)}{U(s)}\right)$ is

(a) $\dfrac{11s+35}{(s-2)(s+4)}$ (b) $\dfrac{11s-35}{(s-2)(s+4)}$

(c) $\dfrac{11s+38}{(s-2)(s+4)}$ (d) $\dfrac{11s-38}{(s-2)(s+4)}$

[2015 : 2 M, Set-3]

7.29 A second-order linear time-invariant system is described by the following state equations

$$\frac{d}{dt}x_1(t) + 2x_1(t) = 3u(t)$$

$$\frac{d}{dt}x_2(t) + x_2(t) = u(t)$$

where $x_1(t)$ and $x_2(t)$ are the two state variables and $u(t)$ denotes the input. If the output $c(t) = x_1(t)$, then the system is

(a) controllable but not observable
(b) observable but not controllable
(c) both controllable and observable
(d) neither controllable nor observable

[2016 : 2 M, Set-3]

7.30 Consider the state space realization:

$$\begin{bmatrix} \dot{x}_1(t) \\ \dot{x}_2(t) \end{bmatrix} = \begin{bmatrix} 0 & 0 \\ 0 & -9 \end{bmatrix} \begin{bmatrix} x_1(t) \\ x_2(t) \end{bmatrix} + \begin{bmatrix} 0 \\ 45 \end{bmatrix} u(t)$$, with the initial

condition $\begin{bmatrix} x_1(0) \\ x_2(0) \end{bmatrix} = \begin{bmatrix} 0 \\ 0 \end{bmatrix}$,

where $u(t)$ denotes the unit step function. The value of $\lim_{t\to\infty}\left|\sqrt{x_1^2(t) + x_2^2(t)}\right|$ is _____.

[2017 : 1 M, Set-2]

7.31 A second order LTI system is described by the following state equations

$$\frac{d}{dt}x_1(t) - x_2(t) = 0$$

$$\frac{d}{dt}x_2(t) + 2x_1(t) + 3x_2(t) = r(t)$$

where $x_1(t)$ and $x_2(t)$ are the two state variables and $r(t)$ denotes the input. The output $c(t) = x_1(t)$. The system is

(a) undamped (oscillatory)
(b) underdamped
(c) critically damped
(d) overdamped

[2017 : 2 M, Set-2]

7.32 The state equation and the output equation of a control system are given below:

$$\dot{x} = \begin{bmatrix} -4 & -1.5 \\ 4 & 0 \end{bmatrix} x + \begin{bmatrix} 2 \\ 0 \end{bmatrix} u,$$

$$y = [1.5 \ \ 0.625]x$$

The transfer function representation of the system is

(a) $\dfrac{3s+5}{s^2+4s+6}$ (b) $\dfrac{3s-1.875}{s^2+4s+6}$

(c) $\dfrac{4s+1.5}{s^2+4s+6}$ (d) $\dfrac{6s+5}{s^2+4s+6}$

[2018 : 2 M]

7.33 Let the state-space representation of an LTI system be $\dot{X}(t) = AX(t) + Bu(t)$, $y(t) = CX(t) + du(t)$ where A, B, C are matrices, d is a scalar, $u(t)$ is the input to the system, and $y(t)$ is its output. Let $B = [0 \ 0 \ 1]^T$ and $d = 0$. Which one of the following options for A and C will ensure that the transfer function of this LTI system is

$$H(s) = \frac{1}{s^3 + 3s^2 + 2s + 1}$$

(a) $A = \begin{bmatrix} 0 & 1 & 0 \\ 0 & 0 & 1 \\ -3 & -2 & -1 \end{bmatrix}$ and $C = [0 \ 0 \ 1]$

(b) $A = \begin{bmatrix} 0 & 1 & 0 \\ 0 & 0 & 1 \\ -1 & -2 & -3 \end{bmatrix}$ and $C = [0 \ 0 \ 1]$

(c) $A = \begin{bmatrix} 0 & 1 & 0 \\ 0 & 0 & 1 \\ -1 & -2 & -3 \end{bmatrix}$ and $C = [1 \ 0 \ 0]$

(d) $A = \begin{bmatrix} 0 & 1 & 0 \\ 0 & 0 & 1 \\ -3 & -2 & -1 \end{bmatrix}$ and $C = [1 \ 0 \ 0]$

[2019 : 2 M]

7.34 For the given circuit, which one of the following is the correct state equation?

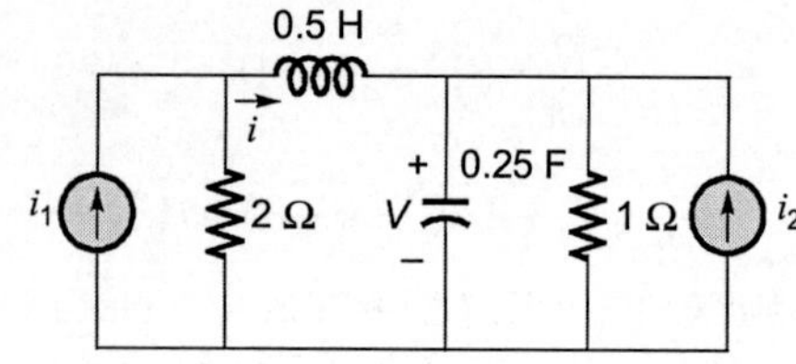

(a) $\frac{d}{dt}\begin{bmatrix} V \\ i \end{bmatrix} = \begin{bmatrix} -4 & 4 \\ -2 & -4 \end{bmatrix}\begin{bmatrix} V \\ i \end{bmatrix} + \begin{bmatrix} 0 & 4 \\ 4 & 0 \end{bmatrix}\begin{bmatrix} i_1 \\ i_2 \end{bmatrix}$

(b) $\frac{d}{dt}\begin{bmatrix} V \\ i \end{bmatrix} = \begin{bmatrix} -4 & -4 \\ -2 & -4 \end{bmatrix}\begin{bmatrix} V \\ i \end{bmatrix} + \begin{bmatrix} 4 & 0 \\ 0 & 4 \end{bmatrix}\begin{bmatrix} i_1 \\ i_2 \end{bmatrix}$

(c) $\frac{d}{dt}\begin{bmatrix} V \\ i \end{bmatrix} = \begin{bmatrix} -4 & -4 \\ -2 & 4 \end{bmatrix}\begin{bmatrix} V \\ i \end{bmatrix} + \begin{bmatrix} 4 & 4 \\ 4 & 0 \end{bmatrix}\begin{bmatrix} i_1 \\ i_2 \end{bmatrix}$

(d) $\frac{d}{dt}\begin{bmatrix} V \\ i \end{bmatrix} = \begin{bmatrix} 4 & -4 \\ -2 & -4 \end{bmatrix}\begin{bmatrix} V \\ i \end{bmatrix} + \begin{bmatrix} 0 & 4 \\ 4 & 4 \end{bmatrix}\begin{bmatrix} i_1 \\ i_2 \end{bmatrix}$

[2020 : 2 M]

7.35 The electrical system shown in the figure converts input source current $i_s(t)$ to output voltage $v_o(t)$.

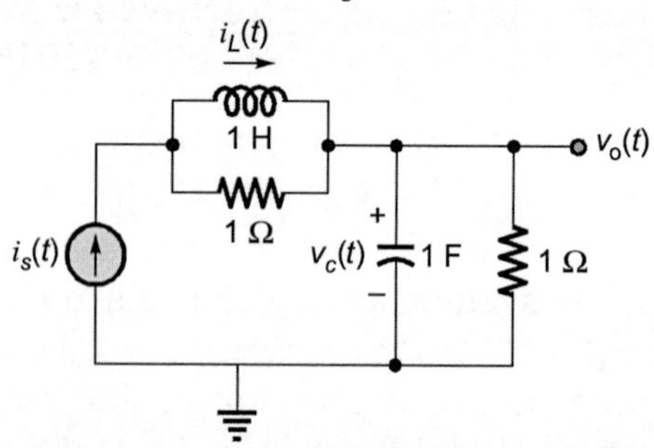

Current $i_L(t)$ in the inductor and voltage $v_c(t)$ across the capacitor are taken as the state variables, both assumed to be initially equal to zero, i.e., $i_L(0) = 0$ and $v_c(0) = 0$. The system is

(a) completely observable but not state controllable

(b) completely state controllable as well as completely observable

(c) neither state controllable nor observable

(d) completely state controllable but not observable

[2021 : 2 M]

7.36 Consider a system S represented in state space as

$$\frac{dx}{dt} = \begin{bmatrix} 0 & -2 \\ 1 & -3 \end{bmatrix}x + \begin{bmatrix} 1 \\ 0 \end{bmatrix}r,\ y = \begin{bmatrix} 2 & -5 \end{bmatrix}x$$

Which of the state space representations given below has/have the same transfer function as that of S?

(a) $\frac{dx}{dt} = \begin{bmatrix} 0 & 1 \\ -2 & -3 \end{bmatrix}x + \begin{bmatrix} 1 \\ 0 \end{bmatrix}r,\ y = \begin{bmatrix} 0 & 2 \end{bmatrix}x$

(b) $\frac{dx}{dt} = \begin{bmatrix} 0 & 1 \\ -2 & -3 \end{bmatrix}x + \begin{bmatrix} 0 \\ 1 \end{bmatrix}r,\ y = \begin{bmatrix} 1 & 2 \end{bmatrix}x$

(c) $\frac{dx}{dt} = \begin{bmatrix} -1 & 0 \\ 0 & -2 \end{bmatrix}x + \begin{bmatrix} -1 \\ 3 \end{bmatrix}r,\ y = \begin{bmatrix} 1 & 1 \end{bmatrix}x$

(d) $\frac{dx}{dt} = \begin{bmatrix} -1 & 0 \\ 0 & -2 \end{bmatrix}x + \begin{bmatrix} 1 \\ 1 \end{bmatrix}r,\ y = \begin{bmatrix} 1 & 2 \end{bmatrix}x$

[2024 : 2 Marks]

■■■

Answers State Space Analysis

7.1 (a)	7.2 (a)	7.3 (a)	7.4 (d)	7.5 (c)	7.6 (d)	7.7 (b)	7.8 (b)	7.9 (a)
7.10 (a)	7.11 (c)	7.12 (d)	7.13 (d)	7.14 (d)	7.15 (c)	7.16 (b)	7.17 (c)	7.18 (c)
7.19 (d)	7.20 (a)	7.21 (a)	7.22 (b)	7.23 (c)	7.24 (a)	7.25 (b)	7.26 (d)	7.27 (d)
7.28 (a)	7.29 (a)	7.30 (5)	7.31 (d)	7.32 (a)	7.33 (c)	7.34 (a)	7.35 (c)	7.36 (b)

Explanations State Space Analysis

7.1 (a)

$$\dot{x}_1 = x_1 - x_2$$
$$\dot{x}_1(0) = x_1(0) - x_2(0)$$
$$\dot{x}_1(0) = 1 - (-1) = 2$$
$$\dot{x}_2 = x_2$$
$$\dot{x}_2(0) = x_2(0) = -1$$
$$y = x_1 + x_2$$
$$\frac{dy}{dt} = \dot{x}_1 + \dot{x}_2$$
$$\left.\frac{dy}{dt}\right|_{t=0} = \dot{x}_1(0) + \dot{x}_2(0) = 2 - 1 = 1$$

7.2 (a)

$$Q_c = \begin{bmatrix} B & AB & A^2B & \dots & A^{n-1}B \end{bmatrix}$$
$$A = \begin{bmatrix} 0 & 1 \\ 2 & -3 \end{bmatrix}$$
$$B = \begin{bmatrix} 0 \\ 1 \end{bmatrix}$$
$$C = [1 \ \ 1]$$
$$AB = \begin{bmatrix} +1 \\ -3 \end{bmatrix}$$
$$\therefore \quad Q_c = \begin{bmatrix} 0 & 1 \\ 1 & -3 \end{bmatrix}_{2\times 2} \neq 0$$

$\therefore$ order 2, rank 2

$\therefore$ controllable

$$Q_0 = \begin{bmatrix} C^T & A^TC^T & A^{T^2}C^T \dots \dots \end{bmatrix}$$
$$C^T = \begin{bmatrix} 1 \\ 1 \end{bmatrix}, \ A^T = \begin{bmatrix} 0 & 2 \\ 1 & -3 \end{bmatrix}$$
$$\therefore \quad Q_0 = \begin{bmatrix} 1 & 2 \\ 1 & -2 \end{bmatrix} \neq 0$$

$\therefore$ rank 2 observable

7.3 (a)

$$\dot{x} = \begin{bmatrix} 0 & 1 & 0 \\ 0 & 0 & 1 \\ 0.5 & 1 & 2 \end{bmatrix} x + \begin{bmatrix} 0 \\ 0 \\ 1 \end{bmatrix} [-0.5 \ \ -3 \ \ -5] x + V$$
$$\therefore \quad \dot{x} = \begin{bmatrix} 0 & 1 & 0 \\ 0 & 0 & 1 \\ 0 & -2 & -3 \end{bmatrix} x + V$$

Characteristics equation

$$= \lambda^3 + 3\lambda^2 + 2\lambda + 0 = 0$$
$$\Rightarrow \lambda = 0, -1, -2$$

7.4 (d)

$$\dot{x}(t) = -2x(t) + 2u(t) \quad \dots(i)$$
$$y(t) = 0.5x(t) \quad \dots(ii)$$

From (i), Taking Laplace transform of (i)

$$sX(s) = -2X(s) + 2U(s)$$
$$X(s)[s+2] = 2U(s)$$
$$\Rightarrow \quad X(s) = \frac{2U(s)}{(s+2)}$$

Taking Laplace transform of (ii)

$$Y(s) = 0.5X(s) = \frac{0.5 \times 2U(s)}{s+2}$$
$$\therefore \quad \frac{Y(s)}{U(s)} = \frac{1}{(s+2)}$$

7.5 (c)

$$(sI - A) = \begin{bmatrix} s & 0 \\ 0 & s \end{bmatrix} - \begin{bmatrix} 1 & 0 \\ 1 & 1 \end{bmatrix} = \begin{bmatrix} s-1 & 0 \\ -1 & s-1 \end{bmatrix}$$
$$(sI - A)^{-1} = \frac{\begin{bmatrix} (s-1) & 0 \\ +1 & (s-1) \end{bmatrix}}{(s-1)^2} = \begin{bmatrix} \frac{1}{s-1} & 0 \\ \frac{+1}{(s-1)^2} & \frac{1}{s-1} \end{bmatrix}$$
$$L^{-1}[sI - A]^{-1} = e^{At} = \begin{bmatrix} e^t & 0 \\ te^t & e^t \end{bmatrix}$$

$$x(t) = e^{At}[x(0)] = \begin{bmatrix} e^t & 0 \\ te^t & e^t \end{bmatrix}\begin{bmatrix} 1 \\ 0 \end{bmatrix} = \begin{bmatrix} e^t \\ te^t \end{bmatrix}$$

7.6 (d)

$$\begin{bmatrix} \dot{x}_1 \\ \dot{x}_2 \end{bmatrix} = \begin{bmatrix} -3 & -1 \\ 2 & 0 \end{bmatrix}\begin{bmatrix} x_1 \\ x_2 \end{bmatrix} + \begin{bmatrix} 1 \\ 0 \end{bmatrix}u$$

$$y = [1\ 0]\begin{bmatrix} x_1 \\ x_2 \end{bmatrix} + \begin{bmatrix} 1 \\ 0 \end{bmatrix}u$$

$$Q_C = [B \quad AB]$$

$$AB = \begin{bmatrix} -3 \\ 2 \end{bmatrix}$$

$$\therefore \quad Q_C = \begin{bmatrix} 1 & -3 \\ 0 & 2 \end{bmatrix} \neq 0$$

$\therefore$ Controllable, $Q_O = [C^T \quad A^T C^T]$

$$C^T = \begin{bmatrix} 1 \\ 0 \end{bmatrix}, \quad A^T = \begin{bmatrix} -3 & 2 \\ -1 & 0 \end{bmatrix}$$

$$A^T C^T = \begin{bmatrix} -3 \\ -1 \end{bmatrix}$$

$$\therefore \quad Q_O = \begin{bmatrix} 1 & -3 \\ 0 & -1 \end{bmatrix} \neq 0$$

$\therefore$ Observable.

7.7 (b)

$$[sI - A] = \begin{bmatrix} s & 0 \\ 0 & s \end{bmatrix} - \begin{bmatrix} 1 & 0 \\ 0 & 1 \end{bmatrix} = \begin{bmatrix} s-1 & 0 \\ 0 & s-1 \end{bmatrix}$$

$$e^{At} = L^{-1}\{[sI - A]^{-1}\}$$

$$= \begin{bmatrix} \frac{1}{s-1} & 0 \\ 0 & \frac{1}{s-1} \end{bmatrix} = \begin{bmatrix} e^t & 0 \\ 0 & e^t \end{bmatrix}$$

7.8 (b)

Eigen values of $A = [\lambda]$

Eigen values of $W = [\mu]$

The eigen values of a system are always unique.

So, $[\lambda] = [\mu]$

But a system can be represented by different state models having different set of state variables.

$$X = W$$
$$X \neq W$$

Both are possible conditions.

7.9 (a)

$$\phi(t) = L^{-1}[sI - A]^{-1}$$

$$= L^{-1}\left[\begin{bmatrix} s & 0 \\ 0 & s \end{bmatrix} - \begin{bmatrix} 0 & 1 \\ -1 & 0 \end{bmatrix}\right]^{-1} = L^{-1}\begin{bmatrix} s & -1 \\ 1 & s \end{bmatrix}^{-1}$$

$$= L^{-1}\begin{bmatrix} \frac{s}{s^2+1} & \frac{1}{s^2+1} \\ \frac{-1}{s^2+1} & \frac{s}{s^2+1} \end{bmatrix}\left[D = s^2 + 1\right] = \begin{bmatrix} \cos t & \sin t \\ -\sin t & \cos t \end{bmatrix}$$

7.10 (a)

$$\begin{bmatrix} \frac{d\omega}{dt} \\ \frac{di_a}{dt} \end{bmatrix} = \begin{bmatrix} -1 & 1 \\ -1 & -10 \end{bmatrix}\begin{bmatrix} \omega \\ i_a \end{bmatrix} + \begin{bmatrix} 0 \\ 10 \end{bmatrix}u$$

$$\Rightarrow \frac{d\omega}{dt} = -\omega + i_a \qquad \text{...(i)}$$

$$\Rightarrow \frac{di_a}{dt} = -\omega - 10\, i_a + 10\, u \qquad \text{...(ii)}$$

Taking Laplace transform (i) & (ii)

$$\Rightarrow s\omega(s) = -\omega(s) + I_a(s)$$

$$\Rightarrow (s+1)\,\omega(s) = I_a(s) \qquad \text{...(iii)}$$

$$\Rightarrow sI_a(s) = -\omega(s) - 10\, I_a(s) + 10\, U(s)$$

$$\Rightarrow \omega(s) = (-10 - s)\, I_a(s) + 10\, U(s)$$

$$= (-10 - s)(s+1)\,\omega(s) + 10\, U(s)$$

$$= -[s^2 + 11s + 10]\omega(s) + 10\, U(s)$$

$$\Rightarrow [s^2 + 11s + 11]\,\omega(s) = 10\, U(s)$$

$$\Rightarrow \frac{\omega(s)}{U(s)} = \frac{10}{(s^2 + 11s + 11)}$$

7.11 (c)

Sum of the eigen value = Trace of the principle diagonal matrix

Sum = –3. Only option (a) satisfies both conditions.

7.12 (d)

Multiplication of the eigen value = determinant of the matrix

$\therefore$ from options it seems determinant should be $\pm$ 2. Only option (d) satisfies as det = 2.

7.13 (d)

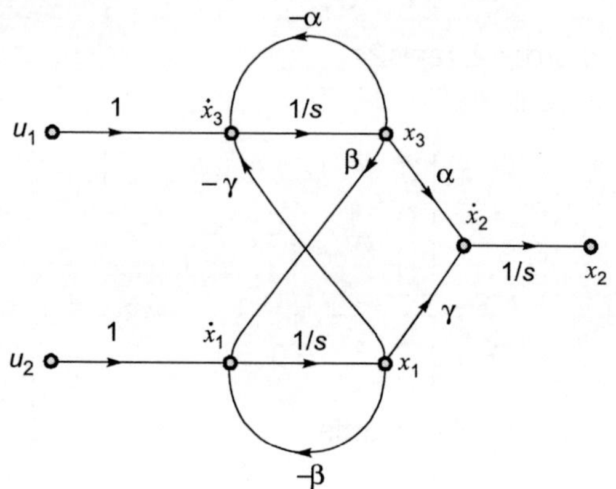

By solving this we get

$$\frac{d}{dt}\begin{pmatrix} x_1 \\ x_2 \\ x_3 \end{pmatrix} = \begin{bmatrix} -\gamma & 0 & \beta \\ \gamma & 0 & \alpha \\ -\beta & 0 & -\alpha \end{bmatrix}\begin{pmatrix} x_1 \\ x_2 \\ x_3 \end{pmatrix} + \begin{bmatrix} 0 & 1 \\ 0 & 0 \\ 1 & 0 \end{bmatrix}\begin{pmatrix} u_1 \\ u_2 \end{pmatrix}$$

7.15 (c)

$$A = \begin{bmatrix} 1 & 0 \\ 0 & 1 \end{bmatrix} \quad ; \quad B = \begin{bmatrix} p \\ q \end{bmatrix}$$

For controllability condition is

$$Q_c = [B, AB, \ldots A^{n-1} B] \neq 0$$

$$A.B = \begin{bmatrix} 1 & 0 \\ 0 & 1 \end{bmatrix}\begin{bmatrix} p \\ q \end{bmatrix} = \begin{bmatrix} p+0 \\ 0+q \end{bmatrix} = \begin{bmatrix} p \\ q \end{bmatrix}$$

So $\quad Q_c = \begin{bmatrix} p & p \\ q & q \end{bmatrix} = 0$

So the system is uncontrollable for all values of p and q.

7.17 (c)

Forward path gain,

$$P_1 = 2\left(\frac{1}{s}\right)\left(\frac{1}{s}\right)(0.5) = \frac{1}{s^2}$$

$$P_2 = 2\left(\frac{1}{s}\right)(1)(0.5) = \frac{1}{s}$$

$$\Delta_1 = 1$$

$$\Delta_2 = 1$$

$$\Delta = 1 - \left\{-\frac{1}{s} - \frac{1}{s^2}\right\} = 1 + \frac{1}{s} + \frac{1}{s^2}$$

Transfer function of the system,

$$\frac{Y(s)}{U(s)} = \frac{P_1\Delta_1 + P_2\Delta_2}{\Delta}$$

$$= \frac{\frac{1}{s^2} + \frac{1}{s}}{1 + \frac{1}{s} + \frac{1}{s^2}} = \frac{s+1}{s^2+s+1}$$

7.18 (c)

$$\frac{Y_1(s)}{U(s)} = \frac{1}{s+2}$$

$$\frac{Y_2(s)}{U(s)} = \frac{2}{s+2}$$

$$\frac{Y_1(s)}{X_1(s)}\frac{X_1(s)}{U(s)} = \frac{1}{s+2}$$

$$\frac{Y_2(s)}{X_2(s)}\frac{X_2(s)}{U(s)} = \frac{2}{s+2}$$

$$\frac{X_1(s)}{U(s)} = \frac{1}{s+2} \text{ and } \frac{Y_1(s)}{X_1(s)} = 1$$

$$\frac{X_2(s)}{U(s)} = \frac{1}{s+2} \text{ and } \frac{Y_2(s)}{U(s)} = 2$$

$$sX_1(s) + 2X_1(s) = U(s)$$

and $\quad Y_1(s) = X_1(s)$

$$sX_2(s) + 2X_2(s) = U(s)$$

and $\quad Y_2(s) = 2X_2(s)$

$$\dot{x}_1(t) + 2x_1(t) = U(t)$$

and $\quad y_1(t) = x_1(t)$

$$\dot{x}_2(t) + 2x_2(t) = U(t)$$

and $\quad y_2(t) = 2x_2(t)$

From Question, $\quad y = [y_1 \;\; y_2]^T = [1 \;\; 2]^T = \begin{bmatrix} 1 \\ 2 \end{bmatrix}$

$$\dot{x}_1(t) = -2x_1(t) + u(t)$$

$$\dot{x}_2(t) = -2x_2(t) + u(t)$$

$$\begin{bmatrix} \dot{x}_1 \\ \dot{x}_2 \end{bmatrix} = \begin{bmatrix} -2 & 0 \\ -2 & 0 \end{bmatrix}\begin{bmatrix} x_1 \\ x_2 \end{bmatrix} + \begin{bmatrix} 1 \\ 1 \end{bmatrix}u(t)$$

or, $\quad \dot{x} = [-2]x + [1]u$

only option (b) is satisfied.

7.19 (d)

$$\begin{bmatrix} \dot{x}_1 \\ \dot{x}_2 \\ \dot{x}_3 \end{bmatrix} = \begin{bmatrix} 0 & a_1 & 0 \\ 0 & 0 & a_2 \\ a_3 & 0 & 0 \end{bmatrix}\begin{bmatrix} x_1 \\ x_2 \\ x_3 \end{bmatrix} + \begin{bmatrix} 0 \\ 0 \\ 1 \end{bmatrix}u$$

$$y = [1 \;\; 0 \;\; 0]\begin{bmatrix} x_1 \\ x_2 \\ x_3 \end{bmatrix}$$

$$Q_c = [B \;\; AB \;\; A^2B] \quad \ldots(1)$$

$$AB = \begin{bmatrix} 0 & a_1 & 0 \\ 0 & 0 & a_2 \\ a_3 & 0 & 0 \end{bmatrix}\begin{bmatrix} 0 \\ 0 \\ 1 \end{bmatrix} = \begin{bmatrix} 0 \\ a_2 \\ 0 \end{bmatrix}$$

$$A^2B = AAB = \begin{bmatrix} 0 & a_1 & 0 \\ 0 & 0 & a_2 \\ a_3 & 0 & 0 \end{bmatrix}\begin{bmatrix} 0 \\ a_2 \\ 0 \end{bmatrix}$$

$$A^2B = \begin{bmatrix} a_1a_2 \\ 0 \\ 0 \end{bmatrix}$$

$$Q_c = \begin{bmatrix} 0 & 0 & a_1a_2 \\ 0 & a_2 & 0 \\ 1 & 0 & 0 \end{bmatrix}$$

For system to be controllable,

$|Q_c| \neq 0$

$\Rightarrow (0 - a_1 a_2^2) \neq 0$

$\Rightarrow a_1 \neq 0$

$a_2 \neq 0$

7.20 (a)

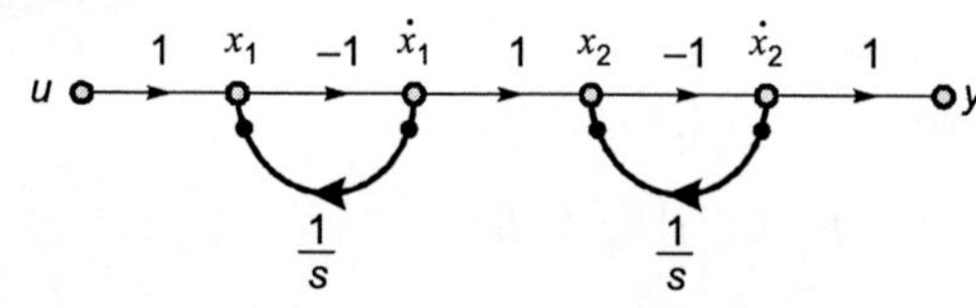

So, $\dot{x}_1 = -x_1 - u$

$$\dot{x}_2 = -(x_2 + \dot{x}_1) = -(x_2 - x_1 - u)$$

$$\dot{x}_2 = x_1 - x_2 + u$$

$$y = \dot{x}_2$$

$$y = x_1 - x_2 + u$$

$$\begin{bmatrix} \dot{x}_1 \\ \dot{x}_2 \end{bmatrix} = \begin{bmatrix} -1 & 0 \\ 1 & -1 \end{bmatrix}\begin{bmatrix} x_1 \\ x_2 \end{bmatrix} + \begin{bmatrix} -1 \\ 1 \end{bmatrix} u$$

$$\dot{x} = \begin{bmatrix} -1 & 0 \\ 1 & -1 \end{bmatrix} X + \begin{bmatrix} -1 \\ 1 \end{bmatrix} u$$

7.21 (a)

$$A = \begin{bmatrix} -1 & 0 \\ 1 & -1 \end{bmatrix}$$

$$sI - A = \begin{bmatrix} s+1 & 0 \\ -1 & s+1 \end{bmatrix}$$

$$[sI - A]^{-1} = \frac{1}{(s+1)\times(s+1)}\begin{bmatrix} s+1 & 0 \\ 1 & s+1 \end{bmatrix}$$

$$[sI - A]^{-1} = \begin{bmatrix} \frac{1}{s+1} & 0 \\ \frac{1}{(s+1)^2} & \frac{1}{s+1} \end{bmatrix}$$

$$\phi(t) = e^{At} = L^{-1}\{(sI - A)^{-1}\}$$

$$e^{At} = \begin{bmatrix} e^{-t} & 0 \\ te^{-t} & e^{-t} \end{bmatrix}$$

7.22 (b)

$$\begin{bmatrix} \dot{x}_1 \\ \dot{x}_2 \\ \dot{x}_3 \end{bmatrix} = \begin{bmatrix} -1 & 1 & 0 \\ 0 & -1 & 0 \\ 0 & 0 & -2 \end{bmatrix}\begin{bmatrix} x_1 \\ x_2 \\ x_3 \end{bmatrix} + \begin{bmatrix} 0 \\ 4 \\ 0 \end{bmatrix} u$$

$$y = [1 \;\; 1 \;\; 1]\begin{bmatrix} x_1 \\ x_2 \\ x_3 \end{bmatrix}$$

$$A = \begin{bmatrix} -1 & 1 & 0 \\ 0 & -1 & 0 \\ 0 & 0 & -2 \end{bmatrix}, B = \begin{bmatrix} 0 \\ 4 \\ 0 \end{bmatrix}, C = [1 \;\; 1 \;\; 1]$$

Check for controllability:

$$Q_c = [B : AB : A^2B] = \begin{bmatrix} 0 & 4 & -8 \\ 4 & -4 & 4 \\ 0 & 0 & 0 \end{bmatrix}$$

For controllable,

$$|Q_c| \neq 0$$

Here, $|Q_c| = 4(0) = 0 \;\therefore$ Uncontrollable.

Check for observability:

$$Q_o = [C^T : A^T C^T : A^{2T} C^T]$$

$$= \begin{bmatrix} 1 & -1 & 1 \\ 1 & 0 & -1 \\ 1 & -2 & 4 \end{bmatrix}$$

For observable,

$$|Q_o| \neq 0$$

Here, $|Q_o| = 1 \;\therefore$ Observable.

7.23 (c)

$$\begin{bmatrix} \dot{x}_1 \\ \dot{x}_2 \end{bmatrix} = \begin{bmatrix} -1 & 0 \\ 0 & -2 \end{bmatrix}\begin{bmatrix} x_1 \\ x_2 \end{bmatrix}$$

$$\dot{x}_1 = -x_1 + 0x_2 \quad \text{...(i)}$$

$$\dot{x}_2 = 0x_1 - 2x_2 \quad \text{...(ii)}$$

Applying Laplace transform on both equations

$\Rightarrow sX_1(s) - x_1(0) = -X_1(s)$

$X_1(s)\,[s + 1] = X_1(0)$ given $x_1(0) = 1$

$$X_1(s) = \frac{1}{s+1}$$

$$\Rightarrow \boxed{e^{-t} = x_1(t)}$$

$\Rightarrow sX_2(s) - X_2(0) = -2X_2(s)$

$(s + 2)X_2(s) = X_2(0)$ $X_2(0) = -1$

$$X_2(s) = -\frac{1}{s+2}$$

$$x_2(t) = -e^{-2t}$$

7.24 (a)

The state equation from the given state diagram is

$$\dot{x}_1 = x_2; \quad \dot{x}_2 = x_3$$

$$\dot{x}_3 = a_3 x_3 + a_2 x_2 + a_1 x_1 + u$$

Also, $y = c_1 x_1 + c_2 x_2 + c_3 x_3$

Thus, state matrix

$$\begin{bmatrix} \dot{x}_1 \\ \dot{x}_2 \\ \dot{x}_3 \end{bmatrix} = \begin{bmatrix} 0 & 1 & 0 \\ 0 & 0 & 1 \\ a_1 & a_2 & a_3 \end{bmatrix}\begin{bmatrix} x_1 \\ x_2 \\ x_3 \end{bmatrix} + \begin{bmatrix} 0 \\ 0 \\ 1 \end{bmatrix} u$$

and $y = \begin{bmatrix} c_1 & c_2 & c_3 \end{bmatrix} \begin{bmatrix} x_1 \\ x_2 \\ x_3 \end{bmatrix}$

Check for controllability:

The system is said to be controllable if, the rank of controllability matrix Q_c is equal to the rank of the state matrix A. However, if the controllability matrix Q_c is a square matrix then the condition for controllability is

$|Q_c| \neq 0$

where,

$$Q_c = [B : AB : A^2B]$$

$$\therefore \quad Q_c = \begin{bmatrix} 0 & 0 & 1 \\ 0 & 1 & a_3 \\ 1 & a_3 & a_2 + a_3^2 \end{bmatrix}$$

$\because |Q_C| \neq 0$ and Rank of Q_c = Rank of $A = 1$

$\therefore$ The system is controllable.

Check for observability:

The system is said to be observable if, the rank of observability matrix Q_o is equal to the rank of the state matrix A. However, if the observability matrix Q_o is a square matrix then the condition for observability is

$|Q_o| \neq 0$

where,

$$Q_o = [C^T : A^TC^T : (A^T)^2 C^T]$$

$$\therefore Q_o = \begin{bmatrix} c_1 & c_2 & c_3 \\ c_3 a_1 & c_1 + c_2a_2 & c_2 + c_2a_3 \\ c_1(c_2 + c_3 a_3) & c_3a_1 + (c_2 + c_3a_3)a_2 & c_1 + c_2a_2 + (c_2 + c_3a_3)a_3 \end{bmatrix}$$

$\because |Q_o|$ depends on value of unknown.

Hence, not observable always.

7.25 (b)

Given $\dot{x}(t) = Ax(t), \quad x(0) = x_0$

Taking the Laplace transform

$$sX(s) - x(0) = AX(s)$$

$$[sI - A]\, X(s) = x(0)$$

$$X(s) = [sI - A]^{-1}\, x(0)$$

$$x(t) = \mathcal{L}^{-1}[sI - A]^{-1}\, x(0) \qquad \text{...(i)}$$

Conditions given are

For $x_0 = \begin{bmatrix} 1 \\ -1 \end{bmatrix}, \; x(t) = \begin{bmatrix} e^{-t} \\ -e^{-t} \end{bmatrix}$

For $x_0 = \begin{bmatrix} 0 \\ 1 \end{bmatrix}, \; x(t) = \begin{bmatrix} e^{-t} - e^{-2t} \\ -e^{-t} + 2e^{-2t} \end{bmatrix}$

Using the linearity property in equation (i)

$$K_1 x_1(t) = \mathcal{L}^{-1}[sI - A]^{-1}\, x_1(0)\, K_1$$

$$K_2 x_2(t) = \mathcal{L}^{-1}[sI - A]^{-1}\, x_2(0)\, K_2$$

Using the linearity property as

$$K_1 x_1(t) + K_2 x_2(t) = \mathcal{L}^{-1}[sI - A]^{-1} [K_1x_1(0) + K_2x_2(0)] \qquad \text{...(ii)}$$

Also $\quad X_3(s) = [sI - A]^{-1}\, x_3(0)$

So, $\quad K_1 \begin{bmatrix} 1 \\ -1 \end{bmatrix} + K_2 \begin{bmatrix} 0 \\ 1 \end{bmatrix} = \begin{bmatrix} 3 \\ 5 \end{bmatrix}$

$$\begin{bmatrix} K_1 + 0K_2 \\ -K_1 + K_2 \end{bmatrix} = \begin{bmatrix} 3 \\ 5 \end{bmatrix}$$

$\Rightarrow \quad K_1 = 3$

$K_2 = 8$

So, from equation (ii), we get $x(t)$

$$x(t) = K_1 x_1(t) + K_2 x_2(t)$$

$$= 3\begin{bmatrix} e^{-t} \\ -e^{-t} \end{bmatrix} + 8\begin{bmatrix} e^{-t} - e^{-2t} \\ -e^{-t} + 2e^{-2t} \end{bmatrix} = \begin{bmatrix} 11e^{-t} - 8e^{-2t} \\ -11e^{-t} + 16e^{-2t} \end{bmatrix}$$

7.26 (d)

Given $\begin{bmatrix} \dot{x}_1 \\ \dot{x}_2 \end{bmatrix} = \begin{bmatrix} 0 & 1 \\ 0 & 0 \end{bmatrix} \begin{bmatrix} x_1 \\ x_2 \end{bmatrix}$

$$A = \begin{bmatrix} 0 & 1 \\ 0 & 0 \end{bmatrix}$$

$$[sI - A] = \begin{bmatrix} s & 0 \\ 0 & s \end{bmatrix} - \begin{bmatrix} 0 & 1 \\ 0 & 0 \end{bmatrix} = \begin{bmatrix} s & -1 \\ 0 & s \end{bmatrix}$$

$$|sI - A| = s^2$$

$$\phi(t) = \mathcal{L}^{-1}[sI - A]^{-1}$$

$$= \frac{1}{s^2}\begin{bmatrix} s & 1 \\ 0 & s \end{bmatrix} = \mathcal{L}^{-1}\begin{bmatrix} \frac{1}{s} & \frac{1}{s^2} \\ 0 & \frac{1}{s} \end{bmatrix} = \begin{bmatrix} 1 & t \\ 0 & 1 \end{bmatrix}$$

7.27 (d)

The response of the system is given by

$$x(t) = L^{-1}\left[(sI - A)^{-1}\, x(0) + (sI - A)^{-1}\, B \cdot U(s)\right]$$

and the complete response

$$y(t) = [C]\, x(t)$$

From the given state model,

$$A = \begin{bmatrix} 0 & 1 \\ 0 & -1 \end{bmatrix}; \; B = 0; C = [0 \;\; 1] \text{ and } x(0) = \begin{bmatrix} 1 \\ 0 \end{bmatrix}$$

$$\therefore \quad (sI - A)^{-1} = \begin{bmatrix} s & -1 \\ 0 & s+1 \end{bmatrix}^{-1} = \frac{1}{s(s+1)} \begin{bmatrix} s+1 & 1 \\ 0 & s \end{bmatrix}$$

$$(sI-A)^{-1}x(0) = \frac{1}{s(s+1)}\begin{bmatrix} s+1 & 1 \\ 0 & s \end{bmatrix}\begin{bmatrix} 1 \\ 0 \end{bmatrix}$$

$$= \frac{1}{s(s+1)}\begin{bmatrix} s+1 \\ 0 \end{bmatrix} = \begin{bmatrix} 1/s \\ 0 \end{bmatrix}$$

$$\Rightarrow \; x(t) = \mathcal{L}^{-1}\left[(sI-A)^{-1}x(0)\right]$$

$$= \begin{bmatrix} 1 \\ 0 \end{bmatrix}u(t)$$

and the complete response

$$y(t) = [0 \;\; 1]\,x(t) = [0 \;\; 1]\begin{bmatrix} 1 \\ 0 \end{bmatrix} = 0$$

7.28 (a)

State matrix form:

$$\dot{x} = \begin{bmatrix} 2 & -1 \\ 0 & -4 \end{bmatrix}x + \begin{bmatrix} 3 \\ -1 \end{bmatrix}u \quad ; \quad y = [3 \;\; -2]x$$

The transfer function from state matrix can be calculated as:

$$T(s) = C[sI-A]^{-1}B + D$$

Here,

$$[sI-A]^{-1} = \begin{bmatrix} s-2 & 1 \\ 0 & s+4 \end{bmatrix}^{-1}$$

$$= \frac{1}{(s-2)(s+4)}\begin{bmatrix} (s+4) & -1 \\ 0 & (s-2) \end{bmatrix}$$

$$\Rightarrow C[sI-A]^{-1}B$$

$$= \frac{1}{(s-2)(s+4)}[3 \;\; -2]\begin{bmatrix} (s+4) & -1 \\ 0 & (s-2) \end{bmatrix}\begin{bmatrix} 3 \\ -1 \end{bmatrix}$$

$$= \frac{1}{(s-2)(s+4)}[3 \;\; -2]\begin{bmatrix} 3s+12+1 \\ -s+2 \end{bmatrix}$$

$$= \frac{1}{(s-2)(s+4)}[9s+39+2s-4]$$

$$= \frac{11s+35}{(s-2)(s+4)}$$

7.29 (a)

$$\dot{x}_1(t) + 2x_1(t) = 3u(t)$$
$$\dot{x}_2(t) + x_2(t) = u(t)$$
$$\dot{x}_1(t) = -2x_1(t) + 3u(t)$$
$$\dot{x}_2(t) = -x_2(t) + u(t)$$

$$\begin{bmatrix} \dot{x}_1 \\ \dot{x}_2 \end{bmatrix} = \begin{bmatrix} -2 & 0 \\ 0 & -1 \end{bmatrix}\begin{bmatrix} x_1 \\ x_2 \end{bmatrix} + \begin{bmatrix} 3 \\ 1 \end{bmatrix}U$$

$$c(t) = x_1(t)$$

$$= [1 \;\; 0]\begin{bmatrix} x_1 \\ x_2 \end{bmatrix}$$

$$A = \begin{bmatrix} -2 & 0 \\ 0 & -1 \end{bmatrix};$$

$$B = \begin{bmatrix} 3 \\ 1 \end{bmatrix};$$

$$C = [1 \;\; 0]$$

For controllability $Q_C = [B \;\; AB] = \begin{bmatrix} 3 & -6 \\ 1 & -1 \end{bmatrix}$

$|Q_C| \neq 0 \rightarrow$ Hence system is controllable.

For observability

$$Q_0 = \begin{bmatrix} C \\ CA \end{bmatrix} = \begin{bmatrix} 1 & 0 \\ -2 & 0 \end{bmatrix}$$

$|Q_0| = 0 \rightarrow$ Hence system is not observable.

7.30 Sol.

$$\begin{bmatrix} \dot{x}_1(t) \\ \dot{x}_2(t) \end{bmatrix} = \begin{bmatrix} 0 & 0 \\ 0 & -9 \end{bmatrix}\begin{bmatrix} x_1(t) \\ x_2(t) \end{bmatrix} + \begin{bmatrix} 0 \\ 45 \end{bmatrix}u(t)$$

By applying Laplace transform on both sides, we get

$$sX_1(s) - x_1(0) = 0$$

$$X_1(s) = \frac{x_1(0)}{s} = 0 \qquad \because x_1(0) = 0$$

So, $\quad x_1(t) = 0$

and $\quad sX_2(s) - x_2(0) = -9X_1(s) + \dfrac{45}{s}$

$$X_2(s) = \frac{45}{s(s+9)} \qquad \because x_2(0) = 0$$

$$\text{Required value} = \lim_{t\to\infty}\left|\sqrt{x_1^2(t) + x_2^2(t)}\right|$$

$$= \left|\lim_{t\to\infty} x_2(t)\right| \qquad \because x_1(t) = 0$$

$$\lim_{t\to\infty} x_2(t) = \lim_{s\to 0} sX_2(s) = \frac{45}{9} = 5$$

So,

Required value $= |5| = 5$

7.31 (d)

Given state variable equations are as follows:

$$\frac{d}{dt}x_1(t) - x_2(t) = 0 \qquad \ldots (i)$$

$$\frac{d}{dt}x_2(t) + 2x_2(t) + 3x_2(t) = r(t) \quad \text{... (ii)}$$

Also given that, input = $r(t)$ and output = $x_1(t)$

By applying Laplace transform to equation (*i*), we get,

$$sX_1(s) = X_2(s) \quad \text{... (iii)}$$

By applying Laplace transform to equation (*ii*), we get,

$$sX_2(s) + 2X_1(s) + 3X_2(s) = R(s)$$

$$X_2(s) = \frac{R(s) - 2X_1(s)}{s+3} \quad \text{... (iv)}$$

By substituting equation (*iv*) in equation (*iii*), we get,

$$sX_1(s) = \frac{R(s) - 2X_1(s)}{s+3}$$

$$(s^2 + 3s + 2)X_1(s) = R(s)$$

So, the transfer function of the given system is,

$$\frac{X_1(s)}{R(s)} = \frac{1}{s^2 + 3s + 2}$$

$$\omega_n^2 = 2$$

$$2\zeta\omega_n = 3$$

$$\xi = \frac{3}{2\omega_n} = \frac{3}{2\sqrt{2}} = 1.06$$

As $\xi > 1$, the given system is overdamped.

7.32 (a)

Transfer function,

$$T(s) = \frac{Y(s)}{U(s)} = C[sI - A]^{-1}B$$

$$A = \begin{bmatrix} -4 & -1.5 \\ 4 & 0 \end{bmatrix}$$

$$B = \begin{bmatrix} 2 \\ 0 \end{bmatrix}$$

$$C = [1.5 \quad 0.625]$$

$$[sI - A] = \begin{bmatrix} s+4 & 1.5 \\ -4 & s \end{bmatrix}$$

$$[sI - A]^{-1} = \frac{1}{(s^2 + 4s + 6)}\begin{bmatrix} s & -1.5 \\ 4 & s+4 \end{bmatrix}$$

$$[sI - A]^{-1}B = \frac{1}{s^2 + 4s + 6}\begin{bmatrix} 2s \\ 8 \end{bmatrix}$$

$$C[sI - A]^{-1}B = \frac{1}{s^2 + 4s + 6}[1.5 \quad 0.625]\begin{bmatrix} 2s \\ 8 \end{bmatrix}$$

$$T(s) = \frac{3s + 5}{s^2 + 4s + 6}$$

7.33 (c)

$$X(t) = Ax(t) + Bu(t)$$

$$y(t) = Cx(t)$$

$$B = \begin{bmatrix} 0 \\ 0 \\ 1 \end{bmatrix}$$

$$\frac{Y(s)}{U(s)} = \frac{Y(s)}{X_1(s)} \times \frac{X_1(s)}{U(s)}$$

$$= 1 \times \frac{1}{s^3 + 3s^2 + 2s + 1}$$

$$X_1(s)[s^3 + 3s^2 + 2s + 1] = U(s)$$

$$x_2 = \dot{x}_1(t)$$

$$\Rightarrow \quad X_2(s) = sX_1(s)$$

$$x_3 = \dot{x}_2(t)$$

$$\Rightarrow \quad X_3(s) = sX_2(s) = s^2X_1(s)$$

So, $\quad sX_3(s) = -X_1(s) - 2X_2(s) - 3X_3(s) + U(s)$

$$\dot{x}_3(t) = -x_1(t) - 2x_2(t) - 3x_3(t) + u(t)$$

$$y(t) = x_1(t)$$

$$\dot{x}(t) = \begin{bmatrix} 0 & 1 & 0 \\ 0 & 0 & 1 \\ -1 & -2 & -3 \end{bmatrix}x(t) + \begin{bmatrix} 0 \\ 0 \\ 1 \end{bmatrix}u(t)$$

$$y(t) = [1 \quad 0 \quad 0]\,x(t)$$

7.34 (a)

From source transformation,

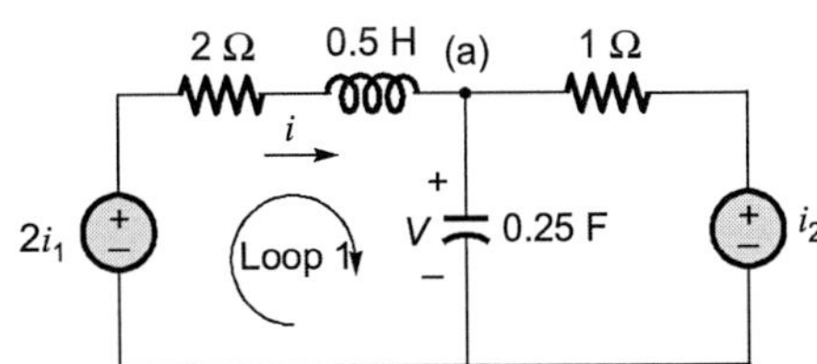

KVL in loop 1,

$$2i_1 = 2i + 0.5\frac{di}{dt} + V$$

$$\frac{di}{dt} = -2V - 4i + 4i_1 \quad \text{...(i)}$$

KCL at node (a),

$$i = 0.25\frac{dV}{dt} + \frac{V - i_2}{1}$$

$$\frac{dv}{dt} = -4V + 4i + 4i_2 \quad \text{...(ii)}$$

$$\frac{d}{dt}\begin{bmatrix} V \\ i \end{bmatrix} = \begin{bmatrix} -4 & 4 \\ -2 & -4 \end{bmatrix}\begin{bmatrix} V \\ i \end{bmatrix} + \begin{bmatrix} 0 & 4 \\ 4 & 0 \end{bmatrix}\begin{bmatrix} i_1 \\ i_2 \end{bmatrix}$$

7.35 (c)

$$i_s = v_i + v_c$$

$$\therefore \quad v_i = -v_c + i_s$$

$$v_L = L\,i_L$$

$$v_L = (i_s - i_L)$$

$$\therefore \quad L\,i_L = i_s - i_L$$

$$\therefore \quad i_L = -i_L + i_s$$

$$v_o = v_c$$

$$X = \begin{bmatrix} -1 & 0 \\ 0 & -1 \end{bmatrix} X + \begin{bmatrix} 1 \\ 2 \end{bmatrix} U$$

$$Y = [0 \quad 1]X + [0]U$$

$$A = -I$$

$$Q_c = [B \quad AB] = \begin{bmatrix} 1 & -1 \\ 2 & -2 \end{bmatrix} \Rightarrow |Q_c| = 0$$

$$Q_o = [C^T \quad A^TC^T] = \begin{bmatrix} 0 & 0 \\ 1 & -1 \end{bmatrix}$$

$$\Rightarrow \quad |Q_o| = 0$$

7.36 (b)

For a system with state-space representation as below:

$$\frac{dx}{dt} = Ax + Br$$

$$y = Cx + D$$

The transfer function is given as,

$$\text{T.F} = C[(sI - A)^{-1}B] + D$$

For the given system,

$$A = \begin{bmatrix} 0 & -2 \\ 1 & -3 \end{bmatrix}, B = \begin{bmatrix} 1 \\ 0 \end{bmatrix}, C = [2 \quad -5], D = [0]$$

$$\therefore \quad \text{T.F} = [2 \quad -5] \cdot \frac{1}{s^2 + 3s + 2} \begin{bmatrix} s+3 & -2 \\ 1 & s \end{bmatrix} \begin{bmatrix} 1 \\ 0 \end{bmatrix}$$

$$\text{T.F} = \frac{2(s+3) - 5}{s^2 + 3s + 2} = \frac{2s+1}{s^2 + 3s + 2}$$

Option (a):

$$\text{T.F}_1 = [0 \quad 2] \frac{1}{s^2 + 3s + 2} \begin{bmatrix} s+3 & 1 \\ -2 & s \end{bmatrix} \begin{bmatrix} 1 \\ 0 \end{bmatrix}$$

$$= \frac{-4}{s^2 + 3s + 2} \neq \text{T.F.}$$

Option (b):

$$\text{T.F}_2 = [1 \quad 2] \frac{1}{s^2 + 3s + 2} \begin{bmatrix} s+3 & 1 \\ -2 & s \end{bmatrix} \begin{bmatrix} 0 \\ 1 \end{bmatrix}$$

$$= \frac{2s+1}{s^2 + 3s + 2} = \text{T.F.}$$

Option (c):

$$\text{T.F}_3 = [1 \quad 1] \frac{1}{s^2 + 3s + 2} \begin{bmatrix} s+1 & 0 \\ 0 & s+2 \end{bmatrix} \begin{bmatrix} -1 \\ 3 \end{bmatrix}$$

$$= \frac{-s - 1 + 3s + 6}{s^2 + 3s + 2} = \frac{2s+5}{s^2 + 3s + 2} \neq \text{T.F.}$$

Option (d)

$$\text{T.F}_4 = [1 \quad 2] \frac{1}{s^2 + 3s + 2} \begin{bmatrix} s+1 & 0 \\ 0 & s+2 \end{bmatrix} \begin{bmatrix} 1 \\ 1 \end{bmatrix}$$

$$= \frac{s + 1 + 2(s+2)}{s^2 + 3s + 2} = \frac{3s+5}{s^2 + 3s + 2} \neq \text{T.F.}$$

Hence, option (b) is the correct answer.

■■■

Electronic Devices and Circuits

UNIT IV

CONTENTS

Electronic Devices and Circuits

Syllabus

Energy bands in intrinsic and extrinsic semiconductors, equilibrium carrier concentration, direct and indirect band-gap semiconductors.

Carrier transport: diffusion current, drift current, mobility and resistivity, generation and recombination of carriers, Poisson and continuity equations.

P-N junction, Zener diode, BJT, MOS capacitor, MOSFET, LED, photo diode and solar cell.

Analysis of Previous GATE Papers

Exam Year	1 Mark Ques.	2 Marks Ques.	Total Marks
1995	11	1	**13**
1996	–	1	**2**
1997	1	4	**9**
1998	2	1	**4**
1999	1	1	**3**
2000	–	–	**–**
2001	2	–	**2**
2002	1	–	**1**
2003	5	5	**15**
2004	3	7	**17**
2005	3	3	**9**
2006	4	4	**12**
2007	2	6	**14**
2008	4	4	**12**
2009	2	3	**8**
2010	2	4	**10**
2011	3	3	**9**
2012	1	3	**7**
2013	2	–	**2**

Exam Year	1 Mark Ques.	2 Mark Ques.	Total Marks
2014 Set-1	2	3	**8**
2014 Set-2	2	3	**8**
2014 Set-3	3	3	**9**
2014 Set-4	3	3	**9**
2015 Set-1	2	4	**10**
2015 Set-2	2	3	**8**
2015 Set-3	2	3	**8**
2016 Set-1	2	4	**10**
2016 Set-2	1	4	**9**
2016 Set-3	3	2	**7**
2017 Set-1	3	3	**9**
2017 Set-2	2	4	**10**
2018	3	3	**9**
2019	3	4	**11**
2020	2	4	**10**
2021	2	2	**6**
2022	3	2	**7**
2023	3	2	**7**
2024	2	5	**12**

1 Basic Semiconductor Physics

1.1 The drift velocity of electrons, in silicon
(a) is proportional to the electric field for all values of electric field
(b) is independent of the electric field.
(c) increases at low values of electric field and decreases at high values of electric field exhibiting negative differential resistance.
(d) increases linearly with electric field at low values of electric field and gradually saturates at higher values of electric field.
[1995 : 1 M]

1.2 The probability that an electron in a metal occupies the fermi level, at any temperature. ($T > 0K$)
(a) 0 (b) 1
(c) 0.5 (d) 1.0 **[1995 : 1 M]**

1.3 In a *p*-type Si simple the hole concentration is $2.25 \times 10^{15}/\text{cm}^3$. The intrinsic carrier concentration is $1.5 \times 10^{10}/\text{cm}^3$ the electron concentration is
(a) zero (b) $10^{10}/\text{cm}^3$
(c) $10^5/\text{cm}^3$ (d) $1.5 \times 10^{25}/\text{cm}^3$
[1995 : 1 M]

1.4 A small concentration of minority carries is injected into a homogeneous semiconductor crystal at one point. An electric field of 10 V/cm is applied across the crystal and this moves the minority carries a distance of 1 cm in 20 μsec. The mobility (in cm^2/V-sec) will be
(a) 1,000 (b) 2,000
(c) 5,000 (d) 500,000 **[1995 : 1 M]**

1.5 The units of $\dfrac{q}{KT}$ are
(a) V (b) V^{-1}
(c) J (d) J/K
[1997 : 1 M]

1.6 The intrinsic carrier density at 300 K is $1.5 \times 10^{10}/\text{cm}^3$, in silicon for *n*-type silicon doped to 2.25×10^{15} atoms/cm^3, the equilibrium electron and hole densities are
(a) $n = 1.5 \times 10^{15}$, $p = 1.5 \times 10^{10}/\text{cm}^3$
(b) $n = 1.5 \times 10^{10}$, $p = 2.25 \times 10^{15}/\text{cm}^3$
(c) $n = 2.25 \times 10^{15}$, $p = 1.0 \times 10^{5}/\text{cm}^3$
(d) $n = 1.5 \times 10^{10}$, $p = 1.5 \times 10^{10}/\text{cm}^3$
[1997 : 2 M]

1.7 The electron concentration in a sample of non-uniformly doped *n*-type silicon at 300 K varies linearly from $10^{17}/\text{cm}^3$ at $x = 0$ to $6 \times 10^{16}/\text{cm}^3$ at $x = 2$ μm. Assume a situation that electrons are supplied to keep this concentration gradient constant with time. If electronic charge is 1.6×10^{-19} Coulomb and the diffusion constant $D_n = 35\text{cm}^2/\text{s}$, the current density in the silicon, if no electric field is present is
(a) zero (b) 120 A/cm^2
(c) +1120 A/cm^2 (d) –1120 A/cm^2
[1997 : 2 M]

1.8 An *n*-type silicon bar 0.1 cm long and 100 μm^2 in cross-sectional area has a majority carrier concentration of $5 \times 10^{20}/\text{m}^3$ and the carrier mobility is 0.13 m^2/V-s at 300 K. If the charge of an electron is 1.6×10^{-19} Coulomb, then the resistance of the bar is
(a) $10^6\ \Omega$ (b) $10^4\ \Omega$
(c) $10^{-1}\ \Omega$ (d) $10^{-4}\ \Omega$ **[1997 : 2 M]**

1.9 The unit of $\dfrac{q}{KT}$ are
(a) V (b) V^{-1}
(c) J (d) J/K **[1998 : 1 M]**

1.10 *n*-type silicon is obtained by doping silicon with
(a) Germanium (b) Aluminium
(c) Boron (d) Phosphorus
[2003 : 1 M]

1.11 The bandgap of silicon at 300 K is
(a) 1.36 eV (b) 1.10 eV
(c) 0.80 eV (d) 0.67 eV **[2003 : 1 M]**

1.12 The intrinsic carrier concentration of silicon sample at 300 K is $1.5 \times 10^{16}/\text{m}^3$. If after doping, the number of majority carriers is $5 \times 10^{20}/\text{m}^3$, the minority carrier density is
(a) $4.50 \times 10^{11}/\text{m}^3$ (b) $3.33 \times 10^{4}/\text{m}^3$
(c) $5.00 \times 10^{20}/\text{m}^3$ (d) $3.00 \times 10^{-5}/\text{m}^3$
[2003 : 1 M]

1.13 The impurity commonly used for realizing the base region of a silicon n-p-n transistor is
(a) Gallium (b) Indium
(c) Boron (d) Phosphorus
[2004 : 1 M]

1.14 The resistivity of a uniformly doped n-type silicon sample is 0.5 Ω – cm. If the electron mobility (μ_n) is 1250 cm²/V–sec and the charge of an electron is 1.6 × 10^{-19} Coulomb, the donor impurity concentration (N_D) in the sample is
(a) $2 \times 10^{16}/cm^3$ (b) $1 \times 10^{16}/cm^3$
(c) $2.5 \times 10^{15}/cm^3$ (d) $2 \times 10^{15}/cm^3$
[2004 : 2 M]

1.15 The longest wavelength that can be absorbed by silicon, which has the bandgap of 1.12 eV, is 1.1 μm. If the longest wavelength that can be absorbed by another material is 0.87 μm, then the bandgap of this material is
(a) 1.416 eV (b) 0.886 eV
(c) 0.854 eV (d) 0.706 eV [2004 : 2 M]

1.16 The neutral base width of a bipolar transistor, biased in the active region, is 0.5 μm. The maximum electron concentration and the diffusion constant in the base are $10^{14}/cm^3$ and D_n = 25 cm²/sec respectively. Assuming negligible recombination in the base, the collector current density is (the electron charge is 1.6×10^{-19} Coulomb)
(a) 800 A/cm² (b) 8 A/cm²
(c) 200 A/cm² (d) 2 A/cm² [2004 : 2 M]

1.17 The bandgap of Silicon at room temperature is
(a) 1.3 eV (b) 0.7 eV
(c) 1.1 eV (d) 1.4 eV [2005 : 1 M]

1.18 The primary reason for the widespread use of Silicon in semiconductor device technology is
(a) abundance of Silicon on the surface of the Earth.
(b) larger bandgap of Silicon in comparison to Germanium
(c) favorable properties of Silicon-dioxide (SiO_2)
(d) lower melting point [2005 : 1 M]

1.19 A Silicon sample *A* is doped with 10^{18} atoms/cm³ of Boron. Another sample *B* of identical dimensions is doped with 10^{18} atoms/cm³ of Phosphorus. The ratio of electron to hole mobility is 3. The ratio of conductivity of the sample *A* to *B* is
(a) 3 (b) 1/3
(b) 2/3 (d) 3/2 [2005 : 2 M]

1.20 The concentration of minority carriers in an extrinsic semiconductor under equilibrium is
(a) directly proportional to the doping concentration
(b) inversely proportional to the doping concentration
(c) directly proportional to the intrinsic concentration
(d) inversely proportional to the intrinsic concentration
[2006 : 1 M]

1.21 Under low level injection assumption, the injected minority carrier current for an extrinsic semiconductor is essentially the
(a) diffusion current
(b) drift current
(c) recombination current
(d) induced current [2006 : 1 M]

1.22 The majority carriers in an *n*-type semiconductor have an average drift velocity *v* in a direction perpendicular to a uniform magnetic field *B*. The electric field *E* induced due to Hall effect acts in the direction
(a) $v \times B$ (b) $B \times v$
(c) along *v* (d) opposite to *v*
[2006 : 2 M]

1.23 A heavily doped *n*-typed semiconductor has the following data:
Hole-electron mobility ratio : 0.4
Doping concentration : 4.2×10^8 atoms/m³
Intrinsic concentration : 1.5×10^4 atoms/m³
The ratio of conductance of the *n*-type semiconductor to that of the intrinsic semiconductor of same material and at the same temperature is given by
(a) 0.00005 (b) 2,000
(c) 10,000 (d) 20,000 [2006 : 2 M]

1.24 The electron and hole concentrations in an intrinsic semiconductor are n_i per cm³ at 300 K. Now, if acceptor impurities are introduced with a concentration of N_A per cm³ (where $N_A >> n_i$) the electron concentration per cm³ at 300 K will be
(a) n_i (b) $n_i + N_A$
(c) $N_A - n_i$ (d) $\frac{n_i^2}{N_A}$ [2007 : 1 M]

1.25 Which of the following is true?
(a) A silicon wafer heavily doped with boron is a p^+ substrate
(b) A silicon wafer lightly doped with boron is a p^+ substrate
(c) A silicon wafer heavily doped with arsenic is a p^+ substrate
(d) A silicon wafer lightly doped with arsenic is a p^+ substrate [2008 : 1 M]

1.26 Silicon is doped with boron to a concentration of 4×10^{17} atoms/cm³. Assume the intrinsic carrier concentration of silicon to be $1.5 \times 10^{10}/cm^3$ and the value of kT/q to be 25 mV at 300 K. Compared to undoped silicon, the Fermi level of doped silicon

(a) goes down by 0.13 eV
(b) goes up by 0.13 eV
(c) goes down by 0.427 eV
(d) goes up by 0.427 eV [2008 : 2 M]

1.27 In an *n*-type silicon crystal at room temperature, which of the following can have a concentration of $4 \times 10^{19} cm^{-3}$?
(a) Silicon atoms
(b) Holes
(c) Dopant atoms
(d) Valence electrons [2009 : 1 M]

1.28 The ratio of the mobility to the diffusion coefficient in a semiconductor has the unit
(a) V^{-1} (b) $cm \cdot V^{-1}$
(c) $V \cdot cm^{-1}$ (d) $V \cdot s$ [2009 : 1 M]

Linked Answer Questions 1.29 and 1.30:

The silicon sample with unit cross-sectional areas shown below is in thermal equilibrium. The following information is given: $T = 300$ K, electronic charge $= 1.6 \times 10^{-19}$ C, thermal voltage = 26 mV and electron mobility = 1350 cm²/V-s.

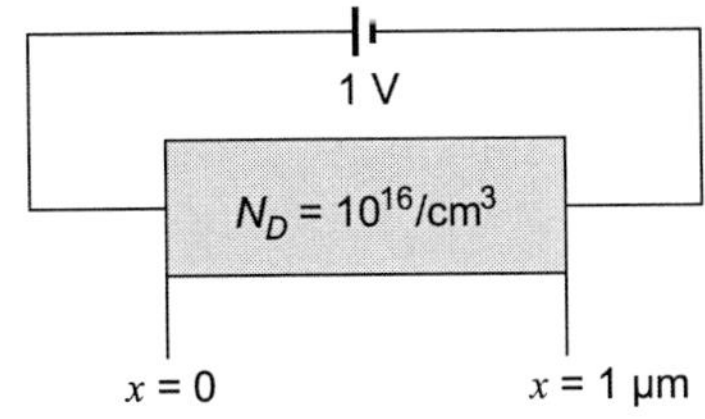

1.29 The magnitude of the electric field at $x = 0.5$ μm is
(a) 1 kV/cm (b) 5 kV/cm
(c) 10 kV/cm (d) 26 kV/cm
[2010 : 2 M]

1.30 The magnitude of the electron drift current density at $x = 0.5$ μm is
(a) 2.16×10^4 A/cm² (b) 1.08×10^4 A/cm²
(c) 4.32×10^3 A/cm² (d) 6.48×10^2 A/cm²
[2010 : 2 M]

1.31 Drift current in semiconductors depends upon
(a) only the electric field
(b) only the carrier concentration gradient
(c) both the electric field and the carrier concentration
(d) both the electric field and the carrier concentration gradient
[2011 : 1 M]

1.32 The doping concentrations on the *p*-side and *n*-side of a silicon diode are 1×10^{16} cm^{-3} and 1×10^{17} cm^{-3}, respectively. A forward bias of 0.3 V is applied to the diode. At $T = 300$ K, the intrinsic carrier concentration of silicon $n_i = 1.5 \times 10^{10}$ cm^{-3} and $\frac{kT}{q} = 26$ mV. The electron concentration at the edge of the depletion region on the *p*-side is
(a) 2.3×10^9 cm^{-3} (b) 1×10^{16} cm^{-3}
(c) 1×10^{17} cm^{-3} (d) 2.25×10^6 cm^{-3}
[2014 : 2 M, Set-1]

1.33 A silicon bar is doped with donor impurities $N_D = 2.25 \times 10^{15}$ atoms/cm³. Given the intrinsic carrier concentration of silicon at $T = 300$ K is $n_i = 1.5 \times 10^{10}$ cm^{-3}. Assuming complete impurity ionization, the equilibrium electron and hole concentrations are
(a) $n_0 = 1.5 \times 10^{16}$ cm^{-3}, $p_0 = 1.5 \times 10^5$ cm^{-3}
(b) $n_0 = 1.5 \times 10^{10}$ cm^{-3}, $p_0 = 1.5 \times 10^{15}$ cm^{-3}
(c) $n_0 = 2.25 \times 10^{15}$ cm^{-3}, $p_0 = 1.5 \times 10^{10}$ cm^{-3}
(d) $n_0 = 2.25 \times 10^{15}$ cm^{-3}, $p_0 = 1 \times 10^5$ cm^{-3}
[2014 : 1 M, Set-2]

1.34 Assume electronic charge $q = 1.6 \times 10^{-19}$C, $kT/q = 25$ mV and electron mobility $\mu_n = 1000$ cm²/V-s. If the concentration gradient of electrons injected into a *p*-type silicon sample is 1×10^{21} cm⁴, the magnitude of electron diffusion current density (in A/cm²) is
[2014 : 2 M, Set-2]

1.35 When a silicon diode having a doping concentration of $N_A = 9 \times 10^{16}$ cm^{-3} on p-side and $N_D = 1 \times 10^{16}$ cm^{-3} on n-side is reverse biased, the total depletion width is found to be 3 μm. Given that the permittivity of silicon is 1.04×10^{-12} F/cm, the depletion width on the *p*-side and the maximum electric field in the depletion region, respectively, are
(a) 2.7 μm and 2.3×10^5 V/cm
(b) 0.3 μm and 4.15×10^5 V/cm
(c) 0.3 μm and 0.42×10^5 V/cm
(d) 2.1 μm and 0.42×10^5 V/cm
[2014 : 2 M, Set-2]

1.36 A thin *P*-type silicon sample is uniformly illuminated with light which generates excess carriers. The recombination rate is directly proportional to
(a) the minority carrier mobility
(b) the minority carrier recombination lifetime
(c) the majority carrier concentration
(d) the excess minority carrier concentration
[2014 : 1 M, Set-3]

1.37 At $T = 300$ K, the hole mobility of a semiconductor $\mu_p = 500$ cm²/V-s and $\frac{kT}{q} = 26$ mV. The hole diffusion constant D_p in cm²/s is ______.
[2014 : 1 M, Set-3]

1.38 At $T = 300$ K, the band gap and the intrinsic carrier concentration of GaAs are 1.42 eV and 10^6 cm^{-3}, respectively. In order to generate electron hole pairs in GaAs, which one of the wavelength (λ_C) ranges of incident radiation, is most suitable? (Given that: Plank's constant is 6.62×10^{-34} J-s, velocity of light is 3×10^{10} cm/s and charge of electron is 1.6×10^{-19} C)

(a) 0.42 μm < λ_C < 0.87 μm
(b) 0.87 μm < λ_C < 1.42 μm
(c) 1.42 μm < λ_C < 1.62 μm
(d) 1.62 μm < λ_C < 6.62 μm

[2014 : 1 M, Set-4]

1.39 In the figure, ln (ρ_i) is plotted as a function of $1/T$, where ρ_i is the intrinsic resistivity of silicon, T is the temperature, and the plot is almost linear.

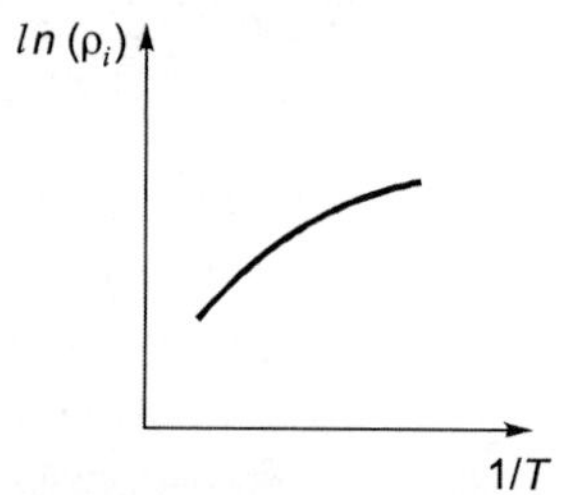

The slope of the line can be used to estimate

(a) band gap energy of silicon (E_g)
(b) sum of electron and hole mobility in silicon ($\mu_n + \mu_p$)
(c) reciprocal of the sum of electron and hole mobility in silicon $(\mu_n + \mu_p)^{-1}$
(d) intrinsic carrier concentration of silicon (n_i)

[2014 : 1 M, Set-4]

1.40 The cut-off wavelength (in μm) of light that can be used for intrinsic excitation of a semiconductor material of bandgap $E_g = 1.1$ eV is _______.

[2014 : 1 M, Set-4]

1.41 Consider a silicon sample doped with $N_D = 1 \times 10^{15}$/cm^3 donor atoms. Assume that the intrinsic carrier concentration $n_i = 1.5 \times 10^{10}$/cm^3. If the sample is additionally doped with $N_A = 1 \times 10^{18}$/cm^3 acceptor atoms, the approximate number of electrons/cm^3 in the sample, at $T = 300$ K, will be _____.

[2014 : 2 M, Set-4]

1.42 An N-type semiconductor having uniform doping is biased as shown in the figure.

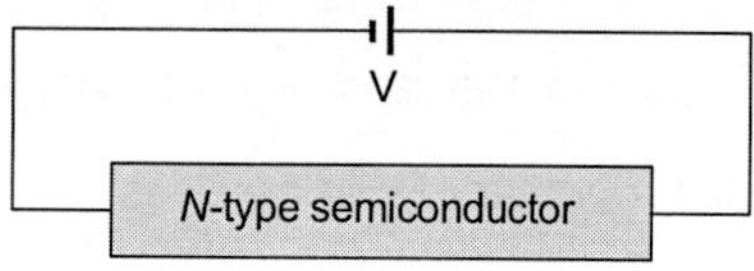

If E_C is the lowest energy level of the conduction band, E_V is the highest energy level of the valance band and E_F is the Fermi level, which one of the following represents the energy band diagram for the biased N-type semiconductor?

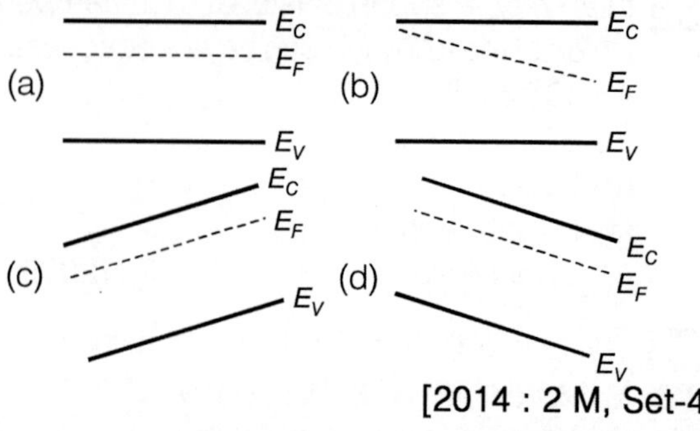

[2014 : 2 M, Set-4]

1.43 A silicon sample is uniformly doped with donor type impurities with a concentration of 10^{16}/cm^3. The electron and hole mobilities in the sample are 1200 cm^2/V-s and 400 cm^2/V-s respectively. Assume complete ionization of impurities. The charge of an electron is 1.6×10^{-19} C. The resistivity of the sample (in Ω-cm) is_______.

[2015 : 1 M, Set-1]

1.44 A piece of silicon is doped uniformly with phosphorous with a doping concentration of 10^{16}/cm^3. The expected value of mobility versus doping versus doping concentration for silicon assuming full dopant ionization is shown below. The charge of an electron is 1.6×10^{-19} C. The conductivity (in S cm^{-1}) of the silicon sample at 300 K is _____.

Hole and Electron Mobility in Silicon at 300K

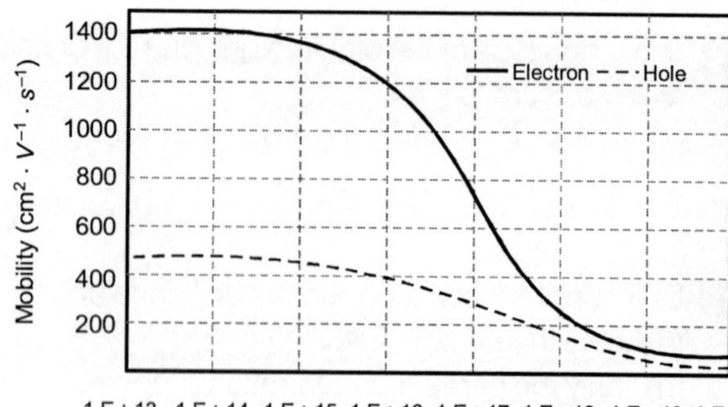

[2015 : 1 M, Set-2]

1.45 An n-type silicon sample is uniformly illuminated with light which generates 10^{20} electron-hole pairs per cm^3 per second. The minority carrier lifetime in the sample is 1 μs. In the steady state, the hole concentration in the sample is approximately 10^x, where x is an integer. The value of x is__________.

[2015 : 1 M, Set-2]

1.46 The energy band diagram and the electron density profile $n(x)$ in a semiconductor are shown in the figures.

Assume that $n(x) = 10^{15} e^{\left(\frac{q\alpha x}{kT}\right)} cm^{-3}$, with $\alpha = 0.1$ V/cm and x expressed in cm. Given $\frac{kT}{q} = 0.026$ V, $D_n = 36$ cm^2 s^{-1}, and $\frac{D}{\mu} = \frac{kT}{q}$. The electron current density (in A/cm^2) at $x = 0$ is

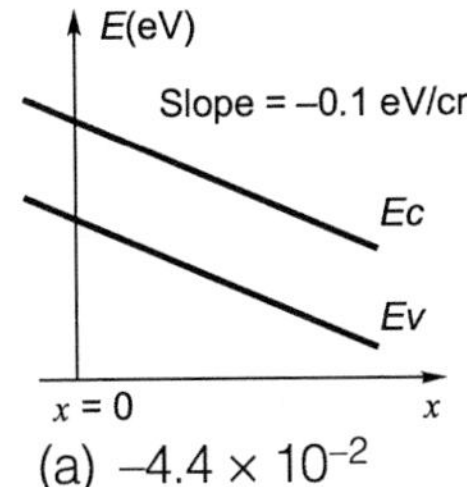

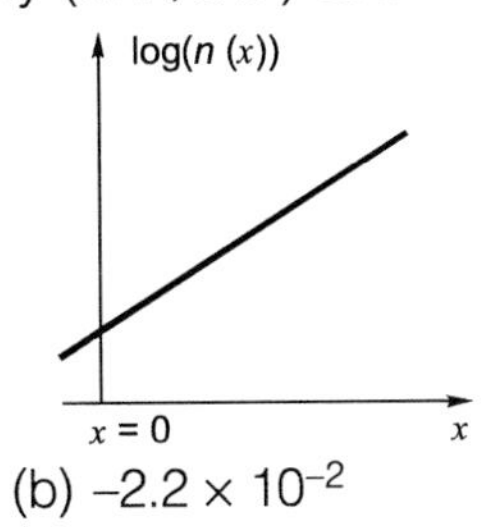

(a) -4.4×10^{-2}
(b) -2.2×10^{-2}
(c) 0
(d) 2.2×10^{-2}

[2015 : 2 M, Set-2]

1.47 A dc voltage of 10 V is applied across an n-type silicon bar having a rectangular cross-section and length of 1 cm as shown in figure. The donor doping concentration N_D and the mobility of electrons μ_n are 10^{16} cm^{-3} and 1000 cm^2 V^{-1}s^{-1}, respectively. The average time (in μs) taken by the electrons to move from one end of the bar to other end is _____.

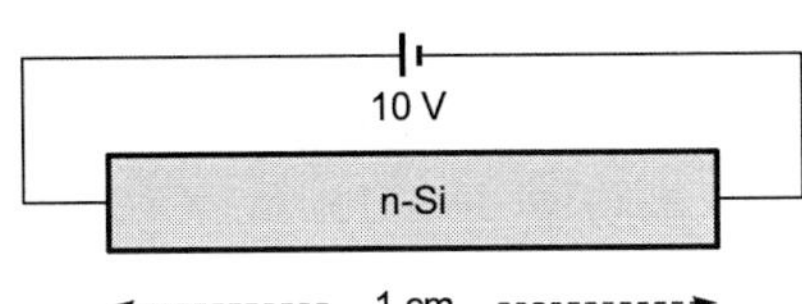

[2015 : 2 M, Set-2]

1.48 A small percentage of impurity is added to an intrinsic semiconductor at 300 K. Which one of the following statements is true for the energy band diagram shown in the following figure?

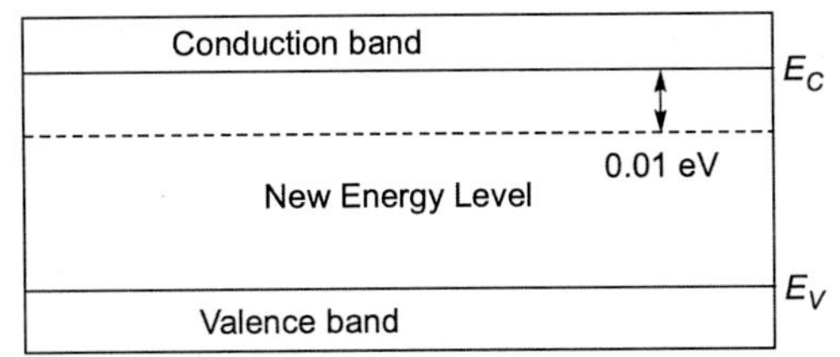

(a) Intrinsic semiconductor doped with pentavalent atoms to form n-type semiconductor
(b) Intrinsic semiconductor doped with trivalent atoms to form n-type semiconductor
(c) Intrinsic semiconductor doped with pentavalent atoms to form p-type semiconductor
(d) Intrinsic semiconductor doped with trivalent atoms to form p-type semiconductor

[2016 : 1 M, Set-1]

1.49 The figure below shows the doping distribution in a p-type semiconductor in log scale.

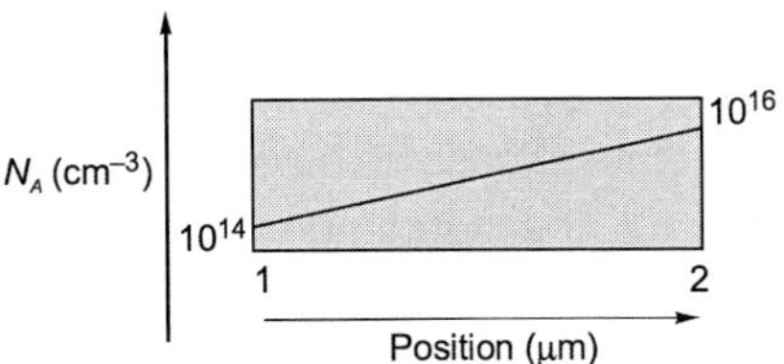

The magnitude of the electric field (in kV/cm) in the semiconductor due to non uniform doping is ____.

[2016 : 2 M, Set-1]

1.50 Consider a silicon sample at $T = 300$ K, with a uniform donor density $N_d = 5 \times 10^{16}$ cm^{-3}, illuminated uniformly such that the optical generation rate is $G_{opt} = 1.5 \times 10^{20}$ cm^{-3} s^{-1} throughout the sample. The incident radiation is turned off at $t = 0$. Assume low-level injection to be valid and ignore surface effects. The carrier lifetimes are $\tau_{p0} = 0.1$ μs and $\tau_{n0} = 0.5$ μs.

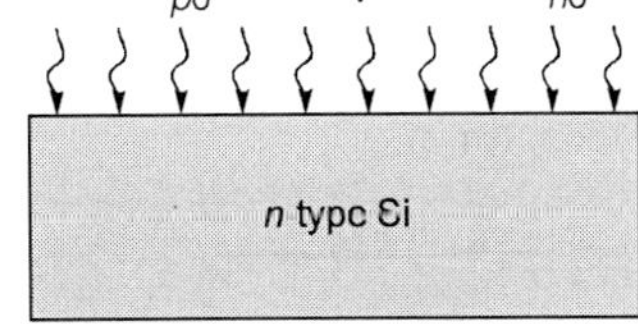

The hole concentration at $t = 0$ and the hole concentration at $t = 0.3$ μs, respectively, are

(a) 1.5×10^{13} cm^{-3} and 7.47×10^{11} cm^{-3}
(b) 1.5×10^{13} cm^{-3} and 8.23×10^{11} cm^{-3}
(c) 7.5×10^{13} cm^{-3} and 3.73×10^{11} cm^{-3}
(d) 7.5×10^{13} cm^{-3} and 4.12×10^{11} cm^{-3}

[2016 : 2 M, Set-1]

1.51 A bar of Gallium Arsenide (GaAs) is doped with Silicon such that the Silicon atoms occupy Gallium and Arsenic sites in the GaAs crystal. Which one of the following statements is true?

(a) Silicon atoms act as p-type dopants in Arsenic sites and n-type dopants in Gallium sites.
(b) Silicon atoms act as n-type dopants in Arsenic sites and p-type dopants in Gallium sites.
(c) Silicon atoms act as p-type dopants in Arsenic as well as Gallium sites.
(d) Silicon atoms act as n-type dopants in Arsenic as well as Gallium sites.

[2017 : 1 M, Set-1]

1.52 As shown, a uniformly doped Silicon (Si) bar of length $L = 0.1$ μm with a donor concentration $N_D = 10^{16}$ cm^{-3} is illuminated at $x = 0$ such that electron and hole pairs are generated at the rate of

$$G_L = G_{L0}\left(1 - \frac{x}{L}\right),\ 0 \le x \le L,$$

where $G_{L0} = 10^{17}$ cm^{-3} s^{-1}. Hole lifetime is 10^{-4} s, electronic charge $q = 1.6 \times 10^{-19}$ C, hole diffusion coefficient D_p = 100 cm^2/s and low level injection condition prevails. Assuming a linearly decaying steady state excess hole concentration that goes to 0 at $x = L$, the magnitude of the diffusion current density at x = L/2, in A/cm^2, is _______.

Light
Si ($N_D = 10^{16}$ cm^{-3})
$x = 0$ $L = 0.1$ μm

[2017 : 2 M, Set-1]

1.53 The dependence of drift velocity of electrons on electric field in a semiconductor is shown below. The semiconductor has a uniform electron concentration of $n = 1 \times 10^{16}$ cm^{-3} and electronic charge $q = 1.6 \times 10^{-19}$ C. If a bias of 5 V is applied across a 1 μm region of this semiconductor, the resulting current density in this region, in kA/cm^2, is _______.

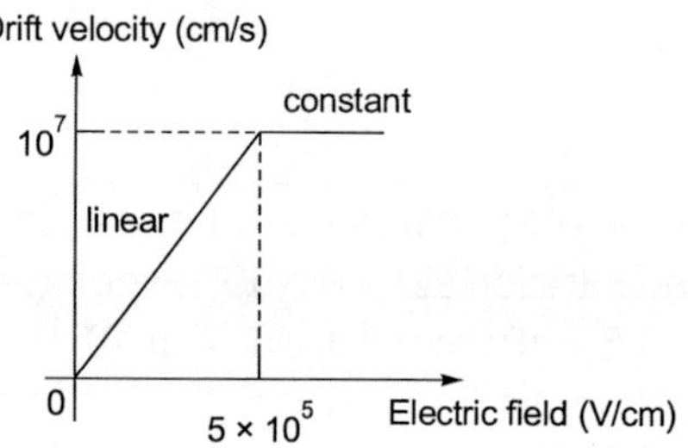

[2017 : 2 M, Set-1]

1.54 A single crystal intrinsic semiconductor is at a temperature of 300 K with effective density of states for holes twice that of electrons. The thermal voltage is 26 mV. The intrinsic Fermi level is shifted from mid-bandgap energy level by

(a) 13.45 meV (b) 18.02 meV
(c) 26.90 meV (d) 9.01 meV

[2020 : 1 M]

1.55 The energy band diagram of a p-type semiconductor bar of length L under equilibrium condition (i.e, the Fermi energy level E_F is constant) is shown in the figure. The valence band E_V is sloped since doping is non-uniform along the bar. The difference between the energy levels of the valence band at the two edges of the bar is Δ.

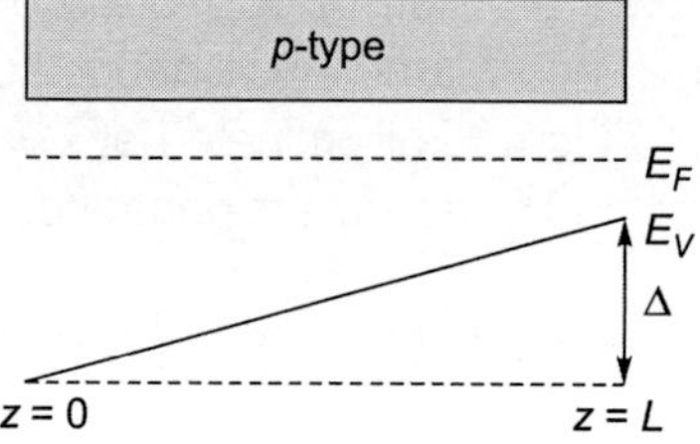

If the charge of an electron is q, then the magnitude of the electric field developed inside this semiconductor bar is

(a) $\frac{2\Delta}{qL}$ (b) $\frac{3\Delta}{2qL}$
(c) $\frac{\Delta}{2qL}$ (d) $\frac{\Delta}{qL}$

[2021 : 1 M]

1.56 A bar of silicon is doped with boron concentration of 10^{16} cm^{-3} and assumed to be fully ionized. It is exposed to light such that electron-hole pairs are generated throughout the volume of he bar at the rate of 10^{20} cm^{-3} s^{-1}. If the recombination lifetime is 100 μs intrinsic carrier concentration of silicon is 10^{10} cm^{-3} and assuming 100% ionization of boron, then the approximate product of steady-state electron and hole concentration due to this light exposure is

(a) 2×10^{20} cm^{-6} (b) 10^{32} cm^{-6}
(c) 10^{20} cm^{-6} (d) 2×10^{32} cm^{-6}

[2021 : 1 M]

1.57 Consider a long rectangular bar of direct bandgap p-type semiconductor. The equilibrium hole density is 10^{17} cm^{-3} and the intrinsic carrier concentration is 10^{10} cm^{-3}. Electron and hole diffusion lengths are 2 μm and 1 μm, respectively. The left side of the bar (x = 0) is uniformly illuminated with a laser having photon energy greater than the bandgap of the semiconductor. Excess electron-hole pairs are generated ONLY at x = 0 because of the laser. The steady state electron density at x = 0 is 10^{14} cm^{-3} due to laser illumination. Under these conditions and ignoring electric field, the closest approximation (among the given options) of the steady state electron density at x = 2 μm, is __________.

(a) 0.37×10^{14} cm^{-3} (b) 0.63×10^{13} cm^{-3}
(c) 3.7×10^{14} cm^{-3} (d) 10^3 cm^{-3}

[2022 : 1 M]

1.58 In a non-degenerate bulk semiconductor with electron density n = 10^{16} cm^{-3}, the value of $E_C - E_{Fn}$ = 200 meV, where E_C and E_{Fn} denote the bottom of the conduction band energy and electron Fermi level energy, respectively. Assume thermal voltage as 26 meV and the intrinsic carrier concentration is 10^{10} cm^{-3}. For $n = 0.5 \times 10^{16}$ cm^{-3}, the closest approximation of the value of (E_C – E_{Fn}), among the given options, is __________.

(a) 226 meV (b) 174 meV
(c) 218 meV (d) 182 meV

[2022 : 1 M]

1.59 Select the CORRECT statement(s) regarding semiconductor devices.

(a) Electrons and holes are of equal density in an intrinsic semiconductor at equilibrium.
(b) Collector region is generally more heavily doped than Base region in a BJT.
(c) Total current is spatially constant in a two terminal electronic device in dark under steady state condition.
(d) Mobility of electrons always increases with temperature in Silicon beyond 300 K.

[2022 : 2 M]

1.60 In a semiconductor, if the Fermi energy level lies in the conduction band, then the semiconductor is known as

(a) degenerate *n*-type.
(b) degenerate *p*-type.
(c) non-degenerate *n*-type.
(d) non-degenerate *p*-type.

[2023 : 1 M]

1.61 For an intrinsic semiconductor at temperature $T = 0\ K$, which of the following statement is true?

(a) All energy states in the valence band are filled with electrons and all energy states in the conduction band are empty of electrons.
(b) All energy states in the valence band are empty of electrons and all energy states in the conduction band are filled with electrons.
(c) All energy states in the valence and conduction band are filled with holes.
(d) All energy states in the valence and conduction band are filled with electrons.

[2023 : 1 M]

1.62 In an extrinsic semiconductor, the hole concentration is given to be $1.5n_i$ where n_i is the intrinsic carrier concentration of 1×10^{10} cm^{-3}. The ratio of electron to hole mobility for equal hole and electron drift current is given as ________.
(rounded off to two decimal places).

[2023 : 2 M]

1.63 In a semiconductor device, the Fermi-energy level is 0.35 eV above the valence band energy. The effective density of states in the valence band at $T = 300$ K is 1×10^{19} cm^{-3}. The thermal equilibrium hole concentration in silicon at 400 K is ________ $\times 10^{13}$ cm^{-3}. (rounded off to two decimal places).
Given kT at 300 K is 0.026 eV.

[2023 : 2 M]

1.64 The free electron concentration profile $n(x)$ in a doped semiconductor at equilibrium is shown in the figure, where the points A, B, and C mark three different positions. Which of the following statements is/are true?

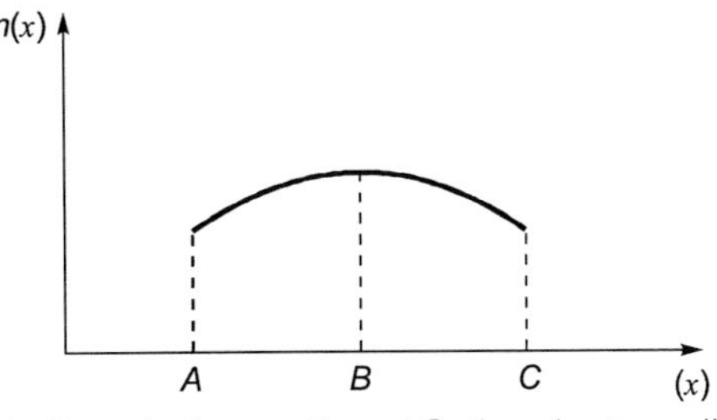

(a) For x between B and C, the electron diffusion current is directed from C to B.
(b) For x between B and A. the electric field is directed from A to B.
(c) For x between B and A, the electron drift current is directed from B to A.
(d) For x between B and C, the electric field is directed from B to C .

[2024 : 1 Mark]

1.65 For non-degenerately doped n-type silicon, which one of the following plots represents the temperature (T) dependence of free electron concentration (n)?

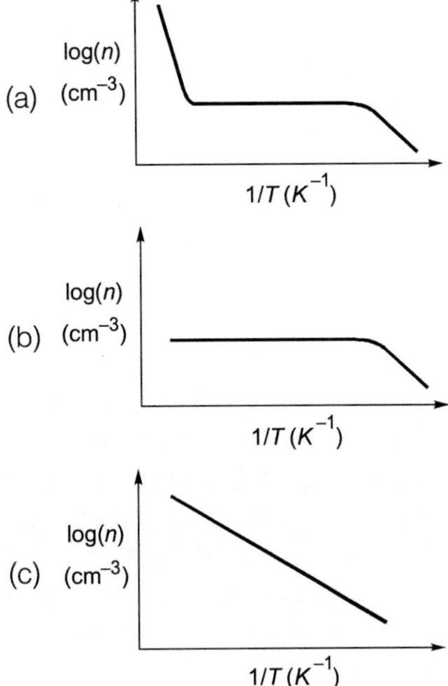

(d) 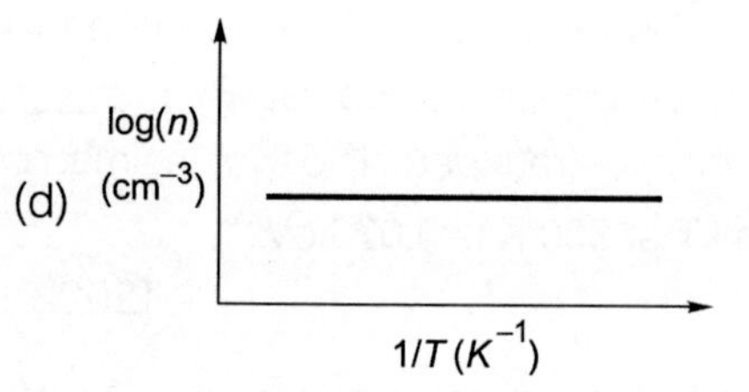

[2024 : 1 Mark]

1.66 A non-degenerate n-type semiconductor has 5% neutral dopant atoms. Its Fermi level is located at 0.25 eV below the conduction band (E_C) and the donor energy level (E_D) has a degeneracy of 2. Assuming the thermal voltage to be 20 mV, the difference between E_C and E_D (in eV, rounded off to two decimal places) is ________.

[2024 : 2 Marks]

■■■

Answers Basic Semiconductor Physics

1.1	(d)	1.2	(c)	1.3	(c)	1.4	(c)	1.5	(b)	1.6	(c)	1.7	(d)
1.8	(a)	1.9	(b)	1.10	(d)	1.11	(b)	1.12	(a)	1.13	(c)	1.14	(b)
1.15	(a)	1.16	(b)	1.17	(c)	1.18	(a)	1.19	(b)	1.20	(b)	1.21	(a)
1.22	(b)	1.23	(d)	1.24	(d)	1.25	(a)	1.26	(c)	1.27	(c)	1.28	(a)
1.29	(c)	1.30	(a)	1.31	(c)	1.32	(a)	1.33	(d)	1.34	(4000)	1.35	(b)
1.36	(d)	1.37	(13)	1.38	(a)	1.39	(a)	1.40	(1.12)	1.41	(225.2)	1.42	(d)
1.43	(0.52)	1.44	(1.92)	1.45	(14)	1.46	(c)	1.47	(100)	1.48	(a)	1.49	(1.198)
1.50	(a)	1.51	(a)	1.52	(16)	1.53	(1.6)	1.54	(d)	1.55	(d)	1.56	(d)
1.57	(a)	1.58	(c)	1.59	(a, c)	1.60	(a)	1.61	(a)	1.62	(2.25)	1.63	(63.36)
1.64	(a, c, d)	1.65	(a)	1.66	(0.18)								

Explanations Basic Semiconductor Physics

1.1 (d)

$$\boxed{v_d = \mu E}$$

where, v_d = Drift velocity ; μ = Mobility

E = Applied electric field

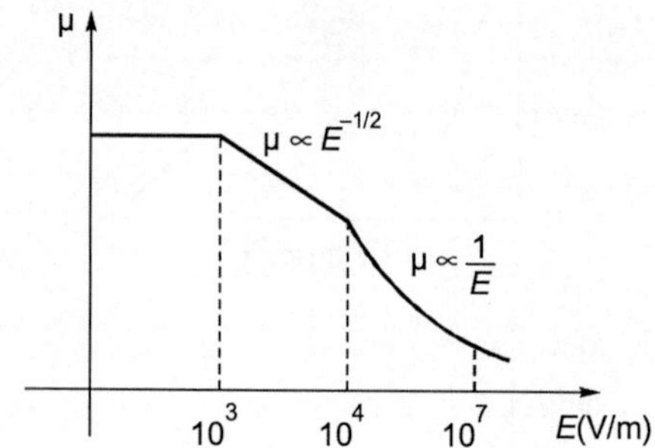

So,

1. For smaller electric field applied, mobility of change carrier will remain almost constant. So for smaller electric field applied, drift velocity (v_d) increases linearly with electric field.
2. For large electric field applied, mobility of charge carrier will be very small, so the drift velocity gradually saturates at higher values of electric field.

1.3 (c)

By Mass Action Law

$$\boxed{n \cdot p = n_i^2}$$

Where, n = electron concentration

p = hole concentration

n_i = intrinsic carrier concentration

$p = 2.25 \times 10^{15}/\text{cm}^3$

$n_i = 1.5 \times 10^{10}/\text{cm}^3$

$$n = \frac{n_i^2}{p} = \frac{(1.5 \times 10^{10})^2}{2.25 \times 10^{15}} = \frac{2.25 \times 10^{20}}{2.25 \times 10^{15}}$$

$n = 10^5/\text{cm}^3$

1.4 (c)

$$\text{Velocity} = \frac{\text{Distance}}{\text{Time}}$$

$$v_d = \frac{1}{20 \times 10^{-6}} = 50{,}000 \text{ cm/sec}$$

Drift velocity $= v_d = \mu E$

Where,

μ = Mobility

E = Electric field

$$\mu = \frac{v_d}{E} = \frac{50{,}000}{10} = 5{,}000 \text{ cm}^2/\text{volt-sec}$$

1.5 (b)

Thermal voltage or volt-equivalent of temperature

$$V_T = \frac{\bar{K}T}{q} = \text{volts}$$

So, $\frac{q}{K\bar{T}} \Rightarrow V^{-1}$

1.6 (c)

$n = 2.25 \times 10^{15}$, $p = 1.0 \times 10^5/\text{cm}^3$

For n-type semiconductor

Electron density = $n = N_D$

$= 2.25 \times 10^{15}$ atoms/cm^3

Given that,

Intrinsic carrier concentration = n_i

$= 1.5 \times 10/\text{cm}^3$

$\because \quad n >> n_i$

By Mass Action Law:

$$\boxed{n \cdot p = n_i^2}$$

$$P = \frac{n_i^2}{n} = \frac{(1.5 \times 10^{10})^2}{2.25 \times 10^{15}} = 10^5 \text{ atoms/cm}^3$$

1.7 (d)

$$\boxed{J_n = nq\mu_n E + D_n q \frac{dn}{dx}}$$

Given $\quad E = 0$

$$J_n = D_n q \frac{dn}{dx}$$

$$\frac{dn}{dx} = \frac{6 \times 10^{16} - 10^{17}}{2 \times 10^{-4} - 0} = -2 \times 10^{20}$$

$J_n = (35)(1.6 \times 10^{-19})(-2 \times 10^{20})$

$J_n = -1120 \text{ A/cm}^2$

1.8 (a)

Given that,

$l = 0.1 \text{ cm} = 10^{-3} \text{ m}$

$A = 100 \text{ μm}^2 = 100 \times 10^{-12} \text{ m}^2$

$n = 5 \times 10^{20}/\text{m}^3$

$\mu_n = 0.13 \text{ m}^2/\text{V-sec}$

$q = 1.6 \times 10^{-19}$ Col.

We know that

$$\boxed{\text{Conductivity} = \sigma = nq\mu_n}$$

$$\text{Resistivity} = \rho = \frac{1}{\sigma}$$

$$\text{Resistance} = \frac{\rho l}{A} = \frac{l}{\sigma A}$$

$$R = \frac{10^{-3}}{(5 \times 10^{20}) \times (1.6 \times 10^{-19})(0.13)(100 \times 10^{-12})}$$

$R = 10^6 \, \Omega$

1.9 (b)

$$\boxed{\text{Thermal voltage} = V_T = \frac{KT}{q}}$$

So, $\frac{q}{KT} \rightarrow V^{-1}$

1.12 (a)

$n_i^2 = np$

n_i = intrinsic concentration

$$p = \frac{n_i^2}{n} = \frac{1.5 \times 10^{16} \times 1.5 \times 10^{16}}{5 \times 10^{20}}$$

$= 45 \times 10^{10} = 4.5 \times 10^{11}/\text{cm}^3$

1.14 (b)

$$\rho = \frac{1}{nq\mu_n}$$

$n = N_D$

$$N_D = \frac{1}{q\mu_n\rho} = \frac{1}{1.6 \times 10^{-19} \times 1250 \times 0.5}$$

$= 10^{16}/\text{cm}^3$

1.15 (a)

$$E_g = \frac{1.24}{\lambda(\mu\text{m})} \text{eV} = \frac{1.24}{0.87\mu\text{m}} \text{eV} = 1.425 \text{ eV}$$

1.16 (b)

$$J_C = qD_n \frac{dn}{dx}$$

$$= 1.6 \times 10^{-19} \times 25 \times \frac{10^{14}}{0.5 \times 10^{-4}}$$

$J_C = 8 \text{ A/cm}^2$

1.19 (b)

$\sigma_n = nq\mu_n$

$$\frac{\sigma_p}{\sigma_n} = \frac{\mu_p}{\mu_n} = \frac{1}{3}$$

1.20 (b)

$$np = n_i^2$$

n_i = constant

For n-type p is minority carrier concentration

$$p = \frac{n_i^2}{n}$$

$$p \propto \frac{1}{n}$$

1.22 (b)

Hall effect; Electric force + magnetic force = 0

$$qE + qv \times B = 0$$

$$E = -v \times B$$

$$E = B \times v$$

1.23 (d)

For n-type semiconductor $\sigma_n = nq\mu_n$

For intrinsic semiconductor,

$$\sigma_i = n_i q(\mu_n + \mu_p)$$

$$\frac{\sigma_n}{\sigma_i} = \frac{n\mu_n}{n_i(\mu_n + \mu_p)} = \frac{4.2 \times 10^8 \times \mu_n}{1.5 \times 10^4 \times \mu_n\left(1 + \frac{\mu_p}{\mu_n}\right)}$$

$$= \frac{4.2 \times 10^8}{1.5 \times 10^4 \times 1.4} = 2 \times 10^4$$

1.24 (d)

By the law of electrical neutrality

$$p + N_D = n + N_A$$

as $N_D = 0$

$$N_A >> n_i \cong 0 \quad p = N_A$$

using mass action law np = n_i^2

So, $$n = \frac{n_i^2}{p} = \frac{n_i^2}{N_A}$$

1.26 (c)

Since, boron is p-type impurity, therefore, fermi level goes down

$$E_i - E_f = kT\,ln\,\frac{N_A}{n_i}$$

$$= 25 \times 10^{-3}\,ln\,\frac{4 \times 10^{17}}{1.5 \times 10^{10}} = 0.427\text{ eV}$$

1.27 (c)

In an N-type silicon crystal at room temperature

$$n \simeq N_d$$

and by mass-action law

$$n \cdot p = n_i^2$$

$$\therefore \quad p = \frac{(1.5 \times 10)^2}{4 \times 10^{19}} = 5.625\text{ cm}^{-3}$$

and $n >> p$

1.28 (a)

$$\frac{D}{\mu} = V_T$$

$$\Rightarrow \quad \frac{\mu}{D} = \frac{1}{V_T} \Rightarrow \text{units : V}^{-1}$$

1.29 (c)

The sample is in thermal equilibrium,

So, $$\boxed{E = \frac{V}{d}} = \frac{1}{1 \times 10^{-6}} = 10^6\text{ V/m} = 10\text{ kV/cm}$$

1.30 (a)

$$\boxed{J = \sigma E} \quad \boxed{\sigma = nq\mu_n = N_D q\mu_n}$$

$$J = N_D q\mu_n E$$

$$J = 10^{16} \times (1.6 \times 10^{-19})(1350)(10 \times 10^3)$$

$$J = 2.16 \times 10^4\text{ A/cm}^2$$

1.31 (c)

$$J = n e v_d$$

Put, $v_d = \mu E$

$\therefore$ $J = n e \mu E$

Hence, $I = n e \mu EA$

So, I depends upon carrier concentration and electric field.

1.32 (a)

The electron concentration at the edge of the depletion region on the p-side is given by

$$n_p = n_{p_o}\exp\left[\frac{qV_f}{kT}\right]$$

where, $$n_{p_o} = \frac{n_i^2}{N_A}$$

$$n_{p_o} = \frac{(1.5 \times 10^{10})^2}{10^{16}} = 2.25 \times 10^4\text{ cm}^{-3}$$

We then have

$$n_p = 2.25 \times 10^4 \exp\left[\frac{0.3}{0.026}\right]$$

$$n_p = 2.25 \times 10^4 \times 1.025 \times 10^5$$

$$n_p = 2.306 \times 10^9\text{ cm}^{-3}$$

1.33 (d)

Since $N_D >> n_i$ therefore equilibrium electron concentration is

$$n \simeq N_D = 2.25 \times 10^5\text{ cm}^{-3}$$

And equilibrium hole concentration is given by mass action law

$$p = \frac{n_i^2}{N_D} = \frac{(1.5\times10^{10})^2}{2.25\times10^{15}} = 1\times10^5\ \text{cm}^{-3}$$

1.34 (4000)

$$J = qD_n\frac{dn}{dx}$$

where, $D_n = \mu_n V_T$

$$D_n = \frac{1000\ \text{cm}^2}{\text{V s}}\times25\times10^{-3}\ \text{V}$$

$$D_n = 25\ \text{cm}^2/\text{s}$$

then, $J = 1.6\times10^{-19}\times25\times1\times10^{21}$

$$J = 4000\ \text{A/cm}^2$$

1.35 (b)

We know that

$$\frac{N_A}{N_D} = \frac{x_n}{x_p}$$

So, $\frac{N_A}{N_D}+1 = \frac{x_n}{x_p}+1$

$$\frac{N_A+N_D}{N_D} = \frac{x_n+x_p}{x_p}$$

$$\frac{10^{17}}{10^{16}} = \frac{3\ \mu\text{m}}{x_p}$$

$\therefore$ $x_p = 0.3\ \mu\text{m}$

and

$$E_{max} = \frac{qN_A x_p}{\epsilon_s} = \frac{qN_D x_n}{\epsilon_s}$$

$$\therefore E_{max} = \frac{1.6\times10^{-19}\times9\times10^{16}\times0.3\times10^{-4}}{1.04\times10^{-12}}$$

$$E_{max} = 4.15\times10^5\ \text{V/cm}$$

1.37 (13)

$$\frac{D_p}{\mu_p} = V_T$$

$$D_p = \mu_P V_T = 500\times26\times10^{-3}$$

$$D_p = 13\ \text{cm}^2/\text{s}$$

1.38 (a)

$$\lambda_c \le \frac{1.24}{E_g(eV)}\ \mu\text{m}$$

$$\lambda_c \le \frac{1.24}{1.42}$$

$$\lambda_c \le 0.87\ \mu\text{m}$$

1.40 (1.12)

$$\lambda_c = \frac{1.24}{E_g(\text{eV})}\ \mu\text{m} = \frac{1.24}{1.1}\ \mu\text{m}$$

$$\lambda_c = 1.12\ \mu\text{m}$$

1.41 (225.2)

For a compensated p-type semiconductor, the minority carrier electron concentration is given by

$$n_o = \frac{n_i^2}{(N_A-N_D)}$$

$$n_o = \frac{2.25\times10^{20}}{(10^{18}-10^{15})} = 225.2\ \text{cm}^{-3}$$

1.43 (0.52)

$$P = \frac{1}{\sigma_N} = \frac{1}{N_D\, q\, \mu_n}$$

$$= \frac{1}{10^{16}\times1.6\times10^{-19}\times1200} = 0.52\ \Omega\text{-cm}$$

1.44 (1.92)

As per the graph, mobility of electrons at the concentration $10^{16}/\text{cm}^3$ is 1200 cm²/V-s.

So, $\mu_n = 1200\ \text{cm}^2/\text{V-s}$

$$\sigma_N = N_D q\mu_n$$

$$= 10^{16}\times1.6\times10^{-19}\times1200 = 1.92\ \text{S cm}^{-1}$$

1.45 (14)

The concentration of hole-electron par in 1 μsec $= 10^{20}\times10^{-6} = 10^{14}/\text{cm}^3$

So, the power of 10 is 14.

$$x = 14$$

1.46 (c)

The concentration of doping is not uniform throughout the semiconductor, thus to maintain equilibrium internal electric field is generated due to which the band of the semiconductor is varying with slope –0.1 eV/cm.

The generated electric field opposes the diffusion of carriers due to concentration gradient and thus there will be no current flow inside the semiconductor.

1.47 (100)

Electric field inside semiconductor bar,

$$E = 10\ \text{V/cm}$$

Velocity of electron, $V = \mu E = 1 \times 10^4$ cm/sec

The average time (in μs) taken by the electrons to move from one end of the bar to other end is

$$\frac{L}{V} = 100\ \mu\text{sec}$$

1.48 (a)

Pentavalent impurity when introduced in Intrinsic SC, a new discrete energy level called Donor energy level is created just below the conduction band.

1.49 (1.198)

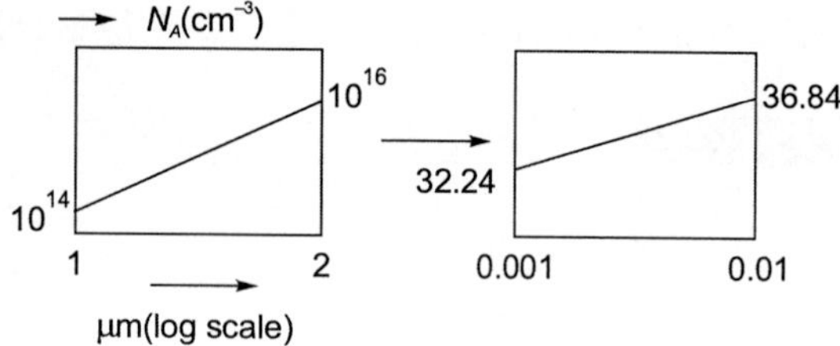

Applying the current density equation

$$J = J_{Drift} + J_{Diff}$$
$$J = 0$$
$$0 = -qD_p\frac{dp}{dx} + q\mu_p pE$$
$$qD_p\frac{dp}{dx} = q\mu_p pE$$
$$E = \frac{V_T}{p(x)}\frac{dp}{dx}$$
$$\therefore \quad |E| = \frac{V_T}{p(x)}\frac{dp}{dx}$$
$$E = V_T\frac{d}{dx}ln[N_A(x)]$$
$$E = V_T\left[\frac{ln(10^{16}) - ln(10^{14})}{(2-1)\times 10^{-4}}\right] = 0.026\left[\frac{36-32.23}{1\times 10^{-4}}\right]$$

$= 1198.6$ V/cm $= 1.198$ kV/cm

1.50 (a)

Given, $G_{opt} = 1.5 \times 10^{20}/\text{cm}^3/\text{sec}$

$$G_{opt} = R = \frac{N_A}{\tau_p}$$
$$\Rightarrow \quad 1.5 \times 10^2 = \frac{N_A}{0.1\times 10^{-6}}$$
$$\therefore \quad N_A = 1.5 \times 10^{13}/\text{cm}^3$$
$$P(t) = P_{n0}e^{-t/\tau_p}$$
$$= 1.5 \times 10^{13}\ e^{\frac{-0.3}{0.1}} = 7.46 \times 10^{11}/\text{cm}^3$$

1.51 (a)

Si acts as p-type dopant in As sites.

Si acts as n-type dopant in GA sites.

1.52 (16)

Net hole density varying in the direction of x is,

$$p_n(x) = p_{no} + \Delta p = p_{no} + G_L\tau_p$$
$$= p_{no} + G_{Lo}\tau_p\left(1-\frac{x}{L}\right)$$
$$J_{p,\text{diff}} = -eD_p\frac{dp}{dx} = -eD_p\left[\frac{-G_{Lo}\tau_p}{L}\right]$$
$$= \frac{1.6\times10^{-19}\times100\times10^{17}\times10^{-4}}{0.1\times10^{-4}}\ \text{A/cm}^2$$
$$= 16\ \text{A/cm}^2$$

1.53 (1.6)

$$E = \frac{V}{d} = \frac{5}{10^{-4}} = 5\times10^4\ \text{V/cm}$$

Slope of the curve,

$$m = \frac{10^7 - 0}{5\times10^5} = 20 \qquad y = mx$$
$$v_d = 20 \times E = 20 \times 5 \times 10^4 = 10^6\ \text{V/cm}$$
$$J = nev_d$$
$$= 1 \times 10^{16} \times 1.6 \times 10^{-19} \times 10^6$$
$$= 1.6 \times 10^3\ \text{A/cm}^2 = 1.6\ \text{kA/cm}^2$$

1.54 (d)

$$\frac{E_C + E_V}{2} - E_{F_i} = \frac{kT}{2}\ln\left(\frac{N_C}{N_V}\right) \qquad \left(\because N_C = \frac{N_V}{2}\right)$$
$$= \frac{0.026}{2}\ln 0.5 = -9.01\ \text{meV}$$

1.55 (d)

The built-in electric field is due to non-uniform doping (the semiconductor is under equilibrium)

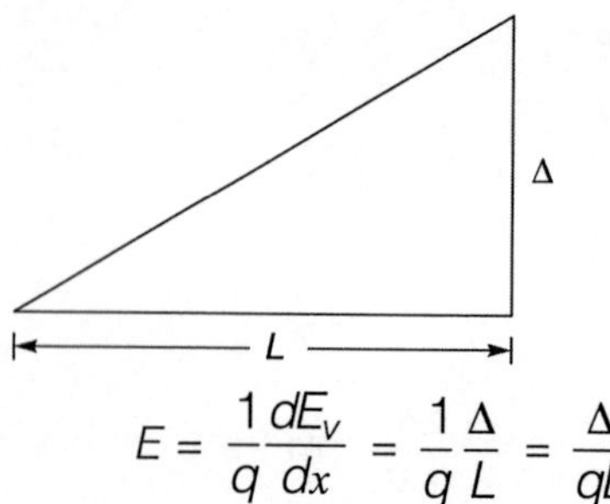

$$E = \frac{1}{q}\frac{dE_V}{dx} = \frac{1}{q}\frac{\Delta}{L} = \frac{\Delta}{qL}$$

1.56 (d)

Boron → Acceptor type doping

$$N_A = 10^{16}\ \text{cm}^{-3}$$
$$g_{op} = 10^{20}\ \text{cm}^{-3}\ \text{s}^{-1}$$
$$\tau = 100\ \mu\text{s}$$
$$n_i = 10^{10}\ \text{cm}^{-3}$$

Product of steady state electron-hole concentration = ?

At thermal equilibrium (before shining light)

Hole concentration,

$$p_o \simeq N_A = 10^{16}\ \text{cm}^{-3}$$

Electron concentration,

$$n_o = \frac{n_i^2}{p_o} = \frac{10^{20}}{10^{16}} = 10^4\ \text{cm}^{-3}$$

After, illumination of light,

Hole concentration,

$$p = p_o + \delta p$$

Electron concentration,

$$n = n_o + \delta n$$

Due to shining light, excess carrier concentration,

$$\delta p = \delta n = g_{op} \cdot \tau = 10^{20} \times 100 \times 10^{-6} = 10^{16}\ \text{cm}^{-3}$$

$$\therefore \quad p = 10^{16} + 10^{16} = 2 \times 10^{16}\ \text{cm}^{-3}$$
$$n = 10^4 + 10^{16} \simeq 10^{16}\ \text{cm}^{-3}$$

So, product of steady state electron-hole concentration

$= np = 10^{16} \times 2 \times 10^{16} = 2 \times 10^{32}\ \text{cm}^{-6}$

1.57 (a)

Given : Equilibrium hole density, $p_o = 10^{17}\ \text{cm}^{-3}$

Intrinsic carrier concentration, $n_i = 10^{10}\ \text{cm}^{-3}$

Diffusion length of electrons, $L_n = 2\ \mu\text{m}$

Diffusion length of holes, $L_p = 1\ \mu\text{m}$

Steady state electron density at $x = 0$

$= \delta n'_{po} = 10^{14}\ \text{cm}^{-3}$ due to laser illumination

Intrinsic electron concentration,

$$n_o = \frac{n_i^2}{p_o} = \frac{10^{20}}{10^{17}} = 10^3\ \text{cm}^{-3}$$

The excess electron concentration generated due to laser illumination varies across the distance x as below:

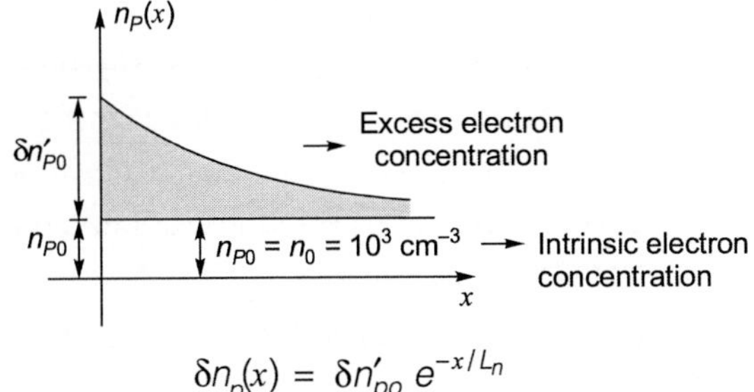

$$\delta n_p(x) = \delta n'_{po}\, e^{-x/L_n}$$

Substituting the given values for $x = 2\ \mu\text{m}$, we get,

$$\delta n_p(2\ \mu\text{m}) = 10^{14} \times e^{-2/2}$$
$$\delta n_p(2\ \mu\text{m}) = 10^{14}\, e^{-1}$$
$$\delta n_p(2\ \mu\text{m}) = 0.367 \times 10^{14}\ \text{cm}^{-3}$$

The steady state electron density at $x = 2\ \mu\text{m}$ is

$$n_p(2\ \mu\text{m}) = n_{po} + \delta n_p(2\ \mu\text{m})$$
$$n_p(2\ \mu\text{m}) = 10^3 + 0.367 \times 10^{14}$$
$$n_p(2\ \mu\text{m}) \approx 0.37 \times 10^{14}\ \text{cm}^{-3}$$

1.58 (c)

Given : For electron density, $N_{D1} = 10^{16}\ \text{cm}^{-3}$,

$$(E_C - E_{Fn})_1 = 200\ \text{meV}$$

Intrinsic carrier concentration, $n_i = 10^{10}\ \text{cm}^{-3}$.

Thermal voltage, $kT = 26\ \text{meV}$

For an n-type semiconductor, the difference between conduction energy level and Fermi energy level is given by

$$(E_C - E_{Fn})_1 = kT \ln\left(\frac{N_c}{N_{D1}}\right) \quad \text{...(i)}$$

For electron concentration, $N_{D2} = 0.5 \times 10^{16}\ \text{cm}^{-3}$,

$$(E_C - E_{Fn})_2 = kT \ln\left(\frac{N_c}{N_{D2}}\right) \quad \text{...(ii)}$$

Subtracting equations (i) and (ii), we get,

$$(E_C - E_{Fn})_1 - (E_C - E_{Fn})_2 = kT \ln\left(\frac{N_C}{N_{D1}}\right) - kT \ln\left(\frac{N_C}{N_{D2}}\right)$$

$$\therefore\ 200\ \text{meV} - (E_C - E_{Fn})_2 = kT \ln\left(\frac{N_{D2}}{N_{D1}}\right)$$

Substituting the given values, we get,

$$200\ \text{meV} - (E_C - E_{Fn})_2 = 0.026 \ln\left(\frac{0.5 \times 10^{16}}{10^{16}}\right)$$

$$200\ \text{meV} - (E_C E_{Fn})_2 = 0.026 \ln(0.5)$$
$$200\ \text{meV} - (E_C - E_{Fn})_2 = -0.01802 = -18.02\ \text{meV}$$
$$\therefore\ (E_C - E_{Fn})_2 = 200 + 18.02 = 218.02\ \text{meV}$$

1.59 (a, c)

- In an intrinsic semiconductor at equilibrium,

  $$n_i = n_o = p_o$$

  i.e. the electron and hole densities are equal at equilibrium and equal to the intrinsic concentration.
- Collector region is generally lightly doped than base region in a BJT to get low collector capacitance. Hence, option (b) is incorrect.

- Total current is spatially constant in a two terminal electronic device in dark under steady state condition as no excess carriers are generated in dark.
- Beyond 300 K, the mobility of electrons decreases with increase in temperature. Hence, option (d) is incorrect.

Therefore, only options (a) and (c) are correct.

1.60 (a)

As the Fermi lies inside the conduction band hence it is degenerate *n*-type semiconductor.

1.61 (a)

Intrinsic semiconductor at T = 0 K behaves as an insulator.
Hence, valence band is completely filled with electrons and conduction band is completely empty.

1.62 (2.25)

Given, intrinsic carrier concentration,

$$n_i = 1 \times 10^{10}\ \text{cm}^{-3}$$

Hole concentration, $p = 1.5 \times n_i$

$$p = 1.5 \times 10^{10}\ \text{cm}^{-3}$$

Given, electron and hole current are equal

$$I_{p\,\text{drift}} = I_{n\,\text{drift}}$$

$$pq\mu_p EA = nq\mu_n EA$$

$$1.5 \times 10^{10}\ \mu_p = n\mu_n \qquad \ldots\text{(i)}$$

But according to mass action law,

$$np = n_i^2$$

$$\therefore \quad n = \frac{n_i}{1.5} = \frac{10^{10}}{1.5}\ \text{cm}^{-3}$$

Put in equation (i)

$$\therefore\ 1.5 \times 10^{10}\ \mu_p = \frac{10^{10}}{1.5} \times \mu_n$$

$$\frac{\mu_n}{\mu_p} = 2.25$$

1.63 (63.36)

Given, $E_F - E_V = 0.35$ eV

[Considering it is given at 400 K]

Also, $V_{T_1} = KT_1 = 0.026$ eV at $T_1 = 300$ K

$$\therefore \quad \frac{V_{T_1}}{V_{T_2}} = \frac{T_1}{T_2} \Rightarrow V_{T_2} = \frac{T_2}{T_1} \times V_{T_1}$$

$$\therefore \quad V_{T_2} = \frac{400}{300} \times 0.026$$

$$V_{T_2} = 0.03466 \text{ eV at } T_2 = 400 \text{ K}$$

Given, effective density of states in the valence band,

$$N_v = 1 \times 10^{19}/\text{cm}^3 \text{ at } T_1 = 300 \text{ K}$$

$$N_v \propto T^{3/2}$$

$$\frac{N_{V_2}}{N_{V_1}} = \left(\frac{T_2}{T_1}\right)^{3/2}$$

$$N_{V_2} = \left(\frac{T_2}{T_1}\right)^{3/2} N_{V_1} \qquad (\because T_2 = 400 \text{ K})$$

$$= \left(\frac{400}{300}\right)^{3/2} N_{V_1}$$

$$N_{V_2} = 1.5396 \times 10^{19}/\text{cm}^3$$

Now, hole concentration at 400 K is given as

$$p = N_v\, e^{-(E_F - E_V)/kT_2}$$

$$= 1.5396 \times 10^{19} \times e^{-0.35\,\text{eV}/0.03466\,\text{eV}}$$

$$p = 63.36 \times 10^{13}\ \text{cm}^{-3}$$

1.64 (a, c, d)

- The diffusion of charge carriers takes place from the region of higher concentration to the region of lower concentration. Hence, the electrons diffuse from B to C. Since, the direction of flow of electrons is opposite to the direction of flow of current. Hence, between B and C, the electron diffusion current is directed from C to B.
- The change in the concentration of the carriers develops a gradient. Due to this gradient, an electric field is produced in the semiconductor. When an electron moves from higher concentration to lower concentration, it leaves an uncompensated positive ion resulting in an electric field from higher electron concentration to lower electron concentration. Hence,
  (i) For x between B and C, the electric field is directed from B to C.
  (ii) For x between B and A, the electric field is directed from B to A.
- Due to the electric field, the force exerted on the electron, $F = -eE$ results in the drift of electrons from A to B. Hence, for x between B and A, the electron drift current is directed from B to A.

Hence, options (a), (c) and (d) are correct.

1.65 (a)

(i) **Ionization Region:** At very low temperatures (large $1/T$), negligible intrinsic electron-hole pairs exist and the donor electrons are bound to the donor atoms. As temperature is raised, ionization occurs leading to increase in the free electron concentration.

(ii) **Extrinsic Region:** In this region, all available dopants has been ionized resulting in $n = N_D$ (donor concentration). In this region, an increase in temperature produces no increase in carrier concentration.

(iii) **Intrinsic Region:** At high temperatures (low $1/T$), the thermally generated carriers outnumber the dopants. In this region, carrier concentration increases with temperature with exponential dependence given by

$$n_i(T) = 2\left[\frac{2\pi kT}{h^2}\right]^{3/2}(m_n^* \, m_p^*)^{3/4}\exp\left[\frac{-E_g}{2kT}\right]$$

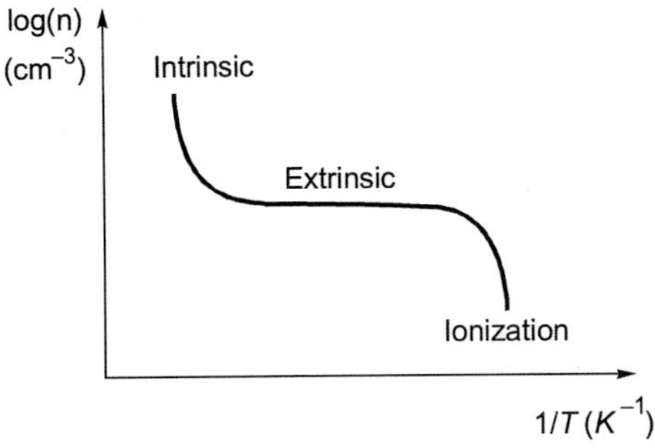

1.66 (0.18)

5% neutral donor atoms/dopant atoms means 95% donor atoms are ionized.

Concentration of electrons occupying the donor level is given as

Conduction band E_c

Discrete donor level E_D

Valence band E_v

Electron energy

$$n_d = \frac{N_d}{1+\frac{1}{g}\exp\left(\frac{E_D - E_F}{kT}\right)}$$

where, g = Degeneracy factor

and $n_d = N_d - N_d^+$ where N_d^+ is ionized donor atoms concentration and N_d is the concentration of donors.

According to question

$$n_d = 5\% \text{ of } N_d$$

Degeneracy factor, $g = 2$

$$\frac{5}{100}N_d = \frac{N_d}{1+\frac{1}{g}\exp\left(\frac{E_D - E_F}{kT}\right)}$$

$$\frac{1}{20} = \frac{1}{1+\frac{1}{g}\exp\left(\frac{E_D - E_F}{kT}\right)}$$

$$1+\frac{1}{g}\exp\left(\frac{E_D - E_F}{kT}\right) = 20$$

$$\exp\left(\frac{E_D - E_F}{kT}\right) = 38$$

$$\frac{E_D - E_F}{kT} = \ln(38)$$

$$\left\{\text{given } V_T = \frac{kT}{q} = 20 \text{ mV}\right\}$$

$$E_D - E_F = \ln(38) \times 0.020$$
$$= 0.07275 \text{ eV}$$

given: $E_C - E_F = 0.25$ eV

$\therefore \; E_C - E_D + E_D - E_F = 0.25$ eV

$$E_C - E_D = 0.25 \text{ eV} - (E_D - E_F)$$
$$= 0.25 \text{ eV} - 0.07275 \text{ eV}$$
$$= 0.17725 \text{ eV} \simeq 0.18 \text{ eV}$$

■■■

2 PN-Junction Diodes and Special Diodes

Note: Tunnel diode, PIN diode, Lasers & Avalanche photodiode topics are out of syllabus for GATE from 2016, it may be useful for PSUs & Interviews.

2.1 The diffusion potential across a *pn*-junction
(a) decreases with increasing doping concentration
(b) increases with decreasing band gap
(c) does not depend on doping concentrations
(d) increases with increase in doping concentration
[1995 : 1 M]

2.2 A Zener diode works on the principle of
(a) tunneling of charge carriers across the junction
(b) thermionic emission
(c) diffusion of charge carriers across the junction
(d) hopping of charge carriers across the junction
[1995 : 1 M]

2.3 The depletion capacitance, C_j of an abruptly *p-n* junction with constant doping on either side varies with R.B. V_R as
(a) $C_J \propto V_R$
(b) $C_J \propto V_R^{-1}$
(c) $C_J \propto V_R^{-1/2}$
(d) $C_J \propto V_R^{-1/3}$
[1995 : 1 M]

2.4 The static characteristic of an adequately forward biased p-n junction is a straight line, if the plot is of
(a) log I vs.log V
(b) log I vs.V
(c) I vs log V
(d) I vs V **[1998 : 1 M]**

2.5 A *pn*-junction in series with a 100 ohms resistor, is forward biased. So that a current of 100 mA flows. If the voltage across this combination is instantaneously reversed at $t = 0$ current through diodes is approximately given by
(a) 0 mA
(b) –100 mA
(c) 100 mA
(d) 50 mA **[1998 : 2 M]**

2.6 In the figure, silicon diode is carrying a constant current of 1 mA. When the temperature of the diode is 20°C, V_D is found to be 700 mV. If the temperature rises to 40°C, V_D becomes approximately equal to
(a) 740 mV
(b) 660 mV
(c) 680 mV
(d) 700 mV

+
V_D
–

[2002 : 1 M]

2.7 Choose proper substitutes for X and Y to make the following statement correct Tunnel diode and Avalanche photodiode are operated in X bias and Y bias respectively.
(a) X: reverse, Y: reverse
(b) X: reverse, Y: forward
(c) X: forward, Y: reverse
(d) X: forward, Y: forward **[2003 : 1 M]**

2.8 Match items in **Group-1** with items in **Group-2**, most suitably.

Group-1	Group-2
P. LED	1. Heavy doping
Q. Avalanche Photodiode	2. Coherent radiation
R. Tunnel diode	3. Spontaneous emission
S. LASER	4. Current gain

(a) P – 1; Q – 2; R – 4; S – 3
(b) P – 2; Q – 3; R – 1; S – 4
(c) P – 3; Q – 4; R – 1; S – 2
(d) P – 2; Q – 1; R – 4; S – 3 **[2003 : 2 M]**

2.9 At 300 K, for a diode current of 1 mA, a certain germanium diode requires a forward bias of 0.1435 V, whereas a certain silicon diode requires a forward bias of 0.718 V. Under the conditions stated above, the closest approximation of the ratio of reverse saturation current in germanium diode to that in silicon diode is
(a) 1
(b) 5
(c) 4×10^3
(d) 8×10^3 **[2003 : 2 M]**

2.10 A particular green LED emits light of wavelength 5490 Å. The energy bandgap of the semiconductor material used there is
(Plank's constant = 6.626×10^{-34} J-s)
(a) 2.26 eV
(b) 1.98 eV
(c) 1.17 eV
(d) 0.74 eV **[2003 : 2 M]**

2.11 In an abrupt *p-n* junction, the doping concentrations on the *p*-side and *n*-side are $N_A = 9 \times 10^{16}/\text{cm}^3$ and $N_D = 1 \times 10^{16}/\text{cm}^3$ respectively. The *p-n* junction is reverse biased and the total depletion width is 3 μm. The depletion width on the *p*-side is
(a) 2.7 μm
(b) 0.3 μm
(c) 2.25 μm
(d) 0.75 μm **[2004 : 2 M]**

2.12 Consider an abrupt *p*-junction. Let V_{bi} be the built-in potential of this junction and V_R be the applied reverse bias. If the junction capacitance (C_j) is 1 pF for $V_{bi} + V_R = 1$ V, then for $V_{bi} + V_R = 4$ V, C_j will be
(a) 4 pF (b) 2 pF
(c) 0.25 pF (d) 0.5 pF [2004 : 2 M]

2.13 A Silicon PN junction at a temperature of 20°C has a reverse saturation current of 10 pico-Amperes (pA). The reverse saturation current at 40°C for the same bias is approximately
(a) 30 pA (b) 40 pA
(c) 50 pA (d) 60 pA [2005 : 1 M]

2.14 A Silicon PN junction diode under reverse bias has depletion region of width 10 μm. The relative permittivity of Silicon, $\varepsilon_r = 11.7$ and the permittivity of free space $\varepsilon_0 = 8.85 \times 10^{-12}$ F/m. The depletion capacitance of the diode per square meter is
(a) 100 μ (b) 10 μ
(c) 1 μ (d) 20 μ [2005 : 2 M]

2.15 The values of voltage (V_D) across a tunnel-diode corresponding to peak and valley currents are V_p and V_v respectively. The range of tunnel-diode voltage V_D for which the slope of its $I - V_D$ characteristics is negative would be
(a) $V_D < 0$ (b) $0 \le V_D < V_p$
(c) $V_p \le V_D < V_v$ (d) $V_D \ge V_v$ [2006 : 1 M]

2.16 In the circuit shown below, the switch was connected to position 1 at $t < 0$ and at $t = 0$, it is changed to position 2. Assume that the diode has zero voltage drop and a storage time t_s. For $0 < t \le t_s$, v_R is given by (all in Volts)

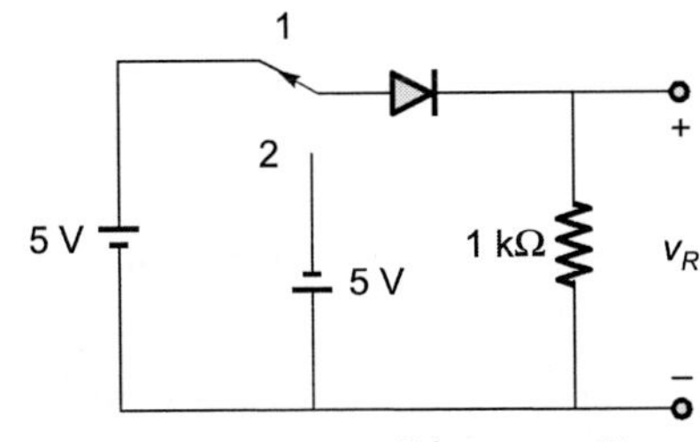

(a) $v_R = -5$ (b) $v_R = +5$
(c) $0 \le v_R < 5$ (d) $-5 < v_R < 0$
[2006 : 2 M]

2.17 Find the correct match between Group 1 and Group 2

	Group 1		Group 2
E.	Varactor diode	1.	Voltage reference
F.	PIN diode	2.	High-frequency switch
G.	Zener diode	3.	Tuned circuits
H.	Schottky diode	4.	Current controlled attenuator

(a) E-4, F-2, G-1, H-3
(b) E-2, F-4, G-1, H-3
(c) E-3, F-4, G-1, H-2
(d) E-1, F-3, G-2, H-4 [2006 : 2 M]

2.18 In a p^+n junction diode under reverse bias, the magnitude of electric field is maximum at
(a) the edge of the depletion region on the *p*-side
(b) the edge of the depletion region on the *n*-side
(c) the p^+n junction
(d) the centre of the depletion region on the *n*-side
[2007 : 1 M]

2.19 Group I lists four types of p-n junction diodes. Match each device in Group I with one of the options in Group II to indicate the bias condition of that device in its normal mode of operation.

	Group I		Group II
P.	Zener Diode	1.	Forward bias
Q.	Solar cell	2.	Reverse bias
R.	LASER diode		
S.	Avalanche Photodiode		

(a) P-1, Q-2, R-1, S-2
(b) P-2, Q-1, R-1, S-2
(c) P-2, Q-2, R-2, S-1
(d) P-2, Q-1, R-2, S-2 [2007 : 2 M]

2.20 Group I lists four different semiconductor devices. Match each device in Group I with its characteristic property in Group II.

	Group I		Group II
P.	BJT	1.	Population inversion
Q.	MOS capacitor	2.	Pinch-off voltage
R.	LASER diode	3.	Early effect
S.	JFET	4.	Flat-band voltage

(a) P-3, Q-1, R-4, S-2
(b) P-1, Q-4, R-3, S-2
(c) P-3, Q-4, R-1, S-2
(d) P-3, Q-2, R-1, S-4 [2007 : 2 M]

2.21 A p^+n junction has a built-in potential of 0.8 V. The depletion layer width at a reverse bias of 1.2 V is 2 μm. For a reverse bias of 7.2 V, the depletion layer width will be
(a) 4 μm (b) 4.9 μm
(c) 8 μm (d) 12 μm [2007 : 2 M]

2.22 Which of the following is NOT associated with a *p-n* junction?
(a) Junction Capacitance
(b) Charge Storage Capacitance
(c) Depletion Capacitance
(d) Channel Length Modulation [2008 : 1 M]

2.23 Consider the following assertions.

S_1: For Zener effect to occur, a very abrupt junction is required

S_2: For quantum tunneling to occur, a very narrow energy barrier is required.

Which of the following is correct?

(a) Only S_2 is true

(b) S_1 and S_2 are both true but S_2 is not a reason for S_1

(c) S_1 and S_2 are both true and S_2 is a reason for S_1

(d) Both S_1 and S_2 are false

[2008 : 2 M]

Common Data for Question 2.24 and 2.25.

Consider a silicon *p-n* junction at room temperature having the following parameters:

Doping on the *n*-side = 1×10^{17} cm^{-3}

Depletion width on the *n*-side = 0.1 μm

Depletion width on the *p*-side = 1.0 μm

Intrinsic carrier concentration = 1.4×10^{10} cm^{-3}

Thermal voltage = 26 mV

Permittivity of free space = 8.85×10^{-14}F·cm^{-1}

Dielectric constant of silicon = 12

2.24 The built-in potential of the junction

(a) is 0.70 V

(b) is 0.76 V

(c) is 0.82 V

(d) cannot be estimated from the data given

[2009 : 2 M]

2.25 The peak electric field in the device is

(a) 0.15 MV·cm^{-1}, directed from *p*-region to *n*-region

(b) 0.15 MV·cm^{-1}, directed from *n*-region to *p*-region

(c) 1.80 MV·cm^{-1}, directed from *p*-region to *n*-region

(d) 1.80 MV·cm^{-1}, directed from *n*-region to *p*-region

[2009 : 2 M]

2.26 Compared to a *p-n* junction with $N_A = N_D = 10^{14}$/cm^3, which one of the following statements is TRUE for a *p-n* junction with $N_A = N_D = 10^{20}$/cm^3?

(a) Reverse breakdown voltage is lower and depletion capacitance is lower

(b) Reverse breakdown voltage is higher and depletion capacitance is lower

(c) Reverse breakdown voltage is lower and depletion capacitance is higher

(d) Reverse breakdown voltage is higher and depletion capacitance is higher

[2010 : 2 M]

2.27 A silicon *PN* junction is forward biased with a constant current at room temperature. When the temperature is increased by 10°C, the forward bias voltage across the *PN* junction

(a) increases by 60 mV

(b) decreases by 60 mV

(c) increases by 25 mV

(d) decreases by 25 mV

[2011 : 1 M]

2.28 A Zener diode, when used in voltage stabilization circuits, is biased in

(a) reverse bias region below the breakdown voltage

(b) reverse breakdown region

(c) forward bias region

(d) forward bias constant current mode

[2011 : 1 M]

2.29 The *i-v* characteristics of the diode in the circuit given below are

$$i = \begin{cases} \dfrac{v - 0.7}{500} A, & v \geq 0.7V \\ 0\ A, & v < 0.7V \end{cases}$$

1 kΩ, 10 V, i, + v −

The current in the circuit is

(a) 10 mA (b) 9.3 mA

(c) 6.67 mA (d) 6.2 mA [2012 : 1 M]

2.30 In a forward biased *pn* junction, the sequence of events that best describes the mechanism of current flow is

(a) injection, and subsequent diffusion and recombination of minority carriers

(b) injection, and subsequent drift and generation of minority carriers

(c) extraction, and subsequent diffusion and generation of minority carriers

(d) extraction, and subsequent drift and recombination of minority carriers

[2013 : 1 M]

2.31 When the optical power incident on a photodiode is 10 μW and the responsivity is 0.8 A/W, the photocurrent generated (in μA) is ____.

[2014 : 1 M, Set-1]

2.32 Consider an abrupt PN junction (at T = 300 K) shown in the figure. The depletion region width x_n on the *N*-side of the junction is 0.2 μm and the permittivity of silicon (ϵ_{si}) is 1.044×10^{-12} F/cm.

At the junction, the approximate value of the peak electric field (in kV/cm) is ________.

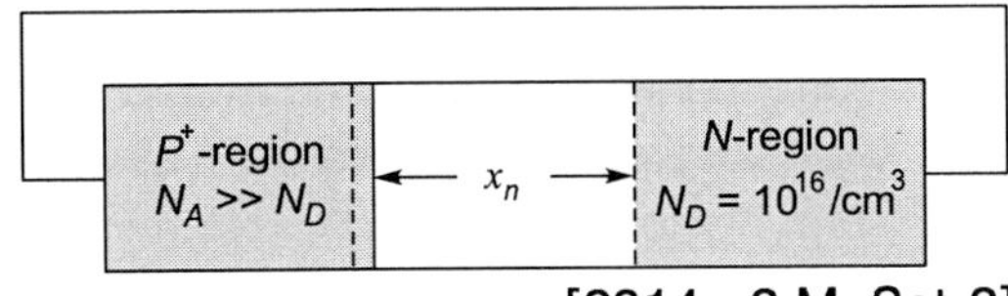

[2014 : 2 M, Set-2]

2.33 The donor and acceptor impurities in an abrupt junction silicon diode are 1×10^{16} cm^{-3} and 5×10^{18} cm^{-3}, respectively. Assume that the intrinsic carrier concentration in silicon $n_i = 1.5 \times 10^{10}$ cm^{-3} at 300 K, $\frac{kT}{q} = 26$ mV and the permittivity of silicon $\epsilon_{si} = 1.04 \times 10^{-12}$ F/cm. The built-in potential and the depletion width of the diode under thermal equilibrium conditions, respectively, are

(a) 0.7 V and 1×10^{-4} cm
(b) 0.86 V and 1×10^{-4} cm
(c) 0.7 V and 3.3×10^{-5} cm
(d) 0.86 V and 3.3×10^{-5} cm [2014 : 2 M, Set-3]

2.34 A region of negative differential resistance is observed in the current voltage characteristics of a silicon *PN* junction if

(a) both the *P*-region and the *N*-region are heavily doped
(b) the *N*-region is heavily doped compared to the *P*-region
(c) the *P*-region is heavily doped compared to the *N*-region
(d) an intrinsic silicon region is inserted between the *P*-region and the *N*-region

[2015 : 1 M, Set-1]

2.35 The built-in potential of an abrupt p-n junction is 0.75 V. If its junction capacitance (C_J) at a reverse bias (V_R) of 1.25 V is 5 pF, the value of C_J (in pF) when $V_R = 7.25$ V is ____.

[2015 : 2 M, Set-1]

2.36 For a silicon diode with long *P* and *N* regions, the accepter and donor impurity concentrations are 1×10^{17} cm^{-3} and 1×10^{15} cm^{-3}, respectively. The lifetimes of electrons in *P* region and holes in *N* region are both 100 μs. The electron and hole diffusion coefficients are 49 cm^2/s and 36 cm^2/s, respectively. Assume $kT/q = 26$ mV, the intrinsic carrier concentration is 1×10^{10} cm^{-3}, and $q = 1.6 \times 10^{-19}$ C. When a forward voltage of 208 mV is applied across the diode, the hole current density (in nA/cm^2) injected from *P* region to *N* region is ______. [2015 : 2 M, Set-1]

2.37 The electric field profile in the depletion region of a *p-n* junction in equilibrium is shown in the figure. Which one of the following statements is NOT TRUE?

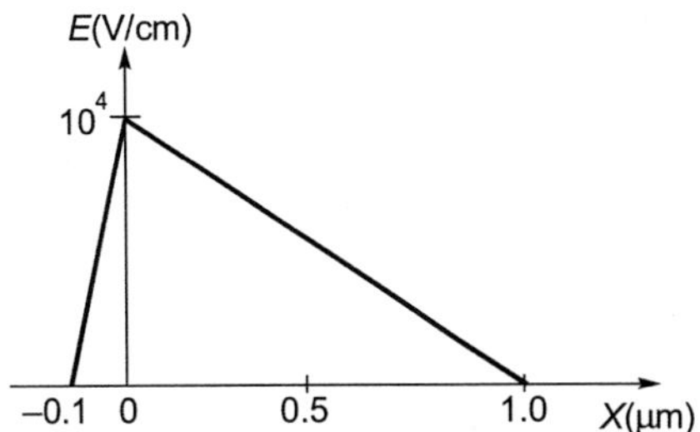

(a) The left side of the junction is *n*-type and the right side is *p*-type
(b) Both the *n*-type and *p*-type depletion regions are uniformly doped
(c) The potential difference across the depletion region is 700 mV
(d) If the *p*-type region has a doping concentration of 10^{15} cm^{-3}, then the doping concentration in the *n*-type region will be 10^{16} cm^{-3}

[2015 : 2 M, Set-3]

2.38 Consider a silicon *p-n* junction with a uniform acceptor doping concentration of 10^{17} cm^{-3} on the p-side and a uniform donor doping concentration of 10^{16} cm^{-3} on the *n*-side. No external voltage is applied to the diode.
Given: $kT/q = 26$ mV, $n_i = 1.5 \times 10^{10}$ cm^{-3}, $\varepsilon_{si} = 12\,\varepsilon_0$, $\varepsilon_0 = 8.85 \times 10^{-14}$ F/cm, and $q = 1.6 \times 10^{-19}$ C.
The charge per unit junction area (nC cm^{-2}) in the depletion region on the *p*-side is ____

[2016 : 2 M, Set-1]

2.39 Consider avalanche breakdown in a silicon p^+n junction. The *n*-region is uniformly doped with a donor density N_D. Assume that breakdown occurs when the magnitude of the electric field at any point in the device becomes equal to the critical field E_{crit}. Assume E_{crit} to be independent of N_D. If the built-in voltage of the p^+n junction is much smaller than the breakdown voltage, V_{BR}, the relationship between V_{BR} and N_D is given by

(a) $V_{BR} \times \sqrt{N_D}$ = constant
(b) $N_D \times \sqrt{V_{BR}}$ = constant
(c) $N_D \times V_{BR}$ = constant
(d) N_D / V_{BR} = constant

[2016 : 2 M, Set-2]

2.40 The I - V characteristics of three types of diodes at the room temperature, made of semiconductors X, Y and Z, are shown in the figure. Assume that the diodes are uniformly doped and identical in all respects except their materials. If E_{gX}, E_{gY}, and E_{gZ} are the band gaps of X, Y and Z, respectively, then

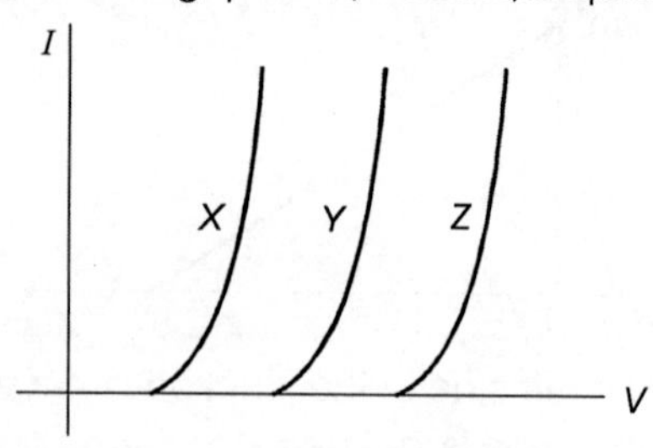

(a) $E_{gX} > E_{gY} > E_{gZ}$
(b) $E_{gX} = E_{gY} = E_{gZ}$
(c) $E_{gX} < E_{gY} < E_{gZ}$
(d) no relationship among these band gaps exists

[2016 : 1 M, Set-3]

2.41 The figure shows the I - V characteristics of a solar cell illuminated uniformly with solar light of power 100 mW/cm^2. The solar cell has an area of 3 cm^2 and a fill factor of 0.7. The maximum efficiency (in %) of the device is ______.

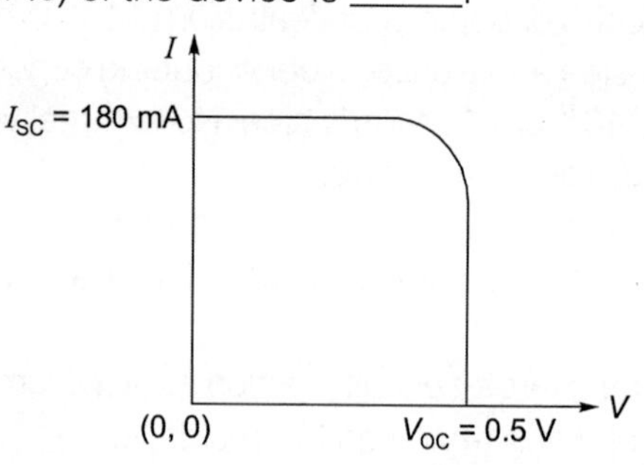

[2016 : 1 M, Set-3]

2.42 Consider a region of silicon devoid of electrons and holes, with an ionized donor density of $N_d^+ = 10^{17}\,\text{cm}^{-3}$. The electric field at $x = 0$ is 0 V/cm and the electric field at $x = L$ is 50 kV/cm in the positive x direction. Assume that the electric field is zero in the y and z directions at all points.

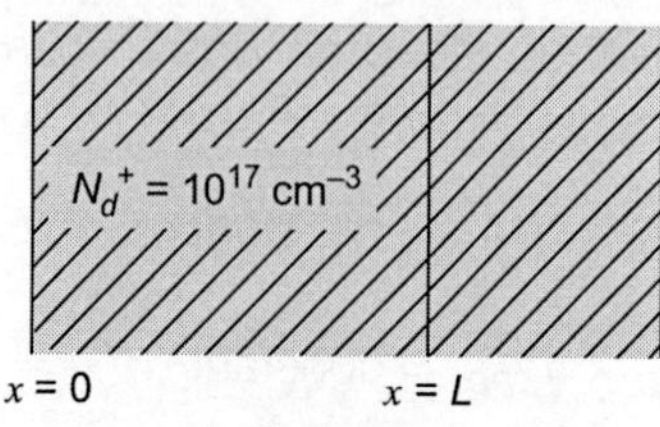

Given $q = 1.6 \times 10^{-19}$ coulomb, $\epsilon_0 = 8.85 \times 10^{-14}$ F/cm, $\epsilon_r = 11.7$ for silicon, the value of L in nm is _____

[2016 : 2 M, Set-2]

2.43 An n^+ - n Silicon device is fabricated with uniform and non-degenerate donor doping concentrations of $N_{D1} = 1 \times 10^{18}\,\text{cm}^{-3}$ and $N_{D2} = 1 \times 10^{15}\,\text{cm}^{-3}$ corresponding to the n^+ and n regions respectively. At the operational temperature T, assume complete impurity ionization, $kT/q = 25$ mV, and intrinsic carrier concentration to be $n_i = 1 \times 10^{10}\,\text{cm}^{-3}$. What is the magnitude of the built-in potential of this device?

(a) 0.748 V (b) 0.460 V
(c) 0.288 V (d) 0.173 V

[2017 : 1 M, Set-1]

2.44 As shown, two Silicon (Si) abrupt p-n junction diodes are fabricated with uniform donor doping concentrations of $N_{D1} = 10^{14}\,\text{cm}^{-3}$ and $N_{D2} = 10^{16}\,\text{cm}^{-3}$ in the n-regions of the diodes, and uniform acceptor doping concentrations of $N_{A1} = 10^{14}\,\text{cm}^{-3}$ and $N_{A2} = 10^{16}\,\text{cm}^{-3}$ in the p-regions of the diodes, respectively. Assuming that the reverse bias voltage is >> built-in potentials of the diodes, the ratio C_2/C_1 of their reverse bias capacitances for the same applied reverse bias, is ________.

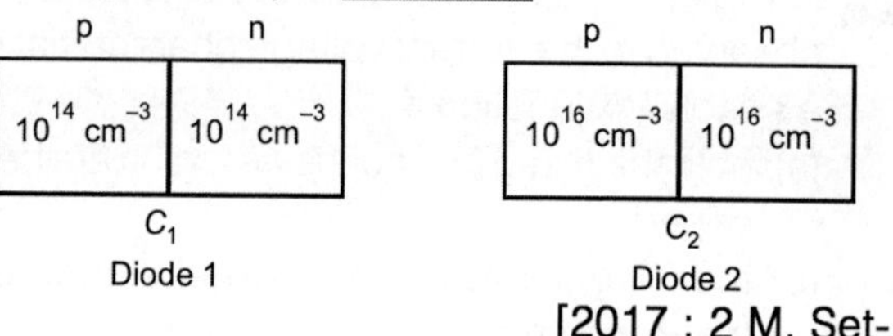

[2017 : 2 M, Set-1]

2.45 For a particular intensity of incident light on a silicon pn junction solar cell, the photocurrent density (J_L) is 2.5 mA/cm^2 and the open-circuit voltage (V_{oc}) is 0.451 V. Consider thermal voltage (V_T) to be 25 mV. If the intensity of the incident light is increased by 20 times, assuming that the temperature remains unchanged, V_{oc} (in volts) will be _____.

[2017 : 2 M, Set-2]

2.46 An abrupt pn junction (located at $x = 0$) is uniformly doped on both p and n sides. The width of the depletion region is W and the electric field variation in the x-direction is $E(x)$. Which of the following figures represents the electric field profile near the pn junction?

(a)

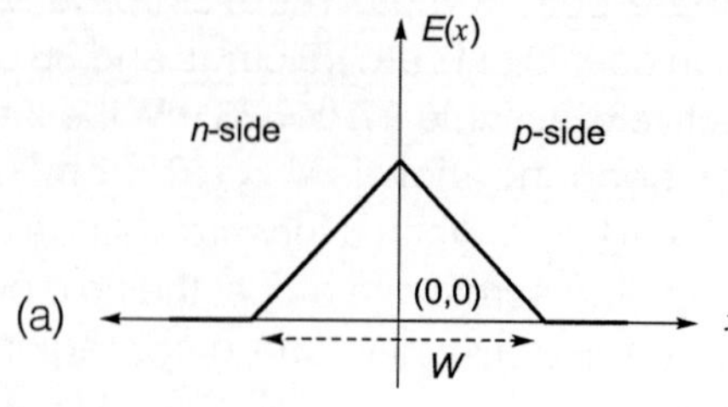

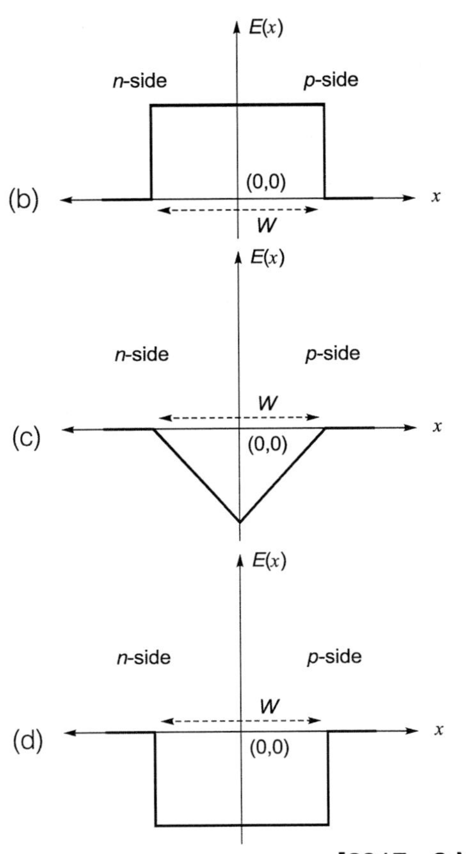

[2017 : 2 M, Set-2]

2.47 In a p-n junction diode at equilibrium, which one of the following statements is NOT TRUE?

(a) The hole and electron diffusion current components are in the same direction.
(b) The hole and electron drift current components are in the same direction.
(c) On an average, holes and electrons drift in opposite direction.
(d) On an average, electrons drift and diffuse in the same direction.

[2018 : 1 M]

2.48 A p-n step junction diode with a contact potential of 0.65 V has a depletion width of 1 μm at equilibrium. The forward voltage (in volts, correct to two decimal places) at which this width reduces to 0.6 μm is ______.

[2018 : 1 M]

2.49 Red (R), Green (G) and Blue (B) Light Emitting Diodes (LEDs) were fabricated using *p-n* junctions of three different inorganic semiconductors having different band-gaps. The built-in voltages of red, green and blue diodes are V_R, V_G and V_B, respectively. Assume donor and acceptor doping to be the same (N_A and N_D respectively) in the *p* and *n* sides of all the three diodes.

Which one of the following relationships about the built-in voltages is TRUE?

(a) $V_R > V_G > V_B$ (b) $V_R < V_G < V_B$
(c) $V_R = V_G = V_B$ (d) $V_R > V_G < V_B$

[2018 : 2 M]

2.50 A junction is made between p^- Si with doping density $N_{A1} = 10^{15}$ cm^{-3} and *p* Si with doping density $N_{A2} = 10^{17}$ cm^{-3}.
Given: Boltzmann constant $k = 1.38 \times 10^{-23}$ J · K^{-1}, electronic charge $q = 1.6 \times 10^{-19}$ C. Assume 100% acceptor ionization.
At room temperature ($T = 300$K), the magnitude of the built-in potential (in volts, correct to two decimal places) across this junction will be ______.

[2018 : 2 M]

2.51 A solar cell of area 1.0 cm^2, operating at 1.0 sun intensity, has a short circuit current of 20 mA, and an open circuit voltage of 0.65 V. Assuming room temperature operation and thermal equivalent voltage of 26 mV, the open circuit voltage (in volts, correct to two decimal places) at 0.2 sun intensity is ______.

[2018 : 2 M]

2.52 Which one of the following options describes correctly the equilibrium band diagram at $T = 300$ K of a Silicon pnn^+p^{++} configuration shown in the figure?

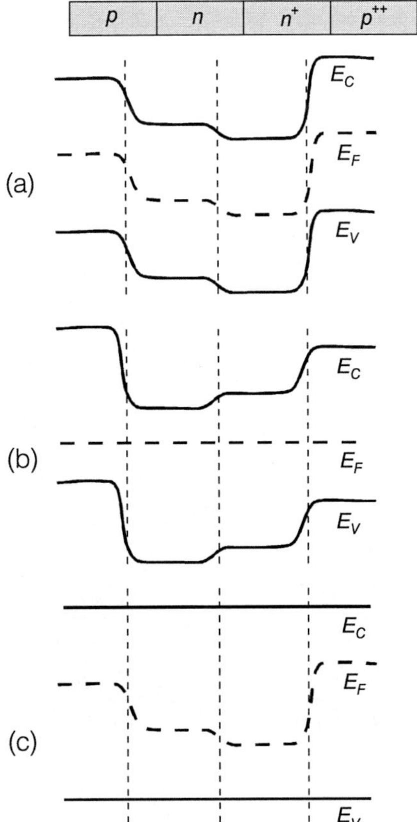

(d) 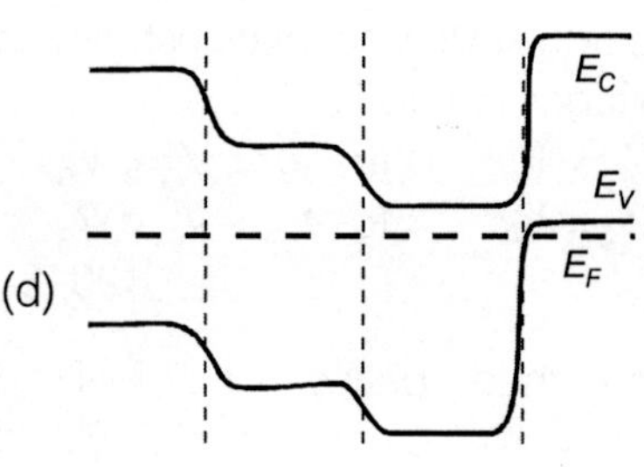

[2019 : 1 M]

2.53 In an ideal *pn* junction with an ideality factor of 1 at $T = 300$ K, the magnitude of the reverse-bias voltage required to reach 75% of its reverse saturation current, rounded off to 2 decimal places, is ______ mV.
[$k = 1.38 \times 10^{-23}$ JK^{-1}, $h = 6.625 \times 10^{-34}$ J-s, $q = 1.602 \times 10^{-19}$C]

[2019 : 2 M]

2.54 A Germanium sample of dimensions 1 cm × 1 cm is illuminated with a 20 mW. 600 nm laser light source as shown in the figure. The illuminated sample surface has a 100 nm of loss-less Silicon dioxide layer that reflects one-fourth of the incident light. From the remaining light, one-third of the power is reflected from the Silicon dioxide-Germanium interface, one-third is absorbed in the Germanium layer, and one-third is transmitted through the other side of the sample. If the absorption coefficient of Germanium at 600 nm is 3×10^4 cm^{-1} and the bandgap is 0.66 eV, the thickness of the Germanium layer, rounded off to 3 decimal places, is ________ μm.

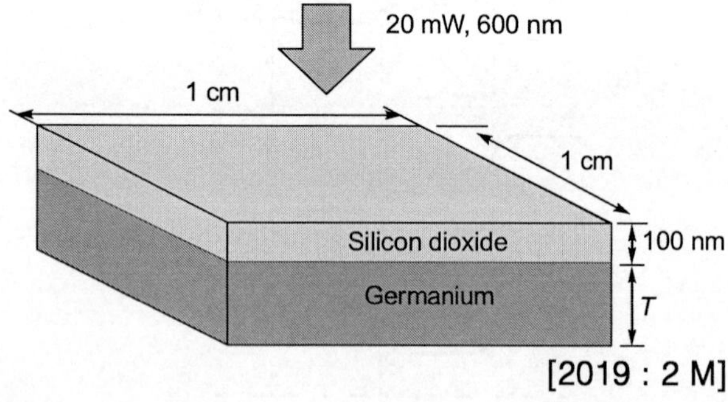

[2019 : 2 M]

2.55 The quantum efficiency (η) and responsivity (R) at a wavelength λ (in μm) in a p-i-n photodetector are related by

(a) $R = \frac{1.24 \times \lambda}{\eta}$ (b) $R = \frac{\eta \times \lambda}{1.24}$

(c) $R = \frac{\lambda}{\eta \times 1.24}$ (d) $R = \frac{1.24}{\eta \times \lambda}$

[2019 : 2 M]

2.56 Consider the recombination process via bulk traps in a forward biased pn homojunction diode. The maximum recombination rate is U_{max}. If the electron and the hole capture cross-section are equal, which one of the following is False?

(a) With all other parameters unchanged, U_{max} decreases if the intrinsic carrier density is reduced.
(b) With all other parameters unchanged, U_{max} increases if the thermal velocity of the carriers increases.
(c) U_{max} occurs at the edges of the depletion region in the device.
(d) U_{max} depends exponentially on the applied bias.

[2020 : 1 M]

2.57 A one-sided abrupt pn junction diode has a depletion capacitance C_D of 50 pF at a reverse bias of 0.2 V. The plot of $1/C_D^2$ versus the applied voltage V for this diode is a straight line as shown in the figure below. The slope of the plot is ____ $\times 10^{20}$ F^{-2} V^{-1}.

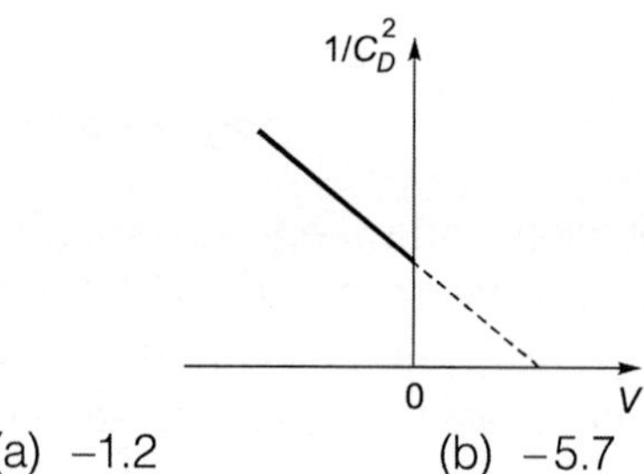

(a) −1.2 (b) −5.7
(c) −3.8 (d) −0.4

[2020 : 2 M]

2.58 A pn junction solar cell of area 1.0 cm^2, illuminated uniformly with 100 mW cm^{-2}, has the following parameters: Efficiency = 15%, open circuit voltage = 0.7 V, fill factor = 0.8, and thickness = 200 μm, The charge of an electron is 1.6×10^{-19} C. The average optical generation rate (in cm^{-3} S^{-1}) is

(a) 1.04×10^{19} (b) 83.60×10^{19}
(c) 0.84×10^{19} (d) 5.57×10^{19}

[2020 : 2 M]

2.59 A silicon P-N junction is shown in the figure. The doping in the P region is 5×10^{16} cm^{-3} and doping in the N region is 10×10^{16} cm^{-3}. The parameters given are
Built-in voltage (Φ_{bi}) = 0.8 V
Electron charge (q) = 1.6×10^{-19} C
Vacuum permittivity (ϵ_o) = 8.85×10^{-12} F/m

Relative permittivity of silicon (ϵ_{si}) = 12

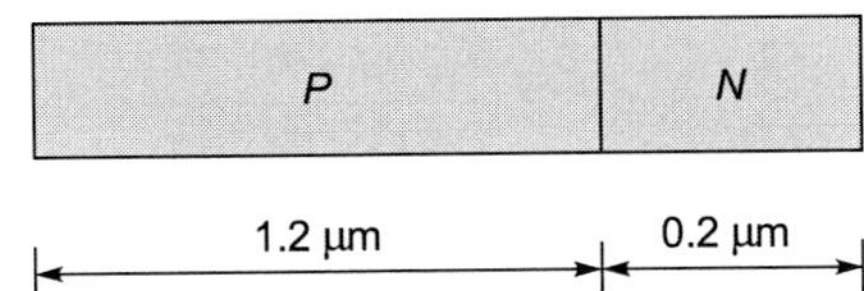

The magnitude of reverse bias voltage that would completely deplete one of the two regions (*P* or *N*) prior to the other (rounded off to one decimal place) is ______ V. [2021 : 2 M]

2.60 A p-type semiconductor with zero electric field is under illumination (low level injection) in steady state condition. Excess minority carrier density is zero at $x = \pm 2l_n$, where $l_n = 10^{-4}$ cm is the diffusion length of electrons. Assume electronic charge, $q = -1.6 \times 10^{-19}$ C. The profiles of photo-generation rate of carriers and the recombination rate of excess minority carriers (R) are shown. Under these conditions, the magnitude of the current density due to the photo-generated electrons at $x = +2l_n$ is ________ mA/cm^2 (rounded off to two decimal places).

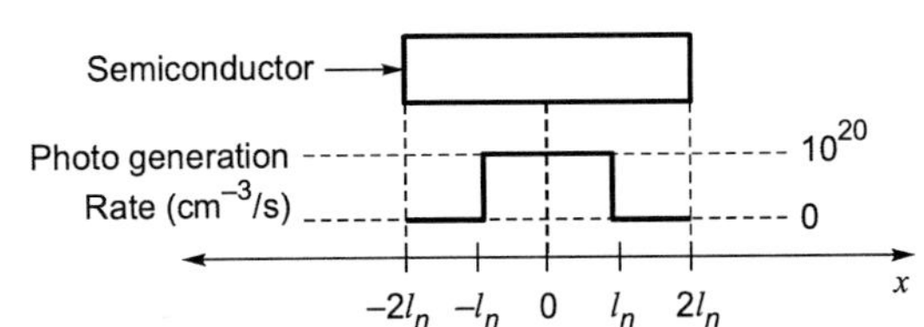

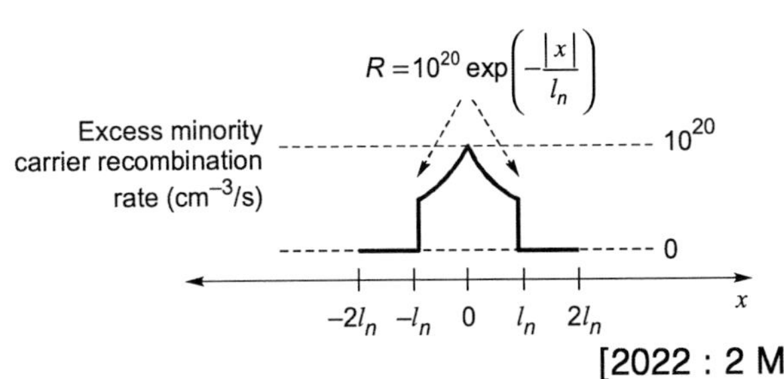

[2022 : 2 M]

2.61 The photocurrent of a PN junction diode solar cell is 1 mA. The voltage corresponding to its maximum power point is 0.3 V. If the thermal voltage is 30 mV, the reverse saturation current of the diode (in nA, rounded off to two decimal places) is ______.

[2024 : 2 Marks]

■■■

Answers PN-Junction Diodes and Special Diodes

2.1 (d)	2.2 (a)	2.3 (c)	2.4 (b)	2.5 (b)	2.6 (b)	2.7 (c)
2.8 (c)	2.9 (c)	2.10 (a)	2.11 (b)	2.12 (d)	2.13 (b)	2.14 (b)
2.15 (c)	2.16 (a)	2.17 (c)	2.18 (c)	2.19 (b)	2.20 (c)	2.21 (a)
2.22 (d)	2.23 (a)	2.24 (b)	2.25 (b)	2.26 (c)	2.27 (d)	2.28 (b)
2.29 (d)	2.30 (a)	2.31 (8)	2.32 (–30.65)	2.33 (d)	2.34 (a)	2.35 (2.5)
2.36 (0.5147)	2.37 (c)	2.38 (–4.83)	2.39 (c)	2.40 (c)	2.41 (21)	2.42 (32.36)
2.43 (d)	2.44 (10)	2.45 (0.526)	2.46 (a)	2.47 (b)	2.48 (0.42)	2.49 (b)
2.50 (0.12)	2.51 (0.61)	2.52 (d)	2.53 (35.83)	2.54 (0.231)	2.55 (b)	2.56 (c)
2.57 (*)	2.58 (c)	2.59 (8.239)	2.60 (0.59)	2.61 (4.13)		

Explanations PN-Junction Diodes and Special Diodes

2.1 (d)

Increases with increase in doping concentration contact potential of diffusion potential across a pn junction

$$V_o = kT\ln\left(\frac{N_A N_D}{n_i^2}\right)$$

at constant temperature

$$V_o \propto N_A N_D$$

2.2 (a)

Tunneling of charge carriers across the junction a Zener diode works on the principle of tunneling of charge carriers across the junction which leads the junction to breakdown.

2.3 (c)

$$C_j \propto V_R^{-1/2}$$

The depletion layer capacitance of a diode is given

$$C_T \propto V^{-n}$$

$n = \frac{1}{2}$ for step graded or abrupt *p-n* junction

$$C_T \propto V_R^{-1/2}$$

2.4 (b)

log (I) V_s V

$$I = I_o[e^{V_d/\eta V_T} - 1] = \text{forward current}$$

$\because \quad e^{V_d/\eta V_T} >> 1$

$$I = I_o e^{V_d/\eta V_T}$$

$$e^{V_d/\eta V_T} = \frac{I}{I_o}$$

$$\frac{V_d}{\eta V_T} = ln\left(\frac{I}{I_o}\right) = ln(I) - ln(I_o)$$

$$V_d = \eta V_T ln(I) - \eta V_T ln(I_o)$$

$$y = mx + C$$

So, $m = \eta V_T$

$x = ln(I)$

$c = -\eta V_T ln(I_o)$

$y = V_d$

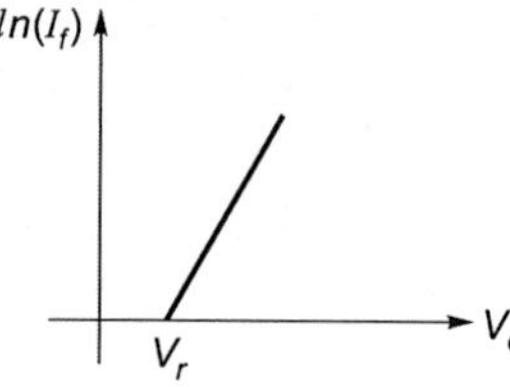

2.5 (b)

Diode is forward bias in the starting.
Diode is reverse bias later.
But at $t = 0$ diode is forward bias
So, $I = -100$ mA

2.6 (b)

For either Si or Ge $\frac{dV}{dT} \simeq -2$ mV /°C

in order to maintain a constant value of I.

$$T_2 - T_1 = 40 - 20 = 20°\text{C}$$

$$-2 \times 20 \text{ mV} = 40 \text{ mV}$$

Therefore, $V_D = 700 - 40 = 660$ mV

2.9 (c)

$\eta = 1$ for Germanium

$\eta = 2$ for Silicon at low value of current

$$I = I_{o_{Si}}\left(e^{\frac{V_{D1}}{\eta V_T}} - 1\right) \quad \text{...(i)}$$

$$I = I_{o_{Ge}}\left(e^{\frac{V_{D2}}{\eta V_T}} - 1\right) \quad \text{...(ii)}$$

L.H.S. of (i) = (ii)

$$\Rightarrow I_{o_{Si}}\left(e^{\frac{V_{D1}}{\eta V_T}} - 1\right) = I_{o_{Ge}}\left(e^{\frac{V_{D2}}{\eta V_T}} - 1\right)$$

$$\frac{I_{o_{Ge}}}{I_{o_{Si}}} = \frac{e^{\frac{V_{D_1}}{\eta V_T}} - 1}{e^{\frac{V_{D_2}}{\eta V_T}} - 1} = \frac{e^{\frac{0.718}{2\times 26\times 10^{-3}}} - 1}{e^{\frac{0.1435}{26\times 10^{-3}}} - 1} = 4 \times 10^3$$

2.10 (a)

$$E_g = \frac{1.24}{\lambda(\mu\text{m})}\text{eV} = \frac{1.24}{5490\times 10^{-4}\mu\text{m}} = 2.26 \text{ eV}$$

2.11 (b)

$$\frac{\omega_n}{\omega_p} = \frac{N_A}{N_D} \qquad \omega \simeq \omega_n \qquad (\because \omega_n >> \omega_p)$$

$$\omega_p = \frac{\omega_n \times N_D}{N_A} = \frac{3\,\mu\text{m} \times 10^{16}}{9\times 10^{16}} = 0.3\,\mu\text{m}$$

2.12 (d)

$$C_j \propto V_j^{-1/2}$$

$$\frac{C_{j2}}{C_{j1}} = \sqrt{\frac{V_1}{V_2}} = \sqrt{\frac{1}{4}} = \frac{1}{2}$$

$$C_{j2} = \frac{C_{j1}}{2} = \frac{1\text{pF}}{2} = 0.5\,\text{pF}$$

2.13 (b)

we know that ,

$$I_{o(T_2)} = I_{o(T_1)} \times 2^{\frac{(T_2 - T_1)}{10}}$$

$$I_{o(40°C)} = I_{o(20°C)} \times 2^2$$

$$I_{o(40°C)} = 40 \text{ pA}$$

2.14 (b)

$$C = \frac{\epsilon_o \epsilon_r A}{d}$$

$$\frac{C}{A} = \frac{8.85\times 10^{-12}\times 11.7}{10\times 10^{-6}} = 10.35\,\mu\text{F}$$

2.15 (c)

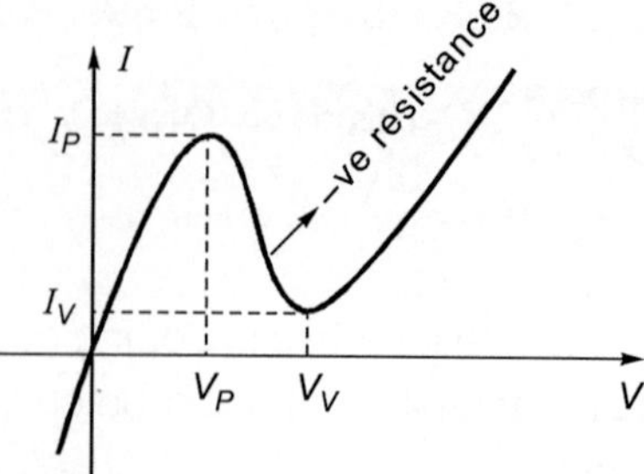

2.16 (a)

When diode is instantaneously switched from a conduction state it needs time to return to non-conduction state, so behaves short circuited for a small time period in reverse direction.

Storage time: The time period for which diode remains in conduction state even in the reverse direction so,

$$V_R = -5\text{ V}$$

For $t < 0$,

5 V, 1 kΩ, $V_R = 5$ volt

For $0 < t < t_s$,

5 V, I, 1 kΩ, $V_R = -5$ volt

2.18 (c)

Electrical field is always maximum at the junction.

2.21 (a)

Junction potential = Built-in potential + reverse bias voltage

$$V_j = V_0 + V_R$$

Now for abrupt *pn*-junction depletion width

$$W \propto V_j^{1/2}$$
$$W = KV_j^{1/2}$$
$$2\ \mu\text{m} = K(0.8 + 1.2)^{1/2} \quad \text{...(i)}$$
$$x = K(0.8 + 7.2)^{1/2} \quad \text{...(ii)}$$

From (i) & (ii) $x = 4\ \mu\text{m}$

2.24 (b)

N side is heavily doped, $W_N = \left(\dfrac{2\epsilon V_j}{eN_D}\right)^{1/2}$

$$V_j = \frac{eN_D \cdot W_N^2}{2\epsilon_0\epsilon_r} \cong 0.76\text{ Volts}$$

2.25 (b)

Electric field, $E = \dfrac{qN_o}{\epsilon}(x - w)$

At the centre $x = 0$

So, $E_{max} = \dfrac{qN_o}{\epsilon}(-w_n)$

$$= \frac{-1.6\times10^{-19}\times10^{17}\times10^{-5}}{12\times8.85\times10^{-14}} = -0.15\text{ MV/cm}$$

2.26 (c)

With increase in doping, the reverse breakdown voltage decreases and depletion width d also decreases. Since depletion capacitance, $C = \epsilon A/d$, so, depletion capacitance increases with increase in doping.

2.27 (d)

$$\frac{dV}{dT} = -2.5\text{ mV/°C}$$

For 10°C voltage across *pn*-junction will decrease by 25 mV.

2.28 (b)

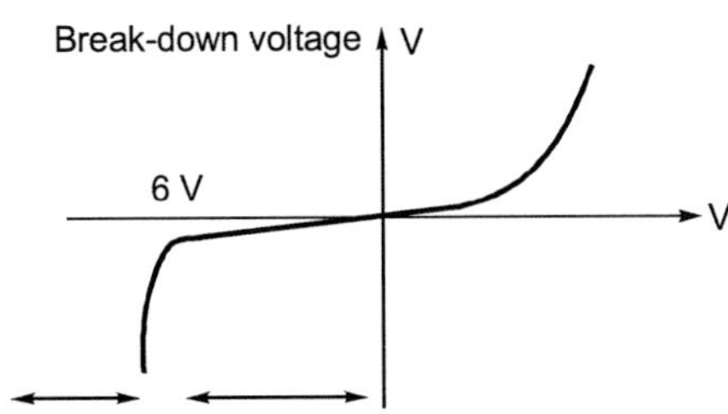

In breakdown region only, Zener diode is useful. In reverse bias region below the breakdown voltage it will behave like an open circuit.

2.29 (d)

Applying KVL in given circuit we have

$$10 - 1000\,i - v = 0 \quad \text{...(1)}$$

since 10 V is greater than 0.7 V so current will flow through diode and is

$$i = \frac{v - 0.7}{500} \quad \text{...(2)}$$

from (1) and (2) we have,

$$10 - 2(v - 0.7) - v = 0$$
$$10 + 1.4 - 3v = 0$$
$$3v = 11.4$$
$$v = \frac{11.4}{3} = 3.8\ v$$

So, $i = \dfrac{v - 0.7}{500} = \dfrac{3.8 - 0.7}{500} = 6.2\text{ mA}$

2.30 (a)

In a forward biased *pn*-junction diode, the current flow is due to diffusion of majority carriers and recombination of minority carriers.

2.31 (8)

Responsivity $(R) = \dfrac{I_p}{P_o}$

where, I_p = Photo current ; P_o = Incident power

$\therefore \quad I_p = R \times P_o = 8\ \mu\text{A}$

2.32 (–30.65)

For abrupt P^+N junction

E_{max} is given:

$$E_{max} = \frac{-qN_D x_n}{\epsilon_s} \qquad \left[\because \begin{array}{c} N_A >> N_D \\ W \simeq x_n \end{array}\right]$$

$$\Rightarrow \quad E_{max} = \frac{-1.6\times10^{-19}\times10^{16}\times0.2\times10^{-4}}{1.044\times10^{-12}}$$

$$E_{max} = -30.65 \times 10^3 \text{ V/cm}$$

$$E_{max} = -30.65 \text{ kV/cm}$$

2.33 (d)

Built-in potential is given by,

$$V_{bi} = V_T \ln\left[\frac{N_A N_D}{n_i^2}\right]$$

$$V_{bi} = 0.026 \ln\left[\frac{10^{16}\times5\times10^{18}}{2.25\times10^{20}}\right]$$

$$V_{bi} = 0.86 \text{ V}$$

and depletion width W is given by,

$$W = \sqrt{\frac{2\in}{q}\left[\frac{1}{N_A}+\frac{1}{N_D}\right]V_{bi}}$$

$$W = \sqrt{\frac{2\times1.04\times10^{-12}}{1.6\times10^{-19}}\left[\frac{1}{5\times10^{18}}+\frac{1}{10^{16}}\right]\times0.86}$$

$W = 3.3 \times 10^{-5}$ cm

2.35 (2.5)

$$C_j \propto \frac{1}{\sqrt{V_{bi}+V_R}}$$

$$\frac{C_{2j}}{C_{1j}} = \sqrt{\frac{V_{bi}+V_{R_1}}{V_{bi}+V_{R_2}}}$$

$$C_{2j} = C_{1j}\sqrt{\frac{2}{8}} = \frac{C_{1j}}{2} = 2.5 \text{ pF}$$

2.36 (0.5147)

The hole current density injected from P region to N region is given by

$$\frac{q\,n_i^2\,D_p}{N_D\,L_p}\left[\exp\left(\frac{V_{FB}}{\eta V_T}\right)-1\right]$$

where,

q = charge on electron

n_i = Intrinsic carrier concentration in silicon

N_D = Donor doping

D_p = Hole diffusion coefficient

L_p = Mean diffusion length of hole

V_{FB} = Forward voltage applied across diode

$V_T = kT/q$ = 26 mV

$$L_p = \sqrt{\tau_p\,D_p} = \sqrt{100\times10^{-6}\times36} = 0.06 \text{ cm}$$

Using the above values, we get hole current density injected from P region to N region is

= 0.5147 nA/cm^2

2.37 (c)

Built in potential,

$$\psi_0 = \frac{1}{2}\times(10^6 \text{ V/m})\times(1.1\times10^{-6}\text{ m}) = 0.55 \text{ Volts}$$

But in question (option (c)) is given as 700 mV.

2.38 (-4.83)

$$V_0 = V_T \ln\frac{N_A N_D}{n_i^2} = 26\times10^{-3}\ln\frac{10^{16}\times10^{17}}{(1.5\times10^{10})^2}$$

$$V_0 = 0.757 \text{ V}$$

$$W = \sqrt{\frac{2\varepsilon}{q}\left(\frac{1}{N_A}+\frac{1}{N_D}\right)V_0}$$

$$= \sqrt{\frac{2\times8.854\times10^{-16}\times12}{1.6\times10^{-19}}\left[\frac{1}{10^{16}}+\frac{1}{10^{17}}\right]0.757}$$

$$W = 3.3255\,\mu\text{cm}$$

$$W_P = \frac{WN_D}{N_A+N_D} = \frac{3.3255\times10^{-6}\times10^{16}}{10^{16}+10^{17}}$$

$$= 0.3023\,\mu\text{cm}$$

Charge per unit junction area in the depletion layer on p side is

$= -qN_A W_P = -1.6\times10^{-19}\times10^{17}\times0.3023\times10^{-6}$

$= -4.8368$ nC/cm^2

2.39 (c)

In any type of PN junction

$$V_{Br} \propto \frac{1}{\text{Doping Concentration}}$$

i.e. $$V_{Br} \propto \frac{1}{N_D} \text{ (or) } V_{Br} = \frac{\varepsilon E^2}{2qN_D}$$

$\therefore V_{Br}\times N_D \rightarrow$ is a constant

2.40 (c)

$X \rightarrow$ Ge Diode ....E_G = 0.2 V

$Y \rightarrow$ Si Diode ....E_G = 0.7 V

$Z \rightarrow$ GaAs Diode (or) LED ....E_G = 1.3 V

2.41 (21)

$$\text{Fill factor} = \frac{\text{Maximum obtained power}}{V_{OC}\cdot I_{SC}}$$

$$0.7 = \frac{\text{Maximum obtained power}}{0.5 \times 180 \times 10^{-3}}$$

$\therefore$ Maximum obtained power, $P_m = 0.063$ W

$$\text{Efficiency } \eta = \frac{P_m}{G \cdot A} = \frac{0.063}{100 \times 10^{-3} \times 3} = 21\%$$

$G \rightarrow$ input light in W/cm^2

2.42 (32.36)

$$|\varepsilon| = \frac{q}{\varepsilon} N_D X_N$$

$$50 \times 10^3 = \frac{1.6 \times 10^{-19}}{8.85 \times 10^{-14} \times 11.7} \times 10^{17} \times X_N$$

On solving, $X_N = 3.2356 \times 10^{-6}$ cm

$= 3.236 \times 10^{-8}$ m

$X_N = L = 32.36$ nm

2.43 (d)

$$V_0 = V_T ln \frac{n_1}{n_2} = V_T ln \frac{N_{D1}}{N_{D2}}$$

$$= 0.25\, ln \frac{10^{18}}{10^{15}} = 0.025\, ln(1000) = 0.173 \text{ V}$$

2.44 (10)

$$W = \sqrt{\frac{2\varepsilon V_R}{e}\left(\frac{1}{N_A} + \frac{1}{N_D}\right)}$$

$\because V_R >> V_{bi}$

Under reverse biased condition,

$$C = \frac{\varepsilon A}{W} \qquad \Rightarrow C \propto \frac{1}{W}$$

So, $$\frac{C_2}{C_1} = \frac{W_1}{W_2}$$

$$= \sqrt{\frac{\frac{2\varepsilon V_R}{e}\left(\frac{2}{10^{14}}\right)}{\frac{2\varepsilon V_R}{e}\left(\frac{2}{10^{16}}\right)}} = 10$$

2.45 (0.526)

$$V_{OC} = V_T ln \frac{J_L}{J_S} \qquad J_L \propto \text{Light intensity}$$

$$V_{OC2} - V_{OC1} = V_T ln \frac{J_{L2}}{J_{L1}}$$

$$V_{OC2} - V_{OC1} = 25 ln\, 20 \cong 75 \text{ mV} = 0.075 \text{ V}$$

$$V_{OC2} = 0.451 \text{ V} + 0.075 \text{ V} = 0.526 \text{ V}$$

2.46 (a)

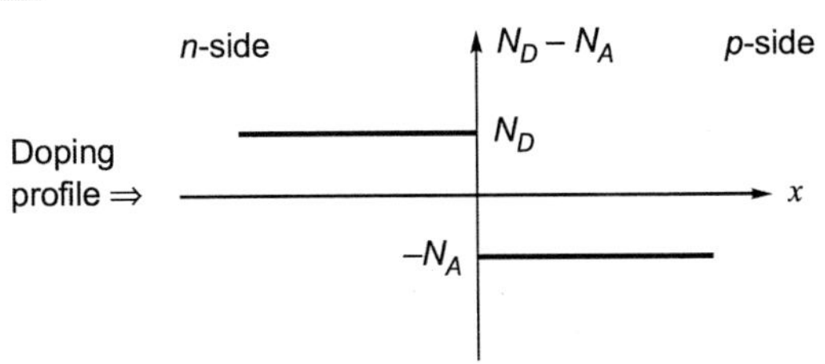

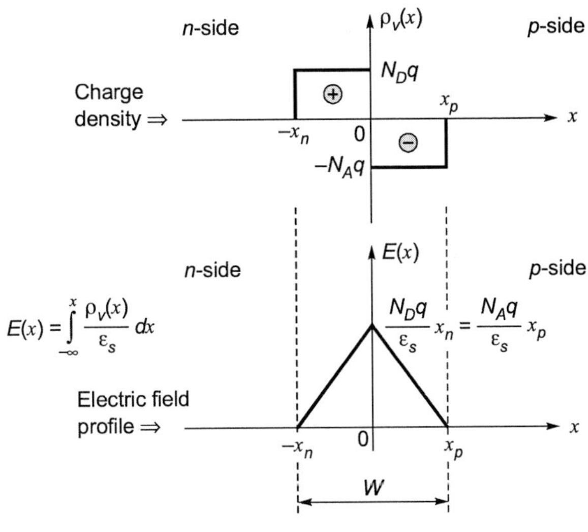

2.47 (b)

p	n

$\longrightarrow$ Hole diffusion

$\longleftarrow$ Electron diffusion

$\longleftarrow$ Hole drift

$\longrightarrow$ Electron drift

$\longrightarrow$ Hole diffusion current

$\longrightarrow$ Electron diffusion current

$\longleftarrow$ Hole drift current

$\longleftarrow$ Electron drift current

2.48 (0.42)

$$W_{dep} = \sqrt{\frac{2\varepsilon}{q}\left(\frac{1}{N_A} + \frac{1}{N_D}\right)(V_{bi} - V_{AK})}$$

$$\frac{\sqrt{(0.65 - V_{AK})}}{\sqrt{0.65}} = \frac{0.6\,\mu\text{m}}{1\,\mu\text{m}} = 0.6$$

$$1 - \frac{V_{AK}}{0.65} = 0.36$$

$$V_{AK} = 0.65\,(1 - 0.36)$$

$$= 0.65 \times 0.64 = 0.416 \text{ V} \simeq 0.42 \text{ V}$$

2.49 (b)

$$\lambda_R > \lambda_G > \lambda_B$$

Energy gap, $E_g \propto \frac{1}{\lambda}$

So, $E_{gR} < E_{gG} < E_{gB}$

Materials with high energy gap will have high built-in voltages, when doping concentrations are same.

So, $V_R < V_G < V_B$

2.50 (0.12)

Built-in potential,

$$V_{bi} = \frac{kT}{q} ln\left(\frac{N_{A2}}{N_{A1}}\right) = \frac{1.38 \times 3}{1.6 \times 100} ln(100) \text{ V} = 0.1192 \text{ V}$$
$$\simeq 0.12 \text{ V}$$

2.51 (0.61)

For solar cell, $V_{OC} = V_T ln\left(\frac{I_{SC}}{I_o}\right)$

$$V_{OC2} - V_{OC1} = V_T ln\left(\frac{I_{SC2}}{I_{SC1}}\right) = V_T ln\left(\frac{0.20}{1.0}\right)$$
$$V_{OC2} = V_{OC1} - 0.026 ln(5)$$
$$= 0.65 - 0.041845 = 0.608 \text{ V}$$

2.53 (35.83)

$$V_T = \frac{kT}{q} = \frac{1.38 \times 10^{-23} \times 300}{1.602 \times 10^{-19}} \text{ V} = 25.843 \text{ mV}$$
$$I = I_0\left(e^{V/V_T} - 1\right) = -\frac{3}{4} I_0$$
$$\Rightarrow \quad V = V_T \ln 0.25 = -35.83 \text{ mV}$$
$$V_R = |V| = 35.83 \text{ mV}$$

2.54 (0.231)

$$P_{absorbed} = P_{incident}(1 - e^{-\alpha T})$$
$$\frac{1}{3} = \frac{2}{3}(1 - e^{-\alpha T}) \quad ; \quad \frac{2}{3} e^{-\alpha T} = \frac{1}{3}$$

where $\alpha = 3 \times 10^4 \text{ cm}^{-1}$, absorption coefficient of Ge sample.

$$\therefore \quad T = \frac{1}{\alpha}\ln(2) = \frac{1}{3 \times 10^4}\ln(2) \text{ cm} = 0.231 \text{ µm}$$

2.55 (b)

$$\eta = \frac{I_{out}}{q} \times \frac{hf}{P_{in}} \ ; \quad R = \frac{I_{out}}{P_{in}}$$

So, $R = \eta \times \frac{q}{hf} = \eta \times \frac{q\lambda}{hc}$

If λ is given in µm, then

$$R = \eta\lambda \times \frac{q \times 10^{-6}}{hc} \quad ; \quad \frac{hc}{q \times 10^{-6}} \simeq 1.24$$

So, $R = \frac{\eta\lambda}{1.24}$

2.57 (*)

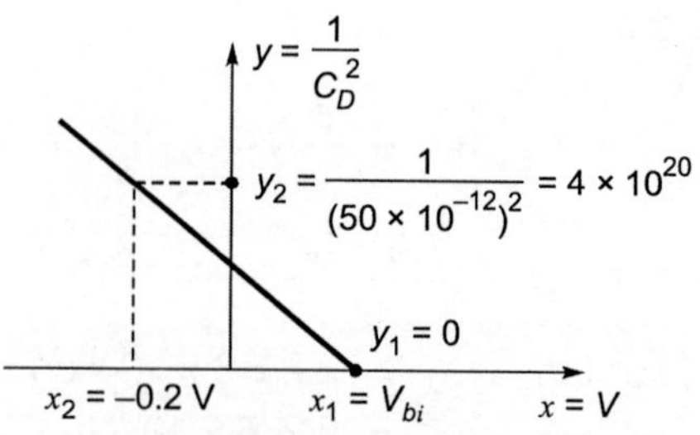

Depletion or transition capacitance is,

$$C_D = \frac{A\epsilon}{W}$$

For one-sided PN junction (Ex : $P^+ N$ junction)

$$W = \sqrt{\frac{2\epsilon V_B}{eN_D}} = \sqrt{\frac{2\epsilon (V_{bi} - V)}{eN_D}}$$

where V is anode to cathode applied potential.

$$\Rightarrow \quad C_D = \frac{A\epsilon}{\sqrt{\frac{2\epsilon (V_{bi} - V)}{eN_D}}}$$

$$\Rightarrow \quad \frac{1}{C_D^2} = \frac{2}{A^2 \epsilon eN_D}(V_{bi} - V)$$

$\frac{1}{C_D^2}$ becomes zero at $V = V_{bi}$

From above graph, $y = \frac{1}{C_D^2} = 0$ at $x_1 = V_{bi}$

And $y_2 = \frac{1}{C_D^2} = 4 \times 10^{20}$ at $x_2 = -0.2$ V

$$\text{Slope} = \frac{y_2 - y_1}{x_2 - x_1} = \frac{4 \times 10^{20} - 0}{-0.2 - V_{bi}}$$

$\because V_{bi}$ is not provided, slope cannot be found.

2.58 (c)

$$\eta = \frac{(FF) V_{OC} I_{SC}}{P_{in}}$$
$$0.15 = \frac{0.8 \times 0.7 \times I_{SC}}{100 \text{ mW}}$$
$$\Rightarrow \quad I_{SC} = \frac{15}{0.56} \text{ mA}$$
$$G_L = \frac{I_{SC}}{q \times \text{Area} \times \text{thickness}}$$

$$= \frac{15\times10^{-3}}{0.56\times1.6\times10^{-19}\times1\times200\times10^{-4}}$$

$$= \frac{15}{0.56\times32}\times10^{19} = 0.837\times10^{19}\ /\text{cm}^3/\text{second}$$

2.59 (8.239)

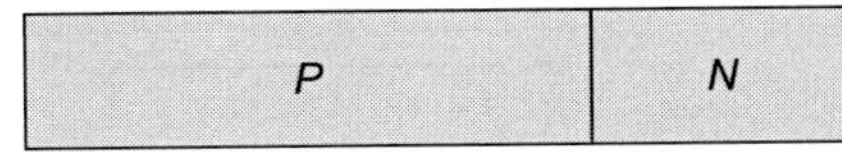

Given : $N_A = 5\times10^{16}$ cm^{-3}

$N_D = 10\times10^{16}$ cm^{-3}

Built-in potential, $\phi_{bi} = 0.8$ V

Electron charge, $q = 1.6\times10^{-19}$ C

Vacuum permittivity,

$\epsilon_o = 8.85\times10^{-12}$ F/m $= 8.85\times10^{-14}$ F/cm

Relative permittivity silicon,

$\epsilon_{si} = 12$

⇒ Doping on both sides is comparable, so smaller region would deplete first.

So, depletion region width on N-side $= x_n = 0.2$ μm

$$\Rightarrow \quad x_n = 0.2\times10^{-4}\text{ cm}$$

$$x_n = \sqrt{\frac{2\epsilon_{si}}{q}\left(\frac{N_A}{N_D}\right)\left(\frac{1}{N_A+N_D}\right)(\phi_{bi}+V_R)}$$

where, $V_R \rightarrow$ Magnitude of reverse bias potential

$$\Rightarrow 0.2\times10^{-4} = \sqrt{\frac{2\times12\times8.85\times10^{-14}}{1.6\times10^{-19}}\times\frac{5\times10^{16}}{10\times10^{16}}\times\frac{1}{(15\times10^{16})}(\phi_{bi}+V_R)}$$

$$\Rightarrow \quad \phi_{bi}+V_R = 9.039$$

$$\Rightarrow \quad V_R = 9.039-0.8$$

$$\Rightarrow \quad V_R = 8.239\text{ V}$$

2.60 (0.59)

The excess minority carrier concentration,

$$\delta n(x) = R\tau_n = 10^{20}\exp\left(\frac{-|x|}{l_n}\right)\times\tau_n$$

for $0 \le x \le l_n$

At $x = l_n$,

$$\delta n(l_n) = 10^{20}e^{-1}\tau_n \qquad \text{...(i)}$$

The minority carrier diffusion equation is given as

$$\frac{\partial\,\delta n(x)}{\partial t} = D_n\cdot\frac{\partial^2\delta n(x)}{\partial x^2} - R + G$$

At steady-state,

$$\frac{\partial\,\delta n(x)}{\partial t} = 0$$

Therefore, $D_n\cdot\dfrac{\partial^2\delta n(x)}{\partial x^2} - R + G = 0$

For $l_n \le x \le 2\,l_n$,

$R = 0$

and $G = 0$

We get,

$$D_n\cdot\frac{\partial^2 n(x)}{\partial x^2} = 0$$

The solution of the above differential equation is

$$\delta n(x) = Ax + B$$

Given, At $x = 2l_n$, $\delta n(x) = 0$. Hence,

$$0 = A(2l_n) + B$$

$$\Rightarrow \quad A = -\frac{B}{2l_n}$$

$$\therefore \quad \delta n(x) = -\frac{B}{2l_n}x + B = B\left(1-\frac{x}{2l_n}\right) \qquad \text{...(ii)}$$

Comparing the excess minority carrier concentration at $x = l_n$.

Using equation (i) and (ii), we get

$$10^{20}e^{-1}\tau_n = B\left(1-\frac{l_n}{2l_n}\right)$$

$$\Rightarrow \quad B = 2\times10^{20}e^{-1}\tau_n$$

$$\therefore \quad \delta n(x) = 2\times10^{20}e^{-1}\tau_n\left(1-\frac{x}{2l_n}\right)$$

for $l_n \le x \le 2l_n$

The electron diffusion current density,

$$|J_n|_{\text{diff}} = qD_n\frac{d\,\delta n(x)}{dx}$$

$$= qD_n\times2\times10^{20}\times e^{-1}\times\tau_n\left(0-\frac{1}{2l_n}\right)$$

$$= \frac{1.6\times10^{-19}\times l_n^2\times2\times10^{20}\times e^{-1}}{2l_n}$$

$\{\because l_n = \sqrt{D_n\tau_n}\}$

$$|J_n|_{\text{diff}} = 1.6\times10^{-19}\times l_n\times10^{20}\times e^{-1}$$

Given, $l_n = 10^{-4}$ cm

$$|J_n|_{\text{diff}} = 1.6\times10^{1}\times1\times10^{-4}\times e^{-1}$$

$$|J_n|_{\text{diff}} = 0.588\text{ mA/cm}^2 \approx 0.59\text{ mA/cm}^2$$

Therefore, the magnitude of the current density due to photo-generated electrons at $x = +2l_n$ is 0.59 mA/cm^2.

2.61 (4.13)

Given: Voltage corresponding to maximum power point, V_m 0.3 V and photocurrent, I_{Ph} = 1 mA

The current flowing through the P-N junction diode solar cell is given as

$$I = I_{Ph} - I_0\left[e^{V/V_T} - 1\right],$$

where V = voltage across diode and I_0 = reverse saturation current.

The power produced by the solar cell is

$$P = V\left[I_{Ph} - I_0\left[e^{V/V_T} - 1\right]\right]$$

To obtain voltage corresponding to maximum power point,

$$\frac{dP}{dV} = 0$$

$$\Rightarrow \quad I_{Ph} - I_0\left[e^{V/V_T} - 1\right] + V\left[\frac{-I_0}{V_T} \cdot e^{V/V_T}\right] = 0$$

$$\Rightarrow \quad I_0\left[e^{V/V_T}\left(1 + \frac{V}{V_T}\right) - 1\right] = I_{Ph}$$

Substituting $V = V_m$, we get

$$I_0 = \frac{I_{Ph}}{\left(1 + \frac{V_m}{V_T}\right)e^{\frac{V_m}{V_T}} - 1}$$

Substituting the values, we get

$$I_0 = \frac{1 \times 10^{-3}}{\left(1 + \frac{0.3}{0.03}\right)e^{\frac{0.3}{0.03}} - 1} = 4.127 \text{ nA}$$

$$I_0 \simeq 4.13 \text{ nA}$$

■■■

3 BJT and FET Basics

Note: JFET topic is out of syllabus for GATE from 2016, it may be useful for PSUs & Interviews.

3.1 The break down voltage of a transistor with its base open is BV_{CEO} and that with emitter open is BV_{CBO}, then
(a) $BV_{CEO} = BV_{CBO}$
(b) $BV_{CEO} > BV_{CBO}$
(c) $BV_{CEO} < BV_{CBO}$
(d) BV_{CEO} is not related to BV_{CBO} [1995 : 1 M]

3.2 A BJT is said to be operating in the saturation Region if
(a) Both the Junctions are reverse biased.
(b) Base-Emitter Junction is reverse biased and Base-Collector Junction is forward biased.
(c) Base-Emitter Junction is forward biased and Base-Collector Junction is reverse-biased.
(d) Both the junctions are forward biased.
[1995 : 1 M]

3.3 A BJT is said to be operating in the saturation region if
(a) both junctions are reverse biased.
(b) base-emitter junction is R.B and base collector junction is forward biased.
(c) base-emitter junction is forward biased and base-collector junction reverse biased.
(d) both the junctions are forward biased.
[1995 : 1 M]

3.4 The early-effect in a bipolar junction transistor is caused by
(a) Fast-turn-on
(b) Fast-turn-off
(c) large collector-base reverse bias
(d) large emitter-base forward bias
[1995 & 1999 : 1 M]

3.5 In a JFET

List-I
A. The pinch-off voltage decreases
B. The transconductance increases
C. The transit time of the carriers in the channel is reduced

List-II
1. The channel doping is reduced
2. The channel length is increases
3. The conductivity of the channel is increased
4. The channel length is reduced
5. The GATE area is reduced

[1995 : 2 M]

3.6 If a transistor is operating with both of its junctions forward biased, but with the collector base forward bias greater than the emitter-base forward bias, then it is operating in the
(a) forward active mode
(b) reverse saturation mode
(c) reverse active mode
(d) forward saturation anode [1996 : 2 M]

3.7 In a bipolar transistor at room temperature, if the emitter current is doubled, the voltage across its base-emitter junction
(a) doubles
(b) halves
(c) increases by about 20 mV
(d) decreases by about 20 mV [1997 : 2 M]

3.8 An *n*-channel JFET has $I_{DSS} = 2$ mA and $V_p = -4$ V. It's transconductance g_m(in mA/V). An applied GATE to source voltage V_{GS} of –2 V is
(a) 0.25 (b) 0.5
(c) 0.75 (d) 1.0 [1999 : 2 M]

3.9 MOSFET can be used as a
(a) current controlled capacitor
(b) voltage controlled capacitor
(c) current controlled inductor
(d) voltage controlled inductor [2001 : 1 M]

3.10 The effective channel length of a MOSFET in saturation decreases with increase in
(a) gate voltage (b) drain voltage
(c) source voltage (d) body voltage
[2001 : 1 M]

3.11 For an n-channel enhancement type MOSFET, if the source is connected at a higher potential than that of the bulk (i.e. $V_{SB} > 0$), the threshold voltage V_T of the MOSFET will
(a) remain unchanged (b) decrease
(c) change polarity (d) increase
[2003 : 1 M]

3.12 When the gate-to-source voltage (V_{GS}) of a MOSFET with threshold voltage of 400 mV, working in saturation is 900 mV, the drain current is observed to be 1 mA. Neglecting the channel width modulation effect and assuming that the MOSFET is operating at saturation, the drain current for an applied V_{GS} of 1400 mV is
(a) 0.5 mA (b) 2.0 mA
(c) 3.5 mA (d) 4.0 mA [2003 : 2 M]

3.13 If for a silicon *n-p-n* transistor, the base-to-emitter voltage (V_{BE}) is 0.7 V and the collector-to-base voltage (V_{CB}) is 0.2 V, then the transistor is operating in the
(a) normal active mode
(b) saturation mode
(c) inverse active mode
(d) cutoff mode [2004 : 1 M]

3.14 Consider the following statements S_1 and S_2.
S_1: The β of a bipolar transistor reduces if the base width is increased.
S_2: The β of a bipolar transistor increases if the doping concentration in the base is increased.
Which one of the following is correct?
(a) S_1 is FALSE and S_2 is TRUE
(b) Both S_1 and S_2 are TRUE
(c) Both S_1 and S_2 are FALSE
(d) S_1 and TRUE and S_2 is FALSE [2004 : 1 M]

3.15 Consider the following statements S_1 and S_2.
S_1: The threshold voltage (V_T) of a MOS capacitor decreases with increase in gate oxide thickness.
S_2: The threshold voltage (V_T) of a MOS capacitor decreases with increase in substrate doping concentration.
Which one of the following is correct?
(a) S_1 is FALSE and S_2 is TRUE
(b) Both S_1 and S_2 are TRUE
(c) Both S_1 and S_2 are FALSE
(d) S_1 is TRUE and S_2 is FALSE [2004 : 2 M]

3.16 The drain of an n-channel MOSFET is shorted to the gate so that $V_{GS} = V_{DS}$. The threshold voltage (V_T) of MOSFET is 1 V. If the drain current (I_D) is 1 mA for V_{GS} = 2 V, then for V_{GS} = 3 V, I_D is
(a) 2 mA (b) 3 mA
(c) 9 mA (d) 4 mA
[2004 : 2 M]

3.17 A MOS capacitor made using *p*-type substrate is in the accumulation mode. The dominant charge in the channel is due to the presence of
(a) holes
(b) electrons
(c) positively charged ions
(d) negatively charged ions [2005 : 2 M]

3.18 The phenomenon known as "Early Effect" in a bipolar transistor refers to a reduction of the effective base-width caused by
(a) electron-hole recombination at the base
(b) the reverse biasing of the base-collector junction
(c) the forward biasing of emitter-base junction
(d) the early removal of stored base charge during saturation-to-cutoff switching
[2006 : 1 M]

Common Data Question 3.19, 3.20 and 3.21:
The figure shows the high-frequency capacitance-voltage (C-V) characteristics of a Metal/SiO_2/silicon (MOS) capacitor having an area of 1×10^{-4} cm^2. Assume that the permittivities ($\varepsilon_0 \varepsilon_r$) of silicon and SiO_2 are 1×10^{-12} F/cm and 3.5×10^{-13} F/cm respectively.

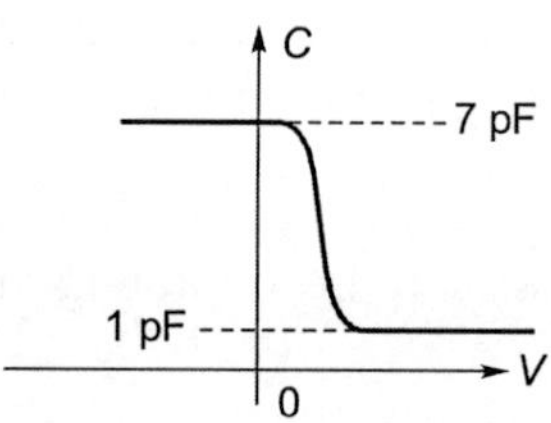

3.19 The gate oxide thickness in the MOS capacitor is
(a) 50 nm (b) 143 nm
(c) 350 nm (d) 1 μm
[2007 : 2 M]

3.20 The maximum depletion layer width in silicon is
(a) 0.143 μm (b) 0.857 μm
(c) 1 μm (d) 1.143 μm
[2007 : 2 M]

3.21 Consider the following statements about the *C-V* characteristics plot:
S_1: The MOS capacitor has as *n*-type substrate.
S_2: If positive charges are introduced in the oxide, the *C-V* plot will shift to the left.
Then which of the following is true?
(a) Both S_1 and S_2 are true
(b) S_1 is true and S_2 is false
(c) S_1 is false and S_2 is true
(d) Both S_1 and S_2 are false [2007 : 2 M]

3.22 The drain current of a MOSFET in saturation is given by $I_D = K(V_{GS} - V_T)^2$ where K is a constant. The magnitude of the trans-conductance g_m is

(a) $\dfrac{K(V_{GS}-V_T)^2}{V_{DS}}$ (b) $2K(V_{GS}-V_T)$

(c) $\dfrac{I_D}{V_{GS}-V_{DS}}$ (d) $\dfrac{K(V_{GS}-V_T)^2}{V_{GS}}$

[2008 : 1 M]

3.23 The measured trans-conductance g_m of an NMOS transistor operating in the linear region is plotted against the gate voltage V_G at a constant drain voltage V_D. Which of the following figures represents the expected dependence of g_m on V_G?

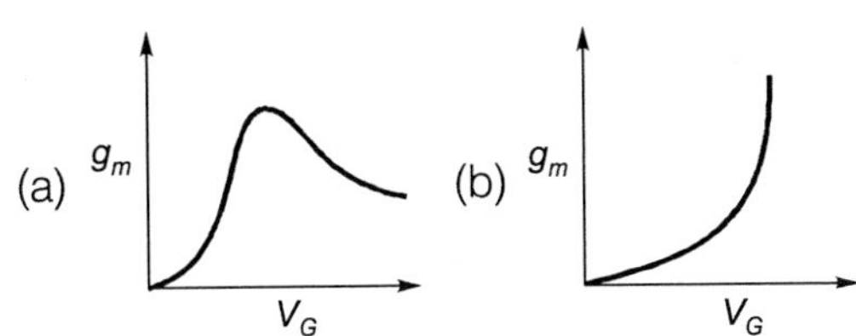

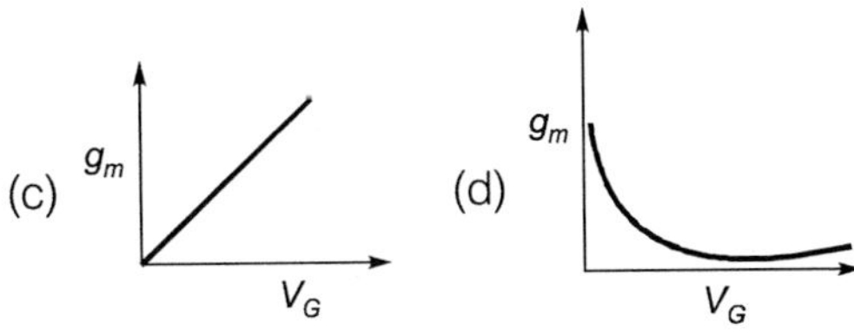

[2008 : 2 M]

3.24 The cross section of a JFET is shown in the following figure. Let V_G be –2 V and let V_p be the initial pinch-off voltage. If the width W is doubled (with other geometrical parameters and doping levels remaining the same), then the ratio between the mutual transconductances of the initial and the modified JFET is

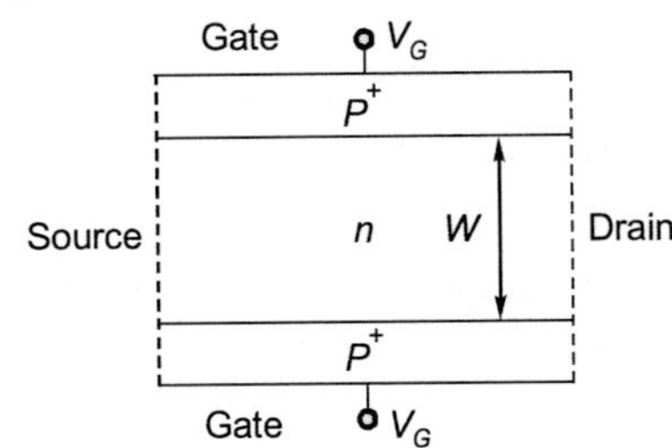

(a) 4 (b) $\dfrac{1}{2}\left(\dfrac{1-\sqrt{2/V_P}}{1-\sqrt{1/(2V_P)}}\right)$

(c) $\dfrac{1-\sqrt{2/V_P}}{1-\sqrt{1/(2V_P)}}$ (d) $\dfrac{1-\left(2/\sqrt{V_P}\right)}{1-\left(1/\left(2\sqrt{V_P}\right)\right)}$

[2008 : 2 M]

3.25 Consider the following two statements about the internal conditions in an n-channel MOSFET operating in the active region

S_1: The inversion charge decreases from source to drain

S_2: The channel potential increases from source to drain

Which of the following is correct?

(a) Only S_2 is true
(b) Both S_1 and S_2 are false
(c) Both S_1 and S_2 are true, but S_2 is not a reason for S_1
(d) Both S_1 and S_2 are true, and S_2 is a reason for S_1

[2009 : 2 M]

3.26 At room temperature, a possible value for the mobility of electrons in the inversion layer of a silicon n-channel MOSFET is

(a) 450 cm^2/V-s (b) 1350 cm^2/V-s
(c) 1800 cm^2/V-s (d) 3600 cm^2/V-s

[2010 : 1 M]

3.27 In a uniformly doped BJT, assume that N_E, N_B and N_C are the emitter, base and collector dopings in atoms/cm^3, respectively. If the emitter injection efficiency of the BJT is close to unity, which one of the following conditions is TRUE?

(a) $N_E = N_B = N_C$
(b) $N_E >> N_B$ and $N_B > N_C$
(c) $N_E = N_B$ and $N_B < N_C$
(d) $N_E < N_B < N_C$ [2010 : 2 M]

3.28 For a BJT, the common-base current gain α = 0.98 and the collector base junction reverse bias saturation current I_{CO} = 0.6 μA. This BJT is connected in the common emitter mode and operated in the active region with a base drive current I_B = 20 μA. The collector current I_C for this mode of operation is

(a) 0.98 mA (b) 0.99 mA
(c) 1.0 mA (d) 1.01 mA [2011 : 2 M]

Common Data for Questions 3.29 and 3.30.

The channel resistance of an N-channel JFET shown in the figure below is 600 Ω when the full channel thickness (t_{ch}) of 10 μm is available for conduction. The built-in voltage of the gate P^+N junction (V_{bi}) is –1. When the gate to source voltage (V_{GS}) is 0 V, the channel is depleted by 1 μm on each side due to the built-in voltage and hence the thickness available for conduction is only 8 μm.

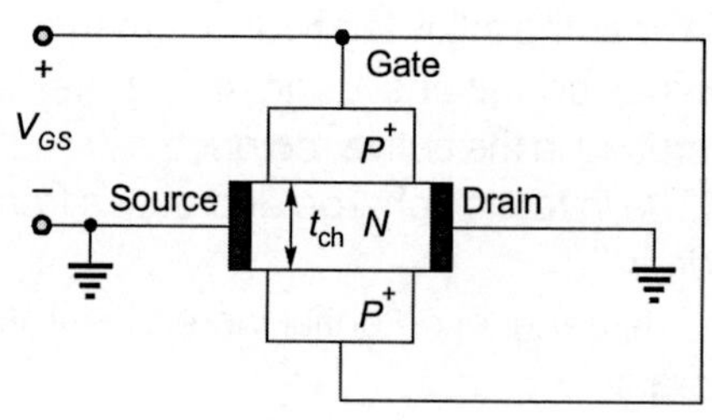

3.29 The channel resistance when V_{GS} = 0 V is
(a) 480 Ω (b) 600 Ω
(c) 750 Ω (d) 1000 Ω **[2011 : 2 M]**

3.30 The channel resistance when V_{GS} = −3 V is
(a) 360 Ω (b) 917 Ω
(c) 1000 Ω (d) 3000 Ω **[2011 : 2 M]**

3.31 The source of a silicon ($n_i = 10^{10}$ per cm³) *n*-channel MOS transistor has an area of 1 sq μm and a depth of 1 μm. If the dopant density in the source is 10^{19}/cm³, the number of holes in the source region with the above volume is approximately
(a) 10^7 (b) 100
(c) 10 (d) 0 **[2012 : 2 M]**

Common Data for Questions 3.32 and 3.33:

In the three dimensional view of a silicon n channel MOS transistor shown below, δ = 20 nm. The transistor is of width 1 μm. The depletion width formed at every p-n junction is 10 nm. The relative permittivities of Si and SiO_2, respectively, are 11.7 and 3.9, and $\varepsilon_0 = 8.9 \times 10^{-12}$ F/m.

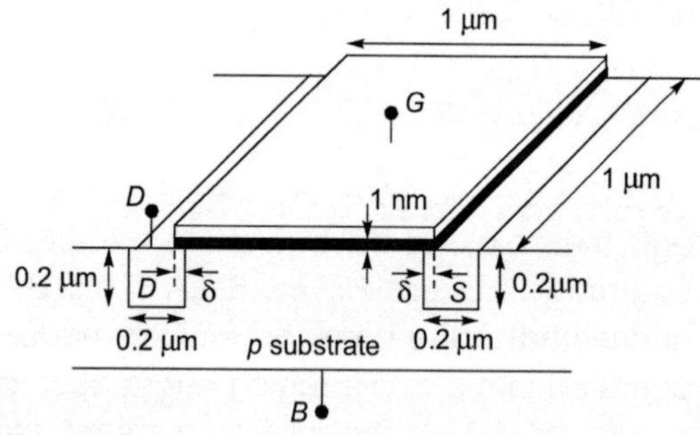

3.32 The source-body junction capacitance is approximately
(a) 2 fF (b) 7 fF
(c) 2 pF (d) 7 pF **[2012 : 2 M]**

3.33 The gate-source overlap capacitance is approximately
(a) 0.7 fF (b) 0.7 pF
(c) 0.35 fF (d) 0.24 pF **[2012 : 2 M]**

3.34 A BJT is biased in forward active mode. Assume V_{BE} = 0.7 V, kT/q = 25 mV and reverse saturation current $I_s = 10^{-13}$ mA. The transconductance of the BJT (in mA/V) is ____.
[2014 : 2 M, Set-1]

3.35 A depletion type N-channel MOSFET is biased in its linear region for use as a voltage controlled resistor. Assume threshold voltage V_{TH} = −0.5 V, V_{GS} = 2.0 V, V_{DS} = 5 V, W/L = 100, $C_{ox} = 10^{-8}$ F/cm² and μ_n = 800 cm²/V-s. The value of the resistance of the voltage controlled resistor (in Ω) is ____.
[2014 : 2 M, Set-1]

3.36 The slope of the I_D vs. V_{GS} curve of an *n*-channel MOSFET in linear region is $10^{-3}\ \Omega^{-1}$ at V_{DS} = 0.1 V. For the same device, neglecting channel length modulation, the slope of the $\sqrt{I_D}$ vs. V_{GS} (in $\sqrt{A}/V$) under saturation region is approximately ________.
[2014 : 2 M, Set-3]

3.37 An ideal MOS capacitor has boron doping-concentration of 10^{15} cm^{-3} in the substrate. When a gate voltage is applied, a depletion region of width 0.5 μm is formed with a surface (channel) potential of 0.2 V. Given that $\varepsilon_0 = 8.854 \times 10^{-14}$ F/cm and the relative permittivities of silicon and silicon dioxide are 12 and 4, respectively, the peak electric field (in V/μm) in the oxide region is _____.
[2014 : 2 M, Set-3]

3.38 Consider two BJTs biased at the same collector current with area A_1 = 0.2 μm × 0.2 μm and A_2 = 300 μm. Assuming that all other device parameters are identical kT/q = 26 mV, the intrinsic carrier concentrations is 1×10^{10} cm^{-3}, and $q = 1.6 \times 10^{-19}$C, the difference between the base-emitter voltages (in mV) of the two BJTs (i.e., $V_{BE1} - V_{BE2}$) is ______.
[2014 : 2 M, Set-4]

3.39 For the NMOSFET in the circuit shown, the threshold voltage is V_{th}, where $V_{th} > 0$. The source voltage V_{SS} is varied from 0 to V_{DD}. Neglecting the channel length modulation, the drain current I_D as a function of V_{SS} is represented by

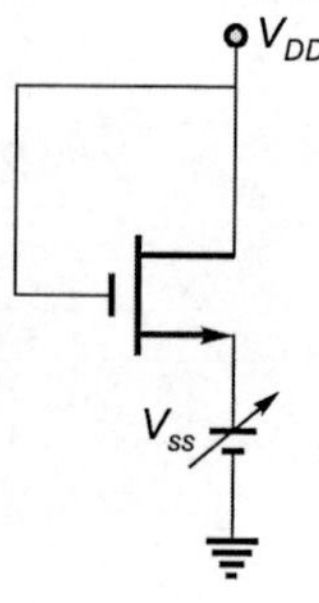

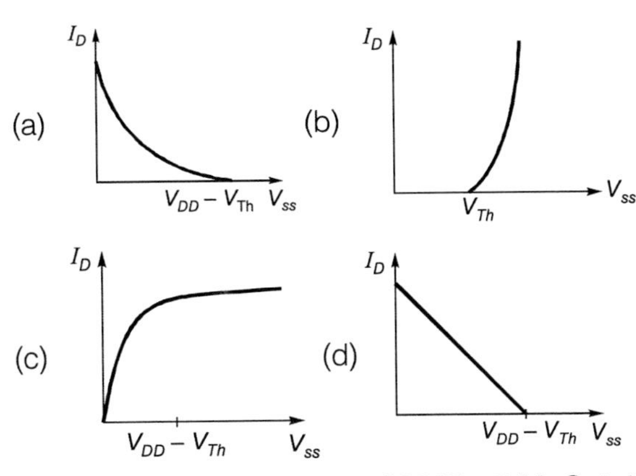

[2015 : 2 M, Set-1]

3.40 A MOSFET in saturation has a drain current of 1 mA for $V_{DS} = 0.5$ V. If the channel length modulation coefficient is 0.05 V^{-1}, the output resistance (in kΩ) of the MOSFET is ______.

[2015 : 2 M, Set-1]

3.41 In MOS capacitor with an oxide layer thickness of 10 mm. The maximum depletion layer thickness is 100 mm. The permittivities of the semiconductor and the oxide layer are ε_s and ε_{ox} respectively. Assuming $\varepsilon_s/\varepsilon_{ox} = 3$ the ratio of the maximum capacitance to the minimum capacitance of this MOS capacitor is______.

[2015 : 2 M, Set-2]

3.42 The current in an enhancement mode NMOS transistor biased in saturation mode was measured to be 1 mA at a drain-source voltage of 5 V. When the drain-source voltage was increased to 6 V while keeping gate-source voltage same, the drain current increased to 1.02 mA. Assume that drain to source saturation voltages is much smaller than the applied drain-source voltage. The channel length modulation parameter λ (in V^{-1}) is _____.

[2015 : 2 M, Set-3]

3.43 If the base width in a bipolar junction transistor is doubled, which one of the following statements will be TRUE?
(a) Current gain will increase
(b) Unity gain frequency will increase
(c) Emitter base junction capacitance will increase
(d) Early voltage will increase

[2015 : 1 M, Set-3]

3.44 An *npn* BJT having reverse saturation current $I_s = 10^{-15}$ *A* is biased in the forward active region with $V_{BE} = 700$ mV. The thermal voltage (V_T) is 25 mV and the current gain (β) may vary from 50 to 150 due to manufacturing variations. The maximum emitter current (in μA) is_____.

[2015 : 2 M, Set-3]

3.45 Consider the following statements for a metal oxide semiconductor field effect transistor (MOSFET):
P: As channel length reduces, OFF-state current increases.
Q: As channel length reduces, output resistance increases.
R: As channel length reduces, threshold voltage remains constant.
S: As channel length reduces, ON current increases.
Which of the above statements are **INCORRECT**?
(a) *P* and *Q* (b) *P* and *S*
(c) *Q* and *R* (d) *R* and *S*

[2016 : 1 M, Set-1]

3.46 Consider an *n*-channel metal oxide semiconductor field effect transistor (MOSFET) with a gate-to-source voltage of 1.8 V. Assume that $\frac{W}{L} = 4$, $\mu_n C_{ox} = 70 \times 10^{-6}\ AV^{-2}$, the threshold voltage is 0.3 V, and the channel length modulation parameter is 0.09 V^{-1}. In the saturation region, the drain conductance (in micro Siemens) is _____

[2016 : 2 M, Set-1]

3.47 A long-channel NMOS transistor is biased in the linear region with $V_{DS} = 50$ mV and is used as a resistance. Which one of the following statements is NOT correct?
(a) If the device width *W* is increased, the resistance decreases.
(b) If the threshold voltage is reduced, the resistance decreases.
(c) If the device length *L* is increased, the resistance increase.
(d) If V_{GS} is increased, the resistance increases.

[2016 : 1 M, Set-2]

3.48 A voltage V_G is applied across a MOS capacitor with metal gate and *p*-type silicon substrate at $T = 300$ K. The inversion carrier density (in number of carriers per unit area) for $V_G = 0.8$ V is 2×10^{11} cm^{-2}. For $V_G = 1.3$ V, the inversion carrier density is 4×10^{11} cm^{-2}. What is the value of the inversion carrier density for $V_G = 1.8$ V?
(a) 4.5×10^{11} cm^{-2} (b) 6.0×10^{11} cm^{-2}
(c) 7.2×10^{11} cm^{-2} (d) 8.4×10^{11} cm^{-2}

[2016 : 2 M, Set-2]

3.49 Consider a long-channel NMOS transistor with source and body connected together. Assume that the electron mobility is independent of V_{GS} and V_{DS}. Given,

$g_m = 0.5\ \mu A/V$ for $V_{DS} = 50$ mV and $V_{GS} = 2$ V,

$g_d = 8\ \mu A/V$ for $V_{GS} = 2$ V and $V_{DS} = 0$ V,

where $g_m = \dfrac{\partial I_D}{\partial V_{GS}}$ and $g_d = \dfrac{\partial I_D}{\partial V_{DS}}$

The threshold voltage (in volts) of the transistor is ______.

[2016 : 2 M, Set-2]

3.50 The figure shows the band diagram of a Metal Oxide Semiconductor (MOS). The surface region of this MOS is in

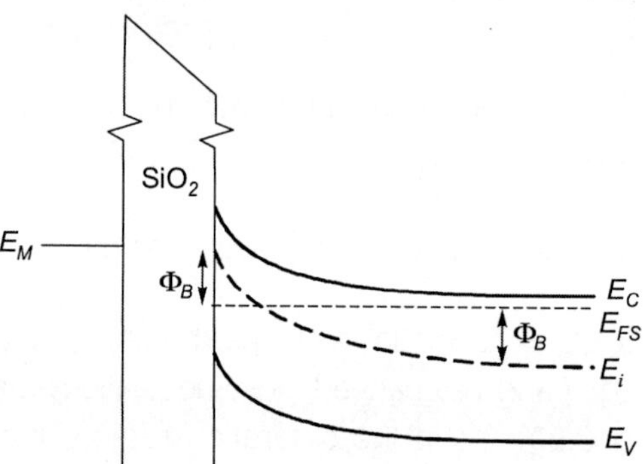

(a) inversion
(b) accumulation
(c) depletion
(d) flat band

[2016 : 1 M, Set-3]

3.51 Figures I and II show two MOS capacitors of unit area. The capacitor in Figure I has insulator materials X (of thickness $t_1 = 1$ nm and dielectric constant $\varepsilon_1 = 4$) and Y (of thickness $t_2 = 3$ nm and dielectric constant $\varepsilon_2 = 20$). The capacitor in Figure II has only insulator material X of thickness t_{Eq}. If the capacitors are of equal capacitance, then the value of t_{Eq} (in nm) is ______.

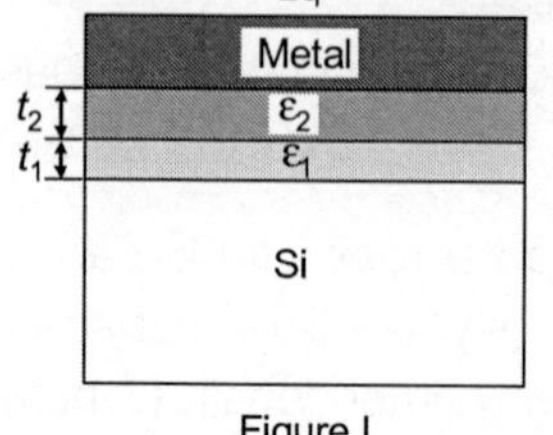

Figure I

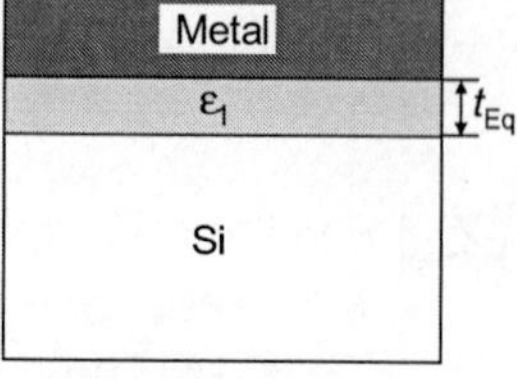

Figure II

[2016 : 2 M, Set-3]

3.52 The injected excess electron concentration profile in the base region of an npn BJT, biased in the active region, is linear, as shown in the figure. If the area of the emitter-base junction is 0.001 cm^2, $\mu_n = 800$ cm^2/(V-s) in the base region and depletion layer widths are negligible, then the collector current I_C (in mA) at room temperature is (Given: thermal voltage $V_T = 26$ mV at room temperature, electronic charge $q = 1.6 \times 10^{-19}$C)

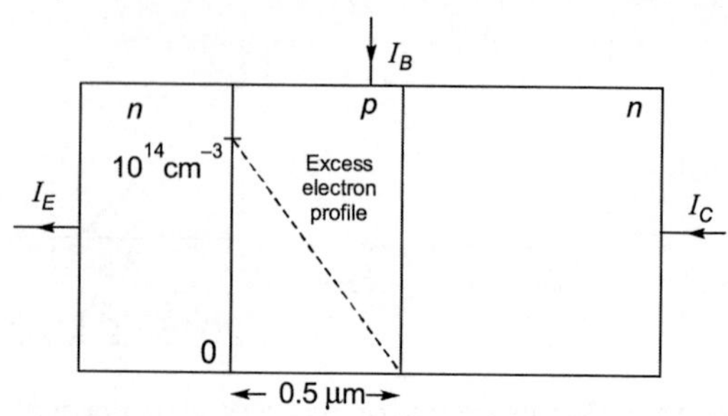

[2016 : 2 M, Set-3]

3.53 For a narrow base PNP BJT, the excess minority carrier concentrations (Δn_E for emitter, Δp_B for base, Δn_C for collector) normalized to equilibrium minority carrier concentrations (n_{E0} for emitter, p_{B0} for base, n_{C0} for collector) in the quasi-neutral emitter, base and collector regions are shown below. Which one of the following biasing modes is the transistor operating in?

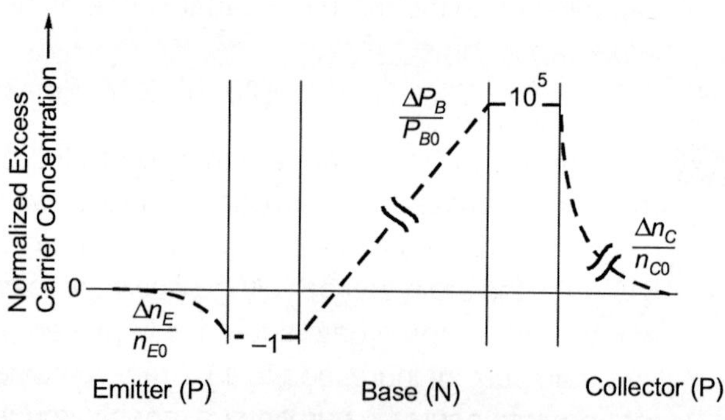

(a) Forward active
(b) Saturation
(c) Inverse active
(d) Cutoff

[2017 : 1 M, Set-1]

3.54 An npn bipolar junction transistor (BJT) is operating in the active region. If the reverse bias across the base-collector junction is increased, then

(a) the effective base width increases and common-emitter current gain increases
(b) the effective base width increases and common-emitter current gain decreases
(c) the effective base width decreases and common-emitter current gain increases
(d) the effective base width decreases and common -emitter current gain decreases

[2017 : 1 M, Set-2]

3.55 Consider an n-channel MOSFET having width W, length L, electron mobility in the channel μ_n and oxide capacitance per unit area C_{ox}. If gate-to-source voltage $V_{GS} = 0.7$ V, drain-to-source voltage $V_{DS} = 0.1$ V. $(\mu_n C_{ox}) = 100\ \mu A/V^2$, threshold voltage $V_{Th} = 0.3$ V and $(W/L) = 50$, then the transconductance g_m (in mA/V) is ______.

[2017 : 1 M, Set-2]

3.56 Two n-channel MOSFETs, T1 and T2, are identical in all respects except that the width of T2 is double that of T1. Both the transistors are biased in the saturation region of operation, but the gate overdrive voltage $(V_{GS} - V_{TH})$ of T2 is double that of T1, where V_{GS} and V_{TH} are the gate-to-source voltage and threshold voltage of the transistors, respectively. If the drain current and transconductance of T1 are I_{D1} and g_{m1} respectively, the corresponding values of these two parameters for T2 are

(a) $8 I_{D1}$ and $2 g_{m1}$ (b) $8 I_{D1}$ and $4 g_{m1}$
(c) $4 I_{D1}$ and $4 g_{m1}$ (d) $4 I_{D1}$ and $2 g_{m1}$

[2017 : 2 M, Set-2]

3.57 A MOS capacitor is fabricated on p-type Si (Silicon) where the metal work function is 4.1 eV and electron affinity of Si is 4.0 eV. $E_C - E_F = 0.9$ eV, where E_C and E_F are the conduction band minimum and the Fermi energy levels of Si respectively. Oxide $\varepsilon_r = 3.9$, $\varepsilon_o = 8.85 \times 10^{-14}$ F/cm, oxide thickness $t_{ox} = 0.1$ µm and electronic charge $q = 1.6 \times 10^{-19}$ C. If The measured flat band voltage of this capacitor is –1 V, then the magnitude of the fixed charge at the oxide-semiconductor interface, in nC/cm² is ______.

[2017 : 2 M, Set-2]

3.58 The figure shows the high-frequency C-V curve of a MOS capacitor (at $T = 300$ K) with $\phi_{ms} = 0$ V and no oxide charges. The flat-band, inversion, and accumulation conditions are represented, respectively, by the points

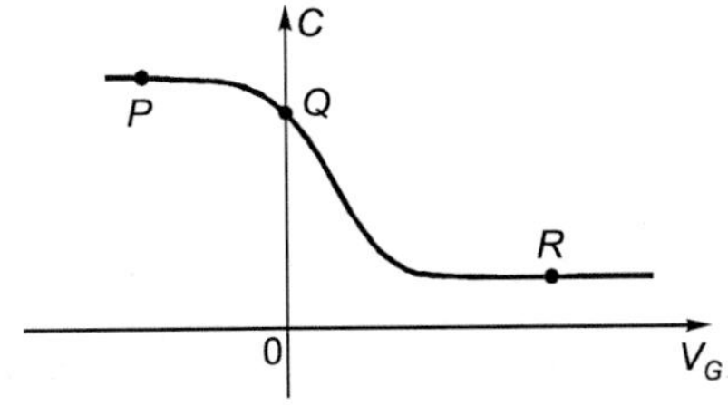

(a) Q, R, P (b) P, Q, R
(c) Q, P, R (d) R, P, Q

[2019 : 1 M]

3.59 Consider a long-channel MOSFET with a channel length 1 µm and width 10 µm. The device parameters are acceptor concentration $N_A = 5 \times 10^{16}$ cm^{-3}, electron mobility $\mu_n = 800$ cm²/V-s, oxide capacitance/area $C_{ox} = 3.45 \times 10^{-7}$ F/cm², threshold voltage $V_T = 0.7$ V. The drain saturation current (I_{Dsat}) for a gate voltage of 5 V is ____ mA (rounded off to two decimal places). [$\varepsilon_0 = 8.854$ X 10^{-14} F/cm, $\varepsilon_{Si} = 11.9$]

[2019 : 2 M]

3.60 The band diagram of a p-type semiconductor with a band-gap of 1 eV is shown. Using this semiconductor, a MOS capacitor having V_{Th} of –0.16 V, C'_{ox} of 100 nF/cm² and a metal work function of 3.87 eV is fabricated. There is no charge within the oxide. If the voltage across the capacitor is V_{Th}. the magnitude of depletion charge per unit area (in C/cm²) is

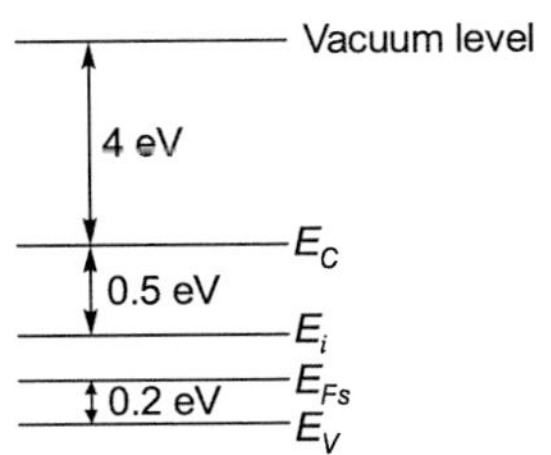

(a) 0.52×10^{-8} (b) 0.93×10^{-8}
(c) 1.41×10^{-8} (d) 1.70×10^{-8}

[2020 : 2 M]

3.61 The base of an npn BJT T1 has a linear doping profile $N_B(x)$ as shown below. The base of another npn BJT T2 has a uniform doping N_B of 10^{17} cm^{-3}. All other parameters are identical for both the devices. Assuming that the hole density profile is the same as that of doping, the common-emitter current gain of T2 is

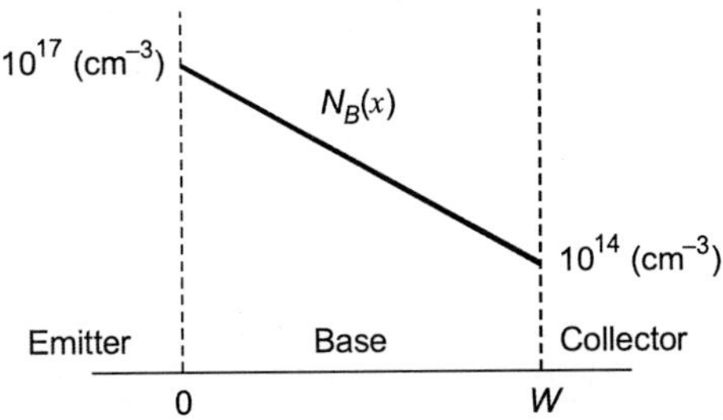

(a) approximately 0.3 times that of T1
(b) approximately 0.7 times that of T1
(c) approximately 2.5 times that of T1
(d) approximately 2.0 times that of T1

[2020 : 2 M]

3.62 For an n-channel silicon MOSFET with 10 nm gate oxide thickness, the substrate sensitivity $\left(\frac{\partial V_T}{\partial |V_{BS}|}\right)$ is found to be 50 mV/V at a substrate voltage $|V_{BS}| = 2$ V, where V_T is the threshold voltage of the MOSFET. Assume that, $|V_{BS}| >> 2\Phi_B$, where $q\Phi_B$ is the separation between the Fermi energy level E_F and the intrinsic level E_i in the bulk. Parameters given are

Electron charge $(q) = 1.6 \times 10^{-19}$ C

Vacuum permittivity $(\epsilon_o) = 8.85 \times 10^{-12}$ F/m

Relative permittivity of silicon $(\epsilon_{si}) = 12$

Relative permittivity of oxide $(\epsilon_{ox}) = 4$

The doping concentration of the substrate is

(a) 4.37×10^{15} cm^{-3} (b) 7.37×10^{15} cm^{-3}

(c) 9.37×10^{15} cm^{-3} (d) 2.37×10^{15} cm^{-3}

[2021 : 2 M]

3.63 An ideal MOS capacitor (p-type semiconductor) is shown in the figure. The MOS capacitor is under strong inversion with $V_G = 2$ V. The corresponding inversion charge density (Q_{IN}) is 2.2 μC/cm². Assume oxide capacitance per unit area as $C_{ox} = 1.7$ μF/cm². For $V_G = 4$ V, the value of Q_{IN} is ______ μC/cm² (rounded off to one decimal place).

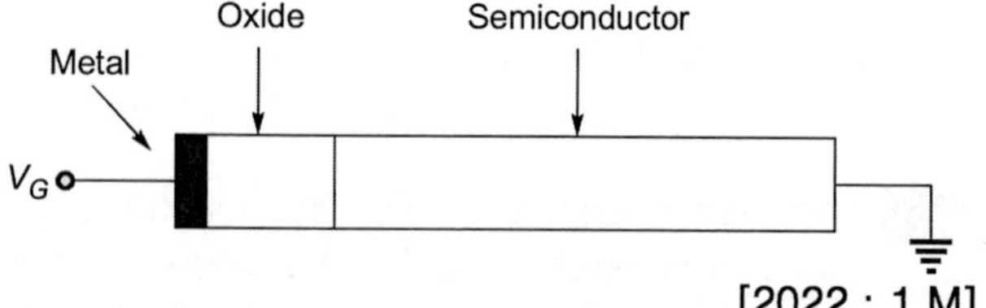

[2022 : 1 M]

3.64 For a MOS capacitor, V_{fb} and V_t are the flat-band voltage and the threshold voltage, respectively. The variation of the depletion width (W_{dep}) for varying gate voltage (V_g) is best represented by

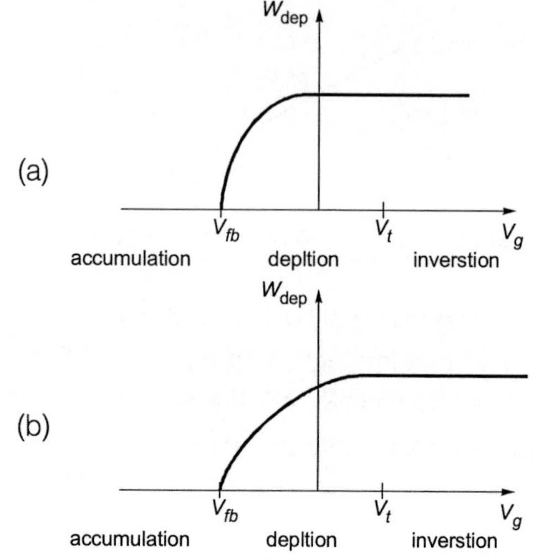

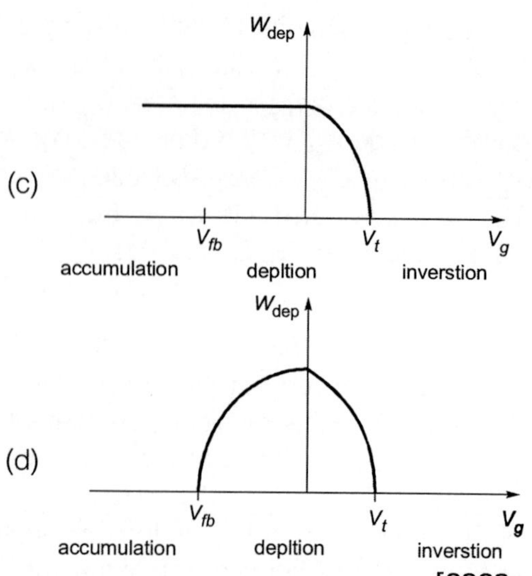

[2023 : 1 M]

3.65 Consider a MOS capacitor made with p-type silicon. It has an oxide thickness of 100 nm, a fixed positive oxide charge of 10^{-8} C/cm² at the oxide-silicon interface, and a metal work function of 4.6 eV. Assume that the relative permittivity of the oxide is 4 and the absolute permittivity of free space is 8.85×10^{-14} F/cm. If the flatband voltage is 0 V, the work function of the p-type silicon (in eV, rounded off to two decimal places) is _____.

[2024 : 2 Marks]

3.66 Which of the following statements is/are true for a BJT with respect to its DC current gain β?

(a) Under high-level injection condition in forward active mode, β will decrease with increase in the magnitude of collector current.

(b) β will be lower when the BJT is in saturation region compared to when it is in active region.

(c) Under low-level injection condition in forward active mode, where the current at the emitter-base junction is dominated by recombination-generation process, β will decrease with increase in the magnitude of collector current.

(d) A higher value of β will lead To a lower value of the collector-to-emitter breakdown voltage.

[2024 : 2 Marks]

3.67 An NMOS transistor operating in the linear region has I_{DS} of 5 μA at V_{DS} of 0.1 V. Keeping V_{GS} constant, the V_{DS} is increased to 1.5 V.

Given that $\mu_n C_{ox} \frac{W}{L} = 50 \mu A/V^2$, the transconductance at the new operating point (in μA/V, rounded off to two decimal places) is ______.

[2024 : 2 Marks]

■■■

Answers BJT and FET Basics

3.1 (c)	3.2 (d)	3.3 (d)	3.4 (c)	3.5 (A-1, B-3, C-4)		3.6 (b)
3.7 (c)	3.8 (b)	3.9 (b)	3.10 (b)	3.11 (d)	3.12 (d)	3.13 (a)
3.14 (d)	3.15 (d)	3.16 (d)	3.17 (a)	3.18 (b)	3.19 (a)	3.20 (b)
3.21 (c)	3.22 (b)	3.23 (*)	3.24 (b)	3.25 (d)	3.26 (b)	3.27 (b)
3.28 (d)	3.29 (c)	3.30 (c)	3.31 (d)	3.32 (a)	3.33 (a)	3.34 (5.76)
3.35 (500)	3.36 (0.0707)	3.37 (2.4)	3.38 (380.28)	3.39 (a)	3.40 (20)	3.41 (4.33)
3.42 (0.02)	3.43 (d)	3.44 (1475)	3.45 (c)	3.46 (28.35)	3.47 (d)	3.48 (b)
3.49 (1.2)	3.50 (a)	3.51 (1.6)	3.52 (6.656)	3.53 (c)	3.54 (c)	3.55 (0.5)
3.56 (b)	3.57 (6.903)	3.58 (a)	3.59 (25.52)	3.60 (d)	3.61 (*)	3.62 (d)
3.63 (5.6)	3.64 (b)	3.65 (4.32)	3.66 (a, b, d)	3.67 (52.5)		

Explanations BJT and FET Basics

3.1 (c)

The relationship between open base breakdown voltage (BV_{CEO}) of BJT with open emitter voltage (BV_{CBO}) is given by

$$BV_{CEO} = BV_{CBO}\sqrt{\frac{1}{\beta}}$$

$\because \quad \beta > 1$

So, $BV_{CEO} < BV_{CBO}$

3.2 (d)

Both the junctions are forward biased.
When emitter base (E-B) junction and collector base (C-B) junction both are forward biased then the BJT is said to be operating in saturation region.

3.3 (d)

In saturation region:
Collector-base (C-B) junction $\rightarrow$ Forward bias
Emitter-base (E-B) junction $\rightarrow$ Forward bias.

3.4 (c)

The process where the effective base width of the transistor is altered by varying the collector junction voltage is called base width modulation or early effect. In a transistor by changing the collector junction voltage, it is seen that the base width of the transistor is altered and this property is called as early effect.

3.5 Sol.

A-1, B-3, C-4

- As the channel doping is reduced then the pinch-off voltage decreases.
- $g_m = \dfrac{I_D}{V_{GS}}$

  So, the transconductance increases as drain current increases and the drain current increases as the conductivity of the channel is increased.
- If the channel length is reduced then the transit time of the carriers in the channel is reduced.

3.6 (b)

In a transistor when both the junction (J_C and J_E) are forward bias and if collector junction voltage is greater than emitter junction voltage then this transistor is in reverse saturation region.

3.7 (c)

$$I_e = I_o\left[e^{V_{be}/\eta V_T} - 1\right]$$

$$\frac{2I_1}{I_1} = \frac{I_o\left[e^{V_{be_2}/\eta V_T} - 1\right]}{I_o\left[e^{V_{be_1}/\eta V_T} - 1\right]}$$

$$\frac{2}{1} = \frac{e^{V_{be_2}/\eta V_T} - 1}{e^{V_{be_1}/\eta V_T} - 1}$$

$$e^{V_{be_2}/\eta V_T} >> 1$$

$$e^{V_{be_1}/\eta V_T} >> 1$$

$$\frac{e^{V_{be_2}/\eta V_T}}{e^{V_{be_1}/\eta V_T}} = 2$$

$$e^{(V_{be_2} - V_{be_1})\eta V_T} = 2$$

$$\frac{V_{be_2} - V_{be_1}}{\eta V_T} = ln2$$

$$V_{be_2} - V_{be_1} = \eta V_T(0.693)$$

Take $\eta = 1$

$$V_{be_2} - V_{be_1} = 1 \times 0.026 \times 0.693$$

$$V_{be_2} - V_{be_1} = 18 \text{ mV}$$

$$V_{be_2} - V_{be_1} \cong 20 \text{ V}$$

3.8 (b)

$$\boxed{g_m = \frac{2I_{DSS}}{|V_p|}\left[1 - \frac{V_{GS}}{V_p}\right]}$$

$$g_m = \frac{2 \times 2 \times 10^{-3}}{|-4|}\left[1 - \frac{(-2)}{(-4)}\right] = 0.5 \text{ mA/V}$$

3.9 (b)

Voltage controlled capacitor.

3.10 (b)

At the edge of saturation i.e. when drain to source voltage reaches $V_{D_{sat}}$ the inversion layer charge at the drain end becomes zero (ideally). The channel is said to be pinched off at the drain end. If the drain to source voltage V_{DS} is increased even further beyond the saturation edge so that $V_{DS} > V_{D_{sat}}$, an even larger portion of the channel becomes pinched off and effective channel length is reduced.

3.11 (d)

$$V_T = V_{To} + \gamma\sqrt{|-2\phi_F + V_{SB}|} - \sqrt{|2\phi_F|}$$

γ = substrate bias coefficient

V_{SB} = substrate bias voltage

3.12 (d)

$$I_D = K(V_{GS} - V_T)^2$$

$$1 \text{ mA} = K(900 - 400)^2$$

$$K = \frac{1 \text{mA}/\text{V}^2}{10^4 \times 25}$$

When, $V_{GS} = 1400$ V ; $I_D = K(V_{GS} - V_T)^2$

$$= \frac{1 \text{mA}}{25 \times 10^4 V^2}(1400 - 400)^2 V^2$$

$$I_D = \frac{1 \text{mA} \times 100}{25 \times 10^4} \times 10^4 = 4 \text{ mA}$$

3.14 (d)

$$\beta = \frac{I_C}{I_B} = \frac{\alpha}{1 - \alpha}$$

When base width increases, recombination in base region increases and α decreases & hence β decreases.

If doping in base region increases, then recombination in base increases and α decreases thereby decreasing β.

3.15 (d)

S_2: V_T of N-MOS is increased by adding extra p-type impurities i.e. V_T of a MOS increases by increase in substrate doping concentration. (True)

S_1: $C_{ox} = \frac{\epsilon_{ox}}{t_{ox}}$ t_{ox} = gate oxide thickeners

$$V_T = \phi_{GC} - 2\phi_F - \frac{Q_R}{C_{ox}} - \frac{Q_{ox}}{C_{ox}}$$

If C_{ox} increases $\frac{Q_B}{C_{ox}}$ and $\frac{Q_{ox}}{C_{ox}}$ decreases and V_T increases.

C_{ox} decreases when t_{ox} increases.

3.16 (d)

$$I_D = K(V_{GS} - V_T)^2$$

$$1 \text{ mA} = K(2 - 1)^2$$

$$K = 1 \text{ mA/V}^2$$

Again, $I_D = K(V_{GS} - V_T)^2 = 4$ mA

3.17 (a)

In accumulation mode for NMOS having p-substrate V_G is –ve. When negative V_G is applied to the gate electrode, the holes in the p-type substrate are attracted to the semiconductor oxide interface. This condition is called carrier accumulation on the surface.

3.19 (a)

$$C = \frac{\varepsilon A}{d}$$

$$d = \frac{\varepsilon A}{C} = \frac{3.5 \times 10^{-13} \times 1 \times 10^{-4} \times 10^{-4}}{7 \times 10^{-12} \times 10^{-2}}$$

$$= 5 \times 10^{-8} \text{ m} = 50 \text{ nm}$$

3.20 (b)

$$\epsilon_{si} = 10^{-12} \text{ F/cm}$$

$$= 10^{-10} \text{ F/m}$$

$$C_{min} = \frac{C_{max} \cdot C_d}{C_{max} + C_d}$$

$$\Rightarrow \quad C_d = \frac{C_{max} \cdot C_{min}}{C_{max} - C_{min}}$$

$$C_d = \frac{7 \times 1}{7-1} = 1.166 \text{ pF}$$

$$C_d = \frac{\in_{si} \cdot A}{w}$$

$$\Rightarrow \quad w = \frac{10^{-10} \times 10^{-4} \times 10^{-4}}{1.166 \times 10^{-12}} = 0.857 \, \mu\text{m}$$

3.22 (b)

$$g_m = \left.\frac{\partial i_d}{\partial V_{GS}}\right|_{V_{DS}} = \frac{\partial K(V_{GS} - V_T)^2}{\partial V_{GS}} = 2K(V_{GS} - V_T)$$

3.23 (*)

I_D is given by,

$$I_D = K_n\left[(V_{gs} - V_t)V_{DS} - \frac{V_{DS}^2}{2}\right] - \text{in linear region}$$

$$\text{and } g_m = \frac{\partial I_D}{\partial V_{gs}} \text{ at constant } V_{DS}$$

$$g_m = \frac{\partial I_D}{\partial V_{gs}} = K_n V_{DS}$$

$\therefore$ g_m is independent of V_{gs}.

3.24 (b)

$$g_m = g_{mo}\left[1 - \left(\frac{-V_G}{V_P}\right)^{\frac{1}{2}}\right]$$

where, $g_{mo} = \dfrac{2az}{\rho L}$

$a \rightarrow$ Half channel height

$L \rightarrow$ Length of channel

$z \rightarrow$ Depth of channel

$\rho \rightarrow$ Resistivity of channel

Since all the parameters are same except width 'W'. So,

$$W \propto a$$

$$\therefore \quad g_{mo} = \frac{2Wz}{\rho L}$$

$$g_m = g_{mo}\left[1 - \left(-\frac{V_G}{V_p}\right)^{1/2}\right]$$

$$g_{initial} = W\left[1 - \left(-\frac{V_{G_1}}{V_{P_1}}\right)^{1/2}\right] \quad \text{...(i)}$$

Now, the width is doubled. So,

$$g_{m\,modified} = 2W\left[1 - \left(-\frac{V_{G_1}}{4V_{P_1}}\right)^{1/2}\right] \quad \text{...(ii)}$$

$$\left[\because V_p = \frac{qN_o a^2}{2\in}\right]$$

$$\frac{g_{m\,modifies}}{g_{m\,initial}} = \frac{1}{2}\left[\frac{1 - \sqrt{\dfrac{2}{V_p}}}{1 - \sqrt{\dfrac{1}{2V_p}}}\right]$$

3.27 (b)

$$\gamma = \frac{1}{1 + \dfrac{N_B}{N_E}}$$

To achieve $\gamma = 1$

$$N_E >> N_B$$

3.28 (d)

$$I_C = \beta I_B + (1 + \beta)I_{CO}$$

$$\beta = \frac{\alpha}{1-\alpha} = \frac{0.98}{1-0.98} = 49$$

$$\therefore \quad I_C = 49 \times 20 + 50 \times 0.6$$

$$= 980 + 30 = 1010 \, \mu\text{A}$$

$$\text{or} \quad I_C = 1.01 \text{ mA}$$

3.29 (c)

Depletion layer width

$$w \propto \sqrt{V_{GS} + V_{bi}}$$

Given, $V_{bi} = -1$

and for, $V_{GS} = 0$ V, $w_1 = 1 \, \mu$m

Now for, $V_{GS} = -3$ V, w_2 can be calculated

$$\text{as} \quad \frac{w_1}{w_2} = \sqrt{\frac{0-1}{-3-1}} = \frac{1}{2}$$

$$\Rightarrow \quad w_2 = 2 \, \mu\text{m}$$

Channel width $\uparrow \Rightarrow$ channel resistance $\uparrow$

For 10 μm channel resistance = 600 Ω

$\therefore$ For 8 μm channel resistance

$$= \frac{600 \times 10}{8} = 750 \, \Omega$$

3.30 (c)

For $(10 - 2w_2 = 6 \, \mu\text{m})$ channel resistance

$$= \frac{600 \times 10}{6} = 1000 \, \Omega$$

3.31 (d)

$$n_i = 10^{10}/\text{cm}^3$$
$$A = 1\times10^{-12}\ \text{m}^2$$
$$d = 10^{-6}\ \text{m}$$
$$V = Ad = 10^{-18}\ \text{m}^3 = 10^{-12}\ \text{cm}^3$$
$$N_D = n = 10^{19}/\text{cm}^3$$
$$p = \frac{n_i^2}{N_D} = \frac{10^{20}}{10^{19}} = 10/\text{cm}^3$$

So, holes in volume V is

$H = pV = 10^{-11}$

Since, holes number can not be a decimal number so $H = 0$.

3.32 (a)

Source body junction capacitor,

$$C_j = \frac{\epsilon A}{d} = \frac{\epsilon_0 \epsilon_r A}{d}$$

here, $\epsilon_r = 11.7$ as there is Si only

$A = 1\ \mu\text{m}\times0.2\ \mu\text{m} = 0.2\times10^{-12}\ \text{m}^2$

d = depletion width of p-n junction

$d = 10\ \text{nm} = 10^{-8}\ \text{m}$

So,
$$C_j = \frac{8.9\times10^{-12}\times11.7\times0.2\times10^{-12}}{10^{-8}}$$
$$C_j = 2.082\times10^{-15}\ \text{F}$$
$$C_j \cong 2\ \text{fF}$$

3.33 (a)

Gate source capacitance C_g is

$$C_g = \frac{\epsilon_1 A_1}{d_1} = \frac{\epsilon_0 \epsilon_{r1} A_1}{d_1}$$

$\epsilon_{r1} = 3.9$ as between gate and source there is SiO_2

$$A_1 = 1\ \mu\text{m}\times\delta = 1\times10^{-6}\times20\times10^{-9} = 2\times10^{-14}\ \text{m}^2$$
$$d_1 = 1\ \text{nm} = 10^{-9}\ \text{m}$$

So,
$$C_g = \frac{8.9\times10^{-12}\times3.9\times2\times10^{-14}}{10^{-9}}$$
$$C_g = 69.42\times10^{-17}\ \text{F}$$
$$C_g = 0.69\times10^{-15} = 0.7\ \text{fF}$$

3.34 (5.76)

We know that,

$$g_m = \frac{I_C}{V_T}$$

where,
$$I_C = I_S e^{\frac{V_{BE}}{V_T}}$$

So,
$$I_C = 10^{-13}\times e^{\frac{0.7}{0.025}}$$
$$I_C = 0.144\ \text{mA}$$

Hence,
$$g_m = \frac{0.144}{0.025} = 5.76\ \text{mA/V}$$

3.35 (500)

Voltage controlled resistance r_{DS} is is given by

$$r_{DS} = \frac{1}{(\mu_n C_{ox})\left(\frac{W}{L}\right)(V_{GS}-V_t)}$$
$$r_{DS} = \frac{1}{800\times10^{-8}\times100\times(2+0.5)}$$
$$r_{DS} = 500\ \Omega$$

3.36 (0.0707)

We known that for linear region,

$$I_D = K_n\left[(V_{GS}-V_t)V_{DS} - \frac{V_{DS}^2}{2}\right]$$
$$\frac{\partial I_D}{\partial V_{GS}} = K_n V_{DS}$$
$$\frac{10^{-3}}{0.1} = K_n$$

i.e. $K_n = 10^{-2}$

and for saturation region

$$I_D = \frac{K_n}{2}(V_{GS}-V_t)^2$$
$$\sqrt{I_D} = \sqrt{\frac{K_n}{2}}(V_{GS}-V_t)$$
$$\frac{\partial\sqrt{I_D}}{\partial V_{GS}} = \sqrt{\frac{K_n}{2}}$$

So
$$\frac{\partial\sqrt{I_D}}{\partial V_{GS}} = \sqrt{\frac{10^{-2}}{2}} = 0.0707\frac{\sqrt{A}}{V}$$

3.37 (2.4)

Peak electric field in the oxide region is

$$E_{ox} = \frac{\epsilon_s}{\epsilon_{ox}}E_s$$

and
$$E_s = \frac{0.2\times2}{0.5}$$
$$E_s = 0.8\ \text{V}/\mu\text{m}$$

$\therefore$
$$E_{ox} = \frac{12}{4}\times0.8$$
$$E_{ox} = 2.4\ \text{V}/\mu\text{m}$$

3.38 (380.28)

I_C is given by

$$I_C = I_S \exp\left[\frac{V_{BE}}{V_T}\right]$$

$$I_{C_1} = I_{S_1} \exp\left[\frac{V_{BE_1}}{V_T}\right] \quad ...(i)$$

$$I_{C_2} = I_{S_2} \exp\left[\frac{V_{BE_2}}{V_T}\right] \quad ...(ii)$$

According to given condition,

$$I_{C_1} = I_{C_2}$$

$$I_{S_1} \exp\left[\frac{V_{BE_1}}{\eta V_T}\right] = I_{S_2} \exp\left[\frac{V_{BE_2}}{\eta V_T}\right]$$

$$V_{BE_1} - V_{BE_2} = V_T \, ln\left[\frac{I_{S_2}}{I_{S_1}}\right]$$

$$V_{BE_1} - V_{BE_2} = 26 \, ln\left[\frac{300 \times 300}{0.2 \times 0.2}\right]$$

$$V_{BE_1} - V_{BE_2} - 380.28 \text{ mV}$$

3.39 (a)

$$V_{GS} = V_{DS}$$

Hence MOS transistor is in saturation.

In saturation,

$$I_D = k(V_{GS} - V_T)^2 = k(V_{DD} - V_{SS} - V_T)^2$$

As, V_{SS} increases I_D decreases

(Not linearly because square factor)

Hence option (a) is correct.

3.40 (20)

Under channel length modulation,

$$I_D = I_{Dsat}(1 + \lambda V_{DS})$$

$$\frac{dI_D}{dV_{DS}} = \frac{1}{r_o} = \lambda I_{Dsat}$$

$$r_o = \frac{1}{\lambda I_{Dsat}} = \frac{1}{0.05 \times 10^{-3}} = 20 \text{ k}\Omega$$

3.41 (4.33)

$$\frac{C_{max}}{C_{min}} = \frac{\frac{\epsilon_{ox}}{t_{ox}}}{\frac{\frac{\epsilon_{ox}}{t_{ox}} \times \frac{\epsilon_S}{X_{dmax}}}{\frac{\epsilon_{ox}}{t_{ox}} + \frac{\epsilon_S}{X_{dmax}}}} = \left[1 + \frac{X_{dmax}}{t_{ox}} \times \frac{\epsilon_{ox}}{\epsilon_S}\right]$$

$$= \left[1 + \frac{100}{10} \times \frac{1}{3}\right] = 4.33$$

3.42 (0.02)

$$\lambda = \frac{\Delta I_D}{I_D \Delta V_{DS}} = \frac{0.02 \text{ mA}}{1 \text{mA} \times 1 \text{V}} = 0.02$$

3.43 (d)

W_B doubled (increased) → early effect is still present but its effect less severe relative to previous W_B. Slope I_C versus V_{CE} decreases.

3.44 (1475)

$$I_C = I_s e^{V_{BE}/V_T}$$

$$I_E = \frac{\beta + 1}{\beta} I_C = \frac{\beta + 1}{\beta} I_s \cdot e^{V_{BE}/V_T}$$

I_E will be maximum when β is 50.

$$= (1.02)(10^{-9} \times 10^{-6}) e^{\frac{700 \times 10^{-3}}{25 \times 10^{-3}}}$$

$$= 1475 \text{ μA}$$

3.46 (28.35)

In the saturation region,

$$g_d = \lambda I_{DS}$$

$$= \lambda\left[\frac{1}{2}\mu_n C_{ox} \frac{W}{L}(V_{GS} - V_T)^2\right]$$

$$= 0.09\left[\frac{1}{2} \times 70 \times 10^{-6} \times 4(1.8 - 0.3)^2\right]$$

$$g_d = 28.35 \text{ μs}$$

3.47 (d)

$$r_{ds} = \frac{1}{\mu_n C_{ox} \frac{W}{L}(V_{GS} - V_T)}$$

r_{ds} = channel resistance

$W \uparrow r \downarrow A \rightarrow$ correct

$V_T \downarrow r \downarrow B \rightarrow$ correct

$L \uparrow r \uparrow C \rightarrow$ correct

$V_{GS} \uparrow r \downarrow \therefore D \rightarrow$ Wrong statement

3.48 (b)

In a MOS – Capacitor

$$(V_{G1} - V_T) \propto Q$$

$$\frac{(V_{G1} - V_T)}{(V_{G2} - V_T)} = \frac{Q_1}{Q_2}$$

$$\frac{0.8 - V_T}{1.3 - V_T} = \frac{2 \times 10^{11}}{4 \times 10^{11}}$$

On solving, $V_T = 0.3 \text{ V}$

Now Consider, $\dfrac{V_{G2}-V_T}{V_{G3}-V_T} = \dfrac{Q_2}{Q_3}$

$$\frac{1.3-0.3}{1.8-0.3} = \frac{4\times10^{11}}{Q_3}$$

Inverse charge density with $V_G = 1.8$ V

$$Q_3 = 6\times10^{11}/\text{cm}^2$$

3.49 (1.2)

Since $V_{GS} > V_{DS}$, MOSFET is in linear operation

$$I_D = K'_N[V_{GS}-V_T]V_{DS}$$

$$\frac{dI_D}{dV_{GS}} = K'_N V_{DS}$$

$$g_m = K'_N V_{DS}$$

$$0.5\times10^{-6} = K'_N[50\times10^{-3}] \quad \therefore K'_N = 10^{-5}$$

$$I_D = K'_N[V_{GS}-V_T]V_{DS}$$

$$\frac{dI_D}{dV_{DS}} = K'_N[V_{GS}-V_T]$$

$$g_d = K'_N[V_{GS}-V_T]$$

$$8\times10^{-6} = 10^{-5}[2-V_T]$$

On solving $V_T = 1.2$ V

3.50 (a)

Substrate is *N*-type SC.

3.51 (1.6)

From the figure I.

Both the capacitors C_1 and C_2 are in series

Total Capacitance, $C = \dfrac{C_1C_2}{C_1+C_2}$

$$C = \frac{\frac{A\varepsilon_1}{t_1}\cdot\frac{A\varepsilon_2}{t_2}}{\frac{A\varepsilon_1}{t_1}+\frac{A\varepsilon_2}{t_2}} = \frac{\frac{A\times4}{10^{-9}}\cdot\frac{A\times20}{3\times10^{-9}}}{\frac{A\times4}{10^{-9}}+\frac{A\times20}{3\times10^{-9}}}$$

$$C = 2.5\times10^9\times A$$

From figure II

$$C = \frac{\varepsilon A}{t_{eq}}$$

Since both capacitors must be equal in figures I and II

$$2.5\times10^9\times A = \frac{A\times4}{t_{eq}}$$

$$\therefore \quad t_{eq} = 1.6\text{ nm}$$

3.52 (6.656)

$$I_C = AqD_n\frac{dn}{dx} \quad \text{where } D_n = \mu_n V_T$$

$$= 0.001(1.6\times10^{-19})800\times26\times10^{-3}\left[\frac{10^{14}}{0.5\times10^{-4}}\right]$$

$$= 66.56\times10^{-4}\text{ A} = 6.656\text{ mA}$$

3.53 (c)

Emitter-base junction (J_E) is in RB

Collector-base junction (J_C) is in FB

Hence, inverse active mode.

3.54 (c)

When a BJT is in active region, as the reverse bias voltage across collector-base junction increased, the width of depletion region increases, which results in decrease of effective base width. This decrease in effective base width reduces the recombinations in base region, hence, common-emitter current gain will increase.

3.55 (0.5)

Given that,

$$V_{GS} = 0.7\text{ V},\ V_{TH} = 0.3\text{ V},\ V_{DS} = 0.1\text{ V}$$

$$V_{GS}-V_{TH} = 0.4 > V_{DS}$$

$\Rightarrow$ MOSFET is in linear region.

In linear region,

$$I_D = K_n\left[2(V_{GS}-V_{TH})V_{DS}-V_{DS}^2\right]$$

Transconductance,

$$g_m = \frac{\partial I_D}{\partial V_{GS}} = 2K_nV_{DS}$$

$$K_n = \frac{K'_n}{2}\left(\frac{W}{L}\right) = \frac{\mu_nC_{ox}}{2}\left(\frac{W}{L}\right)$$

So, $\quad g_m = 2K_nV_{DS} = \mu_nC_{ox}\left(\dfrac{W}{L}\right)V_{DS}$

$= 100\times50\times0.1\ \mu$A/V $= 0.5$ mA/V

3.56 (b)

$$W_2 = 2W,$$

$$V_{GS2}-V_{T2} = (V_{GS1}-V_{T1})$$

$$\frac{I_{D2}}{I_{D1}} = \frac{W_2}{W_1}\times\left(\frac{V_{GS2}-V_{T2}}{V_{GS1}-V_{T1}}\right)^2 = 2\times4 = 8$$

$$I_{D2} = 8I_{D1}$$

$$\frac{g_{m2}}{g_{m1}} = \frac{W_2}{W_1}\times\frac{V_{GS2}-V_{T2}}{V_{GS1}-V_{T1}} = 2\times2 = 4$$

$$g_{m2} = 4g_{m1}$$

3.57 (6.903)

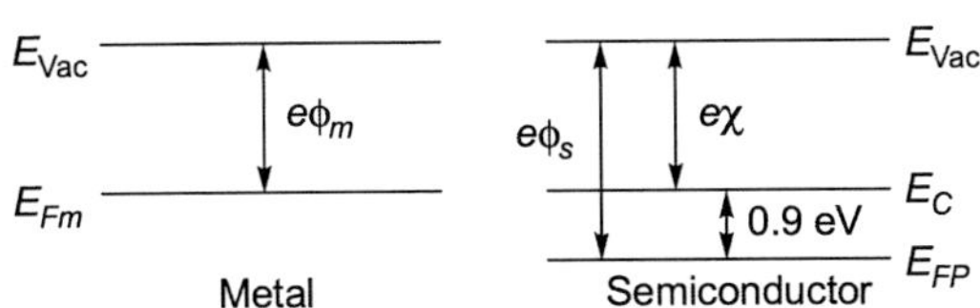

Isolated case

$$e\phi_m = 4.1 \text{ eV}$$
$$e\chi = 4 \text{ eV}$$
$$E_C - E_{FP} = 0.9 \text{ eV}$$
$$e\phi_s = e\chi + 0.9 = 4.9 \text{ eV}$$
$$\phi_m = 4.1 \text{ V}, \quad \phi_s = 4.9 \text{ V}$$
$$\phi_{ms} = \phi_m - \phi_s = 4.1 - 4.9 = -0.8 \text{ V}$$
$$C_{ox} = \frac{\epsilon_{ox}}{t_{ox}} = \frac{3.9 \times 8.85 \times 10^{-14}}{0.1 \times 10^{-4}}$$
$$= 34.515 \times 10^{-9} \text{ F/cm}^2$$
$$V_{FB} = \phi_{ms} - \frac{\phi'_{ox}}{C_{ox}}$$
$$\frac{\phi'_{ox}}{C_{ox}} = -0.8 + 1 = 0.2 \text{ V}$$
$$\phi'_{ox} = 0.2 \times C_{ox}$$
$$= 0.2 \times 34.515 \times 10^{-9} \text{ C/cm}^2$$
$$= 6.903 \text{ nC/cm}^2$$

3.58 (a)

Since $\phi_{ms} = 0$, the MOS-capacitor is ideal.

Point P Represents accumulation

Point Q Represents flat band

Point R Represents Inversion

3.59 (25.52)

$$I_{D(\text{sat})} = \frac{1}{2}\mu_n C_{ox} \frac{W}{L}(V_{GS} - V_T)^2$$
$$= \frac{1}{2} \times 800 \times 3.45 \times 10^{-7} \times \frac{10}{1}(5 - 0.7)^2 \text{ A}$$
$$= 25.5162 \text{ mA} \simeq 25.52 \text{ mA}$$

3.60 (d)

MOS capacitance

$$\phi_m = 3.87, \quad \phi_s = 4.8, \quad \phi_{ms} = -0.93$$
$$V_T = \phi_{ms} - \frac{Q_{ox}}{C_{ox}} - \frac{Q_d}{C_{ox}} + 2\phi_{Fp}$$
$$\phi_{Fp} = E_i - E_F = 0.5 - 0.2 = 0.3$$
$$-0.16 = -0.93 - 0 - \frac{Q_d}{C_{ox}} + 2 \times 0.3$$
$$\frac{Q_d}{C_{ox}} = 0.6 + 0.16 - 0.93 = -0.17$$
$$Q_b = -0.17 \times C_{ox}$$
$$= -0.17 \times 100 \times 10^{-9}$$
$$= -1.7 \times 10^{-8} \text{ C/cm}^2$$

3.61 (*)

$$\frac{\beta_1}{\beta_2} = \frac{\int_0^W N_{A_2}(x)dx}{\int_0^W N_{A_1}(x)dx}$$
$$= \frac{W \times 10^{17}}{\frac{1}{2} \times W \times (10^{17} - 10^{14})} = \frac{2 \times 10^{17}}{10^{17} + 10^{14}} \simeq 2$$
$$\beta_2 = 0.5\beta_1$$

Hence no option is matching.

3.62 (d)

Given, N channel MOSFET

$$t_{ox} = 10 \text{ nm} = 10 \times 10^{-7} \text{ cm}$$
$$\frac{\partial V_T}{\partial |V_{BS}|} = 50 \text{ mV/V}, \quad |V_{BS}| = 2 \text{ V}$$
$$q = 1.6 \times 10^{-19} \text{ C}, \quad |V_{BS}| >> 2\phi_B$$

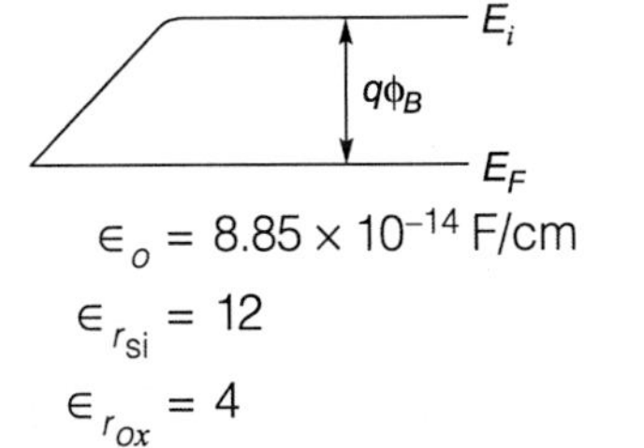

$$\epsilon_o = 8.85 \times 10^{-14} \text{ F/cm}$$
$$\epsilon_{r_{si}} = 12$$
$$\epsilon_{r_{ox}} = 4$$

Threshold voltage, including body effect,

$$V_T = \phi_{ms} + \frac{\sqrt{2\epsilon_{si} qN_A(2\phi_B - V_{BS})}}{C_{ox}} + 2\phi_B$$

In question, we need,

$$|V_{BS}| = |V_{SB}|$$
$$\Rightarrow V_T = \phi_{ms} + \frac{\sqrt{2\epsilon_{si} qN_A(2\phi_B + |V_{SB}|)}}{C_{ox}} + 2\phi_B$$
$$\therefore \quad \frac{\partial V_T}{\partial |V_{BS}|} = 0 + \frac{\sqrt{2\epsilon_{si} qN_A}}{C_{ox}} \cdot \frac{1}{2\sqrt{2\phi_B + |V_{SB}|}} + 0$$
$$50 \times 10^{-3} = \frac{\sqrt{2 \times 8.85 \times 10^{-14} \times 12 \times 1.6 \times 10^{-19} N_A}}{\epsilon_{ox}/t_{ox}} \cdot \frac{1}{2\sqrt{|V_{SB}|}}$$

$$\Rightarrow \left(\frac{50\times10^{-3}\times4\times8.85\times10^{-14}}{10\times10^{-7}}\right)^2 = \frac{2\times8.85\times10^{-14}\times12\times1.6\times10^{-19}}{4\times2}\times N_A$$

$$[\because |V_{SB}| >> 2\phi_B]$$

$$\Rightarrow \quad N_A = 7.375 \times 10^{15}\ \text{cm}^{-3}$$

3.63 (5.6)

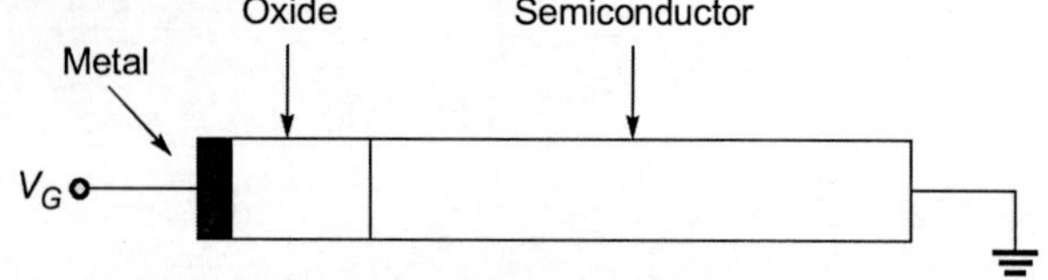

Given: For $V_G = 2$ V,

Inversion charge density, $Q_{IN} = 2.2\ \mu C/cm^2$

The inversion layer charge density is given as

$$Q_{IN} = C_{ox}(V_{GS} - V_T)$$

$$\therefore \quad V_T = V_G - \frac{Q_{IN}}{C_{ox}}$$

For $V_G = 2$ V, substituting the given values, we get,

$$V_T = 2 - \frac{2.2\times10^{-6}}{1.7\times10^{-6}} = 0.706\ \text{V}$$

At $V_G = 4$ V,

$$Q_{IN} = C_{ox}(V_G - V_T)$$

$$Q_{IN} = 1.7 \times 10^{-6} \times (4 - 0.706)$$

$$Q_{IN} = 5.5998 \times 10^{-6}\ \text{C/cm}^2$$

$$Q_{IN} \approx 5.6\ \mu\text{C/cm}^2$$

3.64 (b)

$\because$ We know $V_G < V_{FB}$ then accumulation mode.

$\therefore$ In accumulation mode $\boxed{W_d = 0}$ because there is no depletion charge.

Now, $V_{FB} < V_G < V_T \Rightarrow$ then depletion and inversion mode.

$\therefore$ Depletion width is available.

$\therefore$ $V_G > V_T \Rightarrow$ Strong inversion.

$\therefore$ Depletion width $W_d \Rightarrow$ Constant.

And $W_d = \sqrt{\dfrac{2\epsilon|\phi_S|}{qN_s}}$ and $|\phi_S| \propto V_G$

But after strong inversion, W_d remains constant.

$\therefore$ Correction option is (b).

3.65 (4.32)

The flat-band voltage of a MOS capacitor is defined as:

$$V_{FB} = \phi_{ms} - \frac{Q'_{ss}}{C_{ox}} \quad \text{...(i)}$$

where ϕ_{ms} = work function difference between the metal and the semiconductor

Q'_{ss} = Trapped oxide charge

C_{ox} = Capacitance of the oxide

It is given that: $V_{FB} = 0$ V, $\epsilon_{r(ox)} = 4$, $t_{ox} = 100$ nm, $\epsilon_0 = 8.85 \times 10^{-14}$ F/cm, $Q'_{ss} = 10^{-8}$ C/cm^2

$$\therefore \quad C_{ox} = \frac{\epsilon_{ox}}{t_{ox}} = \frac{\epsilon_{r(ox)}\cdot\epsilon_0}{t_{ox}}$$

$$= \frac{4\times8.85\times10^{-14}}{100\times10^{-7}} = 35.4 \times 10^{-9}\ \text{F/cm}^2$$

Substituting the values in equation (i),

$$0 = \phi_{ms} - \frac{10^{-8}}{35.4\times10^{-9}}$$

$$\Rightarrow \quad \phi_{ms} = \frac{10}{35.4}$$

$$\Rightarrow \quad \phi_m - \phi_s = 0.28\ \text{V}$$

$$\Rightarrow \quad q\phi_m - q\phi_s = 0.28\ \text{eV}$$

Given: Metal work function, $q\phi_m = 4.6$ eV

$$\therefore \quad q\phi_s = 4.6 - 0.28 = 4.32\ \text{eV}$$

Work function of p-type silicon = 4.32 eV

3.66 (a, b, d)

(i) Under high level injection, the recombination of the charge carriers in the base region increases resulting in decrease in current gain i.e. β will decrease.

(ii) We know that,

$$\beta = \frac{I_C}{I_B}$$

In saturation region, the collector saturates at I_{Csat} and does not increase with increase in I_B resulting in a lower value of β compared to active region.

(ii) Under low level injection, β increase with the increase in the magnitude of collector current.

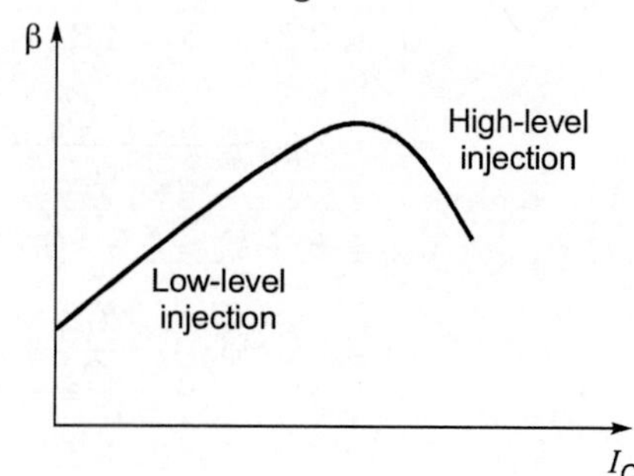

(iv) We have,

$$BV_{CEO} = BV_{CBO}\left(\frac{1}{\beta}\right)^{1/n}$$

Hence, a higher value of β lead to a lower collector to emitter breakdown voltage.

Therefore, options (a), (b) and (d) are correct

3.67 (52.5)

For linear region of MOS,

$$I_D = \mu_n C_{ox}\left(\frac{W}{L}\right)\left[(V_{GS}-V_T)V_{DS}-\frac{1}{2}V_{DS}^2\right]$$

$$5\times 10^{-6} = 50\times 10^{-6}\left[(V_{GS}-V_T)0.1-\frac{1}{2}(0.1)^2\right]$$

$$\frac{1}{10} = (V_{GS}-V_T)0.1-\frac{1}{2}\times 0.1^2$$

$$0.1 + 0.5(0.1)^2 = (V_{GS}-V_T)0.1$$

$$1 + 0.5(0.1) = (V_{GS}-V_T)$$

$$1.05 \text{ volt} = V_{GS}-V_T$$

Keeping V_{GS} constant, V_{DS} is increased to

$$V_{DS} = 1.5 \text{ V}$$

Since, $V_{DS} > V_{GS} - V_T$

$\Rightarrow$ MOSFET is in saturation

$$g_{m(sat)} = \mu_n C_{ox}\left(\frac{W}{L}\right)(V_{GS}-V_T)$$

$$= 50\times 10^{-6}\times 1.05$$

$$= 52.5\times 10^{-6} \text{ A/V} = 52.5\ \mu\text{A/V}$$

■■■

4 IC Fabrication

4.1 If P is Passivation, Q is n-well implant, R is metallization and S is source/drain diffusion, then the order in which they are carried out in a standard n-well CMOS fabrication process, is

(a) P-Q-R-S (b) Q-S-R-P
(c) R-P-S-Q (d) S-R-Q-P

[2003 : 2 M]

4.2 A silicon wafer has 100 nm of oxide on it and is inserted in a furnace at a temperature above 1000° C for further oxidation in dry oxygen. The oxidation rate

(a) is independent of current oxide thickness and temperature
(b) is independent of current oxide thickness but depends on temperature
(c) slows down as the oxide grows
(d) is zero as the existing oxide prevents further oxidation [2008 : 1 M]

4.3 Thin gate oxide in a CMOS process is preferably grown using

(a) wet oxidation (b) dry oxidation
(c) epitaxial deposition (d) ion implantation

[2010 : 1 M]

4.4 In IC technology, dry oxidation (using dry oxygen) as compared to wet oxidation (using stem or water vapour) produces

(a) superior quality oxide with a higher growth rate
(b) inferior quality oxide with a higher growth rate
(c) inferior quality oxide with a lower growth rate
(d) superior quality oxide with a lower growth rate

[2013 : 1 M]

4.5 There are two photolithography systems: one with light source of wavelength λ_1 = 156 nm (System 1) and another with light source of wavelength λ_2 = 325 nm (System 2). Both photolithography systems are otherwise identical. If the minimum feature sizes that can be realized using System1 and System2 are L_{min1} and L_{min2} respectively, the ratio L_{min1}/L_{min2} (correct to two decimal places) is _______. [2018 : 1 M]

4.6 In a uniformly doped BJT, assume that N_E, N_B and N_C are the emitter, base and collector dopings in atoms/cm^3, respectively. If the emitter injection efficiency of the BJT is close to unity, which one of the following conditions is TRUE?

(a) $N_E = N_B = N_C$
(b) $N_E >> N_B$ and $N_B > N_C$
(c) $N_E = N_B$ and $N_B < N_C$
(d) $N_E < N_B < N_C$ [2010 : 2 M]

4.7 If fixed positive charges are present in the gate oxide of an n-channel enhancement type MOSFET, it will lead to

(a) a decrease in the threshold voltage
(b) channel length modulation
(c) an increase in substrate leakage current
(d) an increase in accumulation capacitance

[2014 : 1 M, Set-1]

4.8 In CMOS technology, shallow P-well or N-well regions can be formed using

(a) low pressure chemical vapour deposition
(b) low energy sputtering
(c) low temperature dry oxidation
(d) low energy ion-implantation

[2014 : 1 M, Set-2]

4.9 In MOSFET fabrication, the channel length is defined during the process of

(a) isolation oxide growth
(b) channel stop implantation
(c) poly-silicon gate patterning
(d) lithography step leading to the contact pads

[2014 : 1 M, Set-3]

4.10 Which one of the following processes is preferred to form the gate dielectric (SiO_2) of MOSFETs?

(a) Sputtering
(b) Molecular beam epitaxy
(c) Wet oxidation
(d) Dry oxidation [2015 : 1 M, Set-3]

4.11 The correct circuit representation of the structure shown in the figure is

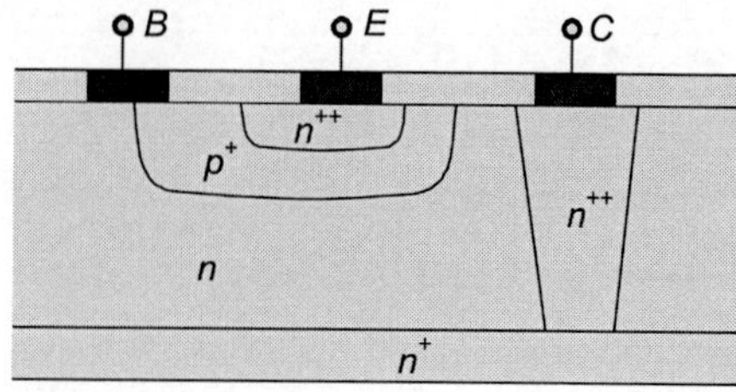

(a) (b)

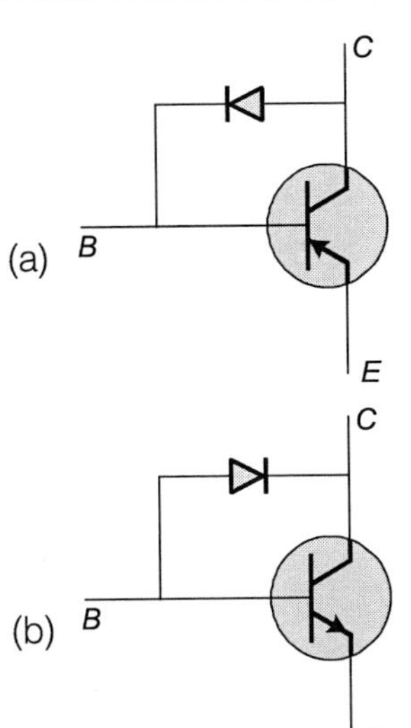

(c)

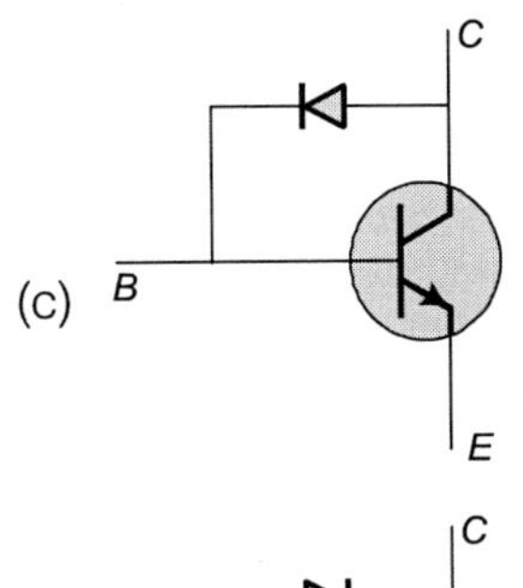

(d)

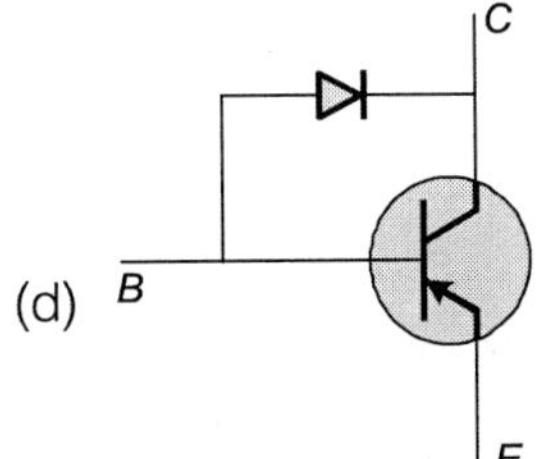

[2019 : 1 M]

■■■■

Answers IC Fabrication

4.1	(b)	4.2	(c)	4.3	(b)	4.4	(d)	4.5	(0.48)	4.6	(b)	4.7	(a)
4.8	(d)	4.9	(c)	4.10	(d)	4.11	(b)						

Explanations IC Fabrication

4.1 (b)

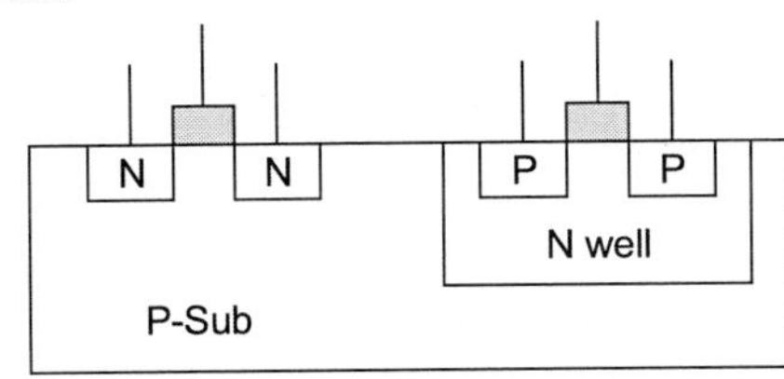

N Well CMOS fabrication process

Q : Well implant

S : Source/drain diffusion

R : Metallisation

P : Passivation

4.2 (c)

$$\text{Growth rate, } \frac{dt_{ox}}{dt} = \frac{B}{A + 2t_{ox}}$$

where A and B are oxidation constants.

as tox increases $\frac{dt_{ox}}{dt}$ decreases.

4.3 (b)

Dry oxidation is preferred for gate oxides.

4.4 (d)

Dry oxidation has better quality over wet oxidation.
Dry oxidation is slower than wet oxidation.

4.5 (0.48)

$$L_{min} \propto \lambda$$

$$\frac{L_{min1}}{L_{min2}} = \frac{\lambda_1}{\lambda_2} = \frac{156 \text{ nm}}{325 \text{ nm}} = 0.48$$

4.7 (a)

Fixed charges reduces threshold voltage.

4.8 (d)

Ion implantation/diffusion is used for well implantation.

4.9 (c)

Channel length is defined during the polysilicon gate pattering.

4.10 (d)

Dry oxidation is preferred for gate oxides.

4.11 (b)

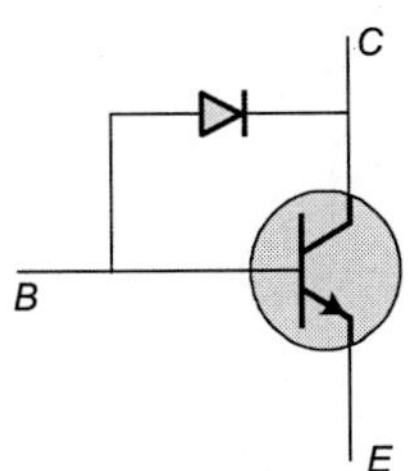

■■■■

Analog Circuits

CONTENTS

Analog Circuits

Syllabus

Diode circuits : Clipping, clamping and rectifiers.

BJT and MOSFET amplifiers : Biasing, ac coupling, small signal analysis, frequency response.

Current mirrors and differential amplifiers.

Op-amp circuits : Amplifiers, summers, differentiators, integrators, active filters, Schmitt triggers and oscillators.

Analysis of Previous GATE Papers

Exam Year	1 Mark Ques.	2 Marks Ques.	Total Marks
1995	10	–	**10**
1996	2	5	**12**
1997	2	2	**6**
1998	13	–	**13**
1999	4	4	**12**
2000	8	2	**12**
2001	3	5	**13**
2002	4	5	**14**
2003	5	8	**21**
2004	4	6	**16**
2005	3	11	**25**
2006	2	7	**16**
2007	2	8	**18**
2008	1	8	**17**
2009	–	6	**12**
2010	3	3	**9**
2011	2	4	**10**
2012	2	3	**8**
2013	3	6	**15**

Exam Year	1 Mark Ques.	2 Mark Ques.	Total Marks
2014 Set-1	4	3	**10**
2014 Set-2	4	3	**10**
2014 Set-3	3	3	**9**
2014 Set-4	3	3	**9**
2015 Set-1	2	3	**8**
2015 Set-2	3	4	**11**
2015 Set-3	3	3	**9**
2016 Set-1	4	2	**8**
2016 Set-2	4	3	**10**
2016 Set-3	3	4	**11**
2017 Set-1	3	4	**11**
2017 Set-2	4	3	**10**
2018	3	4	**11**
2019	1	5	**11**
2020	3	5	**13**
2021	3	3	**9**
2022	3	5	**13**
2023	2	3	**8**
2024	4	3	**10**

1 Operational Amplifiers

1.1 In the given circuit figure, if the voltage inputs V_- and V_+ are to be Amplified by the same Amplification factor, the value of 'R' should be

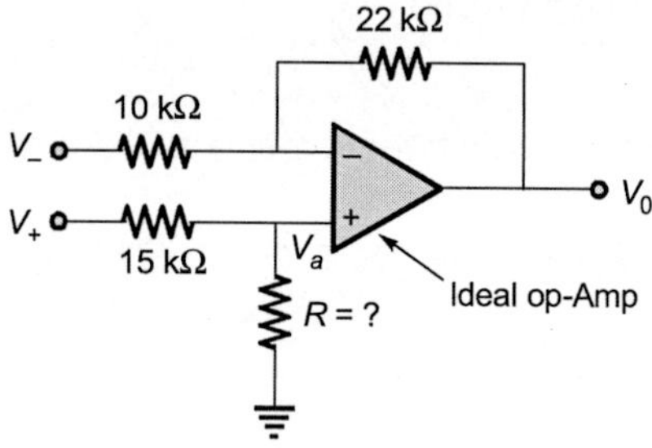

[1995 : 1 M]

1.2 An op-Amp is used as a zero-crossing detector. If maximum output available from the Op-Amp is $\pm 12\ V_{p-p}$ and the slew rate of the Op-Amp is 12 V/μ sec then the maximum frequency of the input signal that can be applied without causing a reduction in the $P-P$ output is

[1995 : 1 M]

1.3 The circuit shown in the figure is that of

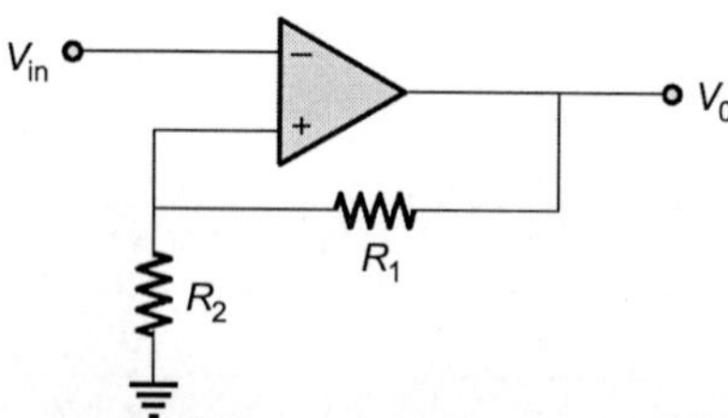

(a) a non-inverting Amplifiers
(b) an inverting Amplifier
(c) an oscillator
(d) a Schmitt trigger [1996 : 1 M]

1.4 The output voltage V_o of the circuit shown in the figure is

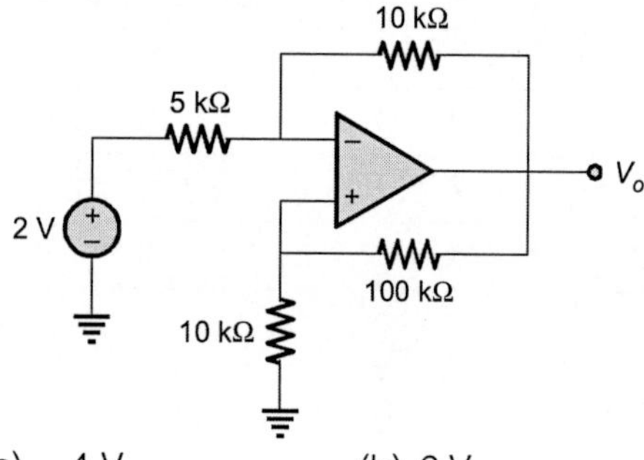

(a) – 4 V (b) 6 V
(c) 5 V (d) –5.5 V [1997 : 2 M]

1.5 One input terminal of high gain comparator circuit is connected to ground and a sinusoidal voltage is applied to the other input. The output of comparator will be
(a) a sinusoid
(b) a full rectified sinusoid
(c) a half rectified sinusoid
(d) a square wave [1998 : 1 M]

1.6 The first dominant pole encountered in the frequency response of a compensated op-amp is approximately at
(a) 5 Hz (b) 10 kHz
(c) 1 MHz (d) 100 MHz

[1999 : 1 M]

1.7 In the circuit of the figure, V_o is

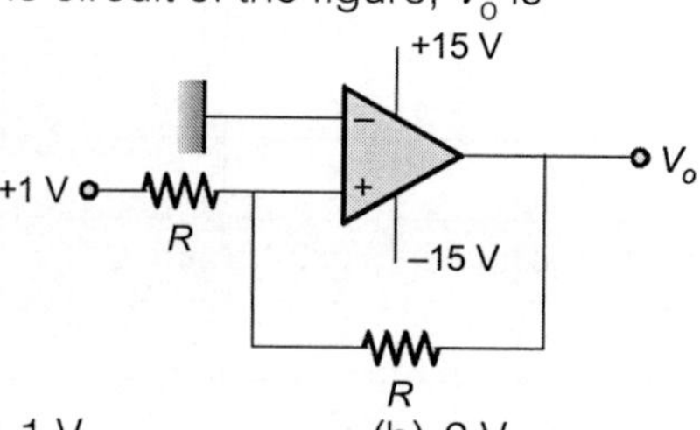

(a) –1 V (b) 2 V
(c) +1 V (d) +15 V [2000 : 1 M]

1.8 If the op-amp in the figure, is ideal, then V_o is

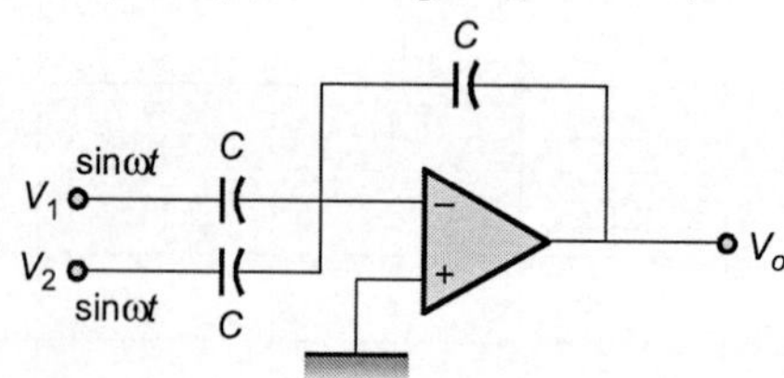

(a) Zero (b) $(V_1 - V_2) \sin \omega t$
(c) $-(V_1 + V_2) \sin \omega t$ (d) $(V_1 + V_2) \sin \omega t$

[2000 : 1 M]

1.9 Assume that the op-amp of the figure is ideal. If V_i is a triangular wave, then V_0 will be

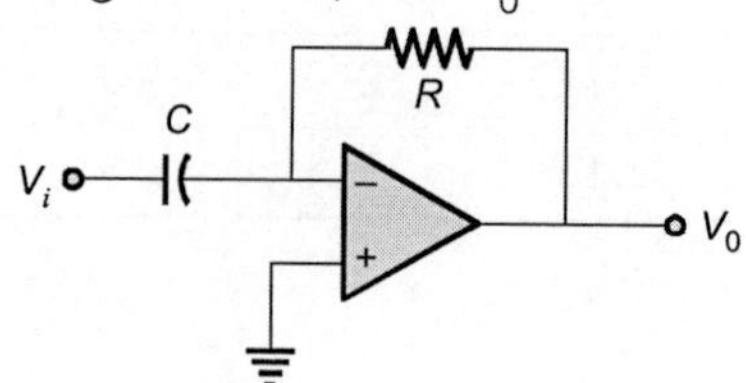

(a) square wave (b) triangular wave
(c) parabolic wave (d) sine wave

[2000 : 1 M]

1.10 The most commonly used amplifier in sample and hold circuits is
(a) a unity gain inverting amplifier
(b) a unity gain non-inverting amplifier
(c) an inverting amplifier with a gain of 10
(d) an inverting amplifier with a gain of 100
[2000 : 1 M]

1.11 If the op-amp in the figure has an input offset voltage of 5 mV and an open-loop voltage gain of 10,000, then V_0 will be

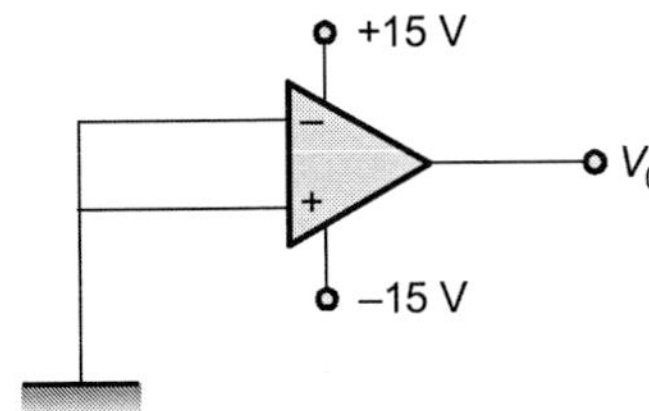

(a) 0 V (b) 5 mV
(c) +15 V or −15 V (d) +50 V or −50 V
[2000 : 2 M]

1.12 The ideal OP-AMP has the following characteristics.
(a) $R_i = \infty, A = \infty, R_0 = 0$
(b) $R_i = 0, A = \infty, R_0 = 0$
(c) $R_i = \infty, A = \infty, R_0 = \infty$
(d) $R_i = 0, A = \infty, R_0 = \infty$ [2001 : 1 M]

1.13 The inverting OP-AMP shown in the figure has an open-loop gain of 100. The closed-loop gain V_0/V_s is

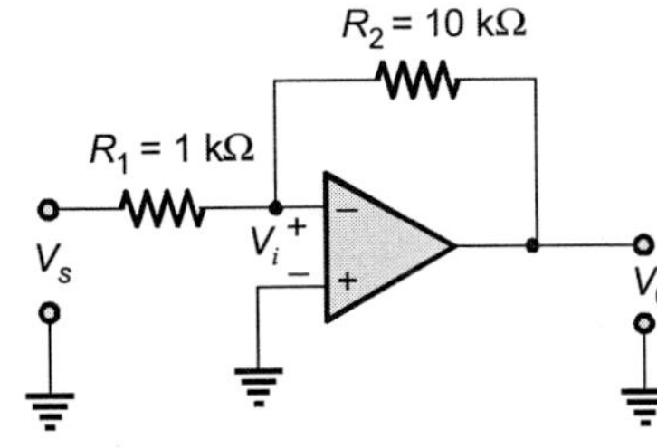

(a) −8 (b) −9
(c) −10 (d) −11 [2001 : 2 M]

1.14 In the figure assume the Op-Amps to be ideal. The output V_0 of the circuit is:

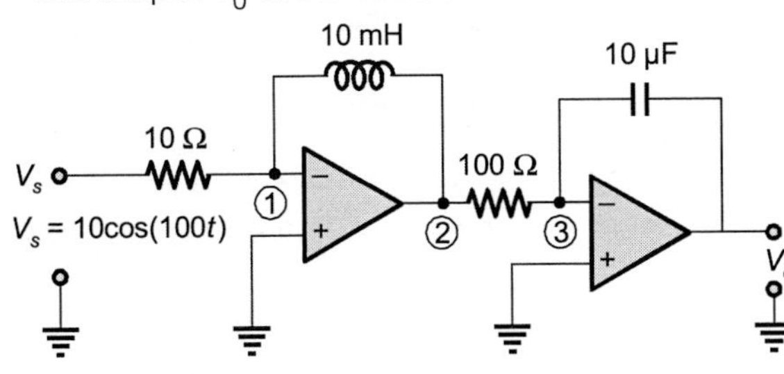

(a) $10\cos(100t)$ (b) $10\int_0^t \cos(100\tau)d\tau$
(c) $10^{-4}\int_0^t \cos(100\tau)d\tau$ (d) $10^{-4}\frac{d}{dt}\cos(100t)$
[2001 : 2 M]

1.15 A 741-type op-amp has a gain-bandwidth product of 1 MHz. A non-inverting amplifier using this op-amp and having a voltage gain of 20 dB will exhibit a −3 dB bandwidth of
(a) 50 kHz (b) 100 kHz
(c) $\frac{1000}{17}$ kHz (d) $\frac{1000}{7.07}$ kHz
[2002 : 1 M]

1.16 An amplifier using an op-amp with a slew-rate SR = 1 V/μsec has a gain of 40 dB. If this amplifier has to faithfully amplify sinusoidal signals from dc to 20 kHz without introducing any slew-rate induced distortion, then the input signal level must not exceed.
(a) 795 mV (b) 395 mV
(c) 79.5 mV (d) 39.5 mV [2002 : 2 M]

1.17 If the input to the ideal comparator shown in the figure is a sinusoidal signal of 8 V (peak to peak) without any DC component, then the output of the comparator has a duty cycle of
(a) 1/2
(b) 1/3
(c) 1/6
(d) 1/12

V_{ref} = 2 V — (−); Input — (+); Output

[2003 : 1 M]

1.18 If the differential voltage gain and the common mode voltage gain of a differential amplifier are 48 dB and 2 dB respectively, then its common mode rejection ratio is
(a) 23 dB (b) 25 dB
(c) 46 dB (d) 50 dB [2003 : 1 M]

1.19 If the op-amp in the figure is ideal, the output voltage V_{out} will be equal to

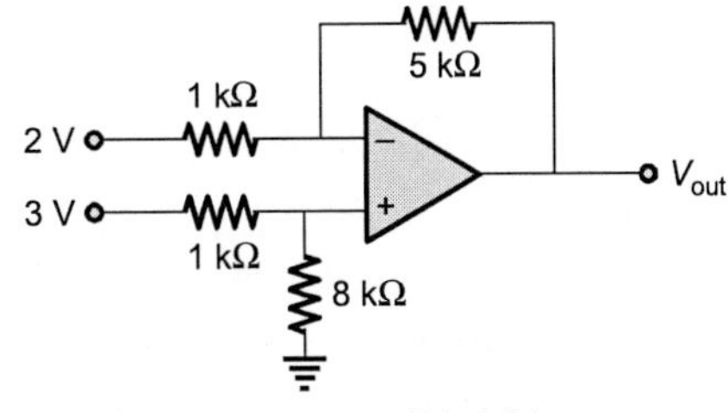

(a) 1 V (b) 6 V
(c) 14 V (d) 17 V

[2003 : 2 M]

1.20 Three identical amplifiers with each one having a voltage gain of 50, input resistance of 1 kΩ and output resistance of 250 Ω, are cascaded. The open circuit voltage gain of the combined amplifier is

(a) 49 dB (b) 51 dB
(c) 98 dB (d) 102 dB

[2003 : 2 M]

1.21 An ideal *op*-amp is an ideal

(a) voltage controlled current source
(b) voltage controlled voltage source
(c) current controlled current source
(d) current controlled voltage source

[2004 : 1 M]

1.22 The circuit in the figure is

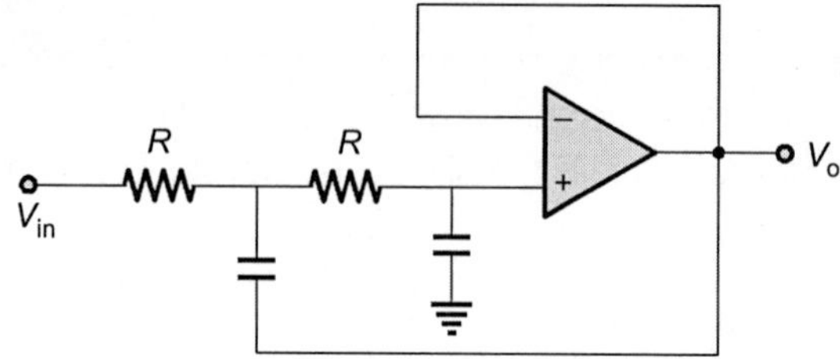

(a) low-pass filter (b) high-pass filter
(c) band-pass filter (d) band-reject filter

[2004 : 1 M]

1.23 In the op-amp circuit given in the figure, the load current i_L is

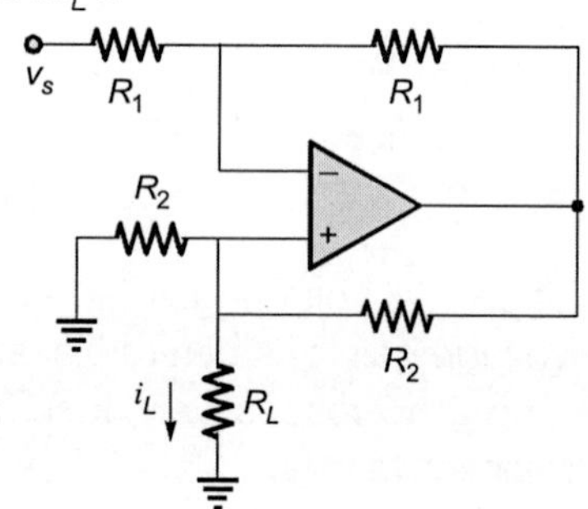

(a) $-\frac{V_s}{R_2}$ (b) $\frac{V_s}{R_2}$

(c) $-\frac{V_s}{R_L}$ (d) $\frac{V_s}{R_1}$

[2004 : 2 M]

1.24 The input resistance R_i of the amplifier shown in the figure is

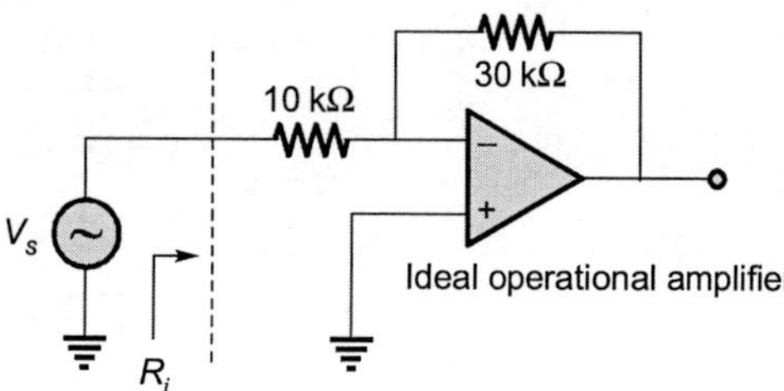

(a) $\frac{30}{4}$ kΩ (b) 10 kΩ
(c) 40 kΩ (d) infinite [2005 : 1 M]

1.25 The voltage e_0 indicated in the figure has been measured by an ideal voltmeter. Which of the following can be calculated?

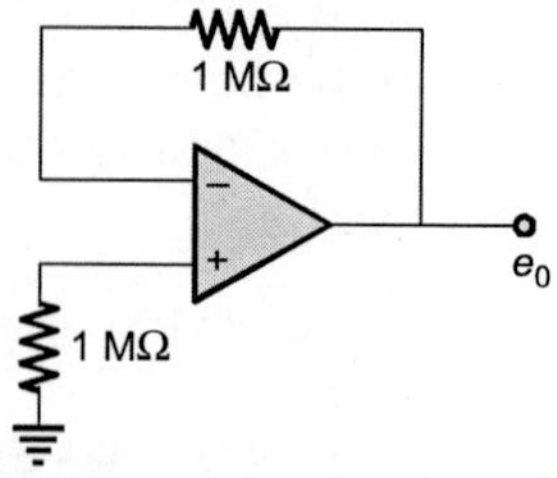

(a) Bias current of the inverting input only
(b) Bias current of the inverting and non-inverting inputs only
(c) Input offset current only
(d) Both the bias currents and the input offset current

[2005 : 2 M]

1.26 The OP-amp circuit shown in the figure is a filter. The type of filter and its cut-off frequency are respectively

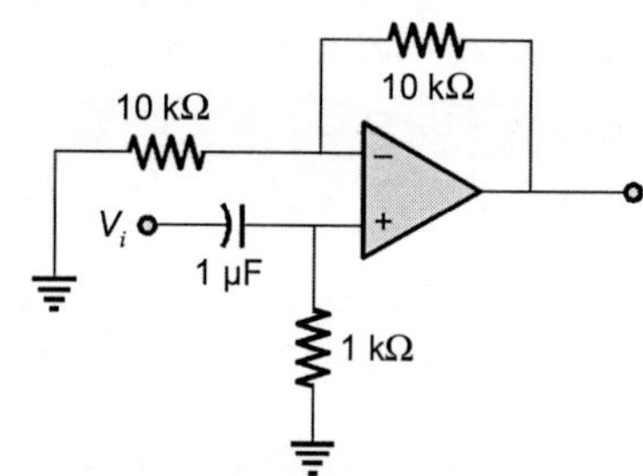

(a) high pass, 1000 rad/sec.
(b) low pass, 1000 rad/sec.
(c) high pass, 10000 rad/sec.
(d) low pass, 10000 rad/sec. [2005 : 2 M]

1.27 For the circuit shown in the following figure, the capacitor C is initially uncharged. At $t = 0$, the switch S is closed. The voltage V_C across the capacitor at $t = 1$ millisecond is

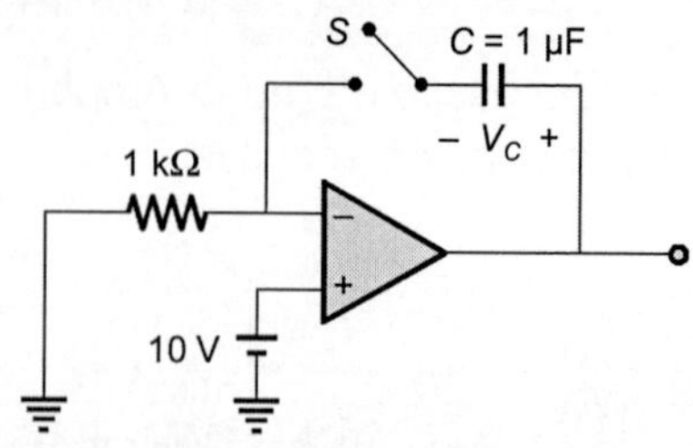

In the figure shown above, the Op-Amp is supplied with ±15 V.

(a) 0 Volt (b) 6.3 Volts
(c) 9.45 Volts (d) 10 Volts

[2006 : 2 M]

1.28 For the Op-Amp circuit shown in the figure, V_0 is

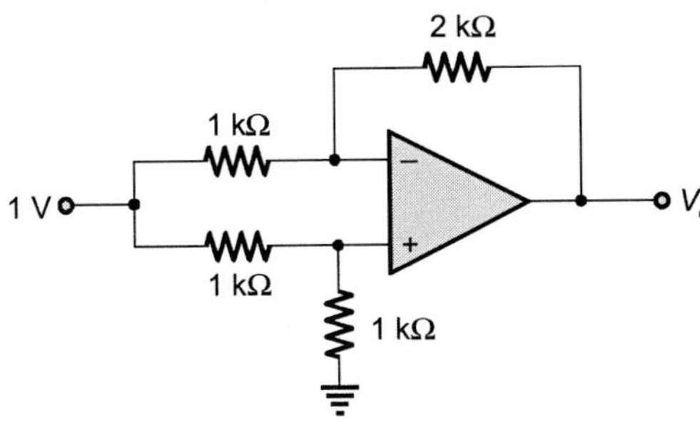

(a) –2 V (b) –1 V
(c) –0.5 V (d) 0.5 V

[2007 : 2 M]

1.29 In the OP-Amp circuit shown, assume that the diode current follows the equation $I = I_s \exp(V/V_T)$. For $V_i = 2$ V, $V_0 = V_{01}$, and for $V_i = 4$ V, $V_0 = V_{02}$. The relationship between V_{01} and V_{02} is

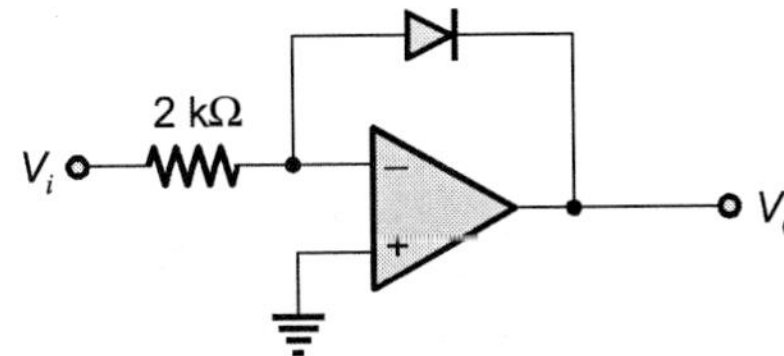

(a) $V_{02} = \sqrt{2}\ V_{01}$ (b) $V_{02} = e^2\ V_{01}$
(c) $V_{02} = V_{01}\ ln\ 2$ (d) $V_{01} - V_{02} = V_T\ ln\ 2$

[2007 : 2 M]

Linked Answer Questions 1.30 and 1.31.

Consider the Op-Amp circuit shown in the figure.

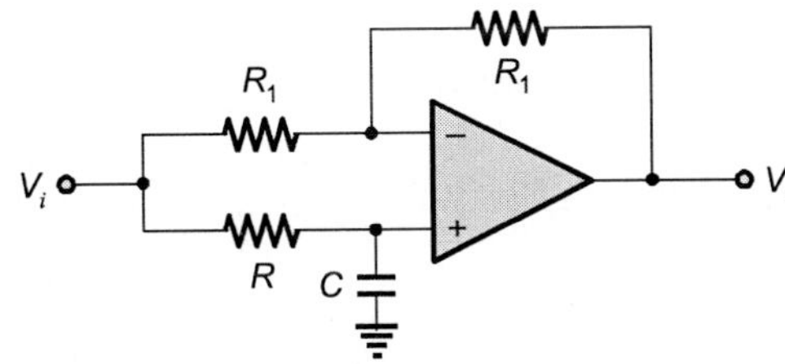

1.30 The transfer function $V_0(s)/V_i(s)$ is

(a) $\frac{1-sRC}{1+sRC}$ (b) $\frac{1+sRC}{1-sRC}$

(c) $\frac{1}{1-sRC}$ (d) $\frac{1}{1+sRC}$

[2007 : 2 M]

1.31 If $V_i = V_1 \sin(\omega t)$ and $V_o = V_2 \sin(\omega t + \phi)$, then the minimum and maximum values of ϕ (in radians) are respectively

(a) $-\pi/2$ and $\pi/2$ (b) 0 and $\pi/2$
(c) $-\pi$ and 0 (d) $-\pi/2$ and 0

[2007 : 2 M]

1.32 Consider the following circuit using an ideal OP-Amp. The I-V characteristics of the diode is described by the relation $I = I_0\left(e^{\frac{V}{V_T}} - 1\right)$ where $V_T = 25$ mV, $I_0 = 1\ \mu$A and V is the voltage across the diode (taken as positive for forward bias).

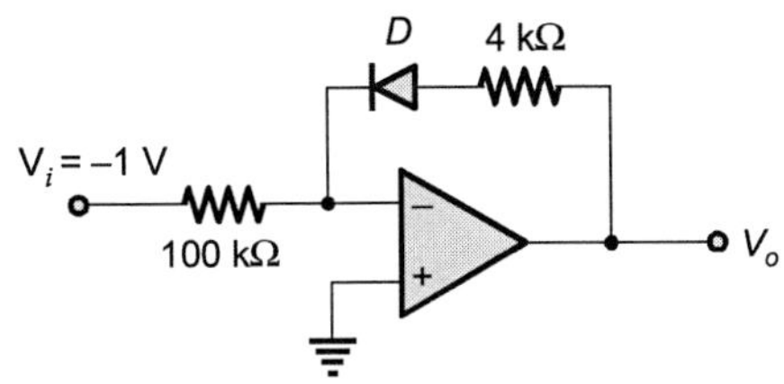

For an input voltage $V_i = -1$ V, the output voltage V_0 is

(a) 0 V (b) 0.1 V
(c) 0.7 V (d) 1.1 V [2008 : 2 M]

1.33 The OP-Amp circuit shown below represents a

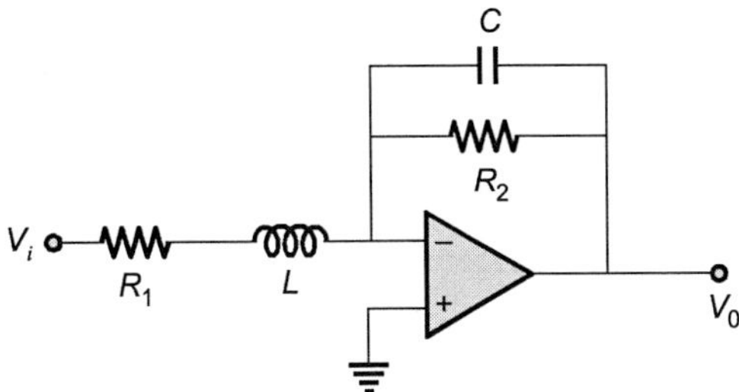

(a) high pass filter (b) low pass filter
(c) band pass filter (d) band reject filter

[2008 : 2 M]

1.34 In the circuit shown below, the op-amp is ideal, the transistor has $V_{BE} = 0.6$ V and $\beta = 150$. Decide whether the feedback in the circuit is positive or negative and determine the voltage V at the output of the op-amp.

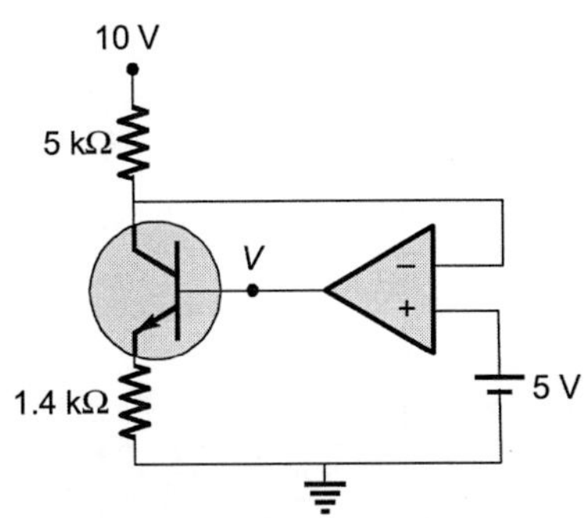

(a) Positive feedback, $V = 10$ V
(b) Positive feedback, $V = 0$ V
(c) Negative feedback, $V = 5$ V
(d) Negative feedback, $V = 2$ V

[2009 : 2 M]

1.35 Assuming the OP-AMP to be ideal, the voltage gain of the amplifier shown below is

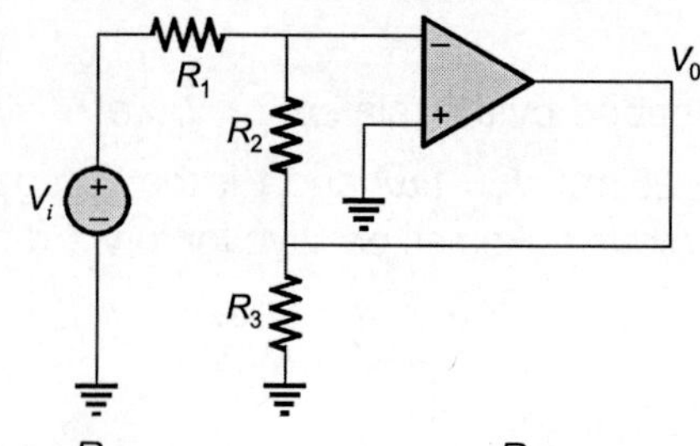

(a) $-\frac{R_2}{R_1}$ (b) $-\frac{R_3}{R_1}$

(c) $-\left(\frac{R_2 \| R_3}{R_1}\right)$ (d) $-\left(\frac{R_2 + R_3}{R_1}\right)$

[2010 : 1 M]

1.36 The transfer characteristic for the precision rectifier circuit shown below is (assume ideal OP-AMP and practical diodes)

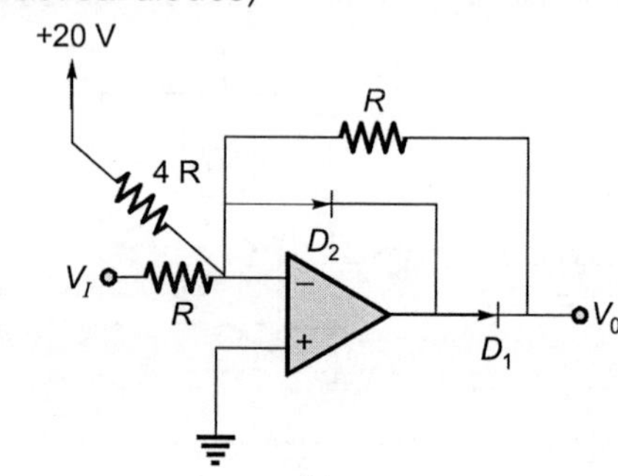

(a)

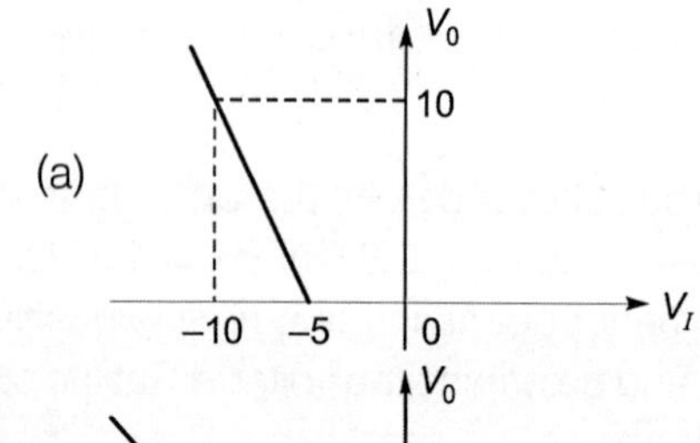

(b)

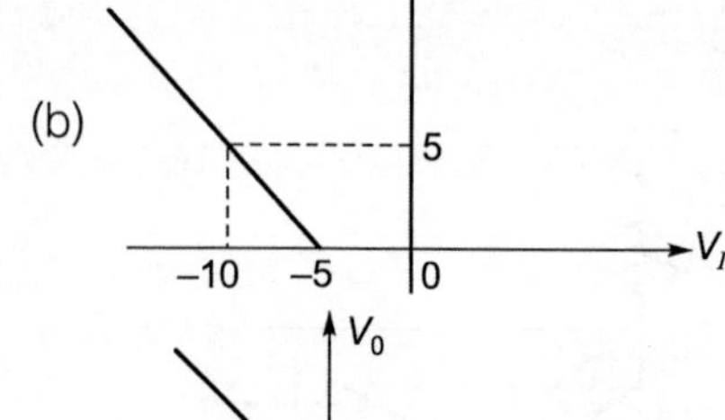

(c)

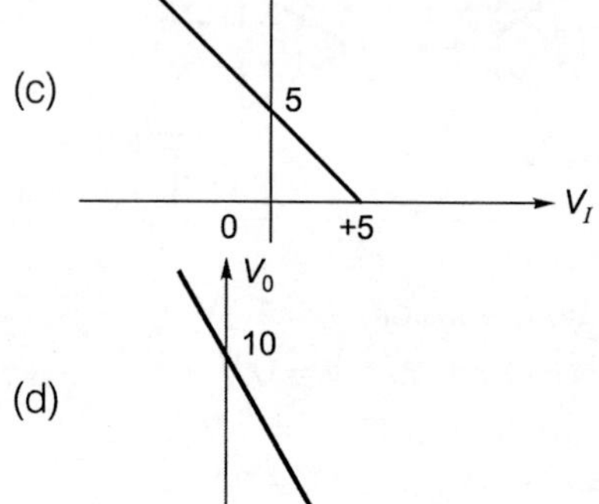

(d)

[2010 : 2 M]

1.37 The circuit below implement a filter between the input current i_i and output voltage v_0. Assume that the op-amp is ideal. The filter implemented is a

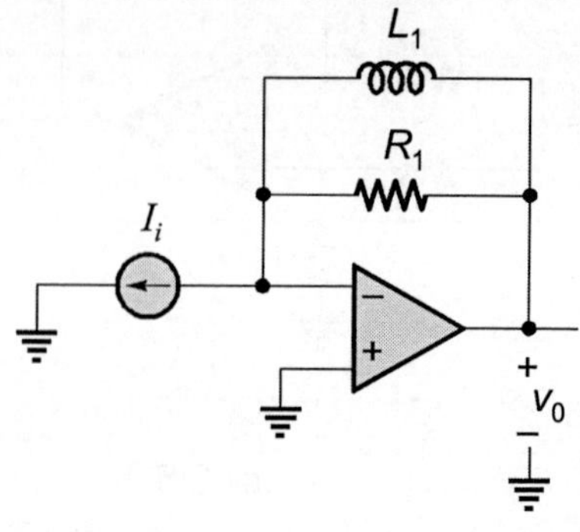

(a) low pass filter (b) band pass filter
(c) band stop filter (d) high pass filter

[2011 : 1 M]

1.38 The circuit shown is a

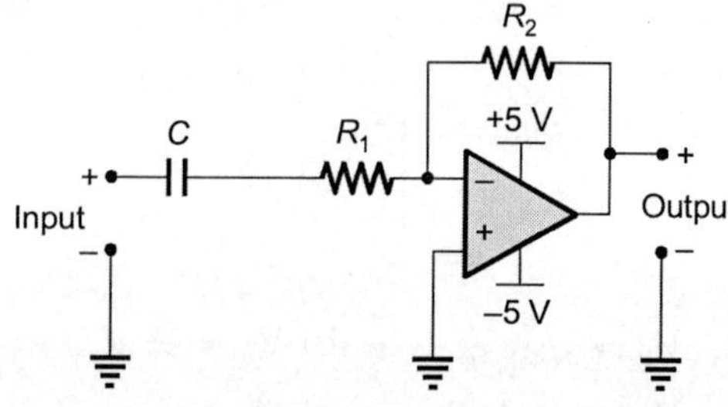

(a) low pass filter with $f_{3dB} = \frac{1}{(R_1 + R_2)C}$ rad/s

(b) high pass filter with $f_{3dB} = \frac{1}{R_1 C}$ rad/s

(c) low pass filter with $f_{3dB} = \frac{1}{R_1 C}$ rad/s

(d) high pass filter with $f_{3dB} = \frac{1}{(R_1 + R_2)C}$ rad/s

[2012 : 2 M]

1.39 In the circuit shown below what is the output voltage (V_{out}) if a silicon transistor Q and an ideal op-amp are used?

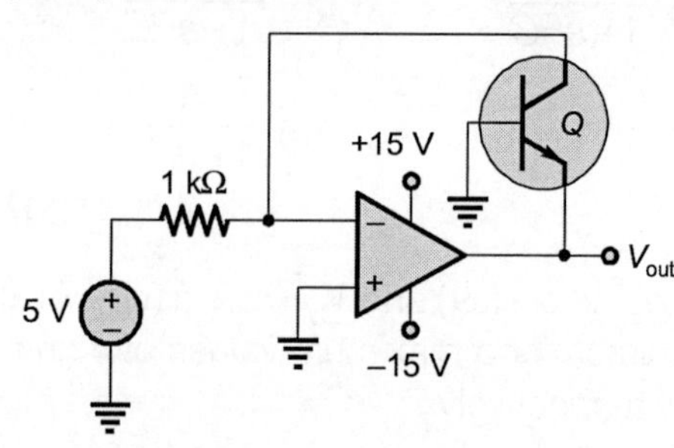

(a) –15 V (b) –0.7 V
(c) +0.7 V (d) +15 V

[2013 : 1 M]

1.40 In the circuit shown below the op-amps are ideal. Then V_{out} in volts is

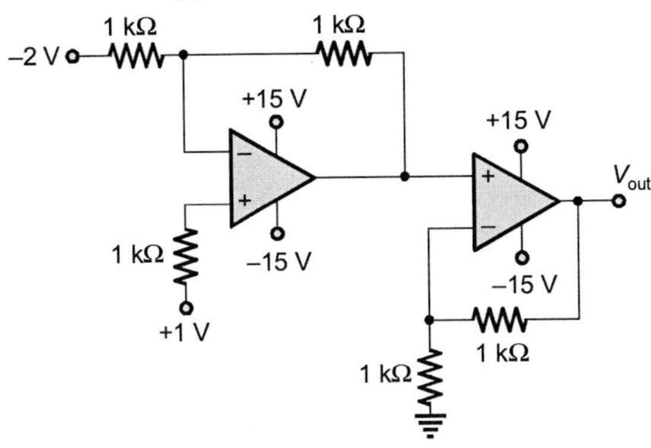

(a) 4 (b) 6
(c) 8 (d) 10 [2013 : 2 M]

1.41 In the low-pass filter shown in the figure, for a cut-off frequency of 5 kHz, the value of R_2 (in kΩ) is ____.

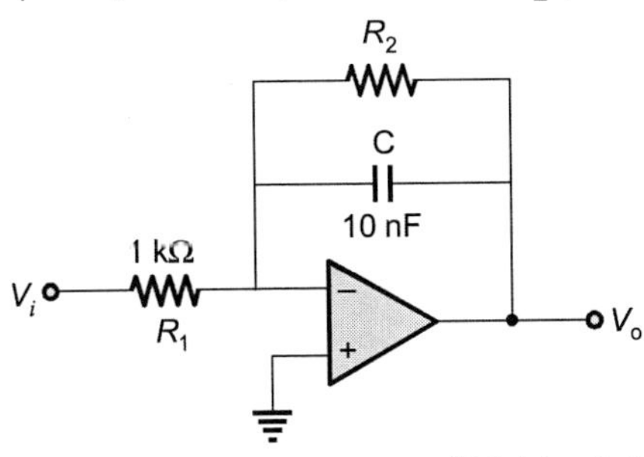

[2014 : 1 M, Set-1]

1.42 In the voltage regulator circuit shown in the figure, the op-amp is ideal. The BJT has V_{BE} = 0.7 V and β = 100, and the Zener voltage is 4.7 V. For a regulated output of 9 V, the value of R (in Ω) is _____.

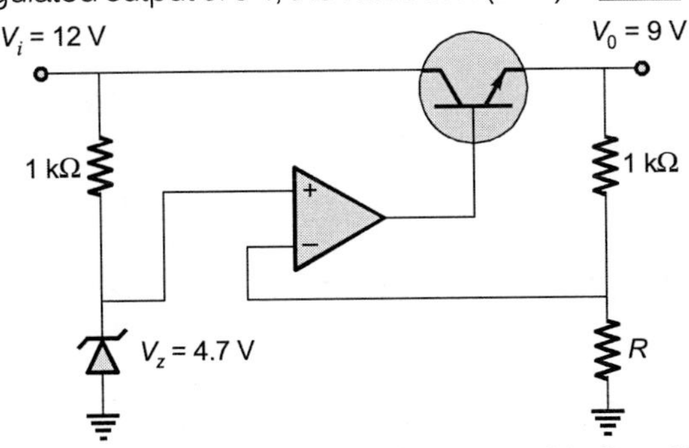

[2014 : 2 M, Set-1]

1.43 In the circuit shown, the op-amp has finite input impedance, infinite voltage gain and zero input offset voltage. The output voltage V_{out} is

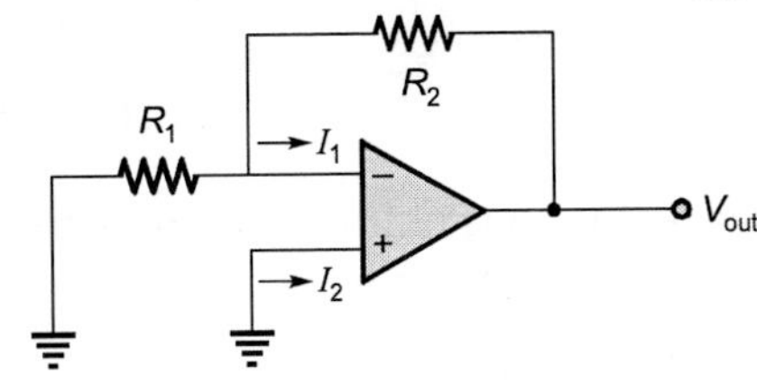

(a) $-I_2(R_1 + R_2)$ (b) $I_2 R_2$
(c) $I_1 R_2$ (d) $-I_1(R_1 + R_2)$
[2014 : 2 M, Set-1]

1.44 In the differential amplifier shown in the figure, the magnitudes of the common-mode and differential-mode gains are A_{cm} and A_d, respectively. If the resistance R_E is increased, then

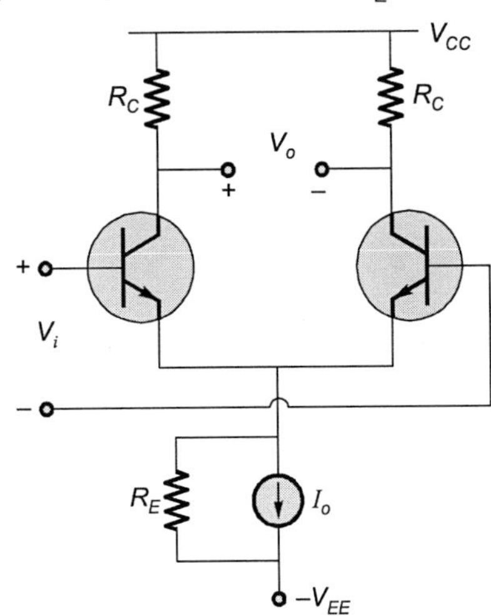

(a) A_{cm} increase
(b) common-mode rejection ratio increases
(c) A_d increase
(d) common-mode rejection ratio decreases
[2014 : 1 M, Set-2]

1.45 Assuming that the Op-amp in the circuit shown is ideal, V_o is given by

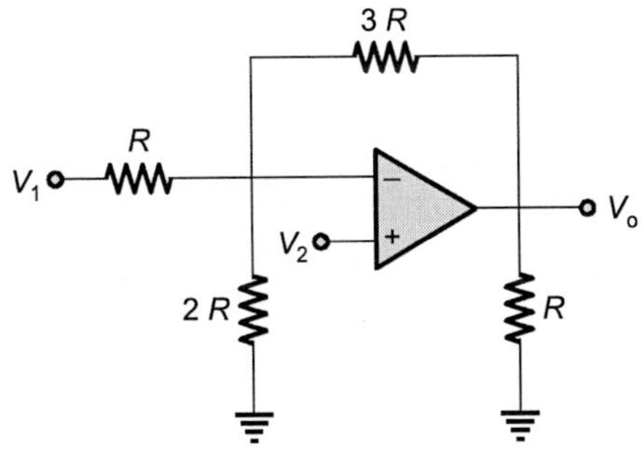

(a) $\frac{5}{2}V_1 - 3V_2$ (b) $2V_1 - \frac{5}{2}V_2$
(c) $-\frac{3}{2}V_1 + \frac{7}{2}V_2$ (d) $-3V_1 + \frac{11}{2}V_2$
[2014 : 2 M, Set-3]

1.46 The circuit shown represents

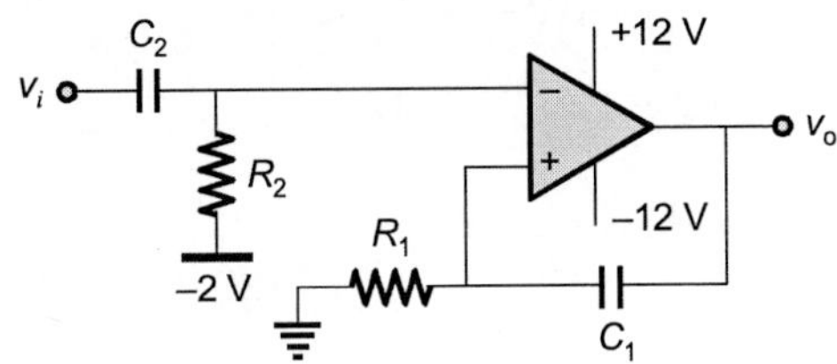

(a) a bandpass filter
(b) a voltage controlled oscillator
(c) an amplitude modulator
(d) a monostable multivibrator

[2014 : 1 M, Set-4]

1.47 In the circuit shown, assume that the op-amp is ideal. The bridge output voltage V_0 (in mV) for $\delta = 0.05$ is ________.

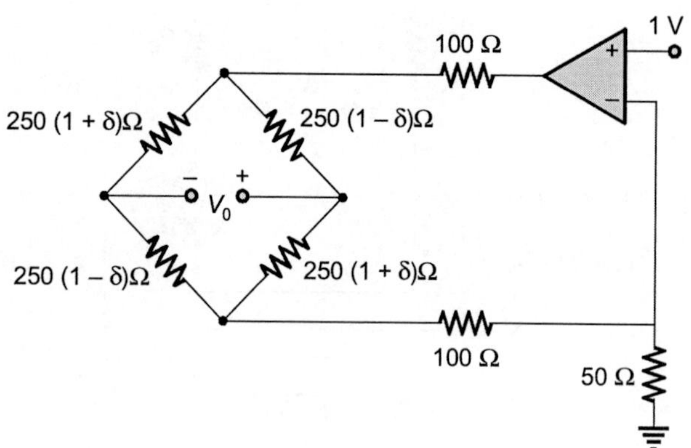

[2015 : 2 M, Set-1]

1.48 In the circuit shown, $V_0 = V_{0A}$ for switch SW in position A and $V_0 = V_{0B}$ for SW in position B. Assume that the op-amp is deal. The value of $\frac{V_{0B}}{V_{0A}}$ is ________.

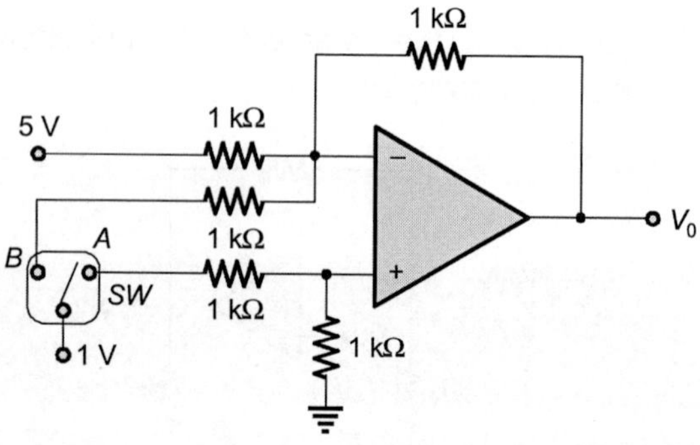

[2015 : 1 M, Set-2]

1.49 In the bistable circuit shown, the ideal op-amp has saturation levels of ± 5 V. The value of R_1 (in kΩ) that gives a hysteresis width of 500 mV is _____.

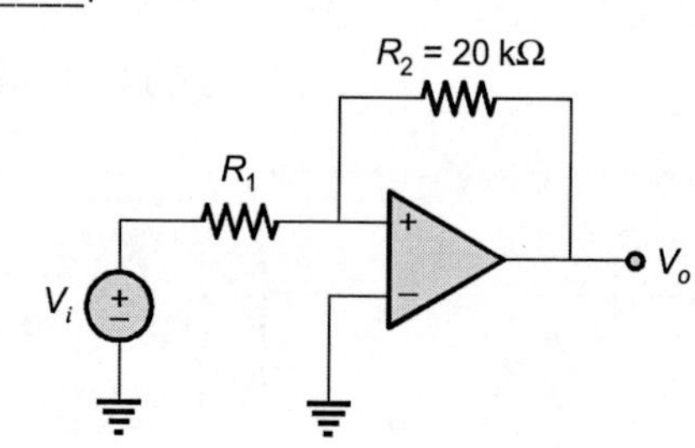

[2015 : 1 M, Set-2]

1.50 For the voltage regulator circuit shown, the input voltage (V_{in}) is 20 V ± 20% and the regulated output voltage (V_{out}) is 10 V. Assume the op-amp to be ideal. For a load R_L drawing 200 mA, the maximum power dissipation in Q_1 (in Watts) is_____.

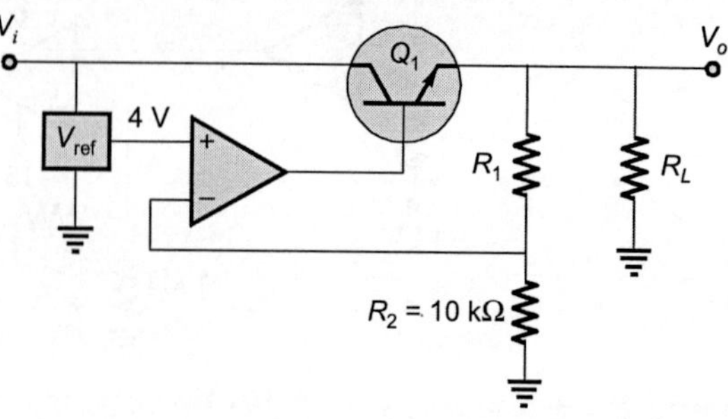

[2015 : 2 M, Set-2]

1.51 Assuming that the op-amp in the circuit shown below is deal, the output voltage V_0 (in volts) is ___.

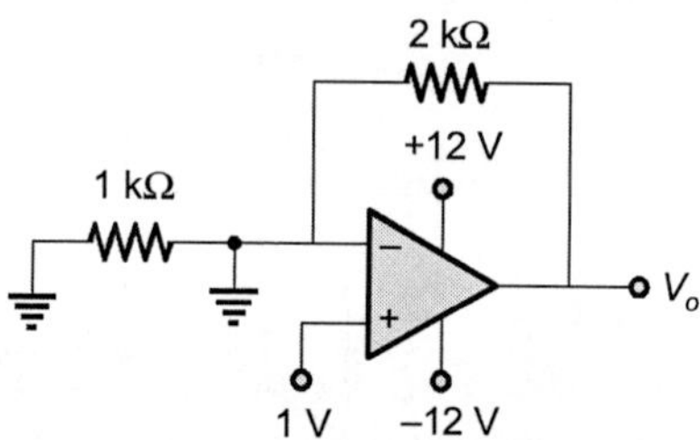

[2015 : 2 M, Set-2]

1.52 In the circuit shown using an ideal op-amp, the 3-dB cut-off frequency (in Hz) is ______.

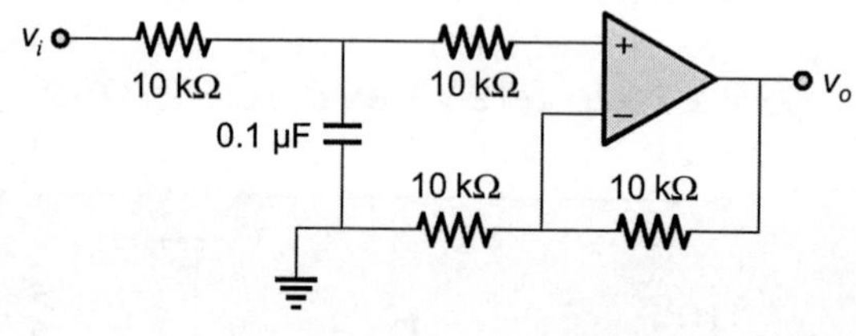

[2015 : 1 M, Set-3]

1.53 In the circuit shown, assume that the op-amp is ideal. If the gain (V_0/V_{in}) is –12, the value of R (in kΩ) is _____.

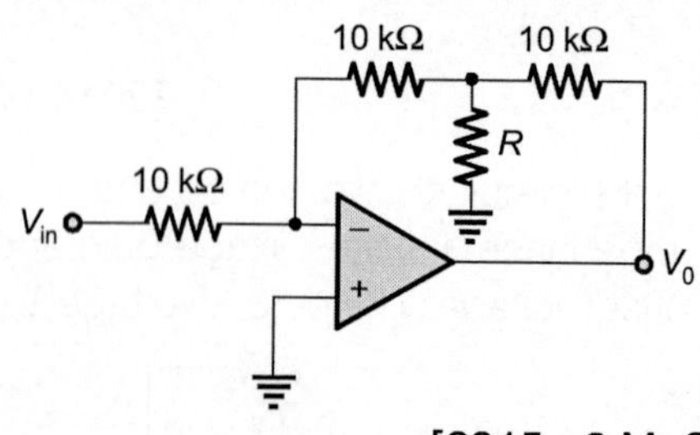

[2015 : 2 M, Set-3]

1.54 Consider the constant current source shown in the figure below. Let β represent the current gain of the transistor.

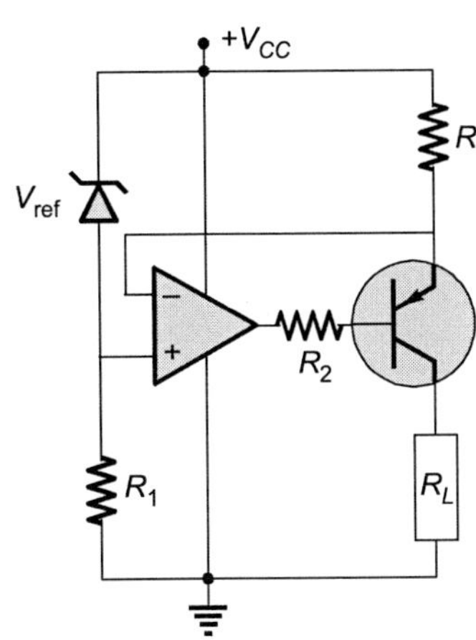

The load current I_0 through R_L is

(a) $I_0 = \left(\dfrac{\beta+1}{\beta}\right)\dfrac{V_{ref}}{R}$ (b) $I_0 = \left(\dfrac{\beta}{\beta+1}\right)\dfrac{V_{ref}}{R}$

(c) $I_0 = \left(\dfrac{\beta+1}{\beta}\right)\dfrac{V_{ref}}{2R}$ (d) $I_0 = \left(\dfrac{\beta}{\beta+1}\right)\dfrac{V_{ref}}{2R}$

[2016 : 1 M, Set-1]

1.55 The following signal V_i of peak voltage 8 V applied to the non-inverting terminal of an ideal op-amp. The transistor has V_{BE} = 0.7 V, β = 100; V_{LED} = 1.5 V, V_{CC} = 10 V and $-V_{CC}$ = −10 V

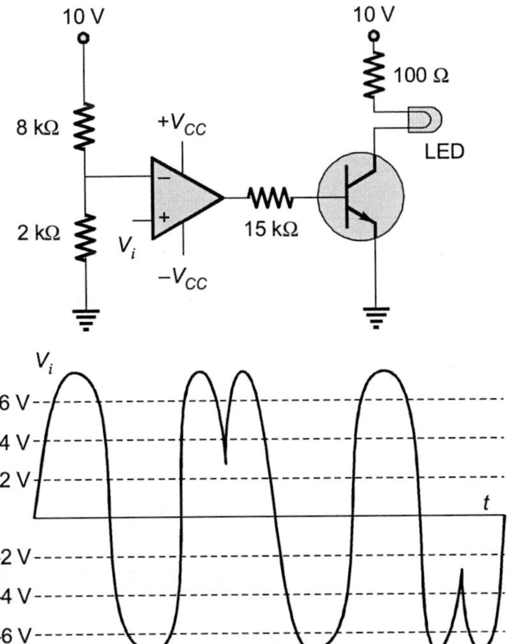

The number of times the LED glows is_____

[2016 : 1 M, Set-1]

1.56 An ideal op-amp has voltage sources, V_1, V_3, V_5, ... V_{N-1} connected to the non-inverting input and V_2, V_4, V_6, ..., V_N connected to the inverting input as shown in the figure below ($+V_{CC}$ = 15 volt, $-V_{CC}$ = −15 volt). The voltages V_1, V_2, V_3, V_4, V_5, V_6, ... are 1, −1/2, 1/3, −1/4, 1/5, −1/6, ... volt, respectively. As N approaches infinity, the output voltage (in volt) is _____.

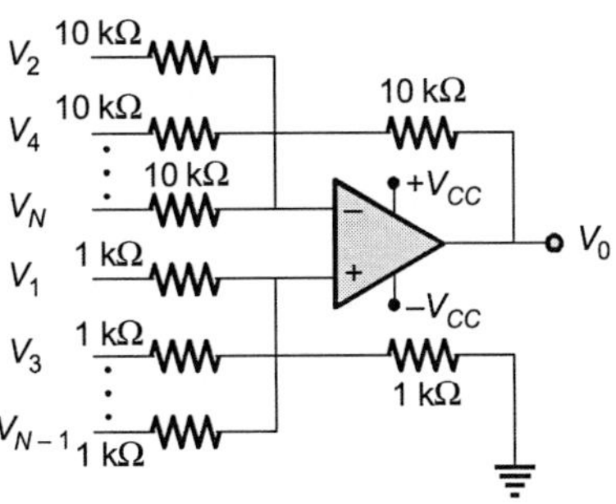

[2016 : 2 M, Set-1]

1.57 A *p-i-n* photodiode of responsivity 0.8 A/W is connected to the inverting input of an ideal op-amp as shown in the figure, $+V_{CC}$ = 15 V, $-V_{CC}$ = −15 V, Load resistor R_L = 10 kΩ. If 10 μW of power is incident on the photodiode, then the value of the photocurrent (in μA) through the load is _____.

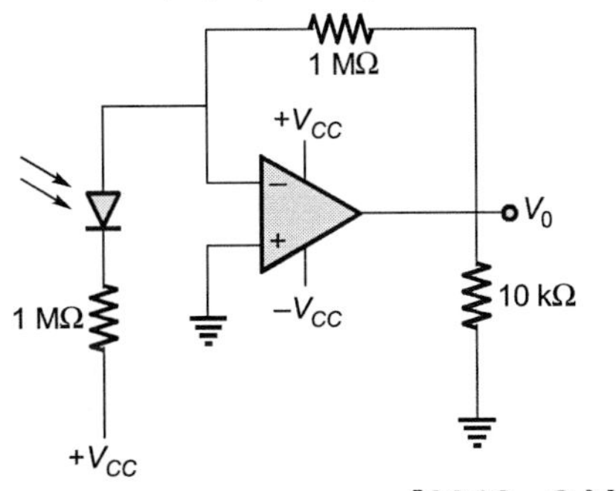

[2016 : 2 M, Set-1]

1.58 In the op-amp circuit shown, the Zener diodes Z_1 and Z_2 clamp the output voltage V_0 to + 5 V or −5 V. The switch S is initially closed and is opened at time t = 0.

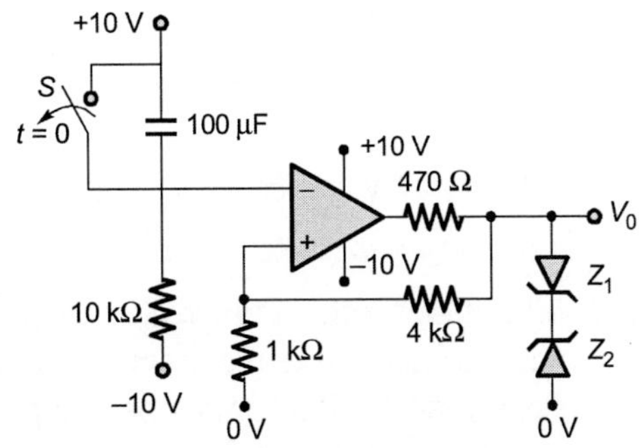

The time $t = t_1$ (in seconds) at which V_0 changes state is ________.

[2016 : 2 M, Set-2]

1.59 An op-amp has a finite open loop voltage gain of 100. Its input offset voltage V_{ios} (= + 5 mV) is modeled as shown in the circuit below. The amplifier is ideal in all other respects. V_{input} is 25 mV.

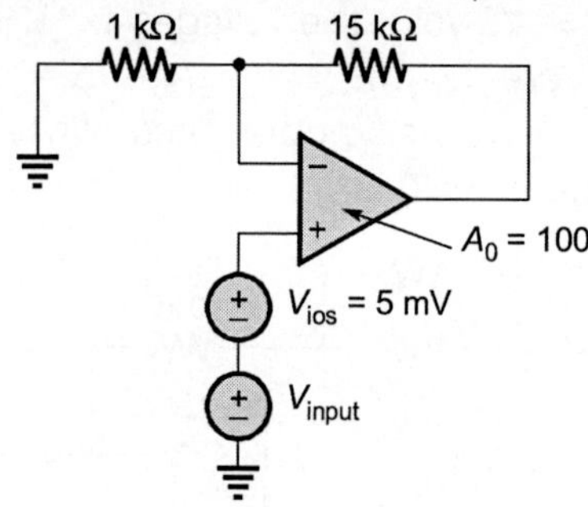

The output voltage (in millivolts) is _______

[2016 : 2 M, Set-2]

1.60 For the circuit shown in the figure, $R_1 = R_2 = R_3 = 1\ \Omega$, $L = 1\ \mu$H and $C = 1\ \mu$F. If the input $V_{in} = \cos(10^6\ t)$, then the overall voltage gain (V_{out} / V_{in}) of the circuit is _______.

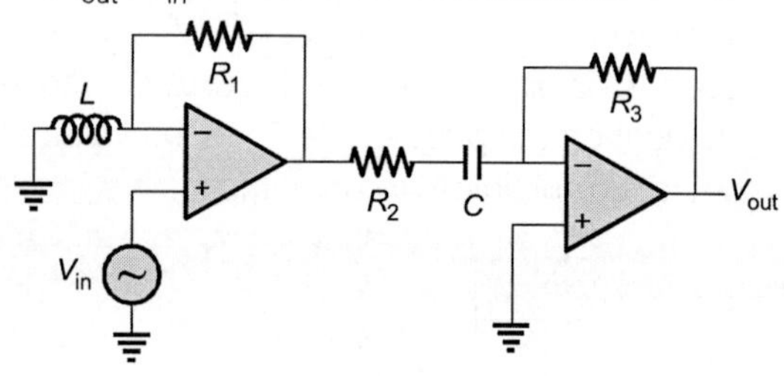

[2016 : 2 M, Set-3]

1.61 For the operational amplifier circuit shown, the output saturation voltages are ± 15 V. The upper and lower threshold voltages for the circuit are, respectively,

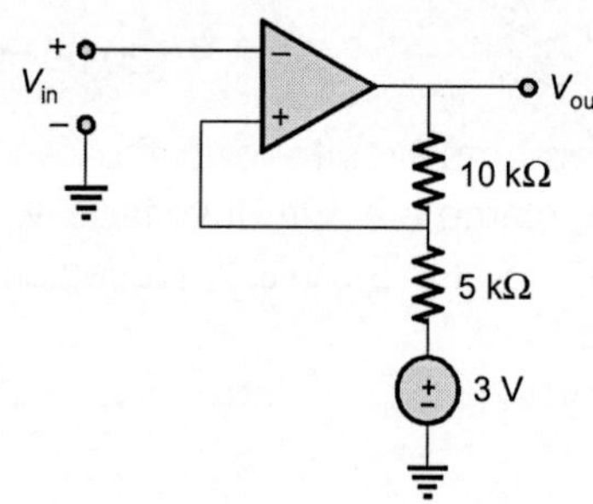

(a) +5 V and –5 V (b) +7 V and –3 V
(c) +3 V and –7 V (d) +3 V and –3 V

[2017 : 1 M, Set-1]

1.62 The amplifier circuit shown in the figure is implemented using a compensated operational amplifier (op-amp), and has an open-loop voltage gain, $A_0 = 10^5$ V/V and an open-loop cut-off frequency, $f_c = 8$ Hz. The voltage gain of the amplifier at 15 kHz, in V/V, is ___________.

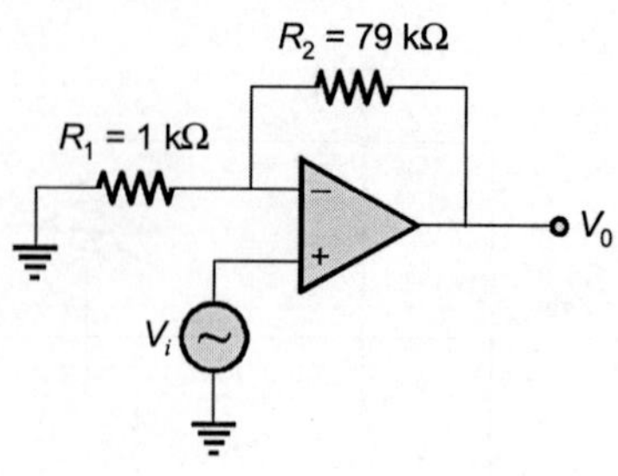

[2017 : 2 M, Set-1]

1.63 In the voltage reference circuit shown in the figure, the op-amp is ideal and the transistors Q_1, Q_2, ... Q_{32} are identical in all respects and have infinitely large values of common-emitter current gain (β). The collector current (I_C) of the transistors is related to their base-emitter voltage (V_{BE}) by the relation $I_C = I_S \exp(V_{BE}/V_T)$, where I_S is the saturation current. Assume that the voltage V_P shown in the figure is 0.7 V and the thermal voltage V_T = 26 mV.

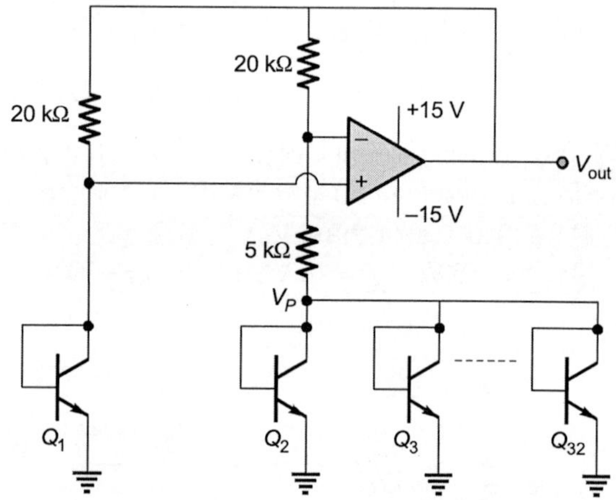

The output voltage V_{out} (in volts) is _____.

[2017 : 2 M, Set-2]

1.64 In the circuit shown below, the op-amp is ideal and Zener voltage of the diode is 2.5 volts. At the input, unit step voltage is applied i.e. $v_{IN}(t) = u(t)$ volts. Also at $t = 0$, the voltage across each of the capacitors is zero

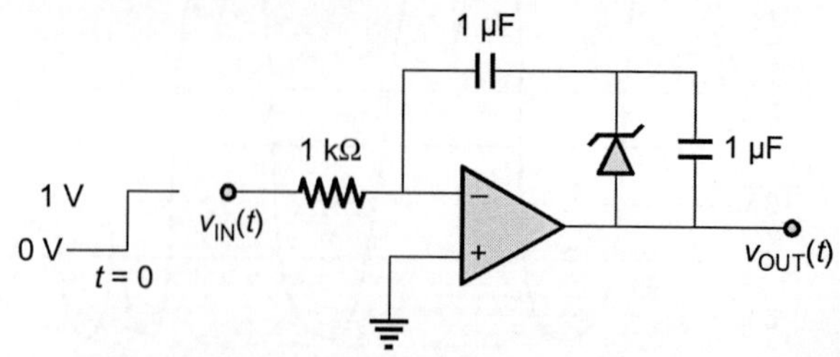

The time t, in milliseconds, at which the output voltage v_{OUT} crosses –10 V is

(a) 2.5 (b) 5
(c) 7.5 (d) 10

[2018 : 1 M]

1.65 An op-amp based circuit is implemented as shown below.

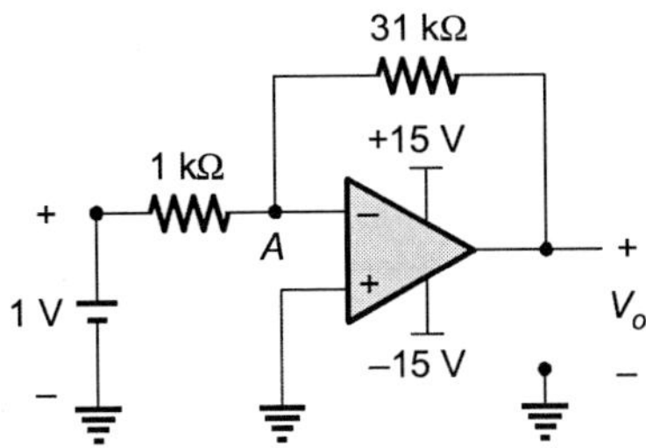

In the above circuit, assume the op-amp to be ideal. The voltage (in volts, correct to one decimal place) at node A, connected to the negative input of the op-amp as indicated in the figure is ____.

[2018 : 2 M]

1.66 In the circuit shown below, all the components are ideal. If V_i is +2 V, the current I_0 sourced by the op-amp is ________ mA.

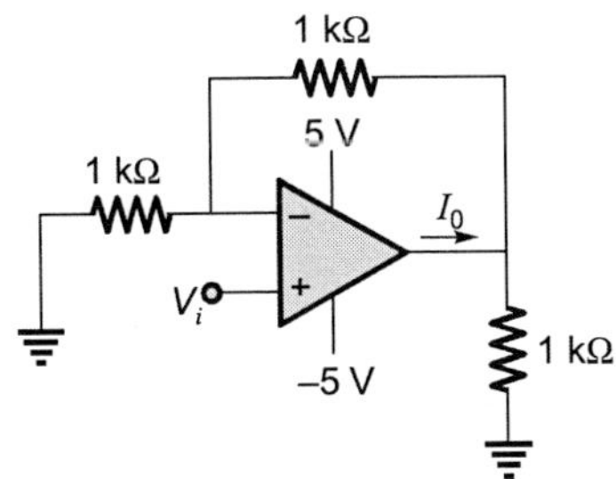

[2020 : 1 M]

1.67 The components in the circuit shown below are ideal. If the op-amp is in positive feedback and the input voltage V_i is a sine wave of amplitude 1 V, the output voltage V_o is

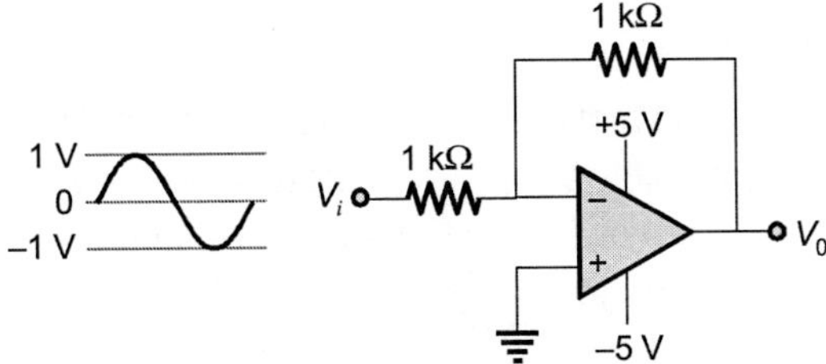

(a) a square wave of 5 V amplitude
(b) an inverted sine wave of 1 V amplitude
(c) a non-inverted sine wave of 2 V amplitude
(d) a constant of either +5 or –5 V

[2020 : 1 M]

1.68 The components in the circuit given below are ideal. If R = 2 kΩ and C = 1 μF, the –3 dB cut-off frequency of the circuit in Hz is

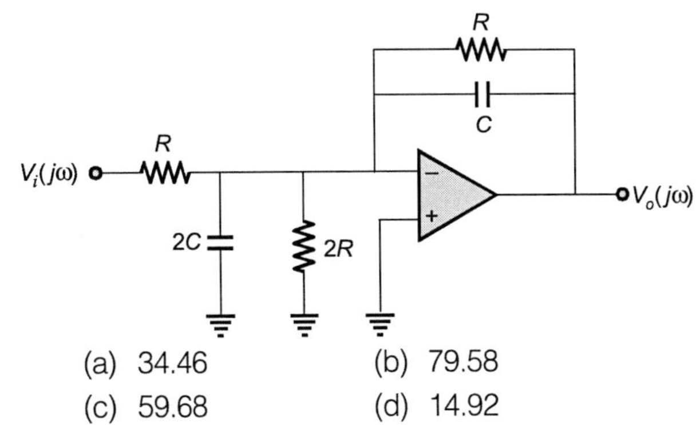

(a) 34.46
(b) 79.58
(c) 59.68
(d) 14.92

[2020 : 2 M]

1.69 For the circuit with an ideal Op-Amp shown in the figure, V_{REF} is fixed.

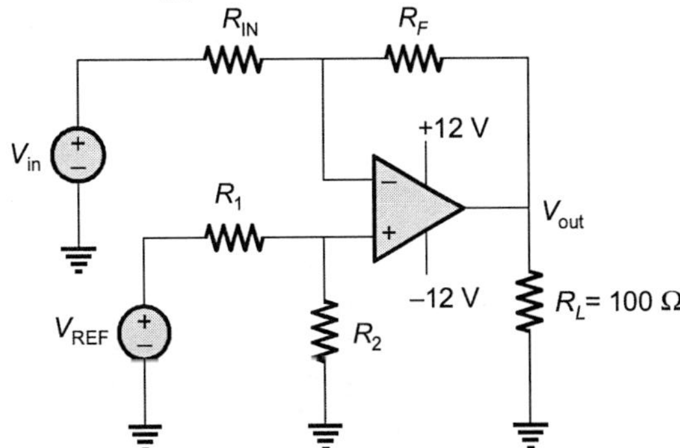

If V_{OUT} = 1 Volt for V_{IN} = 0.1 Volt and V_{OUT} = 6 Volt for V_{IN} = 1 Volt, where V_{OUT} is measured across R_L connected at the output of this Op-Amp, the value of R_F/R_{IN} is

(a) 3.825
(b) 5.555
(c) 2.860
(d) 3.285

[2021 : 1 M]

1.70 Consider the circuit with an ideal Op-Amp shown in the figure.

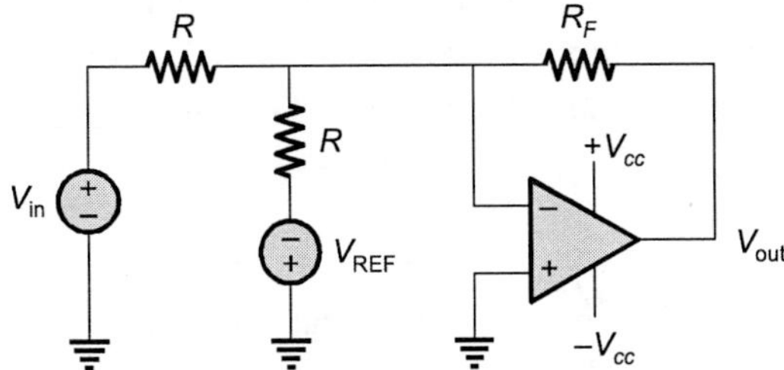

Assuming $|V_{IN}| << |V_{CC}|$ and $|V_{REF}| << |V_{CC}|$, the condition at which V_{OUT} equals to zero is

(a) $V_{IN} = 0.5\ V_{REF}$
(b) $V_{IN} = 2 + V_{REF}$
(c) $V_{IN} = 2\ V_{REF}$
(d) $V_{IN} = V_{REF}$

[2021 : 1 M]

1.71 A circuit with an ideal Op-Amp is shown in the figure. A pulse V_{IN} of 20 ms duration is applied to the input. The capacitors are initially uncharged.

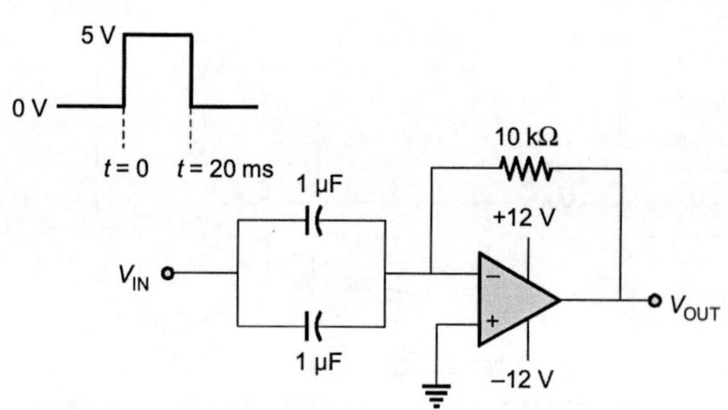

The output voltage V_{OUT} of this circuit at $t = 0^+$ (in integer) is ________ V.

[2021 : 2 M]

1.72 An ideal OPAMP circuit with a sinusoidal input is shown in the figure. The 3 dB frequency is the frequency at which the magnitude of the voltage gain decreases by 3 dB from the maximum value. Which of the options is/are correct?

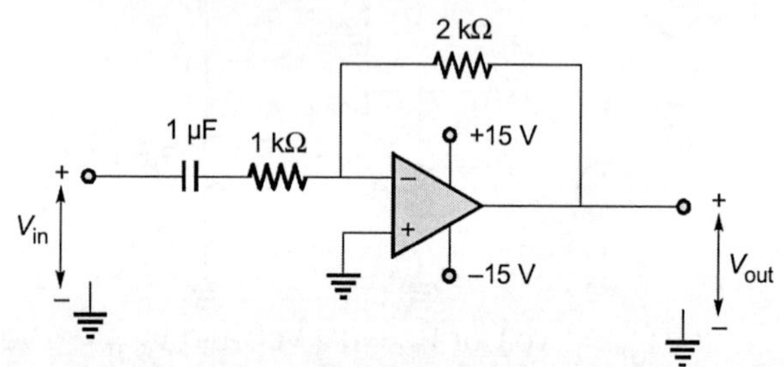

(a) The circuit is a low pass filter.
(b) The circuit is a high pass filter.
(c) The 3 dB frequency is 1000 rad/s.
(d) The 3 dB frequency is $\frac{1000}{3}$ rad/s.

[2022 : 1 M]

1.73 A circuit with an ideal OPAMP is shown. The Bode plot for the magnitude (in dB) of the gain transfer function ($A_V(j\omega) = V_{out}(j\omega)/V_{in}(j\omega)$) of the circuit is also provided (here, ω is the angular frequency in rad/s). The values of R and C are ________.

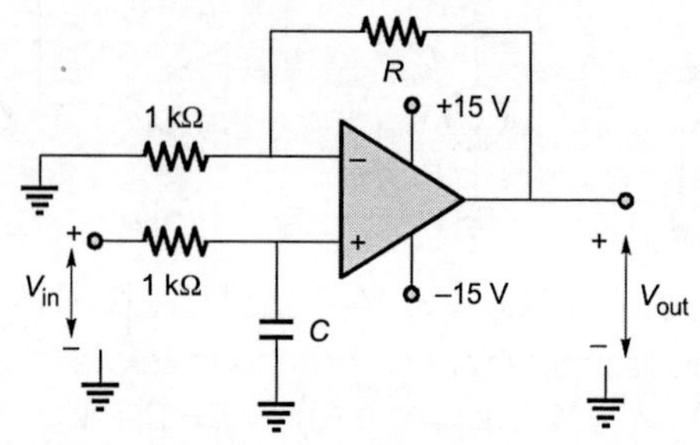

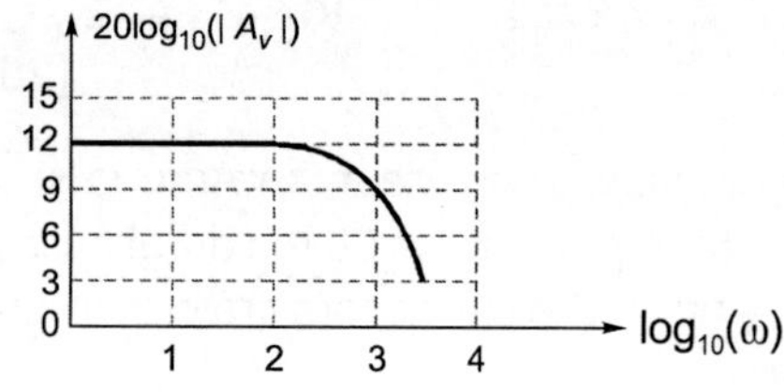

(a) $R = 3\text{ k}\Omega$, $C = 1\ \mu\text{F}$
(b) $R = 1\text{ k}\Omega$, $C = 3\ \mu\text{F}$
(c) $R = 4\text{ k}\Omega$, $C = 1\ \mu\text{F}$
(d) $R = 3\text{ k}\Omega$, $C = 2\ \mu\text{F}$

[2022 : 2 M]

1.74 The V_{OUT}/V_{IN} of the circuit shown below is

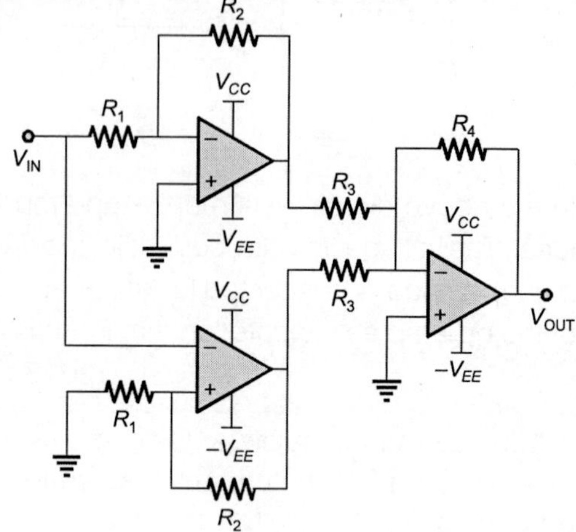

(a) $-\frac{R_4}{R_3}$
(b) $\frac{R_4}{R_3}$
(c) $1+\frac{R_4}{R_3}$
(d) $1-\frac{R_4}{R_3}$

[2023 : 2 M]

1.75 The opamps in the circuit shown are ideal, but have saturation voltages of ±10 V.

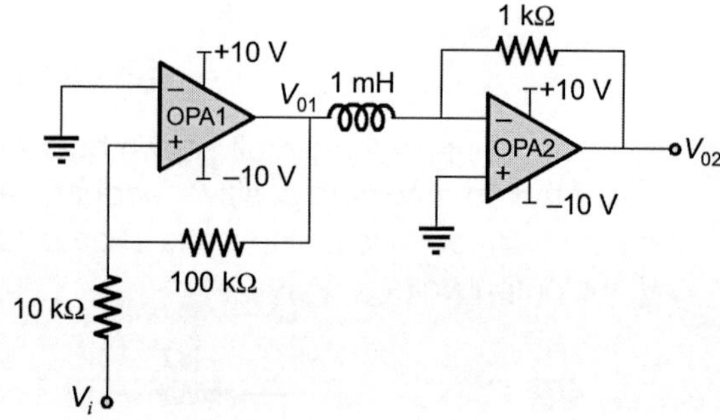

Assume that the initial inductor current is 0 A. The input voltage (V_i) is a triangular signal with peak voltages of ±2 V and time period of 8 μs. Which one of the following statements is true?

(a) V_{01} is delayed by 1 μs relative to V_i, and V_{02} is a trapezoidal waveform.
(b) V_{01} is delayed by 2 μs relative to V_i, and V_{02} is a triangular waveform.
(c) V_{01} is not delayed relative to V_i, and V_{02} is a trapezoidal waveform.
(d) V_{01} is not delayed relative to V_i, and V_{02} is a triangular waveform.

[2024 : 2 Marks]

■■■

Answers Operational Amplifiers

1.1 (33)	1.2 (159)	1.3 (d)	1.4 (d)	1.5 (d)	1.6 (a)	1.7 (d)
1.8 (c)	1.9 (a)	1.10 (b)	1.11 (c)	1.12 (a)	1.13 (b)	1.14 (a)
1.15 (b)	1.16 (c)	1.17 (b)	1.18 (c)	1.19 (b)	1.20 (c)	1.21 (b)
1.22 (a)	1.23 (a)	1.24 (b)	1.25 (c)	1.26 (a)	1.27 (c)	1.28 (c)
1.29 (d)	1.30 (a)	1.31 (c)	1.32 (b)	1.33 (b)	1.34 (d)	1.35 (a)
1.36 (b)	1.37 (d)	1.38 (b)	1.39 (b)	1.40 (c)	1.41 (3.18)	
1.42 (1093.02)	1.43 (c)	1.44 (b)	1.45 (d)	1.46 (d)	1.47 (250)	1.48 (1.5)
1.49 (1)	1.50 (2.8056)	1.51 (12)	1.52 (159.15)	1.53 (1)	1.54 (b)	1.55 (3)
1.56 (15)	1.57 (800)	1.58 (0.798)	1.59 (413.79)	1.60 (–1)	1.61 (b)	1.62 (44.4)
1.63 (1.145)	1.64 (c)	1.65 (0.5)	1.66 (6)	1.67 (d)	1.68 (b)	1.69 (b)
1.70 (d)	1.71 (–12)	1.72 (b, c)	1.73 (a)	1.74 (a)	1.75 (a)	

Explanations Operational Amplifiers

1.1 (33)

$$\boxed{A^- = -\frac{R_f}{R_1}}$$

$$A^- = -\frac{22\text{k}}{10\text{k}} = -2.2$$

$$|A^-| = 2.2$$

$$V_a = V_+\left(\frac{R}{R+15}\right)$$

$$\boxed{A^+ = 1+\frac{R_f}{R_1}}$$

$$\frac{V_o}{V_a} = 1+\frac{R_f}{R_1} = \frac{V_o}{V_+\left(\frac{R}{R+15}\right)}$$

$$\frac{V_o}{V_+} = \left(1+\frac{22}{10}\right)\times\left(\frac{R}{R+15}\right)$$

$$|A^+| = 3.2\left(\frac{R}{R+15}\right)$$

$$\frac{3.2R}{R+15} = 2.2$$

$$3.2\,R = 2.2\,R + 33$$

$$R = 33\text{ k}\Omega$$

1.2 (159)

$$f_{max} = \frac{S.R.}{2\pi V_o}$$

$$f_{max} = \frac{12}{10^{-6}}\times\frac{1}{2\pi\times 12} = 159\text{ kHz}$$

1.3 (d)

The given figure has positive feedback and an input at inverting terminal, so the given figure is of Schmitt trigger.

1.4 (d)

$$\boxed{V_o^- = -\frac{R_f}{R_1}V_{in}} = -\frac{10\text{ k}\Omega}{5\text{ k}\Omega}\times 2\,V = -4\,V$$

$$V_x = V_o\times\frac{10k}{10k+100\text{ k}\Omega} = \frac{V_o}{11}$$

$$\boxed{V_o^+ = \left(1+\frac{R_f}{R_1}\right)V_x} = \left(1+\frac{10\text{ k}\Omega}{5\text{ k}\Omega}\right)\frac{V_o}{11} = \frac{3V_o}{11}$$

Apply superposition principle:

$$\boxed{V_o = V_o^+ + V_o^-}$$

$$V_o = \frac{3V_o}{11} - 4$$

$$V_o - \frac{3V_o}{11} = -4$$

$$V_o = -5.5\text{ V}$$

1.5 (d)

When one input terminal of high gain comparator circuit is connected to ground and a sinusoidal voltage is applied to the other input. Then the output of comparator will be a square wave.

v_a = +ve sinusoidal half $\rightarrow V_o = +V_{sat}$

v_a = –ve sinusoidal half $\rightarrow V_o = -V_{sat}$

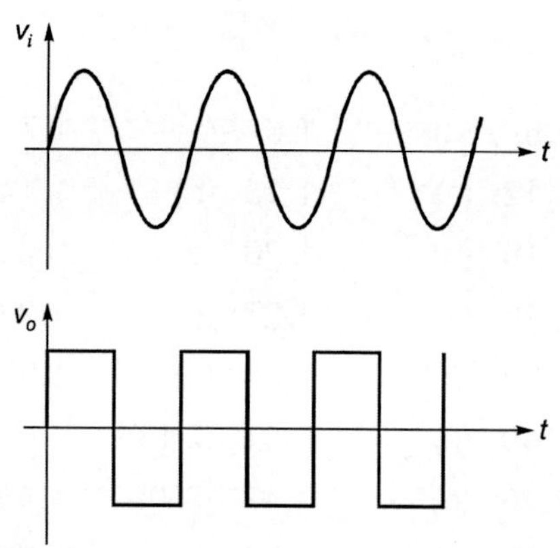

1.7 (d)

In positive feedback op-amp act in its saturation region $\pm V_{sat.}$

Here applied voltage is positive,

$V_0 = +V_{sat} = +15$ V

1.8 (c)

$$V_0 = \frac{-R_f}{R_1} \cdot V_{in}$$

Here $$V_0 = -\left(\frac{\omega C}{\omega C}\right) V_1 \sin\omega t - \frac{\omega C}{\omega C} V_2 \sin\omega t$$

$$= -(V_1 + V_2) \sin \omega t$$

1.9 (a)

This circuit acts as a differentiator and differentiation of triangular wave gives square wave.

$$V_0 = -\frac{R}{1/sC} V_i = -sCRV_i$$

$$V_0 = -RC\frac{dV_i}{dt} \qquad \left(s = \frac{d}{dt}\right)$$

1.10 (b)

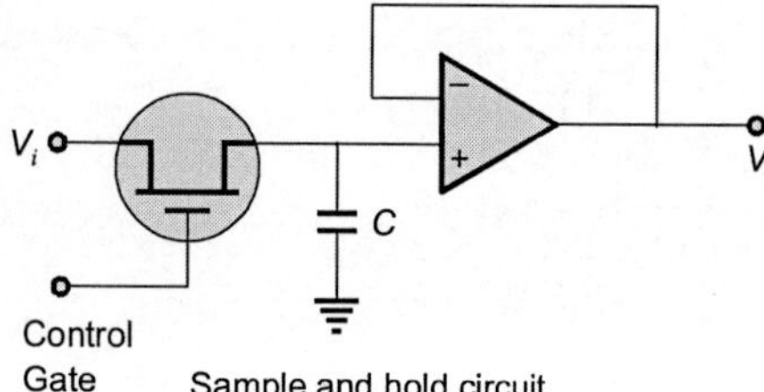

Sample and hold circuit

1.11 (c)

$$V_o = V_{io} \cdot A = 5 \text{ mV} \times 10{,}000 = 50 \text{ V}$$

But $V_o = \pm 15$V,

V_o can never be greater than $\pm V_{sat}$.

1.13 (b)

If open loop gain is not infinite, then

$$\frac{1}{\beta} = -\frac{R_f}{R_1} = -\frac{10\text{ k}}{1\text{k}} = -10$$

$$|\beta| = \frac{1}{10}$$

$$A_f = \frac{-A}{1+A\beta} = \frac{-100}{1+100\times\frac{1}{10}} = -\frac{100}{11}$$

$$A_f \cong -9$$

1.14 (a)

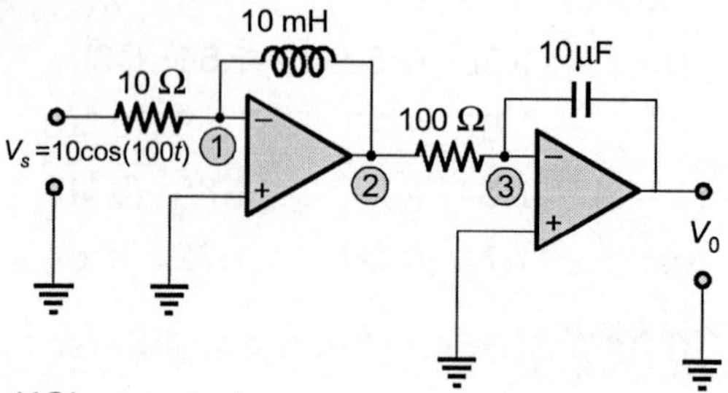

KCL at node 1,

$$\frac{V_2}{V_s} = \frac{-\omega L}{R} = \frac{-100\times 10\times 10^{-3}}{10} = -\frac{1}{10}$$

$$V_2 = -\frac{V_s}{10} = -\cos 100t$$

$$\frac{V_o}{V_2} = \frac{-1/\omega c}{100} = -\frac{1}{100\times 10\times 10^{-6}\times 100}$$

$$= -10$$

$$V_o = -10\ V_2 = -10(-\cos 100\ t)$$

$$V_o = 10 \cos 100\ t$$

1.15 (b)

$$20 \log x = 20 \text{ dB}$$

$$x = 10$$

$$\text{Gain} \times BW = 1 \times 10^6$$

$$BW = \frac{1\times 10^6}{\text{Gain}} = \frac{10^6}{10} = 100 \text{ k} = 10^5$$

1.16 (c)

$$\text{Slew rate} = A\cdot 2\pi f V_m$$

$$V = AV_m \sin \omega t$$

$$\frac{dV}{dt} = AV_m\omega \cos \omega t$$

$$\left.\frac{dV}{dt}\right|_{max} = SR = AV_m 2\pi f_m$$

$$20 \log x = 40$$

$$x = 100 = A$$

$$\Rightarrow V_m = \frac{SR}{A\cdot 2\pi f_m} = \frac{1}{10^{-6}\times 100\times 2\pi\times 20\times 10^3}$$

$$V_m = 79.5 \text{ mV}$$

1.17 (b)

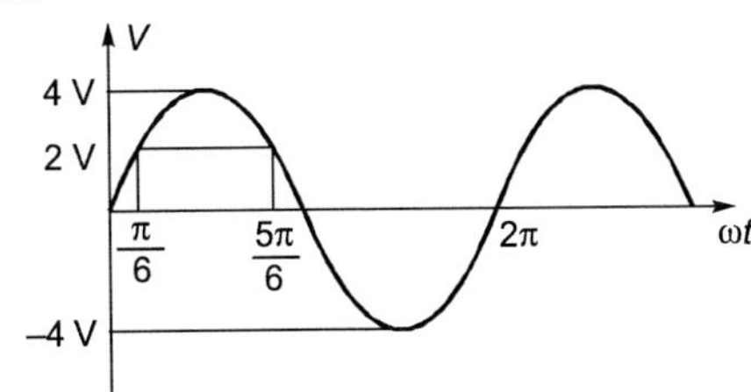

$V_{p\text{-}p} = 8\text{ V}$

So, $V_i = 4\sin\omega t$

At $V_i = 2$

$$\sin\omega t = \frac{1}{2}$$

$$\Rightarrow \quad \omega t = \frac{\pi}{6}$$

Another crossover at $\pi - \frac{\pi}{6} = \frac{5\pi}{6}$

$$\therefore \quad \text{Duty cycle} = \frac{T_{on}}{T} = \frac{\frac{5\pi}{6} - \frac{\pi}{6}}{2\pi} = \frac{\left(\frac{4\pi}{6}\right)}{2\pi} = \frac{1}{3}$$

1.18 (c)

$$\text{CMRR} = 20\log A_d - 20\log A_c$$
$$= 48\text{ dB} - 2\text{ dB} = 46\text{ dB}$$

1.19 (b)

$$V_0 = -\frac{R_F}{R_1}\times 2 + \left(1 + \frac{R_F}{R_1}\right)\cdot\frac{8}{9}\times 3$$

$$= -5\times 2 + 6\times\frac{8}{3}$$

$$V_o = 6\text{ V}$$

1.20 (c)

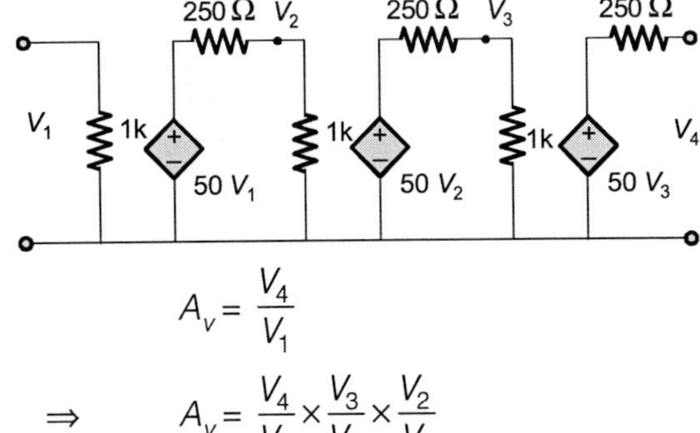

$$A_v = \frac{V_4}{V_1}$$

$$\Rightarrow \quad A_v = \frac{V_4}{V_3}\times\frac{V_3}{V_2}\times\frac{V_2}{V_1}$$

Voltage across 1 k after 1st stage is,

$$V_2 = \frac{1000\times 50\,V_1}{1250}$$

$$\Rightarrow \quad \frac{V_2}{V_1} = 40$$

Similarly $\frac{V_3}{V_2} = 40$

$$\therefore \quad A_v = 40\times 40\times 50 = 8\times 10^4$$

$$A_v\text{ in (dB)} = 20\log(8\times 10^4) = 98\text{ dB}$$

1.22 (a)

At $\omega = \infty$, $\frac{V_o}{V_{in}} = 0$

and at $\omega = 0$, $\frac{V_o}{V_{in}} = 1$

1.23 (a)

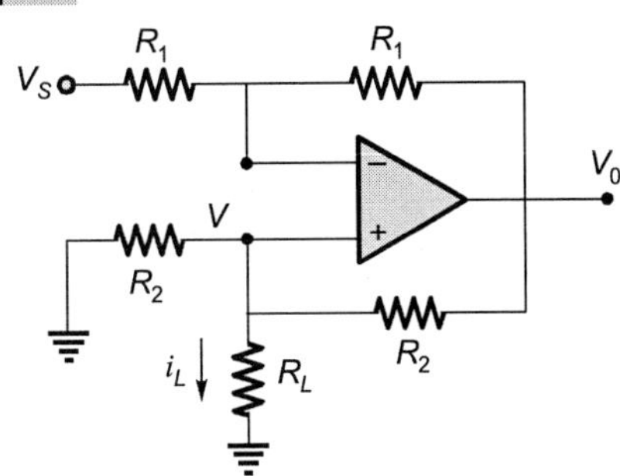

$$\frac{V}{R_2} + \frac{V}{R_L} + \frac{V}{R_2} = \frac{V_o}{R_2}$$

$$\frac{2V}{R_2} + \frac{V}{R_L} = \frac{V_o}{R_2} \quad \text{...(i)}$$

$$\frac{V_s - V}{R_1} = \frac{V - V_o}{R_1}$$

$$V_s - 2V = -V_0 \quad \text{...(ii)}$$

Putting V_o from (i),

$$\frac{-V_s + 2V}{R_2} = +\left(\frac{2V}{R_2} + \frac{V}{R_L}\right)$$

$$i_L = \frac{V}{R_L} = -\frac{V_s}{R_2} = \left[\frac{V}{R_L} = i_L\right]$$

1.24 (b)

Connect a V_s voltage source across inverting terminal of op-amp.

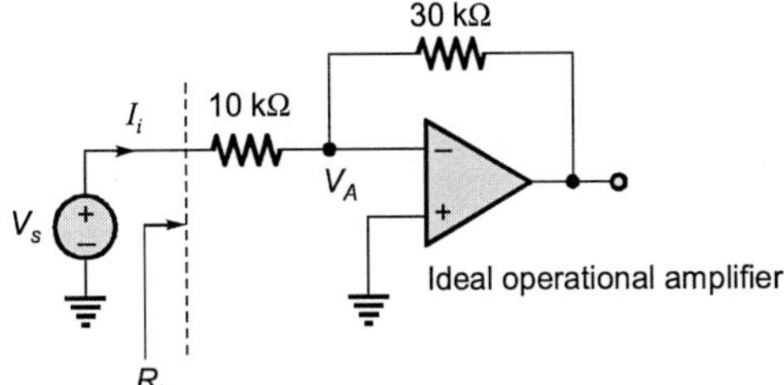

By virtual short

$$V_A = 0\text{ V}$$

$$I_i = \frac{V_s - V_A}{10\text{k}\Omega} = \frac{V_s - 0}{10\text{k}\Omega} = \frac{V_s}{10\text{k}\Omega}$$

So, $R_i = \frac{V_s}{I_i} = 10\text{ k}\Omega$

1.25 (c)

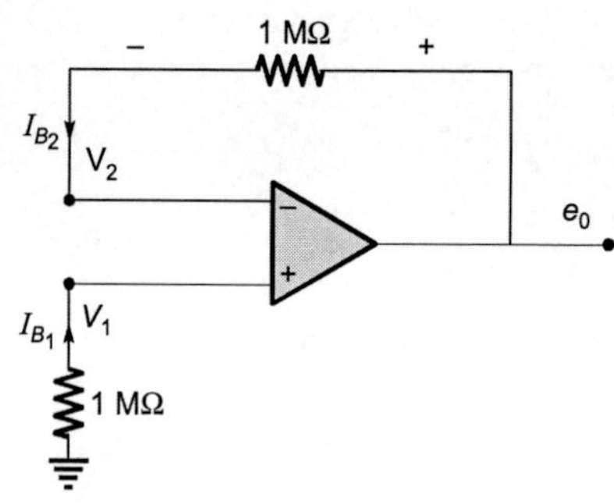

$$V_1 = -I_{B_1} \times 1\text{ M}\Omega,\ V_2 = V_1 = -I_{B_1} \times 1\text{ M}\Omega$$

(due to virtual ground)

Drop in feedback resistor 1 MΩ = $I_{B_2} \times 1\text{M}\Omega$

$$e_0 = V_2 + I_{B_2} \times 1\text{ M}\Omega$$

$$e_0 = -I_{B_1} \times 1\text{ M}\Omega + I_{B_2} \times 1\text{ M}\Omega$$

$$e_0 = (I_{B_2} - I_{B_1}) \times 1\text{ M}\Omega$$

where $(I_{B_2} - I_{B_1})$ is offset current

1.26 (a)

High pass filter,

$$\omega_c = \frac{1}{RC} = \frac{1}{1\times10^3 \times 1\times10^{-6}} = 1000\text{ rad/sec}$$

1.27 (c)

$$V_C = 15\,(1 - e^{-t/RC}) = 15\,(1 - e^{-1}) = 9.45\text{ V}$$

1.28 (c)

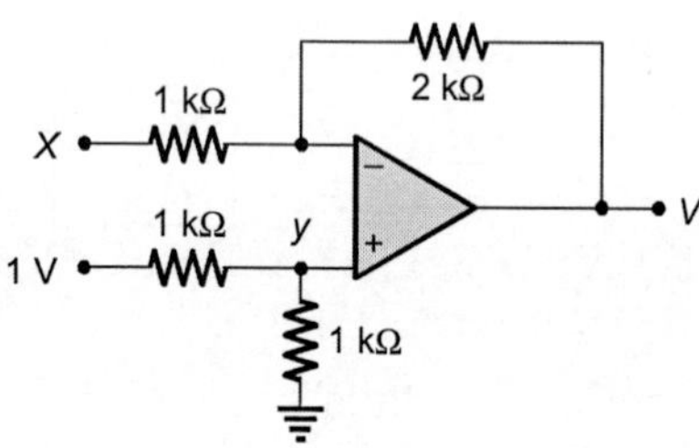

$X = 1$ volt and $y = 0.5$ V

(using voltage division rule)

So $$V_0 = X\left(-\frac{2\text{ K}\Omega}{1\text{K}\Omega}\right) + y\left(1 + \frac{2\text{ K}\Omega}{1\text{K}\Omega}\right)$$

$$= 1 \times (-2) + 0.5 \times (3) = -0.5\text{ V}$$

1.29 (d)

$$V_o = -V_T \ln\left(\frac{V_I}{R_1 I_s}\right)$$

$$V_{o_1} - V_{o_2} = -V_T \ln\frac{2}{R_1 I_s} + V_T \ln\frac{4}{R_1 I_s}$$

$$= V_T \ln\left(\frac{4}{2}\right) = V_T \ln 2$$

1.30 (a)

From the figure given in the question

$$V_0 = V_i\left(-\frac{R_1}{R_1}\right) + V_i \frac{\frac{1}{sC}}{\frac{1}{sC} + R}\left(1 + \frac{R_1}{R_1}\right)$$

$$\frac{V_0}{V_i} = -1 + \frac{1}{1 + RsC} \times 2 = \frac{2 - (1 + RsC)}{1 + RsC}$$

$$\frac{V_0}{V_i} = \frac{1 - RsC}{1 + RsC}$$

1.31 (c)

$$\angle\frac{V_0}{V_i} = \angle\frac{1 - RsC}{1 + RsC}$$

$$\Rightarrow \quad \angle(1 - sRC) - \angle 1 + sCR$$

$$\theta = -\tan^{-1}\omega RC - \tan^{-1}\omega RC$$

$$\theta = -2\tan^{-1}\omega RC$$

minimum value of $\theta = -\pi$ (at $\omega \to \infty$)

maximum value of $\theta = 0$ (at $\omega = 0$)

1.32 (b)

$$I = I_0\left(e^{V/V_T} - 1\right)$$

$$= \frac{0 - (-1)}{100\text{ k}\Omega}$$

$$\Rightarrow \quad 10^{-6}\left(e^{\frac{V}{25\times10^{-3}}} - 1\right) = \frac{1}{10^5}$$

$$\Rightarrow \quad V = 0.06\text{ V}$$

$$\Rightarrow \quad \frac{V_0 - V}{4\text{ k}\Omega} = \frac{1}{100\text{ k}\Omega}$$

$$\Rightarrow \quad \frac{V_0 - 0.06}{4\text{ k}\Omega} = \frac{1}{100\text{ k}\Omega}$$

$$\Rightarrow \quad V_0 = 0.1\text{ V}$$

1.33 (b)

$$\frac{V_i}{R_1 + Ls} = \frac{-V_0}{\frac{R_2}{R_2C_2s + 1}}$$

$$\frac{V_0}{V_i} = -\frac{R_2}{(R_1 + Ls)(R_2C_2s + 1)}$$

which is equivalent to standard form of transfer function of low pass filter, i.e.

$$H(s) = \frac{a}{ps^2 + qs + r}$$

1.34 (d)

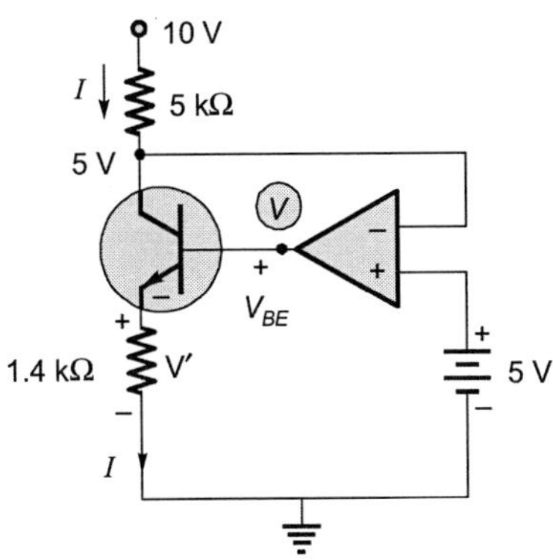

$$I = \frac{10-5}{5} = 1\text{mA}$$

$$V' = 1.4 \times 1 = 1.4 \text{ Volt}$$

$$\therefore \quad V = V' + V_{BE}$$

$$= 1.4 + 0.6 = 2 \text{ Volts}$$

1.35 (a)

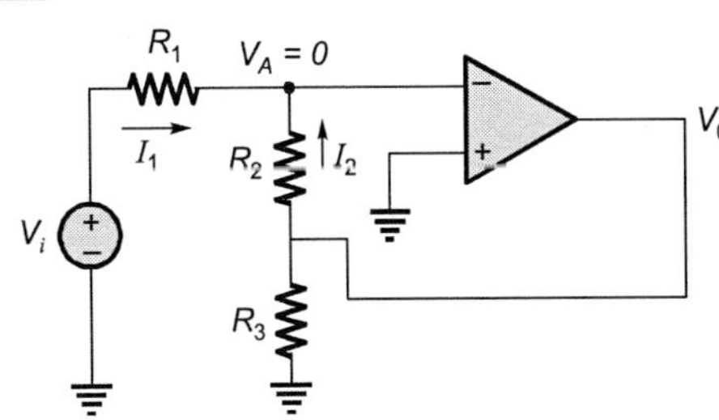

Assuming ideal OP-AMP, voltage at point A is zero. So, the given circuit can be considered as shown below:

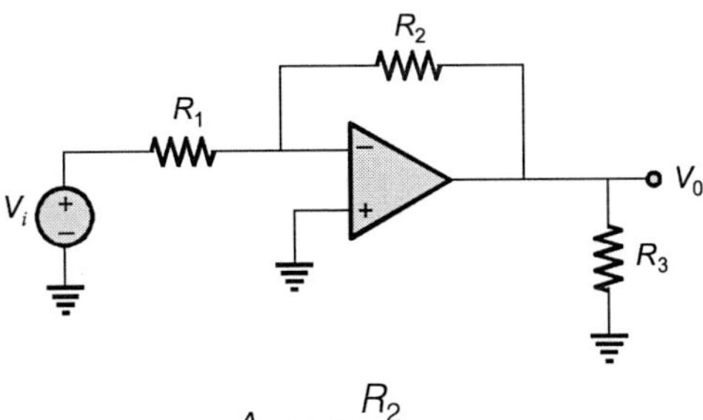

$$A_v = -\frac{R_2}{R_1}$$

1.36 (b)

At, $V_I = -10$ V,

$$\frac{V_0 - 0}{R} = \frac{0-20}{4R} + \frac{0(-10)}{R}$$

$$V_0 = -5 + 10 = 5 \text{ V}$$

At, $V_I = -5$ V,

$$\frac{V_0 - 0}{R} = \frac{0-20}{4R} + \frac{0-(-5)}{R}$$

$$V_0 = -5 + 5 = 0 \text{ V}$$

For $V_I > -5$ V, both diodes are conducting.

So, $V_0 = 0$ V.

1.37 (d)

At $\omega = 0$; $X_L = \omega L = 0$

Hence circuit can be redrawn as below:

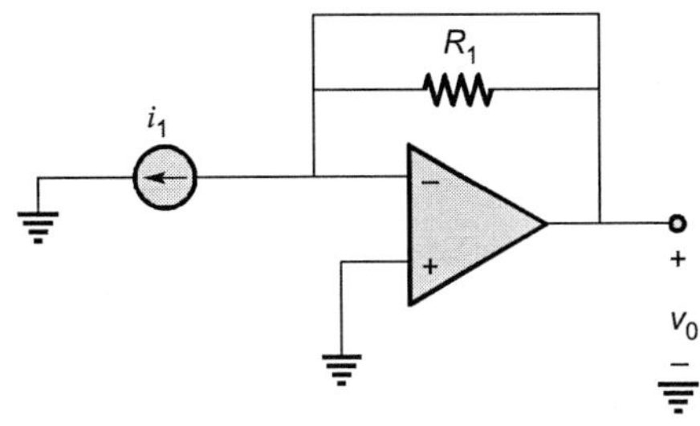

$$\therefore \quad V_0 = 0$$

At $\omega = \infty$; $X_L = \infty$

Hence, circuit can be redrawn as below:

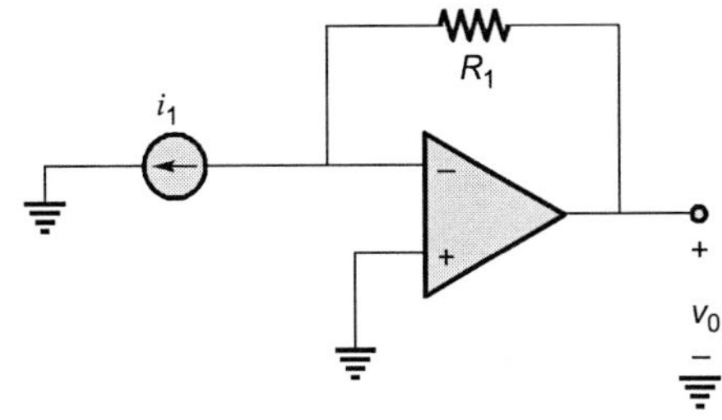

$$\therefore \quad V_0 = R_1 i_L$$

Hence given circuit is a high pass filter.

1.38 (b)

$$T(s) = \frac{(O/P)}{(I/P)} = \frac{-R_2}{\left(R_1 + \frac{1}{Cs}\right)} = \frac{-R_2 sC}{R_1 sC + 1}$$

$$T(s) = \frac{-\left(\frac{R_2}{R_1}\right)s}{s + \frac{1}{R_1 C}}$$

It is the transfer function of high pass filter with cutoff frequency →

$$\omega = \frac{1}{R_1 C} \text{ rad/sec}$$

1.39 (b)

Due to virtual short

$$V_c = V_B = 0 \text{ V}$$

(collector voltage of transistor Q)

and given that

Base voltage of transistor Q

$$V_B = 0 \text{ V}$$

So, $V_C = 0$ V

It means collector to base of transistor Q are short circuited.

If any junction of transistor is short circuited then the junction acts as reverse bias. So, the C-B junction is reverse bias.

The given op-amp is an inverting configuration which have positive voltage as input. So the output voltage of op-amp will be negative voltage.

$$V_{out} = -\text{ve voltage}$$

Emitter voltage of transistor

$$V_E = -\text{ve voltage}$$

given transistor is *n-p-n* transistor and

$$V_B = 0\text{ V}$$

$$V_E = -\text{ve voltage}$$

So, the *E-B* junction will be forward bias. Thus, the transistor is in active region and will behave as closed switch.

So,

$$V_{BE} = 0.7\text{ V} \quad \text{(for silicon transistor)}$$

$$V_B - V_E = 0.7\text{ V}$$

$$V_E = V_B - 0.7$$

$$V_E = 0 - 0.7$$

$$V_E = -0.7\text{ V}$$

1.40 (c)

Output of first op-amp

$$V_{out1} = -2\left[-\frac{1}{1}\right] + 1\left[1 + \frac{1}{1}\right] = 2 + 1(2) = 4\text{ V}$$

$$V_{out} = 4\left(1 + \frac{1}{1}\right) = 4(2) = 8\text{ V}$$

1.41 (3.18)

$$\frac{0 - V_o}{Z_2} + \frac{0 - V_i}{Z_1} = 0$$

where $Z_2 = R_2 || 10\text{ nF}$

$Z_1 = R_1 = 1\text{ k}\Omega$

$$\text{or} \quad \frac{V_o}{V_1} = -\frac{Z_2}{Z_1} = -\frac{R_2 \| \frac{1}{Cs}}{R_1} = \frac{R_2}{R_1(R_2Cs + 1)}$$

cut-off frequency

$$|1 + R_2Cs| = 0$$

$$\text{or} \quad R_2Cs = 1$$

$$\text{or} \quad R_2 = \frac{1}{Cs}$$

$$= \frac{1}{2\pi fC} = 3184.7 \text{ or } 3.18\text{ k}\Omega$$

1.42 (1093.02)

Given circuit is a op-amp series regulator V_o is given by

$$V_o = \left(1 + \frac{R_1}{R_2}\right)V_z$$

$$9\text{ V} = \left(1 + \frac{1\text{k}\Omega}{R}\right)4.7$$

$$\therefore \quad R = 1093.02\ \Omega$$

1.43 (c)

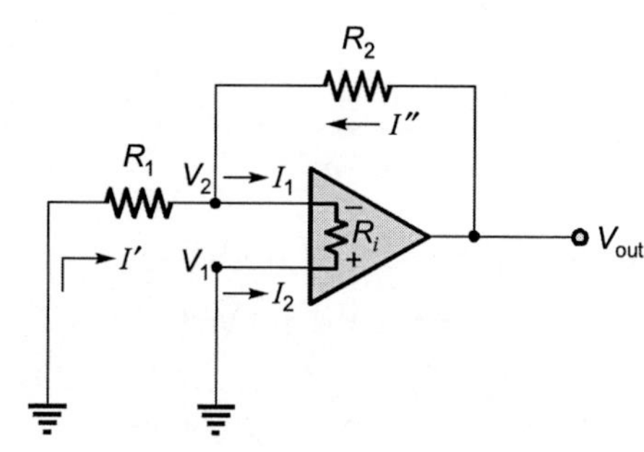

$$I_B = I_1 = I_2$$

$$V_2 = (R_1 \parallel R_2)I_1 \quad \text{...(i)}$$

KCL at node '2'

$$I' + I'' = I_1$$

$$\frac{0 - V_2}{R_1} + \frac{V_o - V_2}{R_2} = \frac{V_2}{R_i}$$

$$\frac{V_o}{R_2} = V_2\left(\frac{R_1 + R_2}{R_1R_2}\right) \quad \text{...(ii)}$$

$\because R_i$ is very high for op-amp.

By using equations (i) and (ii)

$$V_o = I_1 R_2$$

1.44 (b)

If the resistance R_E is increased then A_{cm} decreases.

Hence, CMRR $= \frac{A_d}{A_{cm}}$ increases.

1.45 (d)

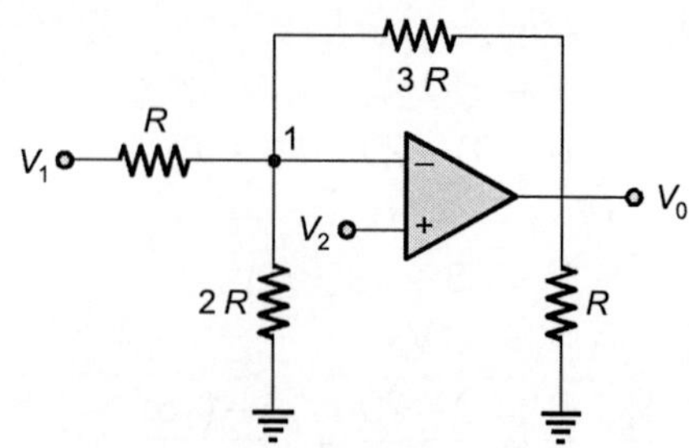

By applying *KCL* at node, we get

$$\frac{V_2 - V_1}{R} + \frac{V_2}{2R} + \frac{V_2 - V_o}{3R} = 0$$

$$\frac{6V_2 - 6V_1 + 3V_2 + 2V_2}{2} = V_o$$

$$V_o = -3V_1 + \frac{11}{2}V_2$$

1.47 (250)

$$I_{50\,\Omega} = \frac{1}{50}A = I_{100\,\Omega}$$

$$V_o = \frac{1}{100}[250(1+\delta) - 250(1-\delta)]$$

$$= \frac{1}{1000} \times 250 = 0.25 \text{ V} = 250 \text{ mV}$$

1.48 (1.5)

$$V_{OB} = -5\left(\frac{1\,\text{k}\Omega}{1\,\text{k}\Omega}\right) - 1\left(\frac{1\,\text{k}\Omega}{1\,\text{k}\Omega}\right) = -6 \text{ V}$$

$$V_{OA} = -5\left(\frac{1\,\text{k}\Omega}{1\,\text{k}\Omega}\right) + \frac{1}{2}\left(1 + \frac{1\,\text{k}\Omega}{1\,\text{k}\Omega}\right)$$

$$= -4 \text{ V} \qquad \left(\because V_+ = \frac{1}{2}V\right)$$

$$\therefore \quad \frac{V_{OB}}{V_{OA}} = 1.5$$

1.49 (1)

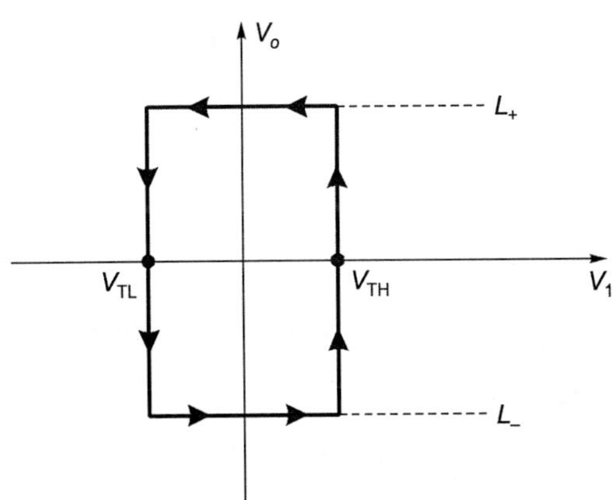

$$\text{Hysteresis} = V_{TH} - V_{TL}$$

$$= -L_-\left(\frac{R_1}{R_2}\right) + L_+\left(\frac{R_1}{R_2}\right)$$

$$500 \text{ mV} = -(-5)\left(\frac{R_1}{20\,\text{k}\Omega}\right) + 5\left(\frac{R_1}{20\,\text{k}\Omega}\right)$$

$$= \frac{R_1}{2\,\text{k}\Omega}$$

$$\therefore \quad R_1 = 500 \times 2 \times 10^3 \times 10^{-3}$$

$$= 1000\ \Omega = 1\text{ k}\Omega$$

1.50 (2.8056)

$$P_{Q_1(\text{max})} = V_{CE(\text{max})} \times I_{c\max} \qquad \text{...(i)}$$

$$V_{CE(\text{max})} = V_{i\max} - V_o = (24 - 10)\text{ V} = 14 \text{ V}$$

$$I_E = I_C = 200 \text{ mA} + 0.4 \text{ mA}$$

$$I_{C\max} = 200.4 \text{ mA} \qquad \left(\because I_{R_2} = \frac{4-0}{10}\text{mA}\right)$$

Put values in equation (i), we get

$$P_{Q_1(\text{max})} = 14 \times 200.4 \times 10^{-3} \text{ Watt} = 2.8056 \text{ Watt}$$

1.51 (12)

$$V_+ > V_-$$

So, $V_o = V_{sat} = 12$ Volts

1.52 (159.15)

$$f_{3\text{ dB}} = \frac{1}{2\pi RC} = 159.15 \text{ Hz}$$

1.53 (1)

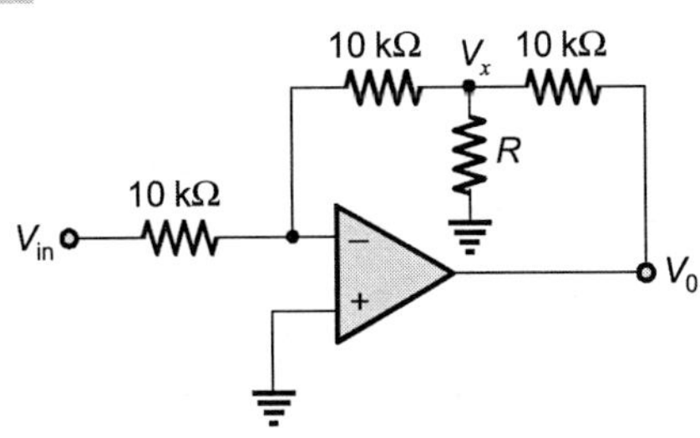

Apply nodal analysis at inverting terminal

$$\frac{V_{in} - 0}{10\,\text{k}\Omega} = \frac{0 - V_x}{10\,\text{k}\Omega}$$

$$\Rightarrow \quad V_{in} = -V_x \qquad \text{...(i)}$$

Again apply nodal analysis at node V_x

$$\frac{0 - V_x}{10\,\text{k}\Omega} = \frac{V_x - 0}{R} + \frac{V_x - V_o}{10\,\text{k}\Omega} \qquad \text{...(ii)}$$

Put the value of V_x from equation (i) in equation (ii), we get

$$R = 1 \text{ k}\Omega$$

1.54 (b)

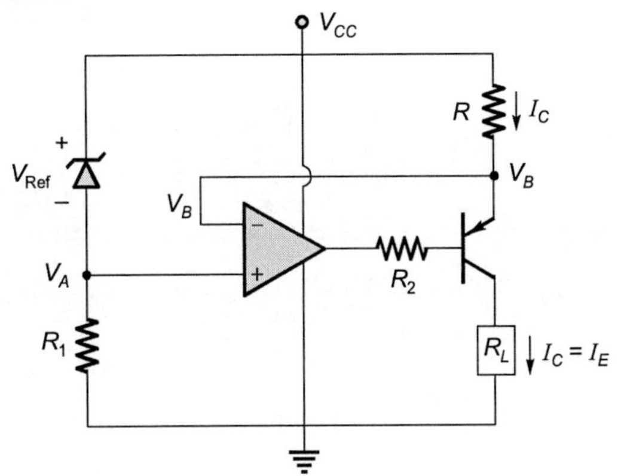

$V_A = V_{CC} - V_{ref}$ and $V_B = V_A$ (since virtual short)

$$I_C = \frac{V_{CC} - V_B}{R} = \frac{V_{CC} - (V_{CC} - V_{Ref})}{R} = \frac{V_{Ref}}{R}$$

$$I_0 = I_E = \frac{I_C}{\alpha} = \left(\frac{\beta}{1+\beta}\right)\frac{V_{Ref}}{R}$$

1.55 (3)

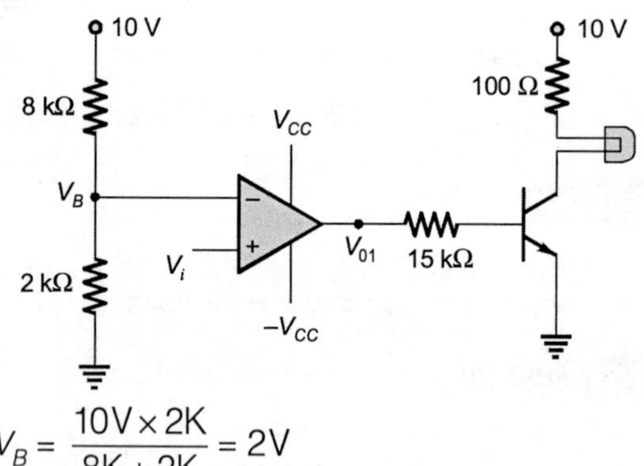

$$V_B = \frac{10\text{V} \times 2\text{K}}{8\text{K} + 2\text{K}} = 2\text{V}$$

When V_1 exceeds 2 V output of op-amp V_{01} goes to V_{CC} and drives BJT into saturation shorted LED will glow,

In the given problem V_i exceeds $2V$ three times and hence output V_{01} of op-amp goes to V_{CC} thrice so that LED glow three times.

1.56 (15)

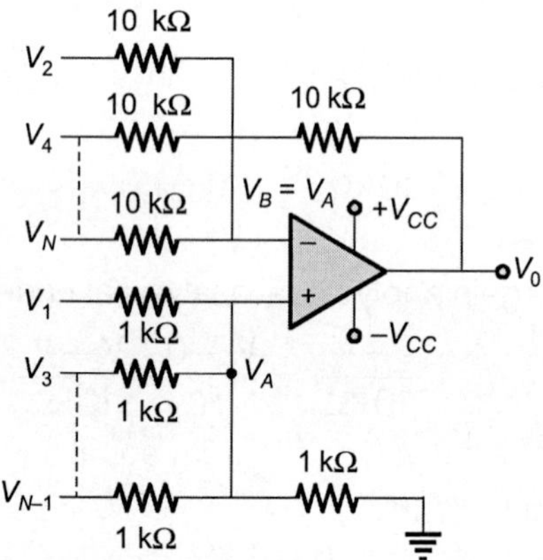

Node A:

$$\frac{V_A - V_1}{1\text{K}} + \frac{V_A - V_3}{1\text{K}} + \ldots\ldots \frac{V_A - V_{N-1}}{1\text{K}} + \frac{V_A}{1\text{K}} = 0$$

$$V_A\left(\frac{N}{2}+1\right) = V_1 + V_3 + \ldots + V_{N-1}$$

$$V_B = V_A \qquad \because \text{ Virtual short}$$

Node B:

$$\frac{V_A - V_2}{10\text{K}} + \frac{V_A - V_4}{10\text{K}} + \ldots + \frac{V_A - V_N}{10\text{K}} + \frac{V_A - V_0}{10\text{K}} = 0$$

$$V_0 = V_A\left(\frac{N}{2}+1\right) - (V_2 + V_4 + V_6 + \ldots + V_N)$$

$$= \left(\frac{N}{2}+1\right)\cdot\frac{(V_1 + V_3 + \ldots + V_{N-1})}{\left(\frac{N}{2}+1\right)} - (V_2 + V_4 + \ldots + V_N)$$

$$= V_1 - V_2 + V_3 - V_4 + \ldots \quad ; \quad = 1 + \frac{1}{2} + \frac{1}{3} + \frac{1}{4} \ldots$$

$$= \Sigma\frac{1}{N} = \infty$$

$\Rightarrow$ Output of op-amp goes to saturation

$$V_0 = V_{sat} = V_{CC} = 15 \text{ V}$$

1.57 (800)

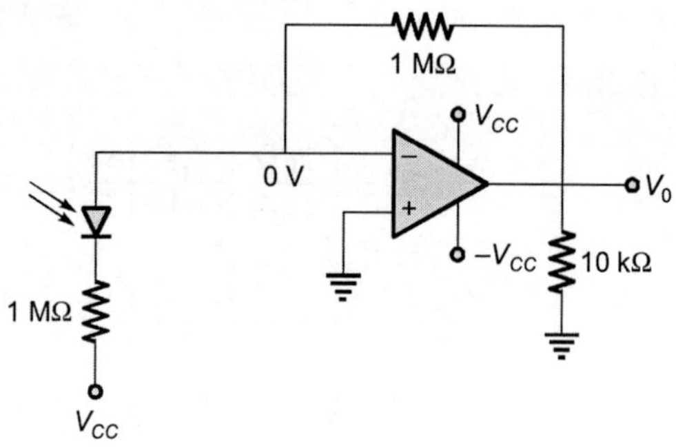

$$\text{Responsivity} = \frac{\text{Generated photo current}}{\text{Incident light power}} = \frac{I_0}{P_i}$$

$$0.8 \text{ A/w} = \frac{I_0}{10 \times 10^{-6}}$$

$$\Rightarrow \quad I_0 = 8\ \mu\text{A}$$

$$V_0 = I_0 \times 1\ \text{M}\Omega$$

$$= 8 \times 10^{-6} \times 1 \times 10^6 = 8 \text{ V}$$

The photo current through load R_L = 10 kΩ is given by

$$I_L = \frac{V_0}{R_L} = \frac{8}{10 \times 10^3}$$

$$= 800\ \mu\text{A (in upward direction)}$$

1.58 (0.798)

Initially switch is closed and V_B = 10 V

$$\Rightarrow \quad V_{01} = -10 \text{ V}$$

$$\Rightarrow \quad V_0 = -V_2 = -5 \text{ V}$$

$$\Rightarrow \quad V_A = \frac{V_0}{4\text{k} + 1\text{k}} \times 1\text{k} = -1 \text{ V}$$

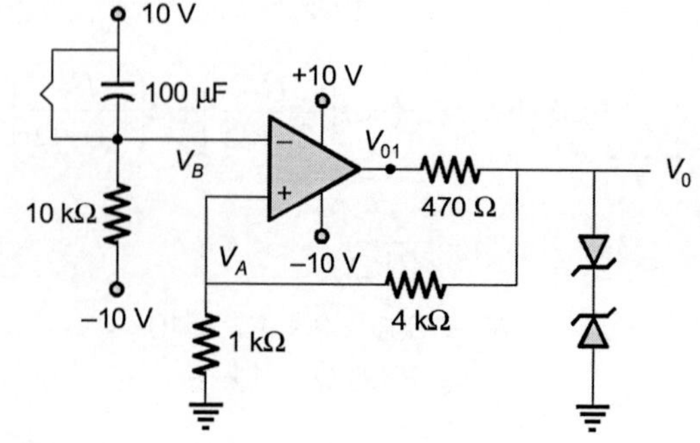

At $t = 0$;

The switch is opened and as $t \to \infty$, V_B approaches -10 V.

Let at $t = T_1$,

V_B exceeds V_A (–1 V) so that V_{01} changes from –10 V to 10 V

$\Rightarrow V_0$ charges from –5 V to 5 V

$$V_B = V_f + (V_i - V_f)e^{-t/\tau}$$
$$= -10 + [10 - (-10)]\, e^{-t/RC}$$

At $t = T_1$ $\quad V_B = -1$

$$-1\text{ V} = -10 + 20\, e^{-T1/RC}$$

$$\Rightarrow \quad T_1 = RC\, ln\frac{20}{9}$$
$$= 10 \times 10^3 \times 100 \times 10^{-6} \times 0.798$$
$$= 0.798 \text{ sec}$$

1.59 (413.79)

$$\text{Overall input} = V_{ios} + V_{input}$$
$$= 5 \text{ mV} + 25 \text{ mV} = 30 \text{ mV}$$

$$V_0 = \frac{\left(1+\frac{R_F}{R_1}\right)}{1+\frac{1}{A_{OL}}\left(1+\frac{R_F}{R_1}\right)} \times \text{Overall input}$$

$$= \frac{1+\frac{15\text{ k}\Omega}{1\text{ k}\Omega}}{1+\frac{1}{100}\left(1+\frac{15\text{ k}\Omega}{1\text{k}\Omega}\right)} \times 30 \times 10^{-3} = 413.79 \text{ mV}$$

1.60 (–1)

$$H_1(s) = \frac{V_{01}}{V_i} = 1+\frac{R_1}{sL}$$

$$= 1+\frac{1}{\frac{sL}{R_1}} = 1+\frac{1}{s\times 10^{-6}}$$

$$H_2(s) = \frac{V_0}{V_{01}} = -\frac{R_3}{\frac{R_2Cs+1}{Cs}}$$

$$= \frac{-sR_3C}{1+sR_2C} = \frac{-Cs}{1+Cs} = \frac{-1}{1+\frac{1}{Cs}}$$

$$H(s) = H_1(s)\, H_2(s)$$

$$= \left(1+\frac{1}{s\times 10^{-6}}\right)\left(\frac{-s\times 1\times 10^{-6}}{1+s\times 10^{-6}}\right) = -1$$

1.61 (b)

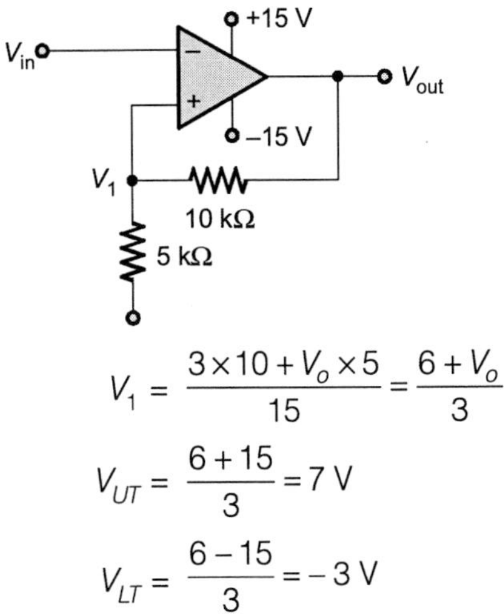

$$V_1 = \frac{3\times 10 + V_o \times 5}{15} = \frac{6+V_o}{3}$$

$$V_{UT} = \frac{6+15}{3} = 7 \text{ V}$$

$$V_{LT} = \frac{6-15}{3} = -3 \text{ V}$$

1.62 (44.4)

- In the given circuit,

  Feedback factor, $\beta = \frac{R_1}{R_1+R_2} = \frac{1}{80}$

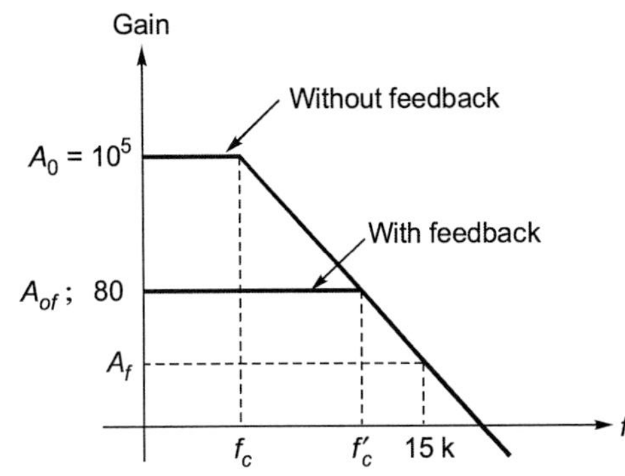

- $A_{of} = \frac{A_o}{1+A_o\beta} \simeq 80$
- $f_c' = f_c(1+A_o\beta) = 8\left(1+\frac{10^5}{80}\right)\text{Hz} = 10{,}008 \text{ Hz}$
- Gain at f = 15 kHz = 15000 Hz is,

$$A_f = \frac{A_{of}}{\sqrt{1+\left(\frac{f}{f_c'}\right)^2}} = \frac{80}{\sqrt{\left(1+\left(\frac{15000}{10008}\right)^2\right)}} \simeq 44.4$$

1.63 (1.145)

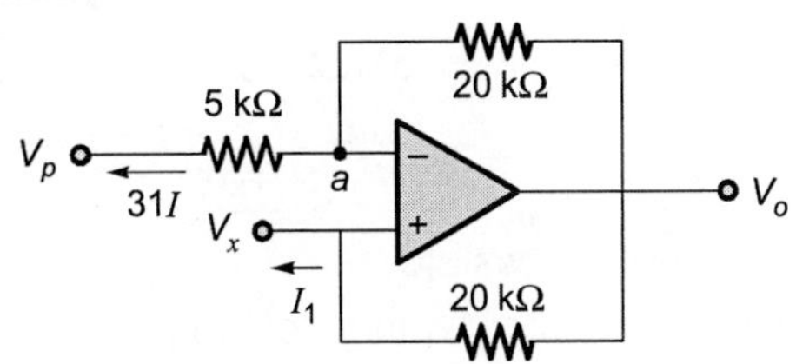

KCL at a

$$\frac{V_o - V_x}{20} = \frac{V_x - 0.7}{5}$$

$$V_o - V_x = 4V_x - 2.8$$

$$V_o = 5V_x - 2.8 \quad \text{...(i)}$$

Now, $I_1 = 31I$

$$I_s e^{V_x/V_T} = 31 I_s e^{V_P/V_T}$$

$$\frac{V_x}{V_T} = ln31 + \frac{V_P}{V_T}$$

$$\frac{V_x - V_P}{V_T} = ln31$$

$\Rightarrow$ $V_x = 0.789\text{ V}$

From equation (i),

$$V_o = 5 \times 0.789 - 2.8 = 1.145\text{ V}$$

1.64 (c)

For $t > 0$,

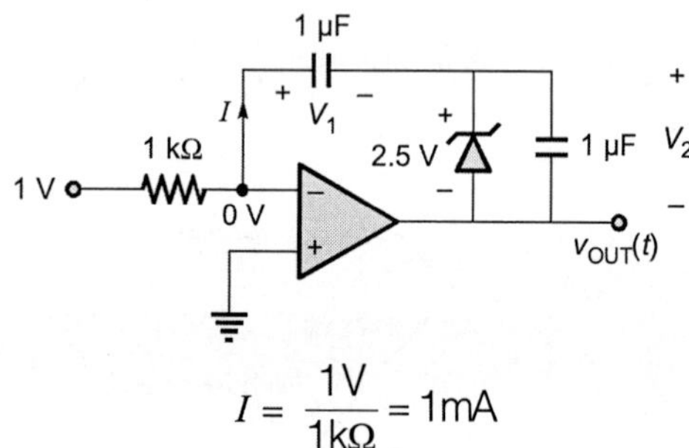

$$I = \frac{1\text{V}}{1\text{k}\Omega} = 1\text{mA}$$

The capacitor charges with constant current I and both V_1 and V_2 will increase till V_2 reaches 2.5 V. Thereafter, $V_2 = 2.5$ V and V_1 increases with time.

When $v_{out}(t) = -10$ V,

$$V_1 = 7.5\text{ V}$$

So,

$$\frac{1}{1\mu\text{F}}\int_0^t (1\text{mA})dt = 7.5\text{ V}$$

$$10^3 t = 7.5$$

$$t = 7.5\text{ msec}$$

1.65 (0.5)

Applying the concept of virtual ground, we get,

$$V_o = -\frac{R_2}{R_1} \cdot V_{in} \qquad [\because \text{non-inverting amplifier}]$$

$$\therefore \qquad V_o = -\frac{31\text{k}\Omega}{1\text{k}\Omega} \times 1\text{V}$$

$$V_o = -31\text{ V} < -15\text{ V}$$

which is not possible

Hence, the output voltage of the op-amp is equal to –15 V.

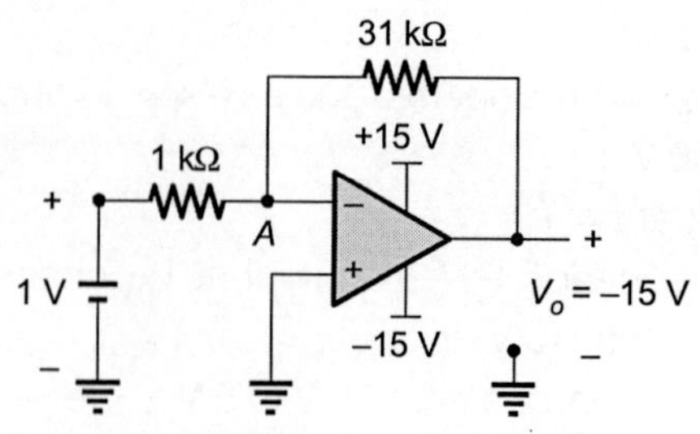

Now applying KCL of node 'A', we get,

$$\frac{V_A - (-15)}{31\text{k}\Omega} + \frac{V_A - 1}{1\text{k}\Omega} = 0$$

$$\frac{V_A}{31\text{k}\Omega} + \frac{V_A}{1\text{k}\Omega} = \frac{-15}{31\text{k}\Omega} + \frac{1}{1\text{k}\Omega}$$

$$V_A\left[\frac{1}{31} + \frac{1}{1}\right] = -\frac{15}{31} + 1$$

$$V_A = 0.5\text{ V}$$

1.66 (6)

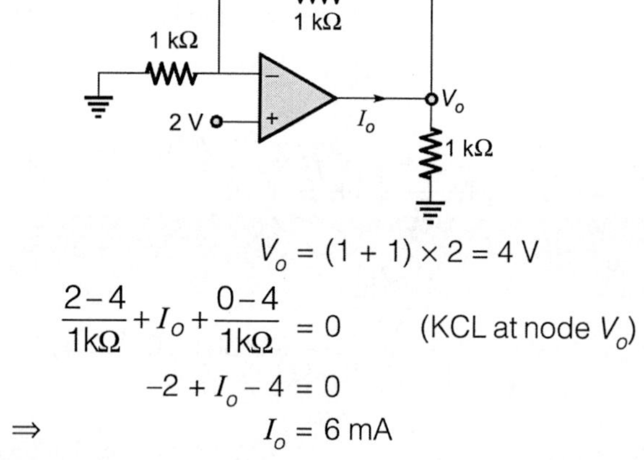

$$V_o = (1 + 1) \times 2 = 4\text{ V}$$

$$\frac{2-4}{1\text{k}\Omega} + I_o + \frac{0-4}{1\text{k}\Omega} = 0 \qquad \text{(KCL at node } V_o\text{)}$$

$$-2 + I_o - 4 = 0$$

$\Rightarrow$ $I_o = 6\text{ mA}$

1.67 (d)

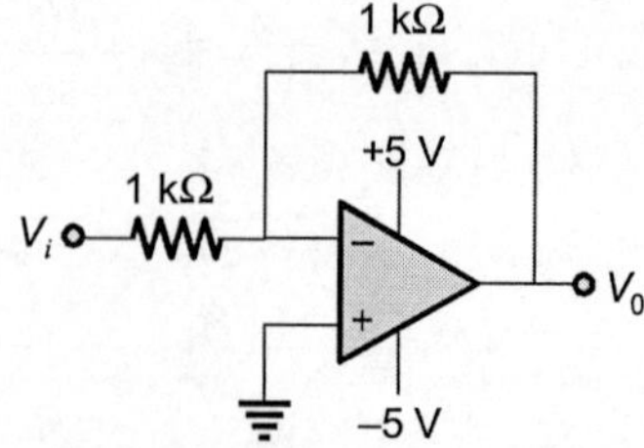

Given circuit is a Schmitt trigger of non-inverting type.

$$V_o = \pm 5\text{ V}$$

$$V^+ = \frac{V_o \times 1 + V_i \times 1}{1+1} = \frac{V_o + V}{2}$$

Let, $V_o = -5$ V,

$$V^+ = \frac{-5 + V_i}{2}$$

V_o can change from –5 V to +5 V if $V^+ > 0$ i.e. $\frac{-5+V_i}{2} > 0 \Rightarrow V_i > 5$ V.

Similarly V_o can change from +5 V to –5 V if $V_i < -5$ V.
But given input has peak value 1 V. Hence output cannot change from +5 V to –5 V or –5 V to +5 V.
∴ Output remain constant at +5 V or –5 V.
Correct answer is option (d)

1.68 (b)

Op-amp active filter (LPF) inverting type 3 dB cut-off frequency,

$$f_c = \frac{1}{2\pi RC} = \frac{1}{2\pi \times 2\times 10^3 \times 10^{-6}}$$

$$= \frac{500}{2\pi} = 79.58 \text{ Hz}$$

1.69 (b)

$$V^- = V^+$$

$$\frac{V_{out}R_{in} + V_{in}R_F}{R_{in}+R_F} = \frac{V_{ref}R_2}{R_1+R_2}$$

$$\frac{1\times R_{in} + 0.1\times R_F}{R_{in}+R_F} = \frac{V_{ref}R_2}{R_1+R_2} \quad \text{...(i)}$$

$$\frac{6R_{in} + 1\times R_F}{R_{in}+R_F} = \frac{V_{ref}R_2}{R_1+R_2} \quad \text{...(ii)}$$

Equate equation (i) and (ii),

$$1 \times R_{in} + 0.1 \times R_F = 6 \times R_{in} + 1 \times R_F$$

$$-5\, R_{in} = 0.9\, R_F$$

$$\therefore \quad \frac{R_F}{R_{in}} = \frac{-5}{0.9} = -5.55$$

(According to the given data magnitude is taken)

1.70 (d)

For ideal op-amp,

$$V^- = V^+ = 0$$

KCL at node V^-:

$$\frac{V_{in}-0}{R} + \frac{(-V_{ref}-0)}{R} + \frac{V_{out}-0}{R_F} = 0$$

$$\frac{V_{out}}{R_F} = \frac{1}{R}(V_{ref}-V_{in})$$

$$V_{out} = \frac{R_F}{R}(V_{ref}-V_{in})$$

We want, $V_{out} = 0$

$\Rightarrow \quad V_{ref} - V_{in} = 0$

$\Rightarrow \quad V_{in} = V_{ref}$

1.71 (–12)

At, $t = 0^+$: Capacitor is short circuit

$\therefore \quad V^- = V_{in} = 5$ V

$V^+ = 0$ V

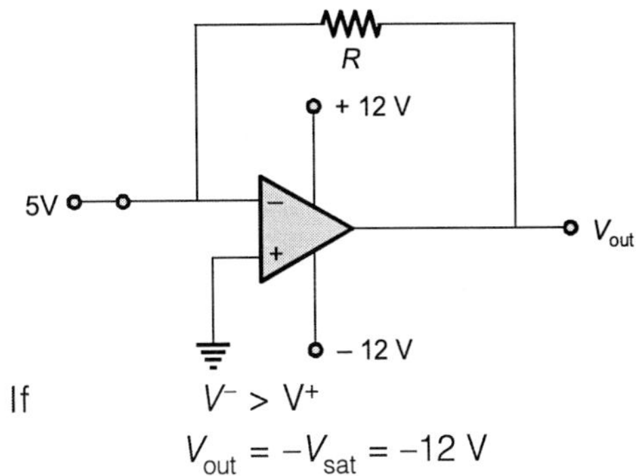

If $\quad V^- > V^+$

$$V_{out} = -V_{sat} = -12 \text{ V}$$

1.72 (b, c)

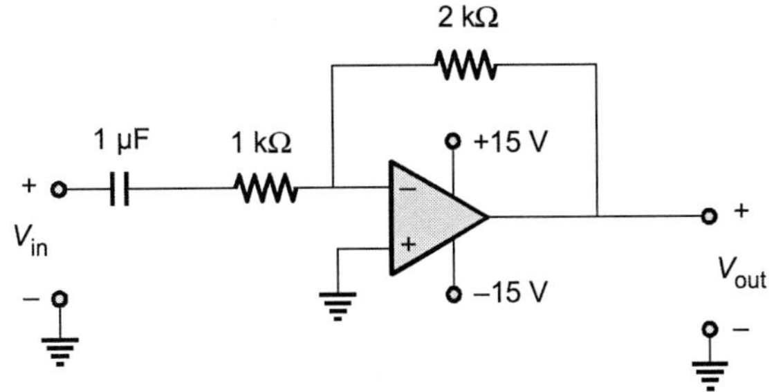

Drawing the circuit in s-domain,

$R_2 = 2$ kΩ

$\frac{1}{Cs} = \frac{10^6}{s}\,\Omega$

$V_{in}(s)$

$R_1 = 1$ kΩ

+15 V

–15 V

$V_o(s)$

$$\frac{V_o(s)}{V_{in}(s)} = \frac{-R_2}{R_1 + \frac{1}{Cs}} = \frac{-2\text{ k}\Omega}{1\text{k}\Omega + \frac{10^6}{s}\,\Omega}$$

$$\frac{V_o(s)}{V_{in}(s)} = \frac{-2s}{s+1000}$$

For $s = 0$, $\frac{V_o(s)}{V_{in}(s)} = 0$

and for $\quad s = \infty$, $\frac{V_o(s)}{V_{in}(s)} = -2$

Hence, the given circuit is a high pass filter.

$$\left|\frac{V_o(s)}{V_{in}(s)}\right|_{max} = 2$$

At 3 dB frequency ω_c, the magnitude of voltage gain decreases by 3 dB from the maximum value i.e.

$$\left|\frac{V_o(s)}{V_{in}(s)}\right|_{s=j\omega_c} = \frac{1}{\sqrt{2}}\times\left|\frac{V_o(s)}{V_{in}(s)}\right|_{max}$$

$$\Rightarrow \quad \frac{2\omega_c}{\sqrt{\omega_c^2+10^6}} = \frac{1}{\sqrt{2}}\times 2$$

$$\Rightarrow \quad \frac{1}{\sqrt{1+10^6/\omega_c^2}} = \frac{1}{\sqrt{2}}$$

$$\Rightarrow \quad \frac{10^6}{\omega_c^2} = 1 \Rightarrow \omega_c^2 = 10^6$$

$$\Rightarrow \quad \omega_c = 10^3 = 1000 \text{ rad/sec}$$

Hence, option (b) and option (c) are the correct answers.

1.73 (a)

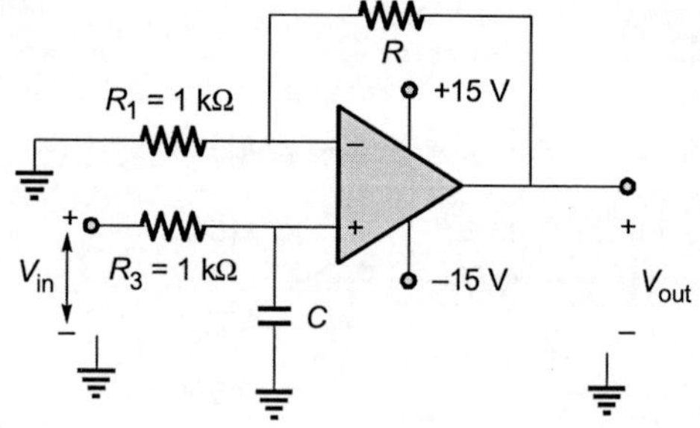

Using voltage division rule, the voltage at non-inverting terminal,

$$V_+ = \frac{\left(\frac{1}{sC}\right)V_{in}}{R_3 + \frac{1}{sC}} = \frac{V_{in}}{1+sR_3C}$$

Using concept of virtual ground,

$$V_- = V_+ = \frac{V_{in}}{1+sR_3C} \quad \text{...(i)}$$

Applying KCL at inverting terminal, we get,

$$\frac{V_- - 0}{R_1} + \frac{V_- - V_{out}}{R} = 0$$

$$\therefore \; V_-\left[\frac{1}{R_1} + \frac{1}{R}\right] = \frac{V_{out}}{R}$$

$$\therefore \quad V_{out} = V_-\left(1+\frac{R}{R_1}\right)$$

Using equation (i),

$$V_{out} = \frac{V_{in}}{1+sR_3C}\left(1+\frac{R}{R_1}\right)$$

$$\therefore \quad A_v = \frac{V_{out}}{V_{in}} = \frac{1+R/R_1}{1+sR_3C} \quad \text{...(ii)}$$

From the given Bode plot as below,

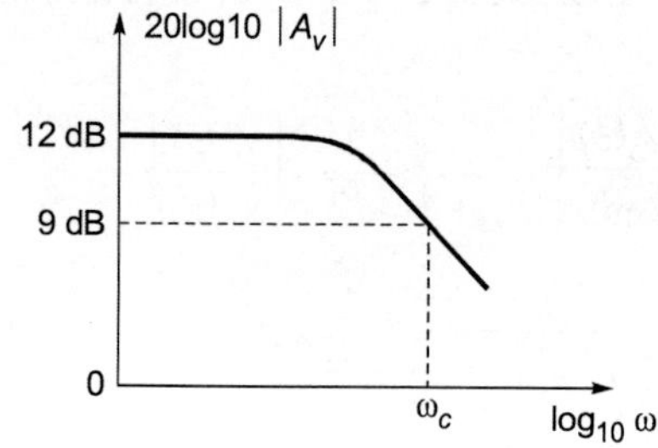

There exists a pole at ω_c for the gain transfer function. Hence,

$$A_v = \frac{K}{s+\omega_c} = \frac{K/\omega_c}{1+s/\omega_c} \quad \text{...(iii)}$$

At $\omega = 0$, $20\log_{10}|A_v| = 12 \text{ dB}$

$$\Rightarrow 20\log_{10}\left(\frac{K}{\omega_c}\right) = 12$$

$$\Rightarrow \log_{10}\left(\frac{K}{\omega_c}\right) = \frac{3}{5}$$

$$\Rightarrow \quad \frac{K}{\omega_c} = 3.98 \simeq 4$$

Comparing equations (ii) and (iii), we get,

$$1+\frac{R}{R_1} = \frac{K}{\omega_c} = 4$$

$$R = 3R_1 = 3 \times 1 = 3 \text{ k}\Omega$$

Also, $\quad \omega_c = \frac{1}{R_3 C}$

$$\therefore \quad C = \frac{1}{\omega_C R_3}$$

From the given Bode plot,

$$\log_{10}(\omega_c) = 3$$

$$\therefore \quad \omega_c = 10^3 = 1000 \text{ rad/sec}$$

Hence, $\quad C = \frac{1}{1000 \times 1000} = 1 \mu\text{F}$

Therefore, $R = 3$ kΩ and $C = 1$ μF.

1.74 (a)

Given circuit,

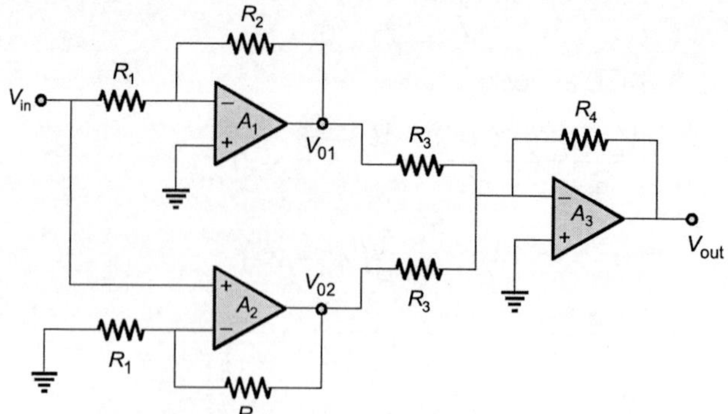

Here, A_1 is an inverting amplifier and A_2 is a non-inverting amplifier.

$$V_{01} = \frac{-R_2}{R_1}V_{in}$$

$$V_{02} = \left(1+\frac{R_2}{R_1}\right)V_{in}$$

Also, A_3 is an inverting summing amplifier,

$$V_{out} = \frac{-R_4}{R_3}V_{01} - \frac{R_4}{R_3}V_{02}$$

$$= \frac{-R_4}{R_3}\left[\frac{R_2}{R_1}V_{in} + \left(1+\frac{R_2}{R_1}\right)V_{in}\right]$$

$$V_{out} = \frac{-R_4}{R_3}V_{in}$$

$$\text{Gain, } \frac{V_{out}}{V_{in}} = \frac{-R_4}{R_3}$$

1.75 (a)

Considering the Op-Amp 1 circuit, it represents a non-inverting Schmitt Trigger.

The Schmitt trigger works by comparing the input voltage V_i with the upper threshold voltage (V_{UT}) and the lower threshold voltage (V_{LT}) given by

$$V_{UT} = \frac{R_1}{R_2}V_{sat} = \frac{10}{100}\times 10 = 1\text{ V}$$

$$V_{LT} = \frac{R_1}{R_2}(-V_{sat}) = \frac{10}{100}\times(-10)$$

$$= -1\text{ V}$$

For an input triangular signal with peak voltages of ±2 V and time period of 8 μs, the output V_{01} is given as below:

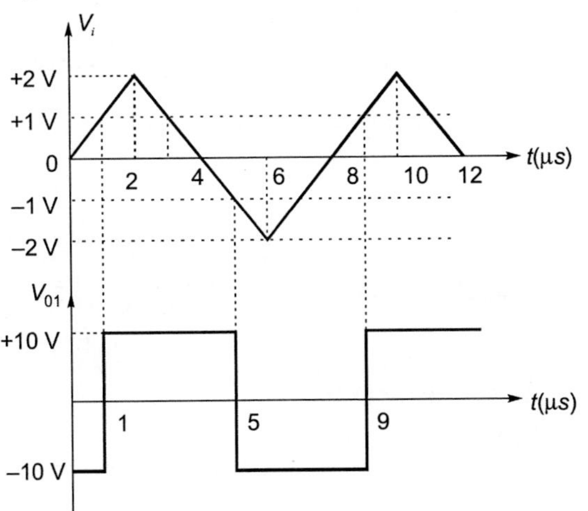

As V_i exceeds V_{UT} i.e., 1 V, V_{01} switches from $-V_{sat}$ to $+V_{sat}$ and when V_i falls below V_{LT} i.e., −1 V, V_{01} switches from $+V_{sat}$ to $-V_{sat}$. Considering the Op-Amp 2 circuit, it represents an integrator.

$$V_{02} = \left[\frac{-1}{L}\int_{-\infty}^{t} V_{01}\cdot dt\right]\cdot R_L$$

$$V_{02} = -\int_{-\infty}^{t} V_{01}dt \times 10^6$$

We can see that V_{02} will be saturated to ±10 V in less than half time period resulting in the waveform as shown below:

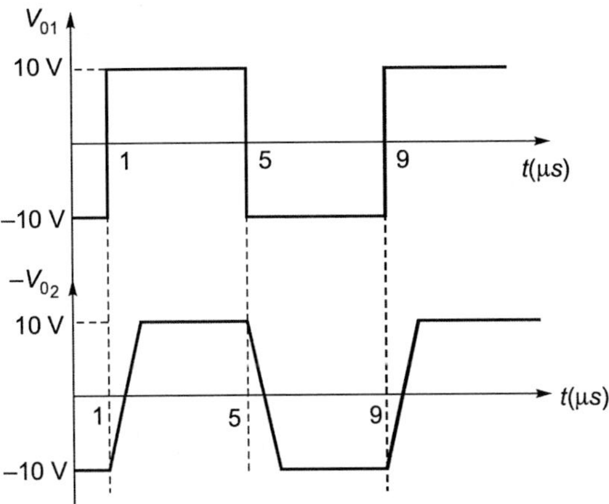

Hence, V_{01} is delayed by 1 μs relative to V_i and V_{02} is a trapezoidal waveform.

2 Diodes Applications

2.1 The Ebers Moll model is applicable to
(a) Bipolar junction transistors
(b) NMOS transistors
(c) Unipolar junction transistors
(d) Junction field-effect [1995 : 1 M]

2.2 A Zener diode in the circuit shown in below figure has a knee current of 5 mA, and a maximum allowed power dissipation of 300 mW. What are the minimum and maximum load currents that can be drawn safely from the circuit, keeping the output voltage V_o constant at 6 V?

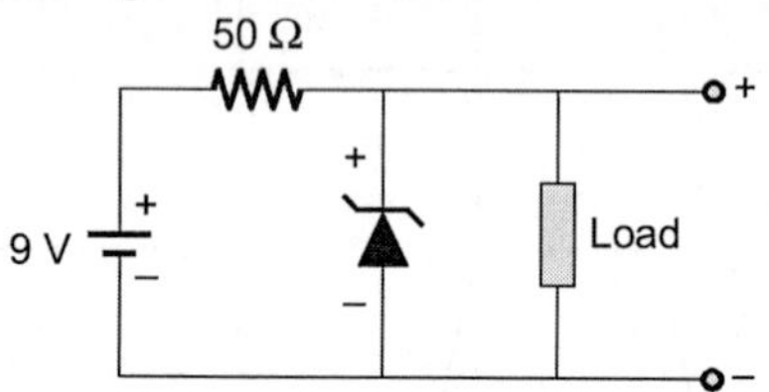

(a) 0 mA, 180 mA (b) 5 mA, 110 mA
(c) 10 mA, 55 mA (d) 60 mA, 180 mA
[1996 : 2 M]

2.3 For small signal ac operation, a practical forward biased diode can be modeled as
(a) a resistance and a capacitance
(b) an ideal diode and resistance in parallel
(c) a resistance and an ideal diode in series
(d) a resistance [1998 : 1 M]

2.4 For full wave rectification, a four diode bridge rectifier is claimed to have the following advantages over a two diode circuit.
1. Less expensive transformer
2. Smaller size transformer and
3. Suitability for higher voltage application of these
(a) only (1) and (2) are true
(b) only (1) and (3) are true
(c) only (2) and (3) are true
(d) (1), (2), as well as (3) are true [1998 : 1 M]

2.5 A dc power supply has a no-load voltage of 30 V, and a full-load voltage of 25 V at a full-load current of 1 A. Its output resistance and load regulation, respectively, are
(a) 5 Ω and 20% (b) 25 Ω and 20%
(c) 5 Ω and 16.7% (d) 25 Ω and 16.7%
[1999 : 2 M]

2.6 The transistor shunt regulator shown in the figure has a regulated output voltage of 10 V, when the input varies from 20 V to 30 V. The relevant parameters for the Zener diode and the transistor are: $V_Z = 9.5$, $V_{BE} = 0.5$ V, $\beta = 99$. Neglect the current through R_B. Then the maximum power dissipated in the Zener diode (P_Z) and the transistor (P_T) are

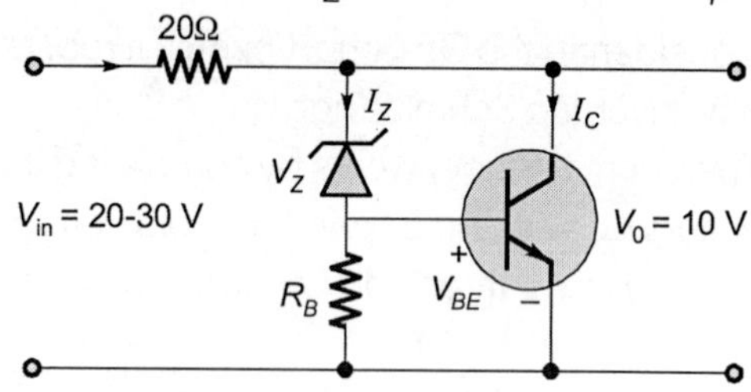

(a) $P_Z = 75$ mW, $P_T = 7.9$ W
(b) $P_Z = 85$ mW, $P_T = 8.9$ W
(c) $P_Z = 95$ mW, $P_T = 9.9$ W
(d) $P_Z = 115$ mW, $P_T = 11.9$ W [2001 : 2 M]

2.7 A Zener diode regulator in the figure is to be designed to meet the specifications: $I_L = 10$ mA, $V_0 = 10$ V and V_{in} varies from 30 V to 50 V. The Zener diode has $V_z = 10$ V and I_{zk} (knee current) = 1 mA. For satisfactory operation

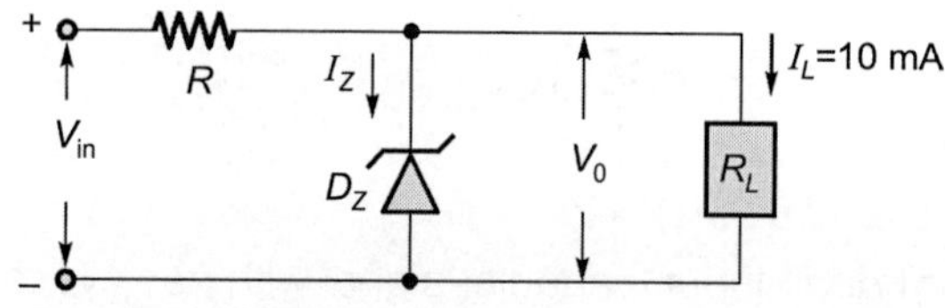

(a) $R \leq 1800\Omega$
(b) $2000\Omega \leq R \leq 2200\Omega$
(c) $3700\Omega \leq R \leq 4000\Omega$
(d) $R > 4000\ \Omega$ [2002 : 2 M]

2.8 The circuit shown in the figure is best described as a

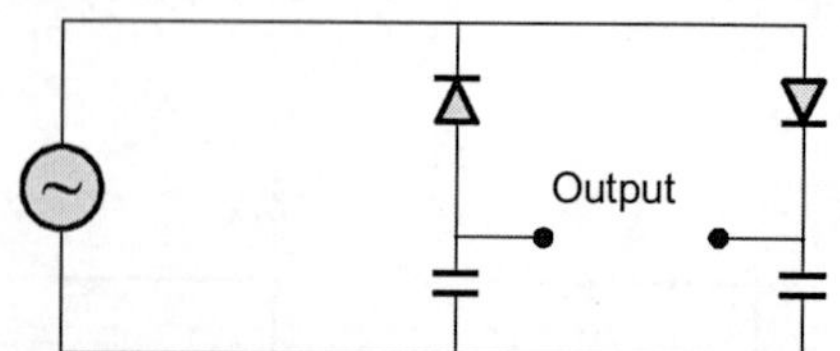

(a) bridge rectifier
(b) ring modulator
(c) frequency discriminatory
(d) voltage doubler [2003 : 1 M]

2.9 The output voltage of the regulated power supply shown in the figure is

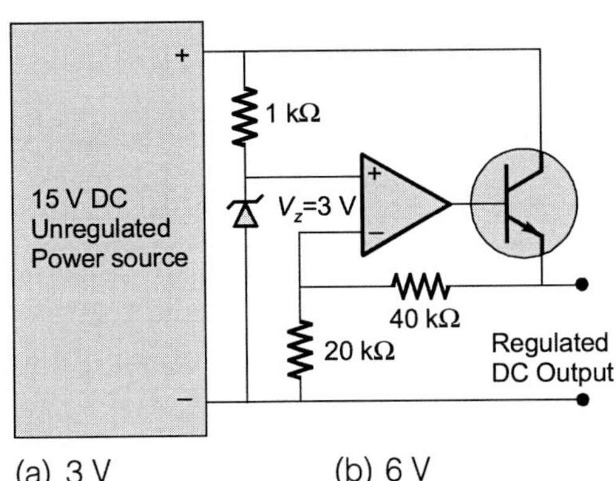

(a) 3 V (b) 6 V
(c) 9 V (d) 12 V

[2003 : 2 M]

2.10 In the voltage regulator shown in the figure, the load current can vary from 100 mA to 500 mA. Assuming that the Zener diode is ideal (i.e., the Zener knee current is negligibly small and Zener resistance is zero in the breakdown region), the value of R is

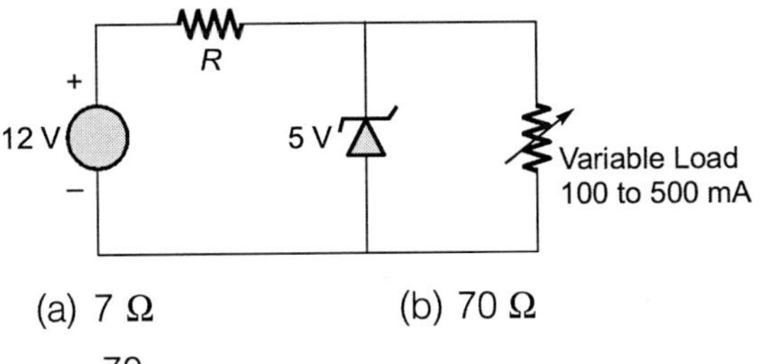

(a) 7 Ω (b) 70 Ω
(c) $\frac{70}{3}$ Ω (d) 14 Ω [2004 : 2 M]

2.11 In a full-wave rectifier using two ideal diodes, V_{dc} and V_m are the dc and peak values of the voltage respectively across a resistive load. If PIV is the peak inverse voltage of the diode, then the appropriate relationships for this rectifier are

(a) $V_{dc} = \frac{V_m}{\pi}, PIV = 2V_m$

(b) $V_{dc} = 2\frac{V_m}{\pi}, PIV = 2V_m$

(c) $V_{dc} = 2\frac{V_m}{\pi}, PIV = V_m$

(d) $V_{dc} = \frac{V_m}{\pi}, PIV = V_m$ [2004 : 2 M]

2.12 The Zener diode in the regulator circuit shown in the figure has a Zener voltage of 5.8 volts and a Zener knee current of 0.5 mA. The maximum load current drawn from this circuit ensuring proper functioning over the input voltage range between 20 and 30 volts, is

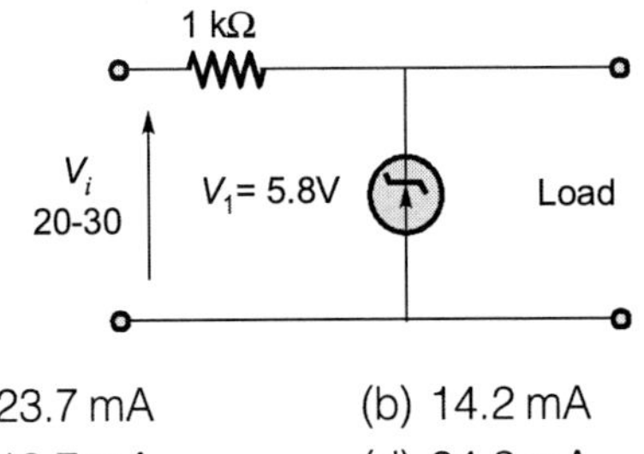

(a) 23.7 mA (b) 14.2 mA
(c) 13.7 mA (d) 24.2 mA

[2005 : 2 M]

2.13 For the circuit shown below, assume that the Zener diode is ideal with a breakdown voltage of 6 volts. The waveform observed across R is

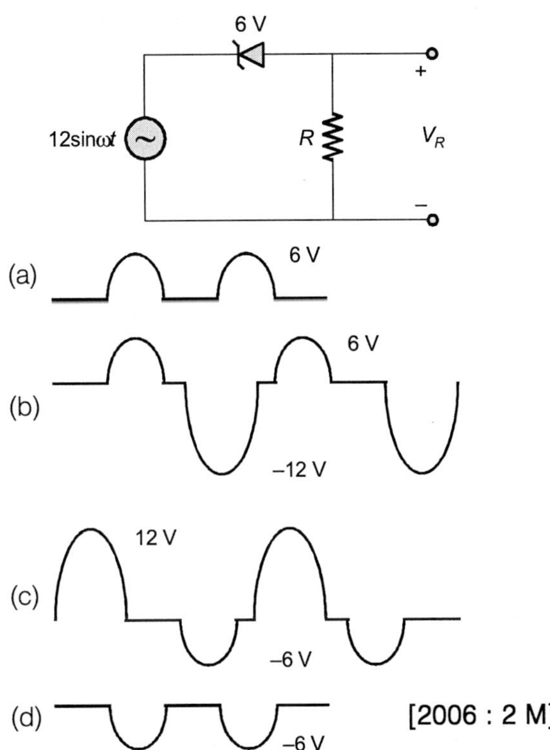

[2006 : 2 M]

Common Data for Questions 2.14 & 2.15.

A regulated power supply, shown in figure below, has an unregulated input (UR) of 15 Volts and generates a regulated output V_{out}. Use the component values shown in the figure

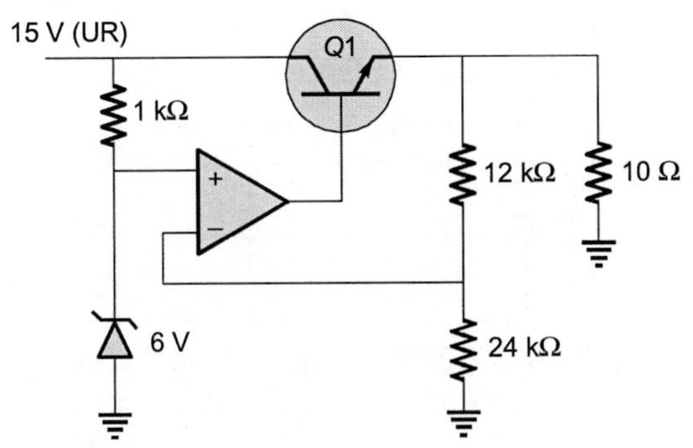

2.14 The power dissipation across the transistor shown in the figure is
(a) 4.8 Watts (b) 5.0 Watts
(c) 5.4 Watts (d) 6.0 Watts
[2006 : 2 M]

2.15 If the unregulated voltage increases by 20%, the power dissipation across the transistor Q_1
(a) increases by 20%
(b) increases by 50%
(c) remains unchanged
(d) decreases by 20% [2006 : 2 M]

2.16 The correct full wave rectifier circuit is

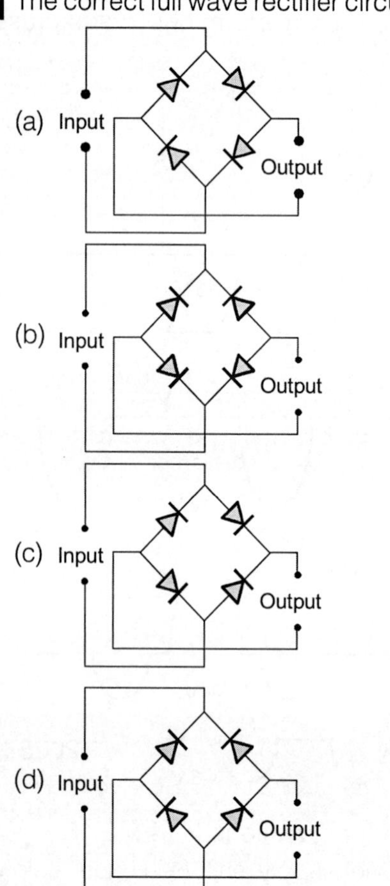

[2007 : 1 M]

2.17 For the Zener diode shown in the figure, the Zener voltage at knee is 7 V, the knee current is negligible and the Zener dynamic resistance is 10 Ω. If the input voltage (V_i) range is from 10 to 16 V, the output voltage (V_0) ranges from

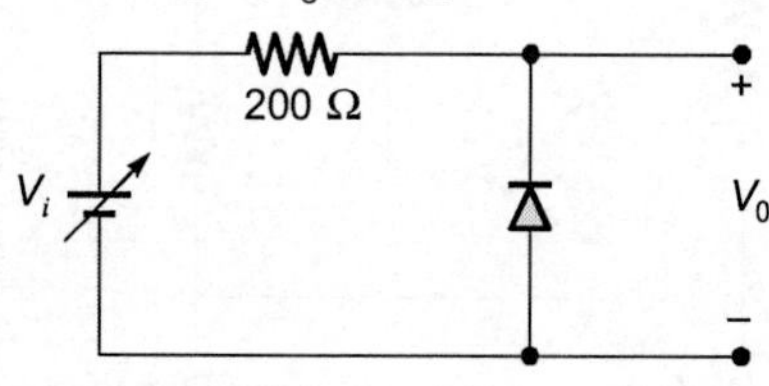

(a) 7.00 to 7.29 V (b) 7.14 to 7.29 V
(c) 7.14 to 7.43 V (d) 7.29 to 7.43 V
[2007 : 2 M]

2.18 In the following limiter circuit, an input voltage $V_i = 10 \sin 100\pi t$ is applied. Assume that the diode drop is 0.7 V when it is forward biased. The Zener breakdown voltage is 6.8 V.

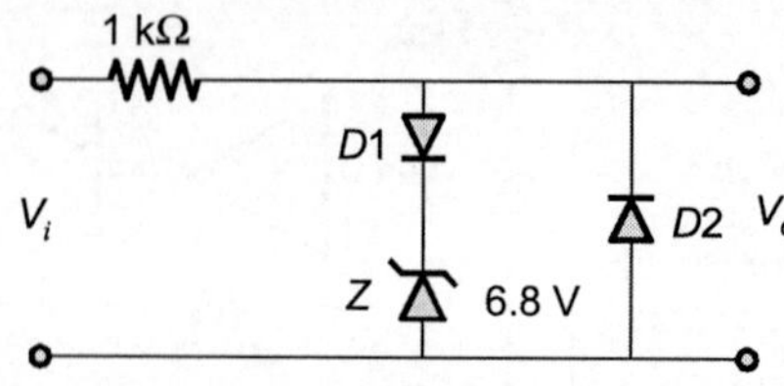

The maximum and minimum values of the output voltage respectively are
(a) 6.1 V, –0.7 V (b) 0.7 V, –7.5 V
(c) 7.5 V, – 0.7 V (d) 7.5 V. – 7.5 V
[2008 : 1 M]

2.19 In the circuit below, the diode is ideal. The voltage V is given by

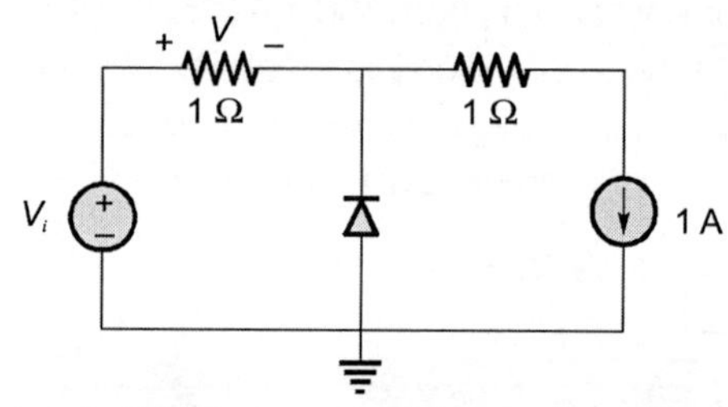

(a) min $(V_i, 1)$ (b) max $(V_i, 1)$
(c) min $(-V_i, 1)$ (d) max $(-V_i, 1)$
[2009 : 2 M]

Statement for Linked Answer Questions 2.20 and 2.21.

In the circuit shown below, assume that the voltage drop across a forward biased diode is 0.7 V. The thermal voltage $V_t = kT/q$ = 25 mV. The small signal input $v_i = V_p \cos(\omega t)$ where V_p = 100 mV.

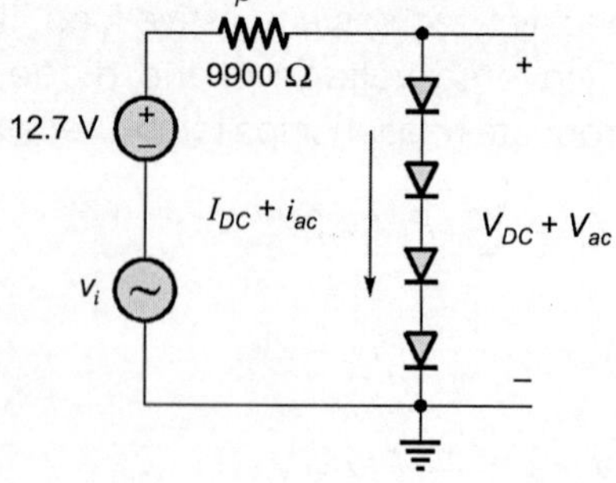

2.20 The bias current I_{DC} through the diodes is
(a) 1 mA (b) 1.28 mA
(c) 1.5 mA (d) 2 mA [2011 : 2 M]

2.21 The ac output voltage v_{ac} is
(a) 0.25 cos (ωt) mV (b) 1 cos (ωt) mV
(c) 2 cos (ωt) mV (d) 22 cos (ωt) mV
[2011 : 2 M]

2.22 The diodes and capacitors in the circuit shown are ideal. The voltage $v(t)$ across the diode $D1$ is

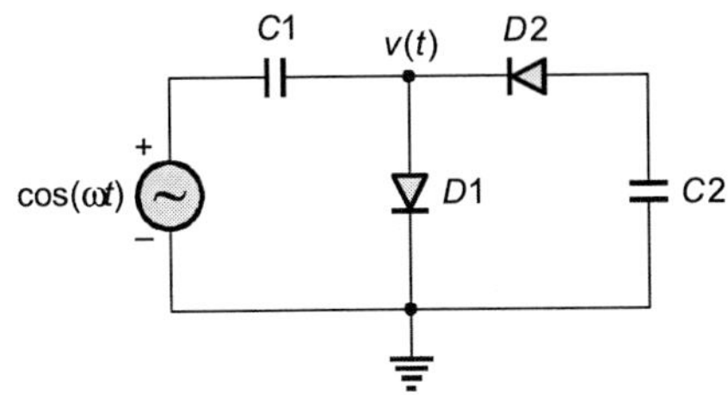

(a) $\cos(\omega t) - 1$ (b) $\sin(\omega t)$
(c) $1 - \cos(\omega t)$ (d) $1 - \sin(\omega t)$

[2012 : 1 M]

2.23 In the circuit shown below, the knee current of the ideal Zener diode is 10 mA. To maintain 5 V across R_L, the minimum value of R_L in Ω and the minimum power rating of the Zener diode in mW, respectively, are

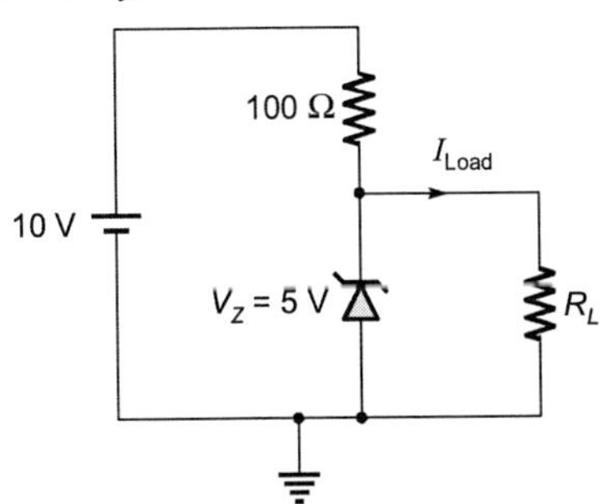

(a) 125 and 125 (b) 125 and 250
(c) 250 and 125 (d) 250 and 250

[2013 : 2 M]

2.24 A voltage 1000 sinωt volts is applied across YZ. Assuming ideal diodes, the voltage measured across WX in volts, is

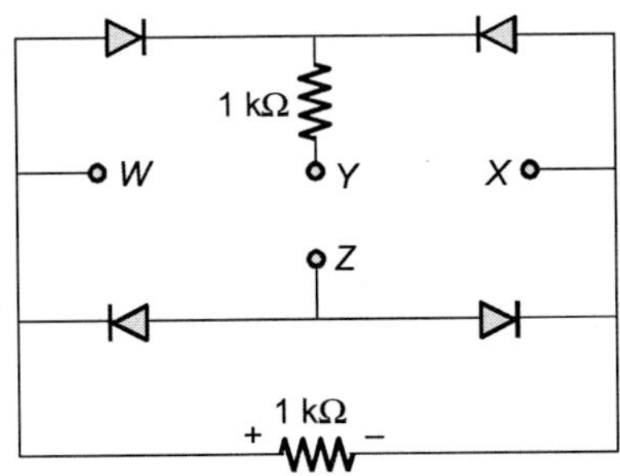

(a) $\sin \omega t$
(b) $(\sin \omega t + |\sin \omega t|)/2$
(c) $(\sin \omega t - |\sin \omega t|)/2$
(d) 0 for all t

[2013 : 2 M]

2.25 In the figure, assume that the forward voltage drops of the PN diode D_1 and Schottky diode D_2 are 0.7 V and 0.3 V, respectively. If ON denotes conducting state of the diode and OFF denotes non-conducting state of the diode, then in the circuit,

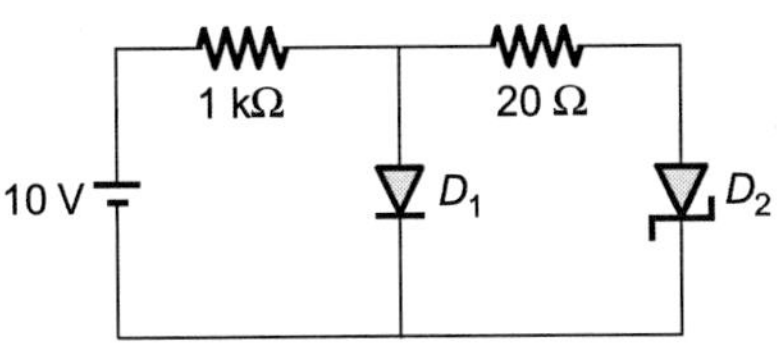

(a) both D_1 and D_2 are ON
(b) D_1 is ON and D_2 is OFF
(c) both D_1 and D_2 are OFF
(d) D_1 is OFF and D_2 is ON

[2014 : 1 M, Set-1]

2.26 The diode in the circuit shown has V_{on} = 0.7 V but is ideal otherwise. If $V_i = 5 \sin(\omega t)$ volts, the minimum and maximum values of V_o (in Volts) are, respectively,

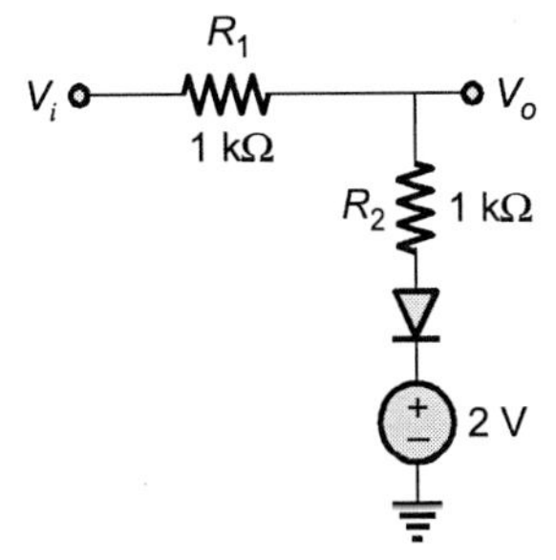

(a) –5 and 2.7 (b) 2.7 and 5
(c) –5 and 3.85 (d) 1.3 and 5

[2014 : 2 M, Set-2]

2.27 The figure shows a half-wave rectifier. The diode D is ideal. The average steady-state current (in Amperes) through the diode is approximately ____.

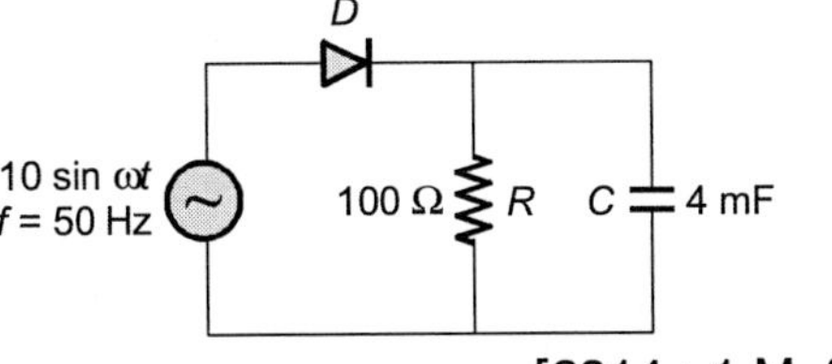

[2014 : 1 M, Set-3]

2.28 Two silicon diodes, with a forward voltage drop of 0.7 V, are used in the circuit shown in the figure. The range of input voltage V_i for which the output voltage $V_o = V_i$, is

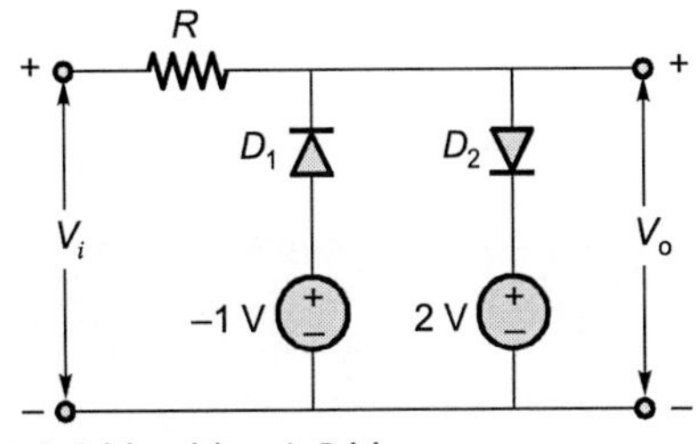

(a) $-0.3\text{ V} < V_i < 1.3\text{ V}$
(b) $-0.3\text{ V} < V_i < 2\text{ V}$
(c) $-1.0\text{ V} < V_i < 2.0\text{ V}$
(d) $-1.7\text{ V} < V_i < 2.7\text{ V}$

[2014 : 1 M, Set-4]

2.29 For the circuit with ideal diodes shown in the figure, the shape of the output (v_{out}) for the given sine wave input (v_{in}) will be

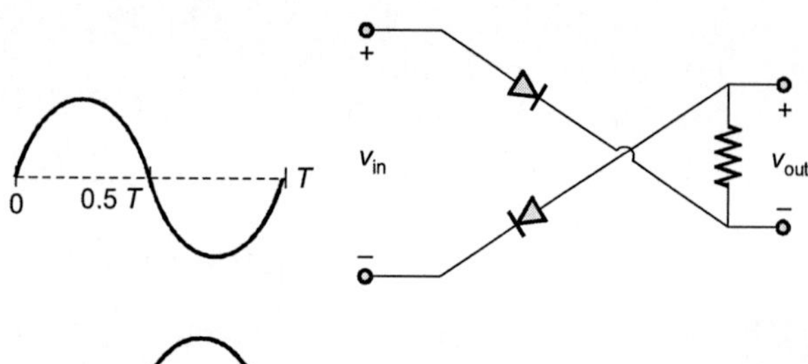

(a)
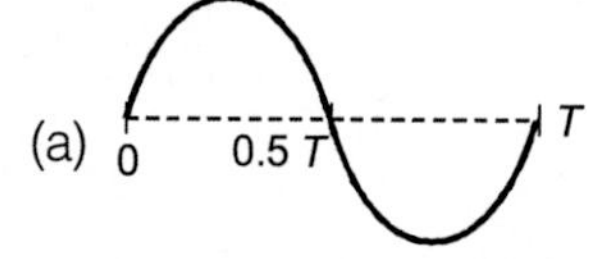

(b)
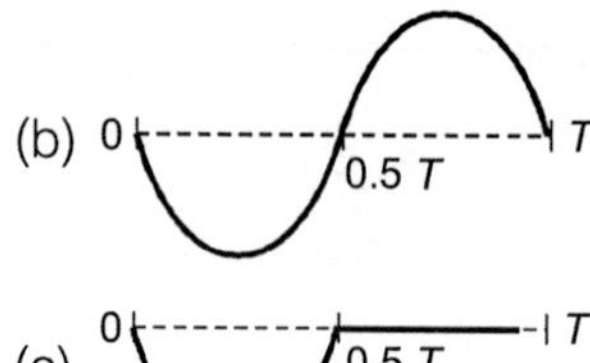

(c)

(d)
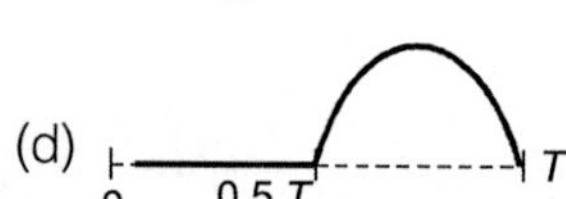

[2015 : 1 M, Set-1]

2.30 In the circuit shown below, the Zener diode is ideal and the Zener voltage is 6 V. The output voltage V_0 (in volts) is__________.

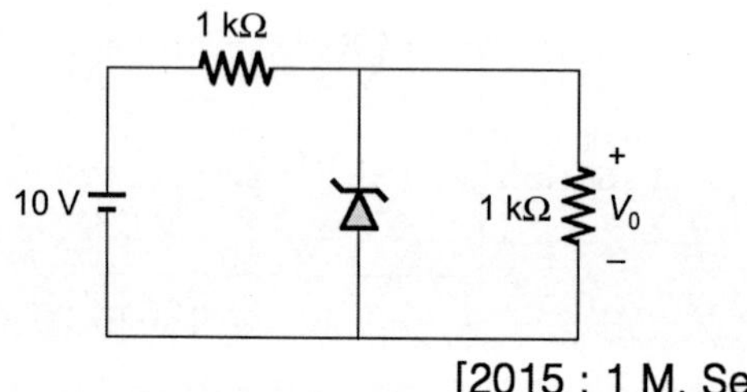

[2015 : 1 M, Set-1]

2.31 If the circuit shown has to function as a clamping circuit, then which one of the following conditions should be satisfied for the sinusoidal signal of period T?

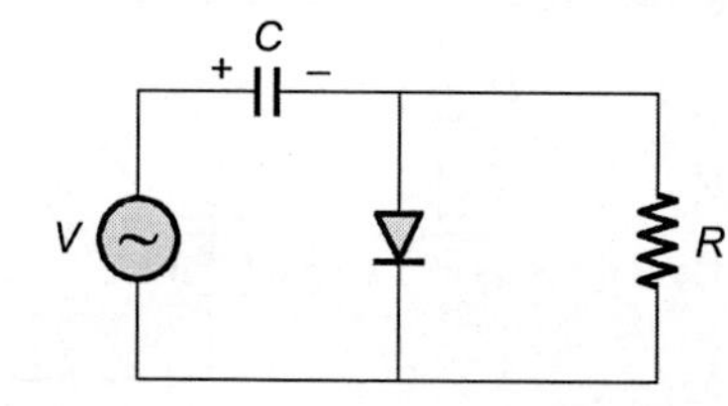

(a) $RC << T$
(b) $RC = 0.35\ T$
(c) $RC \approx T$
(d) $RC >> T$

[2015 : 1 M, Set-2]

2.32 The diode in the circuit given below has $V_{ON} = 0.7$ V but is ideal otherwise. The current (in mA) in the 4 kΩ resistor is_____.

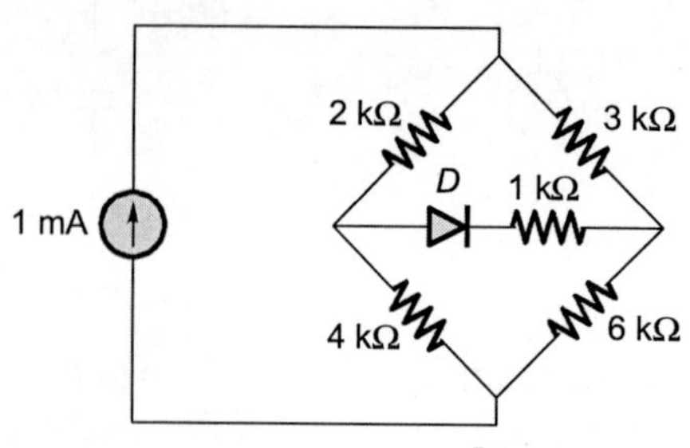

[2015 : 2 M, Set-2]

2.33 In the circuit shown, assume that diodes D_1 and D_2 are ideal. In the steady-state condition the average voltage V_{ab} (in Volts) across the 0.5 μF capacitor is _____.

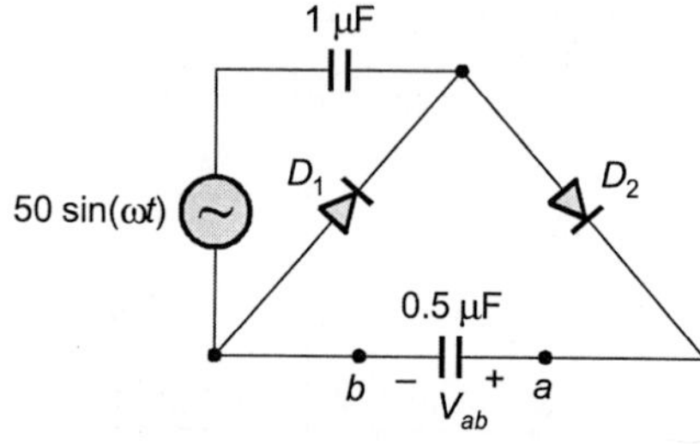

[2015 : 1 M, Set-3]

2.34 In the circuit shown, assume that the diodes D_1 and D_2 are ideal. The average value of voltage V_{ab} (in Volts), across terminals 'a' and 'b' is _____.

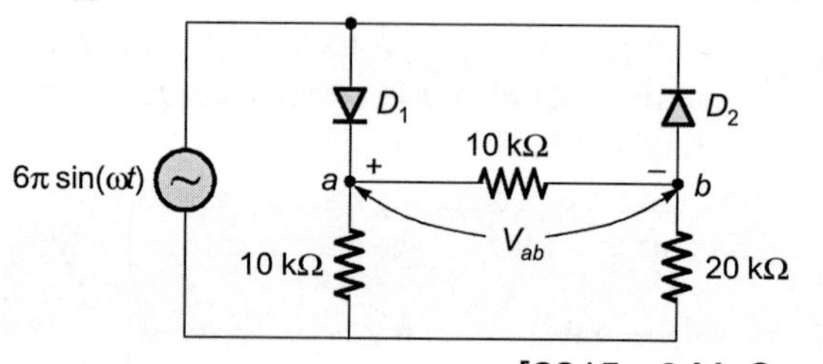

[2015 : 2 M, Set-3]

2.35 The figure shows a half-wave rectifier with a 475 μF filter capacitor. The load draws a constant current I_0 = 1 A from the rectifier. The figure also shows the input voltage V_i, the output voltage V_C and the peak-to-peak voltage ripple u on V_C. The input voltage V_i is a triangle-wave with an amplitude of 10 V and a period of 1 ms.

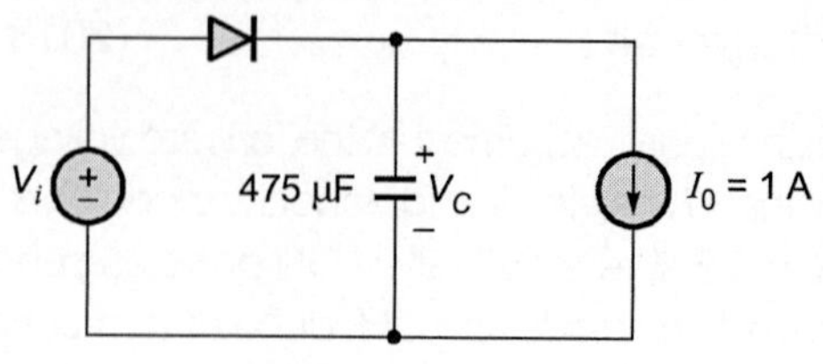

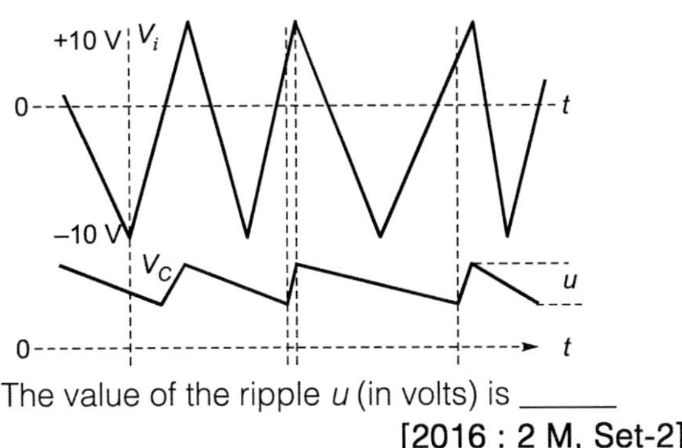

The value of the ripple u (in volts) is ______

[2016 : 2 M, Set-2]

2.36 The diodes $D1$ and $D2$ in the figure are ideal and the capacitors are identical. The product RC is very large compared to the time period of the ac voltage. Assuming that the diodes do not breakdown in the reverse bias, the output voltage V_O (in volt) at the steady state is ________.

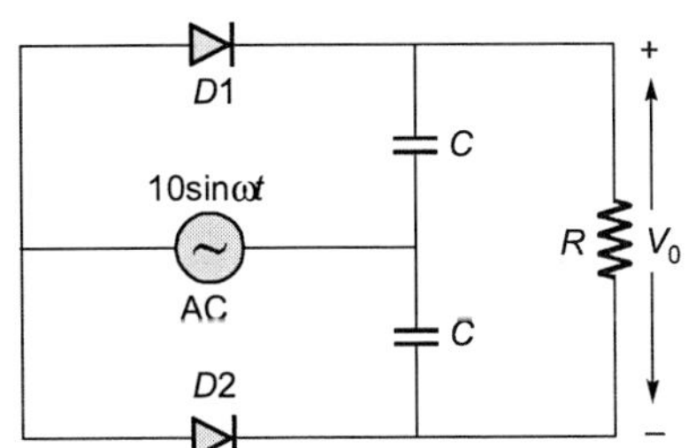

[2016 : 1 M, Set-3]

2.37 Assume that the diode in the figures has V_{on} = 0.7 V, but is otherwise ideal.

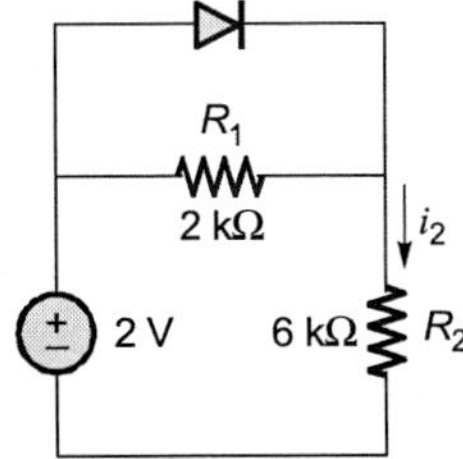

The magnitude of the current i_2 (in mA) is equal to ___.

[2016 : 1 M, Set-2]

2.38 The I-V characteristics of the Zener diodes $D1$ and $D2$ are shown in Figure I. These diodes are used in the circuit given in Figure II. If the supply voltage is varied from 0 to 100 V, then breakdown occurs in

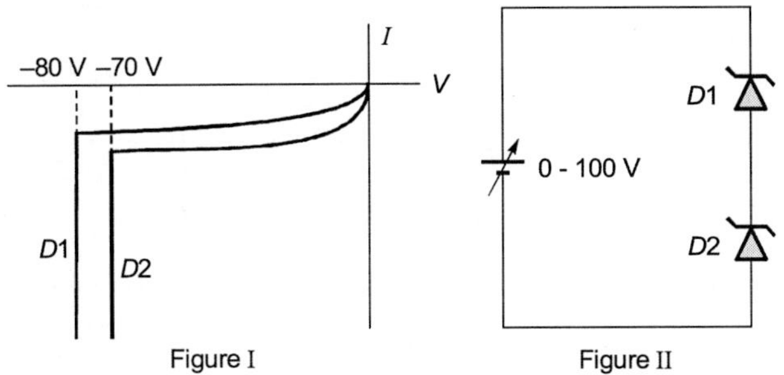

(a) $D1$ only
(b) $D2$ only
(c) both $D1$ and $D2$
(d) none of $D1$ and $D2$

[2016 : 2 M, Set-3]

2.39 The output V_o of the diode circuit shown in the figure is connected to an averaging DC voltmeter. The reading on the DC voltmeter in Volts, neglecting the voltage drop across the diode, is

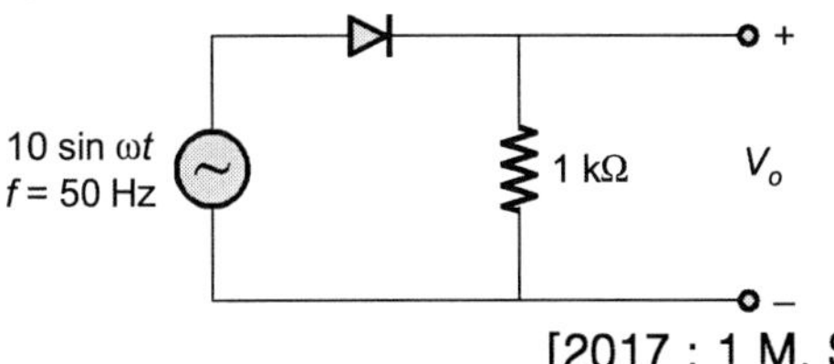

[2017 : 1 M, Set-2]

2.40 In the figure, D1 is a real silicon pn junction diode with a drop of 0.7 V under forward bias condition and D2 is a Zener diode with breakdown voltage of − 6.8 V. The input $V_{in}(t)$ is a periodic square wave of period T, whose one period is shown in the figure.

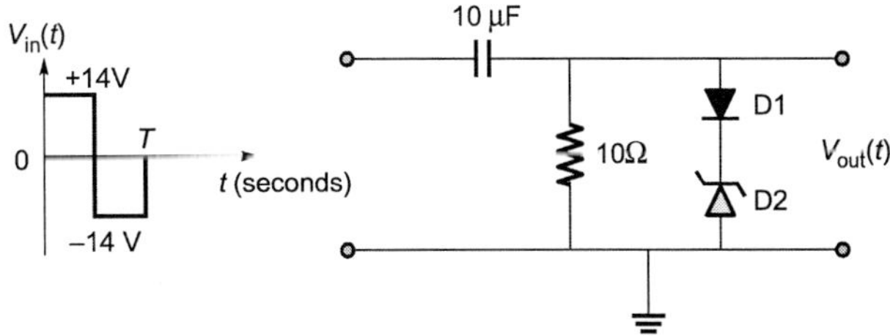

Assuming $10\tau << T$, where τ is the time constant of the circuit, the maximum and minimum values of the output waveform are respectively.

(a) 7.5 V and − 20.5 V
(b) 6.1 V and − 21.9 V
(c) 7.5 V and − 21.2 V
(d) 6.1 V and − 22.6 V

[2017 : 1 M, Set-2]

2.41 A DC current of 26 μA flows through the circuit shown. The diode in the circuit is forward biased and it has an ideality factor of one. At the quiescent point, the diode has a junction capacitance of 0.5 nF. Its neutral region resistances can be neglected. Assume that the room temperature thermal equivalent voltage is 26 mV.

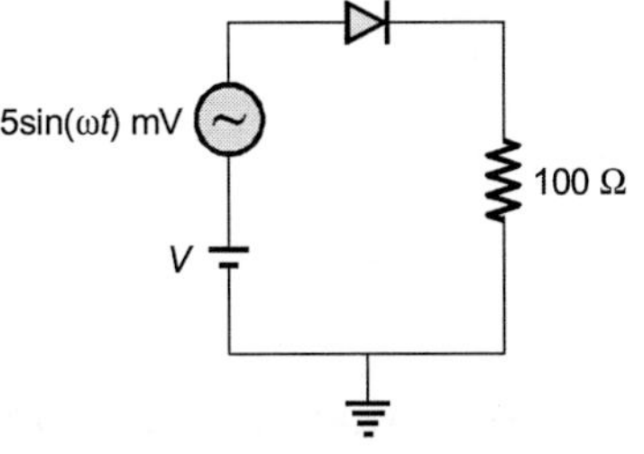

For $\omega = 2 \times 10^6$ rad/s, the amplitude of the small-signal component of diode current (in μA, correct to one decimal place) is ______.

[2018 : 2 M]

2.42 The circuit shown in the figure is used to provide regulated voltage (5 V) across the 1 kΩ resistor. Assume that the Zener diode has a constant reverse breakdown voltage for a current range, starting from a minimum required Zener current, I_{Zmin} = 2 mA to its maximum allowable current. The input voltage V_1 may vary by 5% from its nominal value of 6 V. The resistance of the diode in the breakdown region is negligible.

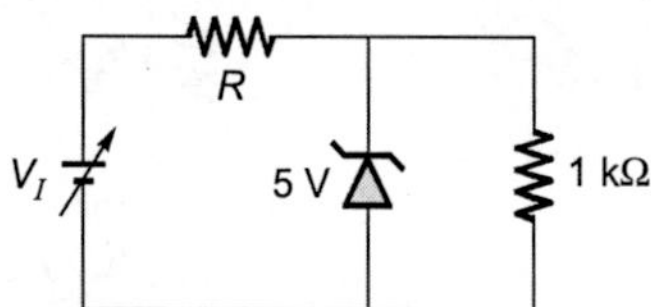

The value of R and the minimum required power dissipation rating of the diode, respectively, are

(a) 186 Ω and 10 mW (b) 100 Ω and 40 mW
(c) 100 Ω and 10 mW (d) 186 W and 40 mW

[2018 : 2 M]

2.43 In the circuit shown, V_s is a square wave of period T with maximum and minimum values of 8 V and –10 V, respectively. Assume that the diode is ideal and $R_1 = R_2 = 50\ \Omega$. The average value of V_L is ________ volts (rounded off to 1 decimal place).

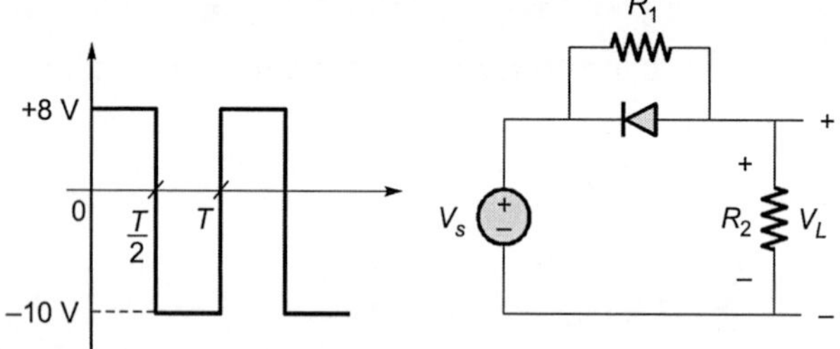

[2019 : 1 M]

2.44 In the circuit shown. V_s is a 10 V square wave of period, T = 4 ms with R = 500 Ω and C = 10 μF. The capacitor is initially uncharged at t = 0, and the diode is assumed to be ideal. The voltage across the capacitor (V_c) at 3 ms is equal to _____ volts (rounded off to one decimal place).

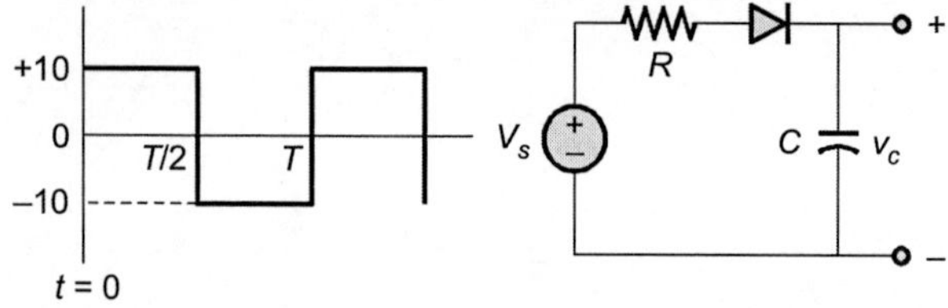

[2019 : 2 M]

2.45 In the circuit shown, the breakdown voltage and the maximum current of the Zener diode are 20 V and 60 mA. respectively. The values of R_1 and R_L are 200 Ω and 1 kΩ. respectively. What is the range of V_i that will maintain the Zener diode in the 'on' state?

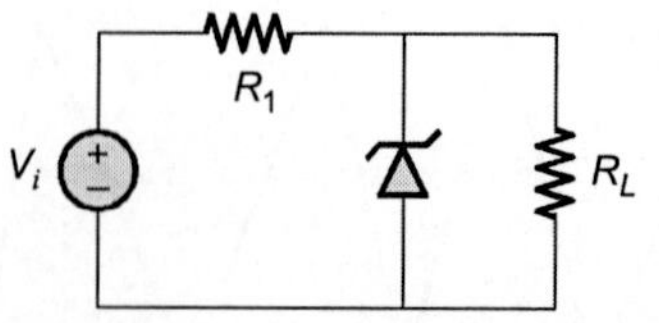

(a) 20 V to 28 V (b) 24 V to 36 V
(c) 18 V to 24 V (d) 22 V to 34 V

[2019 : 2 M]

2.46 In the circuit shown below, all the components are ideal and the input voltage is sinusoidal. The magnitude of the steady-state output V_0 (rounded off to two decimal places) is ________ V.

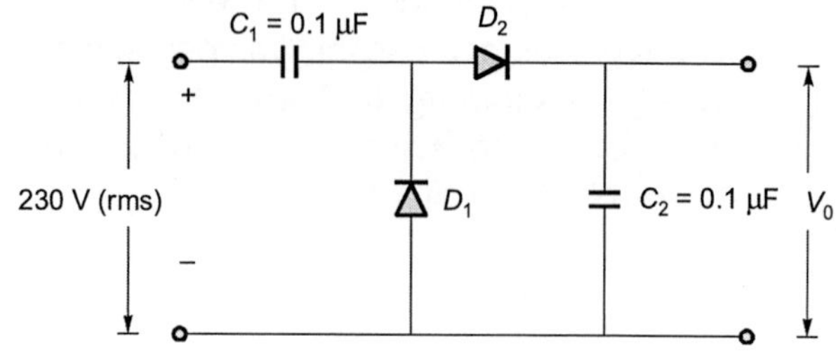

[2020 : 1 M]

2.47 An symmetrical periodic pulse train v_{in} of 10 V amplitude with on-time T_{ON} = 1 ms and off-time T_{OFF} = 1 μs is applied to the circuit shown in the figure. The diode D_1 is ideal.

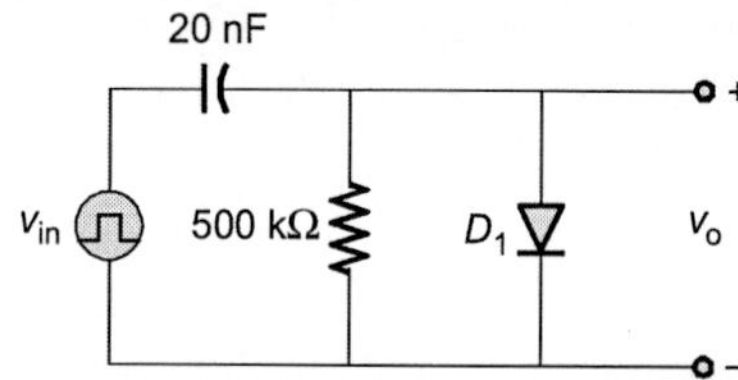

The difference between the maximum voltage and minimum voltage of the output waveform v_o (in integer) is ______ V.

[2021 : 2 M]

2.48 A circuit and the characteristics of the diode (D) in it are shown. The ratio of the minimum to the maximum small signal voltage gain $\frac{\partial V_{out}}{\partial V_{in}}$ is ________ (rounded off to two decimal places).

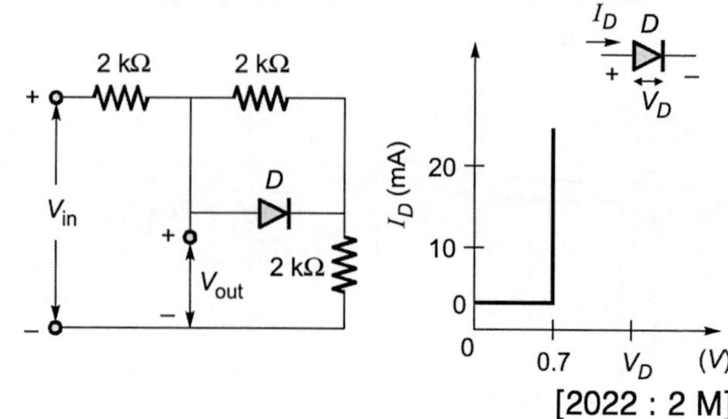

[2022 : 2 M]

2.49 In the circuit shown below, D_1 and D_2 are silicon diodes with cut-in voltage of 0.7 V. V_{IN} and V_{OUT} are input and output voltages in volts. The transfer characteristic is

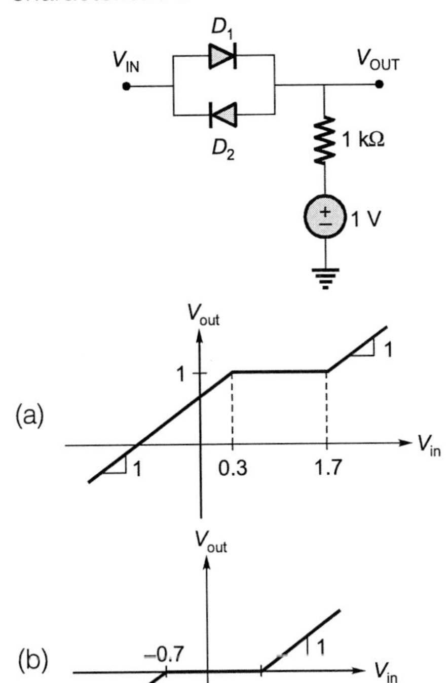

(c)

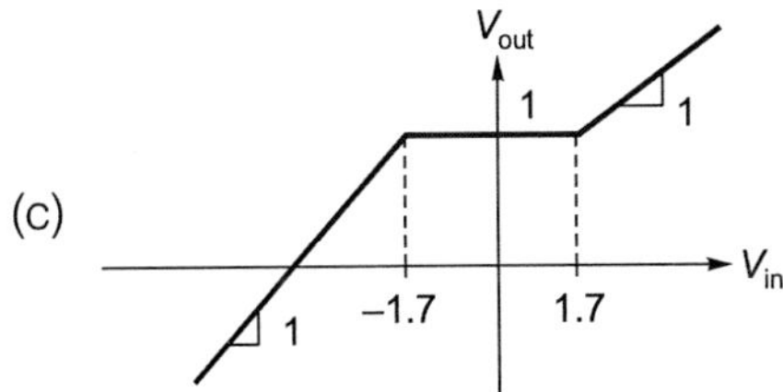

(d)

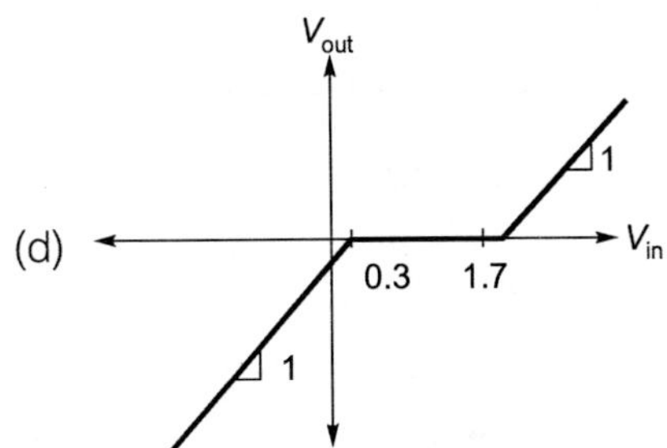

[2023 : 2 M]

2.50 In the circuit shown, the n : 1 step-down transformer and the diodes are ideal. The diodes have no voltage drop in forward biased condition. If the input voltage (in Volts) is $V_s(t) = 10\sin\omega t$ and the average value of load voltage $V_L(t)$ (in Volts) is $2.5/\pi$, the value of n is ________.

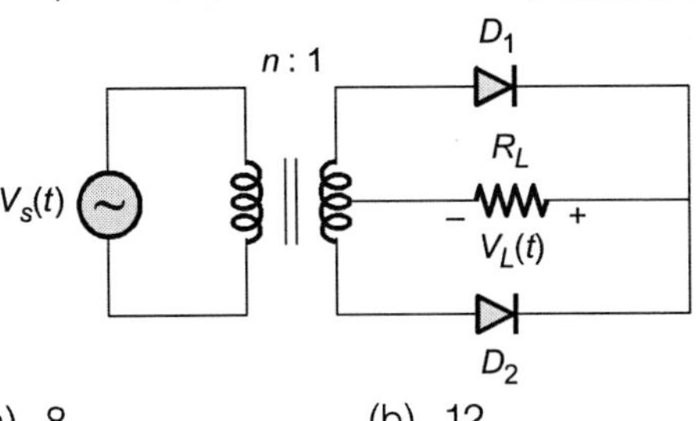

(a) 8 (b) 12
(c) 4 (d) 16

[2024 : 1 Mark]

2.51 As shown in the circuit, the initial voltage across the capacitor is 10 V, with the switch being open. The switch is then closed at t = 0. The total energy dissipated in the ideal Zener diode (V_Z = 5 V) after the switch is closed (in mJ, rounded off to three decimal places) is _______.

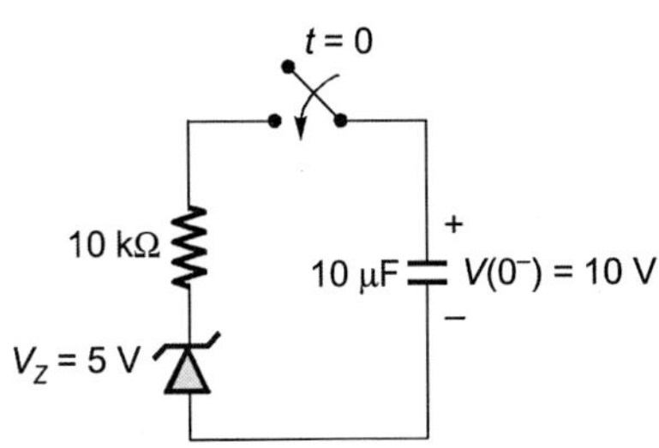

[2024 : 1 Mark]

■■■

Answers Diodes Applications

2.1 (a)	2.2 (c)	2.3 (d)	2.4 (d)	2.5 (b)	2.6 (c)	2.7 (a)
2.8 (d)	2.9 (c)	2.10 (d)	2.11 (b)	2.12 (c)	2.13 (b)	2.14 (d)
2.15 (b)	2.16 (c)	2.17 (c)	2.18 (c)	2.19 (a)	2.20 (a)	2.21 (b)
2.22 (a)	2.23 (b)	2.24 (d)	2.25 (d)	2.26 (c)	2.27 (0.10)	2.28 (d)
2.29 (c)	2.30 (5)	2.31 (d)	2.32 (0.6)	2.33 (100)	2.34 (5)	2.35 (2.105)
2.36 (0)	2.37 (0.25)	2.38 (a)	2.39 (3.183)	2.40 (a)	2.41 (6.40)	2.42 (b)
2.43 (–3)	2.44 (3.3)	2.45 (b)	2.46. (650.40)	2.47 (10)	2.48 (0.75)	2.49 (a)
2.50 (c)	2.51 (0.125)					

Explanations Diodes Applications

2.1 (a)

Ebers Moll model is a composite model and is used to predict the operation of BJT all of its possible modes.

It consists of two ideal diodes placed back to back with reverse saturation currents I_{EO} and I_{CO} and two dependent current controlled sources shunting the diodes.

2.2 (c)

Current through 50 Ω resistance = I

$$I = \frac{9-6}{50} = \frac{3}{50} = 60 \text{ mA}$$

Given that, $I_{z\,min} = 5$ mA

$$\boxed{P_{zmax} = I_{zmax} V_z} = 300 \text{ mW}$$

$$I_{z\,max} = \frac{300\times10^{-3}}{V_z} = \frac{300\times10^{-3}}{6} = 50 \text{ mA}$$

$$\boxed{I = I_{zmin} + I_{Lmax} = I_{zmax} + I_{Lmin}}$$

$$60 = 50 + I_{L\,min}$$

$$I_{L\,min} = 10 \text{ mA}$$

$$60 = 5 + I_{L\,max}$$

$$I_{L\,max} = 55 \text{ mA}$$

2.3 (d)

A resistance: For small signal ac operation, a practical forward biased diode can be modeled as a resistance.

2.4 (d)

A four diode bridge rectifier uses the smaller size of transformer, which is less expensive transformer and these rectifiers are suitable for higher voltage applications, because of low PIV rating required of each diode.

2.5 (b)

$$\text{Regulation} = \frac{V_{NL}-V_{FL}}{V_{FL}} = \frac{30-25}{25} = \frac{1}{5} = 20\%$$

$$R_o = \frac{V_{DCNL}-V_{DCFL}}{I_{DC}} = 5\ \Omega$$

2.6 (c)

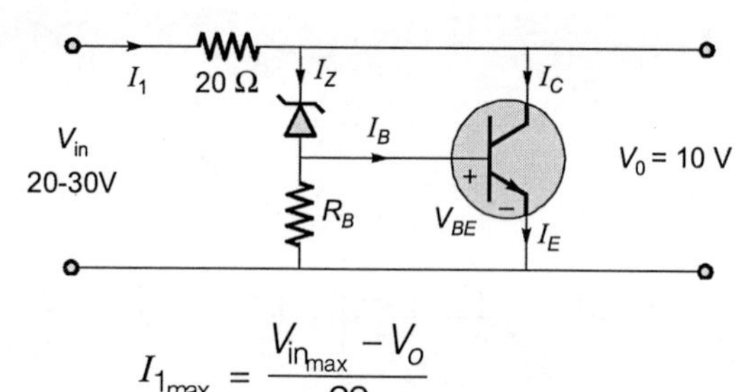

$$I_{1max} = \frac{V_{in_{max}} - V_o}{20}$$

$$I_1 = \frac{30-10}{20} = 1 \text{ A} \quad \text{(i.e. when } I_Z = 0)$$

$$I_E = I_C + I_Z \quad ...(i)$$

$I_B = I_Z$ (as no current flows in R_B)

$$\beta = \frac{I_C}{I_B} = \frac{I_C}{I_Z}$$

$$\Rightarrow \quad I_C = \beta I_Z$$

From (i), $I_E = \beta I_Z + I_Z = (99 + 1)I_Z$

$$I_E = 100\,I_Z$$

$$I_1 = I_E = 100\,I_Z$$

$$I_Z = \frac{I_1}{100} = \frac{1}{100} = 0.01 \text{ A}$$

$$P_Z = V_Z I_Z = 9.5 \times 0.01 = 95 \text{ mW}$$

$$I_C = 99\,I_Z = 99 \times 0.01 = 0.99 \text{ A}$$

$$P_C = V_C I_C = 10 \times 0.99 = 9.9 \text{ W}$$

2.7 (a)

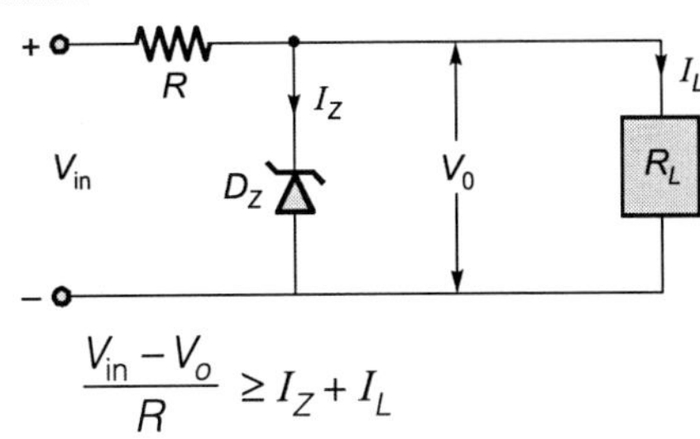

$$\frac{V_{in}-V_o}{R} \geq I_Z + I_L$$

when $V_{in} = 30$ V

$$\frac{30-10}{R} \geq (10+1)\text{mA} \quad ; \quad \frac{20}{R} \geq 11 \text{ mA}$$

$$\Rightarrow \quad R \leq 1818\ \Omega$$

2.9 (c)

As voltage at non inverting terminal is 3 V due to Zener diode, voltage at inverting terminal will be 3 V because of virtual ground.

So, current in 20 k is $\frac{3}{20\text{k}\Omega} = \frac{3}{20}$ mA

$$\therefore \quad V_0 = \frac{3}{20\text{k}\Omega} \times 60\ \text{k}\Omega = 9\text{ V}$$

2.10 (d)

$$I = I_Z + I_L = I_{zmin} + I_{Lmax}$$
$$= 0 + 500 \text{ mA}$$
$$R = \frac{V_{in}-V_Z}{I} = \frac{12-5}{500\text{ mA}} = 14\ \Omega$$

2.12 (c)

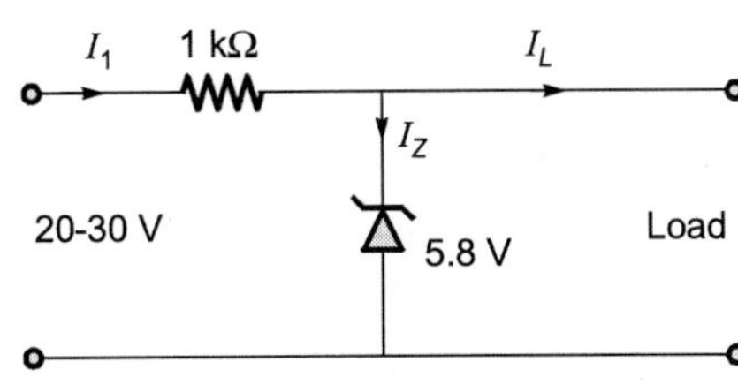

$V_L = 5.8$ V

Maximum load current will be drawn, when

$V_i = V_{min} = 20$ V

$$= \frac{20-5.8}{1\text{ k}\Omega} = I_L + I_Z$$

$$14.2\text{ mA} = I_L + I_Z$$

$$I_L = 14.2\text{ mA} - 0.5\text{ mA} = 13.7\text{ mA}$$

2.13 (b)

Zener diode works as normal diode in *FB*. So, when $V_{in} < 0$, $V_R = V_{in}$

When $0 < V_{in} < 6$, Diode is OFF and $V_R = 0$.

When $V_{in} > 6$, Diode conducts and $V_R = V_{in}$.

2.14 (d)

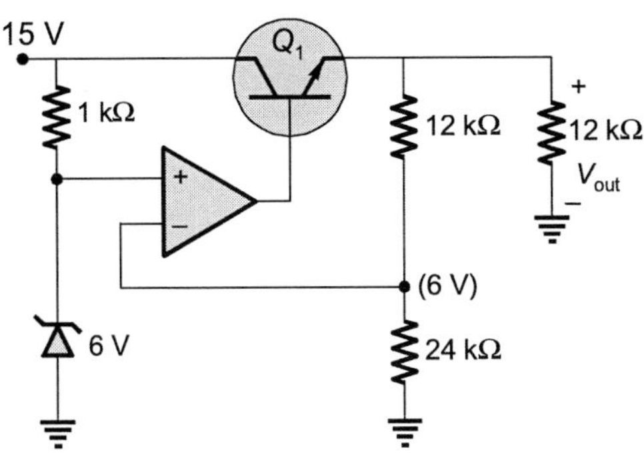

Volt across 24 Ω = 6 V due to virtual ground concept. So volt across 12 Ω is 3 V.

$$V_{out} = V_{12\ \Omega} + V_{24\ \Omega} = 3 + 6 = 9\text{ V}$$
$$V_{CE} = 15 - V_{out} = 15 - 9 = 6\text{ V}$$
$$I_C = \frac{9\text{ V}}{12\ \Omega} + \frac{9\text{ V}}{36\ \Omega} = 1\text{ A}$$
$$\therefore \quad P = V_{CE}I_C = 6\text{ V} \times 1\text{ A} = 6\text{ Watts}$$

2.15 (b)

New unregulated voltage = 18 V

$$\therefore \quad V_{CE} = 18 - 9 = 9\text{ V}$$
$$I_C = 1\text{ A}$$
$$\therefore \quad P = 9 \times 1 = 9\text{ Watts}$$

$$\therefore \text{ \% increase} = \frac{9-6}{6} \times 100 = 50\%$$

2.17 (c)

$$V_o = \frac{V_{in}-V_Z}{R+r_Z} r_Z + V_Z$$

If $V_{in} = 10$ V

$$V_o = \frac{10-7}{200+10} \times 10 + 7 = 7.14\text{ V}$$

If $V_{in} = 16$ V

$$V_o = \frac{16-7}{200+10} \times 10 + 7 = 7.43\text{ V}$$

2.18 (c)

For the positive half of V_i,

*D*1 is forward biased and Zener diode is in breakdown stage

$$V_0 = 0.7 + 6.8 = 7.5\text{ V}$$

For the negative half of V_i, *D*2 is forward biased.

$$V_0 = -0.7\text{ V}$$

2.20 (a)

For d.c. biasing a.c. will be short circuited.

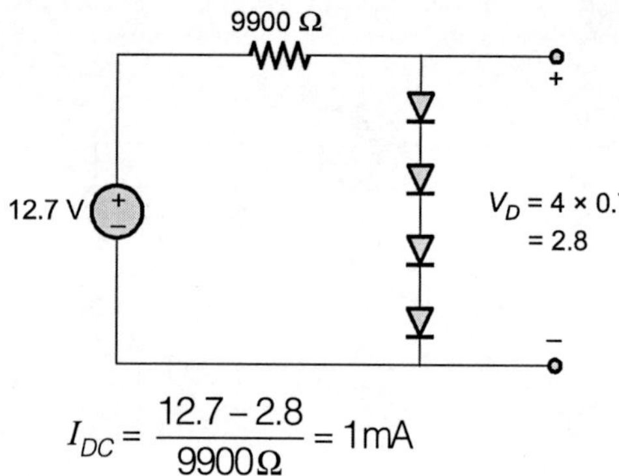

$$I_{DC} = \frac{12.7 - 2.8}{9900\,\Omega} = 1\,\text{mA}$$

2.21 (b)

For a.c. analysis diode will be replaced by its dynamic resistance

$$r_d = \frac{\eta V_T}{I} = \frac{1 \times 25\,\text{mV}}{1\,\text{mA}} = 25\,\Omega$$

$$(v_0)_{ac} = \frac{v_i}{9900 + 100} \times 100$$

$$= \frac{v_i}{100} = \frac{v_p}{100}\cos\omega t$$

$$\Rightarrow \quad (v_0)_{ac} = 1\cos(\omega t)\,\text{mV}$$

2.22 (a)

It is voltage doubler circuit in which C_1 will be charged to maximum value of input that is 1 V.
So $v(t) = (\cos\omega t - 1)$ according to KVL.

2.23 (b)

$$I = \frac{V_{in} - V_z}{R} = \frac{10 - 5}{100} = 50\,\text{mA}$$

$$R_{L\min} = \frac{V_z}{I_{L\max}} = \frac{V_z}{I - I_{z\min}} = 125\,\Omega$$

Minimum power rating of Zener diode

$$P_z = V_z I_{z\max}$$

$$I_{z\max} = 50\,\text{mA}$$

$$Pz = 5 \times 50 = 250\,\text{mW}$$

2.24 (d)

Case 1:

When V_{YZ} is positive,
then all the four diodes are reverse bias

So, $V_w = V_x = 0$

$V_{wx} = 0\,\text{V}$

Case 2:

When V_{YZ} is negative,
then all the four diodes are forward bias

So, all the diodes will be short circuit

$$V_w = V_x$$

$$V_{wx} = V_w - V_x = V_w - V_w = 0$$

$$V_{wx} = 0$$

So, $V_{wx} = 0$ for all t

2.25 (d)

Consider $D_1 \rightarrow$ OFF and $D_2 \rightarrow$ ON then

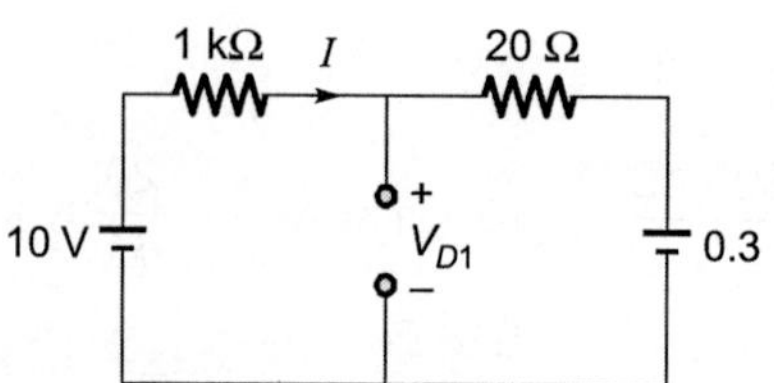

Apply KVL

$$10 = 1000\,I + 20\,I + 0.3$$

$$\therefore \quad I = \frac{9.7}{1020}$$

$$I = 9.5\,\text{mA}$$

Now, we calculate V_{D_1} of

$$10 = 9.5 + V_{D_1}$$

$$\therefore V_{D_1} = 0.5\,\text{V}$$

Since $V_{D_1} < 0.7\,\text{V}$, D_1 is in OFF state i.e. our assumption is correct and hence (d) is the correct option.

2.26 (c)

If $V_{in} = -5\,\text{V} \Rightarrow$ Diode is off

$V_o = V_{in} = -5\,\text{V}$

If $V_{in} = 5\,\text{V} \Rightarrow$ Diode is on

$$V_o = \frac{V_{in} - 0.7 - 2}{1\,\text{k}\Omega + 1\,\text{k}\Omega} \times 1\,\text{k}\Omega + 0.7 + 2$$

$$= \frac{5 - 2.7}{2} + 2.7 = 3.85\,\text{V}$$

2.27 (0.10)

The average current through the diode is approximate,

$$i_{D(avg)} \approx \frac{V_p}{R} = \frac{10}{100} = 0.10\,\text{A}$$

2.28 (d)

For case-i $\quad V_i < -1.7$

$D_1 \rightarrow$ ON

$D_2 \rightarrow$ OFF

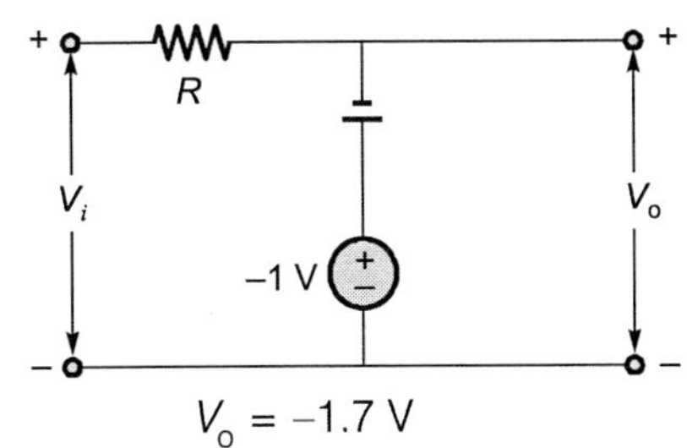

$$V_o = -1.7\text{ V}$$

For case-ii

$$-1.7 < V_i < 2.7$$
$$D_1 \rightarrow \text{OFF}$$
$$D_2 \rightarrow \text{OFF}$$

$$V_o = V_i$$

2.29 (c)

The circuit can be redrawn as

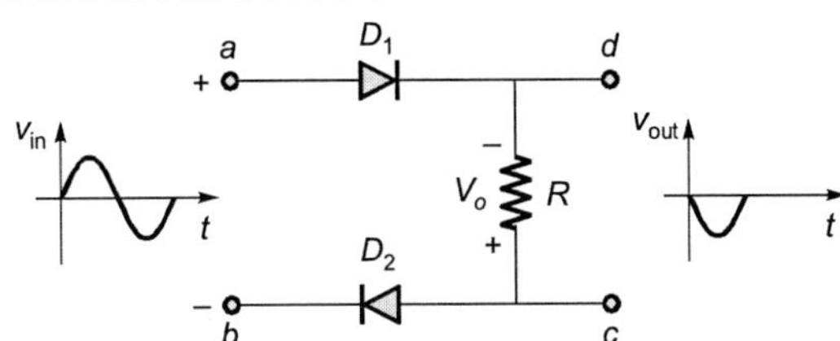

During positive pulse, both diodes are forward biased. During negative pulse, both diodes are reverse biased.

So, $V_o = 0\text{ V}$

2.30 (5)

Zener is not in breakdown region

Hence, $V_o = 10 \times \frac{1}{2} = 5\text{ V}$

2.32 (0.6)

The bridge is balanced bridge, so the diode will be OFF.

Current through 4 kΩ resistor can be obtained by applying current division rule

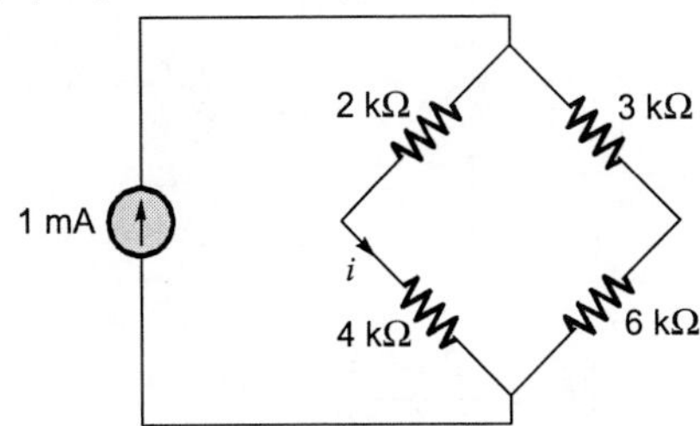

Current through 4 kΩ resistor is 0.6 mA.

2.33 (100)

The circuits work as a voltage doubler.

$\therefore \quad V_{ab} = 2 \times 50 = 100$

2.34 (5)

During positive cycles: $V_1 = \frac{V_{in}}{2}$

During negative cycles: $V_0 = \frac{V_{in}}{3}$

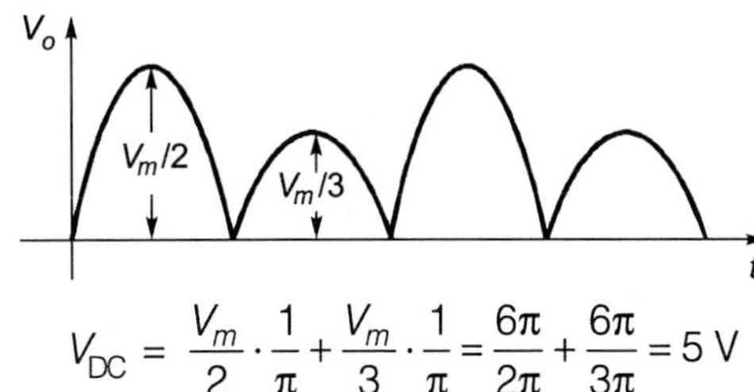

$$V_{DC} = \frac{V_m}{2} \cdot \frac{1}{\pi} + \frac{V_m}{3} \cdot \frac{1}{\pi} = \frac{6\pi}{2\pi} + \frac{6\pi}{3\pi} = 5\text{ V}$$

2.35 (2.105)

$$V_{ripple} = \frac{I_{DC} \cdot T}{C} = \frac{1 \times 1 \times 10^{-3}}{475 \times 10^{-6}} = 2.105\text{ Volts}$$

2.36 (0)

During first positive quarter cycle both diodes $D1$ and $D2$ conduct and at $t = \frac{T}{4}$ both will be OFF.

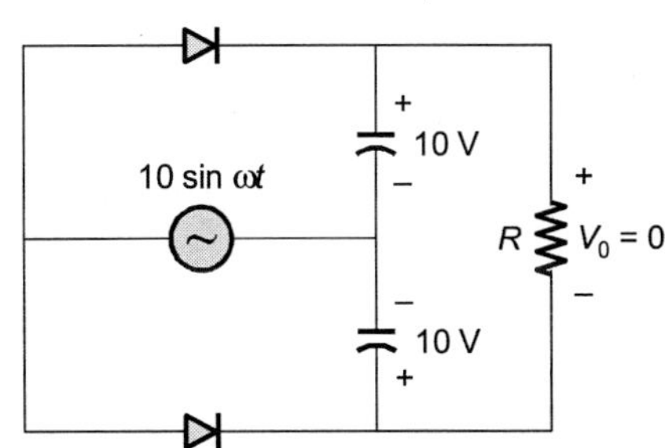

Both the capacitors will charge to 10 V as shown is above figure and $V_0 = 0$ in steady state.

2.37 (0.25)

Let diode : ON

$\Rightarrow V_A = 2 - 0.7$

$= 1.3\text{ V}$

$$i_2 = \frac{1.3}{6\text{ k}\Omega}$$

$$i_1 = \frac{0.7}{2\text{ k}\Omega}$$

$$i_d = i_2 - i_1$$

$$= \frac{1.3}{6}\text{mA} - \frac{0.7}{2}\text{mA} = -\text{ve}$$

not possible

⇒ diode is OFF

⇒ 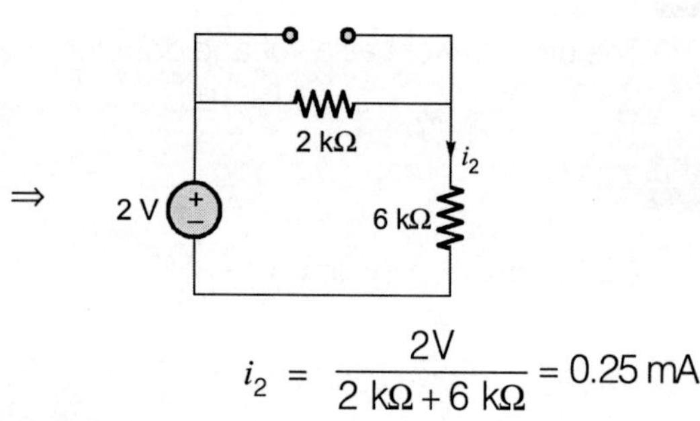

$$i_2 = \frac{2V}{2\text{ k}\Omega + 6\text{ k}\Omega} = 0.25\text{ mA}$$

2.39 (3.183)

The given circuit is a half-wave rectifier.

Voltmeter reads the average value of V_o.

Average value of $V_o = \frac{V_m}{\pi}$

V_m = peak value of the applied sine wave
= 10 V

So, Reading of meter $= \frac{10}{\pi}$ V = 3.183 V

2.40 (a)

Its a negative clamper with V_{ref} = 6.8 V.

In steady state V_{cap} = 14 – 0.7 – 6.8 = 6.5 V

$$\Rightarrow \quad V_{o\,max} = V_{in\,max} - V_{cap} = 14 - 6.5\text{ V}$$

$$V_{o\,min} = V_{in\,in} - V_{cap} = -14 - 6.5 = -20.5\text{ V}$$

2.41 (6.40)

The small-signal equivalent model of the given circuit can be drawn as shown below.

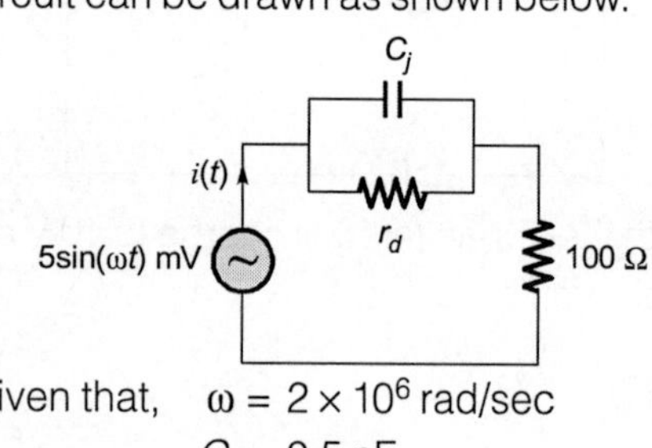

Given that, $\omega = 2 \times 10^6$ rad/sec

C_j = 0.5 nF

I_{DC} = 26 μA

V_T = 26 mV

η = 1

So, $$r_d = \frac{\eta V_T}{I_{DC}} = \frac{26\text{ mV}}{26\text{ μA}} = 1\text{k}\Omega$$

$$\frac{1}{\omega C_j} = \frac{1}{2 \times 10^6 \times 0.5 \times 10^{-9}}\,\Omega = 1\text{k}\Omega$$

So, total impedance of the circuit will be,

$$Z = \left(r_d \| \frac{1}{j\omega C_j}\right) + 100\,\Omega$$

$$\left(r_d \| \frac{1}{j\omega C_j}\right) = \frac{(1000)(-j1000)}{1000 - j1000}\,\Omega$$

$$= \frac{-j(1+j)}{2}\text{ k}\Omega$$

$$= \frac{1}{2}(1-j)\text{ k}\Omega = (500 - j500)\,\Omega$$

$$\therefore \quad Z = 600 - j500\,\Omega$$

$$|Z| = 100\sqrt{36+25} = 100\sqrt{61}\,\Omega$$

$$I_m = \frac{V_m}{|Z|}$$

$$= \frac{5\text{ mV}}{100\sqrt{61}\,\Omega} = \frac{50}{\sqrt{61}}\text{ μA} = 6.40\text{ μA}$$

2.42 (b)

$$V_I = 6\text{ V} \pm 5\% = 6\text{ V} \pm 0.3\text{ V}$$
$$= 5.7\text{ V to } 6.3\text{ V}$$

$$I_L = \frac{5\text{ V}}{1\text{k}\Omega} = 5\text{ mA}$$

$$I_{s(min)} = I_L + I_{Z(min)} = 5\text{ mA} + 2\text{ mA} = 7\text{ mA}$$

$$I_s = \frac{V_I - V_Z}{R}$$

$$I_{s(min)} = \frac{V_{I(min)} - V_Z}{R} = 7\text{ mA}$$

So, $$R = \frac{5.7-5}{7}\text{ k}\Omega = \frac{700}{7}\,\Omega = 100\,\Omega$$

When, $R = 100\,\Omega$,

$$I_{s(max)} = \frac{6.3-5}{100}\text{ A} = 13\text{ mA}$$

$$I_{z(max)} = I_{s(max)} - I_L = 13\text{ mA} - 5\text{ mA} = 8\text{ mA}$$

$$P_{z(min)} = V_z I_{z(max)} = (5 \times 8)\text{ mW} = 40\text{ mW}$$

2.43 (–3)

When V_s = 8 V ⇒ diode is in reverse bias

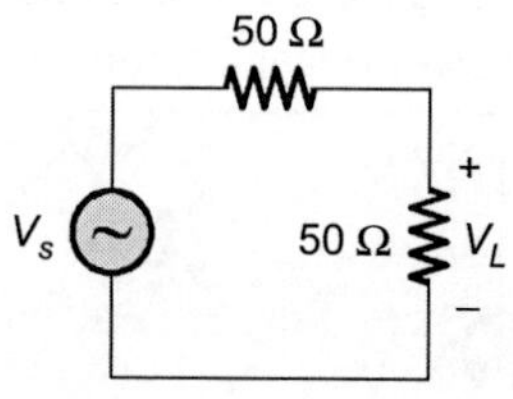

$$V_L = \frac{8 \times 50}{50+50} = 4\text{ V}$$

If $V_s = -10$ V, diode is in forward bias

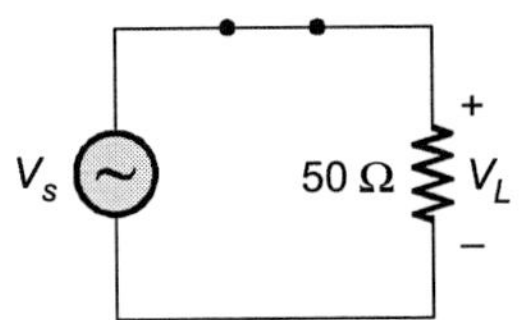

$$\text{Average value of } V_L = \frac{\text{Area}}{\text{Time period}}$$

$$= \frac{4 \times 0.5T + (-10) \times 0.5T}{T} = -3 \text{ V}$$

2.44 (3.3)

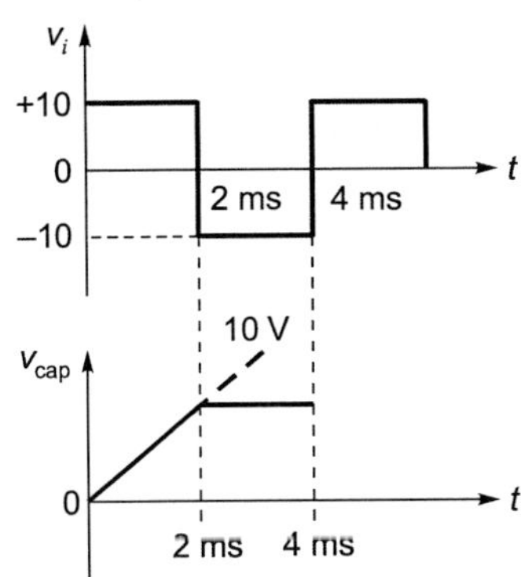

$$\tau = RC = 500 \times 10 \times 10^{-6} = 5 \text{ ms}$$

$$0 < t < \frac{t}{2} : v_{cap} = v_f + (v_i - v_f)e^{-t/\tau}$$

$$= 10 + (0 - 10)e^{-t/RC}$$

At $t = 2$ msec,

$$v_{cap} = 10 - 10e^{\frac{-2\times10^{-3}}{5\times10^{-3}}}$$

$$= 3.296 \text{ V} \simeq 3.3 \text{ V}$$

For 2 ms < t < 4 ms, diode is OFF and capacitor has no path to discharge. Hence, at $t = 3$ ms, $V_{cap} = 3.3$ V.

2.45 (b)

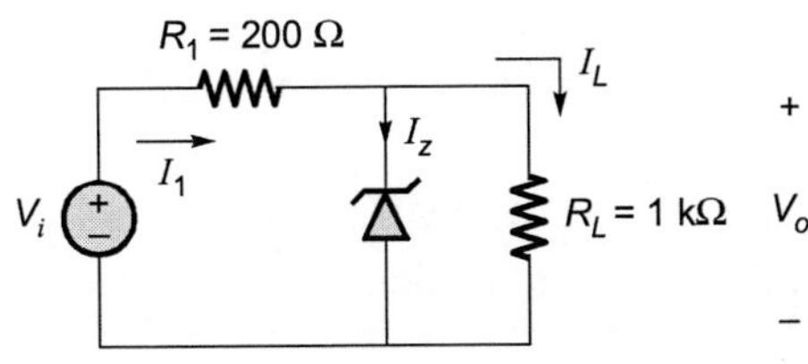

$$V_z = 20 \text{ V}$$

$$I_{z\max} = 60 \text{ mA}$$

Set Zener diode be OFF

$$V_o = \frac{V_i \times 1}{0.2 + 1} = \frac{V_i}{1.2}$$

Zener diode can become ON i.e. it goes into breakdown, when

$$\frac{V_i}{1.2} > 20 \text{ V}$$

$$V_i > 24 \text{ V}$$

When Zener diode is in breakdown region,

$$I_1 = \frac{V_i - 20}{0.2\text{ k}\Omega} = \frac{V_i - 20}{0.2} \text{ mA}$$

$$I_L = \frac{V_o}{R_L} = \frac{20}{1\text{k}\Omega} = 20 \text{ mA}$$

$$I_z = I_1 - I_L = \frac{V_i - 20}{0.2} - 20$$

For safe operation, $I_z \le I_{z\max}$

$$\frac{V_i - 20}{0.2} - 20 \le 60$$

$$\Rightarrow \qquad V_i \le 36 \text{ V}$$

Hence, $\quad 24 < V_i < 36$ V

2.46. (650.40)

Voltage doubles,

$$V_o = 2\,V_m = 2\times230\sqrt{2} \cong 650.4 \text{ V}$$

2.47 (10)

$$V_{in} = 10 \text{ V:} \quad \text{Diode is ON}$$

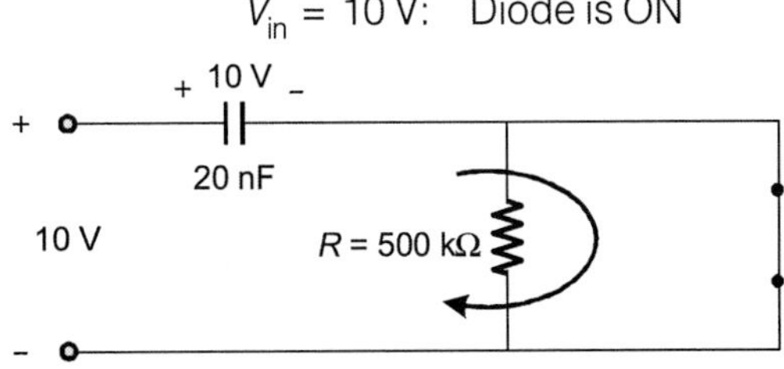

∴ Capacitor charges upto 10 V,

$$\therefore \qquad V_C = 10 \text{ V}$$

$$V_{in} = 0; \text{ Diode is OFF}$$

Discharging time constant = $R \times C$

$$= 10 \text{ m sec}$$

$$\tau_{\text{discharging}} >> \tau_{\text{OFF}}$$

Capacitor discharges negligibly

$$\therefore \qquad V_C = 10 \text{ V}$$

In steady state,

$$V_C = 10 \text{ V}$$

$$V_{out} = V_{in} - V_c = V_{in} - 10 \text{ V}$$

When $\quad V_{in} = 10$ V

$$\Rightarrow \qquad V_{out} = 10 - 10 = 0 \text{ V}$$

$$V_{in} = 10\text{ V}$$

$$\Rightarrow \quad V_{out} = 0 - 10 = -10\text{ V}$$

$$V_{out(max)} - V_{out(min)} = 0 - (-10) = 10\text{ V}$$

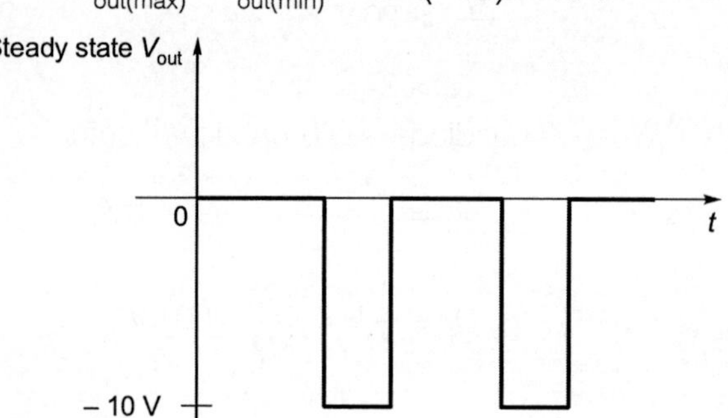

2.48 (0.75)

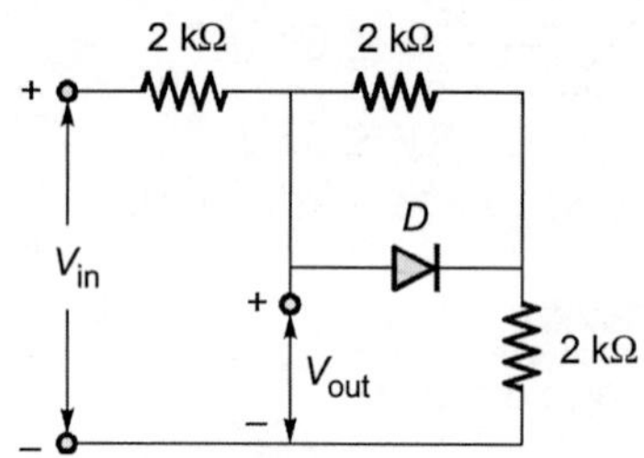

Case-I : Diode D is OFF

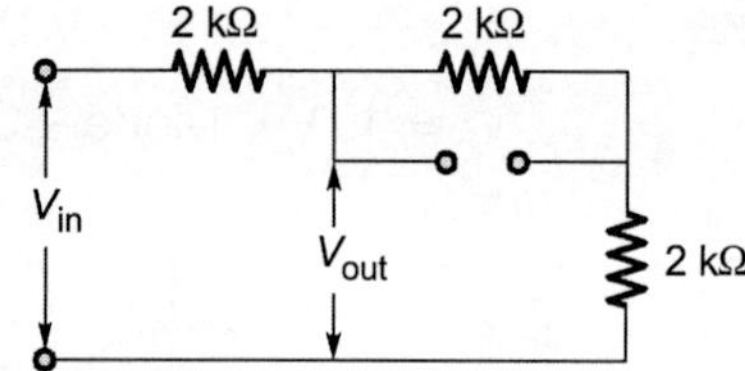

Using the voltage division rule,

$$V_{out} = V_{in}\left(\frac{2+2}{2+2+2}\right) = \frac{2V_{in}}{3}$$

Hence, $$\frac{\partial V_{out}}{\partial V_{in}} = \frac{2}{3} \qquad \text{...(i)}$$

Case-II: Diode D is ON

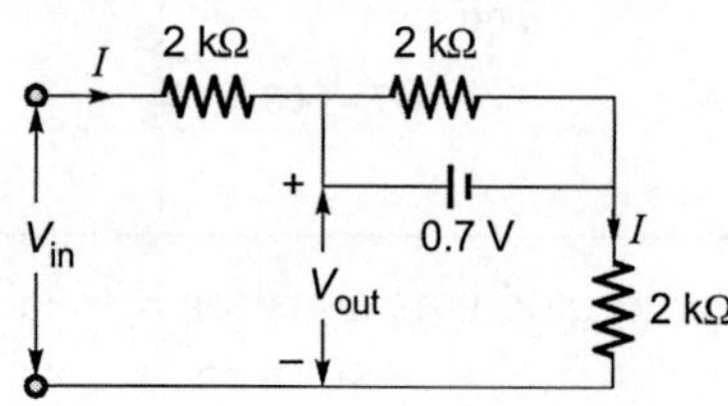

From the given characteristics of the diode, when the diode is forward-biased, voltage drop across the diode is 0.7 V. Applying KVL in the circuit, we obtain

$$V_{out} = 0.7 + (2\text{ k}\Omega)I$$

$$V_{in} = 0.7 + (4\text{ k}\Omega)I$$

$$I = \frac{V_{out} - 0.7}{2\text{k}\Omega}$$

Hence, $$V_{in} = 0.7 + (4\text{k}\Omega)\times\frac{(V_{out} - 0.7)}{2\text{ k}\Omega}$$

$$V_{in} = 0.7 + 2(V_{out} - 0.7)$$

$$V_{in} = 0.7 + 2V_{out} - 1.4$$

$$V_{out} = \frac{V_{in} + 0.7}{2}$$

$$\therefore \quad \frac{\partial V_{out}}{\partial V_{in}} = \frac{1}{2} \qquad \text{...(ii)}$$

From equation (i) and (ii),

Maximum small signal voltage gain

$$\left.\frac{\partial V_{out}}{\partial V_{in}}\right|_{max} = \frac{2}{3}$$

Minimum small signal voltage gain,

$$\left.\frac{\partial V_{out}}{\partial V_{in}}\right|_{min} = \frac{1}{2}$$

$$\therefore \quad \frac{\left(\frac{\partial V_{out}}{\partial V_{in}}\right)_{min}}{\left(\frac{\partial V_{out}}{\partial V_{in}}\right)_{max}} = \frac{\frac{1}{2}}{\frac{2}{3}} = \frac{3}{4} = 0.75$$

2.49 (a)

Case I:

$$V_\gamma = 0.7\text{V}$$

For the +ve half cycle if input V_{in},

$$D_1 \to \text{ON and } D_2 \to \text{OFF}$$

For diode D_1: $V_{in} - 1V > 0.7$

$$V_{in} > 1.7\text{V}$$

$$V_0 = V_{in} - 0.7$$

Case II:

For the +ve half cycle if input V_{in},

$$D_1 \to \text{OFF and } D_2 \to \text{ON}$$

For diode D_2: $1 - V_{in} > 0.7$

$$V_{in} < 0.3\text{V}$$

$$V_0 = V_{in} + 0.7$$

Case III:

$$0.3V < V_{in} < 1.7V$$

$$D_1 \to \text{OFF and } D_2 \to \text{OFF}$$

$$V_0 = 1\text{V}$$

Transfer characteristics,

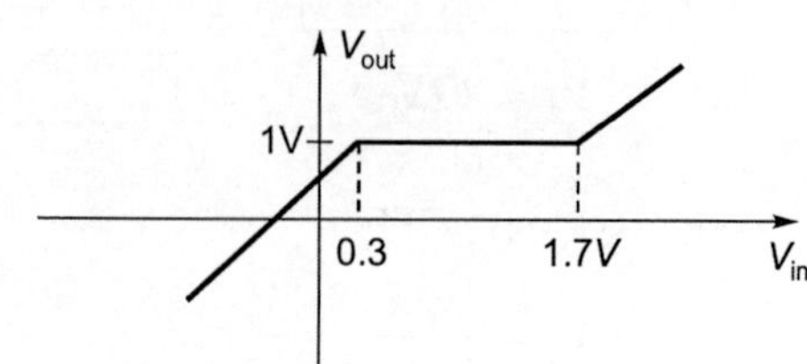

2.50 (c)

The given circuit is a centre-tapped full wave rectifier. Since, the diodes are ideal, cut-in voltage of diode, $V_\gamma = 0$ and forward resistance of diode, $R_f = 0$.

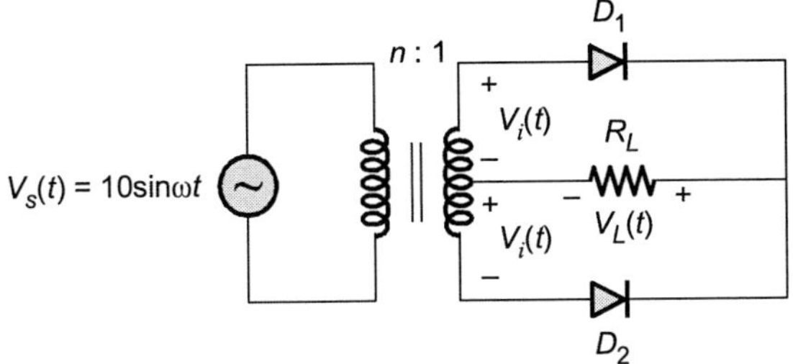

For a n : 1 step-down transformer, the voltage across half of the secondary is given by

$$V_i(t) = \frac{V_s(t)}{2n} = \frac{10}{2n}\sin\omega t \equiv V_m \sin\omega t \quad ...(i)$$

For a full-wave rectifier, the average value of load voltage (i.e. DC value) is given by

$$V_{L(DC)} = \frac{2V_m}{\pi} = \frac{2.5}{\pi} \quad \text{[Given]}$$

$$\Rightarrow \quad V_m = \frac{2.5}{2}\text{ Volts}$$

Using equation (i),

$$\frac{10}{2n} = \frac{2.5}{2}$$

$$\Rightarrow \quad n = \frac{10}{2.5} = 4$$

2.51 (0.125)

For $t > 0$: As the voltage across zener diode is greater than the breakdown voltage, therefore the diode conducts and the capacitor starts discharging through 10 kΩ resistor and zener diode.

$$V_C(t) = 10e^{-t/RC}, \text{ where } RC = 0.1 \text{ sec}$$

Zener diode remains ON till V_C becomes 5 V.

$$5 = 10e^{-t_1/RC}$$

$$\Rightarrow \quad t_1 = RC\ln 2 = 0.0693 \text{ sec}$$

Hence, the Zener diode conducts from $t = 0$ to $t = t_1 = 0.0693$ sec.

Current in the circuit is $i(t) = I_0 e^{-t/RC}$

where $$I_0 = \frac{10-5}{10\text{k}} = 0.5\text{ mA}$$

Total energy dissipated in zener diode is,

$$W = \int_0^{t_1} V_z \times i(t)dt$$

$$W = \int_0^{0.0693} 5 \times 0.5e^{-t/RC}dt$$

$$= 2.5[-0.1]\left[e^{-t/0.1}\right]_0^{0.0693}$$

$$W = 0.125 \text{ mJ}$$

■■■

3 BJT Analysis

3.1 A transistor having $\alpha = 0.99$ and $V_{BE} = 0.7$ V, is used in the circuit of the figure is the value of the collector current will be

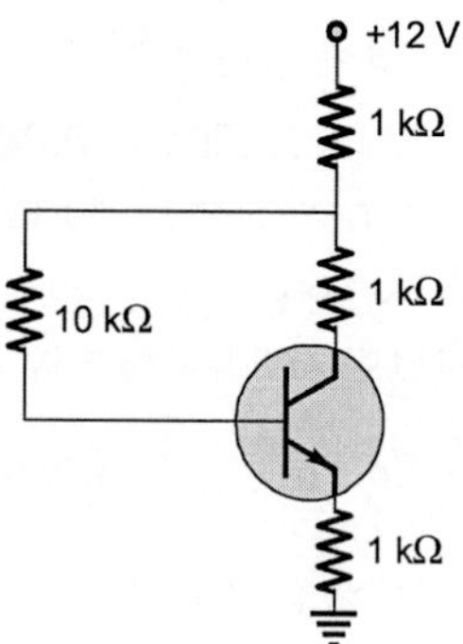

[1995 : 1 M]

3.2 A Darlington stage is shown in the figure is, if the transconductance of Q_1 is g_{m1} and Q_2 is g_{m2}, then the overall transconductance is given by gm is given by

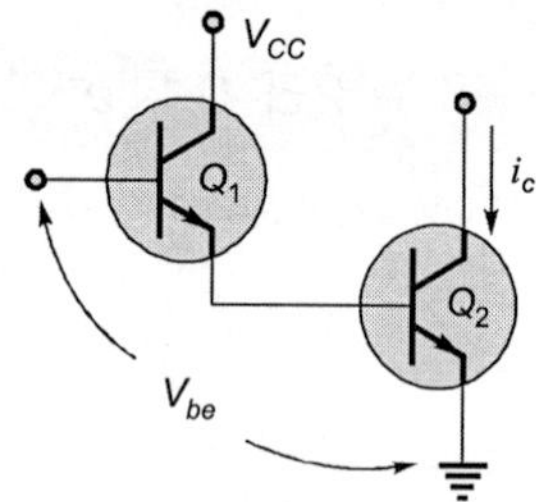

(a) g_{m1}
(b) $0.5\ g_{m1}$
(c) g_{m2}
(d) $0.5\ g_{m2}$

[1996 : 2 M]

3.3 Match the following.
(a) Cascode Amplifier
(b) Differential Amplifier
(c) Darlington pair common-Collector Amplifier
(1) Does not provide current gain
(2) Is a wide band Amplifier
(3) Has very low input impedance Emitter Amplifier and very high current gain
(4) Has very high input impedance and very high current gain
(5) Provides high common mode voltage Rejection

[1996 : 2 M]

3.4 In the BJT amplifier shown in the figure is the transistor is biased in the forward active region putting a capacitor across R_E will

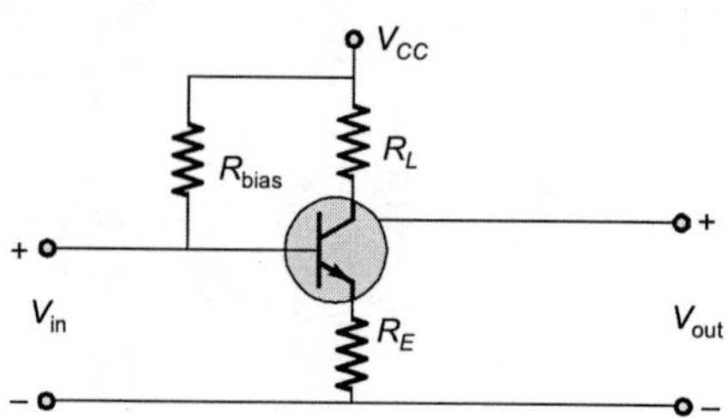

(a) Decrease the voltage gain and decrease the input impedance
(b) Increase the voltage gain and decrease the input impedance
(c) Decrease the voltage gain and increase the input impedance
(d) Increase the voltage gain and increase the input impedance

[1997 : 1 M]

3.5 A cascode Amplifier stage is equivalent to
(a) A common emitter stage followed by a common base stage
(b) A common base stage followed by an Emitter follower
(c) An Emitter follower stage followed by a common base stage
(d) A common base stage followed by a common Emitter stage

[1997 : 1 M]

3.6 The circuit of the figure is an example of feedback of the following type

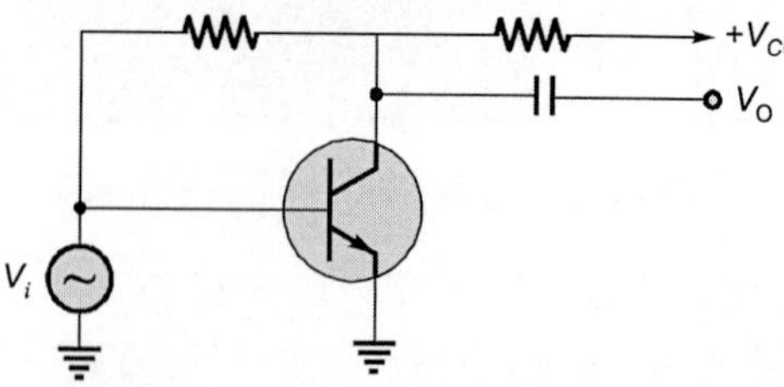

(a) current series
(b) current shunt
(c) voltage series
(d) voltage shunt

[1998 : 1 M]

3.7 From measurement of the rise time of the o/p pulse of an Amplifier whose input is a small amplitude square wave, one can estimate the following parameter of the amplifier
(a) Gain-bandwidth product
(b) Slew Rate
(c) Upper-3-dB frequency
(d) Lower 3-dB frequency

[1998 : 1 M]

3.8 A distorted sinusoid has the Amplitude, A_1, A_2, A_3, ... of the fundamental, second harmonic, third harmonic, ... respectively. The total harmonic distortion is

(a) $\dfrac{A_2 + A_3 + ...}{A_1}$ (b) $\dfrac{\sqrt{A_2^2 + A_3^2 + ...}}{A_1}$

(c) $\dfrac{\sqrt{A_2^2 + A_3^2 + ...}}{\sqrt{A_1^2 + A_2^2 + A_3^2 ......}}$ (d) $\dfrac{(A_2^2 + A_3^2 + ...)}{A_1}$

[1998 : 1 M]

3.9 The Emitter coupled pair of BJT's gives a linear transfer relation between the differential o/p voltage and the differential input voltage V_{id} only. When the magnitude of V_{id} is less 'α' times the thermal voltage, where 'α' is

(a) 4 (b) 3
(c) 2 (d) 1 [1998 : 1 M]

3.10 A multistage Amplifier has a low-pass Response with three Real poles at $s = -\omega_1, -\omega_2$ and ω_3. The approximate overall bandwidth B of the Amplifier will be given by

(a) $B = \omega_1 + \omega_2 + \omega_3$
(b) $\dfrac{1}{B} = \dfrac{1}{\omega_1} + \dfrac{1}{\omega_2} + \dfrac{1}{\omega_3}$
(c) $B = (\omega_1 + \omega_2 + \omega_3)^{1/3}$
(d) $B = \sqrt{\omega_1^2 + \omega_2^2 + \omega_3^3}$ [1998 : 1 M]

3.11 In a series Regulated power supply circuit the voltage gain A_v of the "pass", transistor satisfies the condition

(a) $A_v \to \infty$ (b) $1 << A_v << \infty$
(c) $A_v \approx 1$ (d) $A_v << 1$

[1998 : 1 M]

3.12 In the cascode amplifier shown in the figure, if the common-emitter stage (Q_1) has a transconductance g_{m1}, and the common base stage (Q_2) has a transconductance g_{m2}, then the overall transconductance $g (= i_0/V_i)$ of the cascode amplifier is

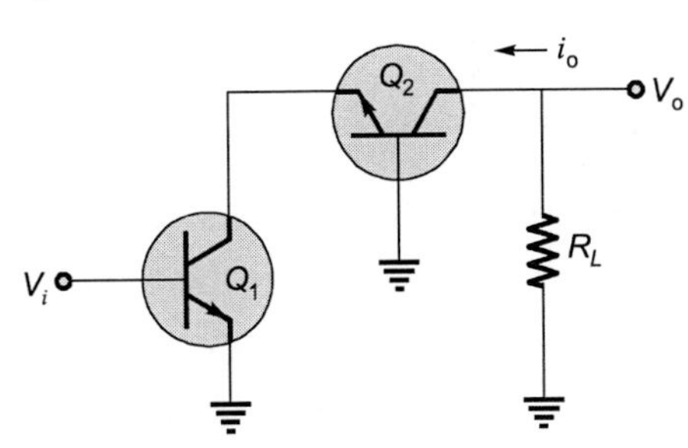

(a) g_{m1} (b) g_{m2}
(c) $\dfrac{g_{m1}}{2}$ (d) $\dfrac{g_{m2}}{2}$ [1999 : 1 M]

3.13 In the differential amplifier of the figure, if the source resistance of the current source I_{EE} is infinite, then the common-mode gain is

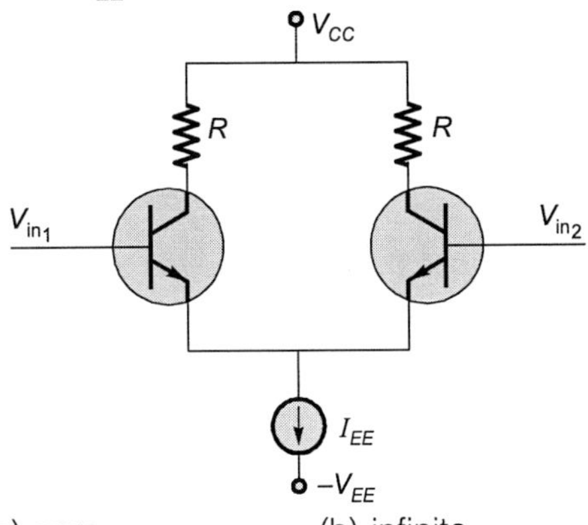

(a) zero (b) infinite
(c) indeterminate (d) $\dfrac{V_{in_1} + V_{in_2}}{2V_T}$

[2000 : 1 M]

3.14 Introducing a resistor in the emitter of a common emitter amplifier stabilizes the dc operating point against variations in

(a) only the temperature
(b) only the β of the transistor
(c) both temperature and β
(d) none of the above [2000 : 1 M]

3.15 The current gain of a bipolar transistor drops at high frequencies because of

(a) transistor capacitances
(b) high current effects in the base
(c) parasitic inductive elements
(d) the Early effect [2000 : 1 M]

3.16 In the circuit of the figure, assume that the transistor is in the active region. It has a large β and its base-emitter voltage is 0.7 V. The value of I_c is

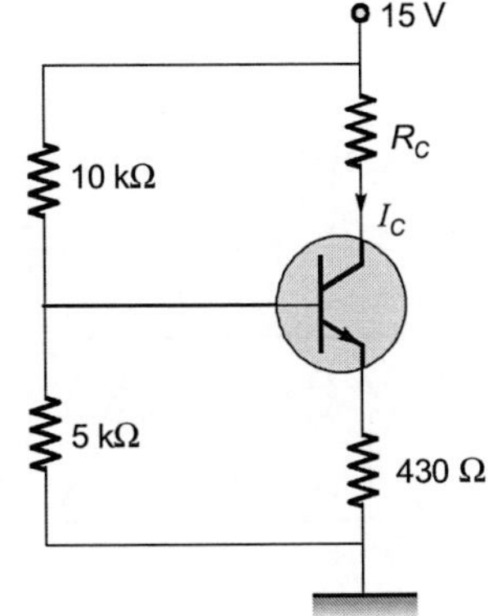

(a) Indeterminate since R_c is not given
(b) 1 mA
(c) 5 mA
(d) 10 mA [2000 : 2 M]

3.17 The current gain of a BJT is

(a) $g_m r_o$
(b) $\frac{g_m}{r_o}$
(c) $g_m r_\pi$
(d) $\frac{g_m}{r_\pi}$

[2001 : 1 M]

3.18 Three identical *RC*-coupled transistor amplifiers are cascaded. If each of the amplifiers has a frequency response as shown in the figure, the overall frequency response is as given in

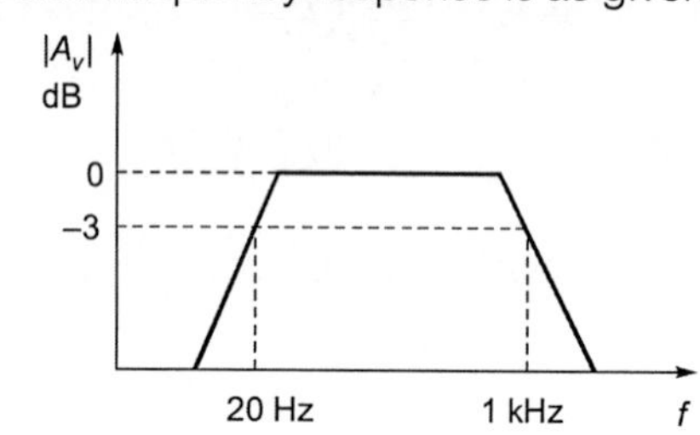

(a)
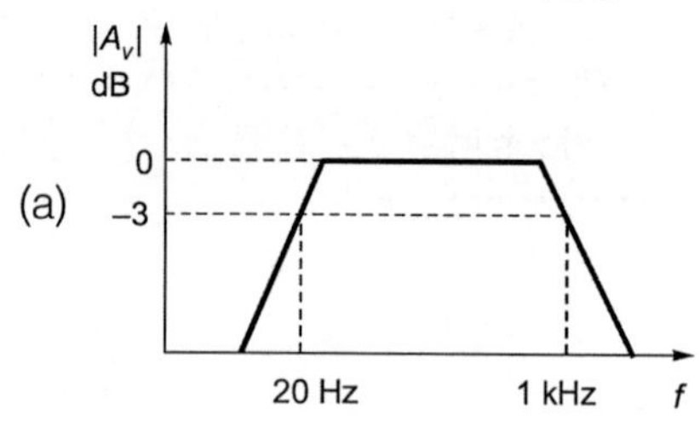

(b)
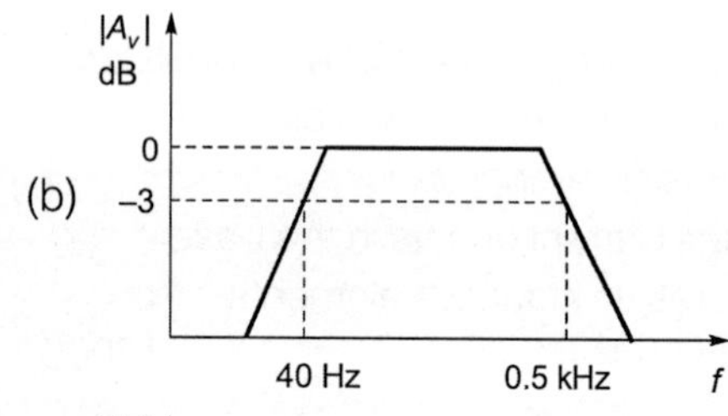

(c)
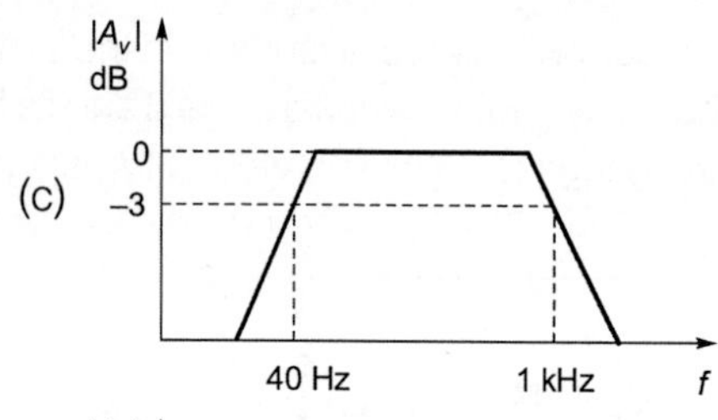

(d)
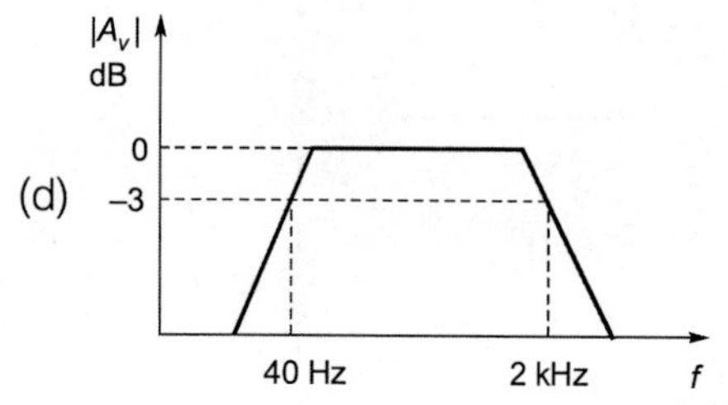

[2002 : 1 M]

3.19 If the transistor in the figure is in saturation, then

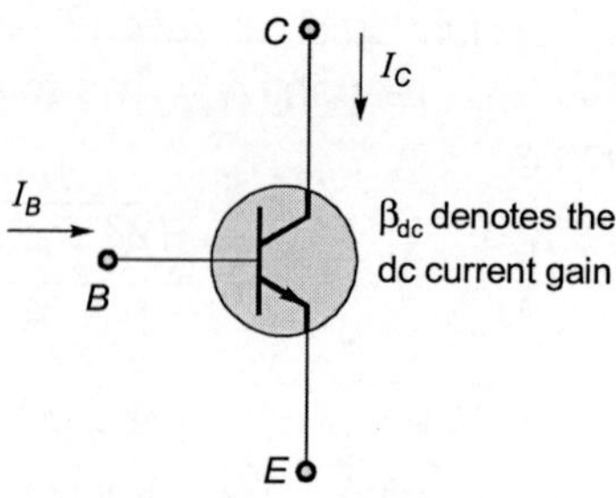

(a) I_C is always equal to $\beta_{dc} I_B$
(b) I_C is always equal to $-\beta_{dc} I_B$
(c) I_C is greater than or equal to $\beta_{dc} I_B$
(d) I_C is less than or equal to $\beta_{dc} I_B$

[2002 : 1 M]

3.20 Choose the correct match for input resistance of various amplifier configurations shown below

Configuration
CB: Common Base
CC: Common Collector
CE: Common Emitter

Input resistance
LO: Low
MO: Moderate
HI: High

(a) CB-LO, CC-MO, CE-HI
(b) CB-LO, CC-HI, CE-MO
(c) CB-MO, CC-HI, CE-LO
(d) CB-HI, CC-LO, CE-MO [2003 : 1 M]

3.21 Generally, the gain of a transistor amplifier falls at high frequencies due to the
(a) internal capacitances of the device
(b) coupling capacitor at the input
(c) skin effect
(d) coupling capacitor at the output

[2003 : 1 M]

3.22 In the amplifier circuit shown in the figure, the values of R_1 and R_2 are such that the transistor is operating at V_{CE} = 3 V and I_C = 1.5 mA when its β is 150. For a transistor with β of 200, the operating point (V_{CE}, I_C) is

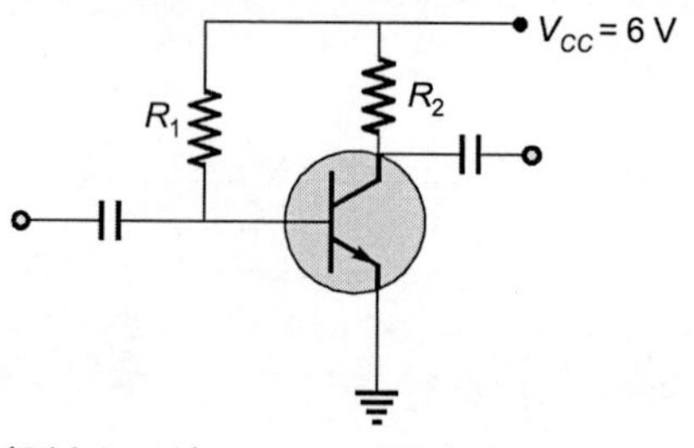

(a) (2 V, 2 mA)
(b) (3 V, 2 mA)
(c) (4 V, 2 mA)
(d) (4 V, 1 mA)

[2003 : 2 M]

3.23 Assuming $V_{CEsat} = 0.2$ V and $\beta = 50$, the minimum base current (I_B) required to drive the transistor in the figure to saturation is

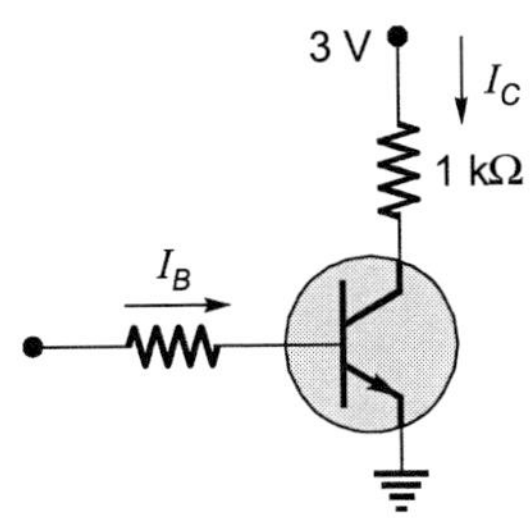

(a) 56 μA (b) 140 μA
(c) 60 μA (d) 3 μA

[2004 : 1 M]

3.24 Assuming that the β of the transistor is extremely large and $V_{BE} = 0.7$ V, I_C and V_{CE} in the circuit shown in the figure are

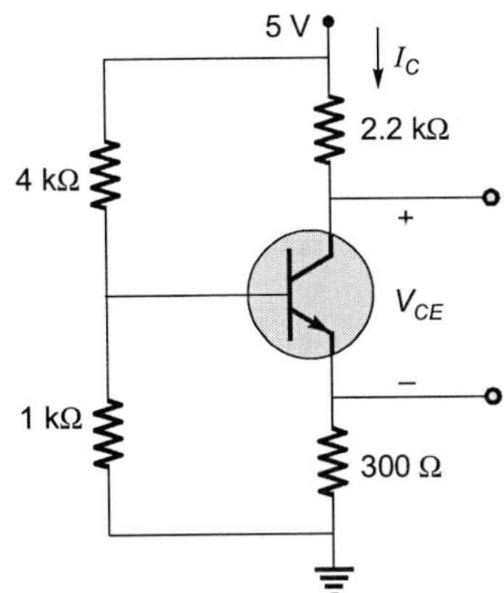

(a) $I_C = 1$ mA, $V_{CE} = 4.7$ V
(b) $I_C = 0.5$ mA, $V_{CE} = 3.75$ V
(c) $I_C = 1$ mA, $V_{CE} = 2.5$ V
(d) $I_C = 0.5$ mA, $V_{CE} = 3.9$ V [2004 : 2 M]

3.25 The cascode amplifier is a multistage configuration of

(a) CC-CB (b) CE-CB
(c) CB-CC (d) CE-CC [2005 : 1 M]

3.26 For an npn transistor connected as shown in the figure, $V_{BE} = 0.7$ volts. Given that reverse saturation current of the junction at room temperature 300°K is 10^{-13} A, the emitter current is

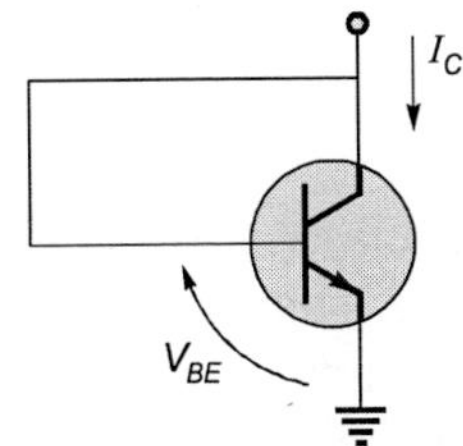

(a) 30 mA (b) 39 mA
(c) 49 mA (d) 20 mA [2005 : 2 M]

3.27 In an ideal differential amplifier shown in the figure, a large value of (R_E).

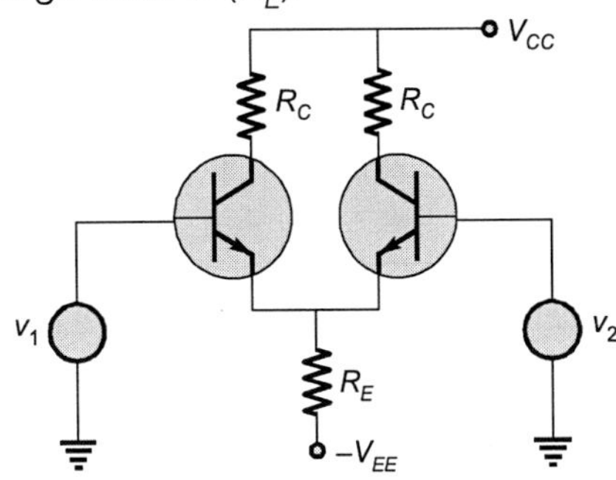

(a) increases both the differential and common-mode gains
(b) increases the common-mode gain only
(c) decreases the differential-mode gain only
(d) decreases the common-mode gain only

[2005 : 2 M]

3.28 The circuit using a BJT with $\beta = 50$ and $V_{BE} = 0.7$ V is shown in the figure. The base current I_B and collector voltage V_C are respectively

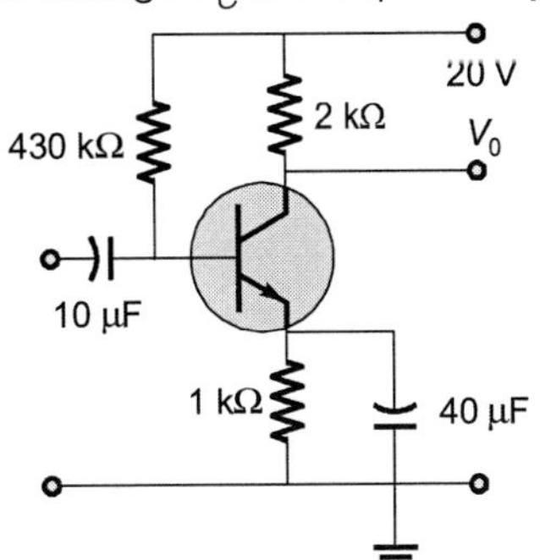

(a) 43 μA and 11.4 Volts
(b) 40 μA and 16 Volts
(c) 45 μA and 11 Volts
(d) 50 μA and 10 Volts [2005 : 2 M]

Common Data for Questions 3.29, 3.30 & 3.31.

In the transistor amplifier circuit shown in the figure below, the transistor has the following parameters:

$\beta_{DC} = 60$, $V_{BE} = 0.7$ V, $h_{ie} \rightarrow \infty$, $h_{fe} \rightarrow \infty$

The capacitance C_C can be assumed to be infinite.

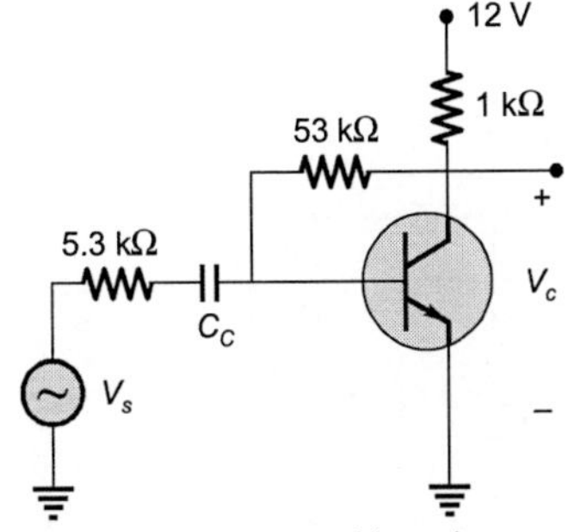

In the figure above, the ground has been shown by the symbol ⏚

3.29 Under the DC conditions, the collector-to-emitter voltage drop is
(a) 4.8 Volts (b) 5.3 Volts
(c) 6.0 Volts (d) 6.6 Volts [2006 : 2 M]

3.30 If β_{DC} is increased by 10%, the collector-to-emitter voltage drop
(a) increases by less than or equal to 10%
(b) decreases by less than or equal to 10%
(c) increases by more than 10%
(d) decreases by more than 10% [2006 : 2 M]

3.31 The small-signal gain of the amplifier V_c/V_s is
(a) –10 (b) –5.3
(c) 5.3 (d) 10 [2006 : 2 M]

3.32 The DC current gain (β) of a BJT is 50. Assuming that the emitter injection efficiency is 0.995, the base transport factor is
(a) 0.980 (b) 0.985
(c) 0.990 (d) 0.995 [2007 : 2 M]

3.33 For the BJT circuit shown, assume that the β of the transistor is very large and V_{BE} = 0.7 V. The mode of operation of the BJT is

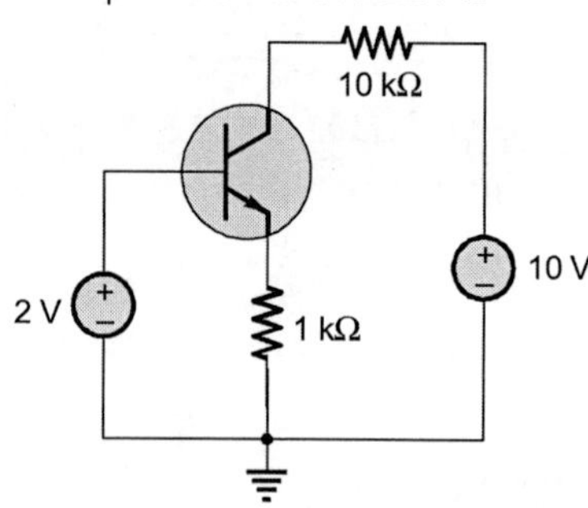

(a) cut-off (b) saturation
(c) normal active (d) reverse active
[2007 : 2 M]

Statement for linked Answer Questions 3.34 and 3.35:

In the following transistor circuit, V_{BE} = 0.7 V, r_e= 25 mV/I_E , and β and all the capacitances are very large.

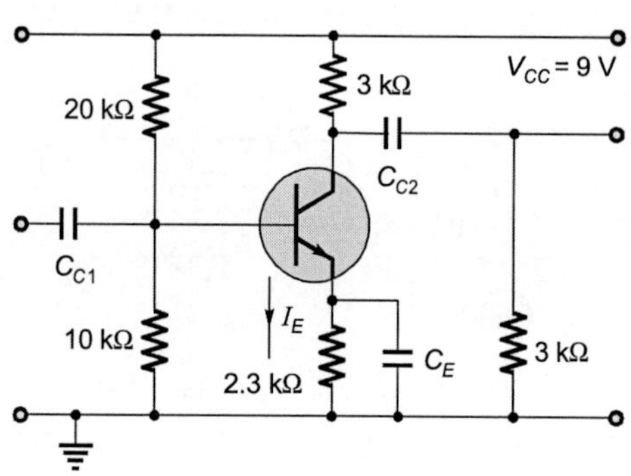

3.34 The Value of DC current I_E is
(a) 1 mA (b) 2 mA
(c) 5 mA (d) 10 mA
[2008 : 2 M]

3.35 The mid-band voltage gain of the amplifier is approximately
(a) –180 (b) –120
(c) –90 (d) –60
[2008 : 2 M]

3.36 A small signal source $v_i(t) = A \cos 20\,t + B \sin 10^6\,t$ is applied to a transistor amplifier as shown below. The transistor has β = 150 and h_{ie} = 3 kΩ. Which expression best approximates $v_0(t)$?

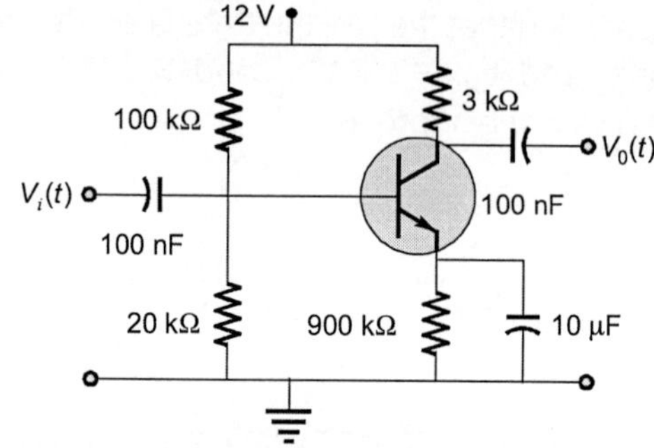

(a) $v_0(t) = -1500(A \cos 20t + B \sin 10^6 t)$
(b) $v_0(t) = -150(A \cos 20t + B \sin 10^6 t)$
(c) $v_0(t) = -1500\, B \sin 10^6 t$
(d) $v_0(t) = -150\, B \sin 10^6 t$
[2009 : 2 M]

3.37 In the silicon BJT circuit shown below, assume that the emitter area of transistor *Q*1 is half that of transistor *Q*2.

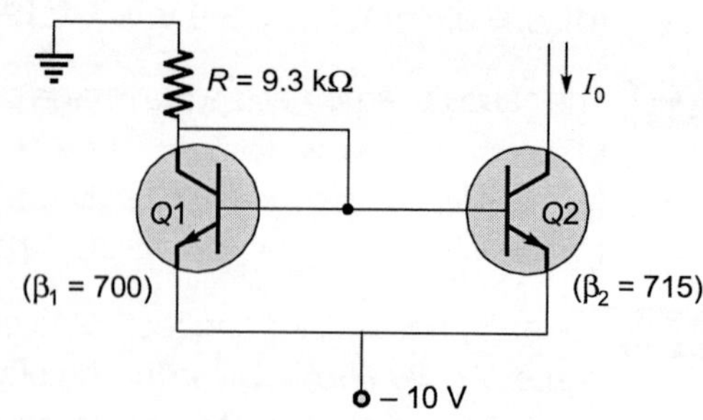

The value of current I_o is approximately
(a) 0.5 mA (b) 2 mA
(c) 9.3 mA (d) 15 mA
[2010 : 1 M]

3.38 The amplifier circuit shown below uses a silicon transistor. The capacitors C_C and C_E can be assumed to be short at signal frequency and the effect of output resistance r_0 can be ignored. If C_E is disconnected from the circuit, which one of the following statements is TRUE?

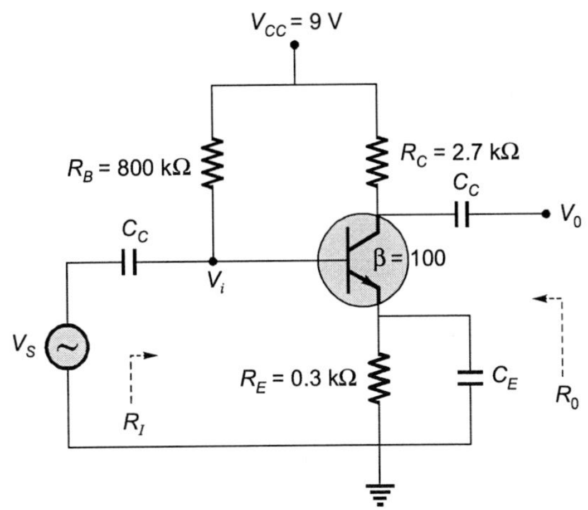

(a) The input resistance R_i increases and the magnitude of voltage gain A_V decreases
(b) The input resistance R_i decreases and the magnitude of voltage gain A_V increases
(c) Both input resistance R_i and the magnitude of voltage gain A_V decrease
(d) Both input resistance R_i and the magnitude of voltage gain A_V increase

[2010 : 1 M]

Common Data for Questions 3.39 and 3.40:

Consider the common emitter amplifier shown below with the following circuit parameters:

$\beta = 100$, $g_m = 0.3861$ A/V, $r_o = \infty$, $r_\pi = 259\ \Omega$, $R_S = 1$ kΩ, $R_B = 93$ kΩ, $R_C = 250\ \Omega$, $R_L = 1$ kΩ, $C_1 = \infty$ and $C_2 = 4.7$ μF.

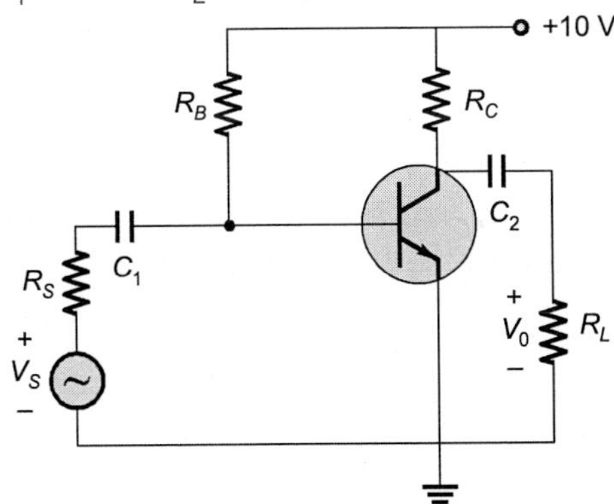

3.39 The resistance seen by the source V_s is
(a) 250 Ω (b) 1258 Ω
(c) 93 kΩ (d) ∞ [2010 : 2 M]

3.40 The lower cut-off frequency due to C_2 is
(a) 33.9 Hz (b) 27.1 Hz
(c) 13.6 Hz (d) 16.9 Hz [2010 : 2 M]

3.41 In the circuit shown below, capacitors C_1 and C_2 are very large and shorts at the input frequency. v_i is a small input. The gain magnitude $|v_0/v_i|$ at 10 M rad/s is

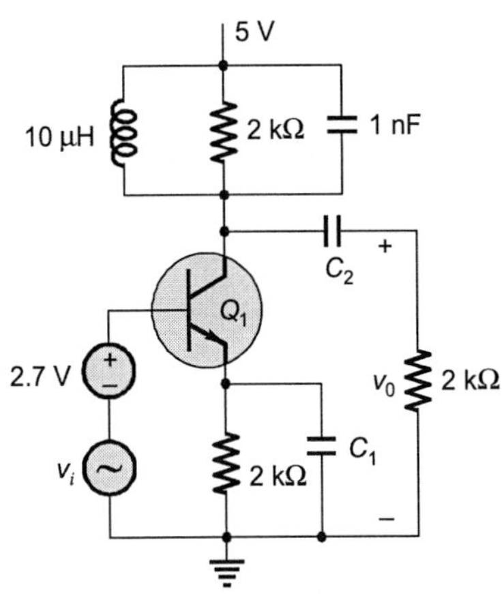

(a) maximum (b) minimum
(c) unity (d) zero [2011 : 1 M]

3.42 For the BJT Q_1 in the circuit shown below, $\beta = \infty$, $V_{BEon} = 0.7$ V, $V_{CEsat} = 0.7$ V. The switch is initially closed. At time $t = 0$, the switch is opened. The time t at which Q_1 leaves the active region is

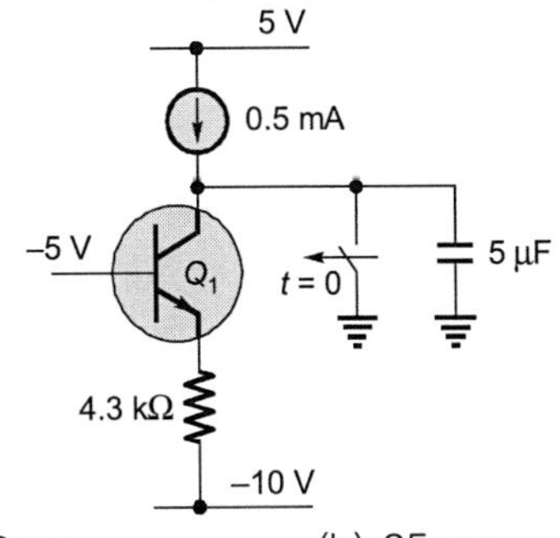

(a) 10 ms (b) 25 ms
(c) 50 ms (d) 100 ms

[2011 : 2 M]

3.43 The voltage gain A_v, of the circuit shown below is

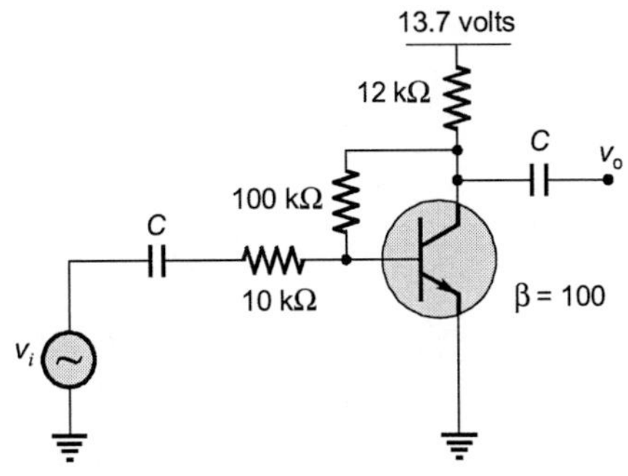

(a) $|A_v| \approx 200$ (b) $|A_v| \approx 100$
(c) $|A_v| \approx 20$ (d) $|A_v| \approx 10$

[2012 : 2 M]

3.44 The current i_b through the base of a silicon npn transistor is $1 + 0.1 \cos(10000\,\pi\, t)$ mA. At 300 K, the r_π in the small signal model of the transistor is

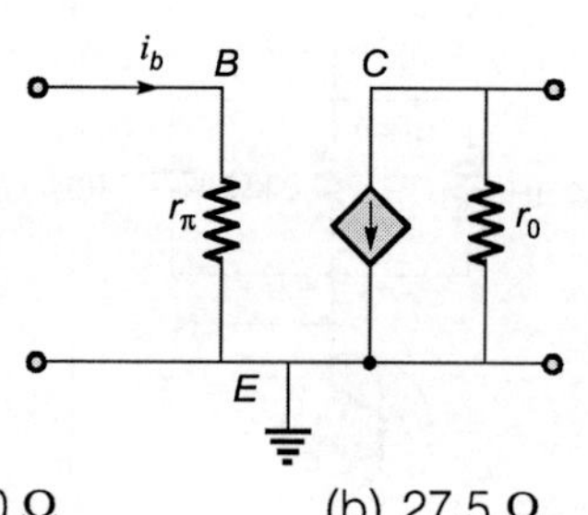

(a) 250 Ω (b) 27.5 Ω
(c) 25 Ω (d) 22.5 Ω

[2012 : 1 M]

3.45 In the circuit shown below, the silicon npn transistor Q has a very high value of β. The required value of R_2 in kΩ to produce I_C = 1 mA is

(a) 20
(b) 30
(c) 40
(d) 50

[2013 : 2 M]

3.46 A good current buffer has
(a) low input impedance and low output impedance
(b) low input impedance and high output impedance
(c) high input impedance and low output impedance
(d) high input impedance and high output impedance

[2014 : 1 M, Set-1]

3.47 For the amplifier shown in the figure, the BJT parameters are V_{BE} = 0.7 V, β = 200, and thermal voltage V_T = 25 mV. The voltage gain (v_o/v_i) of the amplifier is _____.

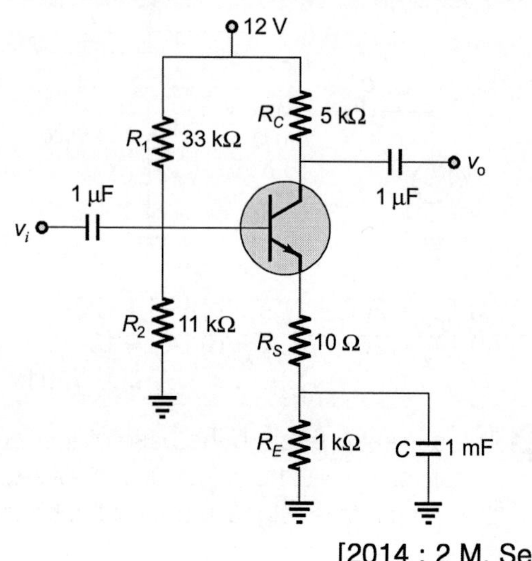

[2014 : 2 M, Set-1]

3.48 A cascade connection of two voltage amplifiers A_1 and A_2 is shown in the figure. The open-loop gain A_{vo}, input resistance R_{in}, and output resistance R_o for A_1 and A_2 are as follows:

$A_1 : A_{vo} = 10,\ R_{in} = 10\ \text{k}\Omega,\ R_o = 1\ \text{k}\Omega$

$A_2 : A_{vo} = 5,\ R_{in} = 5\ \text{k}\Omega,\ R_o = 200\ \Omega$

The approximate overall voltage gain v_{out}/v_{in} is

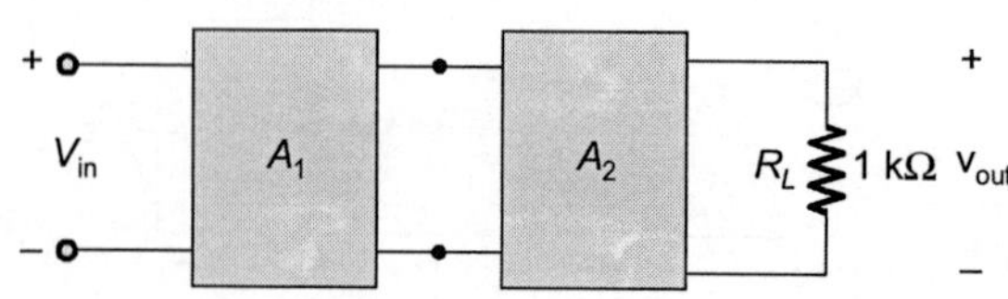

[2014 : 1 M, Set-2]

3.49 In the circuit shown, the PNP transistor has $|V_{BE}| = 0.7\text{ V}$ and $\beta = 50$. Assume that R_B = 100 kΩ. For V_0 to be 5 V, the value of R_C (in kΩ) is _____.

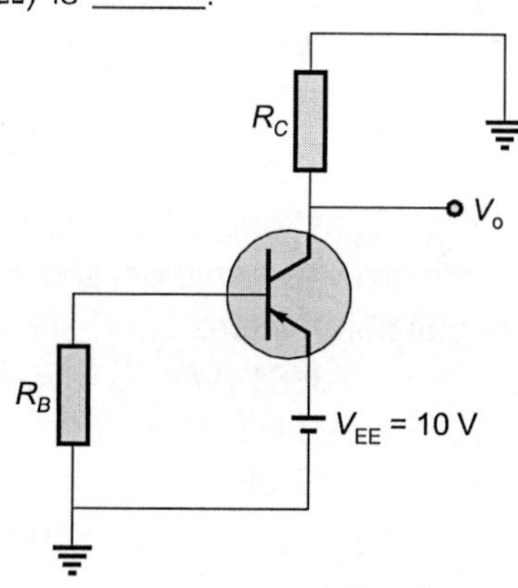

[2014 : 1 M, Set-3]

3.50 In the circuit shown, the silicon BJT has β = 50. Assume V_{BE} = 0.7 V and $V_{CE(sat)}$ = 0.2 V. Which one of the following statements is correct?

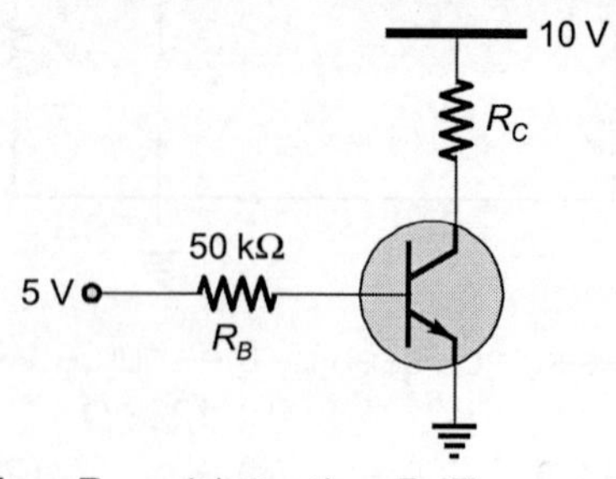

(a) For R_C = 1 kΩ, the BJT operates in the saturation region
(b) For R_C = 3 kΩ, the BJT operates in the saturation region
(c) For R_C = 20 kΩ, the BJT operates in the cut-off region
(d) For R_C = 20 kΩ, the BJT operates in the linear region

[2014 : 2 M, Set-3]

3.51 If the emitter resistance in a common-emitter voltage amplifier is not by passed, it will
(a) reduce both the voltage gain and the input impedance
(b) reduce the voltage gain and increase the input impedance
(c) increase the voltage gain and reduce the input impedance
(d) increase both the voltage gain and the input impedance

[2014 : 1 M, Set-4]

3.52 Consider the common-collector amplifier in the figure (bias circuitry ensures that the transistor operates in forward active region, but has been omitted for simplicity). Let I_C be the collector current, V_{BE} be the base-emitter voltage and V_T be the thermal voltage. Also, g_m and r_o are the small-signal transconductance and output resistance of the transistor, respectively. Which one of the following conditions ensures a nearly constant small signal voltage gain for a wide range of values of R_E?

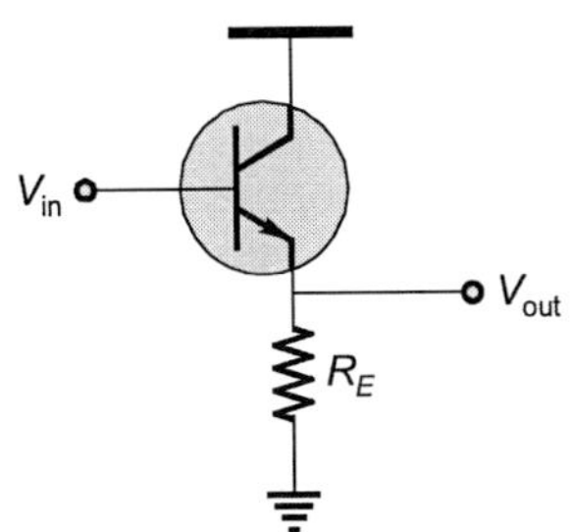

(a) $g_m R_E << 1$
(b) $I_C R_E >> V_T$
(c) $g_m r_o >> 1$
(d) $V_{BE} >> V_T$

[2014 : 2 M, Set-4]

3.53 For the common collector amplifier shown in the figure, the BJT has high β, negligible $V_{CE(sat)}$, and V_{BE} = 0.7 V. The maximum undistorted peak-to-peak output voltage v_o (in Volts) is ______.

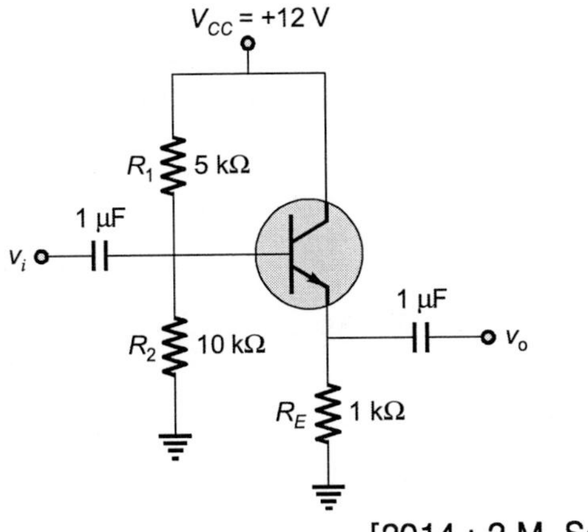

[2014 : 2 M, Set-4]

3.54 An increase in the base recombination of a BJT increase
(a) the common emitter dc current gain β
(b) the breakdown voltage BV_{CEO}
(c) the unity-gain cut-off frequency f_T
(d) the transconductance g_m

[2014 : 1 M, Set-2]

3.55 A BJT in a common-base configuration is used to amplify a signal received by a 50 Ω antenna. Assume kT/q = 25 mV. The value of the collector bias current (in mA) required to match the input impedance of the amplifier to the impedance of the antenna is______.

[2014 : 2 M, Set-4]

3.56 In the circuit shown, I_1 = 80 mA and I_2 = 4 mA. Transistors T_1 and T_2 are identical. Assume that the thermal voltage V_T is 26 mV at 27 °C. At 50 °C, the value of voltage $V_{12} = V_1 - V_2$ (in mV) is

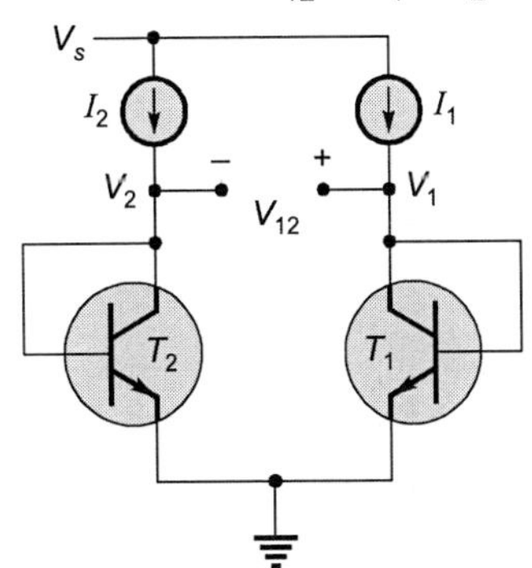

[2015 : 2 M, Set-1]

3.57 In the ac equivalent circuit shown, the two BJTs are biased in active region and have identical parameters with β >> 1. The open circuit small signal voltage gain is approximately _____.

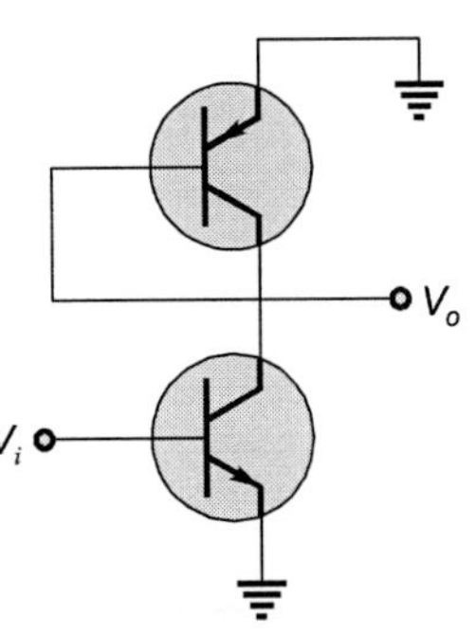

[2015 : 2 M, Set-2]

3.58 In the circuit shown in the figure, the BJT has a current gain (β) of 50. For an emitter base voltage V_{EB} = 600 mV, the emitter collector voltage V_{EC} (in Volts) is ______.

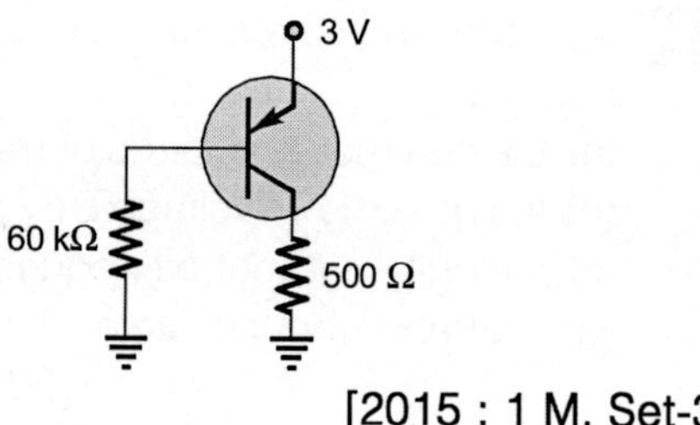

[2015 : 1 M, Set-3]

3.59 The Ebers-Moll model of a BJT is valid
(a) only in a active mode
(b) only in active and saturation modes
(c) only in active and cut-off modes
(d) in active, saturation and cut-off modes

[2016 : 1 M, Set-2]

3.60 Resistor R_1 in the circuit below has been adjusted so that I_1 = 1 mA. The bipolar transistors Q_1 and Q_2 are perfectly matched and have very high current gain, so their base currents are negligible. The supply voltage V_{CC} is 6 V. The thermal voltage kT/q is 26 mV.

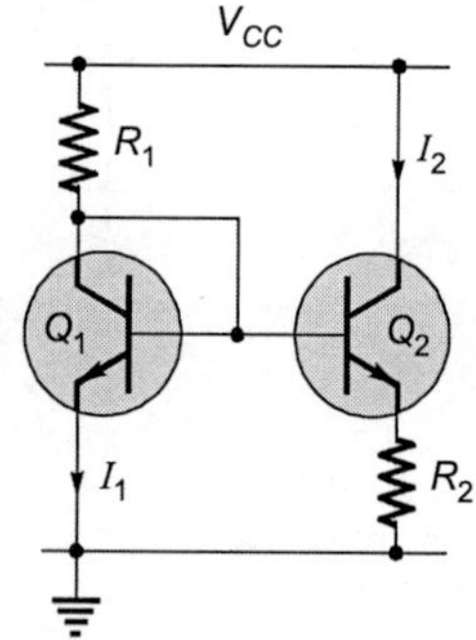

The value of R_2 (in Ω) for which I_2 = 100 μA is ____.

[2016 : 1 M, Set-2]

3.61 Consider the circuit shown in the figure. Assuming $V_{BE1} = V_{EB2}$ = 0.7 volt, the value of the dc voltage V_{C2} (in volt) is ____.

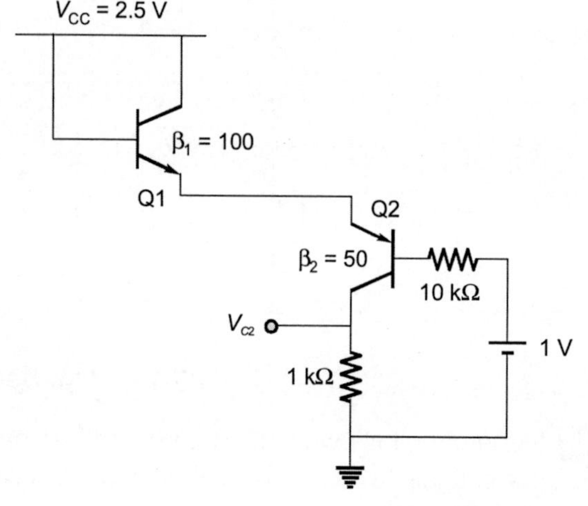

[2016 : 1 M, Set-3]

3.62 In the figure shown, the *npn* transistor acts as a switch

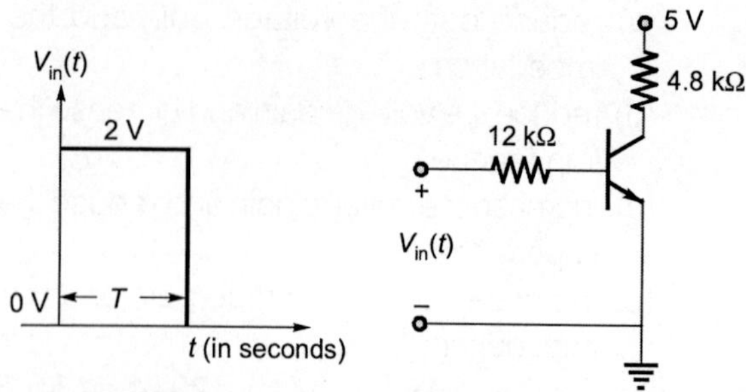

For the input $V_{in}(t)$ as shown in the figure, the transistor switches between the cut-off and saturation regions of operation, when T is large. Assume collector-to-emitter voltage at saturation $V_{CE(sat)}$ = 0.2 V and base-to-emitter voltage V_{BE} = 0.7 V. The minimum value of the common-base current gain (α) of the transistor for the switching should be ________.

[2017 : 2 M, Set-1]

3.63 For the DC analysis of the Common-Emitter amplifier shown, neglect the base current and assume that the emitter and collector currents are equal. Given that V_T = 25 mV, V_{BE} = 0.7 V, and the BJT output resistance r_o is practically infinite. Under these conditions, the midband voltage gain magnitude, $A_v = |v_o/v_i|$ V/V, is ________.

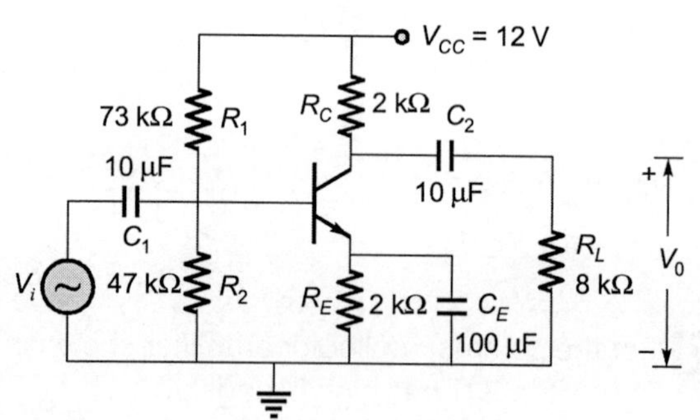

[2017 : 2 M, Set-1]

3.64 Consider the circuit shown in the figure. Assume base-to-emitter voltage V_{BE} = 0.8 V and common-base current gain (α) of the transistor is unity.

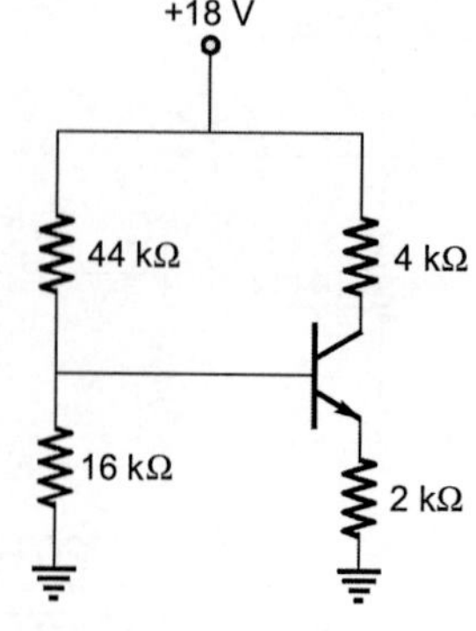

The value of the collector-to-emitter voltage V_{CE} (in volt) is ____.

[2017 : 1 M, Set-2]

3.65 In the circuit shown, transistors Q_1 and Q_2 are biased at a collector current of 2.6 mA. Assuming that transistor current gains are sufficiently large to assume collector current equal to emitter current and thermal voltage of 26 mV, the magnitude of voltage gain V_o/V_s in the mid-band frequency range is _____ (up to second decimal place).

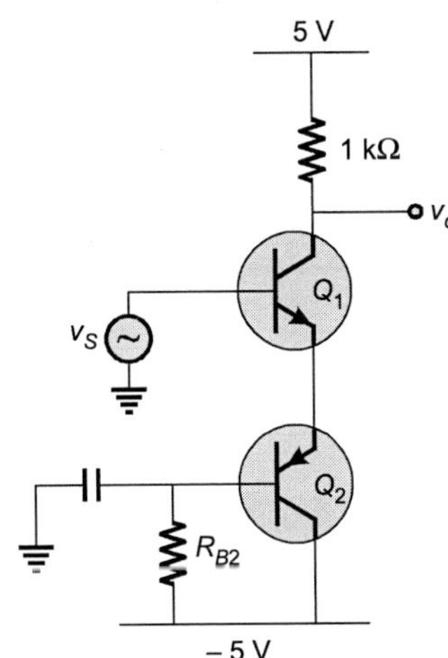

[2017 : 2 M, Set-2]

3.66 For the BJT in the amplifier shown below. V_{BE} = 0.7 V, kT/q = 26 mV. Assume the BJT output resistance (r_o) is very high and the base current is negligible. The capacitors are also assumed to be short circuited at signal frequencies. The input v_i is direct coupled. The low frequency gain v_o/v_i of the amplifier is

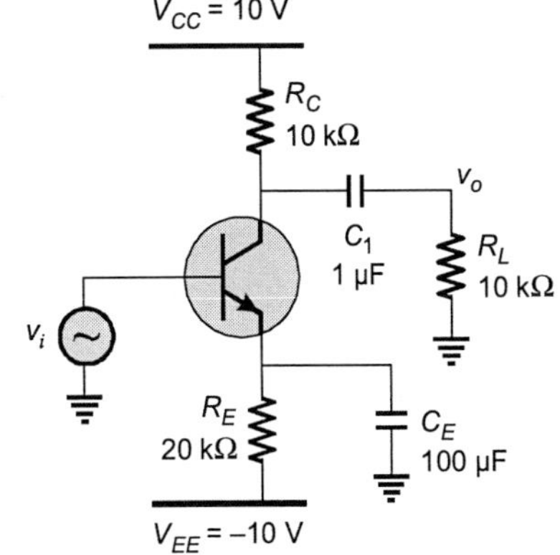

(a) –178.85
(b) –256.42
(c) –128.21
(d) –89.42

[2020 : 2 M]

Answers BJT Analysis

3.1	(3.75)	3.2	(c)	3.3	(a-2, b-5, c-4)			3.4	(b)	3.5	(a)
3.6	(d)	3.7	(c)	3.8	(b)	3.9	(d)	3.10	(b)	3.11	(c)
3.12	(a)	3.13	(a)	3.14	(c)	3.15	(a)	3.16	(d)	3.17	(c)
3.18	(b)	3.19	(d)	3.20	(b)	3.21	(a)	3.22	(a)	3.23	(a)
3.24	(c)	3.25	(b)	3.26	(c)	3.27	(d)	3.28	(b)	3.29	(c)
3.30	(b)	3.31	(a)	3.32	(b)	3.33	(b)	3.34	(a)	3.35	(d)
3.36	(d)	3.37	(b)	3.38	(a)	3.39	(b)	3.40	(b)	3.41	(a)
3.42	(c)	3.43	(d)	3.44	(c)	3.45	(c)	3.46	(b)	3.47	(–233.6)
3.48	(34.72)	3.49	(1.07)	3.50	(b)	3.51	(b)	3.52	(b)	3.53	(9.4)
3.54	(b)	3.55	(0.50)	3.56	(83.15)	3.57	(–1)	3.58	(2.04)	3.59	(d)
3.60	(598.67)	3.61	(0.5)	3.62	(0.902)	3.63	(128)	3.64	(6)	3.65	(50)
3.66	(d)										

Explanations BJT Analysis

3.1 (3.75)

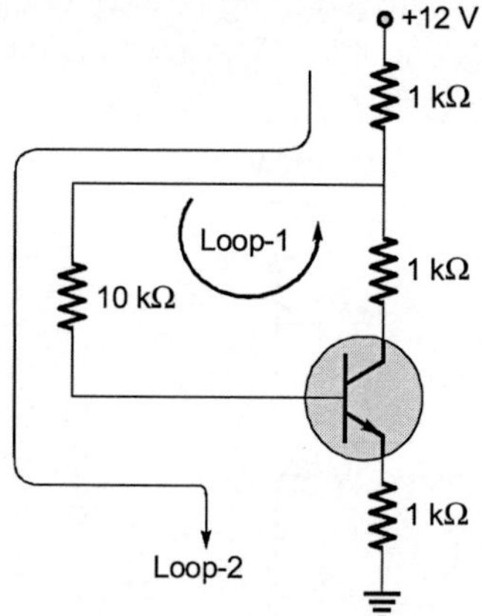

I_C cannot be 5.32 mA because I_C = 5.32 mA will make V_{CE} negative which implies transistor is in saturation. Through KVL,

$12 I_B + 2 I_C = 11.2$ (Loop-1)

$10 I_B - I_C = 0.6$ (Loop-2)

Upon solving

$I_C \cong 3.75$ mA

3.2 (c)

$$g_m = \frac{i_c}{V_{be}} = \frac{i_{c_2}}{V_{be_1} + V_{be_2}} \simeq \frac{i_{c_2}}{V_{be_2}} = g_{m_2}$$

3.3 Sol.

a-2, b-5, c-4

(a) Cascade Amplifier —

(2) It provides a wide band Amplifier

(b) Differential Amplifier —

(5) Provides high common mode voltage Rejection

(c) Darlington pair common-Collector Amplifier —

(4) Has very high input impedance and very high current gain

3.4 (b)

Increase the voltage gain and decrease the input impedance.

The bypass capacitor C across R_E will act as short circuit for ac signal. So, the bypass capacitor C will increase the voltage gain and decrease the input impedance.

3.5 (a)

A common emitter stage followed by common base stage.

3.7 (c)

Upper 3-dB frequency

$$B.W. = f_H = \frac{0.35}{t_r}$$

3.8 (b)

$$\frac{\sqrt{A_2^2 + A_3^2 + \cdots}}{A_1}$$

The total harmonic distortion is

$$\text{T.H.D.} = \frac{\sqrt{A_2^2 + A_3^2 + \cdots}}{A_1}$$

3.10 (b)

$$\frac{1}{B} = \frac{1}{\omega_1} + \frac{1}{\omega_2} + \frac{1}{\omega_3}$$

Cascading of amplifier results in decrease of higher cut-off frequency and increase in lower cut-off frequency.

So, $f_H\downarrow$ and $f_L\uparrow$

B.W. = $f_H - f_L$; So, B.W. $\downarrow$

3.11 (c)

The pass transistor in a series voltage regulator in common collector configuration, the voltage gain is almost equal to 1.

3.12 (a)

$$g_m = \frac{i_o}{V_i}, \qquad i_o \cong I_{E_2} = I_{C_1}$$

$$I_{C_1} = \beta I_{B_1}, \qquad I_{E_2} = I_{C_1}$$

$$i_o \cong \beta I_{B_1}, \qquad V_i = I_{B_1} r_\pi$$

$$\frac{i_o}{V_i} = \frac{\beta I_{B_1}}{I_{B_1} r_\pi} = g_{m_1} = \frac{I_{C_1}}{V_i} \left(as\ I_{C_1} = \beta I_{B_1}\right)$$

3.13 (a)

Common mode gain,

$$V_C = A_C V_i \left(V_{i_1} = V_{i_2} = V_i\right)$$

If R_e is infinite then because of symmetry of fig., V_C becomes zero.

$$i_{e_1} = i_{e_2} = 0$$

$$i_{b_2} << i_{c_2}$$

So, $i_{c_2} \simeq i_{e_2}$

3.15 (a)

$$A_i = \frac{-g_m}{g_{b'e} + j\omega(C_e + C_c)}$$

Where, C_e and C_c are the transistor capacitances so, at high frequencies the current gain of bipolar transistor drops due to the transistor capacitors.

3.16 (d)

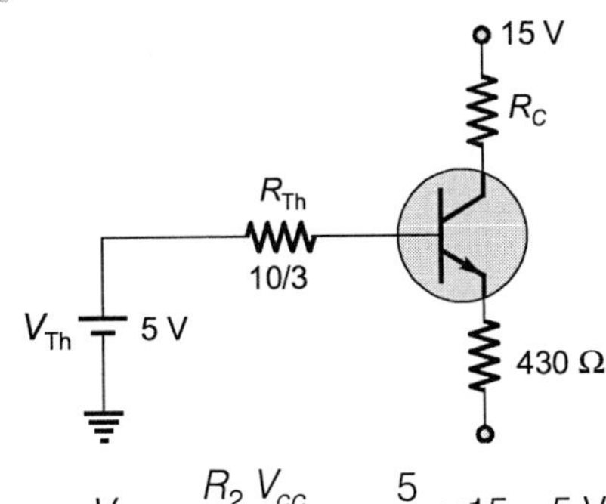

$$V_{Th} = \frac{R_2 V_{CC}}{R_1 + R_2} = \frac{5}{15} \times 15 = 5 \text{ V}$$

Since β is large, $I_B \approx 0$, $R_{Th} = 5 \parallel 10$

$$I_C = \frac{V_{Th} - 0.7}{R_E} = \frac{5 - 0.7}{0.430 \text{ k}\Omega} = \frac{4.3}{0.430 \text{ k}\Omega} = 10 \text{ mA}$$

3.17 (c)

$h_{fe} = g_m \cdot r_\pi$

3.18 (b)

$f_L = 20$ Hz

$f_H = 1$ kHz for single stage.

For cascaded stage

$$f_L^* = \frac{f_L}{\sqrt{2^{1/n} - 1}} = \frac{20}{\sqrt{2^{1/3} - 1}} = 39.2 \text{ Hz}$$

$$f_H^* = f_H\sqrt{2^{1/n} - 1} = 0.5 \text{ k}\Omega$$

3.22 (a)

$$V_{CE} = V_{CC} - I_C R_2$$
$$3 = 6 - 1.5 \text{ mA} \times R_2$$
$$1.5 \text{ mA} \times R_2 = 3$$
$$R_2 = 2 \text{ k}\Omega$$
$$I_B = \frac{I_C}{\beta} = \frac{1.5 \text{ mA}}{150} = 0.01 \text{mA}$$

When $\beta = 200$,

$I_C = \beta I_B$

(as R_1 is same I_B remains same)

$= 0.01 \text{ mA} \times 200$

$I_C = 2$ mA

$$V_{CE} = V_{CC} - I_C R_2 = 6 - 2 \text{ mA} \times 2 \text{ k}\Omega$$
$$V_{CE} = 2 \text{ V}$$

3.23 (a)

$$V_{CE} = V_{CC} - I_C R_C$$
$$0.2 = 3 - I_C \times 1 \text{ k}$$
$$I_C = 2.8 \text{ mA}$$
$$\beta = \frac{I_C}{I_B},\ I_B = \frac{I_C}{\beta} = \frac{2.8 \text{mA}}{50}$$
$$= 0.056 \text{ mA} = 56 \text{ μA}$$

3.24 (c)

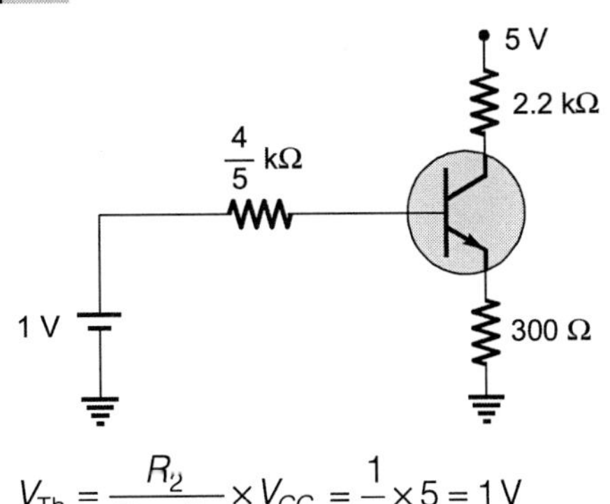

$$V_{Th} = \frac{R_2}{R_1 + R_2} \times V_{CC} = \frac{1}{5} \times 5 = 1 \text{V}$$

$$R_{Th} = R_1 \parallel R_2 = \frac{4}{5} \text{k}\Omega$$

$$I_C \simeq \frac{1 - 0.7}{0.3 \text{k}\Omega} = 1 \text{mA} \quad (\because \beta \text{ is large } I_B \simeq 0)$$

$$V_{CE} = V_{CC} - I_C R_C - I_E R_E \quad (I_E = I_C \text{ as } I_B \simeq 0)$$
$$V_{CE} = 5 - 2.2 \times 1 - 0.3 \times 1 = 2.5 \text{ V}$$

3.26 (c)

When two terminals of a transistor are shorted it acts as diode.

$$I = I_o\left(e^{\frac{V_D}{\eta V_T}} - 1\right) = 10^{-13}\left(e^{\frac{0.7}{1 \times 26 \times 10^{-3}}} - 1\right) = 49 \text{ mA}$$

3.27 (d)

Only common-mode gain depends on R_E and differential mode gain is independent of R_E.

3.28 (b)

$$I_E = I_C + I_B$$
$$= \beta I_B + I_B = (\beta + 1) I_B$$

KVL in input loop gives,

$$V_{CC} - V_{BE} = I_B R_B + I_E R_E$$
$$= I_B R_B + (\beta + 1) I_B R_E$$

$$\therefore I_B = \frac{V_{CC} - V_{BE}}{R_B + (\beta + 1) R_E} = \frac{20 - 0.7}{430 \text{K} + 51 \times 1 \text{K}}$$

$$I_B = 40\,\mu A$$
$$I_C = \beta I_B = 50 \times 40\,\mu A$$
$$= 2000\,\mu A = 2\text{ mA}$$
$$V_C = V_{CC} - I_C R_C = 20 - 2\text{ mA} \times 2\text{ K}$$
$$V_C = 16\text{ V}$$

3.29 (c)

Drawing DC equivalent circuit, the capacitor is treated as open-circuit

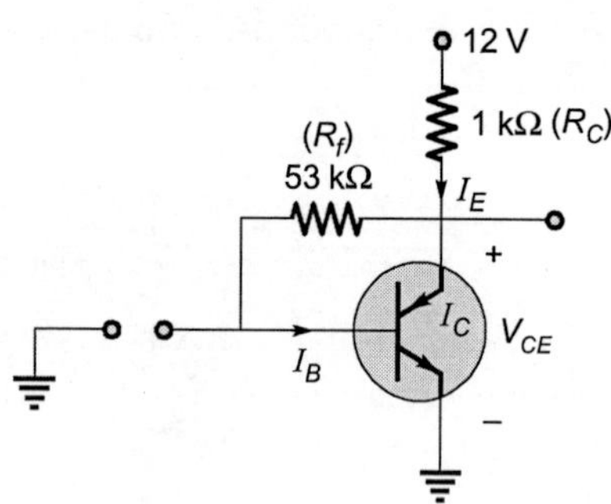

Applying KVL in base-emitter loop,
Applying KVL in base-emitter loop,

$$12 - I_E R_C - I_B \cdot R_f - 0.7 = 0$$

$$12 - 0.7 = I_E \cdot 1\text{k} + \frac{I_E}{(\beta+1)} \cdot 53\,k$$

$$I_E = \frac{11.3}{\left(1+\frac{53}{61}\right)\text{k}} = \frac{11.3}{1.87\text{ k}}$$

$$= 6.046\text{ mA} = 6\text{ mA}$$

$$V_{CE} = V_{CC} - I_E R_C$$
$$= 12 - 6\text{ mA} \times 1\text{ k} = 6\text{ V}$$

3.30 (b)

When β increases by 10%, new $\beta = 66$

$$I_E = \frac{11.3}{\left(1+\frac{53}{67}\right)\text{k}} = 6.31\text{mA}$$

$$V_{CE} = V_{CC} - I_E \,.\, R_C$$
$$= 12 - 6.31\text{ mA} \times 1\text{ k} = 5.7$$

$$\%\text{ change in, } V_{CE} = \frac{6-5.7}{6} \times 100 = 5\%$$

3.31 (a)

Given circuit is a voltage-shunt feedback amplifier. So, the approximate voltage gain is

$$A_{vf} = -\frac{R_f}{R_s} = -\frac{53\text{k}}{5.3\text{k}} = -10$$

3.32 (b)

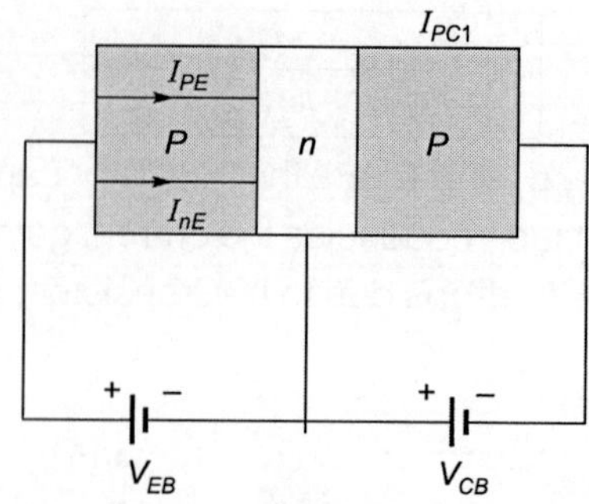

$$\text{Transport factor} = \frac{I_{PC_1}}{I_{PE}} = \beta^*$$

Current in emitter is both due to holes and electrons

Neglect current due to electrons

$$\alpha = \frac{I_{PC_1}}{I_E} = \frac{I_{PC_1}}{I_{PE}} \times \frac{I_{PE}}{I_E} = \beta^* \times \eta$$

$$\left(\eta = \frac{I_{PE}}{I_E} = \text{emitter efficiency}\right)$$

$$\beta^* = \frac{\alpha}{\eta} = \frac{50}{51 \times 0.995} = 0.9853$$

$$\left(\alpha = \frac{\beta}{1+\beta} = \frac{50}{51}\right)$$

3.33 (b)

Given β is large, so $I_B = 0$ and $I_E = I_C$

Assuming BJT is in active

Applying KVL in Base emitter loop

$$2 - 0.7 = 1\text{ k}\Omega \times I_E$$
$$I_E = 1.3\text{ mA}$$

$$\text{Now,}\quad I_{C_{sat}} = \frac{10 - 0.2}{1\text{k} + 10\text{ k}} = \frac{9.8}{11\text{ k}} \approx 0.9\text{ mA}$$

As $I_{C_{active}} > I_{C_{sat}}$

So, BJT is in saturation.

3.34 (a)

The given circuit can be redesigned as shown below:

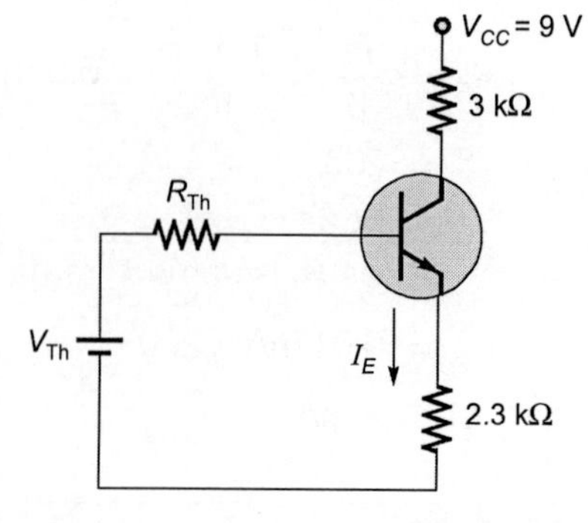

$$V_{Th} = \frac{10}{10+20} \times 9 = 3\text{ V}$$

$$R_{Th} = \frac{10 \times 20}{10 + 20}\text{k}\Omega = 6.67\text{ k}\Omega$$

Since β is large, I_B can be ignored

$$I_E = \frac{3-(0.7)}{2.3\text{ k}\Omega}$$
$$= 1\text{ mA}$$

3.35 (d)

Mid-band voltage gain,

$$A_{V_m} = -\frac{R_C \| R_L}{r_e}$$

$$A_{V_m} = -\frac{R'_L}{r_e} = -\frac{(3 \| 3)}{0.026} = -60$$

$$= -\frac{3 \times 10^3 \| 3 \times 10^3}{25 \times 10^{-3} / 1 \times 10^{-3}} = -60$$

3.36 (d)

The best approx. answer for output voltage V_0 is

$$V_o = A_V \cdot v_i$$

$$\approx \frac{-h_{fe} R_C}{h_{ie}} \times v_i \approx -150\,(A \cos 20t + B \sin 10^6 t)$$

Since the coupling capacitor is large

$$V_o \approx -150\, B \sin 10^6 t$$

3.37 (b)

The given circuit is a current mirror circuit in which the output current is a mirror image of the input current if both the transistors are identical.

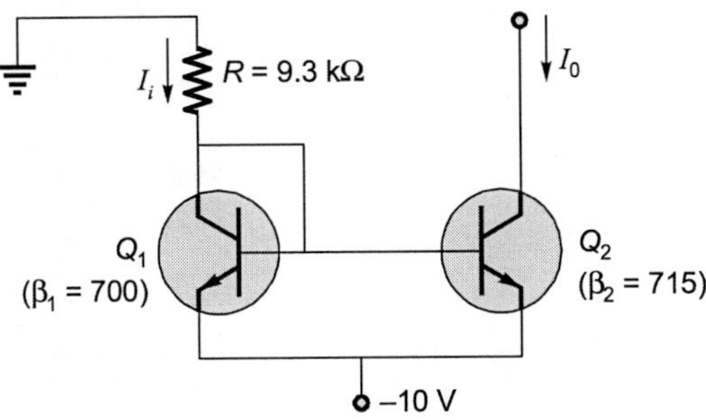

To calculate I_i,

$$9.3\, I_i + 0.7 = 0 - (-10) = 10$$

$$\Rightarrow \quad I_i = 1\text{ mA}$$

Since the emitter area of transistor Q_1 is half that of transistor Q_2,

So $\quad I_i = I_0/2.$

Therefore, I_0 = 2 mA

3.38 (a)

By disconnecting emitter bypass capacitor C_E,

(i) Input impedance increases by a factor of $(1 + \beta R_E)$.

(ii) Magnitude of voltage gain A_V decreases by the same factor.

3.39 (b)

Equivalent model of the given circuit is shown below:

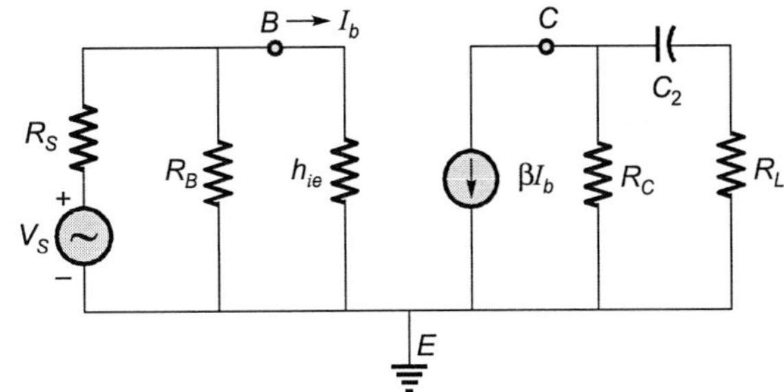

$$h_{ie} = \frac{\beta}{g_m} = \frac{100}{0.3861} = 259\ \Omega$$

The resistance seen by the source V_s,

$$R'_{in} - R_s + (R_B \| h_{ie})$$

$$= 1000 + \frac{93000 \times 259}{(93000 + 259)}$$

$$R'_{in} = 1000 + 258 = 1258\ \Omega$$

3.40 (b)

Lower cut-off frequency due to C_2,

$$f_L = \frac{1}{2\pi(R_C + R_L)C_2} = \frac{1}{2\pi(250 + 1000) \times 4.7 \times 10^{-6}}$$
$$= 27.1\text{ Hz}$$

3.41 (a)

Whenever we use a bypass capacitor in parallel R_E in a BJT biased with "common emitter with R_E" then it always increases voltage gain. As compare to "CE with R_E without capacitor". Hence only answer (a) is possible. And

$$\omega = \frac{1}{\sqrt{LC}} = \frac{1}{\sqrt{10 \times 10^{-6} \times 10^{-9}}}$$

= 10 M rad/sec and gain is maximum at resonance frequency.

3.42 (c)

In active region

$$-5 - 0.7 - 4.3\, I_E = -10$$

$$I_E = \frac{10 - 5.7}{4.3} = \frac{4.3}{4.3} = 1\text{ mA}$$

$$I_C = I_E = I' + 0.5\text{ mA} = 1\text{ mA}$$

$\Rightarrow \qquad I' = 0.5\text{ mA}$

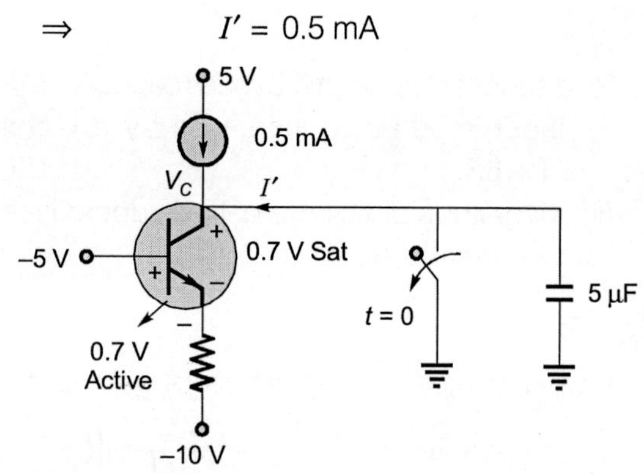

In saturation region ⇒

$V_C - 0.7 - 4.3 \times 1 = -10$

$V_C = -5\text{ V}$

$q = CV_C = -5 \times 10^{-6} \times 5\text{ V} = -25 \times 10^{-6}$

and $q = it$

$I'(0 - t) = -25 \times 10^{-6}$

$$t = \frac{25\times10^{-6}}{0.5\times10^{-3}} = 50\text{ m sec}$$

3.43 (d)

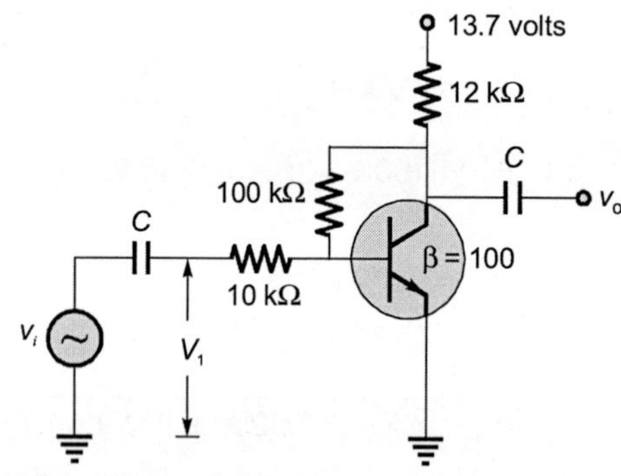

KVL: $-13.7 + 12 \times 101\, I_B \times 100\, I_B + V_{BE} = 0$

$1312\, I_B = B$

$\Rightarrow I_B \equiv 0.01\text{ mA}$

$I_C = \beta I_B = 1\text{ mA}$

$$g_m = \frac{I_C}{V_T} = 38\text{ m℧}$$

$$r_\pi = \frac{\beta}{g_m} = \frac{100}{38\text{ m}} = 2.63\text{ k}\Omega$$

Apply Miller's theorem to 100 k resistor,

$$R_m = \frac{100}{1-A_v} \quad ; \quad R_n = \frac{100}{1-\dfrac{1}{A_v}} \cong 100\text{ k}$$

$$\frac{V_o}{V_1} = -g_m R_L' = -38 \times (12 \,||\, 100) = -407.14$$

$$R_m = \frac{100}{1+407.1} = 0.245$$

$$R_i' = r_\pi \,||\, R_m = 2.63 \,||\, 0.245 = 0.224$$

$$\frac{V_o}{V_i} = \frac{V_o}{V_i} \times \frac{R_i'}{R_s + R_i}$$

$$= -407.14 \times \frac{0.224}{10+0.224} = -8.92$$

$$|A_v| \cong 10$$

3.44 (c)

We know that, $r_\pi = (\beta + 1)r_e = (\beta + 1)\dfrac{V_T}{I_e}$

$$r_\pi = (\beta + 1)\frac{V_T}{(\beta+1)I_b} = \frac{V_T}{I_b}$$

Where I_b is d.c. current through base so

$I_b = 1\text{ mA}$.

$V_T = 25\text{ mV}$ at room temperature.

So, $$r_\pi = \frac{25\times10^{-3}}{1\times10^{-3}} = 25\ \Omega$$

3.45 (c)

Given that, β is very large.

So, $I_c = I_E = 1\text{ mA}$

$V_E = I_E R_E = 1 \times 10^{-3} \times 500 = 0.5\text{ V}$

$V_{BE} = 0.7\text{ V}$

$V_{R2} = V_{BE} + V_E = 0.7 + 0.5 = 1.2\text{ V}$

It is a self bias circuit,

So, $V_{R2} = V_{cc} \times \dfrac{R_2}{R_1 + R_2}$; $1.2 = 3 \times \dfrac{R_2}{60 + R_2}$

$72 + 1.2\, R_2 = 3R_2$

$1.8\, R_2 = 72$

$R_2 = 40\text{ k}\Omega$

3.47 (-233.6)

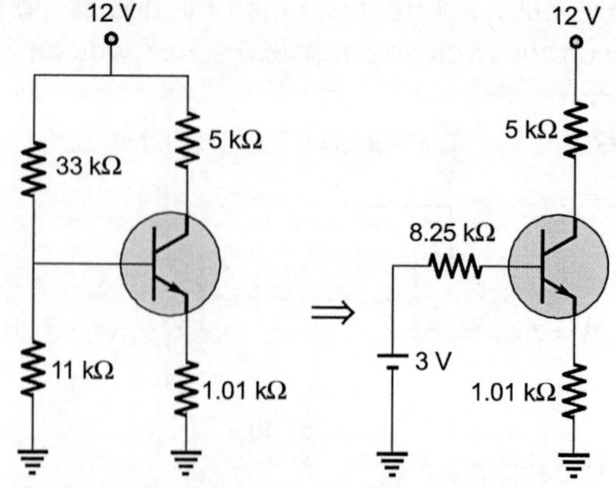

By applying KVL in input loop, we get

$3 = 8.25\text{ k}\, I_B + 0.7 + 1.10\text{ k}\, I_E$

$$2.3 = I_E\left[\frac{8.25\text{ k}}{201} + 1.01\text{k}\right]$$

$2.3 = I_E \times 1.051\text{ k}$

$I_E = 2.2\text{ mA}$

AC equivalent model of amplifier is given as

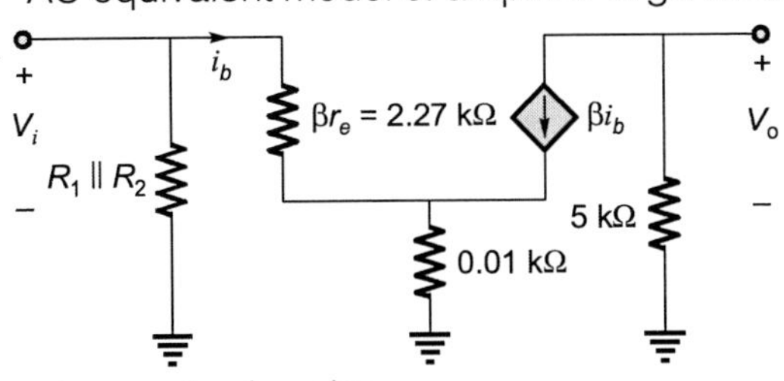

where r_e is given by

$$r_e = \frac{V_T}{I_E}$$

$$r_e = \frac{25\,\text{mV}}{2.2\,\text{mA}} = 11.36\,\Omega$$

$$V_o = -5\beta\, i_b \quad \text{...(i)}$$

$$V_i = 2.27\text{ k}\, i_b + 0.01\text{ k}\,(i_b + \beta i_b) \quad \text{...(ii)}$$

From equations (i) and (ii), we get

$$\frac{V_o}{V_i} = -\frac{5\text{ k}\times 200}{4.28\text{ k}} = -233.6$$

3.48 (34.72)

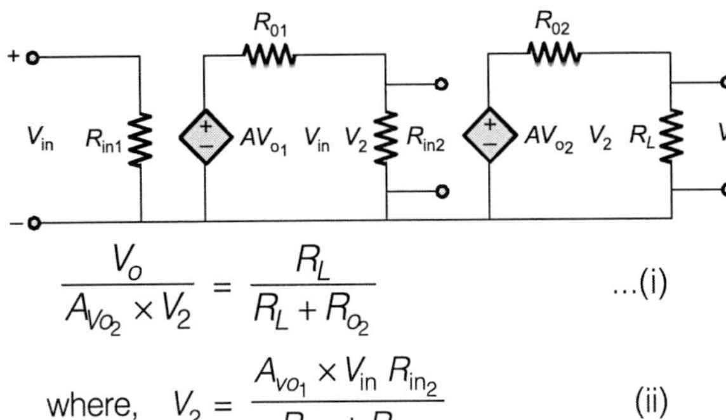

$$\frac{V_o}{A_{Vo_2}\times V_2} = \frac{R_L}{R_L + R_{o_2}} \quad \text{...(i)}$$

where, $$V_2 = \frac{A_{Vo_1}\times V_{in}\, R_{in_2}}{R_{in_2} + R_{o_1}} \quad \text{(ii)}$$

From equations (i) and (ii), we get

$$\frac{V_o}{V_{in}} = A_{V_{o_2}} \times A_{V_{o_1}} \times \frac{R_L}{R_L + R_{o_2}} \times \frac{R_{in_2}}{R_{in_2} + R_{o_1}}$$

$$\Rightarrow \quad \frac{V_o}{V_{in}} = 5\times 10\times \frac{1\text{k}\Omega}{1\text{k}\Omega + 0.2\,\text{k}\Omega}\times \frac{5\,\text{k}\Omega}{5\,\text{k}\Omega + 1\,\text{k}\Omega}$$

$$\Rightarrow \quad \frac{V_o}{V_{in}} = 34.72$$

3.49 (1.07)

Applying KVL in input loop, we get

$$10 = 0.7 + I_B R_B$$

$$I_B = \frac{9.3}{100\,\text{k}\Omega} = 0.93\,\text{mA}$$

Hence $I_C = \beta I_B = 4.65$ mA

where $$I_C = \frac{V_o - 0}{R_C} = \frac{5}{R_C}$$

$$\therefore \quad R_C = \frac{5}{4.65} = 1.07\text{ k}\Omega$$

3.50 (b)

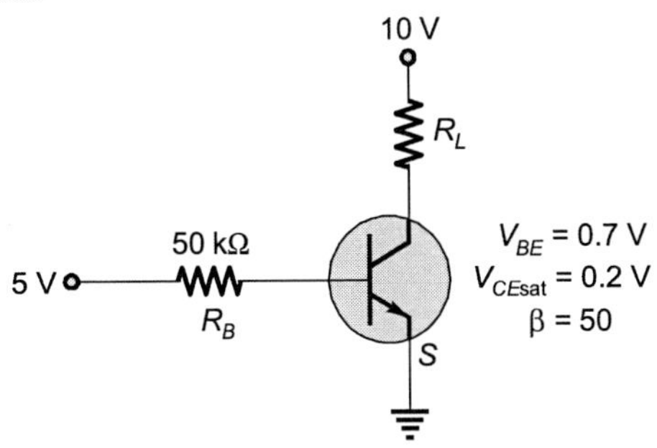

Assume the transistor is in active region

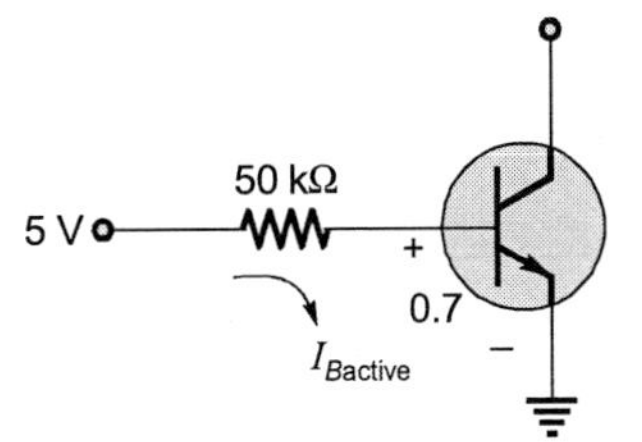

$$I_{B\,\text{active}} = \frac{5 - 0.7}{50\,\text{k}\Omega} = 86\,\mu\text{A}$$

$$I_{C\,\text{active}} = \beta\, I_{B\,\text{active}} = 50 \times 86\,\mu\text{A} = 4.3\text{ mA}$$

Assume $R_e = 1$ kΩ

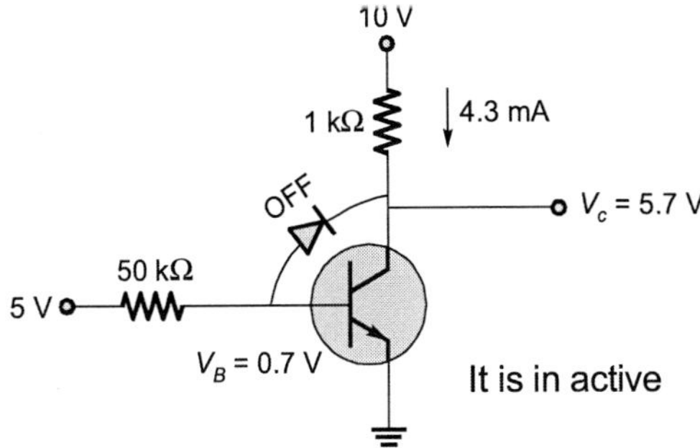

Assume $R_c = 3$ kΩ

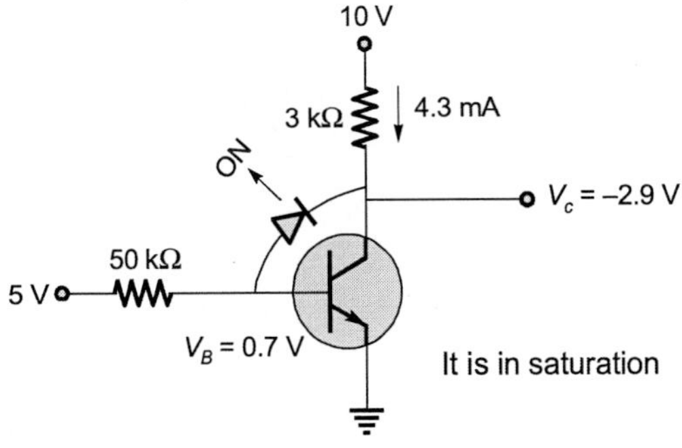

3.51 (b)

For unbypassed R_E,

$$R_i = \beta r_e + (1+\beta)R_E \text{ and}$$

$$A_v = \frac{A_I\, R_L}{R_i}$$

3.52 (b)

AC equivalent circuit for given common collector amplifier is

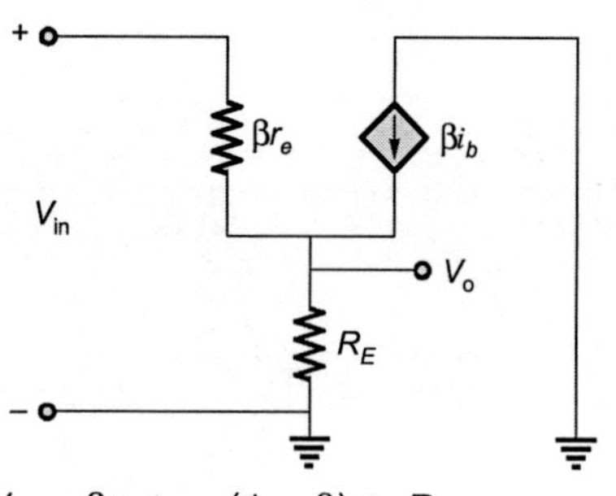

$$V_{in} = \beta r_e i_b + (1+\beta) i_b R_E \quad ...(i)$$
$$V_o = (1+\beta) R_E i_b \quad ...(ii)$$

Then

$$\frac{V_o}{V_{in}} = \frac{(1+\beta) R_E}{\beta r_e + (1+\beta) R_E}$$

$$\frac{V_o}{V_{in}} = \frac{\beta R_E}{\beta r_e + \beta R_E} = \frac{R_E}{r_e + R_E}$$

The condition for small signal voltage gain to be nearly constant is

$$R_E >> r_e$$

$$R_E >> \frac{V_T}{I_C}$$

$$I_C R_E >> V_T$$

3.53 (9.4)

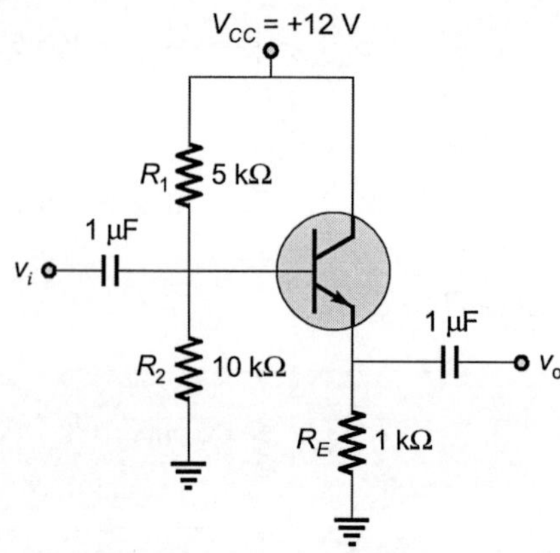

DC analysis

$$(V_B)_q = \frac{V_{CC} \times R_2}{R_1 + R_2} = \frac{12 \times 10\,k\Omega}{15\,k\Omega} = 8\text{ V}$$

$$(V_E)_Q = V_B - 0.7 = 8 - 0.7 = 7.3\text{ V}$$

$$(V_{CE})_Q = (V_C)_Q - (V_E)_Q$$
$$= 12 - 7.3 = 4.7\text{ V}$$

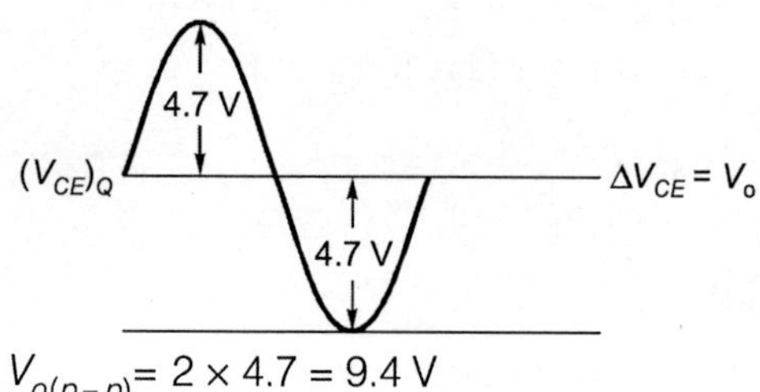

$$V_{o(p-p)} = 2 \times 4.7 = 9.4\text{ V}$$

3.55 (0.50)

The input impedance of CB-amplifier is

$$Z_i = r_e$$

So, according to given condition

$$r_e = 50\ \Omega$$

$$\frac{V_T}{I_C} = 50\ \Omega$$

$$I_C = \frac{25\text{ mV}}{50\ \Omega}$$

$$I_C = 0.50\text{ mA}$$

3.56 (83.15)

$$I_2 = I_S e^{\frac{V_{BE_2}}{\eta V_T}}$$

$$V_{BE_2} = V_2$$

$$I_1 = I_S e^{\frac{V_{BE_1}}{\eta V_T}}$$

$$V_{BE_1} = V_1$$

$$\frac{I_1}{I_2} = e^{\frac{V_1 - V_2}{\eta V_T}}$$

Since, V_T at 27° is 26 mV then V_T at 50° is 27.99 mV. Thus,

$$V_{12} = V_1 - V_2 = 83.15\text{ mV}$$

3.57 (–1)

Both the transistors are biased at the same collector current thus their transconductance will be same, the small signal equivalent circuit is

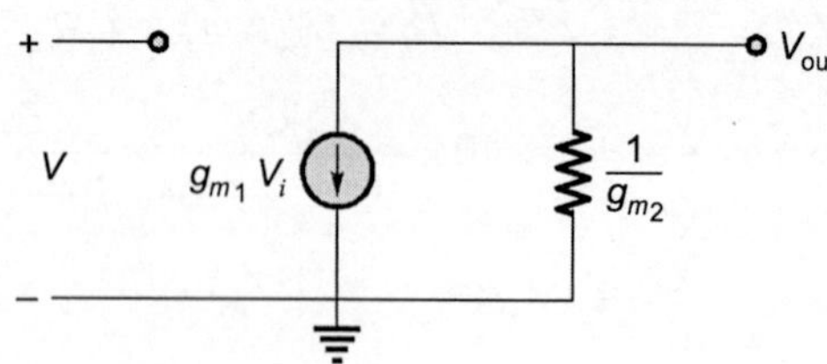

$g_{m_1} = g_{m_2}$, thus $V_o = -V_i$

So, gain equal to –1.

3.58 (2.04)

$$V_{EB} = 0.7\text{ V}$$
$$I_B = 0.0383\text{ mA}$$
$$I_C = 1.916\text{ mA}$$
$$V_{EC} = 3 - I_C R_C$$
$$= 3 - (1.916 \times 0.5) = 2.04\text{ V}$$

3.59 (d)

Ebers-Moll model is valid for all the region of operation.

3.60 (598.67)

$$R_2 = \frac{V_T}{I_2}\ln\left(\frac{I_1}{I_2}\right) = \frac{26\times10^{-3}}{100\times10^{-6}}ln\left(\frac{1\times10^{-3}}{100\times10^{-6}}\right)$$

$= 598.67\ \Omega$

3.61 (0.5)

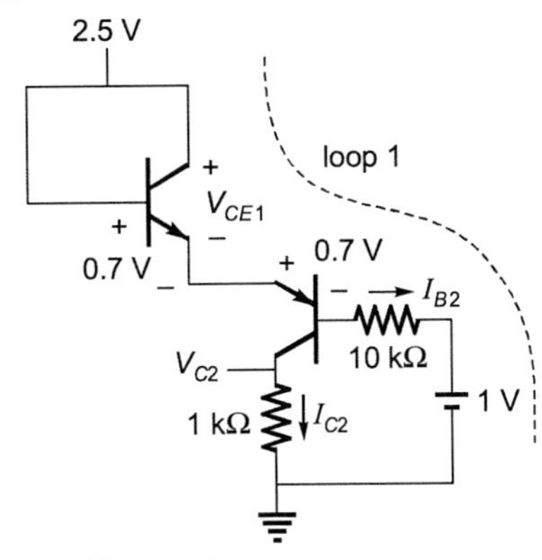

Q_1: $V_{CB} = 0$

We know,

$V_{CB1} + V_{BE1} = V_{CE1}$

$\Rightarrow \quad V_{CE1} = V_{BE1} = 0.7\text{ V}$

KVL to loop 1

$-2.5 + V_{CE1} + 0.7 + I_{B2} \times 10\text{ K} + 1\text{ V} = 0$

$$\Rightarrow \quad I_{B2} = \frac{2.5-1.4-1}{10K} = 0.01\text{mA}$$

$I_{C2} = \beta_2 I_{B2} = 0.5\text{ mA}$

$V_{C2} = I_{C2} \times 1\ K = 0.5 \times 10^{-3} \times 1 \times 10^3$

$= 0.5\text{ V}$

3.62 (0.902)

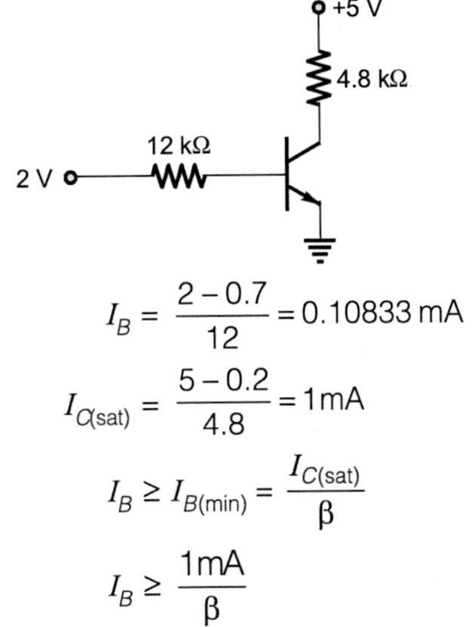

$$I_B = \frac{2-0.7}{12} = 0.10833\text{ mA}$$

$$I_{C(sat)} = \frac{5-0.2}{4.8} = 1\text{mA}$$

$$I_B \geq I_{B(min)} = \frac{I_{C(sat)}}{\beta}$$

$$I_B \geq \frac{1\text{mA}}{\beta}$$

$$\beta \geq \frac{1}{0.10833} \text{ and } \beta_{min} = 9.23$$

$$\alpha_{min} = \frac{\beta_{min}}{1+\beta_{min}} = 0.902$$

3.63 (128)

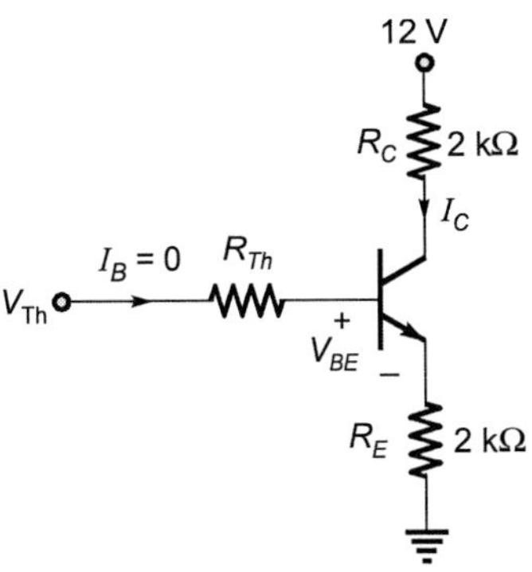

By taking the Thevenin's equivalent between Base and Ground nodes, the given circuit can be reduced as follows for DC analysis:

$$V_{Th} = \frac{12\times47}{73+47} = 4.7\text{ V}$$

Neglecting base current ($I_B = 0$)

$$I_E = I_C = \frac{V_{Th}-V_{BE}}{R_E} = \frac{4.7-0.7}{2}$$

$I_C = 2\text{ mA}$

$$\therefore \quad g_m = \frac{I_C}{V_T} = \frac{2}{25} = 80\text{ m℧}$$

Voltage gain, $A_v =$

$-g_m(R_C \| R_L) = -80(2\|8) = -128$

$\therefore \quad |A_v| = 128$

3.64 (6)

By taking the Thevenin's equivalent between Base and Ground nodes, the given circuit can be reduced as follows:

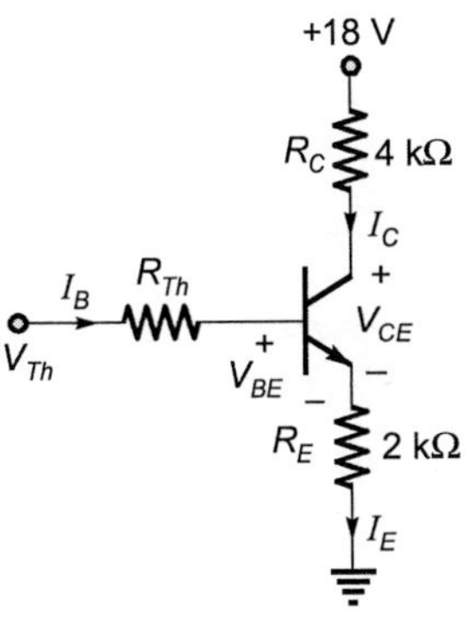

$$V_{Th} = \frac{16}{16+44} \times 18\text{ V} = 4.8\text{ V}$$

$$I_E R_E = V_{Th} - V_{BE} - I_B R_{Th}$$

$$I_B = 0\text{ A} \qquad \because \alpha = 1$$

So, $$I_E R_E = 4.8 - 0.8 = 4\text{ V}$$

$$I_E = \frac{4}{2}\text{ mA} = 2\text{ mA}$$

$$I_C = I_E = 2\text{ mA} \qquad \because \alpha = 1$$

$$V_{CE} = V_{CC} - I_C R_C - I_E R_E$$

$$= 18 - (2 \times 4) - (2 \times 2) = 6\text{ V}$$

3.65 (50)

In AC equivalent circuit, Q_2 becomes diode-connected-transistor because collector and base get shorted.

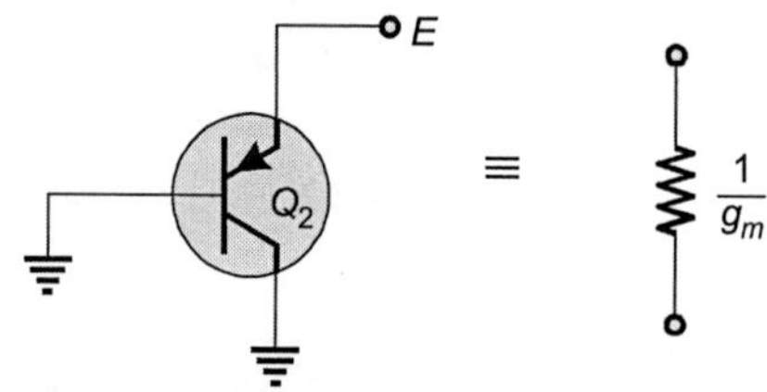

Computer AC equivalent circuit is as shown below:

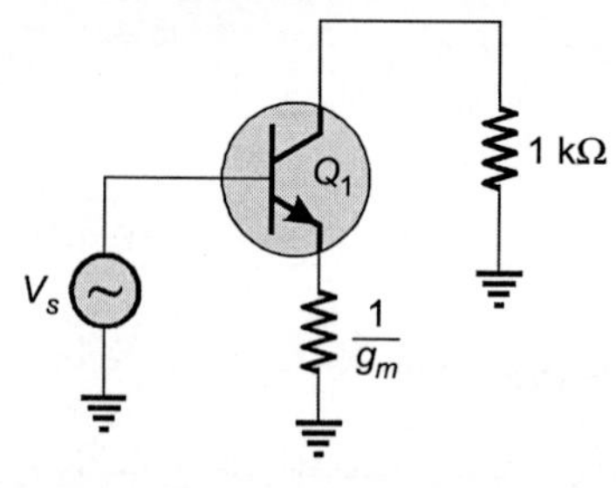

It is a CE amplifier with unbypassed R_E.

$$A_v = \frac{-g_m R_L'}{1 + g_m R_E}$$

where, $$g_m = \frac{I_C}{V_T} = \frac{2.6}{26} = 100\text{ m℧}$$

$$R_L' = 11\ \Omega$$

$$R_E = \frac{1}{g_m}$$

$$A_v = \frac{-100 \times 1}{1+1} = -50$$

$$|A_v| = 50$$

3.66 (d)

$$I_{EQ} = \frac{10 - 0.7}{20} = 0.465\text{ mA}$$

$$g_m = \frac{I_{EQ}}{V_T} = \frac{0.465}{26}\text{ A/V}$$

$$\frac{V_{out}}{V_{in}} = -g_m (R_e \| R_L)$$

$$= \frac{0.465}{26} \times 5000 = -89.423$$

■■■

4 FET and MOSFET Analysis

Note: # JFET topic is out of syllabus for GATE from 2016, it may be useful for PSUs & Interviews.

4.1 An n-channel JFET has $I_{DSS} = 1$ mA and $V_P = -5$ V. Its maximum transconductance is ________

[1995 : 1 M]

4.2 Two identical FETs, each characterized by the parameters g_m and r_d are connected in parallel. The composite FET is then characterized by the parameters.

(a) $\frac{g_m}{2}$ and $2r_d$ (b) $\frac{g_m}{2}$ and $\frac{r_d}{2}$

(c) $2g_m$ and $\frac{r_d}{2}$ (d) $2g_m$ and $2r_d$

[1998 : 1 M]

4.3 In the MOSFET amplifier of the figure is the signal output V_1 and V_2 obey the relationship

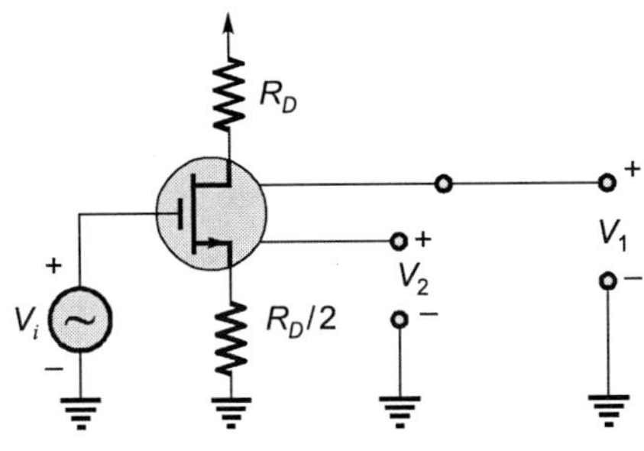

(a) $V_1 = \frac{V_2}{2}$ (b) $V_1 = -\frac{V_2}{2}$

(c) $V_1 = 2\ V_2$ (d) $V_1 = -2\ V_2$

[1998 : 1 M]

4.4# The voltage gain $A_v = \frac{V_0}{V_i}$ of the JFET amplifier shown in the figure is

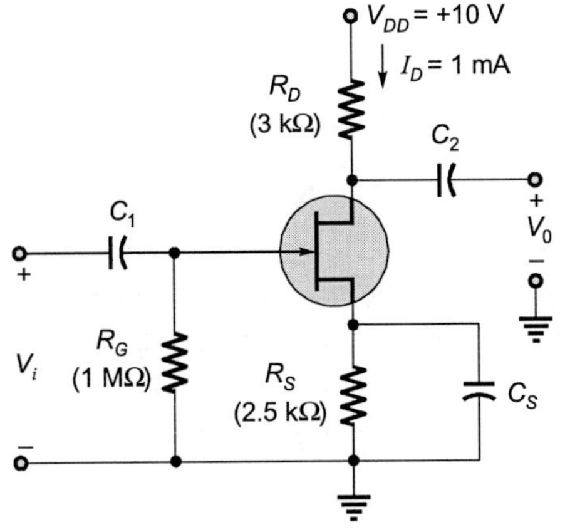

$I_{DSS} = 10$ mA $V_p = -5$ V

(Assume C_1, C_2 and C_s to be very large)

(a) +18 (b) −18

(c) +6 (d) −6

[2002 : 2 M]

4.5 Consider the following statements in connection with the CMOS inverter in the figure, where both the MOSFETs are of enhancement type and both have a threshold voltage of 2 V.

Statement 1: T_1 conducts when $V_i \geq 2$ V.

Statement 2: T_1 is always in saturation when $V_0 = 0$ V.

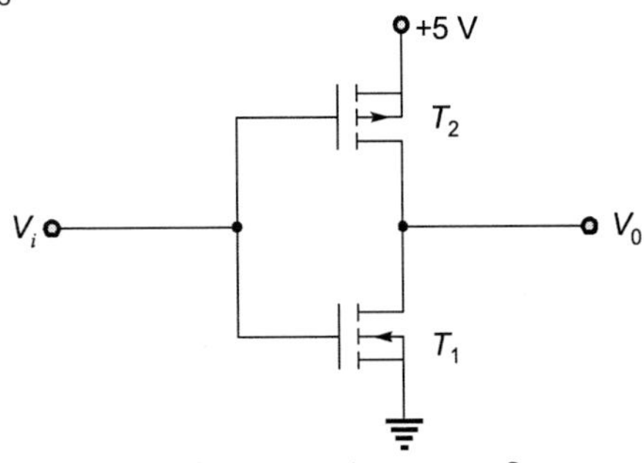

Which of the following is correct?

(a) Only Statement 1 is TRUE
(b) Only Statement 2 is TRUE
(c) Both the statements are TRUE
(d) Both the statements are FALSE

[2002 : 2 M]

4.6 The action of a JFET in its equivalent circuit can best be represented as a

(a) Current Controlled Current Source
(b) Current Controlled Voltage Source
(c) Voltage Controlled Voltage Source
(d) Voltage Controlled Current Source

[2003 : 2 M]

4.7 For an n-channel MOSFET and its transfer curve shown in the figure, the threshold voltage is

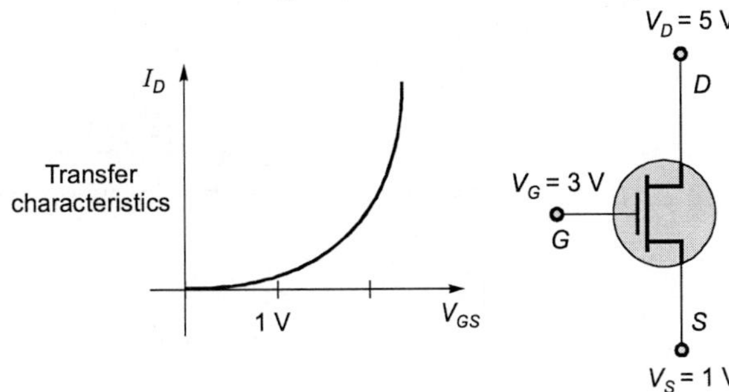

(a) 1 V and the device is in active region
(b) –1 V and the device is in saturation region
(c) 1 V and the device is in saturation region
(d) –1 V and the device is in active region

[2005 : 2 M]

Common Data Questions 4.8, 4.9 & 4.10.

Given, $r_d = 20$ kΩ, $I_{DSS} = 10$ mA, $V_p = -8$ V

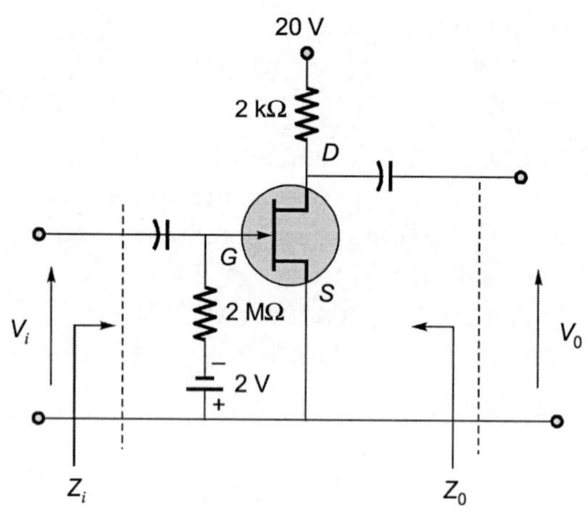

4.8 Z_i and Z_0 of the circuit are respectively
(a) 2 MΩ and 2 kΩ
(b) 2 MΩ and $\frac{20}{11}$ kΩ
(c) infinity and 2 MΩ
(d) infinity and $\frac{20}{11}$ kΩ

[2005 : 2 M]

4.9 I_D and V_{DS} under DC conditions are respectively
(a) 5.625 mA and 8.75 V
(b) 7.500 mA and 5.00 V
(c) 4.500 mA and 11.00 V
(d) 6.250 mA and 7.50 V

[2005 : 2 M]

4.10 Transconductance in milli-Siemens (mS) and voltage gain of the amplifier are respectively
(a) 1.875 mS and 3.41
(b) 1.875 mS and –3.41
(c) 3.3 mS and –6
(d) 3.3 mS and 6

[2005 : 2 M]

4.11 An *n*-channel depletion MOSFET has following two points on its $I_D - V_{GS}$ curve:
(i) $V_{GS} = 0$ at $I_D = 12$ mA and
(ii) $V_{GS} = -6$ Volts at $Z_0 = \infty$

Which of the following *Q*-points will give the highest trans-conductance gain for small signals?
(a) $V_{GS} = -6$ Volts (b) $V_{GS} = -3$ Volts
(c) $V_{GS} = 0$ Volts (d) $V_{GS} = 3$ Volts

[2006 : 1 M]

4.12 In the CMOS inverter circuit shown, if the transconductance parameters of the NMOS and PMOS transistors are $K_n = K_p = \mu_n C_{ox}\frac{W_n}{L_n} = \mu_p C_{ox}\frac{W_P}{L_P} = 40\ \mu A/V^2$ and their threshold voltages are $V_{THn} = |V_{THp}| = 1$ V, the current I is

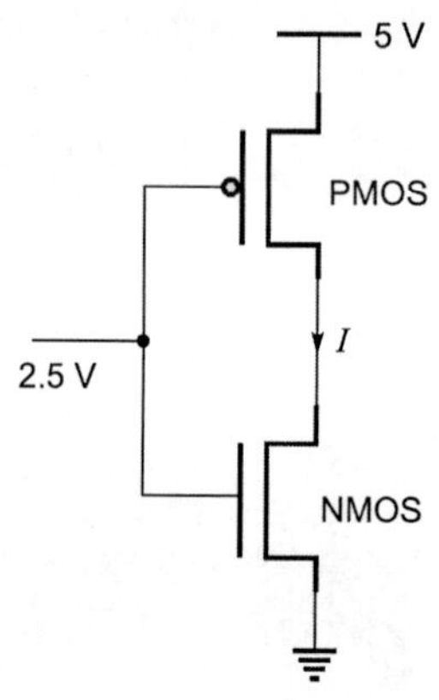

(a) 0 A (b) 25 μA
(c) 45 μA (d) 90 μA [2007 : 2 M]

4.13 Two identical NMOS transistors M_1 and M_2 are connected as shown below. V_{bias} is chosen so that both transistors are in saturation. The equivalent g_m of the pair is defined to be $\frac{\partial I_{out}}{\partial V_i}$ at constant V_{out}.

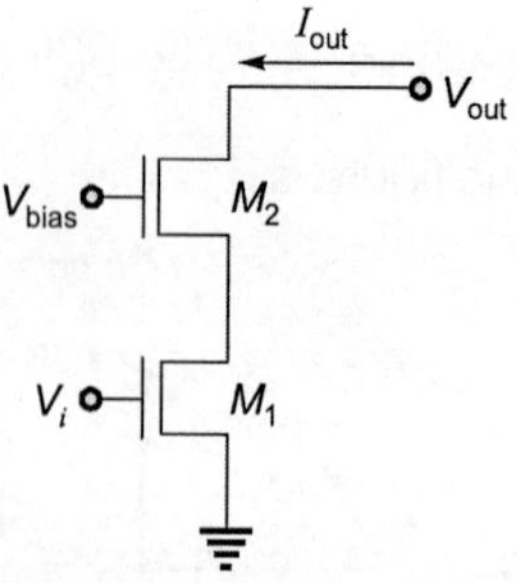

The equivalent g_m of the pair is
(a) the sum of individual g_m's of the transistors
(b) the product of individual g_m's of the transistors
(c) nearly equal to the g_m of M_1
(d) nearly equal to g_m/g_0 of M_2

[2008 : 2 M]

4.14 For the circuit shown in the following figure, transistors M_1 and M_2 are identical NMOS transistors. Assume that M_2 is in saturation and the output is unloaded.

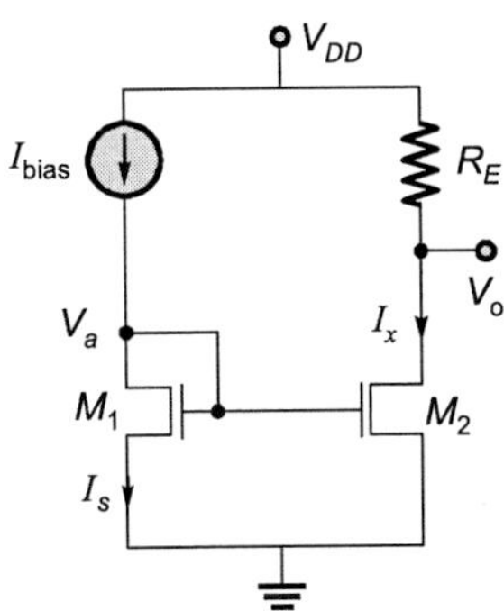

The current I_x is related to I_{bias} as

(a) $I_x = I_{bias} + I_s$

(b) $I_x = I_{bias}$

(c) $I_x = I_{bias} - I_s$

(d) $I_x = I_{bias} - \left(V_{DD} - \frac{V_{out}}{R_E}\right)$ [2008 : 2 M]

Statement for Linked Answer Question 4.15 and 4.16.

Consider the CMOS circuit shown, where the gate voltage V_G of the *n*-MOSFET is increased from zero, while the gate voltage of the *p*-MOSFET is kept constant at 3 V. Assume that, for both transistors, the magnitude of the threshold voltage is 1 V and the product of the transconductance parameter and the (W/L) ratio, i.e. the quantity $\mu C_{ox}(W/L)$, is 1 mA · V^{-2}.

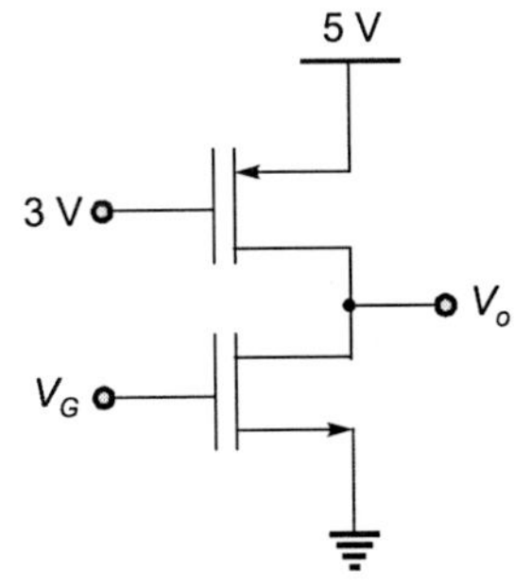

4.15 For small increase in V_G beyond 1 V, which of the following gives the correct description of the region of operation of each MOSFET?

(a) Both the MOSFETs are in saturation region

(b) Both the MOSFETs are in triode region

(c) *n*-MOSFET is in triode and *p*-MOSFET is in saturation region

(d) *n*-MOSFET is in saturation and *p*-MOSFET is in triode region [2009 : 2 M]

4.16 Estimate the output voltage V_0 for V_G = 1.5 V.

(a) $4 - \frac{1}{\sqrt{2}}$ V (b) $4 + \frac{1}{\sqrt{2}}$ V

(c) $4 - \frac{\sqrt{3}}{2}$ V (d) $4 + \frac{\sqrt{3}}{2}$ V

[2009 : 2 M]

4.17 In the circuit shown below, for the MOS transistors, $\mu_n C_{ox} = 100\ \mu A/V^2$ and the threshold voltage V_T = 1 V. The voltage V_x at the source of the upper transistor is

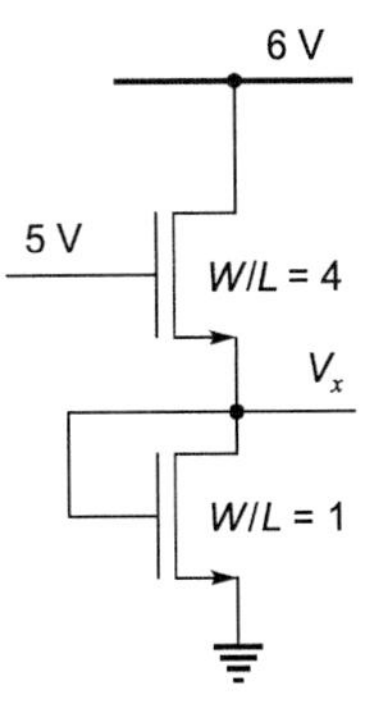

(a) 1 V (b) 2 V

(c) 3 V (d) 3.67 V [2011 : 2 M]

4.18 In the CMOS circuit shown, electron and hole mobilities are equal, and $M1$ and $M2$ are equally sized. The device $M1$ is in the linear region if

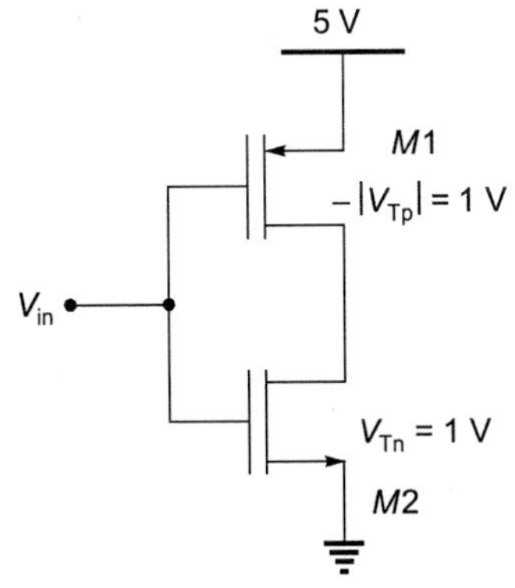

(a) $V_{in} < 1.875$ V

(b) $1.875\text{ V} < V_{in} < 3.125$ V

(c) $V_{in} > 3.125$ V

(d) $0 < V_{in} < 5$ V [2012 : 2 M]

4.19 The small-signal resistance (i.e., dV_B/dI_D) in kΩ offered by the *n*-channel MOSFET M shown in the figure below, at a bias point of V_B = 2 V is (device data for M: device transconductance parameter $k_N = \mu_n C'_{OX}(W/L) = 40\ \mu A/V^2$, threshold voltage V_{TN} = 1 V, and neglect body effect and channel length modulation effects)

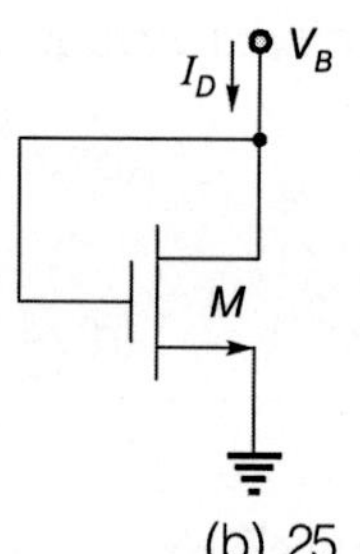

(a) 12.5 (b) 25
(c) 50 (d) 100 [2013 : 2 M]

4.20 In a MOSFET operating in the saturation region, the channel length modulation effect causes
(a) an increase in the gate-source capacitance
(b) a decrease in the transconductance
(c) a decrease in the unity-gain cut-off frequency
(d) a decrease in the output resistance

[2013 : 1 M]

4.21 For the *n*-channel MOS transistor shown in figure, the threshold voltage V_{Th} is 0.8 V. Neglect channel length modulation effects. When the drain voltage V_D = 1.6 V, the drain current I_D was found to be 0.5 mA. If V_D is adjusted to be 2 V by changing the values of R and V_{DD}, the new value of I_D (in mA) is

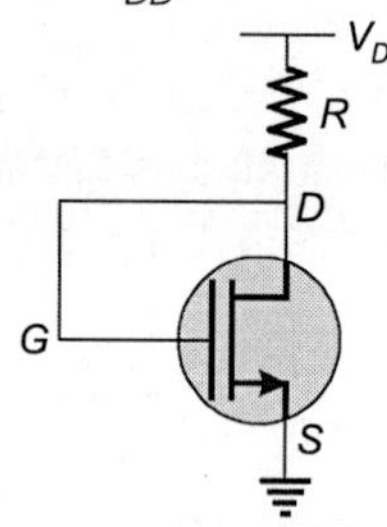

(a) 0.625 (b) 0.75
(c) 1.125 (d) 1.5

[2014 : 2 M, Set-2]

4.22 For the MOSFET shown in the figure, the threshold voltage $|V_t|$ = 2 V and $K = \frac{1}{2}\mu C\left(\frac{W}{L}\right) = 0.1 \text{mA/V}^2$. The value of I_D (in mA) is ______.

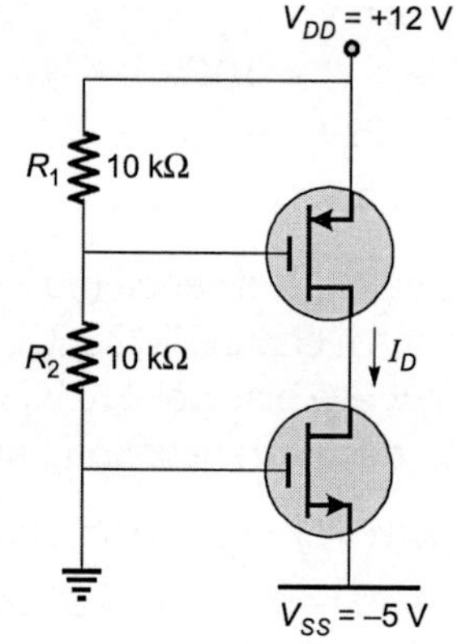

[2014 : 2 M, Set-2]

4.23 For the MOSFET M_1 shown in the figure, assume $W/L = 2$, $V_{DD} = 2.0$ V, $\mu_n C_{ox} = 100\ \mu\text{A/V}^2$ and $V_{TH} = 0.5$ V. The transistor M_1 switches from saturation region to linear region when V_{in} (in volts) is ______.

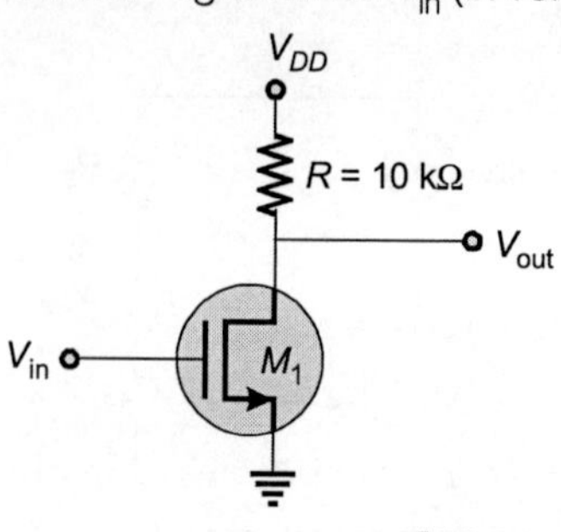

[2014 : 2 M, Set-3]

4.24 In the circuit shown, both the enhancement mode NMOS transistors have the following characteristics: $k_n = \mu_n C_{ox}$ (W/L) = 1 mA/V²; V_{TN} = 1 V. Assume that the channel length modulation parameter λ is zero and body is shorted to source. The minimum supply voltage V_{DD} (in volts) needed to ensure that transistor M_1 operates in saturation mode of operation is ______.

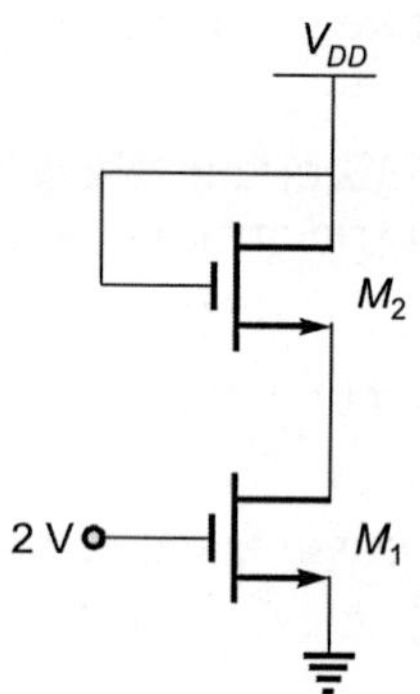

[2015 : 2 M, Set-3]

4.25 What is the voltage V_{out} in the following circuit?

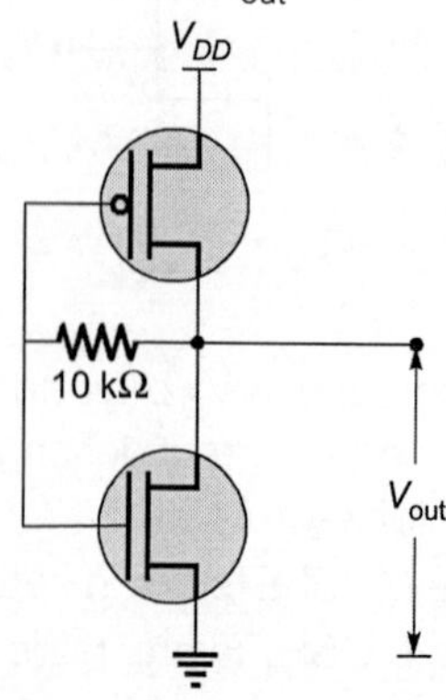

(a) 0 V
(b) ($|V_T$ of PMOS$|$ + V_T of NMOS)/2
(c) Switching threshold of inverter
(d) V_{DD} [2016 : 1 M, Set-1]

4.26 In the circuit shown in the figure, transistor M1 is in saturation and has transconductance g_m = 0.01 Siemens. Ignoring internal parasitic capacitances and assuming the channel length modulation λ to be zero, the small signal input pole frequency (in kHz) is ______.

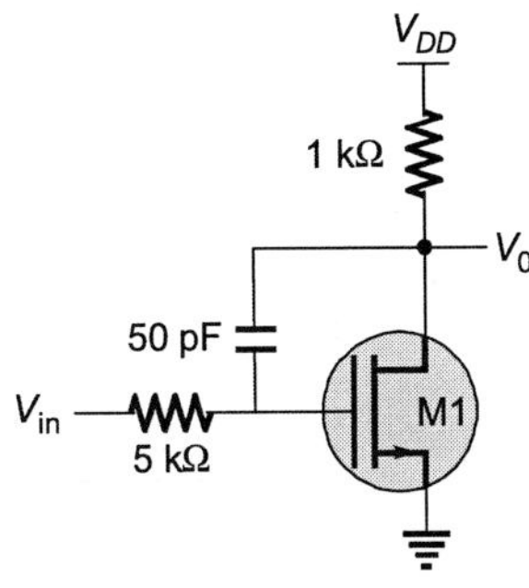

[2016 : 2 M, Set-3]

4.27 In the circuit shown in the figure, the channel length modulation of all transistors is non-zero ($\lambda \neq 0$). Also, all transistors operate in saturation and have negligible body effect. The ac small signal voltage gain (V_o/V_{in}) of the circuit is

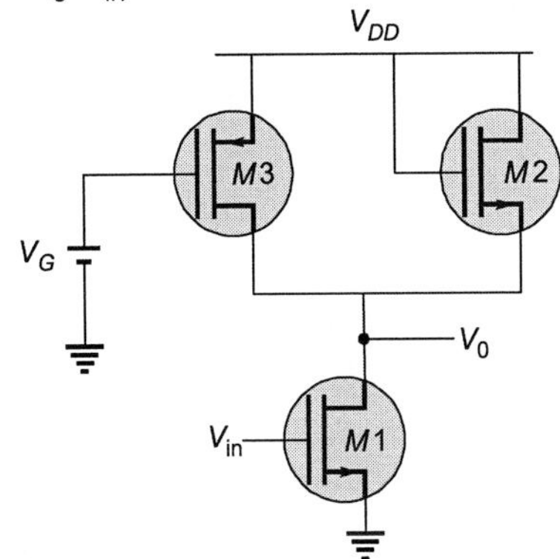

(a) $-g_{m1}(r_{01} \| r_{02} \| r_{03})$

(b) $-g_{m1}\left(r_{01} \left\| \frac{1}{g_{m3}} \right\| r_{03}\right)$

(c) $-g_{m1}\left(r_{01} \left\| \left(\frac{1}{g_{m2}} \| r_{02}\right) \right\| r_{03}\right)$

(d) $-g_{m1}\left(r_{01} \left\| \left(\frac{1}{g_{m3}} \| r_{03}\right) \right\| r_{02}\right)$

[2016 : 2 M, Set-3]

4.28 For the circuit shown, assume that the NMOS transistor is in saturation. Its threshold voltage V_{tn} = 1 V and its transconductance parameter $\mu_n C_{ox}\left(\frac{W}{L}\right) = 1\,\text{mA/V}^2$. Neglect channel length modulation and body bias effects. Under these conditions, the drain current I_D in mA is ______.

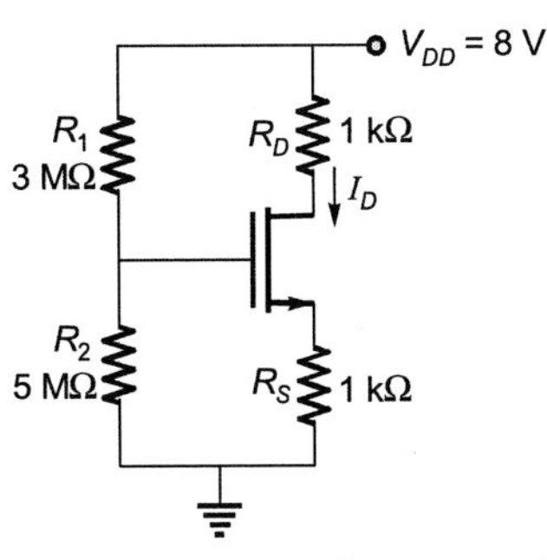

[2017 : 2 M, Set-1]

4.29 An n-channel enhancement mode MOSFET is biased at $V_{GS} > V_{TH}$ and $V_{DS} > (V_{GS} - V_{TH})$, where V_{GS} is the gate-to-source voltage. V_{DS} is the drain-to-source voltage and V_{TH} is the threshold voltage. Considering channel length modulation effect to be significant, the MOSFET behaves as a

(a) voltage source with zero output impedance
(b) voltage source with non-zero output impedance
(c) current source with finite output impedance
(d) current source with infinite output impedance

[2017 : 1 M, Set-2]

4.30 Assuming that transistors M_1 and M_2 are identical and have a threshold voltage of 1 V, the state of transistors M_1 and M_2 are respectively

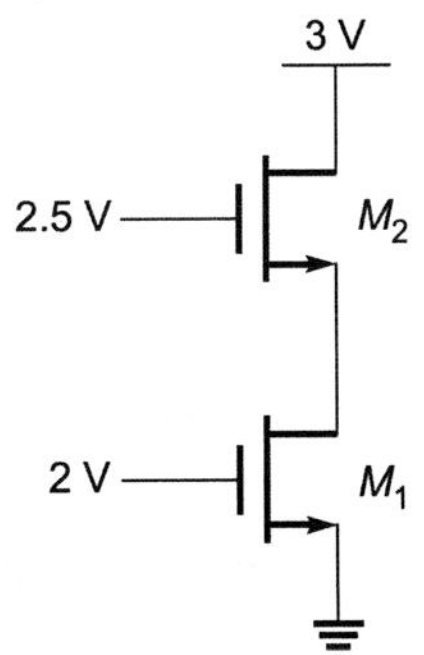

(a) Saturation, Saturation
(b) Linear, Linear
(c) Linear, Saturation
(d) Saturation, Linear

[2017 : 2 M, Set-2]

4.31 Two identical nMOS transistors M_1 and M_2 are connected as shown below. The circuit is used as an amplifier with the input connected between G and S terminals and the output taken between D and S terminals, V_{bias} and V_D are so adjusted that both transistors are in saturation. The transconductance of this combination is defined as $g_m = \frac{\partial i_D}{\partial v_{GS}}$ while the output resistance is $r_o = \frac{\partial v_{DS}}{\partial i_D}$, where i_D is the current flowing into the drain of M_2. Let g_{m1}, g_{m2} be the transconductances and r_{o1}, r_{o2} be the output resistance of transistors M_1 and M_2, respectively

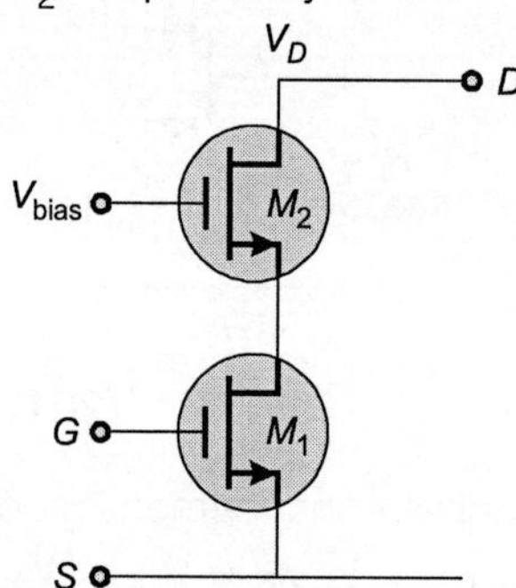

Which of the following statements about estimates for g_m and r_o is correct?

(a) $g_m \approx g_{m1} \cdot g_{m2} \cdot r_{o2}$ and $r_o \approx r_{o1} + r_{o2}$

(b) $g_m \approx g_{m1} + g_{m2}$ and $r_o \approx r_{o1} + r_{o2}$

(c) $g_m \approx g_{m1}$ and $r_o \approx r_{o1} \cdot g_{m2} \cdot r_{o2}$

(d) $g_m \approx g_{m1}$ and $r_o \approx r_{o2}$ **[2018 : 1 M]**

4.32 In the circuit shown below, the (W/L) value for M_2 is twice that for M_1. The two nMOS transistors are otherwise identical. The threshold voltage V_T for both transistors is 1.0 V. Note that V_{GS} for M_2 must be > 1.0 V.

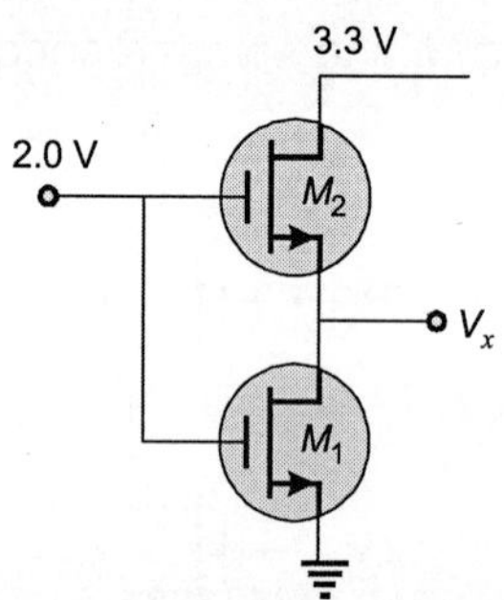

Current through the nMOS transistors can be modeled as

$$I_{DS} = \mu C_{ox}\left(\frac{W}{L}\right)\left((V_{GS} - V_T)V_{DS} - \frac{1}{2}V_{DS}^2\right) \quad \text{for } V_{DS} \le V_{GS} - V_T$$

$$I_{DS} = \mu C_{ox}\left(\frac{W}{L}\right)(V_{GS} - V_T)^2/2 \text{ for } V_{DS} \ge V_{GS} - V_T$$

The voltage (in volts, accurate to two decimal places) at V_x is _____.

[2018 : 2 M]

4.33 A CMOS inverter, designed to have a mid-point voltage V_1 equal to half of V_{dd}, as shown in the figure, has the following parameters:

$V_{dd} = 3$ V

$\mu_n C_{ox} = 100\ \mu A/V^2$; $V_{tn} = 0.7$ V for nMOS

$\mu_n C_{ox} = 40\ \mu A/V^2$; $|V_{tp}| = 0.9$ V for pMOS

The ratio of $\left(\frac{W}{L}\right)_n$ to $\left(\frac{W}{L}\right)_p$ is equal to _______ (rounded off to 3 decimal places).

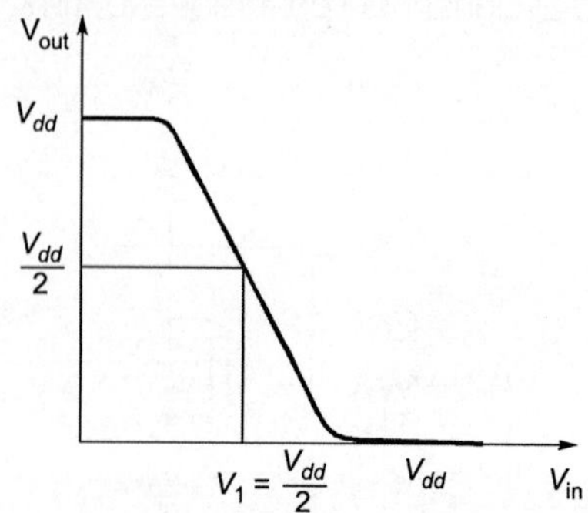

[2019 : 2 M]

4.34 In the circuit shown, the threshold voltages of the pMOS ($|V_{tp}|$) and nMOS (V_{tn}) transistors are both equal to 1 V. All the transistors have the same output resistance r_{ds} of 6 MΩ. The other parameters are listed below:

$$\mu_n C_{ox} = 60\ \mu A/V^2\ ;\ \left(\frac{W}{L}\right)_{nMOS} = 5$$

$$\mu_p C_{ox} = 30\ \mu A/V^2\ ;\ \left(\frac{W}{L}\right)_{pMOS} = 10$$

μ_n and μ_p are the carrier mobilities, and C_{ox} is the oxide capacitance per unit area. Ignoring the effect of channel length modulation and body bias, the gain of the circuit is _____ (rounded off to 1 decimal place).

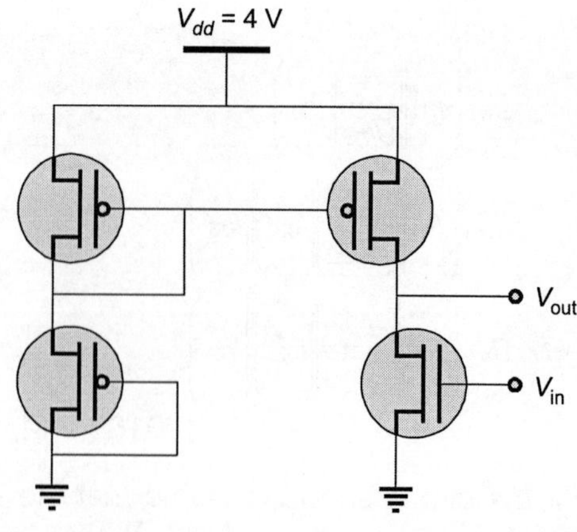

[2019 : 2 M]

4.35 In the circuit shown. $V_1 = 0$ and $V_2 = V_{dd}$. The other relevant parameters are mentioned in the figure. Ignoring the effect of channel length modulation and the body effect, the value of I_{out} is _______ mA (rounded off to 1 decimal place).

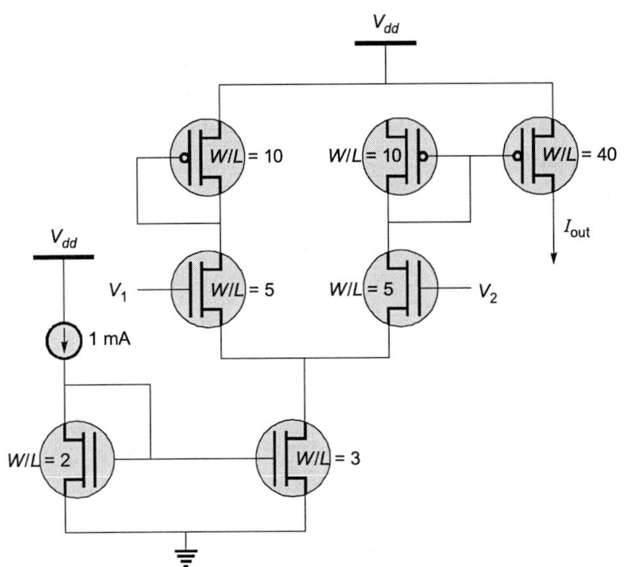

[2019 : 2 M]

4.36 An enhancement MOSFET of threshold voltage 3 V is being used in the sample and hold circuit given below. Assume that the substrate of the MOS device is connected to –10 V. If the input voltage V_i lies between ±10 V, the minimum and the maximum values of V_G required for proper sampling and holding respectively, are

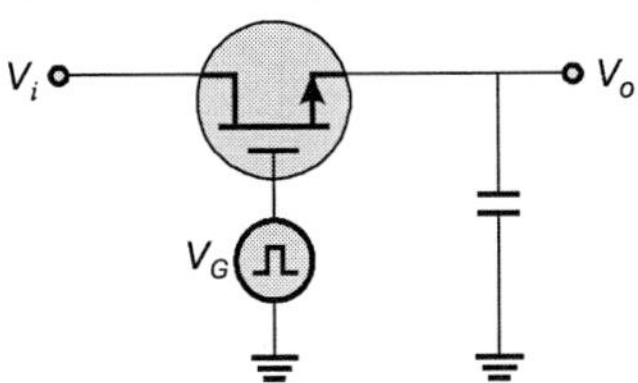

(a) 10 V and –13 V (b) 13 V and –7 V
(c) 10 V and –10 V (d) 3 V and –3 V

[2020 : 2 M]

4.37 Using the incremental low frequency small-signal model of the MOS device, the Norton equivalent resistance of the following circuit is

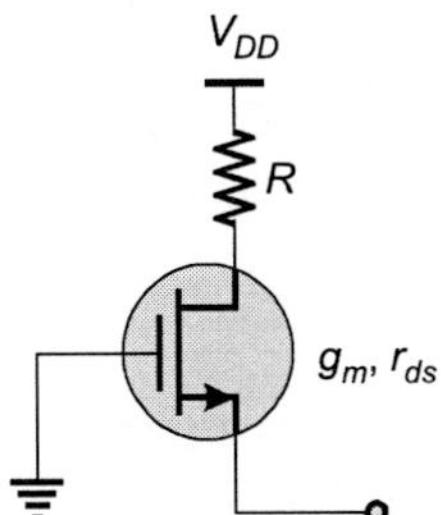

(a) $r_{ds} + R + g_m r_{ds} R$ (b) $r_{ds} + \dfrac{1}{g_m} + R$

(c) $r_{ds} + R$ (d) $\dfrac{r_{ds} + R}{1 + g_m r_{ds}}$

[2020 : 2 M]

4.38 For the transistor M_1 in the circuit shown in the figure, $\mu_n C_{ox} = 100\ \mu A/V^2$ and $(W/L) = 10$, where μ_n is the mobility of electron, C_{ox} is the oxide capacitance per unit area, W is the width and L is the length.

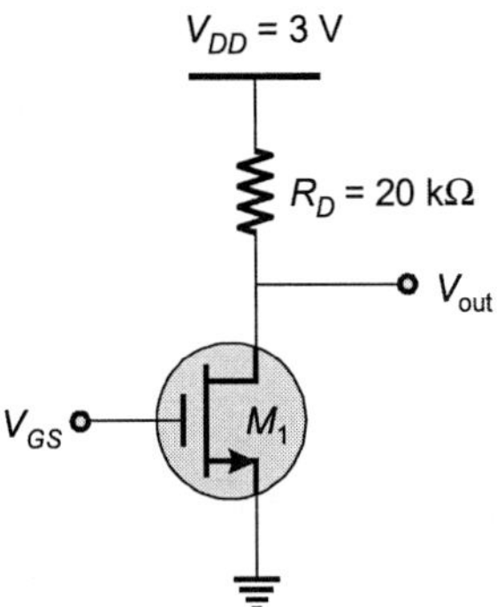

The channel length modulation coefficient is ignored. If the gate-to-source voltage V_{GS} is 1 V to keep the transistor at the edge of saturation, then the threshold voltage of the transistor (rounded off to one decimal place) is _______ V.

[2021 : 2 M]

4.39 In the circuit shown in the figure, the transistors M_1 and M_2 are operating in saturation. The channel length modulation coefficients of both the transistors are non-zero. The transconductance of the MOSFETs M_1 and M_2 are g_{m1} and g_{m2}, respectively, and the internal resistance of the MOSFETs M_1 and M_2 are r_{o1} and r_{o2}, respectively.

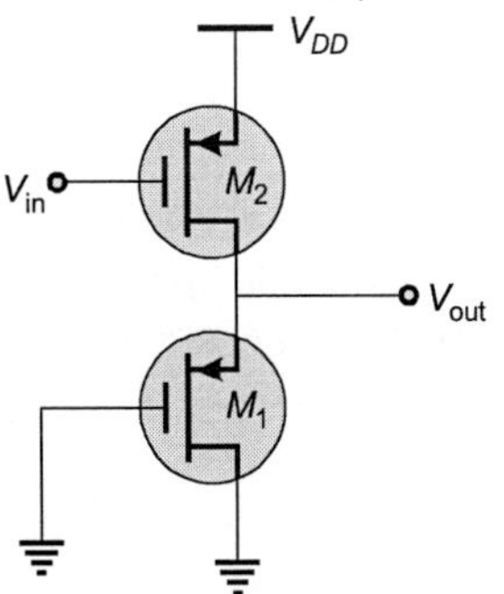

Ignoring the body effect, the AC small signal voltage gain $\left(\dfrac{\partial V_{out}}{\partial V_{in}}\right)$ of the circuit is

(a) $-g_{m1}\left(\dfrac{1}{g_{m2}} \| r_{o_1} \| r_{o_2}\right)$

(b) $-g_{m2}\left(\dfrac{1}{g_{m1}} \| r_{o_1} \| r_{o_2}\right)$

(c) $-g_{m2}\left(\dfrac{1}{g_{m1}} \| r_{o_2}\right)$

(d) $-g_{m2}(r_{o_1} \| r_{o_2})$

[2021 : 1 M]

4.40 Consider the CMOS circuit shown in the figure (substrates are connected to their respective sources). The gate width (W) to gate length (L) ratios $\left(\frac{W}{L}\right)$ of the transistors are as shown. Both the transistors have the same gate oxide capacitance per unit area. For the pMOSFET, the threshold voltage is –1 V and the mobility of holes is $40 \frac{\text{cm}^2}{\text{V.s}}$. For the nMOSFET, the threshold voltage is 1 V and the mobility of electrons is $300 \frac{\text{cm}^2}{\text{V.s}}$. The steady state output voltage V_o is ________.

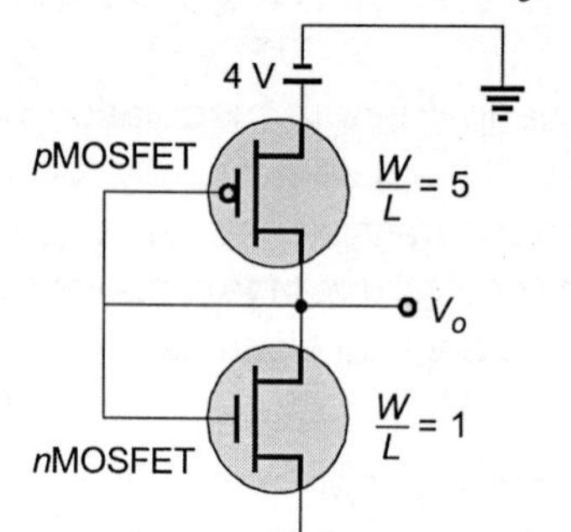

(a) equal to 0 V (b) more than 2 V
(c) less than 2 V (d) equal to 2 V

[2022 : 1 M]

4.41 The ideal long channel nMOSFET and pMOSFET devices shown in the circuits have threshold voltages of 1 V and –1 V, respectively. The MOSFET substrates are connected to their respective sources. Ignore leakage currents and assume that the capacitors are initially discharged. For the applied voltages as shown, the steady state voltages are ________________.

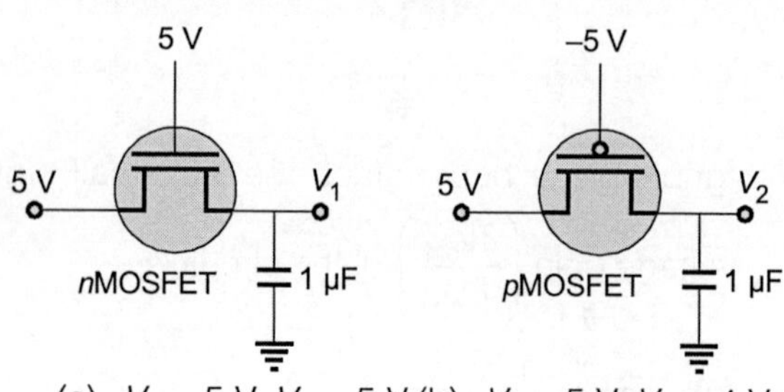

(a) $V_1 = 5$ V, $V_2 = 5$ V (b) $V_1 = 5$ V, $V_2 = 4$ V
(c) $V_1 = 4$ V, $V_2 = 5$ V (d) $V_1 = 4$ V, $V_2 = -5$ V

[2022 : 1 M]

4.42 Consider an ideal long channel nMOSFET (enhancement-mode) with gate length 10 μm and width 100 μm. The product of electron mobility (μ_n) and oxide capacitance per unit area (C_{ox}) is $\mu_n C_{ox} = 1$ mA/V². The threshold voltage of the transistor is 1 V. For a gate-to-source voltage $V_{GS} = [2 - \sin(2t)]$ V and drain-to source voltage $V_{DS} = 1$ V (substrate connected to the source), the maximum value of the drain-to-source current is ________.

(a) 40 mA (b) 20 mA
(c) 15 mA (d) 5 mA

[2022 : 2 M]

4.43 Consider the circuit shown with an ideal long channel nMOSFET (enhancement mode, substrate is connected to the source). The transistor is appropriately biased in the saturation region with V_{GG} and V_{DD} such that it acts as a linear amplifier. v_i is the small-signal ac input voltage. v_A and v_B represent the small-signal voltages at the nodes A and B, respectively. The value of $\frac{v_A}{v_B}$ is ________ (rounded off to one decimal place).

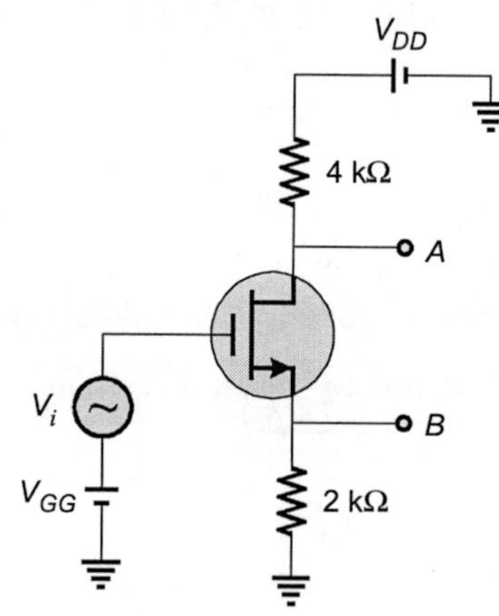

[2022 : 2 M]

4.44 The h-parameters of a two port network are shown below. The condition for the maximum small signal voltage gain $\frac{V_{out}}{V_s}$ is

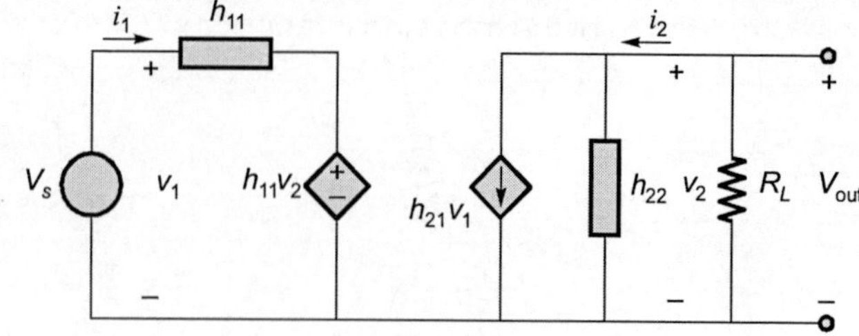

(a) $h_{11} = 0$, $h_{12} = 0$, h_{21} = very high and $h_{22} = 0$
(b) h_{11} = very high, $h_{12} = 0$, h_{21} = very high and $h_{22} = 0$
(c) $h_{11} = 0$, h_{12} = very high, h_{21} = very high and $h_{22} = 0$
(d) $h_{11} = 0$, $h_{12} = 0$, h_{21} = very high and h_{22} = very high

[2023 : 2 M]

4.45 In the circuit below, assume that the long channel NMOS transistor is biased in saturation. The small signal trans-conductance of the transistor is g_m. Neglect body effect, channel length modulation and intrinsic device capacitances. The small signal input impedance $Z_{in}(j\omega)$ is ______.

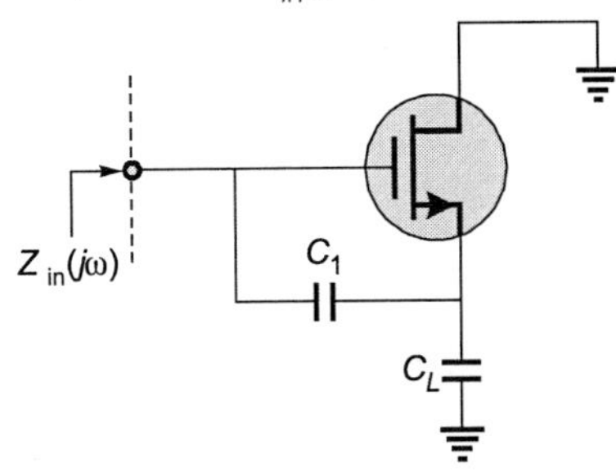

(a) $\dfrac{-g_m}{C_1 C_L \omega^2} + \dfrac{1}{j\omega C_1} + \dfrac{1}{j\omega C_L}$

(b) $\dfrac{-g_m}{C_1 C_L \omega^2} + \dfrac{1}{j\omega C_1 + j\omega C_L}$

(c) $\dfrac{1}{j\omega C_1} + \dfrac{1}{j\omega C_L}$

(d) $\dfrac{g_m}{C_1 C_L \omega^2} + \dfrac{1}{j\omega C_1} + \dfrac{1}{j\omega C_L}$

[2024 : 1 Mark]

4.46 In the circuit shown below, the transistors M_1 and M_2 are biased in saturation. Their small signal transconductances are g_{m1} and g_{m2} respectively. Neglect body effect, channel length modulation and intrinsic device capacitances.

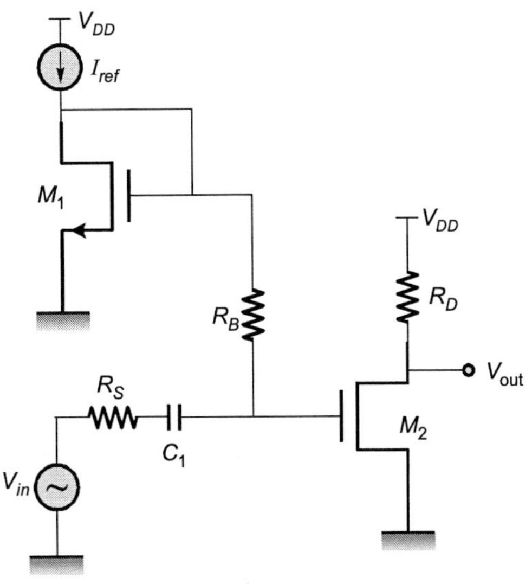

Assuming that capacitor C_1 is a short circuit for AC analysis, the exact magnitude of small signal voltage gain $\left|\dfrac{V_{out}}{V_{in}}\right|$ is ______.

(a) $\dfrac{g_{m2}R_D\left(\dfrac{1}{g_{m1}}\right)}{\dfrac{1}{g_{m1}} + R_s}$

(b) $g_{m2}R_D$

(c) $\dfrac{g_{m2}R_D\left(R_B + \dfrac{1}{g_{m1}}\right)}{R_B + \dfrac{1}{g_{m1}} + R_s}$

(d) $\dfrac{g_{m2}R_D\left(R_B + \dfrac{1}{g_{m1}} + R_s\right)}{R_B + \dfrac{1}{g_{m1}}}$

[2024 : 2 Marks]

■■■

Answers FET and MOSFET Analysis

4.1 (0.4)	4.2 (c)	4.3 (c)	4.4 (d)	4.5 (a)	4.6 (d)	4.7 (c)
4.8 (b)	4.9 (a)	4.10 (b)	4.11 (d)	4.12 (c)	4.13 (c)	4.14 (b)
4.15 (d)	4.16 (d)	4.17 (c)	4.18 (a)	4.19 (b)	4.20 (d)	4.21 (c)
4.22 (0.9)	4.23 (1.5)	4.24 (3)	4.25 (c)	4.26 (57.87)	4.27 (c)	4.28 (2)
4.29 (c)	4.30 (c)	4.31 (c)	4.32 (0.422)	4.33 (0.225)	4.34 (–900)	4.35 (6)
4.36 (b)	4.37 (d)	4.38 (0.5)	4.39 (b)	4.40 (c)	4.41 (c)	4.42 (c)
4.43 (–2)	4.44 (*)	4.45 (a)	4.46 (c)			

Explanations FET and MOSFET Analysis

4.1 (0.4)

$$g_{m(\max)} = \frac{2I_{DSS}}{|V_p|} = \frac{2\times 1\times 10^{-3}}{|-5|} = 0.4\ ms$$

4.2 (c)

$2g_m$ and $\frac{r_d}{2}$

So, $r_d' = \frac{r_d}{2}$

$$\mu' = \mu = g_m' r_d' = g_m' \frac{r_d}{2} = g_m r_d$$

$$g_m' = 2g_m$$

4.3 (c)

$$V_1 = 2V_2$$
$$V_1 = I_D R_D$$
$$I_S = I_D$$
$$V_2 = I_s \frac{R_D}{2} = \frac{I_D R_D}{2} = \frac{V_1}{2}$$

So, $V_1 = 2V_2$

4.4 (d)

$$g_m = \frac{2I_{DSS}}{|V_P|}\left(1-\frac{V_{GS}}{V_P}\right)$$

$V_G = 0$, $V_S = I_D \cdot R_S = 1\text{ mA}\times 2.5$

$V_s = 2.5$ V

$V_{GS} = V_G - V_S = -2.5$ V

$$g_m = \frac{2\times 10\text{ mA}}{5}\left(1-\left(\frac{-2.5}{-5}\right)\right) = 2\text{ mS}$$

$A_v = -g_m R_D$

($\because r_d$ is not given, it is taken as ∞).

$= -2\text{ mS}\times 3\text{ k}\Omega = -6$

4.5 (a)

Statement-2 is false because V_{DS} will be less than $V_{GS} - V_T$, if $V_o = 0$.

4.7 (c)

From the graph its clear that $V_{th} = 1$ V

$V_{GS} = 3 - 1 = 2$ V

$V_{DS} = 5 - 1 = 4$ V

Since $V_{DS} \geq V_{GS} - V_T$.

So MOSFET is in saturation region.

4.8 (b)

$Z_{in} = 2\text{ M}\Omega$

$Z_o = r_d \,||\, R_D = 20\text{ k} \,||\, 2\text{ k}$

$Z_o = \frac{20}{11}$ k

4.9 (a)

$$I_D = I_{DSS}\left(1-\frac{(V_{GS})}{(V_P)}\right)^2 = 10\ mA\left(1-\frac{(-2)}{(-8)}\right)^2$$

$$I_D = 10\text{ mA}\times\frac{9}{16} = 5.625\text{ mA}$$

$V_{DS} = V_{DD} - I_D R_D$

$= 20 - 5.625\text{ mA}\times 2\text{ k}$

$V_{DS} = 8.75$ V

4.10 (b)

$$g_m = \frac{2}{|V_p|}\sqrt{I_D \cdot I_{DSS}}$$

$$= \frac{2}{8}\sqrt{5.625 \text{ mA} \times 10 \text{ mA}} = \frac{1}{4}\sqrt{56.25} \text{ mS}$$

$$g_m = 1.875 \text{ mS}$$

$$A_V = -g_m(r_d \| R_D) \qquad (g_m Z_o)$$

$$= -1.875 \text{ ms} \times \frac{20}{11} \text{k}$$

$$A_v = -3.41$$

4.11 (d)

From the graph it is clear that as V_{GS} increase conductance i.e. slope of graph increase.

Transfer characteristics of n-channel D-MOSFET.

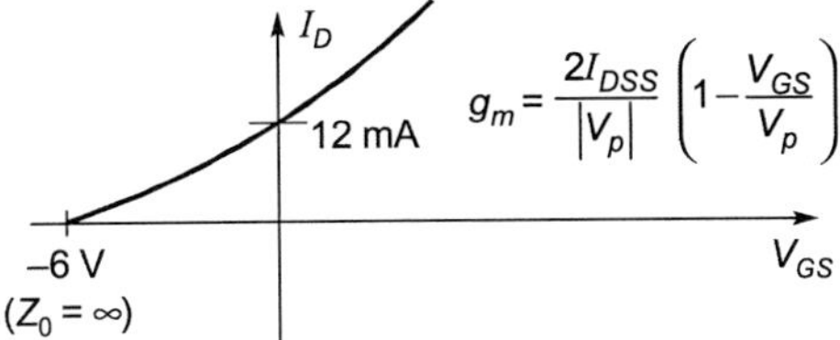

4.12 (c)

V_{GS} for each MOS is 2.5 V.

V_T = 1 Volt, device parameter K = 40 μA/v²

So, $I_D = \frac{K}{2}(V_{GS} - V_T)^2$

$= 20\,(2.5 - 1)^2 = 45\ \mu\text{A}$

4.13 (c)

$$g_{m_1} = \frac{\partial I_{D_1}}{\partial V_i}$$

$$I_{D_1} = I_{D_2}$$

$$g_m = \frac{\partial I_{\text{out}}}{\partial V_i} = \frac{\partial I_{D_2}}{\partial V_i} = \frac{\partial I_{D_1}}{\partial V_i} = g_{m_1}$$

4.14 (b)

It is a current mirror circuit

$$I_{G_1} = I_{G_2} = 0$$

$$I_x = I_s$$

$$I_{\text{bias}} = I_S + I_{G_1} + I_{G_2}$$

$$= I_s + 0 + 0$$

$$I_x = I_{\text{bias}}$$

4.15 (d)

When V_G is little higher than 1 V

- For n-MOSFET,

  $V_{GSn} = V_G = 1^+$ V $(1^+ \Rightarrow$ Little higher than 1)

  $V_{GSn} - V_{Tn} = 0^+$ V $(0^+ \Rightarrow$ Little higher than 0)

- For p-MOSFET,

  $V_{SGP} = V_{DD} - 3\text{ V} = 2\text{ V}$

  $V_{SGP} - |V_{TP}| = 1\text{ V}$

  $(V_{GSn} - V_{Tn}) < (V_{SGP} - |V_{TP}|)$

  So, n-MOSFET will be in saturation region and p-MOSFET will be in triode region.

4.16 (d)

When $V_G = 1.5$ V :

- For n-MOSFET,

  $V_{GSn} = V_G = 1.5\text{ V}$

  $V_{GSn} - V_{Tn} = 0.5\text{ V}$

- For p-MOSFET,

  $V_{SGP} = V_{DD} - 3\text{ V} = 2\text{ V}$

  $V_{SGP} - |V_{TP}| = 2\text{ V} - 1\text{ V} = 1\text{ V}$

  $(V_{GSn} - V_{Tn}) < (V_{SGP} - |V_{TP}|)$

  So, n-MOSFET will be in saturation region and p-MOSFET will be in triode region.

To determine V_o:

$$I_{DSn} = I_{SDP}$$

$$K_n(V_{GSn} - V_{Tn})^2 = K_p\left[2(V_{SGP} - |V_{TP}|)V_{SDP} - V_{SDP}^2\right]$$

Given that, $K_n = K_p$

So, $(0.5)^2 = 2(1\text{ V})(V_{DD} - V_o) - (V_{DD} - V_o)^2$

$$0.25 = 2(5 - V_o) - (5 - V_o)^2$$

$$0.25 = 10 - 2V_o - 25 + 10V_o - V_o^2$$

$$V_o^2 - 8V_o + 15.25 = 0$$

$$V_o = \frac{8 \pm \sqrt{64 - 4(15.25)}}{2} = 4 \pm \frac{\sqrt{3}}{2}$$

$$= 4.866 \text{ V or } 3.134 \text{ V}$$

Check for valid V_o:

- We know that n-MOSFET is in saturation region and p-MOSFET is in triode region.

  So, $V_o \geq V_{GSn} - V_{Tn}$

  $\Rightarrow$ Both the possibilities of V_o satisfies this

  and $[V_{SDP} = (5 - V_o)] < [V_{SGP} - |V_{TP}| = 1\text{ V}]$

- For $V_o = 4.866$ V,

  $[V_{SDP} = 0.314\text{ V}] < [V_{SGP} - |V_{TP}| = 1\text{ V}]$

- For $V_o = 3.134$ V,

  $[V_{SDP} = 0.866\text{ V}] > [V_{SGP} - |V_{TP}| = 1\text{ V}]$

So, the valid value of V_o is 4.866 V or $4 + \frac{\sqrt{3}}{2}$ V.

4.17 (c)

For upper MOS,

$$V_{DS} = 6 - V_x$$
$$V_{GS} - V_T = 5 - V_x - 1$$
$$= 4 - V_x$$

Upper MOS will be in saturation because $V_{DS} > V_{GS} - V_T$.

For lower MOS,

$$V_{DS} = V_x$$
$$V_{GS} - V_T = V_x - 1$$

so, $V_{DS} > V_{GS} - V_T$

Hence both MOS will be in saturation

$$I_{D1} = \mu_n C_{ox}\left(\frac{W}{L}\right)(V_{GS} - V_T)^2$$

$\Rightarrow$ $I_{D1} = \mu_n C_{ox}(4)(4 - V_x)^2$

Similarly, $I_{D2} = \mu_n C_{ox}(1)(V_x - 1)^2$

but, $I_{D1} = I_{D2}$

$\therefore$ $\mu_n C_{ox}(4)(4 - V_x)^2 = \mu_n C_{ox}(1)(V_x - 1)^2$

On solving, $V_x = 3$ volt

4.18 (a)

For PMOS $V_{SG} = V_S - V_G = 5 - V_{in}$

for PMOS to be ON

$$V_{SG} > |V_{TP}|$$

$\Rightarrow$ $5 - V_{in} > 1$

$$\boxed{V_{in} < 4}$$

So, V_{in} must be less than 4 V for MOS to be in linear regions so option *C* and *D* are rejected.

Now we know that for small V_{in} output is high and PMOS is in linear region and NMOS is in cutoff region. Similarly for high V_{in} PMOS is in cutoff and NMOS is in linear region and for V_{in} in between both are in saturation.

So PMOS will be in linear region for $V_{in} < 1.875$ V.

4.19 (b)

$$g_m = \frac{\partial I_D}{\partial V_{GS}}$$
$$V_B = V_G$$
$$V_S = 0$$
$$V_{GS} = V_G - V_S = V_G - 0 = V_G = V_B$$
$$\frac{\partial V_B}{\partial I_D} = \frac{1}{g_m}$$
$$V_D = V_G$$
$$V_D - V_S = V_G - V_S$$
$$V_{DS} = V_{GS}$$
$$V_{DS} > V_{GS} - V_T$$

So, the given MOSFET is in saturation region.

$$g_m = \mu_n C'_{OX}\frac{W}{L}(V_{GS} - V_T)$$
$$g_m = 40 \times 10^{-6} \times (2 - 1) = 4 \times 10^{-5}\text{ S}$$
$$\frac{\partial V_B}{\partial I_D} = \frac{1}{g_m} = \frac{1}{4\times 10^{-5}} = \frac{100\times 10^3}{4} = 25\text{ k}\Omega$$

4.20 (d)

In a MOSFET operating in the saturation region, the channel length modulation effect causes a decrease in the output resistance.

4.21 (c)

We know that

$$I_D = K_n(V_{GS} - V_T)^2 \quad [\because V_{DS} > V_{GS} - V_T]$$

So, $K_n = \dfrac{0.5\text{ mA}}{(1.6 - 0.8)^2}$

$$K_n = 0.78\text{ mA/V}$$

Therefore, for $V_D = 2$ V

$$I_D = 0.78(2 - 0.8)^2 = 1.123\text{ mA}$$

4.22 (0.9)

$$I_{D_1} = I_{D_2} = K_n(V_{GS} - V_T)^2$$
$$= \frac{1}{2}\mu_n C_{ox}\left(\frac{\omega}{L}\right)(V_{GS} - V_T)^2$$
$$= 0.1(5 - 2)^2 = 0.9\text{ mA}$$

4.23 (1.5)

For saturation

$$I_D = K_n(V_{GS} - V_t)^2$$
$$I_D = \frac{\mu_n C_{ox}}{2}\frac{W}{L}(V_{GS} - V_t)^2$$
$$I_D = \frac{1}{2}\times 100\times 10^{-6}\times 2(V_0)^2$$
$$I_D = 100\times 10^{-6} V_0^2 \quad \text{...(i)}$$

Now, KVL in outer loop we get

$$V_{DD} = I_D \times 10\text{ k}\Omega + V_0 \quad \text{...(ii)}$$

From (*i*) and (*ii*), we get

$$2 = 100\times 10^{-6}\times 10\times 10^3 V_0^2 + V_0$$
$$V_0^2 + V_0 - 2 = 0$$

$\therefore$ $V_0 = 1$ V

$$V_{DS} = V_{GS} - V_t$$
$$V_0 = V_{in} - V_t$$

$\therefore$ $V_{in} = 1.5$ V

4.24 (3)

Lower transistor (M_1) to work in saturation

$$V_{DS_1} \geq V_{GS_1} - V_{TN}$$

So, for minimum V_{DD}

$$V_{DS_1} = V_{GS1} - V_{TN}$$

$$V_{DS_1} = 2 - 1 = 1\text{ V}$$

$$\therefore \quad V_{D_1} = 1\text{ V}$$

and $$I_{D_1} = K'(V_{GS_1} - V_{TN})^2$$

$$I_{D_1} = \frac{1\text{mA}}{V^2} \times (2-1)^2 = 1\text{mA}$$

Now transistor M_2, $V_{DG} = 0$ V

So, it will work into saturation region and same current will flow.

$$I_{D_2} = I_{D_1} = K'(V_{GS_2} - V_{TN})^2$$

$$1\text{ mA} = 1\text{ mA/V}^2 \times (V_{DD} - 1 - 1)^2 \qquad (\because V_{S_2} = V_{D_1})$$

$$\therefore \quad V_{DD} = 3\text{ V}$$

4.26 (57.87)

With respect to A.C

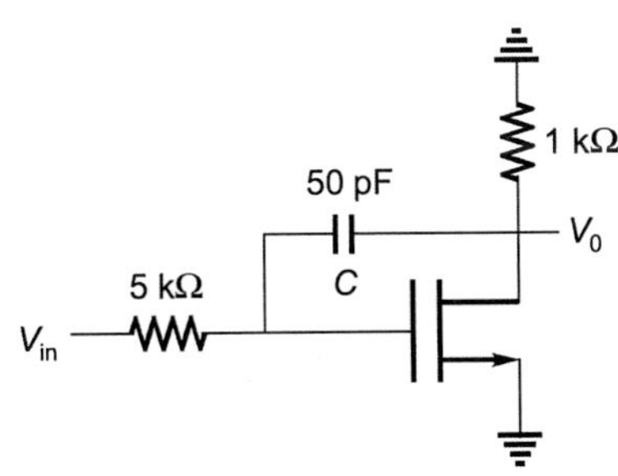

Taking miller's equivalent and assume $r_0 = \infty$

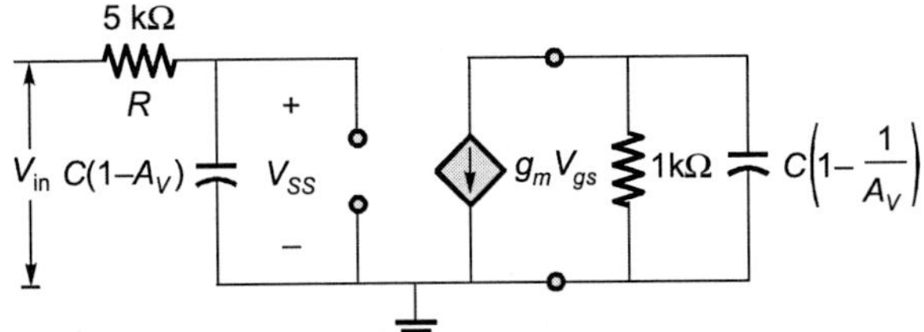

$A_V = -g_m R_D = -0.01 \times 10^3 = -10$

small signal input pole frequency

$$= \frac{1}{2\pi \times 5 \times 10^3 \times 50 \times 10^{-12}(1+10)}$$

$= 57.87$ kHz

4.27 (c)

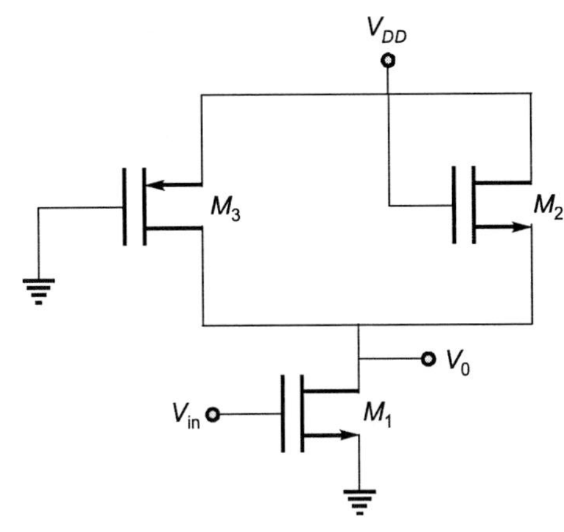

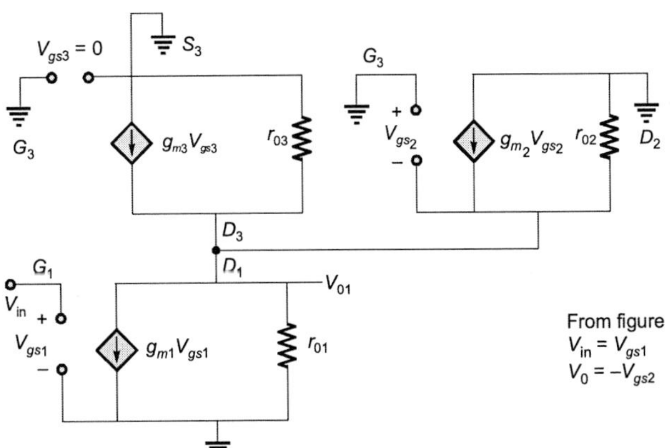

From figure
$V_{in} = V_{gs1}$
$V_0 = -V_{gs2}$

Node equation at P_1:

$$g_{m1}V_{gs1} + \frac{V_0}{r_{01}} + \frac{V_0}{r_{03}} + \frac{V_0}{r_{02}} - g_{m_2}V_{gs_2} = 0$$

$$V_0\left(\frac{1}{r_{01}} + \frac{1}{r_{02}} + g_{m2} + \frac{1}{r_{03}}\right) = -g_{m1}V_{in}$$

$$\Rightarrow \quad A_v = \frac{V_0}{V_{in}} = -g_{m1}\left[r_{01} \,||\, \left[r_{02} \,||\, \frac{1}{g_{m2}}\right] \,||\, r_{03}\right]$$

4.28 (2)

$$V_{GS} = \frac{8 \times 5}{8} - 1 \times I_D = 5 - I_D$$

(Here I_D is numerically in mA)

$$I_D = 5 - V_{GS} \qquad \ldots(i)$$

$$I_D = \frac{\mu_n C_{ox}}{2}\frac{W}{L}(V_{GS} - V_T)^2$$

$$5 - V_{GS} = \frac{1}{2}(V_{GS} - 1)^2$$

$$10 - 2V_{GS} = V_{GS}^2 + 1 - 2V_{GS}$$

$$V_{GS}^2 = 9 \Rightarrow V_{GS} = 3\text{ V}$$

$$I_D = 5 - 3 = 2\text{ mA}$$

4.29 (c)

The small signal equivalent circuit of MOSFET in saturation is as given below.

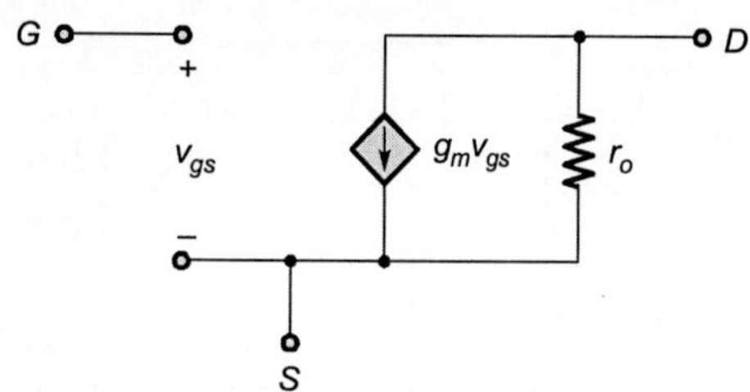

So, when the channel length modulation effect is significant, the MOSFET can be modelled as a current source with finite output impedance.

4.30 (c)

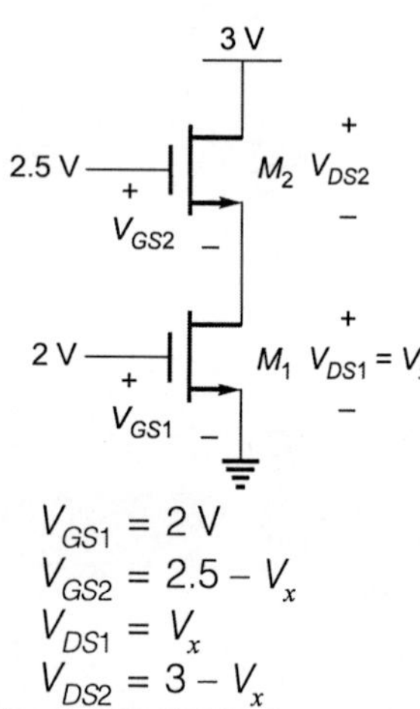

$$V_{GS1} = 2\text{ V}$$
$$V_{GS2} = 2.5 - V_x$$
$$V_{DS1} = V_x$$
$$V_{DS2} = 3 - V_x$$

Assume both MOSFETs in saturation and equate their currents

$$I_{DS1} = I_{DS2}$$

$$\frac{k_n}{2}(2-1)^2 = \frac{k_n}{2}(2.5 - V_x - 1)^2$$

After solving, V_x = 0.5 V, 2.5 V

V_x cannot be 2.5 V. Because this will make M_2 OFF.

Hence V_x may be 0.5 V

⇒ M_1 is in linear region.

⇒ M_2 is in saturation region.

To verify further,

$$I_{DS1} = I_{DS2}$$

$$k_n\left[(2-1)V_x - \frac{V_x^2}{2}\right] = \frac{k_n}{2}(2.5 - V_x - 1)^2$$

$V_{GS2} = 2.5\text{ V} - V_x$ must be greater than $V_{Th} = 1$ V

After solving, V_x = 0.588 V, 1.91 V

Hence correct value of V_x = 0.588 V

This verifies above conclusion i.e.

$M_1 \rightarrow$ Linear region

$M_2 \rightarrow$ Saturation region

4.31 (c)

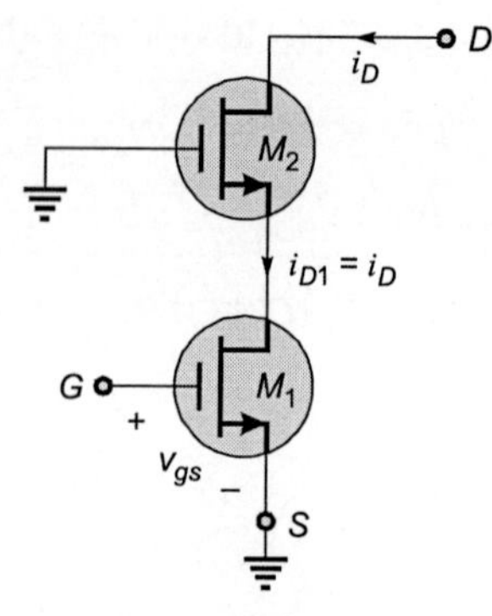

$$g_m = \frac{\Delta I_D}{\Delta V_{in}} = \frac{i_D}{v_{gs}} = \frac{i_{D1}}{v_{gs}} = g_{m1}$$

To calculate r_o :

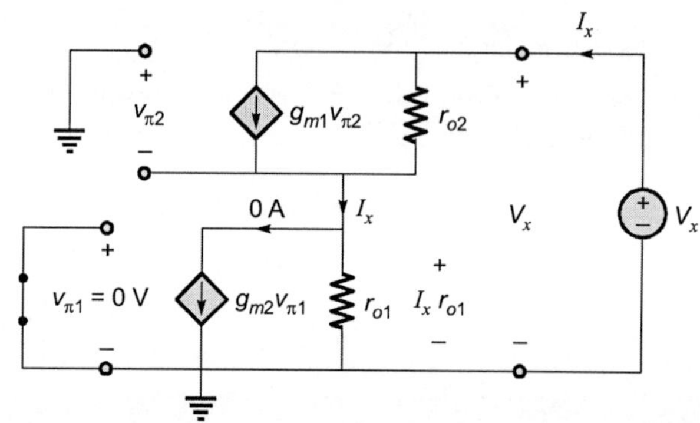

$$v_{\pi 2} = -I_x r_{o1}$$

$$I_x = g_{m2}\, v_{\pi 2} + \frac{(V_x - I_x r_{o1})}{r_{o2}}$$

$$I_x = -g_{m2}\, r_{o1}\, I_x + \frac{V_x}{r_{o2}} - I_x\frac{r_{o1}}{r_{o2}}$$

$$V_x = r_{o2}\left[1 + r_{o1}g_{m2} + \frac{r_{o1}}{r_{o2}}\right]I_x$$

$$r_o = \frac{V_x}{I_x} = r_{o1} + r_{o2} + r_{o1}\, r_{o2}\, g_{m2}$$

$$\approx r_{o1}\, r_{o2}\, g_{m2}$$

4.32 (0.4226)

Let, $K_n = \frac{\mu_n C_{ox}}{2}\left(\frac{W}{L}\right)$

Given that,

$$\left(\frac{W}{L}\right)_2 = 2\left(\frac{W}{L}\right)_1$$

So, $K_{n2} = 2K_{n1}$

For M_1, $V_{GS1} - V_T = 2 - 1 = 1$ V

For M_2, $V_{GS2} - V_T = 2 - V_x - 1 = 1\text{ V} - V_x < 1$ V

$$V_{DS2} = (3.3 - V_x) > (V_{GS2} - V_T)$$

So, M_1 will be in linear region and M_2 will be in saturation region.

$$I_{D_1} = I_{D_2}$$

$$K_{n1}\left[2(V_{GS1}-V_T)V_{DS1}-V_{DS1}^2\right] = K_{n2}(V_{GS2}-V_T)^2$$

$$K_{n1}\left[2(2-1)V_x - V_x^2\right] = 2K_{n1}(2-V_x-1)^2$$

$$2V_x - V_x^2 = 2(1+V_x^2-2V_x)$$

$$= 2V_x^2 - 4V_x + 2$$

$$3V_x^2 - 6V_x + 2 = 0$$

$$V_x^2 - 2V_x + \frac{2}{3} = 0$$

$$V_x = 1 \pm \sqrt{\frac{4-\frac{8}{3}}{4}} = 1 \pm \sqrt{\frac{1}{3}}\ \text{V}$$

$$V_{GS2} = (2-V_x) \geq V_T$$

$$\Rightarrow \quad (1-V_x) \geq 0$$

So, valid value, $V_x = 1 - \sqrt{\frac{1}{3}} = 0.4226$ V

4.33 (0.225)

$I_{Dn} = I_{Dp}$ and both will be in saturation.

If $\quad V_{IN} = \frac{V_{DD}}{2} = 1.5\ \text{V} = V_{GSN} = V_{SGP}$

$$\Rightarrow \frac{1}{2}(\mu_n C_{ox})\left(\frac{W}{L}\right)_n [V_{GSN} - V_{TN}]^2$$

$$= \frac{1}{2}(\mu_p C_{ox})\left(\frac{W}{L}\right)_p [V_{GSP} + V_{TP}]^2$$

$$100 \times 10^{-6}\left(\frac{W}{L}\right)_n [1.5-0.7]^2$$

$$= 40 \times 10^{-6}\left(\frac{W}{L}\right)_p [1.5-0.9]^2$$

$$\Rightarrow \frac{\left(\frac{W}{L}\right)_n}{\left(\frac{W}{L}\right)_p} = \frac{40}{100} \times \frac{(0.6)^2}{(0.8)^2} = 0.225$$

4.34 (-900)

M_3 and M_4 are identical PMOS transistor and they have equal current.

Hence their V_{SG} should be equal.

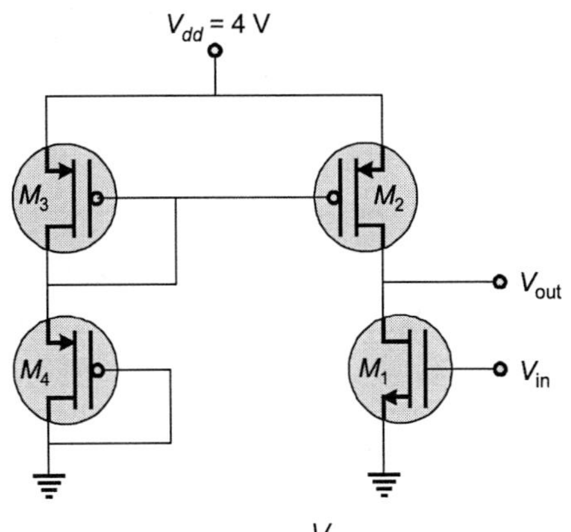

$$V_{SG3} = V_{SG4} = \frac{V_{DD}}{2} = 2\ \text{V}$$

$$I_{SD} = \frac{\mu_p C_{ox}}{2}\left(\frac{W}{L}\right)_p (V_{SG} - |V_T|)^2$$

$$= \frac{30}{2} \times 10(2-1)^2 = 150\ \mu\text{A}$$

now, by using current mirror property all transistor should have equal current.

$$I_{DSN} = I_{SDP} = 150\ \mu\text{A}$$

For M_1 $\quad g_{m1} = \sqrt{2\mu_n C_{ox}\frac{W}{L} \times I_{DS}}$

$$= \sqrt{2\mu_n C_{ox}\frac{W}{L} \times I_{DS}} = \sqrt{2 \times 60 \times 5 \times 150} = 300\ \mu℧$$

M_2, M_3 and M_4 from active load for M_1. This active load in equivalent to resistance r_{ds2} i.e. 6 MΩ.

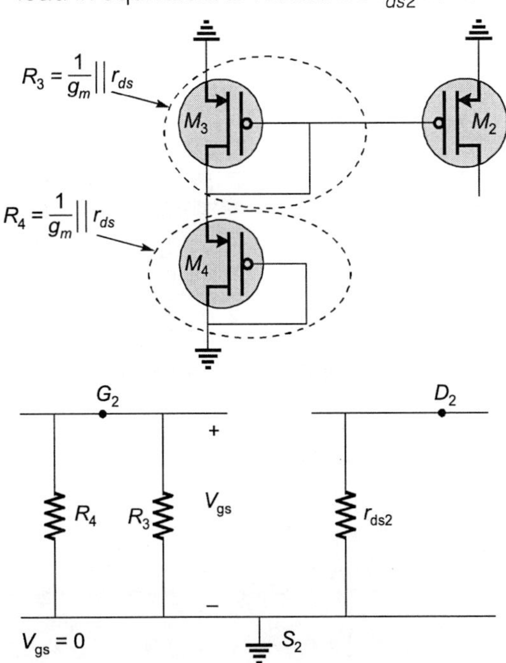

M_1 is common source amplifier.

$$\therefore \quad \frac{V_{out}}{V_{in}} = A_v = -g_{m1} \times (r_{ds2} \parallel r_{ds1})$$

$$= -300\ \mu℧ \times 3\ \text{M}\Omega = -900$$

4.35 (6)

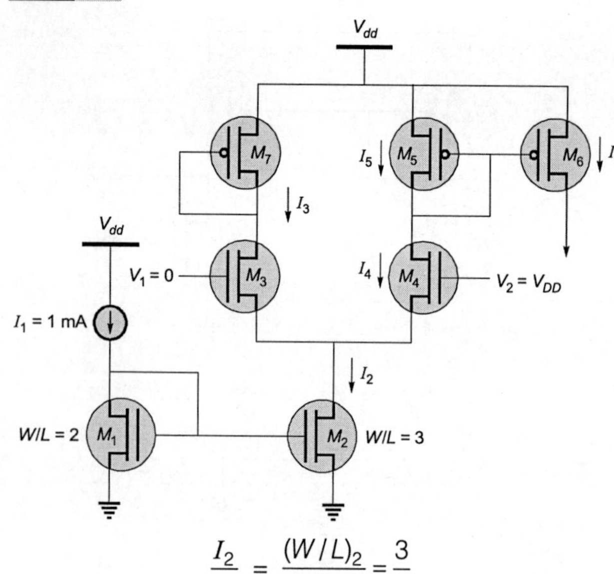

$$\frac{I_2}{I_1} = \frac{(W/L)_2}{(W/L)_1} = \frac{3}{2}$$

$$I_2 = \frac{3}{2} \times I_1 = 1.5\text{ mA}$$

M_3 is OFF because $V_1 = 0 \Rightarrow I_3 = 0$

M_4 is ON because $V_2 = V_{DD}$

$$I_5 = I_4 = I_2 = 1.5\text{ mA}$$

$$\frac{I_6}{I_5} = \frac{(W/L)_6}{(W/L)_5} = \frac{40}{10} = 4$$

$$I_6 = 4I_5 = 4 \times 1.5 = 6\text{ mA}$$

$\therefore$ $I_{out} = I_6 = 6\text{ mA}$

4.36 (b)

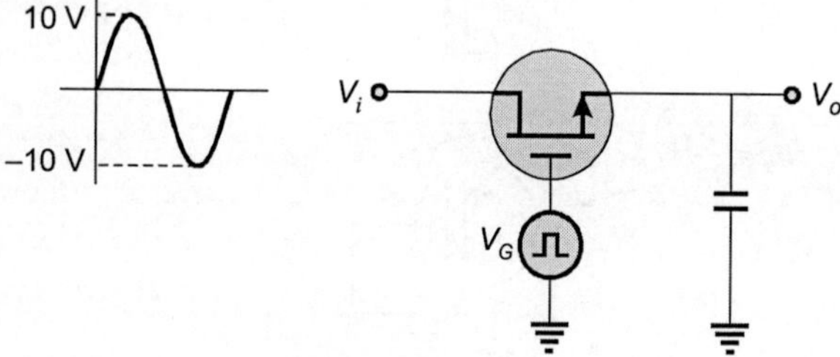

for holding MOSFET should be OFF.

$$V_{1\,min} \rightarrow -10\text{ V}$$

$$V_G - V_{1\,min} < 3$$

$$V_G < 3 - 10\text{ V} \Rightarrow -7\text{ V}$$

For sampling,

$$V_G - V_{i\,max} > 3$$

$$V_G > 3 + V_{i\,max}$$

$$V_G > 13$$

4.37 (d)

$$v_\pi = -V_x$$

$$V_x = (I_x - g_m V_x)\, r_{ds} + I_x R$$

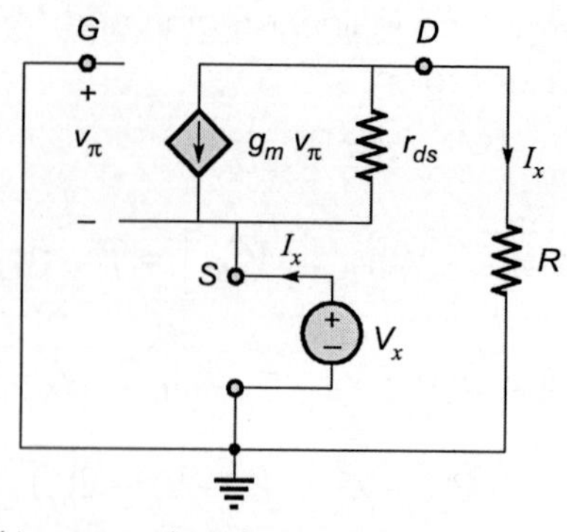

$$V_x(1 + g_m r_{ds}) = (r_{ds} + R)I_x$$

$$R_N = \frac{V_x}{I_x} = \frac{R + r_{ds}}{1 + g_m r_{ds}}$$

4.38 (0.5)

$$I_{DS} = \frac{\mu_n C_{ox}}{2} \times \frac{W}{L}(V_{GS} - V_T)^2$$

Given, $V_{GS} = 1\text{ V}$

$$I_{DS} = \frac{1}{2}(1 - V_T)^2 \quad \text{...(i)}$$

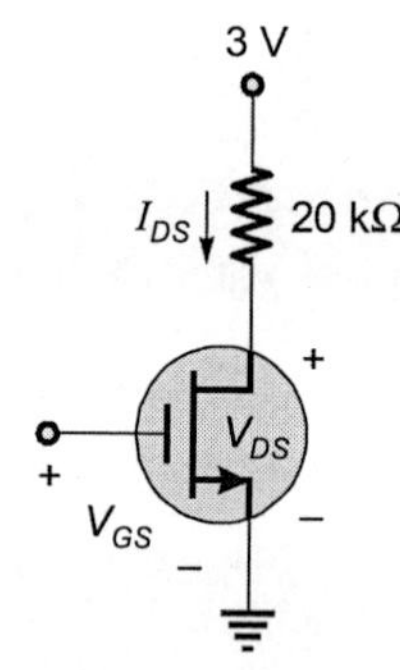

$$V_{DS} = 3 - 20 \times I_{DS}$$

$$V_{DS} = 3 - \frac{20}{2}(1 - V_T)^2$$

$$V_{DS} = 3 - 10(1 - V_T)^2$$

MOSFET operates in saturation if

$$V_{DS} \geq V_{GS} - V_T$$

So, we take, $V_{DS} = V_{GS} - V_T$

$$V_{GS} - V_T = 3 - 10(1 - V_T)^2$$

$$1 - V_T = 3 - 10(1 - V_T)^2$$

Let, $1 - V_T = x$

$$3 - 10x^2 = x$$

$$10x^2 + x - 3 = 0$$

We get, $x = \dfrac{-1 \pm \sqrt{1 + 120}}{20}$

$$x = \frac{-1 \pm 11}{20}$$

$\therefore$ $x = 0.5$ and -0.6

$$x = 0.5$$

$$\Rightarrow \quad 1 - V_T = 0.5$$
$$\Rightarrow \quad V_T = 0.5\text{ V}$$
$$x = -0.6$$
$$\Rightarrow \quad 1 - V_T = -0.6$$
$$\Rightarrow \quad V_T = 1.6\text{ V}$$

But $\quad V_{GS} > V_T$

or $\quad V_T < V_{GS}$

i.e., $\quad V_T < 1$

$\therefore \quad V_T = 0.5\text{ V}$

4.39 (b)

MOSFET M_2 acts as common source amplifier.

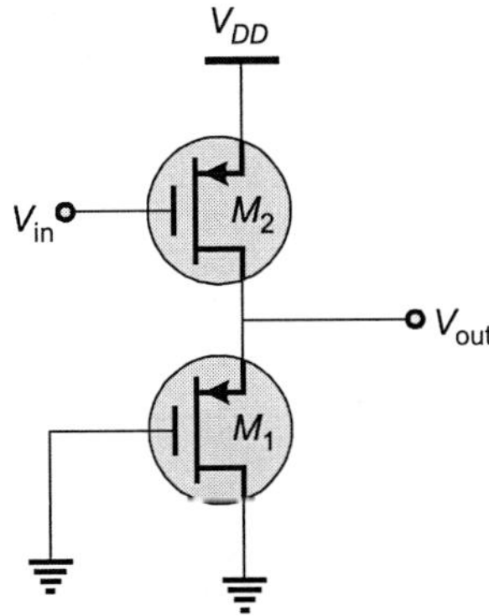

Drain to gate connected MOSFET M_1 acts as load.

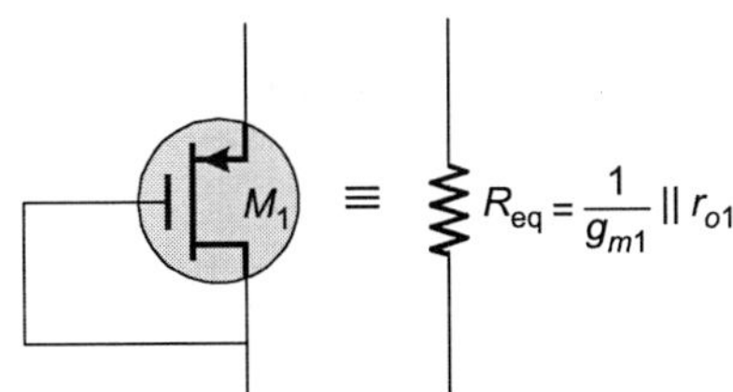

For given circuit, AC equivalent is as shown.

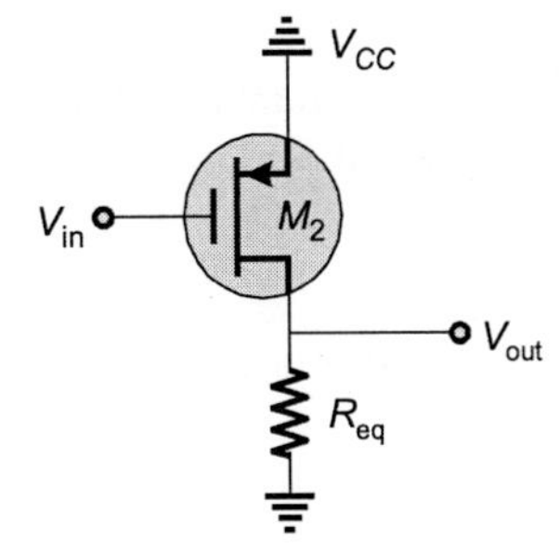

Replace M_2 with small signal model

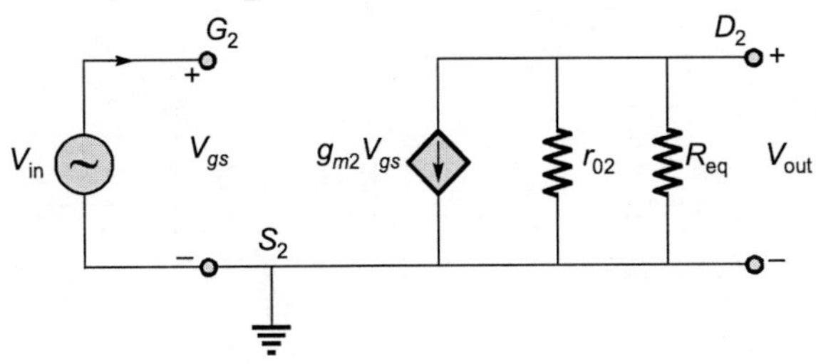

$$\frac{V_{out}}{V_{in}} = \frac{-g_{m2}V_{gs}\left(r_{o_2} \parallel R_{eq}\right)}{V_{gs}}$$

$$A_V = -g_{m2}\left(r_{o_2} \parallel \frac{1}{g_{m1}} \parallel r_{o_1}\right)$$

Hence, option (b) is correct.

4.40 (c)

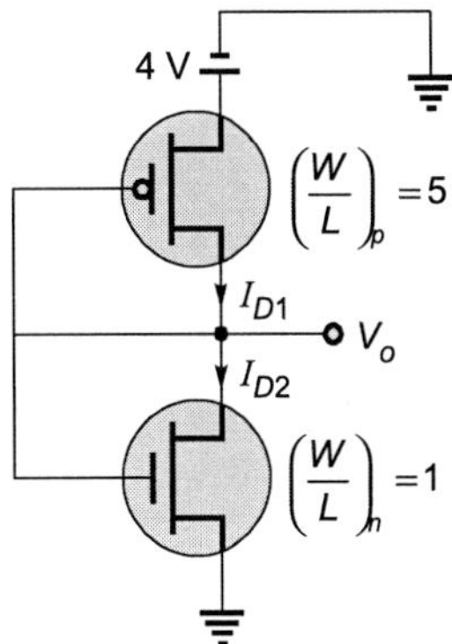

Given :

PMOS : $\quad V_{TP} = -1\text{ V}, \mu_p = 40\text{ cm}^2/\text{V-s}$

NMOS : $\quad V_{TN} = 1\text{ V}, \mu_n = 300\text{ cm}^2/\text{V-s}$

For both the MOSFETs, PMOS and NMOS, drain is shorted to gate.

Therefore, $V_{DS} = V_{GS} > V_{GS} - |V_T|$.

Hence, both the transistors are operating in saturation region. The drain current for both the MOSFETs are equal i.e.

$$I_{D1} = I_{D2}$$

$$\Rightarrow \quad \mu_p C_{ox}\left(\frac{W}{L}\right)_P \left(V_{SGP} - |V_{TP}|\right)^2 = \mu_n C_{ox}\left(\frac{W}{L}\right)_N (V_{GSN} - V_{TN})^2$$

$$\Rightarrow \mu_p\left(\frac{W}{L}\right)_P \left(V_{SGP} - |V_{TP}|\right)^2 = \mu_r\left(\frac{W}{L}\right)_N (V_{GSN} - V_{TN})^2$$

From the figure,

$$V_{SGP} = 4 - V_o \text{ and } V_{GSN} = V_o$$

Substituting the values, we get,

$$40 \times 5\,(4 - V_o - 1)^2 = 300 \times 1 \times (V_o - 1)^2$$

$$\Rightarrow \quad 2(9 + V_o^2 - 6V_o) = 3(V_o^2 - 2V_o + 1)$$

$$\Rightarrow \quad V_o^2 + 6V_o - 15 = 0$$

$$\Rightarrow \quad V_o = \frac{-6 \pm \sqrt{36 + 4 \times 15}}{2}$$

$$\Rightarrow \quad V_o = \frac{-6 \pm \sqrt{96}}{2} = 1.898\text{ V}, -7.89\text{ V}$$

V_o cannot be negative because V_o should lie between 0 and 4 V.

$\therefore \quad V_o = 1.898\text{ V}$

Hence, V_o is less than 2 V.

4.41 (c)

For the nMOSFET shown below,

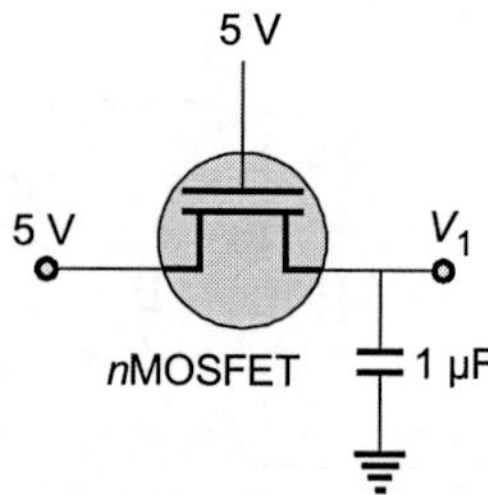

The output node, V_1 gets gradually charged from 0 towards V_{DD}. When V_1 reaches $V_{DD} - V_{Tn}$, then V_{GS} is

$$V_{gate} - V_1 = 5 - (5 - V_{Tn}) = V_{Tn}$$

Which is the minimum voltage required for the NMOS to be ON for current to flow. So, the maximum voltage level that the output node can be charged to is $(5 - V_{Tn})$, hence

$$V_1 = 5 - V_{Tn} = 5 - 1 = 4\text{ V}$$

For the pMOSFET shown below,

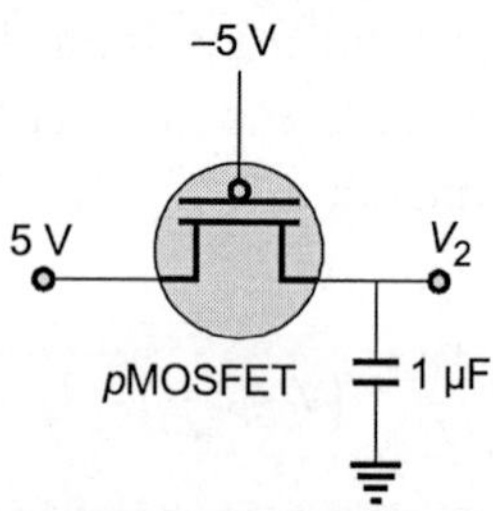

PMOS is good to pass logic 1 and passes weak 0 i.e. Output is $|V_{Tp}|$ in case of 0. In the given circuit, input is logic '1'. So, PMOS passes strong '1' i.e.

$$V_2 = 5\text{ V}$$

4.42 (c)

Given : Gate length, $L = 10\ \mu\text{m}$

Gate width, $W = 100\ \mu\text{m}$

$\mu_n C_{ox} = 1\text{ mA/V}^2$

Threshold voltage, $V_T = 1\text{ V}$

Gate-to-source voltage,

$$V_{GS} = [2 - \sin(2t)]\text{ V}$$

Drain-to-source voltage,

$$V_{DS} = 1\text{ V}$$

Since, $\quad V_{GS} = [2 - \sin(2t)]\text{ V}$

Therefore,

$$V_{GS\min} = 2 - 1 = 1\text{ V}$$

$$V_{GS\max} = 2 + 1 = 3\text{ V}$$

(i) For $1\text{ V} < V_{GS} < 2\text{ V}$,

$$V_{DS} > V_{GS} - V_T$$

MOSFET is in saturation region. The drain current,

$$I_D = \frac{\mu_n C_{ox} W}{L} \cdot \frac{(V_{GS} - V_T)^2}{2}$$

For maximum drain current in saturation region,

$$V_{GS\max} = 2\text{ V}$$

$$I_{D\max} = \frac{1\text{mA/V}^2 \times 100}{10} \cdot \frac{(2-1)^2}{2}$$

$$\therefore \quad I_{D\max} = 5\text{ mA}$$

(ii) For $2\text{ V} < V_{GS} < 3\text{ V}$,

$$V_{DS} < V_{GS} - V_T$$

MOSFET is operating in linear region. The drain current,

$$I_D = \frac{\mu_n C_{ox} W}{L}\left[(V_{GS} - V_T)V_{DS} - \frac{V_{DS}^2}{2}\right]$$

For maximum drain current in linear region,

$$V_{GS\max} = 3\text{ V}$$

$$\therefore \quad I_{D\max} = \frac{\mu_n C_{ox} W}{L}\left[(V_{GS\max} - V_T)V_{DS} - \frac{V_{DS}^2}{2}\right]$$

$$I_{D\max} = \frac{1\text{mA/V}^2 \times 100}{10}\left[(3-1)1 - \frac{1}{2}\right]$$

$$I_{D\max} = 10 \times \frac{3}{2} = 15\text{ mA}$$

Therefore, the maximum value of the drain to source current is 15 mA.

4.43 (–2)

The small signal model of the circuit can be drawn as below:

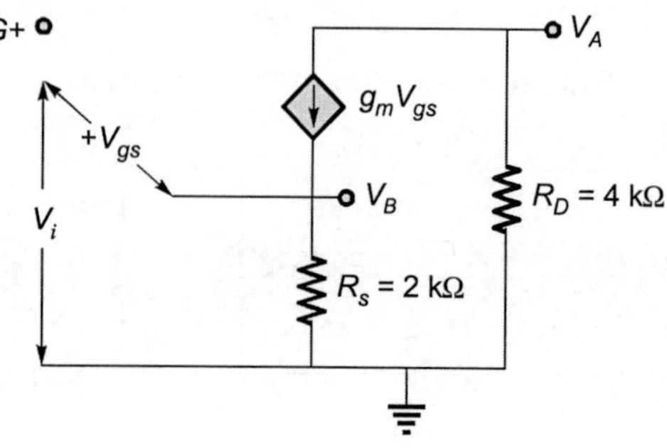

Applying KVL in the circuit, we can write,

$$V_i = V_{gs} + (g_m V_{gs}) \cdot R_s$$

$$V_{gs} = \frac{V_i}{1 + g_m R_s} \quad \text{...(i)}$$

Hence, $V_B = (g_m V_{gs}) R_s$

Using equation (i),

$$V_B = V_{in}\left(\frac{g_m R_s}{1 + g_m R_s}\right) \quad \text{...(ii)}$$

Similarly,

$$V_A = (-g_m V_{gs}) \cdot R_D$$

Using equation (i),

$$V_A = V_{in}\left(\frac{-g_m R_D}{1 + g_m R_s}\right) \quad \text{...(iii)}$$

From equation (ii) and (iii),

$$\frac{V_A}{V_B} = \frac{-R_D}{R_s} = \frac{-4\,\text{k}\Omega}{2\,\text{k}\Omega}$$

$$\therefore \quad \frac{V_A}{V_B} = -2$$

4.44 (*)

Dependent current source should have $h_{21} I_1$ instead of $h_{21} V_1$ according to h-parameter.

$$\text{Voltage gain}, A_v = \frac{V_{out}}{V_s} = \frac{-h_{21} I_1 \times \left(\frac{1}{h_{22}} \| R_L\right)}{h_{11} I_1 + h_{12} V_2}$$

To achieve maximum $\frac{V_{out}}{V_s}$,

$$h_{11} = 0, \quad h_{12} = 0$$

$$h_{21} = \text{Very high}, \quad h_{22} = 0$$

Hence, answer should be (a) according to $h_{21} I_1$.

4.45 (a)

The small-signal equivalent of the circuit can be drawn as below:

$$I_i = \frac{V_i - V_0}{Z_1} = \frac{V_i}{Z_1} - \frac{V_0}{Z_1} \quad \text{...(i)}$$

We have, $V_0 = (I_i + g_m V_{gs}) Z_L$

$$V_0 = I_i Z_L + g_m V_{gs} Z_L \quad [\because V_{gs} = I_i Z_1]$$

$$V_0 = I_i (Z_L + g_m Z_1 Z_L)$$

From equation (i),

$$I_i = \frac{V_i}{Z_1} - \frac{I_i (Z_L + g_m Z_1 Z_L)}{Z_1}$$

$$I_i \frac{[Z_1 + Z_L + g_m Z_1 Z_L]}{Z_1} = \frac{V_i}{Z_1}$$

$$\frac{V_i}{I_i} = Z_1 + Z_L + g_m Z_1 Z_L$$

$$Z_{in}(j\omega) = \frac{1}{j\omega C_1} + \frac{1}{j\omega C_L} - \frac{g_m}{\omega^2 C_1 C_L}$$

4.46 (c)

Neglecting r_0, the small-signal output impedance of diode-connected MOSFET M_1 is $1/g_{m1}$. We obtain the small signal model of the circuit as shown. Hence,

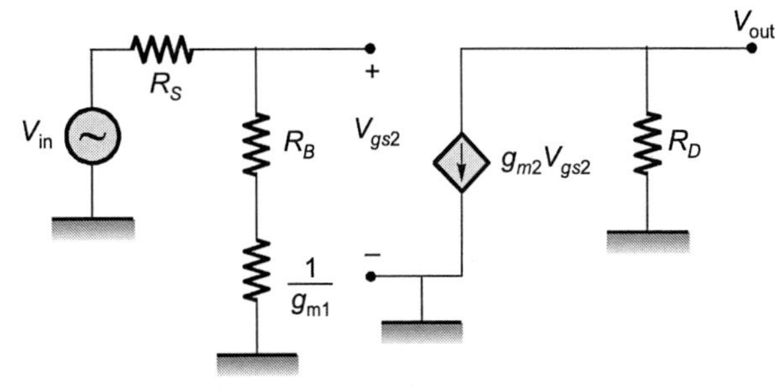

$$V_{out} = -g_{m2} V_{gs2} R_D \quad \text{...(i)}$$

Using the voltage division rule,

$$V_{gs2} = V_{in} \times \frac{R_B + \frac{1}{g_{m1}}}{R_S + R_B + \frac{1}{g_{m1}}} \quad \text{...(ii)}$$

From equations (i) and (ii),

$$\left|\frac{V_{out}}{V_{in}}\right| = \frac{g_{m2} R_D \left(R_B + \frac{1}{g_{m1}}\right)}{R_S + R_B + \frac{1}{g_{m1}}}$$

■■■

5 Frequency Response of Amplifier

5.1 An RC-Coupled Amplifier is assumed to have a single-pole low frequency transfer function. The maximum lower-cutoff frequency allowed for the Amplifier to pass 50 Hz. Square wave with no more than 10% tilt is _______.

[1995 : 1 M]

5.2 An Amplifier has an open-loop gain of 100 and its lower and upper-cut-off frequency of 100 Hz and 100 kHz respectively, a feedback network with a feedback factor of 0.99 is connected to the amplifier. The new lower-and upper-cut-off frequencies are at ...... and ......

[1995 : 1 M]

5.3 An *npn* transistor has a beta cut-off frequency f_β of 1 MHz and Common Emitter short circuit low frequency current gain β_0 of 200 at unity gain frequency f_T and the alpha cut-off frequency f_α respectively are

(a) 200 MHz, 201 MHz (b) 200 MHz, 1999 MHz
(c) 199 MHz, 200 MHz (d) 201 MHz, 200 MHz

[1996 : 1 M]

5.4 In the circuit shown in Fig. is a finite gain amplifier with a gain of *K*, a very large input impedance, and a very low output impedance. The input impedance of the feedback amplifier with the feedback impedance *Z* connected as shown will be

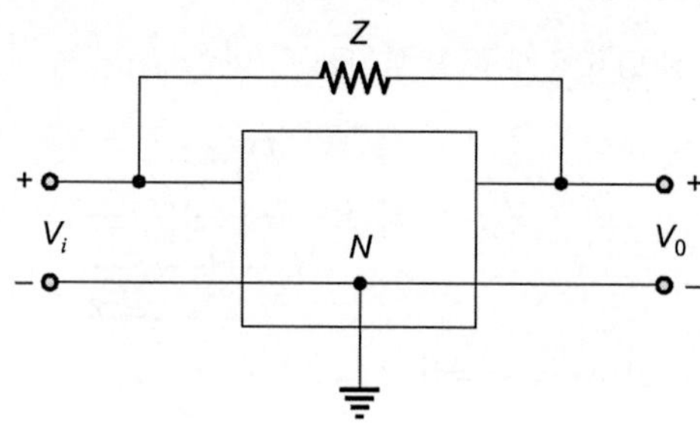

(a) $Z\left[1-\frac{1}{K}\right]$ (b) $Z(1-K)$

(c) $\left[\frac{Z}{K-1}\right]$ (d) $\left[\frac{Z}{1-K}\right]$

[1996 : 2 M]

5.5 The f_T of a BJT is related to its g_m, C_π and $C\mu$ as follows

(a) $f_T = \frac{C_\pi + C_\mu}{g_m}$ (b) $f_T = \frac{2\pi(C_\pi + C_\mu)}{g_m}$

(c) $f_T = \frac{g_m}{C_\pi + C_\mu}$ (d) $f_T = \frac{g_m}{2\pi(C_\pi + C_\mu)}$

[1998 : 1 M]

5.6 An npn transistor (with $C = 0.3$ pF) has a unity-gain cutoff frequency f_T of 400 MHz at a dc bias current $I_c = 1$ mA. The value of its C_μ (in pF) is approximately ($V_T = 26$ mV)

(a) 15 (b) 30
(c) 50 (d) 96

[1999 : 2 M]

5.7 An amplifier is assumed to have a single-pole high-frequency transfer function. The rise time of its output response to a step function input is 35 nsec. The upper −3 dB frequency (in MHz) for the amplifier to a sinusoidal input is approximately at

(a) 4.55 (b) 10
(c) 20 (d) 28.6

[1999 : 2 M]

5.8 An npn BJT has $g_m = 38$ mA/V, $C_\mu = 10^{-14}$ F, $C_\pi = 4 \times 10^{-13}$ F, and DC current gain $\beta_0 = 90$. For this transistor f_T and f_β are

(a) $f_T = 1.64 \times 10^8$ Hz and $f_\beta = 1.47 \times 10^{10}$ Hz

(b) $f_T = 1.47 \times 10^{10}$ Hz and $f_\beta = 1.64 \times 10^8$ Hz

(c) $f_T = 1.33 \times 10^{12}$ Hz and $f_\beta = 1.47 \times 10^{10}$ Hz

(d) $f_T = 1.47 \times 10^{10}$ Hz and $f_\beta = 1.33 \times 10^{12}$ Hz

[2001 : 2 M]

5.9 A bipolar transistor is operating in the active region with a collector current of 1 mA. Assuming that the β of the transistor is 100 and the thermal voltage (V_T) is 25 mV, the transconductance (g_m) and the input resistance (r_π) of the transistor in the common emitter configuration, are

(a) $g_m = 25\text{ mA/V}$ and $r_\pi = 15.625\text{ k}\Omega$
(b) $g_m = 40\text{ mA/V}$ and $r_\pi = 4.0\text{ k}\Omega$
(c) $g_m = 25\text{ mA/V}$ and $r_\pi = 2.5\text{ k}\Omega$
(d) $g_m = 40\text{ mA/V}$ and $r_\pi = 2.5\text{ k}\Omega$

[2004 : 2 M]

5.10 The ac schematic of an NMOS common-source stage is shown in the figure below, where part of the biasing circuits has been omitted for simplicity. For the *n*-channel MOSFET *M*, the transconductance g_m = 1 mA/V, and body effect and channel length modulation effect are to be neglected. The lower cut-off frequency in Hz of the circuit is approximately at

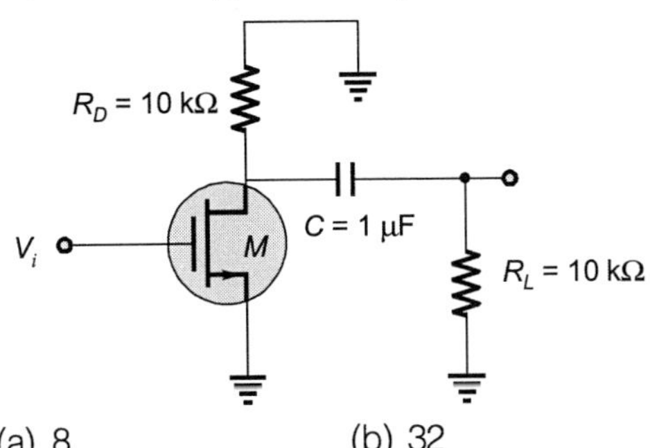

(a) 8
(b) 32
(c) 50
(d) 200

[2013 : 2 M]

5.11 Which one of the following statements is correct about an ac-coupled common-emitter amplifier operating in the mid-band region?

(a) The device parasitic capacitances behave like open circuits, whereas coupling and bypass capacitances behave like short circuits.
(b) The device parasitic capacitances, coupling capacitances and bypass capacitances behave like open circuits.
(c) The device parasitic capacitances, coupling capacitances and bypass capacitances behave like short circuits.
(d) The device parasitic capacitances behave like short circuits, whereas coupling and bypass capacitances behave like open circuits.

[2016 : 1 M, Set-2]

5.12 The Miller effect in the context of a Common Emitter amplifier explains

(a) an increase in the low-frequency cutoff frequency
(b) an increase in the high-frequency cutoff frequency
(c) a decrease in the low-frequency cutoff frequency
(d) a decrease in the high-frequency cutoff frequency

[2017 : 1 M, Set-1]

■■■

Answers Frequency Response of Amplifier

5.1	(1.59)	5.2	(10)	5.3	(a)	5.4	(d)	5.5	(d)	5.6	(a)	5.7	(b)
5.8	(b)	5.9	(d)	5.10	(a)	5.11	(a)	5.12	(d)				

Explanations Frequency Response of Amplifier

5.1 (1.59)

$$\boxed{\%\text{ Tilt} = \frac{\pi f_L}{f} \times 100\%}$$

$$f_L = \frac{f \times \%\text{ Tilt}}{\pi \times 100} = \frac{50 \times 10}{\pi \times 100} = 1.59\text{ Hz}$$

5.2 (10)

$$1 + A\beta = 1 + 100 \times 0.99 = 1 + 99 = 100$$

$$\boxed{f_L^* = \frac{f_L}{1 + A\beta}} = \frac{100}{100} = 1\text{ Hz}$$

$$\boxed{f_H^* = f_H(1 + A\beta)} = 100 \times 10^3 \times (100) = 10\text{ MHz}$$

5.3 (a)

$$\boxed{f_T = \beta f_\beta}$$

$$f_T = 200 \times 1 = 200\text{ MHz}$$

$$\boxed{f_\alpha = \frac{f_\beta}{1 - \alpha} = (1 + \beta) f_\beta}$$

$$f_\alpha = (1 + 200)1 = 201\text{ MHz}$$

5.4 (d)

Miller's Theorem:

Miller's theorem states that if a series impedance Z' is connected between the input and output terminal of a network, then it can be replaced with a shunt impedance Z_1 in the input side and by a shunt impedance Z_2 in the output section of the network.

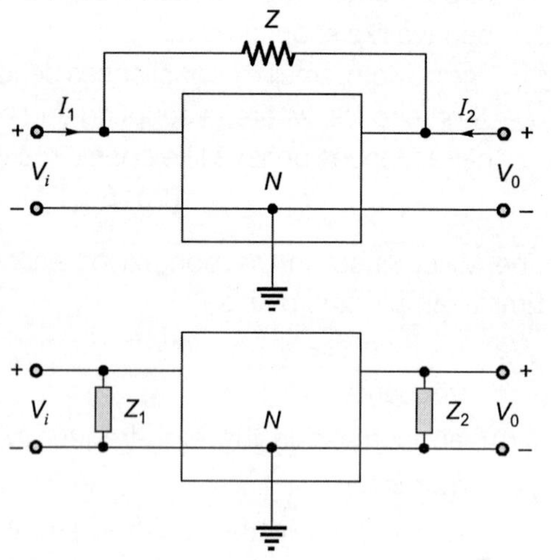

If $V_i > V_2$ then I_1 flows

$$I_1 = \frac{V_1 - V_2}{Z} = \frac{V_1[1 - V_2/V_1]}{Z} = \frac{V_1[1 - A_V]}{Z}$$

$$\frac{V_1}{I_1} = Z_1 = \frac{Z}{1 - A_V}$$

$$Z_1 = \frac{Z}{1 - K}$$

5.5 (d)

Unity gain bandwidth product of the C.E. transistor

$$\boxed{f_T = \frac{g_m}{2\pi(C_\pi + C_\mu)} = \frac{g_m}{2\pi(C_e + C_c)}}$$

5.6 (a)

$$f_T = \frac{1}{2\pi RC}$$

$$\frac{1}{R} = g_m = \frac{I_C}{V_T}$$

$$f_T = \frac{I_C}{2\pi V_T \times C_\mu}$$

$$C_\mu = \frac{1\,\text{mA}}{2\pi \times 26\,\text{mv} \times 400 \times 10^6}\ ; \quad C_\mu = 15\text{ pF}$$

5.7 (b)

$$t_r \times BW = 0.35$$

$$BW = \frac{0.35}{35 \times 10^{-9}} = 10$$

5.8 (b)

$$f_T = \frac{g_m}{2\pi(C_\mu + C_\pi)}$$

$$= \frac{38 \times 10^{-3}}{2\pi \times (10^{-14} + 4 \times 10^{-13})}$$

$$= \frac{38 \times 10^{-3}}{2\pi \times 10^{-13}(0.1 + 4)} = 1.47 \times 10^{10}\text{ Hz}$$

$$f_\beta = \frac{f_T}{\beta_o} = \frac{1.47 \times 10^{10}}{90}$$

$$= 1.64 \times 10^8\text{ Hz} \qquad (\beta_0 = h_{fe})$$

5.9 (d)

$$g_m = \frac{I_C}{V_T} = \frac{1\,\text{mA}}{25\,\text{mV}} = 0.04 = 40\text{ mA/V}$$

$$h_{fe} = g_m \cdot r_\pi,\ h_{fe} = \beta$$

$$r_\pi = \frac{\beta}{g_m} = \frac{100}{40 \times 10^{-3}} = 2.5\text{ k}\Omega$$

5.10 (a)

$$f_L = \frac{1}{2\pi RC} = \frac{1}{2 \times 3.14 \times (10\text{k} + 10\text{k}) \times 1\,\mu\text{F}} = 8\text{ Hz}$$

5.12 (d)

Miller effect increases input capacitance and thereby decreases the higher cut-off frequency.

■■■

6 Feedback Amplifiers

6.1 To obtain very high input and output impedances in a feedback Amplifier, the mostly used is
(a) Voltage - series (b) Current - series
(c) Voltage - shunt (d) Current - shunt
[1995 : 1 M]

6.2 Negative feed back in
(1) voltage series configuration
(2) current shunt configuration
(a) increases input impedance
(b) decreases input impedance
(c) increases closed loop gain
(d) leads to oscillation. [1997 : 2 M]

6.3 In a shunt-shunt negative feedback Amplifier, as compared to the basic Amplifier.
(a) both input and output impedance decreases.
(b) input impedance decreases but output impedance increases
(c) input impedance increases but output impedance decreases
(d) both input and output impedance increases
[1998 : 1 M]

6.4 Negative feedback in an amplifier
(a) reduces gain
(b) increases frequency and phase distortions
(c) reduces bandwidth
(d) increases noise [1999 : 1 M]

6.5 An amplifier has an open-loop gain of 100, an input impedance of 1 kΩ, and an output impedance of 100 Ω. A feedback network with a feedback factor of 0.99 is connected to the amplifier in a voltage series feedback mode. The new input and output impedances, respectively, are
(a) 10 Ω and 1 Ω (b) 10 Ω and 10 Ω
(c) 100 kΩ and 1 Ω (d) 100 kΩ and 1 kΩ
[1999 : 2 M]

6.6 In a negative feedback amplifier using voltage-series (i.e. voltage-sampling, series mixing) feedback.
(a) R_i decreases and R_o decreases
(b) R_i decreases and R_o increases
(c) R_i increases and R_o decreases
(d) R_i increases and R_o increases
(R_i and R_o denote the input and output resistances respectively) [2002 : 1 M]

6.7 An amplifier without feedback has a voltage gain of 50, input resistance of 1 kΩ and output resistance of 2.5 kΩ. The input resistance of the current-shunt negative feedback amplifier using the above amplifier with a feedback factor of 0.2 is
(a) 1/11 kΩ (b) 1/5 kΩ
(c) 5 kΩ (d) 11 kΩ [2003 : 2 M]

6.8 Voltage series feedback (also called series-shunt feedback) results in
(a) increase in both input and output impedances
(b) decrease in both input and output impedances
(c) increase in input impedance and decrease in output impedance
(d) decrease in input impedance and increase in output impedance
[2004 : 1 M]

6.9 The effect of current shunt feedback in an amplifier is to
(a) increase the input resistance and decrease the output resistance
(b) increase both input and output resistances
(c) decrease both input and output resistances.
(d) decrease the input resistance and increase the output resistance
[2005 : 1 M]

6.10 The input impedance (Z_i) and the output impedance (Z_0) of an ideal trans-conductance (voltage controlled current source) amplifier are
(a) $Z_i = 0, Z_0 = 0$ (b) $Z_i = 0, Z_0 = \infty$
(c) $Z_i = \infty, Z_0 = 0$ (d) $Z_i = \infty, Z_0 = \infty$
[2006 : 1 M]

6.11 In a transconductance amplifier, it is desirable to have
(a) a large input resistance and a large output resistance
(b) a large input resistance and a small output resistance
(c) a small input resistance and a large output resistance
(d) a small input resistance and a small output resistance
[2007 : 1 M]

6.12 In a voltage-voltage feedback as shown below, which one of the following statements is TRUE if the gain k is increased?

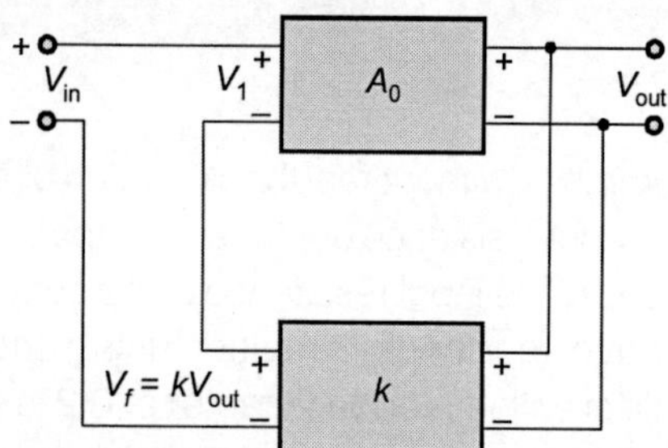

(a) The input impedance increases and output impedance decreases
(b) The input impedance increases and output impedance also increases
(c) The input impedance decreases and output impedance also decreases
(d) The input impedance decreases and output impedance increases **[2013 : 1 M]**

6.13 In the ac equivalent circuit shown in the figure, if i_{in} is the input current and R_F is very large, the type of feedback is

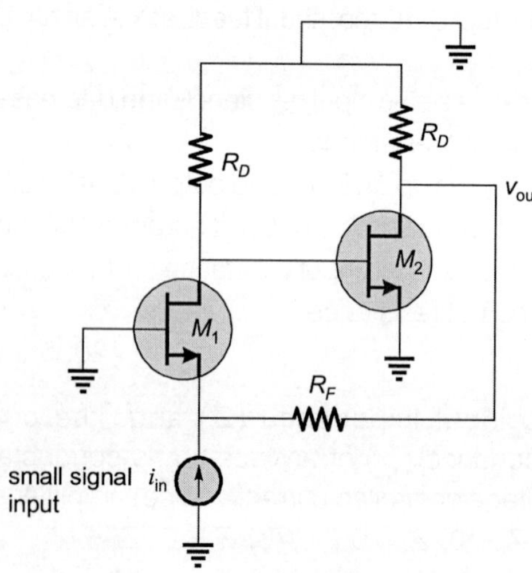

(a) voltage-voltage feedback
(b) voltage-current feedback
(c) current-voltage feedback
(d) current-current feedback

[2014 : 1 M, Set-1]

6.14 The desirable characteristics of a transconductance amplifier are
(a) high input resistance and high output resistance
(b) high input resistance and low output resistance
(c) low input resistance and high output resistance
(d) low input resistance and low output resistance

[2014 : 1 M, Set-3]

6.15 The feedback topology in the amplifier circuit (the base bias circuit is not shown for simplicity) in the figure is

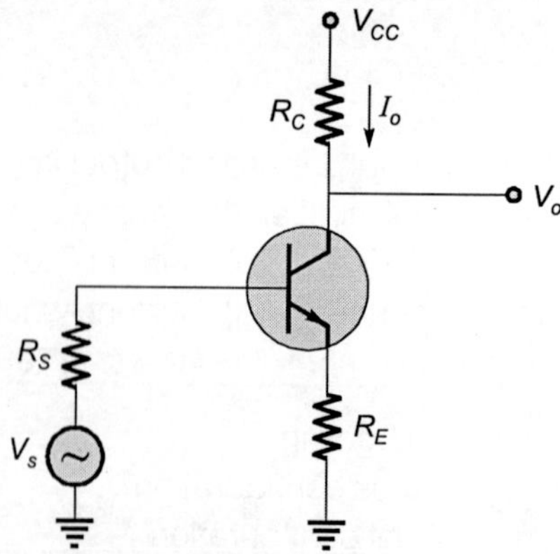

(a) Voltage shunt feedback
(b) Current series feedback
(c) Current shunt feedback
(d) Voltage series feedback

[2014 : 1 M, Set-2]

6.16 A good transconductance amplifier should have
(a) high input resistance and low output resistance
(b) low input resistance and high output resistance
(c) high input and output resistances
(d) low input and output resistances

[2017 : 1 M, Set-1]

6.17 A good transimpedance amplifier has
(a) low input impedance and high output impedance
(b) high input impedance and high output impedance
(c) high input impedance and low output impedance
(d) low input impedance and low output impedance

[2018 : 1 M]

6.18 In the voltage regulator shown below, V_t is the unregulated at 15 V. Assume V_{BE} = 0.7 V and the base current is negligible for both the BJTs. If the regulated output V_o is 9 V, the value of R_2 is ______ Ω.

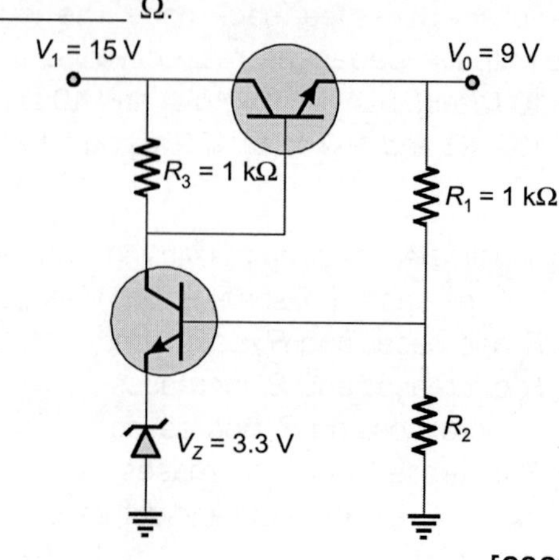

[2020 : 2 M]

6.19 In the circuit shown below, V_1 and V_2 are bias voltages. Based on input and output impedances, the circuit behaves as a

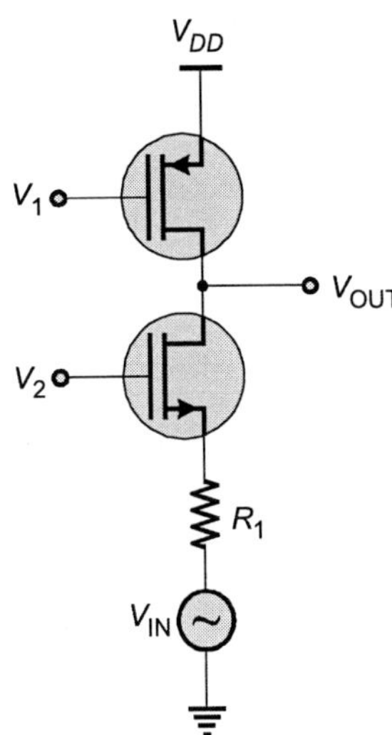

(a) voltage controlled voltage source.
(b) voltage controlled current source.
(c) current controlled voltage source.
(d) current controlled current source.

[2023 : 1 M]

6.20 A cascade of common-source amplifiers in a unity gain feedback configuration oscillates when

(a) the closed loop gain is less than 1 and the phase shift is less than 180°.
(b) the closed loop gain is greater than 1 and the phase shift is less than 180°.
(c) the closed loop gain is less than 1 and the phase shift is greater than 180°.
(d) the closed loop gain is greater than 1 and the phase shift is greater than 180°.

[2023 : 1 M]

6.21 For the closed loop amplifier circuit shown below, the magnitude of open loop low frequency small signal voltage gain is 40. All the transistors are biased in saturation. The current source I_{SS} is ideal. Neglect body effect, channel length modulation and intrinsic device capacitances. The closed loop low frequency small signal voltage gain $\frac{V_{out}}{V_{in}}$ (rounded off to three decimal places) is ______.

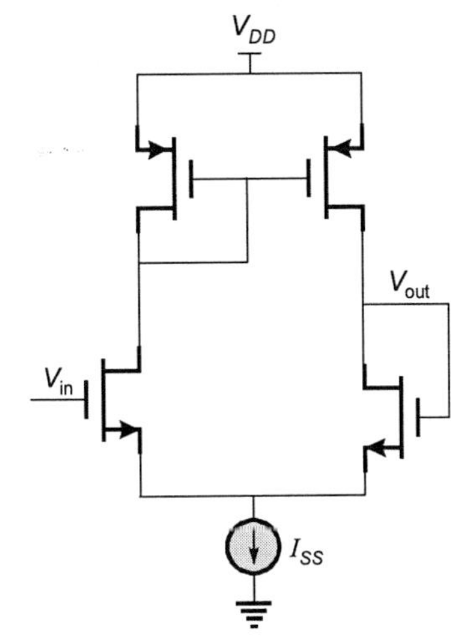

(a) 0.976
(b) 1.000
(c) 0.488
(d) 1.025

[2024 : 1 Mark]

■■■

Answers Feedback Amplifiers

6.1 (b)	6.2 (1-a, 2-b)	6.3 (a)	6.4 (a)	6.5 (c)	6.6 (c)	6.7 (a)
6.8 (c)	6.9 (d)	6.10 (d)	6.11 (a)	6.12 (a)	6.13 (b)	6.14 (a)
6.15 (b)	6.16 (c)	6.17 (d)	6.18 (800)	6.19 (d)	6.20 (*)	6.21 (a)

Explanations Feedback Amplifiers

6.1 (b)

Current-series feedback amplifiers has very high input and very high output impedances.

$$R_{if} = R_i(1+A\beta) = R_i(1+G_m\beta)$$
$$R_{of} = R_o(1+A\beta) = R_o(1+G_m\beta)$$

6.2 Sol.

1-*a*, 2-*b*

Voltage-series configuration-increases the input impedance.

$$R_{if} = R_i(1 + A\beta) = R_i(1 + A_v\beta)$$

Current-shunt configuration-decreases the input impedance.

$$R_{if} = \frac{R_i}{1+A\beta} = \frac{R_i}{1+A_I\beta}$$

6.3 (a)

Shunt-shunt feedback amplifiers has very low input and very low output impedance.

$$R_{if} = \frac{R_i}{1+A\beta} = \frac{R_i}{1+R_m\beta} \; ; \; R_{of} = \frac{R_o}{1+A\beta} = \frac{R_o}{1+R_m\beta}$$

6.4 (a)

$$A_f = \frac{A}{1+A\beta}$$

6.5 (c)

$A = 100, \beta = 0.99$; $1 + A\beta = 100$

For voltage series, $R_i \uparrow$ & $R_o \downarrow$ by $1+A\beta$.

We have, $R_{if} = R_i(1 + A\beta)$ and $R_{of} = \dfrac{R_o}{1+A\beta}$

$$\Rightarrow \quad R_{if} = 100 \times 1\text{ k}\Omega = 100\text{ k}\Omega$$
$$R_{of} = \frac{100\ \Omega}{100} = 1\Omega$$

6.6 (c)

R_i increases by factor of $1 + A\beta$ and R_o decreases by $1 + A\beta$.

6.7 (a)

In current shunt negative feedback amplifier.

$$R_{if} = \frac{R_i}{1+A\beta} \; ; \; R_{if} = \frac{1\times 10^3}{1+50\times 0.2} = \frac{10^3}{1+10}$$
$$R_{if} = \frac{1}{11}\text{k}\Omega$$

6.12 (a)

The given configuration is a voltage-series feedback configuration.

So, the input impedance increases

$$R_{if} = R_i(1 + A_oK)$$

and, the output impedance decreases

$$R_{of} = \frac{R_o}{1+A_oK}$$

6.13 (b)

Sampling → current
Mixing → voltage

6.15 (b)

The feedback topology in the amplifier circuit is current series because

Sampling → Current
Mixing → Series

So current-series feedback.

6.16 (c)

A good transconductance amplifier should have very high input resistance and very high output resistance.

6.17 (d)

A good transimpedance amplifier should have low input impedance and low output impedance.

6.18 (800)

$$9\times\frac{R_2}{R_2+1\text{k}\Omega} = 4$$
$$9R_2 = 4R_2 + 4\text{ k}\Omega$$

$$5R_2 = 4K$$

$$R_2 = \frac{4000}{5} = 800\,\Omega$$

6.19 (d)

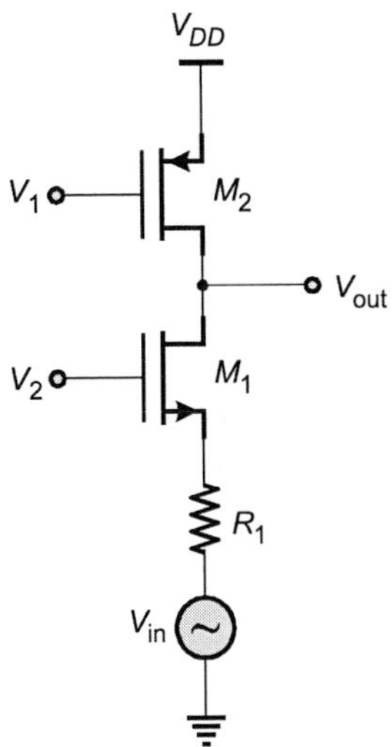

In the given circuit,

Transistor M_1 is common-gate amplifier and transistor M_2 behaves as an active load.

By using properties of common gate (CG) amplifier,

Input impedance (R_i) is low

Output impedance (R_0) is high

So, it is a current amplifier.

Current amplifier is a current controlled current source.

6.20 (*)

For oscillation,

1. Loop gain magnitude ≥ 1.
2. Phase of loop gain = 360°

So, correct answer is option (d).

* In options, closes loop gain is mentioned technically it should be loop gain.

6.21 (a)

Open loop gain, $A_{OL} = 40$

Given circuit is differential amplifier with current mirror active load.

The feedback voltage,

$$V_f = V_{out}$$

∴ Feedback factor,

$$\frac{V_f}{V_{out}} = \beta = 1$$

The closed loop gain,

$$A_{CL} = \frac{A_{OL}}{1+\beta A_{OL}} = \frac{40}{1+(1\times 40)}$$

$$A_{CL} = 0.976$$

■■■

7 Oscillator Circuits

7.1 Value of R in the oscillator shown in the given figure. So chosen that it just oscillates at an angular frequencies of 'ω'. The value of 'ω' and the required value of R will respectively be

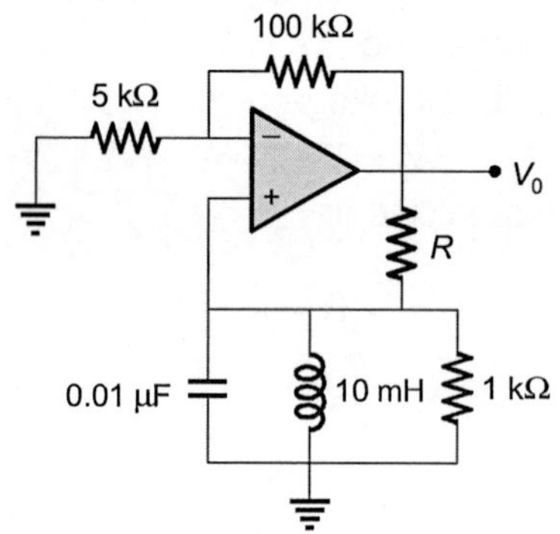

(a) 10^5 rad/sec, 2×10^4 Ω
(b) 2×10^4 rad/sec, 2×10^4 Ω
(c) 2×10^4 rad/sec, 10^5 Ω
(d) 10^5 rad/sec, 10^5 Ω

[1996 : 2 M]

7.2 The configuration of the figure is a

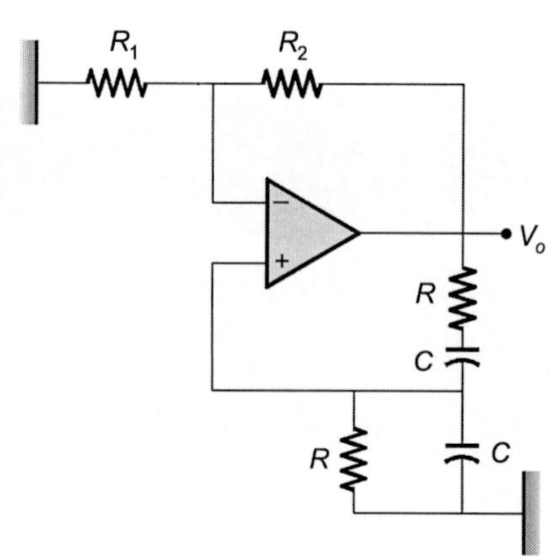

(a) Precision integrator
(b) Hartley oscillator
(c) Butterworth high pass filter
(d) Wien-bridge oscillator

[2000 : 1 M]

7.3 The oscillator circuit shown in the figure is

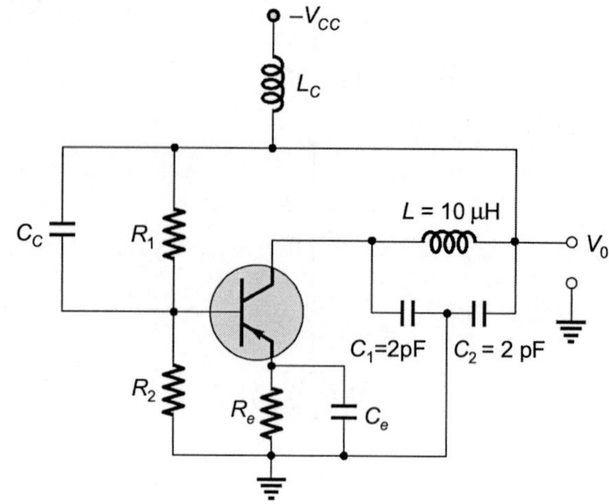

(a) Hartley oscillator with $f_{oscillation}$ = 79.6 MHz
(b) Colpitts oscillator with $f_{oscillation}$ = 50.3 MHz
(c) Hartley oscillator with $f_{oscillation}$ = 159.2 MHz
(d) Colpitts oscillator with $f_{oscillation}$ = 159.2 MHz

[2001 : 2 M]

7.4 The circuit in the figure employs positive feedback and is intended to generate sinusoidal oscillation. If at a frequency f_0, $B(f) = \frac{V_f(f)}{V_0(f)} = \frac{1}{6}\angle 0°$ then to sustain oscillation at this frequency

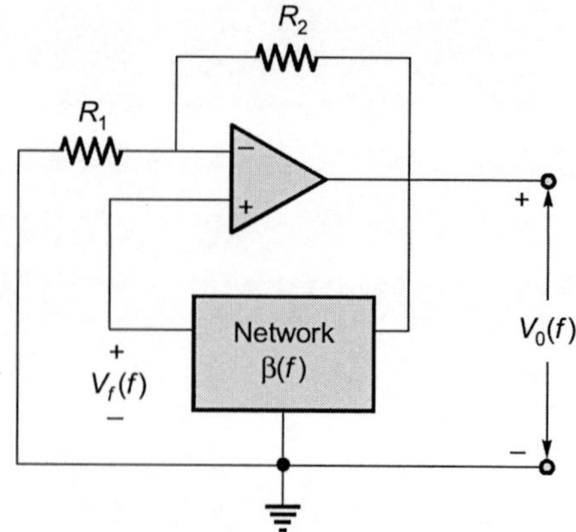

(a) $R_2 = 5R_1$
(b) $R_2 = 6R_1$
(c) $R_2 = \frac{R_1}{6}$
(d) $R_2 = \frac{R_1}{5}$

[2002 : 2 M]

7.5 The oscillator circuit shown in the figure has an ideal inverting amplifier. Its frequency of oscillation (in Hz) is

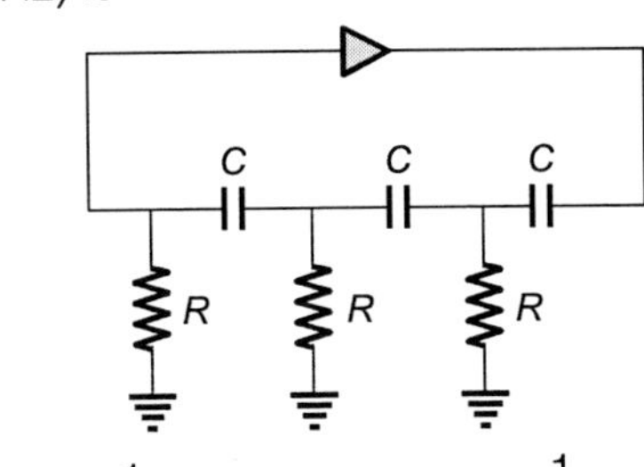

(a) $\frac{1}{(2\pi\sqrt{6}RC)}$ (b) $\frac{1}{(2\pi RC)}$

(c) $\frac{1}{(\sqrt{6}RC)}$ (d) $\frac{1}{\sqrt{6}(2\pi RC)}$

[2003 : 2 M]

7.6 The value of C required for sinusoidal oscillations of frequency 1 kHz in the circuit of the figure is

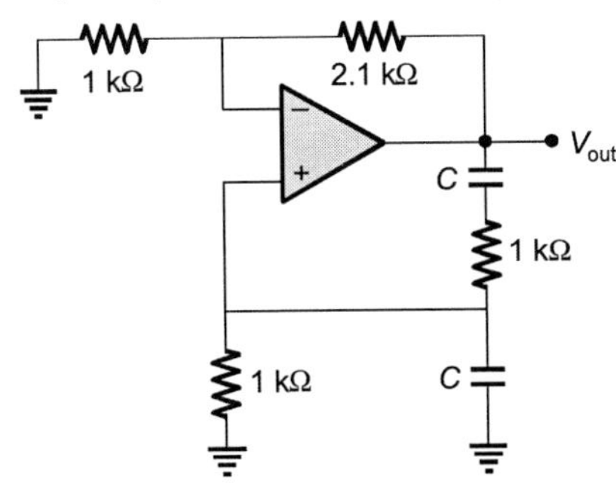

(a) $\frac{1}{2\pi}\mu F$ (b) $2\pi\mu F$

(c) $\frac{1}{2\pi\sqrt{6}}\mu F$ (d) $2\pi\sqrt{6}\mu F$

[2004 : 2 M]

7.7 The circuit shown in the figure has an ideal op-amp. The oscillation frequency and the condition to sustain the oscillations, respectively, are

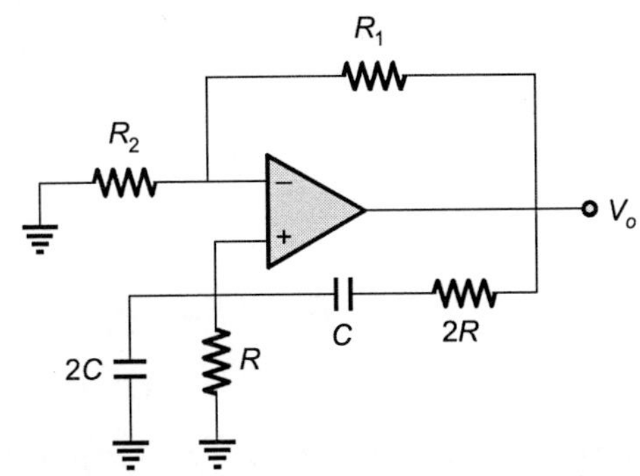

(a) $\frac{1}{CR}$ and $R_1 = R_2$

(b) $\frac{1}{CR}$ and $R_1 = 4R_2$

(c) $\frac{1}{2CR}$ and $R_1 = R_2$

(d) $\frac{1}{2CR}$ and $R_1 = 4R_2$ [2015 : 2 M, Set-1]

7.8 Consider the oscillator circuit shown in the figure. The function of the network (shown in dotted lines) consisting of the 100 kΩ resistor in series with the two diodes connected back-to-back is to:

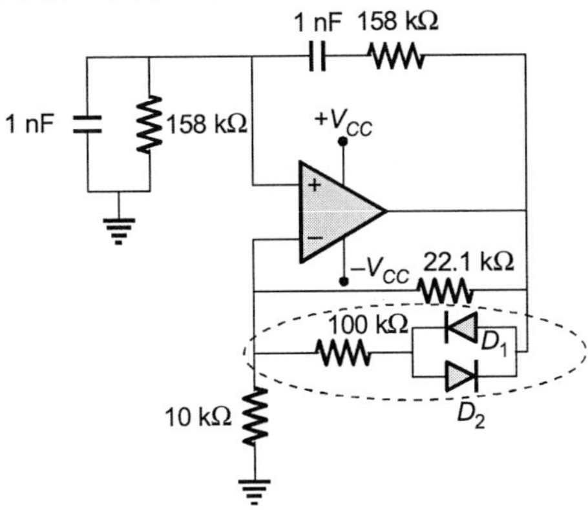

(a) introduce amplitude stabilization by preventing the op amp from saturating and thus producing sinusoidal oscillations of fixed amplitude
(b) introduce amplitude stabilization by forcing the op-amp to swing between positive and negative saturation and thus producing square wave oscillations of fixed amplitude
(c) introduce frequency stabilization by forcing the circuit to oscillate at a single frequency
(d) enable the loop gain to take on a value that produces square wave oscillations

[2016 : 1 M, Set-1]

7.9 In the circuit below, the opamp is ideal.

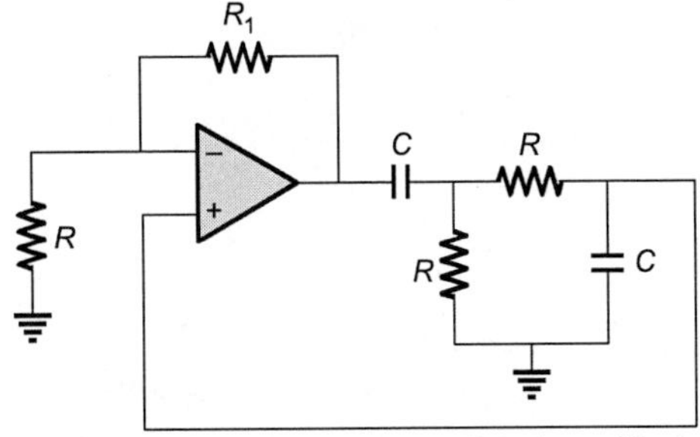

If the circuit is to show sustained oscillations, the respective values of R_1 and the corresponding frequency of oscillation are ______.

(a) $2R$ and $1/(2\pi\sqrt{6}RC)$ (b) $29R$ and $1/(2\pi RC)$

(c) $2R$ and $1/(2\pi RC)$ (d) $29R$ and $1/(2\pi\sqrt{6}RC)$

[2024 : 2 Marks]

■■■

Answers Oscillator Circuits

7.1 (a)	7.2 (d)	7.3 (b)	7.4 (a)	7.5 (a)	7.6 (a)	7.7 (d)
7.8 (a)	7.9 (c)					

Explanations Oscillator Circuits

7.1 (a)

$$\omega = \frac{1}{\sqrt{LC}}$$

$$\omega = \frac{1}{\sqrt{10\times10^{-3}\times0.01\times10^{-6}}}$$

$$\omega = 10^5 \text{ rad/sec}$$

Apply Nodal analysis

$$\frac{V_f - V_o}{R} + \frac{V_f}{1\text{k}} + \frac{V_f}{LS} + V_f CS = 0$$

$$V_f\left[\frac{1}{R}+\frac{1}{1\text{k}}\right] + V_f\left[j\left(\omega C - \frac{1}{\omega L}\right)\right] = \frac{V_o}{R}$$

For oscillation $\omega C - \frac{1}{\omega L} = 0$

$$V_f\left[\frac{1}{R}+\frac{1}{1\text{k}}\right] = \frac{V_o}{R}$$

$$\beta = \frac{V_f}{V_o} = \frac{\frac{1}{R}}{\frac{1}{R}+\frac{1}{1\text{k}}}$$

$$A = 1+\frac{R_f}{R_1} = 1+\frac{100\text{k}}{5\text{k}} = 21$$

$$\beta = \frac{1}{A} = \frac{1}{21} = \frac{\frac{1}{R}}{\frac{1}{R}+\frac{1}{1\text{k}}} = \frac{\frac{1}{R}}{\frac{R+1\text{k}}{R\times1\text{k}}}$$

$$R + 1\,k = 21\times1\,k$$

$$R = 20\text{ k}\Omega = 2\times10^4\ \Omega$$

7.3 (b)

Fig. shown is Colpitts oscillator.

$$f = \frac{1}{2\pi\sqrt{LC_{eq}}}$$

$$C_{eq} = \frac{C_1C_2}{C_1+C_2} = \frac{2\times2}{4} = 1\text{ pF}$$

$$= \frac{1}{2\pi\sqrt{10\times10^{-6}\times10^{-12}}}$$

$$= \frac{1\times10^9}{2\pi\sqrt{10}} = 50.3\text{ MHz}$$

7.4 (a)

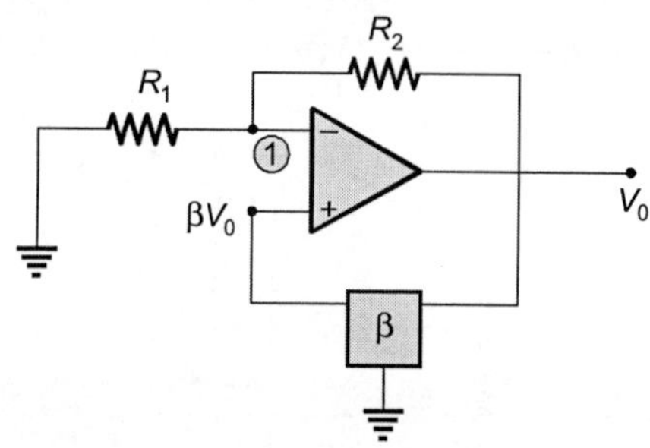

KCL at node 1

$$\frac{\beta V_o - 0}{R_1} + \frac{\beta V_o - V_o}{R_2} = 0$$

$$\beta\left(\frac{1}{R_1}+\frac{1}{R_2}\right) = \frac{1}{R_2}$$

$$\Rightarrow \quad \beta\cdot\frac{R_1+R_2}{R_1} = 1$$

$$\Rightarrow \quad \frac{R_1\left(1+\frac{R_2}{R_1}\right)}{R_1} = \frac{1}{\beta} = 6$$

$$\frac{R_2}{R_1} = 5$$

$$\Rightarrow \quad R_2 = 5R_1$$

7.5 (a)

Frequency of oscillation for RC phase shift oscillator is $\frac{1}{2\pi\sqrt{6}RC}$

7.6 (a)

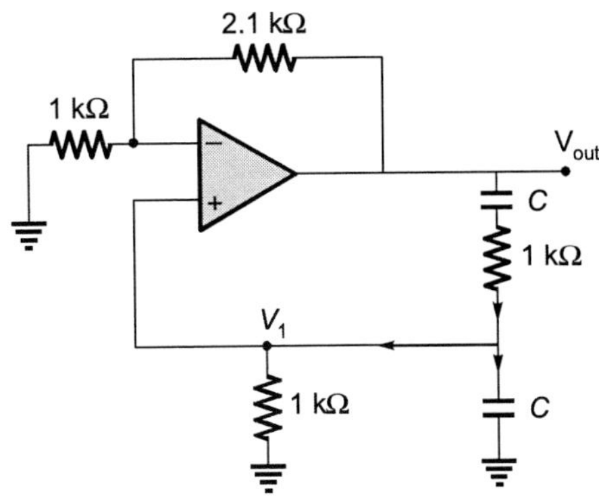

$R = 1$ k

$$\frac{V_o - V_1}{X_C + R} = \frac{V_1}{X_C} + \frac{V_1}{R}$$

$$\frac{V_o - V_1}{X_C + R} = V_1\left(\frac{X_C + R}{X_C \cdot R}\right)$$

$$\frac{V_o}{V_1} = \frac{(X_C + R)^2}{X_C \cdot R} + 1 \qquad \left[X_C = -\frac{j}{\omega C}\right]$$

$$\frac{V_o}{V_1} = \frac{-\frac{1}{\omega^2 C^2} + R^2 - \frac{2jR}{\omega C}}{\frac{R}{j\omega C}} + 1$$

$$\frac{V_o}{V_1} = +\frac{j\omega C}{R}\left(-\frac{1}{\omega^2 C^2} + R^2 - \frac{2jR}{\omega C}\right) + 1$$

For oscillation imaginary part is zero.

i.e. $j\left[\frac{\omega C}{R} \times \frac{-1}{\omega^2 C^2} + \frac{\omega C}{R} \times R^2\right] = 0$

$$\frac{-1}{\omega CR} + \omega CR = 0$$

$$\omega^2 C^2 R^2 - 1 = 0$$

$$\omega^2 = \frac{1}{R^2 C^2}$$

$$C = \frac{1}{\omega R} = \frac{1}{2\pi \times 10^3 \times 10^3}$$

$$C = \frac{1}{2\pi}\mu\text{F}$$

7.7 (d)

Frequency of Wein bridge oscillator is $\omega_o = \frac{1}{RC}$, but in the question time constant is doubled so, frequency becomes half.

$$\omega_o = \frac{1}{2RC}$$

$$Z_1 = 2R + \frac{1}{j\omega C} = 2(R - jR)$$

$$Z_2 = \frac{R \times \frac{1}{2j\omega C}}{R + \frac{1}{2j\omega C}} = \frac{\frac{R^2}{j}}{R - jR}$$

$$\beta = \frac{Z_2}{Z_1 + Z_2} = \frac{1}{5}$$

$$1 + \frac{R_1}{R_2} = 5 \qquad \Rightarrow R_1 = 4R_2$$

7.8 (a)

The given circuit is Wein-bridge oscillator which produced sinusoidal oscillations and the amplitude of output wave is decided by feedback through inverting input terminal of op-amp.

7.9 (c)

The feedback factor of the given oscillator circuit can be obtained as below:

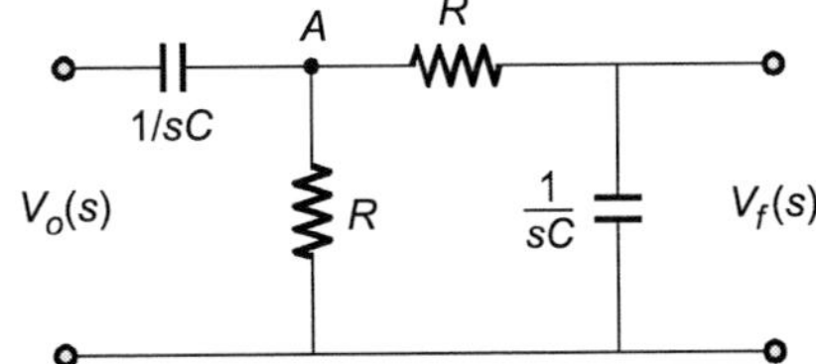

On applying KCL at node A, we get,

$$\frac{V_A(s) - V_o(s)}{\left(\frac{1}{sC}\right)} + \frac{V_A(s)}{R} + \frac{V_A(s)}{R + \frac{1}{sC}} = 0$$

$$\Rightarrow \quad V_A(s)\left[sC + \frac{1}{R} + \frac{sC}{1 + sRC}\right] = sC\, V_o(s)$$

$$\Rightarrow \quad V_A(s) = V_o(s)\left[\frac{sC}{sC + \frac{1}{R} + \frac{sC}{1 + sRC}}\right] \qquad \text{...(i)}$$

Using voltage division rule,

$$V_f(s) = V_A(s) \cdot \frac{\frac{1}{sC}}{R + \frac{1}{sC}} = \frac{V_A(s)}{1 + sRC}$$

Using equation (i),

$$V_f(s) = V_o(s)\left[\frac{sRC}{sRC(1 + sRC) + 1 + sRC + sRC}\right]$$

$$\therefore \qquad \beta = \frac{V_f(s)}{V_o(s)} = \frac{sRC}{1+3sRC+(sRC)^2}$$

$$= \frac{1}{3+\frac{1}{sRC}+sRC}$$

Substituting $s = j\omega$, we get,

$$\beta = \frac{1}{3+j\left(\omega RC - \frac{1}{\omega RC}\right)}$$

For sustained oscillations, imaginary term must be zero at $\omega = \omega_o$. Hence,

$$\omega_o RC - \frac{1}{\omega_o RC} = 0$$

$$\Rightarrow \qquad \omega_o^2 = \frac{1}{R^2C^2} \Rightarrow \omega_o = \frac{1}{RC}$$

$$\Rightarrow \qquad f_o = \frac{1}{2\pi RC}$$

According to Barkhausen criterion,

$$A\beta = 1 \Rightarrow A = \frac{1}{\beta} = 3$$

where A is the gain of non-inverting op-amp amplifier circuit.

We have, $A = 1+\frac{R_1}{R} = 3$

$$\Rightarrow \qquad R_1 = 2R$$

8 Power Amplifiers

8.1 The circuit shown in the figure supplies power to an 8 Ω speaker, LS. The values of I_C and V_{CE} for this circuit will be I_C = ____ and V_{CE} = ______

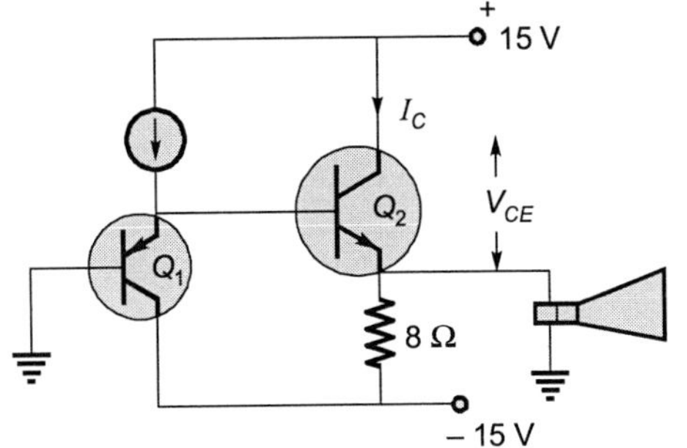

[1995 : 1 M]

8.2 A power Amplifier delivers 50 W output at 50% efficiency. The ambient temperature is 25° C. If the maximum allowable junction temperature is 150° C, then the maximum thermal resistance θg_c that can be tolerated is ______

[1995 : 1 M]

8.3 Crossover distortion behaviour is characteristic of

(a) Class *A* output stage

(b) Class *B* output stage

(c) Class *AB* output stage

(d) Common-base output stage [1999 : 1 M]

■■■

Answers Power Amplifiers

8.1 (15) 8.2 (2.5) 8.3 (b)

Explanations Power Amplifiers

8.1 (15)

$$I_C = \frac{15}{8} \text{ Amp.}$$

$$V_{CE} = 15 \text{ V}$$

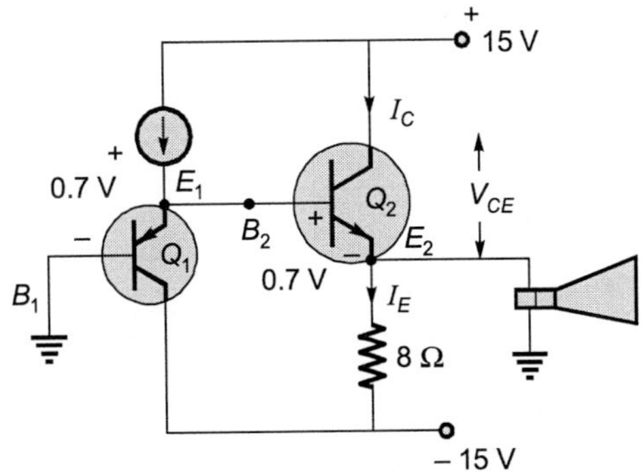

$$V_{E_1B_1} = 0.7 \text{ V}$$

$$V_{E_1} - V_{B_1} = 0.7 \text{ V}$$

$$V_{E_1} - 0 = 0.7 \text{ V}$$

$$V_{E_1} = 0.7 \text{ V}$$

$$V_{E_1} = V_{B_2} = 0.7 \text{ V}$$

$$V_{B_2E_2} = 0.7 \text{ V}$$

$$V_{B_2} - V_{E_2} = 0.7 \text{ V}$$

$$0.7 \text{ V} - V_{E_2} = 0.7 \text{ V}$$

$$V_{E_2} = 0 \text{ V}$$

Apply KVL at emitter terminal

$$0 = I_E R_E - 15$$

$$I_E = \frac{15}{R_e} = \frac{15}{8}$$

$$V_{CE} = V_C - V_E = 15 - 0 = 15 \text{ V}$$

8.2 (2.5)

Power dissipated $(P_D) = 50 \times 50\%$

$$\frac{P_o}{P_{in}} = \frac{1}{2}$$

$$P_{in} = 2P_o = 2 \times 50 = 100 \text{ W}$$

$$P_D = P_{in} - P_o = \frac{T_j - T_A}{\theta}$$

$$\theta = \frac{T_j - T_A}{P_D} = \frac{150 - 25}{50} = \frac{125}{50}$$

$$\theta = 2.5°\text{C/W}$$

■■■

9 Multivibrators and 555 Timer

9.1 Consider the following two statements:
Statement 1: Astable multivibrator can be used for generating square wave.
Statement 2: Bistable multivibrator can be used for storing binary information.
(a) Only statement 1 is correct
(b) Only statement 2 is correct
(c) Both the statements 1 and 2 are correct
(d) Both the statements 1 and 2 are incorrect

[2001 : 1 M]

9.2 An ideal sawtooth voltage waveform of frequency 500 Hz and amplitude 3 V is generated by charging a capacitor of 2 μF in every cycle. The charging requires
(a) constant voltage source of 3 V for 1 ms
(b) constant voltage source of 3 V for 2 ms
(c) constant current source of mA for 1 ms
(d) constant current source of 3 mA for 2 ms

[2003 : 2 M]

9.3 Given the ideal operational amplifier circuit shown in the figure indicate the correct transfer characteristics assuming ideal diodes with zero cut-in voltage.

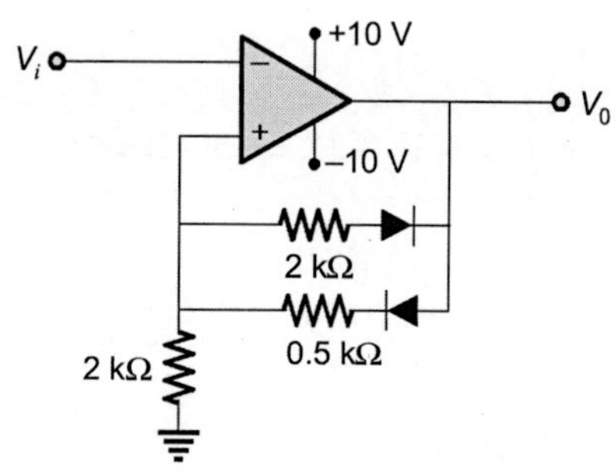

(a)
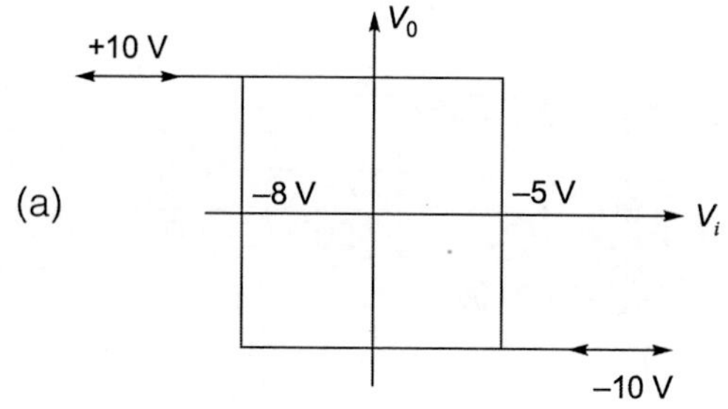

(b)
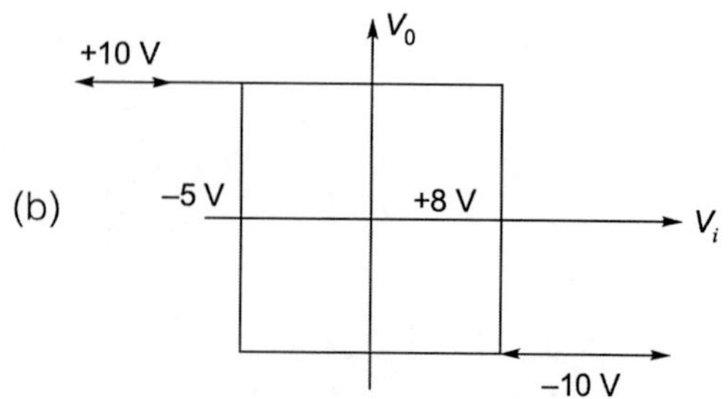

(c)
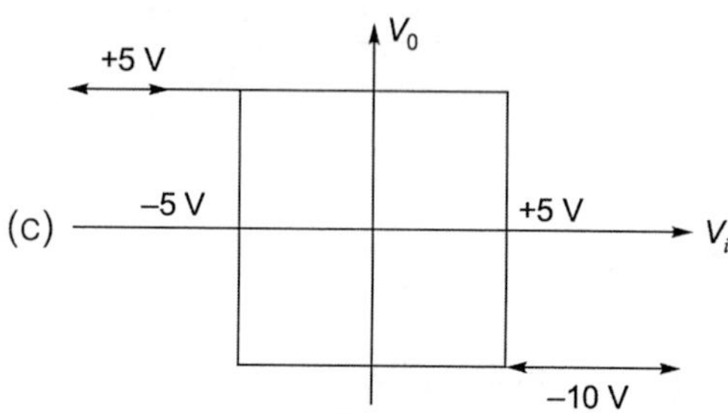

(d)
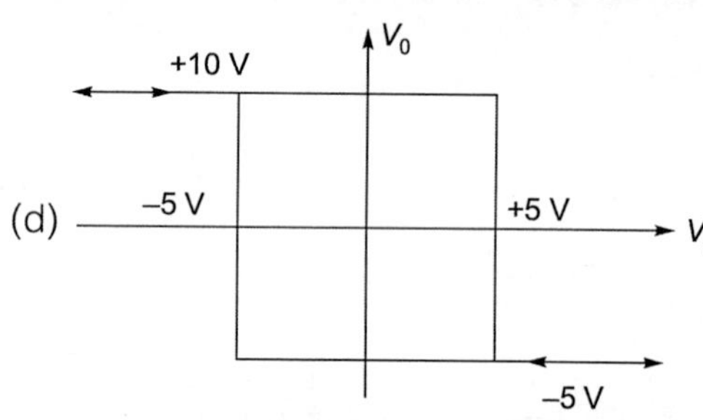

[2005 : 2 M]

9.4 Consider the Schmidt trigger circuit shown below:

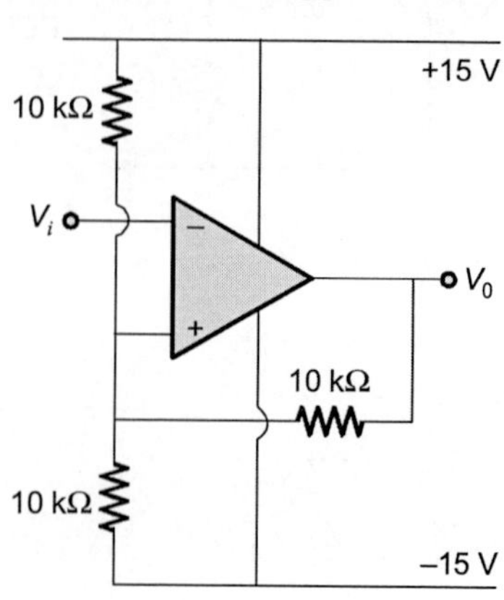

A triangular wave which goes from –12 V to 12 V is applied to the inverting input of the OP-Amp. Assume that the output of the OP-Amp swings from +15 V to –15 V. The voltage at the non-inverting input switches between

(a) –12 V and +12 V (b) –7.5 and +7.5 V
(c) –5 V and +5 V (d) 0 V and 5 V

[2008 : 2 M]

9.5 An astable multi-vibrator circuit using IC 555 timer is shown below. Assume that the circuit is oscillating steadily.

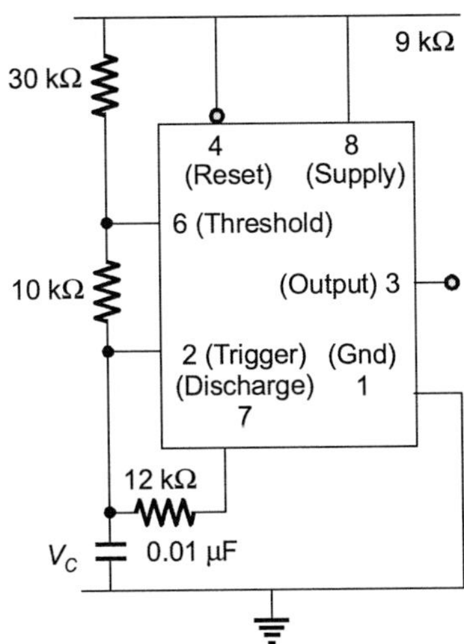

The voltage V_C across the capacitor varies between

(a) 3 V to 5 V (b) 3 V to 6 V
(c) 3.6 V to 6 V (d) 3.6 V to 5 V

[2008 : 2 M]

9.6 In the following astable multivibrator circuit, which properties of $v_0(t)$ depend on R_2?

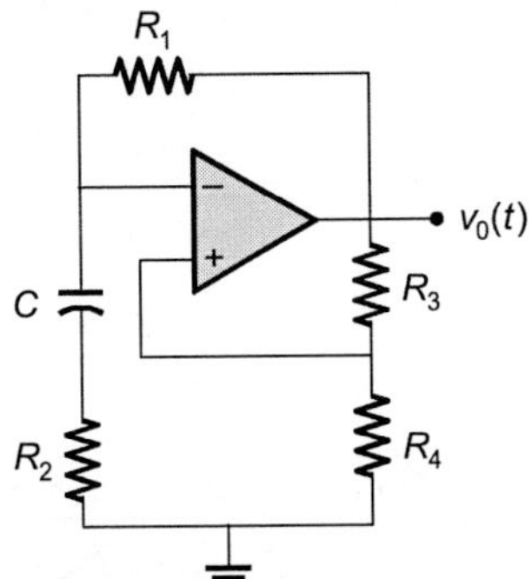

(a) Only the frequency
(b) Only the amplitude
(c) Both the amplitude and the frequency
(d) Neither the amplitude nor the frequency

[2009 : 2 M]

9.7 In the astable multivibrator circuit shown in the figure, the frequency of oscillation (in kHz) at the output pin 3 is ______.

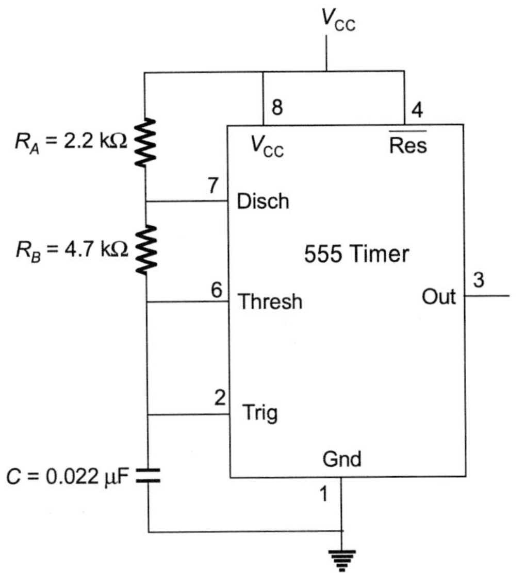

[2016 : 1 M, Set-3]

9.8 For the following circuit with an ideal OPAMP, the difference between the maximum and the minimum values of the capacitor voltage (V_c) is ________.

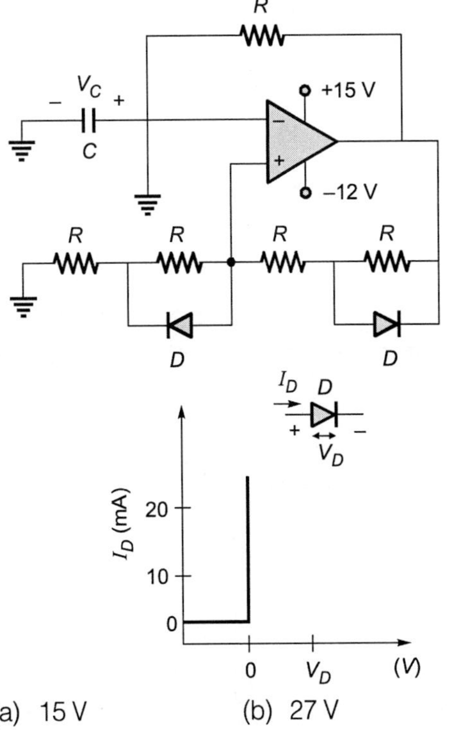

(a) 15 V (b) 27 V
(c) 13 V (d) 14 V

[2022 : 2 M]

■■■

Answers Multivibrators and 555 Timer

9.1 (c) 9.2 (d) 9.3 (b) 9.4 (c) 9.5 (a) 9.6 (a) 9.7 (5.6818)
9.8 (c)

Explanations Multivibrators and 555 Timer

9.2 (d)

$$T = \frac{1}{f} = \frac{1}{500\text{ Hz}} = 2\text{ ms}$$

$$I = \frac{CdV}{dt} = 2\times10^{-6}\times\frac{3}{2\times10^{-3}}$$

$$= 3\times10^{-3}$$

$$I = 3\text{ mA}$$

9.3 (b)

The given circuit is of a Schmitt trigger. The upper and lower transition voltages can be obtained as

$$V_{ut} = \beta_u V_{sat}$$

When lower diode is ON

$$\beta_u = \frac{2\text{ k}\Omega}{2.5\text{ k}\Omega}$$

$$\therefore \quad V_{ut} = \frac{2}{2.5}\times10 = 8\text{ V}$$

$$V_{lt} = -\beta_l V_{sat}$$

(when upper diode is ON, $\beta_l = \frac{2\text{ k}\Omega}{4\text{ k}\Omega} = \frac{1}{2}$)

$$= -\frac{1}{2}\times10 = -5\text{ V}$$

V_{ut} and V_{lt} are upper and lower transition voltage. The transfer characteristics of the circuit is given as below:

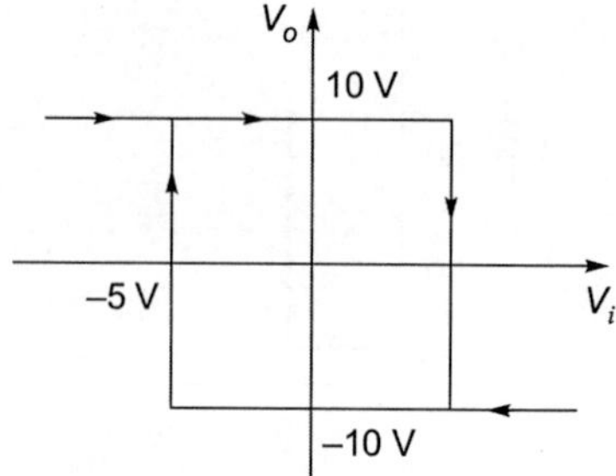

9.4 (c)

Let the voltage at the non-inverting input be V_1. Applying KCL at non-inverting input end,

$$\frac{15-V_1}{10} + \frac{V_0-V_1}{10} = \frac{V_1-(-15)}{10}$$

$$15 - V_1 + V_0 - V_1 = V_1 + 15$$

$$\Rightarrow \quad V_1 = \frac{V_0}{3}$$

Since, V_0 swings from − 15 V to + 15 V,

Therefore, V_1 switches between −5 V & +5 V.

9.5 (a)

V_C varies between 3 V to 5 V.

9.7 (5.6818)

$$f = \frac{1}{0.69(R_A + 2R_B)C}$$

$$= \frac{1}{0.69(2.2\times10^3 + 2\times4.7\times10^3)\times0.022\times10^{-6}}$$

$$= 5.6818\text{ kHz}$$

9.8 (c)

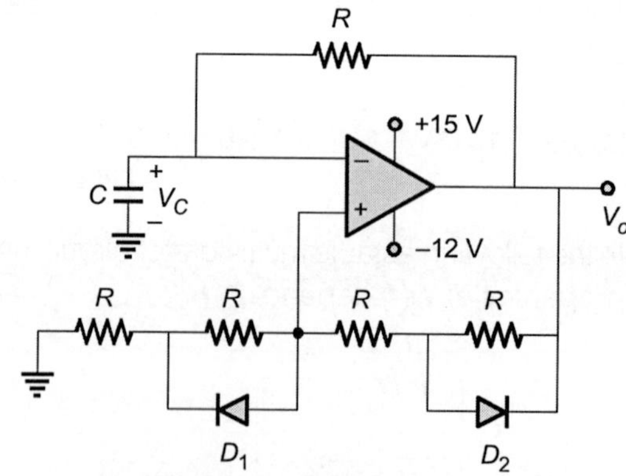

The given circuit is an astable multivibrator.

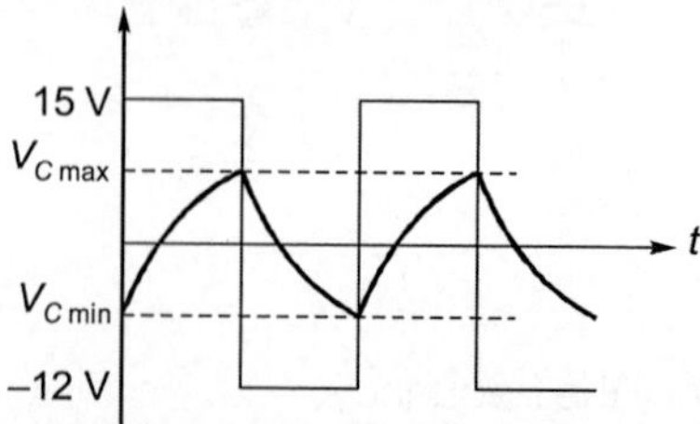

The output is a square wave, with the capacitor charging when V_o = 15 V upto $V_{c\,max}$ and discharging when V_o = − 12 V upto $V_{c\,min}$.

(i) When $V_o = 15$ V,
D_1 is forward-biased (ON) and D_2 is reverse-biased (OFF). The equivalent circuit can be drawn as below:

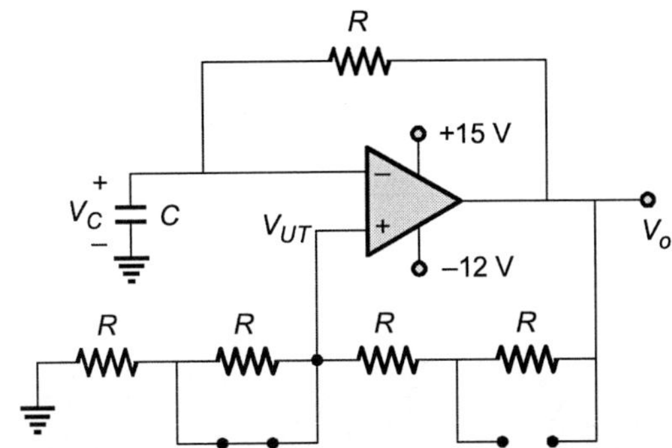

The capacitor charges through resistance R upto $V_{c\,max}$ i.e. the upper threshold voltage, V_{UT} given as

$$V_{UT} = V_{c\,max} = V_o\left(\frac{R}{R+2R}\right)$$

$$\therefore \quad V_{c\,max} = 15\times\frac{1}{3} = 5\text{ V}$$

(ii) When $V_o = -12$ V,
D_1 is reverse-biased (OFF) and D_2 is forward-biased (ON). The equivalent circuit can be drawn as below,

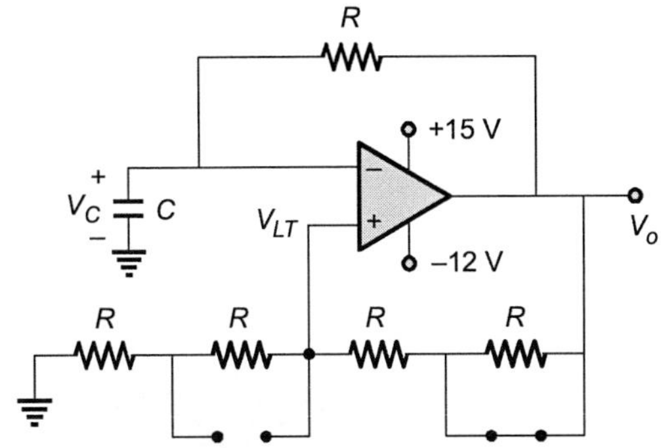

The capacitor discharges through resistance R upto $V_{c\,min}$ i.e. the lower threshold voltage, V_{LT} given as :

$$V_{LT} = V_{c\,min} = V_o\left(\frac{2R}{R+2R}\right)$$

$$V_{c\,min} = -\frac{12\times 2}{3} = -8\text{ V}$$

Hence,

$V_{c\,max} - V_{c\,min} = 5 - (-8) = 13$ V

■■■

Digital Circuits and Computer Organization

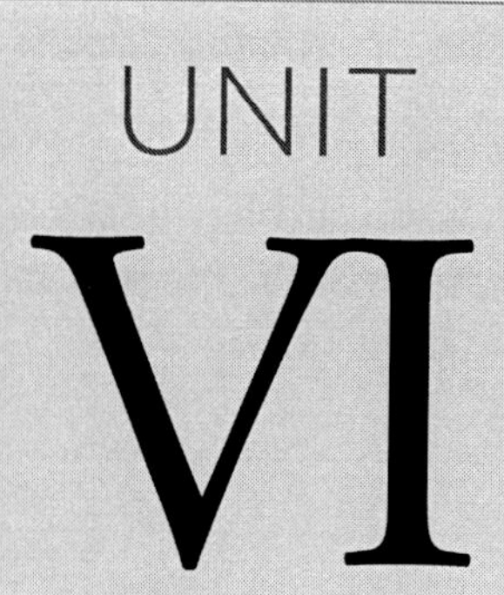

CONTENTS

Digital Circuits and Computer Organization

Syllabus

Number representations : Binary, integer and floating-point- numbers. Combinatorial circuits: Boolean algebra, minimization of functions using Boolean identities and Karnaugh map, logic gates and their static CMOS implementations, arithmetic circuits, code converters, multiplexers, decoders.

Sequential circuits : Latches and flip-flops, counters, shift-registers, finite state machines, propagation delay, setup and hold time, critical path delay.

Data converters : Sample and hold circuits, ADCs and DACs.

Semiconductor memories : ROM, SRAM, DRAM.

Computer organization : Machine instructions and addressing modes, ALU, data-path and control unit, instruction pipelining.

Analysis of Previous GATE Papers

Exam Year	1 Mark Ques.	2 Marks Ques.	Total Marks
1995	6	–	**6**
1996	1	–	**1**
1997	9	–	**9**
1998	9	1	**11**
1999	4	4	**12**
2000	3	4	**11**
2001	2	4	**10**
2002	3	3	**9**
2003	5	7	**19**
2004	6	5	**16**
2005	2	6	**14**
2006	–	5	**10**
2007	2	7	**16**
2008	–	8	**16**
2009	1	6	**13**
2010	2	2	**6**
2011	3	2	**7**
2012	4	1	**6**
2013	1	1	**3**

Exam Year	1 Mark Ques.	2 Mark Ques.	Total Marks
2014 Set-1	3	3	**9**
2014 Set-2	3	2	**7**
2014 Set-3	2	3	**8**
2014 Set-4	2	2	**6**
2015 Set-1	1	3	**7**
2015 Set-2	2	3	**8**
2015 Set-3	2	3	**8**
2016 Set-1	1	2	**5**
2016 Set-2	3	2	**7**
2016 Set-3	2	3	**8**
2017 Set-1	2	3	**8**
2017 Set-2	3	3	**9**
2018	3	4	**11**
2019	4	2	**8**
2020	2	3	**8**
2021	3	3	**9**
2022	3	4	**11**
2023	3	3	**9**
2024	3	3	**9**

1 Number Systems

1.1 A signed integer has been stored in a byte using the 2's complement format. We wish to store the same integer in a 16 bit word. We should

(a) copy the original byte to the less significant byte of the word and fill the more significant byte with zeros.

(b) copy the original byte to the more significant byte of the word and fill the less significant byte with zeros.

(c) copy the original byte to the less significant byte of the word and make each bit of the more significant byte equal to the most significant bit of the original byte.

(d) copy the original byte to the less significant bytes well as the more significant byte of the word.

[1997 : 1 M]

1.2 An equivalent 2's complement representation of the 2's complement number 1101 is

(a) 110100 (b) 001101
(c) 110111 (d) 111101 **[1998 : 1 M]**

1.3 The 2's complement representation of −17 is

(a) 101110 (b) 101111
(c) 111110 (d) 110001 **[2001 : 1 M]**

1.4 4-bit 2's complement representation of a decimal number is 1000. The number is

(a) +8 (b) 0
(c) −7 (d) −8 **[2002 : 1 M]**

1.5 The range of signed decimal numbers that can be represented by 6-bit 1's complement number is

(a) −31 to +31 (b) −63 to + 63
(c) −64 to + 63 (d) −32 to + 31

[2004 : 1 M]

1.6 11001, 1001 and 111001 correspond to the 2's complement representation of which one of the following sets of number?

(a) 25, 9 and 57 respectively
(b) −6, −6 and −6 respectively
(c) −7, −7 and −7 respectively
(d) −25, −9 and −57 respectively

[2004 : 2 M]

1.7 Decimal 43 in Hexadecimal and BCD number system is respectively

(a) B2, 0100 0011 (b) 2B, 0100 0011
(c) 2B, 0011 0100 (d) B2, 0100 0100

[2005 : 1 M]

1.8 A new Binary Coded Pentary (BCP) number system is proposed in which every digit of a base-5 number is represented by its corresponding 3-bit binary code. For example, the base-5 number 24 will be represented by its BCP code 010100. In this numbering system, the BCP code 100010011001 corresponds to the following number in base-5 system

(a) 423 (b) 1324
(c) 2201 (d) 4231

[2006 : 2 M]

1.9 X = 01110 and Y = 11001 are two 5-bit binary numbers represented in two's complement format. The sum of X and Y represented in two's complement format using 6 bits is

(a) 100111 (b) 001000
(c) 000111 (d) 101001

[2007 : 1 M]

1.10 The two numbers represented in signed 2's complement form are

P = 11101101 and Q = 11100110. If Q is subtracted from P, the value obtained in signed 2's complement form is

(a) 100000111 (b) 00000111
(c) 11111001 (d) 111111001

[2008 : 2 M]

1.11 The number of bytes required to represent the decimal number 1856357 in packed BCD (Binary Coded Decimal) form is ________.

[2014 : 1 M, Set-2]

1.12 P, Q, and R are the decimal integers corresponding to the 4-bit binary number 1100 considered in signed magnitude, 1's complement, and 2's complement representations, respectively. The 6-bit 2's complement representation of $(P + Q + R)$ is

(a) 111101 (b) 110101
(c) 110010 (d) 111001

[2020 : 2 M]

1.13 If $(1235)_x = (3033)_y$, where x and y indicate the bases of the corresponding numbers, then
(a) $x = 7$ and $y = 5$ (b) $x = 8$ and $y = 6$
(c) $x = 6$ and $y = 4$ (d) $x = 9$ and $y = 7$

[2021 : 1 M]

1.14 In a number system of base r, the equation $x^2 - 12x + 37 = 0$ has $x = 8$ as one of its solutions. The value of r is ________.

[2024 : 1 Mark]

■■■

Answers Number Systems

1.1 (c) 1.2 (d) 1.3 (b) 1.4 (d) 1.5 (a) 1.6 (c) 1.7 (b) 1.8 (d) 1.9 (c)
1.10 (b) 1.11 (4) 1.12 (b) 1.13 (b) 1.14 (11)

Explanations Number Systems

1.1 (c)

The MSB of the integer in 8 bit format should be repeated to expand the representation of 2's complement form to 16 bit.

1.2 (d)

According to sign bit extension the sign bit can be extended towards left.

1.3 (b)

$$17 = 010001$$
$$-17 = 101111 \text{ (2's complement)}$$

1.4 (d)

1000
MSB is 1 so, –ve number
Take 2's complement for magnitude.

$$\begin{array}{r} 0111 \\ +\quad 1 \\ \hline 1000 \end{array} = 8$$

1.5 (a)

$$\text{Range} = -(2^{n-1} - 1) \text{ to } +(2^{n-1} - 1)$$
$$= -(2^{6-1} - 1) \text{ to } +(2^{6-1} - 1) = -31 \text{ to } +31$$

1.6 (c)

$11001 \rightarrow 00111(+7)$

$1001 \rightarrow 0111(+7)$

$111001 \rightarrow 000111(+7)$

∴ Numbers given in question in 2's complement correspond to –7.

1.7 (b)

	16	43	
$(43)_{10} \rightarrow$	16	2	B
		0	2

∴ $(2B)_H$ $(43)_{10} = (01000011)_{BCD}$

1.8 (d)

$100010011001 \rightarrow 4231$

1.9 (c)

$$\begin{array}{rl} X = & 01110 \\ Y = & 11001 \\ X + Y = 1 & \underline{00111} \end{array}$$

Carry is discarded in the addition of numbers represented in 2's complement form. $X + Y$ in 6 bits is 000111.

1.10 (b)

∵ Signed 2's complement of
$P = 11101101$
∴ No. $P = 00010011$
∵ Signed 2's complement of
$Q = 11100110$

$$P - Q = P + (\text{2's complement of } Q)$$
$$= 00010011$$
$$\underline{+\ 11100110}$$
$$11111001$$

2's complement of
$(P - Q) = 00000111$

1.11 (4)

To represent decimal number into BCD number each decimal number is represented in 4-bits while converting in BCD numbers, as

1 → 0001	6 → 0110
8 → 1000	3 → 0011
5 → 0101	5 → 0101
	7 → 0111

(18 5 6 3 5 7) $\underbrace{0000\ \ 0001}_{\text{1byte}}\ \underbrace{1000\ \ 0101}_{\text{1byte}}$

$\underbrace{0110\ \ 0011}_{\text{1byte}}\ \underbrace{0101\ \ 0111}_{\text{1byte}}$

So total 4 bytes are required.

1.12 (b)

Given, binary number 1100

1's complement of 1100 = –3

Sign magnitude of 1100 = –4

2's complement of 1100 = –4

$\therefore \quad P + Q + R = -4 - 3 - 4 = -11$

The 6 digit 2's complement of (–11) = 110101

1.13 (b)

$$x^3 + 2x^2 + 3x + 5 = 3y^3 + 3y + 3$$

Option (b) will satisfy the equation.

1.14 (11)

Consider the base of the system as 'r'. The equation in the given number system is

$x^2 - 12x + 37 = 0$

Rewriting the equation in decimal number system [Assuming x as single digit]

$x^2 - [r + 2]x + [3r + 7] = 0$

Since $x = 8$ is the root of the equation, substituting $x = 8$, we get

$\Rightarrow 64 - 8(r + 2) + (3r + 7) = 0$

$\Rightarrow \quad 5r = 55$

$\Rightarrow \quad r = 11$

■■■

2 Boolean Algebra

2.1 Two 2's complement numbers having sign bits x and y are added and the sign bit of the result is z. Then, the occurrence of overflow is indicated by the Boolean function.

(a) xyz (b) $\bar{x}\,\bar{y}\,\bar{z}$

(c) $\bar{x}\,\bar{y}\,z + x\,y\,\bar{z}$ (d) $xy + yz + zx$

[1998 : 1 M]

2.2 The *K*-map for a Boolean function is shown in figure. The number of essential prime implicants for this function is

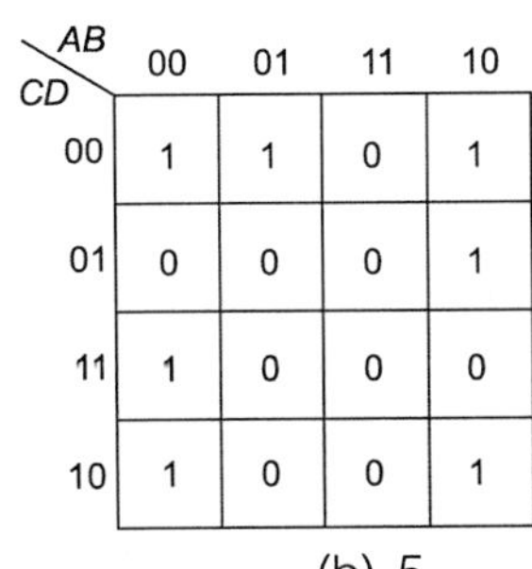

CD \ AB	00	01	11	10
00	1	1	0	1
01	0	0	0	1
11	1	0	0	0
10	1	0	0	1

(a) 4 (b) 5

(c) 6 (d) 8 [1998 : 1 M]

2.3 The logical expression $y = A + \bar{A}B$ is equivalent to

(a) $y = AB$ (b) $y = \bar{A}B$

(c) $y = \bar{A} + B$ (d) $y = A + B$

[1999 : 1 M]

2.4 The minimized form of the logical expression $(\bar{A}\bar{B}\bar{C} + \bar{A}B\bar{C} + \bar{A}BC + AB\bar{C})$ is

(a) $\bar{A}\bar{C} + B\bar{C} + \bar{A}B$ (b) $A\bar{C} + \bar{B}C + \bar{A}B$

(c) $\bar{A}C + \bar{B}C + \bar{A}B$ (d) $A\bar{C} + \bar{B}C + A\bar{B}$

[1999 : 2 M]

2.5 The number of distinct Boolean expressions of 4 variables is

(a) 16 (b) 256

(c) 1024 (d) 65536

[2003 : 1 M]

2.6 If the functions W, X, Y and Z are as follows

$W = R + \bar{P}Q + \bar{R}S$

$X = PQ\bar{R}\bar{S} + \bar{P}\bar{Q}\bar{R}\bar{S} + P\bar{Q}\bar{R}\bar{S}$

$Y = RS + \overline{PR + \bar{P}\bar{Q} + P\bar{Q}}$

$Z = R + S + \overline{PQ + \bar{P}\bar{Q}\bar{R} + P\bar{Q}\bar{S}}$

Then

(a) $W = Z, X = \bar{Z}$ (b) $W = Z, X = y$

(c) $W = Y$ (d) $W = Y = \bar{Z}$

[2003 : 2 M]

2.7 The Boolean expression $AC + B\bar{C}$ is equivalent to

(a) $\bar{A}C + B\bar{C} + AC$

(b) $\bar{B}C + AC + B\bar{C} + \bar{A}C\bar{B}$

(c) $AC + B\bar{C} + \bar{B}C + ABC$

(d) $ABC + \bar{A}B\bar{C} + AB\bar{C} + A\bar{B}C$ [2004 : 2 M]

2.8 The Boolean expression for the truth table shown is

A	B	C	f
0	0	0	0
0	0	1	0
0	1	0	0
0	1	1	1
1	0	0	0
1	0	1	0
1	1	0	1
1	1	1	0

(a) $B(A + C)(\bar{A} + \bar{C})$ (b) $B(A + \bar{C})(\bar{A} + C)$

(c) $\bar{B}(A + \bar{C})(\bar{A} + C)$ (d) $\bar{B}(A + C)(\bar{A} + \bar{C})$

[2005 : 2 M]

2.9 The Boolean function $Y = AB + CD$ is to be realized using only 2-input NAND gates. The minimum number of gates required is

(a) 2 (b) 3

(c) 4 (d) 5 [2007 : 1 M]

2.10 The Boolean expression

$Y = \bar{A}\bar{B}\bar{C}D + \bar{A}BC\bar{D} + A\bar{B}\bar{C}D + AB\bar{C}\bar{D}$

can be minimized to

(a) $Y = \bar{A}\bar{B}\bar{C}D + \bar{A}B\bar{C} + A\bar{C}D$

(b) $Y = \bar{A}\bar{B}\bar{C}D + BC\bar{D} + A\bar{B}\bar{C}D$

(c) $Y = \bar{A}BC\bar{D} + \bar{B}\bar{C}D + A\bar{B}\bar{C}D$

(d) $Y = \bar{A}BC\bar{D} + \bar{B}\bar{C}D + AB\bar{C}\bar{D}$

[2007 : 2 M]

2.11 If X = 1 in the logic equation

$[X + Z\{\bar{Y} + (\bar{Z} + X\bar{Y})\}]\{\bar{X} + \bar{Z}(X + Y)\} = 1$, then

(a) $Y = Z$ (b) $Y = \bar{Z}$

(c) $Z = 1$ (d) $Z = 0$ [2009 : 2 M]

2.12 In the sum of products function $f(X, Y, Z) = \sum(2,3,4,5)$, the prime implicants are

(a) $\bar{X}Y, X\bar{Y}$

(b) $\bar{X}Y, X\bar{Y}\bar{Z}, X\bar{Y}Z$

(c) $\bar{X}Y\bar{Z}, \bar{X}YZ, X\bar{Y}$

(d) $\bar{X}Y\bar{Z}, \bar{X}YZ, X\bar{Y}\bar{Z}, X\bar{Y}Z$ [2012 : 1 M]

2.13 The Boolean expression

$(X+Y)(X+\bar{Y})+\overline{(\bar{X}\bar{Y})+\bar{X}}$ simplifies to

(a) X (b) Y

(c) XY (d) $X+Y$

[2014 : 1 M, Set-1]

2.14 Consider the Boolean function,

$F(w, x, y, z) = wy + xy + \bar{w}xyz + \bar{w}\bar{x}y + xz + \bar{x}\bar{y}\bar{z}.$

Which one of the following is the complete set of essential prime implicants?

(a) $w, y, xz, \bar{x}\bar{z}$ (b) w, y, xz

(c) $y, \bar{x}\bar{y}\bar{z}$ (d) $y, xz, \bar{x}\bar{z}$

[2014 : 2 M, Set-1]

2.15 For an n-variable Boolean function, the maximum number of prime implicants is

(a) $2(n-1)$ (b) $n/2$

(c) 2^n (d) $2^{(n-1)}$

[2014 : 1 M, Set-2]

2.16 The Boolean expression

$F(X, Y, Z) = \bar{X}Y\bar{Z} + X\overline{YZ} + XY\bar{Z} + XYZ$

converted into canonical product of sum (POS) form is

(a) $(X+Y+Z)(X+Y+\bar{Z})(X+\bar{Y}+\bar{Z})(\bar{X}+Y+\bar{Z})$

(b) $(X+\bar{Y}+Z)(\bar{X}+Y+\bar{Z})(\bar{X}+\bar{Y}+Z)(\bar{X}+\bar{Y}+\bar{Z})$

(c) $(X+Y+Z)(\bar{X}+Y+\bar{Z})(X+\bar{Y}+Z)(\bar{X}+\bar{Y}+\bar{Z})$

(d) $(X+\bar{Y}+\bar{Z})(\bar{X}+Y+Z)(\bar{X}+\bar{Y}+Z)(X+Y+Z)$

[2015 : 2 M, Set-1]

2.17 A function of Boolean variables, X, Y and Z is expressed in terms of the min-terms as

$F(X, Y, Z) = \Sigma(1, 2, 5, 6, 7)$

Which one of the product of sums given below is equal to the function $F(X, Y, Z)$?

(a) $(\bar{X}+\bar{Y}+\bar{Z})\cdot(\bar{X}+Y+Z)\cdot(X+\bar{Y}+\bar{Z})$

(b) $(X+Y+Z)\cdot(X+\bar{Y}+\bar{Z})\cdot(\bar{X}+Y+Z)$

(c) $(\bar{X}+\bar{Y}+Z)\cdot(\bar{X}+Y+\bar{Z})\cdot(X+\bar{Y}+Z)$
$\cdot(X+Y+\bar{Z})\cdot(X+Y+Z)$

(d) $(X+Y+\bar{Z})\cdot(\bar{X}+Y+Z)\cdot(\bar{X}+Y+\bar{Z})$
$(\bar{X}+\bar{Y}+Z)\cdot(\bar{X}+\bar{Y}+\bar{Z})$ [2015 : 2 M, Set-2]

2.18 Following is the K-map of a Boolean function of five variables P, Q, R, S and X. The minimum sum-of-product (SOP) expression for the function is

RS \ PQ	00	01	11	10
00	0	0	0	0
01	1	0	0	1
11	1	0	0	1
10	0	0	0	0

$X = 0$

RS \ PQ	00	01	11	10
00	0	1	1	0
01	0	0	0	0
11	0	0	0	0
10	0	1	1	0

$X = 1$

(a) $\bar{P}\bar{Q}S\bar{X}+P\bar{Q}S\bar{X}+Q\bar{R}\bar{S}X+QR\bar{S}X$

(b) $\bar{Q}S\bar{X}+Q\bar{S}X$

(c) $\bar{Q}SX+Q\bar{S}\bar{X}$

(d) $\bar{Q}S+Q\bar{S}$ [2016 : 2 M, Set-3]

2.19 Which one of the following gives the simplified sum of products expression for the Boolean function $F = m_0 + m_2 + m_3 + m_5$, where m_0, m_2, m_3 and m_5 are minterms corresponding to the inputs A, B and C with A as the MSB and C as the LSB?

(a) $\bar{A}B+\bar{A}\bar{B}\bar{C}+A\bar{B}C$ (b) $\bar{A}\bar{C}+\bar{A}B+A\bar{B}C$

(c) $\bar{A}\bar{C}+A\bar{B}+A\bar{B}C$ (d) $\bar{A}BC+\bar{A}\bar{C}+A\bar{B}C$

[2017 : 2 M, Set-1]

2.20 A function $F(A, B, C)$ defined by three Boolean variables A, B and C when expressed as sum of products is given by

$$F = \bar{A}\cdot\bar{B}\cdot\bar{C}+\bar{A}\cdot B\cdot\bar{C}+A\cdot\bar{B}\cdot\bar{C}$$

where, $\bar{A}, \bar{B}$, and $\bar{C}$ are the complements of the respective variables. The product of sums (POS) form of the function F is

(a) $F = (A+B+C)\cdot(A+\bar{B}+C)\cdot(\bar{A}+B+C)$

(b) $F = (\bar{A}+\bar{B}+\bar{C})\cdot(\bar{A}+B+\bar{C})\cdot(A+\bar{B}+\bar{C})$

(c) $F = (A+B+\bar{C})\cdot(A+\bar{B}+\bar{C})\cdot(\bar{A}+B+\bar{C})$
$\cdot(\bar{A}+\bar{B}+C)\cdot(\bar{A}+\bar{B}+\bar{C})$

(d) $F = (\bar{A}+\bar{B}+C)\cdot(\bar{A}+B+C)\cdot(A+\bar{B}+C)$
$\cdot(A+B+\bar{C})\cdot(A+B+C)$

[2018 : 1 M]

2.21 Select the Boolean function(s) equivalent to $x + yz$, where x, y, and z are Boolean variables, and + denotes logical OR operation.

(a) $x + z + xy$ (b) $(x + y)(x + z)$

(c) $x + xy + yz$ (d) $x + xz + xy$

[2022 : 1 M]

2.22 For the Boolean function

$F(A, B, C, D) = \Sigma m(0, 2, 5, 7, 8, 10, 12, 13, 14, 15)$

the essential prime implicants are ______.

(a) $BD, \bar{B}\bar{D}, AB$ (b) $BD, \bar{B}\bar{D}$

(c) $AB, \bar{B}\bar{D}$ (d) BD, AB

[2024 : 1 Mark]

■■■

Answers Boolean Algebra

2.1 (c)	2.2 (a)	2.3 (d)	2.4 (a)	2.5 (d)	2.6 (a)	2.7 (d)
2.8 (a)	2.9 (b)	2.10 (d)	2.11 (d)	2.12 (a)	2.13 (a)	2.14 (d)
2.15 (d)	2.16 (a)	2.17 (b)	2.18 (b)	2.19 (b)	2.20 (c)	2.21 (b, c)
2.22 (b)						

Explanations Boolean Algebra

2.1 (c)

The condition for overflow to occur is

$\bar{x}\,\bar{y}\,z + x\,y\,\bar{z}$

2.2 (a)

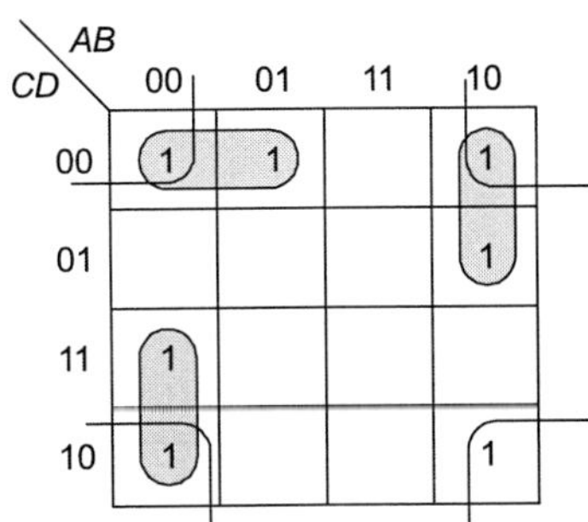

Number of essential prime implicants = 4.

2.3 (d)

$$y = A + \bar{A}B = (A + \bar{A})(A + B) = A + B$$

2.4 (a)

$$\begin{aligned} Y &= \bar{A}\bar{B}\bar{C} + \bar{A}B\bar{C} + \bar{A}BC + AB\bar{C} \\ &= \bar{A}\bar{C}(B + \bar{B}) + \bar{A}BC + AB\bar{C} \\ &= \bar{A}(\bar{C} + BC) + AB\bar{C} \\ &= \bar{A}(C + \bar{C})(\bar{C} + B) + AB\bar{C} \\ &= \bar{A}\bar{C} + \bar{A}B + AB\bar{C} = \bar{A}\bar{C} + B(\bar{A} + A\bar{C}) \\ &= \bar{A}\bar{C} + B(A + \bar{A})(\bar{A} + \bar{C}) = \bar{A}\bar{C} + \bar{A}B + B\bar{C} \end{aligned}$$

2.5 (d)

$$2^{2^n} = 2^{2^4} = 2^{16} = 65536$$

2.6 (a)

(1) W =

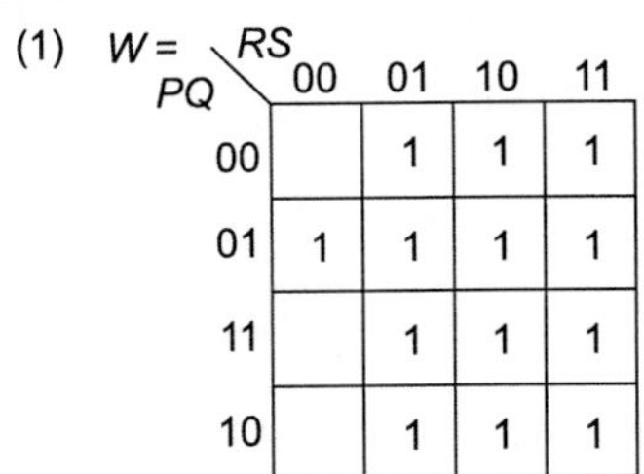

$$W = R + \bar{P}Q + \bar{R}S$$

(2) X =

PQ \ RS	00	01	11	10
00	1			
01				
11	1			
10	1			

$$X = PQ\bar{R}\bar{S} + \bar{P}\bar{Q}\bar{R}\bar{S} + P\bar{Q}\bar{R}\bar{S}$$

(3) Y =

PQ \ RS	00	01	11	10
00			1	
01	1	1	1	1
11	1	1	1	
10			1	

$$\begin{aligned} Y &= RS + \overline{PR + P\bar{Q} + \bar{P}\bar{Q}} \\ &= RS + \overline{PR}.\overline{P\bar{Q}}.\overline{\bar{P}\bar{Q}} \\ &= RS + (\bar{P} + \bar{R})(\bar{P} + Q)(P + Q) \\ &= RS + (\bar{P} + \bar{P}Q + \bar{P}\bar{R} + Q\bar{R})(P + Q) \\ &= RS + \bar{P}Q + \bar{P}Q + \bar{P}Q\bar{R} + PQ\bar{R} + Q\bar{R} \\ &= RS + \bar{P}Q + Q\bar{R}(P + \bar{P}) + Q\bar{R} \\ &= RS + \bar{P}Q + Q\bar{R} \end{aligned}$$

(4) Z =

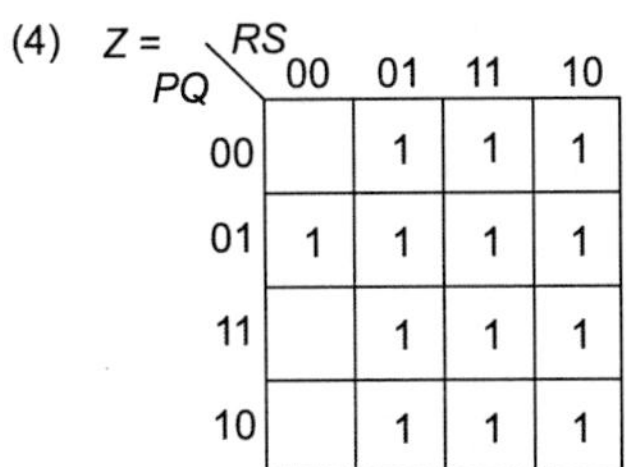

$$\begin{aligned} Z &= R + S + \overline{PQ + \bar{P}\bar{Q}\bar{R} + P\bar{Q}\bar{S}} \\ &= R + S + \overline{PQ}\cdot\overline{\bar{P}\bar{Q}\bar{R}}\cdot\overline{P\bar{Q}\bar{S}} \\ &= R + S + (\bar{P} + \bar{Q})(P + Q + R)(\bar{P} + Q + S) \\ &= R + S + (\bar{P}Q + \bar{P}R + P\bar{Q} + \bar{Q}R)(\bar{P} + Q + S) \\ &= R + S + \bar{P}Q + \bar{P}Q + \bar{P}QS + \bar{P}R + \bar{P}QR \\ &\quad + \bar{P}RS + P\bar{Q} + P\bar{Q}S + \bar{P}\bar{Q}R + \bar{Q}RS \\ &= R + S + \bar{P}Q + \bar{P}R + \bar{P}QS + \bar{P}QR \\ &\quad + \bar{P}RS + P\bar{Q}S + \bar{P}\bar{Q}R + \bar{Q}RS \end{aligned}$$

$$= R+S+\bar{P}Q(1+S)+\bar{P}R(1+Q)$$
$$+\bar{P}RS+P\bar{Q}S+\bar{P}\bar{Q}R+\bar{Q}RS$$
$$= R+S+\bar{P}Q+\bar{P}R+\bar{P}RS+P\bar{Q}S+\bar{P}\bar{Q}R+\bar{Q}RS$$
$$= R+S+\bar{P}Q+\bar{P}R+P\bar{Q}S+\bar{P}\bar{Q}R+\bar{Q}RS$$
$$= R+S+\bar{P}Q+\bar{P}R(1+\bar{Q})+P\bar{Q}S+\bar{Q}RS$$
$$= R+S+\bar{P}Q+\bar{P}R+P\bar{Q}S+\bar{Q}RS$$

$K\cdot$ Map (1) = K.map (4)

$\therefore\ W = Z$

From map (2) & (4), $X = \bar{Z}$

2.7 (d)

$AC+B\bar{C}$

Writing the SOP form of the above expression and filling the *K*-map, we get,

A \ BC	00	01	11	10
0				1
1		1	1	1

Now, writing the SOP form for the given options we get,

(a) $\bar{A}C+B\bar{C}+AC$

A \ BC	00	01	11	10
0		1	1	1
1		1	1	1

(b) $\bar{B}C+AC+B\bar{C}+\bar{A}C\bar{B}$

A \ BC	00	01	11	10
0		1		1
1		1	1	1

(c) $AC+B\bar{C}+\bar{B}C+ABC$

A \ BC	00	01	11	10
0		1		1
1		1	1	1

(d) $ABC+\bar{A}B\bar{C}+AB\bar{C}+A\bar{B}C$

A \ BC	00	01	11	10
0				1
1		1	1	1

Now, comparing it which the expression given in the question, we find option (d) satisfies the given condition.

Alternate method

$ABC+\bar{A}B\bar{C}+AB\bar{C}+A\bar{B}C$

$AC(B+\bar{B})+B\bar{C}(A+\bar{A}) = AC+B\bar{C}$

2.8 (a)

$$f = \bar{A}BC+AB\bar{C} = B(\bar{A}C+A\bar{C}) = B(A+C)(\bar{A}+\bar{C})$$

2.9 (b)

$$y = AB+CD = \overline{\overline{A\cdot B}}+\overline{\overline{C\cdot D}}$$

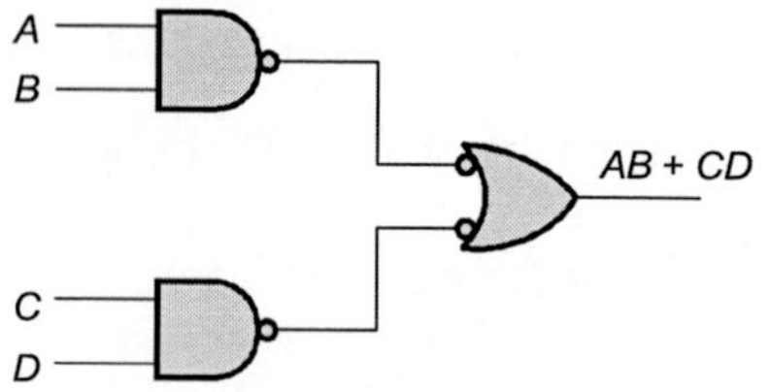

A B C D AB + CD

2.10 (d)

K-map corresponding to given Boolean expression

AB \ CD	00	01	11	10
00		1		
01				1
11	1			
10		1		

simplified expression from the *K*-map

$$Y = \bar{A}BC\bar{D}+ABC\bar{D}+\bar{B}\bar{C}D$$

2.11 (d)

$$[X+Z\{\bar{Y}+(\bar{Z}+X\bar{Y})\}]\ [\bar{X}+\bar{Z}(X+Y)] = 1$$

(first bracket: X = 1, so the bracket = 1; second bracket: $\bar{X}$ = 0, $(X+Y)$ = 1)

$$\Rightarrow \bar{Z}(1+Y) = 1 \quad \Rightarrow \bar{Z} = 1 \quad \Rightarrow Z = 0$$

2.12 (a)

$$f(X, Y, Z) = \Sigma(2, 3, 4, 5)$$

	$\bar{Y}\bar{Z}$	$\bar{Y}Z$	YZ	$Y\bar{Z}$
$\bar{X}$			1	1
X	1	1		

$$f(X, Y, Z) = X\bar{Y}+\bar{X}Y$$

So prime implicants are $X\bar{Y}$ and $\bar{X}Y$.

2.13 (a)

Let $F = (X+Y)(X+\bar{Y})+\overline{(X\bar{Y}+\bar{X})}$

$$F = X+X\bar{Y}+XY+(\overline{X\bar{Y}}\cdot\bar{\bar{X}})$$
$$= X+(\bar{X}+Y)\cdot X = X+XY = X$$

2.14 (d)

$f(w, x, y, z)$

$= wy + xy + \bar{w}xyz + \bar{w}\bar{x}y + xz + \bar{x}\,\bar{y}\,\bar{z}$

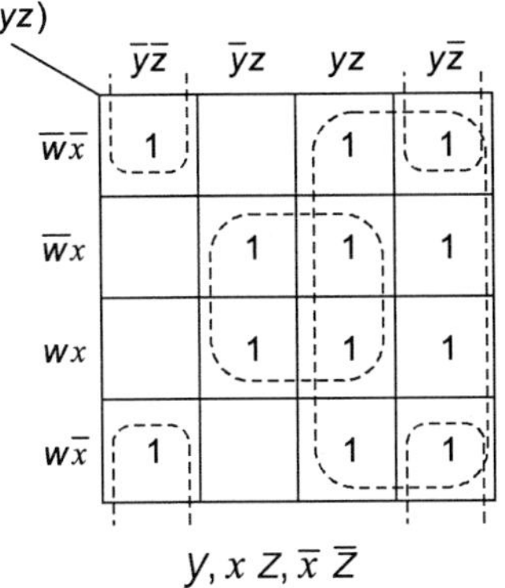

From the above K-map, it can be seen that, the set of essential prime-implicants is

$$y,\ xz,\ \bar{x}\bar{z}$$

2.15 (d)

For a variable Boolean function the maximum number of prime implicants is $2^{(n-1)}$.

2.16 (a)

$F(X, Y, Z) = \bar{X}Y\bar{Z} + X\bar{Y}\bar{Z} + XY\bar{Z} + XYZ$

X \ YZ	00	01	11	10
0	0	0	0	1
1	1	1	0	1

$F = (X + Y + Z)(X + Y + \bar{Z})(X + \bar{Y} + \bar{Z})(\bar{X} + Y + \bar{Z})$

2.17 (b)

$F(X, Y, Z) = \Sigma(1, 2, 5, 6, 7)$

$= \pi[0, 3, 4]$

$= (X + Y + Z)\,(X + \bar{Y} + \bar{Z})\,(\bar{X} + Y + Z)$

2.18 (b)

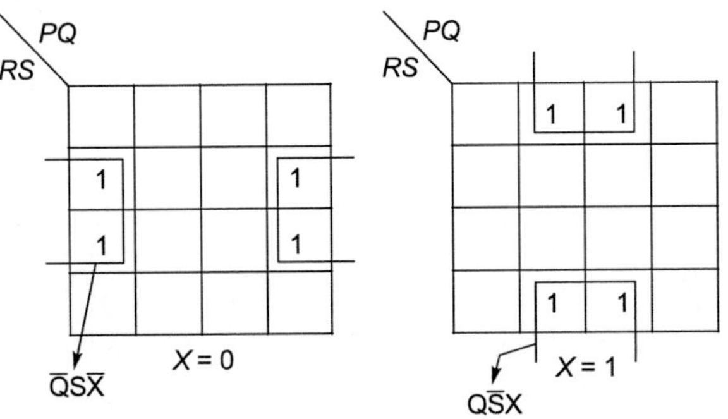

∴ minimum sum of product expression of the function is

$$= \bar{Q}S\bar{X} + Q\bar{S}X$$

2.19 (b)

Given Boolean function is,

$$F = m_0 + m_2 + m_3 + m_5$$

It can be minimized by using K-map as given below.

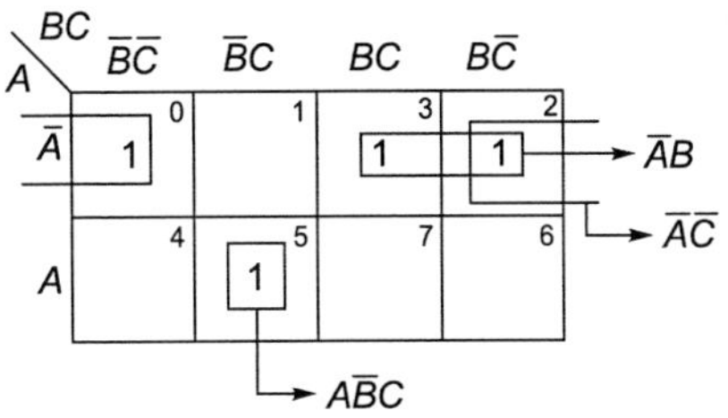

$$F = \bar{A}\bar{C} + \bar{A}B + A\bar{B}C$$

2.20 (c)

$F(A, B, C, D) = \bar{A}\bar{B}\bar{C} + \bar{A}B\bar{C} + A\bar{B}\bar{C}$

$= \Sigma m(0, 2, 4) = \Pi M(1, 3, 5, 6, 7)$

$= (A + B + \bar{C})\,(A + \bar{B} + \bar{C})\,(\bar{A} + B + \bar{C})$
$(\bar{A} + \bar{B} + C)\,(\bar{A} + \bar{B} + \bar{C})$

2.21 (b, c)

Let, $f = x + yz$

Option (a) : $x + z + xy = x(1 + y) + z = x + z$

Option (b) : $(x + y)\,(x + z)$

$= x \cdot x + x \cdot z + y \cdot x + y \cdot z = x + xz + xy + yz$

$= x(1 + z + y) + yz = x + yz$

Option (c) : $x + xy + yz = x(1 + y) + yz = x + yz$

Option (d) : $x + xz + xy = x(1 + z + y) = x$

2.22 (b)

$F(A, B, C, D) = \Sigma m(0, 2, 5, 7, 8, 10, 12, 13, 14, 15)$

The Boolean function can be represented in the K-map as below:

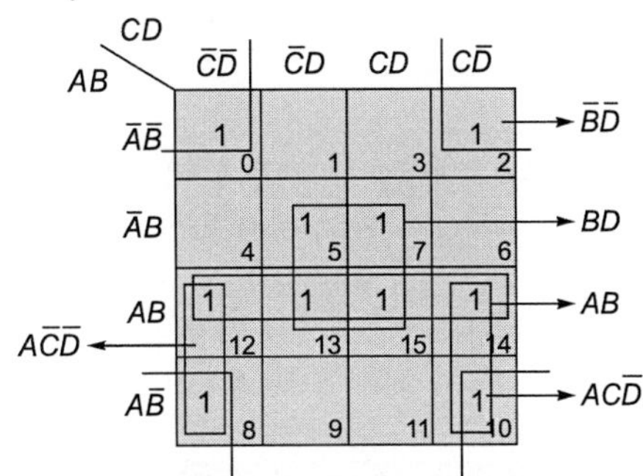

Essential prime implicants are those groups that cover atleast one minterm that can't covered by other prime implicant. They always appear in the final solution.

From the K-map, we can see $\bar{B}\bar{D}$ and BD are the essential prime implicants. We can write,

$$F = BD + \bar{B}\bar{D} + AB \quad \text{or} \quad F = BD + \bar{B}\bar{D} + A\bar{C}\bar{D} + AC\bar{D}$$

■■■

3 Logic Gates

3.1 The minimum number of NAND gates required to implement the Boolean function $A + A\bar{B} + A\bar{B}C$ is equal to
(a) Zero (b) 1
(c) 4 (d) 7 **[1995 : 1 M]**

3.2 The output of the circuit shown in figure is equal to

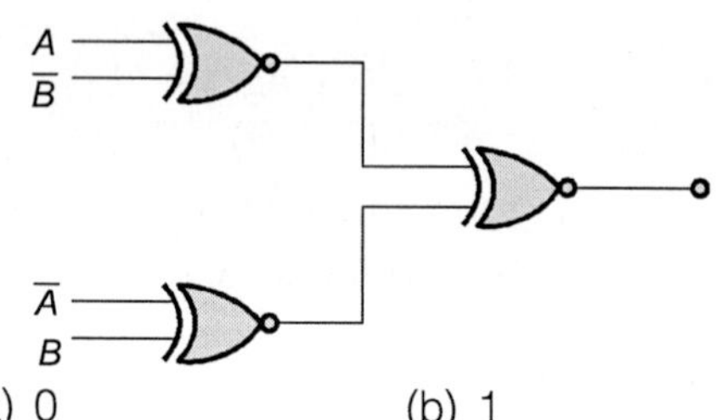

(a) 0 (b) 1
(c) $\bar{A}B + A\bar{B}$ (d) $(\overline{A \oplus B}) \oplus (\overline{A \oplus B})$
[1995 : 1 M]

3.3 The output of the logic gate in figure is

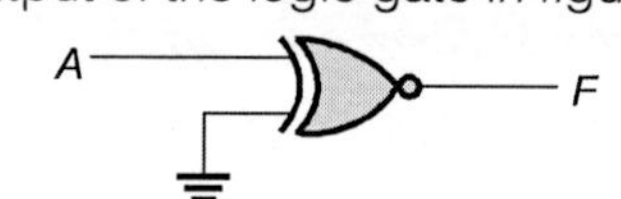

(a) 0 (b) 1
(c) $\bar{A}$ (d) A **[1997 : 1 M]**

3.4 The Boolean function $A + BC$ is a reduced form of
(a) $AB + BC$ (b) $(A + B) \cdot (A + C)$
(c) $\bar{A}B + A\bar{B}C$ (d) $(A + C) \cdot B$
[1997 : 1 M]

3.5 The minimum number of 2-input NAND gates required to implement the Boolean function $Z = A\bar{B}C$, assuming that A, B and C are available, is
(a) two (b) three
(c) five (d) six **[1998 : 1 M]**

3.6 For the identity $AB + \bar{A}C + BC = AB + \bar{A}C$, the dual form is
(a) $(A + B)(\bar{A} + C)(B + C) = (A + B)(\bar{A} + C)$
(b) $(\bar{A} + \bar{B})(A + \bar{C})(\bar{B} + \bar{C}) = (\bar{A} + \bar{B})(A + \bar{C})$
(c) $(A + B)(\bar{A} + C)(B + C) = (\bar{A} + \bar{B})(A + C)$
(d) $\bar{A}\bar{B} + A\bar{C} + \bar{B}\bar{C} = \bar{A}\bar{B} + A\bar{C}$
[1998 : 1 M]

3.7 For the logic circuit shown in the figure, the required input condition (A, B, C) to make the output (X) = 1 is

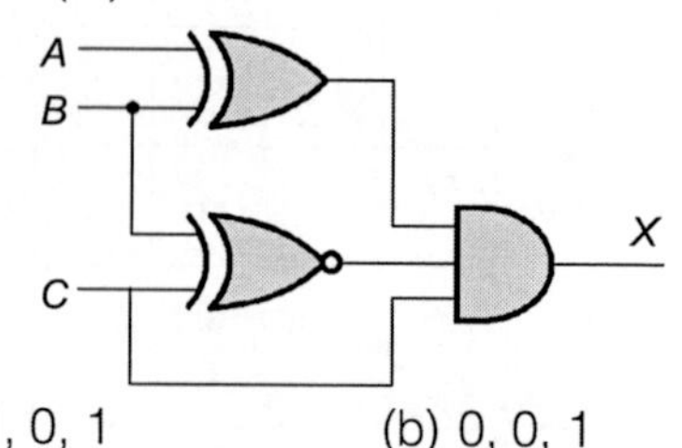

(a) 1, 0, 1 (b) 0, 0, 1
(c) 1, 1, 1 (d) 0, 1, 1
[2000 : 1 M]

3.8 For the logic circuit shown in the figure, the simplified Boolean expression for the output Y is

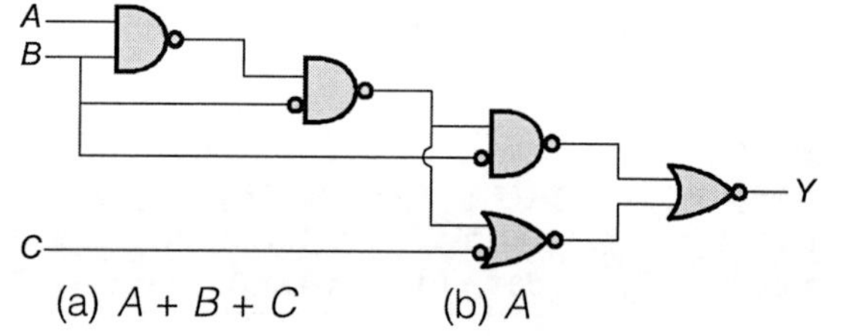

(a) $A + B + C$ (b) A
(c) B (d) C **[2000 : 2 M]**

3.9 In the figure, the LED

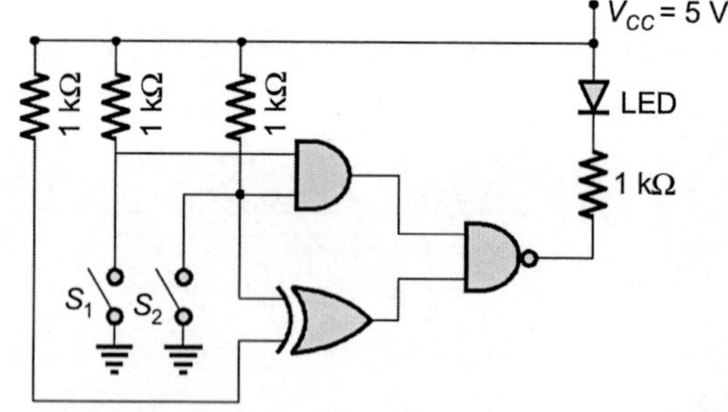

(a) emits light when both S_1 and S_2 are closed.
(b) emits light when both S_1 and S_2 are open.
(c) emits light when only of S_1 and S_2 is closed.
(d) does not emit light, irrespective of the switch positions.
[2001 : 2 M]

3.10 If the input to the digital circuit (in the figure) consisting of a cascade of 20 XOR-gates is X, then the output Y is equal to

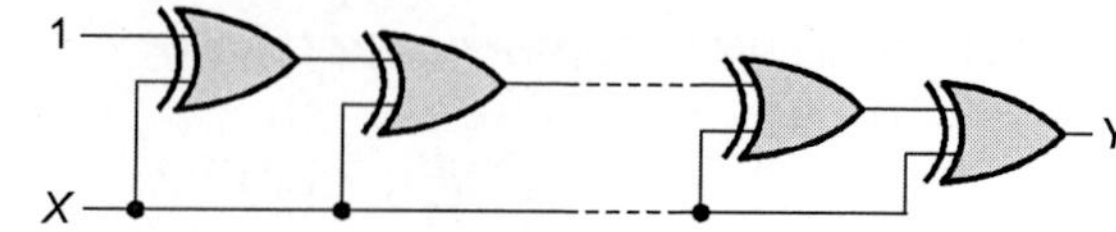

(a) 0 (b) 1

(c) $\overline{X}$ (d) X [2002 : 1 M]

3.11 The gates G_1 and G_2 in the figure have propagation delays of 10 nsec and 20 nsec respectively. If the input V_i makes an abrupt change from logic 0 to 1 at time $t = t_0$, then the output waveform V_0 is

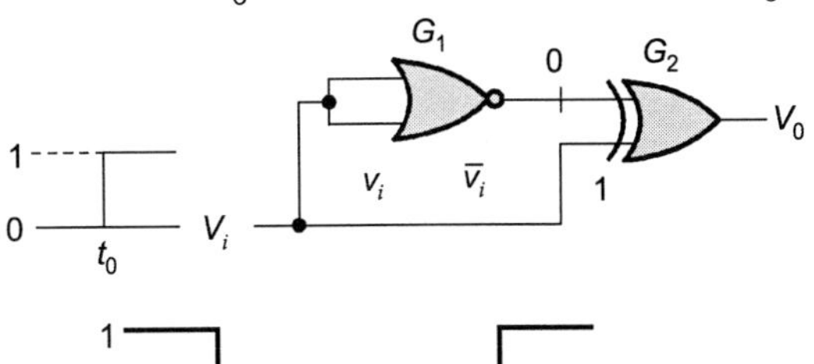

(a)

(b)

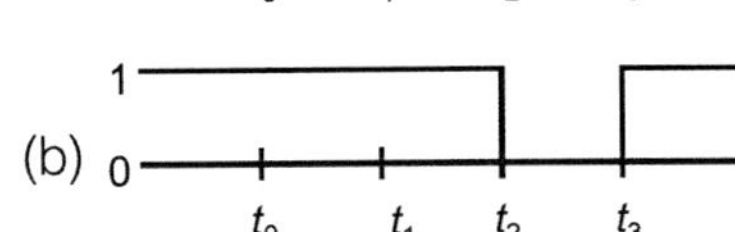

(c)

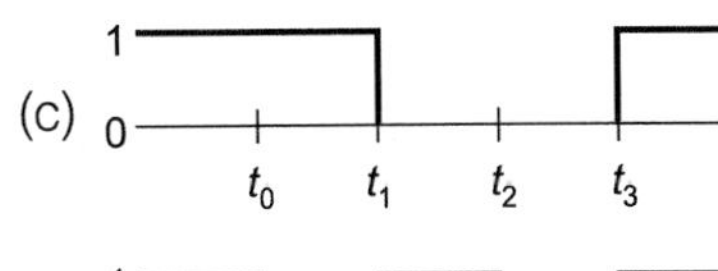

(d)

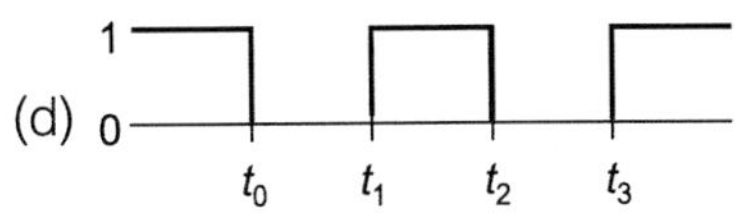

[2002 : 2 M]

3.12 A Boolean function f of two variables x and y is defined as follows:

$f(0, 0) = f(0, 1) = f(1, 1) = 1;\ f(1, 0) = 0$

Assuming complements of x and y are not available, a minimum cost solution for realizing f using only 2-input NOR gates and 2-input OR gates (each having unit cost) would have a total cost of

(a) 1 unit (b) 4 unit

(c) 3 unit (d) 2 unit

[2004 : 2 M]

3.13 The number of product terms in the minimized sum-of-product expression obtained through the following K-map is (where, "d" denotes don't care states)

1	0	0	1
0	d	0	0
0	0	d	1
1	0	0	1

(a) 2 (b) 3

(c) 4 (d) 5 [2005 : 2 M]

3.14 Which of the following Boolean Expressions correctly represents the relation between P, Q, R and M_1?

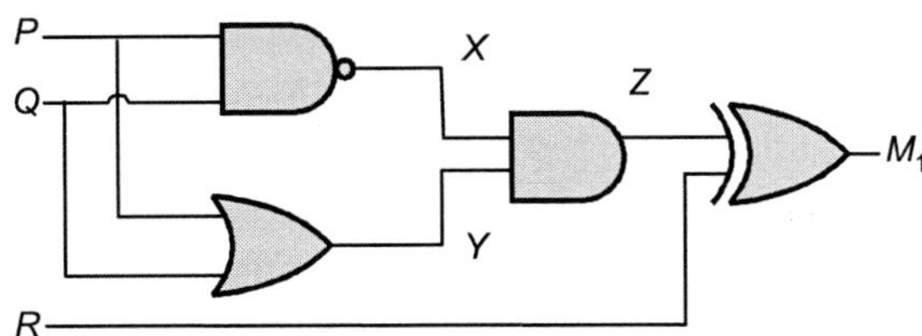

(a) M_1 = (P OR Q) XOR R

(b) M_1 = (P AND Q) XOR R

(c) M_1 = (P NOR Q) XOR R

(d) M_1 = (P XOR Q) XOR R [2008 : 2 M]

3.15 Match the logic gates in Column A with their equivalents in Column B.

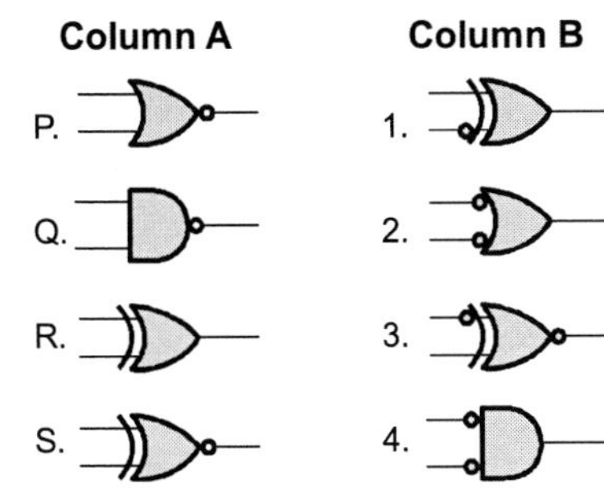

(a) P-2, Q-4, R-1, S-3

(b) P-4, Q-2, R-1, S-3

(c) P-2, Q-4, R-3, S-1

(d) P-4, Q-2, R-3, S-1 [2010 : 1 M]

3.16 For the output F to be 1 is the logic circuit shown, the input combination should be

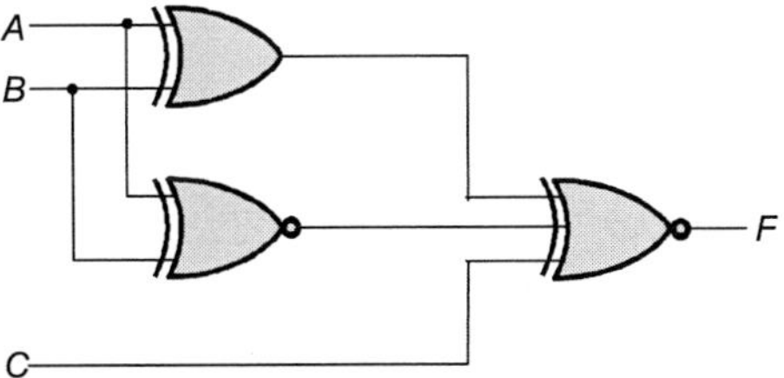

(a) $A = 1, B = 1, C = 0$

(b) $A = 1, B = 0, C = 0$

(c) $A = 0, B = 1, C = 0$

(d) $A = 0, B = 0, C = 1$ [2010 : 1 M]

3.17 The output Y in the circuit below is always "1" when

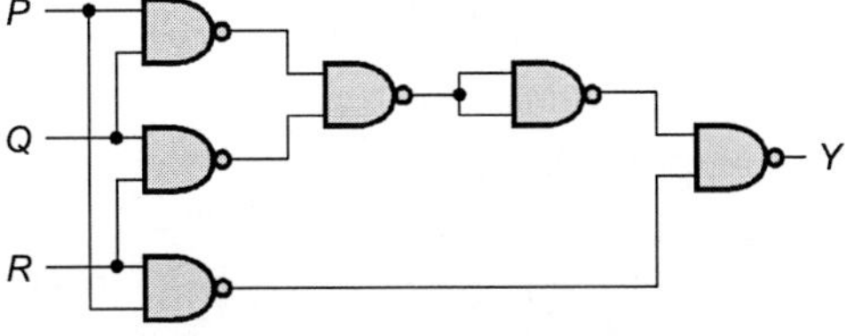

(a) two or more of the inputs P, Q, R are "0"

(b) two or more of the inputs P, Q, R are "1"

(c) any odd number of the inputs P, Q, R is "0"

(d) any odd number of the inputs P, Q, R is "1"

[2011 : 1 M]

3.18 A bulb in a staircase has two switches, one switch being at the ground floor and the other one at the first floor. The bulb can be turned ON and also can be turned OFF by any one of the switches irrespective of the state of the other switch. The logic of switching of the bulb resembles

(a) an AND gate (b) an OR gate
(c) an XOR gate (d) a NAND gate

[2013 : 1 M]

3.19 The output F in the digital logic circuit shown in the figure is

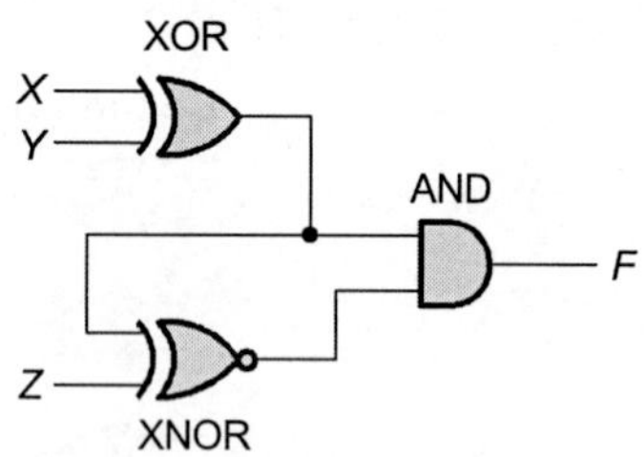

(a) $F = \overline{X}YZ + X\overline{Y}Z$ (b) $F = \overline{X}Y\overline{Z} + XY\overline{Z}$
(c) $F = \overline{X}\overline{Y}Z + XYZ$ (d) $F = \overline{X}\overline{Y}\overline{Z} + XYZ$

[2014 : 2 M, Set-1]

3.20 In the circuit shown in the figure, if $C = 0$, the expression for Y is

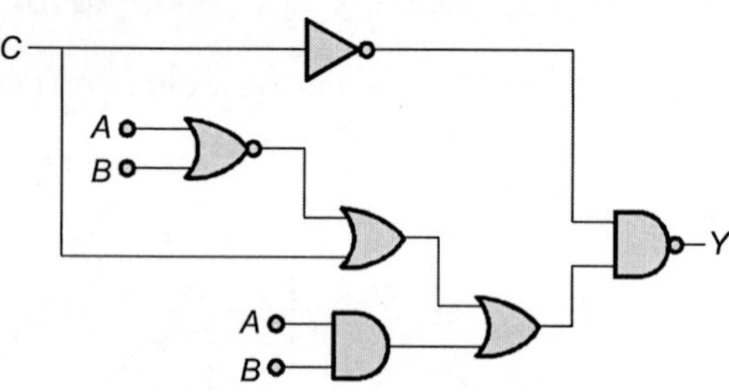

(a) $Y = A\overline{B} + \overline{A}B$ (b) $Y = A + B$
(c) $Y = \overline{A} + \overline{B}$ (d) $Y = AB$

[2014 : 1 M, Set-4]

3.21 A 3-input majority gate is defined by the logic function $M(a, b, c) = ab + bc + ca$. Which one of the following gate is represented by the function

$M(\overline{M(a,b,c)}, M(a,b,\overline{c}), c)$?

(a) 3-input NAND gate
(b) 3-input XOR gate
(c) 3-input NOR gate
(d) 3-input XNOR gate

[2015 : 2 M, Set-1]

3.22 All the logic gates shown in the figure have a propagation delay of 20 ns. Let $A = C = 0$ and $B = 1$ until time $t = 0$. At $t = 0$, all the inputs flip (i.e. $A = C = 1$ and $B = 0$) and remain in that state. For $t > 0$, output $Z = 1$ for a duration (in ns) of

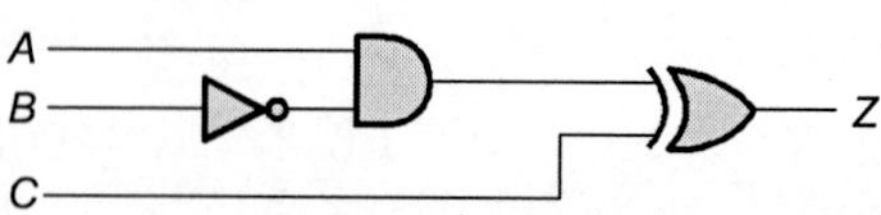

[2015 : 2 M, Set-1]

3.23 In the figure shown, the output Y is required to be $Y = AB + \overline{C}\overline{D}$. The gates $G1$ and $G2$ must be,

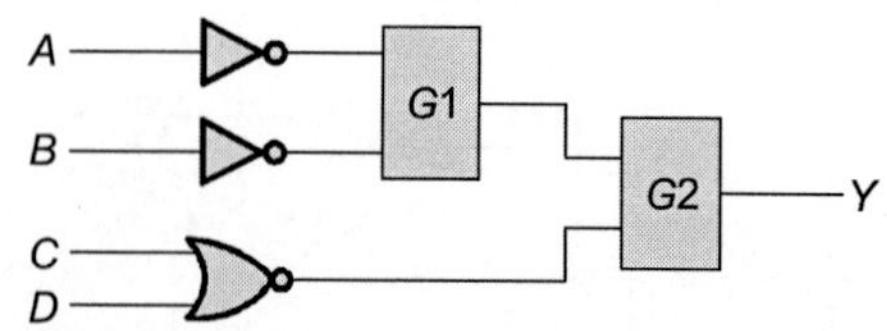

(a) NOR, OR (b) OR, NAND
(c) NAND, OR (d) AND, NAND

[2015 : 1 M, Set-2]

3.24 A universal logic gate can implement any Boolean function by connecting sufficient number of them appropriately. Three gates are shown.

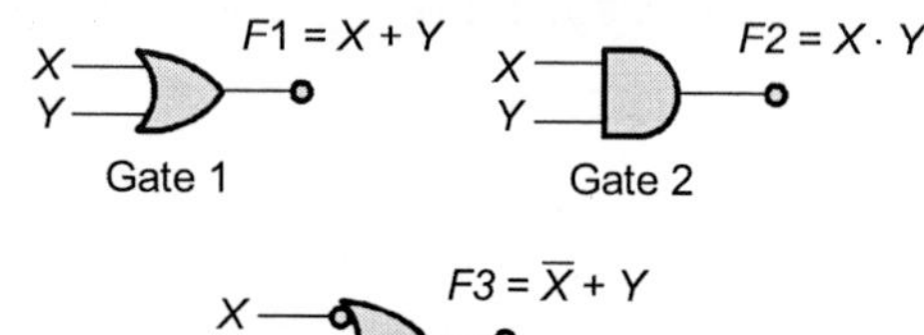

Which one of the following statements is TRUE?

(a) Gate 1 is a universal gate.
(b) Gate 2 is a universal gate.
(c) Gate 3 is a universal gate.
(d) None of the gates shown is a universal gate.

[2015 : 2 M, Set-3]

3.25 The output of the combinational circuit given below is

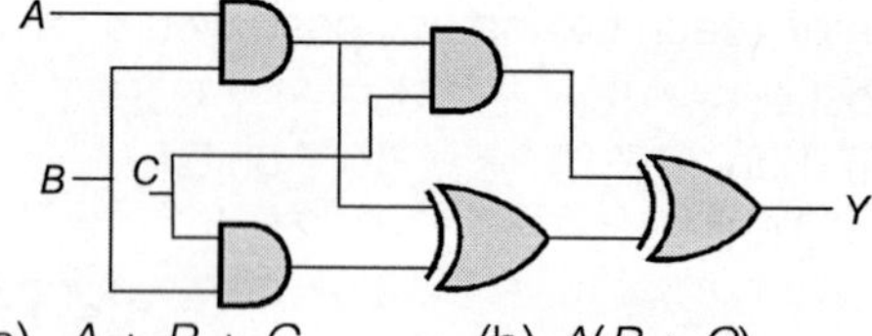

(a) $A + B + C$ (b) $A(B + C)$
(c) $B(C + A)$ (d) $C(A + B)$

[2016 : 1 M, Set-1]

3.26 The minimum number of 2-input NAND gates required to implement a 2-input XOR gate is

(a) 4 (b) 5
(c) 6 (d) 7

[2016 : 1 M, Set-3]

3.27 Consider a Boolean gate (D) where the output Y is related to the inputs A and B as, $Y = A + \overline{B}$, where + denotes logical OR operation. The

Boolean inputs '0' and '1' are also available separately. Using instances of only D gates and inputs '0' and '1', ________ (select the correct option(s)).

(a) NAND logic can be implemented
(b) OR logic cannot be implemented
(c) NOR logic can be implemented
(d) AND logic cannot be implemented

[2022 : 2 M]

Answers Logic Gates

3.1 (a)	3.2 (b)	3.3 (c)	3.4 (b)	3.5 (c)	3.6 (a)	3.7 (d)
3.8 (*)	3.9 (d)	3.10 (b)	3.11 (b)	3.12 (d)	3.13 (a)	3.14 (d)
3.15 (d)	3.16 (d)	3.17 (b)	3.18 (c)	3.19 (a)	3.20 (a)	3.21 (b)
3.22 (40)	3.23 (a)	3.24 (c)	3.25 (c)	3.26 (a)	3.27 (a, c)	

Explanations Logic Gates

3.1 (a)

$F = A + A\bar{B} + A\bar{B}C = A(1 + \bar{B} + \bar{B}C) = A$

So, to implement A, zero NAND gates are required.

3.2 (b)

$Y = (A \odot \bar{B}) \odot (\bar{A} \odot B)$

$Y = A \odot \bar{B} \odot \bar{A} \odot B$

$Y = (A \odot \bar{A}) \odot (\bar{B} \odot B) = 0 \odot 0$

$Y = 1$

3.3 (c)

$F = A \odot 0 = \bar{A} \cdot \bar{0} + A \cdot 0 = \bar{A} \cdot 1 + 0$

$F = \bar{A}$

3.4 (b)

$(A + B)(A + C) = A \cdot A + A \cdot C + A \cdot B + BC$
$= A + AC + AB + BC$
$= A(1 + C + B) + BC = A + BC$

3.5 (c)

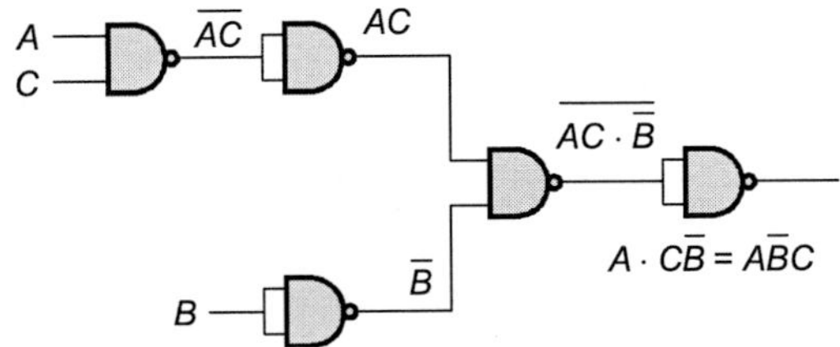

3.6 (a)

AND ↔ OR
OR ↔ AND

3.7 (d)

For $X = 1$

$A \oplus B = 1$ i.e. $A = 0, B = 1$

$B \odot C = 1,$ i.e. $B = 1, C = 1$

3.8 (*)

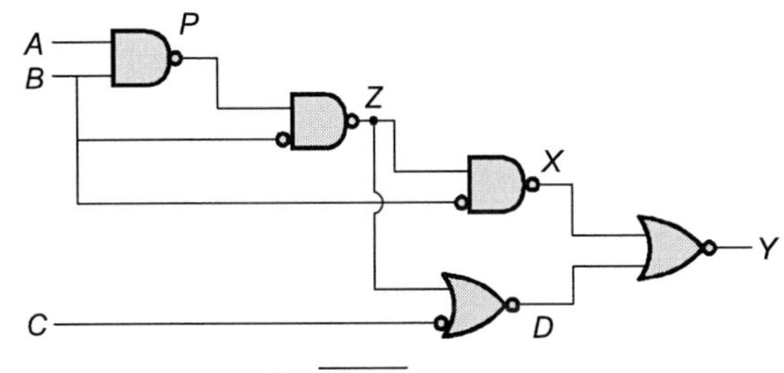

$Y = \overline{X + D}$

$\bar{Y} = X + D$

$D = \overline{(Z + \bar{C})} = \bar{Z}.C$

$\bar{Y} = \bar{Z} + B + \bar{Z}.C$

$X = \overline{(Z.\bar{B})} = \bar{Z} + B$

$\bar{Y} = \bar{Z} + B$

$Z = \overline{P.\bar{B}} = \bar{P} + B$

$\bar{Y} = \overline{(\bar{P} + B)} + B$

$P = \overline{AB}$

$\bar{Y} = P.\bar{B} + B = B + P$

$= B + \overline{AB} = B + \bar{A} + \bar{B} = 1$

$\therefore \quad Y = 0$

3.9 (d)

For LED to be ON, output of NAND gate = 0
No condition of S_1 and S_2 gives output of NAND gate zero. So LED will never glow.

3.10 (b)

Output of 1st XOR gate = $\overline{X}$

Output of 2nd XOR gate = $\overline{X} \oplus X$

$$= (\overline{\overline{X}}) \cdot X + \overline{X} \cdot \overline{X} = X + \overline{X} = 1$$

Output of 20 XOR gates is 1.

3.11 (b)

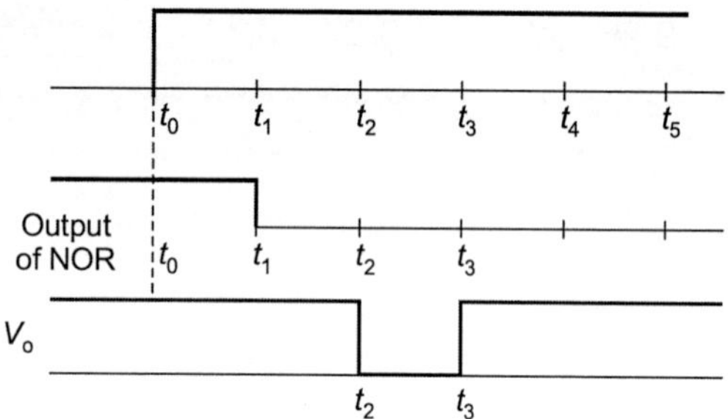

3.12 (d)

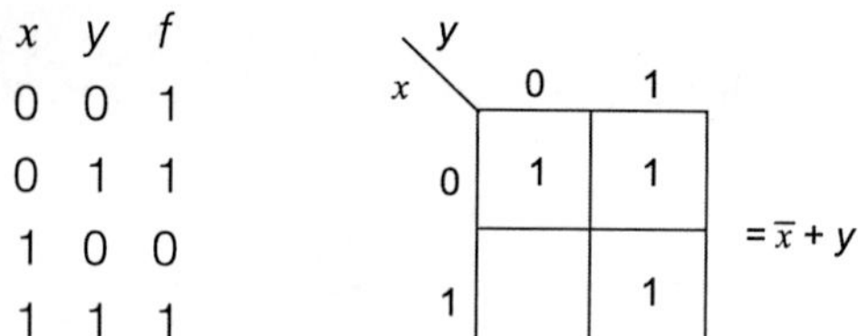

x	y	f
0	0	1
0	1	1
1	0	0
1	1	1

x \ y	0	1
0	1	1
1		1

$= \overline{x} + y$

Since complements are not available.

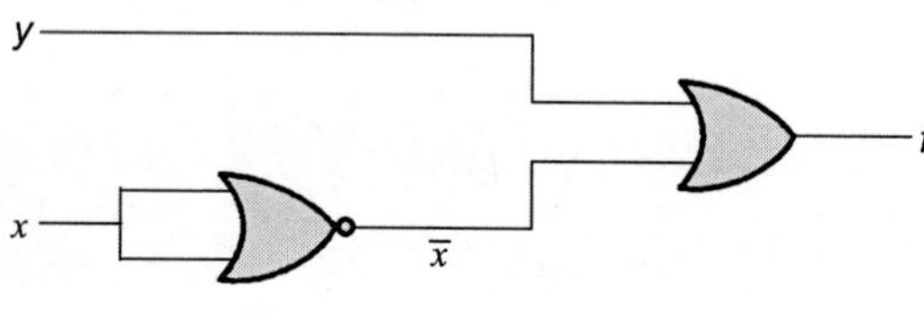

$\therefore$ 2 units.

3.13 (a)

1	0	0	1
0	d	0	0
0	0	d	1
1	0	0	1

3.14 (d)

$$M_1 = \left[\overline{PQ}(P+Q)\right] \oplus R$$

$$= \left[(\overline{P}+\overline{Q})(P+Q)\right] \oplus R$$

$$= (P \oplus Q) \oplus R$$

3.15 (d)

NOR gate is equivalent to INVERT – AND gate.
NAND gate is equivalent to INVERT – OR gate.

R : $F = X\overline{Y} + \overline{X}Y$

3 : $F = \overline{X}Y + \overline{\overline{X}}\,\overline{Y}$

$$F = \overline{X}Y + X\overline{Y}$$

S : $F = XY + \overline{X}\,\overline{Y}$

1 : $F = X\overline{\overline{Y}} + \overline{X}\,\overline{Y}$

$$F = XY + \overline{X}\,\overline{Y}$$

3.16 (d)

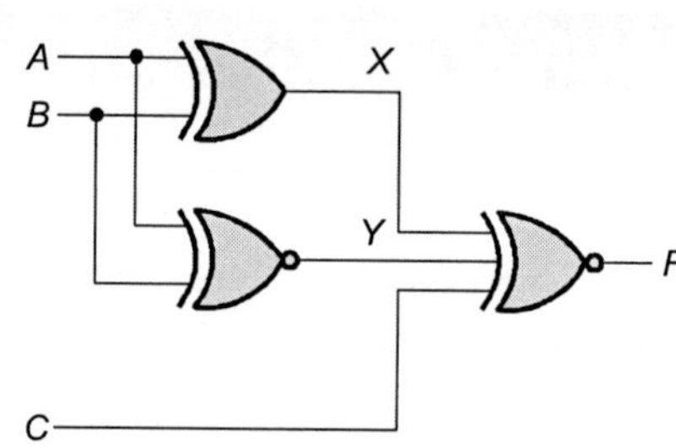

If $A = 0,\ B = 0$ then

$X = \overline{A}B + A\overline{B} = 0$

$Y = AB + \overline{A}\overline{B} = 1$

F will be '1' if even number of inputs to XNOR gate is '1' ; hence option (d) is the correct answer.

3.17 (b)

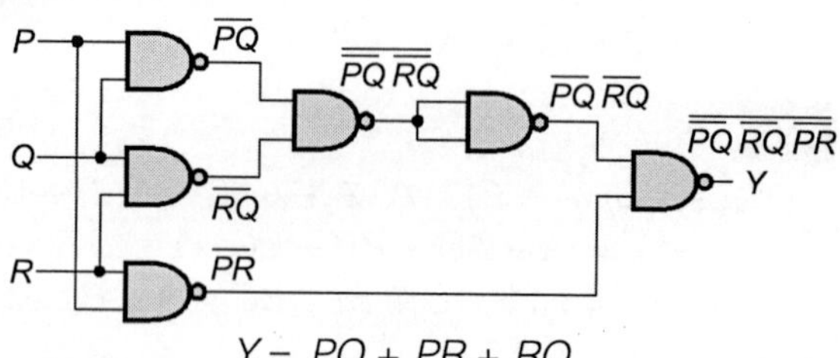

$$Y = PQ + PR + RQ$$

3.18 (c)

Truth table of XOR gate

A	B	Y
0	0	0
0	1	1
1	0	1
1	1	0

So, from the XOR gate truth table it is clear that the bulb can be turned ON and also can be turned OFF by any one of the switches irrespective of the state of the other switch.

3.19 (a)

$A = X \oplus Y$

$B = A \odot Z$

$= A \cdot Z + \bar{A}\bar{Z}$

$= Z(X \oplus Y) + (X \odot Y)\bar{Z}$

$F = AB$

$= (X \oplus Y)\left[Z(X \oplus Y) + \bar{Z}\,(\overline{X \oplus Y})\right]$

$= Z\left[(X \oplus Y)(X \oplus Y) + \bar{A}(\overline{X \oplus Y})(X \oplus Y)\right]$

as we know $A \cdot A = A$

$A \cdot \bar{A} = 0$

$= Z\left[(X \oplus Y) + 0\right] = \bar{X}YZ + X\bar{Y}Z$

3.20 (a)

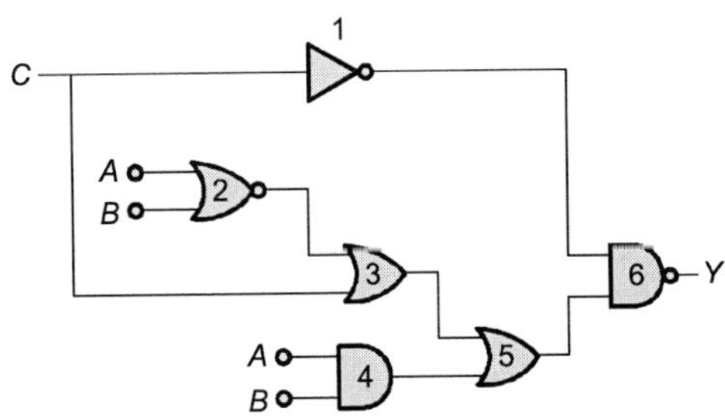

Output of gate 1: $\bar{C}$

Output of gate 2: $\rightarrow (\overline{A+B})$

Output of gate 3: $\rightarrow (\overline{A+B}+C)$

Output of gate 4: $\rightarrow AB$

Output of gate 5: $\rightarrow (\overline{A+B}+C)+AB$

Output of gate 6 is output Y i.e.

$$Y = \overline{\bar{C}\cdot(\overline{A+B}+C)+AB}$$

$$= C + \overline{(\overline{A+B}+C+AB)}$$

Using DeMorgan's theorem

$$= C + (\overline{\overline{A+B}})\cdot\bar{C}\cdot(\overline{AB})$$

$$= C + (A+B)\cdot\bar{C}\cdot(\bar{A}+\bar{B})$$

Given in question $C = 0$,

So, $Y = 0 + (A+B)\cdot\bar{0}\cdot(\bar{A}+\bar{B}) = \bar{A}B + A\bar{B}$

3.21 (b)

$M(a, b, c) = ab + bc + ca$

$\overline{M(a,b,c)} = \overline{ab+bc+ca}$

$= \overline{ab}\cdot\overline{bc}\cdot\overline{ca} = (\bar{a}+\bar{b})(\bar{b}+\bar{c})(\bar{c}+\bar{a})$

$M(a,b,\bar{c}) = ab + b\bar{c} + \bar{c}a$

$M(\overline{M(a,b,c)}, M(a,b,\bar{c})\, c) =$

$$\left[\begin{array}{l}(\overline{ab}\cdot\overline{bc}\cdot\overline{ca})(ab+b\bar{c}+\bar{c}a)\\ +(ab+b\bar{c}+\bar{c}a)(c)+(\overline{ab}\cdot\overline{bc}\cdot\overline{ca})c\end{array}\right]$$

$$= \left[\begin{array}{l}(\bar{a}+\bar{b})(\bar{b}+\bar{c})(\bar{c}+\bar{a})(ab+b\bar{c}+\bar{c}a)\\ +abc+(\bar{a}+\bar{b})(\bar{b}+\bar{c})(\bar{c}+\bar{a})c\end{array}\right]$$

$= (\bar{a}+\bar{b})(\bar{b}+\bar{c})(\bar{c}+\bar{a})\left[ab+b\bar{c}+\bar{c}a+c\right]+abc$

$= (\bar{a}+\bar{b})(\bar{b}+\bar{c})(\bar{c}+\bar{a})\left[a+b+c\right]+abc$

$= (\bar{a}\bar{b}+\bar{b}\bar{c}+\bar{c}\bar{a})\left[a+b+c\right]+abc$

$F = a\bar{b}\bar{c} + b\bar{c}\bar{a} + c\bar{a}\bar{b} + abc$

$F = A \oplus B \oplus C$

3.22 (40)

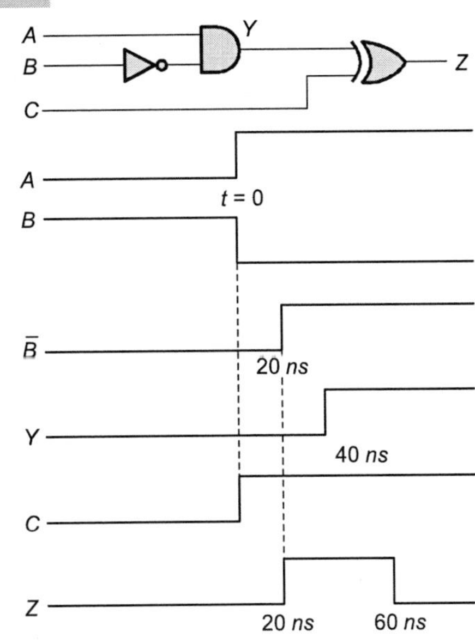

Z is '1' for 40 n-sec.

3.23 (a)

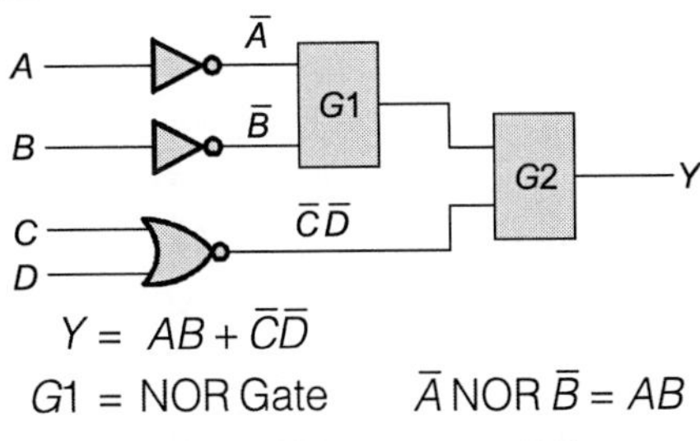

$Y = AB + \overline{CD}$

G1 = NOR Gate $\quad \bar{A}$ NOR $\bar{B} = AB$

G2 = OR GATE $\quad AB + \overline{CD}$

3.24 (c)

Universal gate is a gate by which every other gate can be realized.

Gate 1 and Gate 2 are basic gates and can not be used as universal gates.

3.25 (c)

$y = ABC \oplus AB \oplus BC$

$= \left[\overline{ABC}\cdot AB + ABC\cdot\overline{AB}\right] \oplus BC$

$$= \left[\left(\bar{A}+\bar{B}+\bar{C}\right)\cdot AB + ABC\cdot\left(\bar{A}+\bar{B}\right)\right] \oplus BC$$

$$= (AB\bar{C}) \oplus (BC)$$

$$= \overline{AB\bar{C}}\cdot BC + AB\bar{C}\cdot\overline{BC}$$

$$= (\bar{A}+\bar{B}+C)\cdot BC + AB\bar{C}\cdot(\bar{B}+\bar{C})$$

$$= \bar{A}BC + BC + AB\bar{C}$$

$$= BC(\bar{A}+1) + AB\bar{C} = BC + AB\bar{C}$$

$$= B(C + A\bar{C}) = B(C + A)$$

3.26 (a)

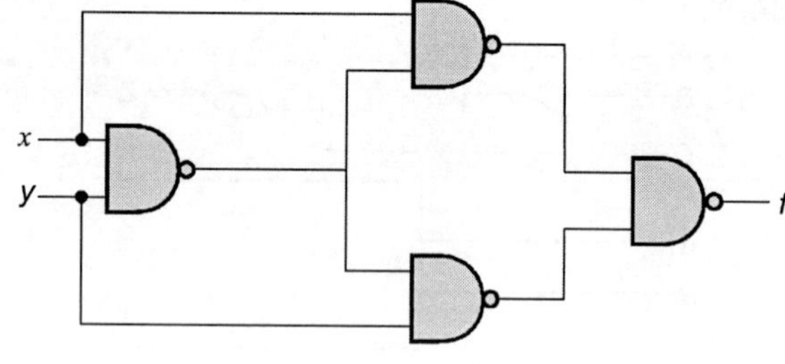

3.27 (a, c)

For the given boolean gate D with inputs *A* and *B*, the output

$$Y(A, B) = A + \bar{B}$$

As the Boolean input 1 and 0 are available separately

$$Y(0, B) = 0 + \bar{B} = \bar{B} \text{ (NOT)}$$

NOT logic can be implemented using *D*-gate

Using the output of NOT Gate, we can obtain

$$Y(A, \bar{B}) = A + \bar{\bar{B}} = A + B \text{ (OR)}$$

∴ OR gate can also be implemented using *D* Gates.

With the combination of OR and NOT, NOR gate can be implemented.

Since NOR gate is universal gate, so all the logic gates can be implemented using D gates.

Hence, option (a) and (c) are correct.

■■■

4 Combinational Circuits

4.1 A 2-bit binary multiplier can be implemented using
(a) 2 input ANDs only
(b) 2 input X-ORs and 4-input AND gates only
(c) Two (2) input NORs and one XNOR gate
(d) XOR gates and shift registers.
[1997 : 1 M]

4.2 For a binary half-subtractor having two inputs A and B, the correct set of logical expressions for the outputs D (= A minus B) and X (= borrow) are
(a) $D = AB + \bar{A}B,\ X = \bar{A}B$
(b) $D = \bar{A}B + A\bar{B} + A\bar{B},\ X = A\bar{B}$
(c) $D = \bar{A}B + A\bar{B},\ X = \bar{A}B$
(d) $D = AB + \bar{A}\bar{B},\ X = A\bar{B}$ [1999 : 2 M]

4.3 Without any additional circuitry, an 8 :1 MUX can be used to obtain
(a) some but not all Boolean functions of 3 variables
(b) all functions of 3 variables but none of 4 variables
(c) all functions of 3 variables and some but not all of 4 variables
(d) all functions of 4 variables
[2003 : 1 M]

4.4 The circuit shown in the figure has 4 boxes each described by inputs P, Q, R and outputs Y, Z with
$Y = P \oplus Q \oplus R$ $\quad Z = RQ + \bar{P}R + Q\bar{P}$

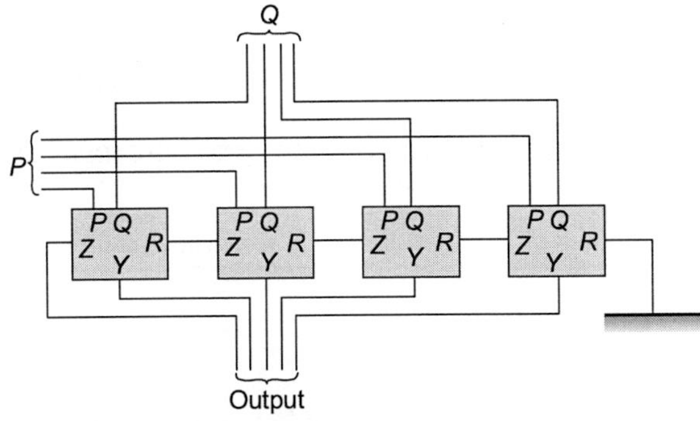

The circuit acts as a
(a) 4 bit adder giving $P + Q$
(b) 4 bit subtractor giving $P - Q$
(c) 4 bit subtractor giving $Q - P$
(d) 4 bit adder giving $P + Q + R$
[2003 : 2 M]

4.5 The circuit shown in the figure converts

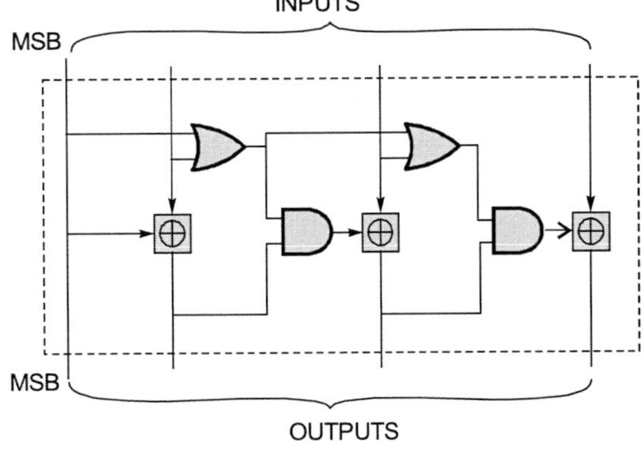

(a) BCD to binary code
(b) Binary to excess - 3 code
(c) Excess - 3 to Gray code
(d) Gray to Binary code
[2003 : 2 M]

4.6 The minimum number of 2-to-1 multiplexers required to realize a 4-to-1 multiplexer is
(a) 1 (b) 2
(c) 3 (d) 4 [2004 : 2 M]

4.7 The Boolean function f implemented in the figure using two input multiplexers is

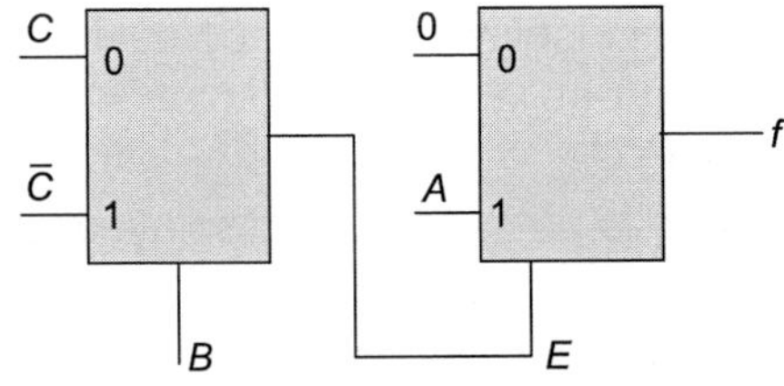

(a) $A\bar{B}C + AB\bar{C}$ (b) $ABC + A\bar{B}\bar{C}$
(c) $\bar{A}BC + \bar{A}\bar{B}\bar{C}$ (d) $\bar{A}\bar{B}C + \bar{A}B\bar{C}$
[2005 : 1 M]

4.8 In the following circuit, X is given by

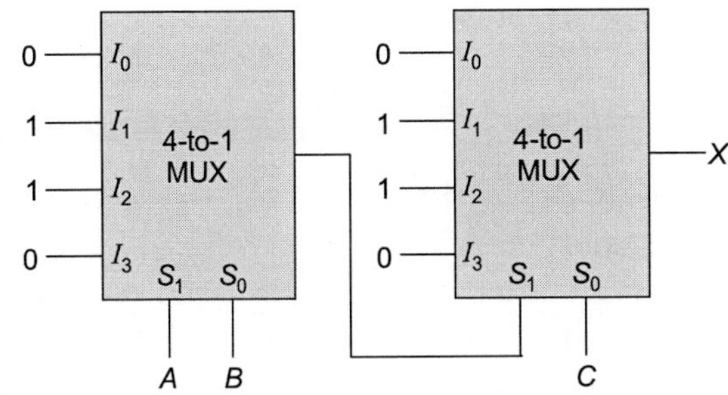

(a) $X = A\bar{B}\bar{C} + \bar{A}B\bar{C} + \bar{A}\bar{B}C + ABC$

(b) $X = \bar{A}BC + A\bar{B}C + AB\bar{C} + \bar{A}\bar{B}\bar{C}$

(c) $X = AB + BC + AC$

(d) $X = \bar{A}\bar{B} + \bar{B}\bar{C} + \bar{A}\bar{C}$ [2007 : 2 M]

4.9 For the circuit shown in the following figure, I_0-I_3 are inputs to the 4 : 1 multiplexer. R(MSB) and S are control bits.

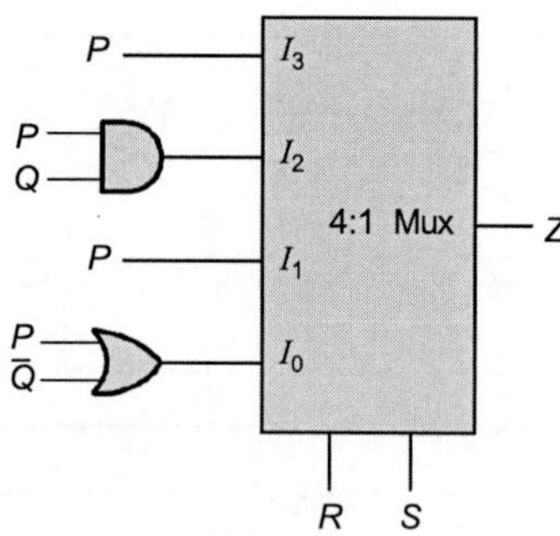

The output Z can be represented by

(a) $PQ + P\bar{Q}S + \bar{Q}\bar{R}\bar{S}$

(b) $P\bar{Q} + PQ\bar{R} + \bar{P}\bar{Q}\bar{S}$

(c) $P\bar{Q}\bar{R} + \bar{P}QR + PQRS + \bar{Q}\bar{R}\bar{S}$

(d) $PQ\bar{R} + PQR\bar{S} + P\bar{Q}\bar{R}S + \bar{Q}\bar{R}\bar{S}$

[2008 : 2 M]

Statement for Linked Answer Question 4.10 and 4.11.

Two products are sold from a vending machine, which has two push buttons P_1 and P_2. When a button is pressed, the price of the corresponding product is displayed in a 7-segment display.

If no buttons are pressed, '0' is displayed, signifying 'Rs. 0'

If only P_1 is pressed, '2' is displayed, signifying 'Rs. 2'

If only P_2 is pressed, '5' is displayed, signifying 'Rs. 5'

If both P_1 and P_2 are pressed, 'E' is displayed, signifying 'Error'

The names of the segments in the 7-segment display, and the glow of the display for '0', '2', '5' and 'E' are shown below.

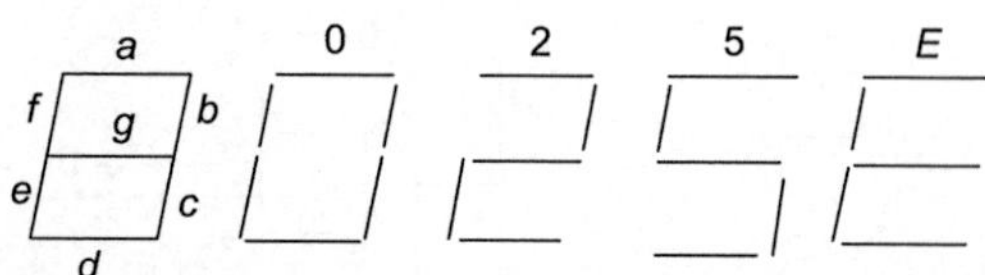

Consider

(i) push button pressed/not pressed in equivalent to logic 1/0 respectively,

(ii) a segment glowing/not glowing in the display is equivalent to logic 1/0 respectively

4.10 If segments a to g are considered as functions of P_1 and P_2, then which of the following is correct?

(a) $g = \bar{P}_1 + P_2, d = c + e$

(b) $g = P_1 + P_2, d = c + e$

(c) $g = \bar{P}_1 + P_2, e = b + c$

(d) $g = P_1 + P_2, e = b + c$ [2009 : 2 M]

4.11 What are the minimum numbers of NOT gates and 2-input OR gates required to design the logic of the driver for this 7-segment display?

(a) 3 NOT and 4 OR (b) 2 NOT and 4 OR

(c) 1 NOT and 3 OR (d) 2 NOT and 3 OR

[2009 : 2 M]

4.12 What are the minimum number of 2-to-1 multiplexers required to generate a 2-input AND gate and a 2-input EX-OR gate?

(a) 1 and 2 (b) 1 and 3

(c) 1 and 1 (d) 2 and 2

[2009 : 2 M]

4.13 The Boolean function realized by the logic circuit shown is

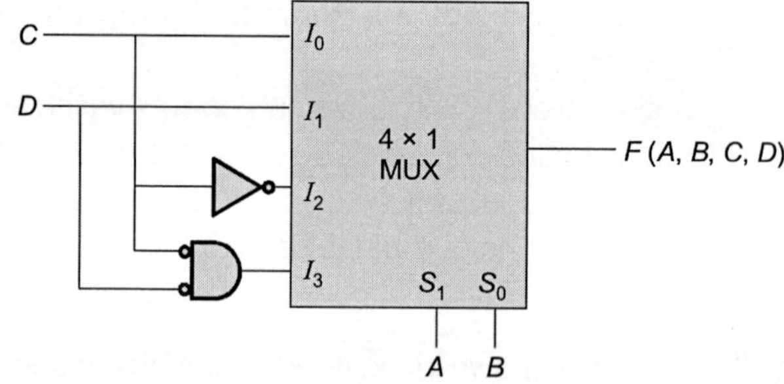

(a) $F = \Sigma m(0, 1, 3, 5, 9, 10, 14)$

(b) $F = \Sigma m(2, 3, 5, 7, 8, 12, 13)$

(c) $F = \Sigma m(1, 2, 4, 5, 11, 14, 15)$

(d) $F = \Sigma m(2, 3, 5, 7, 8, 9, 12)$

[2010 : 2 M]

4.14 The logic function implemented by the circuit below is (ground implies a logic "0")

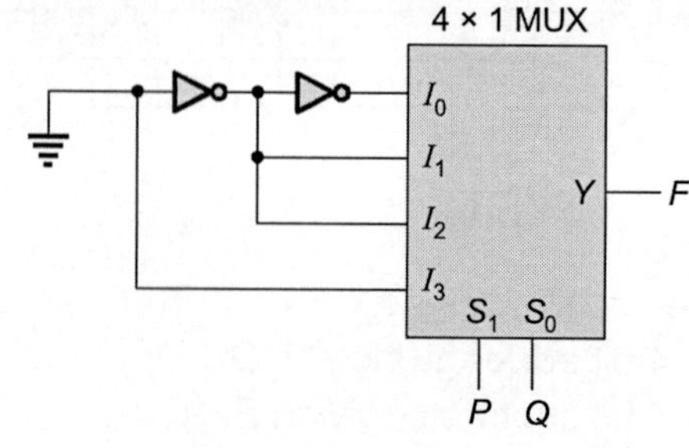

(a) $F = \text{AND}(P, Q)$ (b) $F = \text{OR}(P, Q)$

(c) $F = \text{XNOR}(P, Q)$ (d) $F = \text{XOR}(P, Q)$

[2011 : 1 M]

4.15 The output Y of a 2-bit comparator is logic 1 whenever the 2-bit input A is greater than the 2-bit input B. The number of combination for which the output is logic 1, is

(a) 4
(b) 6
(c) 8
(d) 10

[2012 : 1 M]

4.16 In a half-subtractor circuit with X and Y as inputs, the Borrow (M) and Difference ($N = X - Y$) are given by

(a) $M = X \oplus Y, N = XY$
(b) $M = XY, N = X \oplus Y$
(c) $M = \bar{X}Y, N = X \oplus Y$
(d) $M = X\bar{Y}, N = \overline{X \oplus Y}$

[2014 : 1 M, Set-2]

4.17 Consider the multiplexer based logic circuit shown in the figure.

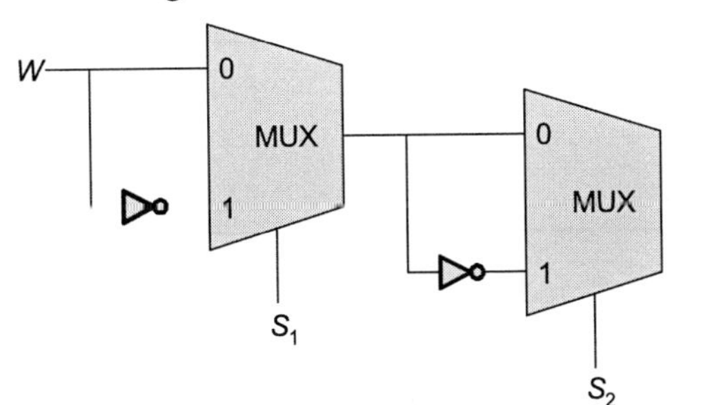

Which one of the following Boolean functions is realized by the circuit?

(a) $F = W\bar{S}_1\bar{S}_2$
(b) $F = WS_1 + WS_2 + S_1S_2$
(c) $F = \bar{W} + S_1 + S_2$
(d) $F = W \oplus S_1 \oplus S_2$

[2014 : 1 M, Set-3]

4.18 In the circuit shown, W and Y are MSBs of the control inputs. The output MSBs is given by

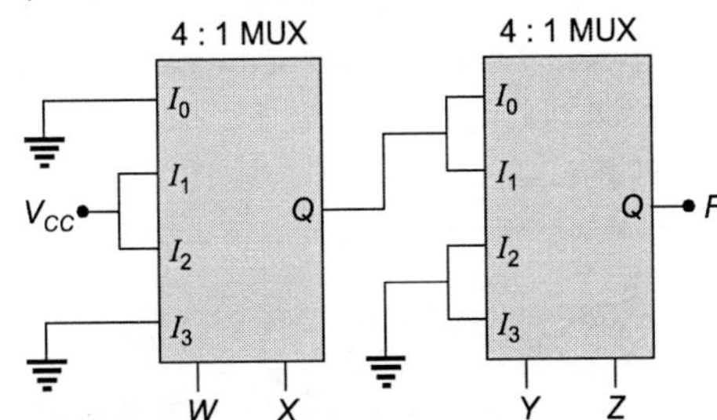

(a) $F = W\bar{X} + \bar{W}X + \bar{Y}\bar{Z}$
(b) $F = W\bar{X} + \bar{W}X + \bar{Y}Z$
(c) $F = W\bar{X}\bar{Y} + \bar{W}X\bar{Y}$
(d) $F = (\bar{W} + \bar{X})\bar{Y}\bar{Z}$

[2014 : 2 M, Set-3]

4.19 If X and Y are inputs and the Difference ($D = X - Y$) and the Borrow (B) are the outputs, which one of the following diagrams implements a half-subtractor?

(a)

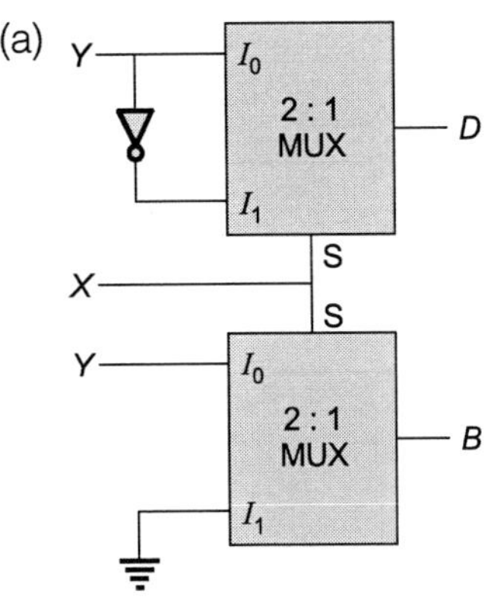

(b)

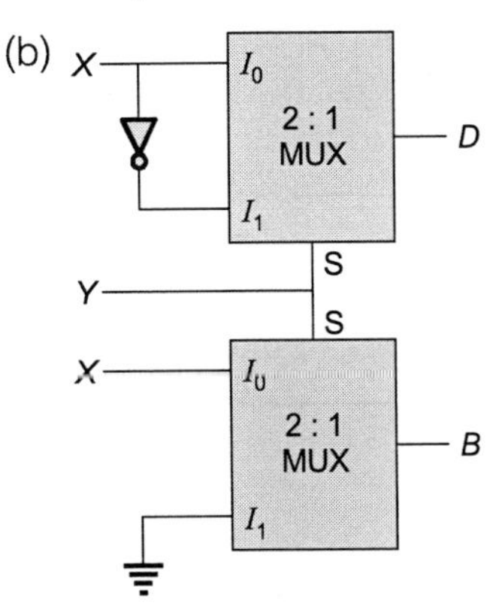

(c)

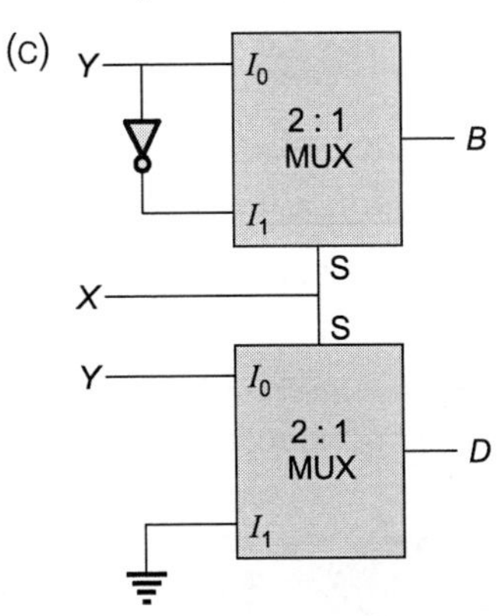

(d)

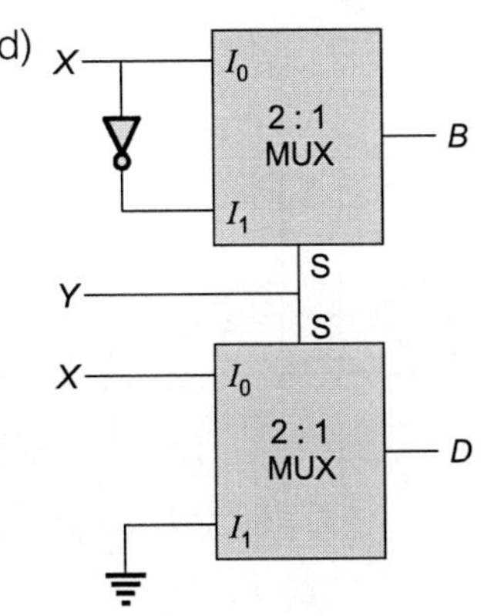

[2014 : 2 M, Set-3]

4.20 An 8-to-1 multiplexer is used to implement a logical function Y as shown in the figure. The output

(a) $Y = A\bar{B}C + A\bar{C}D$

(b) $Y = \bar{A}BC + A\bar{B}D$

(c) $Y = AB\bar{C} + \bar{A}CD$

(d) $Y = \bar{A}\bar{B}D + A\bar{B}C$

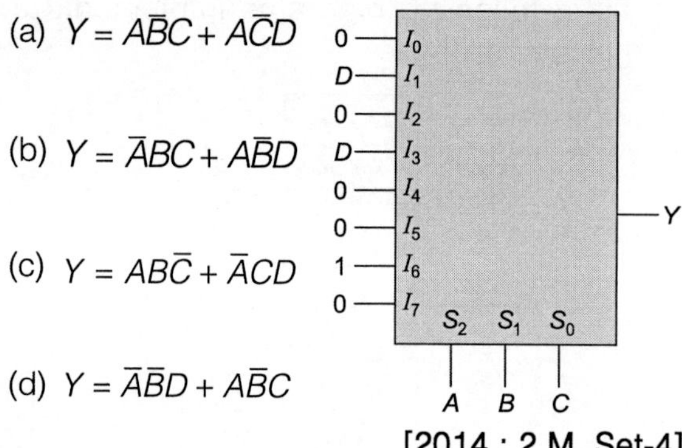

[2014 : 2 M, Set-4]

4.21 A 16-bit ripple carry adder is realized using 16 identical full adders (FA) as shown in the figure. The carry-propagation delay of each FA is 12 ns and the sum-propagation delay of each FA is 15 ns. The worst case delay (in ns) of this 16-bit adder will be ________.

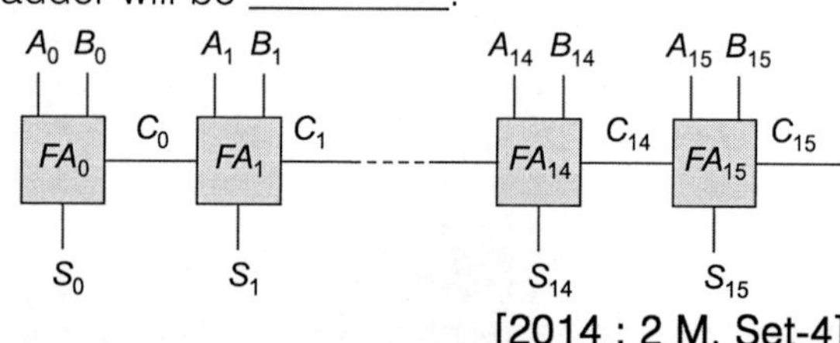

[2014 : 2 M, Set-4]

4.22 A 1-to-8 demultiplexer with data input D_{in}, address inputs S_0, S_1, S_2 (with S_0 as the LSB) and $\bar{Y}_0$ to $\bar{Y}_7$ as the eight demultiplexed outputs, is to be designed using two 2-to-4 decoders (with enable input $\bar{E}$ and address inputs A_0 and A_1) as shown in the figure D_{in}, S_0, S_1 and S_2 are to be connected to P, Q, R and S, but not necessarily in this order. The respective input connections to P, Q, R and S terminals should be

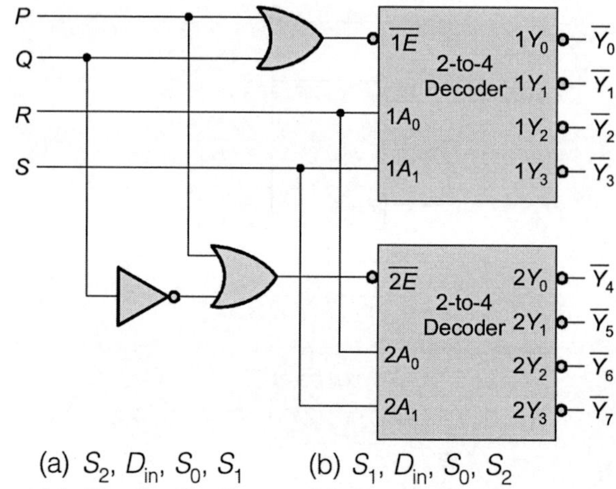

(a) S_2, D_{in}, S_0, S_1
(b) S_1, D_{in}, S_0, S_2
(c) D_{in}, S_0, S_1, S_2
(d) D_{in}, S_2, S_0, S_1

[2015 : 2 M, Set-2]

4.23 Identify the circuit below,

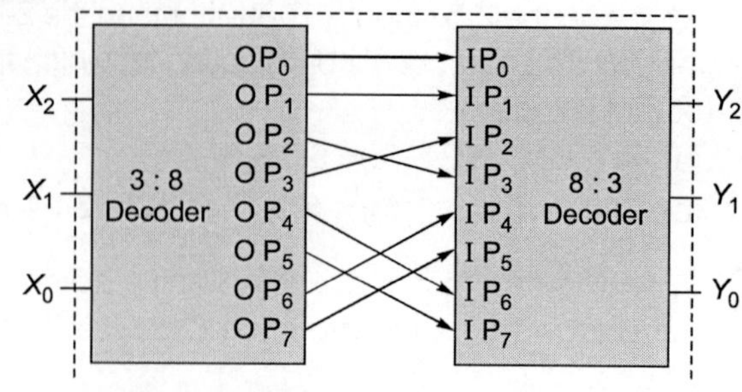

(a) Binary to Gray code converter
(b) Binary to XS3 converter
(c) Gray to Binary converter
(d) XS3 to Binary converter

[2016 : 2 M, Set-1]

4.24 A 4:1 multiplexer is to be used for generating the output carry of a full adder. A and B are the bits to be added while C_{in} is the input carry and C_{out} is the output carry. A and B are to be used as the select bits with A being the more significant select bit.

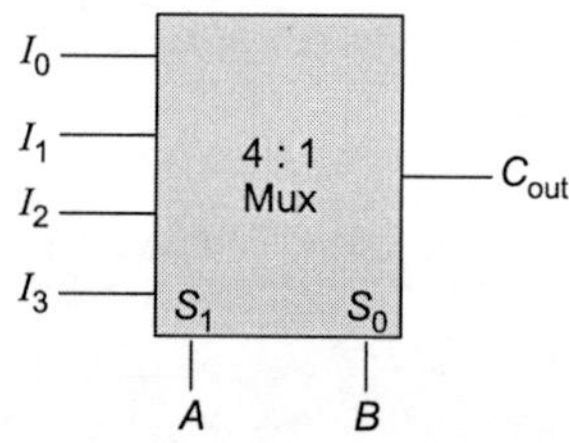

Which one of the following statements correctly describes the choice of signals to be connected to the inputs I_0, I_1, I_2 and I_3 so that the output is C_{out}?

(a) $I_0 = 0$, $I_1 = C_{in}$, $I_2 = C_{in}$ and $I_3 = 1$
(b) $I_0 = 1$, $I_1 = C_{in}$, $I_2 = C_{in}$ and $I_3 = 1$
(c) $I_0 = C_{in}$, $I_1 = 0$, $I_2 = 1$ and $I_3 = C_{in}$
(d) $I_0 = 0$, $I_1 = C_{in}$, $I_2 = 1$ and $I_3 = C_{in}$

[2016 : 1 M, Set-2]

4.25 For the circuit shown in the figure, the delays of NOR gates, multiplexers and inverters are 2 ns, 1.5 ns and 1 ns, respectively. If all the inputs P, Q, R, S and T are applied at the same time instant, the maximum propagation delay (in ns) of the circuit is ________.

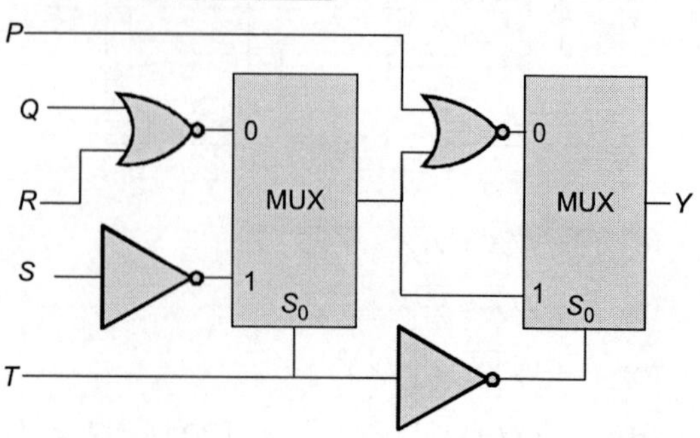

[2016 : 2 M, Set-3]

4.26 Consider the circuit shown in the figure.

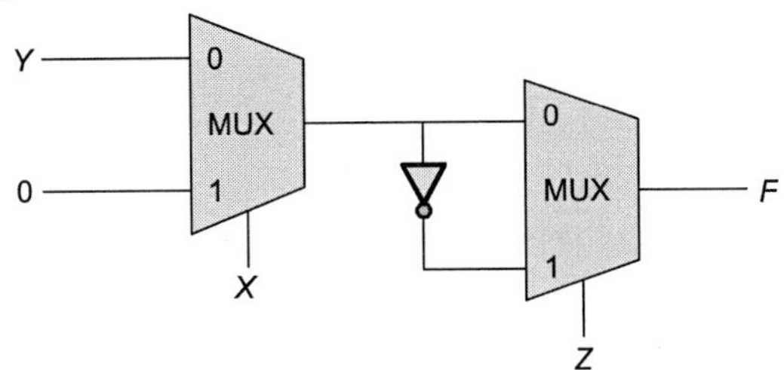

The Boolean expression F implemented by the circuit is

(a) $\bar{X}\bar{Y}\bar{Z}+XY+\bar{Y}Z$ (b) $\bar{X}Y\bar{Z}+XZ+\bar{Y}Z$

(c) $\bar{X}Y\bar{Z}+XY+\bar{Y}Z$ (d) $\bar{X}\bar{Y}\bar{Z}+XZ+\bar{Y}Z$

[2017 : 1 M, Set-2]

4.27 A programmable logic array (PLA) is shown in the figure.

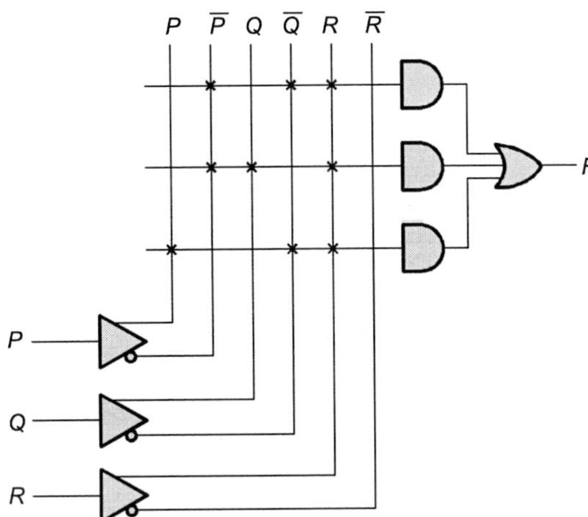

The Boolean function F implemented is

(a) $\bar{P}\bar{Q}R+\bar{P}QR+P\bar{Q}\bar{R}$

(b) $(\bar{P}+\bar{Q}+R)(\bar{P}+Q+R)(P+\bar{Q}+\bar{R})$

(c) $\bar{P}\bar{Q}R+\bar{P}QR+P\bar{Q}R$

(d) $(\bar{P}+\bar{Q}+R)(\bar{P}+Q+R)(P+\bar{Q}+R)$

[2017 : 2 M, Set-2]

4.28 Figure I shows a 4-bit ripple carry adder realized using full adders and Figure II shows the circuit of a full-adder (FA). The propagation delay of the XOR, AND and OR gates in Figure II are 20 ns, 15 ns and 10 ns, respectively. Assume all the inputs to the 4-bit adder are initially reset to 0.

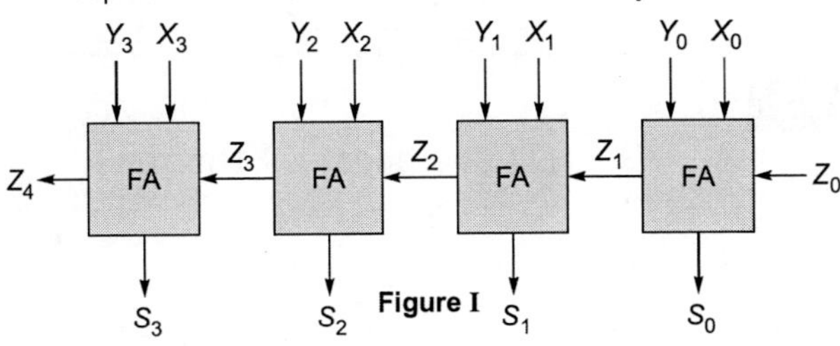

Figure I

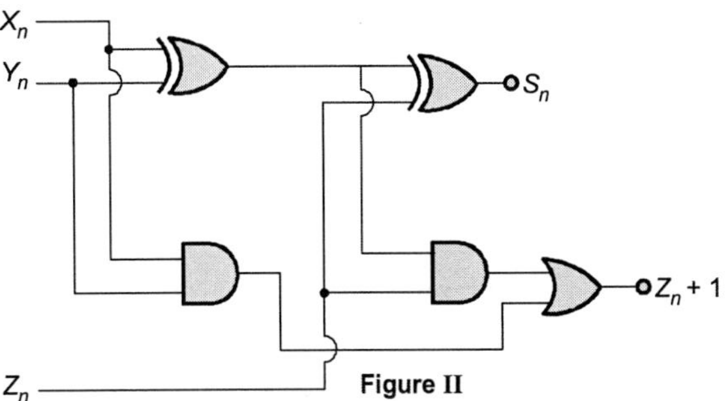

Figure II

At $t = 0$, the inputs to the 4-bit adder are changed to $X_3X_2X_1X_0 = 1100$, $Y_3Y_2Y_1Y_0 = 0100$ and $Z_0 = 1$. The output of the ripple carry adder will be stable at t (in ns) = _____.

[2017 : 2 M, Set-2]

4.29 A four-variable Boolean function is realized using 4×1 multiplexers as shown in the figure.

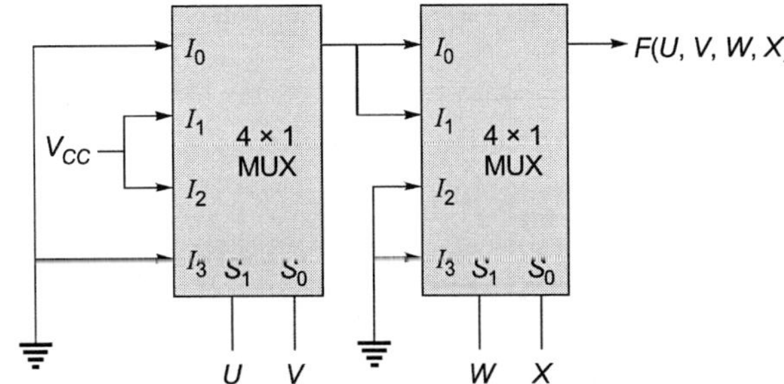

The minimized expression for $F(U, V, W, X)$ is

(a) $(UV+\bar{U}\bar{V})\bar{W}$

(b) $(UV+\bar{U}\bar{V})(\bar{W}\bar{X}+\bar{W}X)$

(c) $(U\bar{V}+\bar{U}V)\bar{W}$

(d) $(U\bar{V}+\bar{U}V)(\bar{W}\bar{X}+\bar{W}X)$ [2018 : 2 M]

4.30 The figure below shows a multiplexer where S_1 and S_0 are the select lines. I_0 to I_3 are the input data lines, EN is the enable line, and $F(P, Q, R)$ is the output. F is

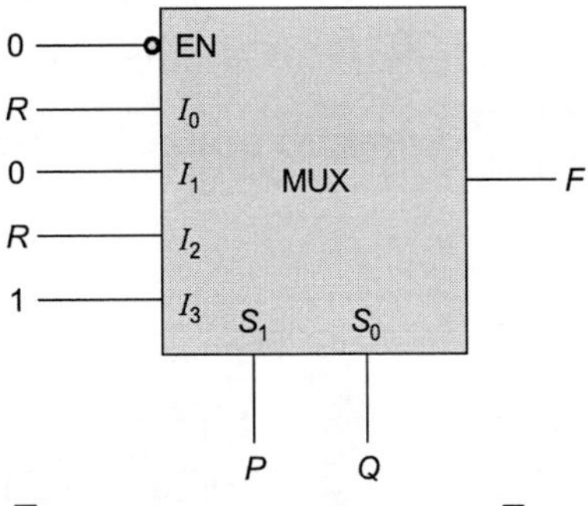

(a) $\bar{Q}+PR$. (b) $P+Q\bar{R}$.

(c) $PQ+\bar{Q}R$. (d) $P\bar{Q}R+\bar{P}Q$.

[2020 : 1 M]

4.31 The propagation delays of the XOR gate, AND gate and multiplexer (MUX) in the circuit shown in the figure are 4 ns, 2 ns and 1 ns, respectively.

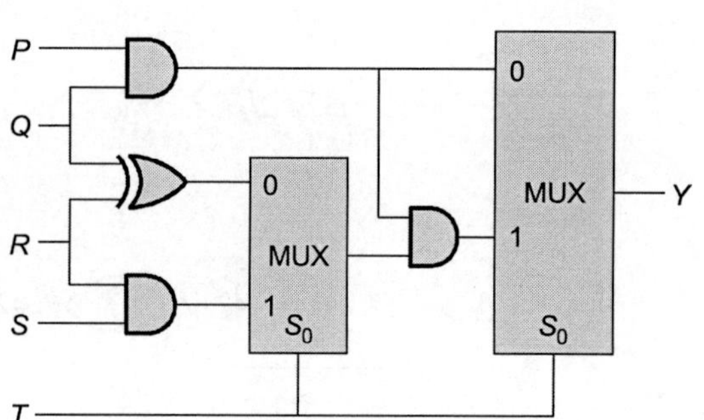

If all the inputs P, Q, R, S and T are applied simultaneously and held constant, the maximum propagation delay of the circuit is

(a) 3 ns (b) 5 ns
(c) 6 ns (d) 7 ns [2021 : 2 M]

4.32 Consider the 2-bit multiplexer (MUX) shown in the figure. For OUTPUT to be the XOR of C and D, the values for A_0, A_1, A_2 and A_3 are ________.

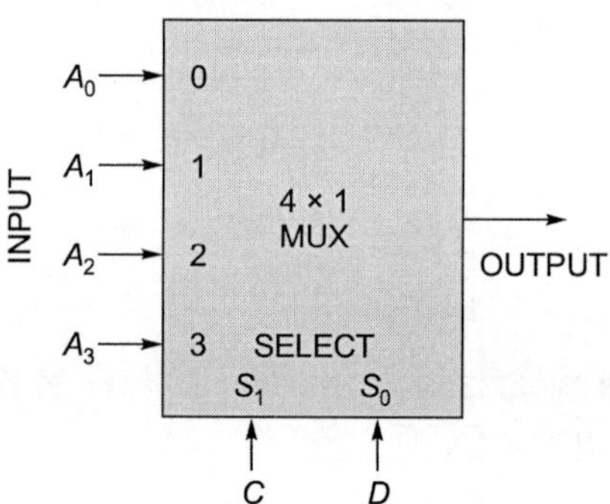

(a) $A_0 = 0, A_1 = 0, A_2 = 1, A_3 = 1$
(b) $A_0 = 1, A_1 = 0, A_2 = 1, A_3 = 0$
(c) $A_0 = 0, A_1 = 1, A_2 = 1, A_3 = 0$
(d) $A_0 = 1, A_1 = 1, A_2 = 0, A_3 = 0$

[2022 : 1 M]

4.33 In the circuit shown below, P and Q are the inputs. The logical function realized by the circuit shown below is

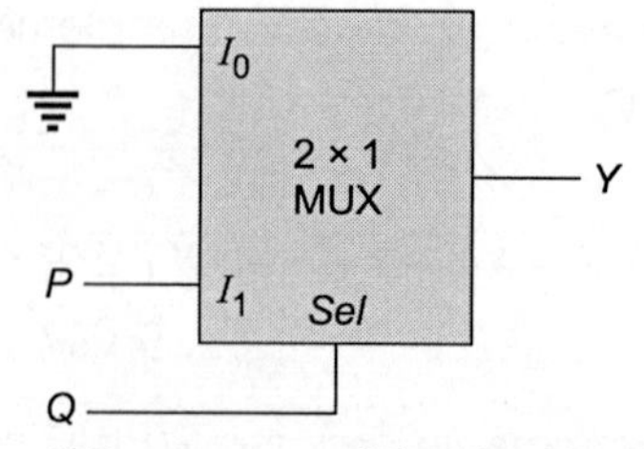

(a) $Y = PQ$ (b) $Y = P + Q$
(c) $Y = \overline{PQ}$ (d) $Y = \overline{P + Q}$

[2023 : 1 M]

4.34 A 4-bit priority encoder has inputs D_3, D_2, D_1 and D_0 in descending order of priority. The two-bit output AB is generated as 00, 01, 10, and 11 corresponding to inputs D_3, D_2, D_1 and D_0, respectively. The Boolean expression of the output bit B is ________.

(a) $\overline{D_3} D_2 + \overline{D_3}\,\overline{D_1}$ (b) $\overline{D_3}\,\overline{D_2}$
(c) $\overline{D_3}\,\overline{D_1}$ (d) $D_3 \overline{D_2} + \overline{D_3} D_1$

[2024 : 2 Marks]

■■■

Answers Combinational Circuits

4.1	(b)	4.2	(c)	4.3	(c)	4.4	(b)	4.5	(d)	4.6	(c)	4.7	(a)
4.8	(a)	4.9	(a)	4.10	(b)	4.11	(d)	4.12	(a)	4.13	(d)	4.14	(d)
4.15	(b)	4.16	(c)	4.17	(d)	4.18	(c)	4.19	(a)	4.20	(c)	4.21	(195)
4.22	(d)	4.23	(*)	4.24	(a)	4.25	(6)	4.26	(b)	4.27	(c)	4.28	(50)
4.29	(c)	4.30	(c)	4.31	(c)	4.32	(c)	4.33	(a)	4.34	(a)		

Explanations Combinational Circuits

4.1 (b)

Two bit binary multiplier

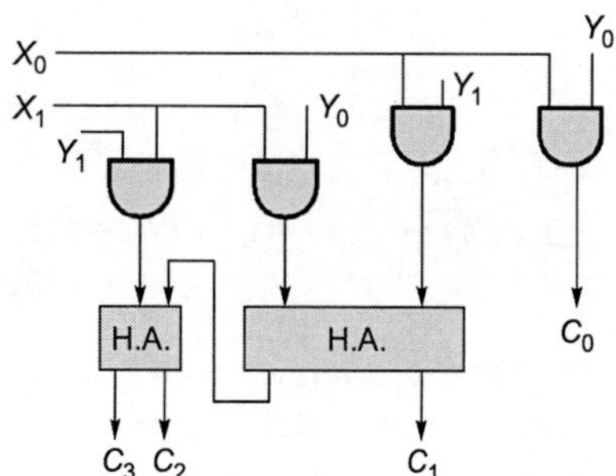

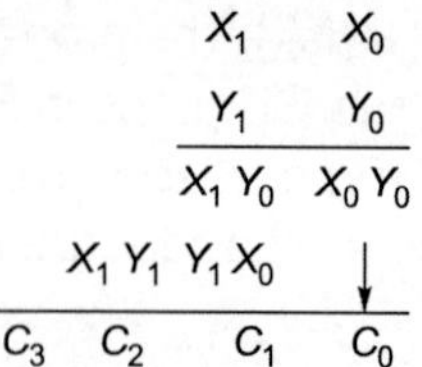

4.2 (c)

A	B	D (difference)	X (borrow)
0	0	0	0
0	1	1	1
1	0	1	0
1	1	0	0

$D = A \oplus B = \bar{A}B + A\bar{B}$

$X = \bar{A}B$

4.4 (b)

$Y = D = P \oplus Q \oplus R$

$Z = \text{Borrow} = RQ + \bar{P}R + Q\bar{P}$

4.5 (d)

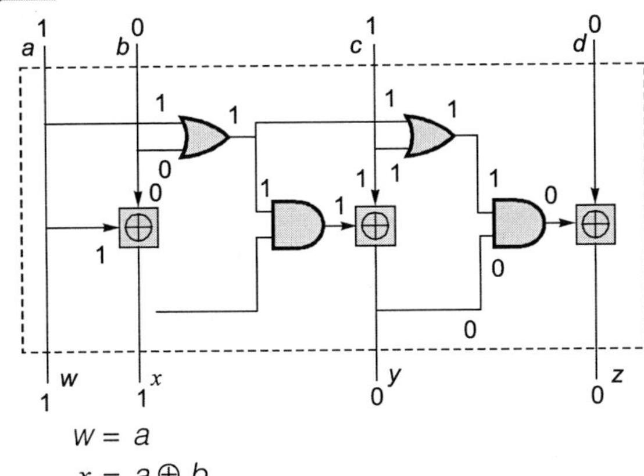

$w = a$

$x = a \oplus b$

$y = c \oplus x(a + b)$

$z = d \oplus y(a + b + c)$

By substituting given option in the Boolean equations of two circuit,: it shows Gray to binary code converter.

The input = 1010 and output = 1100

The circuit is converting Gray code number to Binary code number.

4.6 (c)

$$\text{No. of MUX} = \frac{4}{2} = 2$$

$$\frac{2}{2} = 1$$

$$\therefore \quad \text{Total} = 2 + 1 = 3$$

4.7 (a)

$f = E \cdot A$

$E = \bar{B}C + B\bar{C}$

$\therefore \quad f = A\bar{B}C + AB\bar{C}$

A	B	C	f
1	0	1	1
1	1	0	1

4.8 (a)

Let the output of first MUX is Y

So $Y = \bar{A}B + A\bar{B} = A \oplus B$

$X = \bar{Y}C + Y\bar{C} = Y \oplus C$

So $X = A \oplus B \oplus C$

$= \bar{A}\bar{B}C + \bar{A}B\bar{C} + A\bar{B}\bar{C} + ABC$

4.9 (a)

$Z = PRS + PQR\bar{S} + P\bar{R}S + (P + \bar{Q})\bar{R}\bar{S}$

Mapping above terms in Karnaugh map

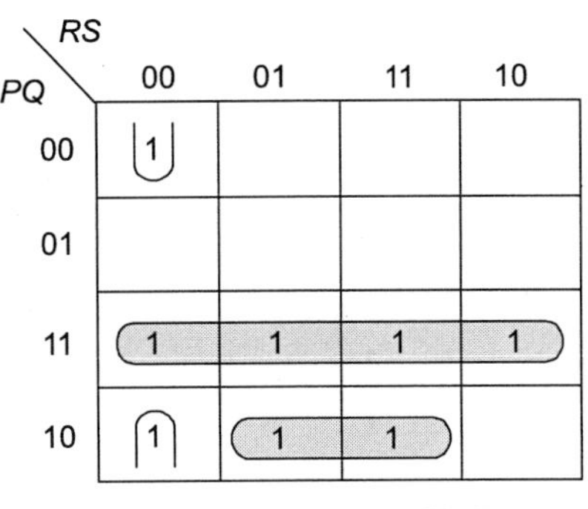

$Z = PQ + P\bar{Q}S + \bar{Q}\bar{R}\bar{S}$

4.10 (b)

P_1	P_2	a	b	c	d	e	f	g
0	0	1	1	1	1	1	1	0
0	1	1	0	1	1	0	1	1
1	0	1	1	0	1	1	0	1
1	1	1	0	0	1	1	1	1

$a = 1$

$b = \bar{P}_2$ ...1 (NOT)

$c = \bar{P}_1$ ...1 (NOT)

$d = 1 = c + e$

$e = P_1 + \bar{P}_2$ ...1 (OR)

$f = \bar{P}_1 + P_2$ ...1 (OR)

$g = P_1 + P_2$ ... 1 (OR)

$\Rightarrow \quad g = P_1 + P_2$

$d = 1 = c + e$

4.11 (d)

2 – NOT gates

3 – OR gates

4.12 (a)

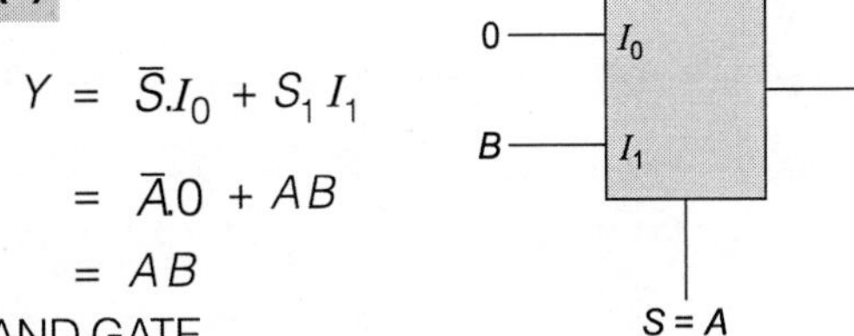

$Y = \bar{S}.I_0 + S_1 I_1$

$= \bar{A}0 + AB$

$= AB$

AND GATE

Similarly EX OR gate required 2 MUX of 2 × 1.

4.13 (d)

$$F(A, B, C, D) = \bar{A}\bar{B}C + \bar{A}BD + A\bar{B}\bar{C} + AB(\bar{C}\bar{D})$$
$$= \bar{A}\bar{B}C(D+\bar{D}) + \bar{A}B(C+\bar{C})D + A\bar{B}\bar{C}(D+\bar{D}) + AB\bar{C}\bar{D}$$

Placing above minterms in Karnaugh map,

AB \ CD	00	01	11	10
00			1	1
01		1	1	
11	1			
10	1	1		

So, $F = \Sigma m(2, 3, 5, 7, 8, 9, 12)$

4.14 (d)

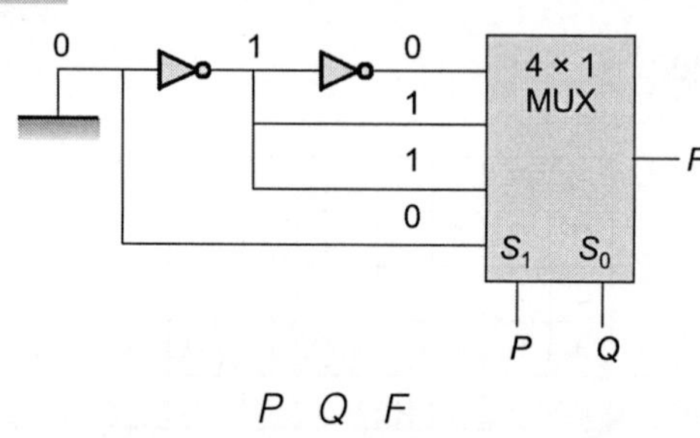

P	Q	F
0	0	0
0	1	1
1	0	1
1	1	0

$F = P \oplus Q$

4.15 (b)

Output will be 1 if $A > B$.

- If $B = 00$ then there will be three combinations for which output will be 1 i.e. when $A = 01$, 10, or 11.
- If $B = 01$ there will be two conditions i.e. $A = 10$ and 11.
- If $B = 10$ there will be one condition i.e. $A = 11$.

So total 6 combinations are there for which output will be 1.

4.16 (c)

X	Y	$N = X - Y$	M
0	0	0	0
0	1	1	1
1	0	1	0
1	1	0	0

$N = X \oplus Y$, $M = \bar{X}Y$

4.17 (d)

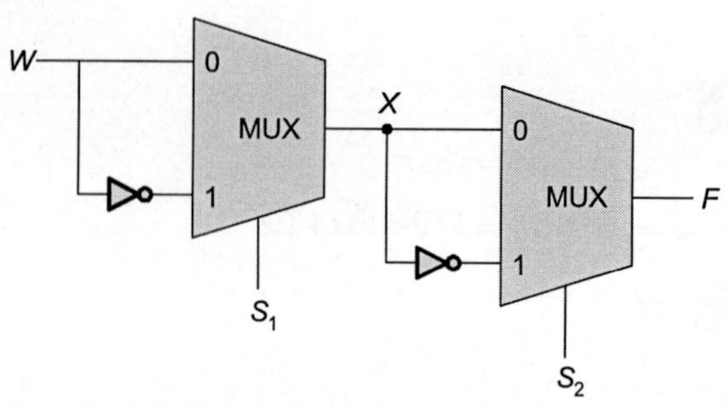

Using the state equation of the multiplexer, we get

$$X = \bar{S}_1 \cdot W + S_1\bar{W}$$

So, $F = \bar{S}_2 \cdot X + S_2\bar{X}$

$$= \bar{S}_2(\bar{S}_1W + S_1\bar{W}) + S_2(\overline{\bar{S}_1W + S_1\bar{W}})$$
$$= \bar{S}_1\bar{S}_2W + S_1\bar{S}_2\bar{W} + S_2\bar{S}_1\bar{W} + S_2S_1W$$
$$F = W \oplus S_1 \oplus S_2$$

4.18 (c)

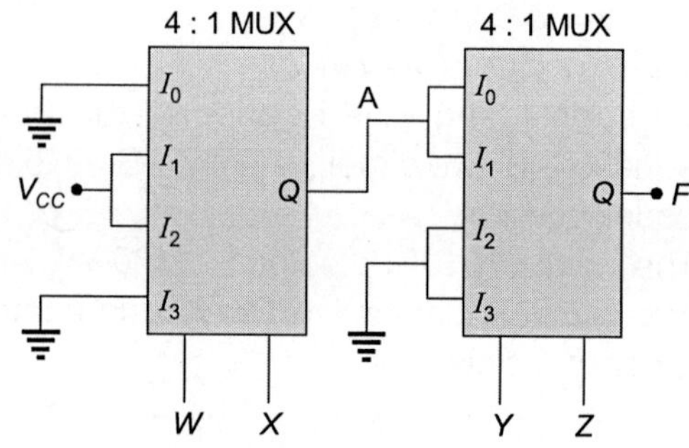

For the state equation of 4 : 1 MUX, we know that

$$A = \bar{W}\bar{X}I_0 + \bar{W}XI_1 + W\bar{X}I_2 + WXI_3$$
$$= \bar{W}X + W\bar{X} \quad \text{(Since } V_{CC} = 1\text{)}$$

Now, output F

$$= \bar{Y}\bar{Z}I_0 + \bar{Y}ZI_1 + Y\bar{Z}I_2 + YZI_3$$
$$= \bar{W}X\bar{Y}\bar{Z} + W\bar{X}\bar{Y}\bar{Z} + \bar{W}X\bar{Y}Z + W\bar{X}\bar{Y}Z$$
$$= \bar{W}X\bar{Y}(\bar{Z} + Z) + W\bar{X}\bar{Y}(\bar{Z} + Z)$$
$$= W\bar{X}\bar{Y} + \bar{W}X\bar{Y}$$

4.19 (a)

X	Y	D	B
0	0	0	0
0	1	1	1
1	0	1	0
1	1	0	0

$\therefore$ Difference $= \bar{X}Y + X\bar{Y} = X \oplus Y$

Borrow $= \bar{X}Y$

Also using the state equation of MUX if X is used as a select line and Y as input, we get

$$D = \bar{X}Y + X\bar{Y}$$

$$B = \bar{X}Y$$

option (a) satisfies

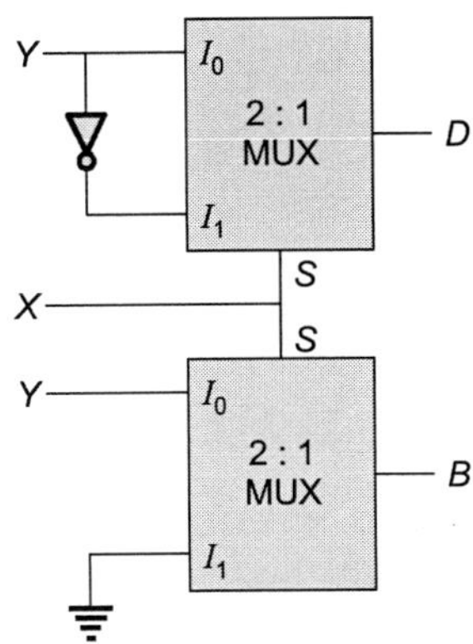

4.20 (c)

Depending upon the select lines one input is transferred to the output in multiplexer.

A	B	C	Y
0	0	0	0
0	0	1	D
0	1	0	0
0	1	1	D
1	0	0	0
1	0	1	0
1	1	0	1
1	1	1	0

$$\therefore \quad Y = \bar{A}\bar{B}CD + \bar{A}BCD + AB\bar{C}$$

$$= \bar{A}CD(\bar{B} + B) + AB\bar{C}$$

$$= \bar{A}CD + AB\bar{C}$$

4.21 (195)

Full adder is a combinational circuit that performs the sum of three bits i.e. add two bits and a carry from previous addition and produces output as sum and carry.

Given carry propagation delay of each FA = 12 nsec.

Sum-propagation delay of each FA = 15 nsec.

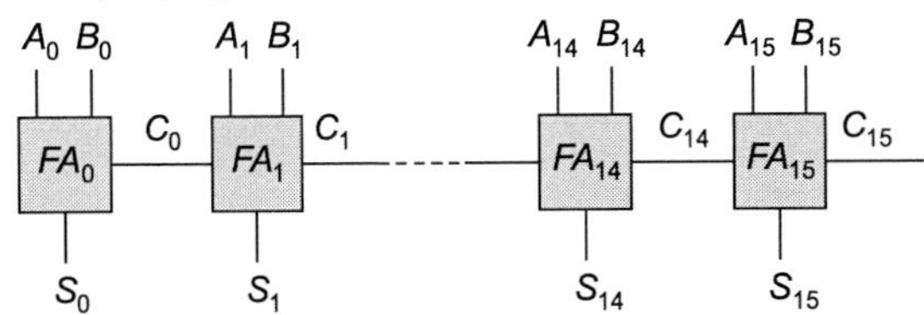

As can be seen from above diagram, full adder is not received from previous full adder.

So, in 16-bit ripple carry adder, worst case delay of this 16-bit adder will be

$$= (15 \times 12) \text{ nsec.} + 15 \text{ nsec.}$$

$$= 195 \text{ nsec.}$$

4.22 (d)

Consider a 1 ×8 demultiplexer

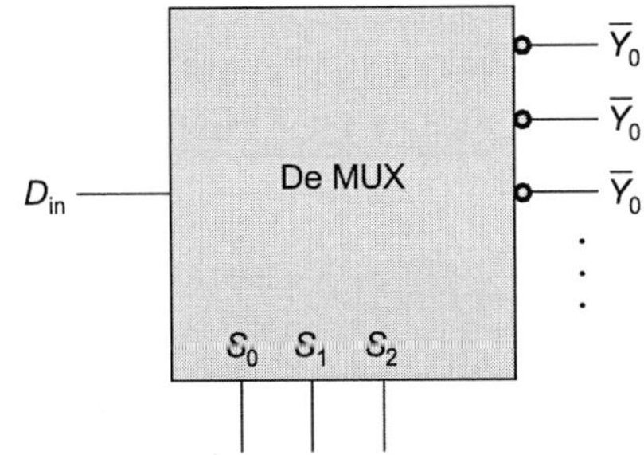

$$\text{So, } \bar{Y}_0 = \overline{(D_{in} \cdot \bar{S}_0 \bar{S}_1 \bar{S}_2)}$$

$$\bar{Y}_0 = (\bar{D}_{in} + S_0 + S_1 + S_2)$$

$$\bar{Y}_1 = (\bar{D}_{in} + \bar{S}_0 + S_1 + S_2)$$

$$\bar{Y}_2 = (\bar{D}_{in} + S_0 + \bar{S}_1 + S_2)$$

$$\bar{Y}_3 = (\bar{D}_{in} + \bar{S}_0 + \bar{S}_1 + S_2)$$

$$\bar{Y}_4 = (\bar{D}_{in} + S_0 + S_1 + \bar{S}_2)$$

$$\bar{Y}_5 = (\bar{D}_{in} + \bar{S}_0 + S_1 + \bar{S}_2)$$

$$\bar{Y}_6 = (\bar{D}_{in} + S_0 + \bar{S}_1 + \bar{S}_2)$$

$$\bar{Y}_7 = (\bar{D}_{in} + S_0 + S_1 + S_2)$$

From the circuit given in question we can see that

$$\bar{Y}_0 = (1A_0 + 1A_1 + \bar{E})$$

$$\bar{Y}_0 = (R + S + P + Q)$$

Similarly,

$$\bar{Y}_1 = (1\bar{A}_0 + 1A_1 + 1\bar{E}) = (P + Q + \bar{R} + S)$$

$$\bar{Y}_4 = (2\bar{A}_0 + 2A_1 + 2\bar{E}) = (R + S + P + \bar{Q})$$

So comparing, we get

$$P = \bar{D}_{in}$$

$$Q = S_2$$

$$R = S_1$$

$$S = S_0$$

4.23 (*)

The truth table of the circuit is shown below,

X_2	X_1	X_0	Y_2	Y_1	Y_0
0	0	0	0	0	0
0	0	1	0	0	1
0	1	0	0	1	1
0	1	1	0	1	0
1	0	0	1	1	0
1	0	1	1	1	1
1	1	0	1	0	0
1	1	1	1	0	1

As per the truth table, none of the options given in the question are correct. However, by making some (minor) changes in the circuit, the answer could be obtained as option (a).

4.24 (a)

In case of a full adder,

$C_{out}(A, B, C_{in}) = \Sigma(3, 5, 6, 7)$

A, B: Applied at select input of MUX; C_{in}: Applied at data input of MUX

	I_0	I_1	I_2	I_3
$\overline{C_{in}}$	0	2	4	(6)
C_{in}	1	(3)	(5)	(7)
	0	C_{in}	C_{in}	1

$\therefore$

$$I_0 = 0$$
$$I_1 = C_{in}$$
$$I_2 = C_{in}$$
$$I_3 = 1$$

4.25 (6)

When, T = logic 0, the path followed by the circuit would be,

NOR gate $\rightarrow$ MUX 1 $\rightarrow$ MUX 2

$\Rightarrow$ 2 ns $\rightarrow$ 1.5 ns $\rightarrow$ 1.5 ns

$\Rightarrow$ 5 ns

When, T = logic 1, the path followed by the circuit would be,

NOR gate $\rightarrow$ MUX 1 $\rightarrow$ NOR gate $\rightarrow$ MUX 2

$\Rightarrow$ 1 ns $\rightarrow$ 1.5 ns $\rightarrow$ 2 ns $\rightarrow$ 1.5 ns

$\Rightarrow$ 6 ns

$\therefore$ Maximum propagation delay is 6 ns

4.26 (b)

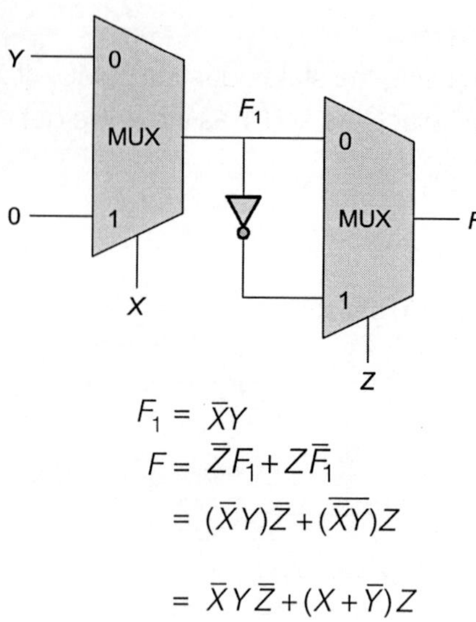

$$F_1 = \bar{X}Y$$
$$F = \bar{Z}F_1 + Z\bar{F_1}$$
$$= (\bar{X}Y)\bar{Z} + (\overline{\bar{X}Y})Z$$
$$= \bar{X}Y\bar{Z} + (X + \bar{Y})Z$$
$$F = \bar{X}Y\bar{Z} + XZ + \bar{Y}Z$$

4.27 (c)

$$F = \bar{P}\bar{Q}R + \bar{P}QR + P\bar{Q}R$$

4.28 (50)

In this question inputs to be added are :

$$X_3X_2X_1X_0 = 1100$$
$$Y_3Y_2Y_1Y_0 = 0100 \text{ and } Z_0 = 1$$

For this combination of addition, total minimum delay depends on the addition of most-significant two bits (since least significant two bits are zeros they does not cause any change in Z_1 and Z_2). So, in the process of addition of given two digits, waveforms at Z_1 and Z_2 become stable at $t = 0$ itself.

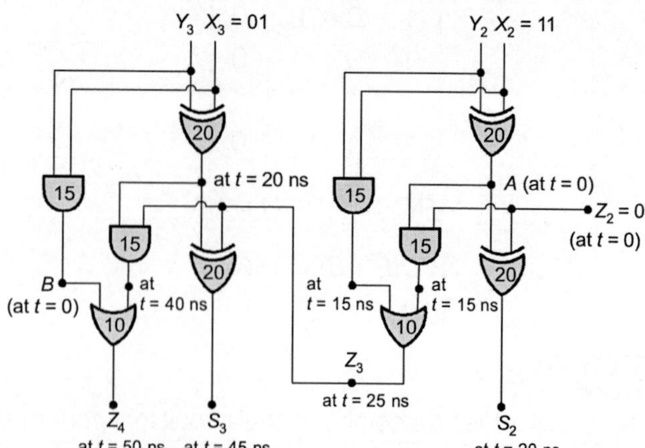

In the above diagram the waveform at A and B become stable at $t = 0$ itself, as the applied input combinations does not cause any change.

So, for he given combination of inputs, output will settle at t = 50 ns.

4.29 (c)

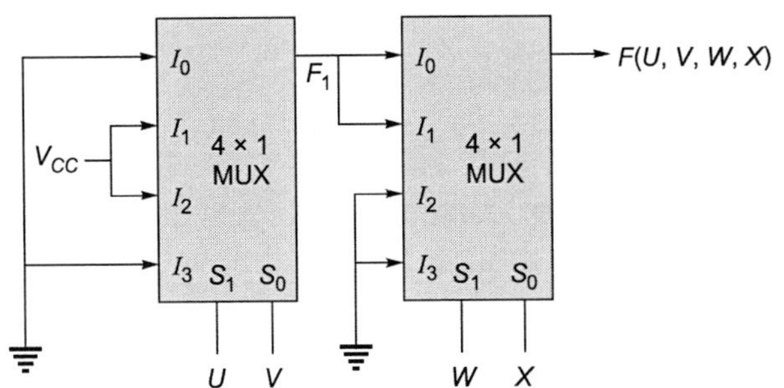

Output of the first multiplexer can be expressed as,

$$F_1 = \bar{U}V + U\bar{V}$$

Output of the second multiplexer can be expressed as,

$$F = \bar{W}\bar{X}F_1 + \bar{W}XF_1 = \bar{W}F_1$$
$$= (\bar{U}V + U\bar{V})\bar{W}$$

4.30 (c)

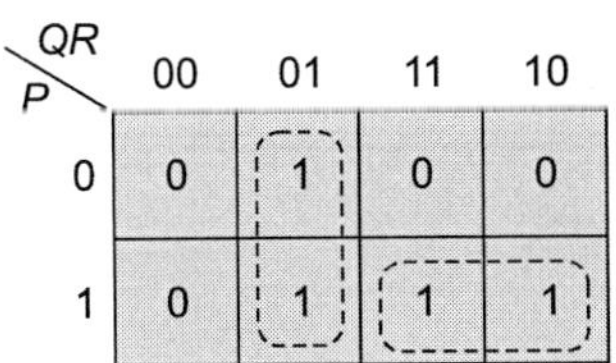

P \ QR	00	01	11	10
0	0	1	0	0
1	0	1	1	1

Output, $F = \bar{P}\bar{Q}R + P\bar{Q}R + PQ$

$$F = \bar{Q}R + PQ$$

4.31 (c)

For $T = 0$

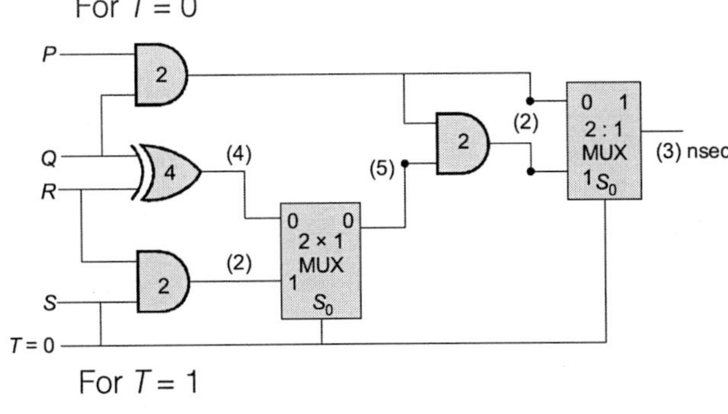

For $T = 1$

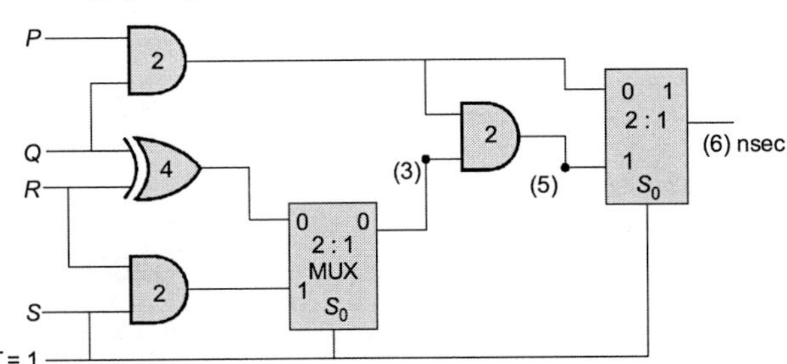

∴ Maximum propagation delay for the circuit is 6ns.

4.32 (c)

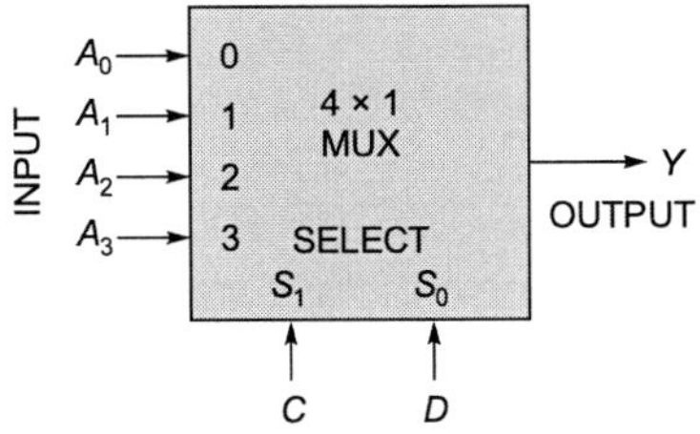

The truth table for the given 4 : 1 MUX is as below:

$S_1(C)$	$S_0(D)$	Y
0	0	A_0
0	1	A_1
1	0	A_2
1	1	A_3

On comparing the above truth table with truth table of XOR gate as shown below:

C	D	$Y = C \oplus D$
0	0	0
0	1	1
1	0	1
1	1	0

We get, $A_0 = 0,\ A_1 = 1,\ A_2 = 1,\ A_3 = 0$

4.33 (a)

From the Y given 2 × 1 MUX,

Output, $Y = \bar{Q}\cdot I_0 + Q\cdot I_1 = \bar{Q}\cdot 0 + Q\cdot P = PQ$

4.34 (a)

For the given 4-bit priority encoder, D_3 has the highest priority followed by D_2, D_1 and D_0. The truth table of the encoder can be drawn as below:

D_3	D_2	D_1	D_0	A	B
0	0	0	0	X	X
0	0	0	1	1	1
0	0	1	X	1	0
0	1	X	X	0	1
1	X	X	X	0	0

(Highest priority) D_3, D_2, D_1, D_0 (Lowest priority) → 4 : 1 Priority encoder → A, B

Drawing the K-Map for the output B,

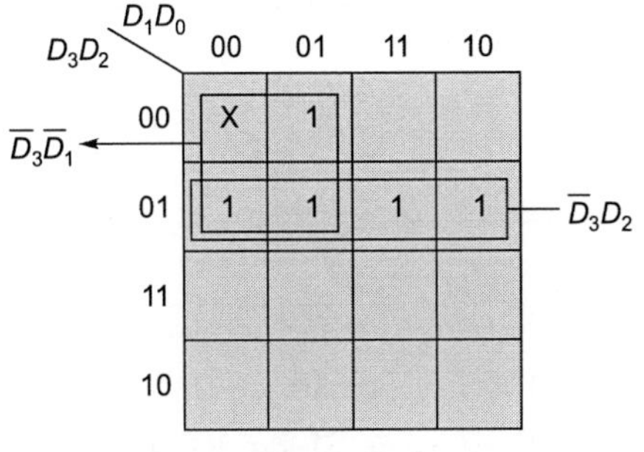

D_3D_2 \ D_1D_0	00	01	11	10
00	X	1		
01	1	1	1	1
11				
10				

$$\therefore \quad B = \overline{D_3}D_2 + \overline{D_3}\,\overline{D_1}$$

5 Sequential Circuits

5.1 A switch-tail ring counter is made by using a single *D* flip-flop. The resulting circuit is a

(a) *SR* flip-flop (b) *JK* flip-flop
(c) *D* flip-flop (d) *T* flip-flop

[1995 : 1 M]

5.2 An *R-S* latch is

(a) combinatorial circuit
(b) synchronous sequential circuit.
(c) one bit memory element
(d) one clock delay element [1995 : 1 M]

5.3 In a *J-K* flip-flop we have $J = \bar{Q}$ and $K = 1$ (see figure) Assuming the flip-flop was initially cleared and then clocked for 6 pulses, the sequence at the *Q* output will be

(a) 010000
(b) 011001
(c) 010010
(d) 010101

[1997 : 1 M]

5.4 Figure shows a mod-*K* counter, Here *K* is equal to

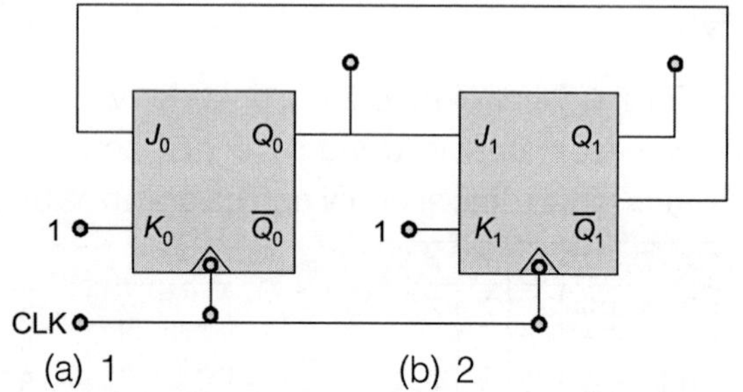

(a) 1 (b) 2
(c) 3 (d) 4 [1998 : 1 M]

5.5 In fig. below, *A* = 1 and *B* = 1. The input *B* is now replaced by a sequence 101010...., the outputs x and *y* will be

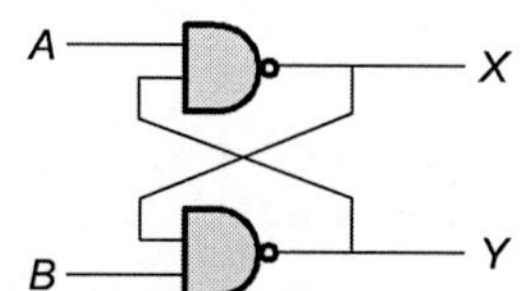

(a) fixed at 0 and 1, respectively
(b) x = 1010 ...... while *y* = 0101 ......
(c) x = 1010 ...... and *y* = 1010 ......
(d) fixed at 1 and 0, respectively

[1998 : 2 M]

5.6 The ripple counter shown in the figure works as a

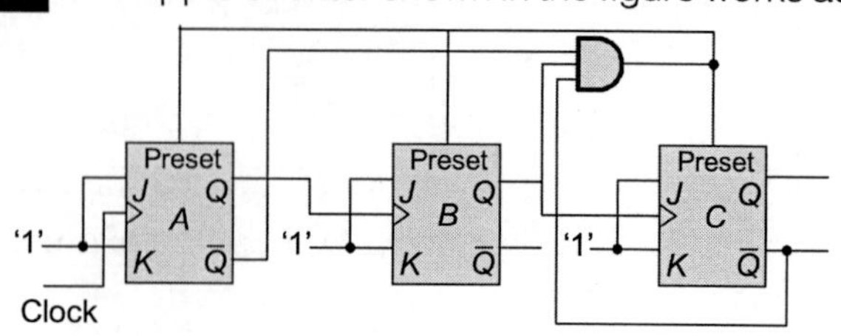

(a) mod-3 up counter
(b) mod-5 up counter
(c) mod-3 down counter
(d) mod-5 down counter

[1999 : 2 M]

5.7 A sequential circuit using *D* Flip-Flop and logic gates is shown in the figure, where *X* and *Y* are the inputs and *Z* is the output. The circuit is

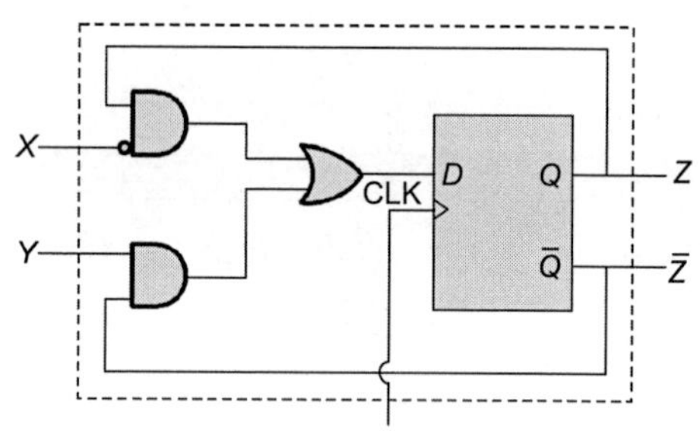

(a) $S-R$ Flip-Flop with inputs $X = R$ and $Y = S$
(b) $S-R$ Flip-Flop with inputs $X = S$ and $Y = R$
(c) $J-K$ Flip-Flop with inputs $X = J$ and $Y = K$
(d) $J-K$ Flip-Flop with inputs $X = K$ and $Y = J$

[2000 : 2 M]

5.8 In the figure, the *J* and *K* inputs of all the four flip-flips are made high The frequency of the signal at output *Y* is

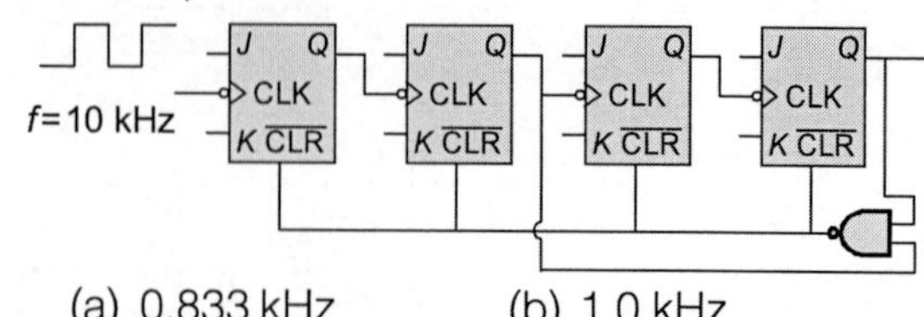

(a) 0.833 kHz (b) 1.0 kHz
(c) 0.91 kHz (d) 0.77 kHz

[2000 : 2 M]

5.9 For the ring oscillator shown in the figure, the propagation delay of each inverter is 100 pico sec. What is the fundamental frequency of the oscillator output?

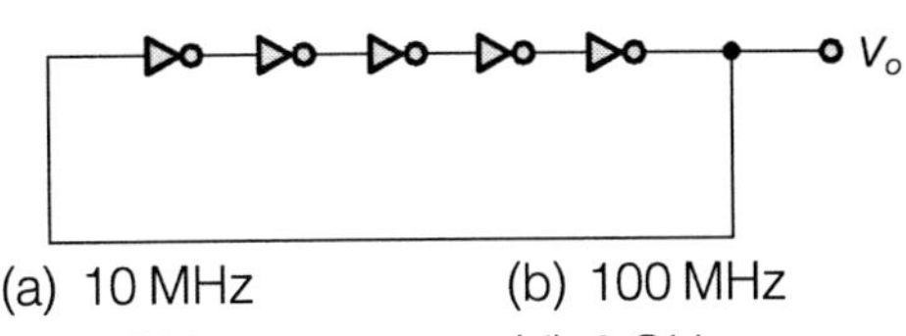

(a) 10 MHz (b) 100 MHz
(c) 1 GHz (d) 2 GHz

[2001 : 1 M]

5.10 The digital block in the figure is realized using two positive edge triggered *D*-flip-flops. Assume that for $t < t_0$, $Q_1 = Q_2 = 0$. The circuit in the digital block is given by:

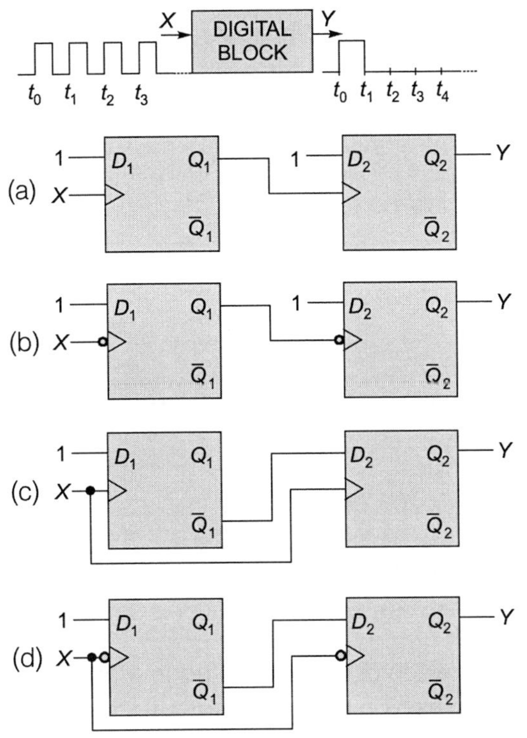

[2001 : 2 M]

5.11 A 0 to 6 counter consists of 3 flip flops and a combination circuit of 2 input gate(s). The combination circuit consists of

(a) one AND gate
(b) one OR gate
(c) one AND gate and one OR gate
(d) two AND gates [2003 : 1 M]

5.12 A 4 bit ripple counter and a 4 bit synchronous counter are made using flip flops having a propagation delay of 10 ns each. If the worst case delay in the ripple counter and the synchronous counter be *R* and *S* respectively, then

(a) R = 10 ns, S = 40 ns
(b) R = 40 ns, S = 10 ns
(c) R = 10 ns, S = 30 ns
(d) R = 30 ns, S =10 ns [2003 : 2 M]

5.13 A master-slave flip-flop has the characteristic that

(a) change in the input immediately reflected in the output
(b) change in the output occurs when the state of the master is affected
(c) change in the output occurs when the state of the slave is affected
(d) both the master and the slave states are affected at the same time [2004 : 1 M]

5.14 Choose the correct one from among the alternatives *A*, *B*, *C*, *D* after matching an item from Group-1 with the most appropriate item in Group-2.

Group-1
P. Shift register
Q. Counter
R. Decoder

Group-2
1. Frequency division
2. Addressing in memory chips
3. Serial to parallel data conversion

(a) P – 3, Q – 2, R–1
(b) P – 3, Q – 1, R – 2
(c) P – 2, Q – 1, R – 3
(d) P – 1, Q – 2, R – 2 [2004 : 1 M]

5.15 In the modulo-6 ripple counter shown in the figure, the output of the 2-input gate is used to clear the *J-K* flip-flops.

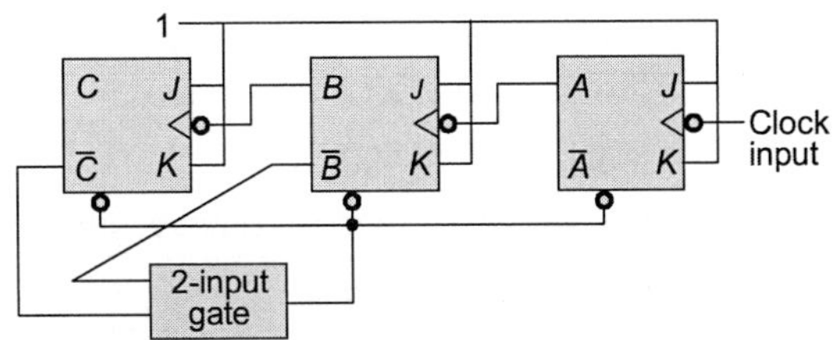

The 2-input gate is

(a) a NAND gate (b) a NOR gate
(c) an OR gate (d) an AND gate

[2004 : 2 M]

5.16 The present output Q_n of an edge triggered *JK* flip-flop is logic 0. If J = 1, then Q_{n+1}

(a) cannot be determined
(b) will be logic 0
(c) will be logic 1
(d) will race around [2005 : 2 M]

5.17 The given figure shows a ripple counter using positive edge triggered flip-flops.
If the present state of the counter is $Q_2Q_1Q_0 = 011$, then its next state $(Q_2Q_1Q_0)$ will be

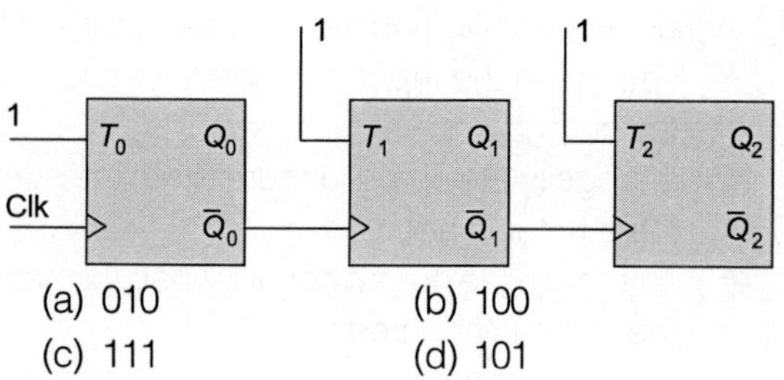

(a) 010 (b) 100
(c) 111 (d) 101

[2005 : 2 M]

5.18 For the circuit shown in the figure below, two 4-bit parallel-in serial-out shift registers loaded with the data shown are used to feed the data to a full adder. Initially, all the flip-flops are in clear state. After applying two clock pulses, the outputs of the full adder should be

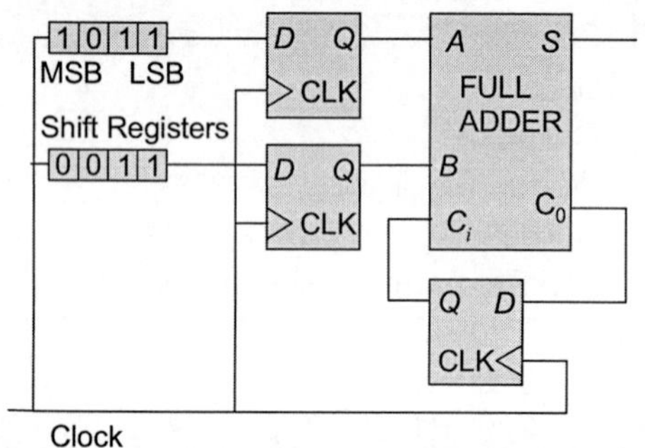

(a) $S = 0$ $C_0 = 0$ (b) $S = 0$ $C_0 = 1$
(c) $S = 1$ $C_0 = 0$ (d) $S = 1$ $C_0 = 1$

[2006 : 2 M]

5.19 Two D-flip-flops, as shown below, are to be connected as a synchronous counter that goes through the following Q_1Q_0 sequence

$00 \rightarrow 01 \rightarrow 11 \rightarrow 10 \rightarrow 00 \rightarrow ......$

The inputs D_0 and D_1 respectively should be connected as

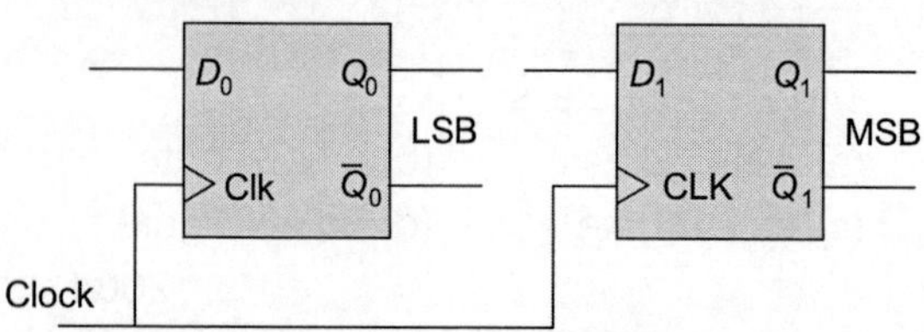

(a) $\overline{Q_1}$ and Q_0 (b) $\overline{Q_0}$ and Q_1
(c) $\overline{Q_1}Q_0$ and $\overline{Q_1}Q_0$ (d) $\overline{Q_1}\,\overline{Q_0}$ and Q_1Q_0

[2006 : 2 M]

5.20 The following binary values were applied to the X and Y inputs of the NAND latch shown in the figure in the sequence indicated below:

$X = 0,\ Y = 1;$
$X = 0,\ Y = 0;$
$X = 1,\ Y = 1.$

The corresponding stable P, Q outputs will be

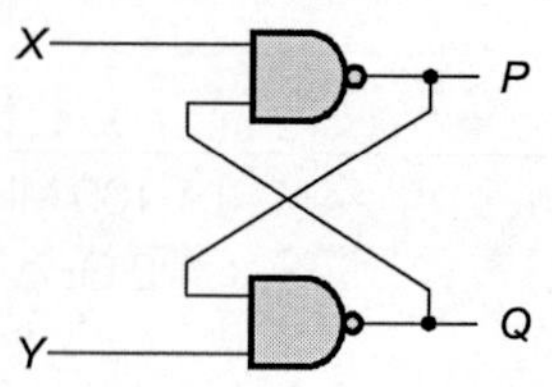

(a) $P = 1,\ Q = 0;\ P = 1,\ Q = 0;\ P = 1,\ Q = 0$ or $P = 0,\ Q = 1$
(b) $P = 1,\ Q = 0;\ P = 0,\ Q = 1$ or $P = 0,\ Q = 1;\ P = 0,\ Q = 1$
(c) $P = 1,\ Q = 0;\ P = 1,\ Q = 1;\ P = 1,\ Q = 0$ or $P = 0,\ Q = 1$
(d) $P = 1,\ Q = 0;\ P = 1,\ Q = 1;\ P = 1,\ Q = 1$

[2007 : 2 M]

5.21 For the circuit shown, the counter state (Q_1Q_0) follows the sequence

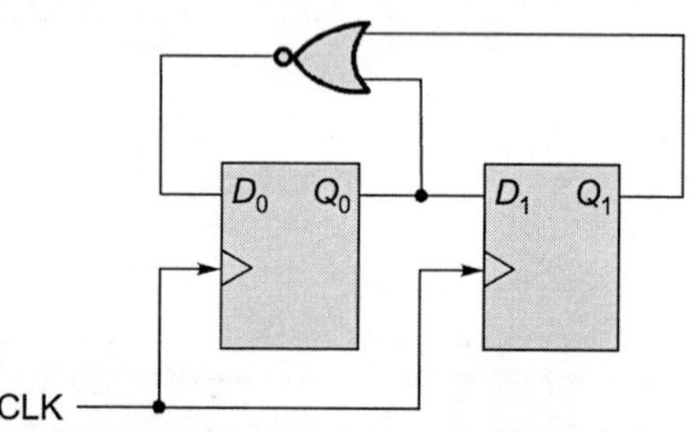

(a) 00, 01, 10, 11, 00 ...
(b) 00, 01, 10, 00, 01 ...
(c) 00, 01, 11, 00, 01 ...
(d) 00, 10, 11, 00, 10 ...

[2007 : 2 M]

5.22 For each of the positive edge-triggered J-K flip flop used in the following figure, the propagation delay is ΔT.

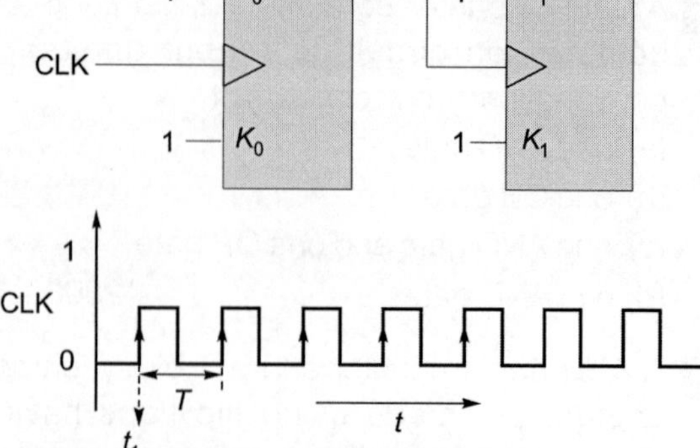

Which of the following waveforms correctly represents the output at Q_1?

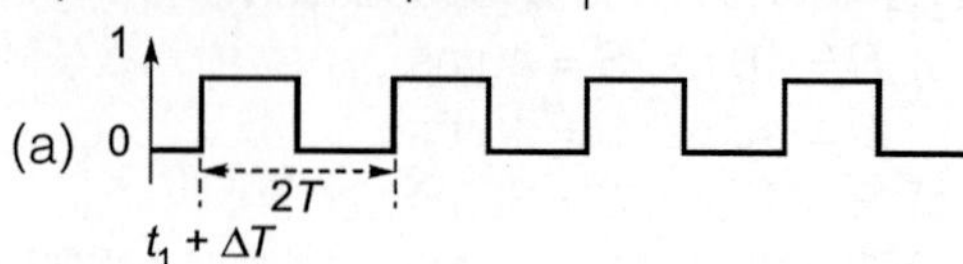

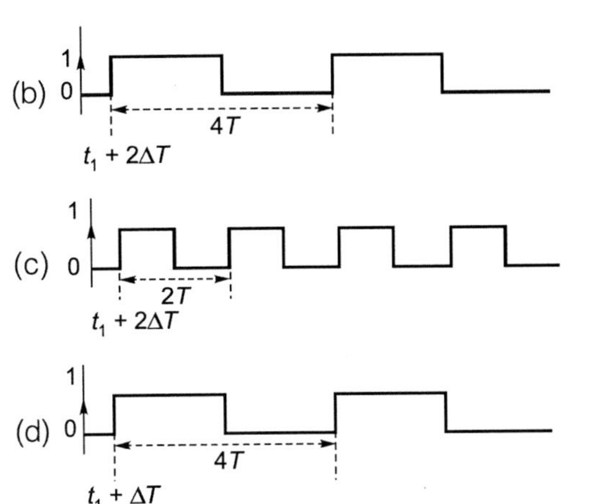

[2008 : 2 M]

5.23 For the circuit shown in the figure, D has a transition from 0 to 1 after CLK changes from 1 to 0. Assume gate delays to be negligible

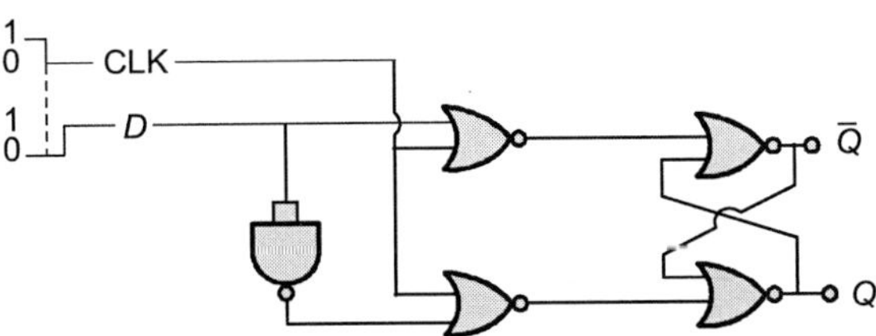

Which of the following statements is true?

(a) Q goes to 1 at the CLK transition and stays at 1.
(b) Q goes to 0 at the CLK transition and stays at 0.
(c) Q goes to 1 at the CLK transition and goes to 0 when D goes to 1.
(d) Q goes to 0 at the CLK transition and goes to 1 when D goes to 1. [2008 : 2 M]

5.24 Refer to the NAND and NOR latches shown in the figure. The inputs (P_1, P_2) for both the latches are first made (0,1) and then, after a few seconds, made (1,1). The corresponding stable outputs (Q_1, Q_2) are

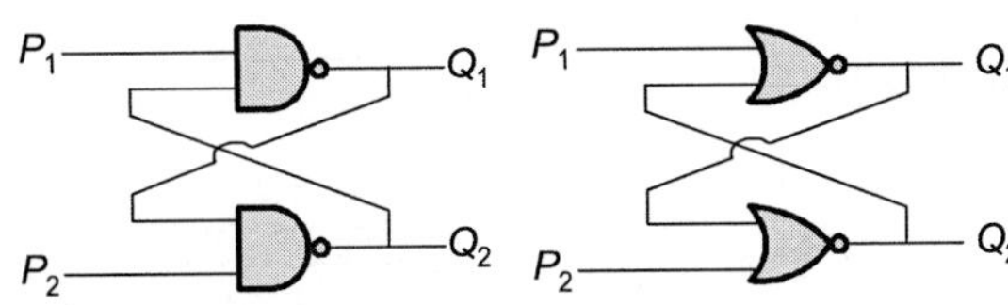

(a) NAND: first (0, 1) then (0, 1) NOR: first (1, 0) then (0, 0)
(b) NAND: first (1, 0) then (1, 0) NOR: first (1, 0) then (1, 0)
(c) NAND: first (1, 0) then (1, 0) NOR: first (1, 0) then (0, 0)
(d) NAND: first (1, 0) then (1, 1) NOR: first (0, 1) then (0, 1) [2009 : 2 M]

5.25 What are the counting states (Q_1, Q_2) for the counter shown in the figure below?

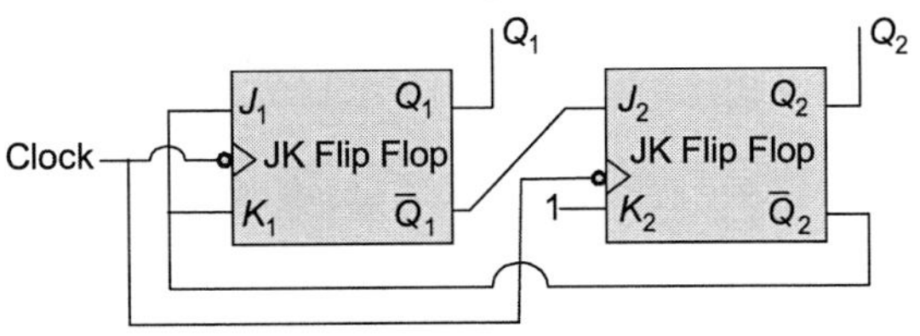

(a) 11, 10, 00, 11, 10, ...
(b) 01, 10, 11, 00, 01 ....
(c) 00, 11, 01, 10, 00, ...
(d) 01, 10, 00, 01, 10, ... [2009 : 2 M]

5.26 Assuming that all flip-flops are in reset conditions initially, the count sequence observed at Q_A in the circuit shown is

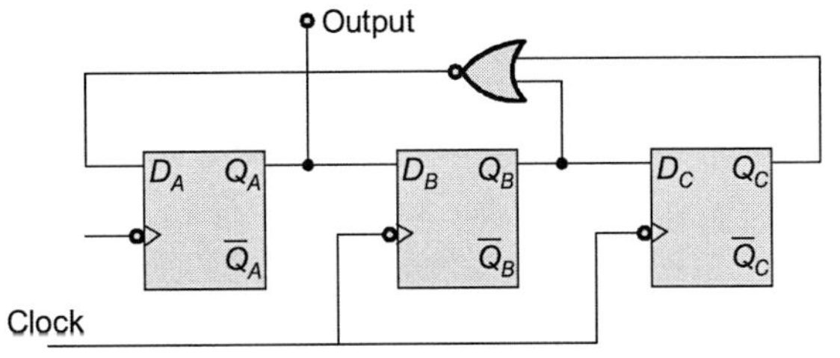

(a) 0010111... (b) 0001011...
(c) 0101111... (d) 0110100...

[2010 : 2 M]

5.27 When the output Y in the circuit below is "1", it implies that data has

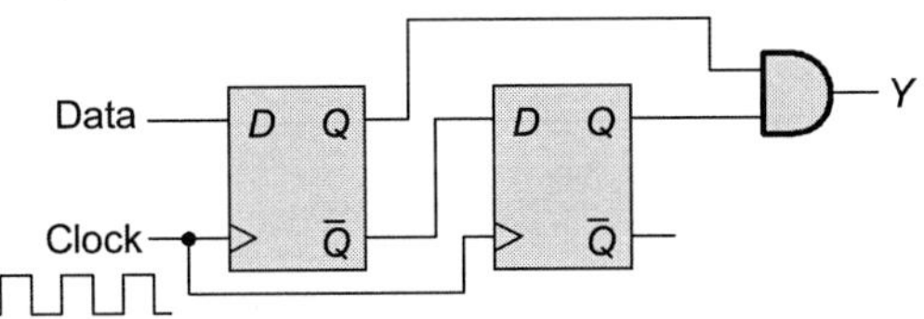

(a) changed from "0" to "1"
(b) changed from "1" to "0"
(c) changed in either direction
(d) not changed [2011 : 1 M]

5.28 Two D flip-flops are connected as a synchronous counter that goes through the following $Q_B\ Q_A$ sequence $00 \to 11 \to 01 \to 10 \to 00 \to ...$
The connections to the inputs D_A and D_B are

(a) $D_A = Q_B, D_B = Q_A$
(b) $D_A = \bar{Q}_A, D_B = \bar{Q}_B$
(c) $D_A = (Q_A\bar{Q}_B + \bar{Q}_A Q_B), D_B = Q_A$
(d) $D_A = (Q_A Q_B + \bar{Q}_A\bar{Q}_B), D_B = \bar{Q}_B$

[2011 : 2 M]

5.29 Consider the given circuit

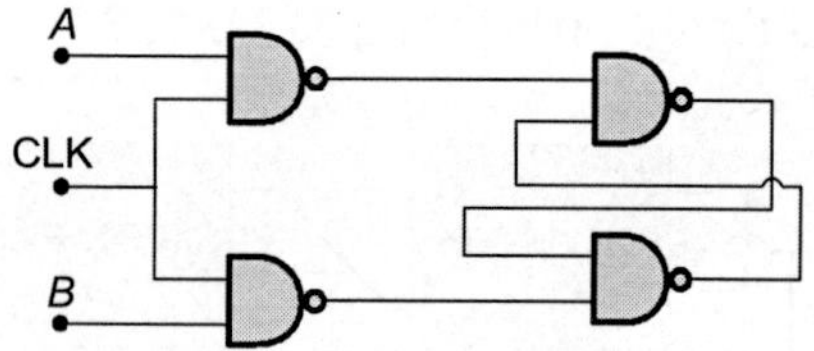

In this circuit, the race around
(a) does not occur
(b) occurs when CLK = 0
(c) occurs when CLK = 1 and $A = B = 1$
(d) occurs when CLK = 1 and $A = B = 0$

[2012 : 1 M]

5.30 The state transition diagram for the logic circuit shown in

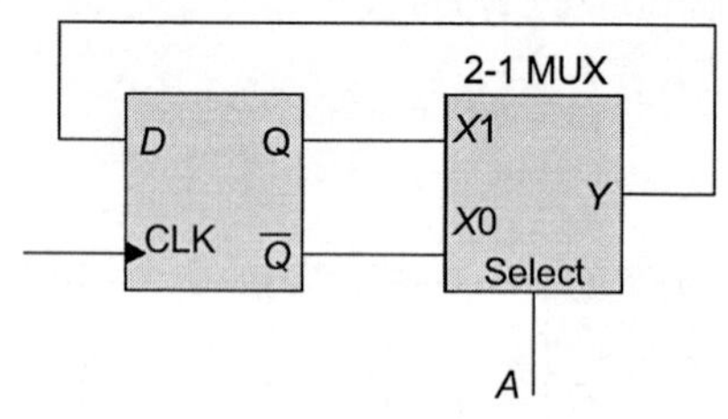

(a)
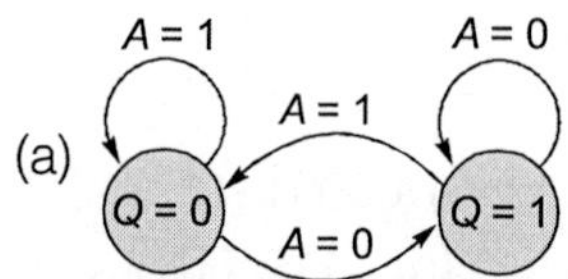

(b)
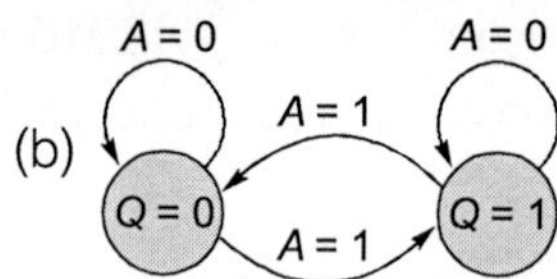

(c)
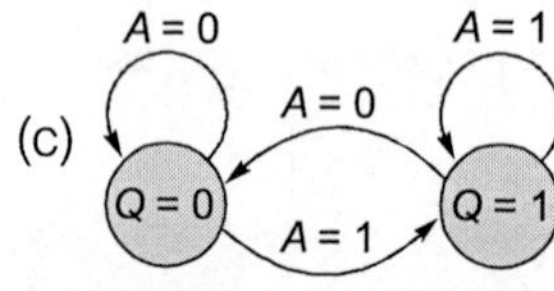

(d)
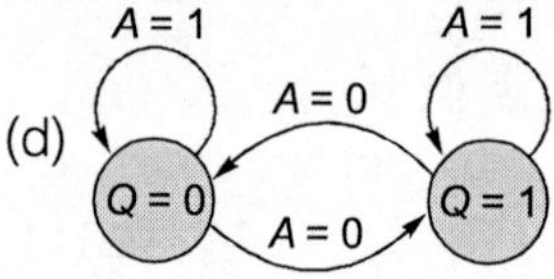

[2012 : 2 M]

5.31 Five JK flip-flops are cascaded to form the circuit shown in Figure. Clock pulses at a frequency of 1 MHz are applied as shown. The frequency (in kHz) of the waveform at Q_3 is ______.

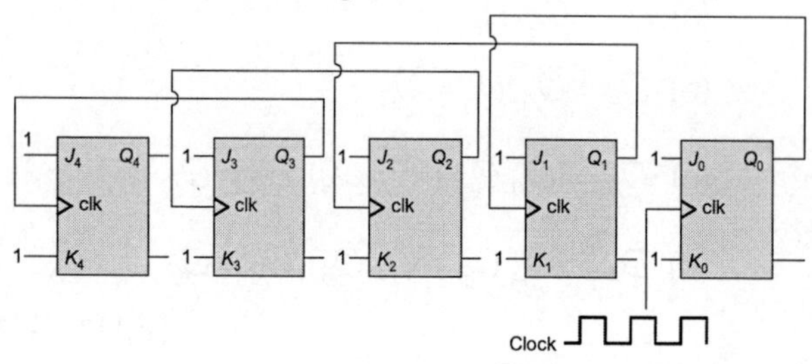

[2014 : 1 M, Set-1]

5.32 In the circuit shown, choose the correct timing diagram of the output (Y) from the given waveforms W_1, W_2, W_3 and W_4.

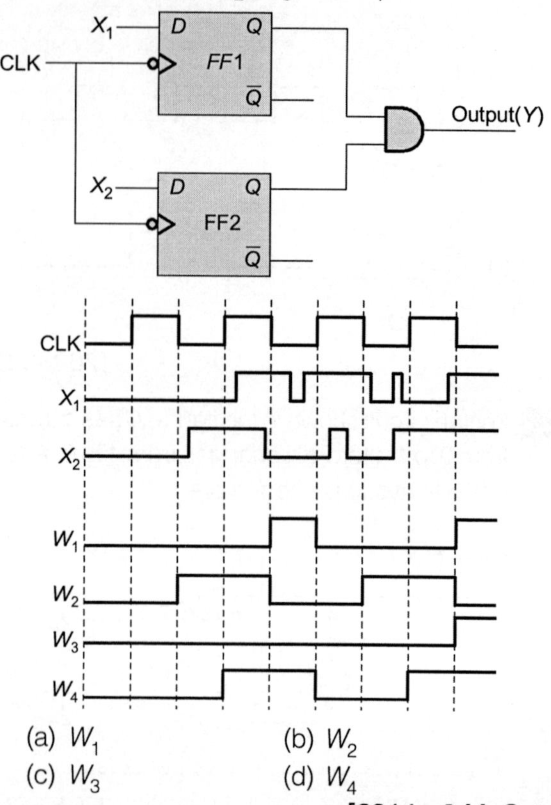

(a) W_1
(b) W_2
(c) W_3
(d) W_4

[2014 : 2 M, Set-2]

5.33 The outputs of the two flip-flops Q_1, Q_2 in the figure shown are initialized to 0, 0. The sequence generated at Q_1 upon application of clock signal is

(a) 01110...
(b) 01010...
(c) 00110...
(d) 01100...

[2014 : 2 M, Set-2]

5.34 The circuit shown in the figure is a

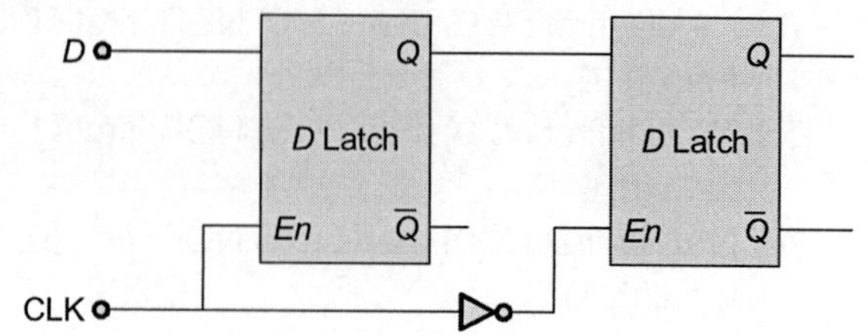

(a) Toggle Flip Flop
(b) *JK* Flip Flop
(c) *SR* Latch
(d) Master-Slave *D* Flip Flop

[2014 : 1 M, Set-3]

5.35 A mod-*n* counter using a synchronous binary up-counter with synchronous clear input is shown in the figure. The value of *n* is _____.

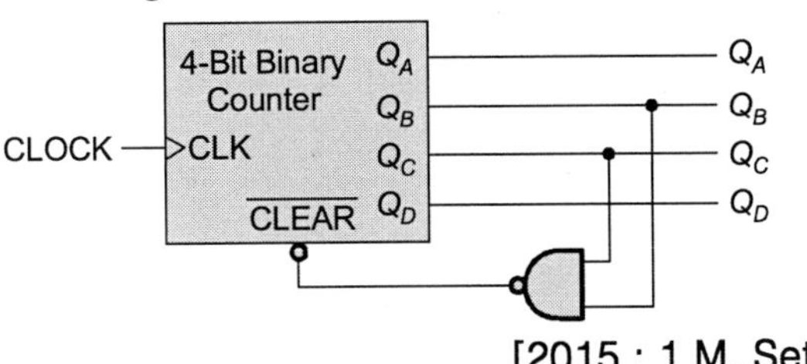

[2015 : 1 M, Set-2]

5.36 The figure shows a binary counter with synchronous clear input. With the decoding logic shown, the counter works as a

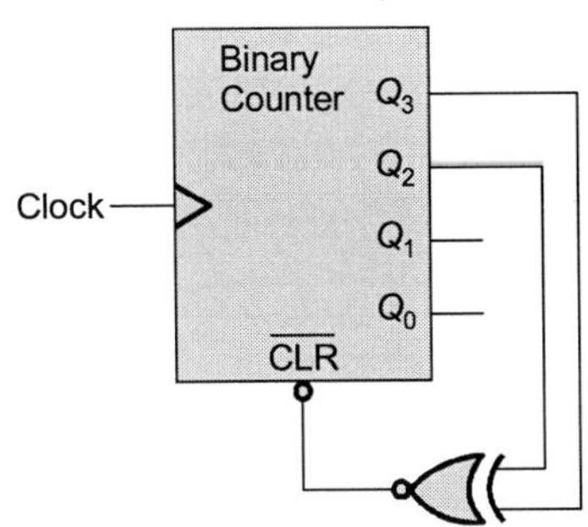

(a) mod-2 counter
(b) mod-4 counter
(c) mod-5 counter
(d) mod-6 counter

[2015 : 2 M, Set-2]

5.37 The circuit shown consists of J-K flip-flops, each with an active low asynchronous reset ($\overline{R_d}$ input). The counter corresponding to this circuit is

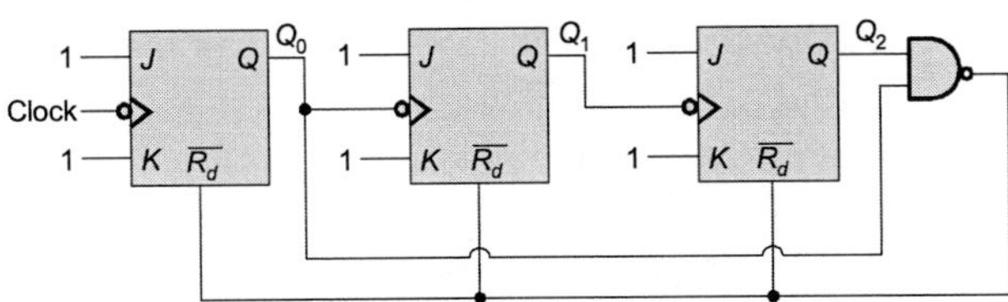

(a) a modulo-5 binary up counter
(b) a modulo-6 binary down counter
(c) a modulo-5 binary down counter
(d) a modulo-6 binary up counter

[2015 : 1 M, Set-3]

5.38 A three bit pseudo random number generator is shown. Initially the value of output $Y = Y_2 Y_1 Y_0$ is set to 111. The value of output *Y* after three clock cycles is

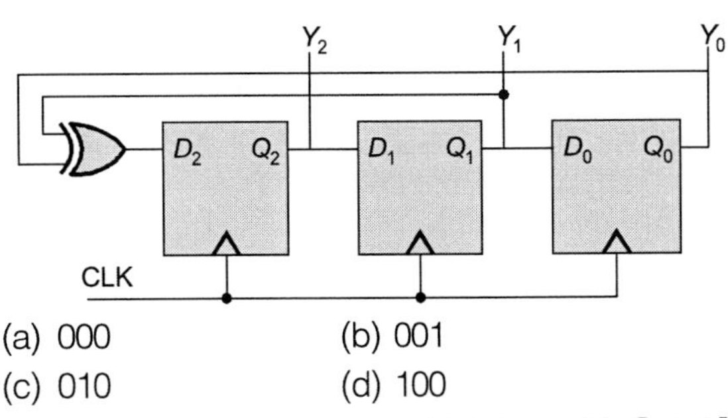

(a) 000
(b) 001
(c) 010
(d) 100

[2015 : 2 M, Set-3]

5.39 Assume that all the digital gates in the circuit shown in the figure are ideal, the resistor $R = 10\ k\Omega$ and the supply voltage is 5 V. The *D* flip-flops D_1, D_2, D_3, D_4 and D_5 are initialized with logic values 0,1,0,1 and 0, respectively. The clock has a 30% duty cycle.

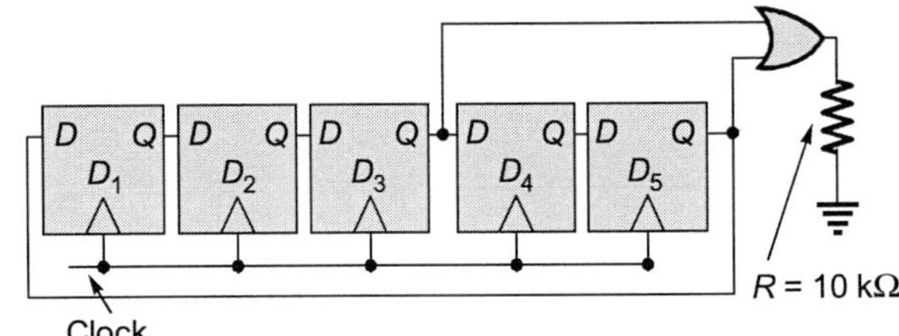

The average power dissipated (in mW) in the resistor *R* is _____.

[2016 : 1 M, Set-2]

5.40 For the circuit shown in the figure, the delay of the bubbled NAND gate is 2 ns and that of the counter is assumed to be zero.

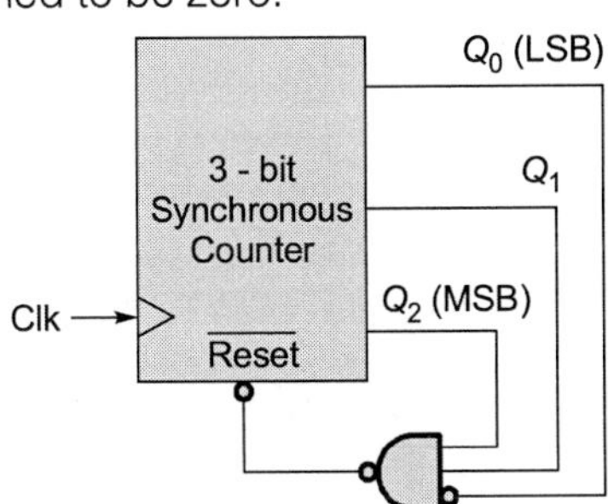

If the clock (Clk) frequency is 1 GHz, then the counter behaves as a

(a) mod-5 counter
(b) mod-6 counter
(c) mod-7 counter
(d) mod-8 counter

[2016 : 2 M, Set-3]

5.41 In the latch circuit shown, the NAND gates have non-zero, but unequal propagation delays. The present input condition is: $P = Q =$ '0'. If the input condition is changed simultaneously to $P = Q =$ '1', the outputs *X* and *Y* are

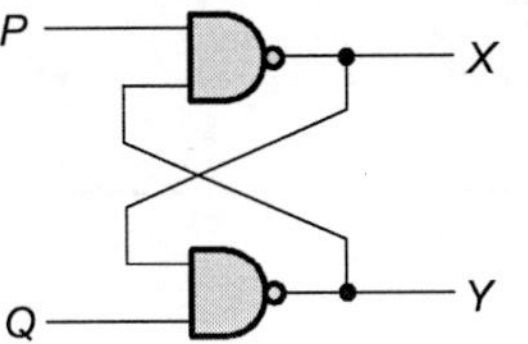

(a) X = '1', Y = '1'
(b) either X = '1', Y = '0' or X = '0', Y = '1'
(c) either X = '1', Y = '1' or X = '0', Y = '0'
(d) X = '0', Y = '0'

[2017 : 1 M, Set-1]

5.42 Consider the D-Latch shown in the figure, which is transparent when its clock input CK is high and has zero propagation delay. In the figure, the clock signal CLK1 has a 50% duty cycle and CLK2 is a one-fifth period delayed version of CLK1. The duty cycle at the output of the latch in percentage is ____.

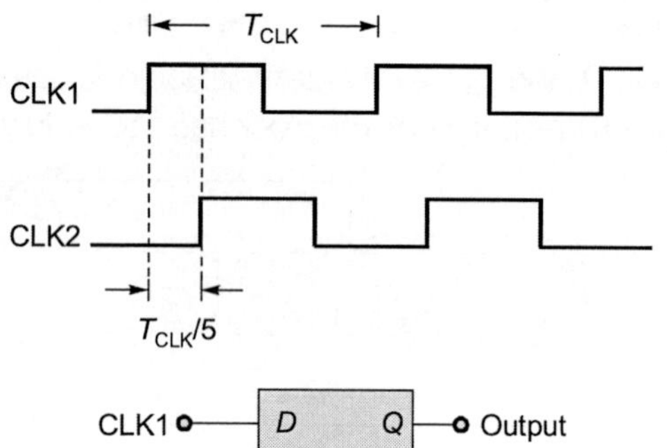

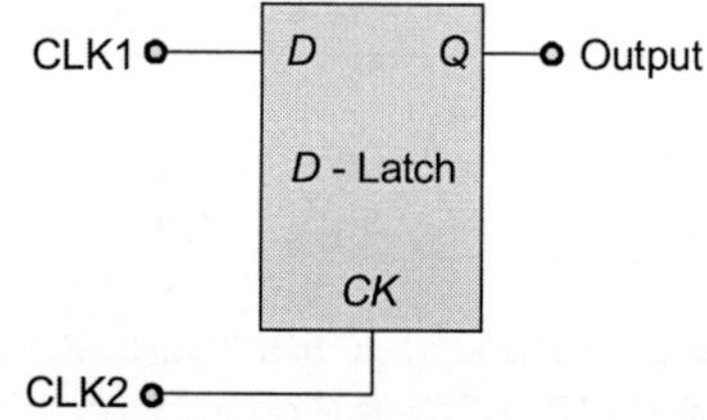

[2017 : 1 M, Set-1]

5.43 A 4-bit shift register circuit configured for right-shift operation, i.e. $D_{in} \rightarrow A, A \rightarrow B, B \rightarrow C, C \rightarrow D$, is shown. If the present state of the shift register is $ABCD$ = 1101, the number of clock cycles required to reach the state $ABCD$ = 1111 is ____.

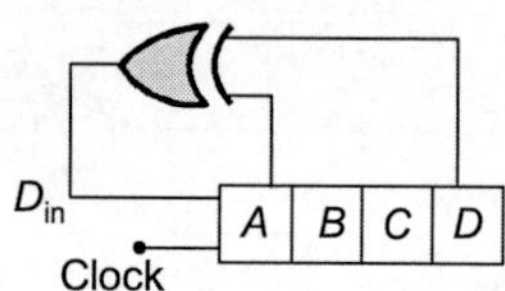

[2017 : 2 M, Set-1]

5.44 In the circuit shown, the clock frequency, i.e., the frequency of the Clk signal, is 12 kHz. The frequency of the signal at Q_2 is ______ kHz.

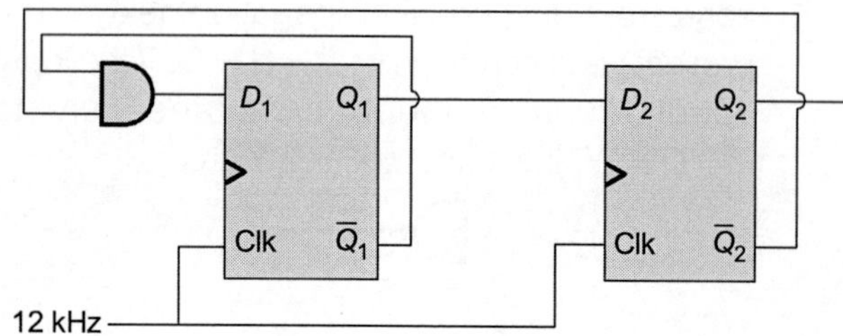

[2019 : 1 M]

5.45 For the components in the sequential circuit shown below, t_{pd} is the propagation delay, t_{setup} is the setup time, and t_{hold} is the hold time. The maximum clock frequency (rounded off to the nearest integer), at which the given circuit can operate reliably, is ____ MHz.

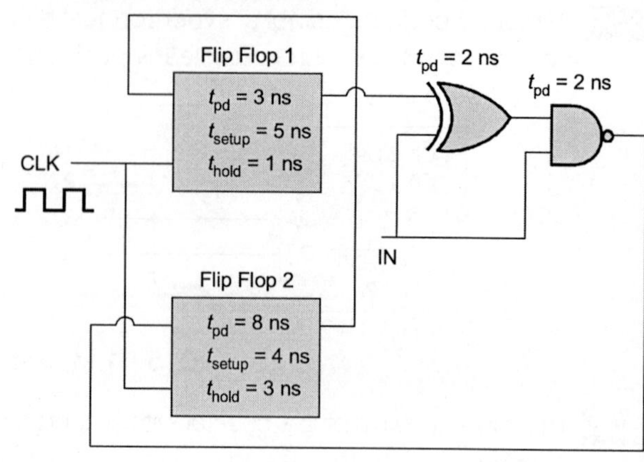

[2020 : 2 M]

5.46 The propagation delay of the exclusive-OR (XOR) gate in the circuit in the figure is 3 ns. The propagation delay of all the flip-flops is assumed to be zero. The clock (Clk) frequency provided to the circuit is 500 MHz.

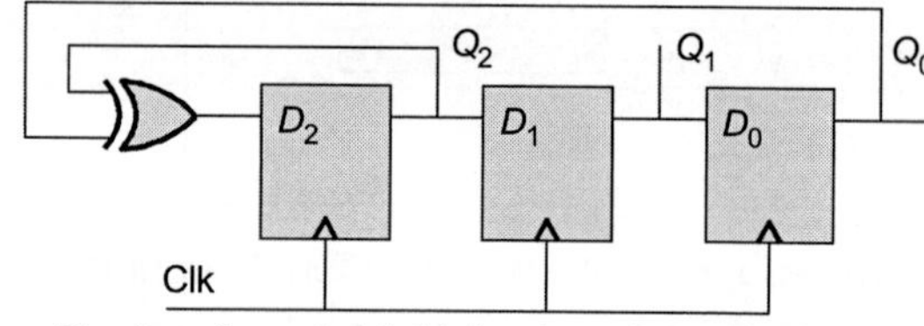

Starting from the initial value of the flip-flop outputs $Q_2Q_1Q_0$ = 111 with D_2 = 1, the minimum number of triggering clock edges after which the flip-flop outputs $Q_2Q_1Q_0$ becomes 100 (in integer) is ___.

[2021 : 2 M]

5.47 For the circuit shown, the clock frequency is f_o and the duty cycle is 25%. For the signal at the Q output of the Flip-Flop, ______.

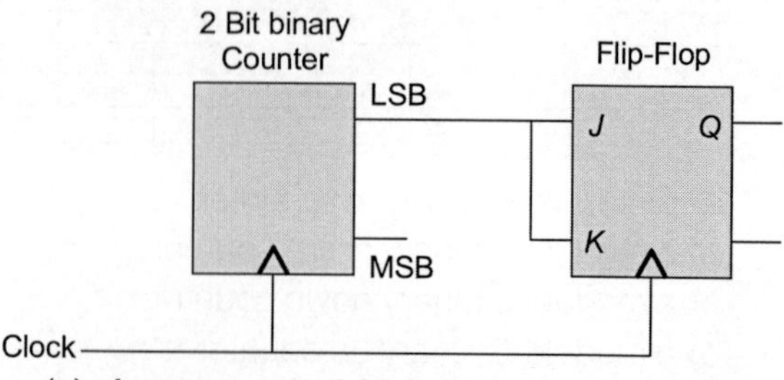

(a) frequency is $f_0/4$ and duty cycle is 50%
(b) frequency is $f_0/4$ and duty cycle is 25%
(c) frequency is $f_0/2$ and duty cycle is 50%
(d) frequency is f_0 and duty cycle is 25%

[2022 : 2 M]

5.48 The synchronous sequential circuit shown below works at a clock frequency of 1 GHz. The throughput, in M bits/s, and the latency, in ns, respectively, are

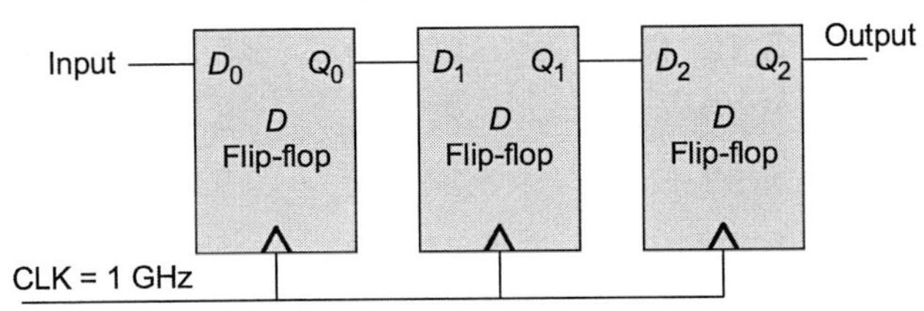

(a) 1000, 3 (b) 333.33, 1
(c) 2000, 3 (d) 333.33, 3

[2023 : 1 M]

5.49 For the circuit shown below, the propagation delay of each NAND gate is 1 ns. The critical path delay, in ns, is ______ (rounded off to the nearest integer).

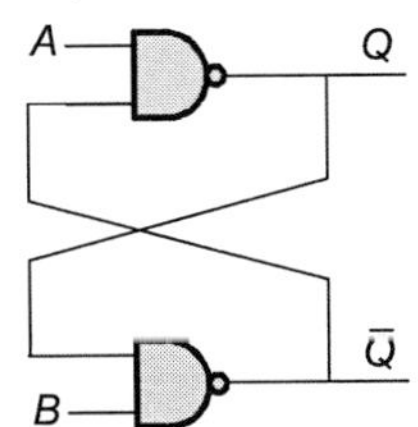

[2023 : 1 M]

5.50 In a given sequential circuit, initial states are $Q_1 = 1$ and $Q_2 = 0$. For a clock frequency of 1 MHz, the frequency of signal Q_2 in kHz, is ______. (rounded off to the nearest integer).

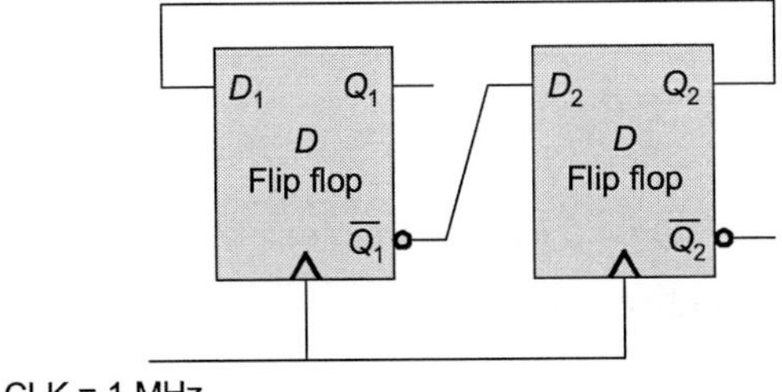

[2023 : 2 M]

5.51 The sequence of states ($Q_1 Q_0$) of the given synchronous sequential circuit is ______.

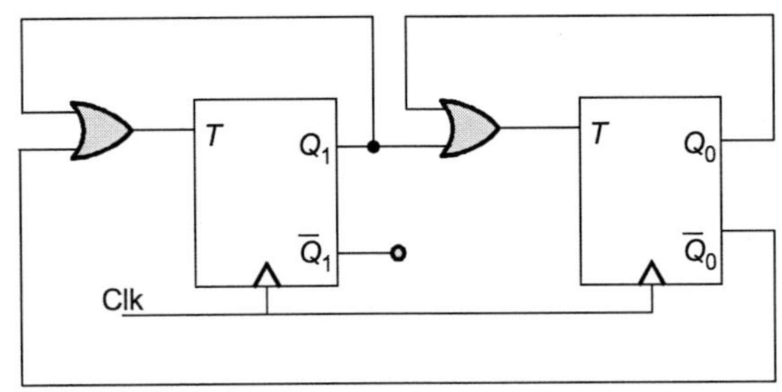

(a) $11 \rightarrow 00 \rightarrow 10 \rightarrow 01 \rightarrow 00$
(b) $00 \rightarrow 01 \rightarrow 10 \rightarrow 00$
(c) $01 \rightarrow 10 \rightarrow 11 \rightarrow 00 \rightarrow 01$
(d) $00 \rightarrow 10 \rightarrow 11 \rightarrow 00$

[2024 : 2 Marks]

5.52 IThe propagation delay of the 2×1 MUX shown in the circuit is 10 ns. Consider the propagation delay of the inverter as 0 ns.

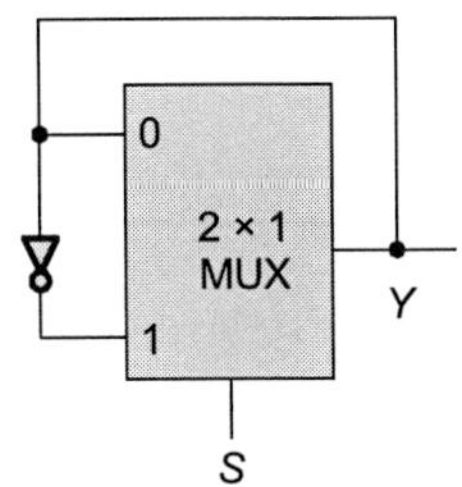

If S is set to 1, then the output Y is ______.
(a) constant at 1
(b) a square wave of frequency 100 MHz
(c) a square wave of frequency 50 MHz
(d) constant at 0

[2024 : 2 Marks]

■■■

Answers Sequential Circuits

5.1 (d)	5.2 (c)	5.3 (d)	5.4 (c)	5.5 (a)	5.6 (d)	5.7 (d)
5.8 (b)	5.9 (c)	5.10 (c)	5.11 (d)	5.12 (b)	5.13 (c)	5.14 (b)
5.15 (c)	5.16 (c)	5.17 (b)	5.18 (d)	5.19 (a)	5.20 (c)	5.21 (b)
5.22 (b)	5.23 (c)	5.24 (c)	5.25 (a)	5.26 (d)	5.27 (a)	5.28 (d)
5.29 (a)	5.30 (d)	5.31 (62.5)	5.32 (c)	5.33 (d)	5.34 (d)	5.35 (7)
5.36 (c)	5.37 (a)	5.38 (d)	5.39 (1.5)	5.40 (d)	5.41 (b)	5.42 (30)
5.43 (10)	5.44 (4)	5.45 (76.92)	5.46 (5)	5.47 (a)	5.48 (a)	5.49 (2)
5.50 (250)	5.51 (a)	5.52 (c)				

Explanations Sequential Circuits

5.1 (d)

If a switch tail ring counter using single D flip-flop, the complementary output $\bar{Q}$ is connected to D, so it becomes a T flip-flop.

5.2 (c)

An R-S latch is a one bit memory element.

5.3 (d)

When, $Q = 0 \rightarrow \bar{Q} = 1$

So, $J = K = 1$

$$Q_{n+1} = \bar{Q}_n = 1$$

When

$$Q = 1 \rightarrow \bar{Q} = 0$$

$$J = 0, K = 1$$

$$Q_{n+1} = 0$$

So, the sequence at the Q output will be 0 1 0 1 0 1.

5.4 (c)

Let initial $Q_1 = Q_0 = 0$

J_0	K_0	J_1	K_1	Q_1	Q_0
				0	0
1	1	0	1	0	1
1	1	1	1	1	0
0	1	0	1	0	0

It is a mod-3 counter

So, $K = 3$

5.5 (a)

When $A = 1$ and $B = 1$

$$X = \bar{Y}$$

$$Y = \bar{X}$$

Now $A = 1$ and $B = 0$

$$Y = 1$$

$$X = 0$$

Now $A = 1$ and $B = 1$

$$X = \bar{Y} = 0$$

$$Y = \bar{X} = 1$$

So, the outputs x and y will be fixed at 0 and 1 respectively.

5.6 (d)

Trick up/down = $CP \oplus Q$, 1 for up and 0 for down.

CP = (clock pulse)

Q = (O/P)

0 = –ve edge; $Q = 1$

1 = +ve edge; $\bar{Q} = 1$

= $1 \oplus 1 = 0$ (down counter)

Counting sequence

1	1	1
1	1	0
1	0	1
1	0	0
0	1	1 (preset state) so Mod 5

5.7 (d)

$$D = \bar{X}Z + Y\bar{Z},\ D = \bar{K}Q + J\bar{Q}$$

$Y = J$, $X = K$, $D = Q$ (for D flip-flop)

5.8 (b)

Given counter is Mod - 10 up counter

$$\therefore \text{ Frequency of output} = \frac{10\text{ k}}{10} = 1\text{k}$$

5.9 (c)

t_{Pd} of all 5 intervals

$$= 5 \times 100 \text{ p sec} = 500 \text{ p sec}$$

∴ fundamental frequency of oscillator

$$\text{Output} = \frac{1}{2 \times 500 \text{ ps}} = 1 \text{ GHz}$$

5.10 (c)

It is given that clock is positive edge triggered than only option (a) and (c) are possible. Out of which only (c) option gives the required result.

5.11 (d)

Decoding logic = 111

Hence, CLR = $Q_2\,Q_1\,Q_0$

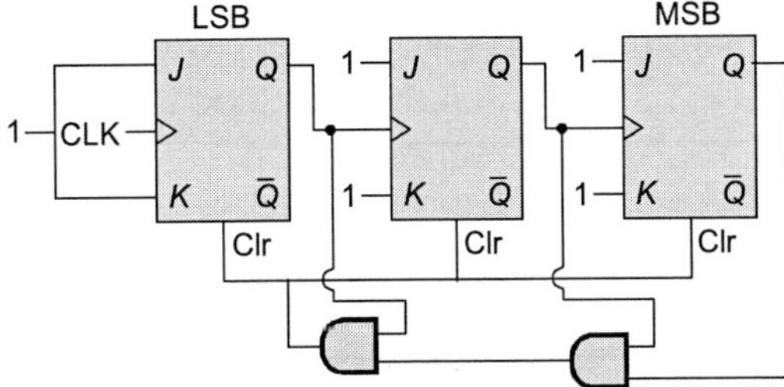

5.12 (b)

Prop. delay of 4 bit ripple counter

$R = 4 \times t_{pd} = 40\,\text{ns}$

In synchronous counter all flip-flops are given clock simultaneously so, S = 10 ns.

5.15 (c)

At the end of 6th pulse all states should be cleared.

$CBA = 110 \qquad \bar{C}\bar{B}X = 00X$

Output of desired gate should be zero as clear is given active low.

So given gate should be OR as OR gate output is zero if both inputs are 0.

5.16 (c)

Since $J = 1$ and $Q_n = 0$ So $Q_{n+1} = 1$

as even if $K = 0$, $Q_{n+1} = 1$ (set)

and if $K = 1$, $Q_{n+1} = \overline{Q_n} = 1$ (toggle)

5.17 (b)

$Q_2Q_1Q_0 = 011$

1st Clk→ $Q_2Q_1Q_0 = 100$

$\bar{Q}_0 = 1$ (triggers T_1)

$\bar{Q}_1 = 1$ (triggers T_2)

5.18 (d)

	A	B	C_i	S	C_o
After 1st CP	1	1	0	0	1
After 2nd CP	1	1	1	1	1

5.19 (a)

Q_1	Q_0	$D_1(Q_0)$	$D_0(\bar{Q}_1)$
0	0	0	1
0	1	1	1
1	1	1	0
1	0	0	0

5.22 (b)

Time period of waveform of output at

$$Q_1 = 2 \times 2 \times T = 4T$$

Delay time at output

$$Q_1 = 2\Delta T$$

Note:

(i) In case of n flip-flops in such case, Time period of last output waveform $= 2^n T$

where,

T = Time period for clock pulse

(ii) Delay time = $n\Delta T$

where,

ΔT = Propagation delay provided by one flip-flop

5.24 (c)

For NAND gates: Inputs [(0,1); (1,1)]

⇒ Output [(1, 0) ; (1, 0)]

For NOR gates : Inputs [(0, 1); (1,1)]

⇒ Output [(1,0); (0,0)]

5.25 (a)

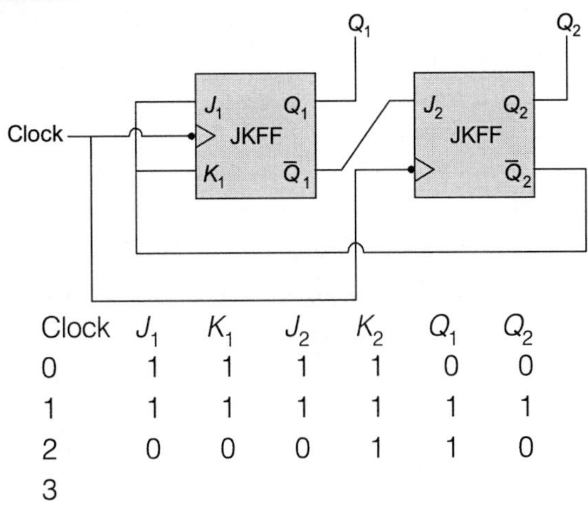

Clock	J_1	K_1	J_2	K_2	Q_1	Q_2
0	1	1	1	1	0	0
1	1	1	1	1	1	1
2	0	0	0	1	1	0
3						

So the sequence is 11, 10, 00, 11, 10, 00...

5.26 (d)

Initially, $Q_A = Q_B = Q_C = 0$

$D_A = Q_B \odot Q_C = 1,\ D_B = Q_A = 0$

$D_C = Q_B = 0$

After one clock pulse,

$Q_A = 1,\ Q_B = 0,\ Q_C = 0$

$D_A = Q_B \odot Q_C = 1$

$D_B = Q_A = 1,\ D_C = Q_B = 0$

After two clock pulse,

$Q_A = 1,\ Q_B = 1,\ Q_C = 0$

$D_A = Q_B \odot Q_C = 0$

$D_B = Q_A = 1,\ D_C = Q_B = 1$

After three clock pulses,

$Q_A = 0,\ Q_B = 1,\ Q_C = 1$

$D_A = Q_B \odot Q_C = 1$

$D_B = Q_A = 0,\ D_C = Q_B = 1$

After four clock pulse,

$Q_A = 1,\ Q_B = 0,\ Q_C = 1$

$D_A = Q_B \odot Q_C = 0$

$D_B = Q_A = 1,\ D_C = Q_B = 0$

After five clock pulse,

$Q_A = 0,\ Q_B = 1,\ Q_C = 0$

$D_A = Q_B \odot Q_C = 0$

$D_B = Q_A = 0,\ D_C = Q_B = 1$

After six clock pulse,

$Q_A = 0,\ Q_B = 0,\ Q_C = 1$

Therefore, the count sequence observed at Q_A is 0110100 ..... .

5.27 (a)

$Y = 1$, it is possible only when both flip-flop outputs are '1'. It means before applying clock both flip-flop input should be '1'.

Before applying clock output of 1st flip-flop should be '0'. (Because input of 2nd flip-flop is connected to $\bar{Q}$) and after applying clock output of 1st flip-flop should be '1'.

And it depends only upon input data when it changes from '0' to '1'.

5.28 (d)

Present State		Next State	
Q_B	Q_A	Q_B	Q_A
0	0	1	1
1	1	0	1
0	1	1	0
1	0	0	0
0	0	1	1

Now using excitation table of D flip-flop

$$D_A = (Q_A Q_B + \bar{Q}_A \bar{Q}_B)$$

$$D_B = \bar{Q}_B$$

5.29 (a)

Given flip-flop is S-R flip-flop with $A = S$ and $B = R$. In S-R flip-flop race around condition does not occur.

5.30 (d)

When $A = 1$, then Q will be selected by MUX and feedback to D flip-flop which gives output Q again.

So at $A = 1$ it holds its state. When $A = 0$, then $\bar{Q}$ will be selected by MUX and feedback to D flip-flop and output will be inverted.

So for $A = 1$ it holds the state and for $A = 0$ it interchange the state i.e. if $Q = 0$ then it will go to $Q = 1$ and if $Q = 1$ then it will go to $Q = 0$.

So option (d) is correct option.

5.31 (62.5)

The circuit or the counter given is the asynchronous counter in which the clock is applied to first flip flop only and the remaining flip flop take the clock from the previous flip flops outputs.

As we know that 1 flip flop divides the frequency by 2. So, output of Q_3 is taken at 4th flip-flop. So, division of frequency be f/16.

$\therefore$ Frequency (in kHz) of the waveform of Q_3

$$= \frac{1\text{MHz}}{16} = 62.5 \text{ kHz}$$

5.32 (c)

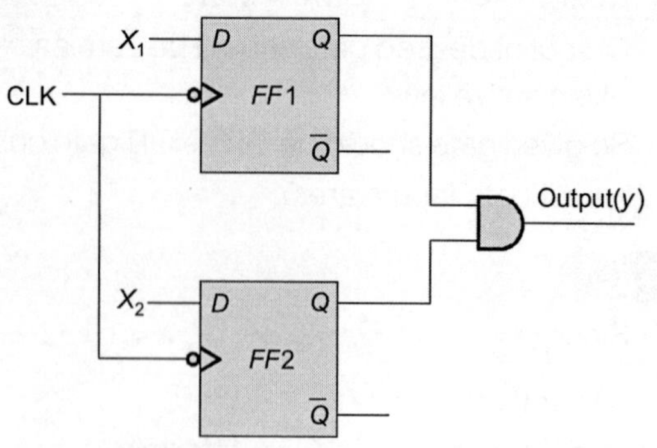

CLK	X_1	X_2	Y
1	X	X	Previous output/ No change
0	0	0	0
0	0	1	0
0	1	0	0
0	1	1	1

So, according to waveform

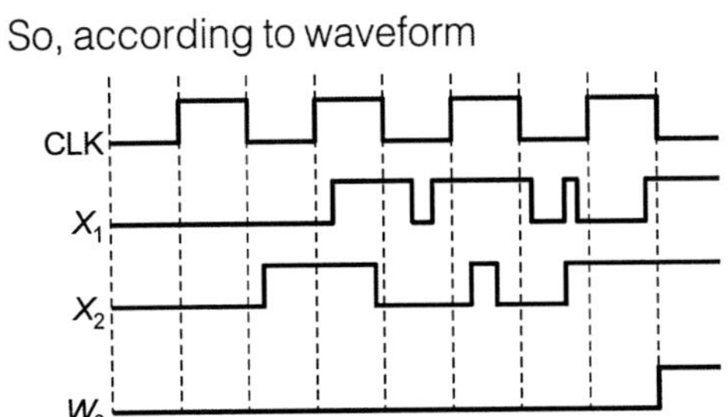

Solution waveform W_3 satisfies it.

5.33 (d)

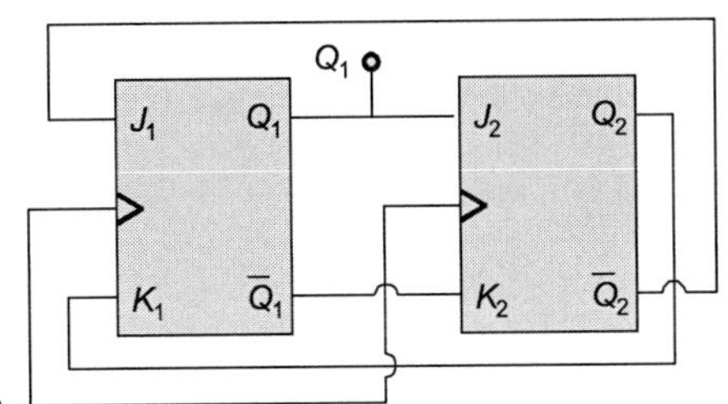

Initially $Q_1 = Q_2 = 0$

Truth table of JK:

J	K	Q_n
0	0	Previous state
0	1	0
1	0	1
1	1	$\bar{Q}_n$

Case-1: 1st clock pulse

$Q_1 = 0,\ \bar{Q}_1 = 1, Q_2 = 0$

$\bar{Q}_2 = 1$

So, $J_1 = 1$, $K_1 = 0$ and $J_2 = 0$, $K_2 = 1$

So $Q_1^+ = 1,\quad Q_2^+ = 0$

Case-2:

$$Q_1 = 1,\ Q_2 = 0,\ \bar{Q}_1 = 0,\ \bar{Q}_2 = 1$$

So, $J_1 = 1,\ J_2 = 1,\ K_1 = 0\quad K_2 = 0$

So, $Q_1^+ = 1,\quad Q_2^+ = 1$

Case-3:

$$Q_1 = 1,\ \bar{Q}_1 = 0, Q_2 = 1, \bar{Q}_2 = 0$$

$$J_1 = 0,\ K_1 = 1,\ J_2 = 1,\ K_2 = 0$$

$$Q_1^+ = 0,\quad Q_2^+ = 1$$

So, the sequence will be 01100.

5.34 (d)

Given circuit diagram is a master slave D flip-flop in which master flip-flop works on the positive cycle of clock and slave flip-flop works on negative cycle of clock.

5.35 (7)

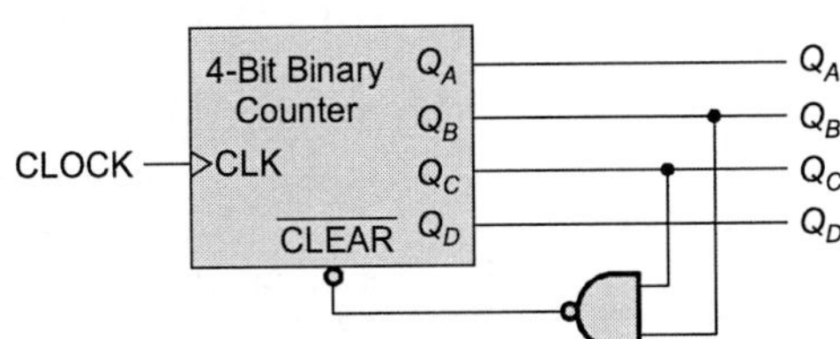

Clear signal is $\overline{Q_B \cdot Q_C}$

Counter is cleared after 0110

$\Rightarrow$ MOD of the counter is 7 as it counts states from 0 to 6.

5.36 (c)

Clear signal = $Q_3 \odot Q_2$

Counter is clear after 0100

So number of states counted are 0 to 4.

$\Rightarrow$ MOD of the counter is 5.

5.37 (a)

Q is applied as negative edge triggering clock.

$\Rightarrow$ Counter is up counter. Reset signal = $\overline{Q_2 \cdot Q_0}$

The counter resets at 101

$\Rightarrow$ Counter is MOD-5

5.38 (d)

The given circuit is similar to

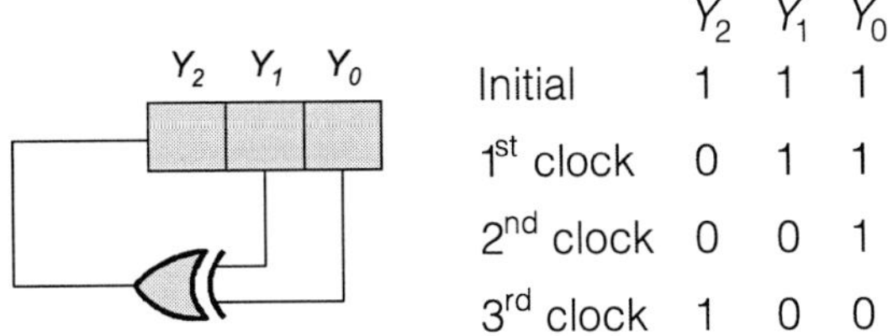

	Y_2	Y_1	Y_0
Initial	1	1	1
1st clock	0	1	1
2nd clock	0	0	1
3rd clock	1	0	0

$\Rightarrow$ After three clock Y is 100.

5.39 (1.5)

CLK	Q_1	Q_2	Q_3	Q_4	Q_5	$Y = Q_3 + Q_5$
0	0	1	0	1	0	0
1	0	0	1	0	1	1
2	1	0	0	1	0	0
3	0	1	0	0	1	1
4	1	0	1	0	0	1
5	0	1	0	1	0	0

The waveform of the gate output

$$Y = Q_3 + Q_5$$

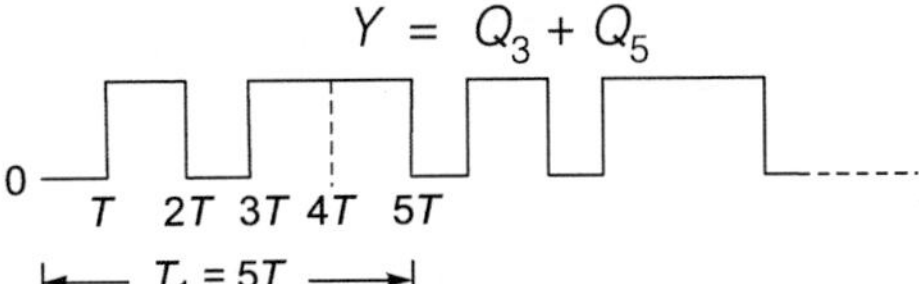

Average power dissipated

$$P = \frac{V^2}{R} \times \frac{T_{ON}}{T} = \frac{5^2}{10\text{ k}} \times \frac{3T}{5T} = 1.5\text{ mW}$$

5.40 (d)

Clock frequency = 1 GHz

$\Rightarrow$ Clock time period = 1 ns

If the propagation delay of the NAND gate were 0 ns, the circuit would have behaved as MOD 6 counter.

However, the delay of NAND gate is 2 ns. During this time, two more clock pulses would reach the counter, and therefore it would count two more states. Hence, it acts as MOD 8 counter.

5.41 (b)

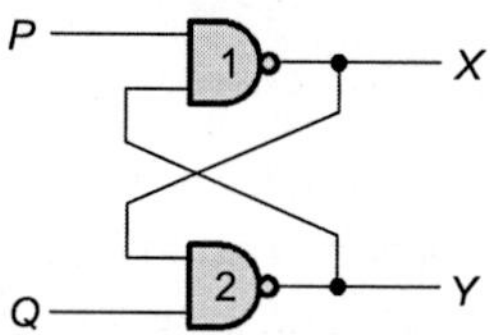

Present input condition : $P = Q = 0$

$\Rightarrow$ Corresponding outputs are $X = Y = 1$

When input condition is changed to $P = Q = 1$ from $P = Q = 0$:

Possibility - 1 :

Let gate-1 is faster than gate-2, then the possible outputs are $X = 0$, $Y = 1$

Possibility - 2 :

Let gate-2 is faster than gate-1, then the possible outputs are $X = 1$, $Y = 0$

5.42 (30)

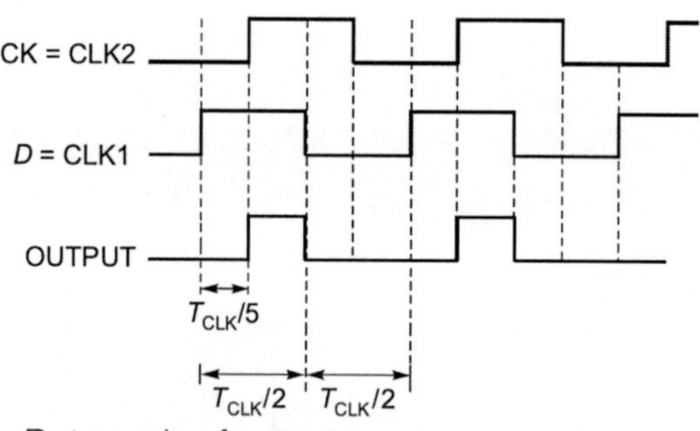

Duty cycle of output

$$= \frac{\frac{T_{CLK}}{2} - \frac{T_{CLK}}{5}}{T_{CLK}} \times 100 = \frac{3}{10} \times 100 = 30\%$$

5.43 (10)

Clock Number	$D_{in} = A \oplus D$ Just before clock	A	B	C	D
Initial	-	1	1	0	1
1	0	0	1	1	0
2	0	0	0	1	1
3	1	1	0	0	1
4	0	0	1	0	0
5	0	0	0	1	0
6	0	0	0	0	1
7	1	1	0	0	0
8	1	1	1	0	0
9	1	1	1	1	0
10	1	1	1	1	1

So, 10 clock cycles are required.

5.44 (4)

$$D_1 = \overline{Q}_2\overline{Q}_1 = \overline{Q_2 + Q_1}$$

PS		D_2	D_1	NS	
Q_2	Q_1			Q_2^+	Q_1^+
0	0	0	1	0	1
0	1	1	0	1	0
1	0	0	0	0	0

MOD = 3

$$f_{Q2} = \frac{f_{clk}}{3} = \frac{12}{3}\text{ kHz} = 4\text{ kHz}$$

5.45 (76.92)

Total propagation delay

$= (t_{pd} + t_{\text{set-up}})_{max} = 8\text{ns} + 5\text{ ns} = 13\text{ ns}$

$\therefore$ Frequency of operations

$$= \frac{1000}{13}\text{ MHz} = 76.92\text{ MHz}$$

5.46 (5)

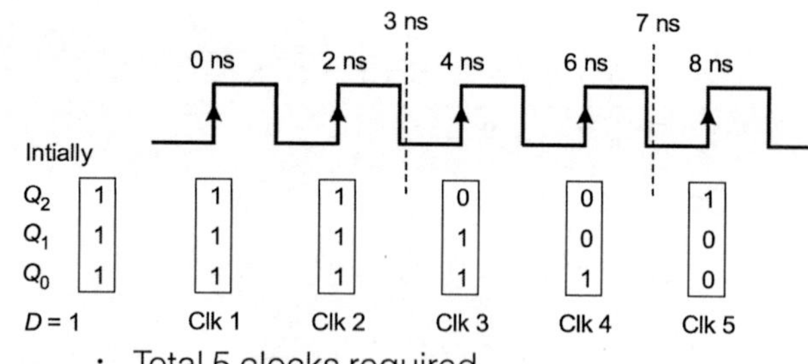

$\therefore$ Total 5 clocks required.

5.47 (a)

The two-bit binary counter counts as below :

CLK	MSB	LSB
0	0	0
1	0	1
2	1	0
3	1	1

For the flip-flop, $J = K$. The waveforms at the input and output of flip-flop can be drawn as below :

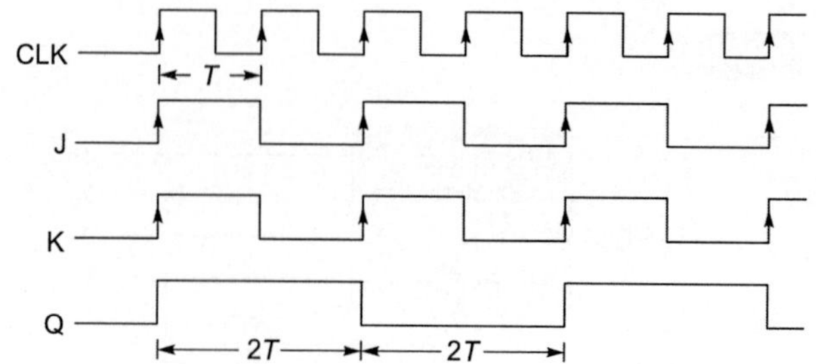

At the output of flip-flop, time period of waveform

$$T_o = 4T$$

$$\therefore \quad f_o = \frac{1}{4T} = \frac{f}{4}$$

Since, $T_o(\text{on}) = T_o(\text{off}) = 2T$. Hence, duty cycle is 50%.

5.48 (a)

The given circuit is a type of serial input serial output.

$\therefore$ Latency $= n \times T_{\text{clk}}$ where, n = number of flip flops

$$= 3 \times 1 \text{ where, } T_{\text{clk}} = \frac{1}{f_{\text{clk}}} = 1\text{ns}$$

$$= 3 \text{ ns}$$

Now, Throughput = Number of bits/sec

$\because$ 1 bit = 1 nsec

$\therefore$ Throughput $= 10^9$ bits/sec

= 1000 Mbps

5.49 (2)

The given circuit can be drawn as;

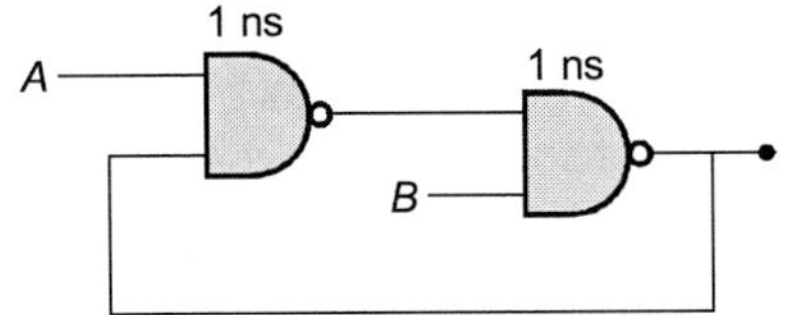

$\therefore$ The critical path delay = 1 ns + 1 ns

= 2 ns

5.50 (250)

Given, initial states, $Q = 1$ and $Q_2 = 0$

Clock frequency, $f = 1$ MHz

Clk	$D_1 = Q_2$	$D_2 = \overline{Q_1}$	Q_1	Q_2
Initial			1	0
1	0	0	0	0
2	0	1	0	1
3	1	1	1	1
4	1	0	1	0

Therefore, the given counter is having MOD-4

$$\therefore \text{ The frequency of signal } Q_2 = \frac{f_i}{4}$$

$$= \frac{1000}{4} \text{ kHz} = 250 \text{ kHz}$$

5.51 (a)

From the given sequential circuit, the excitation inputs to the flip-flops can be obtained as:

$$T_1 = Q_1 + \overline{Q_0}$$

$$T_0 = Q_1 + Q_0$$

The state table of the sequential circuit, thus can be written as below:

Current State		Flip-flop Excitations		Next State	
Q_1	Q_0	T_1	T_0	Q_1^+	Q_0^+
0	0	1	0	1	0
1	0	1	1	0	1
0	1	0	1	0	0
1	1	1	1	0	0

Hence, the sequence of states is:

$00 \rightarrow 10 \rightarrow 01 \rightarrow 00$

State Diagram:

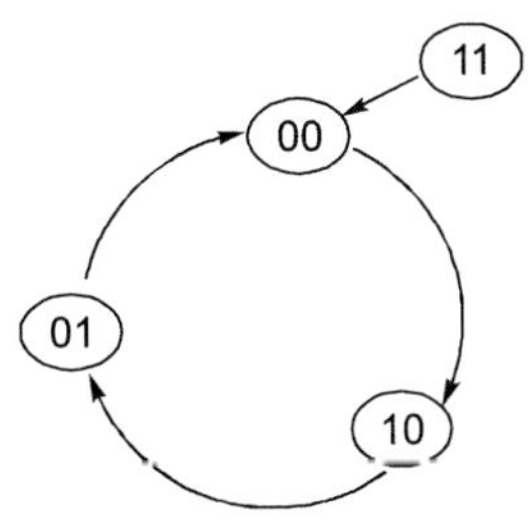

5.52 (c)

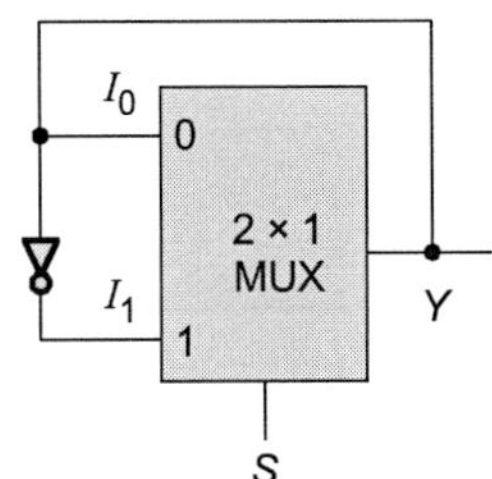

If S is set to 1, the MUX transmits the input line I_1 to the output Y. The propagation delay of the MUX causes the output to change every 10 ns as $I_1 = \overline{Y}$ resulting in a square wave as shown below:

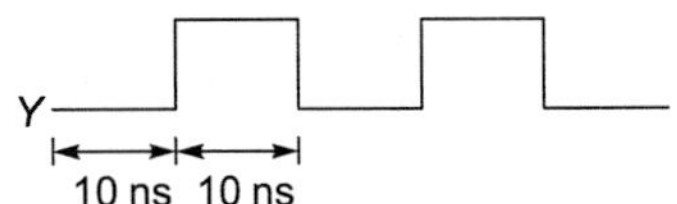

The time period of the square wave,

$$T_o = 10 + 10 = 20 \text{ ns}$$

$$\therefore \text{ Frequency, } f_o = \frac{1}{T_o} = \frac{1}{20 \text{ ns}} = \frac{1}{20 \times 10^{-9}}$$

$$f_o = 50 \times 10^6 \text{ Hz} = 50 \text{ MHz}$$

■■■

6 Finite State Machine and Miscellaneous

6.1 The point P in the following figure is stuck-at-1. The output f will be

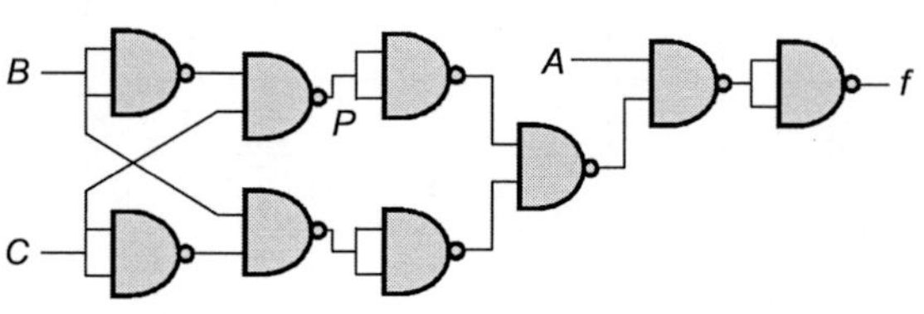

(a) $\overline{AB\overline{C}}$ (b) $\overline{A}$
(c) $AB\overline{C}$ (d) A **[2006 : 2 M]**

6.2 The digital logic shown in the figure satisfies the given state diagram when Q_1 is connected to input A of the XOR gate.

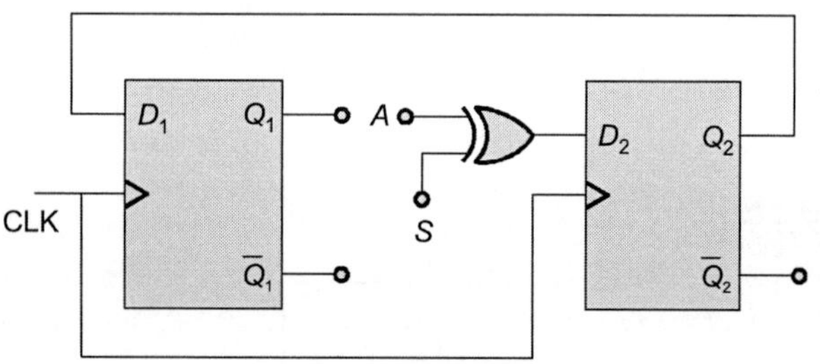

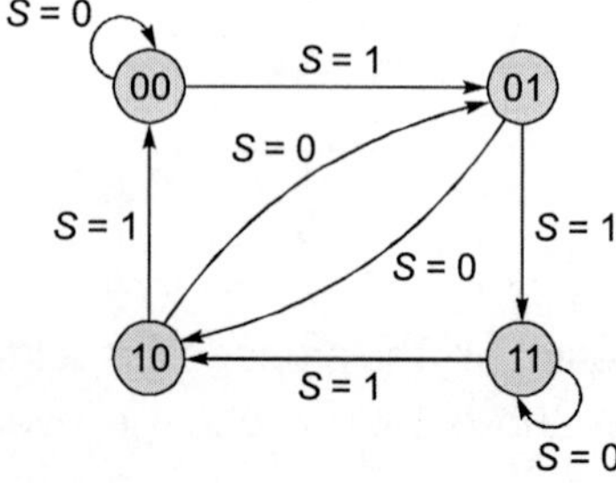

Suppose the XOR gate is replaced by an XNOR gate. Which one of the following options preserves the state diagram?

(a) Input A is connected to $\overline{Q}_2$.
(b) Input A is connected to Q_2.
(c) Input A is connected to $\overline{Q}_1$ and S is complemented.
(d) Input A is connected to $\overline{Q}_1$.

[2014 : 2 M, Set-1]

6.3 The state transition diagram for a finite state machine with states A, B and C, and binary inputs X, Y and Z, is shown in the figure.
Which one of the following statements is correct?

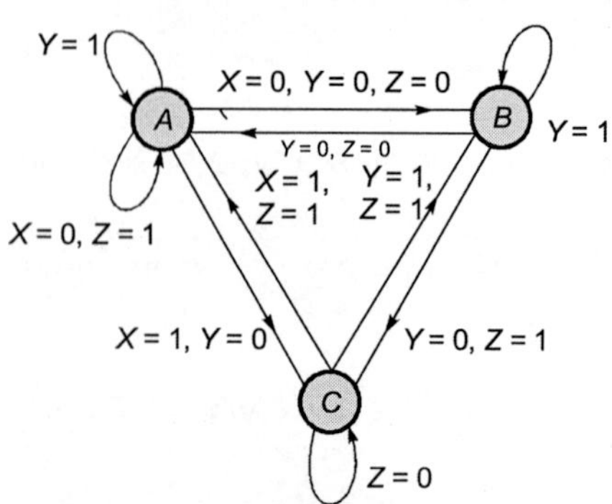

(a) Transitions from State A are ambiguously defined.
(b) Transitions from State B are ambiguously defined.
(c) Transitions from State C are ambiguously defined.
(d) All of the state transitions are defined unambiguously. **[2016 : 2 M, Set-2]**

6.4 Transistor geometries in a CMOS inverter have been adjusted to meet the requirement for worst case charge and discharge times for driving a load capacitor C. This design is to be converted to that of a NOR circuit in the same technology, so that its worst case charge and discharge times while driving the same capacitor are similar. The channel lengths of all transistors are to be kept unchanged. Which one of the following statements is correct?

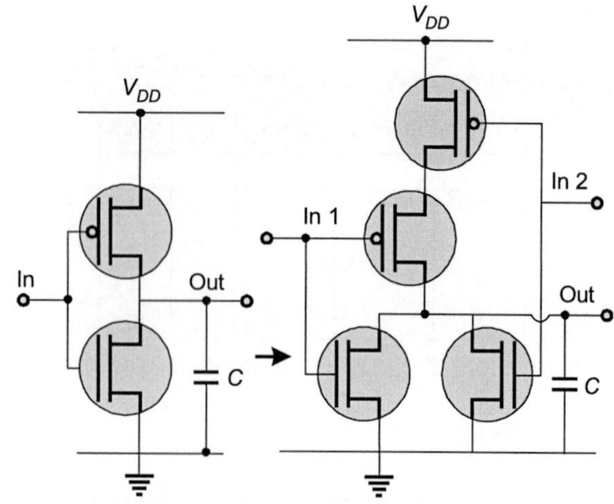

(a) Widths of PMOS transistors should be doubled, while widths of NMOS transistors should be halved.
(b) Widths of PMOS transistors should be doubled, while widths of NMOS transistors should not be changed.

(c) Widths of PMOS transistors should be halved, while widths of NMOS transistors should not be changed.
(d) Widths of PMOS transistors should be unchanged, while widths of NMOS transistors should be halved.

[2016 : 1 M, Set-2]

6.5 A finite state machine (FSM) is implemented using the D flip-flops A and B, and logic gates, as shown in the figure below. The four possible states of the FSM are Q_AQ_B = 00, 01, 10, and 11.

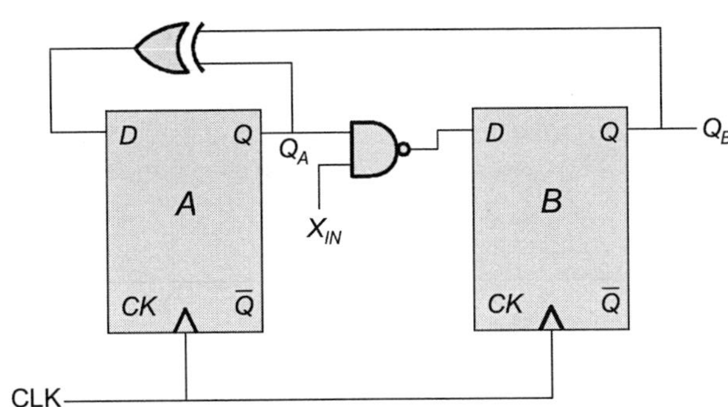

Assume that X_{IN} is held at constant logic level throughout the operation of the FSM. When the FSM is initialized to the Q_AQ_B = 00 and clocked, after a few clock cycles, it starts cycling through

(a) all of the four possible states if X_{IN} = 1
(b) three of the four possible states if X_{IN} = 0
(c) only two of the four possible states if X_{IN} = 1
(d) only two of the four possible states if X_{IN} = 0

[2017 : 2 M, Set-1]

6.6 The state diagram of a finite state machine (FSM) designed to detect an overlapping sequence of three bits is shown in the figure. The FSM has an input 'In' and an output 'Out'. The initial state of the FSM is S_0.

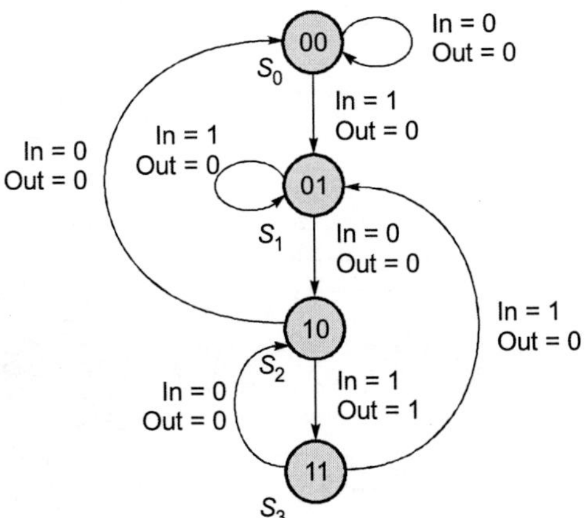

If the input sequence is 10101101001101, starting with the left-most bit, then the number of times 'Out' will be 1 is _____.

[2017 : 2 M, Set-2]

6.7 A traffic signal cycles from GREEN to YELLOW, YELLOW to RED and RED to GREEN. In each cycle, GREEN is turned on for 70 seconds, YELLOW is turned on for 5 seconds and the RED is turned on for 75 seconds. This traffic light has to be implemented using a finite state machine (FSM). The only input to this FSM is a clock of 5 second period. The minimum number of flip-flops required to implement this FSM is ______.

[2018 : 1 M]

6.8 In the circuit shown below, a positive edge-triggered D flip-flop is used for sampling input data D_{IN} using clock CK. The XOR gate outputs 3.3 volts for logic HIGH and 0 volts for logic LOW levels. The data bit and clock periods are equal and the value of $\Delta T/T_{CK}$ = 0.15, where the parameters ΔT and T_{CK} are shown in the figure. Assume that the Flip-Flop and the XOR gate are ideal.

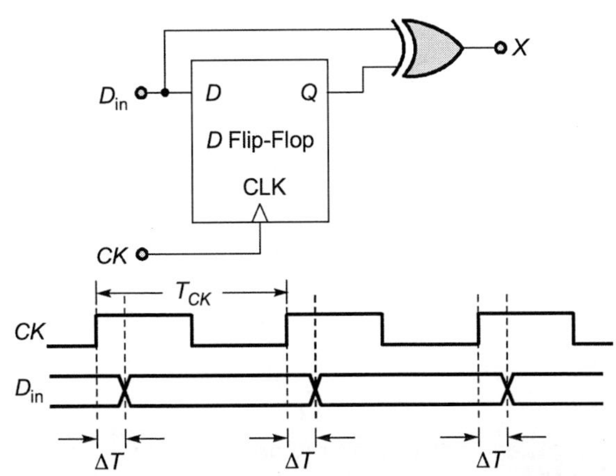

If the probability of input data bit (D_{in}) transition in each clock period is 0.3, the average value (in volts, accurate to two decimal places) of the voltage at node X, is _____.

[2018 : 2 M]

6.9 The state transition diagram for the circuit shown is

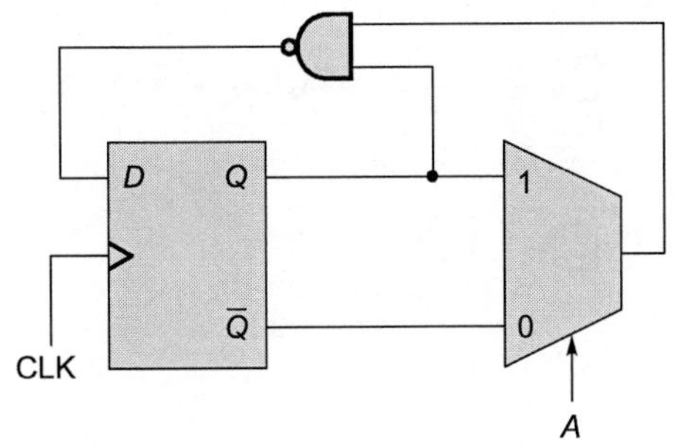

(a) 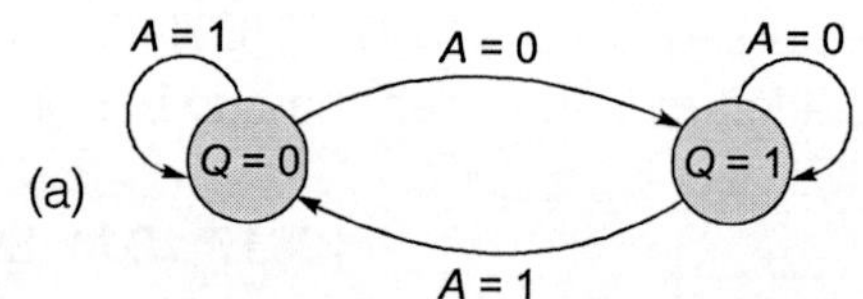

(b) 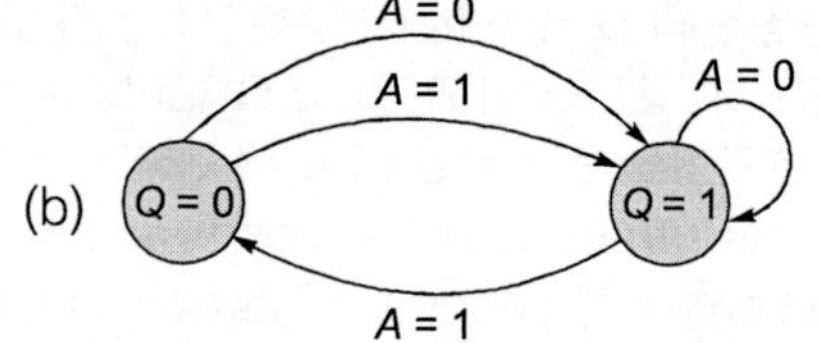

(c)

(d) 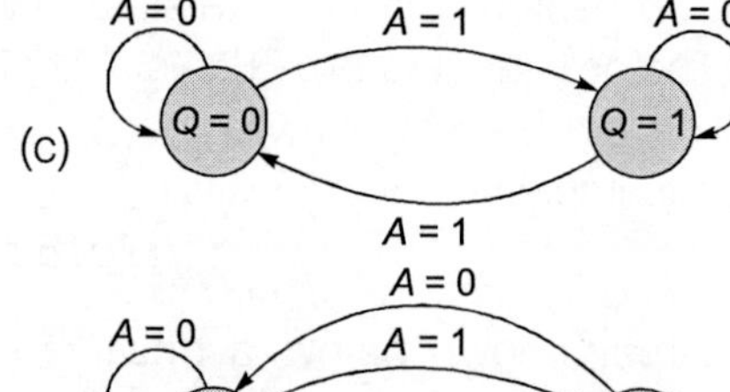

[2019 : 2 M]

6.10 The state diagram of a sequence detector is shown below. State S_0 is the initial state of the sequence detector. If the output is 1, then

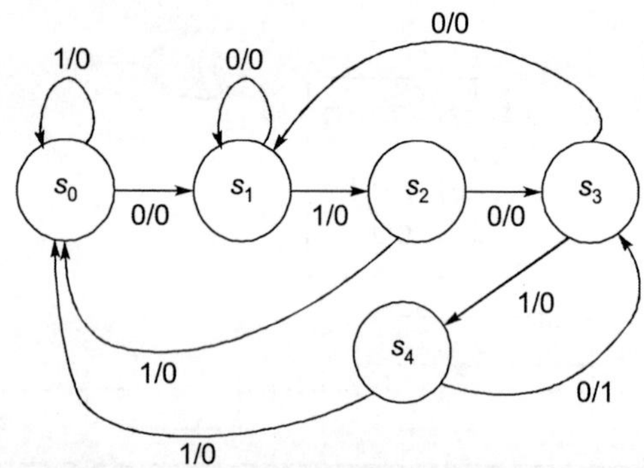

(a) the sequence 01010 is detected
(b) the sequence 01011 is detected
(c) the sequence 01001 is detected
(d) the sequence 01110 is detected

[2020 : 2 M]

6.11 A state transition diagram with states *A*, *B*, and *C*, and transition probabilities $p_1, p_2, ..., p_7$ is shown in the figure (e.g., p_1 denotes the probability of transition from state *A* to *B*). For this state diagram, select the statement(s) which is/are universally true.

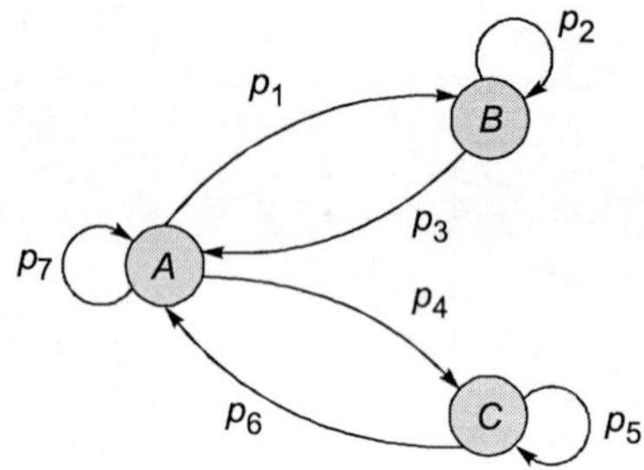

(a) $p_2 + p_3 = p_5 + p_6$ (b) $p_1 + p_3 = p_4 + p_6$
(c) $p_1 + p_4 + p_7 = 1$ (d) $p_2 + p_5 + p_7 = 1$

[2022 : 2 M]

6.12 A sample and hold circuit is implemented using a resistive switch and a capacitor with a time constant of 1 μs. The time for sampling switch to stay closed to charge a capacitor adequately to a full scale voltage of 1 V with 12-bit accuracy is ______ μs.
(rounded off to two decimal places)

[2023 : 2 M]

■■■

Answers Finite State Machine and Miscellaneous

6.1 (d)	6.2 (d)	6.3 (c)	6.4 (b)	6.5 (d)	6.6 (4)	6.7 (5)
6.8 (0.8415)	6.9 (b)	6.10 (a)	6.11 (a, c)	6.12 (8.3177)		

Explanations Finite State Machine and Miscellaneous

6.1 (d)

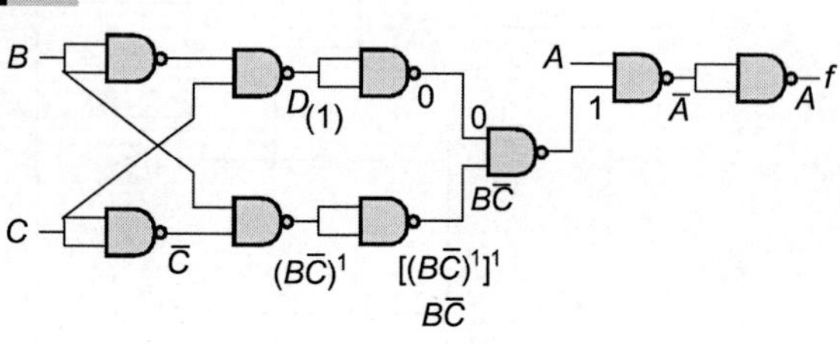

6.2 (d)

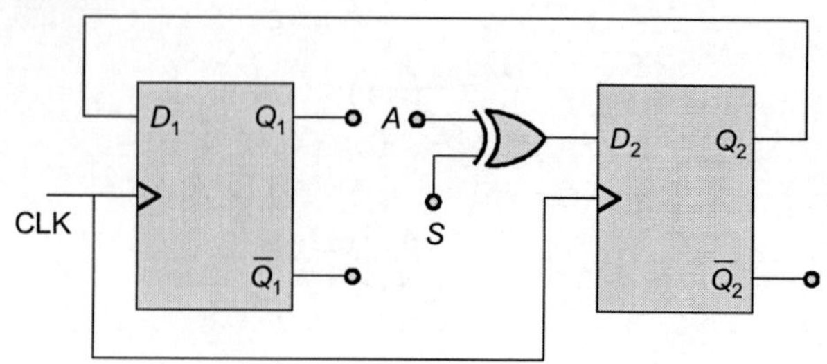

Initially when X-OR gate is connected to the input of flip flop 2 i.e. input of D_2, i.e.

$$D_2 = A \oplus S = Q_1 \oplus S$$

Now, X-OR is replaced by X-NOR, keeping state diagram to be unchanged it means the input to D_2 should not get changed.

When replaced by X-NOR gate

Now $D_2 = A \odot S$

Also, we know that

$\overline{A \oplus B} = \bar{A} \oplus B = A \odot B$

We require output to be as $D_2 = A \oplus S$

So, to keep this output, A should be connected to $\bar{Q}_1$ as

$$D_2 = \bar{Q}_1 \odot S = \overline{Q_1 \odot S} = Q_1 \oplus S$$

6.3 (c)

For State A

X	Y	Z	Present State	Next State
0	0	0	A	B
0	0	1	A	A
0	1	0	A	A
0	1	1	A	A
1	0	0	A	C
1	0	1	A	C
1	1	0	A	A
1	1	1	A	A

For State B

X	Y	Z	Present State	Next State
0	0	0	B	A
0	0	1	B	C
0	1	0	B	B
0	1	1	B	B
1	0	0	B	A
1	0	1	B	B
1	1	0	B	B
1	1	1	B	B

For State C

X	Y	Z	Present State	Next State
0	0	0	C	C
0	0	1	C	C
0	1	0	C	C
0	1	1	C	B
1	0	0	C	C
1	0	1	C	A
1	1	0	C	C
1	1	1	C	AB

In state 'C' when XYZ = 111; the ambiguity occurs. Because from state 'C'

When, $X = 1$, $Z = 1 \Rightarrow$ Next state = A

When, $Y = 1$, $Z = 1 \Rightarrow$ Next state = B

6.5 (d)

In the given diagram,

$$D_A = Q_A \oplus Q_B \text{ and } D_B = \overline{Q_A X_{IN}}$$

For $X_{IN} = 0$:

$$D_A = Q_A \oplus Q_B \text{ and } D_B = 1$$

Present State Q_A	Q_B	D_A	D_B	Next State Q_A^+	Q_B^+
0	0	0	1	0	1
0	1	1	1	1	1
1	1	0	1	0	1
0	1	1	1	1	1

So, for $X_{IN} = 0$, Number of possible states = 2

For $X_{IN} = 1$:

$$D_A = Q_A \oplus Q_B \text{ and } D_B = \bar{Q}_A$$

Present State Q_A	Q_B	D_A	D_B	Next State Q_A^+	Q_B^+
0	0	0	1	0	1
0	1	1	1	1	1
1	1	0	0	0	0
0	0	0	1	0	1

So, for $X_{IN} = 1$, Number of possible states = 3

6.6 (4)

By observing the given state diagram, it is clear that, the FSM can be used to detect the sequence '101'.

It is given in the question that, the FSM detects overlapping sequences also.

The given input sequence is,

10101101001101 $\Rightarrow$ So, output will be 1 for 4 times.

6.7 (5)

GREEN → 70 seconds

YELLOW → 5 seconds

RED → 75 seconds

Clock period → 5 seconds

Total number of unique states required

$$= \frac{70+5+75}{5} = 30$$

Minimum number of flip-flops required is,

$$n = \lceil \log_2(30) \rceil = \lceil 4.91 \rceil = 5$$

6.8 (0.8415)

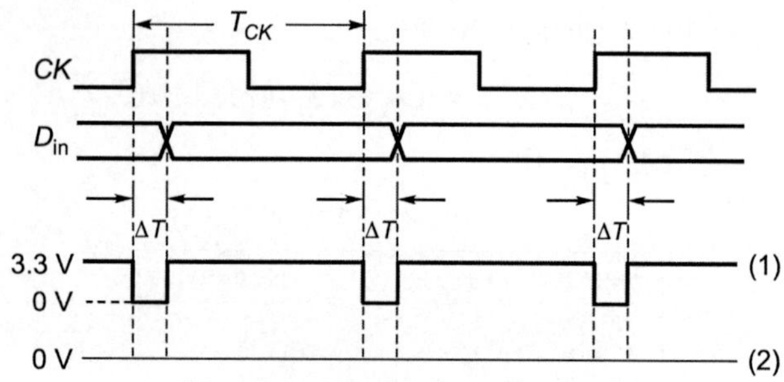

(1) → When input is changed during clock period (with probability 0.3)

(2) → When input is not changed during clock period (with probability 0.7)

$$V_{X(avg)} = \left[0.3 \times 3.3\left(1 - \frac{\Delta T}{T_{CK}}\right)\right] + [0.7 \times 0]\text{ V}$$

$$= 0.3 \times 3.3 \times (1 - 0.15)\text{ V}$$

$$= 0.3 \times 3.3 \times 0.85\text{ V} = 0.8415\text{ V}$$

6.9 (b)

When $A = 0$, $Q_{n+1} = 1$

When $A = 1$, $Q_{n+1} = \bar{Q}_n$

So, the correct state transition diagram is,

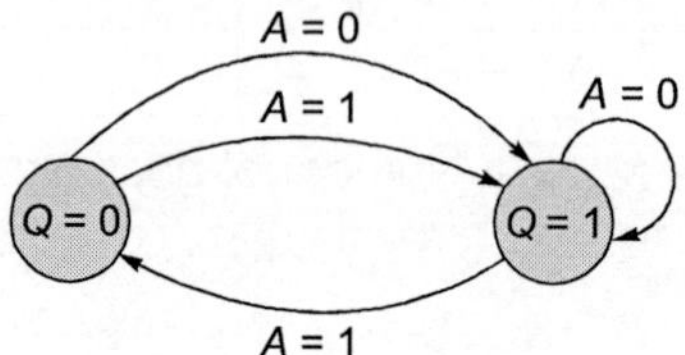

6.10 (a)

The sequence detected is 01010.

6.11 (a, c)

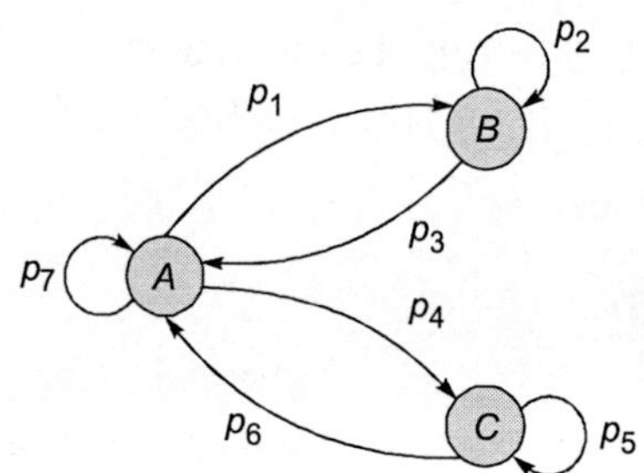

In a state diagram, the output transition expressions for a state are:

(i) Mutually exclusive : For each state, the product of any pair of outgoing transition expressions is "0".

(ii) All-inclusive : For each state, the sum of transition expressions on all outgoing arcs is "1".

Considering the all-inclusive property of states,

For state A,

$$p_1 + p_7 + p_4 = 1$$

For state B,

$$p_2 + p_3 = 1 \quad \text{...(i)}$$

For state C,

$$p_5 + p_6 = 1 \quad \text{...(ii)}$$

From equation (i) and (ii),

$$p_2 + p_3 = p_5 + p_6$$

Hence, options (a) and (c) are correct.

6.12 (8.3177)

Given: Time constant (τ) of 1 μsec.
Full scale voltage = 1 V

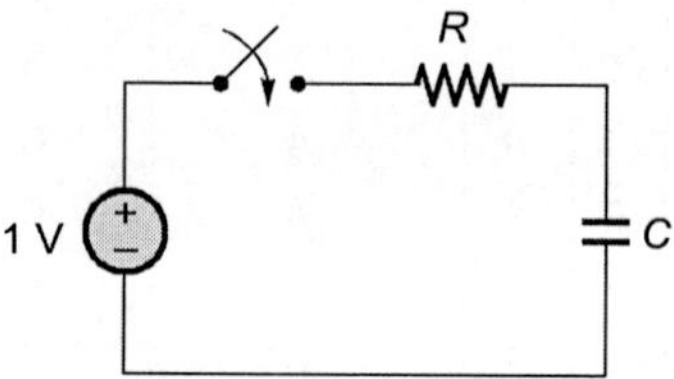

The voltage across capacitor is given as

$$V_c(t) = V_{in}(1 - e^{-t/\tau})$$

$$\therefore \quad V_c(t) = (1 - e^{-t/1\ \mu\text{sec}}) \quad \text{...(i)}$$

To calculate the voltage to stay closed to charge capacitor adequately to a full scale voltage with 12-bit accuracy is given by

$$V_c(t) = V_{ref}\left[1 - \frac{1}{2^n}\right]$$

n = Number of bit = 12

$$\therefore \quad V_c(t) = \left[1 - \frac{1}{4096}\right] \quad \text{...(ii)}$$

Comparing equations (i) and (ii), we get,

$$e^{-t/1\mu\text{sec}} = \frac{1}{4096}$$

$$-\frac{t}{1\mu\text{sec}} = \ln\left\{\frac{1}{4096}\right\}$$

$$t = 8.3177\ \mu\text{sec}$$

7 Logic Families

7.1 In standard TTL the 'totem pole' stage refers to
(a) the multi-emitter input stage.
(b) the phase splitter
(c) the output buffer
(d) open collector output stage [1997 : 1 M]

7.2 The inverter 74 AL S01 has the following specifications:

$I_{OH\,max} = -0.4$ mA, $I_{OL\,max} = 8$ mA, $I_{IH\,max} = 20$ µA, $I_{IL\,max} = -0.1$ mA.

The fan out based on the above will be
(a) 10 (b) 20
(c) 60 (d) 100 [1997 : 1 M]

7.3 For the NMOS logic gate shown in figure, the logic function implemented is

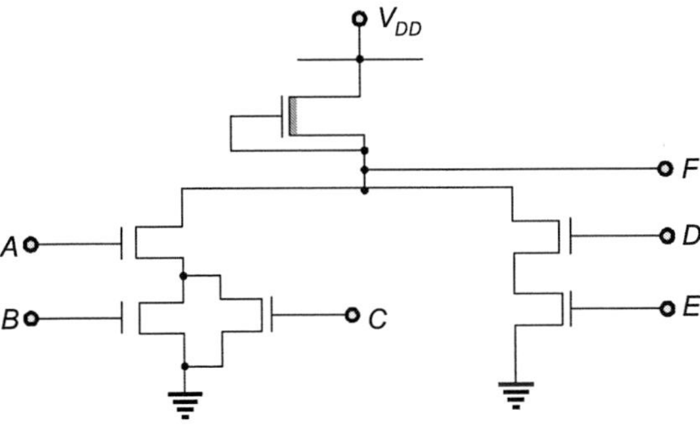

(a) $\overline{ABCDE}$ (b) $(AB+\overline{C})\cdot(\overline{D+E})$
(c) $\overline{A\cdot(B+C)+D\cdot E}$ (d) $\overline{(A+B)}\cdot C+\overline{D}\cdot\overline{E}$
[1997 : 1 M]

7.4 The gate delay of an NMOS inverter is dominated by charge time rather than discharge time because
(a) the driver transistor has a larger threshold voltage than the load transistor
(b) the driver transistor has larger leakage currents compared to the load transistor
(c) the load transistor has a smaller W/L ratio compared to the driver transistor
(d) none of the above.
[1997 : 1 M]

7.5 The noise margin of a TTL gate is about
(a) 0.2 V (b) 0.4 V
(c) 0.6 V (d) 0.8 V [1998 : 1 M]

7.6 The threshold voltage for each transistor in figure is 2 V. For this circuit to work as an inverter, V_i must take the values

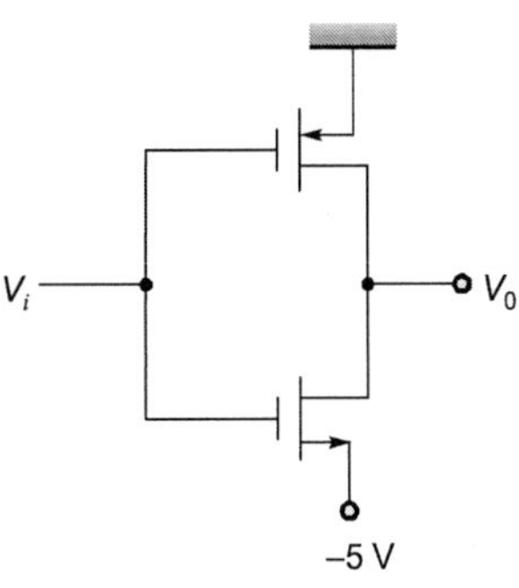

(a) – 5 V and 0 V (b) – 5 V and 5 V
(c) 0 V and 5 V (d) – 3 V and 3 V
[1998 : 1 M]

7.7 A Darlington emitter-follower circuit is sometimes used in the output stage of a TTL gate in order to
(a) increase its I_{OL}
(b) reduce its I_{OH}
(c) increase its speed of operation
(d) reduce power dissipation [1999 : 1 M]

7.8 Commercially available ECL Gates use two ground lines and one negative supply in order to
(a) reduce power dissipation
(b) increase fan-out
(c) reduce loading effect
(d) eliminate the effect of power line glitches or the biasing circuit
[1999 : 1 M]

7.9 In the TTL circuit in the figure, S_2 and S_0 are select lines and X_7 and X_0 are input lines. S_0 and X_0 are LSBs. The output Y is

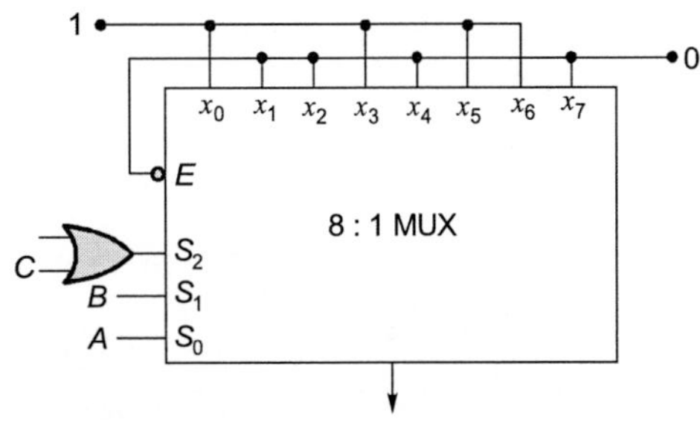

(a) indeterminate (b) $A\oplus B$
(c) $\overline{A\oplus B}$ (d) $\overline{C}(\overline{A\oplus B})+C(A\oplus B)$
[2001 : 2 M]

7.10 The circuit in the figure has two CMOS NOR-gates. This circuit functions as a:

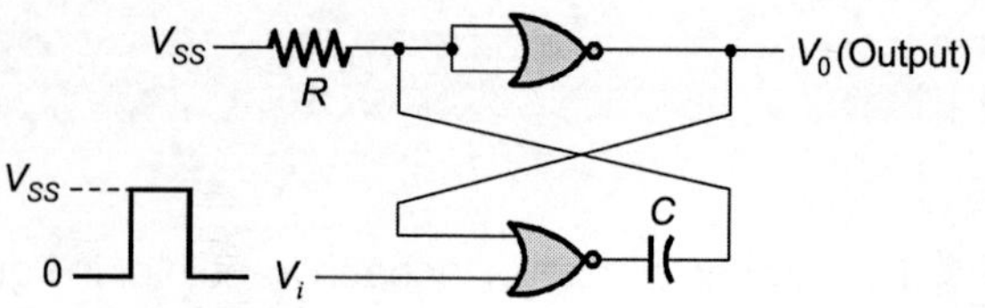

(a) flip-flop
(b) Schmitt trigger
(c) monostable multi-vibrator
(d) astable multi-vibrator [2002 : 2 M]

7.11 The output of the 74 series GATE of TTL gates is taken from a BJT in
(a) totem pole and common collector configuration
(b) either totem pole or open collector configuration
(c) common base configuration
(d) common collector configuration [2003 : 1 M]

7.12 The DTL, TTL, ECL and CMOS famil GATE of digital ICs are compared in the following 4 columns

	(P)	(Q)	(R)	(S)
Fanout is minimum	DTL	DTL	TTL	CMOS
Power consumption is minimum	TTL	CMOS	ECL	DTL
Propagation delay is minimum	CMOS	ECL	TTL	TTL

The correct column is
(a) P (b) Q
(c) R (d) S [2003 : 2 M]

7.13 Given figure is the voltage transfer characteristic of
(a) an NMOS inverter with enhancement mode transistor as load
(b) an NMOS inverter with depletion mode transistor as load
(c) a CMOS inverter
(d) a BJT inverter [2004 : 1 M]

7.14 The figure shows the internal schematic of a TTL AND-OR-Invert (AOI) gate. For the inputs shown in the figure, the output Y is

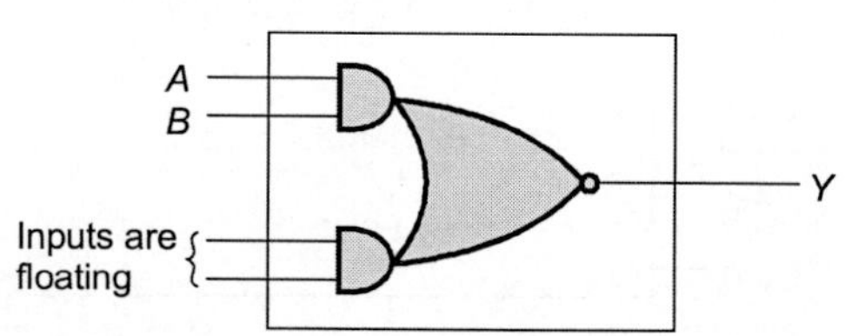

(a) 0 (b) 1
(c) AB (d) $\overline{AB}$ [2004 : 1 M]

7.15 The transistors used in a portion of the TTL gate shown in the figure have a $\beta = 100$. The base-emitter voltage of is 0.7 V for a transistor in active region and 0.75 V for a transistor in saturation. If the sink current $I = 1$ mA and the output is at logic 0, then the current I_R will be equal to

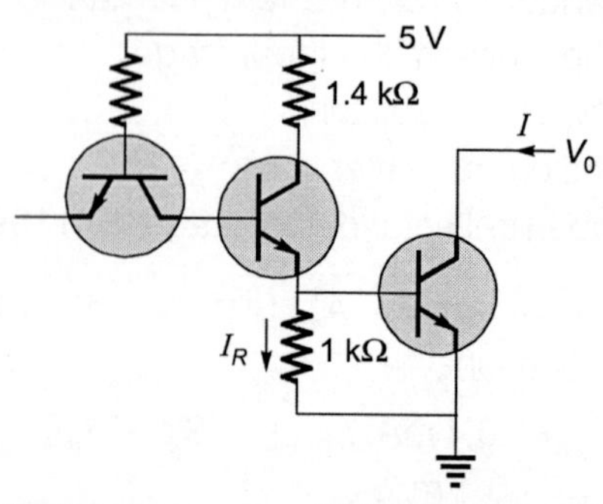

(a) 0.65 mA (b) 0.70 mA
(c) 0.75 mA (d) 1.00 mA [2005 : 2 M]

7.16 Both transistors T_1 and T_2 shown in the figure, have a threshold voltage of 1 Volts. The device parameters K_1 and K_2 of T_1 and T_2 are, respectively, 36 μA/V^2 and 9 μA/V^2. The output voltage V_0 is

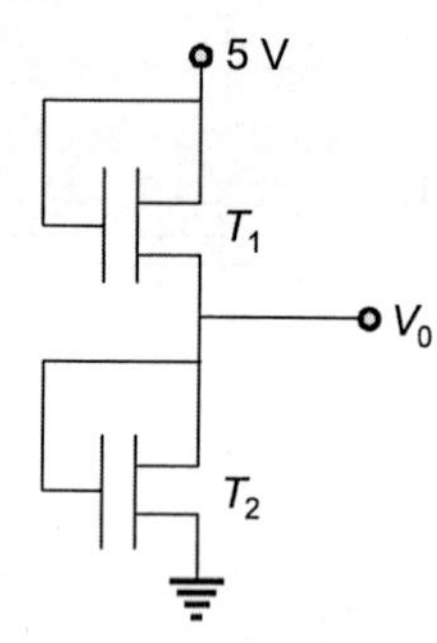

(a) 1 V (b) 2 V
(c) 3 V (d) 4 V [2005 : 2 M]

7.17 The circuit diagram of a standard TTL NOT gate is shown in the figure. When $V_i = 2.5$ V, the modes of operation of the transistors will be

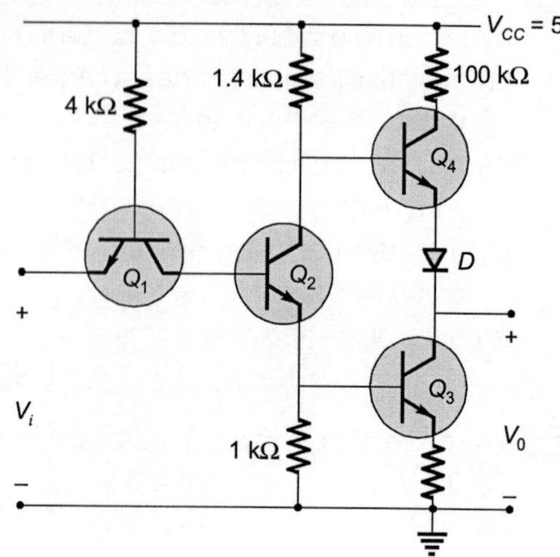

(a) Q_1: reverse active; Q_2: normal active; Q_3: saturation; Q_4: cut-off
(b) Q_1: reverse active; Q_2: saturation; Q_3: saturation; Q_4: cut-off
(c) Q_1: normal active; Q_2: cut-off; Q_3: cut-off; Q_4: saturation
(d) Q_1: saturation; Q_2: saturation; Q_3: saturation; Q_4: normal active

[2007 : 2 M]

7.18 The logic function implemented by the following circuit at the terminal OUT is

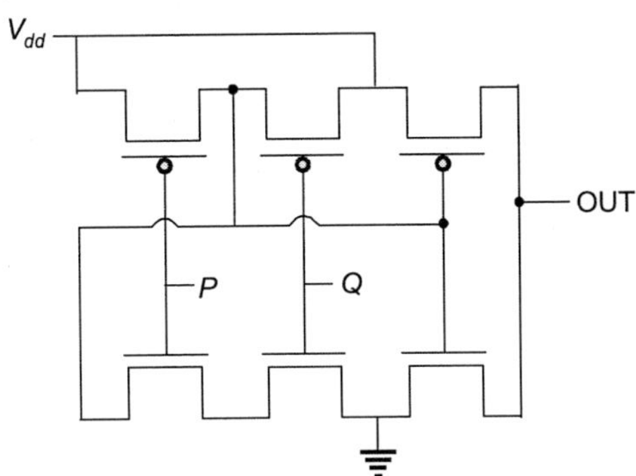

(a) P NOR Q (b) P NAND Q
(c) P OR Q (d) P AND Q

[2008 : 2 M]

7.19 The full forms of the abbreviations TTL and CMOS in reference to logic families are
(a) Triple Transistor Logic and Chip Metal Oxide Semiconductor
(b) Tristate Transistor Logic and Chip Metal Oxide Semiconductor
(c) Transistor Transistor Logic and Complementary Metal Oxide Semiconductor
(d) Tristate Transistor Logic and Complementary Metal Oxide Silicon [2009 : 1 M]

7.20 In the circuit shown

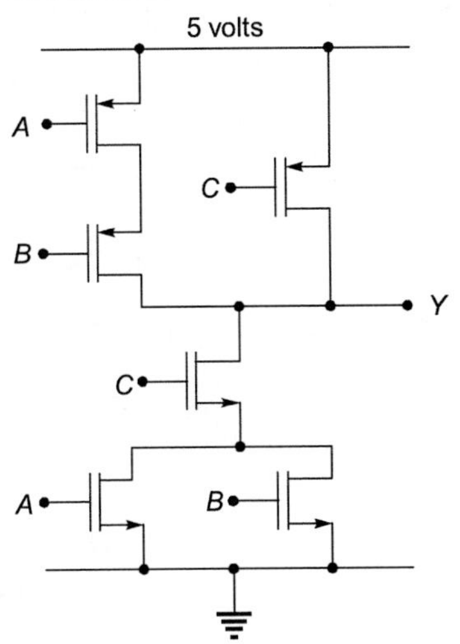

(a) $Y = \overline{A}\overline{B} + \overline{C}$ (b) $Y = (A + B)C$
(c) $Y = (\overline{A} + \overline{B})\overline{C}$ (d) $Y = AB + C$

[2012 : 1 M]

7.21 In the circuit shown below, Q_1 has negligible collector-to-emitter saturation voltage and the diode drops negligible voltage across it under forward bias. If V_{cc} is +5 V, X and Y are digital signals with 0 V as logic 0 and V_{cc} as logic 1, the Boolean expression for Z is

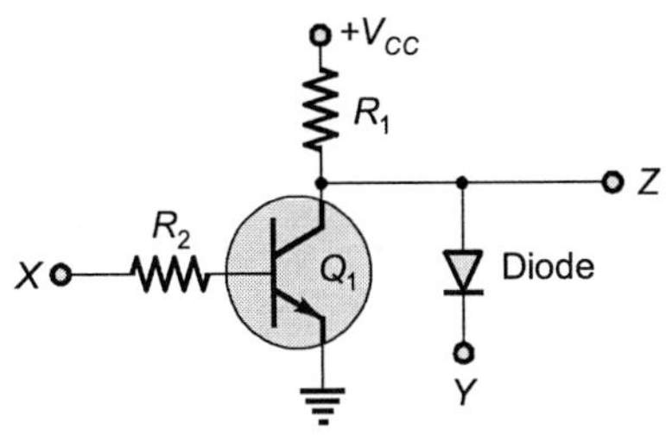

(a) XY (b) $\overline{X}Y$
(c) $X\overline{Y}$ (d) $\overline{XY}$ [2013 : 2 M]

7.22 In the following circuit employing pass transistor logic, all NMOS transistors are identical with a threshold voltage of 1 V. Ignoring the body-effect, the output voltages at P, Q and R are,

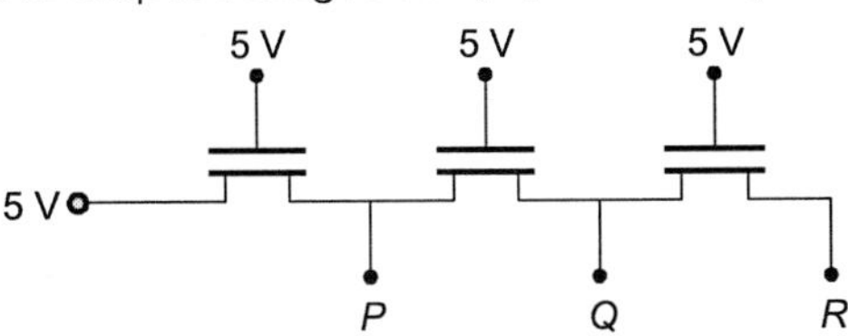

(a) 4 V, 3 V, 2 V (b) 5 V, 5 V, 5 V
(c) 4 V, 4 V, 4 V (d) 5 V, 4 V, 3 V

[2014 : 1 M, Set-1]

7.23 The output (Y) of the circuit shown in the figure is

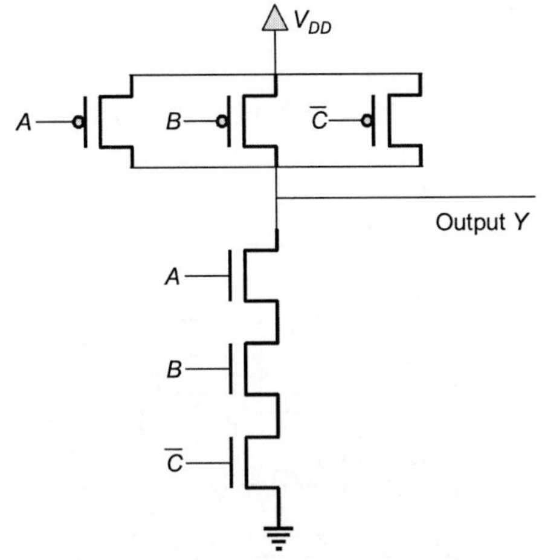

(a) $\overline{A} + \overline{B} + C$ (b) $A + \overline{B} \cdot \overline{C} + A \cdot \overline{C}$
(c) $\overline{A} + B + \overline{C}$ (d) $A \cdot B \cdot \overline{C}$

[2014 : 1 M, Set-4]

7.24 In the circuit shown, diodes D_1, D_2 and D_3 are ideal, and the inputs E_1, E_2 and E_3 are '0 V' for logic '0' and '10 V' for logic '1'. What logic gate does the circuit represent?

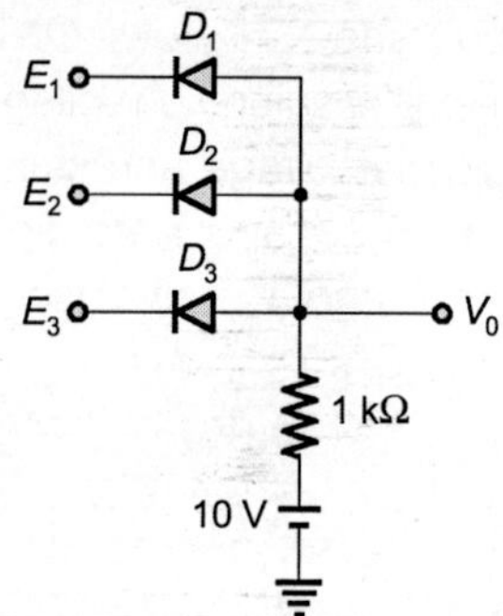

(a) 3 input OR gate (b) 3 input NOR gate
(c) 3 input AND gate (d) 3 input XOR gate

[2015 : 1 M, Set-3]

7.25 An SR latch is implemented using TTL gates as shown in the figure. The set and reset pulse inputs are provided using the push-button switches. It is observed that the circuit fails to work as desired. The SR latch can be made functional by changing

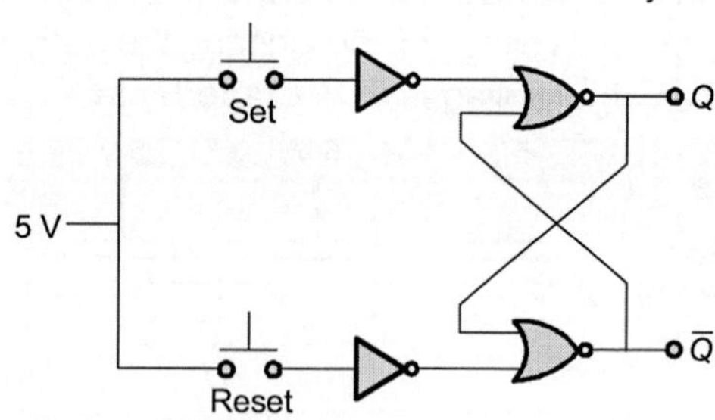

(a) NOR gates to NAND gates
(b) inverters to buffers
(c) NOR gates to NAND gates and inverters to buffers
(d) 5 V to ground

[2015 : 2 M, Set-3]

7.26 The functionality implemented by the circuit below is

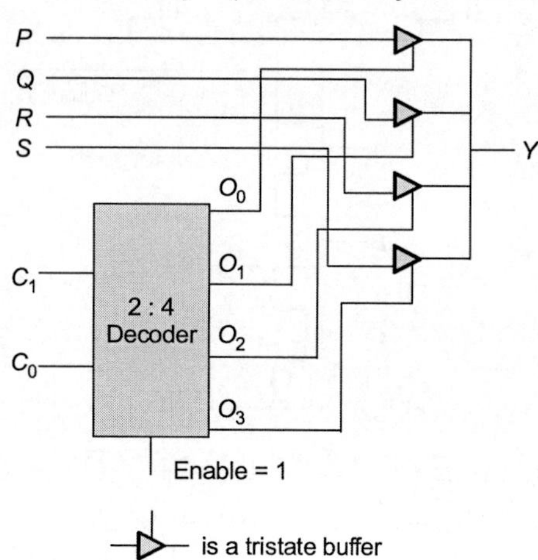

(a) 2-to-1 multiplexer (b) 4-to-1 multiplexer
(c) 7-to-1 multiplexer (d) 6-to-1 multiplexer

[2016 : 2 M, Set-1]

7.27 The logic functionality realized by the circuit shown below is

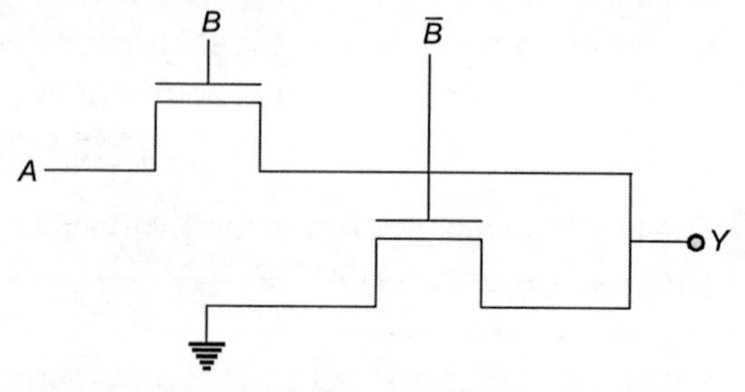

(a) OR (b) XOR
(c) NAND (d) AND

[2016 : 1 M, Set-3]

7.28 For the circuit shown in the figure. P and Q are the inputs and Y is the output.

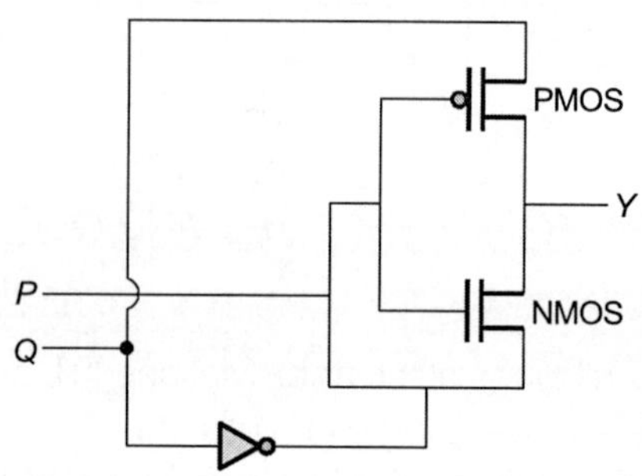

The logic implemented by the circuit is
(a) XNOR (b) XOR
(c) NOR (d) OR

[2017 : 1 M, Set-2]

7.29 The logic $f(X, Y)$ realized by the given circuit is

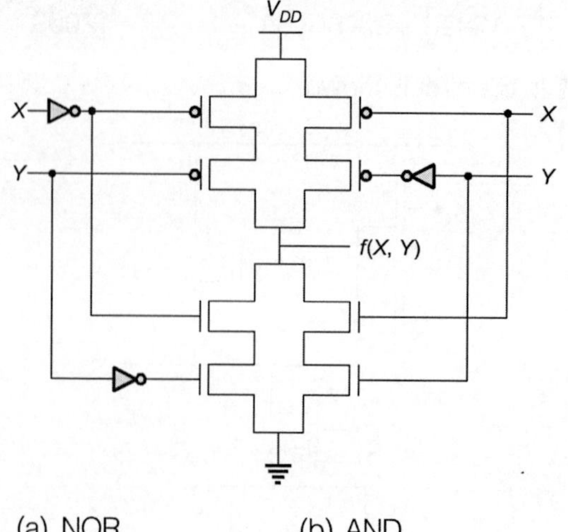

(a) NOR (b) AND
(c) NAND (d) XOR

[2018 : 1 M]

7.30 The logic gates shown in the digital circuit below use strong pull-down nMOS transistors for LOW logic level at the outputs. When the pull-downs are off, high-value resistors set the output logic levels to HIGH (i.e. the pull-ups are weak). Note that some nodes are intentionally shorted to implement "wired logic". Such shorted nodes will be HIGH only if the outputs of all the gates whose outputs are shorted are HIGH.

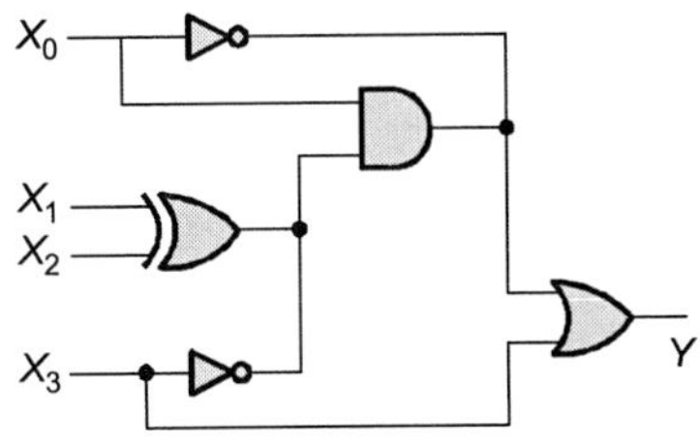

The number of distinct values of $X_3\,X_2\,X_1\,X_0$ (out of the 16 possible values) that give $Y = 1$ is ____.

[2018 : 2 M]

7.31 In the circuit shown, A and B are the inputs and F is the output. What is the functionality of the circuit?

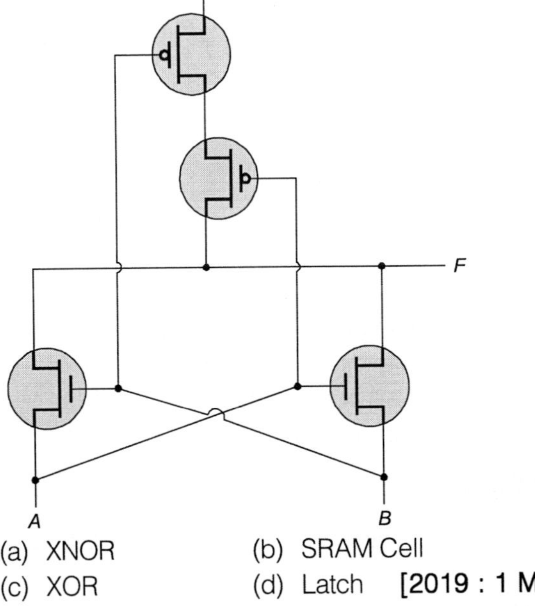

(a) XNOR
(b) SRAM Cell
(c) XOR
(d) Latch

[2019 : 1 M]

7.32 In the circuit shown, what are the values of F for $EN = 0$ and $EN = 1$, respectively?

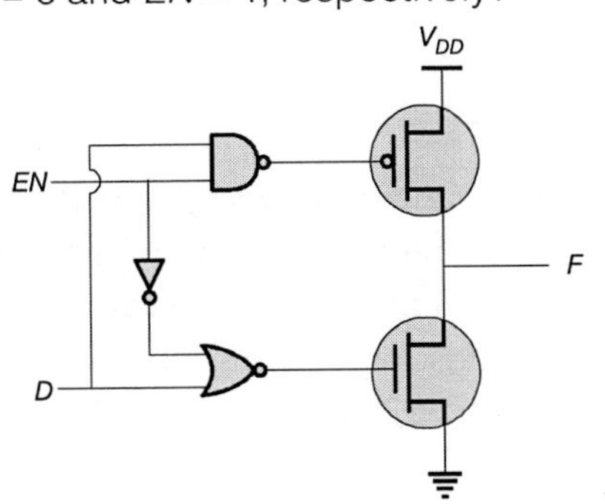

(a) 0 and 1
(b) Hi-Z and D
(c) Hi-Z and $\bar{D}$
(d) 0 and D

[2019 : 1 M]

7.33 A standard CMOS inverter is designed with equal rise and fall times ($\beta_n = \beta_p$). If the width of the pMOS transistor in the inverter is increased, what would be the effect on the LOW noise margin (NM_L) and the HIGH noise margin NM_H?

(a) NM_L decreases and NM_H increases.
(b) No change in the noise margins.
(c) Both NM_L and NM_H increase.
(d) NM_L increases and NM_H decreases.

[2019 : 1 M]

7.34 In the circuits shown, the threshold voltage of each nMOS transistor is 0.6 V. Ignoring the effect of channel length modulation and body bias, the values of V_{out1} and V_{out2}, respectively, in volts, are

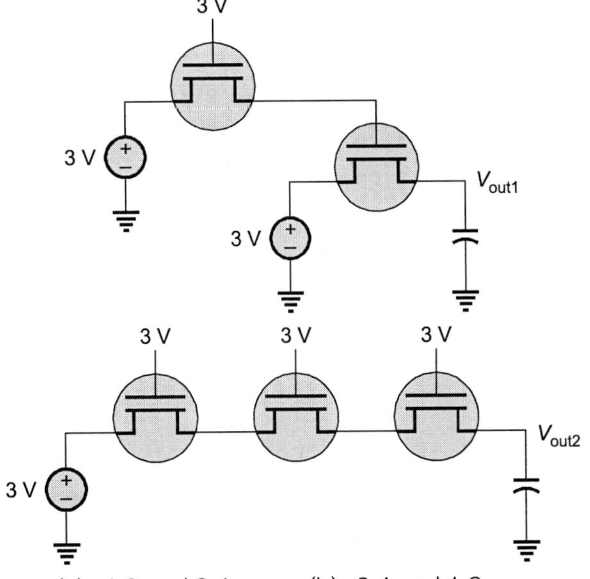

(a) 1.8 and 2.4
(b) 2.4 and 1.2
(c) 1.8 and 1.2
(d) 2.4 and 2.4

[2019 : 2 M]

7.35 Select the correct statement(s) regarding CMOS implementation of NOT gates.

(a) Noise Margin High (NM_H) is always equal to the Noise Margin Low (NM_L), irrespective of the sizing of transistors.
(b) Dynamic power consumption during switching is zero.
(c) For a logical high input under steady state, the nMOSFET is in the linear regime of operation.
(d) Mobility of electrons never influences the switching speed of the NOT gate.

[2022 : 1 M]

7.36 In the circuit below, the voltage V_L is ______ V. (rounded off to two decimal places)

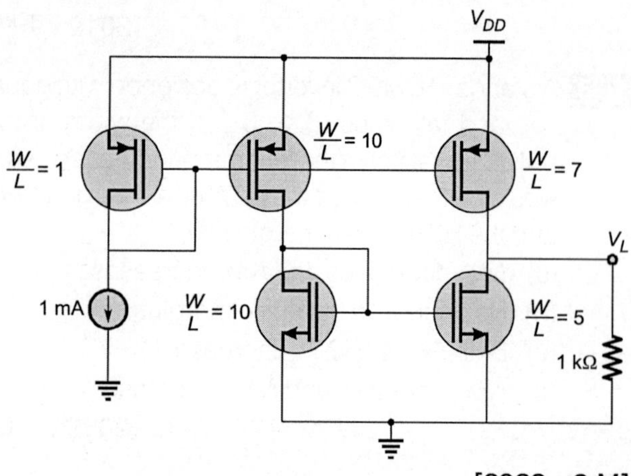

[2023 : 2 M]

Answers Logic Families

7.1 (c)	7.2 (b)	7.3 (c)	7.4 (c)	7.5 (b)	7.6 (a)	7.7 (c)
7.8 (d)	7.9 (b)	7.10 (c)	7.11 (b)	7.12 (b)	7.13 (c)	7.14 (a)
7.15 (c)	7.16 (c)	7.17 (b)	7.18 (d)	7.19 (c)	7.20 (a)	7.21 (b)
7.22 (c)	7.23 (a)	7.24 (c)	7.25 (d)	7.26 (b)	7.27 (d)	7.28 (*)
7.29 (d)	7.30 (8)	7.31 (a)	7.32 (b)	7.33 (d)	7.34 (a)	7.35 (c)
7.36 (2)						

Explanations Logic Families

7.1 (c)

In standard TTL, the totem pole stage is an output stage having two transistors connected in such a way that only one of them is ON at a time. Thus totem pole refers to output buffer.

7.2 (b)

$$(F.O.)_H = \frac{I_{OH(max)}}{I_{IH(max)}} = \frac{0.4 \times 10^{-3}}{20 \times 10^{-6}} = \frac{400}{20} = 20$$

$$(F.O.)_L = \frac{I_{OL(max)}}{I_{IL(max)}} = \frac{8 \times 10^{-3}}{0.1 \times 10^{-3}} = 80$$

$$F.O. = \text{min. of } [(F.O.)_H, (F.O.)_L]$$
$$= \text{min. of } (20, 80) = 20$$

7.3 (c)

$$F = \overline{A.(B+C) + D.E}$$

7.4 (c)

The gate delay of an NMOS inverter is dominated by charge time rather than discharge time because the load transistor has a smaller W/L ratio compared to the driver transistor.

7.5 (b)

The noise margin of a TTL gate is about 0.4 V.

7.6 (a)

$$V_T = 2\text{ V}$$
$$V_i = -5\text{ V} \rightarrow V_o = 0$$
$$V_i = 0\text{ V} \rightarrow V_o = -5\text{ V}$$

7.7 (c)

DEF has low output resistance. The low output resistance provides a short time constant for charging up any capacitive load on the output. This provides very small rise time. Therefore speed of operation is increased.

7.9 (b)

Floating input is accepted by TTL logic gate as logic 1.

Hence $S_2 = 1$

S_2	S_1	S_0	Y
1	B	A	
1	0	1	1
1	0	1	1
1	1	0	1
1	1	1	0

C, $S_2 = 1$

$Y = A \oplus B$

7.14 (a)

For TTL logic floating input = 1

$\therefore \quad Y = (AB + 1)' = \overline{AB}.0 = 0$

7.15 (c)

$I = I_C = 1\text{ mA}$ ($\therefore$ BJT is in saturation).

$V_{BE\,sat} = 0.75$

$\Rightarrow \quad 0.75 = I_R.1\text{ k}\Omega$

$I_R = 0.75\text{ mA}$

7.16 (c)

$$I_{D_1} = I_{D_2}$$

$$K_1(V_{GS1} - V_t)^2 = K_2(V_{GS_2} - V_t)^2$$

$\Rightarrow 36(5 - V_0 - 1)^2 = 9(V_0 - 0 - 1)^2$

$\Rightarrow \quad V_0 = 3\text{ V}$

7.19 (c)

TTL – Transistor-transistor logic

CMOS – Complementary Metal Oxide Semiconductor.

7.20 (a)

Series combination of nMOS is equivalent to AND and parallel combination is equivalent to OR.

So, $Y = \overline{C.(A+B)} = \overline{C} + \overline{(A+B)} = \overline{C} + \overline{A}.\overline{B}$

7.21 (b)

X	Y	Z
0	0	0
0	1	1
1	0	0
1	1	0

X	Y	Q	Z
0	0	OFF	0
0	+5V	OFF	+5V
+5V	0	ON	0
+5V	+5V	ON	0

7.22 (c)

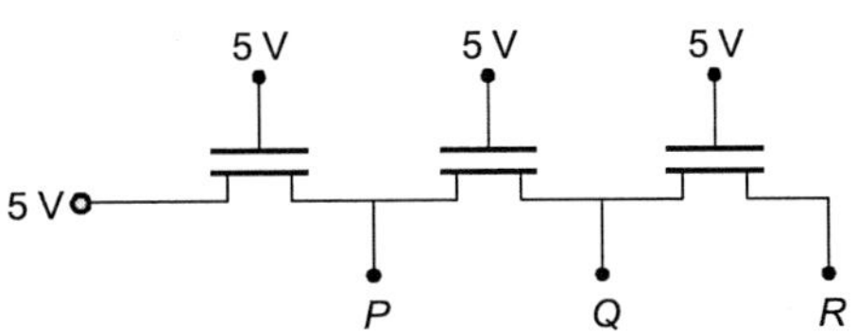

As we know that for proper operation

$V_{DS} = V_{GS} - V_T$

$V_D - V_S = V_G - V_S - V_T$

$\Rightarrow V_D = V_G - V_T$

At P

$\Rightarrow V_D = V_G - V_T$

Given V_G for all transistor = 5 V and

V_T for all transistors = 1 V

$\therefore V_D$ at $P = V_G - V_T = 5 - 1 = 4\text{ V}$

V_D at $Q = V_G - V_T = 5 - 1 = 4\text{ V}$

V_D at $R = V_G - V_T = 5 - 1 = 4\text{ V}$

According to the given options the only condition satisfied for the voltages at P, Q and R is only option (c).

7.23 (a)

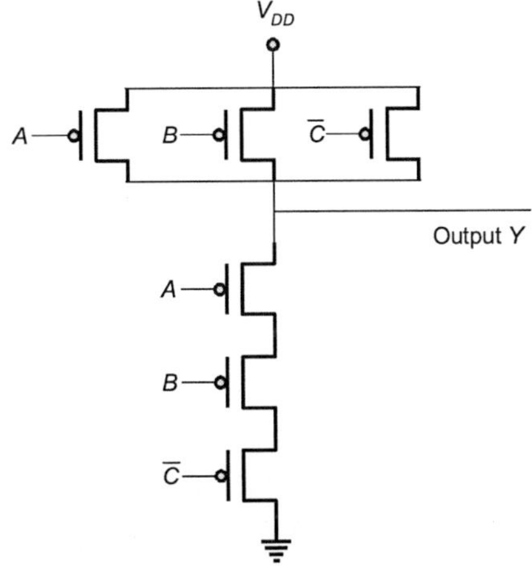

This is the diagram of NAND gates made up of CMOS gates. So output X is given as

$$Y = \overline{A \cdot B \cdot \overline{C}} = (\overline{A} + \overline{B} + C)$$

7.24 (c)

Case (i): If any input is logic 0 (i.e., 0 V) then corresponding diode is "ON" and due to ideal diode output voltage $V_o = 0$ as well as if there is any input logic 1 (i.e., 10 V) corresponding diode will be OFF.

Case (ii): If all the inputs are high (i.e., 10 V) then all the diodes are R.B. (OFF) and output voltage $V_o = 10$ V.

So, it is a positive 3-inputs AND gate.

7.25 (d)

For TTL gate, open end acts as logic '1'.

So if we connect 5 V battery to ground then pressing the switches allow logic '0' while opening the switches allow logic '1'. In case of 5 V battery both opening and closing of switches allow logic '1' and hence circuit can't act as SR latch.

7.26 (b)

When the outputs (O_0, O_1, O_2, O_3) of the decoder are at logic 1, the corresponding tristate buffer is activated. In that case, whatever data is applied at the input of a buffer, becomes its output.

Hence, when

$\Rightarrow \quad C_1C_0 = 00, \quad \text{Then } O_0 = 1,$

$\therefore \quad Y = P$

$\Rightarrow \quad C_1C_0 = 01, \quad \text{Then } O_1 = 1,$

$\therefore \quad Y = Q$

$\Rightarrow \quad C_1C_0 = 10, \quad \text{Then } O_2 = 1,$

$\therefore \quad Y = R$

$\Rightarrow \quad C_1C_0 = 11, \quad \text{Then } O_3 = 1,$

$\therefore \quad Y = S$

$\therefore$ the circuit effectively behaves as a 4 to 1 multiplexer.

7.27 (d)

The output Y will be logic 1, when

$A = 1, B = 1 \text{ and } \overline{B} = 0$

$\Rightarrow \quad Y = A \cdot B \cdot \overline{\overline{B}} = A \cdot B \quad \Rightarrow$ AND operation

7.28 (*)

The given circuit is wrong.

7.29 (d)

From pull-down network,

$$\overline{f(X,Y)} = \overline{X}\,\overline{Y} + XY = X \odot Y$$

$$f(X, Y) = \overline{X \odot Y} = X \oplus Y$$

7.30 (8)

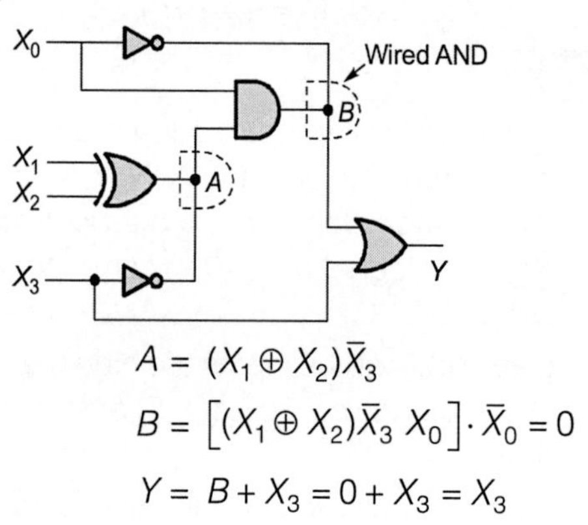

$$A = (X_1 \oplus X_2)\overline{X}_3$$

$$B = \left[(X_1 \oplus X_2)\overline{X}_3 \; X_0\right] \cdot \overline{X}_0 = 0$$

$$Y = B + X_3 = 0 + X_3 = X_3$$

Out of 16 possible combinations of $X_3\, X_2\, X_1\, X_0$, X_3 will be high for 8 combinations. So, Y will be high for 8 combinations.

7.31 (a)

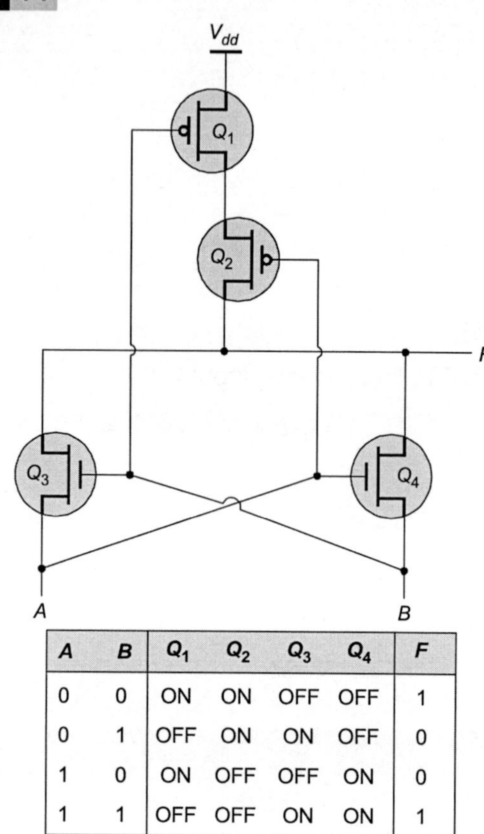

A	B	Q_1	Q_2	Q_3	Q_4	F
0	0	ON	ON	OFF	OFF	1
0	1	OFF	ON	ON	OFF	0
1	0	ON	OFF	OFF	ON	0
1	1	OFF	OFF	ON	ON	1

So, the given logic circuit acts as an XNOR gate.

7.32 (b)

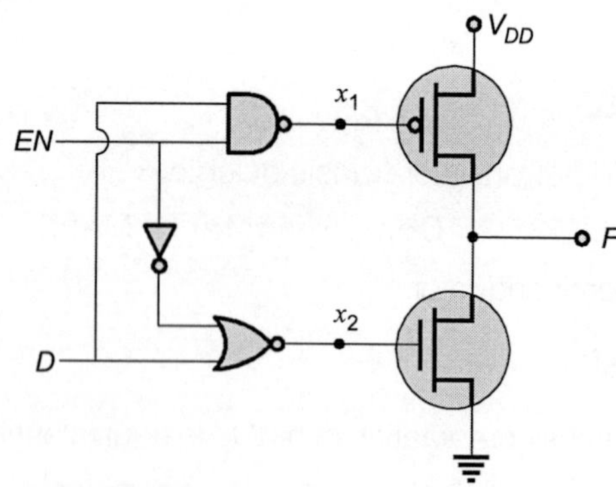

When $EN = 0$:

$x_1 = (\overline{D \cdot 0}) = 1 \Rightarrow$ PMOS is in OFF state

$x_2 = (\overline{1 + \overline{D}}) = 0 \Rightarrow$ NMOS is in OFF state

Both the transistors are in OFF state, which offers high impedance.

When $EN = 1$: $\quad x_1 = (\overline{D \cdot 1}) = \overline{D}$

$x_2 = (\overline{0 + D}) = \overline{D}$

$F = D$

7.33 (d)

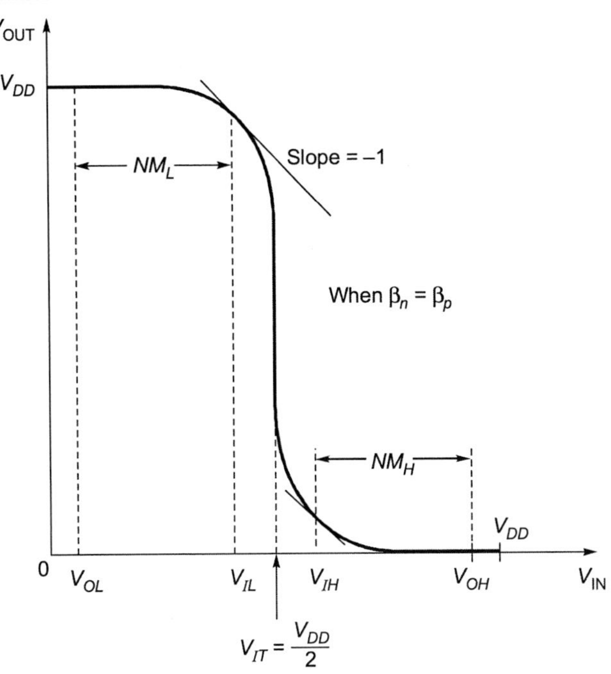

Making PMOS wider, shifts input transition point (V_{IT}) towards V_{DD}

Making NMOS wider, shifts input transition point (V_{IT}) towards zero.

So, as PMOS made wider, NM_L increases and NM_H decreases.

7.34 (a)

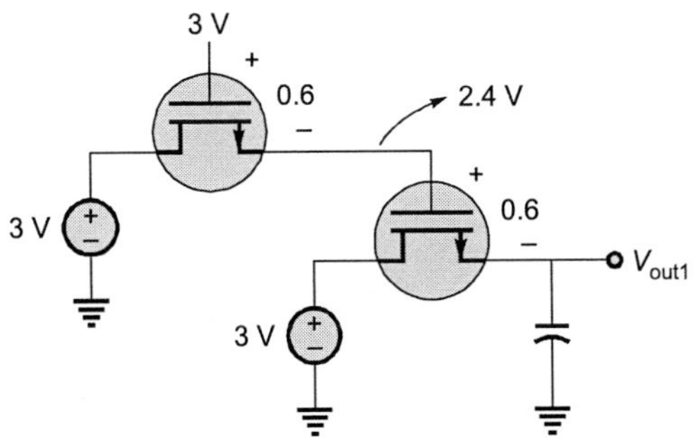

$$V_{out\,1} = 3 - 0.6 - 0.6 = 1.8 \text{ V}$$

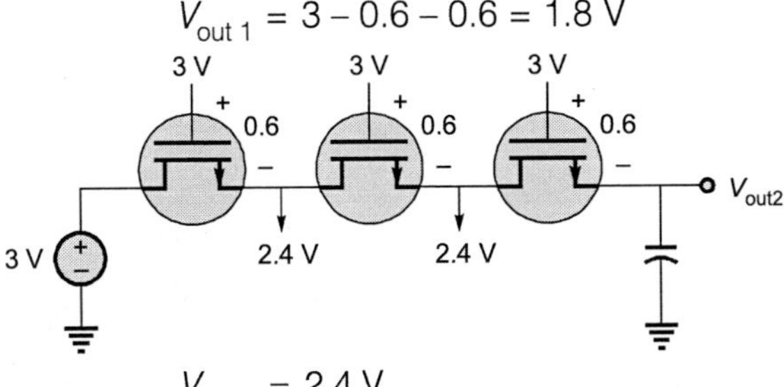

$$V_{out\,2} = 2.4 \text{ V}$$

7.35 (c)

(a) Noise margin is the maximum voltage amplitude of extraneous signal that can exist at the input of a logic gate without it inadvertently switching the output.

High noise margin,

$$NM_H = V_{OH} - V_{IH}$$

Low noise margin,

$$NM_L = V_{IL} - V_{OL}$$

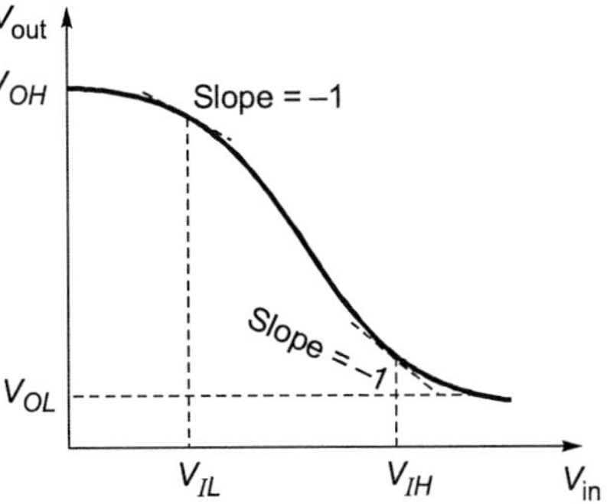

With the change in sizing of the transistors, the voltage transfer characteristics shift as shown below:

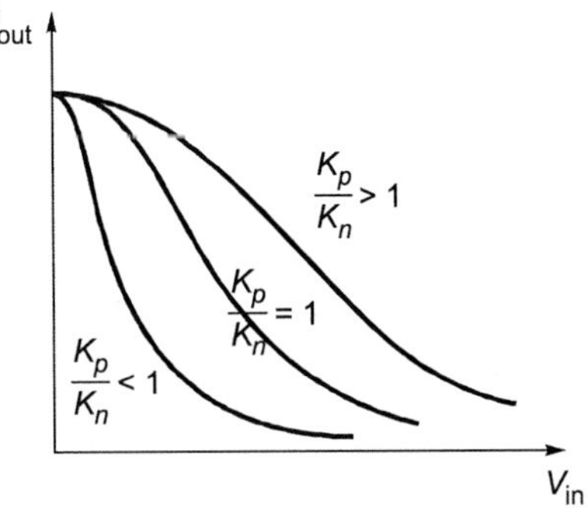

when, $V_{TN} = |V_{TP}|,\ K_N = K_P,\ V_i = \frac{V_{DD}}{2}$

$$NM_L = MM_H$$

when, $\frac{K_p}{K_n} > 1,\quad V_{IL}\uparrow, V_{IH}\uparrow$

$\therefore\ N_{MH}\downarrow$ and $NM_L\uparrow$

when, $\frac{K_p}{K_n} < 1, V_{IL}\downarrow, V_{IH}\downarrow$

$NM_H\uparrow$ and $NM_L\downarrow$

Therefore, NM_H and NM_L depends on transistor sizing and they are equal for certain condition only.

(b) Dynamic power consumption during switching is non-zero due to capacitive loading of next stage.

(c) The transfer characteristics of CMOS NOT Gate is as shown below:

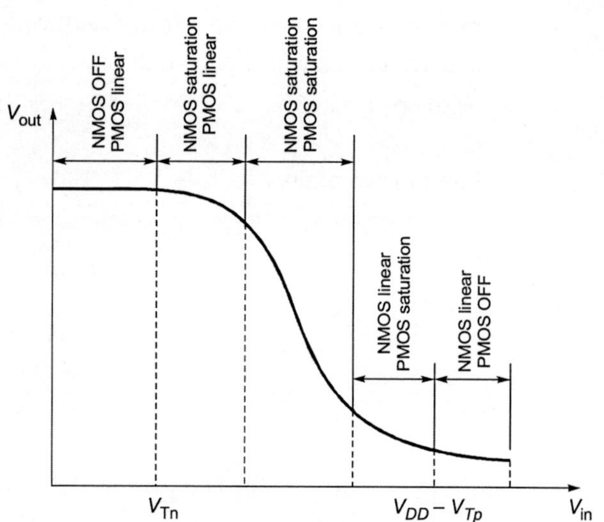

For a logical high input under steady state i.e.

$$V_{in} \geq V_{DD} - |V_{Tp}|$$

NMOS is in linear region of operation and PMOS is in cut-off region.

(d) Switching speed depends on charging and discharging of load capacitor for pull up and pull down of output voltage respectively.

Fast charging and discharging of capacitor depends on mobility of charge carrier.

$\tau_{charging} = R_{PMOS}\, C_L$(to bring output at lotig high)

$\tau_{discharging} = R_{NMOS}\, C_L$ (to bring output at logic low)

Also, propagation delay

$$\tau_p = \frac{\tau_{PLH} + \tau_{PHL}}{2}$$

$$\text{where, } \tau_{PLH} = \frac{C_L V_{DD}}{\mu_p C_{ox}\left(\frac{W}{L}\right)(V_{GS} - V_{TP})^2}$$

$$\tau_{PLH} = \frac{C_L V_{DD}}{\mu_{pn} C_{ox}\left(\frac{W}{L}\right)(V_{GS} - V_{TN})^2}$$

7.36 (2)

Given circuit,

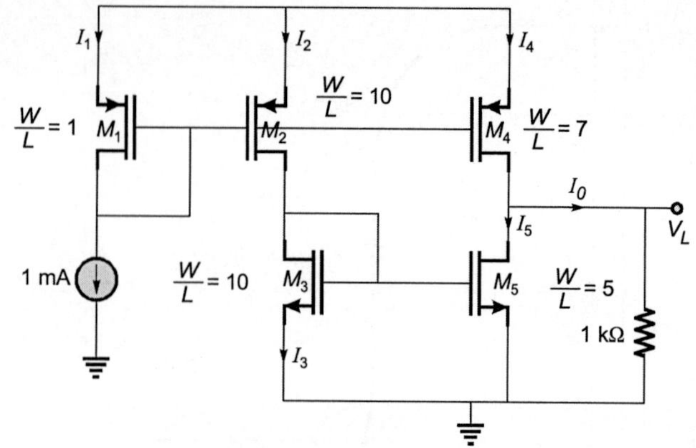

We know that drain current,

$$I_D \propto \left(\frac{W}{L}\right)$$

$$I_1 = 1 \text{ mA}$$

$$I_2 = \frac{10}{1} \times 1 = 10 \text{ mA}$$

$$I_3 = 10 \text{ mA}$$

$$I_4 = 7 \text{ mA}$$

$$I_5 = 5 \text{ mA}$$

$$I_0 = I_4 - I_5 = 7 - 5 = 2 \text{ mA}$$

Load voltage, $V_L = 2 \times 1 = 2\text{V}$

■■■

8 Memories

8.1 The minimum number of MOS transistors required to make a dynamic RAM cell is

(a) 1 (b) 2
(c) 3 (d) 4 **[1995 : 1 M]**

8.2 Each cell of a static Random Access Memory contains

(a) 6 MOS transistors.
(b) 4 MOS transistors and 2 capacitors
(c) 2 MOS transistors and 4 capacitors
(d) 1 MOS transistor and 1 capacitors

[1996 : 1 M]

8.3 In the DRAM cell in the figure, the V_t of the NMOSFET is 1 V. For the following three combinations of WL and BL voltages.

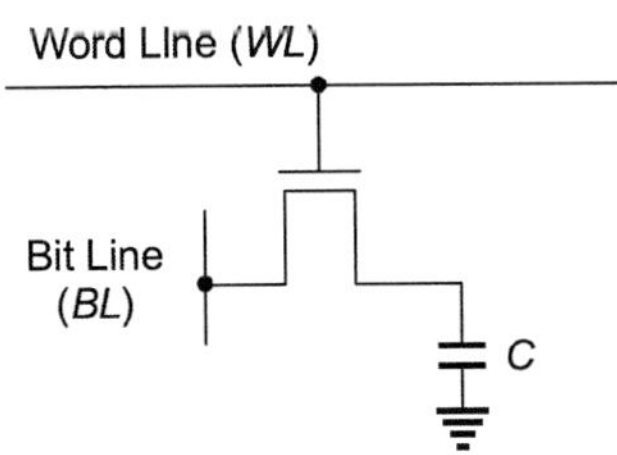

(a) 5 V; 3 V; 7 V (b) 4 V; 3 V; 4 V
(c) 5 V; 5 V; 5 V (d) 4 V; 4 V; 4 V

[2001 : 2 M]

8.4 If the input X_3, X_2, X_1, X_0 to the ROM in the figure are 8 4 2 1 BCD numbers, then the outputs Y_3, Y_2, Y_1, Y_0 are

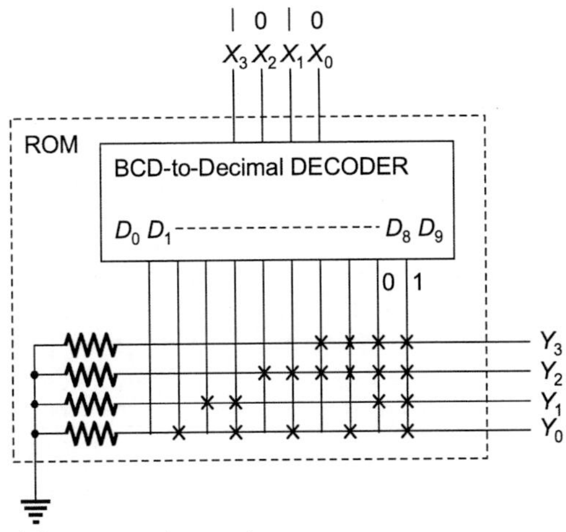

(a) gray code numbers
(b) 2 4 2 1 BCD numbers
(c) excess-3 code numbers
(d) none of the above **[2002 : 2 M]**

8.5 In the circuit shown in the figure, A is parallel-in, parallel-out 4 bit register, which loads at the rising edge of the clock C. The input lines are connected to a 4 bit bus, W. Its output acts as the input to a 16 × 4 ROM whose output is floating when the enable input E is 0. A partial table of the contents of the ROM is as follows

Address	0	2	4	6	8	10	11	14
Data	0011	1111	0100	1010	1011	1000	0010	1000

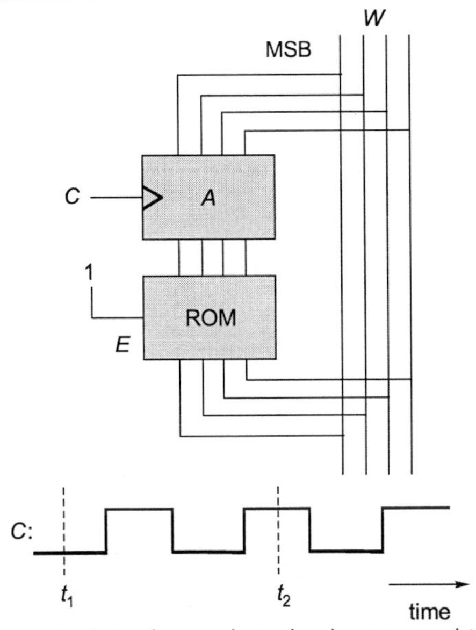

The clock to the register is shown, and the data on the W bus at time t_1 is 0110. The data on the bus at time t_2 is

(a) 1111 (b) 1011
(c) 1000 (d) 0010

[2003 : 2 M]

8.6 If WL is the Word Line and BL the Bit Line, an SRAM cell is shown in

(a)

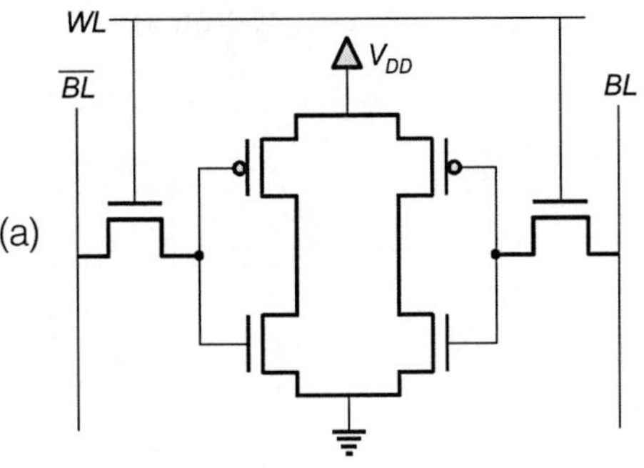

(b)

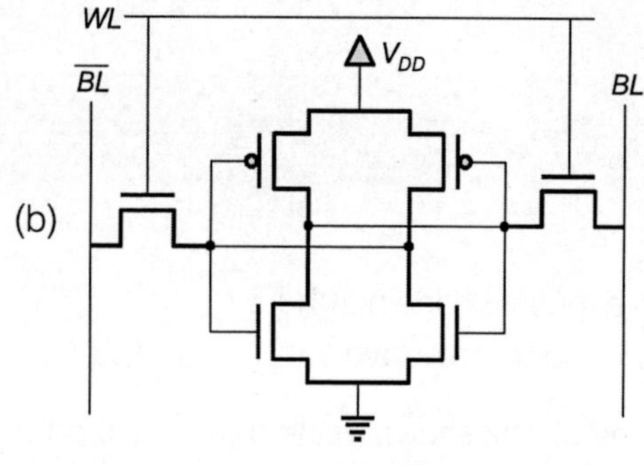

(c)

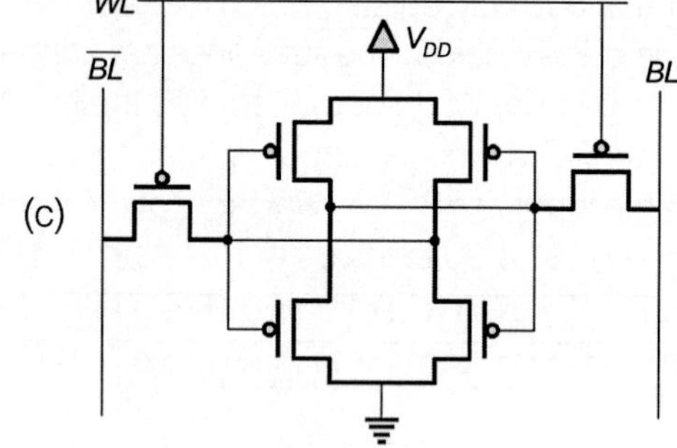

(d)

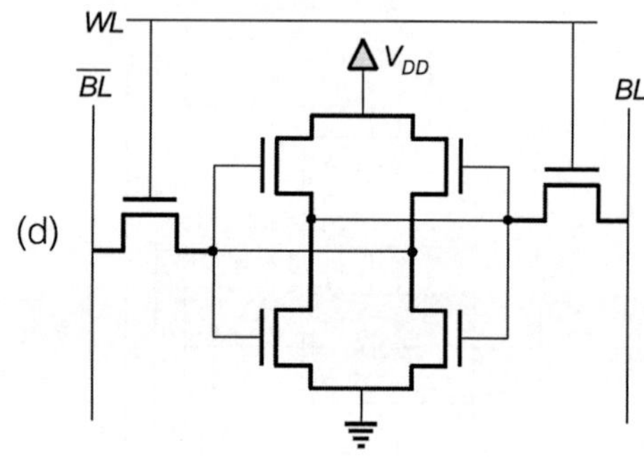

[2014 : 2 M, Set-3]

8.7 In a DRAM,

(a) periodic refreshing is not required
(b) information is stored in a capacitor
(c) information is stored in a latch
(d) both read and write operations can be performed simultaneously

[2017 : 1 M, Set-2]

8.8 A 2 × 2 ROM array is built with the help of diodes as shown in the circuit below. Here W_0 and W_1 are signals that select the word lines and B_0 and B_1 are signals that are output of the sense amps based on the stored data corresponding to the bit lines during the read operation.

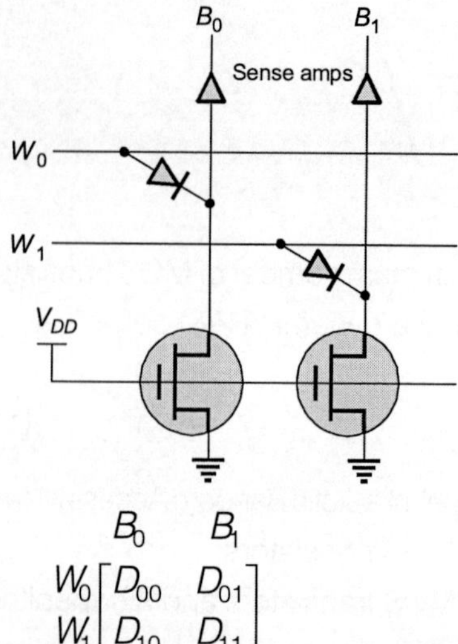

$$\begin{array}{cc} & \begin{array}{cc} B_0 & B_1 \end{array} \\ \begin{array}{c} W_0 \\ W_1 \end{array} & \begin{bmatrix} D_{00} & D_{01} \\ D_{10} & D_{11} \end{bmatrix} \end{array}$$

Bits stored in the ROM Array

During the read operation, the selected word line goes high and the other word line is in a high impedance state. As per the implementation shown in the circuit diagram above, what are the bits corresponding to D_{ij} (where $i = 0$ or 1 and $j = 0$ or 1) stored in the ROM?

(a) $\begin{bmatrix} 1 & 0 \\ 0 & 1 \end{bmatrix}$ (b) $\begin{bmatrix} 0 & 1 \\ 1 & 0 \end{bmatrix}$

(c) $\begin{bmatrix} 1 & 0 \\ 1 & 0 \end{bmatrix}$ (d) $\begin{bmatrix} 1 & 1 \\ 0 & 0 \end{bmatrix}$

[2018 : 2 M]

8.9 Addressing of a 32 K × 16 memory is realized using a single decoder. The minimum number of AND gates required for the decoder is

(a) 2^8 (b) 2^{19}
(c) 2^{15} (d) 2^{32}

[2021 : 1 M]

■■■

Answers Memories

8.1	(b)	8.2	(a)	8.3	(b)	8.4	(b)	8.5	(c)	8.6	(b)	8.7	(b)
8.8	(a)	8.9	(c)										

Explanations Memories

8.2 (a)

Each cell of a static RAM contains six MOS transistors. Each bit on static RAM is stored on four transistors out of which two are PMOS and two are NMOS, that from cross-coupled inverters. Remaining two transistors are used to control reading from or writing into cell.

8.4 (b)

				2	4	2	1	
x_3	x_2	x_1	x_0	y_3	y_2	y_1	y_0	
0	0	0	0	0	0	0	0	$\rightarrow 0$
0	0	0	1	0	0	0	1	$\rightarrow 1$
0	0	1	0	0	0	1	0	$\rightarrow 2$
0	0	1	1	0	0	1	1	$\rightarrow 3$
0	1	0	0	0	1	0	0	$\rightarrow 4$
0	1	0	1	0	1	0	1	$\rightarrow 5$
0	1	1	0	1	1	0	0	$\rightarrow 6$
0	1	1	1	1	1	0	1	$\rightarrow 7$
1	0	0	0	1	1	1	0	$\rightarrow 8$
1	0	0	1	1	1	1	1	$\rightarrow 9$

$\therefore$ It is 8421BCD to 2421BCD.

8.5 (c)

When W has data 0110 i.e. 6 in decimal its data value at that add. is 1010.
Now 1010 i.e. 10 is acting as add. at time t_2 and data at that moment is 1000.

8.6 (b)

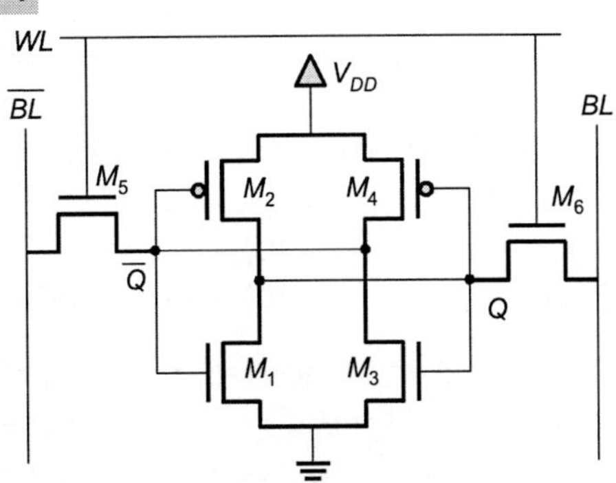

8.7 (b)

In a DRAM, data is stored in the form of charge on capacitor and periodic refreshing is needed to restore the charge on capacitor.

8.8 (a)

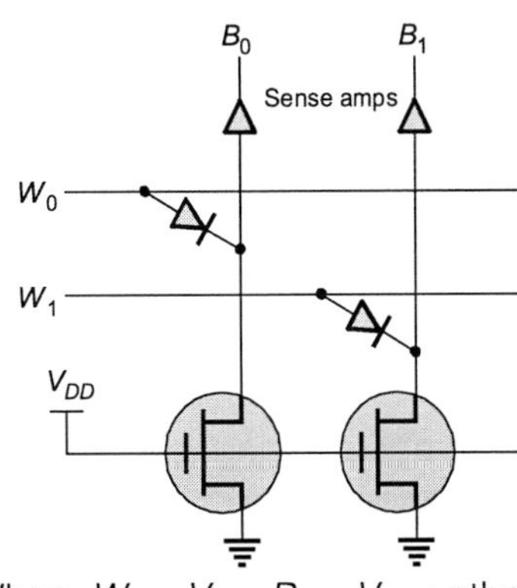

- When, $W_0 = V_{DD}$, $B_0 = V_{DD}$; otherwise $B_0 = 0$
- When $W_1 = V_{DD}$, $B_1 = V_{DD}$; otherwise $B_1 = 0$

So, $B_0 = W_0$ and $B_1 = W_1$

Hence, $\begin{matrix} & B_0 & B_1 \\ W_0 & 1 & 0 \\ W_1 & 0 & 1 \end{matrix}$

8.9 (c)

For $N \times 2^N$, decoder, there are 2^N AND gates required at output.

$\therefore \quad 32K \times 16 = 2^5 \times 2^{10} \times 16 = 2^{15} \times 16$

$\therefore$ 2^{15} AND gates required to realize the given memory address.

■■■

9 ADC and DAC

9.1 For an ADC, match the following: if

List – I

(A) Flash converter

(B) Dual slope converter

(C) Successive Approximation Converter

List – II

(1) requires a conversion time of the order of a few seconds

(2) requires a digital-to-analog converter

(3) minimizes the effect of power supply interference.

(4) requires a very complex hardware

(5) is a tracking A/D converter.

[1995 : 1 M]

9.2 The advantage of using a dual slope ADC in a digital voltmeter is that

(a) its conversion time is small

(b) its accuracy is high

(c) it gives output in BCD format

(d) it does not require a comparator

[1998 : 1 M]

9.3 The resolution of a 4-bit counting ADC is 0.5 Volts. For an analog input of 6.6 Volts, the digital output of the ADC will be

(a) 1011 (b) 1101

(c) 1100 (d) 1110

[1999 : 1 M]

9.4 An 8 bit successive approximation analog to digital converter has full scale reading of 2.55 V and its conversion time for an analog input of 1 V is 20 μs. The conversion time for a 2 V input will be

(a) 10 μs (b) 20 μs

(c) 40 μs (d) 50 μs

[2000 : 1 M]

9.5 The number of comparators in a 4-bit flash ADC is

(a) 4 (b) 5

(c) 15 (d) 16 [2000 : 1 M]

9.6 For the 4 bit DAC shown in the figure, the output voltage V_0 is

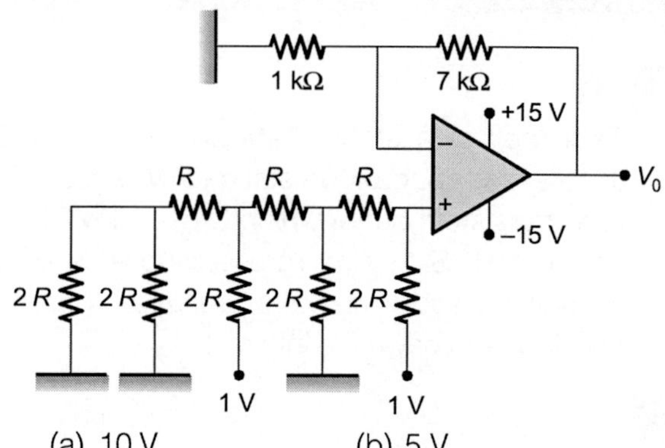

(a) 10 V (b) 5 V

(c) 4 V (d) 8 V [2000 : 2 M]

9.7 The number of comparators required in a 3-bit comparator type ADC is

(a) 2 (b) 3

(c) 7 (d) 8 [2002 : 1 M]

9.8 The minimum number of comparators required to build an 8-bit flash ADC is

(a) 8 (b) 63

(c) 255 (d) 256 [2003 : 1 M]

9.9 The circuit shown in the figure is a 4 bit DAC

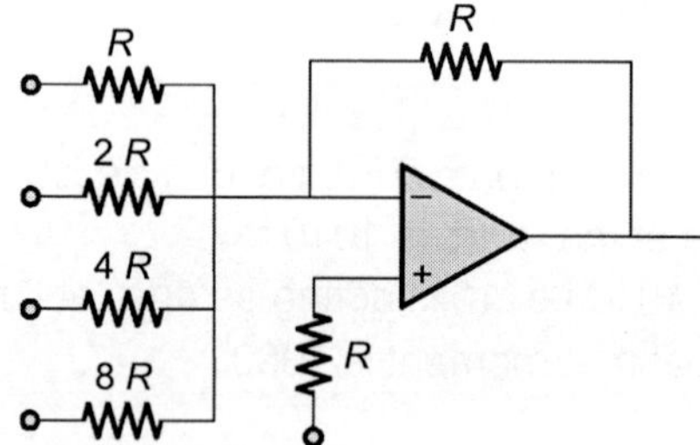

The input bits 0 and 1 are represented by 0 and 5 V respectively. The OP AMP is ideal, but all the resistances and the 5 V inputs have a tolerance of ±10%. The specification (rounded to the nearest multiple of 5%) for the tolerance of the DAC is

(a) ±35% (b) ±20%

(c) ±10% (d) ±5% [2003 : 2 M]

9.10 A digital system is required to amplify a binary-encoded audio signal. The user should be able to control the gain of the amplifier from a minimum to a maximum in 100 increments. The minimum number of bits required to encode, in straight binary, is

(a) 8 (b) 6

(c) 5 (d) 7 [2004 : 1 M]

9.11 A 4-bit D/A converter is connected to a free-running 3-bit UP counter, as shown in the following figure. Which of the following waveforms will be observed at V_o?

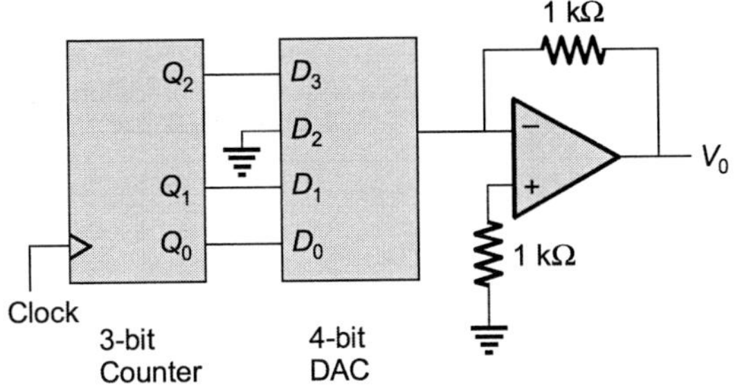

In the figure shown above, the ground has been shown by the symbol ⏚

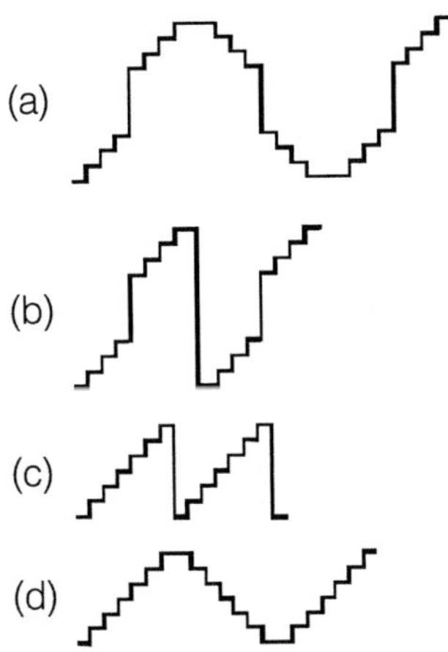

[2006 : 2 M]

Statement for Linked Answer Questions 9.12 and 9.13:

In the Digital-to-Analog converter circuit shown in the figure below, $V_R = 10$ V and $R = 10$ kΩ.

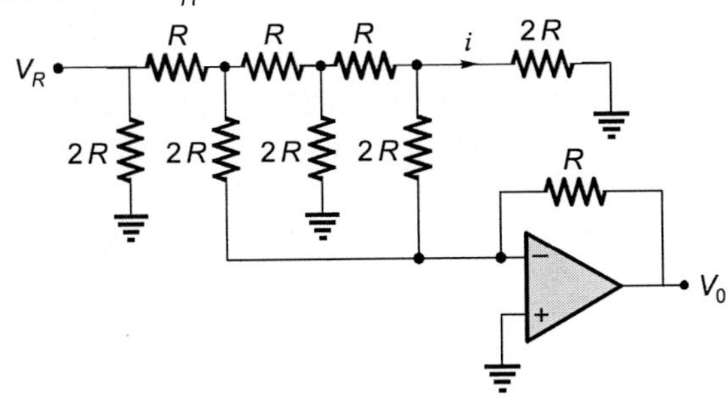

9.12 The current i is

(a) 31.25 μA (b) 62.5 μA
(c) 125 μA (d) 250 μA [2007 : 2 M]

9.13 The voltage V_0 is

(a) –0.781 V (b) –1.562 V
(c) –3.125 V (d) –6.250 V

[2007 : 2 M]

Statement for linked Answer Questions 9.14 and 9.15:

In the following circuit, the comparator output is logic "1" if $V_1 > V_2$ and is logic "0" otherwise. The D/A conversion is done as per the relation

$$V_{DAC} = \sum_{n=0}^{3} 2^{n-1} b_n \text{ Volts, where } b_3\text{(MSB)}, b_2, b_1$$

and b_0(LSB) are the counter outputs.
The counter starts from the clear state.

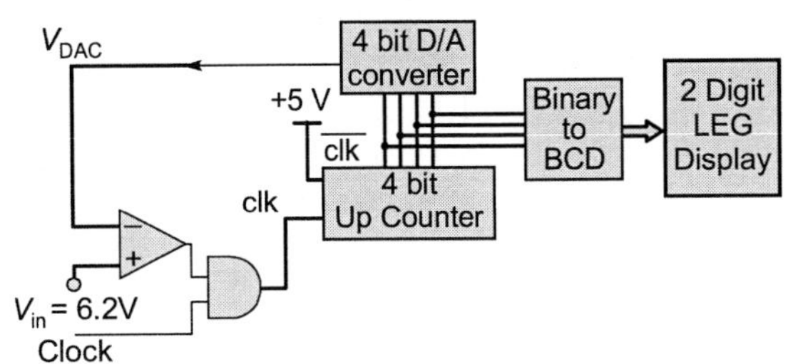

9.14 The stable reading of the LED displays is

(a) 06 (b) 07
(c) 12 (d) 13 [2008 : 2 M]

9.15 The magnitude of the error between V_{DAC} and V_{in} at steady state in volts is

(a) 0.2 (b) 0.3
(c) 0.5 (d) 1.0

[2008 : 2 M]

9.16 The output of a 3-stage Johnson (twisted-ring) counter is fed to a digital-to-analog (D/A) converter as shown in the figure below. Assume all states of the counter to be unset initially. The waveform which represents the D/A converter output V_0 is

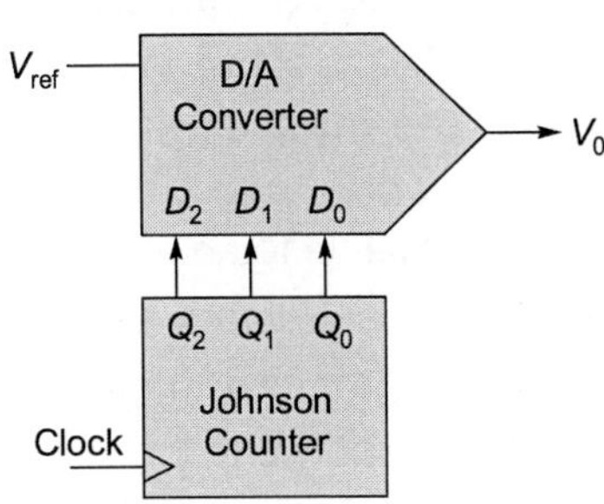

(a)

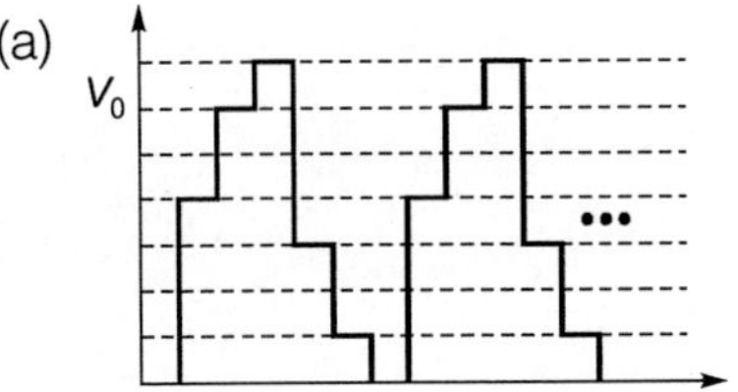

(b)

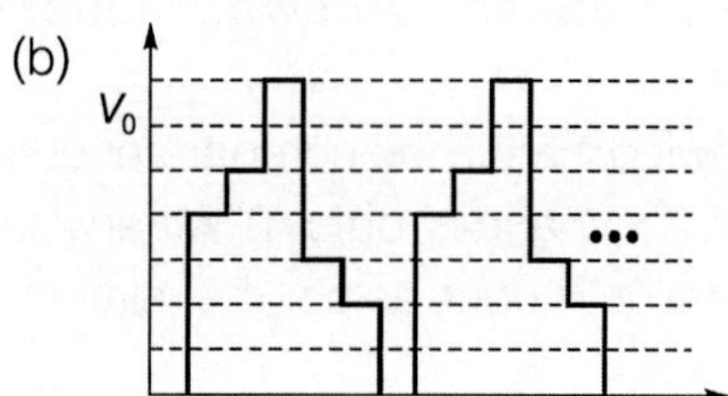

(c)

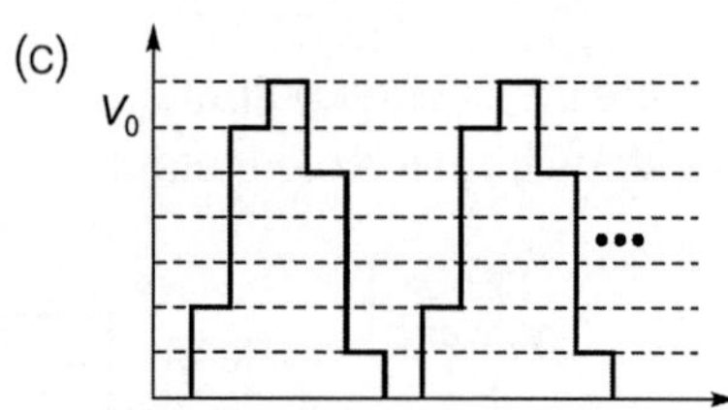

(d)

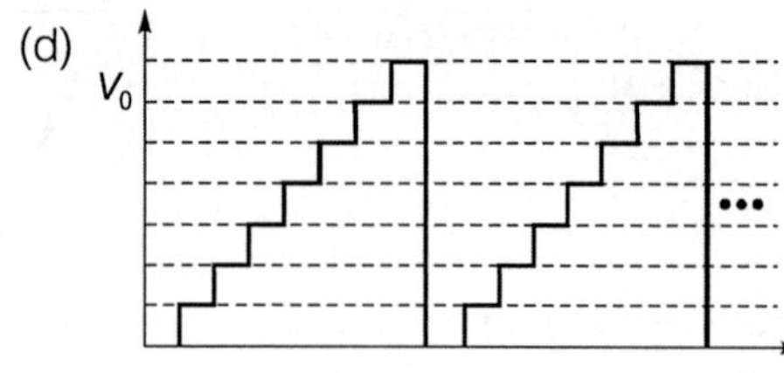

[2011 : 2 M]

9.17 Consider a four bit D to A converter. The analog value corresponding to a digital signals of values 0000 and 0001 are 0 V and 0.0625 V respectively. The analog value (in Volts) corresponding to the digital signal 1111 is ________.

[2015 : 1 M, Set-1]

9.18 In an N bit flash ADC, the analog voltage is fed simultaneously to $2^N - 1$ comparators. The output of the comparators is then encoded to a binary format using digital circuits. Assume that the analog voltage source V_{in} (whose output is being converted to digital format) has a source resistance of 75 Ω as shown in the circuit diagram below and the input capacitance of each comparator is 8 pF.

The input must settle to an accuracy of 1/2 LSB even for a full scale input change for proper conversion. Assume that the time taken by the thermometer to binary encoder is negligible.

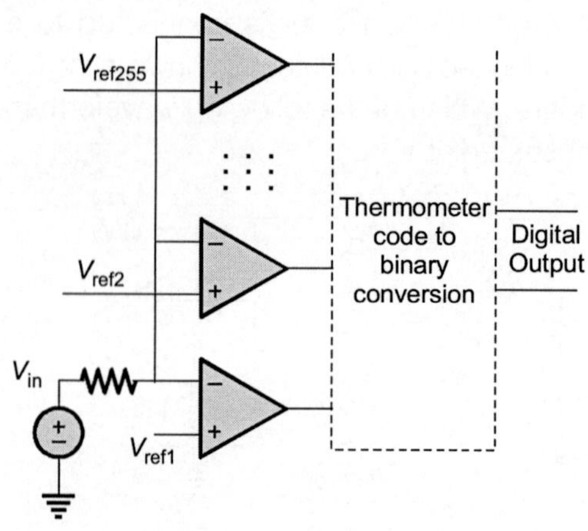

If the flash ADC has 8 bit resolution, which one of the following alternatives is closest to the maximum sampling rate ?

(a) 1 megasamples per second
(b) 6 megasamples per second
(c) 64 megasamples per second
(d) 256 megasamples per second

[2016 : 2 M, Set-2]

9.19 A 10-bit D/A converter is calibrated over the full range from 0 to 10 V. If the input to the D/A converter is 13 A (in hex), the output (rounded off to three decimal places) is ________ V.

[2020 : 1 M]

9.20 An 8-bit unipolar (all analog output values are positive) digital-to-analog converter (DAC) has a full-scale voltage range from 0 V to 7.68 V. If the digital input code is 10010110 (the leftmost bit is MSB), then the analog output voltage of the DAC (rounded off to one decimal place) is ________ V.

[2021 : 1 M]

9.21 Consider the circuit shown with an ideal OPAMP. The output voltage V_0 is ________ V (rounded off to two decimal places).

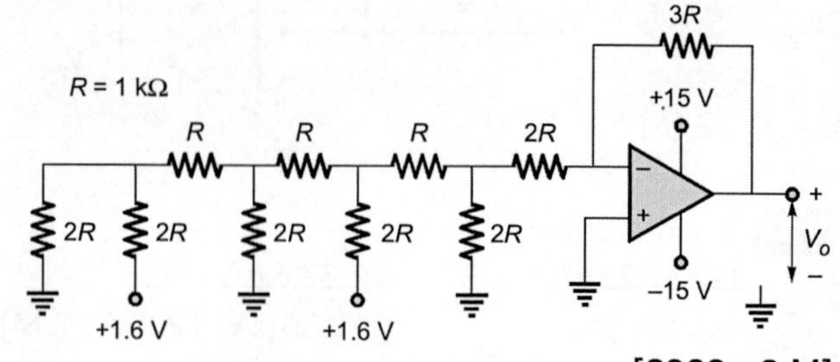

[2022 : 2 M]

Answers ADC and DAC

9.1	(A-4, B-3, C-2.)			9.2	(b)	9.3	(d)	9.4	(b)	9.5	(c)
9.6	(b)	9.7	(c)	9.8	(c)	9.9	(a)	9.10	(d)	9.11	(b)
9.12	(b)	9.13	(c)	9.14	(d)	9.15	(b)	9.16	(a)	9.17	(0.9375)
9.18	(a)	9.19	(3.069)	9.20	(4.5)	9.21	(–0.5)				

Explanations ADC and DAC

9.2 (b)

The advantage of using a dual slope ADC in a digital voltmeter is its high accuracy.

9.3 (d)

$$V_0 = \frac{6.6}{0.5} = 13.2 \simeq 14$$

Binary equivalent = 1110

9.4 (b)

Conversion time of successive approximate ADC depends upon the number of bits only.

9.5 (c)

No. of comparators in a flash ADC is equal to $2^n - 1$ where n = no. of bits.

$2^4 - 1 = 15$

9.6 (b)

$$V_{in} \text{ at non-inverting terminal} = \frac{1}{8} + \frac{1}{2} = \frac{5}{8}$$

$$\therefore \quad V_o = \left(1 + \frac{R_f}{R}\right)V_{in} = 8 \times \frac{5}{8} = 5\text{ V}$$

9.7 (c)

$2^n - 1 = 2^3 - 1$

9.8 (c)

$2^n - 1 = 2^8 - 1 = 255$

9.9 (a)

$$V_0 = -V_R \cdot \left[d_3 \cdot \frac{R}{R} + d_2\frac{R}{2R} + d_1\frac{R}{4R} + d_0\frac{R}{8R}\right]$$

$$\Rightarrow V_0 = -V_R \cdot \frac{R}{R}[\text{constant}]$$

worst case tolerance in

$$V_0 = \frac{1.1 \times 1.1}{0.9} = 35\%$$

9.10 (d)

$$2^n \geq 100$$

$$\therefore \quad n \geq 7$$

9.11 (b)

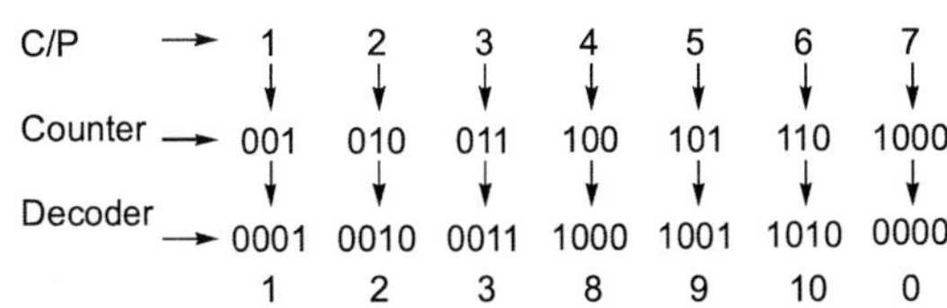

D_2 is connected to ground and Q_2 to D_3.

9.12 (b)

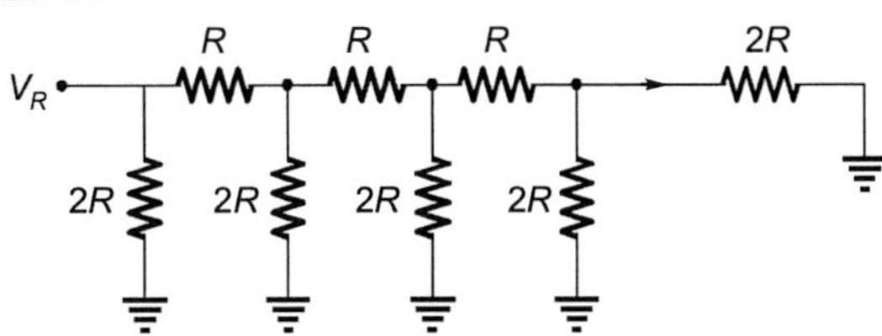

Last both $2R$ resister are in parallel and series with R then after

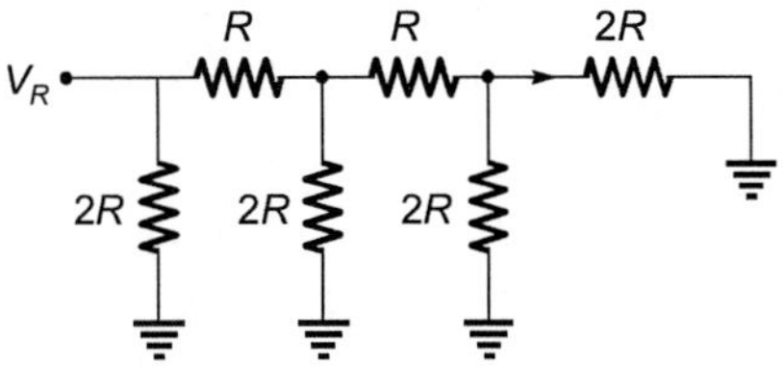

then again similar condition last both $2R$ are in parallel and series with R similarly after solving equivalent circuit is

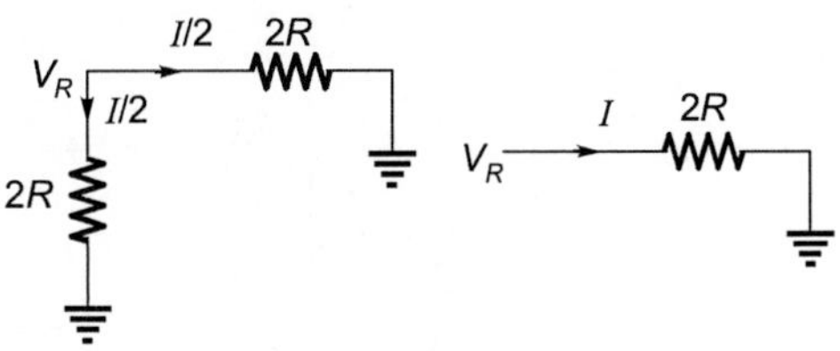

$$I = \frac{V_R}{R} = \frac{10}{10\text{ k}\Omega} = 1\text{ mA}$$

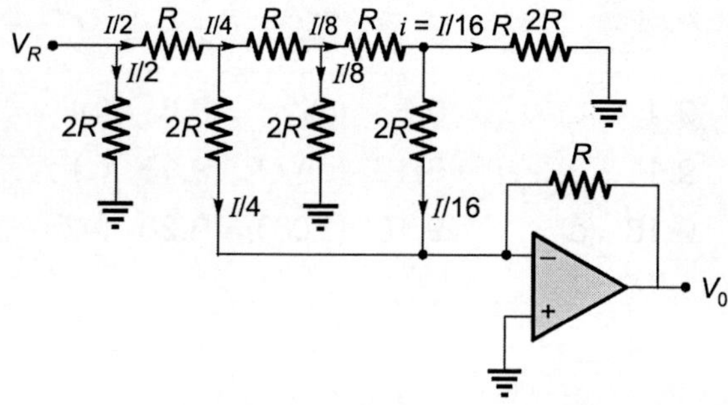

then $i = \frac{I}{16} = \frac{1\times 10^{-3}}{16} = 62.5\ \mu A$

9.13 (c)

Net current in inverting terminal of op-amp

$$= \frac{I}{4} + \frac{I}{16} = \frac{5I}{16}$$

$$V_0 = -R \times \frac{5I}{16}$$

$$= -\frac{10\times10^3 \times 5 \times 1 \times 10^{-3}}{16} = -3.125\ V$$

9.14 (d)

$$V_{DAC} = 2^{-1}b_0 + 2^0 b_1 + 2^1 b_2 + 2^2 b_3$$
$$= 0.5b_0 + b_1 + 2b_2 + 4b_3$$

Counter output will start from 0000 and will increase by 1 at every clock pulse.

Table for V_{DAC} is shown below

b_3	b_2	b_1	b_0	V_{DAC}
0	0	0	0	0
0	0	0	1	0.5
0	0	1	0	1
0	0	1	1	1.5
0	1	0	0	2
0	1	0	1	2.5
0	1	1	0	3
0	1	1	1	3.5
1	0	0	0	4
1	0	0	1	4.5
1	0	1	0	5
1	0	1	1	5.5
1	1	0	0	6
1	1	0	1	6.5
1	1	1	0	7
1	1	1	1	7.5

Counter will increase till $V_{in} > V_{DAC}$. So, when V_{DAC} = 6.5 V, the comparator output will be zero and the counter will be stable at that reading. The corresponding reading of LED display is 13.

9.15 (b)

Magnitude of the error between V_{DAC} and V_{in} at steady state

$$= 6.5 - 6.2 = 0.3\ V$$

9.16 (a)

Sequence of Johnson counter is

Q_2	Q_1	Q_0	D_2	D_1	D_0	V_0
0	0	0	0	0	0	0
1	0	0	1	0	0	4
1	1	0	1	1	0	6
1	1	1	1	1	1	7
0	1	1	0	1	1	3
0	0	1	0	0	1	1
0	0	0	0	0	0	0

9.17 (0.9375)

Step size = 0.0625 V

Decimal equivalent = 15

Analog output = 15 ×0.0625 = 0.9375 Volts

9.19 (3.069)

Given, $n = 10$

$V_{FS} = 10\ V$

Input voltage = $(13A)_{16} = (314)_{10}$

Output voltage

= Resolution × Decimal equivalent of input

$$V_o = \frac{10}{2^{10}-1} \times 314 = 3.069\ V$$

9.20 (4.5)

Given: $V_{ps} = 7.68\ V$

$n = 8$ bit

$$\text{Resolution } (k) = \frac{V_{FS}}{2^n - 1} = \frac{7.68}{2^8 - 1} = 0.03$$

Now, $V_{DAC} = k \times \{\text{Decimal equivalent}\}$

$= 0.03 \times \{150\}$

$= 4.5\ V$

9.21 (–0.5)

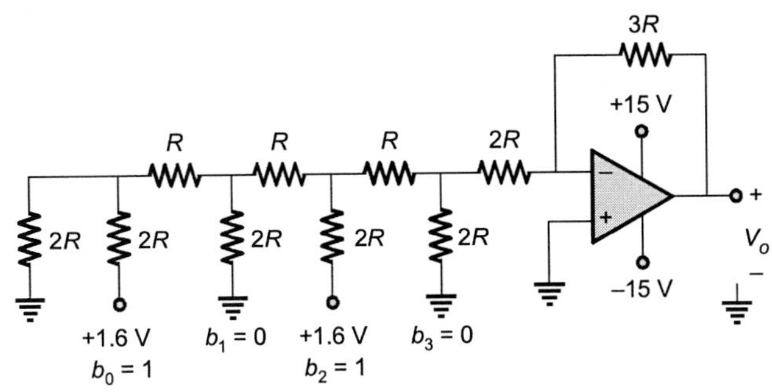

The given circuit is an R-$2R$ ladder type digital to analog converter with digital input

$$(b_3b_2b_1b_0) = (0101)_2 = (5)_{10}$$

$$\text{Resolution of DAC} = \frac{V_r}{2^n} = \frac{1.6}{2^4} = \frac{1.6}{16} = 0.1 \text{ V}$$

The output voltage,

V_0 = Resolution × Gain × Decimal equivalent of binary input

$$\text{Gain} = \frac{-3R}{2R+R} = -1$$

$$\therefore \quad V_0 = 0.1 \times (-1) \times (5)_{10} = -0.5 \text{ V}$$

■■■

10 Computer Organization

10.1 The content of the registers are $R_1 = 25$ H, $R_2 = 30$ H and $R_3 = 40$ H. The following machine instructions are executed,

PUSH{R_1}

PUSH{R_2}

PUSH{R_3}

POP{R_1}

POP{R_2}

POP{R_3}

After execution, the content of registers R_1, R_2, R_3 are

(a) $R_1 = 40$ H, $R_2 = 30$ H, $R_3 = 25$ H

(b) $R_1 = 30$ H, $R_2 = 40$ H, $R_3 = 25$ H

(c) $R_1 = 25$ H, $R_2 = 30$ H, $R_3 = 40$ H

(d) $R_1 = 40$ H, $R_2 = 25$ H, $R_3 = 30$ H

[2021 : 2 M]

10.2 A machine has a 32-bit architecture with 1-word long instructions. It has 24 registers and supports an instruction set of size 40. Each instruction has five distinct fields, namely opcode, two source register identifiers, one destination register identifier, and an immediate value. Assuming that the immediate operand is an unsigned integer, its maximum value is ________.

[2024 : 1 Mark]

■■■

Answers Computer Organization

10.1 (a) 10.2 (2047)

Explanations Computer Organization

10.1 (a)

For PUSH SP is decremented and for POP SP is incremented.

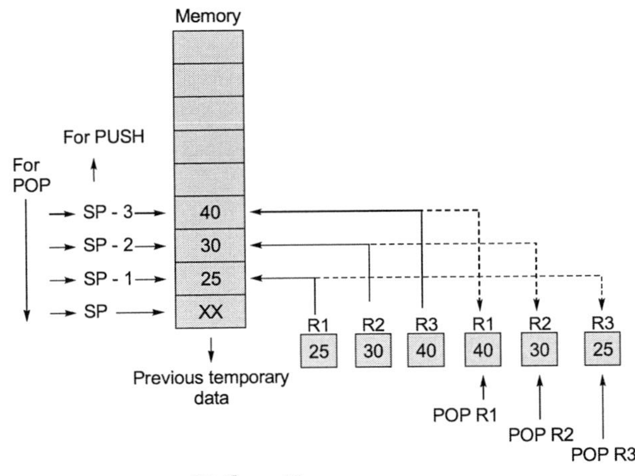

$\therefore$ $[R1] = 40$

$[R2] = 30$

$[R3] = 25$

10.2 (2047)

Given 32-bit architecture {1 word long instructions}

1 word = 32-bits

$\therefore$ The instructions are 32-bit long.

Number of registers = 24. If n bits are used for register identifier,

$\Rightarrow$ $2^n \geq 24$

$n = \lceil \log_2(24) \rceil$

$\therefore$ n = 5-bits

Hence, each register identifier in an instruction requires 5-bits.

Number of instructions = 40 (Given)

$\therefore$ Opcode bits would be $2^p \geq 40$

$\therefore$ $p = \lceil \log_2(40) \rceil$

p = 6-bits

Instruction format would be

← 32-bits →

Opcode	Source Register-1	Source Register-2	Destination Register	Immediate Value
6-bits	5-bits	5-bits	5-bits	?

$\therefore$ Immediate value bits

$= 32 - (6 + 5 + 5 + 5) = 32 - 21 = 11$-bits

The range of n-bit unsigned values = 0 to $2^n - 1$

Range of unsigned immediate value

$= 0$ to $2^{11} - 1 = 0$ to $(2048 - 1)$

$= 0$ to 2047

$\therefore$ Maximum magnitude of immediate value = 2047

■■■

Signals and Systems

CONTENTS

Signals and Systems

Syllabus

Continuous-time signals: Fourier series and Fourier transform, sampling theorem and applications. Discrete-time signals: DTFT, DFT, z-transform, discrete-time processing of continuous-time signals. LTI systems: definition and properties, causality, stability, impulse response, convolution, poles and zeroes, frequency response, group delay, phase delay.

Analysis of Previous GATE Papers

Exam Year	1 Mark Ques.	2 Marks Ques.	Total Marks
1995	5	1	**7**
1996	2	2	**6**
1997	2	2	**6**
1998	12	–	**12**
1999	4	2	**8**
2000	3	3	**9**
2001	4	1	**6**
2002	4	2	**8**
2003	4	4	**12**
2004	3	6	**15**
2005	6	6	**18**
2006	3	3	**9**
2007	1	4	**9**
2008	2	8	**18**
2009	3	5	**13**
2010	3	2	**7**
2011	3	4	**11**
2012	2	3	**8**
2013	6	3	**12**

Exam Year	1 Mark Ques.	2 Mark Ques.	Total Marks
2014 Set-1	3	4	**11**
2014 Set-2	3	3	**9**
2014 Set-3	4	3	**10**
2014 Set-4	4	4	**12**
2015 Set-1	2	3	**8**
2015 Set-2	5	4	**13**
2015 Set-3	4	5	**14**
2016 Set-1	4	4	**12**
2016 Set-2	1	2	**5**
2016 Set-3	3	4	**11**
2017 Set-1	3	4	**11**
2017 Set-2	2	4	**10**
2018	3	3	**9**
2019	3	3	**9**
2020	2	4	**10**
2021	2	3	**8**
2022	2	2	**6**
2023	3	5	**13**
2024	2	4	**10**

1 Basics of Signals and Systems

1.1 A system with an input $x(t)$ and output $y(t)$ is described by the relation: $y(t) = tx(t)$. This system is
(a) linear and time-invariant
(b) linear and time varying
(c) non-linear & time-invariant
(d) non-linear and time-varying

[2000 : 1 M]

1.2 Let $\delta(t)$ denote the delta function. The value of the integral $\int_{-\infty}^{\infty} \delta(t)\cos\left(\frac{3t}{2}\right)dt$ is
(a) 1 (b) –1
(c) 0 (d) $\frac{\pi}{2}$ [2001 : 1 M]

1.3 If a signal $f(t)$ has energy E, the energy of the signal $f(2t)$ is equal to
(a) E (b) $\frac{E}{2}$
(c) $2E$ (d) $4E$ [2001 : 1 M]

1.4 Let P be linearity, Q be time-invariance, R be causality and S be stability. A discrete-time system has the input-output relationship,

$$y(n) = \begin{cases} x(n) & n \geq 1 \\ 0, & n = 0 \\ x(n+1), & n \leq -1 \end{cases}$$

where $x(n)$ is the input and $y(n)$ is the output. The above system has the properties
(a) P, S but not Q, R
(b) P, Q, S but not R
(c) P, Q, R, S
(d) Q, R, S but not P [2003 : 2 M]

1.5 Consider the sequence

$$x[n] = \left[-4 - j5,\ \underset{\uparrow}{1 + j2},\ 4\right]$$

The conjugate antisymmetric part of the sequence is
(a) $[-4 - j2.5,\ j2,\ 4 - j2.5]$
(b) $[-j2.5,\ 1,\ j2.5]$
(c) $[-j5,\ j2,\ 0]$
(d) $[-4,\ 1,\ 4]$ [2004 : 2 M]

1.6 The function $x(t)$ is shown in the figure. Even and odd parts of a unit-step function $u(t)$ are respectively,
(a) $\frac{1}{2}, \frac{1}{2}x(t)$
(b) $-\frac{1}{2}, \frac{1}{2}x(t)$
(c) $\frac{1}{2}, -\frac{1}{2}x(t)$
(d) $-\frac{1}{2}, -\frac{1}{2}x(t)$

x(t)
1
0
t
–1

[2005 : 1 M]

1.7 The power in the signal

$s(t) = 8\cos\left(20\pi t - \frac{\pi}{2}\right) + 4\sin(15\pi t)$ is
(a) 40 (b) 41
(c) 42 (d) 82 [2005 : 1 M]

1.8 The Dirac delta function $\delta(t)$ is defined as
(a) $\delta(t) = \begin{cases} 1 & t = 0 \\ 0 & \text{otherwise} \end{cases}$
(b) $\delta(t) = \begin{cases} \infty & t = 0 \\ 0 & \text{otherwise} \end{cases}$
(c) $\delta(t) = \begin{cases} 1 & t = 0 \\ 0 & \text{otherwise} \end{cases}$ and $\int_{-\infty}^{\infty} \delta(t)dt = 1$
(d) $\delta(t) = \begin{cases} \infty & t = 0 \\ 0 & \text{otherwise} \end{cases}$ and $\int_{-\infty}^{\infty} \delta(t)dt = 1$

[2006 : 1 M]

1.9 A system with input $x[n]$ and output $y[n]$ is given as $y[n] = \left(\sin\frac{5}{6}\pi n\right)x(n)$.

The system is
(a) linear, stable and invertible
(b) non-linear, stable and non-invertible
(c) linear, stable and non-invertible
(d) linear, unstable and invertible [2006 : 2 M]

1.10 A Hilbert transformer is a
(a) non-linear system
(b) non-causal system
(c) time-varying system
(d) low-pass system [2007 : 2 M]

1.11 The input and output of a continuous time system are respectively denoted by $x(t)$ and $y(t)$. Which of the following descriptions corresponds to a casual system?
(a) $y(t) = x(t-2) + x(t+4)$
(b) $y(t) = (t-4)\,x(t+1)$
(c) $y(t) = (t+4)\,x(t-1)$
(d) $y(t) = (t+5)\,x(t+5)$ [2008 : 1 M]

1.12 Let $x(t)$ be the input and $y(t)$ be the output of a continuous time system. Match the system properties P_1, P_2 and P_3 with system relations R_1, R_2, R_3, R_4.
Properties
P_1: Linear but NOT time-invariant
P_2: Time-invariant but NOT linear
P_3: Linear and time-invariant
Relations
R_1: $y(t) = t^2 x(t)$
R_2: $y(t) = t|x(t)|$
R_3: $y(t) = |x(t)|$
R_4: $y(t) = x(t-5)$
(a) $(P_1, R_1), (P_2, R_3), (P_3, R_4)$
(b) $(P_1, R_2), (P_2, R_3), (P_3, R_4)$
(c) $(P_1, R_3), (P_2, R_1), (P_3, R_2)$
(d) $(P_1, R_1), (P_2, R_2), (P_3, R_3)$ [2008 : 2 M]

1.13 The input $x(t)$ and output $y(t)$ of a system are related as $y(t) = \int_{-\infty}^{t} x(\tau)\cos(3\tau)d\tau$. The system is
(a) time-invariant and stable
(b) stable and not time-invariant
(c) time-invariant and not stable
(d) not time-invariant and not stable
[2012 : 2 M]

1.14 For a periodic signal
$v(t) = 30\sin 100t + 10\cos 300t + 6\sin(500t + \pi/4)$,
the fundamental frequency in rad/s is
(a) 100 (b) 300
(c) 500 (d) 1500 [2013 : 1 M]

1.15 The impulse response of a continuous time system is given by $h(t) = \delta(t-1) + \delta(t-3)$. The value of the step response at $t = 2$ is
(a) 0 (b) 1
(c) 2 (d) 3 [2013 : 2 M]

1.16 A discrete-time signal $x[n] = \sin(\pi^2 n)$, n being an integer, is
(a) periodic with period π
(b) periodic with period π^2
(c) periodic with period $\pi/2$
(d) not periodic [2014 : 1 M, Set-1]

1.17 The waveform of a periodic signal $x(t)$ is shown in the figure.

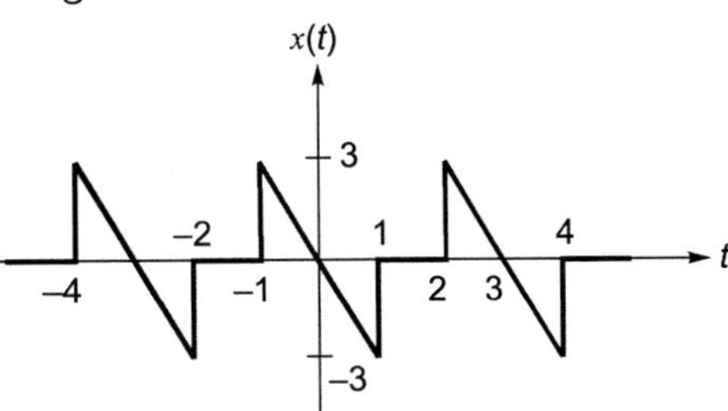

A signal $g(t)$ is defined by $g(t) = x\left(\frac{t-1}{2}\right)$. The average power of $g(t)$ is ________.
[2015 : 1 M, Set-1]

1.18 Two sequences $x_1[n]$ and $x_2[n]$ have the same energy. Suppose $x_1[n] = \alpha\, 0.5^n\, u[n]$, where α is a positive real number and $u[n]$ is the unit step sequence. Assume

$$x_2[n] = \begin{cases} \sqrt{1.5} & \text{for } n = 0, 1 \\ 0 & \text{otherwise} \end{cases}$$

Then the value of α is _____.
[2015 : 2 M, Set-3]

1.19 A continuous time function $x(t)$ is periodic with period T. The function is sampled uniformly with a sampling period T_s. In which one of the following cases is the sampled signal periodic?
(a) $T = \sqrt{2}T_s$ (b) $T = 1.2T_s$
(c) Always (d) Never
[2016 : 1 M, Set-1]

1.20 Consider a single input single output discrete-time system with $x[n]$ as input and $y[n]$ as output, where the two are related as

$$y[n] = \begin{cases} n|x[n]|, & \text{for } 0 \le n \le 10 \\ x[n] - x[n-1], & \text{otherwise.} \end{cases}$$

Which one of the following statements is true about the system?
(a) It is causal and stable
(b) It is causal but not stable
(c) It is not causal but stable
(d) It is neither causal nor stable
[2017 : 1 M, Set-1]

1.21 The input $x(t)$ and the output $y(t)$ of a continuous-time system are related as

$$y(t) = \int_{t-T}^{t} x(u)du$$

The system is

(a) linear and time-variant
(b) linear and time-invariant
(c) non-linear and time-variant
(d) non-linear and time-invariant

[2017 : 1 M, Set-2]

1.22 Let the input be u and the output be y of a system, and the other parameters are real constants. Identify which among the following systems is not a linear system:

(a) $\frac{d^3y}{dt^3} + a_1\frac{d^2y}{dt^2} + a_2\frac{dy}{dt} + a_3y$
$= b_3u + b_2\frac{du}{dt} + b_1\frac{d^2u}{dt^2}$
(with initial rest conditions)

(b) $y(t) = \int_0^t e^{\alpha(t-\tau)}\beta u(\tau)\, d\tau$

(c) $y = au + b, b \neq 0$

(d) $y = au$

[2018 : 1 M]

1.23 Consider the signal

$$f(t) = 1 + 2\cos(\pi t) + 3\sin\left(\frac{2\pi}{3}t\right) + 4\cos\left(\frac{\pi}{2}t + \frac{\pi}{4}\right),$$

where t is in seconds. Its fundamental time period, in seconds, is _______.

[2019 : 1 M]

1.24 Consider a system with input $x(t)$ and output $y(t) = x(e^t)$. The system is
(a) Causal and time invariant
(b) Non-causal and time varying
(c) Causal and time varying
(d) Non-causal and time invariant

[2023 : 1 M]

Answers Basics of Signals and Systems

1.1	(b)	1.2	(a)	1.3	(b)	1.4	(a)	1.5	(a)	1.6	(a)	1.7	(a)
1.8	(d)	1.9	(c)	1.10	(b)	1.11	(c)	1.12	(a)	1.13	(d)	1.14	(a)
1.15	(b)	1.16	(d)	1.17	(2)	1.18	(1.5)	1.19	(b)	1.20	(a)	1.21	(b)
1.22	(c)	1.23	(12)	1.24	(b)								

Explanations Basics of Signals and Systems

1.1 (b)

$$ay_1(t) = atx_1(t)$$
$$ay_2(t) = atx_2(t)$$
$$a[y_1(t) + y_2(t)] = a[tx_1(t) + tx_2(t)]$$

$\therefore$ System is linear.

$$y(t-t_0) = (t-t_0)\,x(t-t_0)$$

Delay is introduced. So, time varying.

1.2 (a)

$$\int_{-\infty}^{\infty} \delta(t)\cos\left(\frac{3t}{2}\right)dt = f(0) = \cos\left(\frac{3\times 0}{2}\right) = \cos 0 = 1$$

1.3 (b)

$$E = \int_{-\infty}^{\infty} f(t)^2\, dt$$

$$E' = \int_{-\infty}^{\infty} f(2t)^2\, dt = \int_{-\infty}^{\infty} f(p)^2 \frac{dp}{2}\left(2t = p; dt = \frac{dp}{2}\right)$$

$$E' = \frac{E}{2}$$

1.4 (a)

$$y(n-n_0) = x(n-n_0+1) \text{(time varying)}$$
$$y(n) = x(n+1) \text{ (depends on future)}$$
i.e. $y(1) = x(2)$ (non causal)

For bounded input, system has bounded output. So it is stable.

$$y(n) = x(n), n \geq 1$$
$$= 0, n = 0$$
$$= x(n+1), n \leq -1$$

So, system is linear.

1.5 (a)

$$x(n) = [-4-j5, \underset{\uparrow}{1+2j}, 4]$$

$$x^*(-n) = [4, \underset{\uparrow}{1-2j}, -4+j5]$$

$$x_{CAS}(n) = \frac{x(n) - x^*(-n)}{2}$$
$$= [-4-2.5j, \underset{\uparrow}{2j}, 4-2.5j]$$

1.6 (a)

Even part $= \dfrac{u(t)+u(-t)}{2}$; Odd part $= \dfrac{u(t)-u(-t)}{2}$

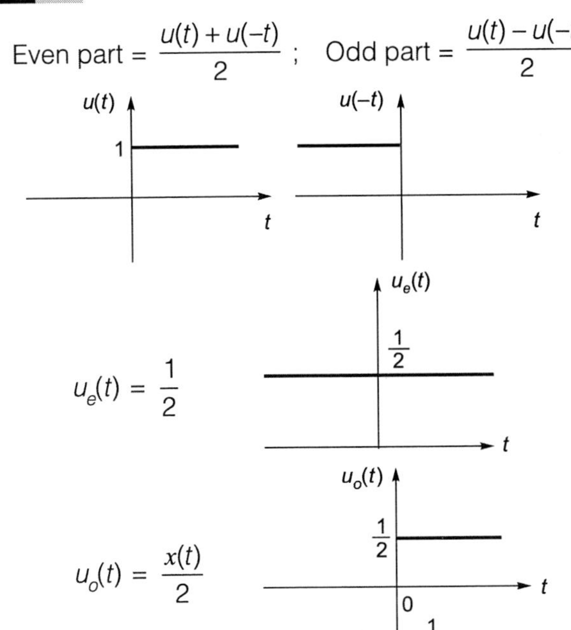

$$u_e(t) = \frac{1}{2}$$

$$u_o(t) = \frac{x(t)}{2}$$

1.7 (a)

$$s(t) = 8\cos\left(\frac{\pi}{2} - 20\pi t\right) + 4\sin 15\pi t$$

$$= 8\sin 20\pi t + 4\sin 15\pi t$$

$$P = \frac{8^2}{2} + \frac{4^2}{2} = 32 + 8 = 40$$

1.8 (d)

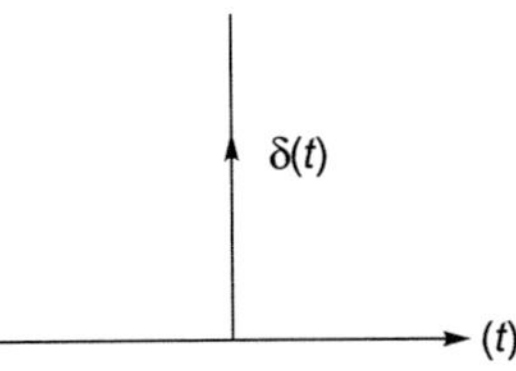

1.9 (c)

$$y(n) = \left(\sin\frac{5}{6}\pi n\right)x(n)$$

Let $x(n) = \delta(n)$

$\therefore$ $y(n) = \sin 0 = 0$ (bounded)

BIBO stable

1.10 (b)

The Hilbert transformer is characterised by the impulse response

$$h(t) = \frac{1}{\pi t} \quad t \in (-\infty, \infty)$$

$$h(t) \neq 0 \text{ for } t < 0$$

Thus, the Hilbert transformer is a non-causal system.

1.11 (c)

A system is casual if the output at any time depends only on values of the input at the present time and in the past.

1.13 (d)

$$y = \int_{-\infty}^{t} x(\tau)\cos(3\tau)\,d\tau$$

$$y(t - t_0) = \int_{-\infty}^{t-t_0} x(\tau)\cos(3\tau)\,d\tau$$

$y'(t)$ for input $x(t - t_0)$ is

$$y'(t) = \int_{-\infty}^{t} x(\tau - t_0)\cos 3\tau\,d\tau$$

$$y'(t) = \int_{-\infty}^{(t-t_0)} x(\tau)\cos 3(\tau + t_0)\,d\tau$$

$y'(t) \neq y(t - t_0)$ so system is not time invariant

for input $x(\tau) = \cos(3\tau)$ (bounded input)

$$y(t) = \int_{-\infty}^{t} \cos^2(3\tau)\,d\tau \to \infty \text{ as } t \to \infty$$

So, for bounded i/p, o/p is not bounded therefore system is not stable.

1.14 (a)

$$\omega_1 = 100$$
$$\omega_2 = 300$$
$$\omega_3 = 500$$

H.C.F. of (ω_1, ω_2 and ω_3) = H.C.F. (100, 300, 500)

$$\omega = 100 \text{ rad/sec.}$$

1.15 (b)

Step response = Integration of impulse response.

$$\int_{-\infty}^{t} \delta(t-1)\,dt = u(t-1)$$

$$\int_{-\infty}^{t} \delta(t-3)\,dt = u(t-3)$$

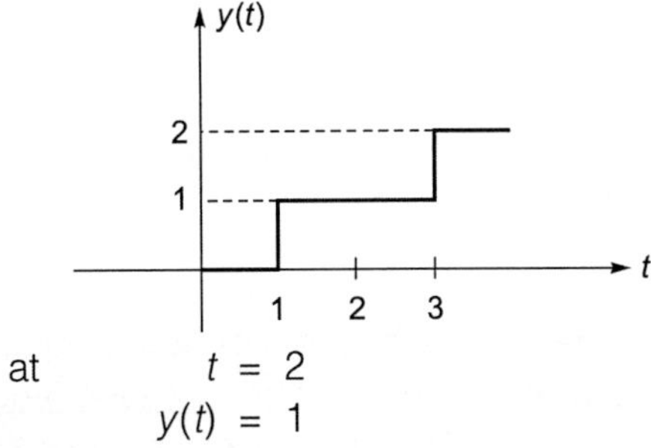

at $t = 2$

$$y(t) = 1$$

1.16 (d)

$$x[n] = \sin(\pi^2 n)$$

$$\omega_0 = \pi^2$$

$$\therefore \quad N = \frac{2\pi}{\omega_o} \cdot m$$

where m is the smallest integer that converts $\frac{2\pi}{\omega_o}$ into a integer value.

$$\therefore \quad N = \frac{2\pi}{\pi^2}\cdot m = \frac{2}{\pi}\cdot m$$

So, there exists no such integer value of m which could make the N integer, so the system is not periodic.

1.17 (2)

$$x(t) = -3t, \quad -1 < t < 1$$

$$x\left(\frac{t-1}{2}\right) = -\frac{3}{2}(t-1) \qquad -1 < t < 3$$

and $T = 6$

$$\text{Average power} = \frac{1}{6}\int_{-1}^{3}\left(-\frac{3}{2}(t-1)\right)^2 dt = 2$$

1.18 (1.5)

$$\text{Energy of } x_1[n] = \sum_{n=-\infty}^{\infty} |x_1[n]|^2$$

$$= \sum_{n=0}^{\infty} \alpha^2\left(\frac{1}{4}\right)^n = \alpha^2\cdot\frac{1}{1-\frac{1}{4}} = \alpha^2\cdot\frac{4}{3}$$

Energy of $x_2[n] = 1.5 + 1.5 = 3$

$$\Rightarrow \quad \alpha^2\frac{4}{3} = 3$$

$$\Rightarrow \quad \alpha^2 = \frac{9}{4}$$

$$\Rightarrow \quad \alpha = 1.5$$

1.19 (b)

A signal is said to be periodic if $\frac{T}{T_s}$ is a rational number.

Here, $T = 1.2\, T_s$

$$\Rightarrow \quad \frac{T}{T_s} = \frac{6}{5} \quad \text{Which is a rational number}$$

1.20 (a)

Since present output does not depend upon future values of input, the system is causal and also every bounded input produces bounded output, so it is stable.

1.21 (b)

Given that, $y(t) = \int_{t-T}^{t} x(u)du$

- Since the given system satisfies both homogeneity and additivity properties, the system is linear.
- **Check for time invariance:**

$$y(t-t_o) = \int_{t-t_o-T}^{t-t_o} x(u)du$$

- When the applied input is $x(t-t_o)$,

$$y_1(t) =$$

$$\int_{t-T}^{t} x(u-t_o)du = \int_{t-t_o-T}^{t-t_o} x(\tau)d\tau$$

$$= y(t-t_o)$$

$\Rightarrow$ System is time invariant

1.22 (c)

$y = au + b$, $b \neq 0$ is a non-linear system.

1.23 (12)

$$f(t) = 1 + 2\cos(\pi t) + 3\sin\left(\frac{2\pi}{3}t\right) + 4\cos\left(\frac{\pi}{2}t + \frac{\pi}{4}\right)$$

$$\omega_1 = \pi$$

$$\omega_2 = \frac{2\pi}{3}$$

$$\omega_3 = \frac{\pi}{2}$$

$$\omega_0 = \text{GCD}\left(\pi, \frac{2\pi}{3}, \frac{\pi}{2}\right) = \frac{\pi}{6}$$

Fundamental period,

$$N = \frac{2\pi}{\omega_o} = \frac{2\pi}{(\pi/6)} = 12$$

1.24 (b)

For causaility,

We have, $y(t) = x(e^t)$

At $t = 0$

$$y(0) = x(1)$$

i.e. present value of output depends on future value of input, hence it is non-causal.

For Time Variant:

Delay the input,

$$y(t) = x(e^t - t_o) \qquad \text{...(i)}$$

Delay the output,

$$y(t-t_o) = x(e^{t-t_o}) \qquad \text{...(ii)}$$

i.e. equations (i) $\neq$ (ii)

Hence, it is time variant system.

■■■

2 LTI Systems Continuous and Discrete (Time Domain)

2.1 Let $h(t)$ be the impulse response of a linear time invariant system. Then the response of the system for any input $u(t)$ is

(a) $\int_0^t h(\tau)u(t-\tau)d\tau$

(b) $\frac{d}{dt}\int_0^t h(\tau)u(t-\tau)d\tau$

(c) $\int_0^t \left[\int_0^t h(\tau)u(t-\tau)d\tau\right]dt$

(d) $\int_0^t h^2(\tau)u(t-\tau)d\tau$ [1995 : 1 M]

2.2 The unit impulse response of a linear time invariant system is the unit step function $u(t)$. For $t > 0$, the response of the system to an excitation $e^{-at}u(t)$, $a > 0$ will be

(a) $a e^{-at}$ (b) $(1/a)(1 - e^{-at})$

(c) $a(1 - e^{-at})$ (d) $1 - e^{-at}$

[1998 : 1 M]

2.3 Let $u(t)$ be the step function. Which of the waveforms in the figure corresponds to the convolution of $u(t)-u(t-1)$ with $u(t)-u(t-2)$?

(a)
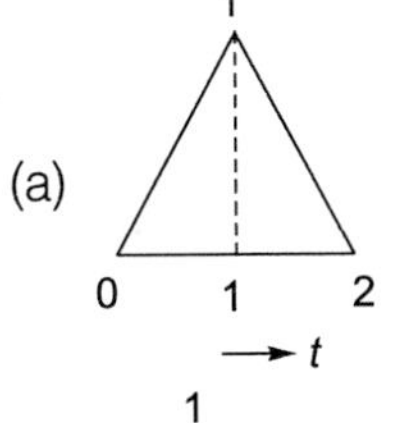

(b)
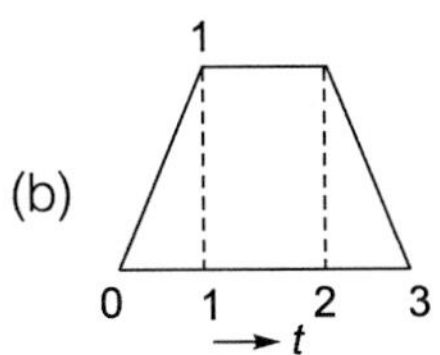

(c)
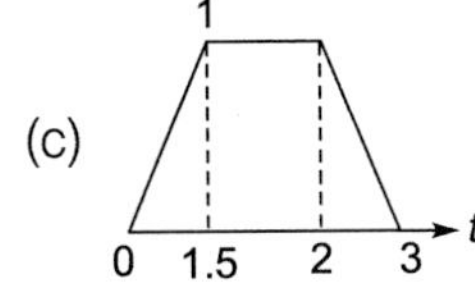

(d)
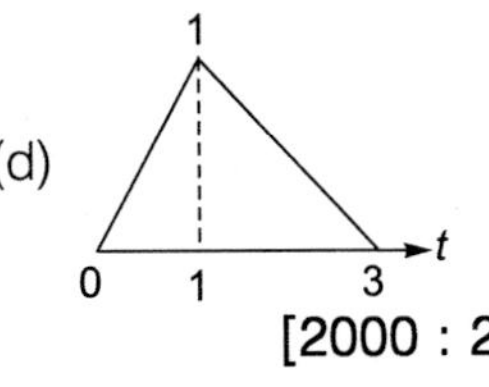

[2000 : 2 M]

2.4 The impulse response functions of four linear systems S_1, S_2, S_3, S_4 are given respectively by

$h_1(t) = 1$ $h_2(t) = u(t)$

$h_3(t) = \frac{u(t)}{t+1}$ $h_4(t) = e^{-3t}u(t)$

Where $u(t)$ is the unit step function. Which of these systems is time invariant, causal, and stable?

(a) S_1 (b) S_2

(c) S_3 (d) S_4 [2001 : 2 M]

2.5 Convolution of $x(t+5)$ with impulse function $\delta(t-7)$ is equal to

(a) $x(t-12)$ (b) $x(t+12)$

(c) $x(t-2)$ (d) $x(t+2)$

[2002 : 1 M]

2.6 The impulse response $h[n]$ of a linear time-invariant system is given by

$h[n] = u[n+3] + u[n-2] - 2u[n-7]$

where $u[n]$ is the unit step sequence. The above system is

(a) stable but not causal

(b) stable and causal

(c) causal but unstable

(d) unstable and not causal [2004 : 1 M]

2.7 Which of the following can be impulse response of a causal system?

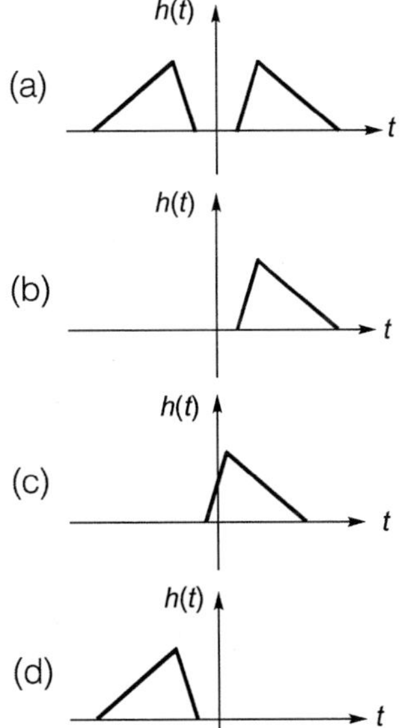

[2005 : 1 M]

2.8 The impulse response $h(t)$ of a linear time-invariant continuous time system is described by $h(t) = \exp(\alpha t)u(t) + \exp(\beta t)\,u(-t)$, where $u(t)$ denotes the unit step function, and α and β are real constants. This system is stable if

(a) α is positive and β is positive

(b) α is negative and β is negative

(c) α is positive and β is negative

(d) α is negative and β is positive

[2008 : 1 M]

2.9 A discrete time linear shift-invariant system has an impulse response $h[n]$ with $h[0] = 1$, $h[1] = -1$, $h[2] = 2$, and zero otherwise. The system is given an input sequence $x[n]$ with $x[0] = x[2] = 1$ and zero otherwise. The number of nonzero samples in the output sequence $y[n]$, and the value of $y[2]$ are, respectively

(a) 5, 2 (b) 6, 2
(c) 6, 1 (d) 5, 3 [2008 : 2 M]

2.10 A system is defined by its impulse response $h(n) = 2^n u(n-2)$. The system is

(a) stable and causal
(b) causal but not stable
(c) stable but not causal
(d) unstable and non-causal [2011 : 1 M]

2.11 Let $y[n]$ denote the convolution of $h[n]$ and $g[n]$, where $h[n] = (1/2)^n u[n]$ and $g[n]$ is a causal sequence. If $y[0] = 1$ and $y[1] = 1/2$, then $g[1]$ equals

(a) 0 (b) 1/2
(c) 1 (d) 3/2

[2012 : 2 M]

2.12 Two systems with impulse responses $h_1(t)$ and $h_2(t)$ are connected in cascade. Then the overall impulse response of the cascaded system is given by

(a) product of $h_1(t)$ and $h_2(t)$
(b) sum of $h_1(t)$ and $h_2(t)$
(c) convolution of $h_1(t)$ and $h_2(t)$
(d) subtraction of $h_2(t)$ from $h_1(t)$

[2013 : 1 M]

2.13 Consider a discrete-time signal

$$x[n] = \begin{cases} n & \text{for } 0 \le 0 \le 10 \\ 0 & \text{Otherwise} \end{cases}$$

If $y[n]$ is the convolution of $x[n]$ with itself, the value of $y[4]$ is ______.

[2014 : 2 M, Set-2]

2.14 The result of the convolution $x(-t) * \delta(-t - t_0)$ is

(a) $x(t + t_0)$ (b) $x(t - t_0)$
(c) $x(-t + t_0)$ (d) $x(-t - t_0)$

[2015 : 1 M, Set-1]

2.15 The impulse response of an LTI system can be obtained by

(a) differentiating the unit ramp response
(b) differentiating the unit step response
(c) integrating the unit ramp response
(d) integrating the unit step response

[2015 : 1 M, Set-3]

2.16 Which one of the following is an eigen function of the class of all continuous-time, linear, time-invariant systems ($u(t)$ denotes the unit-step function)?

(a) $e^{j\omega_0 t}u(t)$ (b) $\cos(\omega_0 t)$
(c) $e^{j\omega_0 t}$ (d) $\sin(\omega_0 t)$

[2016 : 1 M, Set-1]

2.17 Two discrete-time signals $x[n]$ and $h[n]$ are both non-zero only for $n = 0, 1, 2$ and are zero otherwise. It is given that

$x[0] = 1$, $x[1] = 2$, $x[2] = 1$, $h[0] = 1$.

Let $y[n]$ be the linear convolution of $x[n]$ and $h[n]$. Given that $y[1] = 3$ and $y[2] = 4$, the value of the expression $(10y[3] + y[4])$ is ______.

[2017 : 2 M, Set-1]

2.18 Consider the parallel combination of two LTI systems shown in the figure,

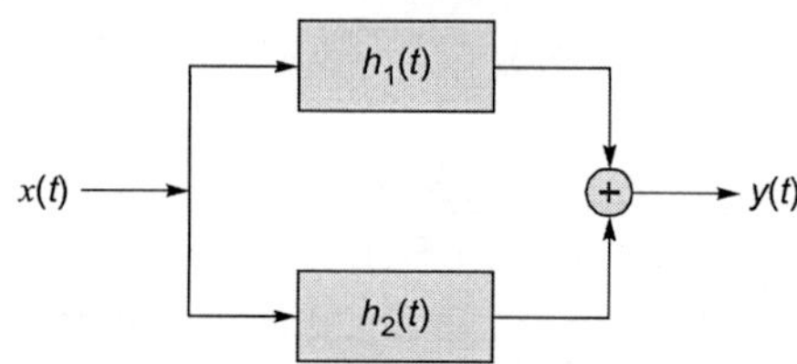

The impulse responses of the systems are

$$h_1(t) = 2\delta(t+2) - 3\delta(t+1)$$
$$h_2(t) = \delta(t-2)$$

If the input $x(t)$ is a unit step signal, then the energy of $y(t)$ is _____.

[2017 : 2 M, Set-2]

2.19 The output $y[n]$ of a discrete-time system for an input $x[n]$ is

$$y[n] = \max_{-\infty \le k \le n} |x[k]|$$

The unit impulse response of the system is

(a) 0 for all n.
(b) 1 for all n.
(c) unit step signal $u[n]$.
(d) unit impulse signal $\delta[n]$.

[2020 : 1 M]

2.20 For a unit step input $u[n]$, a discrete-time LTI system produces an output signal

$$(2\delta[n+1] + \delta[n] + \delta[n-1])$$

Let $y[n]$ be the output of the system for an input $\left[\left(\frac{1}{2}\right)^n u[n]\right]$. The value of $y[0]$ is ____.

[2021 : 2 M]

2.21 Let $x_1(t) = e^{-t}u(t)$ and $x_2(t) = u(t) - u(t-2)$, where $u(\cdot)$ denotes the unit step function.
If $y(t)$ denotes the convolution of $x_1(t)$ and $x_2(t)$, then $\lim_{t\to\infty} y(t) =$ ______ (rounded off to one decimal place).

[2022 : 1 M]

2.22 The outputs of four systems (S_1, S_2, S_3, and S_4) corresponding to the input signal sin(t), for all time t, are shown in the figure.
Based on the given information, which of the four systems is/are definitely NOT LTI (linear and time-invariant)?

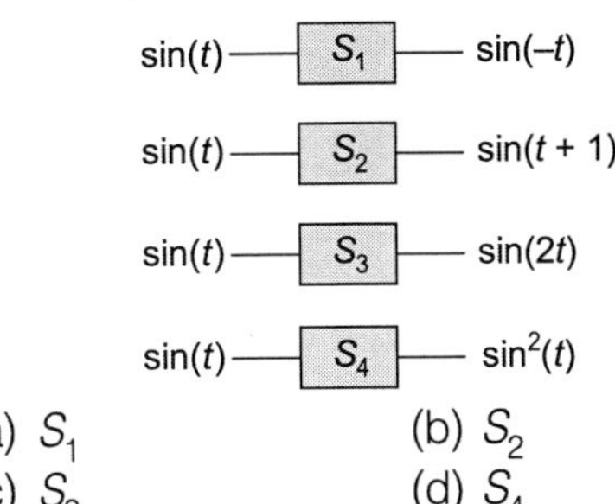

(a) S_1 (b) S_2
(c) S_3 (d) S_4

[2022 : 2 M]

2.23 A causal and stable LTI system with impulse response $h(t)$ produces an output $y(t)$ for an input signal $x(t)$. A signal $x(0.5t)$ is applied to another causal and stable LTI system with impulse response $h(0.5t)$. The resulting output is ______.
(a) $2y(0.5t)$ (b) $4y(0.5t)$
(c) $0.25y(2t)$ (d) $0.25y(0.25t)$

[2024 : 1 Mark]

2.24 The radian frequency value(s) for which the discrete time sinusoidal signal $x[n] = A\cos(\Omega n + \pi/3)$ has a period of 40 is/are ______.
(a) 0.15π (b) 0.3π
(c) 0.45π (d) 0.225π

[2024 : 2 Marks]

Answers LTI Systems Continuous and Discrete (Time Domain)

2.1 (a)	2.2 (b)	2.3 (b)	2.4 (d)	2.5 (c)	2.6 (a)	2.7 (b)
2.8 (d)	2.9 (d)	2.10 (b)	2.11 (a)	2.12 (c)	2.13 (10)	2.14 (d)
2.15 (b)	2.16 (c)	2.17 (31)	2.18 (7)	2.19 (c)	2.20 (0)	2.21 (0)
2.22 (c, d)	2.23 (a)	2.24 (a, b, c)				

Explanations LTI Systems Continuous and Discrete (Time Domain)

2.1 (a)

For LTI system

$$y(t) = u(t) * h(t)$$

$$y(t) = \int_{-\infty}^{\infty} h(\tau)\, u(t-\tau)d\tau$$

$$\because u(t) = 0 \quad \text{for } t < 0$$

$$y(t) = \int_0^t h(\tau)\, u(t-\tau)\, d\tau$$

2.2 (b)

Given that,
Impulse response = $h(t) = u(t)$
So, transfer function

$$H(s) = \frac{1}{s}$$

$$H(s) = \frac{Y(s)}{X(s)}$$

$$Y(s) = H(s) \times X(s)$$

Given that

$$x(t) = e^{-at}u(t)$$

$$X(s) = \frac{1}{(s+a)}$$

$$Y(s) = \frac{1}{s} \times \frac{1}{(s+a)}$$

$$Y(s) = \frac{1}{a}\left[\frac{1}{s} - \frac{1}{s+a}\right]$$

Taking inverse Laplace transform

$$y(t) = \frac{1}{a}\left[1 - e^{-at}\right]$$

2.3 (b)

$$u(t) - u(t-1) = \frac{1}{s}[1-e^{-s}]$$

$$u(t) - u(t-2) = \frac{1}{s}[1-e^{-2s}]$$

Convolution in time domain = Multiplication in s-domain

$$\Rightarrow \frac{1}{s^2}[1-e^{-s}][1-e^{-2s}]$$

$$= \frac{1}{s^2}[1-e^{-2s}-e^{-s}+e^{-3s}]$$

$$= tu(t) - (t-1)\,u(t-1) - (t-2)\,u(t-2) + (t-3)\,u(t-3)$$

$$= r(t) - r(t-1) - r(t-2) + r(t-3)$$

2.4 (d)

$h_1(t) \neq 0$ for $t < 0$

Therefore s_1 is non causal

$h_2(t) = u(t) = 0$ for $t < 0$

$$\int_{-\infty}^{\infty} h_2(t)dt = \int_{-\infty}^{\infty} u(t)dt = \int_{0}^{\infty} dt = \infty$$

Therefore s_2 is unstable and causal

$$h_3(t) = \frac{u(t)}{t+1}$$

at $t = -1$

$h_3(t) = \infty$, $h_3(t) = 0$ for $t < 0$

Therefore s_3 is unstable and causal

$h_4(t) = e^{-3t}\,u(t)$

s_4 is time invariant, causal and stable.

2.5 (c)

$$x(t+5) * \delta(t-7) = x(t+5-7) = x(t-2)$$

2.6 (a)

$$\sum_{k=-\infty}^{\infty} h(k) = \sum_{k=-3}^{\infty} u(k+3) + \sum_{k=2}^{\infty} u(k-2) - 2\sum_{k=7}^{\infty} u(k-7)$$

$$= \sum_{k=-3}^{6} 1 + \sum_{k=2}^{6} 1$$

$$= 10 + 5 = 15 < \infty$$

For bounded input, bounded output. So system is stable.

Response depends on future value of input signal i.e. $u(n+3)$. So, system is not causal.

2.7 (b)

For a causal system

$h(t) = 0$ for $t < 0$

2.8 (d)

$$h(t) = e^{\alpha t}u(t) + e^{\beta t}u(-t)$$

For the system to be stable, $\int_{-\infty}^{\infty} h(t)dt < \infty$

For the above condition, $h(t)$ should be as shown below.

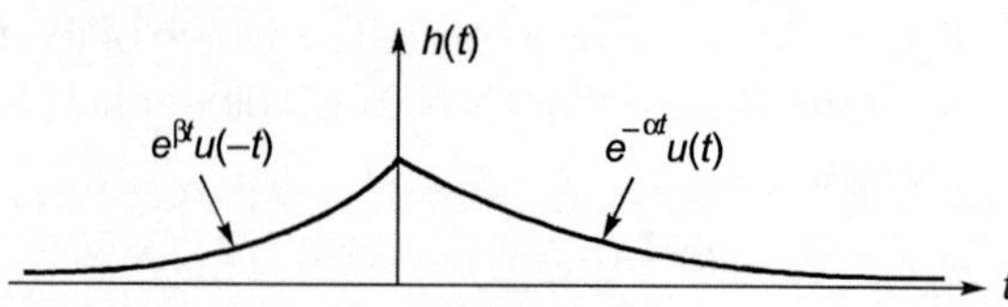

Therefore $\alpha < 0$ & $\beta > 0$.

2.9 (d)

$L_1 = 3$, $L_2 = 3$

$\Rightarrow$ Number of samples in output $= L_1 + L_2 - 1 = 5$

$$y[n] = \sum_{k=-\infty}^{\infty} x[k]\,h[n-k]$$

$$y[2] = \sum_{k=-\infty}^{\infty} x[k]\,h[2-k]$$

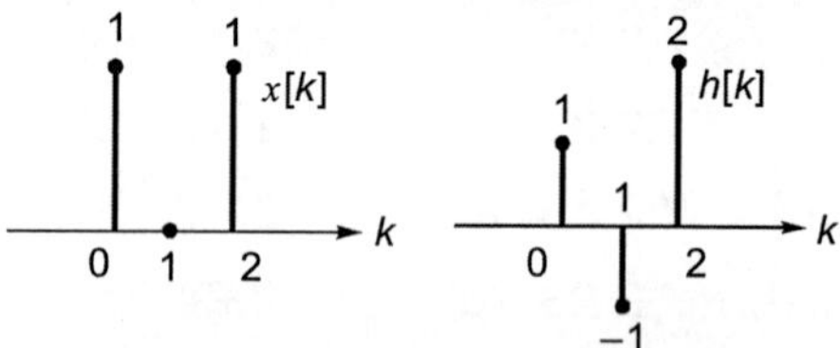

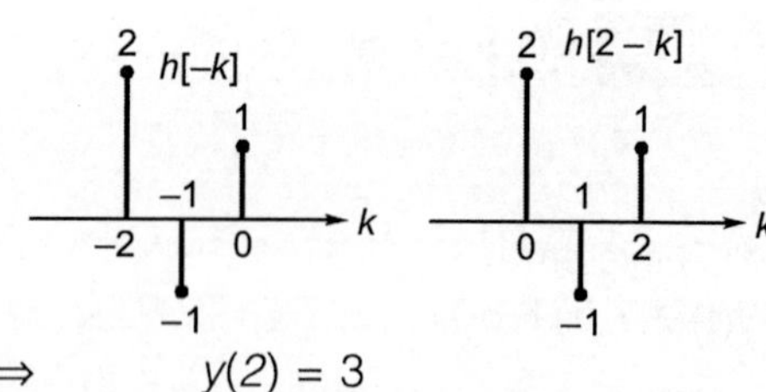

$\Rightarrow$ $y(2) = 3$

2.10 (b)

$$h(n) = 2^n\,u(n-2)$$

For causal system $h(n) = 0$ for $n < 0$.

Hence given system is causal.

For stability:

$\sum_{n=2}^{\infty} 2^n = \infty$, so given system is not stable.

2.11 (a)

$$h[n] = \left(\frac{1}{2}\right)^n u[n]$$

$g[n] = ??$

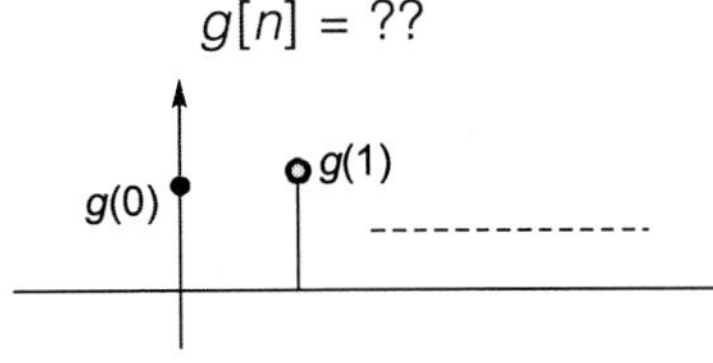

$$y[n] = \sum_{k=-\infty}^{\infty} h[n-k]\,g[k]$$

$$y[0] = \sum_{k=-\infty}^{\infty} h[-k]\,g[k]$$

$$y[0] = h[0]\,g[0]$$

$$1 = 1\,g[0]$$

$$g[0] = 1$$

$$y[1] = \sum_{k=-\infty}^{\infty} h[1-k]\,g[k]$$

$$y[1] = h[1]\,g[0] + h[0]\,g[1]$$

$h[1-k]$ will be zero for $k > 1$ and $g[k]$ will be zero for $k < 0$ as it is causal sequence.

$$\frac{1}{2} = \frac{1}{2}\times 1 + 1\,g[1]$$

$$g[1] = 0$$

2.12 (c)

The overall impulse response $h(t)$ of the cascaded system is given by:

$$h(t) = h_1(t) * h_2(t)$$

2.13 (10)

$$x[n] = \begin{cases} n & \text{for } 0 \le n \le 10 \\ 0 & \text{otherwise} \end{cases}$$

$$y[n] = \sum_{K=\infty}^{\infty} x(k)\,x(n-k)$$

$$y[4] = \sum_{k=0}^{10} x(k)\,x(4-k)$$

$$= x(0)x(4) + x(1)x(3) + x(2)x(2) + x(3)x(1) + 0$$
$$= 0 + (1\times 3) + (2\times 2) + (3\times 1) = 10$$

2.14 (d)

$$x(-t) * \delta(-t-t_0) = x(-t) * \delta(t+t_0)$$
$$= x(-t-t_0)$$

2.15 (b)

$$IR = \frac{d}{dt}(SR)$$

Step response, $y(t) = \int_{-\infty}^{t} h(\tau)\cdot d\tau$

$$h(t) = \frac{d}{dt}y(t)$$

2.16 (c)

If the input to the system is eigen signal then output is also the same eigen signal.

2.17 (31)

$$x[n] = \{1, 2, 1\}$$
$$h[n] = \{1, a, b\}$$
$$y[n] = \{A, 3, 4, B, C\}$$

x \ h	1	a	b
1	1	a	b
2	2	2a	2b
1	1	a	b

$y[0] = 1$; $y[1] = 2 + a$; $y[2] = 1 + 2a + b$;
$y[3] = a + 2b$; $y[4] = b$

Given, $y[1] = 2 + a = 3 \Rightarrow a = 1$

Given, $y[2] = 1 + 2a + b = 4 \Rightarrow b = 1$

So, $y[3] = a + 2b = 3$

and $y[4] = b = 1$

So, $10y[3] + y[4] = 10\times 3 + 1 = 31$

2.18 (7)

Given that,

$$h_1(t) = 2\delta(t+2) - 3\delta(t+1)$$
$$h_2(t) = \delta(t-2)$$

Overall impulse response is,

$$h(t) = h_1(t) + h_2(t)$$
$$= 2\delta(t+2) - 3\delta(t+1) + \delta(t-2)$$

If input, $x(t) = u(t)$, then the output will be,

$$y(t) = x(t) * h(t)$$
$$= u(t) * [2\delta(t+2) - 3\delta(t+1) + \delta(t-2)]$$
$$= 2u(t+2) - 3u(t+1) + u(t-2)$$

By plotting $y(t)$, we get,

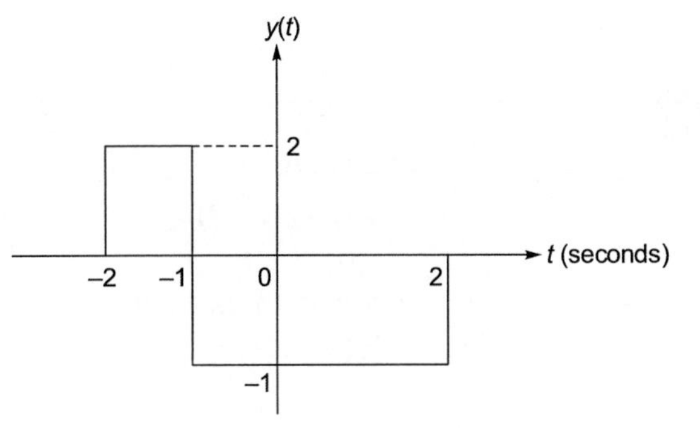

The energy of $y(t)$ can be given as,

$$E_y = \int_{-\infty}^{\infty} y^2(t)dt$$

By plotting $y^2(t)$, we get,

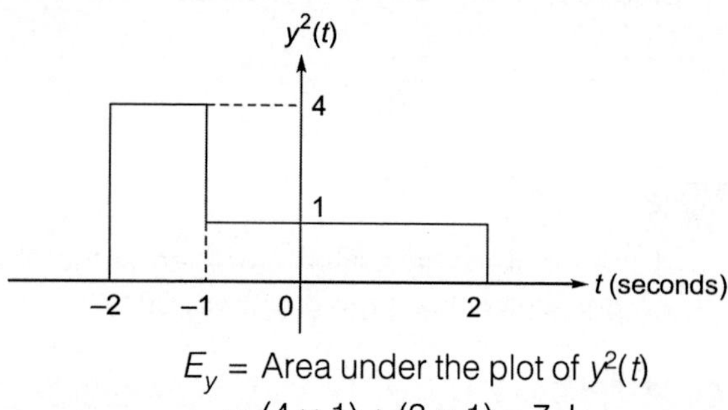

$$E_y = \text{Area under the plot of } y^2(t)$$
$$= (4 \times 1) + (3 \times 1) = 7 \text{ J}$$

2.20 (0)

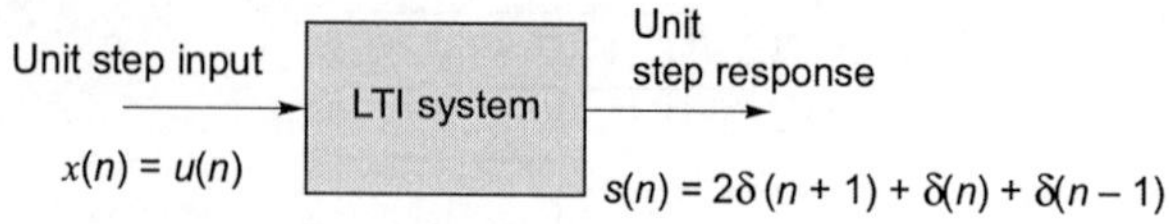

The impulse response $h(n) = s(n) - s(n-1)$

$$h(n) = 2\delta(n+1) + \delta(n) + \delta(n-1)$$
$$- 2\delta(n) - \delta(n-1) - \delta(n-1-1)$$
$$= 2\delta(n+1) + \delta(n) + \delta(n-1)$$
$$- 2\delta(n) - \delta(n-1) - \delta(n-2)$$
$$h(n) = 2\delta(n+1) - \delta(n) - \delta(n-2)$$

For input $x_1(n) = (1/2)^n u(n)$ then output $y_1(n)$

$$= x_1(n) * h(n)$$

$$y_1(n) = \left(\frac{1}{2}\right)^n u(n) * [2\delta(n+1) - \delta(n) - \delta(n-2)]$$

$$y_1(n) = \left(\frac{1}{2}\right)^n u(n) * 2\delta(n+1) - \left(\frac{1}{2}\right)^n u(n) * \delta(n)$$
$$- \left(\frac{1}{2}\right)^n u(n) * \delta(n-2)$$

$$y_1(n) = 2\left(\frac{1}{2}\right)^{n+1} u(n+1) - \left(\frac{1}{2}\right)^n u(n) - \left(\frac{1}{2}\right)^{n-2} u(n-2)$$

$$y_1(n)\Big|_{n=0} = 2\left(\frac{1}{2}\right)^1 u(1) - \left(\frac{1}{2}\right)^0 u(0) - \left(\frac{1}{2}\right)^{-2} u(-2)$$
$$= 1 - 1$$
$$y_1(0) = 0$$

2.21 (0)

Given :
$$x_1(t) = e^{-t} u(t)$$
$$x_2(t) = u(t) - u(t-2)$$
$$y(t) = x_1(t) * x_2(t)$$

Applying Laplace transform, we get,

$$Y(s) = X_1(s) \cdot X_2(s)$$
$$Y(s) = L[e^{-t} u(t)] \cdot L[u(t) - u(t-2)]$$
$$Y(s) = \left(\frac{1}{s+1}\right) \times \left(\frac{1}{s} - \frac{e^{-2s}}{s}\right)$$
$$Y(s) = \frac{1 - e^{-2s}}{s(s+1)}$$

Applying final value theorem,

$$\lim_{t \to \infty} y(t) = \lim_{s \to 0} sY(s)$$
$$\lim_{t \to \infty} y(t) = \lim_{s \to 0} sX\frac{(1 - e^{-2s})}{s(s+1)}$$
$$\therefore \quad \lim_{t \to \infty} y(t) = \frac{1 - e^0}{1} = 1 - 1 = 0$$

2.22 (c, d)

$\sin t \rightarrow$ S_1 $\rightarrow \sin(-t) = -\sin t$

$\sin t \rightarrow$ S_2 $\rightarrow \sin(t+1)$

$\sin t \rightarrow$ S_3 $\rightarrow \sin(2t)$

$\sin t \rightarrow$ S_4 $\rightarrow \sin^2(t) = \frac{1 - \cos 2t}{2}$

The response of an LTI system to a sinusoidal or complex exponential input is a sinusoid or complex exponential output at the same frequency as the input. LTI systems cannot change frequencies. The only change will be in magnitude and phase.

Out of the four given systems, systems S_3 and S_4 are definitely not LTI systems as the input and output sinusoidal frequencies are different.

2.23 (a)

The output $y(t)$ of a causal and stable LTI system with input signal $x(t)$ and impulse response $h(t)$ is given by

$$y(t) = x(t) * h(t)$$

Taking Fourier transform, we get,

$$Y(\omega) = X(\omega) \cdot H(\omega) \qquad \text{...(i)}$$

We have, $x_1(t) = x(0.5t)$ and $h_1(t) = h(0.5t)$. The output of other LTI system is given by

$$y_1(t) = x_1(t) * h_1(t)$$
$$Y_1(\omega) = X_1(\omega) \cdot H_1(\omega) \qquad \text{...(ii)}$$

Using the time-scaling property of Fourier transform,

$$x(t) \longleftrightarrow X(\omega)$$

$$x(at) \longleftrightarrow \frac{1}{|a|} X\left(\frac{\omega}{a}\right)$$

Therefore, we can write

$$X_1(\omega) = 2X(2\omega) \text{ and } H_1(\omega) = 2H(2\omega)$$

From equation (i) and (ii),

$$Y_1(\omega) = 4\,X(2\omega)\,H(2\omega)$$

$$Y_1(\omega) = 4Y(2\omega) = 2\left[\frac{1}{(1/2)} Y\left(\frac{\omega}{1/2}\right)\right]$$

Taking inverse Fourier transform, we get,

$$y_1(t) = 2y\left(\frac{1}{2}t\right) = 2y(0.5t)$$

2.24 (a, b, c)

The time period of the discrete-time sinusoidal signal $x[n] = A\cos(\Omega n + \pi/3)$ is given by

$$N = m\left(\frac{2\pi}{\Omega}\right)$$

where N and m are integers. It is given that $N = 40$, hence

$$\Omega = m\left(\frac{2\pi}{40}\right) = 0.05\,\pi m$$

Since, m is an integer, the possible values of Ω are:

$$\Omega = 0.05\,\pi \quad (m = 1)$$
$$\Omega = 0.10\,\pi \quad (m = 2)$$
$$\Omega = 0.15\,\pi \quad (m = 3)$$
$$\Omega = 0.20\,\pi \quad (m = 4)$$
$$\Omega = 0.25\,\pi \quad (m = 5)$$
$$\Omega = 0.30\,\pi \quad (m = 6)$$
$$\Omega = 0.35\,\pi \quad (m = 7)$$
$$\Omega = 0.40\,\pi \quad (m = 8)$$
$$\Omega = 0.45\,\pi \quad (m = 9)$$

Hence, options (a), (b) and (c) are correct.

■■■

3 Fourier Series

3.1 The trigonometric Fourier series of an even function of time does not have
(a) the dc term
(b) cosine terms
(c) sine terms
(d) odd harmonic terms

[1996 : 1 M]

3.2 The trigonometric Fourier series of a periodic time function can have only
(a) cosine terms
(b) sine terms
(c) cosine and sine terms
(d) dc and cosine terms

[1998 : 1 M]

3.3 A periodic signal $x(t)$ of period T_0 is given by

$$x(t) = \begin{cases} 1, & |t| < T_1 \\ 0, & T_1 < |t| < \frac{T_0}{2} \end{cases}$$

the dc component of $x(t)$ is

(a) $\frac{T_1}{T_0}$ (b) $\frac{T_1}{(2T_0)}$
(c) $\frac{2T_1}{T_0}$ (d) $\frac{T_0}{T_1}$

[1998 : 1 M]

3.4 The Fourier series representation of an impulse train denoted by

$s(t) = \sum_{n=-\infty}^{\infty} \delta(t - nT_0)$ is given by

(a) $\frac{1}{T_0}\sum_{n=-\infty}^{\infty} \exp - \frac{j2\pi nt}{T_0}$

(b) $\frac{1}{T_0}\sum_{n=-\infty}^{\infty} \exp - \frac{j\pi nt}{T_0}$

(c) $\frac{1}{T_0}\sum_{n=-\infty}^{\infty} \exp \frac{j\pi nt}{T_0}$

(d) $\frac{1}{T_0}\sum_{n=-\infty}^{\infty} \exp \frac{j2\pi nt}{T_0}$

[1999 : 2 M]

3.5 One period (0, T) each of two periodic waveforms W_1 and W_2 are shown in the figure. The magnitudes of the nth Fourier series coefficients of W_1 and W_2, for $n \geq 1$, n odd, are respectively proportional to

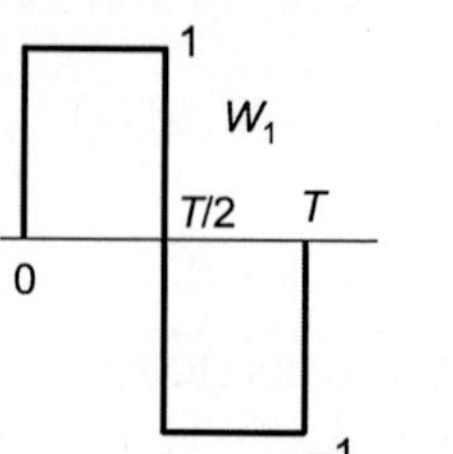

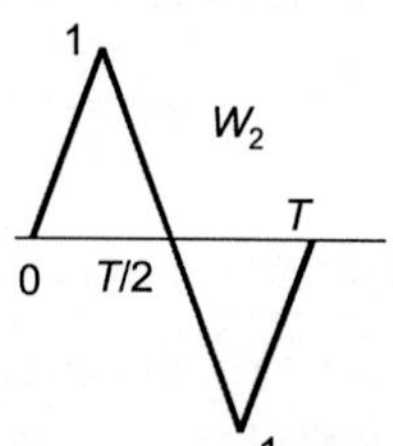

(a) $|n^{-3}|$ and $|n^{-2}|$ (b) $|n^{-2}|$ and $|n^{-3}|$
(c) $|n^{-1}|$ and $|n^{-2}|$ (d) $|n^{-4}|$ and $|n^{-2}|$

[2000 : 2 M]

3.6 Which of the following cannot be the Fourier series expansion of a periodic signal?
(a) $x(t) = 2\cos t + 3\cos 3t$
(b) $x(t) = 2\cos \pi t + 7\cos t$
(c) $x(t) = \cos t + 0.5$
(d) $x(t) = 2\cos 1.5\pi t + \sin 3.5\pi t$

[2002 : 1 M]

3.7 The Fourier series expansion of a real periodic signal with fundamental frequency f_0 is given by

$$g_p(t) = \sum_{n=-\infty}^{\infty} c_n e^{j2\pi n f_0 t}$$

it is given that $c_3 = 3 + j5$. Then c_{-3} is
(a) $5 + j3$ (b) $-3 - j5$
(c) $-5 + j3$ (d) $3 - j5$

[2003 : 1 M]

3.8 Choose the function $f(t)$, $-\infty < t < \infty$, for which a Fourier series cannot be defined.
(a) $3\sin(25t)$
(b) $4\cos(20t + 3) + 2\sin(710t)$
(c) $\exp(-|t|)\sin(25t)$
(d) 1

[2005 : 1 M]

3.9 The Fourier series of a real periodic function has only
P. Cosine terms if it is even
Q. Sine terms if it is even
R. Cosine terms if it is odd
S. Sine terms if it is odd

Which of the above statements are correct?

(a) P and S (b) P and R
(c) Q and S (d) Q and R

[2009 : 1 M]

3.10 The trigonometric Fourier series of an even function does not have the

(a) dc term (b) cosine terms
(c) sine terms (d) odd harmonic terms

[2011 : 1 M]

3.11 Consider the periodic square wave in the figure shown.

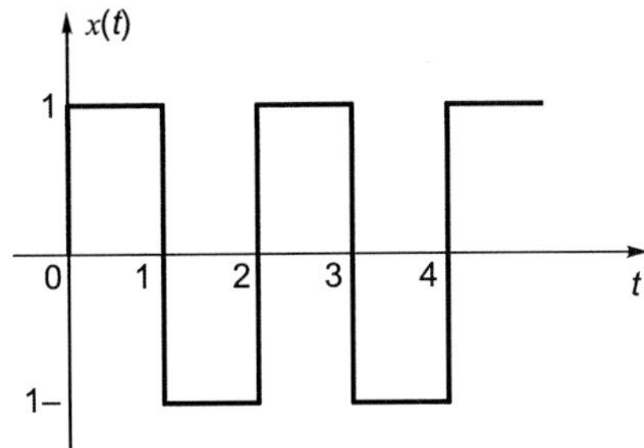

The ratio of the power in the 7th harmonic to the power in the 5th harmonic for this waveform is closest in value to ________.

[2014 : 1 M, Set-2]

3.12 The magnitude and phase of the complex Fourier series coefficient a_k of a periodic signal $x(t)$ are shown in the figure. Choose the correct statement from the four choices given. Notation: C is the set of complex number, R is the set of purely real numbers, and P is the set of purely imaginary numbers.

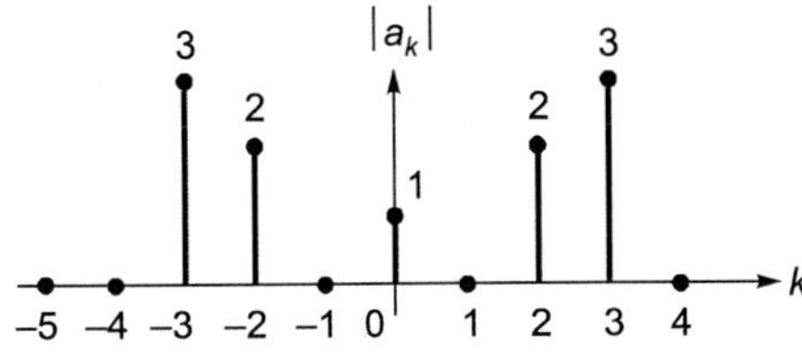

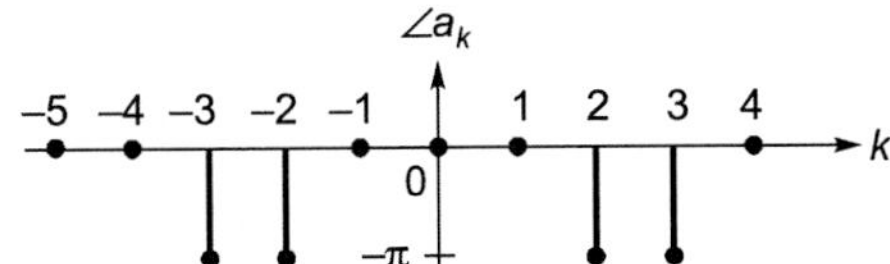

(a) $x(t) \in R$
(b) $x(t) \in P$
(c) $x(t) \in (C\text{–}R)$
(d) the information given is not sufficient to draw any conclusion about $x(t)$

[2015 : 1 M, Set-2]

3.13 A periodic signal $x(t)$ has a trigonometric Fourier series expansion

$$x(t) = a_0 + \sum_{n=1}^{\infty}(a_n \cos n\omega_0 t + b_n \sin n\omega_0 t)$$

If $x(t) = -x(-t) = -x(t - \pi/\omega_0)$, we can conclude that

(a) a_n are zero for all n and b_n are zero for n even
(b) a_n are zero for all n and b_n are zero for n odd
(c) a_n are zero for n even and b_n are zero for n odd
(d) a_n are zero for n odd and b_n are zero for n even

[2017 : 1 M, Set-1]

3.14 Let $x(t)$ be a continuous time periodic signal with fundamental period $T = 1$ seconds. Let $\{a_k\}$ be the complex Fourier series coefficients of $x(t)$, where k is integer valued. Consider the following statements about $x(3t)$:

I. The complex Fourier series coefficients of $x(3t)$ are $\{a_k\}$ where k is integer valued.
II. The complex Fourier series coefficients of $x(3t)$ are $\{3a_k\}$ where k is integer valued.
III. The fundamental angular frequency of $x(3t)$ is 6π rad/s.

For the three statements above, which one of the following is correct?

(a) only II and III are true
(b) only I and III are true
(c) only III is true
(d) only I is true

[2017 : 2 M, Set-1]

3.15 Let $x(t)$ be a periodic function with period $T = 10$. The Fourier series coefficients for this series are denoted by a_k, that is

$$x(t) = \sum_{k=-\infty}^{\infty} a_k e^{jk\frac{2\pi}{T}t}$$

The same function $x(t)$ can also be considered as a periodic function with period $T' = 40$. Let b_k be the Fourier series coefficients when period is taken as T'.

If $\sum_{k=-\infty}^{\infty} |a_k| = 16$, then $\sum_{k=-\infty}^{\infty} |b_k|$ is equal to

(a) 256 (b) 64
(c) 16 (d) 4

[2018 : 1 M]

3.16 The exponential Fourier series representation of a continuous-time periodic signal $x(t)$ is defined as

$$x(t) = \sum_{k=-\infty}^{\infty} a_k e^{jk\omega_0 t}$$

where ω_o is the fundamental angular frequency of $x(t)$ and the coefficients of the series are a_k. The following information is given about $x(t)$ and a_k.

I. $x(t)$ is real and even, having a fundamental period of 6.

II. The average value of $x(t)$ is 2.

III. $a_k = \begin{cases} k, & 1 \le k \le 3 \\ 0, & k > 3 \end{cases}$

The average power of the signal $x(t)$ (rounded off to one decimal place) is _____.

[2021 : 2 M]

3.17 In the table shown below, match the signal type with its spectral characteristics.

Signal Type
(i) Continuous, aperiodic
(ii) Continuous, periodic
(iii) Discrete, aperiodic
(iv) Discrete, periodic

Spectral Characteristics
(a) Continuous, aperiodic
(b) Continuous, periodic
(c) Discrete, aperiodic
(d) Discrete, periodic

(a) (i) → (a), (ii) → (b), (iii) → (c), (iv) → (d)
(b) (i) → (a), (ii) → (c), (iii) → (b), (iv) → (d)
(c) (i) → (d), (ii) → (b), (iii) → (c), (iv) → (a)
(d) (i) → (a), (ii) → (c), (iii) → (d), (iv) → (b)

[2023 : 1 M]

■■■

Answers Fourier Series

3.1	(c)	3.2	(c)	3.3	(c)	3.4	(d)	3.5	(c)	3.6	(b)	3.7	(d)
3.8	(c)	3.9	(a)	3.10	(c)	3.11	(0.51)	3.12	(a)	3.13	(a)	3.14	(b)
3.15	(c)	3.16	(32)	3.17	(b)								

Explanations Fourier Series

3.1 (c)

The trigonometric Fourier series of an even function of time contains only d.c. term and cosine term. The trigonometric Fourier series of an even function of time does not contain sine terms.

3.2 (c)

The trigonometric Fourier series of a periodic time function can have only cosine and sine terms.

3.3 (c)

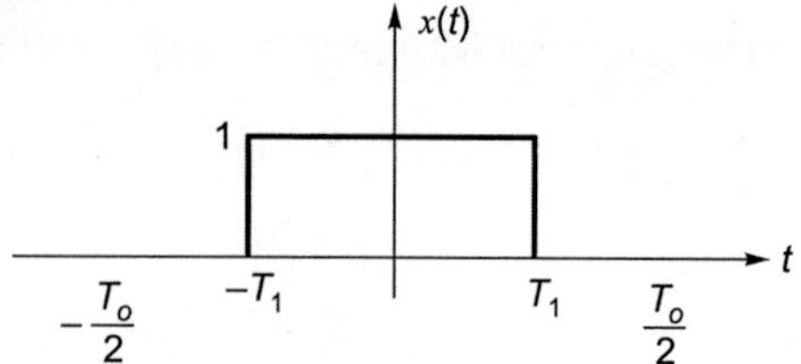

Given periodic signal $x(t)$ has time period T_o

$$x_{dc} = \frac{1}{T_o}\int_{-T_o/2}^{T_o/2} x(t)\,dt = \frac{1}{T_o}\int_{-T_1}^{T_1} 1\,dt = \frac{1}{T_o}[t]_{-T_1}^{T_1}$$

$$x_{dc} = \frac{1}{T_o}[T_1 - (-T_1)] = \frac{2T_1}{T_o}$$

3.4 (d)

$$s(t) = \sum_{K=-\infty}^{\infty} c_k e^{jK\omega_o t}$$

$$c_k = \frac{1}{T_o}\int_{-T_o/2}^{T_o/2} \delta(t) e^{-jK\omega_o t}\,dt = \frac{1}{T_o}$$

$$\therefore \quad s(t) = \frac{1}{T_o}\sum_{K=-\infty}^{\infty} e^{jK\omega_o t}$$

3.5 (c)

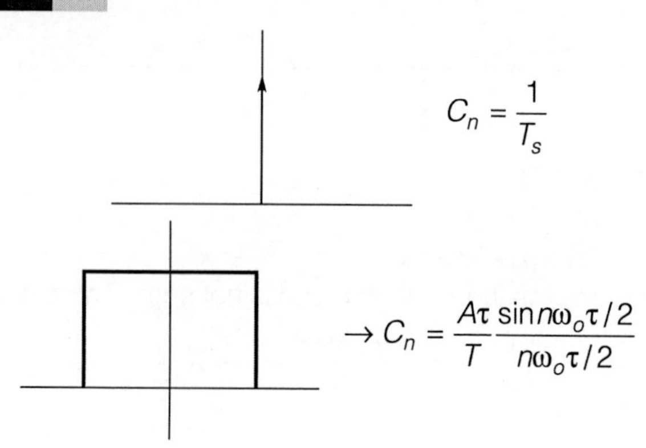

$$= \frac{2A}{Tn\omega_o}\sin\left(n\omega_o\frac{\tau}{2}\right)$$

$$C_n \propto \frac{1}{n}$$

Again integrating

$$C_n \propto \frac{1}{n^2}$$

3.6 (b)

$$x(t) = 2\cos\pi t + 7\cos t$$

$$T_1 = \frac{2\pi}{\omega} = 2$$

$$T_2 = \frac{2\pi}{1} = 2\pi$$

$$\frac{T_1}{T_2} = \frac{1}{\pi} = \text{irrational}$$

$\therefore x(t)$ is not periodic and does not satisfy Dirichlet condition.

3.7 (d)

$$c_{-k} = c_k^*$$

$$c_3 = 3 + j5$$

$$c_{-3} = c_k^* = 3 - j5$$

3.8 (c)

All other functions are either periodic or constant function.

3.9 (a)

The Fourier series of a real periodic function has only cosine terms if it is even and only sine terms if it is odd.

3.10 (c)

Trigonometric Fourier series of an even function has dc and cosine terms only.

3.11 (0.51)

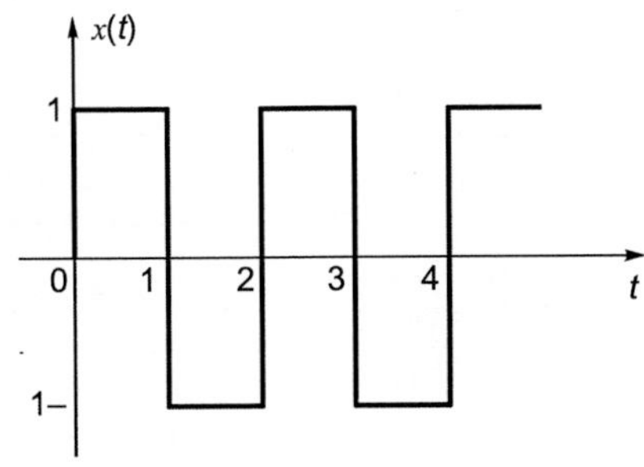

$$X(\omega) = \sum_{K=-\infty}^{\infty} 2\pi\, a_k\, \delta(\omega - K\omega)$$

$$a_K = \frac{1}{2\pi}\int_0^2 x(t)e^{-jK\omega_o t}dt$$

$$= \frac{1}{2\pi}\left\{\int_0^1 1\cdot e^{-jK\omega_o t}d\omega - \int_1^2 1\cdot e^{-jK\omega_o t}d\omega\right\}$$

$$= \frac{1}{2\pi}\left[\frac{1}{(-jK\omega_0)}[e^{(-jK\omega_0)} - 1] - \frac{1}{(-jK\omega_o)}[e^{(-j2K\omega_0)} - e^{(-jK\omega_0)}]\right]$$

$$= \frac{1}{2\pi}\left[\frac{e^{-jK\omega_o/2}}{jK\omega_o}\cdot 2j\sin\frac{K\omega_o}{2} - \frac{1}{jK\omega_o}e^{-\frac{jK\omega_o 3}{2}}\left(2j\sin\frac{K\omega_o}{2}\right)\right]$$

$$= \frac{1}{\pi}\left[\frac{\sin\frac{K\omega_o}{2}}{K\omega_o}(e^{-jK\omega_o})(e^{\frac{jK\omega_o}{2}} - e^{-\frac{jK\omega_o}{2}})\right]$$

$$= \frac{1}{\pi}\left(\frac{\sin K\omega_o/2}{K\omega_o}\right)e^{-jK\omega_o}\times 2j\sin\left(\frac{K\omega_o}{2}\right)$$

$$= \frac{1}{\pi^2}\frac{\left(\sin K\frac{\pi}{2}\right)^2}{K}2j\times\cos K\pi$$

7th harmonic power $= (2\pi\, a_7)^2$

5th harmonic power $= (2\pi\, a_5)^2$

$$\text{Ratio} = \left(\frac{a_7}{a_5}\right)^2 = \left(\frac{5}{7}\right)^2 = \frac{25}{49} = 0.51$$

3.12 (a)

$|a_k|$ – even symmetry

$\angle a_k$ – Odd symmetry $\quad$ (π can be $-\pi$)

$\Rightarrow x(t)$ is real.

3.13 (a)

Signal has odd and half wave symmetries. So, all a_n are zero and b_n are zero for n even.

3.14 (b)

Initially $T = 1$ sec, so $\omega_0 = 2\pi$ rad/sec. When $x(t)$ is compressed by 3, frequency will expand by same factor but there is no change in values of a_k.

So, both statement I and III are correct.

3.15 (c)

Change in only time period or frequency does not change in the value of Fourier series coefficients.

So, $\quad b_k = a_k$

$$\sum_{k=-\infty}^{\infty}|b_k| = \sum_{k=-\infty}^{\infty}|a_k| = 16$$

3.16 (32)

1. $x(t)$ is real and even so a_k is also real and even $a_k = a_{-k}$
2. Average of $x(t)$ is 2 i.e., $a_0 = 2$.
3. $x(t) \rightarrow a_k = k \quad 1 \le k \le 3 \quad a_1 = 1 \quad a_{-1} = 1$
   $\qquad\qquad 0 \qquad k > 3 \quad a_2 = 2 \quad a_{-2} = 2$
   $\qquad\qquad\qquad\qquad a_3 = 3 \quad a_{-3} = 3$
4. $T_0 = 6$

Parseval's Power Theorem

$$\frac{1}{T}\int_0^T \left|x(t)^2\right| dt = \sum_{n=-\infty}^{+\infty} |a_k|^2$$

$$P_{x(t)} = \sum_{n=-\infty}^{+\infty} |a_k|^2$$

$$= |a_{-3}|^2 |a_{-2}|^2 + |a_{-1}|^2 + |a_0|^2 + |a_1|^2 + |a_2|^2 + |a_3|^2$$

$$= 2|a_1|^2 + 2|a_2|^2 + 2|a_3|^2 + |a_0|^2$$

$$= 2 \times 1^2 + 2 \times (2)^2 + 2(3)^2 + (2)^2$$

$$= 2 + 8 + 18 + 4$$

$$P_{x(t)} = 32$$

3.17 (b)

Signal Types	Spectral Characteristics
• Continuous and aperiodic	• Aperiodic and continuous
• Continuous and periodic	• Aperiodic and discrete
• Discrete and aperiodic	• Periodic and continuous
• Discrete and periodic	• Periodic and discrete

4 Fourier Transforms, Frequency Response and Correlation

4.1 Match each of the items, *A*, *B* and *C* with an appropriate item from 1, 2, 3, 4 and 5.

List-I

A. Fourier transform of a Gaussian function
B. Convolution of a rectangular pulse with itself
C. Current through an inductor for a step input voltage

List-II

1. Gaussian function
2. Rectangular pulse
3. Triangular pulse
4. Ramp function
5. Zero

[1995 : 2 M]

4.2 The Fourier transform of a real valued time signal has
(a) odd symmetry (b) even symmetry
(c) conjugate symmetry (d) no symmetry

[1996 : 1 M]

4.3 The function $f(t)$ has the Fourier transform $g(\omega)$.

The Fourier transform of $g(t)=\left(\int_{-\infty}^{\infty} g(t)e^{-j\omega t}dt\right)$ is

(a) $\frac{1}{2\pi}f(\omega)$ (b) $\frac{1}{2\pi}f(-\omega)$
(c) $2\pi f(-\omega)$ (d) none of these

[1997 : 1 M]

4.4 If the Fourier transform of a deterministic signal $g(t)$ is $G(f)$, then
(1) The Fourier transform of $g(t-2)$ is
(2) The Fourier transform of $g(t/2)$ is
(a) $G(f)e^{-j(4\pi f)}$
(b) $G(2f)$
(c) $2G(2f)$
(d) $G(f-2)$

Match each of the items 1, 2 on the left with the most appropriate item *A*, *B*, *C*, or *D* on the right.

[1997 : 2 M]

4.5 The amplitude spectrum of a Gaussian pulse is
(a) uniform (b) a sine function
(c) Gaussian (d) an impulse function

[1998 : 1 M]

4.6 TheFourier Transform of a function $x(t)$ is $X(f)$. The Fourier transform of $\frac{dx(t)}{dt}$ will be

(a) $\frac{dX(f)}{dt}$ (b) $j2\pi fX(f)$
(c) $jfX(f)$ (d) $\frac{X(f)}{jf}$

[1998 : 1 M]

4.7 The Fourier transform of a voltage signal $x(t)$ is $X(f)$. The unit of $|X(f)|$ is
(a) Volt (b) Volt-sec
(c) Volt/sec (d) Volt^2

[1998 : 1 M]

4.8 A signal $x(t)$ has a Fourier transform $X(\omega)$. If $x(t)$ is a real and odd function of *t*, then $X(\omega)$ is
(a) a real and even function of ω
(b) an imaginary and odd function of ω
(c) an imaginary and even function of ω
(d) a real and odd function of ω

[1999 : 1 M]

4.9 The input to a channel is a bandpass signal. It is obtained by linearly modulating a sinusoidal carrier with a single-tone signal. The output of the channel due to this input is given by

$y(t)=(1/100)\cos(100t-10^{-6})\cos(10^{6}t-1.56)$

The group delay (t_g) and the phase delay (t_p) in seconds, of the channel are
(a) $t_g = 10^{-6}$, $t_p = 1.56$
(b) $t_g = 1.56$, $t_p = 10^{-6}$
(c) $t_g = 10^{-8}$, $t_p = 1.56 \times 10^{-6}$
(d) $t_g = 10^{8}$, $t_p = 1.56$

[1999 : 1 M]

4.10 The Fourier Transform of the signal $x(t) = e^{-3t^2}$ is of the following form, where *A* and *B* are constants:

(a) Ae^{Bf^2} (b) Ae^{-Bf^2}
(c) $A+B|f|^2$ (d) Ae^{-Bf}

[2000 : 1 M]

4.11 A linear phase channel with phase delay T_p and group delay T_g must have
(a) $T_p = T_g =$ constant
(b) $T_p \propto f$ and $T_g \propto f$
(c) $T_p =$ constant and $T_g \propto f$
(d) $T_p \propto f$ and $t_g =$ constant
(f denotes frequency)

[2002 : 1 M]

4.12 The Fourier transform $F\{e^{-t}u(t)\}$ is equal to $\frac{1}{1+j2\pi f}$. Therefore, $F\left\{\frac{1}{1+j2\pi t}\right\}$ is

(a) $e^{f}u(f)$ (b) $e^{-f}u(f)$
(c) $e^{f}u(-f)$ (d) $e^{-f}u(-f)$

[2002 : 1 M]

4.13 Let $x(t)$ be the input to a linear, time-invariant system. The required output is $4x(t-2)$. The transfer function of the system should be

(a) $4e^{j4\pi f}$ (b) $2e^{-j8\pi f}$
(c) $4e^{-j4\pi f}$ (d) $2e^{j8\pi f}$

[2003 : 1 M]

Data given below for two questions. Solve the problem and choose the correct answer.

The system under consideration is an RC low-pass filter (RC-LPF) with $R = 1.0\ \text{k}\Omega$ and $C = 1.0\ \mu\text{F}$

4.14 Let $H(f)$ denote the frequency response of the RC-LPF. Let f_1 be the highest frequency such that $0 \le |f| \le f_1; \frac{|H(f_1)|}{H(0)} \ge 0.95$. Then f_1 (in Hz) is

(a) 327.8 (b) 163.9
(c) 52.2 (d) 104.4

[2003 : 2 M]

4.15 Let $t_g(f)$ be the group delay function of the given RC-LPF and $f_2 = 100$ Hz. Then $t_g(f_2)$ in ms, is

(a) 0.717 (b) 7.17
(c) 71.7 (d) 4.505

[2003 : 2 M]

4.16 The Fourier transform of a conjugate symmetric function is always

(a) imaginary
(b) conjugate antisymmetric
(c) real
(d) conjugate symmetric

[2004 : 1 M]

4.17 A rectangular pulse train $s(t)$ as shown in the figure is convolved with the signal $\cos^2(4\pi \times 10^3 t)$. The convolved signal will be a

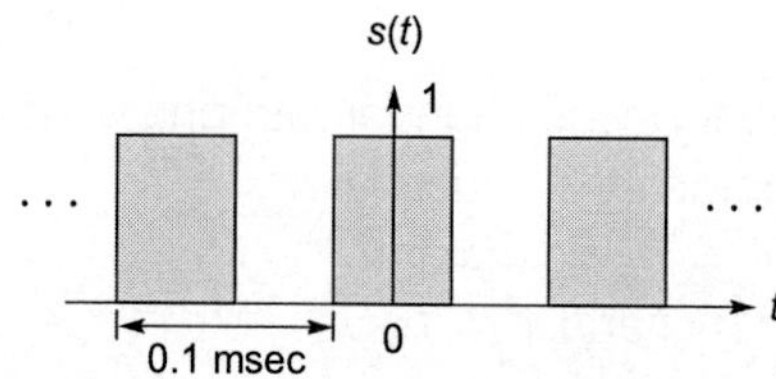

(a) DC (b) 12 kHz sinusoid
(c) 8 kHz sinusoid (d) 14 kHz sinusoid

[2004 : 2 M]

4.18 Let $x(t)$ and $y(t)$ (with Fourier transforms $X(f)$ and $Y(f)$ respectively) be related as shown in the figure. Then $Y(f)$ is

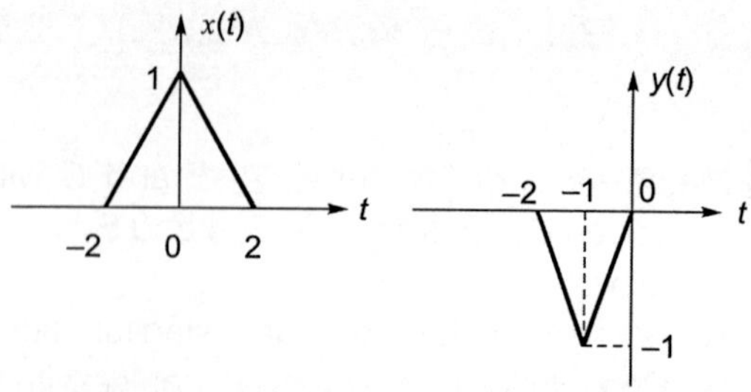

(a) $-\frac{1}{2}X(f/2)e^{-j2\pi f}$ (b) $-\frac{1}{2}X(f/2)e^{j2\pi f}$
(c) $-X(f/2)e^{j2\pi f}$ (d) $-X(f/2)e^{-j2\pi f}$

[2004 : 2 M]

4.19 For a signal $x(t)$ the Fourier transform is $X(f)$. Then the inverse Fourier transform of $X(3f+2)$ is given by

(a) $\frac{1}{2}x\left(\frac{t}{2}\right)e^{j3\pi t}$ (b) $\frac{1}{3}x\left(\frac{t}{3}\right)e^{-j4\pi t/3}$
(c) $3x(3t)e^{-j4\pi t}$ (d) $x(3t+2)$

[2005 : 2 M]

4.20 The output $y(t)$ of a linear time invariant system is related to its input $x(t)$ by the following equation:

$y(t) = 0.5x(t - t_d + T) + x(t - t_d) + 0.5x(t - t_d - T)$

The filter transfer function $H(\omega)$ of such a system is given by

(a) $(1+\cos\omega T)e^{-j\omega t_d}$
(b) $(1+0.5\cos\omega T)e^{-j\omega t_d}$
(c) $(1-\cos\omega T)e^{-j\omega t_d}$
(d) $(1-0.5\cos\omega T)e^{-j\omega t_d}$

[2005 : 2 M]

4.21 Let $x(t) \leftrightarrow X(j\omega)$ be Fourier Transform pair. The Fourier Transform of the signal $x(5t-3)$ in terms of $X(j\omega)$ is given as

(a) $\frac{1}{5}e^{-\frac{j3\omega}{5}}X\left(\frac{j\omega}{5}\right)$ (b) $\frac{1}{5}e^{\frac{j3\omega}{5}}X\left(\frac{j\omega}{5}\right)$
(c) $\frac{1}{5}e^{-j3\omega}X\left(\frac{j\omega}{5}\right)$ (d) $\frac{1}{5}e^{j3\omega}X\left(\frac{j\omega}{5}\right)$

[2006 : 1 M]

4.22 The 3-dB bandwidth of the low-pass signal $e^{-t}u(t)$, where $u(t)$ is the unit step function, is given by

(a) $\frac{1}{2\pi}$ Hz (b) $\frac{1}{2\pi}\sqrt{\sqrt{2}-1}$ Hz
(c) ∞ (d) 1 Hz

[2007 : 2 M]

4.23 The signal $x(t)$ is described by

$$x(t) = \begin{cases} 1 \text{ for } -1 \le t \le +1 \\ 0 \text{ otherwise} \end{cases}$$

Two of the angular frequencies at which its Fourier transform becomes zero are

(a) $\pi, 2\pi$ (b) $0.5\pi, 1.5\pi$
(c) $0, \pi$ (d) $2\pi, 2.5\pi$

[2008 : 2 M]

Statement for Linked Answer Questions 4.24 and 4.25:
The impulse response $h(t)$ of a linear time-invariant continuous time system is given by $h(t) = \exp(-2t)u(t)$, where $u(t)$ denotes the unit step function.

4.24 The frequency response $H(\omega)$ of this system in terms of angular frequency ω, is given by, $H(\omega) =$

(a) $\dfrac{1}{1+j2\omega}$ (b) $\dfrac{sin(\omega)}{\omega}$
(c) $\dfrac{1}{2+j\omega}$ (d) $\dfrac{j\omega}{2+j\omega}$

[2008 : 2 M]

4.25 The output of this system, to the sinusoidal input $x(t) = 2\cos(2t)$ for all time t, is

(a) 0
(b) $2^{-0.25}\cos(2t-0.125\pi)$
(c) $2^{-0.5}\cos(2t-0.125\pi)$
(d) $2^{-0.5}\cos(2t-0.25\pi)$

[2008 : 2 M]

4.26 A function is given by $f(t) = \sin^2 t + \cos 2t$. Which of the following is true?

(a) f has frequency components at 0 and $1/2\pi$ Hz
(b) f has frequency components at 0 and $1/\pi$ Hz
(c) f has frequency components at $1/2\pi$ and $1/\pi$ Hz
(d) f has frequency components at 0, $1/2\pi$ and $1/\pi$ Hz

[2009 : 1 M]

4.27 Consider a system whose input x and output y are related by the equation

$$y(t) = \int_{-\infty}^{\infty} x(t-\tau)h(2\tau)d\tau$$

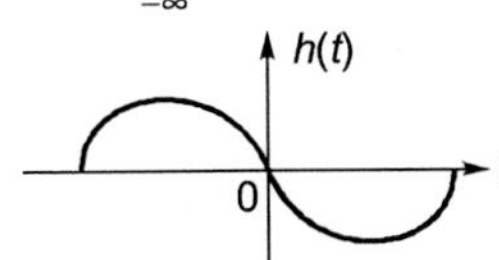

where $h(t)$ is shown in the graph
Which of the following four properties are possessed by the system? BIBO: Bounded input gives a bounded output
Causal: The system is causal
LP : The system is low pass
LTI : The system is linear and time-invariant

(a) Causal, LP (b) BIBO, LTI
(c) BIBO, Causal, LTI (d) LP, LTI

[2009 : 2 M]

4.28 The Fourier transform of a signal $h(t)$ is $H(j\omega) = (2\cos\omega)(\sin 2\omega)/\omega$
The value of $h(0)$ is

(a) 1/4 (b) 1/2
(c) 1 (d) 2

[2012 : 2 M]

4.29 A continuous, linear time-invariant filter has an impulse response $h(t)$ described by

$$h(t) = \begin{cases} 3 & \text{for } 0 \le t \le 3 \\ 0 & \text{otherwise} \end{cases}$$

When a constant input of value 5 is applied to this filter, the steady state output is ____.

[2014 : 1 M, Set-1]

4.30 For a function $g(t)$, it is given that

$\int_{-\infty}^{+\infty} g(t)e^{-j\omega t}dt = \omega e^{-2\omega^2}$ for any real value ω. If

$y(t) = \int_{-\infty}^{t} g(\tau)d\tau$, then $\int_{-\infty}^{+\infty} y(t)dt$ is ____.

(a) 0 (b) $-j$
(c) $-\dfrac{j}{2}$ (d) $\dfrac{j}{2}$

[2014 : 2 M, Set-1]

4.31 The value of the integral $\int_{-\infty}^{\infty} \sin c^2(5t)dt$ is ____.

[2014 : 2 M, Set-2]

4.32 The phase response of a passband waveform at the receiver is given by

$$\varphi(f) = -2\pi\alpha(f-f_c) - 2\alpha\beta f_c$$

where f_c is the centre frequency, and α and β are positive constants. The actual signal propagation delay from the transmitter to receiver is

(a) $\dfrac{\alpha-\beta}{\alpha+\beta}$ (b) $\dfrac{\alpha\beta}{\alpha+\beta}$
(c) α (d) β

[2014 : 1 M, Set-3]

4.33 A real-valued signal $x(t)$ limited to the frequency band $|f| \le \dfrac{W}{2}$ is passed through a linear time invariant system whose frequency response is

$$H(f) = \begin{cases} e^{-j4\pi f}, & |f| \le \dfrac{W}{2} \\ 0, & |f| > \dfrac{W}{2} \end{cases}$$

The output of the system is

(a) $x(t+4)$ (b) $x(t-4)$

(c) $x(t+2)$ (d) $x(t-2)$

[2014 : 1 M, Set-4]

4.34 The energy of the signal $x(t) = \dfrac{\sin(4\pi t)}{4\pi t}$ is _____.

[2016 : 1 M, Set-2]

4.35 A network consisting of a finite number of linear resistor (R), inductor (L), and capacitor (C) elements, connected all in series or all in parallel, is excited with a source of the form

$\sum_{k=1}^{3} a_k \cos(k\omega_0 t)$, where $a_k \neq 0$, $\omega_0 \neq 0$.

The source has nonzero impedance. Which one of the following is a possible form of the output measured across a resistor in the network?

(a) $\sum_{k=1}^{3} b_k \cos(k\omega_0 t + \phi_k)$, where $b_k \neq a_k, \forall k$

(b) $\sum_{k=1}^{3} b_k \cos(k\omega_0 t + \phi_k)$, where $b_k \neq 0, \forall k$

(c) $\sum_{k=1}^{3} a_k \cos(k\omega_0 t + \phi_k)$

(d) $\sum_{k=1}^{2} a_k \cos(k\omega_0 t + \phi_k)$

[2016 : 2 M, Set-1]

4.36 If the signal $x(t) = \dfrac{\sin(t)}{\pi t} * \dfrac{\sin(t)}{\pi t}$ with $*$ denoting the convolution operation, then $x(t)$ is equal to

(a) $\dfrac{\sin(t)}{\pi t}$ (b) $\dfrac{\sin(2t)}{2\pi t}$

(c) $\dfrac{2\sin(t)}{\pi t}$ (d) $\left(\dfrac{\sin(t)}{\pi t}\right)^2$

[2016 : 1 M, Set-3]

4.37 A continuous time signal $x(t) = 4\cos(200\pi t) + 8\cos(400\pi t)$, where t is in seconds, is the input to a linear time invariant (LTI) filter with the impulse response

$$h(t) = \begin{cases} \dfrac{2\sin(300\pi t)}{\pi t}, & t \neq 0 \\ 600, & t = 0 \end{cases}$$

Let $y(t)$ be the output of this filter. The maximum value of $|y(t)|$ is _________.

[2017 : 2 M, Set-1]

4.38 Consider an LTI system with magnitude response

$$|H(f)| = \begin{cases} 1 - \dfrac{|f|}{20}, & |f| \leq 20 \\ 0, & |f| > 20 \end{cases}$$

and phase response

$$\arg\{H(f)\} = -2f$$

If the input to the system is

$$x(t) = 8\cos\left(20\pi t + \frac{\pi}{4}\right) + 16\sin\left(40\pi t + \frac{\pi}{8}\right) + 24\cos\left(80\pi t + \frac{\pi}{16}\right)$$

then the average power of the output signal $y(t)$ is _____.

[2017 : 2 M, Set-2]

4.39 The input $4\text{sinc}(2t)$ is fed to a Hilbert transformer to obtain $y(t)$, as shown in the figure below:

4sinc(2t) → [Hilbert Transform] → y(t)

Here $\text{sinc}(x) = \dfrac{\sin(\pi x)}{\pi x}$. The value (accurate to two decimal places) of $\int_{-\infty}^{\infty} |y(t)|^2 dt$ is _______.

[2018 : 2 M]

4.40 $X(\omega)$ is the Fourier transform of $x(t)$ shown below. The value of $\int_{-\infty}^{\infty} |X(\omega)|^2 d\omega$ (rounded off to two decimal places) is _____.

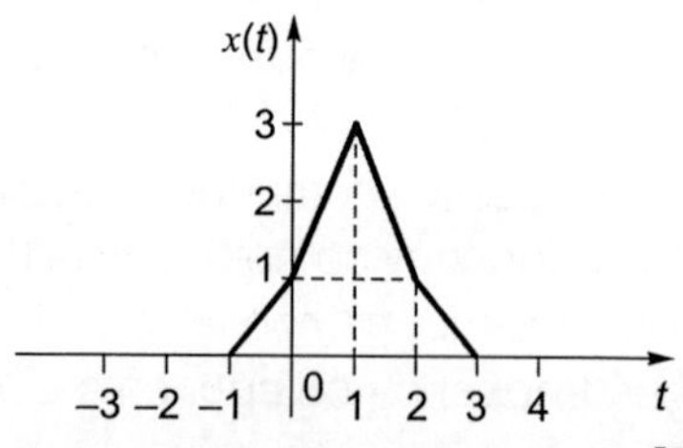

[2020 : 2 M]

4.41 The Fourier transform $X(j\omega)$ of the signal

$$x(t) = \frac{t}{(1+t^2)^2}$$

is _______.

(a) $\dfrac{\pi}{2j}\omega e^{-|\omega|}$ (b) $\dfrac{\pi}{2}\omega e^{-|\omega|}$

(c) $\dfrac{\pi}{2j} e^{-|\omega|}$ (d) $\dfrac{\pi}{2} e^{-|\omega|}$

[2022 : 1 M]

4.42 Let $m(t)$ be a strictly band-limited signal with bandwidth B and energy E. Assuming $\omega_0 = 10B$, the energy in the signal $m(t)\cos\omega_0 t$ is

(a) $\dfrac{E}{4}$ (b) $\dfrac{E}{2}$

(c) E (d) $2E$ [2023 : 1 M]

4.43 Let $x(t) = 10\cos(10.5Wt)$ be passed through an LTI system having impulse response $h(t) = \pi\left(\frac{\sin Wt}{\pi t}\right)^2 \cos 10Wt$. The output of the system is

(a) $\left(\frac{15W}{4}\right)\cos(10.5Wt)$

(b) $\left(\frac{15W}{2}\right)\cos(10.5Wt)$

(c) $\left(\frac{15W}{8}\right)\cos(10.5Wt)$

(d) $(15W)\cos(10.5Wt)$ [2023 : 2 M]

4.44 Let $x_1(t) = u(t+1.5) - u(t-1.5)$ and $x_2(t)$ is shown in the figure below. For $y(t) = x_1(t) * x_2(t)$, the $\int_{-\infty}^{\infty} y(t)\,dt$ is ________.

(rounded off to the nearest integer).

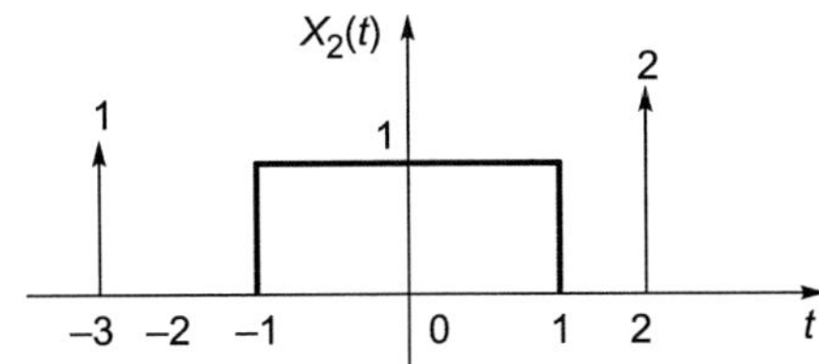

[2023 : 2 M]

4.45 Consider two continuous time signals $x(t)$ and $y(t)$ as shown below

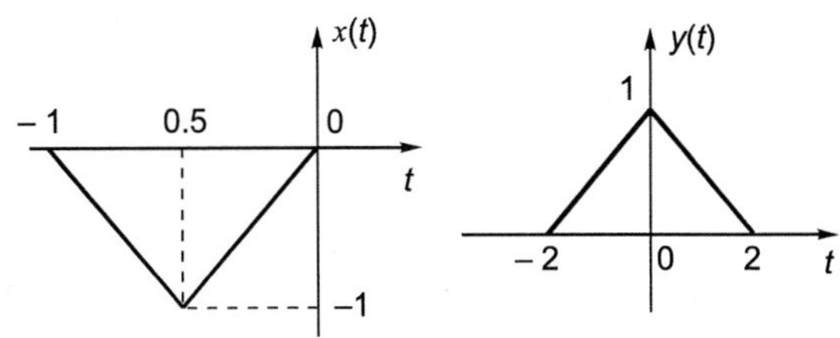

If $X(f)$ denotes the Fourier transform of $x(t)$, then the Fourier transform of $y(t)$ is ________.

(a) $-\frac{1}{4}X(f/4)e^{-j\pi f}$ (b) $-4X(4f)e^{-j\pi f}$

(c) $-4X(4f)e^{-j4\pi f}$ (d) $-\frac{1}{4}X(f/4)e^{-j4\pi f}$

[2024 : 2 Marks]

Answers Fourier Transforms , Frequency Response and Correlation

4.1	(A- 1, B-3, C-4)			4.2	(c)	4.3	(c)	4.4	(sol.)	4.5	(c)
4.6	(b)	4.7	(b)	4.8	(b)	4.9	(c)	4.10	(b)	4.11	(a)
4.12	(c)	4.13	(c)	4.14	(c)	4.15	(a)	4.16	(c)	4.17	(a)
4.18	(b)	4.19	(b)	4.20	(a)	4.21	(a)	4.22	(a)	4.23	(a)
4.24	(c)	4.25	(d)	4.26	(b)	4.27	(b)	4.28	(c)	4.29	(45)
4.30	(b)	4.31	(0.2)	4.32	(c)	4.33	(d)	4.34	(0.25)	4.35	(a)
4.36	(a)	4.37	(8)	4.38	(8)	4.39	(8)	4.40	(58.61)	4.41	(a)
4.42	(b)	4.43	(a)	4.44	(15)	4.45	(c)				

Explanations Fourier Transforms , Frequency Response and Correlation

4.1 Sol.

(A) - 1, (B) - 3, (C) - 4

(A) Fourier transform of a Gaussian function is also a Gaussian function

$$e^{-\pi t^2} \xleftrightarrow{F.T.} e^{-\pi f^2}$$

(B) Convolution of a rectangular pulse with itself is a triangular pulse.

(C) Current through an inductor for a step input voltage is ramp function

$$i = \frac{1}{L}\int V dt = \frac{1}{L}\int u(t)dt = \frac{1}{L}r(t)$$

4.2 (c)

Fourier transform of real valued signal has conjugate symmetry.

$$[X(j\omega)]^* = \left[\int_{-\infty}^{\infty} x(t)e^{-j\omega t}dt\right]^*$$

$$X^*(j\omega) = \int_{-\infty}^{\infty} x(t)e^{-j(-\omega)t}dt$$

$$X^*(j\omega) = X(-j\omega)$$

4.3 (c)

According to the duality property of Fourier transform

$$f(t) \xleftrightarrow{F.T.} g(\omega)$$

$$g(t) \xleftrightarrow{F.T.} 2\pi f(-\omega)$$

4.4 Sol.

1-A, 2-C

$$g(t) \xleftrightarrow{F.T.} G(f)$$

$$g(t-2) \xleftrightarrow{F.T.} e^{-j2\pi 2f}G(f) = G(f)e^{-j(4\pi f)}$$

$$g\left(\frac{t}{2}\right) \xleftrightarrow{F.T} \left(\frac{1}{1/2}\right)G\left(\frac{f}{1/2}\right) = 2G(2f)$$

4.5 (c)

A normalized Gaussian pulse is defined as

$$x(t) = e^{-\pi t^2}$$

$$x(t) = e^{-\pi t^2} \xleftrightarrow{F.T.} e^{-\pi f^2} = X(f)$$

So, $X(f) = e^{-\pi f^2}$ is also a Gaussian pulse in frequency domain.

4.6 (b)

$$x(t) \xleftrightarrow{F.T.} X(f)$$

$$\frac{d}{dt}x(t) \xleftrightarrow{F.T.} (j2\pi f)X(f)$$

4.7 (b)

By definition Fourier transform

$$X(f) = \int_{-\infty}^{\infty} x(t)\,e^{-j2\pi ft}dt$$

$x(t) \Rightarrow$ volts

So, $X(f) \Rightarrow$ Volt-sec

4.8 (b)

Example: $A\sin\omega_0 t = x(t)$

(Real and Odd function)

$$X(\omega) = Aj\pi\left[\delta(\omega+\omega_o) - \delta(\omega-\omega_o)\right]$$

(imaginary and odd function of ω)

4.9 (c)

$$y(t) = \left(\frac{1}{100}\right)\cos(100t - 10^{-6})$$

$$\cos(10^6 t - 1.56)$$

Comparing with

$$y(t) = \cos k(t-t_g)\cos[\omega_c(t-t_p)]$$

$$y(t) = \left(\frac{1}{100}\right)\cos[100(t-10^{-8})]$$

$$\cos\left[10^6\left(t-1.56\times10^{-6}\right)\right]$$

$$\therefore\ t_p = 1.56\times10^{-6}\text{ sec}$$

$$t_g = 10^{-8}\text{ sec}$$

4.10 (b)

Normalized Gaussian function has Gaussian Fourier Transform.

4.11 (a)

$$\theta(\omega) = -\omega t_0$$

$$t_p = \frac{-\theta(\omega)}{\omega} = t_o$$

$$t_g = -\frac{d\theta(\omega)}{d\omega} = t_o$$

$$t_p = t_g = t_0 = \text{constant}$$

4.12 (c)

$$f(t) = e^{-t}u(t)$$

$$F(f) = \frac{1}{1+j2\pi f}$$

From duality, $F(t) \xleftrightarrow{FT} f(-f)$

$$\Rightarrow\quad F(t) = \frac{1}{1+j2\pi t}$$

$$f(-f) = e^{f}u(-f)$$

$$F(t) \xleftrightarrow{FT} e^{f}\,u(-f)$$

4.13 (c)

$$y(t) = 4x(t-2)$$

$$Y(s) = 4e^{-2s}X(s)$$

$$\frac{Y(s)}{X(s)} = 4e^{-2s}$$

$$H(f) = 4e^{-2\times j.2\pi f} = 4e^{-j4\pi f}$$

4.14 (c)

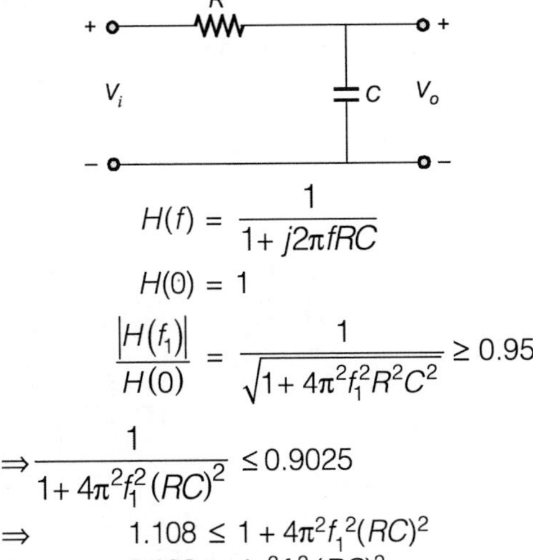

$$H(f) = \frac{1}{1+j2\pi fRC}$$

$$H(0) = 1$$

$$\frac{|H(f_1)|}{H(0)} = \frac{1}{\sqrt{1+4\pi^2f_1^2R^2C^2}} \ge 0.95$$

$$\Rightarrow \frac{1}{1+4\pi^2f_1^2(RC)^2} \le 0.9025$$

$$\Rightarrow\quad 1.108 \le 1+4\pi^2f_1^2(RC)^2$$

$$\Rightarrow\quad 0.108 \ge 4\pi^2f_1^2(RC)^2$$

$$f_1^2 \le \frac{0.108}{4\pi^2(RC)^2}$$

$$f_1 \le \frac{0.329}{2\pi\times10^{-3}}$$

$$\Rightarrow\quad f_{1\,max} = 52.2\text{ Hz}$$

4.15 (a)

$$H(\omega) = \frac{1}{1+j\omega RC}$$

$$\theta(\omega) = -\tan^{-1}RC\omega$$

$$t_g = \frac{10^{-3}}{1+10^{-6}\times4\pi^2\times10^4} = 0.717\text{ ms}$$

4.17 (a)

$$T_0 = 0.1\times10^{-3} = 10^{-4}$$

$$f_0 = \frac{1}{T_o} = 10^4 = 10\text{ kHz}$$

$\cos^2(4\pi\times10^3 t)$ has frequency = 4 kHz

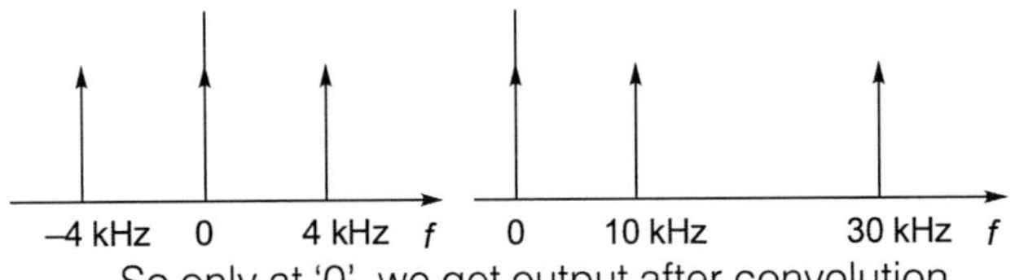

So only at '0', we get output after convolution. (Only odd harmonics are present.

0 f

constant in time domain.

4.18 (b)

$$y(t) = -x[2(t+1)]$$

$$x(t-t_0) \leftrightarrow X(f)e^{-j2\pi ft_o}$$

$$t_0 = -1$$

$$x(at) = \frac{1}{|a|}X\left(\frac{f}{a}\right)$$

$$a = -2$$

$$\therefore\quad Y(f) = -\frac{1}{2}X\left(\frac{f}{2}\right)e^{j2\pi f}$$

4.19 (b)

$$X\left[3\left(f+\frac{2}{3}\right)\right] = \frac{1}{3}x\left(\frac{t}{3}\right)e^{-j4\pi t/3}$$

Applying scaling & shifting property.

4.20 (a)

$$y(t) = 0.5x(t-t_d+T)+0.5x(t-t_d-T)+x(t-t_d)$$

Taking Fourier transform

$$Y(\omega) = \left[0.5e^{j\omega(-t_d+T)}+0.5e^{j\omega(-t_d-T)}+e^{-j\omega t_d}\right]X(\omega)$$

$$\frac{Y(\omega)}{X(\omega)} = e^{-j\omega t_d}\left[0.5e^{j\omega T} + 0.5e^{-j\omega T} + 1\right]$$

$$\frac{Y(\omega)}{X(\omega)} = e^{-j\omega t_d}\left[\cos\omega T + 1\right]$$

$$H(\omega) = (1+\cos\omega T)e^{-j\omega t_d}$$

4.21 (a)

$$x(t) \leftrightarrow X(j\omega)$$

$$x\left[5\left(t-\frac{3}{5}\right)\right] = \frac{1}{5}X\left(\frac{j\omega}{5}\right).e^{-\frac{j3\omega}{5}}$$

Using scaling and shifting property.

4.22 (a)

Laplace transform of $e^{-t}\,u(t) = \dfrac{1}{s+1}$

$\because$ magnitude at 3dB frequency $= \dfrac{1}{\sqrt{2}}$

$$\Rightarrow \quad \frac{1}{\sqrt{2}} = \frac{1}{\sqrt{1+\omega^2}}$$

$$\omega = 1 \text{ rad}$$

$$\therefore \quad f = \frac{1}{2\pi}\text{ Hz.}$$

4.23 (a)

$$X(\omega) = \int_{-\infty}^{\infty} x(t)\,e^{-j\omega t}dt$$

$$= \int_{-1}^{1} e^{-j\omega t}\,dt = \left.\frac{e^{-j\omega t}}{-j\omega}\right|_{-1}^{1}$$

$$= \frac{e^{-j\omega}}{-j\omega} - \frac{e^{j\omega}}{-j\omega} = \frac{1}{j\omega}\left(e^{j\omega} - e^{-j\omega}\right)$$

$$X(\omega) = 0$$

$$\Rightarrow e^{j\omega} - e^{-j\omega} = 0$$

$$\Rightarrow e^{j\omega} - \frac{1}{e^{j\omega}} = 0$$

$$\Rightarrow \quad e^{2j\omega} - 1 = 0$$

$$\Rightarrow \quad e^{2j\omega} = 1$$

$$\Rightarrow \quad e^{j\omega} = \pm 1$$

$$\Rightarrow \quad \omega = \pi, 2\pi$$

4.24 (c)

$$h(t) = \exp(-2t)\,u(t)$$

$$H(\omega) = \int_{-\infty}^{\infty} h(t)e^{-j\omega t}dt$$

$$= \int_{0}^{\infty} e^{-2t}\,e^{-j\omega t}\,dt = \int_{0}^{\infty} e^{-(2+j\omega)t}dt$$

$$= -\frac{1}{2+j\omega}\cdot\left.e^{-(2+j\omega)t}\right|_{0}^{\infty} = \frac{1}{2+j\omega}$$

4.25 (d)

Input $\quad x(t) = 2\cos(2t)$

Frequency response

$$X(\omega) = 2\pi\left[\delta(\omega-2) + \delta(\omega+2)\right]$$

$$H(\omega) = \frac{1}{2+j\omega}$$

Output $\;Y(\omega) = H(\omega)\,X(\omega)$

$$= \frac{1}{2+j\omega}.2\pi\left[\delta(\omega-2)+\delta(\omega+2)\right]$$

$$= \frac{2\pi}{2+j2}\delta(\omega-2) + \frac{2\pi}{2-j2}\delta(\omega+2)$$

$$= \frac{2\pi}{8}\left[(2-j2)\,\delta(\omega-2) + (2+j2)\,\delta(\omega+2)\right]$$

$$= \frac{\pi}{4}\left[2\{\delta(\omega-2)+\delta(\omega+2)\} + j2\{\delta(\omega+2)-\delta(\omega-2)\}\right]$$

$$= \frac{\pi}{2}\left[\delta(\omega-2)+\delta(\omega+2)\right] - j\frac{\pi}{2}\left[\delta(\omega-2)-\delta(\omega+2)\right]$$

$$\therefore \quad y(t) = \frac{\cos 2t}{2} + \frac{\sin 2t}{2}$$

$$= \frac{\sqrt{2}}{2}\left[\frac{1}{\sqrt{2}}\cos 2t + \frac{1}{\sqrt{2}}\sin 2t\right]$$

$$= \frac{1}{\sqrt{2}}\cos(2t - 0.25\pi)$$

$$= 2^{-0.5}\cos(2t - 0.25\pi)$$

4.26 (b)

$$f(t) = \frac{1}{2}(1-\cos 2t) + \cos 2t$$

frequency components are

$$f_1 = 0$$

$$f_2 = \frac{\omega_2}{2\pi} = \frac{2}{2\pi} = \frac{1}{\pi}\text{Hz}$$

4.27 (b)

$$y(t) = \int_{-\infty}^{\infty} x(t-\tau)\,h(2\tau)\,d\tau$$

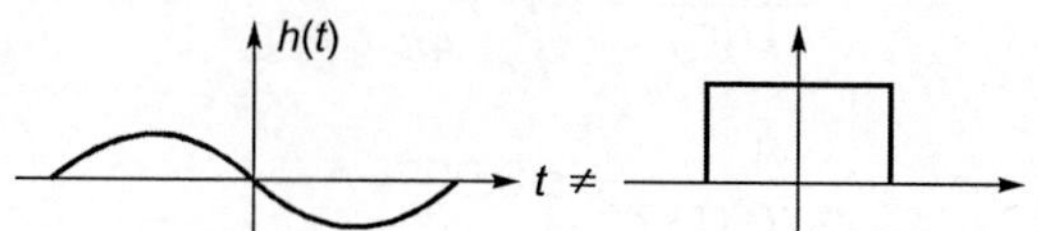

It is not low pass filter.
But the system is LTI and BIBO.

4.28 (c)

$$\frac{2\cos\omega\sin 2\omega}{\omega} = H(j\omega) \qquad ...(1)$$

$$H(j\omega) = 2\cos\omega\left(\frac{2\sin 2\omega}{2\omega}\right) = H_1(j\omega).H_2(j\omega) \qquad ...(2)$$

where, $H_1(j\omega) = 2\cos\omega$

$$H_2(j\omega) = \frac{2\sin 2\omega}{2\omega} = 2\,\text{Sa}(2\omega)$$

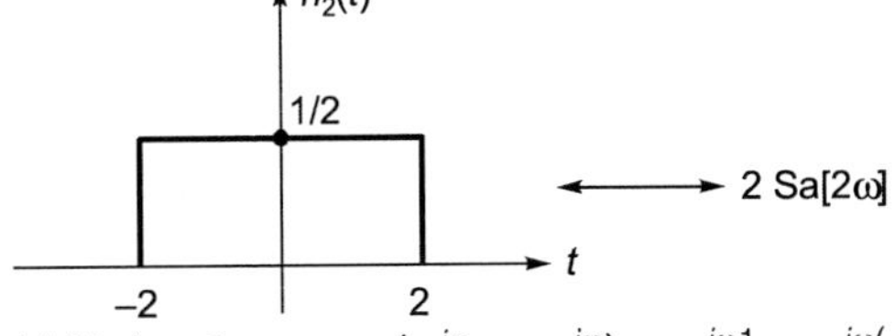

$H_1(j\omega) = 2\cos\omega = (e^{j\omega} + e^{-j\omega}) = e^{j\omega 1} + e^{j\omega(-1)}$

So

$H(j\omega) = 2\cos\omega\, H_2(j\omega) = (e^{j\omega} + e^{-j\omega})\, H_2(j\omega)$

$H(j\omega) = e^{j\omega} H_2(j\omega) + e^{-j\omega} H_2(j\omega)$

$\Rightarrow \quad h(t) = h_2(t+1) + h_2(t-1)$

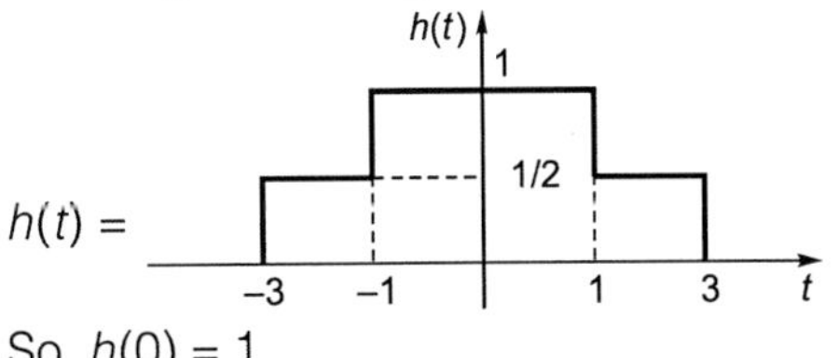

So, $h(0) = 1$

4.29 (45)

$$h(t) = \begin{cases} 3, & \text{for } 0 \le t \le 3 \\ 0, & \text{otherwise} \end{cases}$$

$$h(t) = 3[u(t) - u(t-3)]$$

$$H(s) = 3\left[\frac{1}{s} - \frac{e^{-3s}}{s}\right]$$

$\because \quad x(t) = 5$

Then $X(s) = \dfrac{5}{s}$

We know that

$$Y(s) = H(s)\,X(s) = 3\left[\frac{1}{s} - \frac{e^{-3s}}{s}\right]\cdot\frac{5}{s}$$

Steady state output

$$= \lim_{s\to 0} s\cdot Y(s) = \lim_{s\to 0} s\cdot\frac{15}{s}\frac{(1-e^{-3s})}{s}$$

$$= 15\cdot\lim_{s\to 0} 3e^{-3s} \qquad [\because \text{ L-Hospital's rule}]$$

$$= 45$$

4.30 (b)

$$G(j\omega) = \omega e^{-2\omega^2}$$

$$\int_{-\infty}^{\infty} g(t)dt = G(j0) = 0$$

$$y(t) = g(t) * u(t)$$

$$\Rightarrow \quad Y(j\omega) = G(j\omega)\,U(j\omega)$$

$$\Rightarrow \quad Y(j\omega) = G(j\omega)\left[\frac{1}{j\omega} + \pi\delta(\omega)\right]$$

$$= \frac{\omega e^{-2\omega^2}}{j\omega} + \omega e^{-2\omega^2}\pi\delta(\omega)$$

$$\int_{-\infty}^{\infty} y(t)dt = Y(j0) = \frac{1}{j} = -j$$

4.31 (0.2)

$$\int_{-\infty}^{\infty} \text{sin}c^2(5t)\,dt$$

Let, $f(t) = \sin c(5t)$

$= \text{Sa}(5\pi t) \qquad [\because \text{ sinc}(t) = \text{Sa}(\pi t)]$

$$T\,Sa\left(t\cdot\frac{T}{2}\right) \longleftrightarrow 2\pi\, G_T(\omega)$$

$$\frac{T}{2} = 5\pi \;\Rightarrow\; \boxed{T = 10\pi}$$

$$10\pi\, Sa(5\pi t) \longleftrightarrow 2\pi\, G_{10\pi}(\omega)$$

$$\Rightarrow \quad F(\omega) = \frac{1}{5}G_{10\pi}(\omega)$$

$$\text{ESD}_f = |F(\omega)|^2$$

$$E_f = \int_{-\infty}^{\infty} f^2(t)\,dt = \frac{1}{2\pi}\int_{-\infty}^{\infty} \text{ESD}_f\, d\omega$$

$$\int_{-\infty}^{\infty} \text{sin}c^2(5t)\,dt = \frac{1}{2\pi}\int_{-5\pi}^{5\pi}\left(\frac{1}{5}\right)^2 d\omega$$

$$= \frac{1}{2\pi}\times\frac{1}{25}\times 10\pi = \frac{1}{5}$$

4.32 (c)

Group delay is given by

$$= \frac{-1}{2\pi}\frac{d\phi(f)}{df} = \frac{-1}{2\pi}[-2\pi\alpha]$$

Group delay $= \alpha$

4.33 (d)

$$\text{Given:} \quad H(f) = \begin{cases} e^{-j4\pi f}, & |f| \le \dfrac{\omega}{2} \\ 0, & |f| > \dfrac{\omega}{2} \end{cases}$$

So, $|H(f)| = 1$

$$H(f) = e^{-j2.2\pi f}$$

$$H(\omega) = e^{-j2\omega} \qquad |f| \le \frac{\omega}{2}$$

$$h(t) = \delta(t-2)$$

As we know that

$$y(t) = x(t) * h(t)$$

or $\quad y(t) = X(\omega) \cdot H(\omega)$

$$y(t) = x(t) * \delta(t-2)$$

$$y(t) = x(t-2)$$

4.34 (0.25)

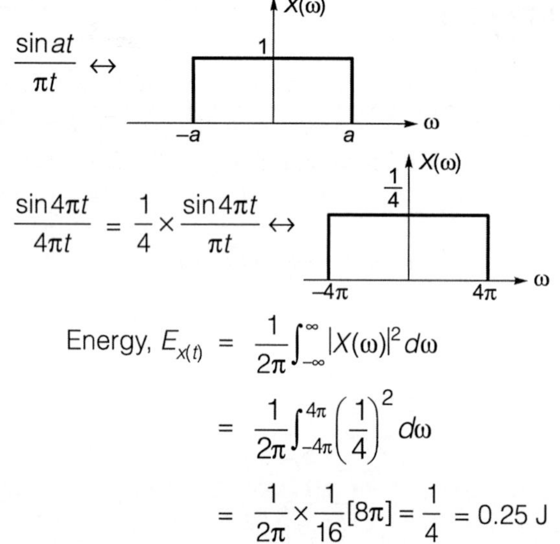

$$\frac{\sin at}{\pi t} \leftrightarrow$$

$$\frac{\sin 4\pi t}{4\pi t} = \frac{1}{4} \times \frac{\sin 4\pi t}{\pi t} \leftrightarrow$$

$$\text{Energy, } E_{x(t)} = \frac{1}{2\pi}\int_{-\infty}^{\infty} |X(\omega)|^2 d\omega$$

$$= \frac{1}{2\pi}\int_{-4\pi}^{4\pi}\left(\frac{1}{4}\right)^2 d\omega$$

$$= \frac{1}{2\pi} \times \frac{1}{16}[8\pi] = \frac{1}{4} = 0.25 \text{ J}$$

4.35 (a)

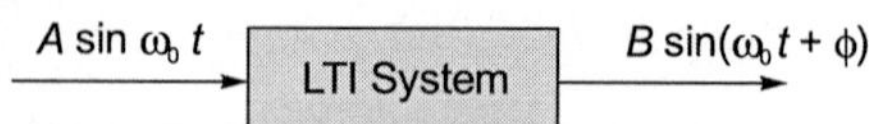

When a sinusoidal input is given to LTI system, the output is also a sinusoid with change in magnitude and the phase shift offered by LTI system.

4.36 (a)

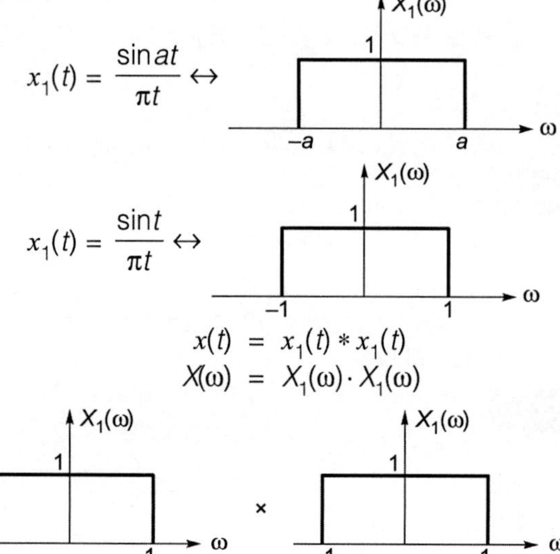

$$x_1(t) = \frac{\sin at}{\pi t} \leftrightarrow$$

$$x_1(t) = \frac{\sin t}{\pi t} \leftrightarrow$$

$$x(t) = x_1(t) * x_1(t)$$

$$X(\omega) = X_1(\omega) \cdot X_1(\omega)$$

$=$

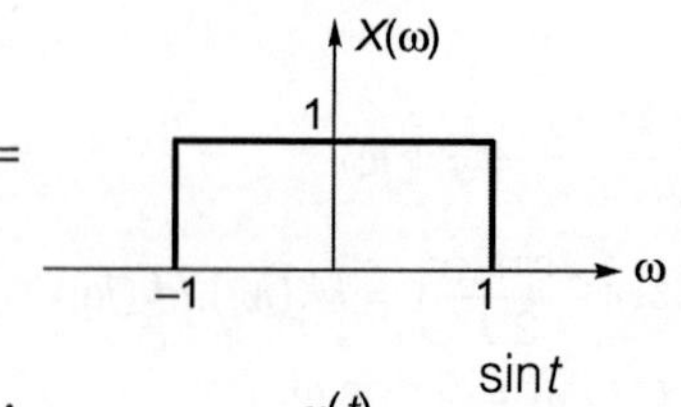

$$\therefore \quad x(t) = \frac{\sin t}{\pi t}$$

4.37 (8)

$$x(t) = 4\cos 200\pi t + 8\cos 400\pi t$$

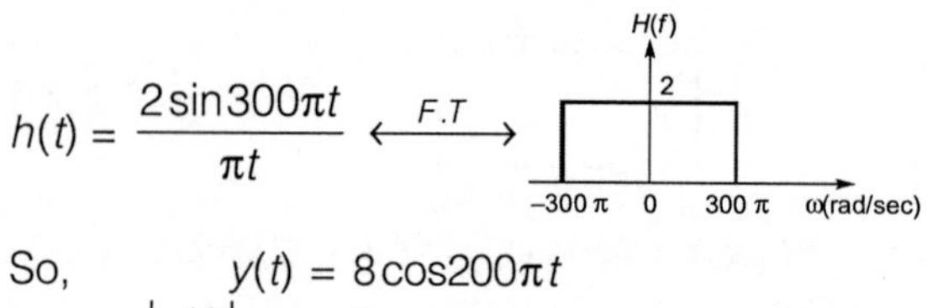

$$h(t) = \frac{2\sin 300\pi t}{\pi t} \xleftrightarrow{F.T}$$

So, $\quad y(t) = 8\cos 200\pi t$

$\therefore \quad |y(t)|_{max} = 8$

4.38 (8)

Since,

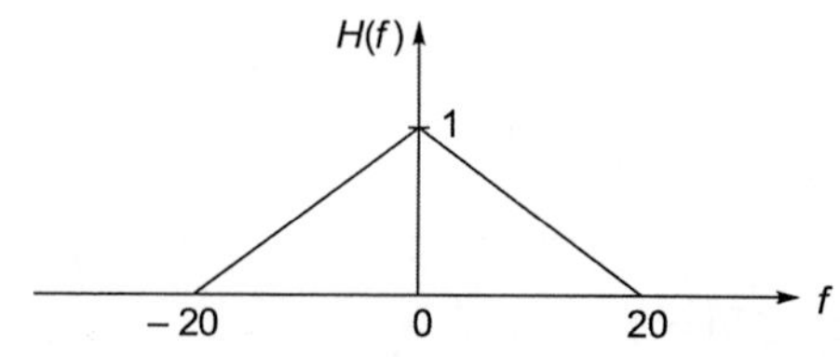

So, $\quad y(t) = \frac{1}{2} \times 8\cos(20\pi t + \phi)$

$$= 4\cos(20\pi t + \phi)\,;\ \phi = \frac{\pi}{4} - 20°$$

So, $\quad P_y = \frac{4^2}{2} = 8 \text{ W.}$

4.39 (8)

Hilbert transform does not alter the amplitude spectrum of the signal.

So, $\int_{-\infty}^{\infty} |y(t)|^2 dt = \int_{-\infty}^{\infty} |x(t)|^2 dt = \int_{-\infty}^{\infty} |X(f)|^2 df$

$$x(t) = 4\text{sinc}(2t)$$

$$\text{sinc}(t) \xleftrightarrow{CTFT} \text{rect}(f)$$

$$4\text{sinc}(2t) \xleftrightarrow{CTFT} \frac{4}{2}\text{rect}\left(\frac{f}{2}\right) = 2\,\text{rect}\left(\frac{f}{2}\right)$$

$$\int_{-\infty}^{\infty} |X(f)|^2 df = 2 \times (2)^2 = 8$$

So,

$$\int_{-\infty}^{\infty} |y(t)|^2 dt = 8$$

X(f): 2 ; –1, 0, 1 ; f(Hz)

4.40 (58.61)

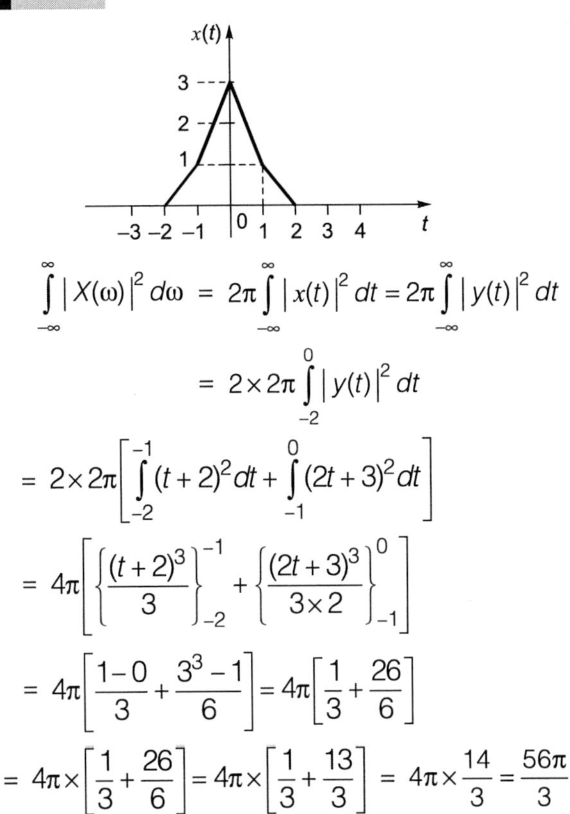

$$\int_{-\infty}^{\infty} |X(\omega)|^2 d\omega = 2\pi \int_{-\infty}^{\infty} |x(t)|^2 dt = 2\pi \int_{-\infty}^{\infty} |y(t)|^2 dt$$

$$= 2\times 2\pi \int_{-2}^{0} |y(t)|^2 dt$$

$$= 2\times 2\pi \left[\int_{-2}^{-1} (t+2)^2 dt + \int_{-1}^{0} (2t+3)^2 dt\right]$$

$$= 4\pi\left[\left\{\frac{(t+2)^3}{3}\right\}_{-2}^{-1} + \left\{\frac{(2t+3)^3}{3\times 2}\right\}_{-1}^{0}\right]$$

$$= 4\pi\left[\frac{1-0}{3} + \frac{3^3-1}{6}\right] = 4\pi\left[\frac{1}{3}+\frac{26}{6}\right]$$

$$= 4\pi\times\left[\frac{1}{3}+\frac{26}{6}\right] = 4\pi\times\left[\frac{1}{3}+\frac{13}{3}\right] = 4\pi\times\frac{14}{3} = \frac{56\pi}{3}$$

4.41 (a)

Consider $z(t) = e^{-|t|}$

By taking the Fourier transform,

$$e^{-|t|} \xleftrightarrow{\text{F.T.}} \frac{2}{1+\omega^2}$$

Using multiplication by 't' property that results in differentiation in frequency domain,

$$tz(t) \xleftrightarrow{\text{F.T.}} j\frac{dZ(\omega)}{d\omega}$$

$$te^{-|t|} \xleftrightarrow{\text{F.T.}} j\left(\frac{d}{d\omega}\left[\frac{2}{1+\omega^2}\right]\right)$$

$$te^{-|t|} \xleftrightarrow{\text{F.T.}} j\left[\frac{0-2\times 2\omega}{(1+\omega^2)^2}\right]$$

$$te^{-|t|} \xleftrightarrow{\text{F.T.}} \frac{-4j\omega}{(1+\omega^2)^2}$$

Applying duality property,

$$x(t) \xleftrightarrow{\text{F.T.}} X(\omega)$$

$$X(t) \xleftrightarrow{\text{F.T.}} 2\pi x(-\omega)$$

We get,

$$\frac{-4jt}{(1+t^2)^2} \xleftrightarrow{\text{F.T.}} 2\pi(-\omega)e^{-|(-\omega)|}$$

$$\frac{t}{(1+t^2)^2} \xleftrightarrow{\text{F.T.}} \frac{-2\pi\omega e^{-|(-\omega)|}}{-4j}$$

$$\Rightarrow \frac{t}{(1+t^2)^2} \xleftrightarrow{\text{F.T.}} \frac{\pi}{2j}\omega e^{-|\omega|}$$

4.42 (b)

$m(t) \overset{\text{FT}}{\Longleftrightarrow}$

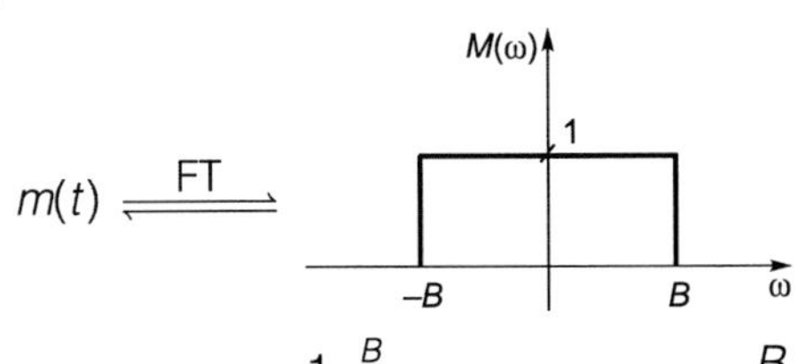

$$\text{Energy }(E) = \frac{1}{2\pi}\int_{-B}^{B} (1)^2 \cdot d\omega \quad ; \quad E = \frac{B}{\pi}$$

Now, let $\quad y(t) = m(t)\cos\omega_0 t$

$$Y(\omega) = \frac{1}{2}[M(\omega-\omega_0) + M(\omega+\omega_0)]$$

Here; $\quad \omega_0 = 10B$

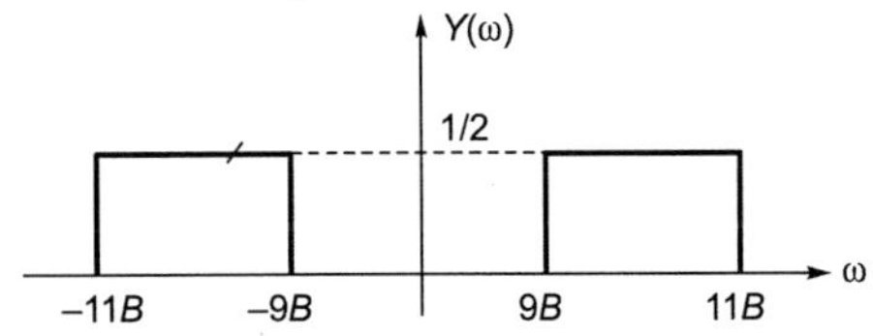

$$\text{Now, Energy }(E') = \frac{1}{2\pi}\int_{-\infty}^{\infty} |Y(\omega)|^2 \cdot d\omega$$

$$= \frac{1}{2\pi}\left[\int_{-11B}^{-9B}\left(\frac{1}{2}\right)^2 \cdot d\omega + \int_{9B}^{11B}\left(\frac{1}{2}\right)^2 \cdot d\omega\right]$$

$$= \frac{1}{2\pi}\left[\frac{1}{4}\times 2B + \frac{1}{4}\times 2B\right]$$

$$E' = \frac{B}{2\pi} = \frac{1}{2}\left(\frac{B}{\pi}\right) = \frac{E}{2}$$

4.43 (a)

Given $h(t)$ is Real and Even. When sinusoidal input applied to LTI system having even impulse response, then output will also be sinusoidal.

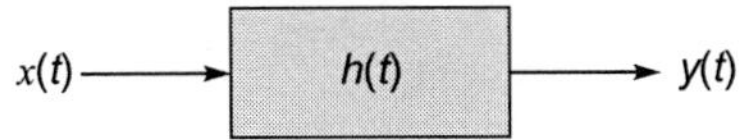

here, $\quad y(t) = H(W)|_{W=10.5W} \cdot 10\cos(10.5Wt)$

let, $\quad h(t) = f(t)\cos 10Wt$

where, $\quad f(t) = \pi\left(\frac{\sin Wt}{\pi t}\right)^2$

$f(t) \overset{\text{FT}}{\Longleftrightarrow}$

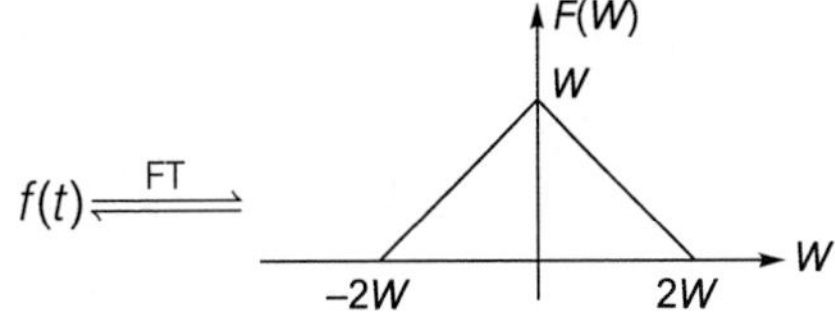

Now; $H(W) = \frac{1}{2}[F(W+10W) + F(W-10W)]$

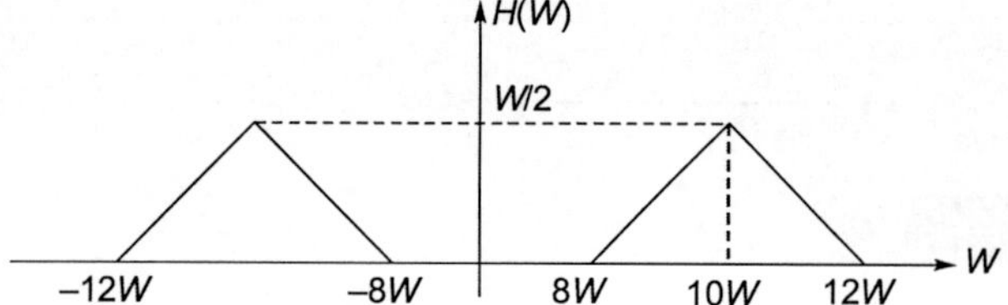

$$\therefore H(W)|_{W=10.5W} = \frac{3}{8}W$$

Hence, $y(t) = \left(\frac{3}{8}W\right)(10\cos 10.5Wt)$

$$= \frac{15}{4}W\cos 10.5Wt$$

4.44 (15)

Given, signal $x_1(t) = u(t+1.5) - u(t-1.5)$

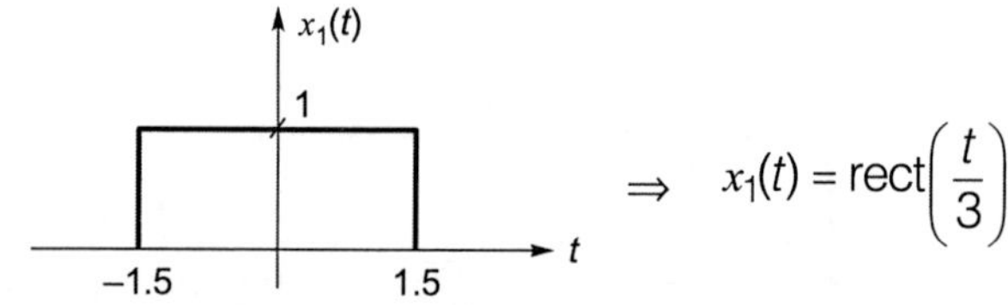

$\Rightarrow \quad x_1(t) = \text{rect}\left(\frac{t}{3}\right)$

$$x_1(t) = \text{rect}\left(\frac{t}{3}\right) \xleftrightarrow{\text{FT}} X_1(\omega)\, 3Sa(1.5\omega)$$

Now, $x_2(t) = \delta(t+3) + \text{rect}\left(\frac{t}{2}\right) + 2\delta(t-2)$

Taking Fourier transform

$$X_2(\omega) = e^{3j\omega} + 2\,Sa(\omega) + 2e^{-2j\omega}$$

$\because \quad y(t) = x_1(t) * x_2(t)$

$$Y(\omega) = X_1(\omega)\cdot X_2(\omega)$$

We know, $Y(\omega) = \int_{-\infty}^{\infty} y(t)\cdot e^{-j\omega t}\cdot dt$

$\therefore \quad \int_{-\infty}^{\infty} y(t) = Y(0)$

$\therefore \quad Y(0) = X_1(0)\cdot X_2(0)$

$$= 3[1+2+2] = 15$$

4.45 (c)

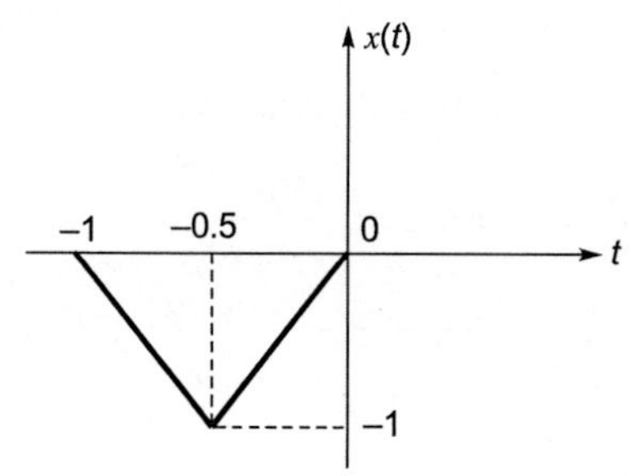

Time scaling the signal by a factor of 4,

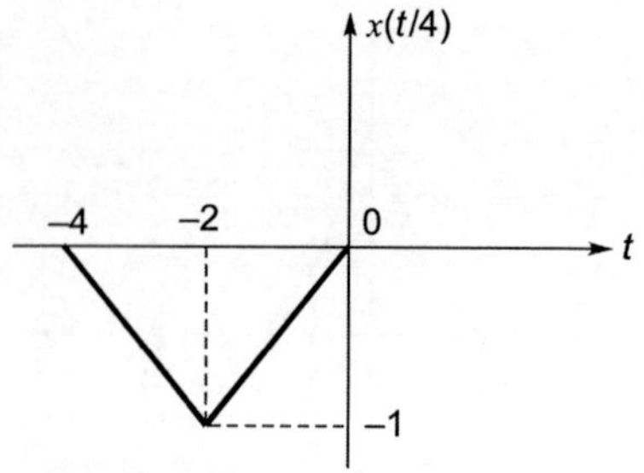

Delaying the signal in time domain by 2 units, we obtain

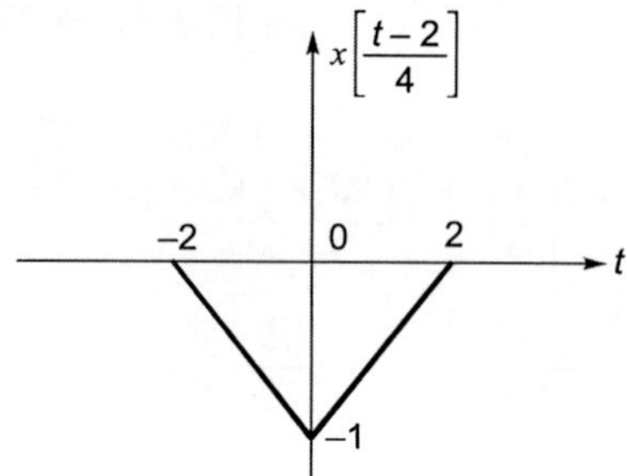

From above, the signal $y(t)$ is obtained as below:

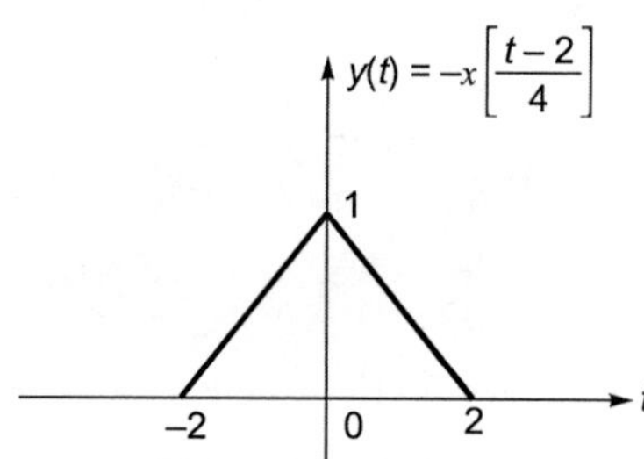

We obtain, $y(t) = -x\left[\frac{t-2}{4}\right]$

We have, $x(t) \longleftrightarrow X(f)$

Using the scaling property of Fourier transform

$$x(at) \longleftrightarrow \frac{1}{|a|}X\left(\frac{f}{a}\right)$$

$$x\left(\frac{t}{4}\right) \longleftrightarrow 4X(4f)$$

Using the time-shifting property of Fourier transform $x(t-t_0) \longleftrightarrow e^{-j2\pi f t_0}X(f)$

$\therefore \quad x\left[\frac{t-2}{4}\right] \longleftrightarrow 4X(4f)\cdot e^{-j4\pi f}$

$\Rightarrow \quad y(t) = -x\left[\frac{t-2}{4}\right] \longleftrightarrow -4X(4f)\cdot e^{-j4\pi f}$

■■■

5 Laplace Transform

5.1 If $L[f(t)] = \frac{2(s+1)}{s^2+2s+5}$, then $f(0^+)$ and $f(\infty)$ are given by
(a) 0, 2 respectively
(b) 2, 0 respectively
(c) 0, 1 respectively
(d) 2/5, 0 respectively
[**Note:** 'L' stands for 'Laplace transform of']
[1995 : 1 M]

5.2 The transfer function of a linear system is the
(a) ratio of the output, $v_0(t)$, and input, $v_i(t)$
(b) ratio of the derivatives of the output and the input
(c) ratio of the Laplace transform of the output and that of the input with all initial conditions zeros
(d) none of these [1995 : 1 M]

5.3 The final value theorem is used to find the
(a) steady state value of the system output
(b) initial value of the system output
(c) transient behaviour of the system output
(d) none of these [1995 : 1 M]

5.4 The inverse Laplace transform of the function $\frac{s+5}{(s+1)(s+3)}$ is
(a) $2e^{-t} - e^{-3t}$ (b) $2e^{-t} + e^{-3t}$
(c) $e^{-t} - 2e^{-3t}$ (d) $e^{-t} + 2e^{-3t}$
[1996 : 2 M]

5.5 The Laplace transform of $e^{\alpha t}\cos(\alpha t)$ is equal to
(a) $\frac{(s-\alpha)}{(s-\alpha)^2+\alpha^2}$ (b) $\frac{(s+\alpha)}{(s+\alpha)^2+\alpha^2}$
(c) $\frac{1}{(s-\alpha)^2}$ (d) none of these
[1997 : 1 M]

5.6 Match each of the items 1, 2 on the left with most appropriate item A, B, C or D on the right.
In the case of a linear time invariant system
(1) Poles in the right half plane implies
(2) Impulse response zero for $t \le 0$ implies
(A) Exponential decay of output
(B) System is causal
(C) No stored energy in the system
(D) System is unstable [1997 : 2 M]

5.7 If $L[f(t)] = \frac{\omega}{(s^2+\omega^2)}$, then the value of $\lim_{t\to\infty} f(t)$
(a) cannot be determined
(b) is zero
(c) is unity
(d) is infinite [1998 : 1 M]

5.8 The transfer function of a zero-order-hold system is
(a) $(1/s)(1 + e^{-sT})$ (b) $(1/s)(1 - e^{-sT})$
(c) $1 - (1/s)e^{-sT}$ (d) $1 + (1/s)e^{-sT}$
[1998 : 1 M]

5.9 If $\mathcal{L}[f(t)] = F(s)$, then $\mathcal{L}[f(t-T)]$ is equal to
(a) $e^{sT}F(s)$ (b) $e^{-sT}F(s)$
(c) $\frac{F(s)}{1+e^{sT}}$ (d) $\frac{F(s)}{1-e^{-sT}}$
[1999 : 1 M]

5.10 Given that

$$L[f(t)] = \frac{s+2}{s^2+1},\ L[g(t)] = \frac{s^2+1}{(s+3)(s+2)},$$

$$h(t) = \int_0^t f(\tau)g(t-\tau)d\tau$$

$L[h(t)]$ is
(a) $\frac{s^2+1}{s+3}$
(b) $\frac{1}{s+3}$
(c) $\frac{s^2+1}{(s+3)(s+2)} + \frac{s+2}{s^2+1}$
(d) None of the above [2000 : 1 M]

5.11 A linear time invariant system has an impulse response e^{2t}, for $t > 0$. If initial conditions are zero and the input is e^{3t}, the output for $t > 0$ is
(a) $e^{3t} - e^{2t}$ (b) e^{5t}
(c) $e^{3t} + e^{2t}$ (d) None of these
[2000 : 2 M]

5.12 The transfer function of a system is given by $H(s) = \frac{1}{s^2(s-2)}$. The impulse response of the system is

(a) $(t^2 * e^{-2t})u(t)$ (b) $(t * e^{2t})u(t)$
(c) $(te^{-2t})u(t)$ (d) $(te^{-2t})u(t)$
($*$ denotes convolution, and $u(t)$ is unit step function)

[2001 : 1 M]

5.13 The Laplace transform of a continuous-time signal $x(t)$ is $X(s) = \frac{5-s}{s^2-s-2}$. If the Fourier transform of this signal exists, then $x(t)$ is

(a) $e^{2t}u(t) - 2e^{-t}u(t)$
(b) $-e^{2t}u(-t) + 2e^{-t}u(t)$
(c) $-e^{2t}u(-t) - 2e^{-t}u(t)$
(d) $e^{2t}u(-t) - 2e^{-t}u(t)$

[2002 : 2 M]

5.14 The Laplace transform of $i(t)$ is given by

$$I(s) = \frac{2}{s(1+s)}$$

As $t \to \infty$, the value of $i(t)$ tends to

(a) 0 (b) 1
(c) 2 (d) ∞

[2003 : 1 M]

5.15 Consider the function $f(t)$ having Laplace transform

$$F(s) = \frac{\omega_0}{s^2+\omega_0^2} \quad \text{Re}[s] > 0$$

The final value of $f(t)$ would be:

(a) 0 (b) 1
(c) $-1 \le f(\infty) \le 1$ (d) ∞

[2006 : 2 M]

5.16 If the Laplace transform of a signal $y(t)$ is $Y(s) = \frac{1}{s(s-1)}$, then its final value is

(a) -1 (b) 0
(c) 1 (d) Unbounded

[2007 : 1 M]

5.17 Given that $F(s)$ is the one-sided Laplace transform of $f(t)$, the Laplace transform of $\int_0^t f(\tau)d\tau$ is

(a) $sF(s) - f(0)$ (b) $\frac{1}{s}F(s)$
(c) $\int_0^s F(\tau)d\tau$ (d) $\frac{1}{s}[F(s) - f(0)]$

[2009 : 2 M]

5.18 A continuous time LTI system is described by

$$\frac{d^2y(t)}{dt^2} + 4\frac{dy(t)}{dt} + 3y(t) = 2\frac{dx(t)}{dt} + 4x(t)$$

Assuming zero initial conditions, the response $y(t)$ of the above system for the input $x(t) = e^{-2t}u(t)$ is given by

(a) $(e^t - e^{3t})u(t)$ (b) $(e^{-t} - e^{-3t})u(t)$
(c) $(e^{-t} + e^{-3t})u(t)$ (d) $(e^t + e^{3t})u(t)$

[2010 : 2 M]

5.19 If the unit step response of a network is $(1 - e^{-\alpha t})$, then its unit impulse response is

(a) $\alpha e^{-\alpha t}$ (b) $\alpha^{-1}e^{-\alpha t}$
(c) $(1-\alpha^{-1})e^{-\alpha t}$ (d) $(1-\alpha)e^{-\alpha t}$

[2011 : 1 M]

5.20 An input $x(t) = \exp(-2t)\,u(t) + \delta(t-6)$ is applied to an LTI system with impulse response $h(t) = u(t)$. The output is

(a) $[1 - \exp(-2t)]\,u(t) + u(t+6)$
(b) $[1 - \exp(-2t)]\,u(t) + u(t-6)$
(c) $0.5[1 - \exp(-2t)]\,u(t) + u(t+6)$
(d) $0.5[1 - \exp(-2t)]\,u(t) + u(t-6)$

[2011 : 2 M]

5.21 If $F(s) = L[f(t)] = \frac{2(s+1)}{s^2+4s+7}$ then the initial and final values of f(t) are respectively

(a) 0, 2 (b) 2, 0
(c) 0, 2/7 (d) 2/7, 0

[2011 : 2 M]

5.22 The unilateral Laplace transform of $f(t)$ is $\frac{1}{s^2+s+1}$. The unilateral Laplace transform of $t\,f(t)$ is

(a) $-\frac{s}{(s^2+s+1)^2}$ (b) $-\frac{2s+1}{(s^2+s+1)^2}$
(c) $\frac{s}{(s^2+s+1)^2}$ (d) $\frac{2s+1}{(s^2+s+1)^2}$

[2012 : 1 M]

5.23 The impulse response of a system is $h(t) = tu(t)$. For an input $u(t-1)$, the output is

(a) $\frac{t^2}{2}u(t)$ (b) $\frac{t(t-1)}{2}u(t-1)$
(c) $\frac{(t-1)^2}{2}u(t-1)$ (d) $\frac{t^2-1}{2}u(t-1)$

[2013 : 1 M]

5.24 Assuming zero initial condition, the response $y(t)$ of the system given below to a unit step input $u(t)$ is

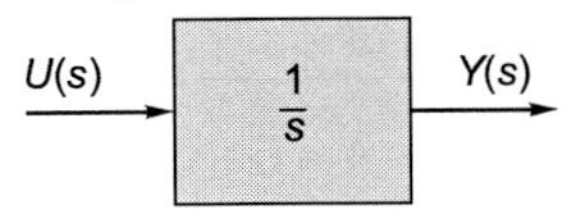

(a) $u(t)$ (b) $tu(t)$

(c) $\frac{t^2}{2}u(t)$ (d) $e^{-t}u(t)$

[2013 : 1 M]

5.25 Which one of the following statements is NOT TRUE for a continuous time causal and stable LTI system?

(a) All the poles of the system must lie on the left side of the $j\omega$ axis.

(b) Zeros of the system can lie anywhere in the s-plane.

(c) All the poles must lie within $|s| = 1$.

(d) All the roots of the characteristic equation must be located on the left side of the $j\omega$ axis.

[2013 : 1 M]

5.26 A system is described by the differential equation

$\frac{d^2y}{dt^2} + 5\frac{dy}{dt} + 6y(t) = x(t)$. Let $x(t)$ be a rectangular pulse given by

$$x(t) = \begin{cases} 1 & 0 < t < 2 \\ 0 & \text{otherwise} \end{cases}$$

Assuming that $y(0) = 0$ and $\frac{dy}{dt} = 0$ at $t = 0$, the Laplace transform of $y(t)$ is

(a) $\frac{e^{-2s}}{s(s+2)(s+3)}$ (b) $\frac{1-e^{-2s}}{s(s+2)(s+3)}$

(c) $\frac{e^{-2s}}{(s+2)(s+3)}$ (d) $\frac{1-e^{-2s}}{(s+2)(s+3)}$

[2013 : 2 M]

5.27 A system is described by the following differential equation, where $u(t)$ is the input to the system and $y(t)$ is the output of the system.

$$\dot{y}(t) + 5y(t) = u(t)$$

When $y(0) = 1$ and $u(t)$ is a unit step function, $y(t)$ is

(a) $0.2 + 0.8e^{-5t}$ (b) $0.2 - 0.2e^{-5t}$

(c) $0.8 + 0.2e^{-5t}$ (d) $0.8 - 0.8e^{-5t}$

[2014 : 2 M, Set-1]

5.28 The input $-3e^{-2t}u(t)$, where $u(t)$ is the unit step function, is applied to a system with transfer function $\frac{s-2}{s+3}$. If the initial value of the output is -2, then the value of the output at steady state is ______.

[2014 : 1 M, Set-3]

5.29 Let $h(t)$ denote the impulse response of a causal system with transfer function $\frac{1}{s+1}$. Consider the following three statements:

S_1 : The system is stable.

S_2 : $\frac{h(t+1)}{h(t)}$ is independent of t for $t > 0$.

S_3 : A non-causal system with the same transfer function is stable.

For the above system,

(a) only S_1 and S_2 are true

(b) only S_2 and S_3 are true

(c) only S_1 and S_3 are true

(d) S_1, S_2 and S_3 are true [2014 : 2 M, Set-3]

5.30 The unilateral Laplace transform of $f(t)$ is $\frac{1}{s^2+s+1}$. Which one of the following is the unilateral Laplace transform of $g(t) = t \cdot f(t)$?

(a) $\frac{-s}{(s^2+s+1)^2}$ (b) $\frac{-(2s+1)}{(s^2+s+1)^2}$

(c) $\frac{s}{(s^2+s+1)^2}$ (d) $\frac{2s+1}{(s^2+s+1)^2}$

[2014 : 2 M, Set-4]

5.31 A stable linear time invariant (LTI) system has a transfer function $H(s) = \frac{1}{s^2+s-6}$. To make this system causal it needs to be cascaded with another LTI system having a transfer function $H_1(s)$. A correct choice for $H_1(s)$ among the following options is

(a) $s + 3$ (b) $s - 2$

(c) $s - 6$ (d) $s + 1$

[2014 : 2 M, Set-4]

5.32 A causal LTI system has zero initial conditions and impulse response $h(t)$. Its output $x(t)$ and input $y(t)$ are related through the linear constant-coefficient differential equation

$$\frac{d^2y(t)}{dt^2} + \alpha\frac{dy(t)}{dt} + \alpha^2 y(t) = x(t)$$

Let another signal $g(t)$ be defined as

$$g(t) = \alpha^2 \int_0^t h(\tau)d\tau + \frac{dh(t)}{dt} + \alpha h(t)$$

If $G(s)$ is the Laplace transform of $g(t)$, then the number of poles of $G(s)$ is _______.

[2014 : 2 M, Set-4]

5.33 Input $x(t)$ and output $y(t)$ of an LTI system are related by the differential equation $y''(t) - y'(t) - 6y(t) = x(t)$. If the system is neither causal nor stable, the impulse response $h(t)$ of the system is

(a) $\frac{1}{5}e^{3t}u(-t) + \frac{1}{5}e^{-2t}u(-t)$

(b) $-\frac{1}{5}e^{3t}u(-t) + \frac{1}{5}e^{-2t}u(-t)$

(c) $\frac{1}{5}e^{3t}u(-t) - \frac{1}{5}e^{-2t}u(t)$

(d) $-\frac{1}{5}e^{3t}u(-t) - \frac{1}{5}e^{-2t}u(t)$ [2015 : 2 M, Set-2]

5.34 Consider the differential equation $\frac{dx}{dt} = 10 - 0.2x$ with initial conduction $x(0) = 1$. The response $x(t)$ for $t > 0$ is

(a) $2 - e^{-0.2t}$
(b) $2 - e^{0.2t}$
(c) $50 - 49e^{-0.2t}$
(d) $50 - 49e^{0.2t}$

[2015 : 2 M, Set-2]

5.35 The bilateral Laplace transform of a function $f(t) = \begin{cases} 1 & \text{if } a \le t \le b \\ 0 & \text{otherwise} \end{cases}$ is

(a) $\frac{a-b}{s}$
(b) $\frac{e^{z}(a-b)}{s}$
(c) $\frac{e^{-as} - e^{-bs}}{s}$
(d) $\frac{e^{-(a-b)}}{s}$

[2015 : 1 M, Set-2]

5.36 Let the signal $f(t) = 0$ outside the interval $[T_1, T_2]$, where T_1 and T_2 are finite. Furthermore, $|f(t)| < \infty$. The region of convergence (ROC) of the signal's bilateral Laplace transform $F(s)$ is

(a) a parallel strip containing the $j\Omega$ axis
(b) a parallel strip not containing the $j\Omega$ axis
(c) the entire s-plane
(d) a half plane containing the $j\Omega$ axis

[2015 : 1 M, Set-2]

5.37 Let $x(t) = \alpha s(t) + s(-t)$ with $s(t) = \beta e^{-4t}u(t)$, where $u(t)$ is unit step function. If the bilateral Laplace transform of $x(t)$ is

$$X(s) = \frac{16}{s^2 - 16} \quad -4 < \text{Re}\{s\} < 4;$$

then the value of β is _____.

[2015 : 2 M, Set-2]

5.38 Consider the function $g(t) = e^{-t}\sin(2\pi t)\,u(t)$ where $u(t)$ is the unit step function. The area under $g(t)$ is _______.

[2015 : 1 M, Set-3]

5.39 The Laplace transform of the causal periodic square wave of period T shown in the figure below is

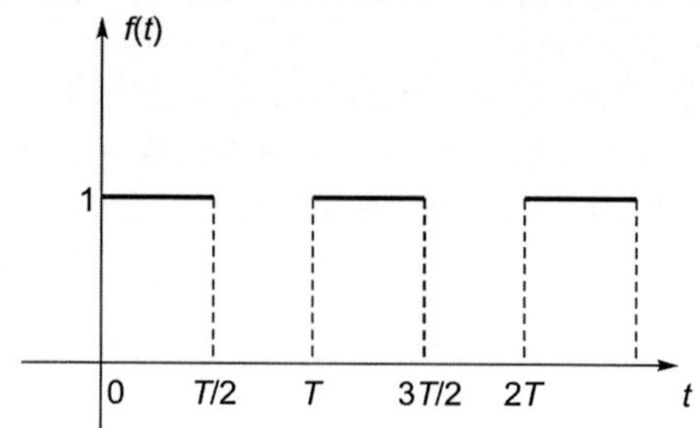

(a) $F(s) = \frac{1}{1 + e^{-sT/2}}$

(b) $F(s) = \frac{1}{s(1 + e^{-sT/2})}$

(c) $F(s) = \frac{1}{s(1 - e^{-sT/2})}$

(d) $F(s) = \frac{1}{1 - e^{-sT}}$

[2016 : 2 M, Set-1]

5.40 A first-order low-pass filter of time constant T is excited with different input signals (with zero initial conditions up to $t = 0$). Match the excitation signals X, Y, Z with the corresponding time responses for $t \ge 0$:

List-I	List-II
X. Impulse	P. $1 - e^{-t/T}$
Y. Unit step	Q. $t - T(1 - e^{-t/T})$
Z. Ramp	R. $e^{-t/T}$

(a) X – R; Y – Q; Z – P
(b) X – Q; Y – P; Z – R
(c) X – R; Y – P; Z – Q
(d) X – P; Y – R; Z – Q

[2016 : 2 M, Set-1]

5.41 A signal $2\cos\left(\frac{2\pi}{3}t\right) - \cos(\pi t)$ is the input to an LTI system with the transfer function

$$H(s) = e^{s} + e^{-s}$$

If C_k denote the kth coefficient in the exponential Fourier series of the output signal, then c_3 is equal to

(a) 0 (b) 1
(c) 2 (d) 3

[2016 : 2 M, Set-3]

5.42 Consider the following statements for continuous-time linear time invariant (LTI) systems.

I. There is no bounded input bounded output (BIBO) stable system with a pole in the right half of the complex plane.

II. There is no causal and BIBO stable system with a pole in the right half of the complex plane.

Which one among the following is correct?

(a) Both I and II are true
(b) Both I and II are not true
(c) Only I is true
(d) Only II is true

[2017 : 1 M, Set-1]

5.43 The transfer function of a causal LTI system is $H(s) = 1/s$. If the input to the system is $x(t) = [\sin(t)/\pi t]\ u(t)$, where $u(t)$ is a unit step function, the system output $y(t)$ as $t \to \infty$ is _____.

[2017 : 2 M, Set-2]

5.44 Let $Y(s)$ be the unit-step response of a causal system, having a transfer function

$$G(s) = \frac{3-s}{(s+1)(s+3)}$$

that is, $Y(s) = \frac{G(s)}{s}$. The forced response of the system is

(a) $u(t)$
(b) $2u(t)$
(c) $u(t) - 2e^{-t}u(t) + e^{-3t}u(t)$
(d) $2u(t) - 2e^{-t}u(t) + e^{-3t}u(t)$

[2019 : 1 M]

5.45 A system with transfer function

$$G(s) = \frac{1}{(s+1)(s+a)},\ a > 0$$

is subjected an input $5\cos 3t$. The steady state output of the system is $\frac{1}{\sqrt{10}}\cos(3t - 1.892)$. The value of a is _____.

[2020 : 2 M]

■■■

Answers Laplace Transform

5.1 (b)	5.2 (c)	5.3 (a)	5.4 (a)	5.5 (a)	5.6 (1-D, 2-B)	5.7 (a)
5.8 (b)	5.9 (b)	5.10 (b)	5.11 (a)	5.12 (b)	5.13 (c)	5.14 (c)
5.15 (c)	5.16 (d)	5.17 (d)	5.18 (b)	5.19 (a)	5.20 (d)	5.21 (b)
5.22 (d)	5.23 (c)	5.24 (b)	5.25 (c)	5.26 (b)	5.27 (a)	5.28 (0)
5.29 (a)	5.30 (d)	5.31 (b)	5.32 (1)	5.33 (b)	5.34 (c)	5.35 (c)
5.36 (c)	5.37 (–2)	5.38 (0.155)	5.39 (b)	5.40 (c)	5.41 (b)	5.42 (d)
5.43 (0.5)	5.44 (a, c)	5.45 (4)				

Explanations Laplace Transform

5.1 (b)

Initial value theorem

$$f(0^+) = \lim_{s\to\infty} \frac{s2(s+1)}{s^2+2s+5}$$

$$f(0^+) = \lim_{s\to\infty} \frac{2s^2+2s}{s^2+2s+5}$$

$$f(0^+) = \lim_{s\to\infty} \frac{s^2(2+2/s)}{s^2\left(1+\frac{2}{s}+\frac{5}{s^2}\right)}$$

$$f(0^+) = \lim_{s\to\infty} \frac{2+2/s}{1+\frac{2}{s}+\frac{5}{s^2}} = \frac{2+0}{1+0+0} = 2$$

Final value theorem

$$f(\infty) = \lim_{s\to 0} sF(s) = \lim_{s\to 0} \frac{s2(s+1)}{s^2+2s+5}$$

$$f(\infty) = 0$$

5.2 (c)

$$H(s) = \frac{Y(s)}{X(s)}$$

5.3 (a)

Final value theorem is used to find the final value or the steady state value of the system output

$$f(\infty) = \lim_{s\to 0} sF(s)$$

5.4 (a)

$$F(s) = \frac{(s+5)}{(s+1)(s+3)} = \frac{2}{s+1} - \frac{1}{s+3}$$

Taking inverse Laplace on both side

$$f(t) = 2e^{-t} - e^{-3t}$$

5.5 (a)

$$\cos(\alpha t)\, u(t) \xleftrightarrow{\text{L.T.}} \frac{s}{s^2+\alpha^2}$$

$$e^{\alpha t}\cos(\alpha t)\, u(t) \xleftrightarrow{\text{L.T.}} \frac{(s-\alpha)}{(s-\alpha)^2+\alpha^2}$$

5.6 Sol.

1-D, 2-B.

(i) Poles lie in the right half of s-plane $\rightarrow$ unstable system.

(ii) $h(t) = 0$ for $t \le 0 \rightarrow$ Causal system.

5.7 (a)

$$F(s) = \frac{\omega}{s^2+\omega^2}$$

$$f(t) = \sin\omega t$$

$$\lim_{t\to\infty} f(t) = \text{lie between } -1 \text{ to } +1$$

So, it can not be determined.

5.8 (b)

The impulse response $h(t)$ of zero-order hold system is given by

$$h(t) = u(t) - u(t-T)$$

Where T is the sampling period

$$H(s) = \frac{1}{s} - \frac{1}{s}e^{-sT} = \frac{1}{s}\left[1-e^{-sT}\right]$$

5.9 (b)

$f(t-T) = F(s)e^{-sT}$

(Time shifting property of Laplace transform)

5.10 (b)

Convolution in time domain is multiplication in s-domain.

$$\therefore \quad L[h(t)] = L[f(t)] \times L[g(t)] = \frac{1}{s+3}$$

5.11 (a)

$$h(t) = e^{2t}\ ;\quad H(s) = \frac{1}{s-2}\ ;\quad x(t) = e^{3t}$$

$$X(s) = \frac{1}{s-3}$$

$$\text{Output} = x(t)*h(t)$$

$$= X(s)\,H(s) = \frac{1}{(s-2)(s-3)}$$

$$\frac{1}{(s-2)(s-3)} = \frac{A}{s-2} + \frac{B}{s-3}$$

$$1 = A(s-3) + B(s-2)$$

$$s = 2,\ 1 = -A$$

$$\Rightarrow \quad A = -1$$

$$s = 3,\ 1 = B$$

$$\Rightarrow \quad B = 1$$

$$\therefore \quad \text{Output} = [e^{3t} - e^{2t}]$$

5.12 (b)

Impulse response of system is $L^{-1}[H(s)]$

$$\frac{1}{s^2(s-2)} = \frac{1}{s^2} \times \frac{1}{s-2} = \left(t * e^{+2t}\right)u(t)$$

5.13 (c)

$$X(s) = \frac{5-s}{s^2-s-2} = \frac{5-s}{(s+1)(s-2)}$$

$$\frac{5-s}{(s+1)(s-2)} = \frac{A}{s+1} + \frac{B}{s-2}$$

$$5 - s = A(s-2) + B(s+1)$$

$$s = 2,\ 3 = 3B$$

$$\Rightarrow \quad B = 1$$

$$s = -1,\ 6 = -3A$$

$$\Rightarrow \quad A = -2$$

$$\therefore \quad x(t) = -2e^{-t}u(t) - e^{2t}u(-t)$$

ROC : $-1 < \sigma < 2$

Note: ROC of Laplace transform should include $\sigma = 0$ line.

5.14 (c)

$$\lim_{t\to\infty} i(t) = \lim_{s\to 0} sI(s) = \lim_{s\to 0} s.\frac{2}{s(1+s)} = 2$$

5.15 (c)

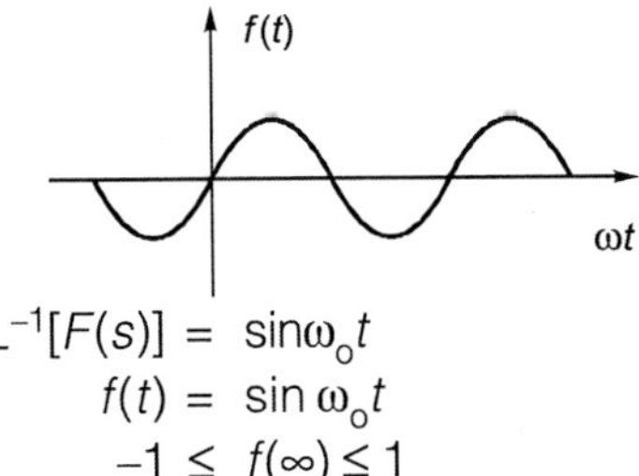

$$L^{-1}[F(s)] = \sin\omega_0 t$$

$$f(t) = \sin\omega_0 t$$

$$-1 \le f(\infty) \le 1$$

5.16 (d)

Final value theorem is applicable only when all the poles of system lies in left half of s-plane.

$\because s = 1$ is right s-plane pole. Therefore, final value theorem cannot be applied.

$$Y(s) = \frac{1}{s-1} - \frac{1}{s}$$

$$y(t) = (e^t - 1)\,u(t)$$

$\therefore$ Unbounded.

5.17 (d)

$$\mathcal{L}\left[\int_0^t f(\tau)\,d\tau\right] = \frac{1}{s}[F(s) - f(0)]$$

5.18 (b)

$$\frac{d^2y(t)}{dt^2} + \frac{4dy(t)}{dt} + 3\,y(t) = \frac{2dx(t)}{dt} + 4\,x(t)$$

Taking Laplace transform on both sides (assuming zero initial conditions),

$$s^2\,Y(s) + 4s\,Y(s) + 3\,Y(s) = 2s\,X(s) + 4\,X(s)$$

or
$$\frac{Y(s)}{X(s)} = \frac{2s+4}{s^2+4s+3} = \frac{2(s+2)}{(s+1)(s+3)}$$

Given that, $x(t) = e^{-2t}\,u(t)$

$$X(s) = \frac{1}{s+2}$$

$$Y(s) = \frac{2(s+2)}{(s+1)(s+3)(s+2)}$$

$$= \frac{2}{(s+1)(s+3)} = \frac{1}{s+1} - \frac{1}{s+3}$$

Taking inverse Laplace transform on both sides,

$$y(t) = (e^{-t} - e^{-3t})\,u(t)$$

5.19 (a)

Unit step response $s(t) = (1 - e^{-\alpha t})$

So, unit impulse response is

$$h(t) = \frac{ds(t)}{dt} = \frac{d}{dt}(1 - e^{-\alpha t})$$

$$= \alpha e^{-\alpha t}$$

5.20 (d)

$$x(t) = e^{-2t}\,u(t) + \delta(t-6)$$

$$X(s) = \frac{1}{s+2} + e^{-6s}$$

$$H(s) = \frac{1}{s}$$

$$\therefore \quad Y(s) = X(s)\cdot H(s)$$

$$\Rightarrow \quad Y(s) = \frac{1}{s(s+2)} + \frac{1}{s}e^{-6s}$$

$$\Rightarrow \quad Y(s) = \frac{1}{2}\left(\frac{1}{s} - \frac{1}{s+2}\right) + \frac{1}{s}e^{-6s}$$

Taking inverse Laplace transform, we have

$$y(t) = 0.5\,(1 - e^{-2t})\,u(t) + u(t-6)$$

5.21 (b)

$$F(s) = L[f(t)] = \frac{2(s+1)}{s^2+4s+7}$$

Initial value,

$$\lim_{t\to 0} f(t) = \lim_{s\to\infty} sF(s) = \lim_{s\to\infty} s\cdot\frac{2(s+1)}{s^2+4s+7}$$

$$= \lim_{s\to\infty} \frac{2s^2\left(1+\frac{1}{s}\right)}{s^2\left(1+\frac{4}{s}+\frac{7}{s^2}\right)} = \frac{2(1+0)}{(1+0+0)} = 2$$

Final value,

$$\lim_{t\to\infty} f(t) = \lim_{s\to 0} sF(s) = \lim_{s\to 0} s\cdot\frac{2(s+1)}{s^2+4s+7}$$

$$= 0$$

5.22 (d)

If $L[f(t)] = \bar{f}(s)$ then

$$L\left[t^n f(t)\right] = (-1)^n \frac{d^n}{ds^n}\bar{f}(s)$$

In this problem,

Given,

$$L[f(t)] = \frac{1}{s^2+s+1} = \bar{f}(s)$$

We need $L[t\,f(t)] = (-1)^1 \dfrac{d^1}{ds^1}\bar{f}(s)$

$$= -\frac{d}{ds}\bar{f}(s) = -\frac{d}{ds}\left[\frac{1}{s^2+s+1}\right]$$

$$= -\left[\frac{-1}{(s^2+s+1)^2}\right]\times(2s+1) = \left[\frac{2s+1}{(s^2+s+1)^2}\right]$$

5.23 (c)

$$h(t) = tu(t)$$

Taking Laplace transform

$$H(s) = \frac{1}{s^2}$$

$$x(t) = u(t-1)$$

Taking Laplace transform

$$X(s) = \frac{e^{-s}}{s}$$

$$\frac{Y(s)}{X(s)} = H(s)$$

$$Y(s) = H(s)\,X(s)$$

$$Y(s) = \frac{1}{s^2}\cdot\frac{e^{-s}}{s} = \frac{e^{-s}}{s^3}$$

Taking the inverse Laplace transform

$$y(t) = \frac{(t-1)^2}{2}u(t-1)$$

5.24 (b)

$$x(t) = u(t)$$

Apply Laplace transform

$$X(s) = \frac{1}{s}$$

$$H(s) = \frac{1}{s} \quad \text{(given)}$$

$$Y(s) = H(s)\,X(s)$$

$$= \frac{1}{s}\times\frac{1}{s} = \frac{1}{s^2}$$

Taking inverse Laplace transform

$$y(t) = tu(t)$$

5.25 (c)

For a causal and stable LTI system the ROC must be right sided and it must include the $s = j\omega$ line ($\sigma = 0$). For all the poles must lie within $|s| = 1$ then all poles will lie between -1 to $+1$.

We know that ROC does not include any pole, so in this case ROC will not be able to include $s = j\omega$ line ($\sigma = 0$). Thus the system will not be stable. Hence this statement is NOT TRUE.

5.26 (b)

$$x(t) = u(t) - u(t-2)$$

$$X(s) = \frac{1}{s} - \frac{e^{-2s}}{s} = \frac{1-e^{-2s}}{s}$$

Given that

$$\frac{d^2y}{dt^2} + 5\frac{dy}{dt} + 6y(t) = x(t)$$

Taking Laplace transform on both the sides

$$\left(y(0) = 0 \text{ and } \frac{dy}{dt} = 0 \text{ at } t = 0\right)$$

$$s^2Y(s) + 5sY(s) + 6Y(s) = X(s)$$

$$Y(s)\,[s^2 + 5s + 6] = \frac{1-e^{-2s}}{s}$$

$$Y(s) = \frac{1-e^{-2s}}{s(s^2+5s+6)} = \frac{1-e^{-2s}}{s(s+2)(s+3)}$$

5.27 (a)

$$\dot{y}(t) + 5y(t) = u(t)$$

Taking the Laplace transform of the above equation

$$sY(s) - y(0) + 5Y(s) = \frac{1}{s}$$

$$(s+5)\,Y(s) - 1 = \frac{1}{s}$$

$$(s+5)\,Y(s) = \frac{s+1}{s}$$

$$Y(s) = \frac{(s+1)}{s(s+5)}$$

$$\frac{A}{s} + \frac{B}{(s+5)} = \frac{(s+1)}{s(s+5)}$$

$$\boxed{A = 0.2}$$

$$B = \frac{-4}{-5} = 0.8$$

$$Y(s) = \frac{0.2}{s} + \frac{0.8}{(s+5)}$$

$$\boxed{y(t) = 0.2 + 0.8e^{-5t}}$$

5.28 (0)

Given, $x(t) = -3e^{-2t}$

Laplace transform of

$$x(t) \longleftrightarrow X(s)$$

where, $$X(s) = \frac{-3}{s+2}$$

Also, $$H(s) = \frac{s-2}{(s+3)}$$

$$\therefore \quad Y(s) = X(s)\,H(s)$$

$$Y(s) = \frac{-3(s-2)}{(s+3)(s+2)}$$

Using final value theorem, the steady state value of $Y(s)$ is

$$y(\infty) = \lim_{s\to 0} sY(s) = 0$$

5.29 (a)

Given $$H(s) = \frac{1}{(s+1)}$$

$$h(t) = e^{-t}u(t)$$

at $t = \infty$, $h(t) = 0$

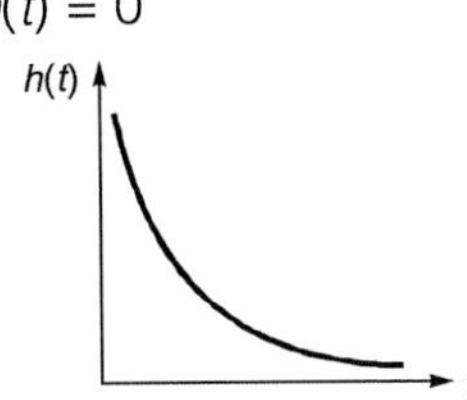

i.e. the system is stable as it converges to zero as $t \to \infty$.

- The system is causal as it can be seen from the graph that $h(t) = 0$ for $t < 0$.
- $\dfrac{h(t+1)}{h(t)} = \dfrac{e^{-(t+1)}}{e^{-t}} = \dfrac{e^{-t}\cdot e^{-1}}{e^{-t}}$
  $= e^{-1}$ i.e. independent of t.

So, statements S_1 and S_2 are true.

5.30 (d)

Given $$F(s) = \frac{1}{s^2+s+1}$$

Laplace transform of $g(t) = t\cdot f(t)$

$$= -\frac{dF(s)}{ds} = \frac{-d}{ds}\left[\frac{1}{(s^2+s+1)}\right] = \frac{(2s+1)}{(s^2+s+1)^2}$$

5.31 (b)

$$H(s) = \frac{1}{(s^2+s-6)}$$

The system is said to be causal if the output at any time depends only on present and/or past values of input or $h(t) = 0$ for $t > 0$ i.e. all poles must lie on LHS of s-plane.

$$H(s) = \frac{1}{(s+3)(s-2)}$$

From the above expression it can be seen that $s = 2$ is a pole which lies on RHS of s-plane. So, $H_1(s)$ system that needs to be cascaded should have one zero at $(s-2)$.

Thus, $$H(s)\,H_1(s) = (s-2)\cdot\frac{1}{(s+3)(s-2)}$$

$$= \frac{1}{(s+3)} \to \text{causal system}$$

5.32 (1)

$$\frac{d^2y(t)}{dt^2} + \alpha\frac{dy}{dt} + \alpha^2 y(t) = x(t)$$

$$\therefore\ (s^2 + \alpha s + \alpha^2)\,Y(s) = X(s)$$

$$H(s) = \frac{Y(s)}{X(s)} = \frac{1}{(s^2+\alpha s+\alpha^2)}$$

Let another signal $g(t)$ be defined as

$$G(s) = \frac{\alpha^2}{s}H(s) + sH(s) + \alpha H(s)$$

$$G(s) = \left(\frac{\alpha^2}{s} + s + \alpha\right)H(s)$$

$$= \left(\frac{\alpha^2+s^2+s\alpha}{s}\right)\cdot\frac{1}{(s^2+\alpha s+\alpha^2)}$$

$$= \frac{(s^2+s\alpha+a^2)}{s(s^2+\alpha s+\alpha^2)} = \frac{1}{s}$$

So, number of poles of $G(s)$ is 1.

5.33 (b)

$$y''(t) - y'(t) - 6y(t) = x(t)$$

$$H(s) = \frac{Y(s)}{X(s)} = \frac{1}{s^2-s-6} = \frac{1}{(s-3)(s+2)}$$

$$H(s) = \frac{\left(\frac{1}{5}\right)}{s-3} + \frac{\left(-\frac{1}{5}\right)}{s+2}$$

If the system is neither causal nor stable, then

$$h(t) = -\frac{1}{5}e^{3t}u(-t) + \frac{1}{5}e^{-2t}u(-t) \quad \text{ROC}: \sigma < -2.$$

5.34 (c)

Applying Laplace transform

$$sX(s) - x(0) = \frac{10}{s} - 0.2\,X(s)$$

$$X(s) = \frac{10}{s(s+0.2)} + \frac{1}{(s+0.2)}$$

Applying inverse Laplace transform, we get

$$x = 50 - 49\,e^{(-0.2)t}$$

5.35 (c)

$$f(t) = \begin{cases} 1 & \text{if } a \le t \le b \\ 0 & \text{otherwise} \end{cases} = u(t-a) - u(t-b)$$

$$\Rightarrow \quad F(s) = \frac{e^{-as} - e^{-bs}}{s}$$

5.36 (c)

ROC of a finite duration signal is entire s-plane.

5.37 (-2)

$$x(t) = \alpha\beta e^{-4t}u(t) + \beta e^{4t}u(-t)$$

Given, $X(s) = \dfrac{16}{s^2 - 16} \qquad -4 < \text{Re}[s] < 4$

$$= \frac{16}{(s+4)(s-4)} = \frac{-2}{s+4} + \frac{2}{s-4}$$

$$x(t) = -2e^{-4t}u(t) - 2e^{4t}u(-t)$$

On comparison $\beta = -2$

5.38 (0.155)

$$g(t) = e^{-t}\sin(2\pi t)\,u(t)$$

$$\text{Area} = \int_{-\infty}^{\infty} g(t)dt$$

$$= \int_{-\infty}^{\infty} \sin(2\pi t)\,u(t)\,e^{-t}dt = G(s)\big|_{s=0}$$

$$G(s) = \frac{2\pi}{(s+1)^2 + 4\pi^2}$$

$$\text{Area} = \left.\frac{2\pi}{(s+1)^2 + 4\pi^2}\right|_{s=0} = 0.155$$

5.39 (b)

$$L(f(t)) = \frac{1}{1-e^{-sT}}\int_0^{T/2} e^{-st}dt$$

$$= \frac{1}{1-e^{-sT}}\left(\frac{e^{-st}}{-s}\right)\Bigg|_0^{T/2}$$

$$= \frac{1}{s(1-e^{-sT})}\cdot(1-e^{-sT/2})$$

$$= \frac{1}{s}\cdot\frac{1-e^{-sT/2}}{(1-e^{-sT/2})(1+e^{-sT/2})}$$

$$= \frac{1}{s}\cdot\frac{1}{1+e^{-sT/2}}$$

5.40 (c)

For 1st order system

$$G(s) = \frac{1}{sT};\; H(s) = 1$$

Impulse response $\qquad R(s) = 1$

$$Y(s) = \left(\frac{G(s)}{1+G(s)H(s)}R(s)\right)$$

$$= \left(\frac{1}{1+sT}\right) = \frac{1}{T}e^{-t/T} \quad \text{for } t \ge 0$$

Unit step response $\qquad R(s) = \dfrac{1}{s}$

$$Y(s) = \frac{1}{s(1+sT)} = \frac{(1+sT)-(sT)}{s(1+sT)}$$

$$= \frac{1}{s} - \frac{T}{(1+sT)} = \frac{1}{s} - \frac{T}{T\left(s+\frac{1}{T}\right)}$$

$$y(t) = 1 - e^{-t/T} \qquad \text{for } t \ge 0$$

Ramp response $\quad R(s) = \dfrac{1}{s^2}$

$$Y(s) = \frac{1}{s^2(1+sT)} = \frac{1}{s^2} - \frac{T}{s} + \frac{T}{s+\frac{1}{T}}$$

$$y(t) = t - T(1-e^{-t/T}) \quad \text{for } t \ge 0$$

5.41 (b)

Given, $\qquad H(s) = e^s + e^{-s}$

$$H(e^{j\omega}) = e^{j\omega} + e^{-j\omega} = 2\cos\omega$$

$x(t)$ —— $H(e^{j\omega})$ —— $y(t)$

If $\qquad x_1(t) = 2\cos\left(\dfrac{2\pi}{3}t\right)$

$$\omega_0 = \frac{2\pi}{3}$$

$$H(j\omega_0) = 2\cos\left(\frac{2\pi}{3}\right) = 2\left(-\frac{1}{2}\right) = -1$$

$$y_1(t) = 2\cos\left(\frac{2\pi}{3}t + 180°\right)$$

if $\qquad x_2(t) = \cos\pi t$

$$\omega_0 = \pi$$

$$H(e^{j\omega_0}) = 2\cos(\pi) = -2$$

$$y_2(t) = 2\cos(\pi t + 180°)$$

$$y(t) = 2\cos\left(\frac{2\pi}{3}t + \pi\right) - 2\cos(\pi t + \pi)$$

$$\omega_1 = \frac{2\pi}{3},\ \omega_2 = \pi$$

$$T_1 = 3 \qquad T_2 = 2$$

$$\therefore \qquad T_0 = 6$$

$$\Rightarrow \qquad \omega_0 = \frac{2\pi}{T_0} = \frac{\pi}{3}$$

$$y(t) = 2\cos(2\omega_0 t + \pi) - 2\cos(3\omega_0 t + \pi)$$

$$y(t) = e^{j(2\omega_0 t+\pi)} + e^{-j(2\omega_0 t+\pi)} - e^{j(3\omega_0 t+\pi)} - e^{-j(3\omega_0 t+\pi)}$$

$$y(t) = -e^{j(2\omega_0 t)} - e^{-j(2\omega_0 t)} + e^{j(3\omega_0 t)} + e^{-j(3\omega_0 t)}$$

$$\therefore \quad C_3 = 1$$

5.42 (d)

- A BIBO stable system can have poles in right half of complex plane, if it is a non-causal system. So, statement-I is wrong.
- A causal and BIBO stable system should have all poles in the left half of complex plane. So, statement-II is correct.
- So, option (d) is correct.

5.43 (0.5)

$$H(s) = \frac{1}{s} \Rightarrow \text{Integrator}$$

$$x(t) = \frac{\sin t}{\pi t} u(t)$$

So, $$y(t) = \int_0^t \frac{\sin\tau}{\pi\tau} d\tau$$

$$y(t)_{t\to\infty} = \frac{1}{\pi}\int_0^\infty \frac{\sin\tau}{\tau} d\tau$$

$$\int_0^\infty \frac{\sin\tau}{\tau} d\tau = \frac{\pi}{2}$$

So, $$y(t)_{t\to\infty} = \frac{1}{\pi}\times\frac{\pi}{2} = \frac{1}{2} = 0.5$$

5.44 (a, c)

Given, $$G(s) = \frac{3-s}{(s+1)(s+3)}$$

$$\therefore \quad Y(s) = \frac{G(s)}{s} = \frac{3-s}{s(s+1)(s+3)}$$

Using partial fractions, we get,

$$Y(s) = \frac{A}{s} + \frac{B}{(s+1)} + \frac{C}{(s+3)}$$

$$A(s^2 + 4s + 3) + B(s^2 + 3s) + C(s^2 + s) = 3 - s$$

$$A + B + C = 0$$

$$4A + 3B + C = -1$$

and $$3A = 3$$

Therefore, we get,

$$A = 1,\ B = -2 \text{ and } C = 1$$

So, $$Y(s) = \frac{1}{s} - \frac{2}{(s+1)} + \frac{1}{(s+3)}$$

and $$y(t) = u(t) - 2e^{-t}u(t) + e^{-3t}u(t)$$

Forced response,

$$y_f(t) = u(t) \Rightarrow \text{option (a)}$$

Some authors also consider the zero state response (ZSR) as the forced response and according to them the correct answer will be option (c).

5.45 (4)

Given that, $$G(j\omega) = \frac{1}{(1+j\omega)(\alpha+j\omega)}$$

$$|G(j\omega)| = \frac{1}{\sqrt{(\omega^2+1)(\omega^2+\alpha^2)}}$$

According to question,

$$|G(j\omega)|_{\omega=3} = \frac{1}{5\sqrt{10}}$$

$$\Rightarrow \quad \frac{1}{\sqrt{(\omega^2+1)(\omega^2+\alpha^2)}} = \frac{1}{5\sqrt{10}}$$

$$\Rightarrow \quad \frac{1}{\sqrt{10(a^2+9)}} = \frac{1}{5\sqrt{10}}$$

$$\alpha^2 + 9 = 25$$

$$\alpha^2 = 16$$

$$\alpha = 4$$

■■■

6 Z-Transform

6.1 The z-transform of the time function $\sum_{k=0}^{\infty} \delta(n-k)$ is

(a) $\frac{z-1}{z}$ (b) $\frac{z}{z-1}$

(c) $\frac{z}{(z-1)^2}$ (d) $\frac{(z-1)^2}{z}$

[1998 : 1 M]

6.2 The z-transform $F(z)$ of the function $f(nT) = a^{nT}$ is

(a) $\frac{z}{z-a^T}$ (b) $\frac{z}{z+a^T}$

(c) $\frac{z}{z-a^{-T}}$ (d) $\frac{z}{z+a^{-T}}$

[1999 : 1 M]

6.3 The z-transform of a signal is given by $C(z) = \frac{1}{4}\frac{z^{-1}(1-z^{-4})}{(1-z^{-1})^2}$. Its final value is

(a) 1/4 (b) zero

(c) 1.0 (d) infinity

[1999 : 2 M]

6.4 The region of convergence of the z-transform of a unit step function is

(a) $|z|>1$

(b) $|z|<1$

(c) (Real part of z) > 0

(d) (Real part of z) < 0

[2001 : 1 M]

6.5 If the impulse response of a discrete-time system is $h[n] = -5^n u[-n-1]$, then the system function $H(z)$ is equal to

(a) $\frac{-z}{z-5}$ and the system is stable

(b) $\frac{z}{z-5}$ and the system is stable

(c) $\frac{-z}{z-5}$ and the system is unstable

(d) $\frac{z}{z-5}$ and the system is unstable

[2002 : 2 M]

6.6 A sequence $x(n)$ with the z-transform $X(z) = z^4 + z^2 - 2z + 2 - 3z^{-4}$ is applied as an input to a linear, time-invariant system with the impulse response $h(n) = 2\delta(n-3)$ where

$$\delta(n) = \begin{cases} 1, & n = 0 \\ 0, & \text{otherwise} \end{cases}$$

The output at $n = 4$ is

(a) –6 (b) zero

(c) 2 (d) –4

[2003 : 1 M]

6.7 The z-transform of a system is $H(z) = \frac{z}{z-0.2}$

If the ROC is $|z| < 0.2$, then the impulse response of the system is

(a) $(0.2)^n u[n]$ (b) $(0.2)^n u[-n-1]$

(c) $-(0.2)^n u[n]$ (d) $-(0.2)^n u[-n-1]$

[2004 : 1 M]

6.8 A causal LTI system is described by the difference equation

$$2y[n] = \alpha y[n-2] - 2x[n] + \beta x[n-1]$$

The system is stable only if

(a) $|\alpha| = 2, |\beta| < 2$

(b) $|\alpha| > 2, |\beta| > 2$

(c) $|\alpha| < 2$, any value of β

(d) $|\beta| < 2$, any value of α

[2004 : 2 M]

6.9 The region of convergence of z-transform of the sequence

$\left(\frac{5}{6}\right)^n u(n) - \left(\frac{6}{5}\right)^n u(-n-1)$ must be

(a) $|z| < \frac{5}{6}$ (b) $|z| > \frac{5}{6}$

(c) $\frac{5}{6} < |z| < \frac{6}{5}$ (d) $\frac{6}{5} < |z| < \infty$

[2005 : 1 M]

6.10 If the region of convergence of $x_1[n] + x_2[n]$ is $\frac{1}{3} < |z| < \frac{2}{3}$, then the region of convergence of $x_1[n] - x_2[n]$ includes

(a) $\frac{1}{3}<|z|<3$ (b) $\frac{2}{3}<|z|<3$

(c) $\frac{3}{2}<|z|<3$ (d) $\frac{1}{3}<|z|<\frac{2}{3}$

[2006 : 1 M]

6.11 The z-transform $X[z]$ of a sequence $x[n]$ is given by $X[z]=\frac{0.5}{1-2z^{-1}}$. It is given that the region of convergence of $X[z]$ includes the unit circle. The value of $x[0]$ is

(a) −0.5 (b) 0

(c) 0.25 (d) 0.5 [2007 : 2 M]

Statement for linked Answer Questions 6.12 and 6.13:

In the following network, the switch is closed at $t = 0^-$ and the sampling starts from $t = 0$. The sampling frequency is 10 Hz.

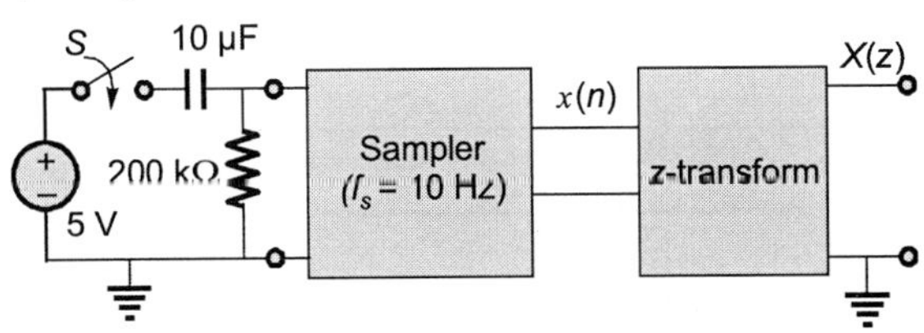

6.12 The sample $x(n)$ (n = 0,1, 2, ...) are given by

(a) $5(1-e^{-0.05n})$ (b) $5e^{-0.05n}$

(c) $5(1-e^{-5n})$ (d) $5e^{-5n}$

[2008 : 2 M]

6.13 The expression and the region of convergence of the z-transform of the sampled signal are

(a) $\frac{5z}{z-e^{5}}, |z|<e^{-5}$

(b) $\frac{5z}{z-e^{-0.05}}, |z|<e^{-0.05}$

(c) $\frac{5z}{z-e^{-0.05}}, |z|>e^{-0.05}$

(d) $\frac{5z}{z-e^{-5}}, |z|>e^{-5}$ [2008 : 2 M]

6.14 The ROC of z-transform of the discrete time sequence

$$x(n)=\left(\frac{1}{3}\right)^n u(n)-\left(\frac{1}{2}\right)^n u(-n-1) \text{ is}$$

(a) $\frac{1}{3}<|z|<\frac{1}{2}$ (b) $|z|>\frac{1}{2}$

(c) $|z|<\frac{1}{3}$ (d) $2<|z|<3$

[2009 : 1 M]

6.15 Consider the z-transform $X(z) = 5z^2 + 4z^{-1} + 3$; $0 < |z| < \infty$. The inverse z-transform $x[n]$ is

(a) $5\delta[n+2]+3\delta[n]+4\delta[n-1]$

(b) $5\delta[n-2]+3\delta[n]+4\delta[n+1]$

(c) $5u[n+2]+3u[n]+4u[n-1]$

(d) $5u[n-2]+3u[n]+4u[n+1]$

[2010 : 1 M]

6.16 Two discrete time systems with impulse responses $h_1[n]=\delta[n-1]$ and $h_2[n]=\delta[n-2]$ are connected in cascade. The overall impulse response of the cascaded system is

(a) $\delta[n-1]+\delta[n-2]$

(b) $\delta[n-4]$

(c) $\delta[n-3]$

(d) $\delta[n-1]\,\delta[n-2]$ [2010 : 1 M]

6.17 The transfer function of a discrete time LTI system is given by

$$H(z)=\frac{2-\frac{3}{4}z^{-1}}{1-\frac{3}{4}z^{-1}+\frac{1}{8}z^{-2}}$$

Consider the following statements:

S_1 : The system is stable and causal for ROC : $|z|>\frac{1}{2}$

S_2 : The system is stable but not causal for ROC : $|z|<\frac{1}{4}$

S_3 : The system is neither stable nor causal for ROC : $\frac{1}{4}<|z|<\frac{1}{2}$

Which one of the following statements is valid?

(a) Both S_1 and S_2 are true

(b) Both S_2 and S_3 true

(c) Both S_1 and S_3 are true

(d) S_1, S_2 and S_3 are all true [2010 : 2 M]

6.18 Two systems $H_1(z)$ and $H_2(z)$ are connected in cascade as shown below. The overall output $y(n)$ is the same as the input $x(n)$ with a one unit delay. The transfer function of the second system $H_2(z)$ is

$x(n) \rightarrow \boxed{H_1(z)=\frac{(1-0.4z^{-1})}{(1-0.6z^{-1})}} \rightarrow \boxed{H_2(z)} \rightarrow y(n)$

(a) $\frac{(1-0.6z^{-1})}{z^{-1}(1-0.4z^{-1})}$ (b) $\frac{z^{-1}(1-0.6z^{-1})}{(1-0.4z^{-1})}$

(c) $\frac{z^{-1}(1-0.4z^{-1})}{(1-0.6z^{-1})}$ (d) $\frac{(1-0.4z^{-1})}{z^{-1}(1-0.6z^{-1})}$

[2011 : 2 M]

6.19 If $x[n] = (1/3)^{|n|} - (1/2)^n u[n]$, then the region of convergence (ROC) of its Z-transform in the Z-plane will be

(a) $\frac{1}{3} < |z| < 3$ (b) $\frac{1}{3} < |z| < \frac{1}{2}$

(c) $\frac{1}{2} < |z| < 3$ (d) $\frac{1}{3} < |z|$

[2012 : 1 M]

6.20 Let $x[n] = \left(-\frac{1}{9}\right)^n u(n) - \left(-\frac{1}{3}\right)^n u(-n-1)$.

The Region of Convergence (ROC) of the z-transform of $x[n]$

(a) is $|z| > \frac{1}{9}$ (b) is $|z| < \frac{1}{3}$

(c) is $\frac{1}{3} > |z| > \frac{1}{9}$ (d) does not exist

[2014 : 2 M, Set-1]

6.21 Let $x[n] = x[-n]$. Let $X(z)$ be the z-transform of $x[n]$. If $0.5 + j0.25$ is a zero of $X(z)$, which one of the following must also be a zero of $X(z)$.

(a) $0.5 - j0.25$ (b) $\frac{1}{(0.5 + j0.25)}$

(c) $\frac{1}{(0.5 - j0.25)}$ (d) $2 + j4$

[2014 : 1 M, Set-2]

6.22 The input-output relationship of a causal stable LTI system is given as

$$y[n] = \alpha y[n-1] + \beta x[n]$$

If the impulse response $h[n]$ of this system satisfies the condition $\sum_{n=0}^{\infty} h[n] = 2$, the relationship between α and β is

(a) $\alpha = 1 - \beta/2$ (b) $\alpha = 1 + \beta/2$

(c) $\alpha = 2\beta$ (d) $\alpha = -2\beta$

[2014 : 2 M, Set-2]

6.23 For an all-pass system $H(z) = \frac{(z^{-1} - b)}{(1 - az^{-1})}$, where $|H(e^{-j\omega})| = 1$, for all ω. If $\text{Re}(a) \neq 0$, $\text{Im}(a) \neq 0$, then b equals

(a) a (b) a^*

(c) $1/a^*$ (d) $1/a$

[2014 : 1 M, Set-3]

6.24 Let $H_1(z) = (1 - pz^{-1})^{-1}$, $H_2(z) = (1 - qz^{-1})^{-1}$, $H(z) = H_1(z) + r\,H_2(z)$. The quantities p, q, r are real numbers. Consider $p = \frac{1}{2}, q = -\frac{1}{4}, |r| < 1$. If the zero of $H(z)$ lies on the unit circle, then $r =$ ________.

[2014 : 2 M, Set-3]

6.25 The z-transform of the sequence $x[n]$ is given by $X(z) = \frac{1}{(1 - 2z^{-1})^2}$, with the region of convergence $|z| > 2$. Then, $x[2]$ is _____.

(a) (b)

(c) (d)

[2014 : 2 M, Set-3]

6.26 For the discrete-time system shown in the figure, the poles of the system transfer function are located at

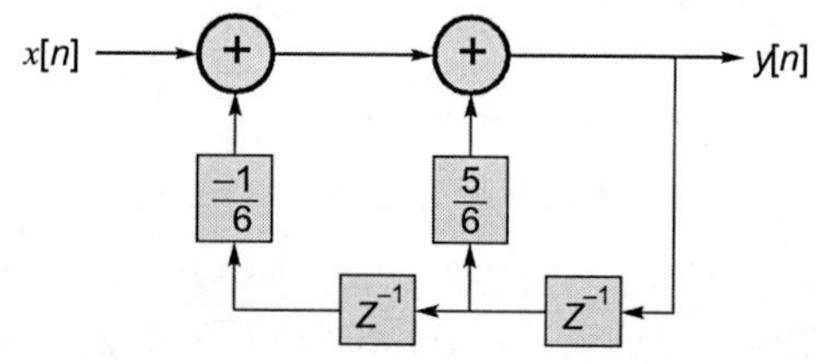

(a) 2, 3 (b) $\frac{1}{2}, 3$

(c) $\frac{1}{2}, \frac{1}{3}$ (d) $2, \frac{1}{3}$

[2015 : 2 M, Set-1]

6.27 The pole-zero diagram of a causal and stable discrete-time system is shown in the figure. The zero at the origin has multiplicity 4. The impulse response of the system is $h[n]$. If $h[0] = 1$, we can conclude

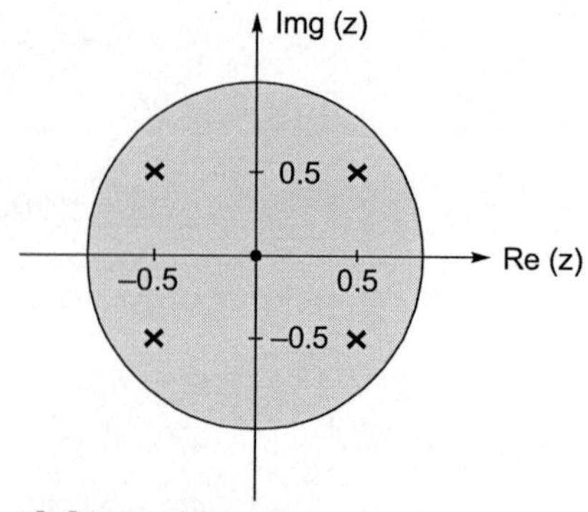

(a) $h[n]$ is real for all n

(b) $h[n]$ is purely imaginary for all n

(c) $h[n]$ is real for only even n

(d) $h[n]$ is purely imaginary for only odd n

[2015 : 2 M, Set-1]

6.28 Two causal discrete-time signals $x[n]$ and $y[n]$ are related as $y[n] = \sum_{m=0}^{n} x[m]$. If the z-transform of $y[n]$ is $\frac{2}{z(z-1)^2}$, the value of $x[2]$ is ______.

[2015 : 1 M, Set-2]

6.29 A realization of a stable discrete time system is shown in figure. If the system is excited by a unit step sequence input $x[n]$, the response $y[n]$ is

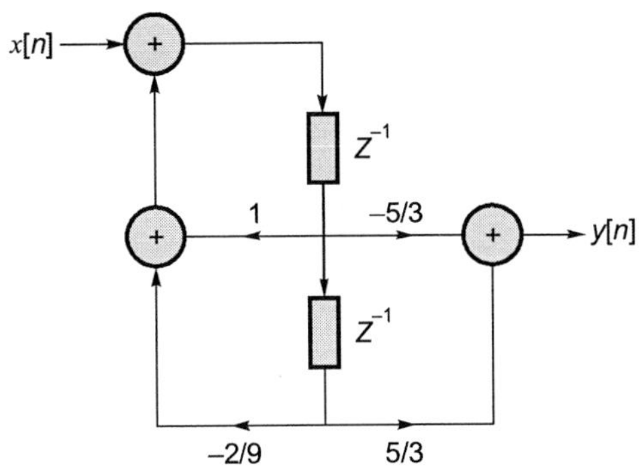

(a) $4\left(-\frac{1}{3}\right)^n u[n] - 5\left(-\frac{2}{3}\right)^n u[n]$

(b) $5\left(-\frac{2}{3}\right)^n u[n] - 3\left(-\frac{1}{3}\right)^n u[n]$

(c) $5\left(\frac{1}{3}\right)^n u[n] - 5\left(\frac{2}{3}\right)^n u[n]$

(d) $5\left(\frac{2}{3}\right)^n u[n] - 5\left(\frac{1}{3}\right)^n u[n]$

[2015 : 2 M, Set-3]

6.30 Consider a four point moving average filter defined by the equation $y[n] = \sum_{i=0}^{3} \alpha_i\, x[n-i]$. The condition on the filter coefficients that results in a null at zero frequency is

(a) $\alpha_1 = \alpha_2 = 0;\ \alpha_0 = -\alpha_3$
(b) $\alpha_1 = \alpha_2 = 1;\ \alpha_0 = -\alpha_3$
(c) $\alpha_0 = \alpha_3 = 0;\ \alpha_1 = \alpha_2$
(d) $\alpha_1 = \alpha_2 = 0;\ \alpha_0 = \alpha_3$

[2015 : 1 M, Set-3]

6.31 Suppose $x[n]$ is an absolutely summable discrete-time signal. Its z-transform is a rational function with two poles and two zeroes. The poles are at $z = \pm 2j$. Which one of the following statements is TRUE for the signal $x[n]$?

(a) It is a finite duration signal.
(b) It is a causal signal.
(c) It is a non-causal signal.
(d) It is a periodic signal.

[2015 : 2 M, Set-3]

6.32 Consider the sequence $x[n] = a^n u[n] + b^n u[n]$, where $u[n]$ denotes the unit-step sequence and $0 < |a| < |b| < 1$. The region of convergence (ROC) of the z-transform of $x[n]$ is

(a) $|z| > |a|$
(b) $|z| > |b|$
(c) $|z| < |a|$
(d) $|a| < |z| < |b|$

[2016 : 1 M, Set-1]

6.33 The ROC (region of convergence) of the z-transform of a discrete-time signal is represented by the shaded region in the z-plane. If the signal

$$x[n] = (2.0)^{|n|}, -\infty < n < +\infty$$

then the ROC of its z-transform is represented by

(a)

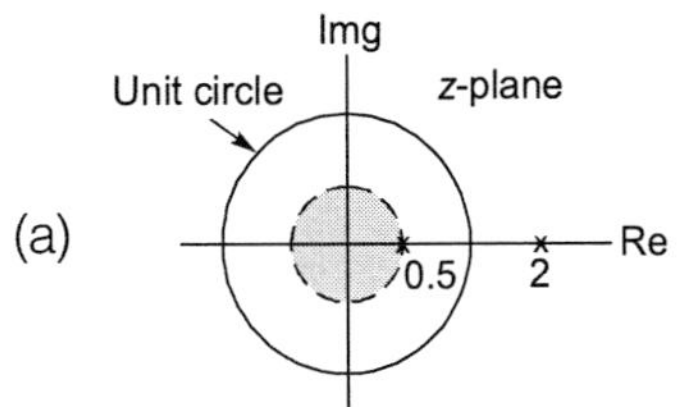

(b)

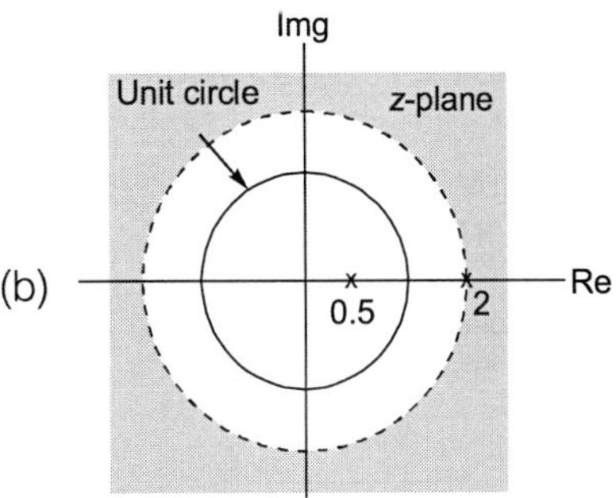

(c)

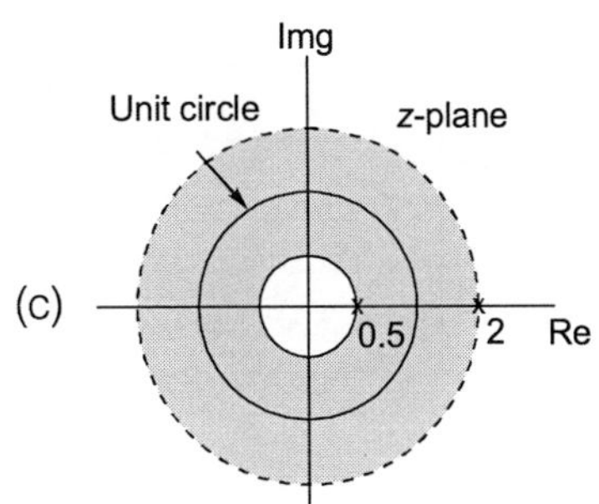

(d)

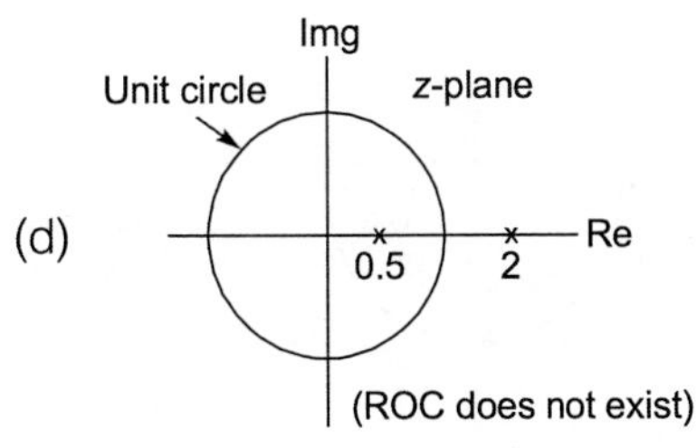

[2016 : 2 M, Set-3]

6.34 A discrete-time signal $x[n] = \delta[n-3] + 2\delta[n-5]$ has z-transform $X(z)$. If $Y(z) = X(-z)$ is the z-transform of another signal $y[n]$, then

(a) $y[n] = x[n]$ (b) $y[n] = x[-n]$

(c) $y[n] = -x[n]$ (d) $y[n] = -x[-n]$

[2016 : 1 M, Set-3]

6.35 A discrete-time all-pass system has two of its poles at $0.25\angle 0°$ and $2\angle 30°$ Which one of the following statements about the system is TRUE?

(a) It has two more poles at $0.5\angle 30°$ and $4\angle 0°$.

(b) It is stable only when the impulse response is two-sided.

(c) It has constant phase response over all frequencies.

(d) It has constant phase response over the entire z-plane.

[2018 : 1 M]

6.36 Let $H(z)$ be the z-transform of a real-valued discrete-time signal $h[n]$. If $P(z) = H(z)H\left(\frac{1}{z}\right)$ has a zero at $z = \frac{1}{2} + \frac{1}{2}j$, and $P(z)$ has a total of four zeros, which one of the following plots represents all the zeros correctly?

(a)

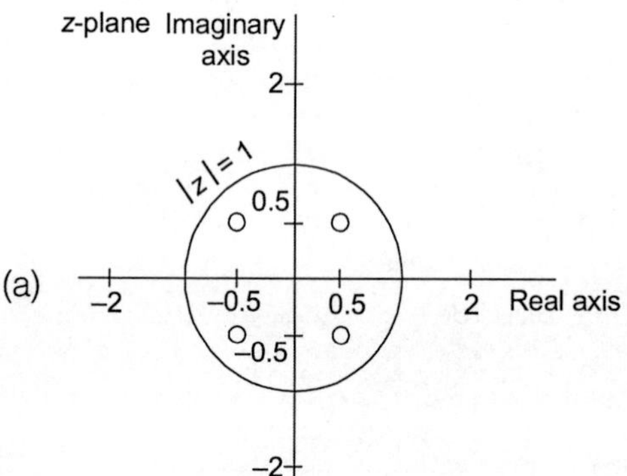

(b)

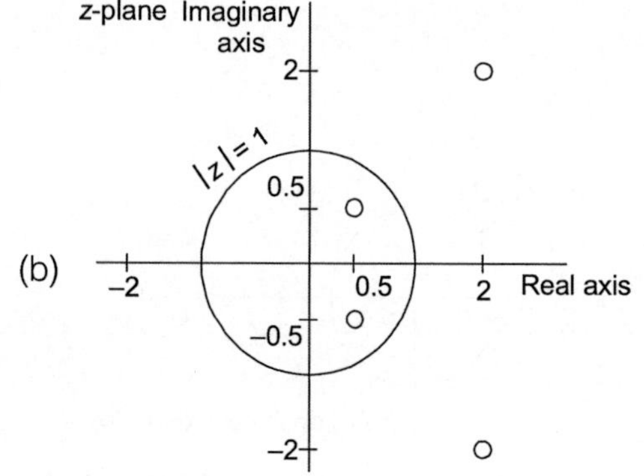

(c)

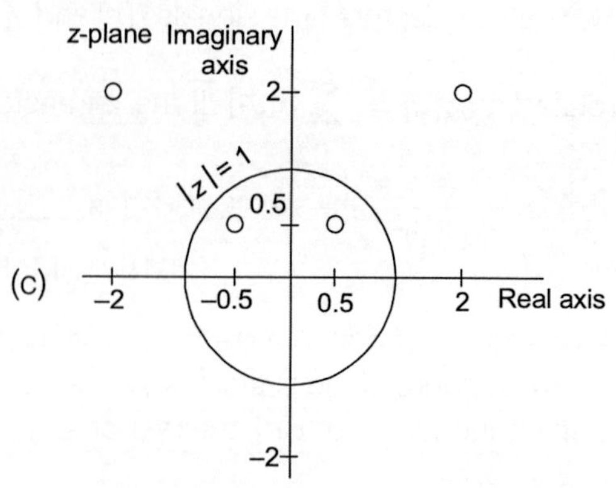

(d)

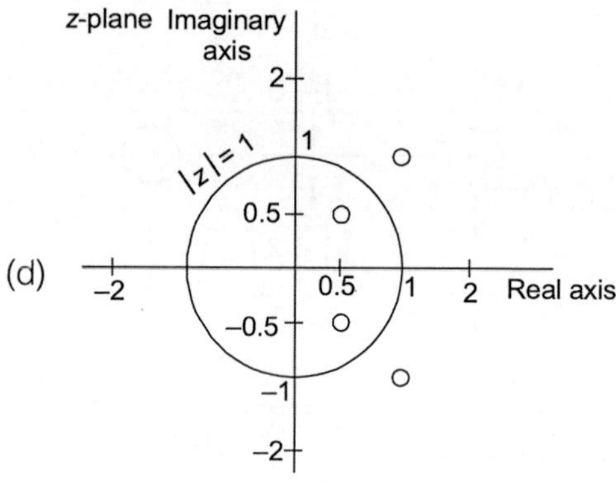

[2019 : 1 M]

6.37 Which one of the following pole-zero corresponds to the transfer function of an LTI system characterized by the input-output difference equation given below?

$$y[n] = \sum_{k=0}^{3} (-1)^k x[n-k]$$

(a)

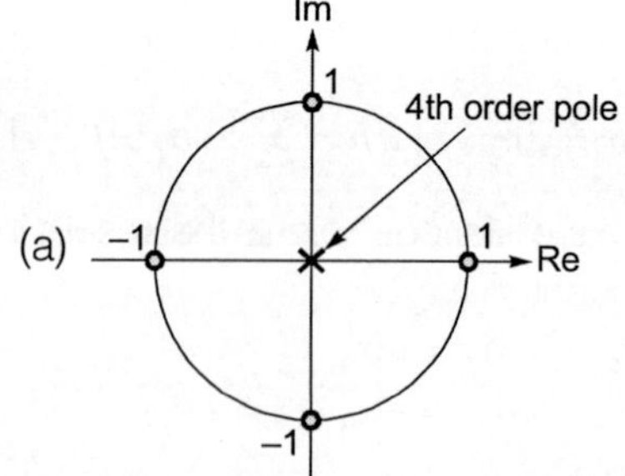

(b)

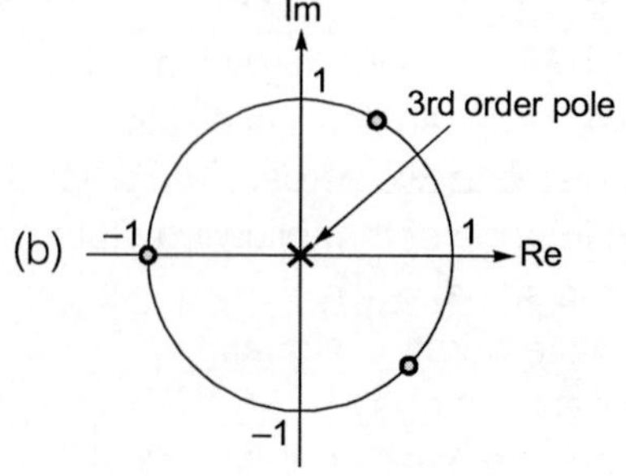

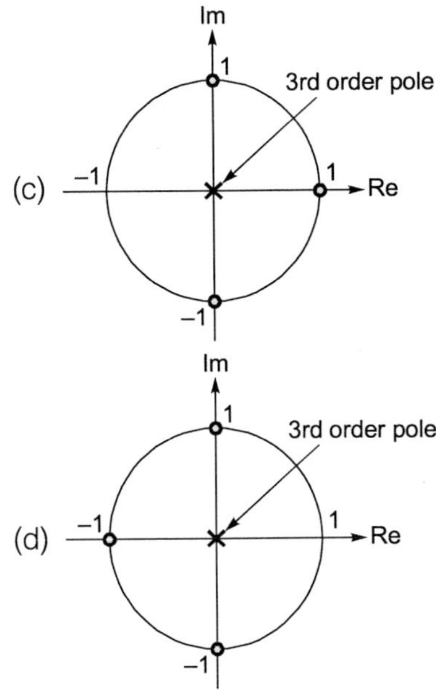

[2020 : 1 M]

6.38 The transfer function of a stable discrete-time LTI system is $H(z) = \dfrac{K(z-\alpha)}{z+0.5}$. where K and α are real numbers. The value of α (rounded off to one decimal place) with $|\alpha| > 1$, for which the magnitude response of the system is constant over all frequencies, is _____.

[2020 : 2 M]

6.39 For a causal discrete-time LTI system with transfer function

$$H(z) = \frac{2z^2+3}{\left(z+\frac{1}{3}\right)\left(z-\frac{1}{3}\right)}$$

Which of the following statements is/are true?

(a) The system is stable.

(b) The final value of the impulse response is 0.

(c) The system is a minimum phase system.

(d) The initial value of the impulse response is 2.

[2024 : 1 Mark]

■■■

Answers Z-Transform

6.1 (b)	6.2 (a)	6.3 (c)	6.4 (a)	6.5 (b)	6.6 (b)	6.7 (d)
6.8 (c)	6.9 (c)	6.10 (d)	6.11 (b)	6.12 (b)	6.13 (c)	6.14 (a)
6.15 (a)	6.16 (c)	6.17 (c)	6.18 (b)	6.19 (c)	6.20 (c)	6.21 (b)
6.22 (a)	6.23 (b)	6.24 (–0.5)	6.25 (12)	6.26 (c)	6.27 (a)	6.28 (0)
6.29 (c)	6.30 (a)	6.31 (c)	6.32 (b)	6.33 (d)	6.34 (c)	6.35 (b)
6.36 (d)	6.37 (c)	6.38 (–2)	6.39 (a, b, d)			

Explanations Z-Transform

6.1 (b)

Given that

$$x(n) = \sum_{k=0}^{\infty} \delta(n-k)$$

$$x(n) = \delta(n) + \delta(n-1) + \delta(n-2) + \cdots$$

$$x(n) = u(n)$$

$$X(z) = \text{Z.T.}[u(n)]$$

$$X(z) = \frac{z}{z-1}$$

6.2 (a)

Z-transform is given by

$$z[f(nT)] = \sum_{n=-\infty}^{\infty} f(nT)z^{-n}$$

$$= \sum_{n=-\infty}^{\infty} a^{nT}z^{-n} = \sum_{n=-\infty}^{\infty} (a^T z^{-1})^n$$

$$F(z) = \frac{1}{1-a^T z^{-1}} = \frac{z}{z-a^T}$$

6.3 (c)

Final value theorem

$$\lim_{n\to\infty} x[n] = \lim_{z\to 1}(1-z^{-1})X(z)$$

$$\lim_{z\to 1}\frac{1}{4}(1-z^{-1})\frac{z^{-1}(1-z^{-4})}{(1-z^{-1})^2}$$

$$= \lim_{z\to 1}\frac{1}{4}\frac{z^{-1}(1-z^{-4})}{(1-z^{-1})}$$

$$= \lim_{z\to 1}\frac{1}{4}\frac{1}{z}\frac{(z^4-1)/z^4}{(z-1)/z}$$

$$= \lim_{z\to 1}\frac{1}{4}\frac{(z^2-1)(z^2+1)}{z^4(z-1)}$$

$$= \lim_{z\to 1}\frac{1}{4}\frac{(z+1)(z-1)(z^2+1)}{z^4(z-1)} = \frac{1}{4}\times 4 = 1$$

6.4 (a)

$$h(n) = u(n)$$

$$H(z) = \sum_{n=0}^{\infty} 1.z^{-n}$$

For the convergence of $H(z)$

$$\sum_{n=0}^{\infty} (z^{-1})^n < \infty$$

$\therefore$ ROC is the range of values of z for which

$$|z^{-1}| < 1 \text{ or } |z| > |1|$$

6.5 (b)

$$h(n) = -5^n u[-n-1]$$

$$H(z) = \sum_{n=-\infty}^{\infty} h(n)z^{-n}$$

$$H(z) = \sum_{n=-\infty}^{-1} -5^n z^{-n}, \text{ Let } n=-k$$

$$= -\sum_{n=\infty}^{1} (5z^{-1})^{-k}$$

$$= 1 - \sum_{K=0}^{\infty} (5^{-1}z)^k$$

$$= 1 - \frac{1}{1-5^{-1}z}, |5^{-1}z| < 1, |z| < |5|$$

$$= 1 - \frac{5}{5-z}$$

$$= -\frac{z}{5-z} = \frac{z}{z-5}$$

The system is stable because ROC includes unit circle.

6.6 (b)

$$Y(z) = H(z)\,X(z)$$

$$H(z) = 2z^{-3}$$

$$\therefore\ Y(z) = 2z^{-3}(z^4 + z^2 - 2z + 2 - 3z^{-4})$$

$= 2(z + z^{-1} - 2z^{-2} + 2z^{-3} - 3z^{-7})$

Taking inverse of z-transform

$y(n) = 2[\delta(n+1) + \delta(n-1) - 2\delta(n-2) + 2\delta(n-3) - 3\delta(n-7)]$

At $n = 4$, $y(4) = 0$

6.7 (d)

$$H(z) = \frac{z}{z-0.2} = \frac{z}{z(1-0.2z^{-1})}$$

$$H(z) = \frac{1}{1-0.2z^{-1}}$$

ROC is $|z| < 0.2$

Comparing with,

$$-a^n u(-n-1) \rightarrow \frac{1}{1-az^{-1}}, |z| < a$$

We get, $h(n) = -(0.2)^n u(-n-1)$

6.8 (c)

$$2y[n] = \alpha y[n-2] - 2x[n] + \beta x[n-1]$$

Taking z-transform

$$2Y(z) = \alpha Y(z)z^{-2} - 2X(z) + \beta X(z)z^{-1}$$

$$Y(z)[2 - \alpha z^{-2}] = X(z)[\beta z^{-1} - 2]$$

$$\frac{Y(z)}{X(z)} = \left(\frac{\beta z^{-1} - 2}{2 - \alpha z^{-2}}\right)$$

For system to be stable, β can be of any value.

$$2 - \alpha z^{-2} = 0$$

$$2z^2 - \alpha = 0$$

$$z = \sqrt{\frac{\alpha}{2}} < 1$$

$$|\alpha| < 2$$

For system to be stable all poles should be inside unity circle.

6.9 (c)

z-transform of $\left(\frac{5}{6}\right)^n u(n) - \left(\frac{6}{5}\right)^n u(-n-1)$

$$= \frac{1}{1-\frac{5}{6}z^{-1}} + \left[1 - \frac{1}{1-\left(\frac{6}{5}\right)^{-1} z}\right]$$

(ROC) $= |z| > \frac{5}{6}$ (ROC) $= |z| < \frac{6}{5}$

$$\frac{5}{6} < |z| < \frac{6}{5}$$

6.10 (d)

ROC remains the same for addition and subtraction in z-domain.

6.11 (b)

$$X(z) = \frac{0.5}{1-2z^{-1}}$$

∵ ROC includes unit circle

⇒ Left handed system

⇒ $x(n) = -(0.5)(2)^n u(-n-1)$

⇒ $x(0) = 0$

6.12 (b)

$$V_R(s) = \left(\frac{200\times 10^3}{200\times 10^3 + \frac{1}{10\times 10^{-6}s}}\right)\frac{5}{s}$$

$$= \frac{5\times 2\times 10^5 \times 10\times 10^{-6}}{2\times 10^5 \times 10^{-5}s + 1} = \frac{10}{2s+1} = \frac{5}{s+0.5}$$

$$v_R(t) = 5e^{-0.5t}$$

Therefore, samples

$$x(n) = 5e^{-0.5n/10} = 5e^{-0.05n}$$

6.13 (c)

$$X(z) = \sum_{n=0}^{\infty} 5e^{-0.05n} z^{-n}$$

$$= 5\sum_{n=0}^{\infty} (e^{-0.05}z^{-1})^n$$

For $|e^{-0.05} z^{-1}| < 1$ or $|z| > e^{-0.05}$

$$X(z) = \frac{5z}{z - e^{-0.05}}$$

6.14 (a)

$$x(n) = (1/3)^n u(n) - (1/2)^n u(-n-1)$$

$(1/3)^n u(n)$ is right sided signal so ROC will be

$$|z| > 1/3 \quad \text{... (i)}$$

$-(1/2)^n u(-n-1)$ is left sided signal so ROC will be

$$|z| < 1/2 \quad \text{... (ii)}$$

from (i) and (ii) we see that ROC of the function will be

$$1/3 < |z| < 1/2$$

6.15 (a)

$$\delta[n + n_0] \xleftrightarrow{Z} z^{n_0}$$

$$X(z) = 5z^2 + 4z^{-1} + 3;\ 0 < |z| < \infty$$

$$x(n) = 5\delta(n+2) + 4\delta(n-1) + 3\delta(n)$$

6.16 (c)

$$h_1[n] = \delta[n-1] \xleftrightarrow{Z} H_1(z) = z^{-1}$$
$$h_2[n] = \delta[n-2] \xleftrightarrow{Z} H_2(z) = z^{-2}$$

Overall impulse response in z-domain,

$$H(z) = H_1(z)\,H_2(z)$$
$$= z^{-1}\,z^{-2} = z^{-3}$$

Overall impulse response in discrete-time domain,

$$h[n] = \delta(n-3)$$

6.17 (c)

1. A discrete-time LTI system is causal if and only if the ROC of its system function is the exterior of a circle, including infinity.
2. A discrete-time LTI system is stable if and only if the ROC of its system function includes the unit circle, $|z| = 1$.

$$H(z) = \frac{\left(1-\frac{1}{4}z^{-1}\right)+\left(1-\frac{1}{2}z^{-1}\right)}{\left(1-\frac{1}{4}z^{-1}\right)\left(1-\frac{1}{2}z^{-1}\right)}$$

or $$H(z) = \frac{1}{1-\frac{1}{4}z^{-1}}+\frac{1}{1-\frac{1}{2}z^{-1}}$$

For ROC : $|z| > \frac{1}{2}$, the system is stable and causal.

For ROC : $|z| < \frac{1}{4}$, ROC does not include unit circle. So, system is not stable.

For ROC : $\frac{1}{4} < |z| < \frac{1}{2}$, ROC does not include unit circle. So, system is not stable.

Also ROC is not the exterior of $|z| = \frac{1}{2}$. So, it is not causal.

6.18 (b)

$$y[n] = x[n-1]$$

Taking z-transform of both sides

$$Y(z) = z^{-1}\,X(z)$$
$$\frac{Y(z)}{X(z)} = z^{-1}$$

For cascaded system

$$H(z) = H_1(z)\cdot H_2(z)$$
$$z^{-1} = \frac{(1-0.4z^{-1})}{(1-0.6z^{-1})}H_2(z)$$
$$\therefore \quad H_2(z) = \frac{z^{-1}(1-0.6z^{-1})}{(1-0.4z^{-1})}$$

6.19 (c)

$$x[n] = \left(\frac{1}{3}\right)^{|n|} - \left(\frac{1}{2}\right)^{n} u[n]$$
$$x[n] = \left(\frac{1}{3}\right)^{n} u[n] + \left(\frac{1}{3}\right)^{-n} u[-n-1] - \left(\frac{1}{2}\right)^{n} u[n]$$
$$\left(\frac{1}{3}\right)^{n} u[n] \longleftrightarrow \frac{1}{1-\frac{1}{3}z^{-1}} \qquad \text{ROC } |z| > \frac{1}{3}$$
$$\left(\frac{1}{3}\right)^{-n} u[-n-1] \longleftrightarrow \frac{-1}{1-3z^{-1}} \qquad \text{ROC } |z| < 3$$
$$\left(\frac{1}{2}\right)^{n} u[n] \longleftrightarrow \frac{1}{1-\frac{1}{2}z^{-1}} \qquad \text{ROC } |z| > \frac{1}{2}$$

So overall ROC will be intersection of there ROCs i.e.

$$\frac{1}{2} < |z| < 3$$

6.20 (c)

$$x(n) = \underbrace{\left(-\frac{1}{9}\right)^{n} u(n)}_{x_1(n)} - \underbrace{\left(-\frac{1}{3}\right)^{n} u(-n-1)}_{x_2(n)}$$
$$x_1(n) = \left(-\frac{1}{9}\right)^{n} u(n) \quad \to \text{Right sided signal}$$
$$X_1(z) = \frac{1}{1-\left(-\frac{1}{9}\right)z^{-1}}$$

ROC $|z| \geq \left|\frac{1}{9}\right|$ or $|z| > \frac{1}{9}$

$$x_2(n) = \left(-\frac{1}{3}\right)^{n} u(-n-1) \quad \to \text{Left sided signal}$$
$$X_2(z) = \frac{1}{1-\left(-\frac{1}{3}\right)z^{-1}} \quad ; \qquad |z| \leq \frac{1}{3}$$

Now, when both the transforms are added using linearity property,

ROC must be between $\frac{1}{3} > |z| > \frac{1}{9}$

6.21 (b)

Given $x[n] = x[-n]$

As we know that,

$$x[n] \xleftrightarrow{Z} X[z]$$
$$x[-n] \xleftrightarrow{Z} X[z^{-1}]$$

So, when $(0.5 + j0.25)$ is a zero of $X(z)$.

Now $x[-n] \longleftrightarrow X(z^{-1})$

Zero of $X(z) = \dfrac{1}{(0.5 + j0.25)}$

6.22 (a)

Input-output relation is given as

$y[n] = \alpha y(n-1) + \beta x[n]$

Taking Z-transform,

$Y(z) = \alpha z^{-1} Y(z) + \beta X(z)$

$Y(z)[1 - \alpha z^{-1}] = \beta X(z)$

$$\frac{Y(z)}{X(z)} = H(z) = \frac{\beta}{1-\alpha z^{-1}}$$

$h[n] = \beta \cdot \alpha^n u[n]$

Also $\sum_{n=0}^{\infty} h[n] = 2 \Rightarrow \sum_{n=0}^{\infty} \beta \cdot \alpha^n u[n] = 2$

$$\frac{\beta}{1-\alpha} = 2 \Rightarrow \beta = 2 - 2\alpha$$

$\Rightarrow 2\alpha = 2 - \beta$

$$\alpha = 1 - \frac{\beta}{2}$$

6.23 (b)

For the given system

$$H(z) = \frac{(z^{-1} - b)}{(1 - az^{-1})} \quad \text{...(i)}$$

The pole lie at $z = a$ and the zero lie at $z = \dfrac{1}{b}$

also 'a' is complex in nature.

Thus, $a = |a| e^{j\angle a}$

For an all pass system where $|H(e^{-j\omega})| = 1$, for all ω, if the pole lies at 'a' then the zero must be present at $\dfrac{1}{a^*}$.

$$\therefore \quad \left(\frac{1}{b}\right) = \frac{1}{|a| e^{-j\angle a}} = \frac{1}{a^*}$$

or $b = a^*$

6.24 (-0.5)

Here, $H_1(z) = \dfrac{1}{1 - pz^{-1}}$

$H_2(z) = \dfrac{1}{1 - qz^{-1}}$

also $p = \dfrac{1}{2}, \; q = -\dfrac{1}{4}$

$\therefore \quad H(z) = H_1(z) + rH_2(z)$

$$= \frac{1}{1 - pz^{-1}} + \frac{r}{1 - qz^{-1}}$$

$$= \frac{1}{1 - \frac{1}{2}z^{-1}} + \frac{r}{1 + \frac{1}{4}z^{-1}}$$

$$H(z) = \frac{\left(1 + \frac{1}{4}z^{-1}\right) + r\left(1 - \frac{1}{2}z^{-1}\right)}{\left(1 - \frac{1}{2}z^{-1}\right)\left(1 + \frac{1}{4}z^{-1}\right)} \quad \text{...(i)}$$

zero of $H(z) \Rightarrow$

$(1 + r) - (q + pr) z^{-1} = 0$ ...(ii)

or $z = \dfrac{q + pr}{1 + r}$

Since, the zero of $H(z)$ lies on the unit circle therefore $|z| = 1$ or $z = \pm 1$

Taking $z = 1$, we get

$$z = \frac{q + pr}{1 + r} = 1 \quad \text{...(iii)}$$

By solving above expression with $p = \dfrac{1}{2}$ and $q = -\dfrac{1}{4}$, we get

$r = -2.5$

But $|r| < 1$ or $-1 < r < 1$

Taking $z = -1$ in equation (iii), we get

$r = -0.5$

6.25 (12)

Given, $X(z) = \dfrac{1}{(1 - 2z^{-1})^2}, \quad |Z| > 2$

$X(z) = (1 - 2z^{-1})^{-2}$

$X(z) = 1 + 4z^{-1} + 12z^{-2} + \cdots$

(using binomial expansion)

Taking inverse z-transform, we get

$\therefore \quad x[n] = \{\underset{\uparrow}{1}, \; 4, \; 12, \; \cdots$

$$\boxed{x[2] = 12}$$

6.26 (c)

The difference equation of the system

$$x(n) - \frac{1}{6}y(n-2) + \frac{5}{6}y(n-1) = y(n)$$

$$H(z) = \frac{Y(z)}{X(z)} = \frac{1}{1 + \frac{1}{6}z^{-2} - \frac{5}{6}z^{-1}}$$

Poles are at $z = \dfrac{1}{2}, \dfrac{1}{3}$

6.27 (a)

$$H(z) = \frac{z^4}{\left[\left(z-\frac{1}{2}\right)^2+\frac{1}{4}\right]\left[\left(z+\frac{1}{2}\right)^2+\frac{1}{4}\right]}$$

$$\text{Poles} = \frac{1}{2} \pm j\frac{1}{2}, -\frac{1}{2} \pm j\frac{1}{2}$$

$$= \frac{z^4}{\left(z^2-z+\frac{1}{2}\right)\left(z^2+z+\frac{1}{2}\right)}$$

$$= \frac{z^4}{z^4+z^2+\frac{1}{4}-z^2}$$

$$H(z) = \frac{z^4}{z^4+\frac{1}{4}}$$

$$= Z^4+\frac{1}{4}\Big)\ Z^4\left(1-\frac{1}{4}z^{-4}\cdots\right.$$

$$\frac{Z^4+\frac{1}{4}}{-\frac{1}{4}}$$

6.28 (0)

$$y[n] = \sum_{m=0}^{n} x[m] = x[n]*u[n]$$

$$Y(z) = X(z)\cdot\left(\frac{z}{z-1}\right)$$

Given

$$Y(z) = \frac{2}{z(z-1)^2} = X(z)\cdot\frac{z}{z-1}$$

$$\Rightarrow X(z) = \frac{2\cdot z^{-2}}{z-1} = 2z^{-3}\left(\frac{z}{z-1}\right)$$

$$\Rightarrow x[n] = 2u[n-3]$$

Therefore, $x[2]=0$

6.29 (c)

$$\frac{Y(z)}{X(z)} = \frac{\frac{5}{3}[1-z]}{z^2-z+\frac{2}{9}}$$

$$\Rightarrow Y(z) = \frac{5}{3}\frac{(1-z)}{\left(z-\frac{1}{3}\right)\left(z-\frac{2}{3}\right)} = \frac{5z}{z-\frac{1}{3}} - \frac{5z}{z-\frac{2}{3}}$$

$$y(n) = 5\left(\frac{1}{3}\right)^n u(n) - 5\left(\frac{2}{3}\right)^n u(n)$$

6.30 (a)

$$y(n) = \sum_{i=0}^{3} \alpha_i\, x(n-i)$$

$$= \alpha_0 x[n]+\alpha_1 x[n-1]+\alpha_2 x[n-2]+\alpha_3 x[n-3]$$

$$\Rightarrow Y(z) = \alpha_0 X(z)+\alpha_1 X(z)z^{-1}+\alpha_2 X(z)z^{-2} +\alpha_3 X(z)z^{-3}$$

$$H(z) = \alpha_0+\alpha_1 z^{-1}+\alpha_2 z^{-2}+\alpha_3 z^{-3}$$

Null at zero frequency

$$\Rightarrow \quad z = e^{j\omega}\Big|_{\omega=0}$$

$$0 = \alpha_0+\alpha_1+\alpha_2+\alpha_3$$

Option (a) satisfies the condition.

6.31 (c)

Poles are at $z = \pm 2j$

Since $x[n]$ is absolutely summable hence, DTFT exists which implies ROC must include unit circle.

$\Rightarrow$ $x(t)$ is a non-causal signal.

6.32 (b)

Given, $x[n] = a^n u[n] + b^n u[n]$

Also given, $0 < |a| < |b| < 1$

ROC $= (|z| > |a|)$ and $(|z| > |b|)$

ROC $= |z| > |b|$

6.33 (d)

Given, $x[n] = (2.0)^{|n|}, -\infty < n < +\infty$

$$= 2^n u[n] + \left(\frac{1}{2}\right)^n u[-n-1]$$

ROC : $|Z| > 2$ and $|Z| < \frac{1}{2}$

$\therefore$ No ROC.

6.34 (c)

Given,

$$x[n] = \delta[n-3]+2\delta[n-5]$$

$$X(z) = z^{-3}+2z^{-5}$$

$$X(-z) = (-z)^{-3}+2(-z)^{-5}$$

$$Y(z) = X(-z) = -z^{-3}-2z^{-5} = -[z^{-3}+2z^{-5}]$$

$$y[n] = -x[n]$$

6.35 (b)

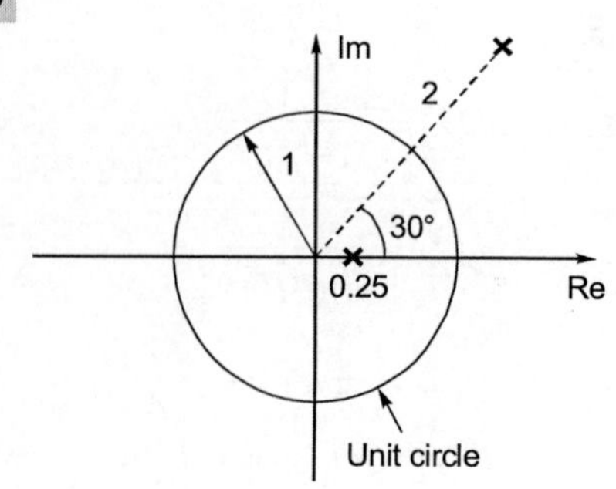

The ROC should include unit circle to make the system stable. From the given pole pattern it is clear that, to make the system stable, the ROC should be finite circular strip or finite annular strip and hence the impulse response of the system should be also two-sided.

6.36 (d)

$$P(Z) = H(z)H\left(\frac{1}{z}\right)$$

(i) $h(n)$ is real. So, $p(n)$ will be also real

(ii) $P(z) = P(z^{-1})$

From (i): If z_1 is a zero of $P(z)$, then z_1^* will be also a zero of $P(z)$.

From (ii): If z_1 is a zero of $P(z)$, then $\frac{1}{z_1}$ will be also a zero of $P(z)$.

So, the 4 zeros are,

$$z_1 = \frac{1}{2}+\frac{1}{2}j$$

$$z_2 = z_1^* = \frac{1}{2}-\frac{1}{2}j$$

$$z_3 = \frac{1}{z_1} = \frac{1}{\frac{1}{2}+\frac{1}{2}j} = 1-j$$

$$z_4 = \left(\frac{1}{z_1}\right)^* = z_3^* = 1+j$$

6.37 (c)

$$y(n) = \sum_{K=0}^{3} (-1)^K x(n-K)$$

$$= x(n) - x(n-1) + x(n-2) - x(n-3)$$

$$\Rightarrow Y(z) = X(z) - z^{-1}X(z) + z^{-2}X(z) - z^{-3}X(z)$$

$$\Rightarrow \quad H(z) = \frac{Y(z)}{X(z)} = 1 - z^{-1} + z^{-2} - z^{-3}$$

$$= \frac{z^3 - z^2 + z - 1}{z^3} = \frac{(z-1)(z^2+1)}{z^3}$$

Pole zero plot:

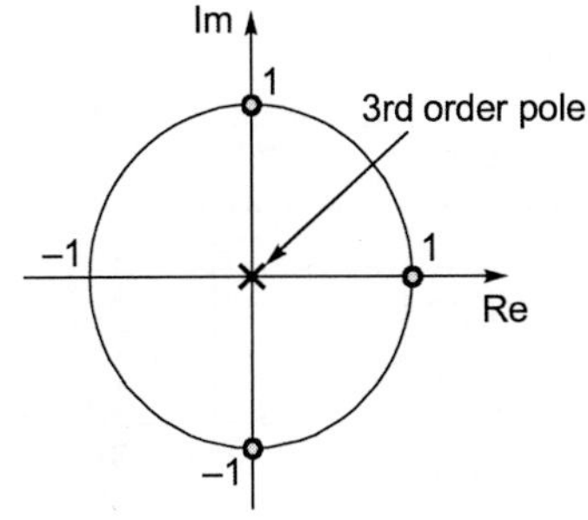

6.38 (–2)

System is all-pass filter.

For digital all-pass filter, condition is

$$\text{Zero} = \frac{1}{\text{Pole}^*} \quad \text{...(i)}$$

By given transfer function,

$$\text{Zero} = \alpha; \quad \text{Pole} = -0.5$$

Using condition (i),

$$\alpha = \frac{1}{-0.5} = -2$$

6.39 (a, b, d)

$$H(z) = \frac{2z^2+3}{\left(z+\frac{1}{3}\right)\left(z-\frac{1}{3}\right)}$$

(i) The poles of the system are located at $|z| = 1/3$. Since, the poles lies inside the unit circuit, hence the system is stable.

(ii) Using Final value theorem,

$$h[\infty] = \lim_{z\to 1}(z-1)\cdot H(z)$$

$$h[\infty] = \lim_{z\to 1}\frac{(z-1)\cdot(2z^2+3)}{\left(z+\frac{1}{3}\right)\left(z-\frac{1}{3}\right)} = 0$$

(iii) The zeros of the system are located at

$$z = \pm j\sqrt{\frac{3}{2}} \Rightarrow |z| = \sqrt{\frac{3}{2}} > 1$$

Hence, the zeros lies outside the unit circle. Thus, the system is a non-minimum phase system.

(iv) Using initial value theorem,

$$h[0] = \lim_{z\to\infty} H(z) = \lim_{z\to\infty}\frac{2z^2+3}{\left(z+\frac{1}{3}\right)\left(z-\frac{1}{3}\right)}$$

$$h[0] = \lim_{z\to\infty}\frac{2+\frac{3}{z^2}}{\left(1+\frac{1}{3z}\right)\left(1-\frac{1}{3z}\right)} = 2$$

Hence, options (a), (b) and (d) are correct.

■■■

7 DTFS, DTFT and DFT

7.1 The impulse response $h[n]$ of a linear time invariant system is given as

$$h[n] = \begin{cases} -2\sqrt{2} & n = 1, -1 \\ 4\sqrt{2} & n = 2, -2 \\ 0, & \text{otherwise} \end{cases}$$

If the input to the above system is the sequence $e^{j\pi n/4}$, then the output is

(a) $4\sqrt{2}e^{j\pi n/4}$ (b) $4\sqrt{2}e^{-j\pi n/4}$

(c) $4e^{j\pi n/4}$ (d) $-4e^{j\pi n/4}$

[2004 : 2 M]

7.2 Let $x(n) = \left(\frac{1}{2}\right)^n u(n)$, $y(n) = x^2(n)$ and $Y(e^{j\omega})$ be the Fourier transform of $y(n)$. Then $Y(e^{j0})$ is

(a) $\frac{1}{4}$ (b) 2

(c) 4 (d) $\frac{4}{3}$

[2005 : 1 M]

7.3 Match the following and choose the correct combination.

Group 1

E. Continuous and aperiodic signal

F. Continuous and periodic signal

G. Discrete and aperiodic signal

H. Discrete and periodic signal

Group 2

1. Fourier representation is continuous and aperiodic
2. Fourier representation is discrete and aperiodic
3. Fourier representation is continuous and periodic
4. Fourier representation is discrete and periodic

(a) E-3, F-2, G-4, H-1

(b) E-1, F-3, G-2, H-4

(c) E-1, F-2, G-3, H-4

(d) E-2, F-1, G-4, H-3

[2005 : 2 M]

Statement of Linked Answer Questions 7.4 and 7.5.

A sequence $x(n)$ has non-zero values as shown in the figure.

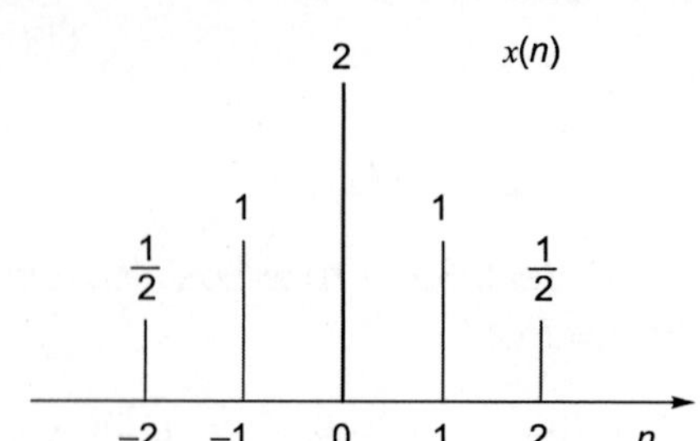

7.4 The sequence

$$y(n) = \begin{cases} x\left(\frac{n}{2} - 1\right) & \text{for } n \text{ even} \\ 0, & \text{for } n \text{ odd} \end{cases}$$

will be

(a)

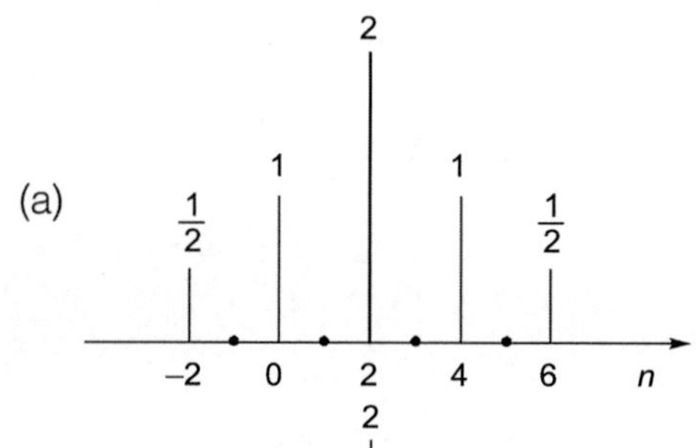

(b)

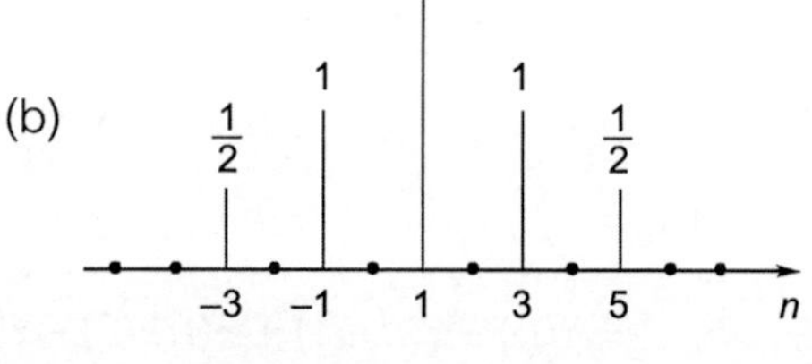

(c)

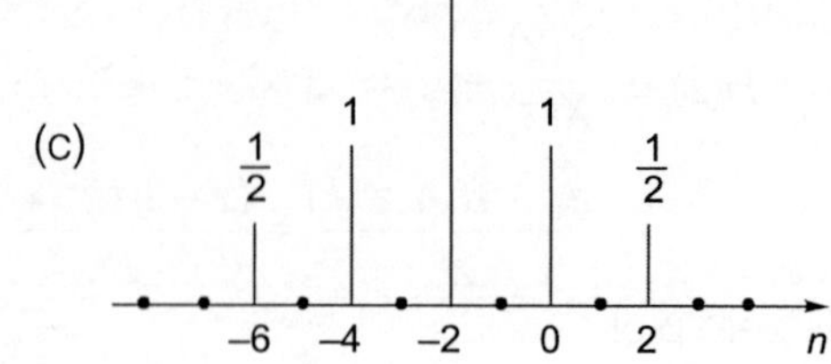

(d)

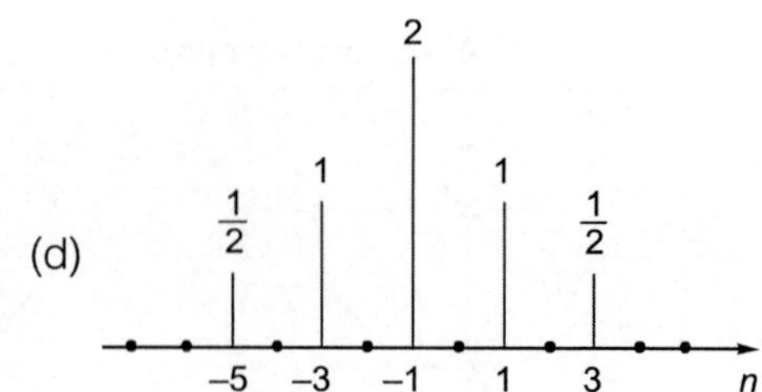

[2005 : 2 M]

7.5 The Fourier transform of $y(2n)$ will be

(a) $e^{-j2\omega}[\cos 4\omega + 2\cos 2\omega + 2]$

(b) $[\cos 2\omega + 2\cos\omega + 2]$

(c) $e^{-j\omega}[\cos 2\omega + 2\cos\omega + 2]$

(d) $e^{\frac{-j\omega}{2}}[\cos 2\omega + 2\cos\omega + 2]$

[2005 : 2 M]

7.6 A signal $x(n) = \sin(\omega_0 n + \phi)$ is the input to a linear time-invariant system having a frequency response $H(e^{j\omega})$. If the output of the system $Ax(n - n_0)$, then the most general form of $\angle H(e^{j\omega})$ will be

(a) $-n_0\omega_0 + \beta$ for any arbitrary real β

(b) $-n_0\omega_0 + 2\pi k$ for any arbitrary integer k.

(c) $n_0\omega_0 + 2\pi k$ for any arbitrary integer k.

(d) $-n_0\omega_0\phi$

[2005 : 2 M]

7.7 A 5-point sequence $x[n]$ is given as $x[-3] = 1, x[-2] = 1, x[-1] = 0, x[0] = 5, x[1] = 1$. Let $X(e^{j\omega})$ denote the discrete-time Fourier transform of $x[n]$. The value of

$\int_{-\pi}^{\pi} X(e^{j\omega})d\omega$ is

(a) 5 (b) 10π

(c) 16π (d) $5 + j10\pi$

[2007 : 2 M]

7.8 $\{x(n)\}$ is a real-valued periodic sequence with a period N. $x(n)$ and $X(k)$ form N-point Discrete Fourier Transform (DFT) pairs. The DFT $Y(k)$ of the sequence

$$y(n) = \frac{1}{N}\sum_{r=0}^{N-1} x(r)\,x(n+r) \text{ is}$$

(a) $|X(k)|^2$

(b) $\frac{1}{N}\sum_{r=0}^{N-1} X(r)X^*(k+r)$

(c) $\frac{1}{N}\sum_{r=0}^{N-1} X(r)X(k+r)$

(d) 0

[2008 : 2 M]

7.9 A system with transfer function $H(z)$ has impulse response $h(n)$ defined as $h(2) = 1$, $h(3) = -1$ and $h(k) = 0$ otherwise. Consider the following statements.

S_1 : $H(z)$ is a low-pass filter

S_2 : $H(z)$ is an FIR filter.

Which of the following is correct?

(a) Only S_2 is true

(b) Both S_1 and S_2 are false

(c) Both S_1 and S_2 are true, and S_2 is a reason for S_1

(d) Both S_1 and S_2 are true, but S_2 is not a reason for S_1

[2009 : 2 M]

7.10 The 4-point discrete Fourier Transform (DFT) of a discrete time sequence {1, 0, 2, 3} is

(a) $[0, -2 + 2j, 2, -2 - 2j]$

(b) $[2, 2 + 2j, 6, 2 - 2j]$

(c) $[6, 1 - 3j, 2, 1 + 3j]$

(d) $[6, -1 + 3j, 0, -1 - 3j]$

[2009 : 2 M]

7.11 For an N-point FFT algorithm with $N = 2^m$, which one of the following statements is TRUE?

(a) It is not possible to construct a signal flow graph with both input and output in normal order

(b) The number of butterflies in the mth state is N/m

(c) In-place computation requires storage of only 2 N node data

(d) Computation of a butterfly requires only one complex multiplication

[2010 : 1 M]

7.12 The first six points of the 8-point DFT of a real valued sequence are 5, $1 - j3$, 0, $3 - j4$, 0 and $3 + j4$. The last two points of the DFT are respectively

(a) 0, $1 - j3$ (b) 0, $1 + j3$

(c) $1 + j3$, 5 (d) $1 - j3$, 5

[2011 : 2 M]

7.13 The DFT of a vector $[a\ b\ c\ d]$ is the vector $[\alpha\ \beta\ \gamma\ \delta]$. Consider the product

$$[p\ q\ r\ s] = [a\ b\ c\ d]\begin{bmatrix} a & b & c & d \\ d & a & b & c \\ c & d & a & b \\ b & c & d & a \end{bmatrix}$$

The DFT of the vector $[p\ q\ r\ s]$ is a scaled version of

(a) $\left[\alpha^2\ \ \beta^2\ \ \gamma^2\ \ \delta^2\right]$

(b) $\left[\sqrt{\alpha}\ \ \sqrt{\beta}\ \ \sqrt{\gamma}\ \ \sqrt{\delta}\right]$

(c) $[\alpha+\beta\ \ \beta+\delta\ \ \delta+\gamma\ \ \gamma+\alpha]$

(d) $[\alpha\ \ \beta\ \ \gamma\ \ \delta]$

[2013 : 2 M]

7.14 Consider a discrete time periodic signal $x[n] = \sin\left(\frac{\pi n}{5}\right)$. Let a_k be the complex Fourier series coefficients of $x[n]$. The coefficients $\{a_k\}$ are non-zero when $k = Bm \pm 1$, where m is any integer. The value of B is _____.

[2014 : 2 M, Set-1]

7.15 An FIR system is described by the system function

$$H(z) = 1 + \frac{7}{2}z^{-1} + \frac{3}{2}z^{-2}$$

The system is

(a) maximum phase (b) minimum phase

(c) mixed phase (d) zero phase

[2014 : 1 M, Set-2]

7.16 The sequence $x[n] = 0.5^n u[n]$ where $u[n]$ is the unit step sequence, is convolved with itself to obtain $y[n]$. Then $\sum_{n=-\infty}^{+\infty} y[n]$ is ______.

[2014 : 1 M, Set-4]

7.17 A Fourier transform pair is given by

$$\left(\frac{2}{3}\right)^n u[n+3] \xleftrightarrow{F.T.} \frac{Ae^{+j6\pi f}}{1-\left(\frac{2}{3}\right)e^{-j2\pi f}}$$

when $u[n]$ denotes the unit step sequence. the value of A is ______.

[2014 : 1 M, Set-4]

7.18 The N-point DFT of a sequence $x[n]$, $0 \le n \le N-1$ is given by

$$X[K] = \frac{1}{\sqrt{N}} \sum_{n=0}^{N-1} x[n]e^{-j\frac{2\pi}{N}nK},\ 0 \le K \le N-1.$$

Denote this relation as $X = \text{DFT}(x)$. For $N = 4$, which one of the following sequences satisfies DFT (DFT $(x)) = x$.

(a) $x = [1\ 2\ 3\ 4]$ (b) $x = [1\ 2\ 3\ 2]$

(c) $x = [1\ 3\ 2\ 2]$ (d) $x = [1\ 2\ 2\ 3]$

[2014 : 2 M, Set-4]

7.19 Two sequences $[a, b, c]$ and $[A, B, C]$ are related as,

$$\begin{bmatrix} A \\ B \\ C \end{bmatrix} = \begin{bmatrix} 1 & 1 & 1 \\ 1 & W_3^{-1} & W_3^{-2} \\ 1 & W_3^{-2} & W_3^{-4} \end{bmatrix} \begin{bmatrix} a \\ b \\ c \end{bmatrix} \quad \text{where, } W_3 = e^{j\frac{2\pi}{3}}$$

If another sequence $[p, q, r]$ is derived as,

$$\begin{bmatrix} p \\ q \\ c \end{bmatrix} = \begin{bmatrix} 1 & 1 & 1 \\ 1 & W_3^1 & W_3^2 \\ 1 & W_3^2 & W_3^4 \end{bmatrix} \begin{bmatrix} 1 & 0 & 0 \\ 0 & W_3^2 & 0 \\ 0 & 0 & W_3^4 \end{bmatrix} \begin{bmatrix} A/3 \\ B/3 \\ C/3 \end{bmatrix}$$

then the relationship between the sequences $[p, q, r]$ and $[a, b, c]$ is

(a) $[p, q, r] = [b, a, c]$

(b) $[p, q, r] = [b, c, a]$

(c) $[p, q, r] = [c, a, b]$

(d) $[p, q, r] = [c, b, a]$

[2015 : 2 M, Set-1]

7.20 Consider two real sequences with time-origin marked by the bold value,

$x_1[n] = \{1, 2, 3, 0\}$, $x_2[n] = \{1, 3, 2, 1\}$

Let $X_1(k)$ and $X_2(k)$ be 4-point DFTs of $x_1[n]$ and $x_2[n]$, respectively.

Another sequence $x_3[n]$ is derived by taking 4-point inverse DFT of $X_3(k) = X_1(k)X_2(k)$.

The value of $x_3[2]$ is _____.

[2015 : 2 M, Set-2]

7.21 The value of $\sum_{n=0}^{\infty} n\left(\frac{1}{2}\right)^n$ is ______.

[2015 : 1 M, Set-3]

7.22 Let $\tilde{x}[n] = 1 + \cos\left(\frac{\pi n}{8}\right)$ be a periodic signal with period 16. Its DFS coefficients are defined by $a_k = \frac{1}{16}\sum_{n=0}^{15} \tilde{x}[n]\exp\left(-j\frac{\pi}{8}kn\right)$ for all k. The value of the coefficient a_{31} is _____.

[2015 : 2 M, Set-3]

7.23 Consider the signal

$x[n] = 6\delta[n+2] + 3\delta[n+1] + 8\delta[n] + 7\delta[n-1] + 4\delta[n-2]$

If $X(e^{j\omega})$ is the discrete-time Fourier transform of $x[n]$, then $\frac{1}{\pi}\int_{-\pi}^{\pi} X(e^{j\omega})\sin^2(2\omega)d\omega$ is equal to ___

[2016 : 2 M, Set-1]

7.24 The Discrete Fourier Transform (DFT) of the 4-point sequence

$x[n] = \{x[0], x[1], x[2], x[3]\} = \{3, 2, 3, 4\}$ is

$X[k] = \{X[0], X[1], X[2], X[3]\} = \{12, 2j, 0, -2j\}$.

If $X_1[k]$ is the DFT of the 12-point sequence $x_1[n]$ = {3, 0, 0, 2, 0, 0, 3, 0, 0, 4, 0, 0}, the value of $\left|\frac{X_1[8]}{X_1[11]}\right|$ is ______.

[2016 : 2 M, Set-2]

7.25 A continuous-time speech signal $x_a(t)$ is sampled at a rate of 8 kHz and the samples are subsequently grouped in blocks, each of size N. The DFT of each block is to be computed in real time using the radix-2 decimation-in-frequency FFT algorithm. If the processor performs all operations sequentially, and takes 20 μs for computing each complex multiplication (including multiplications by 1 and -1) and the time required for addition/subtraction is negligible, then the maximum value of N is _______.

[2016 : 2 M, Set-3]

7.26 Let h[n] be the impulse response of a discrete-time linear time invariant (LTI) filter. The impulse response is given by

$$h[0]=\frac{1}{3};\ h[1]=\frac{1}{3};\ h[2]=\frac{1}{3}$$

and $h[n]=0$ for $n<0$ and $n>2$.

Let $H(\omega)$ be the discrete-time Fourier transform (DTFT) of $h[n]$, where ω is the normalized angular frequency in radians. Given that $H(\omega_0)=0$ and $0<\omega_0<\pi$, the value of ω_0 (in radians) is equal to ___.

[2017 : 2 M, Set-1]

7.27 An LTI system with unit sample response $h[n]=5\delta[n]-7\delta[n-1]+7\delta[n-3]-5\delta[n-4]$ is a

(a) low-pass filter (b) high-pass filter
(c) band-pass filter (d) band-stop filter

[2017 : 1 M, Set-2]

7.28 Let $X[k] = k + 1$, $0 \le k \le 7$ be 8-point DFT of a sequence $x[n]$, where $X[k]=\sum_{n=0}^{N-1} x[n]e^{-j2\pi nk/N}$.

The value (correct to two decimal places) of $\sum_{n=0}^{3} x[2n]$ is ______.

[2018 : 2 M]

7.29 It is desired to find a three-tap causal filter which gives zero signal as an output to an input of the form

$$x[n] = c_1\exp\left(-\frac{j\pi n}{2}\right)+c_2\exp\left(\frac{j\pi n}{2}\right),$$

where c_1 and c_2 are arbitrary real numbers. The desired three-tap filter is given by

$h[0]=1,\quad h[1]=a,\quad h[2]=b$

and $h[n]=0$ for $n<0$ or $n>2$

What are the values of the filter taps a and b if the output $y[n] = 0$ for all n, where $x[n]$ is as given above?

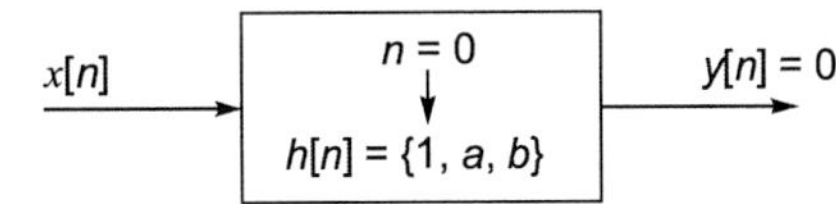

(a) $a = 0, b = -1$ (b) $a = 1, b = 1$
(c) $a = -1, b = 1$ (d) $a = 0, b = 1$

[2019 : 2 M]

7.30 Consider a six-point decimation-in-time Fast Fourier Transform (FFT) algorithm, for which the signal-flow graph corresponding to $X[1]$ is shown hi the figure. Let $W_6=\exp\left(-\frac{j2\pi}{6}\right)$. In the figure, what should be the values of the coefficients a_1, a_2, a_3 in terms of W_6 so that $X[1]$ is obtained correctly?

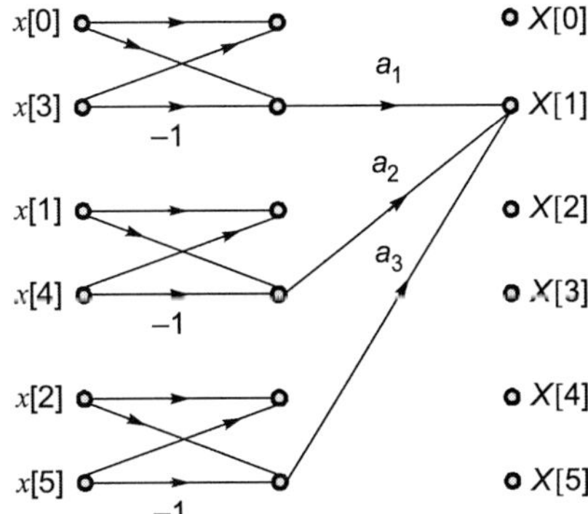

(a) $a_1=1, a_2=W_6^2, a_3=W_6$
(b) $a_1=-1, a_2=W_6^2, a_3=W_6$
(c) $a_1=-1, a_2=W_6, a_3=W_6^2$
(d) $a_1=1, a_2=W_6, a_3=W_6^2$

[2019 : 2 M]

7.31 A finite duration discrete-time signal $x[n]$ is obtained by sampling the continuous-time signal $x(t)=\cos(200\pi t)$ at sampling instants $t = n/400$, $n = 0,1, \ldots, 7$. The 8-point discrete Fourier transform (DFT) of $x[n]$ is defined as

$$X[k]=\sum_{n=0}^{7} x[n]\,e^{-j\frac{\pi kn}{4}},\ k=0,1,\ldots,7$$

Which one of the following statements is true?

(a) Only $X[2]$ and $X[6]$ are non-zero
(b) Only $X[3]$ and $X[5]$ are non-zero
(c) All $X[k]$ are non-zero
(d) Only $X[4]$ is non-zero

[2020 : 2 M]

7.32 Consider two 16-point sequence $x[n]$ and $h[n]$. Let the linear convolution of $x[n]$ and $h[n]$ be denoted by $y[n]$, while $z[n]$ denotes the 16-point inverse discrete Fourier transform (IDFT) of the product of the 16-point DFTs of $x[n]$ and $h[n]$. The value(s) of k for which $z[k] = y[k]$ is/are

(a) $k = 0, 1, 2, \ldots 15$ (b) $k = 15$
(c) $k = 0$ and $k = 15$ (d) $k = 0$

[2021 : 1 M]

7.33 Consider the signals $x[n] = 2^{n-1} u[-n+2]$ and $y[n] = 2^{-n+2} u[n+1]$, where $u[n]$ is the unit step sequence. Let $X(e^{j\omega})$ and $Y(e^{j\omega})$ be the discrete-time Fourier transform of $x[n]$ and $y[n]$, respectively. The value of the integral

$$\frac{1}{2\pi}\int_0^{2\pi} X(e^{j\omega})\, Y(e^{-j\omega})\, d\omega$$

(rounded off to one decimal place) is ______.

[2021 : 2 M]

7.34 For a vector $\bar{x} = [x[0], x[1], \ldots, x[7]]$, the 8-point discrete Fourier transform (DFT) is denoted by $\bar{X} = \text{DFT}(\bar{x}) = [X[0], X[1], \ldots, X[7]]$, where

$$X[k] = \sum_{n=0}^{7} x[n] \exp\left(-j\frac{2\pi}{8}nk\right).$$

Here, $j = \sqrt{-1}$. If $\bar{x} = [1, 0, 0, 0, 2, 0, 0, 0]$ and $\bar{y} = \text{DFT}(\text{DFT}(\bar{x}))$, then the value of $y[0]$ is ______ (rounded off to one decimal place).

[2022 : 2 M]

7.35 Consider a discrete-time periodic signal with period $N = 5$. Let the discrete-time Fourier series (DTFS) representation be $x[n] = \sum_{k=0}^{4} a_k e^{\frac{jk2\pi n}{5}}$, where $a_0 = 1$, $a_1 = 3j$, $a_2 = 2j$, $a_3 = -2j$ and $a_4 = -3j$. The value of the sum $\sum_{n=0}^{4} x[n]\sin\frac{4\pi n}{5}$ is

(a) –10 (b) 10
(c) –2 (d) 2

[2023 : 2 M]

7.36 Let an input $x[n]$ having discrete time Fourier transform.

$X(e^{j\Omega}) = 1 - e^{-j\Omega} + 2e^{-3j\Omega}$ be passed through an LTI system. The frequency response of the LTI system is $H(e^{j\Omega}) = 1 - \frac{1}{2}e^{-j2\Omega}$. The output $y[n]$ of the system is

(a) $\delta[n] + \delta[n-1] - \frac{1}{2}\delta[n-2] - \frac{5}{2}\delta[n-3] + \delta[n-5]$

(b) $\delta[n] - \delta[n-1] - \frac{1}{2}\delta[n-2] - \frac{5}{2}\delta[n-3] + \delta[n-5]$

(c) $\delta[n] - \delta[n-1] - \frac{1}{2}\delta[n-2] + \frac{5}{2}\delta[n-3] - \delta[n-5]$

(d) $\delta[n] + \delta[n-1] + \frac{1}{2}\delta[n-2] + \frac{5}{2}\delta[n-3] + \delta[n-5]$

[2023 : 2 M]

7.37 The relationship between any N-length sequence $x[n]$ and its corresponding N-point discrete Fourier transform $X[k]$ is defined as
$X[k] = F\{x[n]\}$.
Another sequence $y[n]$ is formed as below

$y[n] = F\{F\{F\{F\{x[n]\}\}\}\}$.

For the sequence $x[n] = \{1, 2, 1, 3\}$, the value of $Y[0]$ is ______.

[2024 : 2 Marks]

■■■

Answers DTFS, DTFT and DFT

7.1 (d)	7.2 (d)	7.3 (c)	7.4 (a)	7.5 (c)	7.6 (b)	7.7 (b)
7.8 (a)	7.9 (a)	7.10 (d)	7.11 (d)	7.12 (b)	7.13 (a)	7.14 (10)
7.15 (c)	7.16 (4)	7.17 (3.375)	7.18 (b)	7.19 (c)	7.20 (11)	7.21 (2)
7.22 (0.5)	7.23 (8)	7.24 (6)	7.25 (4096)	7.26 (2.10)	7.27 (c)	7.28 (3)
7.29 (d)	7.30 (d)	7.31 (a)	7.32 (b)	7.33 (8)	7.34 (8)	7.35 (a)
7.36 (c)	7.37 (112)					

Explanations DTFS, DTFT and DFT

7.1 (d)

$$h(n) = 4\sqrt{2}\delta(n+2) - 2\sqrt{2}\delta(n+1) - 2\sqrt{2}\delta(n-1) + 4\sqrt{2}\delta(n-2)$$

$$x(n) = e^{j\pi n/4}$$

$$y(n) = x(n)*h(n)$$

$$= 4\sqrt{2}\delta(n+2)*e^{jn\pi/4} + 4\sqrt{2}\delta(n-2)*e^{jn\pi/4} - 2\sqrt{2}[\delta(n+1)*e^{jn\pi/4} + \delta(n-1)e^{jn\pi/4}]$$

$$x(n)*\delta(n-a) \rightarrow x(n-a)$$

$$y(n) = 4\sqrt{2}\left[e^{j\pi/4(n+2)} + e^{j\pi/4(n-2)}\right] - 2\sqrt{2}\left[e^{j\pi/4(n+1)} + e^{j\pi/4(n-1)}\right]$$

$$= e^{jn\pi/4}[4\sqrt{2}(e^{j\pi/2} + e^{-j\pi/2})] - 2\sqrt{2}\left(e^{j\pi/4} + e^{-j\pi/4}\right)$$

$$= e^{jn\pi/4}\left[0 - 2\sqrt{2}\times 2\cos\pi/4\right]$$

$$y(n) = -4e^{jn\pi/4}$$

7.2 (d)

$$x(n) = \left(\frac{1}{2}\right)^n \cdot u(n)$$

$$y(n) = x^2(n) = \left(\frac{1}{2}\right)^{2n} u^2(n) = \left[\left(\frac{1}{2}\right)^2\right]^n u(n)$$

$$y(n) = \left(\frac{1}{4}\right)^n u(n) \quad ; \quad Y(z) = \frac{1}{1-\frac{1}{4}z^{-1}}$$

Put, $z = e^{j\omega}$

$$Y(e^{j\omega}) = \frac{1}{1-\frac{1}{4}e^{-j\omega}} \quad ; \quad Y(e^{j0}) = \frac{1}{1-\frac{1}{4}} = \frac{4}{3}$$

7.4 (a)

$$y(n) = x\left(\frac{n}{2}-1\right), n \text{ even}$$

$$= 0 \text{ for n odd}$$

$n = 0$, $y(n) = x(-1) = 1$

$n = 2$, $y(n) = x(0) = 2$

$n = 4$, $y(n) = x(1) = 1$

$n = 6$, $y(n) = x(2) = 1/2$

7.5 (c)

$y(2n) = x(n-1)$, seeing in graph

$f(n) = y(2n)$

$$= \frac{1}{2}\delta(n+1) + \delta(n) + 2\delta(n-1) + \delta(n-2) + \frac{1}{2}\delta(n-3)$$

Taking z-transform

$$F(z) = \frac{1}{2}z + 1 + 2z^{-1} + z^{-2} + \frac{1}{2}z^{-3}$$

$$z = e^{j\omega}$$

$$F(e^{j\omega}) = \frac{1}{2}e^{j\omega} + 1 + 2e^{-j\omega} + e^{-2j\omega} + \frac{1}{2}e^{-3j\omega}$$

$$= e^{-j\omega}\left(\frac{1}{2}e^{2j\omega} + e^{j\omega} + 2 + e^{-j\omega} + \frac{1}{2}e^{-2j\omega}\right)$$

$$= e^{-j\omega}\left[\frac{e^{2j\omega}+e^{-2j\omega}}{2} + e^{j\omega} + e^{-j\omega} + 2\right]$$

$$f(n) = e^{-j\omega}\left[\cos 2\omega + 2\cos\omega + 2\right]$$

7.6 (b)

- $\sin(\omega_0 n + \phi) = x(n) \leftrightarrow X(e^{j\omega})$
- $y(n) = Ax(n-n_o)$

  $Y(e^{j\omega}) = Ae^{-jn_o\omega}X(e^{j\omega})$
- $H(e^{j\omega}) = \dfrac{Y(e^{j\omega})}{X(e^{j\omega})} = Ae^{-jn_o\omega}$
- $\angle H(e^{j\omega}) = -n_o\omega_o + 2\pi k \quad (\text{at } \omega = \omega_o)$

7.7 (b)

$$X(e^{j\omega}) = e^{3j\omega} + e^{2j\omega} + 0 + 5 + e^{-j\omega}$$

$$\because \int_{-\pi}^{\pi} e^{aj\omega}d\omega = 0 \text{ if } a\neq 0$$

$$\int_{-\pi}^{\pi} X(e^{j\omega})d\omega = \left|\frac{e^{3j\omega}}{3j} + \frac{e^{2j\omega}}{2j} + 5\omega + \frac{e^{-j\omega}}{-j}\right|_{-\pi}^{\pi}$$

$$= 5\pi + 5\pi = 10\pi$$

7.9 (a)

$h(2) = 1$

$h(3) = -1$

$h(k) = 0$ other wise

It is finite impulse response. It is not low pass filter.

7.10 (d)

4-point DFT of sequence {1, 0, 2, 3} is given as

$$\begin{bmatrix}1 & 1 & 1 & 1\\ 1 & -j & -1 & j\\ 1 & -1 & 1 & -1\\ 1 & j & -1 & -j\end{bmatrix}\begin{bmatrix}1\\0\\2\\3\end{bmatrix} = \begin{bmatrix}1+2+3\\ 1-2+3j\\ 1+2-3\\ 1-2-3j\end{bmatrix} = \begin{bmatrix}6\\ -1+3j\\ 0\\ -1-3j\end{bmatrix}$$

7.11 (d)

For an N-point FFT algorithm with $N = 2^m$, computation of a butterfly requires only one complex multiplication and two complex additions.

7.12 (b)

$$x[n] \rightarrow \text{real}$$
$$X[k] \rightarrow \text{Conjugate symmetric}$$
$$\Rightarrow X[k] = X^*[N-k]$$
$$N = 8$$
$$X[k] = X^*[8-k]$$
$$\therefore X[6] = X^*[8-6] = X^*[2] = 0$$
$$X[7] = X^*[8-7] = X^*[1] = 1 + j3$$

7.13 (a)

$$\text{DFT}[p\ q\ r\ s] = [a\ b\ c\ d] * [a\ b\ c\ d]$$

So, $\text{DFT}[p\ q\ r\ s] = [\alpha^2\ \beta^2\ \gamma^2\ \delta^2]$

7.14 (10)

From discrete Fourier series definition

$$x[n] = \sum_{K=0}^{N-1} a_K e^{jK\left(\frac{2\pi}{N}\right)n}$$

$$= a_{-1} e^{-j\frac{2\pi}{N}n} + a_0 + a_1 e^{j\frac{2\pi}{N}n} \quad \text{...(i)}$$

$$x[n] = \frac{-1}{2j} e^{-j\frac{\pi}{5}n} + \frac{1}{2j} e^{j\frac{\pi}{5}n}$$

Also given

$$x[n] = \sin\left(\frac{\pi n}{5}\right)$$

$$\text{where } N = 10 \quad \left[\frac{2\pi}{N} = \frac{\pi}{5} \Rightarrow N = 10\right]$$

By comparison $a_1 = \dfrac{1}{2j}$ and $a_9 = a_{-1} = \dfrac{-1}{2j}$

As the TFS coefficient a_K is also periodic with period $N = 10$.

$$a_1 = a_{11} = a_{21} \cdots$$
$$a_{-1} = a_9 = a_{19} \cdots$$

Given that a_K is non-zero for $K = Bm \pm 1$.

$$Bm \pm 1 = 1 \qquad 9 \cdots$$
$$-1 \qquad 11 \cdots$$

For $m = 0$ for $m = 1$

To satisfy this condition B must be '10' where 'm' is any integer.

7.15 (c)

Minimum phase $\Rightarrow$ all zeroes are inside the unit circle.

- For maximum phase, all the zeroes are outside the unit circle.
- $H(z) = 1 + \frac{7}{2}z^{-1} + \frac{3}{2}z^{-2} = 0$

$$2z^2 + 6z + z + 3 = 0$$
$$(2z+1)(z+3) = 0$$
$$z_1 = -\frac{1}{2} \text{ (inside the unit circle)}$$
$$z_2 = -3 \text{ (outside the unit circle)}$$

$\therefore$ Mixed phase.

7.16 (4)

Given $x[n] = 0.5^n u[n]$

$$y[n] = x[n] * x[n]$$
$$= \sum_{K=-\infty}^{\infty} x(k)\, h(n-k)$$
$$= \sum_{K=-\infty}^{\infty} (0.5)^k u(k)\, (0.5)^{n-k} u(n-k)$$
$$= (0.5)^n \left[\sum_{k=-\infty}^{\infty} u(k)\, u(n-k)\right]$$

Also we know that $u[n] * u[n] = (n+1)\, u[n]$

$\therefore$ Using the above formula, we get

$$y[n] = (0.5)^n (n+1)\, u[n]$$

Let $\quad p = \sum_{n=0}^{\infty} (0.5)^n (n+1)$

Also let $a = 0.5$

$$\therefore \quad p = \sum_{n=0}^{\infty} a^n (n+1)$$

$$p = a^0(1) + a^1(1+1) + a^2(2+1) + a^3(3+1) + \cdots$$
$$ap = a(1) + a^2(2) + a^3(3) + \cdots$$
$$p - ap = a^0 \cdot 1 + a^1 + a^2 + \cdots$$
$$p(1-a) = \frac{1}{1-a} \quad \Rightarrow p = \frac{1}{(1-a)^2}$$
$$a = 0.5 \text{ or } \frac{1}{2}$$
$$p = \frac{1}{\left(1-\frac{1}{2}\right)^2} = \frac{1}{\left(\frac{1}{2}\right)^2} = 4$$

7.17 (3.375)

$$\left(\frac{2}{3}\right)^n u(n+3) \xleftrightarrow{F.T.} \frac{Ae^{j6\pi f}}{1 - \left(\frac{2}{3}\right)e^{-j2\pi f}}$$

Let, $\quad x[n] = \left(\frac{2}{3}\right)^n u(n+3)$

Discrete Fourier transform is given as

$$X(e^{j\omega}) = \sum_{n=-\infty}^{\infty} x[n]\, e^{-j\omega n} = \sum_{n=-3}^{\infty} \left(\frac{2}{3}\right)^n e^{-j\omega n}$$

Put, $n + 3 = m$

$$\text{So, } X(e^{j\omega}) = \sum_{m=0}^{\infty}\left(\frac{2}{3}\right)^{m-3} e^{-j\omega(m-3)}$$

$$= \left(\frac{2}{3}\right)^{-3}\sum_{m=0}^{\infty}\left(\frac{2}{3}\right)^{m} e^{-j\omega m}\cdot e^{j\omega 3}$$

$$= \left(\frac{3}{2}\right)^{3}\sum_{m=0}^{\infty}\left(\frac{2}{3}e^{-j\omega}\right)^{m}. e^{j\omega 3}$$

$$= \frac{3.375\cdot e^{j2\pi f3}}{1-\frac{2}{3}e^{-j2\pi f}} = \frac{3.375\, e^{j6\pi f}}{1-\frac{2}{3}e^{-j2\pi f}}$$

So, $A = 3.375$

7.18 (b)

DFT (DFT (x)) $= x$

Let, $x = [1\ 2\ 3\ 2]$

$$\text{DFT}(x) = \frac{1}{\sqrt{4}}\begin{bmatrix}1 & 1 & 1 & 1\\ 1 & -j & -1 & j\\ 1 & -1 & 1 & -1\\ 1 & j & -1 & -j\end{bmatrix}\begin{bmatrix}1\\2\\3\\2\end{bmatrix}$$

$$= \frac{1}{2}\begin{bmatrix}8\\-2\\0\\-2\end{bmatrix} = \begin{bmatrix}4\\-1\\0\\-1\end{bmatrix}$$

$$\text{DFT (DFT }(x)) = \frac{1}{\sqrt{4}}\begin{bmatrix}1 & 1 & 1 & 1\\ 1 & -j & -1 & j\\ 1 & -1 & 1 & -1\\ 1 & j & -1 & -j\end{bmatrix}\begin{bmatrix}4\\-1\\0\\-1\end{bmatrix}$$

$$= \frac{1}{2}\begin{bmatrix}2\\4\\6\\4\end{bmatrix} = \begin{bmatrix}1\\2\\3\\2\end{bmatrix}$$

7.19 (c)

$$\begin{bmatrix}A\\B\\C\end{bmatrix} = \begin{bmatrix}a+b+c\\ a+bW_3^{-1}+cW_3^{-2}\\ a+bW_3^{-2}+cW_3^{-1}\end{bmatrix}$$

$$\begin{bmatrix}p\\q\\r\end{bmatrix} = \begin{bmatrix}1 & W_3^2 & W_3^1\\ 1 & W_3^1W_3^2 & W_3^2W_3^1\\ 1 & W_3^2W_3^2 & W_3^1W_3^1\end{bmatrix}\begin{bmatrix}\frac{a+b+c}{3}\\ \frac{a+bW_3^{-1}+cW_3^{-2}}{3}\\ \frac{a+bW_3^{-2}+cW_3^{-1}}{3}\end{bmatrix}$$

where,
$$W_3^1 = e^{j\frac{2\pi}{3}}$$
$$W_3^2 = e^{j\frac{4\pi}{3}}$$
$$W_3^{-1} = e^{-j\frac{2\pi}{3}}$$
$$W_3^{-2} = e^{-j\frac{4\pi}{3}}$$
$$W_3^{-4} = W_3^{-1} = e^{-j\frac{2\pi}{3}}$$

$$\begin{bmatrix}p\\q\\r\end{bmatrix} = \begin{bmatrix}c\\a\\b\end{bmatrix}$$

7.20 (11)

$x_1[n] = \{1, 2, 3, 0\}$, $\quad x_2[n] = \{1, 3, 2, 1\}$

$X(k)_3 = X(k)_1\, X(k)_2$

$\Rightarrow x_3[n] =$ circular convolution of $x_1[n]$ and $x_2[n]$

$= x_1[n] \otimes x_2[n]$

$$\begin{bmatrix}x_3(0)\\x_3(1)\\x_3(2)\\x_3(3)\end{bmatrix} = \begin{bmatrix}1 & 0 & 3 & 2\\ 2 & 1 & 0 & 3\\ 3 & 2 & 1 & 0\\ 0 & 3 & 2 & 1\end{bmatrix}\begin{bmatrix}1\\3\\2\\1\end{bmatrix} = \begin{bmatrix}9\\8\\11\\14\end{bmatrix}$$

$x_3[2] = 11$

7.21 (2)

$$\sum_{n=0}^{\infty}\left(\frac{1}{2}\right)^n = \sum_{n=-\infty}^{\infty} n\left(\frac{1}{2}\right)^n u(n)$$

$$= Z\cdot T\left[n\left(\frac{1}{2}\right)^n u(n)\right]_{z=1}$$

$$= \frac{\frac{1}{2}z^{-1}}{\left(1-\frac{1}{2}z^{-1}\right)^2} = \frac{\frac{z}{2}}{\left(z-\frac{1}{2}\right)^2}\Bigg|_{z=1} = \frac{1/2}{1/4} = 2$$

7.22 (0.5)

$$\tilde{x}[n] = 1+\cos\left(\frac{\pi n}{8}\right)$$

$$= 1+\frac{e^{j\frac{\pi n}{8}}}{2}+\frac{e^{-j\frac{\pi n}{8}}}{2} \qquad (N = 16)$$

$$= 1+\frac{e^{j\left(\frac{2\pi}{16}\right)n}}{2}+\frac{e^{-j\left(\frac{2\pi}{16}\right)n}}{2}$$

$\Rightarrow a_1 = \frac{1}{2}$ and also $a_{31} = a_{15} = a_{-1} = \frac{1}{2}$

7.23 (8)

From the definition of DTFT

$$X(e^{j\omega}) = \sum_{n=-\infty}^{\infty} x[n]e^{-j\omega n}$$

$$x[n] = \frac{1}{2\pi}\int_{-\pi}^{\pi} X(e^{j\omega})e^{j\omega n}d\omega$$

$$x[0] = \frac{1}{2\pi}\int_{-\pi}^{\pi} X(e^{j\omega})d\omega$$

$$\frac{1}{2\pi}\int_{-\pi}^{\pi} X(e^{j\omega})Y(e^{j\omega})d\omega = [x(n)\otimes y(n)]_{n=0}$$

$$Y(e^{j\omega}) = \sin^2(2\omega)$$

$$= \frac{1-\cos 4\omega}{2} = \frac{1}{2}-\frac{1}{4}e^{4j\omega}-\frac{1}{4}e^{-4j\omega}$$

$$y[n] = \frac{1}{2}\delta[n]-\frac{1}{4}\delta[n+4]-\frac{1}{4}\delta[n-4]$$

$$y[n] = \left\{-\frac{1}{4},0,0,0,\underset{\uparrow}{\frac{1}{2}},0,0,0,-\frac{1}{4}\right\}$$

$$\Rightarrow \quad y[0] = \frac{1}{2}$$

$$x[n] = \{6, 3, \underset{\uparrow}{8}, 7, 4\};\quad x[0] = 8$$

$$\frac{1}{\pi}\int_{-\pi}^{\pi} X(e^{j\omega})Y(e^{j\omega})d\omega = 2\sum_{n=-\infty}^{\infty} x[0]\,y[0]$$

$$= 2\times 8\times\frac{1}{2} = 8$$

7.24 (6)

From the given question

$$x_1[n] = x\left[\frac{n}{3}\right]$$

$$X_1[K] = \{12, 2j, 0, -2j, 12, 2j, 0, -2j, 12, 2j, 0, -2j\}$$

$$X_1[8] = 12;\quad X_1(11) = -2j$$

$$\therefore \quad \left|\frac{X_1(8)}{X_1(11)}\right| = \left|\frac{12}{-2j}\right| = 6$$

7.25 (4096)

$$f_s = 8000 \text{ samples/sec}$$

$$\Rightarrow \quad T_s = \frac{1}{f_s} = \frac{1}{8000}\text{ sec.}$$

Time for each multiplication = T_m = 20 μsec.

$$T_m = 20\times 10^{-6}\text{ sec}$$

Suppose block is of size N, then time taken to generate each block = $N\times T_s = N\times\frac{1}{8000}$. No. of multiplications that can be performed by processor in the time taken for each block

$$= \frac{N\times T_s}{T_m}$$

No. of multiplications required to compute DFT by Radix – 2 FFT algorithm

$$= \frac{N}{2}\log_2 N$$

No. of multiplication required by FFT ≤ No. of multiplications that can be performed by the processor in the time taken in each block [Reason : real time]

$$\frac{N}{2}\log_2 N \le \frac{N\times T_s}{T_m}$$

$$\log_2 N \le \frac{2\times\frac{1}{8000}}{20\times 10^{-6}} = 12.5$$

$$N \le 2^{12.5}$$

$$\Rightarrow \quad N \le 2^{12} = 4096$$

7.26 (2.10)

Since $\quad h[n] = \frac{1}{3}\delta[n]+\frac{1}{3}\delta[n-1]+\frac{1}{3}\delta[n-2]$

So, $\quad H(e^{j\omega}) = \frac{1}{3}e^{-j\omega}[1+2\cos\omega]$

$H(e^{j\omega_0}) = 0,\quad$ for $(1 + 2\cos\omega_0) = 0$

or $\quad \cos\omega_0 = -\frac{1}{2}$

or $\quad \omega_0 = \frac{2\pi}{3} = 2.10$ radians

7.27 (c)

$$h[n] = 5\delta[n]-7\delta[n-1]+7\delta[n-3]-5\delta[n-4]$$

Now, $\quad H(e^{j\omega}) = 5 - 7e^{-j\omega} + 7e^{-3j\omega} - 5e^{-4j\omega}$

Now for $\omega = 0$,

$$H(e^{j0}) = 5 - 7 + 7 - 5 = 0$$

and for $\omega = \pi$,

$$H(e^{j\pi}) = 5 - 7(-1) + 7(-1) - 5(1)$$

$$= 5 + 7 - 7 - 5 = 0$$

System is attenuating low and high frequencies whereas passing the mid frequencies. So, its a BPF.

7.28 (3)

$$X(k) = \{1, 2, 3, 4, 5, 6, 7, 8\}$$

$$X(k) = \sum_{n=0}^{7} x[n]\,e^{-j\frac{2\pi}{8}kn} = \sum_{n=0}^{7} x[n]\,e^{-j\frac{\pi}{4}kn}$$

$$\sum_{n=0}^{3} x[2n] = \sum_{n=0,2,4,6} x[n] = \frac{1}{2}\sum_{n=0}^{7}\left(x[n]+(-1)^n x[n]\right)$$

$$\sum_{n=0}^{7} x[n] = X(0) = 1$$

$$\sum_{n=0}^{7}(-1)^n x[n] = \sum_{n=0}^{7} x[n]\, e^{-j\frac{\pi}{4}4n} = X(4) = 5$$

$$\sum_{n=0}^{3} x[2n] = \frac{1}{2}[1+5] = 3$$

7.29 (d)

$$x(n) = c_1 e^{-j\frac{\pi}{2}n} + c_2 e^{j\frac{\pi}{2}n}$$

$$\omega_o = \frac{\pi}{2}\text{ rad/s}$$

$$H(e^{j\omega}) = 1 + ae^{-j\omega} + be^{-j2\omega}$$

To get $y(n) = 0$,

$$H(e^{j\omega_o}) = H(e^{j\omega})\Big|_{\omega=\frac{\pi}{2}} = 0$$

$$1 + ae^{-j\frac{\pi}{2}} + be^{-j2\frac{\pi}{2}} = 0$$

$$1 - ja - b = 0$$

From the given options, $a = 0$ and $b = 1$.

7.30 (d)

$$X(k) = \sum_{n=0}^{N-1} x(n)\, e^{-j\frac{2\pi}{N}kn} \quad ; \quad X(1) = \sum_{n=0}^{5} x(n) W_6^n$$

$$= x(0) + x(1)W_6 + x(2)W_6^2 + x(3)W_6^3 + x(4)W_6^4 + x(5)W_6^5 \quad \text{...(i)}$$

From the given flow graph,

$$X(k) = [x(0) - x(3)]a_1 + [x(1) - x(4)]a_2 + [x(2) - x(5)]a_3$$

By comparing equations (i) and (ii), we get,

$$a_1 = 1,\ a_2 = W_6,\ a_3 = W_6^2$$

7.31 (a)

$$x(t) = \cos 200\pi t$$

$$t = \frac{n}{400}$$

$$x(n) = \cos\left(200\pi\frac{n}{400}\right) = \cos\left(\frac{\pi}{2}n\right)$$

$n = 0, 1, ..., 7$

$$= \begin{Bmatrix} \cos 0, \cos\frac{\pi}{2}, \cos\pi, \cos\frac{3\pi}{2}, \\ \cos 2\pi, \cos\frac{5\pi}{2}, \cos 3\pi, \cos\frac{7\pi}{2} \end{Bmatrix}$$

$$= \{1, 0, -1, 0, 1, 0, -1, 0\} \xleftrightarrow{\text{DFT}} X(k)$$

Suppose, $y(n) = \{1, -1, 1, -1\} \xleftrightarrow{\text{DFT}} Y(k)$

$$[Y(k)] = [W_N]\big|_{N=4}[y(n)]$$

$$= \begin{bmatrix} 1 & 1 & 1 & 1 \\ 1 & -j & -1 & j \\ 1 & -1 & 1 & -1 \\ 1 & j & -1 & -j \end{bmatrix}\begin{bmatrix} 1 \\ -1 \\ 1 \\ -1 \end{bmatrix} = [0,\ 0,\ 4,\ 0]$$

Now, as we know,

If for $\{a, b, c, d\} \xleftrightarrow{\text{DFT}} \{A, B, C, D\}$

Then for $\{a, 0, b, 0, c, 0, d, 0\} \xleftrightarrow{\text{DFT}} \{A, B, C, D, A, B, C, D\}$

Similarly,

for $y(n) = \{1, -1, 1, -1\} \xleftrightarrow{\text{DFT}} Y(k) = \{0, 0, 4, 0\}$

Here, for $x(n) = \{1, 0, -1, 0, 1, 0, -1, 0\}$

$$X(k) = \{\underset{\uparrow}{0},\ 0,\ 4,\ 0,\ 0,\ 0,\ 4,\ 0\}$$

7.32 (b)

If two 'N' point signals $x(n)$ and $h(n)$ are convolving with each other linearly and circularly then

$$y(k) = z(k) \text{ at } k = N-1$$

where, $y(n)$ = Linear convolution of $x(n)$ and $h(n)$

$z(n)$ = Circular convolution of $x(n)$ and $h(n)$

Since, $N = 16$ (Given)

Therefore, $y(k) = z(k)$ at $k = N - 1 = 15$

7.33 (8)

$$x[n] = 2^{n-1}u[-n+2]$$

$$y[n] = 2^{-n+2}u[n+1]$$

$$y[-n] = 2^{n+2}u[-n+1]$$

$$V = \frac{1}{2\pi}\int_0^{2\pi} X(e^{j\omega})\, Y(e^{-j\omega})\,d\omega \quad \text{...(i)}$$

$$z[n] = x[n] * y[-n]$$

$$z[n] \to Z(e^{j\omega})$$

$$z[n] = \frac{1}{2\pi}\int_{-\pi}^{\pi} Z(e^{j\omega})\, e^{j\omega n}\,d\omega$$

Put $n = 0$, $z[0] = \frac{1}{2\pi}\int_0^{2\pi} Z(e^{j\omega})\,d\omega \quad \text{...(ii)}$

Compare equations (i) and (ii)

$$z[0] = V$$

$$Z(e^{j\omega}) = X(e^{j\omega})\, Y(e^{-j\omega})$$

Apply IDTFT,

$$z[n] = x[n] * y[-n] = x[n] * p[n]$$

$\because \quad p[n] = y[-n] = 2^{n+2}\,u[-n+1]$

$$z[n] = \sum_{k=-\infty}^{+\infty} 2^{k-1}u[-k+2]\, 2^{n-k+2}u[-n+k+1]$$

$$= \sum_{k=-\infty}^{2} 2^{k-1} 1\cdot 2^{n-k+2} u[k+1-n]$$

$$= \sum_{k=-\infty}^{2} 2^{k-1+n-k+2} u[k+1-n]$$

$$z[n] = \sum_{k=-\infty}^{2} 2^{n+1} u[k-n+1]$$

Put n = 0, $V = z[0] = \sum_{k=-\infty}^{2} 2^{1}\cdot u[k+1]$

$$= 2\sum_{k=-1}^{2} 1 = 2[1+1+1+1]$$

$$V = 8$$

7.34 (8)

It is given that, $\bar{X} = DFT(\bar{x})$

$\Rightarrow \quad x(n) \xleftrightarrow{DFT} X(k)$

Again applying DFT,

$\bar{y} = DFT(DFT(x)) = DFT(\bar{X})$

$\therefore \quad \bar{y} = \text{DFT}(X(k))$

Using Duality property of DFT,

$x(n) \xleftrightarrow{DFT} X(k) \quad ; \quad X(n) \xleftrightarrow{DFT} Nx(-k)$

$\bar{y}' = N x(-k) = Nx(N-k)$

$\therefore \quad \bar{y} = 8\,[1, 0, 0, 0, 2, 0, 0, 0]$

$\bar{y} = [8, 0, 0, 0, 16, 0, 0, 0]$

Hence, $y[0] = 8$

7.35 (a)

Let, $I = \sum_{n=0}^{4} x(n)\sin\frac{4\pi n}{5}$

$$= \frac{1}{2j}\sum_{n=0}^{4} x(n)\cdot\left[e^{j\frac{4\pi n}{5}} - e^{-j\frac{4\pi n}{5}}\right]$$

$$= \frac{1}{2j}\left[\sum_{n=0}^{4} x(n)\,e^{j\frac{4\pi n}{5}} - \sum_{n=0}^{4} x(n)\cdot e^{-j\frac{4\pi n}{5}}\right] \qquad \text{...(i)}$$

As we know, $a_k = \frac{1}{N}\sum_{n=0}^{4} x(n)\cdot e^{-jK\cdot\frac{2\pi}{N}n}$

$$= \frac{1}{5}\sum_{n=0}^{4} x(n)\cdot e^{-j\frac{2\pi}{N}kn}$$

Put $K = 2$; $a_2 = \frac{1}{5}\sum_{n=0}^{4} x(n)\cdot e^{-j\frac{4\pi n}{5}}$

Put $K = -2$; $a_{-2} = \frac{1}{5}\sum_{n=0}^{4} x(n)\cdot e^{j\frac{4\pi n}{5}}$

From equation (i), $I = \frac{1}{2j}[5a_{-2} - 5a_2]$

$$= \frac{5}{2j}[a_3 - a_2] \quad \left[\because \begin{aligned} a_{-2} &= a_{-2+N} \\ &= a_{-2+5} \\ &= a_3 \end{aligned}\right]$$

$\Rightarrow \quad I = \frac{5}{2j}[-2j - 2j] = -10$

7.36 (c)

For an LTI system, output,

$y[n] = x[n] * h[n]$

By taking DTFT, $Y(e^{j\Omega}) = X(e^{j\Omega})H(e^{j\Omega})$

$$= \left[1 - e^{-j\Omega} + 2e^{-3j\Omega}\right]\left[1 - \frac{1}{2}e^{-2j\Omega}\right]$$

$$= 1 - e^{-j\Omega} + 2.5e^{-3j\Omega} - 0.5e^{-j2\Omega} - e^{-j5\Omega}$$

Taking IDTFT;

$$y[n] = \delta[n] - \delta[n-1] - 0.5\delta[n-2] + 2.5\delta[n-3] - \delta[n-5]$$

7.37 (112)

Using the duality property of N-point DFT,

$$x[n] \xleftrightarrow{N} X[k]$$

$$X[n] \xleftrightarrow{N} Nx[-k]_N$$

We obtain,

$$F\{F\{x[n]\}\} = Nx[-k]_N$$

Again applying DFT operation two times, we get,

$$F\{F\{F\{F\{x[n]\}\}\}\} = N^2 x[n]$$

Hence, we get,

$$y[n] = N^2 x[n]$$

The N-point DFT of $y[n]$ is given as

$$Y[k] = \sum_{n=0}^{N-1} y[n]\cdot e^{-jk\frac{2\pi}{N}n}$$

for $k = 0, 1, \ldots, N-1$

The sequence $x[n] = \{1, 2, 1, 3\}$. Hence, $N = 4$

For $k = 0$, we get,

$$Y[0] = \sum_{n=0}^{3} y[n] = N^2\sum_{n=0}^{3} x[n]$$

$$Y[0] = 16[1 + 2 + 1 + 3]$$

$$Y[0] = 112$$

8 Sampling

8.1 A 1.0 kHz signal is flat-top sampled at the rate of 1800 samples/sec and the samples are applied to an ideal rectangular LPF with cut-off frequency of 1100 Hz, then the output of the filter contains
(a) only 800 Hz component
(b) 800 Hz and 900 Hz components
(c) 800 Hz and 1000 Hz components
(d) 800 Hz, 900 Hz and 1000 Hz components
[1995 : 1 M]

8.2 Flat top sampling of low pass signals
(a) gives rise to aperture effect
(b) implies oversampling
(c) leads to aliasing
(d) introduces delay distortion [1998 : 1 M]

8.3 Let $x(t) = 2\cos(800\pi t) + \cos(1400\pi t)$. $x(t)$ is sampled with the rectangular pulse train shown in the figure. The only spectral components (in kHz) present in the sampled signal in the frequency range 2.5 kHz to 3.5 kHz are

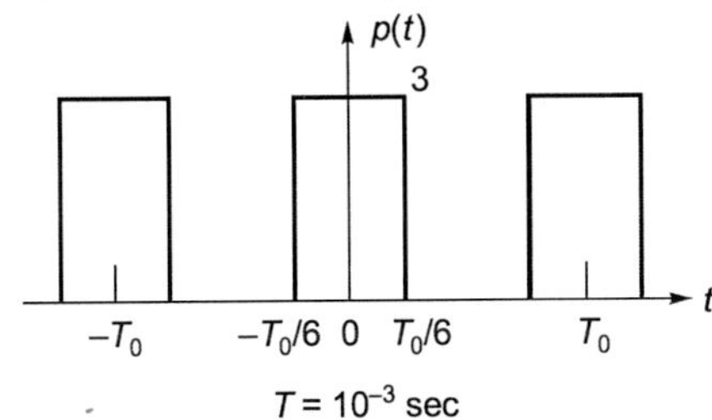

(a) 2.7, 3.4
(b) 3.3, 3.6
(c) 2.6, 2.7, 3.3, 3.4, 3.6
(d) 2.7, 3.3 [2003 : 2 M]

8.4 A 1 kHz sinusoidal signal is ideally sampled at 1500 samples/sec and the sampled signal is passed through an ideal low-pass filter with cut-off frequency 800 Hz. The output signal has the frequency
(a) zero Hz (b) 0.75 kHz
(c) 0.5 kHz (d) 0.25 kHz
[2004 : 2 M]

8.5 A signal $m(t)$ with bandwidth 500 Hz is first multiplied by a signal $g(t)$ where

$$g(t) = \sum_{k=-\infty}^{\infty} (-1)^k \delta(t - 0.5 \times 10^{-4} k)$$

The resulting signal is then passed through an ideal low pass filter with bandwidth 1 kHz. The output of the low pass filter would be
(a) $\delta(t)$ (b) $m(t)$
(c) 0 (d) $m(t)\,\delta(t)$
[2006 : 2 M]

8.6 An LTI system having transfer function $\frac{s^2+1}{s^2+2s+1}$ and input $x(t) = \sin(t+1)$ is in steady state. The output is sampled at a rate ω_s rad/s to obtain the final output $\{y(k)\}$. Which of the following is true?
(a) y is zero for all sampling frequencies ω_s
(b) y is nonzero for all sampling frequencies ω_s
(c) y is nonzero for $\omega_s > 2$, but zero for $\omega_s < 2$
(d) y is zero for $\omega_s > 2$, but nonzero for $\omega_s < 2$
[2009 : 2 M]

8.7 A band-limited signal with a maximum frequency of 5 kHz is to be sampled. According to the sampling theorem, the sampling frequency which is not valid is
(a) 5 kHz (b) 12 kHz
(c) 15 kHz (d) 20 kHz
[2013 : 1 M]

8.8 Consider two real valued signals, $x(t)$ band-limited to [–500 Hz, 500 Hz] and $y(t)$ band-limited to [–1 kHz, 1 kHz]. For $z(t) = x(t)\,y(t)$, the Nyquist sampling frequency (in kHz) is ______.
[2014 : 1 M, Set-1]

8.9 Let $x(t) = \cos(10\pi t) + \cos(30\pi t)$ be sampled at 20 Hz and reconstructed using an ideal low-pass filter with cut-off frequency of 20 Hz. The frequency/frequencies present in the reconstructed signal is/are
(a) 5 Hz and 15 Hz only
(b) 10 Hz and 15 Hz only
(c) 5 Hz, 10 Hz and 15 Hz only
(d) 5 Hz only [2014 : 1 M, Set-3]

8.10 For a given sample-and-hold circuit, if the value of the hold capacitor is increased, then
(a) drop rate decreases and acquisition time decreases

(b) drop rate decreases and acquisition time increases
(c) drop rate increases and acquisition time decreases
(d) drop rate increases and acquisition time increases

[2014 : 1 M, Set-4]

8.11 The signal $\cos\left(10\pi t + \frac{\pi}{4}\right)$ is ideally sampled at a sampling frequency of 15 Hz. The sampled signal is passed through a filter with impulse response $\left(\frac{\sin(\pi t)}{\pi t}\right)\cos\left(40\pi t - \frac{\pi}{2}\right)$. The filter output is

(a) $\frac{15}{2}\cos\left(40\pi t - \frac{\pi}{4}\right)$

(b) $\frac{15}{2}\left(\frac{\sin(\pi t)}{\pi t}\right)\cos\left(10\pi t + \frac{\pi}{4}\right)$

(c) $\frac{15}{2}\cos\left(10\pi t - \frac{\pi}{4}\right)$

(d) $\frac{15}{2}\left(\frac{\sin(\pi t)}{\pi t}\right)\cos\left(10\pi t - \frac{\pi}{2}\right)$

[2015 : 1 M, Set-2]

8.12 Consider a continuous time signal defined as

$$x(t) = \left(\frac{\sin(\pi t/2)}{(\pi t/2)}\right) * \sum_{n=-\infty}^{\infty} \delta(t - 10n)$$

Where '∗' denotes the convolution operation and t is in seconds. The Nyquist sampling rate (in samples/sec) for $x(t)$ is_____.

[2015 : 2 M, Set-3]

8.13 A continuous-time sinusoid of frequency 33 Hz is multiplied with a periodic Dirac impulse train of frequency 46 Hz. The resulting signal is passed through an ideal analog low-pass filter with a cutoff frequency of 23 Hz. The fundamental frequency (in Hz) of the output is _____.

[2016 : 1 M, Set-1]

8.14 Consider the signal $x(t) = \cos(6\pi t) + \sin(8\pi t)$, where t is in seconds. The Nyquist sampling rate (in samples/second) for the signal $y(t) = x(2t + 5)$ is

(a) 8
(b) 12
(c) 16
(d) 32

[2016 : 1 M, Set-3]

8.15 The signal $x(t) = \sin(14000\,\pi t)$, where t is in seconds is sampled at a rate of 9000 samples per second. The sampled signal is the input to an ideal lowpass filter with frequency response $H(f)$ as follows :

$$H(f) = \begin{cases} 1, & |f| \le 12\text{kHz} \\ 0, & |f| > 12\text{kHz} \end{cases}$$

What is the number of sinusoids in the output and their frequencies in kHz?

(a) Number = 1, frequency = 7
(b) Number = 3, frequencies = 2, 7, 11
(c) Number = 2, frequencies = 2, 7
(d) Number = 2, frequencies = 7, 11

[2017 : 2 M, Set-2]

8.16 A band limited low-pass signal $x(t)$ of bandwidth 5 kHz is sampled at a sampling rate f_s. The signal $x(t)$ is reconstructed using the reconstruction filter $H(f)$ whose magnitude response is shown below:

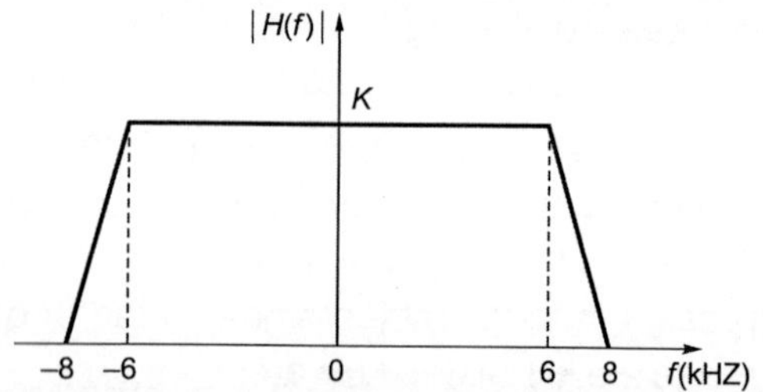

The minimum sampling rate f_s (in kHz) for perfect reconstruction of $x(t)$ is ____.

[2018 : 2 M]

8.17 Consider a real-valued base-band signal $x(t)$, band limited to 10 kHz. The Nyquist rate for the signal $y(t) = x(t)x\left(1 + \frac{t}{2}\right)$ is

(a) 15 kHz
(b) 30 kHz
(c) 20 kHz
(d) 60 kHz

[2021 : 1 M]

8.18 Let $x_1(t)$ and $x_2(t)$ be two band-limited signals having bandwidth $B = 4\pi \times 10^3$ rad/s each. In the figure below, the Nyquist sampling frequency, in rad/s, required to sample $y(t)$, is

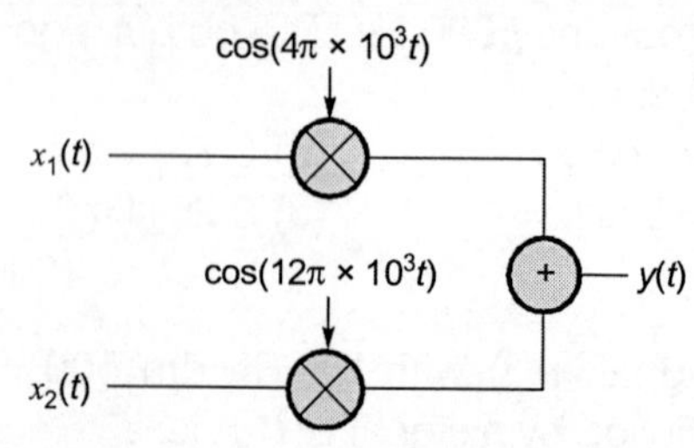

(a) $20\pi \times 10^3$
(b) $40\pi \times 10^3$
(c) $8\pi \times 10^3$
(d) $32\pi \times 10^3$

[2023 : 2 M]

8.19 A continuous time signal $x(t) = 2\cos(8\pi t + \pi/3)$ is sampled at a rate of 15 Hz. The sampled signal $x_s(t)$ when passed through an LTI system with impulse response

$$h(t) = \left(\frac{\sin 2\pi t}{\pi t}\right)\cos(38\pi t - \pi/2)$$

produces an output $x_0(t)$. The expression for $x_0(t)$ is ______.

(a) $15\sin(38\pi t + \pi/3)$ (b) $15\cos(38\pi t - \pi/6)$
(c) $15\sin(38\pi t + \pi/6)$ (d) $15\sin(38\pi t - \pi/3)$

[2024 : 2 Marks]

■■■

Answers Sampling

8.1	(c)	8.2	(a)	8.3	(d)	8.4	(c)	8.5	(b)	8.6	(a)	8.7	(a)
8.8	(3)	8.9	(a)	8.10	(b)	8.11	(a)	8.12	(0.4)	8.13	(13)	8.14	(c)
8.15	(b)	8.16	(13)	8.17	(b)	8.18	(d)	8.19	(a, b)				

Explanations Sampling

8.1 (c)

$f_s = 1800$ samples/sec ; $f_m = 1000$ Hz

The spectrum of sampled signal would have

$nf_s \pm f_m$

So,

1000 Hz, 1800 ± 1000 Hz, 3600 ± 1000 Hz ...

So, 1000 Hz, 800 Hz, 2800 Hz, 2600 Hz, 4600 Hz, ...

The cut-off frequency of LPF is 1100 Hz

So, the output of filter will contains 800 Hz and 1000 Hz components.

8.2 (a)

Flat top sampling of low pass signals gives rise to aperture effect.

8.3 (d)

From Fourier series expansion,

$$C_n = \frac{1}{T_o}\int_{-T_o/6}^{T_o/6} A.e^{-jn\omega_o t}dt = \frac{A}{\pi n}\sin\left(\frac{n\pi}{3}\right)$$

∴ From C_n its clear that 1, 2, 4, 5, 7.... harmonics are present.

Frequency of $p(t)$ corr. to 1, 2, 4, 5, 7...... are 10^3, 2×10^3, 4×10^3....

$x(t)$ has frequency components 0.7 k and 0.4 k

∴ $p(t) \times x(t)$ gives (1 ± 0.7) k, (2 ± 0.7) k, (4 ± 0.7) k, (1 ± 0.4) k, (2 ± 0.4) k....

Frequency present in range of 2.5 k to 3.5 k are 2.7, 3.3.

8.4 (c)

$f_s = 1.5$ kHz ; $f_m = 1$ kHz

Available frequency components are $nf_s \pm f_m$,

So, 1 kHz, 2.5 kHz, 0.5 kHz ......

∵ LPF has $f_c = 0.8$ k

∴ only 0.5 k will appear at output.

8.5 (b)

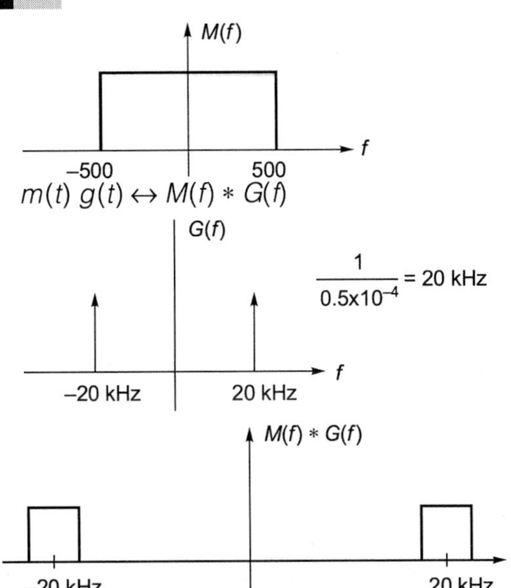

After low pass filtering with $f_c = 1$ kHz

Output is zero.

8.6 (a)

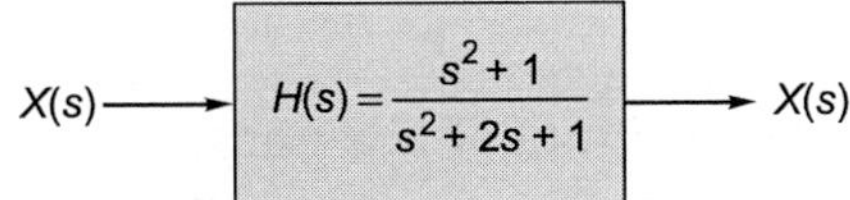

$x(t) = \sin(t + 1)$

$\omega = 1$ rad/sec

$$X(s) = \frac{\omega}{\omega^2 + s^2}e^s = \frac{1}{1^2 + s^2}e^s = \frac{e^s}{s^2 + 1}$$

$$H(s) = \frac{Y(s)}{X(s)}$$

$$Y(s) = H(s)X(s)$$

$$Y(s) = \frac{s^2+1}{s^2+2s+1} \times \frac{e^s}{s^2+1} = \frac{e^s}{s^2+2s+1}$$

$$Y(s) = \frac{e^s}{(s+1)^2}$$

$$y(t) = (t+1)e^{-(t+1)}$$

$$y(\infty) = \lim_{s\to 0} sY(s) = \lim_{s\to 0} \frac{se^s}{(s+1)^2} = 0$$

So, at steady state y remains zero for all sampling frequencies ω_s.

8.7 (a)

$$(f_s)_{min} = 2f_m$$

$$(f_s)_{min} = 2 \times 5 = 10 \text{ kHz}$$

So, $f_s \geq 10$ kHz

8.8 (3)

Multiplication in time domain
= convolution in frequency domain.

$$x_1(t) \cdot x_2(t) \leftrightarrow X_1(\omega) * X_2(\omega)$$

So, highest frequency component contained by the convolved signal $z(t)$ = 1500 Hz

$\therefore$ Nyquist rate = 2 × 1500 = 3000 Hz = 3 kHz

8.9 (a)

$$x(t) = \cos(10\pi t) + \cos(30\pi t)$$

Given sampling frequency, f_s = 20 Hz

$$\omega_s = 40\pi \text{ rad/sec.}$$

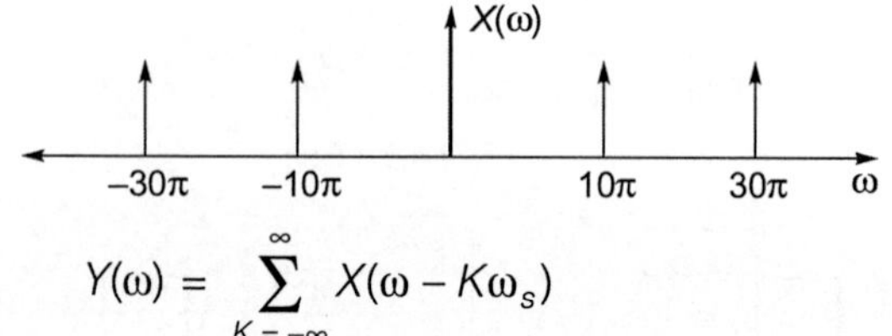

$$Y(\omega) = \sum_{K=-\infty}^{\infty} X(\omega - K\omega_s)$$

After sampling waveform will be

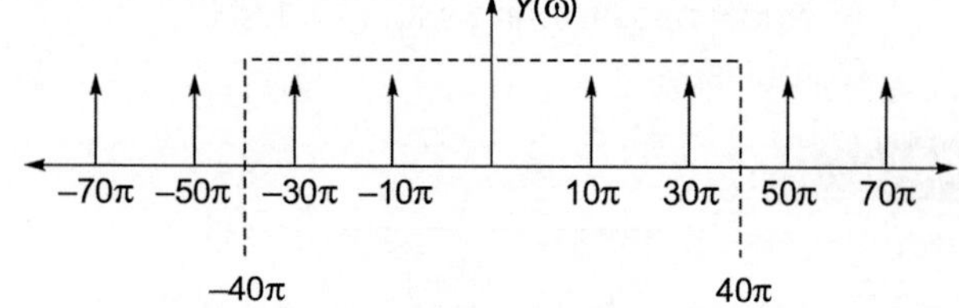

Applying an ideal low-pass filter of cut-off frequency of 20 Hz or 40 π rad/sec, we get the frequencies in reconstructed signal as
10π and 30π rad/sec. or
5 Hz and 15 Hz.

8.10 (b)

We know that

$$Q = CV = i \cdot t \quad \text{...(i)}$$

$$t = \frac{CV}{i} \quad \text{...(ii)}$$

From equation (ii)

$t \propto C$ and it is clear that if value of capacitor increases then the acquisition time increases. For capacitor the drop rate is given as $\frac{dv}{dt}$

$$i = C\frac{dv}{dt}$$

$$\frac{dv}{dt} = \frac{i}{C}$$

$$\frac{dv}{dt} \propto \frac{1}{C}$$

From above relation it is clear that if capacitor value increases drop rate decreases.

8.11 (a)

$$x(t) = \cos\left(10\pi t + \frac{\pi}{4}\right)$$

$$X(f) = \frac{1}{2}[\delta(f-5) + \delta(f+5)]\, e^{j\pi/4}$$

After sampling $x(t)$ by 15 Hz, we get

$$X_s(f) = 15 \sum_{n=-\infty}^{\infty} X(f - 15n)$$

Given, $h(t) = \left[\frac{\sin(\pi t)}{\pi t}\right]\cos\left(40\pi t - \frac{\pi}{2}\right)$

$$H(f) = \frac{1}{2}\left[\text{rect}\frac{(f-20)}{2\pi} + \text{rect}\frac{(f+20)}{2\pi}\right]e^{-j\frac{\pi}{2}}$$

Output $Y(f) = X_s(f) \times H(f)$

$$y(t) = \frac{15}{2}\cos\left(40\pi t - \frac{\pi}{4}\right)$$

8.12 (0.4)

$$x(t) = \frac{\sin\left(\frac{\pi t}{2}\right)}{\left(\frac{\pi t}{2}\right)} * \sum_{n=-\infty}^{\infty} \delta[t - 10n]$$

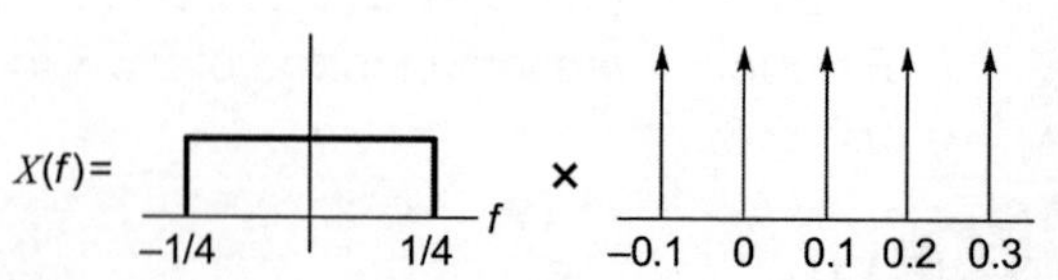

$$f_{max} = 0.2 \text{ Hz}$$

$$\Rightarrow f_s = 2f_{max} = 0.4 \text{ Hz}$$

8.13 (13)

If $x(t)$ is a message signal and $y(t)$ is a sampled signal, then $y(t)$ is related to $x(t)$ as

$$y(t) = x(t)\sum_{n=-\infty}^{\infty}\delta(t-nT_s)$$

$$Y(f) = f_s\sum_{n=-\infty}^{\infty}X(f-nf_s)$$

Spectrum of $X(f)$ and $Y(f)$ are as shown

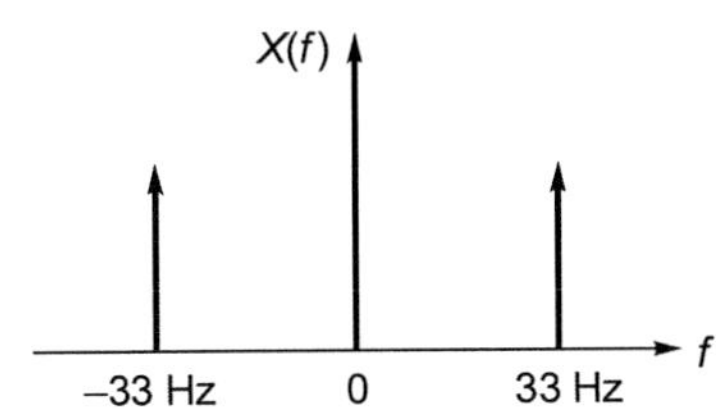

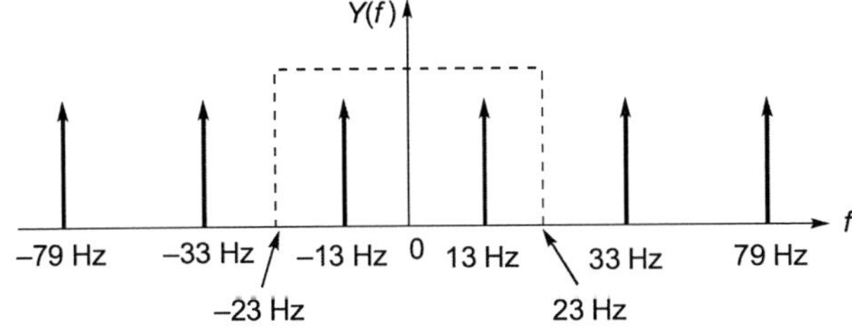

Cut off frequency of LPF = 23 Hz

Hence, frequency at the output is 13 Hz

8.14 (c)

$$X(t) = \cos(6\pi t) + \sin(8\pi t)$$
$$Y(t) = x(2t+5)$$
$$Y(t) = \cos[6\pi(2t+5) + \sin(8\pi(2t+5)]$$
$$= \cos(12\pi t + 30\pi) + \sin(16\pi t + 40\pi)$$
$$f_{m_1} = 6\text{ Hz}; \qquad f_{m_2} = 8\text{ Hz}$$

Nyquist sampling rate,

$$f_s = 2f_{max}$$
$$= 16\text{ samples/second}$$

8.15 (b)

$$x(t) = \sin(14000\pi t)\ ;\ f_m = 7\text{ kHz}$$
$$f_s = 9000\text{ samples per second}$$
$$= 9\text{ kHz}$$

The spectrum of the sampled signal can be given as shown below:

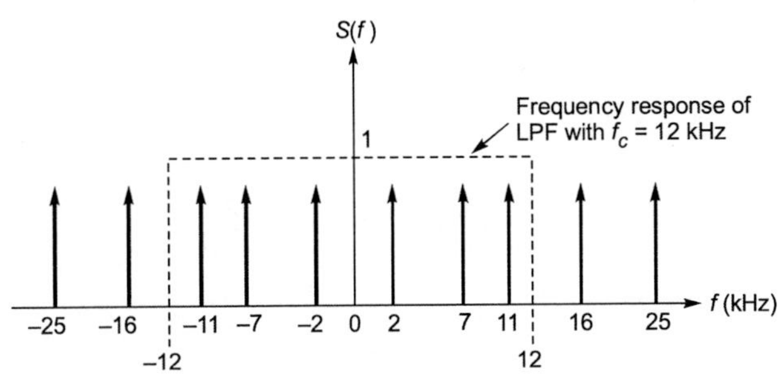

So, three sinusoids will be there at the output of the LPF and the frequencies of those sinusoids are 2 kHz, 7 kHz and 11 kHz.

8.16 (13)

Let us assume an arbitrary spectrum for $x(t)$ as shown below:

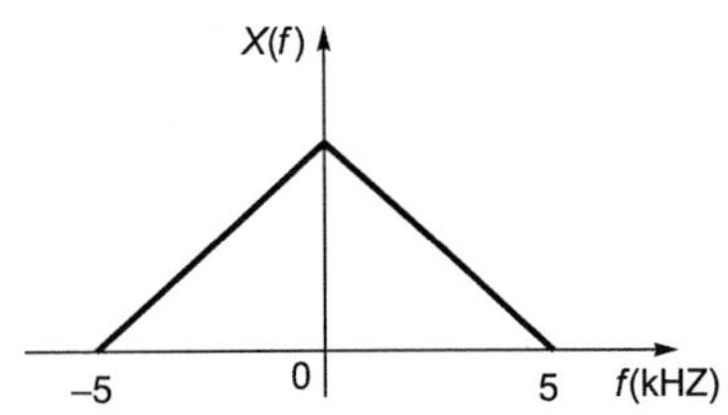

The spectrum of the sampled signal can be given as,

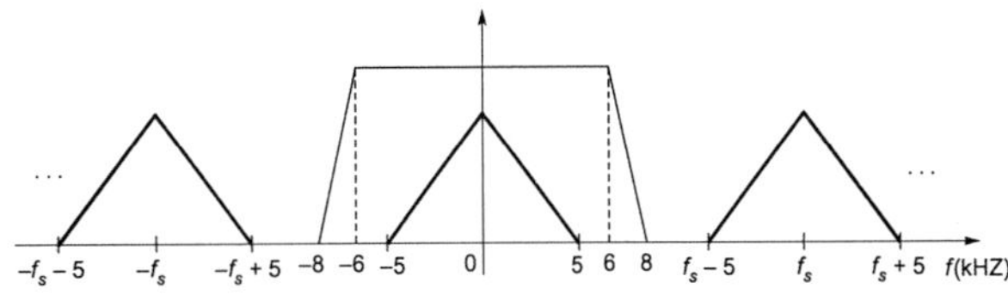

For proper reconstruction of the signal,

$$f_s - 5 \geq 8$$
$$f_s \geq 8 + 5 = 13\text{ kHz}$$

So, $\quad f_{s(\min)} = 13\text{ kHz}$

8.17 (b)

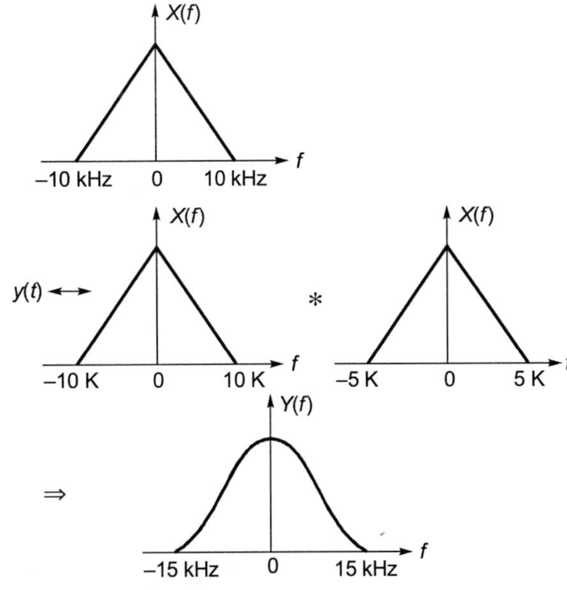

$$\text{NR} = 2 \times f_{max} = 2 \times 15 = 30\text{ kHz}$$

8.18 (d)

Given that, $x_1(t)$ and $x_2(t)$ are two bandlimited signals having bandwidth $B = 4\pi \times 10^3$ rad/sec.

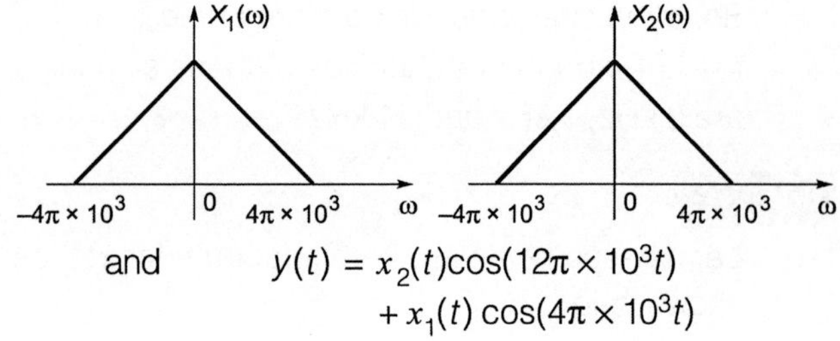

and $\quad y(t) = x_2(t)\cos(12\pi \times 10^3 t) + x_1(t)\cos(4\pi \times 10^3 t)$

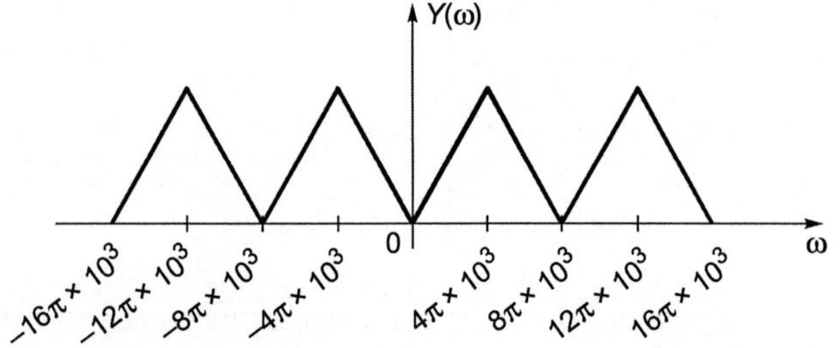

So, Nyquist rate $= 2\omega_{max}$

$= 2[16\pi \times 10^3] = 32\pi \times 10^3$ rad/sec

8.19 (a, b)

$$x(t) = 2\cos\left(8\pi t + \frac{\pi}{3}\right); \; f_0 = \frac{8\pi}{2\pi} = 4 \text{ Hz}$$

After sampling at a rate of $f_s = 15$ Hz, the frequency spectrum of the sampled signal is given by

$$X_s(f) = f_s \sum_{n=-\infty}^{\infty} X(f - nf_s)$$

Hence, the frequencies present in the output of the sampler is $nf_s \pm f_0$ i.e., f_0, $f_s \pm f_0$, $2f_s \pm f_0$, ... = 4 Hz, 11 Hz, 19 Hz, 26 Hz, 34 Hz, ...

The sampled signal is passed through an LTI system with impulse response.

$$h(t) = \left(\frac{\sin 2\pi t}{\pi t}\right) \cdot \cos(38\pi t - \pi/2)$$

We have,

$$P(t) = \frac{\sin 2\pi t}{\pi t}$$

$\xleftrightarrow{\text{F.T.}}$

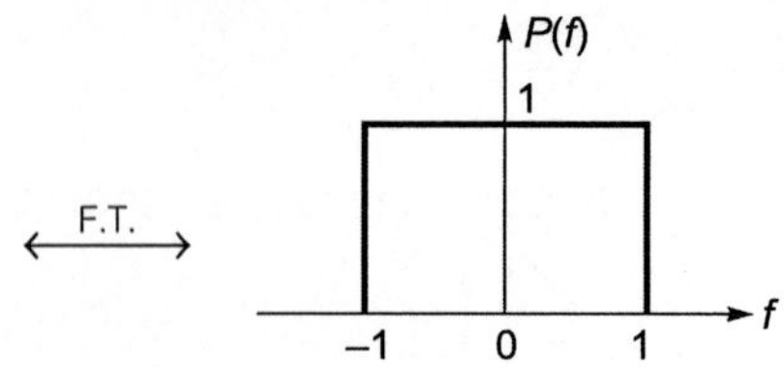

$$h(t) = \left(\frac{\sin 2\pi t}{\pi t}\right)\cos(38\pi t - \pi/2)$$

$\xleftrightarrow{\text{F.T.}}$

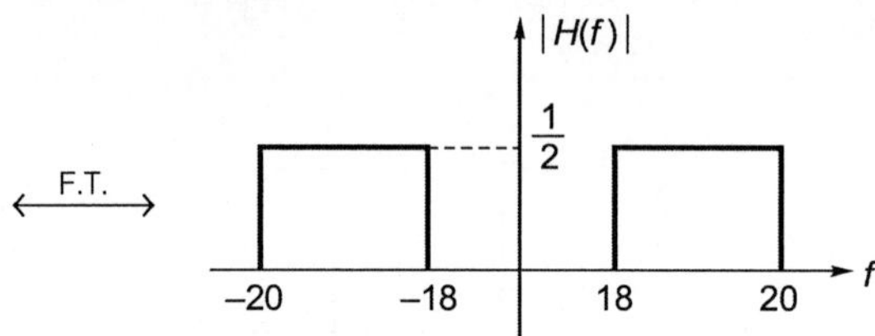

The system will pass only the '19 Hz' component of the sampled signal. The output signal $x_0(t)$ is obtained as below:

$$x_0(t) = 2 \times \frac{f_s}{2}\cos\left[38\pi t + \frac{\pi}{3} - \frac{\pi}{2}\right]$$

$$= 15\cos\left(38\pi t - \frac{\pi}{6}\right)$$

or $\quad x_0(t) = 15\sin\left[38\pi t + \frac{\pi}{3}\right]$

Note: GATE key (marks to all).

■■■

9 Digital Filters

9.1 A continuous-time filter with transfer function $H(s) = \dfrac{2s+6}{s^2+6s+8}$ is converted to a discrete time filter with transfer function $G(s) = \dfrac{2z^2 - 0.5032z}{z^2 - 0.5032z + k}$ so that the impulse response of the continuous-time filter, sampled at 2 Hz, is identical at the sampling instants to the impulse response of the discrete time filter. The value of k is ______.

[2016 : 2 M, Set-2]

9.2 The direct form structure of an FIR (finite impulse response) filter is shown in the figure.

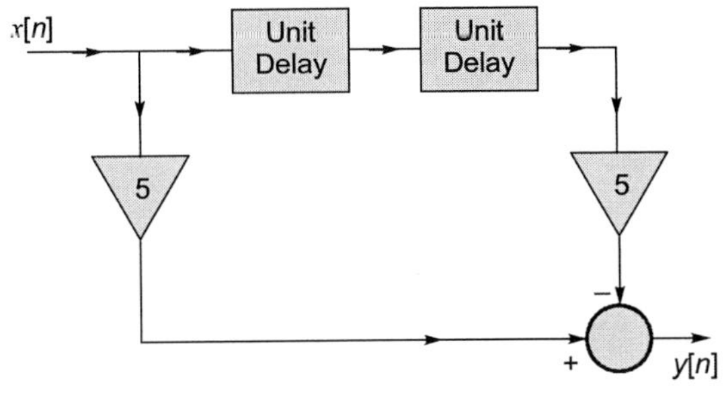

The filter can be used to approximate a

(a) low-pass filter
(b) high-pass filter
(c) band-pass filter
(d) band-stop filter

[2016 : 2 M, Set-3]

9.3 Let $h[n]$ be a length-7 discrete-time finite impulse response filter, given by

$$h[0]=4,\quad h[1]=3,\quad h[2]=2,\quad h[3]=1$$
$$h[-1]=-3,\quad h[-2]=-2,\quad h[-3]=-1,$$

and $h[n]$ is zero for $|n| \geq 4$. A length-3 finite impulse response approximation $g[n]$ of $h[n]$ has to be obtained such that

$$E(h,g) = \int_{-\pi}^{\pi} \left| H(e^{j\omega}) - G(e^{j\omega}) \right|^2 d\omega$$

is minimized, where $H(e^{j\omega})$ and $G(e^{j\omega})$ are the discrete-time Fourier transforms of $h[n]$ and $g[n]$, respectively. For the filter that minimizes $E(h, g)$, the value of $10g[-1] + g[1]$, rounded off to 2 decimal places, is ________ .

[2019 : 2 M]

■■■

Answers Digital Filters

9.1 (0.049) 9.2 (c) 9.3 (–27)

Explanations Digital Filters

9.1 (0.049)

Given, $H(s) = \frac{2s+6}{s^2+6s+8} = \frac{1}{s+2} + \frac{1}{s+4}$

$h(t) = e^{-2t}u(t) + e^{-4t}u(t)$

Given, $f_s = 2\text{ Hz}$

For discrete time, $t = nT_s = \frac{n}{2}$

$$h[n] = (e^{-n} + e^{-2n})\,u[n]$$

$$H(z) = \frac{1}{1-e^{-1}Z^{-1}} + \frac{1}{1-e^{-2}Z^{-1}}$$

$$= \frac{Z}{Z-e^{-1}} + \frac{Z}{Z-e^{-2}}$$

$$= \frac{2Z^2 - 0.5032Z}{Z^2 - 0.5032Z + 0.049}$$

$\therefore$ $K = 0.049$

9.2 (c)

$$y[n] = 5[x[n] - x[n-2]]$$

$$H(z) = \frac{Y(z)}{X(z)} = 5\,[1 - z^{-2}]$$

But, $z = e^{j\omega}$

$$H(e^{j\omega}) = 5\,[1 - e^{-j2\omega}]$$

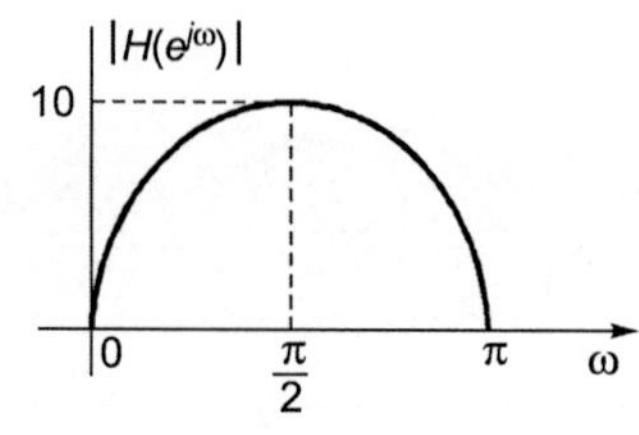

$\omega = 0$ $|H(e^{j\omega})| = 0$

$\omega = \frac{\pi}{2}$ $|H(e^{j\omega})| = 10$

$\omega = \pi$ $|H(e^{j\omega})| = 0$

$\therefore$ The given one is bandpass filter.

9.3 (–27)

From Parseval's theorem,

$$\sum_{n=-\infty}^{\infty} |x[n]|^2 = \frac{1}{2\pi}\int_{-\pi}^{\pi} \left|X(e^{j\omega})\right|^2 d\omega$$

So,

$$\int_{-\pi}^{\pi} \left|H(e^{j\omega}) - G(e^{j\omega})\right|^2 d\omega = 2\pi \sum_{n=-3}^{3} |h(n) - g(n)|^2$$

The solution of $g(n)$ that minimizes $E(h, g)$ also minimizes $\sum_{n=-3}^{3} |h(n) - g(n)|^2$.

$$\sum_{n=-3}^{3} |h(n) - g(n)|^2$$

$$= |4 - g(0)|^2 + |3 - g(1)|^2 + |-3 - g(-1)|^2 + 10$$

The solution of $g(n)$ that minimizes the above equation is

$$g(n) = \{-3, \underset{\uparrow}{4}, 3\}$$

So, $10g(-1) + g(1) = 10(-3) + 3 = -27$

Communication Systems

CONTENTS

Communication Systems

Syllabus

Random processes : autocorrelation and power spectral density, properties of white noise, filtering of random signals through LTI systems.

Analog communications : amplitude modulation and demodulation, angle modulation and demodulation, spectra of AM and FM, superheterodyne receivers.

Information theory : Entropy, mutual information and channel capacity theorem.

Digital communications : PCM, DPCM, digital modulation schemes (ASK, PSK, FSK, QAM), bandwidth, inter-symbol interference, MAP, ML detection, matched filter receiver, SNR and BER.

Fundamentals of error correction, Hamming codes, CRC.

Analysis of Previous GATE Papers

Exam Year	1 Mark Ques.	2 Marks Ques.	Total Marks
1995	8	–	**8**
1996	2	2	**6**
1997	3	1	**5**
1998	9	–	**9**
1999	2	4	**10**
2000	1	4	**9**
2001	2	3	**8**
2002	3	4	**11**
2003	3	13	**29**
2004	5	10	**25**
2005	2	7	**16**
2006	–	14	**28**
2007	3	10	**23**
2008	1	11	**23**
2009	3	5	**13**
2010	3	4	**11**
2011	2	4	**10**
2012	4	3	**10**
2013	2	5	**12**

Exam Year	1 Mark Ques.	2 Mark Ques.	Total Marks
2014 Set-1	3	4	**11**
2014 Set-2	2	4	**10**
2014 Set-3	3	5	**13**
2014 Set-4	2	4	**10**
2015 Set-1	2	4	**10**
2015 Set-2	1	3	**7**
2015 Set-3	2	3	**8**
2016 Set-1	4	4	**12**
2016 Set-2	2	4	**10**
2016 Set-3	3	3	**9**
2017 Set-1	3	2	**7**
2017 Set-2	3	3	**9**
2018	3	3	**9**
2019	2	4	10
2020	2	3	8
2021	5	6	17
2022	3	5	13
2023	2	4	10
2024	5	4	13

1 Analog Communication Systems

1.1 The image (second) channel selectivity of a superheterodyne communication receiver is determined by
(a) antenna and preselector
(b) the preselector and RF amplifier
(c) the preselector and IF amplifier
(d) the RF and IF amplifier [1995 : 1 M]

1.2 A PLL can be used to demodulate
(a) PAM signals
(b) PCM signals
(c) FM signals
(d) DSB - SC signals [1995 : 1 M]

1.3 A PAM signal can be detected by using
(a) an ADC
(b) an integrator
(c) a band pas filter
(d) a high pass filter [1995 : 1 M]

1.4 The image channel rejection in a superheterodyne receiver comes from
(a) IF stages only
(b) RF stages only
(c) detector and RF stages only
(d) detector, RF and IF stages [1996 : 1 M]

1.5 An FM signal with a modulation index 9 is applied to a frequency tripler. The modulation index in the output signal will be
(a) 0 (b) 3
(c) 9 (d) 27 [1996 : 2 M]

1.6 The image channel selectivity of superheterodyne receiver depends upon
(a) IF amplifiers only
(b) RF and IF amplifiers only
(c) Preselector, RF and IF amplifiers
(d) Preselector and RF amplifiers only
[1998 : 1 M]

1.7 A DSB-SC signal is generated using the carrier $\cos(\omega_c t + \theta)$ and modulating signal $x(t)$. The envelop of the DSB-SC signal is
(a) $x(t)$
(b) $|x(t)|$
(c) only positive portion of $x(t)$
(d) $x(t)\cos\theta$ [1998 : 1 M]

1.8 A modulated signal is given by,

$$s(t) = m_1(t)\cos(2\pi f_c t) + m_2(t)\sin(2\pi f_c t)$$

where the baseband signal $m_1(t)$ and $m_2(t)$ have bandwidths of 10 kHz and 15 kHz, respectively. The bandwidth of the modulated signal, in kHz, is
(a) 10 (b) 15
(c) 25 (d) 30 [1999 : 1 M]

1.9 A modulated signal is given by

$$s(t) = e^{-at}\cos[(\omega_c + \Delta\omega)t]u(t)$$

where a, ω_c and $\Delta\omega$ are positive constants, and $\omega_c >> \Delta\omega$. The complex envelope of $s(t)$ is given by
(a) $\exp(-at)\exp[j(\omega_c + \Delta\omega)t]u(t)$
(b) $\exp(-at)\exp(j\Delta\omega t)u(t)$
(c) $\exp(j\Delta\omega t).u(t)$
(d) $\exp[(j\omega_c + \Delta\omega)t]$ [1999 : 1 M]

1.10 The amplitude modulated wave form $s(t) = A_C[1 + K_a m(t)]\cos\omega_C t$ is fed to an ideal envelope detector. The maximum magnitude of $K_a m(t)$ is greater than 1. Which of the following could be the detector output?
(a) $A_c m(t)$
(b) $A_c^2[1 + K_a m(t)]^2$
(c) $[A_c|1 + K_a m(t)|]$
(d) $A_c|1 + K_a m(t)|^2$ [2000 : 1 M]

1.11 A message $m(t)$ bandlimited to the frequency f_m has a power of P_m. The power of the output signal in the figure is

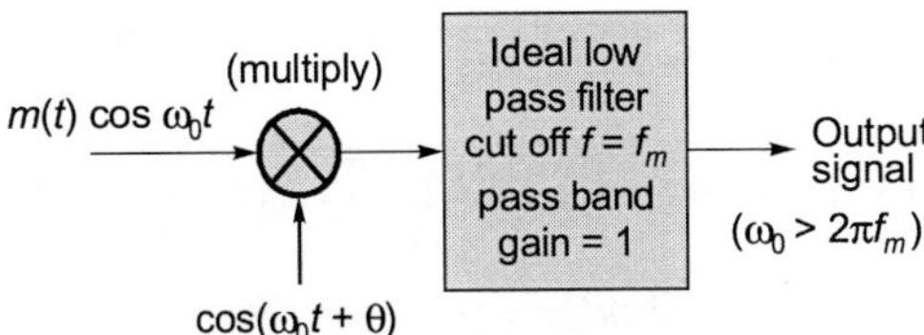

(a) $\frac{P_m\cos\theta}{2}$ (b) $\frac{P_m}{4}$
(c) $\frac{P_m\sin^2\theta}{4}$ (d) $\frac{P_m\cos^2\theta}{4}$
[2000 : 2 M]

1.12 The Hilbert transform of $\cos\omega_1 t + \sin\omega_2 t$ is

(a) $\sin\omega_1 t - \cos\omega_2 t$ (b) $\sin\omega_1 t + \cos\omega_2 t$
(c) $\cos\omega_1 t - \sin\omega_2 t$ (d) $\sin\omega_1 t + \sin\omega_2 t$

[2000 : 2 M]

1.13 In a FM system, a carrier of 100 MHz is modulated by a sinusoidal signal of 5 kHz. The bandwidth by Carson's approximation is 1 MHz. If $y(t)$ = (modulated waveform)^3, then by using Carson's approximation, the bandwidth of $y(t)$ and the spacing of spectral components are, respectively.

(a) 3 MHz, 5 kHz (b) 1 MHz, 15 kHz
(c) 3 MHz, 15 kHz (d) 1 MHz, 5 kHz

[2000 : 2 M]

1.14 A bandlimited signal is sampled at the Nyquist rate. The signal can be recovered by passing the samples through

(a) an RC filter
(b) an envelope detector
(c) a PLL
(d) an ideal low-pass filter with the appropriate bandwidth

[2001 : 1 M]

1.15 A 1 MHz sinusoidal carrier is amplitude modulated by a symmetrical square wave of period 100 μsec. Which of the following frequencies will NOT be present in the modulated signal?

(a) 990 kHz (b) 1010 kHz
(c) 1020 kHz (d) 1030 kHz

[2002 : 1 M]

1.16 An angle-modulated signal is given by
$s(t) = \cos 2\pi\,(2 \times 10^6 t + 30\sin 150t + 40\cos 150t)$.
The maximum frequency and phase deviations of $s(t)$ are

(a) 10.5 kHz, 140π rad
(b) 6 kHz, 80π rad
(c) 10.5 kHz, 100π rad
(d) 7.5 kHz, 100π rad

[2002 : 2 M]

1.17 In the figure $m(t) = \dfrac{2\sin 2\pi t}{t}$, $s(t) = \cos 200\pi t$ and $n(t) = \dfrac{\sin 199\pi t}{t}$. The output $y(t)$ will be

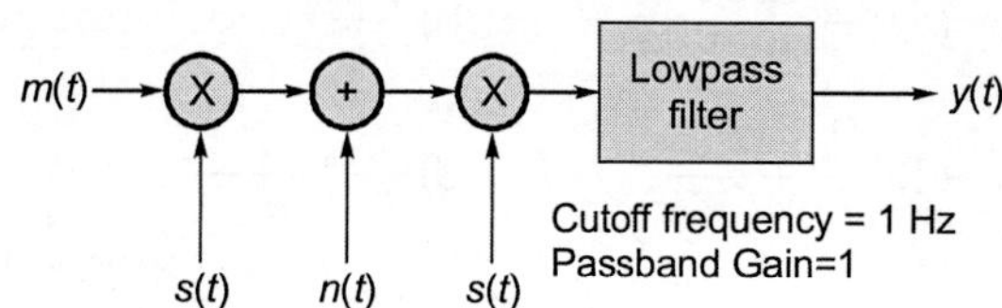

(a) $\dfrac{\sin 2\pi t}{t}$

(b) $\dfrac{\sin 2\pi t}{t} + \dfrac{\sin \pi t}{t}\cos 3\pi t$

(c) $\dfrac{\sin 2\pi t}{t} + \dfrac{\sin 0.5\pi t}{t}\cos 1.5\pi t$

(d) $\dfrac{\sin 2\pi t}{t} + \dfrac{\sin \pi t}{t}\cos 0.75\pi t$

[2002 : 2 M]

1.18 The input to a coherent detector is DSB-SC signal plus noise. The noise at the detector output is

(a) the in-phase component
(b) the quadrature-component
(c) zero
(d) the envelope

[2003 : 1 M]

1.19 A DSB-SC signal is to be generated with a carrier frequency f_c = 1 MHz using a non-linear device with the input-output characteristic
$V_0 = a_0 v_i + a_1 v_i^3$
where, a_0 and a_1 are constants. The output of the non-linear device can be filtered by an appropriate band-pass filter.

Let, $V_i = A_c^i \cos\left(2\pi f_c^i t\right) + m(t)$ where $m(t)$ is the message signal. Then the value of f_c^i (in MHz) is

(a) 1.0 (b) 0.333
(c) 0.5 (d) 3.0

[2003 : 2 M]

Common Data for Questions 1.20 and 1.21

Let $m(t) = \cos\left[(4\pi \times 10^3)t\right]$ be the message signal &

$c(t) = 5\cos\left[(2\pi \times 10^6)t\right]$ be the carrier.

1.20 $c(t)$ and $m(t)$ are used to generate an AM signal. The modulation index of the generated AM signal is 0.5. Then the quantity $\dfrac{\text{Total sideband power}}{\text{Carrier power}}$ is

(a) 1/2 (b) 1/4
(c) 1/3 (d) 1/8

[2003 : 2 M]

1.21 $c(t)$ and $m(t)$ are used to generate an FM signal. If the peak frequency deviation of the generated FM signal is three times the transmission bandwidth of the AM signal, then the coefficient of the term $\cos\left[2\pi\left(1008 \times 10^3 t\right)\right]$ in the FM signal (in terms of the Bessel coefficients) is

(a) $5\,J_4(3)$ (b) $\frac{5}{2}J_8(3)$

(c) $\frac{5}{2}J_8(4)$ (d) $5\,J_4(6)$ [2003 : 2 M]

1.22 Choose the correct one from among the alternative *a, b, c, d* after matching an item in Group 1 with the most appropriate item in Group 2.

Group 1

P. Ring modulator
Q. VCO
R. Foster-Seeley discriminator
S. Mixer

Group 2

1. Clock recovery
2. Demodulation of FM
3. Frequency conversion
4. Summing the two inputs
5. Generation of FM
6. Generation of DSB-Sc

(a) P – 1; Q – 3; R – 2; S – 4
(b) P – 6; Q – 5; R – 2; S – 3
(c) P – 6; Q – 1; R – 3; S – 2
(d) P – 5; Q – 6; R – 1; S – 3 [2003 : 2 M]

1.23 A superheterodyne receiver is to operate in the frequency range 550 kHz-1650 kHz, with the intermediate frequency of 450 kHz. Let, $R = \frac{C_{max}}{C_{min}}$ denote the required capacitance ratio of the local oscillator and I denote the image frequency (in kHz) of the incoming signal. If the receiver is tuned to 700 kHz, then

(a) $R = 4.41, I = 1600$
(b) $R = 2.10, I = 1150$
(c) $R = 3.0, I = 1600$
(d) $R = 9.0, I = 1150$ [2003 : 2 M]

1.24 An AM signal is detected using an envelope detector. The carrier frequency and modulating signal frequency are 1 MHz and 2 kHz respectively. An appropriate value for the time constant of the envelope detector is

(a) 500 μsec (b) 20 μsec
(c) 0.2 μsec (d) 1 μsec

[2004 : 1 M]

1.25 An AM signal and a narrow-band FM signal with identical carriers, modulating signals and modulation indices of 0.1 are added together. The resultant signal can be closely approximated by

(a) broadband FM
(b) SSB with carrier
(c) DSB-SC
(d) SSB without carrier [2004 : 1 M]

1.26 A 100 MHz carrier of 1 V amplitude and a 1 MHz modulating signal of 1 V amplitude are fed to a balanced modulator. The output of the modulator is passed through an ideal high-pass filter with cut-off frequency of 100 MHz. The output of the filter is added with 100 MHz signal of 1 V amplitude and 90° phase shift as shown in the figure. The envelope of the resultant signal is

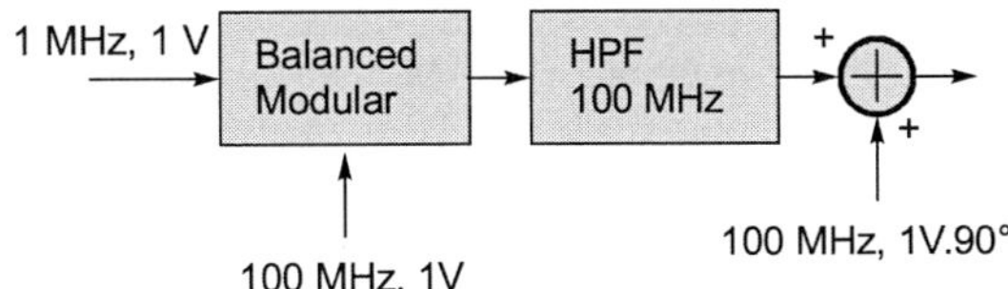

(a) constant
(b) $\sqrt{1+\sin(2\pi\times10^6 t)}$
(c) $\sqrt{5/4-\sin(2\pi\times10^6 t)}$
(d) $\sqrt{5/4+\cos(2\pi\times10^6 t)}$

[2004 : 2 M]

1.27 Two sinusoidal signals of same amplitude and frequencies 10 kHz and 10.1 kHz are added together. The combined signal is given to an ideal frequency detector. The output of the detector is

(a) 0.1 kHz sinusoid
(b) 20.1 kHz sinusoid
(c) a linear function of time
(d) a constant [2004 : 2 M]

1.28 Consider a system shown in the figure. Let $X(f)$ and $Y(f)$ denote the Fourier transforms of $x(t)$ and $y(t)$ respectively. The ideal HPF has the cutoff frequency 10 kHz.

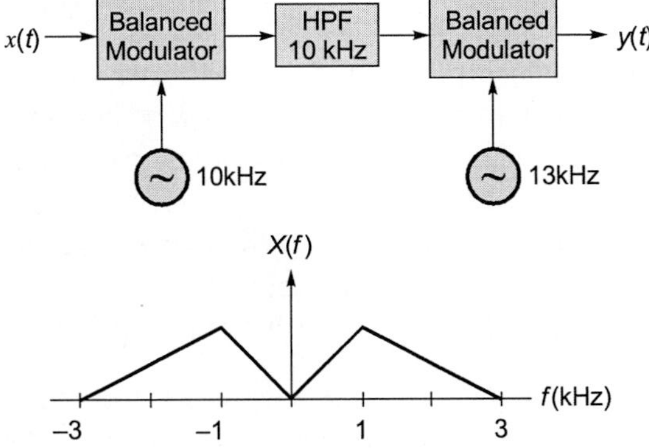

The positive frequencies where $Y(f)$ has spectral peaks are
(a) 1 kHz and 24 kHz
(b) 2 kHz and 24 kHz
(c) 1 kHz and 14 kHz
(d) 2 kHz and 14 kHz

[2004 : 2 M]

1.29 Find the correct match between group 1 and group 2.

Group 1

P. $\{1 + km(t)\}A\sin(\omega_c t)$

Q. $km(t)\,A\sin(\omega_c t)$

R. $A\sin\{\omega_c t + km(t)\}$

S. $A\sin\left[\omega_c t + k\int_{-\infty}^{t} m(t)dt\right]$

Group 2

W. Phase modulation
X. Frequency modulation
Y. Amplitude modulation
Z. DSB-SC modulation

(a) P-Z, Q-Y, R-X, S-W
(b) P-W, Q-X, R-Y, S-Z
(c) P-X, Q-W, R-Z, S-Y
(d) P-Y, Q-Z, R-W, S-X

[2005 : 1 M]

1.30 Which of the following analog modulation scheme requires the minimum transmitted power and minimum channel band-width?
(a) VSB (b) DSB-SC
(c) SSB (d) AM [2005 : 1 M]

1.31 A device with input $x(t)$ and output $y(t)$ is characterized by: $y(t) = x^2(t)$.
An FM signal with frequency deviation of 90 kHz and modulating signal bandwidth of 5 kHz is applied to this device. The bandwidth of the output signal is
(a) 370 kHz (b) 190 kHz
(c) 380 kHz (d) 95 kHz [2005 : 2 M]

1.32 A carrier is phase modulated (PM) with frequency deviation of 10 kHz by a single tone frequency of 1 kHz. If the single tone frequency is increased to 2 kHz, assuming that phase deviation remains unchanged, the bandwidth of the PM signal is
(a) 21 kHz (b) 22 kHz
(c) 42 kHz (d) 44 kHz [2005 : 2 M]

1.33 The diagonal clipping in Amplitude Demodulation (using envelope detector) can be avoided if RC time-constant of the envelope detector satisfies the following condition, (here W is message bandwidth and ω is carrier frequency both in rad/sec)

(a) $RC < \frac{1}{W}$ (b) $RC > \frac{1}{W}$

(c) $RC < \frac{1}{\omega}$ (d) $RC > \frac{1}{\omega}$

[2006 : 2 M]

1.34 A message signal with bandwidth 10 kHz is Lower-Side Band SSB modulated with carrier frequency $f_{c1} = 10^6$ Hz. The resulting signal is then passed through a Narrow-Band Frequency Modulator with carrier frequency $f_{c2} = 10^9$ Hz.
The bandwidth of the output would be
(a) 4×10^4 Hz (b) 2×10^6 Hz
(c) 2×10^9 Hz (d) 2×10^{10} Hz

[2006 : 2 M]

Common Data for Questions 1.35 and 1.36:
Consider the following Amplitude Modulated (AM) signal, where $f_m < B$

$$x_{AM}(t) = 10(1 + 0.5\sin 2\pi f_m t)\cos 2\pi f_c t$$

1.35 The average side-band power for the AM signal given above is
(a) 25 (b) 12.5
(c) 6.25 (d) 3.125

[2006 : 2 M]

1.36 The AM signal gets added to a noise with Power Spectral Density $S_n(f)$ given in the figure below. The ratio of average sideband power to mean noise power would be:

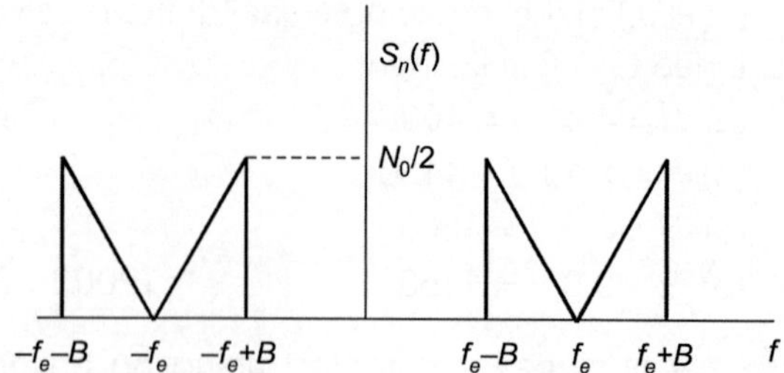

(a) $\frac{25}{8N_0B}$ (b) $\frac{25}{4N_0B}$

(c) $\frac{25}{2N_0B}$ (d) $\frac{25}{N_0B}$

[2006 : 2 M]

1.37 In the following scheme, if the spectrum $M(f)$ of $m(t)$ is as shown, then the spectrum $Y(f)$ of $y(t)$ will be

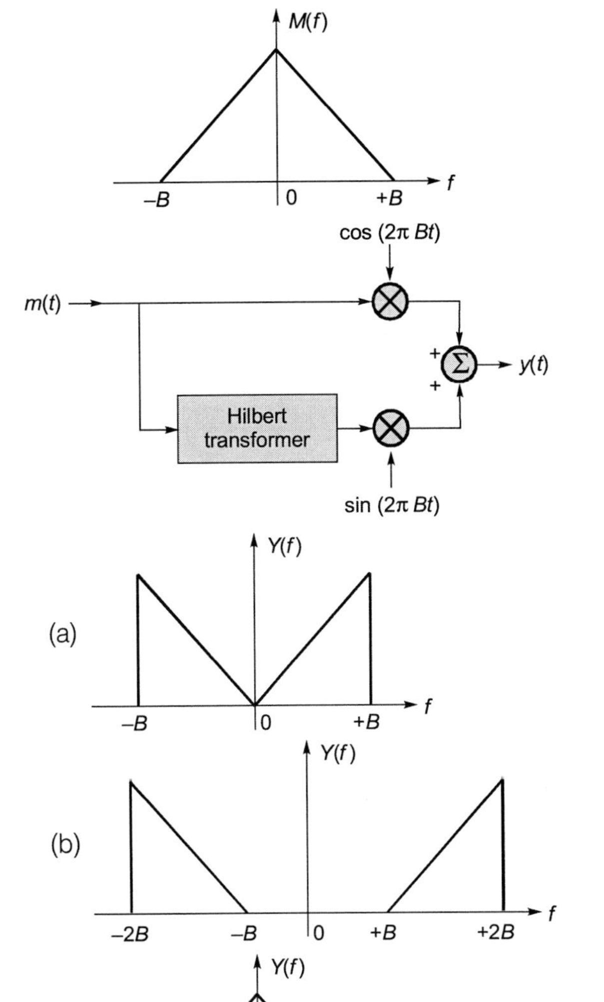

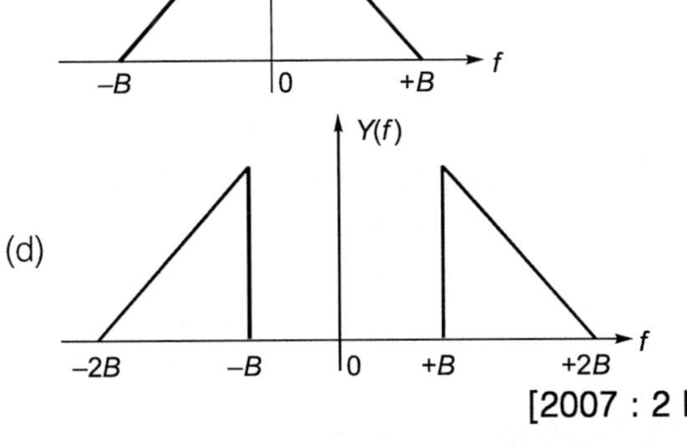

[2007 : 2 M]

1.38 Consider the amplitude modulated (AM) signal $A_c \cos \omega_c t + 2 \cos \omega_m t \cos\omega_c t$. For demodulating the signal using envelope detector, the minimum value of A_c should be

(a) 2 (b) 1
(c) 0.5 (d) 0 [2008 : 1 M]

1.39 Consider the frequency modulated signal $10\cos[2\pi \times 10^5 t + 5\sin(2\pi \times 1500t) + 7.5\sin(2\pi \times 1000t)]$ with carrier frequency of 10^5 Hz. The modulation index is

(a) 12.5 (b) 10
(c) 7.5 (d) 5 [2008 : 2 M]

1.40 The signal $\cos \omega_c t + 0.5\cos \omega_m t \sin \omega_c t$ is

(a) FM only
(b) AM only
(c) both AM and FM
(d) neither AM nor FM [2008 : 2 M]

1.41 For a message signal $m(t) = \cos(2\pi f_m t)$ and carrier of frequency f_c, which of the following represents a single side-band (SSB) signal?

(a) $\cos(2\pi f_m t) \cos(2\pi f_c t)$
(b) $\cos(2\pi f_c t)$
(c) $\cos[2\pi(f_c + f_m)t]$
(d) $[1+\cos(2\pi f_m t)] \cos(2\pi f_c t)$ [2009 : 1 M]

1.42 A message signal given by

$$m(t) = \left(\frac{1}{2}\right)\cos \omega_1 t - \left(\frac{1}{2}\right)\sin\omega_2 t$$

is amplitude-modulated with a carrier of frequency ω_c to generate
$s(t) = [1 + m(t)] \cos \omega_c t$
What is the power efficiency achieved by this modulation scheme?

(a) 8.33 % (b) 11.11 %
(c) 20 % (d) 25 % [2009 : 2 M]

1.43 Suppose that the modulating signal is $m(t) = 2\cos(2\pi f_m t)$ and the carrier signal is $x_C(t) = A_C \cos(2\pi f_c t)$. Which one of the following is a conventional AM signal without over-modulation?

(a) $x(t) = A_C m(t)\cos(2\pi f_c t)$
(b) $x(t) = A_C[1 + m(t)]\cos(2\pi f_c t)$
(c) $x(t) = A_C \cos(2\pi f_c t) + \frac{A_C}{4} m(t)\cos(2\pi f_c t)$
(d) $x(t) = A_C\cos(2\pi f_m t)\cos(2\pi f_c t) + A_C\sin(2\pi f_m t)\sin(2\pi f_c t)$

[2010 : 1 M]

1.44 Consider an angle modulated signal
$x(t) = 6\cos[2\pi \times 10^6 t + 2\sin(8000\pi t) + 4\cos(8000\pi t)]$ V
The average power of $x(t)$ is

(a) 10 W (b) 18 W
(c) 20 W (d) 28 W [2010 : 1 M]

1.45 The **List-I** (lists the attributes) and the **List-II** (lists of the modulation systems). Match the attribute to the modulation system that best meets it.

List-I

A. Power efficient transmission of signals
B. Most bandwidth efficient transmission of voice signals

C. Simplest receiver structure
D. Bandwidth efficient transmission of signals with significant dc component

List-II
1. Conventional AM
2. FM
3. VSB
4. SSB-SC

	A	B	C	D
(a)	4	2	1	3
(b)	2	4	1	3
(c)	3	2	1	4
(d)	2	4	3	1

[2011 : 1 M]

1.46 A message signal

$$m(t) = \cos 2000\,\pi t + 4\cos 4000\,\pi t$$

modulates the carrier $c(t) = \cos 2\pi f_c t$ where $f_c = 1$ MHz to produce an AM signal. For demodulating the generated AM signal using an envelope detector, the time constant RC of the detector circuit should satisfy
(a) 0.5 ms < RC < 1 ms
(b) 1 μs << RC < 0.5 ms
(c) RC << 1 μs
(d) RC >> 0.5 ms

[2011 : 2 M]

1.47 The signal $m(t)$ as shown is applied both to a phase modulator (with k_p as the phase constant) and a frequency modulator with (k_f as the frequency constant) having the same carrier frequency

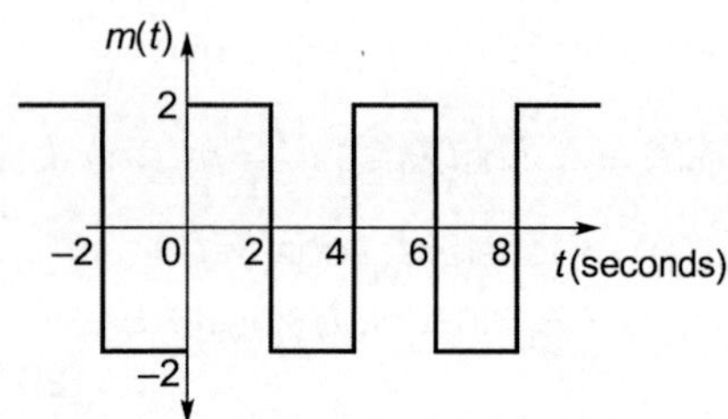

The ratio k_p/k_f (in rad/Hz) for the same maximum phase deviation is
(a) 8π
(b) 4π
(c) 2π
(d) π

[2012 : 2 M]

1.48 Consider sinusoidal modulation in an AM system. Assuming no overmodulation, the modulation index (μ) when the maximum and minimum values of the envelope, respectively, are 3 V and 1 V, is ______.

[2014 : 1 M, Set-2]

1.49 In the figure, $M(f)$ is the Fourier transform of the message signal $m(t)$ where $A = 100$ Hz and $B = 40$ Hz. Given $v(t) = \cos(2\pi f_c t)$ and $w(t) = \cos(2\pi(f_c + A)t)$, where $f_c > A$. The cut-off frequencies of both the filters are f_c.

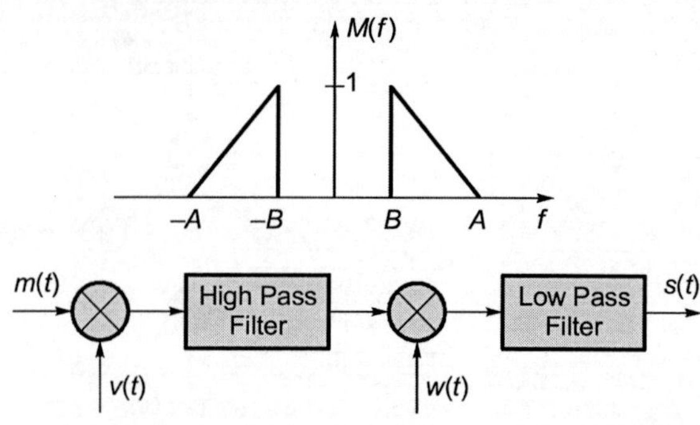

The bandwidth of the signal at the output of the modulator (in Hz) is _____.

[2014 : 2 M, Set-2]

1.50 A modulated signal is $y(t) = m(t)\cos(40000\pi t)$, where the baseband signal $m(t)$ has frequency components less than 5 kHz only. The minimum required rate (in kHz) at which $y(t)$ should be sampled to recover $m(t)$ is _____.

[2014 : 1 M, Set-3]

1.51 Consider an FM signal

$$f(t) = \cos[2\pi f_c t + \beta_1 \sin 2\pi f_1 t + \beta_2 \sin 2\pi f_2 t]$$

The maximum deviation of the instantaneous frequency from the carrier frequency f_c is
(a) $\beta_1 f_1 + \beta_2 f_2$
(b) $\beta_1 f_2 + \beta_2 f_1$
(c) $\beta_1 + \beta_2$
(d) $f_1 + f_2$

[2014 : 1 M, Set-3]

1.52 In a double side-band (DSB) full carrier AM transmission system, if the modulation index is doubled, then the ratio of total sideband power to the carrier power increases by a factor of _____.

[2014 : 1 M, Set-4]

1.53 Consider the signal

$$s(t) = m(t)\cos(2\pi f_c t) + \hat{m}(t)\sin(2\pi f_c t)$$

where, $\hat{m}(t)$ denotes the Hilbert transform of $m(t)$ and the bandwidth of $m(t)$ is very small compared to f_c. The signal $s(t)$ is a
(a) high-pass signal
(b) low-pass signal
(c) band-pass signal
(d) double sideband suppressed carrier signal

[2015 : 1 M, Set-1]

1.54 In the system shown in Figure (a), $m(t)$ is a low-pass signal with bandwidth W Hz. The frequency response of the band-pass filter $H(f)$ is shown in Figure (b). If it is desired that the output signal $z(t) = 10\,x(t)$, the maximum value of W (in Hz) should be strictly less than ________.

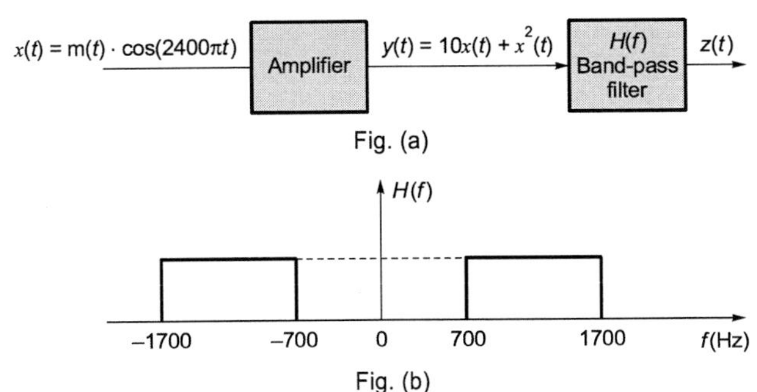

Fig. (a)

H(f)

–1700 –700 0 700 1700 f(Hz)

Fig. (b)

[2015 : 2 M, Set-1]

1.55 A message signal $m(t) = A_m \sin(2\pi f_m t)$ is used to modulate the phase of a carrier $A_c \cos(2\pi f_c t)$ to get the modulated signal $y(t) = A_c \cos(2\pi f_c t + m(t))$. The bandwidth of $y(t)$

(a) depends on A_m but not on f_m
(b) depends on f_m but not on A_m
(c) depends on both A_m and f_m
(d) does not depends on A_m or f_m

[2015 : 1 M, Set-3]

1.56 The complex envelope of the bandpass signal

$$x(t) = -\sqrt{2}\left(\frac{\sin(\pi t/5)}{\pi t/5}\right)\sin\left(\pi t - \frac{\pi}{4}\right),$$ centered

about $f = \frac{1}{2}$ Hz, is

(a) $\left(\frac{\sin(\pi t/5)}{\pi t/5}\right)e^{j\frac{\pi}{4}}$ (b) $\left(\frac{\sin(\pi t/5)}{\pi t/5}\right)e^{-j\frac{\pi}{4}}$

(c) $\sqrt{2}\left(\frac{\sin(\pi t/5)}{\pi t/5}\right)e^{j\frac{\pi}{4}}$ (d) $\sqrt{2}\left(\frac{\sin(\pi t/5)}{\pi t/5}\right)e^{-j\frac{\pi}{4}}$

[2015 : 2 M, Set-3]

1.57 The block diagram of a frequency synthesizer consisting of a Phase Locked Loop (PLL) and a divide-by-N counter (comprising ÷ 2, ÷ 4, ÷ 8, ÷ 16 outputs) is sketched below. The synthesizer is excited with a 5 kHz signal (Input 1). The free-running frequency of the PLL is set to 20 kHz. Assume that the commutator switch makes contacts repeatedly in the order 1 - 2 - 3 - 4.

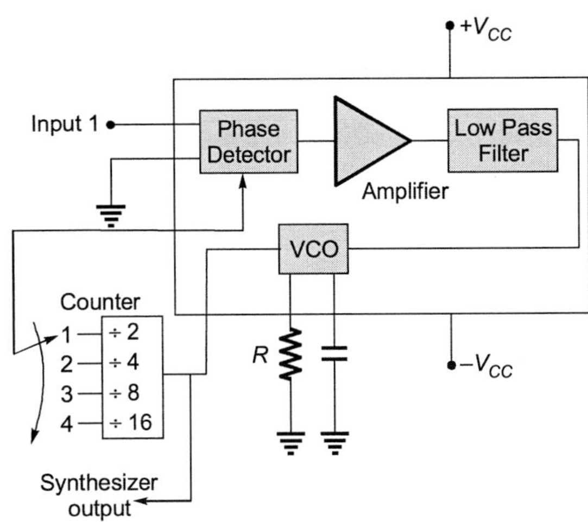

The corresponding frequency synthesized are:

(a) 10 kHz, 20 kHz, 40 kHz, 80 kHz
(b) 20 kHz, 40 kHz, 80 kHz, 160 kHz
(c) 80 kHz, 40 kHz, 20 kHz, 10 kHz
(d) 160 kHz, 80 kHz, 40 kHz, 20 kHz

[2016 : 1 M, Set-1]

1.58 A superheterodyne receiver operates in the frequency range of 58 MHz 68 MHz. The intermediate frequency f_{IF} and local oscillator frequency f_{LO} are chosen such that $f_{IF} \le f_{LO}$. It is required that the image frequencies fall outside the 58 MHz – 68 MHz band. The minimum required f_{IF} (in MHz) is _____.

[2016 : 1 M, Set-1]

1.59 The amplitude of a sinusoidal carrier is modulated by a single sinusoid to obtain the amplitude modulated signal

$$s(t) = 5\cos 1600\pi t + 20\cos 1800\pi t + 5\cos 2000\pi t$$

The value of the modulation index is_____.

[2016 : 1 M, Set-1]

1.60 For a superheterodyne receiver, the intermediate frequency is 15 MHz and the local oscillator frequency is 3.5 GHz. If the frequency of the received signal is greater than the local oscillator frequency, then the image frequency (in MHz) is _______.

[2016 : 1 M, Set-3]

1.61 A modulating signal given by

$$x(t) = 5\sin(4\pi 10^3 t - 10\pi\cos 2\pi 10^3 t)\text{ V}$$

is fed to a phase modulator with phase deviation constant k_p = 5 rad/V. If the carrier frequency is 20 kHz, the instantaneous frequency (in kHz) at t = 0.5 ms is _____.

[2017 : 2 M, Set-2]

1.62 The unmodulated carrier power in an AM transmitter is 5 kW. This carrier is modulated by a sinusoidal modulating signal. The maximum percentage of modulation is 50%. If it is reduced to 40%, then the maximum unmodulated carrier power (in kW) that can be used without overloading the transmitter is _____.

[2017 : 2 M, Set-2]

1.63 Consider the following amplitude modulated signal:

$s(t) = \cos(2000\,\pi t) + 4\cos(2400\,\pi t) + \cos(2800\,\pi t)$

The ratio (accurate to three decimal places) of the power of the message signal to the power of the carrier signal is ________.

[2018 : 1 M]

1.64 Let $c(t) = A_c\cos(2\pi f_c t)$ and $m(t) = \cos(2\pi f_m t)$. It is given that $f_c >> 5f_m$. The signal $c(t) + m(t)$ is applied to the input of a non-linear device, whose output $v_o(t)$ is related to the input $v_i(t)$ as $v_o(t) = av_i(t) + bv_i^2(t)$, where a and b are positive constants. The output of the non-linear device is passed through an ideal band-pass filter with center frequency f_c and bandwidth $3f_m$, to produce an amplitude modulated (AM) wave. If it is desired to have the sideband power of the AM wave to be half of the carrier power, then a/b is

(a) 0.25
(b) 0.5
(c) 1
(d) 2

[2018 : 2 M]

1.65 The baseband signal $m(t)$ shown in the figure is phase-modulated to generate the PM signal $\phi(t) = \cos(2\pi f_c t + km(t))$. The time t on the x-axis in the figure is in milliseconds. If the carrier frequency is f_c = 50 kHz and $k = 10\pi$, then the ratio of the minimum instantaneous frequency (in kHz) to the maximum instantaneous frequency (in kHz) is ______ (rounded off to 2 decimal places).

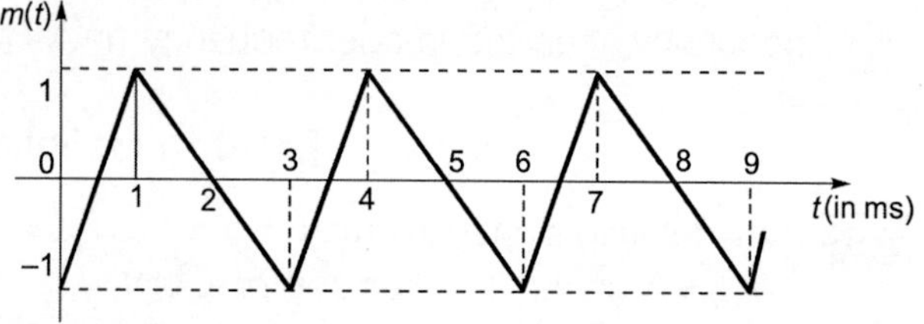

[2019 : 1 M]

1.66 $S_{PM}(t)$ and $S_{FM}(t)$ as defined below, are the phase modulated and the frequency modulated waveforms, respectively, corresponding to the message signal $m(t)$ shown in the figure.

$$S_{PM}(t) = \cos(1000\pi t + K_p\, m(t))$$

and $$S_{FM}(t) = \cos\left(1000\pi t + K_f \int_{-\infty}^{t} m(\tau)\, d\tau\right)$$

where K_p is the phase deviation constant in radians/volt and K_f is the frequency deviation constant in radians/second/volt. If the highest instantaneous frequencies of $S_{PM}(t)$ and $S_{FM}(t)$ are same, then the value of the ratio $\frac{K_p}{K_f}$ is _____ seconds.

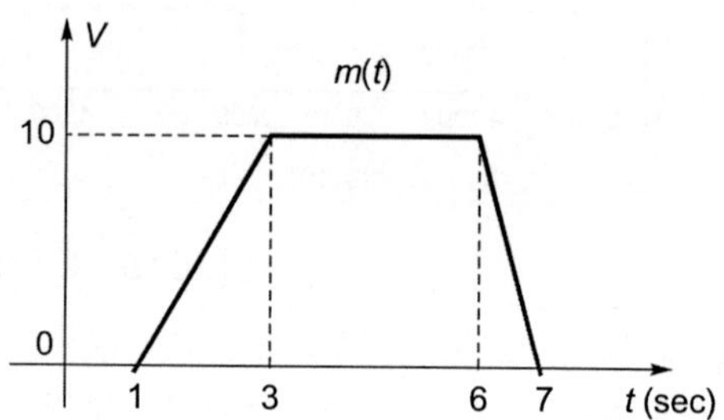

[2020 : 2 M]

1.67 For the modulated signal $x(t) = m(t)\cos\{2nf_c t\}$, the message signal $m(t) = 4\cos(1000\pi t)$ and the carrier frequency f_c is 1 MHz. The signal $x(t)$ is passed through a demodulator, as shown in the figure below. The output $y(t)$ of the demodulator is

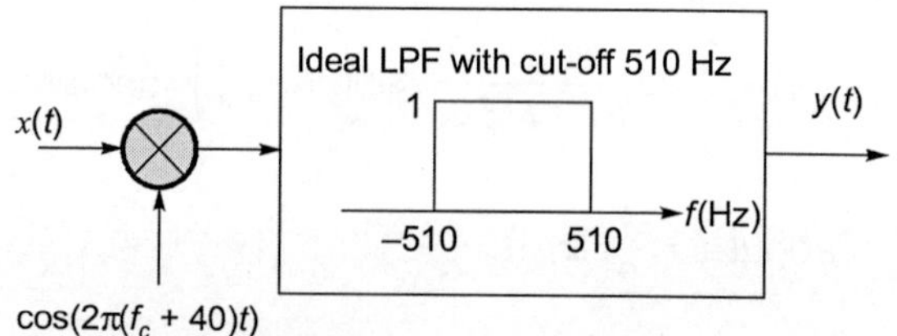

(a) $\cos(1000\pi t)$
(b) $\cos(540\pi t)$
(c) $\cos(920\pi t)$
(d) $\cos(460\pi t)$

[2020 : 2 M]

1.68 Consider a carrier signal which is amplitude modulated by a single-tone sinusoidal message signal with a modulation index of 50%. If the carrier and one of the sidebands are suppressed in the modulated signal, the percentage of power saved (rounded off to one decimal place) is ________.

[2021 : 1 M]

1.69 Consider a superheterodyne receiver tuned to 600 kHz. If the local oscillator feeds a 1000 kHz signal to the mixer, the image frequency (in integer) is ______ kHz.

[2021 : 2 M]

1.70 A sinusoidal message signal having root mean square value of 4 V and frequency of 1 kHz is fed to a phase modulator with phase deviation constant 2 rad/volt. If the carrier signal is $c(t) = 2\cos(2\pi 10^6 t)$, the maximum instantaneous frequency of the phase modulated signal (rounded off to one decimal place) is ______ Hz.

[2021 : 2 M]

1.71 Consider an FM broadcast that employs the pre-emphasis filter with frequency response

$$H_{pe}(\omega) = 1 + \frac{j\omega}{\omega_0},$$

where $\omega_0 = 10^4$ rad/sec.

For the network shown in the figure to act as a corresponding de-emphasis filter, the appropriate pair(s) of (R, C) values is/are ________.

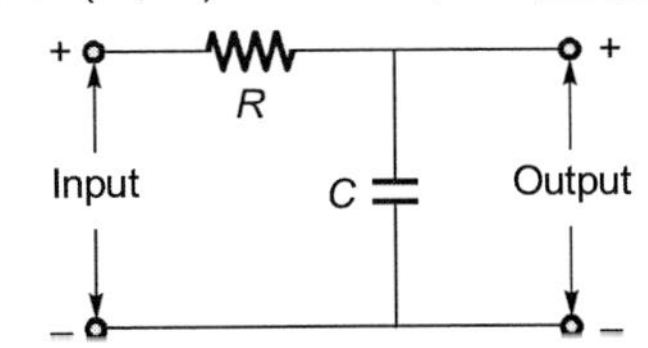

(a) $R = 1$ kΩ, $C = 0.1$ μF
(b) $R = 2$ kΩ, $C = 1$ μF
(c) $R = 1$ kΩ, $C = 2$ μF
(d) $R = 2$ kΩ, $C = 0.5$ μF

[2022 : 2 M]

1.72 Let a frequency modulated (FM) signal $x(t) = A\cos(\omega_c t + k_f \int_{-\infty}^{t} m(\lambda)d\lambda)$, where $m(t)$ is a message signal of bandwidth W. It is passed through a non-linear system with output $y(t) = 2x(t) + 5(x(t))^2$. Let B_T denote the FM bandwidth. The minimum value of ω_c required to recover $x(t)$ from $y(t)$ is

(a) $B_T + W$
(b) $\frac{3}{2}B_T$
(c) $2B_T + W$
(d) $\frac{5}{2}B_T$

[2023 : 2 M]

1.73 An amplitude modulator has output (in Volts)
$s(t) = A\cos(400\pi t) + B\cos(360\pi t) + B\cos(440\pi t)$
The carrier power normalized to 1 Ω resistance is 50 Watts. The ratio of the total sideband power to the total power is 1/9. The value of B (in Volts. rounded off to two decimal places) is ________.

[2024 : 1 Mark]

■■■

Answers Analog Communication Systems

1.1 (b)	1.2 (c)	1.3 (b)	1.4 (b)	1.5 (d)	1.6 (d)	1.7 (b)
1.8 (d)	1.9 (b)	1.10 (c)	1.11 (d)	1.12 (a)	1.13 (a)	1.14 (d)
1.15 (c)	1.16 (d)	1.17 (c)	1.18 (a)	1.19 (c)	1.20 (d)	1.21 (d)
1.22 (b)	1.23 (a)	1.24 (b)	1.25 (b)	1.26 (c)	1.27 (d)	1.28 (b)
1.29 (d)	1.30 (c)	1.31 (a)	1.32 (d)	1.33 (a)	1.34 (b)	1.35 (c)
1.36 (b)	1.37 (a)	1.38 (a)	1.39 (b)	1.40 (a)	1.41 (c)	1.42 (c)
1.43 (c)	1.44 (b)	1.45 (b)	1.46 (b)	1.47 (b)	1.48 (0.50)	1.49 (60)
1.50 (10)	1.51 (a)	1.52 (4)	1.53 (c)	1.54 (350)	1.55 (c)	1.56 (d)
1.57 (a)	1.58 (5)	1.59 (0.5)	1.60 (3485)	1.61 (20)	1.62 (5.208)	1.63 (0.125)
1.64 (d)	1.65 (0.75)	1.66 (2)	1.67 (c)	1.68 (94)	1.69 (1400)	
1.70 (1011313.7)		1.71 (a)	1.72 (b)	1.73 (2.5)		

Explanations Analog Communication Systems

1.1 (b)

The image rejection should be achieved before IF stage because ones it enters into IF amplifier it becomes impossible to remove it from wanted signal. So image channel selectivity depends upon preselector and RF amplifiers only. The IF amplifier helps in rejection of adjacent channel frequency and not image frequency.

1.2 (c)

PLL (Phase Locked Loop) is used to demodulate the FM signals.

1.3 (b)

An integrator.

1.4 (b)

The image rejection should be achieved IF stage because ones it enters into IF amplifier it becomes impossible to remove it from wanted signal. So the image channel rejection in a superheterodyne receiver comes from RF stages only.

1.5 (d)

In a frequency multiplier circuit, the modulation index is multiplied by n.

So, $\boxed{\beta'_{FM} = n\beta_{FM}}$

$\beta' = 3 \times 9 = 27$

1.6 (d)

The image rejection should be achieved before IF stage because ones it enters into IF amplifier it becomes impossible to remove it from wanted signal. So image channel selectivity depends upon preselector and RF amplifiers only. The IF amplifier helps in rejection of adjacent channel frequency and not image frequency.

1.7 (b)

DSB-SC signal= $x(t)\cos(\omega_c t + \theta)$

Envelope of DSB-SC signal = $|x(t)|$

1.8 (d)

$$\text{B.W.} = 2f_m = 2 \times 15 \text{ kHz} = 30 \text{ kHz}$$

1.9 (b)

$$s(t) = e^{-at}\cos[(\omega_c + \Delta\omega)t]u(t)$$

$$= \underbrace{e^{-at}u(t)\cos\Delta\omega t}_{s_i(t)}\cos\omega_c t - \underbrace{e^{-at}u(t)\sin\Delta\omega t}_{s_q(t)}\sin\omega_c t$$

$$s(t) = s_i(t)\cos\omega_c t - s_q(t)\sin\omega_c t$$

$$s_{ce}(t) = s_i(t) + js_q(t)$$

$$= e^{-at}u(t)\cos\Delta\omega t + je^{-at}u(t)\sin\Delta\omega t$$

$$= e^{-at}u(t)e^{j\Delta\omega t}$$

1.10 (c)

When the modulation index of AM wave is less than unity the output of the envelope detector is envelope of the AM wave but when the modulation index is greater than unity then the output of the envelope detector is not envelope but mode of the envelope of the AM wave. Thus the detector output in given case would be $A_c[1 + k_a m(t)]$.

1.11 (d)

Output signal = $\dfrac{m(t)}{2}\cos\theta$

$m(t)$ has power P_m.

Power of a $m(t) \rightarrow a^2.P_m$

Here, $a = \dfrac{1}{2}\cos\theta$

$\therefore$ Power of output signal = $\dfrac{P_m}{4}\cos^2\theta$

1.12 (a)

$$\cos\omega_1 t \xleftrightarrow{\text{H.T.}} \sin\omega_1 t$$

$$\sin\omega_2 t \xleftrightarrow{\text{H.T.}} -\cos\omega_2 t$$

1.13 (a)

$$x(t) = A_c\cos[\omega_c t + \beta\sin\omega_m t]$$

$$y(t) = K\cos[3\omega_c t + 3\beta\sin\omega_m t]$$

[After passing from $y(t) = [x(t)]^3$]

$\therefore$ $\beta^1 = 3\beta$

i.e. $\Delta f^1 = 3 \times \Delta f = \Delta f >> f_m$

BW = $2\Delta f^1 \times 3 = 3$ MHz

f_m remains same

$\therefore$ $f_m = 5$ kHz

1.15 (c)

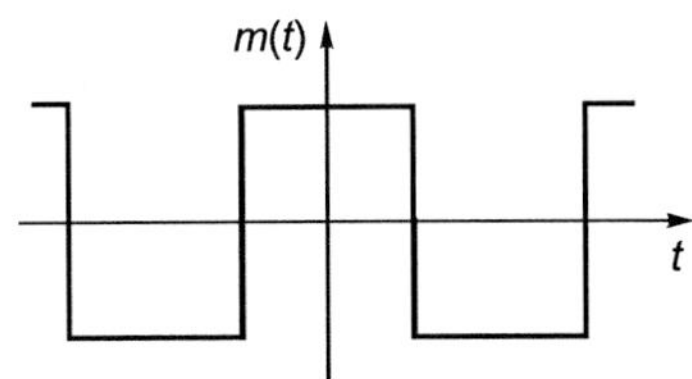

Message sigma $m(t)$ is a symmetrical square wave of period 100 μs.

$$T_o = 100\ \mu s$$

$$f_o = 1/T_o = 10\ \text{kHz}$$

Fourier series representation of $m(t)$ will have only odd harmonics.

So, $m(t)$ contains ⇒ 10 kHz, 30 kHz, 50 kHz, ...

Carrier frequency, f_c = 1 MHz = 1000 kHz

AM signal contains ⇒ ... , 950 kHz, 970 kHz, 990 kHz, 1000 kHz, 1010 kHz, 1030 kHz, 1050 kHz ...

1.16 (d)

$$s(t) = \cos[2\pi\times 2\times 10^6 t + 2\pi\times 30\sin 150t + 2\pi\times 40\cos 150t]$$

$$\frac{d\phi_i}{dt} = 2\pi\times 2\times 10^6 + 2\pi\times 4500\cos 150t + 2\pi\times 6000(-\sin 150t)$$

$$\therefore\ \Delta\omega = \omega_1 - \omega_c = 2\pi(4500^2 + 6000^2)^{1/2}$$

$$= 2\pi\times 7.5\ \text{k rad/sec}$$

$$\therefore\ \Delta f = \frac{\Delta\omega}{2\pi} = 7.5\ \text{kHz}$$

$$\Delta\phi = 2\pi\sqrt{30^2 + 40^2} = 100\pi$$

1.17 (c)

$$m(t)\times s(t) = \frac{2\sin 2\pi t}{t}\times\cos 200\pi t$$

$$= \frac{1}{t}(\sin 202\pi t - \sin 198\pi t)$$

The signal at the input of the LPF will be,

$$x(t) = [m(t)s(t) + n(t)]\,s(t)$$

$$= \frac{1}{t}(\sin 202\pi t - \sin 198\pi t + \sin 199\pi t)\cos 200\pi t$$

$$= \frac{1}{2t}(\sin 402\pi t + \sin 399\pi t - \sin 398\pi t) + \frac{1}{2t}(\sin 2\pi t + \sin 2\pi t - \sin\pi t)$$

After passing $\frac{1}{2t}(\sin 402\pi t + \sin 399\pi t - \sin 398\pi t)$ term through LPF with cut-off frequency of 1 Hz, it results in a component of $\frac{1}{2t}(\sin 2\pi t)$ at the output.

So, after passing $x(t)$ through the LPF, we get,

$$y(t) = \frac{1}{2t}(\sin 2\pi t) + \frac{1}{2t}(\sin 2\pi t + \sin 2\pi t - \sin\pi t)$$

$$= \frac{\sin 2\pi t}{t} + \frac{1}{2t}(2\sin 0.5\pi t\cos 1.5\pi t)$$

$$y(t) = \frac{\sin 2\pi t}{t} + \frac{\sin 0.5\pi t}{t}\cos 1.5\pi t$$

1.18 (a)

The coherent detector rejects the quadrature component of noise and therefore noise at the output has inphase component only. The inphase component of noise and output are additive at output of detector.

1.19 (c)

$$V_o = a_0 V_i + a_1 V_i^3$$

$$= a_0\left[A_c^i\cos 2\pi f_c^i + m(t)\right] + a_1[A_c^{i^3}\cos^3 2\pi f_c^i m^3(t) + 3m^2(t)A_c^i\cos 2\pi f_c^i t + 3m(t)A_c^{i^2}\underbrace{\cos^2 2\pi f_c^i t}_{\frac{1+\cos 4\pi f_c^i t}{2}}]$$

For DSB-SC signal, we have to consider only last term.

$$S_{DSB}(t) = \frac{3a_1 A_c^{i^2}}{2}m(t)\cos 4\pi f_c^i t$$

$$= A_c m(t)\cos 2\pi f_c t$$

$$f_c = 2f_c^i = 1\text{MHz}$$

$$f_c^i = 0.5\ \text{MHz}$$

1.20 (d)

$$P_t = P_c\left(1 + \frac{m_a^2}{2}\right)\quad ;\quad P_c = \frac{A_c^2}{2}$$

$$\text{Sideband power} = \frac{A_c^2}{2}\cdot\frac{m_a^2}{2}$$

$$\therefore\quad \frac{P_s}{P_c} = \frac{\left(\frac{A_c^2 m_a^2}{4}\right)}{\left(\frac{A_c^2}{2}\right)} = \frac{m_a^2}{2} = \frac{1}{8}$$

1.21 (d)

$$x_{FM}(t) = A_c\sum_{n=-\infty}^{\infty} J_n(\beta)\cos(\omega_c + n\omega_m)t$$

$$\text{B.W.} = 2\omega_m$$

$$\Delta\omega = 3\times\text{BW} = 6\omega_m$$

$$\therefore\quad \beta = \frac{\Delta\omega}{\omega_m} = 6$$

$$\omega_c + n\omega_m = 2\pi \times 1008 \times 10^3 \quad \text{(given in question)}$$

$$\therefore 2\pi \times 10^6 + n.4\pi \times 10^3 = 2\pi \times 1008 \times 10^3$$

$$\therefore \quad n = 4$$

$\therefore$ Bessel coefficient $5J_4(6)$

1.23 (a)

$$R = \frac{C_{max}}{C_{min}} = \left(\frac{f_{max}}{f_{min}}\right)^2 = \left(\frac{1650+450}{550+450}\right)^2$$

$$= 4.41$$

$$I = f_s + 2f_{if} = 700 + 2 \times 450 = 1600$$

1.24 (b)

$$\frac{1}{f_c} < RC \le \frac{1}{f_m}$$

(where $1/f_c$ = to avoid fluctuations at recovered output and $1/f_m$ = to avoid diagonal clipping)

$$1\ \mu\text{sec} < RC \le \frac{1}{2\text{ kHz}}$$

$$1\ \mu\text{sec} < RC \le 0.5\text{ms}$$

1.25 (b)

$$s(t) = A_c \cos\omega_c t + \frac{A_c\beta}{2}\cos(\omega_c + \omega_m)t - \frac{A_c\beta}{2}\cos(\omega_c - \omega_m)$$

$$s(t)_{AM} = A_c \cos\omega_c t + \frac{A_c m_a}{2}\cos(\omega_c + \omega_m)t + \frac{A_c m_a}{2}\cos(\omega_c - \omega_m)$$

$s(t) + s(t)_{AM}$ = carrier + USB

= SSB with carrier..

Note: $-s(t) + s(t)_{AM}$ = SSB+ SC

1.26 (c)

1MHz, 1 V → Balance modular (100 MHz, 1 V) → 99 MHz, $\frac{1}{2}$V ; 101MHz, $\frac{1}{2}$V → HPF 100MHz → 101MHz, $\frac{1}{2}$V

101 MHz, $\frac{1}{2}$V → ⊗ (100 MHz, 1 V, 90°) → O/P = $y(t)$

$$y(t) = \frac{1}{2}\cos\{2\pi \times 101 \times 10^6\}t + \sin 2\pi \times 100 + 10^6 t$$

$$= \frac{1}{2}[\cos(2\pi 100 \times 10^6)t \cos 10^6 t \cdot 2\pi - \sin(2\pi \times 100 \times 10^6) t \sin 2\pi 10^6 t + \sin 2\pi \times 100 \times 10^6 t$$

$$= \frac{1}{2}\cos 2\pi \times 100 \times 10^6 t[\cos 2\pi \times 10^6 t] + \frac{1}{2}\sin 2\pi \times 100 \times 10^6 t[-\sin 2\pi \times 10^6 t + 2]$$

$$= A\cos 2\pi \times 10^6 t + B\sin\cos 2\pi \times 10^6 t$$

Envelope

$$= \sqrt{A^2 + B^2}$$

$$= \sqrt{\frac{1}{4}\cos^2 2\pi \times 10^6 t + \frac{1}{4}[\sin 2\pi \times 10^6 t - 2]}$$

Envelope

$$= \sqrt{\frac{5}{4} - \sin 2\pi \times 10^6 t}$$

1.27 (d)

$$x(t) = \cos(2\pi \times 10kt) + \cos(2\pi \times 10.1kt)$$

$$= \cos(2\pi \times 10kt) + \cos[2\pi(10k + 0.1k)t]$$

$$= \cos(2\pi \times 10kt) + \cos(2\pi \times 10kt)\cos(200\pi t) - \sin(2\pi \times 10kt)\sin(200\pi t)$$

$$= \cos(2\pi \times 10kt)[1 + \cos(200\pi t)] - \sin(2\pi \times 10kt)\sin(200\pi t)$$

$$= \sqrt{[1+\cos(200\pi t)]^2 + \sin^2(200\pi t)}\cos[2\pi \times 10kt + \phi(t)]$$

$$\phi(t) = \tan^{-1}\left[\frac{\sin(200\pi t)}{1+\cos(200\pi t)}\right]$$

$A_c\cos[2\pi f_c t + \phi(t)]$ → Frequency detector → Output $\propto \frac{d\phi(t)}{dt}$

$$\text{Output} \propto \frac{d\phi(t)}{dt} = \frac{1}{1+\left[\frac{\sin(200\pi t)}{1+\cos(200\pi t)}\right]^2} \times \left[\frac{[1+\cos(200\pi t)](200\pi\cos 200\pi t)}{[1+\cos(200\pi t)]^2} - \frac{\sin(200\pi t)[-200\pi\sin 200\pi t]}{[1+\cos(200\pi t)]^2}\right]$$

$$\propto \frac{200\pi\cos(200\pi t) + 200\pi}{[1+\cos(200\pi t)]^2 + \sin^2(200\pi t)} = \frac{200\pi[1+\cos(200\pi t)]}{2[1+\cos(200\pi t)]}$$

Output $\propto 100\pi$ (constant)

1.28 (b)

Spectrum at the output of the first balanced modulator,

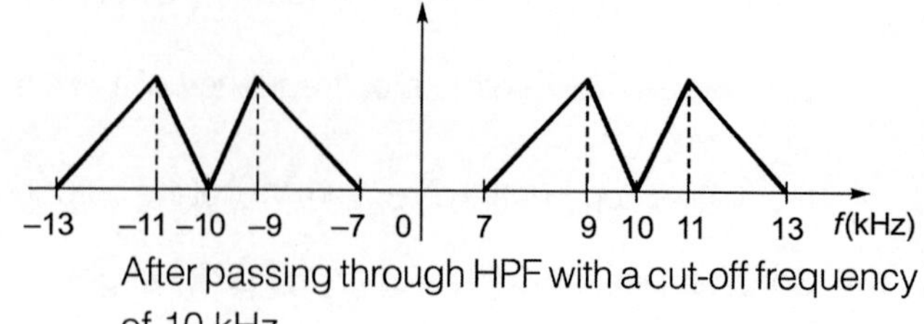

After passing through HPF with a cut-off frequency of 10 kHz,

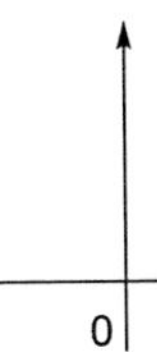
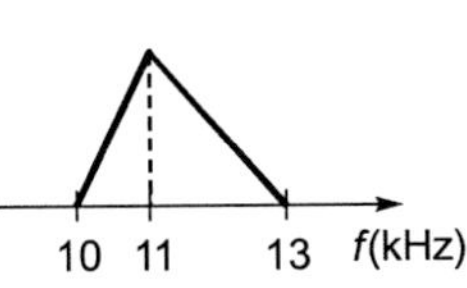

After passing through the second balanced modulator, the spectrum of $y(t)$ for positive frequencies can be drawn as,

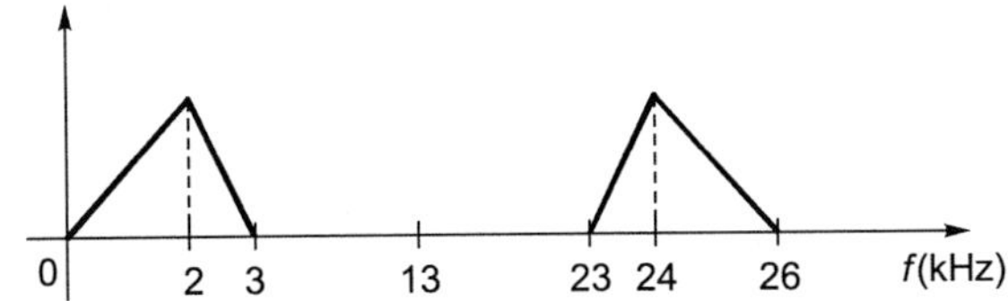

So, the spectral peaks exists at f = 2 kHz and 24 kHz.

1.31 (a)

$$\text{B.W.} = 2(\Delta f + f_m) = 2(180 + 5)$$
$$= 370\text{ K}$$

Note: When FM signal is applied to doubler frequency deviation doubles but f_m remains the same.

1.32 (d)

$$\Delta\phi = m_f$$

m_f remains same for both cases,

$$\Rightarrow \quad \frac{\Delta f}{2} = \frac{10}{1}$$
$$\therefore \quad \Delta f = 20$$
$$\therefore \quad BW = 2(\Delta f + f_m) = 2(20 + 2)$$
$$= 44\text{ K}$$

1.34 (b)

Spectrum of lower sideband SSB signal is,

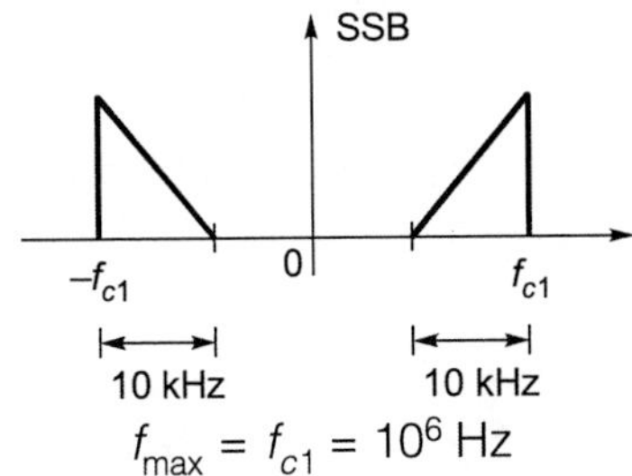

$$f_{max} = f_{c1} = 10^6\text{ Hz}$$

After passing the SSB signal through an NBFM modulator,

$$(BW)_{out} = 2f_{max} = 2 \times 10^6\text{ Hz}$$

1.35 (c)

$$P_s = P_c\frac{m_a^2}{2} = \frac{A_c^2}{2} \times \frac{m_a^2}{2}$$

$$P_s = \frac{100}{2} \times \frac{0.25}{2} = 6.25\text{ Watts}$$

1.36 (b)

$$P_S = \frac{25}{4}\text{Watts}$$
$$P_N = N_O B$$
$$\therefore \quad \text{SNR} = \frac{P_S}{P_N} = \frac{25}{4N_O B}$$

1.38 (a)

Modulated signal,

$$\phi_{AM}(t) = A_c\cos\omega_c t + 2\cos\omega_m t\cos\omega_c t$$
$$= A_c\left[1 + \frac{2}{A_c}\cos\omega_m t\right]\cos\omega_C t$$
$$\mu = \frac{2}{A_c}$$

Condition for envelope detection of an AM signal is

$$\mu \le 1$$
$$\Rightarrow \quad \frac{2}{A_c} \le 1 \text{ or } A_c \ge 2$$

Therefore minimum value of A_C should be 2.

1.39 (b)

Modulation index,

$$m_f = \frac{\delta}{f_m}$$

Where, δ = maximum frequency deviation

f_m = maximum frequency component

Given that, f_m = 1500 Hz

Deviation,

$$\Delta\theta = 5\sin(2\pi \times 1500t) + 7.5\sin(2\pi \times 1000\,t)$$

$$\Delta\omega = \frac{d\theta}{dt} = 5 \times 2\pi \times 1500\cos(2\pi \times 500t)$$
$$+ 7.5 \times 2\pi \times 1000\,\text{Cos}\,(2\pi \times 1000t)$$
$$\Delta\omega_{max} = 2\pi\,(7500 + 7500)$$
$$\delta = \frac{\Delta\omega_{max}}{2\pi} = 15000\text{ Hz}$$
$$m_f = \frac{\delta}{f_m} = \frac{15000}{1500} = 10$$

1.40 (a)

$$s(t) = \cos(\omega_c t) + 0.5\cos\omega_m t\sin\omega_c t$$

In this signal, the carrier components are in phase quadrature.

So, this signal is a narrow-band FM (NBFM) signal.

1.41 (c)

$\cos[2\pi(f_c + f_m)\,t]$ represents only USB of AM-SSB/SC signal.

1.42 (c)

$$\eta = \frac{m_a^2}{2+m_a^2} \times 100\ \%$$

$$m_{a_1} = \frac{V_m}{V_c} = \frac{1}{2} = m_{a_2}$$

$$m_a = \sqrt{m_{a_1}^2 + m_{a_2}^2}$$

$$\eta = \frac{0.25\times 2}{2+0.25\times 2}\times 100\% = 20\%$$

1.43 (c)

Conventional AM signal is

$$x(t) = A_c\,[1 + m(t)] \cos (2\pi\, f_c\, t)$$
$$= A_c \cos (2\pi\, f_c\, t) + A_c\, m(t)\ \cos(2\pi\, f_c\, t)$$

In option (b),

$$\text{Modulation index} = \frac{2}{1} = 2$$

So, it is conventional AM signal but with over-modulation.

In option (c),

$$x(t) = A_c\cos(2\pi f_c t) + \frac{A_c}{4}\, m(t) \cos(2\pi\, f_c\, t)$$

Here, Modulation index

$$= \frac{2}{4} = \frac{1}{2}$$

Therefore, it is conventional AM signal without over-modulation.

1.44 (b)

Average power of angle modulated signal $x(t)$ is $A_c^2/2$. Here, $A_c = 6$.

Therefore,

$$\text{Average power} = \frac{6^2}{2} = 18\ \text{W}$$

1.45 (b)

AM has simple receiver and VSB is bandwidth efficient transmission of signals with significant dc component. SSB-SC is the most bandwidth efficient transmission of voice signals.

1.46 (b)

$$\frac{1}{f_c} << RC << \frac{1}{f_m}$$

$$\frac{1}{1\text{MHz}} << RC << \frac{1}{2\,\text{kHz}}$$

$$\Rightarrow\ 1\ \mu\text{s} << RC << 0.5\ \text{ms}$$

1.47 (b)

For phase modulator

$$\phi(t) = 2\pi f_c t + k_p m(t)$$

maximum phase deviation is

$$(\phi_D)_{\max} = k_p \max[m(t)] = 2k_p \quad ...(1)$$

for frequency modulator

$$\phi(t) = 2\pi f_c t + 2\pi k_f \int_0^t m(t)\, dt$$

$$(\phi'_D)_{\max} = 2\pi k_f \left[\int_0^t m(t)\, dt\right]_{\max}$$

$$(\phi'_D)_{\max} = 2\pi k_f \left[\int_0^2 m(t)\, dt\right]$$

$$(\phi'_D)_{\max} = 2\pi k_f \left[\int_0^2 2\, dt\right]$$

$$(\phi'_D)_{\max} = 8\pi k_f \quad ...(2)$$

given

$$(\phi'_D)_{\max} = (\phi_D)_{\max}$$
$$8\pi k_f = 2\, k_p$$

$$\boxed{\frac{k_p}{k_f} = 4\pi}$$

1.48 (0.50)

$$\mu = \frac{A_{\max} - A_{\min}}{A_{\max} + A_{\min}} = \frac{3-1}{3+1} = \frac{2}{4} = 0.50$$

1.49 (60)

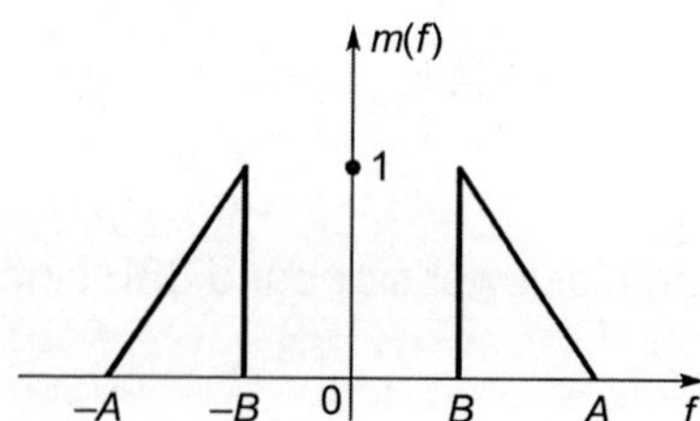

After passing through first multiplier, the output is

$$m(t) \times v(t) = x(t)$$

Therefore,

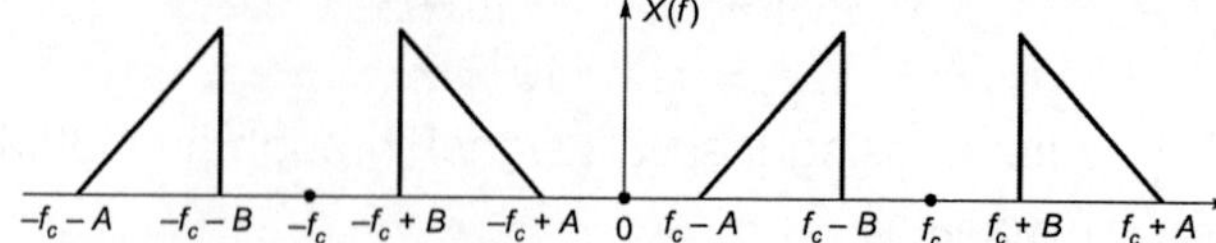

The Fourier transform of output $y(t)$ of the high pass filter centred at f_c is

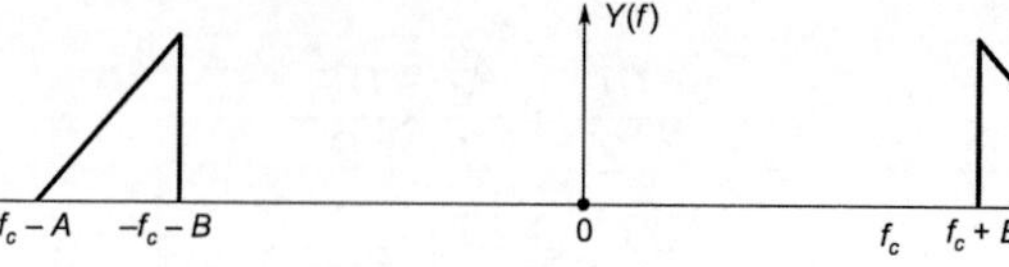

Now, the output of second multiplier is

$$y(t) \cdot w(t) = z(t)$$

Therefore,

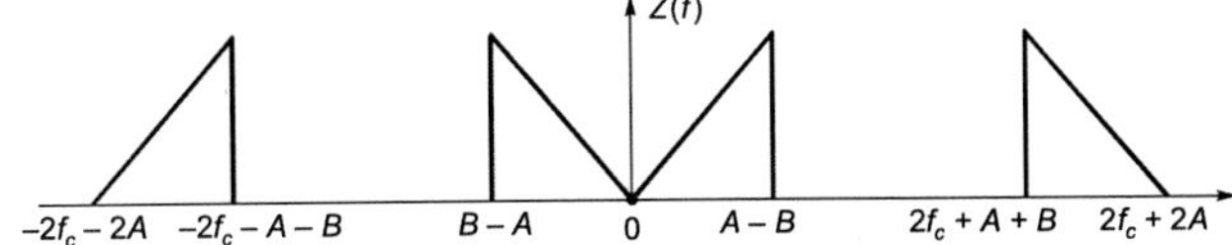

After passing through the low pass filter centred at f_c the Fourier transform of output $s(t)$ is

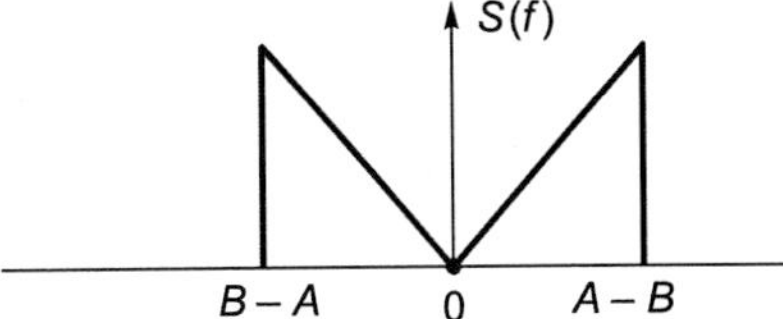

Thus, bandwidth = $(A - B) - 0$

$$= 100 \text{ Hz} - 40 \text{ Hz} = 60 \text{ Hz}$$

1.50 (10)

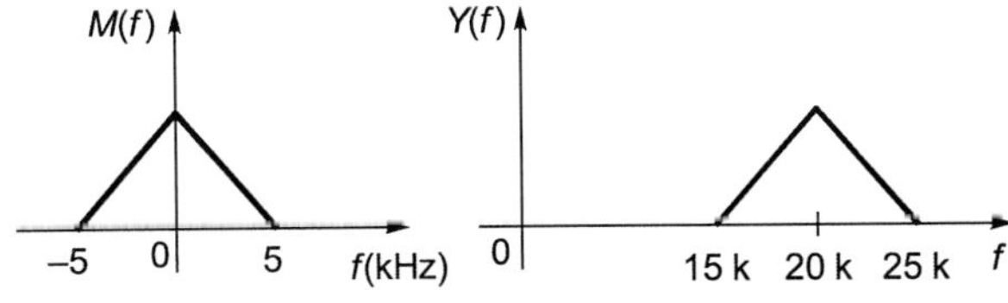

$$Y_s(f) = f_s \sum_{n=-\alpha}^{\infty} Y(f - nf_s)$$

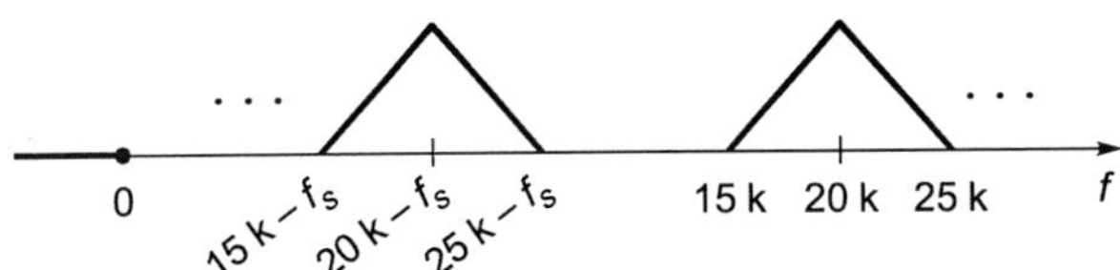

From $Y_s(f)$, to recover $M(f)$ centred about zero

$$20\text{ k} - nf_s = 0$$

$$f_s = \frac{20k}{n}; \quad \text{'}n\text{' can be any integer}$$

$$f_s = 20\text{k (or) } 10\text{k (or) } \frac{20k}{3} \text{ (or) } 5k \quad \text{...(i)}$$

To avoid aliasing $\rightarrow 25\text{k} - f_s \le 15\text{k}$

$$f_s \ge 10 \text{ kHz} \quad \text{...(ii)}$$

From equations (i) and (ii) $\rightarrow (f_s)_{min} = 10$ kHz

1.51 (a)

Maximum frequency deviation is

$$= \frac{1}{2\pi}\left|\frac{d\phi}{dt}\right|_{max}$$

$\phi \rightarrow$ Phase deviation

$$= \frac{1}{2\pi}\left|\beta_1 2\pi f_1 \cos 2\pi f_1 t + \beta_2 2\pi f_2 \cos 2\pi f_2 t\right|_{max}$$

$$= \frac{1}{2\pi}(2\pi\beta_1 f_1 + 2\pi\beta_2 f_2) = \beta_1 f_1 + \beta_2 f_2$$

1.52 (4)

Ratio of P_{SB} to P_C is

$$P_{SB} = \frac{P_C \mu^2}{2}$$

$$\therefore \quad \frac{P_{SB}}{P_C} = \frac{\mu^2}{2}$$

and $$\left(\frac{P_{SB}}{P_C}\right)_1 = \frac{\mu_1^2}{2}$$

$$\left(\frac{P_{SB}}{P_C}\right)_2 = \frac{\mu_2^2}{2}$$

where, $\mu_2 = 2\mu_1$

hence $$\frac{(P_{SB}/P_C)_1}{(P_{SB}/P_C)_2} = \frac{1}{4}$$

$$\left(\frac{P_{SB}}{P_C}\right)_2 = 4\left(\frac{P_{SB}}{P_C}\right)_1$$

Four times.

1.53 (c)

Given $s(t)$ is an SSB modulated signal.

Alternate Solution:

It is the canonical representation of a bandpass signal.

1.54 (350)

Let the spectrum of message signal as shown below.

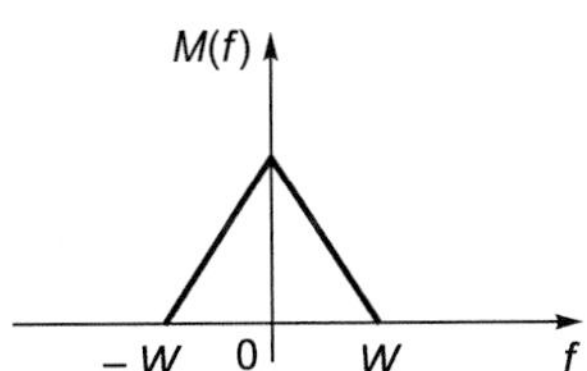

$$x(t) = m(t)\cos 2400\pi t$$

thus $y(t) = 10x(t) + x^2(t)$

$$\therefore\ y(t) = 10m(t)\cos(2400\pi t) + m^2(t)\cos^2(2400\pi t)$$

$$= 10m(t)\cos(2400\pi t) + \frac{1}{2}m^2(t)\frac{[1+\cos(4800\pi t)]}{2}$$

Drawing the spectrum of $x(t)$ be

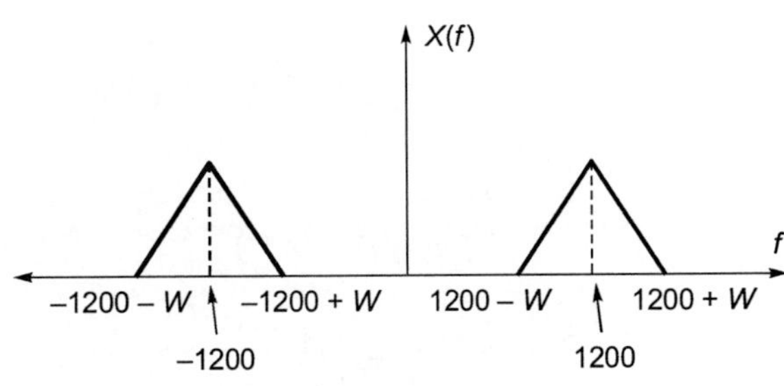

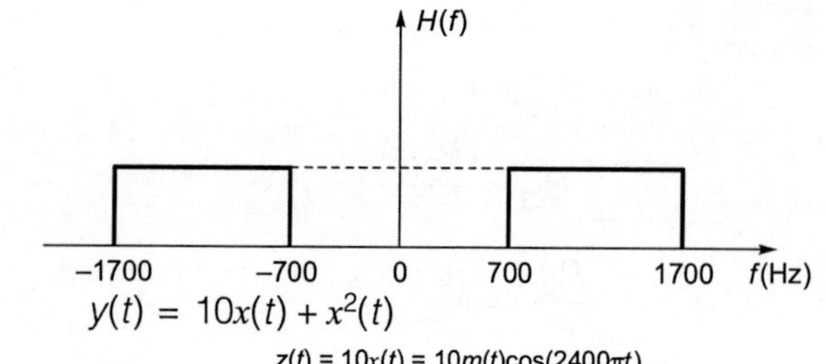

$$y(t) = 10x(t) + x^2(t)$$

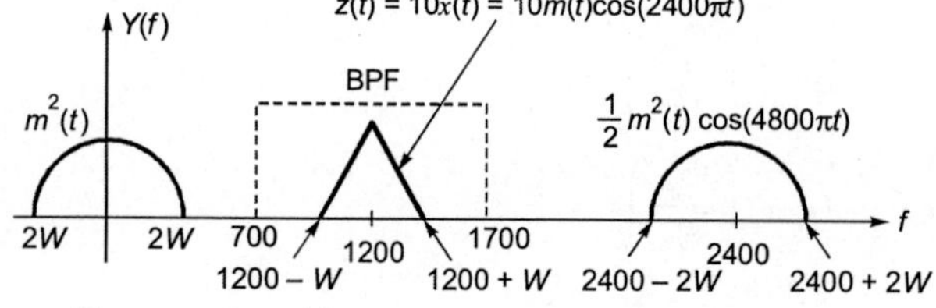

To recover $z(t)$

$$1200 - W > 700 \Rightarrow W < 500$$
$$1200 + W < 1700 \Rightarrow W < 500$$
$$2W < 700 \Rightarrow W < 350$$
$$2400 - 2W > 1700 \Rightarrow W < 350$$

So, that maximum value of W should be less than 350 Hz.

1.55 (c)

The bandwidth of a PM signal is given by

$$B = 2(\Delta f + f_m)$$

Now, for single tone PM signal

$$\Delta f = A_m K_p f_m$$

$$\therefore \quad B = 2(K_p A_m + 1)f_m$$

1.56 (d)

$$x(t) = -\sqrt{2}\left[\frac{\sin\left(\frac{\pi t}{5}\right)}{\frac{\pi t}{5}}\right]\sin\left(\pi t - \frac{\pi}{4}\right)$$

$$= -\sqrt{2}\left[\frac{\sin\left(\frac{\pi t}{5}\right)}{\frac{\pi t}{5}}\right]\left[\sin\pi t\cos\frac{\pi}{4} - \cos\pi t\sin\frac{\pi}{4}\right]$$

$$= \underbrace{\sqrt{2}\left[\frac{\sin\left(\frac{\pi t}{5}\right)}{\frac{\pi t}{5}}\right]\sin\frac{\pi}{4}}_{x_i(t)}\cos\pi t - \underbrace{\sqrt{2}\left[\frac{\sin\left(\frac{\pi t}{5}\right)}{\frac{\pi t}{5}}\right]\cos\frac{\pi}{4}}_{x_q(t)}\sin\pi t$$

$$x(t) = x_i(t)\cos 2\pi f_c t - x_q(t)\sin 2\pi f_c t$$

where, $f_c = \frac{1}{2}$ Hz

$$x_{ce}(t) = x_i(t) + jx_q(t)$$

$$= \sqrt{2}\left[\frac{\sin\left(\frac{\pi t}{5}\right)}{\frac{\pi t}{5}}\right]\sin\frac{\pi}{4} + j\left[\frac{\sin\left(\frac{\pi t}{5}\right)}{\frac{\pi t}{5}}\right]\cos\frac{\pi}{4}$$

$$= j\sqrt{2}\left[\frac{\sin\left(\frac{\pi t}{5}\right)}{\frac{\pi t}{5}}\right]\left[\cos\frac{\pi}{4} - j\sin\frac{\pi}{4}\right] = j\sqrt{2}\left[\frac{\sin\left(\frac{\pi t}{5}\right)}{\frac{\pi t}{5}}\right]e^{-j\pi/4}$$

1.57 (a)

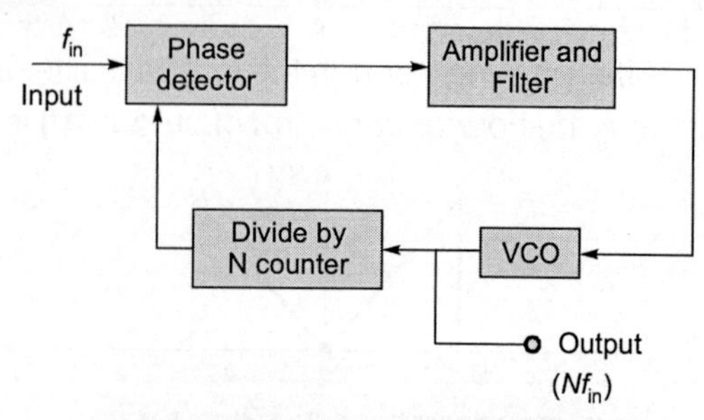

f_{in}	VCO output (Nf_{in})	Divide by N counter
5 kHz	10 kHz	2
5 kHz	20 kHz	4
5 kHz	40 kHz	8
5 kHz	80 kHz	16

1.58 (5)

$$f_s = 58 \text{ MHz} - 68 \text{ MHz}$$

f_{si} should fall outside the range 58 MHz - 68 MHz

Hence $\quad f_{s\min} = 58$ MHz

$$f_{si} = f_s + 2IF > 68 \text{ MHz}$$
$$58 \text{ MHz} + 2IF > 68 \text{ MHz}$$
$$IF > 5 \text{ MHz}$$
$$\Rightarrow \quad (IF)_{\min} = 5 \text{ MHz}$$

1.59 (0.5)

$$s(t) = 5\cos 1600\pi t + 20\cos 1800\pi t + 5\cos 2000\pi t$$
$$s(t) = 20\cos 1800\pi t + 5\cos 1600\pi t + 5\cos 2000\pi t$$
$$s(t) = A_c\cos 2\pi f_c t + \frac{A_c\mu}{2}\cos 2\pi(f_c - f_m)t + \frac{A_c\mu}{2}\cos 2\pi(f_c + f_m)t$$

comparing, we get

$$A_c = 20 \text{ V}; \quad \frac{A_c\mu}{2} = 5 \text{ V}$$
$$\mu = \frac{10}{20} = 0.5$$

1.60 (3485)

$$IF = 15 \text{ MHz}$$
$$f_l = 3.5 \text{ GHz}$$
$$\Rightarrow f_{si} = f_l - \text{IF} = 3500 - 15 \text{ MHz} = 3485 \text{ MHz}$$

1.61 (20)

Given that, the modulating signal is,

$$x(t) = 5\sin(4000\pi t - 10\pi\cos 2000\pi t)\text{V}$$

The standard phase modulated signal can be given as,

$$s(t) = A_c \cos(\omega_c t + k_p x(t))$$

Instantaneous angle of the modulated signal is,

$$\theta(t) = \omega_c t + k_p x(t)$$

Instantaneous frequency is,

$$\omega_i(t) = \frac{d\theta(t)}{dt} = \omega_c + k_p \frac{dx(t)}{dt}$$

$$f_i(t) = f_c + \frac{25}{2\pi}\Big[\cos(4000\pi t - 10\pi\cos 2000\pi t)$$
$$(4000\pi + 20000\pi^2 \sin 2000\pi t)\Big]$$

At $t = 0.5$ ms,

$$f_i(t) = f_c + \frac{25}{2\pi}\left[\cos\left(2\pi - 10\pi\cos(\pi)\right)\right]$$
$$\left(4000\pi + 20000\pi^2 \sin(\pi)\right)$$
$$= f_c + \frac{25}{2\pi}[\cos(12\pi)](4000\pi)$$
$$= f_c + 50 \text{ kHz} = 70 \text{ kHz}$$

$\because$ Given that, $f_c = 20$ kHz

1.62 (5.208)

Given that, $P_C = 5$ kW for $\mu_{(max)} = 0.5$

So, $$P_{t(max)} = P_C\left[1 + \frac{(0.5)^2}{2}\right]$$
$$= 5\left[1 + \frac{0.25}{2}\right] \text{ kW} = 5.625 \text{ kW}$$

For $\mu = 0.4$,

$$P_{C(max)}\left(1 + \frac{\mu^2}{2}\right) = P_{t(max)}$$

$$P_{C(max)} = \frac{5.625}{1 + \frac{(0.4)^2}{2}} \text{ kW} = 5.208 \text{ kW}$$

1.63 (0.125)

The power of the carrier signal is,

$$P_c = \frac{(4)^2}{2} = 8 \text{ W}$$

The power of message signal can be taken as the power carried by the sidebands.

So, the power of the message signal is,

$$P_m = 2\left[\frac{(1)^2}{2}\right] = 1 \text{ W}$$

Required ratio,

$$\frac{P_m}{P_c} = \frac{1}{8} = 0.125$$

1.64 (d)

$$v_i(t) = A_c\cos(2\pi f_c t) + \cos(2\pi f_m t)$$
$$v_o(t) = av_i(t) + bv_i^2(t)$$
$$= \left[aA_c\cos(2\pi f_c t) + a\cos(2\pi f_m t)\right]$$
$$+b\begin{bmatrix} A_c^2\cos^2(2\pi f_c t) + \cos^2(2\pi f_m t) \\ +2A_c\cos(2\pi f_c t)\cos(2\pi f_m t)\end{bmatrix}$$

After passing through the given BPF,

$$y(t) = aA_c\cos 2\pi f_c t + 2bA_c\cos(2\pi f_c t)\cos(2\pi f_m t)$$
$$= aA_c\left[1 + \frac{2b}{a}\cos(2\pi f_m t)\right]\cos(2\pi f_c t)$$

Modulation index,

$$\mu = \frac{2b}{a}$$

$$P_{SB} = \frac{\mu^2}{2}P_c = \frac{1}{2}P_c$$

So, $$\mu^2 = 1 \Rightarrow \mu = 1$$

$$\frac{2b}{a} = 1 \quad \Rightarrow \quad \frac{a}{b} = 2$$

1.65 (0.75)

$$f_i = \frac{1}{2\pi}\frac{d\theta(t)}{dt} = f_c + \frac{k}{2\pi}\frac{dm(t)}{dt} = f_c + 5\frac{dm(t)}{dt}$$

So, $f_{i\,(min)} = 50 \text{ kHz} - (5 \times 1000 \text{ Hz}) = 45 \text{ kHz}$

$f_{i\,(max)} = 50 \text{ kHz} + (5 \times 2000 \text{ Hz}) = 60 \text{ kHz}$

$$\frac{f_{i(min)}}{f_{i(max)}} = \frac{45}{60} = 0.75$$

1.66 (2)

$$S(t)_{pm} = A_c\cos[2\pi f_c t + k_p m(t)]$$

$$S(t)_{Fm} = A_c\cos\left[2\pi f_c t + k_f\int_0^{\infty} m(t)dt\right]$$

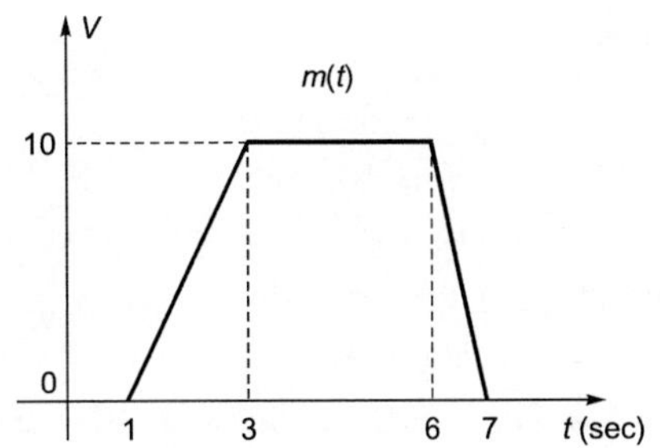

Instantaneous frequency are equal

$$f_i = \frac{1}{2\pi}\frac{d}{dt}\theta(t)$$

$$f_{i\text{PM}} = f_c + \frac{K_p}{2\pi}\frac{d}{dt}m(t) \text{ and } f_{i\text{FM}} = f_c + \frac{K_f}{2\pi}m(t)$$

Given that, $(f_{i\text{PM}})_{\max} = (f_{i\text{FM}})_{\max}$

$$f_c + \frac{k_p}{2\pi}\left[\frac{d}{dt}m(t)\right]_{\max} = f_c + \frac{k_f}{2\pi}[m(t)]_{\max}$$

$$\frac{k_p}{2\pi}\frac{d}{dt}m(t)\bigg|_{\max} = \frac{k_f}{2\pi}m(t)\bigg|_{\max}$$

$$\frac{d}{dt}m(t)\bigg|_{\max} = 5,\ m(t)\big|_{\max} = 10$$

$$5k_p = 10k_f$$

$$\frac{k_p}{k_f} = 2$$

1.67 (c)

Output of multiplier

$$= x(t)\cos 2\pi(f_c + 40)t = m(t)\cos 2\pi f_c t \cdot \cos 2\pi(f_c + 40)t$$

$$= \frac{m(t)}{2}[\cos 2\pi(2f_c + 40)t + \cos 2\pi(40)t]$$

Given, $m(t) = 4\cos 1000\pi t$

So, output of multiplier

$$= 2\cos 2\pi(500)t\,[\cos 2\pi(2f_c + 40)t + \cos 2\pi(40)t]$$

$$= \cos 2\pi(2f_c + 540)t + \cos 2\pi(2f_c - 460)t + \cos 2\pi(540)t + \cos 2\pi(460)t$$

Output of Low pass filter

$$= \cos[2\pi(460)]t = \cos 920\pi t$$

1.68 (94)

If carrier and one of the sidebands are suppressed, then power saved

$$= \frac{\text{Power saved}}{\text{Total power}} = \frac{P_c + \frac{P_c\mu^2}{4}}{P_c\left[1 + \frac{\mu^2}{2}\right]} = \frac{4 + \mu^2}{4 + 2\mu^2}$$

$$= \frac{4 + (0.5)^2}{4 + 2\times(0.5)^2} = 0.944 = 94.4\%$$

1.69 (1400)

$$f_s = 600 \text{ kHz}$$

$$f_l = 1000 \text{ kHz}$$

$$f_{si} = ?$$

Default $f_l > f_s$

$\text{IF} = f_l - f_s = 400$ kHz

$f_{si} = f_s + 2\text{IF} = 600\text{K} + 800\text{K} = 1400$ kHz

1.70 (1011313.7)

Given sinusoidal message signal $\rightarrow V_{\text{rms}} = 4$ V

$$f_m = 1 \text{ kHz}$$

$$V_{\text{peak}} = \sqrt{2}\times V_{\text{rms}} = \sqrt{2}\times 4 \text{ V}$$

$$m(t) = 4\sqrt{2}\sin 2\pi\times 10^3 t$$

$$K_p = 2 \text{ rad/volt}$$

$$c(t) = 2\cos 2\pi\times 10^6 t$$

$$(f_i)_{\max} = f_c + \frac{K_p}{2\pi}\left[\frac{d}{dt}m(t)\right]_{\max}$$

$$\frac{d}{dt}m(t) = 4\sqrt{2}\times(2\pi\times 10^3)\cos 2\pi\times 10^3 t$$

$$\left[\frac{d}{dt}m(t)\right]_{\max} = 4\sqrt{2}\times 2\pi\times 10^3$$

$$(f_i)_{\max} = 10^6 + \frac{2}{2\pi}(4\sqrt{2}\times 2\pi\times 10^3)$$

$$= 1011313.7 \text{ Hz}$$

1.71 (a)

For the de-emphasis filter shown below,

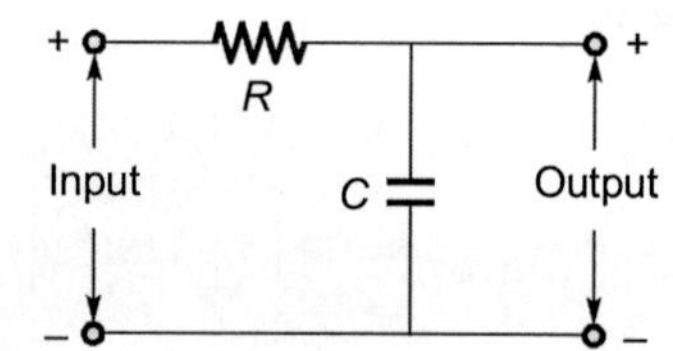

The frequency response

$$H_{De}(\omega) = \frac{\frac{1}{j\omega C}}{R + \frac{1}{j\omega C}} = \frac{1}{1 + j\omega RC}$$

$$\Rightarrow \quad |H_{De}(\omega)| = \frac{1}{\sqrt{1 + \omega^2R^2C^2}} \quad \text{...(i)}$$

Given, the pre-emphasis filter has frequency response,

$$H_{Pe}(\omega) = 1 + \frac{j\omega}{\omega_o},\ \omega_o = 10^4 \text{ rad/sec}$$

$$\Rightarrow \quad |H_{Pe}(\omega)| = \sqrt{1 + \left(\frac{\omega}{\omega_0}\right)^2} \quad \text{...(ii)}$$

The magnitude response of pre-emphasis and de-emphasis filter is related as

$$|H_{Pe}(\omega)| = \frac{1}{|H_{De}(\omega)|}$$

$$\Rightarrow \quad |H_{Pe}(\omega)|^2 = \frac{1}{|H_{De}(\omega)|^2}$$

Using equations (i) and (ii), we get

$$1 + \left(\frac{\omega}{\omega_o}\right)^2 = 1 + (\omega RC)^2$$

$$\Rightarrow \quad \frac{\omega}{\omega_o} = \omega RC$$

$$\Rightarrow \quad \omega_o = \frac{1}{RC} = 10^4 \text{rad/sec}$$

$$\Rightarrow \quad RC = 10^{-4}$$

Option (a) : $RC = 10^3 \times 0.1 \times 10^{-6} = 10^{-4}$

Option (b) : $RC = 2 \times 10^3 \times 10^{-6} = 2 \times 10^{-3}$

Option (c) : $RC = 1 \times 10^3 \times 2 \times 10^{-6} = 2 \times 10^{-3}$

Option (d) : $RC = 2 \times 10^3 \times 0.5 \times 10^{-6} = 10^{-3}$

Therefore, only option (a) satisfies the required condition $RC = 10^{-4}$.

1.72 (b)

$$x(t) = A\cos\left[\omega_c t + K_f \int_{-\infty}^{t} m(\lambda)\, d\lambda\right]$$

Bandwidth $[x(t)] \rightarrow B_T = 2[\Delta f + \omega]$

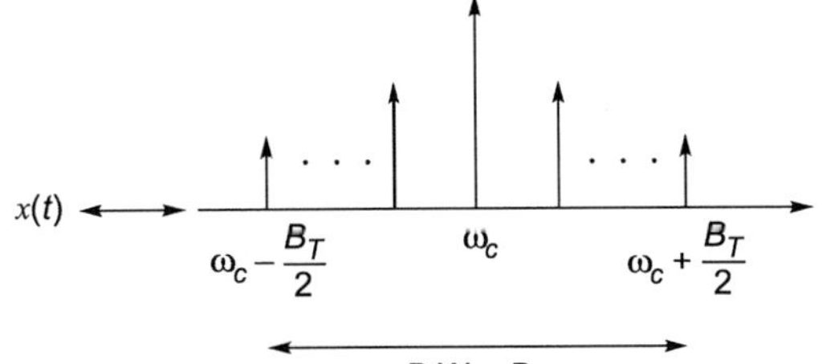

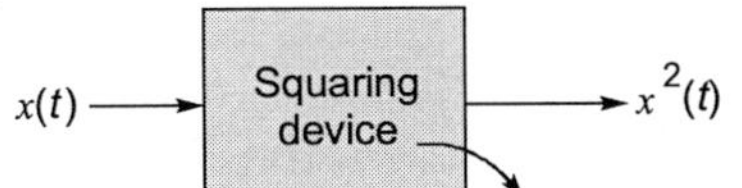

$$x^2(t) \rightarrow \left.\begin{matrix} \Delta f' = 2\Delta f \\ \omega_c' = 2\omega_c \end{matrix}\right\}$$

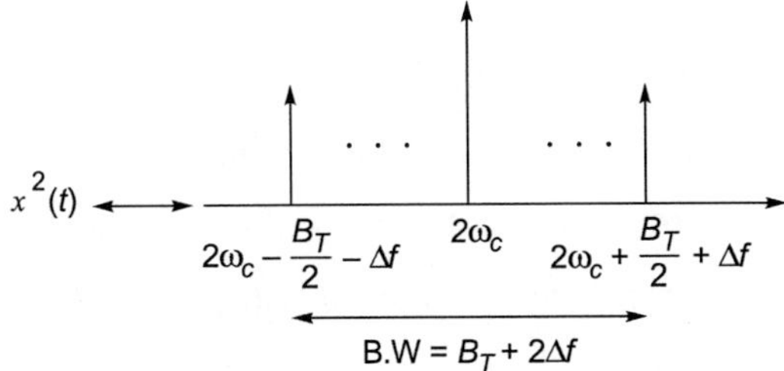

Bandwidth of $[x^2(t)] = 2[\Delta f' + \omega]$

$$= 2[2\Delta f + \omega] = B_T + 2\Delta f$$

$$y(t) = 2x(t) + 5x^2(t)$$

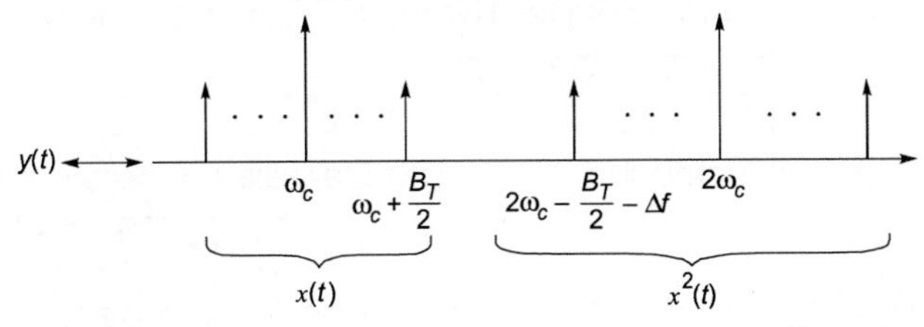

To recover $x(t) \rightarrow 2\omega_C - \frac{B_T}{2} - \Delta f > \omega_c + \frac{B_T}{2}$

$$\omega_c > \Delta f + B_T$$

$$\omega_c > \Delta f + 2\Delta f + 2\omega$$

$$\omega_c > 3\Delta f + 2\omega$$

$$\omega_c > \frac{3}{2}\{2[\Delta f + \omega]\} - \omega$$

$$\omega_c > \frac{3}{2}B_T - \omega$$

Compared to FM BW, message BW is very small. So, that it can be ignored.

$$\omega_C > \frac{3}{2}B_T$$

$$[\omega_c]_{\min} = \frac{3}{2}B_T$$

1.73 (2.5)

The AM signal is given as

$$s(t) = A\cos(400\,\pi t) + B\cos(360\pi t) + B\cos(440\,\pi t)$$

Comparing with the standard AM signal expression,

$$s_{AM}(t) = A_c\cos(2\pi f_c t) + \frac{A_c\mu}{2}\cos[2\pi(f_c\quad f_m)t] + \frac{A_c\mu}{2}\cos[2\pi(f_c + f_m)t]$$

We get,

$$A_c = A \text{ and } \frac{A_c\mu}{2} = B \qquad \text{...(i)}$$

Given:

Carrier power, $P_c = 50$ W

$$\Rightarrow \quad \frac{A_c^2}{2} = 50$$

$$\Rightarrow \quad A_c = 10 \qquad \text{...(ii)}$$

Further, it is given that:

$$\frac{\text{Total sideband power}}{\text{Total power}} = \frac{1}{9}$$

$$\Rightarrow \quad \frac{P_c \cdot \frac{\mu^2}{2}}{P_c\left[1 + \frac{\mu^2}{2}\right]} = \frac{1}{9}$$

$$\Rightarrow \quad \frac{8\mu^2}{2} = 1 \quad \Rightarrow \quad \mu^2 = \frac{1}{4}$$

$$\Rightarrow \quad \mu = \frac{1}{2} \qquad \text{...(iii)}$$

Using equation (i), (ii) and (iii), we get

$$B = \frac{A_c\mu}{2} = \frac{10}{2 \times 2}$$

$$\Rightarrow \quad B = 2.5$$

■■■

2 Random Signals and Noise

2.1 For a narrow band noise with Gaussian quadrature components, the probability density function of its envelope will be
(a) uniform (b) Gaussian
(c) exponential (d) Rayleigh
[1995 : 1 M]

2.2 The auto-correlation function of an energy signal has
(a) no symmetry (b) conjugate symmetry
(c) even symmetry (d) Both (b) and (c)
[1996 : 2 M]

2.3 The power spectral density of a deterministic signal is given by $[\sin(f)/f]^2$, where 'f' is frequency. the autocorrelation function of this signal in the time domain is
(a) a rectangular pulse (b) a delta function
(c) a sine pulse (d) a triangular pulse
[1997 : 2 M]

2.4 An Amplifier A has 6 dB gain and 50 Ω input and output impedances. The noise figure of this Amplifier as shown in the figure is 3 dB. A cascade of two such Amplifiers as in the figure will have a noise figure of

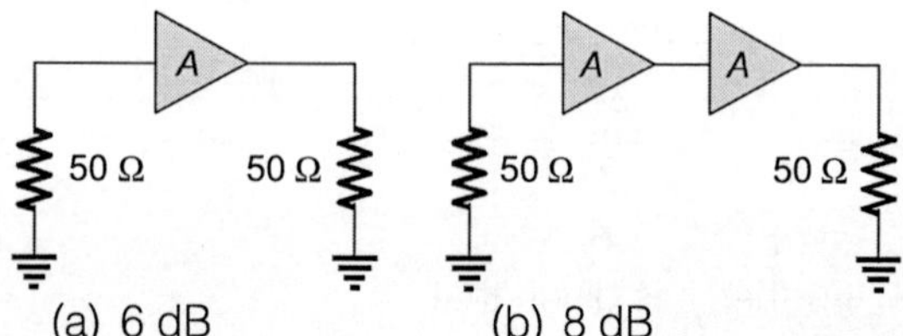

(a) 6 dB (b) 8 dB
(c) 12 dB (d) none of the above
[1997 : 1 M]

2.5 The ACF of a rectangular pulse of duration T is
(a) a rectangular pulse of duration T
(b) a rectangular pulse of duration $2T$
(c) a triangular pulse of duration T
(d) a triangular pulse of duration $2T$ **[1998 : 1 M]**

2.6 A probability density function is given by
$P(x) = K \exp(-x^2/2), -\infty < x < \infty$
The value of K should be
(a) $\frac{1}{\sqrt{2\pi}}$ (b) $\sqrt{\frac{2}{\pi}}$
(c) $\frac{1}{2}\sqrt{\pi}$ (d) $\frac{1}{\pi\sqrt{2}}$ **[1998 : 1 M]**

2.7 The amplitude spectrum of a Gaussian pulse is
(a) uniform
(b) a sine function
(c) Gaussian
(d) an impulse function **[1998 : 1 M]**

2.8 The probability density function of the envelope of narrow band Gaussian noise is
(a) Poisson (b) Gaussian
(c) Rayleigh (d) Rician **[1998 : 1 M]**

2.9 The PDF of a Gaussian random variable X is given by $P_x(x) = \frac{1}{3\sqrt{2\pi}} e^{\frac{-(x-4)^2}{18}}$. The probability of the event $\{X = 4\}$ is
(a) $\frac{1}{2}$ (b) $\frac{1}{3\sqrt{2\pi}}$
(c) 0 (d) $\frac{1}{4}$ **[2001 : 1 M]**

2.10 The PSD and the power of a signal $g(t)$ are, respectively, $S_g(\omega)$ and P_g. The PSD and the power of the signal ag(t) are, respectively,
(a) $a^2S_g(\omega)$ and a^2P_g (b) $a^2S_g(\omega)$ and aP_g
(c) $aS_g(\omega)$ and a^2P_g (d) $aS_g(\omega)$ and aP_g
[2001 : 2 M]

2.11 If the variance σ_X^2 of $d(n) = x(n) - x(n-1)$ is one-tenth the variance σ_X^2 of a stationary zero-mean discrete-time signal $x(n)$, then the normalized autocorrelation function $R_{XX}(k)/\sigma_X^2$ at $k = 1$ is
(a) 0.95 (b) 0.90
(c) 0.10 (d) 0.05 **[2002 : 2 M]**

2.12 The noise at the input to an ideal frequency detector is white. The power spectral density of the noise at the output is
(a) raised-cosine (b) flat
(c) parabolic (d) Gaussian
[2003 : 1 M]

2.13 Let X and Y be two statistically independent random variables uniformly distributed in the ranges (–1, 1) and (–2, 1) respectively. Let

$Z = X + Y$. Then the probability that $(Z \le -2)$ is

(a) zero (b) $\frac{1}{6}$

(c) $\frac{1}{3}$ (d) $\frac{1}{12}$

[2003 : 2 M]

Common Data for Questions 2.14 and 2.15.

Let X be the Gaussian random variable obtained by sampling the process at $t = t_i$ and let

$$Q(\alpha) = \int_{\alpha}^{\infty} \frac{1}{\sqrt{2\pi}} e^{-\frac{y^2}{2}} dy.$$

Auto correlation function $R_{XX}(\tau) = 4\left(e^{-0.2|\tau|} + 1\right)$ and mean = 0

2.14 The probability that $[x \le 1]$ is

(a) $1 - Q(0.5)$ (b) $Q(0.5)$

(c) $Q\left(\frac{1}{2\sqrt{2}}\right)$ (d) $1 - Q\left(\frac{1}{2\sqrt{2}}\right)$

[2003 : 2 M]

2.15 Let Y and Z be the random variables obtained by sampling $X(t)$ at $t = 2$ and $t = 4$ respectively. Let $W = Y - Z$. The variance of W is

(a) 13.36 (b) 9.36

(c) 2.64 (d) 8.00

[2003 : 2 M]

2.16 The distribution function $F_X(x)$ of a random variable X is shown in the figure. The probability that $X = 1$ is

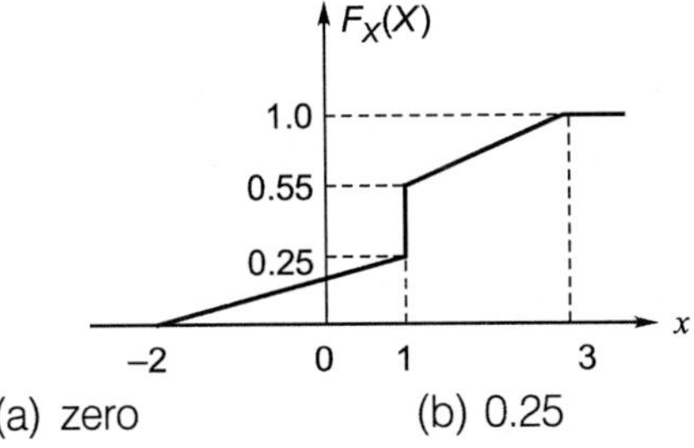

(a) zero (b) 0.25

(c) 0.55 (d) 0.30

[2004 : 1 M]

2.17 A 1 mW video signal having a bandwidth of 100 MHz is transmitted to a receiver through a cable that has 40 dB loss. If the effective one-sided noise spectral density at the receiver is 10^{-20} Watt/Hz, then the signal-to-noise ratio at the receiver is

(a) 50 dB (b) 30 dB

(c) 40 dB (d) 60 dB [2004 : 2 M]

2.18 A random variable X with uniform density in the interval 0 to 1 is quantized as follows:

If $0 \le X \le 0.3$, $x_q = 0$

If $0.3 < X \le 1$, $x_q = 0.7$

where, x_q is the quantized value of X.

The root-mean square value of the quantization noise is

(a) 0.573 (b) 0.198

(c) 2.205 (d) 0.266 [2004 : 2 M]

2.19 Noise with uniform power spectral density of $N_0 W$/Hz is passed through a filter $H(\omega) = 2\exp(-j\omega t_d)$ followed by an ideal low pass filter of bandwidth B Hz. The output noise power in Watts is

(a) $2 N_0 B$ (b) $4 N_0 B$

(c) $8 N_0 B$ (d) $16 N_0 B$ [2005 : 2 M]

2.20 An output of a communication channel is a random variable v with the probability density function as shown in the figure. The mean square value of v is

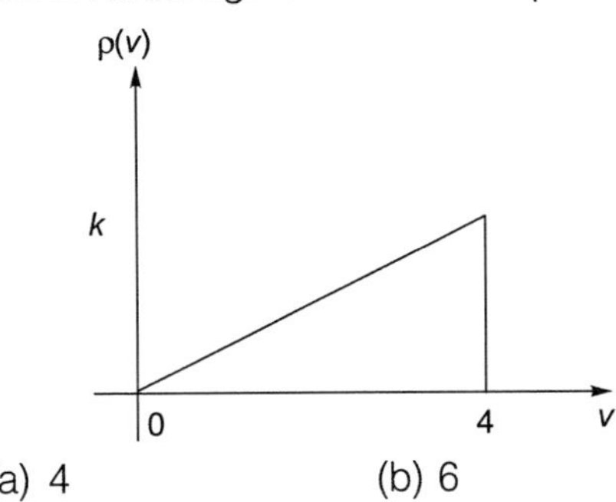

(a) 4 (b) 6

(c) 8 (d) 9

[2005 : 2 M]

Common Data for Questions 2.21 and 2.22.

Asymmetric three-level midtread quantizer is to be designed assuming equiprobable occurrence of all quantization levels.

2.21 If the probability density function is divided into three regions as shown in the figure, the value of a in the figure is

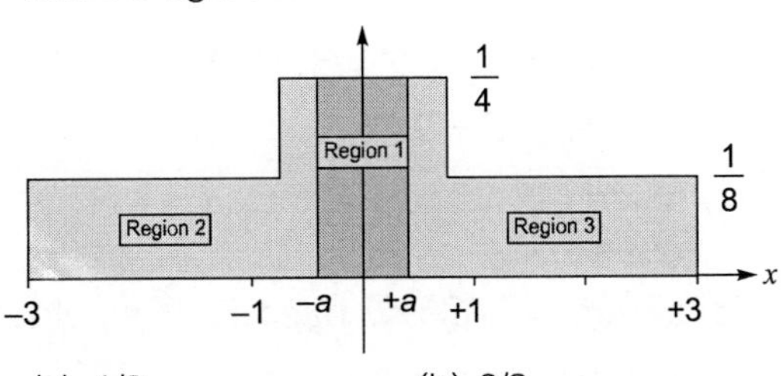

(a) 1/3 (b) 2/3

(c) 1/2 (d) 1/4

[2005 : 2 M]

2.22 The quantization noise power for the quantization region between $-a$ and $+a$ in the figure is

(a) $\frac{4}{81}$ (b) $\frac{1}{9}$

(c) $\frac{5}{81}$ (d) $\frac{2}{81}$

[2005 : 2 M]

2.23 A zero-mean white Gaussian noise is passed through an ideal lowpass filter of bandwidth 10 kHz. The output is then uniformly sampled with sampling period t_s = 0.03 msec. The samples so obtained would be

(a) correlated
(b) statistically independent
(c) uncorrelated
(d) orthogonal

[2006 : 2 M]

Common Data for Questions 2.24 & 2.25:
The following two questions refer to wide sense stationary stochastic processes

2.24 It is desired to generate a stochastic process (as voltage process) with power spectral density

$$S(\omega) = \frac{16}{16+\omega^2}$$

by driving a Linear-Time-Invariant system by zero mean white noise (as voltage process) with power spectral density being constant equal to 1. The system which can perform the desired task could be

(a) first order low-pass *R-L* filter
(b) first order high-pass *R-C* filter
(c) tuned *L-C* filter
(d) series *R-L-C* filter

[2006 : 2 M]

2.25 The parameters of the system obtained in the above question would be

(a) first order *R-L* low-pass filter would have
$R = 4\Omega \quad L = 1\text{ H}$
(b) first order R-C high-pass filter would have
$R = 4\Omega \quad C = 0.25\text{ F}$
(c) tuned *L-C* filter would have $L = 4\text{ H}$
$C = 4\text{ F}$
(d) series *R-L-C* low-pass filter would have
$R = 1\Omega, L = 4\text{H}, C = 4\text{F}$

[2006 : 2 M]

2.26 A uniformly distributed random variable *X* with probability density function

$$f_X(x) = \frac{1}{10}\left(u(x+5) - u(x-5)\right)$$

where $u(\cdot)$ is the unit step function is passed through a transformation given in the figure below. The probability density function of the transformed random variable *Y* would be

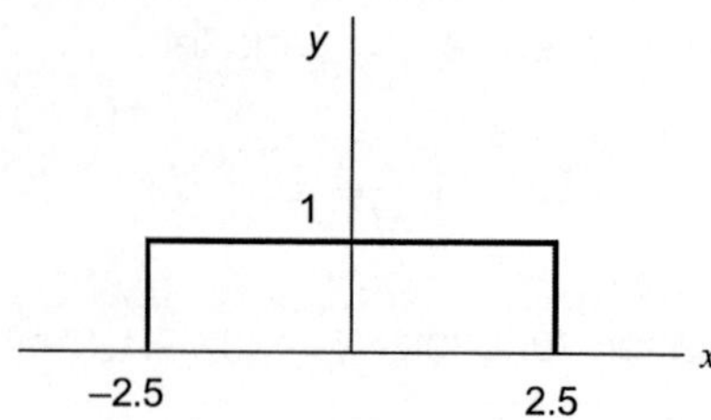

(a) $f_Y(y) = \frac{1}{5}\left(u(y+2.5) - u(y-2.5)\right)$

(b) $f_Y(y) = 0.5\delta(y) + 0.5\delta(y-1)$

(c) $f_Y(y) = 0.25\delta(y+2.5) + 0.25\delta(y-2.5) + 0.5\delta(y)$

(d) $f_Y(y) = 0.25\delta(y+2.5) + 0.25\delta(y-2.5)$
$+\frac{1}{10}\left(u(y+2.5) - u(y-2.5)\right)$

[2006 : 2 M]

2.27 If *E* denotes expectation, the variance of a random variable X is given by

(a) $E[X^2] - E^2[X]$ (b) $E[X^2] + E^2[X]$
(c) $E[X^2]$ (d) $E^2[X]$

[2007 : 1 M]

2.28 If $R(\tau)$ is the auto-correlation function of a real, wide-sense stationary random process, then which of the following is NOT true?

(a) $R(\tau) = R(-\tau)$
(b) $|R(\tau)| \le R(0)$
(c) $R(\tau) = -R(-\tau)$
(d) The mean square value of the process is $R(0)$

[2007 : 1 M]

2.29 If $S(f)$ is the power spectral density of a real, wide-sense stationary random process, then which of the following is ALWAYS true?

(a) $S(0) \ge S(f)$ (b) $S(f) \ge 0$
(c) $S(-f) = -S(f)$ (d) $\int_{-\infty}^{\infty} S(f)df = 0$

[2007 : 1 M]

2.30 The Probability Density Function (PDF) of a random variable X is as shown below.

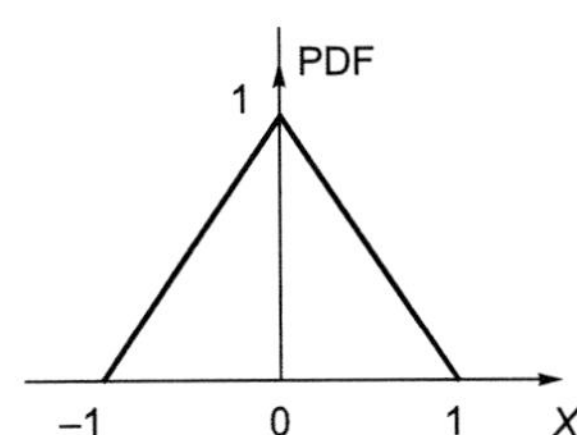

The corresponding Cumulative Distribution Function (CDF) has the form

(a)

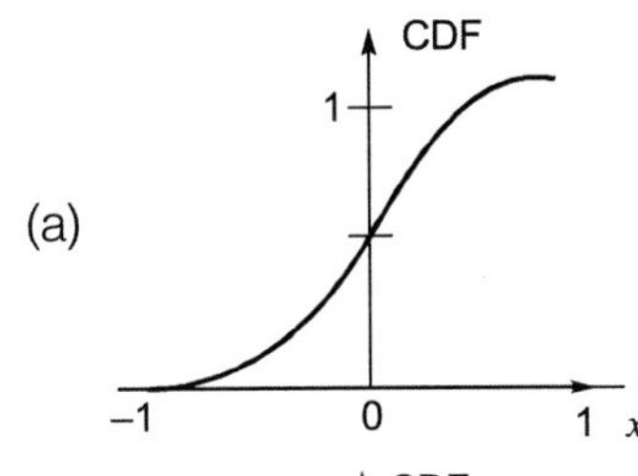

(b)

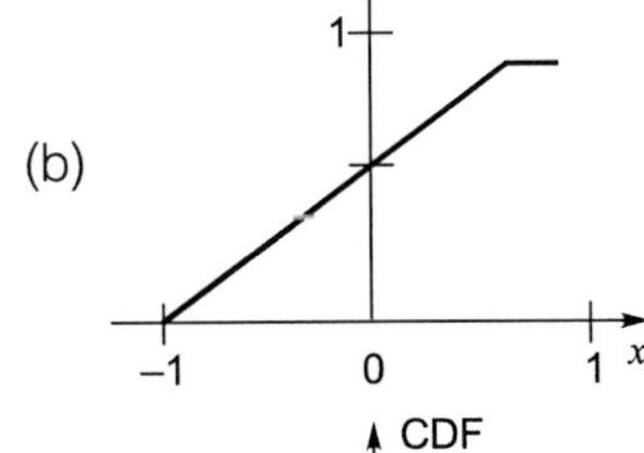

(c)

(d)

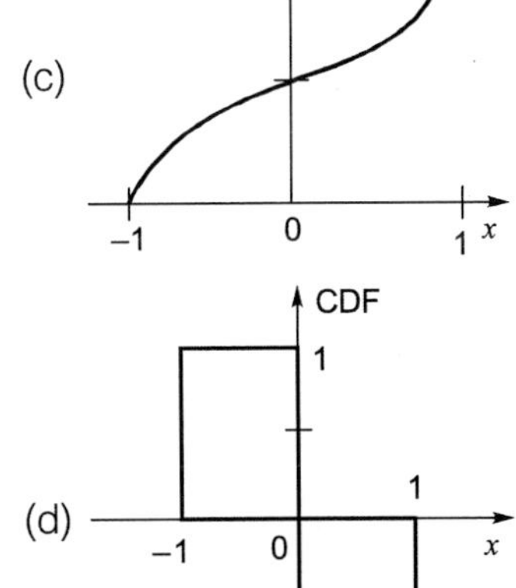

[2008 : 2 M]

2.31 $P_x(x) = M\exp(-2|x|) + N\exp(-3|x|)$ is the probability density function for the real random variable X, over the entire x axis. M and N are both positive real numbers. The equation relating M and N is

(a) $M+\frac{2}{3}N=1$ (b) $2M+\frac{1}{3}N=1$

(c) $M+N=1$ (d) $M+N=3$

[2008 : 2 M]

2.32 Noise with double-sided power spectral density of K over all frequencies is passed through a RC low pass filter with 3 dB cut-off frequency of f_c. The noise power at the filter output is

(a) K (b) Kf_c

(c) $K\pi f_c$ (d) ∞

[2008 : 2 M]

2.33 A white noise process $x(t)$ with two-sided power spectral density 1×10^{-10} W/Hz is input to a filter whose magnitude squared response is shown below.

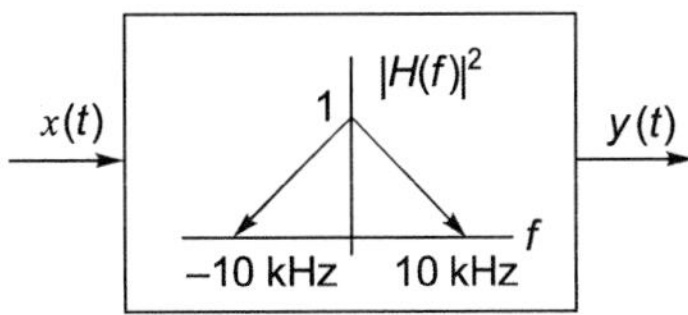

The power of the output process $y(t)$ is given by

(a) 5×10^{-7} W (b) 1×10^{-6} W

(c) 2×10^{-6} W (d) 1×10^{-5} W

[2009 : 1 M]

2.34 If the power spectral density of stationary random process is a sinc-squared function of frequency, the shape of its auto-correlation is

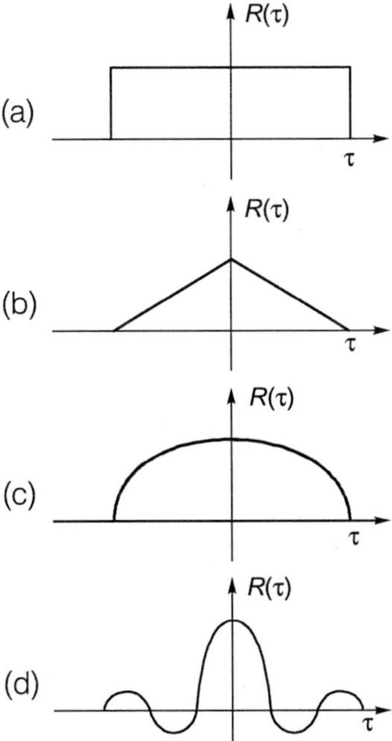

[2009 : 1 M]

2.35 A discrete random variable X takes values from 1 to 5 with probabilities as shown in the table. A student calculates the mean X as 3.5 and her teacher calculates the variance of X as 1.5. Which of the following statements is true?

k	1	2	3	4	5
$P(X = k)$	0.1	0.2	0.4	0.2	0.1

(a) Both the student and the teacher are right
(b) Both the student and the teacher are wrong
(c) The student is wrong but the teacher is right
(d) The student is right but the teacher is wrong

[2009 : 2 M]

2.36 $X(t)$ is a stationary process with the power spectral density $S_X(f) > 0$ for all f. The process is passed through a system shown below.

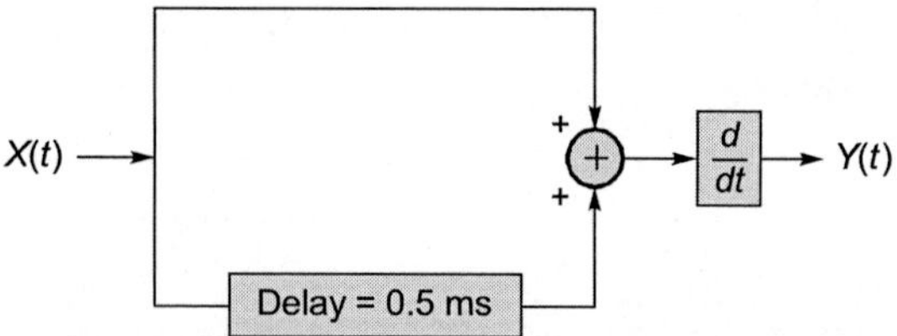

Let $S_Y(f)$ be the power spectral density of $Y(t)$. Which one of the following statements is correct?

(a) $S_Y(f) > 0$ for all f
(b) $S_Y(f) = 0$ for $|f| > 1$ kHz
(c) $S_Y(f) = 0$ for $f = nf_0$, $f_0 = 2$ kHz, n any integer
(d) $S_Y(f) = 0$ for $f = (2n + 1)f_0$, $f_0 = 1$ kHz, n any integer

[2010 : 2 M]

Statement for Linked Answer Questions 2.37 and 2.38:

Consider a baseband binary PAM receiver shown below. The additive channel noise $n(t)$ is white with power spectral density $S_N(f) = N_0/2 = 10^{-20}$ W/Hz. The low-pass filter is ideal with unity gain and cut-off frequency 1 MHz. Let Y_k represent the random variable $y(t_k)$.

$Y_k = N_k$ if transmitted bit $b_k = 0$

$Y_k = a + N_k$ if transmitted bit $b_k = 1$

where N_k represents the noise sample value. The noise sample has a probability density function, $P_{N_k}(n) = 0.5\alpha e^{-\alpha|n|}$ (This has mean zero and variance $2/\alpha^2$). Assume transmitted bits to be equiprobable and threshold z is set to $a/2 = 10^{-6}$ V.

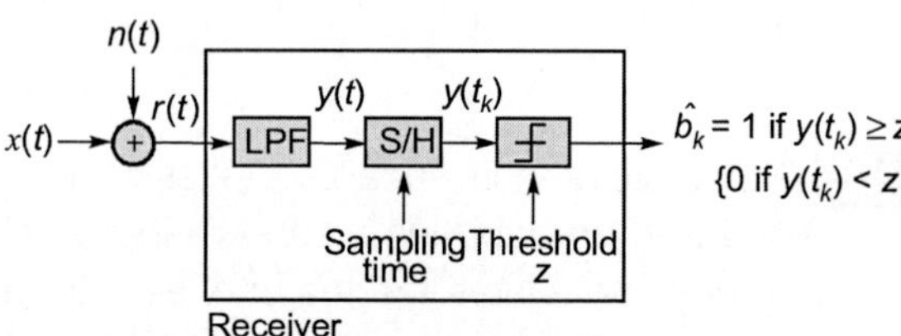

2.37 The value of the parameter α (in V^{-1}) is

(a) 10^{10}
(b) 10^{7}
(c) 1.414×10^{-10}
(d) 2×10^{-20}

[2010 : 2 M]

2.38 The probability of bit error is

(a) $0.5 \times e^{-3.5}$
(b) $0.5 \times e^{-5}$
(c) $0.5 \times e^{-7}$
(d) $0.5 \times e^{-10}$

[2010 : 2 M]

2.39 $X(t)$ is a stationary random process with autocorrelation function $R_X(\tau) = \exp(-\pi\tau^2)$. This process is passed through the system below. The power spectral density of the output process $Y(t)$ is

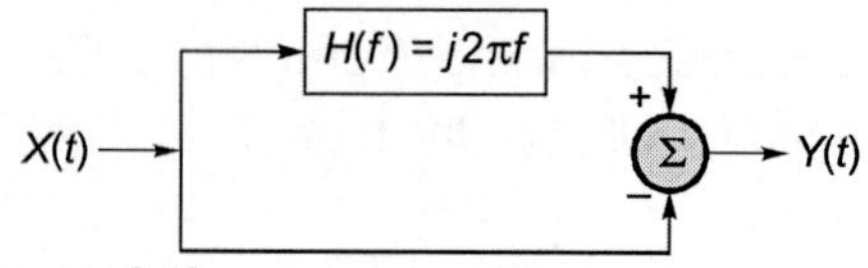

(a) $(4\pi^2 f^2 + 1)\exp(-\pi f^2)$
(b) $(4\pi^2 f^2 - 1)\exp(-\pi f^2)$
(c) $(4\pi^2 f^2 + 1)\exp(-\pi f)$
(d) $(4\pi^2 f^2 - 1)\exp(-\pi f)$

[2011 : 2 M]

2.40 Two independent random variables X and Y are uniformly distributed in the interval $[-1, 1]$. The probability that max $[X, Y]$ is less than 1/2 is

(a) 3/4
(b) 9/16
(c) 1/4
(d) 2/3

[2012 : 1 M]

2.41 A power spectral density of a real process $X(t)$ for positive frequencies is shown below. The values of $E[X^2(t)]$ and $|E[X(t)]|$, respectively are

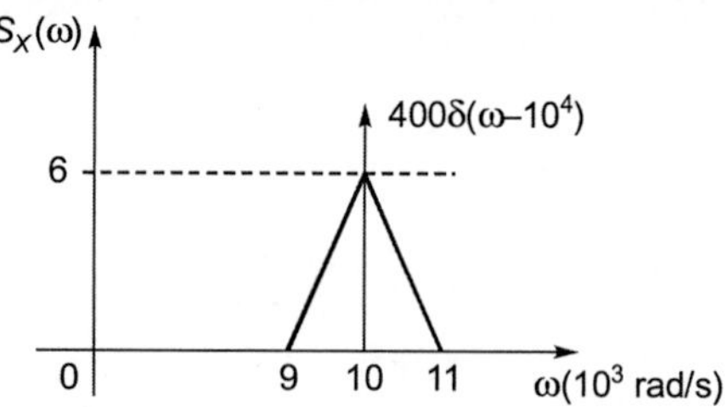

(a) $6000/\pi$, 0
(b) $6400/\pi$, 0
(c) $6400/\pi$, $20/(\pi/\sqrt{2})$
(d) $6000/\pi$, $20/(\pi\sqrt{2})$

[2012 : 1 M]

2.42 Let U and V be two independent zero mean Gaussian random variables of variances $\frac{1}{4}$ and $\frac{1}{9}$ respectively. The probability $P(3V \geq 2U)$ is

(a) 4/9
(b) 1/2
(c) 2/3
(d) 5/9

[2013 : 2 M]

2.43 Consider two identically zero-mean random variables U and V. Let the cumulative distribution functions of U and $2V$ be $F(x)$ and $G(x)$ respectively. Then, for all values of x

(a) $F(x) - G(x) \le 0$ (b) $F(x) - G(x) \ge 0$
(c) $(F(x) - G(x)) \cdot x \le 0$ (d) $(F(x) - G(x)) \cdot x \ge 0$

[2013 : 2 M]

2.44 Let, x_1, x_2, and x_3 be independent and identically distributed random variables with the uniform distribution on [0, 1]. The probability $P\{x_1$ is the largest$\}$ is _____.

[2014 : 1 M, Set-1]

2.45 Let X be a real-valued random variable with $E[X]$ and $E[X^2]$ denoting the mean values of X and X^2, respectively. The relation which always holds

(a) $(E[X])^2 > E[X^2]$ (b) $E[X^2] \ge (E[X])^2$

(c) $E[X^2] = (E[X])^2$ (d) $E[X^2] > (E[X])^2$

[2014 : 2 M, Set-1]

2.46 Consider a random process $X(t) = \sqrt{2}\sin(2\pi t + \phi)$, where the random phase ϕ is uniformly distributed in the interval $[0, 2\pi]$. The autocorrelation $E[X(t_1)\, X(t_2)]$ is

(a) $\cos(2\pi(t_1 + t_2))$ (b) $\sin(2\pi(t_1 - t_2))$

(c) $\sin(2\pi(t_1 + t_2))$ (d) $\cos(2\pi(t_1 - t_2))$

[2014 : 2 M, Set-1]

2.47 The input to a 1-bit quantizer is a random variable X with pdf $f_X(x) = 2e^{-2x}$ for $x \ge 0$ and $f_X(x) = 0$ for $x < 0$. For outputs to be of equal probability, the quantizer threshold should be _____.

[2014 : 2 M, Set-2]

2.48 The power spectral density of a real stationary random process $X(t)$ is given by

$$S_x(f) = \begin{cases} \dfrac{1}{W}, & |f| \le W \\ 0 & |f| > W \end{cases}$$

The value of the expectation

$E\left[\pi x(t) \cdot x\left(t - \dfrac{1}{4W}\right)\right]$ is _____.

[2014 : 2 M, Set-2]

2.49 A real band-limited random process $X(t)$ has two-sided power spectral density

$$S_x(f) = \begin{cases} 10^{-6}(3000 - |f|)\ \text{Watts/Hz} & \text{for } |f| \le 3\,\text{kHz} \\ 0 & \text{otherwise} \end{cases}$$

where f is the frequency expressed in Hz. The signal $x(t)$ modulates a carrier $\cos 16000\pi t$ and the resultant signal is passed through an ideal band-pass filter of unity gain with centre frequency of 8 kHz and band-width of 2 kHz. The output power (in Watts) is _____.

[2014 : 2 M, Set-3]

2.50 Let X_1, X_2 and X_3 be independent and identically distributed random variables with the uniform distribution on [0, 1]. The probability $P\{X_1 + X_2 \le X_3\}$ is _____.

[2014 : 2 M, Set-3]

2.51 A binary random variable X takes the value of 1 with probability 1/3. X is input to a cascade of 2 independent identical binary symmetric channels (BSCs) each with crossover probability 1/2. The output of BSCs are the random variables Y_1 and Y_2 as shown in the figure.

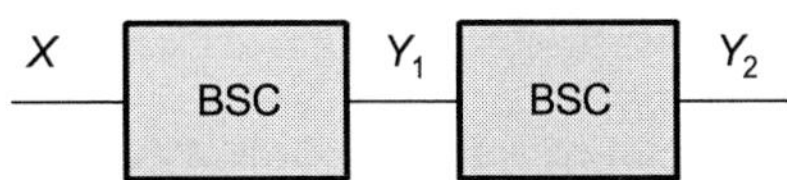

The value of $H(Y_1) + H(Y_2)$ in bits is _____.

[2014 : 2 M, Set-3]

2.52 Let $X(t)$ be a wide sense stationary (WSS) random with power spectral density $S_x(f)$. If $Y(t)$ is the process defined as $Y(t) = X(2t - 1)$, the power spectral density $S_Y(f)$ is

(a) $S_Y(f) = \dfrac{1}{2} S_X\left(\dfrac{f}{2}\right) e^{-j\pi f}$

(b) $S_Y(f) = \dfrac{1}{2} S_X\left(\dfrac{f}{2}\right) e^{-j\pi f/2}$

(c) $S_Y(f) = \dfrac{1}{2} S_X\left(\dfrac{f}{2}\right)$

(d) $S_Y(f) = \dfrac{1}{2} S_X\left(\dfrac{f}{2}\right) e^{j2\pi f}$

[2014 : 2 M, Set-3]

2.53 Let, X be a zero mean unit variance Gaussian random variable. $E[|X|]$ is equal to _____.

[2014 : 1 M, Set-4]

2.54 Consider a communication scheme where the binary values signal X satisfies $P\{X = 1\} = 0.75$ and $P\{X = -1\} = 0.25$. The received signal $Y = X + Z$, where Z is a Gaussian random variable with zero mean and variance σ^2. The received signal Y is fed to the threshold detector. The output of the threshold detector $\hat{X}$ is

$$\hat{X} = \begin{cases} +1, & Y > \tau \\ -1 & Y \le \tau \end{cases}$$

To achieve a minimum probability of error $P\{\hat{X} \neq X\}$, the threshold τ should be

(a) strictly positive
(b) zero
(c) strictly negative
(d) strictly positive, zero, or strictly negative depending on the nonzero value of σ^2

[2014 : 2 M, Set-4]

2.55 Consider the Z-channel given in the figure. The input is 0 or 1 with equal probability.

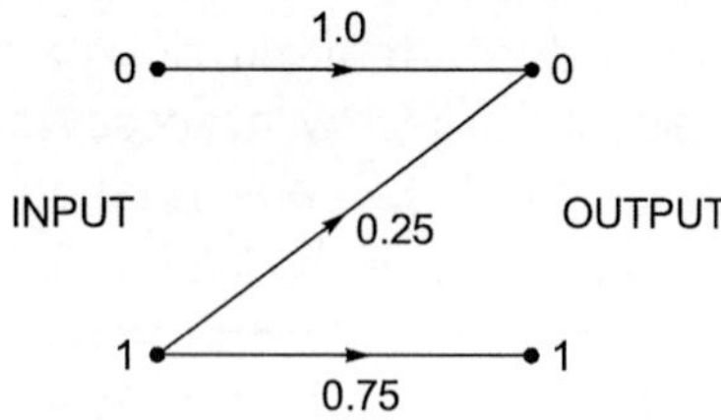

If the output is 0, the probability that the input is also 0 equals ____________.

[2014 : 2 M, Set-4]

2.56 Consider a discrete-time channel $Y = X + Z$, where the additive noise Z is signal-dependent. In particular, given the transmitted symbol $X \in \{-a, +a\}$ at any instant, the noise sample Z is chosen independently from a Gaussian distribution with mean βX and unit variance. Assume a threshold detector with zero threshold at the receiver.

When, $\beta = 0$, the BER was found to be $Q(a) = 1 \times 10^{-8}$

$$\left(Q(v) = \frac{1}{\sqrt{2\pi}} \int_v^\infty e^{-u^2/2} du, \text{ and for } v > 1, \text{ use } Q(v) \approx e^{-v^2/2} \right)$$

When, $\beta = -0.3$, the BER is closest to

(a) 10^{-7}
(b) 10^{-6}
(c) 10^{-4}
(d) 10^{-2}

[2014 : 2 M, Set-4]

2.57 $\{X_n\}_{n=-\infty}^{n=\infty}$ is an independent and identically distributed (i.i.d.) random process X_n equally likely to be + 1 or −1. $\{Y_n\}_{n=-\infty}^{n=\infty}$ is another random process obtained as $Y_n = X_n + 0.5 X_{n-1}$. The autocorrelation function of $\{Y_n\}_{n=-\infty}^{n=\infty}$ denoted by $R_y[k]$, is

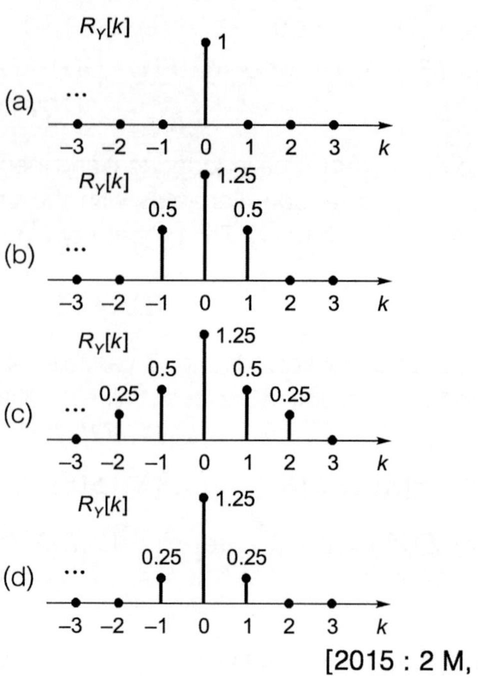

[2015 : 2 M, Set-2]

2.58 A zero mean white Gaussian noise having power spectral density $\frac{N_0}{2}$ is passed through an LTI filter whose impulse response $h(t)$ is shown in the figure. The variance of the filtered noise at $t = 4$ is

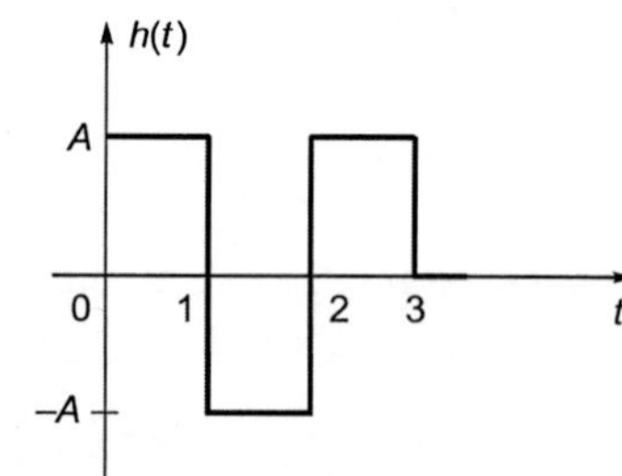

(a) $\frac{3}{2}A^2N_0$
(b) $\frac{3}{4}A^2N_0$
(c) A^2N_0
(d) $\frac{1}{2}A^2N_0$

[2015 : 2 M, Set-2]

2.59 A random binary wave $y(t)$ is given by

$$y(t) = \sum_{n=-\infty}^{\infty} X_n p(t - nT - \phi),$$

where $p(t) = u(t) - u(t - T)$, $u(t)$ is the unit step function and ϕ is an independent random variable with uniform distribution in $[0, T]$. The sequence consists of independent and identically distributed binary valued random variables with $P\{X_n = +1\} = P\{X_n = -1\} = 0.5$ for each n.

The value of the autocorrelation

$R_{yy}\left(\frac{3T}{4}\right) \triangleq E\left[y(t)y\left(t-\frac{3T}{4}\right)\right]$ equals ____.

[2015 : 2 M, Set-3]

2.60 The variance of the random variable X with probability density function $f(x)=\frac{1}{2}|x|e^{-|x|}$ is____.

[2015 : 2 M, Set-3]

2.61 An antenna pointing in a certain direction has a noise temperature of 50 K. The ambient temperature is 290 K. The antenna is connected to a pre-amplifier that has a noise figure of 2 dB and an available gain of 40 dB over an effective bandwidth of 12 MHz. The effective input noise temperature T_e for the amplifier and the noise power P_{a0} at the output of the preamplifier, respectively, are

(a) T_e = 169.36 K and $P_{a0} = 3.73 \times 10^{-10}$ W
(b) T_e = 170.8 K and $P_{a0} = 4.56 \times 10^{-10}$ W
(c) T_e = 182.5 K and $P_{a0} = 3.85 \times 10^{-10}$ W
(d) T_c = 160.62 K and $P_{a0} = 4.6 \times 10^{-10}$ W

[2016 : 2 M, Set-1]

2.62 An information source generates a binary sequence $\{\alpha_n\}$. α_n can take one of the two possible values −1 and +1 with equal probability and are statistically independent and identically distributed. This sequence is precoded to obtain another sequence $\{\beta_n\}$, as $\beta_n = \alpha_n + k\alpha_{n-3}$. The sequence $\{\beta_n\}$ is used to modulate a pulse $g(t)$ to generate the baseband signal

$$X(t)=\sum_{n=-\infty}^{\infty}\beta_n g(t-nT),$$

where, $g(t)=\begin{cases}1, & 0\le t\le T\\ 0, & \text{otherwise}\end{cases}$

If there is a null at $f=\frac{1}{3T}$ in the power spectral density of $X(t)$, then k is ________

[2016 : 2 M, Set-2]

2.63 Consider a random process $X(t) = 3V(t) - 8$, where $V(t)$ is a zero mean stationary random process with autocorrelation $R_V(\tau) = 4e^{-5|\tau|}$. The power in $X(t)$ is ________.

[2016 : 2 M, Set-2]

2.64 A wide sense stationary random process $X(t)$ passes through the LTI system shown in the figure. If the autocorrelation function of $X(t)$ is $R_X(\tau)$, then the autocorrelation function $R_Y(\tau)$ of the output $Y(t)$ is equal to

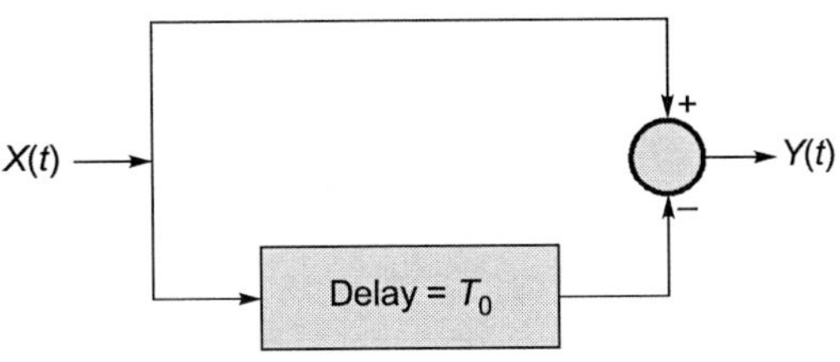

(a) $2R_X(\tau)+R_X(\tau-T_0)+R_X(\tau+T_0)$
(b) $2R_X(\tau)-R_X(\tau-T_0)-R_X(\tau+T_0)$
(c) $2R_X(\tau)+2R_X(\tau-2T_0)$
(d) $2R_X(\tau)-2R_X(\tau-2T_0)$

[2016 : 2 M, Set-3]

2.65 Let $X(t)$ be a wide sense stationary random process with the power spectral density $S_X(f)$ as shown in figure (a), where f is in Hertz (Hz). The random process $X(t)$ is input to an ideal lowpass filter with the frequency response

$$H(f)=\begin{cases}1, & |f|\le\frac{1}{2}\text{ Hz}\\ 0, & |f|>\frac{1}{2}\text{ Hz}\end{cases}$$

as shown in figure (b). The output of the lowpass filter is $Y(t)$.

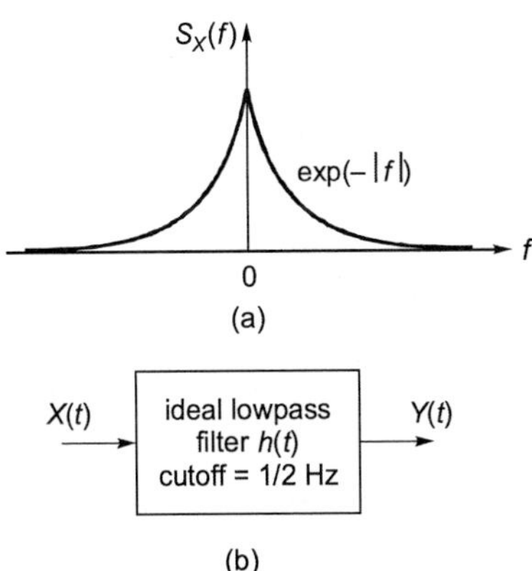

Let E be the expectation operator and consider the following statements:

I. $E(X(t)) = E(Y(t))$
II. $E(X^2(t)) = E(Y^2(t))$
III. $E(Y^2(t)) = 2$

Select the correct option:

(a) only I is true
(b) only II and III are true
(c) only I and II are true
(d) only I and III are true

[2017 : 2 M, Set-1]

2.66 Consider the random process

$$X(t) = U + Vt,$$

where, U is a zero-mean Gaussian random variable and V is a random variable uniformly distributed between 0 and 2. Assume that U and V are statistically independent. The mean value of the random process at $t = 2$ is _____.

[2017 : 1 M, Set-2]

2.67 Consider a white Gaussian noise process $N(t)$ with two-sided power spectral density $S_N(f) = 0.5$ W/Hz as input to a filter with impulse response $0.5e^{-t^2/2}$ (where t is in seconds) resulting in output $Y(t)$. The power in $Y(t)$ in watts is

(a) 0.11 (b) 0.22
(c) 0.33 (d) 0.44

[2018 : 2 M]

2.68 Let a random process $Y(t)$ be described as $Y(t) = h(t) * X(t) + Z(t)$, where $X(t)$ is a white noise process with power spectral density $S_X(f) = 5$ W/Hz. The filter $h(t)$ has a magnitude response given by $|H(f)| = 0.5$ for $-5 \le f \le 5$, and zero elsewhere. $Z(t)$ is a stationary random process, uncorrelated with $X(t)$, with power spectral density as shown in the figure. The power in $Y(t)$, in watts, is equal to W _____ (rounded off to two decimal places).

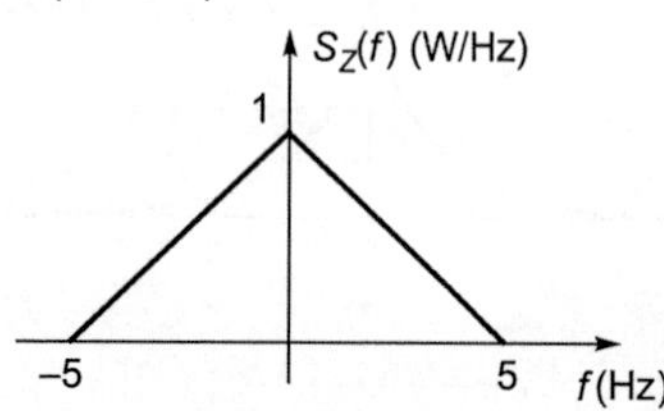

[2019 : 2 M]

2.69 The random variable

$$Y = \int_{-\infty}^{\infty} W(t)\phi(t)dt,$$

where $\phi(t) = \begin{cases} 1; & 5 \le t \le 7 \\ 0; & \text{otherwise} \end{cases}$

and $W(t)$ is a real white Gaussian noise process with two-sided power spectral density $S_W(f) = 3$ W/Hz, for all f. The variance of Y is ________.

[2020 : 1 M]

2.70 The autocorrelation function $R_X(\tau)$ of a wide-sense stationary random process $X(t)$ is shown in the figure.

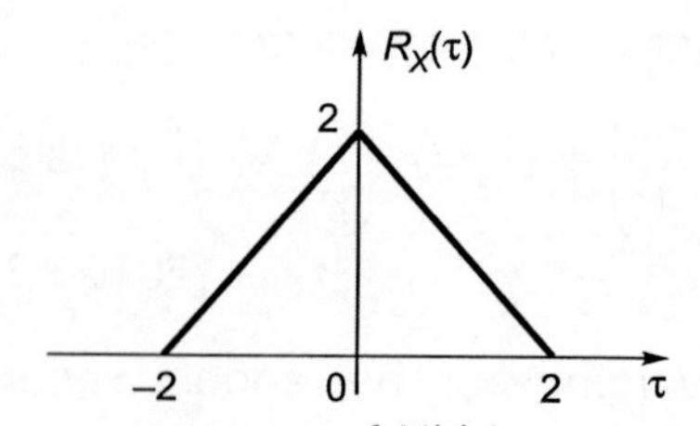

The average power of $X(t)$ is ________.

[2021 : 1 M]

2.71 Two continuous random variables X and Y are related as

$$Y = 2X + 3$$

Let σ_X^2 and σ_Y^2 denote the variance of X and Y, respectively. The variances are related as

(a) $\sigma_Y^2 = 25\sigma_X^2$ (b) $\sigma_Y^2 = 5\sigma_X^2$
(c) $\sigma_Y^2 = 2\sigma_X^2$ (d) $\sigma_Y^2 = 4\sigma_X^2$

[2021 : 1 M]

2.72 Consider a real valued source whose samples are independent and identically distributed random variables with the probability density function, $f(x)$, as shown in the figure.

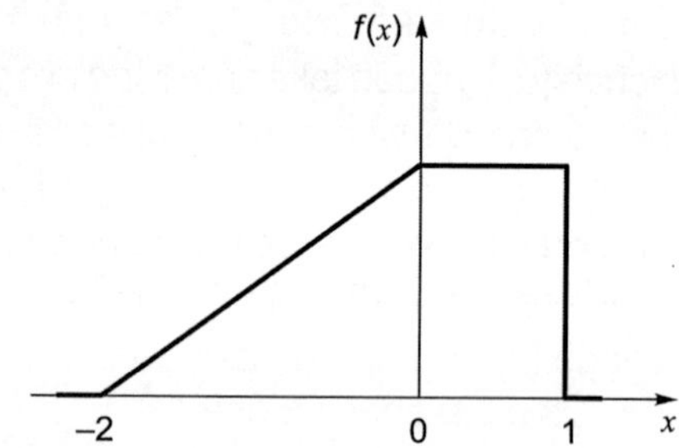

Consider a 1 bit quantizer that maps positive samples to value α and others to value β. If α^* and β^* are the respective choices for α and β that minimize the mean square quantization error, then $(\alpha^* - \beta^*) =$ ________ (rounded off to two decimal places).

[2022 : 2 M]

2.73 The frequency response $H(f)$ of a linear time-invariant system has magnitude as shown in the figure.

Statement I: The system is necessarily a pure delay system for inputs which are bandlimited to $-\alpha \le f \le \alpha$.

Statement II: For any wide-sense stationary input process with power spectral density $S_X(f)$, the output power spectral density $S_Y(f)$ obeys $S_Y(f) = S_X(f)$ for $-\alpha \le f \le \alpha$.

Which one of the following combinations is true?

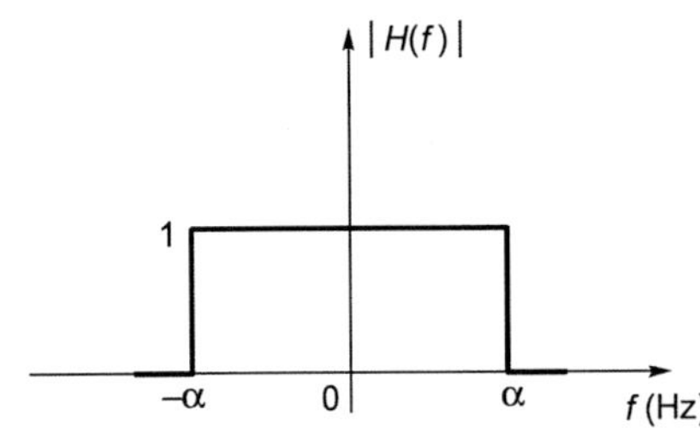

(a) Statement I is correct, Statement II is correct
(b) Statement I is correct, Statement II is incorrect
(c) Statement I is incorrect, Statement II is correct
(d) Statement I is incorrect, Statement II is incorrect

[2022 : 1 M]

2.74 For a real signal, which of the following is/are valid power spectral density/densities?

(a) $S_X(\omega) = \dfrac{2}{9+\omega^2}$

(b) $S_X(\omega) = e^{-\omega^2}\cos^2\omega$

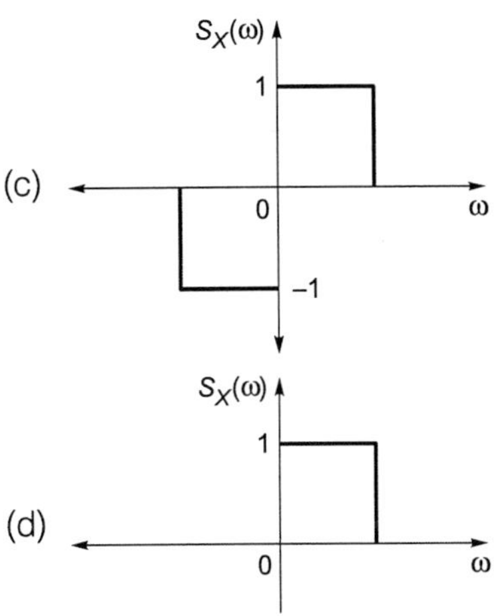

[2023 : 1 M]

2.75 A random variable X, distributed normally as $N(0,1)$, undergoes the transformation $Y = h(X)$, given in the figure. The form of the probability density function of Y is (In the options given below, a, b, c are non-zero constants and $g(y)$ is piece-wise continuous function)

(a) $a\delta(y-1) + b\delta(y+1) + g(y)$
(b) $a\delta(y+1) + b\delta(y) + c\delta(y-1) + g(y)$
(c) $a\delta(y+2) + b\delta(y) + c\delta(y-2) + g(y)$
(d) $a\delta(y+1) + b\delta(y-2) + g(y)$

[2023 : 2 M]

2.76 Let $X(t)$ be a white Gaussian noise with power spectral density $\frac{1}{2}$ W/Hz. If $X(t)$ is input to an LTI system with impulse response $e^{-t}u(t)$. The average power of the system output is _______ W. (Rounded off to two decimal place).

[2023 : 2 M]

2.77 Suppose X and Y are independent and identically distributed random variables that are distributed uniformly in the interval [0, 1]. The probability that $X \geq Y$ is ____.

[2024 : 1 Mark]

2.78 A White Gaussian noise $w(t)$ with zero mean and power spectral density $\frac{N_o}{2}$, when applied to a first-order RC low pass filter produces an output $n(t)$. At a particular time $t = t_k$, the variance of the random variable $n(t_k)$ is _______.

(a) $\dfrac{N_o}{4RC}$
(b) $\dfrac{2N_o}{RC}$
(c) $\dfrac{N_o}{RC}$
(d) $\dfrac{N_o}{2RC}$

[2024 : 1 Mark]

2.79 A source transmits a symbol s, taken from {−4, 0, 4} with equal probability, over an additive white Gaussian noise channel. The received noisy symbol r is given by $r = s + w$, where the noise w is zero mean with variance 4 and is independent of s. Using $Q(x) = \frac{1}{\sqrt{2\pi}}\int_x^{\infty} e^{\frac{-t^2}{2}}\,dt$, the optimum symbol error probability is _______.

(a) $\frac{4}{3}Q(1)$
(b) $\frac{2}{3}Q(1)$
(c) $\frac{4}{3}Q(2)$
(d) $\frac{2}{3}Q(2)$

[2024 : 2 Marks]

■■■

Answers Random Signals and Noise

2.1 (d)	2.2 (d)	2.3 (d)	2.4 (d)	2.5 (d)	2.6 (a)	2.7 (c)
2.8 (c)	2.9 (c)	2.10 (a)	2.11 (a)	2.12 (c)	2.13 (d)	2.14 (d)
2.15 (c)	2.16 (d)	2.17 (a)	2.18 (b)	2.19 (b)	2.20 (c)	2.21 (b)
2.22 (a)	2.23 (a)	2.24 (a)	2.25 (a)	2.26 (b)	2.27 (a)	2.28 (c)
2.29 (b)	2.30 (a)	2.31 (a)	2.32 (c)	2.33 (b)	2.34 (b)	2.35 (b)
2.36 (d)	2.37 (b)	2.38 (d)	2.39 (a)	2.40 (b)	2.41 (a)	2.42 (b)
2.43 (d)	2.44 (0.33)	2.45 (b)	2.46 (d)	2.47 (0.346)	2.48 (4)	2.49 (2.5)
2.50 (0.16)	2.51 (2)	2.52 (c)	2.53 (0.8)	2.54 (c)	2.55 (0.8)	2.56 (c)
2.57 (b)	2.58 (a)	2.59 (0.25)	2.60 (6)	2.61 (a)	2.62 (–1)	2.63 (100)
2.64 (b)	2.65 (a)	2.66 (2)	2.67 (b)	2.68 (17.5)	2.69 (6)	2.70 (2)
2.71 (d)	2.72 (1.16)	2.73 (a)	2.74 (a, b)	2.75 (b)	2.76 (0.25)	2.77 (0.5)
2.78 (a)	2.79 (a)					

Explanations Random Signals and Noise

2.1 (d)

For a narrow band noise with Gaussian quadrature components the probability density function (PDF) of its envelope will be Rayleigh.

2.2 (d)

Auto-correlation function of energy signal has conjugate symmetry

$$\boxed{R_x(\tau) = R_x^*(-\tau)}$$

If the function is real, then the autocorrelation function has even symmetry

$$\boxed{R_x(\tau) = R_x(-\tau)}$$

2.3 (d)

Autocorrelation function and power spectral density makes the Fourier transfer pair.

$$\boxed{R_X(\tau) \xleftrightarrow{\text{F.T.}} G_X(f)}$$

$$R_X(\tau) = F^{-1}\left[\left(\frac{\sin f}{f}\right)^2\right] = F^{-1}\left[\text{sinc}^2\left(\frac{f}{\pi}\right)\right]$$

Inverse Fourier transform of square of sinc function is always a triangular signal in time domain.

2.4 (d)

$$F_1 = F_2 = 10^{0.3} \simeq 2$$
$$G_1 = G_2 = 10^{0.6} = 4$$
$$F_{\text{overall}} = F_1 + \frac{F_2 - 1}{G_1} = 2.25$$
$$= 10\log 2.25 = 3.5 \text{ dB}$$

2.5 (d)

$$R_x(\tau) = \int_{-\infty}^{\infty} x(t+\tau)\, x(t)dt = x(t) \otimes x(-t)\Big|_{t=\tau}$$

The autocorrelation function (ACF) of a rectangular pulse of duration T is a triangular pulse of duration $2T$.

2.6 (a)

Gaussian probability density of random variable 'x' is given by

$$\rho_X(x) = \frac{1}{\sqrt{2\pi}\,\sigma}\exp\left[\frac{-(x-\mu)^2}{2\sigma^2}\right]$$

when $\sigma = 1$ and $\mu = 0$ (zero mean)

$$\rho_X(x) = \frac{1}{\sqrt{2\pi}}\exp\left[-\frac{x^2}{2}\right] \quad \text{...(i)}$$

Given that

$$\rho_X(x) = k \cdot \exp\left[-\frac{x^2}{2}\right] \quad \text{...(ii)}$$

By comparing equation (i) and equation (ii), we have

$$K = \frac{1}{\sqrt{2\pi}}$$

2.7 (c)

The Fourier transform of a Gaussian signal in time domain is also Gaussian signal in the frequency domain.

$$\boxed{e^{-\pi t^2} \xleftrightarrow{\text{F.T.}} e^{-\pi f^2}}$$

2.8 (c)

The probability density function (PDF) of the envelope of narrow band Gaussian noise is Rayleigh.

2.9 (c)

Probability of a Gaussian random variable is defined for an interval and not at a point. So at $X = 4$, it is zero.

2.10 (a)

$$S_X(\omega) \to \boxed{H(\omega)} \to S_Y(\omega)$$

$$S_Y(\omega) = |H(\omega)|^2 S_X(\omega)$$

2.11 (a)

$$E\left[d^2(n)\right] = E\left[\{x(n) - x(n-1)\}^2\right]$$

$$\sigma_d^2 = E[x(n)]^2 + E[x(n-1)]^2 - 2E[x(n)x(n-1)]$$

$$\Rightarrow \frac{\sigma_X^2}{10} = \sigma_X^2 + \sigma_X^2 - 2R_{XX}(1) \to$$

Auto correlation at, $k = 1$

$$2R_{XX}(1) = \frac{19}{10}\sigma_X^2$$

$$\therefore \quad \frac{R_{XX}(1)}{\sigma_X^2} = \frac{19}{20} = 0.95$$

2.12 (c)

Frequency detector can be modelled as a differentiator.

$$S_X(f) = \frac{N_0}{2} \longrightarrow \boxed{H(f) = j2\pi f} \longrightarrow S_Y(f) = S_X(f)|H(f)|^2$$

$$S_Y(f) = \frac{N_0}{2}4\pi^2 f^2 \Rightarrow \text{Parabolic}$$

2.13 (d)

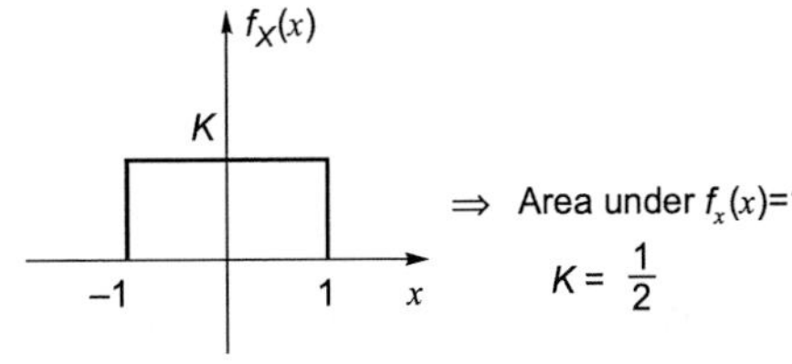

$\Rightarrow$ Area under $f_x(x)=1$

$K = \frac{1}{2}$

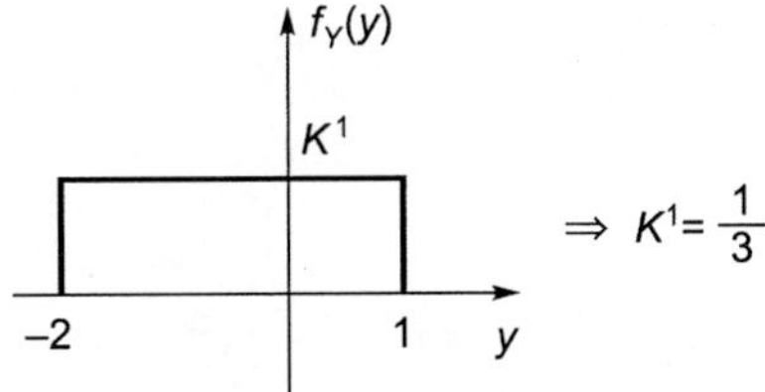

$\Rightarrow K^1 = \frac{1}{3}$

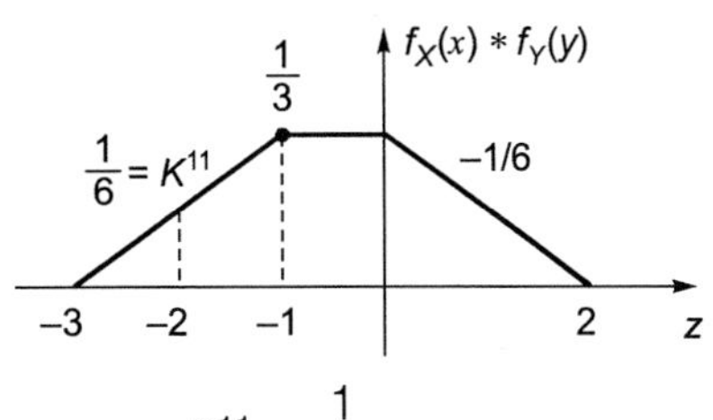

$$\Rightarrow \quad K^{11} = \frac{1}{6}$$

Since it lies in middle of $\frac{1}{3}$.

$\therefore$ Probability $[z \le -2]$

$$= \text{Area of graph}[z \le -2]$$

$$= \frac{1}{2} \times \frac{1}{6} \times 1 = \frac{1}{12}$$

2.14 (d)

$$R_{XX}(\tau) = 4\left[e^{-0.2|\tau|} + 1\right]$$

$$P(X \le 1) = F_X(1)$$

$$= 1 - Q\left(\frac{X-\mu}{\sigma}\right) \text{ at } X = 1$$

$$\mu = 0, \text{ given}$$

$$R_X(0) = \sigma^2$$

$$\sigma^2 = 8$$

$$\Rightarrow \quad \sigma = 2\sqrt{2}$$

$$\therefore \quad P(X \le 1) = 1 - Q\left(\frac{1}{2\sqrt{2}}\right)$$

2.15 (c)

$$W = Y - Z$$

$$\sigma_W^2 = E[Y^2] + E[Z^2] - 2E[Y \cdot Z]$$

Y is sampled at $t = 2$ and Z at $t = 4$

$$\therefore \quad \sigma_W^2 = \sigma_Y^2 + \sigma_Z^2 - 2R_{XX}(2)$$

$$= 8 + 8 - 2[4e^{-0.2121} + 1]$$

$$\Rightarrow \quad \sigma_W^2 = 2.64$$

2.16 (d)

$$P(x = 1) = F_X(x = 1^+) - F_X(x = 1^-)$$

$$= 0.55 - 0.25 = 0.30$$

2.17 (a)

$$\text{SNR} = \frac{P_s}{N_o B} = \frac{10^{-3}}{10^{-20} \times 100 \times 10^6} = 10^9$$

$$(\text{SNR})_{\text{dB}} = 10\log 10^9 = 90 \text{ dB}$$

$$\text{Cable loss} = 40\text{dB}$$

$$\therefore \quad \text{SNR at } R_X = (90 - 40) \text{ dB} = 50 \text{ dB}$$

2.18 (b)

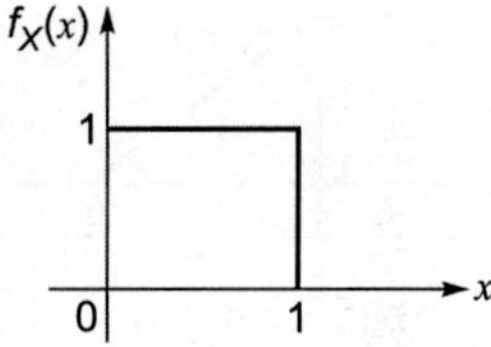

$$Q_e = \text{s.v.} - \text{q.v} = x - x_q$$

Mean square value of quantization noise

$$= E\left[(x - x_q)^2\right] = \int_0^1 (x - x_q)^2 f_X(x)dx$$

$$= \int_0^1 (x - x_q)^2 \cdot 1dx = \int_0^{0.3} (x - 0)^2 dx + \int_{0.3}^1 (x - 0.7)^2 dx$$

$$= 0.039$$

$\therefore$ Root mean square value of the quantization noise

$$= \sqrt{0.039} = 0.198$$

2.19 (b)

$$\text{PSD}_{\text{out}} = |H(\omega)|^2 \text{PSD}_{\text{in}} = 4N_o$$

$$P_N = \text{B.W.} \times SN_{\text{out}} = 4N_oB$$

2.20 (c)

$$E[X^2] = \int_0^4 x^2 f_X(x)dx$$

$$\left[\begin{array}{l} f_X(x) = y = mx + c = \dfrac{x}{8} \\ \dfrac{K \times 4}{2} = \text{Area} = 1 \\ K = \dfrac{1}{2} \end{array}\right]$$

$$\therefore \quad E[X^2] = \int_0^4 x^2\left(\frac{x}{8}\right)dx = \left.\frac{x^4}{4 \times 8}\right|_0^4$$

$$E[X^2] = 8$$

2.21 (b)

Prob. of occurrence of each Region is

$$\int_{-a}^{a} f_X(x)dx = \frac{1}{3}$$

$$\Rightarrow \quad \frac{1}{4}\int_{-a}^{a} xdx = \frac{1}{3}$$

$$\frac{2a}{4} = \frac{1}{3}$$

$$a = \frac{2}{3}$$

2.22 (a)

$$E[X^2] = \int_{-a}^{a} x^2 f_X(x)dx = 2\int_0^a x^2 \frac{1}{4}dx$$

$$E[X^2] = 2\int_0^{2/3} \frac{1}{4}x^2 dx = \frac{1}{2}\left(\frac{x^3}{3}\right)_0^{2/3}$$

$$E[X^2] = \frac{1}{6} \times \frac{8}{27} = \frac{4}{81}$$

2.23 (a)

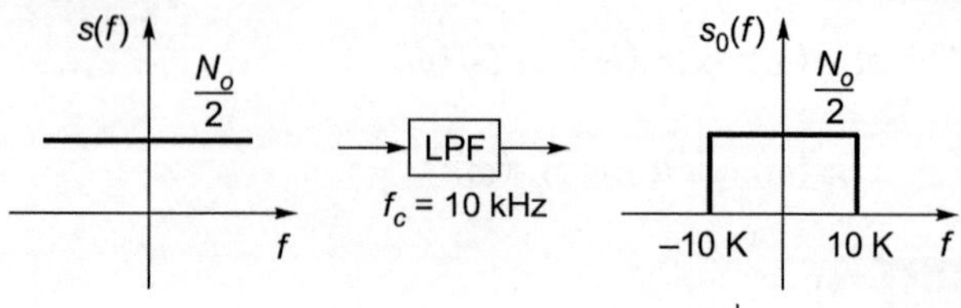

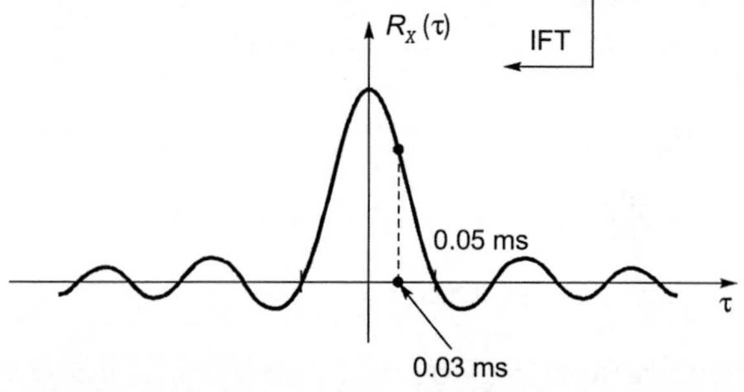

$R_X(\tau) \neq 0$ at $\tau = 0.03$ msec. So correlated.

2.24 (a)

$$S(\omega) = \frac{16}{16 + \omega^2} \quad \text{.... (i)}$$

$$H(s) = \frac{4}{4 + s} \quad \text{.... (ii)}$$

which is a LPF

$\therefore$ It is a low pass *RL* filter.

2.25 (a)

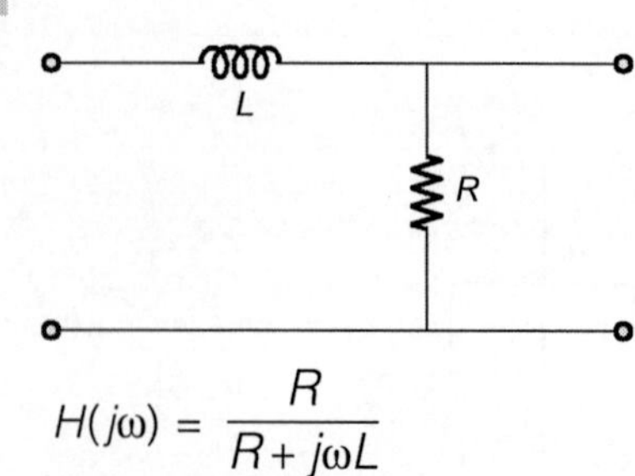

$$H(j\omega) = \frac{R}{R + j\omega L}$$

Comparing with (i) and (ii)

$$R = 4,\ L = 1$$

2.26 (b)

$$f_X(x) = \frac{1}{10}[u(x+5) - u(x-5)]$$

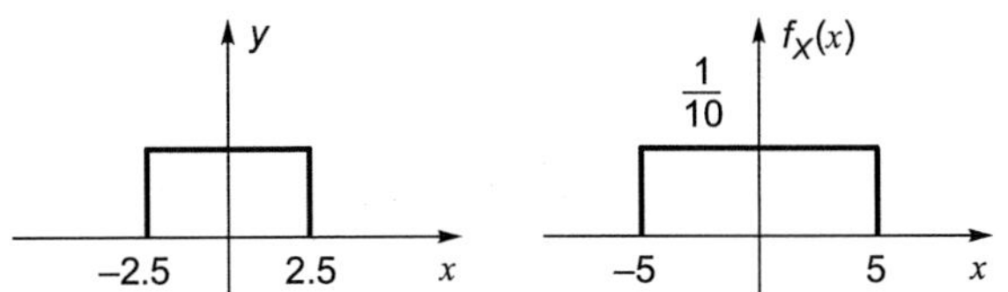

Y is a discrete random variable takes the values of 0 and 1.

$$Y = 1 \ ; \ 2.5 \le x \le 2.5$$

$$Y = 0 \ ; \ |x| \ge 2.5$$

$$P(Y=0) = P(-2.5 \le x \le 2.5) = \frac{1}{2}$$

$$P(Y=1) = P\left(|x| \ge 2.5\right) = \frac{1}{2}$$

$$f_y(y) = \frac{1}{2}\delta(y) + \frac{1}{2}\delta(y-1)$$

2.27 (a)

$$\boxed{\sigma_x^2 = E[x^2] - E^2[x]}$$

$$\boxed{\text{A.C. Power} = \text{Total power} - \text{DC power}}$$

2.28 (c)

Auto-correlation function is an even function.

2.29 (b)

PSD is always a positive quantity.

2.30 (a)

PDF, $f_X(x) = (t+1)u(t+1) - 2tu(t) + (t-1)u(t-1)$

CDF, $F_X(x) = \int_{-\infty}^{x} f_X(x)\,dx$

Integral of increasing ramp signal is increasing parabola and integral of decreasing ramp signal is decreasing parabola.

2.31 (a)

PDF, $P_X(x) = M\exp(-2|x|) + N\exp(-3|x|)$

$$\int_{-\infty}^{\infty} P_X(x)\,dx = 1$$

or $$\int_{-\infty}^{\infty} \left\{Me^{-2|x|} + Ne^{-3|x|}\right\}dx = 1$$

or $$\int_{0}^{\infty} \left(Me^{-2x} + Ne^{-3x}\right)dx = \frac{1}{2}$$

or $$\frac{M}{2} + \frac{N}{3} = \frac{1}{2}$$

or $$M + \frac{2N}{3} = 1$$

2.32 (c)

$$H(f) = \frac{1}{1 + j2\pi fRC} = \frac{1}{1 + j\dfrac{f}{f_c}}$$

$$|H(f)|^2 = \frac{f_c^2}{f^2 + f_c^2}$$

$$\text{Output PSD} = |H(f)|^2 \cdot \text{Input PSD} = \frac{f_c^2}{f^2 + f_c^2} \cdot K$$

$$\text{Output noise power} = \int_{-\infty}^{\infty} (\text{output PSD})df$$

$$= K\int_{-\infty}^{\infty} \frac{f_c^2}{f^2 + f_c^2}df$$

$$= K\pi f_c \ (\text{By substitution } f = f_c \tan\theta)$$

2.33 (b)

PSD of white noise = 1×10^{-10} W/Hz ($\equiv k$)

PSD of output

$$G_0(f) = |H(f)|^2 \cdot G_i(f)$$

$$= k \cdot |H(f)|^2$$

output noise power

$$N_0 = \int_{-f_0}^{+f_0} G_0(f)\,df = k \times (\text{area under } |H(f)|^2 \text{ curve})$$

$$= k \times 2\left(\frac{1}{2}bh\right)$$

$$= k\,f_o \times 1 = 1 \times 10^{-10} \times 10 \times 10^3 = 10^{-6}\ \text{W}$$

2.34 (b)

PSD = $FT[R(\tau)]$

= FT [triangular $R(\tau)$ vs τ]

$\Rightarrow$ sinc^2 function.

2.35 (b)

$$\text{Mean} = \overline{X} = \Sigma x_i \cdot p(x_i)$$

$$= (1 \times 0.1) + (2 \times 0.2) + (3 \times 0.4) + (4 \times 0.2) + (5 \times 0.1) = 3$$

$$\overline{X^2} = \Sigma x_1^2 . P(x_i)$$

$$= (1 \times 0.1) + (4 \times 0.2) + (9 \times 0.4) + (16 \times 0.2) + (25 \times 0.1)$$

$$= 10.2$$

$$\sigma^2 = \overline{X^2} - (\overline{X})^2 = 1.2$$

2.36 (d)

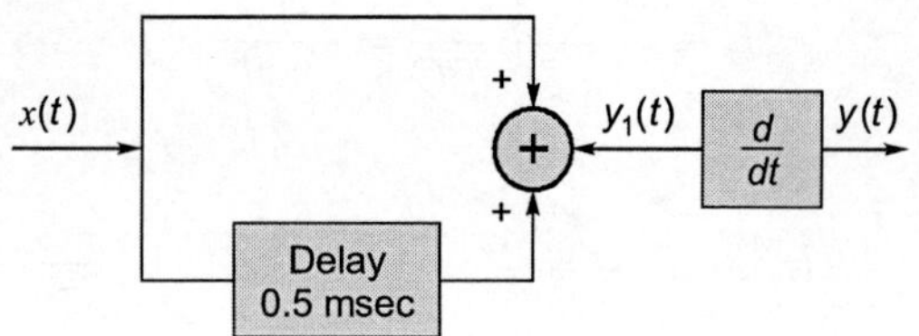

$$y_1(t) = x(t) + x(t - 0.5 \times 10^{-3})$$

$$Y_1(f) = X(f)\left[1 + e^{-j2\pi f(0.5\times 10^{-3})}\right]$$

$$H_1(f) = \frac{Y_1(f)}{X(f)} = 1 + e^{-j\pi f \times 10^{-3}}$$

$$H(f) = H_1(f) \cdot H_2(f)$$

$$= (j2\pi f)\left[1 + e^{-j(\pi f \times 10^{-3})}\right]$$

$$|H(f)|^2 = 4\pi^2 f^2\left[2 + 2\cos(\pi f \times 10^{-3})\right]$$

$$= 8\pi^2 f^2\left[1 + \cos(\pi f \times 10^{-3})\right]$$

$$S_Y(f) = |H(f)|^2 S_X(f)$$

$$= 8\pi^2 f^2\left[1 + \cos(\pi f \times 10^{-3})\right] S_x(f)$$

For $f = (2n+1)f_o$; with $f_o = 1$ kHz

$\pi f \times 10^{-3}$ is an odd multiple of π

$\therefore S_Y(f) = 0$

2.37 (b)

Output noise power = $|H(f)|^2 \cdot$ (input noise PSD)

$$S_{NO}(F) = S_N(f)\,|H(f)|^2$$

$$S_{NO}(f) = 10^{-20}\,|H(f)|^2$$

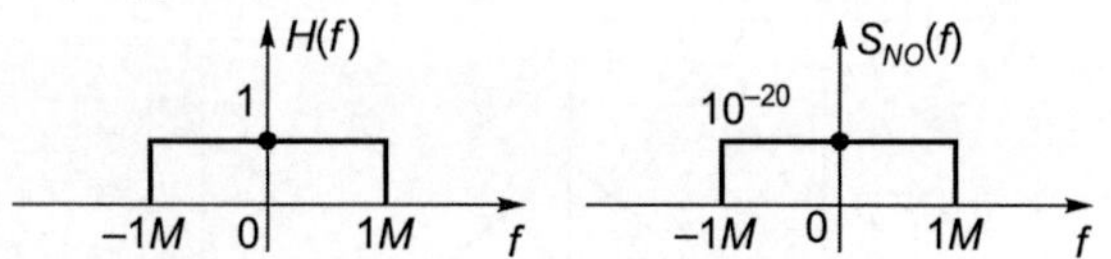

$$\text{Output noise power} = \int_{-\infty}^{\infty} S_{NO}(f)\,df$$

$$= 10^{-20}\frac{\text{W}}{\text{Hz}} \times 2 \times 10^6 \text{ Hz}$$

$$= 2 \times 10^{-14} \text{ W}$$

Since mean square value = power

$$\frac{2}{\alpha^2} = 2 \times 10^{-14} \rightarrow \alpha = 10^7$$

2.38 (d)

When a '1' is transmitted:

$$Y_K = a + N_K$$

$$\text{Threshold } Z = \frac{a}{2} = 10^{-6}$$

$$a = 2 \times 10^{-6}$$

For error to occur, $Y_K < 10^{-6}$

$$2 \times 10^{-6} + N_K < 10^{-6}$$

$$N_K < -10^{-6}$$

$$P(0/1) = \int_{-\infty}^{-10^{-6}} P_{NK}(n)\,dn$$

$$= \int_{-\infty}^{-10^{-6}} 0.5\alpha e^{\alpha n} dn \text{ with } \alpha = 10^7$$

$$P(0/1) = 0.5e^{-10}$$

When a '0' is transmitted:

$$Y_K = N_K$$

For error to occur, $Y_K > 10^{-6}$

$$\therefore\ P(1/0) = \int_{10^{-6}}^{\infty} P_{NK}(n)dn = 0.5 \times e^{-10}$$

Since, both bits are equiprobable;

$$P(0) = P(1) = \frac{1}{2}$$

the probability of bit error

$= P(1) \cdot P(0/1) + P(0) \cdot P(1/0)$

$$= \frac{P(0/1) + P(1/0)}{2} = 0.5 \times e^{-10}$$

2.39 (a)

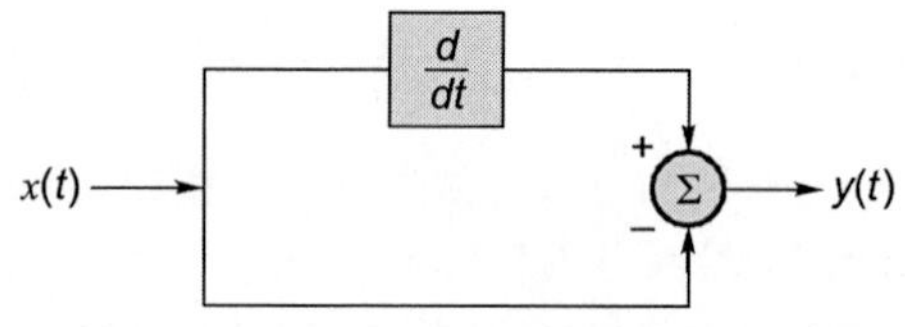

$$y(t) = \frac{d}{dt}x(t) - x(t)$$

FT→ $\quad Y(f) = j2\pi f X(f) - x(f)$

$$Y(f) = [j2\pi f - 1]x(f)$$

PSD→ $\quad S_Y(f) = |j2\pi f - 1|^2 S_X(f)$

$$S_Y(f) = (4\pi^2 f^2 + 1)\, S_X(f)$$

$$S_X(f) = FT[R_X(\tau)]$$

$$S_X(f) = e^{-\pi f^2}$$

$\because$ [Gaussian function $\xrightarrow{\text{FT}}$ Gaussian function]

$$S_Y(f) = \left[4\pi^2 f^2 + 1\right]e^{-\pi f^2}$$

2.40 (b)

$-1 \le x \le 1$ and $-1 \le y \le 1$ is the entire rectangle. The region in which maximum of $\{x, y\}$ is less than $\frac{1}{2}$ is shown below as shaded region inside this rectangle.

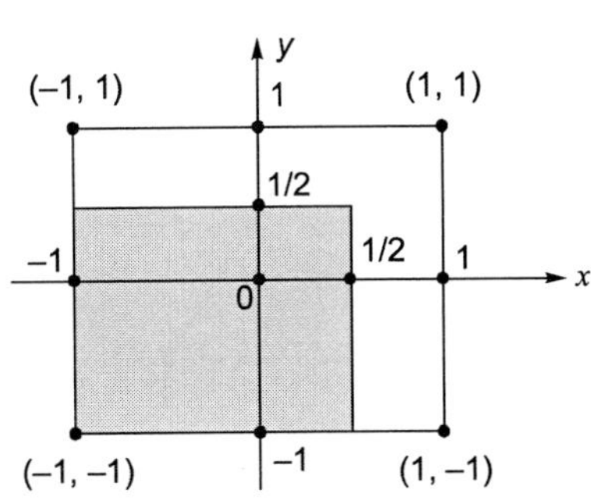

$$p\left(\max\{x,y\}<\frac{1}{2}\right) = \frac{\text{Area of shaded region}}{\text{Area of entire rectangle}}$$

$$= \frac{\frac{3}{2}\times\frac{3}{2}}{2\times 2} = \frac{9}{16}$$

2.41 (a)

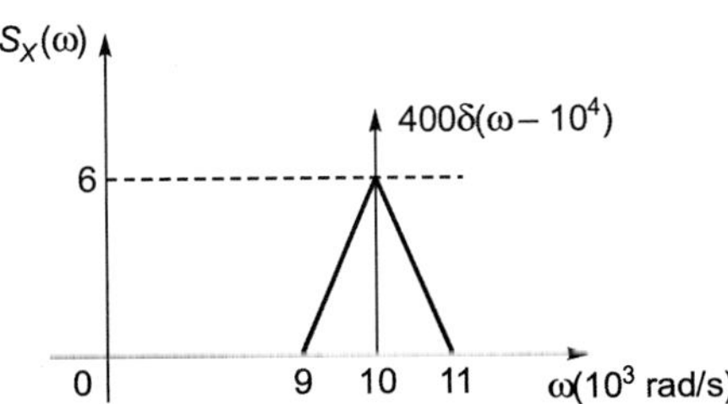

We know that $E[X^2(t)]$ represent the total power in random signal. So

$$P = E[X^2(t)] = 2\times\frac{1}{2\pi}\int_{9\times10^3}^{11\times10^3} S_X(\omega)d\omega$$

$$E[X^2(t)] = \frac{1}{\pi}\left[400+\frac{1}{2}\times6\times2\times10^3\right]$$

$$E[X^2(t)] = \frac{6400}{\pi}$$

At $\omega = 0$, there is no any frequency component present, hence dc value of the process is zero.

2.42 (b)

The probability

$$P(3V-2U) = P(3V-2U\geq 0)$$
$$= P(W\geq 0)$$

where, $W = 3V-2U$

U and V are independent random variables and can be expressed in terms of mean and variance as shown below:

$$U = N\left(0,\frac{1}{4}\right)$$

$$V = N\left(0,\frac{1}{9}\right)$$

$\because\ W = 3V-2U$

$$\therefore\ W = N\left(0, 9\times\frac{1}{9}+4\times\frac{1}{4}\right) = N(0,2)$$

Hence W is Gaussian variable with mean 0 and variance 2.

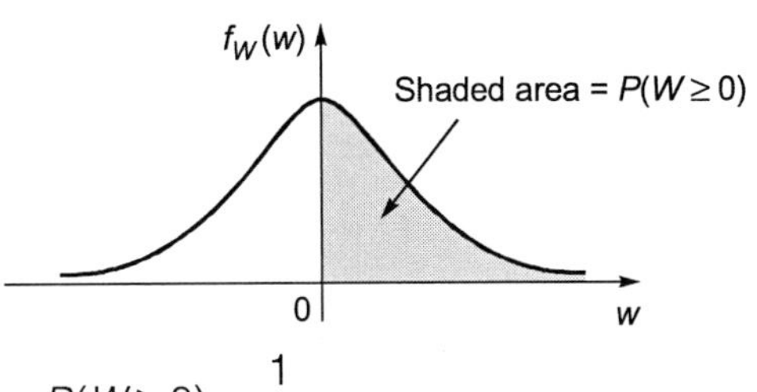

$$\therefore\ P(W\geq 0) = \frac{1}{2}$$

2.43 (d)

$$F(x) = P\{X\leq x\}$$
$$G(x) = P\{2x\leq x\}$$
$$= P\{x\leq x/2\}$$

For positive values of x,

$F(x) - G(x) \geq 0$

For negative values of x,

$F(x) - G(x) < 0$

but $[F(x)-G(x)]\cdot x\geq 0\cdots$ for all values of x.

2.44 (0.33)

If multiple independent random variables are uniformly distributed in the same interval then each random variable will have equal chances to be largest and to be lowest.

$$P(X_1 \text{ is the largest}) = \frac{1}{3}$$

2.45 (b)

Variance $\sigma_X^2 = E(X^2)-[E(X)]^2$

$\because\ \sigma_X^2$ can never be negative.

$\therefore\ E(X^2)\geq[E(X)]^2$

2.46 (d)

$$E[X(t_1)X(t_1)] = E[A\sin(2\pi t_1+\phi)]\times A\sin(2\pi t_2+\phi)]$$

$$= \frac{A^2}{2}E[\cos[2\pi(t_1-t_2)]-\cos[2\pi(t_1+t_2+2\phi)]$$

$$= \frac{A^2}{2}\cos[2\pi(t_1-t_2)$$

2.47 (0.346)

Assume quantization voltage is V_{Th}.

For outputs of quantizer to have equal probability

$\rightarrow P(X \le V_{Th}) = P(X \ge V_{Th})$

$$\int_{-\infty}^{V_{Th}} f_X(x)\,dx = \int_{V_{Th}}^{\infty} f_X(x)\,dx$$

$$\int_{0}^{V_{Th}} 2e^{-2x}dx = \int_{V_{Th}}^{\infty} 2e^{-2x}dx$$

$$e^{-2\times V_{Th}} = \frac{1}{2}$$

$$V_{Th} = 0.346$$

2.48 (4)

$$S_X(f) = \begin{cases} \dfrac{1}{W}, & |f| \le W \\ 0 & |f| > W \end{cases}$$

$\therefore$

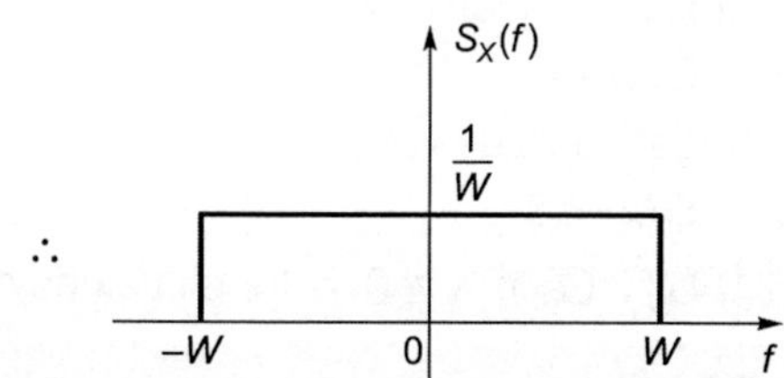

$\because \quad R_X(\tau) \longleftrightarrow [S_X(f)]$

$$= 2\text{sinc}[2W\tau]$$

Also,

$$E\left[\pi x(t) \cdot x\left(t - \frac{1}{4W}\right)\right]$$

$$= \pi E\left[x(t) \cdot x\left(t - \frac{1}{4W}\right)\right]$$

$$= \pi R_X\left(\frac{1}{4W}\right) = \pi \cdot 2\text{sinc}\left(\frac{1}{2}\right) \quad \text{...(i)}$$

We know that,

$$\text{sinc}(x) = \frac{\sin \pi x}{\pi x}$$

From equation (i), we get

$$\therefore \quad 2\pi \frac{\sin\frac{\pi}{2}}{\frac{\pi}{2}} = 4$$

2.49 (2.5)

$$S_X(f) = \begin{cases} 10^{-6}(3000 - |f|) \text{ Watts/Hz} & \text{for } |f| \le 3 \text{ kHz} \\ 0 & \text{otherwise} \end{cases}$$

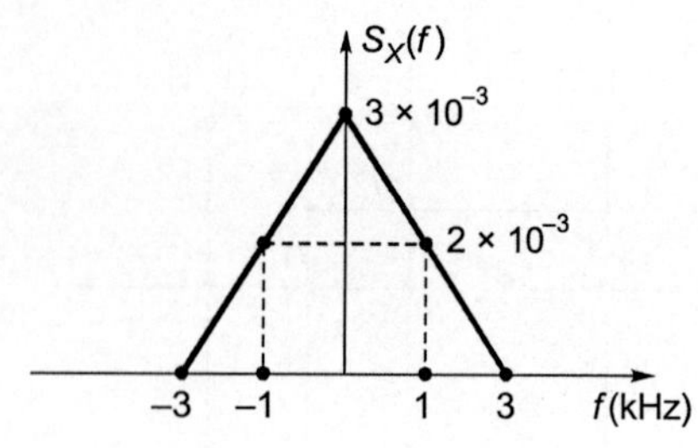

If $\quad x(t) \xleftrightarrow{\text{PSD}} S_X(f)$

Then

$$x(t)\cos 2\pi f_c t \xleftrightarrow{PSD} \frac{S_X(f - f_c) + S_X(f + f_c)}{4}$$

$\therefore$ PSD of $\quad x(t)\cos 2\pi \times 8000t$

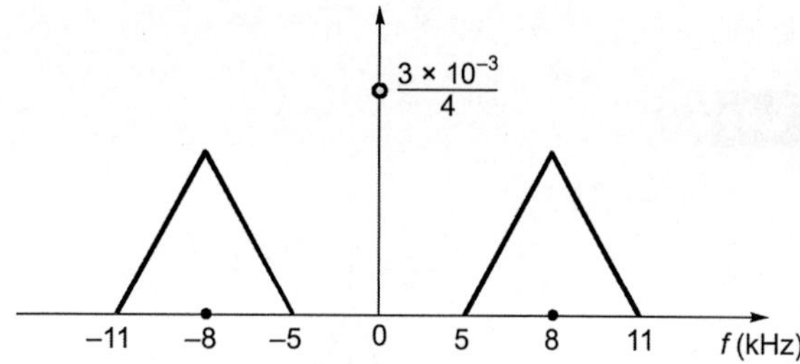

Now, BPF is given by

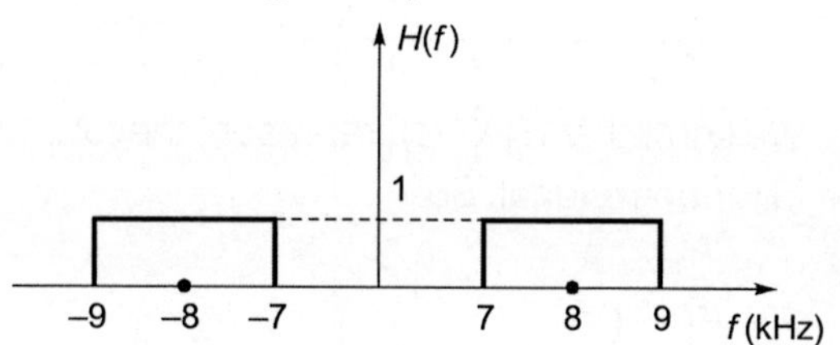

We know that

$S_X(f) \longrightarrow$ [$H(f)$] $\longrightarrow |H(f)|^2 S_X(f)$

$\therefore$ PSD of BPF output

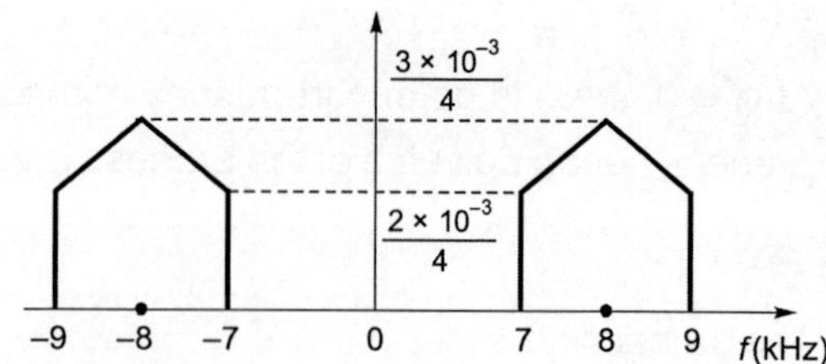

Output power = total area of output PSD

$$= 2\left[(2000)\left(\frac{2\times10^{-3}}{4}\right) + 2\left(\frac{1}{2}\times1000\times\frac{10^{-3}}{4}\right)\right]$$

$$= 2\left[1 + \frac{1}{4}\right] = 2.5 \text{ Watts}$$

2.50 (0.16)

$$X_1 + X_2 \le X_3$$

$$X_1 + X_2 - X_3 \le 0$$

So, $P\{X_1 + X_2 - X_3 \le 0\} = P\{Y \le 0\}$

and Y is another random variable whose probability density function is given by convolution of X_1, X_2, and X_3.

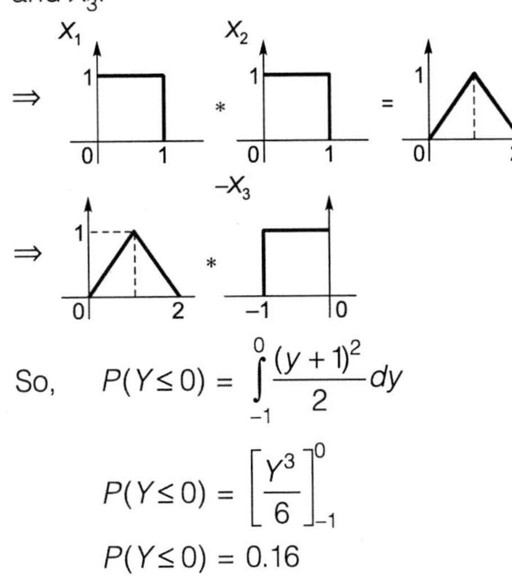

So, $$P(Y \le 0) = \int_{-1}^{0} \frac{(y+1)^2}{2} dy$$

$$P(Y \le 0) = \left[\frac{Y^3}{6}\right]_{-1}^{0}$$

$$P(Y \le 0) = 0.16$$

2.51 (2)

$$P(X = 1) = \frac{1}{3} \text{ and } P(X = 0) = \frac{2}{3}$$

and $$H(Y_1) = P(Y_1)\log_2 \frac{1}{P(Y_1)}$$

$$H(Y_2) = P(Y_2)\log \frac{1}{P(Y_2)}$$

where, $$P(Y_1 = 1) = P(Y_1 = 0) = \frac{1}{3}\times\frac{1}{2}+\frac{2}{3}\times\frac{1}{2} = \frac{1}{2}$$

Similarly, $$P(Y_2 = 1) = P(Y_2 = 0) = \frac{1}{2}$$

Here $$H(Y_1) + H(Y_2) = \frac{1}{2}\log_2 2 + \frac{1}{2}\log_2 2 + \frac{1}{2}\log_2 2 + \frac{1}{2}\log_2 2$$

$$H(Y_1) + H(Y_2) = 2 \text{ bits}$$

2.52 (c)

Shifting operation does not change the power spectral density but scaling operation change the power spectral density

$$y(t) = x(2t-1)$$

$$x(t) \to R_x(z) \longleftrightarrow S_x(f)$$

$$x(2t) \to R_x(2z) \longleftrightarrow \frac{1}{2}S_x\left(\frac{f}{2}\right)$$

$$x(2t-1) \to R_x(2z) \longleftrightarrow \frac{1}{2}S_x\left(\frac{f}{2}\right)$$

So, $$Y(t) \to S_y(f) = \frac{1}{2}S_x\left(\frac{f}{2}\right)$$

2.53 (0.8)

For a Gaussian random variable,

$$f_X(x) = \frac{1}{\sigma_x\sqrt{2\pi}} e^{\left(\frac{\mu - x}{2\sigma^2}\right)^2}$$

For mean (μ) = 0 and Variance (σ_X^2) = 1

$$f_X(x) = \frac{1}{\sqrt{2\pi}} e^{-\frac{x^2}{2}}$$

$$E[|x|] = \int_{-\infty}^{\infty} |x| f_X(x) dx$$

$$= \int_{-\infty}^{\infty} |x| \cdot \frac{1}{\sqrt{2\pi}} e^{\frac{-x^2}{2}} dx$$

$$= \int_{-\infty}^{0} -x \cdot \frac{1}{\sqrt{2\pi}} e^{\frac{-x^2}{2}} dx + \int_{0}^{\infty} x \cdot \frac{1}{\sqrt{2\pi}} e^{\frac{-x^2}{2}} dx$$

$$= 2\int_{0}^{\infty} x \cdot \frac{1}{\sqrt{2\pi}} e^{\frac{-x^2}{2}} dx = \frac{2}{\sqrt{2\pi}} \int_{0}^{\infty} x \cdot e^{\frac{-x^2}{2}} dx$$

Let $$\frac{x^2}{2} = t$$

$$\therefore \quad \frac{2x\,dx}{2} = dt$$

Hence $$\frac{2}{\sqrt{2\pi}} \int_{0}^{\infty} e^{-t} dt = 0.8$$

$$\therefore \quad E[|x|] = 0.8$$

2.54 (c)

The optimum threshold value is

$$\lambda_o = \frac{\sigma^2}{x_1 - x_2}\left[ln \frac{P(x_2)}{P(x_1)} + \frac{x_1^2 - x_2^2}{2\sigma^2}\right]$$

$$x_1 = 1, \quad x_2 = -1$$

$$P(X_1) = 0.75, \; P(X_2) = 0.25$$

$$\lambda_o = \frac{\sigma^2}{2}\left[ln \frac{0.25}{0.75}\right] = -0.55\sigma^2$$

So, λ_o should be strictly negative.

2.55 (0.8)

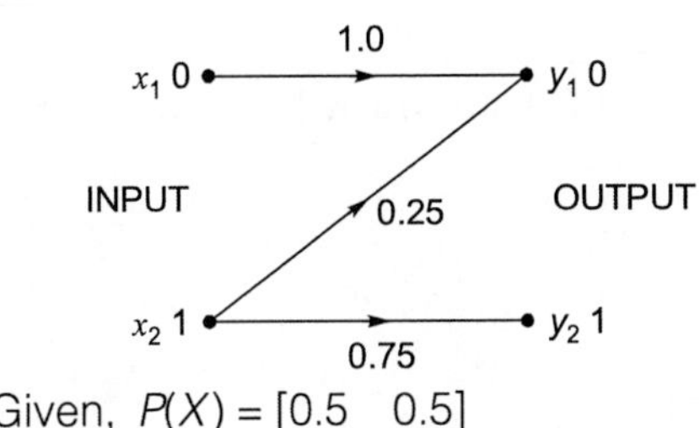

Given, $P(X) = [0.5 \quad 0.5]$

$$P\left[\frac{Y}{X}\right] = \begin{matrix} & y_1 & y_2 \\ x_1 & 1 & 0 \\ x_2 & 0.25 & 0.75 \end{matrix}$$

Then,

$$P[X, Y] = \begin{bmatrix} 0.5 & 0 \\ 0.125 & 0.375 \end{bmatrix}$$

and $P(Y_1) = 0.625$

$P(Y_2) = 0.375$

and the conditional probability matrix.

$P\left[\frac{X}{Y}\right]$ is given by

$$P\left[\frac{X}{Y}\right] = \begin{matrix} & y_1 & y_2 \\ x_1 & 0.8 & 0 \\ x_2 & 0.33 & 1 \end{matrix}$$

Hence $P\left[\frac{x_1}{y_1}\right] = P\left(\frac{0}{0}\right) = 0.8$

2.56 (c)

$$Y = X + Z$$

Z is Gaussian RV with mean βx

$x \in \{-a, +a\}$

when $\beta = 0$ $E[Y] = E[X] + E[Z]$

$E[Y] = E[X] = +a$

$\text{BER} = Q(a) = 1 \times 10^{-8}$

$$Q(v) = \frac{1}{\sqrt{2\pi}} \int_v^{\infty} e^{-\frac{v^2}{2}} du \cong e^{-\frac{v^2}{2}}$$

$$Q(a) = 1 \times 10^{-8} \approx e^{-\frac{a^2}{2}}$$

$a = 6$

when $\beta = -0.3$ mean $= 6 \times -0.3 = -1.8$

So, $E[Y] = E[X] + E[Z] = 6 - 1.8 = 4.2$

So, $\text{BER} = Q(4.2) \cong e^{\frac{-a^2}{2}}$

$\cong 0.0001$

$\cong 10^{-4}$

2.57 (b)

$R_Y(K) = R_Y(n, n + K)$

$= E[Y[n]\, Y[n + K]]$

Given

$$P[x(n) = 1] = \frac{1}{2}$$

$$P[x(n) = -1] = \frac{1}{2}$$

$$Y[n] = X[n] + 0.5\,X[n]$$

$$\therefore R_Y(K) = E[x[n] + 0.5x[n-1]]\,[x[n+K] + 0.5x[n+K-1]]$$

$$= E\big[x[n] \cdot x[n+k]\big] + 0.5E\big[x[n]\, x[n+K-1]\big]$$
$$+0.5E[x[n-1]\, x[n+K]]$$
$$+0.25E\big[x[n-1]\, x[n+K-1]\big]$$

$$= R_X(K) + 0.5R_X(K-1) + 0.5R_X(K+1) + 0.25\,R_X(K)$$

$$= 1.25R_X(K) + 0.5R_X(K-1) + 0.5R_X(R+1)$$

$$= 1.25\,R_X(K) + 0.5\,R_X(K-1) + 0.5\,R_X(K+1)$$

Now, if $K = 0$

$$R_X(0) = E[X^2[n]] = 1 \times \frac{1}{2} + 1 \times \frac{1}{2} = 1$$

$K \neq 0$

$$R_X(K) = E[X(n)\, X(n-k)]$$
$$= E[X(n)]\, E[X(n-k)] = 0$$

Because $\left(E[X(n)] = 1 \times \frac{1}{2} - 1 \times \frac{1}{2} = 0\right)$

$\therefore$ $R_y(0) = 1.25\,R_X(0) + 0.5\,R_X(-1) + 0.5\,R_X(1)$

$= 1.25$

Similarly,

$R_Y(1) = 0.5$

$R_Y(-1) = 0.5$

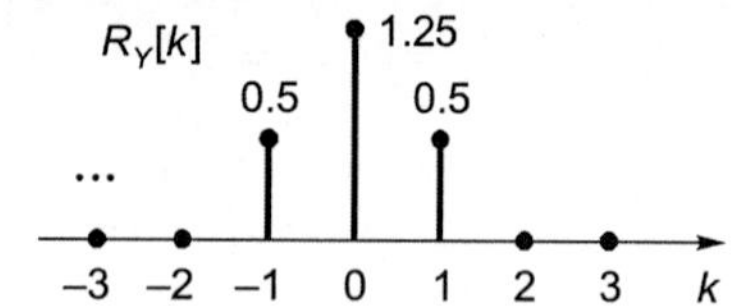

2.58 (a)

Variance of a process $= R_X(0)$

Let variance (output) $= R_Y(0)$

$\because$ It is zero mean

$\therefore$ $R_Y(\tau) = h(\tau) * h(-\tau) * R_N(\tau)$

where, $h(\tau)$ = filter response and $R_N(\tau)$ is the input noise.

$$R_N(\tau) = \frac{N_o}{2}\delta(\tau)$$

$$\therefore R_Y(\tau) = [h(\tau) \times h(-\tau)]\frac{N_o}{2}$$

$$R_Y(\tau) = \frac{N_o}{2}\int_{-\infty}^{\infty} h(\tau)\, h(\tau + z)\, d\tau$$

$$\therefore R_Y(0) = \frac{N_o}{2}\int_{-\infty}^{\infty} h^2(\tau)\, d\tau$$

$$= \frac{N_o}{2}(\text{energy}) = \frac{N_o}{2}(3A^2) = \frac{3}{2}N_o A^2$$

2.59 (0.25)

For a binary symmetric pulse

$$\therefore \quad R_{YY}(t) = \left[1 - \frac{|\tau|}{T}\right]$$

$$\therefore \; R_{YY}\left(\frac{3T}{4}\right) = \left[1 - \frac{3T/4}{T}\right] = \frac{1}{4} = 0.25$$

2.60 (6)

$$f_X(x) = \frac{1}{2}|x|e^{-|x|}$$

$f_X(x)$ is an even function.

$$\text{So,} \quad E(X) = \bar{X} = \int_{-\infty}^{\infty} xf(x)\,dx = 0$$

$$\text{Variance,} \quad \sigma_X^2 = E[X^2] - (\bar{X})^2 = E[X^2]$$

$$\text{So,} \quad \sigma_X^2 = E[X^2] = \int_{-\infty}^{\infty} x^2 f_X(x)\,dx$$

$$= \frac{1}{2}\int_{-\infty}^{\infty} x^2 |x| e^{-|x|} dx = \int_0^{\infty} x^3 e^{-x} dx$$

$$= \left[-e^{-x}(x^3 + 3x^2 + 6x + 6)\right]_0^{\infty} = 6$$

2.61 (a)

(i) $T_e = (F-1)T_0 = \left(10^{2/10} - 1\right)290$

$= 169.6$ K

(ii) $N_i = k(T_{ant} + T_e)B$

$= 1.38 \times 10^{-23} \times (50 + 169.6) \times 12 \times 10^6$

$= 3.63 \times 10^{-14}$ W

$N_o = N_i \times \text{Gain}$

$= 3.63 \times 10^{-14} \times 10^4$

$= 3.63 \times 10^{-10}$ W

2.62 (–1)

Power spectral density of $x(t) = S_X(f)$

$$S_X(f) = \frac{|G(f)|^2}{T}\sum_{n=-\infty}^{\infty} R_b(\tau)e^{j2\pi fn\tau}$$

$$R_b(\tau) = E[\beta_n\beta_{n-\tau}]$$

$= E[(\alpha_n + k\alpha_{n-3})(\alpha_{n-\tau} + k\,\alpha_{n-\tau-3})]$

$= E[\alpha_n\alpha_{n-\tau}] + kE[\alpha_{n-3}\alpha_{n-\tau}] + kE[\alpha_n\alpha_{n-\tau-3}]$

$+ k^2E[\alpha_{n-3}\,\alpha_{n-\tau-3}]$

$= E[\alpha_n\alpha_{n-\tau}] + kE[\alpha_{n-3}\alpha_{n-\tau-3+3}] + kE[\alpha_n\alpha_{n-\tau-3}]$

$+ k^2\,E[\alpha_{n-3}\alpha_{n-\tau-3}]$

$= R(\tau) + kR(\tau - 3) + kR(\tau + 3) + k^2R(\tau)$

$= (1 + k^2)R(\tau) + kR(\tau+3) + kR(\tau - 3)$

Hence autocorrelation function can be defined as:

$$R_b(\tau) = \begin{cases} 1+k^2 & \tau = 0 \\ k & \tau = \pm 3 \\ 0 & \text{otherwise} \end{cases}$$

Power spectral density

$$s_b(f) = 1 + k^2 + 2k\cos(2\pi f 3T)$$

Null will occur at $f = \frac{1}{3T}$

$$\text{at} \quad f = \frac{1}{3T}$$

$$s_b(f) = 1 + k^2 + 2k\cos(2\pi) = 0$$

$\Rightarrow \; 1 + k^2 + 2k = 0$

$\Rightarrow \; (k+1)^2 = 0$

$\therefore \; k = -1$

2.63 (100)

$$X(t) = 3V(t) - 8 \quad \text{and} \quad E[V(t)] = 0$$

$$R_v(\tau) = 4e^{-5|\tau|}$$

$$\text{Power of } X(t) = E[X^2(t)]$$

$$= E[9V^2(t)] + 64 - 48\,E[V(t)]$$

$$= 9E[V^2(t)] + 64 - 48\,E[V(t)]$$

$$E[V^2(t)] = R_v(0) = 4$$

$$\text{Power of } X(t) = ((9 \times 4) + 64)$$

$$= 100$$

2.64 (b)

$$Y(t) = X(t) - (t - T_0)$$

$$R_Y(\tau) = E[Y(t+\tau)\,Y(t)]$$

$$= E\left[(X(t+\tau) - X(t+\tau-T_0))\,(X(t) - X(t-T_0))\right]$$

$$= E[X(t+\tau)X(t) - X(t)X(t+\tau-T_0) - X(t+\tau)X(t-T_0) + X(t+\tau-T_0)\,X(t-T_0)]$$

Since, $X(t)$ is a WSS process,

$R_Y(\tau) = R_X(\tau) - R_X(\tau - T_0) - R_X(\tau + T_0) + R_X(\tau)$

$= 2R_X(\tau) - R_X(\tau - T_0) - R_X(\tau + T_0)$

2.65 (a)

The given input power spectral density is as follows:

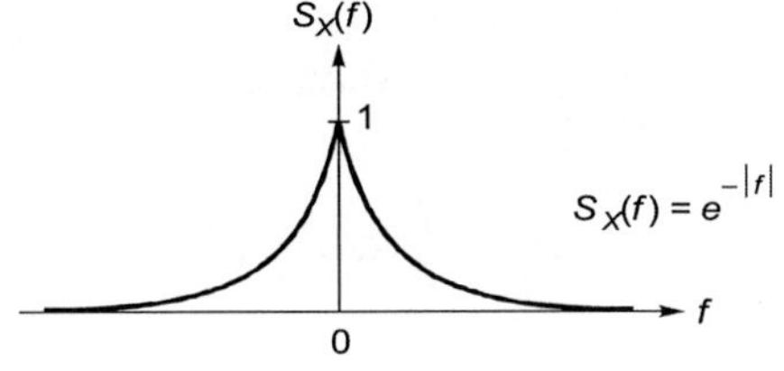

Frequency response of the low pass filter is as follows:

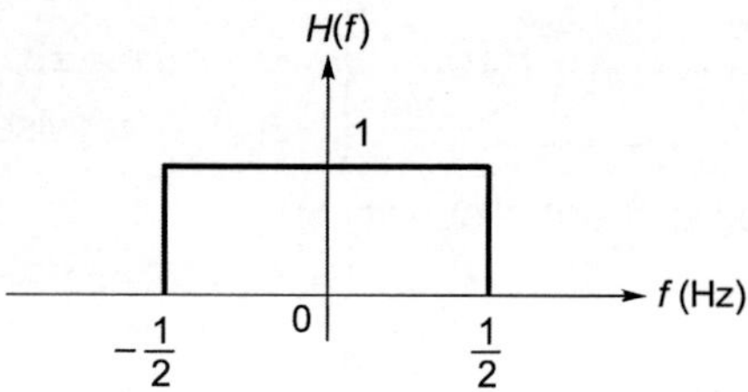

- $E[Y(t)] = H(0)\, E[X(t)]$

$H(0) = 1$

So, $E[Y(t)] = E[X(t)]$

- $E[Y^2(t)] \neq E[X^2(t)]$

Since, LPF does not allow total power from input to output.

- $E[X^2(t)] = \int_0^{\infty} S_X(f)\,df = 2\text{ W}$

As $E[Y^2(t)] \neq E[X^2(t)]$,

$E[Y^2(t)] \neq 2$

- So, only statement - I is correct.

2.66 (2)

$X(t) = U + Vt$

At $t = 2$, $X(t) = X(2) = U + 2V$

$E[X(t)] = E[U + 2V]$

$= E[U] + 2E[V]$

Given that,

$E[U] = 0$

$f_V(v)$

1/2

0 2 v

Also given that,

V is uniformly distributed between 0 and 2

So, $E[V] = \int_{-\infty}^{\infty} f_V(V)\,dv = \int_0^2 \left(\frac{1}{2}\right) dv = 1$

So, $E[X(t)] = 0 + 2(1) = 2$

2.67 (b)

PSD of noise input,

$S_N(f) = 0.50\text{ W/Hz}$

Power of $y(t)$,

$$P_y = \int_{-\infty}^{\infty} S_N(f)\,|H(f)|^2\,df$$

$$= 0.50\int_{-\infty}^{\infty} |H(f)|^2\,df = 0.50\int_{-\infty}^{\infty} |h(t)|^2\,dt$$

Given that,

$$h(t) = \frac{1}{2}e^{-t^2/2}$$

So,
$$P_y = \frac{1}{2}\int_{-\infty}^{\infty}\left(\frac{1}{2}e^{-t^2/2}\right)^2 dt$$

$$= \frac{1}{8}\int_{-\infty}^{\infty} e^{-t^2}\,dt$$

$$= \frac{\sqrt{\pi}}{8} = 0.2215\text{ W} \simeq 0.22\text{ W}$$

2.68 (17.5)

Let, $X_1(t) = h(t) * X(t)$

$S_{X1}(f) = |H(f)|^2\, S_X(f)$

$S_{X1}(f)$

$(0.5)^2\, 5 = 1.25$

-5 5 f(Hz)

$S_Z(f)$

1

-5 5 f(Hz)

Given that, $Z(t)$ and $X(t)$ are uncorrelated.

So, $S_Y(f) = S_{X1}(f) + S_Z(f)$

Power in $y(t)$,

$P_Y =$ [Area under $S_{X1}(f)$] + [Area under $S_Z(f)$]

$= (10 \times 1.25) + (5 \times 1) = 17.5\text{ W}$

2.69 (6)

Given:
$$\phi(t) = \begin{cases} 1 & 5 \le t \le 7 \\ 0 & \text{Otherwise} \end{cases}$$

$S_\omega(f) = 3\text{ Watts/Hz}$

$R_\omega(\tau) = 3\delta(\tau) = 3\delta(t_1 - t_2)$

$\text{Var}[y] = \text{E}[y^2] - \{E[y]\}^2$

$\{E[W(t)]\}^2 =$ DC power = Area under PSD at $f = 0$

$\{E[W(t)]\}^2 = 0$

$E[W(t)] = 0$

$$y = \int_{-\infty}^{\infty} W(t)\,\phi(t)\,dt$$

$$E[Y] = \int_{-\infty}^{\infty} E[W(t)]\,\phi(t)dt = 0$$

$$y = \int_{-\infty}^{\infty} W(t)\,\phi(t)dt \rightarrow E[Y^2]$$

$= S_\omega(f) \cdot \text{Energy}\,[\phi(t)] = 3 \times 2 = 6$

$\text{Var}[Y] = 6 - 0 = 6$

2.70 (2)

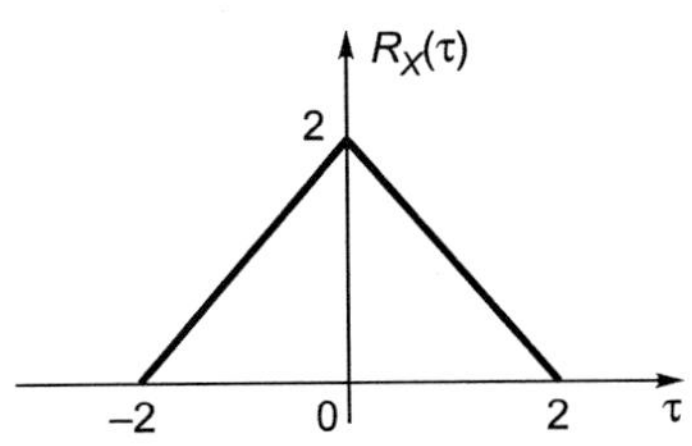

Average power $[X(t)] = R_X(0) = 2$

2.71 (d)

$$Y = 2X + 3$$

$$\text{Var}[Y] = E\left[(Y-\bar{Y})^2\right]$$

$$E[Y] = \bar{Y} = 2\bar{X} + 3$$

$$\text{Var}[Y] = E\left[(2X+3-2\bar{X}-3)^2\right]$$

$$= E\left[4(X-\bar{X})^2\right]$$

$$= 4\cdot E\left[(X-\bar{X})^2\right]$$

$$\sigma_y^2 = 4\cdot\sigma_x^2$$

2.72 (1.16)

The probability density function of the random samples of the source is as given below:

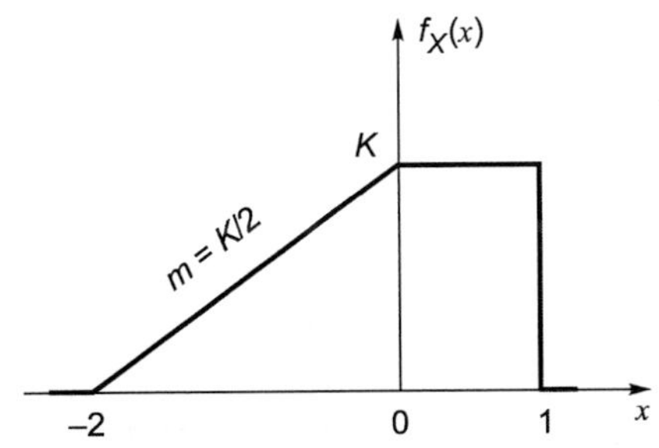

Since, the area of probability density function is unity. Hence,

$$\frac{1}{2}\times K\times 2 + 1\times K = 1$$

$$\Rightarrow \quad K + K = 1$$

$$\Rightarrow \quad K = \frac{1}{2} = 0.5$$

For $-2 \le x \le 0$,

$$f(x) = mx + C$$

$$f(x) = \left(\frac{K}{2}\right)x + C$$

$$f(x) = \frac{x}{4} + C$$

At $x = -2$, $f(x) = 0$. Hence,

$$0 = -\frac{2}{4} + C \Rightarrow C = \frac{1}{2}$$

$$\therefore \quad f(x) = \frac{x}{4} + \frac{1}{2} \quad \text{for } -2 \le x \le 0$$

The probability density function can be written as below:

$$f(x) = \begin{cases} \dfrac{x}{4} + \dfrac{1}{2} & ; \ -2 \le x \le 0 \\ \dfrac{1}{2} & ; \ 0 \le x \le 1 \\ 0 & ; \ \text{otherwise} \end{cases}$$

Given, the quantizer output

$$x_q = \begin{cases} \alpha & ; \ 0 \le x \le 1 \\ \beta & ; \ -2 \le x \le 0 \end{cases}$$

The mean-square quantization error is defined as

$$\text{MSQ}[Q_e] = E[Q_e^2]$$

where, Q_e = Quantization error

Q_e = Sampled value-quantized value

$Q_e = x - x_q$

The mean-square quantization error,

$$N_Q = \text{MSQ}[Q_e] = E[(x - x_q)^2]$$

(i) For $-2 \le x \le 0$

$$N_Q = \int_{-2}^{0} (x-\beta)^2 \cdot \left(\frac{x}{4} + \frac{1}{2}\right) dx$$

$$N_Q = \int_{-2}^{0} (x^2 + \beta^2 - 2\beta x)\left(\frac{x}{4} + \frac{1}{2}\right) dx$$

$$N_Q = \int_{-2}^{0} \left(\frac{x^3}{4} + \frac{1}{2}x^2 + \frac{\beta^2 x}{4} + \frac{\beta^2}{2} - \frac{\beta x^2}{2} - \beta x\right) dx$$

$$N_Q = \int_{-2}^{0} \left[\frac{x^3}{4} + x^2\left(\frac{1-\beta}{2}\right) + x\left(\frac{\beta^2}{4} - \beta\right) + \frac{\beta^2}{2}\right] dx$$

$$N_Q = \left[\frac{x^4}{16} + \frac{x^3}{3}\left(\frac{1-\beta}{2}\right) + \frac{x^2}{2}\left(\frac{\beta^2}{4} - \beta\right) + \frac{\beta^2}{2}x\right]_{-2}^{0}$$

$$N_Q = -\frac{16}{16} + \frac{8}{3}\left(\frac{1-\beta}{2}\right) - \frac{4}{2}\left(\frac{\beta^2}{4} - \beta\right) + \frac{\beta^2}{2}$$

$$N_Q = -1 + \frac{4}{3} - \frac{4\beta}{3} - \frac{\beta^2}{2} + 2\beta + \frac{\beta^2}{2}$$

$$N_Q = \frac{2\beta}{3} + \frac{1}{3}$$

The quantization error will be minimum when $N_Q = 0$ which gives

$$\beta = \beta^* = \frac{-2}{3} \quad \text{...(i)}$$

(ii) For $0 \le x \le 1$,

$$N_Q = \int_0^1 (x-\alpha)^2 \frac{1}{2} dx$$

$$N_Q = \left[\frac{(x-\alpha)^3}{2\times 3}\right]_0^1$$

$$N_Q = \frac{1}{6}\left[(1-\alpha)^3 + \alpha^3\right]$$

To obtain value of α for which N_Q is minimum,

$$\frac{dN_Q}{d\alpha} = 0$$

$$\Rightarrow \frac{1}{6}\left[3(1-\alpha)^2(-1) + 3\alpha^2\right] = 0$$

$$\Rightarrow -\alpha^2 - 1 + 2\alpha + \alpha^2 = 0$$

$$\Rightarrow \quad 2\alpha = 1$$

$$\Rightarrow \quad \alpha = \alpha^* = \frac{1}{2}$$

Hence, $\alpha^* - \beta^* = \frac{1}{2} + \frac{2}{3} + = \frac{7}{6} = 1.167$

2.73 (a)

Statement-I :

Consider a delay system as shown below:

$x(t)$ → Delay T_d → $y(t) = x(t - T_d)$

$$y(t) = x(t - T_d)$$

Taking Fourier transform,

$$Y(f) = X(f) \cdot e^{-j2\pi f T_d}$$

$$\therefore \quad H(f) = \frac{Y(f)}{X(f)} = e^{-j2\pi f T_d}$$

$$|H(f)| = 1 \qquad \text{(Magnitude)}$$

$$\angle H(f) = -2\pi f T_d \qquad \text{(Phase)}$$

If the input is band-limited to $-\alpha \le f \le \alpha$, the magnitude response of the delay system can be represented as below:

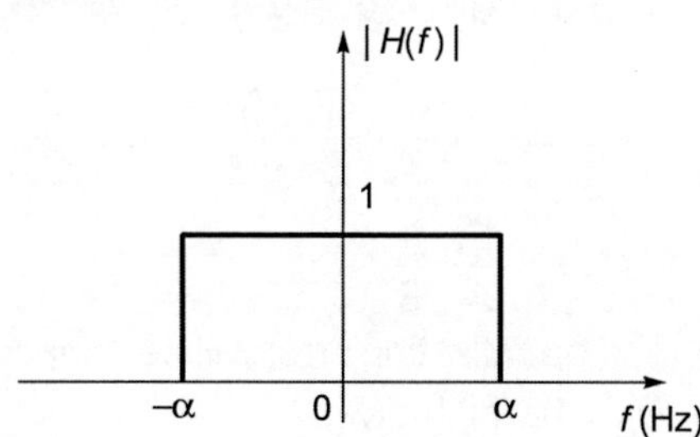

Based on the given response, statement-I is correct.

Note : The phase response of the system is not given in the question. Without knowing the phase response of the system, we can't comment about exact nature of the system.

Statement-II:

For an input process with power spectral density $S_X(f)$ to a system with transfer function $H(f)$, the output power spectral density is given as

$$S_Y(f) = S_X(f) \cdot |H(f)|^2$$

For $-\alpha \le f \le \alpha$, $|H(f)| = 1$ (Given)

Hence, $S_Y(f) = S_X(f)$ for $-\alpha \le f \le \alpha$

Hence, Statement-II is also correct.

2.74 (a, b)

$S_X(\omega) \ge 0$

$S_X(\omega)$ is even function

Hence, options (a) and (b) are valid power spectral densities.

2.75 (b)

Given random variables,

$$X = N(0, 1)$$

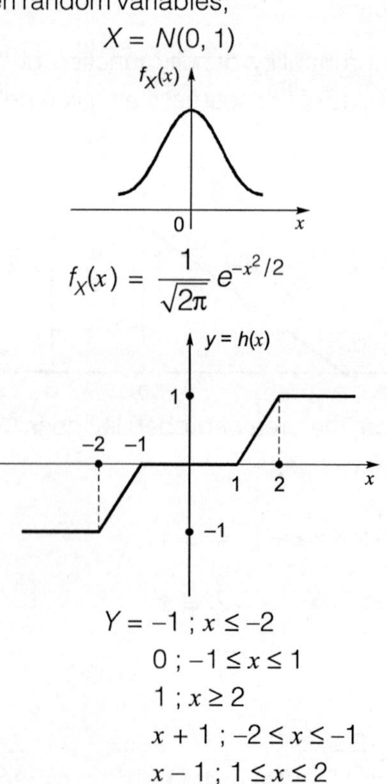

$$f_X(x) = \frac{1}{\sqrt{2\pi}} e^{-x^2/2}$$

$Y = -1 \; ; x \le -2$

$\quad 0 \; ; -1 \le x \le 1$

$\quad 1 \; ; x \ge 2$

$\quad x + 1 \; ; -2 \le x \le -1$

$\quad x - 1 \; ; 1 \le x \le 2$

Y is taking discrete set of values and a continuous range of values, so it is mixed random variable. From the given options, density function of 'Y' will be.

$$f_Y(y) = a\delta(y+1) + b\delta(y) + c\delta(y-1) + g(y)$$

2.76 (0.25)

We know that,

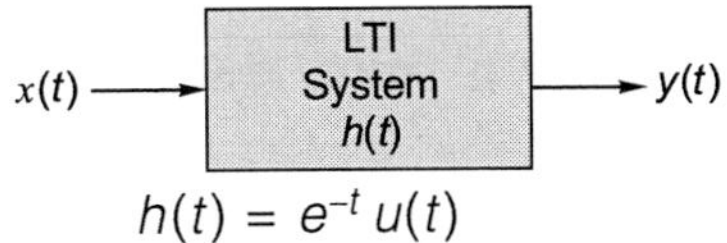

$$h(t) = e^{-t}\, u(t)$$

Given: Input PSD

$$\Rightarrow \qquad S_X(f) = \frac{1}{2}\ \text{W/Hz}$$

We know output PSD,

$$S_Y(f) = S_X(f)\,|H(f)|^2$$

$$S_Y(f) = \frac{1}{2}|H(f)|^2$$

$$\text{Power}\,[y(t)] = \int_{-\infty}^{\infty} S_Y(f)\,df = \int_{-\infty}^{\infty} \frac{1}{2}|H(f)|^2\,df$$

$$= \frac{1}{2}\int_{-\infty}^{\infty} h^2(t)\,dt = \frac{1}{2}\int_{0}^{\infty} e^{-2t}\,dt = \frac{1}{2}\times\frac{1}{2} = \frac{1}{4} = 0.25\ \text{W}$$

2.77 (0.5)

Since, random variables X and Y are independent and distributed uniformly in the interval [0, 1], the probability density function can be drawn as below:

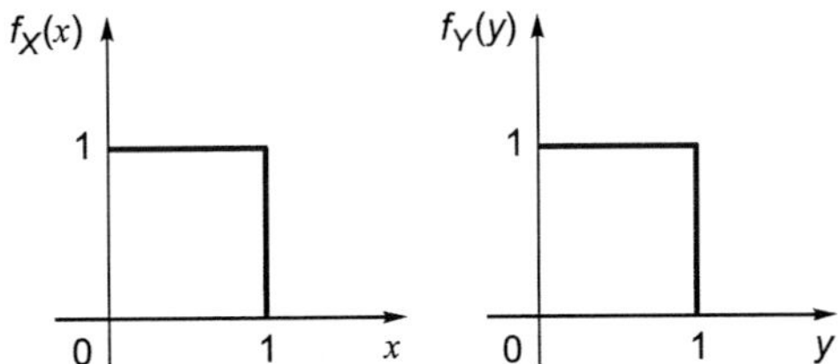

We have to find,

$$P(X \geq Y) = P[X - Y \geq 0]$$

Consider a random variable $Z = X - Y$. Therefore, the required probability is given by $P(Z \geq 0)$.

The probability density function of $Z = X + (-Y)$ is obtained as

$$f_Z(z) = f_X(z) * f_{-Y}(z)$$

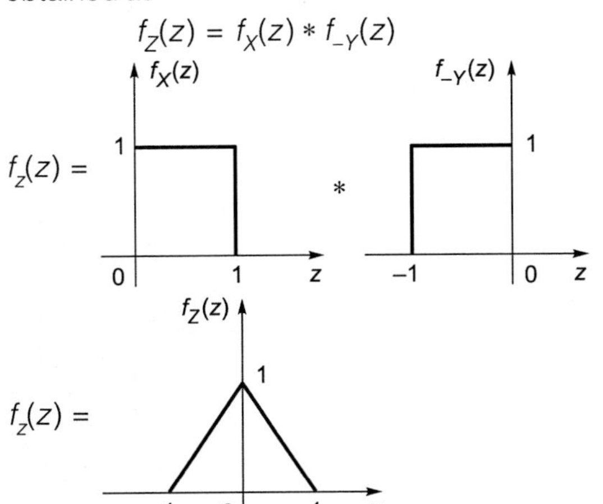

The required probability $P(Z \geq 0)$ is given by

$$P(Z \geq 0) = \frac{1}{2}$$

2.78 (a)

Given: $E[w(t)] = 0$ and $S_W(f) = \frac{N_0}{2}$

The mean of the output random process $n(t)$ after passing through the RC filter with transfer function $H(f)$ is given by

$$E[n(t)] = H(0)\cdot E[w(t)]$$

$$\therefore \qquad E[n(t)] = 0$$

The power spectral density of $n(t)$ is given by

$$S_n(f) = |H(f)|^2 \cdot S_W(f)$$

For an RC low pass filter,

$$H(f) = \frac{1}{1 + j2\pi fRC}$$

$$\therefore \qquad S_n(f) = \frac{1}{1+(2\pi fRC)^2}\cdot\frac{N_0}{2} \qquad \text{...(i)}$$

The variance of the output random process is given by

$$\text{Var}[n(t)] = E[n^2(t)] - [E[n(t)]]^2$$

$$\text{Var}[n(t)] = E[n^2(t)]$$

At $t = t_k$, random variable $n(t_k)$ results with

$$\text{Var}[n(t_k)] = E[n^2(t_k)] = E[n(t_k)\cdot n(t_k)]$$

$$\text{Var}[n(t_k)] = R_n(0)$$

$$[\because R_n(\tau) = E[n(t)\cdot n(t+\tau)]]$$

The inverse Fourier transform of power spectral density gives the auto-correlation function. From equation (i),

$$R_n(\tau) = \text{I.F.T}\left\{\frac{N_0/2}{1+(2\pi fRC)^2}\right\}$$

$$R_n(\tau) = \text{I.F.T}\left\{\frac{N_0/2R^2C^2}{(1/RC)^2+(2\pi f)^2}\right\}$$

We know that $e^{-a|t|} \longleftrightarrow \dfrac{2a}{a^2+(2\pi f)^2}$. Hence,

$$R_n(\tau) = \frac{N_0}{4RC}e^{-|\tau|/RC}$$

$$\therefore \qquad \text{Var}[n(t_k)] = R_n(0) = \frac{N_0}{4RC}$$

2.79 (a)

Received symbol,

$$r = s + w$$

Noise w is a white Gaussian noise with zero mean and variance 4. Hence, the probability density function of noise is given by

$$f_W(w) = \frac{1}{\sqrt{2\pi}} e^{\frac{-w^2}{2\times 4}} = N(0, 4)$$

When –4 is transmitted, $r = -4 + w$

$$f(r/-4) = \frac{1}{\sqrt{2\pi}} e^{\frac{-(r+4)^2}{2\times 4}}$$

$$\Rightarrow \quad r = N(\mu, \sigma^2) = N(-4, 4)$$

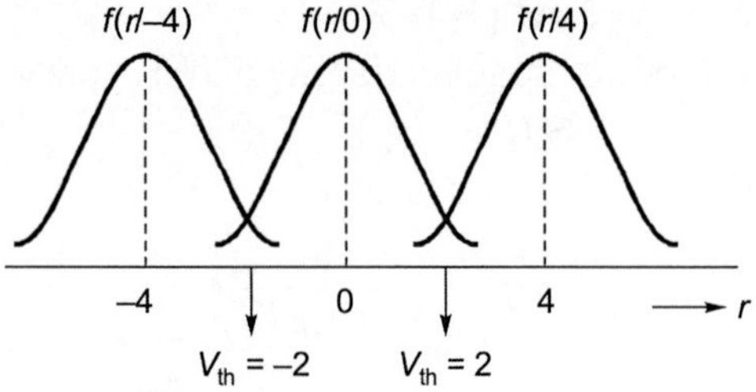

f(r/–4) f(r/0) f(r/4)

–4 0 4 r

$V_{th} = -2$ $V_{th} = 2$

When 0 is transmitted, $r = w$

$$f(r/0) = \frac{1}{\sqrt{2\pi}} e^{-\frac{r^2}{2\times 4}} \Rightarrow r = N(0, 4)$$

When 4 is transmitted, $r = 4 + n$

$$f(r/4) = \frac{1}{\sqrt{2\pi}} e^{\frac{-(r-4)^2}{2\times 4}}$$

$$\Rightarrow \quad r = N(-4, 4)$$

The probability of error,

$$P_e = P(-4)\, P_{e-4} + P(0)\, P_{e0} + P(4) P_{e4}$$

where P_{e-4}, P_{e0} and P_{e4} are the respective errors when –4, 0 and 4 are transmitted.

Given $\quad P(-4) = P(0) = P(4) = \frac{1}{3}$

For the optimum error, using the MAP rule, we obtain the detection threshold as $V_{Th} = -2$ and $V_{Th} = 2$.

$$P_{e-4} = P(r > -2) = \int_{r=-2}^{\infty} f\left(\frac{r}{-4}\right) dr$$

where, $r = N(-4, 4)$

$$P_{e-4} = 1 - F_r(-2)$$

$$= 1 - \left(1 - Q\left(\frac{-2+4}{2}\right)\right) \quad \left[\because F_X(x) = 1 - Q\left(\frac{x-\mu}{\sigma}\right)\right]$$

$$= Q(1)$$

$$P_{e0} = P(r < -2) + P(r > 2)$$

where, $r = N(0, 4)$

$$= F_r(-2) + \{1 - F_r(2)\}$$

$$= \left(1 - Q\left(\frac{-2-0}{2}\right)\right) + Q\left(\frac{2-0}{2}\right)$$

$$= [1 - Q(-1)] + Q(1) = 2Q(1)$$

Similarly, $\quad P_{e4} = P(r < 2) \quad$ where, $r = N(4, 4)$

$$= F_r(2)$$

$$= 1 - Q\left(\frac{2-4}{2}\right)$$

$$= 1 - Q(-1)$$

$$= Q(1)$$

$$\therefore \quad P_e = \frac{1}{3}Q(1) + \frac{2}{3}Q(1) + \frac{1}{3}Q(1)$$

$$= \frac{4}{3}Q(1)$$

■■■

3 Digital Communication Systems

3.1 If the number of bits per sample in a PCM system is increased from a n to $n + 1$, the improvement in signal to quantization noise ratio will be

(a) 3 dB (b) 6 dB
(c) 2 n dB (d) n dB [1995 : 1 M]

3.2 A 1.0 kHz signal is flat top sampled at the rate of 1800 samples/sec and the samples are applied to an ideal rectangular LPF with cut-off frequency of 1100 Hz, then the output of the filter contains

(a) only 800 Hz component
(b) 800 Hz and 900 Hz components
(c) 800 Hz and 1000 Hz components
(d) 800 Hz, 900 Hz and 100 Hz components

[1995 : 1 M]

3.3 The signal to quantization noise ratio in an n-bit PCM system

(a) depends upon the sampling frequency employed
(b) is independent of the value of 'n'
(c) increasing with increasing value of 'n'
(d) decreases with the increasing value of 'n'

[1995 : 1 M]

3.4 For a given data rate, the bandwidth B_p of a BPSK signal and the bandwidth B_0 of the OOK signal are related as

(a) $B_p = B_0/4$ (b) $B_p = B_0/2$
(c) $B_p = B_0$ (d) $B_p = 2B_0$

[1995 : 1 M]

3.5 A rectangular pulse of duration T is applied to a filter matched to this input. The output of the filter is a

(a) rectangular pulse of duration T
(b) rectangular pulse of duration $2T$
(c) triangular pulse
(d) sine function [1996 : 1 M]

3.6 The number of bits in a binary PCM system is increased from n to $n + 1$. As a result, the signal to quantization noise ratio will improve by a factor

(a) $(n + 1)/n$
(b) $2^{(n+1)/n}$
(c) $2^{2(n+1)/n}$
(d) which is independent of n [1996 : 2 M]

3.7 The line code that has zero dc component for pulse transmission of random binary data is

(a) non-return to zero (NRZ)
(b) return to zero (RZ)
(c) alternate mark inversion (AMI)
(d) none of the above [1997 : 1 M]

3.8 A deterministic signal has the power spectrum given in figure. The minimum sampling rate needed to completely represent signal is

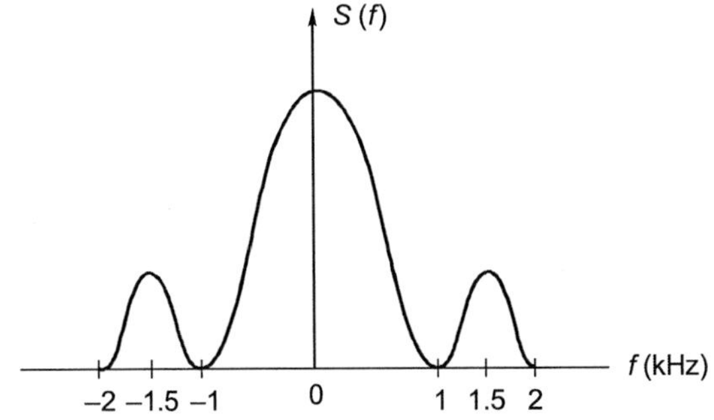

(a) 1 kHz (b) 2 kHz
(c) 3 kHz (d) none of these

[1997 : 1 M]

3.9 In a PCM system with uniform quantization, increasing the number of bits from 8 to 9 will reduce the quantization noise power by a factor of

(a) 9 (b) 8
(c) 4 (d) 2

[1998 : 1 M]

3.10 Flat top sampling of low pass signals

(a) gives rise to aperture effect
(b) implies oversampling
(c) leads to aliasing
(d) introducing delay distortion

[1998 : 1 M]

3.11 Quadrature multiplexing is

(a) the same as FDM
(b) the same as TDM
(c) a combination of FDM and TDM
(d) quite different from FDM and TDM

[1998 : 1 M]

3.12 Compression in PCM refers to relative compression of
(a) higher signal amplitudes
(b) lower signal amplitudes
(c) lower signal frequencies
(d) higher signal frequencies [1998 : 1 M]

3.13 The Nyquist sampling frequency (in Hz) of a signal given by $6 \times 10^4 \sin c^3 (400\, t) * 10^6 \sin c^3 (100\, t)$ is
(a) 200 (b) 300
(c) 1500 (d) 1000 [1999 : 2 M]

3.14 The peak-to-peak input to an 8-bit PCM coder is 2 Volts. The signal power-to-quantization noise power ratio (in dB) for an input of $0.5\cos(\omega_m t)$ is
(a) 47.8 (b) 43.8
(c) 95.6 (d) 99.6
[1999 : 2 M]

3.15 The input to a matched filter is given by

$$s(t) = \begin{cases} 10\sin(2\pi \times 10^6) & 0 < t < 10^{-4} \text{ sec} \\ 0 & \text{Otherwise} \end{cases}$$

The peak amplitude of the filter output is
(a) 10 volts (b) 5 volts
(c) 10 millivolts (d) 5 millivolts
[1999 : 2 M]

3.16 Four independent messages have bandwidths of 100 Hz, 100 Hz, 200 Hz, and 400 Hz, respectively. Each is sampled at the Nyquist rate, and the samples are Time Division Multiplexed (TDM) and transmitted. The transmitted sample rate (in Hz) is
(a) 1600 (b) 800
(c) 400 (d) 200
[1999 : 2 M]

3.17 In a digital communication system employing Frequency Shift Keying (FSK), the 0 and 1 bit are represented by sine waves of 10 kHz and 25 kHz respectively. These waveforms will be orthogonal for a bit interval of
(a) 45 μsec (b) 200 μsec
(c) 50 μ sec (d) 250 μ sec
[2000 : 2 M]

3.18 A video transmission system transmits 625 picture frames per second. Each frame consists of a 400 × 400 pixel grid with 64 intensity levels per pixel. The data rate of the system is
(a) 16 Mbps (b) 100 Mbps
(c) 600 Mbps (d) 6.4 Gbps
[2001 : 2 M]

3.19 The Nyquist sampling interval, for the signal $\text{sin}c(700t) + \sin c(500t)$ is
(a) $\frac{1}{350}$ sec (b) $\frac{\pi}{350}$ sec
(c) $\frac{1}{700}$ sec (d) $\frac{\pi}{175}$ sec
[2001 : 2 M]

3.20 During transmission over a communication channel, bit errors occur independently with probability p. If a block of n bits is transmitted, the probability of at most one bit error is equal to
(a) $1 - (1 - p)^n$
(b) $p + (n - 1)(1 - p)$
(c) $np(1 - p)^{n-1}$
(d) $(1 - p)^n + np(1 - p)^{n-1}$ [2001 : 2 M]

3.21 Consider a sample signal

$$y(t) = 5 \times 10^{-6} x(t) \sum_{n=-\infty}^{+\infty} \delta(t - nT_s)$$

where $x(t) = 10\cos(8\pi \times 10^3)t$ and T_s=100 μsec. When $y(t)$ is passed through an ideal lowpass filter with a cutoff frequency of 5 kHz, the output of the filter is
(a) $5 \times 10^{-6}\cos(8\pi \times 10^3)t$
(b) $5 \times 10^{-5}\cos(8\pi \times 10^3)t$
(c) $5 \times 10^{-1}\cos(8\pi \times 10^3)t$
(d) $10\cos(8\pi \times 10^3)t$ [2002 : 1 M]

3.22 For a bit-rate of 8 Kbps, the best possible values of the transmitted frequencies in a coherent binary FSK system are
(a) 16 kHz and 20 kHz
(b) 20 kHz and 32 kHz
(c) 20 kHz and 40 kHz
(d) 32 kHz and 40 kHz [2002 : 1 M]

3.23 A signal $x(t) = 100\cos(24\pi \times 10^3)t$ is ideally sampled with a sampling period of 50 μsec and then passed through an ideal lowpass filter with cutoff frequency of 15 kHz. Which of the following frequencies is/are present at the filter output?
(a) 12 kHz only
(b) 8 kHz only
(c) 12 kHz and 9 kHz
(d) 12 kHz and 8 kHz
[2002 : 2 M]

3.24 At a given probability of error, binary coherent FSK is inferior to binary coherent PSK by.
(a) 6 dB (b) 3 dB
(c) 2 dB (d) 0 dB
[2003 : 1 M]

3.25 A sinusoidal signal with peak-to-peak amplitude of 1.536 V is quantized into 128 levels using a mid-rise uniform quantizer. The quantization-noise power is

(a) 0.768 V (b) 48×10^{-6} V^2
(c) 12×10^{-6} V^2 (d) 3.072 V

[2003 : 2 M]

3.26 If E_b, the energy per bit of a binary digital signal, is 10^{-5} watt-sec and the one-sided power spectral density of the white noise, $N_0 = 10^{-6}$ W/Hz, then the output SNR of the matched filter is

(a) 26 dB (b) 10 dB
(c) 20 dB (d) 13 dB [2003 : 2 M]

3.27 The input to a linear delta modulator having a step-size $\Delta = 0.628$ is a sine wave with frequency f_m and peak amplitude E_m. If the sampling frequency $f_s = 40$ kHz, the combination of the sine-wave frequency and the peak amplitude, where slope overload will take place is

	E_m	f_m
(a)	0.3 V	8 kHz
(b)	1.5 V	4 kHz
(c)	1.5 V	2 kHz
(d)	3.0 V	1 kHz

[2003 : 2 M]

3.28 If S represents the carrier synchronization at the receiver and ρ represents the bandwidth efficiency, then the correct statement for the coherent binary PSK is

(a) $\rho = 0.5$, S is required
(b) $\rho = 1.0$, S is required
(c) $\rho = 0.5$, S is not required
(d) $\rho = 1.0$, S is not required

[2003 : 2 M]

3.29 A signal is sampled at 8 kHz and is quantized using 8-bit uniform quantizer. Assuming SNR_q for a sinusoidal signal, the correct statement for PCM signal with a bit rate of R is

(a) $R = 32$ kbps, $SNR_q = 25.8$ dB
(b) $R = 64$ kbps, $SNR_q = 49.8$ dB
(c) $R = 64$ kbps, $SNR_q = 55.8$ dB
(d) $R = 32$ kbps, $SNR_q = 49.8$ dB

[2003 : 2 M]

3.30 In a PCM system, if the code word length is increased from 6 to 8 bits, the signal to quantization noise ratio improves by the factor

(a) 8/6 (b) 12
(c) 16 (d) 8 [2004 : 1 M]

3.31 In the output of a DM speech encoder, the consecutive pulses are of opposite polarity during time interval $t_1 \le t \le t_2$. This indicates that during this interval

(a) the input to the modulator is essentially constant
(b) the modulator is going through slope overload
(c) the accumulator is in saturation
(d) the speech signal is being sampled at the Nyquist rate

[2004 : 1 M]

3.32 A source produces binary data at the rate of 10 kbps. The binary symbols are represented as shown in the figure.

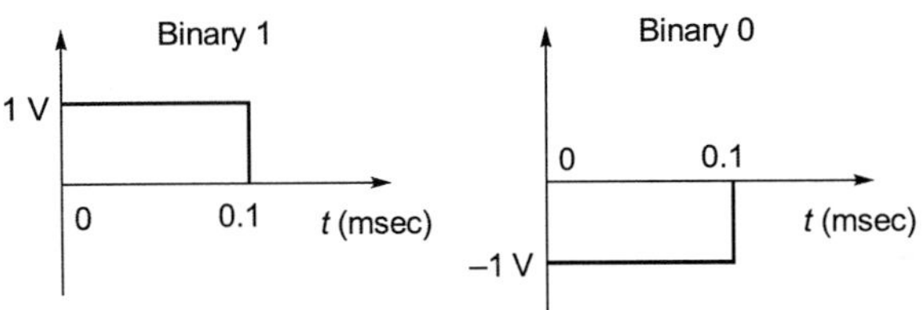

The source output is transmitted using two modulation schemes, namely Binary PSK (BPSK) and Quadrature PSK (QPSK). Let B_1 and B_2 be the bandwidth requirements of BPSK and QPSK respectively. Assuming that the bandwidth of the above rectangular pulses is 10 kHz, B_1 and B_2 are

(a) $B_1 = 20$ kHz, $B_2 = 20$ kHz
(b) $B_1 = 10$ kHz, $B_2 = 20$ kHz
(c) $B_1 = 20$ kHz, $B_2 = 10$ kHz
(d) $B_1 = 10$ kHz, $B_2 = 10$ kHz

[2004 : 2 M]

3.33 Consider a binary digital communication system with equally likely 0's and 1's. When binary 0 is transmitted the detector input can lie between the levels −0.25V and +0.25V with equal probability: when binary 1 is transmitted, the voltage at the detector can have any value between 0 and 1 V with equal probability. If the detector has a threshold of 0.2 V (i.e., if the received signal is greater than 0.2 V, the bit is taken as 1), the average bit error probability is

(a) 0.15 (b) 0.2
(c) 0.05 (d) 0.5

[2004 : 2 M]

3.34 Choose the correct one from among the alternatives *a*, *b*, *c*, *d* after matching an item from Group 1 with the most appropriate item in Group 2.

	Group 1		Group 2
1.	FM	P.	Slope overload
2.	DM	Q.	μ-law
3.	PSK	R.	Envelope detector
4.	PCM	S.	Capture effect
		T.	Hilbert transform
		U.	Matched filter

(a) 1-T, 2-P, 3-U, 4-S
(b) 1-S, 2-U, 3-P, 4-T
(c) 1-S, 2-P, 3-U, 4-Q
(d) 1-U, 2-R, 3-S, 4-Q **[2004 : 2 M]**

3.35 Three analog signals, having bandwidths 1200 Hz, 600 Hz and 600 Hz, are sampled at their respective Nyquist rates, encoded with 12 bit words, and time division multiplexed. The bit rate for the multiplexed signal is

(a) 115.2 kbps (b) 28.8 kbps
(c) 57.6 kpbs (d) 38.4 kbps

[2004 : 2 M]

3.36 Consider the signal $x(t)$ shown in the figure. Let $h(t)$ denote the impulse response of the filter matched to $x(t)$, with $h(t)$ being non-zero only in the interval 0 to 4 sec. The slope of $h(t)$ in the interval $3 < t < 4$ sec is

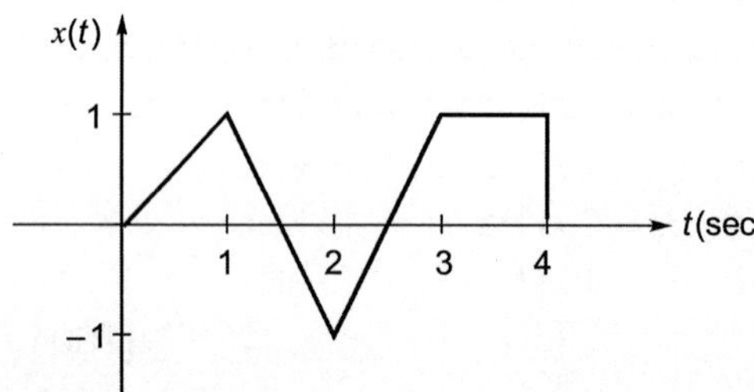

(a) $\frac{1}{2}\text{sec}^{-1}$ (b) $-1\ \text{sec}^{-1}$

(c) $-\frac{1}{2}\text{sec}^{-1}$ (d) $1\ \text{sec}^{-1}$

[2004 : 2 M]

3.37 A signal as shown in the figure is applied to a matched filter. Which of the following does represent the output of this matched filter?

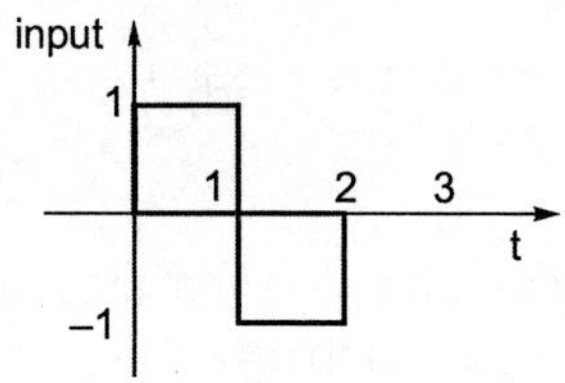

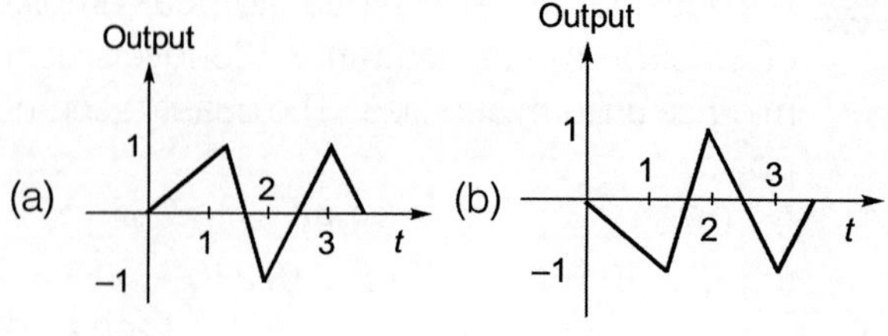

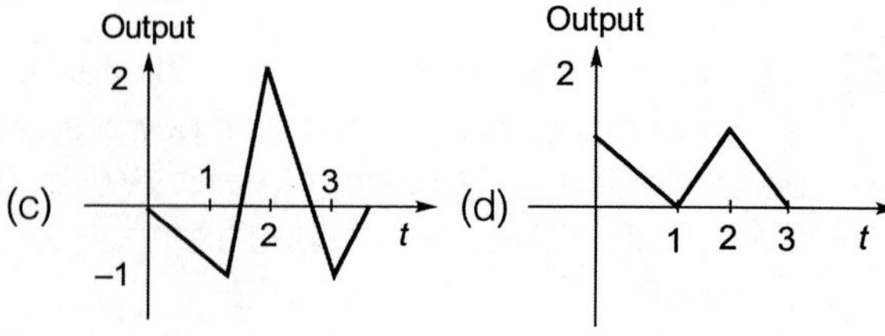

[2005 : 2 M]

3.38 The minimum sampling frequency (in samples/sec) required to reconstruct the following signal from its samples without distortion

$$x(t) = 5\left(\frac{\sin 2\pi 1000t}{\pi t}\right)^3 + 7\left(\frac{\sin 2\pi 1000t}{\pi t}\right)^2$$

would be

(a) 2×10^3 (b) 4×10^3
(c) 6×10^3 (d) 8×10^3 **[2006 : 2 M]**

3.39 The minimum step-size required for a Delta-Modulator operating at 32 K samples/sec to track the signal (here $u(t)$ is the unit-step function)

$$x(t) = 125t(u(t) - u(t-1)) + (250 - 125t)(u(t-1) - u(t-2))$$

so that slope-overload is avoided, would be

(a) 2^{-10} (b) 2^{-8}
(c) 2^{-6} (d) 2^{-4} **[2006 : 2 M]**

3.40 In the following figure the minimum value of the constant "C", which is to be added to $y_1(t)$ such that $y_1(t)$ and $y_2(t)$ are different, is

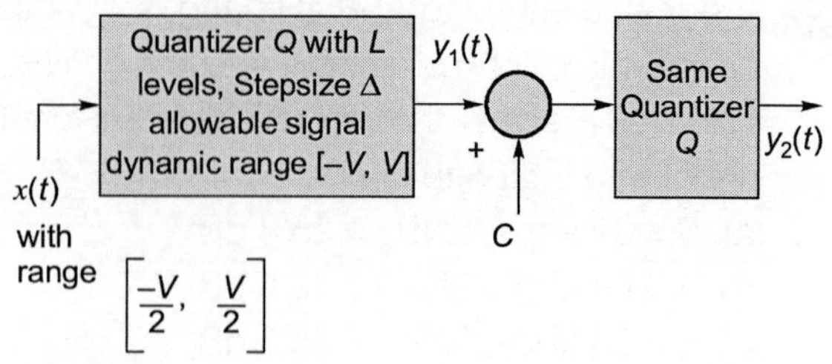

(a) Δ (b) $\Delta/2$
(c) $\Delta^2/12$ (d) Δ/L

[2006 : 2 M]

Common Data for Questions 3.41 and 3.42:

Let $g(t) = p(t)*p(t)$, where '$*$' denotes convolution and $p(t) = u(t) - u(t-1)$ with $u(t)$ being the unit step function.

3.41 The impulse response of filter matched to the signal $s(t) = g(t) - \delta(t-2) * g(t)$ is given as:

(a) $s(1-t)$ (b) $-s(1-t)$
(c) $-s(t)$ (d) $s(t)$

[2006 : 2 M]

3.42 An Amplitude Modulated signal is given as $x_{AM}(t) = 100(p(t) + 0.5g(t))\cos\omega_c t$ in the interval $0 \le t \le 1$. One set of possible values of the modulating signal and modulation index would be

(a) t, 0.5 (b) t, 1.0
(c) t, 2.0 (d) t^2, 0.5

[2006 : 2 M]

3.43 In delta modulation, the slope overload distortion can be reduced by

(a) decreasing the step size
(b) decreasing the granular noise
(c) decreasing the sampling rate
(d) increasing the step size

[2007 : 2 M]

3.44 The raised cosine pulse $p(t)$ is used for zero ISI in digital communications. The expression for $p(t)$ with unity roll-off factor is given by

$$p(t) = \frac{\sin 4\pi Wt}{4\pi Wt\left(1-16W^2t^2\right)}.$$

The value of $p(t)$ at $t = \frac{1}{4W}$ is

(a) –0.5 (b) 0
(c) 0.5 (d) ∞

[2007 : 2 M]

3.45 During transmission over a certain binary communication channel, bit errors occur independently with probability p. The probability of AT MOST one bit in error in a block of n bits is given by

(a) p^n
(b) $1 - p^n$
(c) $np(1-p)^{n-1} + (1-p)^n$
(d) $1-(1-p)^n$

[2007 : 2 M]

3.46 In a GSM system, 8 channels can co-exist in 200 kHz bandwidth using TDMA. A GSM based cellular operator is allocated 5 MHz bandwidth. Assuming a frequency reuse factor of 1/5, i.e. a five-cell repeat pattern, the maximum number of simultaneous channels that can exist in one cell is

(a) 200 (b) 40
(c) 25 (d) 5

[2007 : 2 M]

3.47 In a Direct Sequence CDMA system the chip rate is 1.2288×10^6 chips per second. If the processing gain is desired to be AT LEAST 100, the data rate

(a) must be less than or equal to 12.288×10^3 bits per sec
(b) must be greater than 12.288×10^3 bits per sec
(c) must be exactly equal to 12.288×10^3 bits per sec
(d) can take any value less than 122.88×10^3 bits per sec

[2007 : 2 M]

Common Data for Questions 3.48 & 3.49:

Two 4-ary signal constellations are shown. It is given that ϕ_1 and ϕ_2 constitute an orthonormal basis for the two constellations. Assume that the four symbols in both the constellations are equiprobable. Let $N_0/2$ denote the power spectral density of white Gaussian noise.

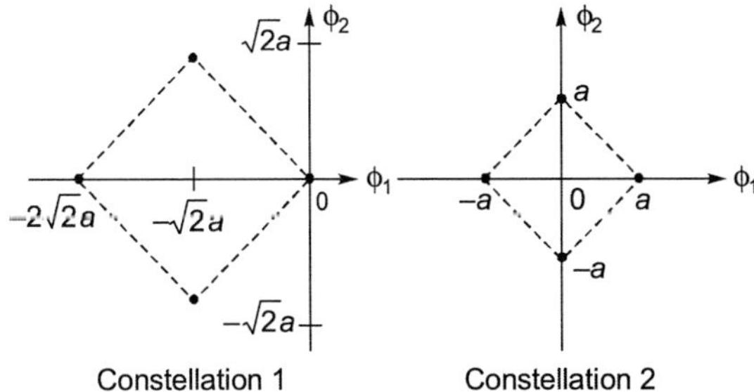

3.48 The ratio of the average energy of Constellation 1 to the average energy of Constellation 2 is

(a) $4a^2$ (b) 4
(c) 2 (d) 8

[2007 : 2 M]

3.49 If these constellations are used for digital communications over an AWGN channel, then which of the following statements is true?

(a) Probability of symbol error for Constellation 1 is lower
(b) Probability of symbol error for Constellation 1 is higher
(c) Probability of symbol error is equal for both the constellations
(d) The value of N_0 will determine which of the two constellations has a lower probability of symbol error

[2007 : 2 M]

Statement for Linked Answer Questions 3.50 and 3.51

An input to a 6-level quantizer has the probability density function $f(x)$ as shown in the figure. Decision boundaries of the quantizer are chosen so as to maximize the entropy of the quantizer output. It is given that 3 consecutive decision boundaries are '–1', '0' and '1'.

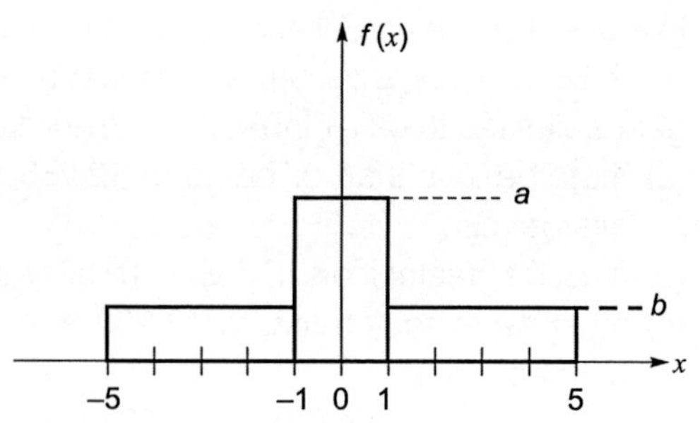

3.50 The values of a and b are
(a) $a = 1/6$ and $b = 1/12$
(b) $a = 1/5$ and $b = 3/40$
(c) $a = 1/4$ and $b = 1/16$
(d) $a = 1/3$ and $b = 1/24$ [2007 : 2 M]

3.51 Assuming that the reconstruction levels of the quantizer are the mid-points of the decision boundaries, the ratio of signal power to quantization noise power is
(a) $\frac{152}{9}$ (b) $\frac{64}{3}$
(c) $\frac{76}{3}$ (d) 28 [2007 : 2 M]

3.52 Consider a Binary Symmetric Channel (BSC) with probability of error being p. To transmit a bit, say 1, we transmit a sequence of three 1s. The receiver will interpret the received sequence to represent 1 if at least two bits are 1. The probability that the transmitted bit will be received in error is
(a) $p^3 + 3p^2(1-p)$ (b) p^3
(c) $(1-p)^3$ (d) $p^3 + p^2(1-p)$
[2008 : 2 M]

3.53 Four messages band limited to W, W, 2 W and 3 W respectively are to be multiplexed using Time Division Multiplexing (TDM). The minimum bandwidth required for transmission of this TDM signal is
(a) W (b) 3 W
(c) 6 W (d) 7 W [2008 : 2 M]

Common Data for Questions 3.54, 3.55 and 3.56.
A speech signal, band limited to 4 kHz and peak voltage varying between +5 V and –5 V, is sampled at the Nyquist rate. Each sample is quantized and represented by 8 bits.

3.54 If the bits 0 and 1 are transmitted using bipolar pulses, the minimum bandwidth required for distortion free transmission is
(a) 64 kHz (b) 32 kHz
(c) 8 kHz (d) 4 kHz
[2008 : 2 M]

3.55 Assuming the signal to be uniformly distributed between its peak to peak value, the signal to noise ratio at the quantizer output is
(a) 16 dB (b) 32 dB
(c) 48 dB (d) 64 dB
[2008 : 2 M]

3.56 The number of quantization levels required to reduce the quantization noise by a factor of 4 would be
(a) 1024 (b) 512
(c) 256 (d) 64
[2008 : 2 M]

Common data for Questions 3.57 and 3.58
The amplitude of a random signal is uniformly distributed between –5 V and 5 V.

3.57 If the signal to quantization noise ratio required in uniformly quantizing the signal is 43.5 dB, the step size of the quantization is approximately
(a) 0.0333 V (b) 0.05 V
(c) 0.0667 V (d) 0.10 V
[2009 : 2 M]

3.58 If the positive values of the signal are uniformly quantized with a step size of 0.05 V, and the negative values are uniformly quantized with a step size of 0.1 V, the resulting signal to quantization noise ratio is approximately
(a) 46 dB (b) 43.8 dB
(c) 42 dB (d) 40 dB
[2009 : 2 M]

3.59 Consider the pulse shape $s(t)$ as shown. The impulse response $h(t)$ of the filter matched to this pulse is

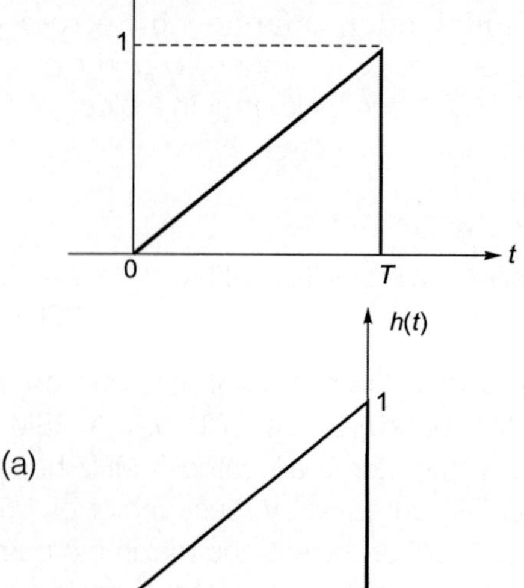

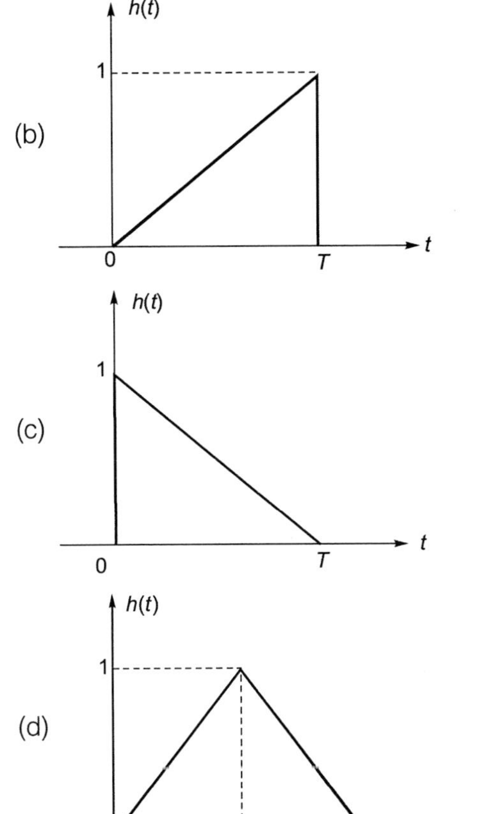

[2010 : 1 M]

3.60 The Nyquist sampling rate for the signal $s(t) = \frac{\sin(500\pi t)}{\pi t} \times \frac{\sin(700\pi t)}{\pi t}$ is given by

(a) 400 Hz (b) 600 Hz
(c) 1200 Hz (d) 1400 Hz [2010 : 2 M]

3.61 An analog signal is band-limited to 4 kHz, sampled at the Nyquist rate and the samples are quantized into 4 levels. The quantized levels are assumed to be independent and equally probable. If we transmit two quantized samples per second, the information rate is

(a) 1 bit/sec (b) 2 bits/sec
(c) 3 bits/sec (d) 4 bits/sec

[2011 : 1 M]

Linked Data Questions 3.62 & 3.63.

A four-phase and an eight-phase signal constellation are shown in the figure below.

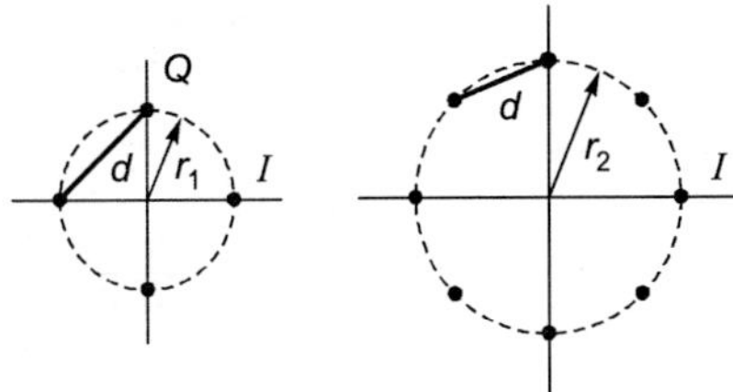

3.62 For the constraint that the minimum distance between pairs of signal points be d for both constellations, the radii r_1 and r_2 of the circles are

(a) $r_1 = 0.707\,d$, $r_2 = 2.782\,d$
(b) $r_1 = 0.707\,d$, $r_2 = 1.932\,d$
(c) $r_1 = 0.707\,d$, $r_2 = 1.545\,d$
(d) $r_1 = 0.707d$, $r_2 = 1.307\,d$ [2011 : 2 M]

3.63 Assuming high SNR and that all signals are equally probable, the additional average transmitted signal energy required by the 8-PSK signal to achieve the same error probability as the 4-PSK signal is

(a) 11.90 dB (b) 8.73 dB
(c) 6.79 dB (d) 5.33 dB

[2011 : 2 M]

3.64 In a baseband communications link, frequencies upto 3500 Hz are used for signaling. Using a raised cosine pulse with 75% excess bandwidth and for no inter-symbol interference, the maximum possible signaling rate in symbols per second is

(a) 1750 (b) 2625
(c) 4000 (d) 5250

[2012 : 1 M]

3.65 A binary symmetric channel (BSC) has a transition probability of 1/8. If the binary transmit symbol X is such that $P(X = 0) = 9/10$, then the probability of error for an optimum receiver will be

(a) 7/80 (b) 63/80
(c) 9/10 (d) 1/10 [2012 : 2 M]

3.66 A BPSK scheme operating over an AWGN channel with noise power spectral density of $N_0/2$, uses equiprobable signals

$$s_1(t) = \sqrt{\frac{2E}{T}}\sin(\omega_c t) \text{ and } s_2(t) = -\sqrt{\frac{2E}{T}}\sin(\omega_c t)$$

over the symbol internal $(0, T)$. If the local oscillator in a coherent receiver is ahead in phase by 45° with respect to the received signal, the probability of error in the resulting system is

(a) $Q\left(\sqrt{\frac{2E}{N_0}}\right)$ (b) $Q\left(\sqrt{\frac{E}{N_0}}\right)$

(c) $Q\left(\sqrt{\frac{E}{2N_0}}\right)$ (d) $Q\left(\sqrt{\frac{E}{4N_0}}\right)$

[2012 : 2 M]

3.67 The bit rate of digital communication system is R kbits/s. The modulation used is 32-QAM. The minimum bandwidth required for ISI free transmission is

(a) $R/10$ Hz (b) $R/10$ kHz
(c) $R/5$ Hz (d) $R/5$ kHz

[2013 : 1 M]

3.68 Let $g(t) = e^{-\pi t^2}$, and $h(t)$ is a filter matched to $g(t)$. If $g(t)$ is applied as input to $h(t)$, then the Fourier transform of the output is

(a) $e^{-\pi f^2}$ (b) $e^{-\pi f^2/2}$
(c) $e^{-\pi|f|}$ (d) $e^{-2\pi f^2}$

[2013 : 1 M]

Common Data For Questions 3.69 and 3.70:

Bits 1 and 0 are transmitted with equal probability. At the receiver, the pdf of the respective received signals for both bits are as shown below:

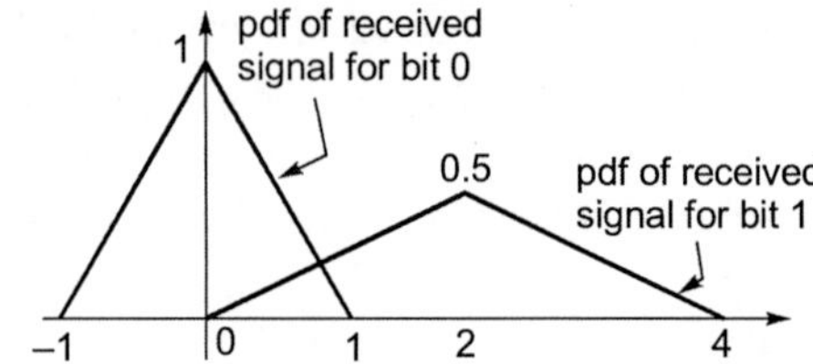

3.69 If the detection threshold is 1, the BER will be

(a) $\frac{1}{2}$ (b) $\frac{1}{4}$
(c) $\frac{1}{8}$ (d) $\frac{1}{16}$

[2013 : 2 M]

3.70 The optimum threshold to achieve minimum bit error rate (BER) is

(a) $\frac{1}{2}$ (b) $\frac{4}{5}$
(c) 1 (d) $\frac{3}{2}$

[2013 : 2 M]

3.71 Let $Q(\sqrt{\gamma})$ be the BER of a BPSK system over an AWGN channel with two-sided noise power spectral density $N_0/2$. The parameter γ is a function of bit energy and noise power spectral density.

A system with two independent and identical AWGN channels with noise power spectral density $N_0/2$ is shown in the figure. The BPSK demodulator receives the sum of outputs of both the channels.

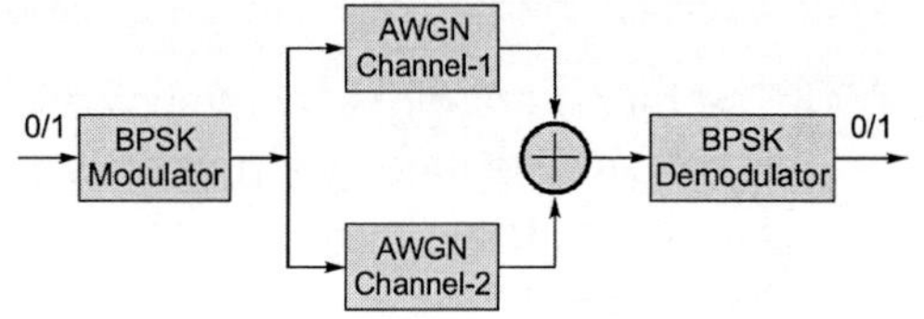

If the BER of this system is $Q(b\sqrt{\gamma})$, then the value of b is ______.

[2014 : 2 M, Set-1]

3.72 Coherent orthogonal binary FSK modulation is used to transmit two equiprobable symbol waveforms $s_1(t) = \alpha\cos 2\pi f_1 t$ & $s_2(t) = \alpha\cos 2\pi f_2 t$, where $\alpha = 4$ mV. Assume an AWGN channel with two-sided noise power spectral density $\frac{N_0}{2} = 0.5 \times 10^{-12}$ W/Hz. Using an optimal receiver and the relation $Q(v) = \frac{1}{\sqrt{2\pi}}\int_v^{\infty} e^{-\frac{u^2}{2}} du$, the bit error probability for a data rate of 500 kbps is

(a) $Q(2)$ (b) $Q(2\sqrt{2})$
(c) $Q(4)$ (d) $Q(4\sqrt{2})$

[2014 : 2 M, Set-2]

3.73 An analog voltage in the range 0 to 8 V is divided in 16 equal intervals for conversion to 4-bit digital output. The maximum quantization error (in V) is ____.

[2014 : 1 M, Set-3]

3.74 In a PCM system, the signal

$$m(t) = \sin(100\pi t) + \cos(100\pi t)$$

is sampled at the Nyquist rate. The samples are processed by a uniform quantizer with step size 0.75 V. The minimum data rate of the PCM system in bits per second is _____.

[2014 : 2 M, Set-3]

3.75 An M-level PSK modulation scheme is used to transmit independent binary digits over a band-pass channel with bandwidth 100 kHz. The bit rate is 200 kbps and the system characteristic is a raised-cosine spectrum with 100% excess bandwidth. The minimum value of M is ________.

[2014 : 2 M, Set-4]

3.76 A sinusoidal signal of 2 kHz frequency is applied to a delta modulator. The sampling rate and step-size Δ of the delta modulator are 20,000 samples per second and 0.1 V, respectively. To prevent slope overload, the maximum amplitude of the sinusoidal signal (in Volts) is

(a) $\frac{1}{2\pi}$ (b) $\frac{1}{\pi}$
(c) $\frac{2}{\pi}$ (d) π

[2015 : 1 M, Set-1]

3.77 The input X to Binary Symmetric Channel (BSC) shown in the figure is '1' with probability of 0.8. The cross-over probability is 1/7. If the received bit $Y = 0$, the conditional probability that '1' was transmitted is ________.

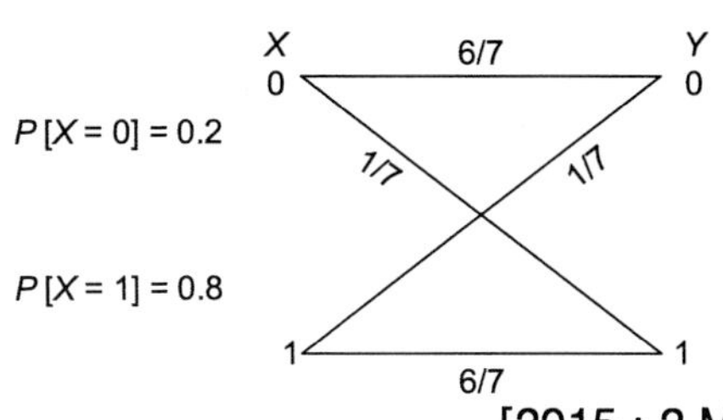

[2015 : 2 M, Set-1]

3.78 The transmitted signal in a GSM system is of 200 kHz bandwidth and 8 users share a common bandwidth using TDMA. If at a given time 12 users are talking in a cell, the total bandwidth of the signal received by the base station of the cell will be at least (in kHz) ________.

[2015 : 2 M, Set-1]

3.79 A source emits bit 0 with probability $\frac{1}{3}$ and bit 1 with probability $\frac{2}{3}$. The emitted bits are communicated to the receiver. The receiver decides for either 0 or 1 based on the received value R. It is given that the conditional density functions of R as

$$f_{R|0}(r) = \begin{cases} \frac{1}{4}, & -3 \le r \le 1 \\ 0, & \text{otherwise} \end{cases} \quad \text{and}$$

$$f_{R|1}(r) = \begin{cases} \frac{1}{6}, & -1 \le r \le 5 \\ 0, & \text{otherwise} \end{cases}$$

The minimum decision error probability is

(a) 0 (b) $\frac{1}{12}$

(c) $\frac{1}{9}$ (d) $\frac{1}{6}$

[2015 : 2 M, Set-1]

3.80 A sinusoidal signal of amplitude A is quantized by a uniform quantizer. Assume that the signal utilizes all the representation levels of the quantizer. If the signal to quantization noise ratio is 31.8 dB, the number of levels in the quantizer is____.

[2015 : 1 M, Set-2]

3.81 Consider a binary, digital communication system which uses pulses $g(t)$ and $-g(t)$ for transmitting bits over an AWGN channel. If the receiver uses a matched filter, which one of the following pulses will give the minimum probability of bit error?

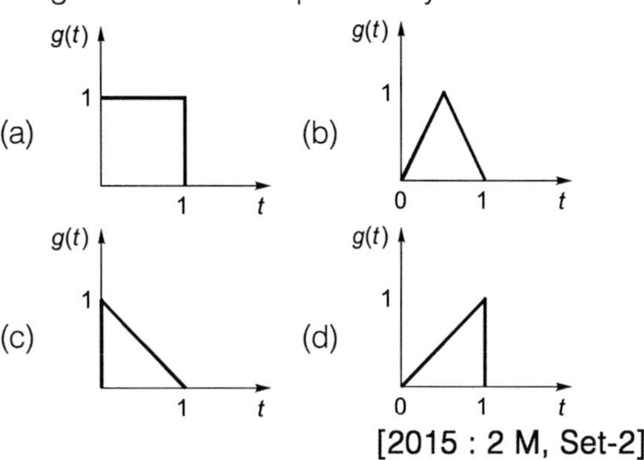

[2015 : 2 M, Set-2]

3.82 The modulation scheme commonly used for transmission from GSM mobile terminals is
(a) 4-QAM
(b) 16-PSK
(c) Walsh-Hadamard orthogonal codes
(d) Gaussian Minimum Shift Keying (GMSK)

[2015 : 1 M, Set-3]

3.83 Consider a binary data transmission at a rate of 56 kbps using baseband binary pulse amplitude modulation (PAM) that is designed to have a raised-cosine spectrum. The transmission bandwidth (in kHz) required of a roll-off factor of 0.25 is _____.

[2016 : 1 M, Set-1]

3.84 An analog pulse $s(t)$ is transmitted over an additive white Gaussian noise (A WGN) channel. The received signal is $r(t) = s(t) + n(t)$ where $n(t)$ is additive white Gaussian noise with power spectral density $\frac{N_0}{2}$. The received signal is passed through a filter with impulse response $h(t)$. Let E_s and E_h denote the energies of the pulse $s(t)$ and the filter $h(t)$, respectively. When the signal to noise ratio (SNR) is maximized at the output of the filter (SNR_{max}), which of the following holds?

(a) $E_s = E_h$; $SNR_{max} = \frac{2E_s}{N_0}$

(b) $E_s = E_h$; $SNR_{max} = \frac{E_s}{2N_0}$

(c) $E_s > E_h$; $SNR_{max} > \frac{2E_s}{N_0}$

(d) $E_s < E_h$; $SNR_{max} = \frac{2E_h}{N_0}$

[2016 : 2 M, Set-1]

3.85 An ideal band-pass channel 500 Hz- 2000 Hz is deployed for communication. A modem is designed to transmit bits at the rate of 4800 bits/s using 16-QAM. The roll-off factor of a pulse with a raised cosine spectrum that utilizes the entire frequency band is ______

[2016 : 2 M, Set-2]

3.86 A speech signal is sampled at 8 kHz and encoded into PCM format using 8 bits/sample. The PCM data is transmitted through a baseband channel via 4-level PAM. The minimum bandwidth (in kHz) required for transmission is _____.

[2016 : 1 M, Set-2]

3.87 A binary baseband digital communication system employs the signal

$$p(t) = \begin{cases} \dfrac{1}{\sqrt{T_s}}, 0 \le t \le T_s \\ 0, \quad \text{otherwise} \end{cases}$$

for transmission of bits. The graphical representation of the matched filter output $y(t)$ for this signal will be

(a)

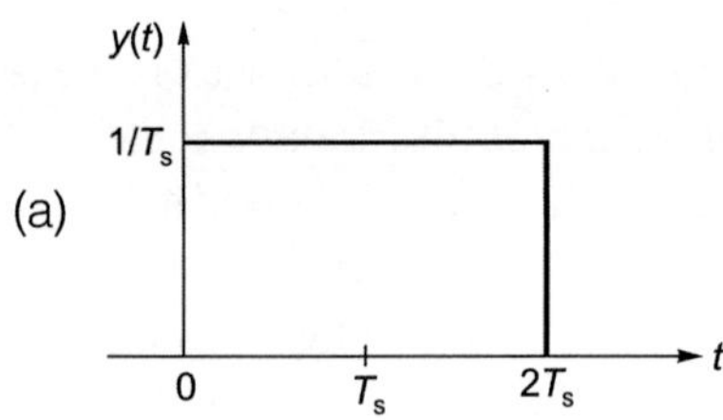

(b)

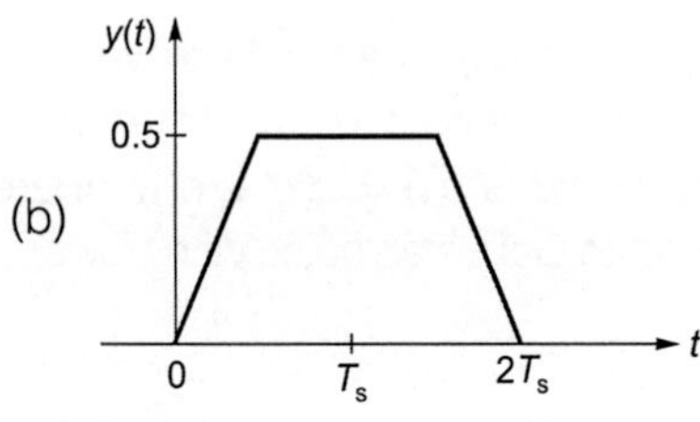

(c)

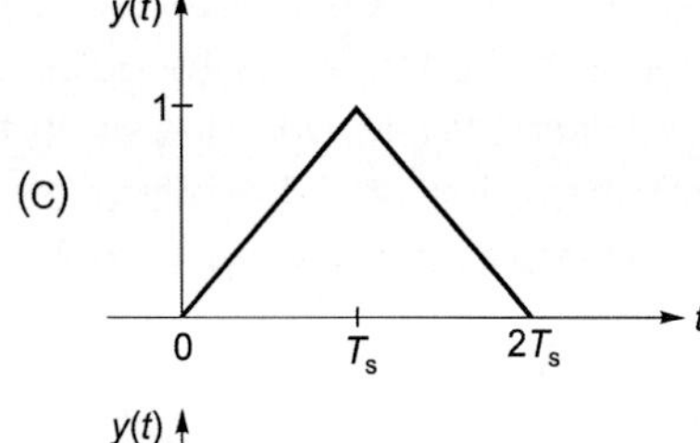

(d)

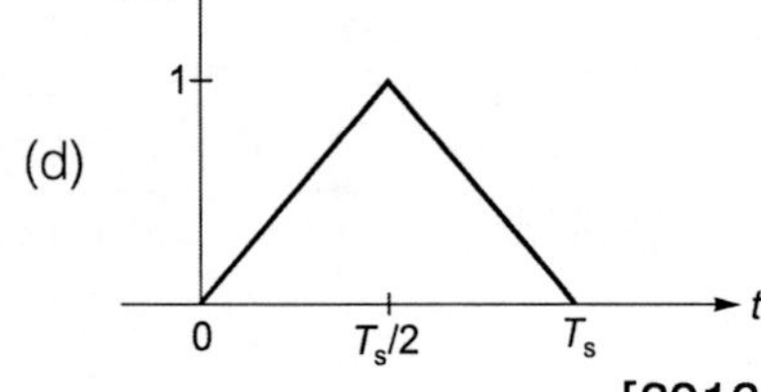

[2016 : 1 M, Set-3]

3.88 The bit error probability of a memoryless binary symmetric channel is 10^{-5}. If 10^5 bits are sent over this channel, then the probability that not more than one bit will be in error is ________.

[2016 : 2 M, Set-3]

3.89 In a digital communication system, the overall pulse shape $p(t)$ at the receiver before the sampler has the Fourier transform $P(f)$. If the symbols are transmitted at the rate of 2000 symbols per second, for which of the following cases is the inter symbol interference zero?

(a)

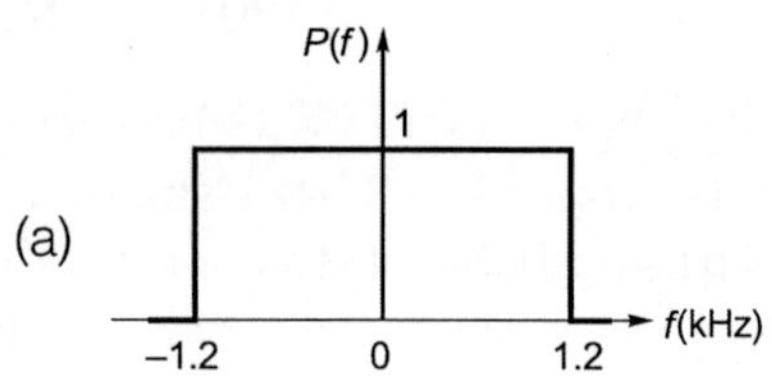

(b)

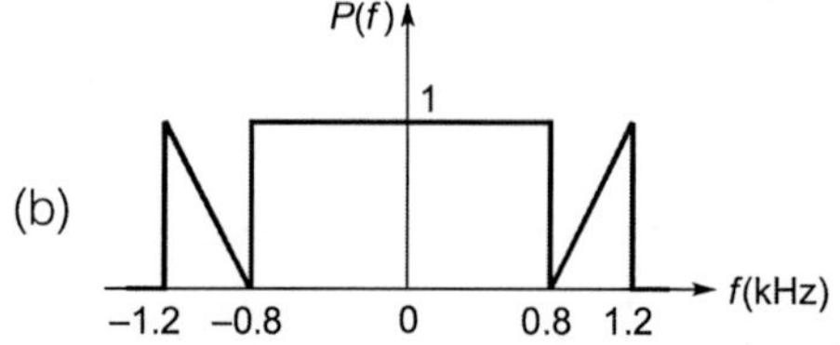

(c)

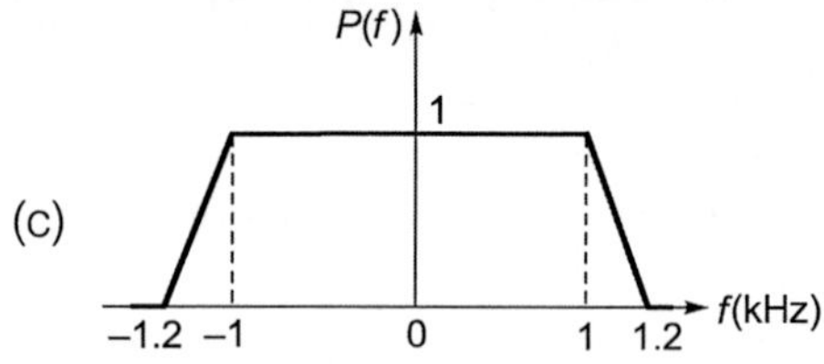

(d)

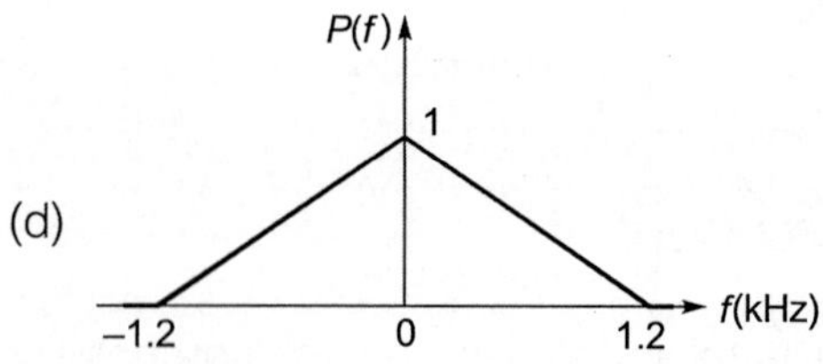

[2017 : 1 M, Set-1]

3.90 Which one of the following statements about differential pulse code modulation (DPCM) is true?

(a) The sum of message signal sample with its prediction is quantized.
(b) The message signal sample is directly quantized, and its prediction is not used.
(c) The difference of message signal sample and a random signal is quantized.
(d) The difference of message signal sample with its prediction is quantized.

[2017 : 1 M, Set-1]

3.91 In binary frequency shift keying (FSK), the given signal waveforms are

$u_0(t) = 5\cos(20000\pi t)$; $0 \le t \le T$ and
$u_1(t) = 5\cos(22000\pi t)$; $0 \le t \le T$,

where T is the bit-duration interval and t is in seconds. Both $u_0(t)$ and $u_1(t)$ are zero outside the interval $0 \le t \le T$. With a matched filter (correlator) based receiver, the smallest positive value of T (in milliseconds) required to have $u_0(t)$ and $u_1(t)$ uncorrelated is

(a) 0.25 ms (b) 0.5 ms
(c) 0.75 ms (d) 1.0 ms

[2017 : 2 M, Set-1]

3.92 A sinusoidal message signal is converted to a PCM signal using a uniform quantizer. The required signal to quantization noise ratio (SQNR) at the output of the quantizer is 40 dB. The minimum number of bits per sample needed to achieve the desired SQNR is ______. [2017 : 1 M, Set-2]

3.93 A random variable X takes values –0.5 and 0.5 with probabilities $\frac{1}{4}$ and $\frac{3}{4}$, respectively. The noisy observation of X is $Y = X + Z$, where Z has uniform probability density over the interval (–1, 1). X and Z are independent. If the MAP rule based detector outputs $\hat{X}$ as

$$\hat{X} = \begin{cases} -0.5, & Y < \alpha \\ 0.5, & Y \ge \alpha, \end{cases}$$

then the value of α (accurate to two decimal places) is ______.

[2018 : 2 M]

3.94 A binary source generates symbols $X \in \{-1, 1\}$ which are transmitted over a noisy channel. The probability of transmitting $X = 1$ is 0.5. Input to the threshold detector is $R = X + N$. The probability density function $f_N(n)$ of the noise N is shown below.

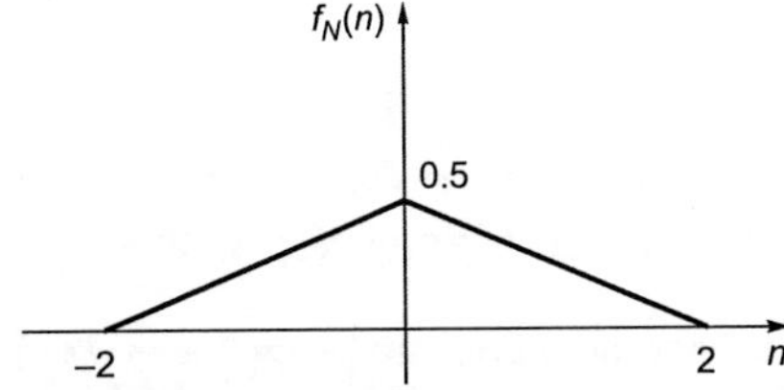

If the detection threshold is zero, then the probability of error (correct to two decimal places) is __________.

[2018 : 1 M]

3.95 A single bit, equally likely to be 0 and 1, is to be sent across an additive white Gaussian noise (AWGN) channel with power spectral density $N_o/2$. Binary signaling, with $0 \to p(t)$ and $1 \to q(t)$, is used for the transmission, along with an optimal receiver that minimizes the bit-error probability.

Let $\varphi_1(t), \varphi_2(t)$ form an orthonormal signal set.

If we choose $p(t) = \varphi_1(t)$ and $q(t) = -\varphi_1(t)$, we would obtain a certain bit-error probability P_b.

If we keep $p(t) = \varphi_1(t)$, but take $q(t) = \sqrt{E}\,\varphi_2(t)$, for what value of E would we obtain the same bit-error probability P_b?

(a) 0 (b) 3
(c) 1 (d) 2

[2019 : 2 M]

3.96 A voice signal $m(t)$ is in the frequency range 5 kHz to 15 kHz. The signal is amplitude- modulated to generate an AM signal $f(t) = A(1 + m(t)) \cos 2\pi f_c t$, where f_c = 600 kHz. The AM signal $f(t)$ is to be digitized and archived. This is done by first sampling $f(t)$ at 1.2 times the Nyquist frequency, and then quantizing each sample using a 256-level quantizer. Finally, each quantized sample is binary coded using K bits, where K is the minimum number of bits required for the encoding. The rate, in Megabits per second (rounded off to 2 decimal places), of the resulting stream of coded bits is ____ Mbps.

[2019 : 2 M]

3.97 A random variable X takes values –1 and +1 with probabilities 0.2 and 0.8. respectively. It is transmitted across a channel which adds noise N, so that the random variable at the channel output is $Y = X + N$. The noise N is independent of X, and is uniformly distributed over the interval [–2 , 2]. The receiver makes a decision

$$\hat{X} = \begin{cases} -1, & \text{if } Y \le \theta \\ +1, & \text{if } Y < \theta \end{cases}$$

where the threshold $\theta \in [-1, 1]$ is chosen so as to minimize the probability of error $\Pr[\hat{X} \ne X]$. The minimum probability of error, rounded off to 1 decimal place, is ____.

[2019 : 2 M]

3.98 In a digital communication system, a symbol S randomly chosen from the set (s_1, s_2, s_3, s_4) is transmitted. It is given that $s_1 = -3$, $s_2 = -1$, $s_3 = +1$ and $s_4 = +2$. The received symbol is $Y = S + W$. W is a zero-mean unit-variance Gaussian random variable and is independent of S. P_i is the conditional probability of symbol error for the maximum likelihood (ML) decoding when the transmitted symbol $S = s_i$. The index i for which the conditional symbol error probability P_i is the highest is _____.

[2020 : 2 M]

3.99 A speech signal, band limited to 4 kHz, is sampled at 1.25 times the Nyquist rate. The speech samples, assumed to be statistically independent and uniformly distributed in the range −5 V to +5 V, are subsequently quantized in an 8-bit uniform quantizer and then transmitted over a voice-grade AWGN telephone channel. If the ratio of transmitted signal power to channel noise power is 26 dB, the minimum channel bandwidth required to ensure reliable transmission of the signal with arbitrarily small probability of transmission error (rounded off to two decimal places) is _______ kHz.

[2021 : 1 M]

3.100 A 4 kHz sinusoidal message signal having amplitude 4 V is fed to a delta modulator (DM) operating at a sampling rate of 32 kHz. The minimum step size required to avoid slope overload noise in the DM (rounded off to two decimal places) is _____ V.

[2021 : 1 M]

3.101 A message signal having peak-to-peak value of 2 V, root mean square value of 0.1 V and bandwidth of 5 kHz is sampled and fed to a pulse code modulation (PCM) system that uses a uniform quantizer. The PCM output is transmitted over a channel that can support a maximum transmission rate of 50 kbps. Assuming that the quantization error is uniformly distributed, the maximum signal to quantization noise ratio that can be obtained by the PCM system (rounded off to two decimal places) is ______.

[2021 : 2 M]

3.102 Consider a polar non-return to zero (NRZ) waveform using +2 V and −2 V for representing binary '1' and '0' respectively, is transmitted in the presence of additive zero-mean white Gaussian noise with variance 0.4 V^2. If the a priori probability of transmission of a binary '1' is 0.4, the optimum threshold voltage for a maximum a posteriori (MAP) receiver (rounded off to two decimal places) is ____ V.

[2021 : 2 M]

3.103 Let $H(X)$ denote the entropy of a discrete random variable X taking K possible distinct real values. Which of the following statements is/are necessarily true?

(a) $H(X) \le \log_2 K$ bits
(b) $H(X) \le H(2X)$
(c) $H(X) \le H(X^2)$
(d) $H(X) \le H(2^X)$

[2022 : 1 M]

3.104 A symbol stream contains alternate QPSK and 16-QAM symbols. If symbols from this stream are transmitted at the rate of 1 mega-symbols per second, the raw (uncoded) data rate is _______ mega-bits per second (rounded off to one decimal place).

[2022 : 1 M]

3.105 Consider a channel over which either symbol x_A or symbol x_B is transmitted. Let the output of the channel Y be the input to a maximum likelihood (ML) detector at the receiver. The conditional probability density functions for y given x_A and x_B are:

$$f_{Y|x_A}(y) = e^{-(y+1)}\,u(y+1),$$

$$f_{Y|x_B}(y) = e^{(y-1)}\,(1-u(y-1)),$$

where $u(\cdot)$ is the standard unit step function. The probability of symbol error for this system is ________ (rounded off to two decimal places).

[2022 : 2 M]

3.106 The signal-to-noise ratio (SNR) of an ADC with a full-scale sinusoidal input is given to be 61.96 dB. The resolution of the ADC is _____ bits. (rounded off to the nearest integer).

[2023 : 1 M]

3.107 TA digital communication system transmits through a noiseless bandlimited channel $[-W\ W]$. The received signal $z(t)$ at the output of the receiving filter is given by $z(t) = \sum_n b[n]\,x(t - nT)$ where $b[n]$ are the symbols and $x(t)$ is the overall system response to a single symbol. The received signal is sampled at $t = mT$. The Fourier transform of $x(t)$ is $X(f)$. The Nyquist condition that $X(f)$ must satisfy for zero intersymbol interference at the receiver is _______.

(a) $\sum_{m=-\infty}^{\infty} X\left(f+\frac{m}{T}\right)=T$

(b) $\sum_{m=-\infty}^{\infty} X\left(f+\frac{m}{T}\right)=\frac{1}{T}$

(c) $\sum_{m=-\infty}^{\infty} X(f+mT)=T$

(d) $\sum_{m=-\infty}^{\infty} X(f+mT)=\frac{1}{T}$ [2024 : 1 Mark]

3.108 A full scale sinusoidal signal is applied to a 10-bit ADC. The fundamental signal component in the ADC output has a normalized power of 1 W, and the total noise and distortion normalized power is 10 μW. The effective number of bits (rounded off to the nearest integer) of the ADC is ______.

(a) 10 (b) 9
(c) 8 (d) 7

[2024 : 2 Marks]

3.109 Let $X(t) = A\cos(2\pi f_0 t + \theta)$ be a random process, where amplitude A and phase θ are independent of each other, and are uniformly distributed in the intervals [–2, 2] and [0, 2π], respectively. $X(t)$ is fed to an 8-bit uniform mid-rise type quantizer. Given that the autocorrelation of $X(t)$ is $R_x(\tau)=\frac{2}{3}\cos(2\pi f_0\tau)$, the signal to quantization noise ratio (in dB, rounded off to two decimal places) at the output of the quantizer is _________.

[2024 : 2 Marks]

■■■

Answers Digital Communication Systems

3.1 (b)	3.2 (c)	3.3 (c)	3.4 (c)	3.5 (c)	3.6 (d)	3.7 (c)
3.8 (d)	3.9 (c)	3.10 (a)	3.11 (d)	3.12 (a)	3.13 (b)	3.14 (b)
3.15 (d)	3.16 (a)	3.17 (b)	3.18 (c)	3.19 (c)	3.20 (d)	3.21 (c)
3.22 (a)	3.23 (d)	3.24 (b)	3.25 (c)	3.26 (d)	3.27 (b)	3.28 (b)
3.29 (b)	3.30 (c)	3.31 (a)	3.32 (c)	3.33 (a)	3.34 (c)	3.35 (c)
3.36 (b)	3.37 (c)	3.38 (c)	3.39 (b)	3.40 (b)	3.41 (c)	3.42 (a)
3.43 (d)	3.44 (c)	3.45 (c)	3.46 (b)	3.47 (a)	3.48 (b)	3.49 (a)
3.50 (a)	3.51 (d)	3.52 (a)	3.53 (d)	3.54 (a)	3.55 (c)	3.56 (b)
3.57 (c)	3.58 (c)	3.59 (c)	3.60 (c)	3.61 (d)	3.62 (d)	3.63 (d)
3.64 (c)	3.65 (d)	3.66 (b)	3.67 (d)	3.68 (d)	3.69 (d)	3.70 (b)
3.71 (1.414)	3.72 (c)	3.73 (0.25)	3.74 (200)	3.75 (17)	3.76 (a)	3.77 (0.4)
3.78 (400)	3.79 (d)	3.80 (32)	3.81 (a)	3.82 (d)	3.83 (35)	3.84 (a)
3.85 (0.25)	3.86 (16)	3.87 (c)	3.88 (0.7357)	3.89 (b)	3.90 (d)	3.91 (b)
3.92 (7)	3.93 (–0.50)	3.94 (0.125)	3.95 (b)	3.96 (11.81)	3.97 (0.10)	3.98 (3)
3.99 (9.25)	3.100 (3.14)	3.101 (30.72)	3.102 (0.0405)	3.103 (a, b, d)	3.104 (3)	3.105 (0.23)
3.106 (10)	3.107 (a)	3.108 (c)	3.109 (45.15)			

Explanations Digital Communication Systems

3.1 (b)

$$\boxed{(\text{SQNR})_{\text{dB}} = (1.76 + 6n)_{\text{dB}}}$$

$$(\text{SQNR})_1 = 1.76 + 6n$$

$$(\text{SQNR})_2 = 1.76 + 6(n+1) = 1.76 + 6n + 6$$

$$(\text{SQNR})_2 - (\text{SQNR})_1 = (1.76 + 6n + 6) - (1.76 + 6n)$$

$$(\text{SQNR})_2 - (\text{SQNR})_1 = 6 \text{ dB}$$

So, for every one bit increase in bits per sample will result in 6 dB improvement in signal to quantization noise ratio.

3.2 (c)

Given that

$$f_m = 1 \text{ kHz}$$

$$f_s = 1.8 \text{ k samples/sec}$$

The frequency components in the sampled signal are

$$\boxed{nf_s \pm f_m}$$

$n = 0 \Rightarrow f_m = 1 \text{ kHz} = 1000 \text{ Hz}$

$n = 1 \Rightarrow 1.8 \pm 1 = 800 \text{ Hz and } 2800 \text{ Hz}$

$n = 2 \Rightarrow 3.6 \pm 1 = 2600 \text{ Hz and } 4600 \text{ Hz}$

Cut-off frequency of LPF 1100 Hz.

So, 800 Hz and 1000 Hz components.

3.3 (c)

The signal to quantization noise ratio in an n-bit PCM system is given by

$$\boxed{(\text{SQNR})_{\text{dB}} = 1.76 + 6n}$$

$$\boxed{\text{SQNR} = \frac{3}{2}2^{2n}}$$

From the above equation it is clear that SQNR increases with increase in value of 'n'.

3.5 (c)

Given that

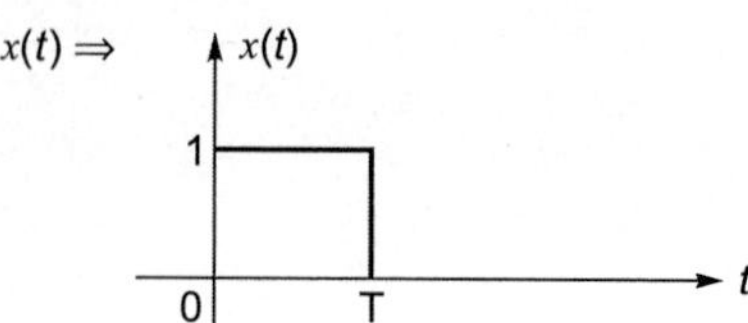

For the matched filter

$$h(t) = x(T - t)$$

$h(t) = x(T - t)$

1

0 T t

$$y(t) = x(t) * h(t)$$

$y(t)$ is a triangular pulse.

3.6 (d)

$$\text{SQNR} = \frac{3}{2}2^{2n}$$

Given that $n_1 = n$

$$n_2 = n + 1$$

$$(\text{SQNR})_1 = \frac{3}{2}2^{2n}$$

$$(\text{SQNR})_2 = \frac{3}{2}2^{2(n+1)} = \frac{3}{2}2^{2n+2}$$

$$= \frac{3}{2}[2^{2n} \cdot 2^2]$$

So, $$\frac{(\text{SQNR})_2}{(\text{SQNR})_1} = \frac{2^2}{1} = 4$$

$$(\text{SQNR})_2 = 4(\text{SQNR})_1$$

Signal to quantization noise ratio increases by a factor of 4. So, this improvement in SQNR is independent of 'n'.

3.7 (c)

Alternate mark inversion (AMI) code has zero dc component for pulse transmission of random binary data.

3.8 (d)

$$f_{s(\min)} = 2f_{(\max)} = 2 \times 2 \text{ kHz} = 4 \text{ kHz}$$

3.9 (c)

QNP of power signal is given by

$$\text{QNP} = \frac{\Delta^2}{12}$$

$$\text{Step size} = \Delta = \frac{V_{p-p}}{2^n}$$

$$\text{QNP} = \frac{V_{p-p}^2}{12 \times 2^{2n}} \; ; \; \text{QNP} \propto \frac{1}{2^{2n}}$$

$$\frac{(\text{QNP})_2}{(\text{QNP})_1} = \frac{2^{2n_1}}{2^{2n_2}} = \frac{2^{2\times 8}}{2^{2\times 9}} = \frac{2^{16}}{2^{18}} = \frac{1}{2^2} = \frac{1}{4}$$

$$(\text{QNP})_2 = \frac{(\text{QNP})_1}{4}$$

So, the QNP reduces by a factor of 4.

3.10 (a)

Flat top sampling of low pass signals gives rise to aperture effect.

3.11 (d)

Quadrature carrier multiplexing utilizes carrier phase shifting and synchronous detection to permit two DSB signals to occupy the same frequency band. It is the scheme where same carrier frequency is used for two different DSB signals. It is also known as quadrature amplitude modulation (QAM). So, quadrature multiplexing is quite different from FDM and TDM.

3.12 (a)

Compression in PCM refers to relative compression of higher signal amplitudes.

3.13 (b)

$\text{sinc}^3(400t) \Rightarrow f_{m1} = 600 \text{ Hz}$

$\text{sinc}^3(100t) \Rightarrow f_{m2} = 150 \text{ Hz}$

$\text{sinc}^3(400t) * \text{sinc}^3(100t) \Rightarrow f_m = \min(f_{m1}, f_{m2}) = 150 \text{ Hz}$

$$f_{s(\min)} = 2f_m = 2 \times 150 = 300 \text{ Hz}$$

3.14 (b)

SNR = 1.76 + 6 × 7 = 1.76 + 42 = 43.8

As input signal is $0.5\cos\omega_m t$, 128 levels are needed to quantize them.

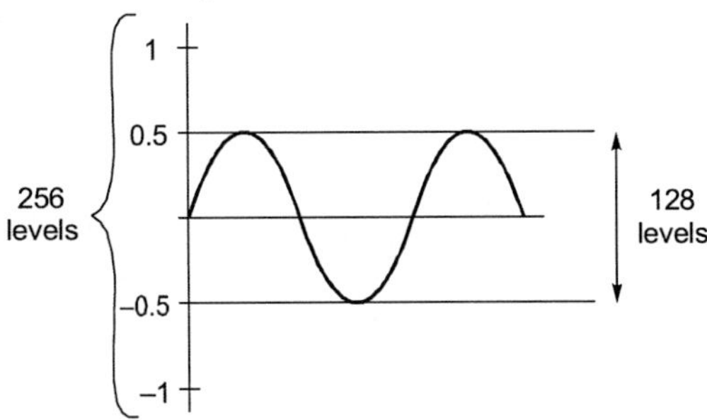

3.15 (d)

Maximum amplitude of matched filter output is

$$\frac{A^2T}{2} = \frac{10^2}{2} \times 10^{-4} = 5 \text{ mV}$$

3.16 (a)

Sample rate = 2(100 + 100 + 200 + 400)Hz
= 1600 Hz

3.17 (b)

$$f_1 = 25 \text{ kHz and } f_0 = 10 \text{ kHz}$$

$$f_1 - f_0 = 15 \text{ kHz}$$

For orthogonality,

$$f_1 - f_0 = \frac{n}{T_b}$$

So, $$T_b = \frac{n}{15k} = \frac{1000n}{15}\ \mu s$$

For $n = 3$, $T_b = 200\ \mu s$

Among the given options, only 200 μs is the integer multiple of (1000/15).

3.18 (c)

Frames per second = 625

Pixels per frame = 400 × 400

64 intensity levels per pixel can be represented by 6 bits per pixel.

∴Date Rate = 625 × 400 × 400 × 6 = 600 Mbps

3.19 (c)

$\sin c(700t) + \sin c(500t)$

Nyquist sampling rate = $2 f_m$ = 700 Hz

$$\therefore \quad \text{Sampling rate} = \frac{1}{700}\text{sec}$$

3.20 (d)

Probability of error = p

Probability of no error = $(1 - p)$

For error in at most one bit there should either be no error or error in only one bit.

$= {}^nC_o(1-p)^n p^0 + {}^nC_1(1-p)^{n-1} p^1$

$=(1-p)^n + np(1-p)^{n-1}$

3.21 (c)

$$Y(f) = \frac{5\times10^{-6}}{100\times10^{-6}} \sum_{n=-\infty}^{\infty} X(f - nf_s)$$

$$f_s = \frac{1}{T_s} = 10\text{ kHz}$$

$x(t)$ is a sinusoidal signal of 4 kHz.

Since $f_s > 8$ kHz and cut-off frequency f_c of the LPF satisfies 4 kHz < (f_c = 5 kHz) < 6 kHz, the output signal of the filter will be,

$$z(t) = \frac{1}{20}x(t) = \frac{10}{20}\cos(8\pi\times10^3 t)$$

$$= 5\times10^{-1}\cos(8\pi\times10^3 t)$$

3.22 (a)

For coherent BFSK system,

$$f_H - f_L = \frac{n}{2T_b} = n\times4\text{ kHz}$$

For best possible set of transmitted frequencies, $f_H - f_L$ should be as minimum as possible. i.e., n can be taken as 1.

$$f_H - f_L = 4\text{ kHz}$$

From option (a),

$$f_H - f_L = 20\text{ kHz} - 16\text{ kHz} = 4\text{ kHz}$$

3.23 (d)

50 μs ⇒ 20 kHz

$$f_m = 12\text{ kHz}$$

$$f_s = 20\text{ k samples/sec}$$

∴ Frequency at filter output with cut-off frequency of 15 k is 20 ± 12 = 8 k and f_m = 12 kHz

3.24 (b)

$$P_{e(\text{BPSK})} = Q\left(\sqrt{\frac{2E_{\text{BPSK}}}{N_0}}\right)$$

$$P_{e(\text{BFSK})} = Q\left(\sqrt{\frac{E_{\text{BFSK}}}{N_0}}\right)$$

To get same probability of error in both the schemes,

$$2E_{\text{BPSK}} = E_{\text{BFSK}}$$

$$\frac{E_{\text{BFSK}}}{E_{\text{BPSK}}} = 2\ \text{(or)}\ 3\text{ dB}$$

So, at a given probability of error, coherent BFSK is inferior to coherent BPSK by 3 dB.

3.25 (c)

$$\Delta = \frac{2M_p}{L} = \frac{1.536}{128}$$

$$P_N = \frac{\Delta^2}{12} = \frac{\left(\frac{1.536}{128}\right)^2}{12} = 12\times10^{-6}V^2$$

3.26 (d)

$$\text{SNR} = \frac{2E_b}{N_0} = \frac{2\times10^{-5}}{10^{-6}} = 20$$

$$(\text{SNR})_{\text{dB}} = 10\log 20 = 13\text{ dB}$$

3.27 (b)

For slope overload to take place,

$$\frac{\Delta}{T_s} \le A_m.\omega_m$$

$$\frac{\Delta}{T_s} = 0.628\times40\times10^3$$

= 25120 = 25.12 K

Matching with options $2\pi \times 1.5 \times 4\text{ K} = 37.7$ which is greater than Δ/T_s.

3.28 (b)

$$\text{Bandwidth efficiency} = \frac{1}{T_b \times \text{BW}}$$

For BPSK, $\text{BW} = \frac{1}{T_b}$

(with Nyquist pulse shaping of baseband data)

So, Bandwidth efficiency = 1

3.29 (b)

$$\text{Bit rate} = nf_s = 8 \times 8\text{ K} = 64\text{ Kbps}$$
$$\text{SNR} = 1.76 + 6n = 1.76 + 48 = 49.8$$

3.30 (c)

For every 1 bit SNR_q improves by a factor of 4. Thus for 2 bits increase in message the SNR_q increases by a factor of 16.

3.31 (a)

In between the two adjacent sampled values, if the base band signal changes an amount less than the step size, the output of the DM is a sequence of alternate positive and negative pulses. This small change in base band signal indicated that the base band is almost constant.

3.32 (c)

$$R_b = 10\text{ k}$$
$$\therefore \quad B_1 = 2R_b = 20\text{ k}$$
$$B_2 = R_b = 10\text{ k}$$

3.33 (a)

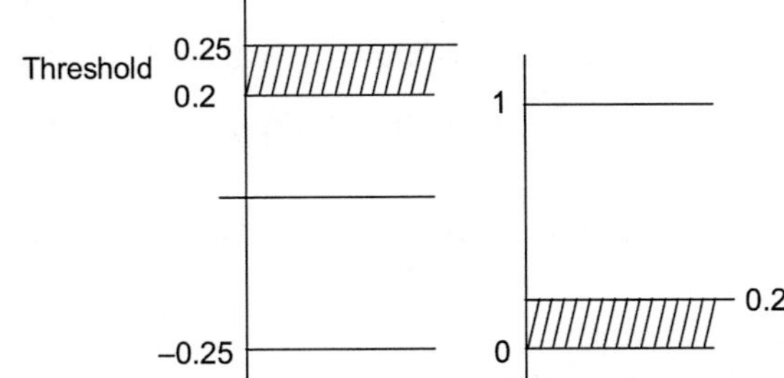

Shaded areas show the probability of error.

$$P_e(0) = \frac{\text{error}}{\text{total}} = \frac{0.25 - 2}{0.25 - (-0.25)} = 0.1$$

$$P_e(1) = \frac{0.2 - 0}{1 - 0} = 0.2$$

Since probability of occurrence of 0 and 1 are equal

$$\text{Avg.} = \frac{P_e(0) + P_e(1)}{2} = 0.15$$

3.35 (c)

$$f_s = 2(1200 + 600 + 600) = 4.8\text{ K}$$
$$R_b = nf_s = 12 \times 4.8\text{ K} = 57.6\text{ Kbps}$$

3.36 (b)

$$h(t) = x(4 - t)$$

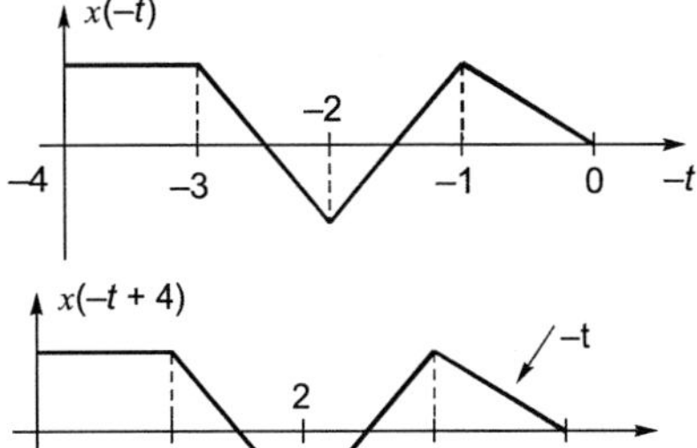

Slope in region $t = 3$ to $4 = -1$

Slope in region $t = 3$ to $4 = -1$

3.37 (c)

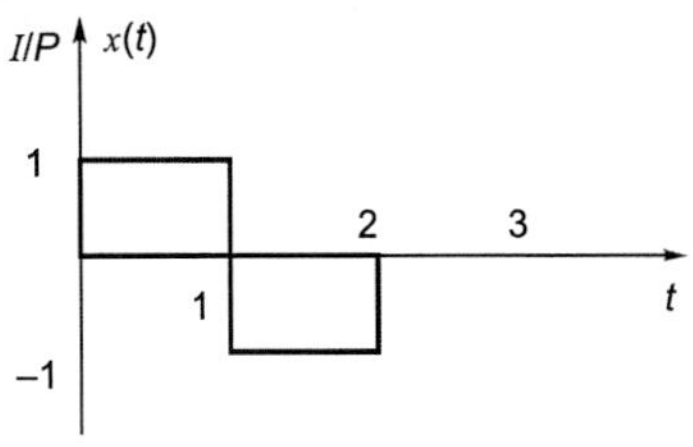

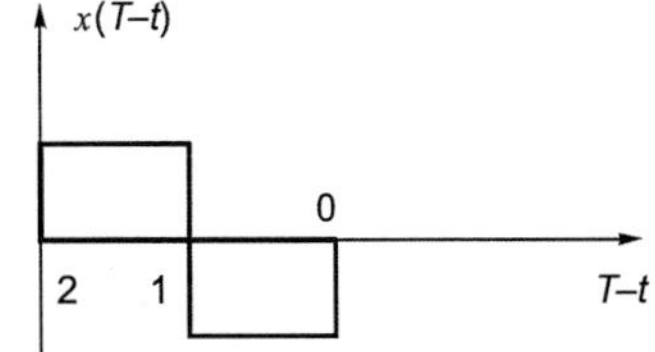

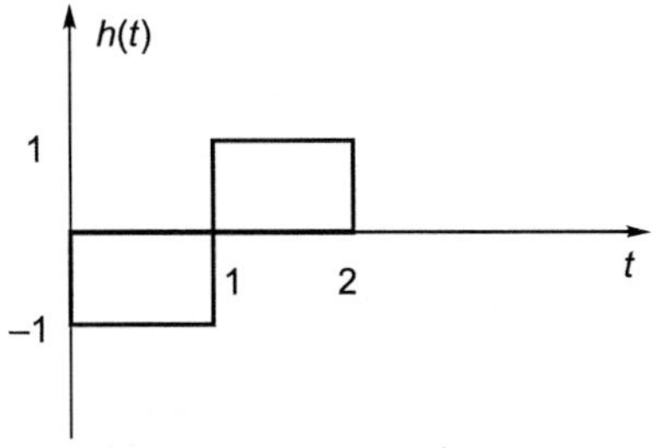

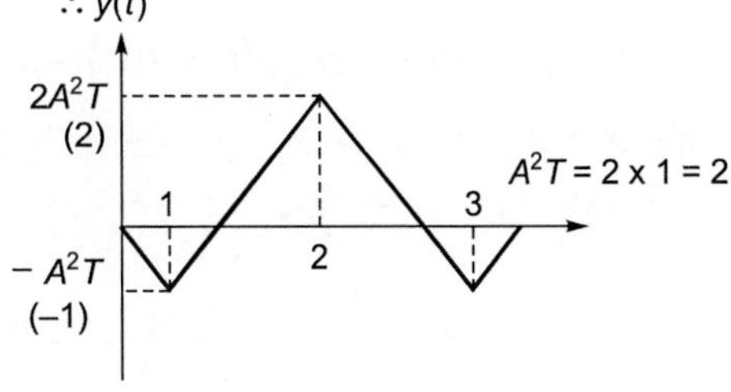

3.38 (c)

$$f_s = (2f_m) \times 3 = (2 \times 1000) \times 3 = 6 \times 10^3 \text{ Hz}$$

3.39 (b)

To avoid slope overload

$$\frac{\Delta}{T_s} \geq m'(t)$$

$$\Delta \times 32 \times 1000 \geq 125$$

$$\Rightarrow \quad \Delta \geq 2^{-8}$$

3.40 (b)

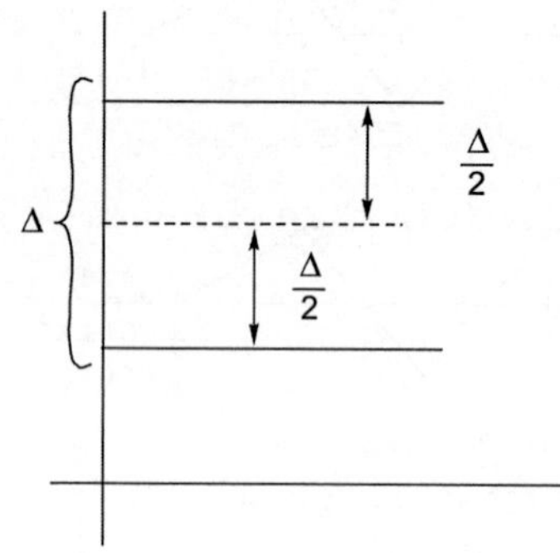

For $y_1(t)$ and $y_2(t)$ to be different minimum step size of $\Delta/2$ is needed else they will be same.

3.41 (c)

$$p(t) = u(t) - u(t-1)$$

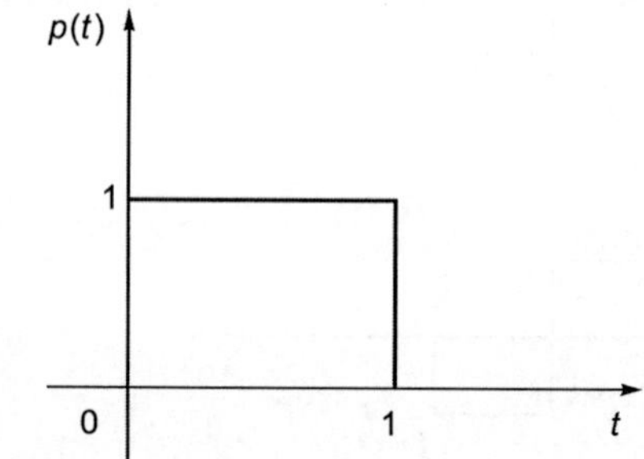

$$g(t) = p(t) * p(t)$$

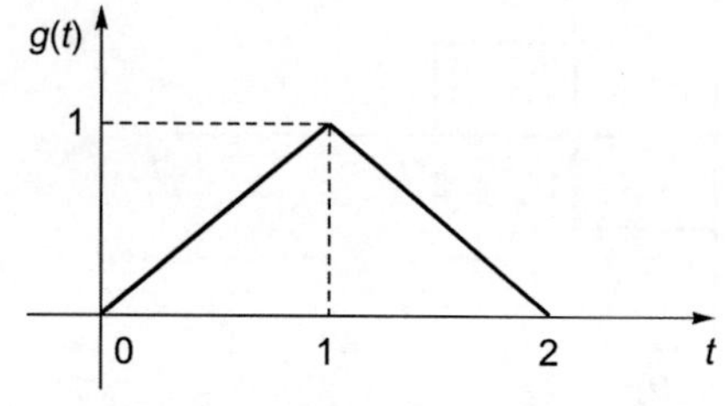

$$s(t) = g(t) - \delta(t-2) * g(t) = g(t) - g(t-2)$$

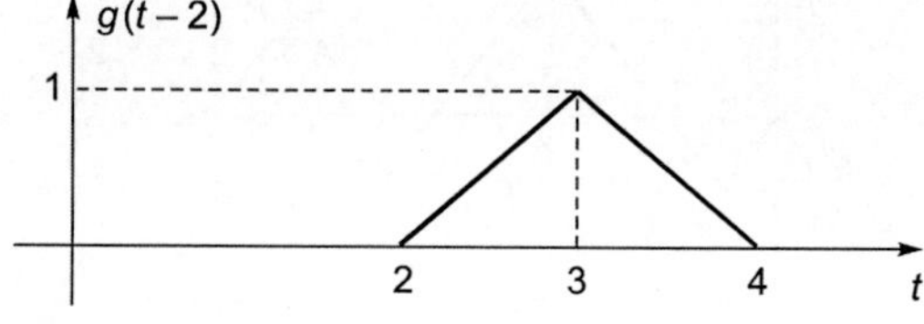

$$s(t) = g(t) - g(t-2)$$

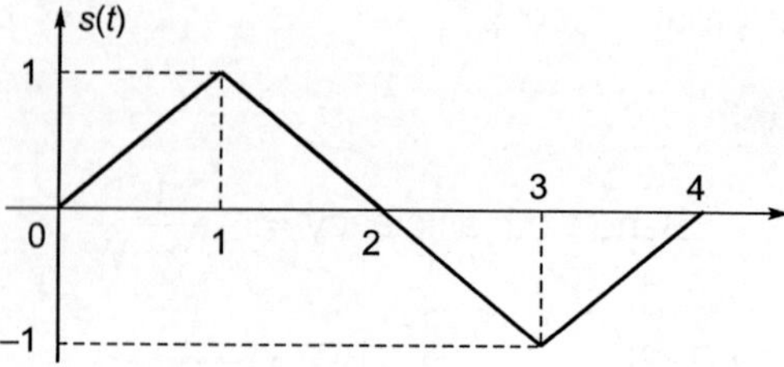

Impulse response of match filter

$h(t) = s(T - t)$

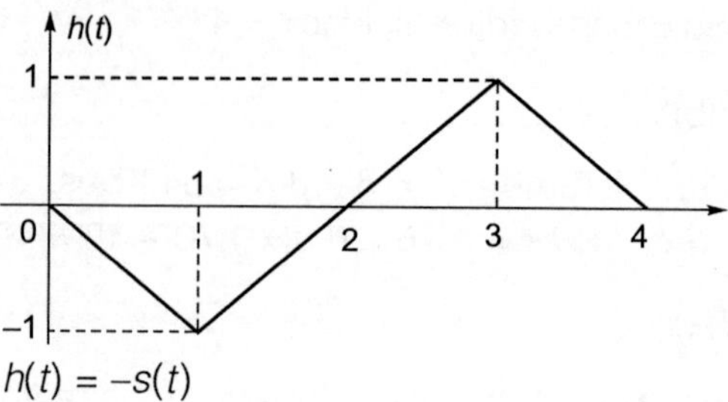

$h(t) = -s(t)$

3.42 (a)

$$X_{AM}(t) = 100[p(t) + 0.5g(t)]\cos\omega_c t \quad \text{...(i)}$$

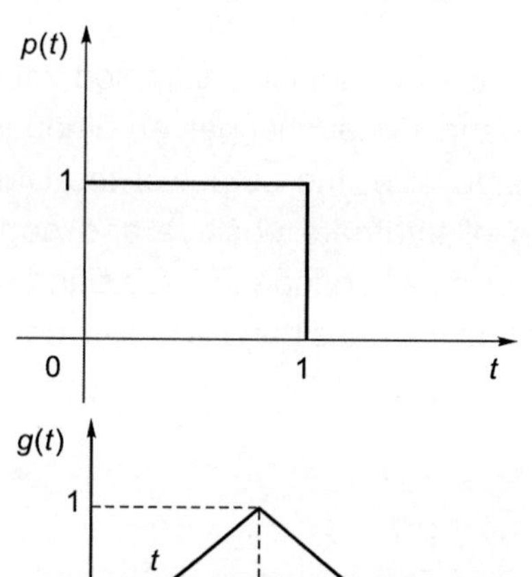

$s(t) = 100(1 + m(t))\cos\omega_c t, \; p(t) = 1$

$m(t) = 0.5\, g(t)$, comparing AM equation with (i)

$= 0.5t$

$\Rightarrow$ Modulating signal = t(from equation of line)

Modulation index = 0.5

3.43 (d)

Condition to avoid slope overload in DM

$$\boxed{\frac{\Delta}{T_s} \geq \frac{d}{dt} m(t)}$$

So, by increasing the step size, slope overload distortion can be avoided.

3.44 **(c)**

$$p(t) = \frac{\sin 4\pi Wt}{4\pi Wt\left(1-16W^2t^2\right)}$$

$$= \frac{\sin 4\pi W \times \frac{1}{4W}}{4\pi W \times \frac{1}{4W}\left(1-16W^2\frac{1}{16W^2}\right)}$$

As it comes in the form of 0/0 so applying LH rule

$$P\left(\frac{1}{4W}\right) = \frac{\frac{d}{dt}\sin 4\pi Wt}{\frac{d}{dt}4\pi Wt\left(1-16W^2t^2\right)} = \frac{\cos 4\pi Wt}{1-48W^2t^2}$$

Putting $t = \frac{1}{4W}$

$$P(t) = \frac{\cos\pi}{1-3} = 0.5$$

3.45 **(c)**

Probability of no error in n bits $= (1-p)^n$

Probability of one error in n bits $= n.p(1-p)^{n-1}$

$\therefore$ Total probability $= np(1-p)^{n-1} + (1-p)^n$

3.46 **(b)**

Allocated Bandwidth = 5 MHz

Frequency reuse factor $= \frac{1}{5}$

Bandwidth allocated for 1 cell

$$= 5 \times \frac{1}{5} = 1 \text{ MHz}$$

To allocate 8 channels 200 kHz bandwidth required.

$$\text{Number of channels in 1 MHz} = \frac{1 \text{ MHz}}{200 \text{ kHz}} \times 8 = 40$$

3.47 **(a)**

$$\text{Processing gain} = \frac{R_c}{R_b}$$

$$\frac{R_c}{R_b} \ge 100$$

$$R_b \le \frac{R_c}{100}$$

$R_b \le 12.288 \times 10^3$ bits per sec.

3.48 **(b)**

Average energy of constellation 1 is

$$E_1 = \frac{0+4a^2+4a^2+8a^2}{4} = 4a^2$$

Average energy of constellation 2 is

$$E_2 = \frac{a^2+a^2+a^2+a^2}{4} = a^2$$

So, $\frac{E_1}{E_2} = \frac{4a^2}{a^2} = 4$

3.49 **(a)**

The probability of error decreases with increase in average energy. As constellation 1 has more average energy than that of constellation 2. So, the probability of symbol error for constellation 1 is lower.

3.50 **(a)**

To maximize the entropy, all the decision boundaries should be equiprobable

$$\int_1^5 \rho_x(x)dx = \frac{1}{3} \quad ; \quad \int_1^5 bdx = \frac{1}{3}$$

$$b[x]_1^5 = \frac{1}{3} \quad ; \quad b[5-1] = \frac{1}{3}$$

$$4b = \frac{1}{3} \quad \Rightarrow b = \frac{1}{12}$$

$$\int_{-1}^1 \rho_x(x)dx = \frac{1}{3} \quad ; \quad \int_{-1}^1 adx = \frac{1}{3}$$

$$a[x]_{-1}^1 = \frac{1}{3} \quad ; \quad a[1-(-1)] = \frac{1}{3}$$

$$2a = \frac{1}{3} \quad \Rightarrow a = \frac{1}{6}$$

3.51 **(d)**

Signal power,

$$S = E[X^2] = \int_{-\infty}^{\infty} x^2 f_X(x)dx$$

$$= 2\left[\int_0^1 ax^2dx + \int_1^5 bx^2dx\right] = 2\left[a\left(\frac{x^3}{3}\right)_0^1 + b\left(\frac{x^3}{3}\right)_1^5\right]$$

$$= 2\left[\frac{a}{3} + \frac{124b}{3}\right] = 2\left[\frac{1}{6\times 3} + \frac{124}{12\times 3}\right] = 7$$

Quantization noise power,

$$N_Q = \sum_{i=1}^{2} \frac{\Delta_i^2}{12} P(\Delta_i)$$

$$\Delta_1 = 1\text{ V} \; ; \; \text{for } 1 < x < 1$$

$$\Delta_2 = 2\text{ V} \; ; \; \text{for } 1 < |x| < 5$$

$$P(\Delta_1) = P\{-1 < x < 1\} = 2a = \frac{1}{3}$$

$$P(\Delta_2) = 1 - P(\Delta_1) = \frac{2}{3}$$

$$\text{So, } N_Q = \frac{1}{3}\left(\frac{1^2}{12}\right) + \frac{2}{3}\left(\frac{2^2}{12}\right) = \frac{1}{4}$$

$$\text{Hence, SQNR} = \frac{S}{N_Q} = 28$$

3.52 (a)

There will be error if all the three received bits are 0 or two of the three received bits are 0.

Therefore, probability of error in output

$$= {}^3C_3.p.p.p + {}^3C_2.p.p.(1-p)$$
$$= p^3 + 3p^2(1-p)$$

3.53 (d)

$$f_{s_1} = 2 \times W = 2\,W$$
$$f_{s_2} = 2 \times W = 2\,W$$
$$f_{s_3} = 2 \times 2\,W = 4\,W$$
$$f_{s_4} = 2 \times 3\,W = 6\,W$$
$$f_s = f_{s_1} + f_{s_2} + f_{s_3} + f_{s_4}$$
$$f_s = 14\,W$$

For minimum bandwidth $n = 1$

$$\boxed{R_b = nf_s}$$

$$R_b = 1 \times 14\,W = 14\,W$$

$$(\text{B.W.})_{min} = \frac{R_b}{2}$$

$$(\text{B.W.})_{min} = \frac{14\,W}{2} = 7\,W$$

3.54 (a)

Binary sequence represented by bipolar pulses

So, B.W. = $R_b = nf_s = 8 \times 8\text{ K} = 64$ kHz

3.55 (c)

Signal to noise ratio

$$\left(\frac{S_0}{N_0}\right)_{dB} \approx 6n\text{ dB}$$

where

n = number of bits per sample quantized

$$\left(\frac{S_0}{N_0}\right)_{dB} \approx 6 \times 8 = 48\text{ dB}$$

3.56 (b)

$$\text{Quantization noise, } N_q = \frac{S^2}{12}$$

where S = Step size of quantization level

$$N_q' = \frac{N_q}{4}$$

$$\Rightarrow \quad \frac{S'^2}{12} = \frac{S^2}{12 \times 4}$$

$$\Rightarrow \quad S' = \frac{S}{2}$$

$$\Rightarrow \quad S = \frac{V_{pp}}{2^n}$$

where, V_{pp} = peak-to-peak value

n = no. of bits/sample

$$\Rightarrow \quad \frac{V_{PP}}{2^{n'}} = \frac{V_{pp}}{2.2^n} = \frac{V_{pp}}{2.2^8}$$

$$\Rightarrow \quad 2^{n'} = 2^9 = 512$$

Therefore, number of quantization levels required to reduce the quantization noise by a factor 4 would be 512.

3.57 (c)

Signal power,

$$S = \int_{-\infty}^{\infty} x^2 f_X(x)dx = \int_{-5}^{5} \frac{1}{10}x^2 dx = \frac{25}{3}$$

$$\text{SQNR} = 43.5\text{ dB (or) } 10^{(4.35)}$$

$$N_Q = \frac{\Delta^2}{12} = \frac{S}{\text{SQNR}}$$

$$\text{So,} \quad \Delta = \sqrt{\frac{12(S)}{\text{SQNR}}} = \sqrt{\frac{12 \times 25}{3(10^{4.35})}} = 0.0668\text{ V}$$

3.58 (c)

$$\Delta_1 = 0.05\text{ V (for positive values)}$$
$$\Delta_2 = 0.1\text{ V (for negative values)}$$

The amplitude of the input signal is symmetrically distributed about zero. Hence,

$$P(\Delta_1) = P(\Delta_2) = 0.50$$

So, the quantization noise power will be,

$$N_Q = \frac{\Delta_1^2}{12}P(\Delta_1) + \frac{\Delta_2^2}{12}P(\Delta_2)$$

$$= \frac{1}{24}(\Delta_1^2 + \Delta_2^2) = \frac{125}{24 \times 10000} = \frac{1}{1920}$$

$$\text{Signal power, } S = \frac{25}{3}$$

$$\text{SQNR} = \frac{S}{N_Q} = \frac{1920 \times 25}{3} = 16000$$

In decilogs,

$$[\text{SQNR}] = 10\log_{10}(\text{SQNR}) = 42\text{ dB}$$

3.59 (c)

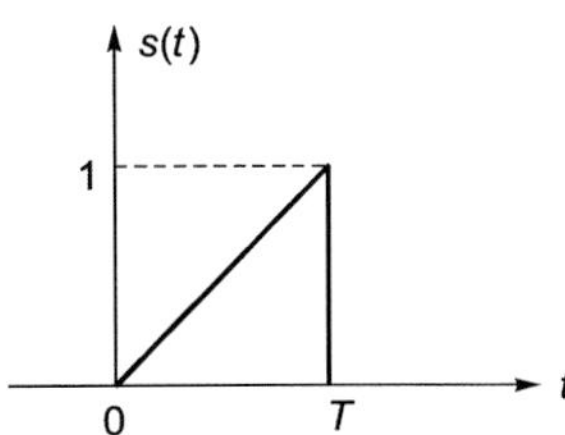

Impulse response of the matched filter,

$$h(t) = s(T-t)$$

$s(T+t)$ is the left-side shifted version of $s(t)$ by T.

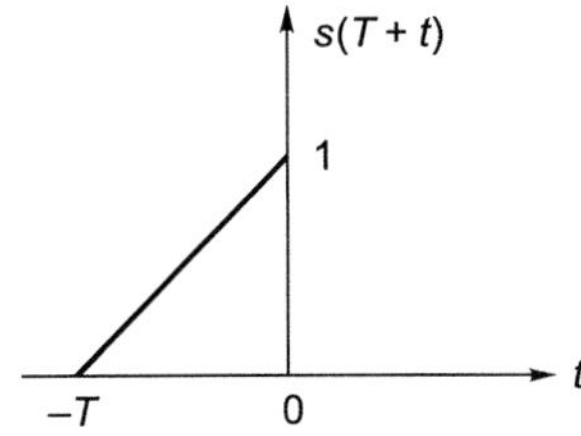

$s(T-t)$ is the mirror image of $s(T+t)$ on y-axis.

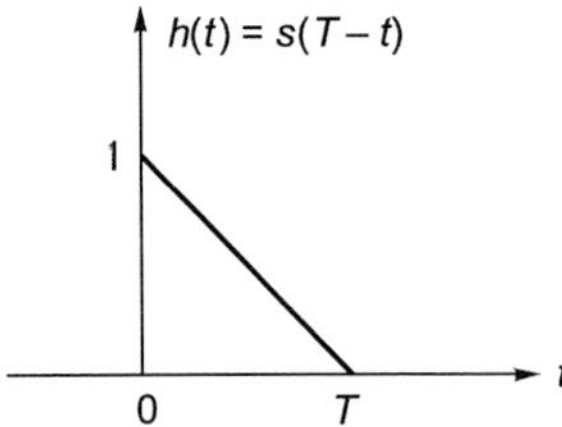

3.60 (c)

$$s(t) = \frac{1}{2}\left[\frac{2\sin(500\,\pi t)\sin(700\,\pi t)}{\pi^2 t^2}\right]$$

$$s(t) = \frac{1}{2\pi^2 t^2}[\cos(700\,\pi t - 500\,\pi t) - \cos(700\,\pi t + 500\,\pi t)]$$

$$s(t) = \frac{1}{2\pi^2 t^2}[\cos(200\,\pi t) - \cos(1200\,\pi t)]$$

Maximum frequency component,

$$f_m = \frac{1200\pi}{2\pi} = 600 \text{ Hz}$$

Nyquist sampling rate,

$$f_{s\,min} = 2\,f_m = 1200 \text{ Hz}$$

3.61 (d)

Quantized levels are equiprobable; hence

$H = \log_2 4 = 2$ bits/sample

$r = 2$ samples/sec

Hence information rate $R = r \cdot H = 2$ samples/sec $\times$ 2 bits/sample

$\Rightarrow$ $R = 4$ bits/sec

3.62 (d)

For M-ary

$$d = 2\sin\left(\frac{\pi}{M}\right)\sqrt{E_s}$$

Distance of any point from origin is $\sqrt{E_s}$

For, 4 – ary, $r_1 = \sqrt{E_{s_1}}$

8 – ary, $r_2 = \sqrt{E_{s_2}}$

For, 4 – ary $M = 4$, $d_1 = 2\sin\left(\frac{\pi}{4}\right) r_1$

8 – ary $M = 8$; $d_2 = 2\sin\left(\frac{\pi}{8}\right) r_2$

if $d_1 = d_2 = \text{d}$ then,

$$2\sin\left(\frac{\pi}{4}\right) r_1 = d \Rightarrow r_1 = \frac{d}{\sqrt{2}} = 0.707\,d$$

$$2\sin\left(\frac{\pi}{8}\right) r_2 = d$$

$$\Rightarrow \quad r_2 = \frac{d}{2\sin\left(\frac{\pi}{8}\right)} = 1.307\,d$$

3.63 (d)

$P_e \propto \sqrt{E_s}$ for both cases

$$\text{so,} \quad \frac{\sqrt{E_{s_1}}}{\sqrt{E_{s_2}}} = \frac{r_1}{r_2} = \frac{0.707\,d}{1.307\,d}$$

$$\Rightarrow \quad \frac{E_{s_2}}{E_{s_1}} = \left(\frac{1.307}{0.707}\right)^2 = 3.42$$

To achieve same error, 2nd must have 3.42 time than 1st.

The value in dB = 10 log(3.42) = 5.33 dB.

3.64 (c)

$$\text{BW} = \frac{R_b}{2}(1+\alpha)$$

$$3500 = \frac{R_b}{2}(1+0.75)$$

$$R_b = 4000 \text{ bits/sec}$$

In base band transmission →

Symbol rate = Bit rate = 4000 symbols/sec

3.65 (d)

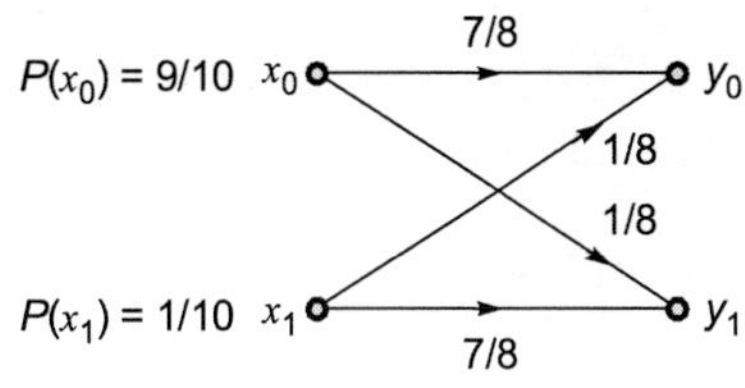

$$P(y_0 \mid x_0)\,P(x_0) = \frac{7}{8}\times\frac{9}{10} = \frac{63}{80}$$

$$P(y_0 \mid x_1)\,P(x_1) = \frac{1}{8}\times\frac{1}{10} = \frac{1}{80}$$

$$P(y_1 \mid x_0)\,P(x_0) = \frac{1}{8}\times\frac{9}{10} = \frac{9}{80}$$

$$P(y_1 \mid x_1)\,P(x_1) = \frac{7}{8}\times\frac{1}{10} = \frac{7}{80}$$

$P(y_0 \mid x_0)\,P(x_0) > P(y_0 \mid x_1)\,P(x_1)$; So, when a symbol is received as y_0, the decision can be made in favour of x_0 in an optimum way.

$P(y_1 \mid x_0)\,P(x_0) > P(y_1 \mid x_1)\,P(x_1)$; So, when a symbol is received as y_1, the decision can be made in favour of x_0 in an optimum way.

So, for the given BSC, with optimum receiver, both the received symbols will be decoded as x_0. Hence, the probability of error is equal to the probability of transmitting x_1.

So, $P_e = P(x_1) = 1/10$

3.67 (d)

Nyquist pulses should be used for baseband pulse shaping to avoid ISI, with minimum bandwidth requirement. In this case the bandwidth of the base band signal will be,

$$(\text{BW})_{\text{baseband}} = \frac{R}{2}\text{ kHz}$$

$$(\text{BW})_{M\text{-QAM}} = \frac{2(\text{BW})_{\text{baseband}}}{\log_2 M}$$

For 32-QAM,

$$(\text{BW})_{32\text{-QAM}} = \frac{2(\text{BW})_{\text{baseband}}}{\log_2(32)} = \frac{R}{5}\text{ kHz}$$

3.68 (d)

$$g(t) = e^{-\pi t^2}$$

$$G(f) = e^{-\pi f^2}$$

$$H(f) = e^{-\pi f^2}$$

$$Y(f) = G(f)\,H(f)$$

$$Y(f) = e^{-\pi f^2}\cdot e^{-\pi f^2} = e^{-2\pi f^2}$$

3.69 (d)

BER is given as

$$P_e = P(0)\,P(1/0) + P(1)\,P(0/1)$$

If detection threshold = 1 then

$$P(0) = P(1) = \frac{1}{2}$$

$$P\left(\frac{Y=1}{X=0}\right) = \int_1^{\infty} f(z/0)dz = 0$$

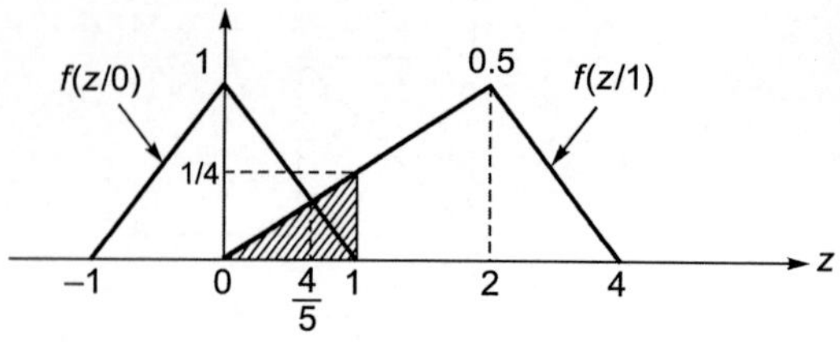

$$P\left(\frac{Y=0}{X=1}\right) = \int_0^{1} f(z/1)dz = \frac{1}{4}\times\frac{1}{4}\times 1 = \frac{1}{8}$$

$$\therefore \quad P_e = \frac{1}{2}\times 0 + \frac{1}{2}\times\frac{1}{8} = \frac{1}{16}$$

3.70 (b)

Optimum threshold is given by the point of intersection of two pdf curves.

$f(z/0) = 1 - |z|$; $|z| \le 1$

$f(z/1) = z/4$; $0 < z < 2$

The point of intersection which decides optimum threshold

$$1 - z = \frac{z}{4}$$

$$1 = z + \frac{z}{4}$$

$$\Rightarrow \quad z = \frac{4}{5}$$

3.71 (1.414)

Since the same signal is passed through two parallel channels and added at the receiver input, the effective distance between the adjacent message points will be $d_{\min} = 4\sqrt{E_b}$ and the effective noise PSD(one-sided) will be $N_{o(\text{eff})} = 2N_o$.

So,

$$\text{BER} = Q\left(\sqrt{\frac{d_{\min}^2}{2N_{o(\text{eff})}}}\right) = Q\left(\sqrt{\frac{16E_b}{4N_o}}\right)$$

$$= Q\left(\sqrt{4\frac{E_b}{N_o}}\right) = Q\left(\sqrt{2}\sqrt{\frac{2E_b}{N_o}}\right)$$

For single channel case,

$$\text{BER} = Q(\sqrt{\gamma}) = Q\left(\sqrt{\frac{2E_b}{N_o}}\right)$$

So, for the given problem,

$$\text{BER} = Q\left(\sqrt{2}\sqrt{\frac{2E_b}{N_o}}\right) = Q(\sqrt{2}\sqrt{\gamma}) = Q(b\sqrt{\gamma})$$

So, $b = \sqrt{2} = 1.414$

3.72 (c)

$$\text{For FSK} \rightarrow \text{Bit error Probability} = Q\left(\sqrt{\frac{A_c^2 T_b}{2N_o}}\right)$$

$$A_c = 4\text{ mV},\ T_b = \frac{1}{R_b} = \frac{1}{500\times10^3}$$

$$\frac{N_o}{2} = 0.5 \times 10^{-12}\text{ W/Hz}$$

$$N_o = 10^{-12}\text{ W/Hz}$$

Bit error Probability = $Q(4)$

3.73 (0.25)

The maximum quantization error is

$$Q_e = \frac{\Delta}{2}$$

where, $$\Delta = \frac{\text{Dynamic range}}{2^n \text{ or } L}$$

$$\Delta = \frac{8}{2^4} = \frac{8}{16} = 0.5$$

$$\therefore \quad Q_e = \frac{0.5}{2} = 0.25$$

3.74 (200)

$$m(t) = \sin(100\pi t) + \cos(100\pi t)$$

$$A\cos2\pi f_o t + B\sin2\pi f_o t = \sqrt{A^2+B^2}\cos(2\pi f_o t - \theta)$$

$$\theta = \tan^{-1}(B/A)$$

$$m(t) = \sqrt{2}\cos(100\pi t - 45)$$

$$\Delta = \frac{V_{p-p}}{2^n} \Rightarrow 0.75 = \frac{2\sqrt{2}}{2^n}$$

$$L = 2^n = 3.8 \rightarrow n = 2$$

$$R_b = nf_s = 2 \times 100 = 200 \text{ bits/sec}$$

3.75 (17)

$$B = \frac{R_b}{\log_2 M}(1+\alpha) \Rightarrow 100 = \frac{200\times 2}{\log_2 M}$$

$\log_2 M = 4$

So, $M = 16$

3.76 (a)

Slope signal = Slope of delta modulator

$$A_m(2\pi f_m) = \Delta f_s$$

$$A_m(2\pi\, 2\times10^3) = 20{,}000 \times 0.1$$

$$A_m = \frac{1}{2\pi}$$

3.77 (0.4)

$$P(X = 1/Y = 0)$$

$$= \frac{P(Y=0/X=1)\,P(X=1)}{P(Y=0/X=1)\,P(X=1) + P(Y=0/X=0)\,P(X=0)}$$

$$= \frac{\left(\frac{1}{7}\right)(0.8)}{\left(\frac{1}{7}\right)(0.8) + \left(\frac{6}{7}\right)(0.2)} = 0.4$$

3.78 (400)

In the question it is given that GSM requires 200 kHz for 8 users and uses TDMA scheme to accommodate them. Thus for the next users we will required an extra of 200 kHz bandwidth. Thus 400 kHz bandwidth is to be used.

3.79 (d)

According to MAP criteria, for optimum threshold,

$$f_R(r|s_0)P(s_0) \underset{H_1}{\overset{H_0}{\gtrless}} f_R(r|s_1)P(s_1)$$

Given that,

$$P(s_0) = \frac{1}{3} \text{ and } P(s_1) = \frac{2}{3}$$

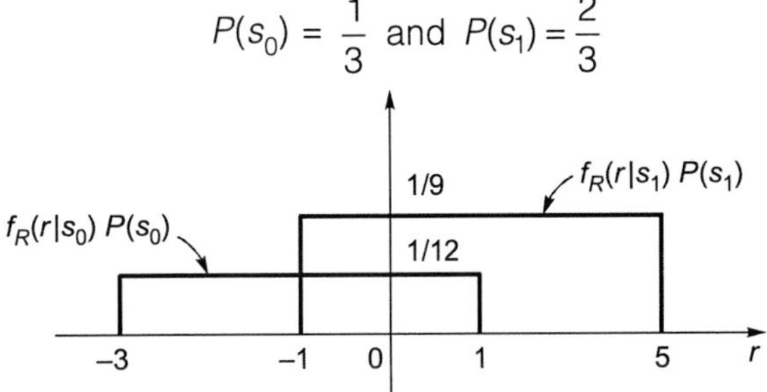

From the above figure, it is clear that,

$$f_R(r|s_0)P(s_0) > f_R(r|s_1)P(s_1) \Rightarrow \text{for } r < -1$$

$$f_R(r|s_1)P(s_1) > f_R(r|s_0)P(s_0) \Rightarrow \text{for } r > -1$$

So, optimum threshold will be, $r_{th} = -1$.

For $r_{th} = -1$,

$$P(r_0|s_1)P(s_1) = 0$$

$$P(r_1|s_0)P(s_0) = \frac{1}{12}[1-(-1)] = \frac{1}{6}$$

$$P_{e(\min)} = P(r_0|s_1)P(s_1) + P(r_1|s_0)P(s_0) = \frac{1}{6}$$

3.80 (32)

SNR of uniform quantized sinusoidal signal is given by

$$\text{SNR} = 1.8 + 6n$$

$$31.8 = 1.8 + 6n$$

$$n = 5$$

Thus total number of levels = $2^5 = 32$.

3.81 (a)

Since $\text{SNR} = \dfrac{2E}{\eta}$

Thus for having minimum error we have to select signal with maximum energy.

Energy of signal 1 = 1

$\because$ Energy of signal 2 = $\dfrac{1}{3}$

Energy of signal 3 = $\dfrac{1}{3}$

Energy of signal 4 = $\dfrac{1}{3}$

3.82 (d)

(GMSK) is used to transmit GSM signals.

3.83 (35)

Bit rate, R_b = 56 kbps

Roll-off factor, α = 0.25

$$\text{Transmission BW} = \frac{R_b}{2}(1+\alpha)$$

$$= \frac{56}{2}(1.25) = 28 \times 1.25$$

$$= 35 \text{ kHz}$$

3.84 (a)

When the signal to Noise ratio is maximum

$$h(t) = s(T-t)$$

but shifting doesn't change the energy

$$\Rightarrow \quad E_h = E_s$$

$$\text{and} (\text{SNR})_{max} = \frac{2E_s}{N_0}$$

3.85 (0.25)

Channel spectrum 500 Hz – 2000 Hz

hence BW = 1500 Hz

$$BW = \frac{R_b}{\log_2 M}(1+\alpha)$$

$$1500 = \frac{R_b}{\log_2 16}(1+\alpha)$$

$$1500 = \frac{4800}{\log_2 16}(1+\alpha)$$

$$1500 = 1200\,(1+\alpha)$$

$$\Rightarrow \quad \alpha = 0.25$$

3.86 (16)

f_s = 8 kHz

n = 8 bits/sample

M = 4

$$BW_{min} = \frac{R_b}{2\log_2 M} = \frac{R_b}{4} = \frac{8\times 8}{4} = 16 \text{ kHz}$$

3.87 (c)

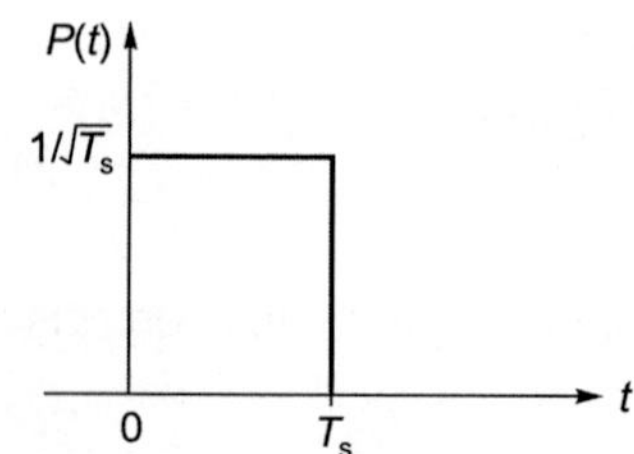

Impulse response of matched filter,

$h(t) = P(T_s - t)$

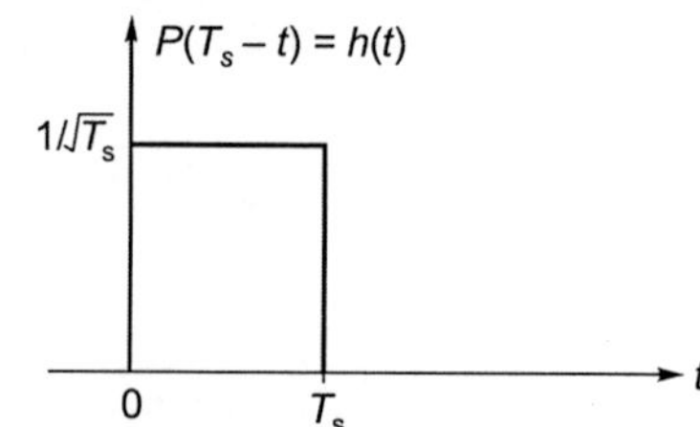

Output of matched filter, $y(t) = P(t) * h(t)$

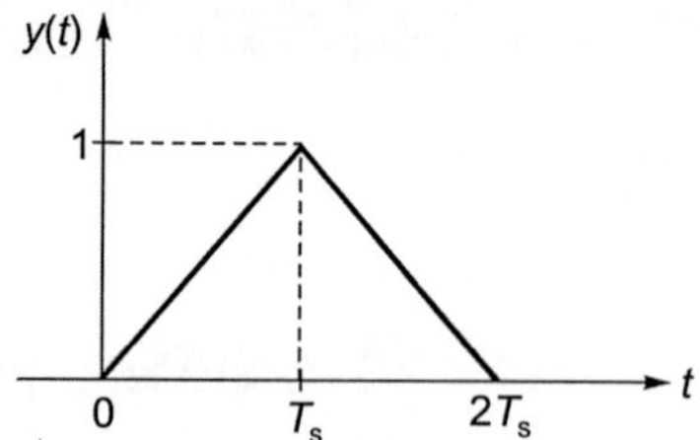

3.88 (0.7357)

bit error probability, $p = 10^{-5}$

Number of bits transmitted,

$n = 10^5$

If n-bits are transmitted, probability of getting error in 'r' bits is given by

$$P_e = {}^nC_r\, p^r (1-p)^{n-r}$$

The probability that not more than one bit will be in error is

$$= {}^nC_0 p^0 (1-p)^{n-0} + {}^nC_1 (p)^1 (1-p)^{n-1}$$

$$= (1-p)^n + np(1-p)^{n-1}$$

$$= (1-10^{-5})^{10^5} + 10^5 \times 10^{-5}(1-10^{-5})^{10^5-1}$$

$$= 0.7357$$

3.89 (b)

Given,

Symbol rate = 2000 symbol/sec

So, $R_b = 2$ kHz

Nyquist's criterion for ISI free transmission is,

$$\sum_{n=-\infty}^{\infty} P(f - nR_b) = \text{Constant}$$

So, only option (b) satisfies this condition for given R_b.

3.90 (d)

In DPCM, difference of message signal sample with its prediction is quantized.

3.91 (b)

$$u_0(t) = 5\cos 20000\pi t$$
$$u_1(t) = 5\cos 22000\pi t$$
$$f_1 = 11000 \text{ Hz and } f_2 = 10000 \text{ Hz}$$

For FSK waveforms to be uncorrelated,

$$f_1 - f_2 = n\frac{R_b}{2} \;;\; n = 1, 2, 3, ...$$

$$R_b = \frac{2(f_1 - f_2)}{n} = \frac{2000}{n} \text{ bits/sec}$$

$$R_{b(\max)} = 2000 \text{ bits/sec}$$

$\because$ minimum value of $n = 1$

$$T_{b(\min)} = \frac{1}{R_{b(\max)}} = 0.5 \text{ ms}$$

3.92 (7)

For sinusoidal input is applied to a PCM system,

SQNR = $6n + 1.8$

n = number of bits per sample

Given that, the required SQNR = 40 dB

so, $6n + 1.8 \geq 40$

$$n \geq \frac{38.2}{6}$$

$n_{\min} = 7$ $\quad\because n$ is always an integer value

3.93 (-0.50)

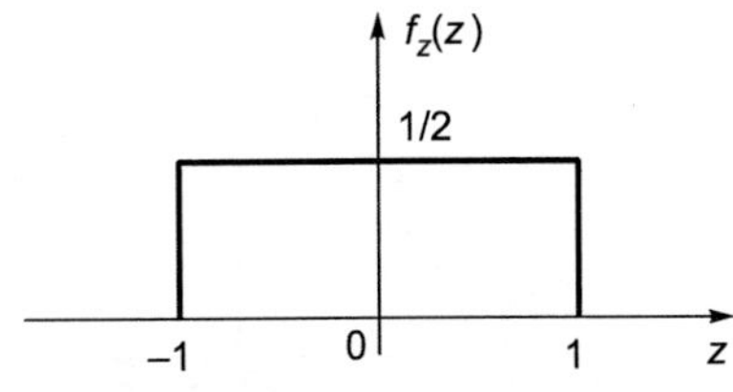

$$P(x_0) = \frac{1}{4} \;;\; P(x_1) = \frac{3}{4}$$

MAP criteria, $f_Y(y \mid x_0) P(x_0) \underset{x_1}{\overset{x_0}{\gtrless}} f_Y(y \mid x_1) P(x_1)$

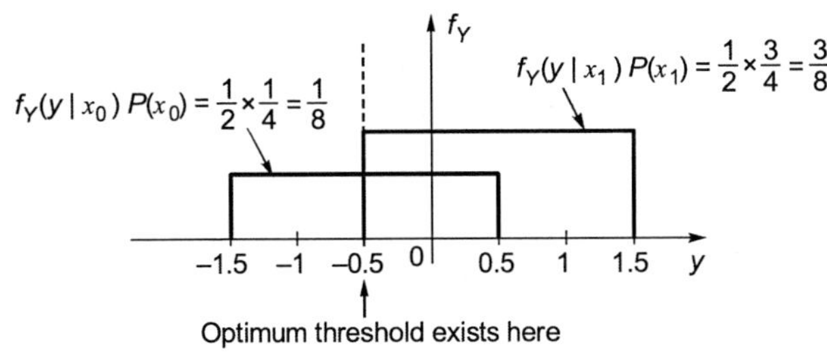

So, $\alpha = -0.50$

3.94 (0.125)

Let s_0 and s_1 be the transmitted symbols representing the transmitted value {-1, 1} respectively and let r_0 and r_1 be the received symbols.

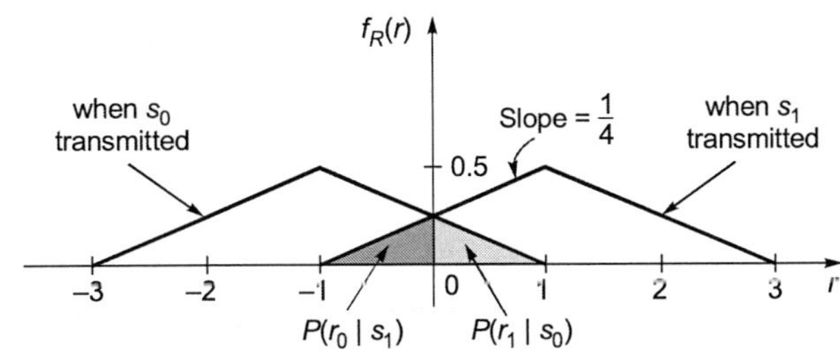

Probability of error,

$$P_e = P(s_1) P(r_0 \mid s_1) + P(s_0) P(r_1 \mid s_0)$$

$$P(r_0 \mid s_1) = P(r_1 \mid s_0) = \frac{1}{2} \times 1 \times \frac{1}{4} = \frac{1}{8}$$

Given that,

$$P(s_0) = P(s_1) = \frac{1}{2}$$

So, $$P_e = \frac{1}{2}\left(\frac{1}{8} + \frac{1}{8}\right) = \frac{1}{8} = 0.125$$

3.95 (b)

When $p(t) = \phi_1(t)$ and $q(t) = -\phi_1(t)$:

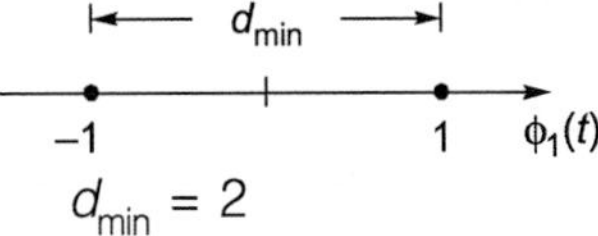

$d_{\min} = 2$

When $p(t) = \phi_1(t)$ and $q(t) = \sqrt{E}\,\phi_2(t)$:

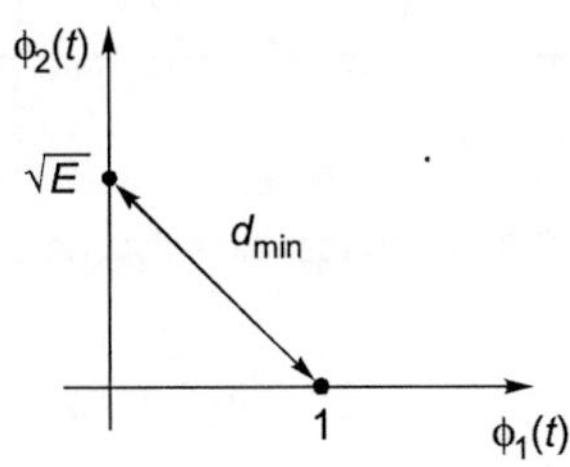

$$d_{min} = \sqrt{(\sqrt{E})^2 + 1} = \sqrt{E+1}$$

To obtain same bit-error probability, d_{min} should be same.

So, $\sqrt{E+1} = 2$

$E = 3$

3.96 (11.81)

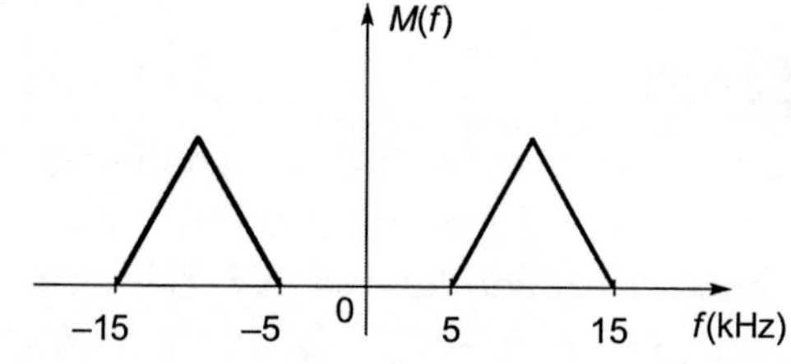

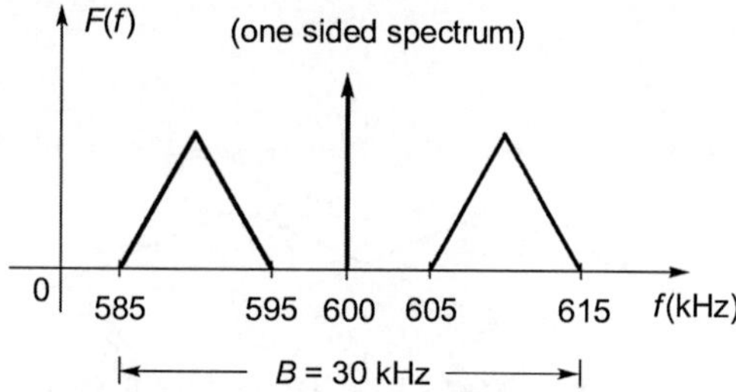

Nyquist rate $= 2 \times 615$ kHz

$= 1230$ kHz $= 1.23$ MHz

$f_s = 1.2 \times 1.23 = 1.476$ MHz

Bits/sample, $n = \log_2(256) = 8$

So, $R_b = nf_s = 8 \times 1.476$

$= 11.808 \simeq 11.81$ Mbps

3.97 (0.10)

MAP criteria should be used to minimise the probability of error.

$$f_Y(y\,|+1)\,P(+1) \underset{-1}{\overset{+1}{\gtrless}} f_Y(y\,|-1)\,P(-1)$$

$P(+1) = 0.80$ and $P(-1) = 0.20$

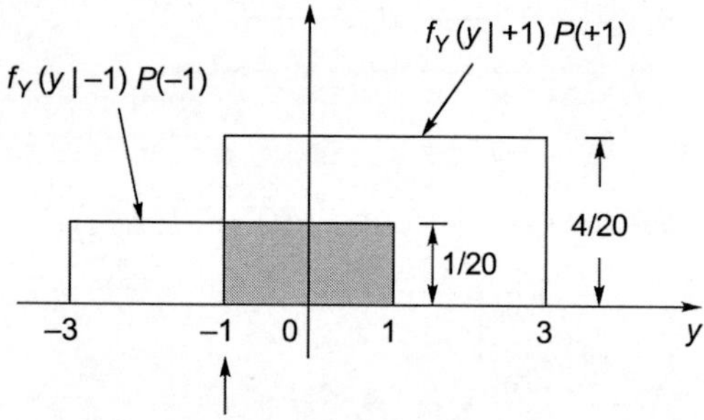

$$P_{e(\min)} = \text{Shaded area} = 2 \times \frac{1}{20} = 0.10$$

3.98 (3)

Since the noise variable is Gaussian with zero mean and ML decoding is used, the decision boundary between two adjacent signal points will be their arithmetic mean.

In the following graphs, the shaded area indicates the conditional probability of decoding a symbol correctly when it is transmitted.

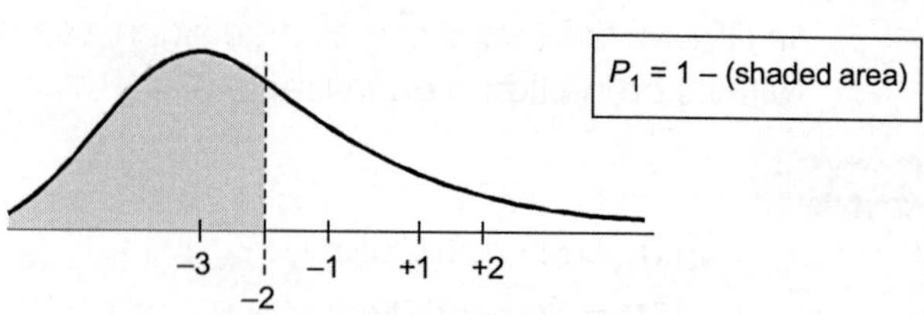

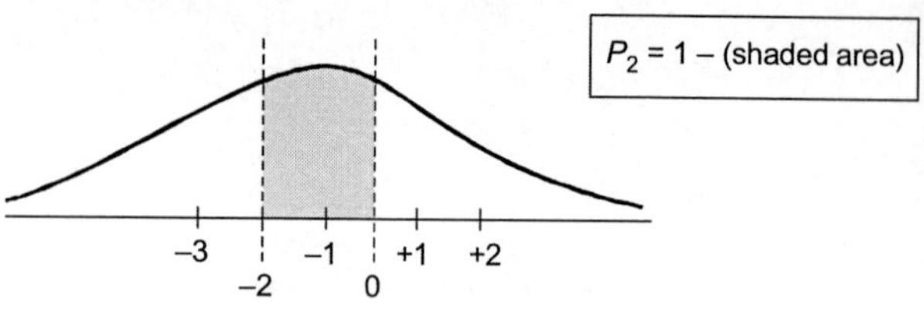

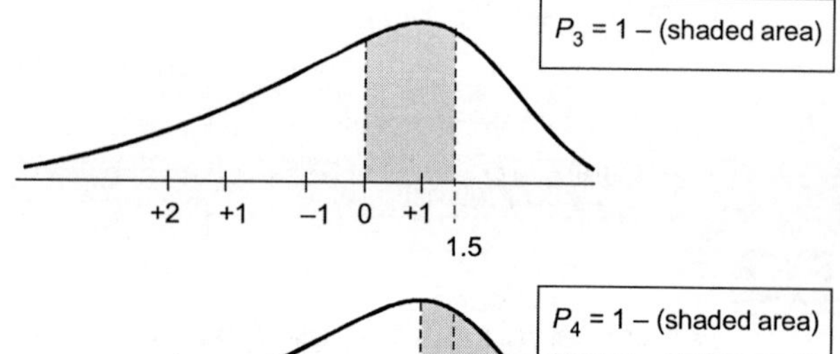

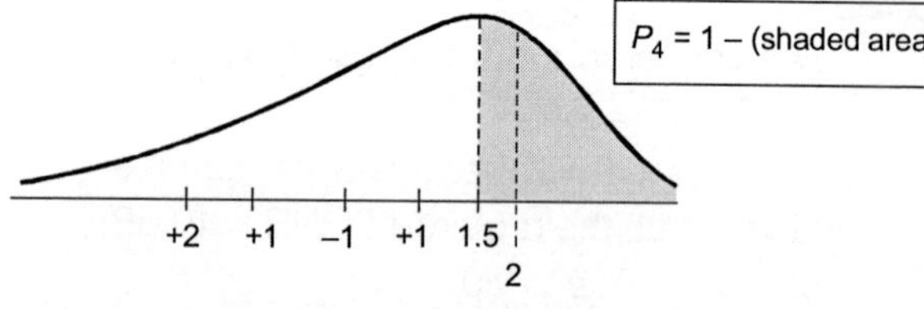

By comparing the above graphs, we can conclude that P_3 is larger among the four.

3.99 (9.25)

$f_m = 4$ kHz

$f_s = 1.25\,\text{NR} = 1.25 \times (2f_m) = 10$ kHz

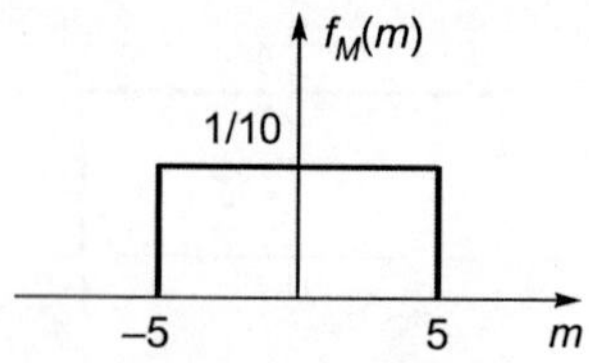

$n = 8$ bits/sample

and $S/N = 26\text{ dB} = 10^{2.6}$

For arbitrarily small probability of transmission error,

$$C \geq R_b \Rightarrow B\log_2\left(1 + \frac{S}{N}\right) \geq nf_s$$

$$B\log_2(1+10^{2.6}) \geq 8\times 10K \Rightarrow B \geq 9.25 \text{ kHz}$$

$$B_{min} = 9.25 \text{ kHz}$$

3.100 (3.14)

Given : $f_m = 4$ kHz ; $A_m = 4$ V ; $f_s = 32$ kHz

To avoid SOE $\rightarrow \dfrac{\Delta}{T_s} \geq 2\pi f_m A_m$

$$\Delta \times 32\,k \geq 2\pi \times 4\,k \times 4 \text{ V}$$

$$\Delta \geq \pi$$

$$(\Delta)_{min} = \pi \text{ volts} = 3.14 \text{ volts}$$

3.101 (30.72)

$$V_{p-p} = 2 \text{ V}$$

Root MSQ$[m(t)]$ = 0.1 V ; $f_m = 5$ kHz

Channel capacity, $C = 50$ kbps

$$\text{Max}\frac{S}{N_Q} = ?$$

Signal power, $S = \text{MSQ}[m(t)] = (0.1)^2 = 0.01$

$$C \geq R_b \Rightarrow 50 \text{ kbps} \geq nf_s$$

$\because$ $f_s = NR = 2f_m = 10$ kHz

$$n \leq 5 \Rightarrow n_{max} = 5$$

$$N_Q = \frac{\Delta^2}{12}$$

$\therefore$ $\Delta = \dfrac{V_{p-p}}{2^n}$

$$\Delta_{min} = \frac{V_{p-p}}{2^{n_{max}}} = \frac{2V}{2^5} = \frac{1}{16}$$

$$(N_Q)_{min} = \frac{\Delta^2_{min}}{12} = 3.25\times 10^{-4}$$

$$\left(\frac{S}{NQ}\right)_{max} = \frac{S}{(N_Q)_{min}} = \frac{0.01}{3.25\times 10^{-4}} = 30.76$$

3.102 (0.0405)

[Block diagram: $Y = X + N$ and V_{Th} inputs → Threshold Comparator → $Y \underset{0}{\overset{1}{\gtrless}} V_{Th}$]

$H_1 : X = +2$ V ; $H_0 : X = -2$ V

$$\text{Var}[N] = \sigma_n^2 = 0.4V^2$$

$$E[N] = 0$$

$$P(1) = 0.4$$

$\Rightarrow$ $P(0) = 0.6$

Opt. V_{Th} by using MAP theorem

$$\frac{V_{Th}[a_1 - a_2]}{\sigma^2} - \frac{a_1^2 - a_2^2}{2\sigma^2} = \ln\frac{P(0)}{P(1)}$$

$$H_1 : a_1 = E[2+N] = E[2] + E[N] = 2$$

$$H_0 : a_2 = -2 \text{ V} = E[-2+N]$$

$$= E[-2] + E[N] = -2$$

$$\sigma^2 = \text{Var}[Y] = \text{Var}[X+N]$$

$$= \text{Var}[X] + \text{Var}[N] = 0 + 0.4 = 0.4$$

$$\frac{V_{Th}[2+2]}{0.4} - \frac{4-4}{2\times 0.4} = \ln\frac{0.6}{0.4}$$

$$V_{Th} = \frac{0.4}{4}\ln\frac{0.6}{0.4} = 0.0405$$

Opt. V_{Th} = 0.0405 Volts

3.103 (a, b, d)

(a) Given that X is a discrete random variable taking K possible distinct real values. The entropy will be maximum when the probability of distinct values of X are equal i.e.

$$P_1 = P_2 = P_3 = \ldots = P_K = \frac{1}{K}$$

$$\therefore H(X)\big|_{max} = -\sum_{i=1}^{K} P_i \log_2 P_i$$

$$= -\sum_{i=1}^{K}\frac{1}{K}\log_2\left(\frac{1}{K}\right)$$

$$= -K\times\frac{1}{K}\log_2\left(\frac{1}{K}\right)$$

$$= \log_2 K$$

$\therefore$ $H(X) \leq \log_2 K$ bits

Hence, option (a) is correct.

(b) Consider Y = $2X$. The distinct K values of X will result in distinct K values of Y. Hence, the probability of the possible values, P_i of X and Y are equal.

Therefore, $H(X) = H(Y) = H(2X)$

Hence, option (b) is correct.

(c) Consider $Y = X^2$. Here, the distinct values of X will not result in distinct values of Y.

For e.g. : For $X = 1$, $Y = 1$ and

For $X = -1$, Y = 1.

Assuming the example as shown below,

X	Y	$P(X)$
–1	1	1/4
0	0	1/2
1	1	1/4

Y	$P(Y)$
0	1/2
1	1/2

$$\therefore \quad H(Y) = H(X^2) = \sum_{i=0}^{1} P_Y(y_i)\log_2\frac{1}{P_y(y_i)}$$

$$H(X^2) = \frac{1}{2}\log_2 2 + \frac{1}{2}\log_2 2$$

$$H(Y) = H(X^2) = 1 \text{ bits/symbol}$$

$$H(X) = \sum_{i=0}^{2} P_X(x_i)\log_2 \frac{1}{P_X(x_i)}$$

$$H(X) = \frac{1}{4}\log_2(4) + \frac{1}{2}\log_2(2) + \frac{1}{4}\log_2(4)$$

$$H(X) = \frac{1}{4}\times 2 + \frac{1}{2}\times 1 + \frac{1}{4}\times 2$$

$$H(X) = 1.5 \text{ bits/symbol}$$

$$\therefore \quad H(X^2) \le H(X)$$

The equility holds when distinct values of X result in distinct values of X^2. Hence, option (c) is not correct.

(d) Consider $Y = 2^X$. Here, the distinct K values of X will result in distinct K values of Y. Hence, the probability of possible values, P_i of X and Y are equal. Therefore,

$$H(Y) = H(2^X) = H(X)$$

Hence, option (d) is correct.

3.104 (3)

Given that symbol stream contains alternate QPSK and 16-QAM signals.

For a QPSK signal, two bits are used per symbol.

For a 16-QAM signal,

Number of bits per symbol = $\log_2 16 = 4$

Since, the symbol stream contains alternate QPSK and 16-QAM signals, the average bits per symbol is

$$n = \frac{2+4}{2} = \frac{6}{2} = 3 \text{ bits/symbol}$$

At the rate of f_s = 1 M symbols per second, the bit rate or data rate is

$$R_b = nf_s$$

$$= (3 \text{ bits/symbol}) \times 1 \times 10^6 \left(\frac{\text{symbol}}{\text{sec}}\right)$$

$$R_b = 3 \times 10^6 \text{ bits/sec}$$

$$R_b = 3 \text{ mega-bits per second}$$

3.105 (0.23)

For the maximum likelihood detector,

$$f_{Y/X_A}(y) \underset{x_B}{\overset{x_A}{\gtrless}} f_{Y/X_B}(y)$$

i.e. if $f_{Y/x_A}(y) > f_{Y/x_B}(y)$, the receiver detects the received symbol as symbol x_A and if $f_{Y/x_A}(y) < f_{Y/x_B}(y)$, the receiver detects the received symbol as symbol x_B.

The given conditional probability density functions, $f_{Y/x_A}(y)$ and $f_{Y/x_B}(y)$ are plotted as below :

$$f_{Y/x_A}(y) = e^{-(y+1)}\, u(y+1)$$

$$f_{Y/x_B}(y) = e^{(y-1)}\,(1-u(y-1))$$

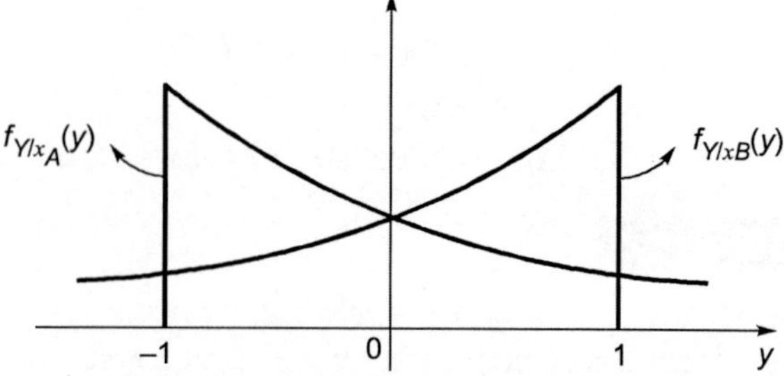

(i) For $-1 < y < 0$ and $1 < y < \infty$,

$$f_{Y/x_A}(y) > f_{Y/x_B}(y)$$

i.e. the receiver detects the received symbol as x_A.

The probability of error for x_B being detected as x_A is

$$P_{e1} = P(x_B)\cdot P(y = x_A / x_B)$$

$$\therefore \quad P_{e1} = P(x_B)\int_{-1}^{0} f_{Y/x_B}(y)\cdot dy$$

$$P_{e1} = P(x_B)\int_{-1}^{0} e^{(y-1)}[1-u(y-1)]\,dy$$

$$P_{e1} = P(x_B)\int_{-1}^{0} e^{y-1}\cdot dy = P(x_B)\left[e^{y-1}\right]_{-1}^{0}$$

$$P_{e1} = P(x_B)(e^{-1} - e^{-2}) = 0.23\,P(x_B) \quad \text{...(i)}$$

(ii) For $-\infty < y < -1$ and $0 < y < 1$

$$f_{Y/x_B}(y) > f_{Y/x_A}(y)$$

i.e. the receiver detects the received symbol as x_B.

The probability of error when x_A is transmitted and detected as x_B is

$$P_{e2} = P(x_A)\cdot P(y = x_B / x_A)$$

$$P_{e2} = P(x_A)\cdot\int_{0}^{1} f_{Y/x_A}(y)\cdot dy$$

$$P_{e2} = P(x_A)\cdot\int_{0}^{1} e^{-(y+1)}u(y+1)\,dy$$

$$P_{e2} = P(x_A)\cdot\int_{0}^{1} e^{-(y+1)}dy$$

$$P_{e2} = P(x_A)\cdot\left[\frac{e^{-(y+1)}}{-1}\right]_0^1$$

$$P_{e2} = P(x_A)\,[e^{-1} - e^{-2}]$$

$$P_{e2} = 0.23\,P(x_A) \quad \text{...(ii)}$$

From equations (i) and (ii), the probability of symbol error for the system is

$$P_e = P_{e1} + P_{e2}$$

$$P_e = 0.23[P(x_A) + P(x_B)]$$

$$P_e = 0.23$$

3.106 (10)

We know that for sinusoidal input, the signal to noise ratio (SNR) is given as,

$$\text{SNR} = 1.76 + 6.02\,n\text{ dB}$$

$$61.96\text{ dB} = 1.76 + 6.02\,n\text{ dB}$$

$$6.02\,n = 61.96 - 1.76$$

$$\therefore \quad n = 10\text{ bits}$$

3.107 (a)

For zero ISI, $x(t)$ should have zero crossings at $\pm T, \pm 2T, \pm 3T$....

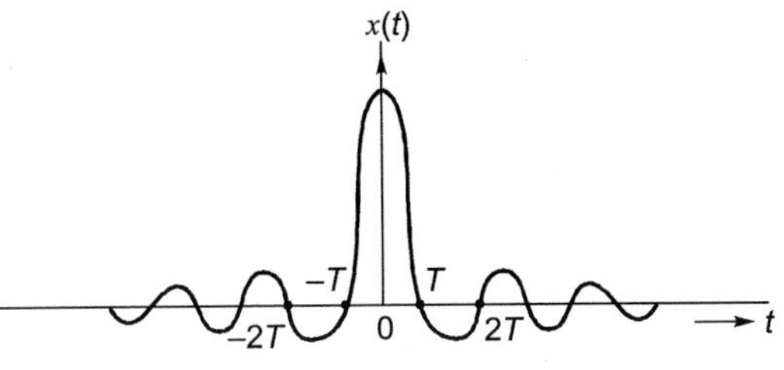

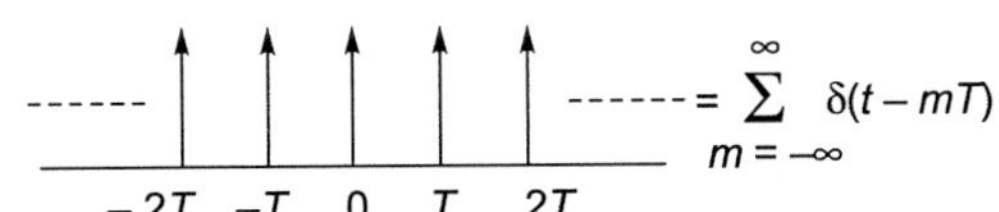

Taking Fourier transform both sides, we get,

$$x(t)\sum_{m=-\infty}^{\infty} \delta(t - mT) = x(0)\delta(t)$$

$$\frac{1}{T}\sum_{m=-\infty}^{\infty} X\left(f - \frac{m}{T}\right) = x(0)$$

$$\sum_{m=-\infty}^{\infty} X\left(f - \frac{m}{T}\right) = T \qquad [\text{Assuming } x(0) = 1]$$

$$\text{(or)} \sum_{m=-\infty}^{\infty} X\left(f + \frac{m}{T}\right) = T$$

Note: Since, $x(0)$ is not mentioned in the question so GATE given marks to all

3.108 (c)

For the given ADC, signal to quantization noise ratio can be calculated as

$$\text{SNR} = \frac{\text{Signal power}}{\text{Noise power}} = \frac{1\text{ W}}{10\,\mu\text{W}}$$

$$\text{SNR} = 10^5$$

In decibels,

$$\text{SNR} = 10\log_{10}(10^5) = 50\text{ dB} \quad \text{...(i)}$$

For an n-bit ADC,

$$\text{SNR} \approx 1.76 + 6.02\,n\text{ dB} \quad \text{...(ii)}$$

From (i) and (ii),

$$1.76 + 6.02n = 50$$

$$\Rightarrow \quad n \approx 8$$

Hence, the effective number of bits of the given *ADC* is 8.

3.109 (45.15)

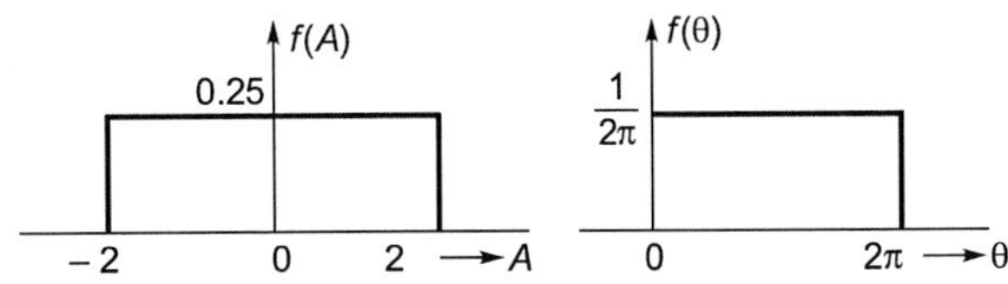

Signal power, $S = R_X(0) = \frac{2}{3}$ W

The quantization noise for a uniform quantizer is given by

$$N_q = \frac{\Delta^2}{12}$$

$$\Delta = \frac{2-(-2)}{2^8} = \frac{4}{2^8}$$

$$N_q = \frac{4^2}{12\times 2^{16}} = \frac{2^4}{12\times 2^{16}} = \frac{1}{12\times 2^{12}}\text{ W}$$

$$\frac{S}{N_q} = \frac{2}{3}\times\frac{12\times 2^{12}}{1} = 32{,}768$$

$$\left(\frac{S}{N_q}\right)_{\text{dB}} = 10\log 32768 = 45.15\text{ dB}$$

■■■

4 Information Theory and Coding

4.1 A source generates three symbols with probabilities 0.25, 0.25, 0.50 at a rate of 3000 symbols per second. Assuming independent generation of symbols, the most efficient source encoder would have average bit rate is

(a) 6000 bits/sec (b) 4500 bits/sec
(c) 3000 bits/sec (d) 1500 bits/sec

[2006 : 2 M]

4.2 A memoryless source emits n symbols each with a probability p. The entropy of the source as a function of n

(a) increases as log n
(b) decreases as log $(1/n)$
(c) increases as n
(d) increases as $n \log n$ [2008 : 2 M]

4.3 A communication channel with AWGN operating at a signal to noise ratio SNR >>1 and bandwidth B has capacity C_1. If the SNR is doubled keeping B constant, the resulting capacity C_2 is given by

(a) $C_2 \approx 2C_1$ (b) $C_2 \approx C_1 + B$
(c) $C_2 \approx C_1 + 2B$ (d) $C_2 \approx C_1 + 0.3B$

[2009 : 2 M]

4.4 A source alphabet consists of N symbols with the probability of the first two symbols being the same. A source encoder increases the probability of the first symbol by a small amount ε and decreases that of the second by ε. After encoding, the entropy of the source

(a) increases
(b) remains the same
(c) increases only if $N = 2$
(d) decreases [2012 : 1 M]

4.5 Let U and V be two independent and identically distributed random variables such that

$P(U = +1) = P(U = -1) = \frac{1}{2}$. The entropy

$H(U + V)$ in bits is

(a) 3/4 (b) 1
(c) 3/2 (d) $\log_2 3$

[2013 : 2 M]

4.6 In a code-division multiple access (CDMA) system with $N = 8$ chips, the maximum number of users who can be assigned mutually orthogonal signature sequences is ______.

[2014 : 1 M, Set-1]

4.7 The capacity of a Binary Symmetric Channel (BSC) with cross-over probability 0.5 is ____.

[2014 : 1 M, Set-1]

4.8 A fair coin is tossed repeatedly until a 'Head' appears for the first time. Let, L be the number of tosses to get this first 'Head'. The entropy $H(L)$ in bits is _______.

[2014 : 2 M, Set-1]

4.9 The capacity of a band-limited additive white Gaussian noise (AWGN) channel is given by

$C = W\log_2\left(1+\frac{P}{\sigma^2 W}\right)$ bits per second (bps),

where W is the channel bandwidth, P is the average power received and σ^2 is the one-sided power spectral density of the AWGN.

For a fixed $\frac{P}{\sigma^2} = 1000$, the channel capacity (in kbps) with infinite bandwidth ($W \to \infty$) is approximately

(a) 1.44 (b) 1.08
(c) 0.72 (d) 0.36

[2014 : 1 M, Set-2]

4.10 Consider a discrete memoryless source with alphabet $S = \{s_0, s_1, s_2, s_3, s_4, ...\}$ and respective probabilities of occurrence

$P = \left\{\frac{1}{2}, \frac{1}{4}, \frac{1}{8}, \frac{1}{16}, \frac{1}{32}, ...\right\}$. The entropy of the source (in bits) is _____. [2016 : 2 M, Set-1]

4.11 A digital communication system uses a repetition code for channel encoding/decoding. During transmission, each bit is repeated three times (0 is transmitted as 000, and 1 is transmitted as 111). It is assumed that the source puts out symbols independently and with equal probability.

The decoder operates as follows: In a block of three received bits, if the number of zeros exceeds the number of ones, the decoder decides in favor of a 0, and if the number of ones exceeds the number of zeros, the decoder decides in favor of a 1. Assuming a binary symmetric channel with crossover probability $p = 0.1$. The average probability of error is _____.

[2016 : 2 M, Set-1]

4.12 A discrete memoryless source has an alphabet $\{a_1, a_2, a_3, a_4\}$ with corresponding probabilities $\left\{\frac{1}{2}, \frac{1}{4}, \frac{1}{8}, \frac{1}{8}\right\}$. The minimum required average codeword length in bits to represent this source for error-free reconstruction is_____

[2016 : 1 M, Set-2]

4.13 A binary communication system makes use of the symbols "zero" and "one". There are channel errors. Consider the following events:

x_0 : a "zero" is transmitted

x_1 : a "one" is transmitted

y_0 : a "zero" is received

y_1 : a "one" is received

The following probabilities are given:

$P(x_0) = \frac{1}{2}, P(y_0|x_0) = \frac{3}{4}$, and $P(y_0|x_1) = \frac{1}{2}$. The information in bits that you obtain when you learn which symbol has been received (while you know that a "zero" has been transmitted) is _______

[2016 : 2 M, Set-2]

4.14 An analog baseband signal, bandlimited to 100 Hz, is sampled at the Nyquist rate. The samples are quantized into four message symbols that occur independently with probabilities $p_1 = p_4 = 0.125$ and $p_2 = p_3$. The information rate (bits/sec) of the message source is ________.

[2016 : 1 M, Set-3]

4.15 A voice-grade AWGN (additive white Gaussian noise) telephone channel has a bandwidth of 4.0 kHz and two-sided noise power spectral density $\frac{\eta}{2} = 2.5 \times 10^{-5}$ Watt per Hz. If information at the rate of 52 kbps is to be transmitted over this channel with arbitrarily small bit error rate, then the minimum bit-energy E_b (in mJ/bit) necessary is ________.

[2016 : 2 M, Set-3]

4.16 Let, (X_1, X_2) be independent random variables. X_1 has mean 0 and variance 1, while X_2 has mean 1 and variance 4. The mutual information $I(X_1 ; X_2)$ between X_1 and X_2 in bits is ________.

[2017 : 1 M, Set-1]

4.17 Which one of the following graphs shows the Shannon capacity (Channel capacity) in bits of a memoryless binary symmetric channel with crossover probability p?

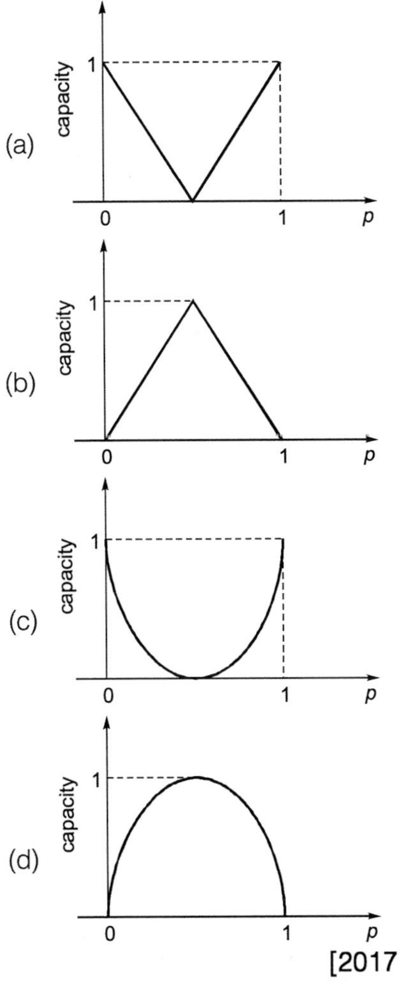

[2017 : 1 M, Set-2]

4.18 Consider a binary memoryless channel characterized by the transition probability diagram shown in the figure.

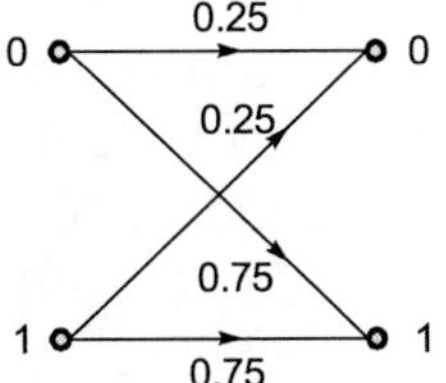

The channel is

(a) lossless
(b) noiseless
(c) useless
(d) deterministic

[2017 : 2 M, Set-2]

4.19 Consider a binary channel code in which each codeword has a fixed length of 5 bits. The Hamming distance between any pair of distinct codewords in this code is at least 2. The maximum number of codewords such a code can contain is _______.

[2018 : 1 M]

4.20 A linear Hamming code is used to map 4-bit messages to 7-bit codewords. The encoder mapping is linear. If the message 0001 is mapped to the codeword 0000111, and the message 0011 is mapped to the codeword 1100110, then the message 0010 is mapped to

(a) 0010011
(b) 1100001
(c) 1111000
(d) 1111111

[2019 : 1 M]

4.21 A digital communication system transmits a block of N bits. The probability of error in decoding a bit is α. The error event of each bit is independent of the error events of the other bits. The received block is declared erroneous if at least one of the its bits is decoded wrongly. The probability that the received block is erroneous is

(a) $N(1-\alpha)$
(b) $1-(1-\alpha)^N$
(c) $1-\alpha^N$
(d) α^N

[2020 : 1 M]

4.22 In a high school having equal number of boy students and girl students, 75% of the students study Science and the remaining 25% students study Commerce. Commerce students are two times more likely to be a boy than are Science students. The amount of information gained in knowing that a randomly selected girl student studies Commerce (rounded off to three decimal places) is ______ bits. [2021 : 2 M]

4.23 A digital transmission system uses a (7, 4) systematic linear Hamming code for transmitting data over a noisy channel. If three of the message-codeword pairs in this code $(m_i ; c_i)$, where c_i is the codeword corresponding to the ith message m_i, are known to be (1100 ; 0101100), (1110 ; 0011110) and (0110 ; 1000110), then which of the following is a valid codeword in this code?

(a) 0110100
(b) 1101001
(c) 0001011
(d) 1011010

[2021 : 2 M]

4.24 The transition diagram of a discrete memoryless channel with three input symbols and three output symbols is shown in the figure. The transition probabilities are as marked.

The parameter α lies in the interval [0.25, 1]. The value of α for which the capacity of this channel is maximized, is ________ (rounded off to two decimal places).

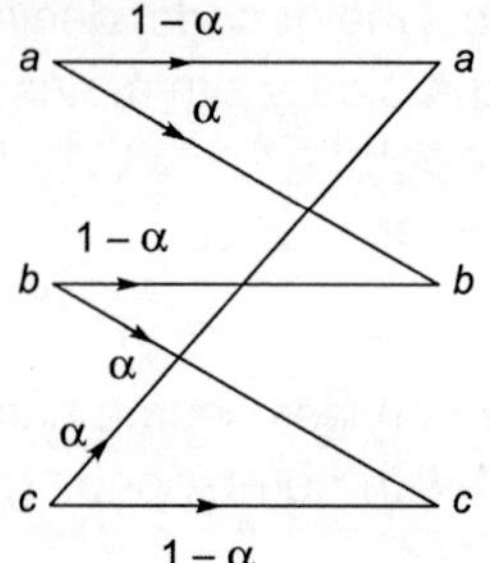

[2022 : 2 M]

4.25 Consider communication over a memoryless binary symmetric channel using a (7, 4) Hamming code. Each transmitted bit is received correctly with probability $(1-\epsilon)$, and flipped with probability ϵ. For each codeword transmission, the receiver performs minimum Hamming distance decoding, and correctly decodes the message bits if and only if the channel introduces at most one bit error.

For $\epsilon = 0.1$, the probability that a transmitted codeword is decoded correctly is ______ (rounded off to two decimal places). [2022 : 2 M]

4.26 The frequency of occurrence of 8 symbols (a-h) is shown in the table below. A symbol is chosen and it is determined by asking a series of "yes/no" questions which are assumed to be truthfully answered. The average number of questions when asked in the most efficient sequence, to determine the chosen symbol, is ______.

(rounded off to two decimal places).

Symbols	a	b	c	d	e	f	g	h
Frequency of occurrence	$\frac{1}{2}$	$\frac{1}{4}$	$\frac{1}{8}$	$\frac{1}{16}$	$\frac{1}{32}$	$\frac{1}{64}$	$\frac{1}{128}$	$\frac{1}{128}$

[2023 : 2 M]

4.27 A source transmits symbols from an alphabet of size 16. The value of maximum achievable entropy (in bits) is ______. [2024 : 1 Mark]

4.28 The information bit sequence (1 1 1 0 1 0 1 0 1) is to be transmitted by encoding with Cyclic Redundancy Check 4 (CRC-4) code, for which the generator polynomial is $C(x) = x^4 + x + 1$. The encoded sequence of bits is ________.

(a) {1 1 1 0 1 0 1 0 1 1 1 0 1}
(b) {1 1 1 0 1 0 1 0 1 1 1 1 0}
(c) {1 1 1 0 1 0 1 0 1 0 1 0 0}
(d) {1 1 1 0 1 0 1 0 1 1 1 0 0} [2024 : 2 Marks]

Answers Information Theory and Coding

4.1	(b)	4.2	(a)	4.3	(b)	4.4	(d)	4.5	(c)	4.6	(8)	4.7	(0)
4.8	(2)	4.9	(a)	4.10	(2)	4.11	(0.028)	4.12	(1.75)	4.13	(0.811)	4.14	(362.2)
4.15	(31.504)	4.16	(0)	4.17	(c)	4.18	(c)	4.19	(16)	4.20	(b)	4.21	(b)
4.22	(3.32)	4.23	(c)	4.24	(1)	4.25	(0.85)	4.26	(1.984)	4.27	(4)	4.28	(d)

Explanations Information Theory and Coding

4.1 (b)

Bit rate = $H.R_b$

$$H = 0.25\log\frac{1}{0.25} + 0.25\log\frac{1}{0.25} + 0.5\log\frac{1}{0.5}$$

$$H = \frac{1}{2} + \frac{1}{2} + \frac{1}{2} = \frac{3}{2} = 1.5$$

Bit rate = 1.5 × 3000 = 4500 bits/sec.

4.2 (a)

Entropy, $H(m) = -\sum_{i=1}^{n} P_i \log P_i$ bits

where, P_i = Probability of individual symbol.
Since probability of each symbol is same, therefore,

$$P_1 = P_2 = \dots = P_n = \frac{1}{n}$$

$$H(m) = -\sum_{i=1}^{n}\frac{1}{n}\log\frac{1}{n} = \log n$$

4.3 (b)

SNR >> 1

$$C_1 = B\log_2\left(1+\frac{S}{N}\right) \cong B\log_2\left(\frac{S}{N}\right)$$

when SNR is doubled

$$C' \cong B\log_2\left(\frac{2S}{N}\right)$$

$$= \underbrace{B\log_2\left(\frac{S}{N}\right)}_{=C_1} + \underbrace{B\log_2 2}_{=B} = C_1 + B$$

4.4 (d)

We know that entropy is maximum when symbols are equal probable so if probability will change from equal to non equal entropy will decrease.

4.5 (c)

$$P(U+V=+2) = \frac{1}{2}\cdot\frac{1}{2} = \frac{1}{4}$$

$$P(U+V=0) = \frac{1}{4} + \frac{1}{4} = \frac{1}{2}$$

$$P(U+V=-2) = \frac{1}{2}\cdot\frac{1}{2} = \frac{1}{4}$$

U	V	$(U+V)$
+1	+1	+2
+1	−1	0
−1	+1	0
−1	−1	−2

$$\therefore\ H(U+V) = \frac{1}{2}\log_2 2 + 2\times\frac{1}{4}\log_2 4$$

$$= \frac{1}{2} + 1 = \frac{3}{2}$$

4.6 (8)

7.99 to 8.01

4.7 (0)

Channel capacity of BSC is

$C = P\log_2 P + (1-P)\log_2(1-P) + 1$

$C = 0.5\log_2 0.5 + 0.5\log_2 0.5 + 1$

$C = 0$

It is the case of channel with independent input and output, hence $C = 0$.

4.8 (2)

If $P = 1$, i.e. if 1 toss is required to gets first head.
Then probability to get the first head

$$\Rightarrow \quad P_1 = \frac{1}{2}$$

If $P = 2$; i.e. if 2 tosses are required to get first head.
Then probability to get the first head

$$\Rightarrow \quad P_2 = \frac{1}{2}\times\frac{1}{2} = \frac{1}{4}$$

If $P = 3$; i.e. if 3 tosses are required to get first head.
Then probability to get the first head

$$\Rightarrow \quad P_3 = \frac{1}{2}\times\frac{1}{2}\times\frac{1}{2} = \frac{1}{8}$$

Entropy $H = \sum_{i=1}^{N} P_i \log_2 \frac{1}{P_i}$

$$= \frac{1}{2}\log_2 \frac{1}{1/2} + \frac{1}{4}\log_2 \frac{1}{1/4} + \log_2 \frac{1}{1/8} + \frac{1}{16}\log_2 \frac{1}{1/16} + \cdots$$

$$= \frac{1}{2}\times 1 + \frac{1}{4}\times 2 + \frac{1}{8}\times 3 + \frac{1}{16}\times 4 + \cdots$$

$$= \frac{1}{2} + \frac{2}{(2)^2} + \frac{3}{(2)^3} + \frac{4}{(2)^4} + \cdots$$

$$H \simeq 2$$

4.9 (a)

$$C = W\log_2\left(1 + \frac{P}{\sigma^2 W}\right)$$

$$C = \frac{P}{\sigma^2}\times\frac{\sigma^2}{P}W\log_2\left(1+\frac{P}{\sigma^2 W}\right)$$

when $W \to \infty$

$$C_\infty = \frac{P}{\sigma^2}\lim_{W\to\infty}\frac{\sigma^2 W}{P}\log_2\left(1+\frac{P}{\sigma^2 W}\right)$$

as $W \to \infty$, $\frac{\sigma^2 W}{P}$ also $\to \infty$

$$\therefore\ C_\infty = \frac{P}{\sigma^2}\lim_{\frac{\sigma^2 W}{P}\to\infty}\frac{\sigma^2 W}{P}\log_2\left(1+\frac{1}{\frac{\sigma^2 W}{P}}\right)$$

$$\left[\because \lim_{x\to\infty} x\log_2\left(1+\frac{1}{x}\right) = \log_2 e\right]$$

$$C_\infty = \frac{P}{\sigma^2}\log_2 e$$

$$C_\infty = 1.44\frac{P}{\sigma^2}$$

4.10 (2)

Entropy of source is given as

$$H = \sum_{i=0}^{N} P_i \log_2 \frac{1}{P_i}$$

(Here, $N = \infty$)

$$= \frac{1}{2}\log_2 2 + \frac{1}{4}\log_2 4 + \frac{1}{8}\log_2 8 + \frac{1}{16}\log_2 16 + \ldots$$

$$H = \frac{1}{2} + 2\times\left(\frac{1}{2}\right)^2 + 3\times\left(\frac{1}{2}\right)^3 + 4\times\left(\frac{1}{2}\right)^4 + \ldots \quad \ldots(i)$$

$$= \sum_{k=0}^{\infty} k\left(\frac{1}{2}\right)^k$$

$$\frac{H}{2} = \left(\frac{1}{2}\right)^2 + 2\times\left(\frac{1}{2}\right)^3 + 3\times\left(\frac{1}{2}\right)^4 + \ldots \quad \ldots(ii)$$

Subtracting (ii) from (i)

$$\frac{H}{2} = \left(\frac{1}{2}\right) + \left(\frac{1}{2}\right)^2 + \left(\frac{1}{2}\right)^3 + \ldots$$

$$\frac{H}{2} = \frac{\left(\frac{1}{2}\right)}{1-\left(\frac{1}{2}\right)} = 1$$

$$\Rightarrow \quad H = 2 \text{ bits/symbol}$$

4.11 (0.028)

Crossover probability,

$$P = 0.1$$

Average probability of error $= 3p^2 - 2p^3$

$$= 3(0.1)^2 - 2(0.1)^3 = 0.028$$

4.12 (1.75)

Minimum required average codeword length in bits for error free reconstruction

$$L_{min} = H \quad \text{(Entropy)}$$

$$H = \frac{1}{2}\log_2 2 + \frac{1}{4}\log_2 4 + \frac{1}{8}\log_2 8 + \frac{1}{8}\log_2 8$$

$$= \frac{1}{2} + \frac{1}{2} + \frac{3}{8} + \frac{3}{8} = 1.75$$

$$\Rightarrow \quad L_{min} = 1.75 \text{ bits/word}$$

4.13 (0.811)

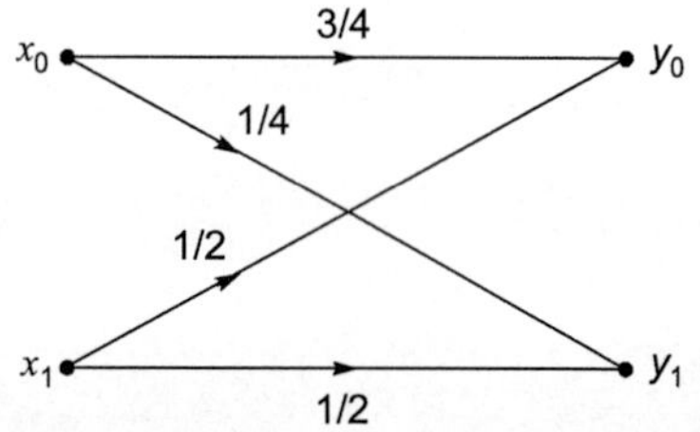

Information in bits when you learn which symbol has been received when zero is transmitted.

Case-I = when y_0 is received and x_0 is transmitted

Case-II = when y_1 is received and x_0 is transmitted

Thus, the overall average information, we can say H.

$$H = P(y_0 | x_0)\log_2\left(\frac{1}{P(y_0|x_0)}\right) + P(y_1|x_0)\log_2\left(\frac{1}{P(y_1|x_0)}\right)$$

$$= \left[\frac{3}{4}\log_2\left(\frac{4}{3}\right) + \frac{1}{4}\log_2(4)\right] \text{bits}$$

$$= 0.811 \text{ bits}$$

4.14 (362.2)

$$f_m = 100 \text{ Hz}$$

$$f_s = 2f_m = 200 \text{ samples/sec}$$

$$P_1 = P_4 = \frac{1}{8}$$

$$P_1 + P_2 + P_3 + P_4 = 1$$

$$\Rightarrow \quad 2P_2 = 2P_3 = 1 - \frac{1}{4} = \frac{3}{4}$$

$$\Rightarrow \quad P_2 = P_3 = \frac{3}{8}$$

$$H = \sum_{i=1}^{N} P_i \log \frac{1}{P_i}$$

$$= \frac{1}{8}\log_2 8 + \frac{1}{8}\log_2 8 + \frac{3}{8}\log_2 \frac{8}{3} + \frac{3}{8}\log_2 \frac{8}{3}$$

$$H = 1.811 \text{ bits/sample}$$

Information rate

$$R = H(\text{bits/sample}) \times r(\text{samples/sec})$$

$$= 1.811 \times 200 = 362.2 \text{ bits/sec}$$

4.15 (31.504)

$$C = B\log_2\left(1 + \frac{S}{N}\right)$$

$$S = E_b \cdot R_b$$

(E_b = bit energy, R_b = information rate bits/sec)

$$C = B\log_2\left(1 + \frac{E_b R_b}{N_o B}\right)$$

For distortionless transmission, C should be atleast of R_b

$$\frac{R_b}{B} = \log_2\left(1 + \frac{E_b R_b}{N_o B}\right)$$

$$E_b = (2^{R_b/B} - 1)\frac{N_o B}{R_b}$$

$$= \frac{(2^{52/13} - 1) \times 2 \times 2.5 \times 10^{-5} \times 4}{52}$$

$$E_b \approx 31.504 \text{ mJ/bit}$$

4.16 (0)

Mutual information of two random variables is a measure of the mutual dependence of the two variables.

Given that, X and Y are independent. Hence, $I(X:Y) = 0$

4.17 (c)

The channel capacity of a memoryless binary symmetric channel can be expressed as,

$$C = 1 + p\log_2 p + (1-p)\log_2(1-p)$$

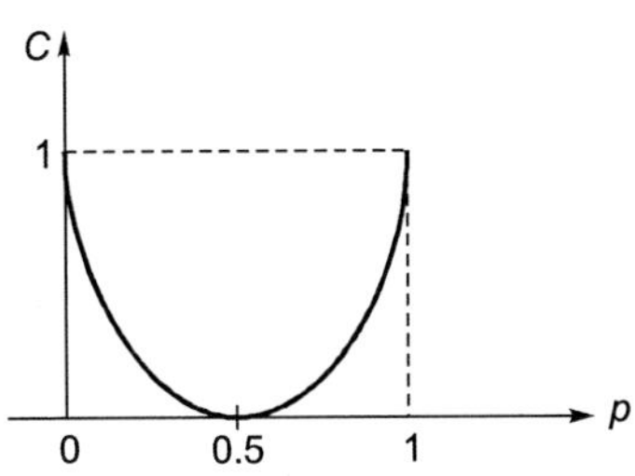

4.18 (c)

Given that, $\left[P\left(\frac{Y}{X}\right)\right] = \begin{bmatrix} 0.25 & 0.75 \\ 0.25 & 0.75 \end{bmatrix}$

If mutual information $I(X;Y) = 0$ for every possible input distribution, then the channel is called as useless (or) zero-capacity channel.

Let $[P(X)] = [\alpha \ (1-\alpha)]$

Then,

$H(X) = -\alpha\log_2\alpha - (1-\alpha)\log_2(1-\alpha)$ bits/symbol

$$[P(Y)] = [P(X)]\left[P\left(\frac{Y}{X}\right)\right] = [0.25 \quad 0.75]$$

$$[P(X,Y)] = \begin{bmatrix} \frac{\alpha}{4} & \frac{3\alpha}{4} \\ \frac{(1-\alpha)}{4} & \frac{3(1-\alpha)}{4} \end{bmatrix}$$

$$\left[P\left(\frac{X}{Y}\right)\right] = \frac{[P(X,Y)]}{[P(Y)]_d} = \begin{bmatrix} \alpha & \alpha \\ (1-\alpha) & (1-\alpha) \end{bmatrix}$$

$$H\left(\frac{X}{Y}\right) = -\sum_i \sum_j P(x_i, y_j)\log_2 P\left(\frac{x_i}{y_j}\right) \text{ bits/symbol}$$

$= -\alpha\log_2\alpha - (1-\alpha)\log_2(1-\alpha)$ bits/symbol

$$I(X;Y) = H(X) - H\left(\frac{X}{Y}\right) = 0$$

So, the given binary memoryless channel is a "useless" channel.

4.19 (16)

Given $n = 5$ and $d_{min} = 2$

From singletone bound $\rightarrow d_{min} \le n - K + 1$

$$K \le 4$$

Number of codewords possible = 2^K

Maximum number of codewords possible

$$= 2^4 = 16$$

4.20 (b)

$$\begin{array}{ccc} 0001 & \longrightarrow & 0000111 \\ \oplus & & \oplus \\ 0011 & \longrightarrow & 1100110 \\ \hline 0010 & \longrightarrow & 1100001 \\ \hline \end{array}$$

4.21 (b)

Probability of error in decoding single bit = α
Then probability of no error will be $1 - \alpha$.
Total N-bits transmitted, so that probability of no error in received block

$$= (1-\alpha)(1-\alpha) \ldots N \text{ times}$$
$$= (1-\alpha)^N$$

The probability of received block is erroneous is $= 1-(1-\alpha)^N$

4.22 (3.32)

Given, $P(S) = \frac{1}{2}$ and $P(B) = \frac{1}{2}$

$P(C) = \frac{1}{4}$ and $P(S) = \frac{3}{4}$

Let probability of selected science student is a boy is $P(B/S) = x$

Given that Commerce students are two time more likely to be a boy than are Science students.

Then, probability of selected Commerce student is a boy

$$P\left(\frac{B}{C}\right) = 2x$$

We have to find probability of randomly selected girl studies Commerce i.e.

$$P\left(\frac{C}{G}\right) = \frac{P(C \cap G)}{P(G)} = \frac{P(C)\,P(G/C)}{P(G)}$$

$$P\left(\frac{C}{G}\right) = \frac{\frac{1}{4} \times P\left(\frac{G}{C}\right)}{\frac{1}{2}} \quad \ldots(i)$$

To find $P(G/C)$, first we have to find $P(B/C)$.

The probability of selected student is a boy

$$P(B) = P(S)\,P(B/S) + P(C) \times P(B/C)$$

$$\frac{1}{2} = \frac{3}{4} \times x + \frac{1}{4} \times 2x$$

$$\Rightarrow \quad x = \frac{2}{5}$$

Then, $P(B/C) = 2x = \frac{4}{5}$

We know that,

$$P\left(\frac{B}{C}\right) + P\left(\frac{G}{C}\right) = 1$$

$$P\left(\frac{G}{C}\right) = 1 - \frac{4}{5} = \frac{1}{5}$$

From equation (i),

$$P\left(\frac{C}{G}\right) = \frac{\frac{1}{4} \times \frac{1}{5}}{1/2} = \frac{1}{10}$$

The amount of information gained in knowing that a randomly selected girl studies Commerce

$$I\left(\frac{C}{G}\right) = \log_2 \frac{1}{P\left(\frac{C}{G}\right)}$$

$$= \log_2 10 = 3.32$$

4.23 (c)

Given codewords:

$$\left.\begin{aligned} C_1 &= 0101100 \\ C_2 &= 0011110 \\ C_3 &= 1000110 \end{aligned}\right\}$$

In given above codewords, first $n - k = 3$ bits are parity bits and last $K = 4$ bits are message bits. Given is (7, 4) systematic linear hamming code. For a linear code, sum of two codewords belong to the code is also a codeword belonging to the code.

$$C_1 \oplus C_2 = 0110010 = C_4$$
$$C_2 \oplus C_3 = 1011000 = C_5$$
$$C_1 \oplus C_3 = 1101010 = C_6$$
$$C_3 \oplus C_4 = 1110100 = C_7$$

Based on codewords C_6 and C_7 options (a) and (d) will incorrect.

Given codewords in the form of

$$\Rightarrow \underbrace{P_1 \; P_2 \; P_3}_{\text{Parity bits}} \; \underbrace{d_1 \; d_2 \; d_3 \; d_4}_{\text{Message bits}}$$

From observation

$$\Rightarrow \quad P_1 = d_1 \oplus d_2 \oplus d_4$$
$$P_2 = d_2 \oplus d_3 \oplus d_4$$
$$P_3 = d_1 \oplus d_2 \oplus d_3$$

From given options, option (c) only satisfying above relations.

4.24 (1)

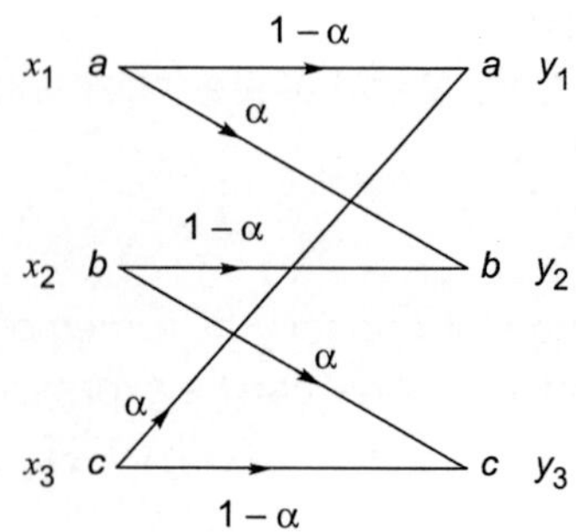

Channel capacity is defined as

$$C_s = \text{Max}[I(X;Y)]$$

where, $I(X;Y)] = H(Y) - H\left(\frac{Y}{X}\right)$

$$H\left(\frac{Y}{X}\right) = -\sum_{i=1}^{3}\sum_{j=1}^{3} P(x_i, y_j)\log_2 P\left(\frac{y_j}{x_i}\right)$$

From the given channel matrix,

$$\left[P\left(\frac{Y}{X}\right)\right] = \begin{matrix} & y_1 & y_2 & y_3 \\ x_1 & 1-\alpha & \alpha & 0 \\ x_2 & 0 & 1-\alpha & \alpha \\ x_3 & \alpha & 0 & 1-\alpha \end{matrix}$$

For simplification convenience, let

$$[P(X)] = [1\ 0\ 0]$$

$$[P(X,Y)] = [P(X)]_d \cdot \left[P\left(\frac{Y}{X}\right)\right]$$

$$[P(X,Y)] = \begin{bmatrix} 1 & 0 & 0 \\ 0 & 0 & 0 \\ 0 & 0 & 0 \end{bmatrix}\begin{bmatrix} 1-\alpha & \alpha & 0 \\ 0 & 1-\alpha & \alpha \\ \alpha & 0 & 1-\alpha \end{bmatrix}$$

$$[P(X,Y)] = \begin{bmatrix} 1-\alpha & \alpha & 0 \\ 0 & 0 & 0 \\ 0 & 0 & 0 \end{bmatrix}$$

$$H\left(\frac{Y}{X}\right) = -\{(1-\alpha)\log_2(1-\alpha) + \alpha\log_2\alpha\}$$

Hence,

$I(X;Y) = H(Y) + (1-\alpha)\log_2(1-\alpha) + \alpha\log_2\alpha$

$C_s = \text{Max}\{I(X:Y)\}$

$C_s = \text{Max}\{H(Y)\} + (1-\alpha)\log_2(1-\alpha) + \alpha\log_2\alpha$

$C_s = \log_2 3 + (1-\alpha)\log_2(1-\alpha) + \alpha\log_2\alpha$

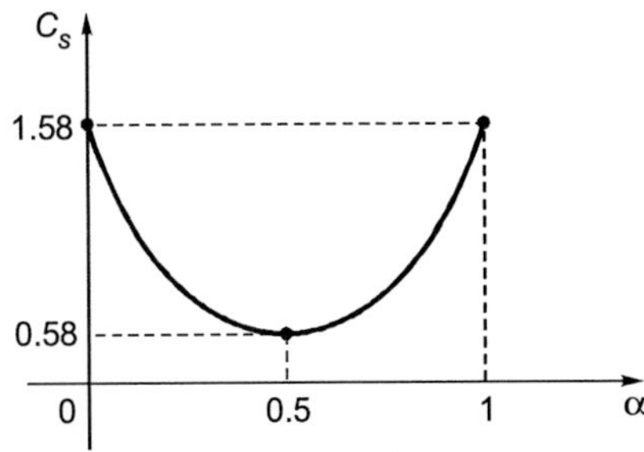

C_S will be maximum at $\alpha = 0$ and 1

Given: $\alpha \in [0.25, 1]$

So, $\alpha = 1$ is the correct answer at which the capacity of the channel is maximum.

4.25 (0.85)

For a (7, 4) Hamming code,

Number of bits in the transmitted codeword = 7

For a binary symmetric channel

$$P(0/1) = P(1/0) = \in = 0.1$$

The receiver correctly decodes the message if it has almost one bit error.

Hence, the probability that a transmitted codeword is decoded correctly is

$$P_c = P(\text{No error}) + P(1-\text{bit error})$$

When 'n' bits are transmitted, the probability of getting error in 'r' bits is ${}^nC_r\, p^r(1-p)^{n-r}$

Where 'p' is bit error probability given as

$$P(1/0) = P(0/1) = p = \in = 0.1$$

Hence, $P_c = P(\text{no error}) + P(1-\text{bit error})$

$$P_c = {}^7C_0(0.1)^0(1-0.1)^7 + {}^7C_1(0.1)^1(1-0.1)^6$$

$$P_c = (0.9)\times 7 + 7\times 0.1\times(0.9)^6$$

$$P_c = 0.8503$$

4.26 (1.984)

The average number of questions when asked in the most efficient sequence, to determine the chosen symbol = min possible number of questions per symbol (H)

$$H = \sum_i P_x(x_i)\log_2\frac{1}{P_x(x_i)}$$

$$= \frac{1}{2}\log_2 2 + \frac{1}{4}\log_2 4 + \frac{1}{8}\log_2 8 + \frac{1}{16}\log_2 16 + \frac{1}{32}\log_2 32 + \frac{1}{64}\log_2 64 + 2\times\frac{1}{128}\log_2 128$$

$$= 1.984\,\frac{\text{Questions}}{\text{Symbol}}$$

4.27 (4)

Consider a source transmits 16 symbols, each with probability P_i. The entropy of the source is given by

$$H = \sum_{i=0}^{15} P_i\log_2\left(\frac{1}{P_i}\right)$$

To achieve maximum entropy, all the symbols must be equiprobable i.e., $P_i = \frac{1}{16}$. Hence,

$$H_{max} = \sum_{i=0}^{15}\frac{1}{16}\log_2(16) = \log_2(16)$$

$$H_{max} = 4 \text{ bits}$$

4.28 (d)

Given information bit sequence → $d = 111010101$

Generator polymial → $C(x) = x^4 + x + 1$

$C = 10011$

Since, generator polynomial is of order 4, hence 4 bits at the end are appended to information sequence.

$d = 1\,1\,1\,0\,1\,0\,1\,0\,1\,0\,0\,0\,0$

```
10011 ) 1110101010000 (
        10011
        ----------------
        Ø111001010000
         10011
        ----------------
         Ø11111010000
          10011
        ----------------
          Ø1100010000
           10011
        ----------------
           Ø101110000
            10011
        ----------------
            ØØ1000000
              10011
        ----------------
              ØØØ1100
```

Encoded sequence = 1 1 1 0 1 0 1 0 1 1 1 0 0

■■■

Engineering Mathematics

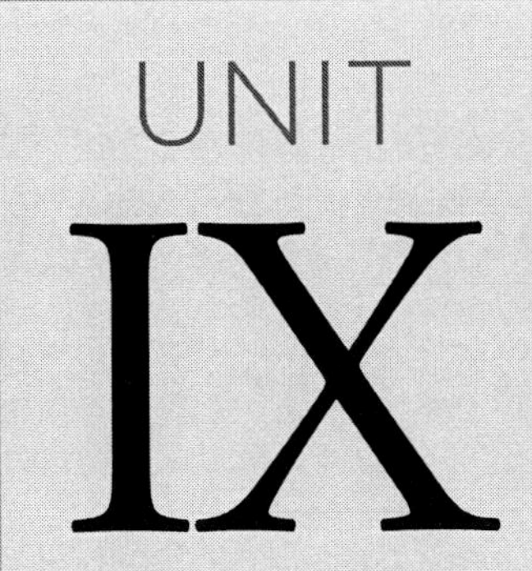

CONTENTS

Engineering Mathematics

Syllabus

Linear Algebra: Vector space, basis, linear dependence and independence, matrix algebra, eigenvalues and eigenvectors, rank, solution of linear equations- existence and uniqueness.

Calculus: Mean value theorems, theorems of integral calculus, evaluation of definite and improper integrals, partial derivatives, maxima and minima, multiple integrals, line, surface and volume integrals, Taylor series.

Differential Equations: First order equations (linear and nonlinear), higher order linear differential equations, Cauchy's and Euler's equations, methods of solution using variation of parameters, complementary function and particular integral, partial differential equations, variable separable method, initial and boundary value problems.

Vector Analysis: Vectors in plane and space, vector operations, gradient, divergence and curl, Gauss's, Green's and Stokes' theorems.

Complex Analysis: Analytic functions, Cauchy's integral theorem, Cauchy's integral formula, sequences, series, convergence tests, Taylor and Laurent series, residue theorem.

Probability and Statistics: Mean, median, mode, standard deviation, combinatorial probability, probability distributions, binomial distribution, Poisson distribution, exponential distribution, normal distribution, joint and conditional probability.

Analysis of Previous GATE Papers

Exam Year	1 Mark Ques.	2 Marks Ques.	Total Marks
1995	–	1	**2**
1996	1	–	**1**
1997	1	–	**1**
1998	2	–	**2**
1999	1	–	**1**
2000	2	–	**2**
2001	–	–	**–**
2002	–	–	**–**
2003	1	–	**1**
2004	–	–	**–**
2005	3	7	**17**
2006	6	8	**22**
2007	5	7	**19**
2008	6	5	**16**
2009	3	4	**11**
2010	3	6	**15**
2011	3	3	**9**
2012	3	5	**13**
2013	5	1	**7**

Exam Year	1 Mark Ques.	2 Mark Ques.	Total Marks
2014 Set-1	4	3	**10**
2014 Set-2	5	4	**13**
2014 Set-3	5	4	**13**
2014 Set-4	5	4	**13**
2015 Set-1	5	4	**13**
2015 Set-2	3	4	**11**
2015 Set-3	3	3	**9**
2016 Set-1	5	4	**13**
2016 Set-2	5	4	**13**
2016 Set-3	5	4	**13**
2017 Set-1	4	5	**14**
2017 Set-2	4	5	**14**
2018	6	4	**14**
2019	7	3	**13**
2020	6	4	**14**
2021	3	3	**9**
2022	5	4	**13**
2023	6	4	**14**
2024	3	3	**9**

1 Linear Algebra

1.1 The eigen values of matrix $A = \begin{bmatrix} 0 & 1 \\ 1 & 0 \end{bmatrix}$ are

(a) 1, 1 (b) −1, −1
(c) $j, -j$ (d) 1, −1 [1998 : 1 M]

1.2 The eigen values of the matrix $\begin{bmatrix} 2 & -1 & 0 & 0 \\ 0 & 3 & 0 & 0 \\ 0 & 0 & -2 & 0 \\ 0 & 0 & -1 & 4 \end{bmatrix}$ are

(a) 2, −2, 1, −1 (b) 2, 3, −2, 4
(c) 2, 3, 1, 4 (d) None [2000 : 1 M]

1.3 Given the matrix $\begin{bmatrix} -4 & 2 \\ 4 & 3 \end{bmatrix}$, the eigenvector is

(a) $\begin{bmatrix} 3 \\ 2 \end{bmatrix}$ (b) $\begin{bmatrix} 4 \\ 3 \end{bmatrix}$

(c) $\begin{bmatrix} 2 \\ -1 \end{bmatrix}$ (d) $\begin{bmatrix} -1 \\ 2 \end{bmatrix}$

[2005 : 2 M]

1.4 Let, $A = \begin{bmatrix} 2 & -0.1 \\ 0 & 3 \end{bmatrix}$ and $A^{-1} = \begin{bmatrix} \frac{1}{2} & a \\ 0 & b \end{bmatrix}$. Then $(a + b) =$

(a) $\frac{7}{20}$ (b) $\frac{3}{20}$

(c) $\frac{19}{60}$ (d) $\frac{11}{20}$ [2005 : 2 M]

1.5 Given an orthogonal matrix

$$A = \begin{bmatrix} 1 & 1 & 1 & 1 \\ 1 & 1 & -1 & -1 \\ 1 & -1 & 0 & 0 \\ 0 & 0 & 1 & -1 \end{bmatrix} \quad \left[AA^T\right]^{-1} \text{ is}$$

(a) $\begin{bmatrix} \frac{1}{4} & 0 & 0 & 0 \\ 0 & \frac{1}{4} & 0 & 0 \\ 0 & 0 & \frac{1}{2} & 0 \\ 0 & 0 & 0 & \frac{1}{2} \end{bmatrix}$ (b) $\begin{bmatrix} \frac{1}{2} & 0 & 0 & 0 \\ 0 & \frac{1}{2} & 0 & 0 \\ 0 & 0 & \frac{1}{2} & 0 \\ 0 & 0 & 0 & \frac{1}{2} \end{bmatrix}$

(c) $\begin{bmatrix} 1 & 0 & 0 & 0 \\ 0 & 1 & 0 & 0 \\ 0 & 0 & 1 & 0 \\ 0 & 0 & 0 & 1 \end{bmatrix}$ (d) $\begin{bmatrix} \frac{1}{4} & 0 & 0 & 0 \\ 0 & \frac{1}{4} & 0 & 0 \\ 0 & 0 & \frac{1}{4} & 0 \\ 0 & 0 & 0 & \frac{1}{4} \end{bmatrix}$

[2005 : 2 M]

1.6 The rank of the matrix $\begin{bmatrix} 1 & 1 & 1 \\ 1 & -1 & 0 \\ 1 & 1 & 1 \end{bmatrix}$ is

(a) 0 (b) 1
(c) 2 (d) 3 [2006 : 1 M]

1.7 The eigenvalues and the corresponding eigenvectors of a 2 × 2 matrix are given by

Eigenvalue	Eigenvector
$\lambda_1 = 8$	$v_1 = \begin{bmatrix} 1 \\ 1 \end{bmatrix}$
$\lambda_2 = 4$	$v_2 = \begin{bmatrix} 1 \\ -1 \end{bmatrix}$

The matrix is

(a) $\begin{bmatrix} 6 & 2 \\ 2 & 6 \end{bmatrix}$ (b) $\begin{bmatrix} 4 & 6 \\ 6 & 4 \end{bmatrix}$

(c) $\begin{bmatrix} 2 & 4 \\ 4 & 2 \end{bmatrix}$ (d) $\begin{bmatrix} 4 & 8 \\ 8 & 4 \end{bmatrix}$ [2006 : 2 M]

1.8 For the matrix $\begin{bmatrix} 4 & 2 \\ 2 & 4 \end{bmatrix}$ the eigenvalue corresponding to the eigenvector $\begin{bmatrix} 101 \\ 101 \end{bmatrix}$ is

(a) 2 (b) 4
(c) 6 (d) 8 [2006 : 2 M]

1.9 It is given that $X_1, X_2, \ldots X_M$ are M non-zero, orthogonal vectors. The dimension of the vector space spanned by the 2 M vectors $X_1, X_2, \ldots X_M$, $-X_1, -X_2, \ldots -X_M$ is

(a) 2 M
(b) $M + 1$
(c) M
(d) dependent on the choice of $X_1, X_2, \ldots X_M$

[2007 : 2 M

1.10 All the four entries of the 2 × 2 matrix $P = \begin{bmatrix} P_{11} & P_{12} \\ P_{21} & P_{22} \end{bmatrix}$ are nonzero, and one of its eigenvalues is zero. Which of the following statements is true?

(a) $P_{11}P_{22} - P_{12}P_{21} = 1$
(b) $P_{11}P_{22} - P_{12}P_{21} = -1$
(c) $P_{11}P_{22} - P_{12}P_{21} = 0$
(d) $P_{11}P_{22} + P_{12}P_{21} = 0$ [2008 : 1 M]

1.11 The system of linear equations

$4x + 2y = 7$
$2x + y = 6$

has

(a) a unique solution
(b) no solution
(c) an infinite number of solutions
(d) exactly two distinct solutions

[2008 : 1 M]

1.12 The eigen values of the following matrix are

$$\begin{bmatrix} -1 & 3 & 5 \\ -3 & -1 & 6 \\ 0 & 0 & 3 \end{bmatrix}$$

(a) $3, 3 + 5j, 6 - j$ (b) $-6 + 5j, 3 + j, 3 - j$
(c) $3 + j, 3 - j, 5 + j$ (d) $3, -1 + 3j, -1-3j$

[2009 : 2 M]

1.13 The eigen values of a skew-symmetric matrix are

(a) always zero
(b) always pure imaginary
(c) either zero or pure imaginary
(d) always real [2010 : 1 M]

1.14 The system of equations

$x + y + z = 6$
$x + 4y + 6z = 20$
$x + 4y + \lambda z = \mu$

has **NO** solution for values of λ and μ given by

(a) $\lambda = 6, \mu = 20$ (b) $\lambda = 6, \mu \neq 20$
(c) $\lambda \neq 6, \mu = 20$ (d) $\lambda \neq 6, \mu \neq 20$

[2011 : 2 M]

1.15 Given that $A = \begin{bmatrix} -5 & -3 \\ 2 & 0 \end{bmatrix}$ and $I = \begin{bmatrix} 1 & 0 \\ 0 & 1 \end{bmatrix}$, the value of A^3 is

(a) $15A + 12I$ (b) $19A + 30I$
(c) $17A + 15I$ (d) $17A + 21I$

[2012 : 2 M]

1.16 The minimum eigen value of the following matrix is

$$\begin{bmatrix} 3 & 5 & 2 \\ 5 & 12 & 7 \\ 2 & 7 & 5 \end{bmatrix}$$

(a) 0 (b) 1
(c) 2 (d) 3 [2013 : 1 M]

1.17 Let A be an $m \times n$ matrix and B an $n \times m$ matrix. It is given that determinant $(I_m + AB)$ = determinant $(I_n + BA)$, where I_k is the $k \times k$ identity matrix. Using the above property, the determinant of the matrix given below is

$$\begin{bmatrix} 2 & 1 & 1 & 1 \\ 1 & 2 & 1 & 1 \\ 1 & 1 & 2 & 1 \\ 1 & 1 & 1 & 2 \end{bmatrix}$$

(a) 2 (b) 5
(c) 8 (d) 16 [2013 : 2 M]

1.18 For matrices of same dimension M, N and scalar c, which one of these properties DOES NOT ALWAYS hold?

(a) $(M^T)^T = M$
(b) $(cM)^T = c(M)^T$
(c) $(M + N)^T = M^T + N^T$
(d) $MN = NM$ [2014 : 1 M, Set-1]

1.19 A real (4×4) matrix A satisfies the equation $A^2 = I$, where I is the (4×4) identity matrix. The positive eigen value of A is ______.

[2014 : 1 M, Set-1]

1.20 Consider the matrix

$$J_y = \begin{bmatrix} 0 & 0 & 0 & 0 & 0 & 1 \\ 0 & 0 & 0 & 0 & 1 & 0 \\ 0 & 0 & 0 & 1 & 0 & 0 \\ 0 & 0 & 1 & 0 & 0 & 0 \\ 0 & 1 & 0 & 0 & 0 & 0 \\ 1 & 0 & 0 & 0 & 0 & 0 \end{bmatrix}$$

which is obtained by reversing the order of the columns of the identity matrix I_6. Let $P = I_6 + \alpha J_6$, where α is a non-negative real number. The value of α for which $\det(P) = 0$ is ______.

[2014 : 2 M, Set-1]

1.21 The determinant of matrix A is 5 and the determinant of matrix B is 40. The determinant of matrix AB is

[2014 : 1 M, Set-2]

1.22 The system of linear equations

$$\begin{bmatrix} 2 & 1 & 3 \\ 3 & 0 & 1 \\ 1 & 2 & 5 \end{bmatrix}\begin{bmatrix} a \\ b \\ c \end{bmatrix} = \begin{bmatrix} 5 \\ -4 \\ 14 \end{bmatrix} \text{ has}$$

(a) a unique solution
(b) infinitely many solutions
(c) no solution
(d) exactly two solutions [2014 : 2 M, Set-2]

1.23 The maximum value of the determinant among all 2×2 real symmetric matrices with trace 14 is ____. [2014 : 2 M, Set-2]

1.24 Which one of the following statements is NOT true for a square matrix A?
(a) If A is upper triangular, the eigenvalues of A are the diagonal elements of it
(b) If A is real symmetric, the eigenvalues of A are always real and positive
(c) If A is real, the eigenvalues of A and A^T are always the same
(d) If all the principal minors of A are positive, all the eigenvalues of A are also positive
[2014 : 2 M, Set-3]

1.25 Consider a system of linear equations:

$$x - 2y + 3z = -1,$$
$$x - 3y + 4z = 1, \text{ and}$$
$$-2x + 4y - 6z = k$$

The value of k for which the system has infinitely many solution is ________.
[2015 : 1 M, Set-1]

1.26 The value of p such that the vector $\begin{bmatrix} 1 \\ 2 \\ 3 \end{bmatrix}$ is an eigenvector of the matrix $\begin{bmatrix} 4 & 1 & 2 \\ p & 2 & 1 \\ 14 & -4 & 10 \end{bmatrix}$ is
[2015 : 1 M, Set-1]

1.27 The value of x for which all the eigen-values of the matrix given below are real is

$$\begin{bmatrix} 10 & 5+j & 4 \\ x & 20 & 2 \\ 4 & 2 & -10 \end{bmatrix}$$

(a) $5 + j$ (b) $5 - j$
(c) $1 - 5j$ (d) $1 + 5j$
[2015 : 1 M, Set-2]

1.28 For $A = \begin{bmatrix} 1 & \tan x \\ -\tan x & 1 \end{bmatrix}$, the determinant of $A^T A^{-1}$ is
(a) $\sec^2 x$ (b) $\cos 4x$
(c) 1 (d) 0
[2015 : 1 M, Set-3]

1.29 Let $M^4 = I$, (where I denotes the identity matrix) and $M \neq I$, $M^2 \neq I$ and $M^3 \neq I$. Then, for any natural number k, M^{-1} equals:
(a) M^{4k+1} (b) M^{4k+2}
(c) M^{4k+3} (d) M^{4k}
[2016 : 1 M, Set-1]

1.30 A sequence $x[n]$ is specified as

$$\begin{bmatrix} x[n] \\ x[n-1] \end{bmatrix} = \begin{bmatrix} 1 & 1 \\ 1 & 0 \end{bmatrix}^n \begin{bmatrix} 1 \\ 0 \end{bmatrix}, \text{ for } n \geq 2.$$

The initial conditions are $x[0] = 1$, $x[1] = 1$, and $x[n] = 0$ for $n < 0$. The value of $x[12]$ is _____
[2016 : 2 M, Set-1]

1.31 The value of x for which the matrix

$$A = \begin{bmatrix} 3 & 2 & 4 \\ 9 & 7 & 13 \\ -6 & -4 & -9+x \end{bmatrix} \text{ has zero as an eigen value}$$

is _____.
[2016 : 1 M, Set-2]

1.32 The matrix $A = \begin{bmatrix} a & 0 & 3 & 7 \\ 2 & 5 & 1 & 3 \\ 0 & 0 & 2 & 4 \\ 0 & 0 & 0 & b \end{bmatrix}$ has $\det(A) = 100$ and trace $(A) = 14$. The value of $|a - b|$ is _______
[2016 : 2 M, Set-2]

1.33 Consider a 2×2 square matrix

$$A = \begin{bmatrix} \sigma & x \\ \omega & \sigma \end{bmatrix}$$

where x is unknown. If the eigenvalues of the matrix A are $(\sigma + j\omega)$ and $(\sigma - j\omega)$, then x is equal to
(a) $+j\omega$ (b) $-j\omega$
(c) $+\omega$ (d) $-\omega$
[2016 : 1 M, Set-3]

1.34 Consider the 5×5 matrix

$$A = \begin{bmatrix} 1 & 2 & 3 & 4 & 5 \\ 5 & 1 & 2 & 3 & 4 \\ 4 & 5 & 1 & 2 & 3 \\ 3 & 4 & 5 & 1 & 2 \\ 2 & 3 & 4 & 5 & 1 \end{bmatrix}$$

It is given that A has only one real eigenvalue. Then the real eigenvalue of A is

(a) –2.5 (b) 0
(c) 15 (d) 25

[2017 : 1 M, Set-1]

1.35 The rank of the matrix $M = \begin{bmatrix} 5 & 10 & 10 \\ 1 & 0 & 2 \\ 3 & 6 & 6 \end{bmatrix}$ is

(a) 0 (b) 1
(c) 2 (d) 3

[2017 : 1 M, Set-1]

1.36 The rank of the matrix

$$\begin{bmatrix} 1 & -1 & 0 & 0 & 0 \\ 0 & 0 & 1 & -1 & 0 \\ 0 & 1 & -1 & 0 & 0 \\ -1 & 0 & 0 & 0 & 1 \\ 0 & 0 & 0 & 1 & -1 \end{bmatrix}$$

is ______.

[2017 : 1 M, Set-2]

1.37 Let M be a real 4×4 matrix. Consider the following statements:

S1: M has 4 linearly independent eigenvectors.
S2: M has 4 distinct eigenvalues.
S3: M is non-singular (invertible).

Which one among the following is TRUE?

(a) S1 implies S2 (b) S1 implies S3
(c) S2 implies S1 (d) S3 implies S2

[2018 : 1 M]

1.38 Consider matrix $A = \begin{bmatrix} k & 2k \\ k^2 - k & k^2 \end{bmatrix}$ and vector $X = \begin{bmatrix} x_1 \\ x_2 \end{bmatrix}$. The number of distinct real values of k for which the equation $AX = 0$ has infinitely many solutions is ______.

[2018 : 1 M]

1.39 The number of distinct eigen values of the matrix

$$A = \begin{bmatrix} 2 & 2 & 3 & 3 \\ 0 & 1 & 1 & 1 \\ 0 & 0 & 3 & 3 \\ 0 & 0 & 0 & 2 \end{bmatrix}$$

is equal to ______.

[2019 : 1 M]

1.40 If $v_1, v_2, \ldots, v_6$ are six vectors in R^4, which one of the following statements is False?

(a) If $\{v_1, v_3, v_5, v_6\}$ spans R^4, then it forms a basis for R^4.
(b) These vectors are not linearly independent.
(c) It is not necessary that these vectors span R^4.
(d) Any four of these vectors form a basis for R^4.

[2020 : 1 M]

1.41 Consider the following system of linear equation.

$$x_1 + 2x_2 = b_1\ ;\quad 2x_1 + 4x_2 = b_2$$
$$3x_1 + 7x_2 = b_3\ ;\ 3x_1 + 9x_2 = b_4$$

Which one of the following conditions ensures that a solution exists for the above system?

(a) $b_3 = 2b_1$ and $3b_1 - 6b_3 + b_4 = 0$
(b) $b_2 = 2b_1$ and $3b_1 - 6b_3 + b_4 = 0$
(c) $b_2 = 2b_1$ and $6b_1 - 3b_3 + b_4 = 0$
(d) $b_3 = 2b_1$ and $3b_1 - 3b_3 + b_4 = 0$

[2020 : 2 M]

1.42 If the vectors (1.0, –1.0, 2.0), (7.0, 3.0, x) and (2.0, 3.0, 1.0) and R^3 are linearly dependent, the value of x is ______.

[2021 : 1 M]

1.43 A real 2×2 non-singular matrix A with repeated eigen value is given as

$$A = \begin{bmatrix} x & -3.0 \\ 3.0 & 4.0 \end{bmatrix}$$

where x is a real positive number. The value of x (rounded off to one decimal place) is ______.

[2021 : 2 M]

1.44 Consider a system of linear equations $Ax = b$, where

$$A = \begin{bmatrix} 1 & -\sqrt{2} & 3 \\ -1 & \sqrt{2} & -3 \end{bmatrix},\ b = \begin{bmatrix} 1 \\ 3 \end{bmatrix}.$$

This system of equations admits ______.

(a) a unique solution for x
(b) infinitely many solutions for x
(c) no solutions for x
(d) exactly two solutions for x

[2022 : 1 M]

1.45 Let α, β be two non-zero real numbers and v_1, v_2 be two non-zero real vectors of size 3×1. Suppose that v_1 and v_2 satisfy $v_1^T v_2 = 0, v_1^T v_1 = 1$, and $v_2^T v_2 = 1$. Let A be the 3×3 matrix given by :

$$A = \alpha v_1 v_1^T + \beta v_2 v_2^T$$

The eigen values of A are _____.

(a) $0, \alpha, \beta$ (b) $0, \alpha+\beta, \alpha-\beta$

(c) $0, \frac{\alpha+\beta}{2}, \sqrt{\alpha\beta}$ (d) $0, 0, \sqrt{\alpha^2+\beta^2}$

[2022 : 2 M]

1.46 Let the sets of eigenvalues and eigenvectors of a matrix B be $\{\lambda_k \mid 1 \le k \le n\}$ and $\{v_k \mid 1 \le k \le n\}$, respectively. For any invertible matrix P, the sets of eigenvalues and eigenvectors of the matrix A, where $B = P^{-1}AB$, respectively, are

(a) $\{\lambda_k \det(A) \mid 1 \le k \le n\}$ and $\{Pv_k \mid 1 \le k \le n\}$

(b) $\{\lambda_k \mid 1 \le k \le n\}$ and $\{v_k \mid 1 \le k \le n\}$

(c) $\{\lambda_k \mid 1 \le k \le n\}$ and $\{Pv_k \mid 1 \le k \le n\}$

(d) $\{\lambda_k \mid 1 \le k \le n\}$ and $\{P^{-1}v_k \mid 1 \le k \le n\}$

[2023 : 1 M]

1.47 Let x be an $n \times 1$ real column vector with length $l = \sqrt{x^T x}$. The trace of the matrix $P = xx^T$ is

(a) l^2 (b) $\frac{l^2}{4}$

(c) l (d) $\frac{l^2}{2}$ [2023 : 2 M]

1.48 The state equation of a second order system is $\dot{x}(t) = Ax(t)$, $x(0)$ is the initial condition.
Suppose λ_1 and λ_2 are two distinct eigenvalues of A and v_1 and v_2 are the corresponding eigenvectors. For constants α_1 and α_2, the solution, $x(t)$, of the state equation is

(a) $\sum_{i=1}^{2} \alpha_i e^{\lambda_i t} v_i$ (b) $\sum_{i=1}^{2} \alpha_i e^{2\lambda_i t} v_i$

(c) $\sum_{i=1}^{2} \alpha_i e^{3\lambda_i t} v_i$ (d) $\sum_{i=1}^{2} \alpha_i e^{4\lambda_i t} v_i$

[2023 : 2 M]

1.49 Let R and R^3 denote the set of real numbers and the three dimensional vector space over it, respectively. The value of α for which the set of vectors

$\{[2 \ \ -3 \ \ \alpha], [3 \ \ -1 \ \ 3], [1 \ \ -5 \ \ 7]\}$

does not form a basis of R^3 is _____.

[2024 : 1 Mark]

1.50 Consider the matrix $\begin{bmatrix} 1 & k \\ 2 & 1 \end{bmatrix}$, where k is a positive real number. Which of the following vectors is/are eigenvector(s) of this matrix?

(a) $\begin{bmatrix} 1 \\ \sqrt{2/k} \end{bmatrix}$ (b) $\begin{bmatrix} 1 \\ -\sqrt{2/k} \end{bmatrix}$

(c) $\begin{bmatrix} \sqrt{2k} \\ -1 \end{bmatrix}$ (d) $\begin{bmatrix} \sqrt{2k} \\ 1 \end{bmatrix}$

[2024 : 2 Marks]

■■■

Answers Linear Algebra

1.1 (d)	1.2 (b)	1.3 (c)	1.4 (a)	1.5 (c)	1.6 (c)	1.7 (a)
1.8 (c)	1.9 (c)	1.10 (c)	1.11 (b)	1.12 (d)	1.13 (c)	1.14 (b)
1.15 (b)	1.16 (a)	1.17 (b)	1.18 (d)	1.19 (1)	1.20 (1)	1.21 (200)
1.22 (b)	1.23 (49)	1.24 (b)	1.25 (2)	1.26 (17)	1.27 (b)	1.28 (c)
1.29 (c)	1.30 (233)	1.31 (1)	1.32 (3)	1.33 (d)	1.34 (c)	1.35 (c)
1.36 (4)	1.37 (c)	1.38 (0, 2)	1.39 (3)	1.40 (d)	1.41 (c)	1.42 (8)
1.43 (10)	1.44 (c)	1.45 (a)	1.46 (c)	1.47 (a)	1.48 (a)	1.49 (5)
1.50 (a, b)						

Explanations Linear Algebra

1.1 (d)

Characteristics equation is $|A-\lambda I| = 0$

$$\begin{vmatrix} -\lambda & 1 \\ 1 & -\lambda \end{vmatrix} = 0$$

$$\lambda^2 - 1 = 0$$

$$\lambda = \pm 1$$

1.2 (b)

Characteristic equation is $|A-\lambda I| = 0$

$$\begin{vmatrix} 2-\lambda & -1 & 0 & 0 \\ 0 & 3-\lambda & 0 & 0 \\ 0 & 0 & -2-\lambda & 0 \\ 0 & 0 & -1 & 4-\lambda \end{vmatrix} = 0$$

$$\begin{vmatrix} 3-\lambda & 0 & 0 \\ 0 & -2\lambda & 0 \\ 0 & -1 & 4-\lambda \end{vmatrix} = 0$$

$(2-\lambda)(3-\lambda)[(-2-\lambda)(4-\lambda)] = 0$

$\lambda = 2, \lambda = 3 \quad -\lambda = 2 \quad 4-\lambda = 0$

$\lambda = -2 \quad \lambda = 4$

eigen values are 2, 3, –2, 4

1.3 (c)

$$A = \begin{bmatrix} -4 & 2 \\ 4 & 3 \end{bmatrix}$$

$$|A-\lambda I| = 0$$

$$\begin{vmatrix} -4 & 2 \\ 4 & 3 \end{vmatrix} - \begin{vmatrix} \lambda & 0 \\ 0 & \lambda \end{vmatrix} = 0$$

$$\Rightarrow \begin{vmatrix} -4-\lambda & 2 \\ 4 & 3-\lambda \end{vmatrix} = 0$$

$$(-4-\lambda)(3-\lambda) - 8 = 0$$

$$-12 + 4\lambda - 3\lambda + \lambda^2 - 8 = 0$$

$$\lambda^2 + \lambda - 20 = 0$$

$$(\lambda+5)(\lambda-4) = 0$$

$$\therefore \quad \lambda = -5, 4$$

Let, eigen vector be $\begin{bmatrix} m_1 \\ m_2 \end{bmatrix}$; Putting, $\lambda_1 = -5$

$$\begin{bmatrix} 1 & 2 \\ 4 & 8 \end{bmatrix}\begin{bmatrix} m_1 \\ m_2 \end{bmatrix} = 0$$

$$\begin{bmatrix} 1 & 2 \\ 1 & 2 \end{bmatrix}\begin{bmatrix} m_1 \\ m_2 \end{bmatrix} = \begin{bmatrix} 0 \\ 0 \end{bmatrix}$$

$$\Rightarrow \begin{bmatrix} 1 & 2 \\ 0 & 0 \end{bmatrix}\begin{bmatrix} m_1 \\ m_2 \end{bmatrix} = \begin{bmatrix} 0 \\ 0 \end{bmatrix}$$

$$m_1 + 2m_2 = 0$$

Taking the value, $m_1 = 2$ and $m_2 = -1$

Thus the eigen vector correspondence to eigen value $\lambda_1 = -5$ is $X_1 = \begin{bmatrix} 2 \\ -1 \end{bmatrix}$.

1.4 (a)

$$[AA^{-1}] = I$$

$$\Rightarrow \begin{bmatrix} 2 & -0.1 \\ 0 & 3 \end{bmatrix}\begin{bmatrix} \frac{1}{2} & a \\ 0 & b \end{bmatrix} = \begin{bmatrix} 1 & 0 \\ 0 & 1 \end{bmatrix}$$

$$\Rightarrow \begin{bmatrix} 1 & 2a-0.1b \\ 0 & 3b \end{bmatrix} = \begin{bmatrix} 1 & 0 \\ 0 & 1 \end{bmatrix}$$

$$\Rightarrow \quad 2a - 0.1b = 0$$

$$3b = 1 \quad \Rightarrow b = \frac{1}{3}$$

$$2a = \frac{0.1}{3} \quad \Rightarrow a = \frac{0.1}{6} = \frac{1}{60}$$

$$a + b = \frac{1}{60} + \frac{1}{3} = \frac{1+20}{60} = \frac{21}{60} = \frac{7}{20}$$

1.5 (c)

$[AA^T]^{-1} = I$

For orthogonal matrix $AA^T = I$ i.e. unity matrix.

Inverse of $I = I$

1.6 (c)

$$R_3 \to R_1 - R_3$$

$$\begin{bmatrix} 1 & 1 & 1 \\ 1 & -1 & 0 \\ 0 & 0 & 0 \end{bmatrix}$$

$\therefore$ Rank = 2

1.7 (a)

$$[\lambda I - A] = 0$$

$$\begin{bmatrix} \lambda & 0 \\ 0 & \lambda \end{bmatrix} - \begin{bmatrix} 6 & 2 \\ 2 & 6 \end{bmatrix} = 1$$

$$(\lambda - 6)^2 - 4 = 0$$

$$\lambda = 8, 4$$

*By property of eigen matrix sum of diagonal elements should be equal to sum of eigen values.

1.8 (c)

$$M = \begin{bmatrix} 4 & 2 \\ 2 & 4 \end{bmatrix}$$

$$M - \lambda I = \begin{bmatrix} 4-\lambda & 2 \\ 2 & 4-\lambda \end{bmatrix}$$

Given eigen vector

$$\therefore \quad \begin{bmatrix} 4-\lambda & 2 \\ 2 & 4-\lambda \end{bmatrix}\begin{bmatrix} 101 \\ 101 \end{bmatrix} = 0$$

$$\Rightarrow \quad (4-\lambda)(101) + 2(101) = 0$$

$$\Rightarrow \quad 4 - \lambda + 2 = 0$$

$$\lambda = 6$$

1.10 (c)

Eigenvalues are the roots of the determinant formed by matrix $[sI-P]$

$$[sI - P] = \begin{bmatrix} s - P_{11} & -P_{12} \\ -P_{21} & s - P_{22} \end{bmatrix}$$

$$[sI - P] = 0$$

$$\Rightarrow \quad (s - P_{11})(s - P_{22}) - P_{12}P_{21} = 0$$

$$\Rightarrow \quad s^2 - (P_{11} + P_{22})s + P_{11}P_{22} - P_{12}P_{21} = 0$$

Since, one of the its eigenvalues is zero, therefore, putting s = 0

$$\boxed{P_{11}P_{22} - P_{12}P_{21} = 0}$$

which is the desired condition.

1.11 (b)

The system can be written in matrix from as

$$\begin{bmatrix} 4 & 2 \\ 2 & 1 \end{bmatrix}\begin{bmatrix} x \\ y \end{bmatrix} = \begin{bmatrix} 7 \\ 6 \end{bmatrix}$$

The Augmented matrix $[A|B]$ is given by

$$\left[\begin{array}{cc|c} 4 & 2 & 7 \\ 2 & 1 & 6 \end{array}\right]$$

Performing Gauss elimination on this $[A|B]$ as follows:

$$\left[\begin{array}{cc|c} 4 & 2 & 7 \\ 2 & 1 & 6 \end{array}\right] \xrightarrow[= R_2 - \frac{1}{2}R_1]{R_2 - \frac{2}{4}R_1} \left[\begin{array}{cc|c} 4 & 2 & 7 \\ 0 & 0 & 5/2 \end{array}\right]$$

Now Rank $[A|B]$ = 2

(The number of non-zero rows in $[A|B]$

Rank $[A]$ = 1

(The number of non-zero rows in $[A]$)

Since, Rank $[A|B] \neq$ Rank$[A]$,

The system has no solution.

1.12 (d)

Sum of the eigen values are the sum of the principle diagonal element of the matrix

sum of the diagonal elements

$$= 3 - 1 - 1 = 1$$

in option (d) sum of the eigen values

$$= 3 - 1 + 3j - 1 - 3j = 3 - 1 - 1 = 1$$

Also since matrix is real, eigen values are either real or complex conjugate.

1.13 (c)

A skew-symmetric matrix is a square matrix whose transpose is equal to its negative value, i.e. $A^T = -A$.

The eigen values of a skew-symmetric matrix always come in pairs $\pm\lambda$ (except in odd-dimensional case where there is an additional unpaired 0 eigen value). For a real skew symmetric matrix the non-zero eigen values are all pure imaginary and thus are of the form $i\lambda_1, -i\lambda_1, i\lambda_2, -i\lambda_2, \ldots$ where each of the λ_K is real.

1.14 (b)

The augmented matrix for the system of equations is

$$[A : B] = \left[\begin{array}{ccc|c} 1 & 1 & 1 & 6 \\ 1 & 4 & 6 & 20 \\ 1 & 4 & \lambda & \mu \end{array}\right]$$

$$[A : B] = \left[\begin{array}{ccc|c} 1 & 1 & 1 & 6 \\ 1 & 4 & 6 & 20 \\ 0 & 0 & \lambda - 6 & \mu - 20 \end{array}\right] \quad [R_3 \to R_3 - R_2]$$

If $\lambda = 6$ and $\mu \neq 20$ then

$$\rho(A : B) = 3 \text{ and } \rho(A) = 2$$

$$\because \quad \rho(A : B) \neq \rho(A)$$

$\therefore$ Given system of equations has no solution for $\lambda = 6$ and $\mu \neq 20$.

1.15 (b)

$$A = \begin{bmatrix} -5 & -3 \\ 2 & 0 \end{bmatrix} \text{ and } I = \begin{bmatrix} 1 & 0 \\ 0 & 1 \end{bmatrix}$$

Characteristic equation of A is

$$\begin{vmatrix} -5-\lambda & -3 \\ 2 & 0-\lambda \end{vmatrix} = 0$$

$$(-5-\lambda)(-\lambda) + 6 = 0$$

$$\Rightarrow \quad \lambda^2 + 5\lambda + 6 = 0$$

So, $A^2 + 5A + 6I = 0$

(by Cayley Hamilton theorem)

$$\Rightarrow \quad A^2 = -5A - 6I$$

Multiplying by A on both sides, we have,

$$A^3 = -5A^2 - 6A$$

$$\Rightarrow \quad A^3 = -5(-5A - 6I) - 6A$$

$$= 19A + 30I$$

1.16 (a)

$$|A - \lambda I| = 0$$

$$\begin{vmatrix} 3-\lambda & 5 & 2 \\ 5 & 12-\lambda & 7 \\ 2 & 7 & 5-\lambda \end{vmatrix} = 0$$

$$(3-\lambda)\left[(12-\lambda)(5-\lambda) - 49\right]$$

$$-5\left[5(5-\lambda) - 14\right] + 2\left[35 - 2(12-\lambda)\right] = 0$$

$$(3-\lambda)\left[60 - 17\lambda + \lambda^2 - 49\right]$$

$$-5(25 - 5\lambda - 14) + 2(35 - 24 + 2\lambda) = 0$$

$$(3-\lambda)(\lambda^2 - 17\lambda + 11) - 5(11 - 5\lambda) + 2(11 + 2\lambda) = 0$$

$$3\lambda^2 - 51\lambda + 33 - \lambda^3 + 17\lambda^2$$

$$-11\lambda - 55 + 25\lambda + 22 + 4\lambda^2 = 0$$

$$-\lambda^3 + 24\lambda^2 - 37\lambda = 0$$

$$\lambda^3 - 24\lambda^2 + 37\lambda = 0$$

$$\lambda(\lambda^2 - 24\lambda + 37) = 0$$

$$\lambda_1 = 0$$

$$\lambda_2, \lambda_3 = \frac{24 \pm \sqrt{(24)^2 - 4(37)}}{2}$$

$$\lambda_2 = 22$$

$$\lambda_3 = 2$$

$$\lambda_{min} = 0$$

1.17 (b)

Take the determinant of given matrix

$$|P| = 2[2(4-1) - 1(2-1) + 1(1-2)]$$
$$-1[1(4-1) - 1(2-1) + 1(1-2)]$$
$$+1[1(2-1) - 2(2-1) + 1(1-1)]$$
$$-1[1(1-2) - 2(1-2) + 1(1-1)]$$

$$= 2[6-1-1] - 1[3-1-1] + 1[1-2+0] - 1[-1+2+0]$$

$$= 2(4) - 1(1) + 1(-1) - 1(1) = 8 - 1 - 1 - 1 = 5$$

1.18 (d)

Matrix multiplication is not commutative.

1.19 (1)

Since, $A^2 = I$, eig(A^2) = eig$(I) = 1$

$$\Rightarrow \quad \text{eig}(A)^2 = 1$$

$$\Rightarrow \quad \text{eig}(A) = \pm 1$$

Therefore, the positive eigen value of A is $+1$.

1.20 (1)

$$I_6 = \begin{bmatrix} 1 & 0 & 0 & 0 & 0 & 0 \\ 0 & 1 & 0 & 0 & 0 & 0 \\ 0 & 0 & 1 & 0 & 0 & 0 \\ 0 & 0 & 0 & 1 & 0 & 0 \\ 0 & 0 & 0 & 0 & 1 & 0 \\ 0 & 0 & 0 & 0 & 0 & 1 \end{bmatrix}$$

$$I_6 + \alpha J_6 = \begin{bmatrix} 1 & 0 & 0 & 0 & 0 & \alpha \\ 0 & 1 & 0 & 0 & \alpha & 0 \\ 0 & 0 & 1 & \alpha & 0 & 0 \\ 0 & 0 & \alpha & 1 & 0 & 0 \\ 0 & \alpha & 0 & 0 & 1 & 0 \\ \alpha & 0 & 0 & 0 & 0 & 1 \end{bmatrix}$$

$$[I_6 + \alpha J_6] = \left[(-1)^2(-1)^4(-1)^6(-1)^8(1-0)\right]$$
$$+\left[(-1)^7(\alpha)(-1)^7(\alpha)(-1)^7(\alpha)(-1)^7(\alpha)(0-\alpha^2)\right]$$

$$1 + \alpha^4(0 - \alpha^2) = 0$$

$$1 - \alpha^6 = 0$$

$$\alpha = 1$$

1.21 (200)

Determinant of $A = 5$

Determinant of $B = 40$

Determinant of $AB = |A|\,|B| = 5 \times 40 = 200$

1.22 (b)

$$\begin{bmatrix} 2 & 1 & 3 \\ 3 & 0 & 1 \\ 1 & 2 & 5 \end{bmatrix}\begin{bmatrix} a \\ b \\ c \end{bmatrix} = \begin{bmatrix} 5 \\ -4 \\ 14 \end{bmatrix}$$

Augmented matrix is $\begin{bmatrix} 2 & 1 & 3 & 5 \\ 3 & 0 & 1 & -4 \\ 1 & 2 & 5 & 14 \end{bmatrix}$

$2R_2 - 3R_1,\ 2R_3 - R_1$

$$\begin{bmatrix} 2 & 1 & 3 & 5 \\ 0 & -3 & -7 & -23 \\ 0 & 3 & 7 & 23 \end{bmatrix}$$

$R_3 + R_2$

$$\begin{bmatrix} 2 & 1 & 3 & 5 \\ 0 & -3 & -7 & -23 \\ 0 & 0 & 0 & 0 \end{bmatrix}$$

Rank of $A = 2$,

Rank of augmented matrix = 2

Number of variables = 3

∴ The system possesses infinitely many solutions.

1.23 (49)

$$A = \begin{bmatrix} a & x \\ x & b \end{bmatrix} \xrightarrow{\text{Trans.}} \begin{bmatrix} a & x \\ x & b \end{bmatrix}$$

Let (a) and (b) are the element of principal diagonal

$$\Rightarrow \quad |A| = ab - x^2$$

$$\frac{d|A|}{dx} = -2x = 0 \qquad (x = 0)$$

So, $|A| = a \cdot b =$ maximum 1 when (a) and (b) should be equal.

$$= (7 \times 7)$$

$$= 49 \qquad \text{(become } 7 + 7 = 14\text{)}$$

1.25 (2)

$x - 2y + 3z = -1$,

$x - 3y + 4z = 1$, and

$-2x + 4y - 6z = k$

$$[A : B] = \begin{bmatrix} 1 & -2 & 3 & : & -1 \\ 1 & -3 & 4 & : & 1 \\ -2 & 4 & 6 & : & k \end{bmatrix}$$

$R_2 \to R_2 - R_1,\ R_3 \to R_3 + 2R_1$

$$\begin{bmatrix} 1 & -2 & 3 & : & -1 \\ 0 & -1 & 1 & : & 2 \\ 0 & 0 & 0 & : & k-2 \end{bmatrix}$$

For infinite may solution

$$\rho(A : B) = \rho(A)$$

$$= r < \text{number of variables}$$

$$\rho(A : B) = 2$$

$$k - 2 = 0$$

$$k = 2$$

1.26 (17)

$$AX = \lambda X$$

$$\begin{bmatrix} 4 & 1 & 2 \\ p & 2 & 1 \\ 14 & -4 & 10 \end{bmatrix} \begin{bmatrix} 1 \\ 2 \\ 3 \end{bmatrix} = \lambda \begin{bmatrix} 1 \\ 2 \\ 3 \end{bmatrix}$$

$$\begin{bmatrix} 12 \\ p+7 \\ 36 \end{bmatrix} = \lambda \begin{bmatrix} 1 \\ 2 \\ 3 \end{bmatrix}$$

$$\frac{p+7}{12} = 2 \Rightarrow p = 17$$

1.27 (b)

For a matrix containing complex number, eigen values are real if and only if

$$A = A^\theta = (\bar{A})^T$$

$$A = \begin{bmatrix} 10 & 5+j & 4 \\ x & 20 & 2 \\ 4 & 2 & -10 \end{bmatrix}$$

$$A^\theta = (\bar{A})^T = \begin{bmatrix} 10 & \bar{x} & 4 \\ 5-j & 20 & 2 \\ 4 & 2 & -10 \end{bmatrix}$$

By comparing these,

$$x = 5 - j$$

1.28 (c)

Long Method:

$$A = \begin{bmatrix} 1 & \tan x \\ -\tan x & 1 \end{bmatrix}$$

$$A^T = \begin{bmatrix} 1 & -\tan x \\ \tan x & 1 \end{bmatrix}$$

$$\text{adj } A = \begin{bmatrix} 1 & \tan x \\ -\tan x & 1 \end{bmatrix}$$

$$A^{-1} = \frac{1}{|A|}[adj\,(A)]^T$$

$$= \frac{1}{1+\tan^2 x}\begin{bmatrix} 1 & -\tan x \\ \tan x & 1 \end{bmatrix}$$

$$= \frac{1}{\sec^2 x}\begin{bmatrix} 1 & -\tan x \\ \tan x & 1 \end{bmatrix}$$

Here,

$$A^T A^{-1} = \begin{bmatrix} 1 & -\tan x \\ \tan x & 1 \end{bmatrix} \frac{1}{\sec^2 x}\begin{bmatrix} 1 & -\tan x \\ \tan x & 1 \end{bmatrix}$$

$$= \frac{1}{\sec^2 x}\begin{bmatrix} 1-\tan^2 x & -2\tan x \\ 2\tan x & 1-\tan^2 x \end{bmatrix}$$

$$A^T A^{-1} = \begin{bmatrix} \dfrac{1-\tan^2 x}{\sec^2 x} & \dfrac{-2\tan x}{\sec^2 x} \\ \dfrac{2\tan x}{\sec^2 x} & \dfrac{1-\tan^2 x}{\sec^2 x} \end{bmatrix}$$

$$|A^TA^{-1}| = \left(\frac{1-\tan^2 x}{\sec^2 x}\right)^2 + \left(\frac{2\tan x}{\sec^2 x}\right)^2$$

$$= \frac{1+\tan^4 x - 2\tan^2 x + 4\tan^2 x}{\sec^4 x}$$

$$= 1$$

(or)

Short Method:

Since, $|AB| = |A|\,|B|$

$$|A^TA^{-1}| = |A^T||A^{-1}|$$

$$= |A| \times \frac{1}{|A|} = 1$$

$$\left(\text{Note} : |A^T| = |A| \text{ and } |A^{-1}| = \frac{1}{|A|}\right)$$

1.29 (c)

Given that $M^4 = I$ or $M^{4k} = I$ or $M^{4(k+1)} = I$

$\therefore \quad M^{-1} \times I = M^{4(k+1)} \times M^{-1}$

$\therefore \quad M^{-1} = M^{4k+3}$

1.30 (233)

For $A = \begin{bmatrix} 1 & 1 \\ 1 & 0 \end{bmatrix}$

equation $\begin{vmatrix} 1-\lambda & 1 \\ 1 & -\lambda \end{vmatrix} = 0$

$$-\lambda + \lambda^2 - 1 = 0$$

$$\lambda^2 - \lambda - 1 = 0$$

By Cayley Hamilton Theorem

$$A^2 - A - I = 0$$

$$A^2 = A + I$$

$$A^4 = A^2 + 2A + I$$

$$= A + I + 2A + I = 3A + 2I$$

$$A^8 = 9A^2 + 12A + 4I$$

$$= 9(A + I) + 12A + 4I$$

$$= 21A + 13I$$

$$A^{12} = A^4 \cdot A^8 = 144A + 89I$$

$$= \begin{bmatrix} 233 & 144 \\ 144 & 89 \end{bmatrix}$$

$$\begin{bmatrix} x[12] \\ x[11] \end{bmatrix} = \begin{bmatrix} 233 & 144 \\ 144 & 89 \end{bmatrix}\begin{bmatrix} 1 \\ 0 \end{bmatrix}$$

$$x[12] = 233$$

1.31 (1)

A has an eigen value is zero

$\therefore \quad |A| = 0$

$$\begin{vmatrix} 3 & 2 & 4 \\ 9 & 7 & 13 \\ -6 & -4 & -9+x \end{vmatrix} = 0$$

$$3(-63 + 7x + 52) - 2(-81 + 9x + 78) + 4(-36 + 42) = 0$$

$$3(7x - 11) - 2(9x - 3) + 4(6) = 0$$

$$21x - 33 - 18x + 6 + 24 = 0$$

$$3x - 3 = 0$$

$$x = 1$$

1.32 (3)

$$\text{Trace of } A = 14$$

$$a + 5 + 2 + b = 14$$

$$a + b = 7 \quad \text{...(i)}$$

$$\det(A) = 100$$

$$5\begin{vmatrix} a & 3 & 7 \\ 0 & 2 & 4 \\ 0 & 0 & b \end{vmatrix} = 100$$

$$5 \times 2 \times a \times b = 100$$

$$10\,ab = 100$$

$$ab = 10 \quad \text{... (ii)}$$

From equation (i) and (ii)

either $a = 5, \quad b = 2$

or $a = 2 \quad b = 5$

$$|a - b| = |5 - 2| = 3$$

1.33 (d)

$$A = \begin{bmatrix} \sigma & x \\ \omega & \sigma \end{bmatrix}$$

$$\text{Trace} = \text{sum of eigen values}$$

$$2\sigma = \sigma + j\omega + \sigma - j\omega$$

$$|A| = \text{product of eigens}$$

$$\sigma^2 - x\omega = (\sigma + j\omega)(\sigma - j\omega) = \sigma^2 + \omega^2$$

which is possible only when, $x = -\omega$

1.34 (c)

$$|A - \lambda I| = 0$$

$$\begin{vmatrix} 1-\lambda & 2 & 3 & 4 & 5 \\ 5 & 1-\lambda & 2 & 3 & 4 \\ 4 & 5 & 1-\lambda & 2 & 3 \\ 3 & 4 & 5 & 1-\lambda & 2 \\ 2 & 3 & 4 & 5 & 1-\lambda \end{vmatrix} = 0$$

Sum of all elements in every one row must be zero.

i.e., $\quad 15 - \lambda = 0$

$$\lambda = 15$$

1.35 (c)

$$M = \begin{bmatrix} 5 & 10 & 10 \\ 1 & 0 & 2 \\ 3 & 6 & 6 \end{bmatrix}$$

$R_1 \leftrightarrow R_2$:

$$\begin{bmatrix} 1 & 0 & 2 \\ 5 & 10 & 10 \\ 3 & 6 & 6 \end{bmatrix}$$

$R_2 \leftarrow R_2 - 5R_1$ and $R_3 \leftarrow R_3 - 3R_1$:

$$\begin{bmatrix} 1 & 0 & 2 \\ 0 & 10 & 0 \\ 0 & 6 & 0 \end{bmatrix}$$

$R_3 \leftarrow R_3 - \frac{6}{10}R_2$:

$$\begin{bmatrix} 1 & 0 & 2 \\ 0 & 10 & 0 \\ 0 & 0 & 0 \end{bmatrix}$$

Which is in Echelon form

Rank of matrix M is,

$$\rho(M) = 2$$

1.36 (4)

$$A = \begin{bmatrix} 1 & -1 & 0 & 0 & 0 \\ 0 & 0 & 1 & -1 & 0 \\ 0 & 1 & -1 & 0 & 0 \\ -1 & 0 & 0 & 0 & 1 \\ 0 & 0 & 0 & 1 & -1 \end{bmatrix}$$

$R_4 \to R_4 + R_1$

$$= \begin{bmatrix} 1 & -1 & 0 & 0 & 0 \\ 0 & 0 & 1 & -1 & 0 \\ 0 & 1 & -1 & 0 & 0 \\ 0 & -1 & 0 & 0 & 1 \\ 0 & 0 & 0 & 1 & -1 \end{bmatrix}$$

$R_4 \to R_4 + R_3$

$$= \begin{bmatrix} 1 & -1 & 0 & 0 & 0 \\ 0 & 0 & 1 & -1 & 0 \\ 0 & 1 & -1 & 0 & 0 \\ 0 & 0 & -1 & 0 & 1 \\ 0 & 0 & 0 & 1 & -1 \end{bmatrix}$$

$R_4 \to R_4 + R_2$

$$= \begin{bmatrix} 1 & -1 & 0 & 0 & 0 \\ 0 & 0 & 1 & -1 & 0 \\ 0 & 1 & -1 & 0 & 0 \\ 0 & 0 & 0 & -1 & 1 \\ 0 & 0 & 0 & 1 & -1 \end{bmatrix}$$

$R_2 \leftrightarrow R_3$ and $R_5 \to R_5 + R_4$

$$= \begin{bmatrix} 1 & -1 & 0 & 0 & 0 \\ 0 & 1 & -1 & 0 & 0 \\ 0 & 0 & 1 & -1 & 0 \\ 0 & 0 & 0 & -1 & 1 \\ 0 & 0 & 0 & 0 & 0 \end{bmatrix}$$

From here,

$\therefore \qquad \rho(A) = 4$

1.37 (c)

Eigen vectors corresponding to distinct eigen values are linearly independent.

Hence, "S2 implies S1".

1.38 (0, 2)

$AX = 0$ has infinitely many solutions

So, $\quad |A| = 0 \quad ; \quad \begin{vmatrix} k & 2k \\ k^2 - k & k^2 \end{vmatrix} = 0$

$k^3 - 2k^3 + 2k^2 = 0 \quad ; \quad k^2(2 - k) = 0$

$k = 0, 2$

$\Rightarrow$ "two" distinct values of k

1.39 (3)

$$A = \begin{bmatrix} 2 & 2 & 3 & 3 \\ 0 & 1 & 1 & 1 \\ 0 & 0 & 3 & 3 \\ 0 & 0 & 0 & 2 \end{bmatrix}$$

Eigen values are 2, 1, 3, 2

Distinct eigen values are 2, 1, 3

$\therefore$ Number of distinct eigen values = 3

1.41 (c)

Given: $x_1 + 2x_2 = b_1$ ...(i)

$2x_1 + 4x_2 = b_2$ ...(ii)

$3x_1 + 7x_2 = b_3$ ...(iii)

$3x_1 + 9x_2 = b_4$ ...(iv)

From equations (ii) and (i)

We can write, $b_2 = 2[x_1 + 2x_2] = 2b_1$

From option (b):

$3b_1 - 6b_3 + b_4 = 3[x_1 + 2x_2] - 6[3x_1 + 7x_2] + 3x_1 + 9x_2 \neq 0$

From option (c):

$b_2 = 2b_1$

and $b_1 - 3b_3 + b_4 = 6[x_1 + 2x_2] - 3[3x_1 + 7x_2] + [3x_1 + 9x_2] = 0$

$6b_1 - 3b_3 + b_4 = 0$

Hence, answer is option (c).

1.42 (8)

$\because$ (1, –1, 2), (7, 3, x) and (2, 3, 1) are linearly dependent.

Then,

$$\begin{vmatrix} 1 & -1 & 2 \\ 7 & 3 & x \\ 2 & 3 & 1 \end{vmatrix} = 0$$

$\Rightarrow 1(3 - 3x) + 1(7 - 2x) + 2(15) = 0$

$\Rightarrow \quad -5x = -40$

$x = 8$

1.43 (10)

$$A = \begin{bmatrix} x & -3 \\ 3 & 4 \end{bmatrix}$$

Characteristic equation

$$|A - \lambda I| = \begin{vmatrix} x-\lambda & -3 \\ 3 & 4-\lambda \end{vmatrix} = 0$$

$\Rightarrow \lambda^2 - (4 + x)\lambda + (4x + 9) = 0$

Roots are repeating

$\Rightarrow \quad b^2 - 4ac = 0$

$(4 + x)^2 - 4(4x + 9) = 0$

$16 + x^2 + 8x - 16x - 36 = 0$

$\Rightarrow \quad x^2 - 8x - 20 = 0$

$$x = \frac{8 \pm \sqrt{64+80}}{2} = \frac{8 \pm 12}{2} = -2, 10$$

Since, x is positive

$\therefore \quad x = 10$

1.44 (c)

The augmented matrix,

$$[A : b] = \begin{bmatrix} 1 & -\sqrt{2} & 3 & \vdots & 1 \\ -1 & \sqrt{2} & -3 & \vdots & 3 \end{bmatrix}_{2\times 4}$$

Since, $\begin{vmatrix} 1 & 1 \\ -1 & 3 \end{vmatrix} = 3 + 1 = 4 \neq 0$

Hence, the rank of augmented matrix

$\rho[A : b] = 2$

For the matrix A,

$$[A] = \begin{bmatrix} 1 & -\sqrt{2} & 3 \\ -1 & \sqrt{2} & -3 \end{bmatrix}$$

The determinant of all 2 × 2 minors is 0.

Hence, $\rho[A] = 1$

Since, $\rho[A : b] \neq \rho[A]$ the system $Ax = b$ is inconsistent and has no solution for x.

1.45 (a)

Given, $A = \alpha v_1 v_1^T + \beta v_2 v_2^T$

(i) Post multiply by v_1 on both sides,

we get, $Av_1 = (\alpha v_1 v_1^T + \beta v_2 v_2^T)v_1$

$Av_1 = \alpha v_1 v_1^T v_1 + \beta v_2 v_2^T v_1$

$Av_1 = \alpha v_1$

$\Rightarrow$ α is an eigen value of A.

(ii) Post multiply A by v_2 on both sides,

we get, $Av_2 = (\alpha v_1 v_1^T + \beta v_2 v_2^T)v_2$

$Av_2 = \alpha v_1 v_1^T v_2 + \beta v_2 v_2^T v_2$

$Av_2 = \beta v_2$

$\Rightarrow$ β is an eigen value of A.

(iii) Post multiply A by $v_1^T v_2$ on both sides, we get,

$Av_1^T v_2 = (\alpha v_1 v_1^T + \beta v_2 v_2^T) v_1^T v_2$

$Av_1^T v_2 = \alpha v_1 v_1^T v_2 + \beta v_2 v_2^T v_1^T v_2$

$Av_1^T v_2 = 0$

$\Rightarrow$ 0 is an eigen value of A.

Therefore, the eigen values of A are 0, α, β.

1.46 (c)

Given, $B = P^{-1}AP$

$\Rightarrow \quad A = PBP^{-1}$

$\Rightarrow$ A, B are called matrices similar.

$\Rightarrow$ Both A, B have same set 7 eigen values

But eigen vectors of A, B are different.

Let $BX = \lambda X$

$\Rightarrow \quad (P^{-1}AP)X = \lambda X$

$\Rightarrow \quad A(PX) = \lambda(PX)$

$\therefore$ Eigen vectors of A are PX.

1.47 (a)

Given, $l = \sqrt{x^T x}, P = \left(xx^T\right)_{n\times n}$

Let
$$(x)_{n\times 1} = \begin{bmatrix} x_1 \\ x_2 \\ x_3 \\ \cdot \\ \cdot \\ x_n \end{bmatrix}$$

$$l = \sqrt{x^T x} = \sqrt{x_1^2 + x_2^2 + x_3^2 + \dots x_n^2}$$

$$P = xx^T$$

$$= \begin{bmatrix} x_1 \\ x_2 \\ x_3 \\ . \\ . \\ x_n \end{bmatrix} [x_1\, x_2\, x_3 \dots x_n]$$

$$P = \begin{bmatrix} x_1^2 & & & & \\ & x_1^2 & & & \\ & & - & & \\ & & & - & \\ & & & & x_n^2 \end{bmatrix}$$

$$\text{Trace of } P = x_1^2 + x_2^2 + \dots + x_n^2 = l^2$$

1.49 (5)

A set of vectors form a basis of R^3 if the set is linearly independent and is a spanning set.

Consider the linear combination,

$$x_1 \begin{bmatrix} 2 \\ -3 \\ \alpha \end{bmatrix} + x_2 \begin{bmatrix} 3 \\ -1 \\ 3 \end{bmatrix} + x_3 \begin{bmatrix} 1 \\ -5 \\ 7 \end{bmatrix} = 0$$

$$\Rightarrow \begin{bmatrix} 2 & 3 & 1 \\ -3 & -1 & -5 \\ \alpha & 3 & 7 \end{bmatrix} \begin{bmatrix} x_1 \\ x_2 \\ x_3 \end{bmatrix} = 0$$

A set of vectors is linearly independent if the only linear combination of the vectors that equals 0 is the trivial linear combination. For trivial solution,

$$\begin{vmatrix} 2 & 3 & 1 \\ -3 & -1 & -5 \\ \alpha & 3 & 7 \end{vmatrix} \neq 0$$

$$\Rightarrow 2[-7 + 15] - 3[-21 + 5\alpha] + 1[-9 + \alpha] \neq 0$$

$$\Rightarrow 16 + 63 - 15\alpha - 9 + \alpha \neq 0$$

$$\Rightarrow -14\alpha \neq -70$$

$$\Rightarrow \alpha \neq 5$$

1.50 (a, b)

The characteristic equation of the matrix $A = \begin{bmatrix} 1 & k \\ 2 & 1 \end{bmatrix}$ is given by

$$|A - \lambda I| = 0$$

$$\Rightarrow \begin{vmatrix} 1-\lambda & k \\ 2 & 1-\lambda \end{vmatrix} = 0$$

$$\Rightarrow (1-\lambda)^2 - 2k = 0$$

$$\Rightarrow \lambda^2 - 2\lambda + (1 - 2k) = 0$$

Solving the equation, we get the eigen values as

$$\lambda = \frac{2 \pm \sqrt{4 - 4(1-2k)}}{2} = 1 \pm \sqrt{2k}$$

For eigen value, $\lambda = 1 + \sqrt{2k}$,

$$[A - \lambda I]X = 0$$

$$\Rightarrow \begin{bmatrix} -\sqrt{2k} & k \\ 2 & -\sqrt{2k} \end{bmatrix} \begin{bmatrix} x_1 \\ x_2 \end{bmatrix} = \begin{bmatrix} 0 \\ 0 \end{bmatrix}$$

$$\Rightarrow -\sqrt{2k}\, x_1 + kx_2 = 0$$

$$2x_1 - \sqrt{2k}\, x_2 = 0$$

It gives $x_1 = \sqrt{\frac{k}{2}} \cdot x_2$

Thus, the possible eigen vectors are

$$\begin{bmatrix} \sqrt{k/2} \\ 1 \end{bmatrix}, \begin{bmatrix} 1 \\ \sqrt{2/k} \end{bmatrix} \quad \dots\text{(i)}$$

For the eigen value $\lambda = 1 - \sqrt{2k}$,

$$[A - \lambda I]X = 0$$

$$\Rightarrow \begin{bmatrix} \sqrt{2k} & k \\ 2 & \sqrt{2k} \end{bmatrix} \begin{bmatrix} x_1 \\ x_2 \end{bmatrix} = \begin{bmatrix} 0 \\ 0 \end{bmatrix}$$

$$\Rightarrow \sqrt{2k}\, x_1 + kx_2 = 0$$

$$2x_1 + \sqrt{2k}\, x_2 = 0$$

It gives, $x_1 = -\sqrt{\frac{k}{2}} \cdot x_2$

Thus, the possible eigen vectors are

$$\begin{bmatrix} -\sqrt{k/2} \\ 1 \end{bmatrix}, \begin{bmatrix} 1 \\ -\sqrt{2/k} \end{bmatrix} \quad \dots\text{(ii)}$$

From equations (i) and (ii), option (a) and option (b) represent the eigen vectors of the matrix.

■■■

2 Calculus

2.1 The value of the integral $I = \frac{1}{\sqrt{2\pi}}\int_0^{\infty} \exp\left(-\frac{x^2}{8}\right)dx$ is

(a) 1 (b) π
(c) 2 (d) 2π [2005 : 2 M]

2.2 The derivative of the symmetric function drawn in given figure will look like

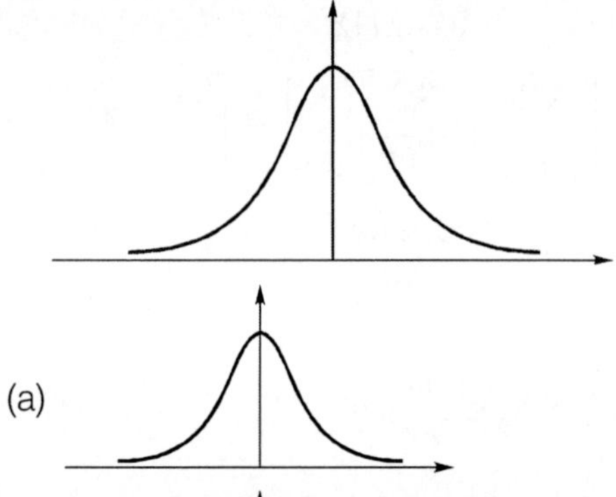

(a)

(b)

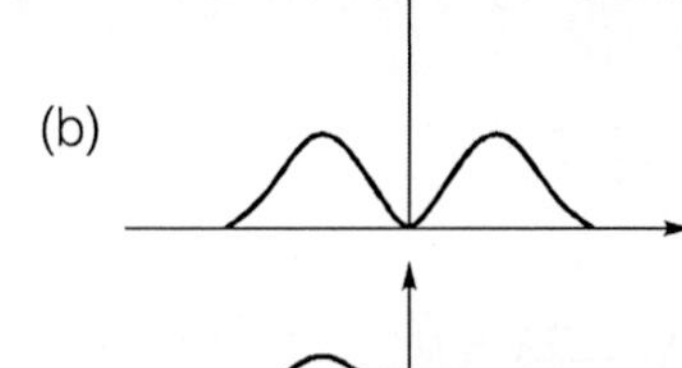

(c)

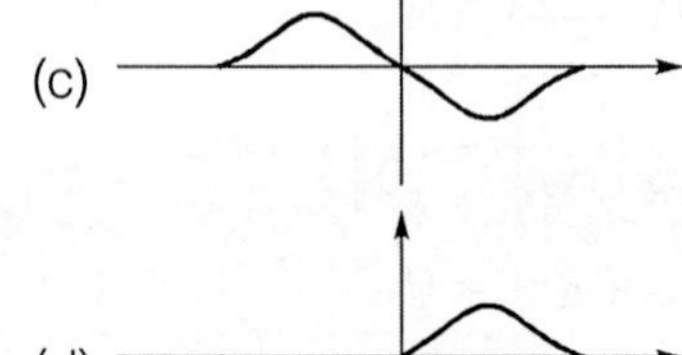

(d)

[2005 : 2 M]

2.3 $\nabla \times \nabla \times P$, where P is a vector is equal to

(a) $P \times \nabla \times P - \nabla^2 P$ (b) $\nabla^2 P + \nabla(\nabla \times P)$
(c) $\nabla^2 P + \nabla \times P$ (d) $\nabla(\nabla \cdot P) - \nabla^2 P$

[2006 : 1 M]

2.4 $\iint (\nabla \times P) \cdot ds$, where P is a vector, is equal to

(a) $\oint P \cdot dl$ (b) $\oint \nabla \times \nabla \times P \cdot dl$
(c) $\oint \nabla \times P \cdot dl$ (d) $\iiint \nabla \cdot P dv$

[2006 : 1 M]

2.5 The integral $\int_0^{\pi} \sin^3\theta \, d\theta$ is given by

(a) 1/2 (b) 2/3
(c) 4/3 (d) 8/3 [2006 : 2 M]

2.6 As x increased from $-\infty$ to ∞, the function

$$f(x) = \frac{e^x}{1+e^x}$$

(a) monotonically increases
(b) monotonically decreases
(c) increases to a maximum value and then decreases
(d) decreases to a minimum value and then increases

[2006 : 2 M]

2.7 The following plot shows a function y which varies linearly with x. The value of the integral

$$I = \int_1^2 y\,dx \text{ is}$$

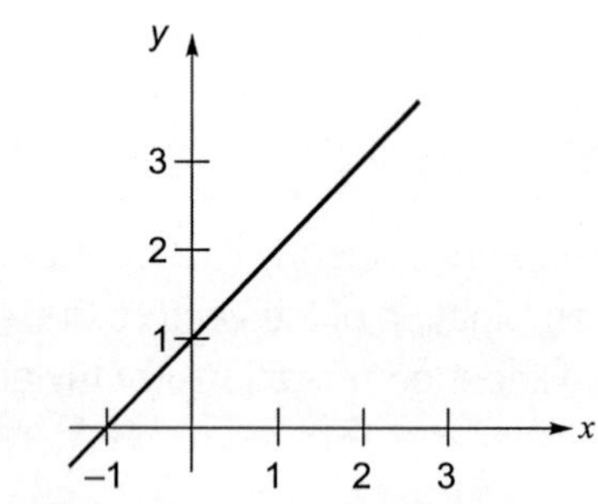

(a) 1.0 (b) 2.5
(c) 4.0 (d) 5.0 [2007 : 1 M]

2.8 For $|x| \ll 1$, cot $h(x)$ can be approximated as

(a) x (b) x^2
(c) $\frac{1}{x}$ (d) $\frac{1}{x^2}$ [2007 : 1 M]

2.9 $\lim_{\theta \to 0} \frac{\sin(\theta/2)}{\theta}$ is

(a) 0.5 (b) 1
(c) 2 (d) not defined

[2007 : 1 M]

2.10 Which one of the following functions is strictly bounded?

(a) $\frac{1}{x^2}$ (b) e^x

(c) x^2 (d) e^{-x^2}

[2007 : 1 M]

2.11 Consider the function $f(x) = x^2 - x - 2$. The maximum value of $f(x)$ in the closed interval [–4, 4] is

(a) 18 (b) 10

(c) –2.25 (d) indeterminate

[2007 : 2 M

2.12 Which of the following functions would have only odd powers of x in its Taylor series expansion about the point $x = 0$?

(a) $\sin(x^3)$ (b) $\sin(x^2)$

(c) $\cos(x^3)$ (d) $\cos(x^2)$

[2008 : 1 M]

2.13 In the Taylor series expansion of $\exp(x) + \sin(x)$ about the point $x = \pi$, the coefficient of $(x-\pi)^2$ is

(a) $\exp(\pi)$ (b) $0.5\exp(\pi)$

(c) $\exp(\pi) + 1$ (d) $\exp(\pi) - 1$

[2008 : 2 M]

2.14 For real values of x, the minimum value of the function $f(x) = \exp(x) + \exp(-x)$ is

(a) 2 (b) 1

(c) 0.5 (d) 0 [2008 : 1 M]

2.15 The value of the integral of the function $g(x, y) = 4x^3 + 10y^4$ along the straight line segment from the point (0, 0) to the point (1, 2) in the x-y plane is

(a) 33 (b) 35

(c) 40 (d) 56

[2008 : 2 M]

2.16 Consider points P and Q in the x-y plane, with $P = (1,0)$ and $Q = (0,1)$. The line integral

$2\int_P^Q (xdx + ydy)$ along the semicircle with the line segment PQ as its diameter

(a) is –1

(b) is 0

(c) is 1

(d) depends on the direction (clockwise or anti-clockwise) of the semicircle

[2008 : 2 M]

2.17 The Taylor series expansion of $\frac{\sin x}{x-\pi}$ at $x = \pi$ is given by

(a) $1+\frac{(x-\pi)^2}{3!}+\ldots$ (b) $-1-\frac{(x-\pi)^2}{3!}+\ldots$

(c) $1-\frac{(x-\pi)^2}{3!}+\ldots$ (d) $-1+\frac{(x-\pi)^2}{3!}+\ldots$

[2009 : 2 M]

2.18 If $e^y = x^{1/x}$, then y has a

(a) maximum at $x = e$

(b) minimum at $x = e$

(c) maximum at $x = e^{-1}$

(d) minimum at $x = e^{-1}$ [2010 : 2 M]

2.19 If $\vec{A} = xy\hat{a}_x + x^2\hat{a}_y$, $\oint_C \vec{A}\cdot d\vec{l}$ over the path shown in the figure is

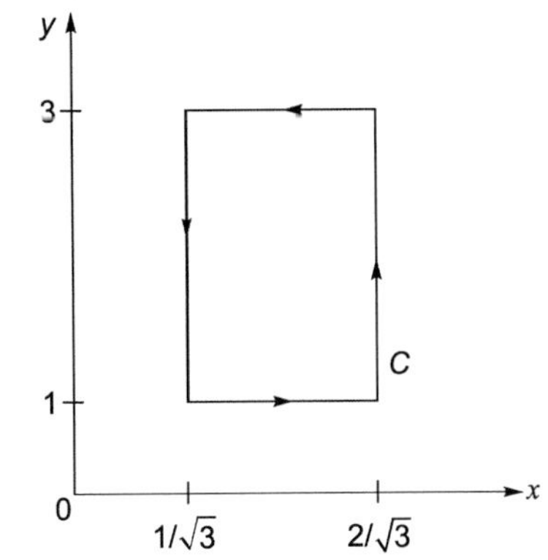

(a) 0 (b) $\frac{2}{\sqrt{3}}$

(c) 1 (d) $2\sqrt{3}$ [2010 : 2 M]

2.20 Consider a closed surface S surrounding a volume V. If $\vec{r}$ is the position vector of a point inside S, with $\hat{n}$ the unit normal on S, the value of the integral $\oiint_S 5\vec{r}\cdot\hat{n}\,dS$ is

(a) 3 V (b) 5 V

(c) 10 V (d) 15 V

[2011 : 1 M]

2.21 The direction of vector A is radially outward from the origin, with $|A| = kr^n$ where $r^2 = x^2 + y^2 + z^2$ and k is a constant. The value of n for which $\nabla \cdot A = 0$ is

(a) – 2 (b) 2

(c) 1 (d) 0 [2012 : 2 M]

2.22 The maximum value of $f(x) = x^3 - 9x^2 + 24x + 5$ in the interval [1, 6] is
(a) 21 (b) 25
(c) 41 (d) 46
[2012 : 2 M]

2.23 Consider a vector field $\vec{A}(\vec{r})$. The closed loop line integral $\oint \vec{A} \cdot \vec{dl}$ can be expressed as
(a) $\oiint (\nabla \times \vec{A}) \cdot \vec{ds}$ over the closed surface bounded by the loop
(b) $\oiiint (\nabla \cdot \vec{A}) dv$ over the closed volume bounded by the top
(c) $\iiint (\nabla \cdot \vec{A}) dv$ over the open volume bounded by the loop
(d) $\iint (\nabla \times \vec{A}) \cdot \vec{ds}$ over the open surface bounded by the loop
[2013 : 1 M]

2.24 The divergence of the vector field
$\vec{A} = x\hat{a}_x + y\hat{a}_y + z\hat{a}_z$ is
(a) 0 (b) 1/3
(c) 1 (d) 3 [2013 : 1 M]

2.25 The Taylor series expansion of $3\sin x + 2\cos x$ is ______.
(a) $2 + 3x - x^2 - \frac{x^3}{2} + \cdots$
(b) $2 - 3x + x^2 - \frac{x^3}{2} + \cdots$
(c) $2 + 3x + x^2 + \frac{x^3}{2} + \cdots$
(d) $2 - 3x - x^2 + \frac{x^3}{2} + \cdots$ [2014 : 2 M, Set-1]

2.26 The volume under the surface $z(x, y) = x + y$ and above the triangle in the x-y plane defined by $\{0 \le y \le x \text{ and } 0 \le x \le 12\}$ is ____.
[2014 : 2 M, Set-1]

2.27 For $0 \le t < \infty$, the maximum value of the function $f(t) = e^{-t} - 2e^{-2t}$ occurs at
(a) $t = \log_e 4$ (b) $t = \log_e 2$
(c) $t = 0$ (d) $t = \log_e 8$
[2014 : 1 M, Set-2]

2.28 The value of $\lim_{x \to \infty} \left(1 + \frac{1}{x}\right)^x$ is
(a) $ln 2$ (b) 1.0
(c) e (d) ∞
[2014 : 1 M, Set-2]

2.29 The maximum value of the function $f(x) = ln(1 + x) - x$ (where $x > -1$) occur at $x =$ ____.
[2014 : 1 M, Set-3]

2.30 The maximum value of $f(x) = 2x^3 - 9x^2 + 12x - 3$ in the interval $0 \le x \le 3$ is
[2014 : 2 M, Set-3]

2.31 The series $\sum_{n=0}^{\infty} \frac{1}{n!}$ converges to
(a) $2\,ln\,2$ (b) $\sqrt{2}$
(c) 2 (d) e
[2014 : 1 M, Set-4]

2.32 For a right angled triangle, if the sum of the lengths of the hypotenuse and a side is kept constant, in order to have maximum area of the triangle, the angle between the hypotenuse and the side is
(a) 12° (b) 36°
(c) 60° (d) 45°
[2014 : 2 M, Set-4]

2.33 If $\vec{r} = x\hat{a}_x + y\hat{a}_y + z\hat{a}_z$ and $|\vec{r}| = r$, then $\text{div}(r^2 \nabla (ln\, r)) =$ ______. [2014 : 2 M, Set-2]

2.34 Given the vector
$\vec{A} = (\cos x)(\sin y)\hat{a}_x + (\sin x)(\cos y)\hat{a}_y$,
where $\hat{a}_x, \hat{a}_y$ denote unit vectors along x, y directions, respectively. The magnitude of curl of $\vec{A}$ is ______. [2014 : 2 M, Set-3]

2.35 The magnitude of the gradient for the function $f(x, y, z) = x^2 + 3y^2 + z^3$ at the point (1,1,1) is ______.
[2014 : 1 M, Set-4]

2.36 The directional derivative of $f(x, y) = \frac{xy}{\sqrt{2}}(x + y)$ at (1, 1) in the direction of the unit vector an angle of $\frac{\pi}{4}$ with y-axis, is given by ______.
[2014 : 1 M, Set-4]

2.37 Given $\vec{F} = z\hat{a}_x + x\hat{a}_y + y\hat{a}_z$. If S represents the portion of the sphere $x^2 + y^2 + z^2 = 1$ for $z \geq 0$, then $\int_s \nabla \times \vec{F} \cdot \vec{ds}$ ______.

[2014 : 2 M, Set-4]

2.38 If $z = xy\,ln(xy)$, then

(a) $x\dfrac{\partial z}{\partial x} + y\dfrac{\partial z}{\partial y} = 0$ (b) $y\dfrac{\partial z}{\partial x} = x\dfrac{\partial z}{\partial y}$

(c) $x\dfrac{\partial z}{\partial x} = y\dfrac{\partial z}{\partial y}$ (d) $y\dfrac{\partial z}{\partial x} + x\dfrac{\partial z}{\partial y} = 0$

[2014 : 1 M, Set-3]

2.39 A function $f(x) = 1 - x^2 + x^3$ is defined in the closed interval [–1, 1]. The value of x in the open interval (–1, 1) for which the mean value theorem is satisfied, is

(a) –1/2 (b) –1/3

(c) 1/3 (d) 1/2

[2015 : 1 M, Set-1]

2.40 A vector $\vec{P}$ is given by $\vec{P} = x^3y\vec{a}_x - x^2y^2\vec{a}_y - x^2yz\vec{a}_z$ Which of the following statements is TRUE?

(a) $\vec{P}$ is solenoidal, but not irrotational

(b) $\vec{P}$ is irrotational, but not solenoidal

(c) $\vec{P}$ is neither solenoidal nor irrotational

(d) $\vec{P}$ is both solenoidal and irrotational

[2015 : 2 M, Set-1]

2.41 Which one of the following graphs describes the function $f(x) = e^{-x}(x^2 + x + 1)$?

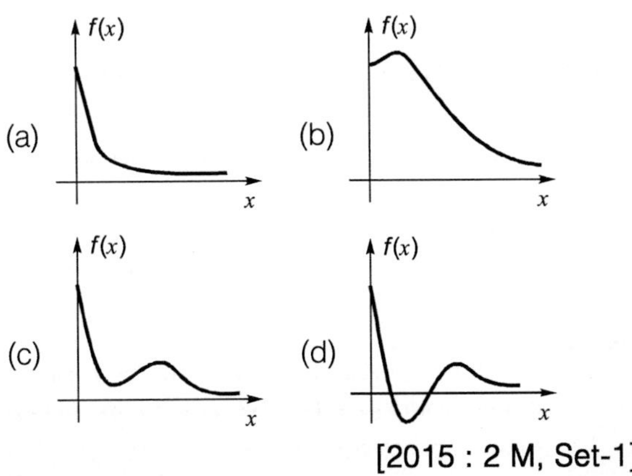

[2015 : 2 M, Set-1]

2.42 The maximum area (in square unit) of a rectangle whose vertices lies on the ellipse $x^2 + 4y^2 = 1$ is______.

[2015 : 2 M, Set-1]

2.43 The value of the integral

$$\int_{-\infty}^{\infty} 12\cos(2\pi t)\frac{\sin(4\pi t)}{4\pi t}dt \text{ is\_\_\_\_\_.}$$

[2015 : 2 M, Set-2]

2.44 The contour on the x-y plane, where the partial derivative of $x^2 + y^2$ with respect to y is equal to the partial derivative of $6y + 4x$ with respect to x, is

(a) $y = 2$ (b) $x = 2$

(c) $x = y = 4$ (d) $x - y = 0$

[2015 : 1 M, Set-3]

2.45 Given the following statements about a function $f: R \rightarrow R$, select the right option:

P: If $f(x)$ is continuous at $x = x_0$, then it is differential at $x = x_0$.

Q: If $f(x)$ is continuous at $x = x_0$, then it may not be differentiable at $x = x_0$.

R: If $f(x)$ is differentiable at $x = x_0$, then it is also continuous at $x = x_0$.

(a) P is true, Q is false, R is false

(b) P is false, Q is true, R is true

(c) P is false, Q is true, R is false

(d) P is true, Q is false, R is true

[2016 : 1 M, Set-1]

2.46 The integral $\dfrac{1}{2\pi}\iint_D (x + y + 10)dxdy$, where D denotes the disc: $x^2 + y^2 \leq 4$, evaluates to _____

[2016 : 2 M, Set-1]

2.47 The region specified by $\{(\rho, \phi, z): 3 \leq \rho \leq 5, \dfrac{\pi}{8} \leq \varphi \leq \dfrac{\pi}{4}, 3 \leq z \leq 4.5\}$ in cylindrical coordinates has volume of _____.

[2016 : 2 M, Set-1]

2.48 Consider the plot $f(x)$ versus x as shown below.

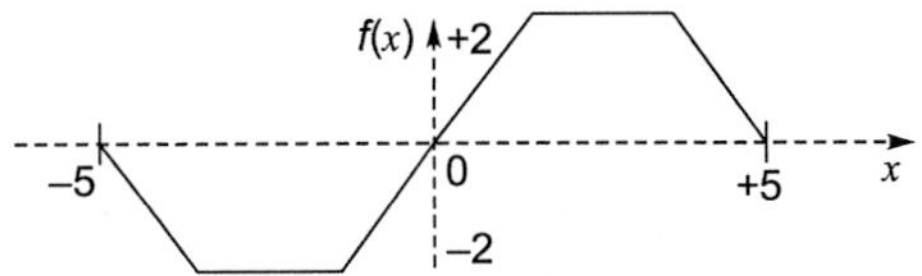

Suppose, $F(x) = \int_{-5}^{x} f(y)dy$. Which one of the following is a graph of $F(x)$?

(a)

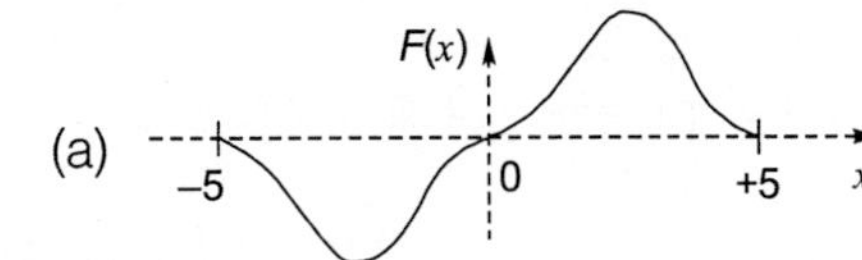

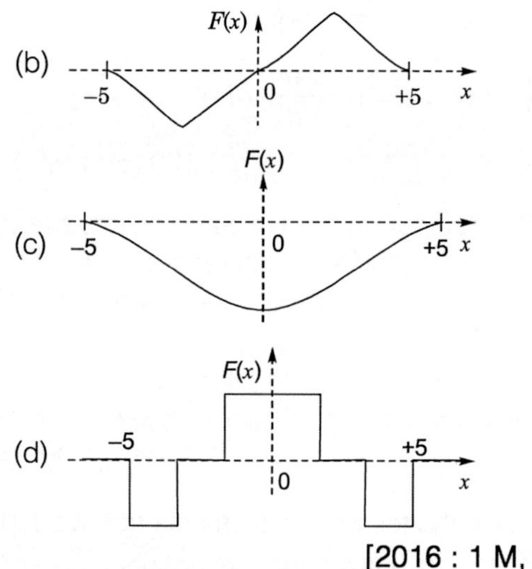

[2016 : 1 M, Set-1]

2.49 Suppose C is the closed curve defined as the circle $x^2 + y^2 = 1$ with C oriented anti clockwise. The value of $\oint(xy^2dx + x^2ydy)$ over the curve C equals ______.

[2016 : 2 M, Set-2]

2.50 As x varies from –1 to +3, which one of the following describes the behaviour of the function $f(x) = x^3 - 3x^2 + 1$?

(a) $f(x)$ increases monotonically.

(b) $f(x)$ increases, then decreases and increases again.

(c) $f(x)$ decreases, then increases and decreases again.

(d) $f(x)$ increases and then decreases

[2016 : 1 M, Set-2]

2.51 How many distinct values of x satisfy the equation $\sin(x) = x/2$, where x is in radians?

(a) 1 (b) 2

(c) 3 (d) 4 or more

[2016 : 1 M, Set-2]

2.52 Consider the time-varying vector $I = \hat{x}15\cos(\omega t) + \hat{y}5\sin(\omega t)$ in Cartesian coordinates, where $\omega > 0$ is a constant. When the vector magnitude $|I|$ is at its minimum value, the angle θ that I makes with the x axis (in degrees, such that $0 \le \theta \le 180$) is _____.

[2016 : 1 M, Set-2]

2.53 The integral $\int_0^1 \frac{dx}{\sqrt{(1-x)}}$ is equal to _____.

[2016 : 1 M, Set-3]

2.54 If the vectors $e_1 = (1, 0, 2)$, $e_2 = (0, 1, 0)$ and $e_3 = (-2, 0, 1)$ form an orthogonal basis of the three-dimensional real space R^3, then the vector $u = (4, 3, -3) \in R^3$ can be expressed as

(a) $u = -\frac{2}{5}e_1 - 3e_2 - \frac{11}{5}e_3$

(b) $u = -\frac{2}{5}e_1 - 3e_2 + \frac{11}{5}e_3$

(c) $u = -\frac{2}{5}e_1 + 3e_2 + \frac{11}{5}e_3$

(d) $u = -\frac{2}{5}e_1 + 3e_2 - \frac{11}{5}e_3$ [2016 : 2 M, Set-3]

2.55 A triangle in the xy-plane is bounded by the straight lines $2x = 3y$, $y = 0$ and $x = 3$. The volume above the triangle and under the plane $x + y + z = 6$ is _________.

[2016 : 2 M, Set-3]

2.56 Consider the following statements about the linear dependence of the real valued functions $y_1 = 1$, $y_2 = x$ and $y_3 = x^2$, over the field of real numbers.

I. y_1, y_2 and y_3 are linearly independent on $-1 \le x \le 0$.

II. y_1, y_2 and y_3 are linearly dependent on $0 \le x \le 1$.

III. y_1, y_2 and y_3 are linearly independent on $0 \le x \le 1$.

IV. y_1, y_2 and y_3 are linearly dependent on $-1 \le x \le 0$.

Which one among the following is correct?

(a) Both I and II are true

(b) Both I and III are true

(c) Both II and IV are true

(d) Both III and IV are true [2017 : 1 M, Set-1]

2.57 Let $I = \int_C (2zdx + 2ydy + 2xdz)$ where x, y, z are real, and let C be the straight line segment from point A: (0, 2, 1) to point B: (4, 1, –1). The value of I is _______

[2017 : 2 M, Set-1]

2.58 Let $f(x) = e^{x + x^2}$ for real x. From among the following, choose the Taylor series approximation of $f(x)$ around $x = 0$, which includes all powers of x less than or equal to 3,

(a) $1 + x + x^2 + x^3$ (b) $1 + x + \frac{3}{2}x^2 + x^3$

(c) $1 + x + \frac{3}{2}x^2 + \frac{7}{6}x^3$ (d) $1 + x + 3x^2 + 7x^3$

[2017 : 2 M, Set-1]

2.59 A three dimensional region R of finite volume is described by

$$x^2 + y^2 \le z^3 \,;\; 0 \le z \le 1,$$

where x, y, z are real. The volume of R (up to two decimal places) is ________.

[2017 : 2 M, Set-1]

2.60 The smaller angle (in degrees) between the planes $x + y + z = 1$ and $2x - y + 2z = 0$ is _____.

[2017 : 1 M, Set-2]

2.61 If the vector function

$$\vec{F} = \hat{a}_x(3y - k_1 z) + \hat{a}_y(k_2 x - 2z) - \hat{a}_z(k_3 y + z)$$

is irrotational, then the values of the constants k_1, k_2 and k_3, respectively, are

(a) 0.3, – 2.5, 0.5 (b) 0.0, 3.0, 2.0
(c) 0.3, 0.33, 0.5 (d) 4.0, 3.0, 2.0

[2017 : 2 M, Set-2]

2.62 The minimum value of the function $f(x) = \frac{1}{3}x(x^2 - 3)$ in the interval $-100 \le x \le 100$ occurs at $x =$ ____.

[2017 : 2 M, Set-2]

2.63 The values of the integrals

$$\int_0^1\left(\int_0^1 \frac{x-y}{(x+y)^3}dy\right)dx \quad \text{and} \quad \int_0^1\left(\int_0^1 \frac{x-y}{(x+y)^3}dx\right)dy$$

are

(a) same and equal to 0.5
(b) same and equal to – 0.5
(c) 0.5 and – 0.5 respectively
(d) – 0.5 and 0.5 respectively

[2017 : 2 M, Set-2]

2.64 Taylor series expansion of $f(x) = \int_0^x e^{-\left(\frac{t^2}{2}\right)}dt$ around $x = 0$ has the form

$$f(x) = a_0 + a_1 x + a_2 x^2 + \ldots$$

The coefficient a_2 (correct to two decimal places) is equal to ______.

[2018 : 1 M]

2.65 Let $f(x, y) = \dfrac{ax^2 + by^2}{xy}$, where a and b are constants. If $\dfrac{\partial f}{\partial x} = \dfrac{\partial f}{\partial y}$ at $x = 1$ and $y = 2$, then the relation between a and b is

(a) $a = \dfrac{b}{4}$ (b) $a = \dfrac{b}{2}$
(c) $a = 2b$ (d) $a = 4b$

[2018 : 1 M]

2.66 Let $r = x^2 + y - z$ and $z^3 - xy + yz + y^3 = 1$. Assume that x and y are independent variables. At $(x, y, z) = (2, -1, 1)$, the value (correct to two decimal places) of $\dfrac{\partial r}{\partial x}$ is ______.

[2018 : 2 M]

2.67 A curve passes through the point $(x = 1, y = 0)$ and satisfies the differential equation $\dfrac{dy}{dx} = \dfrac{x^2 + y^2}{2y} + \dfrac{y}{x}$. The equation that describes the curve is

(a) $ln\left(1 + \dfrac{y^2}{x^2}\right) = x - 1$ (b) $\dfrac{1}{2}ln\left(1 + \dfrac{y^2}{x^2}\right) = x - 1$
(c) $ln\left(1 + \dfrac{y}{x}\right) = x - 1$ (d) $\dfrac{1}{2}ln\left(1 + \dfrac{y}{x}\right) = x - 1$

[2018 : 2 M]

2.68 The value of integral $\int_0^{\pi}\int_y^{\pi} \dfrac{\sin x}{x}\,dx\,dy$, is equal to ____.

[2019 : 1 M]

2.69 Consider a differentiable function $f(x)$ on the set of real numbers such that $f(-1) = 0$ and $|f'(x)| \le 2$. Given these conditions, which one of the following inequalities is necessarily true for all $x \in [-2, 2]$?

(a) $f(x) \le 2|x + 1|$ (b) $f(x) \le \dfrac{1}{2}|x|$
(c) $f(x) \le 2|x|$ (d) $f(x) \le \dfrac{1}{2}|x + 1|$

[2019 : 2 M]

2.70 Consider the line integral

$$\int_C (xdy - ydx)$$

the integral being taken in a counterclockwise direction over the closed curve C that forms the boundary of the region R shown in the figure below. The region R is the area enclosed by the union of a 2×3 rectangle and a semi-circle of radius 1. The line integral evaluates to

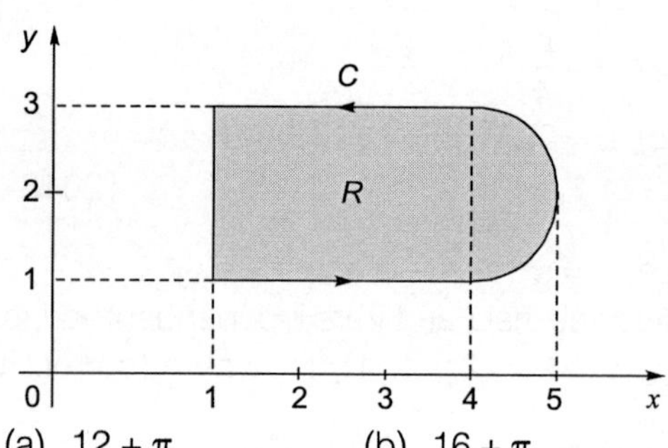

(a) $12 + \pi$
(b) $16 + \pi$
(c) $6 + \pi/2$
(d) $8 + \pi$ [2019 : 2 M]

2.71 The partial derivative of the function

$$f(x,y,z) = e^{1-x\cos y} + xze^{-1/(1+y^2)}$$

with respect to x at the point (1, 0, e) is
(a) −1
(b) 1
(c) 0
(d) $\frac{1}{e}$ [2020 : 1 M]

2.72 For a vector field $\vec{A}$, which one of the following is False?
(a) $\nabla \times \vec{A}$ is another vector field.
(b) $\vec{A}$ is solenoidal if $\nabla \cdot \vec{A} = 0$.
(c) $\vec{A}$ is irrotational if $\nabla^2 \vec{A} = 0$.
(d) $\nabla \times (\nabla \times \vec{A}) = \nabla(\nabla \cdot \vec{A}) - \nabla^2 \vec{A}$ [2020 : 1 M]

2.73 For the solid S shown below, the value of $\iiint_S x\,dx\,dy\,dz$ (rounded off to two decimal places) is ______.

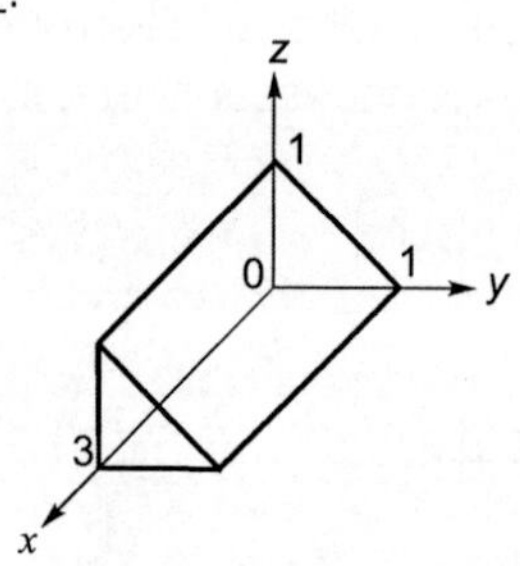

[2020 : 2 M]

2.74 The vector function $\vec{F}(r) = -x\hat{i} + y\hat{j}$ is defined over a circular arc C shown in the figure.

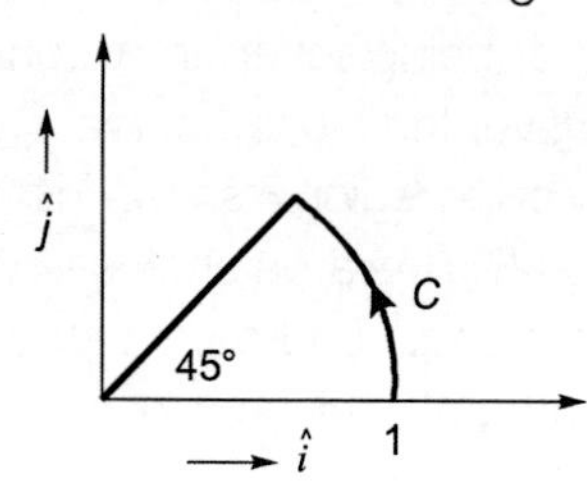

The line integral of $\int_C \vec{F}(r) \cdot d\vec{r}$ is
(a) $\frac{1}{4}$
(b) $\frac{1}{2}$
(c) $\frac{1}{3}$
(d) $\frac{1}{6}$ [2021 : 1 M]

2.75 Consider the two-dimensional vector field $\vec{F}(x, y) = x\vec{i} + y\vec{j}$, where $\vec{i}$ and $\vec{j}$ denote the unit vectors along the x-axis and the y-axis, respectively. A contour C in the x-y plane, as shown in the figure, is composed of two horizontal lines connected at the two ends by two semicircular arcs of unit radius. The contour is traversed in the counter-clockwise sense. The value of the closed path integral

$$\oint_C \vec{F}(x, y) \cdot (dx\vec{i} + dy\vec{j})$$

is ______.

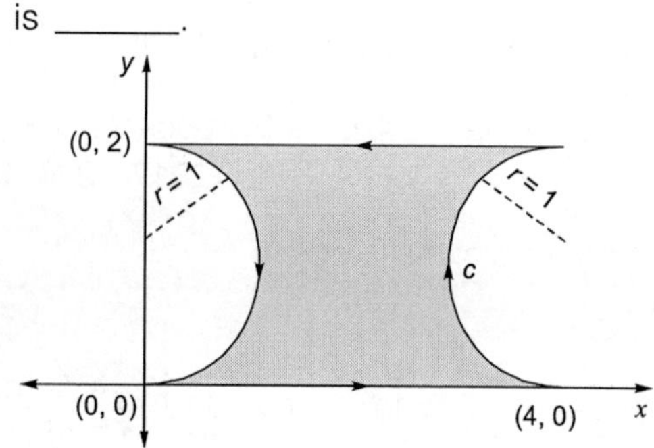

(a) 0
(b) 1
(c) $8 + 2\pi$
(d) −1 [2022 : 1 M]

2.76 The function $f(x) = 8\log_e x - x^2 + 3$ attains its minimum over the interval [1, e] at x = ______. (Here $\log_e x$ is the natural logarithm of x.)
(a) 2
(b) 1
(c) e
(d) $\frac{1+e}{2}$ [2022 : 2 M]

2.77 The value of the integral

$$\iint_D 3(x^2 + y^2)\,dx\,dy,$$

where D is the shaded triangular region shown in the diagram, is ______. (rounded off to the nearest integer).

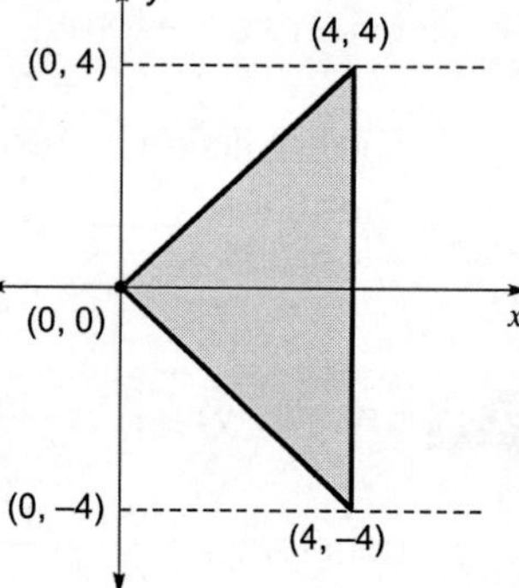

[2022 : 2 M]

2.78 Let $v_1 = \begin{bmatrix} 1 \\ 2 \\ 0 \end{bmatrix}$ and $v_2 = \begin{bmatrix} 2 \\ 1 \\ 3 \end{bmatrix}$ be two vectors. The value of the coefficient α in the expression $v_1 = \alpha v_2 + e$, which minimizes the length of the error vector e, is

(a) $\frac{7}{2}$ (b) $-\frac{2}{7}$

(c) $\frac{2}{7}$ (d) $-\frac{7}{2}$

[2023 : 1 M]

2.79 The rate of increase, of a scalar field $f(x, y, z) = xyz$ in the direction $v = (2,1,2)$ at a point $(0,2,1)$ is

(a) $\frac{2}{3}$ (b) $\frac{4}{3}$

(c) 2 (d) 4

[2023 : 1 M]

2.80 The value of the line integral

$\int_P^Q (z^2 dx + 3y^2 dy + 2xz\, dz)$ along the straight line joining the points $P(1, 1, 2)$ and $Q(2, 3, 1)$ is

(a) 20 (b) 24

(c) 29 (d) –5

[2023 : 2 M]

2.81 The value of the integral $\iint_R xy\, dx\, dy$ over the region R, given in the figure, is ______. (rounded off to the nearest integer).

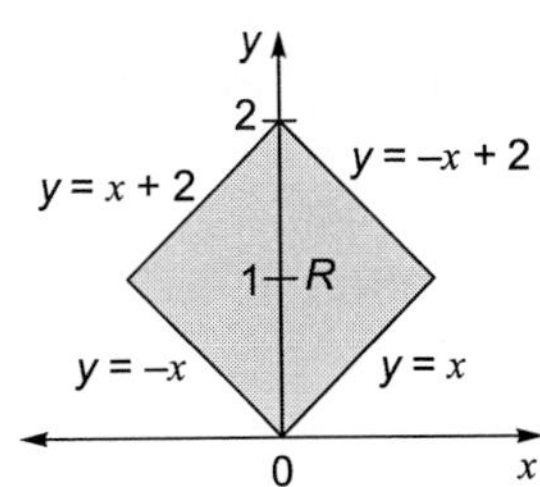

[2023 : 2 M]

2.82 Let $\rho(x, y, z, t)$ and $u(x, y, z, t)$ represent density and velocity, respectively, at a point (x, y, z) and time t. Assume $\frac{\partial \rho}{\partial t}$ is continuous. Let V be an arbitrary volume in space enclosed by the closed surface S and $\hat{n}$ be the outward unit normal of S. Which of the following equations is/are equivalent $\frac{\partial \rho}{\partial t} + \nabla \cdot (\rho u) = 0$?

(a) $\int_V \frac{\partial \rho}{\partial t} dv = \int_V \nabla \cdot (\rho u) dv$

(b) $\int_V \frac{\partial \rho}{\partial t} dv = -\oint_S \rho u \cdot \hat{n} ds$

(c) $\int_V \frac{\partial \rho}{\partial t} dv = \oint_S \rho u \cdot \hat{n} ds$

(d) $\int_V \frac{\partial \rho}{\partial t} dv = -\int_V \nabla \cdot (\rho u) dv$

[2024 : 1 Mark]

2.83 Let F_1, F_2, and F_3 be functions of (x, y, z). Suppose that for every given pair of points A and B in space, the line integral $\int_C (F_1 dx + F_2 dy + F_3 dz)$ evaluates to the same value along any path C that starts at A and ends at B. Then which of the following is/are true?

(a) There exists a differentiable scalar function $f(x, y, z)$ such that $F_1 = \frac{\partial f}{\partial x}, F_2 = \frac{\partial f}{\partial y}, F_3 = \frac{\partial f}{\partial z}$.

(b) $\frac{\partial F_1}{\partial x} + \frac{\partial F_2}{\partial y} + \frac{\partial F_3}{\partial z} = 0$.

(c) $\frac{\partial F_3}{\partial y} = \frac{\partial F_2}{\partial z}, \frac{\partial F_1}{\partial z} = \frac{\partial F_3}{\partial x}, \frac{\partial F_2}{\partial x} = \frac{\partial F_1}{\partial y}$.

(d) For every closed path Γ, we have $\oint_\Gamma (F_1 dx + F_2 dy + F_3 dz) = 0$.

[2024 : 2 Marks]

■■■

Answers Calculus

2.1 (a)	2.2 (c)	2.3 (d)	2.4 (a)	2.5 (c)	2.6 (a)	2.7 (b)
2.8 (c)	2.9 (a)	2.10 (d)	2.11 (a)	2.12 (a)	2.13 (b)	2.14 (a)
2.15 (a)	2.16 (b)	2.17 (d)	2.18 (a)	2.19 (c)	2.20 (d)	2.21 (a)
2.22 (c)	2.23 (d)	2.24 (d)	2.25 (a)	2.26 (864)	2.27 (a)	2.28 (c)
2.29 (0)	2.30 (6)	2.31 (d)	2.32 (c)	2.33 (3)	2.34 (0)	2.35 (7)
2.36 (3)	2.37 (6.28)	2.38 (c)	2.39 (b)	2.40 (a)	2.41 (b)	2.42 (1)
2.43 (3)	2.44 (a)	2.45 (b)	2.46 (20)	2.47 (4.712)	2.48 (c)	2.49 (0)
2.50 (b)	2.51 (c)	2.52 (90)	2.53 (2)	2.54 (d)	2.55 (10)	2.56 (b)
2.57 (–11)	2.58 (c)	2.59 (0.7853)	2.60 (54.73)	2.61 (b)	2.62 (–100)	2.63 (c)
2.64 (0)	2.65 (d)	2.66 (4.50)	2.67 (a)	2.68 (2)	2.69 (a)	2.70 (a)
2.71 (c)	2.72 (c)	2.73 (2.25)	2.74 (b)	2.75 (a)	2.76 (b)	2.77 (512)
2.78 (c)	2.79 (b)	2.80 (b)	2.81 (0)	2.82 (b, d)	2.83 (a, c, d)	

Explanations Calculus

2.1 (a)

$$I = \frac{1}{\sqrt{2\pi}}\int_0^{\infty} e^{-x^2/8}dx$$

Put $\frac{x^2}{8} = t \Rightarrow x^2 = 8t$

$\Rightarrow \quad x = 2\sqrt{2t}$

$\therefore \quad dx = 4\frac{dt}{x} = \frac{4}{2\sqrt{2}}\frac{dt}{\sqrt{t}}$

or $\quad dx = \frac{\sqrt{2}}{\sqrt{t}}dt$

$$I = \frac{1}{\sqrt{2\pi}}\int_0^{\infty} e^{-t}\frac{\sqrt{2}}{\sqrt{t}}dt = \frac{1}{\sqrt{\pi}}\int_0^{\infty} e^{-t}t^{\frac{1}{2}-1}dt$$

$$\left[\because \Gamma n = \int_0^{\infty} e^{-x}x^{n-1}dx \text{ and } \Gamma\tfrac{1}{2} = \sqrt{\pi}\right]$$

$\therefore \quad I = \frac{1}{\sqrt{\pi}}\Gamma\frac{1}{2} = \frac{1}{\sqrt{\pi}}.\sqrt{\pi} = 1$

2.2 (c)

Given function has negative slope in +ve half and +ve slope in –ve half. So its differentiation curve is satisfied by (c).

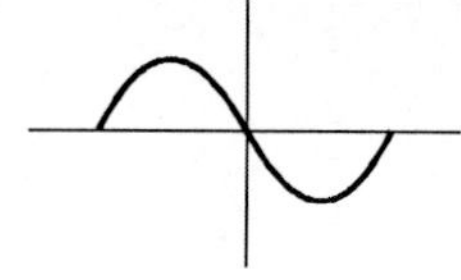

2.3 (d)

From vector triple product.

$$A\times(B\times C) = B(A.C) - C(A.B)$$

$$A = \nabla,\ B = \nabla,\ C = P$$

$$\nabla\times\nabla\times P = \nabla(\nabla\cdot P) - P(\nabla\cdot\nabla) = \nabla(\nabla\cdot P) - \nabla^2 P$$

2.4 (a)

$\iint(\Delta\times P).ds = \oint P.dl$ (Stoke's Theorem)

2.5 (c)

$$I = \int_0^{\pi}\sin^3\theta\cdot d\theta$$

$$\int_0^{\pi}(1-\cos^2\theta)\sin\theta d\theta,$$

Let, $\cos\theta = t$

$$= -\int_{+1}^{-1}(1-t^2)dt$$

$-\sin\theta d\theta = dt,\ \theta = 0,\ \cos 0 = 1 = t$

$\theta = \pi,\ \cos\pi = -1 = t$

$$= \int_{-1}^{+1}(1-t^2)dt$$

$$= \left|t - \frac{t^3}{3}\right|_{-1}^{1} = \left(1-\frac{1}{3}\right) - \left(-1+\frac{1}{3}\right)$$

$$I = \frac{2}{3}+\frac{2}{3} = \frac{4}{3}$$

2.6 (a)

$$f'(x) = \frac{e^x(1+e^x) - e^{2x}}{(1+e^x)^2} = \frac{e^x}{(1+e^x)^2}$$

$\because$ for all x, $f'(x)$ is positive

So, $x \to \infty$, function's value increases.

2.7 (b)

$$y = x + 1$$

$$I = \int_1^2 y\,dx$$

$$= \int_1^2 (x+1)\,dx = \left.\frac{(x+1)^2}{2}\right|_1^2$$

$$= \frac{1}{2}(9-4) = 2.5$$

2.8 (c)

$$\coth hx = \frac{\cos hx}{\sin hx}$$

$$\cot hx = 1 + \frac{x^2}{2!} + \frac{x^4}{4!} + \ldots$$

For $|x| << 1$

$\cosh x \cong 1$ (linear approximation)

and $\sinh x = x + \frac{x^3}{!} + \frac{x^5}{5!} + \ldots$

For $|x| << 1$

$\sinh x \cong x$ (linear approximation)

Hence, $\coth = \frac{\cos hx}{\sin hx} \cong \frac{1}{x}$

2.9 (a)

$$\lim_{\theta \to 0} \frac{\frac{1}{2} \times \sin\left(\frac{\theta}{2}\right)}{\theta \times \frac{1}{2}} = \frac{1}{2} \lim_{\theta \to 0} \frac{\sin\frac{\theta}{2}}{\frac{\theta}{2}} = \frac{1}{2} = 0.5$$

2.10 (d)

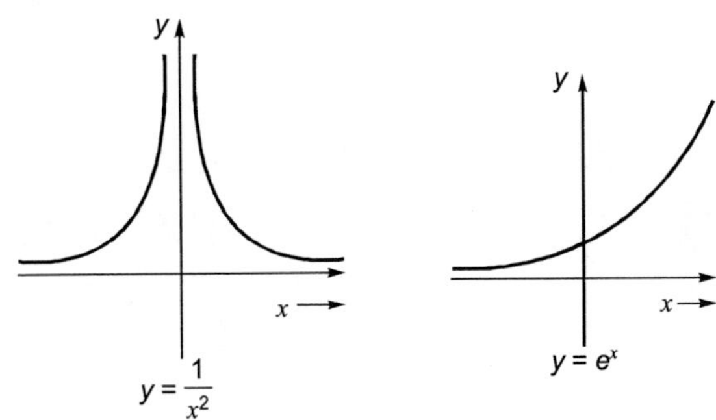

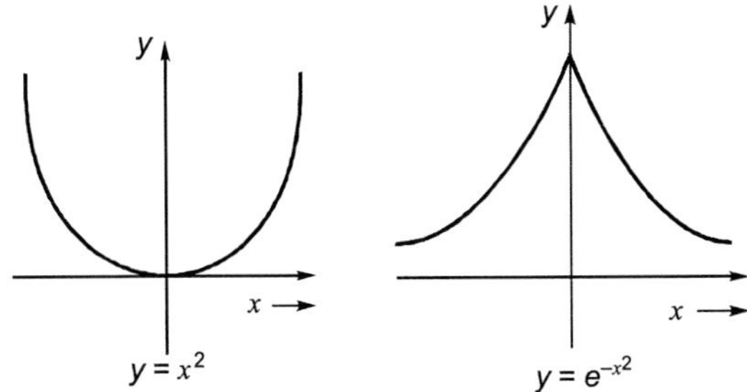

2.11 (a)

$$f(x) = x^2 - x - 2$$
$$= (x+1)(x-2)$$
$$f(-4) = 18$$
$$f(+4) = 10$$

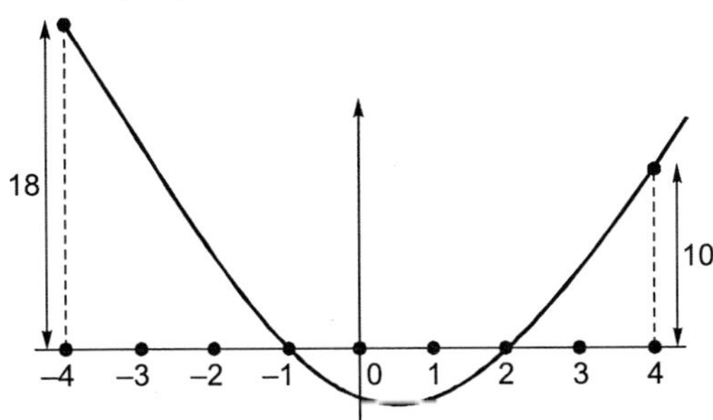

$f(x)$ is maximum in interval $[-4, 4]$ at $x = -4$

2.12 (a)

$$\sin x = x - \frac{x^3}{\lfloor 3} + \frac{x^5}{\lfloor 5} - \frac{x^7}{\lfloor 7} + \ldots.$$

Replace 'x' by x^3

$$\sin(x^3) = x^3 - \frac{(x^3)^3}{\lfloor 3} + \frac{(x^3)^5}{\lfloor 5} \ldots$$

$$= x^3 - \frac{x^9}{\lfloor 3} + \frac{x^{15}}{\lfloor 5} \ldots$$

2.13 (b)

$$f(x) = e^x + \sin x$$

We wish to expand about $x = \pi$

Taylor's series expansion about $x = a$ is

$$f(x) = f(a) + (x-a) f'(a) + \frac{(x-a)^2}{2!} f''(a) + \frac{(x-a)^3}{3!} f'''(a) \ldots.$$

Now about $x = \pi$

$$f(x) = f(\pi) + (x-\pi) f'(\pi) + \frac{(x-\pi)^2}{2!} f''(\pi) + \ldots$$

The coefficient of $(x-\pi)^2$ is $\frac{f''(\pi)}{2!}$

Here $f(x) = e^x + \sin x$

$$f'(x) = e^x + \cos x$$

$$f''(x) = e^x - \sin x$$

$$f''(\pi) = e^{\pi} - \sin\pi = e^{\pi} - 0 = e^{\pi}$$

The coefficient of $(x - \pi)^2$ is therefore

$$\frac{e^{\pi}}{2!} = 0.5\exp(\pi)$$

2.14 (a)

$$f(x) = e^x + e^{-x} = e^x + \frac{1}{e^x}$$

Arithmetic mean of e^x and $\frac{1}{e^x}$ is

$$\text{A.M.} = \frac{e^x + \frac{1}{e^x}}{2}$$

Geometric mean of e^x and $\frac{1}{e^x}$ is

$$\text{G. M.} = e^x \cdot \frac{1}{e^x} = 1$$

It is known that A.M. ≥ G.M.

$$\frac{\left(e^x + \frac{1}{e^x}\right)}{2} \geq 1$$

$$e^x + \frac{1}{e^x} \geq 2$$

Therefore, $(e^x + e^{-x})_{min} = 2$

2.15 (a)

Equation of line joining (0, 0) to (1, 2)

$$C: \quad \frac{x-0}{1-0} = \frac{y-0}{2-0} = t$$

$$C: \quad x = t, \quad y = 2t$$

$\because \quad x = t$, as x varies from 0 to 1

$\therefore$ t also from 0 to 1.

$$\because \quad \int_C 4x^3 + 10y^4 = \int_{t=0}^{1} (4t^3 + 10(2t)^4)dt$$

$$= \left.\frac{4t^4}{4} + 10\times 16\frac{t^5}{5}\right]_{t=0}^{1} = 1 + 32 = 33$$

2.16 (b)

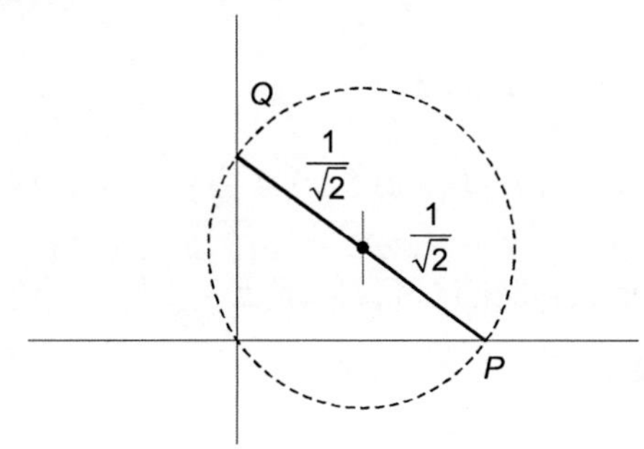

$$= 2\int_P^Q (x\,dx + y\,dx) = 2\int_1^0 x\,dx + 2\int_0^1 y\,dy$$

$$= 2\left[\left.\frac{x^2}{2}\right|_1^0 + \left.\frac{y^2}{2}\right|_0^1\right] = 0$$

2.17 (d)

Put n-π = y $\Rightarrow x = \pi + y$

$$f(x) = \frac{\sin x}{x-\pi} = \frac{\sin(\pi + y)}{y} = \frac{-\sin y}{y}$$

$$= -\frac{1}{y}\left[y - \frac{y^3}{\underline{|3}} + \frac{y^5}{\underline{|5}}\cdots\right]$$

$$= -1 + \frac{(x-\pi)^2}{6} - \frac{(x-\pi)^4}{\underline{|5}} + \ldots$$

2.18 (a)

$$e^y = x^{1/x}$$

Taking natural log on both sides,

$$y = \frac{1}{x} ln x$$

Differentiating on both sides,

$$\frac{dy}{dx} = \frac{1}{x}\cdot\frac{1}{x} - \frac{ln\,x}{x^2} = \frac{1}{x^2}(1 - ln\,x)$$

$$\frac{dy}{dx} = 0$$

$\Rightarrow 1 - ln\,x = 0$

$\Rightarrow \quad ln\,x = 1$

$\Rightarrow \quad x = e$

$$\frac{d^2y}{dx^2} = \frac{-2}{x^3}[1 - ln\,x] + \frac{1}{x^2}\left(\frac{-1}{x}\right)$$

$$= -\frac{1}{x^3}[2 - ln\,x + 1]$$

$$\frac{d^2y}{dx^2} = -\frac{1}{x^3}[3 - ln\,x]$$

Putting, x = e,

$$\left.\frac{d^2y}{dx^2}\right|_{x=e} = \frac{-1}{e^3}[3-1] = \frac{-2}{e^3} < 0$$

Since, second derivative is negative, therefore, y has maxima at x = e.

2.19 (c)

$$\vec{A} = xy\hat{a}_x + x^2\hat{a}_y$$

$$\vec{l} = x\,\hat{a}_x + y\hat{a}_y$$

$$\vec{dl} = dx\,\hat{a}_x + dy\hat{a}_y$$

$\vec{A}\cdot\vec{dl} = xy\,dx + x^2\,dy$

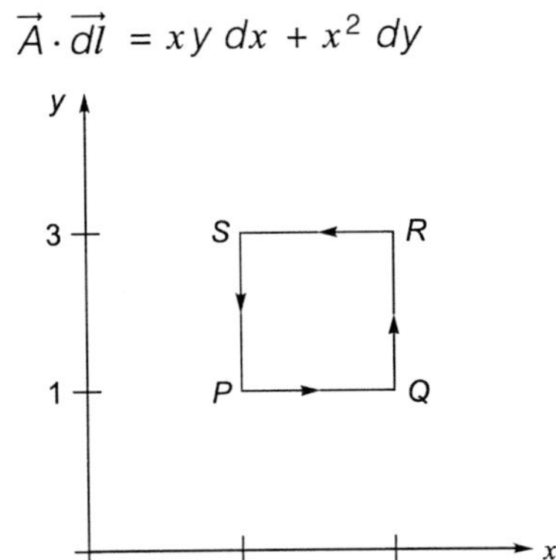

$P-Q$: $y = 1$, $dy = 0$

$$\int_P^Q \vec{A}\cdot\vec{dl} = \int_{1/\sqrt{3}}^{2/\sqrt{3}} x\,dx = \frac{x^2}{2}\Big|_{1/\sqrt{3}}^{2/\sqrt{3}} = \frac{1}{2}$$

$Q-R$: $x = \frac{2}{\sqrt{3}}$, $dx = 0$

$$\int_Q^R \vec{A}\cdot\vec{dl} = \int_1^3 \left(\frac{2}{\sqrt{3}}\right)^2 dy = \frac{4}{3}(3-1) = \frac{8}{3}$$

$R-S$: $y = 3$, $dy = 0$

$$\int_R^S \vec{A}\cdot\vec{dl} = \int_{2/\sqrt{3}}^{1/\sqrt{3}} 3x\,dx = \frac{3}{2}x^2\Big|_{2/\sqrt{3}}^{1/\sqrt{3}}$$

$$= \frac{3}{2}\left(\frac{1}{3}-\frac{4}{3}\right) = \frac{-3}{2}$$

$S-P$:

$$x = \frac{1}{\sqrt{3}}, dx = 0$$

$$\int_S^P \vec{A}\cdot\vec{dl} = \int_3^1 \left(\frac{1}{\sqrt{3}}\right)^2 dy = \frac{1}{3}(1-3) = \frac{-2}{3}$$

So, $$\oint_C \vec{A}\cdot\vec{dl} = \int_P^Q \vec{A}\cdot\vec{dl} + \int_Q^R \vec{A}\cdot\vec{dl} + \int_R^S \vec{A}\cdot\vec{dl} + \int_S^P \vec{A}\cdot\vec{dl}$$

$$= \frac{1}{2} + \frac{8}{3} - \frac{3}{2} - \frac{2}{3} = 1$$

2.20 (d)

$$\oiint_S 5\vec{r}\cdot\hat{n}\,ds = \iiint_V \nabla\cdot 5\vec{r}\,dV$$

$$\nabla\cdot r = \frac{1}{r^2}\frac{\partial}{\partial r}(r^2 r) = \frac{3}{r^2}\cdot r^2 = 3$$

$$5\iiint_V \nabla\cdot\vec{r}\,dV = 5\times 3\,V = 15\,V$$

2.21 (a)

$|A| = kr^n$

$\vec{A} = kr^n\,\hat{a}_r$ (since it is radially outward)

$\nabla\cdot\vec{A}$ in spherical coordinate is

$$\nabla\cdot\vec{A} = \frac{1}{r^2}\frac{\partial}{\partial r}(r^2 A_r) + \frac{1}{r\sin\theta}\frac{\partial}{\partial r}(A_\theta \sin\theta) + \frac{1}{r\sin\theta}\frac{\partial}{\partial\phi}A_\phi$$

$$\nabla\cdot\vec{A} = \frac{1}{r^2}\frac{\partial}{\partial r}(r^2 kr^n) + 0 + 0$$

$$\nabla\cdot\vec{A} = \frac{k}{r^2}\frac{\partial}{\partial r}(r^{n+2})$$

So, $\nabla\cdot\vec{A}$ will be zero if $\frac{\partial}{\partial r}(r^{n+2})$ will be zero, and $\frac{\partial}{\partial r}(r^{n+2})$ will be zero if r^{n+2} will be constant and this is possible if $n + 2 = 0 \Rightarrow n = -2$

2.22 (c)

We need absolute maximum of

$f(x) = x^3 - 9x^2 + 24x + 5$ in the interval [1, 6]

First find local maximum if any by putting $f'(x) = 0$.

i.e. $f'(x) = 3x^2 - 18x + 24 = 0$

i.e. $x^2 - 6x + 8 = 0$

$x = 2, 4$

Now $f''(x) = 6x - 18$

$f''(2) = 12 - 18 = -6 < 0$

(So $x = 2$ is a point of local maximum)

and $f''(4) = 24 - 18 = +6 > 0$

(So $x = 4$ is a point of local minimum)

Now tabulate the values of f at end point of interval and at local maximum point, to find absolute maximum in given range, as shown below:

x	$f(x)$
1	21
2	25
6	41

Clearly the absolute maxima is at $x = 6$ and absolute maximum value is 41.

2.23 (d)

According to Stoke's theorem

$$\oint_C \vec{A}\cdot\vec{dl} = \iint_S (\nabla\times\vec{A})\cdot\vec{ds}$$

2.24 (d)

$$\nabla\cdot\vec{A} = \frac{\partial A_x}{\partial x} + \frac{\partial A_y}{\partial y} + \frac{\partial A_z}{\partial z}$$

$$\nabla\cdot\vec{A} = \frac{\partial}{\partial x}(x) + \frac{\partial}{\partial y}(y) + \frac{\partial}{\partial z}(z) = 1 + 1 + 1$$

$$\nabla\cdot\vec{A} = 3$$

2.25 (a)

The Taylor's series expansion for

$$\sin x = x - \frac{x^3}{3!} + \frac{x^5}{5!} - \frac{x^7}{7!} + \cdots \qquad -\infty < x < \infty$$

and

$$\cos x = 1 - \frac{x^2}{2!} + \frac{x^4}{4!} - \frac{x^6}{6!} \cdots \qquad -\infty < x < \infty$$

$$\therefore\ 3\sin x + 2\cos x = 2 + 3x - \frac{2x^2}{2!} - \frac{3x^3}{3!} + \frac{2x^4}{4!} + \frac{3x^5}{5!} \cdots$$

$$= 2 + 3x - x^2 - \frac{x^3}{2} + \cdots$$

2.26 (864)

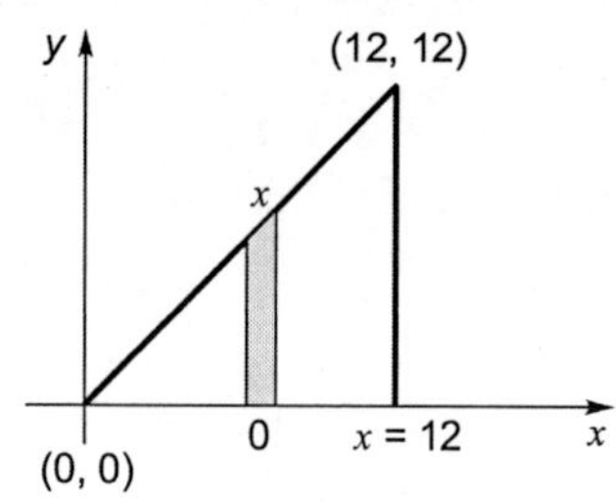

$$\text{Volume} = \iint_R z\,dx\,dy = \iint_R (x+y)\,dx\,dy$$

$$= \int_0^{12}\int_0^{x} (x+y)\,dx\,dy = \int_0^{12} \left(xy + \frac{y^2}{2} \right)\Bigg|_0^x dx$$

$$= \int_0^{12} \left(x^2 + \frac{x^2}{2} \right) dx = \frac{3}{2}\int_0^{12} x^2 dx = \frac{3}{2} \times \frac{x^3}{3}\Bigg|_0^{12}$$

$$= 864 \text{ unit}^3$$

2.27 (a)

$$f(t) = e^{-t} - 2e^{-2t}$$

$$f'(t) = -e^{-t} + 4e^{-2t}$$

For maximum value $f'(t) = 0$

$$f'(t) = 0 = -e^{-t} + 4e^{-2t}$$

$$\Rightarrow \quad 4e^{-2t} = e^{-t}$$

$$4e^{-t} = 1$$

$$\therefore \quad t = \log_e 4$$

2.28 (c)

$$\lim_{x\to\infty}\left(1 + \frac{1}{x}\right)^x = e^{\lim_{x\to\infty} \frac{1}{x}\cdot x} = e^1 = e$$

2.29 (0)

$$f'(x) = \frac{1}{1+x} - 1 = 0$$

$$\frac{1-1-x}{1+x} = 0$$

$$\frac{x}{1+x} = 0$$

$$x = 0$$

$$f''(x) = \frac{-1}{(1+x)^2}$$

$$f''(0) = -1 < 0$$

$f(x)$ have maximum value at $x = 0$

$$f(0) = \ln(1+0) - 0 = 0$$

$$f_{\max} = 0$$

2.30 (6)

$$f(x) = 2x^3 - 9x^2 + 12x - 3$$

$$f'(x) = 6x^2 - 18x + 12$$

$$f'(x) = 0$$

$$6x^2 - 18x + 12 = 0$$

$$x^2 - 3x + 2 = 0$$

$$x = 1, 2$$

Hence critical points are {0, 1, 2, 3,}.

$f(x)$ attains its maximum value at one of these points.

$f(0) = -3$; $f(1) = 2$; $f(2) = 1$; $f(3) = 6$

2.31 (d)

Given: Let, $x(n) = \sum_{n=0}^{\infty} \frac{1}{n!}$

$$= \frac{1}{0!} + \frac{1}{1!} + \frac{1}{2!} + \frac{1}{3!} + \frac{1}{4!} + \cdots$$

$$= 1 + 1 + \frac{1}{2} + \frac{1}{6} + \frac{1}{24} + \cdots$$

Also we know that expression of e^x

$$e^x = 1 + x + \frac{1}{2}x^2 + \frac{1}{6}x^3 + \frac{1}{24}x^4 + \cdots$$

Put $x = 1$ in above expression

$$e^1 = 1 + 1 + \frac{1}{2} + \frac{1}{6} + \frac{1}{24} + \cdots$$

$$e = \sum_{n=0}^{\infty} \frac{1}{n!}$$

2.32 (c)

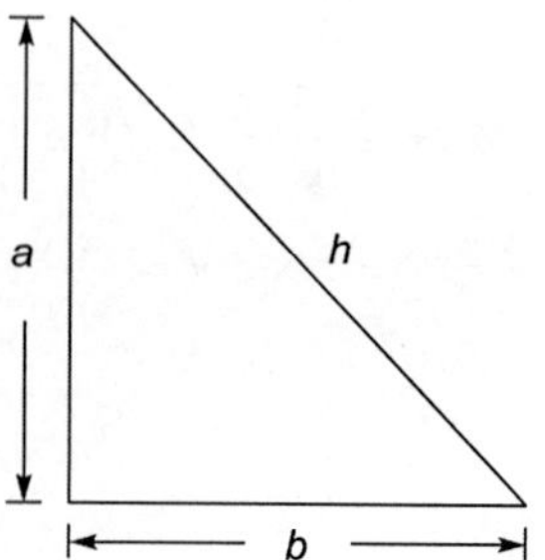

Let a, b and h be the sides of Δ such that

$$h^2 = a^2 + b^2$$

Given,

$$a + h = \text{constant } (K) \text{ say}$$

$$\text{Area of } \Delta = \frac{1}{2} \times a \times b = \frac{1}{2} \times a\sqrt{h^2 - a^2}$$

$$= \frac{1}{2} a\sqrt{K^2 - 2aK}$$

$$\frac{d\Delta}{da} = \frac{1}{2}\sqrt{K^2 - 2aK} + \frac{1}{2} a \times \frac{1}{2} \frac{(-2K)}{\sqrt{K^2 - 2aK}} = 0$$

$$\frac{1}{2}\left[\frac{K^2 - 2aK - aK}{\sqrt{K^2 - 2aK}}\right] = 0$$

$$K(K - 3a) = 0$$

$$K \neq 0 \Rightarrow \quad K = 3a$$

$$h = 2a$$

$$\theta = \cos^{-1}\left(\frac{a}{2a}\right) = \cos^{-1}\left(\frac{1}{2}\right) = 60^\circ$$

2.33 (3)

$$\text{div}(r^2 \nabla (lnr))$$

$$= \text{div}\left(r^2 \cdot \frac{1}{r}\left(\frac{\partial r}{\partial x}\hat{a}_x + \frac{\partial r}{\partial y}\hat{a}_y + \frac{\partial r}{\partial z}\hat{a}_z\right)\right)$$

$$= \text{div}\left(r\frac{\partial r}{\partial x}\hat{a}_x + r\frac{\partial r}{\partial y}\hat{a}_y + r\frac{\partial r}{\partial z}\hat{a}_z\right) \quad ...(i)$$

Putting value of $\bar{r} = x\hat{a}_x + y\hat{a}_y + z\hat{a}_z$ in equation (i)

$$\nabla \cdot (r^2 \nabla(\ln r)) = \nabla \cdot \left[r^2\left(\frac{1}{r}\left(\frac{\partial r}{dx}\hat{i} + \frac{\partial r}{\partial y}\hat{j} + \frac{\partial r}{\partial z}\hat{k}\right)\right)\right]$$

$$= \nabla \cdot \left[r\left(\frac{x}{r}\hat{i} + \frac{y}{r}\hat{j} + \frac{z}{r}\hat{k}\right)\right]$$

$$= \nabla \cdot (x\hat{i} + y\hat{j} + z\hat{k}) = 3$$

2.34 (0)

$$\vec{A} = (\cos x)(\sin y)\hat{a}_x + (\sin x)(\cos y)\hat{a}_y$$

$$\vec{A} = (\cos x)(\sin y)\hat{a}_x + (\sin x)(\cos y)\hat{a}_y + 0\hat{a}_z$$

$$\nabla \times \vec{A} = \begin{vmatrix} \hat{a}_x & \hat{a}_y & \hat{a}_z \\ \frac{\partial}{\partial x} & \frac{\partial}{\partial y} & \frac{\partial}{\partial z} \\ \cos x \sin y & \sin x \cos y & 0 \end{vmatrix}$$

$$= \hat{a}_z[\cos x \cos y - \cos x \cos y] = 0$$

$$\therefore \left|\nabla \times \vec{A}\right| = 0$$

2.35 (7)

$$f(x, y, z) = x^2 + 3y^2 + z^3$$

$$\nabla f = \frac{\partial f}{\partial x}\hat{i} + \frac{\partial f}{\partial y}\hat{j} + \frac{\partial f}{\partial z}\hat{k}$$

$$= 2x\hat{i} + 6y\hat{j} + 3z^2\hat{k}$$

$$(\nabla f)_{(1,1,1)} = 2\hat{i} + 6\hat{j} + 3\hat{k}$$

$$|\nabla f|_{(1,1,1)} = \sqrt{4 + 36 + 9} = \sqrt{49} = 7$$

2.36 (3)

$$f(x, y) = \frac{xy}{\sqrt{2}}(x + y) = \frac{x^2y + xy^2}{\sqrt{2}}$$

$$\nabla f = \frac{\partial f}{\partial x}\hat{i} + \frac{\partial f}{\partial y}\hat{j}$$

$$= \frac{2xy + y^2}{\sqrt{2}}\hat{i} + \frac{(x^2 + 2xy)}{\sqrt{2}}\hat{j}$$

The direction is $\hat{n} = \frac{1}{\sqrt{2}}\hat{i} + \frac{1}{\sqrt{2}}\hat{j}$

Directional derivative in direction of $\hat{n}$ is

$$= \left(\frac{2xy + y^2}{2} + \frac{x^2 + 2xy}{2}\right)_{(1,1)} = \frac{3}{2} + \frac{3}{2} = 3$$

2.37 (6.28)

$$\nabla \times \vec{F} = \begin{vmatrix} \hat{a}_x & \hat{a}_y & \hat{a}_z \\ \frac{\partial}{\partial x} & \frac{\partial}{\partial y} & \frac{\partial}{\partial z} \\ z & x & y \end{vmatrix} = \hat{a}_x + \hat{a}_y + \hat{a}_z$$

Now,

$$\int_s \nabla \times \vec{F} \cdot ds = \int_s (\hat{a}_x + \hat{a}_y + \hat{a}_z) \cdot (\hat{a}_z \cdot dx\,dy)$$

$$= \int_s dxdy = 2\pi(1)^2 = 6.28$$

2.38 (c)

$$\frac{\partial z}{\partial x} = y\ln(xy) + \frac{xy}{xy}y$$

$$\frac{\partial z}{\partial x} = y[\ln(xy) + 1] \quad ...(i)$$

$$\frac{\partial z}{\partial y} = x\ln(xy) + \frac{xy}{xy} \times x$$

$$\frac{\partial z}{\partial y} = x[\ln(xy) + 1] \quad ...(ii)$$

Here, $\boxed{x\frac{\partial z}{\partial x} = y\frac{\partial z}{\partial y}}$

2.39 (b)

Since $f(1) \neq f(-1)$, Rolle's mean value theorem does not apply.

By Lagrange mean value theorem

$$f'(x) = \frac{f(1) - f(-1)}{1-(-1)} = \frac{2}{2} = 1$$

$$-2x + 3x^2 = 1$$

$$x = 1, -\frac{1}{3}$$

x lies in (–1, 1)

$$\Rightarrow \quad x = -\frac{1}{3}$$

2.40 (a)

$$\vec{P} = x^3 y\vec{a}_x - x^2y^2\vec{a}_y - x^2yz\,\vec{a}_z$$

For solenoidal, $\nabla \cdot \vec{P} = 0$

$$\Rightarrow \nabla \cdot \vec{P} = \frac{\partial P_x}{\partial x} + \frac{\partial P_y}{\partial y} + \frac{\partial P_z}{\partial z}$$

$$= 3x^2y - 2x^2y - x^2y$$

$$= 0$$

$\Rightarrow$ $\vec{P}$ is solenoidal

For irrotational, $\nabla \times \vec{P} = 0$

$$\Rightarrow \nabla \times \vec{P} = \begin{vmatrix} \vec{a}_x & \vec{a}_y & \vec{a}_z \\ \dfrac{\partial}{\partial x} & \dfrac{\partial}{\partial y} & \dfrac{\partial}{\partial z} \\ x^3y & -x^2y^2 & -x^2yz \end{vmatrix}$$

$$= \vec{a}_x(-x^2z) + \vec{a}_y(2xyz) + \vec{a}_z(-2x^2 - x^3)$$

$$\neq 0$$

$\Rightarrow$ $\vec{P}$ is not irrotational.

2.41 (b)

$$f(x) = e^{-x}(x^2 + x + 1)$$

$$f'(x) = e^{-x}(2x + 1) - e^{-x}(x^2 + x + 1)$$

$$= e^{-x}(x - x^2) = e^{-x}(x)(1 - x)$$

Putting $f'(x) = 0$, we get

$$x = 0 \quad \text{or} \quad x = 1$$

$$f''(x) = e^{-x}(1 - 2x) - e^{-x}(x - x^2)$$

$$= e^{-x}(1 - 3x + x^2)$$

At $x = 0$, $f''(x) = 1$ (so we have a minimum).

At $x = 1$, $f''(x) = -\frac{1}{e}$

(so we have a maximum).

Only curve (b) shows a single local minimum at $x = 0$ and a single local maximum at $x = 1$.

2.42 (1)

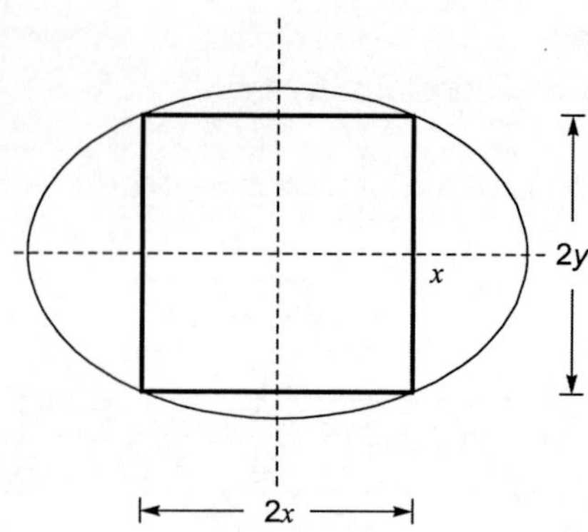

$$x^2 + 4y^2 = 1$$

Area of rectangle

$$= 2x \cdot 2y = 4xy$$

Let $\quad f = (\text{Area})^2$

$$= 16x^2y^2$$

$$= 4x^2(1 - x^2) \quad (\because 1 - x^2 = 4y^2)$$

$$f'(x) = 0$$

$$\frac{d}{dx}\left[4(x^2 - x^4)\right] = 0$$

$$4(2x - 4x^3) = 0$$

We get, $\quad x = \pm\frac{1}{\sqrt{2}}$

$$y = \pm\frac{1}{\sqrt{8}}$$

$$\text{Area} = 4xy = 4 \times \frac{1}{\sqrt{2}} \times \frac{1}{\sqrt{8}} = 1$$

2.43 (3)

$$\int_{-\infty}^{\infty} 12\cos(2\pi t)\frac{\sin(4\pi t)}{4\pi t}dt$$

$$= 12\,F \cdot T\left[\cos(2\pi t)\,\frac{\sin(4\pi t)}{4\pi t}\right]_{\omega=0} = 12 \times \frac{1}{4} = 3$$

2.44 (a)

Partial derivative w.r.t. y

$$\frac{\partial}{\partial y}(x^2 + y^2) = 2y$$

Partial derivative w.r.t. x

$$\frac{\partial}{\partial x}(6y + 4x) = 4$$

From given condition

$$2y = 4$$

$$\Rightarrow \quad y = 2$$

2.45 (b)

P : If $f(x)$ is continuous at $x = x_0$, then it is also differentiable at $x = x_0$

Q : If $f(x)$ is continuous at $x = x_0$, then it may or may not be derivable at $x = x_0$

R : If $f(x)$ is differentiable at $x = x_0$, then it is also continuous at $x = x_0$

P is false

Q is true

R is true   Option (b) is correct

2.46 (20)

Put
$$x = r\cos\theta$$
$$y = r\sin\theta$$
$$dxdy = rdrd\theta$$

$$= \frac{1}{2\pi}\int_0^{2\pi}\int_0^2 (r(\cos\theta + \sin\theta) + 10)r\,drd\theta$$

$$= \frac{1}{2\pi}\int_0^{2\pi}\int_0^2 (r^2(\cos\theta + \sin\theta) + 10r)\,drd\theta$$

$$= \frac{1}{2\pi}\left(\int_0^{2\pi}(\cos\theta + \sin\theta)\left(\frac{r^3}{3}\right)\Bigg|_0^2 d\theta + 10\int_0^{2\pi}\left(\frac{r^2}{2}\right)\Bigg|_0^2 d\theta\right)$$

$$= \frac{1}{2\pi}\int_0^{2\pi}\frac{8}{3}(\cos\theta + \sin\theta)d\theta + \frac{1}{2\pi}\int_0^{2\pi} 5\cdot(4)d\theta$$

$$= \frac{1}{2\pi}\left[\frac{8}{3}(\sin\theta - \cos\theta)\right]_0^{2\pi} + \frac{1}{2\pi}\cdot 20(2\pi)$$

$$= \frac{1}{2\pi}\left(\frac{8}{3}[(0-1)-(0-1)] + 20\right) = 0 + 20 = 20$$

2.47 (4.712)

$$V = \int_{\rho=3}^{5}\int_{\phi=\frac{\pi}{8}}^{\pi/4}\int_{z=3}^{4.5} \rho\,d\rho\,d\phi\,dz$$

$$= \int_3^{4.5}\int_{\pi/8}^{\pi/4}\left(\frac{\rho^2}{2}\right)\Bigg|_3^5 d\phi\,dz$$

$$= \int_3^{4.5}\int_{\pi/8}^{\pi/4} 8\cdot d\phi\,dz = 8\phi\big|_{\pi/8}^{\pi/4}\cdot z\big|_3^{4.5}$$

$$= 8\left(\frac{\pi}{4} - \frac{\pi}{8}\right)(4.5 - 3)$$

$$= 8\cdot\frac{\pi}{8}\cdot(1.5) = 4.712$$

2.48 (c)

$F'(x) = f(x)$ which is density function

$F'(x) = f(x) < 0$ when $x < 0$

$\therefore$ $F(x)$ is decreasing for $x < 0$

$F'(x) = f(x) > 0$ when $x > 0$

$\therefore$ $F(x)$ is increasing for $x > 0$

2.49 (0)

By Green's theorem

$$\int xy^2dx + x^2ydy = \iint_R\left(\frac{d}{dx}(x^2y) - \frac{d}{dy}(xy^2)\right)dxdy$$

$$= \iint_R (2xy - 2xy) = 0$$

2.50 (b)

$$f(x) = x^3 - 3x^2 + 1$$
$$f'(x) = 3x^2 - 6x$$
$$f'(x) = 0$$

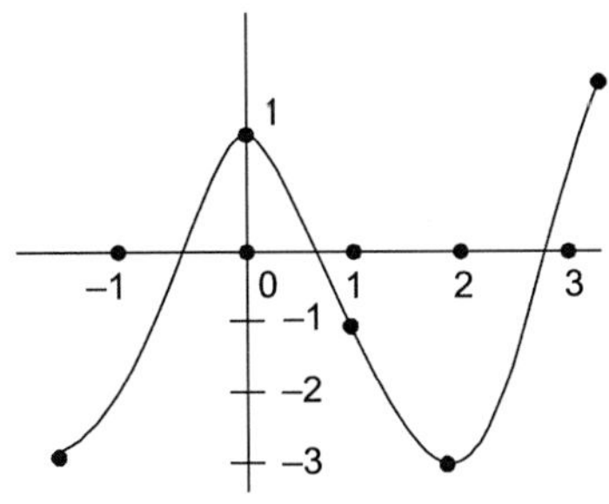

$$3x^2 - 6x = 0$$
$$3x(x-2) = 0$$
$$x = 0, 2$$
$$f''(x) = 6x - 6$$

At $x = 0$ $f''(0) = -6$ maxima

$x = 2$ $f''(2) = 6$ minima

2.51 (c)

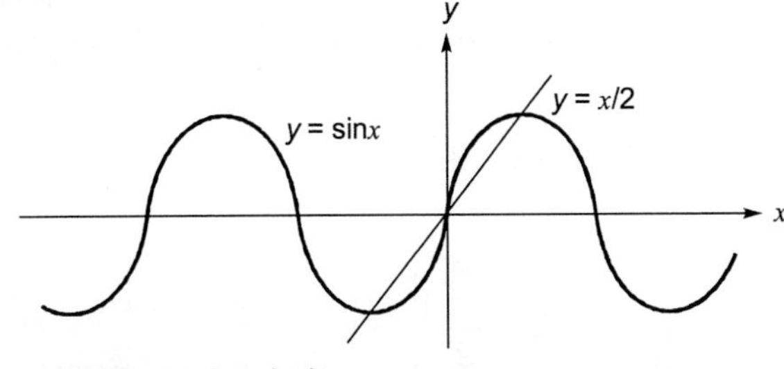

Hence, 3 solutions.

2.52 (90)

$$I = \hat{x}15\cos\omega t + \hat{y}5\sin\omega t$$

$$|I| = \sqrt{(15\cos\omega t)^2 + (5\sin\omega t)^2}$$

$$= \sqrt{225\cos^2\omega t + 25\sin^2\omega t} = \sqrt{25 + 200\cos^2\omega t}$$

$|I|$ is minimum when $\cos^2\omega t = 0$

or $\theta = \omega t = 90°$

2.53 (2)

$$\int_0^1 \frac{1}{\sqrt{1-x}}dx = -2\int_0^1 \frac{1}{2\sqrt{1-x}}dx$$

$$= -2\left(\sqrt{1-x}\right)\Big|_0^1 = -2(0-1) = 2$$

2.54 (d)

$$\begin{bmatrix}4\\3\\-3\end{bmatrix} = a\begin{bmatrix}1\\0\\2\end{bmatrix} + b\begin{bmatrix}0\\1\\0\end{bmatrix} + c\begin{bmatrix}-2\\0\\1\end{bmatrix}$$

$$a - 2c = 4$$
$$b = 3$$
$$2a + c = -3$$

from here $a = -\dfrac{2}{5}$

$$b = 3$$

$$c = -\frac{11}{5}$$

$$u = -\frac{2}{5}e_1 + 3e_2 - \frac{11}{5}e_3$$

2.55 (10)

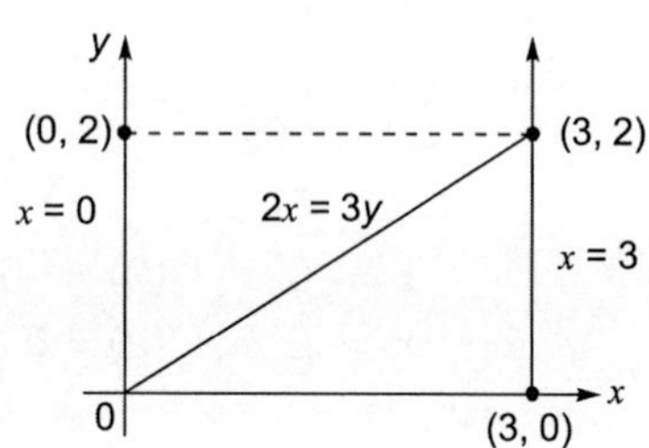

$$\text{Volume} = \iiint dz dx dy = \iint z dy dx$$

$$= \int_0^3 \int_0^{2/3x} (6 - x - y)\,dydx$$

$$= \int_0^3 \left(6y - xy - \frac{y^2}{2}\right)\Bigg|_0^{2/3x} dx = \int_0^3 \left(4x - \frac{8}{9}x^2\right)dx$$

$$= \left(4\frac{x^2}{2} - \frac{8}{9}\cdot\frac{x^3}{3}\right)\Bigg|_0^3 = \left[2x^2 - \frac{8}{9}\left(\frac{x^3}{3}\right)\right]_0^3$$

$$= 18 - 8 = 10$$

2.56 (b)

Any of the given three functions cannot be written as the linear combination of other two functions. Hence, the statements I and III are correct.

2.57 (–11)

$A(0, 2, 1)$ and $B(4, 1, -1)$

The equation of the line AB is

$$\frac{x-0}{4-0} = \frac{y-2}{1-2} = \frac{3-1}{-1-1} = t \text{ say}$$

$$x = 4t\,;\ y = -t + 2\,;\ z = -2t + 1$$

$$dx = 4dt\,;\ dy = -dt\,;\ dz = -2dt$$

t varies from 0 to 1

$$I = \int_0^1 2(-2t+1)\,4dt + 2(-t+2)(-dt) + 2(4t)(-2dt)$$

$$= \int_0^1 (-16t + 8 + 2t - 4 - 16t)dt = \int_0^1 (-30t + 4)dt$$

$$= \left(-30\frac{t^2}{2} + 4t\right)\Bigg|_0^1 = -15 + 4 = -11$$

2.58 (c)

About $x = 0$

$$f(x) = e^x e^{x^2}$$

$$= \left(1 + x + \frac{x^2}{2!} + \frac{x^3}{3!} + \ldots\right)\left(1 + x^2 + \frac{x^4}{2!} + \frac{x^6}{3!} + \ldots\right)$$

$$= 1 + x^2 + \frac{x^4}{2} + \frac{x^6}{6} + \ldots + x + x^3 + \frac{x^5}{2} + \frac{x^7}{6} + \ldots$$

$$+\frac{x^2}{2} + \frac{x^4}{2} + \frac{x^6}{4} + \frac{x^8}{12} \ldots + \frac{x^3}{6} + \frac{x^5}{6} + \frac{x^7}{12} + \frac{x^9}{36} + \ldots$$

$$= 1 + x + \frac{3}{2}x^2 + \frac{7}{6}x^3$$

2.59 (0.7853)

Let, $x^2 + y^2 = t^2$

$$t^2 = z^3$$

Here, revolution is about z axis

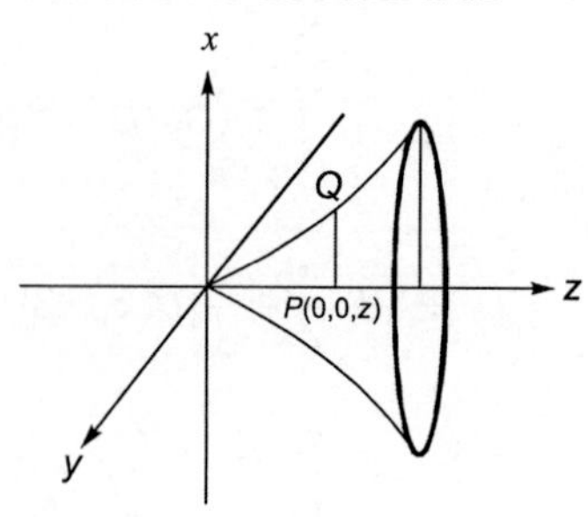

$$\text{volume of region } R = \int_0^1 \pi (PQ)^2 dz$$

Here, PQ is radius of circle at some z, which is given by

$$PQ = \sqrt{x^2 + y^2}$$

$$(PQ)^2 = x^2 + y^2 = z^3$$

so, volume of region R

$$= \int_0^1 \pi t^2 dz = \int_0^1 \pi z^3 dz = \left.\frac{\pi z^4}{4}\right|_0^1 = \frac{\pi}{4} = 0.7853$$

2.60 (54.73)

If θ is the angle between

$a_1 x + b_1 y + c_1 z + d_1 = 0$

$a_2 x + b_2 y + c_2 z + d_2 = 0$ then

$$\cos\theta = \frac{a_1a_2 + b_1b_2 + c_1c_2}{\sqrt{a_1^2 + b_1^2 + c_1^2}\sqrt{a_2^2 + b_2^2 + c_2^2}}$$

$$a_1 = 1 \quad b_1 = 1 \quad c_1 = 1 \quad d_1 = -1$$

$$a_2 = 2 \quad b_2 = -1 \quad c_2 = 2 \quad d_2 = 0$$

$$\cos\theta = \frac{(1)(2) + (1)(-1) + (1)(2)}{\sqrt{(1)^2 + (1)^2 + (1)^2}\sqrt{(2)^2 + (-1)^2 + (2)^2}}$$

$$= \frac{2 - 1 + 2}{\sqrt{3}\sqrt{9}} = \frac{3}{3\sqrt{3}} = \frac{1}{\sqrt{3}}$$

$$\theta = \cos^{-1}\left(\frac{1}{\sqrt{3}}\right) = 54.73^\circ$$

2.61 (b)

$$\vec{F} = \hat{a}_x(3y - k_1 z) + \hat{a}_y(k_2 x - 2z) - \hat{a}_z(k_3 y + z)$$

$$\nabla \times \vec{F} = 0 \quad \text{(irrotational)}$$

$$\nabla \times \vec{F} = \begin{vmatrix} \hat{a}_x & \hat{a}_y & \hat{a}_z \\ \frac{\partial}{\partial x} & \frac{\partial}{\partial y} & \frac{\partial}{\partial z} \\ 3y - k_1 z & k_2 x - 2z & -(k_3 y + z) \end{vmatrix}$$

$$= \hat{a}_x\left[\frac{\partial}{\partial y}[-(k_3 y + z)] - \frac{\partial}{\partial z}(k_2 x - 2z)\right]$$

$$-\hat{a}_y\left[\frac{\partial}{\partial x}[-(k_3 y + z)] - \frac{\partial}{\partial z}(3y - k_1 z)\right]$$

$$+\hat{a}_z\left[\frac{\partial}{\partial x}(k_2 x - 2z) - \frac{\partial}{\partial y}(3y - k_1 z)\right]$$

$$\hat{a}_x[-k_3 + 2] - \hat{a}_y[k_1] + \hat{a}_z[k_2 - 3] = 0$$

$\Rightarrow \quad k_3 = 2,\ k_1 = 0,\ k_2 = 3$

or $\quad k_1 = 0,\ k_2 = 3,\ k_3 = 2$

2.62 (–100)

$$f(x) = \frac{1}{3}x(x^2 - 3) = \frac{1}{3}(x^3 - 3x)$$

$$f'(x) = \frac{1}{3}(3x^2 - 3) = x^2 - 1$$

$$f'(x) = x^2 - 1 = 0$$

$$\Rightarrow \quad x = \pm 1$$

$$f''(x) = 2x$$

At $x = 1$, $\quad f''(1) = 2 = 0 \Rightarrow$ minima

At $x = -1$, $\quad f''(-1) = -2 < 0 \Rightarrow$ maxima

Minimum value of $f(x)$ in [–100, 100] is given by

Minimum $\{f(-100), f(100), f(1)\}$

Minimum $\{-333433.3, 333233.3, -0.666\}$

$= -3335433.3$

Hence the minimum value occurs at $x = -100$

Also graph of the function will be like

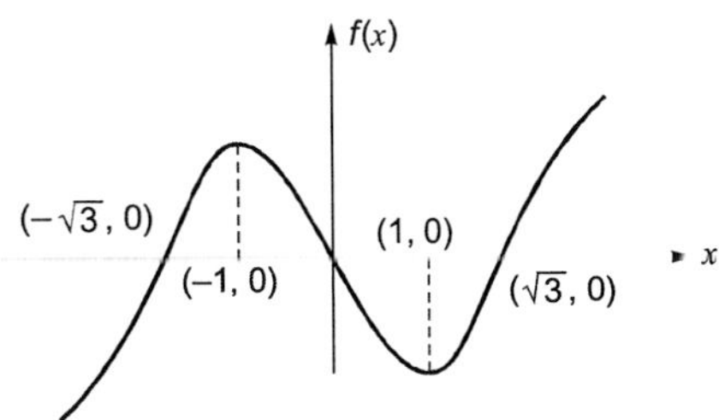

2.63 (c)

$$\text{Integral } I_1 = \int_0^1\int_0^1 \frac{x - y}{(x + y)^3} dy dx$$

$$= \int_0^1\left[\int_0^1\left(\frac{2x}{(x + y)^3} - \frac{1}{(x + y)^2}\right)dy\right]dx$$

$$= \int_0^1\left[2x\left(\frac{-1}{2(x + y)^2}\right)\Bigg|_0^1 + \left(\frac{1}{x + y}\right)\Bigg|_0^1\right]dx$$

$$= \int_0^1 \frac{1}{(x + 1)^2}dx = -\left[\frac{1}{x + 1}\right]_0^1 = 0.5$$

$$\text{and Integral} \quad I_2 = \int_0^1\left(\int_0^1 \frac{x - y}{(x + y)^3}dx\right)dy$$

$$= \int_0^1\left(\int_0^1 \frac{1}{(x + y)^2} - \frac{2y}{(x + y)^3}dx\right)dy$$

$$= \int_0^1\left[\frac{-1}{(x + y)} + \frac{2y}{2(x + y)}\right]_0^1 dy$$

$$= \int_0^1 \frac{-1}{(1 + y)^2}dy = -\left[\frac{-1}{y + 1}\right]_0^1 = -0.5$$

2.64 (0)

$$f(x) = \int_0^x e^{-t^2/2} dt$$

$$f'(x) = e^{-x^2/2} \text{ and } f''(x) = e^{-x^2/2}(-x)$$

$$f''(0) = 0$$

$$a_2 = \frac{f''(0)}{2!} = 0$$

2.65 (d)

$$f(x, y) = \frac{ax^2 + by^2}{xy} = a\left(\frac{x}{y}\right) + b\left(\frac{y}{x}\right)$$

$$\left.\frac{\partial f}{\partial x}\right|_{(1,2)} = \left.\left[\frac{a}{y} - \frac{by}{x^2}\right]\right|_{(1,2)} = \frac{a}{2} - 2b$$

$$\left.\frac{\partial f}{\partial y}\right|_{(1,2)} = \left.\left[-\frac{ax}{y^2} + \frac{b}{x}\right]\right|_{(1,2)} = -\frac{a}{4} + b$$

$$\frac{\partial f}{\partial x} = \frac{\partial f}{\partial y}$$

So, $\frac{a}{2} - 2b = -\frac{a}{4} + b$

$$\frac{3a}{4} = 3b$$

$$a = 4b$$

2.66 (4.50)

$$r = x^2 + y - z \quad \text{...(i)}$$

$$z^3 - xy + yz + y^3 = 1 \quad \text{...(ii)}$$

$$\frac{\partial r}{\partial x} = 2x - \frac{\partial z}{\partial x} \quad \text{...(iii)}$$

$$3z^2 \frac{\partial z}{\partial x} - y + y\frac{\partial z}{\partial x} = 0$$

$$\frac{\partial z}{\partial x} = \frac{y}{3z^2 + y}$$

By substituting $\frac{\partial z}{\partial x}$ in equation (iii), we get,

$$\frac{\partial r}{\partial x} = 2x - \frac{y}{3z^2 + y}$$

At (2, –1, 1),

$$\frac{\partial r}{\partial x} = 2(2) - \frac{(-1)}{3(1)^2 + (-1)}$$

$$= 4 + \frac{1}{2} = 4.50$$

2.67 (a)

$$\frac{dy}{dx} = \frac{x^2}{2y} + \frac{y}{2} + \frac{y}{x}$$

Put, $\frac{y}{x} = t$

$$\frac{dy}{dx} = t + x\frac{dt}{dx}$$

$$t + x\frac{dt}{dx} = \frac{x}{2t} + \frac{tx}{2} + t$$

$$x\frac{dt}{dx} = x\left(\frac{1}{2t} + \frac{t}{2}\right)$$

$$x\frac{dt}{dx} = x\left(\frac{1 + t^2}{2t}\right)$$

$$\int \frac{2t}{1 + t^2} dt = \int dx + C$$

$$ln(1 + t^2) = x + C$$

$$t = \frac{y}{x}$$

So, $ln\left(1 + \frac{y^2}{x^2}\right) = x + C$

At $x = 1$, $y = 0$

$$ln\left(1 + \frac{0}{1}\right) = ln(1) = 0 = 1 + C$$

$$C = -1$$

So, $ln\left(1 + \frac{y^2}{x^2}\right) = x - 1$

2.68 (2)

$$x = y; \quad x = \pi$$

$$y = 0; \quad y = \pi$$

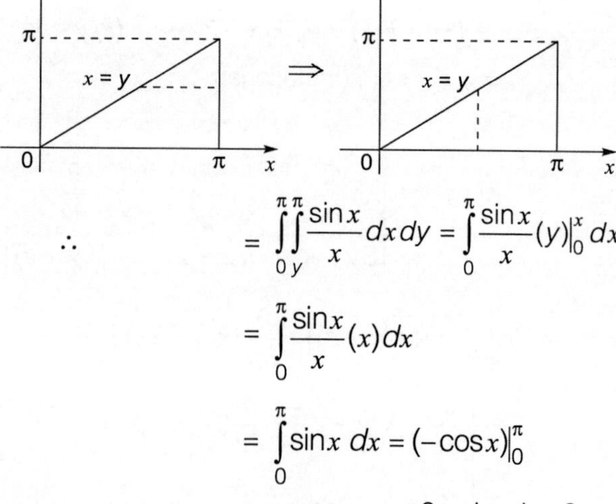

$$\therefore \quad = \int_0^{\pi}\int_y^{\pi} \frac{\sin x}{x} dx\, dy = \int_0^{\pi} \frac{\sin x}{x} (y)\Big|_0^x dx$$

$$= \int_0^{\pi} \frac{\sin x}{x} (x) dx$$

$$= \int_0^{\pi} \sin x\, dx = (-\cos x)\Big|_0^{\pi}$$

$$= -\cos\pi + \cos 0 = 1 + 1 = 2$$

2.69 (a)

Given that, $|f'(x)| \le 2$; $\quad f(-1) = 0$

$$-2 \le f'(x) \le 2$$

$x \in [-2, 2] \rightarrow -2 \le x \le 2$

$\therefore$ By applying mean value theorem in $[-1, 2]$

$$-2 \le f'(x) \le 2$$

$$-2 \le \frac{f(2) - f(-1)}{2-(-1)} \le 2$$

$$-2 \le \frac{f(2) - f(-1)}{3} \le 2$$

$$-6 \le f(2) - 0 \le 6$$

$$-6 \le f(2) \le 6$$

It is satisfied by only option (a).

2.70 (a)

Given, $\int -ydx + xdy$

here, $F_1 = -y$ and $\frac{\partial F_1}{\partial y} = -1$

$F_2 = x$ and $\frac{\partial F_2}{\partial x} = 1$

$$\therefore \int F_1 dx + F_2 dy = \iint \left(\frac{\partial F_2}{\partial x} - \frac{\partial F_1}{\partial y} \right) dx\, dy$$

$$\int -ydx + xdy = \iint 1-(-1)\, dx\, dy$$

$$= 2(\text{Area of region } R)$$

$$= 2\left(6+\frac{\pi}{2}\right) = 12+\pi$$

2.71 (c)

Given $f(x, y, z) = e^{1-x\cos y} + xze^{-1/(1+y^2)}$

$$\frac{\partial f}{\partial x} = e^{1-x\cos y}(0-\cos y) + ze^{-1/1+y^2}$$

$$\left(\frac{\partial f}{\partial x}\right)_{(1,0,e)} = e^0(0-1) + e \cdot e^{-1/(1+0)}$$

$$= -1 + 1 = 0$$

2.73 (2.25)

x : 0 to 3

y : 0 to 1

z : 0 to $1 - y$

$$= \int_{y=0}^{1} \int_{z=0}^{1-y} \int_{x=0}^{3} x\, dx\, dy\, dz$$

$$= \int_{y=0}^{1} \int_{0}^{1-y} \left(\frac{x^2}{2}\right)_0^3 dz\, dy = \int_0^1 \frac{9}{2}(z)_0^{1-y} dy$$

$$= \frac{9}{2}\int_0^1 (1-y)\, dy = \frac{9}{2}\left(y - \frac{y^2}{2}\right)_0^1 = \frac{9}{2}\left(1-\frac{1}{2}\right) = \frac{9}{4}$$

2.74 (b)

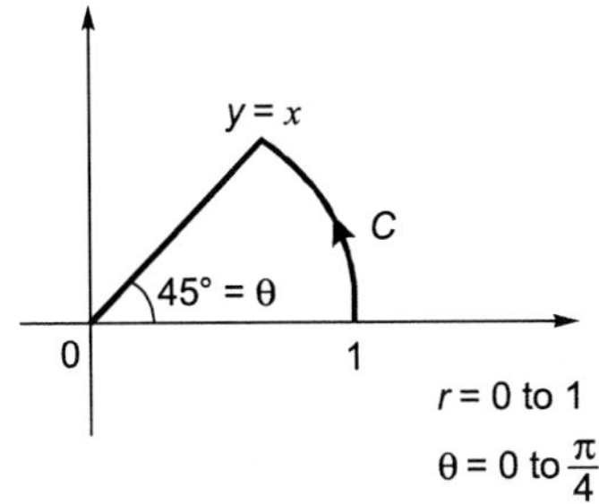

$r = 0$ to 1

$\theta = 0$ to $\frac{\pi}{4}$

$$\bar{F} = -xi + yj$$

$$\int \vec{F} \cdot \vec{dr} = \int_C -xdx + y\,dy$$

In polar form, $x = \cos\theta$, $y = \sin\theta$

$dx = -\sin\theta d\theta$, $dy = \cos\theta d\theta$

$$= \int_{\theta=0}^{45^\circ} (-\cos\theta(-\sin\theta d\theta) + \sin\theta(\cos\theta d\theta))$$

$$= \int_{\theta=0}^{\pi/4} \sin 2\theta\, d\theta = -\frac{\cos 2\theta}{2}\Bigg]_0^{\pi/4}$$

$$= -\frac{1}{2}[0-1] = \frac{1}{2}$$

2.75 (a)

Let, $I = \oint_C \vec{F}(x, y) \cdot (dx\vec{i} + dy\vec{j})$

Given : $\vec{F}(x, y) = x\vec{i} + y\vec{j}$

Hence, $I = \oint_C (x\vec{i} + y\vec{j}) \cdot (dx\vec{i} + dy\vec{j})$

$$I = \oint_C (xdx + ydy)$$

Using Green's theorem,

$$\oint_C Pdx + Qdy = \iint_R \left(\frac{\partial Q}{\partial x} - \frac{\partial P}{\partial y}\right) dx\, dy$$

We get,

$$I = \iint_R \left(\frac{\partial x}{\partial x} - \frac{\partial y}{\partial y}\right) dx\, dy = \iint_R (1-1)dx\, dy = 0$$

2.76 (b)

$$f(x) = 8\ln(x) - x^2 + 3$$

To obtain the stationary points,

$$f'(x) = 0$$

$$\Rightarrow \quad \frac{8}{x} - 2x = 0$$

$$\Rightarrow \quad 2x^2 = 8$$

$\Rightarrow \quad x^2 = 4$

$\Rightarrow \quad x = \pm 2$

In the interval [1, e], $x = 2$.

$$f''(x) = -\frac{8}{x^2} - 2\Big|_{x=2} = -\frac{8}{4} - 2 = -4$$

$\therefore \quad f''(2) < 0$

Hence, $x = 2$ is a maxima.

$\therefore$ Minimum of $f(x)$ in [1, e]

$= \min\{f(1), f(e)\}$

where, $f(1) = 8\log_e 1 - 1 + 3 = 2$

$f(e) = 8\log_e e - e^2 + 3$

$= 8 - e^2 + 3$

$= 11 - 7.39 = 3.61$

$\therefore$ Minimum value of $f(x)$ occurs at $x = 1$.

2.77 (512)

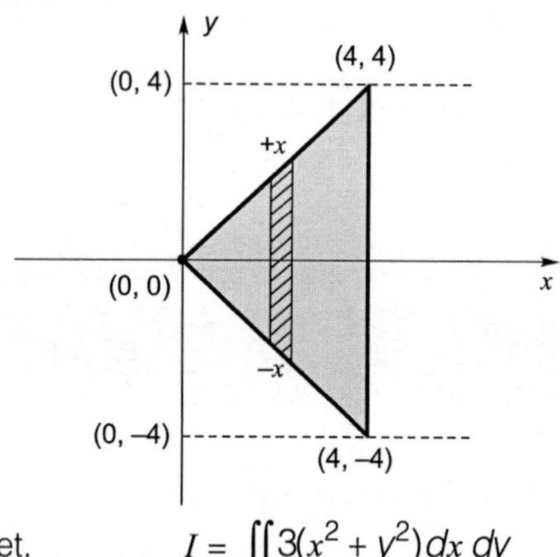

Let, $I = \iint_D 3(x^2 + y^2)dx\,dy$

In the shaded triangular region consider a vertical rectangular strip as shown. To cover the complete region D, y varies from $-x$ to x as x varies from 0 to 4. Therefore,

$$I = \int_{x=0}^{4}\int_{y=-x}^{x} 3(x^2 + y^2)dy\,dx$$

$$I = \int_{x=0}^{4}\left(3x^2y + \frac{3y^3}{3}\right)_{-x}^{x} dx$$

$$I = \int_{x=0}^{4}\left[3x^2(2x) + 2x^3\right]dx$$

$$I = \int_{x=0}^{4}(6x^3 + 2x^3)\,dx = \int_0^4 8x^3 dx$$

$$\therefore \quad I = 8\left[\frac{x^4}{4}\right]_{x=0}^{4} = \frac{8}{4} \times 4^4 = 512$$

$I = 512$

2.78 (c)

From the given expression, we can write

$e = V_1 - \alpha V_2$

$e = (i + 2k + 0k) - \alpha(2i + j + 3k)$

$\hat{e} = (1-2\alpha)\hat{i} + (2-\alpha)\hat{j} + (0-3\alpha)\hat{k}$

$|\hat{e}| = \sqrt{(1-2\alpha)^2 + (2-\alpha)^2 + (-3\alpha)^2}$

$|\hat{e}|^2 = 5 + 14\alpha^2 - 8\alpha$ to be minimum

at $\frac{\partial e^2}{\partial \alpha} = 28\alpha - 8 = 0$

$\therefore \quad \alpha = \frac{2}{7}$ stationary point

2.79 (b)

Given, scalar field,

$f(x, y, z) = xyz$

$\overline{\nabla f} = \hat{i}f_x + \hat{j}f_y + \hat{k}f_z$

Gradient of scalar field,

$= \hat{i}(yz) + \hat{j}(xz) + \hat{k}(xy)$

At point $P(0, 2, 1)$

$\overline{\nabla f}_{(0,2,1)} = \hat{i}(2) + 0\hat{j} + 0\hat{k}$

Directional derivative,

$D.D = \overline{\nabla f}\cdot\frac{\bar{a}}{|\bar{a}|}$

$$= \left(2\hat{i} + 0\hat{j} + 0\hat{k}\right)\cdot\frac{\left(2\hat{i} + \hat{j} + 2\hat{k}\right)}{\sqrt{2^2 + 1^2 + 2^2}} = \frac{4}{\sqrt{9}} = \frac{4}{3}$$

2.80 (b)

Given line integral, $\int_P^Q z^2dx + 3y^2dy + 2xzdz$ along the line joining the points $P(1, 1, 2)$ and $Q(2, 3, 1)$ is

$$= \int_{P(1,2)}^{P(2,1)} z^2dx + 2xydz + \int_{y=1}^{3} 3y^2dy$$

$$= \left(xz^2\right)_{(1,2)}^{(2,1)} + \left(y^3\right)_1^3$$

$= (2 \times 1^2 - 1 \times 2^2) + (3^3 - 1^3)$

$= -2 + 26 = 24$

2.81 (0)

Given, integral,

$$I = \iint_R xy\,dx\,dy$$

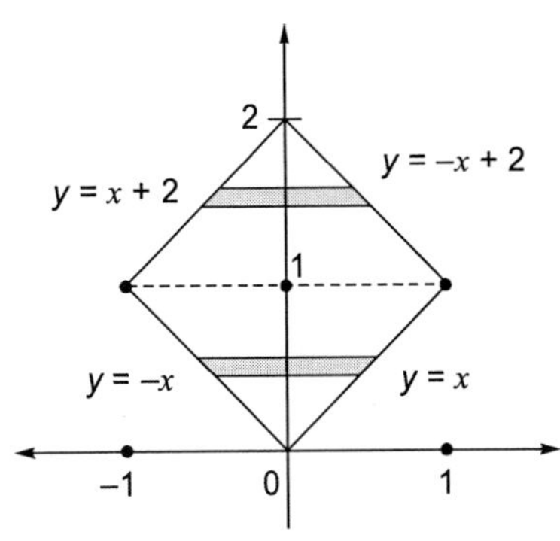

$$= \int_{y=0}^{1}\int_{x=-y}^{y} xy\,dx\,dy + \int_{y=1}^{2}\int_{x=y-2}^{2-y} xy\,dx\,dy$$

$$= \int_0^1 y\left(\frac{x^2}{2}\right)_{-y}^{y} dy + \int_1^2 y\left(\frac{x^2}{2}\right)_{y-2}^{2-y} dy$$

$$= 0 + 0 = 0$$

2.82 (b, d)

Considering the given equation,

$$\frac{\partial \rho}{\partial t} + \nabla\cdot(\rho u) = 0$$

$$\Rightarrow \quad \frac{\partial \rho}{\partial t} = -\nabla\cdot(\rho u)$$

Integrating both sides over the volume V, we get,

$$\int_V \frac{\partial \rho}{\partial t}\cdot dV = -\int_V \nabla\cdot(\rho u)\,dV \quad \text{...(i)}$$

According to Gauss Divergence Theorem, if $\vec{F}$ is a differentiate vector function and V is the volume bounded by a closed surface S, we have

$$\oint \vec{F}\cdot\hat{n}ds = \int_V \nabla\cdot\vec{F}\,dV$$

Using above relation in equation (i), we get,

$$\int_V \frac{\partial \rho}{\partial t}\cdot dV = -\oint_S \rho u\cdot\hat{n}ds$$

Hence, options (b) and (d) are correct.

2.83 (a, c, d)

∵ Given integral is independent of contour 'C'

$$\int_C (F_1dx + F_2dy + F_3dz) = \int_C \overline{F}\cdot\overline{dr} \text{ is independent of '}C\text{'}$$

Hence, $\overline{F}$ is conservative field

$$\therefore \quad \overline{F} = \overline{\nabla}\Phi$$

where, Φ is a differentiable scalar function.

$$\Rightarrow \quad F_1 = \frac{\partial \phi}{\partial x},\ F_2 = \frac{\partial \phi}{\partial y},\ F_3 = \frac{\partial \phi}{\partial z}$$

∴ Option (a) is true.

∵ $\overline{F}$ is conservative

Then, $\overline{\nabla}\times\overline{F} = 0$

$$\Rightarrow \begin{vmatrix} \hat{i} & \hat{j} & \hat{k} \\ \frac{\partial}{\partial x} & \frac{\partial}{\partial y} & \frac{\partial}{\partial z} \\ F_1 & F_2 & F_3 \end{vmatrix} = \vec{0}$$

$$= \hat{i}\left(\frac{\partial F_3}{\partial y} - \frac{\partial F_2}{\partial z}\right) - \hat{j}\left(\frac{\partial F_3}{\partial x} - \frac{\partial F_1}{\partial z}\right) + \hat{k}\left(\frac{\partial F_2}{\partial x} - \frac{\partial F_1}{\partial y}\right)$$

$$= 0\hat{i} + 0\hat{j} + 0\hat{k}$$

Equating both sides,

$$\text{We get,} \quad \frac{\partial F_3}{\partial y} = \frac{\partial F_2}{\partial z},\ \frac{\partial F_3}{\partial x} = \frac{\partial F_1}{\partial z},\ \frac{\partial F_2}{\partial x} = \frac{\partial F_1}{\partial y}$$

∴ Option (c) is true.

If $\overline{F}$ is irrotational then, using Stoke's theorem

$$\oint_C F_1\,dx + F_2\,dy + F_3\,dz = \iint(\overline{\nabla}\times\overline{F})ds = 0$$

∴ Option (d) is also true.

But, $\overline{\nabla}\cdot\overline{F} = \overline{\nabla}\cdot(\overline{\nabla}\phi) \neq 0$

∴ Option (b) is false.

■■■

3 Differential Equations

3.1 The following differential equation has

$$3\left(\frac{d^2y}{dt^2}\right)+4\left(\frac{dy}{dt}\right)^3+y^2+2=x$$

(a) degree = 2, order = 1
(b) degree = 1, order = 2
(c) degree = 4, order = 3
(d) degree = 2, order = 3 **[2005 : 1 M]**

3.2 A solution of the following differential equation is given by $\frac{d^2y}{dx^2}-5\frac{dy}{dx}+6y=0$

(a) $y=e^{2x}+e^{-3x}$ (b) $y=e^{2x}+e^{3x}$
(c) $y=e^{-2x}+e^{3x}$ (d) $y=e^{-2x}+e^{-3x}$

[2005 : 1 M]

3.3 A solution for the differential equation
$\dot{x}(t)+2x(t)=\delta(t)$
with initial condition $x(0^-)=0$ is

(a) $e^{-2t}u(t)$ (b) $e^{2t}u(t)$
(c) $e^{-t}u(t)$ (d) $e^{t}u(t)$ **[2006 : 1 M]**

3.4 For the differential equation $\frac{d^2y}{dx^2}+k^2y=0$ the boundary conditions are

(i) $y=0$ for $x=0$ and (ii) $y=0$ for $x=a$

The form of non-zero solutions of y (where m varies over all integers) are

(a) $y=\sum_m A_m \sin\frac{m\pi x}{a}$

(b) $y=\sum_m A_m \cos\frac{m\pi x}{a}$

(c) $y=\sum_m A_m x^{\frac{m\pi}{a}}$

(d) $y=\sum_m A_m e^{-\frac{m\pi x}{a}}$ **[2006 : 2 M]**

3.5 The solution of the differential equation $k^2\frac{d^2y}{dx^2}=y-y_2$ under the boundary conditions

(i) $y=y_1$ at $x=0$ and
(ii) $y=y_2$ at $x=\infty$, where k, y_1 and y_2 are constants, is

(a) $y=(y_1-y_2)\exp(-x/k^2)+y_2$
(b) $y=(y_2-y_1)\exp(-x/k)+y_1$
(c) $y=(y_1-y_2)\sinh(x/k)+y_1$
(d) $y=(y_1-y_2)\exp(-x/k)+y_2$

[2007 : 2 M

3.6 Which of the following is a solution to the differential equation $\frac{dx(t)}{dt}+3x(t)=0$?

(a) $x(t)=3e^{-t}$ (b) $x(t)=2e^{-3t}$
(c) $x(t)=-\frac{3}{2}t^2$ (d) $x(t)=3t^2$

[2008 : 1 M]

3.7 The order of the differential equation

$$\frac{d^2y}{dt^2}+\left(\frac{dy}{dt}\right)^3+y^4=e^{-t} \text{ is}$$

(a) 1 (b) 2
(c) 3 (d) 4 **[2009 : 1 M]**

3.8 Match each differential equation in Group I to its family of solution curves from Group II

Group I	Group II
A. $\frac{dy}{dx}=\frac{y}{x}$	1. Circles
B. $\frac{dy}{dx}=-\frac{y}{x}$	2. Straight lines
C. $\frac{dy}{dx}=\frac{x}{y}$	3. Hyperbolas
D. $\frac{dy}{dx}=-\frac{x}{y}$	

(a) A-2, B-3, C-3, D-1
(b) A-1, B-3, C-2, D-1
(c) A-2, B-1, C-3, D-3
(d) A-3, B-2, C-1, D-2 **[2009 : 2 M]**

3.9 A function $n(x)$ satisfies the differential equation $\frac{d^2n(x)}{dx^2} - \frac{n(x)}{L^2} = 0$ where L is a constant. The boundary conditions are : $n(0) = K$ and $n(\infty) = 0$. The solution to this equation is

(a) $n(x) = K\exp(x/L)$

(b) $n(x) = K\exp\left(-x/\sqrt{L}\right)$

(c) $n(x) = K^2\exp(-x/L)$

(d) $n(x) = K\exp(-x/L)$ [2010 : 1 M]

3.10 The solution of the differential equation $\frac{dy}{dx} = ky$, $y(0) = c$ is

(a) $x = ce^{-ky}$ (b) $x = ke^{cy}$

(c) $y = ce^{kx}$ (d) $y = ce^{-kx}$

[2011 : 1 M]

3.11 With initial condition $x(1) = 0.5$, the solution of the differential equation, $t\frac{dx}{dt} + x = t$ is

(a) $x = t - \frac{1}{2}$ (b) $x = t^2 - \frac{1}{2}$

(c) $x = \frac{t^2}{2}$ (d) $x = \frac{t}{2}$ [2012 : 1 M]

3.12 If the characteristic equation of the differential equation

$$\frac{d^2y}{dx^2} + 2\alpha\frac{dy}{dx} + y = 0$$

has two equal roots, then the values of α are

(a) ± 1 (b) 0, 0

(c) $\pm j$ (d) $\pm 1/2$

[2014 : 1 M, Set-2]

3.13 Which ONE of the following is a linear non-homogeneous differential equation, where x and y are the independent and dependent variables respectively?

(a) $\frac{dy}{dx} + xy = e^{-x}$ (b) $\frac{dy}{dx} + xy = 0$

(c) $\frac{dy}{dx} + xy = e^{-y}$ (d) $\frac{dy}{dx} + e^{-y} = 0$

[2014 : 1 M, Set-3]

3.14 If a and b are constants, the most general solution of the differential equation

$\frac{d^2x}{dt^2} + 2\frac{dx}{dt} + x = 0$ is

(a) ae^{-t} (b) $ae^{-t} + bte^{-t}$

(c) $ae^{t} + bte^{-t}$ (d) ae^{-2t}

[2014 : 1 M, Set-4]

3.15 With initial values $y(0) = y'(0) = 1$, the solution of the differential equation

$$\frac{d^2y}{dx^2} + 4\frac{dy}{dx} + 4y = 0$$

at $x = 1$ is _____.

[2014 : 2 M, Set-4]

3.16 The solution of the differential equation $\frac{d^2y}{dt^2} + 2\frac{dy}{dt} + y = 0$ with $y(0) = y'(0) = 1$ is

(a) $(2 - t)e^{t}$ (b) $(1 + 2t)e^{-t}$

(c) $(2 + t)e^{-t}$ (d) $(1 - 2t)e^{t}$

[2015 : 2 M, Set-1]

3.17 The general solution of the differential equation

$\frac{dy}{dx} = \frac{1+\cos 2y}{1-\cos 2x}$ is

(a) $\tan y - \cot x = c$ (c is a constant)

(b) $\tan x - \cot y = c$ (c is a constant)

(c) $\tan y + \cot x = c$ (c is a constant)

(d) $\tan x + \cot y = c$ (c is a constant)

[2015 : 1 M, Set-2]

3.18 Consider the differential equation

$\frac{d^2x(t)}{dt^2} + 3\frac{dx(t)}{dt} + 2x(t) = 0$. Given $x(0) = 20$ and $x(1) = 10/e$, where $e = 2.718$, the value of $x(2)$ is_____.

[2015 : 2 M, Set-3]

3.19 The particular solution of the initial value problem given below is

$\frac{d^2y}{dx^2} + 12\frac{dy}{dx} + 36y = 0$ with $y(0) = 3$ and

$\left.\frac{dy}{dx}\right|_{x=0} = -36$

(a) $(3 - 18x)e^{-6x}$ (b) $(3 + 25x)e^{-6x}$

(c) $(3 + 20x)e^{-6x}$ (d) $(3 - 12x)e^{-6x}$

[2016 : 2 M, Set-3]

3.20 Which one of the following is the general solution of the first order differential equation

$$\frac{dy}{dx} = (x + y - 1)^2,$$

where x, y are real?

(a) $y = 1 + x + \tan^{-1}(x + c)$, where c is a constant.
(b) $y = 1 + x + \tan(x + c)$, where c is a constant.
(c) $y = 1 - x + \tan^{-1}(x + c)$, where c is a constant.
(d) $y = 1 - x + \tan(x + c)$, where c is a constant.

[2017 : 2 M, Set-1]

3.21 The general solution of the differential equation

$$\frac{d^2y}{dx^2} + 2\frac{dy}{dx} - 5y = 0$$

in terms of arbitrary constants K_1 and K_2 is

(a) $K_1 e^{(-1+\sqrt{6})x} + K_2 e^{(-1-\sqrt{6})x}$

(b) $K_1 e^{(-1+\sqrt{8})x} + K_2 e^{(-1-\sqrt{8})x}$

(c) $K_1 e^{(-2+\sqrt{6})x} + K_2 e^{(-2-\sqrt{6})x}$

(d) $K_1 e^{(-2+\sqrt{8})x} + K_2 e^{(-2-\sqrt{8})x}$

[2017 : 1 M, Set-2]

3.22 The families of curves represented by the solution of the equation

$$\frac{dy}{dx} = -\left(\frac{x}{y}\right)^n$$

for $n = -1$ and $n = +1$, respectively, are
(a) Hyperbolas and Parabolas
(b) Hyperbolas and Circles
(c) Parabolas and Circles
(d) Circles and Hyperbolas

[2019 : 1 M]

3.23 Consider the homogeneous ordinary differential equation

$$x^2\frac{d^2y}{dx^2} - 3x\frac{dy}{dx} + 3y = 0, \quad x > 0$$

with $y(x)$ as a general solution. Given that

$$y(1) = 1 \text{ and } y(2) = 14$$

the value of $y(1.5)$, rounded off to two decimal places, is ______.

[2019 : 2 M]

3.24 The general solution of $\frac{d^2y}{dx^2} - 6\frac{dy}{dx} + 9y = 0$ is

(a) $y = C_1e^{3x} + C_2e^{-3x}$
(b) $y = (C_1 + C_2x)e^{3x}$
(c) $y = C_1e^{3x}$
(d) $y = (C_1 + C_2x)e^{-3x}$

[2020 : 1 M]

3.25 Which one of the following options contains two solutions of the differential equation $\frac{dy}{dx} = (y - 1)x$?

(a) $\ln|y - 1| = 0.5x^2 + C$ and $y = -1$
(b) $\ln|y - 1| = 2x^2 + C$ and $y = 1$
(c) $\ln|y - 1| = 2x^2 + C$ and $y = -1$
(d) $\ln|y - 1| = 0.5x^2 + C$ and $y = 1$

[2020 : 2 M]

3.26 Consider the differential equation given below :

$$\frac{dy}{dx} + \frac{x}{1 - x^2}y = x\sqrt{y}$$

The integrating factor of the differential equation is

(a) $(1 - x^2)^{-1/4}$
(b) $(1 - x^2)^{-3/4}$
(c) $(1 - x^2)^{-3/2}$
(d) $(1 - x^2)^{-1/2}$

[2021 : 1 M]

3.27 Consider the following partial differential equation (PDE)

$$a\frac{\partial^2 f(x, y)}{\partial x^2} + b\frac{\partial^2 f(x, y)}{\partial y^2} = f(x, y),$$

where a and b are distinct positive real numbers. Select the combination(s) of values of the real parameters ξ and η such that $f(x, y) = e^{(\xi x + \eta y)}$ is a solution of the given PDE.

(a) $\xi = \frac{1}{\sqrt{2a}}, \eta \frac{1}{\sqrt{2b}}$
(b) $\xi = \frac{1}{\sqrt{a}}, \eta = 0$
(c) $\xi = 0, \eta = 0$
(d) $\xi = \frac{1}{\sqrt{a}}, \eta = \frac{1}{\sqrt{b}}$

[2022 : 1 M]

3.28 The general form of the complementary function of a differential equation is given by $y(t) = (At + B)e^{-2t}$, where A and B are real constants determined by the initial condition. The corresponding differential equation is ______.

(a) $\frac{d^2y}{dt^2} + 5\frac{dy}{dt} + 6y = f(t)$

(b) $\frac{d^2y}{dt^2} + 4\frac{dy}{dt} + 4y = f(t)$

(c) $\frac{d^2y}{dt^2} + 4y = f(t)$

(d) $\frac{d^2y}{dt^2} + 3\frac{dy}{dt} + 2y = f(t)$

[2024 : 1 Mark]

■■■

Answers Differential Equations

3.1	(b)	3.2	(b)	3.3	(a)	3.4	(a)	3.5	(d)	3.6	(b)	3.7	(b)
3.8	(a)	3.9	(d)	3.10	(c)	3.11	(d)	3.12	(a)	3.13	(a)	3.14	(b)
3.15	(0.947)	3.16	(b)	3.17	(c)	3.18	(0.8566)	3.19	(a)	3.20	(d)	3.21	(a)
3.22	(b)	3.23	(5.25)	3.24	(b)	3.25	(d)	3.26	(a)	3.27	(a, b)	3.28	(b)

Explanations Differential Equations

3.1 (b)

Order is highest derivative term. Degree is power of highest derivative term.

3.2 (b)

$$AE \Rightarrow \quad D^2 - 5D + 6 = 0$$
$$(D-2)(D-3) = 0$$
$$D = 2, 3$$
$$\therefore \quad y = e^{2x} + e^{3x}$$

3.3 (a)

$$\dot{x}(t) + 2x(t) = \delta(t)$$

Taking L.T. on both sides

$$sX(s) - x(0) + 2X(s) = 1$$
$$X(s)[s+2] = 1$$
$$X(s) = \frac{1}{s+2}$$
$$x(t) = e^{-2t}u(t)$$

3.4 (a)

$$D^2 + K^2 = 0$$
$$D = \pm jK$$
$$\Rightarrow \quad y = A\cos Kx + jB\sin Kx$$

At $x = 0, y = 0$

$$\Rightarrow \quad A = 0, y = jB\sin Kx$$
$$x = a, y = 0$$
$$\Rightarrow \quad 0 = B\sin Ka$$

$\Rightarrow \quad B \neq 0$ else $y = 0$ always

$$\sin Ka = 0$$
$$K = \frac{m\pi x}{a}$$
$$y = \sum_m A_m \sin\left(\frac{m\pi x}{a}\right)$$

3.5 (d)

$$K^2 \frac{d^2y}{dx^2} = y - y_2$$
$$\Rightarrow \frac{d^2y}{dx^2} - \frac{y}{K^2} + \frac{y_2}{K^2} = 0$$
$$m^2 - \frac{1}{K^2} = 0$$
$$m = \pm\frac{1}{K}$$

Complementary function

$$\Rightarrow \quad y = C_1 e^{\frac{x}{K}} + C_2 e^{-\frac{x}{K}} + y_2 \quad \text{...(i)}$$
$$x = 0; y = y_1 = C_1 + C_2 + y_2 \quad \text{...(ii)}$$
$$x = \infty; y = y_2 = C_1 e^{\infty} + 0 + y_2$$

$\Rightarrow \quad C_1 = 0$ and $C_2 = y_1 - y_2$

$$\therefore \quad y = (y_1 - y_2)e^{-x/K} + y_2$$

3.6 (b)

$$(D+3)x(t) = 0$$
$$\Rightarrow \quad D = -3$$

So, $\quad x(t) = Ce^{-3t}$

3.7 (b)

Highest derivative of differential equation is 2.

3.8 (a)

A. $\quad \dfrac{dy}{dx} = \dfrac{y}{x}$

$$\frac{dy}{y} = \frac{dx}{x} \Rightarrow \int\frac{dy}{y} = \int\frac{dx}{x}$$
$$\log y = \log x + \log c$$
$$= \log cx$$

$y = cx \quad$ ... Equation of straight line.

B. $\quad \dfrac{dy}{dx} = \dfrac{-y}{x}$

$$\frac{dy}{y} = \frac{-dx}{x} \Rightarrow \int\frac{dy}{y} = -\int\frac{dx}{x}$$
$$\log y = -\log x + \log c$$
$$\log y + \log x = \log c$$
$$\log yx = \log c$$
$$yx = c$$

$$y = c/x$$ ... Equation of hyperbola.

C. $\frac{dy}{dx} = \frac{x}{y}$, $y\,dy = x\,dx$

$$\Rightarrow \int y\,dy = \int x\,dx$$

$$\frac{y^2}{2} - \frac{x^2}{2} = \frac{c^2}{2} \rightarrow \text{const}$$

$$y^2 - x^2 = c^2$$

$$\frac{y^2}{c^2} - \frac{x^2}{c^2} = 1$$ ... Equation of hyperbola.

D. $\frac{dy}{dx} = \frac{-x}{y}$

$$\Rightarrow \int y\,dy = -\int x\,dx$$

$$\frac{y^2}{2} = -\frac{x^2}{2} + \frac{c^2}{2}$$

$$\frac{y^2}{2} + \frac{x^2}{2} = \frac{c^2}{2}$$

$$x^2 + y^2 = c^2$$ ... Equation of circle.

3.9 (d)

$$\left(D^2 - \frac{1}{L^2}\right) n(x) = 0$$

$$D^2 - \frac{1}{L^2} = 0$$

$$D = \pm\frac{1}{L}$$

So, the solution is

$$n(x) = C_1 e^{x/L} + C_2 e^{-x/L}$$

$$n(0) = K \Rightarrow C_1 + C_2 = K$$

$$n(\infty) = 0 \Rightarrow C_1 = 0$$

$$\Rightarrow C_2 = K$$

Therefore, $n(x) = K \exp(-x/L)$

3.10 (c)

$$\frac{dy}{dx} = ky \quad \Rightarrow \quad \frac{dy}{y} = kdx$$

Integrating both sides

$$\ln y = kx + A$$

Put, $x = 0$

$$\ln y(0) = A$$

$$\therefore \quad A = \ln c \qquad [\because y(0) = c]$$

Hence,

$$\ln y = kx + \ln c$$

$$\Rightarrow \ln y - \ln c = kx$$

$$\Rightarrow \ln\left(\frac{y}{c}\right) = kx$$

$$\Rightarrow \frac{y}{c} = e^{kx}$$

or $y = c\,e^{kx}$

3.11 (d)

The given differential equation is

$t\frac{dx}{dt} + x = t$ with initial condition $x(1) = \frac{1}{2}$ which is same as

$$\frac{dx}{dt} + \frac{x}{t} = 1$$

which is a linear differential equation

$$\frac{dx}{dt} + Px = Q$$

Where, $P = \frac{1}{t}$ and $Q = 1$

Integrating factor $= e^{\int P dt} = e^{\int \frac{1}{t} dt}$

$$= e^{\log_e t} = t$$

Solution is

$$x \cdot (IF) = \int Q \cdot (IF) dt + C$$

$$x \cdot t = \int 1 \cdot t \cdot dt + C$$

$$xt = \frac{t^2}{2} + C$$

$$x = \frac{t}{2} + \frac{C}{t}$$

Put, $x(1) = \frac{1}{2}$

$$\Rightarrow \frac{1}{2} + \frac{C}{1} = \frac{1}{2}$$

$$\Rightarrow C = 0$$

So, $\boxed{x = \frac{t}{2}}$ is the solution.

3.12 (a)

$$\frac{d^2y}{dx^2} + 2\alpha\frac{dy}{dx} + y = 0$$

The characteristic equation is given as

$(m^2 + 2\alpha m + 1) = 0$

$$m_1, m_2 = \frac{-2\alpha \pm \sqrt{4\alpha^2 - 4}}{2}$$

Since both roots are equal i.e.

$$m_1 = m_2$$

$$\frac{-2\alpha + \sqrt{4\alpha^2 - 4}}{2} = \frac{-2\alpha - \sqrt{4\alpha^2 - 4}}{2}$$

$$\sqrt{4\alpha^2 - 4} = -\sqrt{4\alpha^2 - 4}$$
$$2\sqrt{4\alpha^2 - 4} = 0$$
$$4\alpha^2 - 4 = 0$$
$$\alpha^2 = 1$$
$$\alpha = \pm 1$$

3.13 (a)

General form of linear differential equation

$$\frac{dy}{dx} + py = \theta \text{ when } P \text{ and } \theta \text{ can be function of } x.$$

Only option (a) is in this form.

3.14 (b)

The differential equation is given as

$$\frac{d^2x}{dt^2} + 2\frac{dx}{dt} + x = 0$$
$$y = C \cdot F + P \cdot I$$

Since, $Q = 0$, i.e. RHS term is zero, so there will be no particular integral.

$\therefore \quad y = C \cdot F$

Let, $\quad \frac{\partial}{\partial x} = D$

So, $(D^2 + 2D + 1)x = 0$

$\therefore \quad (D + 1)^2 = 0$

$\therefore \quad y = ae^{-t} + bte^{-t}$

3.15 (0.947)

Given,

$y(0) = y'(0) = 1$

$$\frac{d^2y}{dx^2} + \frac{4dy}{dx} + 4y = 0 \quad \text{...(i)}$$
$$D^2 + 4D + 4 = 0$$
$$D = -2, -2$$
$$y = C_1 e^{-2x} + C_2 x e^{-2x}$$
$$y(0) = C_1 = 1$$
$$y' = -2e^{-2x} + C_2 e^{-2x} + C_2(-2)x e^{-2x}$$
$$y'(0) = -2 + C_2 = 1$$
$$\Rightarrow \quad C_2 = 3$$
$$y = e^{-2x} + 3x e^{-2x}$$
$$y(1) = e^{-2} + 3 \times 2e^{-2}$$
$$= 0.947$$

3.16 (b)

$$(D^2 + 2D + 1) = 0$$
$$D = -1, -1$$
$$y(t) = (C_1 + C_2 t)\, e^{-t}$$
$$y'(t) = C_2 e^{-t} + (C_1 + C_2 t)(-e^{-t})$$
$$y(0) = y'(0) = 1$$

From here, $C_1 = 1$, $C_2 = 2$

$\Rightarrow \quad y(t) = (1 + 2t)e^{-t}$

3.17 (c)

$$\frac{dy}{1+\cos 2y} = \frac{dx}{1-\cos 2x}$$
$$\frac{dy}{2\cos^2 y} = \frac{dx}{2\sin^2 x}$$
$$\sec^2 y\, dy = \text{cosec}^2 x\, dx$$

Integrating both sides, we get

$$\tan y = -\cot x + c$$
$$\tan y + \cot x = c$$

3.18 (0.8566)

$$D^2 + 3D + 2 = 0$$
$$D = -1, -2$$
$$x(t) = C_1 e^{-t} + C_2 e^{-2t}$$
$$x(1) = \frac{10}{e} = C_1 e^{-1} + C_2 e^{-2}$$
$$C_1 + C_2 e^{-1} = 10 \quad \text{...(i)}$$
$$C_1 + C_2 = 20 \quad \text{...(ii)}$$

From here, $C_1 = \frac{10e - 20}{e - 1}$; $C_2 = \left(\frac{10e}{e-1}\right)$

$$x(2) = \left(\frac{10e - 20}{e - 1}\right)e^{-2} + \left(\frac{10e}{e-1}\right)e^{-4}$$
$$= 0.8566$$

3.19 (a)

$$(D^2 + 12D + 36) = 0$$
$$(D + 6)^2 y = 0$$
$$D = -6, -6$$
$$y = (C_1 + xC_2)e^{-6x}$$
$$y = C_1 e^{-6x} + C_2 x e^{-6x}$$
$$y(0) = 3$$
$$= C_1 + 0$$
$$\Rightarrow \quad C_1 = 3$$
$$y' = -6C_1 e^{-6x} + C_2 e^{-6x} - 6C_2 x e^{-6x}$$
$$y'(0) = -36$$
$$-36 = -6C_1 + C_2$$
$$-36 = -18 + C_2$$
$$C_2 = -18$$
$$\therefore \quad y = 3e^{-6x} - 18x e^{-6x}$$
$$y = (3 - 18x)e^{-6x}$$

3.20 (d)

$$\frac{dy}{dx} = (x + y - 1)^2 \quad \text{...(i)}$$

Let, $x + y - 1 = t$ ...(ii)

$$1+\frac{dy}{dx} = \frac{dt}{dx}$$

$$\frac{dy}{dx} = \frac{dt}{dx}-1 \quad ...(iii)$$

Substituting equations (ii) and (iii) in equation (i)

$$\frac{dt}{dx}-1 = t^2$$

$$\frac{dt}{t^2+1} = dx$$

Integrating both side

$$\int\frac{1}{t^2+1}dt = \int dx$$

$$\tan^{-1} t = x + c$$

Since, $t = x + y - 1$

$\therefore$ $\tan^{-1}(x + y - 1) = x + c$

$$x + y - 1 = \tan(x + c)$$

$$y = 1 - x + \tan(x + c)$$

3.21 (a)

The general solution of the differential equation

$$\frac{d^2y}{dx^2}+2\frac{dy}{dx}-5y = 0$$

$(D^2 + 2D - 5)y = 0$

Auxiliary equation is

$m^2 + 2m - 5 = 0$

$$m = \frac{-2\pm\sqrt{4+20}}{2} = \frac{-2\pm\sqrt{24}}{2}$$

$$= \frac{-2\pm 2\sqrt{6}}{2} = -1\pm\sqrt{6}$$

Solution is $y = K_1e^{(-1+\sqrt{6})x}+K_2e^{(-1-\sqrt{6})x}$

3.22 (b)

$$\frac{dy}{dx} = -\left(\frac{x}{y}\right)^n$$

$n = -1$, $\dfrac{dy}{dx} = -\dfrac{y}{x}$

$$\frac{dy}{y} = -\frac{dx}{x}$$

$$\int\frac{1}{y}dy = -\int\frac{1}{x}dx$$

$$\ln y = -\ln x + \ln c$$

$$\ln(yx) = \ln c$$

$xy = c$ (Represents rectangular hyperbola)

$n = 1$, $\dfrac{dy}{dx} = -\dfrac{x}{y}$

$$ydy = -xdx$$

$$\int ydy = -\int xdx$$

$$\frac{y^2}{2} = -\frac{x^2}{2}+c$$

$x^2 + y^2 = 2c$ (Represents family of circles)

3.23 (5.25)

$$(x^2D^2 - 3xD + 3) = 0$$

$$(\theta(\theta-1)-3\theta+3)y = 0$$

$$(\theta^2-4\theta+3)y = 0$$

AE is $m^2 - 4m + 3 = 0$

$m = 1, 3$

$CF = C_1e^z + C_2e^{3z}$

Solution is $y = C_1x + C_2x^3$ ...(i)

$y(1) = 1$ $\quad 1 = C_1 + C_2$ ...(ii)

$y(2) = 14$ $\quad 14 = 2C_1 + 8C_2$ ...(iii)

From (ii) and (iii), $C_1 = -1$, $C_2 = 2$

$\therefore$ $y = -x + 2x^3$

$y(1.5) = -1.5 + 2(1.5)^3 = 5.25$

3.24 (b)

Taking $\dfrac{d}{dx} = D$

Given, $D^2 - 6D + 9 = 0$

$(D - 3)^2 = 0$

$D = 3, 3$

So, solution of the given differential equation

$$y = (c_1 + c_2x)\,e^{3x}$$

3.25 (d)

Given differential equation: $\dfrac{dy}{dx} = (y-1)x$

By variable separable method:

$$\int\frac{dy}{y-1} = \int xdx$$

$$\ln|y-1| = \frac{x^2}{2}+c \quad (\text{where } y\neq 1)$$

and the second solution is for $y = 1$.

3.26 (a)

$$\frac{dy}{dx}+\frac{x}{1-x^2}y = x\sqrt{y}$$

Divided by $\sqrt{y}$

$$\frac{1}{\sqrt{y}}\frac{dy}{dx}+\frac{x}{1-x^2}\sqrt{y} = x$$

Put, $\sqrt{y}=u$

$$\frac{1}{2\sqrt{y}}\frac{dy}{dx} = \frac{du}{dx}$$

$$2\frac{du}{dx}+\frac{x}{1-x^2}u = x$$

$$\Rightarrow \quad \frac{du}{dx}+\frac{x}{2(1-x^2)}u = \frac{x}{2}$$

It is a linear differential equation. Hence,

$$\text{I.F} = e^{\int\frac{x}{2(1-x^2)}dx} = e^{-\frac{1}{4}\log(1-x^2)} = e^{\log(1-x^2)^{\frac{-1}{4}}}$$

$$\text{I.F} = \frac{1}{(1-x^2)^{\frac{1}{4}}}$$

3.27 (a, b)

$$f(x, y) = e^{(\xi x+\eta y)}$$

Differentiating $f(x, y)$ two times w.r.t. x, we get,

$$\frac{\partial^2 f(x,y)}{\partial x^2} = \xi^2 e^{(\xi x+\eta y)} \quad \text{...(i)}$$

Differentiating $f(x, y)$ two times w.r.t. y, we get,

$$\frac{\partial^2 f(x,y)}{\partial y^2} = \eta^2 e^{(\xi x+\eta y)} \quad \text{...(ii)}$$

Substituting (i) and (ii) in the given differential equation,

$$a\frac{\partial^2 f(x,y)}{\partial x^2}+b\frac{\partial^2 f(x,y)}{\partial y^2} = f(x, y)$$

$$\Rightarrow \quad a\xi^2 e^{(\xi x+\eta y)}+b\eta^2 e^{(\xi x+\eta y)} = e^{(\xi x+\eta y)}$$

$$\Rightarrow \quad a\xi^2+b\eta^2 = 1$$

Option (a) : For $\xi=\frac{1}{\sqrt{2a}},\ \eta=\frac{1}{\sqrt{2b}}$

$$a\xi^2+b\eta^2 = a\times\frac{1}{2a}+b\times\frac{1}{2b}=1$$

Hence, option (a) is correct.

Option (b) : For $\xi=\frac{1}{\sqrt{a}},\ \eta=0$

$$a\xi^2+b\eta^2 = a\times\frac{1}{a}=1$$

Hence, option (b) is correct.

Option (c) : For $\xi = 0, \eta = 0$

$$a\xi^2+b\eta^2 = 0 \neq 1$$

Hence, option (c) is not correct.

Option (d) : For $\xi=\frac{1}{\sqrt{a}},\ \eta=\frac{1}{\sqrt{b}}$

$$a\xi^2+b\eta^2 = a\times\frac{1}{a}+b\times\frac{1}{b}=2\neq 1$$

Hence, option (d) is not correct.

Therefore, option (a) and option (b) are correct answers.

3.28 (b)

Given: Complementary function

$$= (At+B)e^{-2t} \approx (C_1+C_2t)e^{-2t}$$

i.e. roots of auxiliary equation are $m = -2, -2$

So, AE is $(m + 2)(m + 2) = 0$

$$m^2+4m+4 = 0$$

Replace $m \to D$ we get,

$(D^2+4D+4)y = 0$

i.e. the required differential equation is

$$\frac{d^2y}{dt^2}+4\frac{dy}{dt}+4y = f(t)$$

■■■

4 Complex Variables

4.1 For the function of a complex variable $W = ln\ Z$ (where, $W = u + jv$ and $Z = x + jy$, the u = constant lines get mapped in z-plane as
(a) set of radial straight lines
(b) set of concentric circles
(c) set of confocal hyperbolas
(d) set of confocal ellipses **[2006 : 2 M]**

4.2 The value of the contour integral $\oint_{|z-j|=2} \frac{1}{z^2+4} dz$ in positive sense is
(a) $j\pi/2$ (b) $-\pi/2$
(c) $-j\pi/2$ (d) $\pi/2$ **[2006 : 2 M]**

4.3 If the semicircular contour D of radius 2 is as shown in the figure, then the value of the integral $\oint_D \frac{1}{(s^2-1)} ds$ is

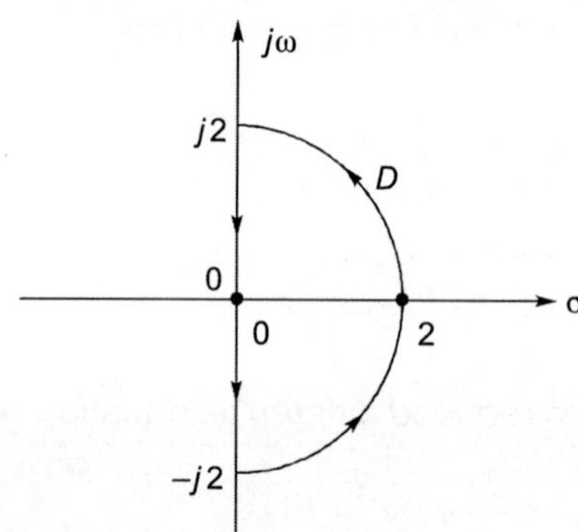

(a) $j\pi$ (b) $-j\pi$
(c) $-\pi$ (d) π **[2007 : 2 M**

4.4 The equation $\sin(z) = 10$ has
(a) no real or complex solution
(b) exactly two distinct complex solutions
(c) a unique solution
(d) an infinite number of complex solutions **[2008 : 1 M]**

4.5 The residue of the function
$f(z) = \frac{1}{(z+2)^2(z-2)^2}$ at $z = 2$ is
(a) $-\frac{1}{32}$ (b) $-\frac{1}{16}$
(c) $\frac{1}{16}$ (d) $\frac{1}{32}$ **[2008 : 2 M]**

4.6 If $f(z) = c_0 + c_1 z^{-1}$, then $\oint_{\text{unit circle}} \frac{1+f(z)}{z} dz$ is given by
(a) $2\pi c_1$ (b) $2\pi(1+c_0)$
(c) $2\pi j c_1$ (d) $2\pi j(1+c_0)$
[2009 : 1 M]

4.7 The residues of a complex function $X(z) = \frac{1-2z}{z(z-1)(z-2)}$ at its poles are
(a) $\frac{1}{2}, -\frac{1}{2}$ and 1 (b) $\frac{1}{2}, \frac{1}{2}$ and -1
(c) $\frac{1}{2}$, 1 and $-\frac{3}{2}$ (d) $\frac{1}{2}$, -1 and $\frac{3}{2}$
[2010 : 2 M]

4.8 The value of the integral $\oint_c \frac{-3z+4}{(z^2+4z+5)} dz$ where c is the circle $|z| = 1$ is given by
(a) 0 (b) 1/10
(c) 4/5 (d) 1 **[2011 : 1 M]**

4.9 If $x = \sqrt{-1}$, then the value of x^x is
(a) $e^{-\pi/2}$ (b) $e^{\pi/2}$
(c) x (d) 1 **[2012 : 1 M]**

4.10 Given $f(z) = \frac{1}{z+1} - \frac{2}{z+3}$.
If C is a counterclockwise path in the z-plane such that $|z + 1| = 1$, the value of $\frac{1}{2\pi j}\oint_C f(z)dz$ is
(a) -2 (b) -1
(c) 1 (d) 2
[2012 : 1 M]

4.11 C is a closed path in the z-plane given by $|z| = 3$. The value of the integral $\oint_c \left(\frac{z^2-z+4j}{z+2j}\right) dz$ is
(a) $-4\pi(1+j2)$ (b) $4\pi(3-j2)$
(c) $-4\pi(3+j2)$ (d) $4\pi(1-j2)$
[2014 : 1 M, Set-1]

4.12 The real part of an analytic function $f(z)$ where $z = x + jy$ is given by $e^{-y}\cos(x)$. The imaginary part of $f(z)$ is

(a) $e^{y}\cos(x)$
(b) $e^{-y}\sin(x)$
(c) $-e^{y}\sin(x)$
(d) $-e^{-y}\sin(x)$

[2014 : 2 M, Set-2]

4.13 Let $z = x + iy$ be a complex variable. Consider that contour integration is performed along the unit circle in anticlockwise direction. Which one of the following statements is **NOT TRUE**?

(a) The residue of $\dfrac{z}{z^2-1}$ at $z = 1$ is 1/2

(b) $\oint_C z^2 dz = 0$

(c) $\dfrac{1}{2\pi i}\oint_C \dfrac{1}{z}dz = 1$

(d) $\bar{z}$ (complex conjugate of z) is analytical function

[2015 : 1 M, Set-1]

4.14 Let $f(z) = \dfrac{az+b}{cz+d}$. If $f(z_1) = f(z_2)$ for all $z_1 \neq z_2$, $a = 2$, $b = 4$ and $c = 5$, then d should be equal to ______.

[2015 : 1 M, Set-2]

4.15 If C denotes the counterclockwise unit circle, the value of the contour integral

$$\frac{1}{2\pi j}\oint_C \text{Re}\{z\}dz \text{ is } \_\_\_\_\_.$$

[2015 : 2 M, Set-2]

4.16 If C is a circle of radius r with centre z_0, in the complex z-plane and if n is a non-zero integer, then $\oint \dfrac{dz}{(z-z_0)^{n+1}}$ equals

(a) $2\pi nj$
(b) 0
(c) $\dfrac{\pi j}{2\pi}$
(d) $2\pi n$

[2015 : 1 M, Set-3]

4.17 In the following integral, the contour C encloses the points $2\pi j$ and $-2\pi j$

$$-\frac{1}{2\pi}\oint_C \frac{\sin z}{(z-2\pi j)^3}dz$$

The value of the integral is _____

[2016 : 2 M, Set-1]

4.18 Consider the complex valued function $f(z) = 2z^3 + b|z|^3$ where z is a complex variable. The value of b for which the function $f(z)$ is analytic is _____

[2016 : 1 M, Set-2]

4.19 The values of the integral $\dfrac{1}{2\pi j}\oint_c \dfrac{e^z}{z-2}dz$ along a closed contour c in anti-clockwise direction for

(i) the point $z_0 = 2$ inside the contour c, and
(ii) the point $z_0 = 2$ outside the contour c, respectively, are

(a) (i) 2.72, (ii) 0
(b) (i) 7.39, (ii) 0
(c) (i) 0, (ii) 2.72
(d) (i) 0, (ii) 7.39

[2016 : 2 M, Set-3]

4.20 For $f(z) = \dfrac{\sin(z)}{z^2}$, the residue of the pole at $z = 0$ is ________.

[2016 : 1 M, Set-3]

4.21 The residues of a function $f(z) = \dfrac{1}{(z-4)(z+1)^3}$ are

(a) $\dfrac{-1}{27}$ and $\dfrac{-1}{125}$
(b) $\dfrac{1}{125}$ and $\dfrac{-1}{125}$
(c) $\dfrac{-1}{27}$ and $\dfrac{1}{5}$
(d) $\dfrac{1}{125}$ and $\dfrac{-1}{5}$

[2017 : 1 M, Set-2]

4.22 An integral I over a counter-clockwise circle C is given by

$$I = \oint_C \frac{z^2-1}{z^2+1}e^z\, dz.$$

If C is defined as $|z| = 3$, then the value of I is

(a) $-\pi i\sin(1)$
(b) $-2\pi i\sin(1)$
(c) $-3\pi i\sin(1)$
(d) $-4\pi i\sin(1)$

[2017 : 2 M, Set-2]

4.23 The contour C given below is on the complex plane $z = x + jy$, where $j = \sqrt{-1}$.

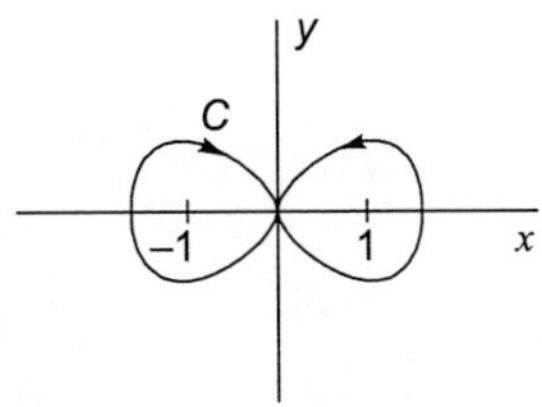

The value of the integral $\dfrac{1}{\pi j}\oint_C \dfrac{dz}{z^2-1}$ is ______.

[2018 : 2 M]

4.24 Which one of the following functions is analytic over the entire complex plane?

(a) $\cos(z)$ (b) $e^{1/z}$

(c) $\ln(z)$ (d) $\frac{1}{1-z}$ [2019 : 1 M]

4.25 The value of the contour integral $\frac{1}{2\pi j}\oint\left(z+\frac{1}{z}\right)^2 dz$ evaluated over the unit circle $|z| = 1$ is________.

[2019 : 1 M]

4.26 Consider the integral

$$\oint_C \frac{\sin(x)}{x^2(x^2+4)}dx$$

where C is a counter-clockwise oriented circle defined as $|x-i| = 2$. The value of the integral is

(a) $\frac{\pi}{8}\sin(2i)$ (b) $-\frac{\pi}{4}\sin(2i)$

(c) $\frac{\pi}{4}\sin(2i)$ (d) $-\frac{\pi}{8}\sin(2i)$

[2021 : 2 M]

4.27 Consider the following series:

$$\sum_{n=1}^{\infty}\frac{n^d}{c^n}$$

For which of the following combinations of c, d values does this series converge?

(a) $c = 1, d = -1$ (b) $c = 2, d = 1$

(c) $c = 0.5, d = -10$ (d) $c = 1, d = -2$

[2022 : 2 M]

4.28 A simple closed path C in the complex plane is shown in the figure. If

$$\oint_C \frac{2^z}{z^2-1}dz = -i\pi A,$$

where $i=\sqrt{-1}$, then the value of A is ______ (rounded off to two decimal places).

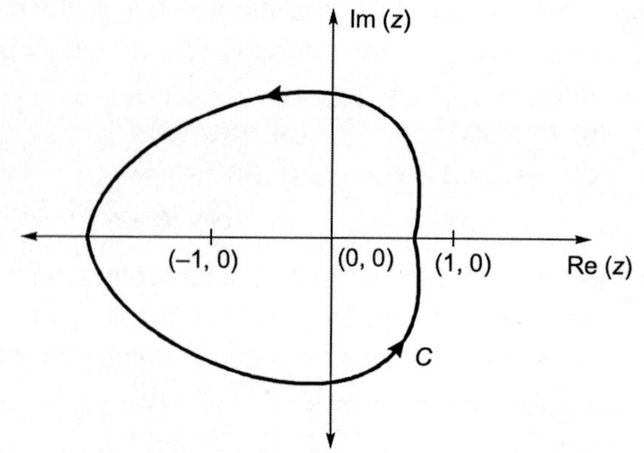

[2022 : 1 M]

4.29 Let $w^4 = 16j$. Which of the following cannot be a value of w?

(a) $2e^{\frac{j2\pi}{8}}$ (b) $2e^{\frac{j\pi}{8}}$

(c) $2e^{\frac{j5\pi}{8}}$ (d) $2e^{\frac{j9\pi}{8}}$ [2023 : 1 M]

4.30 The value of the contour integral,

$\oint_C\left(\frac{z+2}{z^2+2z+2}\right)dz$, where the contour C is $\left\{z:\left|z+1-\frac{3}{2}j\right|=1\right\}$, taken in the counter clockwise direction, is

(a) $-\pi(1+j)$ (b) $\pi(1+j)$

(c) $\pi(1-j)$ (d) $-\pi(1-j)$

[2023 : 1 M]

4.31 Let z be a complex variable. If $f(z) = \frac{\sin(\pi z)}{z^2(z-2)}$ and C is the circle in the complex plane with $|z| = 3$ then $\oint_C f(z)dz$ is ________.T

(a) $-\pi^2 j$ (b) $j\pi\left(\frac{1}{2}-\pi\right)$

(c) $j\pi\left(\frac{1}{2}+\pi\right)$ (d) $\pi^2 j$

[2024 : 2 Marks]

■■■

Answers Complex Variables

4.1	(b)	4.2	(d)	4.3	(a)	4.4	(a)	4.5	(a)	4.6	(d)	4.7	(c)
4.8	(a)	4.9	(a)	4.10	(c)	4.11	(c)	4.12	(b)	4.13	(d)	4.14	(10)
4.15	(0)	4.16	(b)	4.17	(–133.87)	4.18	(0)	4.19	(b)	4.20	(1)	4.21	(b)
4.22	(d)	4.23	(2)	4.24	(a)	4.25	(0)	4.26	(*)	4.27	(b, d)	4.28	(0.50)
4.29	(a)	4.30	(b)	4.31	(a)								

Explanations Complex Variables

4.1 (b)

$$W = lnz = \log_e z$$
$$u + jv = \log_e(x + jy)$$
$$= \frac{1}{2}\log(x^2 + y^2) + i\tan^{-1}\left(\frac{y}{x}\right)$$

u is constant,

$$\therefore \frac{1}{2}\log(x^2 + y^2) = C$$

$x^2 + y^2 = C$ [equation of circle having same centre (0, 0)]

4.2 (d)

$$\frac{1}{z^2+4} = \frac{1}{(z+2j)(z-2j)}$$

Only pole (0, 2j) lies inside the circle $|z - j| = 2$

$\therefore$ By Cauchy's integral formula,

$$I = \oint_{|z-j|=2} \frac{1}{z^2+4}dz = \frac{2\pi j}{2j+2j} = \frac{\pi}{2}$$

4.3 (a)

$$\oint \frac{1}{(s^2-1)}ds = \oint \frac{1}{(s+1)(s-1)}ds$$
$$= 2\pi j \times (\text{Sum of residue})$$

Pole $s = -1$ lies outside 'c'

residue at pole $s = 1$ is

$$= \lim_{s\to 1}\frac{(s-1)}{(s-1)(s+1)} = \frac{1}{2}$$

$$\Rightarrow \oint \frac{1}{(s^2-1)}ds = 2\pi j \times \frac{1}{2} = j\pi$$

4.4 (a)

$$\sin z = 10$$
$$\frac{e^{iz} - e^{-iz}}{2i} = 10$$
$$e^{iz} - \frac{1}{e^{iz}} = 20\,i$$
$$(e^{iz})^2 - 20i\,e^{iz} - 1 = 0$$
$$e^{iz} = \frac{20i \pm \sqrt{-396}}{2} = \frac{20i \pm 6i\sqrt{11}}{2}$$
$$e^{iz} = 10i \pm 3\sqrt{11}\,i$$
$$iz = ln\left[i(10 \pm 3\sqrt{11})\right] = lni + ln(10 \pm 3\sqrt{11})$$
$$iz = ln1 + i\left(\frac{\pi}{2} \pm 2n\pi\right) + ln(10 \pm 3\sqrt{11})$$
$$iz = i\left(\frac{\pi}{2} \pm 2n\pi\right) + ln(10 \pm 3\sqrt{11})$$

Therefore infinite number of complex solutions.

4.5 (a)

Since, $\lim_{z\to 2}\left[(z-2)^2 f(z)\right]$ is finite and non-zero,

$f(z)$ has a pole of order two at, $z = 2$.

The residue at $z = a$ is given for a pole of order n as

$$\text{Res } f(a) = \frac{1}{(n-1)!}\frac{d^{n-1}}{dz^{n-1}}\left[(z-a)^n f(z)\right]_{z=a}$$

Here, $n = 2$ (pole of order 2) & $a = 2$

$$\therefore \text{Res } f(2) = \frac{1}{1!}\frac{d}{dz}\left[(z-2)^2 f(z)\right]_{z=2}$$
$$= \frac{d}{dz}\left[(z-2)^2 \frac{1}{(z+2)^2(z-2)^2}\right]_{z=2}$$
$$= \frac{d}{dz}\left[\frac{1}{(z+2)^2}\right]_{z=2} = \left[-2(z+2)^{-3}\right]_{z=2}$$
$$= \frac{-2}{(2+2)^3} = -\frac{1}{32}$$

4.6 (d)

$$f(z) = c_0 + c_1 z^{-1}$$
$$\oint \frac{1+f(z)}{z}dz = ?$$

unit circle

It has one pole at origin

$$\text{so} \oint_{z\to 0} \frac{z\{1+f(z)\}}{z}dz = 2\pi j\{1 + f(0)\}$$
$$= 2\pi j(1 + c_o)$$

4.7 (c)

Given that,

$$X(z) = \frac{1-2z}{z(z-1)(z-2)}$$

$$\text{Res }(0) = \lim_{z\to 0} 1\cdot\left[\frac{1-2z}{(z-1)(z-2)}\right] = \frac{1}{2}$$
$$\text{Res }(1) = \lim_{z\to 1} 1\cdot\left[\frac{1-2z}{z(z-2)}\right] = \frac{-1}{1(-1)} = 1$$
$$\text{Res }(2) = \lim_{z\to 2} 1\cdot\left[\frac{1-2z}{z(z-1)}\right] = \frac{1-4}{2(2-1)} = \frac{-3}{2}$$

4.8 (a)

$$I = \oint_C \frac{-3z+4}{(z^2+4z+5)}dz = 2\pi i(\text{sum of residues})$$

Poles of $\frac{-3z+4}{(z^2+4z+5)}$ are given by

$$z^2 + 4z + 5 = 0$$

$$z = \frac{-4 \pm \sqrt{16-20}}{2} = \frac{-4 \pm 2i}{2}$$

$$= -2 \pm i$$

Since the poles lie outside the circle $|z| = 1$.

So $f(z)$ is analytic inside the circle $|z| = 1$.

Hence, $\oint_C f(z)dz = 2\pi i(0) = 0$

4.9 (a)

$x = i$, then in polar coordinates,

$$x = \cos\frac{\pi}{2} + i\sin\frac{\pi}{2} = e^{\frac{\pi}{2}i}$$

Now, $x^x = i^i = \left(e^{\pi i/2}\right)^i = e^{i^2\pi/2} = e^{-\pi/2}$

4.10 (c)

Given,

$$f(z) = \frac{1}{z+1} - \frac{2}{z+3}$$

$$= \frac{(z+3)-2(z+1)}{(z+1)(z+3)} = \frac{-z+1}{(z+1)(z+3)}$$

Poles are at –1 and –3 i.e. (–1, 0) and (–3, 0).

From figure below of $|z + 1| = 1$,

we see that (–1, 0) is inside the circle and (–3, 0) is outside the circle.

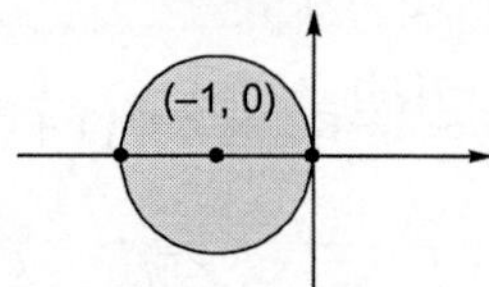

Residue theorem says,

$\frac{1}{2\pi j}\oint_C f(z)\,dz =$ Residue of those poles which are inside C.

So, the required integral $\frac{1}{2\pi j}\oint_C f(z)dz$ is given by the residue of function at pole (–1, 0) (which is inside the circle).

This residue is $= \frac{-(-1)+1}{(-1+3)} = \frac{2}{2} = 1$

4.11 (c)

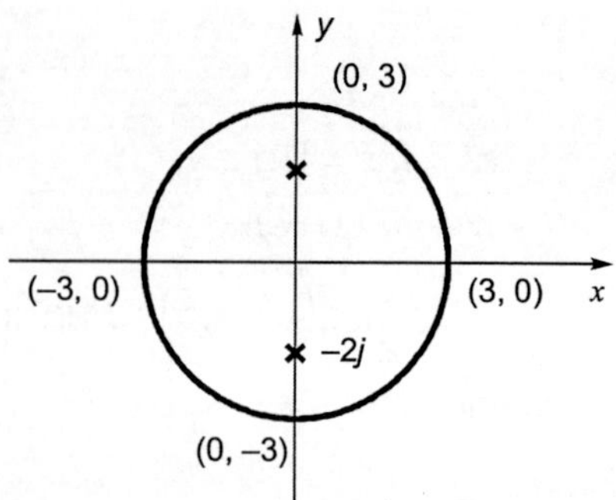

$$\oint_C \frac{z^2 - z + 4j}{z + 2j}$$

Pole = $z = -2j$

which is inside of $|z| = 3$

From Cauchy integral formula

$$\oint \frac{z^2 - z + 4j}{z+2j} = 2\pi i\left[\lim_{z\to -2j} z^2 - z + 4j\right]$$

$$= 2\pi j[-4 + 2j + 4j]$$

$$= 2\pi j[-4 + 6j] = -4\pi[2j + 3]$$

4.12 (b)

$$Z = x + iy$$

$$f(z) = u + iv$$

$$u = e^{-y}\cos(x)$$

$$\frac{\partial u}{\partial x} = -e^{-y}\sin(x)$$

For analytical function

$$\frac{\partial u}{\partial x} = \frac{\partial v}{\partial y}$$

$$\Rightarrow \frac{\partial v}{\partial y} = -e^{-y}\sin(x)$$

Integrating w.r.t. y

$$v = e^{-y}\sin(x)$$

4.13 (d)

$$f(z) = \bar{z} = x - iy$$

$u = x$ $\quad$ $v = -y$

$\Rightarrow u_x = 1$ $\quad$ $v_x = 0$

$u_y = 0$ $\quad$ $v_y = -1$

$u_x \neq v_y$ i.e. $C - R$ not satisfied

$\Rightarrow \bar{z}$ is not analytic function.

4.14 (10)

$$f(z_1) = \frac{az_1 + b}{cz_1 + d} \text{ and } f(z_2) = \frac{az_2 + b}{cz_2 + d}$$

$$\frac{az_1 + b}{cz_1 + d} = \frac{az_2 + b}{cz_2 + d}$$

$$acz_1z_2 + bcz_2 + adz_1 + bd = acz_1z_2 + bcz_1 + adz_2 + bd$$
$$bc(z_2 - z_1) = ad(z_2 - z_1)$$
$$z_2 \neq z_1 \quad \Rightarrow bc = ad$$
$$d = \frac{bc}{a} = \frac{4 \times 5}{2} = 10$$

4.15 (0)

$$\text{Re}\{z\} = \frac{z + \bar{z}}{2}$$

Since there is no pole inside unit circle, so Residue at poles is zero

$$\Rightarrow \frac{1}{2\pi i}\oint_C \text{Re}\{z\}dz = 0$$

4.16 (b)

By Cauchy integral formula

$$\oint \frac{f(z)}{(z - z_o)^{n+1}}dz = \frac{2\pi i\, f^*(z_o)}{n!}$$
$$\oint \frac{dz}{(z - z_o)^{n+1}} = \frac{2\pi i}{n!} \cdot 0 = 0$$

4.17 (–133.87)

$$I = -\frac{1}{2\pi}\int_c \frac{\sin z}{(z - 2\pi j)^3}dz = -\frac{1}{2\pi} \times \frac{2\pi j f''(2\pi j)}{2!}$$
$$f(z) = \sin z$$
$$f'(z) = \cos z$$
$$f''(z) = -\sin z$$
$$I = -\frac{1}{2\pi} \times 2\pi j \frac{-\sin(2\pi j)}{2} = -\frac{1}{2}\sinh 2\pi = -133.87$$

4.18 (0)

$$f(z) = 2z^3 + b_1 |z|^3$$

Given that $f(z)$ is analytic.

which is possible only when $b = 0$

since $|z^3|$ is differentiable at the origin but not analytic.

$2z^3$ is analytic everywhere

$\therefore \quad f(z) = 2z^3 + b|z^3|$ is analytic

only when $b = 0$

4.19 (b)

(i) $Z_0 = 2$ -lies inside C,

So, $$\text{Res } f(z) = \lim_{z \to 2}(z - 2) \cdot \frac{e^z}{z - 2}$$
$$= e^2 = 7.39$$

$$\frac{1}{2\pi i}\int_c \frac{e^z}{z - 2}dz = 2\pi i \cdot \frac{1}{2\pi i}(7.39) = 7.39$$

(ii) $Z_0 = -2$ lies out side C then

$$\text{Res } f(z) = 0$$

So, $$\int_c \frac{e^z}{z - 2}dz = 2\pi i \frac{1}{2\pi i}(0) = 0$$

4.20 (1)

Residue of $\dfrac{\sin z}{z^2}$

$$= \text{coefficient of } \frac{1}{z} \text{ in } \left\{\frac{z - \frac{z^3}{3!} + \frac{z^5}{5!} - - -}{z^2}\right\}$$

$$= \text{coefficient of } \frac{1}{z} \text{ in } \left\{\frac{1}{z} - \frac{z}{3!} + \frac{z^3}{5!} - - -\right\}$$

$= 1$

4.21 (b)

Residue at, $z = 4$ is

$$= \lim_{z \to 4}(z - 4)\frac{1}{(z - 4)(z + 1)^3} = \frac{1}{(4 + 1)^3} = \frac{1}{125}$$

Residue at $z = -1$ is

$$= \lim_{z \to -1} \frac{1}{2!}\frac{d^2}{dz^2}\left((z - 1)^3 \frac{1}{(z - 4)(z + 1)^3}\right)$$

$$= \lim_{z \to -1} \frac{1}{2!}\left(\frac{2}{(z - 4)^3}\right) = \frac{1}{(-1 - 4)^3} = \frac{-1}{125}$$

4.22 (d)

Poles are $z^2 + 1 = 0$

$$z = \pm i$$

$z = i$ lies inside $|z| = 3$

$z = -i$ lies inside $|z| = 3$

Residue at $z = i$ is

$$= \lim_{z \to i}(z - i)\frac{z^2 - 1}{(z - i)(z + i)}e^z$$
$$= \frac{-1 - 1}{2i}e^i = ie^i$$

Residue at $z = -i$ is

$$= \lim_{z \to -i}(z + i)\frac{z^2 - 1}{(z - i)(z + i)}e^z$$
$$= \frac{-1 - 1}{-2i}e^{-i} = \frac{1}{i}e^{-i} = -ie^{-i}$$

By Residues theorem

$$I = 2\pi i(ie^i - ie^{-i})$$
$$= -2\pi(e^i - e^{-i})$$
$$= -2\pi(\cos 1 + i\sin 1 - \cos 1 + i\sin 1)$$
$$= -2\pi(2i\sin 1) = -4\pi i\sin(1)$$

4.23 (2)

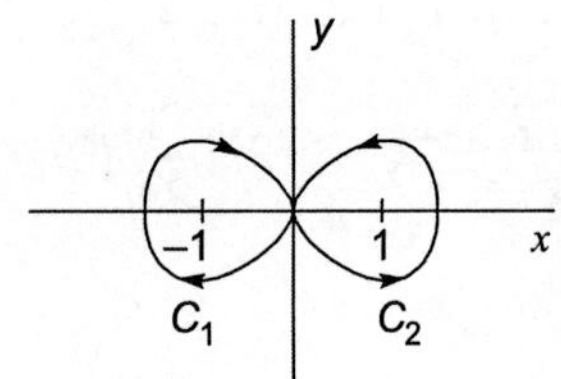

$$\frac{1}{\pi i}\oint_C \frac{dz}{z^2-1} = 2\left[\frac{1}{2\pi i}\oint_{C_1}\frac{dz}{(z+1)(z-1)} + \frac{1}{2\pi i}\oint_{C_2}\frac{dz}{(z+1)(z-1)}\right]$$

$$= 2\left[-\left(\frac{1}{z-1}\right)\bigg|_{z=-1} + \left(\frac{1}{z+1}\right)\bigg|_{z=1}\right]$$

$$= 2\left[-\left(-\frac{1}{2}\right)+\left(\frac{1}{2}\right)\right] = 2$$

4.24 (a)

$f(z) = \cos z$ is analytic every where.

4.25 (0)

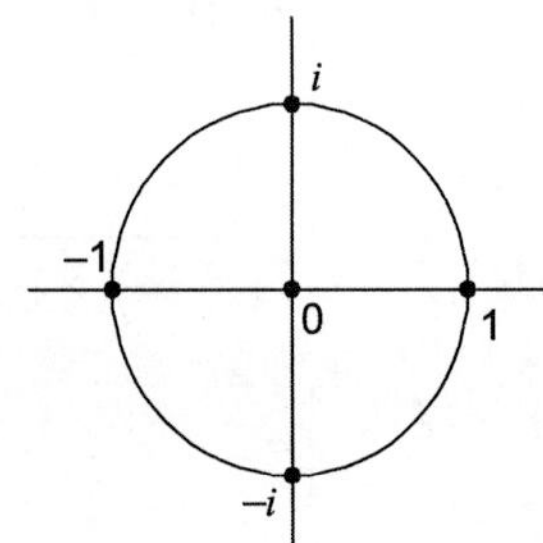

$$\frac{1}{2\pi j}\oint\left(z+\frac{1}{z}\right)^2 dz \text{ where } C \text{ is } |z| = 1$$

$$I = \frac{1}{2\pi j}\int_C \frac{(z^2+1)^2}{z^2}dz$$

$z = 0$ lies inside the circle,

$$I = \frac{1}{2\pi j}\left[\frac{2\pi j}{1!}\frac{d}{dz}(z^2+1)^2\right]_{z=0}$$

$$= \left[\frac{d}{dz}(z^2+1)^2\right]_{z=0} = \left[2(z^2+1)\times 2z\right]_{z=0} = 0$$

4.26 (*)

$$\oint_C \frac{\sin x}{x^2(x^2+4)}dx,\ c: |x-i| = 2$$

Poles are given by

$$x^2 = 0 \text{ and } x^2 + 4 = 0$$

$$\Downarrow$$

$\Rightarrow$ $x = 0$ is a pole of order '2'

$x = i$ are simple nodes

$x = 0$ lies inside 'c'

$x = 2i$ lies inside 'c'

$x = -2i$ lies outside 'c'

Residue at $x = 0$

$$= \frac{1}{(2-1)!}\lim_{x\to 0}\frac{d}{dz}\left[(x-0)^2\frac{\sin x}{x^2(x^2+4)}\right]$$

$$= \lim_{x\to 0}\frac{(x^2+4)\cos x - \sin x(2x)}{(x^2+4)^2} = \frac{1}{4}$$

Residue at $x = 2_i$

$$= \lim_{x\to 2i}(x-2i)\frac{\sin x}{x^2(x-2i)(x+2i)} = \frac{\sin(2i)}{(-4)(4i)}$$

By CRT

$$\oint_C f dx = 2\pi i\left[\text{Res}_0 + \text{Res}_{2i}\right] = 2\pi i\left[\frac{1}{4} - \frac{\sin(2i)}{16}\right]$$

4.27 (b, d)

The given series, $\sum_{n=1}^{\infty}\frac{n^d}{c^n} = \sum_{n=1}^{\infty}u_n$

Option (a) : $c = 1$, $d = -1$

$$\sum_{n=1}^{\infty}u_n = \sum_{n=1}^{\infty}\frac{1}{n}$$

From the p-series test, for the series

$$\sum_{n=0}^{\infty}\frac{1}{n^p}$$

(i) If $p \le 1$, series diverges

(i) If $p > 1$, series converges

In the given case $p = 1$, hence, the series $\sum_{n=1}^{\infty}\frac{1}{n}$ diverge.

Option (b) : $c = 2$, $d = 1$

$$\sum_{n=1}^{\infty}u_n = \sum_{n=1}^{\infty}\frac{n}{2^n}$$

Using the ratio test,

$$\lim_{n\to\infty}\left|\frac{u_{n+1}}{u_n}\right| = \lim_{n\to\infty}\left|\frac{n+1}{2^{n+1}}\times\frac{2^n}{n}\right| = \frac{1}{2} < 1$$

$\therefore$ By ratio test, Σu_n is convergent

Option (c) : $c = 0.5$, $d = -10$

$$\sum_{n=1}^{\infty}u_n = \sum_{n=1}^{\infty}\frac{n^{-10}}{(0.5)^n}$$

Using ratio test,

$$\lim_{n\to\infty}\left|\frac{u_{n+1}}{u_n}\right| = \lim_{n\to\infty}\left|\frac{(n+1)^{-10}}{(0.5)^{n+1}}\times\frac{(0.5)^n}{n^{-10}}\right|$$

$$= \lim_{n\to\infty}\left|\frac{1}{0.5}\times\left(1+\frac{1}{n}\right)^{-10}\right| = \frac{1}{0.5} = 2 > 1$$

$\therefore$ By ratio test, Σu_n is divergent.

Option (d) : $c = 1$, $d = -2$

$$\sum_{n=1}^{\infty} u(n) = \sum_{n=1}^{\infty}\frac{1}{n^2}$$

Using p-test, since $p > 1$. Hence, the series Σu_n is convergent.

4.28 (0.50)

Using Cauchy's residue theorem, if C is a simply closed positively oriented contour (i.e. traversed counterclockwise) in the complex plane and f is analytic except for some points $z_1, z_2, ..., z_n$ inside the contour c, then

$$\oint_C f(z)dz = 2\pi i\sum_{i=1}^{n}\text{Residue at } z_i$$

In the given question,

$$f(z) = \frac{2^z}{z^2-1} = \frac{2^z}{(z-1)(z+1)}$$

The singularity points where $f(z)$ is not analytic lies at $z = 1$ and $z = -1$.

Since, $z = 1$ doesn't lie inside the contour C. Hence,

$$I = \oint_C \frac{2^z}{z^2-1}\cdot dz = \oint_C \frac{\left(\frac{2^z}{z-1}\right)}{z+1}\cdot dz$$

$$I = 2\pi i \quad [\text{Residue at } z = -1]$$

$$I = 2\pi i\times\lim_{z\to(-1)}(z+1)\cdot\frac{2^z}{(z-1)(z+1)}$$

$$I = 2\pi i\times\frac{2^{-1}}{(-1-1)} = -\frac{2\pi i}{4}$$

$$\therefore \quad I = -\frac{\pi i}{2}$$

Given: $I = -\pi i A$. Hence,

$$A = \frac{1}{2} = 0.5$$

4.29 (a)

$w = (2)j^{1/4} = 2(0 + j)^{1/4}$

$$w = 2\left[e^{j(2n+1)\pi/2}\right]^{1/4} = 2\left[e^{j(2n+1)\pi/8}\right]$$

For $n = 0$, $\quad w = e^{j\pi/8}$

For $n = 2$, $\quad w = 2e^{5\pi j/8}$

For $n = 4$, $\quad w = 2e^{9\pi j/8}$

4.30 (b)

Let the contour integral,

$$I = \oint_C \frac{z+2}{z^2+2z+2}dz; \quad c = \left|z+1-\frac{3}{2}i\right| = 1$$

Poles are given $(z+1)^2 + 1 = 0$

$$z + 1 = \pm\sqrt{-1}$$

$$z = -1 + j, -1 - j$$

where $-1 - j$ lies outside 'c'

$z = (-1, 1)$ lies inside 'c'.

by Cauchy's residue theorem,

$$\oint_C f(z)dz = 2\pi i \text{ Res } (f(z), z = -1 + j)$$

$$= 2\pi i\left(\frac{z+2}{2(z+1)}\right)_{z=-1+i} = 2\pi i\left(\frac{-1+j+2}{2(-1+j+1)}\right)$$

$$= \pi(1 + j)$$

4.31 (a)

Poles of $f(z)$ are $z = \underset{\text{simple pole}}{\underset{\downarrow}{2}}$ and $\underset{\text{Double pole}}{\underset{\downarrow}{0}}$

The poles $z = 0$ and $z = 2$ lies inside the complex plane $|z| = 3$. Hence, $f(z)$ is not analytic at $z = 0$ and $z = 2$.

R_1 = Residue of $f(z)$ at $z = 2$,

$$= \lim_{z\to2}\cdot(z-2)f(z) = \lim_{z\to2}\cdot\left(\frac{\sin\pi z}{z^2}\right)$$

$= 0$

R_2 = Residue of $f(z)$ at $z = 0$ (of order $m = 2$),

$$R_2 = \frac{1}{\underline{|2-1}}\left[\frac{d^{2-1}}{dz^{2-1}}(z-0)^2 f(z)\right]_{z=0}$$

$$R_2 = \left[\frac{d}{dz}\left[\frac{\sin\pi z}{z-2}\right]\right]_{z=0} = \left[\frac{(z-2)\cos\pi z(\pi) - \sin\pi z}{(z-2)^2}\right]_{z=0}$$

$$= \frac{(0-2)\cos(0)\cdot\pi - \sin 0}{(0-2)^2}$$

$$= \frac{-2\pi}{4} = \frac{-\pi}{2}$$

By Cauchy Residue Theorem,

$$I = \oint_C f(z)dz = 2\pi j(R_1 + R_2)$$

$$= 2\pi j\left(0 - \frac{\pi}{2}\right) = -\pi^2 j$$

■■■

5 Probability and Statistics

5.1 Events A and B are mutually exclusive and have nonzero probability. Which of the following statement(s) are true?

(a) $P(A \cup B) = P(A) + P(B)$
(b) $P(B^C) > P(A)$
(c) $P(A \cap B) = P(A)\,P(B)$
(d) $P(B^C) < P(A)$

[1988 : 2 M]

5.2 A fair dice is rolled twice. The probability that an odd number will follow an even number is

(a) $\frac{1}{2}$ (b) $\frac{1}{6}$
(c) $\frac{1}{3}$ (d) $\frac{1}{4}$ [2005 : 1 M]

5.3 A probability density function is of the form

$p(x) = Ke^{-\alpha|x|}, x \in (-\infty, \infty)$

The value of K is

(a) 0.5 (b) 1
(c) 0.5 α (d) α [2006 : 1 M]

5.4 Three companies X, Y and Z supply computers to a university. The percentage of computers supplied by them and the probability of those being defective are tabulated below

Company	% of computers supplied	Probability of being defective
X	60%	0.01
Y	30%	0.02
Z	10%	0.03

Given that a computer is defective, the probability that it was supplied by Y is

(a) 0.1 (b) 0.2
(c) 0.3 (d) 0.4 [2006 : 2 M]

5.5 An examination consists of two papers, Paper 1 and Paper 2. The probability of failing in Paper 1 is 0.3 and that in Paper 2 is 0.2. Given that a student has failed in Paper 2, the probability of failing in Paper 1 is 0.6. The probability of a student failing in both the papers is

(a) 0.5 (b) 0.18
(c) 0.12 (d) 0.06

[2007 : 2 M]

5.6 A fair coin is tossed 10 times. What is the probability that ONLY the first two tosses will yield heads?

(a) $\left(\frac{1}{2}\right)^2$ (b) ${}^{10}C_2\left(\frac{1}{2}\right)^2$

(c) $\left(\frac{1}{2}\right)^{10}$ (d) ${}^{10}C_2\left(\frac{1}{2}\right)^{10}$

[2009 : 1 M]

5.7 Consider two independent random variables X and Y with identical distributions. The variables X and Y take values 0,1 and 2 with probabilities $\frac{1}{2}, \frac{1}{4}$ and $\frac{1}{4}$ respectively. What is the conditional probability $P(X + Y = 2 | X - Y = 0)$?

(a) 0 (b) $\frac{1}{16}$
(c) $\frac{1}{6}$ (d) 1

[2009 : 2 M]

5.8 A fair coin is tossed independently four times. The probability of the event "the number of times heads show up is more than the number of times tails show up" is

(a) $\frac{1}{16}$ (b) $\frac{1}{8}$
(c) $\frac{1}{4}$ (d) $\frac{5}{16}$

[2010 : 2 M]

5.9 A fair dice is tossed two times. The probability that the second toss results in a value that is higher than the first toss is

(a) 2/36 (b) 2/6
(c) 5/12 (d) 1/2

[2011 : 2 M]

5.10 A fair coin is tossed till a head appears for the first time. The probability that the number of required tosses is odd, is

(a) 1/3 (b) 1/2
(c) 2/3 (d) 3/4

[2012 : 2 M]

5.11 In a housing society, half of the families have a single child per family, while the remaining half have two children per family. The probability that a child picked at random, has a sibling is

[2014 : 1 M, Set-1]

5.12 Let X be a random variable which is uniformly chosen from the set of positive odd numbers less than 100. The expectation, $E[X]$, is

[2014 : 1 M, Set-2]

5.13 An unbiased coin is tossed an infinite number of times. The probability that the fourth head appears at the tenth toss is

(a) 0.067 (b) 0.073
(c) 0.082 (d) 0.091

[2014 : 1 M, Set-3]

5.14 A fair coin is tossed repeatedly till both head and tail appear at least once. The average number of tosses required is ______ .

[2014 : 2 M, Set-3]

5.15 Parcels from sender S to receiver R pass sequentially through two post-offices. Each post-office has a probability 1/5 of losing an incoming parcel, independently of all other parcels. Given that a parcel is lost, the probability that it was lost by the second post-office is ________.

[2014 : 2 M, Set-4]

5.16 If calls arrive at a telephone exchange such that the time of arrival of any call is independent of the time of arrival of earlier or future calls, the probability distribution function of the total number of calls in a fixed time interval will be

(a) Poisson (b) Gaussian
(c) Exponential (d) Gamma

[2014 : 1 M, Set-4]

5.17 Suppose A and B are two independent events with probabilities $P(A) \neq 0$ and $P(B) \neq 0$. Let $\bar{A}$ and $\bar{B}$ be their complements. Which one of the following statements is FALSE?

(a) $P(A \cap B) = P(A)\,P(B)$
(b) $P(A/B) = P(A)$
(c) $P(A \cup B) = P(A) + P(B)$
(d) $P(\bar{A} \cap \bar{B}) = P(\bar{A})P(\bar{B})$

[2015 : 1 M, Set-1]

5.18 Let the random variable X represent the number of times a fair coin needs to be tossed till two consecutive heads appear for the first time. The expectation of X is_____.

[2015 : 2 M, Set-2]

5.19 Let $X \in \{0, 1\}$ and $Y \in \{0, 1\}$ be two independent binary random variables. If $P(X = 0) = p$ and $P(Y = 0) = q$, then $P(X + Y \geq 1)$ is equal to

(a) $pq(1 - p)(1 - q)$ (b) pq
(c) $p(1 - q)$ (d) $1 - pq$

[2015 : 2 M, Set-2]

5.20 A fair die with faces {1, 2, 3, 4, 5, 6} is thrown repeatedly till '3' is observed for the first time. Let X denote the number of times the die is thrown. The expected value of X is_____.

[2015 : 2 M, Set-3]

5.21 Two random variables X and Y are distributed according to

$$f_{X,Y}(x,y) = \begin{cases} (x+y) & 0 \leq x \leq 1, \quad 0 \leq y \leq 1 \\ 0 & \text{otherwise} \end{cases}$$

The probability $P(X + Y \leq 1)$ is ________

[2016 : 2 M, Set-2]

5.22 The second moment of a Poisson-distributed random variable is 2. The mean of the random variable is _____.

[2016 : 1 M, Set-1]

5.23 The probability of getting a "head" in a single toss of a biased coin is 0.3. The coin is tossed repeatedly till a "head" is obtained. If the tosses are independent, then the probability of getting "head" for the first time in the fifth toss is _______.

[2016 : 1 M, Set-3]

5.24 Three fair cubical dice are thrown simultaneously. The probability that all three dice have the same number of dots on the faces showing up is (up to third decimal place) _______.

[2017 : 1 M, Set-1]

5.25 Passengers try repeatedly to get a seat reservation in any train running between two stations until they are successful. If there is 40% chance of getting reservation in any attempt by a passenger, then the average number of attempts that passengers need to make to get a seat reserved is _____.

[2017 : 2 M, Set-2]

5.26 Let X_1, X_2, X_3 and X_4 be independent normal random variables with zero mean and unit variance. The probability that X_4 is the smallest among the four is ________.

[2018 : 1 M]

5.27 If X and Y are random variables such that $E[2X + Y] = 0$ and $E[X + 2Y] = 33$, then $E[X] + E[Y] =$ ______.

[2019 : 1 M]

5.28 Let Z be an exponential random variable with mean 1. That is, the cumulative distribution function of Z is given by

$$F_Z(x) = \begin{cases} 1 - e^{-x} & \text{if } x \geq 0 \\ 0 & \text{if } x < 0 \end{cases}$$

Then $\Pr(Z > 2 \mid Z > 1)$, rounded off to two decimal places, is equal to ______.

[2019 : 1 M]

5.29 The two sides of a fair coin are labelled as 0 and 1. The coin is tossed two times independently. Let M and N denote the labels corresponding to the outcomes of those tosses. For a random variable X, defined as $X = \min(M, N)$, the expected value $E(X)$ (rounded off to two decimal places) is ________.

[2020 : 1 M]

5.30 A binary random variable X takes the value +2 or −2. The probability $P(X = +2) = \alpha$. The value of α (rounded off to one decimal place), for which the entropy of X is maximum, is ________.

[2020 : 1 M]

5.31 X is a random variable with uniform probability density function in the interval [−2,10]. For $Y = 2X - 6$, the conditional probability $P(Y \leq 7 \mid X \geq 5)$ (rounded off to three decimal places) is _____.

[2020 : 2 M]

5.32 A box contains the following three coins,

I. A fair coin with head on one face and tail on the other face.

II. A coin with heads on both the faces.

III. A coin with tails on both the faces.

A coin is picked randomly from the box and tossed. Out of the two remaining coins in the box, one coin is then picked randomly and tossed. If the first toss results in a head, the probability of getting a head in the second toss is

(a) $\frac{1}{2}$ (b) $\frac{2}{5}$

(c) $\frac{1}{3}$ (d) $\frac{2}{3}$ [2021 : 2 M]

5.33 The bar graph shows the frequency of the number of wickets taken in a match by a bowler in her career. For example, in 17 of her matches, the bowler has taken 5 wickets each. The median number of wickets taken by the bowler in a match is ________ (rounded off to one decimal place).

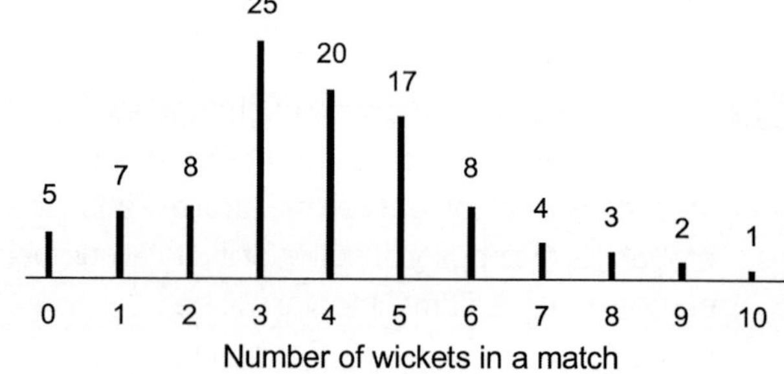

[2022 : 1 M]

■■■

Answers Probability and Statistics

5.1 (a)	5.2 (d)	5.3 (c)	5.4 (d)	5.5 (c)	5.6 (c)	5.7 (c)
5.8 (d)	5.9 (c)	5.10 (c)	5.11 (0.666)	5.12 (50)	5.13 (c)	5.14 (3)
5.15 (0.444)	5.16 (a)	5.17 (c)	5.18 (1.5)	5.19 (d)	5.20 (6)	5.21 (0.33)
5.22 (1)	5.23 (0.072)	5.24 (0.028)	5.25 (3)	5.26 (0.25)	5.27 (11)	5.28 (0.37)
5.29 (0.25)	5.30 (0.5)	5.31 (0.3)	5.32 (c)	5.33 (4)		

Explanations Probability and Statistics

5.1 (a)

$P(A \cup B) = P(A) + P(B)$

For mutually exclusive events A and B

$$\boxed{P(A \cup B) = P(A) + P(B)}$$

5.2 (d)

$$P_o = \frac{3}{6} = \frac{1}{2}$$

$$P_o = \frac{3}{6} = \frac{1}{2}$$

Since both events are independent of each other.

$$P_{(\text{odd/even})} = \frac{1}{2} \times \frac{1}{2} = \frac{1}{4}$$

5.3 (c)

$$\int_{-\infty}^{\infty} p(x)dx = 1$$

$$\int_{-\infty}^{\infty} Ke^{-\alpha|x|}dx = 1$$

$$\int_{-\infty}^{0} Ke^{\alpha x}dx + \int_{0}^{\infty} Ke^{-\alpha x} = 1$$

$$\Rightarrow \frac{K}{\alpha}\left[e^{\alpha x}\right]_{-\infty}^{0} + \frac{K}{-\alpha}\left[e^{-\alpha x}\right]_{0}^{\infty} = 1$$

$$\Rightarrow \frac{K}{\alpha} + \frac{K}{\alpha} = 1$$

$$2K = \alpha$$

$$\Rightarrow K = 0.5\alpha$$

5.4 (d)

$S \rightarrow$ supply by y, $d \rightarrow$ defective

Probability that the computer was supplied by y, if the product is defective

$$P(S/d) = \frac{P(S \cap d)}{P(d)}$$

$$P(S \cap d) = 0.3 \times 0.02 = 0.006$$

$$P(d) = 0.6 \times 0.01 + 0.3 + 0.02 + 0.1 \times 0.03 = 0.015$$

$$P(S/d) = \frac{0.006}{0.015} = 0.4$$

5.5 (c)

$P(A) = 0.2$ $A \rightarrow$ failing in paper 1

$P(B) = 0.3$ $B \rightarrow$ failing in paper 2

$P(A/B) = 0.6$

Probability of failing in both $P(A \cap B)$

$$= P(A/B) \times P(B) = 0.6 \times 0.2 = 0.12$$

5.7 (c)

$$P(A \mid B) = \frac{P(A \cap B)}{P(B)}$$

$$P(X + Y = 2 \mid X - Y = 0) = \frac{P(X + Y = 2, X - Y = 0)}{P(X - Y = 0)}$$

$$P(X - Y = 0) = P(X = 0, Y = 0) + P(X = 1, Y = 1) + P(X = 2, Y = 2)$$

$$= P(X = 0) \cdot P(Y = 0) + P(X = 1) \cdot P(Y = 1) + P(X = 2), P(Y = 2)$$

$$= \frac{3}{8}$$

$$P(X + Y = 2, X - Y = 0) = P(X = 1, Y = 1)$$

$$= P(X = 1) \cdot P(Y = 1) = \frac{1}{16}$$

$$\therefore P(X + Y = 2 \mid X - Y = 0) = \frac{1}{16} \times \frac{8}{3} = \frac{1}{6}$$

5.8 (d)

The only 5 events when the number of times of heads show up is more than the number of times tails show up are given below:

H H H H
H H H T
H H T H
H T H H
T H H H

Since fair coin is tossed independently four times, the probability of each above events is

$$\left(\frac{1}{2}\right)\left(\frac{1}{2}\right)\left(\frac{1}{2}\right)\left(\frac{1}{2}\right) = \frac{1}{16}.$$

All above events are mutually exclusive, so, total probability

$$= \frac{1}{16} + \frac{1}{16} + \frac{1}{16} + \frac{1}{16} + \frac{1}{16} = \frac{5}{16}$$

5.9 (c)

In the first toss, results can be 1, 2, 3, 4, 5
For 1, the second toss results can be 2, 3, 4, 5, 6
For 2, the second toss results can be 3, 4, 5, 6
For 3, the second toss results can be 4, 5, 6
For 4, the second toss results can be 5, 6
For 5, the second toss results can be 6
The required probability

$$= \frac{1}{6}\cdot\frac{5}{6} + \frac{1}{6}\cdot\frac{4}{6} + \frac{1}{6}\cdot\frac{3}{6} + \frac{1}{6}\cdot\frac{2}{6} + \frac{1}{6}\cdot\frac{1}{6} = \frac{5}{12}$$

5.10 (c)

P(no. of tosses is odd)
$= P$(no. of tosses is 1, 3, 5, 7 . . .)
P(no. of toss is 1) $= P$(Head in first toss) $= 1/2$
P(no. of toss is 3) $= P$(tail in first toss, tail in second toss and head in third toss)

$$= \frac{1}{2}\times\frac{1}{2}\times\frac{1}{2} = \frac{1}{8}$$

$$P(\text{no. of toss is 5}) = P(T, T, T, T, H) = \left(\frac{1}{2}\right)^5 = \frac{1}{32}$$

... etc.

$$\text{So } P(\text{no. of tosses is odd}) = \frac{1}{2} + \frac{1}{8} + \frac{1}{32} + \cdots$$

Sum of infinite geometric series with

$$a = \frac{1}{2} \text{ and } r = \frac{1}{4}$$

$$= \frac{\frac{1}{2}}{1-\frac{1}{4}} = \frac{\frac{1}{2}}{\frac{3}{4}} = \frac{2}{3}$$

5.11 (0.666)

Required probability

$$= \frac{\frac{1}{2}\times\frac{2}{3}}{\frac{1}{2}\times\frac{1}{3} + \frac{1}{2}\times\frac{2}{3}} = \frac{2}{3} = 0.666$$

5.12 (50)

$$E[X] = \frac{1+2+3+\cdots 99}{50} = \frac{2500}{50} = 50$$

5.13 (c)

It means 3-head appears in 1st 9 trials.
Probability of getting exactly 3 head in 1st 9 trials
$= \text{Coefficient } p^3 \text{ in } (4 + p)9$

When, $\boxed{p+q=1}$

$$= {}^9C_3\, q^6\, p^3$$

when, p = probability of occur of head
q = probability of occur of tail

$$= {}^9C_3 \times \left(\frac{1}{2}\right)^6 \left(\frac{1}{2}\right)^3 = {}^9C_3 \times \left(\frac{1}{2}\right)^9$$

and in 10th trial head must appears.
So required probability

$$= {}^9C_3\left(\frac{1}{2}\right)^9 \times \frac{1}{2} = \frac{9\times8\times7}{3!}\times\left(\frac{1}{2}\right)^{10} = \frac{84}{1024}$$

$= 0.082$

5.14 (3)

To get both head and Tail atleast once.
We have to toss minimum two times.
The random variable x represents first toss starting with Head.

x	2	3	4	5	...
$P(x)$	$\frac{1}{2^2}$	$\frac{1}{2^3}$	$\frac{1}{2^4}$	$\frac{1}{2^5}$	...

$$E(x) = \Sigma x \cdot P(x)$$

$$= 2\cdot\frac{1}{2^2} + 3\cdot\frac{1}{2^3} + 4\cdot\frac{1}{2^4} + \cdots$$

$$= \frac{1}{2}\left(2\cdot\frac{1}{2} + 3\cdot\frac{1}{2^2} + 4\cdot\frac{1}{2^3} + \cdots\right)$$

$$= \frac{1}{2}\left(1 + 2\cdot\frac{1}{2}\cdot 3\frac{1}{2^2} + 4\cdot\frac{1}{2^3} + \cdots - 1\right)$$

$$= \frac{1}{2}\left[\left(1-\frac{1}{2}\right)^{-2} - 1\right] = \frac{1}{2}\left[\left(\frac{1}{2}\right)^{-2} - 1\right]$$

$$= \frac{1}{2}(4-1) = \frac{3}{2}$$

Y represents first toss starting with Tail.

y	2	3	4	5	...
$P(y)$	$\frac{1}{2^2}$	$\frac{1}{2^3}$	$\frac{1}{2^4}$	$\frac{1}{2^5}$	...

$E(y) = \Sigma y \cdot P(y)$

$$= 2\cdot\frac{1}{2^2} + 3\cdot\frac{1}{2^3} + 4\cdot\frac{1}{2^4} + \cdots$$

$$= \frac{1}{2}\left(2\cdot\frac{1}{2} + 3\cdot\frac{1}{2^2} + 4\cdot\frac{1}{2^3} + \cdots\right)$$

$$= \frac{1}{2}\left(1 + 2\cdot\frac{1}{2}\cdot 3\frac{1}{2^2} + 4\cdot\frac{1}{2^3} + \cdots - 1\right)$$

$$= \frac{1}{2}\left[\left(1-\frac{1}{2}\right)^{-2} - 1\right] = \frac{1}{2}\left[\left(\frac{1}{2}\right)^{-2} - 1\right]$$

$$= \frac{1}{2}(4-1) = \frac{3}{2}$$

$\therefore$ The average number of tosses required

$$= E(x) + E(y) = \frac{3}{2} + \frac{3}{2} = 3$$

5.15 (0.444)

Probability to lost at post-office 1 $= \frac{1}{5}$

Probability to lost at post-office 2 $= \frac{4}{5} \times \frac{1}{5}$

Total probability to lost $= \frac{1}{5} + \frac{4}{25} = \frac{9}{25}$

Required probability $= \dfrac{\frac{4}{25}}{\frac{9}{25}} = \frac{4}{9} = 0.444$

5.17 (c)

$P(A \cup B) = P(A) + P(B) - P(A \cap B)$

Since $P(A \cap B) = p(A)\,p(B)$ (not necessarily equal to zero).

So, $P(A \cup B) = P(A) + P(B)$ is false.

5.18 (1.5)

$$E[X] = \sum_{i=1}^{10} x_i p_i$$

$$= 2\times\left(\frac{1}{4}\right) + 3\left(\frac{1}{8}\right) + 4\left(\frac{1}{16}\right) + \cdots$$

$$= 1.5$$

5.19 (d)

$Z = X + Y$

$p(X = 0) = p, \quad p(Y = 0) = q$

$p(X = 1) = 1 - p, \quad p(Y = 1) = 1 - q$

Now,

X	Y	$Z = X + Y$	
0	0	0	
0	1	1	$z \geq 1$
1	0	1	
1	1	1	

$\therefore\ p(Z \geq 1) = 1 - p(Z < 1) = 1 - pq$

5.20 (6)

Let x = Number of times dice is thrown

Let A = getting '3', $\bar{A}$ = not getting '3'

X	1	2	3	4	5	...
$P(x)$	$\frac{1}{6}$	$\frac{5}{6}\times\frac{1}{6}$	$\left(\frac{5}{6}\right)^2\times\frac{1}{6}$	$\left(\frac{5}{6}\right)^3\times\frac{1}{6}$	-	...

$$E(x) = \sum_{i=1}^{n} x_i P(x = x_i)$$

$$= 1\times\frac{1}{6} + 2\times\frac{5}{6}\times\frac{1}{6} + 3\times\left(\frac{5}{6}\right)^2\frac{1}{6} + \ldots$$

$$= \frac{1}{6}\left[1 + 2\left(\frac{5}{6}\right) + 3\left(\frac{5}{6}\right)^2 + \ldots\right] = \frac{1}{6}\left[\frac{1}{\left(1-\frac{5}{6}\right)^2}\right] = 6$$

5.21 (0.33)

$$P(X + Y \leq 1) = \int_{x=0}^{1}\int_{y=0}^{(1-x)} f_{xy}(x, y)\,dx\,dy$$

$$= \int_{x=0}^{1}\int_{y=0}^{1-x} (x + y)\,dx\,dy$$

$$= \int_{x=0}^{1}\left(xy + \frac{y^2}{2}\right)_0^{1-x}$$

$$= \int_{x=0}^{1}\left(x(1-x) + \frac{(1-x)^2}{2}\right)dx$$

$$= \int_{x=0}^{1}\left(\frac{1}{2} - \frac{x^2}{2}\right)dx = \left(\frac{x}{2} - \frac{x^3}{6}\right)_0^1$$

$$= \frac{1}{2} - \frac{1}{6} = \frac{1}{3} = 0.33$$

5.22 (1)

In Poisson distribution,

Mean = First moment = λ

second moment = $\lambda^2 + \lambda$

Given, that second moment is 2

$$\therefore \quad \lambda^2+\lambda = 2$$
$$\lambda^2+\lambda-2 = 0$$
$$(\lambda+2)(\lambda-1) = 0$$
$$\lambda = 1$$

5.23 (0.072)

$P(H) = 0.3$

$P(T) = 0.7$

since all tosses are independent

so, probability of getting head for the first time in 5th toss is

$= P(T)\,P(T)\,P(T)\,P(T)\,P(H)$

$= 0.7 \times 0.7 \times 0.7 \times 0.7 \times 0.3 = 0.072$

5.24 (0.028)

When three dice are thrown

Total number of possible cases = 6 × 6 × 6 = 216

Favourable cases of all three dice have same number are,

$$\begin{Bmatrix} (1,1,1) & (2,2,2) & (3,3,3) \\ (4,4,4) & (5,5,5) & (6,6,6) \end{Bmatrix}$$

Number of favourable cases = 6

Required probability $= \dfrac{6}{216} = \dfrac{1}{36} = 0.028$

5.25 (3)

Probability of getting success $= \dfrac{4}{10}$

Probability of failure $= \dfrac{6}{10}$

$$E[X] = \sum x_i P(x_i)$$

$$E[X] = 1\times\frac{4}{10}+2\times\frac{4}{10}\times\frac{6}{10}+3\times\frac{4}{10}\left(\frac{6}{10}\right)^2+.... \quad ...(i)$$

$$\frac{6}{10}E[X] = 1\times\frac{4}{10}\times\frac{6}{10}+2\times\frac{4}{10}\times\left(\frac{6}{10}\right)^2 ... \quad ...(ii)$$

subtracting (ii) from (i), we get

$$\frac{4}{10}E[X] = \frac{4}{10}+\frac{4}{10}\times\frac{6}{10}+\frac{4}{10}\times\left(\frac{6}{10}\right)^2 ...$$

$$\frac{4}{10}E[X] = \frac{\frac{4}{10}}{1-\frac{6}{10}} = 1$$

$$E[X] = \frac{10}{4} = 2.5$$

To ensure his success, he requires on an average 3 attempts.

5.26 (0.25)

$$P(X_4 \text{ is smallest}) = \frac{3!}{4!} = \frac{1}{4} = 0.25$$

5.27 (11)

$E[2X+Y] = 0$ and $E[X+2Y] = 33$

then, $2E[X]+E[Y] = 0$ and $E[X]+2E[Y] = 33$

$3E[X]+3E[Y] = 0+33 = 33$

$E[X]+E[Y] = 11$

5.28 (0.37)

Required probability

$$= \frac{P[(z>2)\cap(z>1)]}{P[z>1]} = \frac{P[z>2]}{P[z>1]}$$

$P(z\le 2) = 1-e^{-2} \Rightarrow P(z>2) = e^{-2}$

$P(z\le 1) = 1-e^{-1} \Rightarrow P(z>1) = e^{-1}$

So,Required probability

$$= \frac{e^{-2}}{e^{-1}} = e^{-1} \simeq 0.37$$

5.29 (0.25)

$s = \{(\text{H, H}), (\text{H, T}), (\text{T, H}), (\text{T, T})\}$

$$M = \begin{bmatrix} 1 & 1 & 0 & 0 \\ H & H & T & T \end{bmatrix} \text{ of first toss}$$

$$N = \begin{bmatrix} H & T & H & T \\ 1 & 0 & 1 & 0 \end{bmatrix} \text{ of second toss}$$

Now, $X = \text{Min}\{M, N\}$

$\therefore$ $X = \text{Min}\{\text{H, H}\} = \text{Min}\{(1, 1)\} = 1$

$X = \text{Min}\{\text{H, T}\} = \text{Min}\{1, 0\} = 0$

$X = \text{Min}\{\text{T, H}\} = \text{Min}\{0, 1\} = 0$

$X = \text{Min}\{\text{T, T}\} = \text{Min}\{0, 0\} = 0$

$\therefore \quad P(X=1) = \dfrac{1}{4}, \quad P(X=0) = \dfrac{3}{4}$

We know that,

$$E(X) = \sum_i X_i\, P(x_i)$$

$$= 1\times\frac{1}{4}+0\times\frac{3}{4} = \frac{1}{4} = 0.25$$

5.30 (0.5)

Given that $P(X=2) = \alpha$

Entropy will be maximum; provided probabilities are equal.

i.e. $P(X=2) = P(X=-2) = \alpha = \dfrac{1}{2}$

$$\alpha = \frac{1}{2} = 0.5$$

5.31 (0.3)

x follows uniform distribution over [–2, 10]

$$\therefore \quad f(x) = \frac{1}{b-a} = \frac{1}{10-(-2)} = \frac{1}{12}$$

Given: $y = 2x - 6$

$$\Rightarrow \quad x = \frac{y+6}{2}$$

For $y = 7$

$$x = \frac{7+6}{2} = \frac{13}{2} = 6.5$$

$$P\left[\frac{y<7}{x>5}\right] = P\left[\frac{x<6.5}{x>5}\right]$$

$$= \frac{P[x>5 \text{ and } x<6.5]}{P[x>5]}$$

$$= \frac{P[5<x<6.5]}{P[x>5]} = \frac{\int_5^{6.5} f(x)dx}{\int_5^{10} f(x)dx} = \frac{\int_5^{6.5} \frac{1}{12}dx}{\int_5^{10} \frac{1}{12}dx}$$

$$= \frac{(x)_5^{6.5}}{(x)_5^{10}} = \frac{1.5}{5} = 0.3$$

5.32 (c)

Let,

$P(H_2)$ = Probability of getting head in second toss

$P(H_1)$ = Probability of getting head in first toss

$$P\left(\frac{H_2}{H_1}\right) = \frac{P(H_2 \cap H_1)}{P(H_1)}$$

$$P(H_1) = \frac{1}{3}\times\frac{1}{2}+\frac{1}{3}\times 1+\frac{1}{3}\times 0 = \frac{1}{2}$$

To get head in second toss when head came in first toss, following cases can be made

1. **Fair coin :**

$$\text{Both head coin} = \frac{1}{3}\times\frac{1}{2}\times\frac{1}{2}\times 1 = \frac{1}{12}$$

$$\text{Both tail coin} = \frac{1}{3}\times\frac{1}{2}\times\frac{1}{2}\times 0 = 0$$

2. **Both head coin :**

$$\text{Fair coin} = \frac{1}{3}\times 1\times\frac{1}{2}\times\frac{1}{2} = \frac{1}{12}$$

Both tail coin = 0

$$\therefore \quad P(H_2 \cap H_1) = \frac{1}{12}+\frac{1}{12} = \frac{1}{6}$$

$$\therefore \quad P(H_2/H_1) = \frac{1/6}{1/2} = \frac{1}{3}$$

5.33 (4)

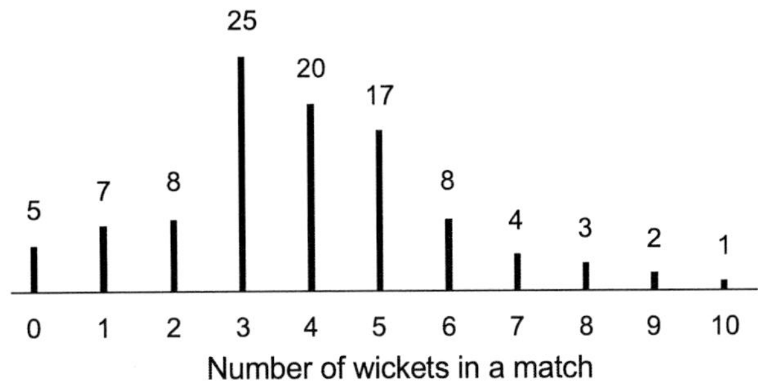

Number of wickets in a match	Number of matches (Frequency)	Cumulative frequency
0	5	5
1	7	12
2	8	20
3	25	45
4	20	65
5	17	82
6	8	90
7	4	94
8	3	97
9	2	99
10	1	100

The cumulative frequency is $N = 100$.

Since, N is even, the median is at the average of the $\left(\frac{N}{2}\right) = 50^{th}$ position and $\left(\frac{N}{2}+1\right) = 51^{st}$ position. The number of wickets for 50th position is 4 and for 51st position is also 4.

$$\therefore \quad \text{Median} = \frac{4+4}{2} = 4$$

■■■

6 Numerical Methods

6.1 Match the following and choose the correct combination:

Group 1
E. Newton-Raphson method
F. Runge-Kutta method
G. Simpson's Rule
H. Gauss elimination

Group 2
1. Solving nonlinear equations
2. Solving linear simultaneous equations
3. Solving ordinary differential equations
4. Numerical integration
5. Interpolation
6. Calculation of Eigenvalues

(a) E-6, F-1, G-5, H-3
(b) E-1, F-6, G-4, H-3
(c) E-1, F-3, G-4, H-2
(d) E-5, F-3, G-4, H-1 [2005 : 2 M]

6.2 For the function e^{-x}, the linear approximation around $x = 2$ is
(a) $(3-x)e^{-2}$
(b) $1-x$
(c) $\left[3+2\sqrt{2}-(1+\sqrt{2})x\right]e^{-2}$
(d) e^{-2} [2007 : 1 M]

6.3 The equation $x^3 - x^2 + 4x - 4 = 0$ is to be solved using the Newton-Raphson method. If $x = 2$ is taken as the initial approximation of the solution, then the next approximation using this method will be
(a) $\frac{2}{3}$ (b) $\frac{4}{3}$
(c) 1 (d) $\frac{3}{2}$ [2007 : 2 M]

6.4 The recursion relation to solve $x = e^{-x}$ using Newton-Raphson method is
(a) $x_{n+1} = e^{-x_n}$
(b) $x_{n+1} = x_n - e^{-x_n}$
(c) $x_{n+1} = (1+x_n)\dfrac{e^{-x_n}}{1+e^{-x_n}}$
(d) $x_{n+1} = \dfrac{x_n^2 - e^{-x_n}(1+x_n)-1}{x_n - e^{-x_n}}$ [2008 : 2 M]

6.5 Consider a differential equation $\dfrac{dy(x)}{dx} - y(x) = x$ with the initial condition $y(0) = 0$. Using Euler's first order method with a step size of 0.1, the value of $y(0.3)$ is
(a) 0.01 (b) 0.031
(c) 0.0631 (d) 0.1 [2010 : 2 M]

6.6 A numerical solution of the equation $f(x) = x + \sqrt{x} - 3 = 0$ can be obtained using Newton-Raphson method. If the starting value is $x = 2$ for the iteration, the value of x that is to be used in the next step is
(a) 0.306 (b) 0.739
(c) 1.694 (d) 2.306
[2011 : 2 M]

6.7 The maximum value of θ until which the approximation $\sin\theta \approx \theta$ holds to within 10% error is
(a) 10° (b) 18°
(c) 50° (d) 90° [2013 : 1 M]

6.8 A polynomial $f(x) = a_4x^4 + a_3x^3 + a_2x^2 + a_1x - a_0$ with all coefficients positive has
(a) no real roots
(b) no negative real root
(c) odd number of real roots
(d) at least one positive and one negative real root
[2013 : 1 M]

6.9 Match the application to appropriate numerical method.

Application
P1: Numerical integration
P2: Solution to a transcendental equation
P3: Solution to a system of linear equations
P4: Solution to a differential equation

Numerical Method
M1: Newton-Raphson Method
M2: Runge-Kutta Method
M3: Simpson's 1/3-rule
M4: Gauss Elimination Method

(a) P1—M3, P2—M2, P3—M4, P4—M1
(b) P1—M3, P2—M1, P3—M4, P4—M2
(c) P1—M4, P2—M1, P3—M3, P4—M2
(d) P1—M2, P2—M1, P3—M3, P4—M4

[2014 : 1 M, Set-3]

6.10 The Newton-Raphson method is used to solve the equation $f(x) = x^3 - 5x^2 + 6x - 8 = 0$. Taking the initial guess as $x = 5$, the solution obtained at the end of the first iteration is_____.

[2015 : 2 M, Set-3]

6.11 Consider the first order initial value problem

$$y' = y + 2x - x^2,\ y(0) = 1,\ (0 \le x < \infty$$

with exact solution $y(x) = x^2 + e^x$. For $x = 0.1$, the percentage difference between the exact solution and the solution obtained using a single iteration of the second-order Runge-Kutta method with step-size $h = 0.1$ is ___.

[2016 : 1 M, Set-3]

6.12 The ordinary differential equation

$$\frac{dx}{dt} = -3x + 2, \text{ with } x(0) = 1$$

is to be solved using the forward Euler method. The largest time step that can be used to solve the equation without making the numerical solution unstable is _______.

[2016 : 2 M, Set-2]

6.13 Starting with $x = 1$, the solution of the equation $x^3 + x = 1$, after two iterations of Newton-Raphson's method (up to two decimal places) is ______.

[2017 : 2 M, Set-1]

6.14 Consider $p(s) = s^3 + a_2 s^2 + a_1 s + a_0$ with all real coefficients. It is known that its derivative $p'(s)$ has no real roots. The number of real roots of $p(s)$ is

(a) 0 (b) 1
(c) 2 (d) 3

[2018 : 1 M]

■■■

Answers Numerical Methods

6.1	(c)	6.2	(a)	6.3	(b)	6.4	(c)	6.5	(b)	6.6	(c)	6.7	(b)
6.8	(d)	6.9	(b)	6.10	(4.2903)	6.11	(0.06)	6.12	(0.66)	6.13	(0.686)	6.14	(b)

Explanations Numerical Methods

6.2 (a)

$f(x) = e^{-x}$ at $x = 2$

The linear approximation of $f(x)$ at $x = a$ is

$$f(x) = f(a) + (x - a)\,f'(a)$$
$$e^{-x} = e^{-2} + (x - 2)(-e^{-2})$$
$$= (3 - x)e^{-2}$$

6.3 (b)

$$f(x) = x^3 - x^2 + 4x - 4$$
$$f'(x) = 3x^2 - 2x + 4$$
$$f(2) = 8$$
$$f'(2) = 12$$

$$\Rightarrow \quad x_1 = x_0 - \frac{f(x_0)}{f'(x_0)} = 2 - \frac{8}{12} = \frac{4}{3}$$

6.4 (c)

The given equation to be solved is

$$x = e^{-x}$$

Which can be rewritten as

$$f(x) = x - e^{-x} = 0$$
$$f'(x) = 1 + e^{-x}$$

The Newton-Raphson iterative formula is

$$x_{n+1} = x_n - \frac{f(x_n)}{f'(x_n)}$$

Here, $f(x_n) = x_n - e^{-x_n}$

$$f'(x_n) = 1 + e^{-x_n}$$

$$\therefore \quad x_{n+1} = x_n - \frac{x_n - e^{-x_n}}{1 + e^{-x_n}}$$

$$= \frac{e^{-x_n}x_n + e^{-x_n}}{1 + e^{-x_n}}$$

$$= (1 + x_n)\frac{e^{-x_n}}{1 + e^{-x_n}}$$

6.5 (b)

Given that $\frac{dy}{dx} = x + y$

x	y	$\frac{dy}{dx}$	$y' = y + \left(\frac{dy}{dx}\right) \times h$
0	0	0	0
0.1	0	0.1	0.01
0.2	0.01	0.21	0.031
0.3	0.031		

From here $y(0.3) = 0.031$

6.6 (c)

$$x_{n+1} = x_n - \frac{f(x_n)}{f'(x_n)}$$

$x = 2,\quad f(x_0) = 2 + \sqrt{2} - 3 = \sqrt{2} - 1$

$$f'(x) = 1 + \frac{1}{2\sqrt{x}}$$

$$f'(x_0) = 1 + \frac{1}{2\sqrt{2}}$$

Then, $x_1 = x_0 - \frac{f(x_0)}{f'(x_0)}$

$$= 2 - \frac{\sqrt{2}-1}{1 + \frac{1}{2\sqrt{2}}}$$

$\Rightarrow \quad x_1 = 1.694$

6.7 (b)

$$10° = \frac{10\pi}{180} = 0.1745$$

$\sin 10° = 0.1736$

So, for $10° \rightarrow \sin \cong \theta$ holds within 10% error

$$18° = \frac{18 \times \pi}{180} = 0.3142$$

$\sin 18° = 0.3090$

So, for $18° \rightarrow \sin\theta \cong \theta$ holds within 10% error

$$50° = \frac{50 \times \pi}{180} = 0.8727$$

$\sin 50° = 0.766$

So, for $50° \rightarrow \sin\theta \cong \theta$ does not hold within 10% error

$$90° = \frac{90 \times \pi}{180} = 1.571$$

$\sin 90° = 1$

So, for $90° \rightarrow \sin\theta \cong \theta$ does not hold within 10% error.

So, the maximum value of θ for the approximation $\sin\theta \cong \theta$ holds to within 10% error is 18°.

6.8 (d)

Using *R-H* criterion

$$\begin{array}{c|ccc} x^4 & a_4 & a_2 & -a_o \\ x^3 & a_3 & a_1 & \\ x^2 & A & & \\ x^1 & a_1 & & \\ x^0 & -a_0 & & \end{array}$$

Where, $A = \frac{a_3 a_2 - a_1 a_4}{a_3}$

So, from the above table it is clear that there is atleast one sign change in the first column. So, at least one positive and one negative real root.

6.10 (4.2903)

$$f(x) = x^3 - 5x^2 + 6x - 8$$
$$x_0 = 5$$
$$f'(x) = 3x^2 - 10x + 6$$
$$x_1 = x_0 - \frac{f(x_o)}{f'(x_o)}$$
$$= 5 - \frac{f(5)}{f'(5)}$$
$$= 5 - \frac{5^3 - 5 \times 5^2 + 6 \times 5 - 8}{3 \times 5^2 - 10 \times 5 + 6}$$
$$= 5 - \frac{22}{31} = 5 - 0.7097$$
$$= 4.2903$$

6.11 (0.06)

$$\frac{dy}{dx} = y + 2x - x^2$$

$y(0) = 1 \qquad 0 \le x \le \infty$

$f(x, y) = y + 2x - x^2$

$x_0 = 0$; $y_0 = 1$; $h = 0.1$

$$k_1 = hf(x_0, y_0)$$
$$= 0.1(1 + 2 \times 0 - 0^2) = 0.1$$

$k_2 = hf(x_0 + h, y_0 + k_1)$

$= 0.1((y_0 + k_1) + 2(x_0 + h) - (x_0 + h)^2)$

$= 0.1((1 + 0.1) + 2(0.1) - (0.1)^2)$

$= 0.129$

$$y_1 = y_0 + \frac{1}{2}(k_1 + k_2)$$

$$= 1 + \frac{1}{2}(0.1 + 0.129)$$

$$= 1.1145$$

Exact solution $y(x) = x^2 + e^x$

$$= (0.1)^2 + e^{0.1}$$

$$= 1.1152$$

$$\text{Error} = 1.1152 - 1.1145 = 0.00069$$

$$\%\text{ error} = 0.06\%$$

6.12 (0.66)

$$\frac{dy}{dx} = -3x + 2, \quad y(0) = 1$$

If $|1 - 3h| < 1$, then solution of differential equation is stable.

$$-1 < 1 - 3h < 1$$

$$-2 < -3h < 0$$

$$0 < h < \frac{2}{3}$$

$$h_{max} = \frac{2}{3} = 0.66$$

6.13 (0.686)

$$f(x) = x^3 + x - 1$$

$$f(1) = 1$$

$$f'(x) = 3x^2 + 1$$

$$f'(1) = 4$$

By Newton-Raphson method,

$$x_{i+1} = x_i - \frac{f(x_i)}{f'(x_i)}$$

For $x_0 = 1$, $x_1 = x_0 - \frac{f(x_0)}{f'(x_0)} = 1 - \frac{1}{4} = 0.75$

For $x_1 = 0.75$,

$$x_2 = x_1 - \frac{f(x_1)}{f'(x_1)} = 0.75 - \frac{f(0.75)}{f'(0.75)}$$

$$= 0.75 - \frac{0.171875}{2.6875} = 0.686$$

6.14 (b)

If $p(s)$ has "r" real roots, then $p'(s)$ will have atleast "$r - 1$" real roots.

■■■

7 Transform Theory

7.1 If $L\{f(t)\} = \dfrac{2(s+1)}{s^2+2s+5}$ then $f(0^+)$ and $f(\infty)$ are given by ______

(a) 0, 2 respectively (b) 2, 0 respectively
(c) 0, 1 respectively (d) 2/5, 0 respectively

[1995 : 2 M]

7.2 The inverse Laplace transform of the function $\dfrac{s+5}{(s+1)(s+3)}$ is ____

(a) $2e^{-t} - e^{-3t}$ (b) $2e^{-t} + e^{-3t}$
(c) $e^{-t} - 2e^{-3t}$ (d) $e^{-t} + 2e^{-3t}$

[1996 : 1 M]

7.3 The Laplace transform of $e^{\alpha t}\cos\alpha t$ is equal to ___.

(a) $\dfrac{s-\alpha}{(s-\alpha)^2+\alpha^2}$ (b) $\dfrac{s+\alpha}{(s+\alpha)^2+\alpha^2}$

(c) $\dfrac{1}{(s-\alpha)^2}$ (d) None

[1997 : 1 M]

7.4 If $L\{f(t)\} = \dfrac{\omega}{s^2+\omega^2}$ then the value of $\lim\limits_{t\to\infty} f(t) =$ ___.

(a) can not be determined
(b) zero
(c) unity
(d) Infinite

[1998 : 1 M]

7.5 If $L\{f(t)\} = F(s)$ then $L\{f(t-T)\}$ is equal to

(a) $e^{sT}F(s)$ (b) $e^{-sT}F(s)$

(c) $\dfrac{F(s)}{1-e^{sT}}$ (d) $\dfrac{F(s)}{1-e^{-sT}}$

[1999 : 1 M]

7.6 If $L\{f(t)\} = \dfrac{s+2}{s^2+1}$, $L\{g(t)\} = \dfrac{s^2+1}{(s+3)(s+2)}$

$h(t) = \int_0^t f(T)g(t-T)dT$ then $L\{h(t)\}$ is ____

(a) $\dfrac{s^2+1}{s+3}$

(b) $\dfrac{1}{s+3}$

(c) $\dfrac{s^2+1}{(s+3)(s+2)} + \dfrac{s+2}{s^2+1}$

(d) None of these

[2000 : 1 M]

7.7 The Laplace transform of $i(t)$ is given by $I(s) = \dfrac{2}{s(1+s)}$. As $t \to \infty$, the value of $i(t)$

(a) 0 (b) 1
(c) 2 (d) ∞

[2003 : 1 M]

7.8 In what range should Re(s) remain so that the Laplace transform of the function $e^{(a+2)t+5}$ exits.

(a) $\text{Re}(s) > a+2$ (b) $\text{Re}(s) > a+7$
(c) $\text{Re}(s) < 2$ (d) $\text{Re}(s) > a+5$

[2005 : 2 M]

7.9 A low-pass filter having a frequency response $H(j\omega) = A(\omega)e^{j\phi(\omega)}$ does not produce any phase distortion if

(a) $A(\omega) = C\omega^2, \phi(\omega) = k\omega^3$
(b) $A(\omega) = C\omega^2, \phi(\omega) = k\omega$
(c) $A(\omega) = C\omega, \phi(\omega) = k\omega^2$
(d) $A(\omega) = C, \phi(\omega) = k\omega^{-1}$

[2006 : 1 M]

7.10 Three functions $f_1(t)$, $f_2(t)$ and $f_3(t)$, which are zero outside the interval [0, T], are shown in the figure. Which of the following statements is correct?

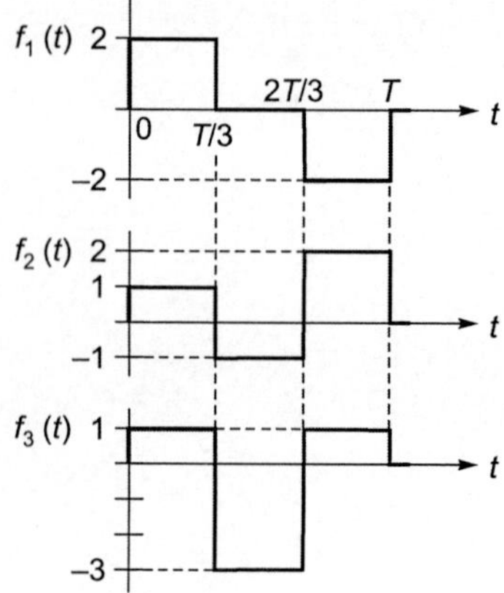

(a) $f_1(t)$ and $f_2(t)$ are orthogonal
(b) $f_1(t)$ and $f_3(t)$ are orthogonal
(c) $f_2(t)$ and $f_3(t)$ are orthogonal
(d) $f_1(t)$ and $f_2(t)$ are orthonormal

[2007 : 2 M

7.11 The trigonometric Fourier series for the waveform $f(t)$ shown below contains

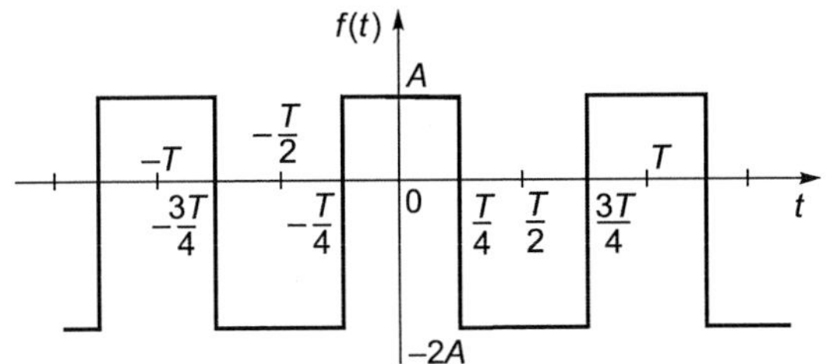

(a) only cosine terms and zero value for the dc component
(b) only cosine terms and a positive value for the dc component
(c) only cosine terms and a negative value for the dc component
(d) only sine terms and a negative value for the dc component

[2010 : 1 M]

7.12 Given $\mathcal{L}^{-1}\left[\dfrac{3s+1}{s^3+4s^2+(K-3)s}\right]$. If $\lim\limits_{t\to\infty} f(t) = 1$, then the value of K is

(a) 1 (b) 2
(c) 3 (d) 4

[2010 : 2 M]

7.13 Consider the differential equation

$$\frac{d^2y(t)}{dt^2} + 2\frac{dy(t)}{dt} + y(t) = \delta(t) \text{ with}$$

$$y(t)\Big|_{t=0} = -2 \text{ and } \frac{dy}{dt}\Big|_{t=0^-} = 0.$$

The numerical value of $\dfrac{dy}{dt}\Big|_{t=0^+}$ is

(a) − 2 (b) − 1
(c) 0 (d) 1

[2012 : 2 M]

7.14 Which one of the following is a property of the solutions to the Laplace equation: $\nabla^2 f = 0$?

(a) The solutions have neither maxima nor minima anywhere except at the boundaries.
(b) The solutions are not separable in the coordinates.
(c) The solutions are not continuous.
(d) The solutions are not dependent on the boundary conditions.

[2016 : 1 M, Set-1]

7.15 The position of a particle $y(t)$ is described by the differential equation:

$$\frac{d^2y}{dt^2} = -\frac{dy}{dt} - \frac{5y}{4}$$

The initial conditions are $y(0) = 1$ and $\dfrac{dy}{dt}\Big|_{t=0} = 0$.

The position (accurate to two decimal places) of the particle at $t = \pi$ is _______.

[2018 : 2 M]

7.16 The Fourier transform $X(\omega)$ of $x(t) = e^{-t^2}$ is

Note: $\int_{-\infty}^{\infty} e^{-y^2} dy = \sqrt{\pi}$

(a) $\sqrt{\pi}\, e^{\frac{\omega^2}{2}}$ (b) $\dfrac{e^{-\frac{\omega^2}{4}}}{2\sqrt{\pi}}$

(c) $\sqrt{\pi}\, e^{-\frac{\omega^2}{4}}$ (d) $\sqrt{\pi}\, e^{-\frac{\omega^2}{2}}$

[2023 : 1 M]

■■■

Answers Transform Theory

7.1 (b)	7.2 (a)	7.3 (a)	7.4 (a)	7.5 (b)	7.6 (b)	7.7 (c)
7.8 (a)	7.9 (b)	7.10 (c)	7.11 (c)	7.12 (d)	7.13 (d)	7.14 (a)
7.15 (–0.21)	7.16 (c)					

Explanations Transform Theory

7.1 (b)

$$F(s) = \frac{2(s+1)}{s^2+2s+5}$$

$$= \frac{2(s+1)}{(s+1)^2+4} = \frac{2(s+1)}{(s+1)^2+2^2}$$

By first shifting theorem

$$f(t) = 2e^{-t}\cos 2t$$

$$f(0^+) = \lim_{t\to 0} f(t) = \lim_{t\to 0} 2e^{-t}\cos 2t = 2$$

$$f(\infty) = \lim_{t\to\infty} f(t) = \lim_{t\to\infty} 2e^{-t}\cos 2t = 0$$

7.2 (a)

$$L^{-1}\left[\frac{s+5}{(s+1)(s+3)}\right] = L^{-1}\left[\frac{2}{s+1} - \frac{1}{s+3}\right]$$

$$= 2e^{-t} - e^{-3t}$$

7.3 (a)

$$L(\cos\alpha t) = \frac{s}{s^2+\alpha^2}$$

By first shifting theorem

$$L(e^{\alpha t}\cos\alpha t) = \frac{(s-\alpha)}{(s-\alpha)^2+\alpha^2}$$

7.4 (a)

$$L(f(t)) = \frac{\omega}{s^2+\omega^2}$$

$\therefore$ $f(t) = \sin\omega t$

$\therefore$ $\underset{t\to\infty}{Lt} f(t) = \underset{t\to\infty}{Lt} \sin\omega t$

= lies between –1 and 1

7.5 (b)

By second shifting theorem

If $L(f(t)) = F(s)$

then $L(f(t-T)) = e^{-sT}F(s)$

7.6 (b)

$$h(t) = \int_0^t f(T)\, g(t-T)\,dT$$

$$= f(t) * g(t)$$

$$L(h(t)) = L(f(t) * g(t))$$

$$= F(s)\cdot G(s)$$

$$= \frac{s+2}{s^2+1}\cdot\frac{s^2+1}{(s+3)(s+2)} = \frac{1}{s+3}$$

7.7 (c)

$$I(s) = \frac{2}{s(1+s)} = \frac{2}{s} - \frac{2}{1+s}$$

$$i(t) = 2 - 2e^{-t}$$

$$\lim_{t\to\infty} i(t) = \lim_{t\to\infty}(2-2e^{-t})$$

$$= 2$$

7.8 (a)

$$f(t) = e^{(a+2)t+5} = e^5 . e^{(a+2)t}$$

$$F(s) = \left[\frac{1}{s-(a+2)}\right]e^5$$

$\therefore$ For LT to exist, Re(s) > a + 2

7.9 (b)

For distortionless transmission,

$$\frac{d\phi(\omega)}{d\omega} = \text{constant}$$

Phase response should be linear

$$\phi(\omega) = K\omega$$

7.10 (c)

Two functions $f(x)$ & $g(x)$ are said to be orthogonal if $\int_{-\infty}^{\infty} f(x)\, g(x)\, dx = 0$

$$\int_{-\infty}^{\infty} f_2(t)\, f_3(t)\, dt = \int_0^T f_2(t)\, f_3(t)\, dt$$

$$= \frac{T}{3} - T + \frac{2T}{3}$$

$$= 0$$

7.11 (c)

Since $f(t)$ is an even function, its trigonometric Fourier series contains only cosine terms.

D.C. component,

$$A_0 = \frac{1}{T}\int_{-T/2}^{T/2} f(t)\,dt = \frac{2}{T}\int_{0}^{T/2} f(t)\,dt$$

$$= \frac{2}{T}\left[\int_0^{T/4} A\,dt + \int_{T/4}^{T/2} (-2A)\,dt\right]$$

$$= \frac{2}{T}\left[\frac{AT}{4} - 2A\left(\frac{T}{2}-\frac{T}{4}\right)\right]$$

$$= \frac{2}{T}\left[-\frac{AT}{4}\right] = -\frac{A}{2}$$

Therefore, the trigonometric Fourier series for the waveform $f(t)$ contains only cosine terms and a negative value for the dc component.

7.12 (d)

$$\lim_{t\to\infty} f(t) = \lim_{s\to 0} sF(s)$$

Given that,

$$F(s) = \left[\frac{3s+1}{s^3+4s^2+(K-3)s}\right]$$

$$\lim_{t\to\infty} f(t) = 1$$

$$\Rightarrow \lim_{s\to 0} s\left[\frac{3s+1}{s^3+4s^2+(K-3)s}\right] = 1$$

$$\Rightarrow \lim_{s\to 0}\left[\frac{3s+1}{s^2+4s+(K-3)}\right] = 1$$

$$\Rightarrow \frac{1}{K-3} = 1$$

$$\Rightarrow K-3 = 1$$

$$\Rightarrow K = 4$$

7.13 (d)

$$\frac{d^2y}{dt^2} + 2\frac{dy}{dt} + y(t) = \delta(t)$$

taking Laplace transform on both the sides we have

$$\left[s^2Y(s) - sy(0^-) - y'(0)\right] + 2\left[Y(s) - y(0^-)\right] + Y(s) = 1$$

$$s^2Y(s) + 2s + 2sY(s) + 4 + Y(s) = 1$$

$$(s^2 + 2s + 1)\,Y(s) = -(2s+3)$$

$$Y(s) = \frac{-(2s+3)}{(s+1)^2}$$

$$Y(s) = -\left[\frac{2}{(s+1)} + \frac{1}{(s+1)^2}\right]$$

$$\Rightarrow \quad Y(t) = -[2e^{-t} + te^{-t}]u(t)$$

$$\frac{dy}{dt} = -[-2e^{-t} + e^{-t} - te^{-t}]u(t)$$

$$\left.\frac{dy}{dt}\right|_{\text{at } t=0^+} = -[-2+1-0]$$

$$\left.\frac{dy}{dt}\right|_{\text{at } t=0^+} = 1$$

7.15 (–0.21)

$$\frac{d^2y}{dt^2} + \frac{dy}{dt} + \frac{5y}{4} = 0$$

$$y(0) = 1$$

$$y'(0) = 0$$

By applying Laplace transform,

$$s^2Y(s) - s(1) + sY(s) - 1 + \frac{5}{4}Y(s) = 0$$

$$Y(s) = \frac{s+1}{s^2+s+\frac{5}{4}} = \frac{s+1}{\left(s+\frac{1}{2}\right)^2+1}$$

$$= \frac{\left(s+\frac{1}{2}\right)}{\left(s+\frac{1}{2}\right)^2+1} + \frac{\frac{1}{2}}{\left(s+\frac{1}{2}\right)^2+1}$$

By taking inverse Laplace transform,

$$y(t) = e^{-t/2}\left[\cos(t) + \frac{1}{2}\sin(t)\right]\ ;\ t>0$$

At $t = \pi$,

$$y(t=\pi) = e^{-\pi/2}[(-1)+(0)]$$

$$= -e^{-\pi/2} = -0.2078 \simeq -0.21$$

7.16 (c)

We know that; $e^{-at^2};\ a>0 \xrightarrow{FT} \sqrt{\frac{\pi}{a}}\cdot e^{-\omega^2/4a}$

Here; $a = 1$

$$\therefore \quad X(\omega) = \sqrt{\pi}\cdot e^{-\omega^2/4}$$

■■■

UNIT X

General Aptitude

CONTENTS

General Aptitude

Syllabus

Verbal Aptitude : Basic English grammar: tenses, articles, adjectives, prepositions, conjunctions, verb-noun agreement, and other parts of speech Basic vocabulary: words, idioms, and phrases in context Reading and comprehension Narrative sequencing.

Quantitative Aptitude : Data interpretation: data graphs (bar graphs, pie charts, and other graphs representing data), 2- and 3-dimensional plots, maps, and tables Numerical computation and estimation: ratios, percentages, powers, exponents and logarithms, permutations and combinations, and series Mensuration and geometry Elementary statistics and probability

Analytical Aptitude : Logic: deduction and induction ; Analogy : Numerical relations and reasoning

Spatial Aptitude : Transformation of shapes: translation, rotation, scaling, mirroring, assembling, and grouping Paper folding, cutting, and patterns in 2 and 3 dimensions

Analysis of Previous GATE Papers

Exam Year	1 Mark Ques.	2 Mark Ques.	Total Marks
2010	5	5	**15**
2011	5	5	**15**
2012	5	5	**15**
2013	5	5	**15**
2014 Set-1	5	5	**15**
2014 Set-2	5	5	**15**
2014 Set-3	5	5	**15**
2014 Set-4	5	5	**15**
2015 Set-1	5	5	**15**
2015 Set-2	5	5	**15**
2015 Set-3	5	5	**15**

Exam Year	1 Mark Ques.	2 Mark Ques.	Total Marks
2016 Set-1	5	5	**15**
2016 Set-2	5	5	**15**
2016 Set-3	5	5	**15**
2017 Set-1	5	5	**15**
2017 Set-2	5	5	**15**
2018	5	5	**15**
2019	5	5	**15**
2020	5	5	**15**
2021	5	5	**15**
2022	5	5	**15**
2023	5	5	**15**
2024	5	5	**15**

1 Numerical Ability

1.1 25 persons are in a room. 15 of them play hockey, 17 of them play football and 10 of them play both hockey and football. Then the number of persons playing neither hockey nor football is

(a) 2 (b) 17
(c) 13 (d) 3

[2010 : 1 M]

1.2 If 137 + 276 = 435 how much is 731 + 672?

(a) 534 (b) 1403
(c) 1623 (d) 1531

[2010 : 2 M]

1.3 5 skilled workers can build a wall in 20 days; 8 semiskilled workers can build a wall in 25 days; 10 unskilled workers can build a wall in 30 days. If a team has 2 skilled, 6 semiskilled and 5 unskilled workers, how long will it take to build the wall?

(a) 20 days (b) 18 days
(c) 16 days (d) 15 days

[2010 : 2 M]

1.4 Given digits 2, 2, 3, 3, 3, 4, 4, 4, 4 how many distinct 4 digit numbers greater than 3000 can be formed?

(a) 50 (b) 51
(c) 52 (d) 54

[2010 : 2 M]

1.5 Hari (H), Gita (G), Irfan (I) and Saira (S) are siblings (i.e. brothers and sisters). All were born on Ist January. The age difference between any two successive siblings (that is born one after another) is less than 3 years. Given the following facts:

1. Hari's age + Gita's age > Irfan's age + Saira's age.
2. The age difference between Gita and Saira is 1 year. However, Gita is not the oldest and Saira is not the youngest.
3. There are no twins.

In what order were they born (oldest first)?

(a) HSIG (b) SGHI
(c) IGSH (d) IHSG

[2010 : 2 M]

1.6 There are two candidates P and Q in an election. During the campaign 40% of the voters promised to vote for P, and rest for Q. However, on the day of election 15% of the voters went back on their promise to vote for P and instead voted for Q. 25% of the voters went back on their promise to vote for Q and instead voted for P. Suppose, P lost by 2 votes, then what was the total number of voters?

(a) 100 (b) 110
(c) 90 (d) 95

[2011 : 1 M]

1.7 The fuel consumed by a motorcycle during a journey while travelling at various speeds is indicated in the graph below

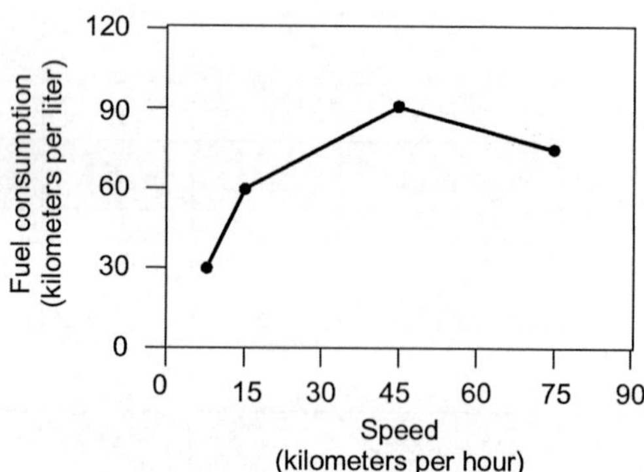

The distance covered during four laps of the journey are listed in the table below

Lap	Distance (kilometers)	Average speed (kilometers per hour)
P	15	15
Q	75	45
R	40	75
S	10	10

From the given data, we can conclude that the fuel consumed per kilometre was least during the lap

(a) P (b) Q
(c) R (d) S

[2011 : 2 M]

1.8 Three friends, *R*, *S* and *T* shared toffee from a bowl. *R* took 1/3rd of the toffees, but returned to the bowl. *S* took 1/4th of what was left but returned three toffees to the bowl. *T* took half of the remainder but returned two back into the bowl. If the bowl had 17 toffees left, how many toffees were originally there in the bowl?

(a) 38 (b) 31
(c) 48 (d) 41 [2011 : 2 M]

1.9 Given that $f(y) = \frac{|y|}{y}$, and q is any non-zero real number, the value of $|f(q) - f(-q)|$ is

(a) 0 (b) –1
(c) 1 (d) 2 [2011 : 2 M]

1.10 The sum of n terms of the series 4 + 44 + 444 + ... is

(a) $(4/81)[10^{n+1} - 9n - 1]$
(b) $(4/81)[10^{n-1} - 9n - 1]$
(c) $(4/81)[10^{n+1} - 9n - 10]$
(d) $(4/81)[10^{n} - 9n - 10]$ [2011 : 2 M]

1.11 If $(1.001)^{1259} = 3.52$ $(1.001)^{2062} = 7.85$, then $(1.001)^{3321}$

(a) 2.23 (b) 4.33
(c) 11.37 (d) 27.64 [2012 : 1 M]

1.12 The data given in the following table summarizes the monthly budget of an average household.

Category	Amount
Food	4000
Clothing	1200
Rent	2000
Savings	1500
Others	1800

The approximate percentage of the monthly budget **NOT** spent on savings is

(a) 10% (b) 14%
(c) 81% (d) 86%

[2012 : 2 M]

1.13 *A* and *B* friends. They decide to meet between 1 PM and 2 PM on a given day. There is a condition that whoever arrives first will not wait for the other for more than 15 minutes. The probability that they will meet on that day is

(a) 1/4 (b) 1/16
(c) 7/16 (d) 9/16

[2012 : 2 M]

1.14 Raju has 14 currency notes in his pocket consisting of only Rs. 20 notes and Rs. 10 notes. The total money value of the notes is Rs. 230. The number of Rs. 10 notes that Raju has is

(a) 5 (b) 6
(c) 9 (d) 10 [2012 : 2 M]

1.15 There are eight bags of rice looking alike, seven of which have equal and one is slightly heavier. The weighing balance is of unlimited capacity. Using this balance the minimum number of weighting required to identify the heavier bag is

(a) 2 (b) 3
(c) 4 (d) 8 [2012 : 2 M]

1.16 In the summer of 2012, in New Delhi, the mean temperature of Monday to Wednesday was 41°C and of Tuesday to Thursday was 43°C. If the temperature on Thursday was 15% higher than that of Monday, then the temperature in °C on Thursday was

(a) 40 (b) 43
(c) 46 (d) 49 [2013 : 1 M]

1.17 A car travels 8 km in the first quarter of an hour, 6 km in the second quarter and 16 km in the third quarter. The average speed of the car in km per hour over the entire journey is

(a) 30 (b) 36
(c) 40 (d) 24 [2013 : 2 M]

1.18 Find the sum to n terms of the series

10 + 84 + 734 +....

(a) $\frac{9(9^n + 1)}{10} + 1$ (b) $\frac{9(9^n - 1)}{8} + 1$

(c) $\frac{9(9^n - 1)}{8} + n$ (d) $\frac{9(9^n - 1)}{8} + n^2$

[2013 : 2 M]

1.19 The set of values of p for which the roots of the equation $3x^2 + 2x + p(p - 1) = 0$ are of opposite sign is

(a) $(-\infty, 0)$ (b) $(0, 1)$
(c) $(1, \infty)$ (d) $(0, \infty)$ [2013 : 2 M]

1.20 What is the change that a leap year, selected at random, will contain 53 Saturdays?

(a) $\frac{2}{7}$ (b) $\frac{3}{7}$

(c) $\frac{1}{7}$ (d) $\frac{5}{7}$

[2013 : 2 M]

1.21 The statistics of runs scored in a series by four batsmen are provided in the following table. Who is the most ***consistent*** batsman of these four?

Batsman	Average	Standard Deviation
K	31.2	5.21
L	46.0	6.35
M	54.4	6.22
N	17.9	5.90

(a) *K* (b) *L*
(c) *M* (d) *N*

[2014 : 1 M, Set-1]

1.22 What is the next number in the series?
12 35 81 173 357 _____.

[2014 : 1 M, Set-1]

1.23 Find the odd one from the following group:
WEKO IQWA FNTX NVBD
(a) WEKO (b) IQWA
(c) FNTX (d) NVBD

[2014 : 2 M, Set-1]

1.24 For submitting tax returns, all resident males with annual income below Rs.10 lakh should fill up Form *P* and all resident females with income below Rs.8 lakh should fill up Form *Q*. All people with incomes above Rs.10 lakh should fill up Form *R*, except non residents with income above Rs.15 lakhs, who should fill up Form *S*. All others should fill Form *T*. An example of a person who should fill Form *T* is
(a) a resident male with annual income Rs.9 lakh
(b) a resident female with annual income Rs.9 lakh
(c) a non-resident male with annual income Rs.16 lakh
(d) a non-resident female with annual income Rs. 16 lakh

[2014 : 2 M, Set-1]

1.25 A train that is 280 metres long, travelling at a uniform speed, crosses a platform in 60 seconds and passes a man standing on the platform in 20 seconds. What is the length of the platform in metres?

[2014 : 2 M, Set-1]

1.26 The exports and imports (in crores of Rs.) of a country from 2000 to 2007 are given in the following bar chart. If the trade deficit is defined as excess of imports over exports, in which year is the trade deficit 1/5th of the exports?

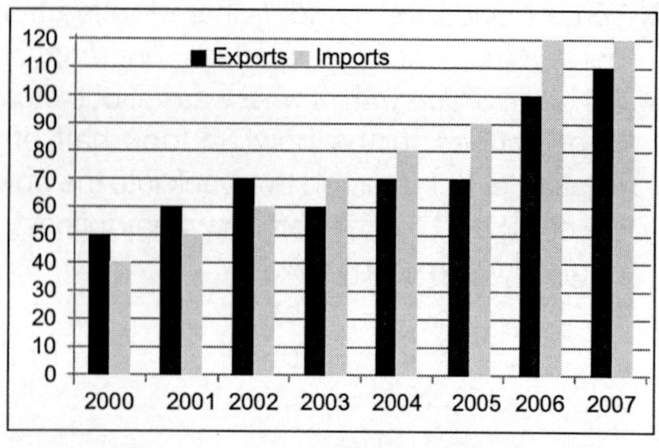

(a) 2005 (b) 2004
(c) 2007 (d) 2006

[2014 : 2 M, Set-1]

1.27 You are given three coins: one has heads on both faces, the second has tails on both faces, and the third has a head on one face and a tail on the other. You choose a coin at random and toss it, and it comes up heads. The probability that the other face is tails is
(a) 1/4 (b) 1/3
(c) 1/2 (d) 2/3

[2014 : 2 M, Set-1]

1.28 A regular die has six sides with numbers 1 to 6 marked on its sides. If a very large number of throws show the following frequencies of occurrence:

$1 \rightarrow 0.167$; $2 \rightarrow 0.167$; $3 \rightarrow 0.152$; $4 \rightarrow 0.166$; $5 \rightarrow 0.168$; $6 \rightarrow 0.180$. We call this die
(a) irregular (b) biased
(c) Gaussian (d) insufficient

[2014 : 1 M, Set-2]

1.29 Fill in the missing number in the series.
2 3 6 15 _?_ 157.5 630

[2014 : 1 M, Set-2]

1.30 Find the odd one in the following group
QWZB, BHKM, WCGJ, MSVX,
(a) QWZB (b) BHKM
(c) WCGJ (d) MSVX

[2014 : 2 M, Set-2]

1.31 Lights of four colors (red, blue, green, yellow) are hung on a ladder. On every step of the ladder there are two lights. If one of the lights is red, the other light on that step will always be blue. If one of the lights on a step is green, the other light on that step will always be yellow. Which of the following statements is not necessarily correct?

(a) The number of red lights is equal to the number of blue lights
(b) The number of green lights is equal to the number of yellow lights
(c) The sum of the red and green lights is equal to the sum of the yellow and blue lights
(d) The sum of the red and blue lights is equal to the sum of the green and yellow lights

[2014 : 2 M, Set-2]

1.32 The sum of eight consecutive odd numbers is 656. The average of four consecutive even numbers is 87. What is the sum of the smallest odd number and second largest even number?

[2014 : 2 M, Set-2]

1.3 The total exports and revenues from the exports of a country are given in the two charts shown below. The pie chart for exports shows the quantity of each item exported as a percentage of the total quantity of exports. The pie chart for the revenues shows the percentage of the total revenue generated through export of each item. The total quantity of exports of all the items is 500 thousand tonnes and the total revenues are 250 crore rupees. Which item among the following has generated the maximum revenue per kg?

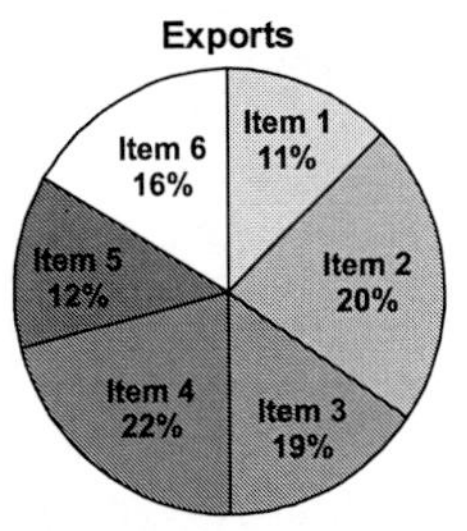

Revenues
Item 1 12%
Item 2 20%
Item 3 23%
Item 4 6%
Item 5 20%
Item 6 19%

(a) Item 2
(b) Item 3
(c) Item 6
(d) Item 5

[2014 : 2 M, Set-2]

1.34 It takes 30 minutes to empty a half-full tank by draining it at a constant rate. It is decided to simultaneously pump water into the half-full tank while draining it. What is the rate at which water has to be pumped in so that it gets fully filled in 10 minutes?
(a) 4 times the draining rate
(b) 3 times the draining rate
(c) 2.5 times the draining rate
(d) 2 times the draining rate

[2014 : 2 M, Set-2]

1.35 "India is a country of rich heritage and cultural diversity." Which one of the following facts best supports the claim made in the above sentence?
(a) India is a union of 28 states and 7 union territories.
(b) India has a population of over 1.1 billion.
(c) India is home to 22 official languages and thousands of dialects.
(d) The Indian cricket team draws players from over ten states.

[2014 : 1 M, Set-3]

1.36 The next term in the series 81, 54, 36, 24, ... is ____.

[2014 : 1 M, Set-3]

1.37 In which of the following options will the expression $P < M$ be definitely true?
(a) $M < R > P > S$
(b) $M > S < P < F$
(c) $Q < M < F = P$
(d) $P = A < R < M$

[2014 : 1 M, Set-3]

1.38 Find the next term in the sequence:
7 *G*, 11 *K*, 13 *M*, _____.
(a) 15 *Q*
(b) 17 *Q*
(c) 15 *P*
(d) 17 *P*

[2014 : 2 M, Set-3]

1.39 The multi-level hierarchical pie chart shows the population of animals in a reserve forest. The correct conclusions from this information are:

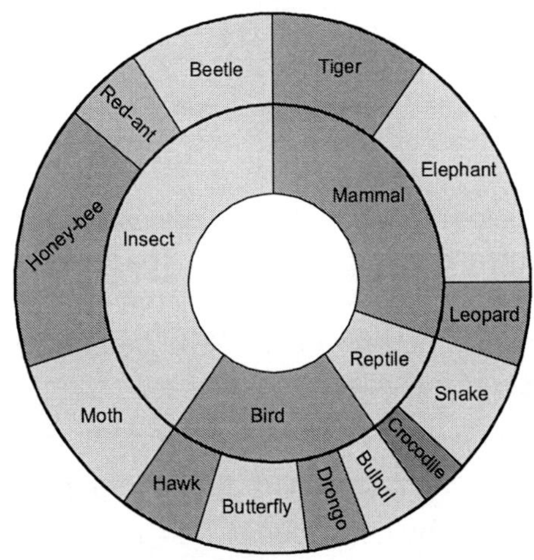

(i) Butterflies are birds
(ii) There are more tigers in this forest than red ants
(iii) All reptiles in this forest are either snakes or crocodiles
(iv) Elephants are the largest mammals in this forest
(a) (i) and (ii) only
(b) (i), (ii), (iii) and (iv)
(c) (i), (iii) and (iv) only
(d) (i), (ii) and (iii) only

[2014 : 2 M, Set-3]

1.40 A man can row at 8 km per hour in still water. If it takes him thrice as long to row upstream, as to row downstream, then find the stream velocity in km per hour.

[2014 : 2 M, Set-3]

1.41 A firm producing air purifiers sold 200 units in 2012. The following pie chart presents the share of raw material, labour, energy, plant & machinery, and transportation costs in the total manufacturing cost of the firm in 2012. The expenditure on labour in 2012 is Rs. 4,50,000. In 2013, the raw material expenses increased by 30% and all other expenses increased by 20%. If the company registered a profit of Rs. 10 lakhs in 2012, at what price (in Rs.) was each air purifier sold?

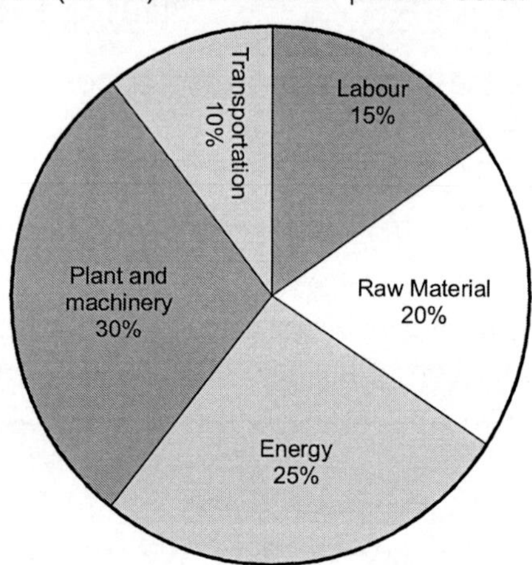

[2014 : 2 M, Set-3]

1.42 A batch of one hundred bulbs is inspected by testing four randomly chosen bulbs. The batch is rejected if even one of the bulbs is defective. A batch typically has five defective bulbs. The probability that the current batch is accepted is ______.

[2014 : 2 M, Set-3]

1.43 Let $f(x, y) = x^n y^m = P$. If x is doubled and y is halved, the new value of f is

(a) $2^{n-m}P$ (b) $2^{m-n}P$
(c) $2(n-m)P$ (d) $2(m-n)P$

[2014 : 1 M, Set-4]

1.44 In a sequence of 12 consecutive odd numbers, the sum of the first 5 numbers is 425. What is the sum of the last 5 numbers in the sequence?

[2014 : 1 M, Set-4]

1.45 Find the next term in the sequence:
13M, 17Q, 19S, ____.

(a) 21 W (b) 21 V
(c) 23 W (d) 23 V

[2014 : 2 M, Set-4]

1.46 If 'KCLFTSB' stands for 'best of luck' and 'SHSWDG' stands for 'good wishes', which of the following indicates 'ace the exam'?

(a) MCHTX (b) MXHTC
(c) XMHCT (d) XMHTC

[2014 : 2 M, Set-4]

1.47 Industrial consumption of power doubled from 2000-2001 to 2010-2011. Find the annual rate of increase in percent assuming it to be uniform over the years.

(a) 5.6 (b) 7.2
(c) 10.0 (d) 12.2

[2014 : 2 M, Set-4]

1.48 A firm producing air purifiers sold 200 units in 2012. The following pie chart presents the share of raw material, labour, energy, plant & machinery, and transportation costs in the total manufacturing cost of the firm in 2012. The expenditure on labour in 2012 is Rs.4,50,000. In 2013, the raw material expenses increased by 30% and all other expenses increased by 20%. What is the percentage increase in total cost for the company in 2013?

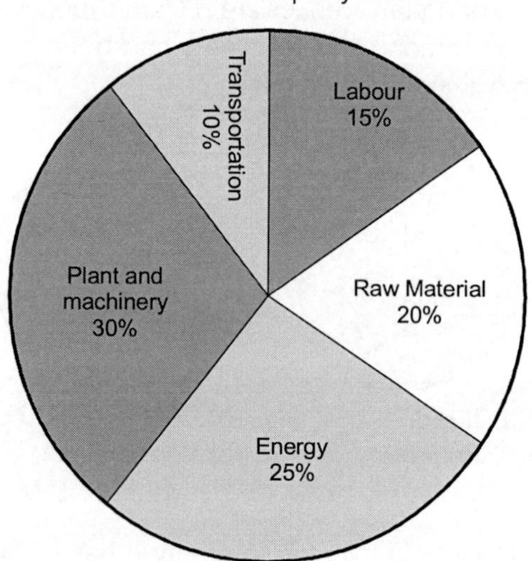

[2014 : 2 M, Set-4]

1.49 A five digit number is formed using the digits 1,3,5,7 and 9 without repeating any of them. What is the sum of all such possible five digit numbers?

(a) 6666660 (b) 6666600
(c) 6666666 (d) 6666606

[2014 : 2 M, Set-4]

1.50 If $\log_x (5/7) = -1/3$, then the value of x is

(a) 343/125 (b) 125/343
(c) –25/49 (d) –49/25

[2015 : 1 M, Set-1]

1.51 Operates □, ◇ and → are defined by:

$a \square b = \frac{a-b}{a+b}$; $a \diamond b = \frac{a+b}{a-b}$; $a \to b = ab$. Find the value of $(66 \square 6) \to (66 \diamond 6)$

(a) –2 (b) –1
(c) 1 (d) 2 [2015 : 1 M, Set-1]

1.52 A cube side 3 units is formed using a set of smaller cubes of side 1 unit. Find the proportion of the number of faces of the smaller cubes visible to those which are NOT visible.

(a) 1 : 4 (b) 1 : 3
(c) 1 : 2 (d) 2 : 3

[2015 : 2 M, Set-1]

1.53 Fill in the missing value

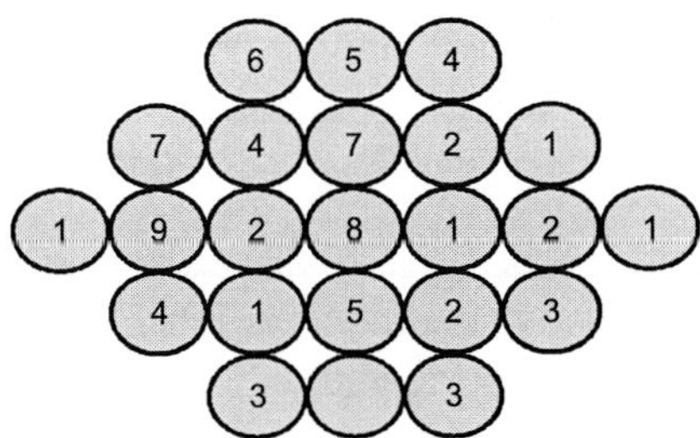

[2015 : 2 M, Set-1]

1.54 An electric bus has onboard instruments that report the total electricity consumed since the start of the trip as well as the total distance covered. During a single day of operation, the bus travels on stretches *M*, *N*, *O* and *P*, in that order. The cumulative distance travelled and the corresponding electricity consumption are shown in the Table below:

Stretch	Cumulative distance (km)	Electricity used (kWh)
M	20	12
N	45	25
O	75	45
P	100	57

The stretch where the electricity consumption per km is minimum is

(a) *M* (b) *N*
(c) *O* (d) *P*

[2015 : 1 M, Set-2]

1.55 Ram and Ramesh appeared in an interview for two vacancies in the same department. The probability of Ram's selection is 1/6 and that of Ramesh is 1/8. What is the probability that only one of them will be selected?

(a) 47/48 (b) 1/4
(c) 13/48 (d) 35/48

[2015 : 1 M, Set-2]

1.56 If $a^2 + b^2 + c = 1$, then $ab + bc + ac$ lies in the interval

(a) [1, 2/3] (b) [–1/2, 1]
(c) [–1, 1/2] (d) [2, –4]

[2015 : 2 M, Set-2]

1.57 A tiger is 50 leaps of its own behind a deer. The tiger takes 5 leaps per minute to the deer's 4. If the tiger and the deer cover 8 metre and 5 metre per leap respectively, what distance in metres will the tiger have to run before it catches the deer?

[2015 : 2 M, Set-2]

1.58 If $x > y > 1$, which of the following must be true?

(i) $\ln x > \ln y$ (ii) $e^x > e^y$
(iii) $y^x > x^y$ (iv) $\cos x > \cos y$

(a) (i) and (ii) (b) (i) and (iii)
(c) (iii) and (iv) (d) (ii) and (iv)

[2015 : 1 M, Set-3]

1.59 Ms. X will be Bagdogra from 01/05/2014 to 20/05/2014 and from 22/05/2014 to 31/05/2014. On the morning of 21/05/2014, she will reach Kochi via Mumbai.

Which one of the statements below is logically valid and can be inferred from the above sentences?

(a) Ms. X will be in Kochi for one day, only in May.
(b) Ms. X will be in Kochi for only one day in May.
(c) Ms. X will be only in Kochi for one day in May.
(d) Only Ms. X will be in Kochi for one day in May.

[2015 : 2 M, Set-3]

1.60 log tan1° + log tan2° +......+ log tan89° is ______.

(a) 1 (b) $\frac{1}{\sqrt{2}}$
(c) 0 (d) –1

[2015 : 2 M, Set-3]

1.61 From a circular sheet of paper of radius 30 cm, a sector of 10% area is removed. If the remaining part is used to make a conical surface, then the ratio of the radius and height of the cone is ______.

[2015 : 2 M, Set-3]

1.62 Find the missing sequence in the letter series below:
A, CD, GHI, ?, UVWXY

(a) LMN (b) MNO
(c) MNOP (d) NOPQ

[2015 : 1 M, Set-3]

1.63 In a huge pile of apples and oranges, both ripe and unripe mixed together, 15% are unripe fruits. Of the unripe fruits, 45% are apples. Of the ripe ones, 66% are oranges. If the pile contains a total of 5692000 fruits, how many of them are apples?

(a) 2029198 (b) 2467482
(c) 2789080 (d) 3577422

[2016 : 1 M, Set-1]

1.64 Michael lives 10 km away from where I live. Ahmed lives 5 km away and Susan lives 7 km away from where *I* live. Arun is farther away than Ahmed but closer than Susan from where *I* live. From the information provided here, what is one possible distance (in km) at which *I* live from Arun's place?

(a) 3.00 (b) 4.99
(c) 6.02 (d) 7.01

[2016 : 1 M, Set-1]

1.65 A person moving through a tuberculosis prone zone has a 50% probability of becoming infected. However, only 30% of infected people develop the disease. What percentage of people moving through a tuberculosis prone zone remains infected but does not show symptoms of disease?

(a) 15 (b) 33
(c) 35 (d) 37

[2016 : 2 M, Set-1]

1.66 Leela is older than her cousin Pavithra, Pavithra's brother Shiva is older than Leela. When Pavithra and Shiva are visiting Leela, all there like to play chess. Pavithra wins more often than Leela does. Which one of the following statements must be TRUE based on the above?

(a) When Shiva plays chess with Leela and Pavithra, he often loses.
(b) Leela is the oldest of three.
(c) Shiva is better chess player than Pavithra.
(d) Pavithra is the youngest of the three.

[2016 : 2 M, Set-1]

1.67 If $q^{-a} = \frac{1}{r}$ and $r^{-b} = \frac{1}{s}$ and $s^{-c} = \frac{1}{q}$, the value of abc is _____.

(a) $(rqs)^{-1}$ (b) 0
(c) 1 (d) $r + q + s$

[2016 : 2 M, Set-1]

1.68 *P*, *Q*, *R* and *S* are working on a project. *Q* can finish the task in 25 days, working alone for 12 hours a day. *R* can finish the task in 50 days, working alone for 12 hours per day. *Q* worked 12 hours a day but took sick leave in the beginning for two days. *R* worked 18 hours a day on all days. What is the ratio of work done by *Q* and *R* after 7 days from the start of the projects?

(a) 10:11 (b) 11:10
(c) 20:21 (d) 21:20

[2016 : 2 M, Set-1]

1.69 Based on the given statements, select the appropriate option with respect to grammar and usage. Statements

I. The height of Mr. *X* is 6 feet.
II. The height of Mr. *Y* is 5 feet.

(a) Mr. *X* is longer than Mr. *Y*.
(b) Mr. *X* is more elongated than Mr. *Y*
(c) Mr. *X* is taller than Mr. *Y*
(d) Mr. *X* is lengthier than Mr. *Y*

[2016 : 1 M, Set-2]

1.70 Given (9 inches)$^{1/2}$ = (0.25 yards)$^{1/2}$, which one of the following statements is TRUE?

(a) 3 inches = 0.5 yards
(b) 9 inches = 1.5 yards
(c) 9 inches = 0.25 yards
(d) 81 inches = 0.0625 yards

[2016 : 1 M, Set-2]

1.71 *S*,*M*,*E* and *F* are working in shifts in a team to finish a project. ***M*** works with twice the efficiency of others but for half as many days as ***E*** worked. ***S*** and ***M*** have 6 hour shifts in a day. whereas ***E*** and ***F*** have 12 hours shifts. What is the ratio of contribution of ***M*** to contribution of ***E*** in the project?

(a) 1 : 1 (b) 1 : 2
(c) 1 : 4 (d) 2 : 1

[2016 : 1 M, Set-2]

1.72 The Venn diagram shows the preference of the student population for leisure activities.

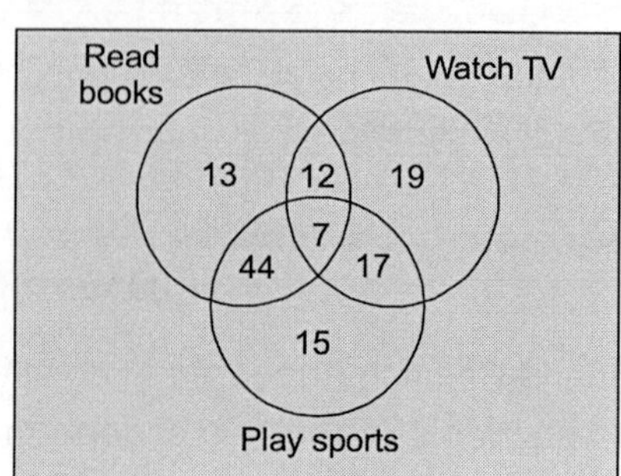

From the data given, the number of students who like to read books or play sports is ____.

(a) 44 (b) 51
(c) 79 (d) 108

[2016 : 2 M, Set-2]

1.73 Two and a quarter hours back, when seen in a mirror, the reflection of a wall clock without number markings seemed to show 1 : 30. What is the actual current time shown by the clock?

(a) 8:15 (b) 11:15
(c) 12:15 (d) 12:45

[2016 : 2 M, Set-2]

1.74 *M* and *N* start from the same location. *M* travels 10 km East and then 10 km North-East. *N* travels 5 km South and then 4 km south-East. What is the shortest distance (in km) between *M* and *N* at the end of their travel?

(a) 18.60 (b) 22.50
(c) 20.61 (d) 25.00

[2016 : 2 M, Set-2]

1.75 A wire of length 340 mm is to be cut into two parts. One of the parts is to be made into a square and the other into a rectangle where sides are in the ratio of 1:2. What is the length of the side of the square (in mm) such that the combined area of the square and the rectangle is a MINIMUM?

(a) 30 (b) 40
(c) 120 (d) 180

[2016 : 2 M, Set-2]

1.76 *M* has a son *Q* and a daughter *R*. He has no other children. *E* is the mother of *P* and daughter-in-law of *M*. How is *P* related to *M*?

(a) *P* is the son-in-law of *M*
(b) *P* is the grandchild of *M*
(c) *P* is the daughter-in law of *M*
(d) *P* is the grandfather of *M*

[2016 : 1 M, Set-3]

1.77 The number that least fits this set: (324, 441, 97 and 64) is ________.

(a) 324 (b) 441
(c) 97 (d) 64

[2016 : 1 M, Set-3]

1.78 It takes 10s and 15s, respectively, for two trains travelling at different constant speeds to completely pass a telegraph post. The length of the first train is 120 m and that of the second train is 150 m. The magnitude of the difference in the speeds of the two trains (in m/s) is ________.

(a) 2.0 (b) 10.0
(c) 12.0 (d) 22.0

[2016 : 1 M, Set-3]

1.79 The velocity *V* of a vehicle along a straight line is measured in m/s and plotted as shown with respect to time in seconds. At the end of the 7 seconds, how much will the odometer reading increase by (in m)?

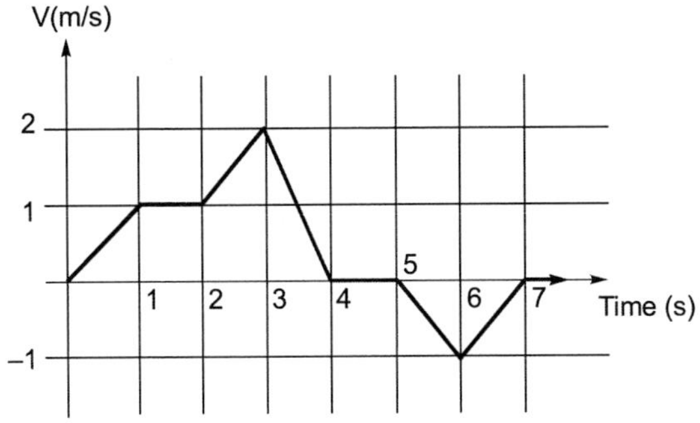

(a) 0 (b) 3
(c) 4 (d) 5

[2016 : 2 M, Set-3]

1.80 A flat is shared by four first year undergraduate students. They agreed to allow the oldest of them to enjoy some extra space in the flat. Manu is two months older than Sravan, who is three months younger than Trideep. Pavan is one month older than Sravan. Who should occupy the extra space in the flat?

(a) Manu (b) Sravan
(c) Trideep (d) Pavan

[2016 : 2 M, Set-3]

1.81 Find the area bounded by the lines $3x + 2y = 14$, $2x - 3y = 5$ in the first quadrant.

(a) 14.95 (b) 15.25
(c) 15.70 (d) 20.35

[2016 : 2 M, Set-3]

1.82 A straight line is fit to a data set ($ln\, x$, y). This line intercepts the abscissa at $\ln x = 0.1$ and has a slope of -0.02. What is the value of y at $x = 5$ from the fit?

(a) -0.030 (b) -0.014
(c) 0.014 (d) 0.030

[2016 : 2 M, Set-3]

1.83 In the summer, water consumption is known to decrease overall by 25%. A Water Board official states that in the summer household consumption decreases by 20%, while other consumption increases by 70%.

Which of the following statements is correct?
(a) The ratio of household to other consumption is 8/17.
(b) The ratio of household to other consumption is 1/17.
(c) The ratio of household to other consumption is 17/8.
(d) There are errors in the official's statement.
[2017 : 1 M, Set-1]

1.84 Some tables are shelves. Some shelves are chairs. All chairs are benches. Which of the following conclusions can be deduced from the preceding sentences?
(i) At least one bench is a table
(ii) At least one shelf is a bench
(iii) At least one chair is a table
(iv) All benches are chairs
(a) only (i) (b) only (ii)
(c) only (ii) and (iii) (d) only (iv)
[2017 : 1 M, Set-1]

1.85 40% of deaths on city roads may be attributed to drunken driving. The number of degrees needed to represent this as a slice of a pie chart is
(a) 120 (b) 144
(c) 160 (d) 212
[2017 : 1 M, Set-1]

1.86 Trucks (10 m long) and cars (5 m long) go on a single lane bridge. There must be a gap of at least 20 m after each truck and a gap of at least 15 m after each car. Trucks and cars travel at a speed of 36 km/h. If cars and trucks go alternately. What is the maximum number of vehicles that can use the bridge in one hour?
(a) 1440 (b) 1200
(c) 720 (d) 600
[2017 : 2 M, Set-1]

1.87 A contour line joins locations having the same height above the mean sea level. The following is a contour plot of a geographical region. Contour lines are shown at 25 m intervals in this plot.

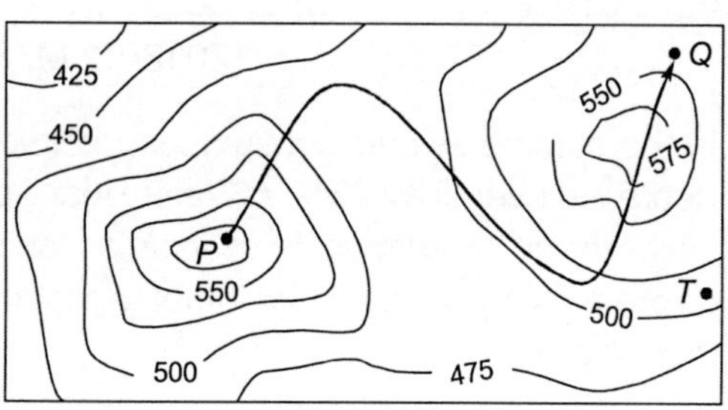

The path from P to Q is best described by
(a) Up-Down-Up-Down
(b) Down-Up-Down-Up
(c) Down-Up-Down
(d) Up-Down-Up
[2017 : 2 M, Set-1]

1.88 There are 3 Indians and 3 Chinese in a group of 6 people. How many subgroups of this group can we choose so that every subgroup has at least one Indian?
(a) 56 (b) 52
(c) 48 (d) 44
[2017 : 2 M, Set-1]

1.89 S, T, U, V, W, X, Y and Z are seated around a circular table. 'T's neighbours are Y and V. Z is seated third to the left of T and second to the right of S. U's neighbours are S and Y; and T and W are not seated opposite each other. Who is third to the left of V?
(a) X (b) W
(c) U (d) T
[2017 : 2 M, Set-1]

1.90 A rule states that in order to drink beer, one must be over 18 years old. In a bar there are 4 people. P is 16 years old, Q is 25 years old, R is drinking milkshake and S is drinking a beer. What must be checked to ensure that the rule is being followed?
(a) Only P's drink
(b) Only P's drink and S's age
(c) Only S's age
(d) Only P's drink. Q's drink and S's age
[2017 : 1 M, Set-2]

1.91 Fatima starts from point P, goes North for 3 km and then East for 4 km to reach point Q. She then turns to face point P and goes 15 km in that direction. She then goes North for 6 km. How far is she from point P and in which direction should she go to reach point P?
(a) 8 km, East (b) 12 km, North
(c) 6 km, East (d) 10 km, North
[2017 : 1 M, Set-2]

1.92 500 students are taking one or more courses out of Chemistry, Physics and Mathematics. Registration records indicate course enrolment as follows: Chemistry (329), Physics (186), Mathematics (295), Chemistry and Physics (83), Chemistry and Mathematics (217) and Physics and Mathematics (63). How many students are taking all 3 subjects?

(a) 37 (b) 43
(c) 47 (d) 53

[2017 : 1 M, Set-2]

1.93 A contour line joins locations having the same height above the mean sea level. The following is a contour plot of a geographical region. Contour lines are shown at 25 m intervals in this plot.

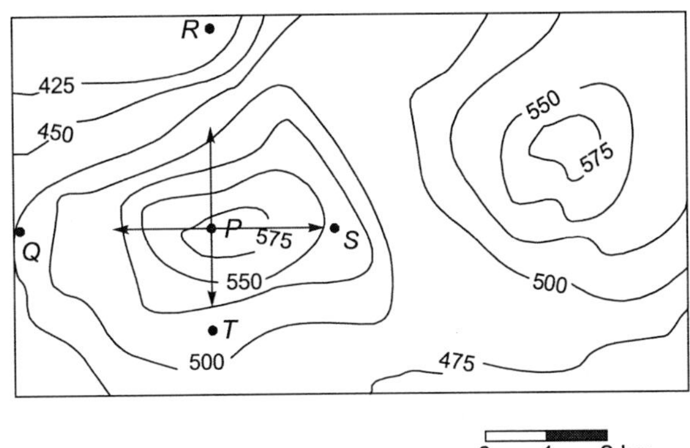

Which of the following is the steepest path leaving from *P*?

(a) *P* to *Q* (b) *P* to *R*
(c) *P* to *S* (d) *P* to *T*

[2017 : 2 M, Set-2]

1.94 1200 men and 500 women can build a bridge in 2 weeks, 900 men and 250 women will take 3 weeks to build the same bridge. How many men will be needed to build the bridge in one week?

(a) 3000 (b) 3300
(c) 3600 (d) 3900

[2017 : 2 M, Set-2]

1.95 Each of *P*, *Q*, *R*, *S*, *W*, *X*, *Y* and *Z* has been married at most once. *X* and *Y* are married and have two children *P* and *Q*. *Z* is the grandfather of the daughter *S* of *P*. Further, *Z* and *W* are married and are parents of *R*. Which one of the following must necessarily be FALSE?

(a) *X* is the mother-in-law of *R*
(b) *P* and *R* are not married to each other
(c) *P* is a son of *X* and *Y*
(d) *Q* cannot be married to *R*

[2017 : 2 M, Set-2]

1.96 The number of 3-digit numbers such that the digit 1 is never to the immediate right of 2 is

(a) 781 (b) 791
(c) 881 (d) 891

[2017 : 2 M, Set-2]

1.97 What is the value of $1+\frac{1}{4}+\frac{1}{16}+\frac{1}{64}+\frac{1}{256}+....?$

(a) 2 (b) $\frac{7}{4}$
(c) $\frac{3}{2}$ (d) $\frac{4}{3}$

[2018 : 1 M]

1.98 A 1.5 m in tall person is standing at a distance of 3 m from a lamp post. The light from the lamp at the top of the post casts her shadow. The length of the shadow is twice her height. What is the height of the lamp post in meters?

(a) 1.5 (b) 3
(c) 4.5 (d) 6

[2018 : 1 M]

1.99 If the number 715 ? 423 is divisible 3 (? denotes the missing digit in the thousandths place), then the smallest whole number in the place of ? is ______.

(a) 0 (b) 2
(c) 5 (d) 6

[2018 : 1 M]

1.100 Two alloys *A* and *B* contain gold and copper in the ratios of 2 : 3 and 3 : 7 by mass, respectively. Equal masses of alloys *A* and *B* are melted to make an alloy *C*. The ratio of gold to copper in alloy *C* is ______.

(a) 5 : 10 (b) 7 : 13
(c) 6 : 11 (d) 9 : 13

[2018 : 2 M]

1.101 Leila aspires to buy a car worth Rs.10,00,000 after 5 years. What is the minimum amount in Rupees that she should deposit now in a bank which offers 10% annual rate of interest, if the interest was compounded annually?

(a) 5,00,000 (b) 6,21,000
(c) 6,66,667 (d) 7,50,000

[2018 : 2 M]

1.102 A cab was involved in a hit and run accident at night You are given the following data about the cabs in the city and the accident.

(i) 85% of cabs in the city are green and the remaining cabs are blue.
(ii) A witness identified the cab involved in the accident as blue.

(iii) It is known that a witness can correctly identify the cab colour only 80% of the time.

Which of the following options is closest to the probability that the accident was caused by a blue cab?

(a) 12% (b) 15%
(c) 41% (d) 80%

[2018 : 2 M]

1.103 It would take one machine 4 hours to complete a production order and another machine 2 hours to complete the same order. If both machines work simultaneously at their respective constant rates, the time taken to complete the same order is ____ hours.

(a) 7/3 (b) 3/4
(c) 2/3 (d) 4/3 [2019 : 1 M]

1.104 Five different books (P, Q, R, S, T) are to be arranged on a shelf. The books R and S are to be arranged first and second, respectively from the right side of the shelf. The number of different orders in which P, Q and T may be arranged is _____.

(a) 12 (b) 2
(c) 120 (d) 6 [2019 : 1 M]

1.105 The bar graph in Panel (a) shows the proportion of male and female illiterates in 2001 and 2011. The proportions of males and females in 2001 and 2011 are given in Panel (b) and (c), respectively. The total population did not change during this period.

The percentage increase in the total number of literates from 2001 to 2011 is ____.

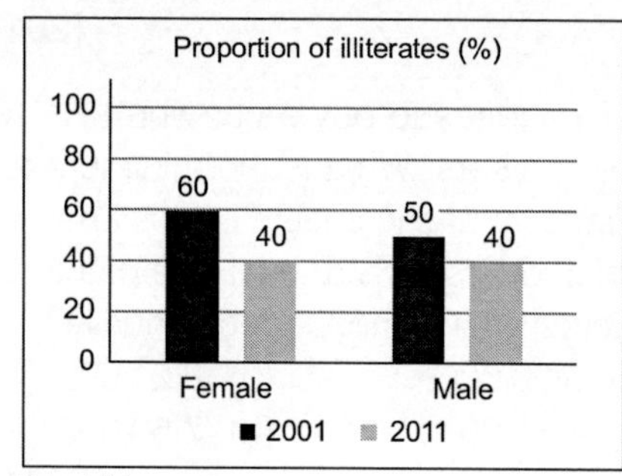

Panel (a)

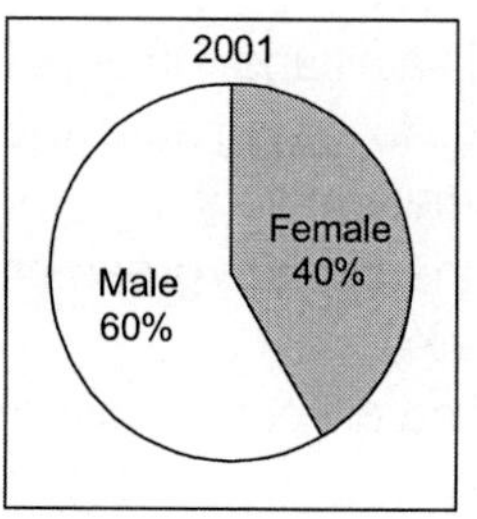

Panel (b)

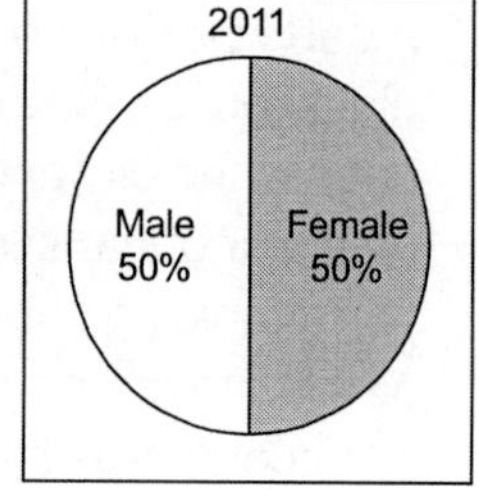

Panel (c)

(a) 35.43 (b) 34.43
(c) 30.43 (d) 33.43 [2019 : 2 M]

1.106 Two design consultants, P and Q, started working from 8 AM For a client. The client budgeted a total of USD 3000 for the consultants. P stopped working when the hour hand moved by 210 degrees on the clock. Q stopped working when the hour hand moved by 240 degrees. P took two lea breaks of 15 minutes each during her shift, but took no lunch break. Q took only one lunch break for 20 minutes, but no tea breaks. The market rate for consultants is USD 200 per hour and breaks are not paid. After paying the consultants, the client shall have USD __________ remaining in the budget.

(a) 433.33 (b) 000.00
(c) 300.00 (d) 166.67 [2019 : 2 M]

1.107 Five people P, Q, R, S and T work in a bank. P and Q don't like each other but have to share an office till T gets a promotion and moves to the big office next to the garden. R, who is currently sharing an office with T wants to move to the adjacent office with S, the handsome new intern. Given the floor plan, what is the current location of Q, R and T?

(O = Office, WR = Washroom)

(a)

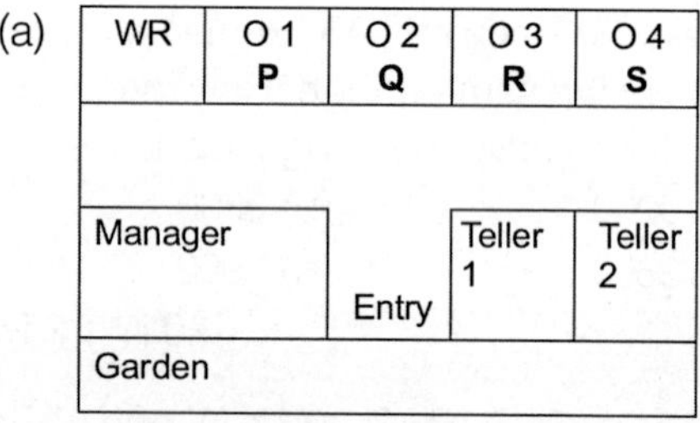

(b)

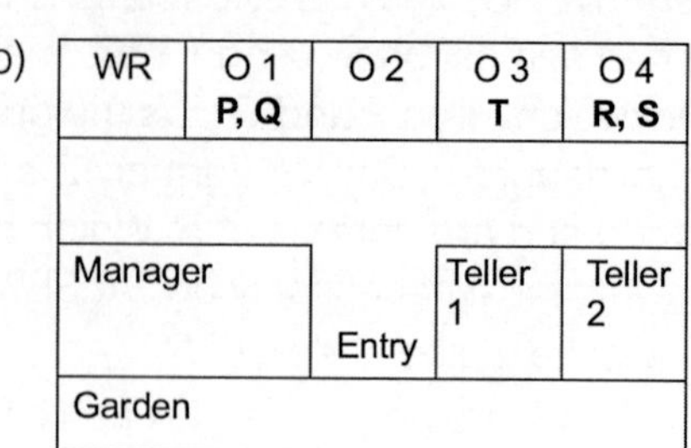

(c)

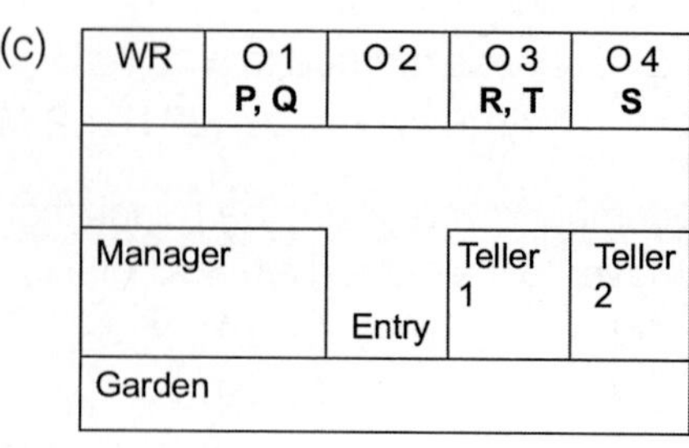

(d)

WR	O 1 P, Q	O 2	O 3 R	O 4 S
Manager T		Entry	Teller 1	Teller 2
Garden				

[2019 : 2 M]

1.108 Four people are standing in a line facing you. They are Rahul, Mathew, Seema and Lohit. One is an engineer, one is a doctor, one a teacher and another a dancer. You are told that:

1. Mathew is not standing nest to Seema.
2. There are two people standing between Lohit and the engineer.
3. Rahul is not a doctor.
4. The teacher and the dancer are standing next to each other.
5. Seema is turning to her right to speak to the doctor standing next to her.

Who among them is an engineer?

(a) Lohit (b) Rahul
(c) Seema (d) Mathew

[2019 : 2 M]

1.109 A superadditive function $f(\cdot)$ satisfies the following property

$$f(x_1 + x_2) \geq f(x_1) + f(x_2)$$

Which of the following functions is a superadditive function for $x > 1$?

(a) $\sqrt{x}$ (b) e^{-x}
(c) e^x (d) $1/x$ [2020 : 1 M]

1.110 The following figure shows the data of students enrolled in 5 years (2014 to 2018) for two schools P and Q. During this period, the ratio of the average number of the students enrolled in school P to the average of the difference of the number of students enrolled in schools P and Q is ____.

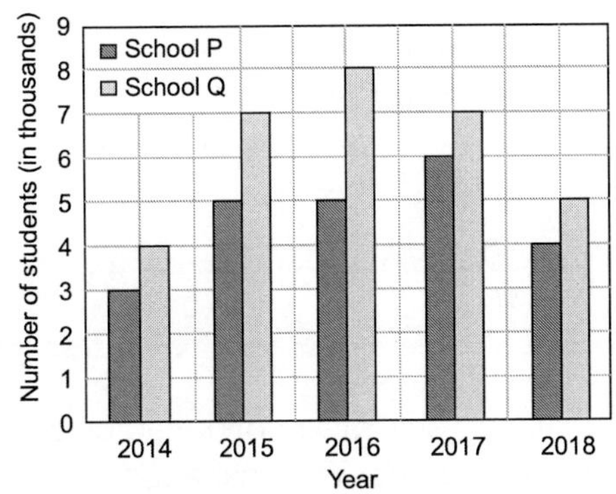

(a) 8 : 23 (b) 23 : 8
(c) 23 : 31 (d) 31 : 23 [2020 : 2 M]

1.111 a, b, c are real numbers. The quadratic equation $ax^2 - bx + c = 0$ has equal roots, which is β, then

(a) $\beta^2 = ac$ (b) $\beta^3 = bc/(2a^2)$
(c) $\beta = b/a$ (d) $b^2 \neq 4ac$

[2020 : 2 M]

1.112 It is quarter past three in your watch. The angle between the hour hand and the minute hand is ________.

(a) 22.5° (b) 7.5°
(c) 0° (d) 15° [2020 : 2 M]

1.113 A circle with centre O is shown in the figure. A rectangle PQRS of maximum possible area is inscribed in the circle. If the radius of the circle is a, then the area of the shaded portion is _____.

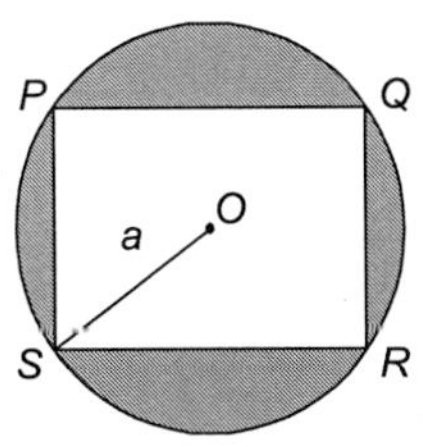

(a) $\pi a^2 - 2a^2$ (b) $\pi a^2 - 3a^2$
(c) $\pi a^2 - \sqrt{2}a^2$ (d) $\pi a^2 - a^2$

[2020 : 2 M]

1.114

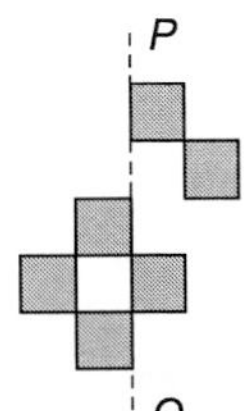

The least number of squares that must be added so that the line P-Q becomes the line of symmetry is _____.

(a) 7 (b) 6
(c) 4 (d) 3 [2021 : 1 M]

1.115 p and q are positive integers and $\frac{p}{q} + \frac{q}{p} = 3$,

then, $\frac{p^2}{q^2} + \frac{q^2}{p^2} =$

(a) 3 (b) 9
(c) 11 (d) 7

[2021 : 1 M]

1.116 The current population of a city is 11,02,500. If it has been increasing at the rate of 5% per annum, what was its population 2 years ago?

(a) 9,92,500 (b) 10,00,000
(c) 9,95,006 (d) 12,51,506

[2021 : 1 M]

1.117 Corners are cut from an equilateral triangle to produce a regular convex hexagon as shown in the figure below.

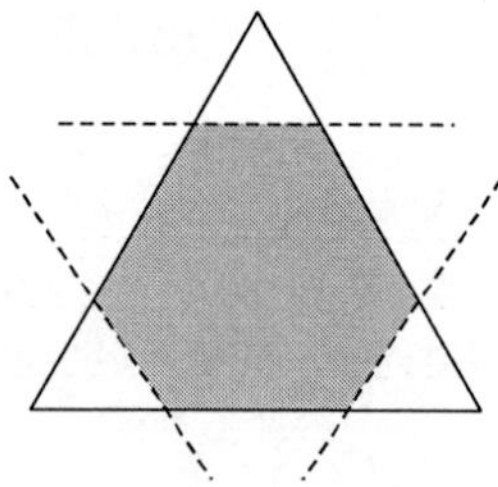

The ratio of the area of the regular convex hexagon to the area of the original equilateral triangle is

(a) 5 : 6 (b) 4 : 5
(c) 2 : 3 (d) 3 : 4

[2021 : 2 M]

1.118 The number of minutes spent by two students, X and Y, exercising every day in a given week are shown in the bar chart below.

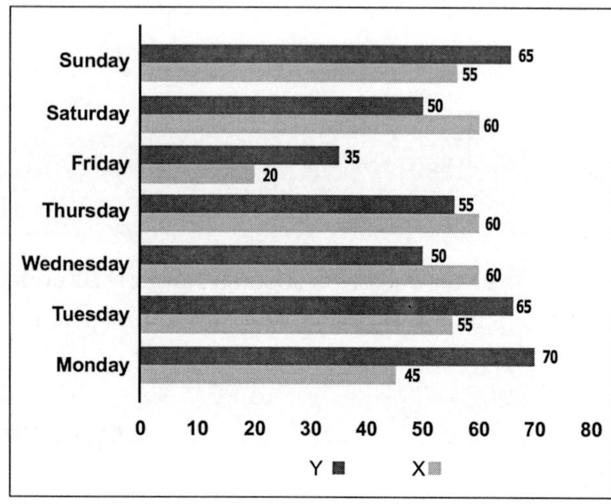

The number of days in the given week in which one of the students spent a minimum of 10% more than the other student, on a given day, is

(a) 4 (b) 7
(c) 5 (d) 6

[2021 : 2 M]

1.119 Given below are two statements and two conclusions.

Statement 1 : All purple are green.
Statement 1 : All black are green.
Conclusion I : Some black are purple.
Conclusion II : No black is purple.

Based on the above statements and conclusions, which one of the following options is logically CORRECT?

(a) Either conclusion I or II is correct.
(b) Both conclusion I and II are correct.
(c) Only conclusion I is correct.
(d) Only conclusion II is correct.

[2021 : 2 M]

1.120 Consider a square sheet of side 1 unit. In the first step, it is cut along the main diagonal to get two triangles. In the next step, one of the cut triangles is revolved about its short edge to form a solid cone. The volume of the resulting cone, in cubic units, is _____.

(a) $\frac{3\pi}{2}$ (b) 3π
(c) $\frac{\pi}{3}$ (d) $\frac{2\pi}{3}$

[2021 : 2 M]

1.121 A sum of money is to be distributed among *P*, *Q*, *R*, and *S* in the proportion 5 : 2 : 4 : 3, respectively.

If *R* gets Rs. 1000 more than S, what is the share of Q (in Rs.)?

(a) 500 (b) 1000
(c) 1500 (d) 2000

[2022 : 1 M]

1.122 A trapezium has vertices marked as *P*, *Q*, *R* and *S* (in that order anticlockwise). The side *PQ* is parallel to side *SR*.

Further, it is given that, *PQ* = 11 cm, *QR* = 4 cm, *RS* = 6 cm and *SP* = 3 cm.

What is the shortest distance between PQ and SR (in cm)?

(a) 1.80 (b) 2.40
(c) 4.20 (d) 5.76

[2022 : 1 M]

1.123 The figure shows a grid formed by a collection of unit squares. The unshaded unit square in the grid represents a hole.

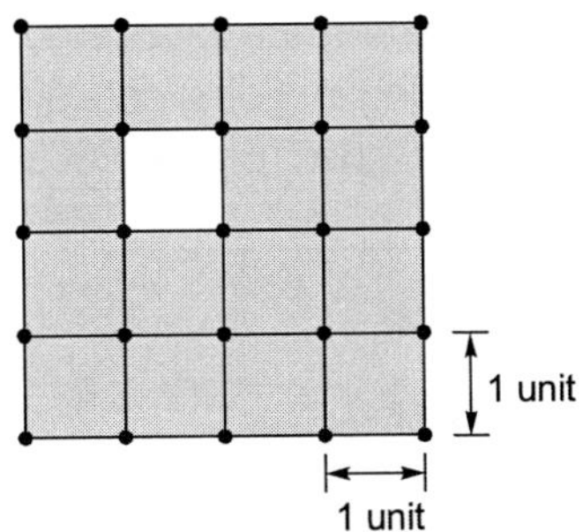

What is the maximum number of squares without a "hole in the interior" that can be formed within the 4×4 grid using the unit squares as building blocks?

(a) 15
(b) 20
(c) 21
(d) 26

[2022 : 1 M]

1.124 Consider the following inequalities.

(i) $2x - 1 > 7$

(ii) $2x - 9 < 1$

Which one of the following expressions below satisfies the above two inequalities?

(a) $x \leq -4$
(b) $-4 < x \leq 4$
(c) $4 < x < 5$
(d) $x \geq 5$

[2022 : 2 M]

1.125 Four points $P(0, 1)$, $Q(0, -3)$, $R(-2, -1)$, and $S(2, -1)$ represent the vertices of a quadrilateral.

What is the area enclosed by the quadrilateral?

(a) 4
(b) $4\sqrt{2}$
(c) 8
(d) $8\sqrt{2}$

[2022 : 2 M]

1.126 Consider the following square with the four corners and the center marked as P, Q, R, S and T respectively.

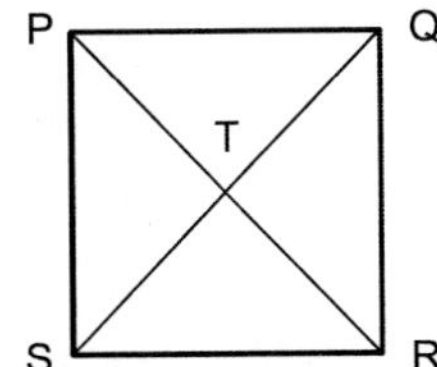

Let X, Y and Z represent the following operations:

X: rotation of the square by 180 degree with respect to the S-Q axis.

Y: rotation of the square by 180 degree with respect to the P-R axis.

Z: rotation of the square by 90 degree clockwise with respect to the axis perpendicular, going into the screen and passing through the point T.

Consider the following three distinct sequences of operation (which are applied in the left to right order).

(1) XYZZ

(2) XY

(3) ZZZZ

Which one of the following statements is correct as per the information provided above?

(a) The sequence of operations (1) and (2) are equivalent

(b) The sequence of operations (1) and (3) are equivalent

(c) The sequence of operations (2) and (3) are equivalent

(d) The sequence of operations (1), (2) and (3) are equivalent

[2022 : 2 M]

1.127 What is the smallest number with distinct digits whose digits add up to 45?

(a) 123555789
(b) 123457869
(c) 123456789
(d) 99999

[2023 : 1 M]

1.128 In a class of 100 students,

(i) there are 30 students who neither like romantic movies nor comedy movies,

(ii) the number of students who like romantic movies is twice the number of students who like comedy movies, and

(iii) the number of students who like both romantic movies and comedy movies is 20.

How many students in the class like romantic movies?

(a) 40
(b) 20
(c) 60
(d) 30

[2023 : 1 M]

1.129 How many rectangles are present in the given figure?

(a) 8
(b) 9
(c) 10
(d) 12

[2023 : 1 M]

1.130 Which one of the following options represents the given graph?

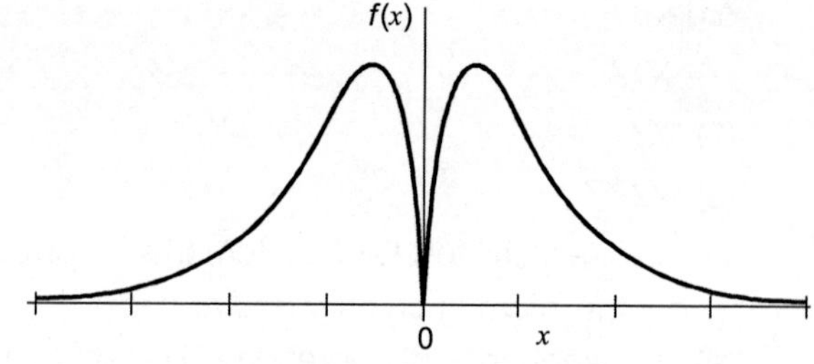

(a) $f(x) = x^2 2^{-|x|}$ (b) $f(x) = x 2^{-|x|}$

(c) $f(x) = |x| 2^{-x}$ (d) $f(x) = x 2^{-x}$

[2023 : 2 M]

1.131 Out of 1000 individuals in a town, 100 unidentified individuals are covid positive. Due to lack of adequate covid-testing kits, the health authorities of the town devised a strategy to identify these covid-positive individuals. The strategy is to:

(i) Collect saliva samples from all 1000 individuals and randomly group them into sets of 5.

(ii) Mix the samples within each set and test the mixed sample for covid.

(iii) If the test done in (ii) gives a negative result, then declare all the 5 individuals to be covid negative.

(iv) If the test done in (ii) gives a positive result, then all the 5 individuals are separately tested for covid.

Given this strategy, no more than ______ testing kits will be required to identify all the 100 covid positive individuals irrespective of how they are grouped.

(a) 700 (b) 600

(c) 800 (d) 1000

[2023 : 2 M]

1.132 100 cm × 32 cm rectangular sheet is folded 5 times. Each time the sheet is folded, the long edge aligns with its opposite side. Eventually, the folded sheet is a rectangle of dimensions 100 cm × 1 cm.

The total number of creases visible when the sheet is unfolded is ______.

(a) 32 (b) 5

(c) 31 (d) 63

[2023 : 2 M]

1.133 The bar chart shows the data for the percentage of population falling into different categories based on Body Mass Index (BMI) in 2003 and 2023.

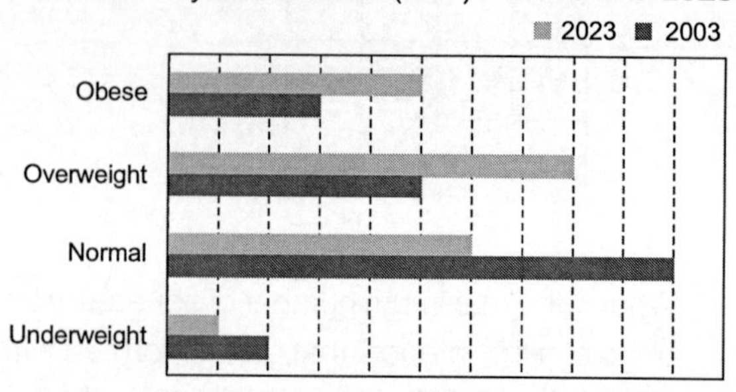

Based on the data provided, which one of the following options is INCORRECT?

(a) The ratio of the percentage of population falling into overweight category to the percentage of population falling into normal category has increased in 20 years.

(b) The ratio of the percentage of population falling into obese category to the percentage of population falling into normal category has decreased in 20 years.

(c) The ratio of the percentage of population falling into underweight category to the percentage of population falling into normal category has decreased in 20 years.

(d) The percentage of population falling into normal category has decreased in 20 years.

[2024 : 2 M]

1.134 Five years ago, the ratio of Aman's age to his father's age was 1:4, and five years from now, the ratio will be 2:5. What was his father's age when Aman was born?

(a) 32 years (b) 35 years

(c) 28 years (d) 30 years

[2024 : 1 M]

1.135 The greatest prime factor of $(3^{199} - 3^{196})$ is

(a) 11 (b) 13

(c) 3 (d) 17 [2024 : 1 M]

1.136 Four identical cylindrical chalk-sticks, each of radius $r = 0.5$ cm and length $l = 10$ cm. are bound tightly together using a duct tape as shown in the following figure

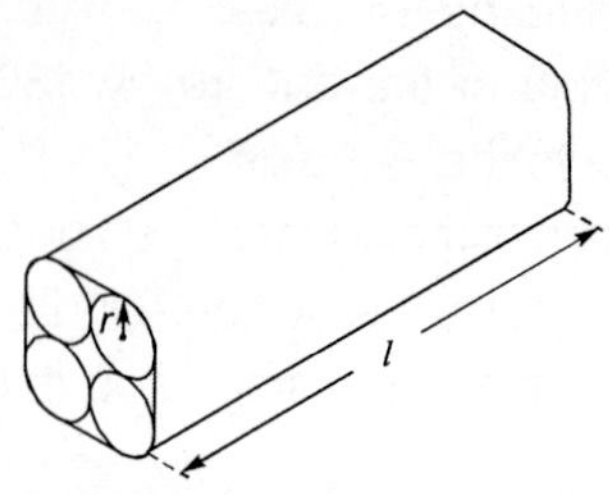

The width of the duct tape is equal to the length of the chalk-stick. The area (in cm^2) of the duct tape required to wrap the bundle of chalk-sticks once, is

(a) $10(8+\pi)$ (b) $20(4+\pi)$
(c) $10(4+\pi)$ (d) $20(8+\pi)$

[2024 : 2 M]

1.137 Two identical sheets A and B, of dimensions 24 cm × 16 cm, can be folded into half using two distinct operations. FO1 or FO2.

In FO1, the axis of folding remains parallel to the initial long edge, and in FO2, the axis of folding remains parallel to the initial short edge.

If sheet A is folded twice using FO1, and sheet B is folded twice using FO2, the ratio of the perimeters of the final shapes of A and B is

(a) 11 : 18 (b) 18 : 11
(c) 11 : 14 (d) 14 : 11

[2024 : 2 M]

1.138 For a real number $x > 1$,

$$\frac{1}{\log_2 x}+\frac{1}{\log_3 x}+\frac{1}{\log_4 x}=1$$

The value of x is

(a) 24 (b) 12
(c) 4 (d) 36 [2024 : 1 M]

1.139 Examples of mirror and water reflections are shown in the figures below:

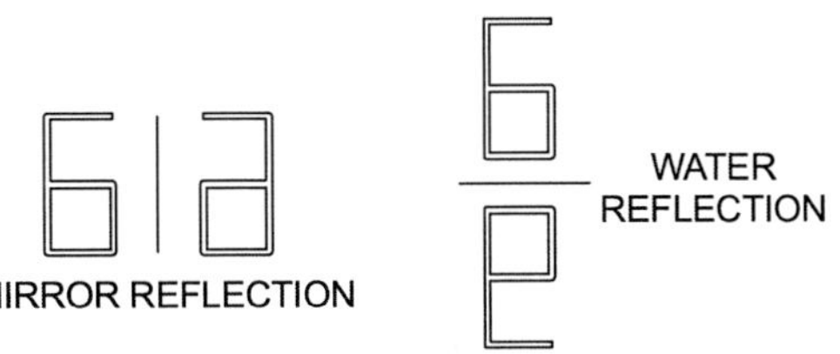

An object appears as the following image after first reflecting in a minor and then reflecting on water.

5·24

The original object is

(a) (b)
(c) (d)

[2024 : 2 M]

■■■

Answers Numerical Ability

1.1	(d)	1.2	(c)	1.3	(d)	1.4	(b)	1.5	(b)	1.6	(a)	1.7	(b)
1.8	(c)	1.9	(d)	1.10	(c)	1.11	(d)	1.12	(d)	1.13	(c)	1.14	(a)
1.15	(a)	1.16	(c)	1.17	(c)	1.18	(d)	1.19	(b)	1.20	(a)	1.21	(a)
1.22	(16)	1.23	(d)	1.24	(b)	1.25	(560)	1.26	(d)	1.27	(b)	1.28	(b)
1.29	(45)	1.30	(c)	1.31	(d)	1.32	(163)	1.33	(d)	1.34	(a)	1.35	(c)
1.36	(16)	1.37	(d)	1.38	(b)	1.39	(d)	1.40	(4)	1.41	(20000)	1.42	(0.81)
1.43	(a)	1.44	(495)	1.45	(c)	1.46	(b)	1.47	(b)	1.48	(22)	1.49	(b)
1.50	(a)	1.51	(c)	1.52	(c)	1.53	(3)	1.54	(d)	1.55	(b)	1.56	(b)
1.57	(800)	1.58	(a)	1.59	(b)	1.60	(c)	1.61	(2.064)	1.62	(c)	1.63	(a)
1.64	(c)	1.65	(c)	1.66	(d)	1.67	(c)	1.68	(c)	1.69	(c)	1.70	(c)
1.71	(b)	1.72	(d)	1.73	(d)	1.74	(c)	1.75	(b)	1.76	(b)	1.77	(c)
1.78	(a)	1.79	(d)	1.80	(c)	1.81	(b)	1.82	(a)	1.83	(d)	1.84	(b)
1.85	(b)	1.86	(a)	1.87	(c)	1.88	(a)	1.89	(a)	1.90	(b)	1.91	(a)
1.92	(d)	1.93	(b)	1.94	(c)	1.95	(b)	1.96	(c)	1.97	(d)	1.98	(b)
1.99	(b)	1.100	(b)	1.101	(b)	1.102	(c)	1.103	(d)	1.104	(d)	1.105	(c)
1.106	(d)	1.107	(c)	1.108	(d)	1.109	(c)	1.110	(b)	1.111	(b)	1.112	(b)
1.113	(a)	1.114	(b)	1.115	(d)	1.116	(2)	1.117	(c)	1.118	(d)	1.119	(a)
1.120	(c)	1.121	(d)	1.122	(b)	1.123	(b)	1.124	(c)	1.125	(c)	1.126	(b)
1.127	(c)	1.128	(c)	1.129	(c)	1.130	(a)	1.131	(a)	1.132	(c)	1.133	(b)
1.134	(d)	1.135	(b)	1.136	(c)	1.137	(d)	1.138	(a)	1.139	(b)		

Explanations Numerical Ability

1.1 (d)

Using the set theory formula

$n(A)$: Number of people who play hockey = 15

$n(B)$: Number of people who play football = 17

$n(A \cap B)$: Persons who play both hockey and football = 10

$n(A \cup B)$: Persons who play either hockey or football or both

Using the formula

$$n(A \cup B) = n(A) + n(B) - n(A \cap B)$$

$$n(A \cup B) = 15 + 17 - 10 = 22$$

Thus people who play neither hockey nor football = 25 – 22 = 3

Alternative Method

Refer to Venn diagram given below:

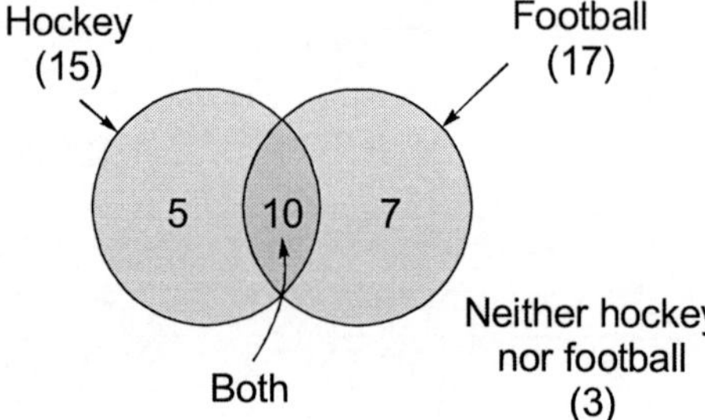

1.2 (c)

7 and 6 added is becoming five means the given two numbers are added on base 8.

$$\begin{array}{r} (137)_8 \\ +(276)_8 \\ \hline (435)_8 \end{array}$$

Hence, we have to add the another two given set of numbers also on base 8.

$$\begin{array}{r} (731)_8 \\ +(672)_8 \\ \hline (1623)_8 \end{array}$$

Hence, the overall problem was based on identifying base which was 8 and adding number on base 8.

1.3 (d)

Let one day work of skilled semi-skilled and unskilled worker be a, b, c units respectively.

$5a \times 20 = 8b + 25 = 10c \times 30$ = Total unit of work

$$100a = 200b = 300c$$

$$a = 2b = 3c$$

$$\Rightarrow \quad b = \frac{a}{2} \text{ and } c = \frac{a}{3}$$

Given that 2 skilled, 6 semi-skilled and 5 unskilled workers are working. Let they finish the work in 'x' days.

$(2a + 6b + 5c)x = 5a + 20$ = Total units of work

$$\left(2a + 3a + \frac{5}{3}a\right)x = 5a \times 20$$

$$\frac{20a}{3}x = 5a \times 20$$

$$x = 15 \text{ days}$$

1.4 (b)

We have to make 4 digit numbers, so the number should be start with 3 or 4, two cases possible;

Case (1) thousands digit is 3

Now other three digits may be any of 2, 2, 3, 3, 4, 4, 4, 4.

(a) Using 2, 2, 3
$\Rightarrow$ 223, 232, 322 -------
$\left(\frac{3!}{2!} = 3 \text{ numbers are possible}\right)$

(b) Using 2, 2, 4 $\Rightarrow$ 224, 242, 422
$\left(\frac{3!}{2!} = 3 \text{ numbers are possible}\right)$

(c) Using 2, 3, 3 $\Rightarrow$ 233, 323, 332
$\left(\frac{3!}{2!} = 3 \text{ numbers are possible}\right)$

(d) Using 2, 3, 4 $\Rightarrow$ 234, 243, 324, 342, 423, 432
(3! = 6 numbers are possible)

(e) Using 2, 4, 4 $\Rightarrow$ 244, 424, 442
$\left(\frac{3!}{2!} = 3 \text{ numbers are possible}\right)$

(f) Using 3, 3, 4 $\Rightarrow$ 334, 343, 433
$\left(\frac{3!}{2!} = 3 \text{ numbers are possible}\right)$

(g) Using 3, 4, 4 $\Rightarrow$ 344, 434, 443
$\left(\frac{3!}{2!} = 3 \text{ numbers are possible}\right)$

(h) Using 4, 4, 4 $\Rightarrow$ 444

$\left(\frac{3!}{3!} = 1 \text{ numbers are possible}\right)$

Total 4 digit numbers in case 1 = 3 + 3 + 3 + 6 + 3 + 3 + 3 + 1 = 25

Case (2) thousands digit is 4 ; Now other three digits may be any of 2, 2, 3, 3, 3, 4, 4, 4.

(a) Using 2, 2, 3 $\Rightarrow$ 223, 232, 322

$\left(\frac{3!}{2!} = 3 \text{ numbers are possible}\right)$

(b) Using 2, 2, 4 $\Rightarrow$ 224, 242, 422

$\left(\frac{3!}{2!} = 3 \text{ numbers are possible}\right)$

(c) Using 2, 3, 3 $\Rightarrow$ 233, 323, 332

$\left(\frac{3!}{2!} = 3 \text{ numbers are possible}\right)$

(d) Using 2, 3, 4 $\Rightarrow$ 234, 243, 324, 342, 423, 432

$\left(\frac{3!}{2!} = 3 \text{ numbers are possible}\right)$

(e) Using 2, 4, 4 $\Rightarrow$ 244, 424, 442

$\left(\frac{3!}{2!} = 3 \text{ numbers are possible}\right)$

(f) Using 3, 3, 3 $\Rightarrow$ 333

$\left(\frac{3!}{3!} = 1 \text{ numbers are possible}\right)$

(g) Using 3, 3, 4 $\Rightarrow$ 334, 343, 433

$\left(\frac{3!}{2!} = 3 \text{ numbers are possible}\right)$

(h) Using 3, 4, 4 $\Rightarrow$ 344, 434, 443

$\left(\frac{3!}{2!} = 3 \text{ numbers are possible}\right)$

(i) Using 4, 4, 4 $\Rightarrow$ 444

$\left(\frac{3!}{3!} = 1 \text{ numbers are possible}\right)$

Total 4 digit numbers in case 2 = 3 + 3 + 3 + 6 + 3 + 3
= 1 + 3 + 1 = 26

Thus total 4 digits numbers using case (1) and case (2)
= 25 + 26 = 51

*** Alternative Method / Shortcut method**

As the number is greater than 3000. So thousand's place can be tiehr 3 or 4.

Let's consider the following two cases

Case (I) When thousand's place is 3.

3 a b c

If there is no restriction on number of two's, three's and four's. Then each of *a*, *b*, *c* can be filled with 2 or 3 or 4 each in 3 ways.

So 3 × 3 × 3 = 27 numbers are there. Out of which 3222, 3333 are invalid as 2 can be used twice & three thrice only so number of such valid numbers beginning with 3 are 27 – 2 = 25.

Case (II) When thousand's place is 4

4 a b c

Without restriction on number of 2's, 3's and 4's *a*, *b*, *c* (as explained in case I) can be filled in 27 ways.

Out of these 27 numbers, 4 2 2 2 is only invalid as two have to be used twice only.

So valid numbers are 27 – 1 = 26.

Total numbers from Case (I) and Case (II) 25 + 26 = 51.

1.5 (b)

Suppose: Hari's age : *H*, Gita's age : *G*,
Saira's age : *S*, Irfan's age : *I*

- $H + G > I + S$
- Using Statement (2) both $G - S = 1$ or $S - G = 1$; *G* can't be oldest and *S* can't be youngest.
- There are no twins thus using statement (2) either GS or SG possible.

(A) HSIG: not possible as there is I between *S* and *G* which is not possible using statement (3)

(B) SGHI: *SG* order is possible, $S > G > H > I$ and $G + H > S + I$ (possible)
Because if { $S = G + 1$; and $G = H+1$ and $H = I + 2$ then $G + (I + 2) > (G + 1) + I$ }

(C) IGSH: according to this $I > G$ and $S > H$ thus adding these both inequalities we get $I + S > G + H$ which is opposite of statement (2) thus not possible.

(D) IHSG: according to this $I > H$ and $S > G$ thus adding both inequalities $I + S > H + G$ which is opposite of statement (2) thus not possible.

1.6 (a)

Let there be overall 'x' candidates. Initially those who decided to vote for *P* and *Q* are. $0.4x$ and $0.6x$ respectively. On the day of election 15% of $0.4x = 0.06x$ went from *P*'s side to *Q*'s side and

25% of $0.6x = 0.15x$ went from Q's side to P's side.
Now, after transfer 'p' has $0.4x - 0.06x + 0.15x = 0.49x$
and after transfer 'q' has $0.6x - 0.15x + 0.06x = 0.51x$
Given, in the question that P lost by 2 votes

$$Q - P = 2 \text{ votes}$$
$$0.51x - 0.49 = 0.02x = 2 \text{ votes}$$

Hence, $(x = 100)$
Total number of votes are 100.

1.7 (b)

Fuel consumed per km will be least when mileage (kilometres per litres) mentioned on 'y' axis of graph will be maximum irrespective of number of kilometres travelled.
From the graph ('y' axis) we can observe that mileage (kilometres per litres) is maximum when vehicle is driven at 45 kilometres per hour. Hence the stretch which 'Q' covered at 45 kmph, mileage was highest and fuel consumption per kilometre was lowest.

1.8 (c)

Shortcut method
The question can be done orally as R took 1/3rd of tofees initially. Hence total number of tofees have to be multiple of 3. There is only one option (c) as 48 which is multiple of 3.
Alternate Solution
Using options:
If we use options for option (c), if total number of tofees are 48. R took 1/3rd if 48 and returned 4. So R has $16 - 4 = 12$ tofees. Out of left over 36 tofees S took 1/4th of 36 and returned 3. So, S has $9 - 3 = 6$ and remaining 30 will be left in bowl. Now T took 1/2 of 30 and returned two back. So, the bowl has 17 left.
*Always solve such questions by options.

1.9 (d)

$$f(y) = \frac{|y|}{y}$$

As we know,

$$|y| = \begin{matrix} y & y \geq 0 \\ -y & y < 0 \end{matrix}\Bigg\} \quad \ldots\text{(i)}$$

So from (i),

$$f(y) = \frac{y}{y} = 1 \quad \text{If } y \geq 0$$

$$f(y) = \frac{-y}{y} = -1 \quad \text{If } y < 0$$

Hence, $|f(q) - f(-q)| = |1 - (-1)| = 2$

1.10 (c)

$4 + 44 + 444$......................

$$\frac{4}{9}[9 + 99 + 999.............]$$

$$= \frac{4}{9}\left[(10-1) + (10^2 - 1) + (10^3 - 1).......\right]$$

$$= \frac{4}{9}\left[(10 + 10^2 + 10^3....10^n) - (1+1+1.....\right]$$

$$= \frac{4}{81}\left[10^{n+1} - 9n - 10\right]$$

1.11 (d)

$(1.001)^{1259} = 3.52$ and $(1.001)^{2062} = 7.85$

$$(1.001)^{1259} \times (1.001) = (1.001)^{3321}$$
$$\downarrow \qquad \downarrow$$
$$3.52 \times 7.85 = (1.001)^{3321} \quad (\text{as } a^m \times a^n = a^{m+n})$$

Hence $(1.001)^{3321} = 27.632 \approx 27.64$

1.12 (d)

Category	Amount
Food	4000
Clothing	1200
Rent	2000
Savings	1500
Others	1800
Total	10500

So total expenses are 10,500.
Expenses excluding savings are 9,000.
Hence required percentage

$$= \frac{9{,}000}{10{,}500} \times 100 = 85.714\% \approx 86\%$$

1.13 (c)

Probability that A & B will meet will be given by the graphical representation where shaded region represents favorable area

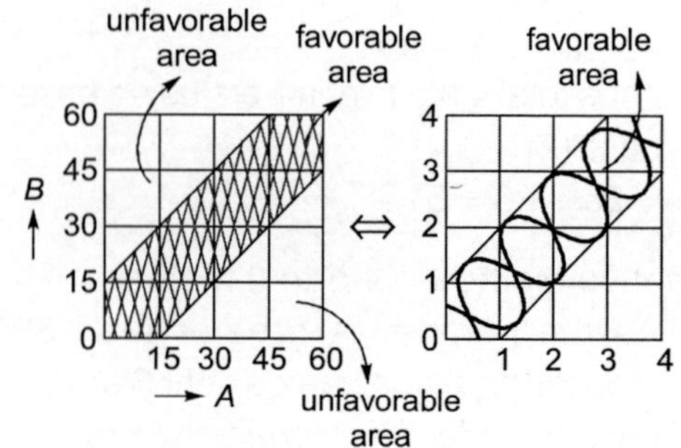

Required probability = $\frac{\text{favourable area}}{\text{total area}}$

$= \frac{7}{16}$

1.14 (a)

Let the number of Rs.20 notes be 'a' and Rs.10 notes be 'b' respectively.

So, $a + b = 14$

$20a + 10b = 230$

on solving $a = 9, b = 5$

Hence number of Rs.10 notes are 5.

Alternative Solution:

If we use option (a)

Number of ten Rs. notes are 5.

Rs.10 × 5 = Rs.50

The remaining nine notes have to be Rs.20 notes

Rs.20 × 9 = 180

and hence 50 + 180 = 230 (total money value)

Hence option (a) number of ten Rs. notes are 5.

1.15 (a)

Divide 8 bags into three parts.

3, 3 and 2 respectively. Now two cases arises.

Case (i):

Firstly, if we compare the 3, 3 bags on pans of a balance. We can identify on which side is the lighter bag placed. And then we will need only one more weighing for identifying faulty bag. So only two weighings are required.

Case (ii):

Secondly, if the 3, 3 compared bags are equal, then to be sure faulty bag is one among the remaing two bags. So, one more weighing is required to find out faulty bag from these two bags. Hence overall two weighings are required. Thus in either cases only two weighings are required.

Alternative Solution:

In case of a weighing balance (i.e. beam balance) the following model can be observed

1 – 3 → 1 weighing required

4 – 9 → 2 weighing required

10 – 27 → 3 weighing required

28 – 81 → 4 weighing required

and the process will go so on.

As there are eight objects. Minimum number of weighings required will be only 2.

1.16 (c)

$$\frac{M + Tue + W}{3} = 41$$

$$\Rightarrow \quad M + Tue + W = 123 \quad ...(i)$$

$$\frac{Tue + W + Th}{3} = 43$$

$$\Rightarrow \quad Tue + W + Th = 129 \quad ...(ii)$$

From eq. (ii) – eq. (i)

$$Th - M = 6$$

Also $\quad Th = 1.15\,M$

$$1.15\,M - M = 6$$

$$\Rightarrow \quad 15\,M = 6$$

$$\Rightarrow \quad M = 40$$

$$Th = 1.15\,m$$

$$\Rightarrow \quad Th = 1.15 \times 40$$

$$\Rightarrow \quad Th = 46$$

1.17 (c)

$$\text{Average speed} = \frac{\text{Total distance}}{\text{Total time}}$$

$$\frac{\text{Total distance}}{\text{Total time}} = \frac{(8+6+16)\text{km}}{\left(\frac{1}{4}+\frac{1}{4}+\frac{1}{4}\right)\text{hr}}$$

$$\frac{30\,\text{km}}{\frac{3}{4}\text{hr}} = 40\text{ km/hr}$$

1.18 (d)

$S = 10 + 84 + 73 + ...$

$S = (9 + 1) + (9^2 + 3) + (9^3 + 5)...$

$S = \{9 + 9^2 + 9^3 + ...9^n\} + \{1 + 3 + 5...(2n - 1)\}$

$$S = \frac{9(9^n - 1)}{(9 - 1)} + n^2$$

$$\left(\begin{array}{l}\text{As } a + ar + ar^2 \,......... \, ar^{n-1} \\ = \frac{a(r^n - 1)}{(r - 1)} \quad \text{for} \quad r > 1\end{array}\right)$$

$$\left(\begin{array}{ll}\text{and } 1 + 3 + 5 \,............\, (2n - 1) & = n^2 \\ \text{Sum of odd natural numbers} & = n^2\end{array}\right)$$

$$\text{So } S = \frac{9(9^n - 1)}{8} + n^2$$

Alternate Method:

Put $n = 2$ in all options and check whether sum = 10 + 84 = 94 is in which option.

1.19 (b)

$3x^2 + 2x + p(p-1) = 0$

If roots are of opposite signs product of the roots will be negative.

$$\frac{p(p-1)}{3} < 0$$
$$p(p-1) < 0$$
$$0 < p < 1$$
$$P \in (0, 1)$$

1.20 (a)

Every leap year will have 366 days.

1 leap year = 366 days

= 52 × 7 + 2 days

Every leap year will have 52 complete weeks. For 53rd Saturday there are two favourable cases from 7 possible cases

Fri	Sat
Sat	Sun
Sun	Mon
Mon	Tue
Tue	Wed
Wed	Thu
Thu	Fri

(Fri Sat, Sat Sun) → 2 favourable cases

Required probability = $\frac{2}{7}$

1.21 (a)

As batsman K have lowest standard deviation, he is the most consistent batsman of these four.

1.22 (16)

12, 35, 81, 173, 357, _____

Difference between consecutive terms of the series are : 13, 46, 92, 184, _____.

35 – 12 = 23 = 23 × 1

81 – 35 = 46 = 23 × 2

173 – 81 = 92 = 23 × 4

357 – 173 = 184 = 23 × 8

So the next difference have to be 23 × 8 = 368, therefore next term have to 357 + 368

= 725

= 23 × 16

725 – 357 = 368 = 23 × 16

1.23 (d)

Difference between 1st and 4th, 2nd and 3rd letters in each series respectively is same except NVBD.

1.24 (b)

A resident female with annual income Rs.9 Lakh should satisfy eligibility for filling form *T*. So answer is option *B* is answer. Option *A*, *C*, *D* corresponds to filling form *P*, *S*, *D* respectively.

1.25 (560)

Let, the length of platform is x and velocity of train is v.

Distance = Vel. × time

$$(280 + x) = V \times 60 \quad \text{... (i)}$$
$$280 = V \times 20 \quad \text{... (ii)}$$

From equation (i) and (ii), we get

$$x = 560 \text{ m}$$

1.26 (d)

Deficit should be equal to (1/5)th of exports which is possible in 2006

$$\text{Deficit} = \frac{\text{Imports} - \text{Exports}}{\text{Exports}} = \frac{120-100}{100} = \frac{20}{100} = \frac{1}{5}$$

1.27 (b)

As one of the face is already shown as head as the possibility of second coin is not there (because second coin has both tail). Now sample space will contain now events (*H*, *H*, *T*). So the probability for another face as tail will be 1/3.

1.28 (b)

As the numbers appearing on the faces are not equally likely because their probability of occurrence over a large number of throws are different (as given) so we can term it to be biased die.

1.29 (45)

2, 3, 6, 15, _____, 157.5, 630

2 × 1.5 = 3

3 × 2 = 6

6 × 2.5 = 15

15 × 3 = 45

45 × 3.5 = 157.5

157.5 × 4 = 630

Hence the answer is 45.

45 is missing term of the series.

1.30 (c)

Adding 6, 3 and 2 to the consecutive letters gives the only (c) does not follow.

1.31 (d)

Sum of the read and green lights as equal to sum of yellow and blue lights are red and blue will be numbered same and green and yellow will be numbered same.

So, $R + H = Y + B$

1.32 (163)

Let eight consecutive odd numbers be

a, a + 2 .... a + 14

and four consecutive over numbers be

$b, b + 2, b + 14, b + 16$

Given $(a) + (a + 2), \ldots\ldots + (a + 14) = 656$

$$8a + 56 = 656$$
$$a = 75$$

Also given

$$(b) + (b + 2) + (b + 4) + (b + 6) = 87 \times 4 = 348$$
$$4b + 12 = 348$$
$$b = 84$$

Sum of smallest odd no (a) and second largest even no ($b + 4$) will be

$= (a) + (b + 4) = 75 + (84 + 4) = 163$

1.33 (d)

For maximum revenue per kg. Let us solve through options

A. $\text{Item 2} = \dfrac{20\% \text{ of } 250}{20\% \text{ of } 500} = 0.5$

B. $\text{Item 3} = \dfrac{23\% \text{ of } 250}{19\% \text{ of } 500} = 0.6$

C. $\text{Item 6} = \dfrac{19\% \text{ of } 250}{16\% \text{ of } 500} = 0.59$

D. $\text{Item 5} = \dfrac{20\% \text{ of } 250}{12\% \text{ of } 500} = 0.83$

Clearly item 4 is largest.

1.34 (a)

Drainage pipe drains half of the tank in 30 minutes or full tank is 68 minutes. Now the tank have to be filled in 10 minutes (when both drainage and filling pipes are in action)

Drainage pipe would drained 1/6th of capacity of tank in 10 minutes. So filling pipe have to fill 1/2 of capacity of 1/6 of capacity more i.e. 2/3rd of the tank in 10 minutes. or complete tank is 15 minutes. That's why the filling tank should have rate 4 times the draining rate.

Alternatively :

Drainage pipe can drain complete tank in 60 min. Let the filling pipe can fill the tank in 'x' min. Now both are operational and tank is half filled. Also remaining half needs to get filled in 10 minute.

$$10\left[\frac{1}{x} - \frac{1}{60}\right] = \frac{1}{2}$$
$$x = 15 \text{ minutes}$$

So filling pipe needs to have rate be times drainage pipe's rate.

1.36 (16)

81, 54, 36, 24, ____

Alternate terms are perfect squares 9^2, 6^2, 4^2 and in between terms are products 9×6, 6×4 m _____ and so on missing terms is

$$4^2 = 16$$

1.37 (d)

$$P = A < R < M$$

P is surely less than M only in option (d) else in any other option it cannot be conveyed surely.

1.38 (b)

Next term will be Q which is 17 letter of English alphabet.

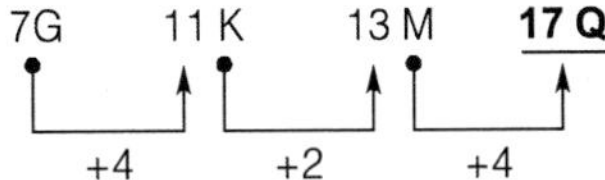

1.39 (d)

On seeing the given figure in question i, ii and iii can be concluded.

1.40 (4)

Let the velocity of stream is 'V' time taken during downstream is 't' and distance is 'x'.

$$t = \frac{x}{8 + V} \quad \ldots(i)$$

During upstream

$$3t = \frac{x}{8 - V} \quad \ldots(ii)$$

Dividing equation (ii) by (i), we get

$$V = 4 \text{ m/s}$$

1.41 (20000)

Since,

$\because \quad 15\% = 4{,}50{,}000$

$\therefore \quad 100\% = \dfrac{4{,}50{,}000 \times 100}{15} = 30{,}00{,}000$

Selling price of purifier

$$= \frac{30,00,000 + 10,00,000}{200} = 20,000$$

1.42 (0.81)

Probability of bulb being defective $= \frac{5}{100}$

$= 0.05$

Probability of bulb being non defective

$= 1 - 0.05 = 0.95$

Probability that the batch is accepted

= None of the four bulb being defective

$= (0.95)^4 = 0.81450625 \approx 0.81$

1.43 (a)

$$f(x, y) = (2x)^n \times \left(\frac{y}{n}\right)^m = 2^{n-m} \times P$$

1.44 (495)

Let the 12 consecutive odd numbers be

$a, a + 2, a + 4, a + 6 ..... a + 22$

Sum of first five

$(a) + (a + 2) + (a + 4) + (a + 6) + (a + 8) = 425$

$5a + 20 = 425 \Rightarrow a = 81$

Sum of last five

$(a + 14) + (a + 16) + (a + 18) + (a + 20) + (a + 22)$

$5a + 90 = 5 \times 81 + 90 = 495$

1.45 (c)

13 M, 17 Q, 19 S

among options only *W* is correct as *W* is 23rd letter in English language.

1.46 (b)

BEST OF LUCK

B–ST –F L–CK

GOOD WISHES

G—D W–SH–S

ACE THE EXAM

–C– TH_ –X–M

So answer is MXHTC (vowel is missing).

1.47 (b)

The value doubled over 10 years

$$A = P\left(1 + \frac{R}{100}\right)^n$$

$$2P = P\left(1 + \frac{R}{100}\right)^{10}$$

$$R = 7.2$$

Alternatively:

Simple growth rate for a quantity to double in 10 years will be 10%. For compounded growth answer have to be less than 10. Among options 7.2 (Ans. (b)) is correct as ((a) (5.6)) will be very much less than 10.

((c) (10)), ((d) (12) cannot be answer as compared growth rate have to be less than 10.

1.48 (22)

In 2012, total cost

$$= \frac{4,50,000 \times 100}{15} = 30,00,000$$

Cost of raw material in 2012

$$= 30,00,000 \times \frac{20}{100} = 6,00,000$$

Cost of raw material in 2013

$$= 600000 + 600000 \times \frac{30}{100}$$

$= 7,80,000$

other expenditure in 2012

$= 3000000 - 600000$

$= 2400000$

Increased cost of other expenditure in 2013

$$= 2400000 + 2400000 \times \frac{20}{100}$$

$= 2880000$

Total cost in 2013

$= 2880000 + 780000 = 3660000$

% increase in total cost in 2013

$$= \frac{3660000 - 3000000}{3000000} \times 100 = 22\%$$

1.49 (b)

Overall there will be 5! = 120 numbers. Digit *I* will appear at unity, tens, hundred, thousand and ten thousand place 4! times each. So sum received due to all 1's will be

$4! \times (1 + 10 + 100 + 1000 + 10000) = 4! \times 11111$

Similarly sum received due to all 3's

$4! \times (3 + 30 + 300 + 3000 + 30000) = 4! \times 3333$

Similarly for digits 3, 4 and 5

So, overall sum received due to all digits

$= 4! \times 11111 [1 + 3 + 5 + 7 + 9]$

$= 24 \times 11111 [25] = 6666600$

1.50 (a)

If $x^a = m$

$$a \log x = \log m \text{ then } a = \frac{\log m}{\log x}$$

$$a = \log_x^m$$

$$\log_x\left(\frac{5}{7}\right) = -\frac{1}{3}$$

$$\Rightarrow \quad (x)^{-1/3} = \frac{5}{7}$$

$$\Rightarrow \quad x = \left(\frac{7}{5}\right)^3$$

$$\Rightarrow \quad x = \frac{343}{125}$$

1.51 (c)

(66 □ 6) → (66 ◇ 6)

$$\left(\frac{66-6}{66+6}\right) \times \left(\frac{66+6}{66-6}\right) = 1$$

1.52 (c)

Number of faces visible = 9 × 6 = 54
(9 on each face × 6 faces)

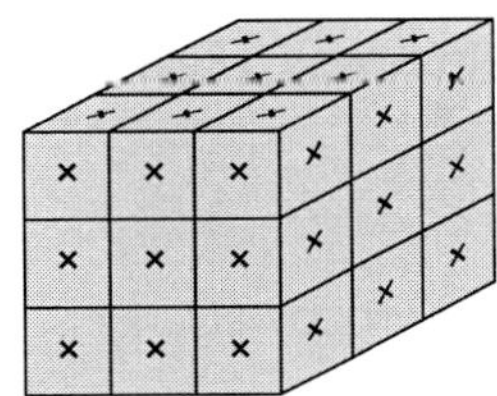

$$\text{Total faces} = 27 \times 6$$

$$\text{Not visible} = \text{Total face} - \text{visible faces}$$

$$= 27 \times 6 - 9 \times 6 = 108$$

$$\text{Required ratio} = \frac{\text{Visible}}{\text{Not visible}} = \frac{54}{108} = \frac{1}{2}$$

1.53 (3)

Look at first row middle term is $\left(\frac{\text{sum}}{2}\right)$ of terms

an either side it $\frac{6+4}{x} = 5.$

Second row middle term is $\left(\frac{\text{sum}}{2}\right)$ of terms an

either side of it $\frac{7+4+2+1}{2} = 7$ is equal to middle term in 2nd row.

Similarly 3rd row middle term

$$\frac{1+9+2+1+2+1}{2} = 8$$

So last row middle term = $\frac{3+3}{2} = 3.$

1.54 (d)

Following chart can be drawn from commulative chart:

Stretch	Distance	Electricity used	
M	20	12	12/20 = 0.6
N	45	25	13/25 = 0.52
O	75	45	20/30 = 0.66
P	100	57	12/25 = 0.48

Hence minimum for *P*.

1.55 (b)

Required probability

$$= \frac{1}{6} \times \left(\frac{7}{8}\right) + \frac{1}{8} \times \left(\frac{5}{6}\right) = \frac{12}{48} = \frac{1}{4}$$

1.56 (b)

We know,

$(a + b + c)^2 = a^2 + b^2 + c^2 + 2[ab + bc + ca]$

$(a + b + c)^2 - 1 = 2[ab + bc + ca]$

[as $a^2 + b^2 + c^2 = 1$]

For $ab + bc + ca$ to be min. $(a + b + c) = 0$

$$\text{Min. } (ab + bc + ca) = -\frac{1}{2}$$

[an putting $a + b + c = 0$]

Only option (b) has min. value as $-\frac{1}{2}$.

1.57 (800)

Tiger is 50 leap of its own behind deer. Given tiger's one leap is 8 m. So initial separation of Tiger and deer = 50 ×8 = 400 m/min

Tiger goes 5 leap in a minute which is equal to 5 ×8 = 40 m/min.

Deer goes 4 leap in a minute which is equal to 4 ×5 = 20 m/min.

Now relative distance of 400 m have to be covered with (40 – 20) = 20 m/min.

$$\text{Required time} = \frac{400\text{ m}}{(40-20)\text{m/min}}$$

$$\frac{\text{(Distance)}}{\text{Relative speed}} = 20 \text{ min.}$$

So distance travelled by Tiger

$$\frac{40\text{ m}}{\text{min}} \times 20 \text{ min} = 800 \text{ m/min.}$$

1.58 (a)

If $1 < y < x$ so $\log x > \log y$ a as well $e^x > e^y$ other two will not always be true.

1.60 (c)

$\log \tan 1° + \log \tan 2° + ... + \log \tan 89°$

$= \log(\tan 1° \times \tan 2° \times \cdots \tan 89°)$

As $[\log m + \log n = \log mn]$

$= \log(\tan 1° \times \tan 2° \cdots \tan(90° - 2°)) \tan(90° - 1°)$

$= \log(\tan 1° \times \tan 2° \times \cdots \tan 45° \times \cot 44 \cdots \cot 2° \cot 1°)$

$= \log(\tan 45°)$

$= \log(1) = 0$

As $\tan(90° - \theta) = \cot\theta$ and

$\tan 1° \times \cot 1° = \tan 2° \times \cot 2° = \cdots 1$

1.61 (2.064)

The shade of part of circular sheet is folded to form a cone (Figure below).

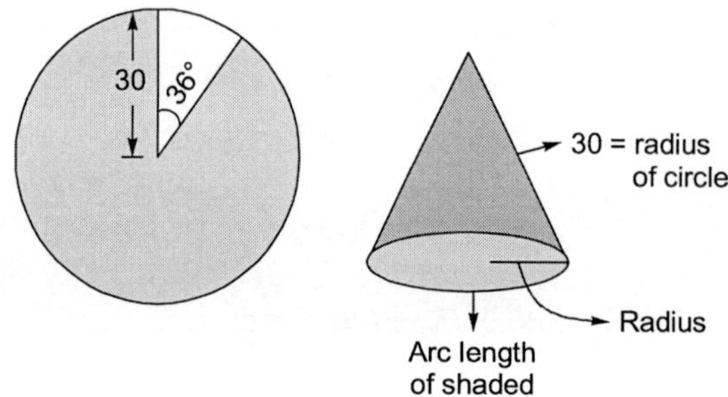

$$2\pi R' = \frac{9}{10} 2\pi \times 30 \qquad R' = 27$$

$$h = \sqrt{30^2 - 27^2} = \sqrt{900 - 729}$$

$$= \sqrt{171}$$

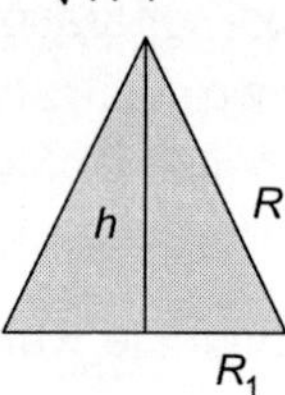

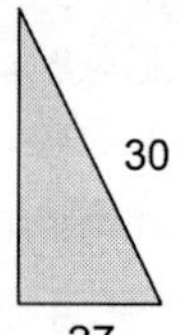

Required ratio $\frac{R'}{h}$

$$= \frac{27}{\sqrt{171}} = 2.064$$

1.62 (c)

A, CD, GHI?, UVWXY

Series is as follows: Write first letter *A*, leave aside nearby one *B*, write next two *CD*. Leave aside net two *EF*. Write next three *GHI* leave aside next *JKL* then we have to write *MNOP*.

1.63 (a)

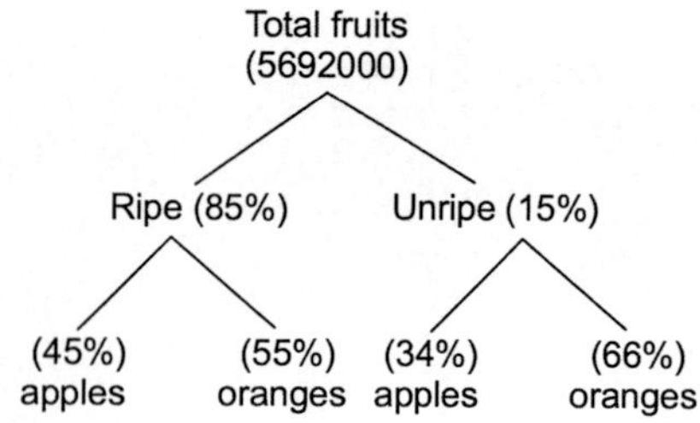

Total number of apples = Ripe apples + Unripe apples

$= (0.85 \times 0.45 + 0.15 \times 0.34)569200 = 2029198$

1.64 (c)

Following line with respective distances can be drawn

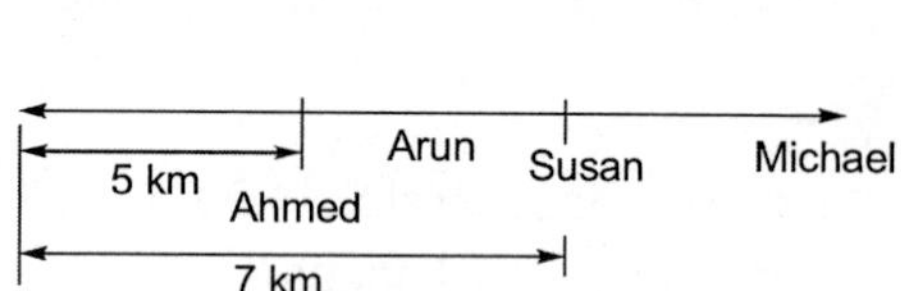

Arun can reside anywhere between Ahmed and Susan i.e. between 5 km and 7 km from I.

$5 < 6.02 < 7$

1.65 (c)

The required probability

$$0.5 \times 0.7 = 0.35 \approx \frac{35}{100} = 35\%$$

1.66 (d)

L > P (Leela is older than Pavithra)

S > L (Shiv is older than Leela)

So, Pavithra is youngest

1.67 (c)

$$q^a = r,\ r^b = s,\ s^c = q$$

$$\Rightarrow \quad a \cdot \log q = \log r,$$

$$b \cdot \log r = \log s,$$

$$c \cdot \log s = \log q$$

So, $$a \times b \times c = \frac{\log r}{\log q} \times \frac{\log s}{\log r} \times \frac{\log q}{\log s} = 1$$

1.68 (c)

Q can do work in $25 \times 12 = 300$ hrs

R can do work in $50 \times 12 = 600$ hrs

So we can say *Q* is twice efficient as *R*

Now *Q* worked only for 5 days at a rate of 12 hrs/day. So, for 60 units of his work (Total work for *Q* i.e. 300 hrs) he will do only $\frac{1}{5}$ of work

$$\left(\frac{60}{300} = \frac{1}{5}\right)$$

While R worked for all 7 days at a rate of 18 hrs/day

So he will do $18 \times 7 = 126$ of his work (Total work for 600 hrs)

He will do $\left(\frac{126}{600} = 0.21\right)$ of his work

So required ratio $\left(\frac{1}{5} : \frac{126}{600}\right) = 120 : 126$

20 : 21

1.71 (b)

'M' works with twice efficiency as E but worked for half as many days. So in this respect they will do equal work if their shifts would have been for same timings. But M's shift is for hrs, while E's shift for 12 hrs. Hence, E will do twice the work as M.

Ratio of contribution of $M : E$ in work, 1 : 2

1.72 (d)

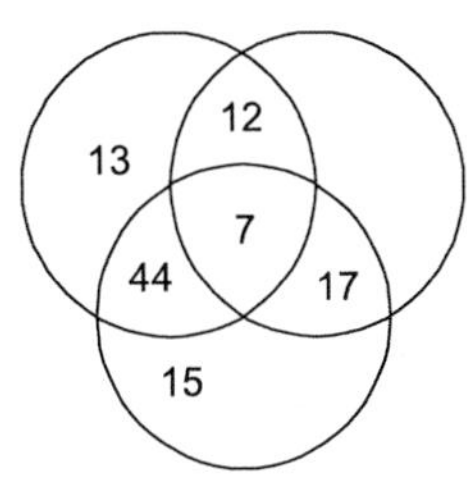

The number of student who like to read books or play sports have been shown.

$= 13 + 12 + 44 + 7 + 15 + 1 = 108$

1.73 (d)

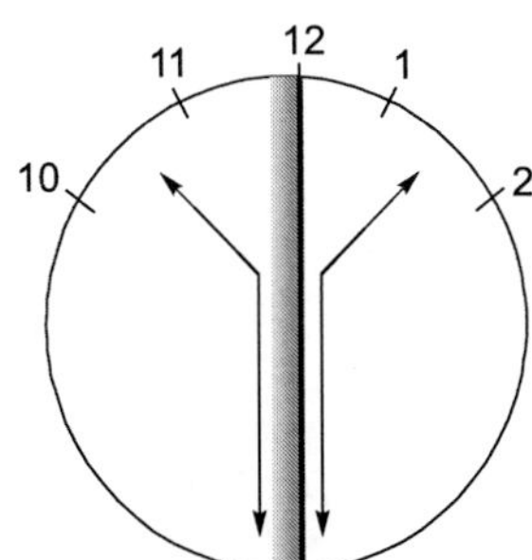

Mirror image of 1 : 20 is 10 : 30

10 : 30 was the time two and quarter hour back so time now will be 12 : 45

1.74 (c)

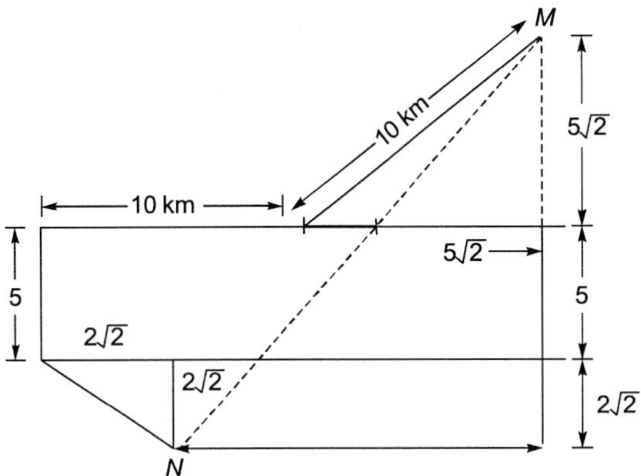

See the adjoining figure for solution

$$MM' = 5\sqrt{2} + 5 + 2\sqrt{2} = 5 + 7\sqrt{2}$$

$$NM' = 10 + 5\sqrt{2} - 2\sqrt{2} = 10 + 3\sqrt{2}$$

$$MN = \sqrt{(MM')^2 + (NM')^2}$$

$$M' = \sqrt{(5+7\sqrt{2}) + (10+3\sqrt{2})^2}$$

$$\approx 20.61$$

1.75 (b)

Let side of square is x and sides of rectangle are y and $2y$

According to given data

$$4x + 2(y + 2y) = 340$$

$$4x + 6y = 340 \quad \text{...(i)}$$

$$\text{Area} = x^2 + y \times 2y$$

$$= x^2 + 2y^2 \quad \text{...(ii)}$$

$x = \left(\frac{340 - 6y}{4}\right)$ by equation (i)

$$\text{Area} = \left(\frac{340 - 6y}{4}\right)^2 + 2y^2$$

For area to be minimum $\frac{d(\text{Area})}{dy} = 0$

$$2\left[\frac{340 - 6y}{4}\right] \times \left(\frac{-6}{4}\right) + 4y = 0$$

From here, $y = 30$

and $x = 40$

$\Rightarrow$ Side of square = 40 mm

1.76 (b)

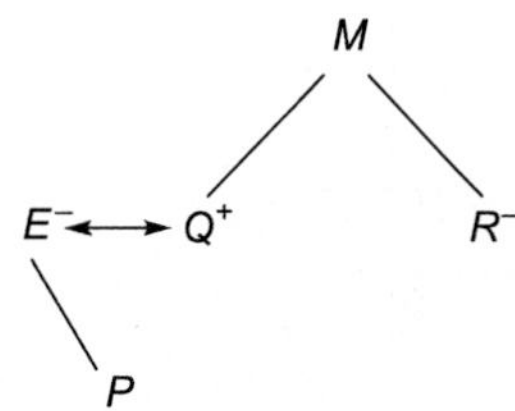

Since gender of M is not known so, from here we can say that P is grandchild of M.

1.77 (c)

97 is not a perfect square.

1.78 (a)

Speed of first train = $\frac{120}{10} = 12$ m/s

Speed of second train = $\frac{150}{15} = 10$ m/s

Difference in their speed = 12 – 10 = 2 m/s

1.79 (d)

The odometer reading increases from starting point to end point.

Magnitude of area of the given diagram = Odometer reading.

Magnitude of area of the velocity and time graph per second.

1st sec ⇒ triangle = $\frac{1}{2} \times 1 \times 1 = \frac{1}{2}$

2nd sec ⇒ square = 1 × 1 = 1

3rd sec ⇒

square + triangle = $1 + 1 + \frac{1}{2} \times 1 \times 1 = 1\frac{1}{2}$

4th sec ⇒ triangle = $\frac{1}{2} \times 1 \times 2 = 1$

5th sec ⇒ straight line = 0

6th sec ⇒ triangle = $\frac{1}{2} \times 1 \times 1 = \frac{1}{2}$

7th sec ⇒ triangle = $\frac{1}{2} \times 1 \times 1 = \frac{1}{2}$

Total Odometer reading at 7 seconds

$= \frac{1}{2} + 1 + 1\frac{1}{2} + 1 + 0 + \frac{1}{2} + \frac{1}{2} = 5$

1.80 (c)

Manu's age = Sravan's age + 2 months

Manu's age = Trideep's age – 3 months

Pavan's age = Sravan's age + 1 month

From this Trideep's age > Manu > Pavan > Sravan

∴ Trideep can occupy the extra spane in the flat.

1.81 (b)

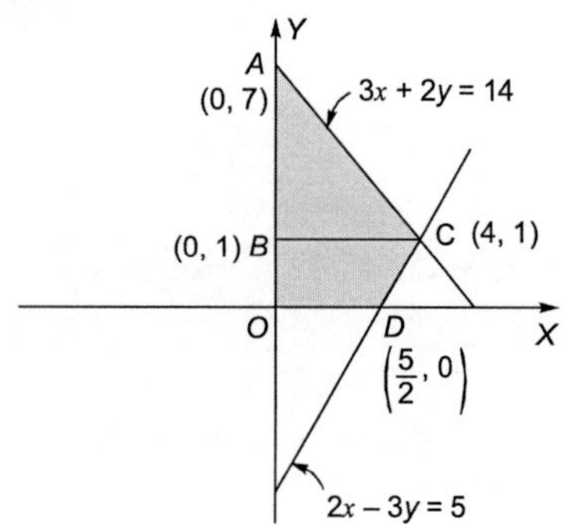

Required area = Area of ΔABC + Area of trapezoid $BCDO$

$$= \frac{1}{2} \times 4 \times 6 + \frac{1}{2} \times (4 + 2.5) \times 1$$

$$= 12 + 3.25 = 15.25 \text{ sq. units}$$

1.82 (a)

Straight line equation $y = mx + c$

m = slope = –0.02

($\ln x$, y)

If $\ln x = X$, then set (x, y) ↓ 0.1, ↓ 0

$y = mX + C$

$o = -0.02 \times 0.1 + C$

∴ $C = 0.002$

$y = mX + C$

$y = -0.02 \times \log x + C$

at $x = 5$

$y = -0.02 \times \log 5 + 0.002$

$= -0.030$

1.83 (d)

Let, H is household consumption and P be other consumption.

$H \times 0.8 + P \times 1.7 = (H + P) \times 0.75$

(According to given condition)

∴ from here ratio is negative.

1.84 (b)

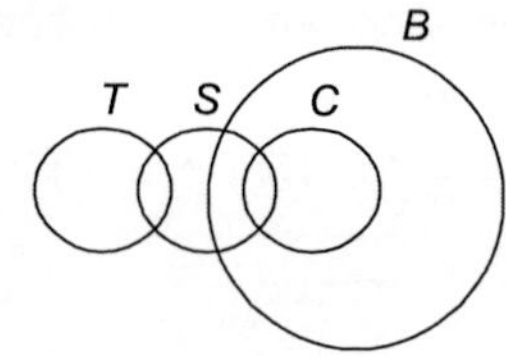

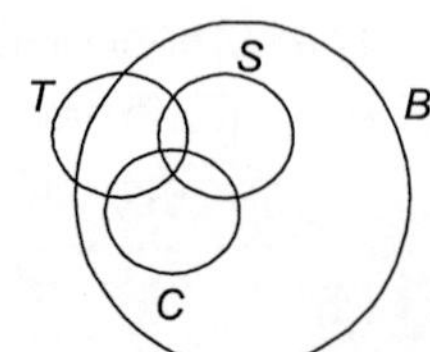

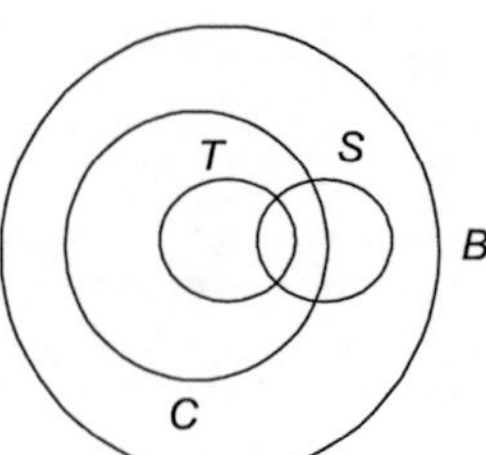

Only conclusion (ii) follows.

1.85 (b)

40% of 360° = $\frac{40}{100} \times 360° = 144°$

So, the angle subtended on pie chart will be = 144°.

1.86 (a)

Length of Truck + gap required = 10 + 20 = 30 m
Length of Car + gap required = 5 + 15 = 20 m
Alternative pairs of Truck and Car needs

30 + 20 = 50 m

Let 'n' be the number of repetition of (Truck + Car) in 1 hour (3600 secs)

Given speed = 36 km/hr = 10 m/sec

$$\frac{50\text{ m} \times n}{3600\text{ secs}} = 36\text{ km/hr}$$

$$\frac{50n}{3600}\text{ m/sec} = 10\text{ m/sec}$$

$$n = \frac{36000}{50} = 720\text{ (Trucks + Cars)}$$

So, = 720 (Trucks + Cars) passes
= 720 × 2 = 1440 vehicles

1.87 (c)

Contour lines can be observed to cross region with height

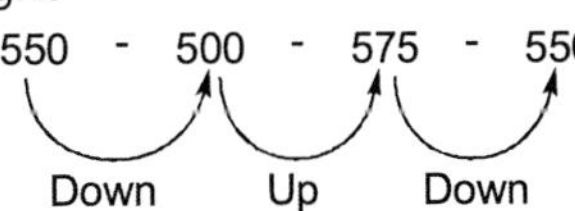

Down-Up-Down satisfies.

1.88 (a)

Subgroups containing only Indians

$= {}^3C_1 + {}^3C_2 + {}^3C_3 = 3 + 3 + 1 = 7$

Subgroups containing one Indian and rest Chinese

$= {}^3C_1\left[{}^3C_1 + {}^3C_2 + {}^3C_3\right]$

$= 3[3 + 3 + 1] = 21$

Subgroups containing two Indian and remaining Chinese

$= {}^3C_2\left[{}^3C_1 + {}^3C_2 + {}^3C_3\right] = 21$

Subgroups containing three Indian and remaining Chinese

$= {}^3C_3\left[{}^3C_1 + {}^3C_2 + {}^3C_3\right] = 7$

Total number of subgroups

= 7 + 21 + 21 + 7 = 56.

1.89 (a)

Following circular seating arrangement can be drawn

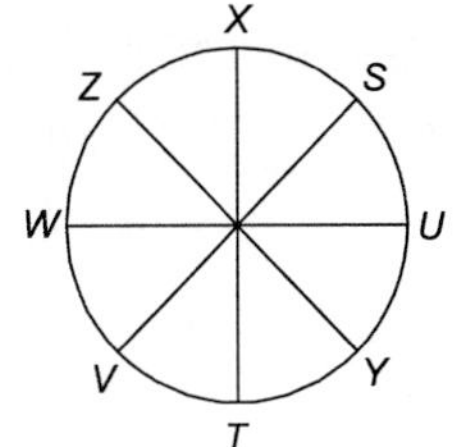

Only one such arrangement can be drawn.
The person on third to the left of V is X.

1.90 (b)

For rules to be followed, we need to check P's drink and S's age.

1.91 (a)

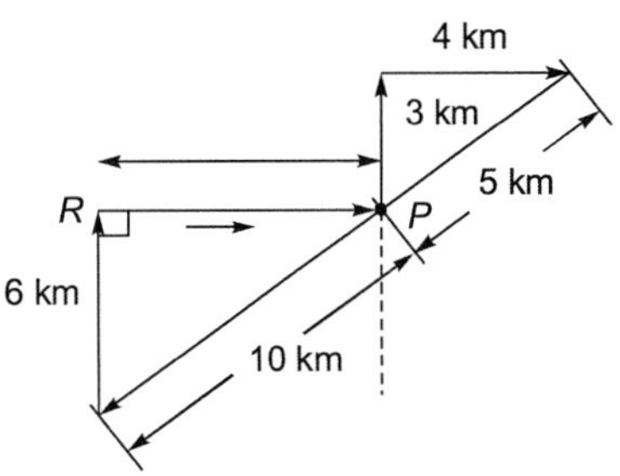

1.92 (d)

$n(C) = 329$
$n(P) = 186$
$n(M) = 295$
$n(C \cap P) = 83$
$n(C \cap M) = 217$
$n(P \cap M) = 63$

$$n(P \cup C \cup M) = n(C) + n(P) + n(M) - n(C \cap P) - n(C \cap M) - n(P \cap M) + n(P \cap C \cap M)$$

$$500 = 329 + 186 + 295 - 83 - 217 - 63 + n(P \cap C \cap M)$$

$$\Rightarrow n(P \cap C \cap M) = 500 - 447 = 53$$

1.93 (b)

The steepest path will be the path which is deepest from see level. So, P to R is the steepest path.

1.94 (c)

Let a man can build the bridge in x weeks and a woman can build the bridge in y weeks.

So, $$\frac{1200}{x} + \frac{500}{y} = \frac{1}{2} \qquad \text{...(i)}$$

$$\frac{900}{x} + \frac{250}{y} = \frac{1}{3} \qquad \text{...(ii)}$$

By equation (i) and (ii), we get

$x = 3600$ and $y = 3000$

⇒ A man build the bridge in 3600 weeks.

1.95 (b)

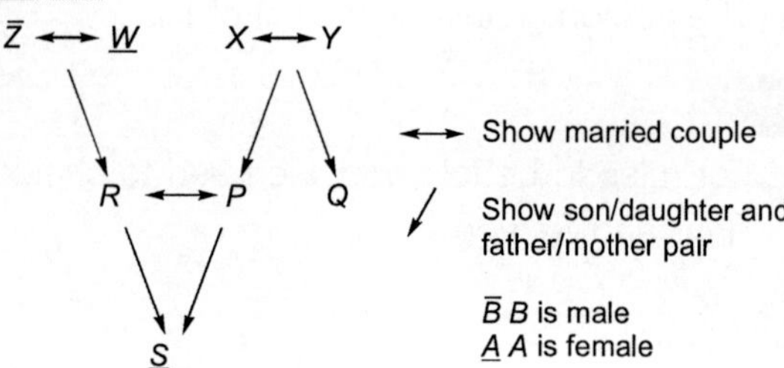

From here,

P and R are married couple. So option (b) is necessarily false.

1.96 (c)

Total number of three digit number are

$$= 9 \times 10 \times 10 = 900$$

Number with 2 is immediate right of 1 are

$$= \boxed{1 \mid 2 \mid x} + \boxed{x \mid 1 \mid 2} = 19$$

$$1 \times 1 \times 10 + 9 \times 1 \times 1$$

So, number with 2 is not immediate right of 1 are

$$= 900 - 19 = 881$$

1.97 (d)

This is a infinite Geometric progression with first term (a) = 1

and common ratio (r) = $\frac{1}{4}$

The formula for finding the sum of infinite G.P is given by $\frac{a}{1-r}$

$$\frac{1}{1-\frac{1}{4}} = \frac{1}{3/4} = \frac{4}{3}$$

1.98 (b)

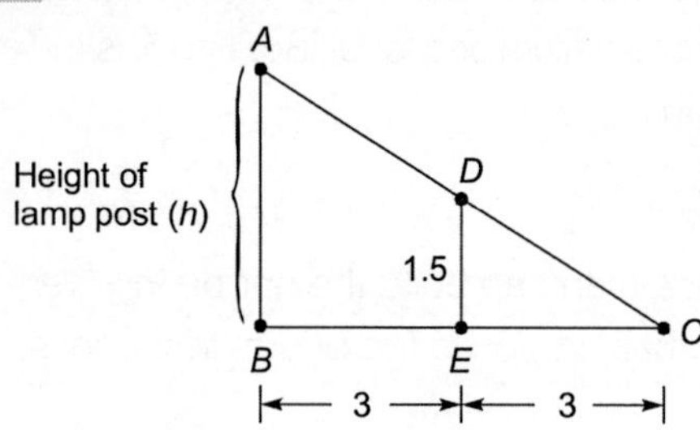

ABC and DEC are similar triangles,

$$\frac{AB}{DE} = \frac{BC}{EC}$$

$$\frac{h}{1.5} = \frac{6}{3}$$

$$\frac{h}{1.5} = 2$$

$$h = 3 \text{ meter}$$

1.99 (b)

715 ? 423

We know that the divisibility rule for 3 is sum of all digits should be divisible by 3.

$7 + 1 + 5 + 4 + 2 + 3 = 22$

So, the next number after sum 22 which are divisible by 3 are 24. 27, 30 etc.

So, $22 + 2 = 24$

$22 + 5 = 27$

But according to minimum condition 2 is right answer.

1.100 (b)

Alloy A contains Gold and Copper.

Let $2x : 3x$

Which is same as $4x : 6x$

Alloy B contains Copper $3x : 7x$

As masses of Alloy A is equal to Alloy of mass B.

i.e.

		Gold		Copper	
i.e.	Alloy 1	$4x$	:	$6x$	$\Rightarrow 10x$ (mass)
	Alloy 2	$3x$	:	$7x$	$\Rightarrow 10x$ (mass)

$\therefore$ Ratio of Gold to Copper is,

$$\frac{\text{Gold from alloy 1} + \text{Gold of alloy 2}}{\text{Copper from alloy 1} + \text{Copper of alloy 2}}$$

$$= \frac{4x + 3x}{6x + 7x} = \frac{7x}{13x} = \frac{7}{13}$$

1.101 (b)

$$A = P\left(1 + \frac{r}{100}\right)^n$$

$$1000000 = P\left(1 + \frac{10}{100}\right)^5$$

$$1000000 = P\left(\frac{11}{10}\right)^5$$

$$1000000 = P\left(\frac{11^5}{10^5}\right)$$

$$P = 1000000 \times \frac{10^5}{11^5} = \frac{10^{11}}{11^5}$$

$$= 620921.323 \approx 621000$$

1.102 (c)

Let us suppose total number of cabs are 100.

85 are green and rest 15 are blue.

Witness is correct 80% of times in identifying.

Total number of blue cabs identified correctly.

$$(80\% \text{ of } 15) \Rightarrow \frac{80}{100} \times 15 = 12$$

Witness is incorrect 20% of times

That incorrectness should happen for green

$(20\% \text{ of } 85) \Rightarrow \frac{20}{100} \times 85 = 17$

Witness identifies 12 + 17 = 29 Cabs in total.

$$\text{Probability} = \frac{\text{Required}}{\text{Total}} = \frac{12}{12+17} = \frac{12}{29}$$
$$= 41.3\% \approx 41\%$$

1.103 (d)

Machine X= 4 hours

Machine Y= 2 hours

In 1 hour X can work = $\frac{1}{4}$

In 1 hour Y can work = $\frac{1}{2}$

In 1 hour $(X + Y)$ can work = $\frac{1}{4} + \frac{1}{2} = \frac{3}{4}$

$(X + Y)$ together = $\frac{4}{3}$ hours

1.104 (d)

Here R and S are fixed in 1st and 2nd places from right.

P Q T [S] [R]

Therefore arrangement of

P, Q, T = $^3P_3 = 3! = 6$

1.105 (c)

Let total population

in 2001 = 100

in 2011 = 100

In 2001, males = 60, females = 40

illiterates,

50% of 60 = 30

60% of 40 = 24

54

literates = 100 – 54 = 46

In 2011, males = 50, females = 50

illiterates,

40% of 50 = 20

40% of 50 = 20

40

literates = 100 – 40 = 60

Percentage of increase literates from 2001 to 2011 = $\frac{60-46}{46} \times 100 = 30.43$.

1.106 (d)

P and Q started 8 am

1 hour = 30

$$P = \frac{210}{30} = 7 \text{ hours}$$

$$Q = \frac{240}{30} = 8 \text{ hours}$$

After break time,

P = 7:30 min = 6 hours 30 min

After break time,

Q = 8:20 min = 7 hours 40 min

Total time = 6 hours 30 min + 7 hour 40 min

= 14 hours 10 min

$= 14 \text{ hour} + \frac{10}{16} \text{ hour}$

$= 14 \text{ hour} + \frac{1}{6} \text{ hour} = \frac{85}{6} \text{ hours}$

In every 1 hour USD 200

$\therefore \frac{85}{6} \text{ hours} = \frac{85}{6} \times 200 = 2833.33$

Remaining= 3000 – 2833.33

= 166.666666 = 166.67

1.107 (c)

Only option (c) shows P and Q in same office, R and T in same office with S being in the adjacent office. Thus, option (c) is correct.

1.108 (d)

Left Right

Given Facing You ↓

Right Left

According to given conditions sequence is

Lohit, Seema, Rahul, Mathew,

↓ ↓

DoctorEngineer

∴Engineer= Mathew

1.109 (c)

Verify with options

Option (a):

$$\frac{1}{x_1 + x_2} > \frac{1}{x_1} + \frac{1}{x_2}$$

$$\frac{1}{2+3} > \frac{1}{2} + \frac{1}{3}$$

$$\frac{1}{5} > \frac{5}{6}$$

0.2 > 0.8 which is wrong.

Option (b):

$$\sqrt{x_1 + x_2} > \sqrt{x_1} + \sqrt{x_2}$$

$$\sqrt{2+3} > \sqrt{2} + \sqrt{3}$$

$$\sqrt{5} > 1.414 + 1.732$$

2.23 > 3.146 which is wrong.

Option (c): $e^{x_1 + x_2} > e^{x_1} + e^{x_2}$

$$e^{1+2} > e^1 + e^2$$

20.085 > 2.718 + 7.389 Satisfying.

1.110 (b)

Average number of students in school,

$$P = \frac{3+5+5+6+4}{5} = \frac{23}{5}$$

Average number of students in school,

$$Q = \frac{4+7+8+7+5}{5} = \frac{31}{5}$$

Difference of the number of students enrolled in school P and Q

$$= \frac{31-23}{5} = \frac{8}{5}$$

Ratio of the average number of the students enrolled in school P to the average of the difference of the number of students enrolled in schools *P* and *Q* is

= 23 : 8

1.113 (a)

Area of shaded portion
= Area of circle – area of rectangle
Maximum possible area of rectangle inscribed in the circle = $2a^2$
So, Required shaded area = $\pi a^2 - 2a^2$

1.114 (b)

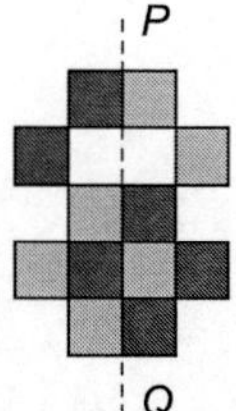

To make *P*-*Q* as symmetric line, minimum number of ■ square added = 6.

1.115 (d)

Given, $\frac{p}{q} + \frac{q}{p} = 3$

Squaring both sides,

$$\left(\frac{p}{q} + \frac{q}{p}\right)^2 = 3^2$$

$$\Rightarrow \frac{p^2}{q^2} + \frac{q^2}{p^2} + 2 = 9$$

$$\Rightarrow \frac{p^2}{q^2} + \frac{q^2}{p^2} = 9 - 2 = 7$$

1.116 (2)

Let the population 20 years ago be x.

$$1102500 = x\left(1 + \frac{5}{100}\right)^2$$

$$1102500 = x\left(\frac{21}{20}\right)^2$$

$$x = 10,00,000$$

1.117 (c)

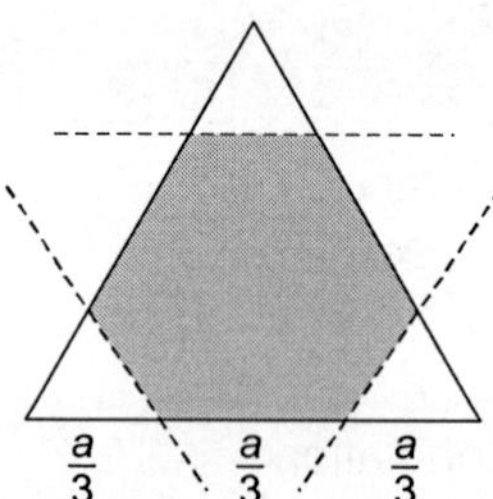

Let the side of the two larger equilateral triangle be a

Then, side of regular hexagon will be $\left(\frac{a}{3}\right)$

Area of regular Hexagon = $6 \times \frac{\sqrt{3}}{4}\left(\frac{a}{3}\right)^2$

Area of triangle = $\frac{\sqrt{3}}{4}(a)^2$

$$\text{Required ratio} = \frac{6\sqrt{3}}{4} \times \frac{a^2}{9} : \frac{\sqrt{3}}{4}a^2$$

$$= 6 : 9 = 2 : 3$$

1.118 (d)

On Sunday, the percentage

$$= \frac{65-55}{55} \times 100 = 18.18\%$$

On Saturday, the percentage

$$= \frac{60-50}{50} \times 100 = 20\%$$

On Friday, the percentage

$$= \frac{35-20}{20} \times 100 = 75\%$$

On Thursday, the percentage

$$= \frac{60-55}{55} \times 100 = 9.09\%$$

On Wednesday, the percentage

$$= \frac{60-50}{50} \times 100 = 20\%$$

On Tuesday, the percentage

$$= \frac{65-55}{55} \times 100 = 18.18\%$$

On Monday, the percentage

$$= \frac{70-45}{45} \times 100 = 55.56\%$$

Total six days are there when one of the students spent a minimum of 10% more than the other student.

1.119 (a)

Possible cases can be,

(i)

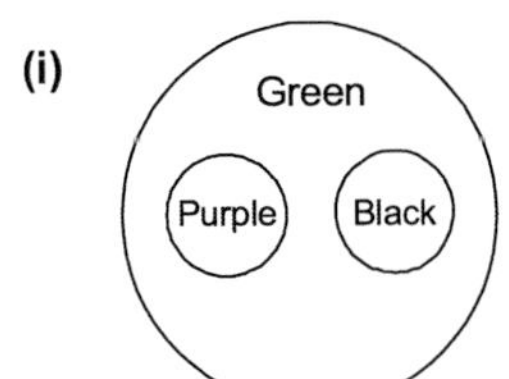

(ii)

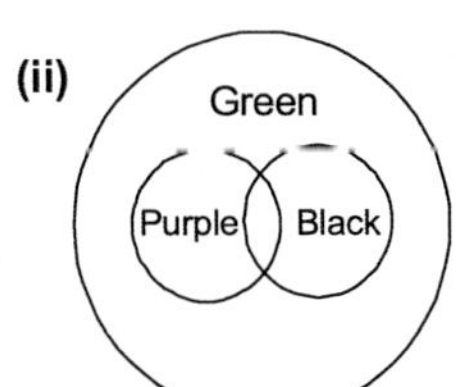

(iii)

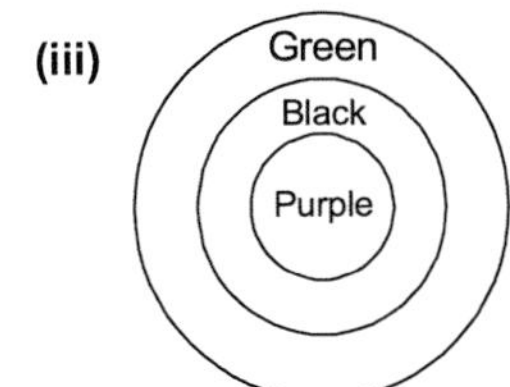

(iv)

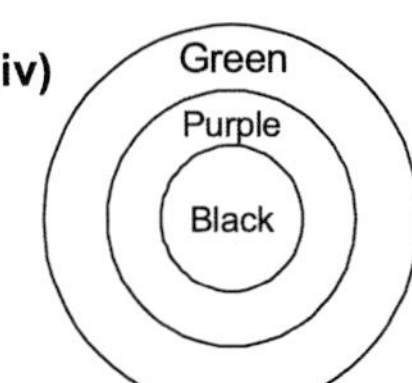

1.120 (c)

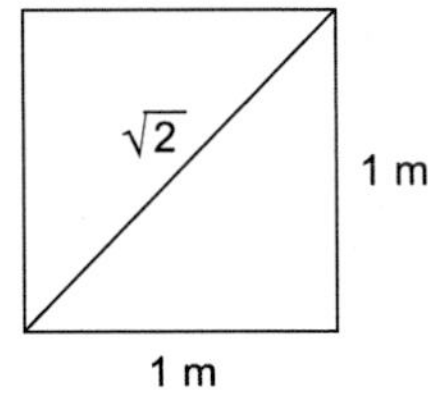

One of the triangle is revolved about its short side, resulting a cone.

Hence, $r = 1$, $H = 1$

Volume of cone, $V = \frac{1}{3}\pi r^2 H = \frac{\pi}{3}(1)^2(1)$

$$V = \frac{\pi}{3}$$

1.121 (d)

Given : $P : Q : R : S = 5 : 2 : 4 : 3$

Assume $P = 5x$, $Q = 2x$, $R = 4x$, $S = 3x$

Given that R gets Rs. 1000 more than S, therefore,

$$4x - 3x = 1000$$

$\Rightarrow \quad x = 1000$

$\therefore$ Share of $Q = 2x = 2 \times 1000 = 2000$

1.122 (b)

Given :

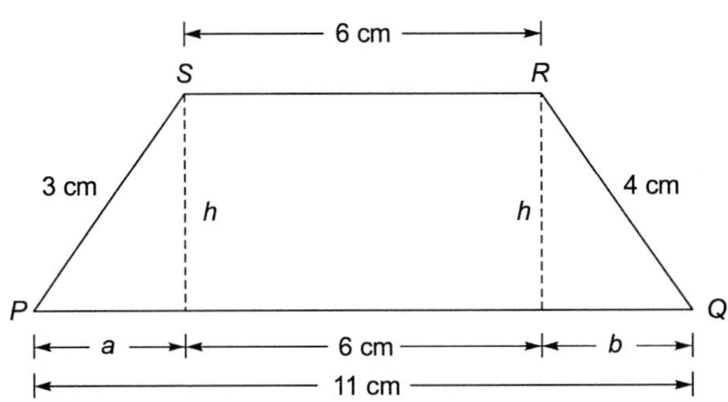

Let h is the shortest (i.e. perpendicular) distance between PQ and SR.

From the trapezium $PQRS$, using Pythagoras theorem,

$$h^2 + a^2 = 9 \quad \text{...(i)}$$

$$h^2 + b^2 = 16 \quad \text{...(ii)}$$

Subtracting equation (i) and (ii), we get,

$$b^2 - a^2 = 16 - 9 = 7$$

$$\Rightarrow (b-a)(b+a) = 7 \quad \text{...(iii)}$$

The length of side PQ,

$$a + 6 + b = 11$$

$$\Rightarrow \quad a + b = 11 - 6 = 5 \quad \text{...(iv)}$$

Using in equation (iii), we get,

$$(b - a) \times 5 = 7$$

$$\Rightarrow \quad b - a = \frac{7}{5} \quad \text{...(v)}$$

Using equation (iv), we get,

$$(5 - a) - a = \frac{7}{5}$$

$$2a = 5 - \frac{7}{5} = \frac{18}{5}$$

$$\Rightarrow \quad a = \frac{9}{5}$$

From equation (i),

$$h^2 = 9 - a^2 = 9 - \left(\frac{9}{5}\right)^2$$

$$h^2 = 9 - \frac{81}{25} = \frac{144}{25}$$

$$h = \frac{12}{5} = 2.40 \text{ cm}$$

1.123 (b)

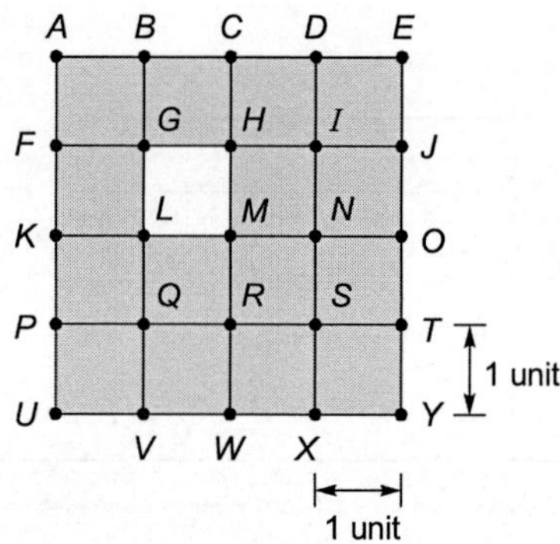

Total number of squares in 4 × 4 grid = 4 × 4 = 16

Number of 1 × 1 squares without a hole in the interior = 16 – 1 = 15

Number of 2 × 2 squares without a hole in the interior

= 5(KMWU, LNXV, MOYW, HJTR, CEOM)

∴ Maximum number of squares that can be formed without a hole in the interior = 15 + 5 = 20.

1.124 (c)

Considering the first inequality,

$$2x - 1 > 7$$

$$\Rightarrow \quad 2x > 8$$

$$\Rightarrow \quad x > 4 \qquad ...(i)$$

From the second given inequality,

$$2x - 9 < 1$$

$$\Rightarrow \quad x < 5 \qquad ...(ii)$$

Combining equations (i) and (ii), we get,

$$4 < x < 5$$

1.125 (c)

According to the given data, the quadrilateral *PSQR* can be plotted as below:

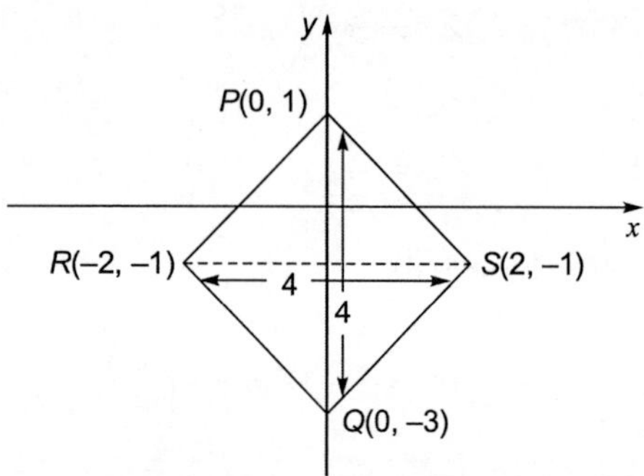

The length of the sides of the given quadrilateral can be calculated as below:

$$PS = \sqrt{(2-0)^2 + (-1-1)^2} = \sqrt{8}$$

$$QS = \sqrt{(2-0)^2 + (-1+3)^2} = \sqrt{8}$$

$$QR = \sqrt{(0+2)^2 + (-3+1)^2} = \sqrt{8}$$

$$PR = \sqrt{(0+2)^2 + (1+1)^2} = \sqrt{8}$$

Hence, all the sides of the quadrilateral are equal. The length of the diagonals,

$$PQ = \sqrt{(0-0)^2 + (1+3)^2} = 4$$

$$RS = \sqrt{(2+2)^2 + (-1+1)^2} = 4$$

$$\therefore \quad PQ = RS = 4$$

Since, the sides of the given quadrilateral are equal. Also, the length of the diagonals are equal and bisect at right angles. Hence, the given quadrilateral is a square.

∴ Area of quadrilateral = (side)^2 = $(PS)^2$ = 8 units

1.126 (b)

The figure obtained after applying the given sequences of operation are as below:

(1) *XYZZ*

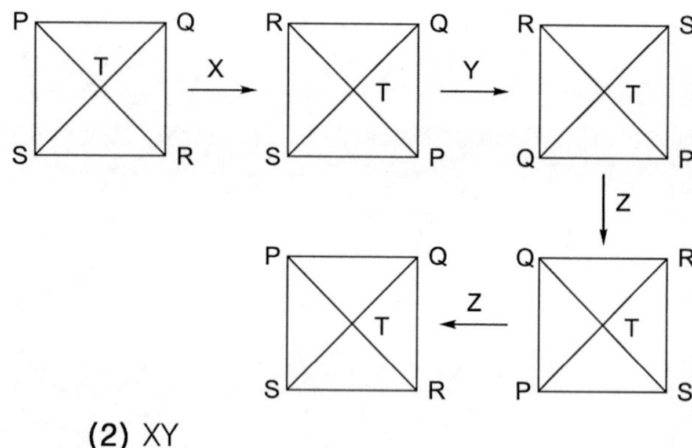

(2) XY

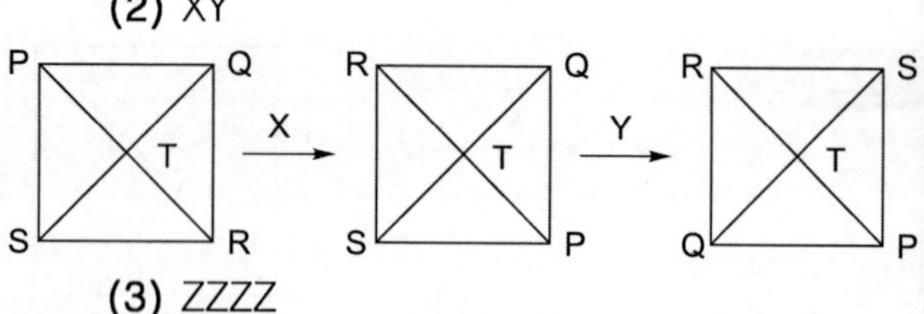

(3) ZZZZ

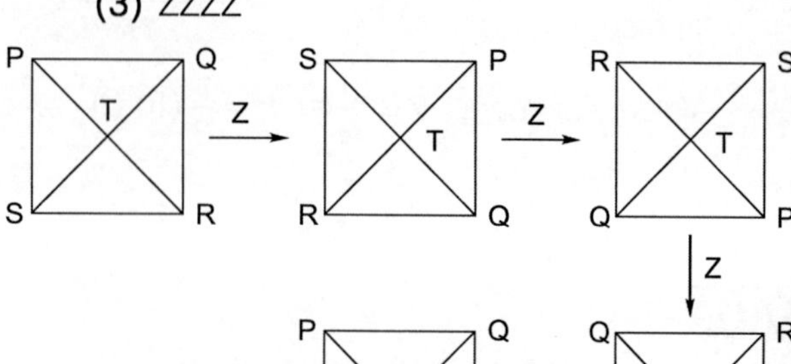

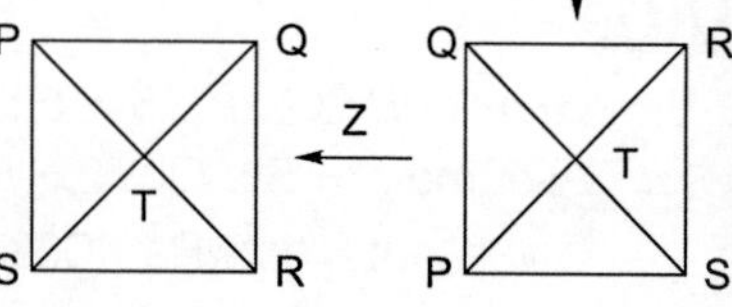

Hence, the sequence of operations (1) and (3) are equivalent.

1.127 (c)

The digits should be distinct and smallest number is 123456789.

1.128 (c)

Let students who like Romantic Movies = R.
Students who like Comedy Movies = C.
Given $R = 2C$
Also, 30 students do not like Romantic and Comedy Movies both.

$$\therefore \quad 100 - 30 = 70 = n(R \cap C)$$

and $n(C \cap R) = 20$

$$n(R \cap C) = n(R) + n(C) - n(C \cap R)$$
$$70 = 2C + C - 20$$
$$3C = 90$$
$$C = 30$$
$$R = 2C = 60$$

1.129 (c)

Number of rectangles = 10 as square is also called as rectangle.

1.130 (a)

Since, the given function is an even function. Option (d) is only represents the even function.

1.132 (c)

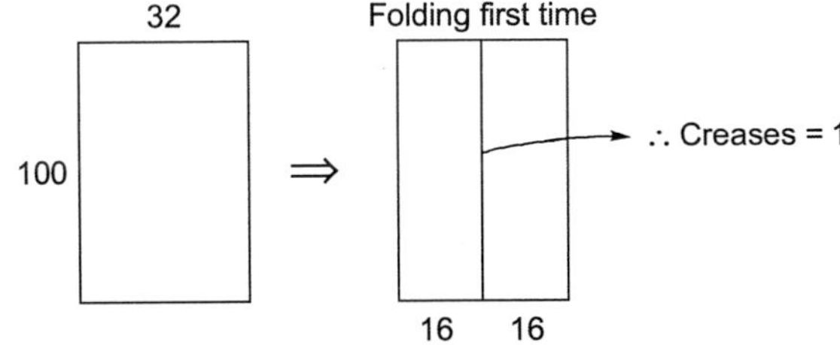

$$\therefore \text{ After folding 5 times} = 1 + 2^1 + 2^2 + 2^3 + 2^4$$
$$= 1 + 2 + 4 + 8 + 16 = 31$$

1.133 (b)

(i) $$\%\left(\frac{\text{Overweight}}{\text{Normal}}\right)_{2003} = \frac{40\%}{30\%} = \frac{4}{3} = 1.33$$

$$\%\left(\frac{\text{Overweight}}{\text{Normal}}\right)_{2023} = \frac{25\%}{50\%} = \frac{1}{2} = 0.5$$

Hence, the ratio of the percentage of population falling into overweight category to the percentage of population falling into normal category has decreased in 20 years. Therefore, option (a) is not correct.

(ii) $$\%\left(\frac{\text{Obese}}{\text{Normal}}\right)_{2003} = \frac{25\%}{30\%} = \frac{5}{6} = 0.83$$

$$\%\left(\frac{\text{Obese}}{\text{Normal}}\right)_{2023} = \frac{15\%}{50\%} = \frac{3}{10} = 0.3$$

Hence, the ratio of the percentage of population falling into obese category to the percentage of population falling into normal category has decreased in 20 years. Therefore, option (b) is correct.

(iii) $$\%\left(\frac{\text{Underweight}}{\text{Normal}}\right)_{2003} = \frac{5\%}{30\%} = \frac{1}{6} = 0.166$$

$$\%\left(\frac{\text{Underweight}}{\text{Normal}}\right)_{2023} = \frac{10\%}{50\%} = \frac{1}{5} = 0.2$$

Hence, the ratio of the percentage of population falling into underweight category to the percentage of population falling into normal category has increased in 20 years. Therefore, option (c) is not correct.

(iv) The percentage of population falling into normal category has increased from 30% to 50% in 20 years. Hence, option (d) is not correct.

1.134 (d)

Let five years ago,
Son's age = x
Father's age = $4x$
After five years,
Son's age = $x + 10$ and Father's age = $4x + 10$

$$\therefore \frac{x+10}{4x+10} = \frac{2}{5}$$
$$5x + 50 = 8x + 20$$
$$3x = 30$$
$$x = 10$$

Son's age five years ago = 10 years
Father's age five years ago = 40 years
At the time of son's birth, father's age
= 40 – 10 = 30 years

1.135 (b)

$$3^{199} - 3^{196} = 3^{196}(3^3 - 1) = 3^{196}(27 - 1)$$
$$= 3^{196} \times 26 = 2 \times 3^{196} \times 13$$

The highest prime factor is 13.

1.136 (c)

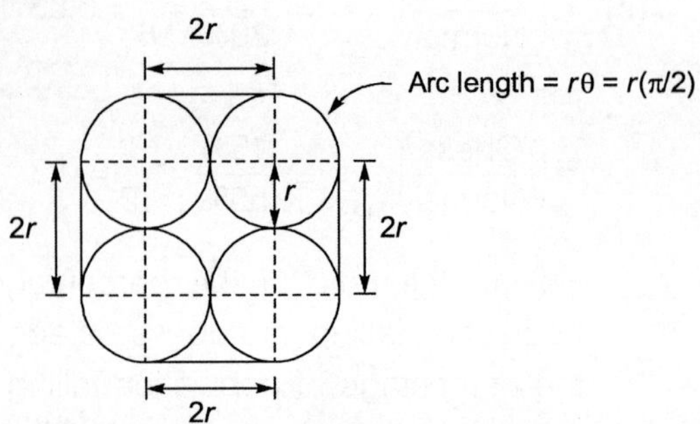

Width of the duct-tape, $W = l = 10$ cm

Length of the duct-tape required,

$$L = 4 \times (2r) + \left(4r \times \frac{\pi}{2}\right)$$

$$L = 8 \times 0.5 + 2 \times 0.5 \times \pi$$

$$L = (4 + \pi) \text{ cm}$$

Hence, the area of duct-tape required,

$$A = W \cdot L = 10(4 + \pi) \text{ cm}^2$$

1.137 (d)

FO1 operation on sheet A:

Folding first time

Folding second time

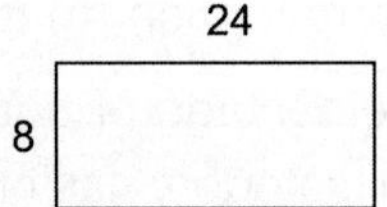

Perimeter $= 2(24 + 4) = 56$ cm

FO2 operation on sheet B:

Folding first time

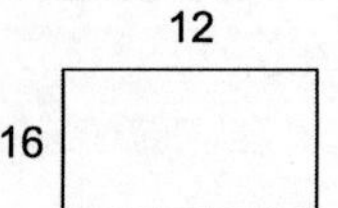

Folding second time

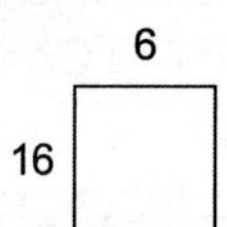

Perimeter $= 2(16 + 6) = 44$ cm

Required ratio = 56 : 44 = 14 : 11

1.138 (a)

$$\frac{1}{\log_2 x} + \frac{1}{\log_3 x} + \frac{1}{\log_4 x} = 1$$

$$\Rightarrow \log_x 2 + \log_x 3 + \log_x 4 = 1$$

$$\Rightarrow \log_x(2 \times 3 \times 4) = 1 \quad \Rightarrow \log_x 24 = 1$$

$$\Rightarrow \quad x = 24$$

1.139 (b)

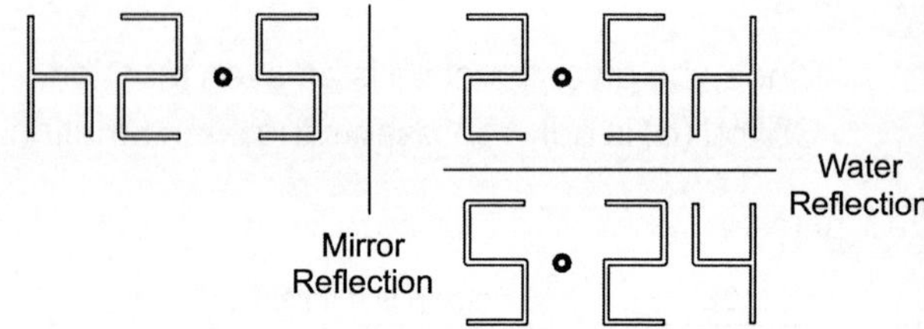

Hence, option (b) is the correct answer.

■■■

2 Verbal Ability

2.1 Which of the following options is the closest in meaning to the word below?

Circuitous

(a) Cyclic (b) Indirect
(c) Confusing (d) Crooked [2010 : 1 M]

2.2 The question below consists of a pair of related words followed by four pairs of words. Select the pair that best expresses the relation in the original pair.

Unemployed: Worker

(a) Fallow : Land (b) Unaware : Sleeper
(c) Wit : Jester (d) Renovated : House

[2010 : 1 M]

2.3 Choose the most appropriate word from the options given below to complete the following sentence:

If we manage to .... our natural resources, we would leave a better planet for our children.

(a) uphold (b) restrain
(c) cherish (d) conserve [2010 : 1 M]

2.4 Choose the most appropriate word from the options given below to complete the following sentence:

His rather casual remarks on politics .... his lack of seriousness about the subject.

(a) masked (b) belied
(c) betrayed (d) suppressed

[2010 : 1 M]

2.5 Modern warfare has changed from large scale clashes of armies to suppression of civilian populations. Chemical agents that do their work silently appear to be suited to such warfare; and regretfully, there exist people in military establishments who think that chemical agents are useful fools for their cause.

Which of the following statements best sums up the meaning of the above passage?

(a) Modern warfare has resulted in civil strife.
(b) Chemical agents are useful in modern warfare.
(c) Use of chemical agents in warfare would be undesirable.
(d) People in military establishments like to use chemical agents in war.

[2010 : 2 M]

2.6 The question below consists of a pair of related words followed by four pairs of words. Select the pair that best expresses the relation in the original pair:

Gladiator : Arena

(a) dancer : stage
(b) commuter : train
(c) teacher : classroom
(d) lawyer : courtroom [2011 : 1 M]

2.7 Choose the most appropriate word from the options given below to complete the following sentence:

It was her view that the country's problems had been ________ by foreign technocrats, so that to invite them to come back would be counter-productive.

(a) identified (b) ascertained
(c) exacerbated (d) analysed

[2011 : 1 M]

2.8 Choose the word from the options given below that is most nearly opposite in meaning to the given word:

Frequency

(a) periodicity (b) rarity
(c) gradualness (d) persistency

[2011 : 1 M]

2.9 Choose the most appropriate word from the options given below to complete the following sentence:

Under ethical guidelines recently adopted by the Indian Medical Association, human genes are to be manipulated only to correct diseases for which ______ treatments are unsatisfactory.

(a) similar (b) most
(c) uncommon (d) available

[2011 : 1 M]

2.10 The horse has played a little known but very important role in the field of medicine. Horses were injected with toxins of diseases until their blood built up immunities. Then a serum was made from their blood. Serums to fight with diphtheria and tetanus were developed this way.

It can be inferred from the passage, that horses were

(a) given immunity to diseases
(b) generally quite immune to diseases
(c) given medicines to fight toxins
(d) given diphtheria and tetanus serums

[2011 : 2 M]

2.11 Which one of the following options is the closest in meaning to the word given below?

Latitude

(a) Eligibility (b) Freedom
(c) Coercion (d) Meticulousness

[2012 : 1 M]

2.12 One of the parts (*A*, *B*, *C*, *D*) in the sentence given below contains an ERROR. Which one of the following is INCORRECT?

I requested that he should be given the driving test today instead of tomorrow.

(a) requested that (b) should be given
(c) the driving test (d) instead of tomorrow

[2012 : 1 M]

2.13 Choose the most appropriate alternative from the options given below to complete the following sentence:

If the tired soldier wanted to lie down, he ____ the mattress out on the balcony.

(a) should take (b) shall take
(c) should have taken (d) will have taken

[2012 : 1 M]

2.14 Choose the most appropriate word from the options given below to complete the following sentence:

Given the seriousness of the situation that he had to face, his ___ was impressive.

(a) beggary (b) nomenclature
(c) jealousy (d) nonchalance

[2012 : 1 M]

2.15 One of the legacies of the Roman legions was discipline. In the legions, military law prevailed and discipline was brutal. Discipline on the battlefield kept units obedient, intact and fighting, even when the odds and conditions were against them.

Which one of the following statements best sums up the meaning of the above passage?

(a) Thorough regimentation was the main reason for the efficiency of the Roman legions even in adverse circumstances.
(b) The legions were treated inhumanly as if the men were animals.
(c) Discipline was the armies inheritance from their seniors.
(d) The harsh discipline to which the legions were subjected to led to the odds and conditions being against them.

[2012 : 2 M]

2.16 Choose the grammatically **CORRECT** sentence:

(a) Two and two add four.
(b) Two and two become four.
(c) Two and two are four.
(d) Two and two make four.

[2013 : 1 M]

2.17 **Statement:** You can always give me a ring whenever you need.

Which one of the following is the best inference from the above statement?

(a) Because I have a nice caller tune.
(b) Because I have a better telephone facility.
(c) Because a friend in need is a friend indeed.
(d) Because you need not pay towards the telephone bills when you give me a ring.

[2013 : 1 M]

2.18 Complete the sentence:

Dare ________ mistakes.

(a) commit (b) to commit
(c) committed (d) committing

[2013 : 1 M]

2.19 They were requested not to **quarrel** with others.

Which one of the following options is the closest in meaning to the word **quarrel**?

(a) make out (b) call out
(c) dig out (d) fall out

[2013 : 1 M]

2.20 **Statement:** There were different streams of freedom movements in colonial India carried out by the moderates, liberals, radicals, socialists, and so on.

Which one of the following is the best inference from the above statement?

(a) The emergence of nationalism in colonial India led to our Independence.
(b) Nationalism in India emerged in the context of colonialism.
(c) Nationalism in India is homogeneous.
(d) Nationalism in India is heterogeneous.

[2013 : 2 M]

2.21 Choose the most appropriate phrase from the options given below to complete the following sentence. The aircraft _______ take off as soon as its flight plan was filed.
(a) is allowed to
(b) will be allowed to
(c) was allowed to
(d) has been allowed to [2014 : 1 M, Set-1]

2.22 Read the statements:
All women are entrepreneurs.
Some women are doctors.
Which of the following conclusions can be logically inferred from the above statements?
(a) All women are doctors
(b) All doctors are entrepreneurs
(c) All entrepreneurs are women
(d) Some entrepreneurs are doctors
[2014 : 1 M, Set-1]

2.23 Choose the most appropriate word from the options given below to complete the following sentence.
Many ancient cultures attributed disease to supernatural causes. However, modern science has largely helped ______ such notions.
(a) impel (b) dispel
(c) propel (d) repel
[2014 : 1 M, Set-1]

2.24 Choose the most appropriate word from the options given below to complete the following sentence.
Communication and interpersonal skills are _______ important in their own ways.
(a) each (b) both
(c) all (d) either
[2014 : 1 M, Set-2]

2.25 Which of the options given below best completes the following sentence?
She will feel much better if she __________.
(a) will get some rest
(b) gets some rest
(c) will be getting some rest
(d) is getting some rest [2014 : 1 M, Set-2]

2.26 Choose the most appropriate pair of words from the options given below to complete the following sentence.
She could not _____ the thought of _______ the election to her bitter rival.
(a) bear, loosing (b) bare, loosing
(c) bear, losing (d) bare, losing
[2014 : 1 M, Set-2]

2.27 The value of one U.S. dollar is 65 Indian Rupees today, compared to 60 last year. The Indian Rupee has ____________.
(a) depressed (b) depreciated
(c) appreciated (d) stabilized
[2014 : 1 M, Set-3]

2.28 'Advice' is ________________.
(a) a verb
(b) a noun
(c) an adjective
(d) both a verb and a noun
[2014 : 1 M, Set-3]

2.29 Which of the following options is the closest in meaning to the word underlined in the sentence below?
In a democracy, everybody has the freedom to **disagree** with the government.
(a) dissent (b) descent
(c) decent (d) decadent
[2014 : 1 M, Set-4]

2.30 After the discussion, Tom said to me, 'Please revert!'. He expects me to _________.
(a) retract (b) get back to him
(c) move in reverse (d) retreat
[2014 : 1 M, Set-4]

2.31 While receiving the award, the scientist said, "I feel vindicated". Which of the following is closest in meaning to the word 'vindicated'?
(a) punished (b) substantiated
(c) appreciated (d) chastened
[2014 : 1 M, Set-4]

2.32 Choose the appropriate word/phrase, out of the four options given below, to complete the following sentence:
Frogs _______________
(a) croak (b) roar
(c) hiss (d) patter
[2015 : 1 M, Set-1]

2.33 Choose the word most similar meaning to the given word:
Educe
(a) Exert (b) Educate
(c) Extract (d) Extend
[2015 : 1 M, Set-1]

2.34 Choose the most appropriate word from the options given below to complete the following sentence.

The principal presented the chief guest with a __________, as token of appreciation.

(a) momento (b) memento
(c) momentum (d) moment

[2015 : 1 M, Set-1]

2.35 The following question presents a sentence, part of which is underlined. Beneath the sentence you find four ways of phrasing the underlined part. Following the requirements of the standard written English, select the answer that produces the most effective sentence.

Tuberculosis, together with its effects, ranks one of the leading causes of death in India.

(a) ranks as one of the leading causes of death
(b) rank as one of the leading causes of death
(c) has the rank of one of the leading causes of death
(d) are one of the leading causes of death

[2015 : 2 M, Set-1]

2.36 Humpty Dumpty sits on a wall every day while having lunch. The wall sometimes breaks. A person sitting on the wall falls if the wall breaks.

Which one of the statements below is logically valid and can be inferred from the above sentences?

(a) Humpty Dumpty always falls white having lunch
(b) Humpty Dumpty does not fall sometimes while having lunch
(c) Humpty Dumpty never falls during dinner
(d) When Humpty Dumpty does not sit on the wall, the wall does not break

[2015 : 2 M, Set-1]

2.37 Read the following paragraph and choose the correct statement.

Climate change has reduced human security and threatened human well being. An ignored reality of human progress is that human security largely depends upon environmental security. But on the contrary, human progress seems contradictory to environmental security. To keep up both at the required level is a challenge to be addressed by one and all. One of the ways to curb the climate change may be suitable scientific innovations, while the other may be the Gandhian perspective on small scale progress with focus on sustainability.

(a) Human progress and security are positively associated with environmental security.
(b) Human progress is contradictory to environmental security.
(c) Human security is contradictory to environmental security.
(d) Human progress depends upon environmental security.

[2015 : 2 M, Set-1]

2.38 Choose the word most similar in meaning to the given word:

Awkward

(a) Inept (b) Graceful
(c) Suitable (d) Dreadful

[2015 : 1 M, Set-2]

2.39 Choose the appropriate word/phase, out of the four options given below, to complete the following sentence:

Dhoni, as well as the other team members of the Indian team _____ present on the occasion.

(a) were (b) was
(c) has (d) have

[2015 : 1 M, Set-2]

2.40 What is the adverb for the given word below?

Misogynous

(a) Misogynousness (b) Misogynity
(c) Misogynously (d) Misogynous

[2015 : 1 M, Set-2]

2.41 Given below are two statements followed by two conclusions. Assuming these statements to be true, decide which one logically follows.

Statements:

I. All film star are playback singers.
II. All film directors are film stars.

Conclusions:

I. All film directors are playback singers.
II. Some film stars are film directors.

(a) Only conclusion I follows.
(b) Only conclusion I nor II follows.
(c) Neither conclusion I nor II follows.
(d) Both conclusions I and II follow.

[2015 : 2 M, Set-2]

2.42 In the following sentence certain parts are underlined and marked *P*, *Q* and *R*. One of the parts may contain certain error or may not be acceptable in standard written communication. Select the part containing an error. Choose *D* as your answer if there is no error.

The student concreted all the errors that the instructor marked on the answer book.
(a) *P* (b) *Q*
(c) *R* (d) No Error

[2015 : 2 M, Set-2]

2.43 Lamenting the gradual sidelining of the arts in school curricula, a group of prominent artists wrote to the Chief Minister last year, asking him to allocate more funds to support arts education in schools. However, no such increase has been announced in this year's Budget. The artists expressed their deep anguish at their request not being approved, but many of them remain optimistic about finding in the future.

Which of the statement(s) below is/are logically valid and can be inferred from the above statements?
(i) The artists expected funding for the arts to increase this year.
(ii) The Chief Minister was receptive to the idea of increasing funding for the arts.
(iii) The Chief Minister is a prominent artists.
(iv) Schools are giving less importance to arts education nowadays.
(a) (iii) and (iv) (b) (i) and (iv)
(c) (i), (ii) and (iv) (d) (i) and (iii)

[2015 : 2 M, Set-2]

2.44 Choose the correct verb to fill in the blank below:
Let us ______.
(a) introvert (b) alternate
(c) atheist (d) altruist

[2015 : 1 M, Set-3]

2.45 Choose the most appropriate word from the options given below to complete the following sentence.
If the athlete had wanted to come first in the race, he ______ several hours every day.
(a) should practice
(b) should have practised
(c) practised
(d) should be practising [2015 : 1 M, Set-3]

2.46 Choose the most suitable one word substitute for the following expression:
Connection of a road or way
(a) Pertinacious (b) Viaticum
(c) Clandestine (d) Ravenous

[2015 : 1 M, Set-3]

2.47 Ram and Shyam shared a secret and promised to each other that it would remain between them. Ram express himself in one of the following ways as given in the choices below. Identify the correct way as per standard English.
(a) It would remain between you and me.
(b) It would remain between I and you.
(c) It would remain between you and I.
(d) It would remain with me.

[2015 : 2 M, Set-3]

2.48 In the following question, the first and the last sentence of the passage are in order and numbered 1 and 6. The rest of the passage is split into 4 parts and numbered as 2, 3, 4 and 5. These 4 parts are not arranged in proper order. Read the sentences and arrange them in a logical sequence to make a passage and choose the correct sequence from the given options.
1. On Diwali, the family rises early in the morning.
2. The whole family, including the young and the old enjoy doing this.
3. Children let off fireworks later in the night with their friends.
4. At sunset, the lamps are lit and the family performs various rituals.
5. Father, mother and children visit relatives and exchange gifts and sweets.
6. Houses looks so pretty with lighted lamps all around.

(a) 2, 5, 3, 4 (b) 5, 2, 4, 3
(c) 3, 5, 4, 2 (d) 4, 5, 2, 3

[2015 : 2 M, Set-3]

2.49 Which of the following is **CORRECT** with respect to grammar and usage?Mount Everest is____ .
(a) the highest peak in the world
(b) highest peak in the world
(c) one of highest peak in the world
(d) one of the highest peak in the world

[2016 : 1 M, Set-1]

2.50 The policeman asked the victim of a theft, "What did you _____?
(a) loose (b) lose
(c) loss (d) louse

[2016 : 1 M, Set-1]

2.51 Despite the new medicine's _____ in treating diabetes, it is not _____ widely.
(a) effectiveness - prescribed
(b) availability - used
(c) prescription - available
(d) acceptance - proscribed

[2016 : 1 M, Set-1]

2.52 In a world filled with uncertainty, he was glad to have many good friends. He has always assisted them in times of need and was confident that they would reciprocate. However, the events of the last week proved him wrong.

Which of the following inference(s) is/are logically valid and can be inferred from the above passage?

I. His friends were always asking him to help them.
II. He felt that when in need of help, his friends would let him down.
III. He was sure that his friends would help him when in need.
IV. His friends did not help him last week.

(a) I and II (b) III and IV
(c) III only (d) IV only

[2016 : 2 M, Set-1]

2.53 The students _____ the teacher on teachers' day for twenty years of dedicated teaching.

(a) facilitated (b) felicitated
(c) fantasized (d) facillitated

[2016 : 1 M, Set-2]

2.54 After India's cricket world cup victory in 1985, Shrotria who was playing both tennis and cricket till then, decided to concentrate only on cricket. And the rest is history.

What does the underlined phrase mean in this context?

(a) history will rest in peace
(b) rest is recorded in history books
(c) rest is well known
(d) rest is archaic [2016 : 1 M, Set-2]

2.55 Social science disciplines were in existence in an amorphous form until the colonial period when they were institutionalized. In varying degrees, they were intended to further the colonial interest. In the time of globalization and the economic rise of postcolonial countries like India, conventional ways of knowledge production have become obsolete.

Which of the following can be logically inferred from the above statements?

I. Social science disciplines have become obsolete.
II. Social science disciplines had a pre-colonial origin.
III. Social science disciplines always promote colonialism.
IV. Social science must maintain disciplinary boundaries.

(a) II only (b) I and III only
(c) II and IV only (d) III and IV only

[2016 : 2 M, Set-2]

2.56 An apple costs ₹10. An onion costs ₹8.

Select the most suitable sentence with respect to grammar and usage.

(a) The price of an apple is greater than an onion.
(b) The price of an apple is more than onion.
(c) The price of an apple is greater than of an onion.
(d) Apples are more costlier than onions.

[2016 : 1 M, Set-3]

2.57 The Buddha said, "Holding on to anger is like <u>grasping</u> a hot coal with the intent of throwing it at someone else; you are the one who gets burnt."

Select the word below which is closest in meaning to the word underlined above.

(a) burning (b) igniting
(c) clutching (d) flinging

[2016 : 1 M, Set-3]

2.58 The overwhelming number of people infected with rabies in India has been flagged by the World Health Organization as a source of concern. It is estimated that inoculating 70% of pets and stray dogs against rabies can lead to a significant reduction in the number of people infected with rabies.

Which of the following can be logically inferred from the above sentences?

(a) The number of people in India infected with rabies is high.
(b) The number of people in other parts of the world who are infected with rabies is low
(c) Rabies can be eradicated in India by vaccinating 70% of stray dogs
(d) Stray dogs are the main source of rabies worldwide

[2016 : 2 M, Set-3]

2.59 I ________ made arrangements had I ________ informed earlier.

(a) could have, been (b) would have, being
(c) had, have (d) had been, been

[2017 : 1 M, Set-1]

2.60 She has a sharp tongue and it can occasionally turn __________.

(a) hurtful (b) left
(c) methodical (d) vital

[2017 : 1 M, Set-1]

2.61 "If you are looking for a history of India, or for an account of the rise and fall of the British Raj, or for the reason of the cleaving of the subcontinent into two mutually antagonistic parts and the effects this mutilation will have in the respective sections, and ultimately on Asia, you will not find it in these pages; for though I have spent a lifetime in the country, I lived too near the seat of events, and was too intimately associated with the actors, to get the perspective needed for the impartial recording of these matters".

Here, the word 'antagonistic' is closest in meaning to

(a) impartial (b) argumentative
(c) separated (d) hostile

[2017 : 2 M, Set-1]

2.62 It is __________ to read this year's textbook __________ the last year's.

(a) easier, than (b) most easy, than
(c) easier, from (d) easiest, from

[2017 : 1 M, Set-2]

2.63 The ninth and the tenth of this month are Monday and Tuesday _____.

(a) figuratively (b) retrospectively
(c) respectively (d) rightfully

[2017 : 1 M, Set-2]

2.64 "If you are looking for a history of India, or for an account of the rise and fall of the British Raj or for the reason of the cleaving of the subcontinent into two mutually antagonistic parts and the effects this mutilation will have in the respective sections, and ultimately on Asia, you will not find it in these pages: for though I have spent a lifetime in the country. I lived too near the seat of events, and was too intimately associated with the actors, to get the perspective needed for the impartial recording of these matters."

Which of the following statements best reflects the author's opinion?

(a) An intimate association does not allow for the necessary perspective.
(b) Matters are recorded with an impartial perspective.
(c) An intimate association offers an impartial perspective.
(d) Actors are typically associated with the impartial recording of matters.

[2017 : 2 M, Set-2]

2.65 "By giving him the last _____ of the cake, you will ensure lasting _____ in our house today."

The words that best fill the blanks in the above sentence are

(a) peas, piece (b) piece, peace
(c) peace, piece (d) peace, peas

[2018 : 1 M]

2.66 "Even though there is a vast scope for its _______, tourism has remained a/an _____ area."

The words that best fill the blanks in the above sentence are

(a) improvement, neglected
(b) rejection, approved
(c) fame, glum
(d) interest, disinterested

[2018 : 1 M]

2.67 A coastal region with unparalleled beauty is home to many species of animals. It is dotted with coral reefs and unspoilt white sandy beaches. It has remained inaccessible to tourists due to poor connectivity and lack of accommodation. A company has spotted the opportunity and is planning to develop a luxury resort with helicopter service to the nearest major city airport. Environmentalists are upset that this would lead to the region becoming crowded and polluted like any other major beach resorts.

Which one of the following statements can be logically inferred from the information given in the above paragraph?

(a) The culture and tradition of the local people will be influenced by the tourists.
(b) The region will become crowded and polluted due to tourism.
(c) The coral reefs are on the decline and could soon vanish.
(d) Helicopter connectivity would lead to an increase in tourists coming to the region.

[2018 : 2 M]

2.68 The Cricket Board has long recognized John's potential as a leader of the team. However, his on-field Temper has always been a matter of concern for them since his junior days. While this aggression has filled stadia with die-hard fans, it has taken a toll on his own batting. Until recently, it appeared that he found it difficult to convert his aggression into big scores. Over the past three

seasons though, that picture of John has been replaced by a cerebral, calculative and successful batsman-captain. After many years, it appears that the team has finally found a complete captain. Which of the following statements can be logically inferred from the above paragraph?
(i) Even as a junior cricketer, John was considered a good captain.
(ii) Finding a complete captain is a challenge.
(iii) Fans and the Cricket Board have differing views on what they want in a captain.
(iv) Over the past three seasons John has accumulated big scores.
(a) (i). (ii) and (iii) only (b) (iii) and (iv) only
(c) (ii) and (iv) only (d) (i), (ii), (iii) and (v)

[2018 : 2 M]

2.69 When he did not come home, she _______ him lying dead on the roadside somewhere.
(a) notice (b) looked
(c) concluded (d) pictured

[2019 : 1 M]

2.70 The strategies that the company ____ to sell its products ______ house-to-house marketing.
(a) used, includes (b) uses, includes
(c) use, includes (d) uses, including

[2019 : 1 M]

2.71 The boat arrived ______ dawn.
(a) on (b) at
(c) in (d) under

[2019 : 1 M]

2.72 "Indian history was written by British historians-extremely well documented and researched, but not always impartial. History had to serve its purpose: Everything was made subservient to the glory of the Union Jack. Latter-day Indian scholars presented a contrary picture."
From the text above, we can infer that:
Indian history written by British historians _______.
(a) was well documented and researched but was sometimes biased
(b) was well documented and not researched but was always biased
(c) was not well documented and researched and was sometimes biased
(d) was not well documented and researched and was always biased

[2019 : 2 M]

2.73 The Canadian constitution requires that equal importance be given to English and French. Last year, Air Canada lost a lawsuit, and had to pay a six-figure fine to a French-speaking couple after they filed complaints about formal in-flight announcements in English lasting 15 seconds, as opposed to informal 5 second messages in French.
The French-speaking couple were upset at ______.
(a) equal importance being given to English and French.
(b) the in-flight announcements being made in English.
(c) the English announcements being longer than the French ones.
(d) the English announcements being clearer than the French ones.

[2020 : 1 M]

2.74 Select the word that fits the analogy:
Explicit : Implicit : : Express : _________
(a) Repress (b) Suppress
(c) Compress (d) Impress

[2020 : 1 M]

2.75 The untimely loss of life is a cause of serious global concern as thousands of people get killed _______ accidents every year while many other die _______ diseases like cardio vascular disease, cancer, etc.
(a) during, from (b) in, of
(c) from, from (d) from, of

[2020 : 1 M]

2.76 The global financial crisis in 2008 is considered to be the most serious world-wide financial crisis, which started with the sub-prime lending crisis in USA in 2007. The sub-prime lending crisis led to the banking crisis in 2008 with the collapse of Lehman Brothers in 2008. The sub-prime lending refers to the provision of loans to those borrowers who may have difficulties in repaying loans, and it arises because of excess liquidity following the East Asian crisis.
Which one of the following sequences shows the correct precedence as per the given passage?
(a) Subprime lending crisis → global financial crisis → banking crisis → East Asian crisis.
(b) Banking crisis → subprime lending crisis → global financial crisis → East Asian crisis.
(c) East Asian crisis → subprime lending crisis → banking crisis → global financial crisis.
(d) Global financial crisis → East Asian crisis → banking crisis → subprime lending crisis.

[2020 : 2 M]

2.77 He was not only accused of theft ________ of conspiracy.

(a) rather (b) rather than
(c) but also (d) but even

[2020 : 1 M]

2.78 Nostalgia is to anticipation as ______ is to ________.

Which one of the following options maintains a similar logical relation in the above sentence?

(a) Past, future (b) Future, past
(c) Future, present (d) Present, past

[2021 : 1 M]

2.79 Consider the following sentences :

(i) I woke up from sleep.
(ii) I woked up from sleep.
(iii) I was woken up from sleep.
(iv) I was wokened up from sleep.

Which of the above sentences are grammatically CORRECT?

(a) (i) and (ii) (b) (ii) and (iii)
(c) (i) and (iv) (d) (i) and (iii)

[2021 : 1 M]

2.80 Computers are ubiquitous. They are used to improve efficiency in almost all fields from agriculture to space exploration. Artificial intelligence (AI) is currently a hot topic. AI enables computers to learn, given enough training data. For humans, sitting in front of a computer for long hours can lead to health issues.

Which of the following can be deduced from the above passage?

(i) Nowadays, computers are present in almost all places.
(ii) Computers cannot be used for solving problems in engineering.
(iii) For humans, there are both positive and negative effects of using computers.
(iv) Artificial intelligence can be done without data.

(a) (ii) and (iii) (b) (i), (iii) and (iv)
(c) (i) and (iii) (d) (ii) and (iv)

[2021 : 2 M]

2.81 Mr. X speaks ________ Japanese ________ Chinese.

(a) neither / or (b) either / nor
(c) neither / nor (d) also / but

[2022 : 1 M]

2.82 An art gallery engages a security guard to ensure that the items displayed are protected. The diagram below represents the plan of the gallery where the boundary walls are opaque. The location the security guard posted is identified such that all the inner space (shaded region in the plan) of the gallery is within the line of sight of the security guard. If the security guard does not move around the posted location and has a 360° view, which one of the following correctly represents the set of ALL possible locations among the locations P, Q, R and S, where the security guard can be posted to watch over the entire inner space of the gallery.

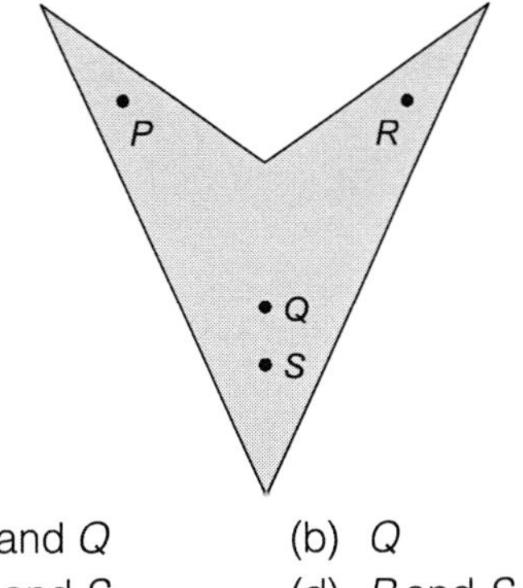

(a) *P* and *Q* (b) *Q*
(c) *Q* and *S* (d) *R* and *S*

[2022 : 1 M]

2.83 Mosquitoes pose a threat to human health. Controlling mosquitoes using chemicals may have undesired consequences. In Florida, authorities have used genetically modified mosquitoes to control the overall mosquito population. It remains to be seen if this novel approach has unforeseen consequences.

Which one of the following is the correct logical inference based on the information in the above passage?

(a) Using chemicals to kill mosquitoes is better than using genetically modified mosquitoes because genetic engineering is dangerous.
(b) Using genetically modified mosquitoes is better than using chemicals to kill mosquitoes because they do not have any side effects.
(c) Both using genetically modified mosquitoes and chemicals have undesired consequences and can be dangerous.
(d) Using chemicals to kill mosquitoes may have undesired consequences but it is not clear if using genetically modified mosquitoes has any negative consequence.

[2022 : 2 M]

2.84 In a class of five students P, Q, R, S and T, only one student is known to have copied in the exam. The disciplinary committee has investigated the situation and recorded the statements from the students as given below.

Statement of P: R has copied in the exam.

Statement of Q: S has copied in the exam.

Statement of R: P did not copy in the exam.

Statement of S: Only one of us is telling the truth.

Statement of T: R is telling the truth.

The investigating team had authentic information that S never lies.

Based on the information given above, the person who has copied in the exam is

(a) R (b) P
(c) Q (d) T

[2022 : 2 M]

2.85 "I cannot support this proposal. My _______ will not permit it."

(a) conscious (b) consensus
(c) conscience (d) consent

[2023 : 1 M]

2.86 Courts : _______ : : Parliament : Legislature

(a) Judiciary (b) Executive
(c) Governmental (d) Legal [2023 : 1 M]

2.87 Forestland is a planet inhabited by different kinds of creatures. Among other creatures, it is populated by animals all of whom are ferocious. There are also creatures that have claws, and some that do not. All creatures that have claws are ferocious.

Based only on the information provided above, which one of the following options can be logically inferred with certainty?

(a) All creatures with claws are animals.
(b) Some creatures with claws are non-ferocious.
(c) Some non-ferocious creatures have claws.
(d) Some ferocious creatures are creatures with claws.

[2023 : 2 M]

2.88 Which one of the following options can be inferred from the given passage alone?

When I was a kid, I was partial to stories about other worlds and interplanetary travel. I used to imagine that I could just gaze off into space and be whisked to another planet.

[Excerpt from The Truth about Stories by T. King]

(a) It is a child's description of what he or she likes.
(b) It is an adult's memory of what he or she liked as a child.
(c) The child in the passage read stories about interplanetary travel only in parts.
(d) It teaches us that stories are good for children.

[2023 : 2 M]

2.89 Sequence the following sentences (P, Q, R, S) in a coherent passage:

P: Shifu's student exclaimed, "Why do you run since the bull is an illusion?"
Q: Shifu said. "Surely my running away from the bull is also an illusion."
R: Shifu once proclaimed that all life is illusion.
S: One day, when a bull gave him chase. Shifu began running for his life.

(a) RSPQ (b) SPRQ
(c) SRPQ (d) RPQS

[2024 : 2 M]

2.90 If '→' denotes increasing order of intensity, then the meaning of the words [charm → enamor → bewitch] is analogous to [bored → ____ → weary]. Which one of the given options is appropriate to fill the blank?

(a) worsted (b) dead
(c) jaded (d) baffled

[2024 : 1 M]

2.91 P, Q, R, S, and T have launched a new startup. Two of them are siblings. The office of the startup has just three rooms. All of them agree that the siblings should not share the same room.

If S and Q are single children, and the room allocations shown below are acceptable to all,

PR	TS	Q

PQ	RT	S

then, which one of the given options is the siblings?

(a) T and R (b) T and Q
(c) P and S (d) P and T

[2024 : 1 M]

■■■

Answers Verbal Ability

2.1 (b)	2.2 (a)	2.3 (d)	2.4 (c)	2.5 (d)	2.6 (a)	2.7 (c)	2.8 (c)	2.9 (c)
2.10 (c)	2.11 (b)	2.12 (a)	2.13 (c)	2.14 (d)	2.15 (a)	2.16 (d)	2.17 (c)	2.18 (a)
2.19 (d)	2.20 (d)	2.21 (c)	2.22 (d)	2.23 (b)	2.24 (b)	2.25 (b)	2.26 (c)	2.27 (b)
2.28 (b)	2.29 (a)	2.30 (b)	2.31 (b)	2.32 (a)	2.33 (c)	2.34 (b)	2.35 (a)	2.36 (b)
2.37 (b)	2.38 (a)	2.39 (b)	2.40 (c)	2.41 (d)	2.42 (b)	2.43 (b)	2.44 (b)	2.45 (b)
2.46 (b)	2.47 (a)	2.48 (b)	2.49 (a)	2.50 (b)	2.51 (a)	2.52 (b)	2.53 (b)	2.54 (c)
2.55 (a)	2.56 (c)	2.57 (c)	2.58 (a)	2.59 (a)	2.60 (a)	2.61 (d)	2.62 (a)	2.63 (c)
2.64 (c)	2.65 (b)	2.66 (a)	2.67 (d)	2.68 (c)	2.69 (d)	2.70 (b)	2.71 (b)	2.72 (a)
2.73 (c)	2.74 (a)	2.75 (b)	2.76 (c)	2.77 (c)	2.78 (a)	2.79 (d)	2.80 (c)	2.81 (c)
2.82 (c)	2.83 (d)	2.84 (b)	2.85 (c)	2.86 (a)	2.87 (d)	2.88 (b)	2.89 (a)	2.90 (c)
2.91 (d)								

Explanations Verbal Ability

2.1 (b)

Circuitous: Deviating from a straight course ⇒ Indirect
(A) Cyclic: Recurring in cycle
(B) Indirect: Not leading by straight line
(C) Confusing: Lacking clarity
(D) Crooked: For shapes (irregular in shape)

2.2 (a)

Unemployed: Worker ⇒ Here one is opposite to other.
(A) Fallow: Land ⇒ Fallow means undeveloped land.
(B) Unaware: sleeper ⇒ Both are same unaware or asleep.
(C) Wit: Jester ⇒ Wit means ability to make jokes and jester is a joker.
(D) Renovated : House ⇒ Renovate means to make better and house can be renovated.

2.3 (d)

(A) Uphold: cause to remain ⇒ not appropriate
(B) Restrain: keep under control ⇒ not appropriate
(C) Cherish: be fond of ⇒ not related
(D) Conserve: Keep in safety and protect from harm, decay, loss, or destruction ⇒ most appropriate.

2.4 (c)

(A) Masked: Hide under a false appearance ⇒ opposite
(B) Belied: Be in contradiction with ⇒ not appropriate
(C) Betrayed: Reveal unintentionally ⇒ most appropriate
(D) Suppressed: To put down by force or authority ⇒ irrelevant

2.5 (d)

(A) Modern warfare has resulted in civil strife: There is no direct consequence of warfare given, so it is not appropriate.
(B) Chemical agents are useful in modern warfare: Passage does not say whether chemical agents are useful or not, so not appropriate.
(C) Use of chemical agents in warfare would be undesirable: Given that people in military think these are useful, undesirable is wrong
(D) People in military establishments like to use chemical agents in war; Correct choice as last statement tells that military people think that chemical agents are useful tools for their cause (work silently in warfare).

2.11 (b)

Latitude refers to freedom of action, freedom of expression from restrictions etc. For example, he allowed his children a fair amount of latitude. Coercion refers to force which is an opposite of the word Latitude. Meticulousness refers to being extremely careful and conscientious.

2.12 (a)

'Requested that' part should be modified. The sentence is in Reported speech so 'should be given' part is correct. Instead of Request, words like 'suggested', 'recommended' can be introduced.

2.13 (c)

If the tired soldier wanted to lie down, he should have taken the mattress out on the balcony.

2.14 (d)

Given the seriousness of the situation that he had to face, his nonchalance was impressive.

Nonchalance refers to casual / non-serious/ indifferent attitude. The author wants to convey that in spite of the seriousness of the situation, he was very casual.

No other word fits in the blank. Beggary is word that is related to extreme poverty (Beggar). Nomenclature refers to the art of naming.

2.15 (a)

Thorough regimentation was the main reason for the efficiency of the Roman legions even in adverse circumstances. As the author says, 'Discipline on the battlefield kept units obedient, intact and fighting, even when the odds and conditions were against them.' So choice (A) emerges as the best answer option.
We have been asked to summarize the paragraph in the best possible manner. Choice (C) presents one-sided picture of the paragraph. Summary is the essence or the gist of the paragraph.
Choice (B) and Choice (D) are wrong as per the information mentioned in the paragraph.

2.16 (d)

This is a famous idiom in English.

2.17 (c)

'You can always give me a ring whenever you need' implies that I am available even at a single call (whenever you need me). 'Ring' has been used as a metaphor here and not literally.

2.20 (d)

As the given argument states that there were different streams of freedom movement during colonialism, it clearly implies that nationalism in India is heterogeneous.
Option (c) is ruled out as it states the contrary.
Options (a) and (b) fail to establish a direct link with the argument.

2.41 (d)

Both the conclusions I and II follow.

2.59 (a)

Could have, been
I could have made arrangements had I been informed earlier. Use of conditional sentence based on past participle form.

2.60 (a)

Hurtful means causing pain or suffering or something that is damaging or harmful.
The expression 'sharp tongue' defines a bitter or critical manner of speaking.

2.61 (d)

Hostile
Antagonist is a adversary or one who opposes/ contends against another.

2.62 (a)

It is easier to read this year's textbook than the last year's.
Use of comparative degree adjectives is made for comparing two nouns.
Comparative adjectives take 'than'.

2.63 (c)

The ninth and tenth of this month are Monday and Tuesday respectively.
Respectively refers to separately or individually in the order already mentioned.

2.64 (c)

An intimate association offers impartial perspective.
Author states that he was too intimately associated with the actors to set the perspective needed for impartial recording.

2.65 (b)

- Peace is a situation when there is no war or fighting and things are calm and quiet.
- Piece is a part of something.

2.66 (a)

Even though there is a vast scope for its improvement, tourism has remained a neglected area.

2.68 (c)

Statement (i) is not true as nowhere it is mentioned that John was a captain in junior team. The introductory line emphasizes on the board recognizing John's potential (Latent quality/ possibility) as leader of the team.

Statement (iii) also manipulates the facts mentioned in the argument.

The 3rd statement of the argument while this aggression has filled stadia with die-hard fans does not indicate fans expectations from John as a caption.

Statement (ii) The concluding statement of the para suggests that finding a completer captain is a tough task as it took John many years to become a successful and calculative batsman - captain.

Statement (iv) can be explicitly concluded from the last 4 lines of the para.

2.69 (d)

When he did not come home, she pictured him lying dead on the roadside somewhere. "pictured" is the most appropriate option in this situation as she is imagining or creating an image of somebody/something in her mind.

2.70 (b)

"Company" is singular, hence, "uses" would be the correct option in blank 1. "products" is plural, hence, "include" would be the correct option in blank 2.

2.71 (b)

Option 1 i.e. "on" is used when one is talking about a day of the week e. g. on Thursday, or a particular part of a date e.g. on Sunday evening, or a particular date e.g. on 9th March.

Option 3 i.e. "in" is used when one is talking about a month e.g. in June, or a season e.g. in winter, or a specific year e.g. in 2019.

Option 4 i.e. "under" is absurd in this situation.

Option 2 i.e. "at" is used when one is referring to a particular time e.g. at dawn, or a calendar festival season e.g. at Christmas.

2.72 (a)

The passage says" extremely well documented and researched, but not always impartial". This suggests that Indian history written by British historians was well documented and researched but was sometimes biased. hence option (a) is the correct answer.

2.78 (a)

Nostalgia refers to a feeling, fondness and slight sadness thinking of past and anticipate is to predicting future.

2.79 (d)

Wake - Woke - Woken are three form of the verb. Hence, option (i) and (iii) are correct.

2.80 (c)

Ubiquitous is the keyword to justify option (i). Positive and negative effect for humans justifies option (iii).

2.81 (c)

Mr. X speaks neither Japanese nor Chinese.

2.82 (c)

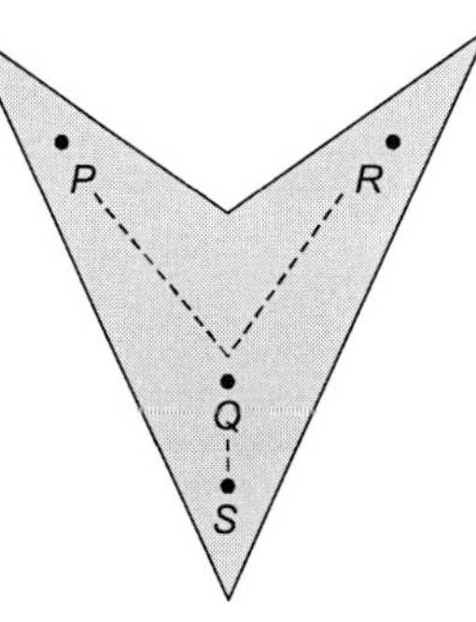

Given that, the plan of the gallery where the boundary walls are opaque, if the security guard will be posted on '*P*' then he doesn't have the visibility of *R* and vice-versa. But if the security guard will be posted at '*Q*' and '*S*', then the security guard can watch over the entire inner space of the gallery.

2.83 (d)

The given passage mentions that it is yet to be seen whether the genetically modified mosquitoes to control the overall mosquito population has any unforseen consequences whereas controlling mosquitoes using chemicals may have undesired consequences.

2.84 (b)

Since, S never lies, therefore the statements of *S* is correct i.e. only one of us telling the truth, which is S only as S never lies.

Therefore, the statements of P, Q, R and T are wrong. Hence, we can infer that

(i) R has not copied in the exam.

(ii) S has not copied in the exam.

(iii) P has copied in the exam.

(iv) R is lying which means P has copied in the exam.

Based on above information, the person who has copied in the exam is P.

2.89 (a)

- Sentence R gives the introduction to passage.
- Sentence Q is the response to a question in sentence P, which is asked on the statement made in sentence S.

Hence, the correct sequence is RSPQ.

2.90 (c)

Jaded means tired or bored with something, because of experiencing it too much.

2.91 (d)

From the given acceptable room allocations, we can infer that:

(i) R and T are not siblings as they share the same room.

(ii) Since, S and Q are single children, the siblings are sharing the rooms with different person and are not allocated single room. Hence, one of the sibling is T and the other is among P and R from the first arrangement of room allocation.

(iii) From the second arrangement, it is clear that R and T can't be siblings. Hence, the only possible option is P and T.

■■■